Pathological Basis of the Connective Tissue Diseases

Pathological Basis of the Connective Tissue Diseases

Dugald Lindsay Gardner

ScD, MA(Cantab); MSc(Manc); MD, PhD(Edin); FRCP; FRCPEd; FRCSEd; FRCPath

Honorary Fellow, Department of Pathology, University of Edinburgh; Emeritus Professor of Histopathology, University of Manchester; Honorary Conservator, Royal College of Surgeons of Edinburgh; *formerly* Musgrave Professor of Pathology, the Queen's University of Belfast; Director of the Kennedy Institute of Rheumatology, London

Edward Arnold
A division of Hodder & Stoughton
LONDON MELBOURNE AUCKLAND

First published in Great Britain 1992

British Library Cataloguing in Publication Data

Gardner, D. L.
Pathological basis of the connective tissue
diseases.—
I. Title
616.7

ISBN 0-7131-4548-X

Whilst the advice and information in this book is believed to be true and accurate at the date of going to press, neither the author nor the publisher can accept any legal responsibility or liability for any errors or omissions that may be made. In particular (but without limiting the generality of the preceding disclaimer) every effort has been made to check drug dosages; however, it is still possible that errors have been missed. Furthermore, dosage schedules are constantly being revised and new side effects recognised. For these reasons the reader is strongly urged to consult the drug companies' printed instructions before administering any of the drugs recommended in this book.

Typeset in Century Old Style by Rowland Phototypesetting Limited, Bury St Edmunds, Suffolk. Printed and bound in Great Britain for Edward Arnold, a division of Hodder and Stoughton Limited, Mill Road, Dunton Green, Sevenoaks, Kent TN13 2YA by Butler and Tanner Limited, Frome, Somerset.

Contents

Contributors

J. C. Anderson, MA PhD
Senior Lecturer in Medical Biochemistry, Department of Biochemistry and Molecular Biology, University of Manchester

C. G. Armstrong, PhD
Reader, Department of Mechanical and Manufacturing Engineering, The Queen's University of Belfast

Maureen M Dawson, MSc PhD
Senior Lecturer in Immunology, Department of Biological Sciences, Manchester Polytechnic; *formerly* Research Fellow, Paterson Institute for Cancer Research, Christie Hospital and Holt Radium Institute, Manchester

A. J. Freemont, BSc MD FRCP(Ed) MRCP(UK) MRCPath
Senior Lecturer in Osteoarticular Pathology, University of Manchester Medical School

I. Isherwood, MD FRCP FRCR FFR RCSI(Hon)
Professor of Diagnostic Radiology, University of Manchester Medical School

J. P. R. Jenkins, MRCP(UK) DMRD FRCR
Consultant Radiologist and Honorary Clinical Senior Lecturer, Department of Clinical Radiology, Manchester Royal Infirmary

J. McClure, BSc MD FRCPath DMJ(Path)
Procter Professor of Pathology, University of Manchester Medical School

M. Moore, DSc PhD FRCPath
Xenova Ltd., Slough; *formerly* Head, Department of Immunology, Paterson Institute for Cancer Research, Christie Hospital and Holt Radium Institute, Manchester and Honorary Reader in Tumour Immunology, University of Manchester Medical School

Patricia O'Connor, PhD
formerly Research Fellow, Department of Histopathology, University of Manchester

E. T. Whalley, PhD
Cortech Inc., Denver, Colorado; *formerly* Lecturer in Pharmacology, Department of Physiological Sciences, University of Manchester

Preface

The purpose of this book is to provide an account of the pathology of the 'rheumatic' or 'connective tissue' diseases. 'Rheumatism' is the most frequent cause of disability and suffering in the Western World and because of the increasing age of this population, the relative importance in medicine and surgery of the connective tissue diseases is likely to continue to rise. In turn, an understanding of the pathology of these disorders will assume ever greater significance. However, unresolved difficulties remain. Spectacular advances in molecular biology and cytogenetics have done much to unravel the mechanisms that can disorganize the connective tissues but the immediate causes of common and important diseases, including osteoarthrosis, rheumatoid arthritis and systemic lupus erythematosus, are still uncertain. Nevertheless, a large body of relevant data has accumulated and I therefore judge that this is an appropriate moment at which to summarize present understanding of the 'natural history' of these diseases.

The book is in three parts: Part I considers the biological structure and functions of the connective tissues. Part II reviews the biochemical, immunological, mechanical and pharmacological basis of the connective tissue diseases and provides an account of fibrosis. Part III describes systematically the pathology of the inherited and acquired diseases of the non-osseous connective tissue system. Throughout, an attempt has been made to relate the observed pathological defects to advancing knowledge of the normal structure and function of the affected tissues. However, the contents of the book are unavoidably selective. Many excellent texts review the pathology of the calcified skeletal tissues and the present volume largely excludes consideration of these disorders. For similar reasons, neoplastic diseases are not discussed. Connective tissue forms the stroma of all the organs of the body but full accounts of visceral abnormalities such as endomyocardial fibrosis and hepatic cirrhosis are beyond the scope of this text.

At first, it seemed that it might still be possible for a single author to cover all aspects of the pathology of the connective tissue diseases. However, it quickly became clear that the necessary degree of special knowledge could not be provided by one person and I decided to seek the assistance of experts in biochemistry, immunology, mechanical engineering, pharmacology and radiology and to ask for help from colleagues with special interests in synovial and bone disease, respectively. This decision accounts for the very valuable contributions made to the book by Dr John Anderson, Dr Maureen Dawson, Dr Michael Moore, Dr Cecil Armstrong, Dr Patricia O'Connor, Dr Eric Whalley, Dr Jeremy Jenkins, Dr Tony Freemont, Professor Ian Isherwood and Professor John McClure, to all of whom I wish to record my thanks. The individual manuscripts prepared by these authors were edited in close collaboration with them and I therefore accept full responsibility not only for the plan, content, scale and arrangement of the text as a whole but also for the style and accuracy of those parts written by my associates. The nature of the book is a reflection of my personal interest and bias.

In many places, I refer to studies made with colleagues in London, Belfast, Manchester and Edinburgh. I wish to acknowledge the generous support of Professors N H F Wilson and C J Kirkpatrick, of Drs S L Carney, R J Elliott, M A H A El-Maghraby, R Fitzmaurice, O Hassan, D M Lawton, R B Longmore, R Mazuryk (Szezcin), J F S Middleton, K Oates, Constance R Orford, D M Salter, Ruth St C Symmers and Maria Warskyj, and of Messrs D S Cunningham, J Pidd, R Simpson and P Sullivan. I express my thanks both to them and, in particular, to Mr A Bradley BVSc, MRCVS and to Dr P Skelton-Stroud MVSc, PhD, FRCVS.

An indirect benefit of a long period as Chairman of the Pathology Committee of the European League Against Rheumatism (EULAR) was the opportunity of annual meetings with European colleagues. I owe much to discussions with Professors M Aufdermauer (Luzern), the late J Ball (Manchester), E G L Bywaters (London), F Eulderink (Leiden), G Geiler (Leipzig), R Lagier (Genève), the late E Maldyk (Warsaw), W Mohr (Ülm) and O Myre-Jensen (Aarhus), and with Drs J E W Ahlquist (Helsinki), M Bely (Budapest), R J Francois (Brussels), P Moller Graabaek (Aarhus), P A Revell (London), S B Refsum (Oslo) and Theresa Wagner (Warsaw).

For many years, my laboratory has enjoyed generous support from the Arthritis and Rheumatism Council for Research, from the Medical Research Council, from the Wellcome Trust and from Ciba Geigy Pharmaceuticals and I am deeply indebted to these organizations. My recent research in Edinburgh would not have been possible without the goodwill of my colleague Professor C C Bird and I wish to record my thanks to him and to the Faculty of Medicine of the University of Edinburgh for making possible both these investigations and the completion of this volume.

In the preparation of the present text, I have drawn extensively upon investigations first published in the *Annals of the Rheumatic Diseases*, *British Journal of Experimental Pathology* (now the *International Journal of Experimental Pathology*), *British Medical Journal*, *Clinical and Experimental Rheumatology*, *Histochemical Journal*, *Journal of Anatomy*, *Journal of Bone and Joint Surgery*, *Journal of Medical Primatology* and *Journal of Pathology* and I am grateful to the publishers of these journals for permission to reproduce material from their columns. I also express my thanks to Academic Press, Blackwell Scientific Publications, Churchill-Livingstone, Pitman Medical, and Wiley and Sons for permission to reproduce illustrations from books published under their auspices.

The volume contains many photographs originally made by Mr James Paul FBIPP, FIMLS and, as on numerous previous occasions, I place on record my appreciation of the tolerance and skill with which he has always responded to my requests. During the final preparation of the manuscript I had the good fortune to be assisted by Miss Ruth Gallacher and I acknowledge the debt that I owe her for much painstaking secretarial help.

Every writer recognizes that sustained composition imposes almost insuperable burdens on the author's family and friends. The completion of this book has been no exception and I shall always be grateful to my wife Helen and my family who have unfailingly supported me through long and often demanding times during which understanding and tolerance have been stretched beyond reasonable limits.

Finally, my thanks are due to my publishers, Edward Arnold, with whom I first joined forces in 1963, and to the unsurpassed skills of my Editor, Mairwen Lloyd-Williams. The publishing house of Edward Arnold has itself undergone change and vicissitude in the years during which this book has been in preparation and the completion of this text is one measure of the success with which they have surmounted their own obstacles.

November 1991 DLG

Historical Introduction

Therefore the moon, the governess of floods,
Pale in her anger, washes all the air,
That rheumatic diseases do abound:
(*A Midsummer Night's Dream*, II: 1
by William Shakespeare)

Progress in understanding the pathology of the connective tissue diseases has been increasingly rapid and it is easy to neglect their historical origins (Fig. 1). The study of pathological anatomy established its roots in the eighteenth century and pathological histology in the nineteenth. Yet it is only in the present era that the majority of the disorders described in this volume have been systematically described and classified. One aim of this Introduction is to seek the reasons for this slow advance—which is partly attributable to the practical difficulties of working with skeletal tissues; a second purpose is to consider the antiquity of connective tissue disorders. In most instances, further historical information is contained in the chapters that describe the individual diseases.

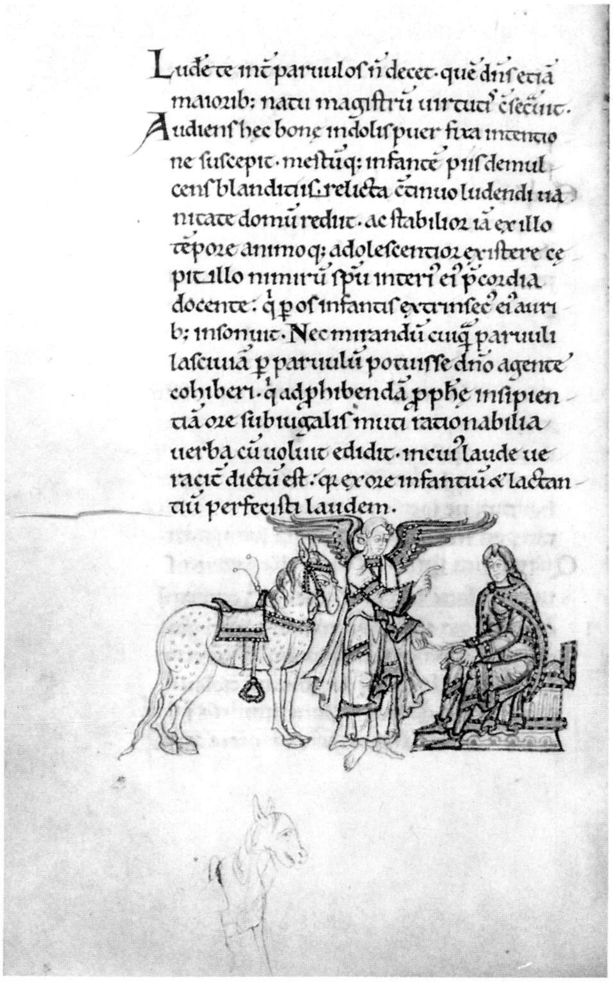

Fig. 1 An angel healing St Cuthbert's knee.
This is one of the earliest known illustrations of connective tissue disease. Reproduced by courtesy of the Master and Fellows of University College, Oxford, and with the assistance of the Department of Western Manuscripts at the Bodleian Library, Oxford.

The foundations of rheumatology

Although the word 'rheumatism'[7] is archaic, it is still in common use and the scientific study of 'rheumatic' diseases is still called rheumatology. The history of rheumatology has been reviewed succinctly (Fig. 2).[63]

Until modern times, the term 'rheumatism' was used in a very broad diagnostic sense. But there were exceptions: it is likely that Hippocrates distinguished gout (see Chapter 10) from acute rheumatism (rheumatic fever) (see Chapter 19). However, before the eighteenth century, European medical thought was constrained by the persistent use of Galenic[32] texts. Gradually, bedside descriptions became more detailed and precise. Scott[63] shows how the development of clinical skills, in parallel with Linnean concepts, encouraged the classification of disease. Sydenham,[67] in

Fig. 2 William Heberden the Elder (1710–1801).
Heberden, physician to King George III, was the 'Father' of British rheumatology. Reproduced by permission of the President and Fellows of the Royal College of Physicians of London from a negative in the possession of the National Portrait Gallery.

Fig. 3 Thomas Sydenham (1624–1689).
Nicknamed the 'English Hippocrates', Sydenham contributed classical clinical accounts of gout, chorea and rheumatic fever during a life devoted to the study of the natural history of disease. From an engraving by Houbraken, after P Lely, in the Wellcome Collection.

particular, gave a unique account of (rheumatic) chorea; he also differentiated gout from other forms of chronic polyarthritis (Fig. 3).[66,67] There were additional individual accounts of rheumatic fever[12,24] and progressive systemic sclerosis[26] (Chapter 16). Cullen's vigorous teachings on rheumatism stimulated a corresponding interest among students of medicine: more than 100 theses on this subject were submitted to the University of Edinburgh between 1755 and 1833. Two examples are those of Trotter[68] and Bell.[9]

The emergence of histology

Xavier Bichat[10,11] can be said to have distinguished the connective tissues (Fig. 4). However, the practice of histology, in the sense in which we now know it, only became

Fig. 4 Marie Francois Xavier Bichat (1771–1802).
Bichat, the founder of histology, scorned the primitive microscopes of his day and classified the body tissues macroscopically. He may justly be regarded as the source of the ideas which led Muller to talk of the 'connective tissues'. From a contemporary stipple after Louis Choquet.

possible with the manufacture of compound microscopes with lenses corrected for chromatic and spherical aberration,[50] and with the development of microtomes to enable the preparation of tissue sections sufficiently thin for critical transillumination. A 'cell theory' originated in botanical studies[61,62] and was extended to the zoology of mammals: the connective, mesenchymal tissues, like those of the epithelia and viscera, were found to be derived from cells and not from the hypothetical tissue fibres of the older anatomists. Cells were formed by the division of parent cells, not by 'free' changes in a germinal matrix.[15] These advances created a new discipline, 'pathological histology'. Much of the evidence used by the Goodsirs[38] and by Virchow[70] to advocate a cell theory of disease was obtained from investigations of the connective tissues, particularly cartilage and bone (Fig. 5). It was argued that just as the cell is the unit of physiological function, an entity formed from previous generations of cells, so all diseases are disturbances of the cells of which the affected tissues are composed.[28]

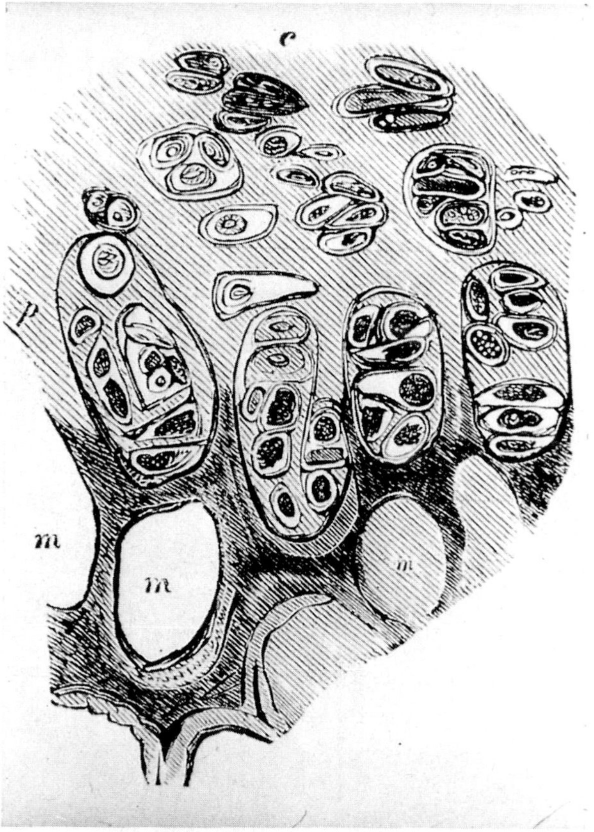

Fig. 5 Perpendicular section through the epiphyseal growth zone of a talus.
Rudolph Virchow (1821–1902) was thoroughly familiar with the histological appearances of many forms of connective tissue and some of his most important arguments in relation to the so-called 'cell theory of disease' are based upon studies of cartilage. Figure 126 from the 1860 English translation, published in America in 1863, of the 2nd German edition of Virchow's *Cellular Pathology*.[70]

Pathology is established

The clinical problems of rheumatism confronted practitioners in all parts of the ancient World including preHispanic Mexico.[3] However, pathology, as a biological science in its own right was a late by-product of Arabian and Renaissance medicine. From the mists of antiquity there had emerged a GraecoRoman era of speculation concerning the nature of disease with writings attributable to Hippocrates,[40] Celsus[17] and Galen.[32] It is a continuing source of surprise that in the earliest Hittite, Babylonian and Egyptian civilizations, anatomical dissection was not practiced, in spite of the custom of embalming. There appears to have been no structural understanding of 'rheumatic' disease, although it is now clear that disorders like tuberculous arthritis (see Chapter 18) and ankylosing spondylitis, which leave permanent bone deformity, were comparatively frequent.[63] That so little was said of the musculoskeletal tissues was a reflection of the fact that anatomical dissection had not become a socially acceptable practice.

Change, however slow, was inevitable. A resurgence of interest in the structure and function of the human body found expression in the early Schools of Anatomy, in the drawings of Leonardo da Vinci and in the *De Humani Corporis Fabrica*, of Vesalius.[69] Instead of a public exercise conducted on the remains of executed criminals, the systematic anatomization of the human body became reputable and even venerated. With the foundation of anatomy, new concepts of disease such as those of Fernel[30] were formulated. Galenic theories yielded to reasoned observation. Physiology emerged: the circulation of the blood was demonstrated[39] and physicians, chemists and microscopists such as Hooke[41] began to present the results of their enquiries to the Royal Society of London and to the European academies.

It is inconceivable that the early anatomists should not have noted the ravages of those connective tissue diseases such as infective arthritis and rheumatic fever which plagued young populations. Yet, for long periods there were very few scientific records of these disorders. The practice of clinicopathological correlation by Morgagni[53] marked a radical change of attitude. Hypotheses concerning the nature of disease were succeeded by finite observations made to explain clinical signs and symptoms. Among the protagonists of this silent revolution was the experimental surgeon, John Hunter (1728–1793). Access to his museum enabled Mathew Baillie not only to compile the first text on pathology in English in 1793[6] and the accompanying atlas of morbid anatomy[5] but also to record a first note on the anatomy of rheumatic heart disease. The pathology of the connective tissue diseases can therefore be said to have originated during the latter part of the eighteenth century.

Early accounts of connective tissue pathology

I have been unable to find any reasonably systematic account of the anatomical changes attributable to connective tissue disease before the time of Morgagni. It is true that Columbo (1559)[20] drew attention to a left ventricular thrombus which, it has been suggested, may have been of rheumatic origin, and that Connor[21] wrote a widely quoted letter to Sir Charles Walgrave describing a skeleton with the signs of what was probably ankylosing spondylitis (Fig. 6). But these were random observations and it seems more reasonable to give priority to John Hunter (1759)[42] and to Morgagni (1761)[53] for their accounts of osteoarthrosis (see Chapter 22) and gout, respectively (Fig. 7). They were followed quickly by Percival Pott (1779)[57] who described vertebral osteitis, without recognizing its tuberculous nature, and by the further account of gout given by Alexander Monro ('Secundus').[52] Wollaston[72] achieved notoriety by demonstrating that tophaceous material from his own gouty

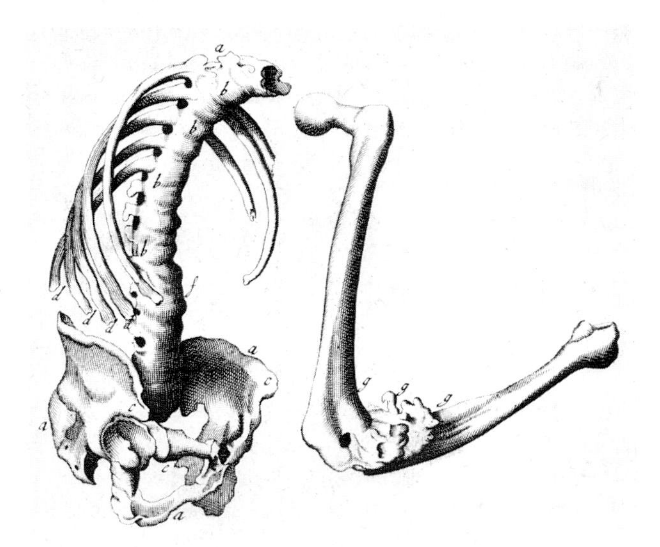

Fig. 6 Illustration from the letter of Bernard Connor (1695).
The figure shows an early representation of the pathological consequences of ankylosing spondylitis. The vertebral column is converted to a rigid structure by a series of fused intervertebral joints; there are numerous exostoses around the knee joint which appear to be incidental to the ankylosing spondylitis.

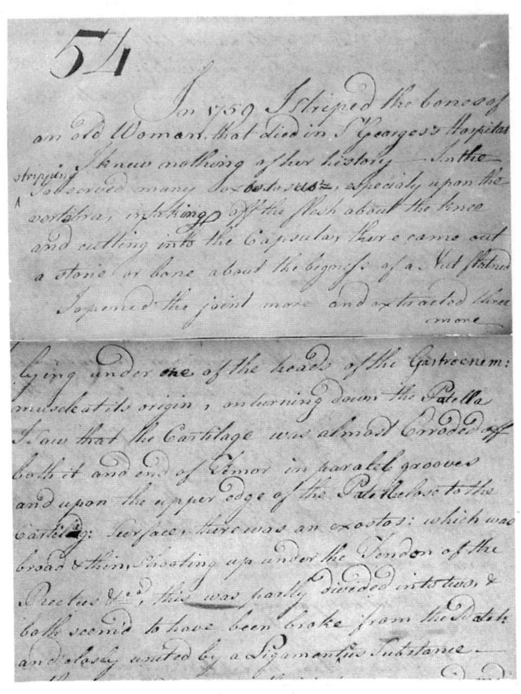

Fig. 7 Extract from John Hunter's manuscript 54[42] in the possession of the Royal College of Surgeons of England.
The notes, recorded by one of Hunter's assistants at his dictation, describe the appearances now recognized as osteoarthrosis (Chapter 22). The notes read as follows:

In 1759 I striped the bones of an old Woman that died in St George's Hospital. I knew nothing of her history. In the stripping I observed many exostoses especialy upon the vertebra, in taking off the flesh about the knee and cutting into the Capsula there came out a stone or bone about the bigness of a Nut flatned. I opened the joint more and extracted three more lying under one of the heads of the Gastrocnem: muscle at its origin, on turning down the Patella I saw that the Cartilage was almost Corroded off both it and (the) end of (the) Femor in paralel grooves and upon the upper edge of the Patella close to the Cartilag: surface, there was an exostos: which was broad and thin, shooting up under the Tendon of the Rectus Fem . . .

Reproduced by permission of the President and Fellows of the Royal College of Surgeons of England.

ear contained uric acid and Benjamin Brodie (1818)[13] recorded the features of both ankylosing spondylitis and of the Reiter-Fiessinger/Leroy syndrome in his book on joint disease.[63] These early reports coincided with the few relevant notes contained in the various editions of Baillie's textbook (1793 *et seq.*)[6]—there are no illustrations of connective tissue diseases in his atlas (1799–1803)[5]—and came shortly before the magnificent colour illustrations of gout and infective arthritis contained in the volumes of Cruveilier's (1829–35)[23] work (Fig. 8). Adams' (1857)[1] engravings appear to be the first proper representation of rheumatoid arthritis (Fig. 9) although there are said to be accounts in the Dutch literature that preceded this date. Soon afterwards, Küssmaul and Maier[49] recorded a de-

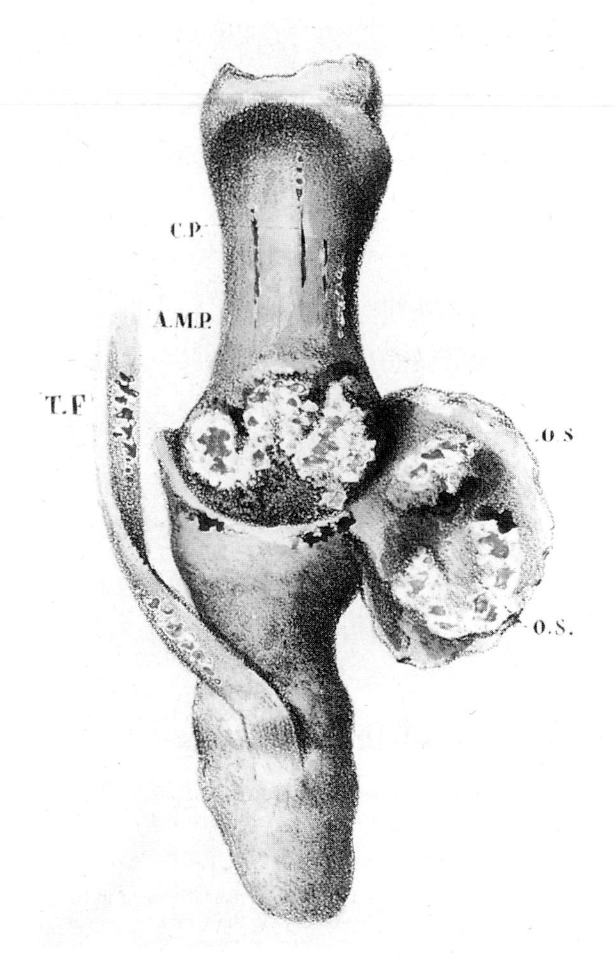

Fig. 8 Part of a plate from *Anatomie Pathologique du Corps Humain.*[23]
Severe gouty arthritis. The gouty tophi disrupt the proximal interphalangeal joint and appear as speckled white flecks because of their calcium content. In the original illustration, which is in colour, the appearances are most striking. Gout of this severity is now rare.

scription of the anatomical lesions of polyarteritis nodosa (Chapter 15), and the Hungarian dermatologist Kaposi, now better known for his identification of a particular form of angiosarcoma,[44] described the disseminated form of lupus erythematosus[43] (Chapter 14), the discoid form of which had been recognized by Cazenave[16] in 1850. In the later years of the nineteenth century, advances in the understanding of new disease entities quickened, catalysed by the emergence of bacteriology.

Disease of collagen

In contemporary terms, the phrases 'disease of collagen' or 'disease of the collagen molecule' signify the many inherited abnormalities (described in Chapters 4 and 9) in which there are defined faults in collagen synthesis, fibrillogenesis, maturation or assembly.

Historically, Klemperer and his colleagues had postulated that systemic lupus erythematosus and systemic sclerosis might be 'diffuse diseases of collagen'.[46] They extended this view to rheumatoid arthritis and related disorders, and the alternative term 'collagenosis' was introduced. 'Fibrinoid',[55] an ill-defined microscopic artefact, had been identified in experimental studies of tissues injured by hypersensitivity reactions.[47,48] Fibrinoid was recognized in systemic lupus erythematosus, systemic sclerosis, rheumatic fever (Fig. 10) and in rheumatoid arthritis; these diseases, it was suggested, also might be of a hypersensitivity nature. The origin of 'fibrinoid' was attributed by Klemperer to injury to collagen. Similarly, it was thought that systemic lupus erythematosus, systemic sclerosis, rheumatic fever and rheumatoid arthritis might originate from disorder(s) of collagenous tissues.[45,46]

This view is now obsolete. Antibodies to collagen are recognizable in arthritis and analogous disorders; and anti-type II collagen arthritis is an effective animal model of human disease. But the production of anti-collagen antibodies in rheumatoid arthritis is likely to be an epiphenomenon, not causally related to the onset of the arthritis, just as the many antibodies synthesized in systemic lupus erythematosus are consequences of this disorder, not causes of it. The hypothesis of 'diffuse collagen disease' should now be abandoned. A further proposal was to distinguish 'disease(s) *of* the collagen molecule' from 'diseases *in* [affecting] collagenous tissues'.[51] This constructive suggestion is of value in emphasizing the distinction between primary abnormalities of collagen such as osteogenesis imperfecta and secondary disorders of collagen such as scurvy, lathyrism and systemic sclerosis.

Early attempts to reproduce connective tissue disease experimentally

The ability to reproduce a disease in an animal has often proved to be an important step in understanding the pathogenesis of the disorder. One of the earliest attempts to create articular disease experimentally was that of Redfern (1852).[58] The first account of an immunological model of arthritis was published by Friedberger (1913).[31] The growth of interest in experimental studies quickened after the classical monographs of Klinge[47,48] and has become increasingly rapid. No synopsis can do justice to a subject of such importance. Gardner[33] analysed the literature published before 1960 and later discussed some of the more

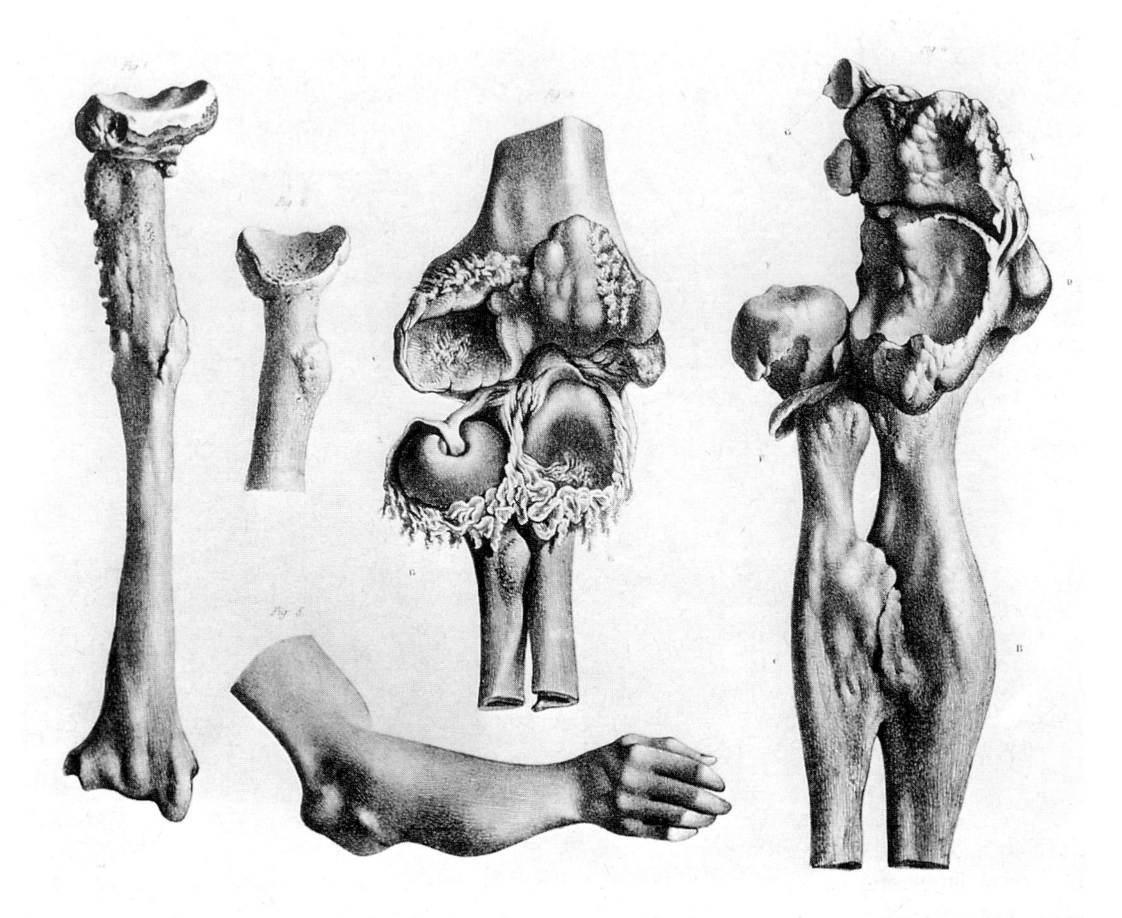

Fig. 9 Plate from *Illustrations of the Effects of Rheumatic Gout or Chronic Rheumatic Arthritis on all the Articulations: with Descriptive and Explanatory Statements.*[1] Not all the changes depicted in this classical work are easy to identify in terms of our present knowledge of the diseases of connective tissue. The centre elbow joint is clearly the site of rheumatoid arthritis. The elbow joint (right) is bounded by numerous marginal osteophytes while the arm (bottom) bears swellings around the elbow which may be gouty tophi. The radius (left) is bounded at its upper end by prominent osteophytes; the periosteum of the upper one-third of the bone has been elevated suggesting the possibility of infection.

useful experimental techniques.[34] In recent years there have been detailed surveys of special topics: the reviews of Adams and Billingham[2] on osteoarthrosis and of Sokoloff[64] on rheumatoid arthritis, merit close attention. The multiplicity of animal models of arthritis is summarized lucidly by Currey.[25] In the present text, both experimental animal diseases and those occurring naturally are discussed in the sections dealing with the aetiology and pathogenesis of individual human disorders.

How old are the connective tissue diseases?

There is speculation concerning the antiquity of the common connective tissue diseases. Evidence has been gained both directly, from the study of fossil remains, mummies and other burials, and indirectly, by the investigation of

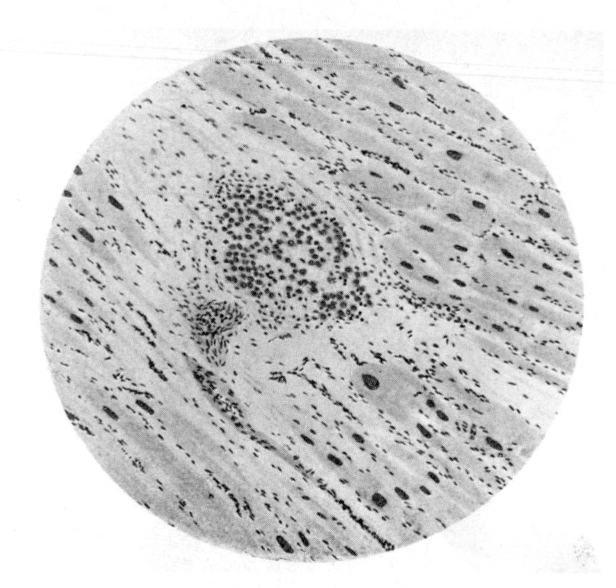

Fig. 10 Illustration from Ludwig Aschoff's text *Pathologische Anatomie.*[4a]
The structure is an 'Aschoff' nodule in the intramyocardial tissue. The illustration should be compared with Figs 19.3 and 19.6. The description by Aschoff of the microscopic lesion which now bears his name was made in 1904; other authors, however, appear to have given earlier but less precise accounts of these changes.

writings, works of art and artefacts that portray the nature of diseases observed in different epochs.

Connective tissue diseases are not confined to modern humans, to hominids or prehominids, and there is evidence that common disorders such as spinal osteophytosis (see Chapter 23) were prevalent among dinosaurs. Palaeopathological studies of human remains have demonstrated lesions attributable to tuberculous vertebral osteitis,[54] gout[29] and osteoarthrosis[60,71] in prehistoric Old and New World man, in an Egyptian mummy and in Saxon and RomanoBritish skeletons, respectively.[63]

Evidence of the prehistoric or early historic presence of diseases of the non-osseous connective tissues is difficult to obtain and validate: these tissues do not survive putrefaction and some, of course, were removed during the preparation of bodies for burial or mummification.[27] Nor do the soft connective tissues fossilize, although tendons, ligaments and cartilage are protected by processes such as tanning that have preserved Grauballe and Lindow men.[65] Under these conditions, the skeleton is demineralized so that bone disease cannot easily be studied.

The question of the antiquity of rheumatoid arthritis has attracted particular comment and discussion. Alfred B Garrod[35] introduced the description 'rheumatoid' arthritis but it was not certain that osteoarthrosis and rheumatoid arthritis were distinct entities until the publication of the fourth edition of Bannatyne's book in 1906.[8] Virchow

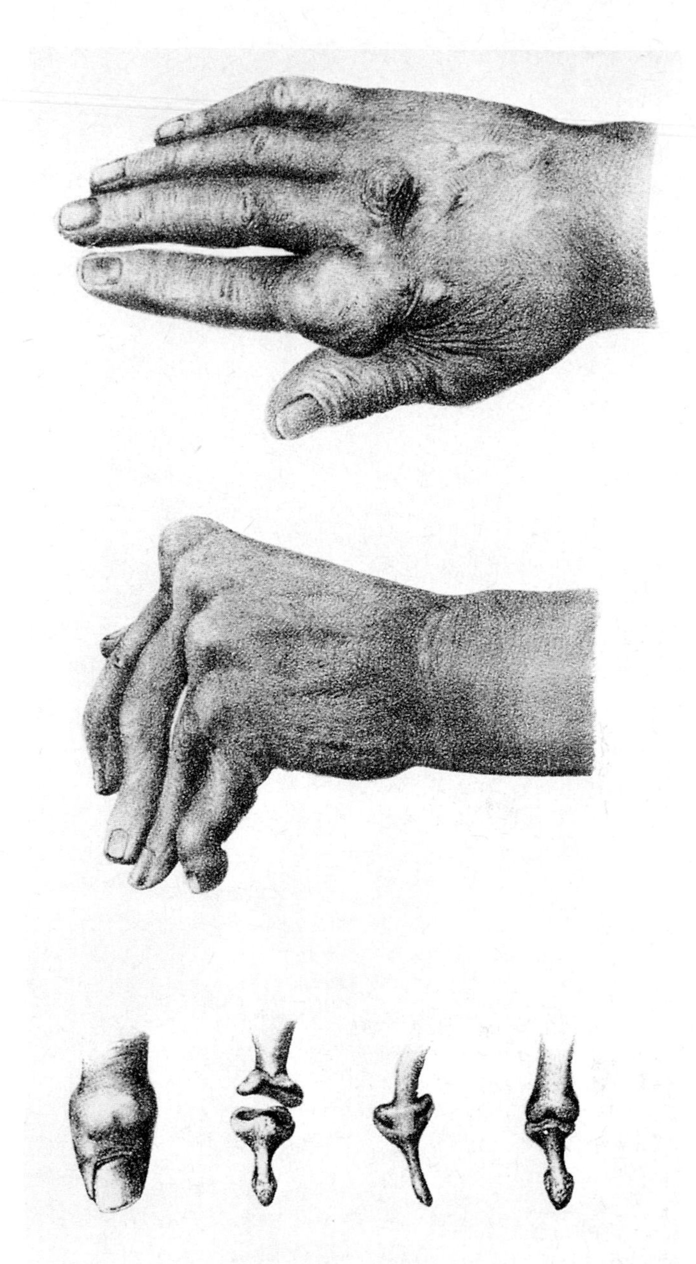

Fig. 11 Nineteenth century illustrations of connective tissue disease.
In this engraving, from his famous book (*Leçons sur les Maladies des Vieillards et les Maladies Chroniques*, 1867) on geriatrics and chronic diseases, Jean Martin Charcot (1825–1893)[18,19] depicts gout (top), rheumatoid arthritis (centre) and Heberden's osteoarthrosic nodes (bottom). Charcot's lectures were published in 1867 and an English translation in 1881.

(quoted by Copeman)[22] and Charcot[18,19] perpetuated confusion by insisting that rheumatoid arthritis and osteoarthrosis were both forms of an 'arthritis deformans' (Figs 11 and 12). By the time Archibald E Garrod[36] wrote (1907) in Albutt's *System of Medicine* however, the two forms of joint disease had come to be accepted as discrete entities and

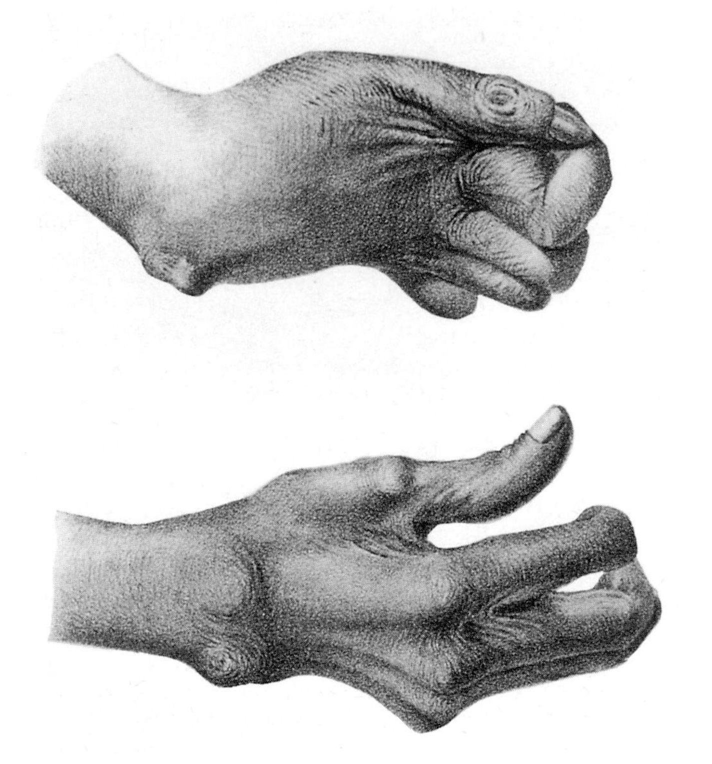

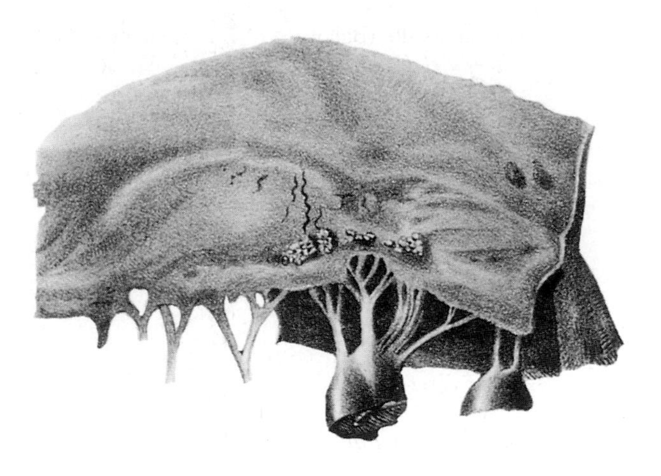

Fig. 12 Further engravings from the work of Jean-Martin Charcot.
Appearances of wrist drop and contracture (top left) and of claw hand or *main-en-griffe* (bottom left) are attributable to long-standing neurological disease. However, the mitral valve (right) displays a form of endocarditis, the appearances of which suggest that it is of rheumatic origin with a superimposed bacterial infection.

this was the view of Goldthwait[37] and of Nichols and Richardson.[56]

A dilemma which confronts modern investigators is this: Did rheumatoid arthritis exist before the 1850s? There are two views. One school[4,14] suggests that rheumatoid-like deformities are identifiable in the hands of the subjects of old Dutch master painters such as Rubens (1577–1640). Rheumatoid arthritis is therefore assumed to have existed long before its first clinical and pathological descriptions. Another view—and one to which I subscribe—is that rheumatoid arthritis and other connective tissue diseases, commonplace in the ageing populations of twentieth century Western society, were rare before 1860–1900 because of the very limited life-span that prevailed until modern times.

The expectation of male life in England and Wales was 39.9 years in 1838–54; 41.3 in 1871–80; 43.7 in 1881–90; and 44.1 in 1891–1900. The corresponding figures for females were 41.8; 44.6; 47.2; and 47.8 years.[59] These figures do not differ conspicuously from those calculated for seventeenth century France or eighteenth century Poland;[73] in turn, the early European death rates and calculated life expectancies resemble those derived for Early Bronze Age, Mesolithic and Upper Palaeolithic man and for *Homo sapiens*, *Homo neanderthalensis* and *Homo erectus*.[73] A propensity to rheumatoid arthritis may have existed for tens of millions of years but a virus or gene capable of causing the disease may only have been expressed when life was prolonged by the conquest of the great bacterial endemics and epidemics.

REFERENCES

1. Adams R. *Illustrations of the Effects of Rheumatic Gout or Chronic Rheumatic Arthritis on all the Articulations: with descriptive and explanatory statements*. London: John Churchill, 1857.

2. Adams M E, Billingham M E J. Animal models of degenerative joint disease. In: Berry C L, ed, *Current Topics in Pathology* (Volume 71). Bone and Joint Disease. Berlin: Springer-Verlag, 1982: 265–97.

3. Alarcon-Segovia D. Descriptions of therapeutic arthrocentesis and of synovial fluid in a Nahuatl text from pre-Hispanic Mexico. *Annals of the Rheumatic Diseases* 1980; **39**: 291–3.

4. Appelboom T, de Boelpaepe C, Ehrlich G E, Famaey J-P. Rubens and the question of antiquity of rheumatoid arthritis. *Journal of the American Medical Association* 1981; **245**: 483–6.

4a. Aschoff K A L. Zur Myocarditisfrage. *Verhandlungen der deutsche pathologischen Gesellschaft* (Berlin, Jena) 1904; **8**: 46–53.

5. Baillie M. *A Series of Engravings, Accompanied with Explanations, to Illustrate the Morbid Anatomy of the Human Body* (1st edition). London: G & W Nichol, 1799–1803.

6. Baillie M. *The Morbid Anatomy of Some of the Most Important Parts of the Human Body* (1st edition). London: J Johnson, 1793.

7. Baillou, Guillaume de (Ballonius). *Opuscula Medica, de Arthritide, de Calculo et Urinarum Hypostasi—Liber de Rheumatismo et Pleuritide Dorsale.* Paris: J Quesnel, 1643. Translated into English by C C Barnard. *British Journal of Rheumatology* 1940; **2**: 141–62.

8. Bannatyne G A. *Rheumatoid Arthritis: its Pathology, Morbid Anatomy and Treatment* (4th edition). Bristol: John Wright, 1906.

9. Bell W. Dissertatio medica inauguralis de arthritide. Edinburgh: 1823.

10. Bichat M F X. *Anatomie Générale, Appliquée à la Physiologie et à la Medicine.* Paris: Brosson, Gabon et Cie, an X, 1802.

11. Bichat M F X. *Anatomie Générale Précédée des Récherches Physiologiques sur la Vie et la Mort.* Paris: Brosson, Gabon et Cie, an VIII, 1800.

12. Boerhaave H. *Academical Lectures on the Theory of Physic.* London: N Innys, 1742–6.

13. Brodie B C. *Pathological and Surgical Observations on the Diseases of the Joints* (1st edition). London: Longman, Hurst, Rees, Orme and Brown, 1818.

14. Buchanan W W, Kean W F. Articular and systemic manifestations of rheumatoid arthritis. In: Scott J T, ed, *Copeman's Textbook of the Rheumatic Diseases* (6th edition). Edinburgh: Churchill Livingstone, 1986: 653–705.

15. Cameron G R. *Pathology of the Cell.* Edinburgh: Oliver and Boyd, 1952.

16. Cazenave P L A. *Des Principales Formés du Lupus et de son Traitement.* Gazette Hôpital (Paris), 3 sér. 1850; **2**: 393.

17. Celsus (AD 30). *De Medicina.* Translated by W G Spencer. Loeb Classical Library. London: Heinemann, 1935–8.

18. Charcot J-M. *Clinical Lectures on Senile and Chronic Diseases.* Translated by W S Tuke. Paris: A Delahaye, 1867.

19. Charcot J-M. *Les Difformés et les Malades dans l'Art.* Paris: Lecrosnier et Babé, 1889.

20. Columbo M R. *De re Anatomica Libris XV.* Venetiis: ex typog. N Bevilacque, 1559.

21. Connor B. An extract from a letter from Bernard Connor, MD, to Sir Charles Walgrave, published in French at Paris: Giving an account of an extraordinary human skeleton, whose vertebrae of the back, the ribs and several bones down to the Os Sacrum were all firmly united into one solid bone, without joynting or cartilage. *Philosophical Transactions of the Royal Society of London* 1695; **29**: 21–7.

22. Copeman W S C. *A Short History of the Gout and Rheumatic Diseases.* Berkeley and Los Angeles: University of California Press, 1964.

23. Cruveilier J. *Anatomie Pathologique de Corps Humain* (Tome 1, Livraisons I A XX). Paris: J-B Baillière, 1829–35.

24. Cullen W. *First Lines of the Practice of Physic for the Use of Students in the University of Edinburgh.* Ms. notes by the author (2nd edition). Edinburgh: William Creech, 1778.

25. Currey H L F. Animal models of arthritis. In: Scott J T, ed, *Copeman's Textbook of the Rheumatic Diseases* (6th edition). Edinburgh: Churchill Livingstone, 1986; 468–85.

26. Curzio C. An account of an extraordinary disease of the skin and its cure. *Philosophical Transactions of the Royal Society of London* 1754; **48**: 579–87.

27. David R. *Mysteries of the Mummies.* London: Cassell, 1978.

28. Doerr W. Jean Cruveilhier, Carl V. Rokitansky, Rudolph Virchow. Fundamente der Pathologie, Gedanken aus Anlass der hundersten Jährung von Rokitanskys Todestag. *Virchows Archiv A. Pathological Anatomy and Histology* 1978; **378**: 1–16.

29. Elliott-Smith G, Dawson W R. *Egyptian Mummies.* London: Allen & Unwin, 1924.

30. Fernel J. *Universa Medicina.* Paris: A Wechel, 1554.

31. Friedberger E. Über aseptisch erregte Gelenkschwellungen beim Kaninchen. *Berliner klinische Wochenschrift* 1913; **50**: 88.

32. Galen of Pergamon. Works, translated by Francis Adams. London: 1844–47.

33. Gardner D L. The experimental production of arthritis: a review. *Annals of the Rheumatic Diseases* 1960; **19**: 297–317.

34. Gardner D L. *Pathology of the Connective Tissue Diseases.* London: Edward Arnold, 1965: 341–62.

35. Garrod A B. *The Nature and Treatment of Gout and Rheumatic Gout.* London: Walton and Maberly, 1859.

36. Garrod A E. Rheumatoid arthritis, osteo-arthritis, arthritis deformans. In: Albutt T C, Rolleston H D, eds, *A System of Medicine* (Volume 3). London: Macmillan, 1907: 3–43.

37. Goldthwait J E. The differential diagnosis and treatment of the so-called rheumatoid disease. *Boston Medical and Surgical Journal* 1904; **151**: 529–34.

38. Goodsir J, Goodsir H D S. *Anatomical and Pathological Observations.* Edinburgh: Myles Macphail, 1845.

39. Harvey W. *Exercitatio Anatomica de Motu Cordis et Sanguinis in Animalibus*, 1628. Translation by G Whitteridge. Oxford: Blackwell, 1976.

40. Hippocrates, *The Genuine Works* of. Translated from the Greek, with a preliminary discourse and annotations by Francis Adams. London: Sydenham Society, 1849. Volume I.

41. Hooke R. *Micrographia*, or *Some Physiological Descriptions of Minute Bodies made by Magnifying Glasses; with Observations and Enquiries Thereupon.* London: J Martyn and J Allestry. (Facsimile reprint, Oxford 1938: *Early Science in Oxford.* R T Gunther, Volume 13. *The Life and Work of Robert Hooke* (Part V). Micrographia, 1665).

42. Hunter J. Manuscript 54 (in the possession of the Royal College of Surgeons of England), 1759.

43. Kaposi M K. Neue Beiträge zur Kennknis des Lupus erythematosus. *Archiv für Dermatologie und Syphilis* (Prague) 1872a; **4**: 36–78.

44. Kaposi M. Idiopathisches multiple ligmentsarkom der Haut. *Archiv für Dermatologie und Syphilis* (Prague) 1872b; **4**: 265–73.

45. Klemperer P. The significance of the intermediate substances of the connective tissue in human disease. *The Harvey Lectures*, Series LXIX. New York: Academic Press, 1955: 100–23.

46. Klemperer P, Pollack A D, Baehr G. Diffuse collagen disease; acute disseminated lupus erythematosus and diffuse scleroderma. *Journal of the American Medical Association* 1942; **119**: 331–2.

47. Klinge F. *Der Rheumatismus; pathologisch anatomische und experimentell-pathologische Tatsachen und ihre Auswertung für arztliche Rheumaproblem.* Munich: Bergmann, 1933.

48. Klinge F. Die rheumatischen Erkrankungen der Knochen und Gelenke und der Rheumatismus. In: Lubarsch O and Henke

F, eds, *Handbuch der speziellen pathologischen Anatomie und Histologie*. Berlin: Springer Verlag, 1934; **9**: 107–251.

49. Küssmaul A, Maier R. Über eine bisher noch nicht beschriebene eigentumliche Arterienerkrankung (Periarteritis nodosa), die mit Morbus Brightii und rapid fortschreitender allgemeiner Muskellahmung einhergeht. *Deutsche Archivs für Klinische Medizin* 1866; **1**: 484–518.

50. Lister J J. On properties in achromatic object-glasses applicable to the improvement of the microscope. *Philosophical Transactions of the Royal Society of London* 1830; **120**: 127–200.

51. Minor R R. Collagen metabolism: a comparison of diseases of collagen and diseases affecting collagen. *American Journal of Pathology* 1980; **98**: 225–80.

52. Monro A, (Secundus). Essays and heads of lectures . . . Edinburgh: 1840.

53. Morgagni G B. *De Sedibus et Causis Morborum per Anatomen Indagatis Libri Quinque*. Venice: typog. Remondiniana, 1761.

54. Morse D. Tuberculosis. In: Brothwell D, Sandison A T, eds, *Diseases in Antiquity*. Springfield, Illinois: Charles C Thomas, 1967.

55. Neumann E. Die Picrocarminfarbung und ihre Anwendung auf die Entzundungslehre. *Archiv für Mikroskopische Anatomie* 1880; **18**: 130–50.

56. Nichols E H, Richardson F L. Arthritis deformans. *Journal of Medical Research* 1909; **21**: 149–222.

57. Pott P. *Remarks on that Kind of Palsy of the Lower Limbs, which is Frequently Found to Accompany a Curvature of the Spine*. London: J Johnson, 1779.

58. Redfern P. *Anormal Nutrition in Articular Cartilages*. Edinburgh: Myles Macphail, 1850.

59. Registrar General. *Supplement to the Sixty-fifth Annual Report of the Registrar General of Births, Deaths and Marriages in England and Wales, 1891–1900*. Part I. London: His Majesty's Stationery Office, 1907.

60. Rogers J, Watt I, Dieppe P. Arthritis in Saxon and Medieval skeletons. *British Medical Journal* 1981; **283**: 1668–70.

61. Schleiden M J. Beiträge zur Phytogenesis. *Archivs der Anatomie, der Physiologie und der Wissenschaftliche Medizin*. (Berlin) 1838; **2**: 137–74.

62. Schwann T. *Mikroscopische Untersuchungen über die Übereinstimmung unter Struktur und dem Wachstum der Tiere und Pflanzen*. Berlin: Sander, 1839.

63. Scott J T. The rheumatic diseases: historical. In: Scott J T, ed, *Copeman's Textbook of the Rheumatic Diseases* (6th edition). Edinburgh: Churchill Livingstone, 1986: 5–18.

64. Sokoloff L. Animal models of rheumatoid arthritis. In: Richter G W, Epstein M A, eds, *International Review of Experimental Pathology* Volume 26. New York and London: Academic Press, 1984: 107–45.

65. Stead I M, Bourke J B, Brothwell D. *Lindow Man. The Body in the Bog*. London: British Museum Publications, 1986.

66. Sydenham T. *Tractatus de Podagra et Hydrope*. London: G Kettilby, 1683.

67. Sydenham T. *Of a Rheumatism: The Whole Works of that Excellent Practical Physician, Dr Thomas Sydenham*. Translated from the Latin by John Pechey (7th edition). London: M Wellington, 1701.

68. Trotter A. . . . *De Rheumatismo*. Edinburgh: 1821.

69. Vesalius A. *De Humani Corporis Fabrica*. Basle: J Oporini, 1543.

70. Virchow R. *Die Cellularpathologie in ihren Begründung auf physiologische und pathologische Gewebelehre*. Berlin: A. Hirschwald, 1858. Translated from the second (1859) German edition by Frank Chance. New York: Dover Publications, 1971.

71. Wells C. The human burials. In: McWhirr A, Viner L, Wells C, eds, *Romano-British Cemeteries at Cirencester*. Cirencester Excavation Committee, 1982: 135–201.

72. Wollaston W H. On gouty and urinary concretions. *Philosophical Transactions of the Royal Society of London* 1797; **87**: 386–400.

73. Young J Z. *An Introduction to the Study of Man*. Oxford: Oxford University Press, 1971.

Part I

Biology of Connective Tissue Disease

Chapter 1
Cells and Matrix

Connective tissue is a multiphase, supporting, interstitial material present throughout mammalian organs and parts. Connective tissue comprises a stroma, distinct from, but inextricably related to, the functioning parenchyma of the viscera; it sustains the organs and parts physically[191] and chemically. Some connective tissues, particularly bone and dentine, are mineralized. With certain exceptions, however, disorders of the bones and teeth are not described in this text since they are analysed extensively in many other contemporary volumes.

Diseases arising in or significantly affecting the connective tissues are, to varying degrees, inherited or acquired. Connective tissue disorders may be anatomically focal, localized or systemic. There is a wide range of primary benign and malignant connective tissue neoplasms but many recent accounts describe the 'soft' and 'hard' tissue tumours and so they are not described in the present work.

Connective tissues often receive a substantial arterial blood supply. These blood vessels are themselves frequent sites for the pathological processes underlying connective tissue disease and they have, moreover, a substantial microskeleton of their own which is also prone to disease. In this text, however, descriptions of vascular abnormalities have been confined to those of immunological and chemical origin and there is no attempt to survey comprehensively the degenerative, infective, traumatic and neoplastic vascular diseases.

Connective tissue is said to be 'loose' or 'compact' (Chapter 2). Some organs are composed predominantly of specialized forms of connective tissue such as cartilage, tendon, synovia, ligament and fat; the majority of these organs have an immediate, mechanical, supportive function. In the viscera and in other parts of the body, such as the skin and skeletal muscle, the connective tissues offer a secondary, structural framework. The phrase 'connective tissue system' can be applied to the aggregate of both organizations.

The connective tissue system comprises cells (Table 1.1), intercellular fibres, intervening matrix materials and the interstitial fluid. More than 70 per cent of the extracellular matrix of the majority of the non-mineralized connective tissues is water. There are wide variations: the nucleus pulposus of the intervertebral disc, Wharton's jelly of the umbilical cord, and the vitreous humour of the eye, are highly hydrated. By contrast, the annulus fibrosus of the intervertebral disc, the tendons, ligaments, labra and menisci have a relatively low water content.

Connective tissue cells

Two groups of cells are found within the mature connective tissues.[4,34,48] The first group is intrinsic (Table 1.1) and includes fibrocytes, fibroblasts, osteocytes, osteoblasts, chondrocytes, chondroblasts, fat cells, type B synoviocytes, mast cells and histiocytes. The majority of the intrinsic cells of the connective tissues are derived from embryonic mesenchyme; they are pluripotential and inherit the genetic information necessary for the differentiation of their progeny to more specialized fibrocytes, chondrocytes, osteocytes and lipocytes. The second group of cells is extrinsic: these cells make their way into the less dense

Table 1.1

The intrinsic connective tissue cells

Mature cell	Precursor	Structural characteristics	Examples of properties
Fibrocyte	Fibroblast	Elongated ovoid cells; much cytoplasm in young cells, less in more mature; rough endoplasmic reticulum; free ribosomes; Golgi apparatus; vesiculate nucleus	Manufacture and export of collagen, proteoglycan and (occasionally) elastic material; repair by granulation and scar tissue formation
Chondrocyte	Chondroblast	Ovoid cells, often in groups of two, three or four, lying within chondrons that have much specialized pericellular proteoglycan and are bounded by a delicate shell of type VI and IX collagens. Cells contain cytoplasmic lipid and glycogen; moderate amount of rough and smooth endoplasmic reticulum; Golgi apparatus; ovoid or round nucleus	Manufacture and export of fibrillar type II, 'minor' types VI, IX, X (at sites of endochondral ossification) and type XI collagens, proteoglycan, glycoproteins and occasionally elastic material (in fibrocartilage and menisci)
Osteocyte	Osteoblast	Elongated ovoid cells with deeply basophilic cytoplasm containing much rough endoplasmic reticulum and many free ribosomes; slender vesiculate nucleus; cytoplasmic processes extend within canaliculi; cell death identified histologically when 'lacunae' of osteons are empty	Synthesis of type I collagen and of proteoglycan; regulation of ossification and mineralization by acid phosphatase; osteocytes lie within lacunae, communicating with the Haversian vascular system via canaliculi
Osteoclast	Pre-osteoclast	Large multinucleate cells with up to 100 nuclei; many infoldings of plasma membranes; numerous large mitochondria, cytoplasmic granules and crystals; sparse endoplasmic reticulum; some free ribosomes	Lacunar bone reabsorption in health and disease; lysosomal enzyme, e.g. tartrate resistant acid phosphatase and protease activity
A(M) synovial cells (macrophage-like)	Synovioblast	Villous cell surface; cytoplasmic vacuoles; dense bodies and lysosomes	Phagocytosis; intracellular digestion by provision of lysosomal enzymes
B(F) synovial cell (fibroblast-like)	Synovioblast	Much endoplasmic reticulum; Golgi apparatus; devoid of surface villi	Synthesis and secretion of hyaluronate-protein of synovial fluid
C Intermediate synovial cell (disputed)	Synovioblast		
Histiocyte	Mesenchymal precursor of loose connective tissue	Large ovoid or polygonal cells with reniform central nucleus; many small cytoplasmic vesicles, lysosomes, dense granules and ingested material within phagosomes	Limited motility; phagocytic; numerous lysosomes with nucleases, proteases, lipases and saccharidases
Fat cell	Lipoblast	Ovoid nucleus displaced to cell margin by cluster of lipid droplets that fuse to form a single inclusion	Support of viscera; storage; and provision of energy

Table 1.1 – continued

The intrinsic connective tissue cells

Mature cell	Precursor	Structural characteristics	Examples of properties
Brown fat cell		Numerous granular cytoplasmic lipoprotein particles; many mitochondria	Analogue of brown fat cell of hibernating animals; responsive to catecholamines that provoke liberation of energy as heat, a mechanism mediated by cyclic adenosine monophosphate
Dendritic macrophage	Histioblast	Long dendritic processes	Binding of antigen in germinal centres of lymph nodes
Mast cell of connective tissue	Mastoblast	Very large ovoid cells; many prominent membrane-bounded cytoplasmic granules; small numbers of mitochondria; little endoplasmic reticulum	Manufacture and storage of heparin, 5-hydroxytryptamine and histamine; binding of IgE antibody to plasma membrane. Contact with specific antigen degranulates cell, releasing chemical mediators of anaphylaxis and inflammation
Pulmonary alveolar septal cell	Endodermal cell	Located on basement membrane of alveolus; cubical, with short microvilli; central round nucleus; inclusion bodies of microlamellar character (cytosomes); many mitochondria	Secretes lipids of alveolar lining (surfactant)
Glomerular mesangial cell	Mesodermal cell	Stellate, with cytoplasmic processes that extend to capillary margins; resembles pericyte or smooth muscle cell	Structural support for capillary tuft of glomerulus; manufacture of mesangial matrix; phagocytosis of foreign material
Ovarian theca externa	Mesenchymal cell	Elongated, fibroblast-like; central ovoid nucleus, free ribosomes, some rough endoplasmic reticulum, lysosomes and many vesicles	Structural support; steroid synthesis
Mesenchymal cell of umbilical cord	Mesenchymal cell	Stellate or spindle-shaped; central oval nucleus; much rough endoplasmic reticulum; clusters of free ribosomes	Manufacture and export of highly hydrated connective tissue matrix of the umbilical cord (Wharton's jelly)
Mesenchymal cell of chorionic villi	Mesenchymal cell	Stellate; round central nucleus; rough endoplasmic reticulum; free ribosomes, glycogen granules, lysosomes and cytoplasmic filaments; occasional lipid droplets	Manufacture and export of loose highly hydrated connective tissue matrix of the maturing villus
Astrocyte	Ectodermal spongioblast	*Protoplasmic*: crenated plasma membranes; irregular shape with cytoplasmic processes; round or oval nucleus; dispersed chromatin; small Golgi apparatus and endoplasmic reticulum; some filaments, mitochondria and glycogen	Located in grey matter; mechanical support of neurones; regulation of passage of extracellular fluid in grey matter
		Fibrous: smooth cell surface; extensive cytoplasmic filaments; crenated central nucleus	Located in white matter; support structure for nerve tissue; involved in healing processes

Table 1.1 – continued

The intrinsic connective tissue cells

Mature cell	Precursor	Structural characteristics	Examples of properties
Oligodendrocyte	Spongioblast	Fine branching cytoplasmic processes. Contains many free ribosomes and numerous microtubules. Three classes: light, medium and dark	Manufacture of myelin membrane lipoprotein; modulation of myelin sheath formation with control of its integrity
Pituicyte		Resembles neuroglial cell; cytoplasmic processes extend from the cells between cylinders of neurosecretory material in the palisade zone of pars posterior of pituitary	Support for structure of pars posterior of pituitary
Pineal interstitial cell	Uncertain	Crenated nucleus; sparse endoplasmic reticulum; free ribosomes, oval mitochondria, cytoplasmic filaments, occasional nerve fibres in cytoplasm	Endings on pinealocytes, for which they provide support; may contain much glycogen
Muller cell of retina (retinal gliocytes)		Irregular shape and very long, round or oval nucleus; long branching processes, cytoplasmic filaments, some rough endoplasmic reticulum; clusters of ribosomes; large glycogen granules; lysosomes	Extend between rod and cone cells of retina; function comparable to astrocytes

connective tissues in the transport or release of metabolites or when an inflammatory reaction occurs. The extrinsic cells include polymorphonuclear neutrophil granulocytes (polymorphs) and eosinophil granulocytes, monocytes (mononuclear phagocytes—macrophages), lymphocytes and plasma cells. Because osteoclasts (see p. 30) and synovial A cells (see p. 32) are believed to be derived from the bone marrow, they may be regarded as extrinsic.

The cells of the connective tissues are functional units of health and disease.[20] In normal connective tissues, the cells are programmed to synthesize and secrete every form of fibrous and non-fibrous connective tissue macromolecule. In practice, differentiated normal connective tissue cells preferentially export substantial quantities of a limited repertoire of molecular species. Thus the mature chondrocyte of adult hyaline articulate cartilage tends to synthesize predominantly type II collagen and chondroitin sulphate proteoglycan. By contrast, the chondrocyte of epiglottic elastic cartilage synthesizes not only collagen and proteoglycan but also elastin, elastic microfibrillar glycoprotein and a protein that acts as a nidus for senile calcification. When environmental conditions change, different genetic programmes are derepressed and the connective tissue cell exports a different series of macromolecules, assuming a different morphology.

Fibroblasts

Fibroblast properties

The prototype of the differentiated mammalian connective tissue cell is the fibroblast or mechanocyte.[2,34,198] The principle functions of these cells, which can be found in almost all tissues, are the synthesis, secretion and modulation of the fibrous and non-fibrous connective tissue proteins; and, through these molecules, the provision of the essential physical attributes and biological properties of the connective tissues. When the cytoskeleton includes much contractile protein, the term 'myofibroblast' (see p. 40) is used.

The fibroblast is readily grown in monolayer culture[2] and can therefore be harvested in abundance. Consequently, the genetics, structure, function, synthetic and secretory characteristics of this cell are well understood. The fibroblast is the precursor of the mature fibrocyte; the young fibroblast is larger and more active and is seen at its most resourceful dividing vigorously in granulation tissue (see Chapter 8). The smaller, more compact, fibrocyte constitutes the cell population of all non-mineralized compact connective tissues and is abundant in loose- and areolar tissues. Connective tissue formed predominantly of fibrocytes and collagen is fibrous tissue: the collagen arrays are orderly by comparison with the less uniform arrangement of collagen in abnormal materials such as those of scars (see p. 300), kelids (see p. 311) and fibromatous masses (see p. 313). Fibrosis is the excessive or inappropriate formation of fibrous tissue. Fibrous tissue and fibrosis are discussed more fully in Chapter 8.

Fibroblast morphology

Fibroblasts in monolayer culture on a plane surface have an elongated, flattened, ovoid or triangular shape: they are approximately 50–100 μm long, 30 μm wide and 3 μm thick (Fig. 1.1a). An isolated, cultured fibroblast retains its shape more constantly and changes its form more slowly than a motile phagocyte such as a mononuclear macrophage. The shape of the fibroblast is maintained by adhesion to the

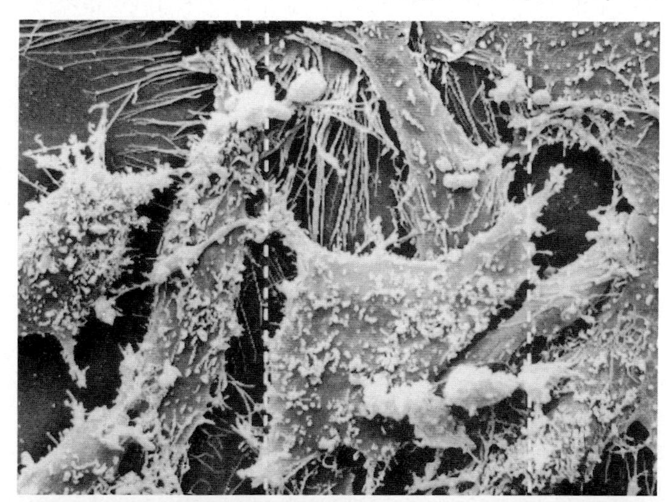

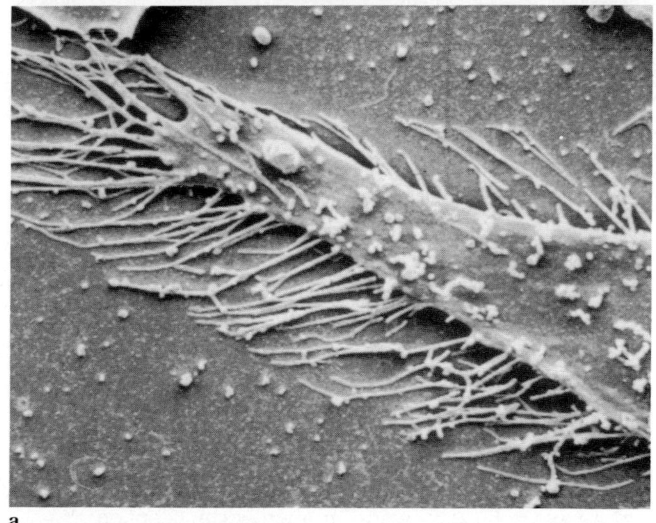

a

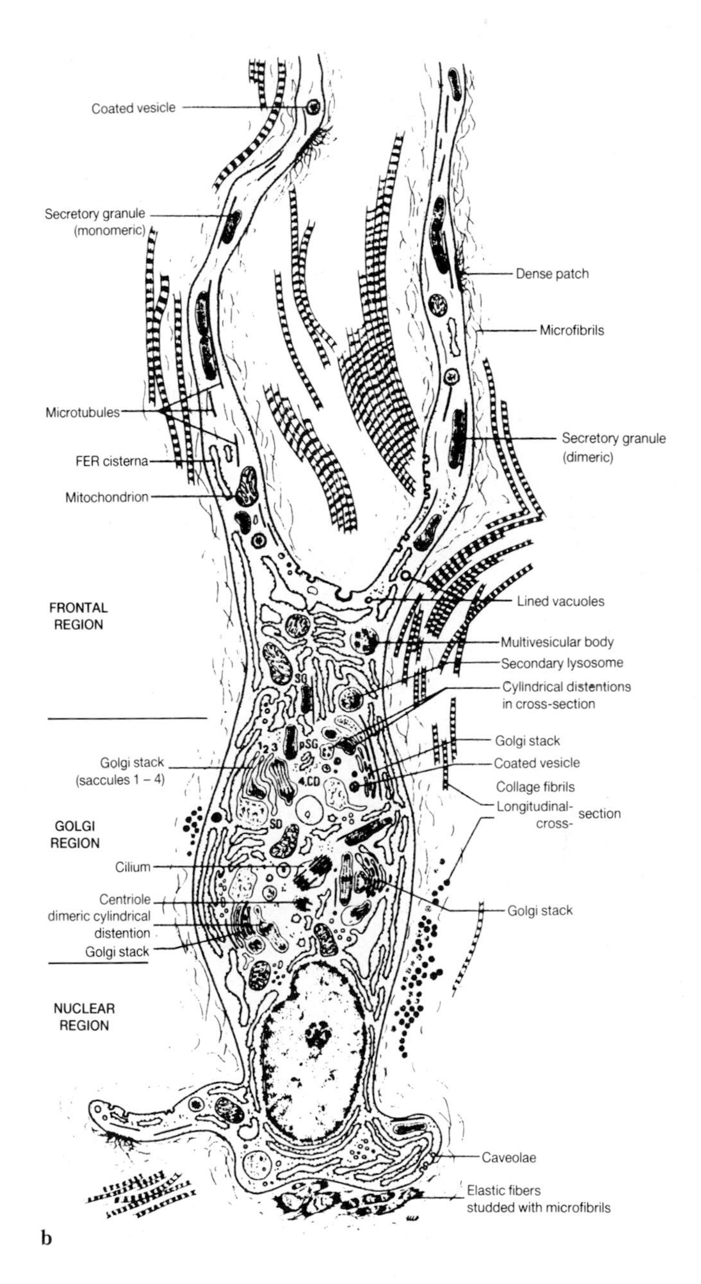

b

Fig. 1.1 Fibroblast.
a. Synovial fibroblasts (type B synoviocytes). Synovial fibroblasts grown in monolayer culture on glass. Note (at top) long, slender cytoplasmic processes, imaged in a dehydrated specimen *in vacuo,* attaching cells to glass surface. For comparison (at bottom), detail of cytoplasmic processes extending from a single cell. Scanning electron microscopy of material, fixed in glutaraldehyde, post-fixed in osmium tetroxide, vacuum dried from propylene oxide. (Top: × 680; bottom: × 1370.) **b.** Diagram of fibroblast-like cell from 20-day-old rat footpad. There are three regions: nuclear, Golgi and frontal. In the nuclear region, note the rough endoplasmic reticulum. In the Golgi region, observe the four stacks of saccules, numbered 1 to 4. The frontal region contains coated vesicles, secretory granules, a mitochondrion, microfibrils and a lysosome. (From Leblond and Laurie,[98] reproduced by courtesy of Professor C P Leblond FRS and S Karger AG.)

substrate on which it is growing.[2] There is no rigid internal skeleton. When the surface attachments are released, the fibroblast, which is contractile, withdraws into an ovoid form. The flattened shape seen when the cells grow in monolayers is very different from the irregular ellipsoidal shape assumed when the cells multiply in a three-dimensional lattice. Under normal, *in vivo* conditions, fibroblasts, in contrast to neighbouring cells, are often arranged in orderly columns or groups separated by increasing amounts of collagen (see pp. 42 and 174) and proteoglycan (see p. 53). The smooth, concave borders of cells in contact in monolayers are the resultants of intercellular forces that are even more complex in compact organized fibrous tissue; they are one expression of an active, intracellular actin-centred system.

In two-dimensional tissue sections the fibroblast is seen to have much cytoplasm. In stained preparations ribosomes, polyribosomes and rough endoplasmic reticulum combine to give a faint basophilia. Mitochondria are numerous and lysosomes are uncommon. Within the cytoplasm are specialized bundles of microfibrils; among them are proteins similar to but distinct from the actin, myosin, tropomyosin, α-actinin and M-protein of striated muscle. There are at least three kinds of fibroblast actinin as well as other proteins such as filamin, not present in myofibrils. The contractile intracellular bundles are transitory: their ends are inserted into the cell surface and at these sites, the plasma membrane forms localized, tight adhesions to the material on which the cell is growing. Similar structures, plaques and attachments appear at zones of cell-to-cell contact. The contracting fibril bundles produce extensive tissue deformation when the substrate on which the cell multiplies is a flexible material like a collagen fibre honeycombe or a mass of fibrin, a phenomenon which strongly influences the maturation and contraction of healing wounds.

Fibroblast adhesion

There is limited understanding of the manner in which fibroblasts interact with the pericellular matrix *in vivo*.[106] In culture systems, fibroblasts adhere to surfaces at adhesion plaques. At the points where adhesion occurs, numerous actin cables are located at the plasma membrane. The actin-binding proteins vinculin and actinin are sited at this locus. The extracellular adhesive glycoprotein fibronectin is concentrated nearby. Heparin sulphate proteoglycan may unite cytoplasmic actin filaments with extracellular collagen, fibronectin and basement membrane laminin. Small chondroitin sulphate proteoglycans may be implicated. That fibronectin (see p. 207) binds to collagen and proteoglycans as well as to the cell surface *in vitro* suggests that this ubiquitous glycoprotein is central to the processes of fibroblast surface adhesion and movement as well as to the control and modulation of cell/matrix interactions *in vivo*. Extracellular adhesive proteins such as fibronectin, chondronectin and anchorin contain a specific tripeptide sequence, Arg-Gly-Asp (R-G-D) recognized by homologous cell surface receptors. The plasma membranes of fibroblasts bear receptors that are alpha–beta heterodimers mediating interactions between the extracellular matrix adhesive glycoproteins and intracellular cytoskeletal proteins such as actin. Other closely related receptors include the integrins (pp. 52 and 208).

Fibroblast migration

Evidence from embryonic tissues shows that fibroblasts actively compress collagen fibre sheets and stretch them into microtendinous cables. Attached to adjacent cells, fibroblasts draw the cells into appropriate positions in the developing organs and tissues.

Fibroblasts move across surfaces like gastropods. Extended cytoplasmic processes, 'lamellipodia', adhere to the surface. The contractile filaments within the cell are activated and the cell body is pulled to a new position. *In vivo*, comparable movements occur as fibroblasts migrate from one connective tissue site to another. Migration may be directed by chemotaxis[138] and many chemoattractants have been demonstrated. They include a lymphocyte-derived chemotactic factor (LDCF-F); a serum-derived factor; collagens, collagen chains and Hyp* peptides; fibronectin; tropoelastin and elastin peptides; and leukotriene B_4. Extracts of macrophages activate a latent T-cell LDCF-F which may be released when a cell-mediated immune response is provoked. Complement (C) activation also liberates a chemotactic product, released from C5 (see p. 237). Curiously C5a is not chemotactic for fibroblasts and C5 is not chemotactic for polymorphs and mononuclear macrophages. Fibroblasts may also be attracted chemotactically by platelet-derived growth factor and transforming growth factor β, and by factors originating in transformed or neoplastic cells. Serum contains a non-cytotoxic protein of approximately 210 kD mol. wt which inhibits the response of fibroblasts to all chemoattractants.

Fibroblast growth

At least two signal sequences are required for fibroblast growth.[137] Competence factors render cells in the G_0 or G_1 phase of the cell cycle capable of growth. Progression factors stimulate DNA synthesis in those fibroblasts that are able to respond (Tables 1.2 and 1.3). The transmission

* Amino acids and sugars are abbreviated in accordance with the recommendations of the Biochemical Society.

Table 1.2

Some factors regulating fibroblast growth (Postlethwaite and Kang[137])

Factors
Competence factors
Platelet-derived growth factor
Fibroblast growth factors
Calcium-containing crystals
Calcium phosphate precipitates
Progression factors
Multiplication-stimulating activity
Somatomedins A and C
Insulin and insulin-like growth factors
Epidermal growth factor
Alveolar macrophage-derived growth factor
Other factors
Interleukin-1α and -1β
Transforming growth factor-β
T-cell-derived fibroblast growth factor
Schistosomal granuloma macrophage-derived growth factor
Vanadate

Table 1.3

Regulation of fibroblast functions by immune reactants and cytokines (Modified from Postlethwaite and Kang[137])

Collagen synthesis	*Collagenase activation*
100–170 kD mol. wt lymphokine	Interleukin-1α and -1β
Transforming growth factor β	
Interleukin-1α and β	
Interferon-γ (inhibits)	
Hyaluronate synthesis	*Multiplication*
60 kD mol wt lymphokine	40–60 kD mol. wt lymphokines
Interleukin-1α and -1β	Interleukin-1α and β
Transforming growth factor β	Transforming growth factor β
Basic fibroblast growth factor	Platelet-derived growth factor
	Interferon-γ (and inhibits)
	Prostaglandin E_2 (inhibits)
Chemotaxis	
22 kD mol. wt lymphokine (lymphocyte-derived chemotactic factor) (LDCF)	Platelet-derived growth factor
	Macrophage-derived fibronectin
Transforming growth factor β	

of mitogenic signals to fibroblasts implies the presence of receptors on the cell surface. Some, such as the receptor for platelet-derived growth factor have been cloned; there is a resemblance to an oncogene product and to the macrophage stimulating factor receptor.

Phosphatidylinositol turnover, cytoskeletal rearrangement and the expression of proto-oncogenes such as *c-myc* and *c-fos* are among the effects of the stimulation of fibroblasts by growth factors.[137] The relationship between these stimuli and mitosis has excited much interest. The introduction of the word 'rheumagene' in the context both of autoimmune connective tissue disease and of connective tissue diseases characterized by mesenchymal cell activation and hyperplasia indicates the direction in which current views are moving.[159]

Fibroblast responsiveness and secretion

Elongated and spindle-shaped in section, flat and stellate in monolayer culture, fibroblasts display a structural orientation clearly related to the property of matrix synthesis and secretion. The macromolecules manufactured and exported by fibroblasts are principally collagen types I and III, elastin, elastin microfibrillar glycoprotein and the proteoglycans; they are described in Chapter 4.

The ultrastructure of fibroblasts engaged in macromolecular synthesis and export is well understood. Immunoelectron microscopic techniques have located the intracellular sites of macromolecular synthesis and radioactive isotopic methods have defined the sequence in which the components of these molecules are formed. Similar sequences probably take place in the other intrinsic connective tissue cells. There are three ultrastructural regions: nuclear, Golgi and frontal.[98] However, serial transmission electron microscopic studies at low temperature suggest that the spatial relationships of these regions are less static than the examination of fixed tissues might suggest. Rough endoplasmic reticulum cisternae and organelles, particularly mitochondria and lysosomes, are present in each region. The Golgi region is of special importance in terms of matrix macromolecular synthesis, the lysosomes in terms of matrix degradation.

The sequence of intracellular, ultrastructural changes that occurs during macromolecular synthesis can be considered in terms of fibrillar collagen (see p. 44). The biosynthesis of collagen is described on p. 174. The Golgi apparatus is a series of stacks of saccules (Fig. 1.1b). The peripheral saccules, located on the forming ('*cis*') side of the Golgi saccules, lie next to cisternae of the rough endoplasmic reticulum. The medial aspect of the Golgi saccules (the 'mature' or '*trans*' side) adjoins a pair of centrioles. Each of four or five Golgi saccules has a distended spherical margin 400–500 nm in diameter: within these distensions, the sequences of collagen synthesis can be followed in serial transmission electron microscopic section and by isotopic labelling. Collagen pro-α-chains are formed in the rough endoplasmic reticulum: the hydroxylation of Pro and Lys is complete after a few minutes. The hydroxylated chains pass to the *cis* aspect of the Golgi stacks. Within 20 min, pro-α-chains, associating as procollagen helices, develop in the spherical distensions of the *cis*-saccules. Some 20–40 min later, parallel procollagen molecules, seen as threads, appear in cylindrical distensions of the Golgi saccules. The distensions are quickly released from the *trans*-Golgi saccules: they form intracellular, prosecretory granules which mature as secretory granules. These granules migrate along the paths of microtubules, reach and fuse with the plasmalemma and escape, as procollagen, into the extracellular tissue space. The processes of propeptide cleavage, fibrillogenesis, cross-linking and fibril maturation are described in Chapter 4. The extracellular assembly of collagen fibres is mentioned on p. 46 and described on p. 302.

Analogous processes take place in proteoglycan synthesis (see p. 200) and the cellular phenomena that occur during the secretion of these macromolecules, by chondrocytes, are summarized on p. 27. Core protein molecules are synthesized within the rough endoplasmic reticulum. Passing to the *cis*-Golgi saccules, sugar residues are added to form the glycosaminoglycan chains. Sulphation occurs and the entire macromolecule moves to the cell surface before secretion.

Less is known of the intracellular mechanisms by which fibroblasts synthesize and secrete elastin (see p. 191). The microfibrils of elastic microfibrillar glycoprotein originate in a manner analogous to collagen. The unique molecules of elastin itself are formed independently and aggregate in relation to elastin-glycoprotein in the extracellular matrix (see p. 47).

Regulation of fibroblast behaviour There is considerable evidence that disordered immune responses (p. 251) influence fibroblast behaviour in progressive systemic sclerosis (see p. 706) and the pathology of many infectious diseases is dominated by fibrosis. Postlethwaite and Kang[137] provide a summary of the properties of fibroblasts that are modulated by immune reactants and cytokines (Table 1.3). Among the cytokines (p. 291) that influence fibroblast behaviour are interferon-γ (1FN-γ), transforming growth factor beta, a collagen production factor and a chemotactic lymphokine. Interleukin-1 (Il-1) modulates fibroblast behaviour, and the fibroblast growth factors (FGF),[8a] particularly basic FGF (bFGF) (p. 296), are particularly important in relation to repair (p. 295) and the formation of granulation tissue (p. 301).

Chondrocytes

The mature chondrocyte[172–174] is seen in typical form in hyaline articular cartilage (see p. 68). Chondrocytes are distinguished from fibroblasts not so much by their ultrastructural morphology as by their secretion of abundant proteoglycan and of varying proportions of type II, VI, IX and XI collagen, structural and enzymic proteins and glycoproteins (see Chapter 4). At sites of endochondral ossification, type X collagen is a unique product. The formation of each of these macromolecular species is independently programmed, their synthesis and secretion being regulated by feedback control. When the quantity of type I collagen synthesized is particularly great, the mature tissue is fibrocartilage (see p. 82). In elastic cartilage (see p. 84), the secretion of both elastin and elastic fibrillar glycoprotein is characteristic.

Chondrocyte morphology

Depending on their functional state, age and microenvironment, chondrocytes are flat and ellipsoidal; ovoid; or spheroidal (Fig. 1.2a). Chondrocytes are larger in embryonic than in infantile cartilage, in younger than in older tissue, and in growing than in mature parts. In *non-articular cartilages* such as those of the bronchi, chondrocytes are relatively uniform in structure and size. When the cells of chondrosarcoma are excluded, the largest cells of non-articular cartilage are those of the hypertrophic zones of the epiphyseal growth plates. (Normal bone growth by endochrondral ossification is not described in this text but endochondral ossification occurs in abnormal sites, for example, in osteophytosis, p. 866). In *hyaline articular cartilage*, the size, morphology and ultrastructure of chondrocytes vary greatly between the five designated zones (p. 69); the variety is as great as that of the zonal extracellular matrix composition and mechanical attributes. Although the ovoid, mature, adult human chondrocytes in, for example, zone III, have gently contoured, crenated or scalloped profiles, features that are strongly influenced by methods of preparation, the much flatter cells of zone I extend long cytoplasmic processes between the orderly laminae of the collagen meshwork. It is believed that a single cilium is a normal feature of each adult articular chondrocyte but cilia are also common in immature cells.

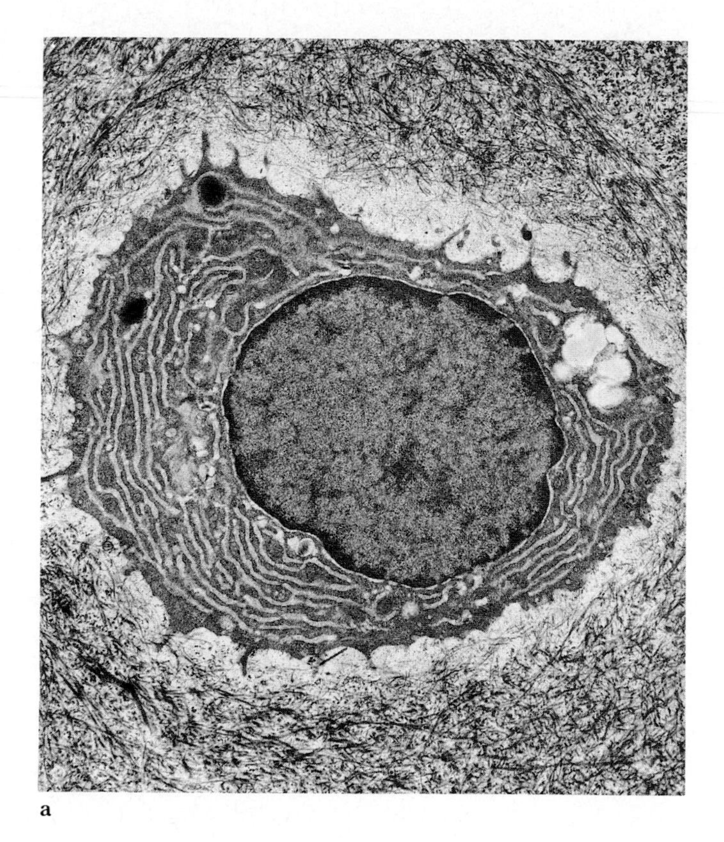

a

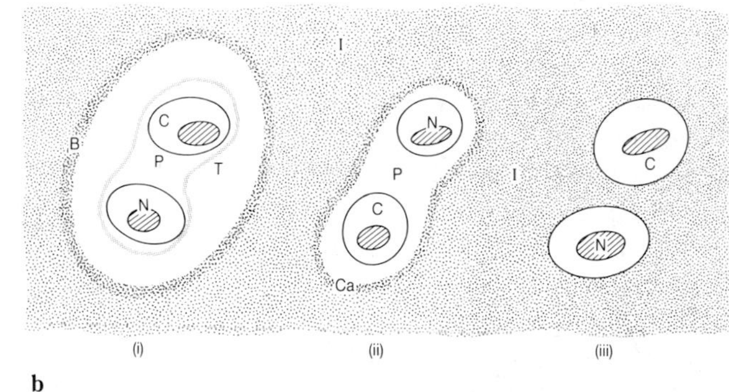

b

Fig. 1.2 Chondrocytes from articular cartilage of rat.
a. Mid-zone (zone III) chondrocyte lying within limited pericellular matrix in which are many small collagen fibrils with cross-striations. The nucleus is almost spherical. Cisternae of the endoplasmic reticulum, in continuity with a perinuclear electron-lucent space, are distended. The cytoplasm contains two dense, residual bodies and the plasma membrane extends short processes into the pericellular matrix. (TEM × 7150) **b.** Chondrocytes and chondrons: the cell–extracellular matrix relationship. The appearances of the chondrocyte and of the nearby extracellular matrix vary according to the technique used in their preparation. (i) Preparations made for light microscopy without precautions to retain proteoglycans *in situ*, reveal artefactual and misleading appearances. They are illustrated by Szirmai.[176a] Metachromatic and cationic dyes tend to reveal a 'specialized', pale-appearing, pericellular matrix, and a territory that is demarcated from the remainder of the 'interterritorial' matrix by an ill-defined, more deeply stained boundary or border. (ii) Preparations made by conventional electron microscopic techniques, without the planned conservation *in situ* of proteoglycans, emphasize the presence of an electron-dense 'capsule' of fibrillar collagen that may be type VI.[134a, 136] An analogous structure is seen when cartilage, fixed without proteoglycan-binding agents, is stained by acridine orange and examined by scanning optical microscopy using light of 488 nm wavelength. (iii) Preparations in which proteoglycan-binding agents such as ruthenium red[43a, 83b] are incorporated with the fixative, do not show the distinct, pericellular, pale zones described in (i) (above). A similar structure is revealed when low temperatures are employed to conserve matrix macromolecules during fixation, embedding and sectioning.[46b] These appearances closely resemble those recognized by Gardner *et al.*[54a] when sections of fresh articular cartilage, cut by cryostat at −25°C, were stained with Alcian blue made up in diethylene glycol. B, boundary; C, chondrocyte cytoplasm; Ca, electron-dense capsule; I, interterritorial matrix; N, chondrocyte nucleus; P, pericellular matrix; T, territorial matrix.

The cilia are thought to be mechanoceptors, acting as sensors of both tensile and compressive stress (Chapter 6).

In adult human hyaline cartilage, the mean diameter of chondrocytes, measured in paraffin sections after aldehyde fixation and dehydration, ranges from 9 to 25 μm. However, these figures may misrepresent measurements made in hydrated specimens. In fixed, hydrated tissue the nuclei measure approximately 6 to 8 μm in diameter. Much more precise measurements have been derived for young animal articular cartilage. In 8-month-old rabbits, for example, the mean cell volume measured in tissue fixed so that proteoglycans are conserved, is 1115 μm^3, the mean cell surface area 670 μm^2.[43a] In rabbit articular cartilage, chondrocytes in the superficial zone, zone I, are much smaller than those in the deeper zones; the cells of zone IV are smaller than those of zones II and III. In the young rabbit, the mean volumes for the cells of 'weight-bearing', femoral condylar cartilage are: zone I: 982 μm^3; zone II: 2142 μm^3; zone III:

2908 μm^3; and zone IV: 1377 μm^3. The corresponding figures for chondrocyte surface areas are: 739 μm^2; 1082 μm^2; 1546 μm^2; and 845 μm^2. For 'less weight-bearing' parts of lapine femoral condylar cartilage, the mean cell volumes are: zone I: 895 μm^3; zone II: 674 μm^3; zone III: 1154 μm^3; and zone IV: 857 μm^3, strongly suggesting that chondrocytes in regions of cartilage subjected to relatively slight compressive stress are substantially smaller than those exposed to normal, diurnal loading.

Chondrocyte cytoplasm is abundant and contains rough endoplasmic reticulum and smooth (Golgi) reticulum (Fig. 1.2a). The modest number of lysosomes in normal chondrocytes include latent enzymes which, when activated, degrade products of metabolism and foreign materials before exocytosis. The lysosomal system is, however, also implicated in the normal metabolic turnover of the extracellular matrix. After matrix macromolecules are ingested, they are degraded within phagolysosomes; they may accumulate in excess in these organelles if digestion and exocytosis are not complete (see Chapter 9).

There are small numbers of mitochondria, less numerous, more compact and denser in mature cells than in the metabolically more active chondroblast. Masses of fine cytoplasmic filaments may accumulate in whorls within older or degenerate cells: their relationship to the delicate actin-like filaments is not certain. The mature chondrocyte contains much glycogen (Fig. 1.3) and lipid. The amount of fat in individual cells varies greatly. So much lipid is recognizable in zone I chondrocytes that its presence has been speculatively related to joint lubrication.[51,176] It is likely that the lipid droplets are stores, influencing energy exchange during oxidative phosphorylation rather than indices of cell injury and obsolescence comparable to the chondrocyte lipid accumulations that form in experimental lipoarthrosis.

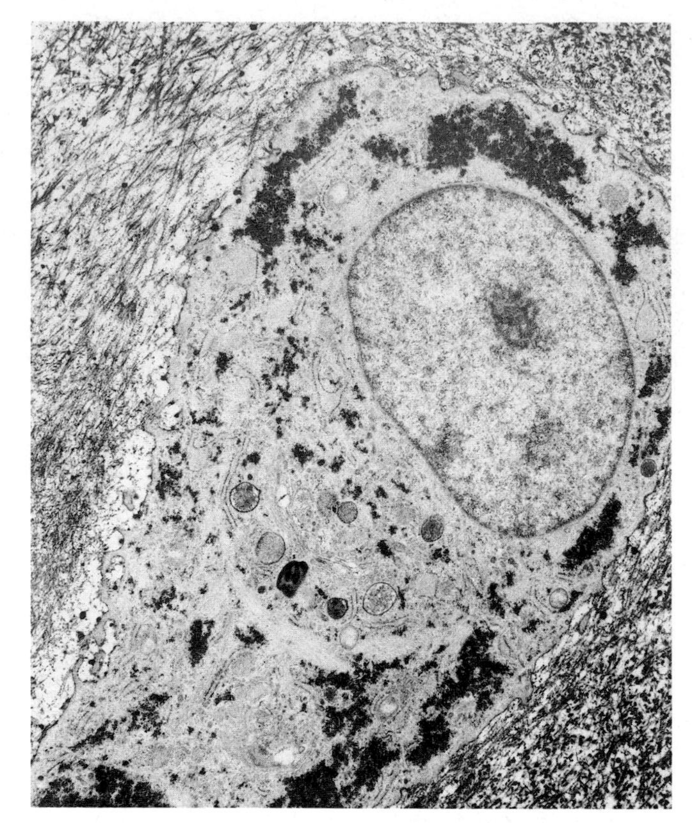

Fig. 1.3 Chondrocyte from articular cartilage of dog. Mid-zone (zone III) cell. The pericellular matrix is of limited extent. An ovoid nucleus contains a well-defined nucleolus. In addition to organelles that are lysosomes and dense bodies, a sparse rough endoplasmic reticulum and poorly defined Golgi stacks, abundant (black-appearing) glycogen is seen. Serial sections of chondrocytes usually reveal a cilium within each cell. (TEM × 5850)

It has proved possible to make indirect measurements of chondrocyte metabolic activity by ultrastructural morphometry. The analysis, for example, of cell size, organelle dimension and lipid and glycogen content, gives considerable insight into cell behaviour. Using such an approach, Paukkonen and Helminen[130a] quantified the synthetic activity of rabbit chondrocytes by electron microscopic stereology. They observed the relative paucity of endoplasmic reticulum in superficial zone chondrocytes and the relatively large amounts of intracytoplasmic filaments in mid- and deep-zone cells. Since the cells with abundant filamentous material contained little rough endoplasmic reticulum, these cells were thought to be 'degenerate'. However, Eggli *et al.*[43a], advanced different views. Thus, in the articular cartilage of young rabbits, the quantity of rough endoplasmic reticulum and the volume occupied by mitochondria are not significantly different between the four cartilage zones. However, the mean volume occupied by glycogen and by intermediate filaments is relatively very large in zones II and III, smaller in zone IV and least in zone I.[43a] These differences account for the differences in cell size recognized between the four zones; there is little variation in nuclear volume. There are comparable differences in the proportions of the cell occupied by intracellular lipid and by vacuoles.

Organization of the extracellular matrix

Throughout life, chondrocytes continue to exercise a regulatory function over the amounts and arrangement of the avascular extracellular matrix. The mechanisms by which matrix changes are signalled to, and perceived by the cell, remain obscure. There are likely to be chemical, mechanical and electrokinetic signals, perhaps channelled through a single receptor pathway at the cell surface.

Whichever efferent/afferent signal mechanism is operative, it is clear that the exact organization of the extracellular matrix adjoining the chondrocyte is of the utmost importance for the normal functioning of the cell and for the maintenance of extracellular matrix integrity.[89a] Historically, the arrangement of the matrix near the chondrocyte has been the centre for debate. However, it is now evident that much of the lack of certainty regarding matrix composition and organization has been the result of limitations in the techniques employed in their study. New methods allow better schemes for the structural relationships of cell and matrix to be proposed. Nevertheless, inspite of these advances, it remains of considerable practical convenience to retain the subjective, relative terms 'pericellular', 'territorial' and 'interterritorial' to describe the zones of extracellular matrix that are, respectively, immediately adjacent to, near, and less near each cell (Fig 1.2b).

Histochemical concepts: pericellular, territorial and interterritorial matrices When sections of aldehyde-fixed, paraffin-embedded cartilage were stained with acidophilic or basophilic dyes, the extracellular matrix did not stain uniformly. The extent and pattern of the zones of variegated acidophilia or basophilia were inconstant in different species and varied during maturation and ageing.[173,174] Moreover, the staining patterns were not the same in different cartilage regions, areas and zones (p. 68). Part of the explanation for these features of artefactual, two-dimensional preparations lay in real differences in extracellular matrix structure, long suspected and increasingly well understood. However, much of the apparent non-homogeneity was apparently artefactual.

An early explanation for the varied staining patterns that are illustrated in standard works[173,174,176a] was the postulated presence in cartilage of populations of non-proteoglycan, anionic proteins which could compete with positively charged dyes, leading to the appearance of areas of false negative staining.[176a] As Stockwell[173,174] points out, when human costal cartilage is stained at pH 2 with azure A, a concentric zone of matrix near to but not contiguous with each chondrocyte is highlighted by an unstained 'border zone' (Fig. 1.2b) which appears, in this way, to separate an internal, densely stained *territorial* matrix, from an external, less densely stained but usually much more extensive, *interterritorial* matrix (Fig. 1.2b). The delineation, on such a basis, of 'territorial' and 'interterritorial' matrices has led to controversy. However, much of this debate appears now to be of diminishing significance. Thus, by staining at neutral pH, a more homogeneous distribution of matrix macromolecules is suggested by an almost uniform staining pattern. In selected cartilages, such as those of the aged, equine nasal septum, this evidence has been substantiated by experiments made with the Alcian blue–critical electrolyte concentration (AB–CEC) techniques of Scott and Dorling.[152] In this procedure, the staining intensity of extracellular matrix macromolecules such as chondroitin sulphate proteoglycan is not susceptible to pH-induced protein competition and appears almost uniform. Nevertheless, there are species and site differences so that the basophilia detected by the application of the AB–CEC technique to human costal and bronchial cartilages demonstrates much chondroitin sulphate proteoglycan in the 'territorial' matrix, whereas, in both equine and human material, high molecular weight keratan sulphate proteoglycan is concentrated near the cells, just as hyaluronate is localized strongly close to the chondrocyte (Fig. 1.2b). The periodic acid–Schiff technique is of value.[92c]

More recent evidence has come from the demonstration that much matrix proteoglycan may be lost or redistributed when articular cartilage is subjected to aldehyde fixation, changes which are exaggerated by fixation in osmium tetroxide for transmission electron microscopy and by ethanol dehydration,[46d]

Conformational alterations, structural collapse and changed antigenicity may also result. These serious technical limitations can be minimized by tissue freezing before microscopy.[83a,83c] When the complex and slow process of freeze substitution, resin infiltration and ultraviolet polymerization at −40°C (233 K) is undertaken, tissue macromolecules can be preserved in an almost native state.[46c,d] When it is wished to avoid low temperature procedures, the loss of matrix molecules can be greatly reduced by combining aldehyde fixation with the addition to the fixative of cetyl pyridinium chloride,[46a,46c] safranin O,[92a] acridine orange, ruthenium hexamine trichloride (RHT)[83b] or cupromeronic blue.[152] In ultrastructural studies, osmication is avoided. However, post-fixation may be undertaken in a 1 per cent (w/v) osmium tetroxide solution containing 1 per cent (w/v) cobalt hexamine trichloride (CHT).[43a,83a]

With the low temperature methods, immunolocalization by double or triple gold labelling is practicable so that the use of poly- and monoclonal antibodies is shedding new light on the organization of the extracellular matrix. Hyaluronate and keratan sulphate are concentrated near the cell. Now there is evidence that some, at least, of the small proportion of cartilage collagen that is type VI is pericellular[134] and type IX collagen is also identifiable at this site.

Ultrastructural concepts: the chondron Ultrastructural studies made after conventional tissue fixation suggest that chondrocytes lie within a specialized pericellular matrix in which a meshwork of fine collagen is enclosed by a condensate ('capsule') of fine, faintly banded collagen fibrils.[134a,135,176a] Cell processes ramify through the pericellular stroma (Fig. 1.4). The structural unit defined by the pericellular capsule is a *chondron*[12a] (Fig. 1.2b).

As Szirmai[176a] had shown with nasal septal cartilage, chondrons can be extracted from articular cartilage in the same way that glomeruli can be separated from renal tissue.[136] However, just as a glomerulus is only part of a physiological unit, the nephron, a chondron is only part of a larger entity that extends to include much territorial and interterritorial matrix.

Concentrated within the chondron, and possibly constituting the capsule, is immunologically demonstrable type VI collagen.[134] Type IX collagen has also been localized to this site. The fine, pericellular collagen fibrils converge and diverge at the superficial ('articular') pole of the capsule and here (the 'cupola') there is an ultrastructural 'channel' (Fig. 1.5). Because of the presence of membrane-bounded matrix vesicles[57] immediately beyond this pericellular channel (Fig. 1.5), the suggestion has been advanced that a flow of materials through such channels during compressive deformation of cartilage (p. 270) may be one mechanism by which chondrocytes excercise homeostatic control over the extracellular matrix.[135,136] The most important matrix component to move in this way must surely be water. The relationship of the chondron as a structural and functional unit to the disposition of pericellular proteoglycan is less clear. Studies made with cartilage fixed in the presence of cationic dyes demonstrate no clear, pericellular ('lacunar') space.[43a] The proteoglycans are fully expanded, the chondrocyte plasma membranes in continuity with the nearby extracellular matrix and the pericellular 'capsule' is inconspicuous.

The pericellular matrix demonstrated by histochemical and lectin-binding techniques is not identical in fetal and adult tissues.[109] It has been claimed that lectins identify keratan sulphate but not chondroitin sulphate. The extent of the pericellular matrix increases with age.[173,174]

Chondrocyte culture

Chondrocytes can be grown in single, two-dimensional layers, in three-dimensional matrices or in free suspension.[170] In free suspension, for example in spinner cultures, the size and metabolic activity of chondrocytes vary widely.[181] Cartilage can also be cultured as a tissue or during explants of organs (see p. 80).

Chondrocytes have been grown successfully in monolayers since the earliest days of cell culture.[171] Many studies have been with chick embryo cells. However, mature mammalian chondrocytes, released by collagenase digestion, can replicate *in vitro*[111] and adult cells are now used as sources of three-dimensional aggregates[10] or to seed agarose gels,[36] collagen gels[11] or porous ceramics (Figs 1.6 and 1.7).[28,94,55,97]

Chondrocytes grow on glass surfaces but cling to them with difficulty: they attach more readily to plastic or coated

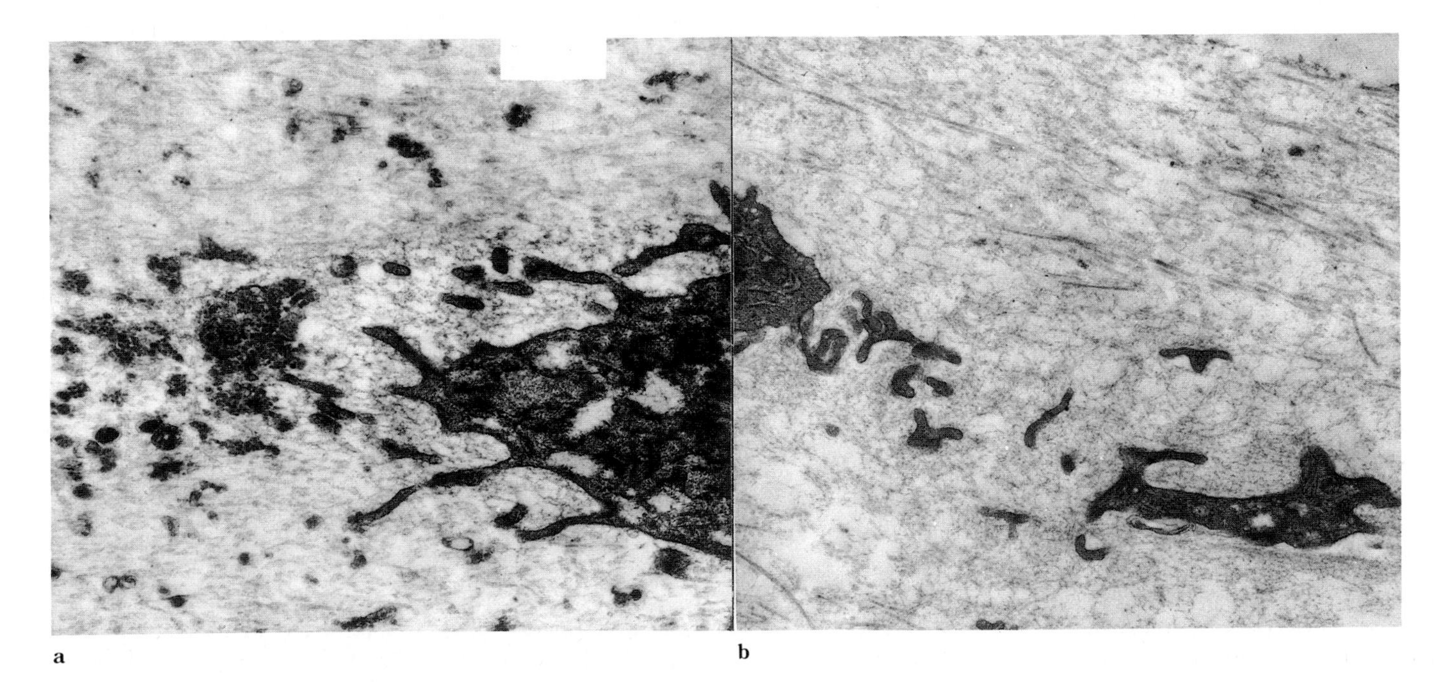

Fig. 1.4 Chondrocytes from articular cartilage of dog.
a. Peripheral margins of zone I chondrocyte. Cytoplasmic processes, cut tangentially and viewed in two-dimensions, appear as vesicular structures in the collagen-rich matrix. Viewed in three-dimensions, these cells are flat and ellipsoidal with amoeboid cytoplasmic extensions. **b.** Compare with **a.** Figure shows opposite margin of cell. (TEM × 27 000.)

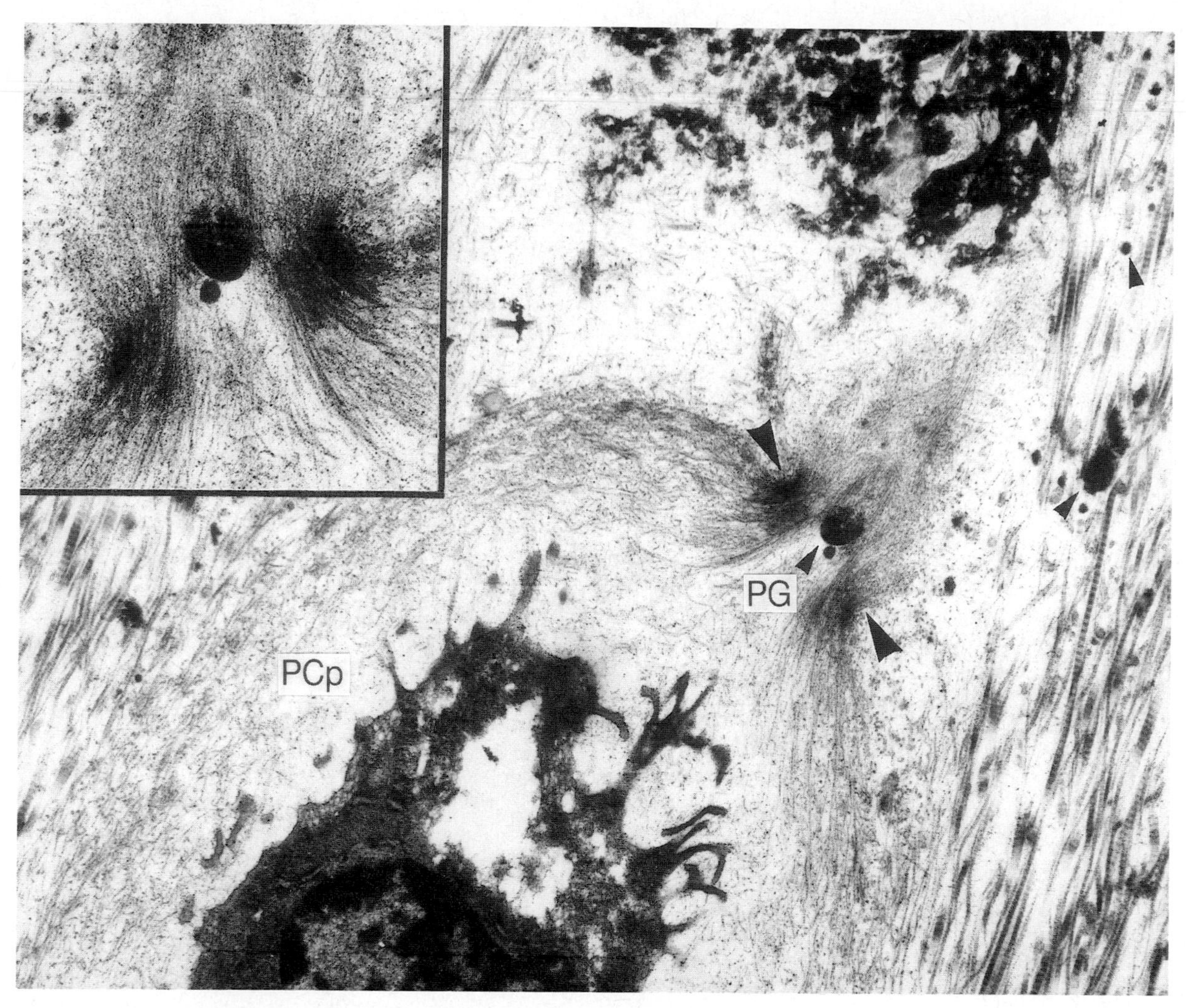

Fig. 1.5 Pericellular matrix of canine chondrocyte.
Pale, pericellular matrix (PCp) surrounds chondrocyte (lower left). Margin of pericellular matrix is defined by 'capsule' of electron-dense fibrils. These fibrils converge and diverge to form electron-dense foci (large arrowheads) on either side of a pericellular channel (PCh) within and above which matrix vesicles (small arrowheads) are often found. (Uranyl acetate × 15 400.) *Inset*: Matrix vesicles within pericellular channel. (× 33 750.) Reproduced by courtesy of Dr C A Poole and the Editor, *Journal of Anatomy*.[134a]

surfaces or to collagen. In monolayers, chondrocytes assume a fibroblast-like morphology and secrete little intercellular matrix material: they express type I rather than type II collagen. The presence of calcium is crucial to the production of type I collagen. Culture, age and cell density influence the nature of the proteoglycans synthesized.[107] The behaviour of cells in suspension is quite different: they multiply slowly but produce much extracellular, metachromatic matrix material and retain the capacity to express type II collagen[124] and proteoglycan. The core proteins that are formed differ from those derived from monolayer cultures.[92]

The growth of cells in three-dimensional agarose,[36] in collagen gels[62a] or on porous ceramic[28] resembles that of cartilage *in vivo*. There is a suggestion that, on calcium phosphate ceramic, the surface cells may express type II collagen, the deeper cells, adjoining the ceramic, type I collagen.[94] The distribution of proteoglycans formed in three-dimensional culture resembles that of the parent tissue, and the cells retain their conventional ultrastructural morphology. Three-dimensional methods are being increasingly used to analyse the behaviour of chondrocytes from diseased tissues.

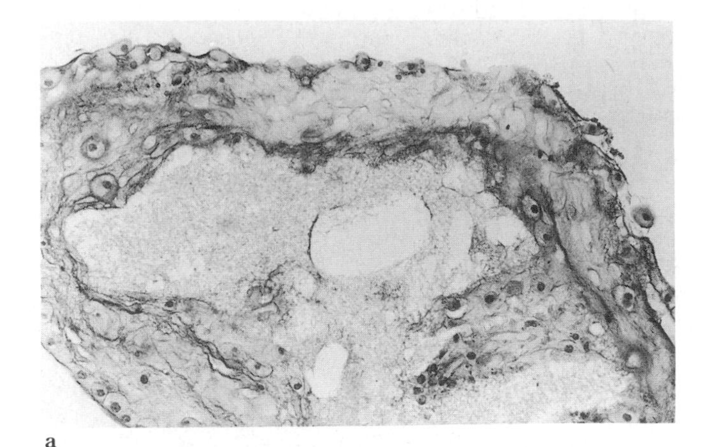

a

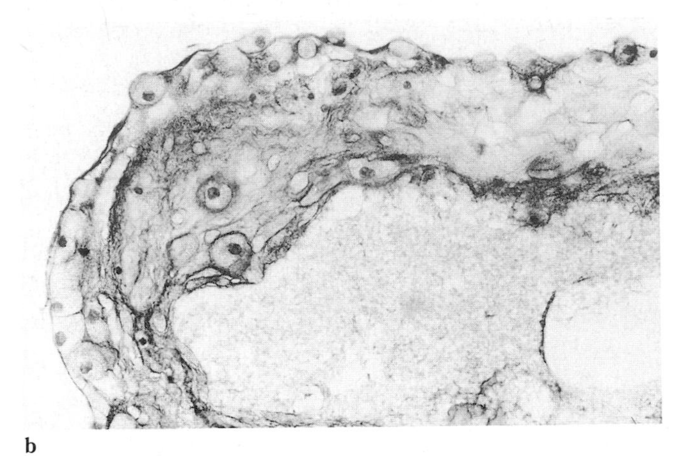

b

Fig. 1.6 a and b. Chondrocytes growing on surface of ceramic granule.
At centre of both fields, granular material is hydroxyapatite ceramic. Chondrocytes cover ceramic surface and have formed extracellular matrix, itself bounded by further layer of plump chondrocytes. Between the two cell layers, matrix material contains type II collagen and chondroitin sulphate proteoglycan. (**a**: × 180; **b**: × 180.)

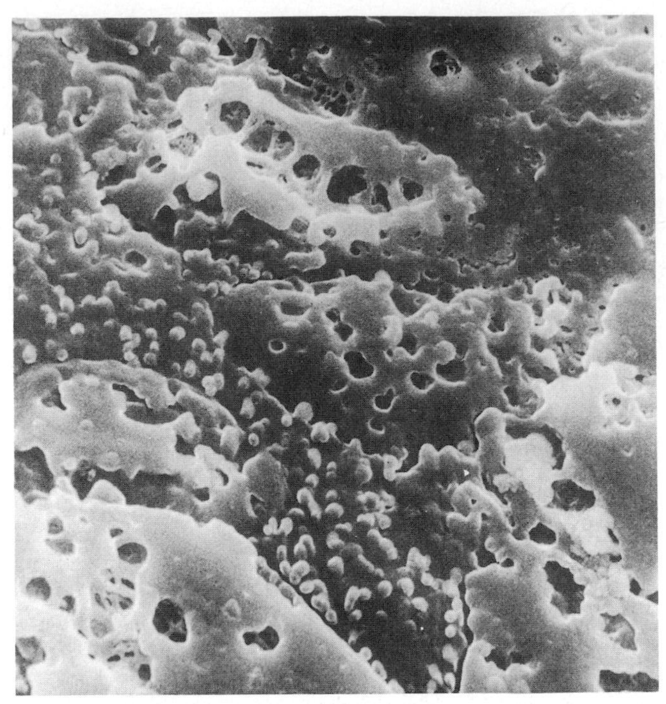

Fig. 1.7 Chondrocytes growing at surface of ceramic granule.
Preparation of chondrocytes multiplying at surface of granule after eight weeks in culture is shown in fully hydrated, unfixed state. Fenestrations represent shallow lamina of extracellular matrix covering cell surfaces and comprise expanded proteoglycan. Note numerous cell processes (at centre). (LTSEM × 1610.)

Chondrocyte division

Cartilage cellularity is discussed in Chapter 2, p. 77.

Chondrocyte nuclei are round or ovoid and, after fixation, often appear irregularly lobated. The extreme activity and degree of cell division in the growing cartilage of embryonic limbs (Fig. 2.15, p. 123) contrasts with the apparently sluggish and inert behaviour of hyaline articular cartilage in the adult. The chondrocytes of mature articular cartilage are very rarely seen to divide by mitosis[74] but deoxyribonucleic acid (DNA) synthesis is demonstrable experimentally, by [^{3}H]-thymidine autoradiography.[73] It may also be possible to demonstrate dividing cells by the application of monoclonal antibodies against markers such as Ki67 and PCNA that define phases of the cell cycle.

The number of cells in a tissue section or cell suspension can be counted and expressed in relative or absolute terms (Fig. 2.14, p. 77). Cartilage chondrocyte DNA can be measured in whole sections by microdensitometry after Feulgen staining or in tissue samples after extraction.[29] By such methods, it has been shown that chondrocytes do not lose their potential for growth *in vivo* or *in vitro* as age advances.[114] After procedures such as the intra-articular injection of papain, chondrocytes very quickly recover their ability to divide[73]: cell division is recognized even before there is demonstrable matrix degradation. Experiments using ultraviolet light to injure chondrocytes have drawn attention to the contribution made by unscheduled DNA synthesis. This process is very slight in resting cells, particularly as they age, but substantial in dividing cells. Cell culture (see p. 24) may be held to introduce artefact either by selecting cells that retain youthful vigour or by activating replicative machinery and obscuring age-related differences in behaviour.[102]

Chondrocytes divide frequently in the embryonic limb, in the growth zones of the epiphyses and, in the adult, more often than normal following cartilage injury (see Chapter 22) and in neoplasia. It is of interest that they are binucleate in auricular but not in normal articular cartilage.[101] Appositional growth is the rule in cartilage other than at epiphyses. Degradation or removal of the matrix provokes cell multiplication. This reaction is observed clearly in early osteoarthrosis (see Chapter 22). Some of the most unusual examples of abnormal chondrocyte multiplication are seen in the disorderly growth of benign and malignant cartilagenous neoplasms and in the chondrodystrophies (see Chapter 9).

Chondrocyte anabolism

Progress has been made in identifying the sites at which intracellular macromolecular synthesis takes place.[144] There are obvious similarities but some differences from the sequences recorded in the fibroblast (see p. 19). Core protein and chondroitin sulphate formation have been analysed by protein A-gold labelling of transmission electron microscopic sections.[143] In normal pig laryngeal cartilage, both proteoglycan core protein and chondroitin sulphate were identified in smooth membraned vesicles: they were galactose-rich medial/*trans*-Golgi cisternae (Fig. 1.8). There was little labelling in the rough endoplasmic reticulum. When monensin was used to treat chondrocytes (and thus to inhibit proteoglycan secretion), much more proteoglycan core protein was found in *trans* and *cis* vesicles than in the distended rough endoplasmic reticulum. Chondroitin sulphate also accumulated after monensin treatment but only in galactose-rich cisternae: chondroitin sulphate synthesis on proteoglycan occurred late and was located only in medial/*trans* Golgi cisternae. The synthesis of glycosaminoglycan on proteoglycan appeared to occur in a compartment in which O-linked and N-linked oligosaccharides were being incorporated onto other glycoproteins. The time for the synthesis and glycosylation of proteoglycan was 22 min; within 6 min, secretion occurred.[143]

Comparable results have been obtained using double immunofluorescence and lectin-localization reactions.[189] The sites of type II procollagen and of chondroitin sulphate proteoglycan can be compared. Type II procollagen is distributed in vesicles throughout the cytoplasm; precursors of chondroitin sulphate proteoglycan accumulate in the perinuclear cytoplasm, a region thought to correspond to the Golgi complex. The differential intracellular localization of the steps in the synthesis of collagen has been reviewed.[142,141,122]

Regulation of chondrocyte metabolism

The heterogeneity of the chondrocyte populations of articular cartilage, and the nature and amounts of the matrix molecules that these cells secrete, are inherent characteristics, retained for periods of up to 6 months in three-dimensional, agarose cultures.[6a,6b]

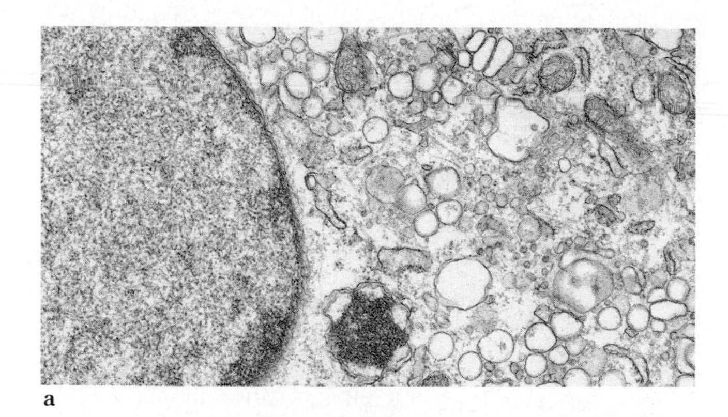

a

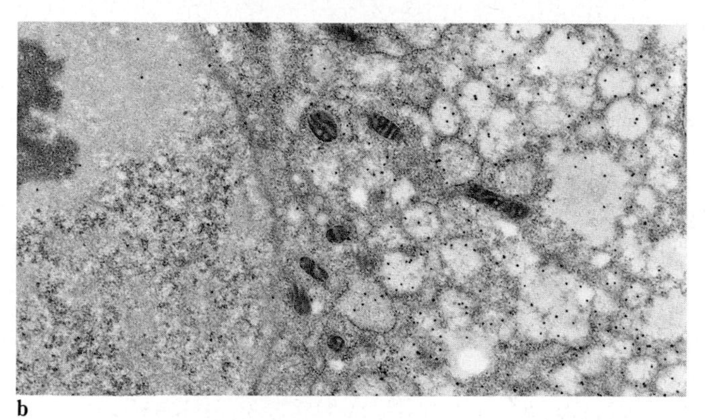

b

Fig. 1.8 Site of proteoglycan synthesis within cell. Figure illustrates use of polyclonal antibody to binding region of proteoglycan core protein in a transmission electron microscopic immunogold labelling procedure, to locate this protein in the smooth-membraned vesicles of a cultured chondrocyte. Further tests, not illustrated, demonstrated that these vesicles were galactose-rich *trans*-Golgi cisternae. There was little labelling of the rough endoplasmic reticulum. The carboxylic ionophore monensin interferes with the intracellular translocation of secretory proteins, thus inhibiting proteoglycan secretion and causing proteoglycan core protein accumulation. Here, monensin was employed to facilitate core protein identification. **a.** Control chondrocytes treated with 1.0 mM monensin only. Distended Golgi cisternae are seen; they are devoid of gold particles. **b.** Chondrocyte from culture treated with monensin and reacted for proteoglycan core protein binding region in immunogold technique. Note gold particles spatially related to Golgi cisternae. (**a**: × 6500; **b**: × 12 850.) Reproduced by courtesy of Drs A Ratcliffe, P R Fryer and T E Hardingham, and the Editor, *Journal of Cell Biology*[143].

The quality and quantity of the macromolecules synthesized and exported by chondrocytes is determined genetically and can be modified by agents that alter DNA structure. Useful information can be gained by *in situ* hybridization.[111a] However, the unique secretions of chondrocytes are also closely affected by the cellular microenvironment.[181] The synthetic properties of chondrocytes are, for example, influenced and regulated by the composition and quantity of nearby matrix proteoglycan and by the mechanical stresses to which the cell and pericellular matrix are periodically subjected.[125] The presence of a normal pericellular matrix acts by negative feedback, to switch off macromolecular synthesis. The delicate feedback mechanisms are susceptible both to inherited anomalies and to the impact of acquired disease.[195] The small proportion of hyaluronate in hyaline cartilage can reduce chondrocyte proteoglycan synthesis and delay proteoglycan export,[196,197] an observation possibly related to the increase of cartilage hyaluronate with age.[44] Other connective tissue cells are not so affected. Hyaluronate interacts with a component of the cell surface where receptors can be destroyed by proteinases.[197] The effects of the physical environment are shown when mechanical loads, in tension or compression[125] (see Chapter 6) change the morphology of ovoid, marginal articular chondrocytes to elongated, fibroblast-like shapes. Collagen secretion increases. In cell culture (see p. 24) contact inhibition is accompanied by a conventional pattern of proteoglycan synthesis. However, when chondrocytes are suspended in spinner culture (see p. 80), the frequency of cell division alters, the amount of proteoglycan synthesized increases greatly and the ratios of the glycosaminoglycans formed alters markedly.[170]

Chondrocyte responses to hormones and growth factors

During chondrocyte proliferation, proteoglycan synthesis declines but hyaluronate synthesis increases. During chondrocyte differentiation, the converse is the case. Studies of fetal bovine cartilage have shown that a somatomedin-like 'cartilage-derived factor' is present.[91] The source of this somatomedin is the liver: its identification in cartilage indicates one of its sites of action. The factor anomalously stimulates cell division and increased proteoglycan synthesis. A cartilage-derived growth factor has also been isolated[69]: unlike cartilage-derived factor and the somatomedins, cartilage-derived growth factor stimulates chondrocyte growth and hyaluronate synthesis but depresses proteoglycan synthesis.

Epiphyseal chondrocytes respond actively to excess or deficient growth hormone.[131,161] *In vivo*, growth hormone acts via the somatomedins (growth hormone-dependent growth factors). They are: insulin-like growth factor-1 (somatomedin-C) (IGF-1), insulin-like growth factor-2 (IGF-2) and multiplication-stimulating activity.[193] The somatomedins are mitogenic. Like the cytokines (see p. 291) they bind to receptors on chondrocyte plasma membranes before increasing DNA synthesis. Articular chondrocytes are also sensitive to many other hormones[164] so that oestrogens, for example, impair cell growth.[162] Insulin enhances the development of the rough endoplasmic reticulum and Golgi apparatus.[163] In diabetes mellitus cartilage formation is diminished and delayed. In the same way, hypothyroidism retards cartilage growth.[160] Testosterone alters epiphyseal cartilage metabolism[177]: the hormone is metabolized before binding to chondrocyte receptors. Cortisone may cause epiphyseal chondrocyte death in experimental rats[163] and triamcinolone hexacetonide, a synthetic analogue of cortisol, depresses DNA synthesis and endochondral bone growth.[165,164] Cortisol itself adversely affects rabbit acetabular cartilage proteoglycan and protein synthesis.[79]

Although purified preparations of thyroid-stimulating hormone and of luteinizing hormone exert no growth-promoting effects on monolayer cultures of rabbit articular chondrocytes, crude preparations do have an effect. The 'contaminant' is a chondrocyte growth factor which suppresses proteoglycan, collagen and protein synthesis.[108,90] Since the pituitary contains a fibroblast growth factor, it seemed possible that the actions on chondrocytes were due to the presence of this factor. This possibility has gained support but the growth promoting actions of fibroblast growth factor are not uniform.[147]

Chondrocyte survival

Adult articular cartilage has no immediate vascular supply and the cells, particularly those of mature cartilage, tolerate low environmental oxygen tensions. They can survive for long periods *in vivo* in circumstances severely damaging to more active mesenchymal cells, such as fibroblasts, and to epithelial cells. Even after long periods of hypoxia or limb ischaemia, chondrocytes can still incorporate ^{35}S-labelled sulphate: they can also survive and proliferate actively under adverse conditions *in vitro*. Nevertheless, the intercellular matrix is readily degraded by proteases *in vitro* (Fig. 1.9) and *in vivo* (see Chapter 12).

Chondrocyte surface molecules

The chondrocyte plasma membrane is antigenic (see Chapter 5) and MHC class I antigens are largely responsible for transplant rejection when cartilage grafts[40] are attempted (see p. 81). The MHC class II antigens are sources of cell activation, for example, in the release of matrix-degrading enzymes. Since the release of mediators from synovial cells and macrophages are stimuli causing chondrocytes to resorb their own matrices (see p. 88), it is of interest that MHC class II antigens are strongly expressed on many of the isolated chondrocytes eluted from damaged articular cartilage in osteoarthrosis, rheumatoid arthritis, traumatic arthritis and the osteochondromata.[19] Similar MHC class II antigens are very uncommon on clustered, normal chondro-

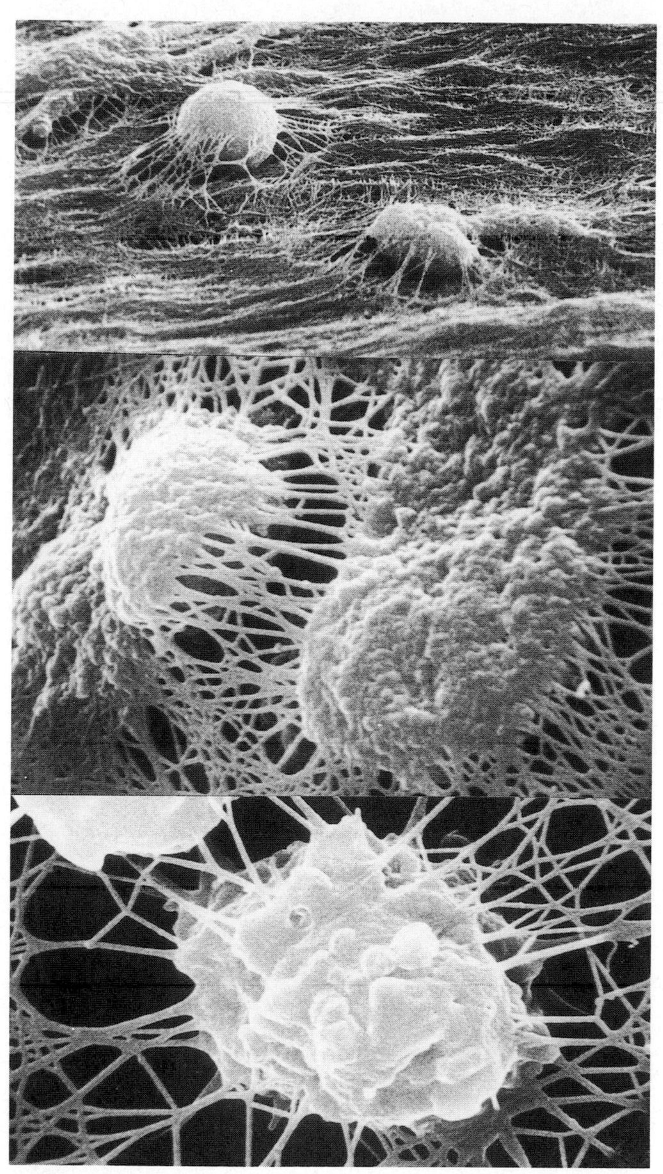

Fig. 1.9 Chondrocytes exposed at articular surface. Femoral condylar cartilage from adult rats was digested with papain before examination by scanning electron microscopy. Three stages in the degradative process are shown. As matrix digestion progresses, proteoglycan-rich matrix is lost first, leaving cells suspended in a spider-like web of type II collagen. Ultimately, after 180–240 min at 37 °C, cells are lost from entire cartilage, leaving calcified cartilage exposed. (SEM × 1100 (top); × 2050 (middle); × 5250 (bottom).) O'Connor, Brereton and Gardner[124a].

cytes. Individual cells from both abnormal and normal patients appear morphologically identical. However, the pericellular matrix is permeable to molecules as large as immunoglobulin G (IgG) and the absence of morphological change is unexplained. Tests with an antibody (83c2) against a surface 85 kD antigen, suggested that cell surface molecules could be shed into the extracellular matrix without provoking recognizable ultrastructural alterations.[19]

Human, canine and bovine chondrocytes have both H_1- and H_2-receptors and, in monolayer culture, respond to histamine by a concentration-dependent increase in intracellular cyclic adenosine monophosphate.[179,178] It is therefore possible that mast cells (see p. 36), which abound in inflammatory joint disease such as rheumatoid arthritis, may also directly influence chondrocyte metabolism.

Osteoblasts

Bone as a tissue is described briefly in Chapter 2. Processes of normal calcification[187] are outlined on p. 107, and abnormal calcification on p. 403.

Osteoblasts[8,13,15,16] line the external (periosteal) and internal (endosteal) surfaces of growing bone (Fig. 1.10). Transmission electron microscopy shows that the diameter of fixed, dehydrated osteoblasts ranges from 15 to 80 μm. The cells are elongated and attenuated in sections, flatter and ellipsoidal when viewed *en face*. Cytoplasmic RNA is abundant; the cisternae of the endoplasmic reticulum are prominent and there are varying numbers of free ribosomes and polyribosomes. The large, ovoid nucleus contains a single, conspicuous nucleolus. Delicate processes of fine cytoplasmic, fibrillar material are seen.

At sites where osteoblasts are of a resting, mature, fibrocyte-like nature, the cells are compact and of a preosteoblastic (osteoprogenitor) form: they react according to genetic programmes that induce and modulate bone formation: they display manifest alkaline phosphatase and latent collagenase activity. In addition to type I collagen, many other glycoproteins and proteoglycans characteristic of mineralizing tissue can be identified. Among these molecular species are the glycosylated bone proteins sialoprotein, osteopontin I and osteopontin II: each is capable of regulating bone cell attachment via internal RGD sequences (p. 18). The mode of adhesion of bone cells is via vitronectin (p. 52). Agents implicated in bone matrix synthesis include the phosphorylated glycoprotein, osteonectin. Provoked to osteoid synthesis (osteogenesis) and to the initiation and regulation of mineralization, the osteoblasts assume a less compact, more open and plump character. Cytoplasmic basophilia increases in stained preparations; the nuclei become more vesicular, the DNA more conspicuous. Slight periosteal trauma, the local growth of a bone neoplasm or the more severe disturbance caused by fracture, in the normal individual, all, rapidly lead to an abrupt resumption of synthetic activity. The role of osteoblasts in osteoclastic bone reabsorption is outlined on page 32.

Osteocytes

Mature osteocytes[180,112] thrive within osteons. From the cell, fine cytoplasmic processes extend within the canaliculi of the haversian systems, defining pathways for respiratory exchange and metabolism. The cells retain the capacity for osteogenesis and those that exist near bone surfaces, whether periosteal or endosteal, incorporate radioactive isotopic precursors during this process. An effete osteocyte is destined to die, crab-like, within its shell of mineralized bone; a histological index of bone death, crude but practicable, is the loss of stainable nuclear chromatin from an osteon.

Although it has been suggested that osteocytes can survive periods of ischaemia of up to two to five days, controlled experiments have shown that, like many other mammalian cells, there is irreversible ultrastructural injury after approximately 2 h.[85] After 24 h of ischaemia, most osteons are devoid of nuclei.

Osteocytes are thought to possess the little understood property of destroying bone in a manner distinct from that of the osteoclast; and this mechanism of insidious osteocytic osteolysis may assume importance in systemic connective tissue—and articular disease. Osteocytes express lysosomal hydrolases including an acid phosphatase, an aminopeptidase and a protease; osteocytic osteolysis may indeed be mediated by lysosomal enzyme activity. This property, like osteoclastic bone reabsorption, is regulated by plasma calcium levels as well as by parathormone, by 1,25-dihydroxyvitamin D and by calcitonin (p. 107).

The pump-like actions of osteocytes, conveying water from periosteal surfaces through the canaliculi to the tissue fluid, transport dissolved calcium for exchange with bone mineral. There is also an exchange of calcium within bone by diffusion. The uptake of plasma proteins into bone matrix can be demonstrated by the administration of radioactive, isotopically labelled protein markers; and the pathways of calcium and phosphate exchange within bone and connective tissue lend themselves to tissue analysis by nuclear magnetic resonance scanning (see p. 162).

Electron microscopy[8] shows that osteocytes have abundant cytoplasm: the cytoplasmic processes of the young cell are short and few; with time, however, they become more numerous and longer processes appear. The processes contain no organelles but establish lines of contact and communication with adjacent cells. Tight junctions are identifiable. Easy metabolic relationships are permitted by the vast exchange areas that exist as vascular surfaces, between blood plasma and the tissue fluid of bone. These areas may be as much as 250 μm^2 per cell but even this large figure is seen to be a considerable underestimate when the interfibrillar and intermolecular spaces within bone matrix are taken into account.

Osteoclasts

Osteoclasts lie on bone surfaces; they are the main agents of bone reabsorption[23,184] in diseases such as rheumatoid arthritis. Osteoclasts are large, multinucleate cells. Because of their size (they occupy volumes of 3–25 × $10^3/\mu m^3$), the number of nuclei seen in a single light-microscopic section may be only five or six and some cells may be erroneously identified as mononuclear. The nuclei in a single cell are of uniform size with fine, evenly dispersed chromatin and one or two nucleoli. The cytoplasm in stained preparations is usually acidophilic and, at the light microscopic level, appears 'frothy'.

Characteristically, osteoclasts lie apposed to bone surfaces (Fig. 1.11) which they digest only after the narrow, intervening lamina of osteoid has been degraded by osteoblast collagenase. The osteoclast plasma membrane bears a variety of integrins (p. 52) and the subjacent cytoskeleton includes talin, vinculin and actin. Along the zone of contact which constitutes a reabsorption front, the osteoclast has a ruffled border where complex folds and projections reach towards the mineralized bone (Fig. 1.11). Vitronectin is detectable over the whole osteoclast surface and among the other ligands present are type I collagen, the osteopontins, bone sialoprotein, fibronectin and fibrinogen. However, in conventional transmission electron microscopic preparations, there is an intervening, extracellular zone where bone mineral is absent but where the cytoplasmic processes of the ruffled border interdigitate with collagen fibres. Deep clefts extend between the villous processes and some are in continuity with cytoplasmic vacuoles which are characteristically numerous. Within these vacuoles, clusters of bone mineral crystals are often seen, derived from

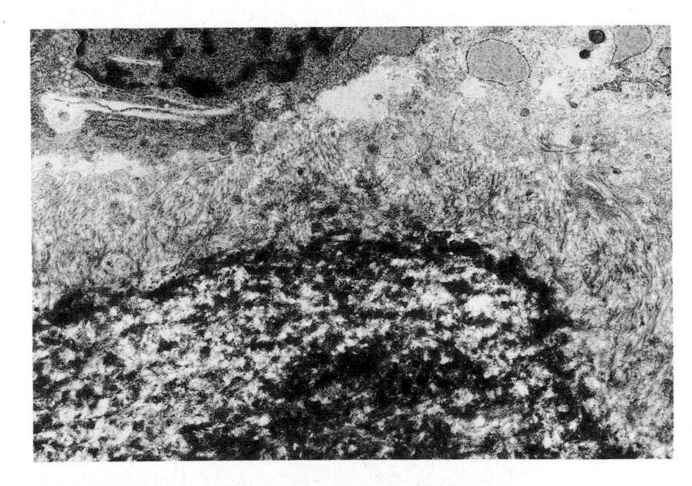

Fig. 1.10 Osteoblast and bone formation. Osteoblast (at top left) adjoins type I collagen-rich osteoid seam (at centre) which forms margin of trabecula of mineralized matrix (bone) (at bottom). (TEM × 4670.)

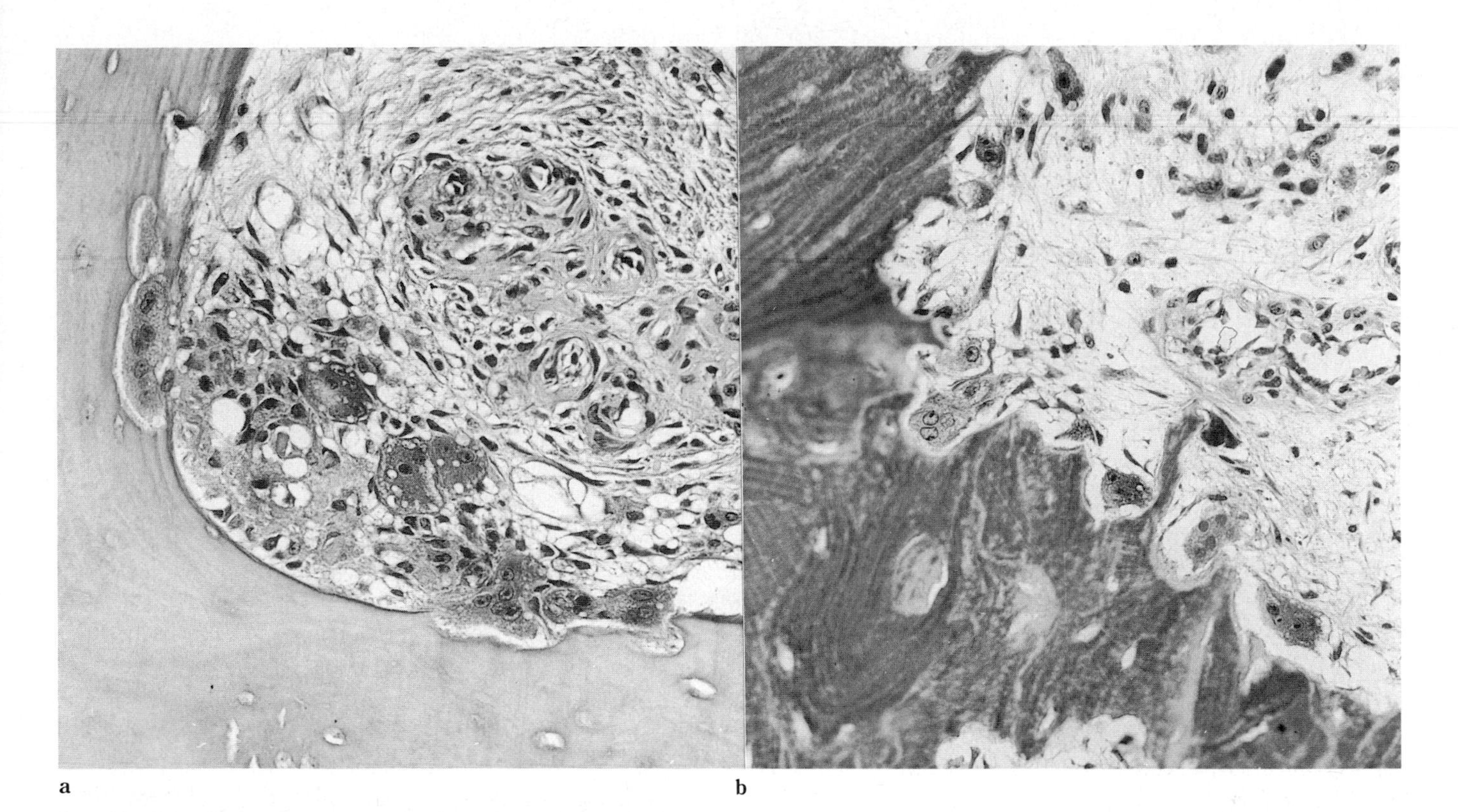

a b

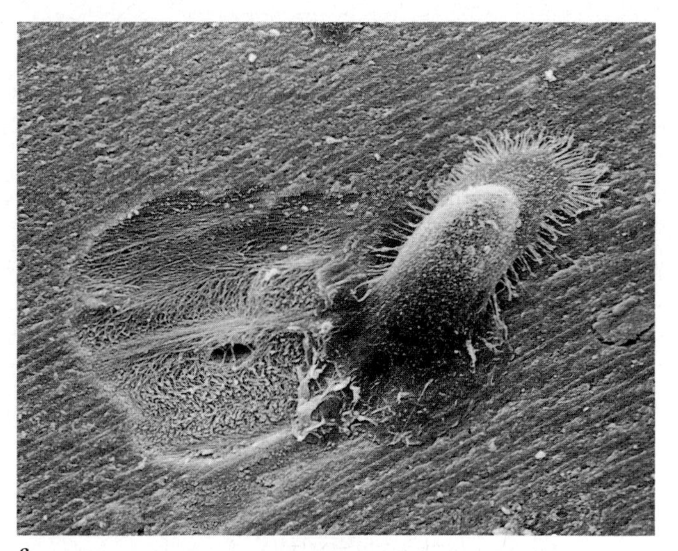

c

Fig. 1.11 Osteoclasts.
a. Vigorous osteoclastic reabsorption of necrotic lamallar bone at site of femoral head necrosis. Osteoclasts degrade calcified bone matrix only after loss of non-mineralized osteoid or its digestion by osteoblast collagenase. Note tangential arrangement of bone lamellae. (HE × 400.) **b.** Compare with 1.11a. Osteoclasts are causing resorption of perpendicularly orientated bone lamellae. (HE × 400.) **c.** Osteoclastic bone reabsorption. Observe the delicate cytoplasmic processes (right) and the lacuna (centre left) where bone reabsorption has taken place. The organization of the collagen fibre bundles of the bone is seen. (SEM × 770.) Courtesy of Professor T J Chambers and the Editor of *Triangle.*[24a]

the bone during degradation and phagocytosis. Isotopic labelling studies confirm that bone matrix material is indeed present within the cell. Cytoplasmic organelles are relatively few in the region of the ruffled border.

An older view that osteoclasts were derived from osteogenic stem cells that cover and line endosteal surfaces was succeeded by the suggestion that the cell of origin was the circulating mononuclear phagocyte (see p. 38). Strong evidence for this opinion came from parabiosis experiments, from studies with osteopetrotic mice and from cross-transplants made with quail and chick embryos;[24] the nuclei of these avian species are individually distinguishable. It was clear that bone-forming and bone-destroying cells had entirely different origins. Now, however, it is apparent that mononuclear phagocytes derived from bone marrow cannot correct osteoclastic defects in osteopetrosis or after irradiation.[23] Osteoclasts display tartrate-resistant acid phosphatase, an enzyme not present in mononuclear macrophages, and lack chloroacetate esterase which is present in the mononuclear cells. Moreover, all the antigenic markers specific for macrophages are absent from osteoclasts. When monoclonal antibodies for myeloid cell markers are tested, osteoclasts are found to be relatively effete in antigenic terms and to have few of the antigens associated with myeloid cells. It appears that the osteoclast must come from a previously unidentified circulating mononuclear cell, the preosteoclast, originating in the bone marrow mononuclear macrophage (see p. 38). Alternatively, an origin from local, non-haematopoietic, perivascular mesenchymal cells is considered possible.[70a] Whether the so-called chondroclast (see Chapter 12) is a form of osteoclast is not clear since morphological evidence suggests that chondroclasts can directly degrade cartilage although it is apparent that their actions are limited to the resorption of calcified cartilage.

Many of the microenvironmental functions of the osteoclast are dependent upon a regulating function expressed by the osteoblast. Osteoblasts, which synthesize and release collagenase, can remove the lining osteoid layer of bone, exposing bone mineral and thus enabling osteoclastic reabsorption to take place. Parathyroid hormone accelerates this process. Calcitonin blocks the capacity of the osteoclast to move and in this way delays the reabsorption of mineralized matrix. Both prostaglandin E_2 and a variety of cytokines (see pp. 288 and 291), induce osteoclastic activity, perhaps by an action on osteoblasts. Osteoclast activating factor and interleukin-1 are related if not identical and there is evidence that other cytokines may be activated. The relationship of this action to the marginal destruction of bone in inflammatory diseases such as rheumatoid arthritis is not fully understood. Pyrophosphate may inhibit this local process but prostaglandins, derived from macrophages, can mediate the response.

Synovial cells

Synovial tissue[59] is described in Chapter 2, p. 84 and synovial fluid on p. 89. The embryology of the joints is briefly outlined on p. 123. The mode of formation of synovial villi is discussed by Edwards *et al.*[42]

Synovial mesenchyme can be identified early in embryogenesis and cavitation is seen in the limbs of both 18-day-old chick embryos and 23 mm human embryos.[128] From the time of cavitation onwards, the inner, non-load-bearing surfaces of all synovial joints are lined by specialized cells.[22,57,62,77,78,145] However, the lining is incomplete and synovial fluid is in continuity with the subsynoviocytic connective tissue matrix.

Synoviocyte morphology and classification

Transmission electron microscopy (TEM)[9,202,57,59,61] confirms the presence of a loose or compact, sometimes discontinuous, two- to four-layer arrangement of synovial cells; in humans, desmosomes and gap junctions are seen.[38] Although type IV collagen and laminin are present,[132a] no structural basement membrane material is seen. Near the synovial cells light microscopy reveals a collagen-containing matrix in direct continuity with the subsynoviocytic connective tissue. A delicate pink lamina covering the surface of the synoviocytes is suggested by picro-Sirius red staining, but no fibrillar collagen is demonstrable in this situation. Conventional scanning electron microscopy has been used to examine the relationship between the long cytoplasmic processes of synovial cells and hydraulic permeability[105] but the constraints imposed by fixation and dehydration make it difficult to draw functional conclusions from such preparations.

At first it seemed that synoviocytes were of two distinct types: A(M) (M = macrophage-like) and B(F) (F = fibroblast-like) (Figs 1.12–1.15).[9,80,81,194] The A cells contain little rough endoplasmic reticulum, well-formed Golgi complexes and numerous vesicles and vacuoles. There are filopodia, intracytoplasmic filaments, many mitochondria and micropinocytotic vesicles. By contrast, the B cells have a well-formed rough endoplasmic reticulum but poorly developed Golgi complexes and few vesicles or vacuoles; their secretory granules, nevertheless, contain glycosaminoglycan and glycoprotein with a protein core.[126]

In spite of this evidence, it was suspected that a third class of intermediate, type C, cell could be found in the normal synoviocytic layer, suggesting a mutable spectrum of ultrastructural morphology, the appearances of the synoviocytes depending upon their functional state. Nevertheless tests made by radioactive sulphate incorporation did not support this view.[14] However, Ghadially[57] argued that both type A and type B cells had secretory and phagocytic potential: in traumatic arthropathy and rheumatoid arthritis, where there was raised phagocytic activity, there was an increased number of type B cells with much rough endoplasmic reticulum and a disappearance of cells of type A morphology. This transformation appeared to be a hypertrophy of rough endoplasmic reticulum in type A cells, not a replacement of type A cells by type B cells and intermediate cells. The concept of mutability of synovial cell morphology also found support in the observations made by Fell *et al.*[49] on porcine synovium in organ culture (see p. 88). Type A cells were normally scanty. During phagocytosis, the other 'intimal' (B) cells withdrew their branched processes, acquired pseudopodia and came to resemble type A cells. The A and B cells were thought to represent different functional

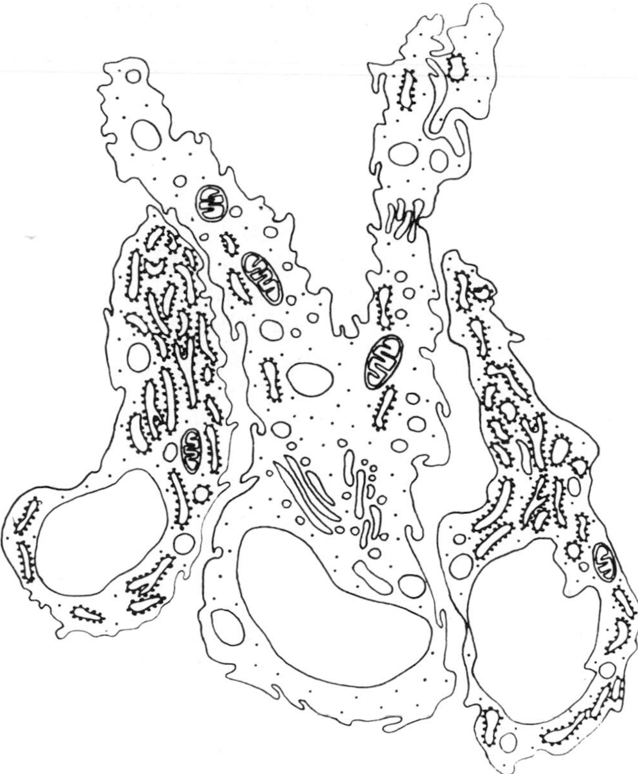

Fig. 1.12 Synovial cells.
Diagrammatic representation of a single type A (centre)and two type B (at margins) synovial cells overlying loose vascular sub-synoviocytic connective tissue. The resemblance of the type A cell to a macrophage and of the type B cells to fibroblasts is emphasized. Redrawn from Barland, Novikoff and Hamerman.[9]

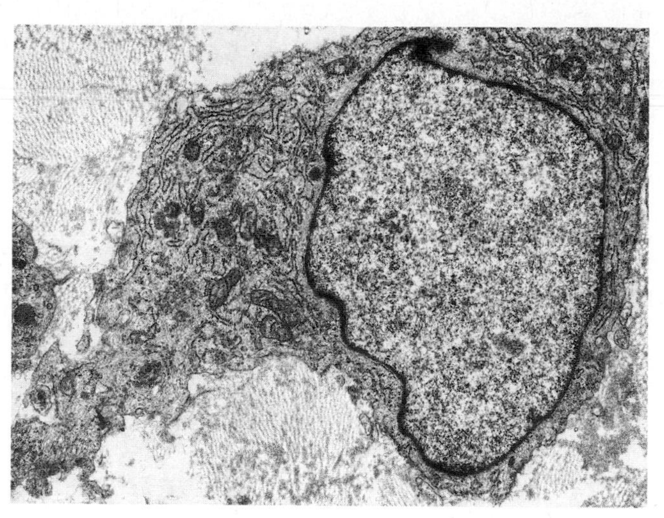

Fig. 1.14 Synovial B cell.
Type B synovial cell from the same tissue. Mitochondria are prominent among the abundant cisternae of the rough endoplasmic reticulum. (TEM × 4300.)

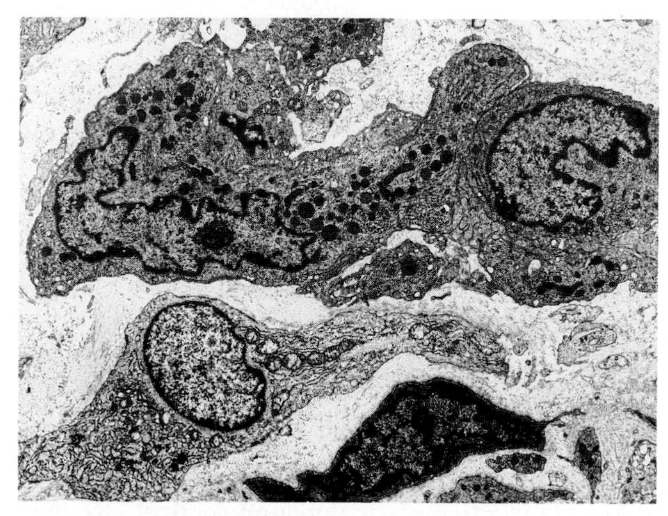

Fig. 1.13 Synovial A cell.
Type A synovial cell from tissue of human right third finger. The two macrophage-like cells contain numerous lysosomes. Beneath them, is a fibroblast-like type B cell. (TEM × 3200.)

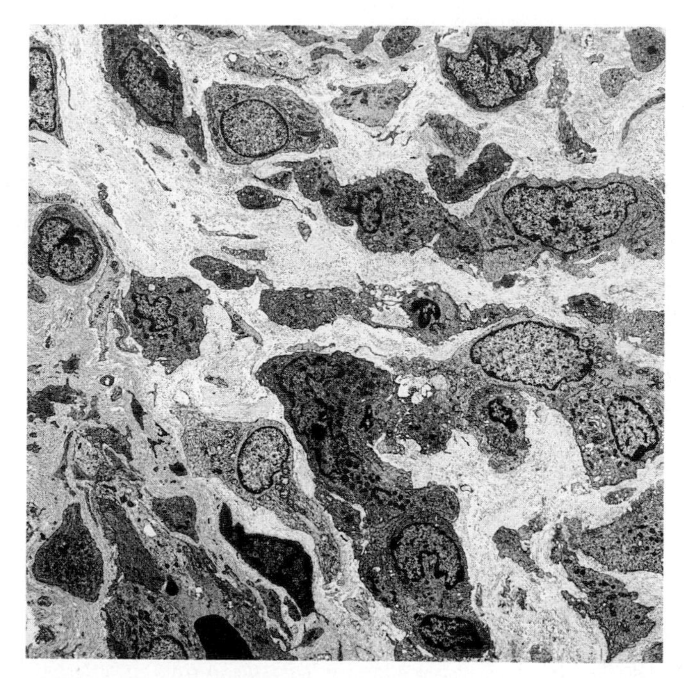

Fig. 1.15 Synovial tissue.
Within this tangential section of synovial connective tissue there are type A and type B cells together with fibrocytes, macrophages and a lymphocyte. A looser, more open intercellular matrix (at right) merges with a more compact connective tissue (lower left) that is part of the joint capsule. (TEM × 1150.)

states of the same cell type and the form assumed was determined by environmental factors.

Further enquiries reinforce the belief that there are indeed distinct populations of synovial cells. The A (macrophage) type of synovial lining cell has been shown to be derived from the bone marrow;[37,41,43] the B cell is a differentiated mesenchymal cell, originating locally. In studies of serial thin sections of rat synovium reconstructed to give three-dimensional models (Fig. 1.16), Graabaek[63,64] claimed a clear-cut definition of two, and only two, cell types, A and S (Figs 1.16–1.18). Type B and intermediate (C) cells were independent profiles of the so-called S cell.

A different organization of the lysosomal enzyme systems of the A and S cells suggested distinct functional attributes for these populations.[65] Intermediate cells, on this evidence, do not exist.[63] The difference in organelle structure between A and S cells is almost complete and the

a

b

Fig. 1.16 Three-dimensional reconstruction of type A synoviocyte derived from a series of equally spaced serial sections.
a. The A cell is elongated and U-shaped. The left and right margins of the cell face the joint cavity. An interrupted line delineates the synovial surface adjoining the joint space. **b.** The identical reconstruction rotated by 90° in relation to the model shown in **a.** Reproduced by courtesy of Dr Pernille Graabaek and the Editor, *Journal of Ultrastructure Research*.[65a]

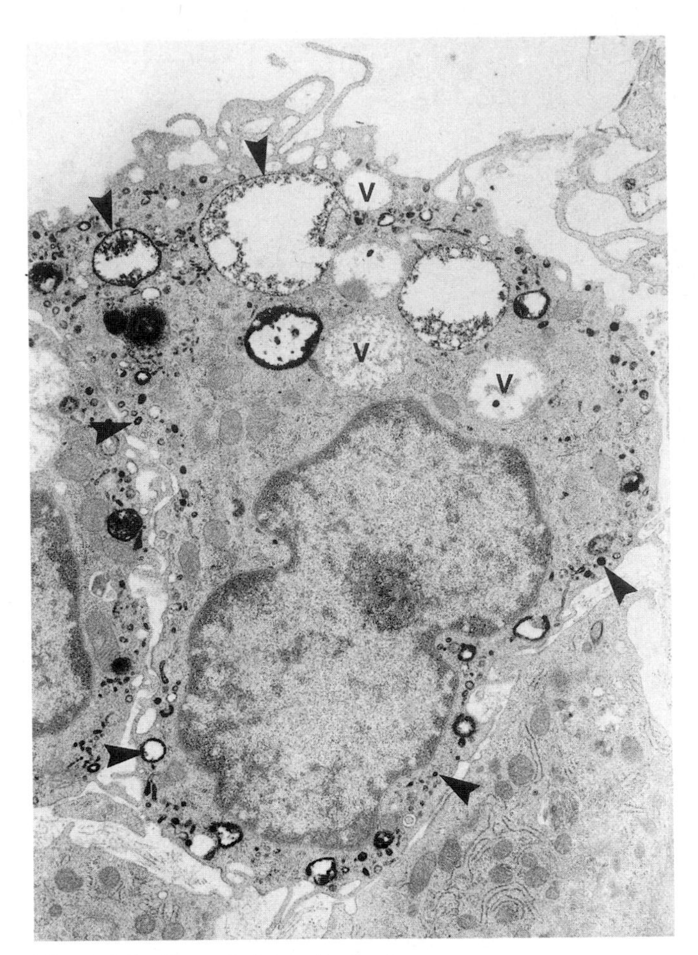

Fig. 1.17 Synovial A cell of rat.
Horseradish peroxidase has been injected intraarticularly. Many vesicles and large vacuoles contain the peroxidase reaction product (arrowheads). Other vacuoles are empty or contain electron-opaque material of low density (**V**). Endoplasmic reticulum is sparse. Portions of S cells are recognized at the bottom corners of the field. (TEM × 8120.) Courtesy of Dr Pernille Graabaek. (See Fig. 1.16.)

organization of the lysosomal system in the two types is quite different.[64,65] The functional distinction between A and S cells is emphasized by the ability of the former to absorb low-dose horseradish peroxidase, a glycoprotein rich in Man groups. S cells do not display this capacity.[65] Type A cells possess Man-specific binding sites; type S cells do not.[65b]

Other functional studies of diseased synovium also favour the presence of discrete synovial cell types. Two varieties of human synovial cell express MHC class II histocompatibility antigens: one is non-phagocytic and stellate/dendritic, the other fibroblastoid but in the mononuclear macrophage lineage.[199] Perhaps surprisingly, the stellate/dendritic cell is found predominantly in patients with rheumatoid arthritis, and the fibroblastic cell in osteoarthrosis. The proportion of these cell types varies. However, the ability to form

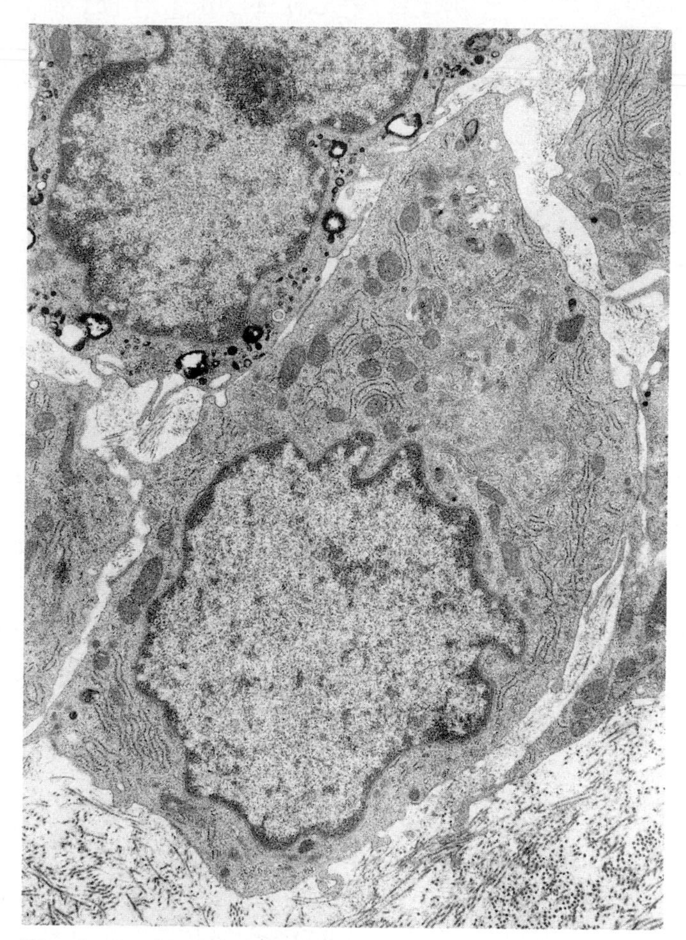

Fig. 1.18 Synovial S cell of rat.
After the same treatment, no peroxidase reaction product is detectable. The adjacent cell (top left) is the A cell seen in Fig. 1.17. (TEM × 8120.) Courtesy of Dr Pernille Graabaek. (See Fig. 1.16.)

rosettes with immunoglobulin G-coated sheep red cells; the presence of MHC class II antigens; the activity of non-specific esterase; and a bone marrow origin, argue in favour of a type A mononuclear macrophage-like cell[43] with surface receptors for Fc and C3, distinct from a locally derived, fibroblast-like B cell and from uncommitted C cells.

Subsequent investigations of synovia from patients with non-inflammatory joint disease demonstrated three cell populations: type I (with MHC class II antigens, Fc receptors, monocyte antigens and capable of phagocytosis); type II (with MHC class II antigens, but without Fc receptors or monocyte markers and non-phagocytic); and type III (with fibroblast antigens but incapable of phagocytosis and without MHC class II antigen or monocyte antigens).[18] It is of particular importance in humans that synoviocytes expressing HLA-DR appear able to present antigens to lymphocytes: they are likely to initiate immune responses to foreign antigens reaching a synovial cavity. On this evidence, the type I synoviocytes of Burmester *et al.*[18] must be assumed to be of high significance in rheumatoid arthritis. Whether the type I (Burmester) cell corresponds to Graabaek's type A cell is not known but systematic studies to examine this possibility have begun.[145]

Synoviocyte division

Synoviocytes divide infrequently: in normal synovial tissue, mitotic figures are very rarely seen and there are suggestions that the renewal of the synovial cell population is entirely by the recruitment of extrinsic precursors.[93a] There is a low labelling index with [^{3}H] thymidine but DNA synthesis and mitotic activity increase quickly and dramatically in response to a great variety of insults ranging from direct injury and surgical trauma to bacterial and viral infection. The lifespan of type A grafted mouse cells is approximately 20 weeks;[43] the potential doubling time, an indication of the time for the regeneration of this cell population, is 24 weeks.

Synoviocyte metabolism

Synovial cells metabolize vigorously.[76] Ribonucleic acid (RNA) formation is active and DNA synthesis is normally at a low rate. There is glycolytic activity, and dehydrogenase enzyme activities are readily demonstrated and measured.[27,26] Synoviocytes synthesize hyaluronate,[118,32] sulphated glycosaminoglycans[70] and link protein;[50] it is probable that they manufacture lubricin (see p. 211). Collagen[21] is a small part of the total protein formed. Lipid metabolism includes eicosanoid synthesis (p. 288) which is often increased in synovitis.

Numerous acid hydrolases, including a phosphatase, β-acetyl-glucosaminidase, cathepsins, β-galactosidase, arylsulphatase A and β-glucuronidase, are characteristically held latent within synovial cell lysosomal walls. The raised activity of cathepsins B and D, neutral collagenase, elastase and acid phosphatase in inflammatory joint disease has attracted particular attention.[201] The secretion of collagenase by synovial fibroblast (B) cells can be induced by iron in the form of ferric nitrilotriacetate[127] and prevented by chelating agents such as desferrioxamine. In addition to the expression of these enzymes, the presence of a plasminogen activator catalyses fibrinolysis.

Synovial cells form many other biologically active substances *in vivo* and *in vitro*. For example, fibronectin (see pp. 51 and 207), is synthesized by explanted normal synoviocytes;[96] and this glycoprotein is a major component of synovial tissue. The variety of molecules is greatly increased in disease (see Chapters 7 and 12). Thus interleukin-1 is released during the enzymatic degradation of cartilage human synovium.[200]

Synoviocyte function

Synoviocytes are potent phagocytes; this property is shared by A and B cells but the macrophage-like type A cells are normally much more active than the more numerous B cells. In practice, almost any solid or fluid substance injected into a synovial joint is avidly removed by phagocytosis or pinocytosis (Fig. 1.19):[158] the red cells in accidental or surgical injury, lipid droplets, contrast media such as thorotrast, elements such as carbon, gold, plutonium or uranium given experimentally, and ^{198}Au, ^{90}Y, or ^{165}Dy administered therapeutically are all seized upon and engulfed. Horseradish peroxidase and iron saccharate are actively taken up by A cells but not, apparently, by S (B + C) cells.[64,65] Crystals, both natural and exogenous, small foreign bodies such as the metabolic particles derived from the abrasive wear of prostheses (see Chapter 24), bacteria, protozoa and fungi are dealt with similarly. Fluids infused into a joint or accumulating as exudates or transudates are removed by pinocytosis;[158] and synovial fluid itself is assimilated in the same way.

So characteristic is the phagocytic potential, so extensive the lining surfaces of the aggregate of synovial joints, that a view has been expressed that the synovium may be regarded as the articular territory of Aschoff's[6] reticuloendothelial system.[58] The reticuloendothelial system[20] (see p. 38) is a collection of cells, widely distributed in vascular and lymphoid tissues, having the common functions of phagocytosis and intravital staining, of a near relationship to blood or lymph channels and the presence of so-called 'reticular' fibres (see p. 50). Synovial phagocytes are closer, anatomically and developmentally, to free connective tissue histiocytes than they are to reticuloendothelial cells. However, the concept of a synovial reticuloendothelial system is valuable since it focuses attention on the observation that the accumulation in the synovia of substances such as haemosiderin and ferritin may contribute to disorders like the anaemia of rheumatoid arthritis.[116] Synovial cells, of course, can synthesize apoferritin.[115] However, the uptake of the radiocolloids ^{198}Au and ^{99}Tc^{m}-sulphide given intravenously to normal rabbits and to animals with antiovalbumin synovitis has been shown to be small in comparison with the activity of the reticuloendothelial cells of the liver, spleen and bone marrow.[67] It appears therefore, that the phagocytic activity of the synovium is essentially a local response which may, nevertheless, have systemic consequences.

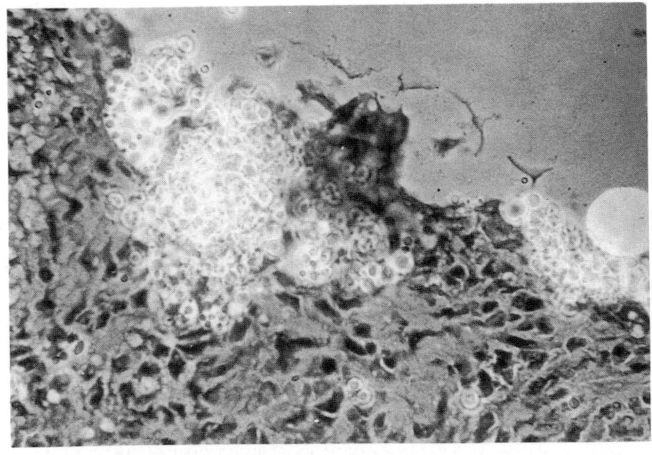

Fig. 1.19 Phagocytosis by synoviocytes.
Latex particles of 5–10 μm diameter were injected under aseptic conditions into the knee joints of rats. Within 30 min, the majority of the particles had been phagocytosed by the synovial type A cells. (Phase contrast × 76.)

Mast cells

Mast cells (gemästete-stuffed) are populations of ovoid or spindle-shaped cells with a rich content of large basophilic granules (Fig. 1.20).[130] They are a normal constituent of nearly every organ but their apparent number varies significantly between different species and, within a single species, between different organs and tissues. Mast cells

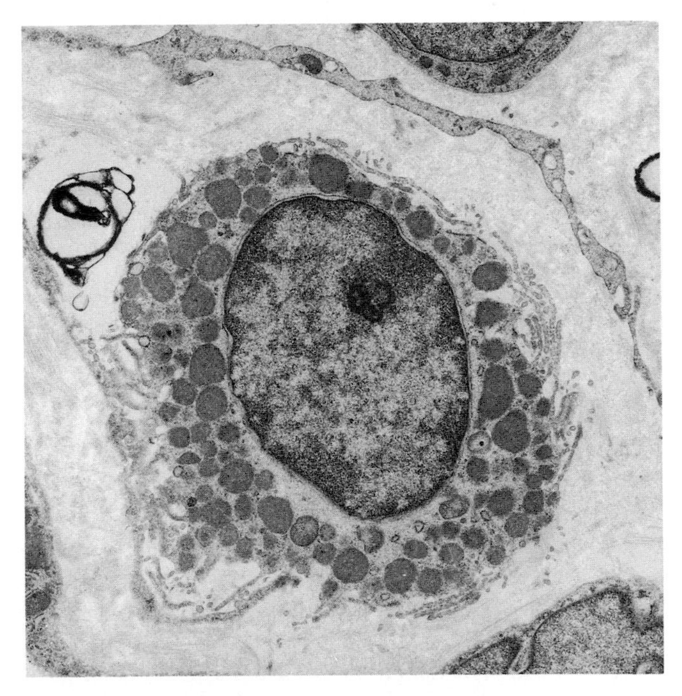

Fig. 1.20 Mast cell.
A connective tissue mast cell is shown within the perineuronal connective tissue of a nerve fibre. The cell contains a myelin body. The mast cell granules occupy a large proportion of the cell. The nuclei (at corners) are those of neurilemmal cells; the attenuated cytoplasm extending diagonally across the field is that of a fibrocyte. (TEM × 4200.)

are of mesodermal origin.[5] Some can be grown in culture in methyl cellulose.[120]

Conventionally, intact mast cells are identified by dyes such as toluidine blue.[167] However, degranulation renders the cell inconspicuous and it may then be necessary to use an antimast cell antibody, transmission electron microscopy or an enzyme histochemical reaction such as chloroacetate esterase to establish their true numbers. To demonstrate mast cell granules reliably by staining techniques, the tissues or cell preparations should be fixed in an alcoholic solution or in the presence of lead acetate.

With such material, it has been established that there are at least two mast cell populations. Using an alcian blue-saffranin O sequence, such as that of Csaba *et al.*[31], applied to tissue fixed in Carnoy's fluid, connective tissues are found to contain mast cells with dark red granules rich in heparin ('typical' mast cells), whereas the lamina propria of the gastrointestinal tract contains mast cells of a smaller size with granules that react briskly for biogenic amines and for chondroitin sulphate proteoglycan but not for heparin ('mucosal' mast cells). Connective tissue mast cells are widely dispersed selectively around small arteries and their nearby nerve bundles. Mucosal mast cells are particularly frequent in the terminal ileum.

Mast cell degranulation

Mast cells readily degranulate, often during preparative procedures. The generation of C5a either by the classic or by the alternate pathway (see p. 238) is the most frequent mechanism. However, elective degranulation is caused by many other agents that interact with the mast cell surface, enhancing calcium ion influx and inhibiting membrane-bound adenylcyclase: the concentration of cyclic adenosine monophosphate falls and there is diminished tubulin polymerization. Among the degranulating agents used experimentally, compound 48/80 is particularly effective.

In response to a stimulus causing degranulation, the granules move to the cell periphery; adjacent granules merge, forming granule complexes. A shallow invagination of the plasma membrane leads to fusion with the granule surface. The disintegration of the fused membranes allows extrusion of the granule contents although some granule matrix remains adherent to the cell surface, slowing the release of the granule constituents and diluting the effector molecules. An elevated histamine concentration is maintained near the cell membrane H_2-receptors. Whether the cell restores its granule content or whether the degranulated mast cell is itself replaced, is not known.[39]

At degranulation, a process blocked by disodium cromoglycate, histamine and heparin are released. Mast cells also liberate eosinophil chemotactic factor, a slow reacting substance of anaphylaxis (see Chapter 5), now known to be a mixture of leukotrienes C4 and D4, and platelet activating factor (see p. 284). Great interest is attached to the mast cell enzyme population: naphthol AS-D chloroacetate esterase is present as it is in polymorphs and mononuclear phagocytes, and myeloperoxidase can be demonstrated. Mast cells can now be obtained from cultures by Ficol gradient concentration in sufficient numbers to allow detailed study of their elastase, chymase, tryptase and collagenase.[139] It is reasonable to suggest that mast cell proteases may play an active part in the degradation of cartilage in rheumatoid arthritis (see Chapter 12) and a role has been attributed to them in osteoarthrosis.[35a]

Mast cells and immune responsiveness

Mast cells play important parts in some humoral and in certain cell-mediated responses. Complement activation, by whatever means (see p. 236), is a potent cause of degranulation. Mast cells bind the Fc component of reaginic (immunoglobulin E) antibody and are of particular significance in those connective tissue disease processes in which anaphylactic (type I) hypersensitivity (see p. 236) is invoked. It is established that, at sites of T-cell aggregation, such as the synovia in rheumatoid arthritis, mast cells increase in numbers,[30,66,86] a response attributable to the liberation by T-cells of a lymphokine, mast cell generating factor.

Mast cells and blood vessels

Mast cells accumulate in the synovial tissues in rheumatoid arthritis and in experimental arthritis as they do around many solid neoplasms; they can be studied in a mouse air pouch model.[166] When the medium from cultured mast cells is tested, it is found to stimulate capillary endothelial cell migration. The mast cell product shown to be responsible for this movement is heparin.[7] It is possible that mast cells influence the proliferation of blood vessels that occurs not only in rheumatoid arthritis[30,36,66] but also in infective arthritis and in the synovitis that complicates osteoarthrosis[35a] and trauma.

The role of mast cells in fibrosis is considered on p. 304.

Histiocytes

Tissue macrophages, histiocytes, are present in almost every loose connective tissue. They cannot normally be distinguished morphologically from occasional blood-derived macrophages (see p. 38) which move into connective tissue spaces. The microglial cells of the central nervous system are extrinsic macrophages of mesodermal origin. Histiocytes of low motility possess a remarkable capacity for the phagocytosis of foreign materials such as silicates, polymers and metals; of cell debris; and of degraded autologous tissue components. Isolated outposts of the reticuloendothelial system empire, the tissue histiocyte becomes laden with non-absorbable materials such as haemosiderin. Haemosiderin accumulates in many parts of the body when, for example, numerous transfusions of blood are given over long periods of time (see p. 971). The transmission electron microscopic appearance of the histiocyte is closely similar to that of the A(M) synoviocyte (p. 32).

Mononuclear phagocytes (macrophages)

Very many of the pathological processes of the connective tissue diseases are executed or influenced by mononuclear phagocytes (Fig. 1.21).[121,146a] Colloidal dyes injected intravenously are avidly phagocytosed by the large mononuclear cells of the reticuloendothelial system.[175,188] The recognition that these cells possess many other properties, particularly the secretion of cytokines regulating messenger, immune, inflammatory, degradative and metabolic functions, led to a new concept of the mononuclear phagocyte.[95] The functions of mononuclear phagocytes centre on ingestion (phagocytosis) and secretion. These properties find expression not only in connective tissue disease processes but also in granulomatous inflammation, atherosclerosis and glomerulonephritis; immediate and delayed hypersensitivity; neoplastic cell death; and amyloidosis.

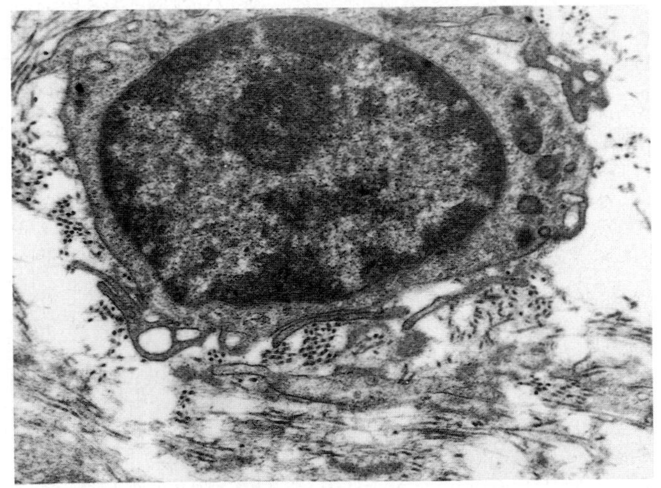

Fig. 1.21 Macrophage.
Mononuclear macrophage within loose connective tissue. Delicate, villous, cytoplasmic processes extend from the cell surface. Cytoplasm contains mitochondria and small numbers of lysosomes. (TEM × 3470.)

Macrophage identification

In human disease, macrophages in cell or tissue preparations are identified by selective stains such as acridine orange; by the application of monoclonal antibodies for macrophage markers such as MAC 387 (calgranulin), MO1 and Y1/82A;[35] and by histochemical tests for enzymes such as chloroacetate esterase. Macrophages can often be recognized by transmission electron microscopy; they have villous, cytoplasmic processes and many lysosomes and phagocytic vacuoles. The uptake of isotopically labelled particles or substrates offers a further means of identification.

Macrophage origins

Mononuclear phagocytes develop from bone marrow precusors. The cells may be liberated into the blood as monocytes which leave the circulation at sites of inflammation or immune reactivity. Re-entry to the blood stream is rare. The population of macrophages in extravascular sites such as the loose subsynoviocytic connective tissue, is augmented by the local formation of further cells.

Macrophage morphology

Although there are differences in size and shape between blood monocytes, bone marrow mononuclear phagocytes and those of the brain (microglial cell), liver (Kupffer cell), kidney (mesangial cell), lung (alveolar and interstitial macrophage), lymph node, spleen and synovium (type A(M) cell) (p. 32), it is convenient to describe the characteristics of a prototype cell. There is great mutability of size and shape. In general, macrophages are large cells with filopodia and an indented reniform nucleus (Fig. 1.16). There is much cytoplasm and the numerous organelles include lyososomes, phagolysosomes and phagocytic vacuoles.

Macrophage activation

Dramatic changes in macrophage behaviour and morphology can be caused by a large variety of exogenous agents. This process of 'activation' is characteristically brought about when macrophages respond to lymphocytes reacting to antigens to which they have been sensitized. The activation is immunologically specific and is induced by lymphokines (cytokines) (see p. 291). This restricted form of activation is highly relevant to the immune responses discussed in Chapter 5. Oxygen-dependent cellular processes and enhanced hydrogen peroxide secretion are involved. A principle agent conferring activation is interferon-γ.

Distinct from immunologically specific and antimicrobial activation is a much wider series of reactions including altered phagocytic potential, increased motility, changed surface receptor behaviour and modified organelle and enzyme content. These non-specific manifestations of activation, oxygen independent, can be provoked by bacterial cell wall components such as muramyl dipeptide (see Chapter 13 and p. 547), by complement, by non-T lymphocytes and by other cells. The non-specific phenomena may be reactions to interleukin-1 or interferon-γ.[121]

Macrophage movement

In response to chemotactic stimuli, circulating macrophages (monocytes) adhere to venular endothelial surfaces and escape from these vessels by passing actively between adjacent cells. Complement component 5a (see p. 238) is the most important chemotactic agent. Penetrating the basement membrane, the macrophage migrates slowly to a site of inflammation. Macrophages adhere to and move particularly quickly on surfaces coated with antigen–antibody immune complexes and on surfaces bearing fibronectin for which macrophages have receptors. The passage of macrophages between cells and through tissues is made possible by a range of enzyme activities of which neutral proteases are the most important. Collagen, proteoglycan, laminin and fibronectin are among the macromolecules that may be degraded by these enzymes. When lymphokines are released by transformed lymphocytes, migration inhibition factor is one cytokine that may act to retain macrophages at sites of tissue injury caused by cell mediated hypersensitivity.

Macrophage ingestion

An important property by which macrophages may be identified is phagocytosis (endocytosis). Normal tissue components are not phagocytosed unless, like red blood cells, they become aged and effete. A recognition system distinguishes old particles from young, self from foreign, injurious from inert. Among the mechanisms that prepare particles for phagocytosis is opsonization. Macrophages have specific receptors for immunoglobulin, complement and fibronectin; they also have non-specific receptors for ligands, glycoproteins and lipoproteins. However, many other materials including fragments of polymers and foreign bodies like glass are vigorously phagocytosed without the mediation of any known receptor. The phagocytosis of monosodium urate crystals is considered on p. 384.

The macrophage, responding to the recognition of 'foreign' surfaces, extends tenuous villi and cytoplasmic processes that move to surround and engulf the foreign material. Drawn into the phagocyte, a particle is held within a phagocytic vacuole (phagosome) until the organelle wall fuses with a nearby lysosome. A change of pH is effected; acid hydrolases are activated and the particle is digested. The degraded products are carried within the phagolysosome to the cell surface and expelled by exocytosis. Alternatively, non-digested material can be retained within residual vacuoles for long periods. In some cases, such as that of the pathogenic mycobacteria, an ingested microorganism may remain viable for months if not years.

Macrophage receptor sites respond most activately to those areas of particle surfaces that have been prepared, coated or opsonized. With the initiation of phagocytosis, there is a burst of cell respiratory activity and a sudden increase in oxygen consumption. Molecular oxygen reduces to superoxide ($O_2^{\cdot -}$) most of which becomes hydrogen peroxide (H_2O_2) (p. 291). The energy for this process derives from the hexose monophosphate shunt. Subsequently, a second respiratory burst releases arachidonic acid and, in turn, numerous pharmacologically active molecules (see p. 288). Nevertheless, phagocytosis and the respiratory burst are dissociable and the energy for the cytokinetics of phagocytosis is thought to come from creatine phosphate and thus, in turn, from the adenosine triphosphate generated by glycolysis.

Macrophage secretion

It is now clear that macrophages synthesize and secrete a great range of molecular species. They are tabulated by Nathan and Cohn.[121] These authors list 14 groups of agent: complement components (e.g. C3); coagulation factors (e.g. Factor VII); other enzymes (e.g. lysozyme, neutral proteases, acid hydrolases); enzyme inhibitors (e.g. α_2-macroglobulin); binding proteins (e.g. fibronectin (see pp. 51 and 207); oligopeptides (e.g. glutathione); bioactive lipids (e.g. many arachidonate metabolites), and platelet-activating factors (see p. 285); nucleosides and metabolites (e.g. thymidine); reactive metabolites of oxygen (e.g. $O_2^{\cdot -}$, H_2O_2); chemotactic factors (e.g. for polymorphs); factors regulating synthetic mechanisms by other cells (e.g. amyloid P component of hepatocytes) (see Chapter 11); collagenase of synoviocytes (see p. 537); factors promoting replication (of lymphocytes (see p. 228) and of fibroblasts); factors inhibiting replication of lymphocytes and of mesangial cells; and other hormone-like factors (e.g. endogenous pyrogen). Many of these groups of substances are briefly referred to on other pages and further consideration of their individual properties is beyond the scope of this text.

Fat cells

Fat cells are encountered in large numbers in connective tissues such as adipose tissue. They occupy retroperitoneal and perivisceral spaces. Mature fat cells are metabolically sluggish and display little oxidative enzyme activity. In the fetus and newborn, several fat droplets can be recognized within a single cell. In the adult, a single large lipid droplet usually displaces the nucleus to one side. The immature human lipoblast also gives rise to the cells of brown fat which are recognizable in the paraaortic and retroperitoneal tissues. Brown fat cells, analogous with those of hibernating animals, are of metabolically high activity. They respond briskly to catecholamines and convert stored energy to heat in reactions mediated by cyclic adenosine monophosphate.

The role of fat in joint lubrication is mentioned on page 22 and the response of the joint to the intraarticular injection of fat in Chapter 24. When adipose tissue is injured directly, as in trauma to the knee synovium or indirectly in ischaemic fat necrosis, degradation of the fat cells liberates products that are locally irritant and chemotactic to polymorphs, plasma cells and histiocytes. The histological consequences are those of fat necrosis.

Smooth muscle cells

Smooth muscle cells (Fig. 1.22) determine the function and behaviour of blood vessels, gut, uterus and other tissues. Smooth muscle cells normally synthesize and secrete the collagen, elastin, proteoglycan and basement membrane which lie between and around them. In important, common disorders such as atherosclerosis (see Chapter 8), the disease is dominated by the excessive formation of these materials. In the uterus, the extent of muscle cell hypertrophy and collagen synthesis during pregnancy is overshadowed only by the speed with which collagen is degraded and lost after parturition.

Endothelial cells also have a capacity to form connective tissue materials such as basement membrane. This ability is well-expressed when diminished blood flow or inflammatory disease culminate in endarterial fibromuscular proliferation, an intimal myofibroblastic response.

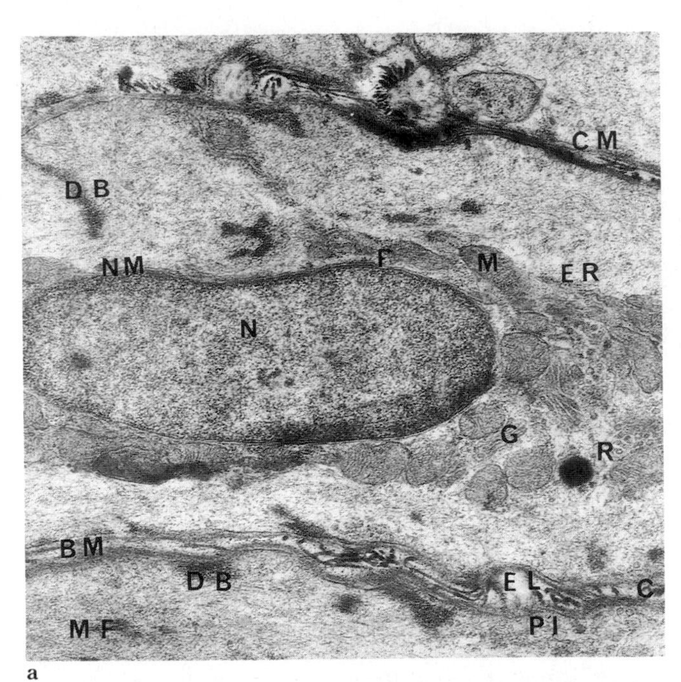

a

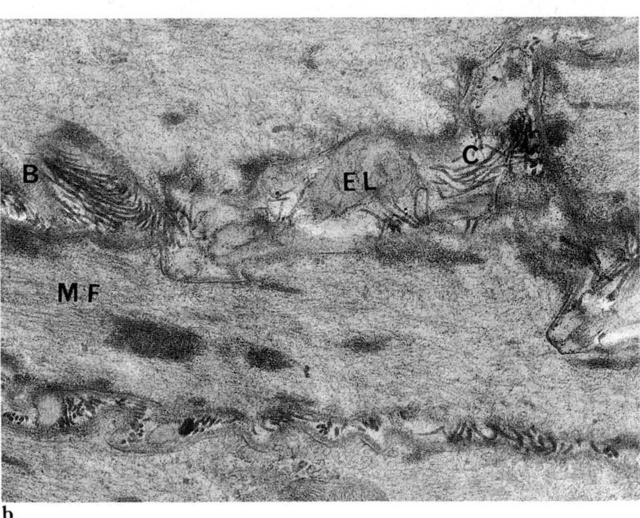

b

Fig. 1.22 Smooth muscle cells.
a. Smooth muscle cells of wall of small artery. Like the elastic chondrocyte, these cells have the programmed capacity simultaneously to manufacture and secrete substantial quantities of matrix collagen, proteoglycan, elastic microfibrillar glycoprotein and elastin in addition to the synthesis of their intrinsic structural, enzymic, contractile and other proteins. *Note*: BM = basement membrane; CM = extracellular matrix; C = collagen fibres; DB = dense body; EL = elastic material; G = Golgi; M = mitochondria; MF = cytoplasmic contractile microfilaments; N = nucleus; NM = nuclear membrane; P = nuclear pore; PL = plasma membrane; R = endoplasmic reticulum. (TEM × 19 240.) **b.** Part of adjacent cell to show island of elastic material in extracellular matrix. (TEM × 18 400.)

Myofibroblasts

There is firm evidence[145a] for the existence of a cell which shares the morphological and synthetic characteristics of the fibroblast with the contractile properties of the smooth muscle cell, the leiomyoblast. The existence of a myofibroblast was suspected by Maximov[111b] but proven only during studies of the contraction and healing of wounds. The part played by the myofibroblast in the maturation of granulation tissue and in fibrosis is outlined on p. 301. The identity and properties of these cells are reviewed by Lipper *et al.*[103] and by Seemayer *et al.*[156]

The myofibroblast, elongated and with a serrated fusiform nucleus, possesses a well-formed Golgi apparatus and rough endoplasmic reticulum. The cytoplasm contains parallel bundles of 4–8 μm diameter actin filaments often located near the plasma membrane where there are attachment sites. Within the cytoplasm, the cytofilaments may be clustered as dense bodies. External to the cell, the filaments may be recognized as microtendinous arrays. Desmosome-like cell junctions appear to facilitate the transmission of forces generated during cell contraction. The myofibroblast responds to vasoconstrictor agents such as prostaglandin F, noradrenaline and angiotensin; cytochalasin B disrupts the contractile filaments and abolishes the contractility of granulation tissue.[156] Myofibroblasts have the ability to manufacture and export type III collagen, elastin, elastic microfibrillar glycoprotein and proteoglycan.

Among the sites in which myofibroblasts have been demonstrated are fetal tendon, duodenal villi and alveolar septa. Myofibroblasts are commonly observed in the stroma of primary and metastatic carcinomata; they are less frequent in sarcomata and are not seen in the lymphomata. When myofibroblasts are the predominant cell type in a diseased tissue, there is a strong likelihood that the process is reactive. This observation accords with the high frequency of myofibroblasts in the neoplasm-like diseases of fibrous tissue, in the fibromatoses and in the fibrous proliferative disorders of infancy and childhood (see Chapter 8). Myofibroblasts are also identifiable in keloids, ganglia, at sites of tendon repair and in the fibrotic reactions to parasites such as *Schistosoma* (Fig. 8.13, p. 323). There remains a possibility that myofibroblasts are implicated in fibrous histiocytoma.

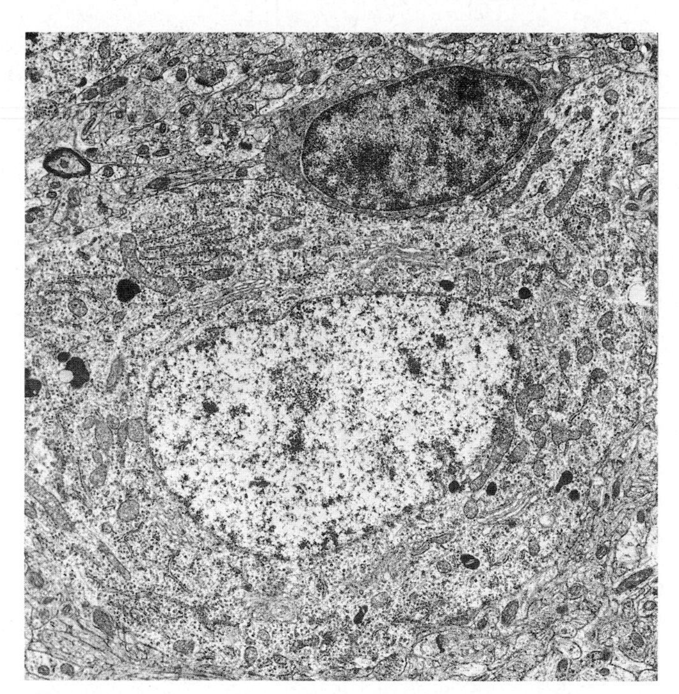

Fig. 1.23 Connective tissue cells of the central nervous system: oligodendrocyte.
Oligodendrocyte from brain of monkey. The cell (at top) is in a satellite position in relation to a nerve cell. (TEM × 5360.)

Other cells that express a connective tissue function

The connective tissues are ubiquitous, and components of the connective tissue system are found in every cellular organ and tissue.[53] In the central nervous system (Figs 1.23, 1.24), glial cells comprise a series of units of neuroectodermal origin, functionally but not developmentally analogous with the fibroblast and its derivatives (Figs 1.23, 1.24). Astrocytes and oligodendrocytes derive from spongioblasts which are themselves progeny of more primitive cells. The astrocyte manufactures glial fibres which are cytoplasmic processes that accumulate in excess in gliosis. The oligodendrocyte maintains a close symbiotic, satellite-

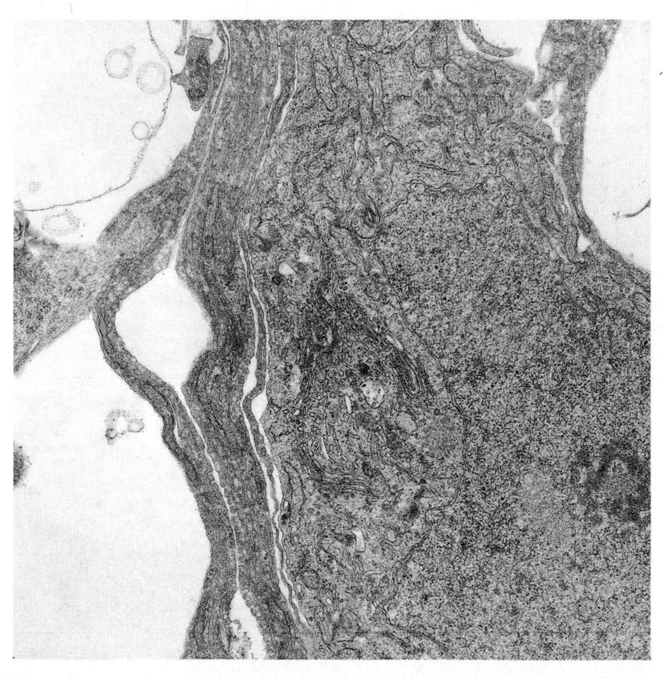

Fig. 1.24 Connective tissue cells of the central nervous system: Muller cell of retina.
The Muller cell stands in the same relationship to the nerve cells of the retina as does the oligodendrocyte to the central nervous system neurones. (TEM × 11 000.)

like relationship with neurones and adjacent small blood vessels.

Connective tissue fibres

Approximately 50 per cent of the dry weight of connective tissues such as articular cartilage is composed of the collagenous proteins which account for 75–85 per cent of the dry weight of the dermis but only 30 per cent of the dry weight of bone.[150] Most of the remaining dry weight of cartilage is proteoglycan (see p. 193). Reticular fibres are rich in type III collagen. In a few sites, connective tissue includes a variable proportion of elastic material (see p. 46). Other non-collagenous extracellular proteins are fibronectin) see p. 207) and chondronectin (see p. 209). Basement membranes, usually comprising non-fibrous, type IV collagen, laminin (see p. 210) and entactin (see p. 211) as well as heparan sulphate proteoglycan, are important connective tissue components (Chapter 4) (Table 1.4).

Collagen

The collagenous proteins (p. 174) form the largest part of the non-aqueous extracellular matrix. Their main function is the provision of a stable tissue scaffolding and a reinforcing and restraining meshwork for cells. They are the main component of bone. The collagens exist in a large variety of molecular and supramolecular forms[93] (Table 4.1, p. 176) (Figs 1.25a, b): their biosynthesis and metabolism are described in Chapter 4. The collagens aggregate extracellularly to comprise an organized macromolecular structure (Fig. 1.26).

There are at least 15 types of known collagen. The collagens of adult articular cartilage are principally type II (95 per cent), type VI (less than 1 per cent) (Fig. 1.27), type IX (1 per cent) (Fig. 1.28) and type XI (3 per cent). Type X collagen is confined to the so-called hypertrophic chondrocytes of epiphyseal growth zones and other sites of cartilage mineralization (p. 107). Bone collagen is type I and this collagen is predominant in tendon and ligament. Menisci (p. 103) such as those of the knee are now known to include collagens types I, II, V and VI while the intervertebral disc (p. 126) collagens comprise at least types I, II, V, VI, IX and XI.

The collagens can be categorized into three classes according to their normal, preferred molecular and supramolecular organization. Class I comprises the 300 nm triple helical, fibril-forming collagens I, II, III, V and XI; the collagens of class II which take the form of interrupted helices found in basement membranes include types IV, VII and VIII; and class III consists of the collagens that are arranged as short helices: types VI, IX, X, XII and XIII. In articular cartilage, 90 per cent of type IX collagen associates with type II and interacts with the type II triple helix in a characteristic way.[203a] Type XI collagen, like the type V collagen found in other tissues, copolymerizes with itself in a 'head-to-tail' arrangement. The extracellular fibrils of types I, II and III collagen display a periodicity of approximately 68 nm. Type VI collagen exists as unusual microfibrils with a periodicity of approximately 100 nm.

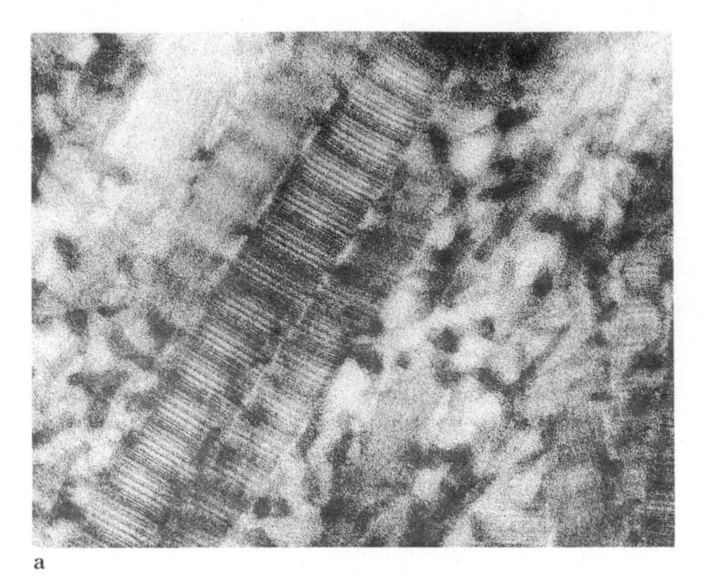

a

b

Fig. 1.25 Collagen and proteoglycan.
a. Collagen fibril, fixed in glutaraldehyde, post-fixed in osmium tetroxide, dehydrated and stained with phosphotungstic acid and uranyl acetate, displaying a characteristic approximately 68 nm cross-banding which is determined by the arrangement, in register, of the individual collagen macromolecules. (TEM × 70 260.)
b. Collagen-proteoglycan relationship. Simultaneous fixation and staining of this collagenous tissue with ruthenium red and by osmium tetroxide allows the demonstration of regularly repeating electron dense foci which are the sites of attachment of proteoglycan molecules to collagen fibrils. In most recorded instances, attachment is at the collagen d band. (TEM × 45 375.)

Table 1.4

Connective tissue fibrous materials
(Modified from Gardner[53])

Fibres	Molecular structure	Site of synthesis	Physical form	Main properties	Staining characteristics
Collagen	Primary amino acid chain has 33 glycine residues per 100; this amino acid occupies every third site in the polypeptide chains; *Hyp* is formed by hydroxylation of many of the *Pro* residues; some lysine is also hydroxylated to give a specifically high hydroxylysine fraction; three chains of amino acids form triple helix of tropo-collagen; the chains vary in composition resulting in the occurrence of at least 15 collagen types	Endoplasmic reticulum; transport within cells to Golgi apparatus; secretion across plasma membrane	Staggered array of tropocollagen molecules forms native collagen fibril; sequential formation of intermolecular cross-links; increasing size of linked fibrils with age. Characteristic 67 nm axial cross-banding on electron microscopy is index of fibrillogenesis	High tensile strength, high elastic modulus; reinforcement of cartilage matrix; nucleation foci for bone mineralization; crystalline translucent structure for lens and cornea	Red with picric acid fuchsin (van Gieson); blue with trichrome methods (e.g. MSB); purple/brown with silver impregnation; doubly refractile (intensified by picro-Sirius red); periodic axial structure revealed by anionic and cationic stains for electron microscopy, or by negative staining
Reticular fibres	Type III collagen content, with associated glycoprotein	As for collagen	As for collagen	Structural support for blood vessels and lymphoreticular system	Blue with Mallory trichrome; black with silver impregnation; not distinctive on electron microscopy
Basement membrane	Type IV procollagen; glycoprotein; heparan sulphate proteoglycan; laminin; fibronectin; entactin	Epithelial cells, mesangial cells, alveolar septal cells, fibroblasts of epithelial lamina	Trilaminar electron-dense structure, continually reformed at molecular level; collagen arranged as mesh, not as fibrils	The ultimate filter between plasma and extracellular fluid	Brown-black to PAS-silver impregnation; electron-dense after uranyl nitrate staining
Elastic	Elastin: protein with little Hyp, unique content of desmosine, isodesmosine and lysinonorleucine; microfibrillar glycoprotein contains hexose and hexosamine; susceptible to chymotrypsin or alkali	Chondrocytes of ear, epiglottis, smooth muscle cells of arteries, septal cells of lung; amino acids essential to polypeptides incorporated on endoplasmic reticulum	Amorphous islands of irregular outline first recognized condensing equidistant from adjacent cells; microfibrils conspicuous at margins of amorphous material	High extensibility; resists large deformations without rupture; support essential for blood vessels in face of pulsatile pressure changes	Dark blue/black with resorcein; pink fluorescence with tetraphenylporphin sulphonate faint brown with orcein; pink with Congo red; according to maturity, stains with potassium permanganate or phosphotungstic acid for electron microscopy

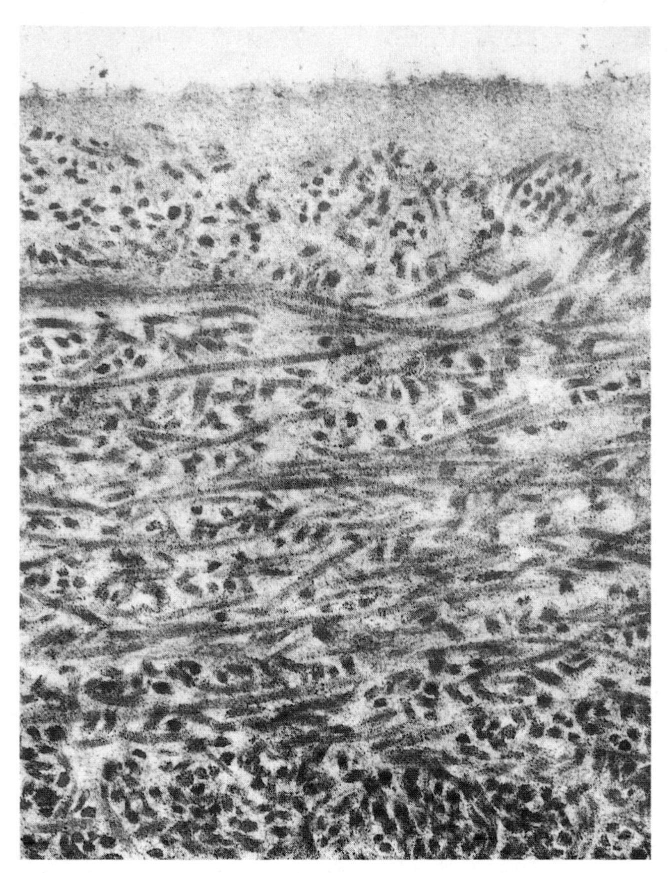

Fig. 1.26 Disposition of collagen fibrils in tissues. In many tissues, such as this adult, articular hyaline cartilage, the fibrils of (type II) collagen, assembled into fibres and fibre bundles, are orientated in preferred directions which relate to the predominant lines of stress to which the structure is subjected. In the most superficial part of this cartilage, a narrow lamina is devoid of fibrillar collagen. Immediately beneath this layer, collagen bundles are arranged in a mat-like array, parallel to the surface. More deeply, the collagen assumes an apparently random structure which is, however, more orderly than had been assumed (see Chapter 2). (TEM × 47 500.)

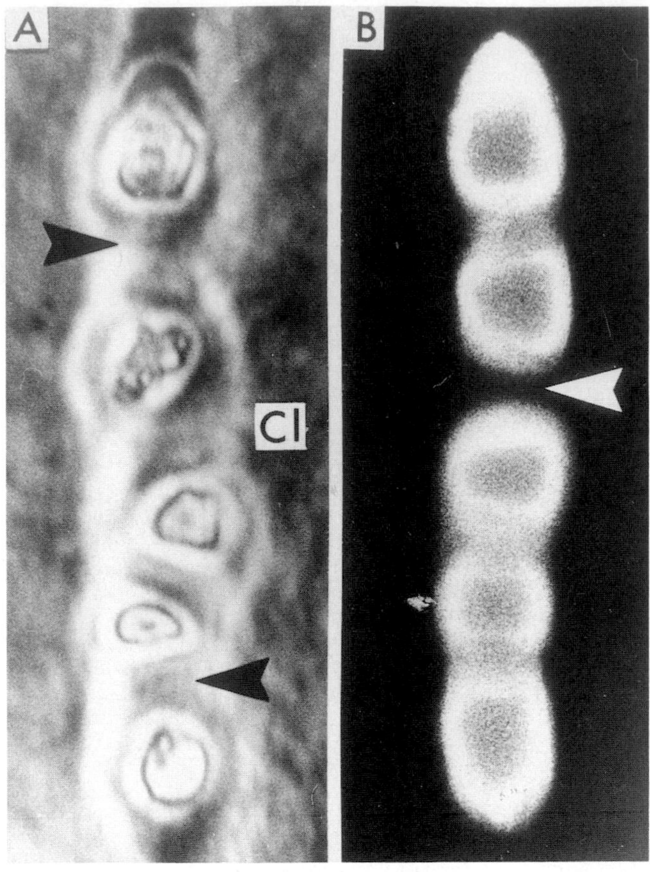

Fig. 1.27 Immunolocalization of type VI collagen. **A.** Columns of chondrocytes within chondrons. Phase contrast microscopy shows connection of five chondrons by common capsular sheath (arrowheads). Frayed collagen bundles (Cl) persistently radiate from the multiple linear arrays. **B.** Following challenge with crude antitype VI collagen antiserum, discrete double and treble chondrones appear to join at a weakly stained interface (arrowhead) to form a column of five chondrons. Radial collagen fibres show no reaction to the antiserum. (**A:** × 705; **B:** × 700.) From Poole *et al.*[134] Reproduced by courtesy of the authors and the Editor, *Journal of Cell Science*.

There is a large and continually evolving literature. For the pathologist, the scale drawings of Trelstad[180a] give a ready impression of the extracellular matrix collagens and of many of the other matrix macromolecules.

Collagens are synthesized by every variety of connective tissue cell except the phagocytic macrophage and the osteoclast; they are also formed by many other cells such as the leiomyocyte (Fig. 1.22). A wide variety of heritable (see Chapter 9) and acquired (see Chapter 10 *et seq.*) diseases is dominated by abnormal, deficient or excessive collagen synthesis. Collagen is used to provide substrates for *in vitro* cell culture[203] and for many other practical purposes such as the covering of burns.

Collagen fibril assembly

Because of its central role as the most abundant protein in metazoan animals, enormous efforts have been directed towards defining the mechanism of assembly of collagen fibres and their extracellular form. The structural molecular biology of collagen is known in fine detail corresponding to that of individual amino acids. The reader is referred to Miller,[113] Chapman and Hulmes,[25] and to Na, Butz and Carrol.[119] Much of the available information has derived from studies of type I (tendon and bone) collagen. However, the ultrastructural organization of types II and III

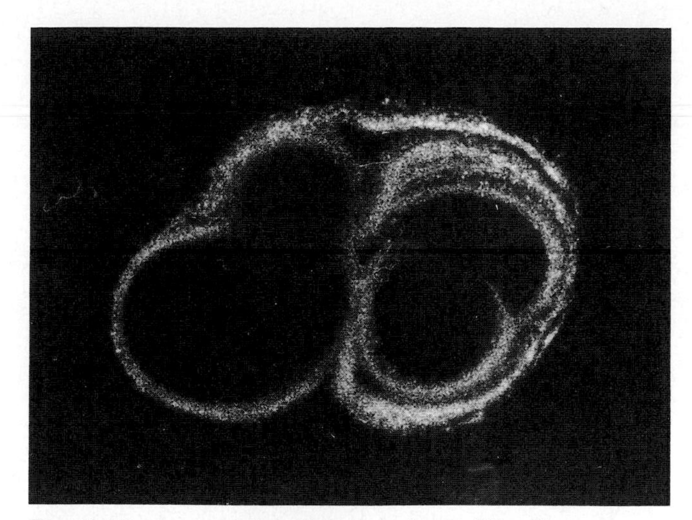

Fig. 1.28 Immunolocalization of type IX collagen.
The capsules of a group of chondrons from a rat chondrosarcoma have been labelled with a monoclonal antibody to type IX collagen and subsequently with a FITC-conjugated anti-mouse IgG. The chondrocytes themselves are not visible within the chondrons. The group of chondrons is ~50 μm in depth and an optical section has been made horizontally through the centre of the group by means of a Bio-Rad Lasersharp MRC500 confocal scanning argon ion laser microscope using light of λ 488 nm. (× 1480.) Courtesy of Mrs S F Wotton, AFRC Institute of Food Research.

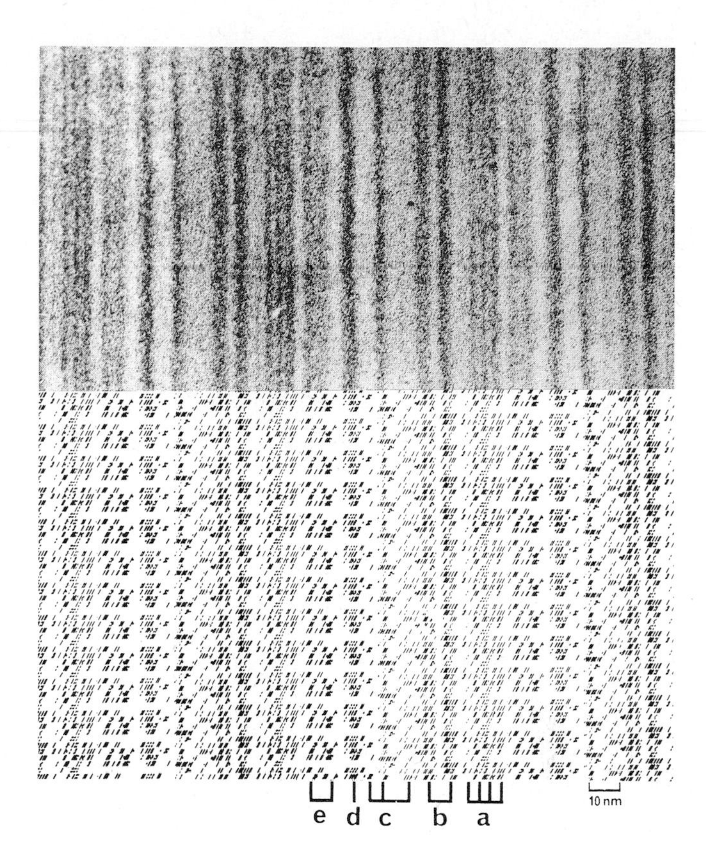

Fig. 1.29 Staining pattern of collagen.
Collagen fibril pattern revealed by staining with phosphotungstic acid and uranyl acetate. The staining pattern is matched with the extended array of charged amino acid residues. The comparison is best made with the page tilted. Courtesy of Dr John Chapman and Dr David Hulmes and Martinus Nijhoff.[25]

collagen is broadly similar. It has proved convenient to employ collagen reconstituted *in vitro* from weak acid or neutral salt solution, and to test embryonic, animal rather than mature, human collagen. Some caution is necessary in extrapolating from these studies of collagen to the changes observed in human disease.

In vivo, collagen fibrils are known often to be several μm long, 8–500 nm thick. In longitudinal array, collagen fibrils display a characteristic periodic structure (Fig. 1.25). Each period is designated D. The structure can be recognized when individual fibrils are teased out on a transmission electron microscopic (TEM) grid and shadow-coated with a heavy metal. A D-periodic structure is also identifiable by TEM when reconstituted fibrils are positively or negatively stained or when fasicles of collagen are surveyed in a thin section (Fig. 1.25). The examination by X-ray diffraction of hydrated rat tail tendon collagen fibrils reveals that the dimension of one D-period is approximately 67 nm. In other tissues this dimension is slightly different. When dehydration precedes TEM examination, the D-period is often approximately 64 nm whereas the study of frozen, hydrated material yields a D-period of approximately 69 nm (Fig. 2.31, p. 102).

The significance of this periodic structure of fibrillar collagen comes from proof that the D-periods are directly determined by the spatial relationship of adjacent collagen molecules (see p. 181). To account for the observed appearances, evidence was advanced for a 'staggering' of collagen molecules within the collagen fibril.[82] A collagen molecule is approximately 300 nm long. When examined in a one-dimensional format (Fig. 1.29), an orderly, periodic structure can be derived when each of a parallel array of molecules is displaced by 68 nm (1-D) with respect to its neighbour. The length of the molecule is 4.48 times the displacement; gaps are therefore seen regularly along the length of the fibril.

The further examination of suitable preparations of collagen fibrils by TEM, after heavy-metal staining, reveals not simply a single electron-dense band at intervals of 1-D, but a series of up to 12 such bands arranged in groups. For historical reasons, these groups are still designated a, b, c, d, e; the bands (Fig. 1.29) are a_{1-4}; $b_{1,2}$; c_{1-3}; d; and $e_{1,2}$. If other techniques such as high resolution neutron diffraction are used, a complex pattern of reflections can be resolved in greater detail, with high precision.[113,17]

The family of bands displayed in an electron micrograph of an appropriate preparation of fibrillar collagen is now known to be an accurate representation of the sequence of amino acids in the collagen polypeptide chains (Fig. 1.29). It is, indeed, a 'molecular staining pattern'.[25] The discrete staining bands represent the positions of charged amino acid residues. The ultrastructural fibril staining pattern can be matched with the molecular sequence of amino acids in the collagen polypeptide chain. A D-period (electron optically) corresponds to 234 amino acid residues.

How are the collagen molecules organized in three-dimension? One explanation was that of Smith.[169] He proposed that the five molecular segments of Hodge and Petruska[82] could be arranged in helical array, around the circumference of a hollow cylinder. The structure thus formed would comprise a filament of five molecular diameters measuring 3.8 nm in thickness overall. Neither this concept, nor modifications of it, could, however, account for the observed data fully[113] and it became clear that the basis of the 3-D organization of the collagen fibril was a six-, not a five-component structure. In cross-section, the six molecular segments appeared to be a slightly flattened hexagon. The arrangement was said to be 'quasi-hexagonal'.[83] The evidence to support this view is examined in detailed by Miller[113] and Chapman and Hulmes.[25] An explanation of the X-ray crystallographic data on which the quasihexagonal model is based is outwith the scope of this text.

The maturation of collagen fibrils necessitates a joining together of adjacent fibrils. The mechanism of cross-linking is discussed in Chapter 4. The union is cemented with increasing strength with advancing age. Stabilization may also be ensured by other recently described mechanisms such as the interactions of the triple helical domains of type IX collagen with nearby type II fibrils.[117] There is an analogy with the co-assembly of types I and V collagen in the cornea (Fig. 1.30).[13a]

Collagen morphology

Mature collagen is recognized by light microscopy as long, straight or wavy, non-branching threads, lying singly or in bundles, as in tendon. The fibres are pale red with haematoxylin and eosin (HE) and can be distinguished from other connective tissue fibres, smooth muscle and polymerized fibrin by appropriate stains (Table 1.4). Mature type I and II collagen fibres cannot be impregnated with silver, i.e. they are not argyrophilic but are readily detected with plane polarized light in which they are birefringent. This property of birefringence can be further exploited with the picro-Sirius red stain after which the fibres display brilliant red double-refractility. It may be possible to use this technique to distinguish type I collagen from type II but the claim that there is a quantitative, linear relationship between the staining intensity and the collagen concentration[87] has been disproved.[142a]

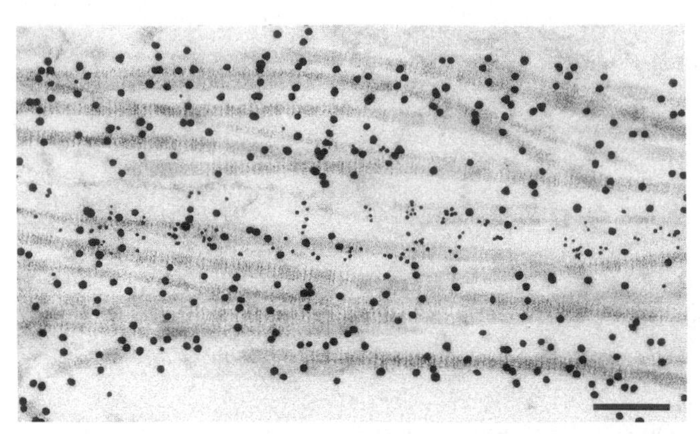

Fig. 1.30 Colocalization of types I and V collagen in chick lathyritic cornea.
Anti-type I collagen antibody labelled with 10 nm gold particles, anti-type V collagen antibody labelled with 5 nm particles. Both collagen types are shown within a single fibril. (Bar = 100 nm.) Courtesy of Dr David E Birk and the Editor, *Journal of Cell Biology*.[13a]

Collagen organization in tissues

The compartmental assembly of extracellular collagen fibrils is discussed in Chapter 8. The distribution and organization of collagen fibres varies widely in different sites. Tendons and ligaments are composed of parallel fibre bundles but the collagen of (superficial) zone I hyaline articular cartilage is arranged in an interlacing meshwork or mat (Fig. 1.26). By contrast, the collagen of (mid-zone) zones II and III articular cartilage is arranged randomly or in a pseudorandom form. The arrangement and organization of collagen fibres in articular cartilage, are considered in Chapter 2.

The suprafibrillar organization of collagen fibre bundles or fasicles is orderly and uniform in ligaments (see p. 99) and tendons (see p. 101). In the lens and cornea, collagen is in crystalline array. The arrangement of collagen fibre bundles in discs (see p. 103), menisci (see p. 103) and labra (see p. 104) follows the radial organization of these tissues. In other structures, for example, in the primate tibial plateau (see Chapter 3), the differential organization of collagen fibre bundles is complex and non-homogeneous.

Elastic material

The term elastic material can be used to describe a composite structure of which the most important part is the protein, elastin.[148] The chemical composition, biosynthesis and degradation of elastin are described in Chapter 4. Three other elastic proteins are known in biology: resilin (from insect wing hinge ligaments), abductin (of mollusc shells), and octopus arterial elastomer. These elastic proteins are rubber-like in their physical properties. They are the only long-range elastic biomaterials that exist as one-phase amorphous polymers.[191] Elastic material is identifiable in almost all organs of the body. However, there are relatively large proportions of elastic material in the intra-articular parts of the temporomandibular joint and in the ear, nasal and epiglottic cartilages; in arteries and veins; in the lungs and bronchi; and in a wide variety of normal and abnormal tissues ranging from the ureter to breast cancers. In the latter, elastic material proliferation may be facilitated by the presence of protease inhibitors.

Elastic material assembly

Transmission electron microscopy[149,183] reveals that the elastic lamellae and fibres recognized, for example, between smooth muscle cells in the walls of arteries, are composed of islands termed elastic aggregates (Fig. 1.31). The aggregates are formed by two morphologically distinct components, well seen in young tissue: a central 'amorphous' substance that accounts for the bulk of the mature fibre or lamella; and a microfibrillar component seen clearly only at the periphery of the elastic aggregate. The amorphous material is elastin, an elastomer. Elastin is arranged in filaments 30–40 nm in diameter, orientated in the long axis of the lamellae and fibres. The filaments display a regular axial periodicity of 4.0 nm. The microfibrillar component, in which individual microfibrils have a tubular structure and are about 11 nm in diameter, is formed of a glycoprotein rich in cysteine.

Histological studies show that the mature elastic aggregate ('fibre') is formed in three stages. Initially, microfibrillar bundles appear which stain with Gomori's aldehyde fuchsin technique only after oxidation by peracetic acid. The resistance of these fibres to acid determined their name, *oxytalan*; they are composed of 10–16 nm diameter microtubules. In a second stage, the microtubules intermingle with amorphous elastin to form *elaunin*.[47,56,151] In the third stage, the mature elastic material is predominantly *elastin*.

Elastic material morphology

Elastic material and elastic fibres can be studied by light microscopic[53,54] (Fig. 1.32), transmission electron microscopic[47,88,89] and scanning electron microscopic techniques[47,123] (Fig. 1.31). Histochemistry,[47,72] immunolocalization,[33,100,47a] and morphometric[182] techniques have been used to assess and measure elastogenesis.

Elastic fibres or laminae seen by light microscopy are

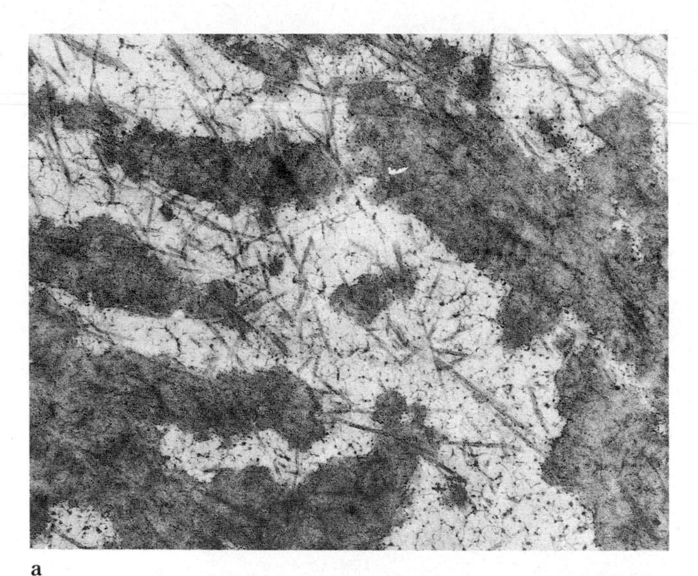

a

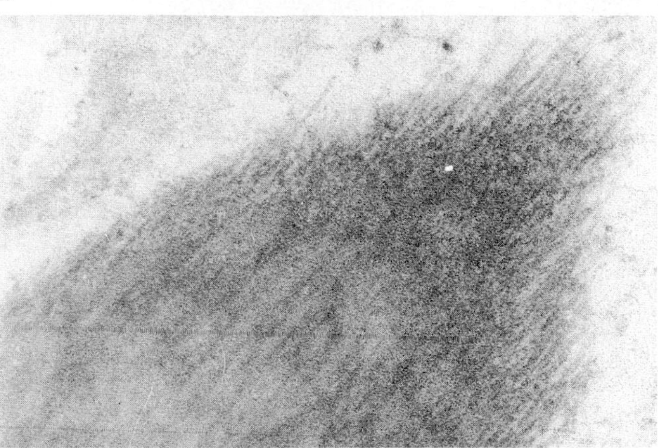

b

Fig. 1.31 Elastic cartilage.
a. Within a matrix in which fibrils of cross-banded collagen are sparsely scattered, islands of elastic material, formed initially from elastic microfibrillar glycoprotein, have accumulated large amounts of elastin. The islands of this maturing material appear grey-black in this section. (TEM × 6200.) **b.** Elastic microfibrillar glycoprotein, seen as arrays of microtubules, within the matrix of elastic cartilage. (TEM × 66 000.) Courtesy of Dr Osama Hassan.[72]

refractile but not birefringent. They are long and branching. They display a faint white fluorescence when viewed by ultraviolet light and become calcified with advancing age. Elastic material stains positively with a large number of dyes; one of the earliest used for this purpose was Congo red. Orcein is a useful reagent for demonstrating the small, delicate elastic fibres of young tissues; Weigert's resorcinol-fuchsin stain, combined with the van Gieson reagents (to demonstrate collagen) is widely employed. Elastic material may also be shown with the fluorescent dye tetraphenylporphin sulphonate which fluoresces a brilliant violet-rose in ultraviolet light. The silver salt of this dye can be employed as an electron-dense stain for transmission

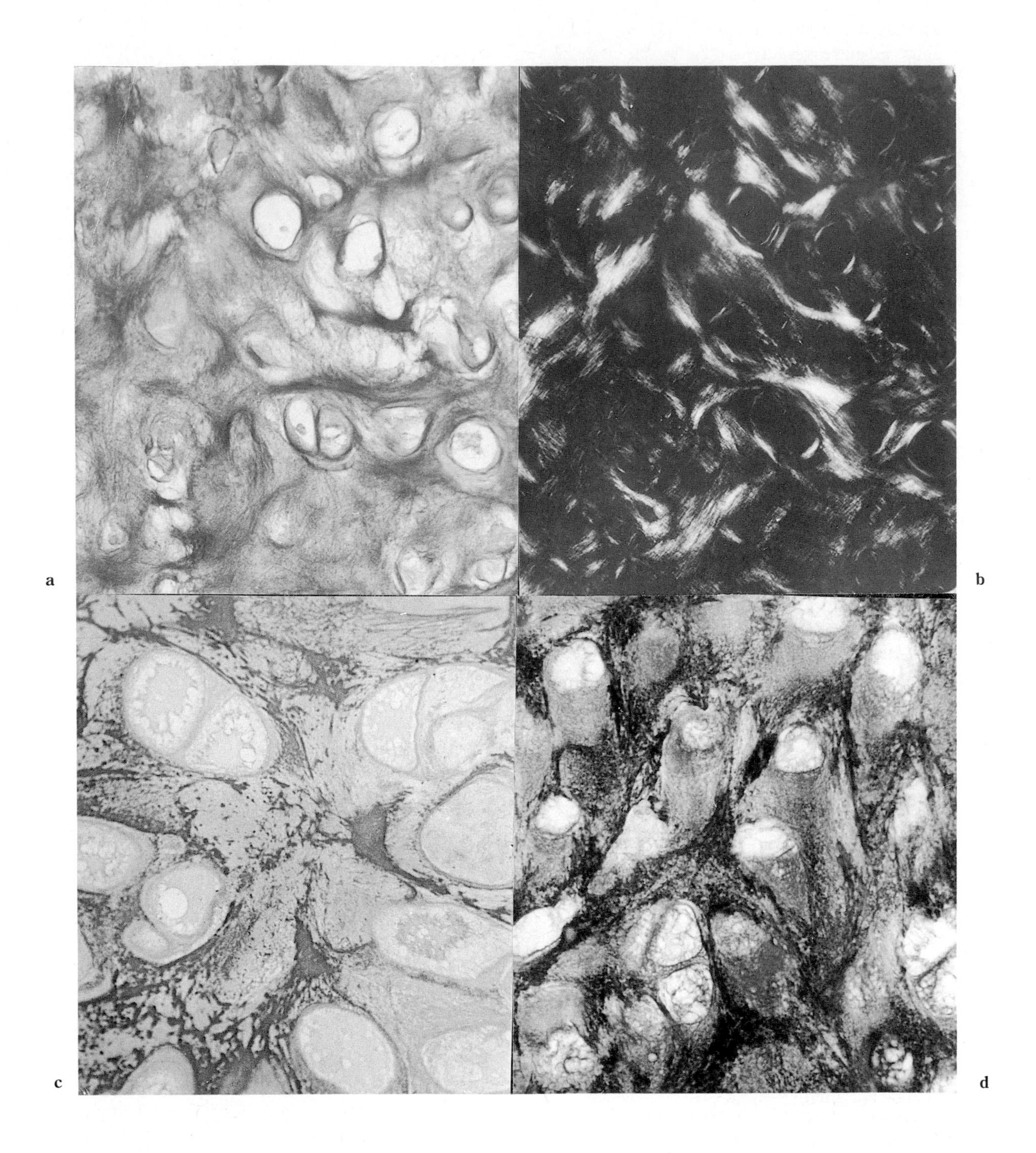

Fig. 1.32 Cells of elastic cartilage.
Adult epiglottic cartilage. Figure shows distribution of collagen and elastic material in relation to chondrocytes. **a.** Collagen fibres, condensed around chondrocyte lacunae, appear dark grey in this paraffin-embedded section. (Picro-Sirius red × 300.) **b.** Same field. (Picro-Sirius red, polarized light × 300.) **c.** Elastic material is organized equidistant from chondrocytes as grey-appearing islands. 1 μm section of Epon-embedded material. (Aldehyde fuchsin × 480.) **d.** Paraffin section of same material. (Orcein × 480.) Reproduced by courtesy of Dr J T Scott and Churchill Livingstone.[155a]

electron microscopy. *In vivo*, tetraphenylporphin sulphonate can cross the placenta; it can therefore be used in studies of embryonic elastogenesis. Immature but not older elastic material reacts with phosphotungistic acid–haematoxylin. The capacity to bind cationic dyes diminishes with time. However, older elastic material continues to be demonstrable with potassium permanganate.

Elastic material is synthesized by the fibroblasts of ligaments, the chondroblasts of elastic cartilage, the smooth muscle cells of elastic arteries (Fig. 1.34) and pulmonary alveolar septal cells. In elastic cartilage, for example, each cell appears capable of the simultaneous synthesis and secretion of elastin, elastic microfibrillar glycoprotein, collagen and proteoglycan, as well as of its own structural and enzymic proteins (Fig. 1.33).

Elastogenesis

The synthesis, secretion and extracellular aggregation of elastic material is programmed genetically; the secretion and maturation of each component of the elastic material begins at exactly defined times in the developing embryo. In the chick vascular system, for example, the onset of elastogenesis is at 5.5 days and is first apparent in the truncus arteriosus. Elastogenesis then extends peripherally in the arterial tree until finally, after 12–14 days, elastic material is recognizable within the smallest arteries (Fig. 1.34).[46] In the pig aorta, there is a four-fold increase in elastin synthesis between the 60th fetal day and birth. In embryonic elastic cartilage, elastin fibres first appear, with collagen, at the margin of the pericellular matrix. In the rat epiglottic cartilage, elastogenesis is not recognizable until 24 h before birth;[72] in human fetal, epiglottic cartilage elastic material formation begins at the fourth month of pregnancy. Elastic fibre formation commences in the outer parts of the walls of the large arteries and extends centripetally towards the endothelium. Elastic material is first recognized in early vascular histogenesis mid-way between adjacent smooth muscle cells, within the abundant proteoglycan matrix (Fig. 1.35). The physiological state of this extracellular matrix is a prerequisite for normal elastogenesis, i.e. defects in elastic material formation can result from an abnormality of the matrix rather than of elastin itself.

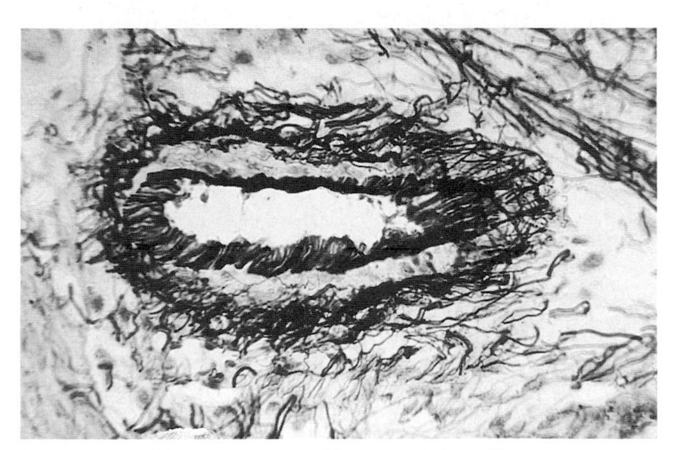

Fig. 1.34 Elastic material in arterial wall.
In this embryonic chick artery, the disposition of the elastic laminae has already attained its adult arrangement. In the chick embryo, elastogenesis is centripetal within the walls of individual arteries such as the aorta but centrifugal in terms of the whole arterial tree. (Weigert–van Gieson × 248.) From El-Maghraby and Gardner[46] by courtesy of the Editor of the *Journal of Pathology*.

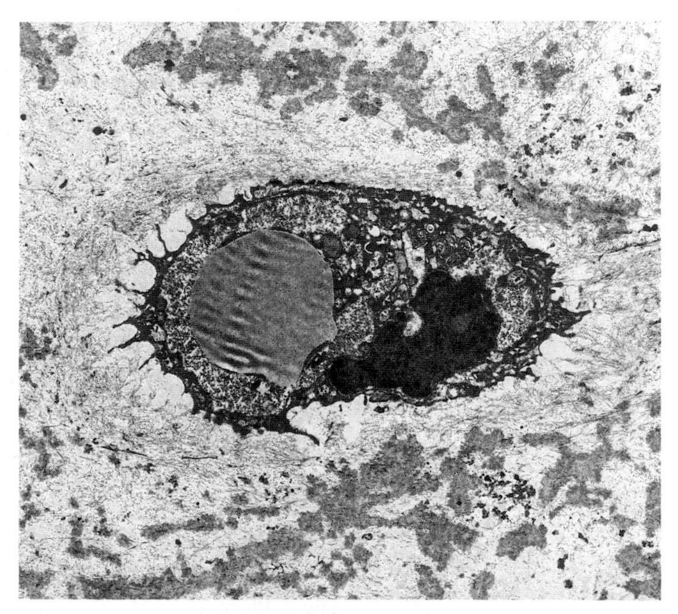

Fig. 1.33 Elastic cartilage chondrocyte.
An ovoid elastic chondrocyte lies in the centre of a matrix rich in islands of elastic material. The dense staining of the cell is in part artefactual, representing a need to enhance the contrast of the adjoining pale-appearing matrix. However, there are signs that the cell may be effete (apoptotic): the nucleus is small and pyknotic and the cytoplasm contains an unusually large lipid droplet. The paler grey-appearing strands and islands of elastic material in the adjoining matrix are granular and their blurred margins still reveal the elastic microfibrillar glycoprotein which has provided a skeleton for elastogenesis (elastin deposition). (TEM × 4300.)

Elastic material

Physical properties Elasticity is discussed in Chapter 6. Physically, the property which characterizes elastin is extreme deformability rather than elasticity. Elasticity is the capacity of a material to recoil perfectly after stretching. Tissue elastic material has a lower Young's modulus (see Chapter 6) than collagen, i.e. it is more extensible than collagenous tissue but less stiff.

Chemical properties Chemically, elastic material is characteristically resistant to hydrolysis by most proteases except the elastases;[88] in inflammatory disorders such as necrotizing vasculitis, elastic material is usually destroyed by polymorph elastase. The synthesis, secretion and biochemistry of elastic material are considered in Chapter 4.

Reticular (reticulin) fibres

The light microscopist distinguishes a delicate, widely dispersed network of non-beaded fibres seen, for example, around the walls of small arteries and renal tubules and intimately related to the microskeleton of the lymphoid and reticuloendothelial systems. The fibres are differentiated from collagen by their non-birefringence in polarized light and by their retention of silver salts after impregnation. There is often a close relationship to the deeper aspects of basement membranes. Reticular fibres are formed of type III collagen: they have 3 α1 (III) component chains. A 67 nm periodic structure is detected by transmission electron microscopy.[93] Reticular fibres are not demonstrable by collagen stains such as picro-Sirius red. Reticular fibres appear very early in embryogenesis, in advance of collagen fibre formation and very far in advance of elastic material synthesis. The recognition and distribution of reticular fibres assist the differential diagnosis of lymphomata and other neoplasms but the identification of reticular fibres is also valuable in the early investigation of connective tissue disorganization.

Basement membranes and basal laminae

Between the epithelial cells of structures such as the intestine and the bronchi and the underlying supporting tissues, is a compact connective tissue layer which provides support, strength and continuity. Basement membranes[3] also regulate important homeostatic functions: renal glomerular excretion, for example, is by ion exchange, a mechanism influenced by glomerular basement membrane polyanionic structure, as well as by molecular sieving. Basement membrane material forms early in the developing yolk sac and embryo.[99] New basement membrane synthesis in later life is very limited: its turnover, which is slow, can be traced by the incorporation of silver as an electron microscopic marker.[192] Glomerular basement membrane collagen has a half-life of approximately 100 days.[140]

The existence of basement membranes, which are difficult to recognize by light microscopical examination of paraffin sections, was first demonstrated by Todd and Bowman.[179a] The faintly eosinophilic basement membranes are well shown by the periodic acid-Schiff and silver impregnation techniques. In inflammatory disease, polymorphs readily penetrate basement membranes, confirming that there is a porous structure. Basement membranes are known to be present at the margins of endothelial, neurilemmal and skeletal muscle cells as well as in epithelia.

Basement membrane morphology

When a basement membrane is examined by transmission electron microscopy, it is found to comprise two parts: there is an inner, 20–100 nm thick basal lamina and an outer reticular lamina. The component of the basal lamina that adjoins epithelial cells is of low electron density but the density increases 30–40 nm from the cell; the inner part is termed the lamina lucida*, the outer the lamina densa (Fig. 1.36). Basal laminae first form as interrupted segments of the lamina densa which are subsequently united by proteoglycan. The external reticular lamina (lamina fibroreticularis) contains varying amounts of fibrillar collagen and differs considerably in thickness in different sites: as Ghadially[60] points out, the reticular lamina of the trachea is 10^2 to 10^3 times thicker than the basal lamina. The reticular lamina of the skin contains elastic material. Where epithelial cells adjoin capillary walls, as in the glomerulus, the basal laminae fuse and the central lamina densa is bounded by two coextensive laminae lucida. The external surfaces of the endothelium of continuous capillaries is covered by a complete basal lamina. The vacuities of fenestrated capillaries are bound in the same way but an incomplete basal lamina surrounds small lymphatic channels and there is no basal lamina in the wall of splenic, bone marrow and lymphoid sinusoids.[60]

The fine structure of basement membranes (Fig. 1.36) is often described as a complex of type IV collagen, laminin and heparan sulphate proteoglycan, arranged as a fibrillar meshwork. However, it may be less misleading to view the arrangement as a closely packed network of cords.[98] The digestion by plasmin of this array reveals a filamentous network of type IV collagen. Laminin and heparan sulphate proteoglycan are bound to the collagen at defined sites along the collagen macromolecule but the demonstration of three further components indicates that basement membranes are more complex than had been suspected. These components are: 1. an 8–10 nm hollow rod or basotubule containing amyloid P component (see p. 432) and formed of piled-up, circular or pentagonal discs, and 2. a loose marginal component. There are also: 3. double pegs of uncertain composition.[98] It is clear that there are large differences in basement membrane ultrastructural architecture between the lamina densa and the lamina rara; there are also marked differences at different sites.

* International Anatomical Nomenclature Committee: *Nomina Anatomica*[123a] quoted by Leblond and Lawrie.[98]

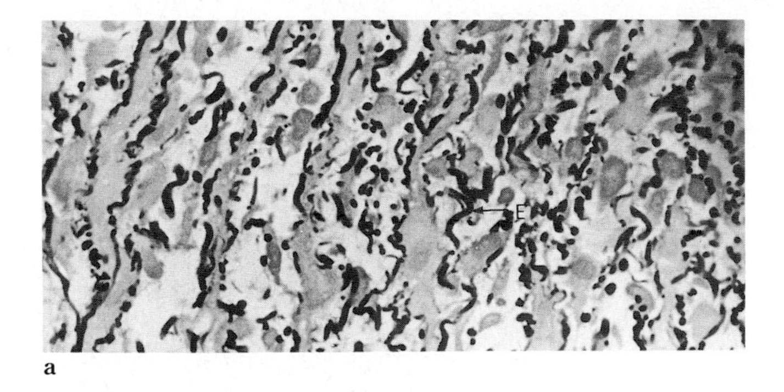

a

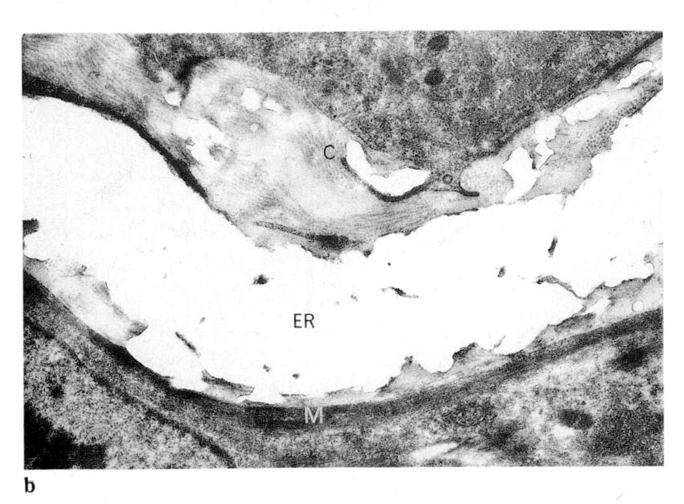

b

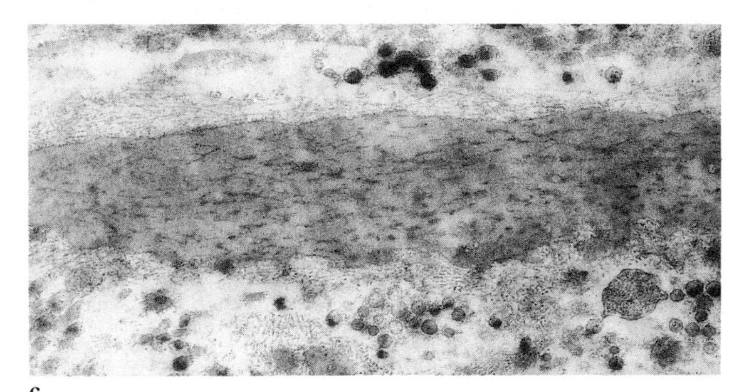

c

Fig. 1.35 Elastic material in arterial walls.
a. Fenestrated laminae in the aortic wall of the rat are seen in two-dimensional paraffin sections as broad, interrupted lines (**E**). (Weigert–van Gieson × 1760.) **b.** After arterial wall tissue has been embedded in a water-soluble polymer, the polymer can be removed prior to digestion by an elastase. In this field, the substantial islands of elastin which accumulate in relation to previously formed elastic glycoprotein, have been removed enzymatically, leaving large, unstained spaces in the connective tissue between the smooth muscle cells. (Durcupan-embedded, elastase-digested, stained with TEM × 11 200.) From Kadar, Bush and Gardner[88] by courtesy of the Editor, *Journal of Pathology*. **c.** For comparison with **b**, intact elastic fibre lying between bundles of transversely sectioned collagen in arterial wall. Elastin has aggregated within an island of elastic microfibrillar glycoprotein the outlines of which can still be recognized. (TEM × 16 060.)

Chemical composition of basement membranes

The chemical composition of basement membranes[75] (Table 1.5) (see Chapter 4) is not yet fully defined. As many as 50 proteins are thought to be present. They must be distinguished from those that lodge in the basement membrane physiologically or in disease. Basement membranes are rich in carbohydrate. The material of basement membranes includes non-fibrillar type IV collagen; heparan sulphate proteoglycan; and the glycoproteins, laminin and entactin. In general, these macromolecules are limited to the lamina densa where fibronectin is also detectable: immunolabelling yields weak reactions in the lamina lucida restricted to that part which joins the lamina lucida to the plasmalemma of nearby epithelial cells; here, membrane receptors such as collagen bind to type IV collagen externally and to the cytoskeleton internally. The role of type VII collagen is not yet fully understood.

Laminin

Laminin (see p. 210) is the principal, non-collagenous basal laminar glycoprotein. It is present in all basement membranes that have been investigated and is a large molecule with a molecular weight of 0.8–1.0 × 10^6 kD. There are two sets of polypeptide chains. Transmission electron microscopy reveals laminin as an asymmetric cross with three short arms and one long arm. There are globular domains at the long-arm ends and in the central region of the short arms (Fig. 4.9). Many cells in culture synthesize laminin. Another polypeptide, *nidogen*, is formed in culture by neoplastic, teratocarcinoma cells. However, *entactin* is also secreted and deposited by cultured fibroblasts; it is recognizable at glomerular epithelial and endothelial cell plasma membranes.

Fibronectin

The fibronectins[84] (p. 207) are a family of closely related glycoproteins with molecular weights of 440–550 kD. They are components of the majority of basement membranes but are ubiquitous and associated with many biological processes ranging from cell adhesion, locomotion and migration to wound healing and tissue repair. Comparable physicochemical properties are exercised by the soluble and insoluble forms of the fibronectins; they display an almost identical amino acid composition and, as a corollary, a similar cross-reactivity to polyclonal antibody. Fibronectins are present in the synovial fluid in rheumatoid arthritis and in pleural exudates as well as in amniotic, cerebrospinal and seminal fluids. There is particular interest in the possibility that fibronectins comprise part of the superficial ultrastructural lamina of cartilage surfaces

Table 1.5

Components of basement membranes
(Modified from Leivo,[99] Abrahamson[3] and Leblond and Laurie[98])

Component	Size	Site	Function
Type IV collagen	450 kD mol. wt assembled as 400 nm, flexible rods	With laminin and HSPG*, as a network of cords ('fibrillar meshwork')	Network formation. Binding to HSPG*, laminin, nidogen and cells
Type VII collagen			Anchoring fibrils projecting from lamina densa into connective tissue
Proteoglycans HSPG* Hyaluronate	HSPG predominates 130–750 kD mol. wt	With laminin, bound at 82 and 206 nm along collagen	Acts as polyanionic negative charge barrier
Laminin	800 kD mol. wt glycoprotein	Throughout lamina densa and lamina rara. Bound at 81, 216 and 291 nm along type II collagen molecule	Cell attachment. Binding of type IV collagen, HSPG, entactin and nidogen
Entactin	150 kD mol. wt sulphated glycoprotein	Associated with laminin	Cell attachment
Fibronectin	V-shaped glycoprotein, each arm, 60 nm long, a 220 kD mol. wt polypeptide	May be extrinsic	Cell–matrix ligand
Hollow rods (basotubule) containing amyloid P	Amyloid P is serum derived glycoprotein with 9 nm pentagonal structure		
Double pegs			

* HSPG = heparan sulphate proteoglycan.

(see Chapter 3) and they are also known to be present in loose subcutaneous and dermal connective tissue, muscle sheaths and organ capsules.

Transmission electron microscopic studies suggest that the fibronectins exist as a pair of interconnected flexible strands devoid of large globular structures.[99] There are multiple domains that exert various biological functions so that the binding to collagen, heparin and heparan sulphate, cell surfaces, fibrinogen and fibrin, staphylococci and actin have been described as fibronectin–fibronectin interactions.

Among the other glycoproteins that act as substrate adhesion molecules are von Willebrand's factor, thrombospondin, haemonectin, chondronectin (p. 209), proteoglycans and vitronectin (p. 209).[40a] *Vitronectin* (serum spreading factor) promotes the attachment and spreading of cells in culture. Localized in the extracellular matrix, vitronectin is associated with platelets; it binds to complement components C5b-9 and takes part in the regulation of complement and coagulation mechanisms by inhibiting membrane insertion and cell lysis. In the tissues, vitronectin is associated with elastic material. Since the quantities of vitronectin are elevated in reactive fibrous tissue, it is thought to contribute to extracellular matrix organization, inflammation and repair.

Integrins

Fibronectin and laminin are attached to cell surfaces by a family of transmembrane α-β heterodimeric phosphoproteins, the integrins (Chapter 4).[84] The integrins are apparently responsible for many cell–matrix and cell–cell interactions. Laminin and fibronectin use different integrins to attach to the cell surface and, depending on which

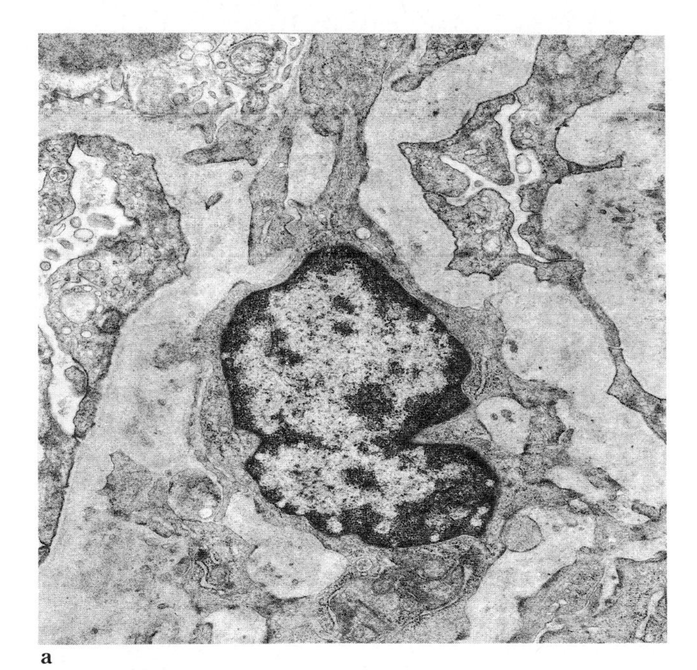

a

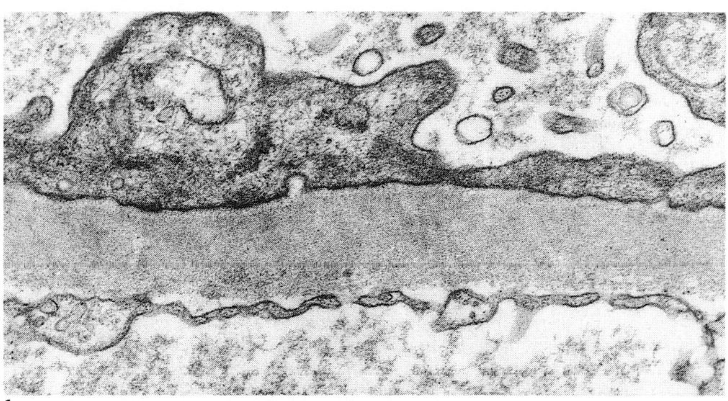

b

Fig. 1.36 Basement membrane.
a. In this glomerulus, a mesangial cell (at centre) lies within ramifying basement membrane material which extends (top left) between glomerular tuft capillaries. (TEM × 9450.) **b.** Higher power view of renal capillary basement membrane. Note the pale, narrow laminae lucida at the margins of the endothelial and epithelial cells, and the broader relatively electron-opaque lamina densa. (TEM × 28 485.)

component is involved, highly different cell behaviour results.[190]

Chondronectin

Chondronectin (p. 209) is a 70 kD mol. wt protein made up of two subunits of 35.5 and 34.5 kD mol. wt, respectively, which form subunits of 77 kD on reduction.[185,186] Chondronectin mediates the attachment of chondrocytes to type II collagen; it can be detected in the cartilage of dogs with the osteoarthrosis that accompanies hip dysplasia (see Chapter 22).

Basal laminae in disease

Basal laminar components may form inadequately or in excess, or they may be metabolized abnormally. Basal laminae may be degraded by proteases activated during inflammation or liberated by cancer cells or microorganisms. A rare disease such as Alports' syndrome, in the early stages of which glomerular basement membranes are thin, centres upon basal laminar defects. A thin basement membrane syndrome has been described in adults with haematuria[1] but the nature of the defect is not understood. More often, as in the spread of cancer cells, in diabetes mellitus and in some immunological disorders, basal laminar changes are secondary consequences of the primary disease. The basal laminae of vascular endothelium thicken with age.

Basal laminae block the escape of neoplastic cells from capillaries and their return to the circulation. Cancer cells bind avidly to basal laminae due to the presence of a laminin-receptor protein. The attached cells secrete enzymes such as heparitinase and type IV collagenase which degrade the basal laminae allowing cancer cells to migrate. It is of interest that metastatic cancer cells are seldom found in synovial tissue.[171a] In diabetes mellitus, basement membrane thickening is an index of renal, retinal and vascular disease. The characteristically thickened basement membrane offers a less effective charge barrier than normal to the passage of serum proteins and they occupy the basement membrane matrix, causing defective filtration. By contrast, insulin deficiency is associated with diminished basement membrane proteoglycan but excess type IV collagen and laminin. In polycystic kidney, the assembly of basement membrane components, an abnormality associated with a gene defect on somatic chromosome 16, is accompanied by increased basement membrane elasticity, tubular dilatation and cyst formation. Some of the renal diseases characterized by the formation of anti-basement membrane antibody or by the lodging in basement membrane of immune complexes, are considered in Chapters 5 and 14. It is not yet certain whether basement membrane components such as laminin can individually provoke autoimmune responses.

Non-fibrous matrix

All connective tissues have an extracellular matrix that is part fibrous, part non-fibrous. The principal component of the non-fibrous matrix is proteoglycan. The proteoglycans are considered in Chapter 4: they are a specialized family of glycoproteins.[71] However, the non-fibrous matrix contains other organic and inorganic materials; in hyaline cartilage, for example, there are glycoproteins such as chondronectin and approximately 1 per cent of the dry weight is non-crystalline calcium phosphate.

Morphology

The proportion of proteoglycan-rich non-fibrous extracellular matrix varies widely between different tissues; it is higher in embryonic and young structures than in old.[45] There is a high but variable proportion of hyaluronate. At one end of the scale, the lens and cornea contain little proteoglycan and these macromolecules are sparse in tendon, osteoid and in bone. At the other end of the scale, there is abundant proteoglycan in the umbilical cord (Wharton's jelly) (Fig. 2.1, p. 66); in the vitreous humour of the eye; in the nucleus pulposus of the young intervertebral disc; and in diseased tissues such as ganglion, myxoma and chordoma. Articular cartilage occupies an intermediate position so that, on average, approximately 50 per cent of the dry weight is proteoglycan; the proportion of proteoglycan in the different forms of cartilage varies inversely with the collagen content and there is relatively little in meniscal fibrocartilage (see p. 82). Chapter 6 explains that where tensile strength and resistance to shear stress is at a premium, collagen predominates. Where the retention of water and resistance to compressive stress is required, proteoglycans abound. However, the distinction between these classes of tissue is not absolute. In most instances, the role of collagen as a microskeleton that acts to retain expanded, hydrated proteoglycan within its domains, means that the proteoglycan–collagen relationship is extremely close.[152,154]

The abundant matrix material among which lie the cells of the loose connective tissues, displays little structure recognizable by light microscopy. Moreover, much proteoglycan is lost during aqueous fixation and in dehydration. Many techniques have been devised to overcome these difficulties (Figs 1.37 and 1.38). Some are discussed on p. 23. Tissue may be rapidly frozen and stained to show matrix proteoglycan after sections have been prepared by cryostat. Alternatively, agents like cetylpyridinium chloride can be added to an aqueous fixative, binding proteoglycans *in situ* so that their loss is minimized.

When these steps have been taken to conserve non-fibrillar connective tissue macromolecules, haematoxylin and eosin staining of an extracellular connective tissue matrix reveals a pale eosinophilic background against which basophilic cell nuclei are prominent. The high proportion of polyanionic extracellular proteoglycan gives the matrix a characteristic metachromasia after staining with cationic dyes in alcoholic solution.[133] The common dyes which stain glycosaminoglycans are thiazines, such as toluidine blue; azins, such as safranin O; the acridines, such as acridine orange; and the phthalocyanins, such as alcian blue 8GX.[44,153] Valuable but less specific staining reactions are given by dialysed ferric hydroxide and colloidal iron or colloidal gold. Ruthenium red,[104,52] and bismuth nitrate[157] yield electron-dense glycosaminoglycan compounds. Important phthalocyanin compounds, cuprolinic blue and cupromeronic blue[152,154] have also been developed for electron histochemistry.

Most of the matrix polysaccharides react slowly with periodate and do not stain with the periodic acid–Schiff (PAS) technique. A positive PAS reaction is, however, given by epithelial mucins, secretions in which distinct oligosaccharides are attached to protein. Nevertheless, in a connective tissue matrix, a positive PAS reaction may be recognizable when sufficient keratan sulphate is present and this may be a basis for the successful use of the technique by Kiviranta *et al.*[92c]

Physical and chemical properties

Many of the most important biological characteristics of the proteoglycans are the consequence of the abundance, in the side-chains, of the polyanionic glycosaminoglycans: they are polyelectrolytes. Each monosaccharide unit bears at least one anionic group, which may be carboxylate, ester sulphate or N-sulphate (see p. 193). These anionic charges are normally associated with counterions, small cationic molecules such as salts of potassium, sodium, magnesium, calcium, strontium and barium that can be exchanged for large molecules like the cationic dyes, under appropriate circumstances (Fig. 1.38).

The polyanionic structure of the glycosaminoglycans accounts for many of the main properties of the proteoglycans[150] and aids understanding of their disorders. The properties include: 1. Counterion binding: proteoglycan can be precipitated from solution by polyvalent cations with which they form insoluble complexes. Proteoglycans consequently react with thiazine dyes such as toluidine blue to display metachromasia, and critically bind phthalocyanin dyes such as alcian blue.[153] 2. Thermodynamic properties: Proteoglycans behave as gels, avidly binding water. The extent of the swelling consequent on water binding is restricted in tissues like cartilage by the collagen network. Other nearby proteoglycans become entangled; 3. Ex-

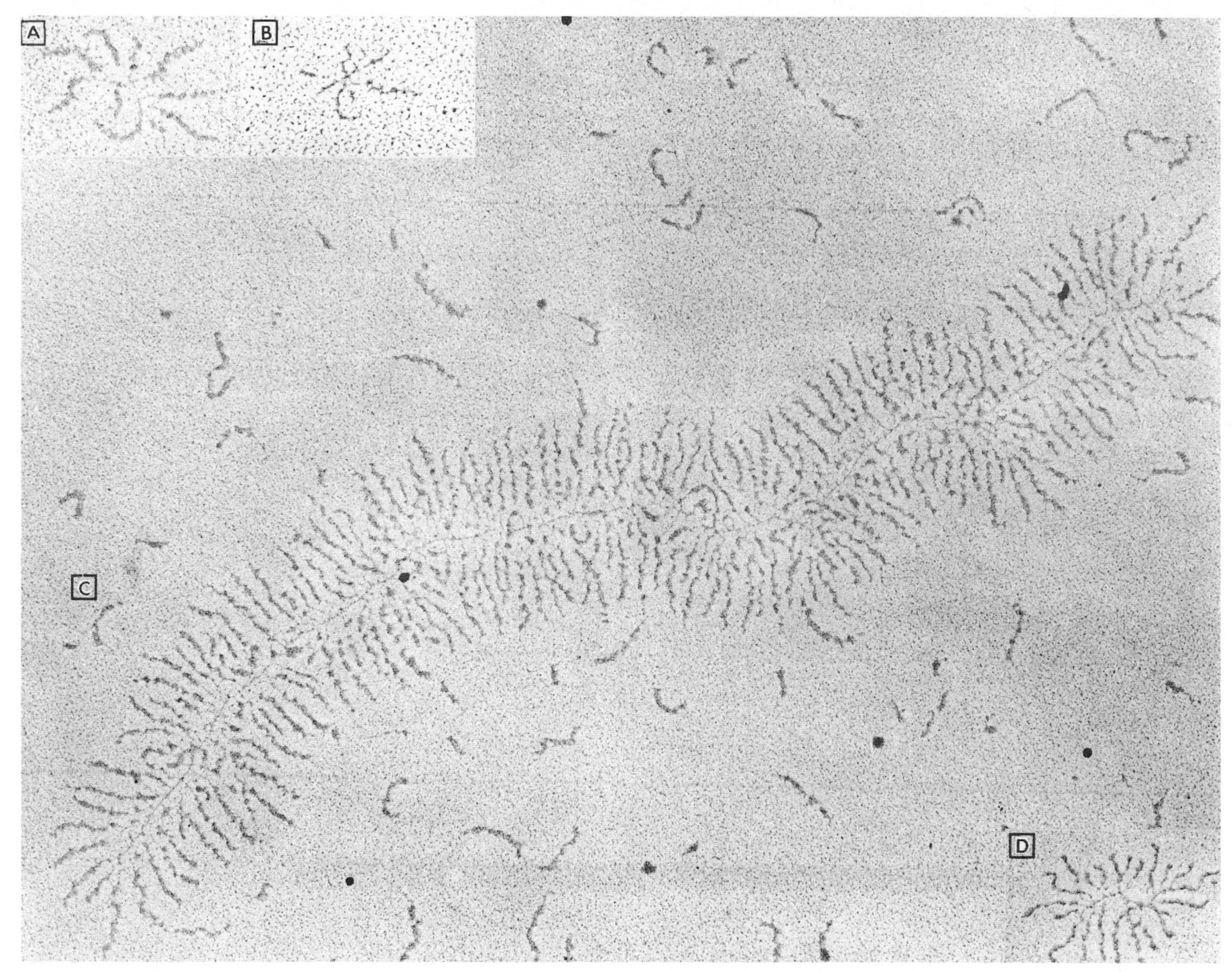

Fig. 1.37 Articular cartilage: age-related changes in proteoglycan aggregates and comparison between articular and cultured chondrocyte material.
Bovine articular cartilage proteoglycan. Spread preparation mixed with cytochrome c on air–liquid interface, followed by low angle rotary shadowing. Transmission electron microscopy shows **A**: proteoglycans synthesized by chondrocytes in culture; **B**: proteoglycans synthesized by adult (steer) chondrocytes in culture; **C**: proteoglycans from calf articular cartilage; and **D**: proteoglycan, from adult (steer) articular cartilage. Length varies from approx. 1130 nm (calf articular cartilage) to approx. 130 nm (adult chondrocytes in culture). (TEM: A, B, D: × 55 800; C: × 60 700.) By courtesy of Dr James A Buckwalter and the Editor, *Electron Microscope Reviews*.[17a]

cluded volume properties. The arrangement of individual proteoglycans within a connective tissue matrix is determined by the anionic groups which act to establish a 'domain' from which other molecules are actively excluded.
4. Flow properties: Proteoglycans tend to be very viscous in solution; they sediment slowly. Viscosity is a particularly important property of synovial fluid (see p. 272) where lubricin provides the unique lubricant in a fluid medium rich in hyaluronate, not in proteoglycan. In this fluid, however, the high viscosity is attributable to hyaluronate protein, not to proteoglycan.

Proteoglycan and collagen relationships

In the fibre-reinforced gel that is articular catrilage, collagen and proteoglycan interact.[173,174,129] *In vitro*, glycosaminoglycan can induce and stabilize fibre formation. Chondroitin sulphate and dermatan sulphate bind to monomeric collagen but hyaluronate and keratan sulphate, lacking the necessary two charges per disaccharide repeat, or unit, do not. The collagen–proteoglycan interactions are temperature-dependent.

Mechanical principles strongly suggest that connective

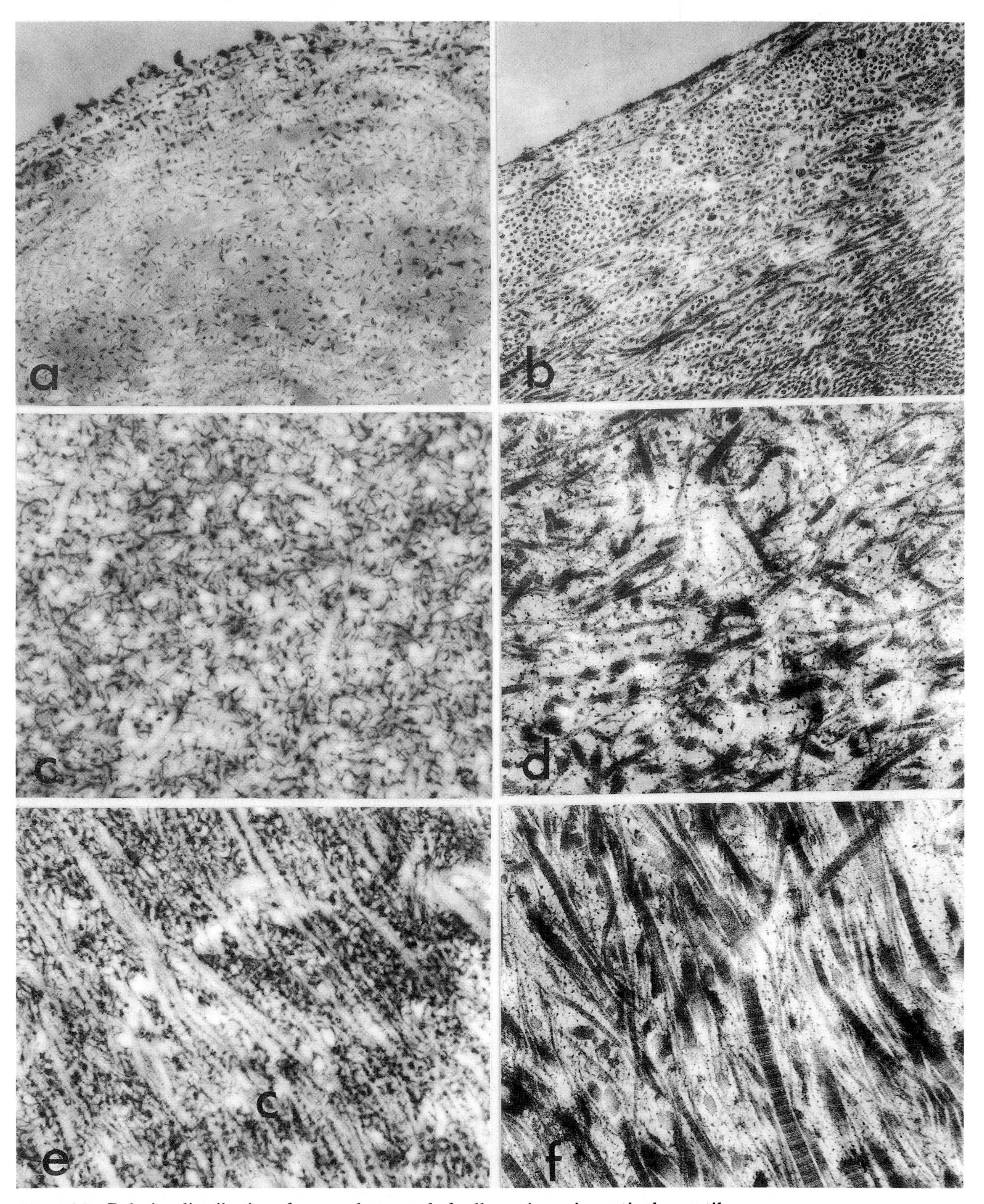

Fig. 1.38 Relative distribution of proteoglycan and of collagen in canine articular cartilage.
Perpendicular sections from central, load-bearing region of femoral condyle to show distribution of proteoglycans (**a**, **c** and **e**) and the organization of fibrillar collagen (**b**, **d** and **f**). **a** and **b** are from zone I (superficial), **c** and **d** from zone II/III (mid) and **e** and **f** from zone IV (deep). **a**, **c** and **e** stained with cupromeronic blue in presence of 0.3 M magnesium chloride; **b**, **d** and **f** with phosphotungstic acid, uranyl acetate and lead citrate. (TEM × 25 000.) From O'Connor, Orford and Gardner[125] courtesy of the Editor, *Annals of the Rheumatic Diseases*.

tissue collagens and proteoglycans are systematically interrelated both in molecular as well as in ultrastructural terms. In rat tail tendon, proteoglycan, 'stained' by the cationic dye cupromeronic blue, is related to collagen at the d-band.[155] A closely similar relationship has been shown in articular cartilage.[129] Comparable evidence has been obtained from studies of the sclera where Young[204] has demonstrated that approximately 50 × 5 nm filaments of proteoglycan are associated with the d-band of collagen. This evidence suggests that a comparable relationship might obtain in the majority of connective tissues. Indeed, Ronziere *et al.*,[146] and others, have confirmed the ultrastructural histochemical evidence that indicates a close and orderly collagen–proteoglycan organization. They employed X-ray diffraction analyses of bovine cartilage. Low-temperature scanning electron microscopic evidence is in accord with these views.[132] The part played by non-fibrillar, 'minor' collagens such as types VI, IX and XI in this organization is being unravelled.[168,134] S-100, a calcium-binding protein expressed in cartilage, may play a part in the mechanisms regulating proteoglycan–collagen relationships.[89a]

There remains the problem of connective tissue heterogeneity discussed further in Chapter 2. Even in the superficially most uniform variety of hyaline cartilage, for example, there is a remarkable variability from one part of the tissue to another.[38a] Different perpendicular cartilage zones contain different proportions of collagen and proteoglycan.[12,92b] The concentration of keratan sulphate varies correspondingly. There are differences between high and low weight-bearing areas of the tibial plateaux and femoral condyles[110] and between the size of proteoglycan monomers in the articular cartilages of high (hip, knee) and low (shoulder) weight-bearing joints.[68]

REFERENCES

1. Abe S, Amagasaki Y, Iyori S, Konishi K, Kato E, Sakaguchi H, Shimoyama K. Thin basement membrane syndrome in adults. *Journal of Clinical Pathology* 1987; **40**: 318–22.

2. Abercrombie M. Fibroblasts. *Journal of Clinical Pathology* 1978; **31**, Suppl. (Royal College of Pathologists) 12: 1–6.

3. Abrahamson D R. Recent studies on the structure and pathology of basement membranes. *Journal of Pathology* 1986; **149**: 257–78.

4. Alberts B, Bray D, Lewis J, Raff M, Roberts K, Watson J D. *Molecular Biology of the Cell* (2nd edition) Differentiated cells and the maintenance of tissue. New York: Garland, 1989: 951–1000.

5. Andrew A, Rawdon B B. The embryonic origin of connective tissue mast cells. *Journal of Anatomy* 1987; **150**: 219–27.

6. Aschoff K A L. Das reticulo-endothelial system. *Ergebnisse für Innere Medizin* 1924; **26**: 1–118.

6a. Aydelotte M B, Greenhill R R, Kuettner K E. Differences between sub-populations of cultured bovine articular chondrocytes. II. Proteoglycan metabolism. *Connective Tissue Research* 1988; **18**: 223–34.

6b. Aydelotte M B, Kuettner K E. Differences between sub-populations of cultured bovine articular chondrocytes. I. Morphology and cartilage matrix production. *Connective Tissue Research* 1988; **18**: 205–22.

7. Azizkhan R G, Azizkhan J C, Zetter B R, Folkmann J. Mast cell heparin stimulates migration of capillary endothelial cells *in vitro*. *Journal of Experimental Medicine* 1980; **152**: 931–44.

8. Bab I, Howlett C R, Ashton B A, Owen M E. Ultrastructure of bone and cartilage formed *in vivo* in diffusion chambers. *Clinical Orthopaedics and Related Research* 1984; **187**: 243–54.

8a. Baird A, Bohlen P. Fibroblast growth factors. In: Sporn M B, Roberts A B, eds, *Handbook of Experimental Pharmacology* (Volume 95/1). *Peptide Growth Factors and their Receptors I.* Berlin: Springer-Verlag, 1990: 369–418.

9. Barland P, Novikoff A B, Hamerman D. Electron microscopy of the human synovial membrane. *Journal of Cell Biology* 1962; **14**: 207–20.

10. Bassleer C, Gysen P, Foidart J M, Bassleer R, Franchimont P. Human chondrocytes in tridimensional culture. In Vitro *Cellular and Developmental Biology* 1986; **22**: 113–9.

11. Bates G P, Schor S L, Grant M E. A comparison of the effects of different substrata on chondrocyte morphology and the synthesis of collagen types IX and X. In Vitro *Cellular and Developmental Biology* 1987; **23**: 374–80.

12. Bayliss M T, Venn M, Maroudas A, Ali S Y. Structure of proteoglycans from different layers of human articular cartilage. *Biochemical Journal* 1983; **209**: 387–400.

12a. Benninghoff A. Ueber den funktionellen Bau des Knorpels. *Anatomische Anzeigung* 1922; **55**: 250–67.

13. Bernard G W, Pease D C. An electron microscopic study of initial intramembranous osteogenesis. *American Journal of Anatomy* 1969; **125**: 271–90.

13a. Birk D E, Fitch J M, Babiarz J P, Linsenmayer T F. Collagen type I and type V are present in the same fibril in the avian corneal stroma. *Journal of Cell Biology* 1988; **106**: 999–1008.

14. Bjorenson J E, Berg N B, Olsen J A, Austin B P. Incorporation of leucine by mouse synovial cells *in vivo*. *Acta Anatomica* 1986; **127**, 237–40.

15. Boivin G, Morel G, Meunier P J, Dubois P M. Ultrastructural aspects after cryoultramicrotomy of bone tissue and sutural cartilage in neonatal mice calvaria. *Biologie de la Cellule* 1983; **49**: 227–30.

16. Boskey A L. Current concepts of the physiology and biochemistry of calcification. *Clinical Orthopaedics and Related Research* 1981; **157**: 225–57.

17. Bradshaw J P, Miller A, Wess T J. Phasing the meridional diffraction pattern of type I collagen using isomorphous derivatives. *Journal of Molecular Biology* 1989; **205**: 685–94.

17a. Buckwalter J A, Rosenberg L C. Electron microscopic

studies of cartilage proteoglycans. *Electron Microscopic Reviews* 1988; **1**: 87–112.

18. Burmester G R, Dimitriu-Bona A, Waters S J, Winchester R J. Identification of three major synovial lining cell populations by monoclonal antibodies directed to Ia antigens and antigens associated with monocytes/macrophages and fibroblasts. *Scandinavian Journal of Immunology* 1983; **17**: 69–82.

19. Burmester R, Menche D, Merryman P, Klein M, Winchester R. Application of monoclonal antibodies to the characterization of cells eluted from human articular cartilage—expression of Ia-antigens in certain diseases and identification of an 85 kD cell surface molecule accumulated in the pericellular matrix. *Arthritis and Rheumatism* 1983; **26**: 1187–95.

20. Cameron G R. *Pathology of the Cell*. Edinburgh: Oliver and Boyd, 1952.

21. Castor C W. The microscopic structure of normal human synovial tissue. *Arthritis and Rheumatism* 1960; **3**: 140–51.

22. Castor C W, Rowe K, Dorstewitz E L, Wright D, Ritchie J C. Regulation of collagen and hyaluronate formation in human synovial fibroblast cultures. *Journal of Laboratory and Clinical Medicine* 1970; **75**: 798–810.

23. Chambers T J. The cellular basis of bone resorption. *Clinical Orthopaedics and Related Research* 1980; **151**: 283–93.

24. Chambers T J. The pathobiology of the osteoclast. *Journal of Clinical Pathology* 1985; **38**, 241–52.

24a. Chambers T J. The effect of calcitonin on the osteoclast. *Triangle* 1988; **27**: 53–60.

25. Chapman J A, Hulmes D J S. Electron microscopy of the collagen fibril. In: Ruggeri A, Motta P M, eds, *Ultrastructure of the Connective Tissue Matrix*. Lancaster: Martinus Nijhoff, 1984: 1–33.

26. Chayen J. Quantitative cytochemistry: a precise form of cellular biochemistry. *Biochemical Society Transactions* 1985; **12**: 887.

27. Chayen J, Bitensky L. Butcher R G, Poulter L W. Redox control of lysosomes in human synovia. *Nature* 1969; **222**: 281–2.

28. Cheung H S. *In vitro* cartilage formation on porous hydroxyapatite ceramic granules. In Vitro *Cellular and Developmental Biology* 1985; **21**: 353–7.

29. Cheung H S, Ryan L M. A method of determining DNA and chondrocyte content of articular cartilage. *Analytical Biochemistry* 1981; **116**: 93–7.

30. Crisp A J, Chapman C M, Kirkham S E, Schiller A L, Krane S M. Articular mastocytosis in rheumatoid arthritis. *Arthritis and Rheumatism* 1984; **27**: 845–51.

31. Csaba G, Surján L, Fischer J, Kiss J, Törö I. On the mechanism of mast cell formation. Effect of glucocorticoids on the mast cells of normal and thymectomized rats. *Acta Biologica Academica Scientifica Hungarica* 1969; **20**: 57–74.

32. Dahl I M S, Husby G. Hyaluronic acid production *in vitro* by synovial lining cells from normal and rheumatoid joints. *Annals of the Rheumatic Diseases* 1985; **44**: 647–57.

33. Damiano V V, Tsang A, Christner P, Rosenbloom J, Weinbaum G. Immunologic localization of elastin by electron microscopy. *American Journal of Pathology* 1979; **96**: 439–56.

34. Darnell J, Lodish H, Baltimore D. *Molecular Cell Biology*. New York: Scientific American Books, 1986.

35. Davey F R, Cordell J L, Erber W N, Pulford K A F, Gatter K C, Mason D Y. Monoclonal antibody (Y1/82A) with specificity towards peripheral blood monocytes and tissue macrophages. *Journal of Clinical Pathology* 1988; **41**: 753–8.

35a. Dean G, Freemont A J. Is osteoarthritis a mast cell-mediated disease? *Journal of Pathology* 1989; **158**: 358A.

36. Delbruck A, Dresow B, Gurr E, Reale E, Schroder H. *In vitro* culture of human chondrocytes from adult subjects. *Connective Tissue Research* 1986; **15**: 155–72.

37. Dreher R. Origin of synovial type-A cells during inflammation: an experimental approach. *Immunobiology* 1982; **161**: 232–45.

38. Dryll A, Lansaman J, Peltier A P, Ryckewaert A. Cellular junctions in normal and inflammatory human synovial membrane revealed by tannic acid and freeze fracture. *Virchows Archiv A* (Pathol. Anat. Histol.) 1980; **386**: 293–302.

38a. Dunham J, Shackleton D R, Billingham M E J, Bitensky L, Chayen J, Muir I H. A reappraisal of the structure of normal canine articular cartilage. *Journal of Anatomy* 1988; **157**: 89–100.

39. Dvorak A M, Galli S J, Schulman E S, Lichtenstein L M, Dvorak H F. Basophil and mast cell degranulation: ultrastructural analysis of mechanisms of mediator release. *Federation Proceedings* 1983; **42**: 2510–5.

40. Editorial. Limb salvage surgery. *Lancet* 1988; **ii**: 662–3.

40a. Editorial. Fibronectins and vitronectin. *Lancet* 1989; **i**: 474–6.

41. Edwards J C W. The origin of type A synovial lining cells. *Immunobiology* 1982; **161**: 227–31.

42. Edwards J C W, Mackay A R, Sedgwick A D, Willoughby D A. Mode of formation of synovial villi. *Annals of the Rheumatic Diseases* 1983; **42**: 585–90.

43. Edwards J C W, Willoughby D A. Demonstration of bone marrow derived cells in synovial lining using giant lysosomal granules as genetic markers. *Annals of the Rheumatic Diseases* 1982; **41**: 177–82.

43a. Eggli P S, Hunziker E B, Schenk R K. Quantitation of structural features characterizing weight- and less-weight-bearing regions in articular cartilage: a stereological analysis of medial femoral condyles in young adult rabbits. *Anatomical Record* 1988; **222**: 217–27.

44. Elliott R J. Biochemical studies of age associated changes in human articular cartilage. PhD thesis, The Queen's University of Belfast, 1976.

45. Elliott R J, Gardner D L. Changes with age in the glycosaminoglycans of human articular cartilage. *Annals of the Rheumatic Diseases* 1979; **38**: 371–7.

46. El-Maghraby M A H A, Gardner D L. Development of connective tissue components of small arteries in the chick embryo. *Journal of Pathology* 1969; **108**: 281–91.

46a. Engfeldt B. Studies on the epiphyseal growth zone. *Acta Pathologica et Microbiologica Scandinavica* 1969; **75**: 201–19.

46b. Engfeldt B, Caterson B, Eklof O, Hultenby K, Muller M. Ultrastructure of hyaline cartilage. 2. Recent developments in preparatory procedures of electron microscopy and immunocytochemistry for the classification of bone dysplasias. *Acta Pathologica et Microbiologica Scandinavica, Section A* 1987; **95**: 371–6.

46c. Engfeldt B, Hjertquist S O. The effect of various fixatives on the preservation of acid glycosaminoglycans in tissues. *Acta Pathologica et Microbiologica Scandinavica* 1967; **71**: 219–32.

46d. Engfeldt B, Hjertquist S O. Studies on the epiphysial growth zone I. The preservation of acid glycosaminoglycans in tissues in some histotechnical procedures for electron microscopy. *Virchows Archiv* (Cell pathol.) 1968; **1**: 222–9.

47. Fanning J C, Cleary E G. Identification of glycoproteins associated with elastin-associated microfibrils. *Journal of Histochemistry and Cytochemistry* 1985; **33**: 287–94.

47a. Farquharson C, Robins S P. The distribution of elastin in developing and adult rat organs using immunocytochemical techniques. *Journal of Anatomy* 1989; **165**: 225–36.

48. Fawcett D W. *The Cell* (2nd edition). Philadelphia, London and Toronto: W B Saunders and Co., 1981.

49. Fell H B, Glauert A M, Barratt M E J, Green R. The pig synovium. 1. The intact synovium *in vivo* and in organ culture. *Journal of Anatomy* 1976; **122**: 663–80.

50. Fife R S, Caterson B, Myers S L. Identification of link proteins in canine synovial cell cultures and canine articular cartilage. *Journal of Cell Biology* 1985; **100**: 1050–5.

50a. Fleming S. Cellular functions of adhesion molecules. *Journal of Pathology* 1990; **161**: 189–90.

51. Freeman M A R, Little T D, Swanson S A V. Lubrication of synovial joints: possible significance of fat. *Proceedings of the Royal Society of Medicine* 1970; **63**: 579–81.

52. Gardner D L. Heberden Oration 1971. The influence of microscopic technology on knowledge of cartilage surface structure. *Annals of the Rheumatic Diseases* 1972; **31**: 235–58.

53. Gardner D L. General pathology of connective tissue and intercellular matrix. In: Passmore R, Robson J S, eds, *A Companion to Medical Studies* (Volume 2, 2nd edition). Oxford: Blackwell Scientific Publications, 1980: 38.1–26.

54. Gardner D L. Structure and function of connective tissue and joints. In: Scott J T, ed, *Copeman's Textbook of the Rheumatic Diseases* (6th edition). Edinburgh: Churchill Livingstone, 1986: 199–250.

54a. Gardner D L, Elliott R J, Gilmore R St C, Longmore R B. Microscopical appearances and organization of articular cartilage surfaces. *Annals of the Rheumatic Diseases* 1975; **34** (Supplement): 2–4.

55. Gardner D L, Lameletie M D J, Lawton D M, Wilson N H F. Response of cultured chondrocytes to porous ceramic. Light- and low temperature scanning electron microscope studies. *Journal of Pathology* 1987; **152**: 189A.

56. Garner A, Alexander R A. Histochemistry of elastic and related fibres in the human eye in health and disease. *Histochemical Journal* 1986; **18**: 405–12.

57. Ghadially F N. Fine structure of joints. In: Sokoloff L, ed, *The Joints and Synovial Fluid* (Volume I). New York: Academic Press, 1978: 105–76.

58. Ghadially F N. The articular territory of the reticuloendothelial system. *Ultrastructural Pathology* 1980; **1**: 249–64.

59. Ghadially F N. *Fine Structure of Synovial Joints.* London: Butterworths, 1983.

60. Ghadially F N. *Ultrastructural Pathology of the Cell and Matrix.* A text and atlas of physiological and pathological alterations in the fine structure of cellular and extracellular components. (3rd edition). London: Butterworths, 1988.

61. Ghadially F N, Roy S. Ultrastructure of rabbit synovial membrane. *Annals of the Rheumatic Diseases* 1966; **25**: 318–26.

62. Ghadially F N, Roy S. *Ultrastructure of Synovial Joints in Health and Disease.* London: Butterworths, 1969.

62a. Gibson G J, Schor S L, Grant M E. Effects of matrix macromolecules on chondrocyte gene expression: synthesis of a low molecular weight collagen species by cells cultured within collagen gels. *Journal of Cell Biology* 1983; **93**: 767–74.

63. Graabaek P M. Characteristics of the two types of synoviocytes in rat synovial membrane. *Laboratory Investigation* 1984; **50**: 690–702.

64. Graabaek P M. Fine structure of the lysosomes in the two types of synoviocytes of normal rat synovial membrane. A cytochemical study. *Cell and Tissue Research* 1985a; **239**: 293–8.

65. Graabaek P M. Absorption of intraarticularly injected horseradish peroxidase in synoviocytes of rat synovial membrane: an ultrastructural-cytochemical study. *Journal of Ultrastructure Research* 1985b; **92**: 86–100.

65a. Graabeck P. Ultrastructural evidence for two distinct types of synoviocytes in rat synovial membrane. *Journal of Ultrastructural Research* 1982; **78**: 321–39.

65b. Graabaek P M. Mannose-specific binding sites in synoviocytes of rat synovial membrane: an ultrastructural-cytochemical study. *Journal of Ultrastructure and Molecular Structure Research* 1986; **94**: 176–87.

66. Godfrey H P, Liard C, Engber W, Graziano F M. Quantitation of human synovial mast cells in rheumatoid arthritis and other rheumatic diseases. *Arthritis and Rheumatism* 1984; **27**: 852–5.

67. Gruhn W B, Devonjee M R, Anderson G S. Comparative study of the reticuloendothelial system pools of the normal and immunologically activated rabbit synovium with colloids. *Journal of Rheumatology* 1980; **7**: 783–7.

68. Gurr E, Mohr W, Pallasch G. Proteoglycans from human articular cartilage: the effect of joint location on the structure. *Journal of Clinical Chemistry and Clinical Biochemistry* 1985; **23**: 811–9.

69. Hamerman D, Sasse J, Klagsbrun M. A cartilage-derived growth factor enhances hyaluronate synthesis and diminishes sulfated glycosaminoglycan synthesis in chondrocytes. *Journal of Cellular Physiology* 1986; **127**: 317–22.

70. Hamerman D, Smith C, Keiser H D, Craig R. Glycosaminoglycans produced by synovial cell cultures. *Collagen and Related Research* 1982; **2**: 313–30.

70a. Hanaoka H, Yabe H, Bun H. The origin of the osteoclast. *Clinical Orthopaedics and Related Research* 1989; **239**: 286–98.

71. Hardinghan T E. Structure and biosynthesis of proteoglycans. In: Kühn K, Krieg T, eds, *Connective Tissue: Biological and Clinical Aspects. Rheumatology, An Annual Review.* Basel: Karger, 1986; **10**: 43–83.

72. Hassan O. Histochemical and electron microscope studies of elastic cartilage under normal and pathological conditions (lathyrism). PhD thesis, The Queen's University of Belfast, 1977.

73. Havdrup T, Henricson A, Telhag H. Papain-induced mito-

sis of chondrocytes in adult joint cartilage. *Acta Orthopaedica Scandinavica* 1982; **53**: 119–24.

74. Havdrup T, Telhag H. Mitosis of chondrocytes in normal adult joint cartilage. *Clinical Orthopaedics and Related Research* 1980; **153**: 248–52.

75. Heathcote J G, Grant M E. The molecular organization of basement membranes. *International Review of Connective Tissue Research* 1981; **9**: 191–264.

76. Henderson B. The contribution made by cytochemistry to the study of the metabolism of the normal and rheumatoid synovial lining cell (synoviocyte). *Histochemical Journal* 1982; **14**: 527–44.

77. Henderson B, Edwards J C W. *The Synovial Lining in Health and Disease*. London: Chapman and Hall, 1987.

78. Henderson B, Pettipher E R. The synovial lining cell: biology and pathobiology. *Seminars in Arthritis and Rheumatism* 1985; **15**: 1–32.

79. Higuchi M, Masuda T, Susuda K, Ishii S, Abe K. Ultrastructure of the articular cartilage after systemic administration of hydrocortisone in the rabbit: an electron microscopic study. *Clinical Orthopaedics and Related Research* 1980; **152**: 296–302.

80. Hirohata K, Mizuhara K, Fujiwara A, Sato T, Imura S, Kobayashi I. Electron microscopic studies on the joint tissues under the normal and pathologic conditions—1. Normal joint tissues (first report). *Journal of the Japanese Orthopaedic Association* 1963a; **36**: 15–9.

81. Hirohata K, Mizuhara K, Fujiwara A, Sato T, Imura S, Kobayashi I. Electron microscopic studies on the joint tissues under the normal and pathologic conditions—2. Normal joint tissues (2nd report). *Journal of the Japanese Orthopaedic Association* 1963b; **37**: 291–301.

82. Hodge A J, Petruska J A. Recent studies with the electron microscope on ordered aggregates of the tropocollagen macromolecule. In Ramachandran G N, ed, *Aspects of Protein Structure*. New York and London: Academic Press, 1963: 289–300.

83. Hulmes D J S, Miller A. Quasi-hexagonal molecular packing in collagen fibrils. *Nature* (London) 1979; **282**: 878–80.

83a. Hunziker E B, Herrmann W. The effect of various cationic dyes, chemically related to ruthenium hexamine trichloride, upon the preservation quality of cartilage during fixation. A light and electronmicroscopic evaluation (quoted by Eggli, Hunziker and Schenk, 1988).

83b. Hunziker E B, Herrmann W, Schenk R K. Ruthenium hexamine trichloride (RHT)-mediated interaction between plasmalemmal components and pericellular matrix proteoglycans is responsible for the preservation of chondrocytic plasma membranes *in situ* during cartilage fixation. *Journal of Histochemistry and Cytochemistry* 1983; **31**: 717–27.

83c. Hunziker E B, Schenk R K. Cartilage ultrastructure after high pressure freezing, freeze substitution and low temperature embedding. *Journal of Cell Biology* 1984; **98**: 277–82.

84. Hynes R D. Fibronectins—a family of complex and versatile adhesive glycoproteins derived from a single gene. *The Harvey Lectures*, series 81. New York: Alan Liss, 1987: 133–52.

85. James J, Steijn-Myagkaya G L. Death of osteocytes. Electron microscopy after *in vitro* ischaemia. *Journal of Bone and Joint Surgery* 1986; **68-B**: 620–4.

86. Janes J, McDonald J R. Mast cells. Their distribution in various human tissues. *Archives of Pathology* 1948: **45**: 622–34.

87. Junquiera L C U, Cossermelli W, Brentani R. Differential staining of collagens type I, II, and III by Sirius red and polarization microscopy. *Archivum Histologicum Japonicum* 1978; **41**: 267–74.

88. Kadar A, Bush V, Gardner D L. Direct elastase treatment of ultrathin sections embedded in water-soluble durcupan. *Journal of Pathology* 1971; **103**: 64–7.

89. Kadar A, Gardner D L, Bush V. Susceptibility of the chick-embryo aorta to elastase: an electron-microscope study. *Journal of Pathology* 1971; **104**: 261–6.

89a. Karabela-Bouropoulou V, Markaki S, Milas Ch. S-100 protein and neuron specific enolase immunoreactivity of normal, hyperplastic and neoplastic chondrocytes in relation to the composition of the extracellular matrix. *Pathology: Research and Practice* 1988; 183: 761–6.

90. Kasper S, Friesen H G. Human pituitary tissue secretes a potent growth factor for chondrocyte proliferation. *Journal of Clinical Endocrinology and Metabolism* 1986; **62**: 70–6.

91. Kato Y, Watanabe R, Hiraki Y, Suzuki F, Canalis E, Raisz L G, Nishikawa K, Adachi K. Selective stimulation of sulfated glycosaminoglycan synthesis by multiplication-stimulating activity, cartilage-derived factor and bone-derived growth factor. *Biochimica et Biophysica Acta* 1982; **704**: 232–9.

92. Keiser H D, Malemud C J. A comparison of the proteoglycans produced by rabbit articular chondrocytesin monolayer and spinner culture and those of bovine nasal cartilage. *Connective Tissue Research* 1983; **11**: 273–84.

92a. Kiviranta I, Jurvelin J, Tammi M, Säämänen A-M, Helminen H J. Microspectrophotometric quantitation of glycosaminoglycans in articular cartilage sections stained with Safranin O. *Histochemistry* 1985; **82**: 249–55.

92b. Kiviranta I, Tammi M, Jurvelin J, Helminen H J. Topographical variation of glycosaminoglycan content and cartilage thickness in canine knee (stifle) joint cartilage. Application of the microspectrophotometric method. *Journal of Anatomy* 1987; **150**: 265–76.

92c. Kiviranta I, Tammi M, Jurvelin J, Säämänen A-M, Helminen H J. Demonstration of chondroitin sulphate and glycoproteins in articular cartilage matrix using the periodic acid–Schiff method. *Histochemistry* 1985; **83**: 303–6.

93. Kühn K. The collagen family—variations in the molecular and supermolecular structure. In: Kuhn K, Krieg T, eds, *Connective Tissue: Biological and Clinical Aspects. Rheumatology, An Annual Review*. Basel: Karger, 1986; **10**: 29–69.

93a. Lalor P A, Mapp P I, Hall P A, Revell P A. Proliferative activity of cells in the synovium as demonstrated by a monoclonal antibody Ki67. *Rheumatology International* 1987; **7**: 183–6.

94. Lameletie M D J. The growth of chondrocytes *in vitro* on porous hydroxyapatite ceramic. BSc thesis, University of Manchester, 1987.

95. Langevoort H C, Cohn Z A, Hirsch J G, Humphrey J H, Spector W G, van Furth R. The nomenclature of mononuclear phagocytes. Proposal for a new classification. In: van Furth R, ed, *Mononuclear Phagocytes*. Oxford: Blackwell Scientific Publications, 1970.

96. Lavietes B B, Carsons S, Diamond H S, Laskin R S. Synthesis, secretion and deposition of fibronectin in cultured human synovium. *Arthritis and Rheumatism* 1985; **28**: 1016–26.

97. Lawton D M, Lameletie M D J, Gardner D L. Biocompatibility of hydroxyapatite ceramic: response of chondrocytes in a test system using low temperature scanning electron microscopy. *Journal of Dentistry* 1989; **17**: 21–7.

98. Leblond C P, Laurie G W. Morphological features of connective tissues. In: Kühn K, Krieg T, eds, *Connective Tissue: Biological and Clinical Aspects. Rheumatology, An Annual Review*. Basel: Karger, 1986; **10**: 1–28.

99. Leivo I. Structure and composition of early basement membranes—studies with early embryos and teratocarcinoma cells. *Medical Biology* 1983; **61**: 1–30.

100. Lethias C, Hartmann D J, Masmajean M, Ravazzola M, Sabbagh I, Ville G, Herbage D, Elay R. Ultrastructural immunolocalization of elastic fibers in rat blood vessels using the protein-A gold technique. *Journal of Histochemistry and Cytochemistry* 1987; **35**: 15–21.

101. Lipman J M, Hicks B J, Sokoloff L. Rabbit chondrocytes are binucleate in auricular but not articular cartilage. *Experientia* 1984; **40**: 553–4.

102. Lipman J M, Sokoloff L. DNA repair by articular chondrocytes. III. Unscheduled DNA synthesis following ultraviolet light irradiation of resting cartilage. *Mechanisms of Ageing and Development* 1985; **32**: 39–55.

103. Lipper S, Kahn L B, Reddick R L. The myofibroblast. In: Sommers S C, Rosen P P, eds (Volume 15), *Pathology Annual*. New York: Appleton-Century-Crofts, 1980; 409–41.

104. Luft J H. The fine structure of hyaline cartilage matrix following ruthenium red fixation and staining. *Journal of Cell Biology* 1965; **27**: 61A.

105. McDonald J N, Levick J R. Morphology of surface synoviocytes *in situ* at normal and raised joint pressure, studied by scanning electron microscopy. *Annals of the Rheumatic Diseases* 1988; **47**: 232–40.

106. Mainardi C L. Localized fibrotic disorders. In: Kelley W N, Harris E D, Ruddy S, Sledge C B, eds, *Textbook of Rheumatology* (3rd edition). Philadelphia: W B Saunders & Co., 1989: 1245–61.

107. Malemud C J, Papay R S. The *in vitro* cell culture age and cell density of articular chondrocytes after sulfated-proteoglycan biosynthesis. *Journal of Cellular Physiology* 1984; **121**: 558–68.

108. Malemud C J, Sokoloff L. The effect of chondrocyte growth factor on membrane transport by articular chondrocytes in monolayer culture. *Connective Tissue Research* 1978; **6**: 1–9.

109. Malinger R, Geleff S, Bock P. Histochemistry of glycosaminoglycans in cartilage ground substance. Alcian-blue staining and lectin-binding affinities in semithin Epon sections. *Histochemistry* 1986; **85**: 121–7.

110. Manicourt D H, Pita J C. Quantification and characterization of hyaluronic acid in different topographical areas of normal articular cartilage from dogs. *Collagen and Related Research* 1988; **8**: 39–43.

111. Manning W K, Bonner W M, Jnr. Isolation and culture of chondrocytes from human adult articular cartilage. *Arthritis and Rheumatism* 1967; **10**: 235–9.

111a. Marles P T, Hoyland T A, Freemont A T. A study of chondrocyte metabolism by *in situ* hybridisation. *Journal of Pathology* 1990; **161**: 360A.

111b. Maximov A. Relation of blood cells to connective tissues and endothelium. *Physiological Reviews*, 1924; **4**: 533–63.

112. Menton D N, Simmons D J, Chang S-L, Orr B Y. From bone lining cell to osteocyte—an SEM study. *Anatomical Record* 1984; **209**: 29–39.

113. Miller A. Collagen: the organic matrix of bone. *Philosophical Transactions of the Royal Society of London B* 1984; **304**: 455–77.

114. Moskalewski S, Golaszewska A, Książek T. *In situ* aging of auricular chondrocytes is not due to the exhaustion of their replicative potential. *Experientia* 1980; **36**: 1294–5.

115. Muirden K D, Fraser J R E, Clarris B. Ferritin formation by synovial cells exposed to haemoglobin *in vitro*. *Annals of the Rheumatic Diseases* 1967; **26**: 251–9.

116. Muirden K D. An electron microscope study of the uptake of ferritin by the synovial membrane. *Arthritis and Rheumatism* 1963; **6**: 289 (abstract).

117. Muller-Glauser W, Humbel B, Glatt M, Strauli P, Winterhalter K H, Bruckner P. On the role of type IX collagen in the extracellular matrix of cartilage: type IX collagen is localized to intersections of collagen fibrils. *Journal of Cell Biology* 1986; **102**: 1931–9.

118. Myers S L, Christine T A. Hyaluronate synthesis by synovial villi in organ culture. *Arthritis and Rheumatism* 1983; **26**: 764–70.

119. Na G C, Butz L J, Carroll R J. Mechanism of *in vitro* collagen fibril assembly. Kinetic and morphological studies. *Journal of Biological Chemistry* 1986; **261**: 12290–9.

120. Nakahata T, Kobayashi T, Ishiguro A, Tsuji K, Naganuma K, Ando O, Yagi Y, Tadokoro K, Akabane T. Extensive proliferation of connective tissue-type mast cells *in vitro*. *Nature* 1986; **324**: 65–7.

121. Nathan C F, Cohen Z A. Cellular components of inflammation: Monocytes and macrophages. In: Kelley W N, Harris E D, Ruddy S, Sledge C B, eds, *Textbook of Rheumatology* (2nd edition). Philadelphia: W B Saunders & Co., 1985: 144–69.

122. Nerlich A G, Poschl E, Voss T, Muller P K. Biosynthesis of collagen and its control. In: Kühn K, Krieg T, eds, *Connective Tissue: Biological and Clinical Aspects. Rheumatology*. Basel: Karger, 1986; **10**: 70–90.

123. Nielsen E H, Bytzer P. High resolution scanning electron microscopy of elastic cartilage. *Journal of Anatomy* 1979; **129**: 823–31.

123a. *Nomina Anatomica* (6th edition). Edinburgh: Churchill Livingstone, 1989.

124. Norrby K, Bergstrom S, Druvefors P. Age-dependent mitogenesis in normal connective tissue cells. *Virchows Archiv B* (Cell Pathol.) 1981; **36**: 27–34.

124a. O'Connor P, Brereton J, Gardner D L. Hyaline articular cartilage dissected by papain: light- and scanning electron microscopy and micromechanical studies. *Annals of the Rheumatic Diseases* 1984; **43**: 320–6.

125. O'Connor P, Orford C R, Gardner D L. Differential res-

ponse to compressive loads of canine hyaline articular cartilage: micromechanical, light and electron microscopic studies. *Annals of the Rheumatic Diseases* 1988; **47**: 414–20.

126. Okada Y, Nakanishi I, Kajikawa K. Secretory granules of B-cells in the synovial membrane. An ultrastructural and cytochemical study. *Cell and Tissue Research* 1981; **216**: 131–42.

127. Okazaki I, Brinckerhoff C E, Sinclair J F, Sinclair P R, Bonkowsky H L, Harris E D. Iron increases collagenase production by rabbit synovial fibroblasts. *Journal of Laboratory and Clinical Medicine* 1981; **97**: 396–402.

128. O'Rahilly R, Gardner E. The embryology of movable joints. In: Sokoloff L, ed, *The Joints and Synovial Fluid* (volume I). New York: Academic Press, 1978: 49–103.

129. Orford C R, Gardner D L. Proteoglycan association with collagen d band in hyaline articular cartilage. *Connective Tissue Research* 1984; **12**: 345–8.

130. Parwaresch M R, Horny H P, Lennert K. Tissue mast cells in health and disease. *Pathology, Research and Practice* 1985; **179**: 439–61.

130a. Paukkinen K, Helminen H J. Rough endoplasmic reticulum and finer intracytoplasmic filaments in articular cartilage chondrocytes of young rabbits; a stereological morphometric study using transmission electron microscopy. *Journal of Anatomy* 1987; **152**: 47–54.

131. Petrovic A, Stutzman J. Growth hormone: mode of action on different varieties of cartilage. *Pathologie et Biologie* 1980; **28**: 42–58.

132. Pidd J G, Gardner D L. Surface structure of baboon (*Papio anubis*) hydrated articular cartilage: study of low temperature replicas by transmission electron microscopy. *Journal of Medical Primatology* 1987; **16**: 301–9.

132a. Pollock L E, Lalor P, Revell P A. Type IV collagen and laminin in the synovial intimal layer: an immunohistochemical study. *Rheumatology International* 1990; **9**: 277–80.

133. Poole A R. The relationship between toluidine blue staining and hexuronic acid content of cartilage matrix. *Histochemical Journal* 1970; **2**: 425–30.

134. Poole C A, Ayad S, Schofield J R. Chondrons from articular cartilage: 1. Immunolocalization of type VI collagen in the pericellular capsule of isolated canine tibial chondrons. *Journal of Cell Science* 1988; **90**: 635–43.

134a. Poole C A, Flint M H, Beaumont B W. Morphological and functional interrelationships of articular cartilage matrices. *Journal of Anatomy* 1984; **138**: 113–38.

135. Poole C A, Flint M H, Beaumont B W. Morphology of the pericellular capsule in articular cartilage revealed by hyaluronidase digestion. *Journal of Ultrastructure Research* 1985; **91**: 13–23.

136. Poole C A, Flint M H, Beaumont B W. Chondrons extracted from canine tibial cartilage: preliminary report on their isolation and structure. *Journal of Orthopaedic Research* 1988; **6**: 408–19.

137. Postlethwaite A E, Kang A H. Fibroblasts. In: Gallin J I, Goldstein I M, Snyderman R, eds, *Inflammation: Basic Principles and Clinical Correlates*. New York: Raven Press, 1988: 577–97.

138. Postlethwaite A E, Snyderman R, Kang A H. Generation of a fibroblast chemotactic factor from serum by the activation of complement. *Journal of Clinical Investigation* 1979; **64**: 1379–85.

139. Powers J C. Serine proteases of leucocyte and mast cell origins: substrate specificity and inhibition of elastase, chymase and tryptases. In: Otterness I, Lewis A, Capetola R, eds, *Therapeutic Control of Inflammatory Diseases: New Approaches to Antirheumatic Drugs* (Volume 11: Advances in Inflammation Research). New York: Raven Press, 1986: 145–57.

140. Price R G, Spiro R D. Studies on the metabolism of the renal glomerular basement membrane. *Journal of Biological Chemistry* 1977; **252**: 8597–602.

141. Prockop D J. Mutations in collagen genes. Consequences for rare and common diseases. *Journal of Clinical Investigation* 1985; **75**: 783–7.

142. Prockop D J, Kivirikko K I, Tuderman L, Guzman A. The biosynthesis of collagen and its disorders. *New England Journal of Medicine* 1979; **301**: 13–23; 77–85.

142a. Puchtler H, Meloan S N, Waldrop F S. Are picro-dye reactions for collagens quantitative? *Histochemistry* 1988; **88**: 243–56.

143. Ratcliffe A, Fryer P R, Hardingham T E. Proteoglycan biosynthesis in chondrocytes: protein A-gold localization of proteoglycan protein core and chondroitin sulfate within Golgi subcompartments. *Journal of Cell Biology* 1985; **101**: 2355–65.

144. Ratcliffe A, Hughes C, Fryer P R, Saed-Nejad F, Hardingham T. Immunochemical studies on the synthesis and secretion of link protein and aggregating proteoglycan by chondrocytes. *Collagen and Related Research* 1987; **7**: 409–21.

145. Revell P A. Synovial lining cells. *Rheumatology International* 1989; **9**: 49–51.

145a. Roche W R. Myofibroblasts. *Journal of Pathology* 1990; **161**: 281–2.

146. Ronziere M-C, Berthet-Colominas C, Herbage D. Low-angle X-ray diffraction analysis of the collagen-proteoglycan interactions in articular cartilage. *Biochimica et Biophysica Acta* 1985; **842**: 170–5.

146a. Roska A K, Lipsky P E. Monocytes and macrophages. In: Kelley W N, Harris E D, Ruddy S, Sledge C B, eds, *Textbook of Rheumatology* (3rd edition). Philadelphia: W. B. Saunders & Co. 1989: 346–66.

147. Sachs B L, Goldberg V M, Moskowitz R W, Malemud C J. Response of articular chondrocytes to pituitary fibroblast growth factor. *Journal of Cellular Physiology* 1982; **112**: 51–9.

148. Sage H. The evolution of elastin: correlation of functional properties with protein structure and phylogenetic distribution. *Comparative Biochemistry and Physiology* 1983; **74B**: 373–80.

149. Sandberg L B, Soskel N T, Leslie J G. Elastin structure, biosynthesis and relation to disease states. *New England Journal of Medicine* 1981; **304**: 566–79.

150. Schubert M, Hamerman D. *A Primer on Connective Tissue Biochemistry*. Philadelphia: Lea & Febiger, 1968.

151. Schwartz E, Fleischmajer P. Association of elastin with oxytalan fibers of the dermis and with extracellular microfibrils of cultured skin fibroblasts. *Journal of Histochemistry and Cytochemistry* 1986; **34**: 1063–8.

152. Scott J E. Proteoglycan histochemistry—a valuable tool

for connective tissue biochemists. *Collagen and Related Research* 1985; **5**: 541–75.

153. Scott J E, Dorling J. Differential staining of acid glycosaminoglycans (mucopolysaccharides) by alcian blue in salt solution. *Histochemie* 1965; **5**: 221–33.

154. Scott J E, Haigh M. Proteoglycan-type I collagen fibril interactions in bone and non-calcifying connective tissues. *Bioscience Reports* 1985; **5**: 71–81.

155. Scott J E, Orford C R. Dermatan sulphate-rich proteoglycan associates with rat tail-tendon collagen at the d-band in the gap region. *Biochemical Journal* 1981; **197**: 213–6.

155a. Scott J T, ed. *Copeman's Textbook of the Rheumatic Diseases* (6th edition). Edinburgh: Churchill Livingstone.

156. Seemayer T A, Lagace R, Schurch W, Thelmo W L. The myofibroblast: biologic, pathologic, and theoretical considerations. In: Somers S C, Rosen P P, eds, *Pathology Annual* (volume 15). New York: Appleton-Century-Crofts, 1980: 443–70.

157. Serafini-Fracassini A, Smith J W. *The Structure and Biochemistry of Cartilage.* Edinburgh: Churchill Livingstone, 1974.

158. Shanon A L, Graham R C. Protein uptake by synovial cells. 1. Ultrastructural study of the fate of intraarticularly injected peroxidases. *Journal of Histochemistry and Cytochemistry* 1971; **19**: 29–42.

159. Sibbitt W L. Oncogenes, normal cell growth, and connective tissue disease. *Annual Review of Medicine* 1988; **39**: 123–33.

160. Silberberg R, Hasler M. Electron microscopy of articular cartilage of mice. *Archives of Pathology* 1969; **87**: 502–13.

161. Silberberg R, Hasler M. Submicroscopic effects of hormones on articular cartilage of adult mice. *Archives of Pathology* 1971; **91**: 241–55.

162. Silberberg R, Hasler M, Silberberg M. Submicroscopic response of articular cartilage of mice treated with estrogenic hormone. *American Journal of Pathology* 1965; **46**: 289–305.

163. Silberberg M, Silberberg R, Hasler M. Fine structure of articular cartilage in mice receiving cortisone acetate. *Archives of Pathology* 1966; **82**: 569–82.

164. Silbermann M, Livne E, Lizarbe M A, von der Mark K. Glucocorticoid hormone adversely affects the growth and differentiation of cartilage cells in neonatal mice. *Growth* 1983; **47**: 77–96.

165. Silbermann M, Maor G. Effect of glucocorticoid hormone on the content and synthesis of nucleic acids in cartilage of growing mice. *Growth* 1979; **43**: 273–87.

166. Sin Y M, Sedgwick A D, Chea E P, Willoughby D A. Mast cells in newly formed lining tissue during acute inflammation: a six day air pouch model in the mouse. *Annals of the Rheumatic Diseases* 1986; **45**: 873–7.

167. Smith E W, Atkinson W B. Simple procedure for identification and rapid counting of mast cells in tissue sections. *Science* 1956; **123**: 941–2.

168. Smith G N, Williams J M, Brandt K D. Interaction of proteoglycans with the pericellular (1α, 2α, 3α) collagen of cartilage. *Journal of Biological Chemistry* 1985; **260**: 10761–7.

169. Smith J W. Molecular patterns in native collagen. *Nature* 1968; **219**: 157–8.

170. Sokoloff L. Articular chondrocytes in culture. *Arthritis and Rheumatism* 1976; **19**: 426–9.

171. Sokoloff L. *In vitro* culture of joints and articular tissues. In: Sokoloff L, ed, *The Joints and Synovial Fluid* (Volume II). New York: Academic Press, 1980: 1–26.

171a. Strachan M W J, Gardner D L. Apparent immunity of synovial joints to metastasis: response of tumour cells to supernates from synovial cell cultures. *Journal of Pathology* 1989; **158**: 355A.

172. Stockwell R A. Chondrocytes. *Journal of Clinical Pathology* 1978, **31**, Supplement (Royal College of Pathologists), 12: 7–13.

173. Stockwell R A. *Biology of Cartilage Cells.* Cambridge: Cambridge University Press, 1979a: 245–7.

174. Stockwell R A. The chondrocytes. In: Freeman M A R, ed, *Adult Articular Cartilage* (2nd edition). Tunbridge Wells: Pitman, 1979b: 69–144.

175. Stuart A E. *The Reticulo-endothelial System.* Edinburgh: Churchill Livingstone, 1970.

176. Swanson S A V. Friction. Wear and lubrication. In: Freeman M A R, ed, *Adult Articular Cartilage* (2nd edition). Tunbridge Wells: Pitman, 1979.

176a. Szirmai J A. Structure of cartilage. In: Engel A, Larsson T, eds, *Ageing of Connective and Skeletal Tissue.* Stockholm: Nordiska Bokhandelns, 1969: 163–84.

177. Tarsoly E. Effect of testosterone administration on the epiphyseal cartilage of hypophysectomized rats. *Acta Histochemica* 1976; **55**: 176–86.

178. Taylor D J, Woolley D E. Evidence for both histamine H_1 and H_2 receptors on human articular chondrocytes. *Annals of the Rheumatic Diseases* 1987; **46**: 431–5.

179. Taylor D J, Yoffe J R, Brown D M, Woolley D E. Histamine H_2 receptors on chondrocytes derived from human, canine and bovine articular cartilage. *Biochemical Journal* 1985; **225**: 315–9.

179a. Todd R B, Bowman W. *The Physiological Anatomy of Man.* London: J W Parker, 1845–59.

180. Tonna E A. An electron microscopic study of skeletal cell aging. II. The osteocyte. *Experimental Gerontology* 1973; **8**: 9–16.

180a. Trelstad R L. Matrix macromolecules: spatial relationships in two dimensions. In: Fleischmajer R, Olsen B R, Kuhn K, eds, *Structure, Molecular Biology, and Pathology of Collagen. Annals of the New York Academy of Sciences* 1990; **580**: 391–420.

181. Trippel S B, Ehrlich M G, Lippiello L, Mankin H J. Characterization of chondrocytes from bovine articular cartilage. I. Metabolic and morphological experimental studies. *Journal of Bone and Joint Surgery* 1980; **62A**: 816–9.

182. Uitto J, Paul J L, Brockley K, Pearce R H, Clark J G. Elastic fibers in human skin: quantitation of elastic fibers by computerized digital image analysis and determination of elastin by radioimmunoassay of desmosine. *Laboratory Investigation* 1983; **49**: 499–505.

183. Urry D W. What is elastin, what is not. *Ultrastructural Pathology* 1983; **4**: 227–51.

184. Vaes G. Cellular biology and biochemical mechanism of bone resorption. A review of recent developments on the forma-

tion, activation and mode of action of osteoclasts. *Clinical Orthopaedics and Related Research* 1988; **231**: 239–71.

185. Varner H H, Furthmayr H, Nilsson B, Fietzek B, Osborne J C, De Luca S, Martin G R, Hewitt A T. Chondronectin: physical and chemical properties. *Archives of Biochemistry and Biophysics* 1985; **243**: 579–85.

186. Varner H H, Horn V J, Martin G R, Hewitt A T. Chondronectin interactions with proteoglycan. *Archives of Biochemistry and Biophysics* 1986; **224**: 824–30.

187. Vaughan J M. *The Physiology of Bone* (3rd edition). Oxford: Clarendon Press, 1981.

188. Vernon-Roberts B. *The Macrophage*. Cambridge: Cambridge University Press, 1972.

189. Vertel B M, Barkman L L, Morrell J J. Intracellular features of type II procollagen and chondroitin sulfate proteoglycan synthesis in chondrocytes. *Journal of Cellular Biochemistry* 1985; **27**: 215–29.

190. Von der Mark K, Goodman S. Adhesive glycoproteins. In: Royce P M and Steinmann B, eds, *Extracellular Matrix and Inheritable Disorders of Connective Tissue*. New York: Alan Liss, 1990.

191. Wainwright S A, Biggs W D, Currey J D, Gosline J M. *Mechanical Design in Organisms*. London: Edward Arnold, 1976.

192. Walker F. The origin, turnover and removal of glomerular basement-membrane. *Journal of Pathology* 1973; **110**: 233–44.

193. Watanabe N, Rosenfeld R G, Hintz R L, Dollar L A, Smith R L. Characterization of a specific insulin-like growth factor-1 somatomedin-C receptor on high density, primary monolayer cultures of bovine articular chondrocytes: regulation of receptor concentration by somatomedin, insulin and growth hormone. *Journal of Endocrinology* 1985; **107**: 275–83.

194. Watanabe H, Spycher M A, Ruttner J R. Ultrastructural study of the normal rabbit synovium. *Pathologie et Microbiologie* 1964; **41**: 283–92.

195. Wiebkin O W, Hardingham T E, Muir H. In: Slavkin H C, Greulich R C, eds, *Extracellular Matrix Influence on Gene Expression*. London: Academic Press, 1975: 209.

196. Wiebkin O W, Muir H. The inhibition of sulphate incorporation in isolated adult chondrocytes by hyaluronic acid. *FEBS Letters* 1973; **37**: 42–6.

197. Wiebkin O W, Muir H. Influence of the cells on the pericellular environment. The effect of hyaluronic acid on proteoglycan synthesis and secretion by chondrocytes of adult cartilage. *Philosophical Transactions of the Royal Society of London B* 1975; **271**: 283–91.

198. Willmer E N. *Cytology and Evolution* (2nd edition). New York: Academic Press, 1970.

199. Winchester R J, Burmester G R. Demonstration of Ia antigens on certain dendritic cells and on a novel elongate cell found in human synovial tissue. *Scandinavian Journal of Immunology* 1981; **14**: 439–44.

200. Wood D D, Ihrie E J, Hamerman D. Release of interleukin-1 from human synovial tissue *in vitro*. *Arthritis and Rheumatism* 1985; **28**: 853–62.

201. Woolley D E. Mammalian collagenases. In: Piez K A and Reddi A H, eds, *Extracellular Matrix Biochemistry*. New York: Elsevier, 1984.

202. Wyllie J C, More R H, Haust M D. The fine structure of normal guinea-pig synovium. *Laboratory Investigation* 1964; **13**: 1254–63.

203. Yang J, Nandi S. Growth of cultured cells using collagen as substrate. In: Bourne G H, Danielli J F, Jeon K W, eds, *International Review of Cytology* 1983; **81**: 249–86.

203a. Yasui N, Benya P D, Nimni M E. Coordinate regulation of type IX and type II collagen synthesis during growth of chick chondrocytes in retinoic acid or 5-bromo-2′-deoxyuridine. *Journal of Biological Chemistry* 1986; **261**: 7997–8001.

204. Young R D. The ultrastructural organization of proteoglycans and collagen in human and rabbit scleral matrix. *Journal of Cell Science* 1985; **74**: 95–104.

Chapter 2
Tissues

The cellular and molecular units that comprise the non-osseous connective tissues are described in Chapters 1 and 4. The integration of these units is now outlined, their arrangement as loose and compact connective tissue, cartilage, synovium, ligament and tendon reviewed. The present account serves as an introduction to Chapter 3 in which the functional components of the connective tissue system are brought together as physiological entities. Because the description of the macromolecules of the normal connective tissues is combined with accounts of their contributions to disease, Chapter 4 is sited immediately before those dealing with connective tissue immunology (Chapter 5), inflammation (Chapter 7) and repair (Chapter 8).

Loose connective tissue

A supportive, loose connective tissue is arranged within and around the viscera and parts (Fig. 2.1).[283,75] The quantity of this tissue is closely related to the state of nutrition of the individual. Surrounding organs such as those of the retroperitoneal and pelvic regions, loose connective tissue acts as a cushion and shock-absorber, minimizing visceral movement when the body accelerates or decelerates, and providing protection against physical injury due to transmitted external forces. Loose connective tissue extends within organs so that it is found, for example, in the renal pelvis, porta hepatis, lung hila and adrenal medulla and, occasionally, in ectopic islands within the viscera.

Loose connective tissue is present in almost every part of the body[154a] with the exception of the intracranial territory. It is formed of intrinsic fibroblasts, fibrocytes and, invariably, less well differentiated mesenchymal cells that retain a pluripotential capacity to develop in any one of a number of directions and to synthesize, according to circumstances, a large variety of fibrous and non-fibrous matrix macromolecules (see Chapter 4). Extrinsic neutrophil granulocytes and mononuclear phagocytes wander into the loose connective tissues in small numbers and the immunological integrity of the tissues is monitored by the presence of sentinal lymphocytes. On average, the quantity of matrix proteoglycan in loose connective tissue tends to be large relative to the quantity of collagen. Elastic fibres are occasional. Blood vessels, particularly capillary channels, are numerous, and venules and arterioles commonplace. There is an abundant lymphatic supply; sensory nerve endings and specialized sensory receptors are regularly encountered.

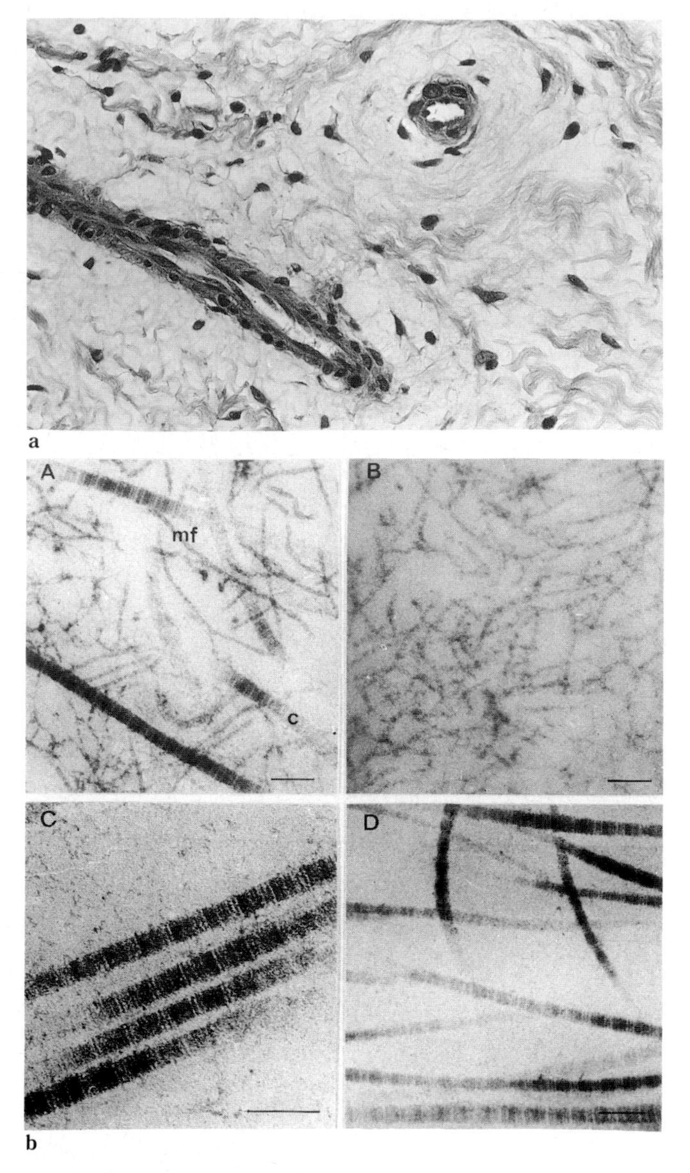

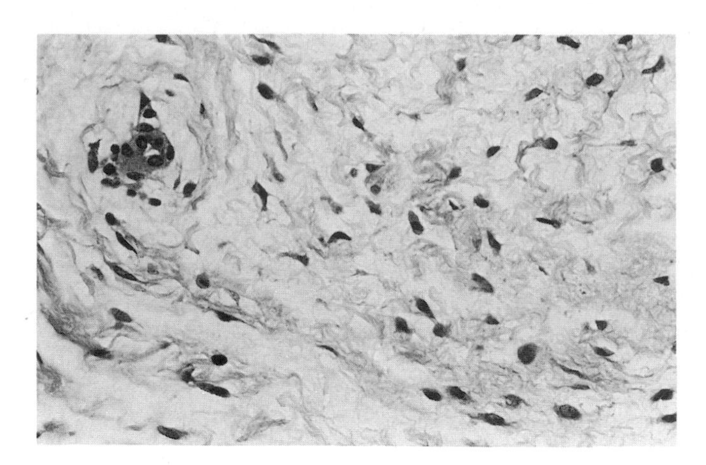

Fig. 2.1 Loose connective tissue.
a. Subcutaneous tissue (light microscopy). **Left** Several small periarticular arterioles lie among a matrix comprising delicate fasciculi of collagen fibres and a sparse, metachromatic extracellular matrix. **Right** Note arteriole (at left). The majority of the cells in the intercellular matrix have the spindle-shaped nuclei of fibrocytes or the plumper nuclei of fibroblasts. Occasional isolated lymphocytes, mast cells and macrophages may be identified. **b.** Ultrastructure. Wharton's jelly of the umbilical cord. **A.** Hyaluronidase-digested tissue with collagen fibrils (c), microfibrils (mf) and fine filaments (arrow) that are often associated with globular bodies. **B.** Hyaluronidase-digested tissue after prolonged treatment with bacterial collagenase. Collagen has been removed; a network of microfibrils and fine filaments remains. **C.** Hyaluronidase-treated tissue after incubation with pronase. The non-collagenous microfibrillar meshwork has been degraded. **D.** Hyaluronidase-treated tissue after pronase digestion. Essentially complete removal of microfibrillar meshwork. (A, B: × 100000; C: × 170 000; D: × 107 000.) Courtesy of Dr Frank A Meyer and the Editor, *Biochimica et Biophysica Acta.*[187]

Areolar tissue

Much of the subcutaneous connective tissue of the adult is areolar or fibroareolar (fibrofatty); in youth, a greater proportion is adipose. Loose connective tissue provides the flexible support for the synovial cell layer of the majority of diarthrodial joints, constituting a lamina propria comparable with, but not analogous to, a submucosa. The areolar tissue supports the synovial intima on delicate islands of fat cells with interlacing bundles of collagen fibres in relation to which blood vessels, lymphatics and nerve fibres extend liberally. The malleable texture of this form of connective tissue is an index of the high ratio of proteoglycan-rich to collagen-rich matrix. Hyaluronate is abundant and the relatively large proportion of proteoglycan determines a large but variable water content. The absence of defined physical limits to the parts occupied by loose connective tissue, despite the fibrous septa that support it and the substantial proportions of hyaluronate and water, has important pathological consequences. Infective processes such as those caused by *Streptococcus pyogenes*, an organism synthesizing hyaluronidase, spread with dramatic speed through extracellular tissue spaces such as those of the subcutis, gaining access quickly to lymphatic channels so that cellulitis and lymphangitis may rapidly advance to bacteraemia and septicaemia.

Adipose connective tissue

Adipose tissue is identified by the very high proportion of fat cells (adipocytes), usually grouped to form lobules. Dividing these islands of fat, there are scanty, stellate, ramifying fibrovascular septa in which run the blood vessels, lymphatics and nerve fibres that supply the fat cells and which contain the associated mast cells. The adipose tissue of the infrapatellar fat pad exemplifies this arrangement. The disposition of grouped adipocytes may determine the three-dimensional microanatomy of a part so that, for example, the hexagonal packing of synovial villi seen *en face* (Fig. 2.22, p. 86) represents the organization not of the synoviocytes but of the underlying adipocytes. In some situations, elastic material in the form of fibres can be identified in adipose connective tissue, but the amount and organization are less than in compact materials like ligaments.

Compact connective tissue

The distinction between 'loose' and 'compact' connective tissue is quantitative. Compact connective tissue describes the arrangement of those forms of connective tissue where strength and mechanical elasticity are the prime needs rather than cushion-like, reversible deformability and flexibility. Non-calcified compact structures such as tendons, ligaments, fascia, aponeuroses, menisci, labra, the dura mater, the lens, the sclera and the annulus fibrosus of the intervertebral disc, all have individual physical requirements. Characteristically, there is much collagen, relatively less proteoglycan and a small number of committed, differentiated cells. Capsules surround synovial joints and form freely moving protective tents that are firmly attached to the opposing bone ends where the capsular collagen merges and is also in continuity with the periosteum. Capsules are not usually continuous sheets: apertures and gaps in their structure allow synovial tissue to protrude outside the joint. There are synovial pockets and bursae which enable the passage, through the joint capsule, of vessels and of collagenous cords such as tendons. In the fascia and aponeuroses, the dispersed collagen fibres, separated by a little proteoglycan, are arranged as interlacing sheets; both strength and flexibility are assured. Bone (see p. 105), dentine and calcified cartilage are special forms of compact connective tissue but many other connective tissues are susceptible to calcification.

Cartilage

Chondrocytes are described on p. 20, and the extracellular matrix on p. 42 and in Chapter 4.

Cartilage is an ubiquitous connective tissue,[116] not always confined to the normal mammalian skeleton, the 'hard' and calcified connective tissues, the nose and respiratory tract and the ears. Cartilage formation[117] is characteristic of a wide range of heritable and acquired disorders (Fig. 2.2). There are many forms of chondrodystrophy and cartilage malformation; they are mentioned in Chapter 10. In humans, ectopic cartilage[20] may form in sites such as the lung when pulmonary hypertension is longstanding. Primary cartilagenous neoplasms are an important category of new growth but cartilage is also found in extraskeletal neoplasms such as ovarian benign cystic teratoma, uterine mixed mesodermal tumour, breast sarcoma and nephroblastoma. Cartilage formation can be provoked experimentally, for example in the aortas of rabbits sensitized against homogenates of heterologous rat or dog aortic proteins.[112,113]

The biochemical distinction between the hyaline cartilage of diarthrodial joints,[53] the young cartilage of some fracture

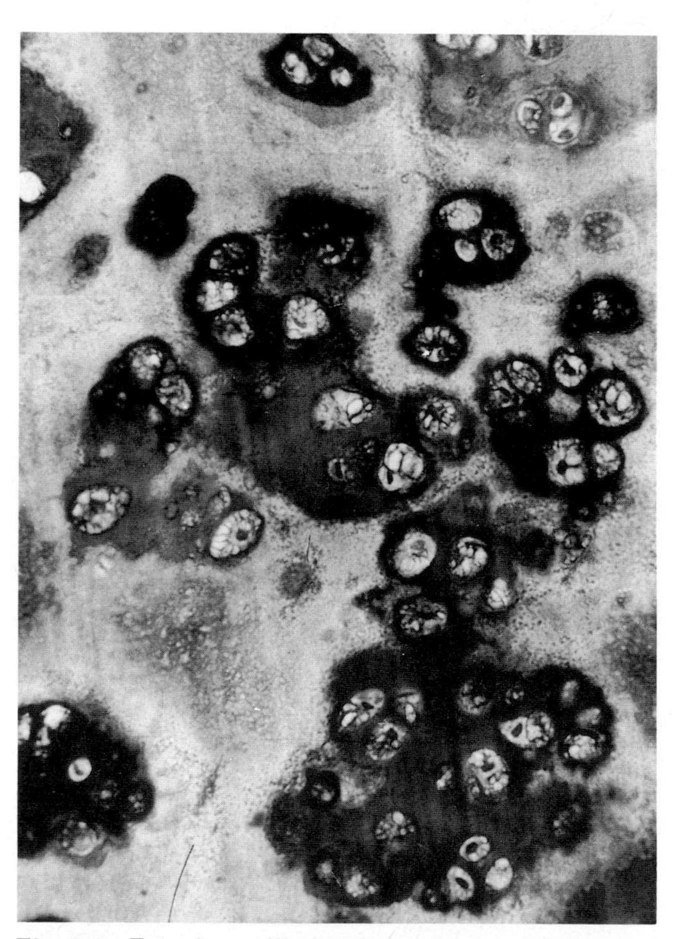

Fig. 2.2 Ectopic cartilage.
Cartilagenous metaplasia is frequent at connective tissue sites of impaired blood flow, hypoxia and old injury. In this example, many deeply metachromatic islands containing large chondrocytes of uniform size are arranged as clusters within a looser tissue that is flecked with groups of calcium hydroxyapatite crystals. (Alcoholic toluidine blue × 400.)

callus, the dysplastic cartilage of the chondrodystrophies and the neoplastic cartilage of the chondromata is not well-understood chemically. However, efforts have begun to identify molecular differences in these distinct but analogous materials and a classification of chondrosarcoma, for example, has been devised, based on the nature of the matrix proteoglycans.[177,178] In view of the heterogeneous structure of mammalian joints, and of the varied arrangement, thickness and extent of the hyaline cartilage covering their load-bearing surfaces, comparable, subtle differences in macromolecular structure and organization are to be expected in these organs.

Cartilage is of three varieties: each differs in the number and morphology of the cells, in the composition and proportions of the extracellular fibrous and non-fibrous macromolecules and in their anatomical location and functions. Hyaline (glass-like) cartilage, in youth translucent and blue-white, in old age yellow-white and opaque, is an avascular material populated by chondrocytes that synthesize and export the components of a highly hydrated, fibre-reinforced gel in which there is much proteoglycan, large quantities of fibrous type II collagen and small amounts of types VI, IX and XI collagen. Fibrocartilage (white fibrocartilage), dull-white, tougher, stronger and less deformable than hyaline cartilage, is manufactured by fibroblast-like cells that synthesize a matrix principally of type I collagen in association with which there is relatively little proteoglycan but, often, some elastic material. The structure is that of an oriented, compact fibrous meshwork, avascular and containing relatively little water. Elastic cartilage (yellow fibrocartilage), like rubber, is highly and reversibly deformable. The morphology and function of elastic cartilage chondrocytes closely resemble those of hyaline cartilage but the simultaneous synthesis by these cells of elastin, collagen and proteoglycan is characteristic.

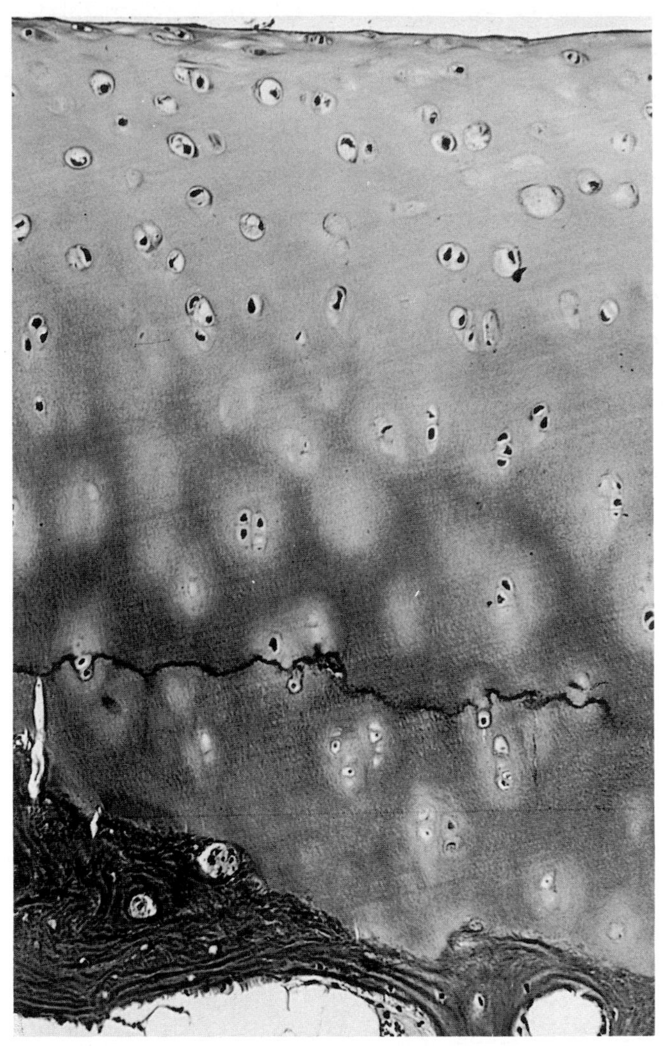

Fig. 2.3 Articular hyaline cartilage.
Perpendicular section through load-bearing surface of synovial joint. Note (at top): non-smooth surface of surface zone I in which dark grey appearance of collagen fibre-rich material contrasts with lighter grey of the immediately subjacent (mid) zones II and III cartilage. Compare the numbers and disposition of the chondrocytes. More deeply, an irregular tide-line delineates the zone IV from the calcified zone V cartilage which is bound to the bony end-plate (at bottom). (HE × 45.)

Hyaline cartilage

This is the material from which the embryonic axial and peripheral skeletons and the bearing surfaces of the diarthrodial joints[53] are formed (Fig. 2.3). Hyaline cartilage is also found in many other normal sites including the nasal septum and bronchi. The bearing surfaces of a diarthrodial joint such as the knee are subject during a 70-year life-span to approximately 1×10^6 loads annually each of 3–4 times body weight. The characteristics that enable articular hyaline cartilage to resist these forces are determined by the components of the extracellular matrix.

The heterogeneity of hyaline cartilage

Descriptions of the pathological changes that may occur in articular hyaline cartilage take account of its complex heterogeneous, three-dimensional character.[72,97] In the sagittal plane (Fig. 2.4) an articular cartilage such as that of the lower end of the human femur can be said to comprise *regions* representing: a. the upper (trochlear) part of the patellar groove; b. the distal part of the patellar groove and the proximal part of the condylar surfaces; c. the main part of the condylar surfaces; and d. the posterior, intercondylar

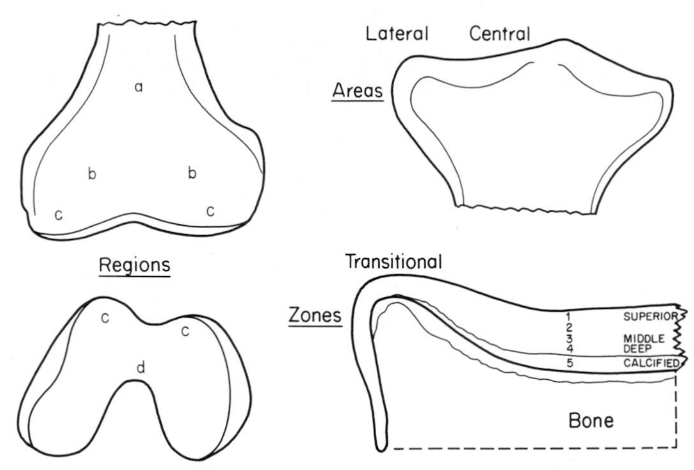

Fig. 2.4 Connective tissue regions, areas and zones. Terms are necessary to distinguish different parts of the connective tissues three-dimensionally. In this example, articular cartilage can be said to comprise sagittal regions, coronal areas and perpendicular zones.

region. In the coronal plane, cartilage can be divided into central and marginal (lateral) *areas*. The latter merge with the transitional collagen-rich tissues between hyaline cartilage and synovium, which, in turn, are in continuity with the periosteum, the loose subsynovial connective tissue, the ligaments, tendons and other articular components. This transitional area is an intermediate territory the principal cells of which are of a fibroblastic character; the scanty intercellular matrix contains less proteoglycan and proportionately more collagen than the nearby hyaline cartilage. In a plane perpendicular to the bearing surface, the layers of cartilage may be described as *zones* of which there can be said to be three (superficial, mid- and deep), four[45] or five.[95,62]

The degree of heterogeneity of articular hyaline cartilage is even greater than these divisions suggest. Articular cartilages are never plane or of uniform thickness. The proportions of collagen, proteoglycan and cells vary greatly. For example, in the central parts of the concave, femoral condylar cartilages, subject to frequent, repetitive, compressive and shearing forces, the chondrocytes are ovoid and single or lie in pairs or triplets within well-defined chondrons* (p. 23); much of the extracellular matrix is proteoglycan with relatively low proportions of the fibrous collagen arranged in a surface mat and a deeper lattice. By contrast, in the marginal, transitional parts of the condyles the cartilage is much thinner,[269] particularly where it is covered by a meniscus: the cells are fibroblast-like, the matrix richly collagenous, the fibre bundles arranged in peripheral circumferential and radial arcades, interspersed with lesser quantities of proteoglycan-rich matrix.

* The old term 'lacuna' is now obsolete (p. 23).

Zonal structure of articular cartilage

Based on the shape, number and arrangement of the cells and on the quantities, molecular composition and macromolecular organization of the collagens and proteoglycans, articular cartilage can therefore be considered as a series of five zones (Table 2.1) (Figs 2.4 and 2.5). This arbitrary division is convenient in analysing the extent of diseases such as

Table 2.1

Cartilage zones and their staining characteristics

Zone	Cells (haematoxylin and eosin)	Collagen content and orientation (picro-Sirius red)	Matrix: proteoglycan concentration (alcoholic toluidine blue)
*Superficial**			
I	Single, ellipsoidal; small	++++ Tangential	+
*Mid**			
II†	Single, ovoid; larger than I; mean diameter 15–20 μm	++ Random (but see p. 74)	++
III	Single, occasional pair; larger than II; mean diameter 20–30 μm	+ Random	+++
*Deep**			
IV A	Single or paired; short columns; larger than III; mean diameter 30–40 μm	+ Random	++++
IV B	As in IV A; short columns of 2–4 cells	+++ Perpendicular	++

* Although convenient, this terminology is highly simplistic.
† Dunham *et al.*[62] suggest that zones IIA and IIB should be designated but they do not distinguish between zones IVA and IVB.

osteoarthrosis (see Chapter 22) but it also reflects the fundamental heterogeneity of articular cartilage so that the intrinsic mechanical properties of the different zones can be shown to be correspondingly diverse.[201,282] The present discussion is centred on appearances recognized in the dog and in the non-human primate but there is much evidence to suggest that comparable heterogeneity prevails in the human tissue.

Zone I The cartilage surface[96] (see Chapter 3) is an artefact, created by disarticulation. In the intact joint, the most superficial part of the articular hyaline cartilage merges with the synovial fluid and the structural sequence, *bone–cartilage–synovial fluid–cartilage–bone* is a developmental, anatomical and functional continuum.

Examined in perpendicular light microscopic section, zone I is, in essence, a dense mat of collagen arranged as orderly, interlacing fibre bundles lying tangential to the articular surface. In the dog, baboon and humans it varies in thickness from 3 to 20 μm. The fasicles of type II collagen are clearly revealed when an exposed surface is briefly digested with papain (Figs 1.9 and 2.6; p. 71); individual bundles or fibres can be shown by scanning electron microscopy or in low temperature carbon replicas (Fig. 3.35, p. 148). The aggregate of zone I collagen is starkly revealed by the polarized light microscopy of sections stained by picro-Sirius red (Fig. 2.7).

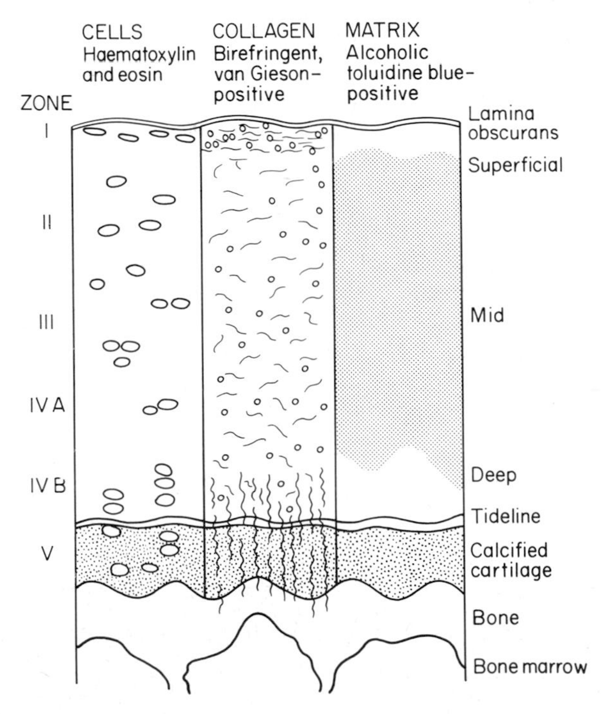

Fig. 2.5 Connective tissue zones.
Perpendicular section of articular cartilage and underlying bone from a synovial joint, indicating the arbitrary zones into which it is convenient to divide the cartilage. At left, distribution of cells displayed in section stained by haematoxylin and eosin (HE); centre, disposition of collagen in section stained by picro-Sirius red; at right, concentrations of proteoglycan indexed by staining with alcoholic toluidine blue. Reproduced from Gardner *et al.*[95]

A *lamina splendens*[171] is now believed to be an optical artefact of phase contrast microscopy.[11] Dunham *et al.*[61] disagree with this interpretation. They confirm the presence of a birefringent cartilage surface and accept Benninghoff's[18] now obsolete view of cartilage collagenous structure. However, the evidence adduced by Dunham *et al.* appears to relate to the whole of the collagen-rich zone I, not simply to the much narrower lamina described by MacConnaill.[171] For this reason, Aspden and Hukin's[11] views must be accepted as correct.

When thin, perpendicular sections of superficial zone I cartilage from a disarticulated joint are examined by transmission electron microscopy, an ultrastructural surface lamina can also be detected (Fig 2.8) (see p. 142). This 'lamina obscurans',[92] is approximately 350–450 nm thick in the normal canine femoral condylar cartilage; it has been identified in low temperature replicas.[207,213] The lamina obscurans has two components. The more superficial, approximately 50 nm thick, is devoid of fibrous collagen and contains no histochemically identifiable chondroitin sulphate proteoglycan.[207] There is no organized fine structure but the layer, which may be glycoprotein, merges imperceptibly with the macromolecules of the synovial fluid (see p. 89). The deeper component, also devoid of fibrous collagen, reacts with cationic dyes such as cupromeronic blue (see p. 54) and includes chondroitin sulphate proteoglycan (see p. 195). The inferior aspect of this component contains the free ends of the banded collagen fibres which are so numerous in the main part of the zone I cartilage.[207]

The chondrocytes of zone I are plate-like or ellipsoidal: their superficial borders lie at or within 1–2 μm of the free, disarticulated surface and their location corresponds to the shallow features seen when non-loaded, disarticulated cartilage is viewed *en face* (see Chapter 3) (Fig. 3.34). The cells extend long cytoplasmic processes between the nearby collagen bundles: thin, ultrastructural sections show apparently separate, cytoplasmic vesicles, remote parts of the chondrocyte cytoplasm, which resemble the matrix vesicles invoked in cartilage calcification (see p. 110) and in the pathogenesis of osteoarthrosis (see Chapter 22). Zone I chondrocytes have thin, disc-shaped nuclei. There is a resemblance to the structure of the fibroblast (see p. 17). Even those cells which lie within 1–2 μm of the surface retain fine structural evidence of active macromolecular synthesis and, in the normal human, primate or dog joint, it is rare to find evidence of cell injury, apoptosis or death.

Zone II The chondrocytes are ovoid or round. Their

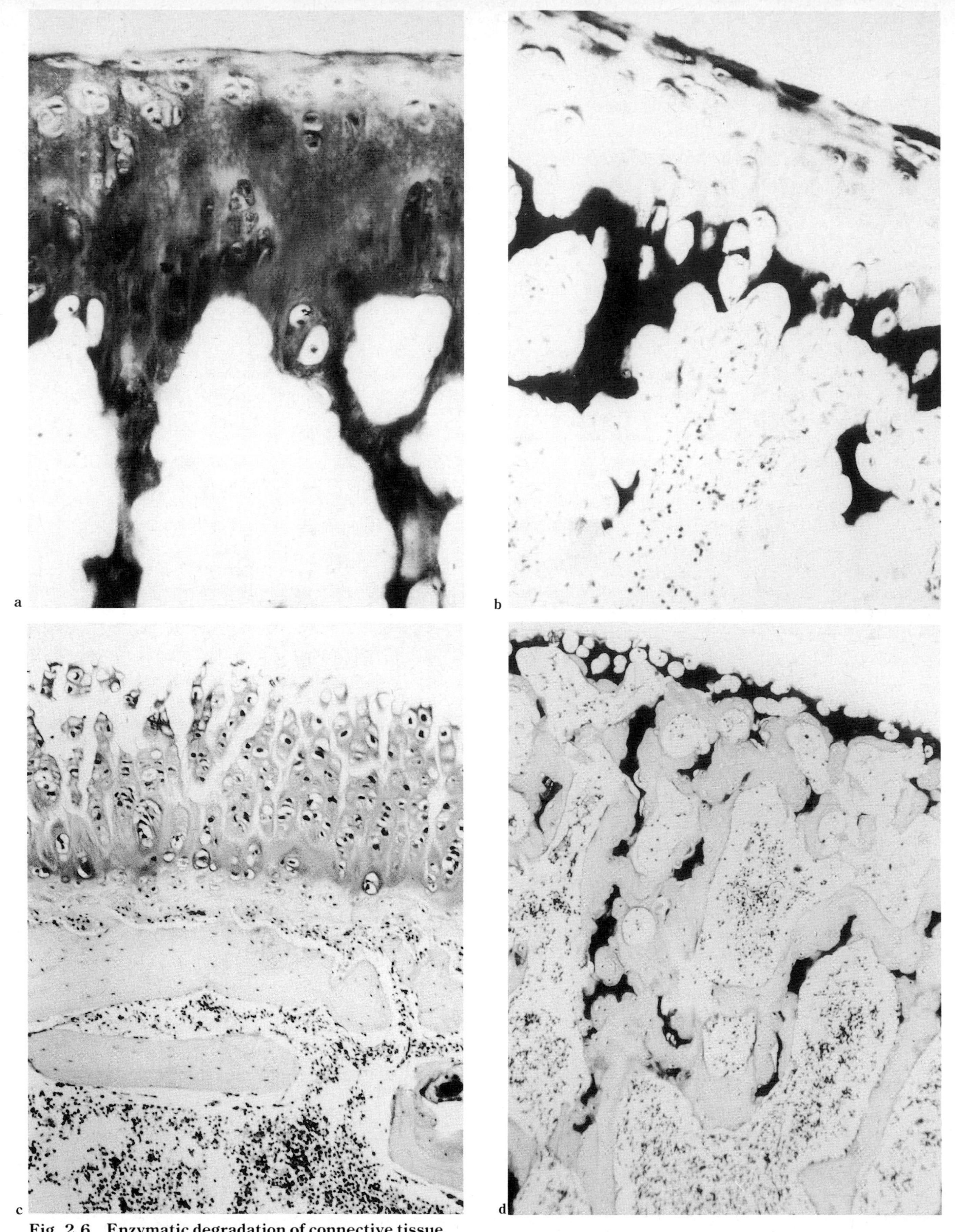

Fig. 2.6 Enzymatic degradation of connective tissue.
Sequential breakdown of rat articular cartilage by papain. **a.** 10 min: early loss of stainable proteoglycan; **b.** 20 min: cartilage matrix now devoid of stainable proteoglycan; **c.** Disintegration of tissue, with fibrillation and loss of cells and chondrons; **d.** Entire loss of tissue, with exposure of calcified cartilage. (**a**, **b** and **d**: toluidine blue × 85; **c**: HE × 85.)

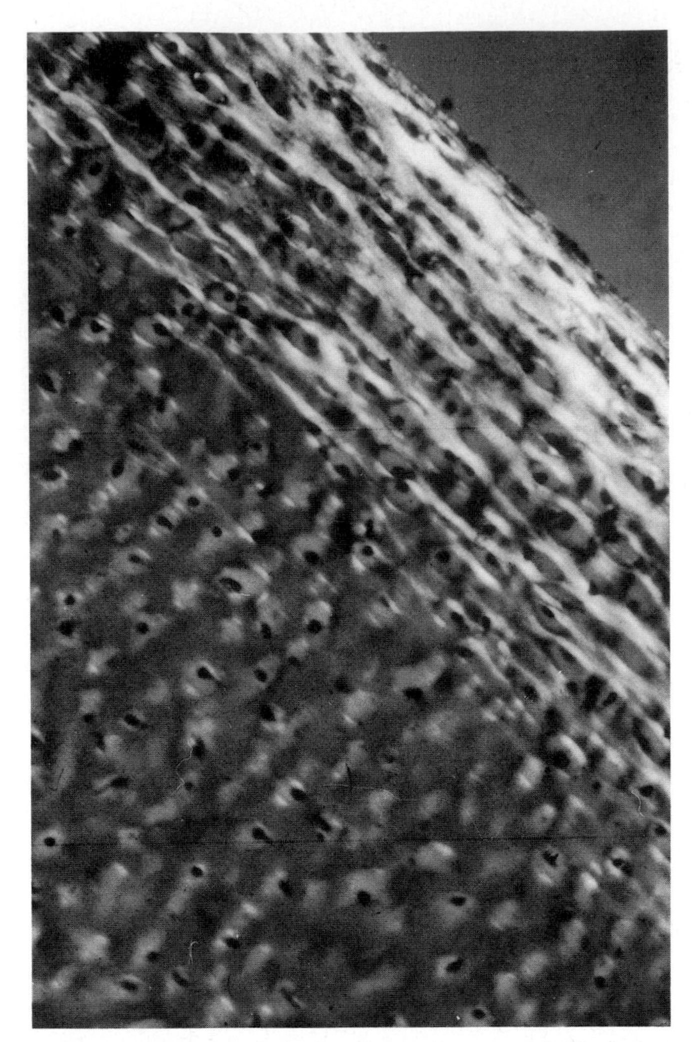

Fig. 2.7 Organization of collagen in non-calcified connective tissues.
In many of the connective tissues, such as this mammalian articular cartilage, collagen fibres and fibre bundles are arranged not at random but in accordance with the prevailing tensile stresses to which the part is subjected. In this articular cartilage, an interlacing, mat-like array of type II collagen fibre bundles can be seen in the zone I and upper zone II cartilage, a less dense and apparently random array of collagen, concentrated around individual chondrons, in the lower zone II and zone III cartilage. (Picro-Sirius red, polarized light × 154.)

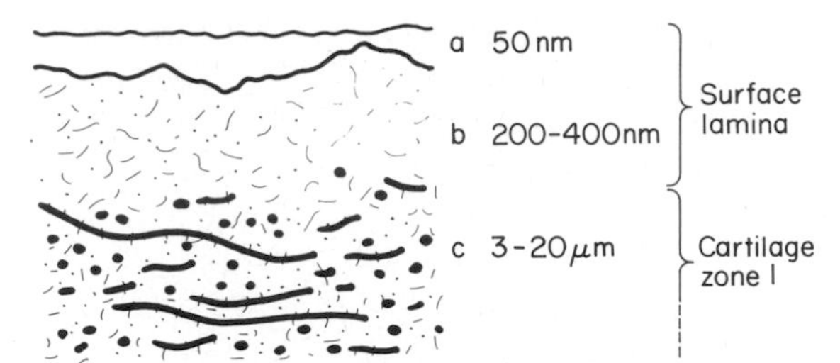

Fig. 2.8 Fine structure of connective tissue surfaces.
In this zone I hyaline articular cartilage, the connective tissue is seen to be organized, at the ultrastructural level, into two surface layers. There is **a.** a superficial fibrillar collagen-free lamina staining with phosphotungstic acid and uranyl acetate but failing to stain for sulphated glycosaminoglycan and resistant to testicular hyaluronidase; and **b.** a deeper lamina with collagen fibrils enmeshed in its deeper aspect. The deeper layer stains weakly with phosphotungstic acid and uranyl acetate but reacts for sulphated glycosaminoglycan; it does not stain for keratan sulphate and is susceptible to testicular hyaluronidase. Deep to this ultrastructural surface lamina, the zone I cartilage is rich in fibrillar type II collagen, arrayed as shown in Fig. 2.7. The interfibrillar matrix resembles that of the less superficial cartilage. In this diagram, the thick lines and heavy dots represent collagen fibres and fibre bundles, the thin lines and light dots, glycosaminoglycan–cupromeronic blue dye complexes. From Orford and Gardner[207] courtesy of the Editor, *Histochemical Journal.*

cytoplasm contains much rough endoplasmic reticulum, clusters of glycogen granules, lipid droplets and a few mitochondria. The most superficial cells display higher levels of oxidative enzyme activity than any other zone. This enzymic heterogeneity[62] contrasts with a selective distribution of alkaline phosphatase activity which, in the rat, is highest in zone IVB, immediately above the calcified cartilage of zone V. On the basis of this diversity of metabolic activity, it has been argued that zone II should be considered to be of two parts, zones IIA and IIB.[62] The zone II cells are single but, progressing more deeply, an increasing number are present as pairs within one chondron.[259] A pericellular matrix (see p. 23) is more abundant than around zone I chondrons (Fig. 3.31, p. 70). The interterritorial matrix includes a meshwork of collagen with fibres extending for varying distances in three dimensions between the chondrons; it constitutes a porous, strong and elastic microskeleton, the interstices of which are occupied by the proteoglycan macromolecules and their water-filled domains. The collagen fibres are finer and more delicate than in zone I; they appear to be arranged randomly, as are those of zones III and IVA but the organization may be pseudorandom (see p. 74).

Zone III The cells, 20–25 μm in diameter, are widely dispersed, plump, ovoid or round and often in pairs; the further they lie from the surface, the larger they appear. The interterritorial matrix contains a collagen array similar to that of zone II and a high proportion of extracellular non-fibrous proteoglycan. The principal glycosaminoglycan is chondroitin sulphate which is more abundant than in zone II; keratan sulphate is scanty.

Zone IV The concentration of proteoglycan is high but variable, the cells, now sometimes in triplets, large. The differentiation between zones IVA and IVB (Fig. 2.5) is subtle: in zone IVB the proportion of chondroitin sulphate diminishes, whereas the proportion of keratan sulphate markedly increases, a change accentuated in advancing age.[261] Finger-like islands of chondroitin sulphate proteoglycan-rich matrix interdigitate with tissue that is less rich in metachromatic material but which contains more stainable collagen which extend upwards from zone IVB.

The tide-line (tide-mark)[76] The tide-line is a faintly basophilic, often reduplicated but sometimes incomplete calcification front that demarcates zone IV from the zone V (calcified) cartilage (Fig. 2.9). The line(s)[219] is 2–5 μm thick. It contains neither neutral fat nor phospholipid and stain techniques such as Bodian's copper-protargol method that demonstrate cement lines, osteocyte lacunae and canaliculi, give no response. The tide-line contains an electron-dense granular material extractable with hyaluronidase or trypsin and undecalcified sections display excess calcium, but not phosphate.[128] The explanation for the basophilia of the tide-line remains uncertain:[259] there is faint toluidine blue metachromasia and alcian blue positivity; the periodic acid-Schiff (PAS) reaction is positive. Glycoproteins, including fibronectin, are present but little glycosaminoglycan. Scanning electron microscopy shows some collagen fibres emerging from the calcified cartilage and passing through the tide-line, but other fibres lie parallel to the irregular surface of the calcified cartilage or are arranged at random.

The tide-line has both biochemical and mechanical significance. Biochemical studies suggest a specific cellular activity at this calcification front, a process associated with the presence of membrane-bound matrix vesicles.[39] The tide-line, like the calcified zone V, is an index of the mechanical behaviour of the cartilage.[219] The mechanical properties of cartilage change with age and the structure of the tide-line alters correspondingly. The junction of the tide-line and the deepest margin of the non-mineralized cartilage constitutes a plane of structural weakness along which microfractures tend to be propagated in the course of osteoarthrosis (see Chapter 22). Cracks form and there is a horizontal splitting and a separation of the cartilage layers.

Calcified zone (zone V) This is the deepest part of the articular cartilage. The superficial boundary is the tide-line(s); the deeper margin of the calcified cartilage is highly irregular, forming an array of processes that interdigitate with the craggy articular aspect of the subchondral bone from which it is delineated by a cement line (Fig. 2.9). The mineralized cartilage front can be directly observed by scanning electron microscopy after the organic tissue is removed. In the case of the rat mandibular condyle, for example, the mineralized component is calcospheritic. The calcospherites are formed of irregular crystalline plates radiating from the centre of the mass. The mineral is hydroxyapatite as in bone.[53,54] Mineralization is dense at the articular edge of the calcified zone, deep to and independent of the tide-line, and on the articular aspect of the cement line. Intermediate, heavily mineralized laminae develop following episodes of arrested calcification, during remodelling of the bone/cartilage interface. The cells of zone V cartilage have much pericellular matrix compared with those of the subchondral osteocytes. There are closely packed collagen fibres perpendicular to the osteochondral junction. It had been assumed that these fibres bind the calcified cartilage and bone together but whether collagen fibres cross the osteochondral junction[203] is still disputed. The extracellular matrix of adult calcified cartilage is PAS-positive and more deeply basophilic than in the child; but the collagen appears to be arranged randomly.

The calcified (zone V) cartilage varies in thickness but is remarkably constant in different joints in a single species, and within different species.[185] However, understanding

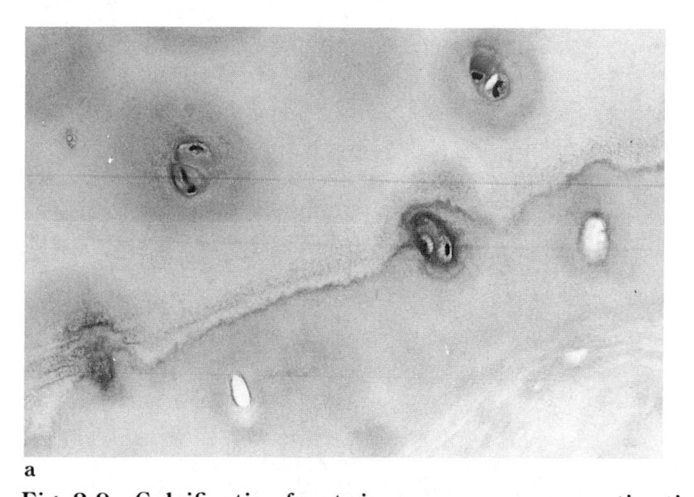

a

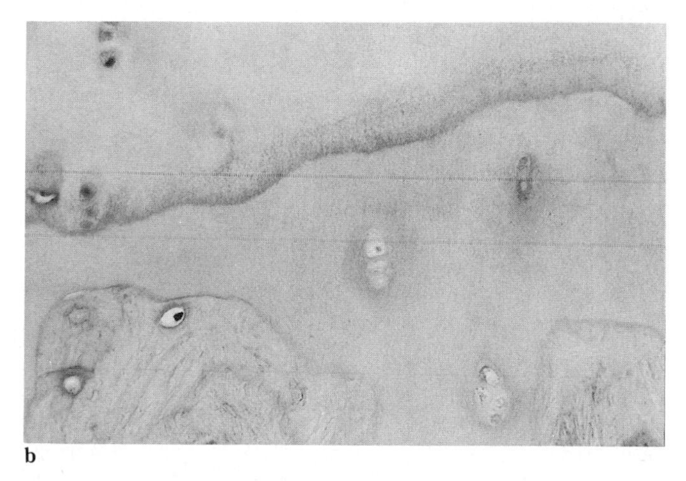

b

Fig. 2.9 Calcification fronts in non-osseous connective tissue.
In articular cartilage, the non-calcified hyaline cartilage is demarcated from the calcified cartilage by a calcification front, the so-called tide-line. The appearances and extent of the tide-line change with age and in disease. (HE × 135)

both of the physiological and pathological responses of the calcified cartilage is not complete. Thus, human articular cartilage appears normally to increase in thickness, at least until the age of about 60 years (see p. 828), whereas the thickness of the calcified zone decreases.[165] Nevertheless, other, conflicting evidence suggests that the ratio of the thickness of the articular cartilage to the thickness of the calcified zone is constant. Both appear to vary in proportion to the mechanical stresses sustained by particular parts of a joint.[195]

Differential distribution and organization of matrix macromolecules A large and rapidly growing literature attests to the efforts being made to elucidate the subcellular and macromolecular organization of cartilage. The extracellular matrix of the placenta has been said to be an integrated unit in terms of collagen types I, III, IV, V and VI;[4] fibronectin; and laminin; and it seems reasonable to view the ultrastructure of cartilage in the same way. New data has emerged from methods centred on the application of monoclonal antibodies in cryotechniques, from low temperature[133,134,132,188] and confocal microscopy (Fig. 1.28, p. 45), and from integrated biochemical, mechanical and ultrastructural studies. However, a significant difficulty in reaching a full understanding of normal cartilage structure remains that of three-dimensional tissue heterogeneity (see p. 68). There are also significant species and age differences so that no single analysis can explain, for example, the molecular organization of the idealized adult human medial tibial condyle.

Collagen Immediately around the articular chondrocyte there is a specialized pericellular (delimited) matrix in which collagen and proteoglycan are intimately associated.[206,229,245] It contains a variable quantity of fine, fibrous collagen and is bounded by a pericellular capsule where type VI collagen is concentrated.[215a] The capsule defines the limit of the chondron (p. 23) (Fig. 1.2b, p. 21). The amount of this material is slight around zone I chondrocytes: instead, they are closely related to the interlacing fasicles of banded, type II collagen. More deeply, in zones II, III and IV, only a little banded collagen immediately surrounds the cells, instead, there is a delicate pericellular corona of fine, non-beaded 5–10 nm diameter filaments that consist largely of type IX collagen. This halo is also circumscribed at its periphery by a discrete 'capsule'.[216] That part of the capsule nearest to the articular surface, the articular pole, is densely compacted as a cupola but has a narrow opening in continuity with the pericellular matrix. Beyond this boundary, external to the capsule, there is an abrupt transition to a surrounding, more coarsely textured matrix.

Between the articular chondrocytes is an interterritorial matrix. Many classical views on the organization of the cartilage matrix were based on the studies of Benninghoff.[18] Using polarized light, he described arcades of collagen bundles originating in the calcified (zone V) cartilage, binding the non-calcified cartilage to this zone and extending in arches towards the articular surface before curving downwards again. In sequence from above downwards, the collagen bundles were 'tangential', 'radial' and 'perpendicular'. It has never been possible for electron microscopists to confirm these observations in fixed material. Instead, they recognize a superficial interlacing mat-like meshwork of collagen fibre bundles in zone I discernible in hydrated tissue by low temperature scanning electron microscopy but an apparently random array of fibres throughout the remainder of the non-calcified matrix. However, interpretation of the organization of fibrous and non-fibrous matrix macromolecules may be influenced by the orientation of the material during the preparation of light microscopic sections.[61] Working on the assumption that elongated proteins such as collagen micelles that can exhibit form-birefringence have to show virtually straight extinction when viewed between crossed polarizing filters, these authors obtained results from canine tibial condylar cartilages which agreed with those of Beninghoff.[18]

Advantage has been taken of techniques that allow hydrated tissue to be viewed microscopically[30,31,34] to re-examine this important question. Articular margins, i.e. transitional areas, are collagen fibre rich: the fibres of these non-load-bearing parts have a macroscopic crimp configuration (see p. 264).[98] More surprising is the identification in hyaline articular cartilage of planes of vertically aligned fibres that have a comparable organization.[34] A similar arrangement of vertical columns of collagen fibres is seen in the central parts of the medial tibial condyles of the baboon.[97] In the interterritorial matrix, these crimped fibrous patterns are always aligned in the direction of mechanical loading. It seems clear therefore that the fibre pattern in hyaline cartilage is, as Broom and Myers[34] say, considerably more 'disciplined' than had been believed.

On the basis of this evidence, it is envisaged that deformable elements, probably hydrated proteoglycan, are trapped in a three-dimensional network of collagen fibres (Fig. 2.10) generated from radial fibre arrays with serial cross-links along their length.[33] Evidence to support this view was obtained by studying the propagation through cartilage of perpendicular notches and of notches made parallel to the surface. The fibrous matrix was apparently highly anisotropic. There was a fundamental difference in cartilage structural configuration in these two main directions.[32] A pseudorandom meshwork was thought to develop from the radial collagen fibre array: each fibre was repeatedly deflected sideways into short-range, oblique orientations along its axis. Overall, this configuration (Fig. 2.11) appeared directional although, viewed in thin electron microscopic section, it was random. The matrix was a braced

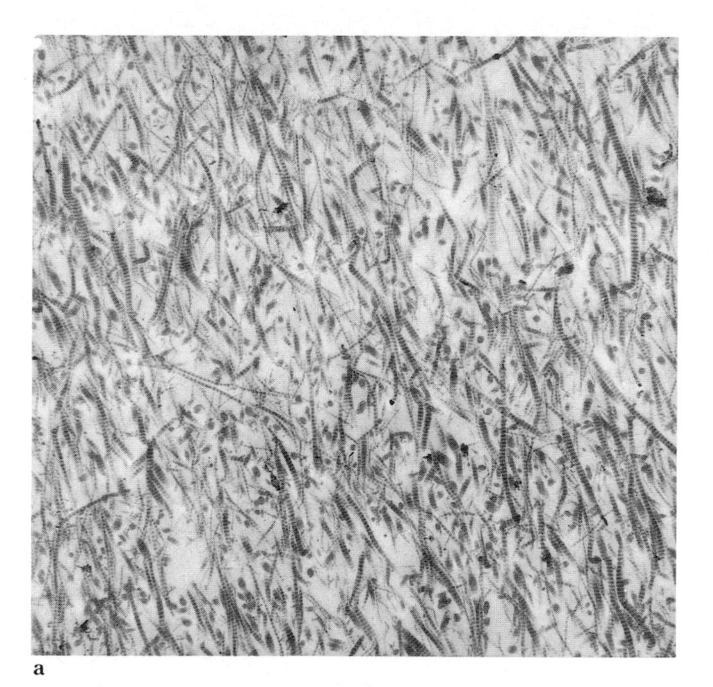

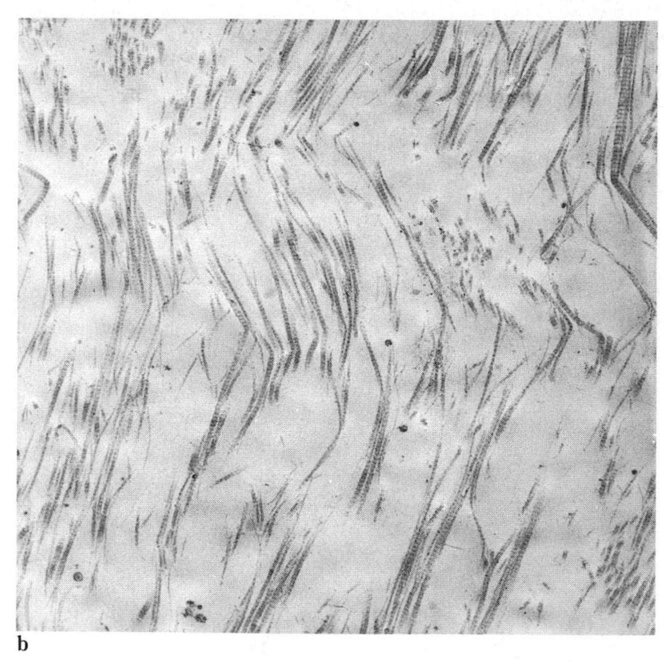

a b

Fig. 2.10 Ultrastructure of zone II/III (midzone) bovine, patellar groove, femoral cartilage.
a. Normal appearance with short fibril segments repeatedly grouped into near-parallel aggregates ('nodes'), with preferential radial orientation. Between 'nodes', structure is more open and irregular. **b.** Structural transformation produced by pre-digestion with hyaluronidase or trypsin followed by bacterial collagenase. Note alignment of fibrillar structure into near-parallel radial arrays in plane of section. Individual fibrils aggregate into fibres and assume a crimped (waveform) organization. (TEM × 7080.) Courtesy of Dr Neil D Broom.

structural system, able to resist the shear forces of direct compression. In non-progressive, age-related cartilage degeneration (see Chapter 21) and in the progressive disorder of osteoarthrosis (see Chapter 22), the greatly increased compressive compliance ('softening') of the cartilage, it was proposed, is immediately associated with a reversion of these sideways deflected collagen fibre segments to a radial configuration of a kind common in disordered cartilage.

COLLAGEN NETWORK TRANSFORMATION
IN ARTICULAR CARTILAGE

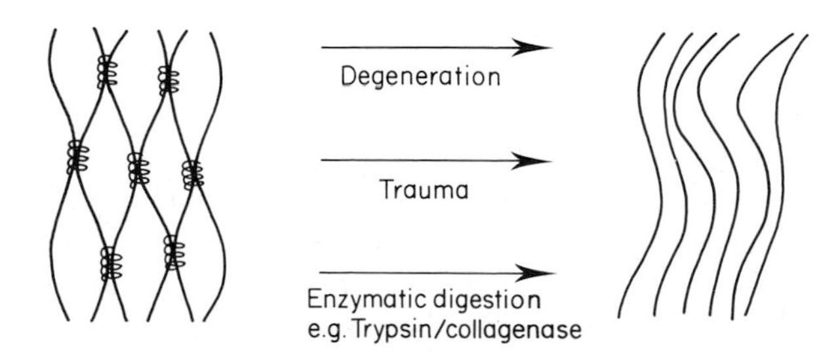

Fig. 2.11 Enzyme-induced cartilage structural transformation.
Diagram of three-fold pathway from a pseudorandom fibrillar arrangement in articular cartilage matrix to a radially aligned, crimped structure. From Broom[31] by courtesy of the Editor, *Arthritis and Rheumatism*.

Proteoglycans It has long been known that there are histochemical differences between the proteoglycan composition of the pericellular and interterritorial matrices. These views have been substantiated by the application of poly- and monoclonal antibodies at light- and transmission electron microscopic levels. Thus, in human material, proteoglycans are deficient in the interterritorial regions of deep zone cartilage, and link protein is deficient in the pericellular regions of deep and mid-zone cartilage.[215] There are analogous differences in the demonstrable proportions of keratan sulphate.

Cartilage water The physical attributes of hyaline cartilage and its essential mechanical properties centre on its water content.[179,185,96] The water content of normal hyaline articular cartilage varies between 60 and 80 per cent; it is high in embryonic and neonatal life and declines with age. The water content of adult articular cartilage is usually between 74 and 76 per cent. The proportion of cartilage water contributed by cells is small. The extracellular water is held within the domains of the very large but incompletely expanded proteoglycan aggregates which are restrained by the fibre bundles of the collagen microskeleton, themselves under tension. As Maroudas[179a] points out, the actual water content of a particular specimen is determined by the balance between the tension in this collagen meshwork and the swelling pressure of the proteoglycan aggregates. The water content in *postmortem* cartilage samples is highest in slices cut from the most superficial part, lowest in the deepest zone.[276a] The reason for these differences is in part variation in proteoglycan concentration, in part the degree to which the collagen microskeleton is under tension.

Water and solutes move freely through normal articular cartilage. Most extracellular water is readily available for the passage of solutes.[180,182] Cartilage water is 'free', not 'trapped'.[179a,181] The water content of cartilage can, of course, be readily altered by changing its physical and chemical composition.[251] By the same token, the ready movement of pericellular and interterritorial water, in response to stress, determines the viscoelastic mechanical characteristics of hyaline cartilage (see Chapter 6).[216]

Cartilage thickness

The cells of hyaline articular cartilage respond quickly to sustained changes in the degree and direction of applied loads. It is not surprising therefore to find that the quantity and organization of cartilage on a bearing surface such as the femoral head is not uniform: the maximum thickness is at points on opposed, contralateral areas,[164] corresponding to parts recognized in lateromedial radiographs to be approximately 20° anterior to the zenith (Fig. 2.12).[9] From these sites, cartilage thickness decreases centripetally. Comparable considerations apply in the human knee (Fig. 2.13) where, as in other erect bipedal primates, the medial part of the medial tibial condylar cartilage is thicker than the lateral (Fig. 3.17, p. 131). The thickness, again, is not uniform and decreases towards the medial and lateral aspects of the

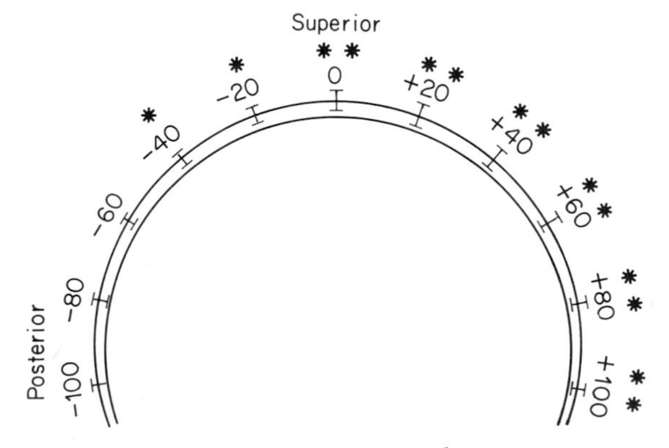

Fig. 2.12 Cartilage thickness: age changes.
Diagram of femoral head as seen in lateromedial radiograph, showing mean cartilage thickness ±3 SD. ** indicates significant regression of cartilage thickness against age alone (simple linear regression). * indicates significant regression of cartilage thickness against age, sex and body height, with age accounting for the greater part of the explained variation. From Armstrong and Gardner.[9]

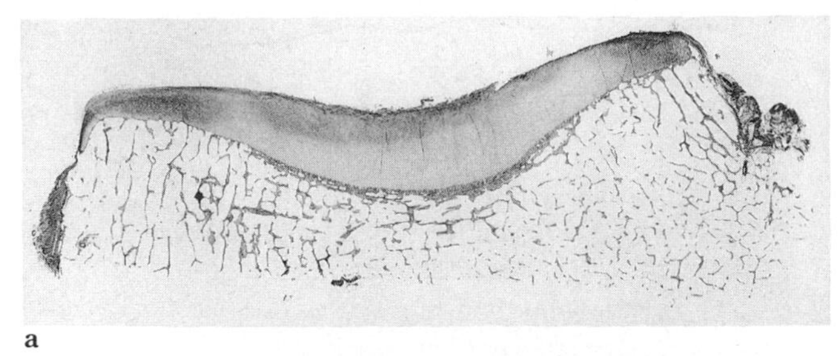
a

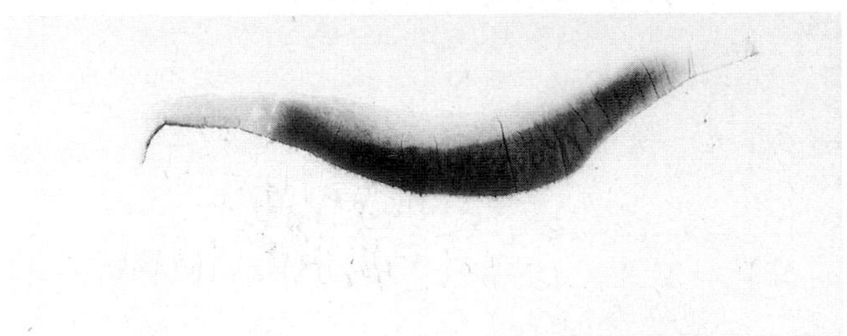
b

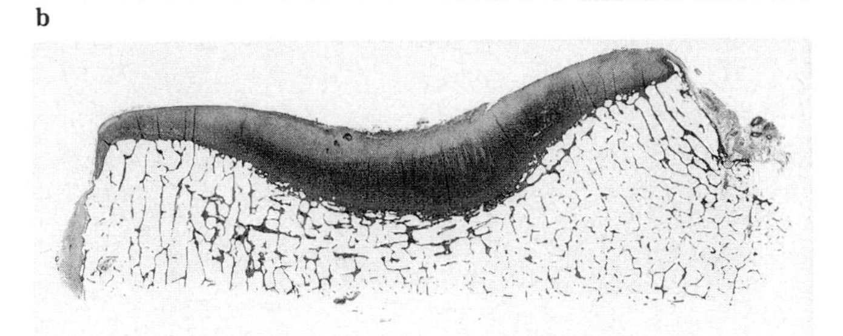
c

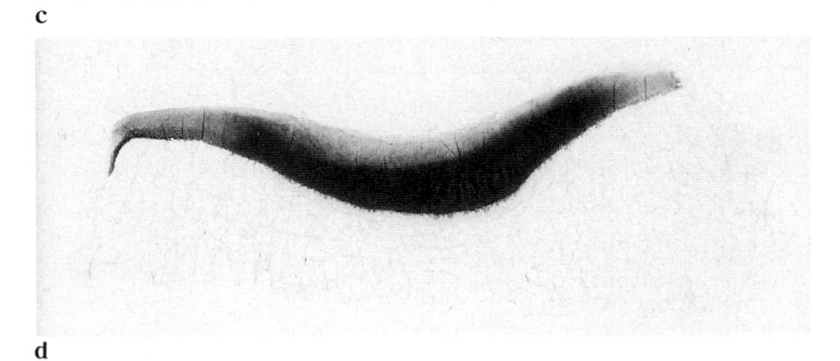
d

Fig. 2.13 Heterogeneity of articular cartilage structure.
Adjacent sections from medial tibial condylar cartilage to show highly varied differential structure. **a.** Picro-Sirius red; **b.** Safranin O; **c.** HE; **d.** Toluidine blue. (× 1.5.)

condyles. Femoral and patellar cartilages are similarly of non-uniform thickness.

Hyaline articular cartilage thickness tends to be greater in males than females; the differences are thought to be a reflection of body mass. It may be difficult to identify a change in cartilage thickness with age; indeed Meachim[183] (Table 2.2) concluded that the thickness of the non-calcified cartilage of older persons was not related to age. However, Armstrong and Gardner[9] found that, between the ages of 20 and 50 years, the femoral head cartilage increased in thickness, the degree of increase being greatest in the thickest part. Their results were derived from amplified lateromedial radiographs. Comparable results have been obtained in studies of the sacroiliac joint in which both sacral and iliac cartilages become thicker as age advances (see Fig. 3.22; p. 949).[94] A method that lends itself to *in vivo* cartilage thickness determinations, is double-contrast arthrography:[84] abnormalities or changes as small as 1 mm diameter can be visualized. However, the technique is much less convenient than high resolution MRI (p. 162) with which variations in cartilage thickness of the order of 1 mm and focal defects of 1 mm diameter can be accurately identified and measured in 3 mm thick MRI slices.[142a] Microfocal radiography (see Chapter 3)[37,38] can enhance this resolution five-to-ten-fold.

Cartilage cellularity

A wide range of cellularity is recognized when the large joints of a single species are compared with the small. These ratios determine a scale effect: the cellularity ('cell density') of the tissue decreases in inverse proportion to the cartilage thickness.[248,249] The number of cells deriving oxygen and metabolites from each unit of load-bearing articular surface is thought to be uniform and the quantity of cartilage covering each unit of load-bearing bone is determined by factors, such as mechanical load, to which the surface is commonly subjected.

Table 2.2

Relationship between articular cartilage thickness and age

Age (yrs)	Number	Mean thickness (mm)
Humeral [183]		
24–44	9	1.48
45–64	13	1.42
65–75	10	1.47

Femoral head [9]
Greatest thickness (approximately 3 mm), 20° anterior to the zenith. Between 20 and 45 years, cartilage thickness increases with age. Results beyond the age of 45 years reveal very large scatter. There is no evident relationship between femoral head diameter, length of femur or body weight

Sacral [94]
Mean thickness of 16 specimens: 1.79 mm (range 1.34–2.16 mm)

Iliac
Mean thickness of 16 specimens: 0.83 mm (range 0.61–1.08 mm)

In both sacral and iliac cartilages, there is a significant trend for thickness to increase as age advances

The overall cellularity of cartilage changes greatly with the stage of development. In the embryonic limb (Fig. 3.5), before bone is formed, the volume proportion of cells to extracellular matrix is approximately 1:3. As the limb grows, the relative volume of the matrix increases. In adult articular cartilage the chondrocytes of zone I tend to be single, the proportions of cells to matrix relatively high. In zones II and III, the chondrocytes, now often paired, each occupy a larger volume but the proportion of matrix is very much greater. In diseased tissue, inconstant cell size and high cell density are characteristic of hyperplasia and neoplasia. By contrast, the cells of normal, mature, differentiated tissue are relatively small, the proportion of extracellular matrix to cell volume relatively high.

Although the cartilage of large synovial joint surfaces, such as the femoral head, tend to become thicker during the period of maturity,[9] there is a reduction in cellularity of the superficial zones.[259] This change is not necessarily detectable in other joints such as the shoulder;[183] it may represent an early index of osteoarthrosis rather than a change attributable solely to senescence (Fig. 2.14). Osteoarthrosis is indeed common in femoral head cartilage, but very uncommon in humeral head cartilage.

There is a tendency for cell density to be greatest at tissue margins, for example, at the perichondrium. There are conspicuous differences in the cell density between cartilage surfaces and deep zone cartilage, and between the cartilages of larger and of smaller species. Thus, the femoral condylar cartilage of the mouse, 60 μm in thick-

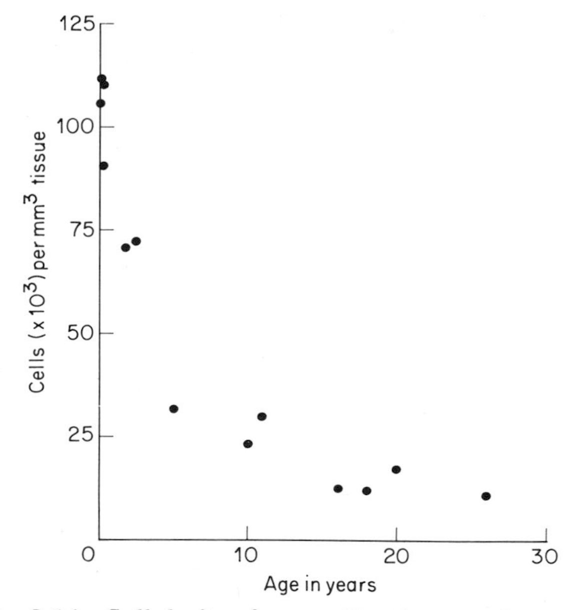

Fig. 2.14 Cellularity of connective tissue and age. Cell density of human femoral condylar cartilage falls during maturation. Other evidence suggests that decline in cellularity continues, at least to age 50–60 years. From Stockwell[258] courtesy of the author and the Editor, *Journal of Clinical Pathology*.

ness, has approximately 1.9×10^4 cells beneath each 1 mm^2 of articular surface, giving a cell density of 33.4.[258] In humans, by contrast, the corresponding cell frequency, in femoral condylar cartilage 2.26 mm in thickness, is 3.1×10^4 beneath each 1 mm^2 of articular surface, giving a cell density of 1.4 (Table 2.3). In sacral cartilage the cell density is approximately half that of the opposed, iliac cartilage which is only half as thick (Fig. 3.23).

Table 2.3

Relationship between cartilage thickness and cell density in eight mammalian species[258]

Source of cartilage	Cartilage thickness (mm)	Cell density (cells $\times$ 10^4 mm^{-3})	No. of cells ($\times$ 10^4) deep to 1 mm^2 articular surface
Man	2.26	1.4	3.1
Cow	1.68	2.0	3.3
Sheep	0.84	5.3	4.2
Dog	0.67	4.4	2.7
Rabbit	0.21	18.8	3.7
Cat	0.33	10.8	3.2
Rat	0.07	26.5	1.9
Mouse	0.06	33.4	1.9

Cartilage pigment

In the fetus and also in infants, hyaline cartilage retains a translucent, gelatinous, blue-white appearance but to a decreasing degree. The opacity advances with time and, in old age, the colour becomes increasingly yellow. The yellow colour is associated with the non-collagen protein fraction of the cartilage but not with the polysaccharide moiety. The nature of the pigment is incompletely understood[274] although it is possible that it is a lipofuscin or bilirubin. It has also been suggested that the yellow pigment of ageing cartilage is a carotenoid.[185] Although relatively large coloured molecules such as haemoglobin (mol. wt 6.8 kD) can penetrate normal cartilage, it is interesting that in synovial haemosiderosis (see p. 971) cartilage remains non-pigmented. The contrast with the remarkable binding of the brown-black oxidation products of homogentisic acid in ochronosis (see p. 351) is striking.

Cartilage response to vascularization

Adult hyaline articular cartilage does not possess a direct blood supply although thin-walled sinusoidal vessels persist in the deep zone cartilage of young porcine and simian cartilages. Avascularity is, by definition, therefore a property of mature hyaline cartilage just as it is of the normal adult cornea and lens. When, because of nearby inflammation or the growth of a vascular neoplasm, blood vessels multiply at a hyaline cartilage margin, there is a tendency for cartilage matrix proteoglycan to be depolymerized. The presence of a normal proportion of matrix proteoglycan in cartilage is indeed incompatible with the ingrowth of blood vessels. Vascular ingrowth is facilitated by angiogenesis factors,[36,162] (see p. 295) catalysed by proteases but normally blocked both by the integrity of proteoglycan and by inhibitors present in the extracellular tissue fluids. The process is accompanied by a fundamental change in the character of the tissue. Chemically, the proportion of polyanions decreases: there is diminished metachromasia. Physically, swelling pressure and ionic gradients change. Mechanically, there is a loss of mechanical strength and increased deformability.

Hyaline cartilage repair

The capacity of adult articular cartilage to repair itself by regeneration has long been doubted. It therefore seemed likely that the indolent nature of osteoarthrosis was, at least in part, attributable to an inability of articular chondrocytes to respond to injury by multiplication or by accelerated matrix synthesis. It was only too easy to accept that articular cartilage could not regenerate under any circumstances. This view is no longer tenable and much evidence has established both the ability of chondrocytes to multiply and of cartilage to repair.[41,253,261,99,100] The growth and proliferation of chondrocytes is outlined in Chapter 1 (see p. 26). In this chapter, the reaction of cartilage to injury is introduced in terms of tissue responsiveness. The effects on cartilage of immobilization and of excess loading are considered in Chapter 22, the adverse effects of minor and major trauma in Chapter 24.

Much of the evidence relating to the repair of cartilage has necessarily come from experimental studies of cartilage wounds in animals. A useful distinction is made between excised and incised wounds and between the effects of sharp and of blunt impact. Conspicuous differences are also noted between superficial injury, affecting cartilage alone, and deeper injuries which extend to the vascular subchondral bone and bone marrow. The former rarely heal; the latter are often capable of repair.

Superficial injury[145]

Excised wounds In the experimental animal, a cartilage defect caused by excising a superficial portion of cartilage remains unaltered over periods of up to two years.[90] The residual, nearby surface cells frequently die and there is a significant loss of proteoglycan from cartilage zones adjoining the defect. The remaining live chondrocytes are rarely provoked to division although they display heightened metabolic activity. An increased content of intracytoplasmic filaments is observed and lysosomes are seen.[99] Cells with the characteristic ellipsoidal shape of zone I chondrocytes do not reform. The superficial lamina of type II collagen is lost. However, in response to continued compressive and shear stresses caused by limb movements there is a displacement of cartilage matrix and cells into the defect, a 'flow formation'.

Incised wounds These superficial injuries also do not repair. There is no difference in response whether the incision is made parallel to or at right angles to the principal axis of alignment of the superficial collagen bundles, those demonstrated by pricking with a round pin (see Chapter 6) (Fig. 6.4a, p. 266).

Blunt impact Contusions of articular cartilage do not cause chondrocyte death or structural damage unless stress maxima of approximately 25 N/m^2 are achieved.[220] The effects of such severe injury persist and fibrillation (see Chapter 22) becomes evident.

Freezing injuries Other modes of cartilage injury have thrown light on the capacity of the tissue to regenerate matrix materials. After localized freezing with a cryoprobe at −40 °C, rabbit articular chondrocytes appear dead. The injured tissue stains poorly for proteoglycan and has little capacity to incorporate [^{35}S]-sulphate.[250] These appearances persist for approximately six months. Thereafter, although the cartilage remains soft, the chondrocyte population gradually increases and new proteoglycan synthesis is accompanied by an irregular pattern of large collagen fibres. If the fibrous microskeleton of cartilage is not destroyed *in situ*, a new colony of chondrocytes can extend into the dead cartilage with the resumption of proteoglycan synthesis.

Radiation injury Cartilage injuries in response to γ-irradiation have been investigated.[66] A loss of metachromatic and alcian blue-positive stainable matrix materials, and a corresponding decrease both in hexose and hexosamine and in Hyp are related to cobalt-60 γ-radiation dosage.

Injuries extending to bone and bone marrow

When wounds in cartilage extend to the vascular, subchondral bone and bone marrow, the reparative sequence is quite different. Early studies such as those of Key[145] were discussed by Meachim and Roberts,[184] by Mitchell and Shepard[191] and by Furukawa *et al.*[91] Many experimental investigations have been reported,[100] and the drilling of cartilage and bone, to encourage repair, in disorders such as osteoarthrosis has led to a correspondingly large number of clinical reports.[136] The cartilage from different parts of normal human femoral condyles appears equally capable of reparative metabolic responses, including proteoglycan synthesis.[169]

The tissue that comes to fill a full-thickness cartilage defect has often been loosely described as fibrous tissue or fibrocartilage, but studies such as those reviewed by Ghadially,[100,101] show that there is a relatively orderly synthesis of a sequence of increasingly differentiated tissues. Shortly after an experimental defect of cylindrical shape, caused by a sharp, sterile instrument like a cork borer and presumably for mechanical reasons, a viscous extension ('flow') of cartilage from the intact wound margins protrudes across the vacuity. A coagulum of fibrin occupies the base of the defect where it penetrates calcified cartilage and bone. Soft, red-brown material becomes increasingly abundant and, in less than a month, largely fills the experimental wound.

Microscopy demonstrates that within a week of injury, the deeper part of the chondroosseous defect is occupied by a vascular connective (granulation) tissue, the more superficial part by a more cellular, loose connective tissue in which there are many spindle-shaped ('fibroblast-like') cells. The deeper part of this reparative tissue is vascular, the more superficial part, related to surrounding cartilage, avascular. Two varieties of tissue evolve from these precursors. The deeper material, destined to differentiate into bone and to fill the bone defect, is a proteoglycan-rich chondroid substance. The more superficial material, weak in proteoglycan until a late stage of repair, is a precursor of the incompletely organized hyaline cartilage into which it will differentiate. It is possible that this distinction between superficial and deeper components of the repair tissue may explain the observations in rabbits of Furukawa *et al.*[91] These authors found that the collagen of early (three-week) repair tissue was type I, but after 6–8 weeks the predominant collagen was type II. Nevertheless, type I collagen persisted in significant amounts for up to a year after injury.

As in superficial wounds, the cartilage matrix which forms the intact edge of a deep chondroosseous injury becomes depleted of stainable proteoglycan and the marginal chondrocytes die. It seems likely that the wounding process initiates the release of interleukins and other cytokines and that the marginal changes are initiated by these messengers. The sequence of cell injury and matrix loss bears similarities to those encountered in cartilage matrix degradation during pannus formation (see Chapter 12).

Whether populations of cells other than those derived from the bone exposed in deep chondroosseous wounds can contribute to their repair has been uncertain. However, it appears likely that mesenchymal cells from the transitional zone can take part in the reconstruction of defects in peripheral cartilage.[192] These authors induced cartilage degradation by the intraarticular injection of papain. Proteoglycan loss quickly followed. The chondrocytes that survived contributed little to repair but marginal (transitional zone) mesenchymal cells migrated medially, differentiated into chondrocytes and reconstituted hyaline cartilage by the secretion of a new extracellular matrix.

Modulation of cartilage repair A large number of procedures has been found to facilitate or enhance cartilage repair. When repair is sought in disorders such as osteoarthrosis, the surgical drilling of residual cartilage at multiple foci has been advocated.[191] Osteoperiosteal grafts have been used for a similar purpose and the value of invoking the reparative capacity of periosteal tissue is clear. New cartilage synthesis by autogenous grafts is an effective form of facilitating repair but immobilization or continuous passive motion have both been claimed to catalyse this process.[202] Value is attached to the intraarticular injection of hyaluronate[263] but the extent of this benefit may be limited to the prevention of water loss from the injured cartilage.[286]

Many materials have been tested in attempts to encourage cartilage to repair quickly and efficiently. Thus, porous composite implants of polytetrafluoroethylene and graphite fibre can successfully stimulate new hyaline cartilage formation.[143] The insertion of a sponge of collagen has a similar influence.[256] Repair has also been attempted by implanting cultured chondrocytes into experimental cartilage wounds[19,110] but it is of course much more difficult to secure pellets of cells *in situ* than it is to implant grafts (see p. 81) that retain continuity with periosteal or synovial tissue. Attempts have also been made to encourage cartilage repair by chemical and endocrine means.[259] Among the procedures tested are: the inhibition of proteases; the stabilization of chondrocyte lysosomal membranes; the modulation of proteoglycan synthesis; and treatment with growth hormone (somatotrophin), somatomedins or cartilage growth factors (see p. 28).

With whatever degree of success repair is attended, the final cartilage formed after deep chondroosseous injury retains some similarity to fibrocartilage. The perpendicular zonation of mature adult articular cartilage clearly does not re-form although a tide-mark delineates zone V (calcified) cartilage from zone IVB. It must be assumed that the mechanical properties of repair cartilage, however completely reconstituted, remain abnormal although there have been few experimental studies to test this view.

Cartilage (organ) culture

Much of our present understanding of cartilage comes from studies made with the techniques of chondrocyte (cell) culture (see p. 24).[254] Cartilage can also be maintained *in vitro* as explants of embryonic chick limb rudiments[78] (Fig. 2.15) or as blocks of excised immature or mature tissue. Maintenance media are employed and the explant can be studied either at the interface between a liquid culture medium and an humidified gaseous environment of controlled composition, or immersed in a fluid medium.[254] Embryogenesis, matrix synthesis and the influence of mediators of inflammation such as the prostaglandins[148–152] can be investigated. Explants of cartilage have proved of value in assessing the influence of synovial tissue on cartilage (see p. 88). Live synovium co-cultured with live or dead cartilage leads to chondroid matrix degradation when the tissues are contiguous. When separated within a single culture dish, live synovium provokes the breakdown of live but not of dead cartilage.[77,78] In classical experiments reviewed by Steinberg, Hubbard and Sledge,[257] it was established that both complement-sufficient antiserum and retinol (vitamin A)[81,83] degrade cartilage matrix by the release and activation of lysosomal enzymes:[83,56] complement injures chondrocyte plasma membranes after antibody has bound to the cell surface; retinol damages chondrocytes directly.

There have been numerous other investigations. For example, the methods of organ culture have been employed to analyse cartilage catabolism and anabolism in experimental models of disease such as the canine cruciate ligament section model of osteoarthrosis (see Chapter 22).[194]

Cartilage can also be studied usefully in the whole chick or mammalian embryo. Among the observations made with

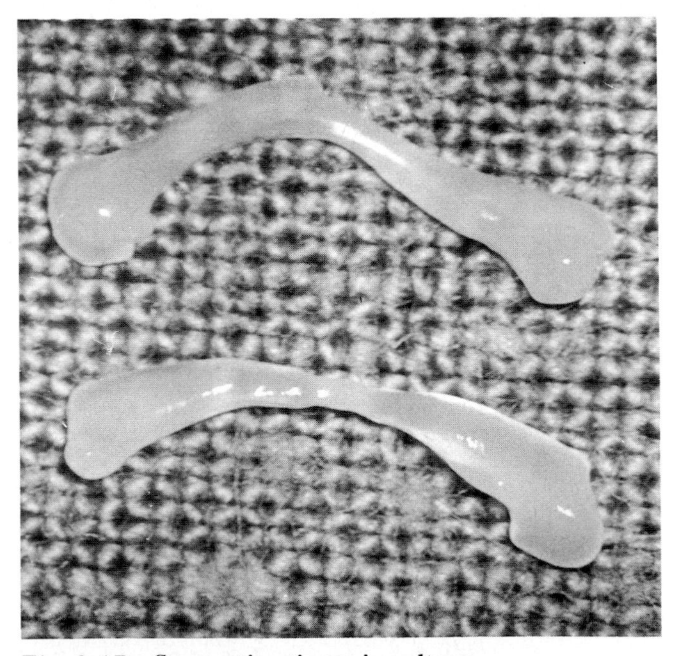

Fig. 2.15 Connective tissue in culture.
Embryonic connective tissues, such as these chick femora, can readily be maintained in organ culture. Cell multiplication may continue. Adult tissues such as synovium, can also be maintained in this way but may not exhibit cell division and growth. Courtesy of Professor C J Kirkpatrick.

this technique is the demonstration that movement is necessary for normal joint formation:[194a] curare or *Clostridium botulinum* toxin cause paralysis and joint development is delayed or inhibited. Limb cartilage regeneration is also impaired by partial denervation[66a].

Cartilage metabolism

The biochemical basis of the connective tissue diseases is reviewed in Chapter 4 and aspects of chondrocyte metabolism in Chapter 1 (p. 27). An extensive appraisal of many aspects of connective tissue metabolism is given by Kühn and Krieg.[163] The components of the extracellular matrix can be investigated *in situ*: their synthesis can be studied by isotopic labelling techniques in which the incorporation of [^{35}S]-SO_4 is used as a guide to glycosaminoglycan (and thus to proteoglycan) synthesis, while the incorporation of [^{3}H]-thymidine or [^{14}C]-Pro provides evidence of DNA and collagen synthesis, respectively. In one example of how these methods can be applied to the study of cartilage metabolism, it was shown that the major proteoglycan synthesized by long-term primary monolayer culture of rabbit articular chondrocytes was similar to the proteoglycan of cartilage matrix.[129] Chondrocytes respire at low oxygen partial pressures. However, the level of activity of the enzymes that catalyse electron transfer differ significantly between different cartilage zones.[62] It is of interest that the differential distribution of other enzyme activities, such as that of alkaline phosphatase, is not the same as that of the oxidoreductases (Fig. 2.16).

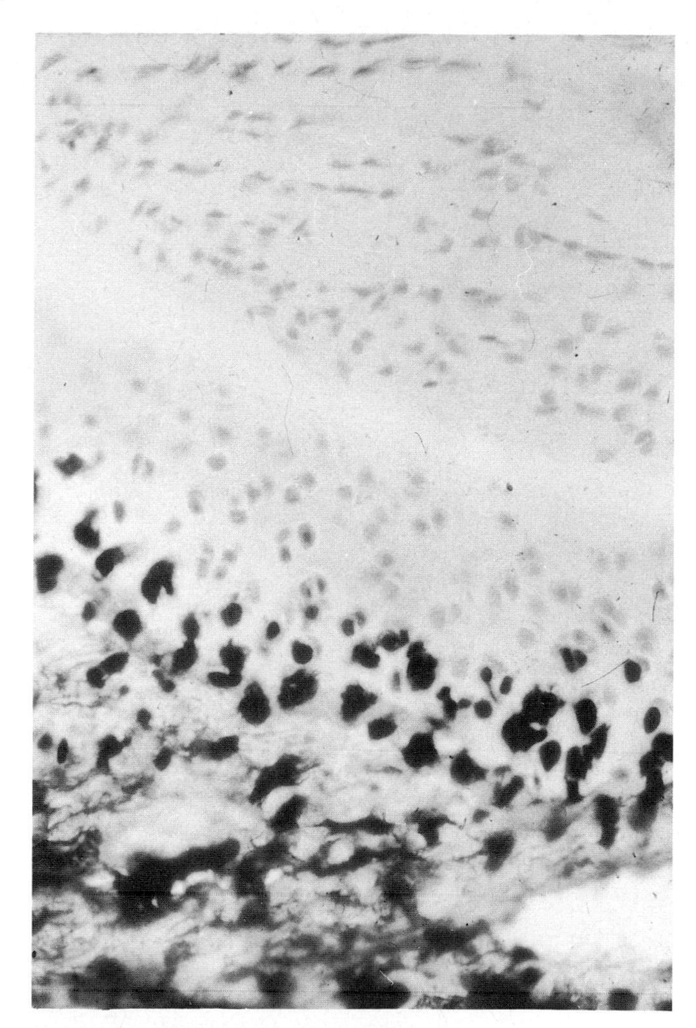

Fig. 2.16 Differential expression of enzyme activities in connective tissue.
Different parts of hyaline articular cartilages, for example, display different levels of enzyme activity. In this figure of rat ankle joint cartilage, alkaline phosphatase activity is high in the deeper zone III, zone IV and calcified (zone V) cartilage, but not detectable in zone I, II and superficial zone III cartilage. Comparable differences have been shown in the activities of important oxidative enzymes.[62]

Cartilage transplantation (grafting)

Whole joints can be transplanted within individual animals, avoiding immune rejection and establishing a viable circulation by microvascular anastomosis.[135] However, this procedure of autografting is rarely of practical value in humans and questions of cartilage rejection have usually been faced when allografts and, occasionally, xenografts, are attempted using whole cartilage preparations or cultured chondrocytes.

Many components of cartilage including the chondrocyte plasma membrane (see p. 28) are antigenic.[70] It had been thought that articular cartilage was almost inert immunologically and might therefore be transplanted with relative ease; cartilage matrix certainly affords some protection against the immunological phenomena of graft rejection.[71] However, allografts are slowly damaged and reabsorbed both because of the formation of host cytotoxic antibodies and through the actions of sensitized lymphocytes. The individual structures of joints are selectively permeable to immunoglobulins.[49] The response of cartilage to antibodies varies in different sites. Foreign antigen is retained in different parts of the joint with an affinity related to charge[273] and charge-dependent antigen retention may be greater than that of antibody-mediated trapping. A mechanism of this kind is likely to influence cartilage graft rejection.

Rats injected with isolated syngeneic chondrocytes develop cell-mediated immunity to these cells and react similarly to grafts of articular cartilage shavings. However, caution must be used in arguing from cells isolated by enzymatic procedures because of possible changes caused in cell surface epitopes. Nevertheless, it appears probable that cartilage can elicit effective, deleterious immune responses that diminish graft effectiveness. The greater part

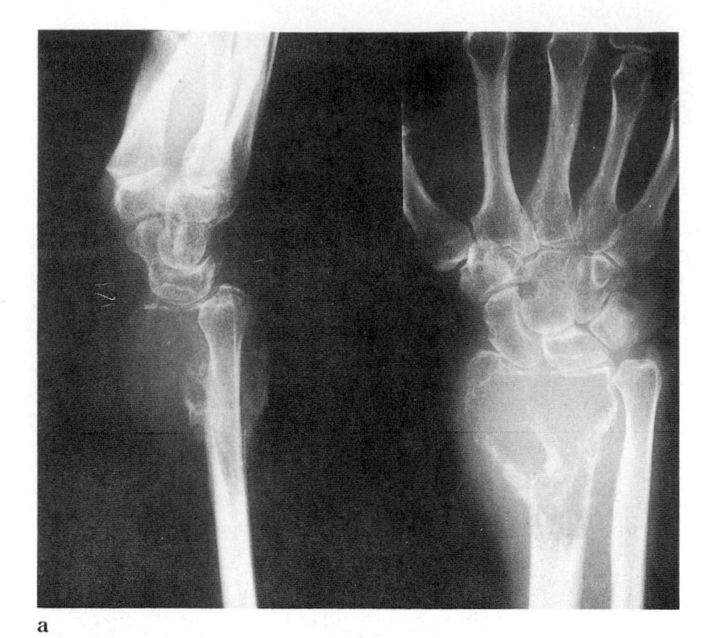

a

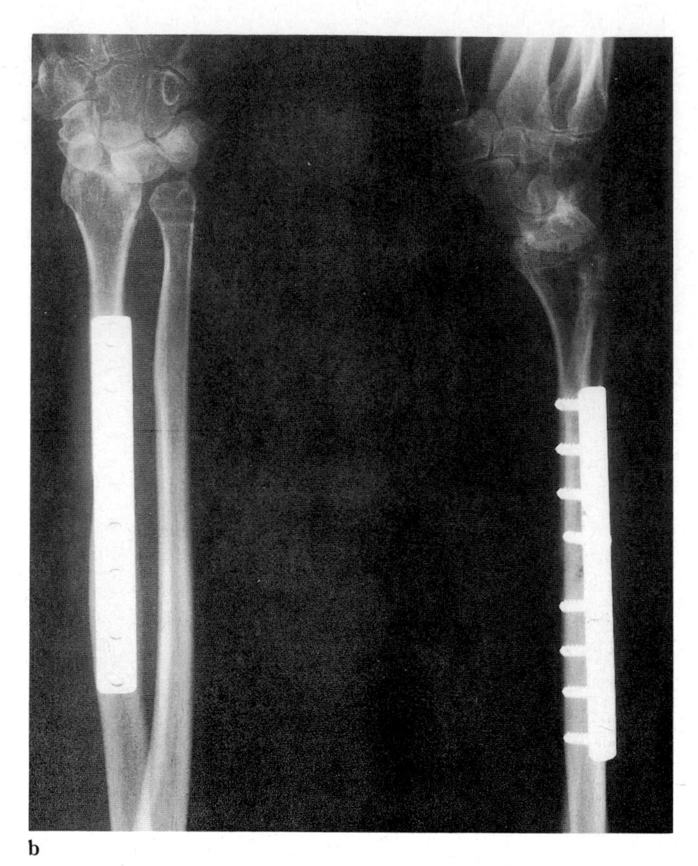

b

Fig. 2.17 Transplantation of connective tissue.
a. Giant cell tumour of lower end of radius. **b.** Replacement of lower end of radius by autograft of lower end of fibula. Note plate securing the graft to the proximal part of the radius. Courtesy of Mr John Chalmers FRCS.

of this adverse reaction is attributable to the antigens of chondrocytes particularly when the antigens have been exposed by preliminary damage to the pericellular matrix. However, extracellular matrix macromolecules can also readily express their antigenicity and the common formation of anticollagen antibody in rheumatoid arthritis (see Chapter 12)[44] and the effectiveness of type II collagen in causing arthritis in the rat and mouse (see Chapter 13) are two examples of the importance of these responses.

In spite of limitations imposed by rejection reactions, massive allografts of bone and cartilage have become a feature of limb salvage surgery.[12,64] Autografts, such as the vascularized fibular graft, replacing the radius, are the optimum materials for transplantation but the necessary source is not always available (Fig. 2.17).

Fibrocartilage

This is an organized, collagen-rich material closer in chemical composition to tendon and ligament than to hyaline cartilage, but differing from the aponeuroses and joint capsules in organization and in its relationship to synovial tissue. Fibrocartilage is exemplified by the intraarticular discs (see p. 103) and the menisci (see p. 103). The discs have a unique developmental origin and a distinctive composition and morphology. The fibrocartilage of menisci (Fig. 2.18) has a very low water content. There are relatively few cells; they are ovoid and elongated. Their fine structure resembles that of fibrocytes (see p. 17). The matrix, which usually has an orientated, structural organization intimately related to the mechanical function of the tissue, contains as much as 80 per cent type I collagen but relatively little metachromatic, non-fibrous proteoglycan.[212,73,1]

In fibrocartilage, individual collagen fibrils are organized into collagen fibres and fibre bundles. The surface structure, for example of menisci, can readily be viewed by scanning electron microscopy.[27,290] There are no intrinsic blood vessels, lymphatics or nerve endings. Fibrocartilage has a low mean oxygen consumption.[260] The tissue therefore tolerates ischaemia well. Disorders like the inflammation of rheumatoid arthritis that are vascular phenomena degrade fibrocartilage margins, via the adjacent synovium. The recognition of matrix vesicles is one indication of cell injury.[104] Because of its location at sites where large forces are transmitted, fibrocartilage is vulnerable to mechanical disruption, wear and abrasion. Invasive neoplastic cells, conveyed to contiguous synovium or nearby bone, destroy fibrocartilage centripetally.

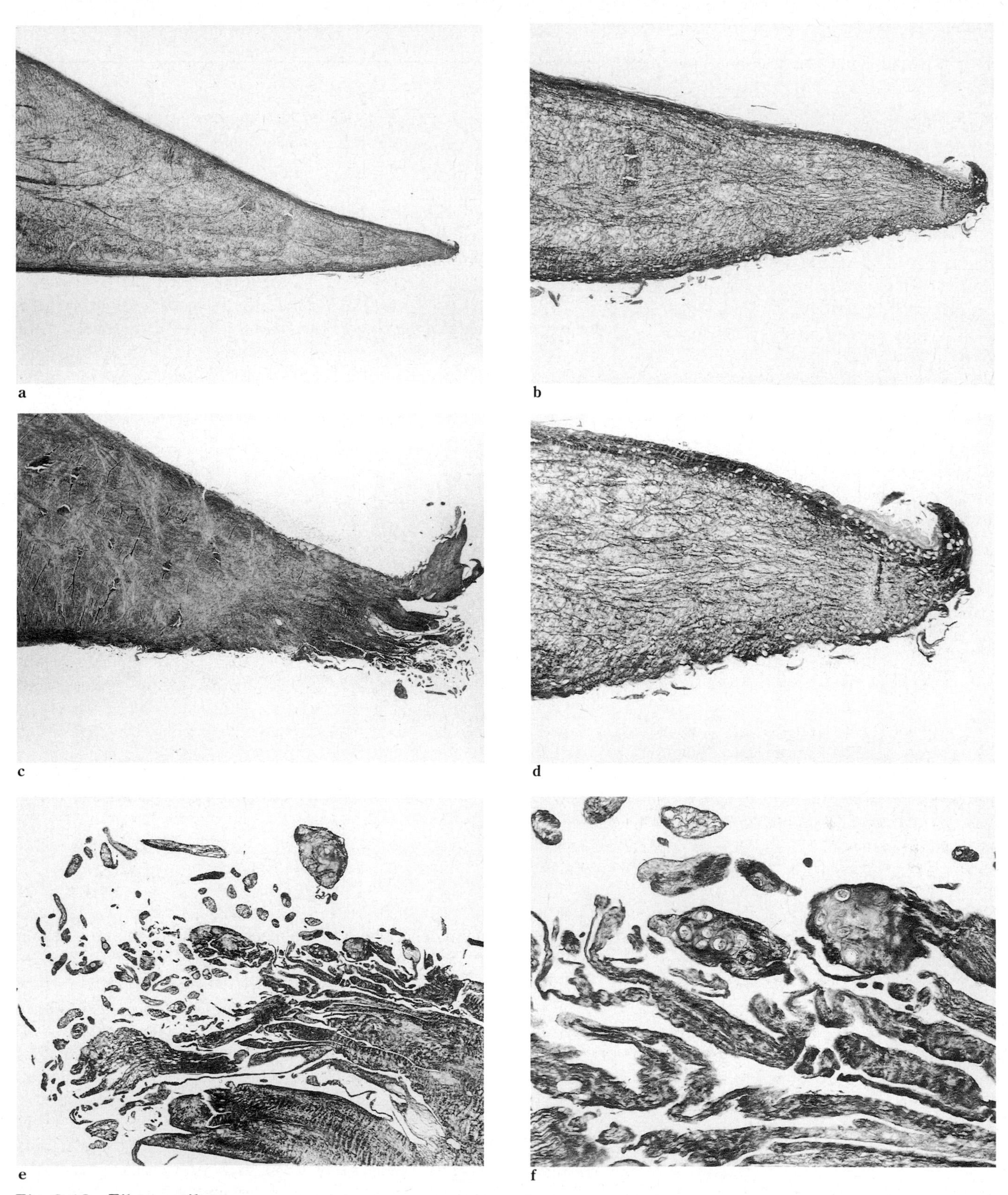

Fig. 2.18 Fibrocartilage.
Perpendicular sections through human knee joint menisci. Dense, intertwining bundles of type I collagen arranged circumferentially comprise the body of the semilunate cartilage; the upper and lower surfaces are formed of orderly, dense collagenous laminae that appear like a skin or shell (see Fig. 2.29). There is a sparse interfasicular, collagen-weak proteoglycan-containing matrix but no blood vessels. However, microcanals are said to be present.[93a, 268] **a.** Normal meniscus; **b.** Higher power view of the same specimen; **c.** Early, age-related marginal fraying; **d.** Similar specimen emphasizing collagenous structure and distribution of cells; **e.** Advanced, age-related fraying of inner rim; **f.** Higher power view of same specimen, demonstrating clusters of chondrocytes forming within residual, fragmenting collagen fibre bundles. (**a.** HE × 12; **b.** HE × 48 (polarized light); **c.** HE × 12; **d.** HE × 48 (polarized light); **e.** HE × 75; **f.** HE × 150.)

Fibrocartilage repair

The repair reactions of fibrocartilage may be considered conveniently in terms of the responses to injury of the knee joint menisci. Clinically, meniscal injuries are usually tears;[291] experimentally, incised wounds have been investigated most frequently.

The pattern of repair in menisci is strongly influenced by the proximity of an injury to the vascular supply which is limited to the peripheral synovium.[52] Among the reasons for analysing meniscal injuries in detail is growing surgical evidence that, after injury, menisci may be encouraged to repair by synovial implantation or by suturing.[105] Total excision is, in any event, more likely to cause secondary, late osteoarthrosis than partial excision, which may be arthroscopic.[197] However, macromolecules such as hyaluronate or semisynthetic sulphated polysaccharides may be protective.[106]

Histological studies suggest that meniscal fibrocartilage undergoes repair when a wound is in continuity with the synovial margin. A form of fibrocartilage is synthesized. The responsible cells may be myofibroblasts rather than meniscal fibroblasts.[102] After experimental meniscetomy, the severity of subsequent hyaline cartilage degeneration is inversely proportional to the extent to which the excised meniscus regenerates.[191a] When a meniscal injury is not in continuity with the synovium, repair is only likely if steps are taken to establish access to a vascular tissue. It is of interest that, after total excision of a meniscus, a flap of fibrous tissue extends centripetally to replace the lost part. This fibrocartilage is less highly organized than that of the original structure and is likely to be mechanically inferior.

Elastic cartilage

Elastic cartilage is distinguished from hyaline cartilage by the presence of elastic material which may comprise as much as 20 per cent of the dry weight of the tissue. The elastic material is arranged as branching fibres, fenestrated sheets or discrete islands. Elastic cartilage lacks the highly orientated fibrous organization of other connective tissue structures, such as the ligamentum nuchae, that are also rich in elastic material; nor does it display the extended, sheet-like morphology of the aponeuroses and joint capsules where elastic fibres are common.

Elastic cartilage (see p. 46) is exemplified by the tissues of the epiglottis and external ear.[120,247] Round or ovoid chondrocytes similar in shape, size and frequency to those of many hyaline cartilages, lie singly or in pairs in chondrons which are rich in proteoglycan and bounded by a permeable, basket-like microcapsule of collagen, possibly type IX (Fig. 1.28, p. 23). Interspersed among the marginal collagen but extending widely through the interterritorial matrix, is elastic material in the form of single or branched fibres. The pericellular and interterritorial collagen is type II. Staining techniques, such as the combined alcian blue-resorcin method, enable the collagen and elastic material to be demonstrated simultaneously. Transmission electron microscopy reveals both the elastic material and the characteristic collagen fibrils and allows their relationships to nearby chondrocytes, from which they derive, to be clearly demonstrated in normal tissue and in disorders such as relapsing polychondritis (see Chapter 17). Scanning electron microscopic studies[27] are facilitated by first removing collagens by enzymatic digestion.

Repair of elastic cartilage

Considerable understanding of the repair of non-articular elastic cartilage (see p. 46) has come from studies of the mammalian ear, and both rabbit and rat ear and epiglottic cartilages have proved valuable models.[116,265] Rabbit skin has a 'cartilage-evoking potential' (see p. 28). Holes punched in the ears of rabbits, hares, cats and insectivorous (but not fruit) bats, repair fully with restoration of the cartilage core; other mammals, including humans, do not have this capacity. Continuity with the adjacent epithelium is a prerequisite for rabbit ear cartilage repair and the participation of an epithelium-derived growth factor is evident.

Synovial tissue

The term synovial, devised by Paracelcus, means synovial fluid and came from the Greek *syn-* (together) and the Latin *ovum* (egg), referring to egg-white (albumin). The synovial tissue is often described as synovium.

The internal, non-load-bearing surfaces of diarthrodial joints are lined by a vascular synovial membrane. A membrane is a thin layer of tissue covering a surface or dividing a space or organ. Synovial cells are described on p. 32. Synovial tissue, in the form of a membrane, extends over the inner, non-weight-bearing parts of all synovial joints, lining the capsules and covering intra-articular structures such as ligaments, tendons and periosteum. Bursae and outpouchings of synovial joints have a comparable internal surface.

Synovial morphology

According to the character of the supporting connective tissue, synovia can be said to be fibrous, fibroareolar, areolar-adipose or adipose (p. 66).

Synovia (synovial membranes) comprise a surface layer of synoviocytes and underlying connective tissue (Fig. 2.19).[205] The layer of synovial cells has been described as an intima, the supporting connective tissue as the subsynovial tissue. In practice, it is more precise to speak of a synoviocyte layer and a subsynoviocytic (connective) tissue.

When a normal synovial joint such as the knee is explored surgically, the adipose synovial tissue of the infrapatellar region is seen as a soft, glistening series of gentle folds, the pale yellow colour of which retains a pink tinge if no tourniquet has been applied. Nearby, areolar synovium is less folded: its faint red colour differs from the pale fibrous synovium that covers adjacent periosteum. The boundary between the synovium and articular cartilage, intra-articular disc or meniscus, a synoviochondral interface, is a precisely delineated zone in which the synovial capillary arcades turn back towards the parent territory. The red colour of the vascular synovium contrasts with the pallor of the avascular cartilage or fibrocartilage. The synoviochondral interface constitutes the site at which all forms of inflammatory arthropathy originate; it is therefore the primary 'attack-zone' for common disorders such as rheumatoid arthritis (Chapter 12) (Figs 2.20 and 2.21). Where synovium covers ligaments or lines tendon sheaths, it is inconspicuous and often transparent.

The structure of the synovial tissue is conveniently viewed by fibreoptic arthroscopy. The areolar surface is a complex series of folds and processes that change in shape as joint movement occurs. The folds may be leaf-shaped, fan-like or villous.[208] It has been suggested that the folds do not form simply as outgrowths but that splitting or, conversely, fusion of adjacent prominences may occur.[65] The area of the folds and villous processes of adipose, areolar and fibroareolar synovium is greatly extended by the presence of numerous blunt villi that resemble those of the small intestine: the villi are packed closely together and thus assume a hexagonal outline viewed *en face* (Fig. 2.22).

The number of synoviocytes lining the synovial membrane varies according to the nature of the tissue. Commonly, the synoviocytic layer is two to three cells thick but, where an attenuated synovium covers a periosteum, the cells are arranged as a squamous lamina, one cell thick, and the covering may be incomplete. There is no underlying basement membrane; there is therefore continuity between the synovial space and the extracellular fluid of the subsynoviocytic connective tissue.

The quantity of this vascular connective tissue varies greatly, ranging from a limited collagen-rich matrix beneath the cells of the thin, fibrous synovia to an abundant, richly vascular matrix in the areolar and adipose synovia. Extend-

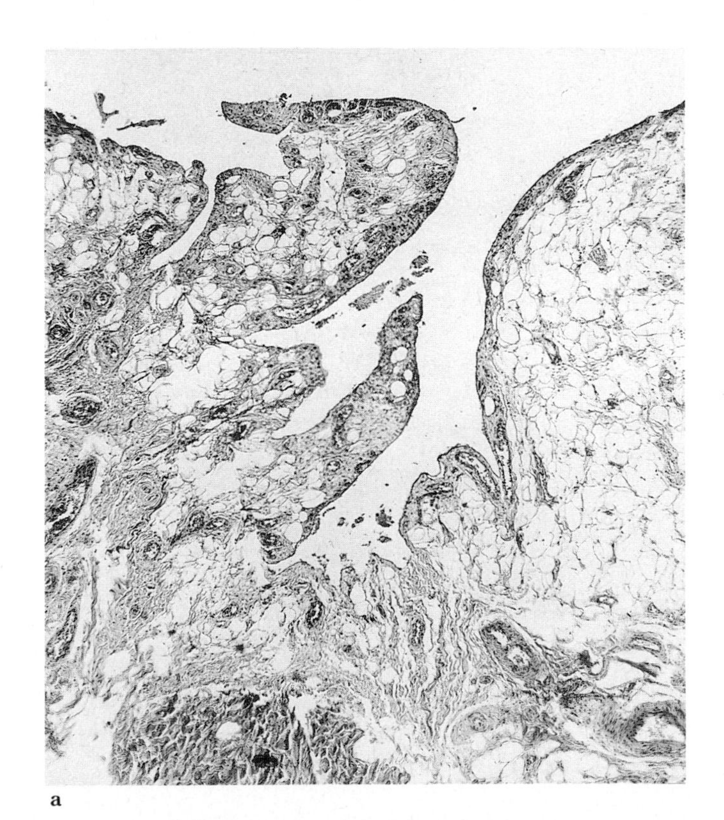

a

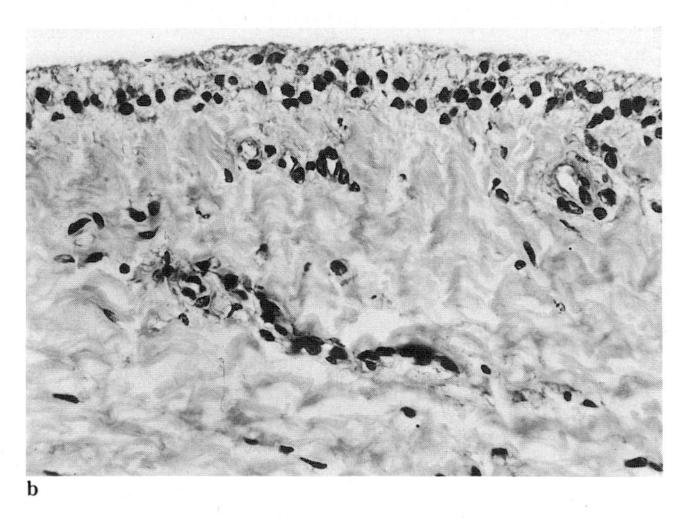

b

Fig. 2.19 Synovial tissue.
a. Finger-like and convex processes of fibrofatty (left) and adipose (right) synovium. Observe considerable numbers of small arteries and veins (left and lower right). (HE × 40.) **b.** Higher power view to show surface arrays of synovial cells (see p. 32) which may constitute a single, double, or triple layer overlying richly collagenous subsynoviocytic connective tissue among which are many small blood vessels, some lymphatics and, occasionally, nerve endings. (HE × 300.)

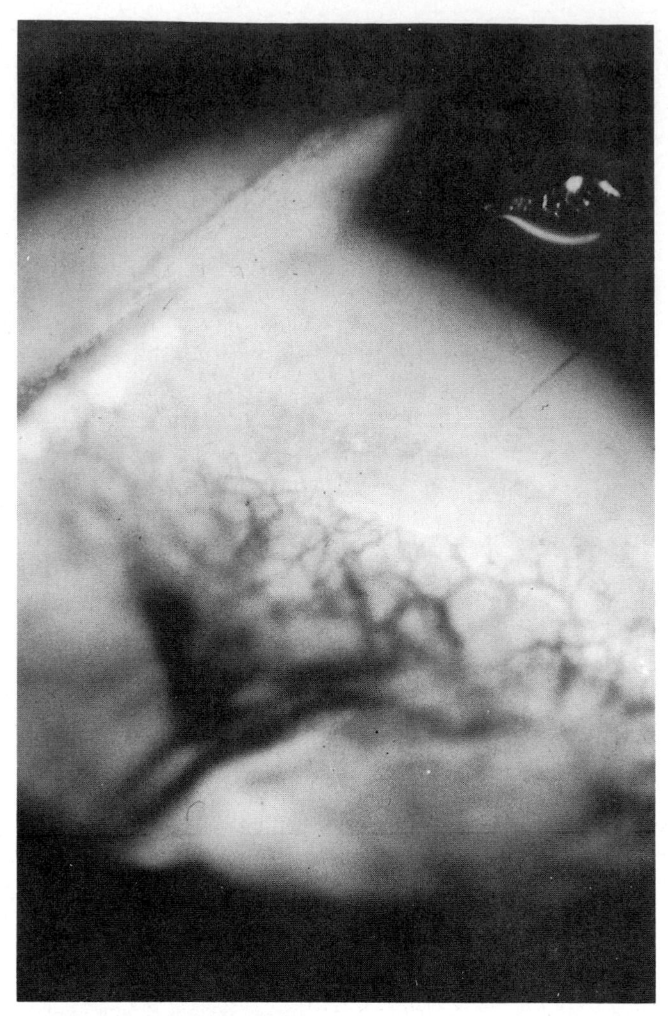

Fig. 2.20 Synovial tissue.
Synoviochondral junction of living rat knee joint. Tissue photographed through glass slide beneath which an air bubble can be seen (top right). Articular cartilage (right) is avascular. Synovial blood vessels terminate as a delicate arcade, the edge of which sharply delineates cartilage margin. (× 5.)

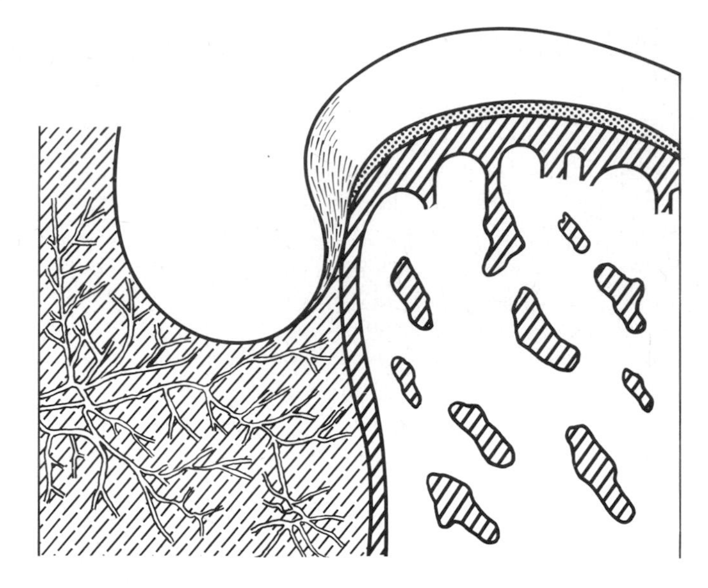

Fig. 2.21 Synoviochondral junction.
The relationship is shown diagrammatically of synovium, dense cortical bone, cancellous bone, transitional area of articular cartilage and principal part of articular cartilage.

ing 2–3 μm deep to the synovial surface of the rabbit knee, the interstitial tissue contains three supramolecular assemblies.[168] There are scattered 32 nm diameter collagen fibrils with a periodic structure, numerous nonperiodic microfibrils and occasional fibrous long spacing fibres. The latter are thought to be type VI collagen and this molecular species may also be represented in the nearby microfibrils. There are many small arteries, arterioles, capillaries, venules and small veins; they are accompanied by lymphatic channels and by periarterial nerve fibres (see p. 87). Mast cells (see p. 36) are scattered in the periarterial territories and the intervening loose connective tissue contains small numbers of histiocytes.

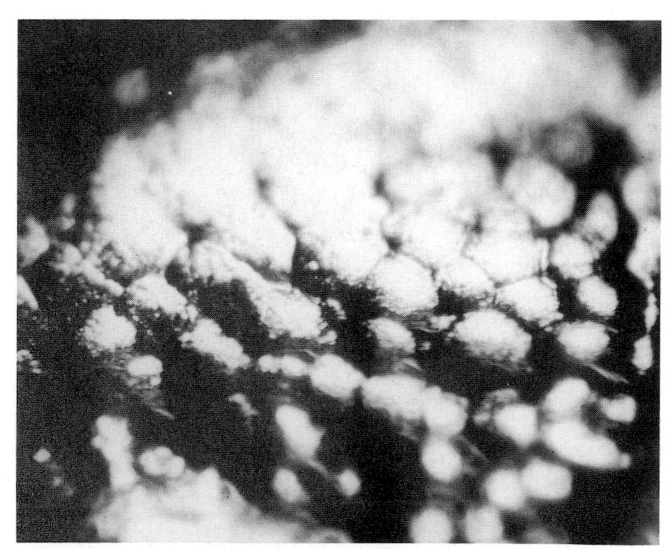

Fig. 2.22 Synovial tissue *en face*.
Finger-like processes amplify the surface area of the synovium; in turn, the presence of microvilli and ultramicroscopic cellular processes increases the surface area still more. The digital processes are closely packed in hexagonal array. (× 22.)

Synovial vasculature

The arterial supply to the synovial joints is relatively large and its distribution well-documented (Fig. 2.23).[283] The anatomical distribution of the circulus vasculosus of the knee joint was indeed clearly demonstrated by William Hunter (1742–43)[131] (Fig. 3.26). The synovial arterioles give rise to an abundant superficial capillary bed that varies in density within synovia of different types. The synovial capillaries are fenestrated.[242] In the rabbit, the areolar or adipose synovium contains 67–83 × 10^3 capillaries cm^{-2} section. Over tendons, the capillary density is only 2000 capillaries cm^{-2} section.[158] The synovial capillaries display a skewed depth distribution with a sharp peak at 6–11 μm. There is a network of related venules the number of which increases in synovitis[63] and a substantial lymphatic drainage.

In the joints of the limbs, synovial arteries and arterioles are characteristically thick-walled (Fig. 2.24). An unusual feature of some normal synovial blood vessels was found to be hyaline sclerosis.[69] The walls of the affected capillaries and venules contained little lipid. The appearance was found in persons with no history or anatomical evidence of arthritis and was seen in small and large joints of upper and lower limbs. The appearances were noted in all age groups from four to 89 years. Nothing is known of any possible association between these arteriovenous changes and the production of angiogenesis factors (see Chapter 7).[36]

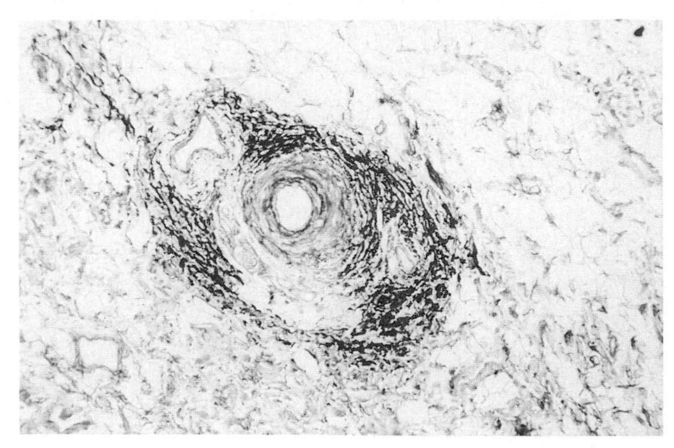

Fig. 2.24 Arteries of the major synovial joints. The small arteries that supply the synovium of large, lower limb joints such as the hip, are thick-walled and, as shown in this figure, may be cuffed by a meshwork of adventitial elastic material. (Weigert's elastic × 64.)

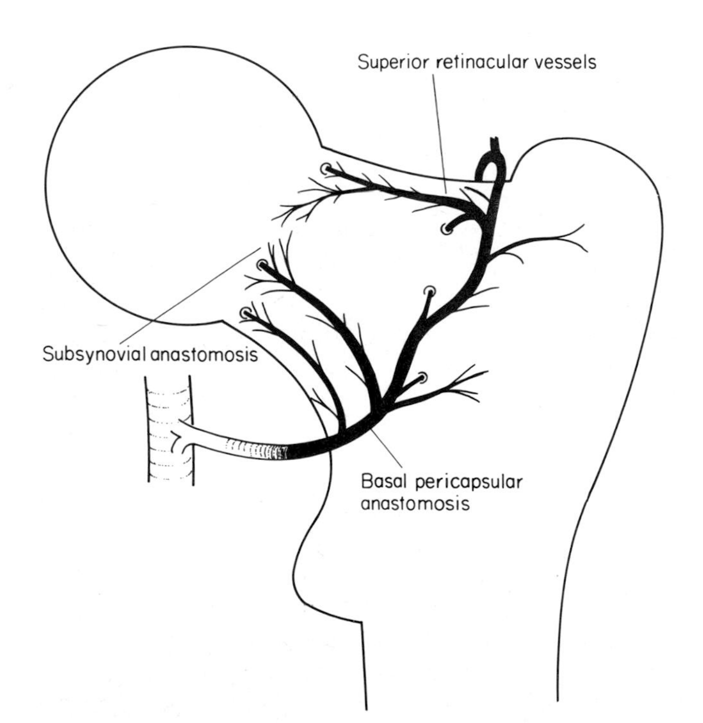

Fig. 2.23 Arterial blood supply to synovial joints. Detailed knowledge of the vascular supply to large joints such as the knee extends back to the early seventeenth century (Fig. 3.26). Here, the arterial blood supply to the vulnerable femoral neck and head is indicated. There is a small additional inflow via the ligamentum teres.

Synovial nerves

Nerve fibre bundles are not seen in the normal subsynoviocytic connective tissue. However, light and electron microscopy reveal autonomic fibres in the arterial adventitia.[115] In addition, nerves conveying pain stimuli are demonstrable by substance P-immunofluorescent reactivity. Encephalin-immunofluorescent nerves are also present.[111]

Synovial tissue regeneration

Synovial cells contribute to regeneration of the joint lining after injury: they readily multiply by mitotic division. Following synovectomy, synovial tissue re-forms within 60–80 days in the rabbit.[144] In the mouse, the synoviocyte layer is restored by the proliferation of synovioblasts from the marginal, transitional zone[204] although the subsynoviocytic layer is repaired by fibrosis. The regenerated synovium in rheumatoid arthritis patients undergoing synovectomy is not only restored but may develop the same histological signs of disease as the original tissue. Ultrastructural studies in rabbits showed that two cell types predominated in the organizing haematoma soon after synovial resection.[190]

In another experiment, three cell types were found.[284] Those that were macrophages became the A cells of the neosynovium; those that were fibroblasts became B cells.[190] After 100 days, the synovium could not be distinguished from normal. One factor that may modulate synovial cell growth is γ-interferon:[28] γ-interferon causes rheumatoid arthritis synovial fibroblasts to express MHC class II (Ia-like) antigen.[5] Thickening of a nuclear fibrous lamina appears to be related to the repair process.[103]

Synovial culture

Although synovial cell culture has been extensively used in biological, immunological, chemical and pharmacological enquiries, synovial tissue (organ) culture has received less attention. In the presence of serum, porcine synovial fragments shrink, losing collagen and releasing collagenase.[272] These changes can be largely inhibited by cortisol and partially by indomethacin. Agents such as sodium fluoride and dibutyryl cyclic adenosine 3′:5′ monophosphate added to synovia in culture, accelerate collagen degradation. A change in intracellular levels of cyclic adenosine monophosphate is a key step in this process.[82]

Synovial tissue and cartilage

Particular interest has centred on the regulation by synovial tissue of cartilage degradation.[57] Live synovial tissue fragments placed in contact with cartilage particles in a culture dish cause a breakdown of the cartilage matrix with the loss of proteoglycan and collagen. The degradation occurs if the cartilage is live or dead (frozen–thawed) although collagen loss is much greater from living cartilage. If the experiment is made with live synovial tissue *not* in contact with the cartilage, there is extensive proteoglycan and collagen loss from the living but not from the dead cartilage (Fig. 2.25).[79,80]

It was deduced that living synovial tissue could exert at least two catabolic effects on cartilage, both invoked as mechanisms of cartilage breakdown in inflammatory joint disease. First, enzymes released from the contiguous synovium can act directly on, and degrade living or dead cartilage; second, synovium exerts an indirect, remote effect causing live chondrocytes (but not dead ones) to break down their own extracellular matrix (Fig. 2.25). Other cells such as mononuclear macrophages, analogous with synovial A cells, can act in the same way if they have been stimulated with substances such as lectins. The remote action of the live synovium on live cartilage is attributable to the liberation of cytokine(s) (see Chapter 7).

The agent released from living synovial tissue was originally termed catabolin[57] and it appeared that there was a family of such substances. Catabolin proved to be a small acidic protein of approximate mol. wt 16–20 000 kD; the activity was localized to the synoviocyte cytosol.[236,237,238] A closely similar catabolin could be obtained from pig leucocytes: it caused the release of proteoglycan from cultured bovine nasal cartilage at concentrations of approximately 20 ppm. Catabolin is now known to be an interleukin: the B synoviocytes may be responsible for its synthesis.[214] Interleukin-1 changes cartilage structure radically in culture so that, ultimately, only a mass of fibroblast-like cells remain. Many other agents can, of course, alter cartilage behaviour: among those of interest is the latent collagenase produced in large amounts by monolayer cultures of rabbit synovial fibroblasts under the influence of phorbol myristate acetate.[29] These effects of phorbol myristate acetate can be inhibited by retinoic acid.

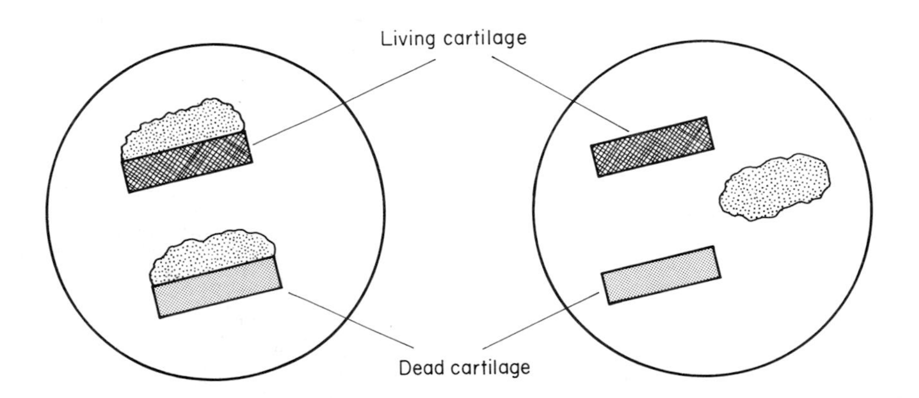

Fig. 2.25 Influence of live, cultured synovium on cartilage.
The experiments of Fell and Jubb.[79,80] demonstrated that live synovial tissue, co-cultured and in contact with cartilage, caused the degradation of this tissue, whether living or dead. When co-cultured but not in contact with cartilage, only living cartilage was degraded, demonstrating that the live cartilage was responsible for its own degradation in response to a messenger, 'catabolin', liberated from the living synovium. Catabolin is interleukin-1.

Synovial tissue ageing

There are few systematic studies. In the rabbit there are reduced numbers of synoviocytes; a relative increase in type A but a reduction in type B cells; the emergence of atrophic cells with few organelles; diminished vascularity; and fibrosis.[139]

Synovial fluid (*Dr A J Freemont*)

The composition of the synovial fluid is now summarized and methods given for its investigation. The changes that occur in synovial fluid in disease are also outlined in the chapters dealing with individual disorders.

Variations in the amount and composition of the synovial fluid reflect not only the physiological and pathological phenomena that take place in the tissues with which the fluid is in continuity[277] but also systemic disease. In the rabbit, the fluid mass is greatest in the least congruent joint, the shoulder, and approximately seven times less in the hip joint.[160] Synovial fluid aspirates may be of clinical significance and occasionally, as in the crystal deposition diseases (see Chapter 10), allow rapid, precise diagnosis. The frequency with which synovial fluid analysis may influence diagnosis is related to the time devoted to its study and the specificity and sensitivity of the procedures used.[262] In hospital practice, synovial fluid analyses have sometimes been found to be neglected and the results inaccurate.[122] However, careful studies and increasingly sophisticated techniques suggest that much more information of value may be gained from the analysis of synovial fluid than has been suspected.

The capillaries of the synovium are fenestrated. In spite of the presence of desmosome-like intercellular junctions, and partly because of the absence of a subsynoviocytic basement membrane, there is direct continuity between the synovial fluid and the extracellular fluid of the synovial tissues. Consequently, both systemic and local inflammatory diseases quickly influence the composition of the synovial fluid which is a particularly sensitive index of inflammatory joint disorders.

Composition

(Table 2.4)

Synovial fluid[166] can be regarded developmentally as a special form of extracellular mesenchymal matrix; it is not analogous with the fluids found in body cavities such as the coelomic spaces, the ventricles of the brain or the anterior chamber of the eye (aqueous humour). Synovial fluid is a dialysate of plasma, enriched by the addition of hyaluronate and by other macromolecules (Table 2.4). Alternatively, synovial fluid can be considered to be a special form of extracellular matrix with a high water, electrolyte and hyaluronate content. Whichever view is taken of this unique, viscous material, it is clear that its composition normally mirrors the state of the vascular synovial tissue and of the avascular cartilage with which it is in continuity. The electrolytes are distributed between blood and synovial fluid in accordance with a Gibbs–Donnan equilibrium. The cell population is small, with approximately 100 mononuclear cells mm^{-3}.

Synovial fluid is highly viscous and contributes to the unique lubrication mechanism that enables most diarthrodial joints, tendons, ligaments and associated structures to function efficiently throughout a normal life-span (see Chapter 6). The physical properties of synovial fluid are non-Newtonian so that there is decreased viscosity with rising shear rates (see Chapter 6). A combination of mechanisms working in concert enables synovial fluid to function as a cartilage lubricant[266] but also to transmit metabolites to and from the avascular articular cartilages, fibrocartilages, discs and menisci. Synovial fluid provides a fluid medium responsive to hydrodynamic change during joint movement (see p. 272). For many years, it had been deduced that synovial fluid hyaluronate was the main determinant of the lubricant. Evidence that hyaluronate protein aggregated on cartilage-bearing surfaces seemed to support the concept of an unusual 'boosted' lubrication mechanism.[279,280,281] However, it is now apparent that both lubrication and the other functions of the synovial fluid are determined by a population of macromolecules which may be derived from the blood, secreted by joint tissues or formed by articular tissue catabolism. These high molecular weight constituents are present in low concentrations: above molecular weights of 1×10^5 kD substances such as α_1-acid-glycoprotein, caeruloplasmin and α_2-macroglobulin are present in inverse proportion to their molecular weight. Other high molecular weight molecules come from the synovial tissues themselves; the most numerous are hyaluronate, the lubricating glycoproteins, including the lubricins[267] (p. 211) and unidentified proteins. A third category of macromolecular substance accumulates in the normal synovial fluid because of tissue catabolism or destruction: these substances include the glycosaminoglycans, derived from cartilage matrix turnover.

In disease, the variety of substances greatly increases and extends from acid hydrolases and neutral proteases to complement, fibrin, cell degradation products and antigens, such as IgG and IgM, synthesized and secreted into the

Table 2.4

Characteristics of normal synovial fluid (From Letner[116])

Synovial fluid is normally present in such small amounts that the values of many normal constituents are not yet reliably established.

Characteristic	Feature/value
General properties	
Appearance	Clear and colourless
Fibrin clot	Absent
Mucin clot	Strong
Glucose	100 mmol l^{-1}
Bacteriological culture	Sterile
Composition	
Volume (knee joint)	< 1.1 ml (0.13–4.0)
Intraarticular pressure	Often negative
Relative density	1.0081–1.015
Relative viscosity (at 37°C)	> 300
Limiting viscosity number	
at shear rate of 500 s^{-1}	$5.23 \times 10^{-3} m^{-3} g^{-1}$ (range 4.5–6.0)
at shear rate of 1300 s^{-1}	$4.36 \times 10^{-3} m^{-3} g^{-1}$ (range 3.8–4.9)
Osmolarity	296 mosm l^{-1} (range 292–300)
pH	7.434 (range 7.31–7.64)
pCO_2	4.7–7.3 kPa
pO_2	< 3.9 kPa
Cells	
Nucleated cells	< 750 ml^{-1}
Neutrophil granulocytes	~ 200 ml^{-1}
Lymphocytes	A few
Mononuclear phagocytes	A few
Synovial cells	Very few
Carbohydrates	
Protein-bound carbohydrate	
Hexosamine	0.15 g/l
Sialic acid	0.10–0.28 g l^{-1}
Proteoglycans	
C4S	~ 42 mg l^{-1}
Hyaluronate	2.03–2.26 g l^{-1}
Lipid	
Phospholipid	Small amounts
Cholesterol	Small amounts
Elements	
Potassium	~ 4.0 mmol l^{-1}
Sodium	~ 136 mmol l^{-1}
Calcium	1.2–2.4 mmol l^{-1}
Magnesium	Lower than serum
Iron	5.19 mmol kg^{-1} (Synovial tissue)
Copper	4.33 mmol kg^{-1} (Synovial tissue)
Zinc	2.69 mmol kg^{-1} (Synovial tissue)

Characteristic	Value
Nitrogenous substances	
Total N	60–286 mmol l^{-1}
Non-protein N	16–31 mmol l^{-1}
Urea	2.5 mmol l^{-1}
Uric acid	0.23 mmol l^{-1}
Proteins	
Total proteins	9–36 g l^{-1} (*post mortem*)
Albumin	~ 8 g l^{-1}
α_1–antitrypsin	0.78 μg l^{-1}
Caeruloplasmin	~ 43 mg l^{-1}
Haptoglobin	~ 90 mg l^{-1}
α_2–macroglobulin	0.31 g l^{-1}
Lactoferrin	0.44 mg l^{-1}
IgG	2.62 g l^{-1}
IgA	0.84 g l^{-1}
IgM	0.14 g l^{-1}
IgE	~ 14 μg l^{-1}
Lubricin	Small amounts
Fibronectin	Small amounts
Enzymes	
Many enzymes are present in small amounts. They include:	
Hydrolytic enzymes	
Acid phosphatase	1.6
Cathepsin C (dipeptidyl peptidase)	2.5
β–D–glucuronidase	< 0.25
β–D–galactosidase	< 0.17
β–N–acetyl–D–glucosaminidase	12.2
Others	
Collagenase	
Lactate dehydrogenase	
Catalase	
Superoxide dismutase	
Alkaline phosphatase	
Hyaluronidase	
Protease inhibitors can also be detected including:	
α_1–antitrypsin	
α_1–antichymotrypsin	
α_2–macroglobulin	
C1 esterase inhibitor	

Cytokine concentrations[285a] (mean of two observations)

IL-1β (pg ml^{-1})	*IL-2* (U ml^{-1})	*TNFα* (ng ml^{-1})	*Interferon-α* (U ml^{-1})	*Interferon-α* (U ml^{-1})
20	15.1	1.38	350	13.7

synovial fluid by cells that enter and multiply in the synovial tissues. The changes in rheumatoid arthritis (see Chapter 12) exemplify these phenomena. The quantity and identity of the proteoglycans present can be used as an index of cartilage metabolism in the arthritides and as a measure of response to treatment[239] and it is of interest that the cytokines interleukin-1 and tumour necrosis factor, invoked in the pathogenesis of inflammatory joint disease, stimulate synovial B cell (fibroblast) synthesis of hyaluronate.[40]

Joint swelling is a common consequence of disease of diarthrodial joints. It is often due to the accumulation of excess synovial fluid. The aspiration of synovial fluid is frequently undertaken therapeutically and it is increasingly apparent that analysis of the fluid can yield important clues to the nature of the disease mechanisms operating in diseased joints as well as data of diagnostic and prognostic value. Disease-related changes in the chemical and cellular composition of the synovial fluid are now discussed.

Collection

Normal synovial fluid is viscous but does not clot, whereas in inflammatory joint disease, where increased vascular permeability results in the leakage of plasma proteins into the fluid, there is a likelihood of fibrin clot formation. It is therefore usual for synovial fluid sent to the laboratory to contain anticoagulant. Oxalate is not used for this purpose since calcium oxalate crystals may form, occasionally resulting in the erroneous diagnosis of crystal arthropathy.[241] It is said that lithium heparin may also crystallize and that sodium heparin should therefore be used. This has not been our experience and we find no problems using 2 ml paediatric lithium heparin bottles* for transporting and storing synovial fluid.

After aspiration, the natural inhibitors of proteolysis in synovial fluid (see below) may be rapidly exhausted, and cytolysis results. Although synovial fluid can be stored overnight at 4 °C without significant cellular deterioration, it is usual to make cytological preparations as soon as possible after aspiration. For serological study it is essential to separate the cells from the supernate before storage.

Gross analysis

The synovial fluid should be examined macroscopically.[222] Gross analysis involves a subjective qualitative or semi-quantitative assessment of colour, clarity, viscosity and mucin clot formation. On the basis of these features and of the cytology of the fluid, it is possible to recognize four distinct categories of arthropathy.[230] The distribution of cell phenotypes within these groups is given in Tables 2.5 and 2.6. The specificity and discriminatory value of the synovial fluid findings in the four groups are summarized in Table 2.7.

Colour

Synovial fluid is normally colourless. In the inflammatory and septic arthropathies, it may appear white or a faint yellow—a result of high concentrations of cells, crystals, haem pigments and/or bacterial chromogens.

Viscosity

The viscosity of normal synovial fluid is due to the presence of hyaluronate which gives it a viscous consistency similar to cervical mucus. In inflammatory joint disease, the viscosity of the fluid falls. Viscosity can be assessed by studying the behaviour of the fluid in the specimen tube and during pipetting; it may be measured (see pp. 99 and 272).

Mucin clot

Mixing synovial fluid and acetic acid[230] leads to the formation of a white precipitate known as the mucin clot. The nature and amount of this precipitate varies from ‘poor’ to ‘good’ and reflects the quantity of hyaluronate-protein present. Poor mucin clots are found in inflammatory disorders. The mechanisms leading to deficient clot formation are unknown but most theories propose variations in the ratio of protein to hyaluronate, the abnormal synthesis of hyaluronate or changes in the enzyme content of the synovial fluid.[46]

Interpretation of gross analysis

Much has been written of the diagnostic value of macroscopic synovial fluid examination. The data given in Table 2.4 are, in broad terms, accurate; however, in practice, there is considerable overlap between the groups of fluid. The importance of the gross analysis of synovial fluid is therefore limited[121,222] but occasionally we have found apparently anomalous macroscopic results which have subsequently proved to be of significance. For instance, the synovial fluid in multicentric reticulohistiocytosis has both a poor mucin clot and a low cell count, a unique combination.[87]

* Teklab.

Table 2.5

Disease distribution of different cell types in synovial fluid (I)

Disease	Cell count 100 mm^{-3} < 1.5	Cell count 100 mm^{-3} > 25	Rago > 65%	Polymorphs < 30%	Polymorphs > 80%	Lymph > 80%	LMC > 60%	CPM > 10%	RNM > 50%	Mast CPM
Infective arthritis										
SA	0	74	44	0	100	0	0	0	0	0
Non-infective arthritis										
Sero + RA	2	12	32	11	46	9	0	0.5	0	0
Sero − RA	4	0	0	8	39	0	0	0	0	0
AS	9	0	0	0	53	0	0	13	0	20
Reiter's	22	5	0	24	38	9	0	19	0	24
Reactive	2	22	0	42	11	7	5	47	0	9
PsA	9	7	0	20	20	0	5	13	0	13
Crystal arthropathy										
Gout	32	4	0	18	36	0	6	4	4	5
CPPD	41	0	0	40	40	0	20	0	10	0
Osteoarthrosis and intervertebral disc disease										
GOA	100	0	0	88	0	0	66	0	33	0
Mon. OA	100	0	0	93	0	14	32	0	11	2
ID	100	0	0	100	0	0	5	0	8	0

Note: Figures underlined represent data of importance in disease differentiation.
SA = Septic arthritis; RA = rheumatoid arthritis; AS = ankylosing spondylitis; PsA = psoriatic arthritis; CPPD = calcium pyrophosphate dihydrate deposition disease; GOA = generalized osteoarthrosis; Mon. OA = Monoarticular arthritis; ID = intervertebral disc disease; Rago = ragocytes; LMC = large mononuclear cells; CPM = cytophagocytic monocytes; RNM = round nuclear mononuclear cells; Mast = mast cells.

Table 2.6

Disease distribution of different cell types in synovial fluid (2)

Disease	Eosin	LE	TC	PC	RC	Mult	MC	Mit	DB
Infective arthritis									
SA	0	0	0	0	0	0	0	0	0
Non-infective arthritis									
Sero + RA	5	3	2	1	6	3	3	4	3
Sero − RA	8	0	0	0	0	0	0	0	0
AS	13	0	0	0	0	0	0	0	0
Reiter's	0	14	0	0	0	0	0	0	0
Reactive	16	0	0	9	0	0	0	9	0
PsA	20	7	0	7	0	7	0	26	0
Crystal arthropathy									
Gout	0	0	0	4	0	4	0	4	0
CPPD	10	0	0	0	0	0	0	0	0
Osteoarthrosis and intervertebral disc disease									
GOA	11	0	0	0	0	23	0	11	0
Mon. OA	14	7	0	0	1	3	0	6	0
ID	0	0	0	0	0	0	0	0	0

Note: Figures underlined represent data of importance in disease differentiation.
SA = Septic arthritis; RA = rheumatoid arthritis; AS = ankylosing spondylitis; PsA = psoriatic arthritis; CPPD = calcium pyrophosphate dihydrate deposition disease; GOA = generalized osteoarthrosis; Mon. OA = Monoarticular arthritis; ID = intervertebral disc disease; Eosin = eosinophils; LE = lupus erythematosus cells; TC = tart cells; PC = phagocytic cells; RC = Reider cells; Mult = multinucleate cells; MC = Mott cells; Mit = cells in mitosis; DB = Döhle bodies.

Table 2.7

Specificity and discriminatory values of synovial fluid features in four categories of disease

Disease	Features specific to the disease	Features of less discriminatory value	Features of least discriminatory value
Septic arthritis	Presence of bacteria		More than 1500 cells per mm^3 *or* More than 1000, but fewer than 1500 cells of which more than 50 per cent are polymorphs
Sero-positive rheumatoid arthritis Sero-negative rheumatoid arthritis Ankylosing spondylitis Reiter's syndrome Reactive arthritis Psoriatic arthropathy	More than 65 per cent ragocytes More than 60 per cent large mononuclear cells	More than 10 per cent cytophagocytic cells *or* Cytophagocytic cells and mast cells	
Gout Chondrocalcinosis	Crystals* identifiable as monosodium biurate Crystals* identifiable as calcium pyrophosphate		
Generalized osteoarthrosis Monoarticular osteoarthrosis Internal derangement disease	Reider cells	Crystals of calcium hydroxyapatite Eosinophil granulocytes Multinucleate cells Cells in motosis	Fewer than 1000 cells per mm^3 *or* More than 1000 but less than 1500 cells per mm^3 with lymphocytes and/or large mononuclear cells comprising more than 50 per cent

*The identification of crystals is subject to the limitations described in Chapter 10.

Microscopic analysis

Microscopy is used to identify cells, crystals, organisms and other particulate material present in the synovial fluid. Crystal-induced and infective diseases are dealt with in Chapters 10 and 18 respectively and are discussed here only briefly.

Crystals

(Figs 2.26 and 2.27)

Three principal classes of crystal are associated pathogenetically with joint disease (see p. 378). Monosodium urate may cause an acute arthritis. Calcium pyrophosphate may be an incidental finding in patients with chondrocalcinosis but can also cause acute inflammatory and chronic destructive arthropathies. Hydroxyapatite is found most frequently in synovial fluid from any joint in which there is damage to cartilage and/or subchondral bone, but rarely, may itself induce a destructive articular or paraarticular disorder.[55] Because of their enhanced solubility when the microenvironment changes, it is essential that these crystals be sought in a 'wet preparation' (see below). Each of the three types of crystal can be seen by phase-contrast microscopy; monosodium urate and calcium pyrophosphate deposition disease can also be demonstrated by polarized light microscopy. Examination in polarized light has the advantage that these crystals can be differentiated by their sign and/or extinction angles (see p. 379).

Hydroxyapatite crystals are too small to be seen individually by polarized light microscopy but they can be identified by transmission or scanning electron microscopy. A simpler method is to instil a drop of alizarin red under the coverslip of a wet preparation where it forms a red, birefringent complex with calcium salts (Fig. 2.27)[55,209] allowing the crystals to be directly visualized by a conventional light microscope.

These are not the only crystalline materials found in joints. A variety of lipids form crystals at room temperature. The most common is cholesterol which has a typical

Fig. 2.26 Crystals of urate in synovial fluid.
(Unstained, wet preparation × 190.)

plate-like appearance with notched corners. Cholesterol plates have been described in rheumatoid arthritis[230,138] and are a common finding in blind, synovium-like sacs and bursae. Steroids, injected into a joint for therapeutic purposes, form crystals which may be confused with calcium pyrophosphate. Steroid crystals can be found up to two months after their injection.[142] Oxalate crystals may form naturally in joints[43] but crystals of calcium oxalate can be present as artefacts if oxalate is used as an anticoagulant.

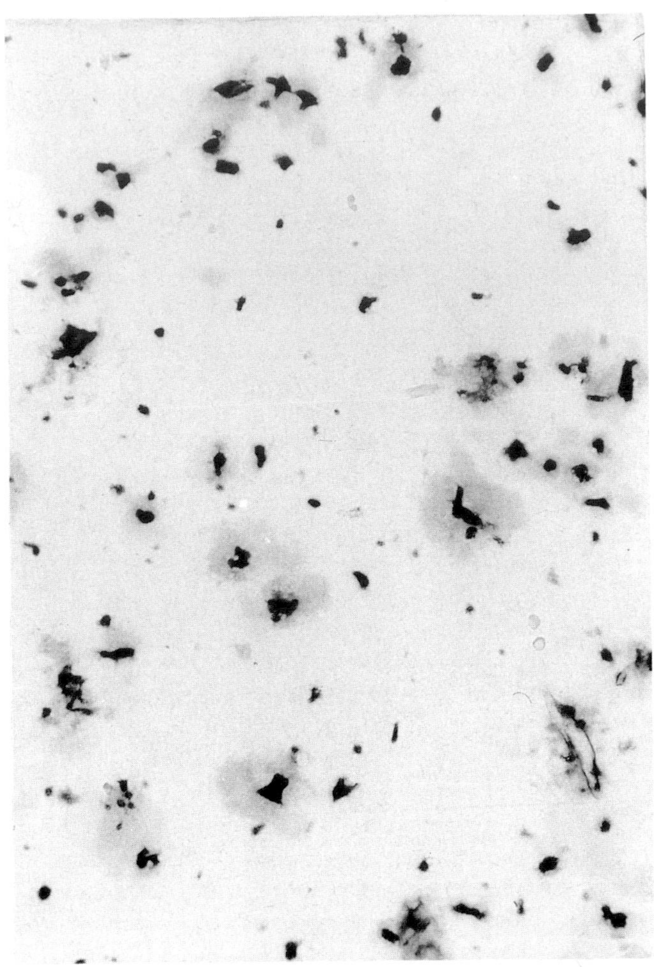

Fig. 2.27 Calcium hydroxyapatite crystals in synovial fluid.
Crystalline aggregates of calcium salts free or within polymorphs, can be demonstrated by light microscopy. Individual crystals are too small to be shown by this method. (Alizarin red × 250.)

Organisms

A large variety of microorganisms can induce infective arthritis (see Chapter 18). Many require special media for optimal growth. For these reasons, and because of the presence of bacteriostatic substances in the synovial fluid, culture is not always diagnostic. Published series suggest that only 30–80 per cent of patients with septic arthritis have positive synovial fluid cultures.[48,172]

We find that the careful microscopic examination of synovial fluid allows microorganisms to be identified in approximately 87 per cent of instances of infective arthritis. The greatest problems are the recognition of Gram-negative organisms and of Gram-positive bacteria rendered Gram-negative by partial antibiotic therapy; and the distinction of contaminants such as throat or skin commensals. After their introduction into a collecting bottle, microorganisms may be rapidly phagocytosed by the synovial fluid leucocytes. They appear intracellularly in a subsequent microscopic preparation.

The mycobacteria and fungi may pose even greater diagnostic problems, particularly in severely ill, immunosuppressed or immunodeficient patients where a cellular reaction to their presence may be atypical or absent. In these patients, the closest co-operation between clinician and pathologist is necessary if a diagnosis is to be made. Polyarthropathies such as rheumatoid arthritis are not infrequently complicated by monoarticular, subclinical, bacterial arthritis. The diagnosis of septic arthritis in this category of patient may be difficult. Moreover, infective and reactive arthritis may coexist particularly in patients with gonorrhoea (see p. 736).

Cells

The majority of cells in synovial fluid are polymorphs, lymphocytes, macrophages and synoviocytes. Although *numerically* important, these four groups represent only a small proportion of the cell *types* present. To appreciate this diversity fully, it is necessary to examine the synovial fluid by a variety of techniques. Although transmission electron microscopic immunocytochemistry and enzyme histochemistry reveal subgroups of cells that cannot be differentiated morphologically, these methods are not easily employed routinely.

For diagnostic purposes we examine 'wet' and centrifuged ('cytospin') preparations by light microscopy.[176] A wet preparation is made by spreading a few drops of agitated, undiluted synovial fluid on a microscope slide. It is examined for 'ragocytes' (see p. 96) in diffused transmitted light with the condensor diaphragm almost closed. We express the number of ragocytes as a proportion of all nucleated cells in the fluid (see below). A measured aliquot of agitated synovial fluid is then diluted in a solution containing methyl violet and the total number of nucleated cells per unit volume of fluid established by counting stained cells in a Fuchs–Rosenthal counting chamber. If appropriate, the fluid is then diluted with normal saline to an optimal concentration of 400×10^6 cells/l. Cell monolayers prepared by cytocentrifugation are fixed in methanol and stained by a standard Jenner–Giemsa technique. The following cells can be identified in these preparations using the criteria detailed below. The most abundant are counted and their number expressed as a percentage of the first 500 leucocytes encountered in random fields of the cytospin preparation. Only the presence or absence of other cell types is noted.

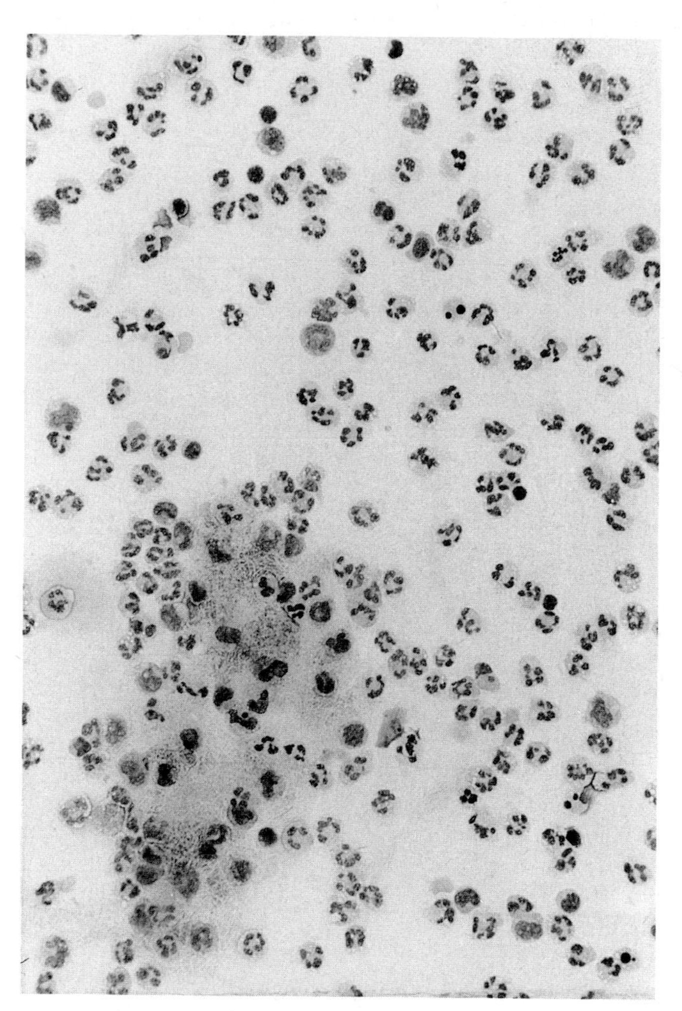

Fig. 2.28 Synovial fluid: polymorph exudate.
In this cytospin preparation, the numerous polymorphs are easily recognized. Although such an appearance could indicate the presence of pyogenic infection, it is commonplace in the sterile fluids of gout and rheumatoid arthritis. (Giemsa × 380.)

Neutrophil polymorphs These cells are recognized by their characteristic nuclear morphology (Fig. 2.28)

Lymphocyte-derived cells There are four morphologically recognizable categories.

Small lymphocytes Cells up to 12 μm in diameter with a nucleocytoplasmic ratio of more than 9:1.

Plasma cells Cells with eccentrically positioned nuclei of typical plasma cell type and blue cytoplasm.

Reider cells These cells are up to 15 μm in diameter with a nucleocytoplasmic ratio of approximately 8:1. The nucleus is multilobed, the lobes showing symmetry about a pale, attenuated central region. The cells are morphologically identical to the Reider cells of blood smears; their appearance is thought to be an artefact induced by oxalate anticoagulant.[287]

Mott cells These cells resemble plasma cells with a large number of Russell body-type, cytoplasmic inclusions.

Large mononuclear cells There are three morphologically defined categories.

Monocytoid mononuclear cells These are cells approximately 13 μm in diameter. They have a nucleocytoplasmic ratio of 6:1.

Cytophagocytic monocytes These are mononuclear cells that contain phagocytosed whole polymorphs or recognizable polymorph nuclei.

Round nuclear, mononuclear cells A subgroup of large mononuclear cells is recognizable. They have smaller, more rounded and denser nuclei than other large mononuclear cells, granular basophilic cytoplasm and a nucleocytoplasmic ratio of approximately 4:1. Some have a pericellular cytoplasmic 'frill'. They resemble synoviocytes.[271]

Eosinophils Cells with orange granular cytoplasm and bilobed nuclei.

Mast cells Cells with metachromatic cytoplasmic granules.

Multinucleate cells These are osteoclast-like polykarions with a variable number of nuclei. Multinucleated plasma cells are not included in this category.

Cells in mitosis Cells in which mitotic figures can be clearly identified.

Phagocytic cells Phagocytes with distinctive cytoplasmic inclusions.

Lupus erythematosus cells (p. 570) Phagocytes containing cytoplasmic inclusions of nuclear material with no recognizable chromatin pattern.

Tart cells Tart cells resemble lupus erythematous cells but have a recognizable chromatin pattern within the inclusions. These cells are identical to the Tart cells seen in peripheral blood preparations.[287]

Döhle body cells and polymorphs These cells, which may be macrophages or polymorphs contain duck-egg blue cytoplasmic inclusions up to 5 μm in diameter. These inclusions resemble Döhle bodies[58]—inclusions found in peripheral blood cells in patients with a variety of systemic disorders.

Ragocytes These are phagocytic cells containing characteristic apple-green, granular inclusions.[127]

Disease distribution

Cells have a characteristic disease distribution illustrated by the following data from our laboratory. Pertinent results from 9000 examples of the most commonly encountered arthropathies are recorded in Tables 2.5 and 2.6. Most of the cell types can be found in the synovial fluid from some patients in each disease group but the absolute number and proportion of each cell type varies significantly between diseases. The diseases comprise four categories: septic arthritis, primary inflammatory arthropathy, crystal-induced arthritis, and non-inflammatory arthropathy. In the tables, these categories are separated by horizontal lines.

On the basis of the cytological examination of the fluids, using the following criteria, it is possible to distinguish between inflammatory and non-inflammatory arthropathies in approximately 97 per cent of cases. Non-inflammatory arthropathies have either synovial fluid counts of less than 1×10^6 cells/l or between 1 and 1.5×10^6 cells/l with a predominance of lymphocytes and/or large mononuclear cells. In synovial fluid from patients with inflammatory disorders, cell counts may be greater than 1.5×10^6/l. Alternatively, if the counts are between 1 and 1.5×10^6/l more than 50 per cent of the cells are usually found to be polymorphs.

Variation in cell number and type is seen between patients in the same clinical categories; thus, whilst most patients with rheumatoid arthritis have a high proportion of polymorphs in their synovial fluids, a small percentage (approximately 9 per cent) have fluids in which lymphocytes predominate. Despite this variability, there are some cytological features specific for one, or a small group of related arthropathies. For instance, excluding crystal-induced and septic arthritis, only in synovial fluids from patients with seropositive rheumatoid arthritis are more than 65 per cent of the leucocytes ragocytes. Similarly, the presence of more than 10 per cent of mononuclear cells with cytophagocytosis is characteristic of the peripheral arthropathy of ankylosing spondylitis, Reiter's disease, psoriasis or reactive arthropathy.[86]

Although these criteria are typical of certain groups of cases, they are encountered in only a proportion of the members of that group. Thus, a ragocyte count of 65 per cent of all leucocytes, although characteristic of seropositive rheumatoid arthritis, is found in only 32 per cent of the patients. In the present series, features typical of a single diagnostic group are found in 31 per cent of cases; of two, three or four diseases in a further 39 per cent. It is concluded that diagnostic criteria, characteristic of a small number of diseases, can be recognized in approximately 70 per cent of cases; in a further 27 per cent, the cytological analysis of the synovial fluid allows an arthropathy to be designated as inflammatory or non-inflammatory.

Special studies of cell types Certain cell types have been studied in detail in particular diseases.

Lymphocytes There has been interest in phenotyping synovial fluid lymphocytes in the T-cell lineage.[88] In rheumatoid arthritis, the ratio of CD4:CD8-positive T lymphocytes is low in the synovial fluid (4:3) by comparison with peripheral blood (10:3). This is not a feature of other arthropathies.[59,23] Moreover, there is an increased expression of MHC class II histocompatibility antigens by cells in rheumatoid synovial fluid, a possible reflection of the numbers of activated CD8-positive cells. However, these cells do not bear other epitopes associated with T-cell activation including transferin receptor and interleukin-2 receptor.[21] In addition, rheumatoid synovial fluid contains fewer natural killer cells than peripheral blood[10] and synovial fluid natural killer cell activity is modulated abnormally.[50]

Large mononuclear cells Mononuclear cells are a mixture of cell types. Immunocytochemical and histochemical methods allow them to be divided into three main groups. *Transformed lymphocytes* are frequent in rheumatoid synovial fluid but rare in septic arthritis and the crystal-induced diseases.[271] *Monocytes* can be distinguished from synoviocytes by their sudanophilia[246] although the type C cells (see p. 32) of the synovia are also said to exhibit sudanophilia.[147] The distribution of these two cells in different diseases has not been thoroughly investigated. In theory, this group should include the antigen-presenting *dendritic cells*. They can be harvested readily from synovial fluid of patients with rheumatoid arthritis and juvenile chronic arthritis but not osteoarthrosis,[119] although we have not been able to identify cells with the dendritic cell phenotype in cytospin preparations, perhaps because of their adherence to and sequestration on the surfaces of glass and plastic.

Ragocytes Ragocytes in rheumatoid arthritis contain diverse polypeptides and proteins including immunoglobulins,[218] rheumatoid factors,[276] fibrin,[15] antinuclear factor,[174] immune complexes[118] and DNA.[210] Ragocytes are also seen in other inflammatory arthropathies.[289,276] The significance of these findings is discussed in Chapters 12, 13 and 14.

Eosinophils Eosinophils are prominent in synovial fluid after arthrography and have been reported in Guinea-worm infestations of the joint (see p. 753), in rheumatoid arthritis and in association with secondary carcinoma.[170]

Other particulate and fibrillar material

Little attention has been paid to the many other microscopic particles that have been found in the synovial fluid. They include the following:

Fat droplets Fat droplets are not present in normal joints or in those of patients with osteoarthrosis or chondromalacia whereas they are found in rheumatoid arthritis and haemarthroses.[240] Fat is conspicuous in the synovial fluid in cases of traumatic arthritis (see p. 955), especially when there is intraarticular fracture.[22] Initially, the globules contain lipid crystals; later, lipid-laden macrophages become prominent.[13] The digestion of articular adipose tissue by lipases released during acute pancreatitis may also release lipid components into a joint, provoking synovitis.

Cartilage Fragments of hyaline cartilage are often seen in the course of osteoarthrosis[126] and, more rarely, in septic arthritis. The fragments may be degraded by endogenous enzymes and remain as collagen fibrils.[153]

'Rice bodies' In various disorders including rheumatoid arthritis (p. 451), tuberculosis and septic arthritis, the synovial fluid contains particles composed of synoviocytes, collagen and fibrin.[42] To the naked eye, they resemble grains of rice. One view is that they represent aggregates of fibrin covered by a layer of mononuclear cells, another that they are derived from synovial villi, perhaps as a result of ischaemia.

Foreign material Probably the commonest form of foreign material in the synovial fluid is derived from prosthetic joint replacements. The particles include metal and plastic (see p. 960).[154]

Ochronosis (see p. 351) Occasionally in ochronosis the synovial fluid contains dark particles coloured by ochronotic pigment. They are said to resemble ground pepper grains.[130]

Amyloid Amyloid fibrils have been described in synovial fluid patients with 'amyloid arthropathy' (see p. 434).[109] Most patients have multiple myeloma but recently a form of the disease has been described in patients on long-term haemodialysis in which the amyloid contains a high proportion of β_2-microglobulin (see p. 437).[85]

Macromolecules

The entry of fluid and macromolecules into the joint is normally balanced by the removal of this fluid via the lymphatics. In the inflammatory arthropathies, enhanced vascular permeability, increased local secretion or release of immunoglobulins, complement and enzymes, and a decreased loss of lymph as a result of lymphatic occlusion, may result in raised protein levels in the synovial fluid.

Antibodies

A range of antibodies is described in rheumatoid synovial fluid. They include rheumatoid factor,[228] anti-collagen antibodies,[186] antinuclear antibodies,[226] antiviral antibodies,[189] antikeratin[294] and β_2-microglobulin.[74]

Immune complexes

Early in rheumatoid arthritis, immune complexes unrelated to rheumatoid factor are formed or trapped in the inflamed joints.[141] Later, in the synovial fluid of established rheumatoid arthritis, immune complexes are present consisting predominantly of IgG and IgM together with C1q, IgA and activated C4 and C3.[175]

Complement

Complement levels are reduced in the synovial fluid in rheumatoid arthritis, gout and infectious arthritis[123,146] and levels of complement breakdown products are increased.[295,211] There is activation of both the classical and alternate pathways (Chapter 5)[235] with the production of mediators, particularly C3a, C3b and C5a. Complement-3b enhances lymphocyte proliferation and causes enzyme release from macrophages[234]. C5a is both a potent chemoattractant for polymorphs and monocytes[234] and may induce the release of interleukin-1 from macrophages.[108]

Lymphokines and monokines

These short-lived, locally active cytokines are reviewed in Chapters 5, 7 and 13. Wood[292] has reported the presence of interleukin-1 in synovial fluid from a variety of arthritides including rheumatoid arthritis, ankylosing spondylitis, psoriatic arthritis, osteoarthrosis and viral arthritis.[199] A similar disease distribution is reported for interleukin-2 (see Chapter 12).[200] New assays allow the distribution of other interleukins to be defined and already the high titre of interleukin-6 in arthritis is exciting interest (Dr S Hopkins, personal communication). One cytokine implicated in the modulation of inflammation in arthritis but not found in synovial fluid is interferon-γ.[225] Other cytokines, particularly tumour necrosis factor are also receiving attention[60] (p. 291).

Eicosanoids

The prostanoids and leukotrienes, the eicosanoids[67] (see Chapter 7) have been identified in rheumatoid synovial fluid;[125] the most common is prostaglandin E_2. It is important to note that the level of prostaglandin E_2 falls in rheumatoid synovial fluids after treatment with non-steroidal anti-inflammatory drugs, but not with steroids.[25] The significance of the remaining eicosanoids found in the synovial fluid is unclear, largely because of uncertainties inherent in their bioassays. Little is known of the concentration of eicosanoids in other arthropathies. In one study, prostaglandin was found in osteoarthrosic synovial fluid; the levels did not fall following non-steroidal anti-inflammatory drug treatment.[270] Leukotriene B_4 has been identified in the synovial fluid of patients with spondyloarthritis[155] and gout.[217]

Platelet activating factor (see p. 284) is a mediator related to the eicosanoids. Although it has yet to be measured in human synovial fluid, in experimentally induced arthritis in rabbits, platelet activating factor is found in biologically significant quantities but only in the first 24 h after induction, implicating it in the earliest stages of this variety of inflammatory arthropathy.[124]

Enzymes

A substantial number of enzymes has been described in synovial fluid, including collagenase,[293] glucuronidase, lipase, pepsin, trypsin, lactate dehydrogenase,[138] amylase, transaminases and phosphatases. An apparent relationship between the total synovial fluid cell count and the activity of the lysosomal enzymes suggests that the enzymes are derived from synovial fluid cells.[252] That tissue damage is relatively limited in most arthropathies despite the presence of many potentially destructive enzymes, is due largely to their inactivity; protease inhibitors are present within the synovial fluid.[114] One such enzyme, the so-called tissue inhibitor of metalloenzymes (see p. 186), which blocks the activity of collagenases, is normally secreted by synovial lining cells (p. 32). In explants of normal synovium the ratio of tissue inhibitor of metalloenzymes to collagenase exceeds unity. In rheumatoid arthritis this ratio is reversed but it can be restored in part by treatment with steroids but not non-steroidal anti-inflammatory drugs.[173]

Proteoglycans and glycoproteins

Because of the availability of many relevant monoclonal antibodies, there is renewed interest in all aspects of these macromolecules in the synovial fluid. The highest levels of cartilage-derived proteoglycans are found in reactive arthritis, calcium pyrophosphate dihydrate deposition arthropathy and juvenile rheumatoid arthritis; the lowest in rheumatoid arthritis and psoriatic arthritis.[239] Patients with rheumatoid arthritis receiving steroids have higher synovial fluid levels of proteoglycans than those given non-steroidal anti-inflammatory drugs or second-line therapy.

Small molecules

Table 2.4 indicates some of the very large number of small molecules that may be found in the normal synovial fluid.[231] The number increases in disease and those listed in Chapter 12 exemplify these changes. Many forms of analysis, including atomic absorption spectrometry, can be used to investigate the elemental composition of synovial fluid but new methods continue to shed light on the small molecules that may be present. Thus, high resolution proton nuclear magnetic resonance (see p. 162) can be used to observe and measure simultaneously a variety of low molecular weight components. The assays are rapid and give direct measurement to levels of approximately 0.1 mM. The samples are treated initially by the addition of a little deuterium oxide, allowing them subsequently to be employed for other tests. Evidence obtained by this method suggests that, unexpectedly, triglycerides and creatinine may be markers of inflammatory activity in rheumatoid arthritis and in traumatic arthritis.[288]

Conclusions

Because of the many advances in methods that allow the identification of cells and their products or contents in synovial fluid, the statement by Galen that: 'The physician must examine the nature of the humor which, in small quantity, envelops the joints . . .'[227] is more pertinent now than ever. Synovial fluid is readily accessible; yet the contribution that can be made to the understanding of articular disease has not yet been fully exploited. Cytoanalysis, in particular, has much to offer in the diagnosis of arthritis and in the understanding of the pathogenesis of joint disease.

Artificial synovial fluids

Artificial synovial fluids have been tested to determine whether they might be used in the treatment or amelioration of osteoarthrosis (Chapter 22) (Table 2.8).[47,140] To be acceptable as a lubricant in the treatment of disease, the lubricant must be biocompatible, of sufficiently high molecular weight, sterile and non-carcinogenic; it should be devoid of acute or chronic irritant qualities and should persist *in situ* for long periods. Many fluids that have been tested cannot meet these criteria.[140] Thus, saline provokes cartilage degeneration with loss of safranin-O stainability. Silicons are effective and inert but, like so many substances introduced into a synovial joint, are removed quickly by macrophages, accumulating in regional lymphoid tissues. Moreover, their physical properties are Newtonian (see Chapter 6). Latex microspheres have been examined but the idea that they might act as micro-ball-bearings was quickly dispelled by the finding that, within a few minutes of injection, the particles had been engulfed by synovial macrophages (see p. 36).[93] Hyaluronate is now available in large volumes and there is renewed interest in its use for the modification of osteoarthrosis.[263]

Table 2.8

Some artificial synovial fluids (After Cooke and Gvozdanovic[47])

Substance	per cent	Surface tension (Nm^{-1})
Polyvinyl alcohol W40/140	5	47
Polyvinyl oxide 2% WSR-301	2	61
Sodium carboxymethyl cellulose	2	63
B-50	2.5	67
Manucol SS/LF	0.2	73
Polyacrylamide 170 H	0.5	84
Carbopol 934 free acid		
Water		73
Synovial fluid		57

Hydrodynamics

(see Chapter 6)

There are extensive reports of the pressure/volume relationships of the synovial fluid in normal animal joints[156,167] and some similar studies in human disease.[137] Intraarticular pressures at a given volume depend on the rate of change of volume, the direction of change, and the history of the joint in terms of inflammation and activity.[159] In physiological terms, joint spaces may be divided into compartments;[157] the distinct knee joint synovial fluid compartments may communicate more effectively when inflammatory disease generates synovial effusions. High pressures can be caused when flexion movements occur.[137] Increased communication may affect the extent of disease and the development of secondary changes such as intraosseous pseudocyst (see Chapter 12) formation.

Ligaments, tendons, discs, menisci and labra

Ligaments

Ligaments are orderly, collagen-rich structures (Fig. 2.29).[283] They are formed of dense, collagen fibre bundles

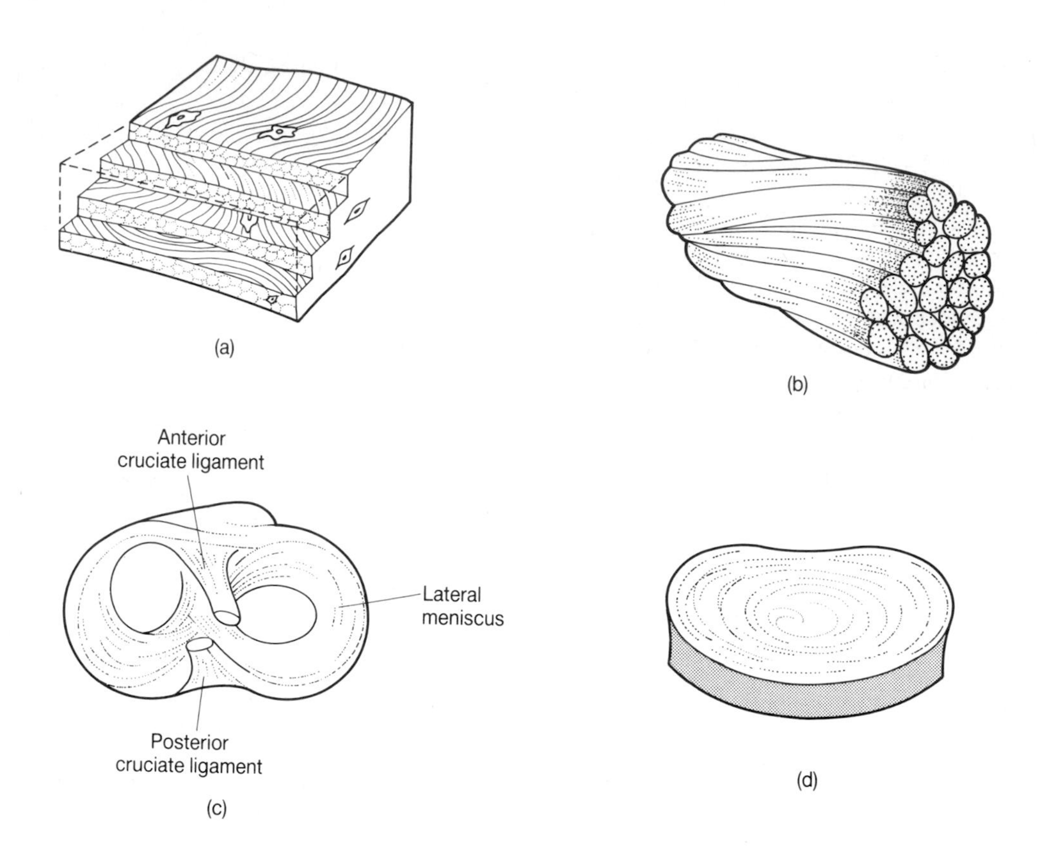

Fig. 2.29 Structure of ligament, tendon and disc.
a. Ligaments (p. 99) are multilayered sheets of collagenous connective tissue, inserted into or originating from bone at entheses. The blood vessels and nerve fibres are situated between the laminae of collagen. The structure of aponeuroses is closely similar to that of ligaments. **b.** Tendons (p. 101) are formed of collagenous fibre bundles, coiled like ropes, originating from skeletal muscle and inserted into bone at characteristic attachment sites. There are specialized nerve endings and a limited vasculature. **c.** Menisci (p. 103) are exemplified by those of the human knee joint. They are dense, fibrocartilagenous semilunate structures, forming an incomplete division between the opposed articular cartilages. Like discs, menisci are avascular but have a system of microscopic canals. They derive their metabolites from a circumferential synovial fringe and have a delicate supply of nerve fibrils, demonstrable by gold impregnation. **d.** In other human synovial joints, and in the knee joint in some non-human primates, the joint may be divided into compartments by a fibrocartilagenous intra-articular disc. Discs also form the main component of each of the principal intervertebral joints (p. 126). The complex structure of the temporomandibular joints (TMJ) (p. 132), with their intra-articular discs, is shown in Fig. 3.18. The TMJ discs have a significant elastic material content.

spread out in sheets but differently arranged within these sheets to provide a broad basis for their mechanical function (see p. 276). Ligamentous collagen bundles are incompletely attached, one to the other; they receive a limited blood supply. This restricted vasculature, and the small number of cells per unit mass of tissue, explains why, after sterile injury, ligaments heal slowly and why healing may be impaired and incomplete. The relatively abundant sensory innervation suggests a reason for the associated pain.

Many ligaments constitute only localized thickenings of a joint capsule. In such instances, the predominant movements of the joint, and the range of these movements, determine the site and extent of the ligaments. Thus, in hinge joints such as the elbow and ankle, the ligaments are collateral. By contrast, in ball-and-socket joints such as the hip, the ligaments are located towards those aspects where maximum movements occur. Other main ligaments secure opposing articulating surfaces. The femoral head and the acetabulum, for example, are firmly united by the insertion of the ligamentum teres near the centre of the joint surfaces.

Some ligaments are wholly intraarticular; they are frequent features of syndemoses (e.g. intervertebral joints) and synchondroses (e.g. the costochondral joints). Where joint use is frequent and load-bearing is constant, ligament size increases; but where there is disuse, due, for example, to limb paralysis, to splinting of fractures or to disorders of other systems such as multiple sclerosis or hemiplegia, ligaments atrophy and their strength decreases.

The insertion of a ligament into a bone is an

enthesis.[196,14] Microscopy shows that immediately before reaching the bone, ligamentous collagen fibres become increasingly compact, the matrix cartilaginous and finally calcified. The attachment zone is defined by a cement line. The extent of calcification varies considerably in depth. There is a vascular communication between the ligament and the bone or bone marrow blood vessels. Biochemical evidence shows that entheses are sites of active molecular exchange and cell metabolism: they are peculiarly susceptible to diseases such as ankylosing spondylitis (see Chapter 19) which is an enthesopathy.

Tendons

Tendons comprise collagen fibre bundles clustered in parallel array; they contain relatively small numbers of fibrocytes[283] (Fig. 2.30). The fibres can readily be separated (Fig. 2.31) and tested by a range of techniques. The mechanical properties of mature collagen ensure that a tendon is strong, only slightly extensible and has a high modulus of elasticity (see p. 275). Movement plays an important role in both the formation and integrity of tendons.[16] In the larger tendons, bundles of collagen fibres, arranged in fascicles (Fig. 2.29), are separated by small quantities of loose, vascular connective tissue within which are nerve fibres and lymphatics. The Golgi tendon organs, at myotendinous junctions, are sprays of non-myelinated nerve terminals. In the non-contracted muscular state, tendons lie relaxed in a crimp-like pattern (see p. 264), a form that has been adduced to explain the disposition of the collagen fibres.[98]

The study of tendinous structure is complicated by uncertain terminology. According to Rowe,[232] a (rat tail) tendon is delineated by an epitenon, continuous with an endotenon which divides the tendon into fasicular units. Smaller units, the tertiary bundles, are delineated by a parietal paratenon, enclosing one or more tendinous bundles, while a single fasicle is enclosed by a visceral paratenon (Fig. 2.32). One alternative nomenclature, more consistent with *Nomina Histologica*[198] is that of Strocchi *et al.*:[264] they describe a paratendineum, an epitendineum, a peritendineum and an endotendineum.

Although there is variability between tendons in different anatomical sites and, presumably, between species a scheme has been devised for rat tail tendon that summarizes tendinous structure.[232] The surface of a tendinous fascicle of collagen fibres reflects the wave pattern of the constituent fibrils which are aggregated side-by-side with a degree of freedom in any direction in the register of the wave patterns.[233]

The attachment of tendons to bone is specialized and has excited interest, particularly because of comparisons with

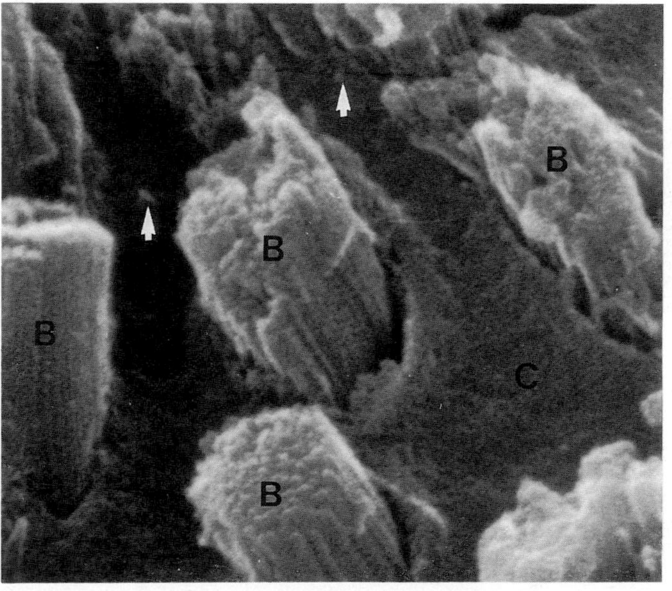

a

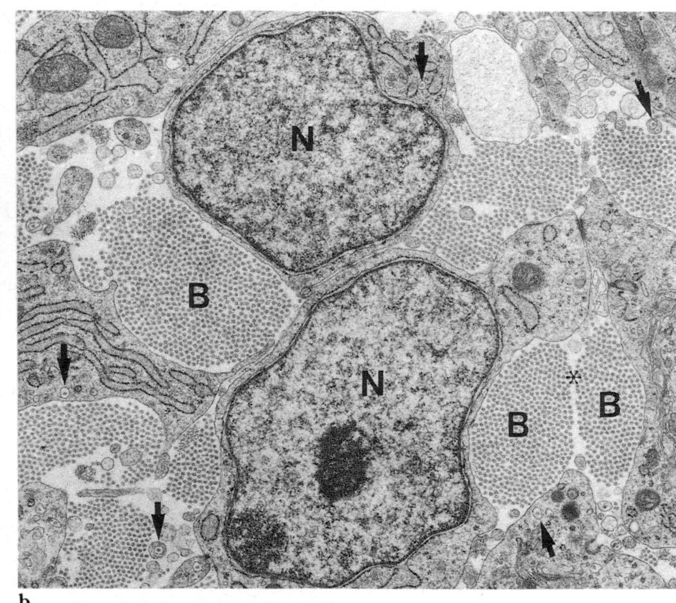

b

Fig. 2.30 Chick embryo tendon, 14 days.
a. Bundles of collagen fibrils collected in compartments formed by a single fibroblast or adjacent fibroblasts. (SEM × 33 150.) **b.** Small channels containing single collagen fibrils are often present (arrow). Note well-defined collagen fibril bundles (B). Because fibroblast cytoplasmic processes are no longer always present, some bundles are coalescing into larger parallel aggregates. (TEM × 10 900.) Courtesy of Drs G C M Yang and D E Birk[293a] who were assisted by PHS grant number RRO1219 supporting the New York State High-Voltage Microscope as a National Biotechnology Resource, awarded by the Division of Research Resources, DHHS, and the Editor, *Journal of Ultrastructure and Molecular Structure Research.*

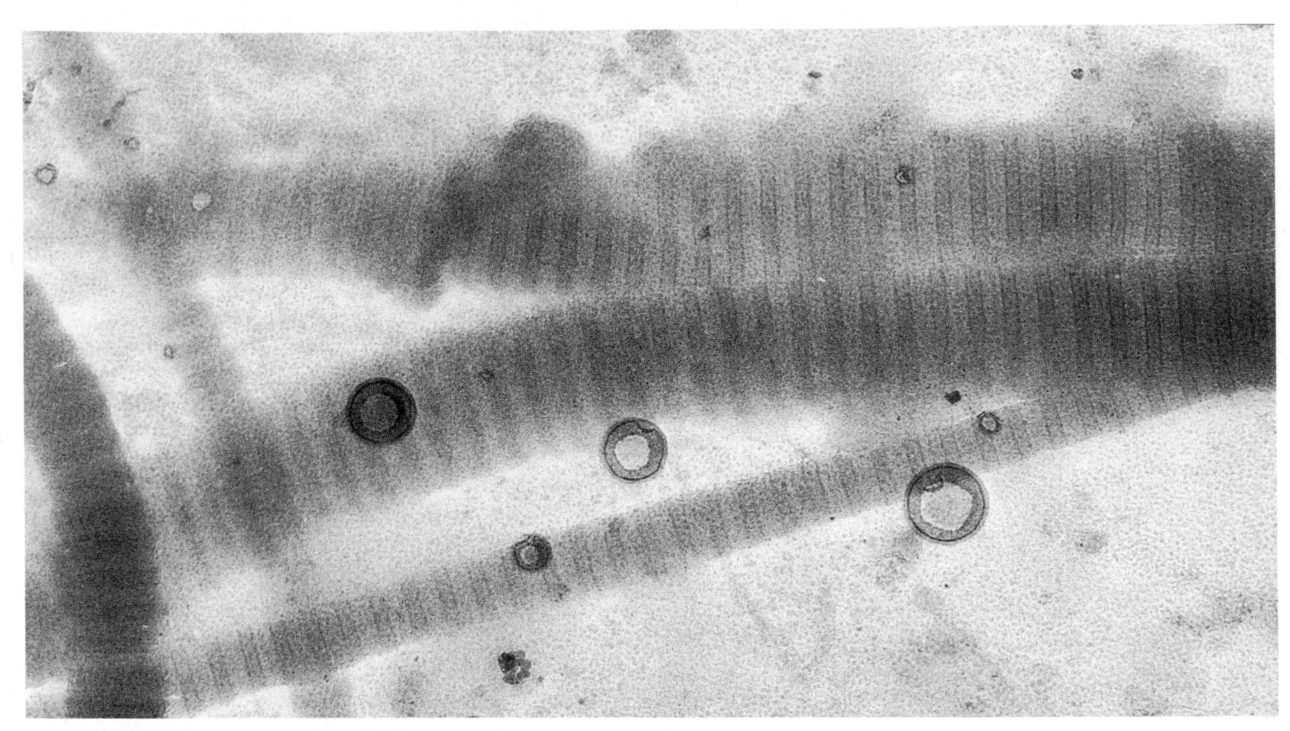

Fig. 2.31 Collagen from tendon: Unfixed, hydrated preparation.
Separated from the main body of a chicken tendon, the individual fibres of the unfixed, hydrated tissue, examined at low temperature, display a periodicity of 69.2 nm. (TEM × 57 800.) Courtesy of Dr Booy.

the ligamentous entheses. At the site of insertion, there are four zones: tendon, fibrocartilage, mineralized fibrocartilage and bone.[51] Movement plays a part in controlling the differentiation and positioning of the fibrocartilage that is associated with tendons. The observation that the calcified cartilage remains attached to bone after a specimen is macerated, explains the smooth surface of the attachment site.[17] These authors observe that: 'blood vessels do not usually traverse the tendon fibrocartilage plugs: the areas are devoid of vascular foramina'—an observation in entire

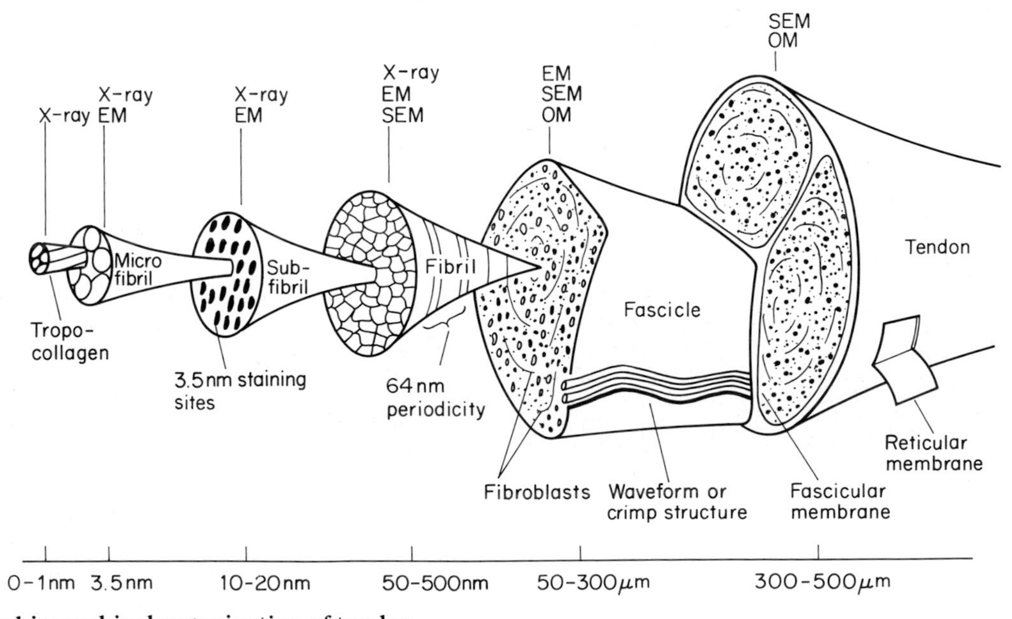

Fig. 2.32 The hierarchical organization of tendon.
Redrawn, courtesy of Betsch and Baer.[24]

contradistinction from present views of the structure and function of entheses (see pp. 126–8).

Because of their accessibility and orderly structure, tendons have long been favoured objects for biochemical and biophysical study. Much of what is known of collagen fibrillogenesis has been learnt from tendons. The synthesis of tendinous collagen and proteoglycan can be correlated readily with mechanical tests[161,278] and the detailed molecular interrelationships between collagen and proteoglycan owe much to experiments conducted with tendons.[243,244]

Tendons, such as the Achilles tendon, have provided a site for the study of ectopic bone formation. After transection of the tendon, repair by silk suture is followed by ectopic calcification in most instances, by ossification in many.[35] This 'paraosteoarthropathy' simulates ankylosing spondylitis although there is diminished vascularity and a lowered content of type V collagen.

Discs

Discs (Fig. 2.33) and menisci, of vestigial evolutionary origin, are present in joints that subserve translational movements. The mechanical properties of discs are outlined on p. 274, the changes of disc chemical composition that occur with age on pp. 126–128. Among the functions attributed to discs are absorption of sudden loads, improved congruence, limitation of translational movements, distribution of forces over increased areas, protection of articular edges and the facilitation of rolling movements and lubrication. Complete discs occur in the inferior radioulnar and sternoclavicular joints. In some sites such as the temporomandibular joint, the disc may be either complete or incomplete. The intervertebral discs are considered in greater detail on p. 126.

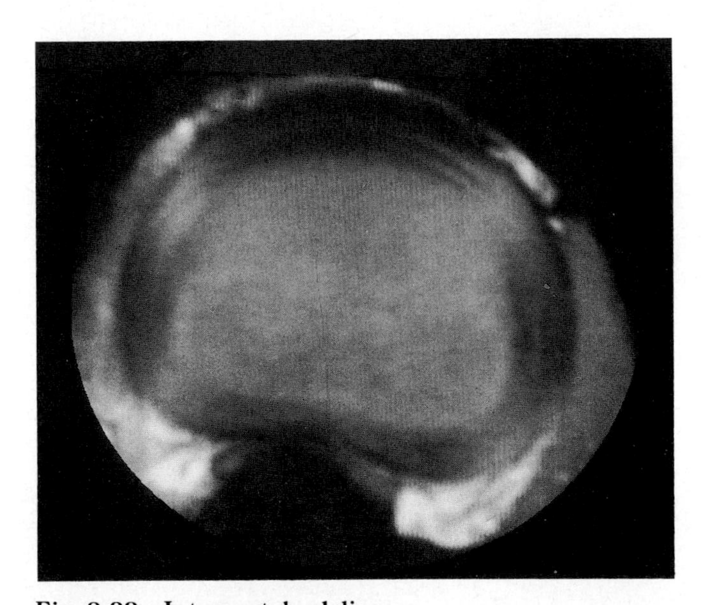

Fig. 2.33 Intervertebral disc.
High resolution transverse T_1-weighted magnetic resonance image of a 16-year-old cadaveric intervertebral disc. Note the laminated structure of the annulus with the fibrous lamellae appearing dark in a light matrix. Such detail becomes less clear in the ageing disc. Reproduced from Hickey *et al.*, by kind permission of the Editor of *Spine*.[124a]

In freely moveable joints discs lie easily between articular surfaces, maintaining two thin lubricating films of synovial fluid. When a synovial joint contains a complete intraarticular disc, it is, in effect, two joints in series: the individual movements of the halves combine to provide a wider range of movement than would be possible for either half alone. Thus, in the forward and backward movement of the temporomandibular joint of omnivores like humans, there is translation of the mandibular condyles together with angulation in a hinge movement. Although there is little translation in the hinge-like joints of carnivores, a disc is still usually present. At their margins, intra-articular discs are linked to the articular capsule via loose vascular connective tissue; in the inferior radioulnar and temporomandibular joints, the links are very substantial and strong. The disc periphery receives a vascular and neural supply from its convex edge; the central and principal part is avascular. Within the disc there are a few cells, flattened peripherally and providing a connective tissue in continuity with the nearby synovium.

Menisci

Menisci[255,197] are incomplete discs. Examples are those of the knee (Fig. 2.33) and acromioclavicular joints. Menisci are largely fibrocartilagenous; their chemical composition is principally of type I collagen with some proteoglycan. Menisci provide supplementary articular surfaces, assisting simultaneous movements of two or more different kinds as in the knee where, for example, rotation accompanies extension (see p. 129). In some species, normal discs are extensively mineralized (Fig. 2.34). Lubrication is facilitated both by the gross shape of the menisci and by their irregular surface structure. The superior and inferior surfaces are characterized by circumferential collagen fibre bundles (see Figs 2.18 and 3.60). There are both concentric and some finer radial external fibres, and the small number of chondrocytes appears at the surface as prominences ('humps') or depressions ('pits'); in the cat, the former are seen in young animals, the latter in the adult.[193]

The menisci of the knee taper towards each end: they fit readily between femorotibial surfaces which are incongruent because of their different radii of curvature. Microscopically, broad bands of collagen, aligned circumferentially, are interspersed with limited amounts of proteoglycan. There are small numbers of fibroblasts. A dense lamina of radially orientated collagen covers the two load-bearing surfaces. The periphery of the meniscus merges with the collagen of the colateral ligaments. At the outer margin of the superior and inferior surfaces, a delicate synovium provides a vascular fringe: this is the 'attack zone' (p. 448) for inflammatory disease. With the exception of this fringe, the menisci are avascular, deriving their limited metabolic needs by diffusion from the synovial fluid. A system of canals has been recognized.[93a,268] The early changes in inflammatory diseases such as rheumatoid arthritis include vascularization of the superficial zone of the meniscal fibrocartilage.

Labra

A labrum is a fibrocartilagenous annular lip, usually triangular in cross-section, like a meniscus, and attached to the margin of an articular surface.[283] Examples are the labra of the acetabulum (Fig. 2.35) and of the glenoid. By means of labra, articular sockets are deepened, the areas of contact between opposing surfaces increased. Labra may therefore spread synovial fluid widely and, like menisci, reduce the volume of the synovial space, limiting drag. Unlike the avascular menisci, labra are not compressed between articular surfaces.

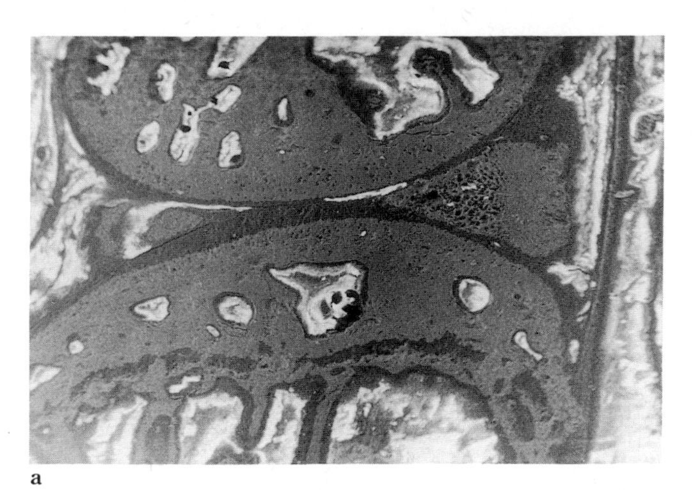

a

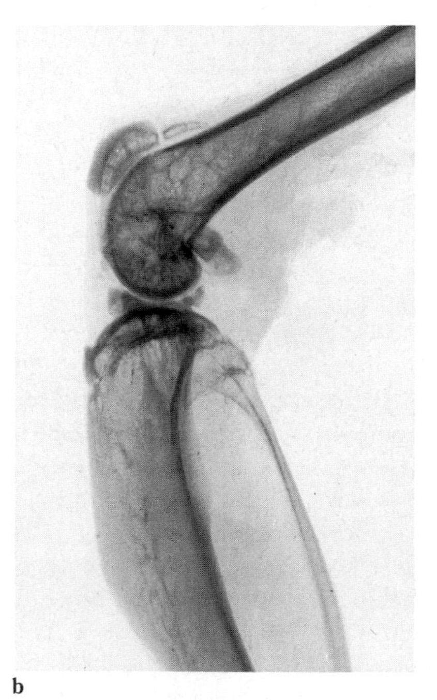

b

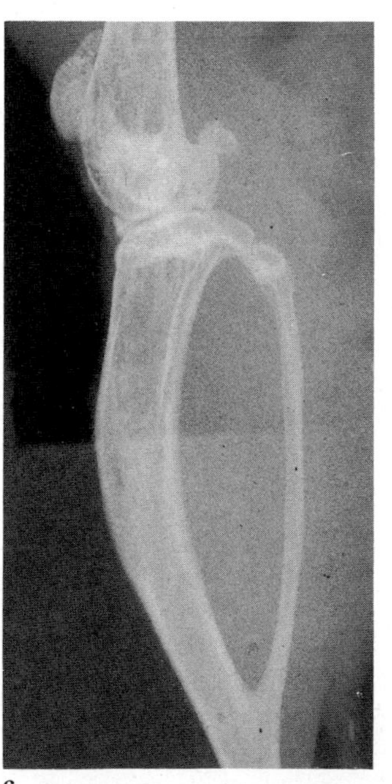

c

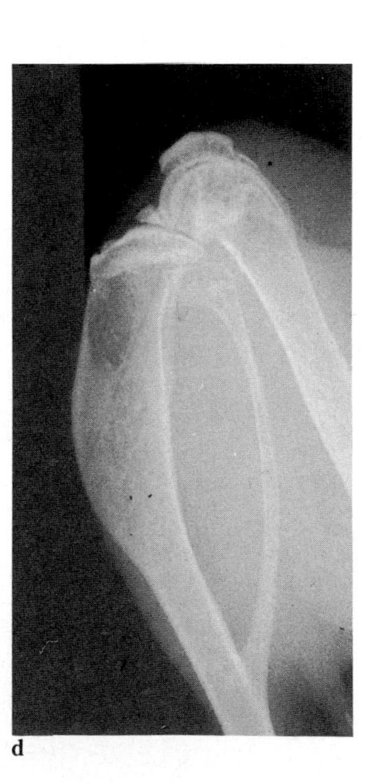

d

Fig. 2.34 Calcification of normal mouse menisci.
a. View of surface block cut sagitally through mouse knee joint. Note large anterior (left) and smaller posterior limbs of knee joint meniscus. The larger, posterior part of the anterior limb is densely calcified. The mineral is calcium hydroxyapatite. The less densely calcified, posterior limb is occupied by calcium pyrophosphate deposition. (× 80.) **b.** Radiograph of normal mouse knee to show the extent of mineralization. Courtesy of Dr M Haddaway. (Microfocal radiograph × 12.) **c.** and **d.** Radiographs of knee of mouse in extension (**c**) and in flexion (**d**) to show that the presence of large quantities of calcium salts in the menisci does not impair knee joint flexion. Courtesy of Dr M Lawton.

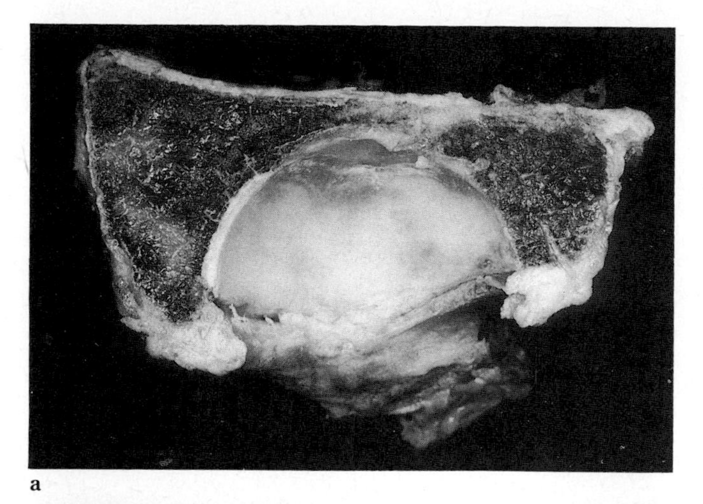

a

b

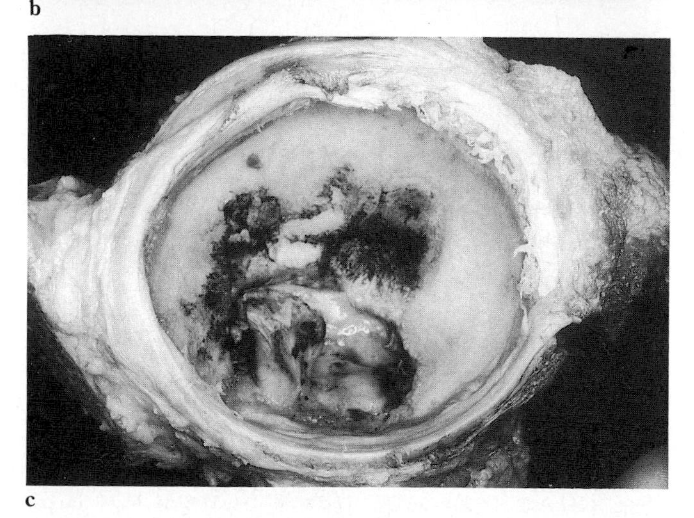

c

Fig. 2.35 Acetabular labrum.
a. Acetabulum in perpendicular section. The labrum has been transected at the inferior margins of the joint surface. Note the asymmetrical thickness of the articular cartilage. **b.** The acetabular labrum forms a rim encircling the concavity of the hip joint. **c.** Extent of age-related fibrillation of the acetabular cartilage (see Chapter 20), highlighted by painting of the surface with Indian ink, for comparison with **b**. (Natural size.)

Bone

Bone (Fig. 2.36) is affected by many of the diseases of non-osseous connective tissues and contributes significantly to the pathogenesis of common disorders such as osteoarthrosis. However, primary bone disease has been the subject of many recent texts and is therefore not described in detail here. For full accounts of the immunology, biochemistry,[285] physiology,[275] anatomy[283] and pathology[224] of bone, the reader is referred to specialized works. Many aspects of bone disease are comprehensively discussed by Resnick and Niwayama.[221]

Bone function

Bones are levers by which mechanical forces are transmitted from muscle during locomotion (see Chapter 6). Bones provide support and protection for the viscera and constitute a reservoir for calcium, phosphate, other minerals, protein and water. The structural and functional unit of bone is the osteon, formed by the laying down of 20–40 lamellae of type I collagen around a vascular channel. The surfaces of this orderly lamellar bone are lined internally by endosteal osteoblasts, externally by those of the periosteum. Both of these layers are potentially osteogenic. Bone shafts, the diplöe of the skull, the vertebral bodies, the ribs and the proximal parts of the shafts of the long bones, accommodate the haemopoetic marrow and may act as selective (strontium) or non-selective (lead, plutonium) reservoirs for extraneous heavy elements that gain access to the body.[275]

The shafts of the mature long bones are formed of compact bone; the interstices are of an open, honeycomb-like meshwork of cancellous bone. The struts and girders of this ramifying structure form a strong but rigid material (Fig. 2.37); their geometrical arrangement is closely related to the main lines of stress to which the bone shaft is subjected, in accordance with Wolf's law.

Bone growth

Bone growth is continuous throughout life although the degree of activity varies greatly in adolescence, during episodes of systemic illness or local trauma, and in response to metabolic, endocrine and pharmacological agents.[275] Bone surfaces that are forming osteoid as a preliminary to mineralization, bind the fluorescent antibiotic tetracycline. This compound, within bone (osteoid) or tooth (dentine), remains permanently linked to the bone precursor; later,

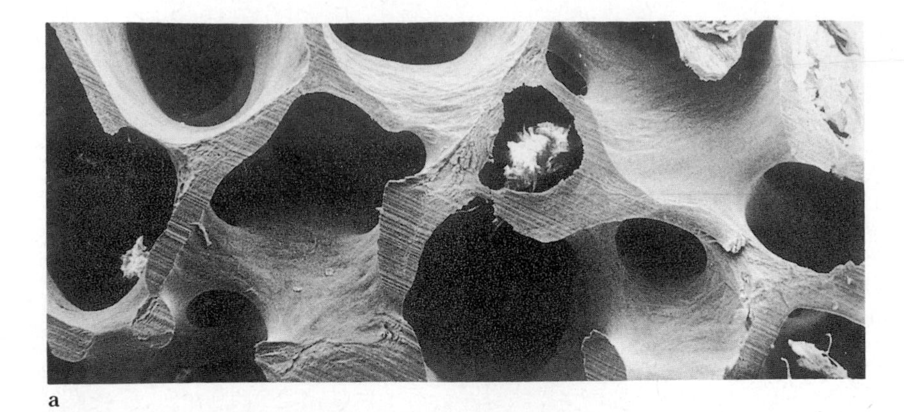

a

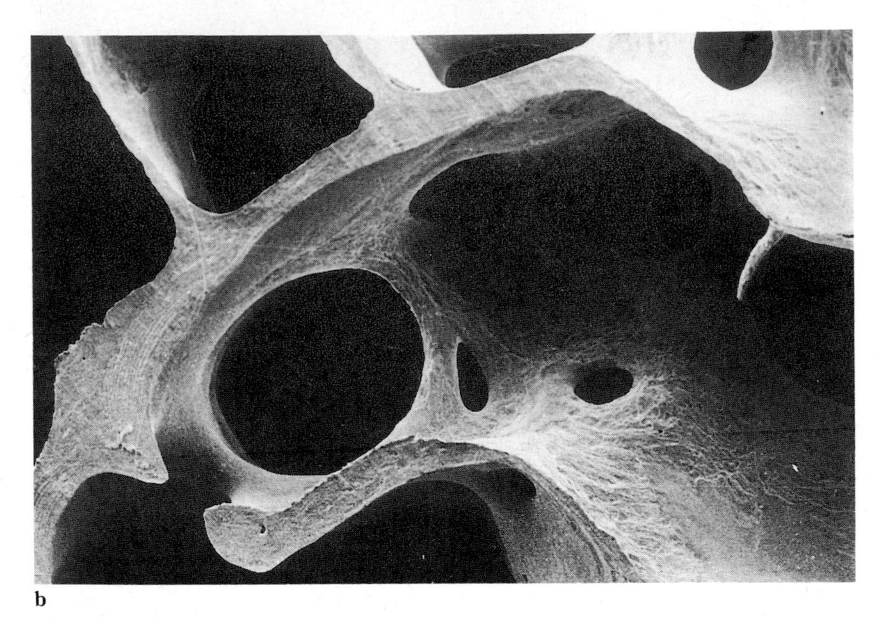

b

Fig. 2.36 Bone: three-dimensional structure.
After fixation and dehydration, cancellous bone has been transected to show its honeycomb-like appearance. (SEM **a.** × 25; **b.** × 50.)

after definitive mineralization, it can be identified by immunofluorescent microscopy. In the adult, tetracycline labelling, undertaken immediately following and again some days after iliac crest biopsy, gives a sensitive measure of bone growth and an important guide to the diagnosis and response to therapy of bone disease (see p. 408). Quantitative analysis of the extent of lacunar reabsorption, of the seams of osteoid, of the extent and width of the mineralization fronts and of the time bands identified by tetracycline labelling, are among the microscopic criteria which can be used in identifying the disorders that accompany calcified tissue disease. The changes can be seen and measured by the techniques of manual and computer-assisted image analysis.[68,223,224]

Both endochondral and intramembranous bone formation require essential, finite amounts of calcium and vitamin D. The incorporation into and the release of calcium from bone during the normal diurnal processes of ossification (see Chapter 1), calcification and reabsorption (see p. 107), are controlled by the interrelated influences of vitamin D, parathormone and calcitonin.

Vitamin D

Vitamin D is essential for calcium absorption and for its use in bone mineralization: it regulates renal tubular phosphate loss. Many of the most important functions of vitamin D_3 (cholecalciferol) are exercised by a metabolic product, 1:25-dihydroxycholecalciferol, formed in the kidney from an intermediary 25-hydroxycholecalciferol, which is itself manufactured in the liver. 1:25 dihydroxycholecalciferol has many of the characteristics of a hormone.

Parathormone

Parathormone is a polypeptide containing 84 amino acids; it acts on gut, kidney and bone. Parathormone increases the intestinal absorption of calcium and reduces urinary excre-

Fig. 2.37 Bone collagen.
Oriented bundles of type I osteocollagen seen at surface of bone trabecula after fixation and dehydration. (SEM × 2250.)

tion; it also increases osteoclastic bone reabsorption, with the liberation of mineral. The actions and interactions of parathormone and 1:25 dihydroxycholecalciferol are intimately related, and since parathormone can lead to a movement of calcium into cells, calcium may act as a 'second messenger' for parathormone.

Calcitonin

Calcitonin, a polypeptide with 32 amino acids, is synthesized by and secreted from the C-(parafollicular) cells of the thyroid; it acts on bone, inhibiting osteoclastic reabsorption. When bone turnover is rapid, as in the young, calcitonin provokes hypocalcaemia. In the adult, calcitonin directly influences bone but there is no reduction in blood calcium. Thus, calcitonin appears to have two main actions: first, to oppose the influence of parathormone so that when calcium is deficient, parathormone-induced bone loss is minimized although the other actions of parathormone continue; second, to control the variations in blood calcium that are likely when there is active bone reabsorption.

Calcification

Calcification,[2,6,8] the deposition of insoluble calcium phosphates in tissues, is an essential part of bone and tooth formation. Calcification is integral to the healing of fractures, the growth of neoplasms such as osteosarcoma, and the restoration of normal skeletal tissues after metabolic disorders, such as rickets and osteomalacia. Calcification is also characteristic of numerous cardiovascular, renal and connective tissue diseases that have little or no direct relationship to skeletal metabolism: examples include atherosclerosis, calcific aortic valve disease, nephrocalcinosis and progressive systemic sclerosis (see Chapter 17). The large volume of information on the nature and regulation of calcification has been reviewed.[7,8,107,2,3,26] Disorders of calcification relevant to the present text are outlined in Chapter 10. The relationship between bone mass and mechanical use is discussed by Frost.[89]

Calcium is present in ionic form in the plasma, extracellular fluid and cells. Much plasma calcium is bound to albumin which acts as a carrier protein. In connective tissue diseases in which there is a change in plasma albumin synthesis, catabolism or concentration, there may be significant alterations in total plasma calcium. In calcification, crystals of insoluble calcium hydroxyapatite form extracellularly. Extracellular fluid calcium and pyrophosphate exist in a metastable condition. In metastatic calcification, their concentrations are sufficient to promote spontaneous calcification; in dystrophic calcification the amounts are insufficient to lead to spontaneous crystallization but sufficient to sustain hydroxyapatite crystal formation once initiated: nucleating crystals act as templates for new crystal formation.

The uncontrolled, further growth of these crystals reflects failure of the delicate mechanisms that regulate normal crystal growth and may be a potent cause of tissue injury.

The factors that initiate calcium hydroxyapatite crystal formation have been vigorously debated. There is now little doubt that in the great majority of normal and abnormal circumstances that have been studied, calcium phosphate is first deposited within matrix vesicles (Figs 2.38, 2.39). These are membrane-bounded, extracellular structures, apparently derived from cells, and containing a substantial number of enzymic and other proteins. The main properties of matrix vesicles are given in Table 2.9. They have now been found in many forms of cartilage, bone, tooth and other tissues,[8] and also in a wide range of abnormal structures (see Chapter 10). Electron spectroscopic imaging can be employed to localize elements such as calcium directly to give fine structural cell detail, enabling matrix vesicles to be measured.[10a]

There are at least three stages in the mineralization theory based on the concept of matrix vesicles. In phase I,[8] hydroxyapatite crystals form within matrix vesicles, located near the inner aspect of the vesicle wall. Calcium is bound by phospholipid, phosphate by the influence of vesicle phosphatases. In phase II, preformed, vesicle-related calcium hydroxyapatite crystals gain exposure to the ex-

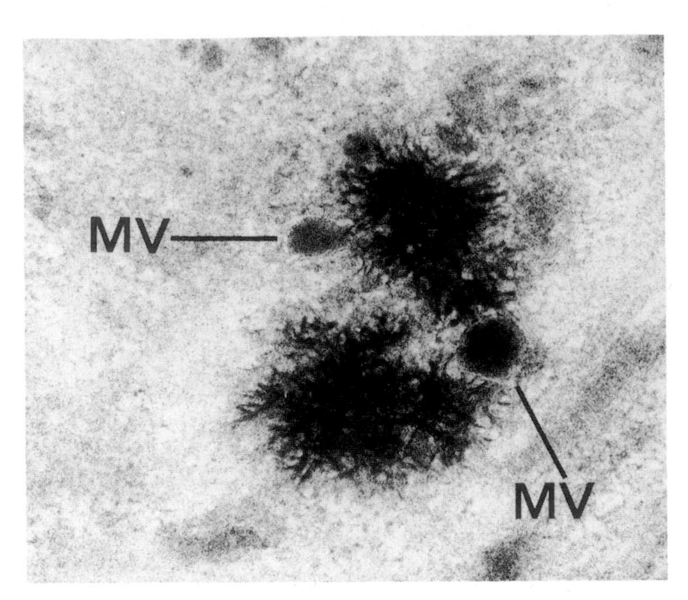

Fig. 2.39 Crystal aggregates and matrix vesicles. Matrix vesicles (MV) adjoin islands of fine crystals of calcium hydroxyapatite. (TEM × 92 000.) Courtesy of Professor S Y Ali.

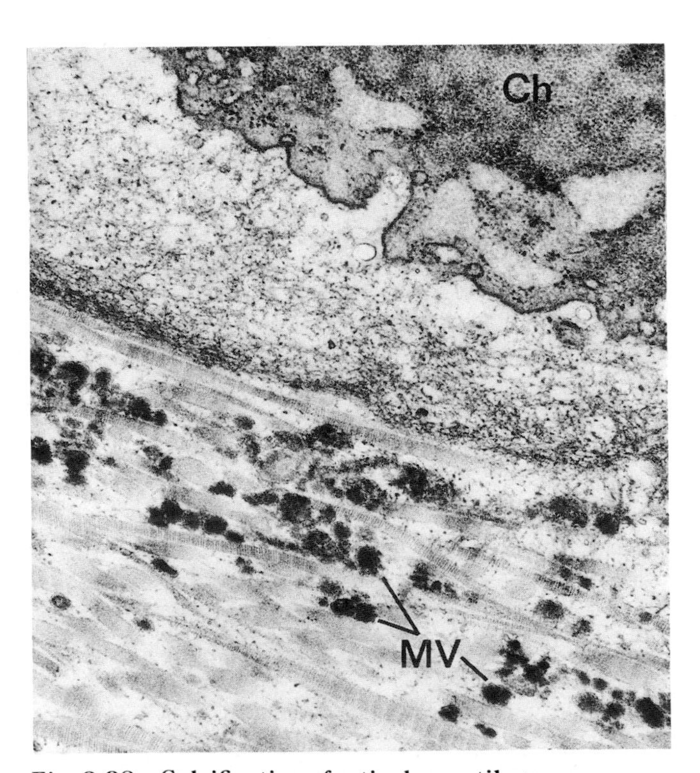

Fig. 2.38 Calcification of articular cartilage. Matrix vesicles (MV) at margin of chondrocyte (Ch). (TEM × 31 000.) Courtesy of Professor S Y Ali.

Table 2.9

Properties of matrix vesicles (Anderson;[8] Ali[3])

Extracellular, cell-derived, membrane-bound 100–200 nm diameter vesicles.
Close association with nascent mineral crystals.
Earliest crystal deposition is in apposition to inner aspect of vesicle membrane; subsequent deposition clusters upon outer surface.
Vesicles contain many enzymes, particularly alkaline phosphatase, adenosine triphosphatase and inorganic pyrophosphatase.
Vesicles are high in phospholipids probably of cell plasma membrane origin. They include much phosphatidyl serine, which has strong affinity for calcium, sphingomyelin and high cholesterol/phospholipid ratio.

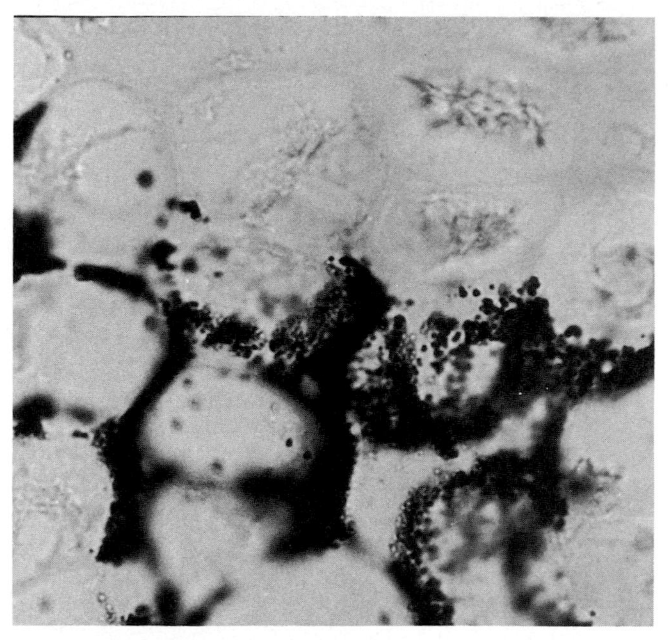

a

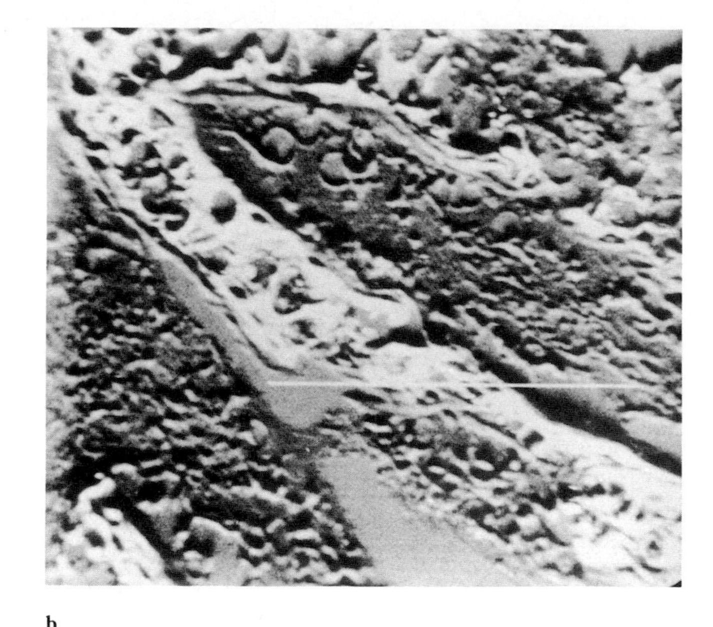

b

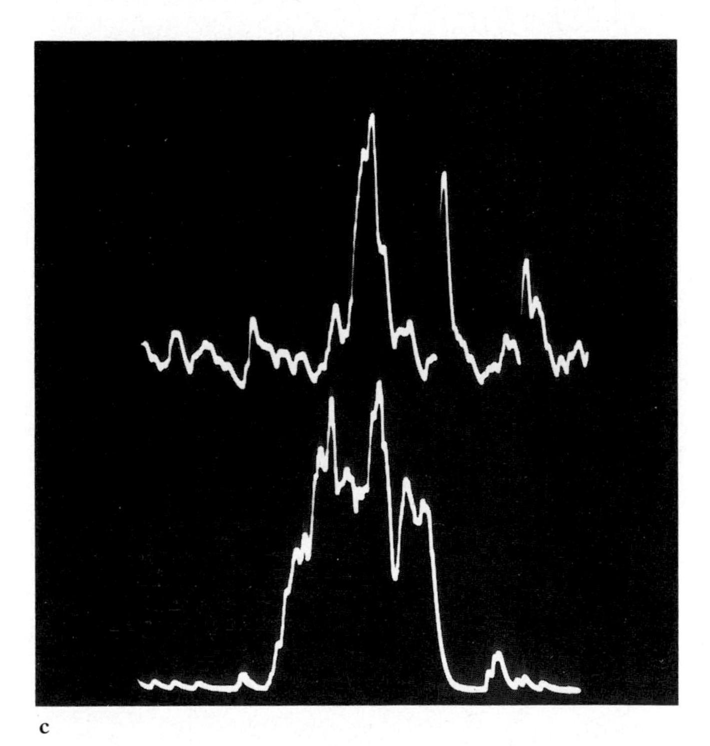

c

Fig. 2.40 Mineralization at sites of endochondral ossification.
a. Calcium hydroxyapatite accumulates in matrix between 'degenerate' chondrocytes at growth plate. **b.** Electron probe scan (indicated by white line) across bone trabecula. **c.** Elemental composition of bone determined by wavelength dispersive X-ray microanalysis. Peaks for calcium (top) and phosphate (bottom) demonstrated along line shown in **b**. (**a**: von Kossa phase contrast × 1160; **b**: × 400.) Reproduced courtesy of Gardner and Hall[94a] and the Editor, *Journal of Pathology*.

tracellular fluid in which there is normally a sufficiently high concentration of calcium and phosphate to facilitate crystal nucleation. In phase III, substances such as collagen, phosphoprotein and pyrophosphate act as promoters and inhibitors to regulate the rate and extent of calcium hydroxyapatite crystal proliferation (Fig. 2.40).

REFERENCES

1. Adams M E, Muir H. The glycosaminoglycans of canine menisci. *Biochemical Journal* 1980; **197**: 385–9.

2. Ali Y, ed. *Cell-Mediated Calcification and Matrix Vesicles*. Amsterdam: Elsevier, 1986.

3. Ali Y. Mechanism of calcification in cartilage and bone. *Bone* 1987; **4**: 18–21.

4. Amenta P S, Gay S, Vaheri A, Martinez-Hernandez A. The extracellular matrix is an integrated unit: ultrastructural localization of collagen types I, III. IV, V, fibronectin and laminin in human placenta. *Collagen and Related Research* 1986; **6**: 125–52.

5. Amento E P, Bhan A K, McCullagh K G, Krane S M. Influences of gamma interferon on synovial fibroblast-like cells. Ia induction and inhibition of collagen synthesis. *Journal of Clinical Investigation* 1985; **76**: 837–48.

6. Anderson H C. Calcification processes. *Pathology Annual* 1980; **15**(2): 45–75.

7. Anderson H C. Calcific diseases—a concept. *Archives of Pathology and Laboratory Medicine* 1983; **107**: 341–8.

8. Anderson H C. Mechanisms of pathologic calcification. *Rheumatic Diseases Clinics of North America* 1988; **14**: 303–19.

9. Armstrong C G, Gardner D L. Thickness and distribution of human femoral head articular cartilage. Changes with age. *Annals of the Rheumatic Diseases* 1977; **36**: 407–12.

10. Armstrong R D, Panayi G S. Natural killer cell activity in inflammatory joint disease. *Clinical Rheumatology* 1983; **2**: 243–9.

10a. Arsenault A L, Ottensmeyer F P, Heath I B. An electron microscopic and spectroscopic study of murine epiphyseal cartilage: analysis of fine structure and matrix vesicles preserved by slam freezing and freeze substitution. *Journal of Ultrastructure and Molecular Structure Research* 1988; **98**: 32–47.

11. Aspden R M, Hukins D W L. The lamina splendens of articular cartilage is an artefact of phase contrast microscopy. *Proceedings of the Royal Society of London B* 1979; **206**: 109–13.

12. Aston J E, Bentley G. Repair of articular surfaces by allografts of articular and growth-plate cartilage. *Journal of Bone and Joint Surgery* 1986; **68**-B: 29–35.

13. Baer A N, Wright E P. Lipid laden macrophages in synovial fluid: a late finding in traumatic arthritis. *Journal of Rheumatology* 1987; **14**: 848–51.

14. Ball J. Enthesopathy of rheumatoid and ankylosing spondylitis. *Annals of the Rheumatic Diseases* 1971; **30**: 213–23.

15. Barnhart M I, Riddle J M, Bluhm G B. Fibrin promotion and lysis in arthritic joints. *Annals of the Rheumatic Diseases* 1967; **26**: 206–18.

16. Beckham C, Dimond R, Greenlee T K. The role of movement in the development of a digital flexor tendon. *American Journal of Anatomy* 1977; **150**: 443–60.

17. Benjamin M, Evans E J, Copp L. The histology of tendon attachments to bone in man. *Journal of Anatomy* 1986; **149**: 89–100.

18. Benninghoff A. Form und Bau der Gelenkknorpel in ihren Beziehungen zur Funktion. II. Der Aufbau des Gelenkknorpels in seinen Beziehungen zur Funktion. *Zeitschrift für Zellforschung und mikroskopische Anatomie* 1925; **2**: 783–862.

19. Bentley G, Greer R B. Homotransplantation of isolated epiphyseal and articular chondrocytes into joint surfaces of rabbits. *Nature* 1971; **230**: 385–8.

20. Beresford W A. Ectopic cartilage, neoplasia and metaplasia. In Hall B K, ed, *Cartilage* (volume III): *Biomedical Aspects*. New York and London: Academic Press, 1983: 1–48.

21. Bergroth V, Konttinen Y T, Nykanen P, von Essen R, Koota K. Proliferating cells in the synovial fluid in rheumatoid disease: an analysis with autoradiography immunoperoxidase double staining. *Scandinavian Journal of Immunology* 1985; **22**: 383–8.

22. Berk R N. Liquid fat in the knee joint after trauma. *New England Journal of Medicine* 1967; **277**: 1411–2.

23. Bertouch J V, Roberts-Thompson P J, Brooks P M, Bradley J. Lymphocyte subsets of inflammatory indices in synovial fluid and blood of patients with rheumatoid arthritis. *Journal of Rheumatology* 1984; **11**: 754–9.

24. Betsch D F, Baer E. Structure and mechanical properties of rat tail tendon. *Biorheology* 1980; **17**: 83–94.

25. Bombardieri S, Cattani P, Ciabottoni R, Di Munno O, Pasero G, Patrono C, Pinca E, Pugliesi F. The synovial prostaglandin system in chronic inflammatory arthritis: Differential effects of steroidal and non-steroidal anti-inflammatory drugs. *British Journal of Pharmacology* 1981; **73**: 891–901.

26. Boskey A L. Current concepts of the physiology and biochemistry of calcification. *Clinical Orthopaedics and Related Research* 1981; **157**: 225–57.

27. Boyde A, Jones S J. Scanning electron microscopy of cartilage. In Hall B K, ed, *Cartilage* (volume I). New York: Academic Press 1983: 105–48.

28. Brinckerhoff C E, Guyre P M. Increased proliferation of human synovial fibroblasts treated with recombinant immune interferon. *Journal of Immunology* 1985; **134**: 3142–6.

29. Brinckerhoff C E, Harris E D. Modulation by retinoic acid and corticosteroids of collagenase production by rabbit synovial fibroblasts treated with phorbol myristate acetate or poly(ethylene glycol). *Biochimica et Biophysica Acta* 1981; **677**: 424–32.

30. Broom N D. Further insights into the structural principles governing the function of articular cartilage. *Journal of Anatomy* 1984; **139**: 275–94.

31. Broom N D. An enzymatically induced structural transformation in articular cartilage. *Arthritis and Rheumatism* 1988; **31**: 210–8.

32. Broom N D, Marra D L. New structural concepts of articular cartilage demonstrated with a physical model. *Connective Tissue Research* 1985; **14**: 1–8.

33. Broom N D, Marra D L. Ultrastructural evidence for fibril-to-fibril associations in articular cartilage and their functional implications. *Journal of Anatomy* 1986; **146**: 185–200.

34. Broom N D, Myers D B. A study of the structural response of wet hyaline cartilage to various loading situations. *Connective Tissue Research* 1980; **7**: 227–38.

35. Brown H, Ehrlich H P, Newberne P M, Kiyoizumi T. Paraosteoarthropathy—ectopic ossification of healing tendon about the rodent ankle joint: histologic and type V collagen changes. *Proceedings of the Society for Experimental Biology and Medicine* 1986; **183**: 214–20.

36. Brown R A, Weiss J B. Neovascularisation and its role in the osteoarthritic process. *Annals of the Rheumatic Diseases* 1988; **47**: 881–5.

37. Buckland-Wright J C, Carmichael I, Walker S R. Quantitative microfocal radiography accurately reflects joint changes in rheumatoid arthritis. *Annals of the Rheumatic Diseases* 1986; **45**: 379–83.

38. Buckland-Wright J C, Walker S R. Incidence and size of erosions in the wrist and hand of rheumatoid patients: a quantitative microfocal radiographic study. *Annals of the Rheumatic Diseases* 1987; **46**: 463–7.

39. Bullough P G, Jagannath A. The morphology of the calcification front in articular cartilage: its significance in joint function. *Journal of Bone and Joint Surgery* 1983; **65-B**: 72–8.

40. Butler D M, Vitti G F, Leizer T, Hamilton J A. Stimulation of the hyaluronic acid levels of human synovial fibroblasts by recombinant human tumor necrosis factor α, tumor necrosis factor β, (lymphotoxin), interleukin-1 α, and interleukin-1 β. *Arthritis and Rheumatism* 1988; **31**: 1281–9.

41. Campbell C J. The healing of cartilage defects. *Clinical Orthopaedics and Related Research* 1969; **64**: 45–63.

42. Cheung H S, Ryan L M, Kozin F, McCarty D J. Synovial origins of rice bodies in joint fluid. *Arthritis and Rheumatism* 1980; **23**: 72–6.

43. Chisholm G D, Heard B E. Oxalosis. *British Journal of Surgery* 1962; **50**: 78–92.

44. Clague R B. Autoantibodies to cartilage collagens in rheumatoid arthritis. Do they perpetuate the disease or are they irrelevant? *British Journal of Rheumatology* 1989; **28**: 1–6.

45. Clarke I C. The microevaluation of articular cartilage surface contours. *Annals of Biomedical Engineering* 1972; **1**: 31–43.

46. Cohen A S, Brandt K D, Krey P R. In: Cohen A S, ed, *Laboratory Diagnostic Procedures in the Rheumatic Diseases* (2nd edition). Boston, Little, Brown Co: 1975.

47. Cooke A F, Gvozdanovic D. Synthetic lubricants for synovial joints. In Doeson D, Wright V, eds, *An Introduction to the Bio-mechanics of Joints and Joint Replacement.* London: Mechanical Engineering Publications, 1981: 139–45.

48. Cooke C L, Owen D S, Irby R, Toone E. Gonococcal arthritis; a survey of 54 cases. *Journal of the American Medical Association* 1971; **217**: 204–5.

49. Cooke T D. Immune pathology in polyarticular osteoarthritis. *Clinical Orthopaedics and Related Research* 1986; **213**: 41–9.

50. Coombe B, Pope R, Darnell B. Regulation of natural killer cell activity by macrophages in the rheumatoid joint and peripheral blood. *Journal of Immunology* 1984; **133**: 709–16.

51. Cooper R R, Misol S. Tendon and ligament insertion. A light and electron microscope study. *Journal of Bone and Joint Surgery* 1970; **52**-A: 1–20.

52. Danzig L, Resnick D, Gonsalves M, Akeson W H. Blood supply to the normal and abnormal menisci of the knee. *Clinical Orthopaedics and Related Research* 1983; **172**: 271–6.

53. Davies D V. The biology of joints. In: Copeman W S C, ed, *Textbook of the Rheumatic Diseases* (4th edition). Edinburgh: E & S Livingstone, 1969: 40–86.

54. Davies D V, Barnett C H, Cochrane W, Palfrey A J. Electron microscopy of articular cartilage in the young adult rabbit. *Annals of the Rheumatic Diseases* 1962; **21**: 11–22.

55. Dieppe P, Calvert P. *Crystals and Joint Disease.* London: Chapman and Hall, 1983.

56. Dingle J T, Page Thomas D P, King B, Bard D R. *In vivo* studies of articular tissue damage mediated by catabolin/interleukin 1. *Annals of the Rheumatic Diseases* 1987; **46**: 527–33.

57. Dingle J T, Saklatvala J, Hembry R, Tyler J, Fell H B, Jubb R. A cartilage catabolic factor from synovium. *Biochemical Journal* 1979; **184**: 177–80.

58. Döhle H. Leukocyteneinschluesse bei Schlach. *Zentralblat für Bakteriologie* 1911; **61**: 63–72.

59. Duclos M, Zeidler H, Liman W, Pichler W J, Rieber P, Peter H N. Characterisation of blood and synovial fluid lymphocytes from patients with rheumatoid arthritis and other joint diseases by monoclonal antibodies (OKT series) and acid α-naphthylesterase staining. *Rheumatology International* 1982; **2**: 75–82.

60. Duff G W, Førre O, Waalen K. Rheumatoid arthritis synovial dendritic cells produce interleukin-1. *British Journal of Rheumatology* 1985; **24**: 94–102.

61. Dunham J, Shackleton D R, Billingham M E J, Bitensky L, Chayen J, Muir I H. A reappraisal of the structure of normal canine articular cartilage. *Journal of Anatomy* 1988; **157**: 89–99.

62. Dunham J, Shackleton D R, Bitensky L, Chayen J, Billingham M E J, Muir I H. Enzymic heterogeneity of normal canine articular cartilage. *Cell Biochemistry and Function* 1986; **4**: 43–6.

63. Dryll A, Lansaman J, Bardin T, Tran Van A, Ryckewaert M, Rabaud M, Brouilhet H. A study of the microvasculature in normal and inflammatory synovial membranes in the rabbit using light and electron microscopy and freeze fracture. *Journal of Submicroscopic Cytology* 1984; **16**: 207–17.

64. Editorial. Limb salvage surgery. *Lancet* 1988; **ii**: 662–3.

65. Edwards J C W, Mackay A, Moore A R, Willoughby D A. The mode of formation of synovial villi. *Annals of the Rheumatic Diseases* 1983; **42**: 585–90.

66. Edwards H E, Moore J S, Philips G O. Effects of ionizing radiations on human costal cartilage and exploration of the procedures to protect the tissue from radiation damage. *Histochemical Journal* 1978; **10**: 389–98.

66a. Egar M, Wallace H, Singer M. Partial denervation effects on limb cartilage regeneration. *Anatomy and Embryology* 1982; **164**: 221–8.

67. Egg D. Concentrations of prostaglandins D_2, E_2, $F_2\alpha$,

6-keto $F_1\alpha$ and thromboxane B_2 in synovial fluid from patients with inflammatory disorders and osteoarthritis. *Zeitschrift für Rheumatologie* 1984; **43**: 89–96.

68. Ellis H A. Metabolic bone disease. In Anthony P P, McSween R N M, eds, *Recent Advances in Histopathology* (Number 1). Edinburgh: Churchill-Livingstone, 1981: 185–202.

69. Elmore S M, Malmgren R A, Sokoloff L. Sclerosis of synovial blood vessels. *Journal of Bone and Joint Surgery* 1963; **45**-A: 318–26.

70. Elves M W. A study of the transplantation antigens on chondrocytes from articular cartilage. *Journal of Bone and Joint Surgery* 1974; **56**-B: 178–85.

71. Elves M W. The immunobiology of joints. In Sokoloff L, ed, *The Joints and Synovial Fluid* (volume 1). New York and London: Academic Press, 1978: 332–93.

72. Engfeldt B, Hutenby K, Müller M. Ultrastructure of hyaline cartilage. I. A comparative study of cartilage from different species and locations using cryofixation, freeze-substitution and low temperature embedding techniques. *Acta Pathologica, Microbiologica et Immunologica Scandinavica A* 1986; **94**: 313–23.

73. Eyre D R, Muir H. The distribution of different molecular species of collagen in fibrous, elastic and hyaline cartilages of the pig. *Biochemical Journal* 1975; **151**: 595–602.

74. Falus A, Meretey K, Glickman G, Svehag S E, Fabian F, Bohm U, Bozsoky S. Beta 2-microglobulin containing IgG complexes in sera and synovial fluids of rheumatoid arthritis and systemic lupus erythematosus patients. *Scandinavian Journal of Immunology* 1981; **13**: 25–34.

75. Fawcett D W. *The Cell* (2nd edition). Philadelphia, London, Toronto: W B Saunders, 1981.

76. Fawns H T, Landells J W. Histochemical studies of rheumatic conditions. I: Observations on the fine structure of the matrix of normal bone and cartilage. *Annals of the Rheumatic Diseases* 1953; **12**: 105–31.

77. Fell H B. The Strangeways Research Laboratory and cellular interactions. In Dingle J T, Gordon J L, eds, *Cellular Interactions*. Amsterdam: Elseiver/North-Holland, 1981.

78. Fell H B, Barratt M E J. The role of soft connective tissue in the breakdown of pig articular cartilage cultivated in the presence of complement-sufficient antiserum to pig erythrocytes. 1. Histological changes. *International Archives of Allergy and Applied Immunology* 1973; **44**: 441–68.

79. Fell H, Jubb R W. The effect of synovial tissue on the breakdown of articular cartilage in organ culture. *Arthritis and Rheumatism* 1977a; **20**: 1359–71.

80. Fell H, Jubb R W. The destructive action of synovial tissue on articular cartilage in organ culture. In Gordon J I, Hasleman B I, eds, *Rheumatoid Arthritis*. Amsterdam: Elsevier/North Holland Biomedical, 1977b: 193–7.

81. Fell H B, Mellanby E M. The effect of hypervitaminosis A on embryonic limb-bones cultivated *in vitro*. *Journal of Physiology* 1952; **116**: 320–49.

82. Fell H B, Reynolds J J, Lawrence C E, Bagga M R, Glauert A M. The promotion and inhibition of collagen-breakdown in organ cultures of pig synovium: the requirement for serum components and the involvement of cyclic adenosine 3′:5′-monophosphate (cAMP). *Collagen and Related Research* 1986; **6**: 51–75.

83. Fell H B, Thomas L. Comparison of the effects of papain and vitamin A on cartilage. II. The effects on organ cultures of embryonic skeletal tissue. *Journal of Experimental Medicine* 1960; **11**: 719–44.

84. Fiedler V von, Schutt H, Beyer D, Roschek H. Zuverlassigkeit der Doppelkontrastarthrographie in der abkarung von Knorpelschaden des Kniegelenkes. *Fortschrift für Rontgenstrahlung* 1979; **131**: 237–43.

85. Freemont A J. Amyloid arthropathy in haemodialysed patients. *Annals of the Rheumatic Diseases* 1986; **45**: 879.

86. Freemont A J, Denton J. The disease distribution of synovial fluid mast cells and cytophagocytic mononuclear cells in inflammatory arthritis. *Annals of the Rheumatic Diseases* 1985; **44**: 312–5.

87. Freemont A J, Jones C J P, Denton J. The synovium and synovial fluid in multicentric reticulohistiocytosis—A light microscopic, electron microscopic and cytochemical analysis of one case. *Journal of Clinical Pathology* 1983; **36**: 860–6.

88. Froland S S, Natvig S S, Husby G. Immunological characterisation of lymphocytes in synovial fluid from patients with rheumatoid arthritis. *Scandanavian Journal of Immunology* 1973; **2**: 67–76.

89. Frost H M. Bone 'mass' and the 'mechanostat': a proposal. *Anatomical Record* 1987; **219**: 1–9.

90. Fuller J A, Ghadially F N. Ultrastructural observations on surgically-produced partial thickness defects in articular cartilage. *Clinical Orthopaedics and Related Research* 1972; **86**: 193–205.

91. Furukawa T, Eyre D R, Koide S, Glimcher M J. Biochemical studies on repair cartilage resurfacing experimental defects in the rabbit knee. *Journal of Bone and Joint Surgery* 1980; **62**-A: 79–89.

92. Gardner D L. Diseases of connective tissue: a consensus. *Journal of Clinical Pathology* 1978; **31**, Supplement (Royal College of Pathologists) 12: 223–38.

93. Gardner D L. Unpublished observations.

93a. Gardner D L. Canal-like structures in menisci. *Annals of the Rheumatic Diseases* 1989; **48**: 175.

94. Gardner D L, Brereton J. Unpublished observations.

94a. Gardner D L, Hall T A. Electron-probe analysis of sites of silver deposition in avian bone stained by the v. Kossa technique. *Journal of Pathology* 1969; **98**: 105–9.

95. Gardner D L, Mazuryk R, O'Connor P, Orford C R. Anatomical changes in pathogenesis of OA in man, with particular reference to the hip and knee joints. In Lock D J, Jasani M K, Birdwood G F B, eds, *Studies in Osteoarthritis: Pathogenesis, Intervention and Assessment*. Chichester: Wiley, 1987: 21–48.

96. Gardner D L, O'Connor P, Middleton J F S, Oates K, Orford C R. An investigation by transmission electron microscopy of freeze replicas of dog articular cartilage surfaces. *Journal of Anatomy* 1983; **137**: 573–82.

97. Gardner D L, Pidd J, Lawton M. Heterogeneity of structure of primate tibial condylar cartilage. Light and electron microscopic studies of *Papio cynocephalus*. Unpublished observations.

98. Gathercole L J, Keller A. Light microscopic waveforms in collagenous tissues and their structural implications. In Atkins E D T, Keller A, eds, *Structure of Fibrous Biopolymers*. London: Butterworths, 1975: 153–87.

99. Ghadially F N. Superficial (partial-thickness) defects and other injuries in articular cartilage. In Ghadially F N, *Fine structure of synovial joints. A Text and Atlas of the Ultrastructure of Normal and Pathological Articular Tissues*. London: Butterworths, 1983a: 261–79.

100. Ghadially F N. Deep (full-thickness) defects in articular cartilage. In Ghadially F N, *Fine structure of synovial joints. A Text and Atlas of the Ultrastructure of Normal and Pathological Articular Tissues*. London: Butterworths, 1983b: 280–306.

101. Ghadially F N, Ailsby R L, Oryschak A F. Scanning electron microscopy of superficial defects in articular cartilage. *Annals of the Rheumatic Diseases* 1974; **33**: 327–32.

102. Ghadially F N, Lalonde J-MA, Yong N K. Myofibroblasts and intracellular collagen in torn semilunar cartilages. *Journal of Submicroscopic Cytology* 1980; **12**: 447–55.

103. Ghadially F N, Oryschak A F, Mitchell D M. Nuclear fibrous lamina in pathological human synovial membrane. *Virchows Archiv Abteilung B: Zellpathologie* 1974; **15**: 223–8.

104. Ghadially F N, Thomas I, Yong N, Lalonde J-MA. Ultrastructure of rabbit semilunar cartilages. *Journal of Anatomy* 1978; **125**: 499–517.

105. Ghadially F N, Wedge J H, Lalonde J-MA. Experimental methods of repairing injured menisci. *Journal of Bone and Joint Surgery* 1986; **68**-B: 106–10.

106. Ghosh P, Taylor T K F. The knee joint meniscus. A fibrocartilage of some distinction. *Clinical Orthopaedics and Related Research* 1987; **224**: 52–63.

107. Glimcher M J. The nature of the mineral component of bone and the mechanism of calcification. *Instructional Course Lectures* 1987; **36**: 49–69.

108. Goodman M G, Chenoweth D E, Weigle W O. Induction of interleukin 1 secretion and enhancement of humoral immunity by binding of human C5a to macrophage surface C5a receptors. *Journal Experimental Medicine* 1982; **156**: 912–7.

109. Gordon D A, Pruzanski W, Orgyzlo M A, Little H A. Amyloid arthritis simulating rheumatoid disease in five patients with multiple myeloma. *American Journal of Medicine* 1973; **55**: 142–54.

110. Grande D A, Singh I J, Pugh J. Healing of experimentally produced lesions in articular cartilage following chondrocyte transplantation. *Anatomical Record* 1987; **218**: 142–8.

111. Gronblad M, Korkala O. Liesi P, Karaharju E. Innervation of synovial membrane and meniscus. *Acta Orthopaedica Scandinavica* 1985; **56**: 484–6.

112. Hadjiisky P, Donev S, Renais J, Scebat L. Cartilage and bone formation in arterial wall. 1. Morphological and histochemical aspects. *Basic Research in Cardiology* 1979; **74**: 649–62.

113. Hadjiisky P, Donev S, Renais J, Scebat L. Cartilage and bone formation in arterial wall. 2. Ultrastructural patterns. *Basic Research in Cardiology* 1980; **75**: 365–77.

114. Hadler N M, Johnson A M, Spitznagel J K, Quinet R J. Protease inhibitors in inflammatory synovial effusion. *Annals of the Rheumatic Diseases* 1981; **40**: 55–9.

115. Halata Z, Groth H-P. Innervation of the synovial membrane of the cats joint capsule. *Cell and Tissue Research* 1976; **169**: 415–8.

116. Hall B K (ed.). *Cartilage* (volumes I–III). New York: Academic Press, 1983a.

117. Hall B K. Tissue interactions and chondrogenesis. In Hall B K, ed, *Cartilage* (volume III) *Development, Differentiation and Growth*. New York: Academic Press, 1983b: 187–222.

118. Hannestad K. Rheumatoid factors reacting with autologous native gamma-G-globulin and joint fluid gamma-G aggregates. *Clinical and Experimental Immunology* 1968; **3**: 671–90.

119. Harding B, Knight S C. The distribution of dendritic cells in the synovial fluids of patients with arthritis. *Clinical and Experimental Immunology* 1986; **63**: 594–600.

120. Hassan O. Histochemical and electron microscope studies of elastic cartilage under normal and pathological conditions (lathyrism). PhD thesis, The Queen's University of Belfast, 1977.

121. Hasselbacher P. Measuring synovial fluid viscosity with a white blood cell diluting pipette. *Arthritis and Rheumatism* 1978; **19**: 1358–65.

122. Hasselbacher P. Variation in synovial fluid analysis by hospital laboratories. *Arthritis and Rheumatism* 1987; **30**: 637–42.

123. Hedberg H. The depressed synovial complement activity in adult and juvenile rheumatoid arthritis. *Acta Rheumatologica Scandinavica* 1964; **10**: 109–18.

124. Henderson B, Edwards J C W. *The Synovial Lining in Health and Disease*. London: Chapman and Hall, 1987: 297.

124a. Hickey D S, Aspden R M, Hukins D W, Jenkins J P, Isherwood I. Analysis of magnetic resonance images from normal and degenerate lumbar intervertebral discs. *Spine* 1986; **11**: 702–8.

125. Higgs G A, Vane J R, Hart F D. Effects of anti-inflammatory drugs on prostaglandins in rheumatoid arthritis. In Robinson H J, Vane J R, eds, *Prostaglandin Synthetase Inhibitors*. New York: Raven Press, 1974: 165–73.

126. Hollander J L. The most neglected differential diagnostic test in arthritis. *Arthritis and Rheumatism* 1960; **3**: 364–7.

127. Hollander J L, McCarty D J, Astorga G, Castro-Murillo E. Studies of the pathogenesis of rheumatoid joint inflammation. I. The 'RA cell' and a working hypothesis. *Annals of Internal Medicine* 1965; **62**: 271–80.

128. Hough A J, Banfield W G, Mottram F C, Sokoloff L. The osteochondral junction of mammalian joints. An ultrastructural and microanalytic study. *Laboratory Investigation* 1975; **31**: 685–95.

129. Hunter G K, Rogakou C C, Pritzker K P H. Extracellular matrix synthesis by articular chondrocytes and synovial fibroblasts in long-term monolayer culture. *Biochimica et Biophysica Acta* 1984; **804**: 459–65.

130. Hunter T, Gordon D A, Ogryzlo M A. The ground pepper sign of synovial fluid; a new diagnostic feature of ochronosis. *Journal of Rheumatology* 1974; **1**: 45–53.

131. Hunter W. Of the structure and diseases of articulating cartilage. *Philosophical Transactions of the Royal Society of London* 1742–43; **42**: 514–21.

132. Hunziker E B, Herrman W. *In situ* localization of cartilage extracellular matrix components by immunoelectron microscopy after cryotechnical tissue processing. *Journal of Histochemistry and Cytochemistry* 1987; **35**: 647–55.

133. Hunziker E B, Herrman W, Schenk R K, Mueller M, Moor H. Cartilage ultrastructure after high pressure freezing, freeze substitution and low temperature embedding. I. Chondrocyte ultrastructure—implications for the theories of mineralization and vascular invasion. *Journal of Cell Biology* 1984; **98**: 267–76.

134. Hunziker E B, Schenk R K. Cartilage ultrastructure after high pressure freezing, freeze substitution, and low temperature embedding. II. Intercellular matrix ultrastructure—preservation of proteoglycans in their native state. *Journal of Cell Biology* 1984; **98**: 277–82.

135. Hurwitz P J. Experimental transplantation of small joints by microvascular anastomosis. *Plastic and Reconstructive Surgery* 1979; **64**: 221–31.

136. Insall J. The Pridie debridement operation on osteoarthritis of the knee. *Clinical Orthopaedics and Related Research* 1974; **101**: 61–7.

137. Jayson M I V, Dixon A St J. Intraarticular pressure in rheumatoid arthritis of the knee. III. Pressure changes during joint use. *Annals of the Rheumatic Diseases* 1970; **29**: 401–8.

138. Jessar R A. The study of synovial fluid. In Hollander J L, McCarty D J, eds, *Arthritis and Allied Conditions*. Philadelphia: Lea & Febiger, 1972: 67–81.

139. Jilani M, Ghadially F N. An ultrastructural study of age-associated changes in the rabbit synovial membrane. *Journal of Anatomy* 1986; **146**: 201–15.

140. Johnson R G, Herbert M A, Wright S, Offierski C, Kellam J, Goodman S, Bobechko W P. The response of articular cartilage to the *in vivo* replacement of synovial fluid with saline. *Clinical Orthopaedics and Related Research* 1983; **174**: 285–92.

141. Jones V, Taylor P C R, Jacoby R K, Wallington T B. Synovial synthesis of rheumatoid factors and immune complex constituents in early arthritis. *Annals of the Rheumatic Diseases* 1984; **43**: 235–9.

142. Kahn C B, Hollander J L, Schumacher H R. Corticosteroid crystals in synovial fluid. *Journal of the American Medical Association* 1970; **211**: 807–9.

142a. Karvonen R L, Negendank W G, Fraser S M, Mayes M D, An T, Fernandez-Madrid F. Articular cartilage defects of the knee: correlation between magnetic resonance imaging and gross pathology. *Annals of the Rheumatic Diseases* 1990; **49**: 672–5.

143. Kessler F B, Homsy C A, Berkeley M E, Anderson M S, Prewitt J M 3rd. Obliteration of traumatically induced articular surface defects using a porous implant. *Journal of Hand Surgery* 1980; **5**: 328–37.

144. Key J A. The reformation of synovial membrane in the knees of rabbits after synovectomy. *Journal of Bone and Joint Surgery* 1925; **7**: 793–813.

145. Key J A. Experimental arthritis: the changes in joints produced by creating defects in the articular cartilage. *Journal of Bone and Joint Surgery* 1931; **13**: 725–39.

146. Kim H J, McCarty D J, Kozin F, Koethe S. Clinical significance of synovial fluid total hemolytic complement activity. *Journal of Rheumatology* 1980; **7**: 143–52.

147. Kinsella T D, Baum J, Ziff M. Studies of isolated synovial lining cells of rheumatoid and non-rheumatoid synovial membranes. *Arthritis and Rheumatism* 1970; **13**: 734–53.

148. Kirkpatrick C J. The effects of prostaglandin A_1 and prostaglandin B_1 on the differentiation of cartilage in the chick embryo. *Cell and Tissue Research* 1980a; **210**: 111–20.

149. Kirkpatrick C J. Cartilage growth inhibition and necrosis *in vitro* caused by prostaglandin A_1. *Virchows Archiv B: Cellular Pathology* 1980b; **33**: 91–105.

150. Kirkpatrick C J, Gardner D L. Chondrocyte growth inhibition by prostaglandin A_1. *Journal of Cellular Biology* 1976; **70**: 168a.

151. Kirkpatrick C J, Gardner D L. Influence of PGA_1 on cartilage growth. *Experientia* 1977; **33**: 504–5.

152. Kirkpatrick C J, Mohr W, Haferkamp O. Effects of prostanoid precursors and indomethacin on chick embryonic cartilage growth in organ culture. *Experimental and Cellular Biology* 1983; **51**: 192–200.

153. Kitridou R, McCarty D J, Prockop D J, Hummeler K. Identification of collagen in synovial fluid. *Arthritis and Rheumatism* 1969a; **12**: 580–8.

154. Kitridou R, Schumacher H R, Sparbaro J L, Hollander J L. Recurrent haemarthrosis after prosthetic knee arthroplasty: Identification of metal particles in the synovial fluid. *Arthritis and Rheumatism* 1969b; **12**: 520–8.

154a. Klein J, Meyer F A. Tissue structure and macromolecular diffusion in umbilical cord. Immobilization of endogenous hyaluronic acid. *Biochimica et Biophysica Acta* 1983; **755**: 400–11.

155. Klickstein L B, Shapleigh C, Goetzl E J. Lipoxygenation of arachadonic acid as a source of polymorphonuclear leukocyte chemotactic factors in synovial fluid and tissue in rheumatoid arthritis and spondylarthritis. *Journal of Clinical Investigation* 1980; **66**: 1166–70.

156. Knight A D, Levick J R. Pressure–volume relationships above and below atmospheric pressure in the synovial cavity of the rabbit knee. *Journal of Physiology* 1982a; **328**: 403–20.

157. Knight A D, Levick J R. Physiological compartmentation of fluid within the synovial cavity of the rabbit knee. *Journal of Physiology* 1982b; **331**: 1–15.

158. Knight A D, Levick J R. The density and distribution of capillaries around a synovial cavity. *Quarterly Journal of Experimental Physiology* 1983a; **68**: 629–44.

159. Knight A D, Levick J R. Time-dependence of the pressure–volume relationship in the synovial cavity of the rabbit knee. *Journal of Physiology* 1983b; **335**: 139–52.

160. Knox P, Levick J R, McDonald J N. Synovial fluid—its mass, macromolecular content and pressure in major limb joints of the rabbit. *Quarterly Journal of Experimental Physiology* 1988; **73**: 33–45.

161. Koob T J, Vogel K G. Proteoglycan synthesis in organ cultures from regions of bovine tendon subjected to different mechanical forces. *Biochemical Journal* 1987; **246**: 589–98.

162. Kuettner K E, Pauli B U. Vascularity of cartilage. In Hall B K, ed, *Cartilage* (volume I): *Structure, Function and Biochemistry*. New York: Academic Press, 1983: 281–312.

163. Kühn K, Krieg T (eds). *Connective Tissue: Biological and Clinical Aspects*. Basel: Karger, 1986.

164. Kurrat H J, Oberlander W. The thickness of the cartilage in the hip joint. *Journal of Anatomy* 1978; **126**: 145–55.

165. Lane L B, Bullough P G. Age-related changes in the thickness of the calcified zone and the number of tidemarks in adult human articular cartilage. *Journal of Bone and Joint Surgery* 1980; **62**-B: 372–5.

166. Letner C (ed.). *Geigy Scientific Tables* (8th edition) (volume 1). Basle: Ciba-Geigy, 1981.

167. Levick J R. Synovial fluid and trans-synovial flow in stationary and moving normal joints. In Helminen H J, Kiviranta I, Säämänen A-M, Tammi M, Paukkonen K, Jurvelin J, eds, *Joint Loading. Biology and Health of Articular Structures*. Bristol: John Wright & Sons, 1987: 149–86.

168. Levick J R, McDonald J N. Microfibrillar meshwork of the synovial lining and associated broad banded collagen: a clue to identity. *Annals of the Rheumatic Diseases* 1990; **49**: 31–6.

169. Luyten F P, Verbruggen G, Veys E M, Goffin E, Pypere H de. *In vitro* repair potential of articular cartilage: proteoglycan metabolism in the different areas of the femoral condyles in human cartilage explants. *Journal of Rheumatology* 1987; **14**: 329–34.

170. Luzar M J, Friedman B M. Acute synovial fluid eosinophilia. *Journal of Rheumatology* 1982; **9**: 961–2.

171. MacConnaill M A. The movement of bones and joints. 4. The mechanical structure of articulating cartilage. *Journal of Bone and Joint Surgery* 1951; **33**B: 251–7.

172. McCord W C, Nies K M, Louie J S. Acute venereal arthritis. *Archives of Internal Medicine* 1977; **137**: 858–63.

173. McGuire M B, Murphy G, Reynolds J J. Production of collagenase and inhibitor (TIMP) by normal, rheumatoid and osteoarthritic synovium *in vitro*: effects of hydrocortisone and indomethacin. *Clinical Science* 1981; **61**: 703–6.

174. MacSween R N M, Dalakos T G, Jasani M K, Boyle J A, Buchanan W W, Goudie R B. A clinical-immunological study of serum and synovial fluid antinuclear factors in rheumatoid arthritis and other arthritides. *Clinical and Experimental Immunology* 1968; **3**: 17–24.

175. Male D K, Roitt I M. Molecular analysis of complement-fixing rheumatoid synovial fluid immune complexes. *Clinical and Experimental Immunology* 1981; **46**: 521–8.

176. Malinin T, Pekin T J, Bauer H, Zfaifler N J. Vacuoles in synovial fluid leukocytes. *American Journal of Clinical Pathology* 1966; **45**: 728–31.

177. Mankin H J, Cantley K P, Lippiello L, Schiller A L, Campbell C J. Biology of human chondrosarcoma. 1. Description of the cases, grading and biochemical analyses. *Journal of Bone and Joint Surgery* 1980; **62**-A: 160–76.

178. Mankin H J, Cantley K P, Schiller A L, Lippiello L. Biology of human chondrosarcoma. 2. Variation in chemical composition among types and subtypes of benign and malignant cartilage tumors. *Journal of Bone and Joint Surgery* 1980; **62**-A: 176–88.

179. Mankin H J, Thrasher A Z. Water content and binding in normal and osteoarthritic human cartilage. *Journal of Bone and Joint Surgery* 1975; **57**-A: 76–80.

179a. Maroudas A. Physicochemical properties of articular cartilage. In: Freeman, M A R, ed, *Adult Articular Cartilage* (2nd edition). Tunbridge Wells: Pitman Medical, 1979: 215–90.

180. Maroudas A, Evans H. Sulphate diffusion and incorporation into human articular cartilage. *Biochimica et Biophysica Acta* 1974; **338**: 265–79.

181. Maroudas A, Schneiderman R. 'Free' and 'exchangeable' or 'trapped' and 'non-exchangeable' water in cartilage. *Journal of Orthopaedic Research* 1987; **5**: 133–8.

182. Maroudas A, Venn M. Swelling of normal and osteoarthritic femoral head cartilage. *Annals of the Rheumatic Diseases* 1977; **36**: 399–406.

183. Meachim G. Effect of age on the thickness of adult articular cartilage at the shoulder joint. *Annals of the Rheumatic Diseases* 1971; **30**: 43–6.

184. Meachim G, Roberts C. Repair of the joint surface from subarticular tissue in the rabbit knee. *Journal of Anatomy* 1971; **109**: 317–27.

185. Meachim G, Stockwell R A. The matrix. In Freeman M A R, ed, *Adult Articular Cartilage* (2nd edition). Tunbridge Wells, Pitman Medical, 1979: 1–67.

185a. Westacott C I, Whicher J T, Barnes I C, Thomson D, Swan A J, Dieppe P A. Synovial fluid concentrations of five different cytokines in rheumatic diseases. *Annals of the Rheumatic Diseases* 1990; **49**: 676–81.

186. Menzel J, Steffen C, Kolarz G. Demonstration of antibodies to collagen and of anti-collagen complexes in rheumatoid arthritis synovial fluids. *Annals of the Rheumatic Diseases* 1978; **35**: 446–52.

187. Meyer F A, Laver-Rudich Z, Tanenbaum R. Evidence for a mechanical coupling of glycoprotein microfibrils with collagen fibrils in Wharton's jelly. *Biochimica et Biphysica Acta* 1983; **755**: 376–87.

188. Middleton J F S, Hunt S, Oates K. Electron probe X-ray microanalysis of the composition of hyaline articular and non-articular cartilage in young and aged rats. *Cell and Tissue Research* 1988; **253**: 469–75.

189. Mims C A, Stokes A, Grahame R. Synthesis of antibodies including antiviral antibodies in the knee joints of patients with arthritis. *Annals of the Rheumatic Diseases* 1985; **44**: 734–41.

190. Mitchell N, Blackwell P. The electron microscopy of regenerating synovium after subtotal synovectomy in rabbits. *Journal of Bone and Joint Surgery* 1968; **50**-A: 675–86.

191. Mitchell N, Shepherd N. The resurfacing of adult rabbit articular cartilage by multiple perforations through the subchondral bone. *Journal of Bone and Joint Surgery* 1976; **58**-A: 230–3.

191a. Moon M-S, Woo Y-K, Kim Y-L. Meniscal regeneration and its effects on articular cartilage in rabbit knees. *Clinical Orthopaedics and Related Research* 1988; **227**: 298–304.

192. Moriizumi T, Yamashita N, Okada Y. Papain-induced changes in the guinea pig knee joint with special reference to cartilage healing. *Virchows Archiv (Cellular Pathology)* 1986; **51**: 461–74.

193. Moschurchak E M, Ghadially F N. A maturation change detected in the semilunar cartilages with the scanning electron microscope. *Journal of Anatomy* 1978; **126**: 605–18.

194. Muir H, Carney S. Pathological and biochemical changes in cartilage and other tissues of the canine knee resulting from induced joint instability. In Helminen H J, Kiviranta I, Säämänen A M *et al.*, eds, *Articular Cartilage and Other Joint Structures in Relation to Loading and Movement of the Joint.* Bristol: John Wright & Sons, 1987: 47–63.

194a. Mitrovic D. Development of the articular cavity in paralyzed chick embryos and in chick embryo limb buds cultured on chorioallantoic membranes. *Acta Anatomica* 1982; **113**: 313–24.

195. Muller-Gerbl M, Schulte E, Putz R. The thickness of the calcified layer of articular cartilage: a function of the load supported? *Journal of Anatomy* 1987; **154**: 103–11.

196. Niepel G A, Kostka D, Kopecky S, Manca S. Enthesopathy. *Acta Rheumatologica et Balnologica Pistiniana* 1966; **1**: 28–64.

197. Noble J, Turner P G. The function, pathology and surgery of the meniscus. *Clinical Orthopaedics and Related Research* 1986; **210**: 62–8.

198. *Nomina Histologica.* In: *Nomina Anatomica* (6th edition). Edinburgh: Churchill Livingstone, 1989.

199. Nouri A M E, Panayi G S, Goodman S M. Cytokines and the chronic inflammation of rheumatic disease. I. The presence of interleukin-1 in synovial fluids. *Clinical and Experimental Immunology* 1984a; **55**: 295–306.

200. Nouri A M E, Panayi G S, Goodman S M. Cytokines and the chronic inflammation of rheumatic disease II. The presence of interleukin-2 in synovial fluids. *Clinical and Experimental Immunology* 1984b; **58**: 402–8.

201. O'Connor P. Orford C R, Gardner D L. Differential response to compressive loads of zones of canine hyaline articular cartilage: micromechanical, light and electron microscopic studies. *Annals of the Rheumatic Diseases* 1988; **47**: 414–20.

202. O'Driscoll S W, Salter R B. The repair of major osteochondral defects in joint surfaces by neochondrogenesis with autogenous osteoperiosteal grafts stimulated by continuous passive motion. An experimental investigation in the rabbit. *Clinical Orthopaedics and Related Research* 1986; **208**: 131–40.

203. Ohnsorge J, Schutt G, Hohn R. Rasterelektronmikroskopischer Untersuchungen des gesunden und arthrotischen Gelenkknorpels. *Zeitschrift für Orthopedie* 1970; **108**: 268–77.

204. Okada Y, Nakanishi I, Kajikawa K. Repair of the mouse synovial membrane after chemical synovectomy with osmium tetroxide. *Acta Pathologica Japonica* 1984; **34**: 705–14.

205. O'Rahilly R, Gardner E. The embryology of movable joints. In Sokoloff L, ed, *The Joints and Synovial Fluid* (volume 1). New York: Academic Press, 1978: 49–103.

206. Orford C R, Gardner D L. Proteoglycan association with collagen d-band in hyaline articular cartilage. *Connective Tissue Research* 1984; **12**: 345–8.

207. Orford C R, Gardner D L. Ultrastructural histochemistry of the surface lamina of normal articular cartilage. *Histochemical Journal* 1985; **17**: 223–33.

208. Palmer D G. Synovial villi: an examination of these structures within the anterior compartment of the knee and metacarpophalangeal joints. *Arthritis and Rheumatism* 1967; **10**: 451–8.

209. Paul H, Reginato A J, Schumacher H R. Alizarin red-S staining as a screening test to detect calcium compounds in synovial fluid. *Arthritis and Rheumatism* 1983; **26**: 191–200.

210. Pekin T T Jr, Malinin T I, Zvaifler N J. The clinical significance of deoxyribonucleic acid particles in synovial fluid. *Annals of Internal Medicine* 1966; **65**: 1229–36.

211. Perrin L H, Nydegger U E, Zubler R H. Correlation between levels of breakdown products of C3, C4 and properdin factor B in synovial fluids from patients with rheumatoid arthritis. *Arthritis and Rheumatism* 1977; **20**: 647–57.

212. Peters T J, Smillie I S. Studies on chemical composition of menisci from the human knee joint. *Proceedings of the Royal Society of Medicine* 1971; **64**: 261–2.

213. Pidd J, Gardner D L. Surface structure of baboon (*Papio anubis*) hydrated articular cartilage: study of low temperature replicas by transmission electron microscopy. *Journal of Medical Primatology* 1987; **16**: 301–9.

214. Pilsworth L M C, Saklatvala J. The cartilage-resorbing protein catabolin is made by synovial fibroblasts and its production is increased by phorbol myristyl acetate. *Biochemical Journal* 1983; **216**: 481–9.

215. Poole A R. Complexity of proteoglycan organization in articular cartilage: recent observations. *Journal of Rheumatology* 1983; **10**: 70–4 (Suppl. 11).

215a. Poole C A, Ayad S, Schofield J R. Chondrons from articular cartilage. 1. Immunolocalization of type VI collagen in the pericellular capsule of isolated canine tibial chondrons. *Journal of Cell Science* 1988; **90**: 635–43.

216. Poole C A, Flint M H, Beaumont B W. Morphological and functional relationships of articular cartilage matrices. *Journal of Anatomy* 1984; **138**: 113–38.

217. Rae S A, Davidson E M, Smith M J H. Leukotriene B4, an inflammatory mediator in gout. *Lancet* 1982; **i**: 677–88.

218. Rawson A J, Abelson N M, Hollander J L. Studies on the pathogenesis of rheumatoid joint inflammation. II. Intracytoplasmic particulate complexes in rheumatoid synovial fluids. *Annals of Internal Medicine* 1965; **62**: 281–4.

219. Redler I, Mow C van, Zimny M L, Mansell J. The ultrastructure and biomechanical significance of the tidemark of articular cartilage. *Clinical Orthopaedics and Related Research* 1975; **112**: 357–62.

220. Repo R U, Finlay J B. Survival of articular cartilage after controlled impact. *Journal of Bone and Joint Surgery* 1977; **59**-A: 1068–76.

221. Resnick D, Niwayama G. *Diagnosis of Bone and Joint Disorders* (volume 2) (2nd edition) *Articular Diseases.* Philadelphia: W B Saunders & Co., 1988.

222. Revell P A. The value of synovial fluid analysis. *Current Topics in Pathology* 1982; **71**: 1–24.

223. Revell P A. Histomorphometry of bone. *Journal of Clinical Pathology* 1983; **36**: 1323–31.

224. Revell P A. *Pathology of Bone*. New York: Springer-Verlag, 1986.

225. Ridley M G, Panayi G S, Nicholas N S, Murphy J. Mechanisms of macrophage activation in rheumatoid arthritis: The role of gamma-interferon. *Clinical and Experimental Immunology* 1986; **63**: 587–96.

226. Robataille P, Zvaifler N J, Tan E. Antinuclear antibodies and nuclear antigens in rheumatoid synovial fluids. *Clinical Immunology and Immunopathology* 1973; **1**: 385–401.

227. Rodnan G P, Benedek T G. Hippocrates, Galen and Synovia. *Annals of Internal Medicine* 1972; **76**: 834–9.

228. Roitt I M, Hay F C, Nineham L J. Rheumatoid arthritis. In Lachman P J, Peters D K, eds, *Clinical Aspects of Immunology*. Oxford: Blackwell Scientific Publications, 1982: 1161.

229. Ronzière M-C, Berthet-Colominas C, Herbage D. Low-angle X-ray diffraction analysis of the collagen-proteoglycan interactions in articular cartilage. *Biochimica et Biophysica Acta* 1985; **842**: 170–5.

230. Ropes M W, Bauer W. *Synovial Fluid Changes In Joint Diseases*. Cambridge, Massachusetts: Harvard University Press, 1953.

231. Ropes M W, Muller A F, Bauer W. The entrance of glucose and other sugars into joints. *Arthritis and Rheumatism* 1960; **3**: 496–514.

232. Rowe R W D. The structure of rat tail tendon. *Connective Tissue Research* 1985a; **14**: 9–20.

233. Rowe R W D. The structure of rat tail tendon fascicles. *Connective Tissue Research* 1985b; **14**: 21–30.

234. Ruddy S. Plasma protein effectors of inflammation: complement. In Kelley W N, Harris E D, Ruddy S, Sledge C B, eds, *Textbook of Rheumatology* (2nd edition). Philadelphia: W B Saunders & Co., 1985: 83.

235. Ruddy S, Fearon D T, Austen K F. Depressed synovial fluid levels of properdin and properdin factor B in patients with rheumatoid arthritis. *Arthritis and Rheumatism* 1975; **18**: 289–95.

236. Saklatvala J. Characterization of catabolin, the major product of pig synovial tissue that induces resorption of cartilage proteoglycan *in vitro*. *Biochemical Journal* 1981; **199**: 705–14.

237. Saklatvala J, Curry V A, Sarsfield S J. Purification to homogeneity of pig leucocyte catabolin, a protein that causes cartilage resorption *in vitro*. *Biochemical Journal* 1983; **215**: 385–92.

238. Saklatvala J, Sarsfield S J. Lymphocytes induce resorption of cartilage by producing catabolin. *Biochemical Journal* 1982; **202**: 275–8.

238a. Saxne T, Heinegard D, Wollheim F A. Therapeutic effects on cartilage metabolism in arthritis as measured by release of proteoglycan structures into the synovial fluid. *Annals of the Rheumatic Diseases* 1986; **45**: 491–7.

239. Saxne T, Heinegård D, Wollheim F A. Cartilage proteoglycans in synovial fluid and serum in patients with inflammatory joint disease: Relation to systemic treatment. *Arthritis and Rheumatism* 1987; **9**: 972–7.

240. Schmid K, MacNair M. Characterisation of the proteins of human synovial fluid in certain disease states. *Journal of Clinical Investigation* 1956; **35**: 814–24.

241. Schumacher H R. Intracellular crystals in synovial fluid anticoagulated with oxalate. *New England Journal Medicine* 1966; **274**: 1372–3.

242. Schumacher H R. The microvasculature of the synovial membrane of the monkey: ultrastructural studies. *Arthritis and Rheumatism* 1969; **12**: 387–404.

243. Scott J E. Collagen-proteoglycan interactions. *Biochemical Journal* 1980; **187**: 887–91.

244. Scott J E, Hughes E W. Proteoglycan-collagen relationships in developing chick and bovine tendons. Influence of the physiological environment. *Connective Tissue Research* 1986; **14**: 267–78.

245. Scott J E, Orford C R. Dermatan sulphate-rich proteoglycan associates with rat tail-tendon collagen at the **d** band in the gap region. *Biochemical Journal* 1981; **197**: 213–6.

246. Sheehan H, Storey G. An improved method of staining leukocyte granules with Sudan black. *British Journal of Pathology and Bacteriology* 1947; **59**: 336–47.

247. Sheldon H. Transmission electron microscopy of cartilage. In Hall B K, ed, *Cartilage* (volume I): *Structure, Function and Biochemistry*. New York: Academic Press, 1983: 87–104.

248. Simon W H. Scale effects in animal joints. 1. Articular cartilage thickness and compressive stress. *Arthritis and Rheumatism* 1970; **13**: 244–55.

249. Simon W H. Scale effects in animal joints. 2. Thickness and elasticity in the deformability of articular cartilage. *Arthritis and Rheumatism* 1971; **14**: 493–502.

250. Simon W H, Richardson S, Herman W, Parsons J R, Lane J. Long-term effects of chondrocyte death on rabbit articular cartilage *in vivo*. *Journal of Bone and Joint Surgery* 1976; **58**-A: 517–26.

251. Simon W H, Wohl D L. Water content of equine articular cartilage: effects of enzymatic degradation and 'artificial fibrillation'. *Connective Tissue Research* 1982; **9**: 227–32.

252. Smith C, Hammerman D. Acid phosphatase in human synovial fluid. *Arthritis and Rheumatism* 1962; **5**: 11–22.

253. Sokoloff L. Cell biology and the repair of articular cartilage. *Journal of Rheumatology* 1974; **1**: 9–16.

254. Sokoloff L. *In Vitro* culture of joint and articular tissues. In: Sokoloff, ed, *The Joints and Synovial Fluid* (volume II). New York: Academic Press, 1980: 1–26.

255. Somer L, Somer T. Is the meniscus of the knee joint a fibrocartilage? *Acta Anatomica* 1983; **116**: 234–44.

256. Speer D P, Chnapil M, Voly R G, Holmes M D. Enhancement of healing in osteochondral defects by collagen sponge implants. *Clinical Orthopaedics and Related Research* 1979; **144**: 326–35.

257. Steinberg J J, Hubbard J R, Sledge C B. *In vitro* models of cartilage degradation and repair. In: Otterness I *et al.*, eds, *Advances in Inflammation Research* (volume II). New York: Raven Press, 1986: 215–41.

258. Stockwell R A. Chondrocytes. *Journal of Clinical Pathol-*

ogy 1978; **31**, Supplement (Royal College of Pathologists) 12: 7–13.

259. Stockwell R A. *Biology of Cartilage Cells.* Cambridge: Cambridge University Press, 1979.

260. Stockwell R A. Metabolism of cartilage. In: Hall B K, ed, *Cartilage* (volume I): *Structure, Function and Biochemistry.* New York: Academic Press, 1983: 253–80.

261. Stockwell R A, Meachim G. The chondrocytes. In: Freeman M A R, ed, *Adult Articular Cartilage* (2nd edition). Tunbridge Wells, Pitman Medical, 1979.

262. Stojan B. Diagnostic potential of synovial fluid testing. *Schweizerische Medizinische Wochenschrift* 1982; **112**: 1514–22.

263. Strachan R, Smith P, Gardner D L. Hyaluronate in therapy. *Annals of the Rheumatic Diseases* 1990; **49**: 949–52.

264. Strocchi R, Leonardi L, Guizzardi S, Marchini M, Ruggeri A. Ultrastructural aspects of rat tail tendon sheaths. *Journal of Anatomy* 1985; **140**: 57–67.

265. Svajger A. Chondrogenesis in the external ear of the rat. *Anatomische Entwicklunggeschichte* 1970; **131**: 236–42.

266. Swann D A, Bloch K J, Swindell D, Shore E. The lubricating activity of human synovial fluids. *Arthritis and Rheumatism* 1984; **27**: 552–6.

267. Swann D A, Silver F H, Slayter H S, Stafford W, Shore E. The molecular structure and lubricating activity of lubricin isolated from bovine and human synovial fluids. *Biochemical Journal* 1985; **225**: 195–201.

268. Sweet M D T, Bird M B E. A system of canals in semilunar menisci. *Annals of the Rheumatic Diseases* 1987; **46**: 670–3.

269. Thompson A M, Stockwell R A. An ultrastructural study of the marginal transitional zone in the rabbit knee joint. *Journal of Anatomy* 1983; **136**: 701–13.

270. Tokunaga M, Ohuchi K, Yoshizawa S. Change of prostaglandin E level in joint fluids after treatment with flurbiprofen in patients with rheumatoid arthritis and osteoarthritis. *Annals of Rheumatic Diseases* 1981; **40**: 462–9.

271. Trachoff R B, Pascual E, Schumacher H R. Mononuclear cells in human synovial fluid. *Arthritis and Rheumatism* 1976; **19**: 743–8.

272. Tyler J A, Fell H B, Lawrence C E. The effect of cortisol on porcine articular tissues in organ culture. *Journal of Pathology* 1982; **137**: 335–51.

273. Van den Berg W B, Van Lent P L E M, Van de Putte L B A, Zwarts W A. Electrical charge of hyaline articular cartilage: its role in the retention of anionic and cationic proteins. *Clinical Immunology and Immunopathology* 1986; **39**: 187–97.

274. Van der Korst J K, Sokoloff L, Miller E J. Senescent pigmentation of cartilage and degenerative joint disease. *Archives of Pathology* 1968; **86**: 40–7.

275. Vaughan J. *The Physiology of Bone* (3rd edition). Oxford: Clarendon Press, 1981.

276. Vaughan J H, Barnett E, Sobel M V. Intracytoplasmic inclusions of immunoglobulins in rheumatoid arthritis and other diseases. *Arthritis and Rheumatism* 1968; **11**: 125–34.

276a. Venn M F, Maroudas A. Chemical composition and swelling of normal and osteoarthrotic femoral head cartilage. I. Chemical composition. *Annals of the Rheumatic Diseases* 1977; **36**: 399.

277. Vernon-Roberts B. Synovial fluid and its examination. In: Scott J T, ed, *Copeman's Textbook of the Rheumatic Diseases* (6th edition). Edinburgh: Churchill Livingstone, 1986: 251–77.

278. Vogel K G, Heinegård D. Characterization of proteoglycans from adult bovine tendon. *Journal of Biological Chemistry* 1985; **260**: 9298–306.

279. Walker P S, Dowson D, Longfield M D, Wright V. 'Boosted lubrication' in synovial joints by fluid entrapment and enrichment. *Annals of the Rheumatic Diseases* 1968; **27**: 512–20.

280. Walker P S, Sikorski J, Dowson D, Longfield M D, Wright V, Buckley T. Behaviour of synovial fluid on surfaces of articular cartilage. A scanning electron microscope study. *Annals of the Rheumatic Diseases* 1969; **28**: 1–14.

281. Walker P S, Unsworth A, Dowson A, Sikorski J, Wright V. Mode of aggregation of hyaluronic acid protein complex on the surface of articular cartilage. *Annals of the Rheumatic Diseases* 1970; **29**: 591–602.

282. Warskyj M, Sullivan P, O'Connor P, Lawton D M, Gardner D L. Micromechanical testing of articular cartilage: recent improvements to test apparatus. *Annals of the Rheumatic Diseases* 1988; **47**: 966–7.

283. Warwick R, Williams P L, (eds). *Gray's Anatomy* (37th edition). Edinburgh: Churchill Livingstone, 1989.

284. Wassilev W. Uber die Ultrastruktur der regenerierten Synovialmembran beim Menschen. *Archiv der orthopedische Unfall-Chirurgie* 1971; **69**: 197–204.

285. Watrous D A, Andrews B S. The metabolism and immunology of bone. *Seminars in Arthritis and Rheumatism* 1989; **19**: 45–65.

286. Wigren A, Falk J, Wik O. The healing of cartilage injuries under the influence of joint immobilization and repeated hyaluronic acid injections. *Acta Orthopaedica Scandinavica* 1978; **49**: 121–33.

287. Williams W J, Beutler E, Ersleu A J, Rundles R W. *Haematology.* New York: McGraw Hill, 1972.

288. Williamson M P, Humm G, Crisp A J. ^{1}H nuclear magnetic resonance investigation of synovial fluid components in osteoarthritis, rheumatoid arthritis and traumatic effusions. *British Journal of Rheumatology* 1989; **28**: 23–7.

289. Willkens R F, Healey L A. The non-specificity of synovial leukocyte inclusions. *Journal of Laboratory and Clinical Medicine* 1966; **68**: 628–35.

290. Wilson N H F, Gardner D L. The microscopic structure of fibrous articular surfaces: a review. *Anatomical Record* 1984; **209**: 143–52.

291. Wirth C R. Meniscus repair. *Clinical Orthopaedics and Related Research* 1981; **157**: 153–60.

292. Wood D D. Interleukin-1 in arthritis. In: Higgs G A, Williams T J, eds, *Inflammatory Mediators.* London: Macmillan, 1985: 183.

293. Woolley D E. Mammalian collagenases. In: Piez K A, Reddi A H, eds, *Extracellular Matrix Biochemistry.* New York: Elsevier, 1984: 119.

293a. Yang G C H, Birk D E, Topographies of extracytoplasmic compartments in developing chick tendon fibroblasts. *Journal of Ultrastructure and Molecular Structure Research* 1986; **97**: 238–48.

294. Youinou P, LeGoff P, Colaco C B. Antikeratin antibodies in serum and synovial fluid shows specificity for rheumatoid arthritis in a study of connective tissue diseases. *Annals of the Rheumatic Diseases* 1985; **44**: 450–5.

295. Zvaifler N J. Breakdown products of C3 in human synovial fluids. *Journal of Clinical Investigation* 1969; **48**: 1532–42.

Chapter 3
Organs

In this chapter, the arrangement of the connective tissues and their assembly into functional units are considered. Because of their importance in the pathology of the connective tissue diseases, particular emphasis is placed on the joints and articulations. Many body structures and organs such as the blood vessels and skin, are rich in connective tissue but their organization is fully reviewed in texts on angiology and dermatology. The crucial role of cartilage in vertebrate evolution has ensured an important part for this material in the formulation of the cell theory of disease,[166,25] and cartilage remains at the centre of current studies of abnormal connective tissue cell behaviour.[44,45,101,155]

Developmental aspects of skeletal connective tissues

Evolution

Components of skeletal connective tissues, such as collagen, are common not only to contemporary animals but also to many species now extinct, including those that can be traced back to very early periods of evolution.[129] There are similar homologies in disease. Little is known of the disorders to which prehistoric creatures were prone although there is evidence of abnormalities such as osteophytosis in dinosaurs. In prehominids and in early man,[99] signs of infective and traumatic disorders can be found in bone fossils but the abnormalities of non-mineralized connective tissue leave little palaeopathological evidence.

Multicellular plants and aninals formed connective tissue for mechanical support during growth and movement. Connective tissue systems, in particular those composed of the extracellular, non-collagenous macromolecules, were stores for water and electrolytes, and bone and mineralized cartilage acted as reservoirs for calcium. The connective tissues also provided protection to the viscera: cells and fibrous and non-fibrous materials became organized around organs and between external and internal body surfaces.

The evolution of the joints is more easily traced in the single subphylum that comprises the vertebrates than among the invertebrates.[115] The terrestial tetrapod limb arose from the pelvic and pectoral fins of crossopterygian fish. Precursors of the humerus/femur, ulna/fibula and radius/tibia were identified. The smaller bones gave origin to the carpals/tarsals but there were too few bones present in these fish to account for the evolution of the metacarpals/metatarsals and an independent origin for these structures and their joints has had to be assumed.

Early tetrapod limbs projected at right angles to the sides of the body which was elevated by bends at the elbow and knee of a form still seen in turtles and many lizards. A comparable arrangement is recognized in extant reptiles (Fig. 3.1). A different, more efficient structure then evolved so that, in the mammals, the hind limbs were rotated forwards, the forelimbs backwards, the knees and

Fig. 3.1 Reptilian locomotion.
Locomotion of the fish-eating Indian Gavial (Gharial), *Ghavialis gangeticus*, exemplifies the problems faced by large reptiles moving on land. Reproduced by permission of the Director, Royal Scottish Museum, Edinburgh.

feet directed forwards, the elbows backwards; To avoid the inconvenience resulting from the consequent backward direction of the arms, the wrist was rotated by 180°, with a crossing of the radius and ulna.

Campbell[26] summarized the manner in which the arboreal life of the lower primates influenced the evolution of the vertebral column. The centre of gravity moved to the hind legs; the tail assumed a special function for balance and elevation, acting as a fifth limb in New World monkeys. Erect sitting and climbing postures developed but the hind limbs maintained considerable flexibility. Other trends appeared: e.g. a reversion to terrestial quadrupedalism and brachiating locomotion with loss of the tail and reduced flexibility of the spine, flattening of the thorax and an erect posture. Later, terrestial bipedalism caused the erect posture to be accompanied by modification of the vertebral column which now acted as a vertical weight bearer. The thorax became flattened, moving the centre of gravity over the pelvis. A modification of the tail formed a floor to the pelvis and the head assumed a balanced posture upon the neck.

The assumption of the erect posture and bipedalism[5,164] culminated in the upright stance of the hominids and of *Homo sapiens* and resulted in fundamental changes to the structure of the pelvis (Fig. 3.2). The pelvis broadened, anchoring the major muscles for forward propulsion and lateral balance, and giving these muscles greater leverage about the hip joint. Pelvic broadening was accompanied by shortening of the ilium and a closer approximation of the sacral articulation to the acetabulum, both points about which the body weight was transmitted.

There were corresponding alterations in the posture of the foot, the whole of which was placed on the ground, and evolutionary changes in the joints of the foot.[103] More weight was borne by the leg bones (Fig. 3.3). The leg extended and the whole body weight was transmitted directly, through the knee joint, from acetabulum to ankle. The leg lengthened in relation to the trunk and limbs. An increase in diameter of the femur had the result that, among primates, the human femur became second only to that of the gorilla. The proportions of the lower limb changed: relative to trunk length, the human leg was now the longest. The bearing surfaces of the acetabulum and femoral head increased by contrast with a relative reduction in the bearing surfaces of the femoral condyles, a consequence of improved weight transmission. The demand for stability and enhanced muscle purchase was associated with the elongated femoral neck. The greater trochanter was placed further from the femoral head.

Comparative anatomy

Classical papers[74] speculated that the formation of tetrapod limbs allowed vertebrates to crawl from water onto land, to

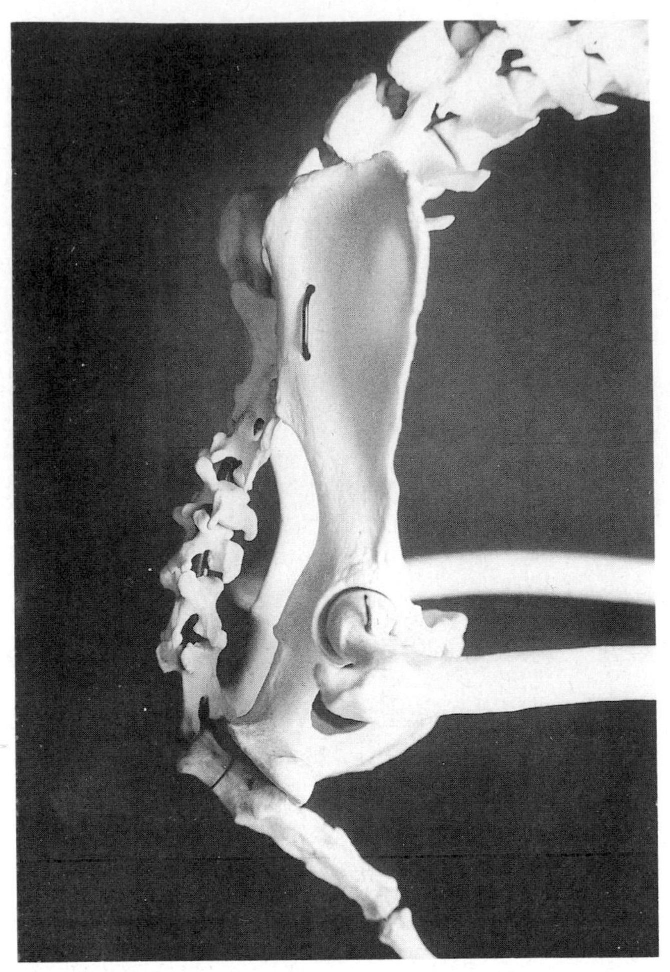

Fig. 3.2 Evolution of the primate skeleton.
With the assumption of the erect posture, radical changes in the shape of the pelvis and in the anatomy of the hip joint and femora were associated with altered mechanics and in changes in susceptibility to disease. This Figure displays the pelvis of a non-human primate, the baboon *Papio cynocephalus*. The narrow, vertical structure may be compared with the broad, flat pelvis of hominids.

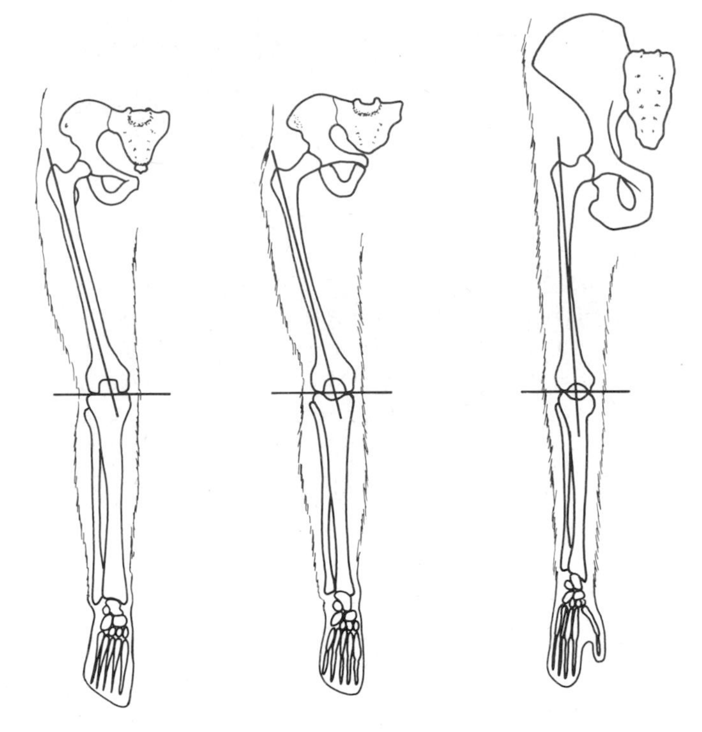

Fig. 3.3 Evolution of human pelvis and leg bones.
Diagrammatic representation of the pelvis and leg bones of man (left), a prehuman, erect primate (centre) and the chimpanzee (right). Note the contrasting pelvic shapes and the alterations in the angles at which loads are transmitted to the knee joints. Redrawn from Johnson and Edey.[99]

eat and reproduce. As D'Arcy Thompson[160] stated: 'the jointed structure of the leg permits one to use it as the shortest possible lever while it is swinging and as the longest possible lever when it is exerting its propulsive force'. Evolutionary modification of this jointed structure led to the diversity of articular organization seen in man today.

Van Sickle and Kincaid[165] traced the variety of joint structure that evolved in the vertebrates. The spectrum ranged from the club-shaped villi of the synovium in the skates and the articulating jaw of the teleosts to the diarthrodial shoulder, hip and elbow joints of amphibia and the true synovial joint cavity of *Salamander maculosa*. These authors note the absence of the patella from the crocodilia, the less massive menisci in the alligators. Birds, also, have single knee-joint cavities; the weaker of the two anterior cruciate ligaments seen in reptiles has disappeared.

The comparative anatomy of the lower limb provides one of many ways in which the primates can be classified; structure and the evolution of function are interrelated. In one approach, an external light source can be used to measure the angles at which interference patterns are generated from a grating fixed onto the anterior surface of the femur, the technique of grating irradiation—moiré topography. According to the measured angles, primates can be divided into two types, with distinct joint structure and the evolution of different functions: Lorisidae, Hylobatidae and Pongidae (type I), and Hominidae and Cercopithecidae (type II).[137]

Differentiation[75]

Cartilage is now known to form the skeleton of many invertebrates;[132] it often becomes mineralized when the body temperature is sustained near 37 °C. Bone and osteoid are recognized only in vertebrates. Whether the individual stem cells from which the connective tissues originate are osteoprogenitor or chondroprogenitor can only be deduced from their proximity to bone or cartilage. The cells cannot be identified morphologically until they have begun to express their genetic potential.

Invertebrates

Until 1972 it was believed that invertebrate endoskeletal tissue did not contain *true* cartilage.[135] Although it is still not certain whether or not invertebrate endoskeletons contain type II collagen, it is clear that they are often cartilaginous. However, the composition of the invertebrate matrix commonly differs from that of vertebrates, and invertebrates do not mineralize an *in vivo* solid phase.

Vertebrates

Cartilage has existed in vertebrates since their origin approximately 5×10^8 years ago[124] but it has undergone important evolutionary changes. The cartilage of coelocanth fish is less cellular than that of modern fish. Among extant fish, the cartilage of hagfish (myxinoids) and lampreys (petromizinoids), representatives of the most primitive vertebrates, has attracted attention.[177] The extracellular territorial matrix of the lingual cartilage of these animals contains cyanogen bromide-insoluble proteins, myxinin (myxinoids) and lamprin (lampreys), respectively, arranged as concentric lamellae. The cartilage of contemporary adult fish and amphibians is less well-differentiated than that of reptiles. Avian and mammalian cartilage have much in common but the growth plate cartilages are very different. In a number of avian species, cartilage is normally present in the viscera including the aorta, the ventricular wall and the papillary muscles.

There are underlying varieties of chondrocyte behaviour. Differentiation and morphogenesis are, of course, influenced both by genetic and by environmental conditions. In the chick embryo, for example, neural crest cells are determined for chondrogenesis while still in the neural tube, before migration to the face and head, whereas in urodele amphibian embryos, stem cells differentiate into cartilage only after migration from the neural folds.[76] Embryonic chick scleral mesenchyme forms cartilage sheets in culture whereas limb bud mesenchyme forms distinct cartilage nodules.

At first, it seemed that morphogenesis was tissue type-specific. However, experimentally, variations in morphogenesis can be shown by varying cell density at culture inoculation, a procedure which controls cell configuration, and by altering the amount of prechondrogenic mesenchyme in the culture.[4] Cartilage morphogenesis *in vitro* depends both on cell form and upon the presence of non-chondrogenic cell types: it is not simply a function of an intrinsic morphogenetic potential of the constituent cells.[4] The intra- and extracellular control of cartilage differentiation are analysed by Hall[75] and by Hunter and Caplan.[89]

Embryology[78]

Within 14 days of fertilization of the human ovum, a primitive extraembryonic mesoderm appears. In the third week, the third of the three germ layers, the intraembryonic mesoderm, is recognized (Table 3.1). At four weeks, a series of segmental, paraxial mesodermal somites develops on either side of the notochord. A non-segmented lateral mesoderm is seen and an intermediate mesoderm forms between the paraxial and lateral components. Limb buds (Fig. 3.4) appear at 4–5 weeks; they are simple, paddle-like protrusions. Forelimb development is ahead of hindlimb. Each limb bud is a loose mass of primitive, gelatinous mesenchyme covered by rudimentary ectoderm. The mesenchymal core gives rise to: 1. the axial tissue of the limbs, including bone and cartilage, the joints, joint capsules and ligaments; 2. the muscles and tendons; 3. dermal and subcutaneous connective tissue; and 4. the blood and lymphatic vessels. In each limb core, a compact scleroblastema foreshadows the shape of the future bone and anticipates the development of a cartilagenous model of the bone and joint.

The morphogenesis of cartilage has been reviewed.[128,161] Joint spaces appear by *in situ* transformation of the cells and matrix of a specialized region of compact mesenchymal tissue that lies mid-way between each bone rudiment (Fig. 3.5). The growth of cartilage is partly

Table 3.1

Timetable of human development (from Scothorne[140])

Age after fertilization (weeks)	Crown-rump length (mm)	Development
3–4		Three germinal layers; first somites
4–5	4–7	Leg buds appear
5–6	11–14	Limb bones, skull base and vertebrae are precartilaginous
6–7	22	First bone formed

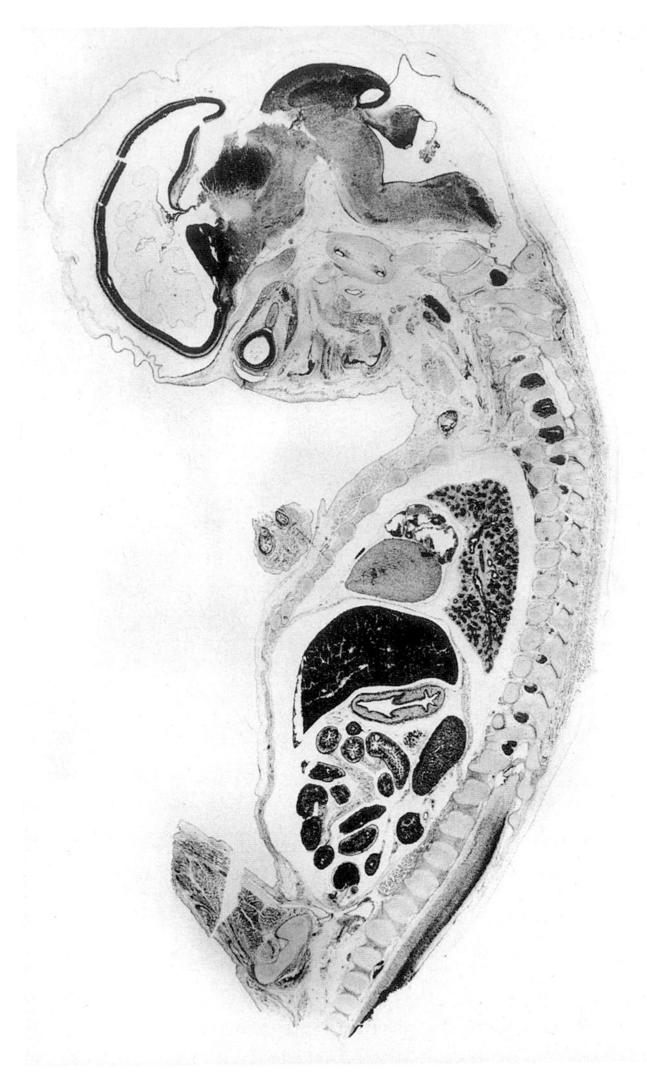

Fig. 3.4 Axial skeleton of the human embryo.
At an early stage of development, the formation of the somites with the segmentation of the spine and the appearance of the intervertebral discs, become evident. Reproduced by courtesy of Dr K Oates.

interstitial, partly appositional.[152] Interstitial growth is central to the continued prolongation of limb bones during childhood and adolescence. Centres of ossification appear and complex cycles of endochondral ossification begin.[140,169] The embryology of the synovial and intervertebral joints is analysed by O'Rahilly and Ernest Gardner.[129]

Joints and articulations

A joint is a junction of two or more bones, especially one which allows relative movement to occur between them.

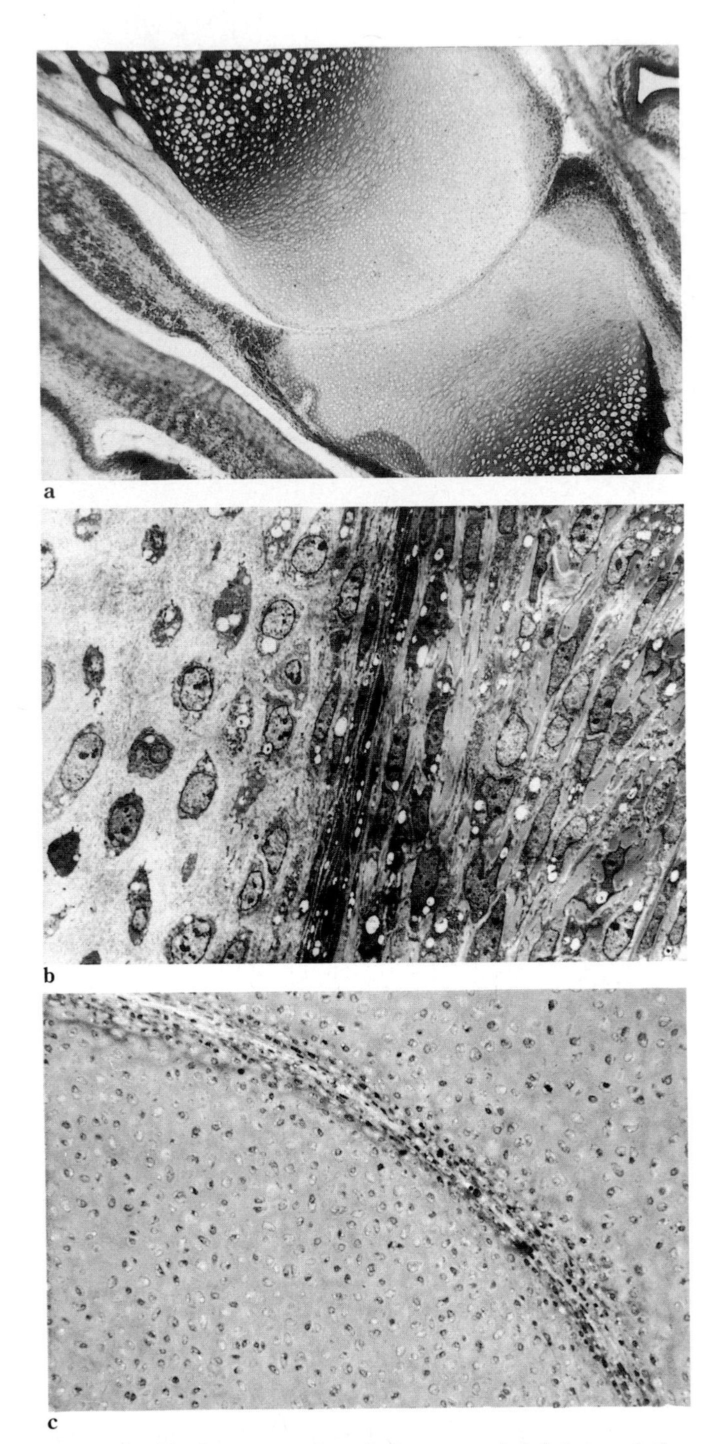

Fig. 3.5 Embryogenesis of the synovial joints of the chick.
A 'condensation' of the mesenchymal tissue located midway between the young bones of the shaft of the leg (**a**) is associated with a change in the morphology and function of the cells at this site: they assume a fibroblast-like appearance (**b**). The joint 'space', later occupied by synovial fluid, is first recognizable in the extracellular matrix that lies between the central array of spindle-shaped cells. The 'space' is first seen peripherally; it extends until there is a complete separation of the opposed cartilages (**c**). (Toluidine blue: **a.** × 29; **b.** × 620; **c.** × 164.) Courtesy of J Pidd.

An articulation is a joint between two skeletal elements, whether of bone or of cartilage. The terms joints, arthroses or articulations are often exchanged indiscriminately and the *Shorter Oxford English Dictionary* (SOED) (1985) does not distinguish between them.

The joints (Table 3.2) can be divided into three categories: fibrous (fixed: synarthroses); cartilagenous (slightly movable: amphiarthroses); and synovial (freely movable: diarthroses).[169] A scheme for the systematic examination of a joint is given in Fig. 3.6. Many aspects of joint structure and function are reviewed by Sokoloff.[147]

Table 3.2

Classification of joints

Fibrous (fixed: synarthrodial) and cartilagenous		
Synchondrosis	*e.g.*	Sphenooccipital (young)
Suture	*e.g.*	Skull sutures
Schindylesis	*e.g.*	Vomer-rostrum
Gomphosis	*e.g.*	Socket of tooth
Syndesmosis	*e.g.*	Dorsal part of sacroiliac
Symphysis	*e.g.*	Intervertebral; manubriosternal
Synovial (freely movable: diarthrodial)		
Plane*	*e.g.*	Carpal; intermetatarsal
Hinge (ginglymus)	*e.g.*	Interphalangeal
Pivot (trochoid)	*e.g.*	Proximal radioulnar; atlantoaxial
Condylar	*e.g.*	Knee; paired temporomandibular
Ellipsoid	*e.g.*	Metacarpophalangeal
Saddle (sellar)	*e.g.*	First carpometacarpal
Ball and socket (spheroidal)	*e.g.*	Hip

* The term is an approximation since no mammalian joint is plane in a strict geometrical sense.

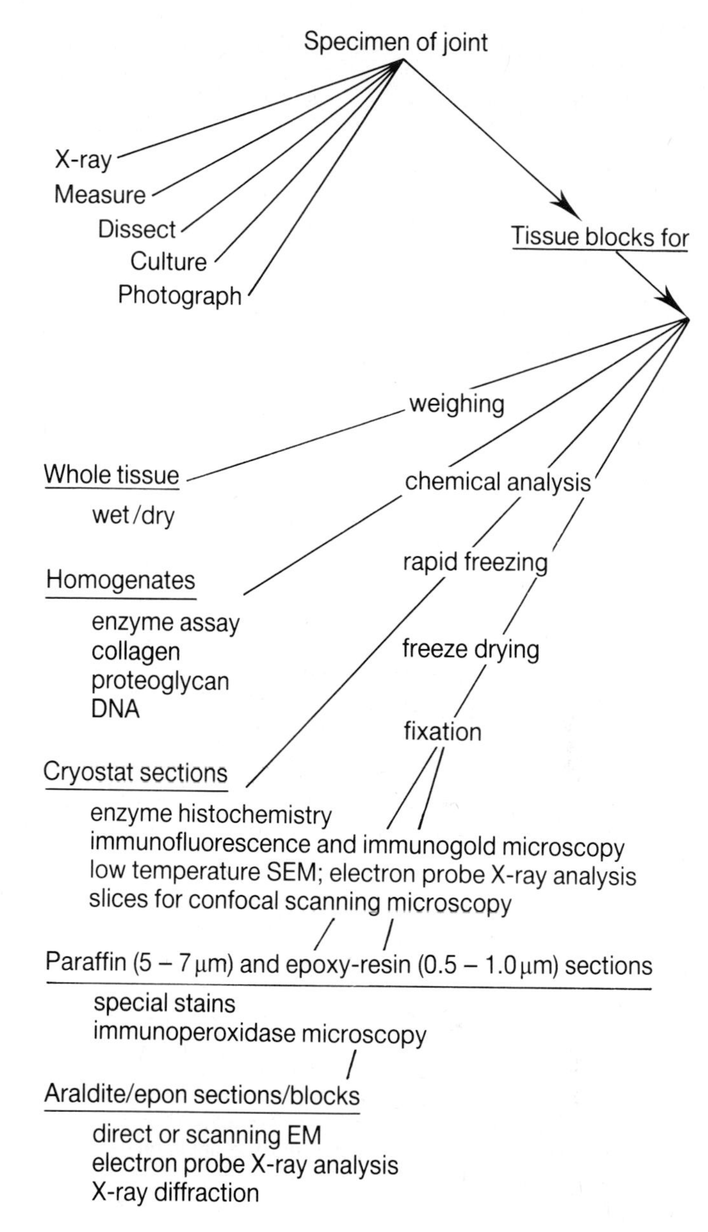

Fig. 3.6 Outline of scheme for the systematic pathological examination of a synovial joint.
A comparable approach may be adopted in the dissection and analysis of any connective tissue structure.

Fibrous and cartilagenous joints

Fibrous tissue or cartilage is invariably present where skeletal components are still being formed. Until growth ceases, sutures and synchondroses always include fibrous and cartilagenous elements. Because of their importance in pathology, particular attention is given here to the main intervertebral joints with their fibrocartilagenous discs.

Synchondroses and other non-synovial joints

Synchondroses are formed of hyaline cartilage: they include parts of the young sacroiliac joints; the temporary articulations between the epiphyses and diaphyses of the immature, postcranial skeleton, and the chondrocranium. *Sutures*, limited to the skull, are separated during growth by sutural periosteum and by a central, loose, vascular, connective tissue in which there may be islands of cartilage. A *schindylesis* is the articulation of a ridged bone where it fits into a groove. The term *gomphosis* is applied to a fibrous articulation, for example between the maxilla or mandible and a tooth. *Syndesmoses*, the unions between bones bound together by an interosseous ligament, are rare in mammals; in man, the inferior tibiofibular joint is the only example.

The syndesmophytes of ankylosing spondylitis (see Chapter 19) are analogous, pathological formations.

Symphyses

Symphyses are fibrocartilagenous joints. The deformation of a connecting pad or disc of fibrocartilage allows limited movement. Although symphyses are permanent, they undergo changes with age: the manubriosternal symphysis, for example, may be transformed to a bony synostosis. Movement is then not possible. The symphyses are important sites for connective tissue disease (see Chapter 19).

Intervertebral joints It is convenient to describe here the characteristics of the intervertebral discs. Fibrocartilage, the principal component of mature discs, is described on p. 82. The intra-articular discs themselves are briefly discussed on p. 103 and diseases of the spine in Chapter 23.

The main intervertebral joints are those between the vertebral bodies. The articulations are fibrocartilagenous. The 24 vertebrae are separated by 23 intervertebral discs the last of which is interposed between the fifth lumbar vertebra and the sacrum. The vertebral bodies and intervertebral discs share a common embryological origin. The shape and thickness of the intervertebral discs vary in different regions and largely determine the cervical and lumbar posterior and the thoracic anterior spinal curves which are respectively convex and concave. The discs are bound to the adjacent residues of the vertebral epiphyseal growth zones. In the adult, a plate of hyaline cartilage separates the disc from the vertebral body and is coextensive with both.

The intervertebral discs are composed of an outer concentric onion-skin-like array of helical type I collagen fibre bundles, the annulus fibrosus, the main part of the fibrocartilage, and a relatively small central or paracentral island of soft, myxoid tissue, the nucleus pulposus (see p. 306) (Fig. 3.7). In the annulus, fibroblast-like cells are scanty but they are more numerous near the superior and inferior surfaces than deeply. In the nucleus, cells are more numerous and often of stellate or polygonal shape.

The intervertebral discs have features in common with other discs such as those of the inferior radioulnar and sternoclavicular joints. By contrast with the complex range of movements characteristic of the synovial joints (see p. 128), those of the main intervertebral joints are limited to anteroposterior and lateral flexion, with restricted rotation. With the exception of the disc periphery, the intervertebral discs have no direct arterial blood supply but receive oxygen and metabolites[114] by diffusion from the vessels of the nearby cancellous vertebral bone, through the adjacent bony endplate and intervening hyaline cartilage.

Mechanical aspects of the intervertebral disc are considered on p. 274. The physical characteristics of the intervertebral discs are very clearly displayed when the spine of a child is divided *post mortem* and examined at once in longitudinal section (Fig. 3.8). Compression or distraction in a vertical plane cause the nuclei pulposus to protrude or intrude, respectively. Bending and rotational forces display the extreme flexibility of the vertebral column. Excised from the spine and placed alone in physiological saline, an intervertebral disc swells. The collagen fibre network in the freed invertebral disc is much looser than in the intact spine and the tension is not sufficiently high to oppose the swelling pressure of the proteoglycans. *In vivo*, a disc is always loaded;[113] the swelling pressure is resisted not only by the collagen fibre network which is very weak in the nucleus pulposus but also by body weight, by muscular contraction and by ligamentous and tendinous collagen. The capacity of the intervertebral discs to swell can be observed when a block of the vertebral column is prepared: immersion in formal saline causes the nuclei pulposus to protrude quickly before fixation takes effect. This nuclear swelling can be observed experimentally and extends to as much as 9 g water per gram dry tissue—i.e. twice the normal volume, a figure very much greater than the 1–2 per cent swelling observed under similar circumstances in hyaline cartilage. Unrestrained swelling of the disc is associated with loss of proteoglycans, which are leached out. Swelling of the divided disc can be minimized by immersion of the free, excised disc in polyethylene glycol.

Intervertebral disc ultrastructure The majority of electron microscopic studies of human intervertebral discs have been by scanning electron microscopy (SEM).[92–94,157] The closely packed collagen fibrillar architecture of the annulus and the chondroosseous endplates of the adult, develop in the seventh month of embryonic life and are complete by term.[82] No fibrillar connection between the hyaline cartilagenous endplate and the underlying vertebral subchondral bone is seen but oblique fibres connect the inner third of the annulus with the hyaline cartilagenous endplate. In the outer third, the lamellar fibrillar bundles are firmly secured to the vertebral bodies. Transmission electron microscopy (TEM)[86] of human fetal annulus fibrosus fibrils reveals, surprisingly, that the collagen fibres do not increase in diameter from the time of first deposition, at 10 weeks, to at least age 24 weeks. The diameter of the collagen fibres is comparable with that of other neonatal tissues in other species. Transmission electron microscopy also confirms that elastic fibres are associated with collagen fibrils in the fetal annulus.

Intervertebral disc composition (see Chapter 4) The biochemical, biophysical[113] and morphological aspects of

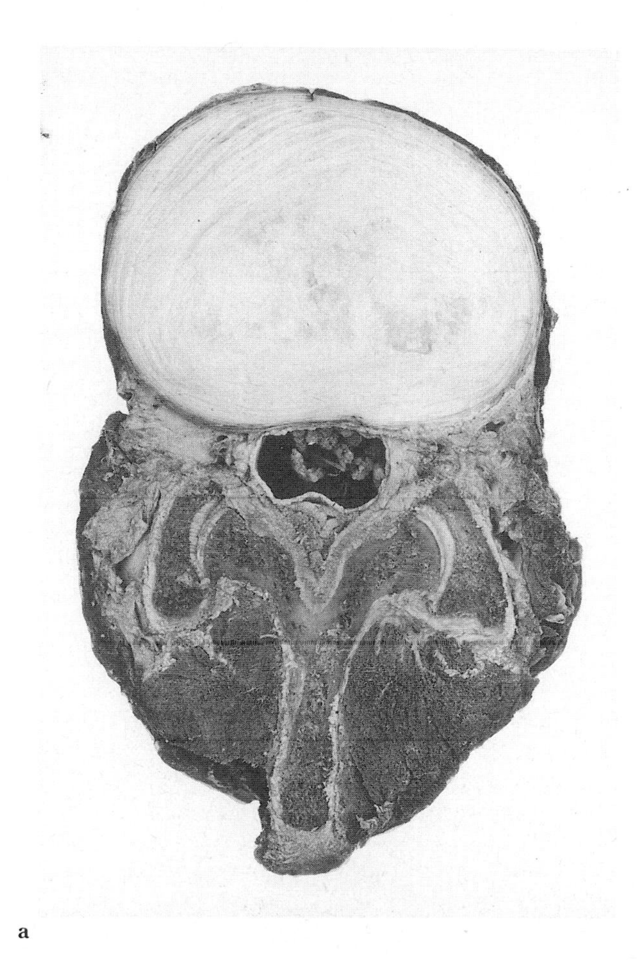

a

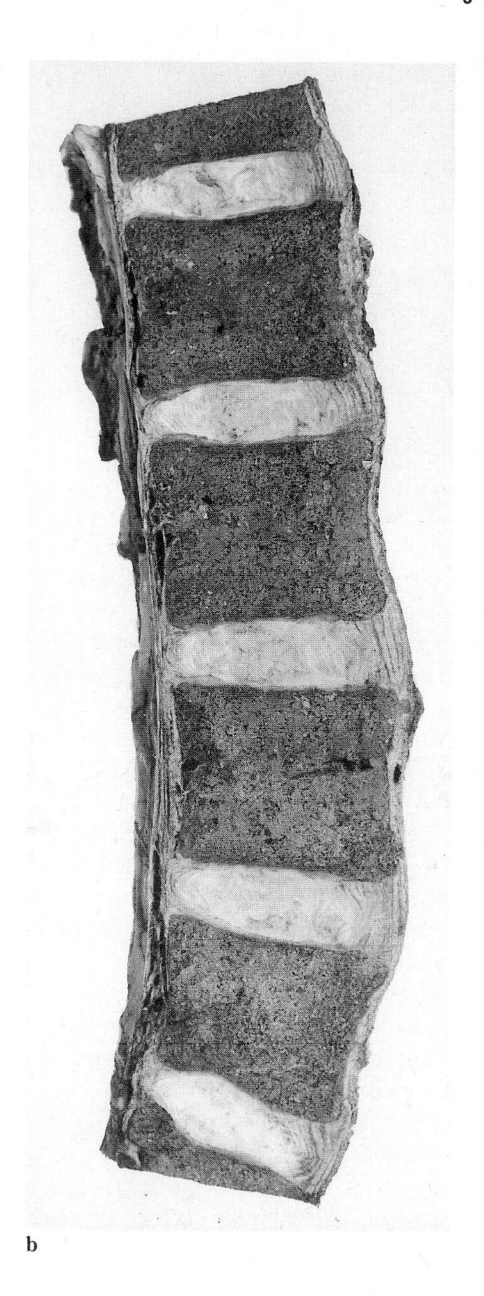

b

Fig. 3.7 Intervertebral disc.
a. A lumbar disc and the pedicles and spinous process of the vertebra are shown in horizontal section. Note the whorled character of the collagen bundles of the annulus fibosus. Immediately posterior to the disc is the spinal canal, posterolateral to which are the synovial facet joints. **b.** Sagittal section of lower thoracic and upper lumbar vertebrae. The anterior (right) and posterior (left) longitudinal ligaments unite the vertebrae and are coextensive with the periphery of the invertebral disc.

intervertebral disc function are still inadequately correlated.[1] Collagen predominates in the annulus fibrosus, and proteoglycan in the nucleus pulposus. Collagen and proteoglycan account for approximately 80 per cent of the dry weight of the disc, the water content of which ranges from 1.8 g water per gram dry weight in the annulus of a 27-year-old individual to 4.3 g water per gram dry weight in the nucleus. The corresponding figures at 74 years are 1.5 and 2.5 g water per gram dry weight.

The intervertebral disc is rich in collagen.[72] The collagen fibrils are associated with proteoglycan.[141] The abundant collagen of the annulus is arranged in interlacing, concentric

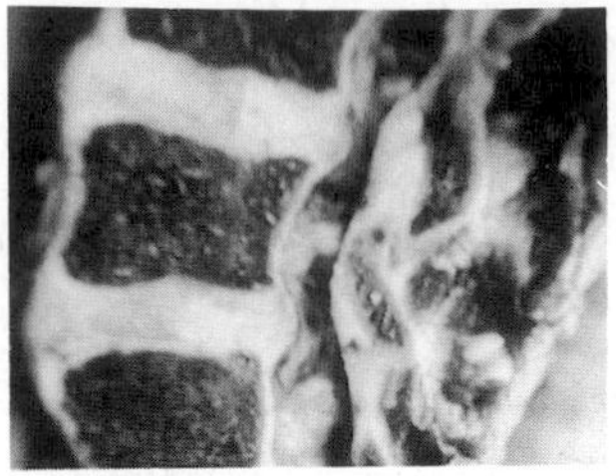
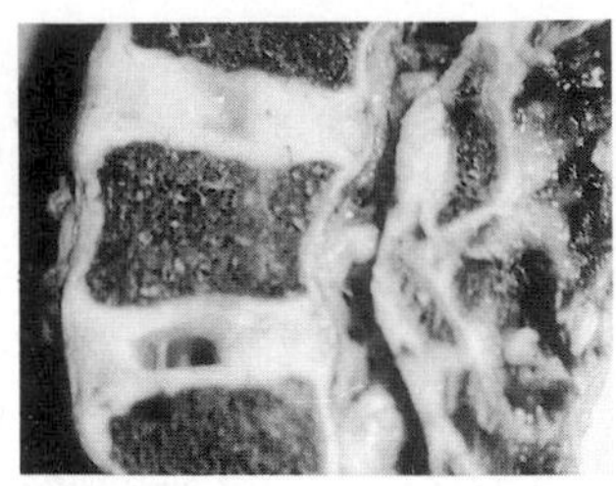

Fig. 3.8 Intervertebral disc—nucleus pulposus. The lower spine of this 7-year-old child is being compressed longitudinally (left) and distracted (right). The applied forces are causing the nuclei pulposus to protrude and recede, respectively.

bundles to give great mechanical strength; the relatively less abundant collagen of the nucleus is arranged randomly. Biochemical and immunofluorescent[10] investigations in the adult reveal principally type I collagen, especially in the outer regions, but little in the nucleus. The outer annulus indeed resembles a perichondrium. Anti-type II collagen antibodies bind to the nucleus pulposus and the whole of the annulus except the outer lamellae.[10] In the newborn, however, type II collagen is restricted to the nucleus and it is of interest that the notochord synthesizes type II collagen. There is therefore an inverse relationship between the amounts of types I and II collagen in the annulus relative to the nucleus. Small amounts of type III collagen, and the 'minor' collagens are also now known to be present in adult discs. The method of preservation of the annulus affects the arrangement of collagen: fixation causes collagen molecules to become more closely packed within the fibrils.[85]

The proteoglycans of intervertebral discs,[19] like those of hyaline cartilage, contain keratan sulphate and chondroitin sulphate attached to a protein core; they may aggregate to hyaluronate. Although the protein core—again, like that of hyaline cartilage—has three regions, one lacking glycosaminoglycan, one rich in keratan sulphate and one rich in chondroitin sulphate, the disc glycosaminoglycan side-chains contain more keratan sulphate and protein and less chondroitin sulphate. They are substantially smaller than those of hyaline cartilage proteoglycans,[151] a difference not attributable to preparative degradation.[150] In the proteoglycans of human intervertebral discs, these features are mainly due to the fact that the region of core protein bearing the chondroitin sulphate chains is shorter. Subtle differences exist in humans between the proteoglycans of the disc nucleus and of the disc annulus: in the annulus a greater proportion of the proteoglycan can bind to hyaluronate. The radial distribution of hyaluronate closely follows the distribution of proteoglycan.[80] The molecular weight of intervertebral disc proteoglycans[21] decreases with age; the disc water content also declines, a change that accompanies an increase in the keratan sulphate/chondroitin sulphate ratio. The altered water content is not a result of changed proteoglycan osmotic properties. Moving from the outer annulus to the centre, there is an increasing gradient in the glycosaminoglycan concentration, an observation confirmed by measured variations in the fixed charge density.[112]

The nucleus contains solvent-extractable lipid approximating to 0.6 per cent of the tissue dry weight for dry extractions and 1.8 per cent for wet extractions. These figures are comparable to those for the vertebral hyaline cartilage endplate, which, however contains less water. Elastic fibres have been identified within the intervertebral disc[18] but the proportion is low. It is assumed that the elastic material within the annulus provides the mechanical property of rubber-like, reversible deformability that complements the strength and resistance to tensile stress of the collagen.

Intervertebral ligaments Although ligaments are described on p. 99, the responses of spinal ligaments in disease cannot be separated from those of the intervertebral discs, and attention is again drawn to them here. The anterior and posterior longitudinal ligaments strongly unite the vertebral bodies and merge with the intervertebral discs. Whereas the longitudinal collagen fibres of the anterior ligament are bound firmly to the intervertebral discs and vertebral bone margins, the attachment to the intervening vertebral bone is relatively weak. The corresponding arrangement of the posterior longitudinal ligament is inside the vertebral canal: basivertebral veins, and veins draining into the anterior, internal vertebral plexuses, emerge between the collagen fibres of the posterior ligament, constituting foci of relative mechanical weakness. A complex criss-cross arrangement of ligaments divides the intervertebral foramina into openings of diverse size.

Synovial joints

The 268* human synovial joints (Figs 3.9; 3.10) are classified by their shape (Table 3.2). Anatomical classifications that are intended to explain joint physiology and mechanics also take account of: 1. the complexity of organization and number of the articular surfaces; 2. the number and distribution of the axes about which movements occur; 3. the geometrical form of the articular surfaces and the gross movements permitted; and 4. definitions of the relationships between surface geometry and the associated movements.

* The number is inconstant. For example, the neurocentral (Luschka) joints are not always present and occasionally cervical or lumbar ribs are recognized. The number of articulating surfaces greatly exceeds the number of joints.

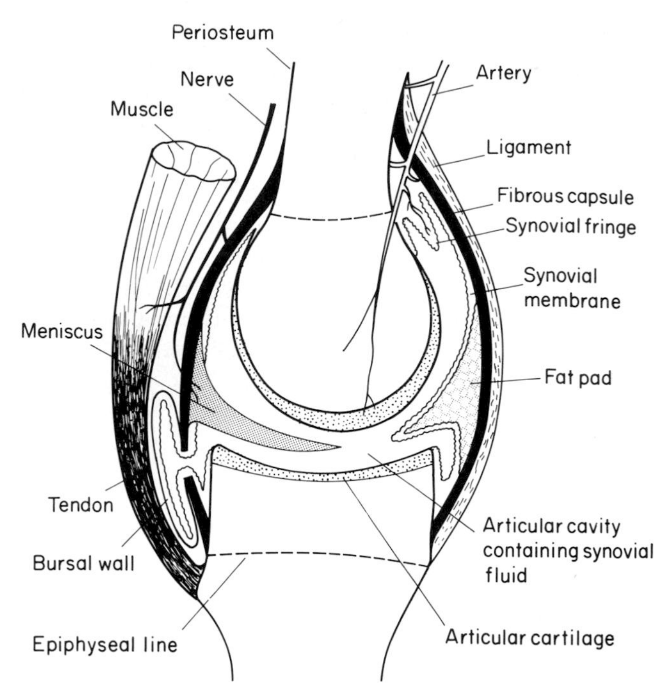

Fig. 3.9 Synovial joint: diagrammatic representation. Redrawn from *Copeman's Textbook of the Rheumatic Diseases*[142] by courtesy of Dr J T Scott and Churchill Livingstone.

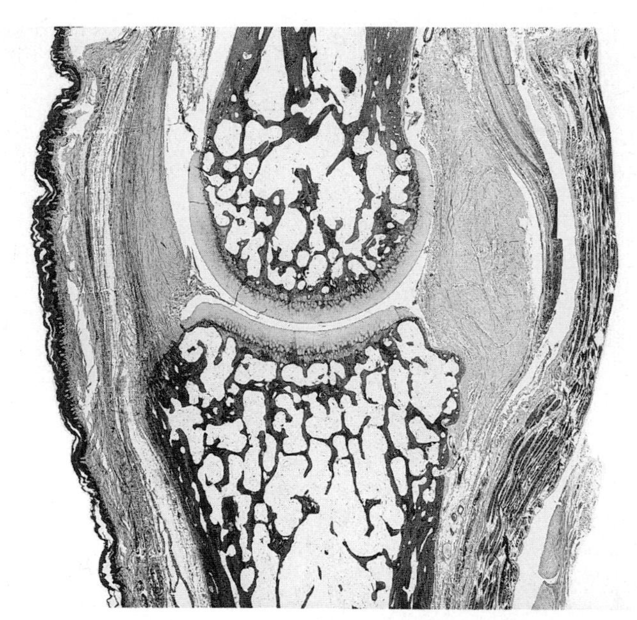

Fig. 3.10 Synovial joint in longitudinal section. Compare with Fig. 3.9. The joint is a proximal interphalangeal articulation; it has been divided in a sagittal plane so that the extensor surface is at left, the flexor surface at right. Observe the tendinous insertions of the corresponding muscles into the periarticular bone margins of the proximal phalanx (bottom). (HE × 3.3.)

Synovial joint articulation

The articular surfaces (see Chapter 3) are apposed but not in continuity. Movement is free, assisted by a very thin film of synovial fluid (see Chapter 2) that lubricates the hyaline cartilages, maintains their metabolism and may be no more than 300 nm in thickness. Many parts of the internal surface of a joint that are not in contact during movement and load bearing, are lined by the synovial membrane (see p. 84). Some synovial joints also contain discs or menisci (see p. 103). Many synovial joints have only two articulating surfaces: the surface with the greater area is 'male' and is often convex in all directions. When a joint has more than one articulating surface it is 'compound'; and where there is an intracapsular disc or meniscus the joint is 'complex'. Computer techniques can be used to construct three-dimensional models of whole, articulating diarthrodial joint surfaces using data obtained by the analysis of serial slices sawn through whole joints.[146]

Synovial joint movement

The movements of joints (Fig. 3.11) are: gliding, angular (flexion and abduction/adduction), circumduction and rotation. Articular surfaces are often ovoid. In these joints, only one 'close-packed' position yields a precise fit; in all others, fit is imperfect or 'loose-packed'. There are, in addition, 'accessory movements' some of which are passive: many movements are limited by the tension of ligaments and by

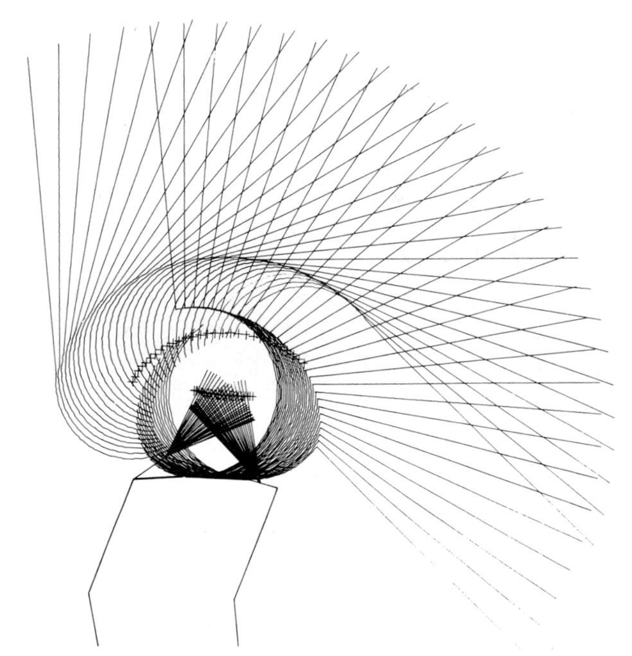

Fig. 3.11 Computer reconstruction of knee joint movements.
Courtesy of Dr J O'Connor.[123]

the approximation of soft tissues as well as by the antagonistic contraction of muscles.

The main movements of one member of the articulating bones of a synovial joint can be considered as rotations around one or more of three mutually perpendicular axes taken to be those of the vertical, transverse and anteroposterior planes of the body. Movement limited to rotation about a single axis is uniaxial; it has one degree of freedom. Movements independently around two axes have 2 degrees of freedom; and so on. At a ball-and-socket joint such as the hip (Figs 3.12; 3.13), movements are apparently multiaxial but they are, of course, within the three convential degrees of freedom. The continuously variable changes of the radius of curvature of all joint surfaces, none are planar, requires that an 'axis' is, in practice, the mean position of a continuously variable axis. Simple movements in one plane are often compound, e.g. rotation often accompanies flexion. In addition, many movements include some translations in which one articular surface slides across its partner. Translational movements are characteristic of the knee and shoulder.

In so-called plane joints, limited gliding takes place without angular movement. In the vertebral arches, however, overall arching and twisting in the long axis are added. Joints that allow angular movements in a single plane only (uniaxial) include the hinge (transverse axis) and pivot (trochoid) (longitudinal axis) articulations. In condylar joints (Figs 3.14; 3.15), movement is principally around one axis but restricted movement around a second axis is possible: a pair of convex, knuckle-shaped condyles is seen to articulate with an opposed pair of appropriately concave condyles. When movement is permitted in two mutually perpendicular axes, a joint is said to be biaxial. In ellipsoid joints, the surfaces are oval, the opposed surfaces convex and concave: flexion/extension and abduction/adduction movements are possible, as at the wrist, but not rotation. In saddle joints, rotary movement is permitted but only when flexion/extension or abduction/adduction occur: the rotary movement is 'conjoint'. The ball and socket joints offer the greatest mobility, allowing flexion/extension, abduction/adduction and rotation.

Given pairs of opposed, articulating surfaces that fit perfectly in only one position, full congruence is achieved when the joint is at the extreme limit of an habitual movement. In the knee, for example, this is achieved at full extension when the tensions within the ligaments and capsule are greatest (Figs 3.16 and 3.40). A final, spiral

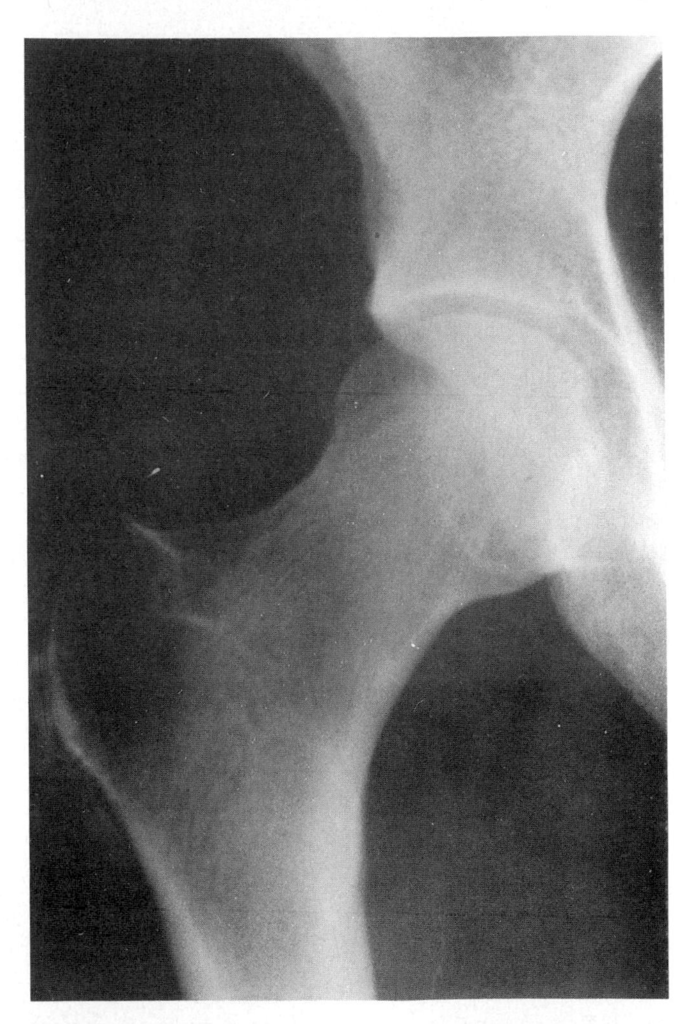

Fig. 3.12 Synovial joint: hip.
The relatively radiolucent space between the acetabulum and the femoral head is often loosely described by the radiologist as the 'joint' space. It is, in fact, the interosseous space. The true intercartilagenous joint space may be no more than 300 nm in width *in vivo*. The site and extent of the true space can only properly be demonstrated by contrast or double-contrast arthrography in which radioopaque media are injected into the cavity.

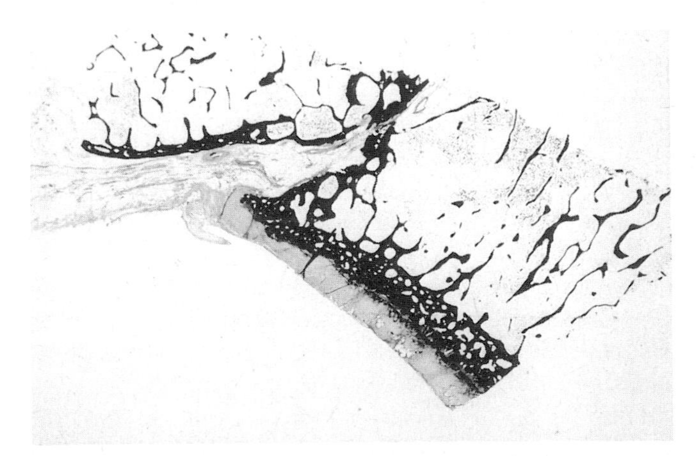

Fig. 3.13 Acetabulum.
The bearing surface of the acetabulum (right) forms a lip adjoining the soft, non-load-bearing connective tissue of the inferior part of the joint. In this example, the high density of the subchondral bony endplate is in striking contrast to the osteoporotic cancellous bone of the pelvis. (HE × 3.)

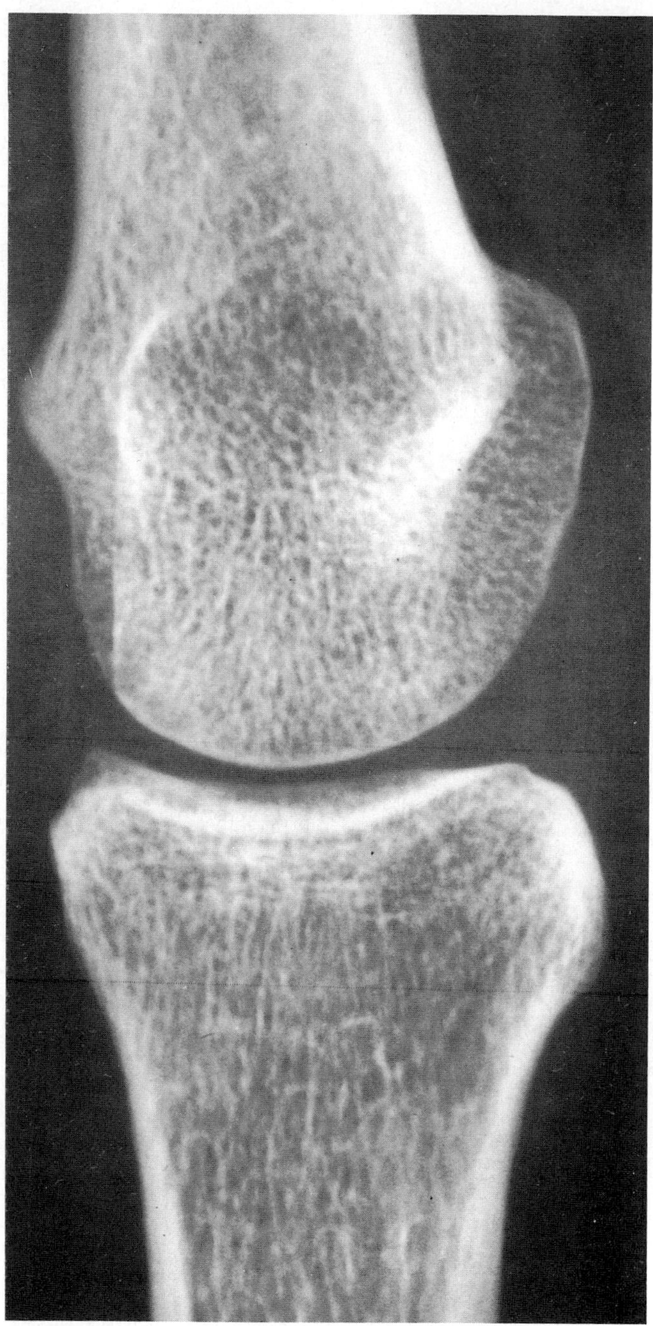

Fig. 3.14 Metacarpophalangeal joints.
The cortical and medullary bone structure and the subchondral bony endplates are shown. (Microfocal radiograph × 5.) Courtesy of Dr C Buckland-Wright.

through conjunct rotation then imposes a maximum congruence and maximum compression in a 'screwed home', close-packed position. In all the remaining joint positions, the opposed surfaces are not congruent and the joint is loose-packed; external forces can readily cause the articular surfaces to separate. Distractive change is not possible when the joint is close-packed; however, the opposing bones are then locked together and the structures comprising the joint become susceptible to trauma.

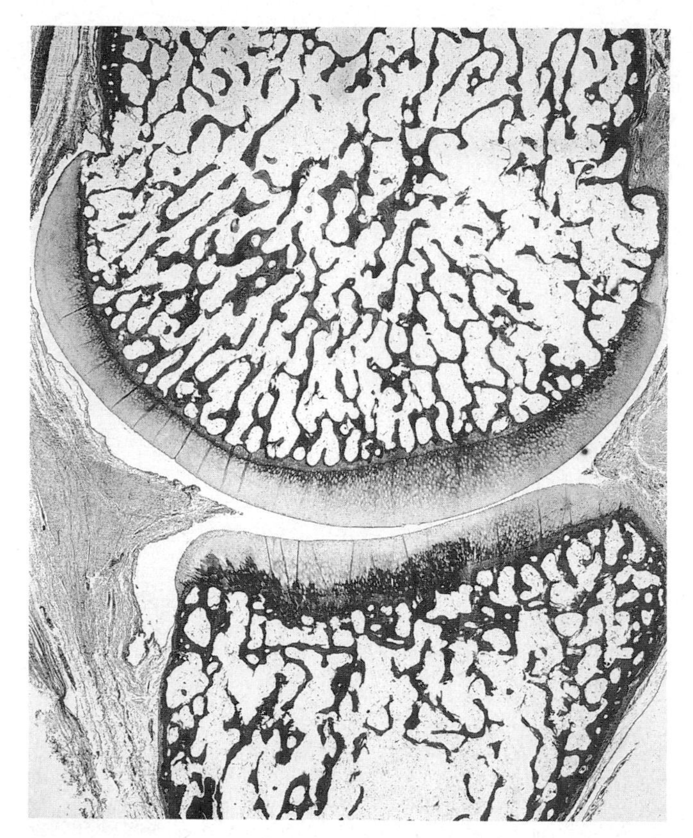

Fig. 3.15 Metacarpophalangeal joint.
Sagittal section through a decalcified paraffin block of finger. Note synovial folds (flexor surface, left; extensor surface, right). (HE × 5.)

Synovial joint size: scale effects

The range in size of synovial joints varies widely within the adults of a single species: the diameter of the human knee joint is approximately 100 times that of the incudomalleolar articulation. Large differences between individual species are evident: the elephant knee is approximately 200 times wider than that of the mouse. Thus the biological properties of hyaline articular cartilage of essentially similar microscopic structure operate efficiently in joints that differ in size by two orders of magnitude. The proportion of an articular surface in contact with its complementary, opposed surface also differs markedly: it is largest in congruent joints such as the ankle, least in condylar joints such as the knee, and ball-and-socket joints such as the hip. It is noticeable that whereas the ankle-joint cartilages are uniform in thickness, those of the hip and knee are unequal, the medial and lateral parts differing widely (Fig. 3.17). Cartilage thickness in a typical lower limb joint is directly related to the body

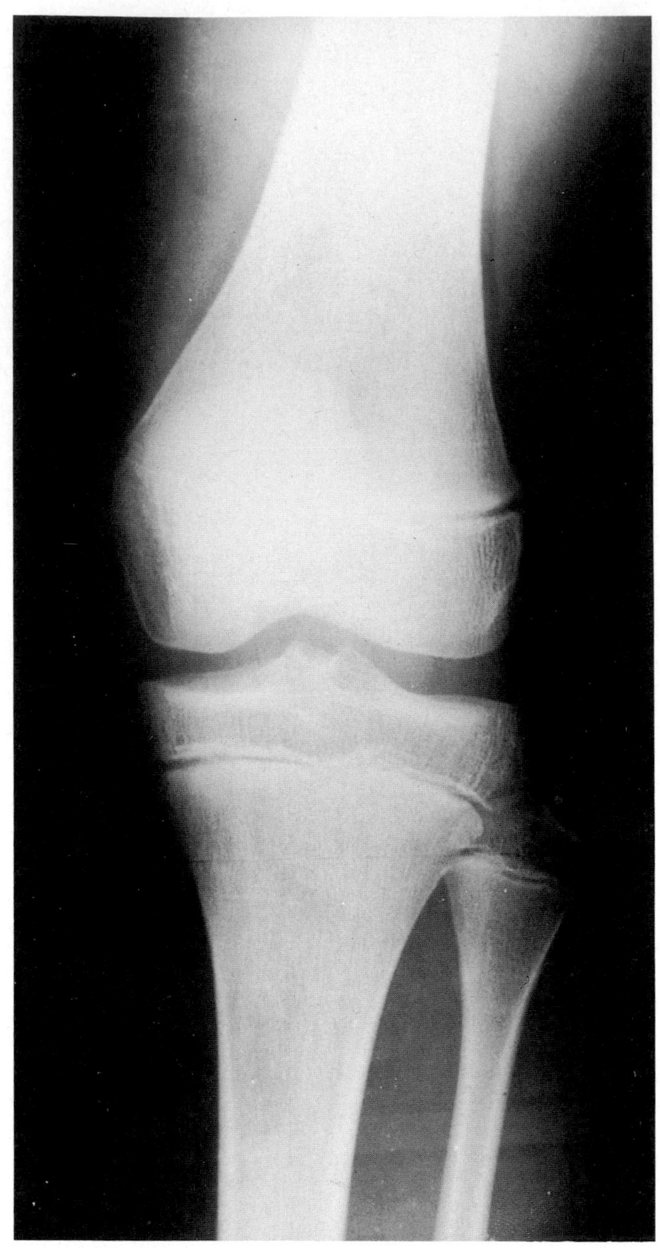

Fig. 3.16 Immature synovial joint.
Knee joint of 9-year-old child demonstrating the unfused growth zones and the epiphyseal plates.

weight, to the hip-to-shoulder distance and to the area of the tibial plateau. A 'law of simple allometry' relates the growth of parts of an organism, so that the thickness and surface area of articular cartilages are 'scaled' from larger to smaller animals.[145] Correcting for scale and comparing large and small animals, it is evident that the compressive stresses on the cartilage of a particular bearing surface generally lie within one order of magnitude of one another; the stresses are not related to the thickness of the cartilage. Limb structure appears to be optimally designed to maintain low static forces on the bearing surfaces and it is of particular interest that, although the volume of extracellular matrix varies considerably, the number of chondrocytes deep to 1 mm^2 of articular cartilage varies little between different joints and between different species (Table 2.3).

Some specialized synovial joints

Whereas the general structure of the larger and smaller limb joints can be understood by reference to a stylized diagram (Fig. 3.9), there are a number of joints of particular importance in pathology which call for individual consideration. They include the temporomandibular; the atlantoaxial; the atlantooccipital; the uncovertebral and the sacroiliac. In addition to the main intervertebral joints, the vertebrae articulate with one another by pairs of posterolateral (apophyseal) joints which are synovial (see p. 84). The typical ribs articulate with the vertebrae by synovial costotransverse joints and by the joints of the heads of the ribs. In the case of the first, tenth, eleventh and twelfth ribs, each articulates with one vertebra: the remainder articulate with two vertebrae and each joint is divided by an intraarticular ligament.

Temporomandibular

The temporomandibular articulation is complex (Fig. 3.18). The left and right joints function together. Opening of the mouth is permitted by anterior movements of each mandibular condyle across the articular surfaces of each temporal bone, movements facilitated first by contact with the lower surfaces of the fibrous articular discs that divide the joints and then by movements of the discs themselves which slide forwards and downwards on the temporal bones. The movement of the discs is limited when the fibroelastic tissue that attaches them posteriorly to the temporal bones reaches the limit of its extensibility.

The condylar process and head of the mandible are often said to be covered by fibrocartilage; however, section reveals a lamina of fibroelastic tissue (Fig. 3.18c) resting on hyaline cartilage that, in turn, is bound to bone. The disc is fibrous. The temporal articular surface is fibrocartilagenous. Although, therefore, the temporomandibular joint is diarthrodial and has an effective marginal synovial membrane, the histopathological changes by which it is affected do not exactly correspond with those in, say, the knee joint. Repetitive injury and forward dislocation result in a disorganization of fibrocartilage that is analogous with but not identical to the fibrillation of hyaline cartilage in osteoarthrosis (see Chapter 22). The synovial lining is, nevertheless, susceptible to the common inflammatory joint diseases including rheumatoid arthritis and juvenile chronic polyarthritis (see Chapter 13).

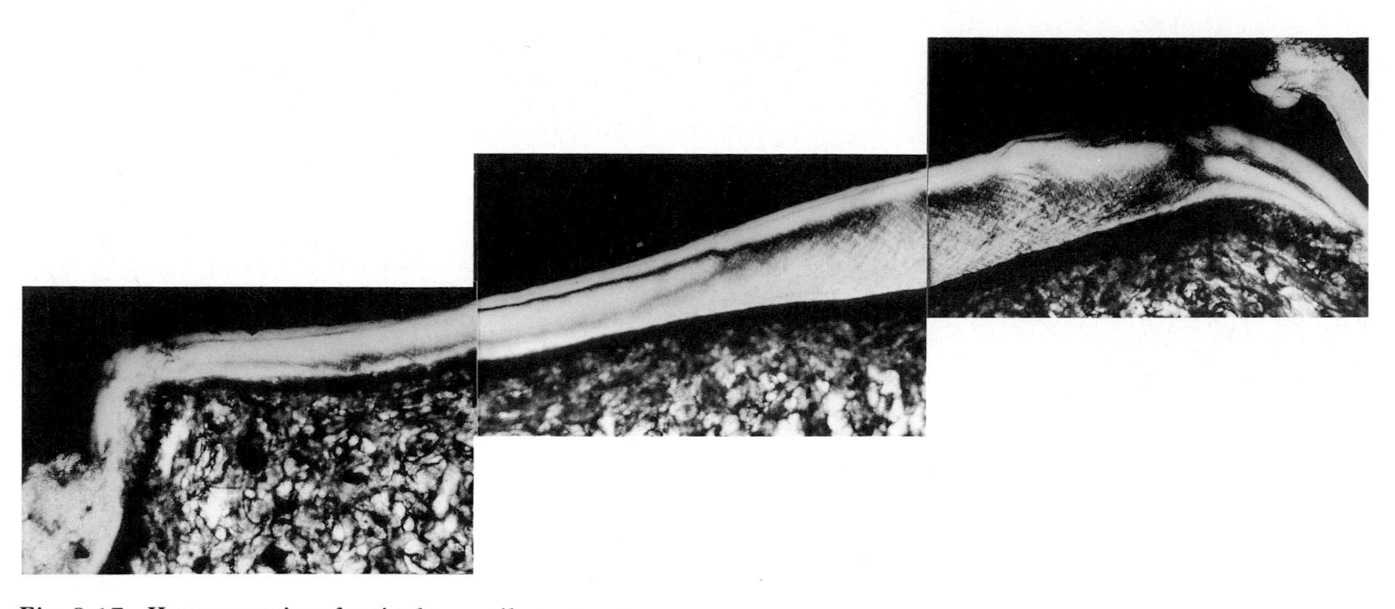

Fig. 3.17 Heterogeneity of articular cartilage structure.
The complex microstructure of this tibial plateau from a young baboon is suggested by the colour patterns, seen in shades of grey, in this 200 μm-thick sawn slice. (Polarized light × 23.5.)

Atlantooccipital The condyles of the occipital bones of the skull rest upon the two superior facets of the lateral mass of the atlas (Fig. 3.19). The shapes of the articular surfaces determine that nodding movements occur around transverse planes, and lateral movements around lateral planes. There are fibrous capsules; the synovial cavities are often in continuity with the synovial bursa between the dens and transverse ligament of the atlas. As in the case of the temporomandibular joints, the two joints behave as one. They are susceptible to both osteoarthrosis and rheumatoid arthritis.

Atlantoaxial Our limited understanding of diseases of the cervical spine is in part due to the inaccessibility of this structure, and in part to the complexity of the joints and ligaments. The atlas and axis articulate via paired synovial joints between the inferior facets of the lateral masses of the atlas and the superior facets of the axis; and, in the median plane, by a third joint between a facet on the anterior aspect of the dens and the anterior arch and transverse ligament of the atlas.[144] There is a thin fibrous capsule lined by synovium around each lateral joint and strong, dense ligaments. The median joint, which is particularly vulnerable to trauma, is pivotal and double: there is a large synovial cavity, a bursa, between the anterior surface of the cartilage-covered transverse ligament of the atlas, and the posterior, grooved surface of the dens. The weight of the head is borne by the lateral atlantoaxial joints. When the head rotates, however, all three joints transmit and regulate this movement, the extent of the rotation being limited by the alar ligaments. The pathological approach to the upper cervical spine and the base of the skull is facilitated by techniques such as those of Eulderink[41] (Fig. 3.20).

Uncovertebral (neurocentral) joints of Luschka

An understanding of these minute, inconstant joints that form between the body of one cervical vertebra and the pedicular base of the subjacent vertebra, provides one key to the pathological anatomy of cervical rheumatoid arthritis. These 'acquired' joints arise during childhood or adolescence as clefts that are bounded by the fibrocartilages of the intervertebral disc annuli. A synovial lining develops: fibro- and hyaline cartilage, meniscus-like folds and a capsule form. The Luschka joints permit gliding and rotation of adjacent vertebral bodies. Since they are just inside the uncinate processes, they are thought to be phylogenetic homologues of the cervical ribs and other earlier costovertebral articulations. The uncovertebral joint synovia are the target for the inflammatory diseases of the cervical spine and they contribute to osteophytosis.

Sacroiliac (Fig. 3.21) These paired joints have a unique structure: in the young individual they are part synovial, part cartilagenous; with advancing age, particularly in the male, the joint space disappears, the surfaces fuse and the

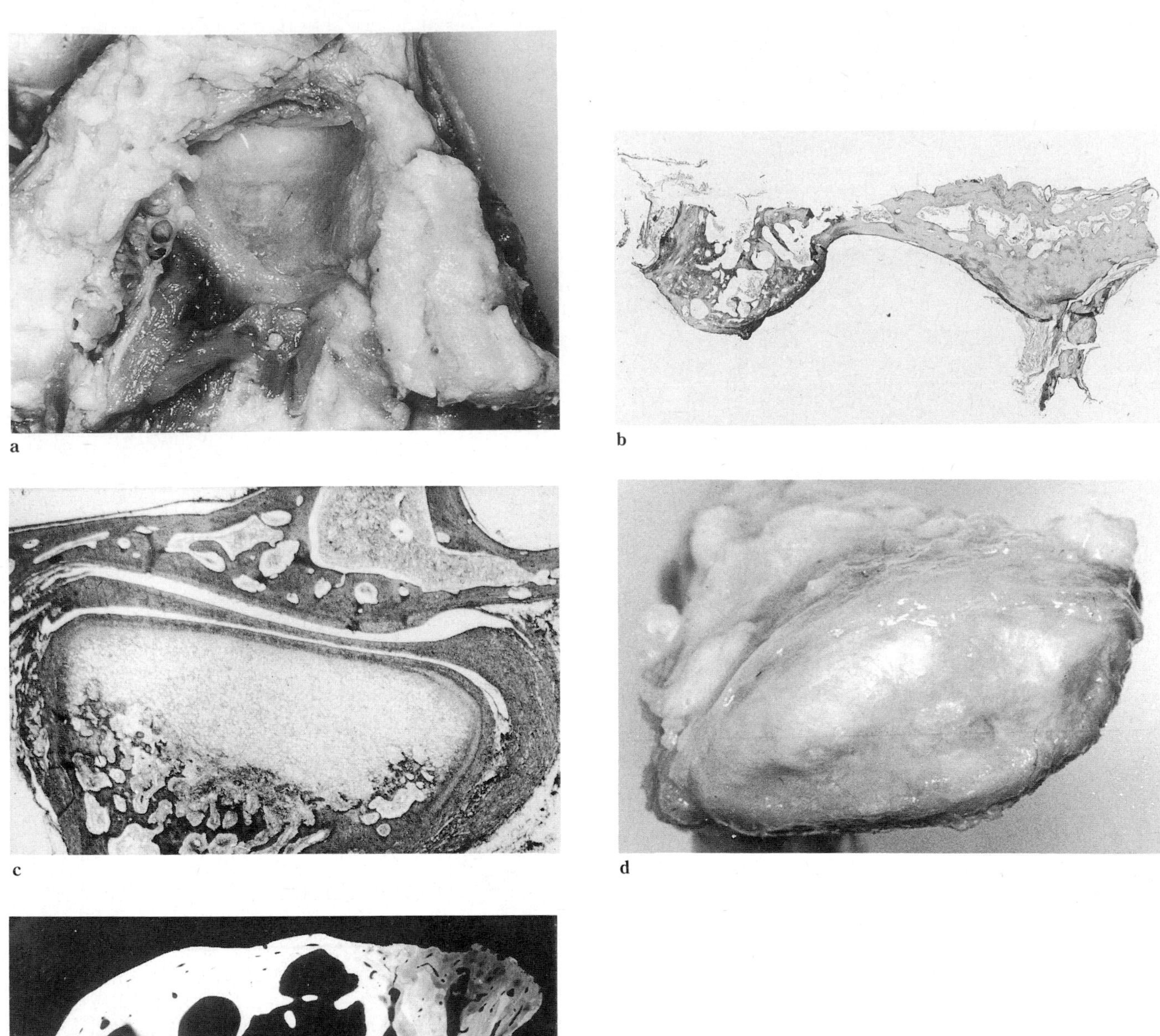

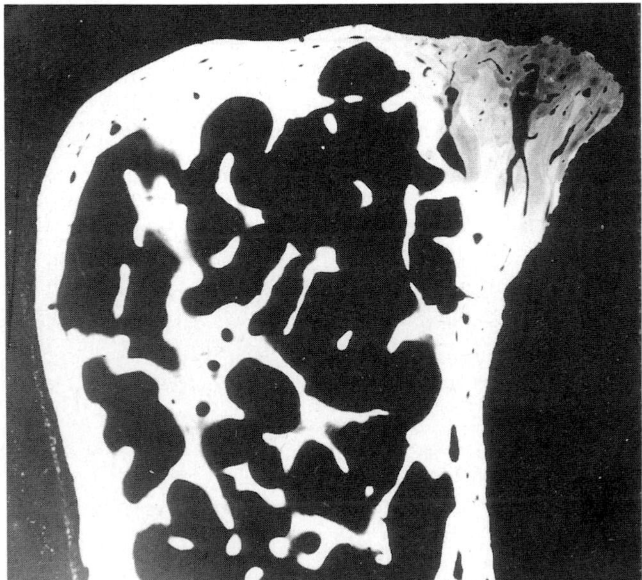

Fig. 3.18 Temporomandibular joint.
a. General view of temporal component of human joint; **b.** Sagittal section of temporal component of human joint; **c.** Section through whole temporomandibular joint of marmoset; **d.** Surface of human mandibular condyle; **e.** Microradiograph of ground section of central part of human mandibular condyle. **a**, **b**, **d** and **e** are reproduced by courtesy of Professor Gunnar E Carlsson; **c** by courtesy of Professor N H F Wilson.

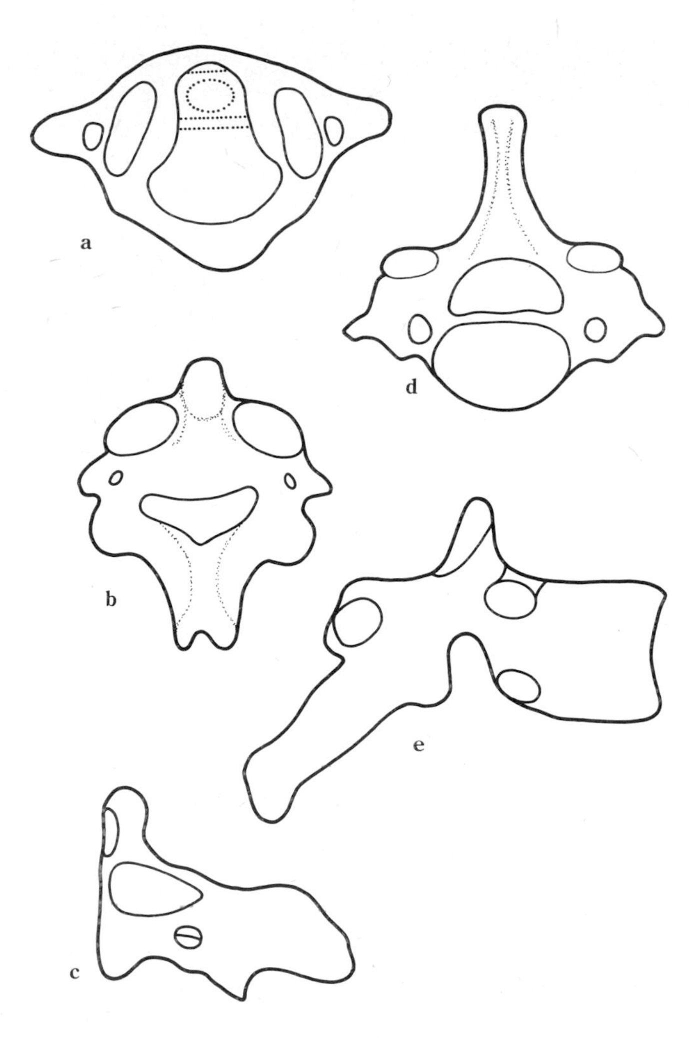

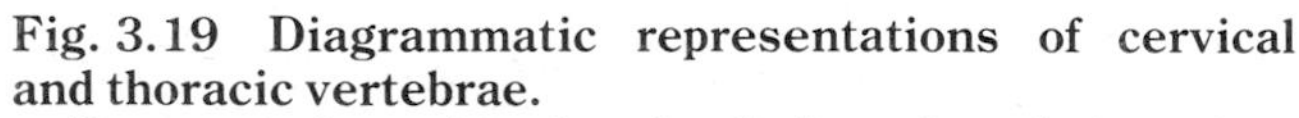

Fig. 3.19 Diagrammatic representations of cervical and thoracic vertebrae.
a. First cervical vertebra, the atlas; **b.** Second cervical vertebra, the axis; **c.** Lateral aspect of axis; **d.** Seventh cervical vertebra; **e.** A thoracic vertebra.

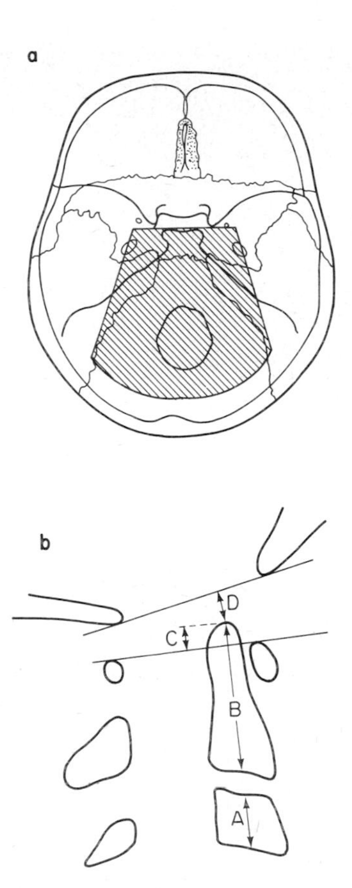

Fig. 3.20 Anatomy of skull base and cervical spine.
a. Schematic drawing of the base of the skull. Cross-hatching indicates the parts excised in preparing tissue for examination. **b.** Diagram of median plane of skull, showing occiput, atlas, axis and the third cervical vertebra. Double-pointed arrows indicate distances measured radiologically. Reproduced from Eulderink and Meijers,[41] with permission of the authors and the Editor of the *Journal of Pathology*.

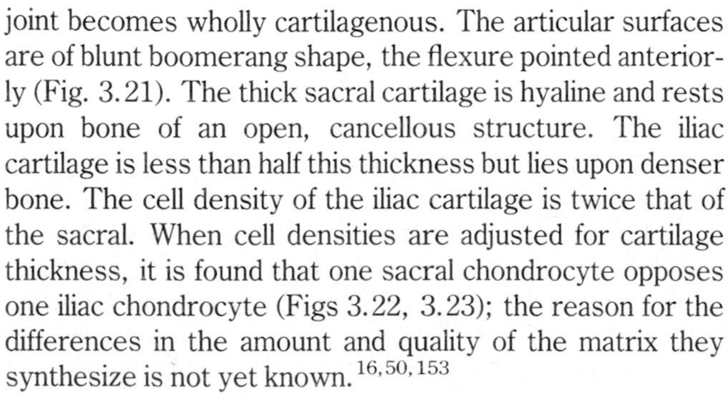

joint becomes wholly cartilagenous. The articular surfaces are of blunt boomerang shape, the flexure pointed anteriorly (Fig. 3.21). The thick sacral cartilage is hyaline and rests upon bone of an open, cancellous structure. The iliac cartilage is less than half this thickness but lies upon denser bone. The cell density of the iliac cartilage is twice that of the sacral. When cell densities are adjusted for cartilage thickness, it is found that one sacral chondrocyte opposes one iliac chondrocyte (Figs 3.22, 3.23); the reason for the differences in the amount and quality of the matrix they synthesize is not yet known.[16,50,153]

The cartilagenous parts of the sacroiliac joints, synchondroses, merge with dense, periarticular collagenous connective tissue that comes together as the ventral, interosseous and dorsal ligaments. The massive, thick interosseous ligament is the principal bond between the sacral and iliac surfaces and occupies the irregular space above and behind the joint. The anterior marginal synovial tissue is sparse and identified with difficulty. However, overtly vascular villous processes are seen to arise from subchondral bone in the main, central parts of the joint, passing in clefts through the cartilage and appearing on the surface (Fig. 3.21) as yellow, papillary islands, presumably in areas where articular contact is slight or intermittent. It is evident that sacroiliac joints are, by definition, susceptible to all disease processes mediated by the vascular circulation (inflammatory, infective, neoplastic) as well as all disorders that are initially not dependent on an active microcirculation (heritable, degenerative or metabolic).

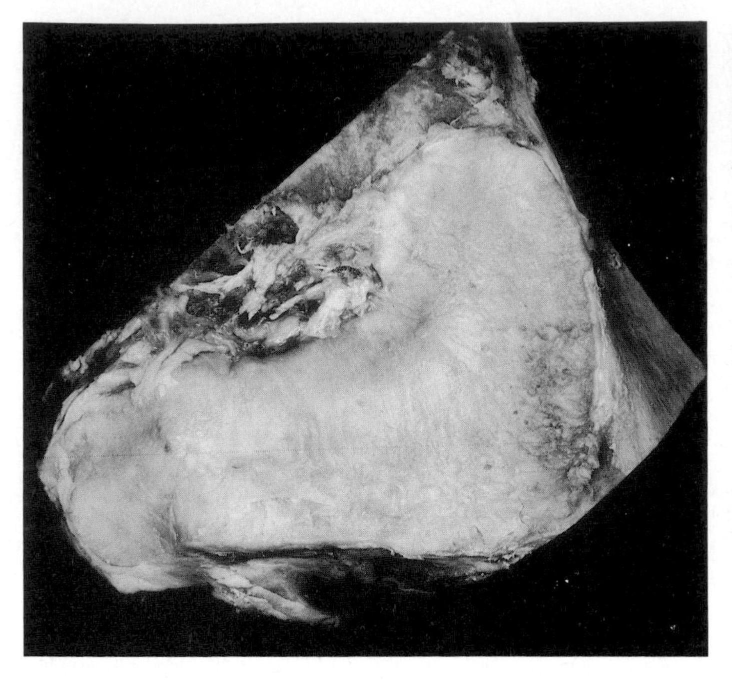

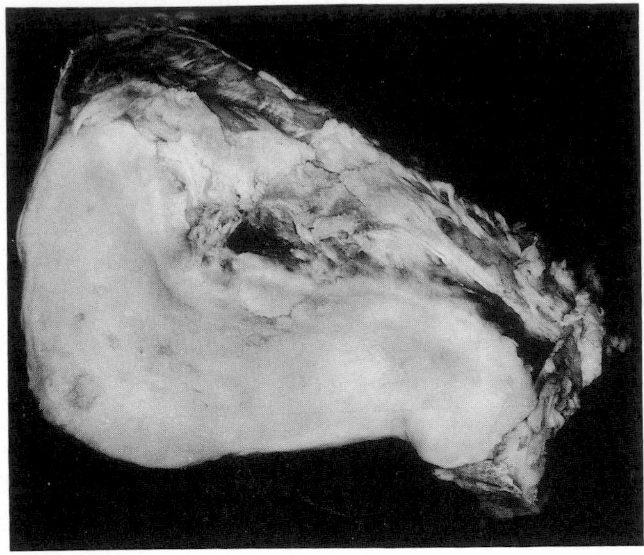

Fig. 3.21 Human Sacroiliac joint.
The iliac cartilage (at left) appears darker than the sacral (at right) because the latter is relatively thick. Observe the boomerang shape of the articulation. The apex is directed towards the pelvic cavity. (Natural size.)

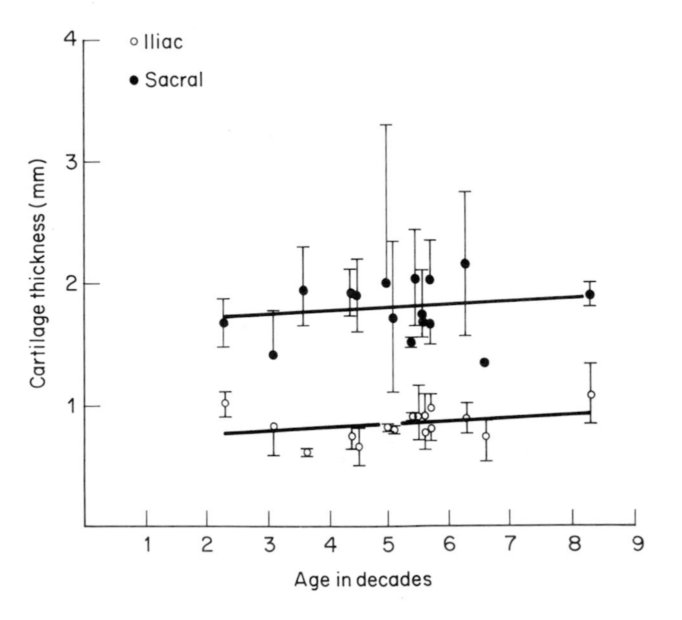

Fig. 3.22 Sacroiliac joint: thickness of cartilages.
On average, the sacral cartilage is twice as thick as the iliac. The thickness of both cartilages tends to increase with age. The articulation is apparently congruent. From Gardner, Brereton and Hollinshead.[51]

Articular surfaces

Synovial joint surfaces

None of the eight classes of diarthrodial joint has an anatomically plane (flat) surface. The high functional efficiency of these joints, facilitated by an approximately 300 nm film of synovial fluid between the opposed cartilages, suggests that under conditions of load-bearing, the contiguous surfaces are likely to be highly smooth and, over the area of contact, flat. So far, little *direct* evidence to confirm or refute this theory has become available and the organization of living, loaded articular cartilage is not fully understood.

When diarthrodial joints are *disarticulated,* the bearing surfaces, freed from loads and contact, undoubtedly change. In the examination of these surfaces, many artefacts can be introduced. For example, when cartilage is released from bone, the inherent forces within the collagen bundles cause a rapid change in shape so that surface features appear more conspicuous and artefactual wrinkles

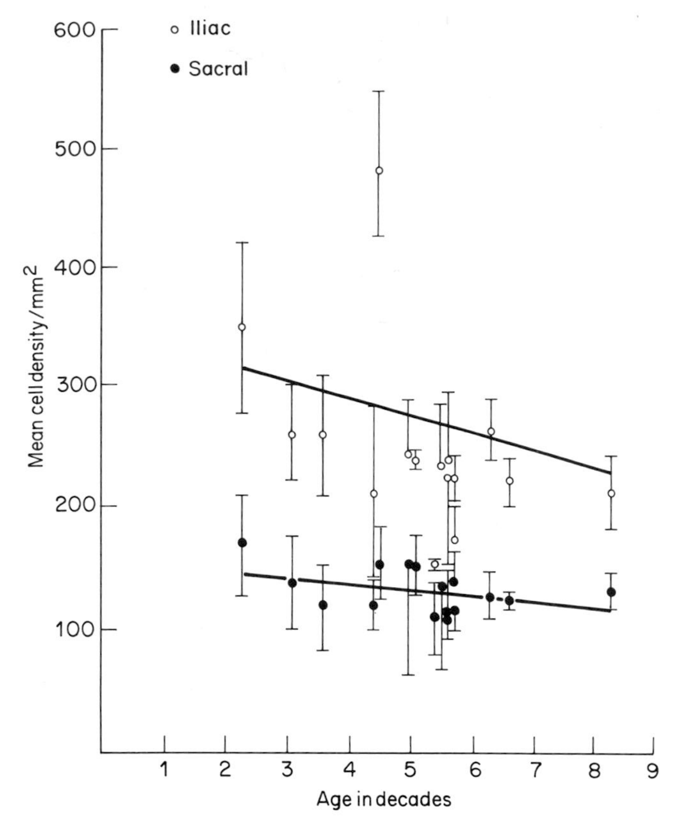

Fig. 3.23 Sacroiliac joint: cell density.
The cell density, i.e. the number of cells/mm^2, is twice as high in the iliac as in the sacral cartilage. Adjusted for cartilage thickness, the evidence shows that one sacral chondrocyte opposes one iliac chondrocyte. Each cell synthesizes a quantitatively and qualitatively distinctive matrix. From Gardner, Brereton and Hollinshead.[51]

are introduced. Excluding artefact, it appears that free, non-loaded concave and planar surfaces, particularly in the young, bear arrays of microscopic contours which are the surface representations of the most superficial chondrocytes (Figs 3.24, 3.25). In the young, the features have been described as prominences ('humps'), in the adult as hollows ('pits'). Similar morphological differences can be recognized when fixed, dry specimens are compared with frozen, hydrated material. Moreover, convex surfaces appear much smoother than concave ones. The numbers, diameters and vertical dimension of these surface features are related to age in the same way as the numbers, size and frequency of articular chondrocytes.

The microscopic shape of *loaded*, living cartilagenous surfaces, a central issue in joint mechanics (see Chapter 6), still remains uncertain. The greater part of the evidence[62,63,58,56,126,49,133] is derived from experiments made with blocks of cartilage, excised from disarticulated mammalian joints *in vitro*, sometimes retained on bone, often

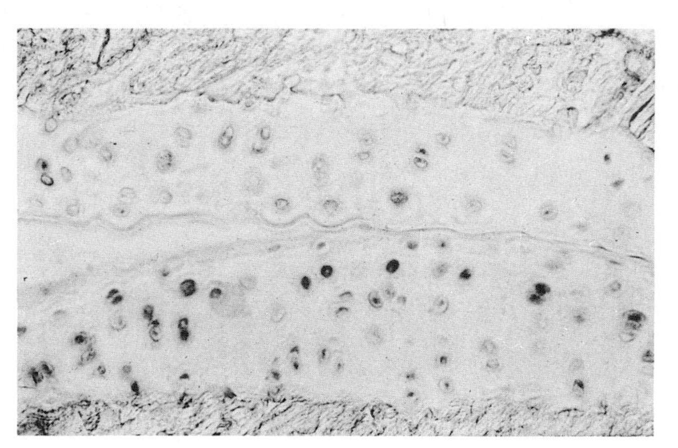

Fig. 3.24 Ankle joint of rat.
In this cryostat section of unfixed, hydrated tissue, is seen a hint of the surface prominences that characterize the planar or concave surface of the lower end of the tibia (top) by contrast with the relatively flatter surface of the calcaneum (bottom). (Tetrazolium salt for lactate dehydrogenase × 169.)

excised from it. It is difficult to extrapolate from these results to the surface structure of living, loaded surfaces.

Disarticulated joints *in vitro*

History Observations of exposed joint surfaces can be traced to William Hunter (1742–43).[90] He believed that the articular cartilage of the diarthrodial joints was vascular, smooth* and covered by a synovial membrane. However, normal adult mammalian cartilage is avascular; indeed the boundary between the arcades of terminal synovial blood vessels and the hyaline cartilage is shown in Hunter's own preparations (Fig. 3.26). Articular cartilage is not covered by a synovial membrane.[102] That the bearing surfaces of disarticulated synovial joints are smooth was a view that was not challenged until 1967–68.†

* The adjective 'smooth' means 'freedom from roughness or irregularity' (Shorter Oxford English Dictionary, 1985). It is a relative expression in the sense that 'absolute smoothness', meaning infinitesimally little roughness, is a theoretical concept analogous with 'absolute velocity' or 'absolute absence of heat'. In molecular terms, therefore, there is no possibility that articular surfaces can be devoid of roughness. If this is accepted, then debate on the 'smoothness' or 'roughness' of the cartilaginous bearing surfaces of diarthrodial joints can be resolved into two paramount questions: to what *degree* are these surfaces rough? and: are the surfaces smooth when loaded *in vivo*?

† There had been early hints that the free, non-loaded surfaces of disarticulated joints might not be 'smooth'.[55] Liston[105] and Birkett[12] made drawings showing approximately 0.5 mm surface irregularities and Hassall[83] depicted 20–30 μm prominences related to underlying chondrocytes. The microscopic irregularity of the non-loaded surface is also indicated in Virchow's[166] engravings but the first to record and comment on these appearances is thought to have been Hammar.[79]

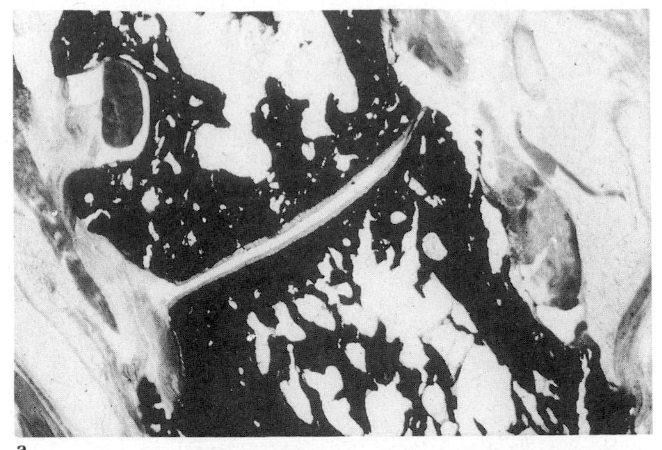

a

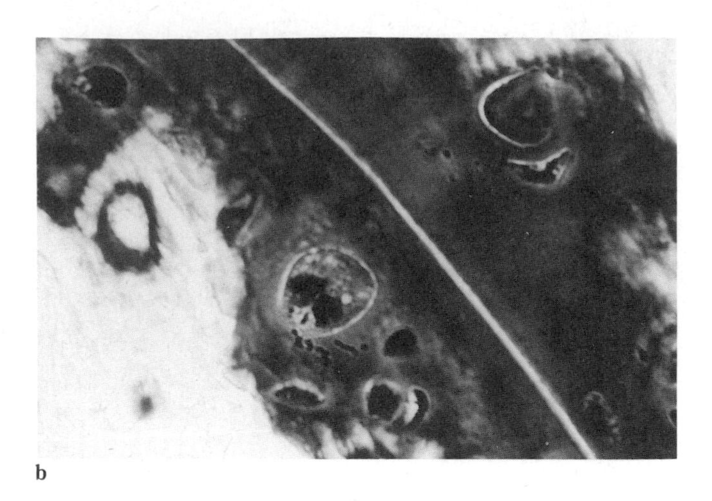

b

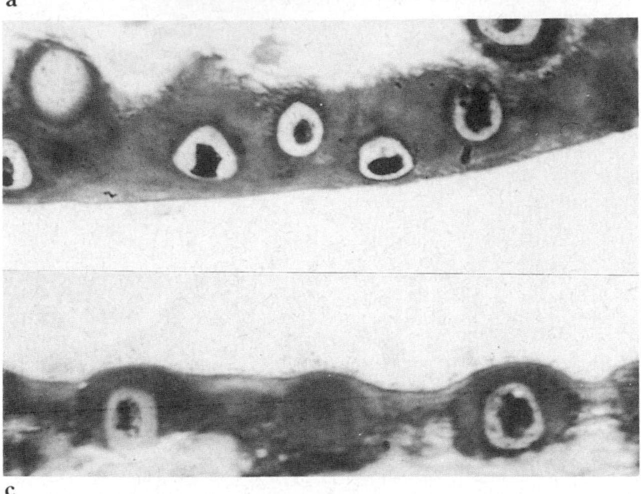

c

Fig. 3.25 Articular surface structure.
a. Tarsal joint of rat. The relatively large contribution made to each synovial joint by bone and the relatively small contribution derived from cartilage are shown. (von Kossa × 10.) **b.** Knee joint of 1-day old mouse. When the joint surfaces are in close apposition and loads are borne, the evidence suggests that the articular surfaces are relatively flat and smooth. (Toluidine blue × 874.) **c.** Knee joint of 1-day old mouse. When the joint is disarticulated, the convex surface retains a flat appearance, the concave surface becomes non-smooth because of the prominences attributable to the most superficial chondrocytes. (Toluidine blue × 752.)

The anatomical contours of joints are the basis for a system of classification (Table 3.2).[169] Superimposed on these contours, which can be said to be 'first order' or 'primary' (1^0) features (Table 3.3) (Fig. 3.19), it is possible to discern with a hand lens, irregularities approximately 0.4–0.5 mm in diameter; they are 'second order' or 'secondary' (2^0) features. When articular, bearing surfaces, and the surfaces of discs and menisci, are viewed more closely, further third order (tertiary, 3^0), fourth order (quarternary, 4^0) and fifth order (quinary, 5^0) irregularities are detectable.

Naked-eye and simple microscopic studies

The primary, anatomical contours of the surfaces of synovial joints have been defined by naked-eye inspection, by contour mapping (see p. 141) and by other imaging techniques (see p. 148). The study of sawn slices and of embedded and cryostat sections is an aid to morphology (Fig. 3.17).

When the anatomical surfaces of a joint are examined with a simple lens or dissecting microscope, they display a pattern of undulating, irregularly shaped features. On average, these second-order features are 0.4–0.6 mm in diameter. They are present in avian and mammalian species and can be detected on living cartilage examined in the operating theatre, on fresh, unfixed material and on the surfaces of fixed, dehydrated cartilage blocks, attached to or detached from the underlying bone. Their functional significance is not known. If they are present *in vivo*, it must be assumed that they influence the distribution of the forces generated in static and dynamic loading (see Chapter 6).

Compound light microscopy

There are numerous records of perpendicular sections of articular cartilage viewed by transillumination.[139,77,14,55] Fixed, dehydrated cartilage blocks were shown to have many, very shallow, third-order surface undulations which were often hollows.[118,52] The surface features could be seen to correspond to the site of zone I (superficial) chondrocytes (e.g. Fig. 1 of Meachim *et al.*[118]) but this relationship was much more easily recognized in sections of cartilage from fetal or young, postnatal animals than in adults, particularly on concave surfaces: non-loaded con-

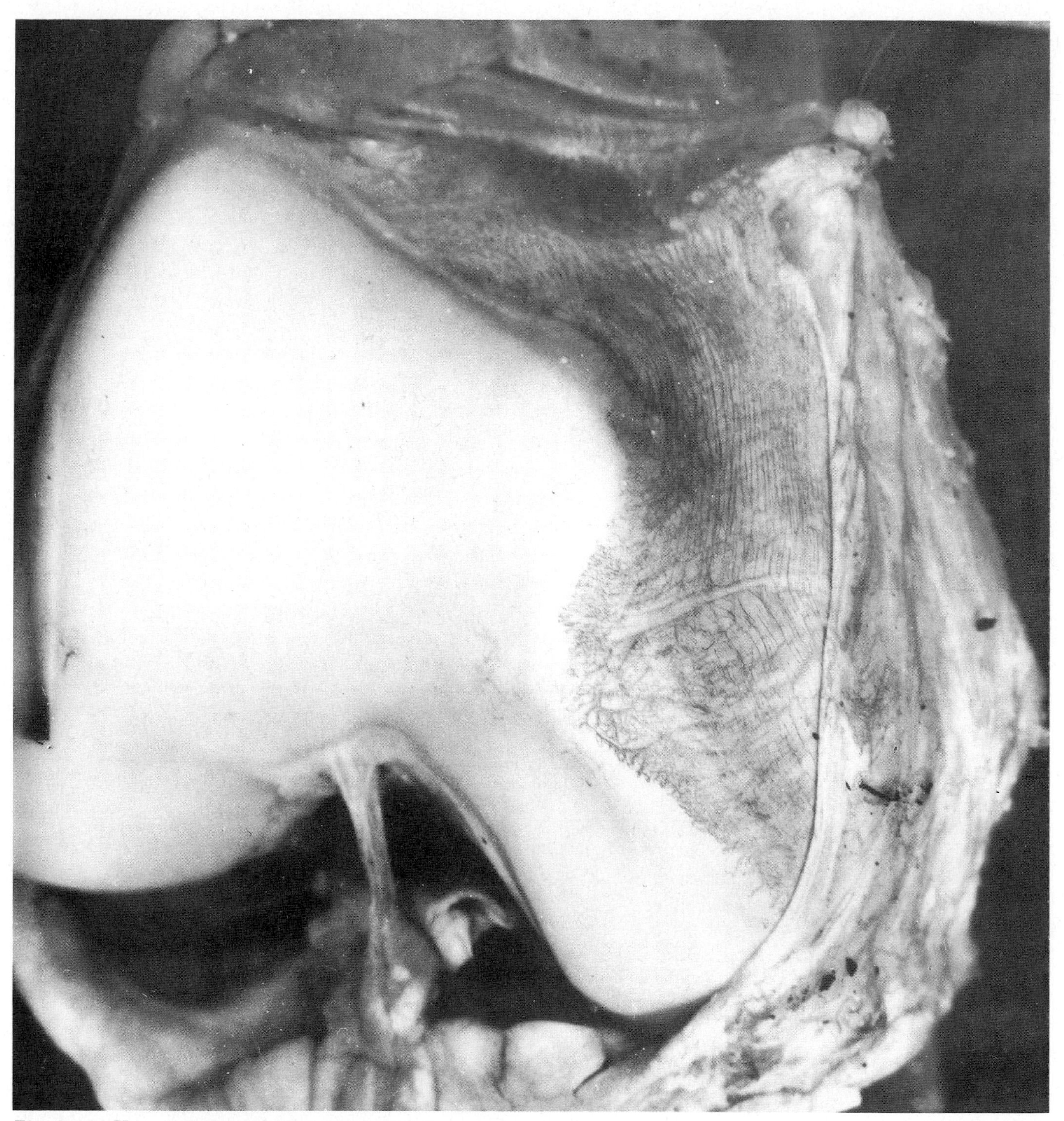

Fig. 3.26 Knee joint vasculature.
The photograph is of one of William Hunter's original preparations. The blood vessels have been injected with red lead and the delicate but sharply defined chondrosynovial junction is clearly seen as is the orderly pattern of small vessels extending over the synovial surface. Courtesy of Professor R J Scothorne, Honorary Curator of the Hunterian Collection, Glasgow.

Table 3.3

Hierarchy of mammalian articular cartilage surface contours from disarticulated joints *in vitro*

Visible by	Feature	Dimension
Naked eye	1st order (1⁰) anatomical contours	1 mm–100 mm
Hand lens	2nd order (2⁰) irregularities	0.3–1.0 mm
Incident light microscopy; conventional scanning electron microscopy; reflected light interference microscopy; one-stage replicas	3rd order (3⁰) hollows, pits; humps, prominences, undulations, figure-of-eight features	Width: 15–35 μm Depth: 0.5–2.5 μm
Low temperature scanning electron microscopy (light microscope sections)	4th order (4⁰) irregularities	0.1–1.0 μm
Transmission electron microscopy of two-stage replicas (transmission electron microscopic sections but see text p. 141)	5th order (5⁰) macromolecular features, including collagen fibrils with 70–80 nm periodicity; expanded proteoglycans	~ 8–10 nm long ~ 150 nm wide

vexities often appeared smooth (Figs 3.24; 3.25). A feature which attracted attention was a superficial 1–2 μm, 'lamina splendens',[110] subsequently proved to be an optical artefact of phase-contrast microscopy (but see p. 70).[6]

Low-angle illumination and incident light microscopy of the surfaces of blocks of cartilage fixed in osmium tetroxide (OsO_4) showed that they were covered by a random series of shallow features sometimes ovoid, sometimes with a 'figure-of-eight' configuration.[60,48] The contours appeared as depressions (pits; hollows) in fixed, dehydrated, adult

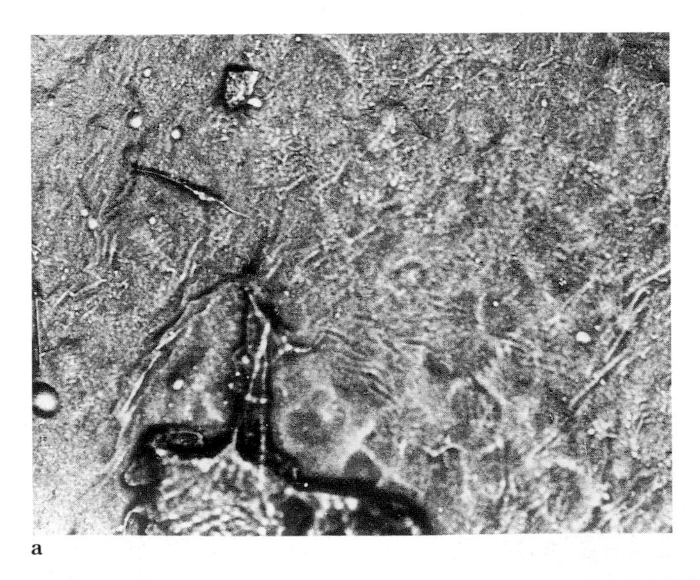
a

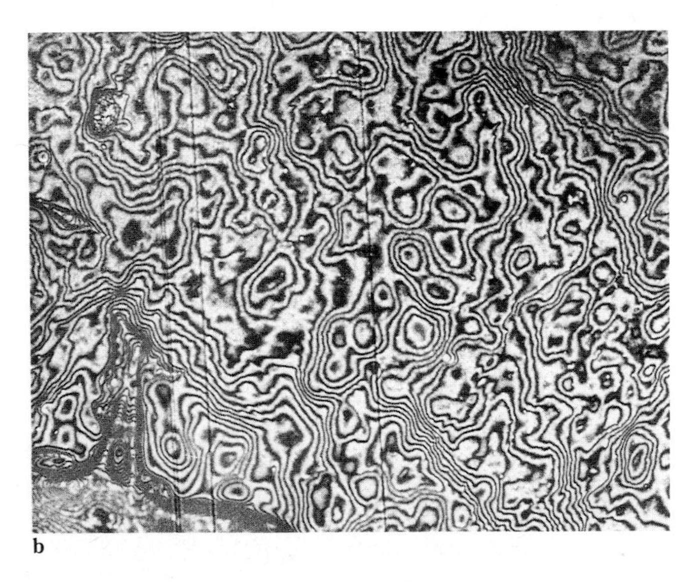
b

Fig. 3.27 Surface of disarticulated human femoral condyle.
a. In this 47-year-old female, appearance of hyaline cartilage surface shows non-smooth structure. (Incident white light × 400.) **b.** Same field, viewed by interference microscopy. The black and white lines, each pair a wavelength, indicate the distance between the objective lens and the surface features, providing a contour guide analogous with a geographic map. Where contours are close together, the slope of the feature is steep. The map allows diameter and frequency of surface features to be determined as well as height/depth. Fluid droplet (lower left) permits initial test to assess which contours are elevations and which are depressions. (Incident 560 nm (green) light in interference mode × 400.) Courtesy of Dr R B Longmore.

material: their centres sometimes displayed elevations like the cone of a volcanic crater. Scattered across the surface of a femoral head, the shallow depressions created a golf ball-like dimpling. A closely similar organization was identified on the hyaline and fibrocartilagenous surfaces of human, baboon, dog, rabbit, rat and mouse synovial joints; the contours were also identifiable on the cartilage surfaces of fetal joints.

Reflected (incident) light interference microscopy can be used to image fresh or fixed, moist cartilage. There is, however, a limited range of resolution and a small working distance so that very rough surfaces cannot be investigated. Within these limitations, white light can be directed onto cartilage surfaces and a pattern of coloured contours can be seen.[49] When green light of λ 560 nm is substituted for white, the coloured patterns become a series of lines with each contour line $\lambda/2$ apart.[53,106] The contour lines (Fig. 3.27) give a measure of distances between asperities on the cartilage surfaces and the objective lens of the interference microscope.

When a survey was made of cadaveric femoral condylar cartilage, the presence of a third order (3^0) of surface features was confirmed. Measurements of these features demonstrated that in the age range 20–50 years, the *number* of features declined with age[107] (Fig. 3.28). The mean depth/height of the third-order features increased linearly with age (Fig. 3.29); and the maximum diameter of the features also increased (Fig. 3.30), but less steeply. Although cartilage blocks, exposed to air during reflected light interference microscopy, undergo drying and distortion, there was reasonable agreement between the number of surface features/mm^2 and the number of chondrocytes detected in radial and tangential cartilage sections (Table 3.6, p. 148). Moreover, analogous evidence was obtained by the interferometric survey of silicon replicas (see p. 148) taken from freshly exposed cartilage.[34] Surveys of fibrous (temporomandibular) joints made by the same method are reviewed on p. 151.

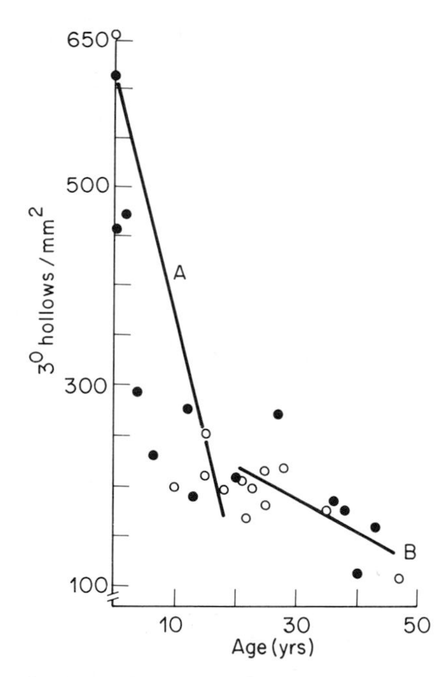

Fig. 3.28 Surface features of disarticulated *postmortem* human femoral condylar cartilage.
Frequency of third-order (3^0) feature, measured by interference microscopy, expressed against age of subject. The shape of the curve defining the decline in surface features with age closely resembles the curve for diminishing cartilage cell frequency obtained by Stockwell,[154] supporting evidence that third-order (3^0) contours seen on unloaded, disarticulated cartilage are representations of superficial (zone I) chondrocytes.

Transmission electron microscopy

Paradoxically, transmission electron microscopic views show that hyaline cartilage-bearing surfaces are 'smooth' although (Fig. 3.31) this is not a universally held opinion. In the rabbit, a relatively cellular zone I was bounded by a 2–3 μm thick acellular lamina.[38] The matrix surface was 'remarkably smooth', the irregularities present of the same order of magnitude as those of polished metal or glass. However, in other hands, the surface of the superficial tangential layer was 'characterized by depressions usually approximately 0.3 μm but occasionally up to 1.5 μm in depth.[171]

Interpretation of transmission electron microscopic evidence There are differences in the importance attached to the ultrastructural evidence obtained from fixed, dehydrated tissue. Bloebaum and Wilson[13] support the contention that the results of studies by plastic embedding, ultramicrotomy and TEM should be the standard against which to judge all other preparative techniques.[15] However, sampling errors and differences in the handling of specimens influence the results. In the human knee there are structural variations in different parts of the cartilage surfaces.[119] A surface which is intact by TEM is 'smooth or gently undulating' (Fig. 3.31), separating the underlying matrix from the joint cavity by a (ultrastructural) membrane-like material. Shallow clefts suggest the origins of fibrillation and a surface which appears intact to the naked eye or by light microscopy, is not always so when examined by TEM. The surface of young, adult rabbit articular cartilage processed attached to bone has 'an amazingly smooth surface'.[70] Whereas free cartilage shavings bear undulations and asperities of up to approximately 0.3 μm deep, TEM blocks with the bone retained show only

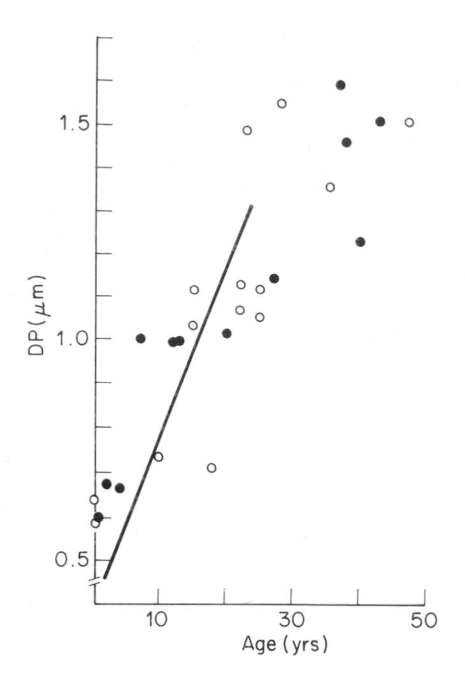

Fig. 3.29 Surface features of disarticulated *postmortem* human femoral condylar cartilage.
Depth/height of third-order (3^0) surface features measured by interference microscopy increases with age. After age 50 years, surface irregularities become so large that incident light interference microscopy ceases to be practicable.

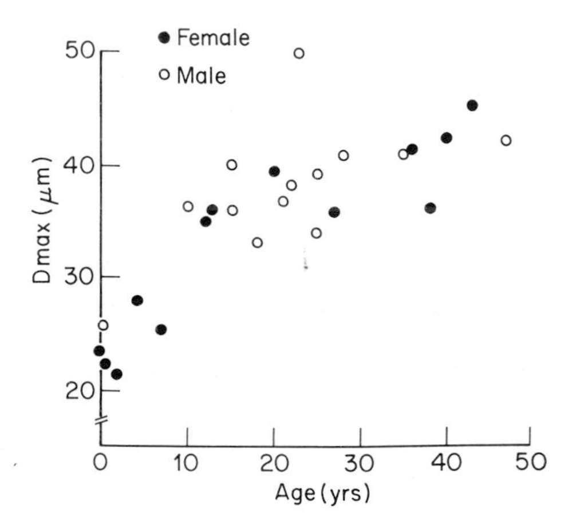

Fig. 3.30 Surface features of disarticulated *postmortem* human femoral condylar cartilage.
Mean diameter of third-order (3^0) surface features increases with age.

features approximately 0.03 μm deep. The significance of this evidence may be confined to adult mammals. After decalcification, neonatal femoral heads fixed in glutaraldehyde (but not dehydrated) yield light microscopic images of chondrocytes protruding from a convex surface and TEM images in which surface depressions are clearly shown.[36]

Ultrastructural surface lamina The most superficial part of the articular cartilage-bearing surface, interfacing with the synovial fluid, is an ultrastructural surface lamina '0.2 μm in thickness, relatively devoid of fibres';[38] it is not visible by light microscopy. The lamina has been described as a 'membrane-like' structure, approximately 200 nm thick[119] or a 'multilaminated 30–100 nm electron-dense surface coat, often rarified or attenuated'.[70] The deeper parts of the coat merge with or lie between the collagen fibres of the extracellular matrix of the superficial, zone I cartilage.

The nature of this delicate surface lamina (Fig. 2.8),[131] and of the comparable layer covering fibrocartilage,[66] fibrous tissue[3] and the marginal synovium[159] is not fully established. Early studies suggested it might be a synovial protein–hyaluronate complex *deposited on* the surface.[8] However, the superficial 20 μm of the cartilage is rich in chondroitin sulphate but poor in hyaluronate. The ultrastructural lamina may be proteinaceous, with some enmeshed collagen fibres[148] but is unlikely to be formed only of proteoglycan. Part of the lamina may be fibronectin which is demonstrable at the cartilage surface.[35,104] The presence of lipid is inconstant.[70] When transmission electron microscopic techniques are combined with the use of the electron-dense cationic dye, cupromeronic blue, in a critical electrolyte concentration technique (see p. 54), the surface lamina can be shown to be of two parts[130] (Fig. 3.32). A 50 nm-thick electron-dense superficial part, corresponding to the 30–100 nm layer of Ghadially *et al.*[70] does not react for sulphated glycosaminoglycan and is neither proteoglycan nor hyaluronate (Fig. 3.32). A 100–400 nm-thick deeper layer reacts with cupromeronic blue at electrolyte concentrations specific for sulphated glycosaminoglycan and is probably chondroitin sulphate proteoglycan; it does not contain keratan sulphate. The deepest part of this layer merges with the superficial parts of the cartilage matrix collagen fibres.

The function of the ultrastructural surface lamina is not yet certain. It is likely to be important in cartilage lubrication and may influence the high permeability of zone I cartilage: the penetration of large molecules may be minimized by the mechanical and charge density barrier created by the lamina.[148] There is little doubt that loss of the lamina plays a part in permitting the early escape of proteoglycan in osteoarthrosis[40] (see Chapter 22). This loss may also

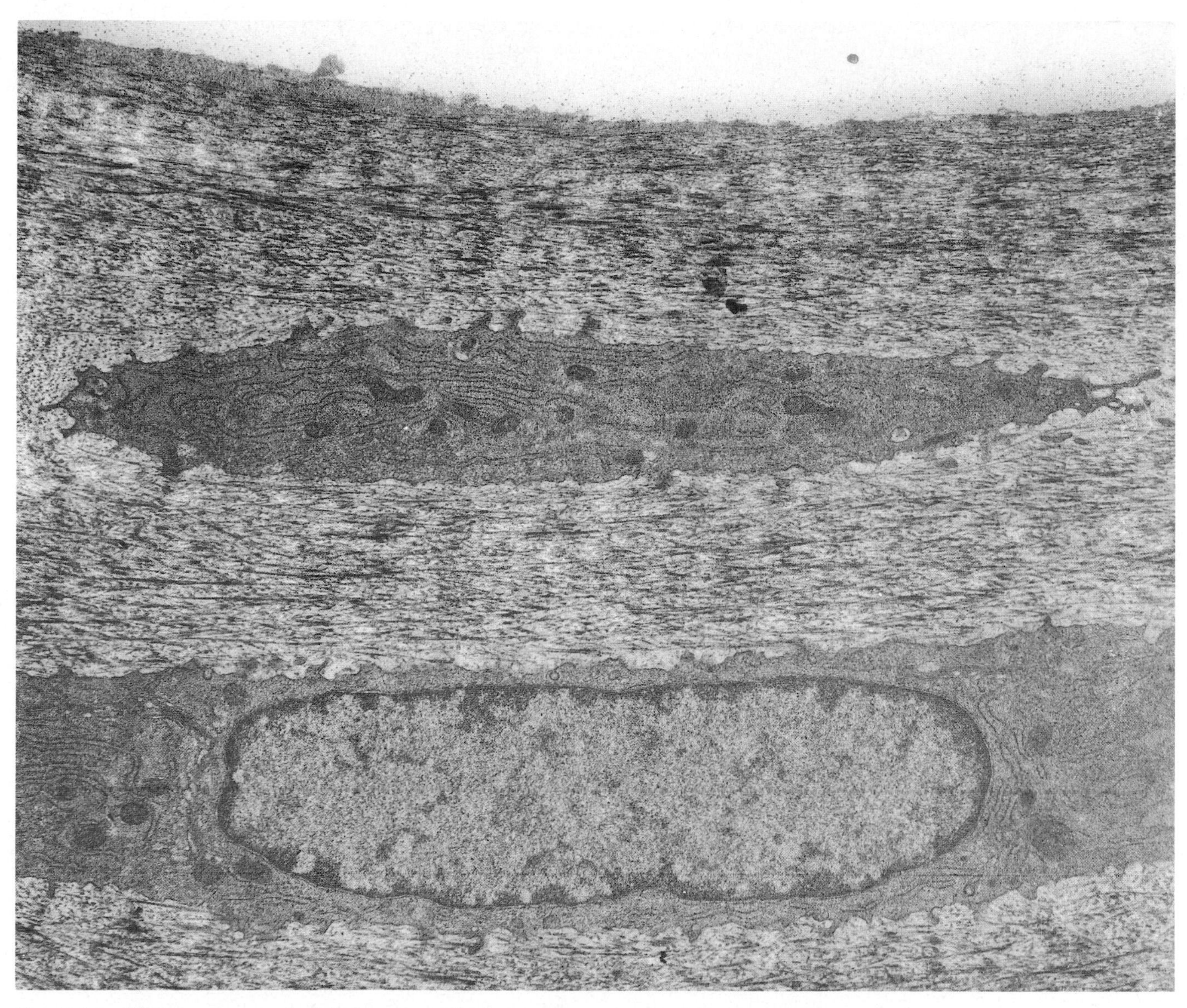

Fig. 3.31 Fine structure of adult human articular cartilage.
Although TEM usually shows the surface of normal hyaline cartilage to be planar ('smooth'), this is not always so. Convexities or concavities are at sites of most superficial chondrocytes. (TEM × 6600.)

contribute to the penetration of enzymes and immune complexes into cartilage and fibrocartilage in early rheumatoid arthritis (see Chapter 13).[127] However, the surface lamina is readily disturbed during preparative procedures and its constant presence throughout normal adult human life is not certain. In spite of this uncertainty, there is evidence that the application of cross-linking agents such as succinimydil propionate offers protection to cartilage surfaces against digestion by enzymes such as collagenase.[148a]

Scanning electron microscopy

Conventional scanning electron microscopy In the material sciences, very precise measurements of surface shapes are practicable by scanning electron microscopy (SEM). Crystalline structures can be subjected to stereoscopic analysis, enabling the determination of the underlying crystallography of three-dimensional planes[100] and true surface topographies can be reconstructed by

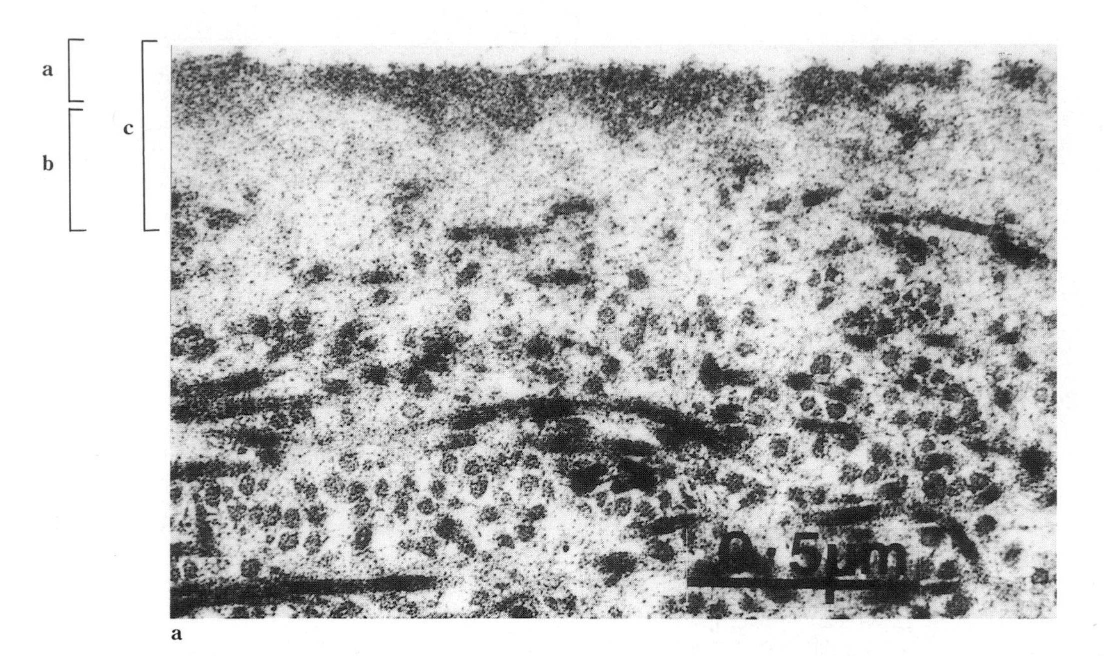

a

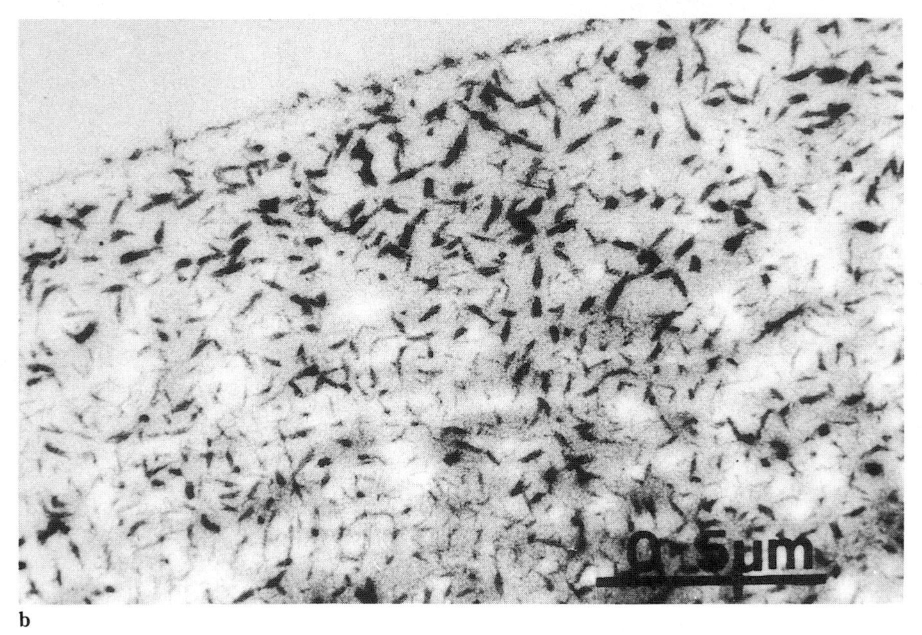

b

Fig. 3.32 Transmission electron microscopic ultrastructure of the load-bearing surface of hyaline articular cartilage.
a. Ultrastructural lamina *c* covering zone I cartilage comprises superficial *a* and deeper *b* layers. Gluteraldehyde and osmium tetroxide fixation; poststained with phosphotungistic acid and uranyl acetate. (TEM × 49 500.) **b.** Deeper portion of surface lamina contains many stained filaments that are proteoglycan. Note periodicity, corresponding to structure of collagen. (TEM × 49 500.)

backscattered imaging.[28] Similar physical procedures can be employed to elucidate subsurface organizations in biological tissues,[95] but in general the interpretation of biological SEM images is subject to many pitfalls. The difficulties centre on tissue collection and preparation. They are therefore less in the case of avascular connective tissues such as cornea and hyaline articular cartilage which tolerate hypoxia well than in epithelia and neural tissues. Nevertheless, mechanical distortion, shrinkage caused by fixation, and dehydration artefact are problems of many SEM preparative procedures as are the thermal effects of coating and the destructive actions of the electron beam itself.

Interpretation of scanning electron microscopic evidence The first SEM evidence of cartilage surface structure was obtained with bulk material, air- or acetone-dried or fixed in formalin, paraformaldehyde or 'osmic acid' before dehydration.[109,59,94,136] The cartilage surfaces did not appear smooth. The second-order (2^0) features comprised single or paired hollows (pits), sometimes with a central, raised prominence (Fig. 3.33). The third-order

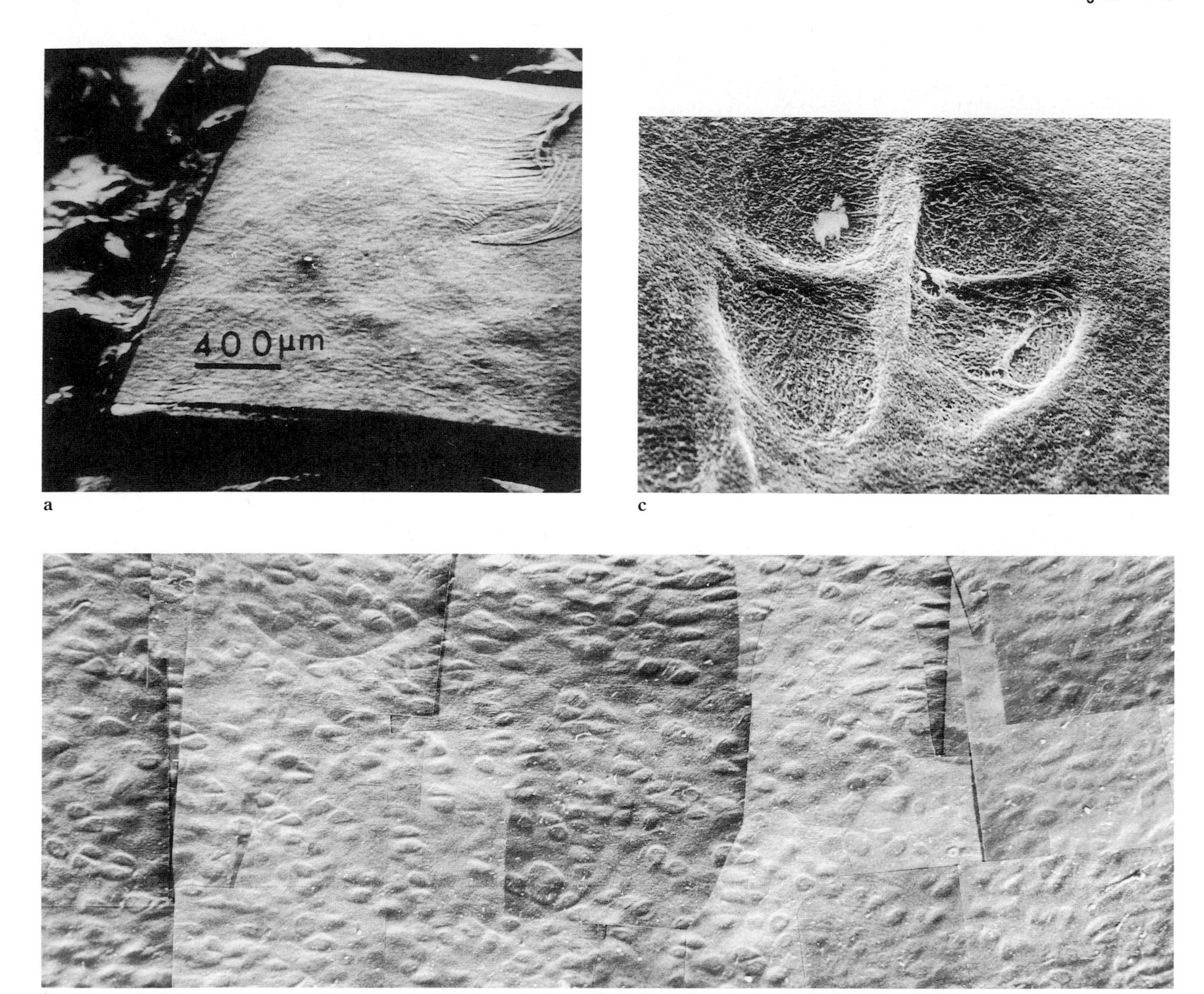

Fig. 3.33 Scanning electron microscopic ultrastructure of load-bearing surface of hyaline articular cartilage.
a. Excised block of fixed, dehydrated mammalian cartilage showing second-order (2^0) 0.4–0.5 mm diameter, and third-order (3^0) 20–40 μm surface features that are recognized on all adult, young, and fetal bearing-surfaces. (SEM × 37.) **b.** Photomontage of fixed, dehydrated adult human cartilage with random arrays of third-order (3^0) surface features. Compare with Fig. 3.34. (SEM × 132.) **c.** Single third-order (3^0) surface feature from fixed dehydrated load-bearing cartilage of young baboon. The quadripartite surface is superficial representation of underlying chondron. (SEM × 3000.)

features corresponded in size and number to the size and structure of irregularities measured by light microscopy (see p. 138). When a comparison was made between the surfaces of the upper half of the human femoral head and the corresponding acetabular cartilage, on the one hand, and the underlying structures revealed by 'peeling off' the cartilage surface material on the other, the 15–30 μm surface hollows (depressions) were found to correspond to sites where the surface layer had collapsed onto underlying chondrocyte lacunae.[29] The study of fractured blocks suggested that the surface layer had a randomly arranged collagen network,[30] but Mital and Millington,[122] using pin-prick tests, Talysurf traces and SEM, confirmed that the most superficial collagen fibrillar layer lay parallel to the surface. A comparison of fresh, moist tissue with formalin-fixed, non-hydrated material and with Araldite replicas showed that undehydrated specimens contained 12–60 μm diameter depressions similar to those identified by the SEM of the replicas. By reflected light microscopy, the surfaces bore 200–323 depressions/mm^2 averaging 18–39 μm in diameter. By SEM, the Araldite replicas displayed 153–700 depressions/mm^2, of average diameter 21–38 μm. The figures corresponded closely to those obtained by reflected or transmitted light microscopy or SEM of tangential sections (Table 3.4).

Species differences There are differences in articular cartilage surface structure between different species. In the (mongrel) dog, for example, glutaraldehyde-fixed, alcohol-dehydrated blocks display a population of numerous prominences (humps) when surveyed by SEM,[65] results in agreement with those of Gardner *et al.*[58] for beagle dog cartilage that had been very rapidly frozen and scanned in the hydrated state at low temperature (see below). Ghadially *et al.*[65] emphasized the contrast with man and rabbit on the cartilage surfaces of which they thought there was a corresponding population of pits (third-order (3^0) hollows/depressions).[64]

Table 3.4

Surface features of excised cartilage: diameter (after Clarke[31])

Technique	Area studied	Diameter of 'lacunae' or surface depressions	
		Range (μm)	Mean (μm)
Reflected light microscopy	Sectioned surface	13–58	25–47
	Articular surface	12–54	18–39
Transmitted light microscopy	Tangential section	12–61	20–38
Scanning electron microscopy	Araldite replica	12–60	21–38
	Acrulite replica	6–60	11–40
	Sectioned surface	10–48	17–30
	Articular surface	7–68	20–35
Overall mean diameter*			20–38

* The Acrulite replica data was excluded from the overall figures.

Note: In this table, the mean diameter of the third-order (3^0) depressions (pits, hollows) displayed on the surfaces of articular cartilage from disarticulated specimens by reflected light microscopy and on replicas of these surfaces and the surfaces themselves by SEM, is compared with the mean diameter of cartilage cell lacunae seen on sections of the cartilage surveyed by reflected light microscopy or transmitted light microscopy or by SEM. The order of size of the natural depressions corresponds to that of the depressions shown on replicas by SEM; both are similar to the mean diameter of cartilage cell lacunae. The evidence supports the view that third-order (3^0) surface features are representations of the shape and size of superficial (zone I) chondrocytes.

Age differences There are also differences between the surface structure of disarticulated, non-loaded cartilage surfaces at different stages of development and at different ages.[68] On the surface of two-day to two-month-old kitten femoral condylar cartilage, as on young human and young rabbit, there are 'innumerable' humps (third-order (3^0) prominences).[67] In 12- and 20-month-old cats, the surface appearance was that of great numbers of pits (third-order (3^0) depressions). In parentheses, the humps seen on young cartilage were noted to be more prominent on air- than on critical point-dried material.

Low temperature scanning electron microscopy

A comparison of preparative procedures for scanning electron microscopic blocks suggested the value of avoiding fixation and dehydration.[24] Low temperature techniques were preferred. Blocks of cartilage freed from or remaining on bone were frozen at −210 °C and scanned in the hydrated state.[134] The presence of the third-order (3^0) surface features detected by light microscopy was confirmed (Fig. 3.34). In contrast to results obtained after dehydration, the surface features appeared as prominences.[57,58] Their three-dimensional structure was validated by the measurement of stereo pairs of micrographs: the prominences were often 20–30 μm in diameter, 1–4 μm in height.

Light and electron microscopic observations had long suggested that the third-order features seen on cartilage surfaces were the representations of superficial chondrocytes.[62] Proof of this assertion was only obtained when it became possible to make electron probe X-ray microanalytic studies of unfixed, hydrated cartilage blocks at low temperature.[121] The X-rays collected from tissue up to 15 μm *below* a third-order feature were identical to those gained by the direct X-ray analysis of perpendicular preparations of mid-zone chondrocytes: there was much phosphorous and potassium. By contrast, the X-rays obtained from surface zones *between* the third-order prominences showed the presence of more sodium, sulphur, calcium and chloride and little potassium. The latter results were comparable to those obtained from femoral condylar cartilage in which there was a pericellular matrix rich in sulphur.[112] The

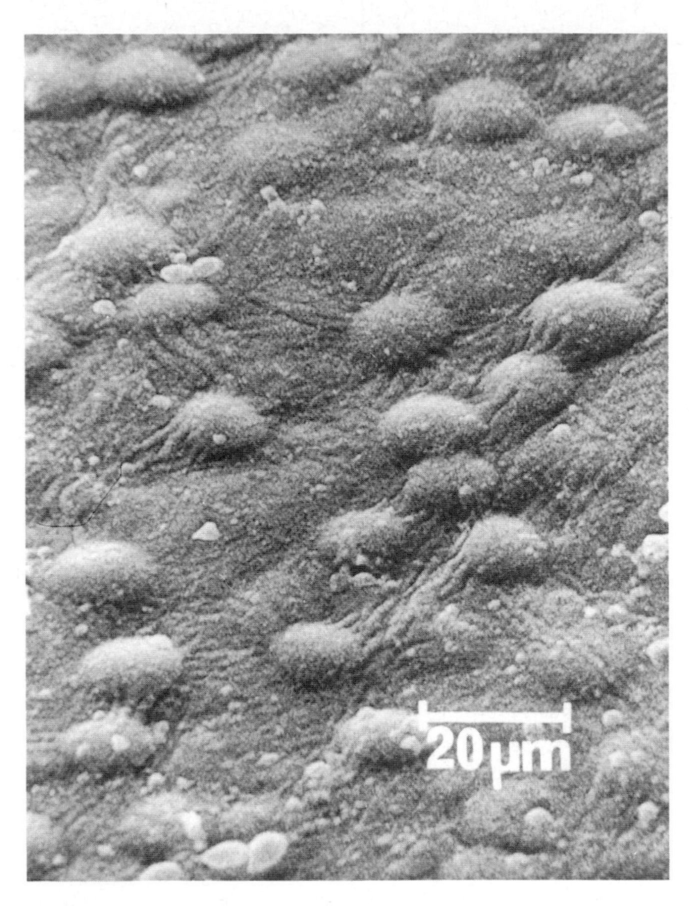

Fig. 3.34 Low temperature scanning electron microscopic ultrastructure of load-bearing surface of hyaline articular cartilage.
Unfixed, hydrated canine articular cartilage with third-order (3^0) surface features shown by the examination of stereo pairs of photomicrographs to be elevations. Although the fine pattern of striations seen at margins of blocks excised from underlying bone may be artefactual, the linear arrays seen in relation to the elevations in this Figure are apparently surface collagen bundles. Compare with Fig. 3.33. (Low temperature SEM at −181°C (82 K) × 900.)

Table 3.5

Surface features of human articular cartilage: frequency (after Clarke [31])

Technique	Area studied	Frequency of 'lacunae' or surface depressions	
		Range (per mm^2)	Mean (per mm^2)
Reflected light microscopy	Sectioned surface	191–366	284
	Free cartilage surface	200–323	244
Transmitted light	Tangential section	237–600	450
Scanning electron microscopy	Replica	153–700	341
	Replica	1040–2530	1385*
	Sectioned surface	185–890	560
	Free cartilage surface	190–837	470
Overall mean frequency*			430

* The Acrulite replica data was excluded from the overall figures.

Note: The table provides a comparison of the frequency of surface, third-order (3^0) depressions assessed by reflected light microscopy of free cartilage surfaces and by SEM of replicas of these surfaces, with the frequency of cartilage cell chondrons ('lacunae') determined by the reflected light microscopy of sections of the surface, the light microscopy of tangential sections and the SEM of replicas of the surfaces and of the surfaces and their tangential sections.

The evidence supports the view that the frequency of cartilage third-order features is similar to the frequency of chondrocyte 'lacunae'.

third-order features seen on disarticulated cartilage-bearing surfaces are therefore formed by superficial chondrocytes, a view that had been suggested by Clarke's[30,31,33] evidence (Tables 3.5, 3.6).

Replicas

Ambient temperature Substances such as latex or a dental vinyl polysiloxane of low viscosity, which are malleable but harden quickly, can be used to 'cast' or replicate cartilage surfaces. The hardened cast, a one-stage replica, can be examined directly by light microscopy or scanning electron microscopy, at resolutions of approximately 0.5–1.0 µm. Alternatively, a two-stage replica can be made with collodion and carbon. The carbon film remaining after the collodion is removed, is sufficiently thin to be examined by transmission electron microscopy. The two-stage replica therefore provides detailed information of surface structure. One-stage replicas confirmed the presence, on free articular cartilage surfaces, of second-order (2^0) irregularities and of third-order features.[60] Two-stage replicas revealed the detailed structure of very small replicated areas.[60]

Table 3.6

Cell frequency of human articular cartilage (modified from Clarke[31])

Source	Material	Section	Cell frequency/ $\times 10^3$ mm^3
Meachim and Collins[117]	Upper end of humerus	Radial 0.22 mm deep	43.5–61.8
Stockwell[152]	Lower end of femur	Radial 0.22 mm deep	37.0–97.0 (mean 62.0)
Stockwell[153]	Lower end of femur	Radial 0.12 mm deep	67.0
Clarke[33]	Upper end of femur	Tangential 10 μm deep	23.7–60.0

Frequency of adult human articular cartilage cell 'lacunae' (obtained from tangential sections 10 μm deep to surface of cartilage) compared to frequency of chondrocyte nuclei derived from sections cut radial (perpendicular) to surface.

The evidence supports the contention that the frequency of articular cartilage chondrons ('lacunae') seen in tangential sections, and used to assess the significance of the third-order (3^0) surface features of cartilage (Tables 3.4 and 3.5) is a reasonable reflection of the overall frequency of cartilage cells.

Low temperature Problems of tissue drying and shrinkage before replication can be avoided by low temperature techniques.[56] Small blocks of cartilage, on or detached from bone, are rapidly frozen. Platinum/carbon replicas of the hydrated, unfixed surface are prepared. The blocks can be examined by TEM. The results demonstrate the presence of two forms of surface structure that cannot be resolved by conventional light microscopy, SEM or low temperature SEM.[56] One part of the surface is amorphous, bears few collagen fibrils and corresponds to the ultrastructural lamina described on p. 142. A second form of fine surface structure is revealed by the loss of this lamina. In this second category, there are many delicate parallel arrays of collagen fibres with diameters of 46–91 nm and periodicities of 69–71 nm. Between these fibrils, which are assumed to be largely type II collagen, are many gently convex, smooth-surfaced elevations, 150–500 nm in diameter; they are two orders of magnitude smaller than the third-order (3^0) prominence (humps) identified, on the non-loaded concave surfaces of disarticulated joints, by light microscopy and SEM. The elevations are thought to be expanded proteoglycan, restrained laterally and deeply by the collagenous microskeleton but free to protrude superficially because of the loss of the ultrastructural surface lamina. Comparable, featureless areas and areas displaying arrays of approximately 50–200 nm diameter collagen fibrils with a periodicity of approximately 90 nm, are identifiable on the cartilage surfaces of non-human primates such as *Papio anubis*[133] (Fig. 3.35).

Profile recording

Some of the limitations of light microscopy, SEM or TEM can be avoided by physical measurements of cartilage surface contours. A sharp, fine stylus can be passed across a surface; vertical movements, attributable to asperities, are amplified and recorded. Clarke[32] concluded that cartilage itself was too soft to test directly. He therefore made acrylic replicas of femoral heads and compared the height of the replicated contours with heights assessed by SEM. The surface features of disarticulated, non-loaded human cartilage *in vitro*, measured approximately 0.19–15 μm (SEM), 0.3–8.9 μm (SEM of replicas) and 0.2–5 μm (profile records of replicas). Sayles *et al.*[138] measured cartilage surfaces directly and assessed the effects of shrinkage and distortion caused by fixation and dehydration. They drew contour maps and recorded roughnesses of approximately 7 μm for the femoral head, with slopes of up to 2° and peak radii of approximately 0.9 μm.

Direct observations of loaded surfaces

The shape of loaded cartilage surfaces has been estimated by applying cartilage to glass lubricated with synovial fluid.[167,168] After separating the frozen, loaded cartilage from the metal, small aggregates of synovial fluid were retained on the metal. They were 'pools' of synovial fluid. The size of the pools was that of the third-order surface features seen by incident light microscopy and by conventional SEM. The evidence gave support to emerging views on cartilage surface organization *in vitro*. However, the concept of boosted lubrication derived from these studies[167] is no longer tenable (see Chapter 6).

Comparable evidence was obtained *in vivo* by the manual application of a glass slide to the surface of an exposed femoral condyle.[54] After Indian ink had been allowed to flow around the loaded zone, filling the space between glass and cartilage, photomicrographs showed a cluster of minute, pale foci in the centre of the zone of cartilage/glass contact. The foci appeared to be third-order prominences remaining under load.

Fig. 3.35 Low temperature replica of natural, load-bearing surface of hyaline articular cartilage.
This carbon replica has been photographed by TEM. The appearance recorded here is therefore a mirror-image of the cartilage surface. It is devoid of the surface lamina illustrated in Figs 2.8 and 3.32 and shows many superficial collagen fibres with a characteristic 70 nm periodicity and, between them, gently convex elevations (fifth-order (5^0) features) thought to be the domains of expanded, hydrated proteoglycan macromolecules. (TEM × 13 100.) Courtesy of Dr K Oates and Mr J Pidd.

Mechanical tests

The collagen-rich surface layer of zone I cartilage is prestressed—i.e. there are inherent, 'interlocked', stresses which give this layer its characteristic mechanical properties. Understanding of microscopic organization of the surface layer has been inferred from studies of its physical behaviour. Some of the tests that have been used to obtain this information are outlined in Chapter 6.

Intact synovial joints *in vitro* The articular surfaces of non-disarticulated diarthrodial joints can be viewed by arthroscope or, directly, through small lateral incisions in the capsule. To see an entire surface, it may be necessary to cut ligaments and tendons with their synovia. The microscopic details obtained from such preparations, either directly by light microscopy, or indirectly by replication, are essentially similar to the results gained from *in vitro* studies of disarticulated joints. However, it is also of value to examine perpendicular sections through the opposed, non-disarticulated joint. For example, cryostat sections through the whole ankle joint of adult rats reveal ovoid chondrocytes at sites of cartilage surface prominences. Although articular surfaces of sawn slices of whole joints are less easy to interpret, sections of whole primate postnatal rat or embryonic chick joints, examined in this way, display prominences (humps) on the non-loaded, concave surfaces (Fig. 3.36). The convex surfaces do not show this appearance.

Disarticulated living synovial joints The surfaces of disarticulated, non-loaded, but living human, rabbit, rat, guinea-pig, mouse and turkey knee joints, and those of neonatal rats and fetal mice, have been examined microsopically.[55] Fresh, unfixed cartilage preparations have been surveyed by conventional incident light microscopy. Immersion techniques are of no advantage: they provide excellent views of the subchondral, bone marrow blood vessels (Fig. 3.36) but not of the articular surface. Because of high reflectivity, dry, incident light microscopy does not clarify the nature of the cartilage surfaces; however, it does suggest that the state of the superficial chondrocytes is intimately related to the integrity of the arterial blood supply to the joint.

There are usually second-order (see p. 138) irregularities. Both in the adult and in the fetal rodent, third-order features are also seen. The appearances change quickly with drying, an observation confirmed in the anaesthetized rat by two-stage replication (see p. 147) and SEM.[173] Second-order irregularities are not usually seen on avian surfaces but the prominent collagen bundles identified in these animals are overlaid by an undulating pattern of third-order hollows and prominences. In similar work, Clarke, Schurman and Amstutz[34] confirmed the presence of approximately 10 μm diameter highlights on the surfaces of dog and rabbit knee joint cartilage surfaces. Silicone replicas were made of the moist, living surfaces and surveyed by differential interference contrast microscopy and SEM. Prominences 7–20 μm in diameter and 0.5–2.5 μm high, corresponding to the highlights seen on the natural surface, covered those parts (approximately 50 per cent) of the replicas that could be interpreted.

Intact living synovial joints Microscopic studies of intact, non-disarticulated living mammalian joint surfaces

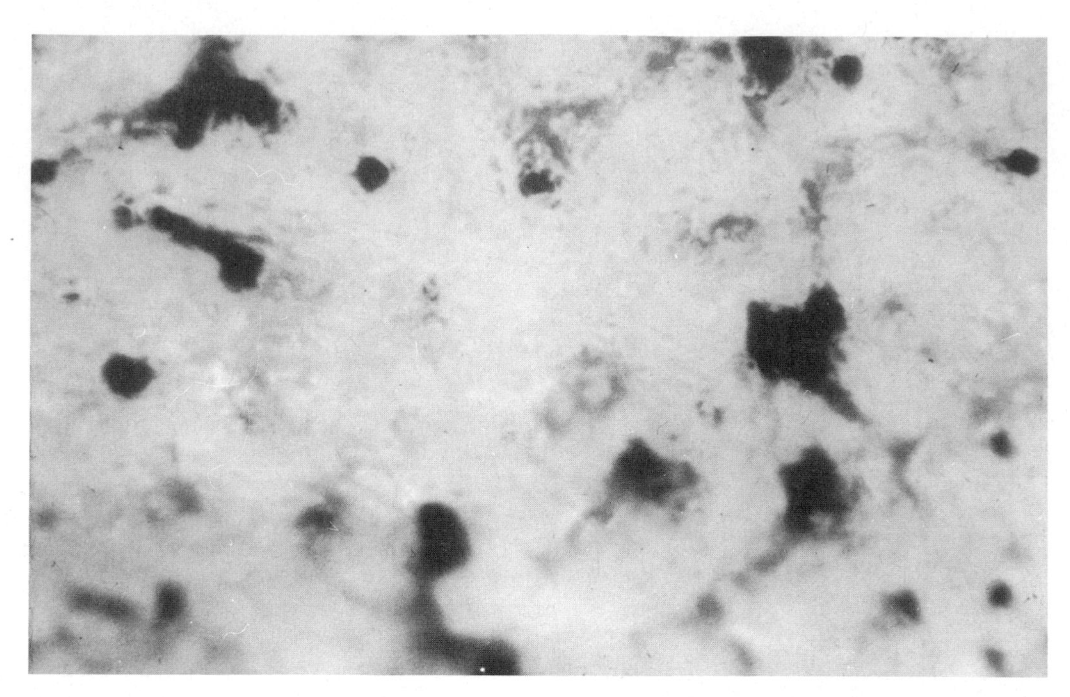

Fig. 3.36 Rat femoral hyaline articular cartilage exposed at surgery and viewed through immersion fluid by incident light microscopy.
Lateral tibial condyle of anaesthetized rat exposed by surgical incision. Microscope objective immersed in 0.9 per cent saline on cartilage surface. Similarity of refractive index of immersion fluid and of cartilage allows observation of subchondral, intraosseous blood vessels. (× 35.)

may be possible when techniques such as laser microholography are more readily available. The bearing surfaces of the main limb joints in man and the larger mammals can be examined at low resolution by contrast and double-contrast imaging methods (see p. 154) which also permit cartilagenous and fibrocartilagenous surfaces to be viewed with the naked eye. Microradiography (see p. 154)[17] allows magnification factors of approximately five-fold to be used for some clinical and experimental purposes.

Fibrocartilagenous and fibrous joint surfaces

Meniscal surfaces

The surfaces of normal human knee joint menisci appear smooth to the naked eye except for occasional fine ridges or furrows.[66] Canal-like openings have been recognized on the surfaces of menisci from young calves, infants and young children; these channels may be related to the distribution of extracellular fluid.[166,11] As age advances, fraying of the inner edge becomes common (Fig. 2.18). Light microscopy demonstrates the upper, bearing surface to be slightly irregular, wavy or even markedly undulating. Electron microscopy shows that there is a gently undulating fine structure; a similarity to hyaline cartilage surfaces is emphasized by the presence of a variable electron-dense surface coat (lamina) approximately 0.8 nm thick.[66]

The examination by SEM of human meniscal material fixed in glutaraldehyde and critical point-dried, suggests a ridged, grooved, fibrous structure[178] (Fig. 3.37) but some of the technical limitations of hyaline cartilage preparations for SEM apply to fibrocartilage. Pin-pricks can also be used to determine the preferred collagen orientation: splits appear at the surface.[7] The splits follow the circumferential orientation of the bulk of the meniscal collagen bundles but the radial orientation of the superficial region (Fig. 2.18). The organization of this part of the human meniscus is similar to that of the rabbit.[69] In both, the cells resemble chondrocytes rather than fibroblasts and, as in hyaline cartilage, some lie within 1–3 μm of the articular surface.

There is a change in the surface structure of the human meniscus with age: in the newborn, there are numerous, randomly arranged linear elevations, folds and cell formations.[120] The folds become longer and wider with time and are most prominent in the inner part of the meniscus.

Another characteristic in common with hyaline cartilage, is said to be the presence of third-order (3^0) prominences ('humps') on the cartilages of young rabbit, cat, monkey and human specimens by comparison with adults in which these surface structures are depressions ('pits').[123]

Non-meniscal fibrous surfaces

Relatively little is known of the structure of non-meniscal, fibrous-bearing surfaces, in spite of their importance in influencing the function of discs and labra.[176] Temporomandibular joint discs can be imaged by computed tomographic scanning,[84] and light microscopic studies of mandibular and temporomandibular joints demonstrate much dense, avascular fibrous tissue containing elastic material. The surfaces, when first exposed, appear smooth and glistening. Washed free from synovial fluid, they display a series of second-order (2^0) features comparable with those present on hyaline cartilagenous surfaces.

Fixation introduces a complex, artefactual pattern on fibrous surfaces which can be seen and measured by reflected light interference microscopy (Fig. 3.38).[174] Discrete or clustered 10–20 μm diameter third-order (3^0) contours, most numerous in younger persons, are superimposed on a second-order (2^0) contour pattern. Incident and reflected light interference microscopic studies of six-month-old guinea-pig fibrous surfaces demonstrate that the third-order features are 13.6 ± 1.3 μm in width, 10.8 ± 1.6 μm in length and 1.1 ± 0.35 μm in height or depth. In baboons, by contrast, the third-order (3^0) features are less well-demarcated, widely spaced and complex: most are elevations, some circumscribed by narrow depressions.

Comparable TEM studies have been made of human,[162,163] baboon, and rat[3,46] fibrous surfaces. There

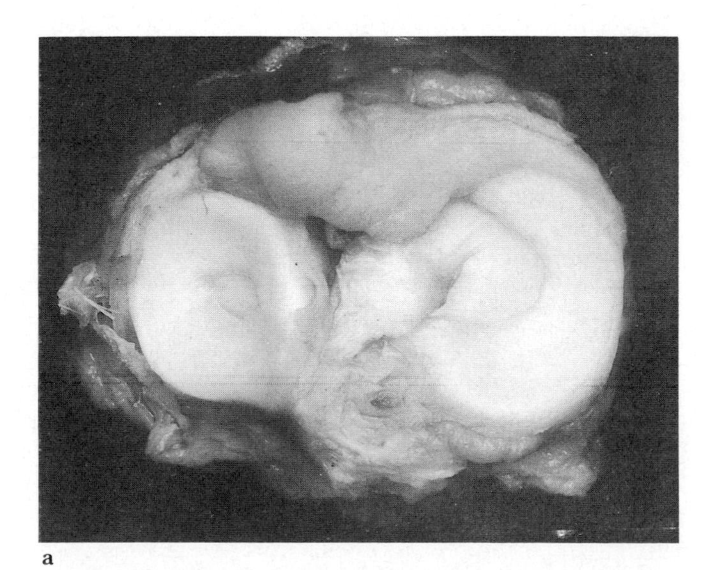

a

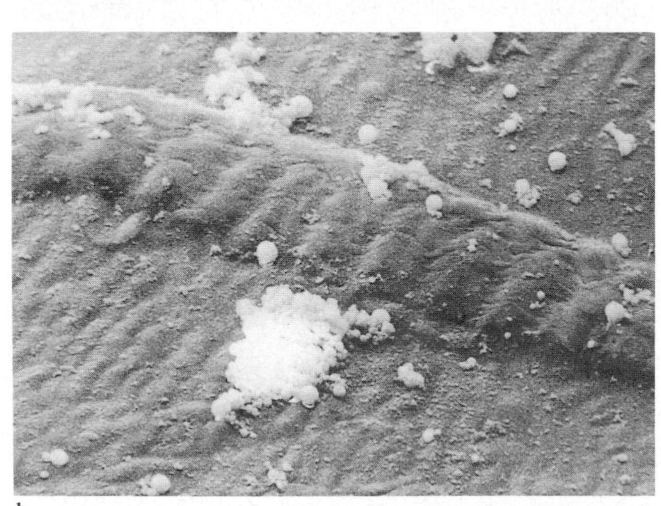

b

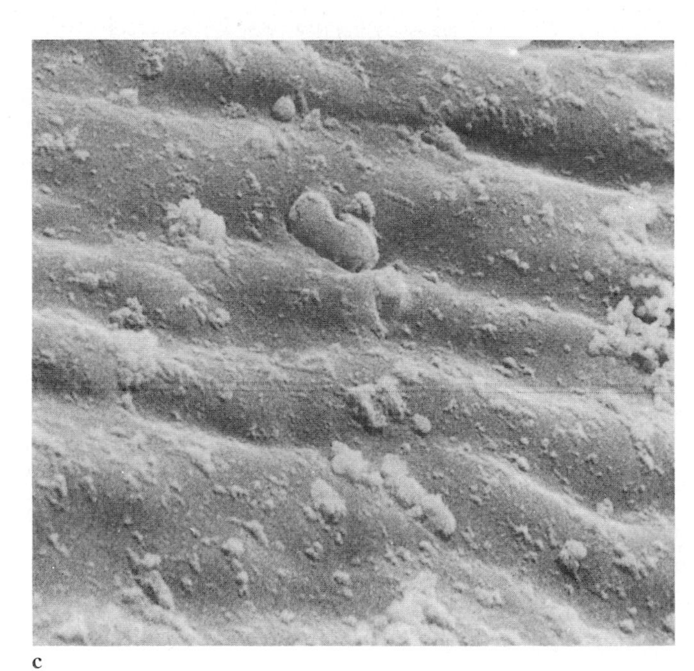

c

Fig. 3.37 Fibrous articular surfaces: menisci.
a. Tibial condyle of young adult baboon. Note the smaller lateral meniscus forming an entire ring upon the condylar surface, the more open semilunate fibrocartilage upon the medial condyle. **b.** Central part of medial tibial condyle of young adult baboon. The bearing surface is fibrocartilagenous. The appearances closely resemble those of the surface of a meniscus. (Low temperature SEM × 200.) **c.** Higher power view of non-human primate tibial condylar fibrocartilage shown in **b.** (Low temperature SEM at −178°C (95 K) × 450.) Courtesy of Dr D M Lawton.

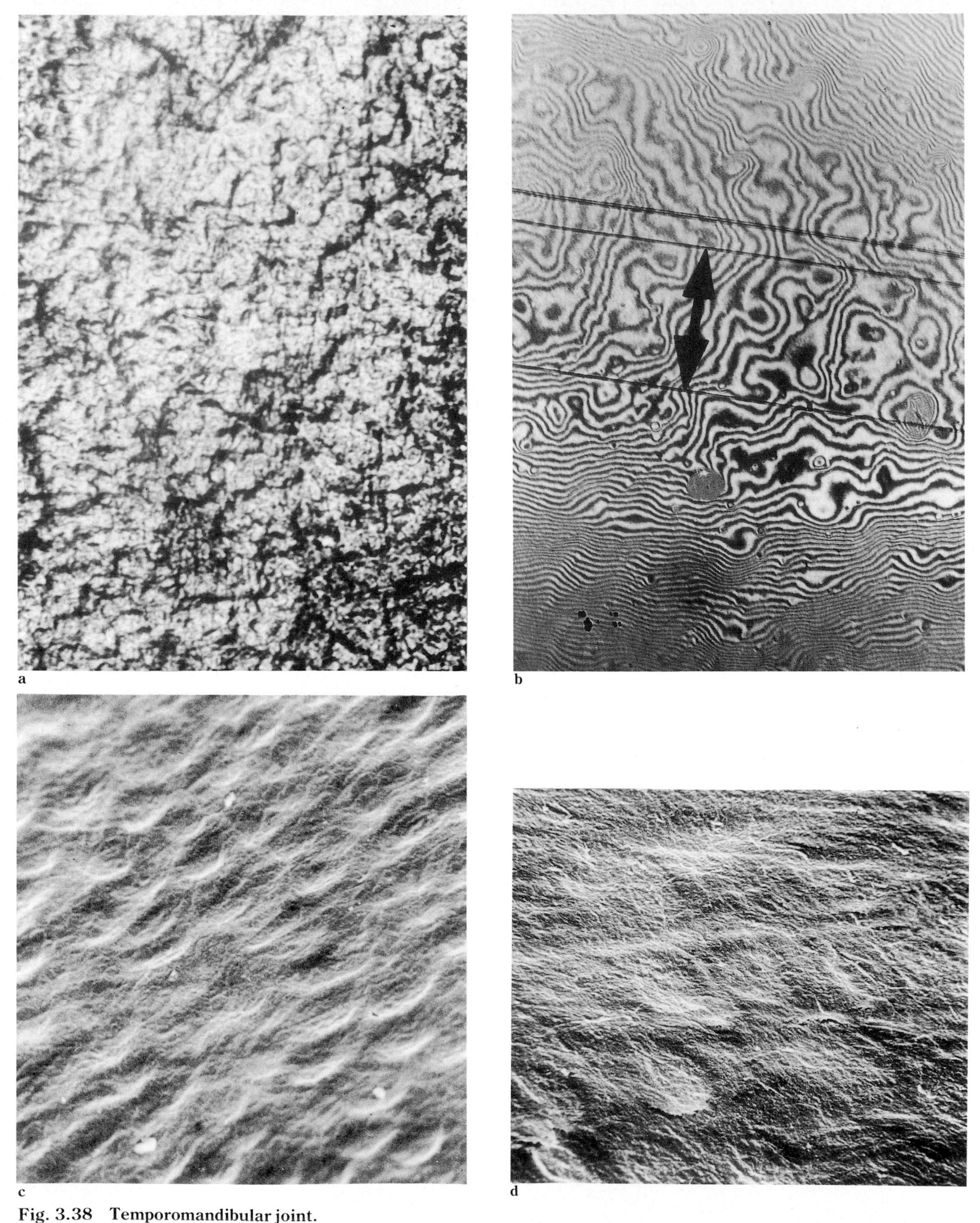

Fig. 3.38 Temporomandibular joint.
a. Surface structure of condyle from baboon temporomandibular joint. (Incident light microscopy × 250.) **b.** Same field to show contour pattern (distance between single reference lines (arrowed) is 100 μm). (Incident light microscopy in interference mode × 200.) **c.** Surface structure of condyle from marmoset temporomandibular joint, showing third-order (3^0) features. (Vacuum dried from propylene oxide: SEM) × 500.) **d.** Surface structure of intraarticular disc from baboon temporomandibular joint. (Critical point dried: SEM × 473.) Courtesy of Professor N H F Wilson.

are many interlacing collagen bundles bounded superficially by an approximately 150 nm ultrastructural lamina (see p. 142). Some collagen fibrils are as much as 200 nm in diameter. In the dog the lamina resembles that of man[131] but in the human temporomandibular joint, Toller[162] found small diameter collagen fibrils and electron-dense cell debris. Appleton[3] proposed that this delicate layer was hyaluronate protein but it seems more probable that this material forms only the most superficial part of the lamina, merging with the synovial fluid. The mid- and deep zone matrix and their cells resemble those of hyaline cartilage. There are many elastic fibres.

Scanning electron microscopic surveys have been made of human, baboon, marmoset, guinea-pig, mouse and rat temporomandibular joints.[176] Fixed, dehydrated human material displays oval or 'figure-of-eight', 20 μm depressions, less frequent in older persons, and resembling those of hyaline cartilage surfaces (Fig. 3.33). Systems of ridges[96] are thought to be the surface expression of superficial collagen fibre bundles but artefact can contribute to the ridges or furrows seen at the temporomandibular joint condylar surface and optimum results are only obtained when the cartilage is held on the underlying bone during SEM.

Imaging techniques in the study of connective tissue disease

J P R Jenkins and I Isherwood

A full understanding of the pathology of connective tissue disease is only possible if advantage is taken of every new method of investigation. Rapid progress in the development and use of new imaging techniques, including digital subtraction angiography (DSA), computed tomography (CT) and magnetic resonance imaging (MRI), together with new contrast media, over the last two decades, has had a profound impact on clinical radiological practice and on pathological knowledge.[47,73,156] The borderline between pathological anatomy and physiology on the one hand, and imaging methods on the other, has never been less sharply defined so that every pathologist investigating connective tissue disease processes requires insight into diagnostic imaging. In view of the large array of imaging techniques now available, it is important that the role of each be kept in perspective as their contribution to the analysis of connective tissue disease is considered.

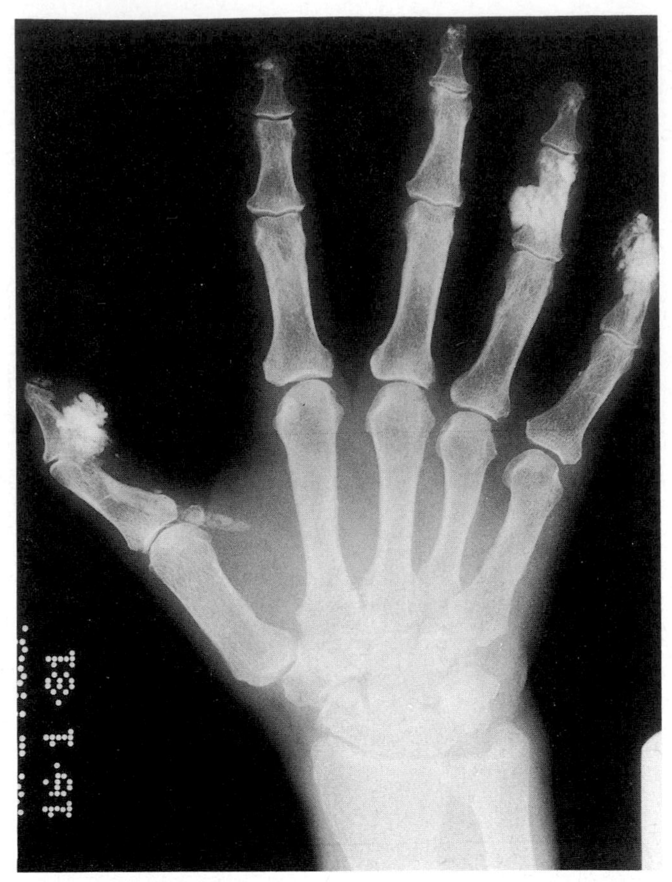

Fig. 3.39 Radiograph in progressive systemic sclerosis.
Right-hand radiograph showing soft-tissue resorption around the terminal phalanges, and calcification (see Chapter 17).

Conventional radiography

Conventional X-ray investigations often allow adequate visualization of bony and some soft tissue structures and abnormalities, including calcification (Fig. 3.39).[42,170] They are, of course, routine in the study of trauma (Fig. 3.40), degenerative, inflammatory (Fig. 3.41) and infective joint disease, and in bone tumours. Plain radiographs are often helpful in localizing a connective tissue abnormality for further investigation, as a screening test (e.g. chest radiograph) (Fig. 3.42 and 3.43)[43] or as an adjunct to other imaging techniques (e.g. radionuclide bone scanning) (see Fig. 3.51). Where fine bony or calcific detail needs to be displayed, magnification studies (macroradiography) or tomography can be used. Tomography is a technique employing linear or complex movements of an X-ray tube and film to obtain a focal plane of tissue (the thickness of which can be varied) with blurring of near and distant structures.

Microfocal radiography (Fig. 3.14) is a recent technique not generally available, that can provide certain advantages over conventional macroradiography.[39] These include the earlier identification of sites of erosions in rheumatoid arthritis and osteoarthrosis and the more precise demonstration of disease progression or response to treatment.[17] Microfocal radiography also offers a useful approach to the problems of assessing the severity of pathological change in experimental joint disease in small animals (see Chapter 13) and in analysing sites of calcium deposition (Fig. 2.34).

Double-contrast barium examinations using barium suspension and air are widely used to outline the hollow viscera of the upper and lower gastrointestinal tracts.[61] These methods are of particular value in progressive systemic sclerosis (see Chapter 17) and enteropathic arthropathy (see Chapter 19). Double-contrast studies are relatively quick but requires a meticulous technique for accuracy in the detection of disease (Fig. 3.44). In the investigation of the small intestine, a barium meal and follow-through examination or the direct infusion of barium (small bowel enema) can be performed.

Arthrography requires the introduction of a contrast medium into the joint space in order to detect congenital or acquired abnormalities. The contrast medium is usually a water-soluble, low osmolar iodinated agent. Most arthrography is performed on synovial joints. For small joints, such as the spinal apophyseal (facet arthrography), a single-contrast technique can be used, but for larger joints,

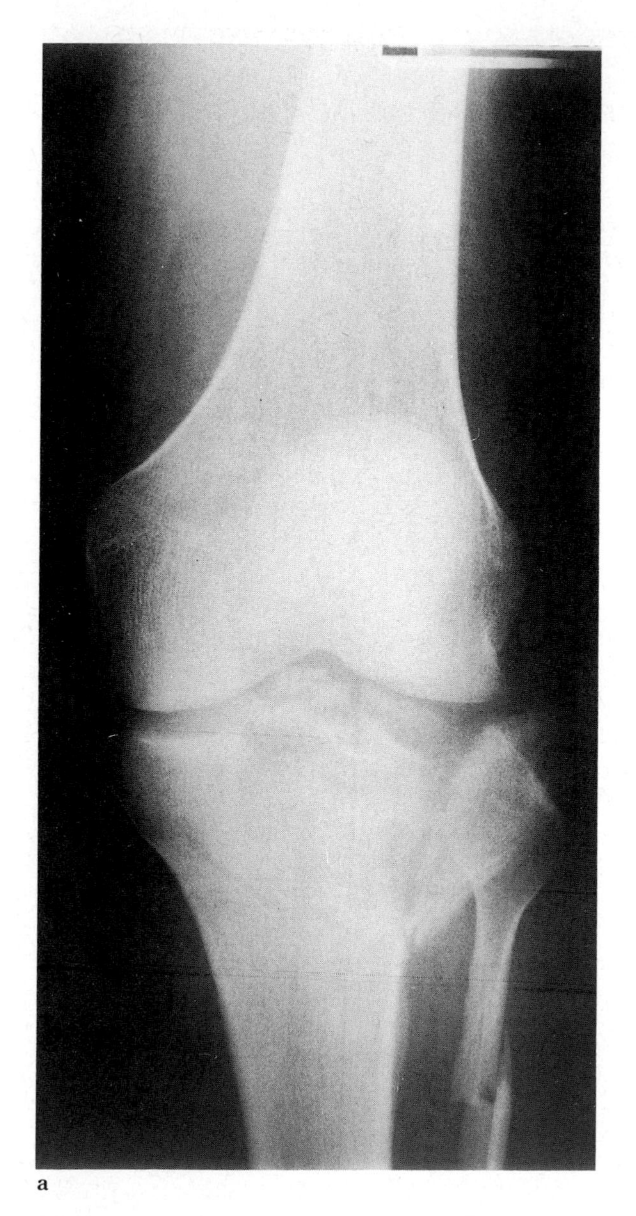

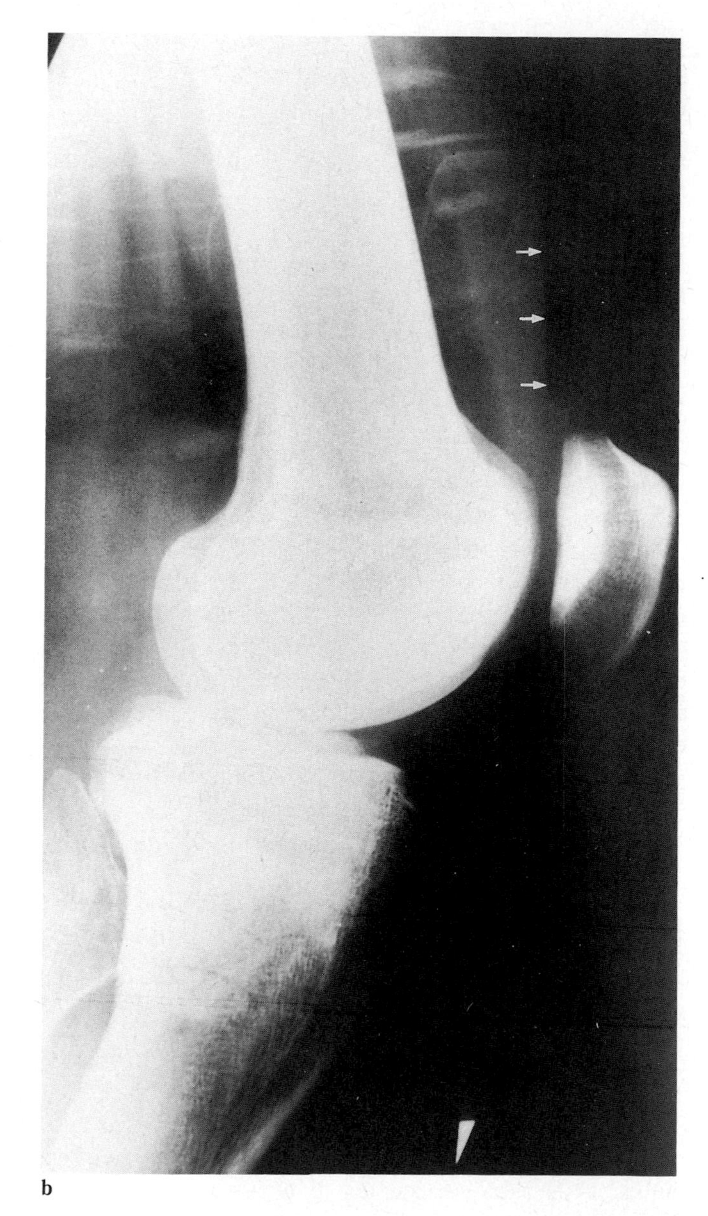

Fig. 3.40 Radiograph of articular fracture.
Left knee radiograph in: **a.** AP and **b.** lateral projections. There are multiple fractures in the upper tibia involving the knee joint associated with a fracture of the fibula shaft. In **b**, taken with a horizontal X-ray beam, there is a fluid level (arrowed) within the suprapatella bursa due to lipohaemarthrosis (see Chapter 24).

such as the knee, a double-contrast method, combining iodinated contrast and air, is usually performed (Fig. 3.45). The procedure is simple but requires expertise, especially in the accurate positioning of the joint and in the interpretation of the resultant radiographs. The technique is not without morbidity, and necessitates the manipulation of an already painful joint. Magnetic resonance imaging is challenging the role of arthrography (compare Fig. 3.45 and Fig. 3.60) and will replace this procedure.

The morphology of an intervertebral disc can be determined by injection of a water-soluble iodinated contrast medium followed by conventional radiography (Fig. 3.46). This technique of discography is invasive but can be used as a provocative test in the assessment of the source of symptomatology in patients with undiagnosed atypical back pain.[108] Magnetic resonance imaging has, however, replaced discography in the initial assessment of the intervertebral disc.

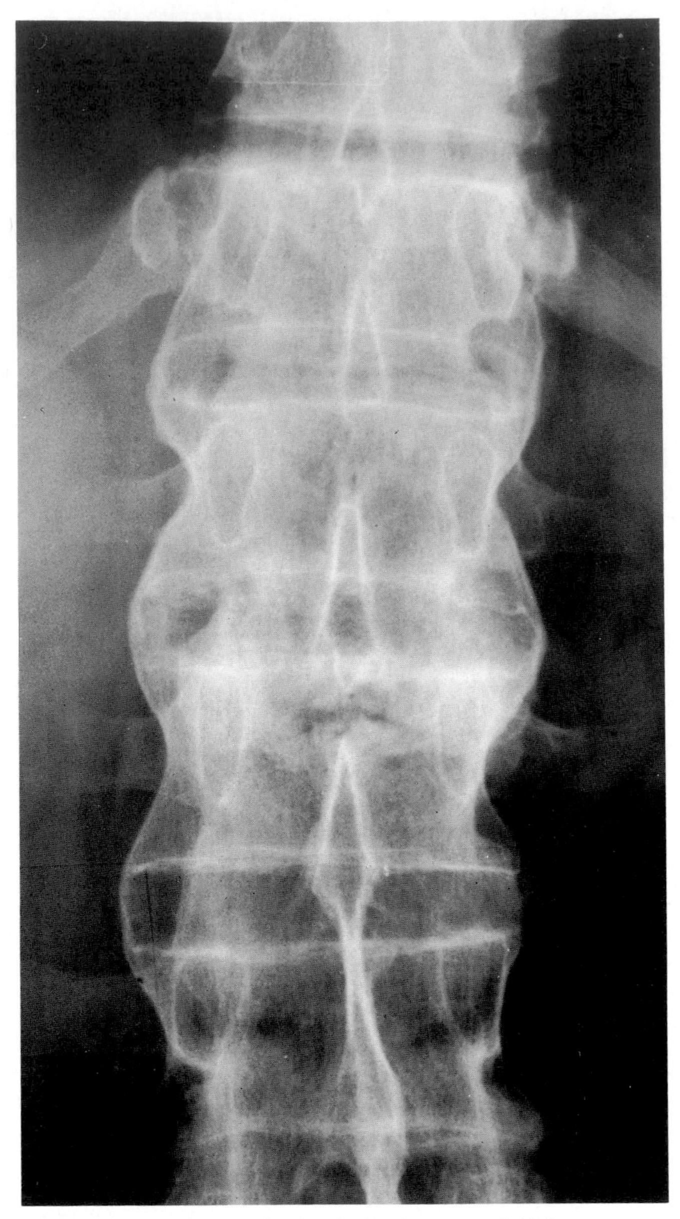

Fig. 3.41 Radiograph of ankylosing spondylitis.
Ankylosing spondylitis involving the lower dorsal and upper lumbar spine. Note ossification of the interspinous ligaments and of the fibres of the annulus fibrosus (syndesmophytes). Courtesy of Dr J T Patton. (See Chapter 19.)

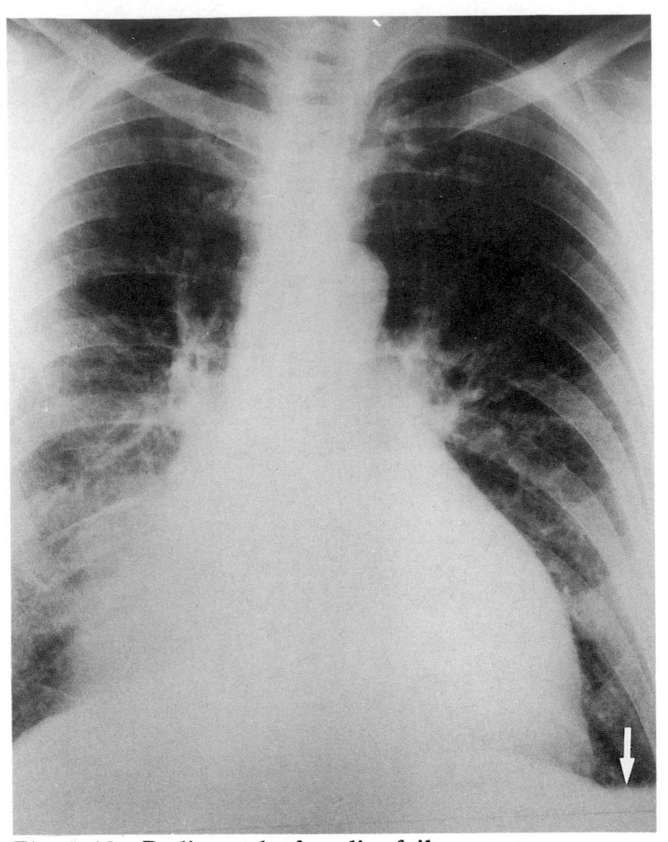

Fig. 3.42 Radiograph of cardiac failure.
Congestive cardiac failure shown on a chest radiograph in a patient with hypertension. The heart is enlarged with a left ventricular configuration. The lungs show evidence of interstitial pulmonary oedema with septal lines due to congested interlobular septa (best seen at the right base) and a left pleural effusion (arrowed).

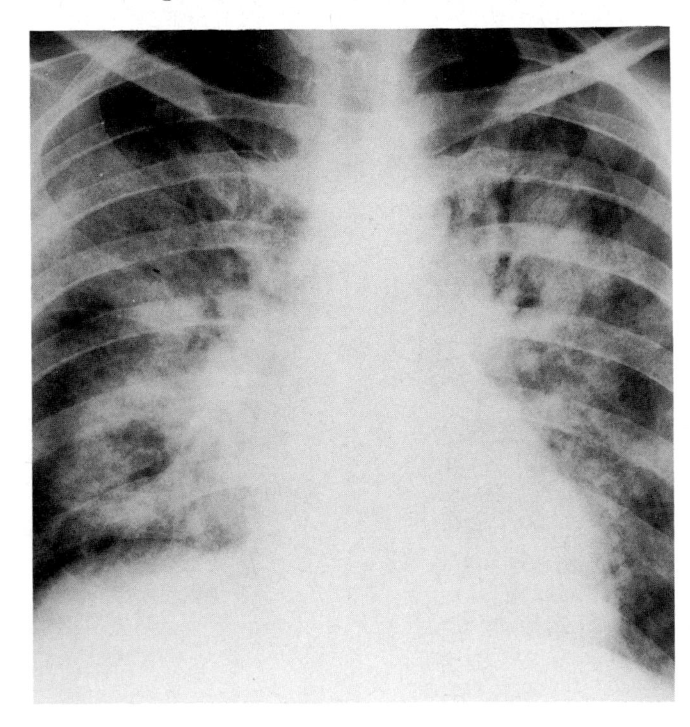

Fig. 3.43 Radiograph of polyarteritis nodosa.
Chest radiograph of a patient with polyarteritis nodosa showing extensive bilateral pulmonary disease characteristic of air-space consolidation with a 'bat's wing' distribution. This appearance is suggestive of bilateral intraalveolar pulmonary oedema but note that the heart size is normal.

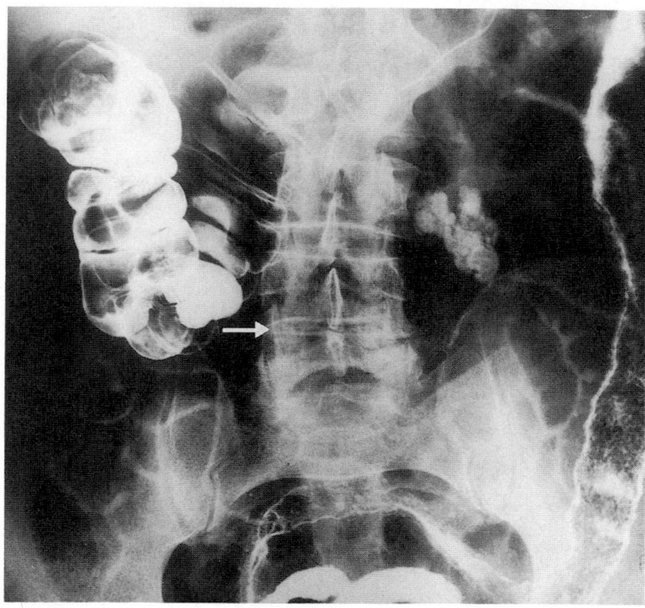

Fig. 3.44 Double-contrast barium enema in ulcerative colitis.
Double-contrast barium enema in a patient with ulcerative colitis involving the rectum, sigmoid and descending colon (see Chapter 19). The affected bowel shows a narrowed and rigid lumen with loss of haustral pattern associated with shallow ulceration ('collar stud' appearance). Note a degenerate and calcified L4/5 disc (arrowed).

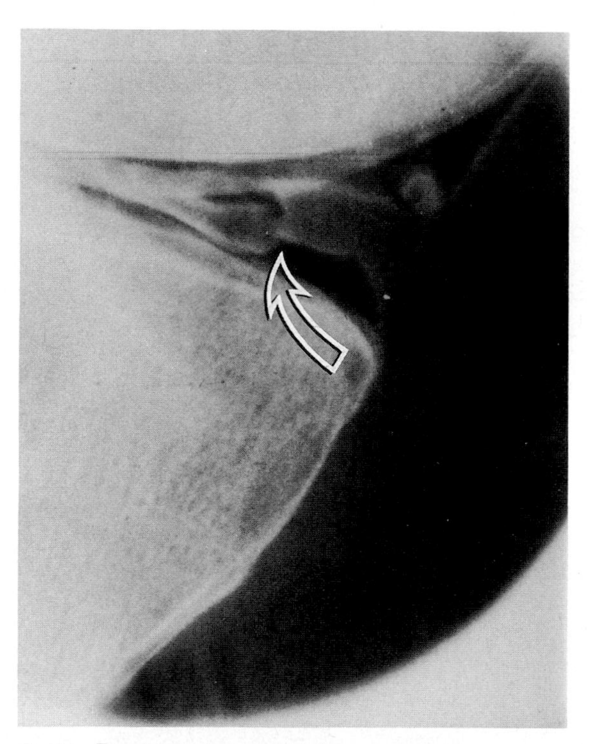

Fig. 3.45 Double-contrast arthrogram after tear of a knee joint meniscus.
Double-contrast knee arthrogram showing a 'bucket-handle' tear of the medial meniscus (arrowed). Courtesy of Dr J T Patton.

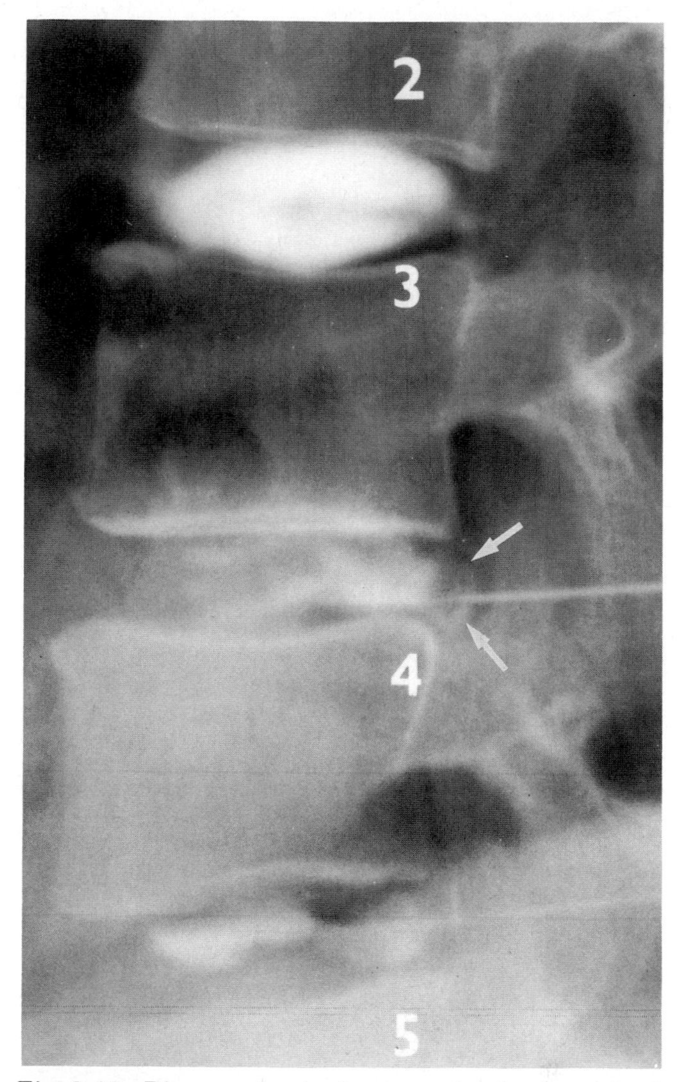

Fig. 3.46 Discography in degenerative intervertebral disc disease.
Lumbar discography at three adjacent levels. The L2/3 disc shows a normal appearance with the nucleus appearing bilocular. Both L3/4 and L4/5 discs show degenerative change. Note the contrast medium bulging beyond the posterior vertebral margin at L3/4 level (arrowed) being contained by the peripheral annular fibres/posterior longitudinal ligament. Discogram needles *in situ* at L3/4 and L4/5. Courtesy of Dr J T Patton. (See Chapter 23.)

Ultrasound

Ultrasound is a technique for imaging soft tissues. Short pulses of high frequency sound are used being produced by a transducer constructed of a piezoelectric material. Transducers act as both transmitter and receiver and focus the ultrasound into a fine beam. When the beam is passed through tissues of differing acoustic impedance some sound is reflected at the boundary zones. These 'reflection

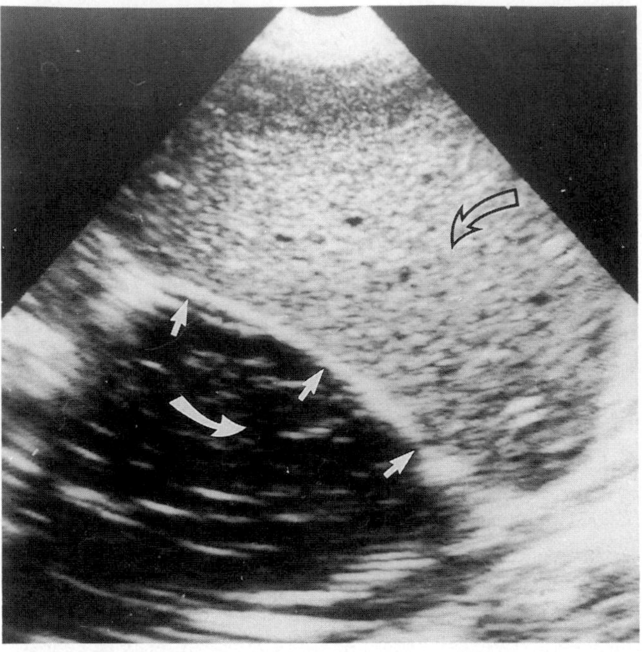

Fig. 3.47 Ultrasound scan in traumatic haematoma and empyema.
A large empyema, secondary to a traumatic haematoma, at the right lung base (solid curved arrow) on a longitudinal ultrasound scan. The diaphragm (straight arrows) is inverted and the liver (open curved arrow) is displaced. Note echoes within the empyema associated with minimal acoustic shadowing posteriorly which differentiates it from a simple pleural effusion.

echoes' are displayed as dots on a monitor screen. The position of each dot indicates the tissue interface that forms the echo. The sectional images so obtained can be displayed on a television monitor or transferred to hard copy film (Figs 3.47 and 3.48).

Ultrasound is safe, fast and inexpensive. It permits a rapid anatomical survey of the abdominal organs in particular (Fig. 3.48a and b), but can be applied to other parts of the body. Both static and real-time scanning systems are available. A real-time display can be particularly useful in areas of tissue motion, e.g. the heart. In addition, by the use of the Doppler principle, blood flow can be measured.[158] New techniques are being developed, including the use of microbubbles from fluids injected intravenously to outline cardiac chambers, to indicate intracardiac shunts and to assess myocardial tissue perfusion.[22]

Difficulties arise at air-to-bone or bone-to-soft-tissue interfaces where total reflection of the ultrasound beam can occur. Large echoes and resultant 'blind' areas can thus be produced. These problems are exemplified by the reflections that occur in abdominal scanning when bowel gas overlies organs, and by the air–bone interface in cardiac scanning. The role of ultrasound has been directed increasingly towards guided needle biopsy and catheter placement.

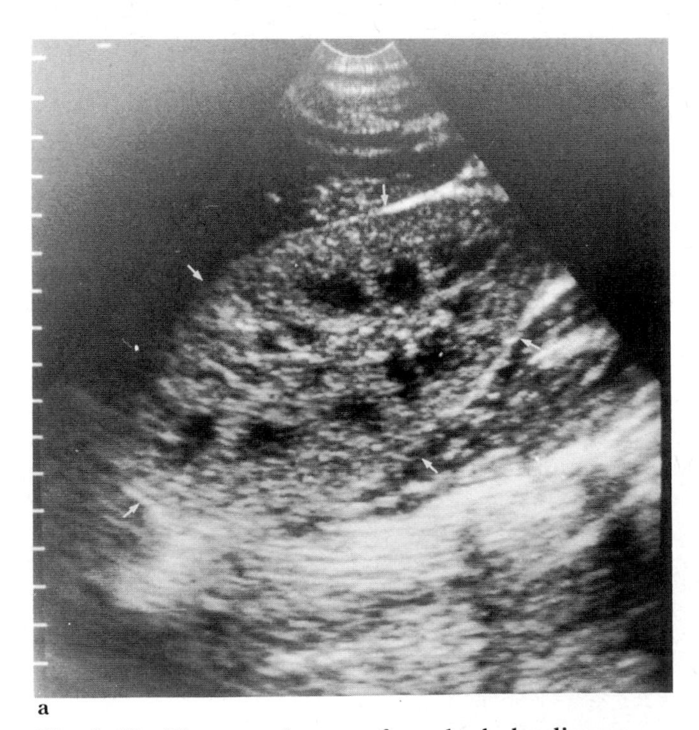

a

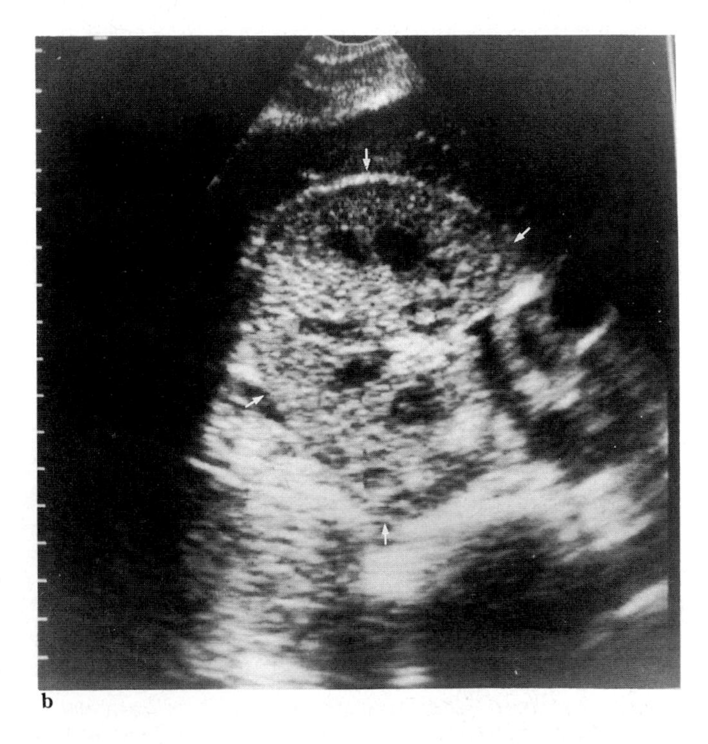

b

Fig. 3.48 Ultrasound scans of renal tubular disease.
a. Longitudinal and **b.** transverse ultrasound scans through a swollen right kidney (arrowed) with diffuse increase in echo pattern. Renal pyramids are shown as echolucent areas. Appearances were due to acute renal disease secondary to ethylene glycol ingestion.

Angiography

Conventional arteriography is invasive. The procedure requires percutaneous puncture of the appropriate vessel or catheterization from a peripheral artery, such as the femoral, and the injection of an iodinated contrast medium. The technique, involving the prior insertion of a guide-wire, is usually performed under local anaesthesia. Special catheters can be employed to facilitate selective catheterization of specific vessels, e.g. renal or craniocervical arteries (Fig. 3.49). The approach is being used with increasing frequency to enable interventional procedures such as therapeutic embolization or dilatational angioplasty.[172]

Digital subtraction angiography (DSA) employes a computer-assisted electronic subtraction and enhancement technique to provide a digital image of contrast medium within the vascular system (Fig. 3.50).[81,88] Contrast material can be injected intravenously (Fig. 3.50) or intraarterially (Fig. 3.49). Each has advantages and disadvantages; in the provision of vascular detail, intraarterial injection is preferred. Both the dose and the rate of administration of contrast medium are lower in intraarterial studies than in either the intravenous DSA technique or in conventional arteriography. In addition, with DSA, the radiation dose, film cost per patient and procedural time are all significantly reduced compared with conventional arteriography. Digital images can be visualized and manipulated at once, a particular advantage during interventional procedures such as angioplasty and in the study of trauma. Disadvantages of DSA compared with conventional arteriography include poorer spatial resolution and occasional misregistration artefacts due to patient movement.

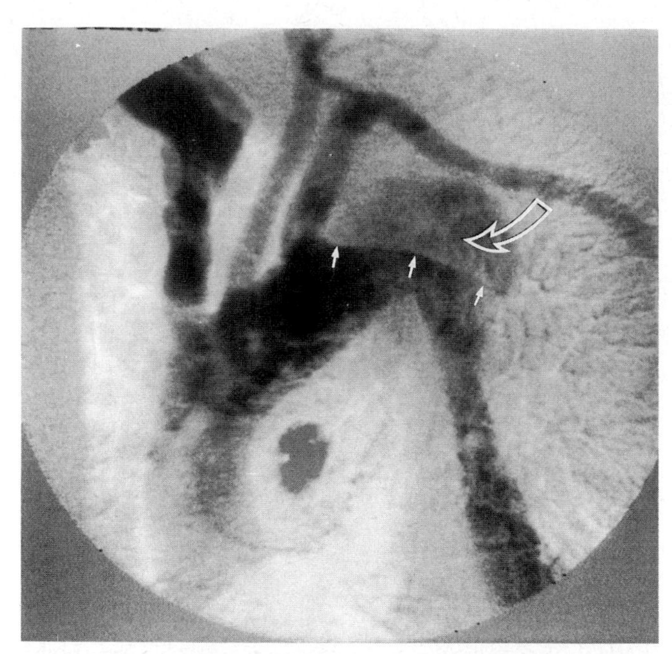

Fig. 3.50 Digital subtraction aortogram in aneurysm. Digital intravenous subtraction arch aortogram showing a dissection distal to the origin of the left subclavian artery with an intimal flap (straight arrows) and contrast medium in the false lumen (curved arrow). Dissection in Marfan's syndrome (see p. 340) begins proximally, extends further.

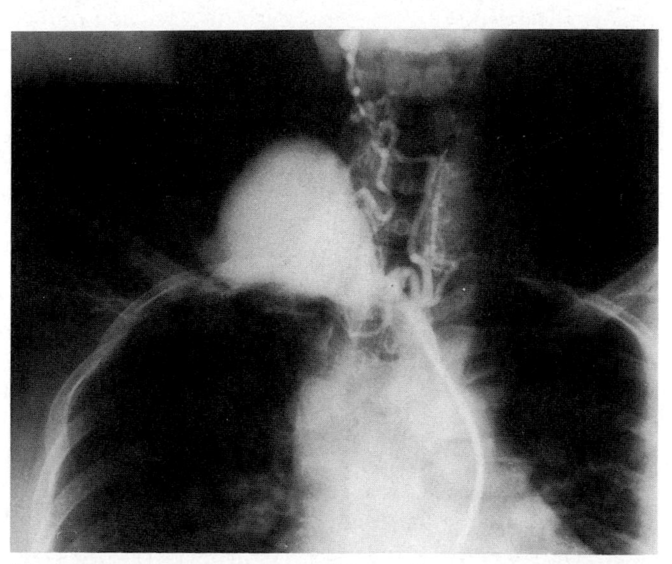

Fig. 3.49 Arteriogram in Ehlers–Danlos syndrome. Traumatic innominate artery aneurysm on a conventional selective arteriogram in a 26-year-old female with Ehlers–Danlos syndrome. Courtesy of Dr V Cope. (See Chapter 9.)

Radionuclide imaging

Radionuclide imaging is made possible by the γ-rays that are emitted by injected radionuclides. The images provide information on the spatial and temporal distribution of radioactivity in the body.[111] A radionuclide is a radioactive species of an atom with a characteristic atomic mass. The mass is described as a superscript of the atomic symbol so that $^{99}Tc^{m}$ refers to technetium-99m, the most frequently used radionuclide. Technetium is an artificial element derived from molybdenum-99 and 'm' designates the metastable variety. Most radionuclides are made artificially and disintegrate spontaneously emitting radiation at a characteristic energy level for each species. A radionuclide can be incorporated into chemical compounds to produce a radiopharmaceutical agent for targetting specific activities. For example, $^{99}Tc^{m}$-methylene diphosphate ($^{99}Tc^{m}$-MDP) is used to tag osteoblastic activity in skeletal scanning (Fig. 3.51a), gallium-67 (^{67}Ga) citrate to detect localized sepsis (Fig. 3.52) and iodine-125 (^{125}I) is incorporated with fibria

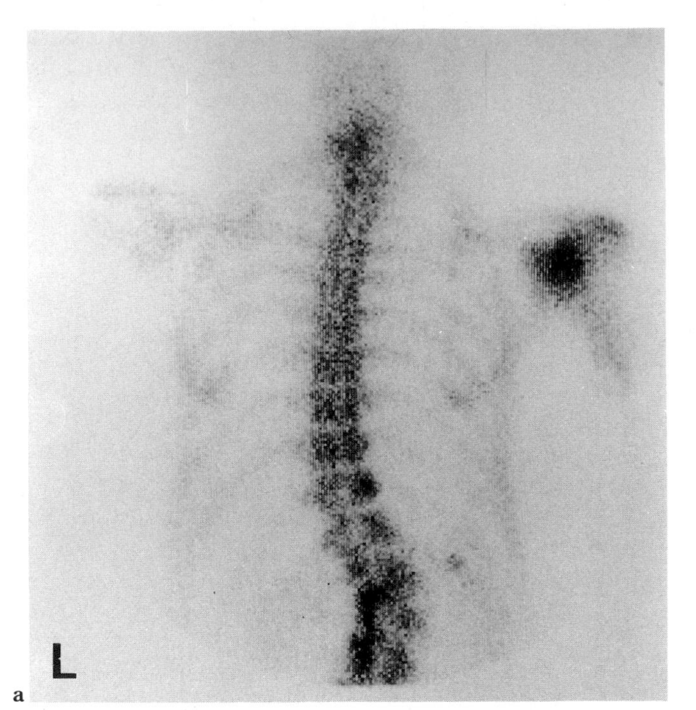

b

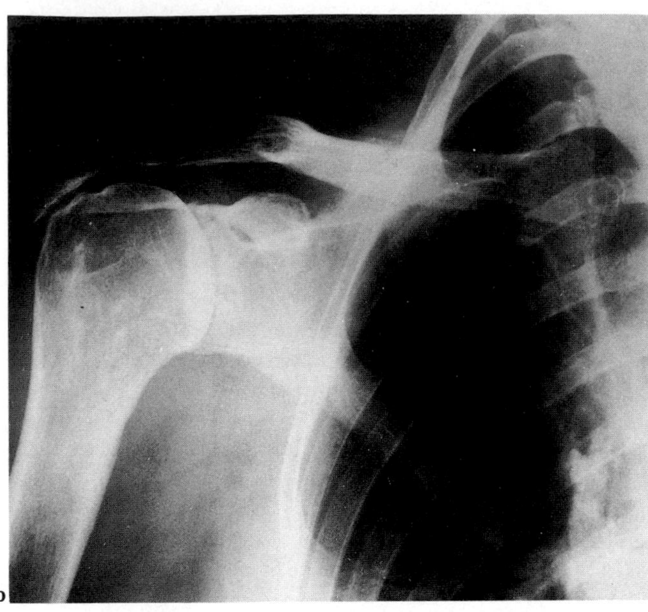

Fig. 3.51 Radionuclide scan and radiograph in shoulder joint osteoarthrosis.
a. Radionuclide bone scan, using technetium-99m-methylene diphosphonate (^{99}Tc^{m}-MDP) demonstrating non-specific increased uptake in the right shoulder joint and lower spine. **b.** Radiograph of the right shoulder demonstrates severe degenerative disease with loss of joint space, eburnation of the humeral head and subchondral cyst formation (see Chapter 22).

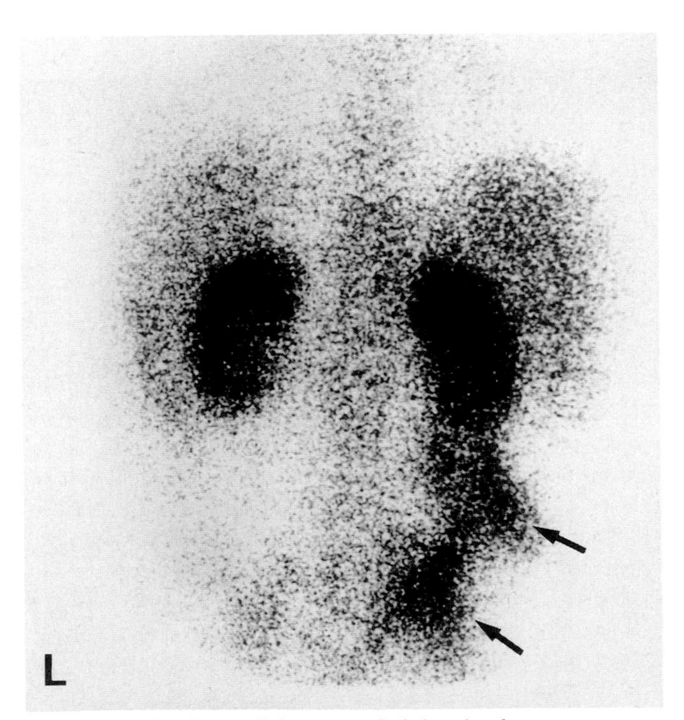

Fig. 3.52 Radionuclide scan of abdominal mass.
Gallium-67 citrate radionuclide scan of an abdomen 96 h after intravenous injection. There is increased uptake of radionuclide in the right lower quadrant (arrowed) in keeping with abscess formation. Note increased uptake of radionuclide in both kidneys due to poor renal function. Courtesy of Dr H J Tesla.

nogen in the investigation of sites of active venous thrombosis such as those encountered in systemic lupus erythematosus (Chapter 14).

Computer processing allows rapid, dynamic radionuclide imaging. Temporal changes in radionuclide concentration can be observed. This has particular relevance in quantitative investigations such as the study of cardiographic or of renal function in ankylosing spondylitis and progressive systemic sclerosis, respectively. Tomographic images of the distribution of radioactivity can also be obtained by emission computed tomography. Single photon emission computed tomography (SPECT) uses conventional radionuclides and a gamma camera.[37,116] Positron emission tomography (PET) requires close proximity to a cyclotron for the production of appropriate short-lived radionuclides such as those of carbon, nitrogen and oxygen. Such techniques can provide unique information on, for example, regional cerebral blood flow and perfusion, and brain function.[97]

Myelography

Myelography is a radiological technique in which the intrathecal injection of contrast medium is used to visualize

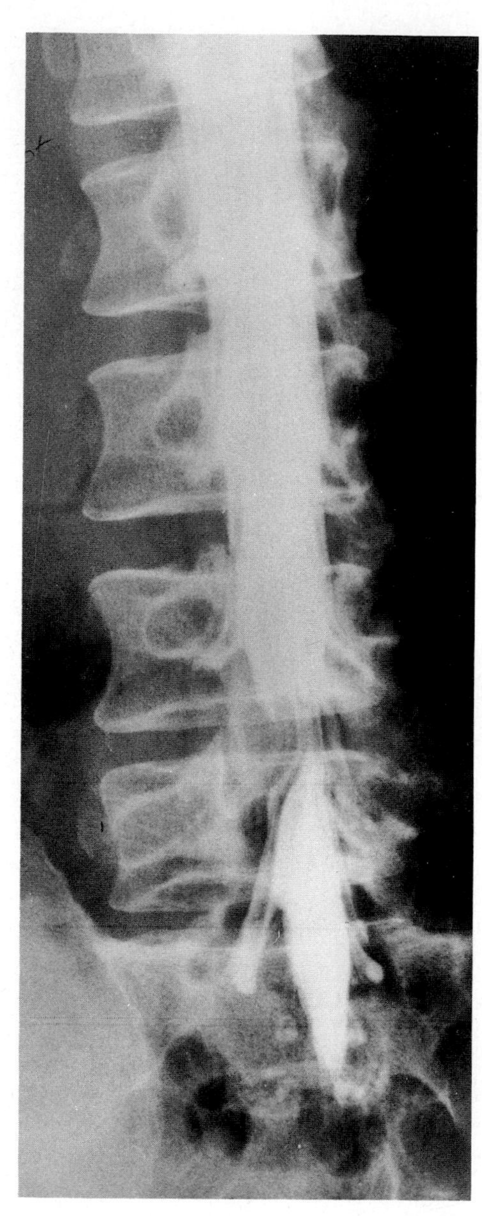

Fig. 3.53 Radiculogram of prolapsed intervertebral disc.
Radiculogram demonstrating a posterolateral prolapsed L4/5 disc compressing and displacing nerve roots. (Same patient as Fig. 3.57). (See Chapter 23.)

the spinal cord, cauda equina, subarachnoid space and nerve root sleeves.[143,9] The preferred contrast media are water-soluble, non-ionic and low osmolar iodinated agents providing good and safe visualization of neural tissue and of pathological processes within the spinal canal. When confined to the lumbar spine, the technique is often referred to as radioculography (Fig. 3.53). Myelography can be combined with CT (CT myelography) enabling spinal disease to be detected with increased sensitivity. Myelography, however, is an invasive technique and its role in the evaluation of spinal disorders is now being superseded by MRI.

Computed tomography

The advent of computed tomography (CT) has revolutionized neuroradiological practice. This technique uses general-purpose body CT scanners. An X-ray source produces a finely collimated beam of X-rays traversing a transaxial section (2–10 mm thickness) of the head or body. An array of sensitive detectors records the attenuation of an X-ray beam leaving the section at intervals round the circumference. A numerical value can be assigned to small volumes of tissue in the cross section; the value corresponds to the ability of a volume to attenuate the X-ray beam. The numbers are referred to as Hounsfield numbers and the scale, based on water as zero, as the Hounsfield scale. The resultant digital image data can be processed for display on a TV monitor, stored on magnetic tape or disc, or transferred to hard copy film.

The cross-sectional, reconstructed image is consequently based on precise X-ray attenuation characteristics of the biological tissue. The image is derived by digital-to-analogue conversion using a fixed grey scale representing bone as white, air as black and biological soft tissues as varying shades of grey. The image is displayed on a TV monitor providing the opportunity for the operator to interact with and interrogate the numeral data. Computed tomography provides information about tissue density differences in uniformly thin slices of tissue. Soft tissues, fat, gas and bone are more clearly differentiated than with conventional radiography. The ability to discriminate changes which accompany connective tissue disease can be further enhanced by the intravenous administration of an iodine-containing contrast medium (Fig. 3.54). Although certain patterns of enhancement are recognized in the normal and in the diseased state, there is significant overlap in the attenuation values. Histological characterization from the numerical data is therefore not possible.

A variety of software facilities are available, including the reformatting of multiple transaxial sections in any chosen plane and true three-dimensional image reconstructions (Fig. 3.55).[71] The latter are of particular value in craniofacial disorders, and in the assessment of the bony spine and pelvis. Quantitative CT data obtained from two different X-ray energies (dual energy CT) can provide improved accuracy in the estimation of bone mass and trabecular bone mineral in both the axial and appendicular skeleton by

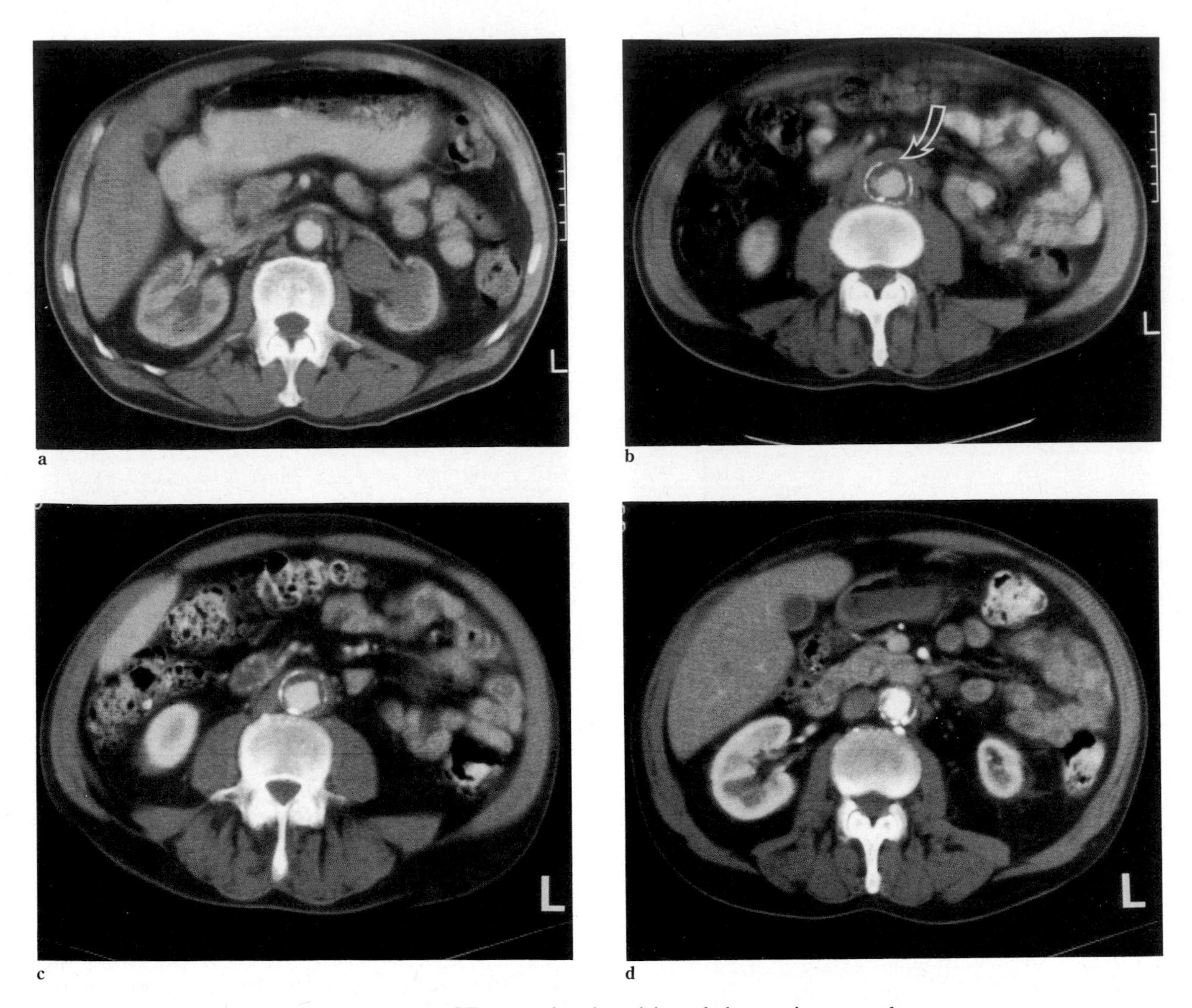

Fig. 3.54 Enhanced computer tomograph (CT) scan of periaortitis and obstructive uropathy.
Periaortitis (arrowed) producing bilateral obstructive uropathy, most marked on the left—**a.** on enhanced CT and **b.** below the renal hila. Repeat examination at comparative levels: **c** and **d** demonstrate resolution of the periaortitis and obstructive uropathy following steroid therapy. Note the ring calcification and thrombus within the abdominal aorta. Periaortitis has been implicated in the origin of retroperitoneal fibrosis (see p. 304).

separating high atomic number calcium from fat in which the elements are of low atomic number.[2,27] Computed tomography is a non-invasive X-ray procedure, although the use of an intravenous contrast medium is not without risk. Computed tomographic scanning can be performed as an outpatient procedure or on ill patients on life-support systems. Data acquisition time per section is usually 2–3 s. Percutaneous biopsy of lesions under CT guidance is now widely practised.

Magnetic resonance imaging

Magnetic resonance imaging is as an increasingly important diagnostic technique which is particularly useful and effective in the investigation of the central nervous system, spine, pelvis, musculoskeletal and cardiovascular systems.[20,87,91,98,149] At present, the image quality of MRI studies of the abdomen is degraded by motion artefacts. However, technical advances related to faster data acquisi-

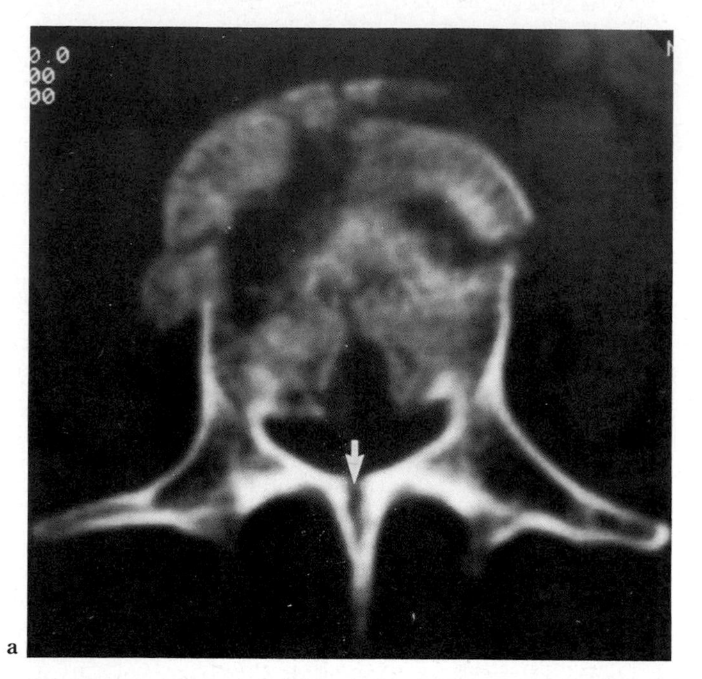

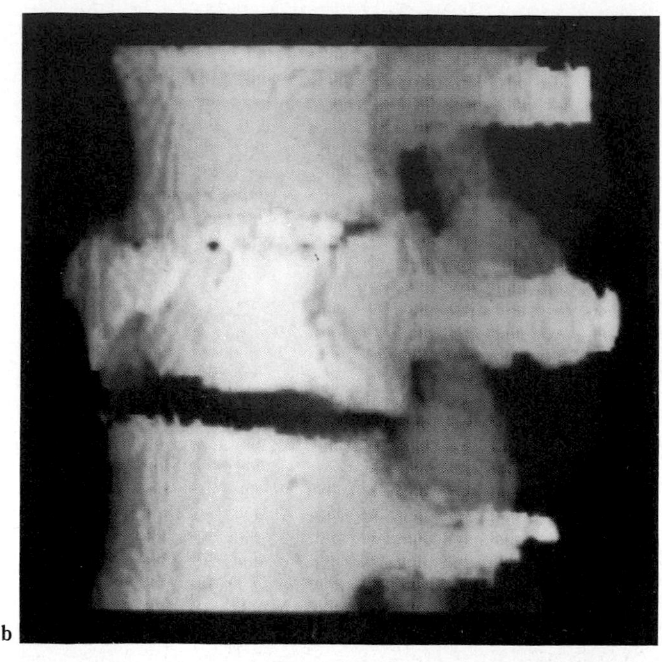

Fig. 3.55 Computed tomographic (CT) scan of vertebral fracture.
Burst fracture of a lumbar vertebra on CT. **a.** High resolution transverse section. **b.** Three-dimensional (3-D) image reconstruction (lateral view). The individual bony fragments are well shown and there is an associated fracture of the spinous process (arrowed).

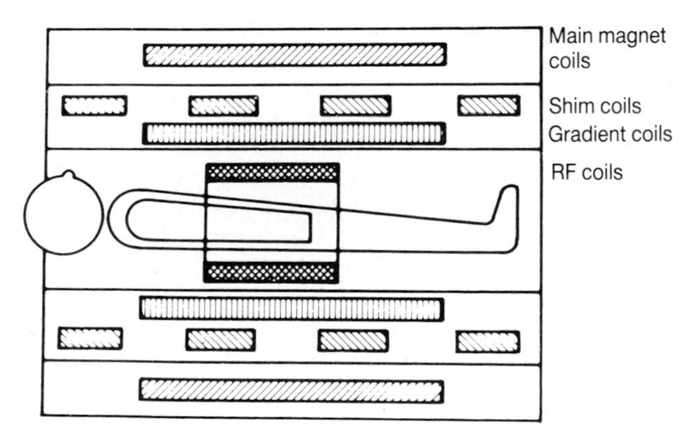

Fig. 3.56 Diagram of magnetic resonance imaging (MRI).
Longitudinal section through an MRI magnet with a patient positioned in the radiofrequency body coil for scanning the abdomen. The main magnetic field is produced by the magnet coils. Homogeneity of the main field, and thus imaging space, is optimized by the use of shim coils which adjust for the effects of iron in the vicinity (e.g. structural steel). A gradient system comprises separate X, Y and Z coils which, when energized, modify the main magnetic field during imaging for short periods. Such a system permits slices of tissue to be identified in any imaging plane. The radiofrequency coils are required to transmit and receive the radiofrequency signals to and from the patient, thereby producing the appropriate signal data for image formation. Reproduced by kind permission from Sutton.[156]

tion and gating techniques are likely to overcome this problem. Further developments await full evaluation.[23]

Magnetic resonance imaging utilizes non-ionizing radiation in the form of a static main magnetic field (currently 0.01–2.0 Tesla (T)) with gradient and pulsed radiofrequency magnetic fields to obtain images directly, in any tomographic plane (Fig. 3.56). Only certain nuclei, i.e. those with an odd number of protons or neutrons, e.g. hydrogen (^{1}H), sodium (^{23}Na) and phosphorus (^{31}P), generate a magnetic moment. The hydrogen nucleus, or proton, of water and fat is the most abundant in biological tissue and also the most sensitive for detection. It is the one most commonly used in human studies, accounting for the term 'proton' magnetic resonance imaging.

The application of appropriate radiofrequency pulses to the part of the body under investigation in the main field induces resonance (i.e. change in aligment) of a specific nuclear species. This change is recorded as a radiofrequency signal and provides the digital data from which images can be formed. An MRI system usually operates at a fixed field strength and radiofrequency. The frequency is determined by the nuclear species under investigation.

Unlike CT in which images are created from a fixed scale of linear attenuation values, the signal intensity of tissue in magnetic resonance images is dependent on a variety of

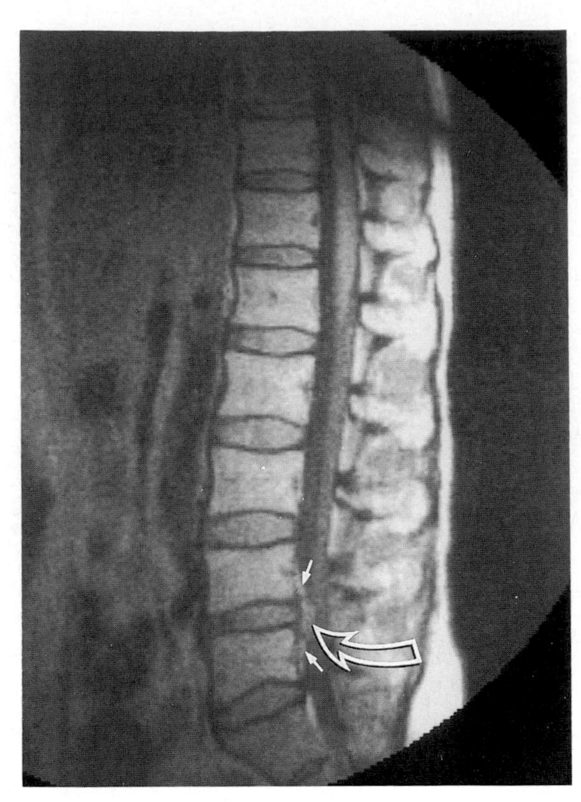

Fig. 3.57 Magnetic resonance (MR) scan of intervertebral disc herniation.
Midline sagittal T_1-weighted MR scan of the lower spine showing a posterior herniated and partially fragmented L4/5 disc (curved arrow) compressing spinal nerve roots and elevating epidural fat (straight arrows). The conus of the spinal cord is opposite T12/L1. (Same patient as Fig. 3.53.) (See Chapter 23.)

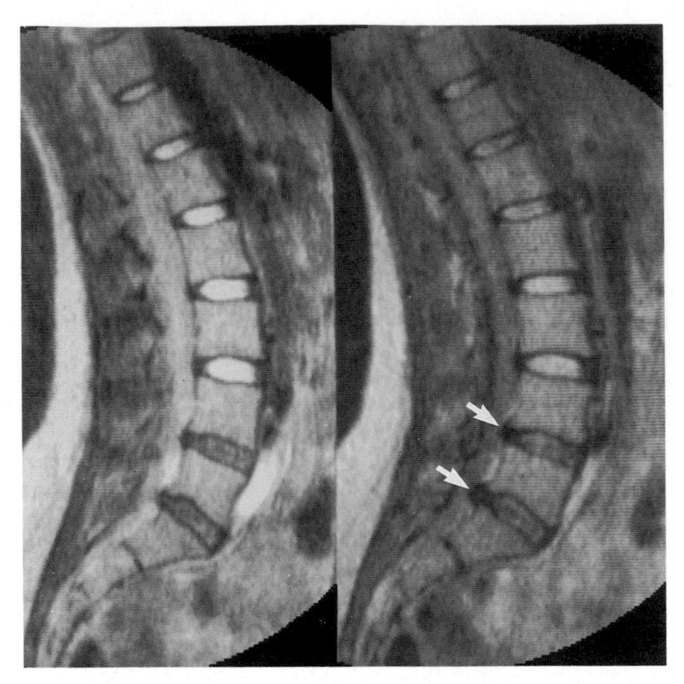

Fig. 3.58 Magnetic resonance (MR) scan of intervertebral disc disease.
Contiguous sagittal T_2-weighted MR images in a young adult showing degenerative and posteriorly herniated L4/5 and L5/S1 intervertebral discs (arrowed). Note the higher signal from the nucleus pulposus of the normal hydrated discs above with a low signal cleft within. Reproduced by kind permission from Galasko and Isherwood.[47]

tissue and technical parameters which provide a correspondingly wide range of tissue contrast (Figs 3.57 and 3.58). Tissue parameters include proton density, relaxation times, blood flow, perfusion and diffusion, all of which can be influenced by the applied radiofrequency pulse sequences. It is important, therefore, that both the imaging technique and the disease process under investigation are well understood for the proper interpretation of MR images. The computer, display unit and digital storage facilities are similar to those employed in CT.

Proton MRI has particular advantages compared with CT. These include a much increased soft-tissue contrast and discrimination capability; an ability to image in any plane; an absence of bone, plastic and some metal artefacts; and the use of non-ionizing radiation. Exquisite morphological detail can be achieved, together with the differentiation of diseases and the provision of functional information. There is good delineation of the fascial planes, subcutaneous fat, vessels, nerves, ligaments, tendons and muscles (Fig. 3.59). The joint capsule, cartilages and menisci, as well as the articulating bone and growth plates, can be distinguished (Fig. 3.60). Although compact bone gives no signal, the haemopoetic and fatty marrow of cancellous bone is well-visualized and the two components of the intervertebral disc (annulus fibrosus and nucleus pulposus) can be differentiated (Figs 3.57; 3.58). Blood velocity can be measured *in vivo* and various disease processes monitored during treatment. The use of paramagnetic tracers such as gadolinium diethylene triamine pentaacetic acid (Gd-DTPA), to enhance soft-tissue contrast and to improve the discrimination of pathological disease, is being evaluated. The current disadvantages of MRI include high cost, limited availability, relatively long scan times (orders of minutes) and, in some patients, a claustrophobic effect. The rapid progress already achieved over only a few years bears witness to the potential future role of MRI as an increasingly powerful diagnostic tool.

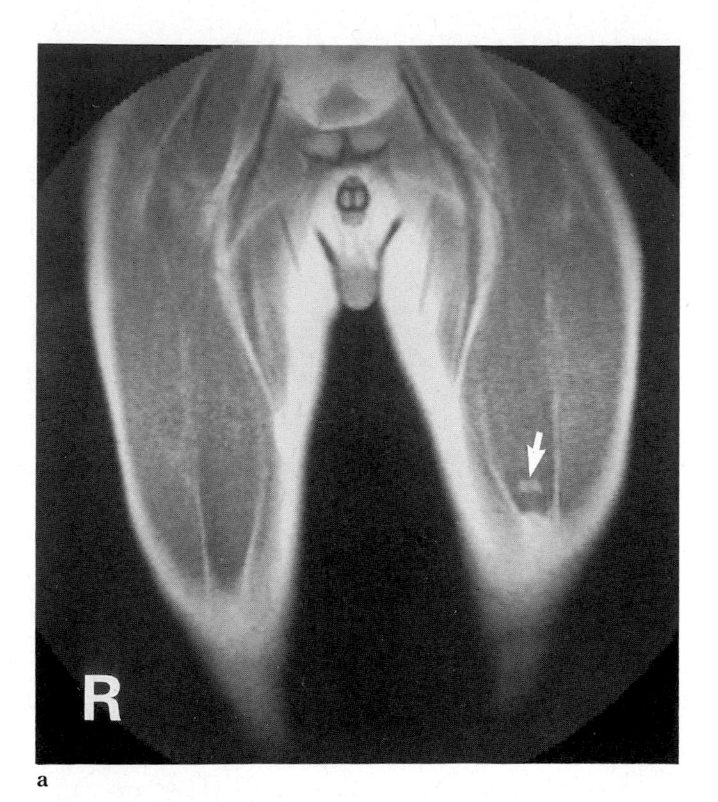

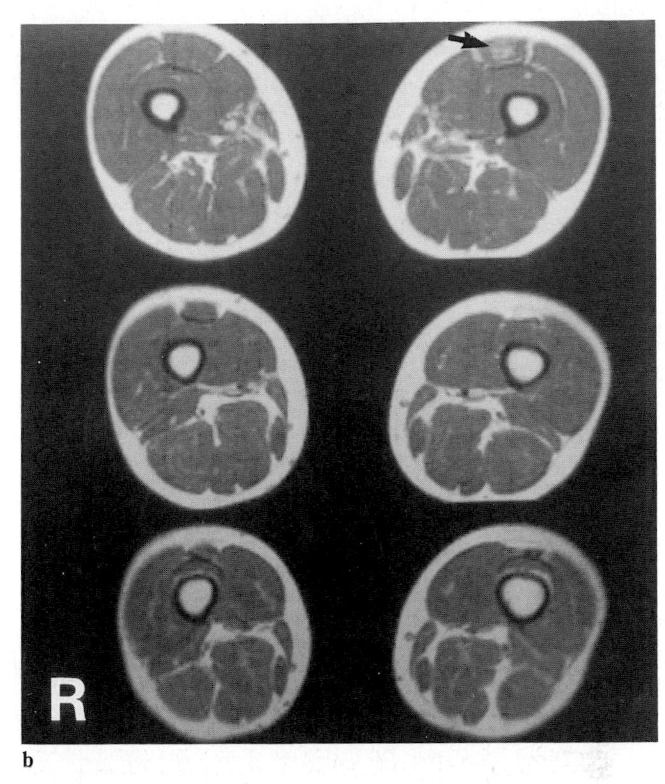

Fig. 3.59 Magnetic resonance scans of ruptured rectus femoris tendon.
Rupture of rectus femoris tendon associated with an intramuscular haematoma (arrow) on: **a.** coronal and **b.** transverse T_1-weighted MR scans. Reproduced by kind permission from Galasko and Isherwood.[47]

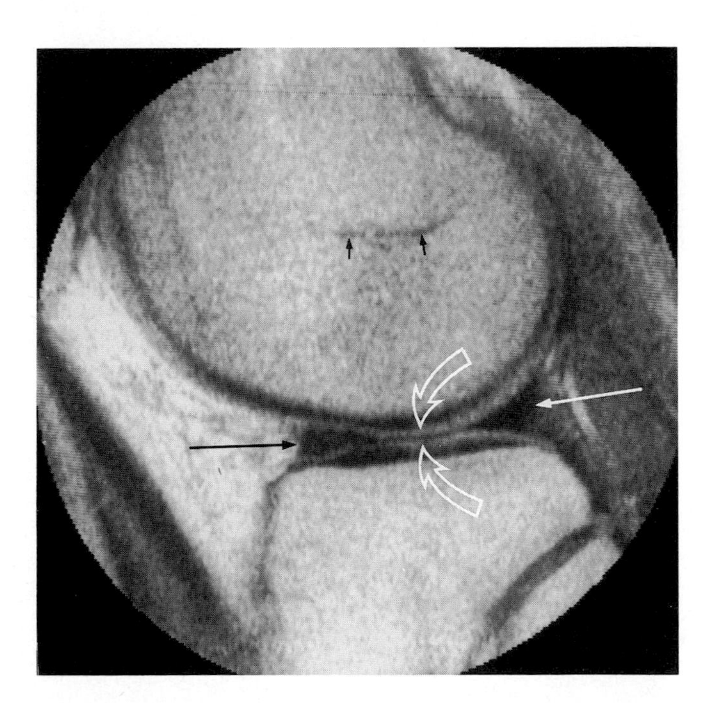

In vivo magnetic resonance spectroscopy in humans is still at an early stage of development and requires magnetic field strengths greater than 1.5 T.

Fig. 3.60 Magnetic resonance scan of knee joint meniscus.
Normal adult knee on a sagittal T_1-weighted MR scan. Menisci are shown as low signal (long straight arrows) with high signal from articular cartilage (curved arrow). A linear low signal in the femoral condyle (small straight arrow), in keeping with trabecular bone, marks the previous location of the epiphyseal growth plate. Reproduced by kind permission from Galasko and Isherwood.[47]

REFERENCES

1. Akeson W H, Woo S L-Y, Taylor T K F, Ghosh P, Bushell G R. Biomechanics and biochemistry of the intervertebral disks: need for correlation studies. *Clinical Orthopaedics and Related Research* 1977; **129**: 133–40.

2. Andersen J, Nielsen H E. Assessment of bone mineral content and bone mass by non-invasive radiologic methods. *Acta Radiologica* 1986; **27**: 609–17.

3. Appleton J. The fine structure of a surface layer over the fibrous articular tissue of the rat mandibular condyle. *Archives of Oral Biology* 1958; **23**: 719–23.

4. Archer C W, Rooney P, Cottrill C P. Cartilage morphogenesis *in vitro*. *Journal of Embryology and Experimental Morphology* 1985; **90**: 33–48.

5. Ashton E H. Primate locomotion: some problems in analysis and interpretation. *Philosophical Transactions of the Royal Society of London B* 1981; **292**: 77–87.

6. Aspden R M, Hukins D W L. The lamina splendens of articular cartilage is an artefact of phase contrast microscopy. *Proceedings of the Royal Society of London B* 1979; **206**: 109–13.

7. Aspden R M, Yarker Y E, Hukins D W L. Collagen orientations in the meniscus of the knee joint. *Journal of Anatomy* 1985; **140**: 371–80.

8. Balazs E A, Bloom G D, Swann D A. Fine structure and glycosaminoglycan content of the surface layer of articular cartilage. *Federation Proceedings* 1966; **25**: 1813–6.

9. Banna M. *Clinical Radiology of the Spine and Spinal Cord.* Maryland: Aspen Systems Corporation, 1985.

10. Beard H K, Roberts S, O'Brien J P. Immunofluorescent staining for collagen and proteoglycan in normal and scoliotic intervertebral discs. *Journal of Bone and Joint Surgery* 1981; **63-B**: 529–34.

11. Bird M D T, Sweet M B E. A system of canals in semilunar menisci. *Annals of the Rheumatic Diseases* 1987; **46**: 670–3.

12. Birkett J. Observations on healthy and morbid articular tissues. *Guy's Hospital Reports* 1848; **6**: 36–52.

13. Bloebaum R D, Wilson A S. The morphology of the surface of articular cartilage in adult rats. *Journal of Anatomy* 1980; **131**: 333–46.

14. Bloom W, Fawcett D W. *A Textbook of Histology*. Philadelphia and London: W B Saunders & Co., 1962: 144–52.

15. Boyde A. Pros and cons of critical point drying for SEM. *Scanning Electron Microscopy* 1978; **II**: 303–14.

16. Brereton J, Gardner D L. Unpublished observations.

17. Buckland-Wright J C, Carmichael I, Walker S R. Quantitative microfocal radiography accurately detects joint changes in rheumatoid arthritis. *Annals of the Rheumatic Diseases* 1986; **45**: 379–83.

18. Buckwalter J A, Cooper R R, Maynard J A. Elastic fibres in human intervertebral discs. *Journal of Bone and Joint Surgery* 1976; **58-A**: 73–6.

19. Buckwalter J A, Pedrini-Mille A, Pedrini V, Tudisco C. Proteoglycans of human infant intervertebral disc. *Journal of Bone and Joint Surgery* 1985; **67-A**: 284–94.

20. Burk D L, Dalinka M K, Schiebler M L, Cohen E K, Kressel H Y. Strategies for musculoskeletal magnetic resonance imaging. *Radiologic Clinics of North America* 1988; **26**: 653–72.

21. Bushell G R, Ghosh P, Taylor T F K, Akeson W H. Proteoglycan chemistry of the intervertebral disks. *Clinical Orthopaedics and Related Research* 1977; **129**: 115–23.

22. Butler B D. Production of microbubbles for use as echo contrast agents. *Journal of Clinical Ultrasound* 1986; **14**: 408–12.

23. Bydder G M. Magnetic resonance imaging: present status and future perspectives. *British Journal of Roentgenology* 1988; **61**: 889–97.

24. Cameron C H D, Gardner D L, Longmore R B. The preparation of human articular cartilage for scanning electron microscopy. *Journal of Microscopy* 1976; **108**: 1–12.

25. Cameron G R. *Pathology of the Cell*. Edinburgh: Oliver and Boyd, 1952.

26. Campbell B G. Body structure and posture. In: *Human Evolution* (2nd edition). Chicago: Aldine Publishing Company, 1974: 123–44.

27. Cann C E. Quantitative CT for determination of bone mineral density: a review. *Radiology* 1988; **166**: 509–22.

28. Carlsen I C. Reconstruction of true surface-topographies in scanning electron microscopes using backscattered electrons. *Scanning* 1985; **7**: 169–77.

29. Clarke I C. A method for the replication of articular cartilage surfaces suitable for the scanning electron microscope. *Journal of Microscopy* 1971a; **93**: 67–71.

30. Clarke I C. Surface characteristics of human articular cartilage—a scanning electron microscope study. *Journal of Anatomy* 1971b; **108**: 23–30.

31. Clarke I C. Human articular surface contours and related surface depression frequency studies. *Annals of the Rheumatic Diseases* 1971c; **30**: 15–23.

32. Clarke I C. The microevaluation of articular surface contours. *Annals of Biomedical Engineering* 1972; **1**: 31–43.

33. Clarke I C. Correlation of SEM, replication and light microscopy studies of the bearing surfaces in human joints. *Scanning Electron Microscopy* 1973; **III**: 660–6.

34. Clarke I C, Schurman D J, Amstutz H C. *In vivo* and *in vitro* comparative studies of animal articular surfaces. *Annals of Biomedical Engineering* 1975; **3**: 100–10.

35. Clemmensen I, Holund B, Johansen N, Andersen R B. Demonstration of fibronectin in human articular cartilage by an indirect immunoperoxidase technique. *Histochemistry* 1982; **76**: 51–6.

36. Cole M B, Narine K R, Ellinger J. Morphological evidence of the shedding of chondrocytes from the articular surface in neonatal rats: relationship to the interlacunar network. *Anatomical Record* 1983; **206**: 439–46.

37. Croft B Y. *Single Photon Emission Computed Tomography*. London: Wolfe Medical Publications, 1988.

38. Davies D V, Barnett C H, Cochrane W, Palfrey A J. Electron microscopy of articular cartilage in young adult rabbit. *Annals of the Rheumatic Diseases* 1962; **21**: 11–21.

39. Ely R V, ed. *Microfocal Radiography*. London: Academic Press, 1980.

40. Erhlich M G, Mankin H J. Biochemical changes in osteoarthritis. In: Nuki G, ed, *The Aetiopathogenesis of Osteoarthrosis*. London: Pitman Medical, 1980: 29–36.

41. Eulderink F, Meijers K A. Pathology of the cervical spine in rheumatoid arthritis: a controlled study of 44 spines. *Journal of Pathology* 1976; **120**: 91–108.

42. Forrester D M, Brown J C, eds. The radiology of joint disease. *Saunders Monographs in Clinical Radiology* (volume 2). Philadelphia: W B Saunders & Co., 1987.

43. Fraser R G, Pare J P A. *Diagnosis of Diseases of the Chest* (2nd edition). Philadelphia: W B Saunders & Co., 1978: 881–980.

44. Freeman M A R. *Adult Articular Cartilage* (2nd edition). Tunbridge Wells: Pitman Medical, 1979.

45. Freeman M A R. *Adult Articular Cartilage* (1st edition). Tunbridge Wells: Pitman Medical, 1973.

46. Fuse S. Scanning and transmission electron microscopic studies on the fiber architecture of the articular disc of rat temporomandibular joint in reference to its functional significance. *Journal of the Stomatology Society of Japan* 1980; **47**: 9–22.

47. Galasko C, Isherwood I, eds. *Imaging Techniques in Orthopaedics*. Berlin: Springer Verlag, 1989.

48. Gardner D L. The influence of microscopic technology on knowledge of cartilage surface structure. *Annals of the Rheumatic Diseases* 1972; **31**: 235–58.

49. Gardner D L. Structure and function of connective tissue and joints. In: Scott J T, ed, *Copeman's Textbook of the Rheumatic Diseases* (6th edition). Edinburgh: Churchill Livingstone, 1986: 199–250.

50. Gardner D L, Brereton J. Unpublished observations.

51. Gardner D L, Brereton J, Holinshed M. Age changes in the cartilage thickness and cell density of the human sacral and iliac cartilages. To be published.

52. Gardner D L, Dodds T C. *Human Histology* (3rd edition). Edinburgh: Churchill Livingstone, 1976.

53. Gardner D L, Longmore R B, Gilmore R St J, Elliott R J. Age-related changes in articular cartilage determined by interferometry. Proceedings of the XIII International Congress of Rheumatology, International Congress Series. *Excerpta Medica* 1973; **299**: 132.

54. Gardner D L, McGillivray D C. Living articular cartilage is not smooth. The structure of mammalian and avian joint surfaces demonstrated *in vivo* by immersion incident light microscopy. *Annals of the Rheumatic Diseases* 1971a; **30**: 3–9.

55. Gardner D L, McGillivray D C. Surface structure of articular cartilage. Historical review. *Annals of the Rheumatic Diseases* 1971b; **30**: 10–14.

56. Gardner D L, O'Connor P, Middleton J F S, Oates K, Orford C R. An investigation by transmission electron miccroscopy of freeze replicas of dog articular cartilage surfaces: the fibre-rich surface structure. *Journal of Anatomy* 1983; **137**: 573–82.

57. Gardner D L, O'Connor P, Oates K. Surface structure of mammalian articular cartilage viewed at low temperature. *Lancet* 1979; **ii**: 538.

58. Gardner D L, O'Connor P, Oates K. Low temperature scanning electron microscopy of dog and guinea-pig hyaline articular cartilage. *Journal of Anatomy* 1981; **132**: 267–82.

59. Gardner D L, Woodward D H. Scanning electron microscopy of articular surfaces. *Lancet* 1968; **ii**: 1246.

60. Gardner D L, Woodward D H. Scanning electron microscopy and replica studies of articular surfaces of guinea-pig synovial joints. *Annals of the Rheumatic Diseases* 1969; **28**: 379–91.

61. Gelfand D W. *Gastrointestinal Radiology*. Edinburgh: Churchill Livingstone, 1984.

62. Ghadially F N. Structure and function of articular cartilage. *Clinics in Rheumatic Diseases* 1981; **7**: 3–28.

63. Ghadially F N. The articular surface. In: *Fine Structure of Synovial Joints*. London: Butterworths, 1983: 80–102.

64. Ghadially F N, Ghadially J A, Oryschak A F, Yong N K. Experimental production of ridges on rabbit articular cartilage: a scanning electron microscope study. *Journal of Anatomy* 1976; **121**: 119–32.

65. Ghadially F N, Ghadially J A, Oryshak A F, Yong N K. The surface of dog articular cartilage. A scanning electron microscope study. *Journal of Anatomy* 1977; **123**: 527–36.

66. Ghadially F N, Lalonde J-M A, Wedge J H. Ultrastructure of normal and torn menisci of the human knee joint. *Journal of Anatomy* 1983; **136**: 773–91.

67. Ghadially F N, Moshurchak E M, Ghadially J A. A maturation change in the surface of cat articular cartilage detected by the scanning electron microscope. *Journal of Anatomy* 1978; **125**: 349–60.

68. Ghadially F N, Moshurchak E M, Thomas I. Humps on young human and rabbit articular cartilage. *Journal of Anatomy* 1977; **124**: 425–35.

69. Ghadially F N, Thomas I, Yong N, Lalonde J-M A. Ultrastructure of rabbit semilunar cartilages. *Journal of Anatomy* 1978; **125**: 499–517.

70. Ghadially F N, Yong N K, Lalonde J-M A. A transmission electron microscopic comparison of the articular surface of cartilage processed attached to bone and detached from bone. *Journal of Anatomy* 1982; **135**: 685–706.

71. Gholkar A, Gillespie J E, Hart C W, Mott D, Isherwood I. Dynamic low-dose three-dimensional computed tomography: a preliminary study. *British Journal of Radiology* 1988; **61**: 1095–9.

72. Ghosh P, Bushell G R, Taylor T F K, Akeson W H. Collagens, elastic and non-collagenous protein of the intervertebral disc. *Clinical Orthopaedics and Related Research* 1977; **129**: 124–32.

73. Grainger R G, Allison D J. *Diagnostic Radiology: an Anglo-American Textbook of Imaging* (4th edition). Edinburgh: Churchill Livingstone, 1986.

74. Haines R W. The development of joints. *Journal of Anatomy* 1947; **81**: 33–55.

75. Hall B K. Intracellular and extracellular control of the differentiation of cartilage and bone. *Histochemical Journal* 1981; **13**: 599–614.

76. Hall B K, Tremaine R. Ability of neural crest cells from the embryonic chick to differentiate into cartilage before their migration away from the neural tube. *Anatomical Record* 1979; **194**: 469–75.

77. Ham A W. *Histology* (7th edition). London: Pitman Medical Publishing, 1974: 454.

78. Hamilton W J, Boyde J D, Mossman H W. *Human Embryology*. Cambridge: Heffer, 1962.

79. Hammar J A. Ueber den feineren Bau den Gelenke. *Archivs für Mikroskopische Anatomie* 1894; **43**: 266–326, 813–85.

80. Hardingham T E, Adams P. A method for the determination of hyaluronate in the presence of other glycosaminoglycans and its application to human intervertebral disc. *Biochemical Journal* 1976; **159**: 143–7.

81. Harrison R M, Isherwood I, eds. *Digital Radiology: Physical and Clinical Aspects.* London: Institute of Physical Sciences in Medicine, 1984.

82. Hashizume H. Three-dimensional architecture and development of lumbar intervertebral discs. *Acta Medica Okayama* 1980; **34**: 301–14.

83. Hassall A H. *The Microscopic Anatomy of the Human Body in Health and Disease.* London: Samuel Highley, 1849: vol. I: 281–93; vol. II: plates 30 and 31.

84. Helms C A, Morrish R B, Kircos L T, Karzberg R W, Dolwick M F. Computed tomography of the meniscus of the temporomandibular joint: preliminary observations. *Radiology* 1982; **145**: 719–22.

85. Hickey D S, Hukins D W L. Effect of methods of preservation of collagen fibrils in connective tissue matrices: an X-ray diffraction study of annulus fibrosus. *Connective Tissue Research* 1979; **6**: 223–8.

86. Hickey D S, Hukins D W L. Collagen fibril diameters of elastic fibres in the annulus fibrosus of human fetal intervertebral disc. *Journal of Anatomy* 1981; **133**: 351–7.

87. Higgins C B, Hricak H. *Magnetic Resonance Imaging of the Body.* New York: Raven Press, 1987.

88. Hillman B J. Newell J D, eds, Digital radiography. *Radiologic Clinics of North America* 1985; **23**: 177–373.

89. Hunter S J, Caplan A I. Control of cartilage differentiation. In: Hall B K, ed, *Cartilage* (volume 2): *Development, Differentiation, and Growth.* New York: Academic Press, 1983: 87–119.

90. Hunter W. Of the structure and diseases of articulating cartilages. *Philosophical Transactions of the Royal Society of London* 1742–3; **42**: 514–21.

91. Hyman R A, Gorey M T. Imaging strategies for MRI of the spine. *Radiologic Clinics of North America* 1988; **26**: 505–33.

92. Inoue H. Three-dimensional observation of collagen framework of intervertebral discs in rats, dogs and humans. *Archivum Histologicum Japonicum* 1973; **36**: 39–56.

93. Inoue H. Three-dimensional architecture of lumbar intervertebral discs. *Spine* 1980; **6**: 139–46.

94. Inoue H, Kodama T, Fujita T. Scanning electron microscopy of normal and rheumatoid articular cartilages. *Archivum Histologicum Japonicum* 1969; **30**: 425–35.

95. Inoue T, Osaka H, Tanaka K. Use of surface tension to enable observation of submembranous structures by scanning electron microscopy. *Journal of Electron Microscopy* 1984; **33**: 258–60.

96. Jagger R G. The surface structure of the temporomandibular joint disk: a scanning electron microscopic study. *Journal of Oral Rehabilitation* 1980; **7**: 225–34.

97. Jamieson D, Alavi A, Jolles P, Chawluk J, Reivich, M. Positron emission tomography in the investigation of central nervous system disorders. *Radiologic Clinics of North America* 1988; **26**: 1075–88.

98. Jenkins J P R, Isherwood I. Magnetic resonance imaging (MRI). In: Galasko C S B, Isherwood I, eds, *Imaging Techniques in Orthopaedics.* Berlin: Springer Verlag, 1989: 159–79.

99. Johnson D C, Edey M A. *Lucy. The Beginnings of Humankind.* London and Toronto: Granada, 1981.

100. Knoesen D, Kristzinger S. Microtopographical analysis of surface structure in a scanning electron microscope. *Journal of Microscopy* 1983; **132**: 87–96.

101. Kühn K, Krieg T, eds, Connective tissue: biological and clinical aspects. *Rheumatology: An Annual Review* (volume 10). Basel: Karger, 1986.

102. Leidy J. On the intimate structure and history of the articular cartilages. *American Journal of Medical Science* 1849; **17**: 277–94.

103. Lewis O J. The joints of the evolving foot. Part I. The ankle joint. *Journal of Anatomy* 1980; **130**: 527–43.

104. Linck G, Stocker S, Grimaud J-A, Porte A. Distribution of immunoreactive fibronectin and collagen (types I, III, IV) in mouse joints. Fibronectin, an essential component of the synovial cavity border. *Histochemistry* 1983; **77**: 323–8.

105. Liston R. On the arrangement of the intermediate vessels on surfaces secreting pus with a note regarding the vascularity of articular cartilages. *Medico-chirurgical Transactions* 1840; **23**: 85–95.

106. Longmore R B, Gardner D L. Development with age of human articular cartilage surface structure. *Annals of the Rheumatic Diseases* 1975; **34**: 26–37.

107. Longmore R B, Gardner D L. The surface structure of ageing human articular cartilage: a study by reflected light interference microscopy (RLIM). *Journal of Anatomy* 1978; **126**: 353–65.

108. McCall I. Radiological investigation of the intervertebral disc. In: Jayson M I V, ed, *The Lumbar Spine and Back Pain* (3rd edition). Edinburgh: Churchill Livingstone, 1987: 236–68.

109. McCall J G. Scanning electron microscopy of articular surfaces. *Lancet* 1968; **ii**: 1194.

110. MacConaill M A. The movements of bones and joints. 4. The mechanical structure of articulating cartilage. *Journal of Bone and Joint Surgery* 1951; **33B**: 251–7.

111. Maisey M N, Britton K E, Gilday D L, eds. *Clinical Nuclear Medicine.* London: Chapman & Hall, 1983.

112. Maroudas A. X-ray microprobe analysis of articular cartilage. *Connective Tissue Research* 1972; **1**: 153–63.

113. Maroudas A. Physical chemistry of articular cartilage and the intervertebral disc. In: Sokoloff L, ed, *The Joints and Synovial Fluid* (volume II). New York and London: Academic Press, 1980: 239–91.

114. Maroudas A, Stockwell R A, Nachemson A, Urban J. Factors involved in the nutrition of the human lumbar intervertebral disc: cellularity and diffusion of glucose *in vitro*. *Journal of Anatomy* 1976; **120**: 113–30.

115. Matthews M B. *Connective Tissue: Macromolecular Structure and Evolution.* Berlin: Springer Verlag, 1975.

116. Maurer A H. Nuclear medicine: SPECT comparisons to PET. *Radiologic Clinics of North America* 1988; **26**: 1059–74.

117. Meachim G, Collins D H. Cell counts of normal and osteoarthritic articular cartilage in relation to the uptake of sul-

phate ($^{35}SO_4$) *in vitro. Annals of the Rheumatic Diseases* 1962; **21**: 45–50.

118. Meachim G, Ghadially F N, Collins D H. Regressive changes in the superficial layer of human articular cartilage. *Annals of the Rheumatic Diseases* 1965; **24**: 23–30.

119. Meachim G, Roy S. Surface ultrastructure of mature adult human articular cartilage. *Journal of Bone and Joint Surgery* 1969; **51B**: 529–39.

120. Merkel K H H. The surface of human menisci and its aging alterations during age. A combined scanning and transmission electron microscopic examination (SEM, TEM). *Archives of Orthopaedic and Traumatic Surgery* 1980; **97**: 185–91.

121. Middleton J F S, Oates K, O'Connor P, Orford C R, Gardner D L. Demonstration by X-ray microprobe analysis of relationship between chondrocytes and tertiary surface structure of hyaline articular cartilage. *Connective Tissue Research* 1984; **13**: 1–8.

122. Mital M A, Millington P F. Surface characteristics of articular cartilage. *Micron* 1971; **2**: 236–49.

123. Moshurchak E M, Ghadially F N. A maturation change detected in the semilunar cartilages with the scanning electron microscope. *Journal of Anatomy* 1978; **126**: 805–18.

124. Moss M L, Moss-Salentijn L. Vertebrate cartilages. In: Hall B K, ed, *Cartilage* (volume 1). *Structure, Function and Biochemistry*. New York: Academic Press, 1983: 1–30.

125. O'Connor J, Shercliff T, Goodfellow J. The mechanics of the knee in the sagittal plane. Mechanical interactions between muscles, ligaments and articular surfaces. In: Muller W, Hackenbruch W P, eds, *Surgery and Arthroscopy of the Knee: 2nd Congress of the European Society*. Berlin: Springer Verlag, 1988: 12–30.

126. O'Connor P, Oates K, Gardner D L. Middleton J F S, Orford C R, Brereton J D. Low temperature and conventional scanning electron microscopic observations of dog femoral condylar cartilage surfaces after anterior cruciate ligament division. *Annals of the Rheumatic Diseases* 1985; **44**: 321–7.

127. Ohno O, Cooke T D. Electron microscopic morphology of immunoglobulin aggregates and their interactions in rheumatoid articular collagenous tissues. *Arthritis and Rheumatism* 1978; **21**: 516–27.

128. Olson M D, Low F N. The fine structure of developing cartilage in the chick embryo. *American Journal of Anatomy* 1971; **131**: 197–216.

129. O'Rahilly R, Gardner E. The embryology of movable joints. In: Sokoloff L, ed, *The Joints and Synovial Fluid* (volume 1). New York: Academic Press, 1978: 49–103.

130. Orford C R, Gardner D L. Ultrastructural histochemistry of the surface lamina of normal articular cartilage. *Histochemical Journal* 1985; **17**: 223–33.

131. Orford C R, Gardner D L, O'Connor P. Ultrastructural changes in dog femoral condylar cartilage following anterior cruciate ligament section. *Journal of Anatomy* 1983; **137**: 653–63.

132. Person P, Philpott D E. On the occurrence and the biologic significance of cartilage tissues in invertebrates. *Clinical Orthopaedics and Related Research* 1967; **53**: 185–212.

133. Pidd J G, Gardner D L. Surface structure of baboon (*Papio anubis*) hydrated articular cartilage: study of low temperature replicas by transmission electron microscopy. *Journal of Medical Primatology* 1987; **16**: 301–9.

134. Potts W T W, Oates K. The ionic concentrations in the mitochondria-rich or chloride cell of Fundulus heteroclitus, *Journal of Experimental Zoology* 1983; **227**: 349–59.

135. Pritchard J J. General histology of bone. In: Bourne G H, ed, *The Biochemistry and Physiology of Bone* (volume 1). New York: Academic Press, 1972: 1–19.

136. Redler I, Zimmy M L. Scanning electron microscopy of normal and abnormal articular cartilage and synovium. *Journal of Bone and Joint Surgery* 1970; **52A**: 1395–1404.

137. Renxiang Z, Zuyun L, Wenji Q, Hongzi Z, Ming L. Comparative study of the femoral articular facies of knees of the primates. *Scientia Sinica* (Series B) 1987; **30**: 960–6.

138. Sayles R S, Thomas T R, Anderson J, Haslock I, Unsworth A. Measurement of the surface microgeometry of articular cartilage. *Journal of Biomechanics* 1979; **12**: 257–67.

139. Schäffer J. Das Knorpelgewebe. In: von Mollendorff W, ed, *Handbuch der mikroskopischen Antomie des Menschen*. Berlin: Springer Verlag, 1930: 210 and 243ff.

140. Scothorne R J. In: Passmore R, Robson J S, eds, *Companion to Medical Studies* (volume 1) (2nd edition). Oxford: Blackwell Scientific Publications, 1976: 19.9–13.

141. Scott J E, Haigh M. Proteoglycan–collagen interactions in intervertebral disc. A chondroitin sulphate proteoglycan associates with collagen fibrils in rabbit annulus fibrosus at the d–e bands. *Bioscience Reports* 1986; **6**: 879–88.

142. Scott J T, ed. *Copeman's Textbook of the Rheumatic Diseases* (6th edition). Edinburgh: Churchill Livingstone, 1985.

143. Shapiro R. *Myelography* (4th edition). London: Wolfe Medical Publication, 1984.

144. Sherke H H, Parke W W. Normal adult anatomy. In: Cervical Spine Research Society Editorial Subcommittee, H H Sherke (secretary), *The Cervical Spine*. Philadelphia: J B Lippincott & Co., 1983: 8–22.

145. Simon W H. Scale effects in animal joints. 1. Articular cartilage thickness and compressive stress. *Arthritis and Rheumatism* 1970; **13**: 244–55.

146. Siu D, Bryant J T, Wevers H W. Three-dimensional reconstruction of joint surfaces using a microcomputer. *Medical and Biological Engineering and Computing* 1986; **24**: 267–74.

147. Sokoloff L, ed, *The Joints and Synovial Fluid*. New York: Academic Press, volume I, 1978; volume II, 1980.

148. Stanescu R, Leibovich S J. The negative charge of articular cartilage surfaces—an electron microscopic study using cationized ferritin. *Journal of Bone and Joint Surgery* 1982; **64A**: 388–98.

148a. Stanescu R, Stanescu V. *In vitro* protection of the articular surface by cross-linking agents. *Journal of Rheumatology* 1988; **15**: 1677–82.

149. Stark D D, Bradley W G, eds. *Magnetic Resonance Imaging*. St Louis, Missouri: C V Mosby, 1988.

150. Stevens R L, Dondi P G, Muir H. Proteoglycans of the intervertebral disc. Absence of degradation during the isolation of proteoglycans from the intervertebral disc. *Biochemical Journal* 1979b; **179**: 573–8.

151. Stevens R L, Ewins R J F, Revell P A, Muir H. Proteo-

glycans of the intervertebral disc. Homology of structure with laryngeal proteoglycans. *Biochemical Journal* 1979a; **179**: 561–72.

152. Stockwell R A. The cell density of human articular and costal cartilage. *Journal of Anatomy* 1967; **101**: 753–63.

153. Stockwell R A. The interrelationship of cell density and cartilage thickness in mammalian articular cartilage. *Journal of Anatomy* 1971; **109**: 411–21.

154. Stockwell R A. Chondrocytes. In: Gardner D L, ed, Diseases of Connective Tissue. *Journal of Clinical Pathology* 1978; **31**, Supplement (Royal College of Pathologists) 12: 7–13.

155. Stockwell R A. *Biology of Cartilage Cells.* Cambridge: Cambridge University Press, 1979.

156. Sutton D, ed. *Textbook of Radiology and Imaging* (4th edition). Edinburgh: Churchill Livingstone, 1987.

157. Takeda T. Three-dimensional observation of collagen framework of human lumbar discs. *Japanese Journal of Orthopaedics* 1975; **49**: 45–57.

158. Taylor K J W, Burns P N, Wells P N T, eds. *Clinical Applications of Doppler Ultrasound.* New York: Raven Press, 1988.

159. Thompson A M, Stockwell R A. An ultrastructural study of the marginal transitional zone in the rabbit knee joint. *Journal of Anatomy* 1983; **136**: 701–13.

160. Thomson D'A W. *On Growth and Form* (abridged edition). Cambridge: Cambridge University Press, 1966.

161. Thorogood T. In: Hall B K, ed, *Cartilage* (volumes I–III). New York: Academic Press, 1983.

162. Toller P A. Ultrastructure of the condylar articular surface in severe mandibular pain-dysfunction syndrome. *International Journal of Oral Surgery* 1977; **6**: 297–312.

163. Toller P A, Wilcox J H. Ultrastructure of the articular surface of the condyle in temporomandibular arthropathy. *Oral Surgery* 1978; **45**: 232–45.

164. Tuttle R H. Evolution of hominid bipedalism and prehensile capabilities. *Philosophical Transactions of the Royal Society of London B* 1981; **292**: 89–94.

165. Van Sickle D C, Kincaid S A. Comparative arthrology. In: Sokoloff L, ed, *The Joints and Synovial Fluid* (volume I). New York: Academic Press, 1978: 1–47.

166. Virchow R. *Die Cellular Pathologie in ihrer Begrundung auf physiologische und pathologische Gewebelehre* Berlin: Hirschwald. English translation (1860) from the 2nd edition (1859) by F. Chance, published (1971) with a new introductory essay by L J Rather. New York: Dover Publications.

167. Walker P S, Dowson D, Longfield M D, Wright V. 'Boosted lubrication' in synovial joints by fluid entrapment and enrichment. *Annals of the Rheumatic Diseases* 1968; **24**: 512–20.

168. Walker P S, Unsworth A, Dowson D, Sikorski J, Wright V. Mode of aggregation of hyaluronic acid protein complex on the surface of articular cartilage. *Annals of the Rheumatic Diseases* 1970; **29**: 591–602.

169. Warwick R, Williams P L. *Gray's Anatomy* (37th edition). Edinburgh: Churchill Livingstone, 1989.

170. Weinstein A S, Kattan K R. Arthritis and other arthropathies. *Radiologic Clinics of North America* 1988; **26** (6).

171. Weiss C, Rosenberg L, Helfet A J. An ultrastructural study of normal young adult human articular cartilage. *Journal of Bone and Joint Surgery* 1968; **50A**: 663–74.

172. White R I. Interventional radiology: reflections and expectations. *Radiology* 1987; **162**: 593–600.

173. Wilson A S, Rees M J C, Bloebaum R D. Studies on the effect of exposing articular cartilage *in vitro* to ambient atmospheric conditions. *Journal of Anatomy* 1980; **130**: 198.

174. Wilson N H F, Gardner D L. Influence of aqueous fixation on articular surface morphology: a reflected light interference microscope study. *Journal of Pathology* 1980; **131**: 333–8.

175. Wilson N H F (personal communication).

176. Wilson N H F, Gardner D L. The microscopic structure of fibrous articular surfaces: a review. *Anatomical Record* 1984; **209**: 143–52.

177. Wright G M, Keeley F W, Youson J H, Babineau D L. Cartilage in the Atlantic hagfish, *Myxine glutinosa. American Journal of Anatomy* 1984; **169**: 407–24.

178. Yasui K. Three-dimensional architecture of human normal menisci. *Journal of the Japanese Orthopaedic Association* 1978; **52**: 391–9.

Part II

Mechanisms of Connective Tissue Disease

Chapter 4

Biochemical Basis of Connective Tissue Disease

JOHN C ANDERSON

A knowledge of biochemistry is essential in order to understand the pathology of many of the connective tissue diseases. More than 3000 single-gene disorders are known in humans, and about 100 involve the connective tissues directly. Some of these disorders are dealt with here, but many, including most of the 70 or so types of chondrodysplasias,[130] are not as little is known of their biochemical basis. However, the most frequent connective tissue disorders such as rheumatoid arthritis (see Chapters 12 and 13) and the infective arthritides (see Chapter 18) are largely attributable to environmental factors.

Few, if any, of the inherited connective tissue diseases can be treated effectively. Currently, management can only be aimed at prevention, not cure; it includes the diagnosis of the disease in a kindred, genetic counselling and prenatal diagnosis. The great progress in human genetics and molecular biology in the last decade makes it possible to analyse the fine structure of human genes in health and in disease. The identification of mutant genes has been facilitated by the development of powerful new techniques,[301,330] some of which are mentioned below.

By contrast, treatment of the acquired connective tissue disorders has advanced significantly so that clinical gout, for example, can be prevented by xanthine oxidase inhibitors, inflammation in rheumatoid arthritis controlled by non-steroidal anti-inflammatory drugs, renal failure in systemic lupus erythematosus regulated by haemodialysis, and hip joint pain in osteoarthrosis eradicated by arthroplasty.

The connective tissue components (see Chapter 1) are collagens, elastin, proteoglycans, glycoproteins, cells, salts, proteins of the extracellular fluid, such as albumin, and water. The composition and organization of each tissue is different and reflects its function. Skin, bone, tendon, ligament, blood vessels, cartilage and intervertebral discs are regarded as the major connective tissues, but connective tissue elements are ubiquitous: muscle, liver and kidney, for example, contain small but significant amounts. Chapter 1 provides a histological introduction to the present account.

Collagens

Collagen[14,46,89,110,158,168,189,190,211,319] provides the essential framework of connective tissues. Many cells lie on a collagenous basal lamina or exist within a collagenous matrix. Cell–collagen interactions are essential to cell movements in inflammation, wound healing, trophoblast implantation, fetal development and cancer.

Structure

The basic unit of the collagen fibre is the α-chain, a polypeptide of 1050 amino acids.[245] The amino acid composition and sequence are unlike those of other proteins. Hydroxyproline* (Hyp) is almost specific to collagen: small amounts also occur in C1q (see p. 237), in acetylcholinesterase and in elastin (see p. 189). Hydroxylysine (Hyl) is found only in collagen where some residues bear sugars (see p. 175). The unique collagen amino acid composition has 33 per cent Gly, 20–25 per cent Pro plus Hyp, and 5–11 per cent Lys plus Hyl. The major sequence in α-chains consists of repeating triplets, Gly–X–Y, where Y is frequently Hyp, X is often Pro and Hyl is always in position Y. This repetitive sequence totals 333 tripeptide units for the major fibrillar collagens. At both ends of the α-chain are telopeptides of more normal amino acid compositions, with 16 and 25 residues in the amino and carboxyl terminal telopeptides, respectively.

The high content of Gly, Pro and Hyp, and the repeating tripeptide sequence, cause the α-chain to assume a left-handed helical secondary structure. The pitch aligns Gly residues on the same side of the helix allowing three chains to aggregate compactly with Gly-bearing faces together, forming a right-handed triple helix. This is the collagen monomer (*tropocollagen*) of molecular weight 285 kD and dimensions 300 × 1.5 nm. The triple helix resists attack by proteinases; however, telopeptides are easily removed.

* Standard abbreviations for amino acids and sugars are used here in accordance with the instructions to authors issued by the *Biochemical Journal*.

Genetically distinct collagens

There are now at least 15 known collagen types, composed of at least 24 α-chains, each a unique gene product. Much of the information given above applies to the commonest collagen, type I; Table 4.1 shows the known characteristics of all types. Types I, II, III, V and XI are referred to as the fibrillar collagens. Most tissues contain several collagen types, their relative proportions being tissue-specific and often changing as the tissue matures.[191] The collagen of adult dermis is predominantly type I (85–90 per cent), with 8–11 per cent type III and 2–4 per cent type V, but in fetal skin these percentages are 70–75, 18–21 and 6–8, respectively.[289] The collagens can be distinguished by electrophoresis and by amino acid analysis: variations occur in Hyp, Pro, Hyl and Lys. The total hexose content and the ratio of Gal–Glc to Gal (see p. 175) also differ between the collagen types and within a single genetic type in different tissues.

Collagen genes

Each of the four pro-α-chains of collagen types I–III requires a 4.5 kilobase (kb) coding sequence of DNA. However, this is split into some 51 exons in total gene lengths of 18 kb (α1[I]), 30 kb (α1[III]) and 38 kb (α2[I] and α1[III]). The helical domain is distinctive. It is coded by exons alternating between 54 and 108 base-pairs (bp), with occasional 45 or 99 bp alternatives, and all 40 exons coding for a triple helix encode a whole number of Gly–X–Y triplets. This complex arrangement may have arisen by duplications of a 54 bp ancestral gene.

Biosynthesis

Because of the complicated structure of the collagen genes, the processing of the primary RNA transcripts to form mRNA is very complex. In addition to the removal of introns, the primary transcripts of some collagen genes (II[262], VI[263] and XIII[307]) are subject to alternative splicing to yield more than one gene product. Two different transcripts are produced from the type IX collagen gene by the use of two widely separated promoters[213] (as has also been reported for the type I collagen gene[29]) and there is also some alternative splicing.

Collagen biosynthesis[207] (Fig. 4.1) is noteworthy for its post-translational processing. The α-chain precursor (prepro-α) is synthesized on the rough endoplasmic reticulum. A short N-terminal polypeptide (the presequence) steers the growing peptide into the cisterna of the rough endoplasmic reticulum (see p. 19), where it is removed, leaving the pro-α-chain (mol. wt 155 kD) with its N- and C-terminal propeptides.

Several modifications are catalysed by membrane-bound enzymes of the rough endoplasmic reticulum. Some Pro and Lys residues are hydroxylated by enzymes needing oxygen, Fe^{2+}, ascorbate and oxoglutarate. Prolyl-4-hydroxylase recognizes most Pro residues in position Y of the sequence Gly–X–Y and prolyl-3-hydroxylase hydroxylates some residues in position X, but only if Y is 4-Hyp. Lysyl hydroxylase hydroxylates C5 of occasional Lys residues in position Y, and some residues in telopeptides. Hydroxylysyl galactosyl transferase catalyses transfer of Gal from uridine diphosphate–Gal to hydroxyl groups of certain Hyl residues, and galactosyl hydroxylysyl glucosyl transferase mediates the transfer of Glc from UDP–Glc to some Gal–Hyl groups to give Glc–Gal–Hyl units. Although the action of these four enzymes begins before synthesis of pro-α-chains is complete, it is finished on free chains. In different tissues, the same α-chain may be hydroxylated and glycosylated to different extents. Other glycosylations occur mainly on C-terminal propeptides where oligosaccharides contain GlcN and Man. In the N–propeptide of α-chains of collagen types III and V, Tyr is sulphated.[148]

As completed pro-α-chains pass down the cisternae of the endoplasmic reticulum, three chains assemble in register to form a triple helix, a process promoted by disulphide bond formation between C-terminal propeptides and catalysed by protein disulphide isomerase, which has 93 per cent homology with the β-subunit of prolyl hydroxylase.[161] Presumably any disulphide bonds linking α-chains, such as those within the triple helix of type III collagen, form simultaneously. Propeptides may also ensure that each trimer contains the correct α-chains. Only Gly–X–Y repeating sequences adopt the triple helical conformation; propeptides (except for a short sequence in the N–propeptide) and telopeptides do not. The action of hydroxylases and glycosyl transferases ceases with triple helix formation. Thus different amounts of Hyl and glycosylated Hyl in various collagens may result from varying rates of triple helix formation. The triple helix is stabilized by its high Gly content and by the Hyp hydroxyl groups which form interchain hydrogen bonds via bridging water molecules. Loss of triple helical structure (referred to as 'melting') if the temperature rises to 39–40 °C shows that stabilization is marginal.

Soluble procollagen passes through the Golgi complex, is packaged for secretion, and leaves the cell by exocytosis for the site of fibrillogenesis (see p. 303). In tendon, fibrillogenesis begins within deep, narrow invaginations in the fibroblast plasma membrane (see Chapter 8).[34] Prior to fibrillogenesis, at least two type-specific proteinases act: N- and C-terminal proteinases cleave their respective propeptides. The action of the latter enzyme alone would leave pN–collagen (see p. 181). N- and C-terminal propeptides from type I procollagen have molecular weights of 20 kD and 35 kD, respectively; the latter is removed as a disulphide-bonded trimer. Proteolysis of types IV and IX procollagen does not occur, the former unit being directly incorporated into basement membranes (see p. 50). Types V and XI procollagens are partially processed, losing half the N-terminal and all of the C-terminal propeptides. For type VII procollagen, the N-terminal propeptide is completely removed but the C-terminal propeptide is retained for its specific function.[183] While recent evidence shows that collagen type VI is not processed,[65] except possibly for α3(VI),[36] the situation for types VIII, X, XII, XIII, XIV and XV is not yet resolved.

Fibril formation

Monomers of collagen types I, II and III resemble flexible rods: under physiological conditions they self-aggregate by alignment so that adjacent monomers are displaced by 67 nm (Fig. 4.1). This 'quarter-stagger' array (see p. 44) maximizes non-covalent interactions by the apposition of regions rich in charged amino acid residues and by the alignment of regions rich in hydrophobic residues. The result is the characteristic banding of interstitial collagen fibres seen by transmission electron microscopy (see Chapter 1).

Non-covalent bonds do not account for tensile strength (see Chapter 6), which is due to the formation of covalent cross-links[83,85,172,251] between α-chains within a single monomer and between α-chains in adjacent monomers. Initially, the oxidative deamination of ε-amino groups of specific Lys or Hyl residues, catalysed by copper-dependent lysyl oxidase, forms aldehydes. These aldehydes produce aldol-type cross-links between α-chains within monomers, via residues in N-terminal telopeptides (Fig. 4.2). In addition, aldehydes combine reversibly with amino groups of Lys, Hyl or glycosylated Hyl residues within the triple helix of an adjacent monomer to form reducible aldimine (Schiff's base) and trivalent 3-hydroxypyridinium (pyridinoline) cross-links (Fig. 4.2), the latter arising predominantly from Hyl. In all tissues except bone and dentin, maturation involves disappearance of reducible cross-links and replacement by more stable derivatives. While pyridi-

Table 4.1

Properties of the different collagen types

Collagen type	No. of different α-chains	Molecular composition	Mol. wt (kD)	Length of triple helix (nm)	Fibre organization	Function	Physicochemical properties	Tissue distribution
I	2	$[\alpha1(I)]_2\alpha2(I)$	285	300	67 nm banded	Fibril formation	Thick fibrils Low glycosylation No S–S bonds	Ubiquitous, except in cartilage, vitreous and nucleus pulposus
I trimer	1	$[\alpha1(I)]_3$	285	300	67 nm banded			Tumours, skin, liver and tendon
II	1	$[\alpha1(II)]_3$	285	300	67 nm banded	Fibril formation	Thin fibrils Moderate glycosylation No S–S bonds	Cartilage, vitreous and nucleus pulposus
III	1	$[\alpha1(III)]_3$	285	300	67 nm banded	Fibril formation	Thin fibrils Low glycosylation S–S bonds present	Fetal skin, gut wall, blood vessels, placenta, uterus, lung, liver and synovia
IV	5	$[\alpha1(IV)]_2\alpha2(IV)$ or $[\alpha1(IV)]_3$ α3(IV)[269]* α4(IV)[113]* α5(IV)*	450	390	'Hexagons'	Network formation	No fibrils formed Highly glycosylated S–S bonds present Interrupted helix	Lamina densa of basement membranes
V	4	$[\alpha1(V)]_2\alpha2(V)$ [α1(V)α2(V) α3(V)] α4(V)?	300+	300	Non-banded filaments	Uncertain: small fibrils or may form core for type I fibrils	Fine fibrils Highly glycosylated No S–S bonds	Ubiquitous
VI[59]	3	[α1(VI)α2(VI) α3(VI)]	500	105	100 nm beaded microfibrils	Microfibrils linking major fibres Anchoring function[152]	Long filaments Highly glycosylated Highly S–S bonded	Present in most tissues

VII	1	[α1(VII)$_3$]	960	424	Segment long spacing crystallites	Anchoring fibrils of basement membranes	S–S bonded, single interruption of helix	Amnion, skin, oesophagus
VIII[144]	2	[α1(VIII)]$_2$ α2(VIII)	180	132	Hexagonal array as seen in Descemer's membrane	Unknown	No S–S bonds	Endothelial cells, Descemet's membrane
IX†	3	[α1(IX)α2(IX) α3(IX)]	300	175 inter-rupted	Associates with type II collagen fibrils	Possibly cross-links type II fibrils and links fibrils to proteo-glycans or to other macro-molecules	No fibrils formed Highly glycosylated plus bound CS/DS S–S bonded, interrupted helix	Cartilage, vitreous, intervertebral disc, cornea
X	1	[α1(X)]$_3$	168	132	?	Associated with matrix vesicles in calcification and with chondrocytes	Glycosylated S–S bonds in mammalian species	Hypertrophic and calcifying cartilage[305]
XI	3	[α1(XI)α2(XI) α3(XI)]	330	300	Non-banded filaments	Cartilage equivalent of type V: α1 can replace α1(V) in type V fibrils	Forms fine fibrils Glycosylated No S–S bonds Similar to type V	Cartilage, vitreous,[30] intervertebral disc
XII†	1	[α1(XII)$_3$][75]	700	75	Possibly associates with type I collagen fibrils	Possibly soft-tissue equivalent of type IX[108]	Interrupted helix	Tendon, bone, cornea
XIII[267,307]	1	?	?	?	?	?	?	?

Type XIV† and XV have been reported at meetings; no doubt further collagen types await discovery

* Trimer composition unknown.

† These collagens are sometimes referred to as Fibril Associated Collagens with Interrupted Triple helix (FACIT) collagens.

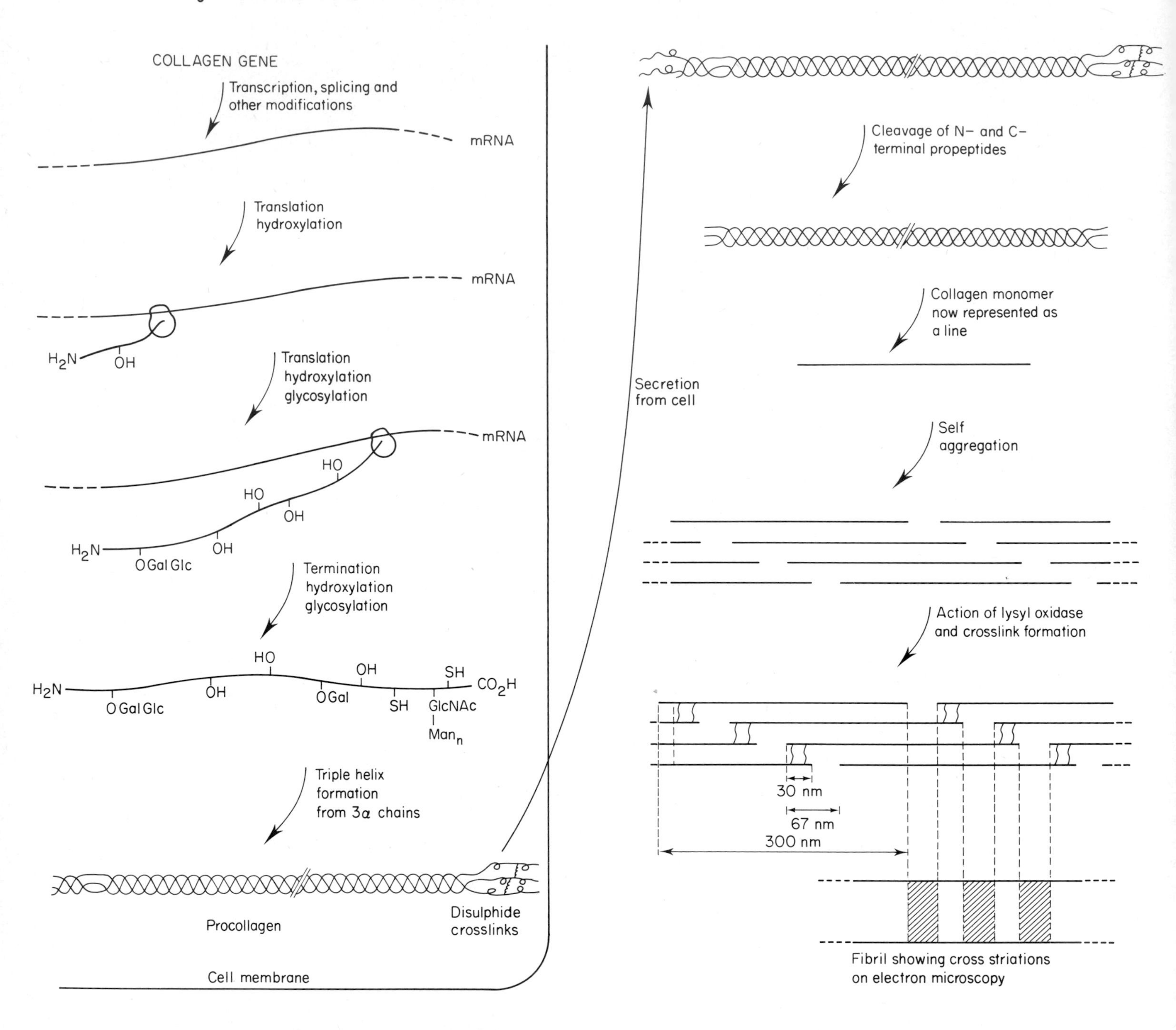

Fig. 4.1 Biosynthesis and assembly of collagen fibrils.
The biosynthesis, post-translational processing and aggregation of the major fibrillar collagens (types I, II and III). This classical scheme does not show the addition of incompletely processed collagen molecules and of other molecules which may limit fibril growth. Modes of aggregation for other collagens are shown in Fig. 4.3.

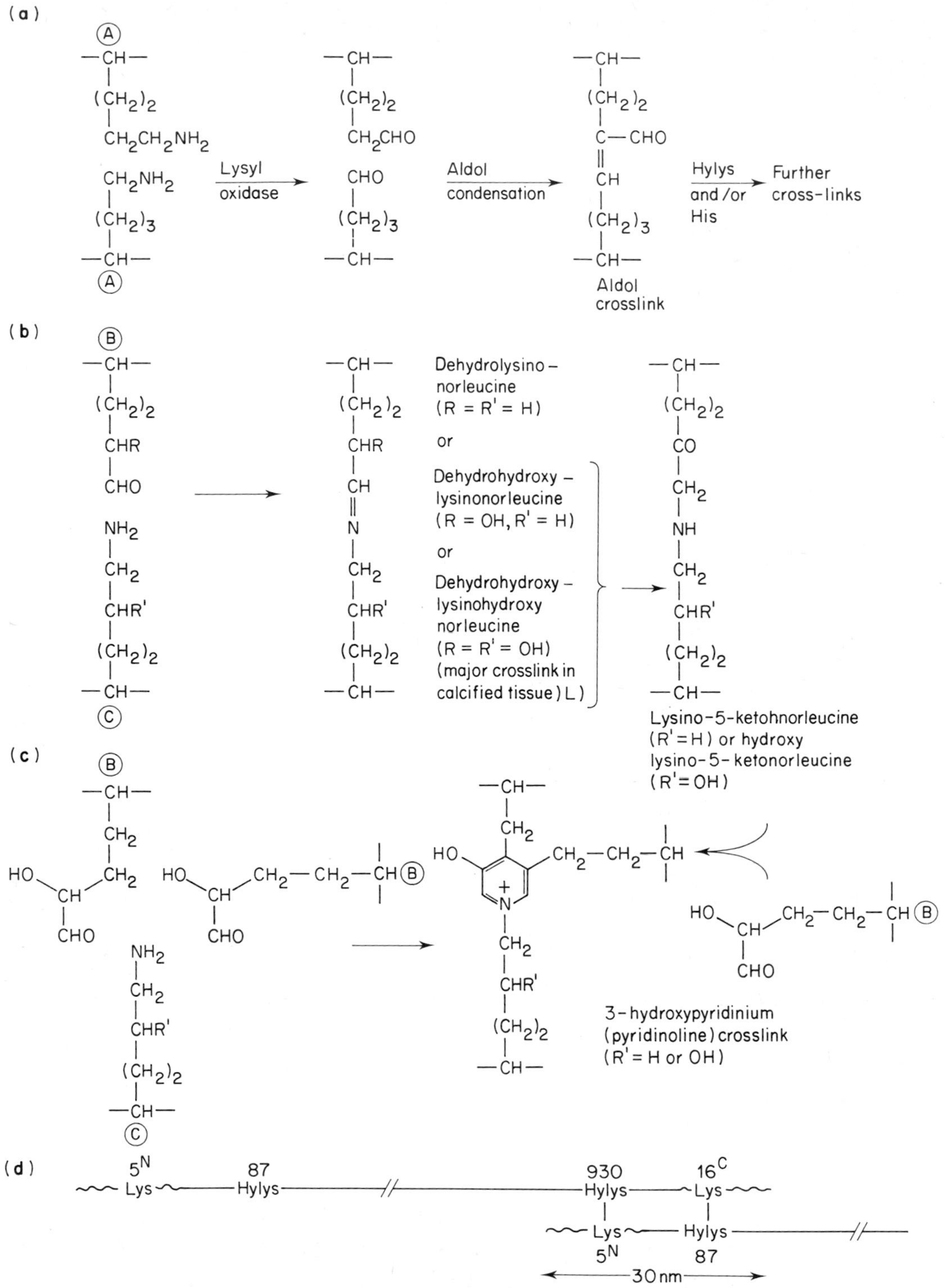

(a)

Ⓐ —CH—(CH$_2$)$_2$—CH$_2$CH$_2$NH$_2$ + CH$_2$NH$_2$—(CH$_2$)$_3$—CH— Ⓐ → Lysyl oxidase → —CH—(CH$_2$)$_2$—CH$_2$CHO + CHO—(CH$_2$)$_3$—CH— → Aldol condensation → —CH—(CH$_2$)$_2$—C(—CHO)=CH—(CH$_2$)$_3$—CH— (Aldol crosslink) → Hylys and /or His → Further cross-links

(b)

Ⓑ —CH—(CH$_2$)$_2$—CHR—CHO + NH$_2$—CH$_2$—CHR'—(CH$_2$)$_2$—CH— Ⓒ → —CH—(CH$_2$)$_2$—CHR—CH=N—CH$_2$—CHR'—(CH$_2$)$_2$—CH—

Dehydrolysinonorleucine (R = R' = H)
or
Dehydrohydroxylysinonorleucine (R = OH, R' = H)
or
Dehydrohydroxylysinohydroxy norleucine (R = R' = OH) (major crosslink in calcified tissue) L)

→ —CH—(CH$_2$)$_2$—CO—CH$_2$—NH—CH$_2$—CHR'—(CH$_2$)$_2$—CH—

Lysino-5-ketohnorleucine (R' = H) or hydroxy lysino-5-ketonorleucine (R' = OH)

(c)

Ⓑ —CH—CH$_2$—CH$_2$—CH(HO)—CHO + HO—CH(CHO)—CH$_2$—CH$_2$—CH Ⓑ + NH$_2$—CH$_2$—CHR'—(CH$_2$)$_2$—CH— Ⓒ →

3-hydroxypyridinium (pyridinoline) crosslink (R' = H or OH) + HO—CH(CHO)—CH$_2$—CH$_2$—CH Ⓑ

(d)

5^N Lys — 87 Hylys — // — 930 Hylys — 16^C Lys
Lys 5^N — Hylys 87 — //
← 30 nm →

Fig. 4.2 Diagram showing established collagen crosslinks.
a. Aldol crosslink formation between α chains in the same tropocollagen unit. **b.** Formation of Schiff's base (aldimine) crosslinks, which link α chains in different tropocollagen units. **c.** Formation of the 3-hydroxypyridinium crosslink, a predominant mature crosslink in many tissues. **d.** The important sites of crosslinking between α chains of tropocollagen units staggered by 4D (D = 67 nm), i.e. overlapping by 30 nm. Both this diagram and that of aggregation shown in Fig. 4.1 are oversimplified two-dimensional representations of a three-dimensional event. In fact the trivalent crosslink shown in **c** joins together two tropocollagen units lying in register with one overlapping by 30 nm, either linking the two C-terminal telopeptides (16^C) to the N-terminal part of the triple helix (87) or two N-terminal telopeptides (5^N) to the C-terminal part of the triple helix (930). A = Lys or Hyl residues in the N-terminal telopeptide (5^N); B = Lys or Hyl residue in the N-terminal telopeptide (5^N) or C-terminal telopeptide (16^C); C = Lys or Hyl in a helical sequence (position 87 or 930).

nolines are regarded as the predominant stable cross-links in most load-bearing tissues (e.g. cartilage, ligament and tendon), the final mature cross-link in skin, sclera and in the cornea, which arises from Lys, Hyl and His, is histidinylhydroxylysinonorleucine.[18,83] Cross-links between adjacent monomers displaced by 67 nm do not produce a cohesive fibril: they must form where the C-terminus of one monomer overlaps the N-terminus of a diagonally related monomer by about 30 nm (Fig. 4.2). Lys and Hyl residues in the telopeptide sequences are therefore essential. The cross-links described above for fibrillar collagens probably occur in other types, but the exact situation has not been established. Intermolecular cross-linking exists between types I and III,[85] and between types II and IX collagens,[321] the latter involving a pyridinoline cross-link.[84] In these various ways, collagen fibrils attain their properties of insolubility, inextensibility, slow turnover and high tensile strength. Increased stability results in a melting temperature well above the value of 40 °C for non-cross linked collagen.

While the situation for types V and IX is unclear, collagen types I, II and III form characteristic 'quarter-stagger' fibrils; types IV, VI–X, XII and XIII do not and have structures distinct from those of fibrillar collagens. Moreover, some have interrupted helices (there are 23 non-helical sections within the helix of type IV procollagen), allowing the formation of flexible structures. In one widely accepted model, type IV procollagen aggregates into an extended 'chicken-wire' network, solely by interactions between the ends of the polypeptides. The N-termini of four monomers overlap by 30 nm and are cross-linked to form a tetrameric domain known as 7S; the C-termini (NC-1 domains) of two monomers are also cross-linked (Fig. 4.3). In both these sites the cross-links are disulphide bonds and those derived from Lys and Hyl.

Pepsinized type V collagen forms fibrils *in vitro*, but its

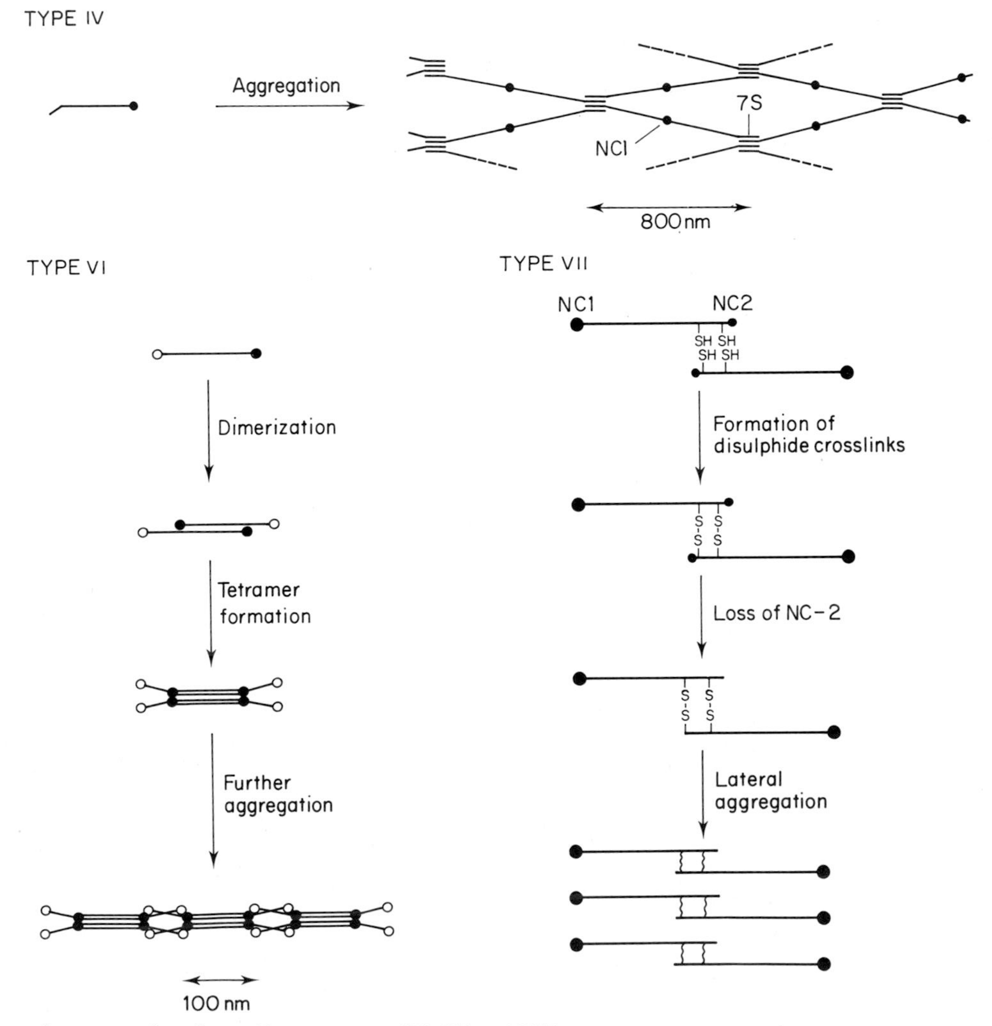

Fig. 4.3 Modes of aggregation for collagen types IV, VI and VII.
NC1 (C-terminal non-collagenous region), NC2 and 7S (N-terminal non-collagenous regions) indicate particular domains.

macromolecular organization *in vivo* remains obscure, although in bone it is seen as non-banded filaments.[41] However, mixed fibrils of types I and V collagens exist in tissues[2,33,276] and type V limits the growth of type I collagen fibrils, especially in cornea.[33] Immunolocalization of type V is impossible without prior removal of type I collagen, suggesting that type V is buried within the fibrils. Type I and III collagens also form mixed fibrils[153] as do types I and II, and types II, IX and XI assemble in a very specific way to form a copolymeric structure.[193] While most type IX in cartilage is present as copolymer, type XI is found primarily as a homopolymer. Type XII is possibly associated with types I and XIV.

In type VI, which may turn out to be a major tissue collagen,[314,344] dimers are formed when two monomers overlap and are stabilized by disulphide bridges (Fig. 4.3). Two dimers then associate side-by-side to form a tetramer which is secreted and then undergoes end-to-end association giving a 100 nm-periodic filament.[45,249] Apparently, type VI collagen lacks covalent cross-links except for disulphide bonds.[344]

Assembly of type VII collagen is initiated by antiparallel association of procollagen as a dimer (Fig. 4.3), by the formation of disulphide bonds between apposed N-termini, followed by proteolysis of N-terminal globular domains and further lateral aggregation.[183] Type VII collagen anchors specialized epithelial basement membranes to the underlying stroma via interaction with the intact C-terminal globular domain. The assembly of collagen types VIII and X is not yet fully understood.

Control of collagen synthesis and fibrillogenesis

The control of collagen synthesis and fibrillogenesis has been extensively reviewed.[131,166,243,246]

There is strong evidence that after cleavage from procollagen, the N- and C-terminal propeptides inhibit collagen synthesis and lead to decreased amounts of procollagen mRNA.[230,343] The N-propeptide appears to be more potent than the C-propeptide. The major influence regulating collagen synthesis is mRNA availability,[5,150] although there is also evidence for translational control.[28] Various factors control mRNA levels, including stimulation by interleukin-1[149] and transforming growth factor-β,[156,322] and depression by retinoids.[216] Increased collagen synthesis in a tissue may be due to hyperplasia: epidermal growth factor stimulates collagen production in granulation tissue by increasing fibroblast proliferation.[170]

Both N- and C-propeptides may have a role in the control of fibrillogenesis.[90,91] Thin fibrils, e.g. for type III collagen, are thought to arise after the initial removal of C-propeptide. Accretion involving pN collagen leaves small fibrils of 40–60 nm diameter with exposed N-propeptides. If these are not cleaved, growth of the fibril is limited. Initial removal of the N-propeptide before the C-propeptide allows the formation of thick fibrils. Thus by regulation of N- and C-proteinase activity, procollagen processing and fibril formation can be controlled. An additional role in calcification has been suggested for the C-terminal propeptide of type II collagen, which is identical to chondrocalcin, a Ca^{2+}-binding protein found in developing fetal epiphyseal cartilage matrix.[110,129] Other factors influencing fibrillogenesis include interaction with proteoglycans[279,280] the collagen type present and interaction of different collagen types (see above).

Biological properties

Apart from mechanical strength and ability to bind other matrix constituents, collagens show additional important properties.

Cell attachment

Many cell types, including human fibroblasts, attach *in vivo* to collagen as their natural substratum and bind equally well to types I, II, III or IV. Attachment is mediated by adhesive glycoproteins (see pp. 51 and 207).

Cell proliferation

Type V collagen inhibits the proliferation of human endothelial cells.[96]

Effect on haemostasis

Exposure of collagen to blood and subsequent aggregation of platelets is important in haemostasis, as is coagulation.[19,199] Adhesion of platelets to collagen fibres triggers release from platelet granules of serotonin, adrenaline, adenosine diphosphate, and thromboxane A_2 which stimulate platelet aggregation (see p. 283). Collagen may also activate the intrinsic coagulation pathway either directly, or indirectly, by release of, or exposure to, coagulation factors bound to platelets. Pathological expression of the haemostatic mechanism probably occurs in thrombosis and atherosclerosis. The ability to induce platelet aggregation seems to be a property of tertiary and quarternary structure rather than of collagen type and the adhesive process is mediated by von Willebrand factor.[88]

Chemotactic properties

Collagen types I–III are chemotactic for fibroblasts, as are fragments generated (see p. 184) by digestion with mammalian collagenases.[6] Collagen type IV is reported to be chemotactic for neutrophils.[282]

Diseases of collagen synthesis

Diseases of collagen synthesis[48,56,236,238,247,300,302,303,317,319] (see Chapter 9) are considered here in the context of steps in collagen biosynthesis and maturation.

Defects in the collagen genes

The chromosomal locations of some of the collagen genes are now known and are: α1(I), chromosome 17; α2(I), 7; α1(II), 12; α1(III), 2; α1(IV) and α2(IV), 13; α5(IV), X; α2(V), 2; α1(VI) and α2(VI), 21; α3(VI), 2, along with the genes for elastin and fibronectin; α1(IX), 6; α1(XI), 1; α2(XI), 6; α1(XIII), 10.[48]

Genes coding for specific pro-α-chains differ slightly in their base sequences in different families. These variations reflect simple Mendelian inheritance and often generate or delete recognition sites for restriction endonucleases. These 'restriction length fragment polymorphisms' can be identified by electrophoresis of restriction enzyme digests of DNA, Southern blotting and hybridization with a suitable probe.[301,330] If this technique allows the two alleles in an affected individual to be distinguished, then inheritance of a mutant gene may be traced. When a particular polymorphism segregates with the disease in a pedigree, causality can be established.[303] Thus a particular type II procollagen genotype appears to be linked to achondroplasia (see p. 365),[80] spondyloepiphyseal dysplasia,[7,176] Stickler syndrome,[308] primary osteoarthrosis associated with mild chondrodysplasia[160] and predisposition to familial osteoarthrosis,[222] but not with diastrophic dysplasia,[204] autosomal dominant spondyloarthropathy[295] or with most heritable chondrodysplasias.[342] Abnormal type II collagens have been isolated from the cartilage of individuals with spondyloepiphyseal and Kniest dysplasias and with achondrogenesis-hypochondrogenesis.[176,308] Polymorphisms in α2(I) and α1(II) genes have been used to study pedigrees with osteogenesis imperfecta[49] (see p. 343) and Ehlers–Danlos syndrome[292] (see p. 334). In one large pedigree with osteogenesis imperfecta (dominant), concordant segregation of the mutant gene close to the α2(I) locus was established. In other pedigrees, α2(I) and α1(II) segregated discordantly from osteogenesis imperfecta and Ehlers–Danlos syndrome and were therefore not the causal mutation.[303] Similarly, discordant segregation between α1(I), α2(I), α1(II), α1(III), α2(V) and α3(VI) genes and Marfan syndrome (see p. 000) has been established,[67,79,215,316] between α1(I), α2(I) and α1(III) genes and scleroderma,[165] and between α1(I), α2(I), α1(III) and α1(V) genes and mitral valve prolapse.[125] The conclusion that defects in neither type I nor type III are involved in Marfan syndrome has been independently confirmed,[200] although one patient lacking α2(I) has been described. Abnormal α2(I) has been associated with osteoporosis and scoliosis.[287] However, failure to find a defect in collagen genes does not eliminate the possibility that the biochemical lesion may lie in post-translational processing.

Many defects in collagen genes have been described and the discovery of many more is probable. Indeed, over 50 mutations in the type I procollagen gene have now been reported.[239] The same phenotype can arise from a number of different genetic changes (e.g. osteogenesis imperfecta type II). The genetic defects of collagen genes may be divided into a number of classes. Some conditions may arise through failure to express a gene, as in an Ehlers–Danlos syndrome variant with no detectable pro-α2(I) chain.[122,270] Cases are known where an allele, e.g. for α1(I), is not expressed at all, resulting in an imbalanced synthesis of chains, giving α1(I) and α2(I) in the ratio 1:1 instead of 2:1. Regulatory mutations with decreased or absent protein production are also known.[236] In addition, numerous partial deletions have been described, of two types. One causes no frameshift (i.e. is a multiple of three bases) and is usually a whole number of exons.[296,303,313,331] Such deletions commonly result in lethal osteogenesis imperfecta.[49] Other deletions lead to a frameshift and thus to a nonsense sequence.[22,337] Further, the insertion of base sequences occurs in diseases including osteogenesis imperfecta and Ehlers–Danlos syndrome.[292] Lastly, point mutations (single-base changes) commonly replace Gly with Cys; these mutations are either lethal or severely disabling.[236] In more than 90 per cent of osteogenesis imperfecta[49] cases there is a point mutation in one of the alleles of this gene.[221] Point mutations[235,312] and exon deletions[296] have also been described in the type III procollagen gene and lead to Ehlers–Danlos syndrome type IV. Mutations in the α5(IV) gene lead to Alport's syndrome.[17,159,205] Point mutations may also lead to deletion of splice sites or generation of ectopic splice sites.[162,313]

A shortened α-chain may be incorporated into the triple helix, giving an unstable structure which is rapidly degraded ('protein suicide'). Since registration occurs at the C-terminus, incorporation of shortened or altered chains into a triple helix may hinder the action of procollagen N-proteinase as is the case in Ehlers–Danlos syndrome type VIIA (altered α1(I)) and VIIB (altered α2(I)).[325,331,338]

Even in the heterozygous state, with one mutant α1(I) allele, only 25 per cent of type I collagen contains two good α1(I) chains; 50 per cent will have one, and 25 per cent none. Instability of 75 per cent of the collagen type I means rapid collagen turnover. Indeed, most mutations lead to helix instability with decreased melting temperature, increased degradation and reduced rate of triple helix formation. Mutations have a particularly striking effect because the process of nucleated growth in the self assembly of fibrils makes collagen especially sensitive to their presence. A reduced rate of triple helix formation allows increased Lys hydroxylation and glycosylation with consequent modification of fibre properties. These changes are usually more marked in the N-terminal half of the collagen molecule. Helix instability has been identified as a reason for lack of secretion of type III collagen in Ehlers–Danlos syndrome type IV.[297]

One possible gene defect in non-fibrillar collagens is in dystrophic epidermolysis bullosa where no anchoring fibrils (type VII collagen) are seen,[183] either due to failure to secrete type VII collagen[178] or to the inability of type VII to assemble into anchoring fibrils.[43,44]

Defective intracellular processes

The level of prolyl hydroxylase reflects the rate of collagen synthesis; this enzyme may catalyse the rate-determining step. No disease due to deficient or defective enzyme protein has been reported, emphasizing the key importance of this enzyme. However, lack of oxygen or the presence of Fe^{2+} chelators limits the activity of prolyl and lysyl hydroxylases. Furthermore, insufficient dietary ascorbate (see p. 414) causes synthesis of underhydroxylated collagen and thus a less stable triple helix, more susceptible to breakdown. Healing wounds and other sites of collagen synthesis are likely to be affected.

Low lysyl hydroxylase activity underlies some cases of Ehlers–Danlos syndrome type VI (see p. 334). Insufficient hydroxylation of Lys impairs glycosylation and decreases the proportion of Hyl-derived cross-links. The abnormal fibre bundles seen in Ehlers–Danlos syndrome type VI suggest that glycosylation is important in fibril organization. Decreased activity has also been implicated in pulmonary fibrosis (see Chapter 8) and in alkaptonuria (see p. 351), where accumulating homogentisate inhibits the enzyme.

Increased levels of hydroxylases and glycosyl transferases have been reported in the kidneys of rats with drug-induced diabetes, in all fibrotic conditions (especially with elevated prolyl hydroxylase) (see Chapter 8), and in wound healing. Indeed, elevated serum levels of prolyl hydroxylase can be detected before there is gross evidence of fibrosis. Glycosylation of Hyl is proportional to the degree of hydroxylation of Lys. A deficiency of galactosyl hydroxylysyl glucosyl transferase has been described in a family with dominant epidermolysis bullosa simplex.[273]

Triple helix formation is impaired if hydroxylation, especially of Pro, is decreased; altered glycosylation may also interfere. A less stable triple helix is seen in several forms of osteogenesis imperfecta (see above).

Defective extracellular collagen processing

Some Ehlers–Danlos syndrome IV (see p. 334) patients have no extracellular type III collagen because type III procollagen remains in the endoplasmic reticulum and is not secreted[12] (see p. 19). Cases of Ehlers–Danlos syndrome type VIIC lack procollagen N-peptidase activity, and so accumulate pN-collagen. The same defect is found in dermatosparaxis, which occurs in cows and sheep. Skin fragility arises because pN-collagen does not form functional fibrils. Impaired self-aggregation probably results from any defect in post-translational processing. Diseases characterized by poor fibril formation are said to have packing defects (e.g. Ehlers–Danlos syndrome types I, II, III; see Chapter 9).

Lysyl oxidase activity appears significantly increased in the lungs of rats with streptozotocin-induced diabetes mellitus. A deficiency of lysyl oxidase was thought to be the cause of Ehlers–Danlos syndrome type V, but enzyme levels proved to be normal. The difference between Ehlers–Danlos syndrome type V and X-linked cutis laxa[318] (see p. 345), in which lysyl oxidase activity is low in skin but not in other tissues, is thus explained.

Inhibition of lysyl oxidase by β-aminopropionitrile occurs in lathyrism (see p. 413). A diet lacking in copper, or faulty copper absorption (the defect in Menkes' syndrome, see p. 355), also inactivates this copper metalloenzyme, as does the treatment of Wilson's disease (see p. 369) with the chelating agent penicillamine. Clearly, decreased lysyl oxidase activity, for whatever reason, results in poorly cross-linked, fragile collagen fibrils.

Collagen cross-linking is blocked by amines: penicillamine and, in homocystinuria (see p. 349), homocystine compete with Lys and Hyl in Schiff's base formation. Faults in cross-linking have been suggested in recessive cutis laxa[318] (see p. 345), Marfan syndrome[241] (see p. 340) and in Ehlers–Danlos syndrome types I, II and III[292] (see p. 334). Electron microscopic studies indicate fibril disorder in Ehlers–Danlos syndrome types I and II, and show that fibrils from Ehlers–Danlos syndrome type II and V tissues are easily disrupted. The rate of collagen turnover dictates the ratio of mature to immature cross-links: the latter appear to be associated with increased Hyl content in osteogenesis imperfecta (see p. 343) and in fibrosis (see the following text and Chapter 8).

Excessive collagen synthesis

Close control of collagen synthesis is crucial; conditions characterized by net collagen deposition include cirrhosis and fibrosis (see Chapter 8),[35,188,254] keloid and hypertrophic scars, the proliferative phase of rheumatoid arthritis (see Chapter 12), progressive systemic sclerosis (see Chapter 16), atherosclerosis (see pp. 40 and 322), and experimental diabetes mellitus. In many of these, elevated procollagen mRNAs have been observed,[9] indicating increased synthesis of collagen.[1] Interestingly, fibrils of type VI collagen are seen most prominently in pathological situations,[194] notably in cutis laxa.[249]

In fibrosis (see Chapter 8) there is an initial, relative increase in type III with later replacement by type I collagen, which is less cross-linked than in mature tissue. Often prolyl hydroxylase activity is increased before there is any evidence of fibrosis. Collagen increases up to sevenfold over the normal liver content of 0.5 per cent dry weight in cirrhosis, with increases in types I, III, IV, V and type I trimer. The intestinal strictures seen in Crohn's disease are due to the proliferation of smooth muscle cells (see Chapter 1) and the accumulation of collagen, with a relative increase in type V;[109] there is also collagen deposition in the liver.[39] Serum levels of fragments such as the N-propeptide of type III collagen may be useful indicators of fibrosis.[272] Reagents that may be used to control fibrosis include synthetic inhibitors of prolyl hydroxylase,[271] colchicine[155] and interferons.[77]

In keloid (see p. 311), increased activities of prolyl hydroxylase and galactosyl hydroxylysyl glucosyl transferase have been observed.[5] Some fibroblasts cultured from keloid tissue show increased synthesis of type I procollagen and its mRNA, but not of type III procollagen mRNA (hence increasing I/III) while others show no changes and normal enzyme activities.[5,319,320] Similarly, skin fibroblasts from some patients with systemic sclerosis synthesize elevated amounts of procollagen I and its mRNA,[150,328] but others do not show this effect.[216] An altered I/III ratio is also seen in idiopathic pulmonary fibrosis and in adult or infantile idiopathic respiratory distress syndrome.[250] An increased content of type V collagen is apparent in inflammatory and proliferative disease, hypertrophic scars and carcinomata.[206]

The atherosclerotic plaque has a higher collagen content than the normal vessel wall, containing collagen types I, III, IV, V and VI.[20] There is evidence that the proportions of type III and type V are decreased and increased, respectively.[189] These changes may be initiated by endothelial cell loss, stimulation of platelet aggregation by underlying collagens followed by release of platelet factors inducing migration and proliferation of smooth muscle cells or myofibroblasts.

Complications in diabetes include thickening of basement membranes in the lens, retina, peripheral nerves and kidney (paralleled by nephropathy). A key chemical change is increased covalent attachment of glucose to many proteins including type IV collagen and fibronectin. Glucosylation of these proteins decreases their ability to bind together[283] and[265] leads to a significant reduction in the stimulation of heparin (and probably of heparan sulphate proteoglycan) binding to fibronectin by type IV collagen.[304] These changes may provoke increased synthesis of basement membrane components, including type IV collagen, and/or decreased degradation, resulting in basement membrane thickening. It has been observed that glucosylation increases with age[212] and leads to increased cross-linking.[128]

Autoimmunity to collagen

Autoimmunity to collagens as a contributory factor in rheumatoid arthritis is discussed on pp. 245, 533 *et seq*, and 547. Goodpasture's syndrome (see p. 233) is an autoimmune disorder characterized by progressive glomerulonephritis and linear deposition of antiglomerular basement membrane antibody, with or without lung haemorrhage. The NC-1 carboxyl terminal domain of the α3(IV) chain[269] IV collagen is the principal, but perhaps not the sole, antigen.[47,154] Patients with bullous pemphigoid, who show IgG autoantibodies to epidermal basement membrane, both in the circulation and deposited linearly along the epidermal basement membrane, may develop Goodpasture's syndrome.[70]

Collagen degradation

Collagen degradation[146,187,201–203,315,341] occurs as part of normal tissue turnover, where collagen has a half-life of between 50 and 300 days.[173] It is increased in development, growth, tissue remodelling, wound healing, in some disease processes and is elevated in the involuting uterus and parturient cervix.[244]

Complete collagen degradation is usually the result of the synergistic action of several matrix metalloproteinases (MMPs) active at neutral pH. These enzymes (Table 4.2) require both Zn^{2+} and Ca^{2+}, and are secreted in latent proenzyme forms by cells such as fibroblasts, chondrocytes, osteoblasts and endothelial cells; they show considerable homology. While animal collagenases (MMP1) have molecular weights of between 30 and 80 kD, the active form in most mammalian tissues has a molecular weight of 45 kD. Collagenases (MMP1) from various connective tissue cells are similar[121] but are different from polymorph

Table 4.2

Characteristics of the known matrix metalloproteinases

Designation	Name	Proenzyme molecular weight (kD)	Degrades
MMP1	Interstitial collagenase	52	Type I, II, III, VII[280], X[97] collagens
MMP2	Type IV collagenase (gelatinase)	72	Type IV, V, VII collagens, gelatins, fibronectin, elastin
MMP3	Stromelysin	57	Type III, IV, V, IX[217] collagens, gelatins, fibronectin, laminin, proteoglycans
MMP7	PUMP-1	28	Gelatins, fibronectin
MMP8	PMN collagenase	75	Type I, II, III collagens
MMP9	Type IV collagenase (gelatinase)	92	Type IV, V collagens, gelatins
MMP10	Stromelysin-2	53	Type III, IV, V collagens, gelatins, fibronectin

collagenase (MMP8, Table 4.2). The former cleave specific Gly–X bonds, where X is a hydrophobic amino acid residue. Animal collagenases cleave fibrillar collagens at a single point in the triple helix, giving two fragments (TC_A and TC_B). The bond Gly_{775}–X_{776} is cleaved specifically, where X is Ile for $\alpha1(I)$ and Leu for $\alpha2(I)$, even though similar bonds occur elsewhere in the molecule. Bacterial collagenases are distinct enzymes and split the collagen monomer into many tripeptide (Gly–X–Y) fragments.

Fate of fragments

While non-cross-linked collagen is most susceptible to collagenase action, this enzyme can eventually digest insoluble cross-linked collagen. However, it is considerably assisted by non-specific proteinases such as gelatinases (also secreted by polymorphs), stromelysins, polymorph elastase[99] and cathepsins B, G, L and D,[40] the latter probably acting in a local acidic pericellular environment. These enzymes initially act on telopeptides, disrupting cross-links. The triple helical structure of TC_A and TC_B melts out below 37 °C,[68] yielding gelatins, so even if linked to the fibre via cross-linked telopeptides, these fragments become susceptible to non-specific proteinases. The products of collagen degradation are Hyp peptides, which are excreted in the urine, as Hyp cannot be re-used in collagen biosynthesis; this excretion is an index of collagen turnover. A greater role for non-specific proteinases in type III collagen catabolism is likely as trypsin, elastase and thermolysin cleave its triple helix near the collagenase site; several non-specific proteinases also degrade basement membrane collagens.

A secondary intracellular pathway of collagen degradation can occur after preliminary extracellular breakdown. Collagen fragments enter the cell by endocytosis and are degraded within secondary lysosomes and within the cisternae of the endoplasmic reticulum and Golgi apparatus by the collagenolytic cathepsins B, D, L and N. Both routes of degradation are shown in Fig. 4.4.

Differential susceptibility of collagens to collagenases

Interstitial collagenases digest types I, II, III, VII[281] and X[275] (Table 4.2),[99] while types IV, V, VI, IX and XI are resistant. Some collagenases digest type III collagen more rapidly than type I; for others the converse is true.[120] Most collagenases attack type II collagen most slowly because of its greater degree of glycosylation. Collagens like type IV with interrupted helices are susceptible to cleavage by type IV collagenases (MMPs 2 and 9) and by non-specific proteinases including gelatinase, stromelysin and polymorph elastase.[232]

A high degree of cross-linking makes collagen degradation more difficult. Often differences in relative rates of degradation of collagen monomers disappear with polymeric collagen substrates.[120] The thicker a collagen fibre, the more slowly it is degraded as enzymes only act at the fibre surface. The association of proteoglycans and glycoproteins can shield collagen from collagenase and fibronectin may be rather effective as it binds to collagen at the collagenase cleavage site. Cartilage proteoglycans protect reconstituted collagen fibres from human synovial collagenase.

Regulation of metalloproteinase activity

Collagenase, gelatinase and stromelysin are regulated at the level of synthesis, activation of latent proenzymes and

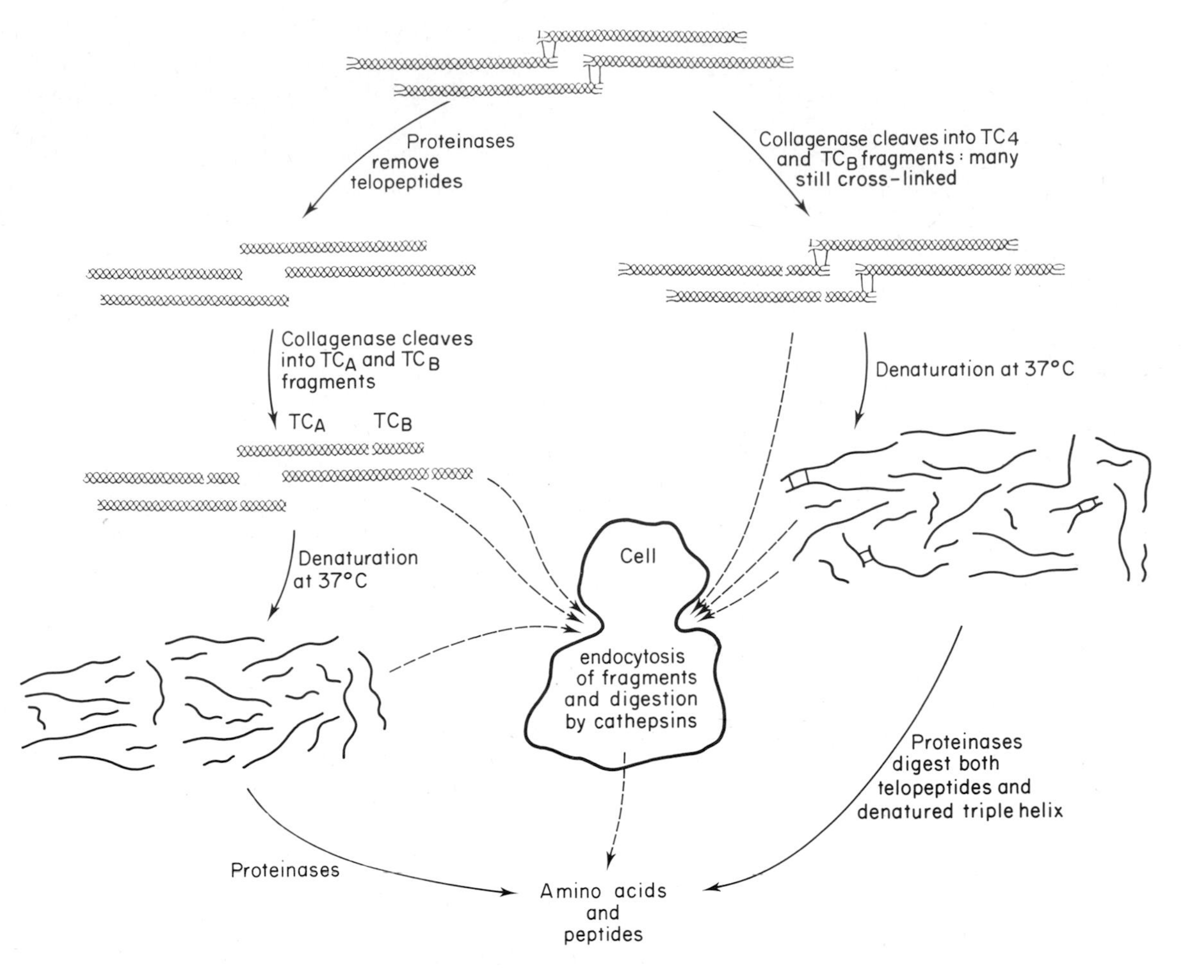

Fig. 4.4 Degradation of collagen.
The diagram shows the degradation of polymeric collagen by the sequential action of collagenase and proteinases (right-hand pathway) and vice versa (left-hand pathway). Presumably both pathways occur simultaneously. Also shown is the pathway of intracellular digestion of fragments produced by the other two mechanisms.

inhibition of active enzymes by tissue inhibitor of metalloproteinase (TIMP, see below).[117] In many cells there is coordinated production of these metalloproteinases, often paralleled by TIMP synthesis (Fig. 4.5).

Cytokines (p. 291) and growth factors (p. 18) known to stimulate synthesis of metalloproteinases by connective tissue cells include interleukins, tumour necrosis factor, epidermal growth factor, fibroblast growth factor and interferons. While many of these are produced by mononuclear cells, undoubtedly some are also synthesized by connective tissue cells themselves, demonstrating important autocrine regulatory mechanisms. Control is exerted at the transcriptional level: exposure of fibroblasts to epidermal growth factor, fibroblast growth factor or embryonal carcinoma-derived growth factor leads to increased collagenase, stromelysin and TIMP mRNAs. In contrast, transforming growth factor β, in the presence of other growth factors, reduces collagenase mRNA but increases TIMP mRNA. Retinoids and dexamethasone decrease, and phorbol esters increase, synthesis of collagenase mRNA; retinoids also increase TIMP mRNA but dexamethasone is without effect.[62] Dexamethasone diminishes the level of stromelysin gene induction caused by interleukin-1, epidermal growth factor, phorbol esters and cAMP.[95] There is evidence that growth factors include expression of proto-oncogene products C-*Fos* and C-*Jun* which regulate MMP gene transcription.[187]

Activation of the latent proenzymes can be effected *in vitro* by proteinases such as trypsin and by organomercurials (e.g. 4-aminophenyl mercuric acetate). The process *in vivo* is probably initiated by plasmin although kallikrein and cathepsin B, which has some activity at neutral pH, have also been implicated. Plasmin is formed from plasminogen by cell-derived plasminogen activators, serine proteinases themselves synthesized as proenzymes, and inhibited by plasminogen activator inhibitor, also secreted from the cell.

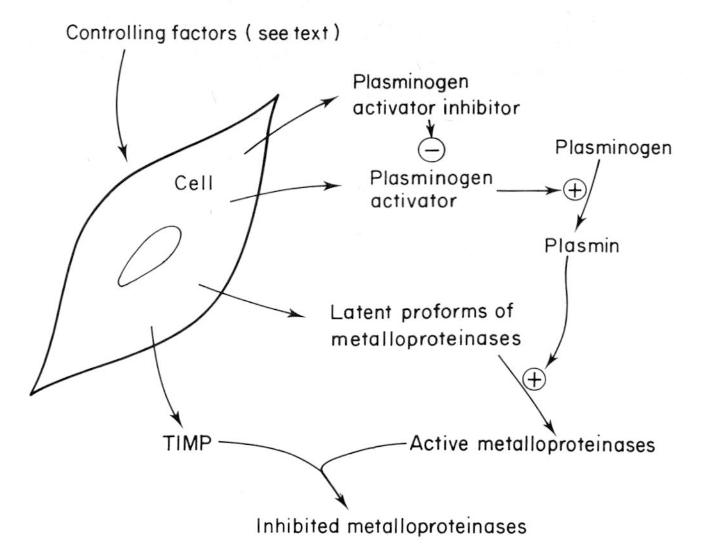

Fig. 4.5 Coordinate production, activation and inhibition of metalloproteinases.
The metalloproteinases collagenase, gelatinase and stromelysin are activated by endogenous proteinases, principally plasmin. In turn, plasmin is formed from plasminogen by the action of plasminogen activator. Besides secreting plasminogen activator, plasminogen activator inhibitor and the latent proforms of metalloproteinases, the cell secretes tissue inhibitor of metalloproteinase, which irreversibly inhibits all three metalloproteinases. Factors stimulating secretion of these components from the cell are discussed in the text.

There is a decrease in molecular weight of about 10 kD on activation of metalloproteinases, but further proteolysis can occur by autolysis and/or the action of proteinases in a complex process yielding several active forms. The process of activation is thus controlled by the relative concentrations of plasmin, plasminogen activator, and plasminogen activator inhibitor. The secretion of the latter two is stimulated by transforming growth factor β differentially in different cell types:[156] in lung fibroblasts plasminogen activator inhibitor predominates while in carcinoma cells plasminogen activator secretion exceeds that of plasminogen activator inhibitor.

The activity of the matrix metalloproteinases is also regulated by inhibition. Plasma α_2-macroglobulin is a potent inhibitor, but its high molecular weight (750 kD) precludes extravascular action. Tissue inhibitor of metalloproteinase is the most important inhibitor of metalloproteinases in the extracellular matrix.[103] Inhibition by TIMP is irreversible. There is no precursor form of TIMP, a glycoprotein of molecular weight 28 kD produced by most connective tissue cells; it is also found in the plasma, where it has been known as β_1-anticollagenase. Synthesis of TIMP is increased in cells stimulated by interleukin-1, phorbol esters or corticosteroids. More details of regulation of TIMP synthesis are given above. A second TIMP, known as TIMP-2, has been described.[106]

Metalloproteinases are also controlled by hormones. Corticosteroids are strong inhibitors of metalloproteinase synthesis (see dexamethasone above) and also decrease secretion of plasminogen activator by several cell types and enhance production of plasminogen activator inhibitor.[156] These actions are probably significant in the anti-inflammatory action of corticosteroids and in their use in cancer therapy. Plasminogen activator activity in cell cultures is raised by sex steroids, vasopressin, prolactin, parathyroid hormone and thyrotropin. Collagen degradation in uterine involution is undoubtedly a response to changes in the levels of oestradiol and progesterone. Prolactin and insulin decrease type IV collagenase activity in mammary epithelium. Osteoblasts (p. 29) produce collagenase in response to parathyroid hormone but other cells do not.

Other factors increasing collagenase secretion are lectins and lymphokines (see Chapter 5), (which stimulate macrophages and monocytes to secret cytokines such as interleukin-1), phagocytosis of poorly digestible particles, such as latex, exposure to proteinases, collagens, colchicin, cytochalasin B, phorbol derivatives, ionic iron (see p. 291), and mycoplasmas (see Chapter 18). Immune complexes (see p. 238) and F_c fragments probably also stimulate collagenase production via an interaction with mononuclear cells. Physical factors, such as attainment of confluency, low pH, increased pO_2 and mechanical stress, all increase collagenase secretion from cells.

Collagenase activity is enhanced by small rises above normal body temperature: inflammation may thus promote collagenolysis. Human skin collagenase is especially remarkable in this respect, increasing its rate three-fold for a 2 °C rise in temperature. However, activity in normal skin, which can be up to 10 °C colder than body core temperature, may be low.

Tissue location of collagenase

Immunofluorescence studies have demonstrated that collagenase is localized on collagen bundles, reticular fibres and basement membranes. Collagenase has also been located pericellularly around the dendritic cells of rheumatoid synovia and intracellularly in human skin fibroblasts and in several murine cell types. Whether the active enzyme is localized depends on antibody specificity: some antibodies recognize both the latent proenzyme and the active enzyme, others bind only to the active enzyme.

Collagen degradation and disease[146,341]

Excess collagenolytic activity

Elevated tissue collagenase activity can result from an imbalance in latent proenzyme, active enzyme and tissue inhibitor of metalloproteinase (TIMP); increased concentration of active enzyme and/or decreased latent enzyme and/or TIMP. Increased collagenase activity may be the primary event leading to the disease or may be secondary to some other event such as hormone imbalance. Therapy based on collagenase inhibition may be possible in some diseases, e.g. corneal ulceration.

Corneal ulceration Injury to the corneal surface epithelium causes eventual destruction and ulceration of the underlying stroma. Subsequent scar tissue formation leads to loss of vision. Involvement of collagenase is suggested by the observations that cultured explants from ulcerated corneas secrete more collagenase than do normal corneal explants and that collagenase can be immunolocalized in stromal tissue from ulcerated corneas but not in normal tissue.

The epithelial cells, fibroblasts and polymorphs involved in corneal ulceration can produce collagenases, and their role is demonstrable by prevention of ulceration if stromal re-epithelialization is blocked by an attached lens. Epithelial cells may secrete a factor chemotactic for polymorphs. Movement of these cells into the damaged area with secretion of collagenases and proteinases upsets the balance of active and latent proenzyme, leading to further ulceration.

Treatment of corneal ulceration with hormones has been suggested as progesterone prevents collagenase production by macrophages and blocks collagenolysis in the postpartum uterus. Medroxyprogesterone more than halves the collagenolytic activity of corneas burned by alkali. In addition, collagenases that cause ulceration in the rabbit cornea can be inhibited by application of calcium ethylene diamine tetra-acetic acid (EDTA), cysteine or acetyl cysteine.

Rheumatoid arthritis (see p. 536) The immunopathology of rheumatoid arthritis is discussed in Chapter 5 and its pathogenesis is covered in Chapter 13. The initial event is likely to be tissue injury by virus with resulting inflammation. The inflammatory sequence, including free radical damage, is outlined in Chapter 7. Degradation of the extracellular matrix components with exposure of previously hidden epitopes follows[74] and elicits the production of autoantibodies. The formation of antigen–antibody complexes in rheumatoid joints stimulates macrophages to secrete interleukin-1 and polymorphs to secrete interleukin-1 and proteinases, especially elastase. Interleukin-1 elicits further interleukin-1 secretion from synovial cells and fibroblasts, and causes chondrocytes, synovial cells, fibroblasts and osteoclasts to liberate proteinases. Thus further breakdown can occur and the cycle of degradation is perpetuated. Since secreted proteinases can include metalloproteinases, activation of these enzymes by processes described on p. 185 would allow degradation of collagens and proteoglycans. Destruction of collagen can be detected by the elevated amount of the cross-linking compound pyridinoline in urine from both rheumatoid arthritis and osteoarthrosis.[252] Many other regulatory factors are likely to be involved in the modulation of cell behaviour, as suggested by the studies summarized below.

The secretion of collagenase and prostaglandin E_2 has been demonstrated in cultures of synovial explants from rheumatoid patients. The purified enzyme, of molecular weight 32 kD, degrades cartilage collagen, insoluble collagen fibres and slices of human articular cartilage. Monospecific antiserum to pure enzyme reacts only weakly with latent proenzyme: active enzyme immunolocalizes in 40 per cent of rheumatoid specimens at the junction between the advancing pannus and articular cartilage. No immunoreactive enzyme has been found elsewhere, although latent proenzyme is present in rheumatoid synovial fluid. The only cell type in the inflammatory cell mass that produces collagenase detectable by immunolocalization techniques is the dendritic or stellate cell and not all of these do so. However, such cells in culture produce high levels of collagenase and prostaglandin E_2, a process stimulated by interleukin-1.

Some rheumatoid patients have a second collagenase of molecular weight 50 kD, probably originating from polymorphs, which is less susceptible than the other enzyme to inhibition by α_2-macroglobulin and tissue inhibitor of metalloproteinase. Neutral proteinases have also been detected in rheumatoid synovium, but their extracellular function awaits proof.

Collagenase action in rheumatoid arthritis may expose and/or generate epitopes which could explain the observed increase in IgG and IgM anticollagen antibodies in sera and synovial fluid, the intraarticular antigen–antibody complexes[61,145] and the perpetuation of inflammation. However, sera from rheumatoid patients not only contain antibodies to both native and denatured type II collagen, but also to native and denatured types I, IX and XI,[55,196,197] suggesting an action of other proteinases on the latter two types. Autoimmunity to collagens was thought to be a possible initiator of the inflammatory process but this is no longer considered likely.

Epidermolysis bullosa The recessive dystrophic

form of this disorder is characterized by the formation of cutaneous blisters (bullae) on the skin and mucous membranes as a result of slight injuries. The blisters show increased collagenase activity when compared with normal controls, but so does uninvolved skin from the same patient. It has been suggested that the increased collagenase may destroy the type VII collagen of anchoring fibrils (see p. 51).[23]

Periodontal disease

Periodontal disease is an inflammatory condition in which periodontal ligament and alveolar bone are slowly destroyed. Secretion of collagenase by gingival explants in culture is greater in periodontitis cases than in normal controls. Collagenase is secreted by epithelial cells and fibroblasts from inflamed gingiva and by macrophages and polymorphs as latent proenzyme. Crevicular fluid collagenase activity is increased with disease severity in gingivitis and periodontitis,[324] and there is less inhibitor than normal in periodontally diseased roots.[198] In immunolocalization studies of collagenase on eight specimens of gingival tissue from patients with untreated periodontal disease, at least half showed intense fluorescence on disintegrating collagenous structures.

Neoplastic invasion (see p. 980)[157,182,225,315,335,341]

Metastasis of malignant neoplasms involves localized loss of basement membranes so that neoplastic cells may pass into the bloodstream and invade the target organ. Many neoplastic cells secrete metalloproteinases, including type IV collagenases, or cause other cells to secrete these enzymes: several types of neoplastic cells produce cytokines probably related to interleukin-1. Elevated matrix metalloproteinases in tumours may be due to the presence of an activated oncogene such as Ha-*ras*, known to be a potent inducer of MMPs 1, 2, 3, 9 and 10.[187] There is evidence of increased cathepsin B and D[40] and plasminogen activator and decreased tissue inhibitor of metalloproteinase secretion by highly invasive neoplastic cells. In some human cancer cells, collagenolytic, gelatinolytic and cysteine proteinase activities are significantly increased[346] but in other cases, such as human colorectal carcinoma, secretion of collagenase is not above normal.[141] Inhibitors of metalloproteinases block neoplastic invasion, and the resistance of cartilage and arteries to this process is probably due to high inhibitor concentrations. Neoplasm-derived collagenases that degrade types I, II and III collagens have been characterized, and have been immunolocalized to some, but not all, cancer cells. Failure to identify collagenase secretion by neoplastic cells may reflect either absence of secretion by the cell itself, or sporadic enzyme production. A type IV collagenase, of molecular weight 62–65 kD, has been identified in metastatic cancer cells and endothelial cells, and antibodies to this enzyme recognize invading breast carcinoma cells and breast carcinoma lymph node metastases.[195] Type IV collagenase cleaves type IV collagen a quarter of the way along from the 7S domain (Fig. 4.3), but different neoplastic cell lines produce different breakdown patterns, probably due to the action of different metalloproteinases.[291] However, it has been proposed that the loss of basement membranes in some neoplasms may reflect the reduced ability of the transformed cells to synthesize and secrete basement membrane components.[329]

Decreased collagenolytic activity

Excessive deposition of collagen occurs if synthesis exceeds breakdown. Normally, synthesis predominates in growth and repair but lack of control may result in disease. Decreased collagenolytic activity may be due to decreased concentration of active enzyme and/or to increased latent proenzyme and/or tissue inhibitor of metalloproteinase. Impairment of collagen degradation may play a role in hypertrophic scar (see p. 312) and keloid (see p. 311) formation.[146]

Cirrhosis of the liver (see p. 323)

Cultured tissue explants from the liver in the early stages of carbon tetrachloride-induced cirrhosis secret collagenase, which declines as the disease advances; immunofluorescence studies show collagenase located on collagen and reticular fibres. In the later, irreversible stages, no collagenase is present and fibrosis begins.

Diabetes mellitus

The excessive deposition of basement membrane material implies a defect in turnover. One explanation is that the type IV collagen has become less susceptible to collagenase because of non-enzymatic glucosylation (see p. 175).

Elastin

Elastin[71,256,266] is a rubber-like fibrous protein found in virtually every organ of the body (see p. 46). The elastic material, which may be organized as sheets or lamellae, can be recognized microscopically because of its characteristic morphology (see p. 47). Tissues whose function demands elasticity include elastic cartilage (see p. 84), skin, ligaments, lung, intestine and blood vessels, particularly the thoracic aorta. Elastin is an exceedingly intractable protein to study because of its great insolubility. Fortunately, copper deficiency leads to inhibition of cross-linking, so that the soluble precursor, tropoelastin, can be extracted and studied.

Structure

Newly synthesized soluble elastin, tropoelastin (mol. wt 72 kD), has a distinctive composition, being rich in hydrophobic amino acids. Thus, Gly, Pro, Ala, Val, Phe, Ile and Leu predominate. Asp, Glu, Lys and Arg make up less than 5 per cent. Gly constitutes one-third of all amino acid residues, but there is no regularly repeated sequence Gly–X–Y. In tropoelastin, Hyp constitutes about 1 per cent of all residues, but Try, His, Met and Hyl are absent and there is no attached carbohydrate.

The human tropoelastin molecule consists of 786 amino acid residues, and its complete primary structure is known.[140] It contains repeated hydrophobic sequences such as Val-Pro-Gly-Val-Gly and Pro-Gly-Val-Gly-Val-Ala, which produce a β-spiral conformation, a structure not found in other mammalian proteins. There are also α-helical segments that include eight Ala-rich sequences in which two Lys residues, separated by two or three Ala residues, are preceded by a sequence of between three and nine Ala residues. These peptides are the sites of cross-linking that transform tropoelastin into insoluble elastin. The action of lysyl oxidase converts Lys to allysine residues (lysyl oxidase oxidizes both collagen and and elastin substrates). One Lys and three allysine residues react together to form the cross-links desmosine and isodesmosine (Fig. 4.6). Since pairs of Lys residues occur close together, each cross-link unites two chains only. However, each chain probably has four or five cross-links joining it to other chains. On hydrolysis, desmosine and isodesmosine can be quantitated by amino acid analysis. There are 1–1.5 residues per 1000. Other cross-links, such as those involving aldol condensation and dehydrolysinonorleucine (Fig. 4.2), are common to elastin and collagen.[85]

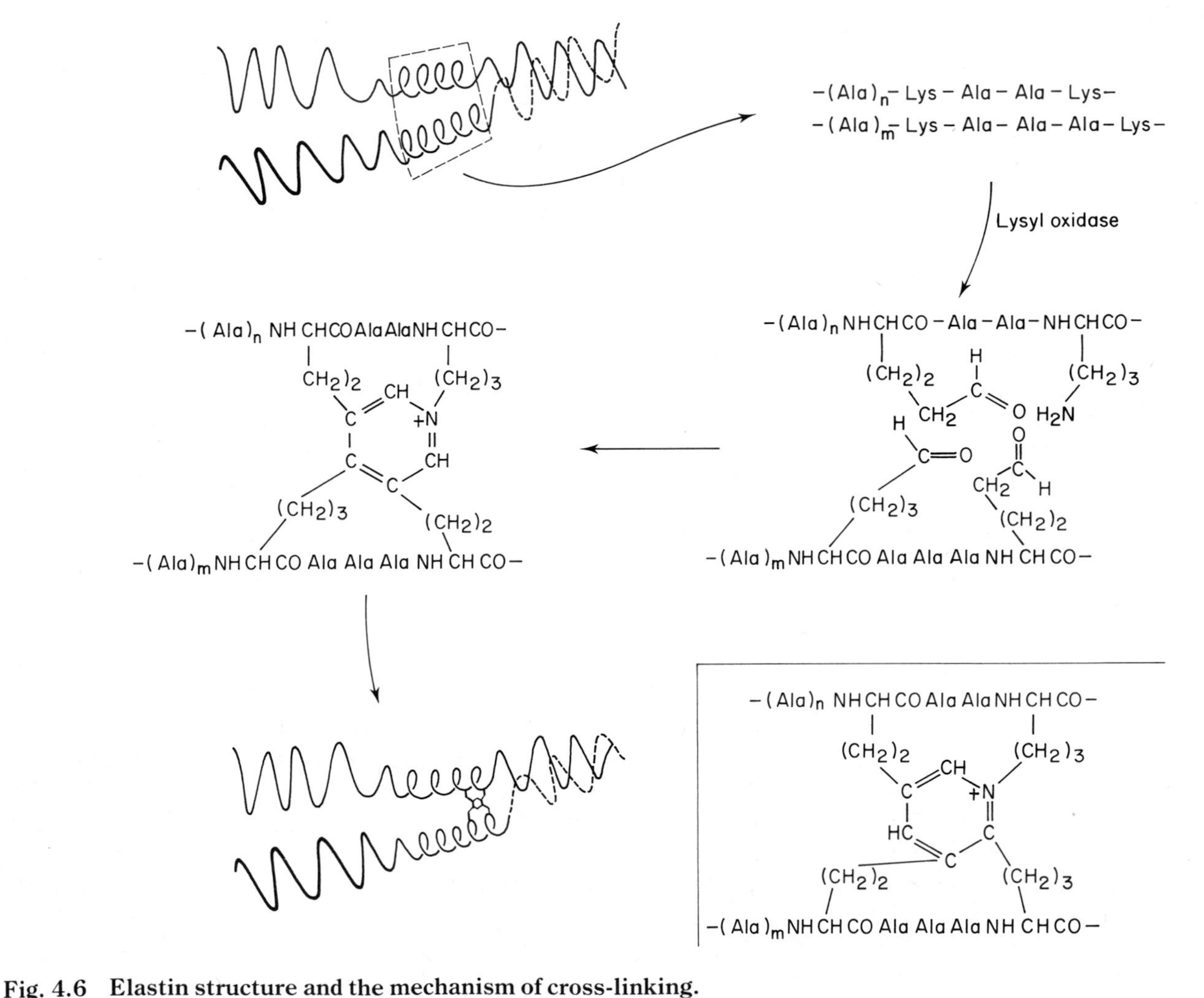

Fig. 4.6 Elastin structure and the mechanism of cross-linking.
The cross-linking of two adjacent tropoelastic molecules occurs within segments which are thought to be α-helical, and which each contain two Lys residues close together. Condensation of three allysine residues with one unmodified Lys shows the formation of the desmosine cross-link. A similar mechanism produces isodesmosine (inset).

In the accepted three-dimensional elastin structure, the α-helical segments in tropoelastin become cross-linked to other tropoelastin units as the insoluble material forms (Fig. 4.6). The α-helical sections are separated by coiled hydrophobic sequences containing β-spiral structures, formerly designated 'random coil'. The elasticity of the material or fibre is a property of this hydrophobic structure; under tension it uncoils, but when the stress is released it regains its original structure and length (see Chapter 6).

Like tropocollagen, tropoelastin reversibly aggregates into fibrils under physiological conditions, a process known as coacervation. Coacervation must involve non-covalent interactions, i.e. electrostatic, hydrogen and hydrophobic bonds; these bonds are therefore important in the formation of elastic fibres and lamellae in tissues.

Biosynthesis

The tropoelastin gene is over 40 kb long, and consists of 17 exons separated by some enormous introns; elastin messenger RNA (mRNA) is only 3.5 kb in length. Although all the evidence shows that there is only one human elastin gene, multiple elastin mRNAs with different sequences have been identified.[140] The most likely explanation is alternative splicing of the single primary transcript.[345] Translation of these mRNAs will give appreciable variation in amino acid sequence and length of tropoelastin, which explains why at least two forms of tropoelastin have been found in several species.[223] There could thus be significant variation in elastin molecular structure in the human population. Moreover, changes in splicing might explain the observed increase in the proportion of polar amino acids in elastin with age[290] and in atherosclerosis (see p. 322) and other diseases.[139] There is evidence of a decrease in elastin gene expression with ageing.[86].

The biosynthesis of elastin proceeds by the normal mechanism of protein synthesis (see p. 19). A signal peptide whose function is to steer the growing peptide into the cisternae of the rough endoplasmic reticulum is removed at that site. The existence of a proelastin precursor has been proposed, but most experimental evidence does not support this view.

Post-translational modifications include hydroxylation of Pro, oxidative deamination of Lys and cross-link formation. The hydroxylation of Pro is catalysed by the same prolyl hydroxylase as for collagen; the Pro must be in the sequence –Gly-X-Pro-Gly–. However, Pro hydroxylation is not the key step that it is in collagen biosynthesis: there is no evidence of defective elastin synthesis or structural organization in scurvy, perhaps because failure to hydroxylate Pro, does not affect the secretion of tropoelastin. Moreover, incubation of connective tissue cells in the presence of the iron chelator αα'dipyridyl does not inhibit the synthesis and secretion of elastin, although no triple helical collagen is found in the medium.

Tropoelastin collects in vacuoles 'budding' from the rough endoplasmic reticulum. As the vacuole moves to the plasma membrane and secretes its contents from the cell, coacervation and limited cross-linking may occur, to be completed at the site of the growing elastic fibres and lammelae.

Elastin synthesis is stimulated by glucocorticoids and insulin-like growth factor 1 (IGF-1 = somatomedin C). In addition to being responsive to circulating IGF-1, both fibroblasts and smooth muscle cells themselves can produce IGF-1, suggesting that IGF-1 could be important in the regulation of elastogenesis in the arterial matrix.[92]

Microfibrillar glycoproteins (see pp. 47 and 211) are a second important component of elastic fibres.[63] The microfibrils are a prominent morphological feature, especially during fibre formation, when these elements form a framework to aid the deposition of elastin.

Diseases of elastin synthesis

Excess elastin deposition may be the result of increased synthesis and/or decreased degradation. Conversely, decreased elastin deposition can ensue from decreased synthesis and/or increased degradation. Diseases characterized by the accumulation of elastin are sometimes described as elastoses, and include pseudoxanthoma elasticum[135] (see p. 355), actinic (solar) elastosis (see Chapter 15), fibrotic breast and liver disease (see p. 323), elastosis perforans serpiginosa, elastofibroma (see p. 310), atherosclerosis (see p. 322) and progeria.[285] Many are characterized by fragmentation and thickening of elastic fibres, but in none is the biochemical cause known. Increased elastin deposition in large arteries is a characteristic of hypertension.

In the atheroma, the link between the development of the plaque and changes in the connective tissue matrix is not fully understood.[20,177] A constant early event is splitting and the movement of smooth muscle cells into the intima. Proliferation of these cells and the accumulation of their secreted extracellular matrix, together with the ingress of plasma lipoproteins, increases intimal thickness and forms the intimal lesion. Smooth muscle cell proliferation may be caused by factors released from platelets following their aggregation on exposed collagen after endothelial injury. Alternatively, smooth muscle cell hyperplasia may be a result of altered cell phenotype, leading to secretion of a changed intimal matrix, with increased proteoglycans, elastin and collagen, and an elevated type I/type III collagen ratio. Elastin isolated from atheromatous plaques contains a higher proportion of polar amino acids than normal elastin, perhaps accounting for its increased ability to bind calcium, heparan sulphate and dermatan sulphate, and also for the ingress of plasma lipid, since it also binds lipids more strongly than normal.

Loss of mechanical strength and elastic lamellae from aortas in Marfan syndrome[241] patients is associated with a decreased elastin concentration and degree of cross-linking.[229] Nevertheless, it is thought unlikely that the defect lies in the elastin gene.[79] Decreased elastin gene expression has been reported in some cases of cutis laxa.[284]

The impairment of elastin cross-linking by low lysyl oxidase activity occurs in X-linked cutis laxa (see p. 345), lathyrism (see p. 413) and copper deficiency (see p. 412), and has been suggested in lethal perinatal osteogenesis imperfecta (see p. 343).[224] However, additional genetic anomalies must be involved since each of these conditions is distinctive. Interference with cross-link formation probably also occurs in homocystinuria (see p. 349).

Degradation of elastin

In normal tissues, degradation[32,203] occurs in the course of very slow elastin turnover. Elastin is degraded by elastases isolated from polymorphs, macrophages, platelets and the pancreas. Elastases are serine proteinases specific for small hydrophobic amino acid residues such as Val, Leu and Ile. Elastases are thus not specific for elastin; they also degrade type III and IV collagens, fibronectin and proteoglycans. Elastases can hydrolyse insoluble elastin fibres, usually breaking peptide bonds next to Gly, Ala or Val. Once soluble elastin fragments are generated, hydrolysis can be completed by other neutral proteinases like stromelysin and gelatinase. As for collagenases, the activity of elastases can be controlled by inhibitors. Known inhibitors are α_1-antiproteinase (α_1-antitrypsin) and α_2-macroglobulin.

Defects in elastin degradation

Pulmonary emphysema is the best documented example of a disease resulting from deranged elastin degradation.[32,340] Indications that this might be the cause of this disease were obtained from experiments in which papain, trypsin or elastases originating from pancreas, polymorphs or macrophages were injected intratracheally into animals. The lungs of treated animals resembled those of patients with emphysema: total collagen was unchanged, but within a few hours the elastin content decreased to be slowly replaced over the ensuing one to two months. However, when degraded elastin was replaced, the lungs did not return to a normal morphology: the changes of emphysema were permanent.

Emphysema can originate in several ways.[98] In hereditary pulmonary emphysema, the substitution of a Lys for a Glu residue by gene mutation results in decreased solubility of α_1-antiproteinase so that it precipitates in hepatocytes and cannot be secreted. Presumably the balance of enzyme and inhibitor is upset in favour of elastase, resulting in increased elastin degradation. Pulmonary emphysema is also found in (Indian) childhood cirrhosis (see Chapter 8), in which α_1-antiproteinase is again not secreted from hepatocytes. In this case, the defect appears to be lack of glycosylation, and inclusion bodies are seen in the hepatocytes, containing unglycosylated α_1-antiproteinase. Deficiency of α_1-antiproteinase can be treated with recombinant α_1-antiproteinase.[53]

The commonest form of emphysema arises from cigarette smoking; other environmental causes are fumes, dusts and cadmium. Lungs of smokers contain more polymorphs and macrophages than normal lung. As both these cell types secrete elastases, the amount of elastase in smokers' lungs is increased, especially when the enzyme is released from cells in response to smoke. Plasma levels of neutrophil polymorph elastase-derived fibrinopeptides are five times higher in smokers than normal.[332] Although the elastase content of macrophages is very low compared with that of polymorphs, their role cannot be disregarded because of their longer life-span, larger number, and the striking increase in their secretion of proteinases in inflammation. Moreover, of the three macrophage elastases, only one is inhibited by α_1-antiproteinase. Macrophages also act by internalizing polymorph elastase, possibly protecting it from inhibitors, and carrying it to sites of tissue degradation. A further neutral proteinase from human alveolar macrophages degrades type V collagen and denatured collagens and is cross-reactive with neutrophil polymorph gelatinase.[127] Another significant observation is that α_1-antiproteinase from smokers' lungs only exerts 62 per cent of its normal inhibitory activity, although its concentration is the same as in normal lung. The decreased activity is probably caused by oxidation or proteolytic attack. Thus smokers' emphysema is multifactorial. Moreover, smoke inhibits lysyl oxidase-catalysed oxidation of Lys residues. As in hereditary pulmonary emphysema, it appears that the elastin content does not decrease, but turnover increases and this, by some process that is not understood, results in tissue deformity.

Pulmonary emphysema is the most common internal defect in cutis laxa (see p. 345). In X-linked cutis laxa, where the defect is a deficiency of lysyl oxidase, the condition could be explained by a failure to form insoluble elastin. The aetiology is thus quite distinct from that of pulmonary emphysema.

Acute oedematous lung injury is seen in pulmonary oxygen toxicity and in adult respiratory distress syndrome. It is proposed[253] that the activity of xanthine oxidase in lung endothelial cells in hyperoxia generates toxic oxygen metabolites which enhance the susceptibility of elastin to polymorph elastase, probably by inactivating elastase inhibitor.

Increased elastase activity has also been identified as a cause of saccular aneurysms of the abdominal aorta, in which the elastin content is greatly reduced,[50] and is also observed in rheumatoid arthritis,[3] diabetes,[64] unstable angina pectoris and acute myocardial infarction,[73] in cystic fibrosis,[147] and in cutix laxa and actinic elastosis.[278]

Derangements in elastases have also been implicated in leucocytoclastic vasculitis (see p. 628), neoplastic invasion (see p. 980), atherosclerosis (see pp. 322 and 340) and Marfan syndrome (see Chapter 9).

Proteoglycans

Proteoglycans[21,99,115,116,124,233,234,259] are macromolecular glycoconjugates consisting of specialized polysaccharide chains called glycosaminoglycans attached covalently to a protein core. Most proteoglycans also bear covalently attached O- and N-linked oligosaccharides typical of glycoproteins. The term ‘proteoglycan’ has replaced the old term ‘(acid) mucopolysaccharide’, which was unsatisfactory because it was used to denote both proteoglycans and glycosaminoglycans, as well as partially degraded proteoglycans. Hyaluronate is included here under the heading proteoglycans for convenience, not accuracy (see below).

An important property of proteoglycans and hyaluronate, due to their highly hydrophilic structures, is their ability to bind water. With associated water molecules, these components of the extracellular matrix tend to occupy large volumes called domains. This property strongly influences the mechanical characteristics of many connective and other tissues (see Chapter 6).

Glycosaminoglycans

The general structures of the glycosaminoglycans are shown in Table 4.3. These polysaccharides differ from the oligosaccharides of glycoproteins and glycolipids as follows:

1. In principle glycosaminoglycans are repeating disaccharides of a hexuronic acid and an N-acetylhexosamine sulphate. Their content of hexuronate is unique in animal tissues so that hexuronate analysis provides a simple marker for proteoglycans and glycosaminoglycans.
2. They are highly negatively charged, usually having one carboxylate and one sulphate monoester group per repeating disaccharide. Thus proteoglycans and glycosaminoglycans form precipitates with organic cationic compounds such as cetyl pyridinium chloride, and bind a wide range of cationic dyes (see Chapter 1).
3. Glycosaminoglycans are usually much larger than oligosaccharides of glycoproteins and glycolipids, and are never branched (except for keratan sulphate); oligosaccharides often are.

There are exceptions to these generalizations (Table 4.3). Hyaluronate has no sulphate groups, is of a much higher molecular weight than other glycosaminoglycans and is not found covalently linked to protein. Keratan sulphate has Gal residues instead of hexuronate, many sulphated at C6, and small but measurable amounts of sialic acid, Fuc and Man, forming a branched structure. Heparan sulphate and especially heparin carry a number of N-sulphate groups in place of N-acetyl groups, contain IdA in place of some GlcA residues with some of the former sulphated in the 2 position. Generally, heparin has a higher proportion of N-sulphate and IdA than heparan sulphate. These variations mean that heparan sulphate and heparin contain unique oligosaccharide sequences with specific properties (see pp. 198–200). Lastly, dermatan sulphate has GlcA and IdA residues; the former occur in one or two clusters per chain. Again, some of the IdA residues are sulphated in the 2 position, allowing some sequence variability. Glycosaminoglycans possess specialized oligosaccharide sequences which link them to their protein cores (Table 4.3).

Hyaluronate

Hyaluronate[82,107,174,293,310] is a ubiquitous component of the extracellular matrix. Large amounts are also present in synovial fluid (see p. 89). Its large hydrodynamic volume, together with entanglement of chains, gives hyaluronate solutions properties such as pronounced viscoelasticity. Hyaluronate is part of large proteoglycan aggregates (see Fig. 4.7) or exists free as in synovial fluid and vitreous humour. Attempts have been made to use hyaluronate as a synovial fluid substitute. It occurs pericellularly around chondroblasts, fibroblasts, myofibroblasts, synovial cells and myoblasts. Pericellular coats of hyaluronate shield cells

Table 4.3

Typical repeating sequences of the glycosaminoglycans and their linkage to core protein (modified from Poole[233])

Glycosaminoglycan	Linkage to core protein
Hyaluronate $\xrightarrow{\beta1,4}$ GlcA $\xrightarrow{\beta1,3}$ GlcNAc $\xrightarrow{\beta1,4}$ GlcA $\xrightarrow{\beta1,3}$	Not linked to a core protein
Chondroitin sulphate $\xrightarrow{\beta1,4}$ GlcA $\xrightarrow{\beta1,3}$ GalNAc $\xrightarrow{\beta1,4}$ GlcA $\xrightarrow{\beta1,3}$ (GalNAc: 4 or 6–SO_3^-)	$\xrightarrow{\beta1,3}$ GalNAc $\xrightarrow{\beta1,4}$ GlcA $\xrightarrow{\beta1,3}$ Gal $\xrightarrow{\beta1,3}$ Gal $\xrightarrow{\beta1,4}$ Xyl $\xrightarrow{\beta1,0}$ Ser
Dermatan sulphate $\xrightarrow{\beta1,4}$ IdA $\xrightarrow{\alpha1,3}$ GalNAc $\xrightarrow{\beta1,4}$ GlcA $\xrightarrow{\beta1,3}$ (IdA: ± 2–SO_3^-; GalNAc: 4 or 6–SO_3^-)	$\xrightarrow{\alpha1,3}$ GalNAc $\xrightarrow{\beta1,4}$ GlcA $\xrightarrow{\beta1,3}$ Gal $\xrightarrow{\beta1,3}$ Gal $\xrightarrow{\beta1,4}$ Xyl $\xrightarrow{\beta1,0}$ Ser
Heparan sulphate and heparin $\xrightarrow{\alpha1,4}$ IdA $\xrightarrow{\alpha1,4}$ GlcN $\xrightarrow{\alpha1,4}$ GlcA $\xrightarrow{\beta1,4}$ (IdA: ± 2–SO_3^-; GlcN: ± 3* or 6–SO_3^- and SO_3^- or Ac†; GlcA: ± 2*SO_3^-)	$\xrightarrow{\beta1,4}$ GlcNAc $\xrightarrow{\alpha1,4}$ GlcA $\xrightarrow{\beta1,3}$ Gal $\xrightarrow{\beta1,3}$ Gal $\xrightarrow{\beta1,4}$ Xyl $\xrightarrow{\beta1,0}$ Ser (Xyl: P‡)
Keratan sulphate $\xrightarrow{\beta1,4}$ GlcNAc $\xrightarrow{\beta1,3}$ Gal $\xrightarrow{\beta1,4}$ GlcNAc $\xrightarrow{\beta1,3}$ (GlcNAc: 6–SO_3^-)	*Cartilage* $\xrightarrow{\beta1,3}$ Gal $\xrightarrow{\beta1,4}$ GlcNAc $\xrightarrow{\beta1,6}$ GalNAc $\xrightarrow{\alpha1,0}$ Ser (Thr) Sialic acid $\xrightarrow{\alpha2,3}$ Gal $\xrightarrow{\beta1,3}$ GalNAc *Cornea* $\xrightarrow{\beta1,3}$ Gal $\xrightarrow{\beta1,4}$ GlcNAc $\xrightarrow{\beta1,2}$ Man $\xrightarrow{\alpha1,6}$ Man $\xrightarrow{\beta1,4}$ GlcNAc $\xrightarrow{\beta1,4}$ GlcNAc $\xrightarrow{\beta1,N}$ Asn $\xrightarrow{\beta1,3}$ Gal $\xrightarrow{\beta1,4}$ GlcNAc $\xrightarrow{\beta1,2}$ Man $\xrightarrow{\alpha1,3}$ Man (GlcNAc ← Fuc, $\alpha1,6$)

* Rare, but may confer very specific properties.

† N-sulphation is variable in heparan sulphate (~ 50%), but extensive in heparin (> 80%).

‡ Xylose is sometimes phosphorylated in heparan sulphate.

Ac = acetyl; P = phosphate.

from virus infection and lymphocytic attack; in particular neoplastic cells are protected from host cell cytotoxicity. Further, pericellular hyaluronate coats impede adhesion of migrant or proliferating cells to stationary components, preventing inappropriate cell immobilization. A distinct interaction of hyaluronate with cells is by a cell surface receptor protein. The function here may be cell aggregation, but activation of macrophages and granulocytes, and uptake of hyaluronate by liver endothelial cells are probably also receptor-mediated processes. The effect of hyaluronate on cell behaviour depends on its size and concentration and on the cell type. Hyaluronate fragments of between three and 25 disaccharide units stimulate capillary angiogenesis;[334] native hyaluronate is not angiogenic. High hyaluronate concentrations inhibit movement, adherence and phagocytosis by leucocytes, and aggregation of several cell types. At low concentrations hyaluronate stimulates these processes. In embryonic development, hyaluronate becomes prominent and stimulates cell proliferation and inhibits cytodifferentiation. Changes in hyaluronate concentration may generate forces causing translocation of cells or create spaces between cells and fibrous elements for routes of cell migration. In contrast, hyaluronate inhibits endothelial cell migration and the proliferation of fibroblasts and synovial cells.

Variability in proteoglycan structure

The number and type of glycosaminoglycan chains attached to a protein core is variable, and influenced by the size of the core. Interestingly, a number of interstitial proteoglycans have common core protein structures (Table 4.4).[123] Another variable is the length of glycosaminoglycan chains: cartilage proteoglycan has chondroitin sulphate chains of average molecular weight 19 kD, while the heparin chains of mast cell heparin proteoglycan (see below) have molecular weights of between 80 and 100 kD.

Chondroitin sulphate proteoglycans

Chondroitin sulphate proteoglycans[116,233] may bear chondroitin sulphate as the sole glycosaminoglycan, or may have, in addition to chondroitin sulphate, either keratan sulphate or heparan sulphate chains. A different variation is seen in proteoglycans with copolymeric chondroitin-/dermatan sulphate chains (p. 198 and Table 4.4).

Since cartilage is the richest source of proteoglycan (up to 50 per cent dry weight and 50–80 mg/ml), its aggregating chondroitin sulphate proteoglycan, aggrecan (Table 4.4), is the best known and most thoroughly studied proteoglycan of all. It bears 100–140 chondroitin sulphate chains (mol. wt 19 kD), and 30–60 keratan sulphate chains (mol. wt 7 kD). The protein core (mol. wt 350 kD) is rich in Glu/Gln, Ser and Gly, with clusters of Ser-Gly sequences;[218] however, the core constitutes less than 10 per cent of the total mass of the proteoglycan. The protein core has several domains (Fig. 4.7). At the N-terminus there is a globular domain (G1 50 kD) which binds hyaluronate. Next to this is a second globular domain (G2 40 kD) of unknown function. There follows an extended keratan sulphate-rich region (E1 15 kD), an extended chondroitin sulphate attachment region (E2 180 kD) and sometimes a final globular domain at the C-terminus, G3, which can contain sequence homologies with vertebrate lectins and complement regulatory protein.[167,218] The whole molecule has a molecular weight of 2000 kD. Aggregation of these proteoglycan subunits with hyaluronate (Fig. 4.7) has been extensively studied; the minimum requirement for binding is a decasaccharide fragment of hyaluronate. The aggregate of hyaluronate and chondroitin sulphate proteoglycan is locked firmly together by link glycoprotein (see p. 211) (mol. wt 47 kD), which binds to both hyaluronate and chondroitin sulphate proteoglycan protein core. The final aggregated chondroitin sulphate proteoglycan–hyaluronate complex may be as much as 200 000 kD in molecular weight.

Cartilage chondroitin sulphate proteoglycans occupy a solution volume 30–50 times their own dry weight. However, in cartilage they are not fully hydrated: further swelling is prevented by the type II collagen network. The resultant swelling pressure confers on cartilage its compressive stiffness and its ability to withstand repeated loading (see p. 265). Similar aggregating chondroitin sulphate proteoglycans, albeit at much smaller concentrations, are found in the aorta and in tendon, where, presumably, they make an essential contribution to mechanical properties. A further property of chondroitin sulphate proteoglycans is their ability to inhibit the attachment of cells to type I collagen and fibronectin.

As polyanions, proteoglycans have a significant affinity for cations, and are probably responsible for the high sodium concentration in cartilage. However, their affinity for sodium is much less than that for calcium, and the presence of chondroitin sulphate proteoglycan-bound calcium in cartilage allows the solubility product of calcium phosphate to be exceeded.[185] Cartilage proteoglycans may thus control calcification by concentrating calcium without precipitating calcium phosphate. The pH of cartilage is 0.3 lower than that of synovial fluid, indicating that proteoglycans bind hydrogen ions.[185]

Table 4.4

Properties of some well-characterized proteoglycans. The listing is not exhaustive but is meant to define the names in current use for the most well-known proteoglycans

New name	Previous name or synonym	Molecular weight of protein core (kD)	Glycosamino-glycans and number of chains if known	Function	Distribution
Extracellular matrix					
Aggrecan	Aggregating proteoglycan of cartilage	207	>100 CS 20 KS	Compressive resilience of cartilage	Cartilage, bone, skin, sclera aorta
Versican	Human fibroblast proteoglycan	263	14(?) CS	Cell attachment and migration	Skin, lung, aorta
Biglycan	PGSI, PGI	36	2CS or DS		Cartilage and other tissues
Decorin	PGS2, PGII, PG40	36	1CS or DS	Associated with collagen fibrils. Control of fibrillogenesis	Wide distribution, especially skin and tendon
Perlecan*	Engelbreth–Holm–Swarm Sarcoma HS proteoglycan	400	3 or 4HS	Provision of aniomic permeability barrier	Basement membranes
–	Type IX collagen	100	1CS or DS	See Table 4.1	Cartilage, cornea, vitreous, intervertebral disc
Membrane-intercalated					
Betaglycan	Type III transforming growth factor beta receptor	110		Cell receptor for transforming growth factor beta	Most cells
Syridecan	–	31	3HS, 2CS variable	Cell receptor for components of extracellular matrix	Epithelial cells
Lymphocyte homing receptor	H-CAM, Hermes antigen, CD44		5 or 6CS	Lymphocyte receptor	High endothelial venules
Intracellular					
Serglycin(e)	Secretory granule proteoglycan	10	Heparin (mast cells) CS (natural killer cells) CSE and CSdiB (mucosal mast cells)	Packaging of basic proteinases and histamine? Regulation of proteinase activity?	Mast cells, lymphocytes, haemopoietic cells

CS = chondroitin sulphate, DS = dermatan sulphate, HS = heparan sulphate, KS = keratan sulphate, CSE = chondroitin 4,6 disulphate, CSdiB = dermatan 4,6 disulphate (chondroitin sulphate B is the old name for dermatan sulphate).

* Name proposed – may not gain wide acceptance.

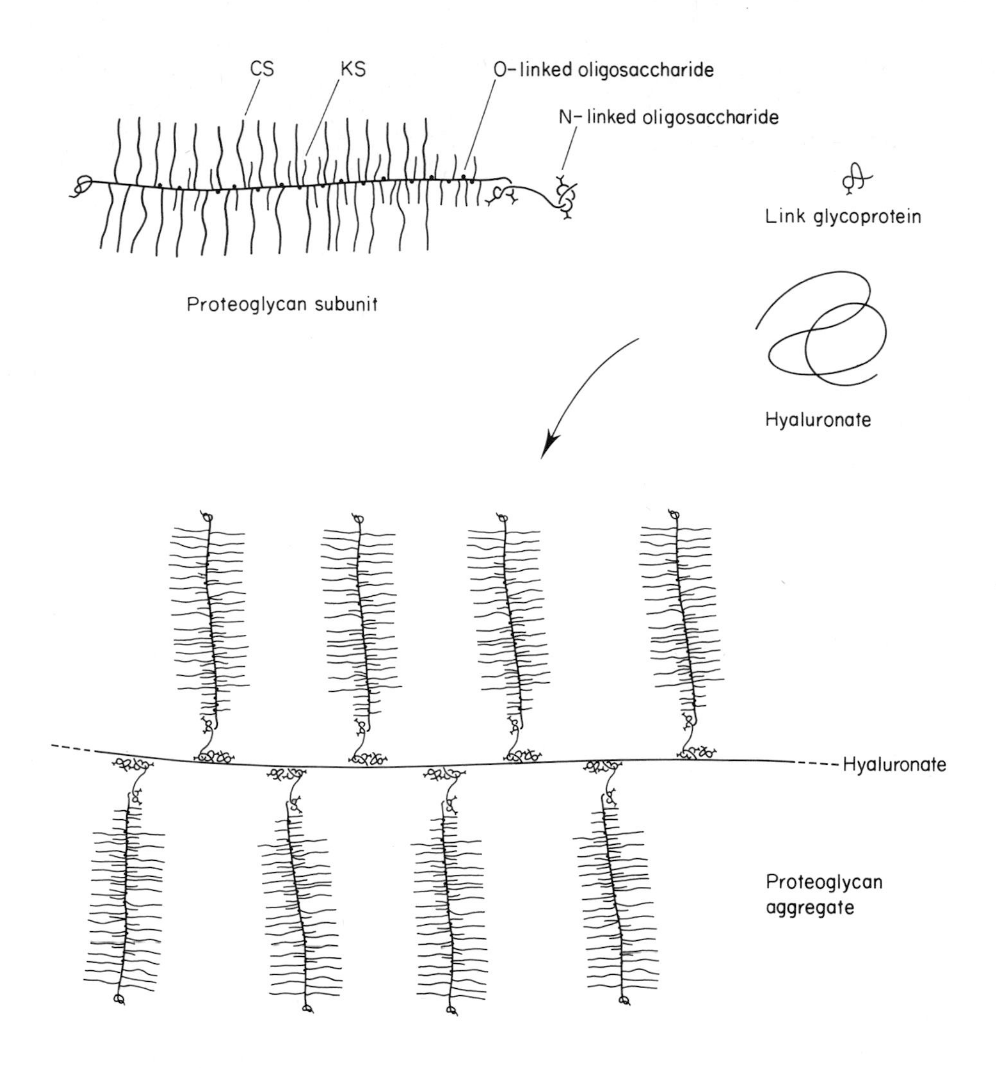

Fig. 4.7 Formation of proteoglycan aggregates typical of cartilage.
This schematic diagram shows how the proteoglycan subunit, bearing chondroitin sulphate, keratan sulphate, O-linked and N-linked oligosaccharides, aggregates with hyaluronate and link glycoprotein to form a supramolecular aggregate.

A large chondroitin sulphate proteoglycan (versican) with somewhat different properties is synthesized by fibroblasts.[100] Similarities to aggrecan include a hyaluronate binding region at the N-terminus and hepatic lectin and complement regulatory protein (CRP) regions at the C-terminus of the protein core. Between these two C-terminal domains, versican has two epidermal growth factor-like regions. Possibly these domains of aggrecan and versican assist in cell adhesion and organization during development. Cell surface proteoglycans, bearing chondroitin sulphate chains, with protein cores intercalated in the plasma membrane include syndecan,[31] betaglycan[8] and lymphocyte homing receptor[100] (Table 4.4). These proteoglycans are potentially able to interact with the cytoskeleton via the cytoplasmic C-terminal parts of their protein cores. The lymphocyte homing receptor, which is present in all high endothelial venules, is particularly interesting as it has a globular hyaluronate-binding region homologous with that in aggrecan and link protein. Lastly, it should not be forgotten that type IX collagen is a proteoglycan, bearing a chondroitin or dermatan sulphate chain attached to the NC3 domain of the $\alpha 2$ chain.

Dermatan sulphate proteoglycans

Chondroitin sulphate chains are the precursors of dermatan sulphate chains, since after the synthesis of chondroitin sulphate proteoglycan,[233,255,279] some GlcA residues are epimerized to IdA residues. The dermatan sulphate chains of dermatan sulphate proteoglycans always contain some GlcA, which indeed may still be the major hexuronate species present. However, these proteoglycans are often referred to as dermatan sulphate proteoglycans.

Dermatan sulphate proteoglycans are commonly found in the extracellular matrix of many normal and neoplastic tissues, and associated with plasma membranes and basement membranes. The small interstitial dermatan sulphate proteoglycans from many tissues, including sclera, skin, tendon, ligament, cartilage, and the small chondroitin sulphate proteoglycans from bone and cornea, have similar, if not identical, protein cores (Table 4.4).[123] Many of these proteoglycans may differ merely in the degree of epimerization of GlcA during synthesis. These small proteoglycans have molecular weights of 70–100 kD, and a protein core of 36 kD with 1 (decorin) or 2 (biglycan) dermatan sulphate (or chondroitin sulphate) chains (15–45 kD).

In vitro interactions of glycosaminoglycans and proteoglycans with collagens have been much studied. However, the demonstration of actual interactions in tissue, which may influence fibrillogenesis and calcification, has only been possible with the development of adequate techniques. Decorin, the major proteoglycan secreted by skin fibroblasts, binds to the d- and e-bands of collagen fibrils in cartilage, tendon, skin, sclera and cornea.[280] This interaction may inhibit the radial growth of collagen fibrils by interferring with the addition of collagen monomers to the growing collagen fibrils. Moreover, the presence of dermatan sulphate proteoglycan in the gap region may interfere with cross-link formation. Tendon dermatan sulphate proteoglycan inhibits fibrillogenesis of both type I and II collagen *in vitro*. Indeed, in adult cartilage, a small dermatan sulphate proteoglycan is localized mainly near the articular surface where collagen fibres are thin: in fetal tissue it is distributed throughout the cartilage matrix. Decorin also binds to fibronectin and membrane receptors via its protein core.

Dermatan sulphate proteoglycans from articular cartilage strongly inhibit the attachment and spreading of fibroblasts, and a large dermatan sulphate proteoglycan (mol. wt 400–800 kD) from rat yolk sac tumour inhibits the attachment of neoplastic cells to fibronectin and type I collagen, binding to fibronectin and type I collagen via its core protein. The complete amino acid sequence of this core protein (104 amino acids) is known and includes a central 49 amino acid sequence of alternating Ser and Gly residues, where the glycosaminoglycan chains are bound.

Dermatan sulphate chains with a high IdA content can self-aggregate strongly. This ability is related to the distribution of GlcA and IdA residues. Some dermatan sulphate proteoglycans from articular cartilage and sclera show this property, which may mediate the association of collagen fibrils via bound dermatan sulphate proteoglycans.

A GlcA-rich dermatan sulphate proteoglycan (mol. wt 100–150 kD) intercalates between collagen fibrils in cornea, possibly regulating fibril spacing and thus maintaining transparency. Some dermatan sulphate proteoglycans also bind hyaluronate and include cell-associated dermatan sulphate proteoglycans from fibroblasts, a very large dermatan sulphate proteoglycan from glial and glioma cells, and one from sclera that may be important in maintaining eye shape. Another very large dermatan sulphate proteoglycan endows ovarian follicular fluid with a high viscosity, presumably maintaining follicular shape. Lastly, type IX collagen can itself be a dermatan sulphate proteoglycan (see above).

Keratan sulphate proteoglycans

The large aggregating proteoglycans contain keratan sulphate chains (see p. 193) and are therefore keratan sulphate proteoglycans[111,233,279] as well as chondroitin sulphate proteoglycans. Keratan sulphate proteoglycans in which keratan sulphate is the sole glycosaminoglycan are found in cornea. Here the protein cores have molecular weights of 30–40 kD with attached keratan sulphate chains of 7 kD linked via Asn (Table 4.3). These keratan sulphate proteoglycans interact with collagen at the a- and c-bands specifically. Dermatan sulphate proteoglycans, which aggregate with the keratan sulphate proteoglycans, can also be isolated from cornea (see p. 201).

Until recently, keratan sulphate was thought to be confined to cartilage and cornea. However, it is now known to be widespread and present in many tissues. A new keratan sulphate proteoglycan present in cartilage is related to the collagen-binding protein fibromodulin, which has homology with decorin and biglycan.[219]

Heparan sulphate proteoglycans

Heparan sulphate proteoglycans[93,94,101,102,233] have excited much interest: first because there is strong evidence for their importance in the regulation of cell–cell and cell–matrix interactions,[258] and second because they are key components of basement membranes (see p. 50), which *inter alia* control the process of glomerular ultrafiltration.

Heparan sulphate proteoglycans and heparin have the most diverse polysaccharide structures of all glycosaminoglycans. This is due to the presence of hexuronate as IdA or GlcA; to the N-sulphation or N-acetylation of GlcN; to the possible 2-sulphation of hexuronates, especially IdA; and to the possible 3- and 6-sulphation of GlcN. Altogether 24 different disaccharide units may be present in many different sequences. Thus heparan sulphate and heparin chains can possess highly specific oligosaccharide sequences, exemplified by the binding of a unique pentasaccharide with antithrombin III. An additional feature of heparan sulphate is the presence of phosphorylated Xyl in the linkage region (Table 4.3).

There may be many different protein cores associated with different heparan sulphate proteoglycans. In some, heparan sulphate is attached to Ser residues in a sequence of Ser-Gly repeats.

Heparan sulphate proteoglycans are found in the extracellular matrix and on the surfaces of most adherent mammalian cells, either intercalated in the plasma membrane or associated via a plasma membrane receptor.

Heparan sulphate proteoglycans can be intercalated into the plasma membrane either via a glycosylphosphatidylinositol tail or via a protein core which extends through the membrane to the cytoplasmic side where it can interact with cytoskeletal elements. Therefore such molecules have the potential to act as receptors in cell–matrix and cell–cell interactions. Plasma membrane receptors for heparan sulphate proteoglycans recognize either heparan sulphate chains or the inositol phosphate exposed after removal of a glycosylphosphatidylinositol tail by phospholipase C. Heparan sulphate proteoglycans with an intercalated protein core include syndecan[31] and betaglycan[8] (Table 4.4). Both are conjugate proteoglycans bearing heparan and chondroitin sulphate chains with the former predominating. Syndecan is present on simple and stratified epithelia and also on fibroblasts. It interacts with collagen types I, III and V and with fibronectin and thrombospondin via its heparan sulphate chains. These interactions are modulated on stimulation by transforming growth factor beta (TGFβ) which increases the synthesis of syndecan molecules with a higher proportion of chondroitin sulphate chains. The type III TGFβ receptor (betaglycan) occurs in most tissues. Its interaction with TGFβ is a function of the protein core, so the role of its heparan sulphate chain is obscure. Heparan sulphate proteoglycan anchored to the cell membrane by a covalent glycosylphosphatidylinositol link has been reported in hepatoma cells, Schwann cells and in fibroblasts. A further function of the cell surface (and matrix) heparan sulphate proteoglycans is the binding of growth factors, particularly of the fibroblast growth factor family. Heparan sulphate proteoglycan–growth factor complexes may assist in providing locally high concentrations of growth factors able to stimulate cell proliferation.

Heparan sulphate proteoglycan at the surface of endothelial cells can interact specifically with a number of plasma proteins, including lipoproteins, lipoprotein lipase, thrombin and antithrombin III. The latter effect is important in endowing the endothelial cell surface with an anticoagulant property.

Heparan sulphate produced by confluent smooth muscle cells in monolayer culture inhibits the growth of these cells, but heparan sulphate from subconfluent cells does not. Smooth muscle cell growth is also inhibited by heparan sulphate released from endothelial cells. Hepatocyte nuclei contain a dynamic pool of a unique heparan sulphate, rich in the disaccharide GlcA-2-SO_4-GlcNSO_3-6-SO_4, which undergoes changes in structure and increases in amount when the cells reach confluency. Heparan sulphate is commonly undersulphated in transformed cells. Coupled with the finding that heparan sulphate from some cell lines inhibits DNA polymerase, these observations strongly suggest a role for heparan sulphate in the control of hyperplasia.[142]

Self-association of heparan sulphate occurs by the interaction of specific disaccharide sequences, thus possibly mediating cell–cell interaction and adhesion, as in postconfluent fibroblasts. In the growth phase, heparan sulphate chains show little association due to a reduced concentration at the cell surface and also because different oligosaccharide sequences are probably present. In neoplastic cells, heparan sulphate proteoglycan is also deficient at cell surfaces. Heparan sulphate proteoglycans are involved in the adhesion of cells to the substratum via 'footpad' adhesion sites on the plasma membrane, which are enriched in heparan sulphate proteoglycan. Attachment can be to collagen (types I, III and V), laminin, fibronectin (covalently in teratocarcinoma cells) and to platelet factor IV. When heparan sulphate proteoglycan is intercalated in the plasma membrane, the cytoplasmic end of its protein core can bind to actin. Heparan sulphate proteoglycan is thus a potential means of communication between the cytoskeleton and the extracellular matrix.

Heparan sulphate proteoglycans are the major proteoglycans of mammalian nervous tissue. They are strongly implicated in the development of membrane specialization at synaptic junctions, being major components of cholinergic synaptic vesicles and of the synaptic basal lamina after fusion with the synaptic plasma membrane. Neurite outgrowth in culture is stimulated by heparan sulphate proteoglycans, and neuronal cell surface heparan sulphate proteoglycan is required for Schwann cell proliferation.

Heparan sulphate proteoglycans are the predominant proteoglycan species in basement membranes, in which type IV collagen and the glycoproteins laminin and entactin are also present, creating support for the attachment of epithelial and endothelial cells. Classic studies have defined the structure of the heparan sulphate proteoglycan from Engelbreth–Holm–Swarm sarcoma, a basement membrane tumour (Table 4.4). This proteoglycan, species closely related to it or generated from it by limited proteolysis are now thought to be widely distributed in basement membranes. Studies on human glomerular basement membrane heparan sulphate proteoglycans have revealed two species of molecular weights 350 and 210 kD, with core proteins of molecular weights 140 and 110 kD, respectively.[288] In the renal glomerulus, the basement membrane lies between two cell layers where it constitutes the primary component of the filtration apparatus. The glomerular basement membrane (see p. 50) consists of a lamina densa bounded on each side by laminae rarae. The lamina densa mostly contains type IV collagen, while the heparan sulphate proteoglycan lies predominantly in the laminae rarae, giving the observed lattice-like network of anionic sites. Heparan sulphate chains play an important role in maintaining the charge-selective permeability properties of the glomerular basement membrane to large molecules such as proteins. Kidney glomerular basement membrane becomes permeable to ferritin and albumin after treatments which specifically degrade heparan sulphate chains. A deficiency of heparan sulphate proteoglycan would therefore be expected to lead to increased glomerular permeability. Such a deficiency has been observed in nephrotic kidneys, diabetic nephropathy and congenital nephrotic syndrome. In diabetes, it appears that the synthesis of glomerular basement membrane heparan sylphate proteoglycan is decreased, but there is also evidence for increased breakdown, as elevated urinary heparan sulphate has been reported.[37]

Heparin and heparin proteoglycan

Heparin is secreted by mast cells (see p. 36), and is generated from a precursor heparin proteoglycan[102, 117, 233] (see below and Table 4.4). Hexuronate residues are 70–90 per cent IdA; most are sulphated. Over 80 per cent of GlcN residues are N-sulphated, but O-sulphation exceeds N-sulphation (Table 4.2).

Heparin occurs as a single glycosaminoglycan chain of 7–25 kD molecular weight. However, heparin proteoglycan of molecular weight 750 kD has been isolated from rat mast cells, and has heparin chains of molecular weight 80–100 kD. Indeed heparin proteoglycan is now recognized as one of a group of intracellular proteoglycans known as secretory granule proteoglycans or serglycin(e). These occur in mast cells, white blood cells and haematopoietic cells. Short protein cores are characterized by predominant Ser–Gly repeats and carry a dense cluster of glycosaminoglycan chains conferring resistance to proteolysis. A common or highly homologous protein core is present in all these cell types, variously substituted, e. g. with heparin, in connective tissue mast cells or highly sulphated chondroitin sulphate in mucosal mast cells (Table 4.4).

Heparin binds to fibronectin and to blood coagulation factors IXa, Xa and thrombin as well as antithrombin III. It also binds to the surfaces of many cells; thus hepatocyte membrane proteins show an affinity for heparin. Antithrombin III inhibits the serine proteinases of the coagulation cascade, especially thrombin, on the formation of a 1:1 complex. Heparin potentiates this inhibition by about 1000-fold by binding Lys residues on antithrombin to its specific pentasaccharide sequence. Heparin is therefore a powerful anticoagulant. Heparin also suppresses the proliferation of vascular smooth muscle cells and inhibits DNA and RNA synthesis.

Biosynthesis of proteoglycans
(see Chapter 1)

The biosynthesis of proteoglycans[102, 115, 116, 233] begins with the synthesis of the protein core by normal mechanisms of protein synthesis. In the cisternae of the rough endoplasmic reticulum, N-linked high-Man oligosaccharides are added by transfer from a dolichol diphosphate intermediate to an Asn residue. These oligosaccharides are later modified in the Golgi apparatus (see p. 27). Here also the glycosaminoglycan chains are built up on the protein core by transfer of successive sugar residues from their uridine diphosphate derivatives to the non-reducing end of the growing chain. Every addition is catalysed by a specific glycosyl transferase. Thus synthesis of chondroitin sulphate chains requires the sequential action of xylosyl transferase, galactosyl transferases I and II, glucuronosyl transferase I, N-acetylgalactosaminyl transferase and glucuronosyl transferase II. Synthesis of the repeating sequence requires the continued, alternate action of the latter two enzymes. Finally, sulphate groups are transferred from 3′-phosphoadenosyl-5′-phosphosulphate to the appropriate sites on the glycosaminoglycans in a process also catalysed by specific transferases. O-linked oligosaccharides are attached in the Golgi, also by stepwise transfer of single sugar residues to a Ser or Thr residue. Next, these oligosaccharides are modified, in some cases exposing the recognition site for keratan sulphate chain synthesis. The proteoglycan is then secreted from the cell, in the case of aggregating proteoglycans as a complex with link glycoprotein. However, the ability of the latter proteoglycans to bind hyaluronate develops slowly, presumably due to a gradual conformational change.[24]

The biosynthesis of heparin proteoglycans and heparan sulphate proteoglycans follows the same mechanism as for chondroitin sulphate proteoglycans using the appropriate uridine diphosphate sugars. There are subsequently several polymer modifications: 1. removal of acetyl groups from some GlcNAc residues; 2. sulphation of the resultant free amino groups; 3. epimerization of many β-D-GlcA residues to α-L-IdA residues; 4. O-sulphation at C2 of some IdA residues; and 5. O-sulphation at C6 of some GlcN residues. In addition, occasional O-sulphation occurs at C2 of GlcA and at C3 of GlcN residues. For the generation of heparin, the long heparin chains of heparin proteoglycan are degraded by a specific endoglucuronidase.

Dermatan sulphate proteoglycans are derived biosynthetically from chondroitin sulphate proteoglycans by the operation of modifications 3 and 4 above.

Whether the final glycosaminoglycan chain length is precisely determined or is the result of a haphazard addition of sugars is unknown. One suggestion is that the glycosyl transferases are arranged on the membrane as a multienzyme complex which controls the final chain length. On the other hand, glycosylation may halt when the proteoglycan passes beyond the active sites on the membrane-bound enzymes. A further uncertainty relates to the factors that determine the type of glycosaminoglycan chain to be added to a protein core.

In young cartilage, proteoglycan synthesis is stimulated by insulin, insulin growth factor, ascorbate, platelet-derived growth factor, pituitary-derived fibroblast growth factor, hydrocortisone and dexamethasone. Inhibition is caused by prostaglandins E_1, E_2, A_1 and B_1. Proteoglycan synthesis in human arterial smooth muscle cells is stimulated by transforming growth factor-β, which has an effect more than five times that of epidermal growth factor, platelet-derived growth factor or heparin-binding growth factor.[57]

Biosynthesis of hyaluronate

The site of hyaluronate biosynthesis[174] is the plasma membrane, where the enzyme hyaluronate synthetase is situated on the cytoplasmic face. In contrast to the synthesis of proteoglycan-bound glycosaminoglycans, residues from uridine diphosphate sugars are transferred to the reducing end of the growing chain, which protrudes through a pore in the membrane into the extracellular space.

Diseases of proteoglycan synthesis

Few examples are known of diseases where the basic defect lies in proteoglycan biosynthesis:[233] without doubt more await discovery.

Corneal macular dystrophy

This disorder is characterized by corneal clouding and the accumulation of extracellular material. Affected patients synthesize normal corneal chondroitin sulphate proteoglycans but no normal keratan sulphate proteoglycan. Instead, a glycoprotein is synthesized with oligosaccharides smaller than normal keratan sulphate chains; this may be an incompletely or abnormally processed precursor of mature keratan sulphate proteoglycan (see p. 198). Probable failure to synthesize cartilage keratan sulphate causes no clinical symptoms.[306]

Core protein expression

Symptoms of Coffin–Lowry syndrome, a rare condition, include mental retardation, soft, thick skin, tapering fingers and skeletal abnormalities. The probable defect lies in the expression, biosynthesis or post-translational processing of a dermatan sulphate proteoglycan core protein. The defective proteoglycan lacks the recognition marker for endocytosis and accumulates extracellularly, attracting water and producing soft, thick skin.[27]

There is evidence that abnormally high glucose levels in insulin-dependent diabetic mothers increase the incidence of congenital malformations, especially of the skeleton, to 3–4 times the normal level. The possible mechanism for these teratogenic effects has been identified as inhibition of core protein gene expression.[180] Deficiency of the core protein of dermatan sulphate proteoglycans has been reported in a variant of Ehlers–Danlos syndrome.[97]

Galactosyl transferase I deficiency

Fibroblasts from a patient with progeroid appearance and symptoms of Ehlers–Danlos syndrome showed a deficiency of galactosyl transferase I.[242] Thus there was a decreased ability to convert the core protein of decorin into the proteoglycan.

Undersulphation

In one form of spondyloepiphyseal dysplasia, there is evidence for the undersulphation of chondroitin sulphate and for a deficiency of the sulphotransferase which transfers sulphate to glycosaminoglycan chains. There may also be an increased content of undersulphated glycosaminoglycan, predominantly chondroitin sulphate, in neoplastic liver.[164]

Degradation of proteoglycans

Since proteoglycans are conjugates of protein and sulphated polysaccharides, complete proteoglycan degradation[203, 233, 234] requires the action of proteinases, glycosidases and sulphatases. Endoglycosidases cleave non-terminal glycosidic bonds within polysaccharide chains while exoglycosidases are able to remove only single terminal sugar residues.

Extracellular events

Degradation is initiated by limited extracellular digestion. Although proteoglycans can be degraded by all known tissue proteinases, most proteoglycans probably suffer extracellular proteolysis by the action of neutral metalloproteinases such as stromelysins and gelatinases (see p. 185 and Table 4.2).

Although glycosaminoglycan chains do not normally undergo extracellular degradation, heparan sulphate is an exception as it can be cleaved by specific endoglycosidases produced by platelets and activated T lymphocytes (see p. 228). Also, some cells secrete hyaluronidase, which is an endo-β-N-acetyl hexosaminidase able to degrade chondroitin sulphate (especially chondroitin-4-sulphate, C4S) as well as hyaluronate. Partially degraded proteoglycans then enter the cell by receptor-mediated endocytosis. The structure recognized by the receptors may include the small oligosaccharides which are part of most proteoglycan structures, but in some instances such as the small dermatan sulphate proteoglycans, it is the protein core.

Intracellular events

Intracellular digestion is carried out by lysosomal proteinases (cathepsins), glycosidases and sulphatases. Initial glycosaminoglycan cleavage is brought about by endohexosaminidase or endoglucuronidase activity, followed by sequential digestion by a battery of exoglycosidases and sulphatases acting in turn. For example, digestion of the chondroitin sulphate unit shown in Table 4.3, assuming a free GlcA at the non-reducing end, would require the sequential action of β-glucuronidase, N-acetyl galactosamine 4- or 6-sulphatase and β-N-acetyl galactosaminidase. More varied structures require more enzymes for degradation. Thus degradation of heparan sulphate, after endoglucuronidase digestion, requires the sequential action of iduronate-2-sulphatase, α-L-iduronidase, N-sulphatase, acetyl transferase, N-acetyl glucosamine sulphatase, α-N-acetyl glucosaminidase and β-glucuronidase. The common trisaccharide linking glycosaminoglycans to protein core is degraded by the action of β-galactosidase and a β-xylosidase.

Control of proteoglycan degradation[233]

Control of metalloproteinase activity has been dealt with in detail (see p. 185). Little is known of the control of lysosomal enzymes.

Diseases of proteoglycan degradation

The arthritides

Any study of the arthritides[24, 118, 234, 274, 339] is made against a background of the normal ageing processes of cartilage proteoglycans (see Chapter 21). These include increases in: the protein/hexuronate ratio; the GlcN/GalN ratio, and hence keratan sulphate; the C6S/C4S ratio (where C6S = chondroitin-6-sulphate); hyaluronate; and the relative amount of lower molecular weight proteoglycan species. Composition may also vary at different points on the cartilage surface, and with depth where increases are observed in protein/hexuronate and GlcN/GalN ratios.

While degradation of collagen releases a large proportion of proteoglycans from cartilage, an early event in joint disease is also proteoglycan degradation with the liberation of relatively large fragments into the synovial fluid. The smallest and most commonly detected fragment is the hyaluronate-binding region (see p. 195). These fragments pass into the circulation via the lymphatic system. The concentration of degradation products in plasma and synovial fluid may be a useful indicator of abnormal catabolism of cartilage proteoglycans. Indeed, total urinary glycosaminoglycan excretion is significantly increased in rheumatoid arthritis, but not in osteoarthrosis (see Chapters 12 and 22).[60]

Fragmentation of proteoglycan is primarily caused by elevated activities of proteinases. Possible mechanisms in rheumatoid arthritis (see Chapter 13) are discussed on pp. 535–539. Increased metalloproteinase activity has been reported in osteoarthrosic lesions (see Chapter 22)[227] and is likely to originate from chondrocytes[214] stimulated by interleukin-1 from synovial cells. Osteoarthrosic cartilage is never normal: the most frequent change is loss of up to 20 per cent of proteoglycan. Most studies have also found changes in proteoglycan composition in severely fibrillated cartilage from human osteoarthrosic joints, the most consistent being decreased keratan sulphate. However, opinions are divided on the precise nature of these changes, for the reason that few studies on normal human cartilage proteoglycan structure have been carried out. It has therefore been impossible to propose a mechanism which would explain all the observations.

Hyaluronate found in rheumatoid synovial fluid is of significantly lower molecular weight than that of normal synovial fluid.[107] Native hyaluronate, at normal synovial fluid concentration, inhibits cell proliferation. Hyaluronate of lower molecular weight is less inhibitory or even stimulatory. Thus the changes in hyaluronate, in addition to modifying the viscoelastic properties of synovial fluid, may favour the growth of pannus.

In contrast to normal chondrocytes, cells from osteoarthrosic patients synthesize fibronectin and may form collagen types I and III; they also show increased synthesis of types IX and XI. Thus the resulting altered matrix may well have inferior mechanical properties.

The mucopolysaccharidoses (see p. 359)

The heritable defect in each of the mucopolysaccharidoses[181,192] is decreased activity of one or more of the lysosomal enzymes necessary for glycosaminoglycan degradation (see p. 202). In mild forms of the diseases, enzymic activity is almost sufficient to allow normal glycosaminoglycan breakdown, and patients may lead a relatively normal life. In contrast, the enzymic activity in severe forms is very low, and patients manifest prominent signs of the diseases.

If an enzyme is absent or demonstrates reduced activity, the substrate accumulates intralysosomally. The sequential action of other exoglycosidases and sulphatases required for complete glycosaminoglycan degradation is thus prevented because the terminal group which they recognize is not exposed. The enzymic deficiencies responsible for the mucopolysaccharidotic syndromes are shown in Table 4.5. Many were identified by Neufeld *et al.*[209] They found that affected cells in culture would take up $^{35}SO_4^{2-}$ from the medium, incorporate it into proteoglycan and accumulate radioactivity intracellularly at a linear rate. By contrast, within normal cells radioactivity stabilized at a low, constant level. The defect could be corrected by co-culture with normal fibroblasts or with fibroblasts having a different defect. Thus, the defect in cells from a case of Hunter syndrome was corrected by the presence of cells from a case of Hürler syndrome and vice versa. Fibroblasts from Sanfilippo A and Sanfilippo B patients behaved similarly. However, no correction resulted from co-culturing fibroblasts from Hürler and Scheie patients, since both have the same enzyme deficiency: the latter is a mild variant of the former. The conclusion from these cross-correction experiments is that some active lysosomal enzymes leave the cell and can be taken up by other cells. Factors that could also correct the abnormal accumulation of sulphate by cultured fibroblasts from mucopolysaccharidotic patients were detected in normal urine. After purification they were identified as lysosomal enzymes.

The inability of affected cells to degrade proteoglycan completely results in metachromatic staining with toluidine blue. Cells have large lysosomal inclusions of partially degraded proteoglycans, leading to hypertrophy and hence to organomegaly. When cells become sufficiently engorged, the material begins to escape by an unknown mechanism and to appear in the urine. For most mucopolysaccharidoses, the urinary content of glycosaminoglycans exceeds the normal range and is the basis of diagnostic urinary tests. Some of the syndromes can be differentiated by establishing the urinary distribution of glycosaminoglycans (see Table 4.5). Thus, if the urine contains only heparan sulphate or only dermatan sulphate the diagnosis is likely to be Sanfilippo or Maroteaux–Lamy syndrome, respectively; if both heparan sulphate and dermatan sulphate are found the patient may be suffering from the Hürler, Scheie or Hunter syndromes or from β-glucuronidase deficiency.

Patients with the Sanfilippo syndrome develop no skeletal deformities but become severely mentally retarded. There is therefore a strong inferred connection between this fact, the inability to degrade heparan sulphate and the observation that heparan sulphate proteoglycans are the major proteoglycan component of mammalian nervous tissue (see p. 199).

Attempts to provide normal enzyme in cases of mucopolysaccharidosis by treatments including plasma infusion, fibroblast implants and bone marrow transplants have at best only decreased the progression of the disease. However, it has recently been demonstrated that correction of the gene defect is possible: murine mucopolysaccharidsosis VII was corrected by a human transgene.[169]

I-cell disease (mucolipidosis II)

I-cell disease (p. 363),[209,294] so-called because of large cellular inclusions of undegraded material, is a severe disorder. Whereas all cells in mucopolysaccharidotic patients display the inherited defect, in I-cell disease the abnormality is largely confined to connective tissue cells; parenchymal cells such as hepatocytes appear normal. In I-cell disease and pseudo-Hürler polydystrophy (a mild

Table 4.5

The mucopolysaccharidoses (modified from McKusick and Neufeld[192] and Glew *et al.*[105])

Designation		Clinical features	Tissue*	Genetics	GAG stored and excreted	Enzyme deficient
MPS IH	Hürler syndrome (severe)	Early clouding of cornea, grave manifestations of dysostosis multiplex, death usually before age 10	LE, F	Homozygous for MPS IH gene	HS, DS	All MPSI show varying activity of α-L-iduronidase from very low (IH) to appreciable (IS)
MPS IS	Scheie syndrome (mild)	Stiff joints, cloudy cornea, aortic regurgitation, normal intelligence, consistent with normal life-span	LE, F	Homozygous for MPS IS gene	HS, DS	
MPS IH/S	Hürler/Scheie compound (intermediate)	Phenotype intermediate between Hürler and Scheie	LE, F	Genetic compound of above genes	HS, DS	
MPS II	Hunter syndrome (severe)	No clouding of cornea, milder course than in MPS IH, but death usually before age 15		Hemizygous for X-linked gene	HS, DS	Iduronate-2-sulphatase
MPS II	Hunter syndrome (mild)	Survival to 30–50 years, fair intelligence	S, LY, F	Hemizygous for X-linked allele for mild form	HS, DS	Iduronate-2-sulphatase
MPS IIIA	Sanfilippo syndrome A	All Sanfilippo variants show identical phenotype: mild somatic, severe CNS effects	F, LE	Homozygous for Sanfilippo A gene	HS	Heparan N-sulphatase
MPS IIIB	Sanfilippo syndrome B		F, S, LE	Homozygous for Sanfilippo B gene (at different locus)	HS	N-acetyl-α-D-glucosaminidase
MPS IIIC	Sanfilippo syndrome C		F, LE	Homozygous for Sanfilippo C gene	HS	Acetyl CoA: α-glucosaminide N-acetyl-transferase
MPS IIID	Sanfilippo syndrome D		F, LE	Homozygous for Sanfilippo D gene	HS	N-acetyl-α-D-glucosamine-6-sulphatase
MPS IVA	Morquio syndrome A	Severe bone changes of distinctive type, cloudy cornea, aortic regurgitation	F, LE	Homozygous for Morquio A gene	KS, CS	N-acetyl-galactosamine-6-sulphatase *and* galactose-6-sulphatase

Table 4.5 – continued

MPS IVB	Morquio syndrome B	Mild bone changes, cloudy cornea	F, LE	Homozygous for Morquio B gene: allele of mutant gene GM_1 gangliosidosis?	KS	β-galactosidase
Type V MPS now reclassified as: MPS IS (Scheie)						
MPS VI	Maroteaux–Lamy syndrome					
	A severe	Severe bone and corneal changes; normal intellect	F	Homozygous for M–L gene	DS	Arylsulphatase (N-acetylgalactosamine-4-sulphatase)
	B mild	Mild bone and corneal changes; normal intellect		Homozygous for allele at M–L locus		
	C intermediate			Homozygous for allele at M–L locus		
MPS VII	Sly syndrome	Hepato-splenomegaly, dysostosis multiplex, white cell inclusions, mental retardation	LE, F, S	Homozygous for mutant gene at β-glucuronidase locus	DS, HS	β-glucuronidase
Multiple sulphatase deficiency			LE, F	Homozygosity for multiple sulphatidosis genes	HS, DS	All MPS sulphatases deficient to varying degree

* Tissue diagnosis of biochemical defect using leucocytes (LE); fibroblasts (F); serum (S); lymphocytes (LY). MPS = mucopolysaccharidosis, HS = heparan sulphate, DS = dermatan sulphate, KS = keratan sulphate, CS = chondroitin sulphate.

form of I-cell disease), lysosomal enzymes are secreted from the cell but are not taken up again. Lysosomal enzymes are glycoproteins and part of their oligosaccharide is a phosphorylated Man residue (Fig. 4.10). This residue is recognized in receptor-mediated endocytosis, but it is missing from the enzymes secreted by I-cells (see p. 363). These enzymes are, however, fully active and lead to elevated enzymic activities in body fluids. Enzymes from normal cells are taken up perfectly well by I-cells. The normal process of directing lysosomal enzymes from the Golgi into primary lysosomes also requires the Man phosphate residue, so this is why active lysosomal enzymes in I-cell disease are secreted from the cell and are not transported to lysosomes.

Other diseases

Diseases affecting the permeability of the glomerular basement membrane (p. 53) have been discussed. Increased renal activity of N-acetyl-β-D-glucosaminidase and β-galactosidase has been reported in diabetics.[15] Neoplastic cell invasion may involve the degradation of proteoglycans by glycosidases[210] as well as by proteinases (see p. 202).

In psoriasis, urinary excretion of chondroitin sulphate and dermatan sulphate is elevated. Moreover, the extent of skin involvement is significantly correlated with the excretion of dermatan sulphate.[237]

Glycoproteins

Glycoproteins[124,143,179,327] are widely distributed in the body. Most plasma proteins, with the notable exception of albumin, are glycoproteins. Some plasma proteins, for example, albumin and the immunoglobulins, are small enough to pass through capillary walls and are found in connective tissues. Glycoproteins are important components of the mucus of the gastrointestinal, respiratory and reproductive tracts. A carbohydrate content of up to 85 per cent is undoubtedly important in endowing these glycoproteins with their mucoid properties. Cell-surface glycoproteins are integral proteins which float in the lipid bilayer and expose their oligosaccharides exclusively on the external surface of the cell. Plasma membrane glycoproteins, together with glycolipids, are crucial to cell recognition phenomena. Lastly, glycoproteins are important components of connective tissues.

Structure

Glycoproteins consist of one or more oligosaccharide units attached covalently to a protein core. Proteoglycans are thus a particular form of glycoprotein, and collagens are glycoproteins because of their Glu-Gal moieties (see p. 175). Typical monosaccharide constituents of glycoproteins are GlcNAc, Man, Gal, Fuc, sialic acid, occasionally GalNAc and, rarely, Glc. Fuc and sialic acid always occupy non-reducing terminal positions in the oligosaccharide, and Gal is often the next sugar residue.

There are two main groups of mammalian oligosaccharides: those in which the linkage to protein is O-glycosidic via GalNAc and Ser or Thr, and those in which the linkage is N-glycosidic via GlcNAc and the amide group of Asn. The latter group is characterized by the presence of a core, common to many glycoproteins, containing Man and GlcNAc. Examples of both groups are shown in Fig. 4.8.

Both types of oligosaccharide often exist in the same glycoprotein, for example, in proteoglycans. Occasionally, oligosaccharides may bear a sulphate or phosphate group.

Whereas knowledge of a peptide sequence establishes almost uniquely its chemical structure, the sequence of monosaccharides in an oligosaccharide by no means defines the precise structure. This is because different glycosidic bonds can be formed between two monosaccharides. Thus, for the disaccharide Man → Man the number of possible linkages is 8 ($\alpha 1 \rightarrow 2$, $\beta 1 \rightarrow 2$, $\alpha 1 \rightarrow 3$, $\beta 1 \rightarrow 3$, $\alpha 1 \rightarrow 4$, $\beta 1 \rightarrow 4$, $\alpha 1 \rightarrow 6$, $\beta 1 \rightarrow 6$). When this number of possibilities is multiplied by the number of glycosidic bonds, the number of different structures possible for one oligosaccharide sequence is evident. This explains why oligosaccharides can confer much greater specificity than peptides of comparable size: the antigenic properties of oligosaccharides are well-known (e.g. blood group determinants).

The chemical study of connective tissue glycoproteins is more difficult than that of collagens or proteoglycans because glycoproteins contain no unique constituent such as Hyp or hexuronate that aids their detection. A further complication is the presence of plasma glycoproteins in connective tissues. Although many glycoproteins can be demonstrated by immunocytochemical techniques the only absolute criterion available to establish a glycoprotein as a true connective tissue component is the demonstration of its synthesis by cells of the extracellular matrix.

Soluble glycoproteins

Most extracellular enzymes, such as those concerned with post-translational modifications of collagen, are glycoproteins. Even though some, like prolyl hydroxylase, act intracellularly, they can be found outside the cell. Lysyl oxidase binds tightly to collagen, and is not easily extractable. Lysosomal enzymes, such as those concerned with proteoglycan degradation (see p. 202), are glycoproteins, and are known to appear extracellularly (see p. 98).

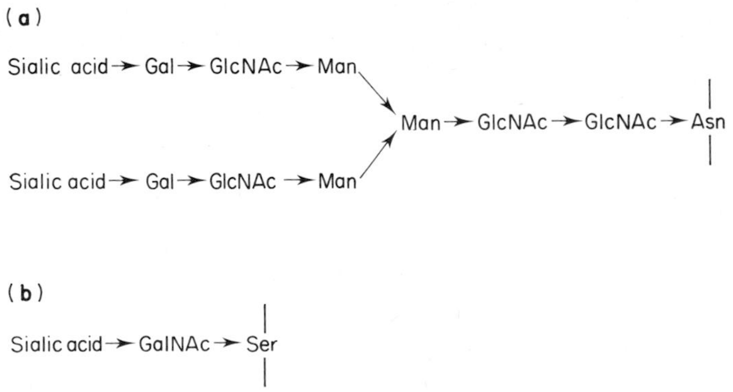

(a)

Sialic acid → Gal → GlcNAc → Man
Man → GlcNAc → GlcNAc → Asn
Sialic acid → Gal → GlcNAc → Man

(b)

Sialic acid → GalNAc → Ser

Fig. 4.8 Typical oligosaccharides of glycoproteins.
a. N-linked oligosaccharide of fibronectin. Glycosidic linkages are not shown but note that although identical oligosaccharide sequences are found in transferrin, the two oligosaccharides differ in some of their glycosidic linkages. Other N-linked oligosaccharides are more complex than this, with more branches, and often a Fuc residue substituted on the GlcNAc adjacent to the Asn residue. **b.** O-linked oligosaccharide of the simplest type, found in ovine submaxillary mucin. Most are much more complex than this.

Fibronectins

The term fibronectin[52,133,136,137,138,260,327] (see p. 51) has been given to a family of glycoproteins found on cell surfaces, and in most extracellular matrices and basement membranes. The functions of fibronectin include cell adhesion and migration, cytoskeletal organization, embryological development and morphogenesis, haemostasis, thrombosis, wound healing and malignant transformation. A similar molecule occurs in plasma (0.3 g/l) and other body fluids, where it is known as cold-insoluble globulin and by other synonyms such as opsonic protein.

Structure

Fibronectin has a molecular weight of 440 kD, and consists of two subunits of molecular weight 220 kD connected by disulphide bonds (Fig. 4.9). The amino acid sequence has been deduced from the base sequence of cDNA. Fibronectin contains 4 per cent carbohydrate as four to six oligosaccharide units connected to a protein core via Asn linkages. The major monosaccharide components are Man, Gal, GlcN and sialic acid (Fig. 4.8) with lesser amounts of Fuc and possibly Glc.

In each fibronectin subunit, the protein is organized in tightly folded globular domains, resistant to proteolytic attack, and connected by polypeptides susceptible to proteolysis. Controlled digestion by proteinases therefore 'dissects' fibronectin into its constituent globular domains, which can then be separated to allow study of their individual properties. Domains with binding sites for collagens, heparin, cell surfaces, fibrin and bacteria are now recognized (Fig. 4.9). Within the domains, three types of short amino acid sequence are repeated many times, together accounting for some 95 per cent of the total sequence of fibronectin. The type I and type II sequences, each of about 50 amino acid residues, are repeated 12 times and twice respectively, in each fibronectin subunit, while the type III sequence, of 90–95 amino acids, is repeated 15–17 times (Fig. 4.9). Blood clotting factor XIIa also contains type I and type II sequences, and type I is also found in plasminogen activator. A single exon in the fibronectin gene usually encodes each type I and type II repeat, but type III repeats are encoded by pairs of exons. Predictably, the fibronectin gene is complex, containing 50 exons. From the single fibronectin gene, up to 20 different versions of fibronectin can be generated by variable RNA splicing. Thus exons encoding ED-A and ED-B (Fig. 4.9) sequences can either be spliced in or out. Five variants can arise from alternative splicing of the IIICS exon (Fig. 4.9). Two occur when IIICS is either spliced in or out, but a further three result when different sections of IIICS are spliced in. Thus fibronectins vary in size from 2145 to 2445 amino acids. Plasma fibronectin contains neither ED-A nor ED-B.[231] ED-A is expressed in normal tissues, e.g. lung,[220] while ED-B expression is more restricted. However, ED-B is preferentially expressed in foetal and malignant tissues.[221,51] Differences in IIICS expression have also been observed among different tissues and cell types, and between normal and transformed cells.[126,231] The mechanisms which regulate alternative splicing are unknown. However, there is evidence that, besides elevating fibronectin production,[156] transforming growth factor β alters the expression of fibronectin isoforms,[16] increasing relative content of mRNA encoding ED-A and ED-B.[38]

Interaction with collagens

Fibronectin binds equally well to denatured collagen types I, II and III but less strongly to type V. Interaction with native collagens is generally much weaker, although type III appears to bind fibronectin more strongly than do other collagens. Fibronectin binds to collagenous sequences of C1q and acetylcholinesterase. Collagens have several binding sites for fibronectin; in the α1(I)-chain that of highest affinity is found in the sequence 757–791, which also contains the collagenase-sensitive bond. There are hydrophobic sequences in this peptide, and it is known that the fibronectin–collagen interaction is largely hydrophobic.

Interaction with proteoglycans

Most studies have been made with glycosaminoglycans. Fibronectin shows little or no binding to chondroitin sulphate or dermatan sulphate, but does interact with hyaluronate, heparan sulphate and heparin. However, the hyaluronate binding site is distinct from the dual sites for binding heparan sulphate and heparin. One of the latter has its affinity modulated by calcium. Binding of these glycosaminoglycans appears to strengthen the association of fibronectin with collagen and with fibrinogen. Fibronectin may thus function as a cross-link between collagen and heparan sulphate or hyaluronate. As it is well known that fibronectin associates with proteoglycans at cell surfaces, the most likely candidate is heparan sulphate proteoglycan.

Inflammation and wound healing (see Chapters 7 and 8)[112]

Fibronectin and fibronectin fragments, from degradation by polymorph proteinases, are chemotactic for monocytes, polymorphs, fibroblasts and endothelial cells. Fibronectin released from these cells enhances their adherence at the inflammatory site. In wound healing,[333] fibronectin promotes migration of keratinocytes,[228] platelet adhesion and

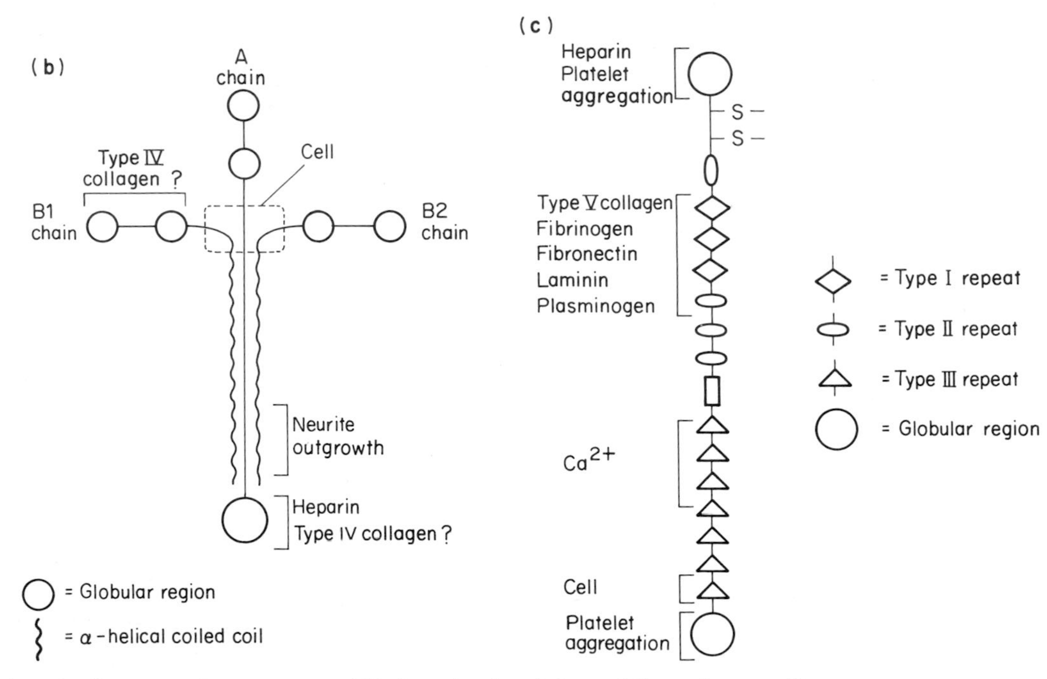

Fig. 4.9 Schematic diagram of structures of fibrinectin, laminin and thrombospondin.
Domains containing binding sites for various other components are indicated. Localization of other functions to specific regions is also shown. **a.** A fibronectin subunit, showing the different repeating structures. Shaded units are variably expressed to generate multiple forms. **b.** The typical cruciform structure of laminin. **c.** A subunit of thrombospondin.

stabilization of the clot. Fibronectin binds to fibrin by non-covalent interactions which are reinforced by peptide cross-links formed between fibronectin and fibrin, bridging side chains of Glu and Lys, a process catalysed by blood clotting factor XIIIa (transglutaminase). Factor XIIIa also catalyses cross-link formation between fibrin monomers, between fibronectin and collagen, and between collagen chains. Immobilization of fibronectin promotes the adhesion of platelets and fibroblasts within the clot and is important for wound healing. In clot removal, fibronectin binds both plasminogen activator and plasminogen, enhancing plasmin formation and fibrinolysis. Fibronectin is thus important for the formation and dissolution of thrombi. The opsonizing role of fibronectin in facilitating phagocytosis, both in the circulation and in tissues, is not fully understood but no doubt its ability to bind complement factors such as C1q and C3, and to promote the activation of C3 and IgFc activation is important here as are the fibronectin receptors on monocytes and polymorphs.[42]

Interaction with cell surfaces via integrins

Many types of eukaryotic cells attach preferentially to surfaces coated with fibronectin, although not all cells do so.[134] Thus, while the attachment of platelets to collagen is mediated by fibronectin, chondrocytes use chondronectin for attachment to type II collagen and epidermal cells attach to type IV collagen via laminin. Soluble fibronectin does not bind directly to cell surfaces unless it is first complexed to a collagen substratum or some other surface. However, fibronectin extracted from cell surfaces attaches to trypsinized normal cells (cell surface fibronectin is removed by proteinases).

The fibronectin receptor on fibroblasts has now been

identified as one member of a family of structurally related receptors, the integrins (see p. 52).[4,261] Other members of the integrin family serve as receptors for other glycoproteins (see below). The fibronectin receptor consists of an α-subunit (mol. wt 160 kD) and a β-subunit (130 kD) both of which span the plasma membrane, connecting the extracellular matrix to the cytoskeleton, via the cytoskeletal proteins talin and vinculin. A striking observation is the alignment of cell surface fibronectin fibrils with the intracellular actin matrix. The fibronectin binding site on the outside of the cell recognizes the amino acid sequence Arg-Gly-Asp-Ser (*also known as* RGDS) in fibronectin; here the Arg and Asp residues are critical. However, other structural features are necessary, since RGD (see above) sequences exist in other proteins (e.g. laminin, collagen, entactin and tenascin) which are not recognized by fibroblast receptors. A second cell receptor for fibronectin, also an integrin, is a complex IIb/IIIa of platelets, endothelial cells, monocytes and polymorphs,[4,42] which also recognizes RGD sequences in fibrinogen, vitronectin and von Willebrand factor.

Cell surface fibronectin is organized into an insoluble fibrillar matrix by extensive disulphide bonding. Long fibrils make contact with the cell membrane. Fibronectin is unevenly distributed when one cell adheres to another or to a substratum and is localized in the area of contact. The amount of cell surface fibronectin on a cell is reduced in cells undergoing mitosis and in some malignantly transformed cells. Neoplastic cells display loss of adhesion, hyperplasia and a disordered cytoskeleton; some make less fibronectin than normal; others do not bind it or assemble it into fibrils. These defects may result from the synthesis of a mutant fibronectin or increased degradation after synthesis: transformation leads to increased secretion of proteinases, particularly plasminogen activator. The defect could also lie in the integrin complex or the cytoskeletal binding site.[72,228] Defects within such a critical nexus could strongly influence the behavioural properties of neoplastic cells. The addition of fibronectin to these cells corrects the cytoskeletal disarray but does not rectify the hyperplasia. The indirect connection between fibronectin and the cytoskeleton is substantiated by the observation that cytochalasin B disrupts microfilaments causing the release of surface-associated fibronectin. The intracellular influence of fibronectin is also apparent on its addition to chondrocytes in culture. The cells assume a fibroblastic appearance and synthesize less type II collagen and proteoglycans.

Fibronectin and morphogenesis[134]

Fibronectin appears in tissues prior to and during cell migration, assisting in the direction of cell differentiation and morphogenetic movement.[76] Thus fibronectin appears in the migratory pathways of neural crest cells. Myoblasts also express fibronectin, but lose this capacity before fusion into myotubes. The addition of fibronectin to myoblast cultures inhibits fusion. Foetal tissues are distinctive in expressing the fibronectin isoform containing the ED-B sequence (see above).[51,220]

Fibronectin and disease

There is increased fibronectin production in both congenital and Duchenne muscular dystrophies,[114,248,257] in diabetes and in myelofibrosis,[119] and a suggestion of mutant fibronectin in a mild Ehlers–Danlos syndrome variant.[10] Fibronectin is important in infectious diseases: many pathogenic microorganisms attach to their host cell via its cell surface fibronectin. In contrast, others attach to host cells only after the fibronectin has been removed by proteolysis.[240] Antibodies to fibronectin have been detected in patients suffering from syphilis.[240]

Chondronectin

Chondronectin[133,323] is a serum glycoprotein located pericellularly in cartilage and vitreous humour. It is synthesized by chondrocytes and stimulates their attachment to type II collagen. Chondronectin is a globular glycoprotein of molecular weight 176 kD, consisting of three subunits each of molecular weight 55 kD linked by disulphide bonds. Chondronectin bears oligosaccharides containing Fuc, Man, GlcNAc, sialic acid and Gal. It binds specifically to chondroitin sulphate rather than to other glycosaminoglycans; it is thought to organize the matrix close to the cell surface by forming complexes with cartilage chondroitin sulphate proteoglycans and type II collagen.

Vitronectin

Vitronectin,[11,66,133,298] otherwise known as S protein or serum spreading factor is a glycoprotein found in plasma (200–500 μg/ml), in basement membranes of aorta and glomerulus, in elastic tissue of blood vessel walls and in dermis. It promotes cell adhesion and spreading and plays a role in platelet aggregation. Vitronectin consists of species of molecular weights 65 and 75 kD. It has an N-terminal 44-amino acid sequence identical to that of somatomedin B, a cell attachment region containing an RGD sequence, and a C-terminal glycosaminoglycan-binding region of 34 amino acids, with preference for heparin.

Thrombospondin

Thrombospondin[175] is an adhesive glycoprotein involved in the formation of extracellular matrices elaborated in culture by endothelial cells, smooth muscle cells and fibroblasts. It is found in macrophages, in monocytes and in the α-granules of platelets, which release it on exposure to thrombin. Thrombospondin aids platelet aggregation by calcium-dependent association with the platelet surface and by binding to fibrinogen. Thus, after vascular injury, thrombospondin probably assists in platelet–platelet interactions, clot formation and subsequent wound repair.

Thrombospondin is a disulphide-bonded aggregate of three identical chains, each of molecular weight 150–180 kD (Fig. 4.9). Globular N- and C-terminal domains are connected by a central extended region, containing three types of repeating sequences. The adhesive properties of thrombospondin result from its interaction with heparin, fibronectin,[69] fibrinogen, laminin, type V collagen and its cooperative binding to calcium.

Laminin

Laminin[26,133,182,186,226,309,326] (see p. 51) is a glycoprotein found only in basement membranes, where it comprises 30–50 per cent of the protein. Although laminin is detected throughout the basement membrane, it may be most concentrated in the lamina lucida. As a major basement membrane component able to bind type IV collagen, heparan sulphate proteoglycan and entactin, laminin is important in basement membrane matrix organization. Laminin also influences the behaviour of epithelial cells, including adhesion, spreading, growth, morphology, differentiation and migration and the outgrowth of neurites. Indeed laminin is one of the earliest matrix components to appear during embryonic development.

Structure

Laminin is a disulphide-bonded aggregate of molecular weight 900 kD, made up of one 400 kD subunit (the A chain) and two 200 kD subunits (the B chains). The two B chains, B1 and B2, are distinct gene products. Carbohydrate constitutes about 15 per cent of laminin, mostly as N-linked oligosaccharides. All subunits are glycosylated. Laminin can be visualized by transmission electron microscopy as a flexible, cruciform structure (Fig. 4.9) with one long arm and three short arms. Each short arm has two globular regions, one at the end of the arm. The long arm has a single larger globular end. It is not clear how the chains assemble to form this structure but it is known that they are linked by 50–60 disulphide bonds in the central proteinase-resistant part of the three short arms.

It has recently become apparent that different isoforms of laminin exist in different tissues.[268] The isoform described above, besides its presence in the Engelbreth–Holm–Swarm tumour, is found in tubular basement membrane of kidney. A second isoform in which the A chain is replaced by the homologous merosin (M) chain occurs in extrasynaptic basement membranes of muscle and endoneurial nerve trunk. The third isoform, in which the B1 chain is replaced by the S chain, is present in synaptic basement membranes of muscle and in glomerular and arterial basement membranes. Lastly, a fourth isoform, with A and B1 chains replaced by M and S chains, respectively, has been identified in synaptic muscle, endometrial nerve trunk and placenta. All forms contain the B2 chain.

Interaction with basement membrane components

The role of laminin depends on its ability to interact with cells as well as with matrix components. Laminin does not bind to gelatin, and of the native collagens, will only interact with type IV. Indeed epithelial cells bind to type IV collagens in preference to other collagens provided laminin is present. Laminin binds to heparin and heparan sulphate proteoglycans, and to entactin (see below). Localization of binding sites on the cruciform structure of laminin is shown in Fig. 4.9.

Interaction with cells

Laminin has a profound effect on the aspects of cell behaviour mentioned above. Some of these effects are mediated via the laminin receptor, also an integrin (see above),[104] which binds to an RGD (see p. 18) sequence near the intersection of the arms. One or more of the short-arm globular end-regions promote cell spreading; the long arm contains the site-stimulating neurite outgrowth.

Laminin and disease

An increased content of laminin receptor has been noted in malignant neoplastic cells.[336] Presumably this facilitates the attachment of cells to the basement membrane prior to its proteolytic destruction (see p. 53).[151,335]

Circulating antibodies to laminin in Chagas' disease, American cutaneous leishmaniasis and normal individuals are not specific to laminin, but are directed against a Gal disaccharide found in several other glycoproteins and glycolipids.[311] Similar epitopes are found on the surface of parasites.

Entactin

Entactin[133,186,226,309] is a sulphated glycoprotein of molecular weight 158 kD found in basement membranes. It is localized with other basement membrane components in prominent chord-like structures. The two globular domains of entactin are connected by a rod-like section which contains cys-rich epidermal growth factor (EGF)-like repeats. These repeats display a Ca^{2+}-binding sequence and contain the RGD cell binding motif.[54] Laminin–entactin complexes have been isolated, suggesting that entactin binds to laminin, and probably also to the NC1 domain of type IV collagen. Entactin has also been implicated in cell adhesion. Nidogen, a glycoprotein from mouse Ehler–Danlos syndrome tumours, is very similar, if not identical, to entactin.[78]

Lubricin

(see Chapter 2)

The boundary lubricating properties of synovial fluid are not due to hyaluronate, but to a glycoprotein, lubricin[299] (see p. 273). In human synovial fluid, lubricin has a molecular weight of 166 kD. Similar glycoproteins are found in other joints and other species. Bovine lubricin has a molecular weight of 227 kD, with 150 oligosaccharide chains, 110 of which have the structure sialic acid-Gal-GalNAc-Thr.

Link glycoproteins

Link glycoproteins[116] present in cartilage stabilize the aggregate of proteoglycan subunit and hyaluronate (see p. 195) preventing its disruption by competitive binding, e.g. by hyaluronate oligosaccharides. Link glycoproteins have also been identified, in other tissues, such as aorta and tendon. Link glycoprotein-1 (mol. wt 51 kD) has a higher carbohydrate content than link glycoprotein-2 (mol. wt 47 kD). Link glycoprotein-2 is derived from link glycoprotein-1 by proteolysis, for in tissues just beginning proteoglycan synthesis only link glycoprotein-1 is found. Where there is considerable tissue proteolytic activity only link glycoprotein-2 occurs. In the normal ageing process, link glycoproteins are degraded to fragments of molecular weights 26–30 kD.[234]

Microfibrillar proteins

Microfibrillar proteins[63] are major constituents of the microfibrils that surround the core of elastin in elastic tissue (p. 46). Little is known of their structure. Two microfibrillar proteins have been isolated from cultures of bovine ligamentum nuchae fibroblasts. Microfibrillar protein-1 (mol. wt 150 kD) is partially collagenous, containing Hyp and Hyl and being sensitive to bacterial collagenase. It is distinct from the 140 kD chains of type VI collagen.[13] Microfibrillar protein-2 is non-collagenous (mol. wt 300 kD). A further microfibrillar protein named fibrillin is a 350 kD glycoprotein isolated from the medium of cultured human fibroblasts. It is localized periodically along 10 nm diameter microfibrils in many tissues,[264] especially in dermis.[66] There is evidence of fibrillin deficiency in Marfan patients, who lack fibrillin in the dermis.[79] Moreover, fibrillin deficiency segregates with the Marfans condition and distribution of microfibrils correlates closely with tissues affected by Marfans syndrome. Hence lens dislocation is explained because microfibrils constitute the main component of the suspensory ligament.

Since microfibrils appear long before elastin, microfibrillar proteins may aid the deposition and organization of elastin. Although elastogenesis always involves microfibrils, microfibrils are also formed in situations where deposition of amorphous elastic tissue does not occur. Microfibrillar protein is also closely associated with collagen fibrils where the collagenous sequence of microfibrillar protein-1 may interact with collagen.

Structural glycoproteins

The name structural glycoprotein[63] was coined many years ago when it was recognized that connective tissues contained glycoproteins only extractable by disruptive methods or by removal of collagen. The glycoproteins now characterized as fibronectin, laminin and microfibrillar protein would doubtless have been designated structural glycoproteins, so this term has now rather outlived its usefulness. However, it is likely that other glycoproteins of this class await characterization or discovery. It is interesting to reflect that a morphogenetic role was proposed for structural glycoproteins before fibronectin and laminin were characterized.

Other glycoproteins

Tenascin,[58,184] originally called myotendinous antigen, hexabrachion, glioma mesenchymal extracellular matrix protein and cytotactin, is a glycoprotein of the extracellular matrix. Subunits of molecular weight 190, 200 and 220 kD are linked by disulphide bonding into a hexameric structure.

The tenascin structure contains 50 per cent fibronectin type III repeats as well as regions homologous to epidermal growth factor and fibrinogen sequences. There is evidence of alternative splicing. Tenascin colocalizes with fibronectin but is more restricted than fibronectin in its distribution in adult tissues. It is found in skin and myotendinous junctions. However, it is more prominent in embryonic and developing tissues, in tumours, and during inflammation and repair. These observations, coupled with the presence of an RGD sequence recognized by an integrin cell receptor, suggest roles for tenascin in cell adhesion, migration, growth and differentiation.

Cartilage matrix glycoprotein[87] occurs in hyaline cartilage and fibrocartilage. It consists of disulphide-bonded subunits of mol. wt 116 kD. A new extracellular matrix glycoprotein, undulin,[277] has recently been isolated from skin and placenta. It is an aggegate (mol. wt $>10^6$) of subunits of 270, 190 and 180 kD and is localized mainly between densely packed collagen fibrils.

Glycoproteins of calcified tissue[287]

Secreted protein and rich in cysteine (SPARC), BM-40 and osteonectin are identical glycoproteins of mol. wt 43 kD, widely distributed in connective tissues. Osteonectin is one of the most abundant non-collagenous proteins produced by bone cells. It has significant calcium-binding ability, located in domains at the C-terminus and N-terminus, the latter being rich in Glu residues.[81,132] Osteonectin has a high affinity for type I collagen and thrombospondin. Its synthesis is increased in regenerating and remodelling tissues and it is especially associated with bone growth repair and remodelling.

Two distinct sialoproteins are known: osteopontin (sialoprotein I, mol. wt 50 kD) and bone sialoprotein (sialoprotein II, mol. wt 75 kD). The latter is made by osteoblasts and is the major non-collagenous protein of bone. Since it contains an RGD sequence its putative function is cell adhesion. Calcium binding to γ-carboxyglutamyl (Gla) residues is implicated in the properties of osteocalcin, also known as bone Gla protein (BGP, mol. wt 6 kD) and in matrix Gla protein (MGP, mol. wt 9 kD).

Biosynthesis of glycoproteins

The biosynthesis of glycoproteins[163,294] is more complex than that of proteoglycans. Oligosaccharide units are built up by a stepwise addition of monosaccharide residues but there are differences in the mechanisms of biosynthesis of O- and N-linked oligosaccharides. The biosynthesis of O-linked oligosaccharide units resembles that of glycosaminoglycan chains: they are assembled on the protein core by the successive transfer of monosaccharide residues from their nucleotide derivatives under the influence of specific transferases. Thus ovine submaxillary mucin oligosaccharide sialic acid-GalNAc-Ser (Fig. 4.8) requires the transfer of GalNAc from its uridine diphosphate derivative followed by transfer of sialic acid from its cytidine monophosphate derivative. Sialic acid is not alone among monosaccharides in not using the uridine diphosphate derivative: Man and Fuc are activated as GDP-Man and GDP-Fuc.

The biosynthesis of N-linked oligosaccharides is more complex still. Initially, the oligosaccharide is built up by stepwise transfer of monosaccharide residues from nucleotide derivatives to a lipid acceptor. This lipid is dolichol phosphate which inserts into the lipid bilayer of the endoplasmic reticulum with its hydrophilic phosphomonoester extending into the cisterna. A unit, $Glc_3Man_9GlcNAc_2$ is first assembled (Fig. 4.10) and is then transferred *en bloc* from the dolichol phosphate to the Asn receptor. The antibiotics bacitracin and tunicamycin block the sequence at the steps shown. The core oligosaccharide is now modified. While still in the endoplasmic reticulum, four terminal residues (three Glc and one Man) are removed. In the Golgi, the remaining Glc residue and five further Man residues are stripped off and GlcNAc, Gal, sialic acid and Fuc residues added (Fig. 4.10).

Defects in glycoprotein synthesis

Defects in the protein core of glycoproteins could, of course, arise by mutation. While inherited defects in the protein would be specific for that protein because a mutation would affect a single gene, defects in post-translational processing may be more extensive because many glycoproteins require common enzyme systems for oligosaccharide synthesis. A case in point is I-cell disease (see p. 363), where the enzyme catalysing the formation of the phosphodiester bond between GlcNAc and Man (Fig. 4.10) is defective.

An indirect affect is the indiscriminate glucosylation of proteins in diabetes mellitus due to high circulating levels of glucose (see p. 417).

Degradation of glycoproteins

The previous description of proteoglycan degradation[203] (see p. 202) applies in general to glycoprotein degradation.

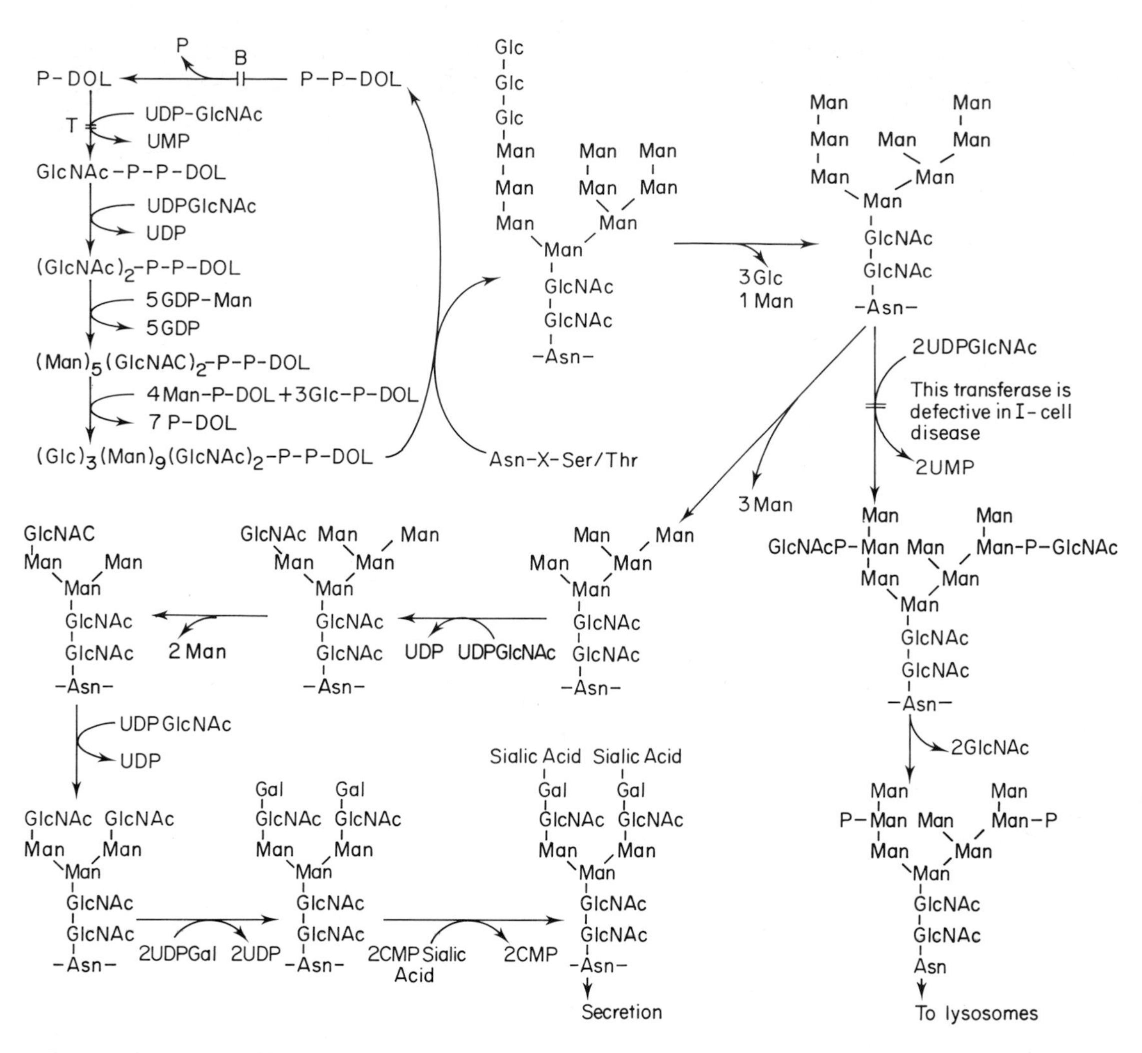

Fig. 4.10 Biosynthesis of an N-linked oligosaccharide typical of fibronectin.
Other N-linked oligosaccharides of different structure can be synthesized by variations of this pathway.[163,294] Abbreviations: DOL = dolichol; P = phosphate; B = inhibition by bacitracin; T = inhibition by tunicamycin; UDP = uridine diphosphate.

Breakdown requires the action of proteinases and glycosidases. Proteinases have been discussed on pp. 184–189. Any proteinase that can attack the core protein of a proteoglycan will almost certainly attack a glycoprotein, although appreciable glycosylation may confer some resistance.

The classical studies of Ashwell[208] established that extracellular glycoproteins are rapidly taken up by cells when terminal sialic acid residues are removed. Specific cell receptors recognize terminal Gal or Fuc (hepatocytes), or GlcNAc or Man residues (reticuloendothelial cells). The first step in glycoprotein degradation is probably the removal of sialic acid residues by extracellular neuraminidases, followed by specific uptake of the desialated macromolecules prior to their degradation within lysosomes.

The degradation pathway for N-linked oligosaccharides has been precisely established from studies of the glycoproteinoses (see p. 214). After the action of proteinases, the Asn oligosaccharide is split between the two GlcNAc residues, a reaction catalysed by endo-N-acetylglucosaminidase. The smaller GlcNAc-Asn fragment may have an attached Fuc residue which is removed by fucosidase. The bond between GlcNAc and Asn is then split in a reaction catalysed by asparaginyl-N-acetylglucosaminidase. The large oligosaccharide, with GlcNAc at its reducing terminus, is degraded by the stepwise action of exoglycosidases. Thus degradation of the oligosaccharide unit shown in Fig. 4.8, with the sialic acid residues removed, requires the successive actions of β-galactosidase, N-acetyl-β-glucosaminidase, α-mannosidase and β-mannosidase.

Less is known of the catabolism of glycoproteins with

O-linked oligosaccharides. In the light of present knowledge of degradation of glycoproteins with N-linked oligosaccharides and of proteoglycans, there is reason to suppose that breakdown of O-linked oligosaccharides is accomplished by similar pathways.

Diseases of glycoprotein degradation (glycoproteinoses)
(see p. 212)

Several deficiencies of lysosomal enzymes concerned in glycoprotein degradation[25] have been identified (Table 4.6). Clinical phenotypes resemble mild mucopolysaccharidoses and generalized symptoms include facial dysmorphia, mental retardation and dysostosis multiplex. Other frequent symptoms are cataracts and corneal opacities in mannosidosis; increased sodium chloride content of sweat in fucosidosis; blindness or reduced visual acuity and cherry-red spot—myoclonus syndrome in sialidosis; and lens opacity in aspartylglycosaminuria. Partially degraded oligosaccharides accumulate in the tissues and are excreted in the urine. The structural analysis of these urinary fragments has elucidated the breakdown pathway described above. Table 4.6 shows the deficient enzymes associated with various disease states. It is noteworthy that in I-cell disease and pseudoHürler polydystrophy (mucolipidoses II and III; see p. 203) patients lack lysosomal enzymes including fucosidase, galactosidase and mannosidase. These diseases are thus also diseases of glycoprotein breakdown. It is also worth noting that deficiency of β-galactosidase in G_{M1} gangliosidosis and of β-N-acetylhexosaminidases in G_{M2} gangliosidoses affects glycoprotein degradation: oligosaccharides of glycolipids and glycoproteins are similar structurally. However, it appears that the α-neuraminidase activity lacking in sialidosis is distinct from that active in hydrolysis of glycolipid substrates.

No definitive treatment is available for the glycoproteinoses; however, prenatal diagnosis is possible.

Table 4.6

Diseases of defective glycoprotein degradation

Disease	Defective enzyme	Material excreted
Sialidosis	α-Neuraminidase	Sialyl oligosaccharides*
Fucosidosis	α-Fucosidase	Fucosyl glycolipids Fucosyl oligosaccharides* Fucosyl glycoasparagines
Mannosidosis	α-Mannosidase	Mannosyl oligosaccharides*
Asparaginyl-glycosaminuria	Asparaginyl-N-acetyl glucosaminidase	Glycoasparagines
I-cell disease, pseudoHürler polydystrophy	Lysosomal glycosidases	Normal
GM_1 gangliosidosis	β-Galactosidase	GM_1 ganglioside and galactosyl oligosaccharides*
Sandhoff's disease	β-N-Acetyl hexosaminidases A and B	GM_2 ganglioside and N-acetylglucosaminyl oligosaccharides*

* The name of the monosaccharide residue at the non-reducing terminus prefixes the word oligosaccharide. The monosaccharide at the reducing terminus is GlcNAc.

Concluding remarks

Over the past decade, the study of connective tissues has burgeoned, and with it our understanding of the structure and metabolism of the macromolecular components of the extracellular matrix and the cells that secrete them. Inevitably, a significant part of this research has striven for elucidation of disease processes affecting connective tissues. In most cases there is still a long way to go before the precise association can be established between the basic biochemical defect and the resultant clinical syndrome. The gap reflects our ignorance of the processes by which elements of the extracellular matrix are laid down to establish a specific tissue morphology.

While recognizing that environmental factors can dominate or contribute to the causes of disease processes, in most cases there is an underlying genetic defect. The altered gene may simply be that for a connective tissue component such as a collagen chain. On the other hand, it may code for a defective protein that may malfunction in processes such as transport, regulation of biosynthesis, post-translational modification, degradation or regulation of degradation. Some of these processes are poorly understood, and it may take many years of research to define them completely.

Lastly, the value of genetic analysis in future clinical investigation cannot be overemphasized. Techniques now exist for gene localization and rapid sequencing, so that genetic abnormalities can be recognized. Genetic registers of the defects in the population are being constructed; they can aid genetic counselling and antenatal diagnosis.

Acknowledgements

The author is indebted to S Ayad, R Boot-Handford, M F Dean, J T Gallagher, D L Gardner, M E Grant, M J Humphries, G Murphy, M A Whittle, D E Woolley and C A Shuttleworth for valuable advice in the preparation of this chapter.

REFERENCES

1. Abergel R P, Chu M L, Bauer E A, Uitto J. Regulation of collagen gene expression in cutaneous diseases with dermal fibrosis: evidence for pretranslational control. *Journal of Investigative Dermatology* 1987; **88**: 727–31.

2. Adachi E, Hayashi T. *In vitro* formation of hybrid fibrils of type V collagen and type I collagen. *Connective Tissue Research* 1986; **14**: 257–66.

3. Adeyemi E D, Campos L B, Loizou S, Walport M J, Hodgson H J F. Plasma lactoferrin and neutrophil elastase in rheumatoid arthritis and systemic lupus erythematosus. *British Journal of Rheumatology* 1990; **29**: 15–20.

4. Akiyama S K, Nagata K, Yamada K M. Cell surface receptors for extracellular matrix components. *Biochemica et Biophysica Acta* 1990; **1031**: 91–110.

5. Ala-Kokko L, Rintala A, Savolainen E-R. Collagen gene expression in keloids: analysis of collagen metabolism and type I, III and V procollagen mRNAs in keloid tissue and keloid fibroblast cultures. *Journal of Investigative Dermatology* 1987; **89**: 238–44.

6. Albini A, Adelmann-Grill B C, Muller P K. Fibroblast chemotaxis. *Collagen and Related Research* 1985; **5**: 283–96.

7. Anderson I J, Goldberg R B, Marion R W, Upholt W B, Tsipouras P. Spondyloepiphyseal dysplasia congenita: genetic linkage to type II collagen (COL2A1). *American Journal of Human Genetics* 1990; **46**: 896–901.

8. Andres J L, Stanley K, Cheifetz S, Massague J. Membrane-anchored and soluble forms of betaglycan, a polymorphic proteoglycan that binds transforming growth factor β. *Journal of Cell Biology* 1989; **109**: 3137–45.

9. Arakawa M, Hatamochi A, Takeda K, Ueki H. Increased collagen synthesis accompanying elevated mRNA levels in cultured Werner's syndrome fibroblasts. *Journal of Investigative Dermatology* 1990; **94**: 187–90.

10. Arneson M A, Hammerschmidt D E, Furcht L T, King R A. A new form of Ehlers Danlos Syndrome. *Journal of the American Medical Association* 1980; **244**: 144–7.

11. Asch E, Podack E. Vitronectin binds to activated human platelets and plays a role in platelet aggregation. *Journal of Clinical Investigation* 1990; **85**: 1372–8.

12. Aumailley M, Poschl E, Martin G R, Yamada Y, Muller P K. Low production of procollagen III by skin fibroblasts from patients with Ehlers–Danlos syndrome type IV is not caused by decreased levels of procollagen III mRNA. *European Journal of Clinical Investigation* 1988; **18**: 207.

13. Ayad S, Chambers C A, Berry L, Shuttleworth C A, Grant M E. Type VI collagen and glycoprotein MFP1 are distinct components of the extracellular matrix. *Biochemical Journal* 1986; **236**: 299–302.

14. Ayad S, Weiss J B. Biochemistry of the intervertebral disc. In: Jayson M I V, ed, *The Lumbar Spine and Back Pain*. Edinburgh: Churchill Livingstone, 1987: 100–37.

15. Baggio B, Briani G, Cicerello E, Gambaro G, Bruttomesso D, Tiengo A, Borsatti A, Crepaldi G. Urinary glycosaminoglycans, sialic acid and lysosomal enzymes increase in nonalbuminuric diabetic patients. *Nephron* 1986; **43**: 187–90.

16. Balza E, Borsi L, Allemanni G, Zardi L. Transforming growth factor β regulates the levels of different fibronectin isoforms in normal human cultured fibroblasts. *FEBS Letters* 1988; **228**: 42–4.

17. Barker D F, Hostikka S L, Zhou J, Chow L T, Oliphant A R, Gerken S C, Gregory M C. Skolnick M H, Atkin C L, Tryggvason K. Identification of mutations in the COL4A5 collagen gene in the Alport syndrome. *Science* 1990; **248**: 1224–7.

18. Barnard K, Light N D, Sims T J, Bailey A J. Chemistry of collagen cross-links. Origin and partial characterisation of a putative mature cross-link of collagen. *Biochemical Journal* 1987; **244**: 303–9.

19. Barnes M J. The collagen-platelet interaction. In: Weiss J B, Jayson M I V, eds, *Collagen in Health and Disease*. Edinburgh: Churchill Livingstone, 1982: 179–97.

20. Barnes M J. Collagens in atherosclerosis. *Collagen and Related Research* 1985; **5**: 65–97.

21. Barry F, ed. Genetic and structural organization of proteoglycans. *Transactions of the Biochemical Society* 1990; **18**: 197–214.

22. Bateman J F, Lamande S R, Dahl H-HM, Chan D, Mascara T, Cole W G. A frameshift mutation results in a truncated non-functional carboxyl terminal pro alpha 1(I) propeptide of type I collagen in osteogenesis imperfecta. *Journal of Biological Chemistry* 1989; **264**: 10960–4.

23. Bauer E A, Tabas M. A perspective on the role of collagenase in recessive dystrophic epidermolysis bullosa. *Archives of Dermatology* 1988; **124**: 734–6.

24. Bayliss M T. Proteoglycan structure in normal and osteoarthritic human cartilage. In: Kuettner K E, Schleyerbach R, Hascall V C, eds, *Articular Cartilage Biochemistry*. New York: Raven Press, 1986: 295–310.

25. Beaudet A L. Disorders of glycoprotein degradation: mannosidosis, fucosidosis, sialidosis and aspartylglycosaminuria. In:

Stanbury J B, Wyngaarden J B, Fredrickson D S, Goldstein J L, Brown M S, eds, *The Metabolic Basis of Inherited Disease.* New York: McGraw Hill, 1983: 788–802.

26. Beck M, Glossl J, Ruter R, Kresse H. Abnormal proteodermatan sulphate in three patients with Coffin–Lowry syndrome. *Pediatric Research* 1983; **17**: 926–9.

27. Beck K, Hunter I, Engel J. Structure and function of laminin: anatomy of a multidomain glycoprotein. *FASEB Journal* 1990; **4**: 148–60.

28. Bennet V D, Adams S L. Characterization of the translational control mechanism preventing synthesis of α2(I) collagen in chicken vertebral chondroblasts. *Journal of Biological Chemistry* 1987; **262**: 14806–14.

29. Bennet V D, Adams S L. Identification of a cartilage specific promoter within intron 2 of the chick α2(I) collagen gene. *Journal of Biological Chemistry* 1990; **265**: 2223–30.

30. Bernard M, Yoshioka H, Rodriguez E, van der Rest M, Kimura T, Ninomiya Y, Olsen B R, Ramirez F. Cloning and sequencing of the pro α1(XI) collagen cDNA demonstrates that type XI belongs to the fibrillar class of collagen and reveals that the expression of the gene is not restricted to cartilaginous tissue. *Journal of Biological Chemistry* 1988; **263**: 17159.

31. Bernfield M, Sanderson R D. Syndecan a developmentally regulated cell surface proteoglycan that binds extracellular matrix and growth factors. *Philosophical Transactions of the Royal Society of London B* 1990; **327**: 171–86.

32. Bieth J G. Elastases: catalytic and biological properties. In: Mecham R P, ed, *Regulation of Matrix Accumulation.* New York: Academic Press, 1986: 217–320.

33. Birk D E, Fitch J M, Babiarz J P, Doane K J, Linsenmayer T F. Collagen fibrillogenesis *in vitro*: interaction of types I and V collagen regulates fibril diameter. *Journal of Cell Science* 1990; **95**: 649–57.

34. Birk D E, Trelstad R L. Extracellular compartments in tendon morphogenesis: collagen fibril bundle and macroaggregate formation. *Journal of Cell Biology* 1986; **103**: 231–40.

35. Bissell D M. Connective tissue metabolism and hepatic fibrosis: an overview. *Seminars in Liver Disease* 1990; **10**: iii and following articles.

36. Bonaldo P, Russo V, Bucciotti F, Doliana R, Colombatti A. Structural and functional features of the α3 chain indicate a bridging role for chicken collagen VI in connective tissues. *Biochemistry* 1990; **29**: 1245–54.

37. Bonavita M, Reed P, Donelly P V, Hill L L, DiFerrante N. The urinary excretion of heparan sulfate by juvenile and adult-onset diabetic patients. *Connective Tissue Research* 1984; **13**: 83–7.

38. Borsi L, Castellani P, Risso A M, Leprini A, Zardi L. Transforming growth factor beta regulates the splicing pattern of fibronectin messenger RNA precursor. *FEBS Letters* 1990; **261**: 175–8.

39. Bosma A, Menuissen S G M, Stricker B H C, Brouwer A. Massive pericellular collagen deposition in the liver of a young female with severe Crohn's disease. *Histopathology* 1989; **14**: 81–90.

40. Briozzo P, Morriset M, Capaony F, Rougeot C, Rochefort H. *In vitro* degradation of extracellular matrix with M_r 52 000 cathepsin D secreted by breast cancer cells. *Cancer Research* 1988; **48**: 3688–92.

41. Broek D L, Madri J, Eikenberry E F, Brodsky B. Characterization of the tissue form of type V collagen from chick bone. *Journal of Cellular Biochemistry* 1985; **260**: 555–62.

42. Brown E J, Goodwin J L. Fibronectin receptors of phagocytes. *Journal of Experimental Medicine* 1988; **167**: 777–93.

43. Bruckner-Tudermann L, Niemi K M, Kero M, Schnyder V W, Reunala T. Type VII collagen is expressed but anchoring fibrils are defective in dystrophic epidermolysis bullosa inversa. *British Journal of Dermatology* 1990; **122**: 383–90.

44. Bruckner-Tudermann L. Epidermolysis bullosa. In: Royce P M, Steinmann B, eds, *Extracellular Matrix and Inheritable Disorders of Connective Tissue.* New York: Alan R. Liss, 1990: chapter 17.

45. Bruns R R, Press W, Engvall E, Timpl R, Gross J. Type VI collagen in extracellular, 100 nm periodic filaments and fibrils: identification by immunoelectron microscopy. *Journal of Cell Biology* 1986; **103**: 393–404.

46. Burgeson R E. New collagens, new concepts. *Annual Review of Cell Biology* 1988; **4**: 551–77.

47. Butkowski R J, Langeveld J P M, Wieslander J, Hamilton J, Hudson B G. Localisation of the Goodpasture epitope to a novel chain of basement membrane collagen. *Journal of Biological Chemistry* 1987; **262**: 7874–77.

48. Byers P H. Brittle bones – fragile molecules: disorders of collagen gene structure and expression. *Trends in Genetics* 1990; **6**: 293–300.

49. Byers P H. Osteogenesis imperfecta. In: Royce P M, Steinmann B, eds, *Extracellular Matrix and Inheritable Disorders of Connective Tissue.* New York: Alan R. Liss, 1990: chapter 10.

50. Campa J S, Greenhalgh R M, Powell J T. Elastin degradation in abdominal aortic aneurysms. *Atherosclerosis* 1987; **65**: 13–21.

51. Carnemolla B, Balza E, Siri A, Zardi L, Nicotra M B, Bigotti A, Natali P G. A tumour-associated fibronectin isoform generated by alternative splicing of messenger RNA precursors. *Journal of Cell Biology* 1989; **108**: 1139–48.

52. Carson S E. *Fibronectin in Health and Disease.* Philadelphia: CRC Press, 1990.

53. Casolaro M A, Fells G, Wewers M, Pierce J E, Ogushi F, Hubbard R, Sellers S, Fostrom J, Lyons D, Kawasaki G, Crystal R G. Augmentation of lung antineutrophil elastase capacity with recombinant α1-antitrypsin. *Journal of Applied Physiology* 1987; **63**: 2015–23.

54. Chakravarti S, Tam M F, Chung A E. The basement membrane glycoprotein entactin promotes cell attachment and binds calcium ions. *Journal of Biological Chemistry* 1990; **265**: 10597–603.

55. Charriere G, Hartmann D J, Vignon E, Ronziere M-C, Herbage D, Ville G. Antibodies to types I, II, IX and XI collagen in the serum of patients with rheumatic diseases. *Arthritis and Rheumatism* 1988; **31**: 325–32.

56. Cheah K S E. Collagen genes and inherited connective tissue disease. *Biochemical Journal* 1985; **229**: 287–303.

57. Chen J-K, Hoshi H, McKeenan W L. Transforming growth factor type β specifically stimulates synthesis of proteoglycan in human adult arterial smooth muscle cells. *Proceedings of the National Academy of Sciences of the United States of America* 1987; **84**: 5267–91.

58. Chiquet-Ehrisman R. What distinguishes tenascin from fibronectin? *FASEB Journal* 1990; **4**: 2598–604.

59. Chu M L, Zhang R Z, Pan T C, Stokes D. Conway D, Kuo H J, Glanville R, Mayer U, Mann K, Deutzmann R, Timpl R. Mosaic structure of globular domain in the human type VI collagen alpha 3 chain: similarity to von Willebrand factor, fibronectin, actin, salivary proteins and aprotinin type protease inhibitors. *EMBO Journal* 1990; **9**: 385–93.

60. Chuck A J, Murphy J, Weiss J B, Grennan D M. Comparison of urinary glycosaminoglycan excretion in rheumatoid arthritis, osteoarthritis, myocardial infarction and controls. *Annals of the Rheumatic Diseases* 1986; **45**: 162–66.

61. Clague R B, Moore L J. IgG and IgM antibody to native type II collagen in RA serum and synovial fluid: evidence for the presence of collagen-anticollagen immune complexes in synovial fluid. *Arthritis and Rheumatism* 1984; **27**: 1370–7.

62. Clark S D, Kobayashi D K, Welgus H G. Regulation of the expression of tissue inhibitor of metalloproteinase and collagenase by retinoids and glucocorticoids in human fibroblasts. *Journal of Clinical Investigation* 1987; **80**: 1280–8.

63. Cleary E G, Gibson M A. Elastin-associated microfibrils and microfibrillar proteins. In: Hall D A, Jackson D S, eds, *International Review of Connective Tissue Research.* New York: Academic Press, 1983: 97–210.

64. Collier A, Jackson M, Bell D, Patrick A W, Matthews D M, Young R J, Clarke B F, Davies J. Neutrophil activation by increased neutrophil elastase activity in type I (insulin-dependent) diabetes mellitus. *Diabetes Research and Clinical Experimentation* 1989; **10**: 135.

65. Colombatti A, Bonaldo P. Biosynthesis of chick type VI collagen. II Processing and secretion in fibroblasts and smooth muscle cells. *Journal of Biological Chemistry* 1987; **262**: 14461–6.

66. Dahlback K, Ljungquist A, Loftberg H, Dahlback B, Engvall E, Sakai L Y. Fibrillin immunoreactive fibers constitute a unique network in the human dermis: immunohistochemical comparison of distributions of fibrillin, vitronectin, amyloid P component and orcein stainable structures in normal skin and elastosis. *Journal of Investigative Dermatology* 1990; **94**: 284–91.

67. Dalgleish R, Hawkins J R, Keston M. Exclusion of the α2(I) and α1(III) collagen genes as the mutant loci in a Marfan syndrome family. *Journal of Medical Genetics* 1987; **24**: 148–51.

68. Danielsen C C. Thermal stability of human fibroblast collagenase cleavage products of type I and type III collagens. *Biochemical Journal* 1987; **247**: 725–9.

69. Dardik R, Lahar J. Multiple domains are involved in the interaction of endothelial cell thrombospondin with fibronectin. *European Journal of Biochemistry* 1989; **185**: 581–8.

70. Davenport A, Verbov J L, Goldsmith H J. Circulating anti-skin basement membrane zone antibodies in a patient with Goodpasture syndrome. *British Journal of Dermatology* 1987; **117**: 125–7.

71. Davidson J M, Giro M G. Control of elastin synthesis. In: Mecham R P, ed, *Regulation of Matrix Accumulation.* New York and London: Academic Press, 1986: 178–216.

72. Dedhar S, Saulnier R. Alterations in integrin receptor expression on chemically transformed human cells; specific enhancement of laminin and collagen receptor complexes. *Journal of Cell Biology* 1990; **110**: 481–9.

73. Dinerman J L, Mehta J L, Saldeen T G P, Emerson S, Wallin R, Davda R, Davidson A. Increased neutrophil elastase release in unstable angina pectoris and acute myocardial infarction. *Journal of the American College of Cardiology* 1990: **15**: 1559.

74. Dodge G R, Poole A R. Immunohistochemical detection and immunochemical analysis of type II collagen degradation in human normal, rheumatoid and osteoarthritic articular cartilages and in explants of bovine articular cartilage cultured with IL-1. *Journal of Clinical Investigation* 1989; **83**: 647–61.

75. Dublet B, Oh S, Sugrue S P, Gordon M K, Gerecke D R, Olsen B R, van der Rest M. The structure of avian type XII collagen. Alpha 1(XII) chains contain 190 kDa non-triple helical amino terminal domains and form homotrimeric molecules. *Journal of Biological Chemistry* 1989; **264**: 13150–6.

76. Dufour S, Duband J L, Kornblitt A R, Thiery J P. The role of fibronectins in embryonic cell migrations. *Trends in Genetics* 1988; **4**: 189–203.

77. Duncan M R, Berman B. Persistence of a reduced collagen-producing phenotype in cultured scleroderma fibroblasts after short term exposure to interferons. *Journal of Clinical Investigation* 1987; **79**: 1318–24.

78. Durkin M E, Carlin B E, Vergnes J, Bartos B, Merlie J, Chung A E. Carboxy-terminal sequence of entactin deduced from a cDNA clone. *Proceedings of the National Academy of Sciences of the United States of America* 1987; **84**: 1570–4.

79. Editorial: Fibrillin and Marfan's syndrome: a real clue. *Lancet* 1990; **336**: 973–4.

80. Eng C E L, Pauli R M, Strom C M. Non-random association of a type II procollagen genotype with achondroplasia. *Proceedings of the National Academy of Sciences of the United States of America* 1985; **82**: 5465–9.

81. Engel J, Taylor W, Paulsson M, Sage H, Hogan B. Calcium binding domains and calcium-induced transition of SPARC/BM-40/osteonectin, an extracellular glycoprotein expressed in mineralised and non-mineralised tissues. *Biochemistry* 1987; **26**: 6958–65.

82. Evered D, Whelan J, eds. *The Biology of Hyaluronan.* Chichester: Academic Press, 1989.

83. Eyre D R. Collagen cross-linking amino acids. In: Cunningham L W, Frederiksen D W, eds, Biochemistry of the Major Components of the Extracellular Matrix. *Methods in Enzymology.* New York: Academic Press, 1987; **144**: 115–39.

84. Eyre D R, Apon S, Wu J J, Ericsson L H, Walsh K A. Collagen type IX: evidence for covalent linkages to type II collagen in cartilage. *FEBS Letters* 1987; **220**: 337–41.

85. Eyre D R, Paz M A, Gallop P M. Cross-linking in collagen and elastin. *Annual Review of Biochemistry* 1984; **53**: 717–48.

86. Fazio M J, Olsen D R, Kuivaniemi H, Chu M L, Davidson J M, Rosenbloom J, Uitto J. Isolation and characterization of human elastin cDNAs and age-associated variation in elastin gene expression in cultured skin fibroblasts. *Laboratory Investigation* 1988; **58**: 270–7.

87. Fife R S. Identification of cartilage matrix glycoprotein in synovial fluid in human osteoarthritis. *Arthritis and Rheumatism* 1988; **31**: 553–6.

88. Fitzsimmons C M, Cockburn C G, Hornsey V, Prowse C V, Barnes M J. The interaction of von Willebrand factor (vwf) with collagen: investigation of vwf binding sites in the collagen molecule. *Thrombosis and Haemostasis* 1988; **59**: 186–92.

89. Fleischmajer R, Olsen B R, Kuhn K. Structure, molecular biology and pathology of collagen. *Annals of the New York Academy of Sciences* 1990; **580**.

90. Fleischmajer R, Perlish J S, Olsen B R. Amino and carboxyl propeptides in bone collagen fibrils during embryogenesis. *Cell and Tissue Research* 1987; **247**: 105–9.

91. Fleischmajer R, Perlish J S, Olsen B R. The carboxyl-propeptide of type I procollagen in skin fibrillogenesis. *Journal of Investigative Dermatology* 1987; **89**: 212–15.

92. Foster J, Rich C B, Florini J R. Insulin like growth factor-I, somatomedin C, induces the synthesis of tropoelastin in aortic tissue. *Collagen and Related Research* 1987; **7**: 161–9.

93. Fransson L A, Carlstedt I, Coster L, Malmstrom A. The functions of the heparan sulphate proteoglycans. In: Evered D, Whelan J, eds, *CIBA Foundation Symposium 124: Functions of the Proteoglycans*. Chichester: John Wiley & Sons, 1986: 125–42.

94. Fransson L-A. Heparan sulphate proteoglycans: structure and properties. In: Lane D A, Lindahl U, eds, *Heparin, Chemical and Biological Properties, Clinical Applications*. London: Edward Arnold, 1989: 115–33.

95. Frisch S M, Ruley H E. Transcription from the stromelysin promoter is induced by interleukin 1 and repressed by dexamethasone. *Journal of Biological Chemistry* 1987; **262**: 16300–4.

96. Fukuda K, Koshihara Y, Oda H, Ohyama M, Ooyama T. Type V collagen selectively inhibits human endothelial cell proliferation. *Biochemical and Biophysical Research Communications* 1988; **151**: 1060.

97. Fushimi H, Kameyama M, Shinkai H. Deficiency of the core protein of dermatan sulphate proteoglycans in a variant form of Ehlers Danlos syndrome. *Journal of Internal Medicine* 1989; **226**: 409–16.

98. Gadek J E. α_1 Antitrypsin deficiency. In: Royce P M, Steinmann B, eds, *Extracellular Matrix and Inheritable Disorders of Connective Tissue*. New York: Alan R. Liss, 1990: chapter 19.

99. Gadher S J, Eyre D E, Duance V C, Wotton S F, Heck L W, Schmid T M, Woolley D E. Susceptibility of cartilage collagens types II, IX, X and XI to human synovial collagenase and neutrophil elastase. *European Journal of Biochemistry* 1988; **175**: 1–7.

100. Gallagher J T. The extended family of proteoglycans: social residents of the pericellular zone. *Current Opinion in Cell Biology* 1989; **1**: 1201–18.

101. Gallagher J T, Lyon M. Molecular organization and functions of heparan sulphate. In: Lane D A, Lindahl U, eds, *Heparin, Chemical and Biological Properties, Clinical Applications*. London: Edward Arnold, 1989: 135–58.

102. Gallagher J T, Lyon M, Steward W P. Structure and function of heparan sulphate proteoglycans. *Biochemical Journal* 1986; **236**: 313–25.

103. Gavrilovic J, Hembry R M, Reynolds J J, Murphy G. TIMP regulates extracellular collagen degradation by chondrocytes and endothelial cells. *Journal of Cell Science* 1987; **87**: 357–62.

104. Gehlsen K R, Dickerson K, Argraves W S, Engvall E, Ruoslahti E. Subunit structure of a laminin-binding integrin and localisation of its binding site on laminin. *Journal of Biological Chemistry* 1989; **264**: 19034–8.

105. Glew R H, Basu A, Prence E M, Remaley A T. Lysosomal storage diseases. *Laboratory Investigation* 1985; **53**: 250–69.

106. Goldberg G I, Marmer B L, Grant G A, Eisen A Z, Wilhelm S, He C. Human 72-kilodalton type IV collagenase forms a complex with a tissue inhibitor of metalloproteinases designated TIMP-2. *Proceedings of the National Academy of Sciences of the USA* 1989; **86**: 8207–11.

107. Goldberg R L, Toole B P. Hyaluronate inhibits cell proliferation. *Arthritis and Rheumatism* 1987; **30**: 769–78.

108. Gordon M K, Gerecke D R, Dublet B, van der Rest M, Olsen B R. Type XII collagen. A large multidomain molecule with partial homology to type IX collagen. *Journal of Biological Chemistry* 1989; **264**: 19772–8.

109. Graham M F, Diegelmann R F, Elson C O, Lindblad W J, Gotschalk N, Gay S, Gay R. Collagen content and types in the intestinal strictures of Crohn's disease. *Gastroenterology* 1988; **94**: 257–65.

110. Grant M E, Ayad S, Kwan A P L, Bates G P, Thomas J T, McClure J. The structure and synthesis of cartilage collagens. In: Glauert A G, ed, *The Control of Tissue Damage*. Amsterdam: Elsevier, 1988: 3–28.

111. Greiling H, Scott J E, eds. *Keratan Sulphate*. London: The Biochemical Society, 1989.

112. Grinnell F. Fibronectin and wound healing. *Journal of Cellular Biochemistry* 1984; **26**: 107–16.

113. Gunwar S, Saus J, Noelken M E, Hudson B G. Glomerular basement membrane. Identification of a fourth chain, alpha 4, of type IV collagen. *Journal of Biological Chemistry* 1990; **265**: 5466–9.

114. Hantai D, Labat-Robert J, Grimaud J-A, Fardeau M. Fibronectin, Laminin and type I, III, IV collagens in Duchenne's muscular dystrophy, congenital muscular dystrophies and congenital myopathies: an immunocytochemical study. *Connective Tissue Research* 1985; **13**: 273–81.

115. Hicks M, Delbridge L, Yue D K, Reeve T S. Increase in cross-linking of non-enzymatically glycosylated collagen induced by products of lipid peroxidation. *Archives of Biochemistry and Biophysics* 1989; **268**: 249–54.

116. Hardingham T E. Structure and biosynthesis of proteoglycans. In: Krieg T, Kuhn K, eds, Connective tissue in normal and pathological states. *Rheumatology* 1986; **10**: 143–83.

117. Hardingham T E, Beardmore-Gray M, Dunham D G, Ratcliffe A. Cartilage proteoglycans. In: Evered D, Whelan J, eds, *Ciba Foundation Symposium 124: Functions of the Proteoglycans*. Chichester: John Wiley & Sons, 1986: 30–46.

118. Harris E D. Regulation of collagenolysis in synovial cell systems. In: Evered D, Whelan J, eds, *Ciba Foundation Symposium 124: Functions of the Proteoglycans*. Chichester: John Wiley & Sons, 1986.

119. Hascall V C, Glant T T. Proteoglycan epitopes as potential markers of normal and pathologic cartilage metabolism. *Arthritis and Rheumatism* 1987; **30**: 541–8.

120. Hasselbalch H, Clemmensen I. Plasma fibronectin in idiopathic myelofibrosis and related chronic myeloproliferative

disorders. *The Scandinavian Journal of Clinical and Laboratory Investigation* 1987; **47**: 429–33.

121. Hasty K A, Jeffrey J J, Hibbs M S, Welgus H G. The collagen substrate specificity of human neutrophil collagenase. *Journal of Biological Chemistry* 1987; **262**: 10048–52.

122. Hasty K A, Stricklin G P, Hibbs M S, Mainardi C L, Kang A H. The immunologic relationship of human neutrophil and skin collagenases. *Arthritis and Rheumatism* 1987; **30**: 695–9.

123. Hata R, Kurata S, Shinkai H. Existence of malfunctioning pro $\alpha 2(I)$ collagen genes in a patient with a pro $\alpha 2(I)$ chain defective variant of Ehlers–Danlos syndrome. *European Journal of Biochemistry* 1988; **174**: 231–7.

124. Heinegard D, Franzen A, Hedbom E, Sommarin Y. Common Structures of the core proteins of interstitial proteoglycans. In: Evered D, Whelan J, eds, *Ciba Foundation Symposium 124: Functions of the Proteoglycans*. Chichester: John Wiley & Sons, 1986: 69–88.

125. Heinegard D, Oldberg A. Proteoglycans and glycoproteins. In: Royce P M, Steinmann B, eds, *Extracellular Matrix and Inheritable Disorders of Connective Tissue*. New York: Alan R. Liss, 1990: chapter 5.

126. Henney A M, Tsipouras P, Schwartz R C, Child A H, Devereux R B, Leech G J. Genetic evidence that mutations in the COL1A1, COL1A2, COL3A1 or COL5A2 collagen genes are not responsible for mitral valve prolapse. *British Heart Journal* 1989; **61**: 292–9.

127. Hershberger R P, Culp L A. Cell-type-specific expression of alternatively spliced human fibronectin IIICS mRNAs. *Molecular and Cellular Biology* 1990; **10**: 662–71.

128. Hibbs M S, Hoidal J R, Kang A H. Expression of a metalloproteinase that degrades native type V collagen and denatured collagens by cultured alveolar macrophages. *Journal of Clinical Investigation* 1987; **80**: 1644–50.

129. Hinek A, Reiner A, Poole A R. The calcification of cartilage matrix in chondrocyte culture. Studies of the C-propeptide of type II collagen (chondrocalcin). *Journal of Cell Biology* 1987; **104**: 1435–41.

130. Horton W A, Chondrodysplasias. In: Royce P M, Steinmann B, eds, *Extracellular Matrix and Inheritable Disorders of Connective Tissue*. New York: Alan R. Liss, 1990: chapter 24.

131. Horton W E. Regulation of the collagen II gene in development and disease. *Pathological and Immunopathological Research* 1988; **7**: 90.

132. Hughes R C, Taylor A, Sage H, Hogan B L M. Distinct patterns of glycosylation of colligin, a collagen-binding glycoprotein, and SPARC (osteonectin), a secreted Ca^{2+}-binding glycoprotein. *European Journal of Biochemistry* 1987; **163**: 57–65.

133. Humphries M J, Yamada K M. Non-collagenous glycoproteins. In: Krieg T, Kuhn K, eds, Connective tissue in normal and pathological states. *Rheumatology* 1986; **10**: 104–42.

134. Humphries M J, Obara M, Olden K, Yamada K M. Role of fibronectin in adhesion, migration and metastasis. *Cancer Investigation* 1989; **7**: 373–93.

135. Hurwitz S A, Goodman R M. Pseudoxanthoma elasticum. In: Royce P M, Steinmann B, eds, *Extracellular Matrix and Inheritable Disorders of Connective Tissue*. New York: Alan R. Liss, 1990: chapter 13.

136. Hynes R O. Fibronectins—a family of complex and versatile adhesive glycoproteins derived from a single gene. *The Harvey lectures, series 81*. New York: Alan R Liss, 1987: 133–52.

137. Hynes R O. *Fibronectins*. Heidelberg, London: Springer-Verlag, 1989.

138. Hynes R O. Fibronectins. *Scientific American* 1986; **254**: 32–41.

139. Indik Z, Yeh H, Ornstein-Goldstein N, Kucich U, Abrams W, Rosenbloom J C, Rosenbloom J. Structure of the elastin gene and alternative splicing of elastin mRNA: implications of human disease. *American Journal of Medical Genetics* 1989; **34**: 81–90.

140. Indik Z, Yeh H, Ornstein-Goldstein N, Sheppard P, Anderson N, Rosenbloom J C, Peltonen L, Rosenbloom J. Alternative splicing of human elastin mRNA indicated by sequence analysis of cloned genomic and complementary DNA. *Proceedings of the National Academy of Sciences of the United States of America* 1987; **84**: 5680–4.

141. Irimura T, Yamori T, Bennett S C, Ota D M, Cleary K R. The relationship of collagenolytic activity to the stage of human colorectal carcinoma. *International Journal of Cancer* 1987; **40**: 24–31.

142. Ishihara M, Fedarko N S, Conrad H E. Involvement of phosphatidyl inositol and insulin in the coordinate regulation of proteoheparan sulphate metabolism and hepatocyte growth. *Journal of Biological Chemistry* 1987; **262**: 4708–16.

143. Ivatt R J. *The Biology of Glycoproteins*. New York: Plenum Press, 1984.

144. Jander R, Korsching E, Rauterberg J. Characteristics and occurrence of type VIII collagen. *European Journal of Biochemistry* 1990; **189**: 601–7.

145. Jasin H E. Autoantibody specificities of immune complexes sequestered in articular cartilage of patients with RA and OA. *Arthritis and Rheumatism* 1985; **28**: 241–8.

146. Jeffrey J J. The biological regulation of collagenase activity. In: Mecham R P, ed, *Regulation of Matrix Accumulation*. New York: Academic Press, 1986: 53–98.

147. Jones M M, Seilheimer D K, Pier G B, Rossen R D, Increased elastase secretion by peripheral blood monocytes in cystic fibrosis patients. *Clinical and Experimental Immunology* 1990; **80** 344–9.

148. Jukkola A, Risteli J, Niemela O, Risteli L. Incorporation of sulphate into type III procollagen by cultured human fibroblasts. *European Journal of Biochemistry* 1986; **154**: 219–24.

149. Kahari V M, Heino J, Vuorio E. IL-1 increases collagen production and mRNA levels in cultured skin fibroblasts. *Biochimica et Biophysica Acta* 1987; **929**: 142–7.

150. Kahari V M, Multimaki P, Vuorio E. Elevated pro $\alpha 2(I)$ collagen mRNA levels in cultured scleroderma fibroblasts result from an increased transcription rate of the corresponding gene. *FEBS Letters* 1987; **215**: 331–4.

151. Kanemoto T, Reich R, Royce L, Greatorex D, Adler SH, Shiraishe N, Martin G R, Yamada Y, Kleinmann H K. Identification of an amino acid sequence from the laminin A chain that stimulates metastasis and collagenase IV production. *Proceedings of the National Academy of Sciences of the USA* 1990; **87**: 2279–83.

152. Keene D R, Engvall E, Glanville R W. Ultrastructure of type VI collagen in human skin and cartilage suggests an anchoring function for this filamentous network. *Journal of Cell Biology* 1988; **107**: 1995–2006.

153. Keene D R, Sakai L Y, Bachinger H P, Burgeson R E. Type III collagen can be present on banded collagen fibrils regardless of fibril diameter. *Journal of Cell Biology* 1987; **105**: 2393–2402.

154. Kefalides N A. The Goodpasture antigen and basement membranes: the search must go on. *Laboratory Investigation* 1987; **56**: 1–3.

155. Kershenobich D, Vargas F, Carcia-Tsao G, Perez Tamayo R, Gent M, Rojkind M. Colchicine in the treatment of cirrhosis of the liver. *New England Journal of Medicine* 1988; **318**: 1709–13.

156. Keski-Oja J, Raghow R, Sawdey M, Loskutoff D J, Postlethwaite A E, Kang A H, Moses H L. Regulation of mRNAs for type I plasminogen activator inhibitor, fibronectin, and type I procollagen by transforming growth factor β. *Journal of Biological Chemistry* 1988; **263**: 3111–115.

157. Khoka R, Denhardt D T. Matrix metalloproteinases and tissue inhibitors of metalloproteinases: a review of their role in tumorigenesis and tissue invasion. *Invasion and Metastasis* 1989; **9**: 391.

158. Kielty C M, Hopkinson I, Grant ME. The collagen family: structure, assembly and organisation in the extracellular matrix. In: Royce P M, Steinmann B, eds, *Extracellular Matrix and Inheritable Disorders of Connective Tissue.* New York: Alan R. Liss, 1990: chapter 3.

159. Kleppel M M, Kashtan C E, Butkowski R J, Fish A J, Michael A F. Alport familial nephritis. Absence of 28 kilodalton non-collagenous monomers of type IV collagen in glomerular basement membrane. *Journal of Clinical Investigation* 1987; **80**: 263–6.

160. Knowlton R G, Katzenstein P L, Moskowitz R W, Weaver E J, Malemud C J, Pathria M N, Jimenez S A, Prockop D J. Genetic linkage of a polymorphism in the type II procollagen gene (COL2A1) to primary osteoarthritis associated with mild chondrodysplasia. *New England Journal of Medicine* 1990; **322**: 526–30.

161. Koivu J, Myllyla R, Helaakoski T, Pihlajaniemi T, Tasanen K, Kivirikko K I. A single polypeptide acts both as the β subunit of prolyl-4-hydroxylase and as a protein disulphide isomerase. *Journal of Biological Chemistry* 1987; **262**: 6447–9.

162. Kontusaari S, Tromp G, Kuivaniemi H, Ladda R L, Prockop D J. Inheritance of an RNA splicing mutation (G + 1 IVS 20) in the type III procollagen gene (COL3A1) in a family having aortic aneurysms and easy bruisability: phenotypic overlap between familial arterial aneurysms and Ehlers Danlos syndrome type IV. *American Journal of Human Genetics* 1990; **47**: 112.

163. Kornfeld R, Kornfeld S. Assembly of Asparagine-linked oligosaccharides. *Annual Review of Biochemistry* 1985; **54**: 633–64.

164. Kovalszky I, Pogamy G, Molnar G, Jeney A, Lapis K, Karacsonyi S, Szecseny A, Iozzo R V. Altered glycosaminoglycan composition in reactive and nephrotic human liver. *Biochemical and Biophysical Research Communications* 1990; **167**: 883–90.

165. Kratz L E, Boughman J A, Needleman B W. Lack of association between scleroderma and type I and type III procollagen gene restriction length polymorphisms. *Arthritis and Rheumatism* 1989; **32**: 1597.

166. Kream B, Harrison J, Bailey R, Petersen D, Rowe D, Lichtler A. Hormonal regulation of collagen gene expression in osteoclastic cells – overview and new findings. *Connective Tissue Research* 1989; **20**: 187–92.

167. Krusius T, Gehlsen K R, Ruoslahti E. A fibroblast chondroitin sulfate proteoglycan core protein contains lectin-like and growth factor-like sequences. *Journal of Biological Chemistry* 1987; **262**: 13120–5.

168. Kuhn K. The collagen family—variations in the molecular and supermolecular structure. In: Krieg T, Kuhn K, eds, Connective tissue in normal and pathological states. *Rheumatology* 1986; **10**: 29–69.

169. Kyle J W, Birkenmeier E H, Gwynn B, Vogler C, Hoppe PC, Hofmann J W, Sly W S. Correction of murine mucopolysaccharidosis VII by a human transgene. *Proceedings of the National Academy of Sciences of the USA* 1990; **87**: 3914–8.

170. Laato M, Kahari V M, Niinikoski J, Vuorio E. Epidermal growth factor increases collagen production in granulation tissue by stimulation of fibroblast proliferation and not by activation of procollagen genes. *Biochemical Journal* 1987; **247**: 385–8.

171. Lane D A, Lindahl U, eds. *Heparin, Chemical and Biological Properties, Clinical Applications.* London: Edward Arnold, 1989.

172. Last J A, Armstrong L G, Reiser K M. Biosynthesis of collagen crosslinks. *International Journal of Biochemistry* 1990; **22**: 559.

173. Laurent G J. Dynamic state of collagen: pathways of collagen degradation *in vivo* and their possible role in regulation of collagen mass. *American Journal of Physiology* 1987; **252**: C1–C9.

174. Laurent T C, Fraser J R E. The properties and turnover of hyaluronan. In: Evered D, Whelan J, eds, *Ciba Foundation Symposium 124: Functions of the Proteoglycans.* Chichester: John Wiley & Sons, 1986: 9–29.

175. Lawler J, Hynes R O. Structural organization of the thrombospondin molecule. *Seminars in Thrombosis and Hemostasis* 1987; **13**: 245–54.

176. Lee B, Vissing H, Ramirez, F. Rogers D, Rimoin D. Identification of the molecular defect in a family with spondyloepiphyseal dysplasia *Science* 1989; **244**: 978–80.

177. Lee K T. Atherosclerosis. *Annals of the New York Academy of Sciences* 1985; **454**.

178. Leigh I M, Eady R A, Heagerty H M, Purtzis P E, Whitehead P A, Burgeson R E. Type VII collagen is a normal component of epidermal basement membrane which shows altered expression in recessive dystrophic epidermolysis bullosa. *Journal of Investigative Dermatology* 1988; **90**: 639–42.

179. Lennarz W J. *The Biochemistry of Glycoproteins and Proteoglycans.* New York: Plenum Press, 1980.

180. Leonard C M, Bergmann M, Frenz D A, Macreery L A, Newman S A. Abnormal ambient glucose levels inhibit proteoglycan core protein gene expression and reduce proteoglycan accumulation during chondrogenesis: possible mechanism for teratogenic effects of diabetes. *Proceedings of the National Academy of Sciences of the USA* 1989; **86**: 10113–7.

181. Leroy J G, Weismann U. Disorders of lysosomal enzymes. In: Royce PM, Steinmann B, eds, *Extracellular Matrix and Inheritable Disorders of Connective Tissue.* New York: Alan R. Liss, 1990: chapter 23.

182. Liotta L A, Rao C N, Wewer U M. Biochemical interactions of tumor cells with the basement membrane. *Annual Review of Biochemistry* 1986; **55**: 1037–57.

183. Lunstrum G P, Kuo H J, Rosenbaum L M, Keene D R, Glanville R W, Sakai L Y, Burgeson R E. Anchoring fibrils contain the carboxyl terminal globular domain of type VII procollagen, but lack the amino terminal globular domain. *Journal of Biological Chemistry* 1987; **262**: 13706–12.

184. Mackie E J, Tucker R P, Halfter W, Chiquet-Ehrismann R. The distribution of tenascin coincides with pathways of neural crest cell migration. *Development* 1988; **102**: 237–50.

185. Maroudas A. Physicochemical properties of articular cartilage. In: Freeman M A R, ed, *Adult Articular Cartilage*. London: Pitman, 1979: 215–90.

186. Martin G R, Timpl R. Laminin and other basement membrane components. *Annual Review of Cell Biology* 1987; **3**: 57–85.

187. Matrisian L M. Metallöproteinases and their inhibitors in matrix remodelling. *Trends in Genetics* 1990; **6**; 121–5.

188. Mauch C, Krieg T. Pathogenesis of fibrosis—introduction and general aspects. In: Krieg T, Kuhn K, eds, Connective tissue in normal and pathological states. *Rheumatology* 1986; **10**: 372–84.

189. Mayne R. Collagenous proteins of blood vessels. *Arteriosclerosis* 1986; **6**: 585–93.

190. Mayne R, Burgeson R E, eds, *Structure and Function of Collagen Types*. New York: Academic Press, 1987.

191. Mays P K, Bishop J E, Laurent G J. Age-related changes in the proportions of types I and III collagen. *Mechanisms of Ageing and Development* 1988; **45**: 203–12.

192. McKusick V A, Neufeld E F. The mucopolysaccharide storage diseases. In: Stanbury J B, Wyngaarden J B, Fredrickson D S, Goldstein J L, Brown M S, eds, *The Metabolic Basis of Inherited Disease*. New York: McGraw Hill, 1983: 751–77.

193. Mendler M, Eron-Bender S G, Vaughan L, Winterhalter K H, Bruckner P. Cartilage contains mixed fibrils of collagens types II, IX and XI. *Journal of Cell Biology* 1989; **108**: 191–7.

194. Mohan P S, Carter W G, Spiro R G. Occurrence of type VI collagen in extracellular matrix of renal glomerulus and its increase in diabetes. *Diabetes* 1990; **39**: 31–7.

195. Monteagudo C, Merino M J, Sanjuan J, Liotta L A, Stetlerstevenson W G. Immunohistochemical distribution of type IV collagenase in normal, benign and malignant breast tissue. *American Journal of Pathology* 1990; **136**: 585–92.

196. Morgan K. What do anti-collagen antibodies mean? *Annals of Rheumatic Diseases* 1990; **49**: 62–5.

197. Morgan K, Clague R B, Collins I, Ayad S, Phinn S, Holt P J L. Incidence of antibodies to native and denatured cartilage collagens (types II, IX and XI) and to type I collagen in rheumatoid arthritis. *Annals of the Rheumatic Diseases* 1987; **46**: 902–7.

198. Morris M L, Harper E. The presence of an inhibitor of human skin collagenase in the roots of healthy and periodontally diseased teeth. *Journal of Periodontal Research* 1987; **22**: 78–80.

199. Morton L F, Fitzsimmons C M, Rauterberg J, Barnes M J. Platelet reactive sites in collagen. Collagen I and III prossess different aggregatory sites. *Biochemical Journal* 1987; **248**: 373–81.

200. Muller K P, Nerlich A G, Kunze D, Muller P K. Study of collagen metabolism in the Marfan syndrome. *European Journal of Clinical Investigation* 1987; **17**: 218–25.

201. Murphy G, Docherty A J P. Molecular studies on the connective tissue metalloproteinases and their inhibitor TIMP. In: Glauert A G, ed, *The Control of Tissue Damage*. Amsterdam: Elsevier, 1988: 223–41.

202. Murphy G, Reynolds J J. Current views of collagen degradation. *BioEssays* 1985; **2**: 55–60.

203. Murphy G, Reynolds J J. Extracellular matrix degradation. In: Royce P M, Steinmann B, eds, *Extracellular Matrix and Inheritable Disorders of Connective Tissue*. New York: Alan R. Liss, 1990: chapter 9.

204. Murray L W, Hollister DW, Rimoin D L. Diastrophic dysplasia–evidence against a defect in type II collagen. *Matrix* 1990; **9**: 459–67.

205. Myers J C, Jones T A, Pohljolainen E R, Kadri A S, Goddard A D, Sheer D, Solomon E, Pihlajaniemi T. Molecular cloning of alpha 5(IV) collagen and assignment of the gene to the region of the X chromosome containing the Alport syndrome. *American Journal of Human Genetics* 1990; **46**: 1024–33.

206. Narayanan A S, Page R C. Synthesis of type V collagen by fibroblasts derived from normal, inflamed and hyperplastic human connective tissues. *Collagen and Related Research* 1985; **5**: 297–304.

207. Nerlich A G, Poschl E, Voss T, Muller P K. Biosynthesis of collagen and its control. In: Krieg T, Kuhn K, eds, Connective tissue in normal and pathological states. *Rheumatology* 1986; **10**: 70–90.

208. Neufeld E F, Ashwell G. Carbohydrate recognition systems for receptor-mediated pinocytosis. In: Lennarz W J, ed, *The Biochemistry of Glycoproteins and Proteoglycans*. New York: Plenum Press, 1980: 241–66.

209. Neufeld E F, McKusick V A. Disorders of lysosomal enzyme synthesis and localisation: I-cell disease and pseudo-Hurler polydystrophy. In: Stanbury J B, Wyngaarden J B, Frederickson D S, Goldstein J L, Brown M S, eds, *The Metabolic Basis of Inherited Disease*. New York: McGraw Hill, 1983: 778–87.

210. Niedbala M J, Madiyalakan R, Matta K, Crickard K, Sharma M, Bernacki R J. Role of glycosidases in human ovarian carcinoma cell mediated degradation of subendothelial extracellular matrix. *Cancer Research* 1987; **47**: 4634–41.

211. Nimni M E. Collagen: biochemistry, biomechanics, biotechnology. *Critical Reviews in Chemistry*. Philadelphia: CRC Press, 1988.

212. Nishimoto S, Oimomi M, Baba S. Glycation of collagen in the aorta and the development of aging. *Clinica Chimica Acta* 1989; **182**: 235–8.

213. Nishimura I, Muragaki M, Olsen B R. Tissue specific forms of type IX collagen-proteoglycan arise from the use of two widely separated promoters.*Journal of Biochemistry* 1989; **264**: 20033–41.

214. Nojima T, Towle C A, Mankin H J, Treadwell B V. Secretion of higher levels of active proteoglycanases from human osteoarthritis chondrocytes. *Arthritis and Rheumatism* 1986; **29**: 292–5.

215. Ogilvie D J, Wordsworth B P, Priestley L M, Dalgleish R, Schmidtke J, Zoll B, Sykes B C. Segregation of all four major

fibrillar collagen genes in the Marfan syndrome. *American Journal of Human Genetics* 1987; **41**: 1071–82.

216. Ohta A, Uitto J. Procollagen gene expression by scleroderma fibroblasts in culture. *Arthritis and Rheumatism* 1987; **30**: 404–11.

217. Okada Y, Konomi H, Yada T, Kimata K, Nagase H. Degradation of type IX collagen by matrix metalloproteinase 3 (stromelysin) from human rheumatoid synovial cells. *FEBS Letters* 1989; **244**: 473–6.

218. Oldberg A, Antonsson P, Heinegard D. The partial amino acid sequence of bovine articular cartilage proteoglycan, deduced from a cDNA clone, contains numerous Ser Gly sequences, arranged in homologous repeats. *Biochemical Journal* 1987; **243**: 255–9.

219. Oldberg A, Antonsson P, Lindblom K, Heinegard D. A collagen binding 59 kDa protein (fibromodulin) is structurally related to the small interstitial proteoglycans PG-S1 and PG-S2 (decorin). *EMBO Journal* 1989; **8**: 2601–4.

220. Oyama F, Hirohashi S, Shimosato Y, Titani K, Sekiguchi K. Oncodevelopmental regulation of the alternative splicing of fibronectin pre-messenger RNA in human lung tissues. *Cancer Research* 1990; **50**: 1075–8.

221. Pack M, Constantinou C D, Kalia K, Nielsen K B, Prockop D J. Substitution of serine for α1(I) glycine 844 in a severe variant of osteogenesis imperfecta minimally destabilises the triple helix of type I procollagen. The effects of glycine substitutions on thermal stability are either position or amino acid specific. *Journal of Biological Chemistry* 1989; **264**: 19694–9.

222. Palotie A, Vaisanen P, Ott J, Ryhanen L, Elima K, Vikkula M, Cheah K, Vuorio E, Peltonen L. Predisposition to familial osteoarthrosis linked to type II collagen gene. *Lancet* 1989; **i**: 924–7.

223. Parks W C, Secrist H, Wu L C, Mecham R P. Developmental regulation of tropoelastin isoforms. *Journal of Biological Chemistry* 1988; **263**: 4416–23.

224. Pasquali-Ronchetti I, Quaglio D, Baccarani-Contri M, Tenconi R, Bressan G M, Volpin D. Aortic elastin abnormalities in OI type II. *Collagen and Related Research* 1986; **6**: 409–21.

225. Pauli B U, Knudson W. Tumour invasion: a consequence of destructive and compositional matrix alterations. *Human Pathology* 1988; **19**; 628–39.

226. Paulsson M. Non-collagenous proteins of basement membranes. *Collagen and Related Research* 1987; **7**: 443–61.

227. Pelletier J P, Martel-Pelletier J, Cloutier J M, Woessner J F. Proteoglycan-degrading acid metalloprotease activity in human osteoarthritic cartilage, and the effect of intra-articular steroid injections. *Arthritis and Rheumatism* 1987; **30**: 541–8.

228. Peltonen J, Larjava H, Jaakola S, Gralnick H, Akiyama S K, Yamada S S, Yamada K M, Uitto J. Localisation of integrin receptors for fibronectin, collagen and laminin in human skin – variable expression in basal and squamous cell carcinomas. *Journal of Clinical Investigation* 1989; **84** 1916–23.

229. Perejda A J, Abraham P A, Carnes W H, Coulson W F, Uitto J. Marfan's syndrome: structural, biochemical and mechanical studies of the aortic media. *Journal of Laboratory and Clinical Medicine* 1985; **106**: 376–83.

230. Perlish J S, Timpl R, Fleischmajer R. Collagen synthesis regulation by the aminopropeptide of procollagen I in normal and scleroderma fibroblasts. *Arthritis and Rheumatism* 1985; **28**: 647–51.

231. Peters J H, Sporn L A, Ginsberg M H, Wagner D D. Human endothelial cells synthesize, process and secrete fibronectin molecules bearing an alternatively spliced type III homology (ED1). *Blood* 1990; **75**: 1801–8.

232. Pipoly D J, Crouch E C. Degradation of native type IV procollagen by human neutrophil elastase. Implications for leukocyte-mediated degradation of basement membranes. *Biochemistry* 1987; **26**: 5748–54.

233. Poole A R. Proteoglycans in health and disease: structures and functions. *Biochemical Journal* 1986; **236**: 1–14.

234. Poole A R. Changes in the collagen and proteoglycan of articular cartilage in arthritis. In: Krieg T, Kühn K, eds, Connective tissue in normal and pathological states. *Rheumatology* 1986; **10**: 316–71.

235. Pope F M, Nicholls A C, Narcisi P, Temple A, Chiu Y, Fryer P, De Paepe A, De Groote W P, McEwan J R, Compston D A, Oorthyus H, Davies J, Dinwoodie D L. Type III collagen mutations in Ehlers–Danlos syndrome type IV and other related disorders. *Clinical and Experimental Dermatology* 1988; **13**: 285–302.

236. Pope F M, Nicholls A. Molecular abnormalities of collagen in human disease. *Archives of Disease in Childhood* 1987; **62**: 523–8.

237. Poulsen J H, Cramers M K. Dermatan sulphate in urine reflects the extent of skin affection in psoriasis. *Clinica Chimica Acta* 1982; **126**: 119–26.

238. Prockop D J, Kuivaniemi H. Inborn errors of collagen. In: Krieg T, Kühn K, eds. Connective tissue in normal and pathological states. *Rheumatology* 1986; **10**: 246–71.

239. Prockop D J. Mutations that alter the primary structure of type I collagen. *Journal of Biological Chemistry* 1990; **265**: 15349–52.

240. Proctor R A. Fibronectin: a brief overview of its structure, function and physiology. *Reviews of Infectious Diseases* 1987; supplement **4**: S317–S321. (See also other articles in this issue.)

241. Pyeritz R. The Marfan Syndrome. In: Royce P M, Steinmann B, eds, *Extracellular Matrix and Inheritable Disorders of Connective Tissue*. New York: Alan R. Liss, 1990: chapter 14.

242. Quentin E, Gladen A, Roden L, Kresse H. A genetic defect in the biosynthesis of dermatan sulphate proteoglycan: galactosyl transferase I deficiency in fibroblasts from a patient with a progeroid syndrome. *Proceedings of the National Academy of Sciences of the USA* 1990; **87**: 1342–6.

243. Raghow R, Thompson J P. Molecular mechanisms of collagen gene expression. *Molecular and Cellular Biochemistry* 1989; **B6**: 5–18.

244. Rajabi M R, Dean D D, Beydoun S N, Woessner J F. Elevated tissue levels of collagenase during dilation of the uterine cervix in human parturition. *American Journal of Obstetrics and Gynecology* 1988; **159**: 971–6.

245. Ramachandran G N. Stereochemistry of collagen. *International Journal of Peptide and Protein Research* 1988; **31**: 1–16.

246. Ramirez F, Di Liberto M. Complex and diversified regulatory programs control the expression of vertebrate collagen types. *FASEB Journal* 1990; **4**: 1616–23.

247. Ramirez F, Boast S, D'Alessio M, Prince J, Su M W,

Vissing H. Molecular pathology of human collagens. *Connective Tissue Research* 1989; **21**: 79–89.

248. Rasmussen L M, Heickendorff L. Accumulation of fibronectin in aortae from diabetic patients. A quantitative immunohistochemical and biochemical study. *Laboratory Investigation* 1989; **61**: 440–6.

249. Rauterberg J, Jander R, Troyer D. Type VI collagen. A structural glycoprotein with a collagenous domain. In: Labat-Robert J, Timpl R, Robert L, eds, *Structural Glycoproteins in Cell-Matrix Interactions*. Basel: Karger, 1986: 90–109.

250. Reiser K M, Last J A. A molecular marker for fibrotic collagen in lungs of infants with respiratory distress syndrome. *Biochemical Medicine and Metabolic Biology* 1987; **37**: 16–21.

251. Ricard-Blum S, Ville G. Collagen cross-linking. *Cellular and Molecular Biology* 1988; **34**: 581.

252. Robins S P, Stewart P, Astbury C, Bird H A. Measurement of the cross-linking compound pyridinoline in urine as an index of collagen degradation in joint disease. *Annals of the Rheumatic Diseases* 1986; **45**: 969–73.

253. Rodell T C, Cheronis J C, Ohnemus C L, Piermattei D J, Repine J E. Xanthine oxidase mediates elastase-induced injury to isolated lungs and endothelium. *Journal of Applied Physiology* 1987; **63**: 2159–63.

254. Rojkind M, Perez-Tamayo R. Liver fibrosis. In: Hall D A, Jackson D S, eds, *International Review of Connective Tissue Research*. New York: Academic Press, 1983: 333–94.

255. Rosenberg L C, Choi H U, Poole A R, Lewandowska K, Culp L A. Biological roles of dermatan sulphate proteoglycans. In: Evered D, Whelan J, eds, *Ciba Foundation Symposium 124: Functions of the Proteoglycans*. Chichester: John Wiley & Sons, 1986: 47–68.

256. Rosenbloom J. Elastin. In: Royce P M, Steinmann B, eds, *Extracellular Matrix and Inheritable disorders of Connective Tissue*. New York: Alan R. Liss, 1990: chapter 4.

257. Roy S, Sala R, Cagliero E, Lorenzi M. Overexpression of fibronectin induced by diabetes or high glucose: phenomena with a memory. *Proceedings of the National Academy of Sciences of the USA* 1990; **87**: 404–8.

258. Ruoslahti E. Proteoglycans in cell regulation. *Journal of Biological Chemistry* 1989; **264**: 13369–72.

259. Ruoslahti E. Structure and biology of proteoglycans. *Annual Review of Cell Biology* 1988a; **4**: 229–255.

260. Ruoslahti E. Fibronectin and its receptors. *Annual Review of Biochemistry* 1988b; **57**: 375–413.

261. Ruoslahti E, Pierschbacher M D. New perspectives in cell adhesion: RGD and integrins. *Science* 1987; **238**: 491–7.

262. Ryan M C, Sandell L J. Differential expression of a cysteine-rich domain in the amino terminal propeptide of type II (cartilage) procollagen by alternative splicing of mRNA. *Journal of Biological Chemistry* 1990; **265**: 10334–9.

263. Saitta B, Stokes D G, Vissing H, Timpl R, Chu M-L. Alternative splicing of the human alpha-2(VI) collagen gene generates multiple messenger RNA transcripts which predict 3 protein variants with distinct carboxyl termini. *Journal of Biological Chemistry* 1990; **265**: 6473–80.

264. Sakai L Y, Keene D R, Engvall E. Fibrillin, a new 350 kilodalton glycoprotein is a component of extracellular microfibrils. *Journal of Cell Biology* 1986; **103**: 2499–2509.

265. Sandberg L B, Soskel N T, Leslie J G. Elastin structure, biosynthesis, and relation to disease states. *New England Journal of Medicine* 1981; **304**: 566–79.

266. Salmela P I, Oikarainen A, Pirttiaho H, Knip M, Niemi M, Ryhanen L. Increased non-enzymatic glycosylation and reduced solubility of skin collagen in insulin-dependent diabetic patients. *Diabetes Research and Clinical Experimentation* 1989; **11**: 115.

267. Sandberg M, Tamminen M, Hirvonen H, Vuorio E, Pihlajeniemi T. Expression of mRNAs coding for the α1 chain of type XIII collagen in human fetal tissues: comparison with expression of mRNAs for collagen types I, II and III. *Journal of Cell Biology* 1989; **109**: 1371–9.

268. Sanes J R, Engvall E, Butkowski R, Hunter D D. Molecular heterogeneity of basal laminae: isoforms of laminin and collagen IV at the neuromuscular junction and elsewhere. *Journal of Cell Biology* 1990; **111**: 1685–99.

269. Sans J, Wieslander J, Langeveld J P M, Quinones S, Hudson B G. Identification of the Goodpasture antigen as the α3(IV) chain of collagen IV. *Journal of Biological Chemistry* 1988; **263**: 13374–80.

270. Sasaki T, Arai K, Ono M, Yamaguchi T, Furuta S, Nagai Y. Ehlers–Danlos syndrome: a variant characterised by a deficiency of pro-α2 chain of type I procollagen. *Archives of Dermatology* 1987; **123**: 76–9.

271. Sasaki T, Majamaa K, Uitto J. Reduction in collagen production in keloid fibroblast cultures by ethyl 3,4 dihydroxybenzoate. Inhibition of prolyl hydroxylase activity is a mechanism of action. *Journal of Biological Chemistry* 1987; **262**: 9397–9403.

272. Savolainen E R, Brocks D, Ala-Kokko L, Kivirikko K I. Serum concentrations of the N-terminal propeptide of type III procollagen and two type IV collagen fragments and gene expression of the respective collagen types in liver in rats with dimethylnitrosamine-induced hepatic fibrosis. *Biochemical Journal* 1988; **249**: 753–7.

273. Savolainen E R, Kero M, Pihlajaniemi T, Kivirikko K I. Deficiency of galactosylhydroxylysylglucosyl transferase, an enzyme of collagen synthesis, in a family with dominant epidermolysis bullosa simplex. *New England Journal of Medicine* 1981; **304**: 197–204.

274. Saxne T, Heinegard D, Wollheim F A. Cartilage proteoglycans in synovial fluid and serum in patients with inflammatory joint disease—relation to systemic treatment. *Arthritis and Rheumatism* 1987; **30**: 972–9.

275. Schmid T M, Mayne R, Jeffrey J J, Linsenmeyer T F. Type X collagen contains two cleavage sites for a vertebrate collagenase. *Journal of Biological Chemistry* 1986; **261**: 4184–9.

276. Schuppan D, Becker J, Boehm H, Hahn E G. Immunofluorescent localisation of type V collagen as a fibrillar component of the interstitial connective tissue of human oral mucosa, artery and liver. *Cell and Tissue Research* 1986; **243**: 535–43.

277. Schuppan D, Cantaluppi M C, Becker J, Veit A Bunte T, Troyer D, Schuppan F, Schmid M, Ackermann R, Hahn E G. Undulin, an extracellular matrix glycoprotein associated with collagen fibrils. *Journal of Biological Chemistry* 1990; **265**: 8823–32.

278. Schwartz E, Cruickshank F A, Lebwohl M G. Elastase-like protease and elastolytic activities expressed in cultured dermal fibroblasts derived from lesional skin of patients with pseudox-

anthomata elasticum, actinic elastosis and cutis laxa. *Clinica Chemica Acta* 1988; **176**: 219–24.

279. Scott J E. Proteoglycan-collagen interactions. In: Evered D, Whelan J, eds, *CIBA Foundation Symposium 124: Functions of Proteoglycans*. Chichester: John Wiley & Sons, 1986: 104–24.

280. Scott J E. Proteoglycan collagen interactions and subfibrillar structure in collagen fibrils. Implication in the development and ageing of connective tissues. *Journal of Anatomy* 1990; **169**: 23.

281. Seltzer J L, Eisen A Z, Bauer E A, Morris N P, Glanville R W, Burgeson R E. Cleavage of type VII collagen by interstitial collagenase and type IV collagenase (gelatinase) derived from human skin. *Journal of Biological Chemistry* 1989: **264**: 3822–6.

282. Senior R M, Hinek A, Griffin G L, Pipoly D J, Crouch E C, Mecham R P. Neutrophils show chemotaxis to type IV collagen and its 7S domain and contain a 67 kDa type IV collagen binding protein with lectin properties. *American Journal of Respiratory Cell Molecular Biology* 1989; **109**: 1371.

283. Sensi M, Tanzi, P, Bruns M R, Pozzilli P, Mancuso M, Gambardella S, DiMario U. Non-enzymic glycation of isolated human glomerular basement membrane changes its physicochemical characteristics and binding properties. *Nephron* 1989; **52**: 222–6.

284. Sephel G C, Byers P H, Holbrook K A, Davidson J M. Heterogeneity of elastin expression in cutis laxa fibroblast strains. *Journal of Investigative Dermatology* 1989; **93**: 147–53.

285. Sephel G C, Giro M G, Davidson J M. Increased elastin production by progeria skin fibroblasts is controlled by the steady state levels of elastin mRNA. *Journal of Investigative Dermatology* 1988; **90**: 643–7.

286. Seyedin S M, Rosen D M. Matrix proteins of the skeleton. *Current Opinion in Cell Biology*. 1990; **2**: 914–9.

287. Shapiro J R, Burn V E, Chipman S D, Velis K P, Bansal M. Osteoporosis and familial idiopathic scoliosis association with abnormal alpha 2(I) collagen. *Connective Tissue Research* 1989; **21**: 117–24.

288. Shimomura H, Spiro R G. Studies on macromolecular components of human glomerular basement membrane and alterations in diabetes. *Diabetes* 1987; **36**: 374–81.

289. Smith L T, Holbrook K A, Madri J A. Collagen types I, III and V in human embryonic and foetal skin. *American Journal of Anatomy* 1986; **175**: 507–21.

290. Spina M, Garbisa S, Hinnie J, Hunter J C, Serafini-Fracassini A. Age-related changes in the composition and mechanical properties of the tunica media of the upper thoracic human aorta. *Arteriosclerosis* 1983; **3**: 64–76.

291. Starkey J R, Stanford D R, Magnuson J A, Hamner S, Robertson N P, Gasic G J. Comparison of BM matrix degradation by purified proteinases and metastatic tumour cells. *Journal of Cellular Biochemistry* 1987; **35**: 31–49.

292. Steinmann B, Superti-Furga A, Royce P M. The Ehlers–Danlos Syndrome. In: Royce P M, Steinmann B, eds, *Extracellular Matrix and Inheritable Disorders of Connective Tissue*. New York: Alan R. Liss, 1990: chapter 11.

293. Strachan R, Smith P, Gardner D L. Hyaluronic acid in rheumatology and orthopaedics. Is there a role? *Annals of the Rheumatic Diseases* 1990; **49**: 949–52.

294. Stryer L. *Biochemistry* (3rd edition). San Francisco: W H Freeman, 1988: Chapter 31.

295. Superti-Furga A, Steinmann B, Lee B, Ramirez F, Lehner T, Ott J, Gaucher A, Moreau P, Weryha G. Autosomal dominant spondyloarthropathy; no linkage to the type II collagen gene. *New England Journal of Medicine* 1990; **322**: 552–3.

296. Superti-Furga A, Gugler E, Gitzelman R, Steinmann B. Ehlers–Danlos syndrome type IV; a multiexon deletion in one of the two COL3A1 alleles affecting structure stability and processing of type III procollagen. *Journal of Biological Chemistry* 1988; **263**: 6226–32.

297. Superti-Furga A, Steinmann B. Impaired secretion of type III procollagen in Ehlers–Danlos syndrome type IV fibroblasts: correction of the defect by incubation at reduced temperature and demonstration of subtle alterations in the triple helical region of the molecule. *Biochemical and Biophysical Research Communications* 1988; **150**: 140–7.

298. Suzuki S, Oldberg A, Hayman E G, Piersbacher M D, Ruoslahti F. Complete amino acid sequence of human vitronectin deduced from cDNA. Similarity of cell attachment sites in vitronectin and fibronectin. *EMBO Journal* 1985; **4**: 2519–24.

299. Swann D A, Silver F H, Slayter H S, Stafford W, Short H. The molecular structure and lubricating activity of lubricin isolated from bovine and human synovial fluids. *Biochemical Journal* 1985; **225**: 195–201.

300. Sykes B. Inherited collagen disorders. *Molecular Biology and Medicine* 1989; **6**: 19.

301. Sykes B. Principles of medical genetics. In: Royce P M, Steinmann B, eds, *Extracellular Matrix and Inheritable Disorders of Connective Tissue*. New York: Alan R. Liss, 1990: chapter 1.

302. Sykes B. Genetics cracks bone disease. *Nature* 1987; **330**: 607–8.

303. Sykes B, Smith R. Collagen and collagen gene disorders. *Quarterly Journal of Medicine* 1985; **56**: 533–47.

304. Tarsio J F, Reger L A, Furcht L T. Decreased interaction of fibronectin, type IV collagen and heparin due to non-enzymic glycation. Implications for diabetes mellitus. *Biochemistry* 1987; **26**: 1014–20.

305. Thomas T, Boot-Handford R P, Grant M E. Modulation of type X collagen gene expression by calcium beta-glycerophosphate and levamisole: implications for a possible role for type X collagen in endochondral bone formation. *Journal of Cell Science* 1990; **95**: 639–48.

306. Thonar E J M A, Meyer R F, Dennis R F, Lenz M E, Maldonado B, Hassell J R, Hewitt A T, Stark W J, Stock E L, Kuettner K E, Klintworth G K. Absence of normal keratan sulfate in the blood of patients with macular corneal dystrophy. *American Journal of Ophthalmology* 1986; **102**: 561–9.

307. Tikka L, Pihlajaniemi T, Henttu P, Prockop D J, Tryggvason K. Gene structure for the $\alpha 1$ chain of a human short chain collagen type XIII with alternatively spliced transcript and translation termination codon at the 5′ end of the last exon. *Proceedings of the National Academy of Sciences of the USA* 1988; **857**: 7491–5.

308. Tiller G E, Rimoin D L, Murray L W, Cohn D H. Tandem duplication within a type II collagen gene (COL2A1) exon in an individual with spondyloepiphyseal dysplasia. *Proceeding of the National Academy of Sciences of the USA* 1990; **87**: 3889–93.

309. Timpl R. Structure and biological activity of basement

membrane proteins. *European Journal of Biochemistry* 1989; **180**: 487–502.

310. Toole B P, Goldberg R L, Chi-Rosso G, Underhill C B, Orkin R W. Hyaluronate-cell interactions. In: Trelstad R L, ed, *42nd Symposium of the Society for Developmental Biology: The Role of Extracellular Matrix in Development.* New York: Alan R Liss, 1984: 43–66.

311. Towbin H, Rosenfelder G, Wieslander J, Avila J L, Rojas M, Szarfman A, Esser K, Nowack H, Timpl R. Circulating antibodies to mouse laminin in Chagas disease, American cutaneous leishmaniasis and normal individuals recognize terminal galactosyl(alpha 1–3)-galactose epitopes. *Journal of Experimental Medicine* 1987; **166**: 419–32.

312. Tromp G, Kuivaniemi H, Stolle C, Pope F M, Prockop D J. Single base mutation in the type III procollagen gene that converts the codon for glycine 883 to aspartate in a mild variant of Ehlers–Danlos syndrome IV. *Journal of Biological Chemistry* 1989; **264**: 19313–7.

313. Tromp G, Prockop D J. Single base mutation in the pro α2(I) collagen gene that causes efficient splicing of RNA from exon 27 to exon 29 and synthesis of a shortened but in frame pro α2(I) chain. *Proceedings of the National Academy of Sciences of the USA* 1988; **85**: 5254–8.

314. Trueb B, Schreier T, Bruckner P, Winterhalter K H. Type VI collagen represents a major fraction of connective tissue collagens. *European Journal of Biochemistry* 1987; **160**: 699–703.

315. Tryggvason K, Hoyhtya M, Salo T. Proteolytic degradation of extracellular matrix in tumour invasion. *Biochimica et Biophysica Acta* 1987; **907**: 191–217.

316. Tsipouras P. A workshop on Marfan syndrome. *Journal of Medical Genetics* 1990; **27**: 139–40.

317. Tsipouras P, Ramirez F. Genetic disorders of collagen. *Journal of Medical Genetics* 1987; **24**: 2–8.

318. Uitto J. Cutis Laxa and premature ageing syndromes. In: Royce P M, Steinmann B, eds, *Extracellular Matrix and Inheritable Disorders of Connective Tissue.* New York: Alan R. Liss, 1990: chapter 12.

319. Uitto J, Murray L W, Blumberg B, Shamban A. Biochemistry of collagen in diseases. *Annals of Internal Medicine* 1986; **105**: 740–56.

320. Uitto J, Perejda A J, Abergel R P, Chu R-L, Ramirez F. Altered steady-state ratio of type I/III procollagen mRNAs correlates with selectively increased type I procollagen biosynthesis in cultured keloid fibroblasts. *Proceedings of the National Academy of Sciences of the United States of America* 1985; **82**: 5935–9.

321. Van der Rest M, Mayne R. Type IX collagen proteoglycan from cartilage is covalently cross-linked to type II collagen. *Journal of Biological Chemistry 1988*; **263**: 1615–8.

322. Varga J, Rosenbloom J, Jimenez S A. Transforming growth factor beta (TGF-β) causes a persistent increase in steady state amounts of type I and type III collagen and fibronectin mRNAs in normal human dermal fibroblasts. *Biochemical Journal* 1987; **247**: 597–604.

323. Varner H P, Furthmayr H, Nilsson B, Fietzek P P, Osborne J C, de Luca S, Martin G R, Hewitt A T. Chondronectin: physical and chemical properties. *Archives of Biochemistry and Biophysics* 1985; **243**: 579–85.

324. Villela D, Cogan R B, Bartolucci A A, Birkedal-Hansen H. Crevicular fluid collagenase activity in healthy, gingivitis, chronic adult perioaortitis and localised juvenile periodontitis patients. *Journal of Periodontal Research* 1987; **22**: 209–11.

325. Vogel B E, Dolez R, Kadler K E, Hojima Y, Engel J, Prockop D J. A substitution of cysteine for glycine 748 of the α1 chain produces a kink at this site in the procollagen I molecule and an altered N-proteinase cleavage site over 225nm away. *Journal of Biological Chemistry* 1988; **263**: 19249–55.

326. Von der Mark K, Kühn U. Laminin and its receptor. *Biochemica et Biophysica Acta* 1985; **823**: 147–60.

327. Von der Mark K, Goodman S. Adhesive glycoproteins. In: Royce P M, Steinmann B, eds, *Extracellular Matrix and Inheritable Disorders of Connective Tissue.* New York: Alan R. Liss, 1990: chapter 6.

328. Vuorio T, Makela J K, Vuorio E. Activation of type I collagen genes in cultured scleroderma fibroblasts. *Journal of Cellular Biochemistry* 1985; **28**: 105–13.

329. Warburton M J, Ferns S A, Kimbell R, Rudland P S, Monoghan P, Gusterson B A. Loss of BM deposits and development of invasive potential by virally-transformed rat mammary cells are independent of collagenase production. *International Journal of Cancer* 1987; **40**: 270–7.

330. Weatherall D J. *The New Genetics and Clinical Practice.* Oxford: Oxford University Press, 1985.

331. Weil D, D'Alessio M, Ramirez F, Steinmann B, Wirtz M K, Glanville R W, Hollister D W. Temperature-dependent expression of a collagen-splicing defect in the fibroblasts of a patient with Ehlers–Danlos syndrome type VII. *Journal of Biological Chemistry* 1989; **264**: 16804–9.

332. Weitz J I, Crowley K A, Landman S L, Lipman B I, Yu J. Increased neutrophil elastase activity in cigarette smokers. *Annals of Internal Medicine* 1987; **107**: 680–2.

333. Welch M P, Odland G F, Clark R A F. Temporal relationships of F-actin bundle formation, collagen and fibronectin receptor expression to wound contraction. *Journal of Cell biology* 1990; **110**: 133–45.

334. West D C, Hampson I N, Arnold F, Kumar S. Angiogenesis induced by degradation products of hyaluronic acid. *Science* 1985; **228**: 1324–6.

335. Wewer U M, Albrechtsen R, Rao C N, Liotta L A. The extracellular matrix in malignancy. In: Krieg T, Kuhn K, eds, Connective tissue in normal and pathological states. *Rheumatology* 1986; **10**: 451–78.

336. Wewer U M, Taraboletti S, Sobel M E, Albrechtsen R, Liotta L A. Role of laminin receptor in tumour cell migration. *Cancer Research* 1987; **47**: 5691–8.

337. Willing M C, Cohn D H, Byers P H. Frameshift mutation near the 3′ end of the COL1A1 gene of type I collagen predicts an elongated pro alpha1(I) chain and results in osteogenesis imperfecta type I. *Journal of Clinical Investigation* 1990; **85**: 282–290.

338. Wirtz M K, Keene D R, Hori H, Glanville R W, Steinmann B, Rao V H, Hollister D W. *In vivo* and *in vitro* association of excised alpha1(I) amino terminal propeptides with mutant pN alpha2(I) collagen chains in native mutant collagen in a case of Ehlers–Danlos syndrome type VII. *Journal of Biological Chemistry* 1990; **265**: 6312–7.

339. Witter J, Roughley P J, Webber C, Roberts N, Keystone

E, Poole A R. The immunologic detection and characterization of cartilage proteoglycan degradation products in synovial fluids of patients with arthritis. *Arthritis and Rheumatism* 1987, **30**: 519–29.

340. Woodward S C. Elastin breakdown and regeneration in emphysema. *Human Pathology* 1989; **20**: 613–4.

341. Woolley D E. Mammalian collagenases. In: Piez K A, Reddi A H, eds, *Extracellular Matrix Biochemistry*. Amsterdam: Elsevier, 1984: 119–51.

342. Wordsworth P, Ogilvie D, Priestley L, Smith R, Wynne-davies R, Sykes B. Structural and segregation analysis of the type II collagen gene (COL2A1) in some heritable chondrodysplasias. *Journal of Medical Genetics* 1988; **25**: 521–7.

343. Wu C H, Donovan C B, Wu G Y. Evidence for pretranslational regulation of collagen synthesis of procollagen propeptides. *Journal of Biological Chemistry* 1986; **261**: 10482–4.

344. Wu J J, Eyre D R, Slayter H S. Type VI collagen of the intervertebral disc. Biochemical and electronmicroscopic characterization of the native protein. *Biochemical Journal* 1987; **248**: 373–81.

345. Yeh H, Anderson N, Ornstein-Goldstein N, Bashir M M, Rosenbloom J C, Abrams W, Indik Z, Yoon K, Parks W, Mecham R, Rosenbloom J. Structure of the bovine elastin gene and S1 nuclease analysis of alternative splicing of elastin mRNA in bovine nuchal ligament. *Biochemistry* 1989; **28**: 2365–70.

346. Zucker S, Wieman J M, Lysik R M, Wilkie D, Ramamurthy N S, Golub L M, Lane B. Enrichment of collagen and gelatin degrading activities in the plasma membranes of human cancer cells. *Cancer Research* 1987; **47**: 1608–14.

Chapter 5

Immunological Basis of Connective Tissue Disease

M MOORE and M DAWSON

Introduction

An understanding of the immune system[127,170,171] is a necessary prerequisite to an account of the pathology of the connective tissue diseases and this chapter summarizes many current views. Fuller, more detailed surveys of the immunology of these conditions have been included in the references.[70,144] The articles cited are intended to be an entrée into a much wider literature.

The present chapter assumes a general familiarity with the basic principles of immunology. While recapitulation is inevitable, the chapter is in other respects necessarily selective and seeks to focus on some recent conceptual advances in our understanding of the complexity of immunological processes in three major connective tissue diseases. Facets of the immunology of the other, less common disorders are discussed in the relevant chapters of the book.

The immune response

The major lymphocyte populations are B-cells, T-cells and 'null' cells, so-called because they lack distinctive T- and B-cell markers.[166] B-cells recognize free antigen via membrane-associated immunoglobulin and produce antibodies of the same specificity.[28] In most cases B-cells require different signals from T-cells to make an adequate response. By contrast, T-cells recognize fragments of the mature antigen which are submitted on the surface of antigen-presenting cells (APC). Some antigen-presenting cells are of the monocyte/macrophage series, others are peripheral blood dendritic cells which develop into interdigitating cells in the lymph nodes.[67] T-cells (Fig. 5.1) only recognize antigen on the surface of the APC when antigen is in conjunction with major histocompatibility complex (MHC) class II molecules, the allelic forms of which determine whether a given response takes place (MHC restriction). This recognition is effected by two types of antigen-specific receptors (αβ or γδ) on T-cells.[213] These receptors, which are closely associated with proteins comprising the CD3 complex (where CD = cluster of determinants) through which intracellular activation signals are transduced, resemble the variable regions of immunoglobulin molecules; their molecular genetics have recently been elucidated.[2,71,133,137,15a] Compared with 'classical' $\alpha\beta^+$ T-cells, the immune functions of $\gamma\delta$ cells, which comprise a small subset of the T-cell pool, are not well understood. T-cells can be classified into different functional subsets by CD markers.[187] Direct contact of APC with antigen-specific $\alpha\beta^+$ T helper (T_H) cells induces activation via secreted molecules such as the cytokines interleukin-1 (IL-1) and IL-6.[47,151] Additional activation signals induced by cell:cell interaction involve the accessory molecules CD4, CD8, lymphocyte function-associated antigen (LFA-1) and CD2. These bind to MHC class II, MHC class I, intercellular adhesion molecule (ICAM-1) and LFA-3 on APC, respectively. T-cells respond to these signals by the production and secretion of cytokines (the 'cytokine cascade'). They include IL-2, B-cell growth- and differentiation-factors (IL-4, IL-5) and interferon-γ as well as factors which arrest, recruit and activate cells of the mononuclear phagocytic system and others which influence haematopoietic colony formation (Fig. 5.1).

Recognition of an antigen by specific receptors on the surface of B-cells or T-cells results in the elaboration of receptors for IL-2 and other cytokines on their surfaces.[121,146,169] Interleukin-2 is a hormone-like growth factor which promotes the proliferation and differentiation of cells which have receptors for it. Consequently, any cell recognizing an antigen undergoes clonal expansion and

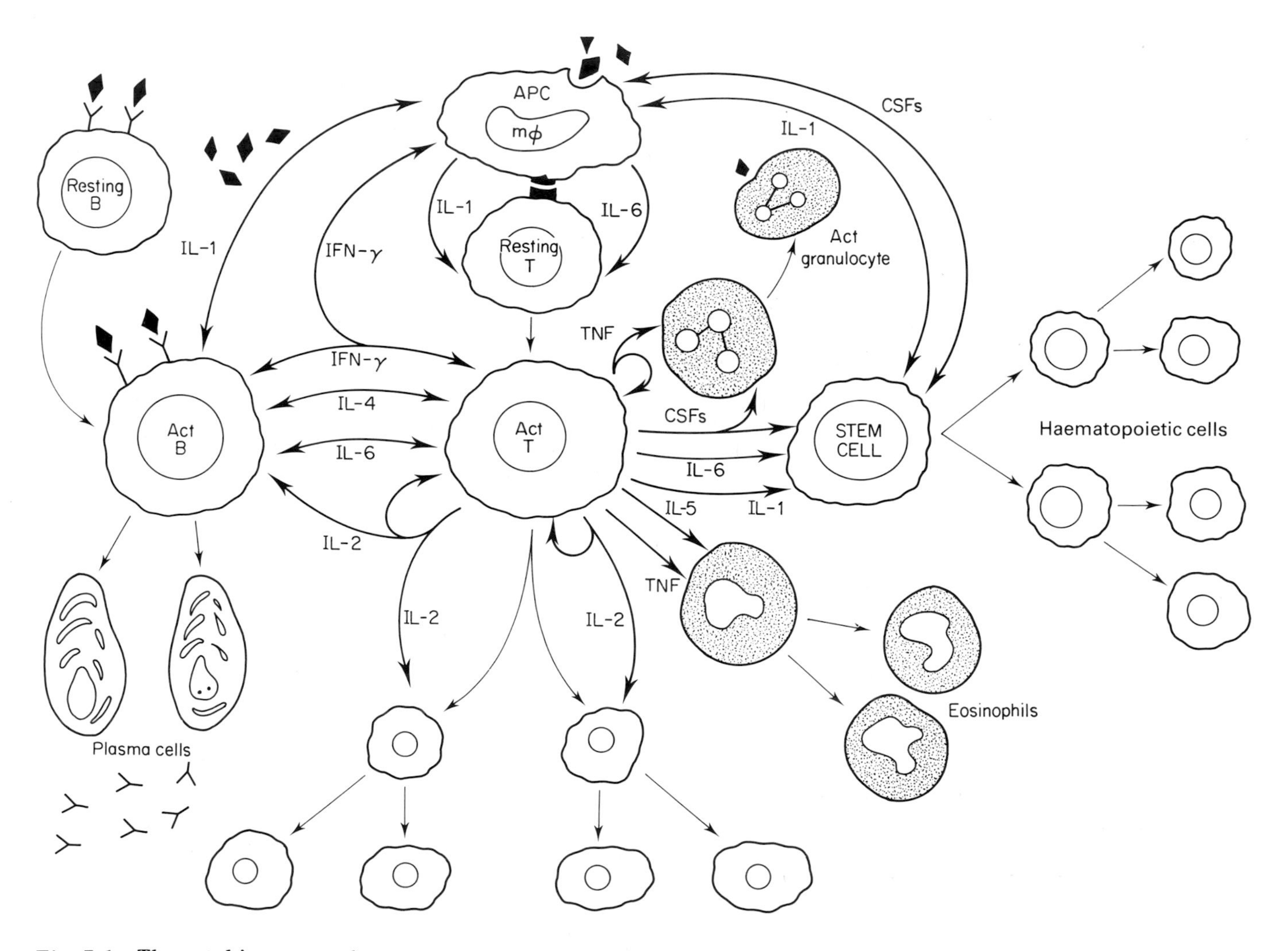

Fig. 5.1 The cytokine network.
Antigen-presenting cells (APC) process antigen and present it in association with class II MHC molecules to a T-helper cell (T_H). Activation of T_H by antigen, and by interleukins produced by the APC, results in the production and secretion of a number of cytokines which interact in a complex network to regulate the effector function of a variety of cells in the immune response. From Balkwill[9] with permission. ACTT = activated T cell; CSFs = colony-stimulating factors; IFN = interferon; IL = interleukin; Mø = macrophage; TNF = tumour necrosis factor; T = T cell; B = B cell; Act B = activated B cell.

develops into a fully functional effector cell. Interleukin-4 and IL-7 are also involved in T-cell activation. B-cells also require additional signals (IL-4, IL-5, interferon-γ) which determine the isotype of antibody produced and regulate differentiation into plasma cells.[135] Several other cytokines interact in the complex network controlling immune responses to 'foreign' antigens. They are illustrated in Fig. 5.1. The recent availability of many cytokines as genetically engineered recombinant molecules has facilitated the elucidation of their numerous biological activities.[42,72,128]

It is important to appreciate that in addition to release by lymphoid cells undergoing antigenic stimulation, many non-antigenic stimuli can induce cytokine synthesis and liberation. Moreover, many non-lymphoid cells can both produce cytokines and/or express receptors for them. This has particular implications for the pathogenesis of important connective tissue disorders. It will also be realized that many of the biological characteristics of so-called 'factors' reported before 1984–86 are now attributable to the pleiotropic behaviour of a single cytokine, interleukin-1. Equally, numerous properties of individual cytokines such as IL-1 and tumour necrosis factor overlap substantially; these cytokines are structurally distinct molecules and they react with unique receptors on individual target cells.

In defined experimental conditions, certain $\alpha\beta^+$ T-cell subsets have distinct functions: T-cell populations can have helper activity essential for an antibody response; proinflammatory (delayed type hypersensitivity, DTH), cytotoxic or suppressor activity. These functions are to some extent distinguishable by CD markers. For example, cytotoxic cells responsible for killing virus-infected cells are $CD8^+$ and helper cells express CD4. However, within these two major subclasses of T-cells, whether functionally exclusive subpopulations exist is controversial. In particular, although there are many examples of T-cells down-regulating immune responses, i.e. acting as suppressor cells, it is not yet clear whether this represents the activity of a separate subset or the same subset which has received a stimulus to secrete suppressive cytokines, such as interferon-γ or IL-10.[140] To a large extent, the quality of response evoked is a function of the immunization schedule.

Cytotoxic T-lymphocytes are the effector cells of cell-mediated immunity. They are $CD8^+$. They bind to epitopes and peptides associated with MHC class I antigens on the surface of target cells (see p. 241). Destruction of a target is mediated by cytolytic proteins of short range and brief half-life.

The 'null' cells comprise 5–10 per cent of peripheral blood lymphocytes. They contribute to 'innate' as distinct from 'acquired' immunity and are phenotypically heterogeneous.[115] The major population is $CD3^-$, $CD4^-$ and $CD8^-$. These 'natural killer' cells can destroy a variety of tumour cell lines and virus-infected cells *in vitro*, without prior sensitization, in an antigen non-specific (i.e. non-MHC restricted) manner. Natural killer cells are morphologically identifiable as Fc-receptor-positive, large granular lymphocytes. This population is, however, heterogeneous and other functions attributed to large granular lymphocytes include the regulation of haematopoiesis and of B-cell function; lymphokine production; and the killing of IgG-coated target cells (*antibody-dependent cellular cytotoxicity*). Peripheral blood natural killer cells do not have rearranged T-cell receptor genes. They represent immature lymphoid cells in the T-cell lineage. The nature of the receptors by which they interact with their targets is unknown.

Large granular lymphocytes can produce and respond to IL-1 and IL-2 and produce interferon in response to the latter. Some large granular lymphocytes constitutively express the IL-2 receptor and respond to this lymphokine by displaying cytotoxicity towards a variety of cell types, normal as well as neoplastic. This phenomenon is lymphokine-activated killing.[68] Current concepts of natural killer and lymphokine-activated killing cytotoxicity suggest that killing is mediated by cells that have overlapping functional properties. The target cell repertoire reflects a differing degree of activation of the respective effector populations.

$\gamma\delta$ T-cells are also cytotoxic for various targets without prior sensitization. In addition, they produce lymphokines (IL-4) and suppress B-cell responses.

Immunological tolerance, autoimmunity and idiotypy

Several common connective tissue diseases are characterized by the presence of antibodies to tissue components to which an individual is normally 'tolerant'. Immunological tolerance is the absence of a specific immune response to epitopes present on tissue components of the individual. An outline of the mechanism(s) of immunological tolerance is therefore essential in order to gain insight into the production of untoward 'autoimmune' responses.

Tolerance and autoimmunity

A central property of the immune system is the ability to distinguish foreign antigens from the self-components of the host. Recognition of the former focusses the destructive mechanisms of immunity on the invading pathogen whereas mistaken recognition of the latter predisposes to autoimmune disease.

There is a broad spectrum of disease with associated

autoimmune phenomena. Autoimmune diseases range from disorders in which humoral and cell-mediated reactions are directed selectively against a single organ to diseases in which antibodies and sensitized cells react against widely distributed, non-organ-specific antigen. The local changes of the thyroid gland in Hashimoto's thyroiditis epitomize the former group, the disseminated tissue changes of systemic lupus erythematosus (SLE), the latter. In both SLE and rheumatoid arthritis, in which antibodies to immunoglobulin are frequent, the production of immune complexes results in inflammatory reactions at sensitive sites.

Tolerance is an acquired characteristic and a property of antigen-specific lymphocytes. The subject has recently been the focus of several in-depth reviews.[167,191,195] Discrimination between self and non-self components can arise during lymphocyte development: On contact with antigen immature T- or B-cells either die ('clonal deletion') or become unresponsive ('clonal anergy'). This mechanism does not wholly account for immunological tolerance, however, since lymphocyte development is restricted to discrete anatomical compartments (thymus for T-cells and bone marrow for B-cells). It is difficult to conceive how antigens apparently expressed only in selected tissues would also be present in the thymus in quantities sufficient to induce clonal deletion or anergy. Thus these are unlikely to be the only mechanisms of tolerance induction, a conclusion which accords with experimental observations of almost 30 years ago in which tolerance was induced in mature animals by intravenous administration of antigen.

Clonal deletion or anergy was more readily demonstrable as a property of immature B-cells, than immature T-cells, because B-cell activity was technically easier to monitor. It was recently established in experiments involving transgenic mice that mature peripheral B-lymphocytes, in common with immature B-cells, can be tolerized by interaction with self-antigen in the absence of antigen-specific helper T-cells, provided receptor occupancy exceeds a critical threshold.[61] This tolerant state which is most readily induced in high-affinity B-cells is closely correlated with down-regulation of membrane IgM but not IgD antigen-receptors. It could thus be an important mechanism for decreasing autoantibody production.

The advent of mice expressing transgene-encoded T-cell receptors and of antibodies capable of recognizing the products of certain gene segments (V_{β}) of the T-cell receptor have more recently facilitated the analysis of T-cells with a given antigen specificity. On intrathymic encounter with antigen in association with MHC products, immature T-cells are clonally deleted. Mature helper T-cells can also become unresponsive on contact with antigen. This phenomenon has also been observed *in vitro* with helper T-cells at the clonal level. If these T-cells encounter their antigen–MHC complex on the surface of chemically modified APC, a state of antigen unresponsiveness results. However, such cells can be induced to proliferate on exposure to IL-2, indicating that anergy may be a reversible state. The distinction between the induction of anergy and the induction of a positive response, including antibody production and lymphokine secretion, is the provision of a second signal (currently undefined) by the APC. The scenario is depicted in Fig. 5.2.

The importance of additional signals in determining the outcome of peptide–MHC recognition by T-cells is illustrated in experiments using monoclonal antibodies to the CD4 molecule.[13] CD4 is preferentially expressed on helper T-cells and possesses both cell binding (via MHC Class II molecules) and cell signalling functions. In certain circumstances *in vivo* treatment leads to the induction of tolerance to antigens administered around the same time. Cumulative data indicate that peripheral $CD4^+$ T-cells can be turned off (tolerized) if presented with antigen in circumstances where the function of cell surface CD4 has been perturbed by appropriate antibodies. Moreover, similar effects can be achieved with anti-LFA-1 as well as with CD8 antibodies (on $CD8^+$ T-cells).

The following model has been advanced to account for these observations.[220] Two basic assumptions are made: first, that T-cell receptor occupancy and the signal provided by recognition of the peptide–MHC antigen complex is equivalent for both tolerance and immunity; and second, that the critical decision (ON or OFF) is determined by secondary signals. Co-stimulatory signals are deemed to arise through multiple cell interactions, each contributing some level of help which ultimately reaches a threshold and activates the system. By inference a T-cell which has adequately bound antigen but is isolated from collaborating partners cannot receive co-stimulatory signals and is unresponsive. In other words whether immunity or tolerance is induced or not is determined by the frequencies of antigen-reactive T-cells and the probability of these being brought together in a collaborative unit. For a potent antigen with many foreign epitopes the chances would be relatively high. However, any situation where T-cells are isolated from each other, either in space or time, would predispose to tolerance. Classical routes to experimental tolerance induction can be explained by this model.

The threshold concentration of antigen for tolerance induction in B-cells is greater than for T-cells. It has been shown for several self-antigens that there is incomplete B-cell tolerance but complete tolerance in the T-cell compartment. Under normal circumstances this does not lead to the breakdown to self-tolerance, because B-cells are dependent on T-cell help. However, if antigen is able to stimulate B-cells without T-cell help (T-cell independent antigen), or an alternate source of T-cell help, autoantibody production can occur. Cross-reaction between a bacterial

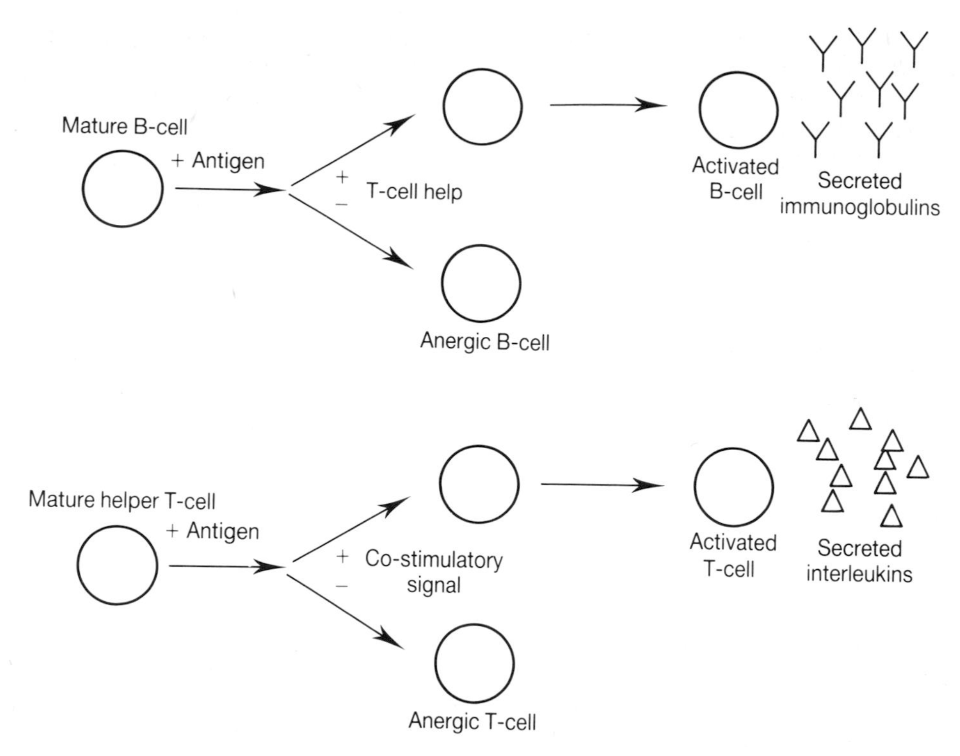

Fig. 5.2 Lymphocyte–antigen interaction.
Lymphocytes which elude inactivation by recognition of self-components at an immature stage, may subsequently come into contact with self- or foreign antigens through their antigen-specific receptors. This leads to cellular proliferation and immune responses provided appropriate second signals are also received. For B-cells such second signals can come from helper T-cells (in the form of interleukins), or in the case of T-cell independent antigens, the second signals derive from macrophages. For helper T-cells, the antigen-presenting cell provides a co-stimulatory signal. In either case if mature lymphocytes come into contact with antigen in the absence of the second signal, the cell enters an extended period of unresponsiveness. From DeFranco with permission.[46a]

antigen and a self-component to which B-cells have not been tolerized, would lead to the generation of anti-self-antibodies. This apparently occurs in rheumatic fever (p. 811) where antibody to streptococcal antigen cross-reacts with cardiac tissue to destructive effect.

Induction and maintenance of the natural unresponsive state would occur only if processed peptides of relevant antigens were operationally visible to T-cells, i.e. associated with MHC on cell types that could 'present' for tolerance; and if certain tissue peptides were not exposed to the immune system.

The majority of 'visible' peptides include the processed products of those molecules, extra- and intra-cellular, which are abundant in the body: gamma globulins, albumin, complement components, haemoglobin, a range of intracellular enzymes and receptors and even MHC molecules themselves. By contrast tissue-specific peptides confined to their respective anatomical compartments are essentially 'cryptic'. For such peptides ever to be visible they would need to become associated with MHC on APC at a level sufficient to induce a response. Such a response could induce either tolerance or immunity, depending on the signals received by the responding T-cells as discussed above.

To induce immunity to foreign antigens, exogenous peptides would need to compete with a plethora of self-peptides. In common with self-peptides, the factors which determine the efficacy of such competition would be antigen concentration, localization, antigen processing and the ability of individual peptides to bind MHC molecules. Viruses are well-equipped in these respects since they furnish multiple repeats of any given protein from which multiple peptide fragments are generated. They also have special means of accessing intracellular entry and privileged processing pathways. Likewise bacteria express a range of natural sugar receptors or are coated with pre-existing antibodies which facilitate entry, processing and presentation in cells of the mononuclear phagocyte series.

Certain peptides associated with MHC are not necessarily capable of inducing a primary response, though they can still be targets for effector systems in other ways. For instance, a potential target such as the beta-cell of the islets of Langerhans in the pancreas, may release sufficient antigen for recognition by *primed* effector cells, but insufficient for initiation of an immune response. Such differential sensitivity in the induction and effector arms of the immune response may, in rare circumstances, lead to breakdown of non-responsiveness by molecular mimicry. In these circumstances a foreign peptide (in association with host MHC molecules) may induce a response leaving host effector systems to detect and attack target cells expressing a low level of a cross-reacting cryptic self-peptide in association with MHC determinants. If this response were sufficient to evoke damage then inflammation and subsequently a cascade process might expose a range of other peptides that would perpetuate damage and thus sustain a chronic autoimmune condition.

In autoimmunity where cryptic antigens are presumed to become visible through tissue damage, the situation is less likely to be one of breakdown of tolerance than one in which particular self-peptide fragments successfully compete to associate with sufficient MHC molecules on APC capable of initiating a response. True tolerance to these peptides may never have existed. Alternatively, the association of self-peptide with MHC components may be on target cells capable of being recognized by a response induced through molecular mimicry. Where the resulting inflammatory reaction is not self-limiting autoimmune processes will continue.

A further factor in the development of autoimmunity may be the inappropriate expression of class II molecules of the major histocompatibility complex (MHC). For example, both epithelial and endothelial cells are able to produce and express cell surface class II molecules, especially when induced with cytokines. A local immune response may result in the release of interferon-γ, often in association with other lymphokines (Fig. 5.1) causing the expression of class II molecules by cells normally devoid of this ability. These cells may act as presenters of 'self-antigen' to T-cells; they may then initiate an autoimmune response.[20,21] In one instance, Sjögren's syndrome (p. 529) is associated with aberrant class II molecules on salivary duct cells.[120] In another, many class II positive macrophages and dendritic cells are found in affected joints in rheumatoid arthritis (p. 455), stimulating both excessive T-cell activation (Fig. 5.1) and the sequelae that activation entails.[102]

Idiotypy

The concept of idiotype recognition was formulated by Jerne.[90] Jerne's network theory proposed that the ability of antibodies both to recognize and to be recognized, creates an interacting system capable of regulating the immune response. Since the receptor on a given B-lymphocyte and the antibody secreted by that B-cell express the same antigen recognition site, antiidiotype antibodies produced against the latter could bind to and regulate antibody production by the former. T-cells would also possess characteristic idiotypes — antigen-specific receptors — which would be recognized and regulated by antiidiotype antibody.

The concept of idiotypy is illustrated in Fig. 5.3. Whether idiotype–antiidiotype recognition can assume a role in autoimmunity has recently been reviewed.[103] The issue is complex because in theory, anti-idiotypy can induce autoimmune disease, be involved in its regulation or be of value in its prevention or treatment.

The pertinent facts are as follows.[30] The capacity of antiidiotype antibody to recognize idiotypes on antigen-induced antibody (Ab_1 in Fig. 5.3) applies to a situation in which the antigen is a self-determinant, the antibody (Ab_1) an autoantibody. In some diseases the magnitude and/or duration of an immune response can be modified by anti-idiotypes. This is particularly so when the antigen-induced

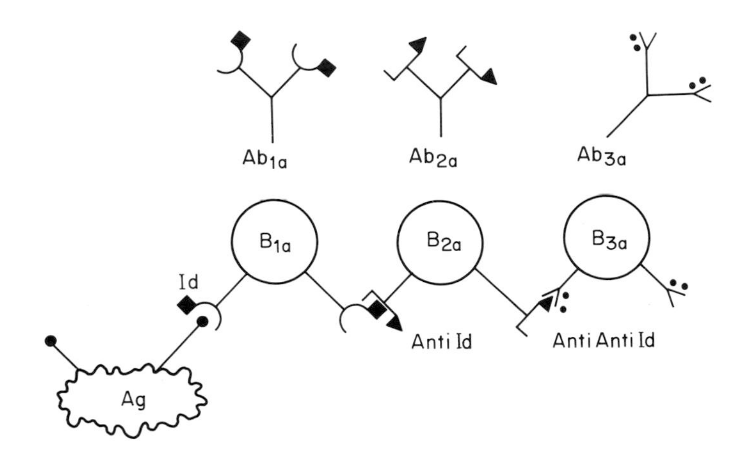

Fig. 5.3 The concept of idiotypy.
Antigens (Ag) contain one or more determinants (solid circles) that induce B-cells to produce antibodies (Ab) which recognize those determinants. The antibody combining site contains a structure which is complementary with the antigenic determinant so that a reasonably good 'fit' between the two can be achieved. Antibody to antigen is called Ab_{1a} and is produced by the B_{1a} cell. In addition to having a structure which recognizes antigen (semicircle), Ab_{1a} also has unique determinants which themselves can be immunogenic. Thus, the solid square on the receptor of Ab_{1a} is a structure (idiotype [Id]) which is recognized by the immune system, and antibodies to that structure can be produced. Such antibodies to idiotypes are called antiidiotype antibodies (anti-Id). Since they are the second antibodies in a series, they are also called Ab_{2a}. The antigen combining site of Ab_{2a} does not recognize the original antigen: instead, it recognizes the idiotype on Ab_{1a}. Ab_{2a}, however, also has a unique structure (solid triangle) on its combining site. This structure is capable of inducing an immune response which leads to the production of the antiantiidiotype or Ab_{3a}. Reproduced with permission of D M Klinman and the Editor, *Arthritis and Rheumatism*.[103]

response is dominated by a cross-reactive idiotype. Many autoantibodies express cross-reactive idiotypes and anti-idiotype administered exogenously can suppress a developing immune response. Such antiidiotype antibodies are detectable in the sera of patients in remission from autoimmune disease. Mechanisms by which antiidiotypic antibodies may promote the development of the autoimmune state can be postulated. One hypothesis[103] proposes that antiidiotypic antibodies interfere with the process by which autoreactive B-cells are tolerized. Immature B-cells are tolerized if they bind with high affinity to antigen. However, this interaction can be competitively inhibited by excessive amounts of a second antigen which binds with low affinity to the B-cell. The low affinity interaction prevents the formation of the stable antigen–B-cell complexes necessary to induce tolerance.

Conventional antibodies may exist which cross-react by chance with antigen receptors on autoreactive B-cells so that antiidiotypes can inhibit the tolerization of these B-cells. By such a mechanism, it is possible to conceive of environmental antigens, unrelated to self, contributing to the pathogenesis of an autoimmune process. The process might be facilitated if some of the antibodies induced by exogenous antigens recognized the idiotype on autoreactive immature B-cells, by cross-reactivity, preventing their tolerization.

Many B-cell clones contribute to an autoimmune response and not all express a common idiotype. This means that no single antiidiotype can suppress autoantibody production. Moreover, it is possible that when antiidiotype therapy reduces idiotypic antibody levels, a compensatory increase in idiotype-negative antibody production may occur, leaving the overall level of the antigen-specific immune response unchanged.

In view of the frequency with which cross-reactive idiotypes are present on autoantibodies, some *in vivo* regulation by antiidiotypic antibody must be contemplated. Only certain autoantibodies of particular affinity, charge, isotype and other immunochemical properties are pathogenic; their selective elimination could partly ameliorate autoimmune disease even if the concentration of non-pathogenic autoantibodies of identical specificity were increased.

Autoantibodies in connective tissue disease

Several connective tissue diseases are characterized by the presence of autoantibodies. The role of autoantibodies in pathogenesis is not always clear but they may be of diagnostic value. Rheumatoid factor (RFA) and the many antinuclear factors,[14] typical of rheumatoid arthritis (see Chapters 12 and 13) and systemic lupus erythematosus (see Chapter 14), respectively, are the best known. Other disorders are characterized by the formation of antibodies against connective tissue components but, by convention, are not usually classed as connective tissue diseases. In Goodpasture's syndrome, for example, autoantibodies are directed mainly against the type IV collagen of glomerular basement membrane.[164] In poststreptococcal glomerulonephritis autoantibodies react with the heparan sulphate proteoglycan and laminin of this basement membrane as well as with the type IV collagen.[1]

Rheumatoid factors

Rheumatoid factors are antibodies, frequently present in the sera of patients with rheumatoid arthritis; they react with antigenic sites on the CH2 domain of the Fc portion of IgG.[148] It is likely that the reactive site is carbohydrate rather than polypeptide, and that the glycosylation pattern of serum IgG is altered in rheumatoid arthritis and certain arteritides (see also p. 246).[154] Some reactive sites are also found on IgG from other species; others are unique to human IgG. The RFAs show various specificities. Rheumatoid factors have been found to be IgG, IgA and, rarely IgD. The most common is, however, IgM. This immunoglobulin forms 22S complexes with circulating IgG.[153] Rheumatoid factors are not exclusive to rheumatoid arthritis; they are also but less frequently found in patients with systemic lupus erythematosus. Whether RFAs play a significant part in the genesis of rheumatoid arthritis is uncertain. Hypo- or agammaglobulinaemic patients sometimes develop rheumatoid arthritis. By contrast, high RFA titres are associated with progressive disease, systemic complications, subcutaneous granulomata and a poor prognosis. Rheumatoid factors may play an active role in the exacerbation of rheumatoid arthritis. Some seronegative patients may have IgG RFA not detectable in a standard assay. Alternatively, all the RFAs may be present as immune complexes so that free RFAs are no longer demonstrable.[4,39]

IgG/RFA complexes are often deposited in the synovium and articular cartilage. They are released into the synovial fluid where they activate complement leading to the chemotaxis of polymorphs, monocytes and macrophages into the joint and the depletion of synovial complement. The initial stimulus for the synthesis of RFAs is unknown. Rheumatoid factors are often found in the sera of patients with bacterial infections such as leprosy and syphilis. It is possible that their origin in connective tissue disorders such as systemic lupus erythematosus and progressive systemic sclerosis has an analogous explanation and this has led to a number of hypotheses. First, the stimulus to RFA formation might be an aetiological agent with determinants that cross-react with those found on IgG. Second, a polyclonal activator of B-cells might be the original stimulus for RFA production. One putative agent is Epstein–Barr virus, a non-specific stimulator of B-cells which can induce the synthesis *in vitro* of RFAs by lymphocytes from both healthy controls and rheumatoid patients. Third, the production of immune complexes between IgG and a hypothetical agent may result in the effective aggregation of IgG (self-aggregation) which could, in turn, be autostimulatory.[50]

Hypotheses that polyclonal RFAs present in rheumatoid arthritis are triggered by autologous IgG must account for the fact that anti-allotypic specificities in this disease do not correlate with the patient's own immunoglobulin allotype.[69]

Much of the pathological significance of RFAs lies in their ability to complex with IgG and to trigger the activation of complement.[205] The range of infectious disease with which RFAs are associated is extensive and RFA production may represent a normal immune response possibly concerned with the elimination of idiotype/antiidiotype immune complexes. However, in almost all cases, RFA production is qualitatively and quantitatively different when associated with infectious disease and ceases to be detectable with recovery from infection.

Rheumatoid factors can be assayed by the Rose–Waaler (SCAT) test or by a latex agglutination test. The Rose–Waaler test relies on the ability of 'classical' IgM RFA to agglutinate sheep red blood cells coated with subagglutinating rabbit antisheep antibody.[174] Heterophile activity is removed by prior adsorption with uncoated sheep red blood cells. Latex particles coated with human IgG may be used but this test gives less specific results. These reactions measure IgM RFA but may fail to detect RFAs of other immunoglobulin classes. 'Hidden' antibody, caused by high avidity RFA bound *in vivo* to circulating IgG, may also remain undetected unless immune complexes are dissociated by treatment at a pH below 2.5. More recently, radioimmune assays and enzyme-linked immunosorbent assays (ELISA) have been used to measure IgM, IgG and IgA RFAs, taking advantage of the ability of patients' sera to bind to human IgG attached to plastic surfaces.[32,107] Binding can be detected indirectly with labelled antihuman immunoglobulin of an appropriate class.

Plasma cells producing RFAs have been detected by immunofluorescence microscopy. In diseased joints (see p. 458) cells producing IgG RFA are more frequent than those secreting IgM but cells secreting IgA[194] are also common.

Much attention has been directed towards the definition of the epitopes recognized by human RFA. Such studies are likely to provide insights into mechanisms governing the induction of RFAs in disease. Disclosure of the diversity of RFAs in rheumatoid patients is in progress. For instance, germ-line genes corresponding to cross-reactive idiotypes present on V_κIIIa and V_κIIIb light chains have been identified. Four germ-line V_κ genes from two unrelated subjects were shown to be highly homologous to V_κIIIa ($6B6.6^+$ RFLes L chain). Three of the germ-line genes are potentially functional and differ by only one to six bases.[122]

Autoreactive antibodies in systemic lupus erythematosus

The lupus erythematosus cell phenomenon (see p. 570) is due to the opsonization of cell nuclei coated with autoantibody to chromatin.[80] The presence of circulating antibodies against double-stranded (ds) DNA still provides the most reliable evidence for the diagnosis of systemic lupus erythematosus (SLE) although these antibodies may also be detected in Sjögren's syndrome, polymyositis/dermatomyositis, mixed connective tissue disease and progressive systemic sclerosis.[33] The detection of autoantibodies to nuclear antigens is most easily achieved by immunofluorescence tests on cryostat sections. Many different patterns have been observed ranging from rim fluorescence or speckled patterns to antimitotic spindle fluorescence. In SLE, antibodies to single-stranded (ss) DNA are frequently present alongside those to dsDNA. Both are implicated in SLE nephritis (p. 584). Antibodies to ribonucleoprotein are found in up to 50 per cent of patients together with a number of antibodies against non-histone proteins (5–10 per cent). Anti-DNA antibodies exist in IgG and IgM classes. They display a range of epitope specificities. Complexes of DNA and specific antibodies are frequently found in the sera of patients with SLE. A typical complex is a single IgG molecule bound to a dsDNA fragment of 20–1200 base pairs. IgG antibodies are more often associated with nephritis and clinically active SLE than IgM antibodies. There is evidence to suggest that high-avidity, cationic antibodies are deposited more often on polyanionic sites of the glomerular basement membrane. Antilymphocyte antibodies are another important class of autoantibodies detectable in most patients with SLE at some stage of the disease.[224] These antibodies are not truly disease-specific; several commonly occur in various immunological, infectious and neoplastic disorders, as well as in normal individuals after prophylactic immunization. In this respect, antilymphocyte antibodies are analogous to RFAs and cold agglutinins. The mechanism of their formation is unknown.

Origin of autoantibodies in systemic lupus erythematosus

Progress towards explaining the origin and diversity of SLE autoantibodies has been achieved with the development of monoclonal antibodies. Hydridomas that secrete monoclonal SLE autoantibodies have been developed from (NZB × NZW) F1 and MRL *1pr/1pr* mice, as well as from SLE patients.[188] Studies of anti-DNA antibodies have shown that a single autoantibody can bind to multiple nucleic acid antigens of widely different base composition. Thus, a monoclonal autoantibody which binds to denatured DNA can also bind to dsDNA, poly (I) and poly (dT). In such cases, the epitope must reside in the sugar–phosphate backbone of the DNA and not in any of the bases of the polynucleotide. The putative epitope is a phosphodiester-linked phosphate structure that occurs in all polynucleotides. Structural differences in polynucleotide backbones due to variations in helical configuration and interphosphate distances can explain individual differences in binding specificities among the monoclonal autoantibodies.

Phosphodiester-containing epitopes also account for the ability of some monoclonal anti-DNA autoantibodies to bind to certain phospholipids, including cardiolipin, accounting for SLE anticoagulant activity and a false-positive microflocculation assay for syphilis (p. 570). Phospholipids resemble polynucleotides; both have phosphodiester-linked phosphate groups. Displayed on micellar surfaces, these groups form repeating units that promote immunochemically specific interactions with anti-DNA antibodies.

Other cross-reactions of anti-DNA antibodies with non-nuclear components have been reported: a monoclonal anti-DNA antibody from an (NZB × NZW) F1 mouse has been shown to react with several distinct polypeptides in the plasma membrane of a variety of mammalian cells. The triggering antigen in SLE could be protein rather than DNA.[87] Additional cross-reactions occur with negatively charged glycosaminoglycans such as chondroitin sulphate and hyaluronate.[51] Several types of molecule can act as targets for anti-DNA antibodies provided that their structures contain repeating negatively charged groups. The reactivity of anti-DNA autoantibodies with heparan sulphate, features strongly in a postulated new mechanism of SLE nephritis.[48]

The wide cross-reactivity of anti-DNA antibodies in SLE raises the question whether any of the target molecules is implicated in the aetiology of the disease. Sensitive radioimmune assays for the detection of anti-DNA antibodies have brought the proportion of positive SLE sera to approximately 70 per cent. The fluorescent antinuclear antibody test is positive in virtually all SLE patients and is recognized as an independent criterion in the 1982 revised criteria for the classification of SLE by the American Rheumatism Association (now the American College of Rheumatology) (see Table 14.1, p. 568).[204] Many molecular constituents of the nucleus may serve as antigens;[203] they include RNA, histones and a variety of ribonucleoprotein complexes. The conclusion is inescapable that the major immunological event in SLE is a change in the immunogenicity of the nucleus. The mechanism by which this change occurs is obscure. That negatively charged molecules such as cardiolipin and hyaluronate are incapable of eliciting antibodies to basic nuclear proteins such as histones means that they cannot play a significant role in pathogenesis.

Several conclusions follow. First, the spectrum of autoantibodies in SLE may not be as complex as the wide catalogue of antigen specificities suggests. Second, DNA is not necessarily an immunogen in SLE; indeed, nucleic acids in general are poorly immunogenic. Third, the multiple specificities of some monoclonal antibodies indicate that the origin of spontaneously produced autoantibodies cannot be inferred from analyses of serum.

In the mouse, SLE autoantibodies have been subject to idiotypic analysis. The reader is referred to pp. 598–603. Antibodies raised in rabbits or appropriate strains of mice against the idiotypes of monoclonal anti-DNA autoantibodies of (NZB × NZW) F1 hybrid mice, have shown that a high proportion of the latter with similar ligand-binding specificities share a common idiotype.[182] These antibodies appear to originate from a restricted family of germ-line genes which, far from being limited to autoimmune mice, are widely dispersed throughout the species. Normal mice produce anti-DNA antibodies when their B-lymphocytes undergo polyclonal maturation. Thus, populations of lymphocytes that can secrete anti-DNA autoantibodies occur in normal, as well as in SLE-prone mice. The spontaneous production of anti-DNA autoantibodies appears therefore to involve impairment of a regulatory mechanism that operates either at the level of immunological circuits or at that of the expression of the immunoglobulin genes. The issue is, however, complex.[104]

There are therapeutic implications. The occurrence of high frequency idiotypes of anti-DNA autoantibodies in murine SLE indicates the feasibility of suppressing the principal serological abnormality of the disease by the administration of antiidiotypic sera. This approach, impracticable if the diversity of serological abnormalities is matched by a corresponding diversity of idiotypes, may lead to the development of new forms of treatment.

Immunological hypersensitivity

Hypersensitivity is present when a state of humoral or cell-mediated immunity causes tissue damage. Coombs and Gell[38] described four types of hypersensitivity in each of which the tissue injury was attributable to different facets of the immune response. The bacterial or viral immunogens involved in many connective tissue diseases are not potent allergens and type I (anaphylactic or immediate) hypersensitivity is rarely evoked. However, many manifestations of hypersensitivity in the connective tissue diseases involve more than one of the other three mechanisms.

Mechanisms of antibody and immune complex-mediated tissue injury

Type II (cytotoxic) hypersensitivity occurs where there is an antigen/antibody interaction at a target cell surface. Complement is activated via the classical pathway. Tissue damage results, either from the inflammatory effects of complement by-products or from substances released from the target cells. In complement-independent cytotoxicity, cell-bound antibody of an appropriate class interacts with the corresponding Fc receptors on killer cells (antibody-dependent cellular cytotoxicity) (see p. 229). Cytotoxic hypersensitivity may be involved in connective tissue diseases where autoantibodies are produced but the damage is usually a result of immune complex formation.

Type III (complex-mediated) hypersensitivity is frequently a component of connective tissue disease. Characteristically, as in rheumatoid arthritis, foci of inflammation result. Immune complexes (p. 238) induce inflammation principally through complement activation. Complexes, in antibody excess or at antibody/antigen equivalence, are precipitated and cause local inflammation. However, immune complexes in antigen excess are soluble, circulate in the blood and accumulate in selected sites. Serum-sickness typifies this reaction. Arthralgia, arteritis and glomerulonephritis are common consequences.

Mechanisms of cell-mediated tissue injury

Delayed hypersensitivity (cell-mediated, type IV, hypersensitivity) occurs in response to immunogens which stimulate cell-mediated rather than humoral immunity. Delayed hypersensitivity is a feature of some connective tissue diseases particularly those in which autoimmunity plays a part. Typically, there is a slow onset after challenge by immunogen.[214]

The lesions of delayed hypersensitivity are exemplified by the response at the site of intradermal injection of tuberculoprotein into the skin of an individual sensitized to *Mycobacterium tuberculosis*. A typical lesion results, the onset of which occurs some 18 h after challenge. Thereafter, the lesion may enlarge and reach its maximum size after 36–48 h and take weeks or months to resolve. Sensitized T-cells react with immunogen and release lymphokines locally. These lymphokines are chemotactic for mononuclear phagocytes; they also increase the permeability of blood vessels and amplify the reaction. There is inhibition of the egress of cells from affected areas by macrophage inhibition factor. In some ways, the sterile synovial granuloma in rheumatoid arthritis (see Chapter 12) resembles a tuberculin reaction and it is of interest to note new views on the relationship of sensitivity to *M. tuberculosis* and injury to proteoglycan (see p. 531). The injection of lymphokines such as IL-1 and IL-2 into the joints of animals can result in inflammatory reactions akin to those seen in the joints of rheumatoid patients. Type IV reactions may also be important in the pathogenesis of progressive systemic sclerosis.[152]

Role and activation of complement

Complement (C) is a complex of 22 proteins, present in the plasma and some tissue fluids, that act in a sequence or pathway to promote phagocytosis, cell and bacterial lysis, and some of the phenomena of inflammation (see Chapter 7).[53] The complement pathway is activated classically when antigens bind to antibodies. Alternatively, C3 can be activated in the absence of antigen/antibody binding, by exposure to proteins such as properdin or lipoproteins such as Gram-negative bacterial cell walls. The onset or sequelae of some connective tissue diseases may be substantially determined by the activation of complement or by the acquired or heritable deficiency of complement components (see p. 244).

The classical pathway

The classical complement (C) pathway involving components C1, C2, C3 and C4 is usually stimulated by the combination of antigen with antibody (Fig. 5.4a). It is not affected by the location of antigen. The combination of some classes of immunoglobulin (namely IgG and IgM) with specific epitopes causes cryptic sites on the Fc regions of these immunoglobulins (CH3 for IgG, CH4 for IgM) to be revealed. C1 can bind to these sites. C1 consists of three subcomponents. The first, C1q, binds to the cryptic sites. Clq is distinctive: it has been compared structurally to a bunch of six tulips, the 'stalks' fused while the 'heads' containing the binding sites, are free. Activation of C1 requires that a molecule of C1q must bind to two Fc region sites on the surface of the immunogen. IgM, with five sites per molecule, is the most efficient. A satisfactory spacing of IgG molecules is necessary to permit the binding of C1q to C2. The lysis of one red blood cell via complement may require 1000 molecules of IgG; a single molecule of IgM can achieve the same result.

C1s acquires enzymatic activity via C1r after C1q is bound. C1s is an esterase able to cleave C4 and C2 each into two fragments: C4a and C4b, C2a and C2b. The larger fragments, C4b and C2a, combine to produce a new enzyme, C3 convertase, which cleaves C3 into fragments, C3a and C3b.

Newly formed C4b and C3b have short-term, hydrophobic binding sites. Subsequent reactions are therefore likely to take place on the membrane of a target cell. If either molecule fails to bind, the complement sequence continues in fluid. Subsequent molecules with short-term binding sites may recruit further complement to the cell surface.

The last enzymic step in classical activation is the combination of a molecule of C3b with classical C3 convertase to produce C5 convertase; C5 is changed to C5a and C5b. Many C5a and C5b molecules may be produced as the result of the activation of a single C1q molecule since each enzyme can act on many substrate molecules, producing an enzymic cascade. Subsequent reactions are not enzymic and there is no further amplification of response. Each C5b molecule which binds to the surface of a target cell now binds each of the following components: C6, C7, C8 and 6 molecules of C9. The result is a membrane attack complex with a relative molecular weight of approximately 1×10^6 kD which comes to be inserted as a tubule into the cell membrane to produce a lesion. Many thousands of such lesions, produced by the activation of a single C1 molecule, result in the leakage of essential ions and small metabolites from the cell, leading to the death of the cell through osmotic lysis.

Although lysis of foreign cells is undoubtedly a critical function of complement, its action in promoting inflammation and phagocytosis is thought to be more important *in vivo* (p. 288). Complement molecules with pharmacological activities include C3a and C5a which are anaphylatoxins, i.e. they stimulate the degranulation of basophils and mast

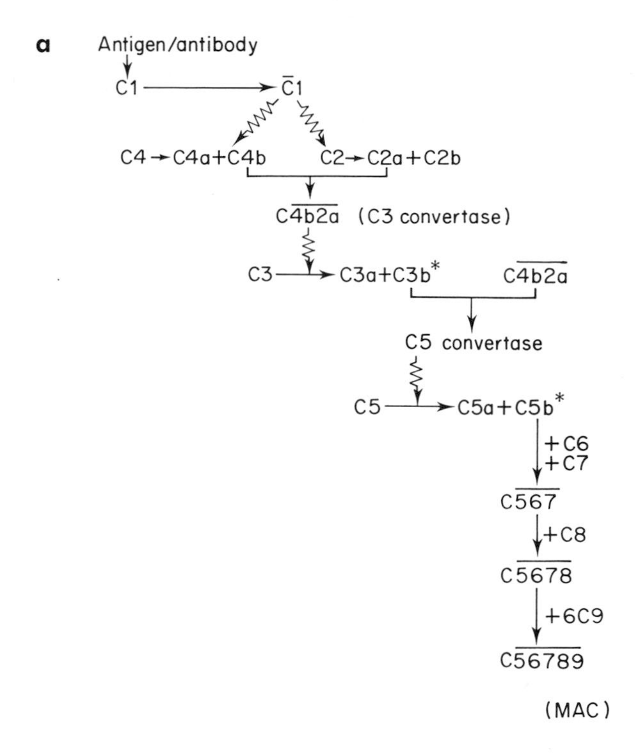

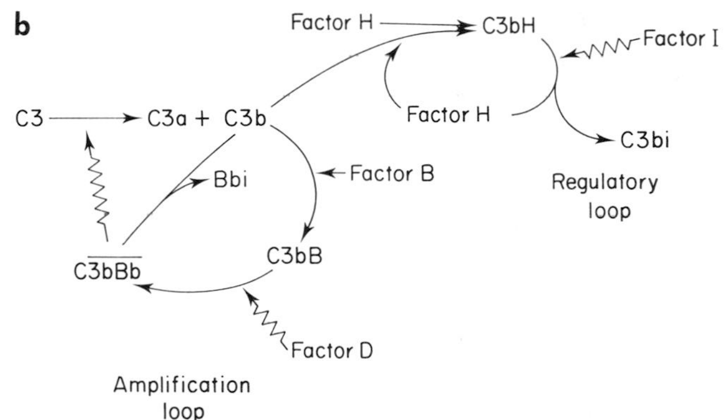

Fig. 5.4 Activation of complement.
a. Classical activation of complement. C1 is activated by complexes of IgG or IgM with appropriate antigen. Two products (C4b and C2a) combine to form the classical C3 convertase. C3b combines with a molecule of the C3 convertase to yield a new enzymatic activity, C5 convertase, which splits C5 into C5a and C5b. C5b binds to membranes of target cells. Subsequent addition of C6, C7, C8 and multiple molecules of C9 produces the membrane attack complex (MAC). (Symbols: (–) activated complement component; (⌇) enzymic conversion; (*) hydrophobic binding site.) **b.** Alternate pathway for complement activation. In the alternate pathway, the amplification loop is balanced by the regulatory loop. C3b, produced by C3 convertase (from either pathway), binds to serum factor B to produce C3bB. C3bB is in turn cleaved by factor D to yield C3bBb, the alternate pathway C3 convertase. C3bBb may cleave more C3 or may spontaneously decay to regenerate C3b. Substances such as bacterial lipopolysaccharide stabilize C3bBb, prevent its decay and thus act as alternate pathway activators. The regulatory loop removes C3b by combining it with factor H, the complex being cleaved by factor I to yield factor H and the inactive form of C3b, C3bi.

cells so that factors with known vasodilatatory activity, such as histamine, are released locally. C3a and C5a are chemotactic for polymorphs.[84,143]

The alternate pathway

This pathway can be activated *in vivo* by bacterial endotoxins or *in vitro* by components of yeast cell walls such as zymosan. Antibody is not required although aggregated IgA may activate complement in the absence of specific antigen. The alternate pathway is a positive feedback loop in which C3b, produced either as a result of the classical pathway or from a natural slow 'ticking over' of C3, combines with factor B to produce C3bB, a compound with weak C3 convertase activity (Fig. 5.4b). In turn, C3bB is acted on by serum factor D to produce C3bBb, a powerful C3 convertase which completes the cycle. Thus an amplification loop is established. The loop is controlled by the spontaneous breakdown of C3bBb to yield C3b and inactive Bbi. Alternate pathway activators may stabilize convertase perhaps by providing a suitable surface for reactions to take place. Convertase breakdown and the decay of the amplification loop are prevented. C3b produced via this pathway can feed into the lytic cycle with all the consequences that this implies.

The classical pathway may have significance in systemic connective tissue diseases in which autoantibodies to tissue components are produced or in which antibodies to unknown causative agents trigger activation at sensitive sites. A causative agent of this kind may invoke the alternate pathway by direct stimulation. In either event, the resultant inflammation may cause or perpetuate severe tissue injury.

Immune complexes and immune complex disease

Immune complex disease is a frequent local or systemic complication of autoimmune disease: the continued production of autoantibody to a self-antigen leads to sustained immune complex formation and overload of the mononuclear phagocytic system[63] which is responsible for the clearance of immune complexes and the deposition of immune complexes in tissues.[131,207] Other categories of immune complex disease arise from the inability of weak antibodies to eliminate persistent infective agents or repeated exposure to extrinsic antigen. Immune complexes are not in themselves harmful but they may initiate damaging inflammatory processes (see p. 627).

Detection of immune complexes

Immune complexes are often detected in the circulation by their physical or biological properties which are antigen-non-specific, i.e. by characteristics not capable of distinguishing true immune complexes from non-specifically aggregated immunoglobulin.[206] The presence of circulating immune complexes is not specific for any immune complex disease. Thus, complexes are not infrequently found in the sera of healthy individuals and lesions induced by immune complexes can exist without detectable circulating immune complexes. Much greater significance therefore attaches to the detection of immune complexes in the affected parts. For this purpose, biopsies are examined for the presence of immunoglobulin and complement. The composition, pattern and microanatomical site of deposition can provide valuable information on the severity, behaviour and prognosis of a disease. This advantage is exemplified in glomerulonephritis (p. 588) in which continuous, granular subepithelial deposits of IgG herald a poor prognosis, by comparison with deposits confined to the mesangium.

Induction of tissue damage

Immune complexes tend to be deposited selectively; they activate complement and induce inflammation in susceptible tissues (Fig. 5.5).[207] Antigen complexed with IgG and IgM activates complement by the classical pathway; aggregated IgA and IgE activate the alternate pathway. In either event, complement activation has important consequences. First, immune complexes containing complement-fixing antibodies may become coated with C3b. Macrophages have receptors for C3b but so do tissue cells such as those of the glomerular epithelium. The glomerulus is particularly susceptible, therefore, to injury attributable to immune complexes. Second, complement activation can cause solubilization of immune complexes by the intercalation of C3b on to the complexes; the response is dependent on alternate pathway convertase initially assembled on the complex. The process is a physiological reaction that can modulate overwhelming inflammatory reactions. In hypocomplementaemic patients this process may be inadequate and can predispose to prolonged immune complex deposition. Such a defect has been detected in the sera of patients with immune complex disease (see p. 244). Third, C3a and C5a have anaphylatoxic and chemotactic properties that can cause the release of vasoactive amines from mast cells (see p. 36) and basophils, increasing vascular permeability and attracting polymorphs. Fourth, factor 3B of the alternate complement pathway can stimulate macrophage activation.

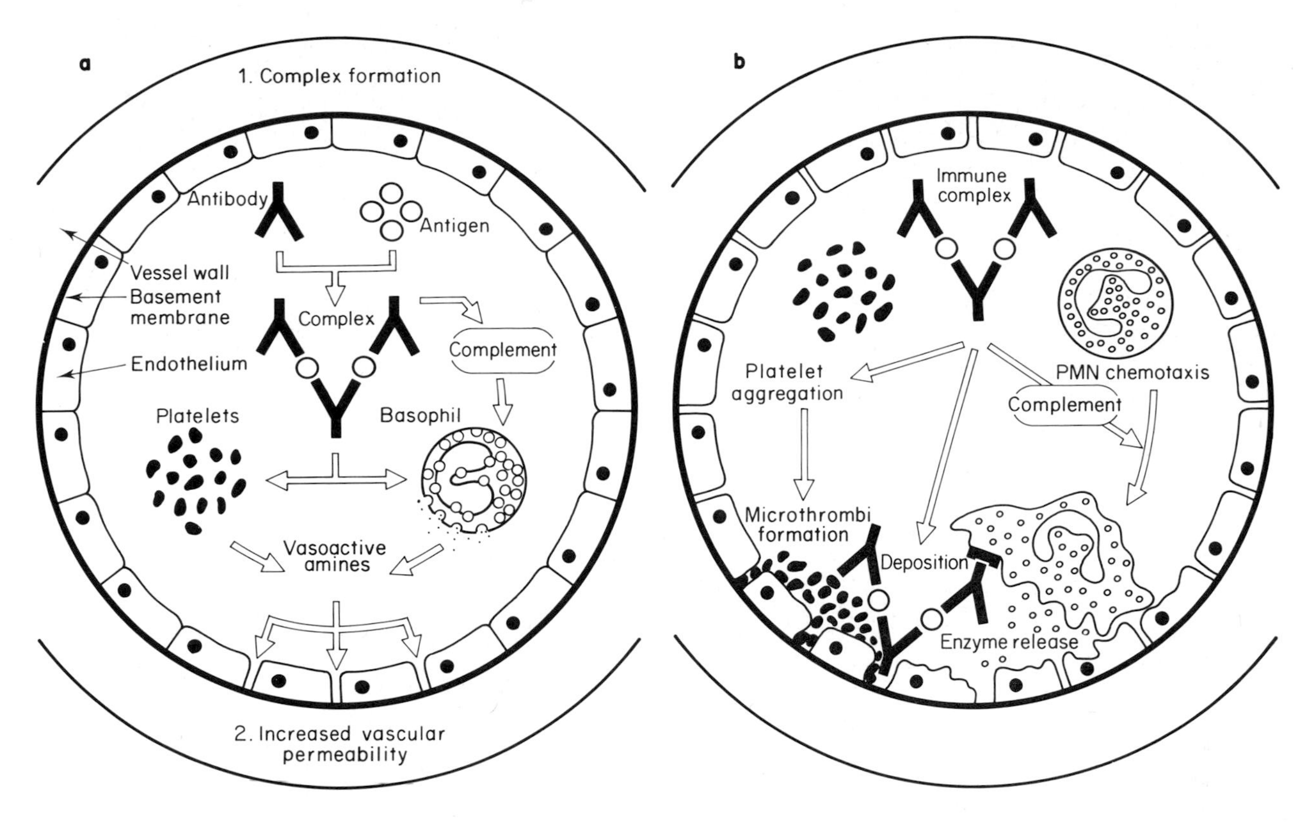

Fig. 5.5 Deposition of immune complexes in blood vessel walls.
a. Antibody and antigen combine to form immune complexes (1). The complexes act on complement (to release C3a and C5a), which in turn acts on basophils to release vasoactive amines. The complexes act directly on basophils to release vasoactive amines and also platelets (in humans) to promote amine release. The amines released include histamine and 5-hydroxytryptamine, which cause endothelial cell retraction and thus increased vascular permeability (2). **b.** With increased vascular permeability, complexes become deposited in the vessel wall. The complexes induce platelet aggregation and complement activation. The platelets aggregate to form microthrombi on the exposed collagen of the basement membrane of the endothelium. Polymorphs, attracted to the site by chemotactic complement peptides, cannot phagocytose the complexes and so release their lysosomal enzymes to the exterior of the cell causing damage to the vessel wall. Reproduced by permission of I Roitt and Churchill Livingston.[171]

Fifth, B-cells, some activated T-cells, and a variety of other cells, such as those of the glomerular epithelium, unconnected with the immune response, can bind immune complexes through Fc and/or complement receptors. The interaction of immune complexes with platelets via their Fc receptors may lead to platelet aggregation and microthrombus formation. There is a further increase in vascular permeability due to the release of vasoactive amines.

Attempts by attracted polymorphs to phagocytose tissue-entrapped immune complexes are often thwarted. The cells release their lyzosomal enzymes into the external milieu causing local tissue damage.

There are experimental models of immune complex disease; those of (NZB × NZW) F1 and MLR-*1pr*/*1pr* mice (see p. 598) are the most important in discussions of autoimmunity.

Persistence of immune complexes

In health, immune complexes are removed from the circulation by cells of the mononuclear phagocytic system (see p. 38) particularly those of liver, lung and spleen. Complex size is very important in determining clearance: large complexes are removed much more rapidly than small. The most important factor governing the clearance of large immune complexes is liver blood flow. Factors which affect immune complex size clearly influence their removal. A genetic defect, for example, leading to enhanced production of low affinity antibody, can lead theoretically to the formation of small immune complexes. Small complexes tend to persist in the circulation and these circumstances predispose to immune complex disease. Antibodies to self-antigen usually recognize only a few epitopes and this situation also favours the production of small immune complexes by limiting the formation of cross-links.

The physical form of immune complexes determines the extent to which complement is implicated in their degradation. Both complement (C3) and IgG are apparently necessary for the destruction of particular immune complexes but complement does not seem to be involved in the removal of soluble complexes.

In the presence of large amounts of immune complexes, the mononuclear phagocytic system may become overloaded. In experimental situations, *in vivo* blockade of the system can lead to persistence of immune complexes in the circulation, accompanied by deposition in glomeruli. However, in human immune complex disease it is difficult to determine whether defective clearance is a function of primary defect or secondary overload.

Tissue deposition

There are many known causes for the tissue deposition of immune complexes. The most important initiating factor may be increased vascular permeability. This increase can be initiated by a variety of mechanisms thought to vary in importance in different diseases. In experimental serum sickness and in autoimmune NZB/NZW mouse disease (p. 599), the long-term administration of vasoactive amine antagonists or platelet depletion have pronounced ameliorating effects on immune complex deposition. Deposition is also favoured in vascular sites where blood pressure is high or where there is turbulence; glomerular capillaries and arterial walls are particularly susceptible.

There is evidence that some immune complexes 'home' to particular organs e.g. DNA/anti-DNA complexes in systemic lupus erythematosus favour the kidney whereas immune complexes in rheumatoid arthritis, containing RFA, are rarely found there. In rheumatoid arthritis, immune complexes are generated in the synovium; large aggregates are prone to be deposited at that site. The deposition of immune complexes in particular regions of an organ such as the kidney also depends on the size of the complexes: large immune complexes do not pass through the glomerular basement membrane and are found in subendothelial sites; small immune complexes are found on the epithelial side of the basement membrane.

Lupus nephritis

In systemic lupus erythematosus (SLE) (see Chapter 14), anti-DNA antibody titres rise with the degree of activity of the renal disease and anti-DNA antibody can be isolated from the affected kidneys.[223] Reflecting the deposition of immune complexes, there is complement activation so that serum haemolytic complement levels fall in the active disorder. There are inconsistencies: for example, some patients may have renal disease but no untoward serological abnormalities and *vice versa*. One possible explanation is that immune complex formation occurs preferentially with antibodies of particular binding properties, or with a high ability to fix complement; another is that antibodies other than those reactive with DNA, are implicated in immune complex formation.

The interesting suggestion has been made that negatively charged, cross-reactive antigens such as heparan sulphate proteoglycan (see p. 198) can localize tissue damage by serving as 'planted antigens', binding anti-DNA antibody in glomerular capillary walls. Negatively charged chemical groups are now known to be important components of glomerular and probably of the majority of basement membranes. Proteoglycans, particularly heparan sulphate, are glomerular anionic sites, closely involved in barrier functions.

A mechanism for tissue damage in SLE nephritis may thus be envisaged: anti-DNA antibodies are produced as part of an abnormal SLE response towards entire nuclei. A fraction of these antibodies cross-react with negatively charged molecules in basement membranes, possibly with proteoglycans, and are deposited in glomerular tissue. The existence of this step has been established by experiments in mice in which it has been shown that anti-DNA antibodies in the glomerular eluates from SLE-affected (WEB × WEW) F1 mice, are restricted to alkaline antibodies with high isoelectric points. In SLE, anti-DNA antibody may, it seems, bind to anionic sites in glomerular basement membrane or to the glycosaminoglycans at cell surfaces in other tissues. Tissue damage can be expected to follow. The proposed mechanism may be by direct blockage of the anionic sites and/or by complement-mediated inflammation.

Theoretically, the proposed situation could be exacerbated by the 'autocatalytic' opening-up of new anionic sites in damaged tissue and by the binding of IgM and IgG rheumatoid factor to the membrane-immobilized anti-DNA, increasing immune complex size and complement fixation. Rheumatoid factor antibodies reactive with anti-DNA IgG have been detected in the sera of all diseased (NZB × NZW) F1 mice,[48] as well as in the sera of MRL mice. This mechanism resembles the two-phase experimental model of Masugi nephritis.[132,223] Masugi-type nephritis is initiated by challenging an animal with heterologous antibodies against renal antigen; it is further developed by the formation of host antibody against the heterologous immunoglobulin. Chains of events analogous with this model can account for some of the highly variable pathological changes in SLE nephritis.[40] If the existence of this mechanism is substantiated, new means of treating SLE nephritis and similar disorders may emerge.

Major histocompatibility complex

The human leucocyte antigen system

The human leucocyte antigen (HLA) system, part of the major histocompatibility complex (MHC) of genes, is located on the short arm of chromosome 6. The gene complex encodes cell-surface molecules which are intimately involved in the regulation of the immune response and which serve as target antigens in transplantation reactions.[209] Other products of the MHC region relevant to the connective tissue diseases include the complement components C2, C4 and Bf (p. 237). Complement polymorphism can influence antibody and immune complex-mediated reactions.

The HLA system is a closely linked series of polyallelic genes (Fig. 5.6). The specific combinations of the alleles at all loci within the MHC constitute the haplotype, inherited as a single co-dominant trait. This helps genetic analyses of serological reaction patterns and the demonstration of allelic relationships.

The genes HLA-A, -B and -C encode the class I antigens present on virtually all nucleated cells. They are detectable serologically with 'typing sera' from transfused or multiparous donors or with monoclonal antibody. Class I products comprise a heavy (α) chain of molecular weight 44 kD in non-covalent association with a light chain (β_2-microglobulin; mol. wt 12 kD) encoded on chromosome 12. The protein domain of an HLA molecule distal to the plasma membrane constitutes a groove or cleft, the dimensions of which have been shown to be approximately 2.5 by 1.0 by 1.1 nm. This groove, with a 'floor' comprising β-strands and 'walls' of α-helical strands, is the site of antigen binding and the MHC molecule and bound peptide form a composite ligand for the T-cell receptor.[16,17] The polymorphic residues are concentrated in this area and are responsible for the difference in binding specificity among different MHC molecules (Fig. 5.7).

The HLA class I molecules are the target antigens in transplantation reactions; they also act as restriction ele-

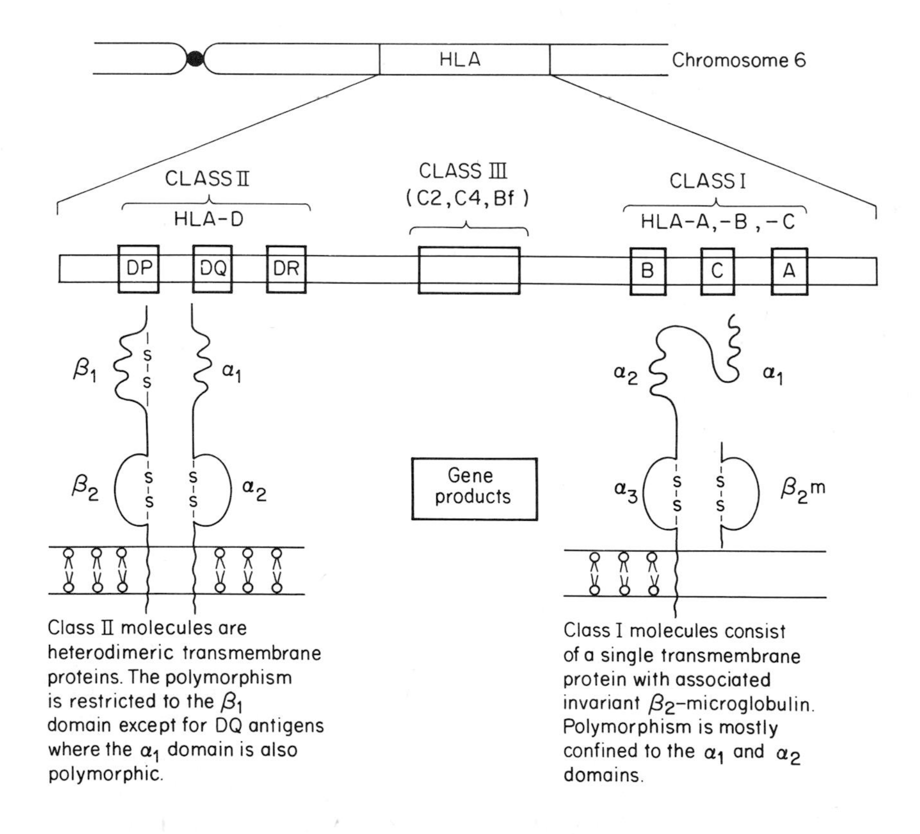

Fig. 5.6 Simplified genetic map and gene products of the HLA-region located on the short arm of chromosome 6.

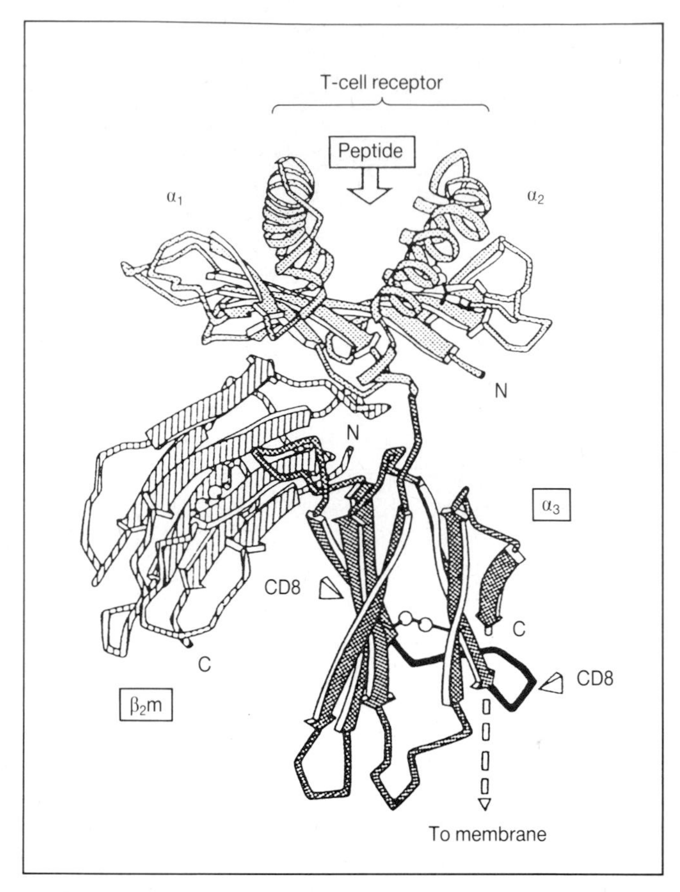

Fig. 5.7 Ternary complex of interaction between an MHC class I molecule, peptide and T-cell receptor on $CD8^+$ T-cells.
The beta strands of HLA-A2 are shown as arrows and the alpha helices as helical ribbons. The $alpha_1$ and $alpha_2$ domains (dotted area) form the peptide-binding site, with the groove in which the peptide binds and the face with which the T-cell receptor is believed to interact indicated at the top. The two immunoglobulin-like domains are beta-2-microglobulin (hatched area) and $alpha_3$ (heavily shaded area). The loop that has been implicated in the interaction with CD8 is shown as a solid line. The membrane anchor and cytoplasmic domain, which are on the carboxy terminal side of the $alpha_3$ domain are not shown. N and C denote the amino and carboxy terminals, respectively. A similar trimolecular complex is formed between MHC class II molecules, peptide and the T-cell receptor on helper T-cell ($CD4^+$) cells. Reproduced by permission of Bjorkman *et al.*[17] and the publisher.

ments in the recognition and killing of autologous virus-infected cells by cytotoxic T-lymphocytes (p. 229).

Class II antigens are encoded by three sub-loci (DP, DQ, DR) which comprise the HLA-D region (Fig. 5.6).[8,18] Their tissue distribution is more restricted than the class I antigens. Expression is mainly a property of cells of the immune system: B-lymphocytes, activated T-cells, macrophages and dendritic cells. However, the class II antigens can be induced in other tissues, including epithelia. Induction is by physiological mediators such as interferon-γ and hormones. Although inducible class II expression may be a normal property of epithelia, the phenomenon has been invoked in the induction of autoimmunity.[21] The cell surface molecules, consisting of heterodimeric heavy (α) and light (β) chain complexes, are structurally different from the class I gene products. Together with antigen presented on the surface of antigen-presenting cells, they stimulate T_H-cells to provide 'help' in the production of antibody and other functions (see p. 229). Products of the D locus are detected by a mixed lymphocyte reaction. Monoclonal antibodies to D-region allotypes as well as to monomorphic determinants are now available. Although each individual possesses two alleles for each gene, one maternally and the other paternally derived, the total number of specificities in the population is large.[18]

Complement component C4 is the most polymorphic of the HLA-linked complement components. There are two loci (C4A or C4F, and C4B or C4S); each expresses two co-dominant alleles. Eight different alleles exist at the F (electrophoretically *F*ast) locus and seven at the S (*S*low) locus. The incidence of null alleles amounts to 13–19 per cent of the total frequency. Linkage disequilibrium (see below) between HLA antigen and C4 alleles is often strong. Factor B has two common alleles, S and F, inherited codominantly. Rare alleles and null alleles also exist. There are six allotypes of C2. The most common occurs with a frequency of 94 per cent: a null allele is associated with the HLA A25, B18, Dw2 haplotype. Complement levels may be controlled by MHC genes other than the structural complement component genes. C3 is also polymorphic and encoded outwith the MHC, on chromosome 19. There are two common electrophoretic variants controlled by codominant autosomal alleles and about 20 rare alleles.

Subtle differences in the functional expression of the various C components may be important for the initial response to antigen, for the subsequent fate of immune complexes and for the efficacy of lysis.

Linkage disequilibrium

Major histocompatability complex (MHC) genes are inherited in accordance with Mendelian laws; recombination is rare. That certain genes exist together at frequencies significantly higher than the products of their individual frequencies (as predicted by the Hardy/Weinberg equilibrium), implies that some combinations of genes confer selective advantage. This is 'linkage disequilibrium'. One selective advantage could be the better ability of some phenotypes to withstand certain diseases. Much has been learnt from the mouse: immune response genes, which confer the ability to make good (or poor) responses to particular artificial antigens, have been mapped to a region equivalent to the DR region of humans.

Linkage disequilibrium complicates the interpretation of serological data, e.g. the susceptibility to SLE associated with HLA B8 DR3. However, a null complement allele is almost always found in linkage disequilibrium with B8 DR3 so that it is difficult to determine the relative importance of this complement 'abnormality', the HLA type, or both.

Immunogenetics of the connective tissue diseases

Ankylosing spondylitis

(see Chapter 19)

The first association between a disease and the inheritance of a specific HLA genotype was recorded by Brewerton.[25] This association was the link between ankylosing spondylitis (see p. 780) and the B27 antigen.[12] Pooled data for Caucasian populations demonstrated this HLA antigen in approximately 70 per cent of ankylosing spondylitic patients. Links have now been found between many other diseases and particular HLA specificities. Such a link is conveniently expressed as the *relative risk* for the development of the disease compared with a control population which does not display this specificity. In some American studies, possession of B27 carries a relative risk of 90, i.e., those who inherit B27 are 90 times more likely to develop ankylosing spondylitis than those who do not. No other HLA allele has proved to be so certain in the prediction of disease; many alleles are associated with disease susceptibility but with much smaller relative risks.

There are several possible explanations for the nature of the increased risk incurred by the inheritance of particular HLA genotypes. First, the presence of a particular allele may be related to the possession of 'poor' immune response genes directed to a specific causative agent. This is an attractive hypothesis when the disease is linked to an allele of the DR region. Similarly, an antigen such as B27 may be specified by a gene not in itself an immune response gene but possibly in linkage disequilibrium with another, as yet undisclosed, immune response gene. The B region maps close to the D (DR) region (Fig. 5.6).[209] Second, MHC class II antigens are important in the presentation of antigen by antigen-presenting cells to T_H-cells[67] and some haplotypes may be associated with poor presentation of antigen. Third, HLA antigens may act as receptors for particular pathogens, directly influencing susceptibility to a disease.[78,88] With the exception of B27 and ankylosing spondylitis, relative risks are, however, much lower than would be expected if this were the case. Other factors are undoubtedly involved. Fourth, diseases which correlate with the presence of a particular HLA allele may be caused by infectious agents bearing cell surface antigens similar to MHC products coded by that particular allele. Thus, an individual, tolerant to self-antigen, may be 'cross-tolerant' to an infectious agent. This 'molecular mimicry' has been seen with other pathogens, notably parasitic worms such as the Schistosomes. It has been suggested that cross-reactivity between the HLA B27 allele and some strains of *Klebsiella* may account for Reiter's disease (see p. 764).[97] Molecular mimicry has also been invoked in the causation of polymyositis/dermatomyositis (see Chapter 16).

Rheumatoid arthritis

(see Chapter 13)

There is much evidence that an immunogenetic element contributes to the development of rheumatoid arthritis.[66] First, haplotype sharing in affected sib pairs shows linkage of a susceptibility gene for rheumatoid arthritis to the HLA region. Second, the frequency of rheumatoid arthritis among monozygotic twins is increased compared to dizygotic twins one of whom is affected. Third, the HLA antigens, DR4 and Dw4 are highly disease-associated.[117] Moreover, seropositive juvenile rheumatoid arthritis is linked both to Dw4 and Dw14[149] and the prevalence of Dw14 in juvenile rheumatoid arthritis is very high (36 per cent) compared with the normal population (5 per cent).

C2 and C4 allotypes do not facilitate the identification of individuals 'at risk'. Although an unusual C4F variant has been reported in patients with rheumatoid arthritis, it is confined to a small subset of DR4 patients. Neither C3 allotypes nor C3b receptor numbers appear to be useful additional markers.

HLA-DR4 is a serological specificity present on multiple allelic products of a single gene within the HLA-DR-locus. Rheumatoid arthritis has been described as a Dw4-associated disease. However, recent DNA studies have disclosed an association between an uncommon DR4 genetic variant, Dw14, in six of seven rheumatoid arthritis patients homozygous for HLA-DR4.[150] These investigations have used allele-specific oligonucleotide probes capable of distinguishing between two closely related but distinct alleles encoding the HLA-DR4 specificity. Five of these patients were heterozygous for Dw4 and Dw14 genes, suggesting that features of both haplotypes may be important in determining susceptibility. The structural differences between Dw4 and Dw14 correspond to amino acids 71 and 86 of the 234 residue polypeptide. These differences trigger alloreactive recognition in the mixed lymphocyte reaction which defines the Dw4 and Dw14 phenotypes. There is an analogy to the variable region which accounts for immune response function changes in murine histocompatibility antigen. It is thought likely that this region of the DR-molecule underlines a susceptibility to rheumatoid arthritis.[126] Disease susceptibility is a reflection of the polymorphism of the HLA class II genes coding for the variable β chains which comprise one of the α helical walls of the groove (p. 241) in the HLA molecule. As Maini[126] points out, susceptibility to rheumatoid arthritis correlates with the amino acid sequence of the DRβ1 third hypervariable region from position 70 to 75 which falls within the α helical structure of the groove. Glutamine is present at position 70, arginine or lysine at position 71. The sequence is identified in HLA DR1 and in HLA DR4, subtypes Dw4, Dw14 and Dw15, all recognized with significantly increased frequency in patients with rheumatoid arthritis. Since the frequency of Dw14 is higher among $DR4^{+}$ patients with rheumatoid arthritis than in healthy subjects, the small nucleotide differences that distinguish the Dw14 allele may be critical.

New HLA restriction fragment length polymorphisms have been described in association with autoimmune diseases, including rheumatoid arthritis and insulin-dependent diabetes mellitus.[54] There is also a highly significant association of sodium aurothiomaleate-induced proteinuria with HLA-B8 and DR3 in patients with rheumatoid arthritis. This association is consistent with the hypothesis that this drug complication is caused by mechanisms involving complement in the renal glomeruli (Fig. 5.7).[226]

Systemic lupus erythematosus
(see Chapter 14)

Seventy per cent of patients with systemic lupus erythematosus (SLE) are female but the genetic implications of this observation have yet to be elucidated. Among the genetic factors implicated in the aetiology of SLE, the inherited complement (C) deficiency states are the most relevant.[31,55,168] Of these, C2 deficiency is the most frequent and SLE has been recorded in approximately 30 per cent of subjects with homozygous C2 deficiency. Partial C2 deficiency, observed in subjects with one C2 null allele (heterozygous C2 deficiency) may also be a risk factor. In addition, defects in C1 esterase inhibition, the classical components and the membrane attack complex components, have been reported.[178] The most consistent associations of HLA antigen with SLE in white populations involve A1, B5, B7, B8, DR2 and DR3. The significance of these associations is obscure; recent evidence suggests that they reflect linkage disequilibrium with other loci. HLA region-encoded complement polymorphisms of C2, C4A, C4B have been disclosed; the majority are located on MHC haplotypes encoding DR3. Earlier studies of HLA associations with SLE showed the strongest to be with HLA-B8 and DR3,[221] and HLA-B8 is known to be in strong linkage disequilibrium with the C4A null allele. Combined HLA and IgG heavy-chain markers (Gm allotypes) can be used to identify SLE patients at high risk for manifesting renal abnormalities (p. 584).[199]

Among the mechanisms by which complete or partial deficits of complement components predispose to SLE, are the persistence of immune complexes or of an infective agent, through defective clearance. Protracted immune stimulation may result.[178] However, other diseases presumed to be mediated by immune mechanisms are also associated with HLA-B8 or DR3 in white populations. These include chronic active hepatitis, idiopathic Addison's disease, Grave's disease, type I diabetes mellitus, gold-induced nephropathy and coeliac disease.[11]

The capacity to acetylate drugs with an aromatic amine structure exhibits a genetically controlled polymorphism. Slow acetylation is a consequence of homozygosity of the recessive allele. Rapid acetylators are coded for by an autosomal dominant gene. Hydralazine-induced lupus[118] (see p. 594) is an example of how different genetic markers enable accurate forecasts of disease susceptibility to be made. HLA-DR4 females are almost certain to develop the complication if they are slow acetylators; DR4-negative, rapid acetylator males display no risk.[11]

Progressive systemic sclerosis

Immunogenetic studies have revealed a familial association between progressive systemic sclerosis (see p. 694) and systemic lupus erythematosus (SLE); an autoimmune component (see p. 229) is more prominent in the latter than in the former. Early studies suggested an association between HLA-B8 and the haplotype B8/DR3;[82,94] both are associated with a variety of autoimmune diseases. This view has been challenged. Nevertheless in the largest series studied, an increased frequency of DR3 has been observed in patients with progressive systemic sclerosis and pulmonary fibrosis[124] and there is association with the A9/DR3 haplotype.[49] Although progressive systemic sclerosis is characterized by increased deposition of collagen in skin and viscera it is unlikely that the genetic influence in this disease is on account of any MHC control of anti-collagen responses.

Other genetic markers evaluated in the connective tissue diseases include α-1-protease inhibitor and superoxide dismutase.[117]

Immune mechanisms in common disorders of the connective tissue system

The extent to which the immune system determines the aetiology and influences the pathogenesis of individual diseases of the connective tissue system varies greatly. In SLE (see Chapter 14), the prototype autoimmune disorder, the diversity of abnormalities of both cellular and humoral immune function is pathognomonic. At the other end of the spectrum of immunological anomaly it is clear that immune disorders play only small parts in the pathology of primary gout (see p. 380) and idiopathic osteoarthrosis (see p. 843).[152]

Rheumatoid arthritis

Abnormalities of the immune system

Most contemporary data accord with a view that rheumatoid arthritis (see Chapters 12 and 13) is a chronic T-lymphocyte/macrophage-dependent response to foreign antigen or to autoantigen expressed on synovial tissue.[74,101,126] Rheumatoid arthritis is often perceived as an immune complex disease because of the frequency with which IgM-anti-IgG rheumatoid factors (RFA) are detected. However, the formation of RFA in many forms of persistent human and animal infections and the possibility of their induction experimentally by repeated immunization, suggest that they take part in normal immunoregulatory responses or in mechanisms of inflammation amplification (see p. 284). In rheumatoid arthritis, IgM-anti-IgG complexes could reflect the persistence of an unidentified microbial agent which might be a consequence rather than a cause of the arthritis. That rheumatoid arthritis may develop in patients with hypogammaglobulinaemia strongly suggests that cell-mediated, not humoral immune phenomena play relatively important parts in the natural history of the disease.

Inability to identify foreign antigen

Attempts to identify a disease-specific antigen in rheumatoid arthritis have been unsuccessful. Many bacteria, mycoplasmas, and viruses have been implicated without proof of their pathogenicity (see p. 531). Rheumatoid synovial tissue, tissue extracts or synovial fluid stimulate autologous peripheral blood leucocytes to a variable degree, but the interpretation of the data is complicated by the coexistence of non-specific stimulating factors, immune complexes and lymphokines which are often present in crude preparations as are commensal bacteria such as *Corynebacteria* spp. Attention has been devoted to the significance of immune responsiveness to Epstein–Barr virus infections (see p. 532). An unusual antibody directed against a unique, Epstein–Barr virus associated antigen (rheumatoid arthritis-associated nuclear antigen-RANA) is detectable with greater frequency in rheumatoid arthritis patients than in healthy controls;[5] Epstein–Barr virus 'early antigen' is found in approximately 30 per cent of rheumatoid synovia.[216]

Epstein–Barr virus (EBV) causes peripheral B-lymphocyte transformation more rapidly in rheumatoid patients than in healthy subjects.[10] This response is secondary to a T-cell deficiency: rheumatoid patients generally lack the peripheral blood suppressor T-cell function for EBV-induced, B-cell, immunoglobulin synthesis that is detectable in normal individuals.[211] T-cell induced retardation of the outgrowth of EBV-infected B-cells is mediated by interferon.[208] Interferon production by T-cells from rheumatoid patients is impaired under certain conditions, e.g. in the autologous mixed lymphocyte reaction. It has been suggested that this diminished activity could contribute to a defective regulation of EBV-lymphoblast transformation by patients' lymphocytes.[75] The defect in interferon production may be related to the increased susceptibility of rheumatoid patients' lymphocytes to proteoglycan. Deficiency of interferon-γ production could also be secondary to decreased interleukin-2 formation, a property of rheumatoid lymphocytes infected *in vitro* with EBV.[215] There is cumulative evidence for an abnormal immune response to EBV in rheumatoid arthritis although it falls short of directly implicating this virus in the aetiology of the disease.

Disturbances of humoral immunity

There are provisional data to show heightened B-cell function in patients who are anergic, as determined by lack of T-cell responsiveness *in vitro* to soluble 'recall' antigen to which patients could be presumed to be sensitized. The hyperactivity is expressed as increased spontaneous immunoglobulin production.[46] Although B-cell hyperfunction with raised antibody titres to immunoglobulin, rheumatoid factor, nucleic acid and collagen[201] have often been reported, hypergammaglobulinaemia has not been correlated previously with T-cell aberrations in rheumatoid arthritis. This apparent inversion in T-cell/B-cell function has been described in other chronic inflammatory diseases, including leprosy[147] and AIDS.[114] The co-existence of T-cell dysfunction and B-cell hyperfunction may represent a migration of activated mononuclear cells from the synovial fluid to the peripheral circulation. There may also be selective homing, possibly regulated by a distinct lymphocyte-endothelial recognition system.[89] The T_H-cells may be retained locally, at sites of synovial inflammation, causing depletion of this lymphocyte subset from the blood.

Defective glycosylation

Other abnormalities of the humoral immune mechanism have been identified. It has been proposed that defective glycosylation of IgG could be of importance[154] although this has been disputed. The defect could contribute to the pathogenesis of rheumatoid arthritis by heightened IgG autoantigenicity.[186] Alternatively, the persistence of the abnormal IgG could be an epiphenomenon.

Glycosylation of IgG takes place at the CH2 domain of each gamma chain (Fig. 5.8). In healthy individuals the majority of IgG molecules are galactosylated in this way, although some always lack Gal. The proportion of agalactosyl oligosaccharides is age-related, the number increasing significantly between the ages of 1 and 15 years and between 40 and 70 years. Even allowing for variations in the number of agalactosyl oligosaccharides with age, it is clear that patients with adult and with juvenile rheumatoid arthritis have significantly higher numbers of these agalactosyl oligosaccharides than healthy persons.[155] Moreover, in individuals with adult rheumatoid arthritis, the number of such oligosaccharides has been related to the extent of the disease. However, the phenomenon is not restricted to connective tissue disease and in a large screening of sera, significantly increased numbers of such oligosaccharides were found in patients with Crohn's disease or with tuberculosis but not with osteoarthrosis as had been previously reported.[154] Defective galactosylation of the oligosaccharides at the CH2 domains of IgG may reveal amino acids which would normally be hidden by the sites of longer oligosaccharides. Monoclonal antibodies, obtained by immunizing mice with a peptidoglycan/polysaccharide complex from group A streptococci, bind to agalactosyl IgG from rheumatoid arthritis patients.[173] This observation supports the argument that agalactosyl IgG contains epitopes that are undetectable on normal galactosylated IgG and that abnormally galactosylated IgG might thus contribute to the induction of RFA in rheumatoid arthritis, systemic lupus erythematosus and other diseases.[210]

An increase in the number of agalactosyl oligosaccharides has been attributed to decreased activity of the enzyme β-galactosyl transferase which catalyses the transfer of uridine diphosphate-galactose to the asialo-agalactosyl oligosaccharide of IgG. The specific activity of this enzyme, isolated from the B-cells of patients with rheumatoid arthritis, is significantly lower than that of the enzyme obtained from healthy individuals.[7]

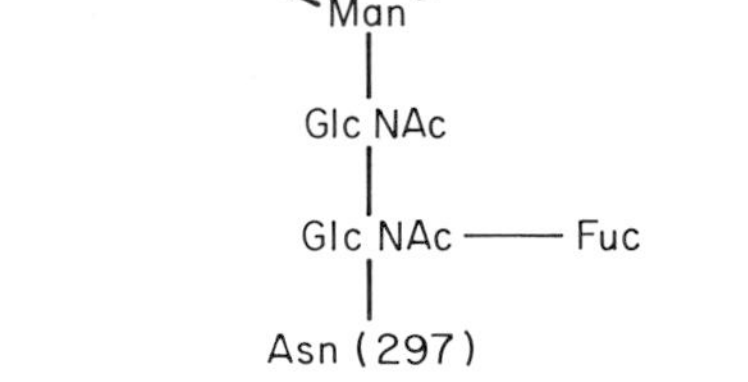

α1-3 arm
α1-6 arm
Neu 5Ac
Neu 5Ac
Gal
Gal
Glc NAc
Glc NAc
Man
Glc NAc
Man
Man
Glc NAc
Glc NAc
Fuc
Asn (297)

Fig. 5.8 Oligosaccharide side chains of human IgG. Individual CH2 domains are linked to the oligosaccharide comprising a 'core' chain itself linked by N-acetyl glucosamine (Glc NAc) to the asparagine (Asn) residue at position 297. The penultimate saccharide of the core is mannose (Man) and this molecule is linked to two oligosaccharide 'arms' via α1-3 and α1-6 linked residues. A single galactose (Gal) residue is found on each arm of one of the oligosaccharides attached to one of the CH2 domains and on the α1-5 arm only of the second oligosaccharide. In healthy subjects the majority of IgG molecules are galactosylated in this way, although some always lack Gal. Neu 5Ac = sialic acid; Fuc = fucose.

B-cell activation

The $CD5^+$ B-cell subset is enlarged and activated in rheumatoid arthritis patients. The cells secrete several antibodies including polyreactive, low-affinity RFA (cross-reactive with single-stranded DNA, thyroglobulin and insulin among other antigens) and high-affinity monoreactive RFA. This latter activity is not detected in $CD5^+$ B-cells from normal controls.[26] $CD5^+$ cells could also have a role in antigen presentation within the inflamed joint.

Disturbances of cell-mediated immunity

The evidence implicating non-specific aberrations of cellular immune function in the pathogenesis of rheumatoid arthritis is more substantial for synovial fluid than for peripheral blood where studies of mitogen responsiveness, of T_H:$T_{C/S}$ cell ratios and suppressor cell function have not always been consistent.

There are more activated T-cells (DR^+) in the peripheral blood of rheumatoid patients than of normal subjects.[46,229] Active disease correlates positively with DR^+ cells and anti-DR antibody and negatively with anti-idiotypic antibodies[183] suggestive of network regulation. There is impairment of the response to, and the absorption and production of interleukin-2,[136] features which do not correlate with the activity or duration of the disease. There is nevertheless, a relationship between synovial immunopathology (see p. 453) and the responsiveness to 'recall' antigen.[46,129] Patients with extensive lymphocytic infiltration are often anergic, as determined by lack of T-cell responsiveness to soluble recall antigens *in vitro*; those with few infiltrating lymphocytes are not. The infiltrates of the former resemble the changes of cell-mediated hypersensitivity (see p. 236): there are T-cells, plasma cells and macrophages and the cells are often aggregated. By comparison with the latter group, there is greater synoviocytic hyperplasia, more DR^+ cells and higher T_H-/T_S-cell ($CD4^+/CD8^+$) ratios. Lymphoid cells are often juxtaposed to DR^+ accessory cells in the subsynoviocytic tissues. These local conditions may favour the generation of factors which support Ig production (cf Fig. 5.1). They also suggest an interrelationship between the peripheral blood response to antigen and the traffic of activated mononuclear cells between the synovial compartment and the periphery.

The cellular immunological abnormalities of anergic patients can be reversed when leucophoresis is used to cause mild lymphopenia. Part of the explanation may be a migration of activated T-cells from the synovium to the circulation. Responses to recall antigens return to normal levels and there is a transient decline in disease activity. Non-anergic patients with synovial tissue of low cellularity do not show this clinical improvement.[219] These observations must be interpreted in the light of an awareness that rheumatoid arthritis patients with similar clinical signs, display a wide spectrum of immunopathological features. It is not known whether anergic and non-anergic patients reflect different stages of a single disease entity, different types or amounts of initiating stimuli, or different types and degrees of host response.

Estimates of non-specific cellular immune function within the synovial tissue apparently disclose more consistent changes than do peripheral blood determinations. Despite interpatient variability, it appears that T-lymphocytes predominate.[217] By several criteria, a substantial proportion of synovial fluid T-cells are in an activated state, i.e. they express antigen DR and the receptor for interleukin-2; this may account, in part, for their poor responsiveness to mitogens and in the autologous mixed lymphocyte reaction.[190] Most of these T-cells suppress the α and β chains of the T-cell receptor. However, T-cells expressing $\gamma\delta$ chains are also present.

In situ activation of T-cells is also most likely to be reflected in the changes observed in CD4 subsets: $CD4^+$ cells of the 'helper inducer' ($4B4^+$) phenotype are significantly increased, whereas those of the 'suppressor inducer' ($2H4^+$) phenotype are substantially reduced relative to peripheral blood.[116] 2H4 (CD45R) is a marker for the stage of differentiation of $CD4^+$. $2H4^+$ T-cells are virgin T-cells while $2H4^-$ T-cells are memory T-cells.

Synovial lymphocytes from rheumatoid arthritis patients exhibit greater expression of cell adhesion molecules [leucocyte function-associated 1(LFA-1) and very late activation-1(VLA-1) antigens] than peripheral blood lymphocytes.[41] These molecules are likely to be important for the binding of T-cells from inflamed synovium to fibroblast-like synovial cells.[76]

The presence of increased numbers of $CD8^+$ cells in the synovial fluid of rheumatoid patients[113] is not associated with increased T_S-cell function. Indeed, it appears that rheumatoid synovial fluid lymphocytes not only exhibit *decreased* suppressor cell activity, but may actually augment *in vitro* immunoglobulin production in concert with low numbers of T_H-cells.[172] The cytotoxic lymphocytes of synovial fluid appear to have a different target cell repertoire from 'classical' natural killer cells; their behaviour may be analogous to the lymphokine-activated killer cells observed when peripheral blood lymphocytes are cultivated in the presence of interleukin-2. Activated synovial fluid T-cells may arise by *in vivo* antigen stimulation; they are a potential source of lymphokines contributing to exacerbations of the disease (see Fig. 5.1 and below).[43,230]

Investigations of cellular immune reactivity toward self-antigen that might play a part in initiating and/or perpetuating rheumatoid arthritis have concentrated on the role of collagen and on determinants on light chains and/or the Fab fragment of IgG. The study of experimental arthritis caused by a response to collagen and incomplete Freund's adjuvant, gives some support to suggestions that reactions to collagen may contribute to the pathogenesis of rheumatoid arthritis (see p. 547). However, experiments made to measure collagen sensitivity in humans have produced inconclusive results.[152] In terms of T-cell responses to immunoglobulin determinants, it appears[165] that peripheral blood T-cells from a large majority of rheumatoid patients are activated on exposure to pooled IgG; control peripheral blood lymphocytes are unresponsive. The responding cells 'recognize' determinants on monoclonal light chains and/or Fab fragments, with preferential responses to either kappa or lambda light chains or equal responses to the respective light chains. The responding T-cells may influence the generation of rheumatoid factors by cooperating with B-cells specific for determinants on Fc fragments.

The role of cytokines

Cytokines (p. 291) can modulate the growth and function of connective tissue.[181,218] Hyperplasia of synoviocytes and fibroblasts, in association with the appearance and persistence of lymphocytes, is an early, characteristic feature of rheumatoid arthritis synovitis.[112,176] Synovial tissue with its associated mononuclear cells, spontaneously generates mediators of fibroblastic growth,[130] suggesting that cell: cell interactions are involved. They may promote and control the function of pannus. The release of soluble factors by the inflammatory cells mediates synovial proliferation, contributing to the high levels of collagenase and eicosanoids (see Chapter 7) generated in synovitis.[44] In turn, these enzymes catalyse connective tissue destruction. There is nearby erosion of bone, cartilage and soft connective tissue (see p. 467).

The amplified population of synovial fibroblasts can be further modulated by the monocyte product interleukin-1 (IL-1)[151,225] to release enhanced amounts of prostaglandins and collagenase.[138] There is an association between IL-1 production and the severity of inflammation as assessed arthroscopically and radiographically. Analysis of synovial cell mRNA reveals both IL-1α and IL-1β as well as tumour necrosis factor-α (TNF-α) and TNF-β (lymphotoxin). Both cytokines are detected in synovial cell tissue culture fluids[22a]. Interleukin-1β induces enhanced secretion of prostaglandin E_2, thromboxane B2 (TxB_2) and 6-keto prostaglandin $F_{1\alpha}$ by rabbit chondrocytes.[35] Induction of prostanoid synthesis may occur via activation of phospholipase A_2.[60]1 Suppression of chondrocyte growth by IL-1 is also observed in the rabbit model; in rheumatoid arthritis such an effect could interfere with cartilage repair mechanisms.

Other cytokines detectable in the synovial fluid of rheumatoid arthritis patients include IL-6, macrophage colony-stimulating factor (M-CSF) and transforming growth factor β (TGF-β). Interleukin-2 and IL-3 are absent and interferon-γ is present only at minimal levels. This situation is difficult to reconcile with the notion that rheumatoid arthritis is predominantly T-cell driven; more recent evidence is consistent with a disorder which is predominantly macrophage driven.[56] Neuropeptides also induce inflammatory cytokines in monocytes.[123] Interleukin-1 can also stimulate plasminogen activator from synovial fibroblasts, resulting in the formation of plasmin. This enzyme can degrade connective tissue elements.[73] Activated monocytes also secrete collagenase and prostaglandins, extending tissue destruction. Synovial lymphocytes and monocytes consequently generate mediators of fibroblast growth which enhance the production of enzymes that destroy connective tissue, and mediators that augment inflammation.

Inflammatory cells, in close proximity to which cartilage undergoes substantial loss of proteoglycan and collagen, provide an alternative mechanism to chondrocyte-mediated proteolytic enzyme destruction. The active participation of polymorphs has been demonstrated[139] but the significance of these cells in disputed. In addition to their cartilage-erosive properties, inflammatory cells probably also influence subchondral bone resorption. Monocytes and fibroblasts can mediate this change, through the elaboration of prostaglandins[145] and activated T-cells secrete osteoclast-activating factor (OAF) (see p. 32), with similar properties.[81] Interestingly, production of OAF by T-cells is controlled by monocyte prostaglandin,[227] imputing to the latter a central role in bone resorption, either directly or by modulation of osteoclast activity. Thus, hyperplasia of synovial tissue as well as the production and release of the degradative enzymes that contribute to tissue destruction and its perpetuation may all be modulated by factors generated by mononuclear cells.

As previously discussed the cytokine-driven chronic synovial inflammatory reaction of rheumatoid arthritis may be a response to foreign or to autologous stimuli. Among the exogenous agents that have been considered are corynebacteria, the mycoplasmas (see p. 731), rubella (see p. 727), Epstein–Barr virus (see p. 724) and *Borrelia burgdorferi* (see p. 741). However, none has been established as a cause of rheumatoid arthritis. There is preliminary evidence that the oncogenes *myb, myc, ras* and *fos* are activated in rheumatoid synovial cells but rarely in osteoarthrosic synovial cells.[65]

Clonal analyses of T-cells from the synovial tissue and synovial fluid may help to determine whether there are any strictly antigen-specific responses in rheumatoid arthritis.[196] T-cells bearing $V_{\beta}5$ or $V_{\beta}8$ T-cell receptor gene products are occasionally increased in synovium relative to peripheral blood suggesting restricted clonal diversity in rheumatoid synovium.[22] However, other studies based on analysis of restriction fragment length polymorphisms of the T-cell receptor beta chain genes, have failed to reveal a dominant TCR β gene rearrangement.[100] This argues for polyclonality of the T-cell response.

Currently much interest is focussed on heat shock proteins (hsp) as potential autoantigens in rheumatoid arthritis.[96] T-cells from synovial fluid respond to mycobacterial hsp 65, though to a lesser extent than the synovial T-cells of patients with reactive arthritis. In one patient with acute arthritis T-cell recognition was mapped to an epitope which is not conserved between bacteria and humans. Thus T-cells reactive to hsp 65 that contribute to autoimmunity need not necessarily be directed against epitopes common to bacteria and host. They may also comprise epitopes unique to microbes. T-cells recognizing hsp 65 express γδ receptors.[18a]

Vaccination of rheumatoid arthritis patients with subpathogenic doses of T-cells reactive with autoantigen(s) could theoretically ameliorate autoimmunity through the generation of anti-clonotypic T-cells which inhibit the patient's autoreactive T-cells.

Although the nature of the critical antigen(s) to which

T_H-cells are exposed within a joint remains to be clarified, the following summary hypothesis may account for the local, clinical features of rheumatoid arthritis. After stimulation by DR^+ accessory cells such as macrophages, dendritic cells, and by B-cells, T_H-cells proliferate. They release lymphokines into the microenvironment, contributing to B-cell hyperactivity. The lymphokines amplify and perpetuate the inflammatory response (Figs 5.1; 5.9). A variety of cytokines is known to be present in the synovia.[22a,134] They are predominantly products of activated macrophages. Many of these cytokines act on other macrophages to stimulate increased phagocytic activity and the secretion of biologically active mediators including chemotactic factors and inflammatory proteins. Cytokines also contribute to the recruitment and activation of inflammatory cells. T-cell lymphokines stimulate macrophages to secrete collagenase and plasminogen activator and these enzymes may contribute to tissue damage and to the vascular phenomena of synovitis.

Animal analogues

A great variety of experimental models can be compared with rheumatoid arthritis and used to throw light on this disease (see p. 542). Type II collagen-induced arthritis and bacterial cell wall-induced arthritis are of special interest. Most of the anticollagen activity in rheumatoid arthritis is directed against covalent structural determinants present on denatured collagen, suggesting that they develop secondarily in response to tissue injury. Antibodies to bacterial cell walls are common in normal subjects, as well as in rheumatoid patients.[201] Their association with rheumatoid arthritis is thus neither evidence for nor against an aetiological involvement.

Experimental type II collagen-induced rat arthritis (see p. 547) is seen as an antibody-mediated and complement-dependent model disease: antibodies to native type II collagen bind to articular cartilage collagen, activating complement and initiating the lesion.[98,200] Bacterial cell wall-induced arthritis (see p. 543), on the other hand, develops after cell walls have been deposited in synovial tissues where they persist. The initial acute inflammatory reaction reflects a pro-inflammatory toxic effect of the cell walls; the development of chronic disease results from a T-lymphocyte reaction to the cell walls, or even to autologous antigen.

Systemic lupus erythematosus

Systemic lupus erythematosus (SLE) (see Chapter 14) is characterized by the disturbed regulation of immunoglobu-

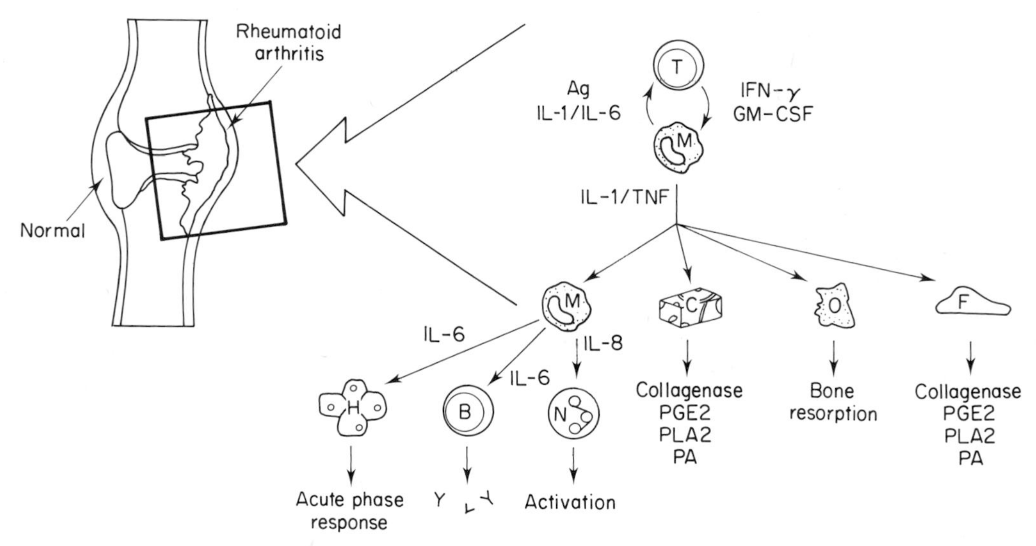

Fig. 5.9 Schematic representation of inflammatory cell-mediated destruction of synovial connective tissue.[46] Macrophages (M) and other antigen-presenting cells process and present antigen (Ag) to T-lymphocytes (T) in conjunction with interleukin-1 (IL-1) and IL-6. These activated T-lymphocytes produce several cytokines (e.g. interferon-γ, IFN-γ) and colony-stimulating factors (e.g. GM-CSF) that recruit and activate macrophages. Activated macrophages secrete high levels of IL-1 and tumour necrosis factor (TNF) which induce chondrocytes (C) and fibroblasts (F) to produce prostaglandin E2 (PGE2), collagenase, phospholipase A2 (PLA2) and plasminogen activator (PA) which contribute directly to connective tissue destruction. They also produce IL-6 (B-cell stimulation/differentiation factor) which contributes to plasma cell infiltration and autoantibody production including rheumatoid factors and which is a major regulator of acute phase protein synthesis in hepatocytes (H). Another important cytokine released by activated macrophages is IL-8 which is chemotactic for, and an activator of, neutrophils (N) (IL-1 and TNF possess no intrinsic chemotactic activity). Activated T-lymphocytes produce TNFβ which activates osteoclasts (O) to induce bone resorption and fibroblasts to proliferate. Thus, mononuclear cell signals may modulate synovial cell hyperplasia and the release of enzymes and mediators that are responsible for the tissue destruction characteristic of rheumatoid arthritis.

lin production. Increased levels of common antibodies and a wide spectrum of autoantibodies result. To elucidate this hyperactive state, the function of immune cell populations involved in immunoglobulin expression has been analysed. These cells include the antibody-producing B-cells, the T_H-cells and T_S-cells which modulate B-cell activity, and the macrophages and other accessory cells responsible for antigen presentation and for initiating an immune response. The interactions between these cells are complex and several immunoregulatory defects can lead to the same disorder. Thus, intrinsic B-cell disturbance, decreased T_S-cell function and increased T_H-cell function can all promote immunoglobulin production. It is probable that B-cell hyperactivity in SLE is caused by more than one mechanism.

Murine systemic lupus erythematosus-like diseases

In mice, several different inbred strains (NZB, NZB/W Fl; MRL-*lpr*/*lpr*; BXSB) exhibit a disease-state characterized by B-cell hyperactivity, antinuclear antibody production, immune complex deposition and glomerulonephritis (see p. 598). The strains differ in important clinical features. This heterogeneity is reflected in varied patterns of genetic control, with the genes responsible for disease differing among the various strains. In NZ mice, the situation is especially complicated; different functional serological abnormalities are determined by independent genetic defects. By contrast, the *lpr* gene can be seen as a single-gene model for autoimmunity.[6] MRL-*lpr*/*lpr* mice spontaneously develop an autoimmune disease manifest as arthritis, vasculitis, immune complex glomerulonephritis and autoantibody production. They also exhibit profound lymphoproliferation. In the face of massive T-cell growth *in vivo*, *lpr* spleen and lymph node cells display paradoxical and severely impaired interleukin-2 production; they proliferate poorly in response to mitogens. It is therefore believed that abnormalities of T-cell function play a fundamental aetiologic role in the autoimmune disease of MRL *lpr*/*lpr* mice.[175]

The regulatory action of the *lpr* gene on mouse autoantibody production is apparently generalized since strains of other genetic backgrounds with *lpr* also develop high levels of anti-DNA. However the latter do not succumb to the severe MRL-*lpr*/*lpr* nephritis. This suggests that, while a single gene may be sufficient for autoantibody production, other genes or factors predispose to renal disease.[86]

The study of cellular immunity in these mice has revealed similarly diverse abnormalities.[192,198] NZB mice manifest disturbances of both B- and T-cell populations, providing at least two mechanisms for B-cell hyperactivity. By contrast, MRL-*lpr*/*lpr* mice exhibit intact B- and T_S-cell function; the mouse disease is caused by excessive T_H-cell function, driving otherwise normal B-cells to high levels of antibody production. One hypothesis to account for this behaviour suggests that the T-cell subset which recognizes self MHC class II (Ia) antigen is stimulated to produce lymphokines; in turn, the lymphokines induce further Ia expression on non-T-cells and proliferation of the T-cell pool. This cycle ultimately results in lymphadenopathy and stimulation of autoantibody and IgG production by B-cells.[175]

The different behaviour of mouse strains is reflected in their diverse responses to immunotherapeutic manipulation. For example, thymectomy ameliorates disease in MRL *lpr*/*lpr* mice but not in NZB mice. These findings suggest that the nature of the underlying immunoregulatory disturbance determines not only the pattern of disease but also the response to treatment.

Human systemic lupus erythematosus

Systemic lupus erythematosus in humans is also characterized by B-cell hyperactivity in association with disturbances of T-cell immunoregulation.[24,45,52,228] Paradigms from NZB mice have concentrated on the demonstrable impairment of T-cell mechanisms. However, it is now widely appreciated that the T-cell defects in SLE are heterogeneous.[179,193] No consistent abnormality is detectable in T-cell subsets in respect of phenotypic markers or function. The immunoregulatory aberrations may be dynamic, varying with the severity of the disease. It is important to note that these abnormalities can be found in disorders other than SLE, such as primary biliary cirrhosis and multiple sclerosis.

Patients with active SLE are frequently lymphopenic: their lymphocytes are hyporesponsive to mitogens and to skin-test antigen.[152] An increased proportion of peripheral T-cells is activated,[229] sometimes in association with enhanced numbers of B-cells spontaneously secreting immunoglobulin.[37] T_S-cell function is defective and there is a selective deficit of lymphocytes of $CD5^+/8^+$ phenotype in patients with active disease in comparison with those with inactive disease and healthy controls.[142] The autologous mixed lymphocyte reaction is depressed in active disease[180] and the cytotoxic functions of T lymphocytes and natural killer (NK)-cells impaired.[34,95] The defect in NK activity may be intrinsic: SLE NK cells are hyporesponsive to interferon.[212] This may be mediated by antilymphocytic autoantibody[64].

Monocyte/macrophage function is normal in SLE but there is evidence for decreased tissue clearance of immune complexes, possibly associated with impaired Fc-receptor expression.[58] Since much tissue damage in SLE is attributed to the deposition of immune complexes, this defective clearance, if it exists, may potentiate the disease.[103a]

Many, if not all, T-cell defects in SLE are believed to be

secondary to the antilymphocyte antibodies directed against T-cells.[224] There are several major types of antilymphocytic antibody, with specificities for different lymphocyte subpopulations but the full degree of heterogeneity is not yet known. Those antibodies most fully studied are the cold reactive lymphocytotoxic antibodies which vary in titre with the activity of disease in the same way as anti-dsDNA antibodies.[27] High levels are associated with lymphocytopenia. This indicates a probable role for antibody in T-cell depletion and in the alteration of subset ratios. However, the relevant structural and functional molecular relationships have not yet been elucidated.

Antilymphocyte antibodies are mostly IgM and are detectable by indirect immunofluorescence and complement-dependent cytotoxicity assays. Antibodies to T-cell subsets are of great interest because of their probable role in the pathophysiology of SLE. They are also of importance because of their capacity to reproduce the immunoregulatory abnormalities characteristic of SLE lymphocytes, by effects exerted on normal cells *in vitro*. The main targets are cells which mediate suppressor function ($CD8^+$)[180] and the T_H-cell subset (suppressor-inducer cell) required for the function of the T_S-cell population.[141] There are also antibodies specific for activated T-cells that have the ability to diminish the expression of interleukin-2 receptors and, in this way, the responsiveness to antigen- or mitogen-induced cell proliferation. These antibodies therefore have the potential to control both the amplification of specifically reactive T-cells and the induction of lymphokines. This suggests that the behaviour of T-cells in SLE is impaired in the presence of antilymphocyte antibody, altering the mechanism of interleukin-2 production, but that the function of these lymphocytes may be intrinsically normal.[189]

In addition to complement-mediated lysis, antilymphocytic IgG antibodies may co-operate in antibody-dependent cellular cytotoxicity to bring about the lysis of T-cells by non-T-cells. Additional non-cytotoxic mechanisms by which antilymphocyte antibodies may affect the immune system include the modulation of cell surface determinants and ligand/receptor triggering.[224]

Progressive systemic sclerosis (scleroderma)

In progressive systemic sclerosis (see Chapter 17) the pathological evidence supports a role for interactions between immune, vascular and connective tissue processes.[29,156,158] In the skin and, to some extent, in the viscera, perivascular mononuclear cell infiltrates are prevalent.[77] In the early stages of the disease the dense inflammatory cell infiltrate, which contributes to skin thickening,[177] is associated with deposition of newly formed Type III collagen. Later, there are fewer inflammatory cells and more mature (Type I) collagen.[57] Inflammatory cells play a role in *in vivo* collagen deposition; mononuclear cells *in vitro* secrete factors which stimulate collagen synthesis by fibroblasts (see p. 302).

The view that vascular and fibrotic features of progressive systemic sclerosis have an immunological basis is further strengthened by observations on graft-versus-host disease[197] in animals and humans. Dermal fibrosis similar to that of progressive systemic sclerosis develops after neonatal rats are tolerized to allogeneic bone marrow cells. The animals succumb to a graft-versus-host reaction characterized by cutaneous mononuclear cell infiltration. Comparable observations have been made in humans grafted with bone marrow.[79] In some long-term human survivors, the cutaneous changes of progressive systemic sclerosis are accompanied by the development of Sjögren's syndrome and, in a smaller proportion, of Raynaud's phenomenon.[59] There is early pulmonary and renal arterial proliferation and pulmonary interstitial inflammation and fibrosis.[79]

Disordered cell-mediated immunity

Aberrant cellular immunity in progressive systemic sclerosis is milder than in rheumatoid arthritis or SLE.[91] This may in part account for discrepancies in the literature which also reflect patient selection, inappropriate or unmatched controls and the general state of health. It is not clear whether the abnormalities that are consistently detectable are secondary to the inflammatory disease or antedate this condition.[79,92,160] Mitogen responses mediated, in some cases, by monocytes diminish in proportion to the severity of the disease.[83] Monocytes from progressive systemic sclerosis patients can also depress primary *in vitro* antibody responses to trinitrophenyl,[184] an effect attributable to prostaglandin E-mediated lymphocyte regulation.[62] This results from excessive monocyte numbers or increased sensitivity of lymphocytes to prostaglandin E-mediated effects.

T-cells numbers are depressed in the blood of patients with progressive systemic sclerosis,[222] particularly those of $CD8^+$ phenotype though suppressor cell activity remains intact. Compared with normal T-cells, those from patients with progressive systemic sclerosis demonstrate increased functional help for B-cell proliferation, differentiation and Ig synthesis.[3,85] Mitogen-induced suppression, spontaneous suppressor cell activity and antigen-specific suppression appear normal[99] although there may be impaired responses to phytohaemagglutinin. A high proportion of patients with systemic sclerosis of recent onset have elevated serum levels of IL-2.[93] There is a positive correlation between serum level and skin progression index (skin score/disease duration).

The autologous mixed lymphocyte reaction is defective in progressive systemic sclerosis, as in rheumatoid arthritis and SLE. The non-T-lymphocytes of patients with progressive systemic sclerosis are poor stimulators in the allogeneic mixed lymphocyte reaction. However, T-lymphocytes respond normally in the latter. The data on lymphocyte cytotoxicity are conflicting.[152]

Disordered humoral immunity

There is evidence of both humoral and cellular immunity against connective tissue elements in progressive systemic sclerosis. Autoantibodies to collagen, observed in other systemic connective tissue disorders, particularly rheumatoid arthritis, are also frequently detectable in progressive systemic sclerosis.[125] Antibodies to fibroblast cell surface proteins can be demonstrated.[23]

Signs of humoral autoimmunity are much less prominent in progressive systemic sclerosis than in SLE or in organ-specific diseases such as autoimmune thyroiditis. The autoantibodies in systemic sclerosis are mostly directed against nuclear constituents. Antinucleolar antibodies are detected in greater frequency than in certain other connective tissue diseases. However, antinuclear antibodies are also demonstrable: the anti-Scl-70 and anticentromere antibodies appear to be the most specific, the former enriched in the minority of patients with a substantial mortality in the first decade of an illness that begins with a close association between Raynaud's phenomenon and early skin disease, the latter in those in whom life-threatening visceral lesions develop late.[119] The relatively frequent detection of immune complexes is rarely associated with either hypocomplementaemia or with the clinical features of immune complex disease.[185]

Responses to collagen

There is cell-mediated immunity to collagen.[202] Skin extracts stimulate the release of macrophage inhibition factor from patients' lymphocytes.[105] The low molecular weight cutaneous antigen which may mediate this effect appears to be absent from organ extracts other than those of skin.[106] Progressive systemic sclerosis lymphocytes are cytotoxic toward fibroblasts but the lytic interaction is not specific, and other cells such as those of epithelia and monocytes are affected. In the presence of progressive systemic sclerosis serum, normal peripheral blood mononuclear cells are cytotoxic for vascular endothelium.[157] Although the factor elutes with IgG and is present in approximately 25 per cent of patients, direct evidence that this represents an antibody to endothelial cells is lacking. At present there is little evidence to associate injury to vascular endothelium in progressive systemic sclerosis with cell-mediated immunity toward endothelial cells.[152]

Differences in the profile of cell-surface proteins in fibroblasts from patients with progressive systemic sclerosis compared to normal subjects have been reported. The relationship, if any, to the humoral and cellular autoreactivity against fibroblasts is unclear.

The association between cell-mediated immunity and collagen synthesis in progressive systemic sclerosis, is complex.[108] Factors produced by lymphocytes and monocytes are capable both of stimulating and suppressing collagen synthesis by fibroblasts *in vitro*. The opposing effects are mediated by distinct molecules in the case of lymphocytes.[161] Interferons α and γ can suppress collagen synthesis in fibroblasts[92] but these cytokines are not identical with the lymphocyte-suppressive mediators that are generated in activated T-cells.[161] Macrophages suppress collagen synthesis by the stimulation of fibroblast prostaglandin E synthesis[36] but have separate direct stimulatory actions.

The role of cytokines

Regulation of fibroblast proliferation by cytokines is also bi-directional. Interleukin-1 can directly stimulate fibroblast proliferation[159] but may also stimulate fibroblast prostaglandin E synthesis with the suppression of fibroblast proliferation.[110] Macrophage-derived growth factor, a separate monocyte product, stimulates fibroblast proliferation.[15] Cytokines with the capability both to suppress and stimulate fibroblast proliferation have been described.[163] It is also clear that lymphocytes can recruit fibroblasts to inflammatory sites by the secretion of chemotactic factors.[162] That fibroblasts are heterogeneous in their ability to synthesize collagen[19] may assist understanding of the mechanisms operative in the dermal and visceral fibrosis of progressive systemic sclerosis (p. 706). Dermal hyperplasia may in this way be accompanied by the clonal selection of subpopulations that secrete excess collagen.[119] Clones of fibroblasts are also heterogeneous in their growth responses to cytokines[111] and the action of these mediators could lead to overgrowth of subpopulations producing large amounts of collagen. Similar preferential proliferation has been shown to occur in the presence of progressive systemic sclerosis serum.[19] Transient exposure of normal fibroblasts to cytokines can induce protracted phenotypic change.[109] Consequently, activation of the immune system may not only promote the fibroblast proliferation and increased collagen synthesis of progressive systemic sclerosis but also result in permanent alterations in the fibroblast population.

REFERENCES

1. Abrahamsson D R. Recent studies on the structure and pathology of basement membranes. *Journal of Pathology* 1986; **149**: 257–78.

2. Acuto O, Reinherz E L. The human T-cell receptor, structure and function. *New England Journal of Medicine* 1985; **312**: 1100–11.

3. Alarçon-Segovia D, Palacios R, Ibanez de Kasep G. Human post thymic precursor cells in health and disease. VII: immunoregulatory circuits of the peripheral blood mononuclear cells from patients with progressive systemic sclerosis. *Journal of Clinical and Laboratory Immunology* 1981; **5**: 143–8.

4. Allen J C, Kunkel H G. Hidden rheumatoid factors with specificity for native γ-globulins. *Arthritis and Rheumatism* 1966; **99**: 758–68.

5. Alspaugh M A, Jensen F C, Rabin H, Tan E M. Lymphocytes transformed by Epstein–Barr virus: induction of nuclear antigen reactive with antibody in rheumatoid arthritis. *Journal of Experimental Medicine* 1978; **147**: 1018–27.

6. Andrews B S, Eisenberg R A, Argyrios N, Theofilopoulos A N, Izui S, Wilson C B, McConahey P J, Murphy E D, Roths J B, Dixon F J. Spontaneous murine lupus-like syndrome. Clinical and immunopathological manifestations in several strains. *Journal of Experimental Medicine* 1978: **148**: 1198–1215.

7. Axford J S, Mackenzie L, Lydyard P M, Hay F C, Isenberg D A, Roitt I M. Reduced B cell galactosyl transferase activity in rheumatoid arthritis. *Lancet* 1987; **ii**: 1486–8.

8. Bach F H. The HLA Class II genes and products: the HLA-D region. *Immunology Today* 1985; **6**: 89–94.

9. Balkwill F R. *Cytokines in Cancer Therapy*. Oxford: Oxford University Press, 1989.

10. Bardwick P A, Bluestein H G, Zvaifler N J, Depper J M, Seegmiller J E. Altered regulation of Epstein–Barr virus induced lymphoblast proliferation in rheumatoid arthritis lymphoid cells. *Arthritis and Rheumatism* 1980; **23**: 626–32.

11. Batchelor J R, Welsh K I. Association of HLA antigens with disease. In: Lachmann P J, Peters D K, eds, *Clinical Aspects of Immunology* (4th edition). Oxford: Blackwell Scientific Publications, 1982: 283–306.

12. Benjamin R, Parham P. Guilt by association: HLA-B27 and ankylosing spondylitis. *Immunology Today* 1990; **11**: 137–42.

13. Benjamin R J, Qin S, Wise M P, Cobbold S P, Waldmann M. Mechanisms of monoclonal antibody–facilitated tolerance induction: a possible role for the CD4 (L3T4) and CD11a (LFA-1) molecules in self-non-self discrimination. *European Journal of Immunology*, 1988; **18**: 1079–88.

14. Bernstein R M. Humoral autoimmunity in systemic rheumatic disease. A review. *Journal of the Royal College of Physicians of London* 1990; **24**: 18–25.

15. Bitterman P B, Rennard S I, Hunninghake G W, Crystal R G. Human alveolar macrophage growth factor for fibroblasts. Regulation and partial characterization. *Journal of Clinical Investigation* 1982; **70**: 806–22.

15a. Bjorkman P J, Davis M M. T-cell antigen receptor genes and T-cell recognition. *Nature* 1990; **334**: 395–402.

16. Bjorkman P J, Saper M A, Samraoui B, Bennett W S, Strominger J L, Wiley D C. Structure of the human class I histocompatibility antigen, HLA-A2. *Nature* 1987; **329**: 506–12.

17. Bjorkman P J, Saper M A, Samraoui B, Bennett W S, Strominger J L, Wiley D C. The foreign antigen binding site and T-cell recognition regions of class I histocompatibility antigens. *Nature* 1987; **329**: 512–18.

18. Bodmer J, Bodmer W. Histocompatibility 1984. *Immunology Today* 1984; **5**: 251–4.

18a. Born W, Happ M P, Dallas A, Reardon C, Kubo R, Shinnick T, Brennan P, O'Brien R. Recognition of heat shock proteins and γδ cell function. *Immunology Today* 1990; **11**: 40–3.

19. Botstein G R, Sherer G K, Leroy E C. Fibroblast selection in scleroderma. An alternative model of fibrosis. *Arthritis and Rheumatism* 1982; **25**: 189–95.

20. Bottazzo G F, Pujol-Borrell R, Hanafusa T, Feldmann M. Role of aberrant HLA-DR expression and antigen presentation in induction of endocrine autoimmunity. *Lancet* 1983; **ii**: 1115–9.

21. Bottazzo G F, Todd I, Mirakian R, Belfiore A, Pujol-Barrell R. Organ-specific autoimmunity: A 1986 overview. *Immunological Reviews* 1986; **94**: 137–69.

22. Brennan F M, Allard S, Londei M, Savill C, Boylston A, Carrel S, Maini R N, Feldmann M. Heterogeneity of T cell receptor idiotypes in rheumatoid arthritis. *Clinical and Experimental Immunology* 1988; **73**: 417–23.

22a. Brennan F M, Chantry D, Jackson A, Maini R, Feldmann M. Inhibitory effect of TNF alpha antibodies on synovial cell interleukin-1 production in rheumatoid arthritis. *Lancet* 1989; **ii**: 244–7.

23. Brentnall T J, Kenneally D, Barnett A J, De Aizpurua H J, Lolait S J, Ashcroft R, Toh B H. Autoantibodies to fibroblasts in scleroderma. *Journal of Clinical and Laboratory Immunology* 1982; **8**: 9–12.

24. Bresnihan B, Jasin H E. Suppressor function of peripheral blood mononuclear cells in normal individuals and in patients with systemic lupus erythematosus. *Journal of Clinical Investigation* 1977; **59**: 106–16.

25. Brewerton D A, Hart F D, Nicholls A, Caffrey M, James D C O, Sturrock R D. Ankylosing spondylitis and HL-A27. *Lancet* 1973; **i**: 904–7.

26. Burastero S E, Casali P, Wilder R L, Notkins A L. Monoreactive high affinity and polyreactive low affinity rheumatoid factors are produced by $CD5^+$ B cells from patients with rheumatoid arthritis. *Journal of Experimental Medicine* 1988; **168**: 1979–92.

27. Butler W T, Sharp J T, Rossen R D. Relationship of the clinical course of systemic lupus erythematosus to the presence of circulating lymphocytotoxic antibodies. *Arthritis and Rheumatism* 1972; **15**: 251–8.

28. Calvanico N J. Structure and function of immunoglobulins. In: Zouhair Atassi M, Van Oss C J, Absolom D R, eds, *Molecular Immunology*. New York: Marcel Dekker Inc, 1984: 141–74.

29. Campbell P M, LeRoy E C. Pathogenesis of systemic sclerosis: a vascular hypothesis. *Seminars in Arthritis and Rheumatism* 1975; **4**: 351–68.

30. Capra J D, Kehoe J M. Hypervariable regions, idiotypy and the antibody combining site. *Advances in Immunology* 1975; **20**: 1–40.

31. Carbonara A O, De Marchi M. Immunogenetics of systemic lupus erythematosus. *Contributions to Nephrology* 1985; **45**: 157–74.

32. Carson D A, Lawrance S, Catalano M A, Vaughan J H, Abraham G. Radioimmunoassay of IgG and IgM rheumatoid factors reacting with human IgG. *Journal of Immunology* 1977; **119**: 295–300.

33. Chapel H, Haeney M. *Essentials of Clinical Immunology* (1st edition). Oxford: Blackwell Scientific Publications, 1984: 188–219.

34. Charpentier B, Carnaud C, Bach J F. Selective depression of the xenogeneic cell-mediated lympholysis in systemic lupus erythematosus. *Journal of Clinical Investigation* 1979; **64**: 351–60.

35. Chin J C, Lin Y. Effects of recombinant human interleukin-1 beta on rabbit articular chondrocytes: stimulation of prostanoid release and inhibition of cell growth. *Arthritis and Rheumatism* 1988; **31**: 1290–6.

36. Clark J G, Kostal K M, Marino B A. Bleomycin-induced pulmonary fibrosis in hamsters. An alveolar macrophage product increases fibroblast prostoglandin E_2 and cyclic adenosine monophosphate and suppresses fibroblast proliferation and collagen production. *Journal of Clinical Investigation* 1983; **72**: 2082–91.

37. Cohen P L, Litvin D A, Winfield J B. Association between endogenously activated T cells and immunoglobulin-secreting B cells in patients with active systemic lupus erythematosus. *Arthritis and Rheumatism* 1982; **25**: 168–73.

38. Coombs R R A, Gell P G H. Classification of allergic reactions responsible for clinical hypersensitivity and disease. In: Gell P G H, Coombs R R A, Lachmann P J, eds, *Clinical Aspects of Immunology* (3rd edition). Oxford: Blackwell Scientific Publications, 1975: 761–82.

39. Cracchiolo A, Bluestone R, Goldberg L S. Hidden antiglobulins in rheumatic disorders. *Clinical and Experimental Immunology* 1970; **7**: 651–5.

40. Culpepper R M, Andreoli T E. The pathophysiology of the glomerulonephropathies. *Advances in Internal Medicine* 1983; **28**: 161–206.

41. Cush J J, Lipsky P E. Phenotypic analysis of synovial tissue and peripheral blood lymphocytes isolated from patients with rheumatoid arthritis. *Arthritis and Rheumatism* 1988; **31**: 1230–8.

42. Dawson M M. *Lymphokines and interleukins*. Open University Press: Milton Keynes, 1991.

43. Dayer J-M, Demczuk S. Cytokines and other mediators in rheumatoid arthritis. *Springer Seminars in Immunopathology* 1984; **7**: 387–413.

44. Dayer J-M, Krane S M, Russell R E G, Robinson D R. Production of collagenase and prostaglandins by isolated adherent rheumatoid synovial cells. *Proceedings of the National Academy of Science of the USA* 1976; **73**: 945–9.

45. Decker J L. Systemic lupus erythematosus: evolving concepts. Proceedings of an NIH Conference. *Annals of Internal Medicine* 1979; **91**: 587–604.

46. Decker J L, Malone D G, Haraovi B *et al.* Rheumatoid arthritis: evolving concepts of pathogenesis and treatment. Proceedings of an NIH Conference. *Annals of Internal Medicine* 1984; **101**: 810–24.

46a. DeFranco D E F. Tolerance; a second mechanism. *Nature* 1989; **342**: 340.

47. Dinarello C A. Interleukin 1. *Reviews of Infectious Disease* 1984; **6**: 51–95.

48. Eilat D. Cross-reactions of anti DNA antibodies and of the central dogma of lupus nephritis. *Immunology Today* 1985; **6**: 123–7.

49. Ercilla M G, Arriaga F, Gratacos M R, Coll J, Lecha V, Vives J, Castillo R. HLA antigens and scleroderma. *Archives of Dermatological Research* (Berlin) 1981; **271**: 381–5.

50. Espinoza L R. Rheumatoid arthritis: etiopathogenetic considerations. *Clinics in Laboratory Medicine*, 1986; **6**: 27–40.

51. Faaber P, Capel P J, Rijke G P M, Vierwinden G, Van de Putte L B A, Koene R A P. Cross-reactivity of anti-DNA antibodies with proteoglycans. *Clinical and Experimental Immunology* 1984; **55**: 502–8.

52. Fauci A S, Steinberg A D, Haynes B F, Whalen G. Immunoregulatory aberrations in systemic lupus erythematosus. *Journal of Immunology* 1978; **121**: 1473–9.

53. Fearon D T. The structure and function of complement. In: Zouhair Atassi M, Van Oss C J, Absolom D R, eds, *Molecular Immunology*. New York: Marcel Dekker Inc, 1984: 511–25.

54. Festenstein H, Awad J, Hitman G A, Cutbush S, Groves A V, Cassell P, Ollier, W, Sachs J A. New HLA DNA polymorphisms associated with autoimmune diseases. *Nature* (London) 1986; **322**: 64–7.

55. Fielder A H L, Walport M J, Batchelor J R, Rynes R I, Black C M, Dodi I A, Hughes G R V. Family study of the major histocompatibility complex in patients with systemic lupus erythematosus: importance of null alleles of C4A and C4B in determining disease susceptibility. *British Medical Journal* 1983; **286**: 425–8.

56. Firestein G S, Xu W D, Townsend K, Bride D, Alvaro-Garcia J, Glasbrook A, Zvaifler N J. Cytokines in chronic inflammatory arthritis: failure to detect T-cell lymphokine (interleukin 2 and interleukin 3) and presence of colony-stimulating factor (CSF-1) and a novel mast cell growth factor in rheumatoid synovitis. *Journal of Experimental Medicine* 1988; **168**: 1573–86.

57. Fleishmajer R, Gay S, Meigel W N, Perlish J S. Collagen in the cellular and fibrotic stages of scleroderma. *Arthritis and Rheumatism* 1978; **21**: 418–28.

58. Frank M M, Hamburger M I, Lawley T J, Kimberley R P, Plotz P H. Defective reticuloendothelial system Fc receptor function in systemic lupus erythematosus. *New England Journal of Medicine* 1979; **300**: 518–23.

59. Furst D E, Clements P J, Graze P, Gale R P, Roberts N. A syndrome resembling progressive systemic sclerosis after bone marrow transplantation. A model for scleroderma? *Arthritis and Rheumatism* 1979; **22**: 904–10.

60. Gilman S C, Chang J, Zeigler P R, Uhl J, Mochan E. Interleukin-1 activates phospholipase-A2 in human synovial cells. *Arthritis and Rheumatism* 1988; **31**: 127–30.

61. Goodnow C C, Crosbie J, Jorgensen H, Brink R A, Basten A. Induction of self-tolerance in mature peripheral B lymphocytes. *Nature* 1989; **342**: 385–91.

62. Goodwin J S, Bankhurst A D, Messner R P. Suppression of human T-cell mitogenesis by prostaglandin: existence of a

prostaglandin-producing suppressor cell. *Journal of Experimental Medicine* 1977; **146**: 1719–34.

63. Goren M B. Phagocyte lysosomes: interactions with infectious agents, phagosomes and experimental perturbations in functions. *Annual Review of Microbiology* 1977; **31**: 507–33.

64. Goto M, Tanimoto K, Horiuchi Y. Natural cell mediated cytotoxicity in systemic lupus erythematosus: suppression by antilymphocyte antibody. *Arthritis and Rheumatism* 1980; **23**: 1274–81.

65. Gray S, Huang G, Ziegler B, Fassbender H-G, Gay R E. Expression of *myb*, *myc*, *ras* and *fos* oncogenes in synovial cells of patients with rheumatoid arthritis (RA) or osteoarthritis (OA). *Arthritis and Rheumatism* 1989; **32** (suppl): 39S.

66. Grennan D M, Dyer P A. Immunogenetics and rheumatoid arthritis. *Immunology Today* 1988; **9**: 33–4.

67. Grey H M, Chesnut R W. Antigen processing and presentation to T cells. *Immunology Today* 1985; **6**, 101–6.

68. Grimm E A, Mazumder A, Zhang H Z, *et al.* Lymphokine-activated killer-cell phenomenon. Lysis of natural killer-resistant fresh solid tumor cells by interleukin 2-activated autologous human peripheral blood lymphocytes. *Journal of Experimental Medicine* 1982; **155**: 1823–41.

69. Grubb R, Matsumoto H, Sattar M A. Incidence of anti-human Ig with restricted specificity in Japanese, Kuwaiti and Swedish patients with rheumatoid arthritis. *Arthritis and Rheumatism* 1988; **31**: 60–2.

70. Gupta S, Talal N. *Immunology of Rheumatic Diseases*. New York: Plenum Medical Book Company, 1985.

71. Haas W, Kaufman S, Martinez-A. The development and function of γδ T cells. *Immunology Today* 1990; **11**: 340–3.

72. Hamblin A S. *Lymphokines*. Washington: IRL Press, 1988.

73. Hamilton J A, Zabriskie J B, Lachman L B, Chen Y S. Streptococcal cell walls and synovial cell activation. Stimulation of synovial fibroblast plasminogen activity by monocytes treated with group A streptococcal cell wall sonicates and muramyl dipeptide. *Journal of Experimental Medicine* 1982; **155**: 1702–18.

74. Harris E D. Rheumatoid arthritis: pathophysiology and implications for therapy. *New England Journal of Medicine* 1990; **323**: 1277–89.

75. Hasler F, Bluestein H G, Zvaifler N J, Epstein L B. Analysis of the defects responsible for the impaired regulation of EBV-induced B cell proliferation by rheumatoid arthritis lymphocytes. II. Role of monocytes and the increased sensitivity of rheumatoid arthritis lymphocytes to prostaglandin E. *Journal of Immunology* 1983; **131**: 768–72.

76. Haynes B F, Grover B J, Whichard L P, Hall L P, Nunby J A, McCollum D E, Singer K H. Synovial microenvironment T cell interactions in human T-cells bind to fibroblast-like synovial cells *in vitro*. *Arthritis and Rheumatism* 1988; **31**: 947–55.

77. Haynes D C, Gershwin M E. The immunopathology of progressive systemic sclerosis (PSS). *Seminars on Arthritis and Rheumatism* 1982; **11**: 331–51.

78. Helenius A, Morein B, Fries E, Simons K, Robinson P, Schirrmacher V, Terhorst C, Strominger J L. Human (HLA-A and HLA-B) and murine (H-2K and H-2D) histocompatibility antigens are cell surface receptors for Semliki Forest Virus. *Proceedings of the National Academy of Sciences of the USA* 1978; **75**: 3846–50.

79. Herzog P, Clements P J, Roberts N K, Furst D E, Johnson C E, Feig S A. Case report: progressive systemic sclerosis-like syndrome after bone marrow transplantation. Clinical, immunologic and pathologic findings. *Journal of Rheumatology* 1980; **7**: 56–64.

80. Holman H R, Robbins W C. Anti-nuclear antibodies in lupus erythematosus. *Science* 1957; **126**: 1232.

81. Horton J E, Raisz L G, Simmons H A, Oppenheim J J, Mergenhagen S E. Bone resorbing activity in supernatant fluid from cultured human peripheral blood leukocytes. *Science* 1972; **177**: 793–5.

82. Hughes P, Gelsthorpe K, Doughty R W, Rowell N R, Rosenthal F D, Sneddon I B. The association of HLA-B8 with visceral disease in systemic sclerosis. *Clinical and Experimental Immunology* 1978; **31**: 351–6.

83. Hughes P, Holt S, Rowell N R, Allonby I D, Janis K, Dodd J K. The relationship of defective cell-mediated immunity to visceral disease in systemic sclerosis. *Clinical and Experimental Immunology* 1977; **28**: 233–40.

84. Hugli T E, Muller-Eberhard H J. Anaphylatoxins: C3a and C5a. *Advances in Immunology* 1978; **26**: 1–53.

85. Inoshita T, Whiteside T L, Rodnan G P, Taylor F H. Abnormalities of T lymphocyte subsets in patients with progressive systemic sclerosis (PSS, scleroderma). *Journal of Laboratory and Clinical Medicine* 1981; **97**: 264–77.

86. Izui S, Kelley V E, Masuda K, Yoshida M, Roths J B, Murphy E D. Induction of various autoantibodies by mutant gene lpr in several strains of mice. *Journal of Immunology* 1984; **133**: 227–33.

87. Jacob L, Tron F, Bach J F, Louvard D. A monoclonal anti-DNA antibody also binds to cell-surface protein(s). *Proceedings of the National Academy of Sciences of the USA* 1984; **81**: 3843–5.

88. Jacobsen E, Biddison W E. Major histocompatibility complex molecules as virus receptors. *Immunology Today* 1984; **5**: 262–3.

89. Jalkanen S, Steere A C, Fox R I, Butcher E C. A distinct endothelial cell recognition system that controls lymphocyte traffic into inflamed synovium. *Science* 1986; **233**: 556–8.

90. Jerne N K. Towards a network theory of the immune system. *Annales d'Immunologie* (Paris) 1974; **125**: 373–89.

91. Jimenez S A. Cellular immune dysfunction and the pathogenesis of scleroderma. *Seminars in Arthritis and Rheumatism* 1983; **13** (Suppl. 1): 104–13.

92. Jimenez S A, Freundlich B, Rosenbloom J. Selective inhibition of human diploid fibroblast collagen synthesis by interferons. *Journal of Clinical Investigation* 1984; **74**: 1112–6.

93. Kahaleh M B, LeRoy E C. Interleukin-2 in scleroderma: Correlation of serum level with extent of skin involvement and disease duration. *Annals of Internal Medicine* 1989; **110**: 446–50.

94. Kallenberg C G M, Vander Voort-Beelen J M, D'Amaro J, The Th. Increased frequency of B8/DR3 in scleroderma and association of the haplotype with impaired cellular immune response. *Clinical and Experimental Immunology* 1981; **43**: 478–85.

95. Karsh J, Dorval G, Osterland C K. Natural cytotoxicity in

stimulation and inhibition of collagen production by different effector molecules. *Journal of Immunology* 1984; **132**: 2470–7.

162. Postlethwaite A E, Snyderman R, Kang A H. The chemotactic attraction of human fibroblasts to a lymphocyte-derived factor. *Journal of Experimental Medicine* 1976; **144**: 1188–203.

163. Potter S R, Bienenstock J, Lee P. Clinical associations of fibroblast growth promoting factor in scleroderma. *Journal of Rheumatology* 1984; **11**: 43–7.

164. Price R G, Wong M. Heterogeneity of Goodpastures' antigen. *Journal of Pathology* 1988; **156**: 97–9.

165. Radoiu N, Cleveland R P, Leon M A. Peripheral T lymphocytes from rheumatoid arthritis patients recognize determinants on light chains and/or Fab fragments of immunoglobulin G. *Journal of Clinical Investigation* 1982; **70**: 329–34.

166. Raff M C. T and B lymphocytes and immune responses. *Nature* (London) 1973; **242**: 19–23.

167. Ramsdell F, Fowlkes B J. Clonal deletion versus clonal anergy: The role of the thymus in inducing self-tolerance. *Science* 1990; **248**: 1342–8.

168. Reveille J D, Bias W B, Winkelstein J A, Provost T T, Dorsch C A, Arnett F C. Familial systemic lupus erythematosus: Immunogenetic studies in eight families. *Medicine* (Baltimore) 1983; **62**: 21–35.

169. Robb R J. Interleukin-2: The molecule and its function. *Immunology Today* 1984; **5**: 203–9.

170. Roitt I. *Essential Immunology* (6th edition). Oxford: Blackwell Scientific Publications, 1988.

171. Roitt I, Brostoff J, Male D. *Immunology*. Edinburgh: Churchill Livingstone, 1985.

172. Romain P L, Burmester G R, Enlow R W, Winchester R J. Multiple abnormalities in immunoregulatory function of synovial compartment T-cells in patients with rheumatoid arthritis. Recognition of a helper augmentation effect. *Rheumatoid International* 1982; **2**: 121–7.

173. Rook G A W, Steele J, Rademacher T A. A monoclonal antibody raised by immunising mice with group A streptococci birds to agalactosyl IgG from rheumatoid arthritis. *Annals of the Rheumatic Diseases* 1988; **47**: 247–50.

174. Rose H M, Ragan C, Pearce E, Lipman M O. Differential agglutination of normal and sensitized sheep erythrocytes by sera of patients with rheumatoid arthritis. *Proceedings of the Society of Experimental Biology and Medicine* 1948; **68**: 1–6.

175. Rosenberg Y J, Steinberg A D, Santoro T J. The basis of autoimmunity in MRL-lpr/lpr mice: a role for self Ia-reactive T-cells. *Immunology Today* 1984; **5**: 64–6.

176. Ross R. Platelet-derived growth factor. *Lancet* 1989; **i**: 1179–82.

177. Roumm A D, Whiteside T L, Medsger T A Jnr, Rodnan G P. Lymphocytes in the skin of patients with progressive systemic sclerosis. Quantification, subtyping and clinical correlations. *Arthritis and Rheumatism* 1984; **27**: 645–53.

178. Rynes R I. Inherited complement deficiency states in SLE. *Clinics in Rheumatic Disease* 1982; **8**, 29–47.

179. Sakane T, Steinberg A D, Arnett F L, Reinertsen J L, Green I. Studies of immune functions in patients with systemic lupus erythematosus. *Arthritis and Rheumatism* 1979; **22**: 770–6.

180. Sakane T, Steinberg A D, Reeves J P, Green I. Studies of immune functions of patients with systemic lupus erythematosus. Complement-dependent immunoglobulin M anti-thymus-derived cell antibodies preferentially inactivate suppressor cells. *Journal of Clinical Investigation* 1979; **63**: 954–65.

181. Schmidt J A, Mizel S B, Cohen D, Green I. Interleukin 1: A potential regulator of fibroblast proliferation. *Journal of Immunology* 1982; **128**: 2177–82.

182. Schwartz R S. Monoclonal lupus autoantibodies. *Immunology Today* 1983; **4**: 68–9.

183. Searles R P, Savage S M, Brozek C M, Marnell L L, Hoffman C L. Network regulation in rheumatoid arthritis: studies of DR$^+$ T cells anti-DR, antiidiotypic antibodies and clinical disease activity. *Arthritis and Rheumatism* 1988; **31**: 834–43.

184. Segond P, Salliere D, Galanaud P, Desmottes R M, Massias P, Fiessinger J N. Impaired primary *in vitro* antibody response in progressive systemic sclerosis patients: role of suppressor monocytes. *Clinical and Experimental Immunology* 1982; **47**: 147–54.

185. Seibold J R, Medsger T A Jnr, Winkelstein A, Kelly R H, Rodnan G P. Immune complexes in progressive systemic sclerosis (scleroderma). *Arthritis and Rheumatism* 1982; **25**: 1167–473.

186. Sharif M, Rook G, Wilkinson L S, Worrall J G, Edwards J C W. Terminal N-acetylglucosamine in chronic synovitis. *British Journal of Rheumatology* 1990; **29**: 25–31.

187. Shaw S. Characterisation of human leukocyte differentiation antigens. *Immunology Today* 1987; **8**: 1–3.

188. Shoenfeld Y, Rauch J, Massicotte H, Datta S K, André-Schwartz J, Stollar B D, Schwartz R S. Polyspecificity of monoclonal lupus autoantibodies produced by human–human hybridomas. *New England Journal of Medicine* 1983; **308**: 414–20.

189. Sibbitt W L Jnr, Kenny C, Spellman C W, Ley K D, Bankhurst A D. Lymphokines in autoimmunity: relationship between interleukin-2 and interferon-gamma production in systemic lupus erythematosus. *Clinical Immunology and Immunopathology* 1984; **32**: 166–73.

190. Silver R M, Redelman D, Zvaifler N J. Studies of rheumatoid synovial fluid lymphocytes II. A comparison of their behaviour with blood mononuclear cells in the autologous mixed lymphocyte reaction and response to TCGF. *Clinical Immunology and Immunopathology* 1983; **27**: 15–27.

191. Sinha A A, Lopez M T, McDevitt H O. Autoimmune diseases: The failure of self-tolerance. *Science* 1990; **248**: 1380–8.

192. Smith H R, Steinberg A D. Autoimmunity—a perspective. *Annual Review of Immunology* 1983; **1**: 175–210.

193. Smolen J S, Chused T M, Leiserson W M, Reeves J P, Alling D, Steinberg A D. Heterogeneity of immunoregulatory T-cell subsets in systemic lupus erythematosus. Correlation with clinical features. *American Journal of Medicine* 1982; **72**: 783–90.

194. Solari R, Kraehenbuhl J P. The biosynthesis of secretory component and its role in the transepithelial transport of IgA dimer. *Immunology Today* 1985; **6**: 17–21.

195. Sprent J, Gao E-K, Webb S R. T cell reactivity to MHC molecules: Immunity versus tolerance. *Science* 1990; **248**: 1357–63.

196. Stamenkovic I, Stegagno M, Wright K A, Krane S M, Amento E P, Colvin R B, Duquesnoy R J, Kurnick J T. Clonal dominance among T-lymphocyte infiltrates in arthritis. *Proceedings of the National Academy of Sciences of the USA* 1988; **85**: 1179–83.

197. Stastny P, Stembridge V A, Ziff M. Homologous disease in the adult rat, a model for autoimmune disease I. General features and cutaneous lesions. *Journal of Experimental Medicine* 1963; **118**: 635–48.

198. Steinberg A D, Raveche E S, Laskin C A, Smith H R, Santoro T, Miller M L, Plotz P H. Systemic lupus erythematosus: insights from animal models. *Annals of Internal Medicine* 1984; **100**: 714–27.

199. Stenszky V, Kozma L, Szegedi G, Farid N R. Interplay of immunoglobulin G heavy chain markers (Gm) and HLA in predisposing to systemic lupus nephritis. *Journal of Immunogenetics* 1986; **13**: 11–17.

200. Stuart J M, Cremer M A, Townes A S, Kang A H. Type II collagen-induced arthritis in rats: passive transfer with serum and evidence that IgG anticollagen antibodies can cause arthritis. *Journal of Experimental Medicine* 1982; **155**: 1–16.

201. Stuart J M, Huffstutter E H, Townes A S, Kang A H. Incidence and specificity of antibodies to types I, II, III, IV and V collagen in rheumatoid arthritis and other rheumatic diseases as measured by 125 I radioimmunoassay. *Arthritis and Rheumatism* 1983; **26**: 832–40.

202. Stuart J M, Postlethwaite A E, Kang A H. Evidence for cell-mediated immunity to collagen in progressive systemic sclerosis. *Journal of Laboratory and Clinical Medicine* 1976; **88**: 601–7.

203. Tan E M. Autoantibodies to nuclear antigens (ANA): their immunobiology and medicine. *Advances in Immunology* 1982; **33**: 167–240.

204. Tan E M, Cohen A S, Fries J F, Masi A T, McShane D J, Rothfield N F, Schaller J G, Talal N, Winchester R J. The 1982 revised criteria for the classification of systemic lupus erythematosus. *Arthritis and Rheumatism* 1982; **25**: 1271–7.

205. Tanimoto K, Cooper H R, Johnson J S, Vaughan J H. Complement fixation by rheumatoid factor. *Journal of Clinical Investigation* 1975; **55**: 437–45.

206. Theofilopoulos A N, Dixon F J. The biology and detection of immune complexes. *Advances in Immunology* 1979; **28**: 89–220.

207. Theofilopoulos A N, Dixon F J. Immune complexes in human diseases. A review. *American Journal of Pathology* 1980; **100**: 531–91.

208. Thorley-Lawson D A. The transformation of adult but not newborn human lymphocytes by Epstein–Barr virus and phytohemagglutinin is inhibited by interferon: the early suppression by T-cells of Epstein–Barr infection is mediated by interferon. *Journal of Immunology* 1981; **126**: 829–33.

209. Tiwari J L, Terasaki P I. *HLA and Disease Associations.* Berlin: Springer-Verlag, 1985.

210. Tomana M, Schrohenlohr K E, Koopman W J, Alarçon G S, Paul W A. Abnormal glycosylation of serum IgG from patients with chronic inflammatory diseases. *Arthritis and Rheumatism* 1988; **31**: 333–8.

211. Tosato G, Steinberg A D, Blaese R M. Defective EBV-specific suppressor T cell function in rheumatoid arthritis. *New England Journal of Medicine* 1981; **305**: 1238–43.

212. Tsokos G C, Rook A H, Djeu J Y, Balow J E. Natural killer cells and interferon responses in patients with systemic lupus erythematosus. *Clinical and Experimental Immunology* 1982; **50**: 239–45.

213. Tsoukas C D, Valentine M, Lotz M, Vaughan J H, Carson D A. The role of the T3 molecular complex in antigen recognition and subsequent activation events. *Immunology Today* 1984; **5**: 311–3.

214. Turk J L. *Delayed Hypersensitivity* (2nd edition). Amsterdam: North Holland, 1975.

215. Vaughan J H. Immune system in rheumatoid arthritis: possible implications in neoplasms. *American Journal of Medicine* 1985; **78** (Suppl. A): 6–11.

216. Vaughan J H, Carson D A, Fox R I. The Epstein–Barr virus and rheumatoid arthritis. *Clinical and Experimental Rheumatology* 1983; **1**: 265–72.

217. Von Boxel J A, Paget S A. Predominantly T cell infiltrate in rheumatoid synovial membranes. *New England Journal of Medicine* 1975; **293**: 517–20.

218. Wahl S, Gateley C L. Modulation of fibroblast growth by a lymphokine of human T-cell and continuous T-cell line origin. *Journal of Immunology* 1983; **130**: 1226–30.

219. Wahl S M, Wilder R L, Katona I M, Wahl L M, Allen J B, Scher I, Decker J L. Leukapheresis in rheumatoid arthritis: association of clinical improvement with reversal of anergy. *Arthritis and Rheumatism* 1983; **26**: 1076–84.

220. Waldmann H, Cobbold S, Benjamin R, Qin S. A theoretical framework for self-tolerance and its relevance to therapy of autoimmune disease. *Journal of Autoimmunity* 1988; **1**: 623–9.

221. Walport M J, Black C M, Batchelor J R. The immunogenetics of SLE. *Clinics in Rheumatic Diseases* 1982; **8**: 3–21.

222. Whiteside T L, Kumagai Y, Roumm A D, Almendinger R, Rodnan G P. Suppressor cell function and T lymphocyte subpopulations in peripheral blood of patients with progressive systemic sclerosis. *Arthritis and Rheumatism* 1983; **26**: 841–7.

223. Wilson C B. Nephritogenic antibody mechanisms involving antigens within the glomerulus. *Immunology Review* 1981; **55**: 257–97.

224. Winfield J B. Antilymphocyte antibodies in systemic lupus erythematosus. *Clinics in Rheumatic Diseases* 1985; **11**: 523–49.

225. Wood D D, Ihrie E J, Dinarello C A, Cohen P L. Isolation of an interleukin-1-like factor from human joint effusions. *Arthritis and Rheumatism* 1983; **26**: 975–83.

226. Wooley P H, Griffin J, Panayi G S, Batchelor J R, Welsh K I, Gibson T J. HLA-DR antigens and toxin reaction to sodium aurothiomalate and D-penicillamine in patients with rheumatoid arthritis. *New England Journal of Medicine* 1980; **303**: 300–2.

227. Yoneda T, Mundy G. Monocytes regulate osteoclast-activating factor production by releasing prostaglandins. *Journal of Experimental Medicine* 1979; **150**: 338–50.

228. Yu C L, Chang K L, Chiu C C, Chiang B N, Han S H, Wang S R. Alteration of mitogenic responses of mononuclear cells by

anti-ds DNA antibodies resembling immune disorders in patients with systemic lupus erythematosus. *Scandinavian Journal of Rheumatology* 1989; **18**: 265–76.

229. Yu D T Y, Winchester R J, Fu S M, Gibofsky A, Ko H S, Kunkel H G. Peripheral blood Ia-positive T-cells. Increases in certain diseases and after immunization. *Journal of Experimental Medicine* 1980; **151**: 91–100.

230. Zvaifler N J. Pathogenesis of the joint disease of rheumatoid arthritis. Proceedings of oral gold symposium. *American Journal of Medicine* 1983; **75**: 3–8.

Chapter 6

Mechanical Basis of Connective Tissue Disease

C G ARMSTRONG, P O'CONNOR and D L GARDNER

Methodology of mechanics

Attempts to understand the pathogenesis of connective tissue disease often invoke the laws of mechanics. These principles are not generally described in pathology texts. The present Chapter therefore brings together some physical and biological aspects of the connective tissues with particular regard to ways in which common diseases may alter their mechanical properties. There are two aims: first, to outline methods by which tissue mechanics can be assessed and defined; second, to examine how these properties may be disordered. A glossary of terms is provided.

Bone is not considered in the present text and the reader is referred to Carter and Spengler,[36] Frankel and Nordin,[50] Currey[39] and Currey *et al.*[40] for reviews of its mechanical properties. The mechanical responses of the skin and of blood vessels are mentioned in Chapters 9 and 15, respectively. For reviews of their properties, Fung[54] provides a full account.

The principles of mechanics are described in standard works.[160] For entertaining introductions to the mechanical properties of materials the reader may consult Gordon.[60–62] Wainwright *et al.*[175] discuss the loading characteristics of a broad range of biological tissues. Comprehensive descriptions of the mechanics of the connective tissue matrix[54] and of the tissue matrix[77] are recommended.

Mechanics is the study of the response of matter to forces

The principles of biomechanics rest on the need to measure the physical properties of tissues and organs. The tests employed by mechanical engineers are best suited to the study of materials that have a natural, defined, *in vitro* resting shape and size. The biological connective tissues are not like this[53]: their shape and size vary in response to the forces exerted on them. Inconveniently, the same biological tissue may behave differently at different times of day, at different temperatures and in distinct ways in the young and old, the male and female. One solution to this problem is to precondition a material before testing by preliminary preloading or repetitive stressing until a 'steady state' is achieved.[53,171] The justification for this preliminary treatment is that connective tissue is often prestressed *in vivo*: tendons retract when cut; cartilage splits when punctured with a round needle. When tissue is dissected out, the support provided by the surrounding material is removed; fluid movement occurs. Pressure, drying and osmotic changes cause swelling or shrinkage and the restoration of an *in vivo*-like state is a desirable preliminary to testing.

Glossary of Terms

Force tends to cause acceleration of a body from a state of rest or uniform motion. **Mechanics** is the study of the response of matter to force systems, and is based upon Newton's Laws of Motion. The SI unit of force is the Newton (N), which is the force required to accelerate a mass of 1 kg at 1 m/s^2, that is, in one second the velocity of the mass will have increased by 1 m/s. Less precisely, 1 N is the approximate weight of an apple. A force of one pound equals 4.45 N.

Stress is a measure of the force per unit area acting on a surface. **Tensile stress** tends to cause an increase in length of a body, **compressive stress** tends to cause a decrease in length. **Shear stress** results when the applied force lies in the plane of the surface. Friction causes shear stress on a surface. When a round bar is twisted shear stress is developed on its ends. The SI unit of stress is the N/m^2 or Pascal (Pa). 1 MPa = 10^6 Pa = 145 psi, 1 kPa = 10^3 N/m^2 = 7.5 mmHg.

Strain is a measure of deformation per unit length of material. **Tensile strain** is the increase in length per unit length of material along a given line, while **compressive strain** is the decrease in length per unit length. **Shear strain** is conventionally the change in angle in the deformed state between two lines which were perpendicular in the undeformed state. Since strain is a ratio of two lengths (a deformation and an initial length) it is dimensionless. It is often given as a percentage rather than a ratio.

The **mechanical properties** of a material can be divided into two main groups: how stiff the material is (how much does it deform under a given stress); and how strong it is (how much stress can the material support before it fractures or permanently deforms).

Elastic material is one which immediately recovers its undeformed shape when the load is removed (engineering usage). When a rectangular strip or round bar of material is loaded in tension the ratio of tensile stress:tensile strain is called the **elastic modulus**. This material property, which has the units of stress, is independent of the size of specimen used to make the experimental measurement.

Viscoelastic material is one in which the strain depends not only on the stress but on the length of time for which the stress is applied. The deformation caused by a rubber band round a finger becomes greater the longer the band remains. This time-dependent increase in strain under constant load is known as **creep**. In engineering terms, bone is an elastic material, while elastic fibres exhibit viscoelastic properties.

The **ultimate tensile strength** of a material is the largest tensile stress which can be supported before the material fractures. In fibre-reinforced materials such as connective tissue the properties of the material are usually different when tested in different directions, a phenomenon known as **anisotropy**. Materials can also fail because of repetitive stressing at a level below the ultimate strength of the material. This is **fatigue failure**.

The testing of whole parts

Test systems can be developed that allow whole composite structures such as synovial joints to be studied. There are problems in such analyses. In the case of the hip joint tested in compression, it is, for example, exceedingly difficult to record the magnitude and direction of forces throughout the joint *in vivo* and impossible to measure deformation in all parts of each of the many structures of which the joint is composed. It is therefore often necessary to approach the mechanical attributes of a whole organ by analysing those of individual parts.

The testing of individual components

In many connective tissue organs and structures, molecular heterogeneity is reflected in mechanical complexity, and the testing of these parts or organs is no less difficult than the testing of the whole structure. In other instances, of which hyaline articular cartilage is the most important example, a superficial appearance of homogeneity facilitates the examination of selected specimens in the shapes (Fig. 6.1) preferred by mechanical engineers. The samples can be examined and tested with reasonable ease: applied and resultant forces can be measured and the observations extrapolated to the whole cartilage. The selection of specimens of uniform shape enables anomalous stress concentrations such as those caused by retention within a clamp during testing, to be minimized. In turn, the capacity to use standardized tests on specimens of predetermined, uniform shape and size permits both the prediction of intrinsic material properties and the comparison of different samples.

A disadvantage imposed by the assumption of molecular homogeneity is that it is easy to overlook intrinsic structural heterogeneity. Hyaline articular cartilage (see p. 68) is, in fact, extremely heterogeneous; the problems encountered in analysing its mechanical properties are considered in detail on p. 265.

In any event, whereas the microscopist records topographical descriptions of organs or tissues, and supplements these observations with histochemical, immunocytochemical and other techniques, the mechanical engineer assembles quantitative data, describing the properties of an individual organ or tissue in terms of well-established physical tests that have often been developed for the analysis of inanimate materials, metals or plastics. There may be considerable difficulty in relating the changed topography of a diseased tissue to its altered mechanical properties.

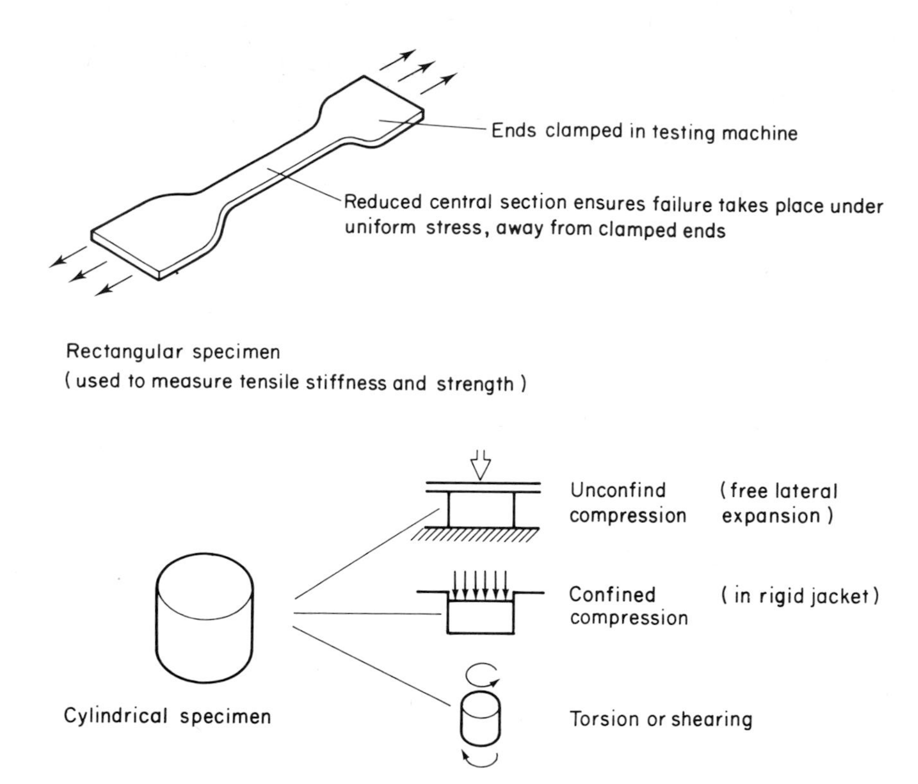

Fig. 6.1 Testing of connective tissue components. Geometry of different species used to determine mechanical properties of material. Information derived from these tests is: 1. relationship between stress and strain, i.e. the modulus or stiffness of the material provided that the stresses do not cause permanent damage; 2. strength of the material, i.e. what stress or force/unit area can the material support before permanent deformation or fracture occurs.

Mechanical properties of connective tissue components

Collagen

The fibrous collagens (see pp. 42, 174 and 302) are the main determinants of the mechanical behaviour in tension of tendon, ligament, meniscus, labra and aponeurosis; they contribute greatly to the properties of discs and bone. The properties of such tissues are not simply the product of the mechanical characteristics of the collagen-rich material; they are also strongly influenced by the spatial relationships of individual fibres to each other and of individual fibre bundles to other fibre bundles. For example, Akeson *et al.*[4] emphasize that proteoglycan and water have a lubricating function within fibrous connective tissue, separating fibres and allowing independent gliding of the microfibrils; without this independent fibre motility, there is limited tissue mobility. In tissues such as hyaline cartilage, fibrocartilage and fibrous tissue in which compressive forces predominate, the fibrous collagen alone is ineffective in resisting these major stresses; its function in these tissues is to form a three-dimensional network to stabilize and restrain the flow of the proteoglycan gel which resists the hydrostatic pressure. In calcified cartilage, these matrix materials are reinforced by the deposition of calcium hydroxyapatite crystals and the orderly arrangement of these crystals in relation to the lamellae of bone affords particular strength.[15,36] Both the non-mineralized and mineralized compact connective tissues are much stiffer and less extensible than the loose connective tissues such as those of the skin.

Collagen resists tensile stress

Individual collagen molecules are highly organized; their geometrical arrangement is of overriding importance in influencing the mechanics of fibrous connective tissue.[20,42] The mechanical properties of collagen are dependent on cross-links (see p. 175) between the molecules; without these cross-links, there is loss of mechanical strength, a structural failure of a kind encountered in the inherited disorder of dyspraxia and in the metabolic defects of copper

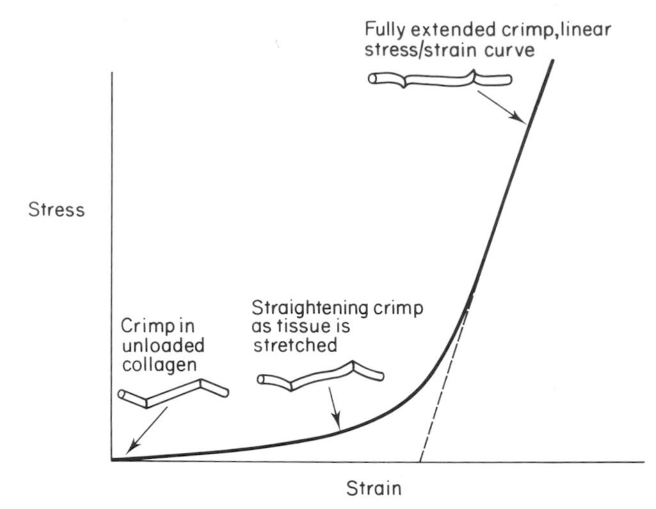

Fig. 6.2 Stress/strain behaviour of tendon.
Plot of force/unit area (stress) applied to tendon against resulting strain or increase in length per unit length. Initially, tendon is relatively flexible due to crimped collagen fibres. As stress increases, crimp is extended and tendon becomes stiffer.

deficiency, lathyrism and scurvy (see Chapter 10). Collagen strength decreases with increasing fibre diameter.[67] The large diameter fibres are made up of smaller fibrils with a greater chance of failure because of slippage between fibrils, rather than because of the breakage of collagen molecules. In native collagen, the initial mechanical response to tensile forces is linearly elastic. In collagen derived from connective tissue such as tendon[42] or skin,[93] the stress/strain curve has a characteristic shape (Fig. 6.2). Initially, relatively large strains develop in response to low levels of stress. Subsequently, there is an increase in stiffness until an increase in length or strain of approximately 4 per cent is achieved. The stress/strain response then remains linear until a point is reached at which the fibres come apart (dislocate): this is the yield point.[20] The appearance of the stress/strain curve in Fig. 6.2 indicates that large extensions of the collagen fibrous structure can occur in response to small loads: when very high loads are applied, extreme deformation is prevented.

Collagen in tendon, ligament and loose connective tissue has a natural zig-zag structure called 'crimp'

The collagen fibres of unstressed compact connective tissues such as tendon and ligament, and of loose connective tissue such as areolar synovial tissue, have a natural zig-zag, crimped structure (Fig. 6.3).[144,159] This arrangement determines the mechanical behaviour of these tissues in response to low tensile loads. Because a crimped structure offers little initial resistance to loading, there is a relatively large deformation as the zig-zag straightens out. Beyond the point at which straightening is complete, the tissue is much stiffer.[126] During straightening, other fibrous materials with different physical properties, such as elastin, may also come to carry part of the load. Since the crimp has a variety of wavelengths, not all the collagen fibres are straightened at the same time; a load is assumed by fibres one-by-one, the stiffness increasing gradually.[174]

Graphs showing the stress/strain response of collagenous tissue such as tendon and ligament can be divided into three regions. In the first, collagen fibres are crimped; in the second, there is progressive straightening of the crimp; and in the third, elongation of the fibres occurs (Figs 6.2 and 6.3). Most physiological strains are within the first two regions. If a tissue is strained in excess of approximately 4 per cent, permanent, non-recoverable strains may be incurred.[37,47,67] These severe strains may be associated with tendon or ligament rupture, joint dislocation or the evulsion of bone.

Betsch and Baer[20] claim that crimp is only seen in tissues that form tensile units with little matrix intervening between the collagen fibres. However, Broom and Myers[23] and Broom[22] found localized islands of crimped collagen fibres in hyaline cartilage; they employed differential interference contrast microscopy (see p. 74).

The molecular explanation for crimping is unknown. In a theoretical analysis of flat collagenous tissues, Lanir[93,94] proposed two models: the first assigned crimp to a pres-

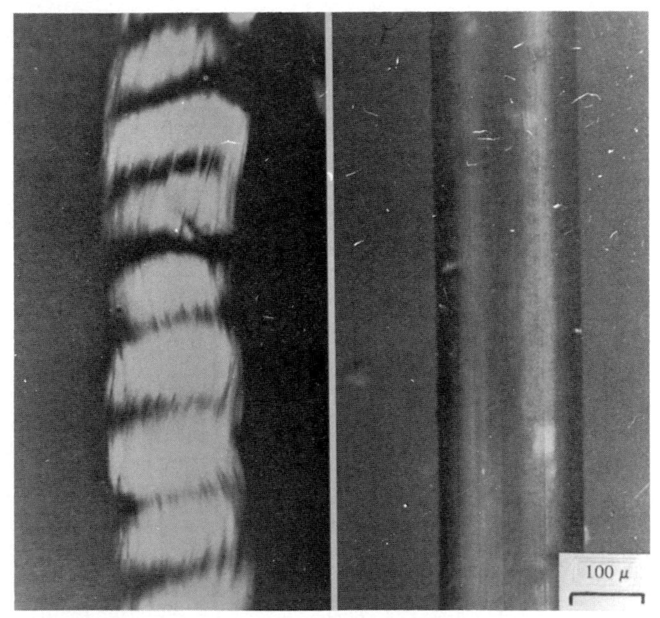

Fig. 6.3 Crimped, zig-zag structure of tendon.
Polarized light optical micrograph without applied tension (left) and with 1 per cent extension to straighten crimp in fibrils (right). Reproduced with permission of D F Betsch and the editor, *Biorheology*.[20])

tress in the accompanying elastic fibres which, attached to collagen fibres at regular intervals, caused collagen to 'buckle'. The second identified the submolecular organization of collagen itself; elastin existed independently. In this model the elastic fibres were straight at equilibrium stress while the collagen fibres undulated. Both theories assumed that there was no interaction between the fibres and the proteoglycan matrix. This now seems unlikely to be true.[25,128,151,152] In osteoarthrosis (see p. 858), crimp is often observed, by light microscopy, in the collagen fibre bundles that become incompletely detached from the surface as fibrillation develops.

The effect of crimp on the mechanics of cartilage

The abnormal appearance of a crimped structure in hyaline articular cartilage is associated with a reduction in the stiffness of these localized areas. It has been argued that crimp in the pericellular fibre arrays may act as a strain-sensing system, translating matrix deformation into chemical information comprehensible to the chondrocyte.[24]

Proteoglycan

Hydrated proteoglycan aggregates determine the compressive resistance of cartilage

The strong tendency of proteoglycan to imbibe water creates a swelling pressure within cartilage which forms the basis of compression resistance in cartilage. If the molecular subunits of proteoglycan (see p. 193), the glycosaminoglycans, are degraded by enzymes such as trypsin and bacterial chondroitinase,[90] the compressive stiffness is severely reduced. Degradation of proteoglycans also influences the tensile behaviour of articular cartilage, at least at low stress levels[90] where abnormally large strains are incurred. This was thought to be due to the removal of the resistance to fibre realignment that proteoglycans have at low stress levels.

Proteoglycans bind water within their domains. Restrained within a molecular framework of collagen fibrils, the hydrated proteoglycans are the main determinants of the mechanical behaviour in compression of hyaline articular cartilage, nucleus pulposus, dermis and arterial intima.

Mechanical properties of connective tissue: hyaline articular cartilage

The mechanical properties of hyaline articular cartilage are best considered against a background in which the overall, mechanical characteristics of synovial joints are reviewed.

Synovial joints (see p. 128) permit a wide range of movement. They are very efficient mechanically and, subjected to approximately 10^6 loads annually (each about 3–4 times body weight), function satisfactorily for 70 years or so. Articular cartilage gives synovial joints many of their remarkable mechanical attributes. In engineering terms, articular cartilage can be regarded as a fibre-reinforced, hydrated gel. The fibrous microskeleton is collagen, the gel largely proteoglycan. The proteoglycans (see p. 193) are strongly hydrophilic because of their high negative fixed charge density. They retain much water within their domains, restricting fluid flow and acting as the compression-resisting component during load bearing. A fibre component, principally of type II collagen, forms a three-dimensional, tension-resisting network which restrains and entraps the expanded proteoglycan molecules. Articular cartilage is heterogeneous and anisotropic; it can be considered as a layered (zoned) material (see p. 68). The most superficial zone (zone I) is rich in collagen fibres arranged parallel to the surface. Within this zone the fibres in different areas of the joint have preferred orientations that influence the split pattern produced when a round pin punctures the cartilage surface (Fig. 6.4).[78,111,124,125] In zones II and III, collagen fibres are arranged randomly or pseudo-randomly; in zone IV they form arrays that anchor the cartilage to the underlying calcified cartilage and bone.

Hyaline cartilage serves two main functions

Articular cartilage has mechanical properties which enable it to serve two main functions within diarthrodial joints. First, the cartilage reduces the stresses applied to the subchondral bone; second, it provides a bearing surface of low friction which is also resistant to wear. Beneath the cartilage, articulating bone ends are not congruent. They have an irregular shape. The bone is hard and not easily deformed. If bone is loaded directly, as it is when cartilage is lost in advanced osteoarthrosis (see Chapter 22), extremely large local stresses are created. Hyaline cartilage pro-

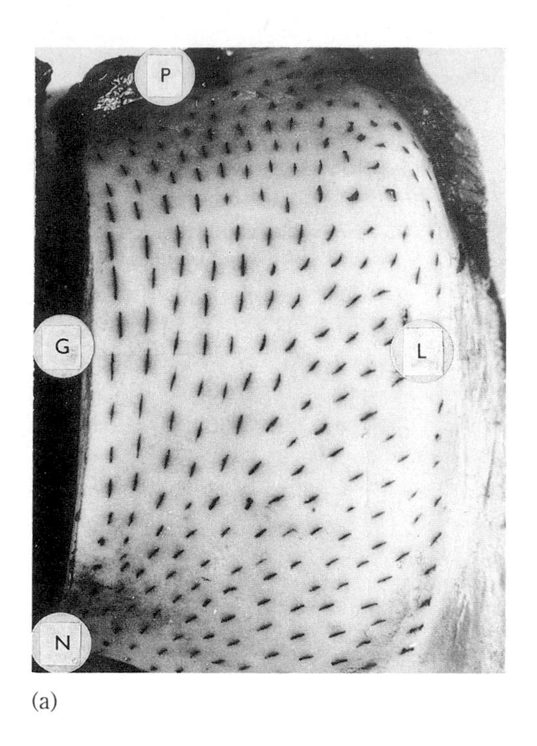

(a)

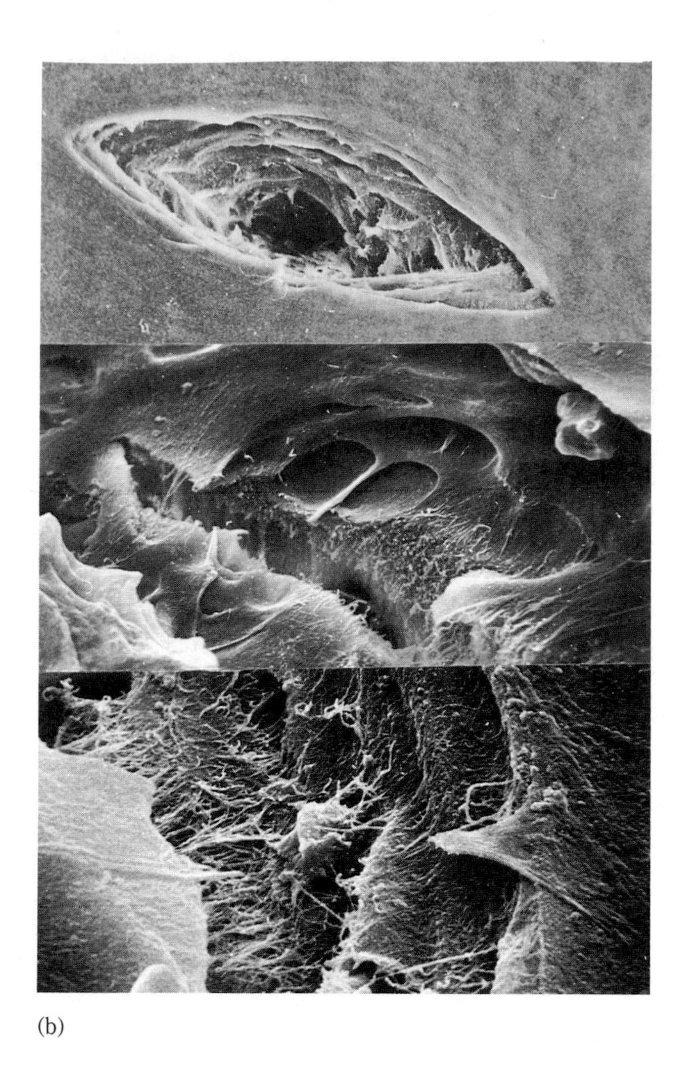

(b)

Fig. 6.4 Split pattern in articular cartilage.
a. Artificial splits made by pin-pricks display orientation of surface collagen fibre bundles on lateral surface of femoral patellar groove. From male aged 12 years. **G**: midline of groove; **L**: lateral edge; **N**: apex of intercondylar notch; **P**: proximal edge. By courtesy of Dr G. Meachim[111] and the Editor, *Journal of Anatomy*.) **b. top**: detail of unidirectional layered split; **centre**: deep crack in floor of layered split; **bottom**: part of unidirectional layered split to indicate orientation of collagen fibres. From O'Connor, Bland and Gardner,[125] by courtesy of the Editor, *Journal of Pathology*.

tects subchondral bone from mechanical damage by increasing the contact area, spreading the load and reducing stress concentrations. Provided its macromolecular composition is unchanged, cartilage is effective in distributing loads over a joint surface even if the cartilage has undergone thinning as it does in advanced osteoarthrosis. A critical point is reached however, when the cartilage thickness diminishes so that bony asperities and the irregularly shaped underlying bone are exposed; bone-to-bone contact then begins, resulting in high stress concentrations in the remaining, nearby cartilage and in the exposed bone, damaging both.[41]

The friction in synovial joints is extremely low. The coefficient of friction is approximately 0.01–0.04 (Table 6.1). The mode of lubrication is still debated (see p. 272 *et seq.*) but the rates of physiological wear are negligible. The relationship between defective lubrication, i.e. excess friction and wear, and the fibrillation of osteoarthrosis is not fully understood, but impaired lubrication is no longer thought to be important in the pathogenesis of this disorder. The degeneration of cartilage with age is most noticeable on surfaces that are not in habitual contact (see Chapter 21).[29]

Cartilage is subject to dynamic forces

In order to understand the mechanical function of a biological material, the magnitude and direction of physiological stresses must be known. The resultant stresses depend on

Table 6.1

Typical coefficients of friction*

	Coefficient of friction
Non-biological systems	
Rubber tyre on dry road	1.0
Nylon against steel	0.3
Fluid-film bearing	0.001
Entire synovial joints	
Human hip	0.01–0.04

* The coefficient of friction is the ratio of the force required to cause motion in a bearing to the force the bearing is supporting.
(From Unsworth *et al.*[172], O'Kelly *et al.*[127] and Roberts *et al.*[145])

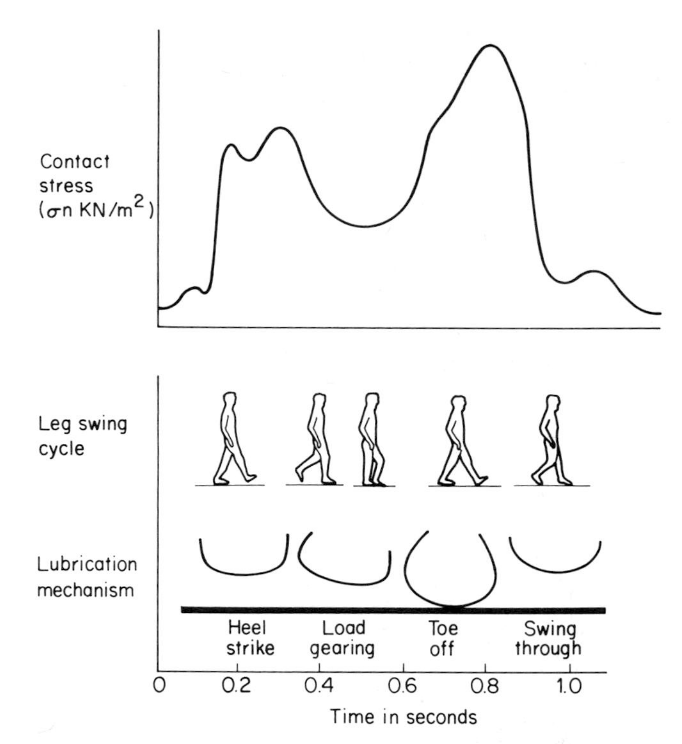

Fig. 6.5 Walking cycle, load and lubrication of human hip joint.
Diagrammatic relationship between: **a** total load on hip joint calculated by gait analysis; **b** phases of walking cycle; and **c** lubrication mechanisms. Peak loads of 4–5 times body weight occur when heel strikes ground and at 'toe-off' position when weight is pushed forward onto other leg. From Paul[135] and Unsworth *et al.*[172]

all the external forces acting on the system. In a synovial joint, the total force transmitted during articulation is a resultant of external forces acting on the extremity, the effects of gravity and acceleration due to movement and the internal forces resulting from muscular contraction. For simple static cases such as the hip joint reaction while standing on one leg, the analysis is straightforward. Frankel and Nordin[50] describe the forces in large joints under these conditions. For dynamic loading, when several muscle groups are active simultaneously, the analysis is much more complex.

Lower limb

Indirect measurements have been made by means of systems that permit aspects of gait to be filmed while loads are recorded (dynamic gait analysis). Paul[135–138] measured the relative movements of the joint components of the hip and knee during normal walking and stair-climbing. Cine-filming, force plate analyses and electromyograms allowed the measurement of ground reactions and the calculation of limb segment dynamics. From these results, the magnitude and direction of the resultant forces imposed on these joints were determined. Similar methods were used to analyse the forces during movement in the human hip[38,57] and knee[154] and in the surgical assessment of abnormal knee and hip joints.[65,68,83] These studies indicate that, even in everyday activities such as walking, forces of the order of 4–5 times body weight are generated at the hip joint (Fig. 6.5).

Indirect measurements make assumptions about muscle activity and lines of action which are difficult to verify. Telemetry is a direct method of monitoring forces at joint surfaces;[35,146–148] it has been used to transmit loading information of compressive forces from transducers positioned in prosthetic joint replacements. However, this technique can only afford information about joint loads after surgical replacement rather than in the normal articulation. One example of evidence obtained in this way is the demonstration that the most effective reduction of load on a hip joint after surgical replacement is by supporting the body weight using a walking stick on the opposite side.

Upper limb

The measurement of forces in the upper limb is more difficult. The range of movements is more extensive and under more precise neural control. Much of the limited data is concerned with the forces in finger joints[19,129,140] during activities such as tap-turning. The elbow[118,119] and shoulder[46,49] have been less fully studied. Calculations by Amis and co-workers[6] show forces of up to 3 kN (670 lb) during strenuous isometric effort at the humeroulnar and humeroradial joints.

Hyaline articular cartilage sustains stresses

If the total force imposed on a joint and the size of the contact area are known, surface pressures can be estimated. Many methods have been used to measure contact areas: tests have been made under different conditions of loading *in vitro*. Dyes such as safranin O have been injected into intact, unloaded or loaded cadaver joints positioned in selected orientations. Contact areas remain unstained.[59,63,64] Quick-setting acrylic cements[2,44,156,176] and radioopaque contrast media[105] have been used with radiography to define these areas. Maximum contact in the hip joint is achieved at approximately 21 per cent of body weight.[64] A correlation has been drawn between the extent of regular contact and non-contact areas and the presence of progressive and non-progressive disease (Chapter 22).[31–33] Cartilage degeneration in non-contact areas is believed to be non-progressive; in contact areas degeneration, when present, is progressive.

To investigate regional variations in loading over a joint surface, Day *et al.*[41] removed parts of the articular cartilage in sequence. More direct measurements of pressure distribution have been attempted by interposing a pressure-sensitive layer[79] or film[3,52,76] between the articular surfaces. Armstrong *et al.*[9,10] estimated the distribution of pressure on hip joint cartilage from radiographic measurements of cartilage deformation. Miniature pressure transducers inset into the cartilage either from beneath[1] or above a surface[26] have also been used.

Measurements have been made *in vivo* in patients into whom artificial joint prostheses incorporating pressure transducers, have been fitted.[146,147] The results of measurements from these various studies suggest that even in a geometrically simple joint such as the hip, local variations in 1. joint surface shape (geometry),[147] 2. cartilage thickness,[11,146] and 3. stiffness,[89] lead to an irregular distribution of pressures, with peak values of the order of 4–8 MPa (600–1200 psi).[26,76]

The stress required to cause articular cartilage to fail in tension is in the range 3–35 MPa.[87] However, the development of a surface pressure of, say, 8 MPa does not imply that the material will necessarily fail: a hydrostatic pressure applied equally from all directions will not cause mechanical damage. The most simple, useful criterion for assessing the failure of ductile engineering materials (with which cartilage can be compared) expresses the difference in the principal stresses in a material or the equivalent maximum shear stress. If an articular surface is subjected to a uniform pressure, then any given unit of cartilage is supported laterally by its neighbouring units (Fig. 6.6). The resulting lateral compressive stress ensures that the principal stress difference is negligible, i.e. there is no tendency to cause failure of the material. In a loaded joint, pressure cannot be uniform at all sites. Where pressure gradients exist, the support for a unit of cartilage will be less on the side subjected to lower pressures. Cartilage will tend to be

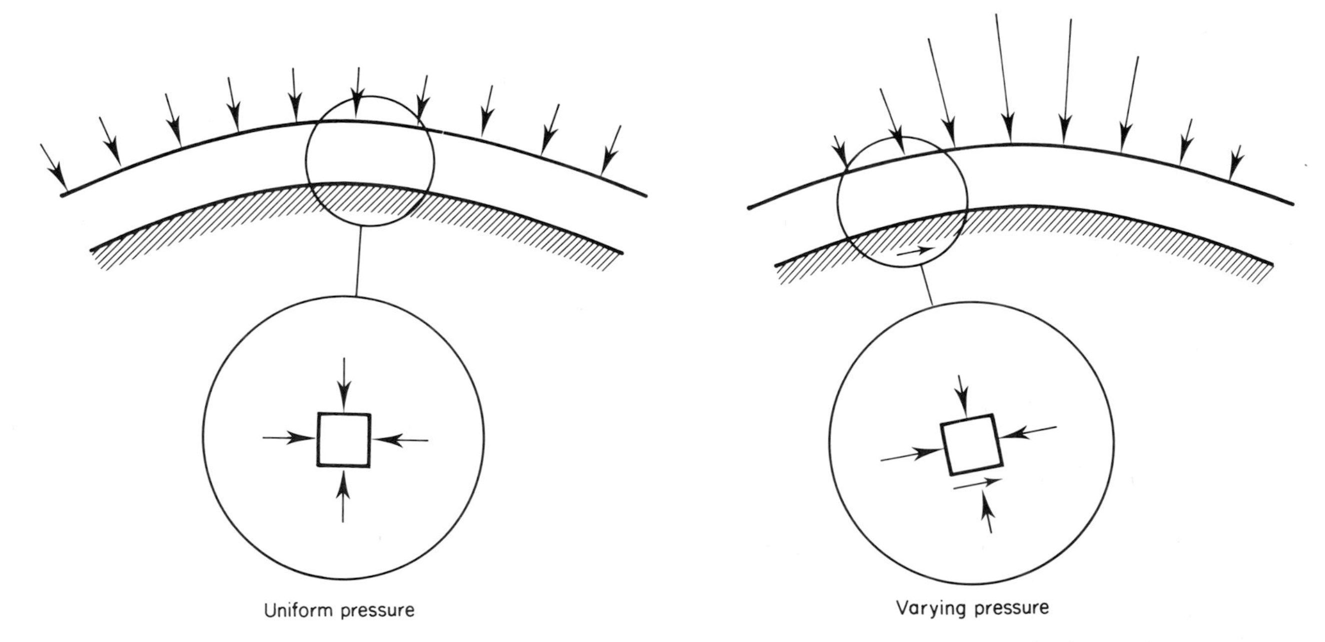

Fig. 6.6 Stresses in articular cartilage in response to uniform and varying pressure on articular surface.
Uniform pressure creates a state of hydrostatic pressure in cartilage which will not be damaging. In the varying pressure case, pressure to right of element identified in the figure is greater than pressure to left creating lateral pressure difference across element. Pressure difference must be balanced by shear stress at subchondral bone junction.

pushed in this direction (Fig. 6.6) causing large shear stresses at the subchondral bone boundary.[8] These shear stresses may explain the horizontal splitting observed *post-mortem* at the cartilage–subchondral bone junction of the patella.[110] Defects of this kind could account for the basal degeneration (blistering) of patellar cartilage observed by Goodfellow, Hungerford and Woods.[58]

At an articular surface, a tendency for the mechanical failure of cartilage depends on the thickness of the material and on the rate at which the pressure varies with position along the surface.[8] In a comparative analysis of cartilage thickness and congruence in joints from different species, Simon[163–165] found osteoarthrosis to be more prevalent in thick cartilage from incongruent joints than in thin cartilage from congruent joints, where the pressure variations are presumably less severe. Cartilage degeneration in areas devoid of habitual contact could also be caused by occasional contact: the pressure variations would be particularly rapid near the edge of the contact areas.

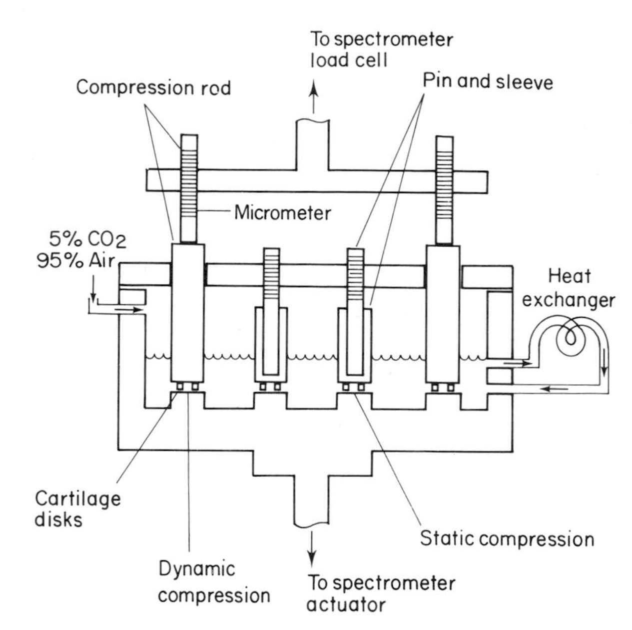

Fig. 6.7 Mechanical testing in compression of cartilage explants under *in vitro* culture conditions.
Chamber designed to permit biosynthetic properties of cartilage to be measured in response to unconfined dynamic loading under living conditions. Cartilage discs can be dynamically compressed between base of chamber and rods. Simultaneously, control discs in same reservoir of culture medium, can be statically compressed between chamber base and steel pins. Culture medium is recirculated through heat exchanger. Temperature is maintained at 37°C while a 95% air/5% CO_2 gas mixture is passed into the chamber. From Sah *et al.*[149] by courtesy of Professor A J Grodzinsky and the Editor, *Journal of Orthopaedic Research.*

Hyaline cartilage can be tested by many different techniques

Information of very great value in explaining the behaviour of cartilage in disease can be obtained by the mechanical testing of this tissue. In spite of its apparent uniformity, it is difficult to study the mechanical properties of articular cartilage directly; in molecular terms, it is a complex and composite material. Tension, compression, indentation and shear tests have been performed to determine both the instantaneous and the time-dependent (viscoelastic) properties. Experiments have been made with whole joints; on single components such as the femoral head; on intact surfaces, by indentation; on isolated pieces of cartilage or cartilage-on-bone, of uniform geometry; and on individual tissue constituents such as collagen or proteoglycan (see p. 262).

Techniques have also been developed which allow some of the mechanical properties of cartilage to be tested *in vitro* under conditions which permit the biosynthesis of proteoglycan to be measured. By these means, the effects of compressive stress on cell behaviour can be estimated (Fig. 6.7).

The response to tensile loading

Tensile testing may be performed by pulling uniform, usually rectangular or dumbbell-shaped specimens between grips or clamps in a direction parallel to the long axis of the specimen.

In a normal synovial joint, articular cartilage is not subjected directly to tensile force. However, tensile strains or stretching deformations result at the articular surface because of the non-uniform distribution of pressure.[8,16] Furthermore, most collagen fibres are under tension due to the swelling pressure of the proteoglycans enmeshed within their network.

It is clear that any change in the alignment of collagen influences the tensile properties of the cartilage. For example, the anisotropy[124,125,187] of the superficial layer, in a plane parallel to the surface, is responsible for the split (Fig. 6.4) caused by pricking the surface of articular cartilage with a sharp, round pin.[78,111,124,125] Each prick produces an elliptical elongated split: the split has a preferred direction (see p. 265). The pattern revealed when a series of such pricks is made on a main bearing surface such as that of a femoral condyle is characteristic and shows that, generally, splits run at right angles to the commonest movements of the opposed bearing surfaces. The direction of the splits can be correlated with the orientation of surface collagen fibres, shown by polarized light microscopy,[28] transmission electron microscopy[111] and by scanning electron microscopy.[125] The arrangement of collagen fibres is,

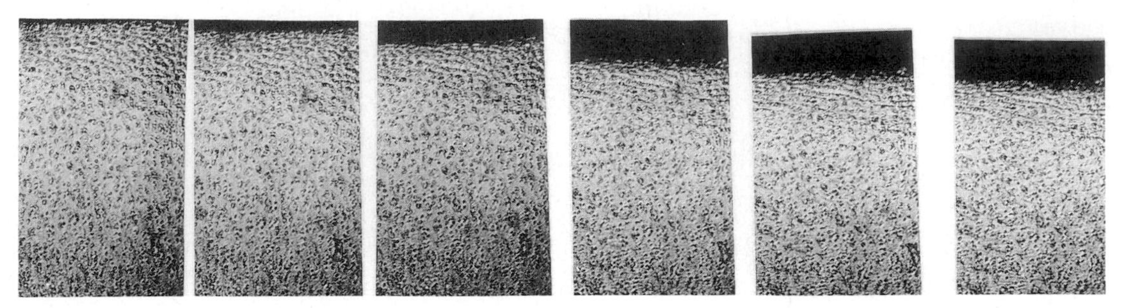

Fig. 6.8 Compressive loading of cartilage.
100 μm section of fresh articular cartilage under compression in surface to bone direction while simultaneously viewed by differential interference contrast microscopy. Deformation response is different in each layer: the mid zone is stiffest, the layer closest to the articular surface the softest.[125]

however, difficult to discover by the examination of sections. Nevertheless, much progress has been made in understanding this complex organization (see p. 74).

Anisotropy has also been observed during the tensile testing of dumbbell-shaped slices cut tangential to the surface of human femoral condylar cartilage[88] and of bovine humeral cartilage.[183,185] Specimens cut parallel to split lines have greater stiffness and strength than others cut perpendicularly; a similar, smaller difference is seen with specimens from deeper cartilage layers. Anisotropy is not confined to surface material. Tensile stiffness and strength, in directions parallel to the split lines, decrease with depth, indicating changes to a random and then to a radial arrangement of collagen fibres. Local differences in mechanical behaviour at different depths have also been measured during the compressive loading of thin (~100 μm) slices of articular cartilage (Fig. 6.8).[126]

The experimental degradation or removal of collagen by enzymes such as clostridial collagenase, adversely affects the tensile properties of articular cartilage.[90] The consequences of disrupting proteoglycan are less predictable (see p. 202).

The response to compression forces

Indentation A beguilingly simple method of applying a compressive load to an exposed articular cartilage surface or to pieces of excised cartilage is to use an indentor of known shape. The resulting deflection or deformation can be measured. The technique was applied to articular cartilage by Bar[18] and by Gocke[48] to test viscoelastic behaviour; an initial instantaneous deformation was recorded followed by a phase in which deformation increased with time towards an asymptotic, equilibrium value. In early studies, performed in air, complete recovery was not achieved when a load was removed. If, however, the cartilage was prevented from drying, the deformation recovered fully.[56,89,133]

A band of cartilage, stiffer than the surrounding material and corresponding to the area occupied by the opposed acetabular cartilage, has been mapped on the human femoral head by indentation tests.[89,34] Although easy and quick, indentation testing produces complicated stress fields which are difficult to analyse,[70,103] so that experimental results need to be carefully interpreted.[85,114] It is also important to note that the stresses under a small indentor used to cause localized deformation are substantially different from those due to joint loading *in vivo*.

Compression of isolated pieces of cartilage

The extent of the deformation produced in an isolated piece of cartilage under a compressive load depends on whether or not the tissue is allowed to expand laterally, i.e. whether the cartilage is 'unconfined'[12,27] or 'confined'.[14] In the intact joint, cartilage forms a continuous sheet across the subchondral bone. Under load, lateral expansion can most easily occur at the margins of the articular surfaces or around discontinuities such as the insertion of the ligamentum teres of the femoral head, sites where age-related fibrillation is observed (see p. 829). Zones of local abnormal softening, degeneration or thinning also accommodate lateral expansion and influence the deformation of the nearby cartilage.

The difference between the mechanical responses of tissues tested in confined and in unconfined modes can be very large. If a cylinder of tissue is compressed in a rigid chamber by an impervious piston, very large stresses are generated. Freedom for the tissue to expand laterally dramatically reduces its apparent stiffness. Both 'confined' and 'unconfined' tests can be used to assess the intrinsic mechanical properties of a tissue matrix.

A second major influence on the mechanical response of cartilage during compression testing is the extent to which interstitial fluid motion is allowed to take place (see p. 76 *et seq.*). When a static load is applied to an articular surface in confined compression, via a porous filter, deformation increases rapidly with time due to the flow of fluid out of the sample.[14,98] If however rapid cyclic (high frequency) loading is applied, the tissue appears to be much stiffer, since the resistance to the rapid cyclic movement of water is large.[74]

Differential interference contrast microscopy has proved very useful for observing the behaviour of thin slices of cartilage in tests made both in compression (Fig. 6.8) and in tension.[22–25,163,176a] In compression tests of slices taken perpendicular to an articular surface in which lateral expansion and fluid flow are restricted to a single plane (plane strain compression), O'Connor *et al.*[126] observed that midzone cartilage was stiffer than either superficial or deep cartilage.

A further consequence of confined compression is a mechanical-to-electrical transduction.[49a,b] Oscillatory compression with physiological loads produces electrical potentials that are a result of an electrokinetic transduction mechanism. Sinusoidal mechanical compression causes a sinusoidal streaming current density; conversely, a sinusoidal current density generates sinusoidal mechanical stress. These phenomena, demonstrable in living tissue, are related to the high fixed charge density of the extracellular matrix.[107] For this reason, electric streaming potentials can be used as indices of matrix degradation.[49c] Because of the intimate relationship between cell synthetic mechanisms and matrix macromolecular composition, it is highly likely that electrical, mechanical and chemical signals to and from chondrocytes converge via common pathways.[66]

The response to shearing stress

A shearing deformation involves a distortion or change in shape of a material without a change in its volume. In connective tissue this means that a small shearing deformation should not be accompanied by the gross movement of fluid. This allows the shear properties of the tissue matrix to be assessed without any contribution to load support from fluid pressure gradients in the matrix. Shear stresses can be generated either by the twisting of a cylindrical specimen,[71] or by direct shearing of a small rectangular specimen.[69] A shear deformation is equivalent to an extension along one diagonal of the specimen with an equal and opposite compression along the other diagonal. Both collagen and proteoglycan influence the shear properties of cartilage: collagen resists extension of the cartilage, proteoglycan resists compression and stabilizes the fibrous network. These same micromechanical functions for collagen and proteoglycan have recently been identified during indentation[84] and unconfined compression.[95]

Deformation and failure under repetitive loading

Fatigue failure of an engineering structure may occur when the structure is subjected to repetitive loading at stresses well below the ultimate strength of the materials involved: fracture of a paper clip caused by repeated flexing is a familiar example. This mode of failure may occur in connective tissue structures. Under the highly artificial conditions of loading produced by repeated indentation of femoral head articular cartilage, splitting is caused.[179] These authors concluded that cartilage is susceptible to fatigue damage. Similar damage to femoral head cartilage has been produced in cyclic compression testing.[157] Repetitive impulse loading of tibial cartilage and bone plugs reveals an increase in deformability of bone with the time of cycling. This has led to the hypothesis that the damage to cartilage may be due either to direct fatigue of cartilage itself or to softening of the underlying bone, causing increased strain in the overlying cartilage.[157] Cyclic tensile tests of isolated pieces of femoral head cartilage may also lead to fatigue failure. The numbers of test cycles necessary to cause failure decreases with increasing stress.[177,178] Fatigue resistance is independent of the content of proteoglycan and collagen although it decreases with age,[178] presumably because of the accumulation of microscopic defects in the material. *In vitro* fatigue testing can, of course, be criticized: first, because the stresses must be very large to permit failure within a reasonable time, and second, because tissue repair is not possible *in vitro*.

Testing of cartilage *in vivo*

Repetitive loading is the proximate cause of cartilage failure in animal models of human disease and in naturally occurring diseases of animals. The stifle joint of the dog and the femorotibial joint of the rabbit have attracted particular attention (see Chapter 22). Few direct studies have been made of the mechanical changes although there is frequent comment on cartilage softening, wear, bone remodelling and osteophyte formation.[139] Few measurements are recorded of the extent to which perturbations of joint mechanics affect the cartilage stress.

Instability of the rabbit knee joint after the operation of medial meniscetomy with division of the collateral and cruciate ligaments, leads to cartilage softening, increased total creep deformation and increased rate of deformation within 24 h. The changes, measured by indentation tests,[92] are ultimately reversible.[75] Mechanical tests can be sensitive measures of early cartilage change although alterations to underlying bone structure may contribute to these effects.

When rabbit heel pads are repeatedly subjected to sudden loads, microfractures occur. The fractures heal and lead to increased stiffness of the subchondral bone.[141] Paradoxically, either immobilization (disuse) or impulse loading and continuous compression (excess use) can cause cartilage degeneration (see Chapter 22). More than one mechanism of tissue breakdown is implied. The mechanisms may include: mechanical insults that cause damage directly; defective nutrition such as that resulting from failure of the intermittent loading that pumps fluid round the joint space; or alterations in endocrine and humoral functions that lead to changes in growth control mechanisms.

The flow of fluid through hyaline articular cartilage determines its mechanical behaviour

Since approximately 72–78 per cent by weight of cartilage is water, it is not surprising that interstitial fluid pressure and flow through the tissue matrix play important roles in tissue mechanics. The water content is attributable to the proteoglycans: they generate a swelling pressure of approximately 0.2 MPa (Chapter 4).[107] Expansion of the proteoglycan domains is resisted by the collagen network so that collagen fibres are under tensile stress even when no external load is applied. Depending on the precise way in which the fibres resist swelling, the resulting tensile stresses will be in the range 0.2–2.5 MPa.[108,116]

Mechanical loading[107] and changes in the external ionic environment[117,134] both perturb the equilibrium between collagen tension and proteoglycan swelling. When the material is subject to an external load, there is an immediate change in shape without significant change in volume (Fig. 6.9). The change in shape is accomplished by alterations in the form of proteoglycan domains and reorientation of the collagen fibrils. If a compressive load has been applied, a rise in interstitial fluid pressure occurs. Fluid begins to flow from the loaded site. As fluid is lost, the proteoglycan concentration is increased. The proteoglycan swelling pressure increases. Equilibrium is eventually reached between the swelling pressure and the externally applied compressive stress, and fluid motion ceases.[116] The rate of fluid flow, and thus the rate of tissue deformation, is dependent on tissue permeability and the pressure gradient. The permeability of cartilage is very low: it may take several hours for a specimen a few millimetres thick to reach equilbrium.[13]

Permeability is determined by the size and relation of the proteoglycan domains; an increase in glycosaminoglycan concentration, representing *more* proteoglycans or proteoglycans of different structure, results[106] in a decrease in permeability to water and small electrolytes. Cartilage

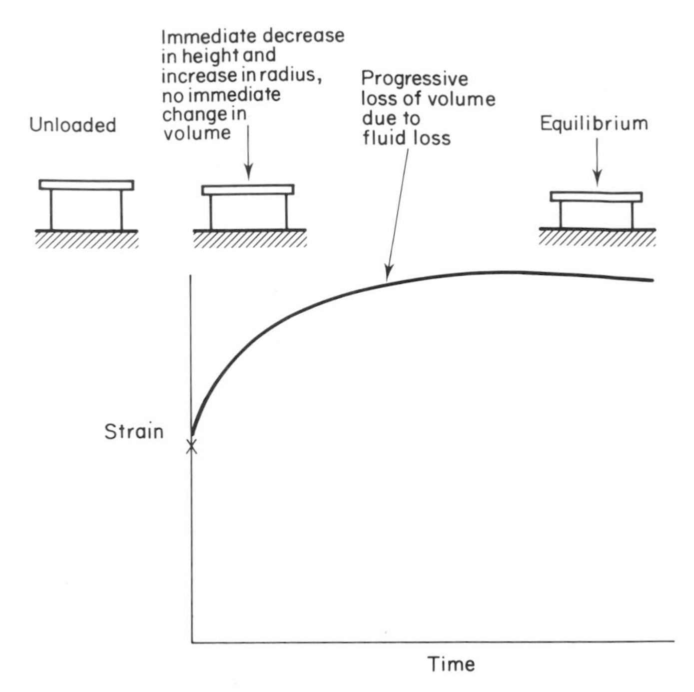

Fig. 6.9 External loading of cartilage cylinder. Response of cylinder of cartilage in unconfined compression under constant load. Initial elastic response is followed by transient phase due to fluid movement. Eventually, equilibrium state is reached. After Armstrong *et al.*[12]

permeability decreases at high compressive strains: the glycosaminoglycan concentration greatly increases because of water loss.[104,106,112] When cartilage samples containing different amounts of proteoglycan are subjected to the same compressive stress, deformation at equilibrium varies with the proteoglycan content. Increased proteoglycan leads to decreased total deformation and rate of fluid loss.[13,107] In cartilage treated experimentally with enzymes such as papain or trypsin to remove proteoglycan, fluid flow and compressive deformation greatly increase. The same conditions apply in disease. Proteoglycan is lost or degraded in cartilage zones subject to the fibrillation of ageing or osteoarthrosis: the degree of degradation is in proportion to the increased manifest activity of a metalloproteinase but not a collagenase.[109] The very low permeability of normal cartilage means that much of the functional compressive stress in the material is carried by the hydrostatic pressure of the entrapped interstitial fluid.

Hyaline articular cartilage is efficiently lubricated

Lubrication of cartilage-bearing surfaces during movement is one of the important properties of synovial fluid (see p. 89). The failure of lubrication often exacerbates diseases such as osteoarthrosis and rheumatoid arthritis. An understanding of the physiological lubricating mechanisms assists the analysis of joint function in health and disease. Central to this understanding is knowledge of the normal properties of synovial fluid (see p. 89): the mechanical functions of synovial fluid are to minimize friction during movement and to protect against excessive wear.

The nature of the bearing surfaces themselves is also of crucial importance. All cartilage-bearing surfaces that have been investigated in the disarticulated, non-loaded state (see p. 137) display orders of irregularities, the presence of which may influence lubrication (see p. 99). Many studies of these surfaces have been made by mechanical techniques which have contributed to an understanding of the role of the asperities and their size and frequency.[150,170] One factor relevant to the efficiency of joint lubrication is the congruence of an individual joint (see p. 129); another is the extent to which the surfaces are loaded at different points in the cycles of movements such as walking. Thomas and his colleagues[170] showed that in the human hip joint during heel strike the real area of cartilage contact was approximately 1.3 cm^2, the mean gap between the surfaces approximately 60 μm and the volume of fluid trapped between the surfaces less than 80 per cent of that in the standing position.

The forms of lubrication in any bearing are of two main categories: boundary and fluid film. Boundary lubrication is afforded by chemical adsorption of a monolayer of lubricant molecules on to the articulating surfaces; the surfaces are allowed to slide over one another, preventing the adhesion and abrasion of natural asperities. This mechanism is usually independent both of the physical properties of the fluid and surfaces and of the speed and magnitude of loading. In fluid-film lubrication a load is supported by pressure in a fluid film that is maintained externally (hydrostatic) or by the relative motions of the two surfaces (hydrodynamic or squeeze (fluid) film). Fluid-film lubrication is dependent on film thickness, load, viscosity and speed of movement. It is most likely to be effective during rapid, relative motion of the surfaces at light load. If the surfaces deform under load, in elastohydrodynamic lubrication, the load capacity is increased. Fluid-film and boundary lubrication can co-exist, resulting in mixed lubrication: part of the load is supported by pressure in the lubricant and part by the contact of asperities on the bearing surfaces.

Detailed reviews of joint lubrication mechanisms are available.[102,115,168,186] The coefficient of friction in synovial joints is extremely low, approximately 0.01–0.04 (Table 6.2), implying the existence of fluid-film lubrication. However, the fluid-film thicknesses predicted by any of the theoretical lubrication mechanisms (above) are 20 nm to 2 μm; except perhaps for relatively short-lived elastohydrodynamic squeeze films,[73] they are substantially less than the unloaded roughness (see p. 136) of the articular surfaces. This has led a number of workers[45,100] to propose novel modes of lubrication in which fluid motion to or from the porous, permeable cartilage contributes to lubrication. The motion of the interstitial fluid during joint movement has been the subject of debate.[45,73,100] A comprehensive account has been given by Mow and Lai.[113] As an area of joint surface moves into the contact zone, and, if there is no pressurized lubricant film already in existence, the interstitial fluid is expressed into the gap between the articular surfaces to form one. As the area moves off the contact zone the elastic solid matrix rebounds and resorbs the fluid expressed. This circulation of fluid is also potentially beneficial to cartilage nutrition.

Another mechanism by which fluid film lubrication could be maintained in unfavourable conditions has recently been proposed by Dowson and Jin.[43] They calculate that microelastohydrodynamic action largely smooths out the initial roughness of the soft articulating surfaces. Thus fluid films of the order of 0.5 μm may well be sufficient to maintain surface separation, even if the unloaded roughness is of the order of 5 μm.

Synovial fluid also acts as a boundary lubricant: it retains a lubricating ability even when its viscosity has been artificially reduced.[101,142,180] Swann *et al.*[167] have isolated a potent lubricant from synovial fluid; it is a family of glycoproteins called lubricin (see p. 211).

Synovial joints are subject to loading and sliding conditions which range from lightly loaded, high-speed articulation during the swing phase of walking (Fig. 6.5) to fixed static loads of long duration in standing. It is unlikely that these varied demands upon the bearing can be satisfied by a single mode of lubrication. It is probable however that elastohydrodynamic and microelastohydrodynamic fluid mechanisms of both the sliding and squeeze film type play important roles in joint lubrication, at least under favourable conditions. As the conditions of load and sliding become more severe, the fluid-film thickness decreases, contact of surface asperities occurs and the contribution of fluid circulating from the cartilage becomes dominant. At worst, the articular surface is protected by the adsorbed boundary lubricant in the synovial fluid.

Mechanical properties of connective tissue: fibrocartilage

Fibrocartilage (see p. 82) has a sparse cell population: it is a fibre-reinforced composite, rich in collagen. The principles by which fibrocartilage is tested mechanically are identical to those employed to test hyaline cartilage. The main forms of fibrocartilage subject to disease are the 23 intervertebral discs, and the menisci; labra are less often affected.

Intervertebral discs

(see pp. 103 and 126)

Due to the very short moment (lever) arm of the erector spinae muscles, large compressive forces are supported by the spinal column.[51] Forces up to 4 kN can be generated in the lumbar discs while lifting a load of only 20 kg or 200 N.[97] The function of an intervertebral disc is to permit spinal flexibility while transmitting these large loads from one vertebra to the next. The nucleus pulposus acts hydrostatically; it transmits loads radially to the surrounding fibrous lamellae of the annulus fibrosus and vertically to the vertebrae. The annulus fibrosus restrains the lateral bulging of the nucleus during compression, and carries the loads that bending, twisting and shear impose on the disc.

The behaviour of segments of the spine in response to loads has been studied both experimentally[132,143,169] and by theoretical stress analysis.[161,166] A simple theoretical model,[72,91] considers that each disc is a fibre-reinforced composite, the fibres having the same properties as those of tendon. This model has given insight into disc function and mechanical failure. Compressive loading of an intervertebral disc raises the internal pressure of the nucleus, stretching the annular fibres and causing the disc to bulge. With anterior, posterior or lateral bending of the spine, tensile stresses occur on the convex side of the annulus. Compressive stresses are incurred on the concave side; maximum stresses are at the outer margin.

Torsion or rotation of the spine about its long axis produces shearing stresses in the annulus. Because of the helical arrangement of the fibres, these stresses are resisted by tension in the fibres of the annulus. The magnitude of strain incurred at particular locations in a disc is proportional to the distance from the axis of rotation: the distance is maximal in the posterolateral position of the annulus. A combination of twisting and bending with rotation under load increases both stresses and strains and may cause disruption of the annulus and protrusion and prolapse of the disc (see Chapter 23).

Isolated specimens of intervertebral disc swell when immersed in 0.9% saline: much of the proteoglycan then leaches out of the tissue matrix.[173] This observation confirms that the collagen network is unable to retain the proteoglycan gel; the matrix is much less cohesive than, for example, that of hyaline articular cartilage.[75a] In extensive investigations of the mechanical properties of human intervertebral discs, Galante[55] and Panagiotacopulos *et al.*[130,131] used conditions of controlled humidity rather than immersion to regulate swelling. The tensile strength of annulus tissue was greatest in the direction of the collagen fibres; it was almost as great as that of tendon when the values were expressed in terms of collagen content.[55] The water content of the disc specimens had a large effect on the viscoelastic properties of the tissue,[131] presumably due to changes in proteoglycan swelling pressure.

A severely degenerate disc has impaired hydraulic and viscoelastic properties; there is both a reduction in water content and a loss of proteoglycan.[99] The annulus is prone to rupture, the cartilagenous endplate to degenerate. The loss of hydraulic capacity can be shown by injecting saline into the disc;[7] it is not retained in the degenerate tissue. Because the annulus is loaded directly in compression, the fibres of a degenerate disc separate: the nucleus is extruded and protrudes.[158] However, herniation is not caused by direct compressive stress alone: the experimental vertical loading of sections of human spine leads to failure of the bony endplates before herniation occurs.

The lumbar disc is often thought to be a cause of human back pain, a view supported by biochemical and biomechanical evidence of degeneration.[80,99] Although disc degeneration appears to be one price paid for the erect, bipedal stance of hominids, it also occurs in quadrupeds. A common factor in back pain is disc herniation associated with sites of maximum resultant mechanical stress.

Menisci

(see p. 103)

The knee menisci, in combination with the ligaments and capsule surrounding the joint, extend the tibial surface to accommodate the femoral condyles and stabilize the joint[153] by distributing loads over large areas. Loss of integrity of any of these structures leads to an increase in joint laxity.

The role of the menisci in transmitting load in the knee joint has been described by Shrive *et al.*[162] When the knee is lightly loaded there is little direct contact between the femoral and tibial articular cartilages; most contact is via the menisci. With increasing load, pressures compress the wedge-shaped menisci; they migrate radially, allowing femur and tibia to come into contact and distributing loads over larger areas (Fig. 6.10). Seedhom,[153] and Seedhom and Hargreaves[155] estimate that the menisci account for 44 per cent of load transmission across the medial compartment of the knee and 65 per cent across the lateral compartment.

The circumferential collagen fibres of the menisci, bound to the tibia at their periphery, resist the tensile stresses set

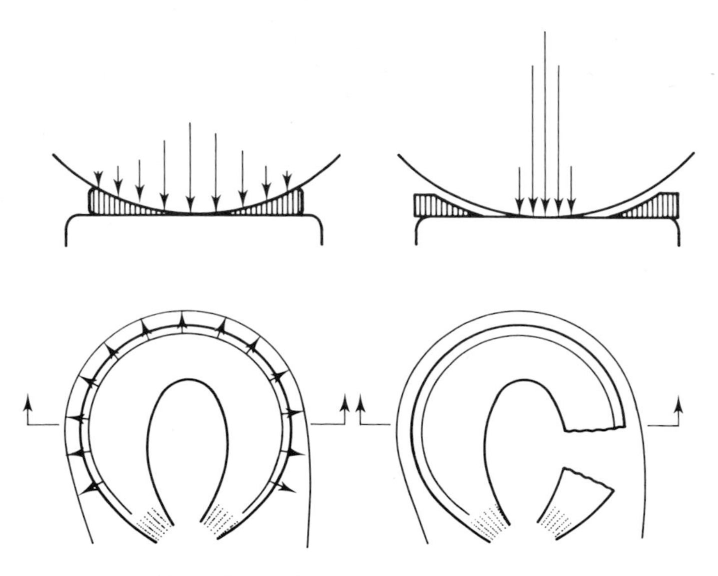

Fig. 6.10 Functions of menisci.
Menisci distribute a joint load over a larger contact area, reducing applied pressure on cartilage (left). Rupture or excision of meniscus (right) grossly disorganizes joint function and increases contact pressure.

up when the menisci are forced to migrate radially under joint loading.[30] Separation of the circumferential fibres is prevented by a surface lamina of radially oriented fibres.[17] In compression, meniscal cartilage creeps more than articular cartilage, due to the lower proteoglycan content of the former.[116]

Mechanical properties of connective tissues: tendons and ligaments

Tendons

(see p. 101)

The primary function of tendon is to transmit tensile (see p. 262) forces from muscle to bone or fascia, producing motion in joints or maintaining joint position under load (Fig. 6.11a). The tensile stiffness and strength of tendon are greater than those of ligament, cartilage or skin tissues in which the collagen content is lower, the fibres less highly orientated.[181,182,184] Despite the high modulus of elasticity and strength of tendon in tension (Table 6.2), tendons are flexible in bending and torsion: the interfaces between fascicles and fibrils allow mobility. The interfaces contribute

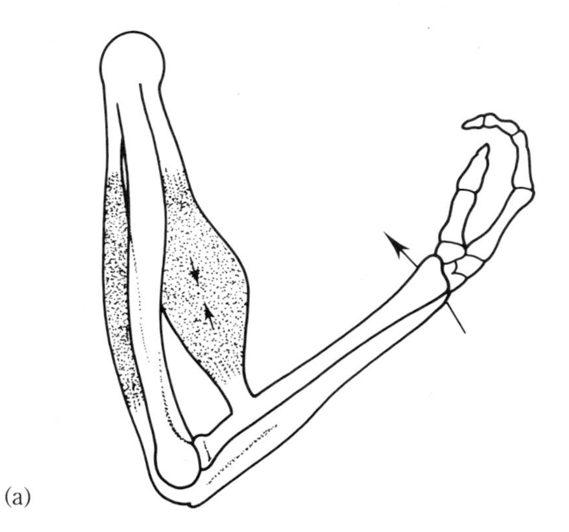

(a)

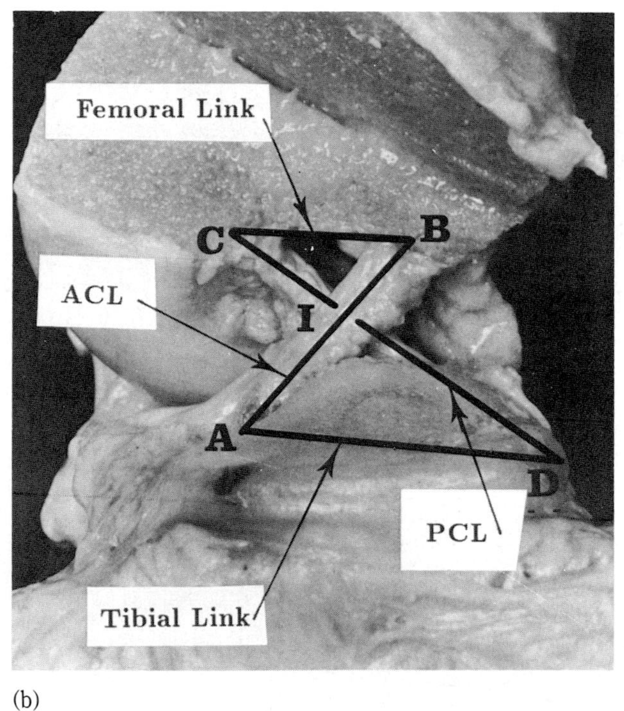

(b)

Fig. 6.11 Function of tendons and ligaments.
a. Function of tendons. Tendons transmit tensile forces from muscle to bone. **b.** Function of ligaments. Human knee joint. Removal of the lateral femoral condyle has exposed the cruciate ligaments. The cruciate linkage (**A,B,C,D**) is shown. **AB**: anterior cruciate ligament. **CD**: posterior cruciate ligament. From O'Connor *et al.*[123]. Reproduced by courtesy of J. O'Connor and Springer-Verlag.

to the fracture toughness of the tissue by preventing the build-up of stress concentrations near cracks or defects.[82]

As the load and extension of a tendon are increased, crimp[20,42] (Fig. 6.2) is lost and the structure becomes progressively stiffer.[42,86] Once crimp has straightened, the stress/strain relationship remains linear until stress levels

Table 6.2

Typical mechanical properties

	Elastic modulus (MPa)	Ultimate tensile strength (MPa)
Collagen	4000	150–300
Tendon	1000	50–100
Polyethylene	200–700	20–40
Nylon	2000–4000	50–100
Mild steel	200 000	400

sufficient to cause permanent damage or failure are reached.

Tendon and ligament remodel in response to mechanical demand.[5,120] In animal models, disuse or immobilization decrease strength and stiffness[121]; exercise increases the mass and enhances the material properties of the tissue.[184] Exercise also increases the strength of the ligament–bone junctions (entheses) (p. 101) and the tendon–bone unions.

Ligaments

(see p. 99)

The primary functions[96] of a ligament are to maintain proper joint kinematics in normal activity and to protect the joint against external loads that are too sudden or too large to be stabilized by the muscles (Fig. 6.11b). Recently Brand[21] has suggested that ligaments also function as sensors for the nervous system. If this is so, they may play a larger role in normal joint function than has been realized and may make a correspondingly large contribution to the pathological consequences of injuries.

In the majority of ligaments there is a predominance of collagen fibres. However, two-thirds of the ligamentum flavum and ligamentum nuchae of the spine is composed of elastic fibres. It has been suggested[120] that these ligaments have a specialized role: they prestress intervertebral discs, protect nerve roots from mechanical injury and provide intrinsic stability to the spine.

Human anterior cruciate ligaments undergo changes with age.[122] There is a decrease in strength, stiffness and energy absorption. The way in which bone–ligament–bone complexes fail mechanically depends on the rate of loading. Under slowly increasing load, failures occur at the bony insertion of a ligament. Under rapid loading, the load required to cause failure is high and most failures occur within the ligament body, not at its insertion. The strength of the ligamentous insertion is stronger under sudden loads.[5]

Mechanical properties of connective tissues: elastic tissue

Elastic fibres (see pp. 46, 84 and 189) have rubber-like properties. The fibres lie close to collagen: the mechanical behaviour of elastic material is therefore likely to be overshadowed by those of this protein. By comparison with collagen, elastic fibres are much more extensible. The modulus of elasticity is 0.30 MPa.[81] In the relaxed state, elastic fibres are highly retractile. Four mechanical functions are attributable to elastic fibres: first, fibre networks disseminate stress by distributing loads and reducing stress concentrations; second, rhythmic movements of body parts are co-ordinated; third, energy is conserved by maintaining tone during muscle relaxation; and fourth, the shape and size of organs is maintained by retractility.

REFERENCES

1. Adams D, Kempson G E, Swanson S A V. Technical note: direct measurement of local pressures in the cadaveric hip joint. *Medical and Biological Engineering and Computing* 1978; **16**: 113–5.

2. Afoke N Y P, Byers P D, Hutton W C. The incongruous hip joint: a casting study. *Journal of Bone and Joint Surgery* 1980; **62-B**: 511–4.

3. Ahmed A M, Burke D L, Tencer A, Miller J, Stachiewicz J W. A method for the *in vitro* measurement of pressure distribution at articular interfaces of synovial joints. *Transactions of the 23rd Orthopaedic Research Society Meeting* 1977; **178**.

4. Akeson W H, Amiel D, Woo S L-Y. Immobility effects on synovial joints: the pathomechanics of joint contracture. *Biorheology* 1980; **17**: 95–110.

5. Akeson W H, Frank C B, Amiel D, Woo S L-Y. Ligament biology and biomechanics. In: Finerman G, ed, *American Academy of Orthopaedic Surgeons Symposium on Sports Medicine. The Knee*. St Louis, Missouri: The C V Mosby Company, 1985: 111–51.

6. Amis A A, Dowson D, Wright V. Elbow joint force prediction for some strenuous isometric actions. *Journal of Biomechanics* 1980; **13**: 765–76.

7. Andersson G, Shultz A. Effects of fluid injection on mechanical properties of intervertebral discs. *Journal of Biomechanics* 1979; **12**: 453–8.

8. Armstrong C G. An analysis of the stresses in a thin layer of articular cartilage in a synovial joint. *Engineering in Medicine* 1986; **15**: 55–61.

9. Armstrong C G, Bahrani A S, Gardner D L. *In vitro* measurement of articular cartilage deformations in the intact human hip joint under load. *Journal of Bone and Joint Surgery* 1979; **61-A**: 744–55.

10. Armstrong C G, Bahrani A S, Gardner D L. Changes in the deformational behaviour of human hip cartilage with age. *Journal of Biomechanical Engineering* 1980; **102**: 214–20.

11. Armstrong C G, Gardner D L. Thickness and distribution of human femoral head articular cartilage. *Annals of the Rheumatic Diseases* 1977; **36**: 407–12.

12. Armstrong C G, Lai W M, Mow V C. An analysis of the unconfined compression of articular cartilage. *Journal of Biomechanical Engineering* 1984; **106**: 165–73.

13. Armstrong C G, Mow V C. Biomechanics of normal and osteoarthrotic articular cartilage. In: Wilson P D, Straub L R, eds, *Clinical Trends in Orthopaedics*. New York: Thieme-Stratton, 1982; 189–97.

14. Armstrong C G, Mow V C. Variations in the intrinsic mechanical properties of human articular cartilage with age, degeneration and water content. *Journal of Bone and Joint Surgery* 1982; **64A**: 88–94.

15. Ascenzi A, Bonucci E. The mechanical properties of the osteon in relation to its structural organisation. In: Balazs E A, ed, *Chemistry and Molecular Biology of the Intercellular Matrix*. New York: Academic Press, 1970; **3**: 1341–59.

16. Askew M J, Mow V C. The biomechanical function of the collagen fibril ultrastructure of articular cartilage. *Journal of Biomechanical Engineering* 1978; **100**: 105–15.

17. Aspden R M, Yarker Y E, Hukins D W L. Collagen orientations in the meniscus of the knee joint. *Journal of Anatomy* 1985; **140**: 371–80.

18. Bar E. Elastizätsprüfungen der Gelenkknorpel. *Archives of Engineering Mechanical Organisation* 1926; **108**: 739–60.

19. Berme N, Paul J P, Purves W K. A biomechanical analysis of the metacarpophalangeal joint. *Journal of Biomechanics* 1977; **10**: 409–12.

20. Betsch D F, Baer E. Structure and mechanical properties of rat tail tendon. *Biorheology* 1980; **17**: 83–94.

21. Brand R A. Knee ligaments: a new view. *Journal of Biomechanical Engineering* 1986; **108**: 106–10.

22. Broom N D. Abnormal softening in articular cartilage. *Arthritis and Rheumatism* 1982; **25**: 1209–16.

23. Broom N D, Myers D B. Fibrous waveforms or crimp in surface and subsurface layers of hyaline cartilage maintained in its wet functional condition. *Connective Tissue Research* 1980; **7**: 165–75.

24. Broom N D, Myers D B. A study of the structural response of wet hyaline cartilage to various loading situations. *Connective Tissue Research* 1980b; **7**: 227–37.

25. Broom N D, Poole C A. Articular cartilage collagen and proteoglycans: their functional interdependency. *Arthritis and Rheumatism* 1983; **26**: 1111–9.

26. Brown T D, Shaw D T. *In vitro* contact stress distributions in the natural human hip. *Journal of Biomechanics* 1983; **116**: 373–84.

27. Brown T D, Singerman R J. Experimental determination of the linear biphasic constitutive coefficients of human fetal proximal femoral chondroepiphysis. *Journal of Biomechanics* 1986; **19**: 597–605.

28. Bullough P, Goodfellow J. The significance of the fine structure of articular cartilage. *Journal of Bone and Joint Surgery* 1968; **50-B**: 852–7.

29. Bullough P, Goodfellow J, O'Connor J. The relationship between degenerative changes and load-bearing in the human hip. *Journal of Bone and Joint Surgery* 1973; **55-B**: 746–58.

30. Bullough P G, Munuera L, Murphy J, Weinsten A M. The strength of the menisci of the knee as it relates to its fine structure. *Journal of Bone and Joint Surgery* 1970; **52-B**: 564–70.

31. Byers P D, Contepomi C A, Farkas T A. A post-mortem study of the hip joint. *Annals of the Rheumatic Diseases* 1970; **29**: 15–31.

32. Byers P D, Contepomi C A, Farkas T A. Post-mortem study of the hip joint. II. Histological basis for limited and progressive cartilage alterations. *Annals of the Rheumatic Diseases* 1976a; **35**: 114–21.

33. Byers P D, Contepomi C A, Farkas T A. Post-mortem study of the hip joint. III. Correlations between observations. *Annals of the Rheumatic Diseases* 1976b; **35**: 122–6.

34. Cameron H, Pillar R M, MacNab I. The microhardness of articular cartilage. *Clinical Orthopaedics and Related Research* 1975; **108**: 275–8.

35. Carlson C E, Mann R W, Harris W H. A radio telemetry device for monitoring cartilage surface pressures in the human hip. *Biomedical Engineering* 1974; **21**: 257–64.

36. Carter D R, Spengler D M. Mechanical properties and composition of cortical bone. *Clinical Orthopaedics and Related Research* 1978; **135**: 192–217.

37. Crisp D C. Properties of tendon and skin. In: Fung Y C, Perrone N, Anliker M, eds, *Biomechanics—Its Foundations and Objectives*. Englewood Cliffs: Prentice Hall, 1972: 141–79.

38. Crowninshield R D, Johnston R C, Andrews J G, Brand R A. A biomechanical investigation of the human hip. *Journal of Biomechanics* 1978; **11**: 75–85.

39. Currey J D. Properties of bone, cartilage and synovial fluid. In: Dowson D, Wright V, eds, *Introduction to the Biomechanics of Joints and Joint Replacement*. London: Mechanical Engineering Publications, 1981: 103–7.

40. Currey J D, Carter D R, Viidik A, eds. F Gaynor Evans anniversary issue on bone biomechanics. *Journal of Biomechanics* 1987; **20**: 1013–50.

41. Day W H, Swanson S A V, Freeman M A R. Contact pressures in the loaded human cadaver hip. *Journal of Bone and Joint Surgery* 1975; **57-B**: 302–13.

42. Diamant J, Keller A, Baer E, Litt M, Arridge R G C. Collagen; ultrastructure and its relation to mechanical properties as a function of ageing. *Proceedings of the Royal Society of London B* 1972; **180**: 293–315.

43. Dowson D, Jin Z-M. Micro-elastohydrodynamic lubrication of synovial joints. *Engineering in Medicine* 1986; **15**: 63–5.

44. Dowson D, Longfield M D, Walker P S, Wright V. An investigation of the friction and lubrication in human joints. *Pro-*

ceedings of the Institute of Mechanical Engineering 1967–68; **182**:3N: 70–8.

45. Dowson D, Unsworth A, Wright V. Analysis of 'boosted lubrication' in human joints. *Journal of Mechanical Engineering Science* 1970; **12**: 364–9.

46. Dvir Z. Biomechanics of the shoulder. PhD Thesis, University of Strathclyde, Glasgow 1978.

47. Elliott D H. Structure and function of mammalian tendon. *Biological Reviews* 1965; **40**: 392–421.

48. Gocke E. Elastizitatsstudien am jungen und alten Gelenkknorpel. *Verhandlungen der deutsche Orthopädische Gesellschaft* 1927; **22**: 130–47.

49. Engin A E. On the biomechanics of the shoulder complex. *Journal of Biomechanics* 1980; **13**: 575–90.

49a. Frank E H, Grodzinsky A J. Cartilage electromechanics – I. Electrokinetic transduction and the effects of electrolyte pH and ionic strength. *Journal of Biomechanics* 1987a; **20**: 615–27.

49b. Frank E H, Grodzinsky A J. Cartilage electromechanics – II. A continuum model of cartilage electrokinetics and correlation with experiments. *Journal of Biomechanics* 1987b; **20**: 629–39.

49c. Frank E H, Grodzinsky A J, Koob T J, Eyre D R. Streaming potentials: a sensitive index of enzymatic degradation in articular cartilage. *Journal of Orthopaedic Research* 1987; **5**: 497–508.

50. Frankel V H, Nordin M. *Basic Biomechanics of the Skeletal System*. Philadelphia: Lea & Febiger, 1980.

51. Frymoyer J N, Stokes I A F. Mechanical function of the low back. In: Owen R, Goodfellow J, Bullough P, eds, *Scientific Foundations of Orthopaedics and Traumatology*. London: Heinemann Medical Books, 1980: 90–6.

52. Fukubayashi T, Kurosawa H. The contact area and pressure distribution pattern of the knee. A study of normal and osteoarthrotic joints. *Acta Orthopaedica Scandinavica* 1980; **51**: 871–9.

53. Fung Y C. Biorheology of soft tissue. *Biorheology* 1973; **10**: 199–212.

54. Fung Y C. Biomechanics. *Mechanical Properties of Living Tissues*. Berlin: Springer-Verlag, 1981.

55. Galante J O. Tensile properties of the human lumbar annulus fibrosus. *Acta Orthopaedica Scandinavica* 1967; Suppl. 100: 1–83.

56. Elmore S M, Sokoloff L, Norris G, Carmeci P. Nature of 'imperfect' elasticity of articular cartilage. *Journal of Applied Physiology* 1963; **18**: 393–6.

57. Goel V K, Svensson N L. Forces on the pelvis. *Journal of Biomechanics* 1977; **10**: 195–200.

58. Goodfellow J, Hungerford D S, Woods C. Patello-femoral joint mechanics and pathology 2. Chondromalacia patellae. *Journal of Bone and Joint Surgery* 1976; **58-B**: 291–9.

59. Goodfellow J, Hungerford D S, Zindel M. Patello-femoral joint mechanics and pathology. I. Functional anatomy of the patello-femoral joint. *Journal of Bone and Joint Surgery* 1976; **58-B**: 287–90.

60. Gordon J E. *The New Science of Strong Materials or Why You Don't Fall through the Floor*. London: Pelican, 1976.

61. Gordon J E. *Structures or Why Things Don't Fall Down*. London: Pelican, 1978.

62. Gordon J E. *The Science of Structures and Materials*. New York: Scientific American Books Inc, 1988.

63. Greenwald A S, Haynes D W. Weight-bearing areas in the human hip joint. *Journal of Bone and Joint Surgery* 1972; **54-B**: 157–63.

64. Greenwald A S, O'Connor J J. The transmission of load through the human hip joint. *Journal of Biomechanics* 1971; **4**: 507–28.

65. Grieve D W, Leggett D, Wetherstone B. The analysis of normal stepping movements as a possible basis for locomotion assessment of the lower limbs. *Journal of Anatomy* 1979; **127**: 515–32.

66. Grodzinsky A J. Electromechanical and physicochemical properties of connective tissue. *CRC Critical Reviews in Biomedical Engineering* 1983; **9**: 133–99.

67. Harkness R D. Mechanical properties of collagenous tissues. In: Gould B S, ed, *Treatise on Collagen* (volume IIa). New York: Academic Press, 1968: 248–310.

68. Harrington I J. A bioengineering analysis of force actions at the knee in normal and pathological gait. *Biomedical Engineering* 1976; **11**: 167.

69. Hayes W C, Bodine A J. Flow-independent viscoelastic properties of articular cartilage matrix. *Journal of Biomechanics* 1978; **11**: 407–19.

70. Hayes W C, Keer L M, Herrman G, Mockros L F. A mathematical analysis for indentation tests of articular cartilage. *Journal of Biomechanics* 1972; **5**: 541–51.

71. Hayes W C, Mockros L F. Viscoelastic properties of human articular cartilage. *Journal of Applied Physiology* 1971; **31**: 562–8.

72. Hickey D S, Hukins D W L. Relationship between the structure of the annulus fibrosus and the function and failure of the intervertebral disc. *Spine* 1980; **5**: 106–16.

73. Higginson G R, Norman R. A model of investigation of squeeze film lubrication in animal joints. *Physics in Medicine and Biology* 1974; **19**: 785–92.

74. Higginson G R, Snaith J E. The mechanical stiffness of articular cartilage in confined oscillating compression. *Engineering in Medicine* 1979; **8**: 11–14.

75. Hoch D H, Grodzinsky A J, Koob T J, Albert M L, Eyre D R. Early changes in material properties of rabbit articular cartilage after menisectomy. *Journal of Orthopaedic Research* 1983; **1**: 4–12.

75a. Holm S H, Urban J P G. The intervertebral disc: factors contributing to its nutrition and matrix turnover. In: Helminen H J, Kiviranta I, Saamanen A-M, Tammi M, Paukkonen K, Jurvelin J, eds, *Joint Loading*. Bristol: John Wright & Sons, 1987: 187–226.

76. Huberti H H, Hayes W C. Patellofemoral contact pressures. The influence of Q-angle and tendofemoral contact. *Journal of Bone and Joint Surgery* 1984; **66-A**: 715–24.

77. Hukins D W L. *Connective Tissue Matrix*. London: Macmillan, 1984.

78. Hültkrantz J W. Ueber die spaltrichtungen der gelenkknorpel. *Verhandlungen der Anatomischen Gesellschaft* 1898; **12**: 248–56.

79. Ingelmark B E, Blomgren E. An apparatus for the measurement of pressure, especially in human joints. *Upsala Lakareforennings Forhandlungen* 1947; **53**: 75–94.

80. Jayson M I V. Back pain, spondylosis and disc disorders. In: Scott J T, ed, *Copeman's Textbook of the Rheumatic Diseases* (6th edition). Edinburgh: Churchill Livingstone, 1986: 1407–34.

81. Jenkins R B, Little R W. A constitutive equation for parallel-fibred elastic tissue. *Journal of Biomechanics* 1974; **7**: 397–402.

82. Jeronimidis G, Vincent J F V. Composite materials. In: Hukins D W L, ed, *Connective Tissue Matrix*. London: Macmillan, 1984: 187–210.

83. Johnson F, Leitl S, Waugh W. The distribution of load across the knee. *Journal of Bone and Joint Surgery* 1980; **62B**: 346–9.

84. Jurvelin J, Kiviranta I, Arokoski J, Tammi M, Helminen H J. Indentation study of the biomechanical properties of articular cartilage in the canine knee. *Engineering in Medicine* 1987; **16**: 15–22.

85. Jurvelin J, Säämänen A-M, Arokoski J, Helminen H J, Kiviranta I, Tammi M. Biomechanical properties of the canine knee articular cartilage as related to matrix proteoglycans and collagen. *Engineering in Medicine* 1988; **17**: 157–62.

86. Kastelic J, Palley I, Baer E. A structural mechanical model for tendon crimping. *Journal of Biomechanics* 1980; **13**: 887–93.

87. Kempson G E. Mechanical properties of articular cartilage. In: Freeman M A R, ed, *Adult Articular Cartilage* (2nd edition). Tunbridge Wells: Pitman Medical, 1979: 333–414.

88. Kempson G E, Muir H, Pollard C, Tuke M. The tensile properties of cartilage of human femoral condyles related to the content of collagen and glycosaminoglycans. *Biochimica et Biophysica Acta* 1973; **297**: 456–72.

89. Kempson G E, Spivey C J, Swanson S A V, Freeman M A R. Patterns of cartilage stiffness on normal and degenerate human femoral heads. *Journal of Biomechanics* 1971; **4**: 597–609.

90. Kempson G E, Tuke M A, Dingle J T, Barrett A J, Horsfield P H. The effect of proteolytic enzymes on the mechanical properties of adult human articular cartilage. *Biochimica et Biophysica Acta* 1976; **428**: 741–60.

91. Klein J A, Hickey D S, Hukins D W. Radial bulging of the annulus fibrosus during compression of the intervertebral disc. *Journal of Biomechanics* 1983; **16**: 211–17.

92. Lane J M, Chisena E, Black J. Experimental knee instability: early mechanical property changes in articular cartilage in a rabbit model. *Clinical Orthopaedics and Related Research* 1979; **140**: 262–5.

93. Lanir Y. A structural theory for the homogeneous biaxial stress-strain relationships in flat collapsed tissues. *Journal of Biomechanics* 1979; **12**: 423–36.

94. Lanir Y. Constitutive equations for fibrous connective tissues. *Journal of Biomechanics* 1983; **16**: 1–12.

95. Lanir Y. Biorheology and fluid flux in swelling tissues. II. Analysis of unconfined compressive response of transversely isotropic cartilage disc. *Biorheology* 1987; **24**: 189–205.

96. Lewis J L, Lew W D, Shybut G T, Jasty M, Hill J A. Biomechanical function of knee ligaments. In: Finerman G, ed, *American Academy of Orthopaedic Surgeons Symposium on Sports Medicine: The Knee*. St Louis Missouri: The C V Mosby Company, 1985: 152–67.

97. Lindl M. Biomechanics of the lumbar spine. In: Frankel V H, Nordin M, eds, *Basic Biomechanics of the Skeletal System*. Philadelphia: Lea & Febiger, 1980: 255–90.

98. Linn F C, Sokoloff L. Movement and composition of interstitial fluid of cartilage. *Arthritis and Rheumatism* 1965; **8**: 481–94.

99. Lipson S J, Muir H. Proteoglycans in experimental intervertebral disc degeneration. *Spine* 1981; **6**: 194–210.

100. McCutchen C W. Sponge hydrostatic and weeping bearings. *Nature* 1959; **184**: 1284–5.

101. McCutcheon C W. Physiological lubrication. *Proceedings of the Institute of Mechanical Engineers* 1966–67; **181**: 55–62.

102. McCutchen C W. Lubrication of joints. In: Sokoloff L, ed, *The Joints and Synovial Fluid* (volume I). New York: Academic Press, 1978: 437–83.

103. Mak A F, Lai W M, Mow V C. Indentation of articular cartilage: a biphasic analysis. In: Thibault L, ed, *Advances in Bioengineering*. New York: American Society of Mechanical Engineers, 1982: 71–4.

104. Mansour J, Mow V C. The permeability of articular cartilage under compressive strain and at high pressures. *Journal of Bone and Joint Surgery* 1976; **58A**: 509–16.

105. Maquet P G, Van de Berg A J, Simonet J C. Femorotibial weight-bearing areas: experimental determination. *Journal of Bone and Joint Surgery* 1975; **57A**: 766–71.

106. Maroudas A. Biophysical chemistry of cartilaginous tissue with special refererence to solute and fluid transport. *Biorheology* 1975; **12**: 233–48.

107. Maroudas A. Physiochemical properties of articular cartilage. In: Freeman M A R, ed, *Adult Articular Cartilage* (2nd edition). Tunbridge Wells: Pitman Medical, 1979: 215–90.

108. Maroudas A, Baylis M T, Venn M F. Further studies on the composition of human femoral head cartilage. *Annals of the Rheumatic Diseases* 1980; **39**: 514–23.

109. Martel-Pelletier J, Martel-Pelletier J-P. Neutral metalloproteases and age related changes in human articular cartilage. *Annals of the Rheumatic Diseases* 1987; **46**: 363–9.

110. Meachim G, Bentley B. Horizontal splitting in patellar articular cartilage. *Arthritis and Rheumatism* 1978; **21**: 669–74.

111. Meachim G, Denham D, Emery I H, Wilkinson P H. Collagen alignments and artificial splits at the surface of human articular cartilage. *Journal of Anatomy* 1974; **118**: 101–18.

112. Mow V C, Holmes M H, Lai W M. Fluid transport and mechanical properties of articular cartilage: a review. *Journal of Biomechanics* 1984; **17**: 377–94.

113. Mow V C, Lai W M. Recent development in synovial joint biomechanics. *Society for Industrial and Applied Mathematics Review* 1980; **22**: 275–317.

114. Mow V C, Lai W M, Holmes M H. Advanced theoretical and experimental techniques in cartilage research. In: Huiskes R, van Campen D, DeWijn J, eds, *Biomechanics: Principles and Applications*. The Hague: Martinus Nijhoff, 1982: 47–74.

115. Mow V C, Mak A. Lubrication of diarthrodial joints. In: Skalak R, Chien S, eds, *Handbook of Bioengineering*. New York: McGraw-Hill, 1987.

116. Myers E R, Armstrong C G, Mow V C. Swelling pressure and collagen tension. In: Hukins D W L, ed, *Connective Tissue Matrix*. London: Macmillan, 1984: 161–86.

117. Myers E R, Lai M W, Mow V C. A continuum theory and

an experiment for the ion-induced swelling behaviour of articular cartilage. *Journal of Biomechanical Engineering* 1984; **106**: 151–8.

118. Nicol A C. Elbow joint prosthesis design—biomechanical aspects. PhD Thesis, University of Strathclyde, Glasgow 1977.

119. Nicol A C, Berme N, Paul J P. A biomechanical analysis of elbow joint function. In: *Joint Replacement in the Upper Limb*. London: Institution of Mechanical Engineers, 1977: 45–51.

120. Nordin M, Frankel V H. Biomechanics of collagenous tissues. In: Frankel V H, Nordin M, eds, *Basic Biomechanics of the Skeletal System*. Philadelphia: Lea & Febiger, 1980: 87–110.

121. Noyes F R. Functional properties of knee ligaments and alterations induced by immobilization: a correlative biomechanical and histological study in primates. *Clinical Orthopaedics and Related Research* 1977; **123**: 210–42.

122. Noyes F R, Grood E S. The strength of the anterior cruciate ligament in humans and Rhesus monkeys. Age-related and species-related changes. *Journal of Bone and Joint Surgery* 1976, **56A**: 1074–82.

123. O'Connor J, Shercliff T, Goodfellow J. The mechanics of the knee in the sagittal plane. Mechanical interactions between muscles, ligaments and articular surfaces. In: Muller W, Hackenbruch W P, eds, *Surgery and Arthroscopy of the Knee: 2nd Congress of the European Society*. Berlin: Springer-Verlag, 1988: 12–30.

124. O'Connor P, Bland C, Bjelle A, Gardner D L. Production of split patterns on the articular cartilage surfaces of rats. *Journal of Pathology* 1980a; **130**: 15–21.

125. O'Connor P, Bland C, Gardner D L. Fine structure of artificial splits in femoral condylar cartilage of the rat: a scanning electron microscopic study. *Journal of Pathology* 1980b; **132**: 169–79.

126. O'Connor P, Orford C R, Gardner D L. Differential response to compressive loads of zones of canine hyaline articular cartilage: micromechanical, light and electron microscopic studies. *Annals of the Rheumatic Diseases* 1988; **47**: 414–20.

127. O'Kelly J, Unsworth A, Dowson D, Hall D A, Wright V. A study of the role of synovial fluid and its constituents in the friction and lubrication of human hip joints. *Engineering in Medicine* 1978; **7**: 73–83.

128. Orford C R, Gardner D L. Proteoglycan association with collagen d band in hyaline articular cartilage. *Connective Tissue Research* 1984; **12**: 345–8.

129. Pagowski S, Piekarski K. Biomechanics of metacarpophalangeal joint. *Journal of Biomechanics* 1977; **10**: 205–9.

130. Panagiotacopulos N D, Block R, Knauss W G, Harvey P, Patzakis M. *On the Mechanical Properties of the Human Intervertebral Disc*. Report No. AFOSR-TR-78-0054. Springfield: National Technical Information Service, 1978.

131. Panagiotacopulos N D, Knauss W G, Block R. On the mechanical properties of human intervertebral disc material. *Biorheology* 1979; **16**: 317–30.

132. Panjabi M M, Brand R A, White A A. The three dimensional flexibility and stiffness properties of the thoracic spine. *Journal of Biomechanics* 1976; **9**: 185–92.

133. Parsons J R, Black J. The viscoelastic shear behaviour of normal rabbit articular cartilage. *Journal of Biomechanics* 1977; **10**: 21–30.

134. Parsons J R, Black J. Mechanical behaviour of articular cartilage; quantitative changes with alteration of ionic environment. *Journal of Biomechanics* 1979; **12**: 765–72.

135. Paul J P. Forces transmitted by joints in the human body. *Proceedings of the Institute of Mechanical Engineers* 1967; **181**:3J: 8–15.

136. Paul J P. The effect of walking speed on the force actions transmitted at the hip and knee joints. *Proceedings of the Royal Society of Medicine* 1970; **63**: 200–4.

137. Paul J P. Techniques of gait analysis. *Proceedings of the Royal Society of Medicine* 1974; **67**: 401–4.

138. Paul J P. Force actions transmitted by joints in the human body. *Proceedings of the Royal Society of London B* 1976; **192**: 163–72.

139. Pond M J, Nuki G. Experimentally-induced osteoarthritis in the dog. *Annals of the Rheumatic Diseases* 1973; **32**: 387–8.

140. Purves W K, Berme N, Paul J P. Finger joint mechanics. In: Kenedi R M, Paul J P, Hughes J, eds, *Disability*. London: Macmillan, 1978: 318–32.

141. Radin E L, Ehrlich M G, Chernack R, Abernethy P, Paul I L, Rose R M. Effect of repetitive impulsive loading on the knee joints of rabbits. *Clinical Orthopaedics and Related Research* 1978; **131**: 288–93.

142. Radin E L, Swann D A, Weisser P A. Separation of a hyaluronate-free lubricating fraction from synovial fluid. *Nature* 1970; **228**: 377–8.

143. Reuber M, Schultz A, Denis F, Spencer D. Bulging of lumbar intervertebral disks. *Journal of Biomechanical Engineering* 1982; **104**: 187–92.

144. Rigby B J, Hirac N, Spikes J D, Eyring H. The mechanical properties of rat tail tendon. *Journal of General Physiology* 1959; **43**: 285–83.

145. Roberts B J, Unsworth A, Mian N. Modes of lubrication in human hip joints. *Annals of the Rheumatic Diseases* 1982; **41**: 217–24.

146. Rushfeldt P D, Mann R W, Harris W H. Improved techniques for measuring *in vitro* the geometry and pressure distribution in the human acetabulum. I. Ultrasonic measurement of acetabular surfaces, sphericity and cartilage thickness. *Journal of Biomechanics* 1981a; **14**: 253–60.

147. Rushfeldt P D, Mann R W, Harris W H. Improved techniques for measuring *in vitro* the geometry and pressure distribution in the human acetabulum. II. Instrumented endoprosthesis measurement of articular surface pressure distribution. *Journal of Biomechanics* 1981b; **14**: 315–23.

148. Rydell N. Biomechanics of the hip joint. *Clinical Orthopaedics and Related Research* 1973; **92**: 6–15.

149. Sah R L-Y, Kim Y-J, Doong J-Y H, Grodzinsky A J, Plaas A H K, Sandy J D. Biosynthetic response of cartilage explants to dynamic compression. *Journal of Orthopaedic Research* 1989; **7**: 619–36.

150. Sayles R S, Thomas T R, Anderson J, Haslock I, Unsworth A. Measurement of the surface microgeometry of articular cartilage. *Journal of Biomechanics* 1979; **12**: 257–67.

151. Scott J E, Haigh M. Proteoglycan-collagen interactions in intervertebral disc. A chondroitin sulphate-proteoglycan associates with collagen fibrils in rabbit annulus fibrosus at the d–e bands. *Bioscience Reports* 1986; **6**: 213–6.

152. Scott J E, Orford C R. Dermatan sulphate-rich proteoglycan associated with rat tail tendon collagen at the d band in the gap region. *Biochemical Journal* 1981; **197**: 213–6.

153. Seedhom B B. Transmission of the load in the knee joint with special reference to the role of the menisci. I. Anatomy, analysis and apparatus. *Engineering in Medicine* 1979; **8**: 207–19.

154. Seedhom B B. Biomechanics of the lower limb (knee). In: Dowson D, Wright V, eds, *Introduction to Biomechanics of Joint Replacement*. London: Mechanical Engineering Publications, 1981: 73–81.

155. Seedhom B B, Hargreaves D J. Transmission of the load in the knee joint with special reference to the role of the menisci. II. Experimental results, discussion and conclusions. *Engineering in Medicine* 1979; **8**: 220–8.

156. Seedhom B B, Takeda T, Tsubuku M, Wright V. Mechanical factors and patellofemoral osteoarthrosis. *Annals of the Rheumatic Diseases* 1979; **38**: 307–16.

157. Serink M T, Nachemson A, Hansson G. The effect of impact loading on rabbit knee joints. *Acta Orthopaedica Scandinavica* 1977; **48**: 250–62.

158. Shah J S. Experimental stress analysis of the lumbar spine. In: Jayson M I V, ed, *The Lumbar Spine and Back Pain*. London: Sector Publishing Ltd, 1976: 271–92.

159. Shah J S, Jayson M I V, Hampson W G J. Mechanical implications of crimping in collagen fibres of human spinal ligaments. *Engineering in Medicine* 1979; **8**: 95–102.

160. Shames I H. *Engineering Mechanics. Statics and Dynamics*. New Jersey: Prentice-Hall, 1980.

161. Shirazi-Adl A, Shrivastava S C, Ahmed A M. Stress analysis of the lumbar disc-body unit in compression: a three dimensional non-linear finite element study. *Spine* 1984; **9**: 120–34.

162. Shrive N G, O'Connor J J, Goodfellow J W. Load-bearing in the knee joint. *Clinical Orthopaedics and Related Research* 1978; **131**: 279–87.

163. Simon W H. Scale effects in animal joints. I. Articular cartilage thickness and compressive stress. *Arthritis and Rheumatism* 1970; **13**: 244–55.

164. Simon W H. Scale effects in animal joints. II. Thickness and elasticity in the deformability of articular cartilage. *Arthritis and Rheumatism* 1971; **14**: 493–502.

165. Simon W H, Friedenberg S, Richardson S. Joint congruence. A correlation of joint congruence and thickness of articular cartilage in dogs. *Journal of Bone and Joint Surgery* 1973; **55A**: 1614–20.

166. Spilker R L, Daugirda D M, Schultz A B. Mechanical response of a simple finite element model of the intervertebral disc under complex loading. *Journal of Biomechanics* 1984; **17**: 103–12.

167. Swann D A, Block K J, Swindell D, Shore E. The lubricating activity of human synovial joints. *Arthritis and Rheumatism* 1984; **27**: 552–6.

168. Swanson S A V. Lubrication. In: Freeman M A R, ed, *Adult Articular Cartilage*. Tunbridge Wells: Pitman Medical, 1979: 415–60.

169. Tencer A F, Ahmed A M, Burke D L. Some static mechanical properties of the lumbar intervertebral joint, intact and injured. *Journal of Biomechanical Engineering* 1982; **104**: 193–201.

170. Thomas T R, Sayles R S, Haslock I. Human joint performance and the roughness of articular cartilage. *Journal of Biomechanical Engineering* 1980; **102**: 50–6.

171. Tong P, Fung Y-C. The stress-strain relationship for the skin. *Journal of Biomechanics* 1976; **9**: 649–57.

172. Unsworth A, Dowson D, Wright W. Some new evidence on human joint lubrication. *Annals of the Rheumatic Diseases* 1975; **34**: 277–85.

173. Urban J P G, Maroudas, A. Swelling of the intervertebral disc *in vitro*. *Connective Tissue Research* 1981; **9**: 1–10.

174. Viidik A. On the rheology and morphology of soft collagenous tissue. *Journal of Anatomy* 1969; **105**: 184.

175. Wainwright S A, Biggs W D, Currey J D, Gosline J M. *Mechanical Design in Organisms*. Princeton: Princeton University Press, 1982.

176. Walker P S, Hajek V J. The load bearing area in the knee joint. *Journal of Biomechanics* 1972; **5**: 581–9.

176a. Warskyj M, Sullivan P, O'Connor P, Lawton M, Gardner D L. Micromechanical testing of articular cartilage: recent improvements to test apparatus. *Annals of the Rheumatic Diseases* 1988; **47**: 966–7.

177. Weightman B. Tensile fatigue of human articular cartilage. *Journal of Biomechanics* 1976; **9**: 193–200.

178. Weightman B, Chappell D J, Jenkins E A. A second study of tensile fatigue properties of human articular cartilage. *Annals of the Rheumatic Diseases* 1978; **37**: 58–63.

179. Weightman B, Freeman M A R, Swanson S A V. Fatigue of articular cartilage. *Nature* 1973; **244**: 303–4.

180. Wilkins J F. Proteolytic destruction of synovial boundary lubrication. *Nature* 1968; **219**: 1050–1.

181. Woo S L-Y. Biorheology of soft tissues: the need for interdisciplinary studies. *Biorheology* 1980; **17**: 39–43.

182. Woo S L-Y. Mechanical properties of tendons and ligaments. I. Quasi-static and nonlinear viscoelastic properties. *Biorheology* 1982; **19**: 385–96.

183. Woo S L-Y, Akeson W H, Jemmott G. Measurements of nonhomogenous directional mechanical properties of articular cartilage in tension. *Journal of Biomechanics* 1976; **9**: 785–91.

184. Woo S L-Y, Gomes M A, Wood Y-K, Akeson W H. Mechanical properties of tendons and ligaments. II. The relationships of immobilization and exercise on tissue remodelling. *Biorheology* 1982; **19**: 397–408.

185. Woo S L-Y, Lubock P, Gomez M A, Jemmott G F, Kuei S C, Akeson W H. Large deformation nonhomogeneous and directional properties of articular cartilage in uniaxial tension. *Journal of Biomechanics* 1979; **12**: 437–46.

186. Wright V, Dowson D. Lubrication and cartilage. *Journal of Anatomy* 1976; **121**: 107–18.

187. Yarker Y E, Aspden R M, Hukins D W L. Birefringence of articular cartilage and the distribution of collagen fibril orientations. *Connective Tissue Research* 1983; **11**: 207–13.

Chapter 7

Inflammatory and Vascular Basis of Connective Tissue Disease

E T WHALLEY

Diseases of vascular connective tissue are often dominated by the phenomena of inflammation.[28] The inflammatory process is extensively described in standard texts[18,65,77] and the present account is a brief survey of some features of inflammation designed to link outlines of the biochemical and immunological aspects of connective tissue disease (see Chapters 4 and 5) to descriptions of their systematic pathology (see Chapters 9 *et seq*).

The cardinal signs of inflammation are the outward, visible manifestations of molecular and cellular events mediated by small blood vessels. Where inflammation characterizes a connective tissue disease, the disorder centres on tissues such as synovia that have a rich vasculature. Inflammatory disease, by definition, cannot begin in avascular connective tissues such as articular cartilage or cornea. In synovial disorders such as rheumatoid arthritis (see Chapter 12), cartilage destruction is best viewed as a secondary phenomenon. Matrix degradation is enzymatic. The enzymes responsible derive from inflammatory cells originating in the synovium. The cells respond to cytokines released during the expression of immune phenomena (see Chapter 5).

The processes of acute inflammation comprise vasodilatation, increased vascular permeability, exudation, raised lymph flow, cellular infiltration and, subsequently, phagocytosis. These local phenomena are responses to irritants, microbial agents, immune complexes or injury. When severe or widespread, the local changes are accompanied by systemic effects.[18,45]

Cellular phenomena are of critical importance both in acute and in sustained inflammation. Many aspects of inflammation, in particular the early destruction of microorganisms, the digestion of foreign material or cell debris and the degradation of components of contiguous tissues, are the prerogative of polymorphonuclear neutrophil (polymorph) and, occasionally, eosinophil granulocytes. Subsequent changes such as the phagocytosis of bacterial products, the processing of antigens for transmission to lymphoid tissues and the manufacture and release of cytokines, are attributable to mononuclear macrophages. Lymphocytes exert critical roles in inflammatory reactions in which delayed hypersensitivity is invoked; they also regulate antibody production. Mast cells (see also p. 36) liberate chemical mediators, play a part in coagulation and modulate repair. Foci of connective tissue and vascular injury in systemic connective tissue diseases may first be detected as 'fibrinoid' (p. 294). Ultimately, tissue injury is followed by healing which may be regenerative or substitutive (p. 295).

Vascular changes

Altered blood flow

Many of the early vascular changes that immediately follow a physical or chemical stimulus causing inflammation, simulate the artefactual triple response. Transient arteriolar vasoconstriction is followed by vasodilatation. Altered arteriolar blood flow influences the capillary circulation: the capillaries become suffused with blood. Additional capillary channels open and there is a consequential increased blood flow through postcapillary venules. However, within 30–60 min of injury, blood flow through the capillary bed begins to slow and in the most central zone of injury, to oscillate or cease. The extent of this zone of stagnation increases for some hours. The reasons for capillary blood flow stasis remain obscure but haemoconcentration, with decreased velocity (sludging); an increase in interstitial fluid pressure; increased venular blood pressure; and raised venous tone are explanations that have been advanced.

Exudation

After injury, the venules soon display increased permeability. A separation of the endothelial cells[15a] occurs: gaps appear that are large enough to allow the escape of 800 nm-diameter structures. The gaps enable the passive movement from the vessel of the deformable red blood cells and facilitate the active migration of marginated leucocytes; they also encourage the passage through the venules of fluid, rich in protein, derived from the blood plasma. Although the venular basement membrane offers resistance to this movement of fluid and cells, the accumulation of extravascular fluid is aided and accelerated by the local increase in venular blood pressure and by a local fall in the plasma colloid osmotic pressure.

Cellular phenomena

Polymorphonuclear granulocytes

Both neutrophil and eosinophil granulocytes play important parts in many inflammatory processes.

Polymorphs, stimulated by chemotactic factors, move quickly from the bloodstream towards inflammatory foci where they can be identified within 20–30 min. The numbers are greatest after 4–6 h. In certain situations, of which eosinophilic fasciitis is one, polyarteritis nodosa another, eosinophil granulocytes are similarly stimulated; an eosinophil chemoattractant has been identified. Less is understood of basophil granulocytes but, in general, their properties resemble those of mast cells (see pp. 36 and 304).

Polymorphs release substances which play a fundamental role in inflammation. These products are contained in cytoplasmic granules. There are two main forms of granule: the lysosomes which are azurophilic, and the specific granules which are neutrophilic. The former contain many proteases; the latter contain non-enzymic cationic proteins such as lactoferrin and other enzymes including lysozyme (muramidase; N-acetyl glucosaminidase). The four main proteolytic enzymes which have effects on extracellular structures are collagenase, elastase, cathepsin G and cathepsin D. These enzymes degrade the collagen of tendon and cartilage and the elastic material in ligaments, elastic cartilage and blood vessels. Other actions include the modulation of inflammatory mediators. The activity of these enzymes is regulated by plasma proteinase inhibitors including α_1-antitrypsin and α_2-macroglobulin (see p. 185).

Mononuclear phagocytes

Macrophages move much more slowly than granulocytes. They accumulate in most sustained inflammatory foci but are characteristically numerous in granulomatous diseases such as tuberculous synovitis[97] (see p. 743). Macrophages are discussed in Chapter 1; their properties in relation to the immune diseases of connective tissue are considered on p. 38 and in Chapter 5.

Lymphocytes

The role of lymphocytes is preeminent in inflammation originating in immunologically mediated reactions such as rheumatoid arthritis (p. 453). Lymphocytes are discussed on p. 227.

Mast cells

Mast cells play particular roles in anaphylaxis, in atopic allergy, in neovascularization in rheumatoid arthritis synovia and in fibrosis (see p. 304).

Platelets

Platelets exert an important influence in the early phases of inflammatory responses: they can release many inflammatory mediators. The platelet and its products are represented in Fig. 7.1. Platelets form three main types of granule; in turn, the granules contain different mediators. The first category of granule is formed of lysosomes and contains acid hydrolases and cathepsins. The second type are α-granules: these granules contain platelet factor IV,

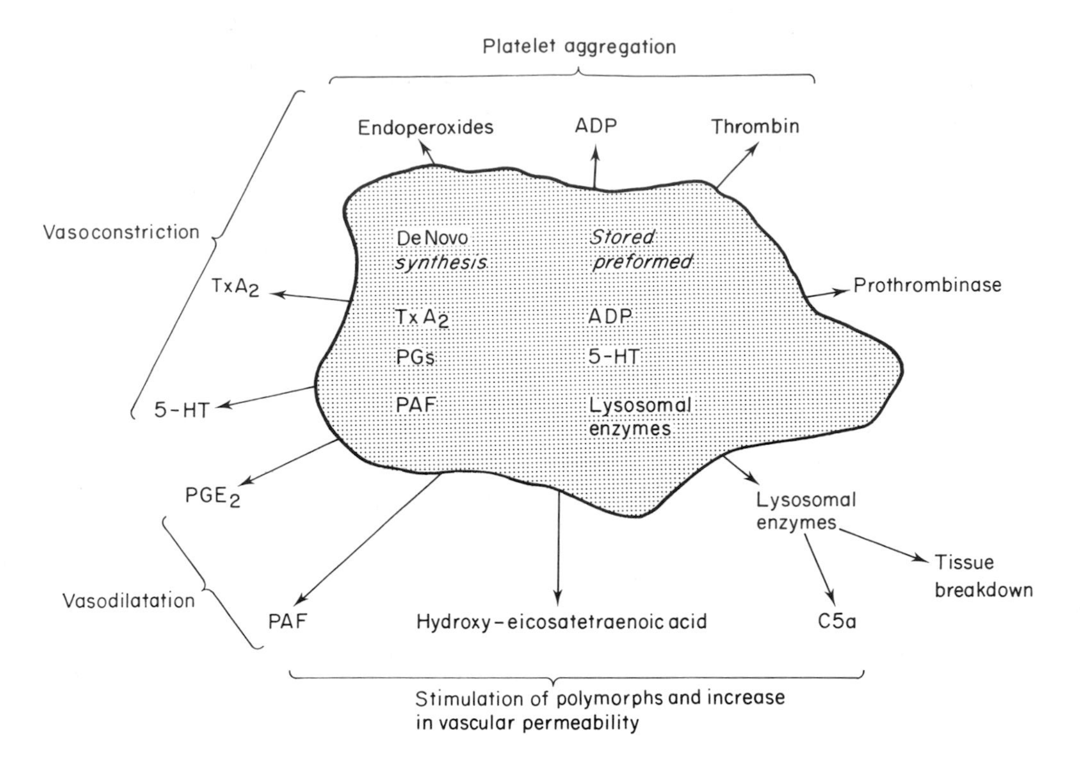

Fig. 7.1 **Blood platelets and inflammation.**
Diagrammatic representation of a blood platelet and the inflammatory mediators produced when activation occurs. The inflammatory effects of these mediators are shown. Some of the mediators are stored preformed whereas others are produced *de novo*. PG = prostaglandin; 5-HT = 5′-hydroxytryptamine; ADP = adenosine disphosphate; PAF = platelet-activating factor.

fibrinogen, a permeability enhancing factor, a platelet-derived growth factor and β-thromboglobulin. The third form is the dense granule. These granules release calcium adenosine diphosphate and 5-hydroxytryptamine. Many other mediators including membrane-derived products such as platelet-activating factor and arachidonic acid derivatives (p. 288) such as PGE_2, thromboxane A_2 and hydroperoxy products, are released when platelets aggregate. There is little doubt that the platelet plays a crucial role in many, if not all of the inflammatory diseases of connective tissue. It appears likely that platelets exert particularly important functions in systemic lupus erythematosus (see Chapter 14) and in the vasculitides (see Chapter 15).

Mediators and modulators

A wide variety of chemical substances causes or influences the inflammatory response. These substances can be designated as exogenous (from outside the body) or endogenous (from inside the body). Of the *exogenous* mediators, bacterial products are important and are responsible, for example, for vascular leakage and the attraction, by chemotaxis, of leucocytes. Of greater significance, however, are the *endogenous* mediators which can be derived from the plasma or from the tissues. The most important mediators are shown in Table 7.1

Definition and classification

For a substance to be accepted as a mediator of inflammatory and immune reactions, well-defined criteria have to be satisfied. These criteria were laid down by Dale[17] for substances classified as neurotransmitters. The three main criteria now suggested for the classification of an inflammatory mediator are: the induction by the putative mediator of some or all of the signs of inflammation; the release of the proposed mediator during an inflammatory reaction; and the reduced release of the agent, or its inactivation or antagonism, by known anti-inflammatory agents and specific receptor-antagonists.

The majority of the plasma and tissue-derived sub-

Table 7.1

Sites of origin of some of the main endogenous inflammatory mediators/modulators

Plasma-derived	Tissue-derived
Kinins	*Arachidonic acid products*
Bradykinin	Prostaglandins E_2, I_2
Des-Arg^9-bradykinin	Thromboxane
	Leukotrienes
Complement factors	
(Anaphylatoxins)	*Vasoactive amines*
C3a, C3b, C4a	Histamine
C5a, $\overline{C567}$	5-hydroxytryptamine
Fibrin degradation products	*Cytokines*
FDP-6A, FDP-6D	e.g. Interleukin-1
	Interleukin-2
Proteinase inhibitors	
e.g. α_2-macroglobulin	*Free radicals*
α_1-antitrypsin	
	Platelet-activating factor
Others	
e.g. C-reactive protein	
Haptoglobulin	

stances shown in Table 7.1 fulfil most of these criteria. These mediators can be subdivided further into substances that influence vascular tone or permeability reversibly, affecting the functions of cells involved in inflammation; and enzymes that cleave substrates either in cells or in the circulation or that irreversibly destroy the lipid membranes of cells.

The chemical mediators can be further classified by origin, function or molecular size, by their contributions to acute or prolonged inflammation and by the extent to which they mediate the whole response. Mediating substances can be divided into those such as histamine and the acute-phase proteins, involved in the early phases of inflammation and those with a role in sustained inflammation. Many inflammatory mediators can in addition be listed according to their molecular weight. There are low molecular weight substances such as histamine and the prostanoids and high molecular weight agents such as bradykinin and many of the anaphylatoxins (see Chapter 5). Finally, some mediators appear to participate in all aspects of inflammation whereas others produce only one or some of the macroscopic or microscopic inflammatory responses.

Nearly all the known mediators and modulators of inflammation appear to have the ability to activate other pathways whose products cause or prolong inflammation. For example, complement (see p. 236) can be activated by lysosomal proteases. Free radicals (see p. 291) can be produced during the respiratory burst when phagocytosis induces the formation of chemoattractant. Prostaglandins can activate immunological systems and complement, Hageman factor (Factor XII), kinin generation and clotting are intricately interconnected. There is an obvious need for complementary modulator and inhibitory feedback systems to prevent inflammatory reactions from unrestrained progression. The activation of kinin and associated systems by exposed and altered collagen is an example of this requirement.[78]

Only the main mediator systems shown in Table 7.1 are outlined here; the evidence for their involvement and importance in the inflammatory reactions of connective tissue is presented. Lesser mediators are mentioned with respect to their interactions with other systems and their sites of origin, particularly when they are derived from the cells involved in inflammation.

Plasma-derived mediators of inflammation

There are four important systems, all plasma enzyme cascades, activation of which results in the formation of potent inflammatory mediators. The systems are: the *kallikrein–kinin system,* which initiates vascular changes along with other systems and mediators; the *complement system* (see p. 236) which influences the vascular changes once initiated and catalyses cellular infiltration; the *blood coagulation system* which arrests bleeding, thus indirectly limiting the spread of inflammatory mediators and inflammation; and the *fibrinolytic system,* which results in the breakdown of thrombi and the formation of fibrinopeptides and other fibrin-degradation products which promote vascular changes. The four systems are interlinked. When they are activated, a series of conversions of pro-enzymes to enzymes occurs in which there is progressive multiplication of the numbers of molecules and of the involved, biologically active mediators. The most important focus for interaction between the four systems is at the site of activation of Hageman factor (Factor XII). Factor XII can be activated by acidification; by dilution; by treatment with organic solvents and by contact with materials such as collagen as well as by surfaces such as those of sodium urate crystals (see p. 383), aggregates of immune complexes, bacterial cell wall polysaccharides and vascular basement membranes. Activated Hageman factor is capable of directly triggering the coagulation, fibrinolytic and kinin-generating systems, and indirectly the complement system and the generation of histamine. The systems activated by Hageman factor are outlined in Fig. 7.2. There are numerous positive and

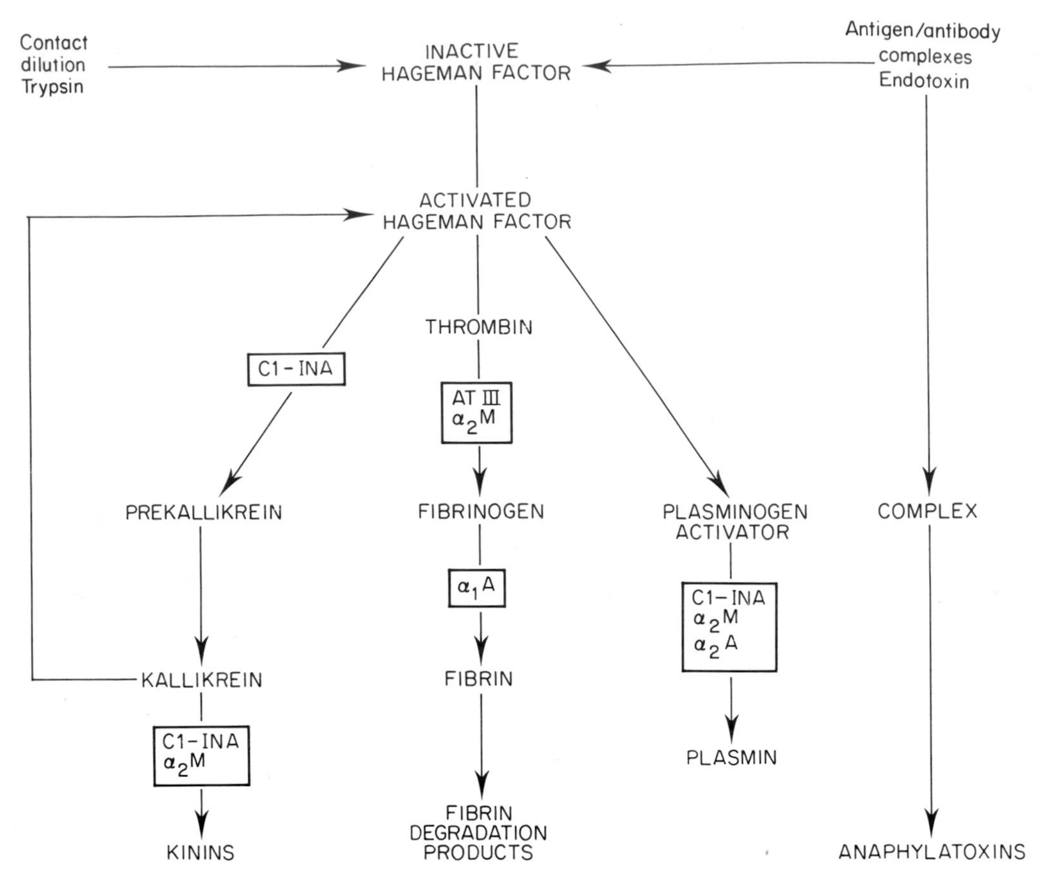

Fig. 7.2 Proteolytic enzymes and inflammatory mediators.
An outline of the proteolytic enzyme cascade systems and their relationships to the inflammatory mediators produced when activation of Hageman Factor (Factor XII) occurs. The pathways are interconnected and negative and positive feedback mechanisms operate. The sites of action of the various plasma proteinase inhibitors are shown: C1-inactivator (C1-INA); antithrombin III (AT-III); α_2-macroglobulin (α_2M); α_1-antitrypsin (α_1A).

negative feedback mechanisms and the pathways are regulated by a variety of plasma proteinase inhibitors such as α_2-macroglobulin, C1-inactivator, antithrombin III and α_1-antitrypsin.

The kallikrein–kinin system

The exact role of the kallikrein–kinin system in inflammation is unclear. However, it is postulated that the system may play a role in the arthritides, in allergic and anaphylactic reactions and in bacteraemic shock.[15,62,64] The results of *in vitro* and *in vivo* studies of models of inflammation show that bradykinin and related kinins possess actions which qualify them as mediators of inflammation.[75,86]

Bradykinin is formed from a precursor, high molecular weight kininogen, and α_2-globulin, through the action of the plasma enzyme kallikrein. Trypsin and plasmin also have the capacity to generate bradykinin from the precursor. A second kinin, lysyl–bradykinin (kallidin) can be generated from another precursor, low molecular weight kininogen, under the influence of tissue kallikreins. Both kinins produce qualitatively similar effects. The association between this group of peptides and the phenomena of inflammation rests predominantly on their demonstrated ability to enhance the vascular permeability of postcapillary venules. This increase in permeability may be the result of endothelial cell contraction and a widening of intercellular junctions, changes which can be prevented by β-adrenergic agonists. The enhancement of permeability is an energy-requiring process: it appears to be mediated via interaction with specific kinin receptors and ultimately by the activation of acyl hydrolases such as phospholipase A_2 which liberate arachidonic acid from membrane-bound phospholipids, leading to the formation of prostanoids.

Some workers claim that infiltration of leucocytes occurs at sites where kinins are injected and hence, by inference, at sites of their proposed formation.[48] Kinins are, moreover, vasodilatator substances and increase blood flow; they include some of the most potent naturally occurring, pain-producing species. Pain is elicited when they are injected intradermally or when they are applied to the base of an exposed blister on human skin.[3] Bradykinin increases local lymph flow, another characteristic of local inflammation.

Kinins are therefore capable of eliciting the classical signs of inflammation. However, it has not been easy to demonstrate directly that kinins are important mediators of inflammation *in vivo*. This problem reflects an inability to measure the rate of activation and inactivation of kinins at a target zone of inflammation; that, in most inflammatory events, many mediator systems are involved simultaneously; and the lack of stable kinin antagonists for use *in vivo*, in models of inflammation. Moreover, recent work has demonstrated an antagonism between the inflammatory effects of bradykinin in models which reproduce the classical signs of inflammation[75,83–85] and other systems.[61,59]

One of the largest problems for those investigating the role of kinins in inflammation is that these peptides are quickly broken down by kininases. There are two main degradative enzymes. The first, kininase II, is identical to angiotensin-converting enzyme, the enzyme responsible for the conversion of angiotensin I to the potent pressor agent, angiotensin II. This zinc-containing metalloenzyme is a peptidyl dipeptidase which cleaves the two C-terminal amino acids (Phe-Arg) from bradykinin (BK) and related kinins and inactivates them (Fig. 7.3). Kininase II is present in the plasma and in several tissues: its activity is concentrated at the luminal surface of endothelial cells. Kininase II can be inhibited by compounds such as captopril. The second enzyme, kininase I, is also a zinc-containing metalloenzyme; it is variously termed carboxypeptidase N, carboxypeptidase B and anaphylatoxin inactivator. Kininase I cleaves C-terminal basic amino acids from peptides involved in inflammation: they include bradykinin, kallidin, fibrinopeptides and complement components, C3a, C4a and C5a. The actions of these enzymes are shown in Fig. 7.3: the breakdown products can be active or inactive. For example, the breakdown product of bradykinin is des-Arg^9-BK which acts on a specific kinin receptor, B_1.[64,85] This receptor is not normally present. It has been termed a 'pathological' kinin receptor; it can be induced in a variety of tissues *in vitro* and *in vivo* by inflammatory insults[64,52,83] such as the response to lipopolysaccharide.

The mechanism of action of captopril is believed to be by the inhibition of kininase II. Although captopril is a very effective antihypertensive agent, it causes many side-effects which are themselves inflammatory in nature.[50] It is possible that captopril can not only elevate endogenous bradykinin levels by inhibiting kininase II but can also expose bradykinin to kininase I.[58] Inflammatory phenomena occur in susceptible patients. This effect and the results of animal studies which show the potentiation of responses to inflammatory stimuli by captopril, constitute confirmatory evidence that the kinins are involved in aspects of the inflammatory response.

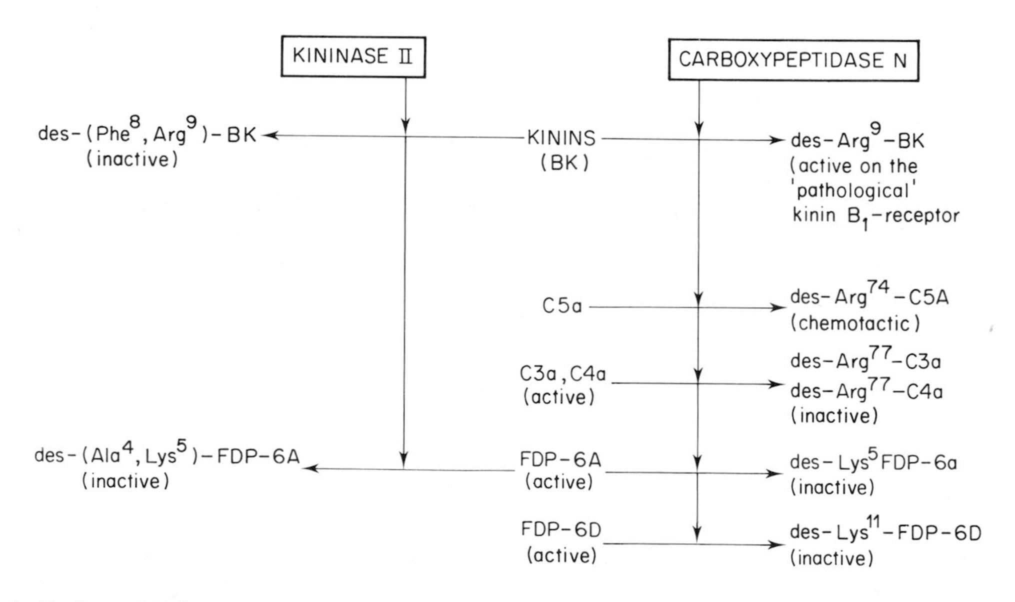

Fig. 7.3 Metabolic fate of inflammatory mediators.
The metabolic fate of mediators produced from the kinin, complement (C3a, 4a and 5a) and fibrinolytic (FDP-6A and 6D) systems. The enzyme carboxypeptidase N metabolizes bradykinin (BK) to des-Arg^9-BK which is active on pathological kinin-B_1-receptors. Similarly, des-Arg^{74}-C5a, the breakdown product of C5a, has chemotactic properties. Carboxypeptidase N inactivates C3a, C4a, FDP-6A and FDP-6D to inactive products. BK is also metabolized by kininase II as in FDP-6A which competes with the enzyme and can thus potentiate the effect of BK. The final breakdown products of BK and FDP-6A by kininase II are inactive.

The complement system

Complement (C) and its activation are described in Chapter 5 (see Figs 5.3a, b, p. 236). The complement products of greatest interest here are C3a, C4a and C5a: some cause a lethal respiratory response when injected *in vivo*: they are anaphylatoxins. Among their effects are erythema and increased vascular permeability. The vascular changes are in part a consequence of the release of histamine and other mediators from mast cells (see p. 36) and basophils.[39,38] In terms of chemotaxis and vascular permeability, the most potent anaphylatoxin is C5a.[37]

The anaphylatoxins are large peptides. They have 77 (C3a and C4a) or 74 (C5a) amino acids, respectively. A common property shared with the kinins and fibrin degradation products (FDP 6A and 6D) is a susceptibility to degradation by carboxypeptidase N. The breakdown products of C3a and C4a are inactive. Although des-Arg[74]-C5a, the breakdown product of C5a, is less potent than C5a in releasing tissue amines, it combines with a plasma factor to form a potent and stable chemotactic agent. Thus, the transient peptide C5a may be converted to a longer lived mediator of the later phases of inflammation together with the formation of des-Arg[9]-BK from bradykinin.

Two of the anaphylatoxins, C3a and C5a may have immunoregulatory properties (see p. 237). C3a suppresses, and C5a enhances polyclonal and specific antibody responses. C3a is, however, unable to suppress cellular proliferative responses such as the T-cell response to phytohaemagglutinin or T- and B-cell-proliferative responses to pokeweed mitogen. C5a enhances antigen induced immune cell proliferation but has no effect on nonspecific proliferation.[26]

The role of complement-mediated responses in the pathogenesis of rheumatoid arthritis is considered on p. 238 *et seq*. The influence of complement deficiency on the origins of systemic lupus erythematosus is reviewed on p. 244 *et seq*.

The fibrinolytic system

The low molecular weight fibrin degradation products 6A and 6D which may play a part in inflammation, are known to increase vascular permeability.[8] Like bradykinin and the anaphylatoxins, both are inactivated by carboxypeptidase N. Substrate competition may potentiate their proinflammatory action by reducing the overall activity of the inactivator. This is particularly relevant for the fibrin degradation product peptides formed by extravascular fibrinolysis or in similar stagnant environments.[71] Fibrin degradation product 6A also competes for kininase II.[71] Fibrin degradation products may therefore potentiate and prolong the inflammatory actions of bradykinin in two ways, first, by competing for kininase II and second, by acting as a preferential substrate for carboxypeptidase N.

The plasma proteinase inhibitors

The processes involved in the enzyme cascade systems described above are complex and interrelated. Fully activated, they are potentially lethal. Full activation, however, rarely occurs: the production of the inflammatory mediators is regulated by at least six plasma-derived proteinase inhibitors. These include α_1-antitrypsin (α_1A), α_2-macroglobulin (α_2M), antithrombin III (ATIII), C1-inactivator (C1-INA), α_1-antichymotrypsin and inter-α-trypsin inhibitor. A wide range of proteinases is inhibited, including elastase, collagenase, plasmin, trypsin, chymotrypsin, Hageman factor, kallikrein and thrombin. Most of the plasma inhibitors interact with several enzymes. The site of action of some are shown in Fig. 7.2.

Tissue-derived mediators of inflammation

Arachidonic acid products—prostanoids and leukotrienes

The products of arachidonic acid include the prostaglandins, the thromboxanes and the leukotrienes. Together, they constitute the eicosanoids[87] (Fig. 7.4). The precursors of these substances are polyunsaturated fatty acids, derived from essential fatty acids and including dihomo-γ-linolenic acid, arachidonic acid and eicosapentaneoic acid. They are found mainly in the cell membrane phospholipids. Arachidonic acid is the most abundant.[55]

The mechanisms by which arachidonic acid is metabolized is shown in Fig. 7.4. Arachidonic acid itself is formed from membrane phospholipids by acylhydrolases such as phopholipase A_2, present, for example, in the lysosomes of polymorphs. Oxidation of arachidonic acid then occurs. Two pathways are involved.

The first pathway is via a cyclo-oxygenase enzyme system (Fig. 7.4), leading to the formation of short-lived intermediate endoperoxides which are converted, via three further enzyme pathways, to:

1. Stable prostaglandins (prostaglandins E_2, F_2 and D_2); *or to*
2. Short-lived prostacyclin (PGI_2) which is broken down predominantly to the inactive 2, 6-keto-prostaglandin $F_1\alpha$. Prostacyclin I_2 (PGI_2) is synthesized mainly by the vascular endothelium; *or* to:
3. Thromboxane A_2 (TxA_2), which is highly unstable and breaks down to its inactive metabolite TxB_2. One of

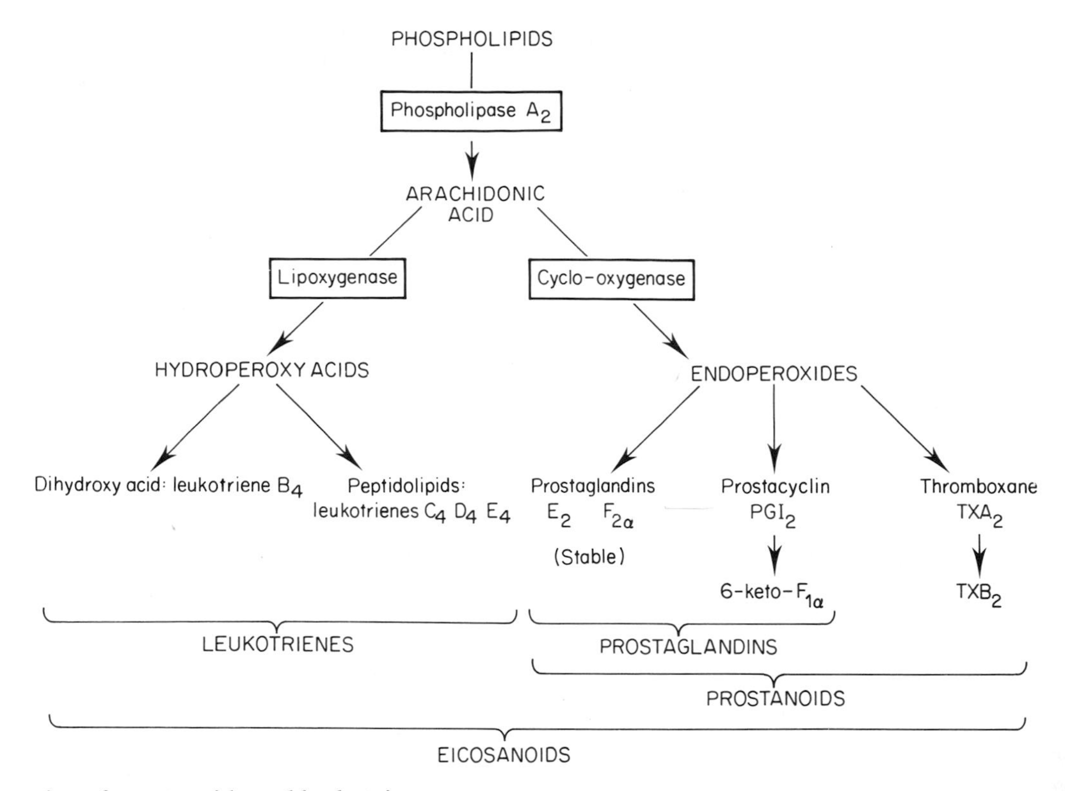

Fig. 7.4 Formation of prostanoids and leukotrienes.
The metabolic pathways leading to the formation of prostanoids and leukotrienes from arachidonic acid.

the major sites of production of TxA_2 is the blood platelet.

The second pathway for the metabolism of arachidonic acid involves the lipoxygenase enzymes. The initial products are hydroperoxy acids which can be converted to substances which include monohydroxy acids, a dihydroxy fatty acid, leukotriene B_4 (LTB_4); and peptidolipids such as LTC_4, LTD_4 and LTE_4.

Prostanoids The prostaglandins are 20-carbon fatty acids that have a central cyclopentane ring and two side-chains. The thromboxanes have a similar structure but the cyclopentane ring is replaced with an oxane ring. Virtually all mammalian cells have the capacity to manufacture prostaglandins and they are synthesized *de novo* when required. The only sites where prostaglandins are stored are the seminal vesicles. During an inflammatory reaction, phospholipases are released from an inflammatory focus where there are leucocytes. Arachidonic acid is liberated. Since this fatty acid is present in all membrane phospholipids, it is not surprising that prostaglandins are synthesized whenever cells are damaged or disordered.

The phenomena of inflammation are induced by the prostaglandins[87] as a consequence of their ability to dilate small blood vessels[49] and to increase vascular permeability. Intradermal injections of prostaglandin E_2 into normal human skin in concentrations of less than 10 ng/ml, cause pronounced and sustained *erythema* with a rise of temperature. Although prostaglandins induce wheal and flare responses in the human skin,[16] it is not clear whether the permeability effects in humans are due to a direct effect on the microvasculature[92] or to the release of vasoactive substances.[44]

Experimental evidence indicates that prostaglandins E_2 and I_2 are probably more effective at potentiating the actions of other mediators of inflammation than they are at inducing inflammation directly. Williams and Peck[92] suggest that the capacity of the products derived from arachidonic acid to provoke modest *oedema* and to potentiate oedema induced by other mediators such as bradykinin and histamine, reflects enhanced vasodilatation. They also demonstrate that other substances such as isoprenaline (which increases skin blood flow) and angiotensin (which reduces skin blood flow when injected locally) increase and decrease plasma exudation, respectively, in response to histamine and bradykinin. Thus, the quantity of an inflammatory exudate appears to be partially determined by the level of vasodilator substance produced when generated simultaneously with agents that increase vessel wall permeability.[89]

Prostaglandins contribute to the *pain* of inflammation. In

humans, the intradermal injection of the hydroperoxide of arachidonic acid, of acetylcholine, of histamine or of bradykinin causes transitory pain. By contrast, the pain produced by prostaglandin E_1 may last more than 2 h. The prostaglandins may sensitize pain-receptors to the effects of other pain-producing substances such as bradykinin and histamine.[24] The E-series prostaglandins have been implicated in this phenomenon but prostacyclin is also an effective hyperalgesic agent.[25] The non-steroidal anti-inflammatory drugs produce their effects by inhibiting cyclo-oxygenase and, consequently, prostaglandin synthesis. These drugs also inhibit the pain-provoking action of bradykinin,[79] emphasizing the modulatory role of the prostaglandins on this response.

The prostaglandins *impair function*. The injection of prostaglandin E_1 and prostaglandin E_2 into rat paw or dog knee joints interferes with ambulation[93] and may even cause debilitating arthritis. Prostaglandin E_2 is a potent stimulator of bone resorption *in vitro*. The addition of serum to culture media stimulates bone resorption,[76] a process that is complement-dependent and possibly prostaglandin-mediated.[63] This mechanism may contribute to bone resorption in the chronically inflamed joint where complement is activated and prostaglandin concentrations high. Diseases in which the cyclo-oxygenase products are elevated include rheumatoid arthritis, osteoarthrosis, psoriatic arthropathy, ulcerative colitis and gout.[36] Synovial tissue from patients with rheumatoid arthritis, maintained in culture, produces larger amounts of the prostaglandins than synovium from patients with osteoarthrosis.[67]

Leukotrienes The major products of the lipoxygenase pathways are the leukotrienes. They are a heterogeneous group of cyclic eicosapolyanionic acids with a common triene structure. Early studies had demonstrated that a substance could be released from sensitized guinea-pig lungs which produced a slow contraction of guinea-pig ileum.[23] This *s*low *r*eacting *s*ubstance released during *a*naphylaxis was called SRS-A.[10] It is now known that SRS-A is a mixture of leukotrienes. Leukotrienes C_4 and D_4 are derived from mast cells and basophils whereas leukotriene B_4 appears to be formed mainly in polymorphs. The leukotrienes can be released from their storage sites by immunological (via bound IgE) or non-immunological stimuli.

Although leukotrienes C_4 and D_4 produce an increase in vascular permeability, they also cause vasoconstriction which tends to mask the former activity. Leukotriene B_4 is the most active chemotactic and chemokinetic arachidonic acid-product causing polymorph infiltration. This effect, together with an early flare response, is seen when leukotriene B_4 is injected into human skin.[74] Finally, evidence suggests that leukotriene B_4, like prostaglandin E_2 and prostacyclin, has a suppressive effect on T-lymphocyte activity. Thus it is apparent that leukotriene B_4 is the most potent lipoxygenase product with a role relevant to the pathophysiology of inflammation. The main actions of the leukotrienes are shown in Table 7.2.

Table 7.2

Effects of leukotrienes relevant to inflammation

Leukotriene B_4	Leukotrienes C_4, D_4, E_4
Aggregation of polymorphs	Leakage from postcapillary venules
Chemotaxis (polymorphs)	Oedema formation
Chemokinesis of polymorphs	Vasoconstriction
Plasma exudation	Phospholipase A_2 stimulation
Phospholipase A_2 stimulation	

Vasoactive amines

Histamine (β-imidozolylethylamine) and 5-hydroxytryptamine (serotonin) are vasoactive amines that mediate the vascular phenomena of acute inflammation. The main sources of these amines are mast cells (see p. 36), basophil granulocytes and platelets.

The present discussion is confined to histamine which has attracted most attention. The actions of histamine on the microcirculation are of particular importance. After an inflammatory insult or injury, mast cells accumulate and degranulate near the arteriolar bed. Vasodilatation occurs. There is an increased escape of fluid from the postcapillary venules because of endothelial cell contraction. Vasodilatation is mediated by the activation of histamine H_1- and H_2-receptor types, effects which can be blocked by mepyramine and cimetidine, respectively. By contrast, the increase in vascular permeability is a response to histamine H_1-receptor activation alone.[7,43] Many of the vascular effects of histamine are enhanced by the prostaglandins.

Histamine may function as a regulator, modifying the responses of cells to inflammatory and immune stimuli.[68] Histamine can inhibit its own release by negative feedback and may reduce basophil and eosinophil granulocyte and polymorph chemotaxis[47] and enhance chemokinesis. Histamine may serve as a negative feedback regulator of cellular immune reactions. The generation of suppressor T-lymphocytes induced by concanavalin A (see p. 229) is inhibited by this amine.[72]

The role of histamine in inflammation is complex and its analysis difficult. It is likely that histamine, like other inflammatory mediators has proinflammatory and anti-inflammatory properties. It may also influence immunological responses, promoting immunosuppression.

Cytokines

Many of the mediators and modulators released during inflammation affect both vascular permeability and cell infiltration.[56] There is increasing evidence that the pathogenesis of chronic inflammatory joint diseases such as rheumatoid arthritis involves persistent lymphocyte activation (see p. 246). The activated leucocytes release mediator molecules that exert profound, often remote effects on other leucocytes as well as on non-lymphoid cells such as synoviocytes. The macromolecular products of lymphocytes that influence other cells were called lymphokines, those of monocytes, monokines. Collectively, these messenger molecules are now cytokines (see pp. 228 and 247).

At first, many cytokines were shown to be generated by the *in vitro* incubation of sensitized lymphocytes with antigen. The products were termed lymphokines. However, macrophages, synoviocytes and other cells also proved to be sources of similar proteins so that additional names such as 'monokine' were adopted. The principal role of the majority of the cytokines is to influence the cells of the immune system and their part in the activation of T- and of B-cells and of macrophages, is outlined in Chapter 5. Interleukin-1[19,53] and tumour necrosis factor are among the cytokines known to act on connective tissue cells; their role in cartilage degradation is considered on pp. 88 and 537 *et seq*, and their part in fibrosis on p. 302 *et seq*. The contribution of cytokines to bone reabsorption is mentioned on p. 474.

Cytokines (Table 7.3) are proteins of molecular weights ranging from 14 to 76 kD produced by cells in response to induction signals generated at cell surfaces. An increasing number of cytokines has been obtained in homogeneous preparation by gene cloning, allowing detailed understanding of their structure and permitting their manufacture in quantities sufficient for experimental and therapeutic use. Much less is known of the receptors to which cytokines bind on target cells. However, it is believed that the ligand attaches to an extracellular domain and that there are both a hydrophobic transmembrane region and an intracellular domain. Among the second messengers produced in response to cytokine-receptor interaction at the cell membrane are cyclic nucleotides; there is hydrolysis of phosphatidylinositol 4,5-biphosphate, activation of protein kinase C and an elevation of intracytoplasmic calcium.[5,34]

In acute inflammation, interleukin-1 and tumour necrosis factor exert both local and systemic effects. Interleukin-1, for example, provokes the release of polymorph granules and lysosomal enzymes, and of synoviocyte collagenase; the production of acute-phase proteins is increased.[19,53] Both interleukin-1 and tumour necrosis factor may activate vascular endothelial cells, increasing leucocyte adhesion and promoting diapedesis. Macrophage inhibition factor impedes the movement of mononuclear phagocytes, facilitating the inflammatory sequence.[97] It is of interest that interleukin-1 has been identified in rheumatoid synovial fluid[95] and this cytokine is firmly believed to contribute to the cascade of local inflammatory changes in this disease. Interleukin-2 causes the induction and proliferation of rheumatoid synovial T-cells. The actions of interleukin-2 are reviewed in Chapter 5 and the relationships between the interleukins and lymphocyte activation summarized in Fig. 5.1.

Free radicals

In response to particulate stimuli such as bacteria or bacterial fragments, and to non-particulate agents and molecular species such as aggregated IgG, complement components or interleukin-1,[31] polymorphs initiate the complex sequence of events termed phagocytosis. Phagocytosis is an energy-dependent process. As the cell moves and enzymes are activated, oxygen is quickly consumed. The response is exceedingly rapid and constitutes a 'respiratory burst'. During oxygen consumption, free radicals[20,53a] are produced. Free radicals are atoms or molecules with one or more unpaired electrons; the radicals are capable of an independent, if transient, existence. The most important free radical formed during a respiratory burst is the superoxide anion, $\dot{O}_2^-$. Superoxide plays a part in the production of other reactive derivatives, particularly the hydroxyl radical $OH^{\cdot}$, a potent agent capable of interacting with any tissue component within several nanometers of its release; and hydrogen peroxide, H_2O_2. The arachidonic acid cascade (p. 288) also involves the formation of free radicals[22] which may then have a potentiating effect on prostaglandin synthesis.[46] Free radicals have the capacity to kill microbial pathogens.[4] By the same token, however, they are liable to cause tissue injury.

The role of free radicals as promoters of inflammation has attracted great interest in rheumatoid arthritis. The rheumatoid joint contains both acute and chronic inflammatory cells. A polymorph leucocytosis is invariably present in the synovial cavity and synovial fluid. In isolation, these cells release low levels of both $\dot{O}_2^-$ and H_2O_2.[9] Free radicals can inhibit the actions of antiproteineases, enhancing the destructive effect of lysosomal proteineases.[33] Superoxide radicals not only react with a component in human plasma forming a product which is chemotactic for polymorphs[82]

Table 7.3

Some cytokines that have been purified, their sources and their targets (modified from Hamblin[34])

Cytokine	Source	Target
Interferon-γ (IFN-γ)	T-cells, natural killer (NK) cells	Macrophages, T-cells, B-cells, NK cells
Interleukin-1α (IL-1α) Interleukin-1β (IL-1β)	Macrophages, endothelial cells, large granular lymphocytes, B-cells, fibroblasts, epithelial cells, astrocytes, keratinocytes, osteoblasts	Thymocytes, polymorphs, hepatocytes, chondrocytes, myocytes, epithelial cells, epidermal cells, osteocytes, macrophages, T-cells, B-cells, fibroblasts
Interleukin-2 (IL-2)	T-cells	T-cells, B-cells, macrophages
Interleukin-3 (IL-3)	T-cells	Multipotential stem cells, mast cells
Interleukin-4 (IL-4)	T-cells	T-cells, mast cells, B-cells, macrophages, haematopoietic progenitor cells
Interleukin-5 (IL-5)	T-cells (mouse)	Eosinophil granulocytes, B-cells (mouse)
Interleukin-6 (IL-6)	Fibroblasts, T-cells	B-cells, thymocytes, haemato poietic cells
Interleukin-7 (IL-7)	Stromal cells, thymus	Pre-B cells, thymocytes
Interleukin-8 (IL-8)	Macrophages	Neutrophil polymorphs
Interleukin-9 (IL-9)	T-cells	Fetal, adult haematopoietic cells
Interleukin-10 (IL-10)	Th2-T-cells	T-cells, B-cells, mast cells
Interleukin-11 (IL-11)		Haematopoletic cells, megakaryocytes
Granulocyte macrophage colony-stimulating factor (GM-CSF)	T-cells, endothelial cells, fibroblasts, macrophages	Multipotential stem cells
Macrophage colony-stimulating factor (M-CSF)	Fibroblasts, monocytes, endothelial cells	Multipotential stem cells
Granulocyte colony-stimulating factor (G-CSF)	Macrophages, fibroblasts	Multipotential stem cells
Tumour necrosis factor-α (TNF-α)	Macrophages, T-cells thymocytes, B-cells, NK cells	Tumour cells, transformed cell lines, fibroblasts, macrophages, osteoclasts, polymorphs, adipocytes, eosinophils, endothelial cells, chondrocytes, hepatocytes
Lymphotoxin (TNF-β)	T-cells	Tumour cells, transformed cell lines, polymorphs, osteoblasts

but also promote platelet aggregation[35] with the consequent release of inflammatory mediators such as 5-hydroxytryptamine, prostanoids, platelet-activating factor and lysosomal enzymes.

Free radicals can damage protein resulting in the formation of factors which act as indices and prognostic markers of the rheumatoid process. When damaged in such a fashion, IgG develops a characteristic autofluorescence[51] identical to that produced by fluorescent complexes present within synovial fluid and serum from rheumatoid arthritis patients.

The extent to which these actions occur in connective tissue disease is unclear. It is likely that free radicals are 'mopped up' by a complex inhibitor system which includes superoxide dismutase and catalase which are free radical 'scavengers'. It has been suggested that chronic synovitis may be an example of reperfusion injury.[96] After a period of hypoxia, tissue injury occurs during the 'reperfusion' phase as a consequence of generation of oxygen and free radicals.[53a] Temporary intestinal and myocardial ischaemia can be inhibited by superoxide dismutase and catalase. Indeed both these free radical scavengers have proved to be very effective in suppressing acute and chronic inflammation in animal models.

It is therefore highly likely that free radicals play a role in maintaining inflammatory reactions. The efficacy of free radical scavengers in arresting or reversing the destructive processes seen in rheumatoid arthritis in humans remains to be assessed.

Interactions of inflammatory mediators

Many inflammatory mediators interact on the effector tissue so that the responses are potentiated or modified.[90] In one example, the capacity of some agents to cause vasodilatation or exudation can be distinguished: the responses are probably determined by different effector cells. Vasodilatation is a property of smooth muscle cell relaxation whereas exudation is a consequence of endothelial cell contraction. A dual mechanism has been proposed by Williams[88] to explain the potentiation of the exudation-promoting activity of kinins and histamine by the vasodilators prostaglandins E_2 and I_2. Both these prostaglandins have no significant capacity to enhance permeability.[91] However, both enhance blood flow to an inflamed zone.

According to Williams[88] and Wedmore and Williams[80] inflammatory mediators can therefore be divided into three groups:

1. *Auxiliary vasodilators* which are inactive in terms of vascular permeability but markedly enhance the exudate produced by agents acting on the endothelium. Prostaglandins of the E and I series are the prototypes of this group.
2. *Mediators acting purely on vascular permeability.* These mediators are almost inactive in terms of permeability when applied alone but become active when given together with a vasodilator. This group includes platelet-activating factor and C5a and its breakdown product des-Arg^{74}-C5a.
3. *Mediators which increase vascular permeability and cause vasodilatation directly.* This group includes the kinins and histamine whose effects are potentiated by prostaglandins.

Mediators such as bradykinin, histamine and platelet-activating factor increase vascular permeability by a direct action; those such as leukotriene B_4 and complement factor C5a act indirectly. The latter appear to be dependent upon the presence of polymorphs. When these cells are absent, no response can be observed (Fig. 7.5), a situation prevailing in urate arthropathy (see p. 389).

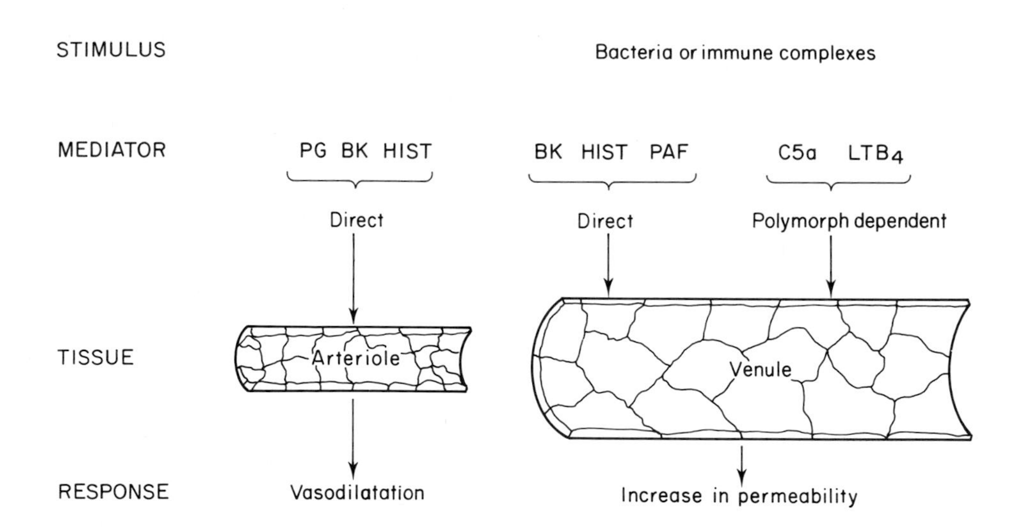

Fig. 7.5 Effects of inflammatory mediators.
Diagrammatic representation of the effects of various inflammatory mediators. The prostaglandins, bradykinin (BK) and histamine (HIST) act directly on arteriolar smooth muscle producing vasodilatation. BK, HIST and platelet activating factor (PAF) can act directly on the venules causing an increase in permeability. An increase in permeability is also produced by C5a and leukotriene B. However, the effect of these two agents depends upon the presence of polymorphs, suggesting an indirect mechanism of action. The mediator released from polymorphs which increases permeability is unknown. By increasing blood flow, the E-series prostaglandins (PG) can enhance the vascular permeability effects of all the other mediators shown. LTB_4 = leukotriene B_4.

Finally, the prostaglandins, particularly those of the E-series, are potent hyperalgesic agents. By themselves, these mediators are relatively ineffective pain-producing agents. However, they sensitize pain receptors to the actions of other pain-producing substances such as histamine and bradykinin.

Fibrinoid

'Fibrinoid' is an old descriptive term still widely used in diagnostic histopathology[1,69,70] and often mentioned in contemporary works on general pathology.[2,66,77] 'Fibrinoid' implies a microscopic appearance that is a non-specific, shared and common result of a wide variety of pathological processes including hypersensitivity (e.g. systemic lupus erythematosus); vascular disease (e.g. accelerated hypertension); ulceration (e.g. peptic ulcer) and injury (e.g. burns). Although the processes that underlie 'fibrinoid' are neither always inflammatory nor necessarily succeeded by inflammation, in many instances there is either altered vessel permeability or the accumulation in the extracellular matrix of macromolecules derived from the circulation.

The term fibrinoid has often been used to describe eosinophilic zones of abnormal connective tissue matrix noted in serous inflammation, in the ruptured walls of aneurysms and in verrucous endocarditis. Neumann[57] deduced from the morphology and staining characteristics of fibrinoid that it was the result of a chemical transformation of connective tissue collagen fibres into a substance resembling fibrin and Klemperer[40] believed that the formation of fibrinoid (in progressive systemic sclerosis) reflected a chemical abnormality of the intercellular matrix 'ground substance'. There were unique staining characteristics, suggesting a transformation into a material derived from collagen. When there was associated cell death, e.g. in the media of arterioles in accelerated hypertension, the phrase 'fibrinoid necrosis' was applicable, indicating a more advanced phase of the same process.

These older views were seminal in the formulation of the hypothesis of the 'diffuse diseases of collagen'[41] (see the Historical Introduction) but they are now obsolete. It is clear that the same histological appearances can be presented in different situations by distinct materials of varied composition.[29] Light microscopy shows that 'fibrinoid' does indeed resemble fibrin although, in haematoxylin and eosin-stained sections, it has a more amorphous, granular character than fibrin deposits; it is refractile but not birefringent nor does it fluoresce in ultraviolet light. Fibrinoid is Schiff-positive, bright red with the picro-Mallory and Martius-Scarlet-Blue methods and slightly metachromatic. Silver impregnation yields a nondescript yellow-brown colour.

One explanation of the problem that accords with these observations is that of Kaplan.[39a] In a discussion of the coagulation pathway, Kaplan states 'Once fibrin is formed, the process for the degradation of fibrin (fibrinolysis) is also set in motion. The final product, plasmin, degrades fibrin to a variety of intermediate products. These degradation products, to which cold-insoluble globulin (fibronectin) is attached, account for the homogeneous staining material in histological sections of inflamed tissues that is known as fibrinoid'.

Early transmission electron microscopic studies demonstrated the integrity of collagen fibres at sites where fibrinoid was seen[14,94] although the X-ray diffraction pattern of the collagen differed from normal. The remaining non-collagen, dense, filamentous material resembled or was identical with fibrin: part of it was probably proteoglycan-rich.[14]

Material with the classical microscopical characteristics of fibrinoid can be identified in many of the lesions of the immune disorders of the connective tissue system. Fibrinoid is recognized in rheumatic fever, progressive systemic sclerosis and in systemic lupus erythematosus[41] and is prominent in polyarteritis nodosa; it is demonstrable in the Aschoff body (p. 798) and in the articular, subcutaneous and visceral lesions of rheumatoid arthritis. The presence of fibrinoid is not, however, pathognomonic for this group of diseases since material with closely similar microscopic appearances is found not only under physiological circumstances, e.g. around the villi of mature placentas, but in lesions as varied as the afferent arteriole in accelerated hypertension and the small blood vessels in the generalized Shwartzman reaction[29] (Table 7.4).

It is no longer justifiable therefore to regard fibrinoid as a specific end-product of tissue injury in one class of disease. It is not a single substance. As immunocytochemical and electron microscopic techniques have advanced, it has become clear that fibrinoid may be mainly fibrin, DNA, altered matrix material or disorganized collagen. Fibrinoid is still a convenient broad term for use in diagnostic histopathology, setting the scene for further, more exact tests. However, fibrinoid can originate in sites rich in collagen, proteoglycan, muscle, plasma protein or immune complexes. Moreover it may be heterogeneous, arising either from alteration in normal mesenchymal tissues or by the accretion or insudation of substances not normally present at the affected site. These observations show that the presence of fibrinoid cannot be used reliably either to prove common characteristics shared by a group of diseases or to show that a single mode of injury prevails. It is of interest that the significance of fibrinoid in the understanding of the systemic connective tissue diseases has so diminished that the material is neither indexed in Ghadially's[30] text on the ultrastructural pathology of the cell and matrix nor in large, contemporary works on rheumatology.[73]

Table 7.4

The sites of occurrence and probable chemical nature of the materials called 'fibrinoid'

Sites	Chemical nature
Inflammatory reactions	
Bacterial endocarditis	Fibrin
Active peptic ulcers	Fibrin
Miscellaneous processes	
Generalized Shwartzman reaction	Fibrin and other proteins
Around mature placental villi	Fibrin
Tissue death	
Arteriolar injury in accelerated (malignant) hypertension	Fibrin; other plasma proteins
Aneurysmal sacs	Fibrin
Immunological reactions	
Rheumatoid arthritis nodule	Fibrin; low hydroxyproline content
Rheumatic fever	
Aschoff body	Immunoglobulin
Pericarditis	Fibrin; albumin; immunoglobulin
Nodule	Degraded collagen; proteoglycan
Systemic lupus erythematosus	Immunoglobulin; complement; DNA
Polyarteritis nodosa	Fibrin; other plasma proteins

Healing

To heal is to make whole. Healing, in pathology, is synonymous with repair. Repair is only possible in living tissues. The repair of connective tissues may be by regeneration of the original tissue, or by the substitution for the injured part, of a mesenchymal or fibrous tissue the presence of which constitutes a scar (p. 300). Particular interest centres on the capacity to which avascular connective tissues such as articular cartilage or cornea, or connective tissues with a limited direct blood supply such as tendon, are capable of regeneration. Where injury has been limited to the degradation or loss of matrix proteoglycan, without disrupting the collagenous microskeleton or irreversibly damaging the cells, the proteoglycans can be effectively and swiftly replaced. The balance of evidence suggests that any effective regeneration extending beyond new macromolecular synthesis is closely related to the proximity of an arterial blood supply. The problems of cartilage and fibrocartilage repair are addressed in Chapter 2. To a varying extent, the absence or limitation of an apparent blood flow can be overcome by the growth, into injured connective tissue, of a new vasculature. Interest has therefore centred on the mechanisms of neovascularization.

New blood vessel formation

Neovascularization strongly influences healing. In addition, the ingrowth of endothelial cell buds and new vessel formation (angiogenesis) are important in modulating the tissue changes at the margin of synovial joints in chronic inflammatory diseases such as rheumatoid arthritis (see p. 540) and tuberculous arthritis (see p. 743). Angiogenesis may also contribute secondarily to the natural history of non-inflammatory disorders such as osteoarthrosis (see p. 892), through an influence on cartilage growth and mineralization.

Angiogenesis

The mechanisms of angiogenesis have been reviewed.[26,27] There are functional differences between the endothelial cells of capillaries and those of large arteries and veins.[15a] Following a stimulus to vessel formation of the kind that, it is thought, occurs in the origin of granulation tissue (see p. 301) and in the synovia in rheumatoid arthritis (see p. 461), vascular buds begin to arise from small venules. Proximal basement membranes are degraded, a step attributable to the secretion by endothelial cells of enzymes such as collagenase (see p. 184) and of plasminogen activator. The cytoplasmic processes of endothelial cells extend through the gaps in the basement membranes. Other cells follow; they are directionally orientated towards the angiogenic stimulus. Endothelial cells begin to curve and a lumen appears. Mitotic activity is recognized in the median part of the new vascular cell extension but not at its leading edge. Individual finger-like processes join and blood flow slowly starts. The acquisition of pericytes, not of endothelial cell origin, and of new basement membrane material, are later steps in this sequence.

Angiogenesis can be initiated by local events such as sustained inflammation or hypoxia[54,92] (p. 322). The processes of new vessel formation are stimulated by angiogenesis factors, and controlled and regulated by other molecular species that act as modulators. Angiogenesis can be blocked by antagonists (see below), and avascular

Michinson[118] analysed 40 cases. The mean age was 53 years; 30 were males, 10 females. Twelve were studied at necropsy. Fibrosis encircled the lower abdominal aorta, seldom extending below the pelvic rim but following the course of the iliac vessels. The vena cava was less completely surrounded and fibrosis extended a small distance laterally, drawing the ureters medially but not involving them directly. Superiorly, fibrosis did not approach the renal arteries although, in three of 12 cases, periaortic thoracic fibrosis was detected. Forward extension to the small bowel was seen once; displacement of the duodenum was recognized three times; infiltration of the pelvic mesocolon twice.

Histologically, two patterns were described.[118] Eleven of the 40 cases displayed avascular, acellular and often calcified fibrous tissue; 29 revealed many inflammatory cells among a vascular, myxoid* matrix. Lymphocytes and plasma cells were present in large numbers in 23 cases, eosinophil granulocytes in 15. In cases of an idiopathic nature, the polyclonality of the plasma cell population helps to distinguish the fibrotic reaction from the connective tissue response in the sclerosing lymphomas.[130] Histiocytes and mast cells were identified but polymorphs were invariably absent. A progression of this active, recent inflammatory response into the more mature collagenous connective tissue, especially under the influence of corticosteroids, was seen. Veins became obliterated. Nearby adipose tissue contained a lymphocytic infiltrate and skeletal muscle included much fibrous tissue. Lymphatics became blocked.

Retroperitoneal fibrosis and atherosclerosis

The coexistence of aortic atherosclerosis attracted particular attention. In three instances[119] aortitis or the extrusion of atheromatous debris into the inflamed adventitia was recognized. The suggestion was therefore made that aortic wall damage, succeeded by adventitial inflammation due to a hypersensitivity reaction to a component of the atherosclerotic plaque, could be one sequence in the development of retroperitoneal and of mediastinal fibrosis. The plaques contained insoluble lipid and immunoglobulin and a view emerged, linking adventitial periaortic inflammation in aortic atheroma with a hypersensitivity response.[120] The idea of chronic periaortitis could apparently unite views on the nature of retroperitoneal fibrosis and inflammatory aortic aneurysms:[121] early removal of the fibrosing stimulus might allow the regression of the retroperitoneal disorder as could the administration of corticosteroids or immunosuppression.[80]

* The term 'myxoid' is commonly used to describe a loose, vascular, highly hydrated, proteoglycan-rich, collagen-poor connective tissue in which stellate fibroblasts or their precursors, less differentiated mesenchymal cells, resemble those of the 'Wharton's jelly' of the umbilical cord connective tissue. The glycoaminoglycan moiety is rich in hyaluronate (see also p. 193).

Propranolol fibrosis

A study of propranolol led to the suggestion that this drug could increase the tensile strength of aortic tissue in turkeys. Propranolol acts partly by stimulating lysyl oxidase to produce greater amounts of reactive aldehydes for intermolecular elastin cross-linking, partly by enhancing the formation of stable, intermolecular elastin cross-links, and partly by reducing the density of the age-related intermolecular cross-links of collagen.[24] Another mechanism, not yet clarified, may be upon the encoding, translation and transcription of collagen amino acid chains within the fibroblast itself. When inflammation can be demonstrated, it is reasonable to invoke the role of peripheral blood leucocytes. Stimulated by T-cell antigen or mitogen, these cells can release soluble factors, inducing fibroblast mitogenesis factor (see p. 18).[147]

Peritoneal fibrosis

The majority of instances of peritoneal fibrosis are the late consequences of infection, surgery, dialysis, perforation of a viscus, or irradiation. However, silicates in the form of talc, used in glove powder or by drug addicts[30] may provoke peritoneal fibrosis.

Practolol fibrosis

Exceptionally, peritoneal fibrosis is caused by other pharmaceutical components. Read[153] described 16 patients with abdominal fibrosis following practolol therapy. Fibrosis has also been reported after sotalol.[98] Postpractolol fibrosis is *peritoneal*, not retroperitoneal. Dilatation of the duodenum and of the small intestine with concertina change, fixation, delays in transit and sacculation were seen; they were accompanied by a separation of the loops of gut due to the thickening of the peritoneum. The visceral peritoneum was grossly thickened, the loops of intestine adherent and, sometimes, the whole small gut, but not the colon, encased within a cocoon-like shell of fibrous tissue (Fig. 8.4). Microscopically, dense fibrous tissue was accompanied by a slight inflammatory reaction; the collagen cross-links were unusually stable. Skin, pleural and lung fibrosis were also recognized (Fig. 8.5).

Peritoneal sclerosis in dialysis

Recurrent, continued inflammation of the peritoneal cavity may induce sclerosis, and patients subjected to long-term peritoneal dialysis for renal disorders can develop symptomatic peritoneal disease.[55] It is of interest to contrast this local peritoneal reaction with the deposits of amyloid and of iron-rich tissue formed in sites such as the carpal tunnel and shoulder.[26a] The dialysing fluid is hypertonic and acidic and has been found to include irritants ranging from talc and lactic acid to glucuronic acid, endotoxin and even formaldehyde (Fig. 8.5). Added antibiotics, β-adrenergic blocking agents and silicone catheters have also contributed to peritoneal disease. Recurrent bacterial infection may complicate the peritoneal response, terminating dialysis or precipitating intestinal obstruction.[173]

Neoplasm-like lesions of fibrous tissue

In addition to the numerous forms of reactive and reparative fibrosis, some of which are described on pp. 301 *et seq*, there is a large number of local or generalized proliferative disorders of fibroblasts and myofibroblasts in which the abnormal cell growth results in the formation of a neoplasm-like mass (tumour) (Table 8.1). The pathological characteristics of this incompletely understood but important category of disease of myofibroblastic cells have been fully described by Enzinger and Weiss.[49] The present summary adopts this classification but takes account of the earlier views of MacKenzie,[111] Allen[7,8] and Hajdu.[71]

In essence, the 'tumour-like' lesions of fibrous tissue display the neoplastic property of inappropriate and excessive growth but not of metastasis. Some tend to recur after resection. They are closely related to but distinct from fibrosarcoma, an uncommon malignant neoplasm with a propensity for both recurrence and metastasis. Although it is widely assumed that the cell responsible for these 'tumour-like' lesions is the fibroblast, cells with the phenotypic characteristics of myofibroblasts (see pp. 40 and 301) are very often identifiable within the proliferative tissue.

Nodular fasciitis

The term 'fasciitis' is used to describe a self-limited, non-neoplastic proliferative mass of subcutaneous or deep fas-

Fig. 8.4 Practolol fibrosis.
A broad lamina of fibrous tissue (top) has formed beneath the surface of the peritoneum. (HE × 60.) Courtesy of Dr J D Davis.

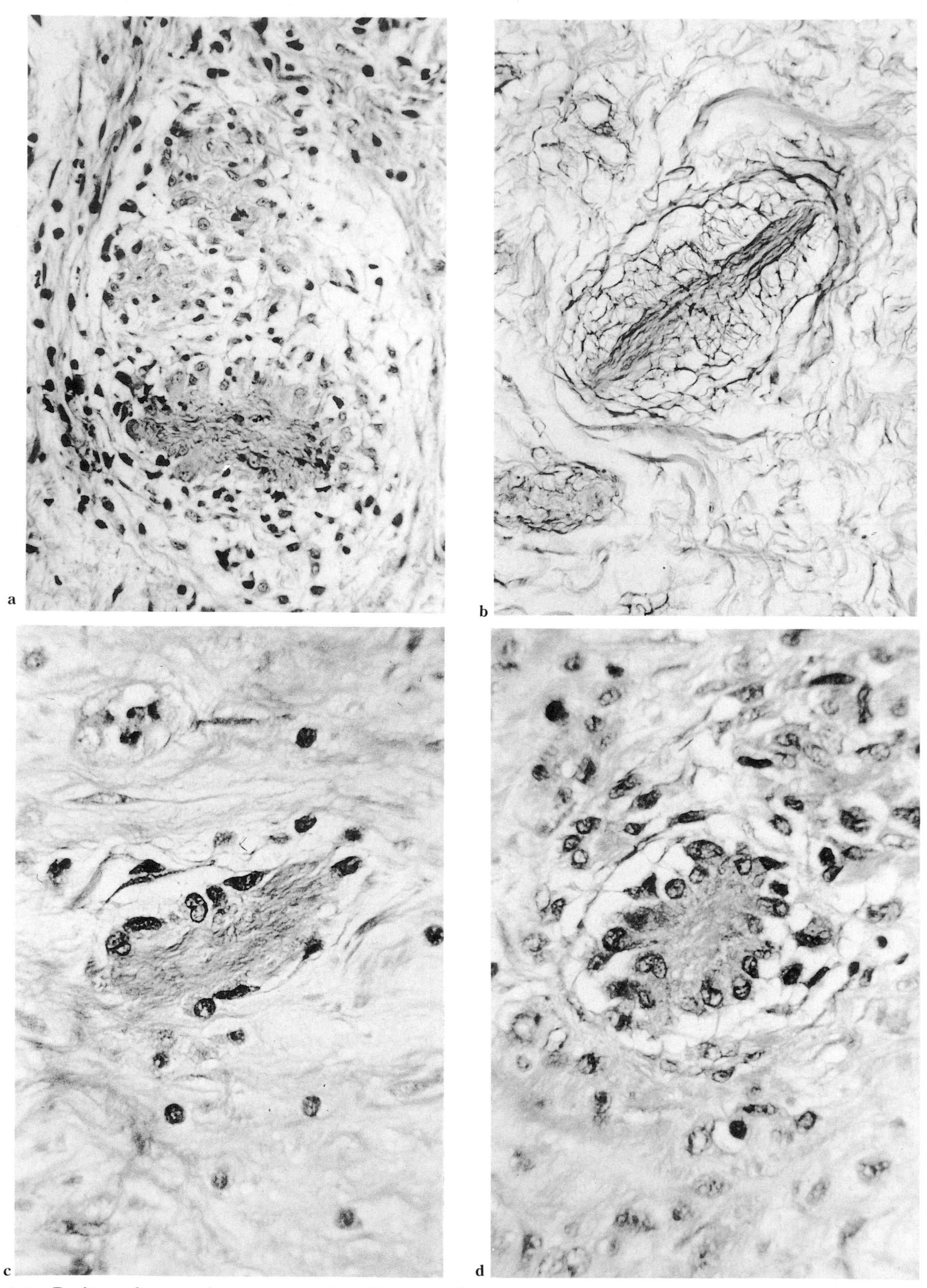

Fig. 8.5 Peritoneal macrophage response to collagenous foci.
Although no doubly refractile material is present in these microscopic peritoneal granulomata, reticular and collagen fibres have elicited a mononuclear macrophage reaction of the kind encountered in response to foreign bodies or talc. **a.** Disposition of collagen; **b.** Reticular fibres; **c.** Macrophages surround collagenous focus; **d.** Further, nearby focus. (**a.** Elastic-van Gieson × 240; **b.** Silver × 240; **c** and **d.** Masson trichrome × 400.)

Table 8.1

Neoplasm-like lesions of fibrous tissue (adapted from Enzinger and Weiss[49])

I Benign fibroblastic proliferation
 Nodular fasciitis
 Proliferative fasciitis
 Proliferative myositis
 Elastofibroma
 Keloid
 Hypertrophic scar

II Fibromatoses (desmoid 'tumours')
 Superficial (fascial)
  Palmar (Dupuytren's contracture)
  Plantar (Ledderhose's disease)
  Penile (Peyronie's disease)

 Deep (musculoaponeurotic)
  Extraabdominal
  Abdominal
  Intraabdominal
   – pelvic
   – mesenteric
   – Gardner's syndrome

III Fibrous proliferation of infancy and childhood
 Non-recurrent
  Fibrous hamartoma
  Infantile myofibromatosis
  Fibromatosis colli

 Recurrent
  Digital fibromatosis
  Hyaline fibromatosis (juvenile)
  Desmoid fibromatosis (infantile)
  Calcifying aponeurotic fibroma

cial, fibrous connective tissue.[71,81] There is a component of fibroblasts and inflammatory cells and their presence justifies the retention of a term suggesting an inflammatory origin. Incomplete excision allows 10 per cent of cases to recur. Fasciitis is distinguished from the fibromatoses (see p. 313) because of the absence from the latter of a vascular, granulomatous response.

Fasciitis is now diagnosed with increased frequency but it is not a new disease. The first full report was that of Konwaler, Keasbey and Kaplan.[97] By 1970, more than 600 cases had been described and Enzinger and Weiss[49] refer to 1000 cases reviewed in a 20-year period. The male: female frequency is approximately 1:1. The lesion may occur at any age although the largest number of cases is recognized between 20 and 40 years.[111]

Macroscopically, fasciitis presents as a poorly defined, tender 10–50 mm diameter mass that has grown quickly. Some 46 per cent of instances arise in the upper extremity, 20 per cent in the head and neck, 18 per cent in the trunk, and 16 per cent in the lower extremities.[49] The lesion is a nodular infiltrate rather than an integral structure and, rarely, may be multiple. It is free from attachment to the skin but often in continuity with either the superficial or deep fascia and may be intramuscular. Exceptionally, the mass is parosteal and new bone or cartilage formation can be found. Depending on the quantity of fat, the cut surface is grey-white, with yellow or brown foci, or a more uniform grey.

The fasciitic mass can be considered to have four main components. The most important part is fibroblastic. Numerous, large stellate fibroblasts with vesiculate nuclei are arranged at random. Mitotic activity is common but variable. Occasional multinucleate forms may also be seen, perhaps derived by aneuploidy: alternatively they may be formed by the fusion of macrophages. The nuclei, as in the mononucleate fibroblast, are pale and ovoid with conspicuous nucleoli. Together with the faintly basophilic, myxoid matrix, the large, scattered, stellate fibroblasts give an appearance recalling that of fibroblasts in monolayer culture. An important feature distinguishing fasciitis from connective tissue neoplasms such as fibrosarcoma, is the presence, at the margin, of radial columns of fibroblasts, a feature resembling the vascular, finger-like processes of young granulation tissue at sites of repair or at the edges of granulomata. Shimizu, Hashimoto and Enjoji[179] have used the classification of Price, Silliphant and Shuman[148] to define three histological categories which adequately cover the spectrum of change in their 250 cases. The categories are type 1 (myxoid); type 2 (cellular); and type 3 (fibrous). The histological patterns roughly correlated with the duration of the nodules. In this series, follow-up studies over a period averaging 5–7 years revealed one recurrence and no deaths attributable to the connective tissue disease.

The fasciitic nodule or mass is vascular. Although capillary channels are often numerous, they may, nevertheless, be sparse and the endothelial cell component may amount to no more than an island of cells without a lumen. It appears as though a proliferating bud of syncitial endothelial cells, extending radially, may have been cut in transverse section near its tip. The loose texture of the matrix within which lie the many fibroblasts and the vascular channels, is characteristic. The lesion extends into and implicates subcutaneous fat but the intercellular material is largely proteoglycan: it stains avidly with methods for anionic glycosaminoglycan. There is a widespread, sometimes substantial reticular fibre meshwork. By comparison with scar tissue, fibroma, fibromatosis, desmoid tumour and elastofibroma, the quantity of collagen is limited. Numerous lymphocytes, some large mononuclear cells and, occasionally, other inflammatory cells, are constantly present. Mast cells are few, plasma cells commonplace.

In *proliferative fasciitis*, the analogue of proliferative myositis, a firm, subcutaneous nodule arises in older adults *de novo* or after local trauma in the forearm or thigh. Among a population of immature spindle cells are many basophilic giant cells, sometimes binucleate. A collagenous, myxoid or hyaline matrix may resemble osteoid. The lesion does not recur.

Proliferative myositis

This inflammatory lesion occurs in men slightly more often than women.[49,91,111] It is often seen in middle-life although an age distribution of 22–82 years is recorded.[48] There is a history of trauma. Clinically, there is a rapid growth of a muscular lesion, often in the trunk, shoulder or arm; the mass, painless and not tender, develops within a few weeks. The lesion is unattached to the skin overlying the affected muscle within which there is found a poorly circumscribed, deep-seated grey-white mass 10–60 mm in diameter. Excision effects cure without recurrence.

The ill-defined mass extends to involve all parts of the muscle. There are two main components, proliferating fibroblasts and loose aggregates of basophilic, multinucleate giant cells of pleomorphic form.[71] The appearance of these cells recalls those of ganglion cells but they have been said to have a myoblastic,[91] a fibroblastic,[48] or histiocytic[71] origin. The fibroblastic proliferation involves perimysium, epimysium and endomysium. Muscle bundles are disrupted without a sarcolemmal response and muscle is not wholly replaced; in this respect the lesion contrasts with fasciitis and with myositis ossificans. Fibroblast proliferation is conspicuous in the muscle zones adjoining fascial planes and within connective tissue septa: mitotic activity is commonplace but the mitotic figures are not abnormal. In spite of the nature of the disease, inflammatory cell infiltration is often very slight. The presence of small, radioopaque foci of bone formation is occasionally confirmed but the principal matrix structure is collagenous and myxoid.

Elastofibroma

Elastofibroma[2,85,86,111,124] is an uncommon non-neoplastic mass that usually arises in the inferior part of the subscapular space between the latissimus dorsi and rhomboid muscles.[71,124] Occasionally, there are infraolecranon, hip or ischial masses. Rare instances have been described in which as many as seven separate elastofibromata have developed in one individual.

The classical features of elastofibroma are illustrated by large Japanese series but the condition is also relatively frequent in other parts of Asia including Papua New Guinea. Of the 170 cases reported by Nagamine *et al.*[124], from the islands of Okinawa, 158 were women, giving a female: male ratio of 8.1:1.0. The mean age at onset was 70 years, the range 35–94 years. There were a few cases in young adults. The patients were predominantly farmers or labourers. In New Guinea, the development of a mass often follows the habitual carriage of backpacks. In 32.3 per cent of the numerous Japanese cases, the disease was familial. It was of great interest that a disproportionately large number of cases came from mainland Okinawa and two only of the 60 nearby islands; in one island, Tonalki, 66 per cent of cases were familial. The masses were usually subscapular (139: 70 per cent) but in 27 cases (16 per cent) they were also found inferior to the olecranon. The subscapular masses varied in diameter from 20 to 140 mm; the largest infraolecranon mass measured up to 60 mm. In all, 122 were bilateral.

Macroscopically, elastofibroma is an ill-defined, firm or semisolid mass. There is no true capsule and the mass extends into nearby tissue. The microscopic structure (Fig. 8.6) reveals much collagen of low cellularity arranged in broad bands, together with small numbers of blood vessels and a little fat. The cells are fibroblastic. Many branched and unbranched eosinophilic fibres are scattered among the collagen: the fibres, together with non-fibrous aggregates of granular material, stain for elastic material. The staining affinity of the fibres varies. The elastin-positive component may appear beaded. There is often a darker, elastin-staining central core and a delicate covering which stains for collagen.

Nagamine *et al.*[124] recognized five classes of elastic fibre in elastofibroma, according to the degree to which a central core was present in the elastic material; more core material was present in cases of longer duration. A serrated edge can be detected so that the elastic fibres have the appearance of delicate, coiled springs. However, with phosphotungstic acid-haematoxylin, the elastic fibres are orange with occasional dark blue or brown centres.[111] Reticular fibres are detected near and within the elastic fibres; this material can be removed by digestion with pancreatic

Fig. 8.6 Elastofibroma.
Painless but inconvenient mass in left subscapular region in 64-year-old female, excised surgically. Broad bands of collagen of low cellularity interspersed with small islands of fat cells are punctuated by darker bands of elastic material that can be recognized without an elastic stain. (HE × 60.)

elastase and with bacterial elastase. Trypsin exerts no effect but there is degradation by prolonged exposure to acid pepsin.

Transmission electron microscopic studies support the view that the elastic material is produced in excess by the sparse fibroblasts but it has also been suggested that elastofibroma arises by denaturation of collagen.[191] This seems improbable. There is no reason to suspect any source for the collagen, elastic and reticular fibres other than the fibroblasts and myofibroblasts that comprise the cell population. The most likely initial cause for elastofibroma appears to be trauma or physical irritation: experimentally, trauma to dermal connective tissue is an effective cause of elastogenesis.[60] The sequence of fibre formation seems likely to follow that characteristic of embryogenesis: reticular fibrils, elastic fibres and, later, collagen.[109a]

A histological comparison of elastofibroma with senile elastosis and with the lesions of pseudoxanthoma elasticum (see p. 355) is to be expected. However, aetiologically, anatomically and histologically, these three forms of elastic material disease appear entirely distinct.

Keloid and hypertrophic scar

From time-to-time scar tissue may form in excess[62]. When there is an insidious, outward growth of the scar tissue the resultant structure is described as a keloid (*Gk*: claw).[5,6] It is a non-neoplastic, indolent proliferation of dermal or subcutaneous connective tissue. When the overgrowth of the collagenous connective tissue is confined to the contours of the initial scar, the term 'hypertrophic scar' is often used. The distinction between keloid and hypertrophic scar is blurred; the value of attempting to make this distinction is related to the need to assess the behaviour of the lesion and to judge the efficacy of treatment.

Keloid

These are unsightly, painful or irritating, compact but never flattened exophytic masses that form at the sites of deep burns, injections, surgical wounds, tattoos, insect bites or vaccination marks. They occur much more frequently in Negro than in Caucasian races, estimates of the relative frequency ranging from 12:1[71] to 3:1.[27] Keloids cause cosmetic anxiety since they often arise in or from the ear lobe, presternal or shoulder regions, although no part is immune (Fig. 8.7a). The masses vary in size from 20 to 150 mm (ear lobe) and from 40 to 110 mm (abdominal). Keloids are generally single but may be multiple, particularly when there is evidence of a familial predisposition as well as a racial trend. Rarely, keloids may be congenital. They reach a final size after 6–18 months[111] but are likely to recur after excision or after treatment by ionizing radiation. By this time the early firm consistence has given way to a softer texture. The spontaneous disappearance of keloid after maturation and regression is very rare and keloids with a characteristic microscopic structure recurred in more than 60 per cent of 135 cases.[23]

Structure The gross structure of a keloid is that of a very firm mass of white or grey-white, compact connective tissue, devoid of recognizable blood vessels but covered by epidermis. The overlying epidermis is often thin and atrophic. Whether dermal appendages persist or not is inconstant and of no diagnostic significance in distinguishing keloid from scar tissue. Occasionally, the epidermis is acanthotic. Microscopically, the overwhelming impression is that of a mass of dense, collagen-rich tissue arranged as irregularly structured fibre bundles. There may be very few cells (Fig. 8.7b). The collagen processes interdigitate and extend into the adjacent dermal and subcutaneous tissue without a circumscribed margin; there is, therefore, anatomical continuity. There are few blood vessels and lymphatics cannot be identified. Elastic material is absent so

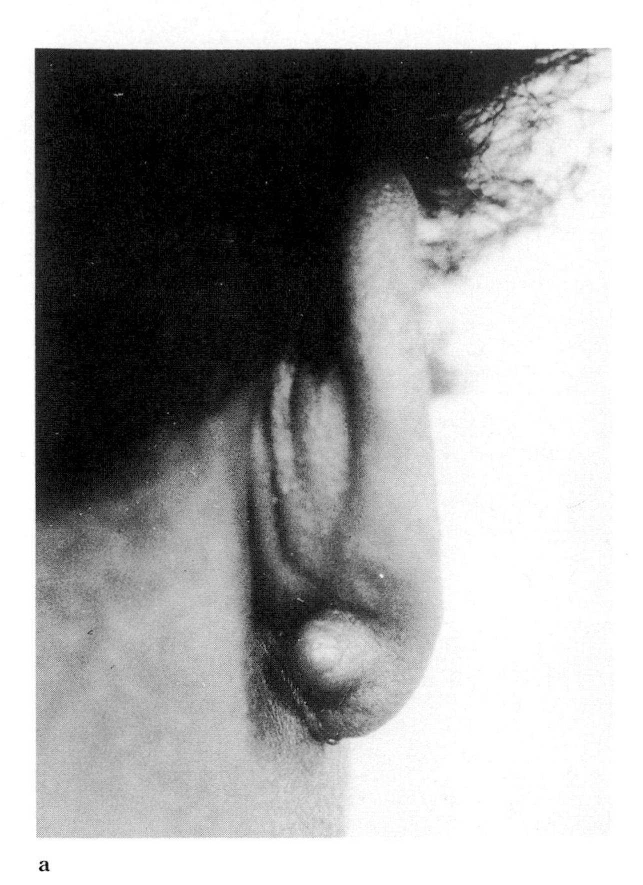

a

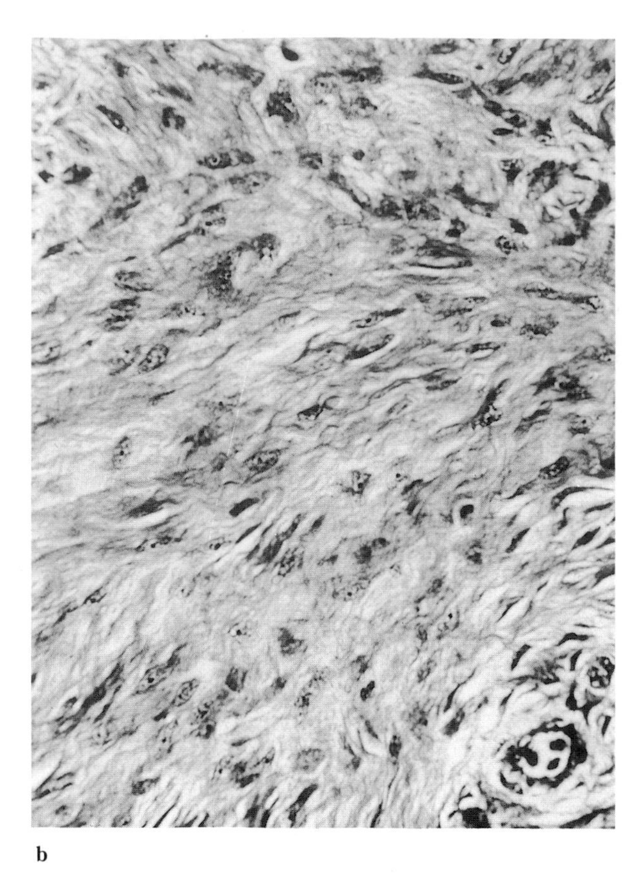

b

Fig. 8.7 Keloid.
a. Recurrent keloid at site of ear piercing in young Negro woman. The nodule has twice been excised but, each time, has recurred. **b.** The cellular response is unusually cellular. The large numbers of fibroblasts are an index of increased growth rate as recurrence occurs. (HE × 220.)

that the stroma of a keloid is in striking contrast with the common appearance of the stroma of elastofibroma (p. 310). The distinction between keloid and non-recurrent hypertrophic scar centres on the presence, in keloid, of thick, glassy, faintly refractile, pale staining collagen bundles with very little faintly basophilic matrix. There is no contiguous foreign body giant cell reaction.[23]

Biochemical structure There is increased collagen synthesis, prolonged beyond the period of normal repair. The presence of an excessive proportion of delicate collagen fibrils with a preponderance of the type III amino acid chain, supports the concept that there is also a fault in the regulation of collagen synthesis, in its secretion and in its cross-linking. Degradation is also active so that prolyl hydroxylase and collagenase activities are increased. The matrix is relatively rich in chondroitin sulphate proteoglycan and an increased interaction between collagen and proteoglycan[105] is demonstrable. Both hypertrophic scars and keloids contain much fibronectin.[95]

Causes Many causes of keloid have been suspected but no single agent has been shown to be responsible. A genetic and racial predisposition is accepted. The persistence of a foreign body or of keratin at the site of an original injury is not thought significant. There is no proof that the fibroblasts from keloids behave differently in culture from those of hypertropic scar, nor that they originate as mononuclear macrophages. Curiously, keloid tissue transplanted from one part of an individual's skin to another does not survive. This evidence is held to show that keloid may be a local phenomenon and that systemic causes are unimportant.

Hypertrophic scar

These scars differ from keloids both in structure and in behaviour. They follow uncomplicated surgery but occur where tension in wounds is high.[128] Excess scar tissue is formed within the anatomical outlines of the scar. The nearby tissue is not affected. The scar tissue is initially erythematous, raised and firm; it regresses with time, becoming paler, flatter and less conspicuous. Contractures occur and can be disabling and disfiguring. Recurrence does not take place and repetitive surgical excision is not necessary. However, if tension is re-applied and the forces to which the scar tissue is subjected once more increase, a further new growth of scar tissue may develop.

A hypertrophic scar is formed of compact collagen fibre bundles which do not show the glassy, refractile appearance seen in keloid. Fibroblasts are numerous. In the young, exuberant scar, thin-walled blood vessels persist but gradually become fewer. There is a scanty, proteoglycan-containing basophilic matrix. Depending on the nature of the injury that caused the scar, a nearby foreign body reaction may be seen and is of value in diagnosis.

Collagen synthesis and metabolism are changed.[36,39] Hypertrophic scars include much type III (fetal) collagen. The characteristics of embryonal collagen are retained[12] and it is of interest that even after many years, the collagen of hypertrophic scars remains type III.[16] There is also a relatively high proportion of chondroitin sulphate proteoglycan.[96] The type III collagen is represented by its large procollagen precursor: it is less easily extractable than native type I collagen. There have been advances in understanding of the non-reducible cross-links in ageing collagen[12,14] but the nature of those that determine the stability of hypertrophic scars is not yet fully understood. Relevant evidence has been obtained from experimental sponge implants.

Fibromatosis

The lesions of fibromatosis[7,8,49,70,71,111,126,160] are a group of non-inflammatory, dysplastic disorders of connective tissue in which the extension and infiltration of fibroblastic elements into adjacent structures, and a tendency for local recurrence after excision, are neoplasm-like. Nevertheless, for reasons given below, the diagnosis of neoplasia appears inappropriate. The fibromatoses are 'benign, fibroblastic proliferative lesions of the soft tissues distinguished from the self-limiting scars';[186] they may occur at any site, are of widely variable extent, and may be fatal or harmless.[111] Little is known of the cause of fibromatosis although it is of interest to note that a peculiar retroperitoneal fibromatosis of macaque monkeys is often followed by the development of an immunodeficiency syndrome, recalling the characteristics of Kaposi's sarcoma.[59]

The principal proliferating cell of fibromatosis is the myofibroblast (see p. 40).[126] The subcategories of fibromatosis are identified by clinical pattern rather than by histopathological criteria. The fibromatoses have many shared, microscopic characteristics. The component cells are well-differentiated; mitotic division is infrequent. A resemblance to fibrosarcoma is suggested by infiltration and by an ill-defined margin. A distinction (Table 8.1) between the fibromatoses of congenital, infantile or childhood origin, and those that occur at an adult age is made although this division has little practical meaning.[160]

Superficial fibromatosis

Palmar fibromatosis (Dupuytren's contracture)

This is a disorder of middle-aged and older males; Caucasian rather than Negro races are affected. The male:female ratio is 7:1, and in one case in four there is a familial trait. Many individuals who develop palmar fibromatosis[46,71,109,111,194] have been engaged in heavy manual work, and an association with alcoholic cirrhosis is recognized.[28] The contracture is frequently bilateral and may occur in individuals with plantar fibromatosis or with Peyronie's disease. Three clinical phases are defined,[109] corresponding to different stages of development of the fibroblastic lesion.

One or more painless nodules appear within the ulnar half of the palm or the corresponding aspect of the fingers (Fig. 8.8). The nodules are found between the skin and the underlying palmar fascia. There is a random proliferation of immature fibroblasts but little collagen secretion. Fibroblasts begin to be oriented in the lines of stress already indicated by the nodules; the proportion of cells decreases and more collagen is formed. Collagen shrinkage marks the onset of contracture;[54] typically, the fifth or the fourth and fifth fingers are drawn across the closing palm. Ultimately, the nodules are replaced by a dense, tendon-like collagen cord which creates mechanical disability and cosmetic embarrassment. The entire palmar aponeurosis may be affected and correction of the deformity by serial reconstructive operations may be only partly successful.

Plantar fibromatosis (Ledderhose's disease)

Plantar fibromatosis[101] shares many features with the palmar disease but contracture of the foot due to collagen maturation and cross-linking is unlikely.[145] The nodules are initially larger than those of the hand. The medial plantar fascia is affected more severely than the lateral. The microscopic features are closely similar to those of palmar fibromatosis. The response to surgery is not readily predictable.

Penile fibromatosis (Peyronie's disease)

A fibroproliferative disease of the corpora cavernosa of the penis bears the name of Peyronie who described three

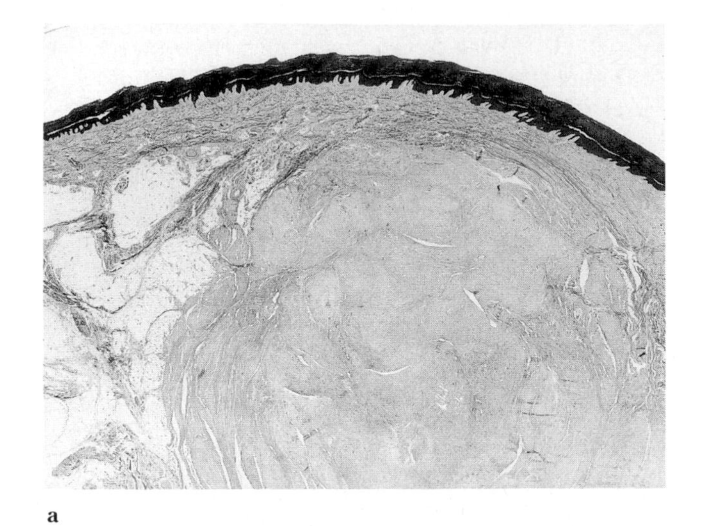

a

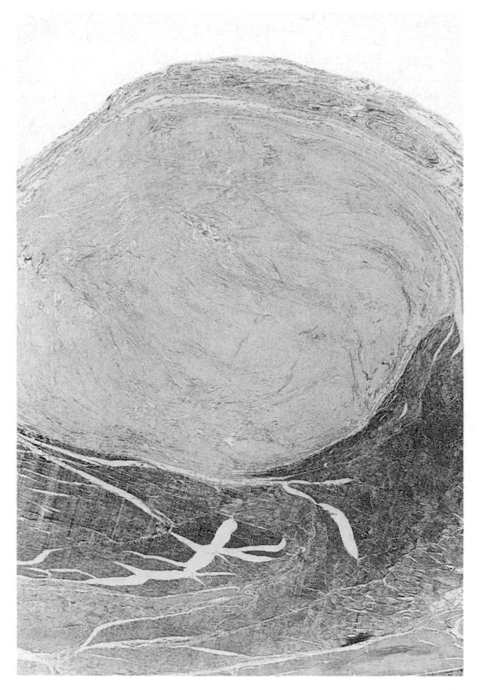

b

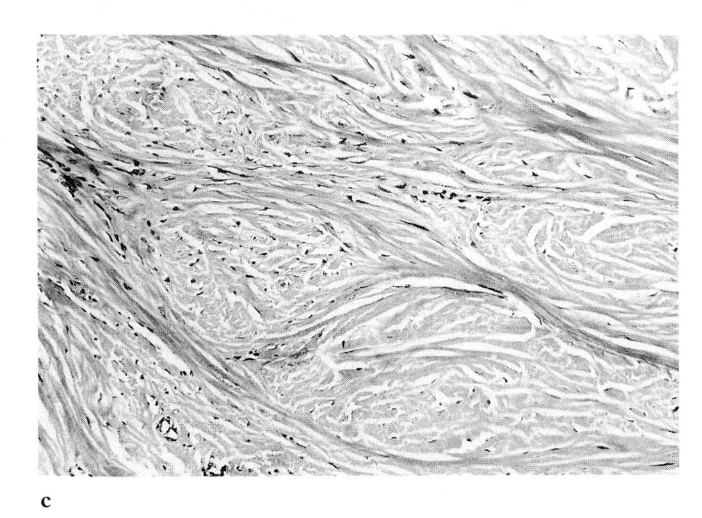

c

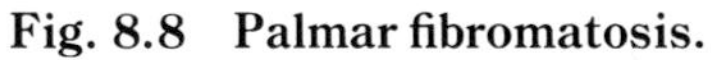

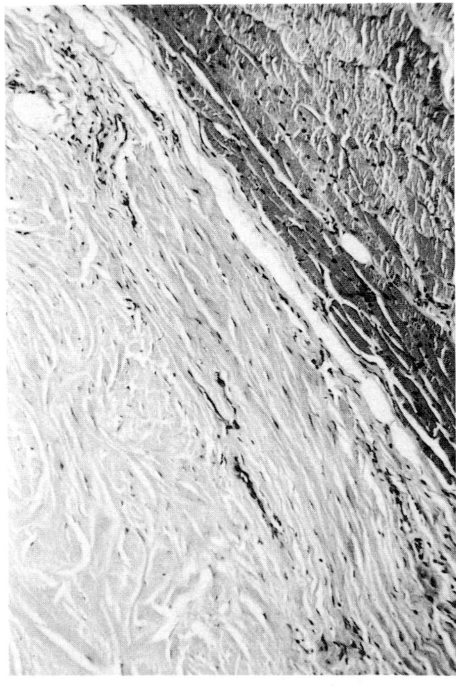

d

Fig. 8.8 Palmar fibromatosis.
a. Male aged 57 years with painless, ill-defined nodule slowly increasing in size within the palmar fascia. The mass of cellular connective tissue is bounded (at top) by displaced normal dense and fibrofatty connective tissue. **b.** Further zone where abnormal, cellular connective tissue (at bottom) adjoins vascular tissue of palm. **c.** Whether the small blood vessels of the fibromatous process are derived from host vessels, as the fibromatous mass grows, or whether they are incorporated in the lesion, is uncertain. **d.** Almost all the cells of the fibromatotic tissue are compact with spindle-shaped nuclei. They are assumed to be fibroblasts. (HE **a.** × 24; **b.** × 85; **c.** × 113; **d.** × 220.)

cases in 1743.[141] Among the proposed causes for this uncommon, non-neoplastic, localized mesenchymal disorder are: local trauma, infection and vasculitis. Associations with venereal disease, gout, arthritis, diabetes and atherosclerosis have been alleged[111] but all are now thought to be coincidental. The histological lesions display two components, first an inflammatory cell infiltrate of plasma cells and lymphocytes, often with a perivascular distribution, leading to, second, a perivascular fibrosis not associated with vasculitis but sometimes accompanied by bone formation and muscle destruction. There may be an underlying myofibroblastic dysplasia. In this context, the observations that Peyronie's disease develops most often in middle-aged adults and that there is an association with Dupuytren's contracture, are significant.

Deep fibromatosis

Extra-abdominal and abdominal fibromatosis (desmoid)

Extra-abdominal desmoid 'tumours' occur as slowly growing, painless masses at a mean age of 25–35 years although their occurrence in children and adolescents is not rare and they may be congenital. There is a male preponderance. The back, chest wall, head and neck, and lower limb are affected (Fig. 8.10)[71] and the fibromatosis may occasionally be multiple.

The so-called 'desmoid tumours' characteristically involve the abdominal wall but may occur in many other sites. Whereas Mackenzie[111] classified the abdominal wall and extra-abdominal 'desmoid tumours' as fibromatoses, Hadju,[71] on the basis of the experience of the Sloan–Kettering Cancer Center, considered these lesions to be low-grade fibrosarcomata. Hajdu, nevertheless, accepted the unique natural history of 'desmoid tumours' and analysed them separately from 'fully malignant fibrosarcoma'. Desmoid tumours have a remarkable tendency to recur locally at the site of previous surgical excision. In spite of their innocent, banal microscopic structure, there is therefore agreement that it is the natural history and clinical behaviour of desmoid tumours and not their morphology that must determine the attitude towards them of pathologists and surgeons.

Anterior abdominal wall desmoid tumours are not common and there were only 11 of 162 (7 per cent) in Hajdu's[71] series. They are unencapsulated, infiltrating lesions, more than 50 mm in diameter, occurring in the deep connective tissue and more common in females than males. They develop at all ages but a mean age of 30 years and a relationship to pregnancy are recognized, and there is evidence that analogous experimental masses are hormone-dependent.

The macroscopic appearance of abdominal and of extraabdominal desmoid tumours is closely similar. On section, the solitary masses, ranging in size from 20–30 to 150–200 mm and weighing 10 g to 10 kg, are found to be uniformly firm or hard and rubber-like in consistency. There may be compression of adjacent tissues and the false capsule formed in this way may mask the infiltrating character of the desmoid. The colour is grey-white; there is neither haemosiderin nor fat. The appearance of the cut surface is dominated by an abundance of collagen arranged in broad fibre bundles orientated in the direction of nearby muscle but often random. Vascularity is low.

The growth is infiltrating. In the pseudocapsule well-differentiated fibroblasts and fibrocytes synthesize and secrete much collagen. The cells, ovoid and elongated with vesiculate nuclei containing normal proportions of chromatin, have been said to be 'shorter, less pyknotic and more vesicular than those of high-grade fibrosarcoma'.[71] Mitotic figures are not present. The cellularity is variable and a myxoid stroma is occasional; a secondary, peripheral, mononuclear cell reaction occurs. The intrinsic blood vessels are capillaries but thicker walled channels also form, recalling those of hamartomata.

The aetiology of desmoid tumours is contentious. Endocrine factors and trauma are invoked. Genetic predisposition is considered probable. The coincidence of desmoid tumour with the other characteristics of Gardner's* syndrome (see below) suggests the formation of a heritable 'fibrous tissue-forming factor', perhaps a fibroblast growth factor (p. 18), in individuals with this familial variety of polyposis.

Intra-abdominal (mesenteric) fibromatosis

Fibroblastic masses within the mesentery or omentum may occur as isolated abnormalities, or as part of Gardner's syndrome.[57]

In the isolated condition, a mass or masses 100–200 mm in diameter develop in the mesentery of the small intestine of individuals of any age and of either sex.[203] Mechanical disorders of the gut, the formation of a mass, or weight loss are common signs. Exceptionally, the neoplastic-like process occurs in other sites such as the gastrosplenic ligament. The microscopic structure is fibroblastic with infrequent zones of myxoid appearance. The cell density varies; mitotic activity is slight. Removal of the mass is rarely followed by recurrence.

In Gardner's syndrome[57] the formation of numerous, ill-defined fibroblastic masses is one part of a widespread

* E J Gardner is an American geneticist.

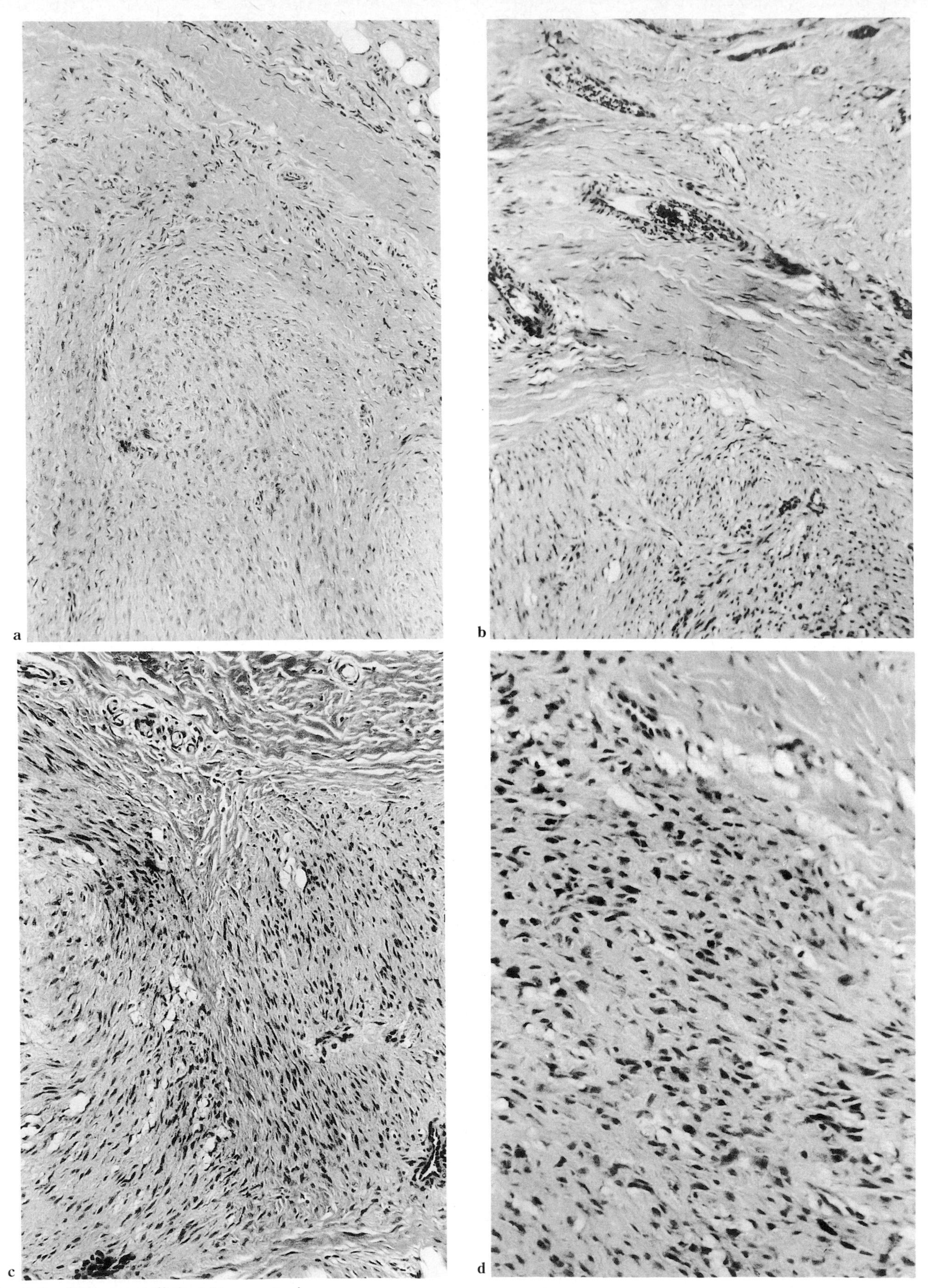

Fig. 8.9 Recurrent digital fibromatosis.
Although the shoulder and upper arm are most often affected, in this case the lesion was of a whole first finger. The enlargement of the finger resembled the appearance of so-called 'local acromegaly', clinically stimulating the change encountered in neurofibromatosis. **a.** Location of mass deep to dermis. **b.** Cross-section of mass. **c.** The mass is formed of orderly, interlacing collagen fibre bundles, of low cellularity. **d.** Margin of region of fibromatosis, with sharp delineation from nearby tendon and skeletal muscle. (HE **a.** × 60; **b.** × 60; **c.** × 120; **d.** × 280.) Courtesy of Professor D Hourihane.

disorder, inherited as an autosomal dominant characteristic. There is intestinal polyposis, osteomata, epidermoid cysts and lipomatosis.[198] The fibroblastic masses do not recur but excision of the mesenteric lesion may be complicated by the effects of previous intra-abdominal surgery. Appendicectomy or cholecystectomy, for example, may be a stimulus predisposing to further fibrosis. The abdominal masses are poorly defined, irregular but firm; the nearby peritoneal surface is commonly adherent. Microscopically, delicate, finger-like processes of fibrous tissue extend through adjacent adipose tissue. The relative quantities of collagen and the number of cells vary considerably but there are few mitotic figures. The cell and nuclear characteristics of malignancy are not seen.

Fibrous dysplasia of bone

An analogy has been drawn between the proliferation of fibroblastic tissue in localized (monostotic) fibrous dysplasia of bone, a locally destructive lesion of cancellous bone, and the fibromatoses; fibrous dysplasia has been termed a 'fringe fibromatosis'. The condition is not considered more fully here and the reader is referred to Dahlin[40a] and Revell.[155]

Polyostotic fibrous dysplasia[103] is an analogous, rare, generalized skeletal disease; it may be associated with skin pigmentation, endocrine disorder and precocious puberty in the female.[4]

Fibrous proliferations of infancy and childhood

Several categories have been defined.[7,8,49,70,71,111] The most common in infancy are: subdermal, colli, digital recurrent (Fig. 8.9), fascial and multifocal (generalized).[71] The

a

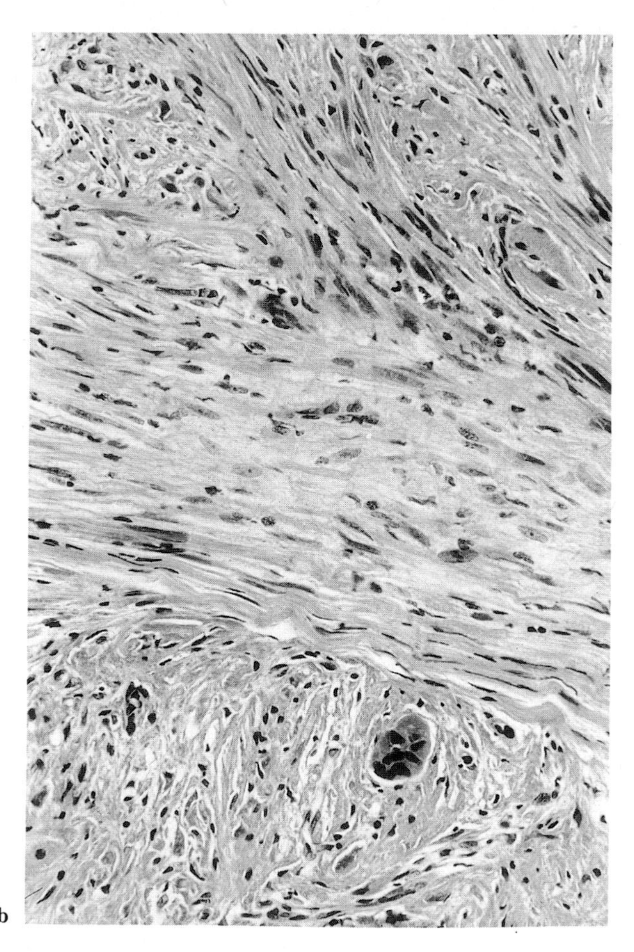

b

Fig. 8.10 Extraabdominal (musculoaponeurotic) fibromatosis.
a. Within this slowly enlarging mass of the shoulder, spindle-shaped fibroblasts are seen infiltrating nearby skeletal muscle (at top). **b.** The component cells are devoid of obvious mitotic activity. A single multinucleated giant cell is likely to be of skeletal muscle origin. (**a.** HE × 25; **b.** × 120.)

most frequent in childhood are: gingival, plantar, aponeurotic and visceral (mesenteric).

Fibrous hamartoma of infancy

This disorder predominates in males and is often present at birth.[47] The shoulder, axilla and upper arm are affected: recurrence after surgical excision is unusual. The lesion is an ill-defined mass less than 100 mm in diameter, grey-white, with yellow foci and occasional cystic spaces. Fibroblasts, adipose, muscular and vascular components are seen. Islands of collagen-rich fibroblastic tissue are arranged as strata of varying cellularity. There is a resemblance to angiomyolipoma. Elastic fibres are scanty. The cells, like those of the embryonic mesenchyme, are often stellate, the matrix loose, open and myxoid. However, collagen may predominate, as in keloid, and adipose tissue comprises a large proportion of the lesion. Transmission electron microscopy demonstrates that the cells are myofibroblasts.

Fibromatosis colli

The rare *congenital torticollis* (sternomastoid tumour; fibromatosis colli) affects the sternomastoid muscles, principally the inferior third.[31] Within two weeks of birth, a small intramuscular mass appears. The mass regresses in size but, by three to four years of age, as growth proceeds, the effects of the deformity are evident. There may be an associated congenital dislocation of the hip. Microscopically, muscle cells and bundles are interrupted by proliferating fibrous tissue: muscle cell atrophy takes place.

Infantile and juvenile fibromatosis

There are three clinical categories of infantile and juvenile fibromatosis.[7–9,156] Two are outlined. In *infantile dermal fibromatosis*, the extensor surfaces of the distal phalanges of the fingers and toes of very young male infants are occupied by circumscribed nodules of fibroblastic and collagen structure, not eroding bone. Recurrence is likely. Transmission electron microscopy[126] reveals large cytoplasmic masses of dense granulofibrillary material corresponding to the eosinophilic inclusions seen by light microscopy. The centre of the inclusion is composed of a granular substance intermingled with membrane fragments, vesicles and a few glycogen particles. In *diffuse infantile fibromatosis*, a number of muscle groups can be affected after birth by proliferating fibroblastic tissue. The head, neck and upper limb are particularly susceptible.

Juvenile (calcifying) aponeurotic fibromatosis

This appears in children or adolescents as an excessive growth of fibroblastic tissue. Foci of calcification, ossification and cartilagenous differentiation occur (Fig. 8.11). Palmar and plantar sites are most common. There is an obvious analogy with palmar and plantar fibromatosis of adults. The plump, oval fibroblasts are very numerous and there is wide infiltration and replacement of skeletal muscle, fat, blood vessels, nerves and dermal appendages. The main difference from other fibromatoses relates to the extracellular matrix where transmission electron microscopy shows chondroid areas and cells, thought to be myofibroblasts, with an irregular shape due to frequent stellate cytoplasmic projections and filopodia. The cytoplasm often contains small pools of glycogen.[126] Recurrence is likely after limited excision. The recurring lesion continues to display masses of closely packed collagen fibres, with cells oriented in a single, preferred direction and a fine, scattered deposition of radioopaque calcium salts.

Infantile myofibromatosis: congenital, generalized (multifocal) fibromatosis

This rare, generalized disease occurs in newborn infants; it may be quickly fatal.[34,180,185] Sixty-one cases were analysed by Chung and Enzinger.[34] The affected child is born with a single or numerous subcutaneous or soft-tissue nodules 2–15 mm in diameter.[111] By the time surgery is performed, up to 12 years of age, a lesion may be as large as 70 mm diameter and the size and location of this solitary nodule dominates the clinical disease. The lesions occur in the head and neck (35 per cent), trunk (33 per cent), upper limb (13 per cent) and lower limb (18 per cent). Among the viscera found at necropsy to be similarly affected have been skeletal muscle, bone, lung, pancreas, heart, kidney and lymph node. In spite of a resemblance to metastases, the lesions are multicentric, primary foci.

The nodules comprise many plump myofibroblasts together with a central vascular structure resembling haemangiopericytoma.[34] Nodules, often lobulated, are well

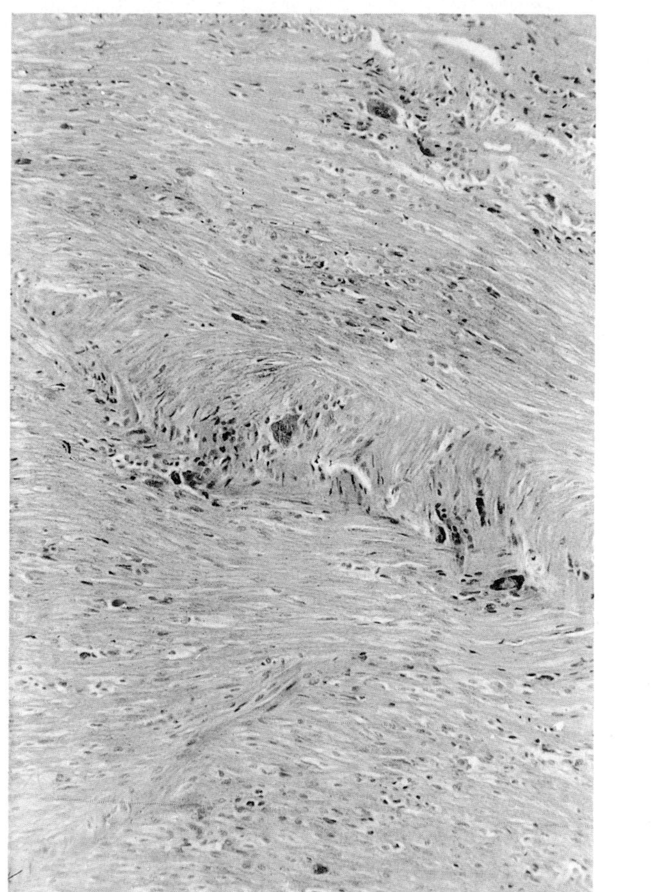

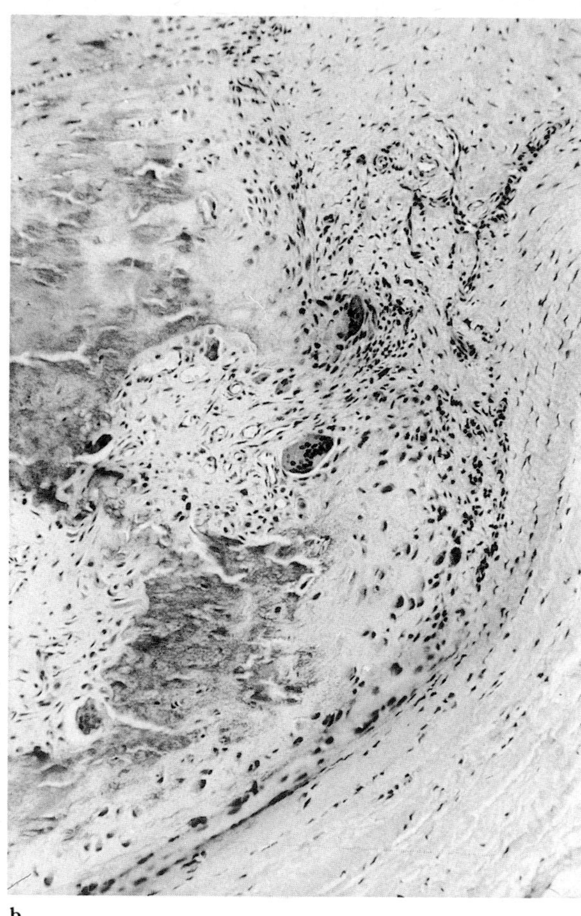

a b

Fig. 8.11 Calcifying aponeurotic fibroma.
a. Broad, angled sheets of richly collagenous connective tissue interspersed with very occasional small blood vessels lie within the subcutis. Note unusual presence of small numbers of multinucleated giant cells. (HE × 60.) **b.** Subcutaneous tissue from palmar surface of metatarsophalangeal joint. A narrow zone of cartilagenous differentiation bounds a region where ill-defined bone is interrupted by loose, vascular connective tissue. (HE × 100.)

circumscribed. Delicate bundles of collagen separate the cellular whorls and islands. Centrally, a rich vascular structure is associated with many polyhedral cells with large, slightly pleomorphic hyperchromatic nuclei and faintly eosinophilic or amphophilic cytoplasm. Hyalinization, focal haemorrhage, cystic degeneration, coagulation necrosis and peripheral foci of calcification are described. Mitotic figures average 3/0.4 mm^2. Visceral lesions display a similar structure; bronchial obliteration and polypoidal intestinal protrusion with ulceration are recognized.

The designation infantile myofibromatosis was chosen because of the myofibroblastic appearance of the component cells. The lesion may be an analogue of neurofibromatosis, haemangiomatosis or lipoblastomatosis or a hamartoma with a mainly fibroblastic composition. The association with neurofibromatosis, in which the responsible gene is located on somatic chromosome 17, suggests that this chromosome may bear genes encoding for myofibromatosis. Patients with solitary or multiple lesions confined to the soft tissue and bone do well[34] but those with multicentric visceral lesions do poorly, dying at birth or shortly thereafter. Recurrence of solitary lesions takes place in 7 per cent of cases; all are cured by re-excision. There is occasional spontaneous regression but it is noteworthy that the behaviour of the lesion is not influenced by the degree of mitotic activity or by the presence of intravascular growth.

Gingival fibromatosis

In this rare, inherited, dominant disorder, there is a painless gingival nodule that begins to form when the permanent teeth erupt.[151] There is an association with hypertrichosis, mental retardation, neurofibromatosis and cherubism. The teeth may be obscured by or engulfed within a mass composed of proliferating fibroblasts and collagen. The connective tissue components extend beyond the clinical margins of the lesion and the overlying epithelium is hyperplastic, the rete processes long and thin.

Fibrosis and the viscera

Lung

The lung[38,44,142] is particularly susceptible to infective, carcinogenic and fibrogenic materials inhaled during occupational exposure or by random environmental contamination.[78,184] Intravascular silicates may also provoke lung fibrosis.[40] Much of what is known of the mammalian response to dusts, and in particular silicates, other minerals, metals and inorganic elements, has been derived from the investigation of human lung disease, from the administration of these minerals to small animals, and from other experimental studies.[28a] Bleomycin, for example, is widely used to provoke pulmonary fibrosis in the rat or hamster.[99,174] Epithelial and endothelial injury precede fibroblast activation.[64] Immunofluorescence microscopy has allowed collagen types to be identified in a miscellaneous group of fibrotic lung diseases.[113] A large increase in type I collagen is often found in the thickened alveolar septa; type III collagen is much reduced and is confined to a perivascular location. Type V collagen may be increased in the interstitial tissue in miscellaneous forms of lung fibrosis in areas of smooth muscle proliferation, but no change in the amount or distribution of type IV collagen is recorded.

Fibrosing alveolitis

Fibrosing alveolitis without apparent cause ('cryptogenic') is a complication of rheumatoid arthritis, polymyositis/dermatomyositis and other systemic connective tissue diseases (Fig. 8.12) (see Chapter 12). Immunoglobulins and complement have been identified in alveolar and capillary walls. Circulating immune complexes have been found in patients with active interstitial disease, both in instances where the lung disorder occurs in isolation[76] and where it complicates rheumatoid arthritis.[43] The ratio of type I to type III collagen varies as the connective tissue organization is seen to be loose or dense.[188] The presence of type III collagen correlates not only with active, unstable lung disease but also with the response to corticosteroid treatment.[19] Although glucocorticoids can suppress fibroblasts in culture,[178] the clinical response is slight. Cytotoxic drugs may be beneficial.[175]

Fibrogenic dusts

The pulmonary reaction to asbestos is by alveolar macrophages; asbestos bodies are ultimately formed. Fibrosis[157] and the formation of pleural and peritoneal plaques are characteristic. Initially, free alveolar macrophages detoxify inhaled asbestos fibres, coating them with cell membrane components, proteoglycan and other unidentified compounds into which haemosiderin is incorporated.[112] The diffuse fibrosis that may ultimately affect the lung parenchyma represents a failure to incorporate and coat the asbestos fibres. Experimentally, all asbestos types injected into the pleura or peritoneum can provoke neoplasia[41] although in humans, crocidolite is responsible for the majority of mesotheliomata. Crocidolite is more carcinogenic in the peritoneal than in the pleural cavity.[41] An investigation into the reaction of the synovia to the fibrogenic dusts would be of interest. Silica induces raised levels of enzymes of collagen biosynthesis.[146] Whether fibrogenic dusts have the same effect on synovia is not known. Fibrous glass dusts have a peribronchiolar fibrotic action[63] just as they may cause synovitis when they gain access to a joint accidentally.[35a]

Other fibrogenic agents

The reaction of the lung to other forms of physical agent has long been a focus of attention and the reactions to ionizing radiation are of particular note. The development of fibrosis is related to the route by which either particulate or waveform radiation reaches the lungs. It is also related to the dose, the exposure level, and the efficiency of deposition and clearance. Pulmonary fibrosis is one consequence of X- and γ-radiation used to treat pulmonary neoplasms. When yttrium-90 was given as an aerosol to hamsters in fused, dry montmorillonite particles, pulmonary collagen synthesis and degradation were both increased within 14 days.[144] Scarring advanced, and continued radiation was not necessary to sustain radiation injury. Aerodynamic, 1.1 μm diameter monodisperse aerosols of plutonium-238 oxide were tested in the same way.[143] Diffuse alveolar thickening, interstitial fibrosis and dense fibrous scars resulted: interstitial fibrosis resolved spontaneously whereas the dense fibrous scars did not. Colchicine is effective in inhibiting hydroxyproline accumulation by experimentally irradiated lung but it is not yet known whether the beneficial effects demonstrable in the treatment of hepatic cirrhosis (see p. 323) are mirrored by similar actions on lung tissue.

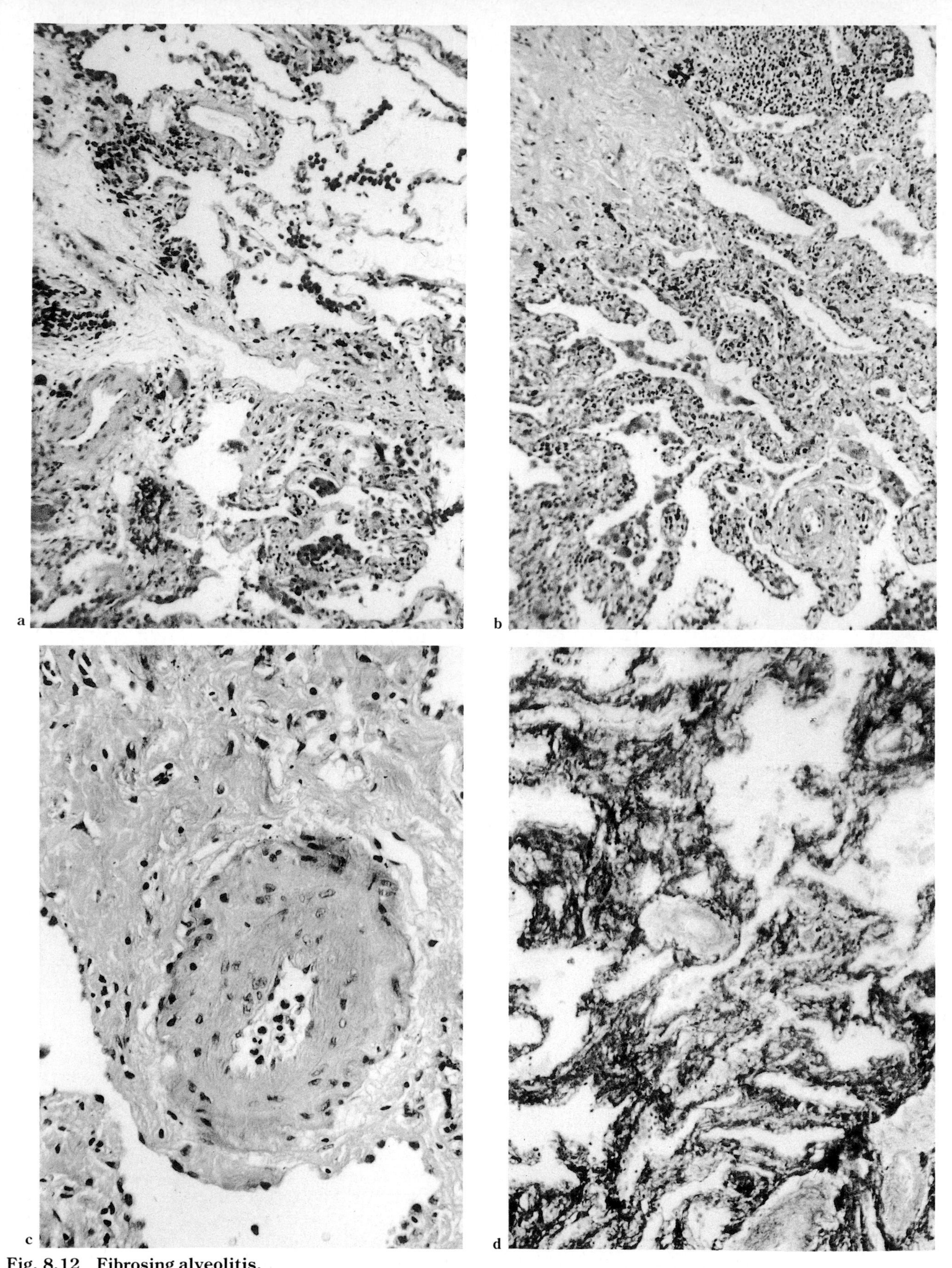

Fig. 8.12 Fibrosing alveolitis.
a. The walls of the alveolar ducts and alveoli are thickened by the presence of excess collagen. A broad band of fibrous tissue extends horizontally across the field, marking the boundary between two lung segments. **b.** The thickened interalveolar septa are infiltrated by lymphocytes. A microscar (top left) marks the site of earlier inflammatory disease. **c.** Within the excess pulmonary fibrous tissue, thick-walled pulmonary arteriolar branches create high resistance to pulmonary blood flow. **d.** The thickened interalveolar septa contain an excess of elastic material. (**a.** HE × 100; **b.** HE × 100; **c.** HE × 250; **d.** Elastic-van Gieson × 100.)

Eye

Fibrosis of the cornea is a common consequence of injury. Particular attention has been given to the effects on the retina of high concentrations of oxygen administered to premature infants.

Corneal wounds

Corneal wounds provoke the transformation of marginal keratocytes into fibroblasts. These cells have close similarities to myofibroblasts (see p. 40) and display large quantities of filamentous actin.[89] The cellular changes begin very quickly after injury. Within 1 h, structural abnormalities are recognizable. There is new protein and RNA synthesis in 6 h, DNA synthesis in 12–18 h and new glycosaminoglycan formation within 48–72 h.[26] Changes in collagen synthesis and organization occur later but are characteristically irreversible.

Retrolental fibroplasia

Retrolental fibroplasia[181] was identified as a cause of blindness in premature infants in 1941.[190] The disorder commences two to three months after premature birth in which high concentrations of oxygen have been used to promote survival of the infant. The retinal blood vessels become dilatated and tortuous; a retinal swelling occurs and detachment begins. The abnormality represents an effect on developing vessels; nearby choroidal arteries are not affected. Atrophy of the young vessels is followed by a delayed, disorganized vascular development and fibrous tissue formation. Collagen synthesis and maturation result in fibre maturation and shrinkage and retinal detachment is promoted.

Systemic connective tissue disease

Ocular abnormalities are characteristic of connective tissue disorders such as Marfan's syndrome, storage diseases and ochronosis, and the eye is frequently affected in inflammatory diseases such as rheumatoid arthritis, systemic lupus erythymatosus, ankylosing spondylitis and vasculitis. These pathological changes are considered elsewhere (see Chapters 9 *et seq.*).

Heart

Connective tissue changes have been closely investigated both in natural and experimental pericardial[100] and myocardial disease.[19a] In cardiac ischaemia, infection or hypersensitivity reactions, the injury of a critical proportion of myocardial cells leads to cardiac failure and death.[53,187] The irreversible contraction of myocardial cells occurs as they lose adenosine triphosphate.[53] Necrosis results. If, however, a proportion of the cell population less than this critical number dies or is injured, survival of the organism is possible and the heart can undergo substitutive repair. The regeneration of cardiac muscle cells by mitotic division is slight and ineffectual; repair is therefore by fibrosis, leading to scar formation. At the margin of the infarct, fibroblasts multiply; within a few days, capillaries grow in from the viable tissue outside the infarct edge. Fibrosis occurs in a widespread patchy fashion in zones that have been most severely disorganized. Cardiac collagen matures and contracts and the heart diminishes in size and 'withers'. New connective tissue fibres and matrix are actively synthesized but scar formation may not be complete for two to three months.

Cardiac fibrosis can be a manifestation of congenital disease (endocardial fibroelastosis) and its formation may reflect intrauterine viral infection. Intrinsic connective tissue abnormalities also contribute to cardiomyopathy.[37] In childhood, rheumatic fever (see Chapter 20) may lead to scarring of valves, pericardial fibrosis and adhesions, but there is little myocardial fibroblastic reaction. In adult life, cardiac muscle is found to be resistant to the injury and subsequent scarring associated with ionizing radiation but susceptible to metabolic diseases, bacteria, viruses and protozoa, and to immunological insults. Recovery from Coxsackie, influenza and rubella virus infections can be a preliminary to fibrosis, as can South American trypanosomiasis, typhus and diphtheria.

Artery

Atherosclerosis evolves through a series of stages in which fatty streaks form as a preliminary to mural fibrosis.[165] The disease affects the arterial intima but as medial thinning and aneurysm develop, fibrosis extends to the media and adventitia as well as the intima. Many of the tissues affected by the common inflammatory and non-inflammatory diseases of joints, such as rheumatoid arthritis, osteoarthrosis, ankylosing spondylitis and gout, are also affected by atherosclerosis. Yet there is little evidence that the coexistence of more or less severe vascular insufficiency influences the natural history of the articular connective tissue diseases. Nevertheless, few experiments have been made to examine this relationship. In a study of the influence of limb ischaemia on the natural history of adjuvant arthritis in the rat (see p. 544), I found that neither femoral artery, internal iliac artery nor lower (subrenal) aortic ligation prevented the onset of lower limb arthritis. However, lower aortic ligation delayed the secondary synovitis that is characteristic of this adjuvant disease.

The grey-white, fibrous plaque of atherosclerosis comprises an accumulation of fat-laden, intimal myofibroblasts. The cells synthesize, and come to be increasingly surrounded by extracellular lipid, collagen, elastic material and proteoglycan. Whether fatty streaks are strictly precursors of fibrous plaques is unclear: the two components may develop simultaneously in different parts of the same large artery. The formation of a collagenized fibrous plaque is preceded by a local fibromuscular proliferation initially free from lipid. The smooth muscle cell response and increased collagen and elastic material synthesis is similar to the vascular responses to mechanical injury, homocystinaemia and experimental immunological injuries. Endothelial cell loss is an essential preliminary to platelet activation. In turn, smooth muscle cell proliferation is provoked by the activated platelets and can be prevented by their inactivation. Other serum factors, particularly low-density lipoproteins, have a synergistic effect and hormones such as insulin sustain the response.

Arterial smooth muscle cells in culture synthesize type I and type III collagen.[162] Relatively more type III than type I collagen is formed but in the atherosclerotic plaque the proportion of type I to type III collagen becomes 2:1. Much new elastogenesis occurs in the plaque and the proteoglycan is rich in dermatan sulphate with much less chondroitin-4- and chondroitin-6-sulphates, and very little hyaluronate. Interaction between dermatan sulphate and low-density lipoprotein may be coincidental.

Liver

The liver responds to a wide variety of toxins, chemicals, viruses and immunological insults by the synthesis of excess collagen and the formation of new fibrous tissue (Fig. 8.13). The volume density of this collagen can be assessed by histophotometry,[84] and its extent by computer-assisted morphometry. When liver cell injury is accompanied by regeneration and when the new fibrous tissue disorganizes the liver architecture with altered portal blood flow, the term 'cirrhosis' is used. Hepatic fibrosis and cirrhosis are 'collagen-formative diseases'.[110] The collagen deposition in cirrhosis alters vascular flow, causing shunting of blood from the portal to the hepatic veins, and from the portal venous system to the hepatic arterial, bypassing the parenchyma. The collagen content of the regenerating nodules is modified leading to closure of the sinusoids and their conversion to capillaries with a new, subjacent basement membrane, within which type VI as well as types I and/or III collagen[168] are deposited. A serious and adverse effect on plasma–hepatocyte metabolic exchange results.

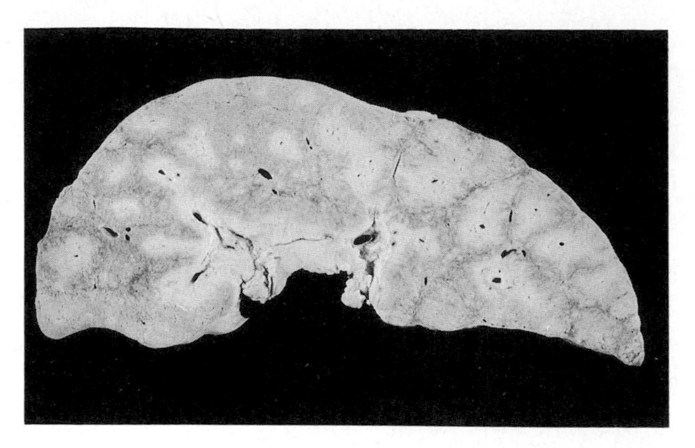

Fig. 8.13 Hepatic cirrhosis in schistosomiasis. The so-called (clay) 'pipe-stem' cirrhosis of infestation with *Schistosoma mansoni* results from the irritant properties of the eggs, carried to the liver in the portal venous system.

The increased collagen content of the liver in cirrhosis may be attributed to an increased rate of collagen synthesis, to a decreased rate of collagen degradation or to both mechanisms.[110] Much of the available data demonstrates increased collagen synthesis. There is both direct evidence of this change, assessed on biopsy material and indirect, suggested by raised levels of prolyl hydroxylase activity.[20,83] Decreased collagen breakdown has been implicated[65] but this is disputed.[106] Collagenase can be shown to be present in the earlier phases of experimental cirrhosis if not in the later ones[123] and the experimental production of fibrosis by carbon tetrachloride can be enhanced by acute-phase protein via an exaggeration of prolyl hydroxylase activity.

Among the agents responsible for stimulating prolonged new collagen synthesis are a series of cytokines and growth factors similar to those present in healing wounds and in experimental pulmonary fibrosis. There is therefore particular interest in the role of the macrophage in hepatic fibrosis. Macrophages are linked with fibroblast proliferation (see p. 39): they secrete a factor that has this effect in hepatic and cardiac tissue as well as in skin and in schistosomal granulomata.[202] Hepatocytes and hepatic mesenchymal cells, themselves, synthesize much new collagen and express prolyl hydroxylase activity. Collagen-stimulating factors do not appear to derive from necrotic hepatocytes. The role of the perisinusoidal cell (Ito cell) in hepatic fibrosis is, however, relevant.[127] In cirrhosis, there is a raised number of fibroblasts in the portal tracts. The perisinusoidal cells are also increased in number but the reason for this response is not clear. Penicillamine and colchicine modify the increased collagen biosynthesis of cirrhosis, and colchicine has been shown to prolong survival greatly in patients given long-term treatment.[92]

Intestine: collagenous colitis

Endoscopic biopsy continues to reveal new pathological processes. In 1976, Lindstrom[107] described the case of a patient with chronic watery diarrhoea associated with an excessive formation of collagenous connective tissue in the lamina propria of the distal colonic and rectal mucosa. In analogy with the disorder 'collagenous sprue',[199] the newly recognized abnormality was called collagenous colitis. Occasionally, the disease coexists with rheumatoid arthritis.[50] At first it was thought to be rare, but increasing numbers of cases are being reported;[58] 36 have been reviewed by Rams *et al.*[152] The patients are almost always middle-aged women. In one account, there is reference to 24 women and 3 men;[50] in another, the female: male sex ratio is given as 10:1 and the mean age as 54.7 years.[82]

Pathological changes

The diagnostic changes are microscopic. Any part of the large gut may be affected, but the severity of the disorder is greatest in the descending colon and rectum. The sub-epithelial collagen band in the normal colon measures approximately 5 μm.[94] A slightly thickened collagen lamina is a rare but non-specific feature of several colonic diseases.[61] By contrast, the thickness of the collagen layer in collagenous colitis is approximately 35–60 μm.[32,56,82] Regression can occur[42] and in the case reported by these authors the lamina diminished from 24 to 2.5 μm over a 15-month period. Rarely, collagenous colitis has been found to antedate Crohn's disease.[32]

In addition to the excess collagen, there is a reduction in numbers of pericryptal fibroblasts, an increased number of lamina propria lymphocytes and plasma cells, occasionally excess histiocytes, and, rarely, multinucleated giant cells.[81] No foreign material is recognizable, nor is there any microbiological abnormality. Transmission electron microscopy confirms occasional focal loss of surface epithelial cells, variable changes in the distance between these cells, goblet cell diminution, and a reduction in epithelial cell microvilli.[50] Scanning electron microscopy can be used to view the mucosa.[50]

The collagen fibres of the abnormal lamina are 30–40 nm thick; their periodicity is approximately 60–70 nm or 64 nm.[51,56] The collagen fibres accumulate as bundles external to a basement membrane that varies only slightly in thickness but which is occasionally lacking. The bundles are interlacing and may be perivascular; their orientation may be tangential to the epithelial surface but this organization is not constant. Immunofluorescence microscopy demonstrates that the collagen fibres are type III[51,52] although type I collagen is also present. Where immunoreactivity to type IV collagen has been shown, it is evidently related to the basement membrane.[61] Many mast cells are seen in the outer parts of the collagen deposits.[51] They are close to the capillaries which have been displaced externally by the collagen aggregates.

Aetiology and pathogenesis

The cause(s) of collagenous colitis is unknown. There is no known, genetic basis. An autoimmune, inflammatory disorder has been proposed. In view of the advances in the knowledge of microorganisms, such as the spirochaetes recently found on gastric and colonic epithelial surfaces, the possibility of bacterial infection cannot be discarded. Nevertheless, the extreme female preponderance of cases argues against a purely infective cause. Furthermore, no virus has been found. A disorganization of intestinal mesenchyme cell turnover time[135] has been postulated. The immature colonic fibroblast secretes little collagen until it has migrated, with the basal epithelial cells, up the intestinal crypt. In collagenous colitis (Fig. 8.14), the fibroblasts may spend longer in their mature phase, secreting excess fibrillar collagen.[94] A vascular factor such as plasmatic vasculosis has been invoked by these authors: a possible stimulus for this change could be the action of an exogenous chemical or toxin, or the presence of microbial exotoxin. It is difficult to see how the transient change in vascular permeability characteristic of plasmatic vasculosis could provoke collagenosis, although a similar hypothesis was advanced by Lendrum[102] to explain the collagenization of glomeruli in diabetes mellitus.

Fibrositis

The fibrositis syndrome is 'the most common condition diagnosed in new patients seen by practicing rheumatologists'.[183] The disorder centres on the clinical recognition of local sites of deep tenderness. There is little histological evidence to validate the nature of these tender foci and the very concept of 'fibrositis' was dismissed by Kellgren[90] as 'a false localization of pain and tenderness over muscles, referred from the joints of the spine and limb girdles and felt over the muscles of the trunk and (proximal) limbs'. In the absence of material for analysis, pathologists can contribute little to this debate. They are, however, aware that entrapment or herniation of incarcerated fat can

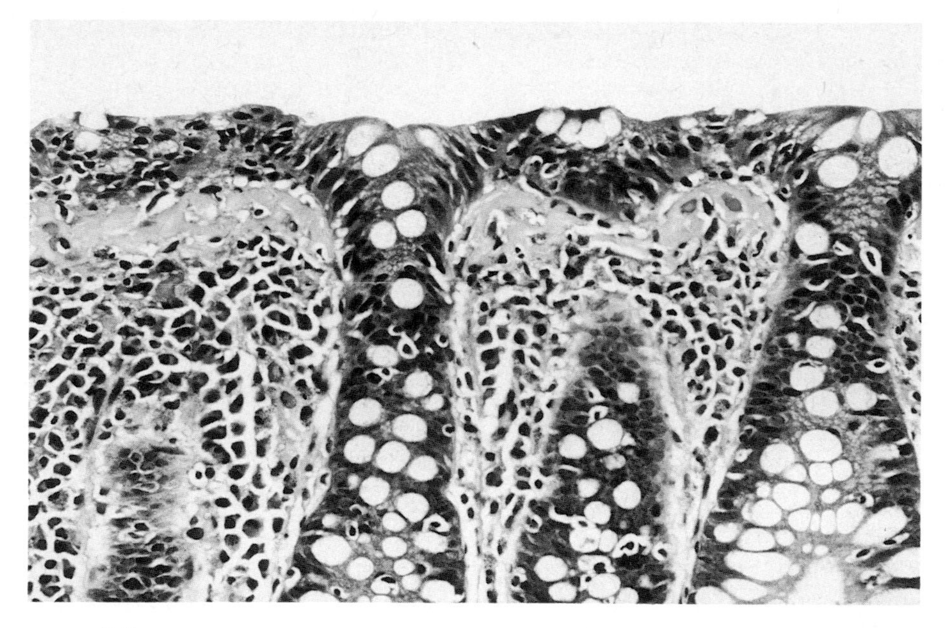

Fig. 8.14 Collagenous colitis.
Immediately beneath the most superficial cells of the intestinal epithelium is an abnormally broad band of acellular collagen. Depending upon the plane of section, the band may appear continuous or discontinuous. (HE × 230.) Courtesy of Dr H M Gilmour.

occur through the inferior lumbar space[88,104] and know that fat, herniated through foramina such as those of the inguinal canal and obturator foramen can undergo ischaemic necrosis with signs of local inflammation or peritonitis. There is some evidence for the presence of a metabolic disorder accompanied by increased catecholamine synthesis[166] and it is relevant to record that of the 17 other causes of soft-tissue pain tabulated by Smythe,[183] four were related to forms of inflammation such as tendinitis and bursitis, and four to vasculitis or vasculopathy (see Chapter 15).

REFERENCES

1. Adams D O. The granulomatous inflammatory response. *American Journal of Pathology* 1976; **84**: 163–91.
2. Akhtar M, Miller R M. Ultrastructure of elastofibroma. *Cancer* 1977; **40**: 728.
3. Albarran J. Retention rénale par pericuretèrité libération extern de l'uretère. *Association Française d'Urologie* 1905; **9**: 511.
4. Albright F, Butler A M, Hampton A O, Smith P. Syndrome characterised by osteitis fibrosa disseminata, areas of pigmentation and endocrine dysfunction with precocious puberty in females. *New England Journal of Medicine* 1937; **216**: 727.
5. Alibert J L M. *Description des Maladies de la Peau Observées à l'Hôpital Saint-Louis et Exposition des Meilleures Méthodes Suivés pour leur Traitement* (2nd edition). Brussels: Wablen, 1806.
6. Alibert J L M. Nôte sur la keloide. *Journal Universel des Sciences Médicales*, Paris, 1816; **2**: 207.
7. Allen P W. The fibromatoses: A clinicopathologic classification based on 140 cases. Part 1. *American Journal of Surgical Pathology* 1977; **1**: 255–70.
8. Allen P W. The fibromatoses: A clinicopathologic classification based on 140 cases. Part 2. *American Journal of Surgical Pathology* 1977; **1**: 305–21.
9. Allen P W. Recurring digital fibrous tumors of childhood. *Pathology* 1972; **4**: 215–23.
10. Antonowitz I, Kodicek E. The effect of scurvy on glycosaminoglycans of granulation tissue and costal cartilage. *Biochemical Journal* 1968; **110**: 609–16.
11. Bailey A J. Tissue and species specificity in the crosslinking of collagen. *Pathologie Biologie* (Paris) 1974; **22**: 675–80.
12. Bailey A J. Age related changes during the biosynthesis and maturation of collagen fibres. *Biochemical Society Transactions* 1975; **3(1)**: 46–8.
13. Bailey A J. Some aspects of collagen maturation and fibrosis. *Annals of the Rheumatic Diseases* 1977; **36** (Suppl.): 5–7.
14. Bailey A J. Collagen and elastin fibres. *Journal of Clinical Pathology* 1978; **31**: Suppl. (Royal College of Pathologists) 12: 49–58.
15. Bailey A J, Bazin S, Delaunay A. Changes in the nature of the collagen during development and resorption of granulation tissue. *Biochemica et Biophysica Acta* 1973; **328**: 383–90.

16. Bailey A J, Bazin S, Sims T J, Le Lous M, Nicholatis C, De Cauney A. Characterization of the collagen of human hypertrophic and normal scars. *Biochimica et Biophysica Acta* 1975; **405**: 412–21.

17. Bailey A J, Robins S P, Balian G. Biological significance of the intermolecular crosslinks of collagen. *Nature* 1974; **251**: 105–9.

18. Bailey A J, Sims T J, Le Lous M, Bazin, S. Collagen polymorphism in experimental granulation tissue. *Biochemical and Biophysical Research Communications* 1975; **66**: 1160–5.

19. Bateman E D, Turner-Warwick M, Haslam P L, Adelmann-Grill B C. Cryptogenic fibrosing alveolitis. Prediction of fibrogenic activity from immunohistochemical studies of collagen types in lung biopsy specimens. *Thorax* 1983; **38**: 93–101.

19a. Becker A E, Anderson R H. *Cardiac Pathology: An Integrated Text and Colour Atlas*. Edinburgh: Churchill Livingstone, 1984.

20. Benjamin I S, Than T, Ryan S, Rodger M C, McGee J O'D, Blumgart L H. Prolyl hydroxylase and collagen biosynthesis in rat liver following varying degrees of partial hepatectomy. *British Journal of Experimental Pathology* 1978; **59**: 333–6.

21. Birk D E, Trelstad R L. Fibroblasts compartmentalize the extracellular space to regulate and facilitate collagen fibril, bundle, and macro-aggregate formation. *Extracellular Matrix: Structure and Function*. New York: Alan R Liss 1985: 373–82.

22. Birk D E, Trelstad R L. Extracellular compartments in tendon morphogenesis: collagen fibril, bundle, and macroaggregate formation. *Journal of Cell Biology* 1986; **103**: 231–40.

23. Blackburn W R, Cosman B. Histologic basis for keloid and hypertrophic scar differentiation: clinicopathologic correlation. *Archives of Pathology* 1966; **82**: 65–71.

24. Boucek R J, Gunja-Smith Z, Noble N L, Simpson C F. Modulation by propranolol of the lysyl-cross-links in aortic elastin and collagen of the aneurysm-prone turkey. *Biochemical Pharmacology* 1983; **32**: 275–80.

25. Bourne G H. Nutrition and wound healing. In: Glynn L E, ed, *Handbook of Inflammation* (volume 3) *Tissue Repair and Regeneration*. Amsterdam: Elsevier/North-Holland Biomedical Press, 1981: 211–42.

26. Bracher R. Radioautographic analysis of the synthesis of protein, RNA, DNA, and sulfated mucopolysaccharides in the early stages of corneal wound healing. *Investigative Ophthalmology and Visual Science* 1967; **6**: 565.

26a. Cary N R, Sethi D, Brown E A, Erhardt C C, Woodrow D F, Gower P E. Dialysis arthropathy: amyloid or iron? *British Medical Journal* 1986; **293**: 1392–4.

27. Calnan J. *Recent Advances in the Surgery of Trauma*. London: Churchill, 1963: 104.

28. Calnan J. Keloid and Dupuytren's contracture. *Annals of the Rheumatic Diseases* 1977; **36**, Suppl. 2: 18–22.

28a. Cantor J O. Experimental pulmonary fibrosis. In: Greenwald R A, Diamond H S, eds, CRC handbook of animal models for the rheumatic diseases (volume I). Boca Raton, Florida: CRC Press Inc., 1988: 205–18.

29. Carini M, Selli C, Rizzo M, Durval A, Constantini A. Surgical treatment of retroperitoneal fibrosis with omentoplasty. *Surgery* 1982; **91**: 137–41.

30. Castelli M J, Armin A-R, Husain A, Orfei E. Fibrosing peritonitis in a drug abuser. *Archives of Pathology and Laboratory Medicine* 1985; **109**: 767–9.

31. Chandler A. Muscular torticollis. *Journal of Bone and Joint Surgery* 1948; **30-A**: 566–9.

32. Chandratre S, Bramble M G, Cooke W M, Jones R A. Simultaneous occurrence of collagenous colitis and Crohn's disease. *Digestion* 1987; **36**: 55–60.

33. Choi K L, Claman H N. Mast cells, fibroblasts and fibrosis. New clues to the riddle of mast cells. *Immunology Research* 1987; **6**: 145–52.

34. Chung E B, Enzinger F M. Infantile myofibromatosis. *Cancer* 1981; **48**: 1807–18.

35. Claman H N. Mast cell depletion in murine chronic graft-versus-host disease. *Journal of Investigative Dermatology* 1985; **84**: 246–8.

35a. Cleland L G, Vernon-Roberts B, Smith K. Fibre glass induced synovitis. *Annals of the Rheumatic Diseases* 1984; **43**: 530–4.

36. Cohen I K, Diegelman R F, Bryant C P, Keiser H R. Collagen metabolism in keloid and hypertrophic scar. In: Gibson T, van der Meulen J C, eds, International Symposium on Wound Healing: International Foundation for Cooperation in the Medical Sciences 69, 1974.

37. Cohen-Gould L, Robinson T F, Factor S M. Intrinsic connective tissue abnormalities in the heart muscle of cardiomyopathic Syrian hamsters. *American Journal of Pathology* 1987; **127**: 327–34.

38. Colby T V, Churg A C. Patterns of pulmonary fibrosis. In: Sommers S C, Rosen P P, Fechner R E, eds, *Pathology Annual* (volume 21, part 1). Norwalk, Connecticut: Appleton-Century-Crofts, 1986: 277–309.

39. Craig R D P, Schofield J D, Jackson D S. Collagen biosynthesis in normal human skin, normal and hypertrophic scar and keloid. *European Journal of Clinical Investigation* 1975; **5**: 69–74.

40. Crouch E, Churg A. Progressive massive fibrosis of the lung secondary to intravenous injection of talc. A pathologic and mineralogic analysis. *American Journal of Clinical Pathology* 1983; **80**: 520–6.

40a. Dahlin D C. *Bone Tumors* (3rd edition). Springfield, Illinois: Charles C Thomas, 1978: 362–7.

41. Davis J M G. The histopathology and ultrastructure of pleural mesotheliomas produced in the rat by injections of crocidolite asbestos. *British Journal of Experimental Pathology* 1979; **60**: 642–52.

42. Debongnie J C, De Galocsy C de, Caholessur M O, Haot J. Collagenous colitis: a transient condition? *Diseases of the Colon and Rectum* 1984; **27**: 672–6.

43. Dehoratius R J, Abruzzo J L, Williams R C. Immunofluorescent and immunologic studies of rheumatoid lung. *Archives of Internal Medicine* 1972; **129**: 441–6.

44. Dunhill M S. Pulmonary Pathology (2nd edition). Edinburgh: Churchill Livingstone, 1987.

45. Dunphy J E, van Winkle W. *Repair and Regeneration*. New York: McGraw-Hill, 1969.

46. Dupuytren G, le baron. De la rétraction des doigts par suite d'une affection de l'aponeurose palmaire, opération chirurgicale

qui convient dans ce cas. *Journal Universel et Hebdomadaire de Médecine et de Chirurgie Pratiques* 1831; **5** (2 ser.): 352–65.

47. Enzinger F M. Fibrous harmatoma of infancy. *Cancer* 1965; **18**: 241–8.

48. Enzinger F M, Dulcey F. Proliferative myositis. Report of thirty-three cases. *Cancer* 1967; **20**: 2213–23.

49. Enzinger F M, Weiss S W. *Soft Tissue Tumors* (2nd edition). St Louis, Missouri: The C V Mosby Company, 1988.

50. Fausa O, Foerster A, Hovig T. Collagenous colitis—a clinical, histological and ultrastructural study. *Scandinavian Journal of Gastroenterology* 1985; **20**: Suppl. 107: 8–23.

51. Flejou J F, Grimaud J A, Molas G, Baviera E, Potet F. Collagenous colitis: ultrastructural study and collagen immunotyping of four cases. *Archives of Pathology and Laboratory Medicine* 1984; **108**: 977–82.

52. Foerster A, Fausa O. Collagenous colitis. *Pathology, Research and Practice* 1985; **180**: 99–104.

52a. Forrest M J, Brooks P M, Takagi T, Kowanko I. The subcutaneous air-pouch model of inflammation. In: Greenwald, R A, Diamond H S, eds, CRC handbook of animal models for the rheumatic diseases (volume I). Boca Raton, Florida: CRC Press Inc., 1988: 125–34.

53. Fox A C. Infarction and rupture of the heart. *New England Journal of Medicine* 1983; **309**: 551–3.

54. Gabbiani G, Majno G. Dupuytren's contracture: fibroblast contraction? *American Journal of Pathology* 1972; **66**: 131–46.

55. Gandhi V C, Humayun H M, Ing T S, Daugirdas J T, Jablokow V R, Iwatsuki S, Geis W P, Hano J E. Sclerotic thickening of the peritoneal membrane in maintenance peritoneal dialysis patients. *Archives of Internal Medicine* 1980; **140**: 121–2.

56. Gardiner G W, Goldberg R, Currie D, Murray D. Colonic carcinoma associated with an abnormal collagen table. *Cancer* 1984; **54**: 2973–7.

57. Gardner E J, Richards R C. Multiple cutaneous and subcutaneous lesions occurring simultaneously with hereditary polyposis and osteomatosis. *American Journal of Human Genetics* 1953; **5**: 139.

58. Giardiello F M, Bayless T M, Jessurun J, Hamilton S R, Yardley J H. Collagenous colitis: physiologic and histopathologic studies in seven patients. *Annals of Internal Medicine* 1987; **106**: 46–9.

59. Giddens W E, Tsai C-C, Morton W R, Ochs H D, Knitter G H, Blakely G A. Retroperitoneal fibromatosis and acquired immunodeficiency syndrome in macaques. Pathologic observations and transmission studies. *American Journal of Pathology* 1985; **119**: 253–63.

60. Gillman T, Penn J, Bronks D, Roux M. Abnormal elastic fibres. Appearance in cutaneous carcinoma, irradiation injuries and arterial and other degenerative connective tissue lesions in man. *Archives of Pathology* 1955; **59**: 733–49.

61. Gledhill A, Cole F M. Significance of basement membrane thickening in the human colon. *Gut* 1984; **25**: 1085–8.

62. Glynn L E. The pathology of scar tissue formation. In: Glynn L E, ed, *Handbook of Inflammation* (volume 3). Amsterdam: Elsevier/North Holland Biomedical Press, 1981: 211–42.

63. Goldstein B, Rendall R E G, Webster I. A comparison of the effects of exposure of baboons to crocidolite and fibrous-glass dusts. *Environmental Research* 1983; **32**: 344–59.

64. Goldstein R H, Fine A. Fibrotic reactions in the lung: the activation of the lung fibroblast. *Experimental Lung Research* 1986; **11**: 245–61.

65. Gool J van, Nie I de, Zuyderhoudt F M J. Mechanisms by which acute phase proteins enhance development of liver fibrosis: effects on collagenase and prolyl-4-hydroxylase activity in the rat liver. *Experimental and Molecular Pathology* 1986; **45**: 160–70.

66. Graham J R. Methysergide for prevention of headache: experience in five hundred patients over three years. *New England Journal of Medicine* 1964; **270**: 67–72.

67. Graham J R. Localised systemic scleroses. In: Kelley W N, Harris E D, Ruddy S, Sledge C B, eds, *Textbook of Rheumatology*. Philadelphia: W B Saunders & Co., 1981: 1235–52.

68. Graham J R, Suby H I, LeCompte P R, Sadowsky N L. Fibrotic disorders associated with methysergide therapy for headache. *New England Journal of Medicine* 1966; **274**: 359–68.

69. Gupta A, Saibil F, Kassim O, McKee J. Retroperitoneal fibrosis caused by carcinoid tumour. *Quarterly Journal of Medicine* 1985; **56**: 367–75.

70. Hajdu S. In: *Pathology of Soft Tissue Tumours*. Philadelphia: Lea & Febiger. 1969: 305.

71. Hajdu S. *Pathology of Soft Tissue Tumours*. Philadelphia: Lea & Febiger, 1979.

72. Hanley P C, Shub C, Lie J T. Constrictive pericarditis associated with combined idiopathic retroperitoneal and mediastinal fibrosis. *Mayo Clinic Proceedings* 1984; **59**: 300–4.

73. Harris E D, Welgus H G, Krane S M. Regulation of the mammalian collagenases. *Collagen and Related Research* 1984; **4**: 493–512.

74. Hart P H, Powell L W, Cooksley H G, Halliday J W. Mononuclear cell factors that inhibit fibroblast collagen synthesis. I. *In vitro* conditions determining their production and expression. *Scandinavian Journal of Immunology* 1983; **18**: 41–9.

75. Hart P H, Powell L W, Cooksley W G E, Halliday J W. Mononuclear cell factors that inhibit fibroblast collagen synthesis. II. Properties of the factors. *Scandinavian Journal of Immunology* 1983; **18**: 51–8.

76. Haslam P L. Circulating immune complexes in patients with cryptogenic fibrosing alveolitis. *Clinical and Experimental Immunology* 1979; **37**: 381–90.

77. Hellstrom H R, Perez-Stable E C. Retroperitoneal fibrosis with disseminated vasculitis and intrahepatic sclerosing cholangitis. *American Journal of Medicine* 1966; **40**: 184–7.

78. Heppleston A G. Environmental lung disease. In: Thurlbeck W M, ed, *Pathology of the Lung*. Stuttgart, New York: Georg Thieme, 1988: 591–685.

79. Hering T M, Marchant R E, Anderson J M. Type V collagen during granulation tissue development. *Experimental and Molecular Pathology* 1983; **39**: 219–29.

80. Hollingworth P, Denman A M, Gumpel J M. Retroperitoneal fibrosis and polyarteritis nodosa successfully treated by intensive immunosuppression. *Journal of the Royal Society of Medicine* 1980; **73**: 61–4.

81. Hutter R V P, Stewart F W, Foote F W. Fasciitis. A report

of 70 cases with follow-up proving the benignity of the lesion. *Cancer* 1962; **15**: 992–1003.

82. Hwang W S, Kelly J K, Shaffer E A, Hershfield N B. Collagenous colitis: a disease of pericryptal fibroblast sheath? *Journal of Pathology* 1986; **149**: 33–40.

83. Jain S, Scheuer P J, McGee J O'D, Sherlock S. Hepatic collagen proline hydroxylase activity in primary biliary cirrhosis. *European Journal of Clinical Investigation* 1978; **8**: 15–7.

84. James J, Bosch K S, Zuyderhoudt F M J, Houtkooper J M, Gool J van. Histophotometric estimation of volume density of collagen as an indication of fibrosis in rat liver. *Histochemistry* 1986; **85**: 129–33.

85. Järvi O, Saxén E. Elastofibroma dorsi. *Acta Pathologica et Microbiologica Scandinavia* 1961; **51** (Suppl. 144): 83–4.

86. Järvi O H, Saxén E, Hopsu-Havu V K, Wartiovaara J J, Vaissalo V T. Elastofibroma—a degenerative pseudotumor. *Cancer* 1969; **23**: 42–63.

87. Jayson M I V, ed. Symposium on the fibrotic process. *Annals of the Rheumatic Diseases* 1977; **36** (Suppl. 2): 1–79.

88. Jayson M I V. Back pain, spondylosis and disc disorders. In: Scott J T, ed, *Copeman's Textbook of the Rheumatic Diseases* (6th edition). Edinburgh: Churchill Livingstone, 1986; 1423–4.

89. Jester J V, Rodrigues M M, Herman I M. Characterization of avascular corneal wound healing fibroblasts. New insights into the myofibroblast. *American Journal of Pathology* 1987; **127**: 140–8.

90. Kellgren J H. Pain. In: Scott J T, ed, *Copeman's Textbook of the Rheumatic Diseases* (5th edition). Edinburgh: Churchill Livingstone 1978; 61–77.

91. Kern W H. Proliferative myositis: a pseudosarcomatous reaction to injury: a report of seven cases. *Archives of Pathology* 1960; **69**: 209–16.

92. Kershenobich D, Vargas F, Carcia-Tsao G, Perez Tamayo R, Gent M, Rojkind M. Colchicine in the treatment of cirrhosis of the liver. *New England Journal of Medicine* 1988; **318**: 1709–13.

93. Kinder C H. Retroperitoneal fibrosis. *Journal of the Royal Society of Medicine* 1979; **72**: 485–6.

94. Kingham J G C, Levison D A, Morson B C, Dawson A M. Collagenous colitis. *Gut* 1986; **27**: 570–7.

95. Kischer C W, Hendrix M J C. Fibronectia (FN) in hypertrophic scars and keloids. *Cell and Tissue Research* 1983; **231**: 29–37.

96. Kischer C W, Shetlar M R. Collagen and mucopolysaccharides in the hypertrophic scar. *Connective Tissue Research* 1974; **2**: 205–13.

97. Konwaler B E, Keasbey L, Kaplan L. Subcutaneous pseudosarcomatous fibromatosis (fasciitis). *American Journal of Clinical Pathology* 1955; **25**: 241–52.

98. Laakso M, Arvala I, Tervonen S, Sotaranta M. Retroperitoneal fibrosis associated with sotalol. *British Medical Journal* 1982; **285**: 1085–6.

99. Last J A. Changes in the collagen pathway in fibrosis. *Fundamental and Applied Toxicology* 1985; **5**: 210–8.

100. Leak L V, Ferrans V J, Cohen S R, Eidbo E E, Jones M. Animal model of acute pericarditis and its progression to pericardial fibrosis and adhesions: ultrastructural studies. *American Journal of Anatomy* 1987; **180**: 373–90.

101. Ledderhose G. Zur Pathologie der Aponeurose des Fusses und der Hand. *Archiv für Klinische Chirurgie* 1897; **55**: 694–712.

102. Lendrum A C. The hypertensive diabetic kidney as a model of the so-called collagen diseases. *Canadian Medical Association Journal* 1963; **88**: 442–52.

103. Lichtenstein L O. Polyostotic fibrous dysplasia. *Archives of Surgery* 1938; **36**: 874–98.

104. Light H G. Hernia of the inferior lumbar space. *Archives of Surgery* 1983; **118**: 1077–80.

105. Linares H A. Measurement of collagen-proteoglycan interaction in hypertrophic scars. *Plastic and Reconstructive Surgery* 1983; **71**: 818–20.

106. Lindblad W J, Fuller G C. Hepatic collagenase activity during carbon tetrachloride induced fibrosis. *Fundamental and Applied Toxicology* 1983; **34**: 34–40.

107. Lindström C G. 'Collagenous colitis' with watery diarrhoea—a new entity? *Pathologica Europea* 1976; **11**: 87–9.

108. Lipper S, Kahn L B, Reddick R L. The myofibroblast. In: Sommers S C, Rosen P P, eds, *Pathology Annual* (volume 15). New York: Appleton-Century-Crofts, 1980: 409–41.

109. Luck J V. Dupuytren's contracture. *Journal of Bone and Joint Surgery* 1959; **41-A**: 635.

109a. El-Maghraby M A H A, Gardner D L. Development of connective-tissue components of small arteries in the chick embryo. *Journal of Pathology*. 1972; 108: 281–91.

110. McGee J O'D, Fallon A. Hepatic cirrhosis: a collagen formative disease? *Journal of Clinical Pathology* 1978; **31**, Suppl. (Royal College of Pathologists) 12: 150–70.

111. Mackenzie D H. *The Differential Diagnosis of Fibroblastic Disorders*. Oxford: Blackwell Scientific Publications, 1970: 44–9.

112. McLemore T L, Mace M L, Roggli V, Marshall M V, Lawrence E C, Wilson R K, Martin R R, Brinkley B R, Greenberg S D. Asbestos body phagocytosis by human free alveolar macrophages. *Cancer Letters* 1980; **9**: 85–93.

113. Madri J A, Furthmayr H. Collagen polymorphism in the lung. *Human Pathology* 1980; **11**: 353–66.

114. Mainardi C L, Kang A H. Localized fibrotic disorders. In: Kelley W N, Harris E D, Ruddy S, Sledge C B, eds, *Textbook of Rheumatology* (2nd edition). Philadelphia: W B Saunders & Co., 1985: 1209–23.

115. Martin B M, Gimbrone M A, Majeau G R, Unanue E R, Cotran R S. Stimulation of human monocyte/macrophage-derived growth factor (MDGF) production by plasma fibronectin. *American Journal of Pathology* 1983; **111**: 367–73.

116. Meshorer A, Prionas S D, Fajardo L F, Meyer J L, Hahn G M, Martinez A A. The effects of hyperthermia on normal mesenchymal tissues. Application of a histologic grading system. *Archives of Pathology and Laboratory Medicine* 1983; **107**: 328–34.

117. Minor R R. Collagen metabolism: a comparison of diseases of collagen and disease affecting collagen. *American Journal of Pathology* 1980; **98**: 225–80.

118. Mitchinson M J. The pathology of idiopathic retroperitoneal fibrosis. *Journal of Clinical Pathology* 1970; **23**: 681–9.

119. Mitchinson M J. Aortic disease in idiopathic retroperitoneal and mediastinal fibrosis. *Journal of Clinical Pathology* 1972; **25**: 287.

120. Mitchinson M J. Macrophages, oxidised lipids and atherosclerosis. *Medical Hypotheses* 1983; **12**: 171–8.

121. Mitchinson M J. Retroperitoneal fibrosis revisited. *Archives of Pathology and Laboratory Medicine* 1986; **110**: 784–6.

122. Mitchinson M J, Wright D G D, Arno J, Milstein B B. Chronic coronary periarteritis in two patients with chronic periaortitis. *Journal of Clinical Pathology* 1984; **37**: 32–6.

122a. Mohr W, Doms E, Wessinghage D. Ultrastruktur des Reparationskollagens und seine Beziehungen zum residualen hyalinen Gelenkknorpel bei der entzundlichen Knorpeldestruktion. *Zeitschrift für Rheumatologie* 1989; **48**: 30–8.

123. Montfort I, Perez-Tamayo R. Collagenase in experimental carbon tetrachloride cirrhosis of the liver. *American Journal of Pathology* 1978; **92**: 411–20.

124. Nagamine N, Nohara Y, Ito E. Elastofibroma in Okinawa. A clinicopathologic study of 170 cases. *Cancer* 1982; **50**: 1794–1805.

125. Nakagawa H, Kitagawa H, Aikawa Y. Tumor necrosis factor stimulates gelatinase production by granulation tissue in culture. *Biochemical and Biophysical Research Communications* 1987; **142**: 791–7.

126. Navas-Palacios J J. The fibromatoses—an ultrastructural study of 31 cases. *Pathology: Research and Practice* 1983; **176**: 158–75.

127. Okanoue T, Burbige E J, French S W. The role of the Ito cell in perivenular and intralobular fibrosis in alcoholic hepatitis. *Archives of Pathology and Laboratory Medicine* 1983; **107**: 459–63.

128. Ordman L J, Gillman T. Studies in the healing of cutaneous wounds. III. A critical comparison in the pig of the healing of surgical incisions closed with sutures or adhesive tape based on tensile strength and clinical and histological criteria. *Archives of Surgery* 1966; **93**: 911–28.

129. Ormond J K. Bilateral ureteral obstruction due to envelopment and compression by an inflammatory retroperitoneal process. *Journal of Urology* 1948; **59**: 1072–9.

130. Osborne, B M, Butler J J, Bloustein P, Sumner G. Idiopathic retroperitoneal fibrosis (sclerosing retroperitonitis. *Human Pathology* 1987; **18**: 735–9.

131. Page R C, Benditt E P. Molecular diseases of connective and vascular tissues: I. The source of lathyritic collagen. *Laboratory Investigation* 1966; **15**: 1643–51.

132. Page R C, Benditt E P. Molecular diseases of connective and vascular tissues. II. Amine oxidase inhibition by the lathyrogen beta-aminopropionitrile. *Biochemistry* 1967; **6**: 1142–8.

133. Page R C, Benditt E P. Molecular diseases of connective and vascular tissues. III. The aldehyde content of normal and lathyritic soluble collagen. *Laboratory Investigation* 1968; **18**: 124–30.

134. Pardo A, Perez-Tamayo R. The collagenase of carrageenin granuloma. *Connective Tissue Research* 1974; **2**: 243–51.

135. Pascal R R, Kaye G I, Lane N. Colonic pericryptal fibroblast sheath: replication, migration and cytodifferentiation of a mesenchymal cell system in adult tissue: I. Autoradiographic studies of normal rabbit colon. *Gastroenterology* 1968; **54**: 835–51.

136. Pérez-Tamayo R. Collagen resorption in carrageenin granulomas. I. Collagenolytic activity in *in vitro* explants. *Laboratory Investigation* 1970; **22**: 137–41.

137. Pérez-Tamayo R. Collagen resorption in carrageenin granulomas. II. Ultrastructure of collagen resorption. *Laboratory Investigation* 1970; **22**: 142–59.

138. Pérez-Tamayo R. Las 'verdaderas' enfermedades de la colagena. *Tres Variaciónes Sobre la Muerte*. México: La Prensa Médica Méxicana, 1974.

139. Pérez-Tamayo R. Pathology of collagen degradation. *American Journal of Pathology* 1978; **92**: 509–66.

140. Pérez-Tamayo R. Mechanisms of disease. *An Introduction to Pathology* (2nd edition). Chicago: Year Book Publishers, Inc, 1985.

141. Peyronie de la F. Sur quelques obstacles qui s'opposent à l'éjaculation naturelle de la semence. Section III, Chapter 19(4). *Mémoires de l'Academie de Chirurgie* 1743; 425: 1745; 317.

142. Phan S H. Fibrotic mechanisms in lung disease. In: Ward P A, ed, *Handbook of Inflammation* (volume IV) *Immunology of Inflammation*. Amsterdam: Elsevier, 1983: 121–62.

143. Pickrell J A, Diel J H, Slavson D O, Halliwell W H, Mauderly J L. Radiation-induced pulmonary fibrosis resolves spontaneously if dense scars are not formed. *Experimental and Molecular Pathology* 1983; **38**: 22–32.

144. Pickrell J A, Harris D V, Benjamin S A, Cuddihy R G, Pfleger R C, Mauderly J L. Pulmonary collagen metabolism after lung injury from inhaled ^{90}Y in fused clay particles. *Experimental and Molecular Pathology* 1976; **25**: 70–81.

145. Pickren J W, Smith A G, Stevenson T W, Stout A P. Fibromatosis of the plantar fascic. *Cancer* 1951; **4**: 846–56.

146. Poole A. Measurements of enzymes of collagen synthesis in rats with experimental silicosis. *British Journal of Experimental Pathology* 1985; **66**: 89–94.

147. Postlethwaite A E, Kang A H. Induction of fibroblast proliferation by human mononuclear leukocyte derived proteins. *Arthritis and Rheumatism* 1983; **26**: 22–7.

148. Price E B Jr, Silliphant W M, Shuman R. Nodular fasciitis: a clinico-pathologic analysis of 65 cases. *American Journal of Clinical Pathology* 1961; **35**: 122–36.

149. Pun K K, Lui F S, Wong K L, Yeung C K. Retroperitoneal fibrosis associated with *Schistosoma japonicum* infestation —an immunologically mediated disease? *Tropical and Geographical Medicine* 1984; **36**: 281–3.

150. Quie P G, Kaplan E L, Page A R, Gruskay F L, Malawista S E. Defective polymorphonuclear-leukocyte function and chronic granulomatous disease in two female children. *New England Journal of Medicine* 1968; **278**: 976–80.

151. Ramon Y, Berman W, Bubis J J. Gingival fibromatosis combined with cherubism. *Oral Surgery* 1967; **24**: 435–8.

152. Rams H, Rogers A I, Ghandur-Mnaymneh L. Collagenous colitis. *Annals of Internal Medicine* 1987; **106**: 108–13.

153. Read A E. Practolol peritonitis. *Annals of the Rheumatic Diseases* 1977; **36** (Suppl. 2): 37–8.

154. Rennard S I, Chen Y-F, Robbins R A, Gader J E, Crystal R G. Fibronectin mediates cell attachment to C1q: a mechanism for the localization of fibrosis in inflammatory disease. *Clinical and Experimental Immunology* 1983; **54**: 239–47.

155. Revell P A. *Pathology of Bone.* New York: Springer-Verlag, 1986.

156. Reye R D K. Recurring digital fibrous tumours of childhood. *Archives of Pathology* 1965; **80**: 228–31.

157. Richards R J, Morris T G. Collagen and mucopolysaccharide production in growing lung fibroblasts exposed to chrysotile asbestos. *Life Sciences* 1973; **12**: 441–51.

158. Roberts J D, Newman R A, Kimberley P J, Hacker M P. Regional fibrosis after intraperitoneal administration of mafosfamide. *Investigational New Drugs* 1986; **4**: 61–5.

159. Roche W R, Du Boulay C E H. A case of ovarian fibromatosis with disseminated intra-abdominal fibromatosis. *Histopathology* 1989; **14**: 101–7.

160. Rosenberg H S, Stenback W A, Spjut H J. The fibromatoses of infancy and childhood. In: Rosenberg H S, Bolande R P, eds, *Perspectives in Pediatric Pathology.* Chicago: Year Book Medical Publications, 1978: 269–348.

161. Ross R. The fibroblast and wound repair. *Biological Reviews* 1968; **43**: 51–96.

162. Ross R. Connective tissue cells, cell proliferation and synthesis of extracellular matrix—a review. *Philosophical Transactions of the Royal Society of London B* 1975; **271**: 247–59.

163. Ross R, Benditt E P. Wound healing and collagen formation. I. Sequential changes in components of guinea-pig skin wounds observed in the electron microscope. *Journal of Biophysical and Biochemical Cytology* 1961; **11**: 677–700.

164. Ross R, Benditt E P. Wound healing and collagen formation. V. Quantitative electron microscope radiographic observations of proline-^{3}H utilization by fibroblasts. *Journal of Cell Biology* 1965; **27**: 83–106.

165. Ross R, Glomset J A. The pathogenesis of atherosclerosis. *New England Journal of Medicine* 1976; **295**: 420–5.

166. Russell I J, Vipraio G A, Morgan W W, Bowden C L. Is there a metabolic basis for the fibrositis syndrome? *American Journal of Medicine* 1986; **81**: 50–6.

167. Ryan G B, Cliff W J, Gabbiani G, Irle C, Statkov P R, Majno G. Myofibroblasts in an avascular fibrous tissue. *Laboratory Investigation* 1973; **2**: 197–206.

168. Sakakibara K, Ooshima A, Igarashi S, Sakakibara J. Immunolocalization of type III collagen and procollagen in cirrhotic human liver using monoclonal antibodies. *Virchows Archiv* A (*Pathological Anatomy and Histopathology*) 1986; **409**: 37–46.

169. Saxton H M, Kilpatrick F R, Kinder C H, Lessof H H, McHardy-Young S, Wardle D F H. Retroperitoneal fibrosis. A radiological and follow-up study of fourteen cases. *Quarterly Journal of Medicine* 1969; **38**: 159–81.

170. Schilling J A. Wound healing. *Physiological Reviews* 1968; **48**: 374–423.

171. Schilling J A. Wound healing. *Surgical Clinics of North America* 1976; **56**: 859–74.

172. Schilling J A. Advances in knowledge related to wounding, repair, and healing: 1885–1984. *Annals of Surgery* 1985; **201**: 268–77.

173. Schmidt R W, Blumenkrantz M. Peritoneal sclerosis. A 'sword of Damocles' for peritoneal dialysis? *Archives of Internal Medicine* 1981; **141**: 1265–6.

174. Schraufnagel D E, Mehta D, Harshbarger R, Treviranus K, Wang N-S. Capillary remodeling in bleomycin-induced pulmonary fibrosis. *American Journal of Pathology* 1986; **125**: 97–106.

175. Scott D G I, Bacon P A. Responses to methotrexate in fibrosing alveolitis associated with connective tissue disease. *Thorax* 1980; **35**: 725–38.

176. Scott P G, Chambers M, Johnson B W, Williams H T. Experimental wound healing: increased breaking strength and collagen synthetic activity in abdominal fascial wounds healing with secondary closure of the skin. *British Journal of Surgery* 1985; **72**: 777–9.

177. Seemayer T A, Lagacé R, Schürch W, Thelmo W L. The myofibroblast: biologic, pathologic and theoretical considerations. In: Sommers S C, Rosen P P, eds, *Pathology Annual* (volume 15). New York: Appleton-Century-Crofts, 1980: 443–70.

178. Seidmann J C, Castor C W. Connective tissue activation in guinea pig lung fibroblast cultures: regulatory effects of glucocorticoids. *In Vitro* 1981; **17**: 133–8.

179. Shimizu S, Hashimoto H, Enjoji M. Nodular fasciitis. An analysis of 250 patients. *Pathology* 1984; **16**: 161–6.

180. Shnitka T K, Asp D M, Horner R H. Congenital generalized fibromatosis. *Cancer* 1958; **11**: 627–39.

181. Silverman W A. The lesson of retrolental fibroplasia. *Scientific American* 1977; **236**: 100–7.

182. Singer I I, Kawka D W, Kazakis D M, Clark R A F. *In vivo* co-distribution of fibronectin and actin fibers in granulation tissue: immunofluorescence and electron microscope studies of the fibronexus at the myofibroblast surface. *Journal of Cell Biology* 1984; **98**: 2091–106.

183. Smythe H A. Non-articular rheumatism and psychogenic musculoskeletal syndromes. In: McCarty D J, ed, *Arthritis and Allied Conditions. A Textbook of Rheumatology* (10th edition). Philadelphia: Lea & Febiger, 1985: 1083–94.

184. Spencer H. *Pathology of the Lung (excluding pulmonary tuberculosis)* (4th edition). Oxford: Pergamon Press, 1984.

185. Stout A P. Juvenile fibromatoses. *Cancer* 1954; **7**: 953–78.

186. Stout A P, Lattes R. Tumours of the soft tissues. In: *Atlas of Tumour Pathology* (2nd series). Washington DC: Armed Forces Institute of Pathology, 1967.

187. Strehler B L, Wilder R, Raychaudhuri A, Gee M, Press G. Studies on the mechanism of cellular death. III. Changes in proteins and connective tissue elements during early and late cardiac necrosis. *Journal of Gerontology* 1967; **22**: 52–8.

188. Takiya C, Peyrol S, Cordier J-F, Grimaud J-A. Connective matrix organization in human pulmonary fibrosis. Collagen polymorphism analysis in fibrotic deposits by immunohistological methods. *Virchows Archiv B* (Cell. Pathol.) 1983; **44**: 223–40.

189. Taylor P E, Tejada C, Sanchez M. The effect of malnutrition on the inflammatory response. As exhibited by the granuloma pouch of the rat. *Journal of Experimental Medicine* 1967; **126**: 539–56.

190. Terry T L. Extreme prematurity and fibroblastic overgrowth of persistent vascular sheath behind each crystalline lens. I. Preliminary report. *American Journal of Ophthalmology* 1942; **25**: 203–4.

191. Tighe J R, Clarke A E, Turvey D J. Elastofibroma dorsi. *Journal of Clinical Pathology* 1968; **21**: 463–9.

192. Trelstad R L, Birk D E. The fibroblast in morphogenesis and fibrosis: cell topography and surface-related functions. In: Bailey A J, ed, *Fibrosis. Ciba Foundation Symposium* 1985; **114**: 4–19.

193. Turck C W, Dohlman J G, Goetzl E J. Immunological mediators of wound healing and fibrosis. *Journal of Cellular Physiology* (Suppl.) 1987; **5**: 89–93.

194. Viljanto J A. Dupuytren's contracture: a review. *Seminars in Arthritis and Rheumatism* 1973; **3**: 155–76.

195. Wagner B M. Fibrosis and cell metabolism. *Human Pathology* 1987; **18**: 1083–4.

196. Walker M A, Harley R A, LeRoy E C. Inhibition of fibrosis in TSK mice by blocking mast cell degranulation. *Journal of Rheumatology* 1987; **14**: 299–301.

197. Warren K S. The cell biology of granulomas (aggregate of inflammatory cells) with a note on giant cells. In: Weissman G, ed, *Handbook of Inflammation* (volume II) *The Cell Biology of Inflammation.* Amsterdam: Elsevier/North Holland Biomedical Press, 1980: 543–57.

198. Weary P E, Linthicum A, Cawley E P, Coleman C C, Graham G F. Gardner's syndrome: a family group study and review. *Archives of Dermatology* 1964; **90**: 20–30.

199. Weinstein W M, Saunders D R, Tytgat G N, Rubin C E. Collagenous sprue: an unrecognized type of malabsorption. *New England Journal of Medicine* 1970; **283**: 1297–301.

200. Williams G. The pleural reaction to injury: a histological and electron optical study with special reference to elastic-tissue formation. *Journal of Pathology* 1970; **100**: 1–7.

201. Willis R A. Hamartomas and hamartomatous syndromes. *The Borderline of Embryology and Pathology.* London: Butterworth, 1958: 341–81.

202. Wyler D J, Wahl S M, Wahl L M. Hepatic fibrosis in schistosomiasis: Egg granulomas secrete fibroblast stimulating factor *in vitro. Science* 1978; **202**: 438–40.

203. Yannopoulos K, Stout A P. Primary solid tumors of the mesentery. *Cancer* 1963; **16**: 914–27.

Part III

Pathology of Connective Tissue Disease

Chapter 9

Heritable Diseases of Connective Tissue

Primary defects of structural proteins
- Ehlers–Danlos syndrome
- Marfan's syndrome
- Osteogenesis imperfecta
- Cutis laxa
- Floppy mitral valve

Secondary diseases of fibrous structural proteins
- Homocystinuria
- Alkaptonuria and hereditary ochronosis
- Menkes' kinky (steely) hair syndrome
- Pseudoxanthoma elasticum (Grönblad–Strandberg syndrome)

Disorders of macromolecular degradation: lysosomal storage diseases
- Mucopolysaccharidoses
- I-cell disease
- Mannosidosis

Other hereditary disorders
- Dwarfism
- Synphalangism
- Epidemic familial arthritis of Malnad
- Congenital pseudoarthrosis
- Achondroplasia (chondrodystrophia fetalis)
- Recurrent polyserositis (familial Mediterranean fever)
- Haemochromatosis
- Wilson's disease
- Cystic fibrosis

Understanding of the nature and consequences of the genetic abnormalities of humans is now advancing extremely rapidly but details of the localization, isolation and characterization of the individual gene defects is beyond the scope of this text. The reader is referred to Scriver *et al.*[209a] and to the annual publication, Human Gene Mapping.[117a]

The synthesis, maturation and organization of the connective tissues is often disturbed because of genetic abnormalities.[153,156] The disorders may be of individual molecules, their manufacture, export, maturation or metabolism; of particular cells; of the assembly of tissues such as cartilage, dermis and bone; or of the form of whole limbs, organs; or of whole skeletons or whole individuals. Occasionally, as in trisomy-21 (Down's syndrome) the defect is attributable to a chromosomal anomaly: in other instances, as in some forms of osteogenesis imperfecta and chondrodystrophia fetalis (achondroplasia), the disease is inherited as a Mendelian dominant characteristic. However, the inherited abnormalities of connective tissue which have attracted the greatest interest are those in which the disorders are inherited either as autosomal or sex-linked Mendelian recessives or on the basis of multiple gene (multifactorial) defects.

Heredity may play an overwhelming part in connective tissue disease; alternatively, its role may be contributory or trivial. When there is an inherited defect in the formation of an enzyme participating in collagen synthesis or cross-linking, as in the Ehlers–Danlos syndrome, or responsible for glycosaminoglycan degradation, as in the lysosomal storage disorders, the role of heredity is pre-eminent. When an individual inherits particular patterns of immune responsiveness, e.g. those indexed by MHC class I, II or III (HLA) antigens (see p. 241) the role of heredity is contributory. This is the background to diseases such as ankylosing spondylitis and Reiters' syndrome, and, in part, to rheumatoid arthritis and systemic lupus erythematosus (SLE). When there is no more than the inheritance of particular social, occupational and behavioural patterns as in the case of those injured accidentally or infected by *Mycobacteria*, *Neisseria* or *Streptococci* spp., then the role of heredity may be slight.

A large number of heritable disorders of the connective tissues is known;[223] they may coexist with acquired connective tissue disease.[253] Many are characteristic of individual species although some, such as chondrodysplasia,[88a] are recognized in several species. There are many reviews.[218a] Fairbank[69] described the known skeletal disorders, and their genetic basis was analysed by Carter and Fairbank.[31] McKusick compiled extensive accounts of the heritable disorders of connective tissue[153–156] and further descriptions were given by Beighton,[12] Cremin and Beighton[34] and Horan and Beighton.[105] The heritable disorders of particular relevance to orthopaedic surgery were analysed by Wynne-Davies,[254] the skeletal dysplasias (Table 9.1) (Fig. 9.1) by Rimoin and Sillence,[197] the constant bone dysplasias (Table 9.2) by Spranger *et al.*[224] and the patterns of human malformation by Smith.[219] Shapiro,[210] reviewing the advances in surgical approaches to the epiphyseal disorders, examined the contribution to

Table 9.1

Pathophysiologic classification of the skeletal dysplasias[199]

Site of defect	Mechanism	Possible examples
Limb bud development	Arrest in fetal bone development	Grebe disease
Cartilage anlage development	Cartilage anlage from which bone arises develops abnormally in size or shape or both	Achondrogenesis: Parenti–Fraccaro type; some forms of mesomelic dwarfism
Chondrocyte metabolism	A metabolic defect may lead to abnormalities in the matrix secreted or reduced survival of the resting chondrocyte	Diastrophic dysplasia; Kniest dysplasia; achondrogenesis: Langer–Saldino type mucopolysaccharidoses
Chondrocyte proliferation	Reduced cell division in the proliferative zone results in diminished linear bone growth	Adenosine deaminase deficiency; achondroplasia
Chondrocyte maturation and degeneration	Abnormalities of the normal chondrocyte maturation degeneration sequence (column development) lead to irregular metaphyseal vascular invasion and bone formation	Metaphyseal chondrodysplasias, thanatophoric dysplasia
Epiphyseal ossification	Abnormality in development of epiphyseal ossification centres (secondary ossification)	Multiple epiphyseal dysplasias
Membranous ossification	Failure of fibroblast cells in periosteum to transform into osteoblasts or produce normal bone matrix (osteoid) results in irregular cortical ossification	Melnick–Needles syndrome
Ossification—generalized	Abnormality in bone matrix composition or mineralization	Osteogenesis imperfecta, hypophosphatasia
Calcified cartilage resorption	Failure to resorb calcified cartilage spicules leads to increased skeletal density	Osteopetrosis, dysosteosclerosis
Bone resorption	Failure to resorb cortical bone results in poorly modelled bones of increased density	Hyperostotic bone dysplasias (craniometaphyseal dysplasia, Engelmann disease)
Aberrant chondrocyte growth	Erratic and excessive growth of chondrocytes disturbs the normal growth plate and extends into (enchondroma) or outside (exostosis) the metaphysis	Ollier disease. Multiple exostosis syndrome
Premature epiphyseal fusion	Early closure of growth plate produces shortening of tubular bones	Tricho-rhino-phalangeal syndrome

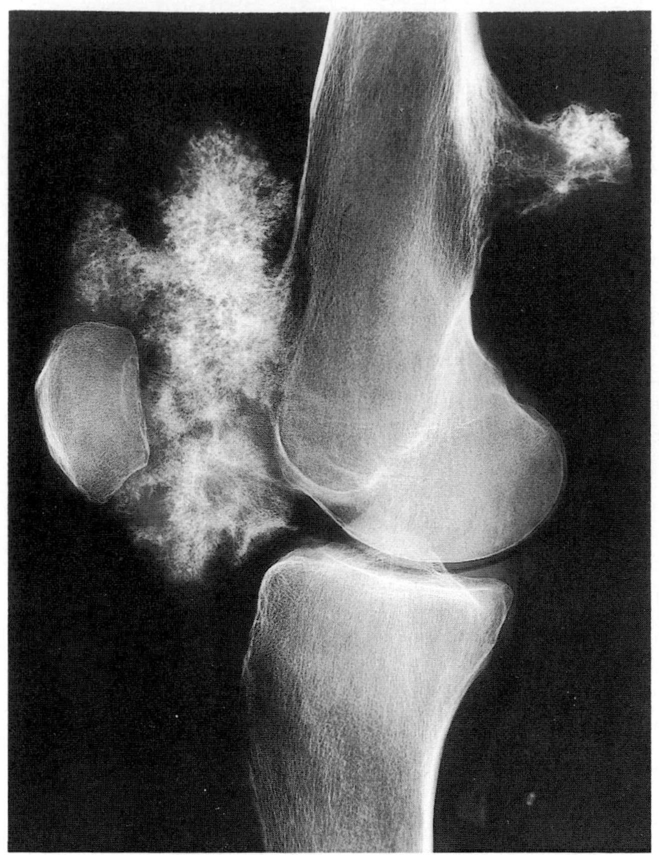

Fig. 9.1 Skeletal dysplasia—diaphyseal aclasis (multiple ecchondromatosis).
At the growing end of the femur two osteochondromatous processes are seen: the larger (at left) extends from the cortex of the femur towards the patella; a smaller process is seen (upper right). Each process comprises a core of bone capped by hyaline cartilage (see Table 9.1).

prenatal diagnosis made possible by techniques such as ultrasonography (see Chapter 3). Siegel and Pinnell,[215] Prockop and Kivirikko,[188] Minor [166] and Pope and Nichols[187] summarized the diseases of structural proteins, while Legum, Schorr and Berman,[133] Kelly[117] and McKusick and Neufeld[157] reviewed the heritable disorders of glycosaminoglycan and lipid metabolism and the 'mucopolysaccharide' storage diseases.

Because of the difficulty of correlating the increasing number of clinical syndromes with the rapid advances in knowledge of the underlying biochemical disorders, Maroteaux, Frezal and Cohen-Selal[162] have proposed a new classification for the genetic disorders of collagen metabolism. They categorize the diseases according to whether skin, joint, bone or blood vessels are mainly involved (Table 9.3). This 'logical clinical framework' is of value to the pathologist. Comparable classifications could be devised for disorders of the proteoglycans and for those of elastic material.

The present chapter centres on the heritable connective tissue disorders of most interest in pathological practice. Many other connective tissue disseases have a heritable basis but because their pathological effects are manifested as immunological, inflammatory or metabolic disorders, they are considered in other chapters. Examples are: rheumatoid arthritis (see Chapters 12 and 13), systemic lupus erythematosus (see Chapter 14) and reactive arthritis (see Chapter 19).

Primary defects of fibrous structural proteins

The reader is referred to Chapter 4 for an outline of collagen synthesis and metabolism, and to Weatherall[245] for a summary of the recent genetic studies that are leading to improved understanding of the heritable connective tissue diseases.[118,130,135,167]

Ehlers–Danlos syndrome

Clinical presentation

This rare group of disorders[104,156] has been known since the seventeenth century.[241] The eponymous title of the disease derives from the descriptions by Ehlers[60] of joint laxity and subcutaneous haemorrhages, and by Danlos[44] of the subcutaneous tumours. Through the ages, circus sideshows have exhibited adults such as the 'India rubber man' who can hyperextend his fingers and wrists, draw out the skin of his neck into bat-like webs (Fig. 9.2), touch the tip of his nose with his tongue, and who has lax skin (hyperelastosis cutis) and double-jointedness (Fig. 9.3).[86] The defective wound-healing with coarse scarring (Fig. 9.4), the tendency to bleed excessively from minor injury, acrocyanosis, the lipomatous tumours of soft tissues and the risk of death from arterial tears are less obvious. In some instances, bowel and uterine rupture are the presenting clinical findings; in others the early signs are of joint pain and swelling.[178] In children, dislocation of lax articulations after physical activity, myopia and midsystolic clicks or murmurs are occasional signs. Blue sclerotics are common; diaphragmatic hernia may occur and the lipomatous nodules may become calcified or ossified.

There are animal disorders of a similar kind[167,239] and studies of Ehlers–Danlos syndrome-like disorders of dogs, cats and mink; of dermatosparaxis in cattle, sheep, cats and

Table 9.2

Systematic classification of bone dysplasias[199]

I *Epiphyseal dysplasia*
  A. Epiphyseal hypoplasia
    1. Failure of articular cartilage: spondyloephiphyseal dysplasia
    2. Failure of ossification of centre: multiple epiphyseal dysplasia
  B. Epiphyseal hyperplasia
    1. Excess of articular cartilage: dysplasia epiphysealis hemimelica

II *Physeal dysplasia*
  A. Cartilage hypoplasia
    1. Failure of proliferating cartilage: achondroplasia
    2. Failure of hypertrophic cartilage: metaphyseal dysostosis; cartilage-hair hypoplasia
  B. Cartilage hyperplasia
    1. Excess of proliferating cartilage: hyperchondroplasia (Marfan syndrome)
    2. Excess of hypertrophic cartilage: enchondromatosis

III *Metaphyseal dysplasia*
  A. Metaphyseal hypoplasia
    1. Failure to form primary spongiosa: hypophosphatasia
    2. Failure to absorb primary spongiosa: osteopetrosis
    3. Failure to absorb secondary spongiosa: craniometaphyseal dysplasia
  B. Metaphyseal hyperplasia
    1. Excessive spongiosa: multiple exostoses

IV *Diaphyseal dysplasia*
  A. Diaphyseal hypoplasia
    1. Failure of periosteal bone formation: osteogenesis imperfecta
    2. Failure of endosteal bone formation: idiopathic osteoporosis, *congenita et tarda*
  B. Diaphyseal hyperplasia
    1. Excessive periosteal bone formation: progressive diaphyseal dysplasia (Engelmann's disease)
    2. Excessive endosteal bone formation: hyperphosphatasemia (including juvenile Paget's disease and van Buchem's disease)

dogs; of the aneurysm-prone mouse; and of sporadic diseases of horses and cats are among those that have been the object of detailed genetic and biochemical enquiry.

It is now apparent that the Ehlers–Danlos syndrome is a group of at least 10 heritable conditions. Affected families have been investigated.[141] The Ehlers–Danlos syndrome may co-exist with the Marfan syndrome.[84]

Inheritance

The mode of inheritance is usually autosomal dominant but exceptional cases are recessive and type V is X-linked recessive (Table 9.4).

Biochemical changes

The biochemical basis of the Ehlers–Danlos syndrome (Table 9.4) is described in Chapter 4. The changes are summarized here.

The biochemical abnormalities in Ehlers–Danlos syndrome types I, II and III are not yet understood. In Ehlers–Danlos syndrome type IV, a heterogeneous group of disorders results from disordered type III collagen structure and processing.[24] In the dominant forms, the defective molecules are retained within the cell. Occasionally, linkage has been shown between the disease and restriction length polymorphisms closely associated with the gene coding for type III collagen, located at 2q23. In one of the recessive forms there may be decreased collagen synthesis and fibroblasts in culture produce relatively small quantities of type III procollagen. In Ehlers–Danlos syndrome type V, some patients were thought to have a defect in collagen cross-linking; however, this evidence has not been substantiated and the biochemical defect remains uncertain. Abnormal copper metabolism, determined genetically, had been discussed as a possible cause of a suspected reduction in lysyl oxidase activity.[122] Ehlers–Danlos syndrome type VI displays lysyl hydroxylase deficiency; the Hyl content of the skin is almost nil. In Ehlers–Danlos syndrome type VII,

Table 9.3

A clinical classification of genetic disorders of collagen metabolism[162]

Disorder	Transmission	Defect
Disorders with predominant skin involvement		
Ehlers–Danlos		
Type I	AD	?
Type II	AD	?
Type V	Xr	Lysyloxidase deficiency
Type IX	Xr	Lysyloxidase deficiency
Type VI	AD	Lysylhydroxylase deficiency
Type VIII	AD	?
Cutis laxa	AD	?
Cutis laxa	Ar	?
Disorders with predominant ligamentous involvement		
Larsen syndrome	Ar	?
	AD	?
Desbuquois syndrome	Ar	?
Ehlers-Danlos Type VII	AD	Procollagen α_2 (I) anomaly
(Arthrochalasis multiplex)	Ar	Procollagen N-peptidase deficiency
Ehlers-Danlos Type XI	AD	?
(Familial joint instability syndrome)		
Ehlers-Danlos Type III	AD	?
(Benign hypermobility syndrome)		
Familial simple joint laxity syndrome	AD	?

Disorder		Inheritance	Defect
Disorders with predominant skeletal involvement			
Perinatally symptomatic osteogenesis imperfecta			
Lethal	II A	D	Short pro-α (I) collagen chains
		R(r)	—
	II C	r	?
Severe	II B	r(D)	Pro-α_2 (I) chains
	III	r(D)	Pro-α_1 (I) trimers
'Regressive'	IV	D	Pro-α chains
	I	D	Pro-α chains (IA)
Disorders with predominant vascular involvement			
Ehlers-Danlos	Severe form	Ar	Absence of collagen III
Type IV	Benign form	Ar	
	Severe form	AD	Deficit of collagen III
	Benign form (acrogeria)	AD	

AD, autosomal dominant; Xr, X-linked recessive; Ar, autosomal recessive; D, dominant; R(r), rarely recessive; r, recessive; r(D), recessive, but possible dominant mutations.

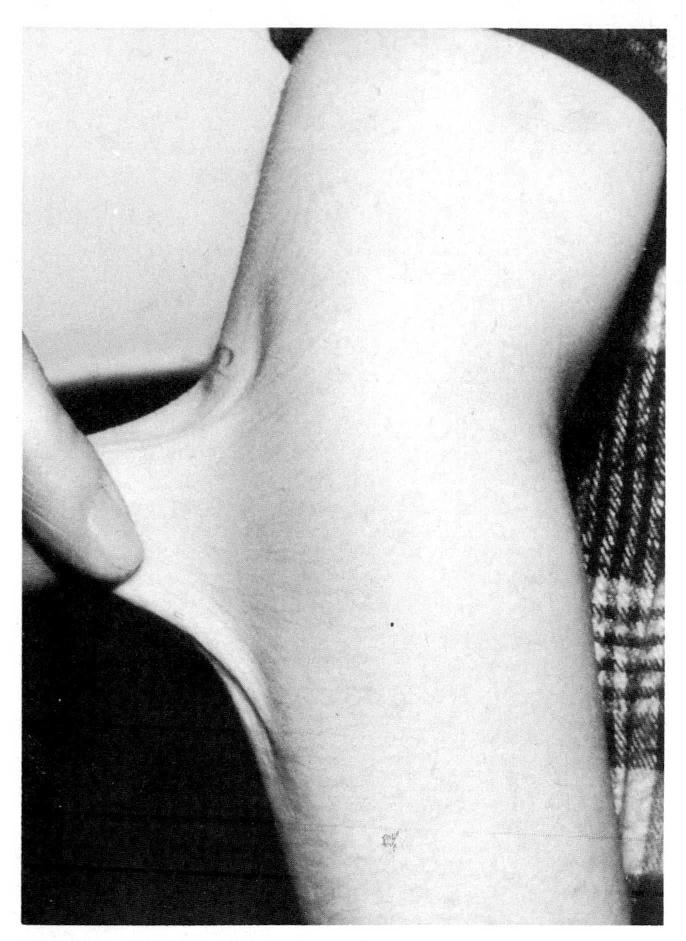

Fig. 9.2 Ehlers–Danlos syndrome.
The excessive extensibility of the skin is shown.

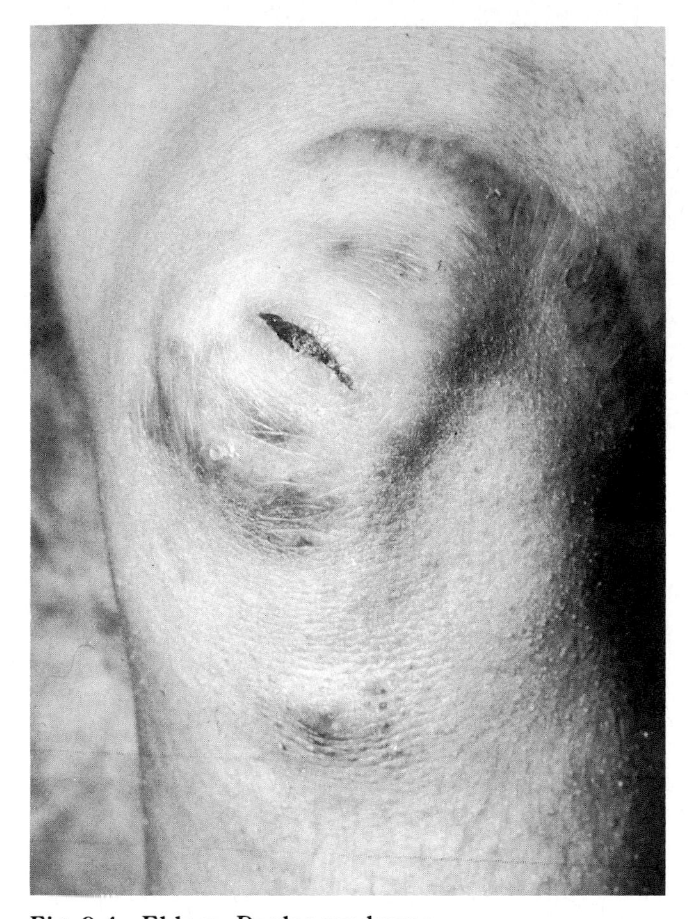

Fig. 9.4 Ehlers–Danlos syndrome.
The skin heals with difficulty: wound gape and scars are poorly formed.

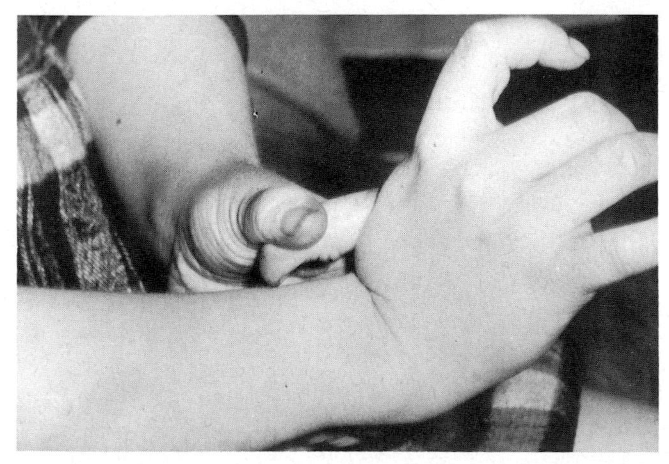

Fig. 9.3 Ehlers–Danlos syndrome.
Hyperextension of the fingers and wrists is one feature of the disorder which may share common signs with other heritable diseases, such as the Marfan syndrome.

a recessive form (VIIa) has an abnormality of the NH_2-terminal procollagen peptidase whereas dominant forms (VIIb) display structural anomalies in both pro α_2 chains that inhibit enzymatic cleavage of the N-terminal propeptides. No specific biochemical disorders are yet known in Ehlers–Danlos syndrome type VIII. In Ehlers–Danlos syndrome type IX defective collagen cross-linking again corresponds to a deficiency of lysyl oxidase. In Ehlers–Danlos syndrome type X, a recessive variant, a defect in fibronectin has been proposed,[7] coinciding with faulty platelet function.

Pathological changes

Less is known of the systemic lesions in the Ehlers–Danlos syndromes than of the biochemical defects. The histological changes in the skin were first investigated by Unna.[240] An account, presumably of Ehlers–Danlos syndrome types I, II and III, was given by Jansen.[112] The histological abnormalities in the skin (Fig. 9.5) are not commensurate in

Table 9.4

Types of Ehlers-Danlos syndrome and the defects in collagen[36,103,188,239]

Type	Genetic defect	Consequences	Inheritance	Collagen fibril diameter	Collagen fibril density	Fraying of fibres	Elastic changes
I	Not known		AD	Large with 'flowers' or small and disorganized	Normal	Present	Fragmented or unaltered
II	Not known		AD	Large or variable; collagen 'flowers'	Normal	Present	Fragmented, excessive or unaltered
III	Not known		AD	Large with collagen 'flowers', or variable	Normal	Present	Unaltered
IV A	Abnormally functioning alleles for pro-α chains	Marked decrease in type III collagen					
B	Altered pro α (III) migrating slowly on electro-phoresis	Unstable triple helix	Ar or AD	Hetero-geneous: at least 2 populations of small or variable diameter	Normal	Present	Excessive amount, branched, fragmented, contains inclusions
C	Poorly defined mutations altering structure of pro α chains	Decreased rate of secretion of type III collagen					
D	Others (unidentified)						
V		? defective collagen cross-linking	XLR	Small homogeneous, collagen 'flowers', serrated	Normal	Present	?fragmented

Table 9.4 – continued

VI	Lysine hydroxylase deficiency	Hydroxylysine-deficient collagen and defective cross-linking	Ar	Normal with flowers, or small	Normal	Present	Electron-dense or unaltered; irregular distribution
VII A	Procollagen N-proteinase deficiency	Persistence of pN collagen, intermediate in conversion of procollagen	AD	Heterogeneous	Decreased	?	Small and unevenly stained or large and unaltered: 'moth-eaten'
B	Deletion or RNA splitting defect altering structure of pro-α_2(I) chains	Resistance to procollagen N-proteinase and persistence of pN collagen	Ar				
VIII		Not known	AD	Variable	?	?	Unaltered
IX	Defective copper metabolism	Lysyl oxidase activity deficiency with defective collagen cross-linking	XLR	Increased (135 nm), collagen 'flowers'	Increased	Not present	Unaltered
X	Functional fibronectin deficiency		Ar	?	?	?	

AD = Autosomal dominant; Ar = Autosomal recessive; XLR = X-linked.

severity with the observed clinical signs. Dermal elastic tissue is seen in relative excess but the collagen density, measured by light microscopic histomorphometry after picro-Sirius red staining,* is reduced.[114] An increase in metachromatic extracellular matrix material has been observed and a medial, basophilic 'mucoid' change of the aortic and large arterial walls is thought to contribute to aneurysm formation,[212] dissection and rupture.[109] The small nodules found subcutaneously in zones exposed to trauma are fibrofatty; the cholesterol content suggests that they may originate as haematomata. The presence of these nodules and the dystrophic calcification and ossification that they undergo, appear to be incidental sequelae of the disease. In Ehlers–Danlos syndrome type IV, there is a relative increase in the elastic material content of the thin skin. There is hyperextensibility of this fragile skin and joint hypermobility and the Marfan-like characteristics are accompanied by keratoconus, scoliosis and ocular globe fragility. The signs of Ehlers–Danlos syndrome type V are similar to those of Ehlers–Danlos syndrome type II. Many of the changes in Ehlers–Danlos syndrome type VI are also ocular; the skin is hyperextensible, the joints excessively mobile. In Ehlers–Danlos syndrome type VIII, no specific histological changes are known. Among the unclassified examples of Ehlers–Danlos syndrome, sinus of Valsalva

* Puchtler[190] denies that picro-Sirius red distinguishes between type I and type II collagen.

Fig. 9.5 Ehlers–Danlos syndrome.
Beneath the well-formed epidermis the dermal papillae are intact but there is an excess of dermal elastic material arranged as black-staining fibres. (Weigert's elastic-van Gieson × 110.)

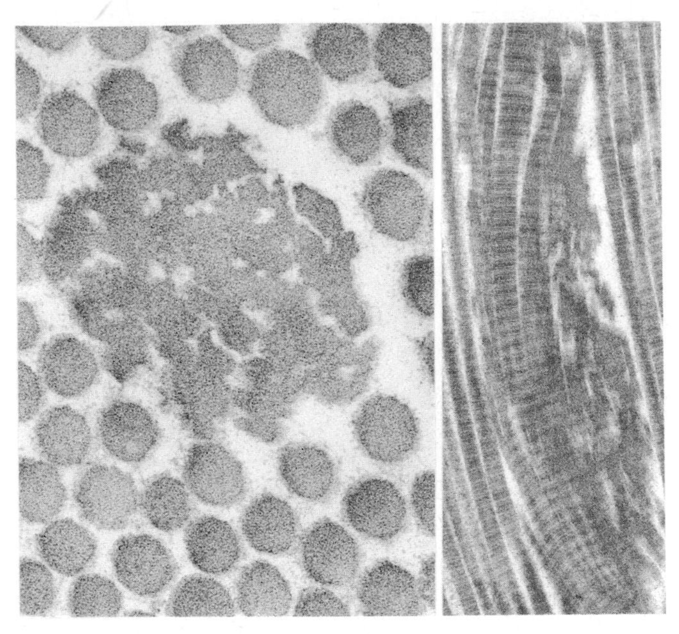

Fig. 9.6 Ehlers–Danlos syndrome.
Collagen 'flowers' from skin of patient with dominant form of Ehlers–Danlos syndrome. Normally shaped fibres surround abnormal. Left: transmission electron microscopy, transverse section, × 79 200; Right: transmission electron microscopy, longitudinal section, × 18 900. By courtesy of Dr Karen A Holbrook and the Editor, *Journal of Investigative Dermatology.*[103]

aneurysms, myocardial infarction and cerebral heterotopia are described.

The ultrastructural changes of the Ehlers–Danlos syndromes are beginning to be recorded.[103] Transmission electron microscopy reveals large, irregular dermal collagen fibres in Ehlers–Danlos syndrome types I, II and III, and the collagen bundles display abnormalities when viewed by scanning electron microscopy.[113] The elastic tissue is fragmented or unaltered. Comparable abnormalities have been identified in affected cats, dogs, horses and mink.[239] The large fibres may aggregate into clusters but the flower-like* groups so formed[243] have been recognized in the skin in pseudoxanthoma elasticum and in the aorta in the Marfan syndrome and are therefore not pathognomonic[24] (Fig. 9.6) although they can be measured by the computer-assisted image analysis of electron photomicrographs.[185] In type IV Ehlers–Danlos syndrome, the collagen fibres are small, frayed and of reduced diameter[24] but there is a relatively large elastin content. Defective type III collagen secretion is manifest as a retention of type III procollagen in the rough endoplasmic reticulum (Table 9.4). The deficiency of type III collagen has been confirmed by transmission electron microscopy in a patient with pulmonary bullae and pneumothoraces: there were dilated cisternae of the rough endoplasmic reticulum but individual extracellular collagen fibres appeared normal,[32] small, or of variable diameter.[103] The ultrastructural changes in Ehlers–Danlos syndrome type V are varied; the collagen fibres are small and homogeneous and collagen 'flowers' are seen. In Ehlers–Danlos syndrome type VI, collagen fibres may be normal but their arrangement in bundles appears to be altered with some smaller bundles and fibres resembling those of Ehlers–Danlos syndrome type I. The collagen fibril architecture in Ehlers–Danlos syndrome type VII is heterogeneous and in Ehlers–Danlos syndrome type VIII the appearances are variable. Platelet changes have been identified by transmission electron microscopy.[116] In Ehlers–Danlos syndrome type IX, collagen fibril density and diameter are increased and 'flowers' are present but little is known of any collagen fibril abnormalities in Ehlers–Danlos syndrome type X. In the many cases of Ehlers–Danlos syndrome in which no type can yet be specified, there may be normal skin collagen cross-banding but irregular fibril size, the fibril diameter varying from 25 to 130 nm (mean 83 nm). The fibrils are loosely packed and frayed. In one instance, the abnormal dermal collagen fibres were heterogeneous in size, reduced in number, irregular and disperse.[36] The collagen of the intervertebral disc may be normal and homogeneous.

* Variously called cauliflowers;[185] wires, ropes, stars, hieroglyphs.

Marfan's syndrome

The Marfan syndrome[160] (arachnodactyly, dolicostenomelia, dystrophia mesodermalis congenita) is a heterogeneous, inherited group of connective tissue disorders distinct from homocystinuria (see p. 349). The presence of spider-like fingers (arachnodactyly) was noted by Achard.[2] The mode of inheritance was first discussed by Weve.[251] Males and females are affected equally.

Clinical presentation

The clinical characteristics have often been reviewed.[150,155,218] There is skeletal deformity, in particular an abnormally high span-to-height ratio, long fingers and toes (Fig. 9.7), pigeon chest and hyperextensibility of the joints. These changes may be associated with dislocation of the lenses of the eyes and medial mucoid change of the aorta which can lead to dissecting aneurysm. Four entities have been distinguished:[155]

1. Aesthenic
2. Non-aesthenic (Lincoln-like) (see below)
3. 'Marfanoid' hypermobility syndrome
4. Contractural arachnodactyly, possibly the form originally described by Marfan.[160]

Many patients with the syndrome are first seen in ophthalmological clinics on account of changes resulting from upward displacement of a lens which is usually small and round. Others, in particular males, are recognized when fatal dissection of the aorta occurs. Diaphragmatic hernia has been described. Relatively small numbers of patients have the entire range of classical clinical features with large stature, increased span-to-height ratio, long fingers and flat feet. In other patients, incomplete expression results in a single clinical defect such as recurrent dislocation of the hip. The recognition of kyphoscoliosis may lead to the detection of multiple abnormalities. Investigation of a family can reveal both normal persons and others with every variety and severity of the known deformities, singly or in combination. The condition was demonstrated in a distant relative of Abraham Lincoln and the suggestion was made that he himself displayed evidence of the trait,[131,208] perhaps predisposing the arteries to the effects of injury. Could the virtuosity of Paganini also be attributed to arachnodactyly?[204]

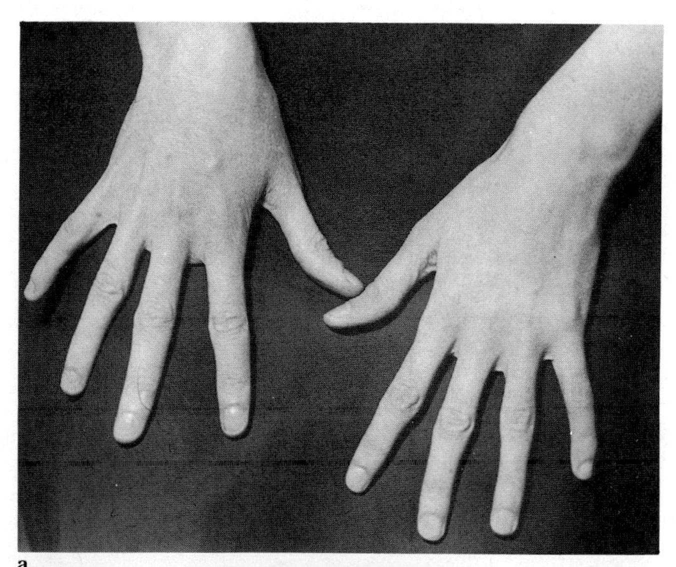

a

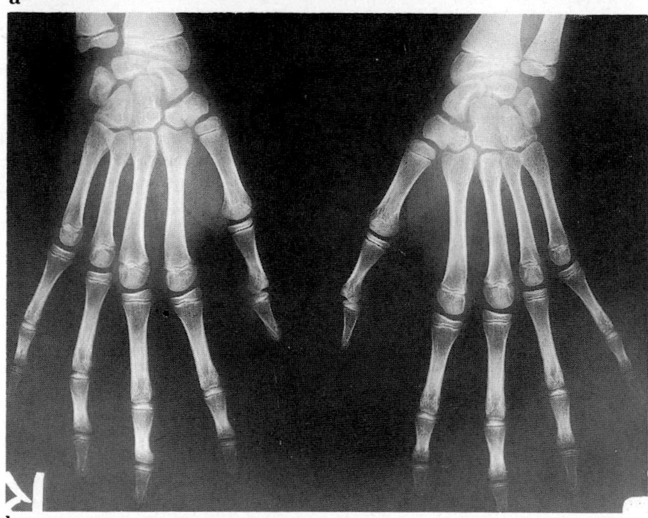

b

Fig. 9.7 Marfan syndrome.
a. Arachnodactyly: the fingers are abnormally long and flexible, and the span very large. **b.** Radiograph of hands similar to those shown in **a**. Bone and joint structure are normal in microscopic terms but their organization is abnormal.

Inheritance

Inheritance is usually as a Mendelian autosomal-dominant characteristic. The original account emphasized the features of fully developed cases but incomplete expression of the genetic defect is common.

Since the diagnostic localization of the abnormality, in some cases of the Marfan syndrome, to the long arm of chromosome 15,[114a] the isolation and characterization of the responsible mutation can safely be predicted. Fibrillin is a candidate protein for the fundamental abnormality.

The frequency of the syndrome has been underestimated. Additional probands have been found as a result of family surveys[218] while the occasional occurrence of families with apparently normal parents and several affected children has raised the question of whether a recessive form may occur.

Biochemical changes

The dominant mode of inheritance suggests that the defects involve structural proteins (see p. 174).[174] However, no agreed collagen abnormality has yet been identified and a recent account has eliminated genetic defects in collagen types I, II and III. A single example of a defect in α_2(I) chain production has been reported.[26,203] Alterations in proteoglycan[129] and in collagen metabolism, including a lack of aortic type I collagen, have been described. The phenotypic similarities to homocystinuria (see p. 349) suggest that impaired collagen cross-link formation may be the critical abnormality.[73] However, there is likely also to be a defect in elastic material since aortic elastic material cross-linking has been found to be reduced,[1] an observation which helps to explain the vascular fragility in this syndrome.

Pathological changes

In the few cases studied systematically, the pathological investigation of the skeleton has been unrewarding. A deformity such as genu recurvatum is not easily overlooked but the search for the precise underlying cause of such an abnormality generally fails to reveal any diagnostic histological feature. In a typical case death took place quickly in a young male following dissection of the aorta; a comprehensive investigation of the skeleton revealed no conventional microscopic change in the connective tissues. No identifiable light-microscopic abnormalities were found in the disturbed joints, in the deformed bones or in the displaced lens ligaments.

Cardiovascular anomalies are well recognized.[149,151] The majority give little clinical evidence of their presence. Aortic incompetence, with supravalvular dilatation,[10] interatrial septal defect and dissection of the aorta are easily detected. Type II DeBakey dissections are the most frequent but occasional examples of type I and III are encountered. Dissection may take place during pregnancy.[170] Coarctation also occurs.[62] The abnormalities are particularly likely to be encountered in young persons. A floppy mitral valve (see p. 347) is often present and severe mitral valve regurgitation may result. Secondary renovascular hypertension has been identified.[102]

Microscopically, the appearances are those of 'mucoid change' of the aorta (cystic medial necrosis).[67] The musculoelastic wall of the aorta is replaced, to a varying extent, by an excess of faintly basophilic proteoglycan (Fig. 9.8). The patchy aggregation of this material partly replaces the normal aortic media and leads to weakness, dilatation and a tendency to rupture or dissection; in the young males in whom this is most common, associated atherosclerotic changes are slight. The matrix material of the aortic wall is not cystic and cystic medial necrosis is, indeed, an inaccurate term: 'mucoid medial change' is more precise. Similar changes have been discovered in the mitral valve. The microscopic lesions in the other varieties of congenital cardiac disease which accompany the Marfan syndrome are poorly defined: there is no evidence that 'mucoid change' in the extracellular matrix accounts for cardiovascular abnormalities such as patent interatrial septum.

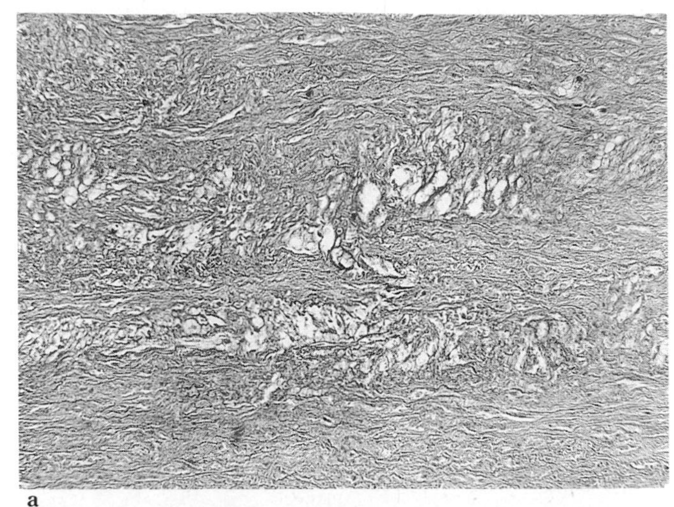
a

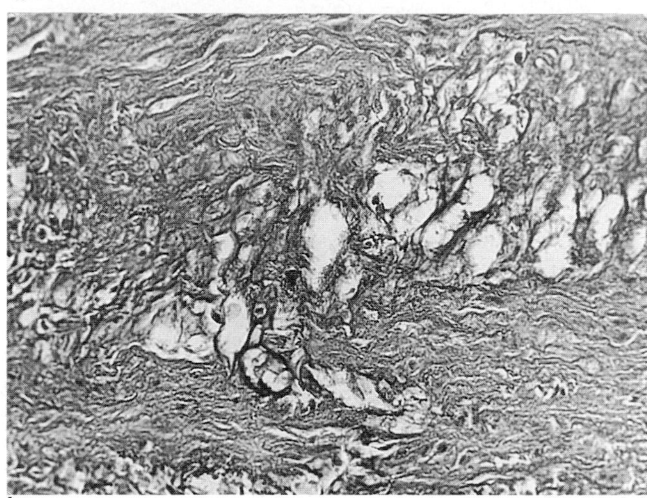
b

Fig. 9.8 Marfan's syndrome.
a. Aortic media: extensive replacement of elastic, collagen and smooth muscle by metachromatic extracellular matrix material, rich in proteoglycan which has been dissolved from formalin-fixed, paraffin section during preparation, leaving empty spaces. **b.** Higher-power view of same field. This is a site susceptible to dissection. (Toluidine blue **a.** × 62; **b.** × 157.)

In affected connective tissue, dissociated bundles of dense, angular collagen fibrils have been seen with an excessive amount of diffuse matrix material.[68] At the light microscopic level, the elastic fibrils in these instances appear 'moth-eaten'. Transmission electron microscopy shows small diameter collagen fibrils; the elastic material is fragmented.[103]

Speculation about the possible cause of the Marfan syndrome is rife. Systematic light and transmission electron microscopic studies of human Marfan aortas, of the aortas of chicks fed a copper-deficient diet, and of rats fed β-aminopropionitrile,[186] have confirmed that the structural alterations in all three are similar (see p. 413).[217] This similarity has suggested that the inherited fault in the Marfan syndrome may be one of connective tissue metabolism, related to deficiency of a substance necessary for normal aortic intercellular matrix formation, and that the mode of genetic action may be via an enzyme controlling the synthesis of collagen molecular species.

There have been advances in therapy, particularly in those cases where there is aortic regurgitation and dilatation. Propranolol diminishes the force of ventricular ejection and slows the progress of aortic dilatation, postponing the need for surgery. In young girls, oestradiol and progesterone treatment may control the ultimate stature and diminish the hazards of scoliosis.

Osteogenesis imperfecta

This is an uncommon class of hereditary disorder of connective tissue.[29,69,150,221,222] There are four main groups (Table 9.5). In the most frequent examples, abnormal bone fragility, blue sclerotics and nerve deafness attract attention. The terms osteogenesis imperfecta and dentinogenesis imperfecta are applied to the disorders of bone and tooth, respectively. Males and females are affected in equal numbers. Survival to adult life is not rare although severely affected cases may die in infancy or even *in utero*. The distinction between the clinical groups is not clearcut.

History

The classical description of osteogenesis imperfecta is that of Ekman.[61] In a doctoral thesis he outlined the clinical characteristics of the adult condition and gave examples of families in which the disease was transmitted as an hereditary disorder. The pathological anatomy of osteogenesis imperfecta was described by Lobstein[138] and Vrolik;[244] the histological appearances were studied by Stilling[227] and analysed by Looser.[139] In 1912, Adair-Deighton reported the occurrence of deafness in the Ekman–Lobstein syndrome;[3] Ruttin[200] recognized that this was due to otosclerosis. Similar histological changes in the teeth, with hyperostosis of periodontal bone, were identified by Bauer.[9] Osteogenesis imperfecta was considered to be a generalized hypoplasia of the mesenchyme.[57] Subsequently, biochemical and genetic studies led to the identification of the four main clinical types. Classically, the fully developed adult syndrome (osteogenesis imperfecta tarda) is termed Ekman–Lobstein disease, the infantile syndrome (osteogenesis imperfecta congenita) Vrolik's disease; but at least 28 different names are listed.[29]

Clinical presentation

There is a wide range of signs and symptoms. In the most severe form of the congenital disease, numerous fractures are present at birth and can be detected radiologically *in utero*. The affected child may be stillborn. Examination shows the entire range of clinical signs with ear and eye involvement and congenital anomalies of the viscera. In the delayed or adult form of the disease, there may be only slight evidence of a systemic connective tissue disorder: blue sclerae may be found in isolation or fractures occur first at a late age. When the skeleton is affected in a surviving patient, there is dwarfism, due partly to fractures involving epiphyseal growth zones. Cranial Wormian bones may be abnormally numerous and osteoporosis leads to vertebral collapse and kyphoscoliosis. The bones of the face (formed in cartilage) remain normal but the bones of the skull (formed in membrane) are defective; the skull becomes large and thin with an overhanging occiput.

Clinically, deafness resembling otosclerosis is frequent; it can be relieved by a fenestration operation. Cardiovascular anomalies have been described but, with the exception of peripheral arterial calcification, are probably not related to the underlying connective tissue disorder. It is unlikely that the cardiovascular changes are due to a defect in vascular connective tissue. The importance of endocrine factors in the behaviour of the disease is shown by its improvement after puberty and by the occasional onset of deafness in pregnancy.

Inheritance

Some forms are transmitted as autosomal-dominant, others as autosomal-recessive characteristics.[230]

Underlying the four main groups of osteogenesis imperfecta, there are a very large number of recognized genetic defects, all changes in the base sequences of genes coding for collagen molecules.[220,230a] These defects can be caused by deletions, insertions or mutations but the majority are deletions.

Biochemical changes

The reader is referred to reports by Wenstrup, Hunter and Byers,[250] to the Proceedings of the Third International Conference on Osteogenesis Imperfecta[230] and to p. 182.

Table 9.5

Types of osteogenesis imperfecta and the underlying inherited defects[24,187,188]

Type	Examples of the underlying genetic defects	Biochemical consequences	Inheritance	Clinical features
I. Dominant with blue sclerae				
I A	Abnormal gene function of alleles for pro-α chains	Half normal amount of type I collagen	AD	Blue sclerae, bone fragility, hearing impairment, onset of fractures after birth. No dentinogenesis imperfecta
B	Shortened pro-α_2(I) chains	Abnormal collagen fibrils		As A, but accompanied by dentinogenesis imperfecta
C	Unidentified			Similar to A but greater severity with reduced stature and deformity
II. Lethal, perinatal				
II A	Shortened pro-α(I) chains	Unstable triple helix; increased synthesis of pro α_1(III)	AD, rarely Ar	
B	Shortened pro-α_2(I) chains and abnormal function of alleles for pro-α_2(I) chains	Uncertain	Ar (AD)	Intrauterine growth retardation, rhizomelic limb shortening, limb bowing. Dark blue sclerae. Broad, 'concertina-like' femurs, continuous beading of ribs, minimal calvarial mineralization
III. Progressive, deforming				
III A	Mutations altering the structure of the C-propeptides of pro-α_2(I) chains	Synthesis of pro α_1(I) trimers		Severe disability. Fractures at birth, progressive limb deformation and kyphoscoliosis; normal sclerae and hearing. Dentinogenesis imperfecta common
B	Mutations altering the structure of the C-propeptides of pro-α_1(I) or pro-α_2(I) chains	Increased mannose in C-propeptide and decreased solubility of type I procollagen	Ar	
C	Unidentified			

Table 9.5 – continued

IV. Dominant with normal sclerae				
IV A	Shortened pro-α_1(I)-chains with 500 bp deletion in this collagen gene	Alterations in triple helix resembling those in osteogenesis imperfecta type I	AD	These are mild disorders that are much less frequent than osteogenesis imperfecta IA. Fragile bones, blue sclerae, normal hearing. No dentinogenesis imperfecta
B	Shortened pro-α_2(I) chains	Resistance to procollagen N-proteinase and persistence of pN procollagen		B. Similar to IVA, but accompanied by dentinogenesis imperfecta. Variable deformity

AD = Autosomal dominant; Ar = Autosomal recessive.

Pathological changes

The pathological lesions are beautifully illustrated by Bullough and Vigorita.[23] The disorders are of mesenchymal development and are attributable to the inherited defects of collagen.

In the less severe, adult disease, bone matrix is insufficiently and abnormally formed (Fig. 9.9); the consequence is frequently a variety of osteoporosis. The abnormal collagen fibres fail to assume their adult configuration. The matrix of cartilage is not involved. The bone which is ultimately formed is fully mineralized in the normal way. No abnormality in calcification is present but, because matrix is deficient in amount and quality, the total calcium content is inevitably reduced. Periosteal bone formation is defective (Fig. 9.10) and bone trabeculae originating at the periosteal surfaces are sparse, slender and incomplete. Osteoblasts are few but the collagenous skeleton of lamellar bone appear normal. Where bone is formed in cartilage, the definitive trabeculae are distorted and slender; microfractures may be seen. In the bones of the face, these changes are less easy to identify.

In the more severe, infantile forms of the disease many fractures are found at birth; they are both gross and microscopic and involve cartilage which has undergone provisional calcification during epiphyseal growth. At this early point of development, abnormal bone matrix is laid down. The matrix material resembles osteoid in being basophilic, argyrophilic and PAS-positive. The birefringent collagen fibres are disorganized, an observation confirmed by X-ray diffraction and microradiography. However, in the less severe adult syndromes, the bone displays no collagen abnormality recognizable with polarized light.

Bone sections from an early, severe case show small, poorly formed trabeculae, occasionally with microfractures. There are increasing numbers of transmission electron microscopic studies;[89] they demonstrate small collagen fibrils and fragmented elastic fibres.[103] In the milder cases which survive longer, microscopic bone changes may be very slight. Eventually, the deformities are likely to lead to secondary osteoarthrosis. In every instance of the disease the primary histological picture may be complicated by the occurrence of varying numbers of fractures which begin to heal normally, sometimes with excess callus, but which culminate in further deformity as remodelling progresses.

Cutis laxa

This very rare syndrome, in which loose skin folds hang from the face, around the eyes and in other dependent parts, was originally confused with the Ehlers–Danlos syndrome (see p. 334) and, in early reports, with neurofibromatosis.[153] Cutis laxa is distinct from pseudoxanthoma elasticum (see p. 355). The abnormality is present from birth.

Inheritance

The fundamental defect[25] is believed to lie in the cross-linking of collagen and of elastic fibrils. There are two principal disease forms;[12] a dominant variety which involves only the dermis, and a recessive form in which other generalized abnormalities include emphysema; pulmonary infection and cor pulmonale; and gastrointestinal diverticulitis, hernia and genitourinary tract diverticula.

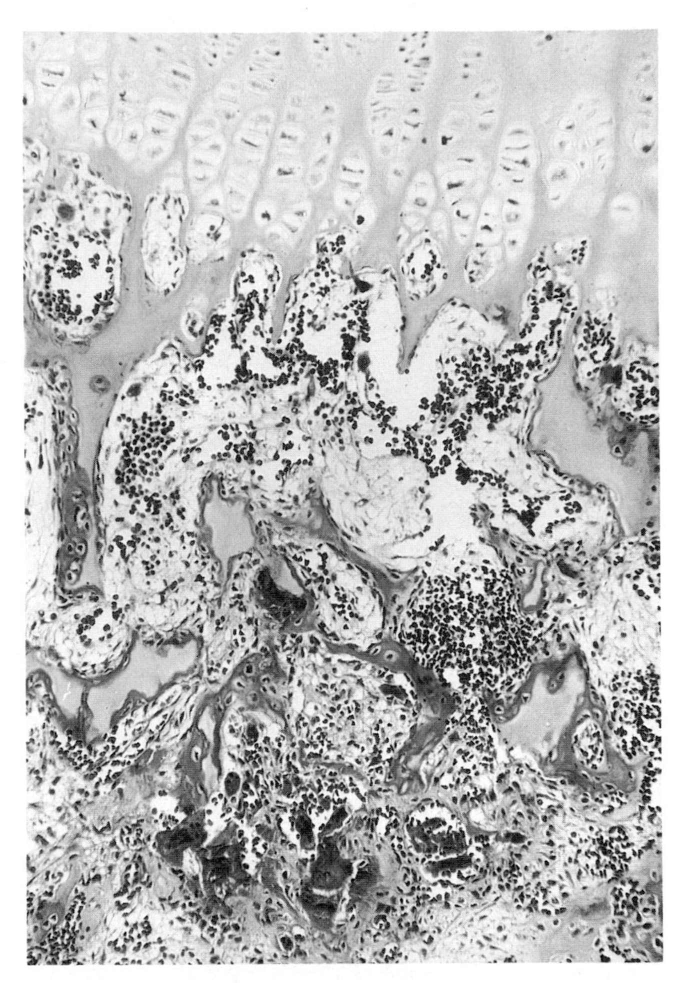

Fig. 9.9 Osteogenesis imperfecta.
Defective bone formation. Regular proximal edge of epiphyseal cartilage is preserved but metaphyseal bone trabeculae are poorly organized, small, slender and, in some instances, fractured. (HE × 140.)

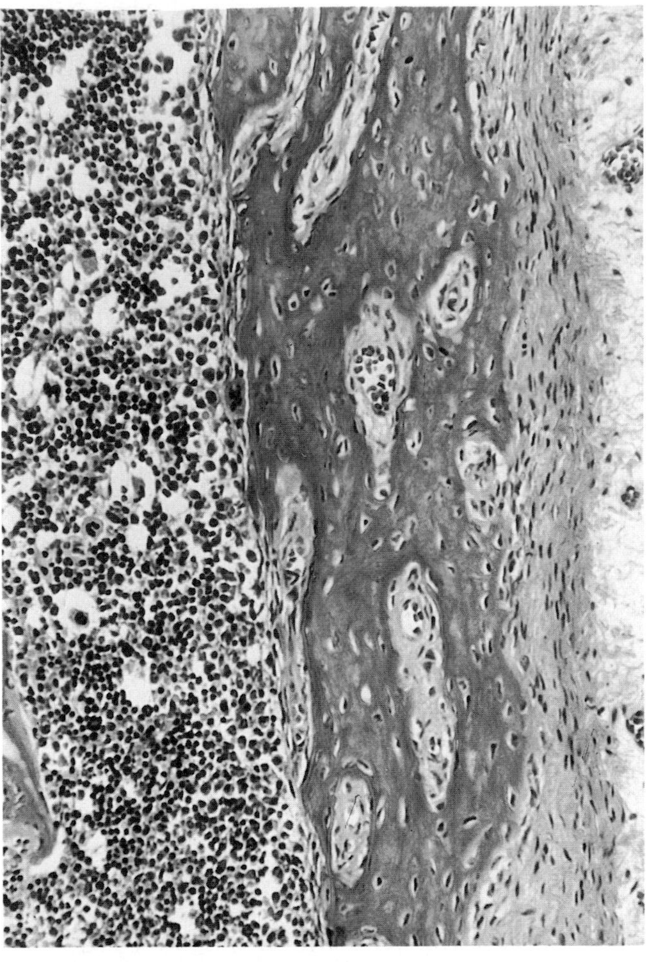

Fig. 9.10 Osteogenesis imperfecta.
Disorganized periosteal bone formation. The compact bone of the cortex is less well defined than normal, the periosteal margin irregular. (HE × 170.)

Biochemical changes

Although the biochemical abnormalities determined by these heritable anomalies are not understood, a family has been described[25] in which low serum copper and caeruloplasmin levels accompanied skin, skeletal and genitourinary disease inherited as X-linked characteristics. There was a reduction in lysyl oxidase activity in the skin only; however, the enzyme activity was also reduced in the medium retrieved after fibroblast culture.

Pathological changes

The pathological changes are incompletely understood. Holbrook and Byers[103] have tabulated the ultrastructural characteristics (Figs 9.11–9.13). In the dominant and recessive autosomal forms, collagen was found to be unaltered but there was a decreased quantity of elastic material and small 'dust-like' elastic fibres. There were collagen fibrils of varying diameter, with collagen and elastic fibrils intermingled. The elastic fibres, globular and deficient in amount but reticular in appearance, contained excessive numbers of tangled microfilaments. Holbrook and Byers[103] showed that non-specific collagen 'flowers' (p. 340) might be present, or that collagen was decreased in amount with a small fibril diameter and fragmented elastic. In the rare X-linked form of cutis laxa, there was also abnormal collagen fibrillogenesis. In the reticular dermis, collagen fibrils were larger than normal, with a mean diameter of 135 nm: their shape was normal but the packing tight. Light and

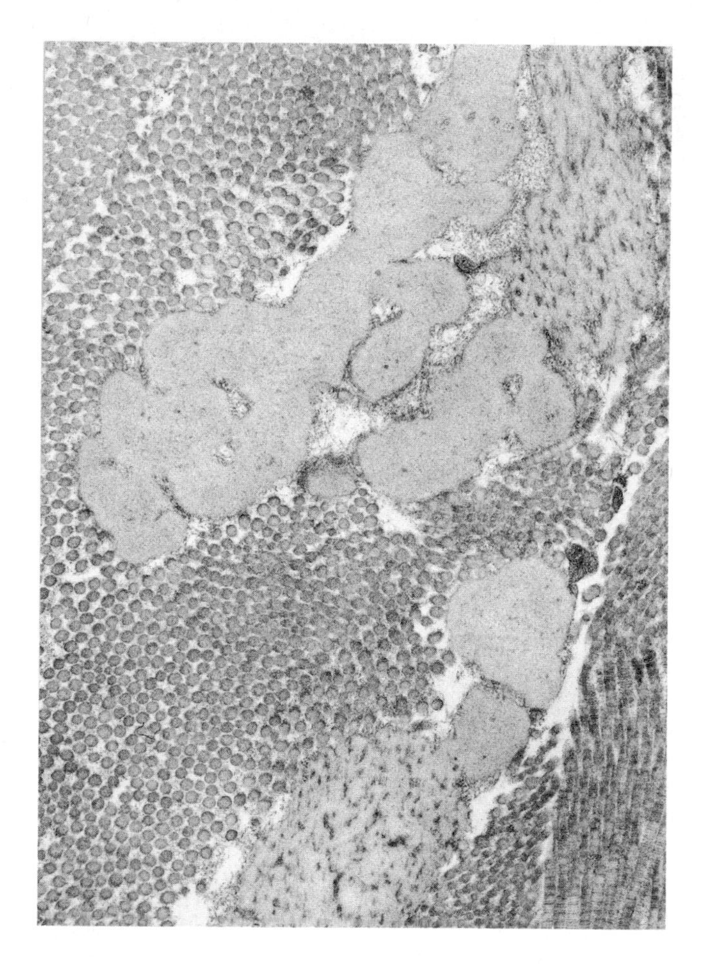

Fig. 9.11 Cutis laxa.
Abnormal elastic material within bundle of collagen in dermis of patient with cutis laxa. (Transmission electron microscopy × 14 140.) Courtesy of Dr Karen A Holbrook and the Editor, *Journal of Investigative Dermatology*.[103]

transmission electron microscopic studies suggested that elastic fibres were normal with intact associated microfibrils. The remaining epidermal and dermal structures were not recognizably altered.

Floppy mitral valve

Expansion of the cusp region of the mitral valve, with elongation of the chordae tendineae, permits prolapse of the valve into the left atrium during ventricular systole. Mitral regurgitation results. There are many underlying causes.[154] The valve abnormality is encountered in heritable disorders of collagen such as type I osteogenesis

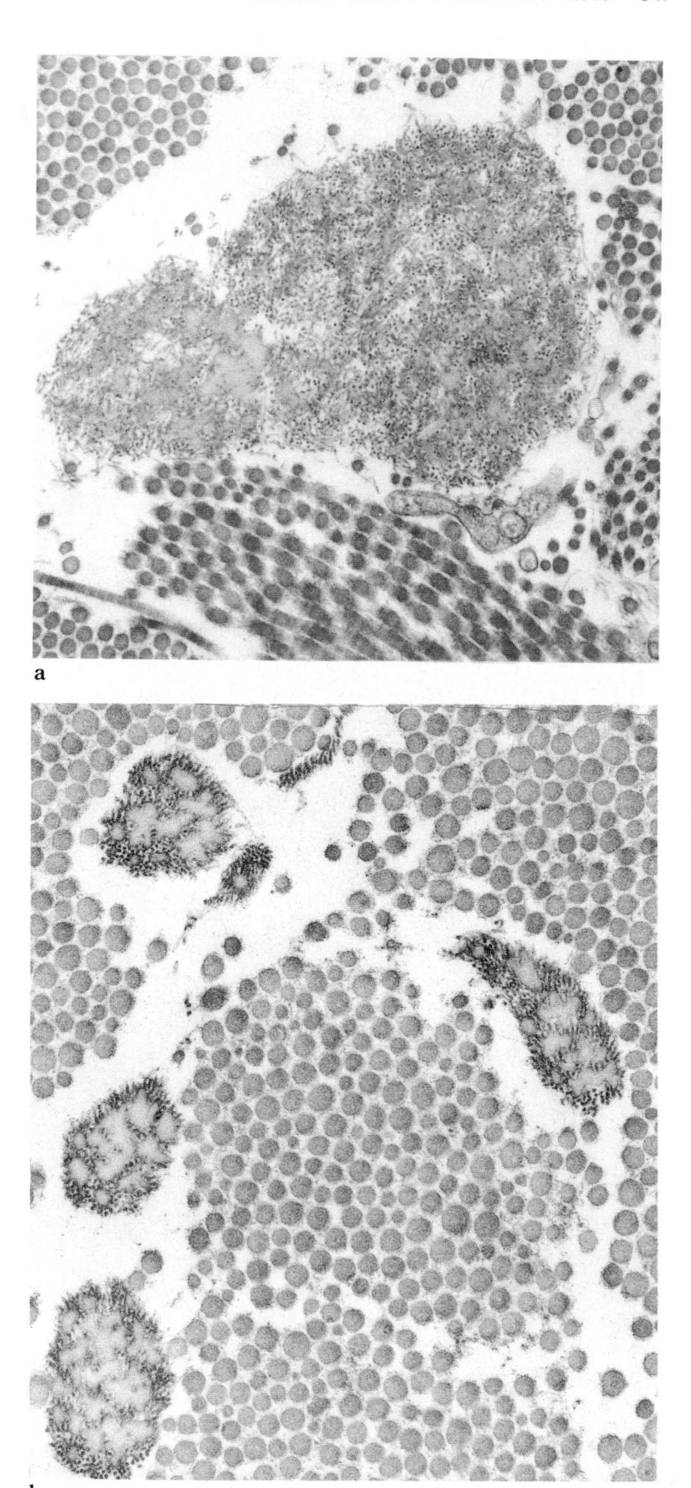

Fig. 9.12 Cutis laxa.
Examples of abnormal elastic material in dermis of patients with dominant and recessive forms of the disorder. **a.** Microfibrils appear tangled; lack of normal elastin. (Transmission electron microscopy × 18 000.) **b.** Small, globular islands of elastin associated with microfibrillar glycoprotein. (Transmission electron microscopy × 25 000.) Courtesy of Dr Karen A Holbrook and the Editor, *Journal of Investigative Dermatology*.[103]

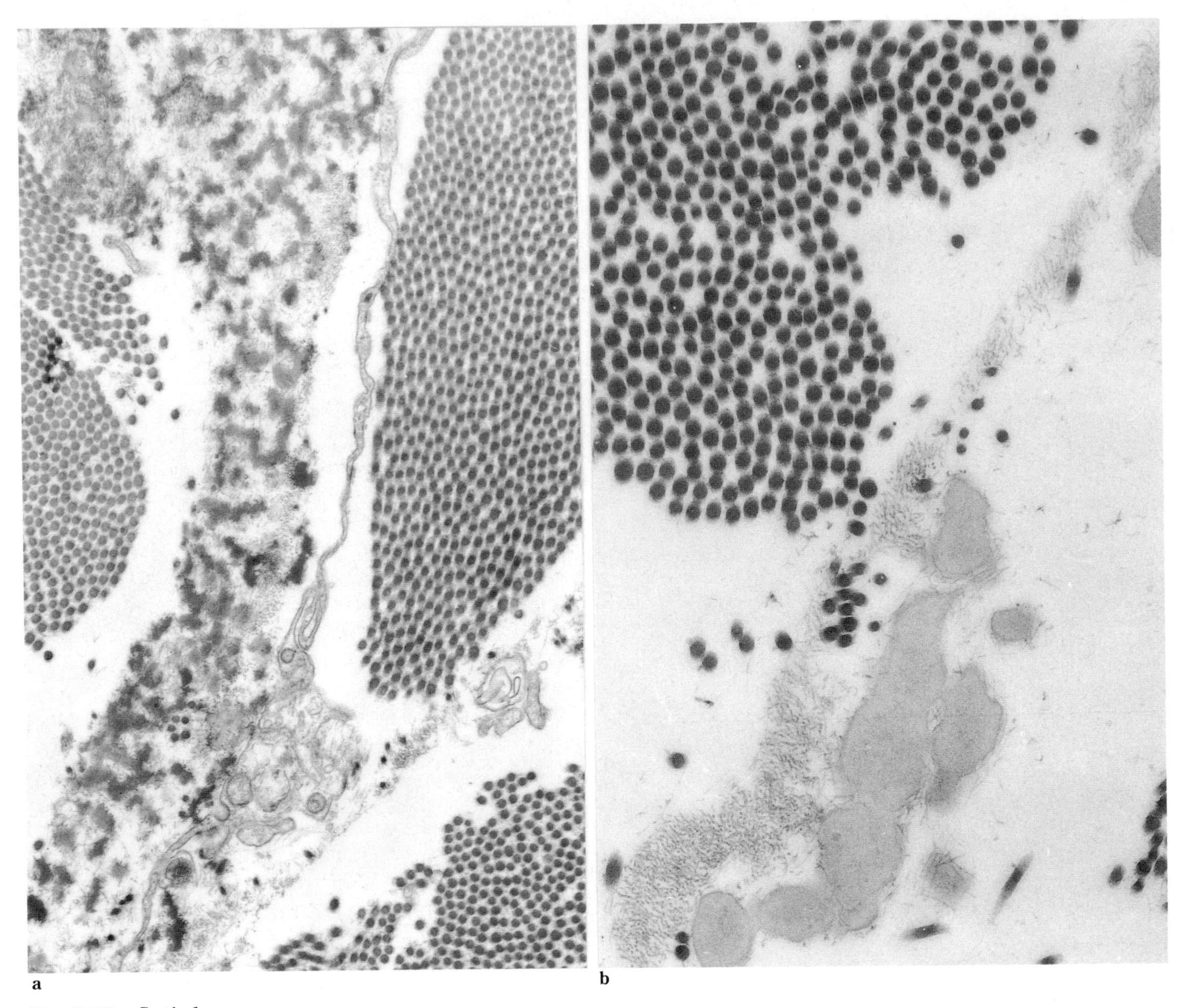

Fig. 9.13 Cutis laxa.
a. Abnormal dermal elastic material. Reticular appearance of elastic fibre structure due small relative proportion of elastin. (Transmission electron microscopy × 12 000.) **b.** Apparent segregation of elastin deposition from elastic microfibrils. (Transmission electron microscopy × 10 500.) Courtesy of Dr Karen A Holbrook and the Editor, *Journal of Investigative Dermatology.*[103]

imperfecta, type VII Ehlers–Danlos syndrome and the Marfan syndrome but it is also frequently found as an isolated anomaly.

Biochemical changes

The histological structure and biochemical composition of individual floppy valves is very variable.[137] An apparent accumulation of proteoglycan can be demonstrated in a central fibrous core. The excess proteoglycan is not thought to be specific but to reflect a change in cardiac collagen organization.[252] Accompanying the high proteoglycan content is a large increase in proteoglycan extractability[137] attributable to defective collagen integrity and accompanied by a rise in the elastin content. In pooled, isolated valves, the disorder of collagen synthesis and maturation has been supported by the demonstration of excess procollagen and single-chain collagen,[16] in analogy with inefficient cleavage of aminopropeptides of the kind encountered in type VII Ehlers–Danlos syndrome and in dermatosparaxis. A deficiency of procollagen peptidase has been suggested. However, proof of a collagen defect is difficult to obtain and it appears that an observed change in the ratios of type I and III collagen reflects the extent of secondary fibrosis rather than the primary, underlying biochemical defect.[137]

Secondary diseases of fibrous structural proteins

Many of the rare heritable defects in connective tissue synthesis, maturation and function are secondary results of defects in Mendelian recessive inheritance. A E Garrod first postulated the concept of the 'Inborn Errors of Metabolism'.[79,80] Since then, biochemical investigation has helped to elucidate not only the true nature of alkaptonuria (see p. 351) but also of numerous other secondary disorders affecting connective tissues,[30] ranging from homocystinuria to the glycosaminoglycan storage diseases ('mucopolysaccharidoses') and Menkes' kinky hair syndrome.[225] In the present account, clinical, morphological and histological changes are emphasized.

Although the 'one-gene, one-enzyme' concept explains many metabolic diseases inherited as recessive characteristics, there is a growing awareness of defects in enzyme regulation as opposed to enzyme activity.[76] Classical inborn errors of metabolism exert their effects at the active site of an enzyme or carrier protein, or the enzymes are simply absent; defects in enzyme regulation, by contrast, are operative at sites modulating enzyme or carrier protein actions. Defective enzyme regulation may also help to explain metabolic disorders such as gout (see p. 380) and lipomatosis.

Homocystinuria

Among the numerous inherited disorders of amino acid metabolism is a group of at least nine abnormalities of the transsulphuration pathway. Of the nine, one, homocystinuria, is particularly likely to prejudice connective tissue.

Clinical presentation

The lesions that result affect the ocular, skeletal, central nervous and vascular tissues.[171,171a] There are characteristic pathological changes; they culminate in clinical disorders which may resemble the Marfan and Ehlers–Danlos syndromes.[154] Joint tightness rather than laxity,[214] kyphoscoliosis, pigeon breast, genu valgum, osteoporosis and ectopia lentis are described. Mental retardation and psychiatric disorder result from a lack of cystathionine in the central nervous system.[30] Vascular thrombosis and dilatation affect medium-sized arteries and veins. The thrombotic tendency may be a result of increased platelet adhesiveness but this is disputed. Other explanations for thrombosis include vascular intimal defects, changes in mural glycosaminoglycan structure and activation of Hageman factor (p. 285).

Inheritance

Homocystinuria is the designation for a biochemical anomaly, not a disease entity, and, in fact, the excessive excretion of urinary homocystine, homocystinuria, may be the consequence of not less than four different genetic faults.[171]

The most frequent of the rare causes of homocystinuria is the defect, inherited as an autosomal recessive characteristic, of cystathionine β-synthase. It is distinguishable from the other categories of homocystinuria by the accumulation of homocysteine and methionine in the plasma. An 'overflow aminoaciduria' follows.[30]

Biochemical changes

Cystathionine β-synthase is absent from or deficient in cells such as fibroblasts (Fig. 9.14).[171,171a] Cultures of these cells can be used to confirm a diagnosis suspected on the basis of a positive urinary cyanide nitroprusside test for homocystine. Heterozygous individuals can be distinguished from homozygous by the recognition of intermediate levels of enzyme activity in fibroblasts, lymphocytes or hepatocytes. As a result of the enzyme defect, the cross-linking of collagen fibres is impaired and there is enhanced platelet aggregatability. Features in common with osteolathyrism (see p. 413) are emphasized. It has indeed been established that homocysteine interferes with the formation of the intermolecular cross-links required to stabilize the collagen macromolecular network via its reversible binding to aldehyde functional groups (see p. 175).[115] Collagen stability is decreased and it appears likely that these changes contribute to the connective tissue anomalies.[215]

Pathological changes

There is widespread arterial disease.[146,147,171,171a] Episodes of thrombosis occur; the thrombi become organized. The arterial media is thinned but intimal fibromuscular hyperplasia develops. Although transverse striations have

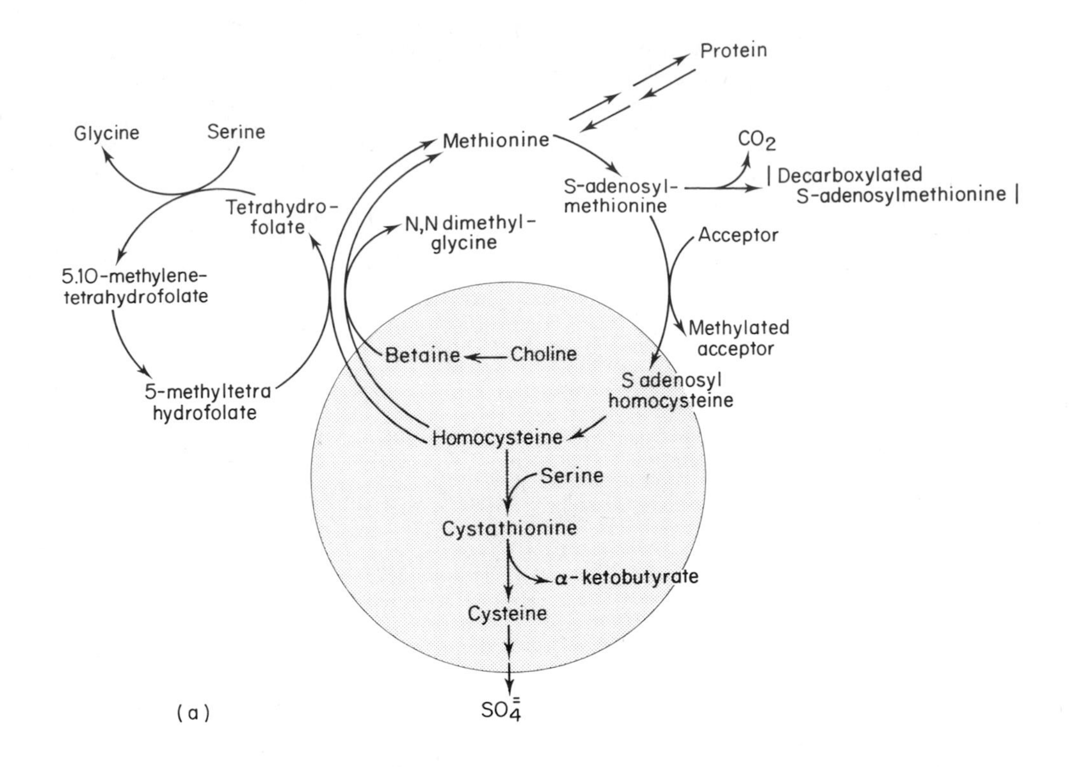

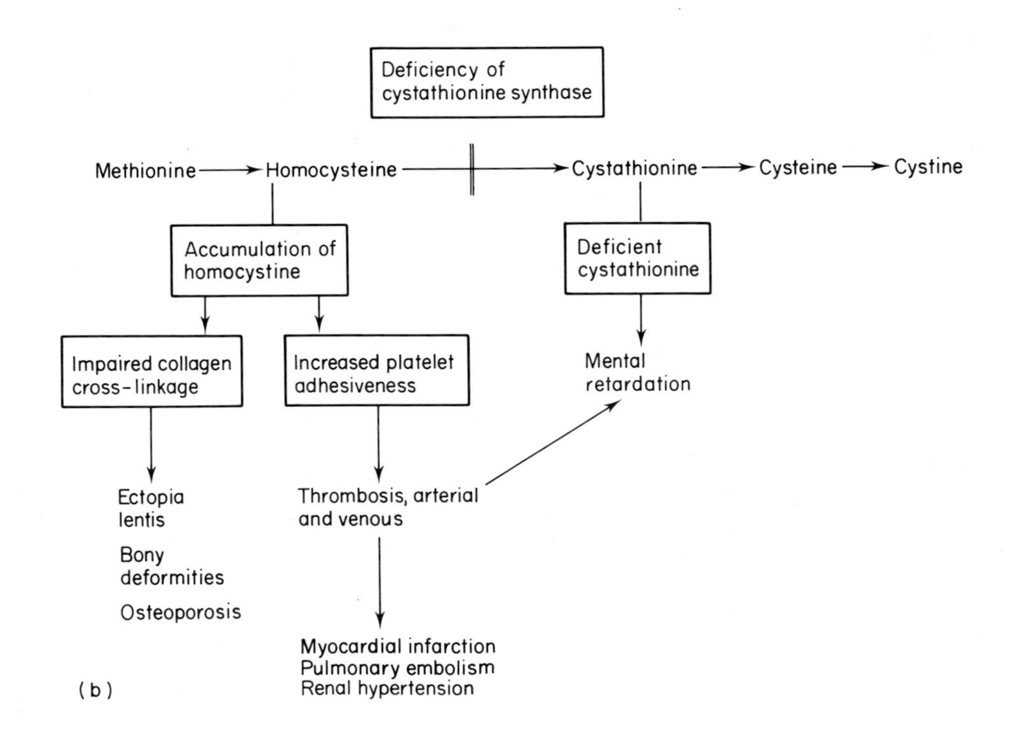

Fig. 9.14 Homocystinuria.
a. Diagram shows consequences of heritable deficiency of cystathione synthase modified from Mudd and Levy.[171] **b.** Sequence of pathogenetic events in homocystinuria. Reproduced from McKusick.[154]

been found in the intima of the descending aorta, and abdominal aortic thrombosis has been seen, aortic aneurysmal dilatation is not a feature of this syndrome and dissection does not result. There are conspicuous ocular changes.[100,192] The brain undergoes focal necrosis, spongiosus and subsequent gliosis. Pulmonary emboli may occur, and together with portal venous thrombosis, they are important causes of death. The liver is often enlarged and fat-laden. The skeleton is porotic and there is a tendency to fracture. Kyphoscoliosis, chest deformity and excessively long limb bones combine with disordered ossification to give a Marfanoid habitus. There are associated abnormalities of the shoulders, wrist joints and metacarpals.

Alkaptonuria and hereditary ochronosis

Alkaptonuria is a rare, heritable disorder of amino acid metabolism attributable to a deficiency of homogentisic acid oxidase.[123,124,207] The accumulation within connective tissues of homogentisic acid precedes the formation of a deep brown-grey pigment, the condition of ochronosis. Arthropathy is a common consequence.

Ochronotic pigment has been identified in Egyptian mummies[132] and the disease is of great antiquity. The review of ochronosis by O'Brien, La Du and Bunim[175] drew attention to the description by Scribonius[209] of a schoolboy who passed dark urine. An early account of this condition was that of Marcet (1822–23).[159] Virchow[242] noted the pathological features and used the term 'ochros' (sallow) to describe the complexion. Welkow and Baumann[249] identified the urinary pigment as an oxidation product of homogentisic acid. Alkaptonuria led Garrod[78] to the use of the term 'Inborn errors of metabolism'.[79,80]

There are relatively few pathological reports[136,169] but an increasing interest in transmission,[75] and scanning electron microscopic[74] and electron probe X-ray microanalytical[145] studies. Comprehensive reviews have been made.[111,119,123,124,168] Six hundred and four cases of this disorder had been described up to 1964,[20] and I have had the opportunity to study the pathological changes in four examples.

Clinical presentation

In the early years of life, alkaptonuria is symptomless. The disorder may be identified in infancy when the urine becomes discoloured on exposure to air. However, the recognition of ochronosis is often coincidental: in one case, degenerative intervertebral disc disease with calcification and secondary knee osteoarthrosis drew attention to alkaptonuria in a 50-year-old man who was found to have small brown-grey pigment deposits in the lobes of the ears and in the sclerae; in another, the disease was diagnosed when hip arthroplasty was required after a fall had caused femoral neck fracture; and in a third, the diagnosis was first made when black costal cartilages were seen *postmortem*.

Inheritance

There is a deficiency of homogentisic acid oxidase inherited as an autosomal recessive characteristic. The formation of fumarylacetoacetic acid from homogentisic acid is impaired because this enzyme is absent or is defective in the nature or degree of its activity.

Biochemical changes

In alkaptonuria, homogentisic acid, a normal intermediary in the principal pathway for the degradation of the amino acids tyrosine and phenylalanine, is excreted in great excess in the urine as a consequence of a defect in the enzymic mechanism which normally accounts for complete homogentisic acid catabolism (Figs 9.15 and 9.16). Homogentisic acid, given in known amounts to patients with alkaptonuria, is quantitatively excreted[125,126] whereas the acid is completely metabolized by normal persons. The fault lies in an inability to break the benzene ring with –OH groups in the 2:5 position. Additional evidence exists, however, to show that the high urinary concentration of homogentisic acid may be attributed in part to its formation by the kidney or to active renal secretion. The renal threshold, as in normal persons, is low.

The blackening of the exposed urine on oxidation gives rise to the term alkaptonuria. Oxidation products, often yellow-brown or black, are slowly liberated from homogentisic acid. Deposited in connective tissues, they cause organ pigmentation and premature degenerative changes. The chemical explanation for the predilection of ochronotic pigment for skeletal connective tissue is not fully understood; it does not appear that there is a simple interaction with proteoglycan.

Inheritance

Because of the blackening of the urine, the disease, equally common in males and in females, is recognizable in infancy. This is particularly likely when a sib with the disorder has already been diagnosed. The disease is transmitted by a single autosomal gene, usually recessive; in the parents of an affected sib there is commonly a history of consanguin-

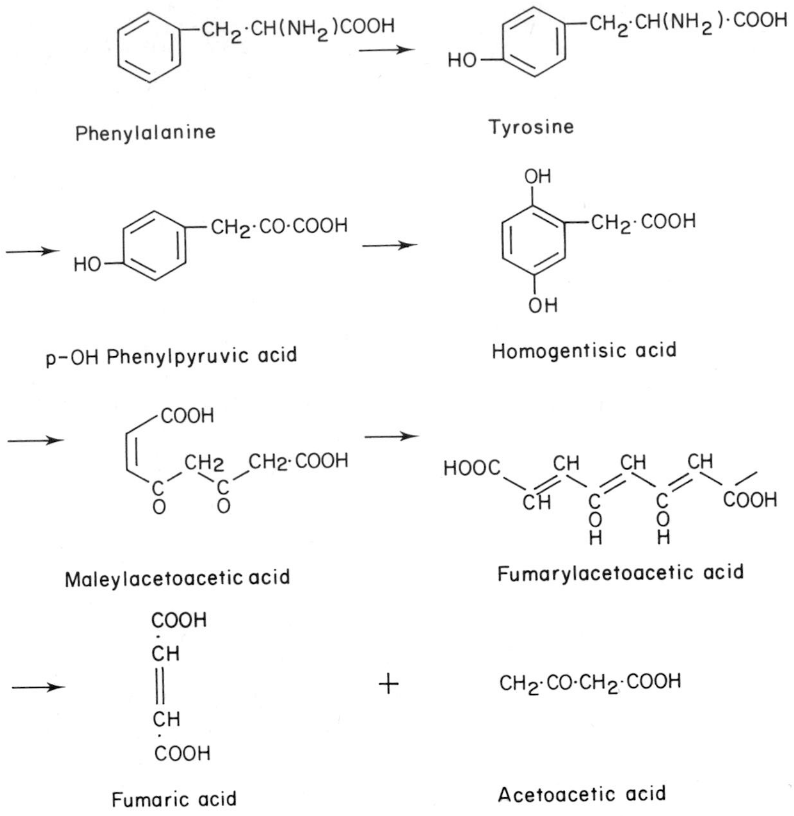

$CH_2 \cdot CH(NH_2)COOH$ → HO– $CH_2 \cdot CH(NH_2) \cdot COOH$

Phenylalanine Tyrosine

→ HO– $CH_2 \cdot CO \cdot COOH$ → OH, $CH_2 \cdot COOH$, OH

p-OH Phenylpyruvic acid Homogentisic acid

→ COOH, CH_2, $CH_2 \cdot COOH$, C, C, O, O → HOOC, CH, CH, CH, CH, C, C, COOH, O, O, H, H

Maleylacetoacetic acid Fumarylacetoacetic acid

→ COOH · CH ‖ CH · COOH + $CH_2 \cdot CO \cdot CH_2 \cdot COOH$

Fumaric acid Acetoacetic acid

Fig. 9.15 Alkaptonuria.
Synthesis and degradation of homogentisic acid.

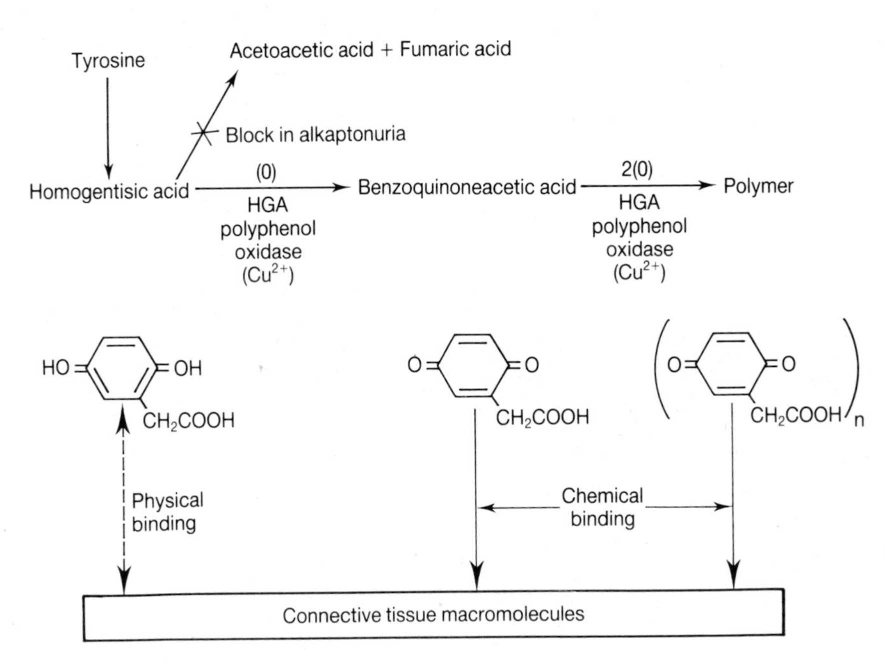

Fig. 9.16 Ochronosis.
Scheme suggested for ochronotic pigment formation as result of heritable deficiency of homogentisic acid oxidase. After La Du.[126]

ity. Less often, a dominant gene with incomplete penetrance may be involved.

Pathological changes

For many years, the patient is symptom-free. Ultimately, ochronosis develops, the result of the cumulative deposition of the oxidation products of homogentisic acid.

The accumulation of homogentisic acid or of a pigmented derivative of this compound is limited principally to the connective tissues (Figs 9.17; 9.18). The manner of this accumulation has been suggested by work with guinea pigs into which homogentisic acid was injected.[125] It was shown that the main connective tissue components have a low affinity for homogentisic acid but that benzoquinonacetic acid, an oxidation product of homogentisic acid, was strongly bound to connective tissue polymers. The suggestion was made therefore that the pigment deposited in the connective tissue of affected persons was a homogentisic acid derivative of this type. Not only are the normal connective tissues of the body involved but so are others such as fibrous scar and granulation tissue, formed in the course of coexistent diseases.

The distribution of ochronotic pigment has been detailed by Lichtenstein and Kaplan.[136] It can be distinguished histochemically from melanin. Microscopically, pigment lies in the walls of arteries; in cartilage; in skeletal connective tissue, such as tendons and ligaments; in other collagenous connective tissue, such as cardiac valve cusps or healed myocardial infarcts; in excretory organs such as sweat glands and the kidney; in endocrine glands such as the islets of Langerhans and the pituitary; and in the reticuloendothelial system.

There are frequent secondary, degenerative changes as a consequence of pigment accumulation. The arcuate arteries of the kidneys stand out as black rings against the cut surface of the organ. Where medium-sized arteries such as those of the coronary tree are the site of atherosclerosis,

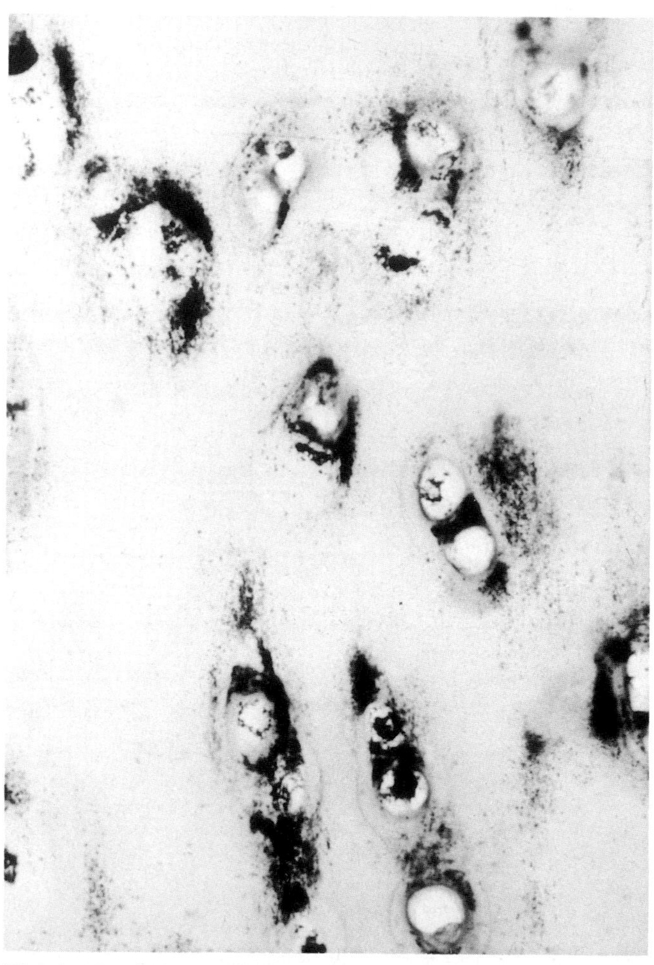

Fig. 9.17 Ochronosis.
Accumulation of ochronotic pigment in pericellular matrix principally within chondrons. (HE × 510.)

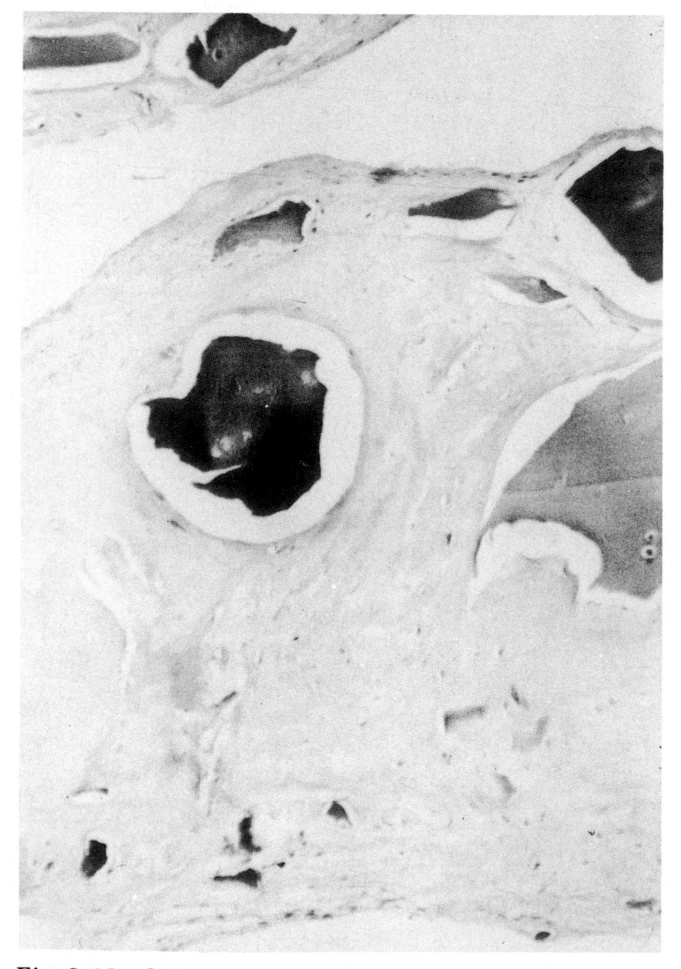

Fig. 9.18 Ochronosis.
Fragments of cartilage impregnated with ochronotic pigment lie within the synovium of a knee joint. (HE × 94.)

Fig. 9.19 Ochronosis.
Vertebral column, bisected in the coronal plane to show the grey-black pigmentation of the intervertebral discs. Early osteophytosis is seen.

the fibrous plaques become pigmented. In the heart, mottled pigmentation of the pericardium may accompany grey-black discoloration of the pulmonary and aortic valve commissures. Secondary calcification leads to distortion and stenosis of the aortic valve. A generalized blue-grey pigmentation of the cartilages of the larynx, trachea and bronchi is recognized but the organs of the alimentary system show little change. The brain is spared but the dura mater and cerebral blood vessels may be discoloured. However, where gliosis occurs within the central nervous tissue, a localized grey-black pigmentation is seen.

The most conspicuous systemic changes are those in the skeletal system. Intervertebral discs and the associated ligaments become uniformly pigmented (Fig. 9.19). A remarkable degeneration of the discs results in their progressive and sometimes total loss, the condition of ochronotic spondylosis. Marginal disc calcification accentuates the reduced intervertebral margins radiologically. The ultimate result is a deforming kyphoscoliosis in which rigidity follows disc disorganization. Destruction of the nearby vertebrae may follow. Degenerative cartilage loss in the apophyseal joints is an associated change.[127]

The intervertebral disc calcium aggregation is in the form of calcium pyrophosphate dihydrate deposition (see p. 393) and it is possible that the accumulation of this crystalline material, rather than ochronotic pigment, is the principal explanation both for the characteristic intervertebral disc degeneration and for the secondary osteoarthrosis of synovial joints.[145] Ochronotic joint disease, like the corresponding disorders found in haemochromatosis (see p. 369) and Wilson's disease (see p. 369), may indeed be a calcium pyrophosphate dihydrate arthropathy.

In diarthrodial joints such as the knee (Fig. 9.20) and hip[127] in which the effects of ochronosis are most frequent, pigment deposited in articular cartilage results in a change in the physical characteristics of this tissue which becomes brittle and fragments. The underlying bone is exposed, as in severe osteoarthrosis, and particles ('shards') of pigmented cartilage are seen by light microscopy to accumulate within nearby synovial villi (Fig. 9.18) causing a low-grade secondary synovitis and progressive disability, ochronotic arthritis.[81,168] Transmission electron microscopic studies show that the fragments of ochronotic cartilage in the synovium are accompanied by much smaller 'microshards' consisting of large collagen fibres encrusted with ochronotic pigment.[75] The collagen fibres are ingested by synovial macrophages and held within phagolysosomes.

Light and transmission electron microscopic observations demonstrate the accumulation of ochronotic pigment around chondrocytes within the pericellular matrix.[145,169] The presence of cytoplasmic pigment is, however, sometimes recognizable. Mohr[169] reported that the quantity of pigment increased with increasing distance from the cells. The suggestion was made that ochronotic pigment binds to

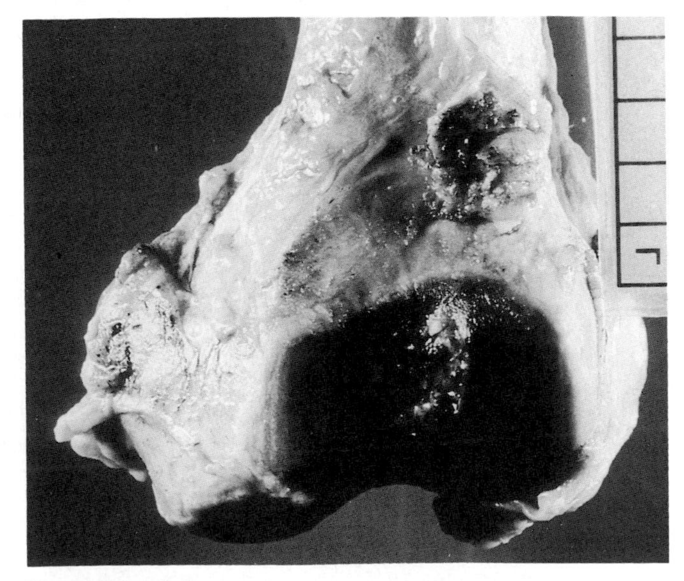

Fig. 9.20 Ochronosis.
Deep pigmentation of the hyaline articular cartilage of the patellar surface of the femur. Courtesy of Professor B Vernon-Roberts.

proteoglycan, causing decreased cartilage strength and consequential fragmentation. Cracks presumably propagate from foci of impaired strength. Subsequent release of pigmented cartilage fragments leads to their phagocytosis by synovial cells; incomplete degradation then permits residual ochronotic bodies to remain within the synoviocytes.

Exogenous ochronosis

Ochronosis also exists in an acquired form as a consequence of the accidental or deliberate ingestion of phenol or phenolic compounds over prolonged periods. The prolonged ingestion or inhalation of phenols may result in the deposition in the connective tissues of a pigment, 2,5-dihydroxyphenylacetic acid, which, like homogentisic acid, is also excreted in the urine. In persons frequently exposed to phenol, as were surgeons in the Listerian era of antisepsis or patients to whom carbolic acid was applied as dressings, there might therefore develop a progressive, focal pigmentation of ear cartilages, articular cartilage and sclerae similar to that encountered in hereditary ochronosis.[11] Picric acid (trinitrophenol) may have the same effects. Fortunately, the deposition of these exogenous derivatives of phenol ceases when intake is arrested and pigmentation slowly regresses. Black costal cartilages are one consequence of the prolonged administration of levodopa for Parkinson's disease.[33a]

Menkes' kinky (steely) hair syndrome

(see p. 412)

In this rare X-linked recessive disorder of copper metabolism, neurological, cardiac and vascular changes develop together with skeletal deformity and curious, kinky hair that is poorly keratinized and weakly pigmented.[155,165] The bone changes resemble those of scurvy (see p. 414). The abnormal hair, sparse and wiry, is described histologically as 'pili torti': copper-dependent enzymes catalyse keratinization and the keratin of Menkes' syndrome is abnormally formed. The affected males rarely survive more than three years.

There is an inherited defect in copper absorption[42] but the molecular defect is still not fully understood.[118] There is copper deficiency associated with overproduction of a copper-binding metallothionein. Copper accumulates selectively in many cell types including cells cultured *in vitro*; the accumulation is caused by the ability of low copper concentrations to induce metallothionein in RNA synthesis in Menkes' but not in normal fibroblasts.[134] The affected males can be identified by demonstrating the increased uptake of copper-64 in fibroblast culture. Muscle copper measurements can be used in confirmatory tests and this method has been applied to fibroblasts from female relatives in *postmortem* diagnosis of an affected boy.[235,236] The copper-dependent enzyme lysyl oxidase invokes the initial phase of cross-linking in elastogenesis and influences collagen cross-linking. It is not surprising, therefore, that when collagen is deficient, there should be weakness of the large, elastic arteries which tend to rupture as they do in lathyrism (see p. 413), in the Marfan syndrome (see p. 340), and in natural and experimental copper deficiency. The vessels form abnormally. Tortuous blood vessels on the surface of the brain and meningovertebral angiodysplasia are almost diagnostic of Menkes' syndrome.[4]

In experimental studies, it was found that sheep grazed on soil deficient in copper developed signs of the defective action of copper-dependent enzymes. This evidence led to the demonstration[41–43] that there was reduced transport of copper through the intestinal wall and probably across all cell walls. Mutated Brindled (mottled) mice inherit a defect similar to that of humans with Menkes' disease.[51] The distribution of copper-64 can be followed by whole-body autoradiography.

Pseudoxanthoma elasticum (Grönblad–Strandberg syndrome)

Pseudoxanthoma elasticum (elastorrhexis) is an inherited disorder of the elastic and vascular tissues of the skin, eyes and cardiovascular system, more common in females than in males.[6] Scattered, pale-yellow, intradermal plaques, chorioretinitis and gastrointestinal haemorrhage are associated with arterial disease. Grönblad[88] and Strandberg[228] described the skin and retinal lesions; the vascular changes were recognized by Touraine.[237] An early account of the pathological findings was given by Balzer[8] although the non-xanthomatous nature of the lesions was only made clear by Darier.[45] That the disease may be primarily one of collagen fibres rather than of elastic material was suggested by Hannay[92] and by Tunbridge *et al.*[238] The condition was reviewed by McKusick.[152]

Clinical presentation

Yellow papules and plaques are present along the skin folds of the neck, groin, axillae and popliteal fossae (Fig. 9.21); they may remain undiagnosed for many years or attract attention for cosmetic reasons only. The disease is, however, not confined to the skin, and careful examination reveals

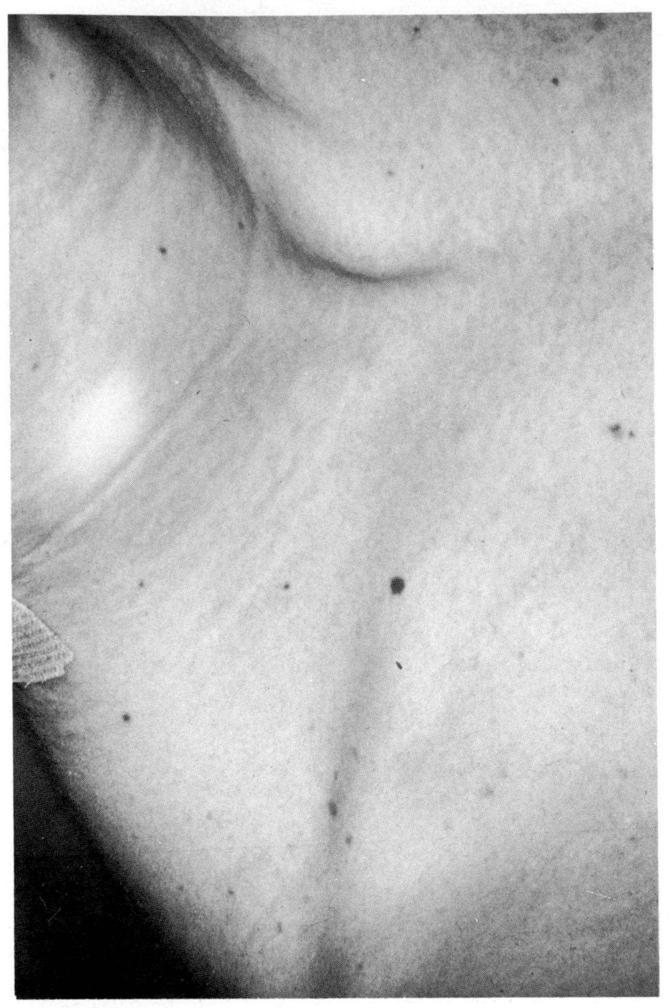

Fig. 9.21 Pseudoxanthoma elasticum.
Yellow papules and plaques lie in the planes of the skin folds of the axilla.

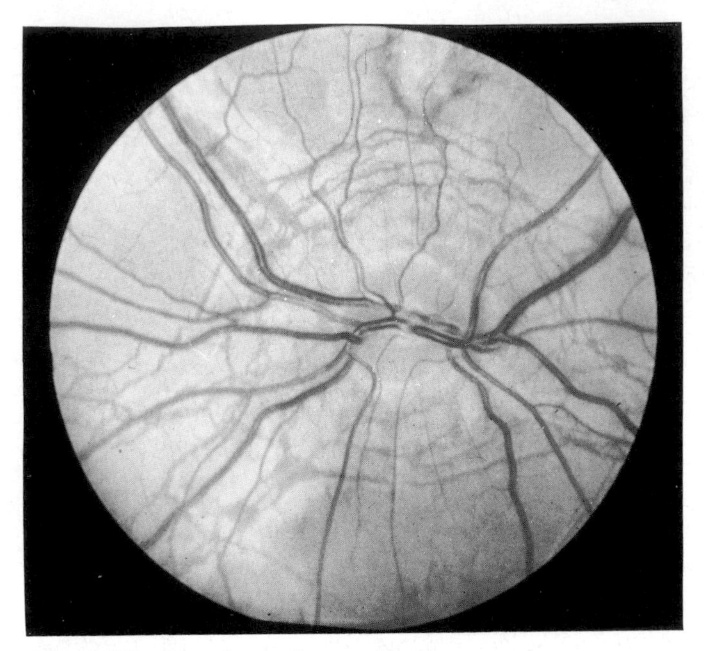

Fig. 9.22 Pseudoxanthoma elasticum.
Angioid streaks can be seen on the surface of the retina.

frequent systemic manifestations. Angioid streaks (breaks in Bruch's membrane) (Fig. 9.22) are a hallmark and chorioretinitis or retinal haemorrhage cause impairment of vision. Vascular involvement is shown by peripheral circulatory insufficiency and by unexplained arterial calcification. Hypertension is common and leads to cerebral haemorrhage and myocardial infarction. Patients with this form of the disease have a comparatively short life expectancy. Intermittent claudication, angina pectoris, peripheral vascular calcification, mental disturbances, epilepsy, thyrotoxicosis and diabetes mellitus are encountered. In one case, diabetes insipidus was present. Telangiectases are visible at the edges of lesions and gastrointestinal bleeding is the result of similar changes in the gut.[229] Articular and connective tissue disorders are uncommon[215] but joint hypermobility has been described. Pseudoxanthoma elasticum may coexist with progressive systemic sclerosis, Paget's disease of bone, tumoral calcinosis (p. 405) and hyperphosphataemia.

Inheritance

There are two forms of autosomal dominant disorder and two of autosomal recessive.[6,152]

Pathological changes

The yellow colour of the skin nodules is due to the presence in the dermis of large numbers of small, basophilic, curved and apparently fragmented fibres (Fig. 9.23) which stain for elastic material and which often bear calcium deposits (Fig. 9.24). Transmission electron microscopy demonstrates the origin of calcification in the centre of the elastic material, spreading to the periphery.[15] Multinucleated giant cells accumulate at affected sites. The nature of the fragmented substance is disputed. It is variously suggested that it is degenerate elastic tissue or that it is fundamentally collagenous, a physical alteration to the dermal collagen resulting in positive staining for elastic material. The consensus of opinion appears to be against the view that there is a primary disorder of elastic tissue, although elastic fibres are involved and atrophic. The affected collagen fibres retain their characteristic 64 nm striation in spite of susceptibility to digestion by elastase; there may be an unusually high proportion of type III collagen. Although the affected zones appear to be separated from the epidermis by a band of normal collagen and elastic fibrils, thick frozen sections

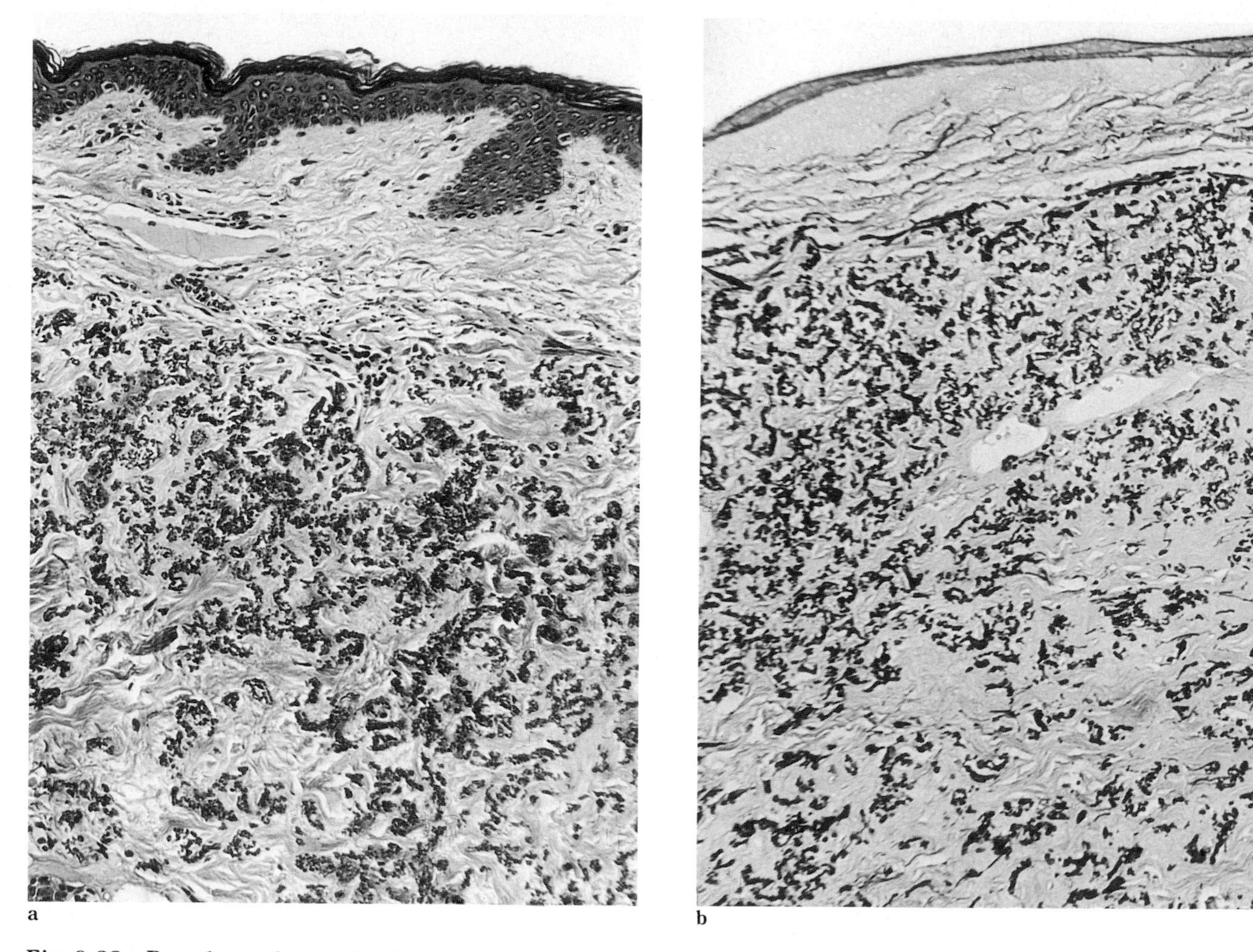

Fig. 9.23 Pseudoxanthoma elasticum.
a. Numerous small fragments of elastic material are scattered throughout the deeper dermis. (HE × 100.) **b.** Compare with Fig. 9.23a. (Weigert's elastic-van Gieson × 100.)

show that the fine elastic fibres which normally form a network in the dermal papillae are widened, increased in number and more twisted than normal. Fibres thought to be collagen and which do not retain an elastic stain are not affected in this way. The blood vessels of the skin are intact.

In the deeper parts of the dermis, circumscribed zones of more complete fibre disruption are present. Here, there is granular, basophilic material resembling tissue debris and staining patchily with Weigert's method. The fibres display axial splitting analogous with the arrangement of collagen in tendons. No fat is present and these brokendown elastic fibres accumulate copper. The reason for the deposition of different amounts of copper in different sites is not yet obvious: the deposits are not associated with age or the duration of the illness.

One explanation for the connective tissue changes may be inappropriate protease activation.[85] Because of the age of onset of the disease or the age at which clinical presentation occurs, it is also possible that the disorder may be due to a change in fibre maturation. Finally, there may be altered proportions of type I and type III collagens.

The viscera display lesions similar to those of the skin. The affected retinal and gastrointestinal arteries show changes in elastic material structure.[229] These vascular alterations account for the resulting calcification and the subsequent angina, claudication and hypertension[70] and their complications. Endomyocardial biopsy permits the diagnosis of vasculopathy of the small coronary arterial branches.[189] The cardiomyopathy that results from these lesions may lead to myocardial ischaemia and cause sudden unexpected death.[106] In the limb vessels, vascular changes account for ulceration and gangrene. Rupture of smaller visceral arteries results from dystrophic calcification.

Skin biopsy facilitates diagnosis. Elastic-like fibres impregnated with calcium are present in the mid- and lower dermis. Transmission electron microscopy confirms this

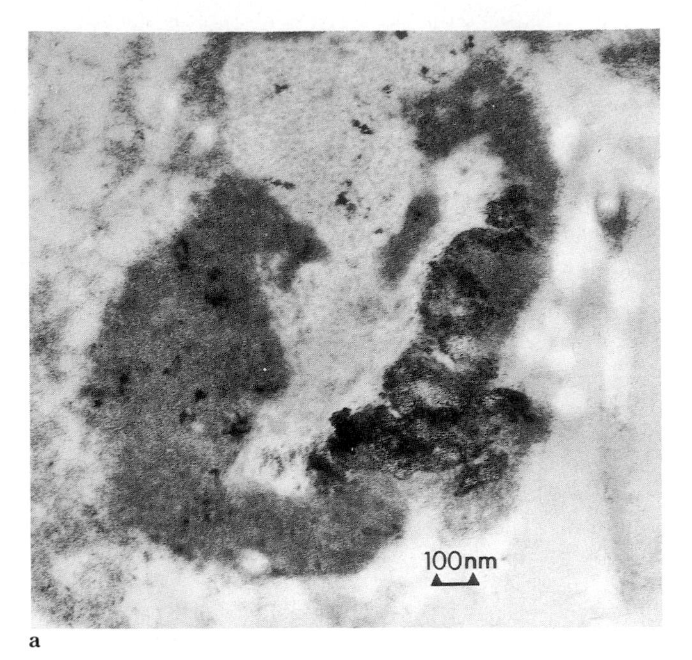

a

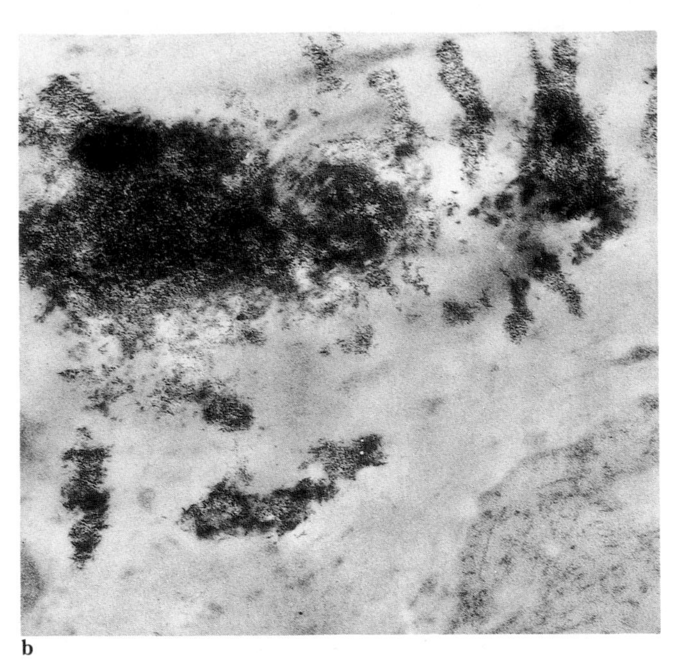

b

Fig. 9.24 Pseudoxanthoma elasticum.
a. Disorganized elastic material in dermis. Calcification is minimal. **b.** Dense, focal dystrophic calcification of abnormal elastic material. (Transmission electron microscopy × 49 500.) Courtesy of Dr Philip McKee.

association[15,198] (Fig. 9.24). The presence of calcium and of phosphate in the centre of the elastic material is demonstrable by X-ray microanalysis. As calcification extends in each island of elastic material, the central core of elastin is degraded and ultimately lost. Comparable findings have been described in the skin of clinically normal heterozygotes.

Disorders of macromolecular degradation: lysosomal storage diseases

This is a group of about 30 rare, inherited diseases each attributable to the deficiency of one or more acid lysosomal hydrolases.[117,133,157] Each disease is the result of the accumulation of a substrate normally degraded by the deficient enzyme(s); their biochemical basis[83] is outlined in Chapter 4. They are distinguishable from lysosomal transport defects such as cystinosis and Salla disease.[58]

The characteristic features of a lysosomal storage disease are: 1. the presence of a membrane-bound nondegraded substrate in the form of cellular inclusions displaying acid phosphatase activity, a lysosomal marker enzyme; 2. multiple organ involvement; 3. the storage of heterogeneous material;[101,117] and 4. an unremitting, progressive course. Although in a typical disorder, such as the Hürler syndrome, there may be many systemic lesions, here the skeletal and connective tissue system changes are of principal interest.

The lysosomal storage diseases are tabulated according to the main, incompletely degraded substrate that accumulates.[83] They are:

Mucopolysaccharidoses: the stored substrate is mainly glycosaminoglycan. There are at least 18 varieties (Table 4.3, p. 203).
Mucolipidoses: the stored substrates include both glycosaminoglycan and sphingolipids (not mucolipids). There are at least five varieties.
Glycoproteinoses: the stored substrates are mannose-rich oligosaccharides or fucose-containing glycolipids, glycoproteins and keratan sulphate.
Sphingolipidoses: the accumulated substrates include GM_1 ganglioside, sialofeturin and keratan sulphate or GM_2 ganglioside.
Others: these include mucosulphatidosis, Farber disease and teleophysic dwarfism.[117]

The greater part of this account is devoted to the mucopolysaccharidoses since connective tissue and skeletal lesions are commonplace. A brief reference is made to one example of the mucolipidoses (I-cell disease) and one glycoproteinosis (mannosidosis). The reader is referred to reviews of the other lysosomal storage diseases.[83]

Mucopolysaccharidoses

The prototypes of the mucopolysaccharidoses were described by Hunter[107] and Hürler.[108] Dwarfing, a grotesque bodily configuration, mental retardation, deafness, cardiac anomalies and hepatomegaly are among the signs noted in more severely disabled children; survival into adult life is likely only in those less affected. There is a great range of phenotypic expression, however, and as techniques of genetic and biochemical analysis have advanced, the complexity of the syndrome has become apparent (Table 4.3, p. 204).

Clinical presentation

The autosomal recessive class of cases described by Hürler[108] may be taken to exemplify the mucopolysaccharidoses which mainly affect connective and skeletal tissues (Fig. 9.25). The newborn infant appears normal although there is already a cellular accumulation of glycosaminoglycan. The disease is progressive and without remission. The accumulation of glycosaminoglycan disorganizes to varying degrees the skeleton, the central nervous system, the heart, eyes, skin and reticuloendothelial system; the descriptive term 'dysostosis multiplex' is used. After a period of accelerated growth, internal hydrocephalus and neurosensory deafness are found to accompany mental deterioration. Corneal clouding, repetitive lung infections and restricted limb movement become evident. Hirsutism develops together with a coarseness of appearance, enlargement of the spleen and liver, umbilical hernia and macroglossia. Growth and development diminish, and death between the ages of 4 and 8 years results from cardiac failure, internal hydrocephalus or pulmonary infection. Disorders of the cardiovascular system include cardiomyopathy and conduction defects. Hepatosplenomegaly and bone marrow foam cells are indices of substrate accumulation; hepatic fibrosis may result.[179] The skin is coarse and thickened and there is gingival hyperplasia resembling that produced by antiepileptic agents.[117]

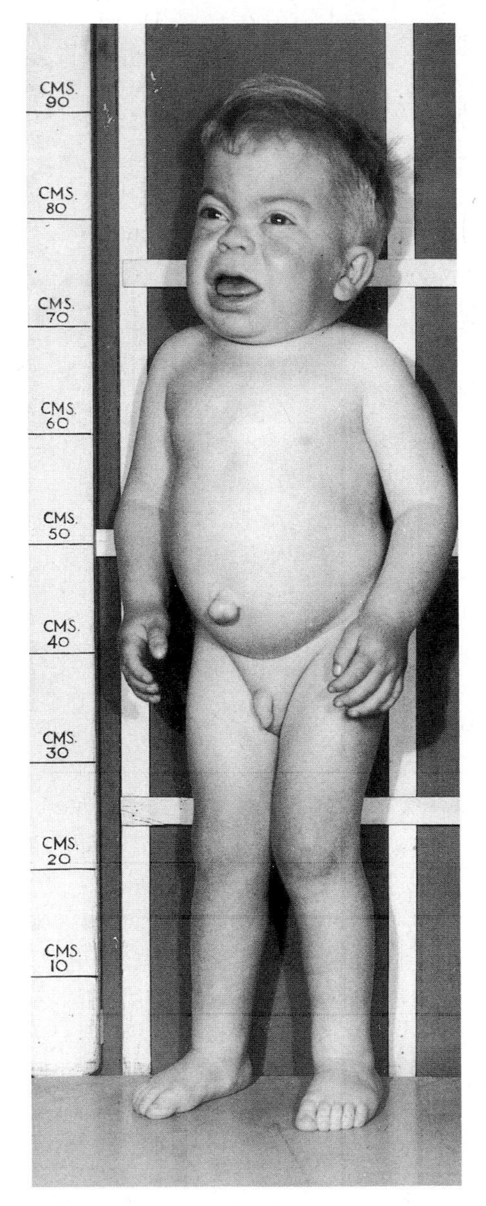

Fig. 9.25 Mucopolysaccharidosis.
Child with bodily habitus of the Hunter–Hürler syndrome.

Inheritance

There is not usually a family history. However, all lysosomal storage diseases are inherited recessively, and, very occasionally, inbreeding is revealed. Genetic analyses demonstrate heterogeneity that may be manifested through three mechanisms: first, closely similar phenotypes may be the result of different enzymic deficiencies; second, different phenotypes may be consequences of a single enzymic defect: a single incompletely degraded macromolecule accumulates in different amounts, determining different clinical manifestations; and third, deficiency of a single enzyme may influence the catalytic degradation of different substrates so that again, distinct clinical phenotypes result.[173,173a]

Biochemical changes

(Table 4.3, p. 203)

The chemical basis of the mucopolysaccharidoses[173,173a] is described in Chapter 4 (p. 203).

Pathological changes

In principle, the macromolecules that accumulate in the mucopolysaccharidoses can aggregate in all organs and in all tissues. The deposits may contain heparan sulphate, dermatan sulphate or chondroitin sulphate. Routine histochemical processing is inadequate to demonstrate the microscopic foci but the use of a technique modified from Dorling[55] together with transmission electron microscopic studies of glutaraldehyde-fixed sections, allows the extent of intra- and extracellular changes to be assessed.[35] Much of the glycosaminoglycan is lost from routinely processed transmission electron microscopic sections but the technique, particularly valuable in fetal studies, has greater sensitivity than light microscopy for the recognition of intracellular material. The sites of accumulation of glycosaminoglycan in Hürler-like disease (MPS IH) have been fully described. They include the hepatocytes,[27] splenic cells (Fig. 9.26), bone marrow cells,[181] thyroid follicular cells, Sertoli cells, neurones,[50] pituitary chromophobe cells and the fibroblast-like connective tissue cells of the meninges (Fig. 9.27), cardiac valves, and periosteum.[172] Transmission electron microscopy confirms that, where clear vacuolated cells are recognized, the glycosaminoglycan is contained within lysosomes.[128] The liver has been studied[27,28] and the delicate, membrane-bound vacuolar contents shown to have histochemical properties similar to those of heparan sulphate.[120]

In the connective tissues, glycosaminoglycan deposits have been observed both as electron-lucent and as electron-dense inclusions in the vacuolated keratocytes of the cornea as well as in the extracellular spaces, isolating large collagen fibres.[231] Less is known of the bone deposits (Fig. 9.28). In the iliac crest, and in the costal cartilages in Hürler-like disease (MPS IH) chondrocytes contain many single, membrane-bound vacuoles[174] (Fig. 9.29). They have delicate filaments or electron-opaque flakes or granules among an electron-lucent background material. Other vacuoles have lipid inclusions: electron-dense lipid particles are attached to the inner aspect of the single vacuolar membrane and there is a delicate fibrillar intercellular matrix.[216]

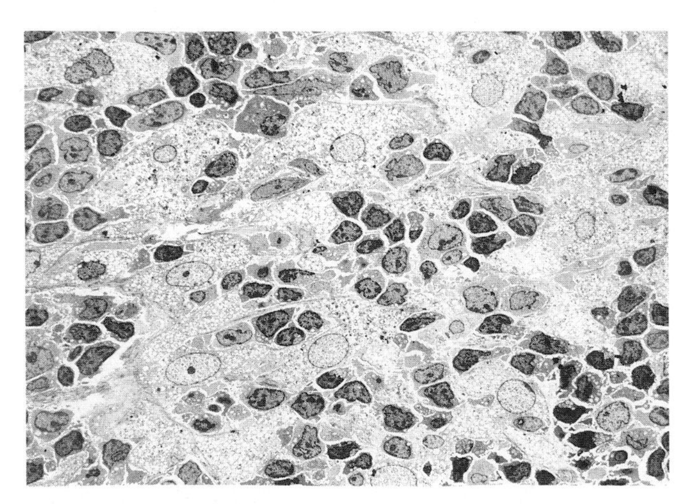

Fig. 9.26 Mucopolysaccharidosis.
Numerous vacuolated cells in nasopharyngeal tissue from child aged 4.3 years. (Transmission electron microscopy × 785.) Courtesy of Dr Julie Crow and the Editor, *Journal of Clinical Pathology*.[35]

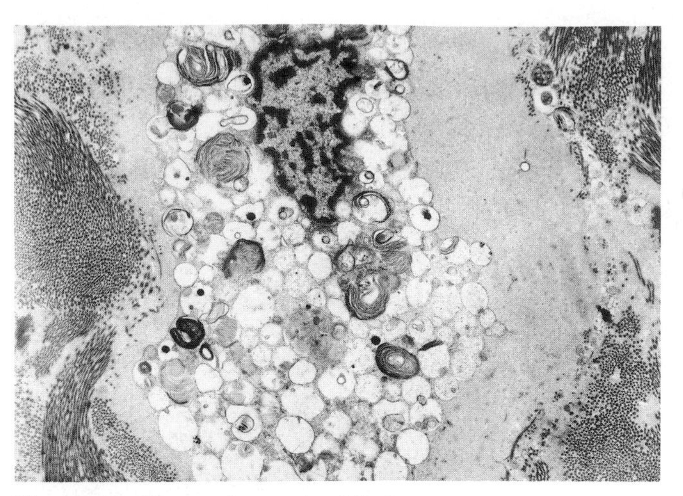

Fig. 9.27 Mucopolysaccharidosis.
Cell in dura mater of 21-year-old man with Scheie disease. Cytoplasm contains clear vacuoles and many membranous, myelin structures resembling the Zebra bodies found in Hürler disease. (Transmission electron microscopy × 3950.) Courtesy of Dr Julie Crow and the Editor, *Journal of Clinical Pathology*.[35]

The fibroblasts and macrophages of the dermis in Hürler-like disease (MPS IH) are often vacuolated. As in the viscera and cartilage, the vacuoles[46] are single membrane-bounded. Schwann cells contain similar vacuoles and a single large vacuole is present in many epidermal cells. The stored material in the lysosomes of keratinocytes, fibroblasts, macrophages and Schwann cells is accompanied by laminated, membranous cytoplasmic structures resembling gangliosides.[13]

The cardiac abnormalities have been extensively reviewed.[176] In 27 male and 13 female patients aged 4 months to 37 years there was usually cardiomegaly. Of the atrioventricular valves, the mitral was most frequently affected, the tricuspid and aortic less often and the pulmonary seldom. In a typical valve, there was nodular thickening and an array of zones of dilatation at the free margin giving a scalloped appearance. Mitral stenosis resulted in some, and regurgitation in others. The chordae tendineae and their insertions were short, thick and chondroid. Thickening of the left atrial and left ventricular myocardium was occasionally accompanied by diffuse endocardial fibroelastosis.

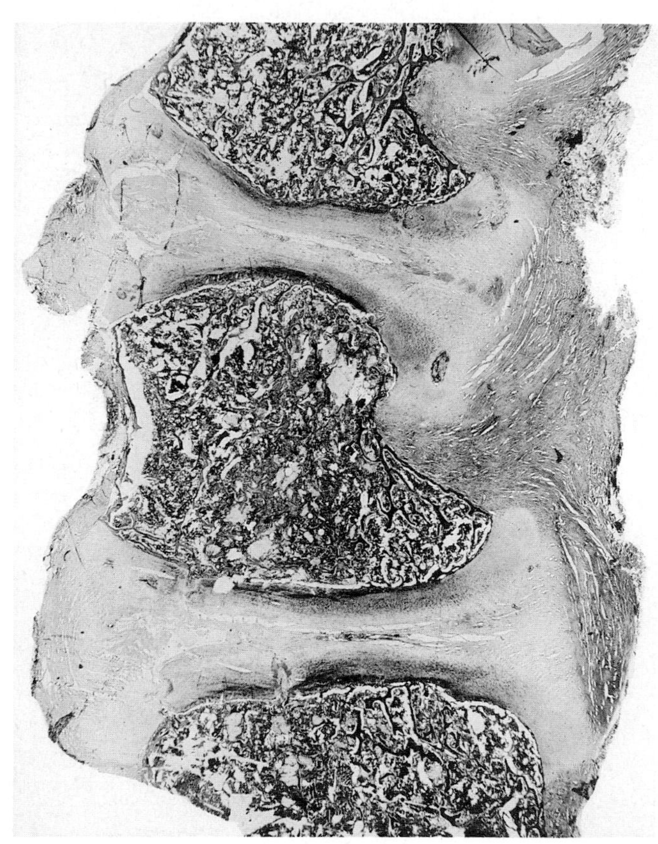

Fig. 9.28 Mucopolysaccharidosis.
In this example of the Hunter–Hürler syndrome in a child, the vertebral bodies are poorly formed, the cartilagenous plates biconcave. The discs are of fish-tail shape. There are no recognizable islands of vacuolated cells in the adjacent bone. (HE × 2.8.)

Histologically, the myocardial connective tissue has a hyaline appearance: the interfibrillar matrix material is amorphous or finely granular. Collagen fibres are said to be dense, 'oedematous', thick and afibrillar. The most characteristic appearance is that of the many large foamy cells with basophilic vacuolated cytoplasm ('gargoyle' cells): some are fibroblast-like; others resemble the rheumatic fever Anitschkow 'myocyte' (macrophage) (see p. 799). Muscle cells may also be vacuolated. The excess connective tissue and numerous vacuolated cells are causes of the partial or complete heart block that, like the cardiomyopathy, contributes to cardiac failure.

Electron microscopic investigations have led to the suggestion that excess dermatan sulphate in Hürler-like disease provokes excessively high rates of collagen synthesis.[195] In mitral and aortic valves, collagen fibres have been identified within membrane-bounded cytoplasmic dense bodies in small granular cells; the dense bodies are distinct from the clear vacuoles present in the fibroblasts and macrophages of the same cases. The presence in myocardial cells of organelles of three types is recognized: zebra bodies;[82] membranous cytoplasmic bodies; and granulomembranous bodies. The suggestion has been made that, since zebra bodies and membranous cytoplasmic bodies are known to accumulate gangliosides, these substances rather than glycosaminoglycan may be stored within cardiac muscle cells in Hürler-like disease. It is possible that the glycosaminoglycans are confined to the cells of the muscular connective tissues.

Laboratory diagnosis

(Fig. 9.30)

Biochemical diagnosis can be made by identifying the enzyme defect in isolated fibroblasts, serum or leucocytes, or by characterization of the excreted glycosaminoglycan. Proton nuclear magnetic resonance (^{1}H-n.m.r.) spectroscopy may facilitate this recognition.[201]

In early studies, it was found that skin fibroblasts cultured from patients with the mucopolysaccharidoses display metachromasia.[38,40,233] Mucopolysaccharidosis type IV is an exception: cell abnormalities are restricted to the skeletal system. Skin fibroblast culture allows cellular abnormalities to be directly correlated with biochemical analysis,[202] cellular metachromasia with fibroblast intracellular glycosaminoglycan.

Heterozygotes can be identified.[39,54,164] However, it is now clear that the culture method does not distinguish between the mucopolysaccharidoses and ganglioside storage diseases.[142] Further, changes in the pH of the culture medium can result in similar microscopic appearances in normal cells.[133] Conjunctival biopsy has therefore taken the place of skin biopsy.[35] Material is collected for transmission electron microscopy enabling measurements of ultrastructural changes, their distribution and appearance. These observations can be correlated with the biochemical identification of the inherited enzyme defect. Diagnosis can also be made by the recognition of characteristic inclusions in lymphocytes or other leucocytes.

Treatment

Initially, it had been thought possible to mobilize an increased glycosaminoglycan degradation by infusing plasma from normal persons into patients with mucopolysaccharidosis type III[49] or type I and II,[53] or with Fabry's disease.[158] Attempts were also made to supplement deficient enzymes by transplanting skin from a girl to a brother with the Hunter syndrome. A pronounced increase in the degradation of glycosaminoglycans was evident for nine weeks, six months after graft rejection[47,48] but the long-term effects proved disappointing. Fibroblasts have also been transplanted to provide a mechanism of long-term

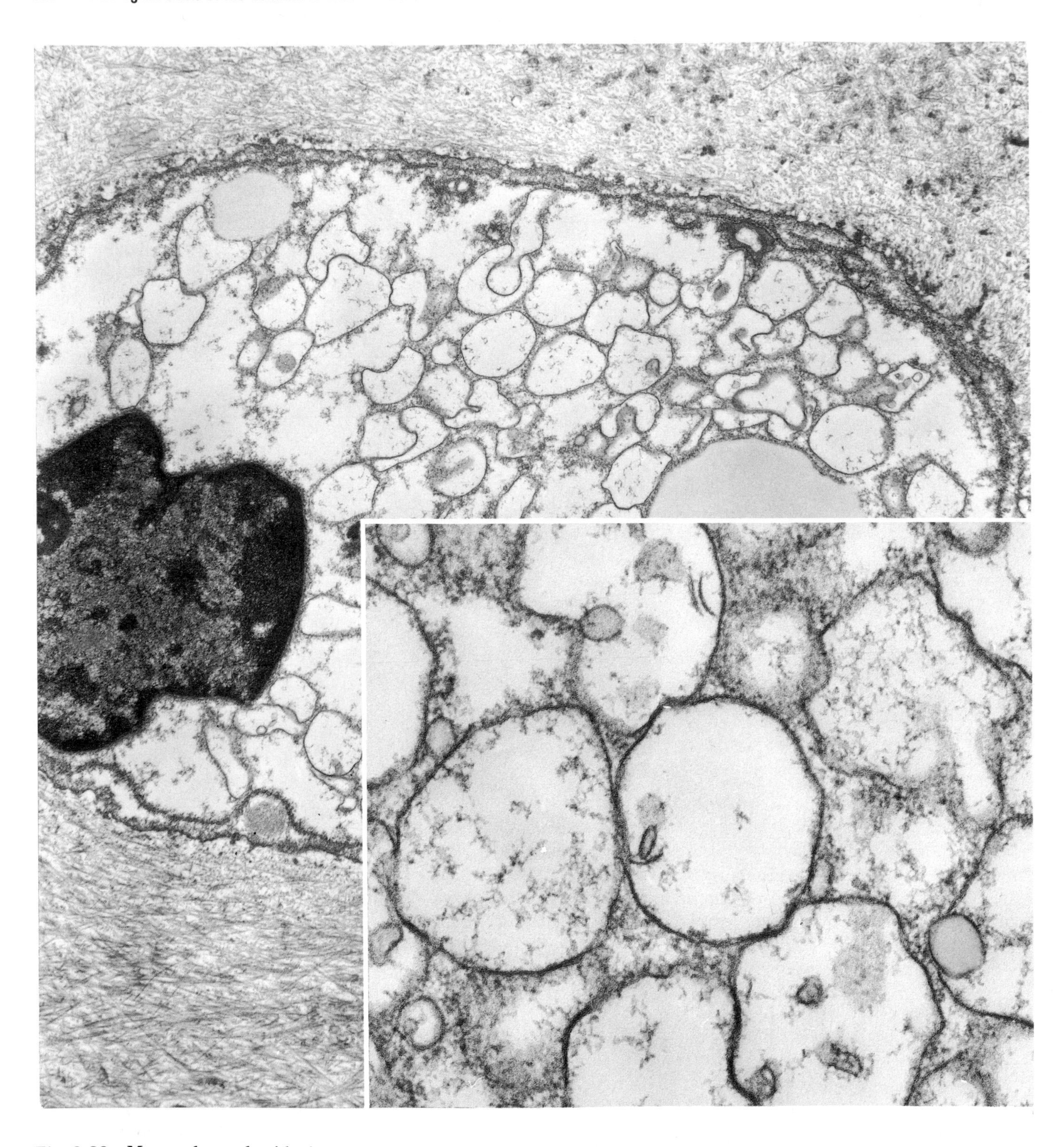

Fig. 9.29 Mucopolysaccharidosis.
Chondrocyte from articular cartilage showing numerous single membrane-bounded vacuoles. (Transmission electron microscopy × 6350.) Inset: Detail from main section. (Transmission electron microscopy × 22 000.) Courtesy of Dr Julie Crow.

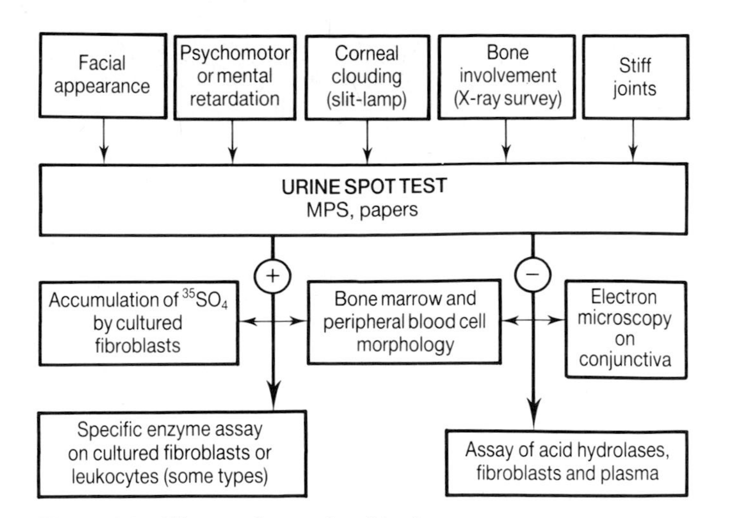

Fig. 9.30 Mucopolysaccharidosis.
A diagnostic approach to mucopolysaccharidoses which may be phenotypically indistinguishable. From Legum, Schorr and Berman.[133]

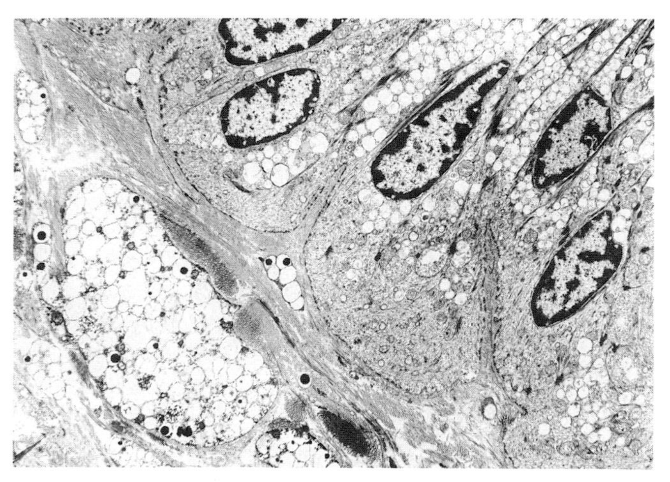

Fig. 9.31 Mucopolysaccharidosis.
Site of fibroblast transplant into skin of an 8.5-year-old child with Hürler disease into whom cells had been grafted on three occasions. Note vacuoles in host epithelial and hair follicle cells and in adjacent connective tissue cells of dermis. (Transmission electron microscopy × 3020.) Courtesy of Dr Julie Crow and the Editor, *Journal of Clinical Pathology*.[35]

enzyme replacement but there is little evidence that there is any resulting change in the activity of enzymes such as α-L-iduronidase in the brain, liver or kidney[35] or in the urinary glycosaminoglycan composition[191] (Fig. 9.31). More success has attended the use of bone marrow transplants.

Animal studies

Spontaneous mucopolysaccharidoses occur in the domestic cat. Mucopolysaccharidosis types I and VI have been identified in the short-haired animal.[94,96–98] 'Zebra' bodies are found in dorsal horn neurones and membrane-bound cytoplasmic inclusions containing granular material in hepatocytes. A deficiency of α-L-iduronidase is demonstrable in cultured cat fibroblasts and peripheral leucocytes. Pathological studies of animals with feline mucopolysaccharidosis type VI[95] confirm the presence of membrane-bound cytoplasmic inclusions in hepatocytes, bone marrow granulocytes, smooth muscle cells and dermal, corneal and cardiac valve fibroblasts. It is interesting that the relatively slight neurological changes are accompanied by spinal cord compression due to vertebral exostoses that, in one illustration, distantly recall the appearances of the syndesmophytes of ankylosing spondylitis (see p. 771).

An analogous condition has been recognized in the dog: the morphological and biochemical changes show similarities to human mucopolysaccharidosis type I.[213] An experimental animal model of mucopolysaccharidosis[52] has also been produced by the intracerebral injection of the trypanocidal drug suramin into rats[193] and mice.[161] The intravenous injection of suramin causes a large increase in liver glycosaminoglycan within 10 days.[33] The activity of some hepatic lysosomal enzymes including iduronate sulphatase is decreased and that of others increased. The biochemical and pathological changes are reversible.[194] An analogous condition, showing features both of mucopolysaccharidosis and of lipidosis, can be produced in the rat by the prolonged oral administration of tilorone, a compound with antiviral and antitumour activities which stimulates interferon production.[140]

I-cell disease

(see pp. 203 and 927)

I-cell disease[173] and pseudo-Hürler polydystrophy are biochemically related, rare disorders which resemble the classical mucopolysaccharidoses clinically. For historical reasons, both diseases were formerly termed mucolipidoses. The inherited defect in I-cell disease, transmitted as an autosomal recessive characteristic, is in the localization of some acid hydrolases which are secreted rather than retained within lysosomes. The biochemical abnormality is described in Chapter 4 (see p. 203). The pathological consequences are similar to those of the mucopolysaccharidoses. The term I-cell disease was derived from the very large number of single membrane-bounded inclusions found in the cytoplasm of cultured skin fibroblasts. The inclusions comprise fibrogranular and lamellar material, similar to that found in the Hürler syndrome. The deposition is selective, so that the axons of peripheral nerves, Schwann cells and perivascular cells are severely affected, as are renal and connective tissue cells generally, whereas neurones and hepatocytes are relatively unaffected. Joint contractures and lumbar kyphosis[167a] are components of a widespread disorder of growth. Mental retardation results from neuronal disease and both the muscular and connective tissue cells of the heart and aorta are affected leading to valvular thickening and cardiomyopathy.

Diagnosis rests upon the demonstration of a large excess of enzymes such as β-hexosaminidase or iduronate sulphatase in the plasma. In fibroblast culture, the cells lack these enzymes whereas they are present in excess in the culture medium.

Mannosidosis

(see p. 206)

Mannosidosis is a rare heritable disorder in which absence of the lysosomal form of α-D-mannosidase results in the inability to degrade various glycoproteins. The disorder is relatively frequent in black Angus cattle. In humans, there are multiple skeletal anomalies, psychomotor retardation, hepatomegaly and a susceptibility to infection. Serum immunoglobulin levels are low and blood lymphocytes may be vacuolated. Weiss and Kelly[248] described the case of a 13-year-old female with a destructive bilateral ankle joint synovitis, resembling pigmented villonodular synovitis (see Chapter 24). The synovial villous folds contained many histiocytes with PAS-positive material. Transmission electron microscopy confirmed that this material lay within membrane-bounded vacuoles. There is an analogy with the synovial response in histiocytosis and Whipple's disease (see p. 788).

Other hereditary disorders

There are many other inherited diseases in which the organized connective tissues, the joints and the skeleton are affected; some are familial. Hereditary skeletal disease may coexist, for example, with immune deficiency.[224] Chromosomal anomalies have sometimes been identified, as in the case described by Bühler *et al.*[22]: a child aged 13 years had short fifth fingers, hypoplastic toes, habitual patellar luxation and severe valgus deformity of the feet. There were multiple exostoses, and a deletion of about half of the G-negative band q24 of somatic chromosome 8.

Dwarfism

Dwarfism[196] can contribute to articular, connective tissue abnormality. The causes range from familial factors and constitutional growth delay to emotional deprivation, malnutrition, chronic disease and intrauterine retardation. There may be a chromosomal anomaly and disorders of growth hormone secretion and skeletal dysplasia. There is overlap with diseases such as osteogenesis imperfecta as discussed above (see p. 343). Few systematic pathological studies of joint disease have been made in these heritable syndromes. Occasionally, short stature and other skeletal disease is accompanied by joint stiffness[183] and extreme limitation of articular movement of the hands was described in a male with osteogenic sarcoma and multiple visceral anomalies.[180] Bone enlargement affecting the synovial joints, with premature cartilage degeneration, joint stiffness and synovitis, particularly of the knees and hips, has been identified in a family with myopia,[226] the syndrome of hereditary progressive arthroophthalmopathy. Proteoglycan structure and metabolism may be disordered in individuals with congenital absence of the tibia[177] and in nanomelia.[148]

Synphalangism

This disorder[37] has attracted attention over the years, encouraging Elkington and Huntsman[63] to review the extraordinary history of the Talbot Fingers. Stiff fingers, with hereditary absence of one or more proximal interphalangeal joints, may have been present in as many as 14 generations of this family, from the fourteenth to the twentieth century. The causes of absence of individual bones must be viewed in the light of evidence of the possible presence or absence of trauma and infection. The existence of a form of severe congenital amputation involving both limbs, acheiropody, limited to Brazil, exemplifies the difficulty in reaching the truth in such cases.[234]

Epidemic familial arthritis of Malnad

During the past 25 years, a circumscribed area of Karnataka, India, has been subject to a previously unrecognized disorder in which epidemic degenerative hip joint disease (osteoarthrosis—Chapter 22) has occurred at all ages except in preschool children. Forty villages have been affected.[14] The area is one where both nutritional and genetic connective tissue diseases are common. In an attempt to disentangle the predisposing and causal factors of the epidemic arthritis, Bhat and Krishnamachari[14] investigated the 60 families (of 206) in which members were involved. The joint disease displayed a 'strong genetic component'. Dwarfism was common. The possible role of pesticides in provoking the epidemic arthritis was considered. An analogy with Kashin–Beck disease (see Chapter

22) is suggested, and the possibility of the action of mycotoxins formed during the fungal contamination of grain crops was reviewed. However, as with similar disorders identified in Russia and South Africa, community studies, including surveys of families and of identical versus non-identical twins, are handicapped by the frequent coexistence of congenital malformations, osteomalacia or rickets, and of both infective and non-suppurative polyarthritis.

Congenital pseudarthrosis

NF1, the most common form of neurofibromatosis, is transmitted as an autosomal recessive characteristic. The gene coding for NF1 has been located on chromosome 17.

Individuals with neurofibromatosis occasionally display pseudarthrosis (Figs 9.32 and 9.33).[56,72,99,247] The anomaly, found in approximately 1 in 140 000 newborn children, is almost always unilateral, affects the tibia, is more frequent on the left than the right and more common in males than females. Although the medullary canal may be normal, bone sclerosis with obliteration of the canal is recognized. The non-united site may be cystic in character but the tibia, which is reduced in length, is usually recognized to be the site of local neurofibromatosis[87] or non-specific fibrosis near a discrete focus of neurofibromatosis.[72,110] Intraosseous neurofibromata or schwannomata have been found. Nevertheless, ultrastructural evidence[19,21] is not in favour of the presence of Schwann cells and the spindle-shaped cells at the non-united site may be fibroblasts, myofibroblasts or perineural cells.

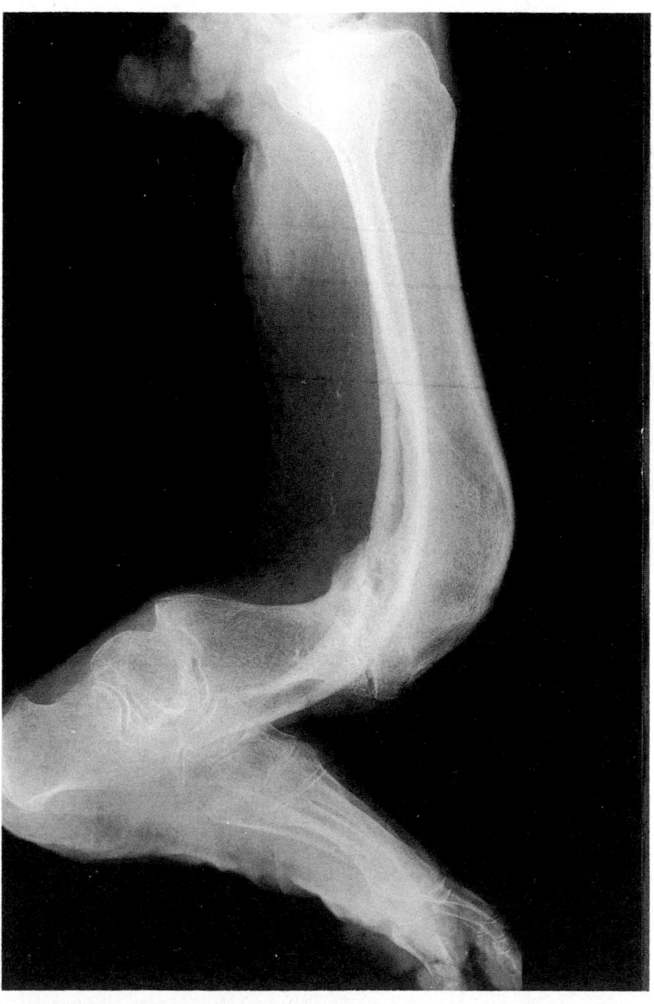

Fig. 9.32 Pseudarthrosis of tibia.
Lower leg of elderly female, one of identical twins. Note articulation towards lower end of tibia. The other twin did not display this anomaly.

Achondroplasia (chondrodystrophia fetalis)

This uncommon autosomal dominant disorder is remarkably consistent in its expression. The characteristic dwarfing is easily recognized in later life; X-ray changes may allow the disease to be identified at birth.[93]

Clinical presentation

The deformity is of the 'short-limb'-type by contrast with, say, the Morquio syndrome (mucopolysaccharidosis type IV) which is 'short-trunk'. Nevertheless, with the exception of the bones of the calvarium, formed in membrane, there are widespread abnormalities and the excess cartilage in the vertebral bodies, separating the vertebral ossification centres, is one sign that permits neonatal diagnosis. Eighty per cent of cases of achondroplasia are sporadic but achondroplastics intermarry and children homozygous for the dominant gene can survive birth.

Achondroplasia must be distinguished from thanatophoric dwarfism.[163] Achondroplasia is a well-recognized disorder of cattle, rabbits and dogs[5,77] and can be a considerable inconvenience to breeders seeking to encourage the expression of desirable somatic features by unnatural selection.

Biochemical changes

Proteoglycan Observations made with the fibular growth plate of newborn infants and young children suggest that an abnormally high proportion of matrix material is in the form of proteoglycan aggregates. Individual glycosaminoglycans do not differ from normal. The principal anomaly in achondroplasia, expressed during chondrocyte matura-

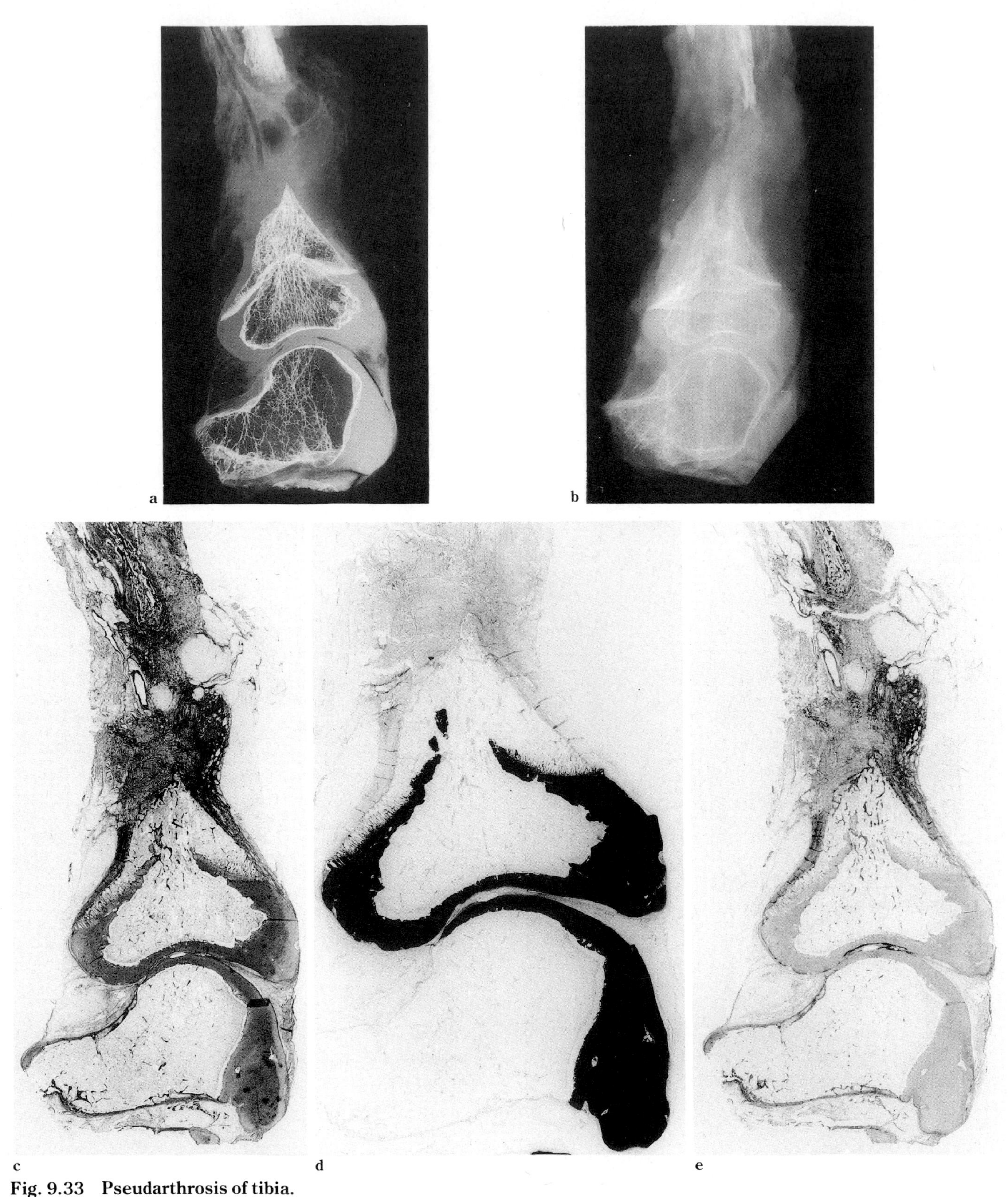

Fig. 9.33 Pseudarthrosis of tibia.
Male aged 6 years. Crippling deformity of lower leg had required long-term use of callipers and immobility was one cause of severe osteoporosis. **a.** Radiograph exposed to show bone structure of resected specimen; **b.** Radiograph exposed to show soft-tissue structure of resected specimen; **c.** Decalcified specimen demonstrating deficiency of bone structure in lower part of tibia; **d.** High concentration of proteoglycan in ankle joint cartilages, low concentrations in connective tissue at site of defect; **e.** Zone of pseudoarthrosis devoid of bone or osteoid, contains abundant, irregularly dispersed collagen. (**a**, **b**, **c**, **e**: Natural size; **d**: × 1.5; **c**: HE; **d**: toluidine blue; **e**: picro-Sirius red.) Reproduced by courtesy of Mr M Macnicol FRCS.

tion, hypertrophy and degeneration, may therefore relate to a qualitative or quantitative disorder of these proteoglycan aggregates.[182]

Collagen There is evidence for the association with achondroplasia, of a particular type II procollagen genotype.[66]

Pathological changes

There is a primary defect in the development of bone formed in cartilage (Fig. 9.32). The limbs are most severely affected; the disorder is of proximal limb bones (rhizomelic) but minor abnormalities of the face, base of skull, vertebrae and digits are customary. There is disorganization of epiphyseal bone growth. The columns of cartilage cells, normally slender, regular and orderly, are abbreviated and disorderly (Figs 9.34 and 9.35). The even, horizontal margins of cartilage which define the normal growth zone are irregular, but vascular ingrowth and calcification are not impaired. Because of the defect in cartilage replacement, the epiphyseal centres of ossification come to lie excessively near the diaphyses. An abundance of cartilage forms at the bone ends giving an abnormally wide tissue X-ray shadow between the opposing articular margins of diarthrodial joints. The growth of bone longitudinally is slow and irregular. Consequently, the child, in whom the skull and central nervous system are intact, has short limbs, broadened at the growing ends, and a deformed spine: kyphosis is frequent (Fig. 9.36). The soft tissues are thrown into folds, giving a 'Michèlin-tyre man' appearance.

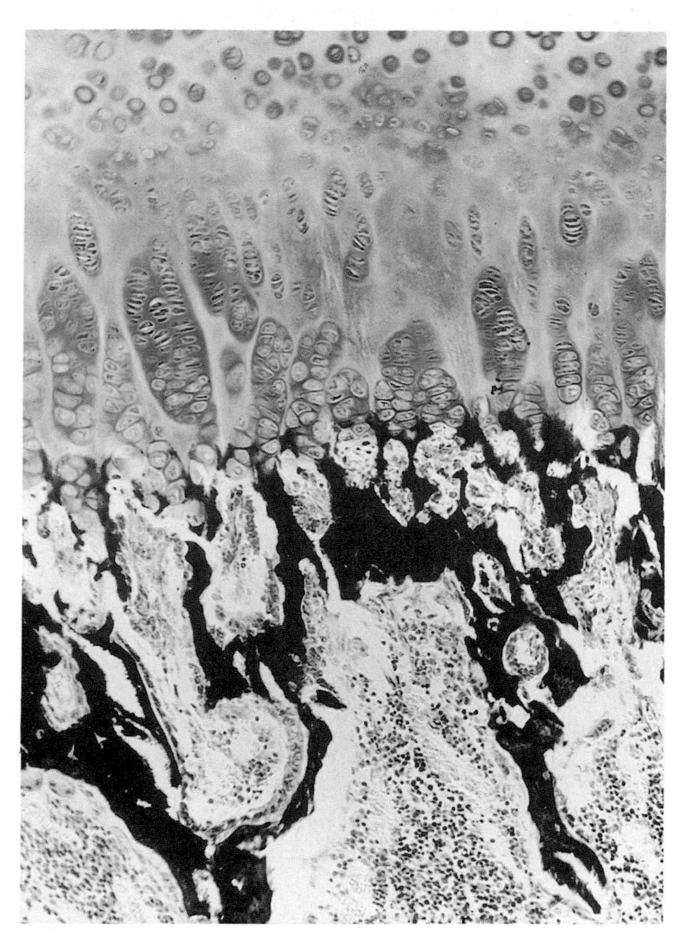

Fig. 9.35 Achondroplasia in the dog.
Endochondral bone formation is characterized by the presence of short, thick cartilage columns but an orderly boundary between the zone of provisional calcification and the non-mineralized cartilage. (von Kossa × 125.) Courtesy of the Editor, *Journal of Pathology*.

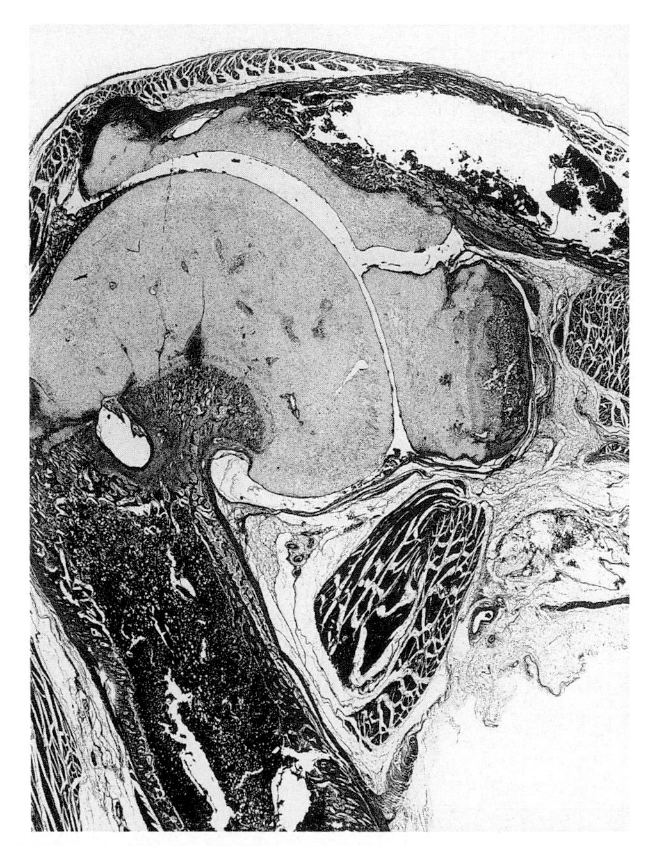

Fig. 9.34 Achondroplasia in the dog.
Both acetabular and femoral head cartilage are excessive in amount. The femur is abnormally short. (HE × 6.)

Experimental studies

The possibility has been considered that the development of achondroplasia is mediated by endocrine rather than by purely genetic factors. In experimental studies, however, it has been shown that neither deficient pituitary growth

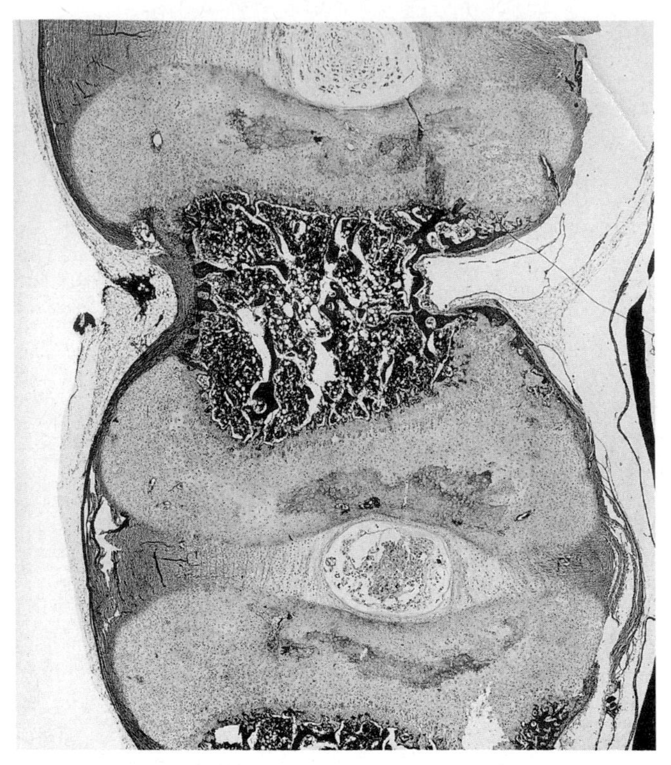

Fig. 9.36 Achondroplasia in the dog.
An excess of epiphyseal cartilage combines with bone deformity to give a 'collar-stud' appearance. (HE × 8.7.)

hormone nor a diminished responsiveness of endocrine end-organs plays a part in the pathogenesis of the disease in achondroplastic (*ac*/*ac*) or dwarf (*Dw*/*Dw*) rabbits.[246] Webber *et al.*[246] interpret the failure of vascular canals to penetrate the zone of proliferating cartilage normally as due to either a generalized retardation of limb development; or a defect in the penetration of cartilage vascular canals into the cartilage of the epiphysis.

Recurrent polyserositis (familial Mediterranean fever)

This systemic, inflammatory disease is prevalent among Jews, Turks, Armenians and Arabs. The disorder affects young people, and males more often than females. The large majority of the affected Jews are of Sephardic or Iraqi origin.[64]

Clinical presentation

The clinical manifestations are dominated by febrile peritonitis which resolves without surgery. Skin rashes, pleurisy and pericarditis may develop. The connective tissues are affected in two ways: by arthritis and amyloidosis (see Chapter 11). On average, 55 per cent of those with recurrent polyperositis develop an asymmetrical, non-destructive, mono- or oligoarthritis of large joints.[65] Knee and ankle joints are affected three times more frequently than hips, shoulders, feet and wrists. The signs are those of aseptic inflammation: the synovial fluid is sterile but contains many polymorphs. A migratory behaviour of the arthritis sometimes resembles that of rheumatic fever (Chapter 20). In five per cent of cases the acute disease does not resolve quickly and in two per cent of chronic cases, destructive arthritis develops.

Inheritance

Familial Mediterranean fever is inherited as an autosomal recessive characteristic. There is no association with any known MHC genotype.

Pathological changes

The pathogenesis is not well understood. There is a similarity to rheumatic fever, glomerulonephritis, Henoch–Schönlein purpura and atopic allergy and these conditions, together with polyarteritis nodosa, occur significantly often in association with familial Mediterranean fever. Raised serum immunoglobulin levels, circulating immune complexes and a possible C5a-inhibitor deficiency are among features that suggest that the widespread inflammatory process has its origin in a disorder of immune regulation.

The pathological features were tabulated by Kaushansky, Finerman and Schwabe.[116a] The hip joints were most often involved but a destructive arthropathy also affected knees, ankles, sacroiliac, shoulder and temporomandibular joints. The gross abnormalities were synovial thickening, cartilage destruction, osteophyte formation and cystic bone changes. They closely resembled the abnormalities of osteoarthrosis (see Chapter 22).

Microscopically, there was synovial vascular proliferation, chronic inflammatory changes, synovial cellular hypertrophy and perivascular lymphocytic infiltration. The histological changes centre on the venules and arterioles: their walls are thickened and surrounded by varying numbers of polymorphs, lymphocytes, macrophages, plasma cells and eosinophil granulocytes. Within affected synovial joints there may be limited evidence of cartilage destruction. Synovial venules display concentric, multilayer basement membrane thickening attributable to recurrent episodes of vascular cell death and regeneration. The explanation for these changes is unknown and there is not yet an animal model that might throw light on the disease process.

Haemochromatosis

In this rare inborn error of iron metabolism, iron accumulates in very many of the tissues of the predominantly male patients. Exceptionally, postmenopausal women are affected. A familial incidence has been recorded.[211]

Clinical presentation

Instances of a severe disabling haemochromatotic joint disease were described by Schumacher[205] and by Kra *et al.*[121] An analysis of the clinical features of 32 cases of this iron storage disorder revealed 16 with arthropathy;[90] some presented with episodes of inflammatory arthritis. In 10 of these 16 patients chondrocalcinosis (see p. 393) was found simultaneously. Articular disease may precede the diagnosis of haemochromatosis by many years. The association of the joint disorder, a form of osteoarthrosis (see Chapter 22), with the diabetes of haemochromatosis may be significant. However, there is a suspicion that synovial and chondrocytic accumulations of iron may be the proximate cause of articular injury.[91] Loss of radiographic joint space with cartilage fragmentation and erosion, were features of Hamilton's[91,90] cases, recalling those of aseptic necrosis rather than secondary osteoarthrosis. Involvement of the metatarsophalangeal joints confirmed a distribution similar to that of rheumatoid arthritis although the haemochromatotic patients were seronegative and did not develop granulomata.

Inheritance

Idiopathic haemochromatosis is believed to be an autosomal recessive characteristic the full expression of which is only likely in adult males.

Pathological changes

At arthroscopy or necropsy the synovia are found to be dull brown and hypertrophic. Microscopically, synovial sections give a strongly positive reaction for iron. The iron is located in synovial B rather than in A cells. There is little sign of inflammation. The quantity of iron in the synovia is related to the duration and severity of the disease. However, effective treatment by venesection or chelation does not diminish the severity of the joint disease: either synovial iron accumulation is not the main determinant of articular disease, or articular disease becomes established in advance of therapy and cannot be reversed.

A conspicuous feature in many cases is the presence—at first in superficial, later in deeper cartilage—of radioopaque deposits of calcium pyrophosphate dihydrate crystals. Extensive white powdery accumulations of this insoluble material are found obscuring articular surfaces when joints are opened at necropsy. Chondrocalcinosis (see p. 393) is most common in the knee menisci but the tibial condylar cartilage may be affected and in diminishing order of frequency, the triangular ligament of the wrist, the symphysis pubis, the hip, metacarpophalangeal, ankle, elbow, shoulder and metatarsophalangeal joints. Although the sacroiliac joints are radiologically normal, calcification of the longitudinal ligaments of the lumbar spine mimics ankylosing spondylitis (p. 768). The mechanisms for calcium pyrophosphate dihydrate deposition may lie in an inhibition of synovial pyrophosphatase by synovial ferrous iron. Ferritin[206] and haemosiderin, an insoluble form of storage iron formed by the denaturation of ferritin, the loss of apoferritin and micellar aggregation, are found by transmission electron microscopy in chondrocyte lacunae and in the cells themselves. By analogy with haemophilia (p. 967), it may be argued that iron, retained as siderosomes, provokes cartilage breakdown directly by lysosomal enzyme activation, or indirectly, via interleukin-1 release. However, some cases with joint disease and calcium pyrophosphate dihydrate deposition display no ferritin iron. Since iron and calcium pyrophosphate dihydrate are not in a constant morphological relationship,[206] iron probably contributes to cartilage breakdown by an indirect mechanism and not by crystal deposition.

Wilson's disease

Wilson's disease is a rare autosomal recessive disorder in which copper accumulates in the liver, brain (hepatolenticular degeneration), kidney, cornea and bone. The gene for Wilson's disease has been mapped to somatic chromosome 13.[59] Skeletal abnormalities are well-recognized. In particular, there may be calcification of articular cartilage, joint capsules and the sites of tendinous insertions.[17] Among the cases of Feller and Schumacher[71] were calcified loose bodies at the wrist, premature degenerative arthropathy of the knee, with chondromalacia and, occasionally, chondrocalcinosis. Since red blood cell pyrophosphatase activity can be inhibited by copper and ferrous iron,[143] it was proposed that the radioopaque deposits in Wilson's disease were calcium pyrophosphate dihydrate (p. 393). This view was substantiated by McClure and Smith[144] in a study of intervertebral discs. Staining techniques, radiography and X-ray microanalysis showed the presence in both the annulus and the nucleus of crystalline deposits with a phosphate: calcium ratio of 0.96, strongly suggestive of the composition of calcium pyrophosphate dihydrate (p. 394).

Cystic fibrosis

Cystic fibrosis is the most common lethal or semilethal heritable disease of Caucasians.[15a,232] The responsible gene has been identified on somatic chromosome 18. The consequent metabolic defect is not fully understood although molecules of molecular weight of approximately 10 kD are among 'markers' present in the plasma and extracellular fluid. These substances inhibit sodium transport, induce mucus secretion and change the behaviour of ciliated cells. The secretion of excess, tenacious mucus leads to chronic obstructive lung disease, pancreatic insufficiency, cirrhosis and other disorders and sets the scene for chronic respiratory tract infection, usually with *Staphylococcus aureus* or *Pseudomonas aeruginosa.*

Improved care determines that increased proportions of patients with cystic fibrosis survive to adult life. Some develop an episodic arthropathy in addition to their hypertrophic pulmonary osteoarthropathy (see Chapter 24) which complicates the chronic lung disease.[18,53a] Arthritis developed in 8.5 per cent of the 59 patients studied by these authors. The male: female ratio was 3:2. The joint disorder is polyarticular but self-limiting and does not progress to joint destruction. Synovial biopsy has rarely been undertaken. The limited evidence is of vascular congestion and oedema without inflammation but with IgM or IgG and complement but not IgA, deposition. In one longstanding case, treated by corticosteroids, synovial granulomata were demonstrated.[184] Their pathogenesis is not understood.

REFERENCES

1. Abraham P A, Perejda A J, Carnes W H, Uitto J. Marfan syndrome. Demonstration of abnormal elastin in aorta. *Journal of Clinical Investigation* 1982; **70**: 1245–52.
2. Achard C. Arachnodactyly. *Bulletin de la Societé Médicale des Hôpitaux de Paris* 1902; **19**: 834–40.
3. Adair-Deighton C A. Four generations of blue sclerotics. *Ophthalmoscope* 1912; **10**: 188–9.
4. Ahlgren P, Vestermark S. Menkes' kinky hair disease. *Neuroradiology* 1977; **13**: 159–63.
5. Almlöf J. On achondroplasia in the dog. *Zentralblatt für Veterinärmedizin* 1961; **8**: 45–56.
6. Altman L K, Fialkow P J, Parker F, Sagebiel R W. Pseudoxanthoma elasticum. An underdiagnosed genetically heterogeneous disorder with protean manifestations. *Archives of Internal Medicine* 1974; **134**: 1048–54.
7. Arneson M A, Hammerschmidt D E, Furcht L T, King R A. A new form of Ehlers–Danlos syndrome. *Journal of the American Medical Association* 1980; **244**: 144–7.
8. Balzer F. Récherches sur les charactères anatomiques du xanthelasma. *Archives de Physiologie* 1884; **4**: 65–80.
9. Bauer K H. Über identität und Wesen der sogenannten osteopsathyrosis idiopathica und osteogenesis imperfecta. *Deutsche Zeitschrift für Chirurgie* 1920; **160**: 289–351.
10. Becker H. Der typische supravalvuläre aortenabriss bei Marfan-syndrom. *Zeitschrift für Kreislaufforschung* 1967; **7**: 658–78.
11. Beddard A P. Ochronosis associated with carboluria. *Quarterly Journal of Medicine* 1909–10; **3**: 329–36.
12. Beighton P. The inherited disorders of connective tissue. *Bulletin on the Rheumatic Diseases* 1972–3; **23**: 696–707.
13. Belcher R. Ultrastructure of the skin in the genetic mucopolysaccharidoses. *Archives of Pathology* 1972; **94**: 511–8.
14. Bhat R V, Krishnamachari K A V R. Endemic familial arthritis of Malnad—an epidemiological study. *Indian Journal of Medical Research* 1977; **66**: 777–86.
15. Blümcke S, Langness V, Liesegang B, Fooke-Achterrath M, Thiel H J. Light and electron microscopic and element analysis of pseudoxanthoma elasticum (Darier–Gronblad–Strandberg syndrome). *Beiträge zür Pathologie* 1974; **152**: 179–99.

15a. Boat T F, Welsh M J, Beaudet A L. Cystic fibrosis. In: Scriver C R, Beaudet A L, Sly W S, Valle D, eds, *The Metabolic Basis of Inherited Disease* (6th edition). New York, London: McGraw-Hill Information Services Company, 1989: 2649–80.

16. Bonella D, Parker D J, Davies M J. Accumulation of procollagen in human floppy mitral valves. *Lancet* 1980; **i**: 880–1.
17. Boudin G, Pepin B, Hubault A. Les arthropathies de la maladie de Wilson. *Sociétié Médicale de Hôpitaux de Paris* 1963; **114**: 617–22.
18. Bourke S, Rooney M, Fitzgerald M, Bresnihan B. Episodic arthropathy in adult cystic fibrosis. *Quarterly Journal of Medicine* 1987; **64(NS)**: 651–9.
19. Briner J, Yunis E. Ultrastructure of congenital pseudoarthrosis of the tibia. *Archives of Pathology* 1973; **95**: 97–9.
20. Brookler M I, Martin W J, Underdahl L O, Worthington J W, Mathieson D R. Alkaptonuria and ochronosis: further experiences. *Proceedings of the Mayo Clinic* 1964; **39**: 107–17.
21. Brown G A, Osebold W R, Ponseti IV. Congenital pseudoarthrosis of long bones: a clinical, radiographic, histologic and ultrastructural study. *Clinical Orthopaedics and Related Research* 1977; **128**: 228–42.
22. Bühler E M, Bühler U K, Stadler G R, Jani L, Jurik L P. Chromosome deletion and multiple cartilaginous exostoses. *European Journal of Pediatrics* 1980; **133**: 163–6.
23. Bullough P G, Vigorita V J. *Atlas of Orthopaedic Pathology with Clinical and Radiologic Correlations.* London: Butterworths, 1984: 3.2–3.8.
24. Byers P H, Barsch G S, Holbrook K A. Molecular pathology in inherited disorders of collagen metabolism. *Human Pathology* 1982; **13**: 89–95.
25. Byers P H, Siegel R C, Holbrook K A, Narayanan A S, Bornstein P, Hall J G. X-linked cutis laxa: defective cross-link formation in collagen due to decreased lysyl oxidase activity. *New England Journal of Medicine* 1980; **303**: 61–5.
26. Byers P H, Siegel R C, Peterson K E, Rowe D W, Hol-

brook K A, Smith L T, Chang Y H, Fu J C. Marfan syndrome —abnormal alpha-2 chain in type I collagen. *Proceedings of the National Academy of Sciences of the USA* 1981; **78**: 7745–9.

27. Callahan W P, Hackett R L, Lorincz A E. New observations by light microscopy on liver histology in the Hürler's syndrome. *Archives of Pathology* 1967; **83**: 507–12.

28. Callahan W P, Lorincz A E. Hepatic ultrastructure in the Hürler syndrome. *American Journal of Pathology* 1966; **48**: 277–98.

29. Caniggia A, Stuart C, Guideri R. Fragilitas ossium hereditaria tarda: Ekman–Lobstein disease. *Acta Medica Scandinavica* 1958; **162** (Suppl.): 1–172.

30. Carson N A J, Neill D W. Metabolic abnormalities detected in a survey of mentally backward individuals in northern Ireland. *Archives of Diseases in Childhood* 1962; **37**: 505–13.

31. Carter C O, Fairbank T J. *The Genetics of Locomotor Disorders*. Oxford: Oxford University Press, 1974.

32. Clark J G, Kuhn C, Uitto J. Lung collagen in type IV Ehlers–Danlos syndrome: ultrastructural and biochemical studies. *American Review of Respiratory Disease* 1980; **122**: 971–8.

33. Constantopoulos G, Rees S, Cragg B G, Barranger J A, Brady R O. Experimental animal model for mucopolysaccharidosis—suramin-induced glycosaminoglycan and sphingolipid accumulation in the rat. *Proceedings of the National Academy of Sciences of the USA* 1980; **77**: 3700–4.

33a. Connolly C E, O'Reilly U, Donlon J. Black cartilage associated with levodopa. *Lancet* 1986; **i**: 690.

34. Cremin B J, Beighton P. *Bone Dysplasias of Infancy*, Berlin: Springer-Verlag, 1978.

35. Crow J, Gibbs D A, Cozens W, Spellacy E, Watts R W E. Biochemical and histopathological studies on patients with mucopolysaccharidoses, two of whom had been treated by fibroblast transplantation. *Journal of Clinical Pathology* 1983; **36**: 415–30.

36. Cupo L N, Pyeritz R E, Olson J L, McPhee S J, Hutchins G M, McKusick V A. Ehlers–Danlos syndrome with abnormal collagen fibrils, sinus of Valsalva aneurysms, myocardial infarction, panacinar emphysema and cerebral heterotopias. *American Journal of Medicine* 1981; **71**: 1051–8.

37. Cushing H. Hereditary anchylosis of the proximal phalangeal joints (symphalangism). *Proceedings of the National Academy of Sciences of the USA* 1915; **1**: 621.

38. Danes B S, Bearn A G. Hürler's syndrome: effect of retinol (vitamin A alcohol) on cellular mucopolysaccharides in cultured human skin fibroblasts. *Journal of Experimental Medicine* 1966a; **124**: 1181–98.

39. Danes B S, Bearn A G. Hürler's syndrome, a genetic study in cell culture. *Journal of Experimental Medicine* 1966b; **123**: 1–16.

40. Danes B S, Bearn A G. Cellular metachromasia, a genetic marker for studying the mucopolysaccharidoses. *Lancet* 1967; **i**: 241–3.

41. Danks D M. Steely hair, mottled mice and copper metabolism. *New England Journal of Medicine* 1975; **293**: 1147–9.

42. Danks D M, Campbell P E, Stevens B J, Mayne V, Cartwright E. Menkes' kinky-hair syndrome. An inherited defect in copper absorption with widespread effects. *Pediatrics* 1972; **80**: 188–201.

43. Danks D M, Cartwright E. Menkes' kinky-hair disease; further definition of the defect in copper transport. *Science* 1973; **179**: 1140–1.

44. Danlos M. Un cas de cutis laxa avec tumeurs par contusion chronique des condes et des génoux (xanthoma juvenile pseudo-diabetique de MM Hallopean et Macc de Lepinay). *Bullétin de la Société français de Dermatologie et de Syphilologie* 1908; **19**: 70–2.

45. Darier J. Pseudoxanthoma elasticum. *Monatschrift für praktische Dermatologie* 1896; **23**: 609–17.

46. DeCloux R J. Ultrastructural studies of the skin in Hürler's syndrome. *Archives of Pathology* 1969; **88**: 350–8.

47. Dean M F. Replacement therapy in the mucopolysaccharidoses. *Journal of Clinical Pathology* 1978; **31**, Suppl. (Royal College of Pathologists) 12: 120–7.

48. Dean M F, Muir H, Benson P F. Mobilization of glycosaminoglycans by plasma infusion in mucopolysaccharidosis type III—two types of response. *Nature* 1973; **243**: 143–6.

49. Dean M F, Muir H, Benson P F, Button L R, Boylston A, Mowbray J. Enzyme replacement therapy by fibroblast transplantation in a case of Hunter syndrome. *Nature* 1976; **261**: 323–5.

50. Dekaban A S, Patton V M. Hürler's and Sanfilippo's variants of mucopolysaccharidosis. *Archives of Pathology* 1971; **91**: 434–43.

51. Delhez H, Prins H W, Prinsen L, Vanderhamer C J A. Autoradiographic demonstration of the copper-accumulating tissues in mice with a defect homologous to Menkes' kinky hair disease. *Pathology—Research and Practice* 1983; **178**: 48–50.

52. Desnick R J, Patterson D F, Scarpelli D G. *Animal Models of Inherited Metabolic Diseases*. New York: Alan R Liss Inc, 1982.

53. Di Ferrante N, Nichols B L, Donnelly P V, Neri G, Hrgovcic R, Berglund R K. Induced degradation of glycosaminoglycans in Hürler's and Hunter's syndromes by plasma infusion. *Proceedings of the National Academy of Sciences of the USA* 1971; **68**: 303–7.

53a. Dixey J, Redington A N, Butler R C, Smith M J, Batchelor J R, Woodrow D F, Hodson M E, Batten J C, Brewerton D A. The arthropathy of cystic fibrosis. *Annals of the Rheumatic Diseases* 1988; **47**: 218–23.

54. Dorfman A, Matalon R. The mucopolysaccharidoses (a review). *Proceedings of the National Academy of Sciences of the USA* 1976; **73**: 630–7.

55. Dorling J. Localisation of sulphated glycosaminoglycans in the mucopolysaccharidoses by a simple technique using cryostat sections. *Journal of Clinical Pathology* 1980; **33**: 897–8.

56. Ducroquet R. A propos des pseudarthroses et inflexions congenitales du tibia. *Mémoires Academie de Chirurgie* 1937; **63**: 863.

57. Eddowes A. Blue sclerotics, fragile bones and deafness. *British Medical Journal* 1900; **II**: 222.

58. Editorial. Lysosomal storage disease. *Lancet* 1986; **ii**: 898–9.

59. Editorial. Homing in on Wilson's disease. *Lancet* 1989; **i**: 822–7.

60. Ehlers E. Cutis laxa, Neigung zu Haemorrhagien in der Haut, Lockerung mehrerer Artikulationen. *Dermatologische Zeitschrift.* 1901; **8**: 173–4.

61. Ekman O J. *Dissertatio Medica Descriptione et Casus ali-*

polysaccharidoses and mucolipidoses: review and comment. *Advances in Pediatrics* 1976; **22**: 305–47.

134. Leone A, Pavlakis G N, Hamer D H. Menkes disease: abnormal metallothionein gene regulation in response to copper. *Cell* 1985; **40**: 301–9.

135. Levene C I. Diseases of the collagen molecule. *Journal of Clinical Pathology* 1978; **31**, Supplement, (Royal College of Pathologists) 12: 82–94.

136. Lichtenstein L, Kaplan L. Hereditary ochronosis: pathological changes observed in two necropsied cases. *American Journal of Pathology* 1954; **30**: 99–125.

137. Lis Y, Burleigh M C, Parker D J, Child A H, Hogg J, Davies M J. Biochemical characterization of individual normal, floppy and rheumatic human mitral valves. *Biochemical Journal* 1987; **244**: 597–603.

138. Lobstein J. *Lehrbuch der Pathologischen Anatomie*. Stuttgart: Brodhag, 1835: 179.

139. Looser E. Zur Kenntnis der osteogenesis imperfecta congenita et tarda (sogenannte idiopathische osteopsathyrosis). *Mitteilungen aus den Grenzgebieten der Medizin und Chirurgie* 1905; **15**: 161–206.

140. Lüllman-Rauch R. Tilorone-induced lysosomal storage mimicking the features of mucopolysaccharidoses and of lipidosis in rat liver. *Virchows Archiv B (Cell Pathology)* 1983; **44**: 355–68.

141. Lynch H T, Larsen A L, Wilson R, Magnuson C L. Ehlers–Danlos syndrome and 'congenital' arteriovenous fistulae. A clinicopathologic study of a family. *Journal of the American Medical Association* 1965; **194**: 1011–14.

142. Lyon G, Hors-Caryla M C, Jonsson V, Maroteaux P. Aspects ultrastructuraux et signification biochemique des granulations métachromatiques et autres inclusions dans les fibroblastes en culture provenant de lipidoses et de mucopolysaccharidoses. *Journal of Neurological Sciences* 1973; **19**: 235–53.

143. McCarty D J, Pepe P F, Solomon S D *et al.* Inhibition of human erythrocyte pyrophosphatase activity by calcium, cupric and ferrous ions. *Arthritis and Rheumatism* 1971; **13**: 336.

144. McClure J, Smith P S. Calcium pyrophosphate dihydrate deposition in the intervertebral discs in a case of Wilson's disease. *Journal of Clinical Pathology* 1983; **36**: 764–8.

145. McClure J, Smith P S, Gramp A A. Calcium pyrophosphate dihydrate (CPPD) deposition in ochronotic arthropathy. *Journal of Clinical Pathology* 1983; **36**: 894–902.

146. McCully K S. Vascular pathology of homocysteinemia; implications for the pathogenesis of arteriosclerosis. *American Journal of Pathology* 1969; **56**: 111–28.

147. McCully K S. Homocysteinemia and arteriosclerosis. *American Heart Journal* 1972; **83**: 571–3.

148. McKeown-Longo P J, Goetinck P F. Characterization of the tissue-specific proteoglycans synthesized by chondrocytes from nanomelic chick embryos. *Biochemical Journal* 1982; **201**: 387–94.

149. McLeod M, Wynn-Williams A. The cardiovascular lesions in Marfan's syndrome. *Archives of Pathology* 1956; **61**: 143–8.

150. McKusick V A. Heritable disorders of connective tissue. III. The Marfan syndrome. *Journal of Chronic Diseases* 1955; **2**: 609–44.

151. McKusick V A. The cardiovascular aspects of Marfan's syndrome: a heritable disorder of connective tissue. *Circulation* 1955; **11**: 321–42.

152. McKusick V A. Heritable disorders of connective tissue. VI. Pseudoxanthoma elasticum. *Journal of Chronic Diseases* 1956; **3**: 263–83.

153. McKusick V A. *Human Genetics* (2nd edition). New Jersey: Prentice-Hall Inc., 1969.

154. McKusick V A. Homocystinuria. In: *Heritable Disorders of Connective Tissue* (4th edition). St Louis, Missouri: The C V Mosby Company, 1972: 224–81.

155. McKusick V A. Heritable disorders of connective tissue: new clinical and biochemical aspects. In: Peters D K, ed, *Twelfth Symposium on Advanced Medicine*. Tunbridge Wells: Pitman Medical, 1976: 170–91.

156. McKusick V A. *Mendelian Inheritance in Man* (6th edition). Baltimore: Johns Hopkins University Press, 1983.

157. McKusick V A, Neufeld E F. The mucopolysaccharide storage diseases. In: Stanbury J B, Wyngaarden J B, Fredrickson D S, Goldstein J L, Brown M S, eds, *The Metabolic Basis of Inherited Disease* (5th edition). New York: McGraw-Hill, 1983: 751–77.

158. Mapes C A, Anderson R L, Sweeley C C, Desnick R J, Krivit W. Enzyme replacement in Fabry's disease; an inborn error of metabolism. *Science* 1970; **169**: 987–9.

159. Marcet A J G. Account of singular variety of urine which turned black soon after being discharged; with some particulars respecting its chemical properties. *Medico-chirurgical Transactions London* 1822–23; **12**: 37–45.

160. Marfan A B. Un cas de déformation congénitale des quatre membres plus prononcé aux extrémités characterisé par l'allongement des os avec un certain degré d'amincissement. *Bulletins et Mémoires de la Société Médicale des Hôpitaux de Paris* 1896; **13**: 220–6.

161. Marjomäki V, Salminen A. Morphological and enzymatic heterogeneity of suramin-induced lysosomal storage disease in some tissues of mice and rats. *Experimental and Molecular Pathology* 1986; **45**: 76–83.

162. Maroteaux P, Frézal J, Cohen-Solal L. The differential symptomatology of errors of collagen metabolism: a tentative classification. *American Journal of Medical Genetics* 1986; **24**: 219–30.

163. Maroteaux P, Lamy M, Robert J M. Le nanisme thanatophore. *La Presse Médicale* 1967; **75**: 2519–24.

164. Matalon R, Dorfman A. Occasional survey—acid mucopolysaccharides in cultured human fibroblasts. *Lancet* 1969; **ii**: 838–40.

165. Menkes J H, Alter M, Steigleeder G K, Weakley D R, Sung J H. An X-linked recessive disorder with retardation of growth, peculiar hair, and focal cerebral and cerebellar degeneration. *Pediatrics* 1962; **29**: 764–79.

166. Minor R R. Collagen metabolism: a comparison of diseases of collagen and diseases affecting collagen. *American Journal of Pathology* 1980; **98**: 225–80.

167. Minor R R, Wootton J A M, Prockop D J, Patterson D F. Genetic diseases of connective tissues in animals. In: Wuepper K D, Geddedahl T, eds, *Biology of Heritable Skin Diseases. Current Problems in Dermatology*. Basel: Karger, 1987; **17**: 199–215.

167a. Mogle P, Amitai Y, Rotenberg M, Yatziv S. Calcification of intervertebral disks in I-cell disease. *European Journal of Pediatrics* 1986; **145**: 226–7.

168. Mohr W. Gelenkkrankheiten. *Diagnostik und Pathogenese Makroskopischer und Histologischer Strukturveranderungen.* Stuttgart: Georg Thieme Verlag, 1984: 213–26.

169. Mohr W, Wessinghage D, Lenschow E. The ultrastructure of hyaline cartilage and articular capsular tissue in alcaptonuria (ochronosis). *Zeitschrift für Rheumatologie* 1980; **39**: 55–73.

170. Moore H C. Marfan syndrome, dissecting aneurysm of the aorta, and pregnancy. *Journal of Clinical Pathology* 1965; **18**: 277–81.

171. Mudd S H, Levy H L. Disorders of transsulfuration. In: Stanbury J B, Wyngaarden J B, Frederickson D S, Goldstein J L, Brown M S, eds, *The Metabolic Basis of Inherited Disease* (5th edition). New York: McGraw-Hill, 1983; 522–59.

171a. Mudd S H, Levy H L, Skovby F. Disorders of transsulfuration. In: Scriver C R, Beaudet A L, Sly W S, Valle D, eds, *The Metabolic Basis of Inherited Disease* (6th edition). New York, London: McGraw-Hill Information Services Company, 1989: 693–734.

172. Nagashima K, Endo H, Sakakibara K, Konishi Y, Miyachi K O, Wey J J, Suzuki Y, Onisawa J. Morphological and biochemical studies of a case of mucopolysaccharidosis II (Hunter's syndrome). *Acta Pathologica Japonica* 1976; **26**: 115–32.

173. Neufeld E F, McKusick V A. Disorders of lysosomal enzyme synthesis and localization: I-cell disease and pseudo-Hürler polydystrophy. In: Stanbury J B, Wyngaarden J B, Frederickson D S, Goldstein J L, Brown M S, eds, *The Metabolic Basis of Inherited Disease* (5th edition). New York: McGraw-Hill, 1983: 778–87.

173a. Neufeld E, Muenzer J. The mucopolysaccaridoses. In: Scriver C R, Beaudet A L, Sly W S, Valle D, eds, *The Metabolic Basis of Inherited Disease* (6th edition). New York, London: McGraw-Hill Information Services Company, 1989: 1565–87.

174. Nogami H, Oohira A, Ozeki K, Oki T, Ogino T, Murachi S. Ultrastructure of cartilage in heritable disorders of connective tissue. *Clinical Orthopaedics and Related Research* 1979; **143**: 251–9.

175. O'Brien W M, La Du B N, Bunim J J. Biochemical, pathologic and clinical aspects of alkaptonuria, ochronosis and ochronotic arthropathy: review of world literature (1584–1962). *American Journal of Medicine* 1963; **34**: 813–38.

176. Okada R, Rosenthal I M, Scaravelli G, Lev M. A histopathologic study of the heart in gargoylism. *Archives of Pathology* 1967; **84**: 20–30.

177. Oohira A, Tamaki K, Ozeki K, Takamatsu K, Nogami H. Human cartilage proteoglycans isolated from normally ossifying and congenitally malformed leg bones. *Calcified Tissue International* 1980; **30**: 183–9.

178. Osborn T, Lichtenstein J R, Moore T L, Weiss T, Zuchner J. Ehlers–Danlos syndrome presenting as rheumatic manifestations in the child. *Journal of Rheumatology* 1981; **8**: 79–85.

179. Parfrey N A, Hutchins G M. Hepatic fibrosis in the mucopolysaccharidoses. *American Journal of Medicine* 1986; **81**: 825–9.

180. Parry D M, Safyer A W, Mulvihill J J. Waardenburg-like features with cataracts, small head size, joint abnormalities, hypogonadism, and osteosarcoma. *Journal of Medical Genetics* 1978; **15**: 66–9.

181. Pearson H A, Lorincz A E. A characteristic bone marrow finding in the Hürler syndrome. *Pediatrics* 1964; **34**: 281–2.

182. Pedrini-Mille A, Pedrini V. Proteoglycans and glycosaminoglycans of human achondroplastic cartilage. *Journal of Bone and Joint Surgery* 1982; **64A**: 39–46.

183. Pfeiffer R A, Palm D, Teller W. Syndrome of short stature, amimic facies, enamel hypoplasia, slowly progressive stiffness of joints and high pitched voice in 2 siblings. *Journal of Pediatrics* 1977; **91**: 955–7.

184. Philips B M, David T N. Pathogenesis and management of arthropathy in cystic fibrosis. *Journal of the Royal Society of Medicine* 1986; **25** (Suppl.): Abstr. 85.

185. Piérard G E, Le T, Hermanns J-F, Nusgens B V, Lapière C M. Morphometric study of cauliflower collagen fibrils in dermatosparaxis of calves. *Collagen and Related Research* 1986; **6**: 481–92.

186. Ponseti I V, Baird W A. Scoliosis and dissecting aneurysm of the aorta in rats fed with *Lathyrus odoratus* seeds. *American Journal of Pathology* 1952; **28**: 1059–78.

187. Pope F M, Nicholls A C. Molecular abnormalities of collagen in human disease. *Archives of Diseases in Childhood* 1987; **62**: 523–8.

188. Prockop D J, Kivirikko K I. Heritable diseases of collagen. *New England Journal of Medicine* 1984; **311**: 376–86.

189. Pryzbojewski J Z, Maritz F, Tiedt F A C, Van der Walt J J. Pseudoxanthoma elasticum with cardiac involvement. A case report and review of the literature. *South African Medical Journal* 1981; **59**: 268–75.

190. Puchtler H, Meloan S N, Waldrop F S. Are picro-dye reactions for collagens quantitative? *Histochemistry* 1988; **88**: 243–56.

191. Purkiss P, Gibbs D A, Watts R W E. Studies on the composition of urinary glycosaminoglycans and oligosaccharides in patients with mucopolysaccharidoses who were receiving fibroblast transplants. *Clinica Chimica Acta* 1983; **131**: 109–21.

192. Ramsey M S, Yanoff M, Fine B S. The ocular histopathology of homocystinuria. A light and electron microscopic study. *American Journal of Ophthalmology* 1972; **74**: 377–85.

193. Rees S. Membranous neuronal and neuroglial inclusions produced by intracerebral injection of suramin. *Journal of Neurological Sciences* 1978; **36**: 97–109.

194. Rees S, Constantopoulos G, Brady R O. The suramin-treated rat as a model of mucopolysaccharidosis. Variation in the reversibility of biochemical and morphological changes among different organs. *Virchows Archiv B (Cell Pathology)* 1986; **52**: 259–72.

195. Renteria V G, Ferrans V J. Intracellular collagen fibrils in cardiac valves of patients with the Hürler syndrome. *Laboratory Investigation* 1976; **34**: 263–72.

196. Rimoin D L, Horton W A. Short stature. *Journal of Pediatrics* 1978; **92**: 523–8, 697–704.

197. Rimoin D L, Sillence D O. The skeletal dysplasias: nomenclature, classification and clinical evaluation. In: Akeson W H, Bornstein P, Glimcher M J, eds, *Heritable Disorders of*

Connective Tissue. St Louis, Missouri: The C V Mosby Company, 1982; 325–32.

198. Ross R, Fialkow P J, Altman L K. Fine structure alterations of elastic fibres in pseudoxanthoma elasticum. *Clinical Genetics* 1978; **13**: 213–23.

199. Rubin P. *Dynamic Classification of Bone Dysplasias*. Chicago: Yearbook Medical Publishers, 1964.

200. Ruttin E. Osteopsathyrosis und Otosklerose. *Zeitschrift für Hals-, Nasen-, und Ohrheilkunde* 1922; **3**: 263–79.

201. Savage A V, Applegarth D A. Diagnosis of mucopolysaccharidoses using ^{1}H-n.m.r. spectroscopy of glycosaminoglycans. *Carbohydrate Research* 1986; **149**: 471–4.

202. Schafer I A, Sullivan J C, Svejcar J, Kofoed J, Robertson W van B. Study of the Hürler syndrome using cell culture: definition of the biochemical pheno-type and the effect of ascorbic acid on the mutant cell. *Journal of Clinical Investigation* 1968; **47**: 321–8.

203. Scheck M, Siegel R C, Robert C, Parker J, Chang Y H, Fu J C C. Aortic aneurysm in Marfan's syndrome: changes in the ultrastructure and composition of collagen. *Journal of Anatomy* 1979; **129**: 645–57.

204. Schoenfeld M R, Nicolo Paganini: musical magician and Marfan mutant? *Journal of the American Medical Association* 1978; **239**: 40–2.

205. Schumacher H R. Haemochromatosis and arthritis. *Arthritis and Rheumatism* 1964; **7**: 41–50.

206. Schumacher H R. Articular cartilage in the degenerative arthropathy of haemochromatosis. *Arthritis and Rheumatism* 1982; **25**: 1460–8.

207. Schumacher H R. Ochronosis, hemochromatosis and Wilson's disease. In: McCarty D J, ed, *Arthritis and Allied Conditions*. Philadelphia: Lea & Febiger, 1985: 1565–78.

208. Schwartz H. Abraham Lincoln and the Marfan syndrome. *Journal of American Medical Association* 1964; **187**: 473–9.

209. Scribonius G A. De Inspectione Urinarum. Germany: Lemgo, 1584: 50.

209a. Scriver C R, Beaudet A L, Sly W S, Valle D. *The Metabolic Basis of Inherited Disease* (6th edition). New York, London: McGraw-Hill Information Services Company, 1989.

210. Shapiro F. Epiphyseal disorders. *New England Journal of Medicine* 1987; **317**: 1702–10.

211. Sheldon J H. *Haemochromatosis*. Oxford: Oxford University Press, 1935.

212. Shohet I, Rosenbaum I, Frand M, Duksin D, Engelberg S, Goodman R M. Cardiovascular complications in the Ehlers–Danlos syndrome with minimal external findings. *Clinical Genetics* 1987; **31**: 148–52.

213. Shull R M, Munger R J, Spellay E, Hall C W, Constantopoulos G, Neufeld E F. Animal models of human disease: canine α-l-iduronidase deficiency: a model of mucopolysaccharidosis. *American Journal of Pathology* 1982; **109**: 244–8.

214. Siegel R C. The connective tissue defect in homocystinuria. *Clinical Research* 1975; **23**: 263A.

215. Siegel R C, Pinnell S R. Diseases associated with abnormalities of structural proteins. In: Kelly W N, Harris E D, Ruddy S, Sledge C B, eds, *Textbook of Rheumatology*. Philadelphia: W B Saunders & Co., 1981; 1675–83.

216. Silberberg R, Rimoin D L, Rosenthal R E, Hasler M B. Ultrastructure of cartilage in the Hürler and Sanfilippo syndromes. *Archives of Pathology* 1972; **94**: 500–10.

217. Simpson C F, Boucek R J, Noble N L. Similarity of aortic pathology in Marfan's syndrome, copper deficiency in chicks and β-aminopropionitrile toxicity in turkeys. *Experimental and Molecular Pathology* 1980; **32**: 81–90.

218. Sinclair R J G, Kitchin A H, Turner R W D. The Marfan syndrome. *Quarterly Journal of Medicine* 1960; **29**: 19–46.

218a. Slavkin H C, Greulich R C. *Extracellular Matrix Influences on Gene Expression*. New York, San Francisco, London: Academic Press, 1975.

219. Smith D W. *Recognizable Patterns of Human Malformations*. Philadelphia: W B Saunders & Co., 1970.

220. Smith R. Collagen and disorders of bone. *Clinical Science* 1980; **59**: 215–23.

221. Smith R, Francis M J O, Houghton G R. *The Brittle Bone Syndrome*. London: Butterworths, 1983.

222. Smith R, Sykes B. Osteogenesis imperfecta (the brittle bone syndrome): advances and controversies. *Calcified Tissue International* 1985; **37**: 107–11.

223. Smithwick E M, Finelt M, Pahwa S, Good R A, Naspitz C K, Mendes N F, Kopersztyk S, Spira T J, Nahmias A J. Cranial synostosis in Job's syndrome. *Lancet* 1978; **i**: 826.

224. Spranger J W, Langer L O, Wiedmann H R. *Bone Dysplasias: An Atlas of Constant Disorders of Skeletal Development*. Philadelphia: W B Saunders & Co., 1974.

225. Stanbury J B, Wyngaarden J B, Fredrickson D S, Goldstein J L, Brown M S. *The Metabolic Basis of Inherited Disease* (5th edition). New York: McGraw-Hill, 1983.

226. Stickler G B, Belau P G, Farrell F J, Jones J D, Pugh D G, Steinberg A G, Ward L E. Hereditary progressive arthroophthalmopathy. *Mayo Clinic Proceedings* 1965; **40**: 433–55.

227. Stilling H. Osteogenesis imperfecta. *Virchows Archivs für Pathologische Anatomie und Physiologie und für Klinische Medizin* 1889; **115**: 357–70.

228. Strandberg J. Pseudoxanthoma elasticum. *Zeitschrift für Haut und Geschlechtskrankheiten* 1929; **31**: 689.

229. Strole W E, Margolis R J. Pseudoxanthoma elasticum. *New England Journal of Medicine* 1983; **308**: 579–85.

230. Sykes B. Genetics cracks bone disease. *Nature* 1987; **330**: 607–8.

230a. Sykes, B. Mapping collagen gene mutations. In: Fleischmajer R, Olsen B R, Kuhn K, eds, *Structure, Molecular Biology and Pathology of Collagen. Annals of the New York Academy of Sciences* 1990; **580**: 385–9.

231. Tabone E, Grimaud J-A, Peyrol S, Grandperret D, Durrand L. Ultrastructural aspects of corneal fibrous tissue in the Scheie syndrome. *Virchows Archives B (Cell Pathology)* 1978; **27**: 63–7.

232. Talamo R C, Rosenstein B J, Berninger R W. Cystic fibrosis. In: Stanbury J B, Wyngaarden J B, Fredrickson D S, Goldstein J L, Brown M S, eds, *The Metabolic Basis of Inherited Disease*. New York: McGraw-Hill, 1983: 1889–917.

233. Taysi K, Kistenmacher M L, Punnett H H, Mellman W J. Limitations of metachromasia as a diagnostic aid in pediatrics. *New England Journal of Medicine* 1969; **281**: 1108–10.

234. Toledo S P A, Saldanha P H. A radiological and genetic investigation of acheiropody in a kindred including six cases.

Journal de Génétique Humaine 1966; **15**: 81–9.

235. Tønnesen T, Muller-Schauenburg G, Damsgaard E, Horn N. Copper-measurement in a muscle-biopsy. A possible method for postmortem diagnosis of Menkes' disease. *Clinical Genetics* 1986; **29**: 258–61.

236. Tønnesen T, Silango M, Gerdes A-M, Hansen J C, Reske-Nielsen E, Franceschini P, Horn N. Postmortem Menkes' diagnosis from carrier testing of female relatives. *Clinical Genetics* 1987; **32**: 393–7.

237. Touraine A. L'élastorrhexie systématisée. *Bulletin de la Société français de Dermatologie et de Syphilologie* 1940; **47**: 255–73.

238. Tunbridge R E, Tattersall R N, Hale D A, Astbury W T, Reed R. The fibrous structure of normal and of abnormal skin. *Clinical Science* 1952; **11**: 315–31.

239. Uitto J, Minor R R. Animal models of the Ehlers–Danlos syndrome. In: Maibach H I, Lowe N J, eds, *Models in Dermatology*. Basel: Karger, 1987; **3**: 148–58.

240. Unna P G. *The Histopathology of the Diseases of the Skin* (translated from the German with the assistance of the author by N Walker). New York: Macmillan, 1896: 984–8.

241. Van Makeeren J A. De dilatabilitate extraordinaria cutis. *Observationes Medico-chirurgiae*. Amsterdam: 1682.

242. Virchow R. Ein Fall von allgemeiner Ochronose der Knorpel und Knorpelahnliche Theile. *Virchows Archives für Pathologische Anatomie und Physiologie und für Klinsche Medizin* 1866; **37**: 212–9.

243. Vogel A, Holbrook K A, Steinmann B, Gitzelmann R, Buyers P H. Abnormal collagen fibril structure in the gravis form (type I) of the Ehlers–Danlos syndrome. *Laboratory Investigation* 1979; **40**: 201–6.

244. Vrolik W. Tabulae ad illustrandum embryogenesis hominis et mammalium, tam naturalem quam abnormen. *Treasury of Human Inheritance* (quoted by Bell J, 1928). London: University of London Press, 1849: Vol. 2, part 3.

245. Weatherall D J. Molecular pathology of single gene disorders. *Journal of Clinical Pathology* 1987; **40**: 959–70.

246. Webber R J, Fox R R, Sokoloff L. *In vitro* culture of rabbit growth plate chondrocytes. 2. Chondrodystrophic mutants. *Growth* 1981; **45**: 269–78.

247. Weiland A J, Daniel R K. Congenital pseudarthrosis of the tibia—treatment with vascularized autogenous fibular grafts—a preliminary report. *Johns Hopkins Medical Journal* 1980; **147**: 89–95.

248. Weiss S W, Kelly W D. Bilateral destructive synovitis associated with alpha mannosidase deficiency. *American Journal of Surgical Pathology* 1983; **7**: 487–94.

249. Welkow M, Baumann E. Ueber das Wesen de Alkaptonurie. *Hoppe-Seylers Zeitschrift für physiologische Chemie* 1891; **15**: 228–85.

250. Wenstrup R J, Hunter A G W, Byers P H. Osteogenesis imperfecta type IV: evidence of abnormal triple helical structure of type I collagen. *Human Genetics* 1986; **74**: 47–53.

251. Weve H. Über Arachnodaktylie (Dystrophia mesodermalis congenita, typus Marfanis). *Archiv für Augenheilkunde* 1931; **104**: 1–46.

252. Whittaker P, Boughner D R, Perkins D G, Canham P B. Quantitative structural analysis of collagen in chordae tendineae and its relation to floppy mitral valves and proteoglycan infiltration. *British Heart Journal* 1987; **57**: 264–9.

253. Wicks I P, Fleming A. Chondrosarcoma of the calcaneum and massive soft tissue calcification in a patient with hereditary and acquired connective tissue diseases. *Annals of the Rheumatic Diseases* 1987; **46**: 346–8.

254. Wynne-Davis R. *Heritable Disorders in Orthopaedic Practice*. Oxford: Blackwell Scientific Publications, 1973.

Chapter 10

Metabolic, Nutritional and Endocrine Diseases of Connective Tissue

D L GARDNER and J McCLURE

The extent of the connective tissues means that they are directly or indirectly affected by many metabolic disorders. There is often a hereditary predisposition, and the diseases in which genetic rather than environmental causes are most important are described in Chapter 9. An important class of connective tissue disease is characterized by the formation of insoluble, extracellular crystals or by inappropriate, extraosseous mineralization.

The normality or abnormality of the calcified skeleton also strongly influences patterns of disease expressed in the extraskeletal connective tissues. Other disorders result from the adverse effects of exogenous organic or inorganic compounds or from the influences of dietary deficiencies or excesses. There remains a category of connective tissue disease attributable to endocrine dysfunction.

Crystal deposition disease

Insoluble crystals (Fig. 10.1; Table 10.1) and non-crystalline mineral deposits often provoke connective tissue disease.[47,132,149,239] The clinical signs, investigation, aetiology and pathogenesis of the crystal deposition diseases have been fully described.[47,132,163a,108,128] There are relatively few comprehensive pathological accounts.[101,159,160] Within articular tissues, monosodium urate monohydrate, calcium pyrophosphate dihydrate (CPPD) and calcium hydroxyapatite are deposited in acute and chronic gouty arthritis; in acute and chronic pyrophosphate arthropathy;[43] and in acute calcific periarthritis and osteoarthrosis, respectively[47] (Tables 10.2 and 10.8). Among other crystals that may be detected in joints are dicalcium phosphate dihydrate (Brushite) and basic calcium phosphate; cholesterol; and calcium oxalate. Corticosteroids may be present as a result of therapeutic procedures.

The larger crystals are distinguishable by light microscopy from fragments of bone and cartilage; metals, plastics, fibres and vegetable material; and cements, polymers and radioopaque media (see Chapter 24). They can also be differentiated histologically from artefacts such as squamous cells, atmospheric dust and crystals formed from mountants. The smaller crystals such as those of calcium hydroxyapatite may only be recognizable when techniques such as scanning electron microscopy are employed. Within the connective tissues, crystal deposits may be found by chance or when symptoms, such as those of acute arthritis, lead to systematic investigation. Individual clinical signs indicate the probability of a crystal disease but further procedures are necessary before the diagnosis can be confirmed pathologically. A first step is the demonstration

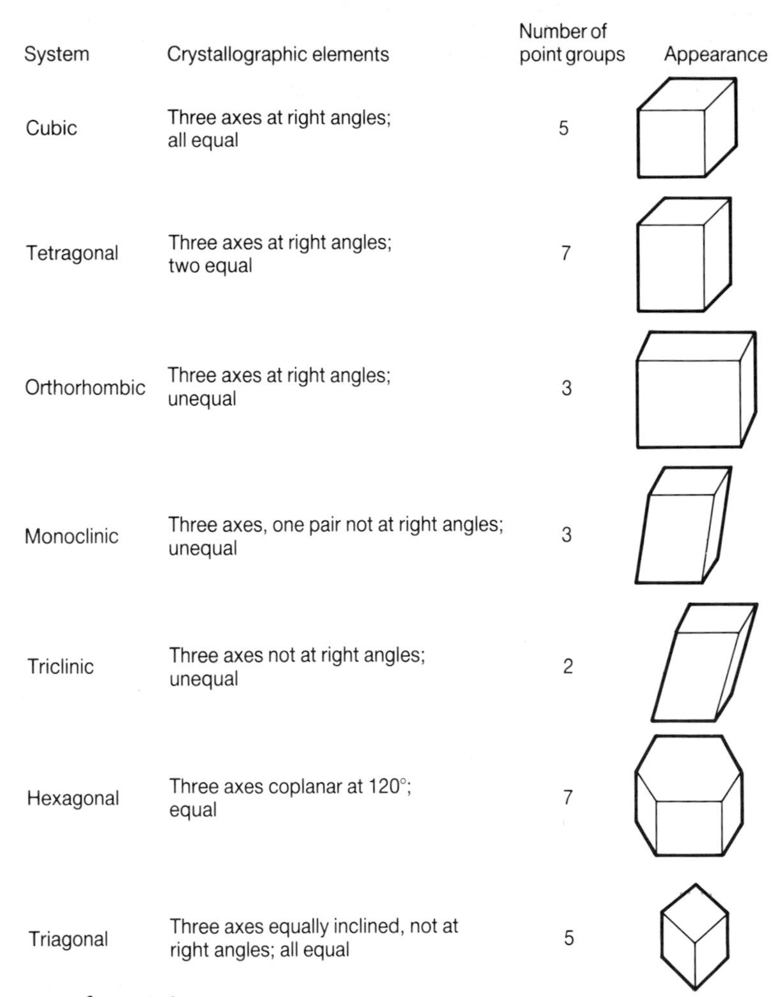

Fig. 10.1 The seven classes of crystals.
Redrawn from Dieppe and Calvert.[47]

of crystals in the affected tissues; a second step is the identification of the crystals.

The variety of techniques which can be used in the pathological diagnosis of the crystal deposition diseases is considerable (Table 10.3).[27,42,47] There are difficulties in the interpretation of samples received for pathological study.[47] The collection of synovial and other fluids by joint aspiration may dislodge crystals from skin, subcutaneous tissue, articular capsule or cartilage. After aspiration, pH and ambient temperature affect crystal formation and dissolution. The number of crystals is increased by cooling, decreased by warming. Samples can be covered with oil to reduce carbon dioxide loss and pH change but brushite crystals, for example, can first form at room temperature,[49] and increased crystal formation may occur when the pH of a gouty synovial fluid is increased to 7.4. The analysis of histological specimens may also introduce artefact. Without buffering, fixatives and staining reagents can promote pH change and crystal dissolution. Crystals may fall out of light and transmission electron microscopic preparations. Measurements of crystal size and shape in tissues and the determination of optical characteristics,[27,42,259] so important for exact diagnosis, may be imprecise.

For these reasons, the reliable identification of crystals in histological preparations is often difficult. Crystals in synovial fluid preparations are exposed to fewer circumstances likely to cause artefact than are those in tissue sections. The collection and study of synovial fluids (p. 89) therefore plays a particularly important part in the investigation of the crystal deposition diseases.

Foremost among the light microscopic techniques employed in the search for and classification of crystals in synovial fluids and tissue sections is polarization microscopy. Laboratories likely to be confronted with the need to find and identify crystals therefore require a microscope with lenses of appropriate design, a rotating stage, easily manipulated polarizing and analysing plates ('filters'), and a lambda plate. The use of polarization microscopy in patholo-

Table 10.1

Crystals* that have been identified in human articular and periarticular tissues[47]

Monosodium urate monohydrate
'Urate' spherulites[66]
Ultra-small urate crystals
Monoclinic calcium pyrophosphate dihydrate
Triclinic calcium pyrophosphate dihydrate
Ultra-small pyrophosphate crystals
Hydroxyapatite
Octacalcium phosphate
Apatite spherulites
Dicalcium phosphate dihydrate
Tricalcium phosphate
Calcium triphosphate
Calcium carbonate
Calcium oxalate
Cholesterol
Liquid lipid crystals
(mixtures of crystals)

* Materials like diamond crystallize in a cubic system (Fig. 10.1): they are singly refracting.

There are six other crystal systems; in each, the structure is asymmetrical and light incident upon them is generally split into two rays which take different paths and for which the refractive indices are different: they are doubly refracting.

Table 10.2

Crystals containing calcium identified in human synovial fluid, synovium or cartilage[84,130]

Crystal	Chemical formulæ	Molar ratios of calcium to phosphate
Basic calcium phosphates		
Hydroxyapatite	$Ca_5(PO_4)_3OH.2H_2O$	1.67
Octacalcium phosphate	$Ca_8H_2(PO_4)_6.5H_2O$	1.33
Tricalcium phosphate (Whitlockite)	$Ca_3(PO_4)_2$	1.5
Others		
Calcium phosphate dihydrate (Brushite)	$CaHPO_4.2H_2O$	1.0
Calcium pyrophosphate dihydrate	$Ca_2P_2O_7.2H_2O$	1.0
Calcium oxalate	$CaC_2O_4.H_2O$	—

Table 10.3

Some histopathological techniques in the diagnosis of crystal deposition diseases

Stains
von Kossa (for calcium salts)
Methenamine silver (for urates)
Alizarin (for calcium)

Histochemical test for uric acid/urates
Uricase digestion

Light microscopic methods
Polarized light microscopy[259] without or with first-order red compensator (lambda plate)

Electron optical methods
Transmission electron microscopy
Scanning electron microscopy
Electron probe X-ray microanalysis[83]

Other methods applicable if adequate samples are available
Infrared spectroscopy[42]
X-ray diffraction (requires substantial pure samples of crystals)[27]

gical histology is comprehensively described by Wolman[259] whose account is amplified by Dieppe and Calvert[47] and by Vernon-Roberts.[247a]

Urate deposition disease (gout)

Gout[261] is a clinical syndrome. It is an occasional consequence of persistently elevated serum urate levels, hyperuricaemia. Hyperuricaemia is commonplace in Western societies (Table 10.4).[22,128,209,210] Why gout does not develop more often remains unexplained. In gout, monosodium biurate monohydrate crystals accumulate in selected sites. Acute gout affects mainly synovial joints, cartilage, tendon sheaths and bursae but the local aggregation of monosodium biurate crystals also occurs in non-articular cartilage. Articular deposits culminate in a destructive arthropathy; local aggregates of monosodium biurate may also appear as disfiguring granulomata. Visceral and cardiovascular disease are frequent and renal calculi form.[209,236] Hyperuricaemia is usually unexplained; and gout occurring in these circumstances is said to be primary. Gout may also arise as a secondary result of a distinct disorder such as lymphoproliferative disease.[171] Some drugs such as diuretics promote raised serum urate levels

Table 10.4

Factors causing hyperuricaemia[210]

Increased uric acid formation
Specific enzyme abnormalities, e.g. decreased hypoxanthine-guanine phosphoribosyl transferase
Increased nucleoprotein turnover
Diet
Exogenous chemicals
Decreased uric acid excretion
Alterations in renal function; reduced extracellular fluid volume
Drugs
Lactic acidaemia
Starvation and ketosis
Essential hypertension
Lead poisoning
Hypercalcaemia
Myxoedema
Other factors
Race, sex and age
Genetic
Body weight
Social class, intelligence
Alcohol
Cardiovascular disease
Diabetes mellitus
Haemoglobin and plasma proteins

while excess ethanol, obesity, hyperlipidaemia, and lead intoxication are other causes.

History[29,41,86,195]

Gout is a disease of antiquity; its features were noted by Hippocrates. The name *gutta* (Latin = a drop) came from the belief that a noxious substance was distilled drop-by-drop into affected joints. Sydenham,[232] a chronic sufferer, depicted in graphic detail the agonies of the patient. Uric acid was identified in urinary calculi.[204] Gouty tophi (see p. 387) were then found to contain urate,[258] and Garrod[69] demonstrated unusual amounts of uric acid in the blood. In recent years, understanding of hereditary enzyme defects[261] and the use of uricosuric agents[211] have greatly advanced knowledge of the disease.

Clinical presentation

Asymptomatic hyperuricaemia may persist for more than 20 years before arthritis or urinary calculi are recognized. Acute attacks of gout are associated with either a rising or a diminishing serum urate concentration, rather than with a sustained, constantly high urate level. Although acute gout often appears spontaneously, it may be provoked by local trauma, systemic illness or surgical operation, or by drug therapy. Allopurinol treatment may cause xanthine or hypoxanthine crystals to accumulate in the tissues.[250]

Gout begins with explosive inflammation. The first metatarsophalangeal joint is characteristically attacked:

> The victim goes to bed and sleeps in good health. About two o'clock in the morning he is awakened by a severe pain in the great toe; more rarely, in the heel, ankle or instep. The pain is like that of a dislocation, and yet the parts feel as if cold water were poured over them . . . now it is a violent stretching and tearing of the ligaments, now it is a gnawing pain and now a pressure and tightening
>
> Sydenham, 1683[232]

The signs mimic acute purulent infection and there is fever and polymorph leucocytosis. Rarely, pyoarthrosis may indeed complicate gouty arthritis.[93] An effusion appears. The first metatarsophalangeal joint is affected in approximately 70 per cent of cases but the ankles, other foot joints, knees, fingers, elbows and wrists are attacked in this order of frequency; the disease is distal rather than proximal. The carpal tunnel syndrome may be a result of gouty tenosynovitis.[102] Untreated gout may persist without remission. The duration of attacks increases with age. Episodes of severe acute arthritis recur with increasing frequency but they are separated by intervals in which the patient is symptom-free and capable of physical activity without restraint. Although 90 per cent of first attacks affect only one joint, 5–10 per cent of cases are polyarticular and systemic connective tissue disease (see Chapter 12) may be simulated.

Recurrence is likely. Recurrent acute inflammation culminates in the progressive local accumulation of monosodium biurate crystals. This aggregate is a *tophus* (see p. 387) (Latin = a loose porous stone). With time, tophi become larger. They develop in soft connective tissue and in cartilages, including those of the helix of the ear and within subchondral bone. Secondary calcification occurs and the lesion ulcerates through overlying skin surfaces from which fragments of white, crystalline material may escape. When changes of this type occur within joint tissue, a chronic destructive articular disease develops which elicits secondary inflammation. Gouty arthritis becomes more deforming as repetitive acute attacks are succeeded by the increasing local deposition of urates; ultimately, the result is a severe, degenerative arthropathy which can be regarded as a form of secondary osteoarthrosis (see Chapter 22). Gout may coincide with rheumatoid arthritis[170,193,234] or with one of the other less common systemic connective tissue diseases of immunological origin. As many as 6 per cent of those with gout also display chondrocalcinosis (see p. 393).[231] An association with diabetes mellitus is suspected.[17,255]

Radiological changes include zones 'punched-out' from the bone of, for example, the first metatarsal. The erosions recall those of rheumatoid arthritis, osteoarthrosis and granulomatous infection but an overhanging margin extending from the bone is diagnostic. Comparable changes, with the formation of monosodium biurate tophi, develop asymmetrically in the bones of the fingers and toes, but less often in the long bones and axial skeleton. Extensive joint disorganization may occasionally be followed by bony ankylosis and by pathological fracture.

Frequency

Gout is a disorder of older men. The disease is uncommon in women, particularly before the menopause. Thereafter, the frequency increases. The Framingham study suggested that gout was seven times more frequent in men than women.

Gout accounts for 5–6 per cent of rheumatology clinic cases in the UK[209] but the prevalence varies in different populations. The prevalence of the disease in a European population was 0.3 per cent, in a North American 0.27 per cent. There are underlying geographical differences in serum urate concentrations in single ethnic groups. The prevalence of gout among Filipinos, for example, varies according to whether they live in the Philippines or in the USA.[91]

Inheritance[228,90]

In rare instances genetic abnormalities are responsible for single enzyme defects leading to gout (see below). In the great majority of cases of gout, the role of heredity remains uncertain. Hyperuricaemia and gout may be familial. Whether hyperuricaemia is then determined by a single autosomal dominant gene or on a polygenic, multifactorial basis, is disputed. The operation of a number of genes regulating a series of enzymes may determine hyperuricaemia but in a single population, the influence of one gene may predominate.[28,169,82]

Biochemical basis

The biochemical basis[130,211,236,251] of gout is still not fully understood. In man, uric acid is the normal end-product of purine metabolism (Fig. 10.2). Uric acid is excreted by glomerular filtration but much of the filtrate is reabsorbed in the proximal renal tubule, to be lost subsequently by active tubular secretion (Fig. 10.3). In addition, one third of the

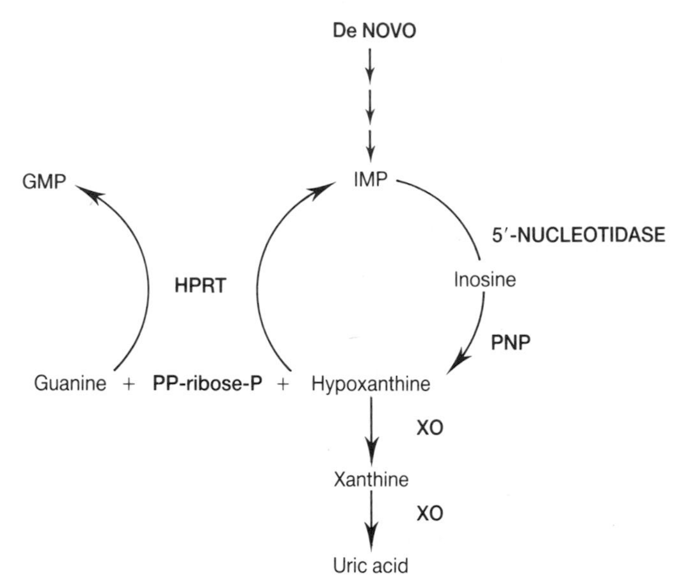

Fig. 10.2 Uric acid formation.
Catabolism of purine ribonucleotides, re-use of purine base and oxidation of purine to uric acid. GMP, guanosine monophosphate; IMP, inosine 5′-monophosphate; HPRT, hypoxanthine-guanine phosphoribosyl transferase; PNP, purine nucleoside phosphorylase; PP-ribose-P, phosphoribosyl pyrophosphate; XO, xanthine oxidase. Redrawn from Levinson.[128]

uric acid produced daily is excreted into the gastrointestinal tract. There is no evidence that hyperuricaemia results from excess purine absorption, and the contribution of purine metabolism to hyperuricaemia must come from decreased purine excretion, diminished purine destruction, increased urate formation or from more than one of these mechanisms. Many patients with gout excrete a great excess of uric acid. Others, with normal urinary loss, may in part dispose of the excess urate by increased destruction in organs other than the kidney; the body pool of urate is increased. In patients with gout, the rate of breakdown of uric acid outside the kidney is not reduced: it may even be increased. Those excreting excess uric acid may also synthesize uric acid more quickly than normal but this phenomenon is detectable in others who excrete normally.

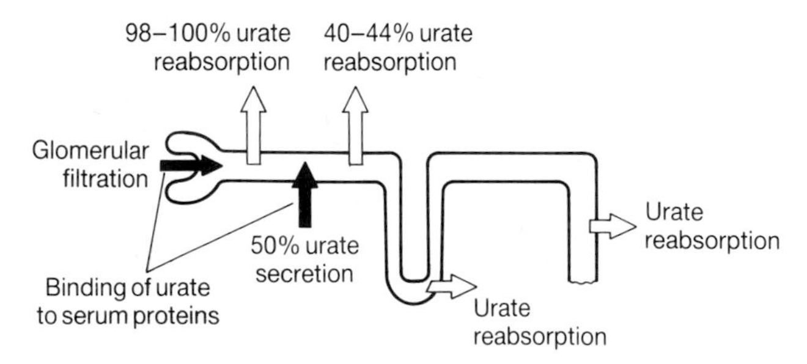

Fig. 10.3 Renal processing of urate.
Three aspects of the handling of urate by the kidney: glomerular filtration, tubular reabsorption and tubular secretion. Redrawn from Levinson.[128]

There have been extensive investigations of purine metabolism. Thus, purine synthetic activity can be measured reliably in lymphocytes prepared from blood, using carbon-14 formate.[79] Phosphoribosyl-pyrophosphate and the purine ribonucleotides are important components of the metabolic path. The role of particular enzymes in catalysing different parts of this pathway for purine synthesis and the ways in which components of this pathway control synthesis have been fully described;[128] in studies of lymphocytes from patients with clinical gout no subgroup was found with intrinsic, *de novo* abnormalities.

Rare heritable enzymatic defects that cause excess uric acid production are well known.[15,213,214] Understanding of these anomalies has helped to elucidate the pathogenesis of gout; the pathological features of these cases do not differ from those of idiopathic gout. The heritable enzyme defects are divisible into three groups: 1. those causing an increased concentration or availability of phosphoribosyl-pyrophosphate; 2. those resulting in a diminished intracellular pool of purine ribonucleotides; and 3. those leading to a raised glutamine concentration. The abnormalities that may result in increased phosphoribosyl-pyrophosphate are shown in Fig. 10.4. No enzymatic abnormalities are yet known in which a diminished pool of purine ribonucleotides results. The theoretical possibility that glutaminase- and glutamate dehydrogenase activity may lead to increased levels of available glutamine and increased purine synthesis has not yet been substantiated.

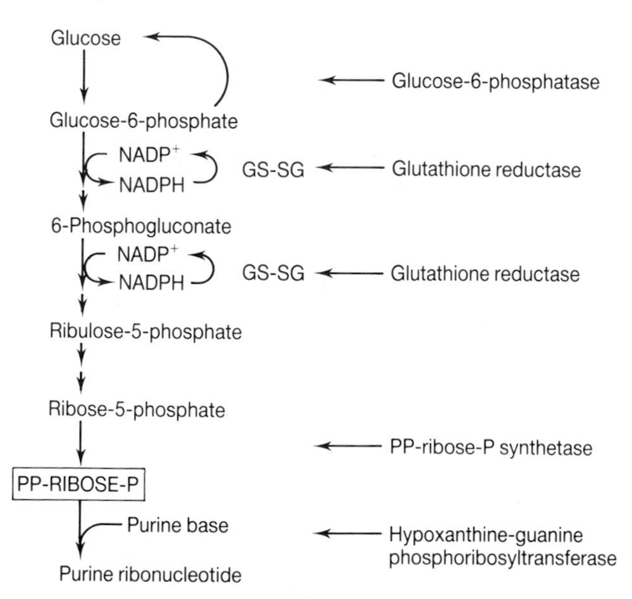

Fig. 10.4 Abnormal PP-ribose-P production.
At right, enzymes, excess or reduced activity of which might promote raised PP-ribose-P formation. NADP, nicotinamide-adenine dinucleotide phosphate; NADPH, reduced (hydrogenated) form of NADP; GS-SG, glutathione; PP-ribose-P, phosphoribosyl pyrophosphate. Redrawn from Levinson.[128]

The heritable biochemical abnormalities that attract the greatest interest are those associated with hypoxanthine–guanine phosphoribosyl transferase or glucose-6-phosphatase deficiencies, or with phosphoribosyl-pyrophosphate synthetase overactivity. Hypoxanthine-guanine phosphoribosyl transferase deficiency, an X-linked disorder, may be inherited in partial or complete forms. Complete deficiency is manifested as the Lesch–Nyhan syndrome. Hyperuricaemia and renal calculus formation are accompanied by neurological abnormality, mental retardation, self-mutilation, choreoathetotic movements and spasticity. Glucose-6-phosphatase deficiency underlies glycogen storage disease type I, von Gierke's disease; gout develops prematurely and contributes to increased morbidity. In X-linked phosphoribosyl-pyrophosphate synthetase overactivity, there are no distinct clinical or pathological features but gout and renal calculi develop early in life.

Pathological characteristics[129,208]

There can be said to be five phases.[259b]

1. Crystal formation and recognition The origin of gout centres on crystal formation[53,128] In principle, crystals of monosodium biurate monohydrate form when the solute concentration of urate is high (Fig. 10.5). Nucleation slowly occurs, promoted by macromolecules such as

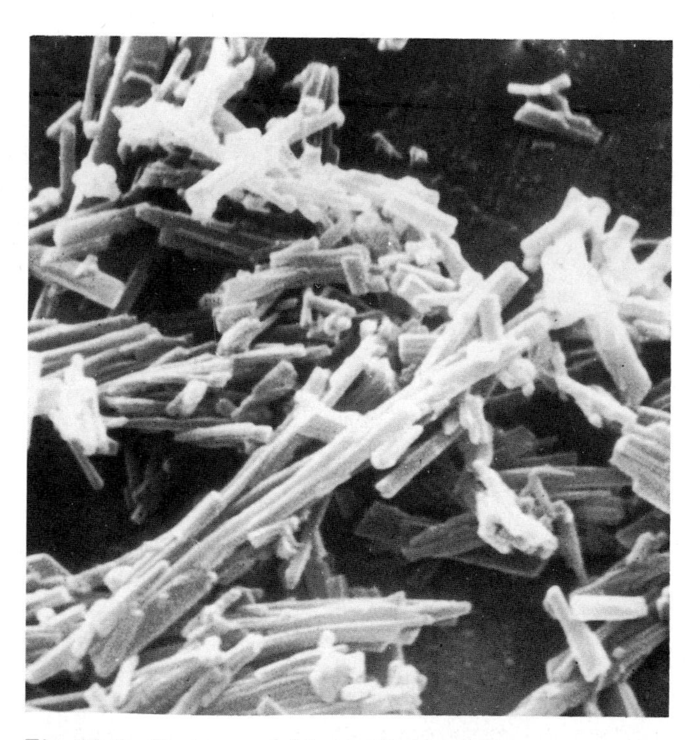

Fig. 10.5 Urate crystal formation *in vitro*.
Urate crystals formed in a supersaturated solution of sodium urate. (SEM × 6250.) By courtesy of Professor Paul Dieppe.

fibrillar collagen and the associated proteoglycan. Single crystals facilitate the growth of others and the release of crystals from the synovium or articular cartilage into the joint may contribute to this response. The presence of other particulate material such as fragments of bone may also catalyse crystal growth and it is accepted that trauma is one factor predisposing to crystal formation.

Why monosodium biurate crystals excite inflammation much more readily than those of calcium pyrophosphate or calcium hydroxyapatite has excited debate.[259b] In general terms the size and shape of crystals are regarded as relevant factors;[66,172] crystals within the range 2–20 μm in length appear most active. Roughness, in atomic terms, contributes to the inflammatory potential, as does the net surface negative charge. However, the surfaces of active and inert crystals appear equally rich in oxygen and hydroxy groups. The influence of surface ionic charge may, in addition, be indirect so that crystals carrying high net negative charges, bind proteins such as the Fab fragment of immunoglobulin avidly.

The surface structure of monosodium biurate crystals plays a crucial part in initiating tissue responses.[182] Although the internal architecture of these crystals comprises a complex, interconnecting but electron-lucent network, the external ultrastructure is superficially smooth. Naked monosodium biurate crystals have the property of lysing cell membranes but they achieve potency as phlogistic agents only when proteins bind to the crystal surface.[87,131,133] Many serum and matrix molecules, for example, acute phase proteins, lipoproteins and fibronectin have this capacity.[38] However, the most important, in terms of predisposition to enhanced biological reactivity, are the immunoglobulins of which the binding capacity of immunoglobulin G is very much greater than that of immunoglobulins A and M.

As Woolf and Dieppe[259b] emphasize, interactions between the crystal surface and the many molecular species to which the crystal is exposed *in vivo*, are certain to be highly complex. Nevertheless, numerous *in vitro* studies indicate that the binding of immunoglobulin G to the monosodium biurate crystal is an important step in a sequence culminating in inflammation: crystal–cell interactions are promoted and pathways, such as those of complement activation,[200] initiated. It is not surprising that a preliminary coating of monosodium biurate with a polymer such as polyvinyl pyridine-N-oxyide, inhibits these effects as do some serum proteins (see below).

Immunoglobulin-G-coated crystals can be vigorously phagocytosed by polymorphs,[176,243] mononuclear macrophages and synoviocytes. The positively charged component Fab of the immunoglobulin molecule binds with the negatively charged monosodium biurate crystal surface, leaving the Fc portion exposed. A complex but rapid sequence of responses is promoted. Of these responses, leukocyte chemotaxis and phagocytosis are the most important.

Chemotaxis, in this pre-inflammatory phase of gout, is effected by mediators such as leukotriene B_4 (p. 288) but also by less well-defined molecules such as 'chemotaxin'. Chemotaxin(s) is variously described as a 8.5 or 11.5 kD mol. wt glycoprotein which can be identified in polymorph lysosomes. This agent(s) is synthesized immediately before its release. Since chemotaxin(s) is synthesized by, released from and acts on polymorphs, it is said to have autocrine properties.

In response to chemotactic agents, and subsequent to a rise in intra-arterial blood pressure, the secretion of prostaglandin E_2 and increased vascular permeability, polymorphs quickly leave the blood stream and move actively towards the monosodium biurate clusters. The key role of polymorphs in initiating the inflammatory phase of urate arthritis is shown by experimental evidence that neutropenia greatly reduces the biological response to these crystals. Many of the subsequent changes now depend on phagocytosis of the urate crystals. By contrast with the reactions to calcium pyrophosphate and calcium hydroxyapatite, crystals which evoke relatively less phagocytosis and which may remain inert within intracellular phagolysosomes, monosodium biurate crystals are vigorously phagocytosed[254] and incorporated into phagolysosomes. Here, it is thought, the immunoglobulin G crystal coat is removed. As one result, organelle membranes are disrupted[220] ('perforation from within'), leading to cell death. This process of autolysis is accompanied by the liberation from the dying cells of both lysosomal and cytoplasmic enzymes.

By reason of mobility and numbers, polymorphs are the most prominent agent in the *in vivo* response to monosodium biurate crystals. They are not, however, the only cell lineage implicated in crystal processing. Platelets are activated and their constituents released.[75] Mononuclear macrophages assemble in significant numbers. Monosodium biurate, calcium pyrophosphate and calcium hydroxyapatite appear less injurious to these cells than do polymorphs. For this reason, and because of the demonstrated release from urate-laden macrophages of lysosomal enzymes, prostaglandin E_2 and interleukin-1, the role of the mononuclear macrophage in promoting gouty inflammation is receiving growing recognition. In parenthesis, it is of interest that interleukin-1 secretion is not stimulated under these circumstances by crystals of calcium pyrophosphate or calcium hydroxyapatite in spite of the demonstrable presence of interleukin-1 in the serum of patients with gout and with pseudogout. The levels of interleukin-1 may be sufficient to explain the acute phase response and fever of acute crystal arthritis.

Synoviocytes also play an active part in mediating the biological response to intra-articular crystals. Synovial A cells are, in effect, mononuclear macrophages. Although monosodium biurate crystals are seldom identified within synoviocytes in biopsy material, these cells are capable of phagocytosing urate crystals given experimentally. Evidence, principally from cell culture studies, has established the release from the crystal-laden cells, of the same variety of inflammatory mediators and cytokines shown to be liberated under these conditions from mononuclear macrophages.

Crystal formation is closely related to the properties of the connective tissue matrix. The tissues in which monosodium biurate is deposited are usually rich in proteoglycan.[224] Heparin and chondroitin sulphate bind to monosodium biurate particularly in the presence of calcium[64] but it is not clear whether this mechanism is significant in the localization of tissue deposits. Although monosodium biurate crystals appear to form early in articular cartilage, cartilage proteoglycans actually increase urate solubility[106,112] a property dependent on proteoglycan aggregation by hyaluronate.[174,175] Serum proteoglycan levels are greatly raised in those with gouty arthritis but not in those with uncomplicated hyperuricaemia, suggesting that enzymatic degradation of proteoglycan may be a preliminary to monosodium biurate crystallization. Like calcium pyrophosphate dihydrate and other crystals, however, monosodium biurate selectively adsorbs IgG,[115] so that an immunological mechanism may amplify the phlogistic properties that are a consequence of the shape of the monosodium biurate crystals. Calcification may be an associated disorder whether chondrocalcinosis develops or not.[50]

2. Inflammation (see Chapter 7) This second phase begins abruptly.[238] The diagnosis of acute gouty arthritis is made by the identification of monosodium biurate crystals in a sterile joint effusion that is rich in polymorphs.[136] The crystals are recognized by polarized light microscopy[25,262] (p. 379). The long, needle-shaped crystals (Fig. 10.6) measuring approximately 1×0.01 μm, display strong negative birefringence. Diagnosis is possible with quantities of synovial fluid as small as those aspirated from a single metatarsophalangeal joint.[2] Crystals of calcium pyrophosphate dihydrate may coexist as may calcium hydroxyapatite. As a sequel to the changes described above, many other inflammatory mediators (see Chapter 7) are also released; they include prostaglandins, leukotriene B_4 and oxygen radicals. Hageman factor (Factor XII) is activated,[107] although acute arthritis has been observed with Factor XII deficiency.[81] Vascular permeability is enhanced. The classical complement pathway (see p. 236) is triggered whether immunoglobulin is present or not, and C3a and C5a contribute to the inflammatory response.

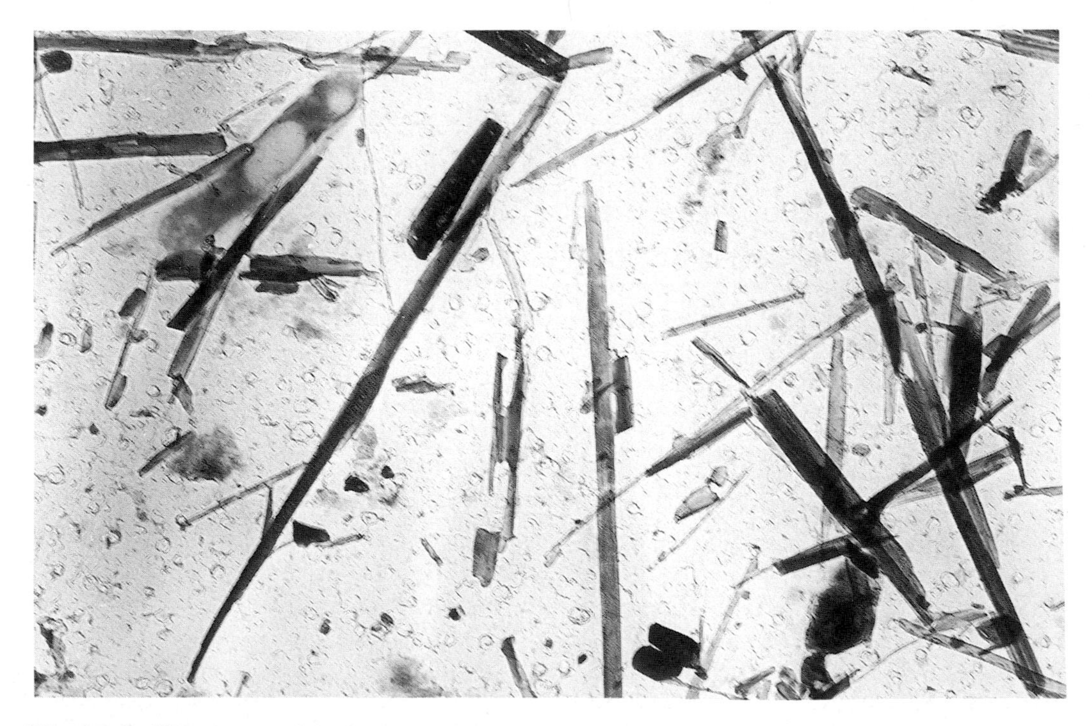

Fig. 10.6 Urate crystal formation ***in vitro.***
Urate crystals formed in the laboratory. (TEM × 11 550.)

Garrod[70] recognized that urate crystals (Figs 10.7, 10.8) may exist in joints without provoking gout. This was confirmed by Moens, Moens and Moens[158] and by Wall *et al.*[248] They showed, respectively, that the synovial fluid of as few as 4 per cent or as many as 18 per cent of hospital patients examined routinely *post mortem* contained monosodium biurate crystals. These studies, and the knowledge that acute gouty arthritis is usually self-limited[128] has encouraged studies of agents that regulate the inflammatory reaction. For example, a low molecular weight glycoprotein, crystal chemotactic factor, is generated, suppressing phagocytosis. Many proteins and polypeptides bind to monosodium biurate (and other) crystals *in vivo* and *in vitro*,[87,89] and lipoproteins are abundant in the crystal protein coat. The very low and low density apo-β-containing lipoproteins can inhibit some of the reactions usually provoked in polymorphs, by monosodium biurate crystals.[240] It is suggested[238] that apo-β-containing lipoproteins may, indeed, be the major specific regulators of the polymorph response to crystals (Fig. 10.9).

Fig. 10.7 Urate crystals in synovial fluid.
Polarized light with lambda plate. (× 560.)

The overall concept (Fig. 10.10) is therefore as follows. Monosodium biurate crystals, liberated into synovial fluid, interact with complement components and bind messenger proteins. Neutrophil polymorphs, attracted by chemotaxis, generate inflammatory mediators. Vascular permeability is increased, and large molecules, such as the apo β-lipoproteins, enter the synovial fluid. These molecules form an additional protein coat; they suppress the monosodium biurate–polymorph interaction and the inflammatory response subsides. A cycle of action and reaction is established. Monosodium biurate crystals from articular cartilage surfaces have indeed been found to be smooth[182] whereas those from gouty synovial fluid have been seen by scanning electron microscopy to have an amorphous coat. Against this concept, is evidence from transmission electron microscopic surveys in which Paul, Reginato and

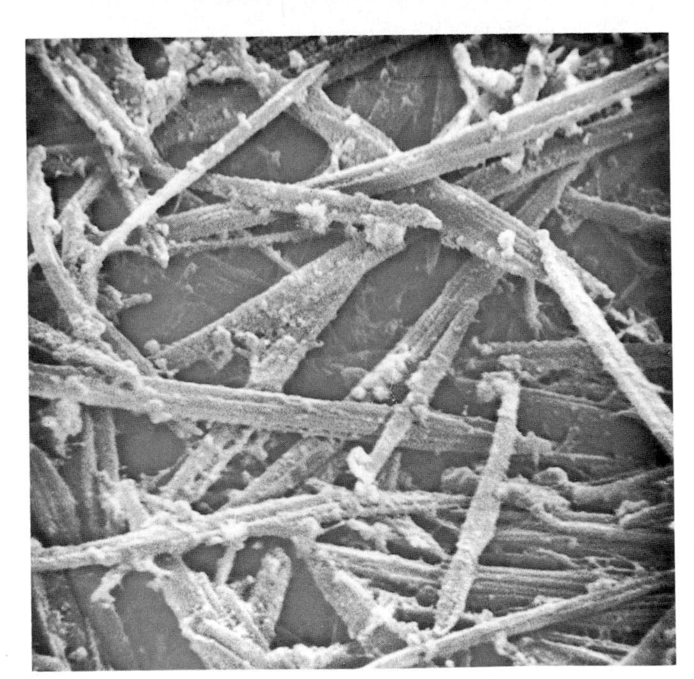

Fig. 10.8 Urate crystals in synovial fluid.
(SEM × 2 700.)

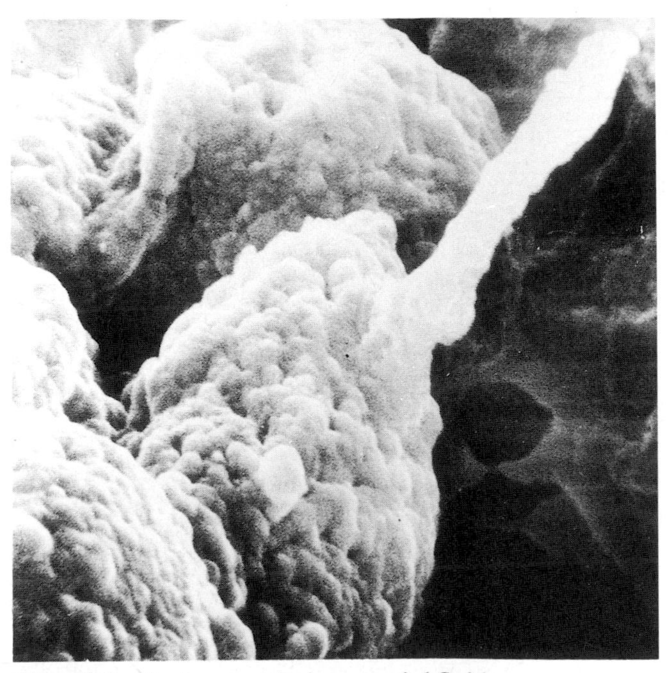

Fig. 10.9 Urate crystals in synovial fluid.
Cluster of polymorphs near protein-coated urate crystal. (SEM × 6956.)

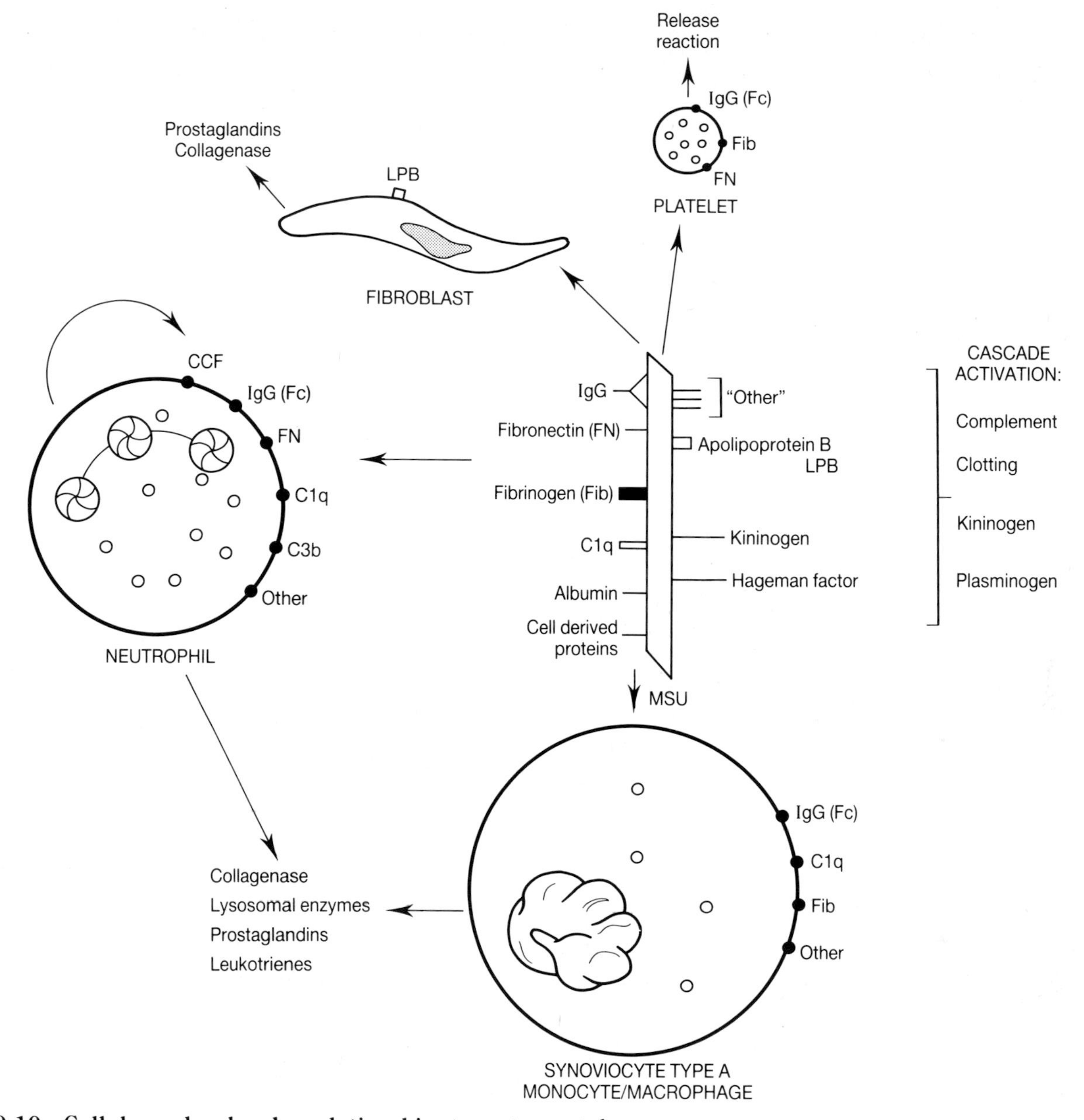

Fig. 10.10 Cellular and molecular relationships to urate crystals.
Schema of some pathways that may be invoked in the onset of acute gout and its perpetuation. IgG(Fc), Fc component of immunoglobulin G molecule; Fib, fibrinogen; FN, fibronectin; LPB, lipoprotein B; CCF, crystal chemotactic factor; C1q, C1q component of complement; C3b, C3b component of complement; MSU, monosodium biurate. Redrawn from McCarty.[133a]

Schumacher[172] detected a protein-like surface covering in only a single asymptomatic patient in whose synovial fluid were many free monosodium biurate crystals but no inflammatory cells.

The inflammatory phenomena of gout can be reproduced by the injection of urate crystals into joints[57,58] or serous cavities *in vivo*.[76] A variety of animal models has been used. Any doubt that the inflammation of gout is caused by monosodium biurate crystals themselves was, indeed, dispelled when such tests showed that the sterile crystals caused immediate violent inflammation.[57] The phenomena of gouty inflammation can, therefore, be studied experimentally although there are differences in species reactivity. Uric acid nephropathy (see p. 392) can also be reproduced.[227] The arthritis produced by urates in dog stifle joints can be suppressed by the use of antipolymorph serum to destroy leucocytes.[37] By contrast, bacterial endotoxin adsorbed to monosodium biurate crystals intensifies the inflammatory response.

3. Chronic inflammation

The third phase is characterized by repetitive episodes of inflammation. The process becomes chronic either as a result of a change in the form of the monosodium biurate deposits, or of renewed crystal formation, or the activation of a crystal-associated immunoglobulin-mediated response. The crystal deposits merge and become larger; tissue necrosis occurs. Fibrosis is stimulated and granulomata (see p. 301) form (Figs 10.11 and 10.12).[98]

Urate crystals, soluble in water, are said not always to be demonstrable in light microscopic material fixed in 10 per cent neutral buffered formalin. In practice, formalin-fixed tissue is often found to contain crystalline deposits; they may not be seen at the periphery of a lesion, whence they have been dissolved, but are identifiable in the centre. Crystals located in this site are not necessarily removed during dehydration and staining.

By the same token, crystals can often be identified in transmission electron microscopic sections;[208] alternatively, only crystal-shaped clefts remain. Other materials, including lipids, may leave spaces that can be mistaken for urate: the clefts caused by cholesterol (see p. 408), cystine or oxalates may also cause confusion. Calcium pyrophosphate dihydrate and calcium hydroxyapatite are insoluble; their calcium content is readily demonstrable and they can therefore be identified by electron probe X-ray microanalysis.

Crystalline monosodium biurate deposits form stellate bundles in tissues (Fig. 10.11). They are colourless in sections stained with haematoxylin and eosin. The crystals elicit a low-grade foreign-body cellular response and are surrounded by histiocytes and smaller numbers of multinucleated giant cells (Fig. 10.12). The deposits are often multicentric. They form simultaneously in several sites or may arise in irregular sequence at long time intervals. The connective tissues surrounding a deposit become increasingly vascular. Fibroblasts proliferate locally; collagen is deposited, delineating the lesion which now constitutes a granulomatous tophus (Figs 10.13 and 10.14). Central necrosis does not occur but necrotic tissue may be found at the margin. Granular and amorphous material may coexist.

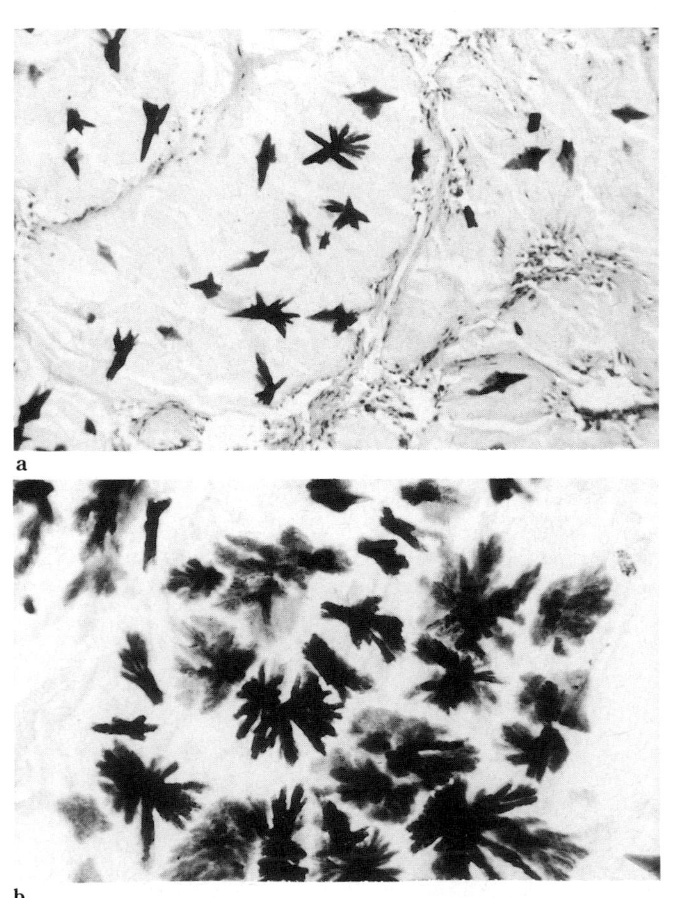

Fig. 10.11 Urate crystals in gouty tophus.
Crystals aggregate as stellate clusters. Form of clusters is varied and is influenced by methods of tissue preservation, fixation and dehydration. (HE **a:** × 190; **b:** × 380.)

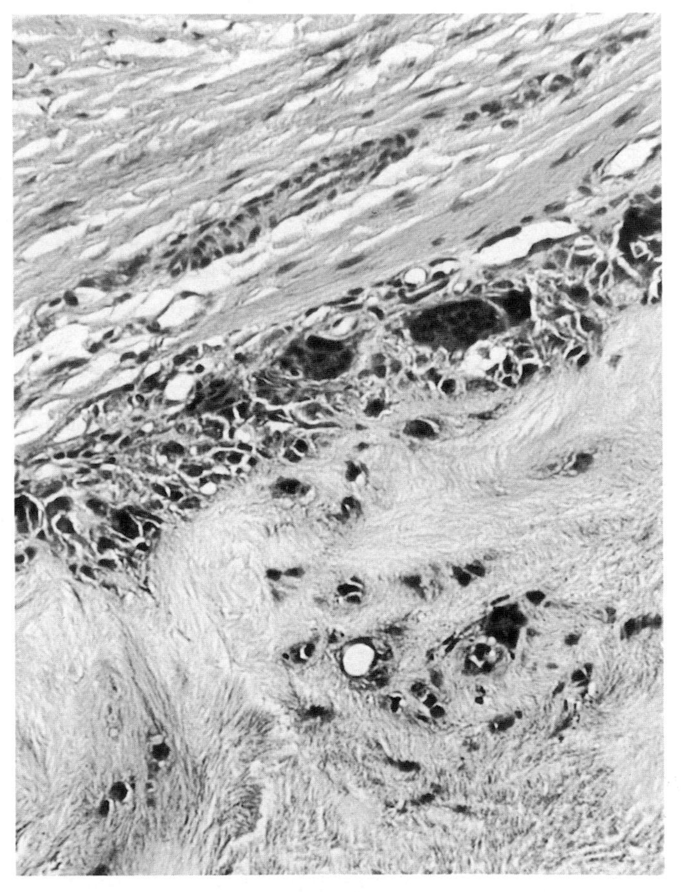

Fig. 10.12 Giant-cell reaction at margin of gouty tophus.
Crystal clusters (at bottom) separated from fibrous margin (at top) of long-standing tophus by array of multinucleated giant cells derived from fusion of macrophages. (HE × 190.)

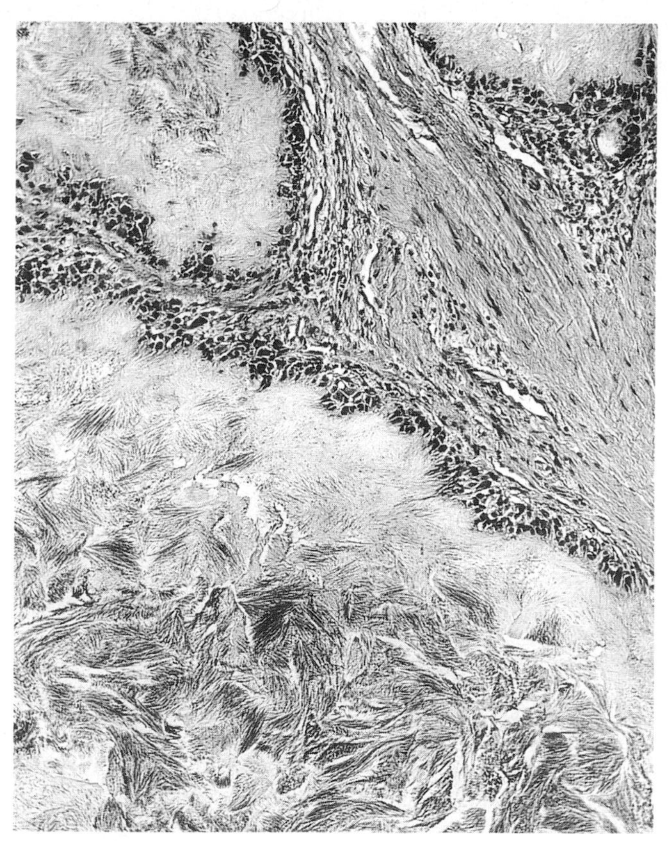

Fig. 10.13 Urate crystal clusters in gouty tophus. Although many crystals tend to dissolve in aqueous fixatives, a phenomenon which can be avoided by fixing gouty material in ethanol, numerous crystals often still remain and are available for diagnosis. (HE, polarized light × 82.)

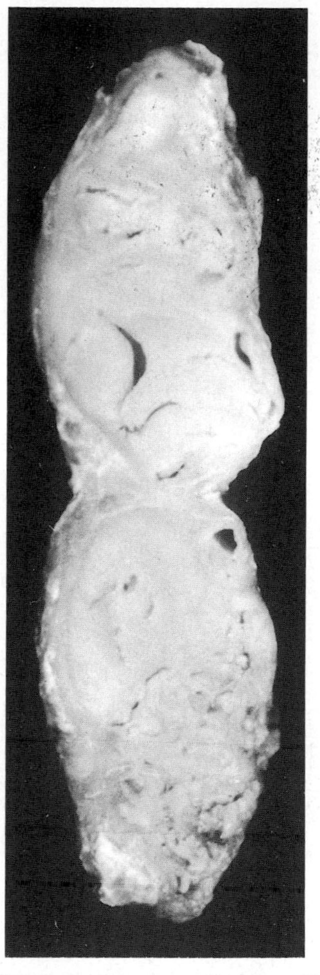

Fig. 10.14 Gouty tophus from human elbow. Note the pale flecks of calcium the accumulation of which accounts for the radioopacity of many tophi. The tophus is multiloculated. It was excised both because of pain and inconvenience. (× 2.)

Dystrophic calcification (see p. 403) is frequent so that the tophus becomes radioopaque.

Tophi are often located in the central part of the elastic cartilage of the helix of the ear, occasionally in the antehelix, in the subcutaneous tissues overlying elbows or fingers, in tendon sheaths and bursae, or in relation to the first metatarsophalangeal joints. The tendinitis may mimic the changes of chondrocalcinosis (see p. 393).[241] Tophi have been described in the skin of the finger tips, palms or soles. They form much less commonly in other cartilages, such as those of the larynx and nose or in the tarsal plates of the eyelids and in the cornea and sclera. Tophi appear as raised, firm or hard subcutaneous nodules; ulceration often occurs.

4. Arthritis The development of arthritis constitutes a fourth phase. A raised concentration of urates within synovial fluid is accompanied by the presence of a fine, white, amorphous snow-like deposit upon the surface of articular cartilage. Urates also aggregate, presumably by diffusion, within the superficial layers of the hyaline cartilage and initiate a sequence of degenerative change, a form of secondary osteoarthrosis (see Chapter 22). The degenerative response is complicated by a brisk inflammatory reaction in which the synovia are involved. The synovial tissues are congested, swollen and inflamed while the synovial fluid comes to contain a great excess of polymorphs. Progressive destruction and loss of articular cartilage follows, accompanied by articular and periarticular inflammation. The reaction is gouty arthritis (Figs 10.15, 10.16 and 10.17).

Articular cartilage Gouty arthritis of the first metatarsophalangeal joint is 'podagra'. The elbows, fingers, knees, ankles and vertebrae may also be affected ('cheiragra') but no joint is spared and the sacroiliac and manubriosternal[110] articulations are occasionally involved. Just as urates are deposited in fibrous tissue and fibrocartilage, with tophus formation, so they are laid down in the

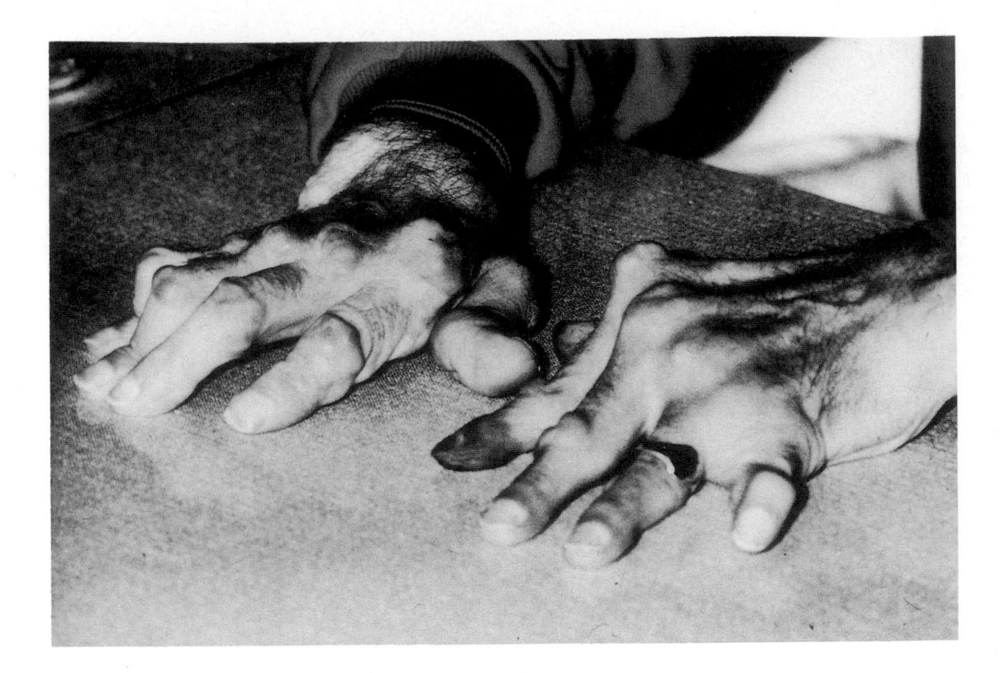

Fig. 10.15 Gouty arthritis. Before the introduction of xanthine oxidase inhibitors, it was not rare for the arthritis of gout to progress to severe deformity. Calcium was extruded through the tophaceous masses, onto the skin surface.

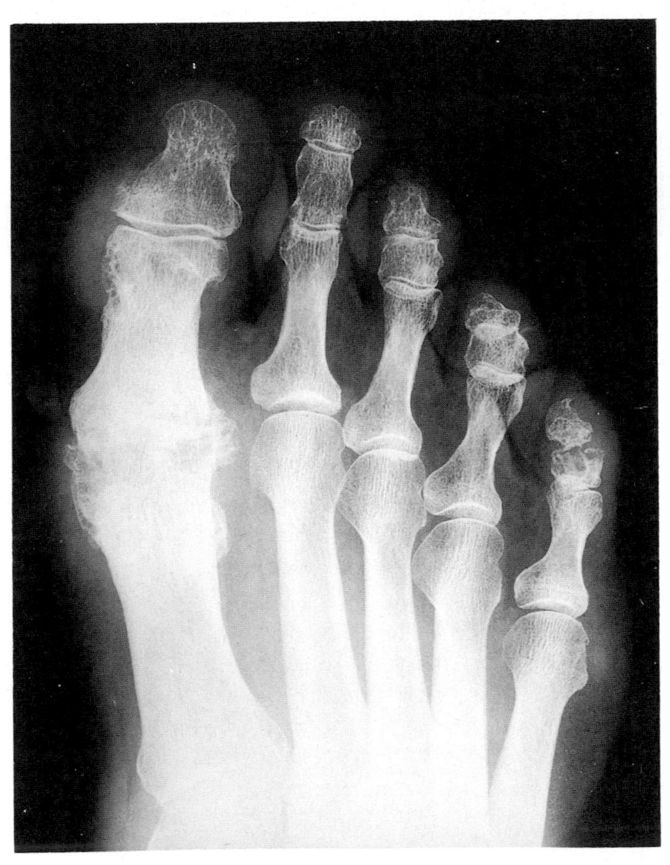

Fig. 10.16 Radiograph of foot with gouty arthritis. A frequent site for gout is the first metatarsophalangeal joint. The reason for this selective localization is still not fully understood. Here, the X-ray shows destruction of the joint space, zones of periarticular radiotranslucency and osteophytosis.

deeper layers of articular cartilage. The white granular deposits found on the surface of the cartilage protrude from the joint when the synovial cavity is opened. In other cases, the amorphous material has a yellow colour due to the presence of haemosiderin pigment. There is a similar colouration of synovial and periarticular tissues. Cartilage begins to break away from subarticular bone. The destructive process excites a secondary inflammatory synovitis. Urates are simultaneously deposited within the synovia. The replacement of cartilage by fibrous tissue culminates in fibrous ankylosis. The changes of osteoarthrosis are added to those of gout and can lead to an extreme degree of deformity exaggerated by the presence of urate deposits in subchondral bone and in periosteal connective tissue (Fig. 10.18). Ultimately, in long-standing and untreated cases,[129] the local destruction of bone and cartilage can be said to constitute a form of arthritis mutilans. Necrosis of the femoral head, for example, may develop but the radiological appearances can be masked by nearby urate and calcium deposits. When arthroplasty has been undertaken for osteoarthrosis, the local accumulation of urate crystals may be one factor leading to loosening of the femoral component.[169a]

Other articular tissues The connective tissues of periarticular bursae (Fig. 10.19) and of the fibrocartilagenous intervertebral discs are occasionally sites for urate accumulation. In the latter case, adjacent vertebral bone is involved in a manner similar to the lesions of vertebral bone in hereditary ochronosis (see p. 351). All parts of the spine

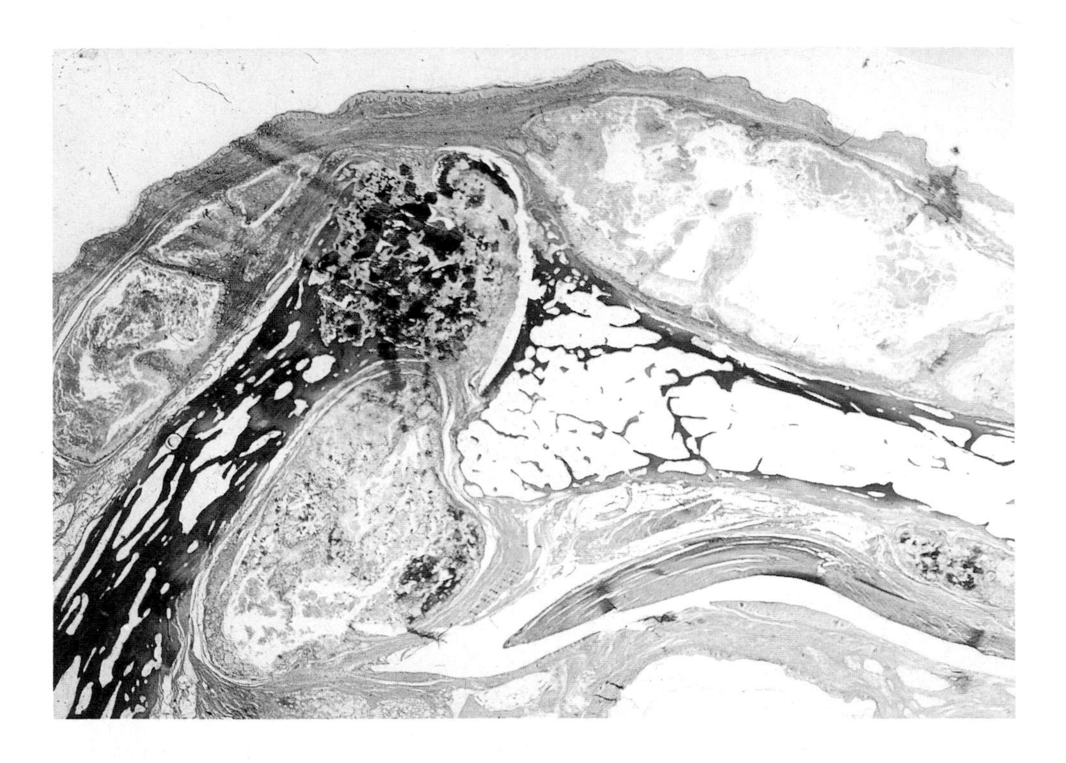

Fig. 10.17 Gouty arthritis. In this interphalangeal joint, urate deposits have led to extensive destruction. An ovoid mass of urate crystals impinges on the inferior margin of the proximal phalanx (lower left) and a large mass of crystals lies above the middle phalanx (upper right). Loss of articular cartilage is a feature of the secondary osteoarthrosis that has developed and there is subluxation. (HE × 3.5.) By courtesy of the late Professor Dr H Zollinger.

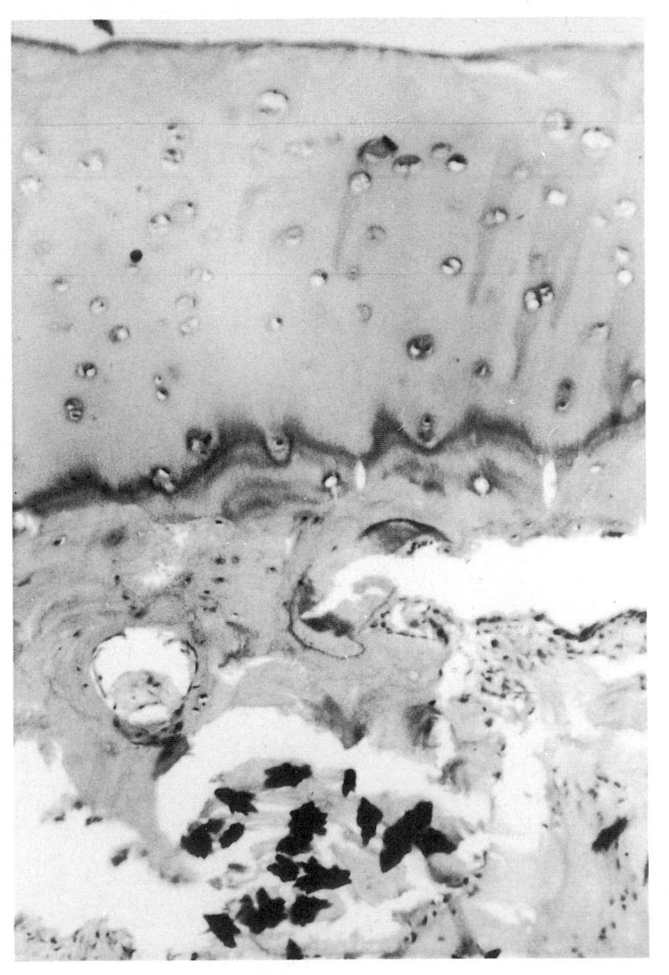

may be affected but symptomatic disease of the cervical spine[3a] and sacroiliac joints is rare. Paraplegia has resulted, but spondylitis is very rare.

There is no direct pathological evidence to show that gouty arthritis can be arrested or ameliorated by effective, prolonged uricosuric therapy but clinically urate deposits and the reactions they cause, slowly diminish during treatment.

Synovia Lesions closely similar to tophi are seen in the synovia.[3,208] In the acute lesions, the ultrastructural changes of inflammation can be identified. Type A and intermediate type synovial cells predominate. Tophi form near the synovial surface; the crystal deposits are acicular and comprise small, radial clusters. A rim of fibrocytes, a few polymorphs, lymphocytes and multinucleated giant cells are characteristic and small numbers of plasma cells may be recognized. The fibrocytes often contain lipid droplets; some have monosodium biurate crystals within phagosomes.

5. Visceral disease In this fifth stage, monosodium biurate crystals accumulate in the subcutaneous tissues,

Fig. 10.18 Gouty arthritis. Urate deposits in bone may act as foci for the concentration of lines of force and the propagation of cracks. The site of the urate crystals is clearly related to foci of tissue disintegration despite the presence of technical artefacts. (HE × 113.)

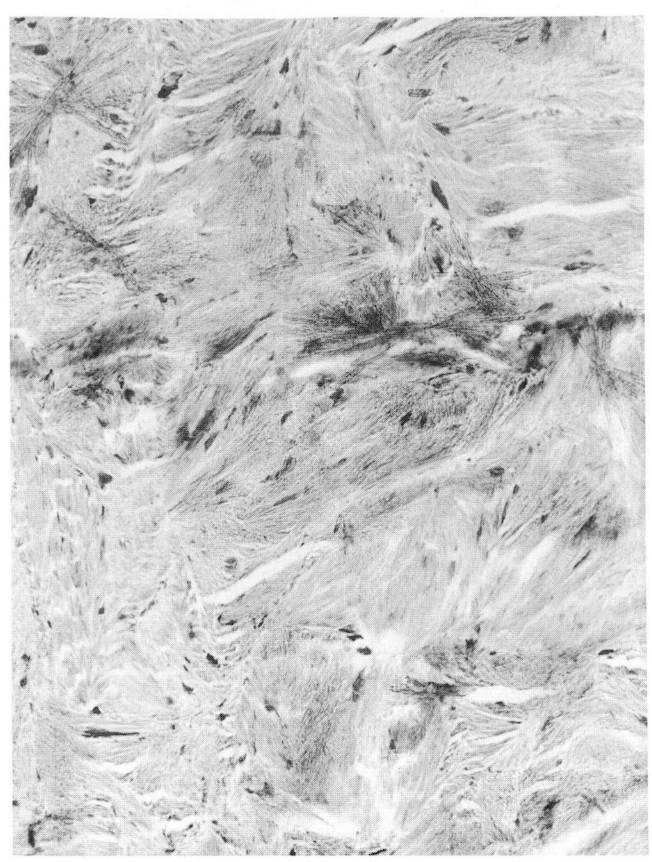

Fig. 10.19 Gouty bursitis.
Urates are deposited in many connective tissue sites in addition to synovial joint. Here, a tophus lies in the wall of an adventitious bursa. (HE × 83.)

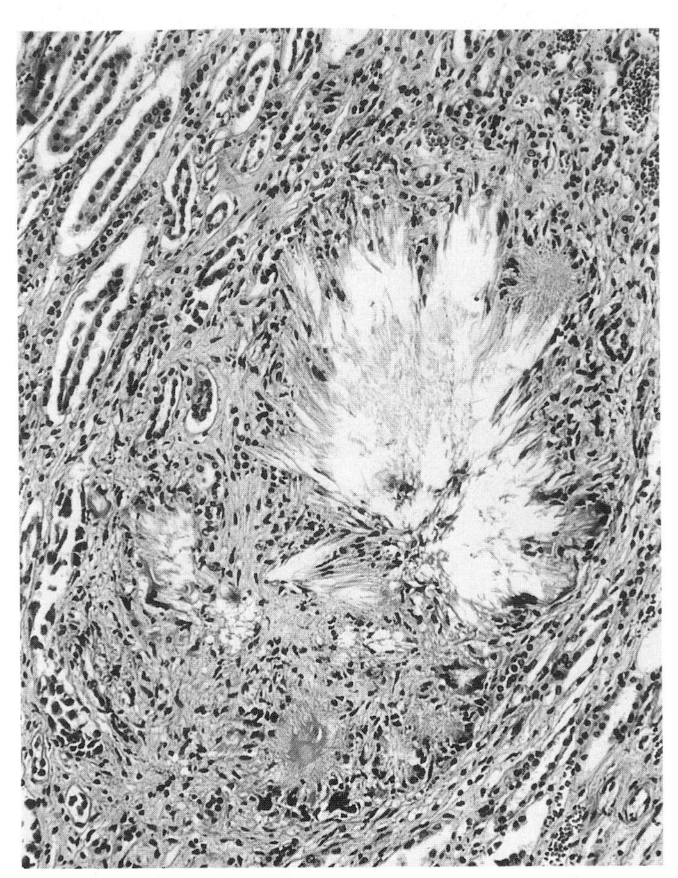

Fig. 10.20 Urate granuloma in kidney.
A stellate group of urate crystals is present within the renal cortex. (HE × 120.)

cornea, sclera and respiratory tract cartilages. Monosodium biurate crystals may also be recognized within the kidney, heart and blood vessels, less commonly the tongue and penis. Extradural deposits may lead to spinal cord compression.[124,147]

Renal failure is an important cause of death.[249] In this series, only three normal kidneys were found in the first 191 cases examined. Pyelonephritis was very frequent; arterial disease common. In approximately 20 per cent of cases, renal insufficiency was of critical importance. In a smaller proportion, malignant hypertension was encountered.[199]

Three categories of renal disease are recognized: urate nephropathy, uric acid nephropathy and uric acid calculi. Urate nephropathy results when monosodium biurate crystals form in the interstitial tissues (Fig. 10.20). Renal failure may indeed be provoked by hyperuricaemia.[113] Uric acid nephropathy, a condition not specific to gout and seen commonly in febrile disorders in childhood and neoplastic disease in the adult, represents the crystallization of uric acid in the collecting tubules, collecting ducts and ducts of Bellini. In later life, both in gout and, very often, in its absence, uric acid calculi form in the renal calyces or pelves.

The term 'gouty kidney' is usually confined to urate nephropathy. The kidney is small and finely scarred; upon the cut surfaces pale, linear aggregates of monosodium biurate crystals are recognized. Occasionally, calcification is an added change. The formation of renal crystals leads to tubular atrophy, granulomatous macrophage and foreign body giant-cell reactions, glomerular hyalinization, and fibrosis. The hypertension so often found in gouty subjects may therefore be of secondary, renal origin.

In an extensive necropsy survey, Talbott and Terplan[235] divided the histological changes into four categories: 1. kidneys with no abnormality; 2. those with pyelonephritis; 3. those with vascular changes, including arterio- and arteriolonephrosclerosis; and 4. those with amyloidosis. Amyloid A (see Chapter 11) protein accumulates occasionally in patients with polyarticular gout.[199]

In the vascular system, monosodium biurate has been described within the walls of visceral arteries, in the pericardium, myocardium and endocardium of the mitral and aortic valves. Rarely, myocardial deposits have been reported as the cause of heart block.

Secondary gout

In diseases of the haemopoietic system such as polycythaemia, myeloid metaplasia, myelofibrosis, myelocytic leukaemia, the sickle cell diseases,[198] Mediterranean anaemia[171] and the lymphomata, a syndrome of secondary gout may develop. The mechanism is an increase in nucleoprotein metabolism with the production of more uric acid than normal. Simultaneously, the normal red cell destruction of urates may diminish. The pathological findings of secondary gout are almost identical with those of the primary disease: tophi and gouty arthritis may result. The skeletal lesions are indistinguishable from those that accompany the primary disorder. Sometimes renal disease may be a combination of an underlying arteriosclerosis with progressive renal excretory failure caused by the mechanical and inflammatory consequences of renal deposition of urate crystals. Hypertension is commonly associated with hyperuricaemia[32] and may be accompanied by clinical gout. This complication of hypertension may be a result of an antihypertensive regimen in which diurietics are given over a prolonged period. Secondary gout of this variety is encountered with increasing frequency in ageing Western societies.

Saturnine gout

Gout is occasionally associated with lead poisoning.[128,209] The articular lesions of saturnine gout are identical with those of primary gout. The description 'saturnine' refers not to the 'sluggish, cold and gloomy' temperament of the patient but to lead intoxication. In Roman times, wines were heavily contaminated with lead when boiled-down grape syrup was used to enhance the colour, sweetness, bouquet and preservation of the wine; but lead was introduced into Roman wines in many other ways.[168] It has been suggested that the epidemic character of gout among the Romans and among the English of the eighteenth and nineteenth centuries is attributable to this custom. In modern times, lead paint in Australia, home-brewed liquor ('moonshine') in the southern United States and leaded petroleum in Spain are among other possible causes of saturnine gout. Lead is thought to act upon renal tubular uric acid excretion. Uric acid is retained but, simultaneously, there may be a diminution in plasma volume due to defects in the renin–angiotensin–aldosterone system.[209,210] However, the role of lead nephropathy in causing the hyperuricaemia of saturnine gout is disputed[192] and other factors such as heredity, may be contributory.

Calcium pyrophosphate deposition disease (chondrocalcinosis; pseudogout)

Calcium pyrophosphate dihydrate (CPPD) deposition disease (CPPDD)[47,132,163a,201] is a spectrum of connective tissue disorder resulting from the accumulation in the tissues of calcium pyrophosphate dihydrate ($Ca_2O_2O_7.2H_2O$) crystals.[24] Acute arthritis, chronic inflammatory joint disease, osteoarthrosis- and rheumatoid arthritis-like changes, and chondrocalcinosis of the spinal joints and of the ligaments, tendons and menisci, are components of the syndrome. A frequent sign of articular chondrocalcinosis is inflammation: the arthritis mimics gout and this form of the disorder was originally termed 'pseudogout' to indicate one clinical manifestation.[59]

Pyrophosphates (Fig. 10.21) are physiological, intermediary sources of high-energy phosphate. For reasons not yet fully understood, excess pyrophosphate may accumulate and crystallize as calcium pyrophosphate in tissues and body fluids. This process is thought to be common in otherwise normal persons. Consequently, calcium pyrophosphate is often detectable in the synovia and

Fig. 10.21 Calcium pyrophosphate dihydrate crystals. The broad, rhomboidal character of the crystals is clearly seen. (SEM × 4000.)

synovial fluid of diarthrodial joints in symptomless individuals. The onset of signs or symptoms of calcium pyrophosphate deposition heralds CPPD deposition disease. In approximately 1 per cent of cases, the disorder is familial. Familial or hereditary chondrocalcinosis has been identified in many parts of the world, rarely in Ashkenazi Jews.[55a] The inheritance is likely to be as an autosomal dominant characteristic. A hereditary predisposition, associated with the inheritance of HLA-2 and HLA-W18, affects particular populations and is geographically selective. A further 10 per cent of symptomatic cases results from metabolic diseases that include hyperparathyroidism, haemochromatosis, Wilson's disease,[141] primary ochronosis,[143] hypophosphatasia, diabetes mellitus or gout. There is an unexplained association with amyloidosis (see p. 437). The histological recognition of CPPD deposition disease is simpler than the clinical: the insoluble crystals may be readily seen and identified. However, the frequency of such deposits is not established. Although there may be as many as 20–30 000 symptomatic cases of CPPD deposition disease in the UK, the prevalence of the histological disorder remains uncertain.

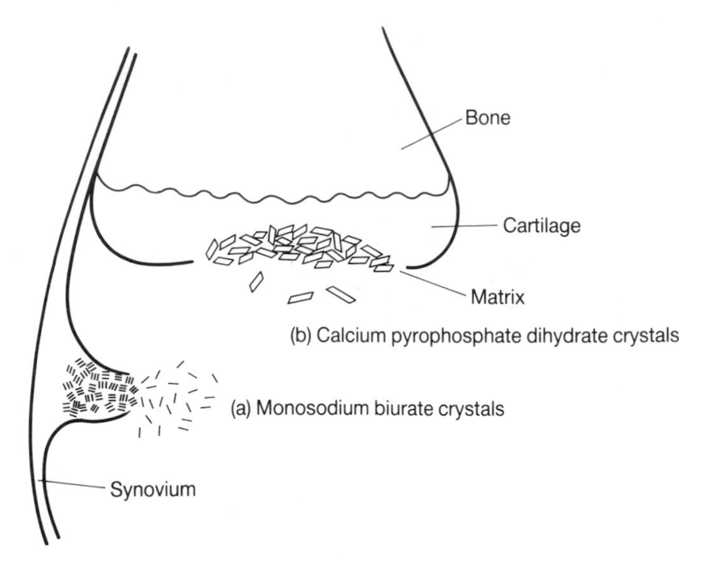

Fig. 10.22 Hypothetical mechanism of escape of crystals into synovial joints.
Defective enzymatic regulation of pyrophosphate dihydrate metabolism, perhaps inherited, allows accumulation of calcium pyrophosphate dihydrate crystals in articular hyaline and fibrocartilages. Subsequently, crystals escape into joint space and evoke inflammation. Redrawn with permission from McCarty.[133a]

History

The familial occurrence of articular chondrocalcinosis was recognized by Žitňan and Sităj.[263–265] When McCarty and his colleagues identified calcium pyrophosphate crystals in the synovial fluid of patients with the associated syndrome of pseudogout, the cause of the disorder became clearer.[33,34,114,132,134]

Biochemical basis

As in gout, an attack of CPPD deposition synovitis may be precipitated by surgery or by acute medical disorders such as myocardial infarction and cerebral infarction, as well as by congestive cardiac failure and the administration of diuretics. The episodes occur when serum calcium levels are at their nadir. They are caused by the liberation of crystals of CPPD into synovial cavities, following the solution of the margin of crystalline tissue deposits, the process of 'crystal shedding' (Fig. 10.22).

The proximate cause of CPPD deposition disease is obscure. In sporadic CPPD deposition disease, a genetic disturbance of midzone cartilage metabolism, subclinical hyperparathyroidism and apoptosis of zone II chondrocytes are among phenomena that may cause CPPD deposition. When there is an underlying systemic metabolic disorder it is possible that a product, like homogentisic acid in ochronosis (see p. 351), can lead to abnormal chondrocyte function with altered proteoglycan–collagen relationships and nucleation of mineral at exposed collagen sites. An alternative mechanism that can explain both the high concentrations of CPPD in chondrocalcinosis and the calcium hydroxyapatite-rich matrix vesicles of osteoarthrosis (see p. 852) has come from Howell and his colleagues.[97,166] Pyrophosphate ions can be generated by cartilage extracts. The ectoenzyme nucleoside triphosphate pyrophosphohydrolase, associated with plasma membranes, has been identified in normal articular cartilage and high activity demonstrated in CPPD deposition disease and osteoarthrosic cartilage. The enzyme activity can be related to processes like altered matrix synthesis or damaging physical forces: it functions effectively on adenosine triphosphate but also on oridine triphosphate, cytidine triphosphate and guanosine triphosphate. If the raised concentrations of pyrophosphate liberated by this enzyme are not eliminated by pyrophosphatase, sufficient CPPD can accumulate to cause the aggregates characteristic of CPPD deposition disease.

Clinical presentation (Fig. 10.23)

Diagnostic criteria exist (Table 10.5).[47,163a,201] CPPD deposition disease is 'a great mimic'. The clinical signs may simulate gout, rheumatoid arthritis, osteoarthrosis or other connective tissue disease[188] (Table 10.6). There are seven patterns. In type I (pseudogout), acute or subacute attacks of arthritis lasting from one day to four weeks recall the syndrome of gout. In type II, the local and systemic disease

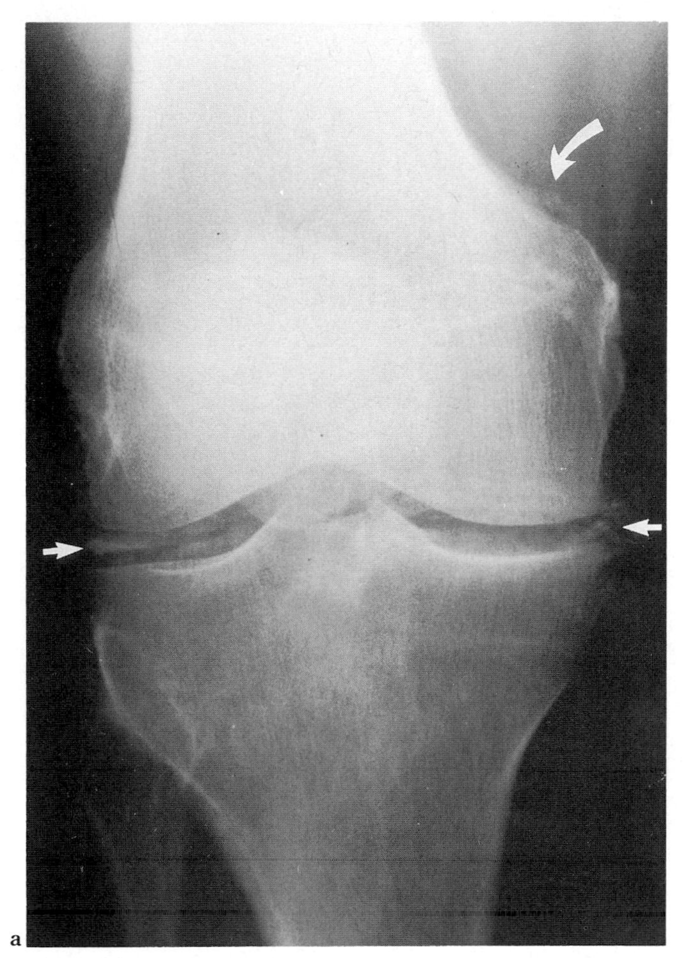

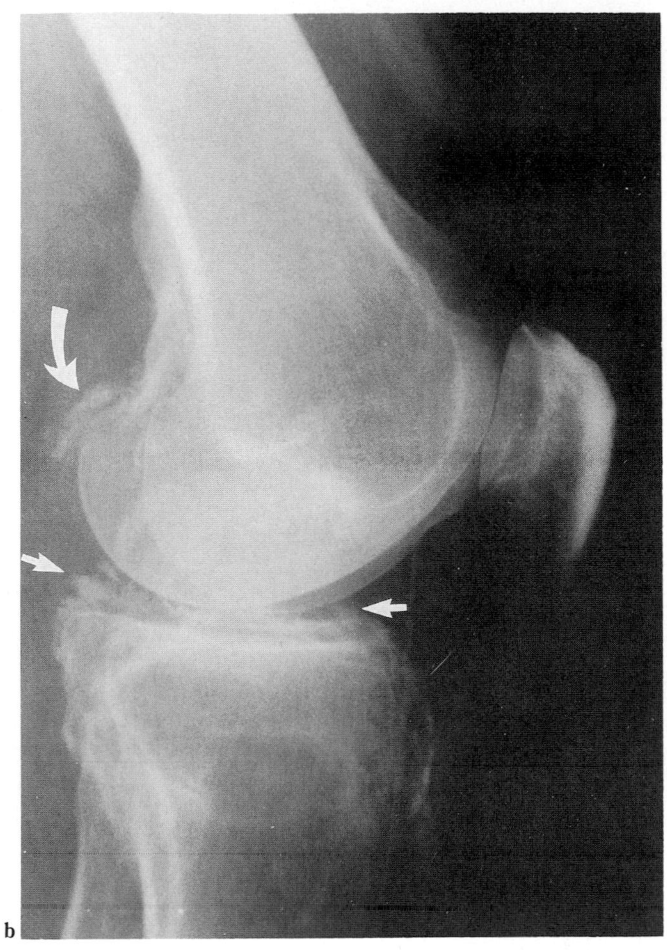

Fig. 10.23 Chondrocalcinosis of right knee.
There is calcification of both the hyaline articular cartilage of the femoral and tibial condyles and menisci (straight arrows) on (a) antero-posterior and (b) lateral radiographs. Note calcification in the adductor tendon (curved arrow). Courtesy of Department of Diagnostic Radiology, University of Manchester.

resembles rheumatoid arthritis. In type III, progressive degenerative disease of many joints, particularly the knees, wrists, metacarpophalangeal, hips and spine, resembles osteoarthrosis; those without inflammatory signs are categorized as type IV. In patients coming to knee arthroplasty because of osteoarthrosis, the relative risk for chondrocalcinosis in one series was six times that of control cases.[224a] In type V, the slow, progressive deposition of CPPD is asymptomatic; in type VI, a neuropathic-like arthropathy is found in the absence of neurological disease. There is a remaining category (type VII) in which the clinical signs range from those of ankylosing spondylitis-like disease or hyperostosis to monoarticular disorder, meningeal irritation or nerve root signs. The pattern of clinical disease may change during its evolution.

There is usually no evidence of parathyroid dysfunction. Raised levels of plasma parathyroid hormone are found in sporadic cases but similar rises are detected in control osteoarthrosic patients. The disorder of pyrophosphate metabolism is local rather than systemic. Synovial fluid levels of pyrophosphate are high by comparison with other joint disorders with effusion.[201] Acute episodes of pseudogout seem to occur when crystalline pyrophosphate is mobilized; the crystals escape into the joint when alkaline phosphatase levels are diminished. These circumstances occur both in the sporadic disease and in those rare cases where CPPD deposition disease is secondary to hypophosphatasia.[52]

The synovial fluid volume increases. In acute CPPD deposition disease, there are approximately 7–35 × 10^6 leucocytes per litre, mainly polymorphs; in the chronic disease, 0.2 is an average figure; of these, many are mononuclear phagocytes. The crystals that lie free within the synovial fluid, or within synovial fluid cells, are much more easily studied and identified than those present in islands within synovial tissue or in fibro- or hyaline cartilage.

Table 10.5

Diagnostic criteria for calcium pyrophosphate deposition disease (modified from Ryan and McCarty[201])

Criteria

I Demonstration of crystals by definitive means

II (A) Identification of monoclinic and/or triclinic crystals showing no, or a weakly positive birefringence
(B) Presence of typical calcifications in X-rays

III (A) Acute arthritis, especially of knees or other large joints, with or without hyperuricaemia
(B) Chronic arthritis, especially if accompanied by acute exacerbations. The chronic arthritis shows the following features:
1. Uncommon site for primary osteoarthrosis—wrist, metacarpophalangeal, elbow, shoulder
2. Radiological: radiocarpal or patellofemoral joint space narrowing, especially if isolated (patella 'wrapped around' the femur)
3. Subchondral cysts
4. Severity of degeneration—progressive, with subchondral bony collapse and fragmentation and intraarticular radiodense bodies
5. Osteophyte formation—variable and inconstant
6. Tendon calcification
7. Axial skeleton involvement

Categories

A Definite—I or II(A) plus (B)
B Probable—II(A) or II(B)
C Possible—III(A) or (B)

Table 10.6

Metabolic diseases with which calcium pyrophosphate deposition disease is associated and the strength of the association[163a]

Diabetes mellitus	8–73 per cent
Gout	0–8
Rheumatoid arthritis	8–28
Hyperparathyroidism	20
(frequency of hyperparathyroidism in calcium pyrophosphate deposition disease	5–15
Haemochromatosis	65–80
Hypothyroidism	Insufficient numbers to calculate
Neuropathic arthropathy	
Wilson's disease	
Hypophosphatasia	
Ochronosis	

Identification of free calcium pyrophosphate dihydrate crystals

It may be necessary to identify small numbers of crystals in the synovial fluid where they are often intracellular, to analyse aggregates of white crystalline material lying free on articular surfaces or to examine larger or smaller clusters of crystals in sections examined *post mortem* or at biopsy.

Light microscopy Calcium pyrophosphate dihydrate crystals can very readily be mistaken for urate. Polarization microscopy (see p. 379) is therefore important[253] especially when in an intercurrent phase of CPPD deposition disease, there is little effusion, slight inflammation and little calcification; the majority of the crystals appear as needles and rods, particularly when there is coincident hyperuricaemia; or pyrophosphate and urate crystals coexist. The synovial fluid CPPD crystals appear as small rods, needles or prisms,[47] or as rods, rhomboids and plates.[253] Needle-shaped crystals and rods (Fig. 10.24) are frequent in acute CPPD deposition disease, while the less soluble rhomboids and rectangular plates are more frequent in the chronic disease. The crystals range in length from 1.0 to 25.0 μm, in width from 0.2 to 2.0 μm. They exhibit weak, positive birefringence and often appear to have a chip out of one corner because of the apposition of two crystals of unequal length, the process of 'twinning'.

In synovial fluid, both CPPD and calcium hydroxyapatite crystals can be recognized by alizarin red S staining and by polarization light microscopy (Table 10.7). In paraffin sections, the clusters are faintly basophilic and react positively with the von Kossa technique (Fig. 10.28). Polarized light reveals the shape of the crystals. Those of CPPD are short, broad and many-sided (Fig. 10.32). They are seen in the articular capsule as well as in the subsynoviocytic tissues from which, it is assumed, they are carried peripherally.

Electron microscopy Transmission and scanning electron microscopy yield additional information of the shapes and sizes of crystals which may be necessary for clinical diagnosis. Scanning electron microscopic preparative techniques are straightforward. Images can be quickly obtained. Electron-probe X-ray microanalysis can then be used to assess atomic composition and chemical identity. Artefactual particles, for example, from glass (silicate) surfaces, may contaminate synovial fluid samples[14] but can be identified by X-ray microanalysis.

The rectangular and pyramidal crystals of calcium pyro-

Fig. 10.24 Calcium pyrophosphate dihydrate crystals from synovial fluid.
There is considerable variability of shape. Urate crystals may coexist and there may be a third population of unidentifiable crystals. Polarization microscopy is often, but not always reliable in the distinction between urate and calcium pyrophosphate dihydrate. Courtesy of Professor Paul Dieppe.

Table 10.7

Some histochemical techniques for the demonstration of calcium[144]

Metal substitution techniques	
von Kossa	Black
von Kossa with substitution of cobalt; lead; iron; copper or mercury	Brown-black, dark brown, vivid blue or blue
Dye-lake reactions	
Haematoxylin	Purple-red
Alizarin	Purple-red
Alizarin red S	Purple-red
Purpurin	Purple-red
Nuclear Fast Red	Purple-red
Gallamine blue	Blue-purple
Sodium rhodizonate	Red-orange
Others	
Murexide	Orange-brown
Chloranilic acid	Red-brown, with possible enhanced birefringence
Transmission electron microscopy	
Potassium pyroantimonate-osmium	
Other techniques for demonstrating calcium in tissues	
Electron probe X-ray microanalysis	
Electron spin resonance	
Autoradiography	
Neutron activation analysis	

phosphate contrast with those of corticosteroids.[31] Gaucher *et al.*[71,72] described two groups of crystals in CPPD deposition disease, the first up to 100 μm long and shaped like arrowheads, the second, no more than 20 μm in length and prism- or lozenge-shaped. X-ray spectrometry, X-ray powder diffraction and monocrystalline studies with the K_α radiation of a cobalt-60 source were used to confirm the identity of the pyrophosphate. Nagahashi[167] studied calcified menisci, articular cartilages and synovial fluid. He demonstrated deposits of CPPD in different layers of a calcified knee meniscus and in different parts of the calcified articular cartilage. The three-dimensional ultrastructure of CPPD could be imaged at the cut surfaces of calcified knee menisci. X-ray microanalysis made it possible to examine the phosphate/calcium ratios in the menisci, proving that this ratio, 0.7, was the same as that of commercially available CPPD (Tables 10.2 and 10.8).

Pathological changes

The descriptive pathological evidence is limited,[101,160] perhaps because of the relatively recent interest in the identification of crystals in tissues. Many cases are diagnosed solely on the basis of clinical,[62] radiological and synovial fluid studies, without biopsy.

The diagnostic histological feature of CPPD deposition disease is the local formation of crystalline aggregates. In some cases acute inflammation, in others chronic insidious tissue destruction,[119] are associated with or caused by the deposits. CPPD crystals are most likely to form in fibrocartilages such as those of the knee menisci, symphysis pubis, wrist and intervertebral discs. Less often (13.5 per cent) crystals are identified within tendons such as those of the tendo achilles and quadriceps and in the plantar fascia.[74] Hyaline cartilage midzones are prone to CPPD crystal deposition. The knees, shoulders, elbows, wrists, hips, metacarpophalangeal and other diarthrodial joints are likely to be affected. In addition, islands of CPPD are recognizable in the synovial tissues and, in synovial fluid effusions, CPPD crystals are usually identifiable within polymorphs.

Why calcium pyrophosphate crystals preferentially form within fibrocartilage is not known.[46] As in hyaline cartilage, there is a peripheral blood supply. The contrast with gout is evident; in gout, hyperuricaemia culminates in monosodium biurate crystal formation and inflammation is excited in selected sites. Local factors influence crystallization. In idiopathic CPPD deposition disease, there is frequently no systemic disorder of calcium metabolism. Crystal formation, except in instances of CPPD deposition secondary, for example, to hyperparathyroidism, is solely on the basis of local changes in tissue composition or physiology.

Synovia Chalky white deposits are seen at arthroscopy or biopsy on synovial surfaces. The appearances resemble 'snow-covered whiskers'[153] (Fig. 10.25). Microscopically, tophaceous deposits, 0.1–1.5 mm in diameter, are constant. The crystal clusters (Fig. 10.26) lie near synovial surfaces, often in zones of low vascularity. Synovial cell hyperplasia is slight. Individual type A and B synovial cells also contain CPPD crystals, as do polymorphs within the synovial fluid.

The CPPD deposits appear faint blue in haematoxylin and eosin-stained, non-decalcified sections. There is an eosinophilic or pale purple margin. The greater part of the deposit may be composed of crystals, 2.0–4.0 μm in length, 1.0 to 1.0 μm wide. Exceptionally, in chronic cases, individual needle-shaped crystals are recognized. The crystals are disorderly, randomly arranged and stain strongly with

Fig. 10.25 Fragment of articular cartilage in calcium pyrophosphate dihydrate arthropathy.
Particles of disorganized hyaline cartilage break away from the articular surface. A fragment is seen in this figure with several nearby calcium pyrophosphate dihydrate crystal aggregates. (× 6.) Courtesy of Professor O. Myre Jensen.

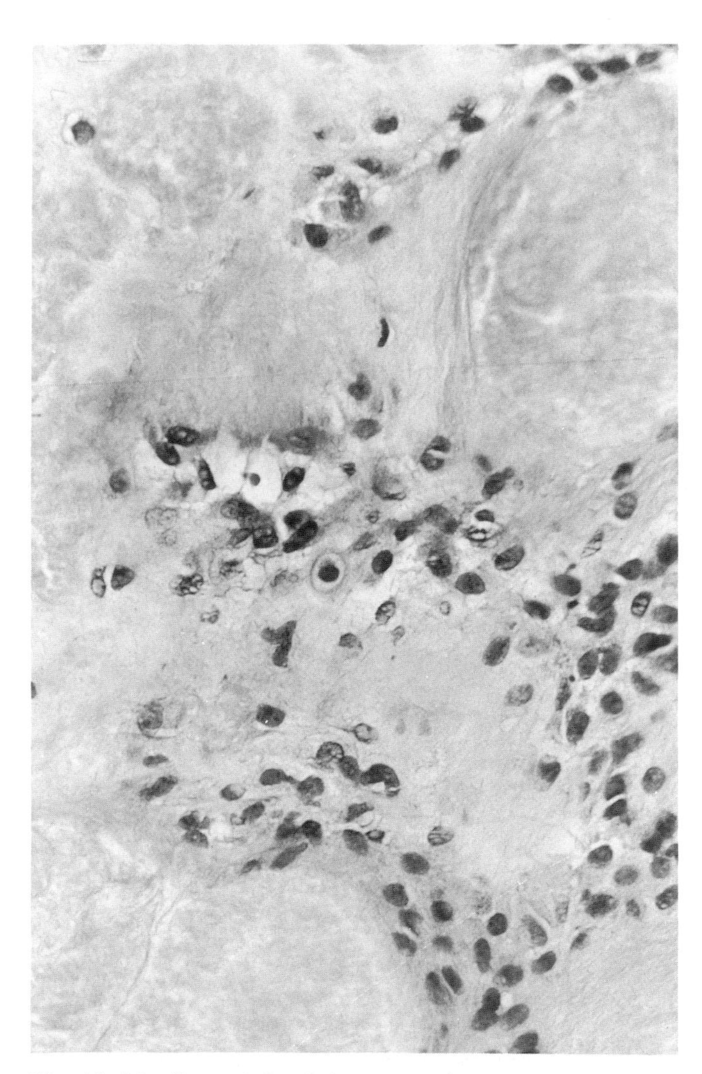

Fig. 10.26 Synovial calcium pyrophosphate dihydrate deposits.
Crystalline aggregates of calcium pyrophosphate dihydrate may be very small and inconspicuous. Polarization microscopy may allow recognition of small numbers of the crystals but a search for their presence should be accompanied by stains for calcium. (HE × 330.)

the von Kossa technique (Figs 10.27, 10.28). The number of crystals remaining in a deposit after a section has been processed is inconstant; this may indicate a transient state during life. When only a very few remain, serious difficulties may arise in their identification and tissue may then require to be searched by scanning electron microscopy. Because the islands of CPPD crystals are faintly coloured, focal and haphazard, they are easily overlooked during light microscopy.[153] They are, however, conspicuous in polarized light. Occasionally, however, crystals are lost from light microscopic sections as they are from transmission electron microscopic preparations[207] and only tissue clefts remain.

Around the crystalline deposits in chronic CPPD deposition disease are variable numbers of mononuclear phagocytes with fewer lymphocytes and plasma cells. However, the number of phagocytes may be large. Multinucleate, foreign body type giant cells are recognized. There are not normally any CPPD crystals within the mononuclear phagocytes or giant cells but urates may be deposited as well and these crystals may appear intracellularly. When absolute identification of tissue crystals is required, blocks of formalin-fixed, wet tissue or of deparaffinized material can be examined by infrared spectrophotometry, by chemical analysis and by spectrographic methods. Preparations may sometimes be available for X-ray diffraction.[153]

Hyaline articular cartilage The appearances are often indistinguishable from those of severe osteoarthrosis with which CPPD deposition disease coexists (see p. 860).

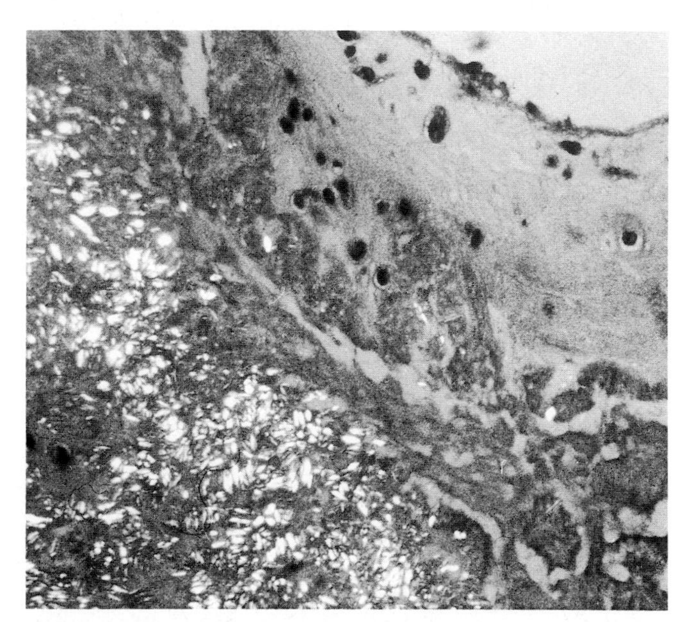

Fig. 10.27 Synovial chondrocalcinosis.
Calcium pyrophosphate dihydrate crystals are seen at a joint margin. (HE, polarized light × 120.)

There is fibrillation and loss of articular cartilage, eburnation and exposure of subchondral bone, bone sclerosis, cyst formation and osteophytosis.[153,160] Foci of faintly basophilic CPPD crystals occur characteristically in mid-zone cartilage or in the superficial part. Microscopically, free groups of crystals or isolated crystals of characteristic appearance lie between the processes of fibrillated cartilage, often adjoining chondrocyte clusters with which they appear to be closely associated. Scanning electron microscopy amplifies this information and displays the crystals on the cartilage surfaces where they can be analysed by electron probe X-ray microanalysis.

Transmission electron microscopic studies[19,185] have been used to display CPPD crystals in hyaline articular cartilage. The crystals are rhomboid or rod-shaped clefts in thin, decalcified sections;[19] they are present as clusters in

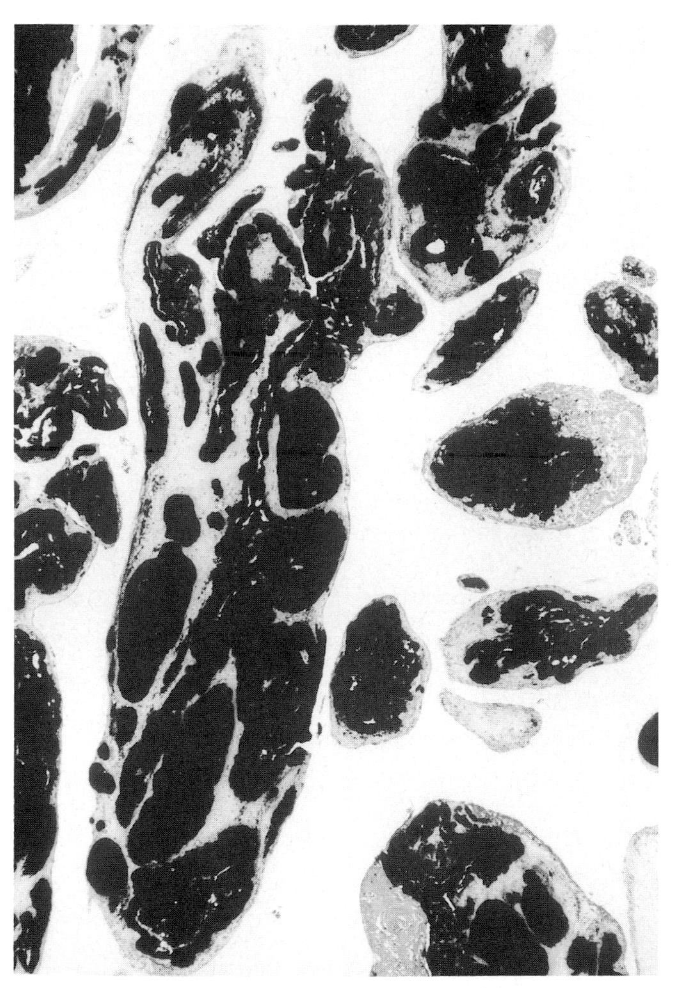

Fig. 10.28 Synovial calcium pyrophosphate dihydrate deposits.
Stains for calcium (Fig. 2.40, Table 10.7) offer a useful means of defining small calcium pyrophosphate dihydrate deposits in articular tissues. Urate aggregates which may appear morphologically similar, rarely stain for calcium. (von Kossa × 35.)

the interterritorial matrix, in mid-zonal (zones II and III) cartilage layers and as isolated, small extracellular structures.[185] The surrounding matrix is either homogeneous or displays faint cross-striations. Increased matrix electron density, with the presence of longitudinally fragmented collagen fibrils, is seen, usually without associated crystal deposition. No specific relationship is detectable between the location of the crystals and the collagen fibrils or non-collagen matrix material in spite of chemical evidence that impaired proteoglycan metabolism is associated with, precedes and predisposes to crystal deposition.[18]

Fibrocartilage

Menisci The knee meniscus (Fig. 10.29) may be intact or show tearing and incomplete repair. X-ray reveals symmetrical midzone radioopacity (Fig. 10.30). Histologically, islands of crystalline material staining for calcium lie in the poorly cellular meniscal fibrocartilagenous laminae. There is often microscopic splitting in the axes of the collagen bundles. The sites of crystal deposition are avascular and neither mononuclear phagocytes nor foreign body giant cells are seen.

Discs Intervertebral disc tissue is prone to symptomless calcification[13] and this is a particular feature of disorders such as Wilson's disease[141] (p. 369). The deposits are often CPPD[134] (Fig. 10.31, 10.32 and 10.33). Severe disc degeneration may result. Microscopy of the tissue excised at laminectomy may give the first evidence of the condition. When CPPD crystals are found, the patient is screened for occult, predisposing hyperparathyroidism.[13] Again, there may be no cellular reaction.[161] Degenerative changes in the affected disc tissue are invariable. The crystal deposits stain with the von Kossa technique but the demonstration of birefringence may be difficult. Scanning electron microscopy of the affected fibrocartilage allows the X-ray microanalytical demonstration of calcium and phosphate (K_α).[160] The characteristic bands of CPPD are shown by X-ray diffraction analysis.

Erosive osteoarthrosis of the articular processes indicates CPPD deposition disease of the apophyseal joints. The synovial crystal deposits are identical with those found in the knee, wrist and hip synovia. In a case of sciatica, biopsy was used to confirm disc CPPD crystal deposition.[233] Mohr *et al.*[161] discovered crystalline deposits in approximately 4 per cent of 2000 discs removed surgically for prolapse: polarization microscopy and X-ray analysis confirmed that the crystals were indeed usually CPPD. Their presence was related to age so that intervertebral disc chondrocalcinosis is an age-related phenomenon, particularly in men. These figures compare with those of Lagier and Wildi[121] who reported CPPD crystal deposition in 3.1 per cent of 1000 cases subjected to surgery for spinal disease. This frequency was higher than for CPPD deposition disease of knee menisci from subjects of the same age but lower than that for menisci from subjects of a higher mean age examined *post mortem*. In a case of ankylosing hyperostosis, Ferrerroca *et al.*[65] describe CPPD deposition in the nuclei pulposus of the intervertebral discs as well as CPPD deposits in the annulus fibrosus and numerous other fibrocartilages.

Articular capsule

The response to CPPD deposition in the vascular parts of the capsule is cellular.[160] There is much less evidence of cellular infiltration in those zones that are of low vascularity. The cellular response, seen in 80 per cent of instances, is a foreign body reaction to deposits of faintly basophilic clusters of CPPD crystals identical with those described in the synovium (see p. 398). There is a mononuclear phagocytic infiltrate, although, exceptionally,

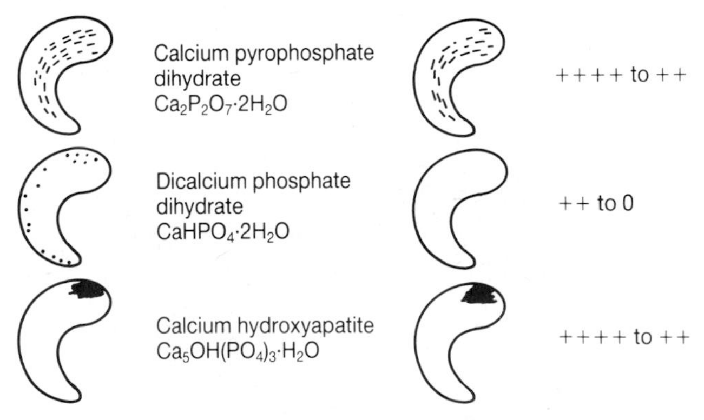

Fig. 10.29 Sites of deposits of calcium in knee menisci. Calcification due to calcium hydroxyapatite appears focal by contrast with accumulations of calcium pyrophosphate dihydrate and calcium phosphate dihydrate. Redrawn from Ryan and McCarty.[201a]

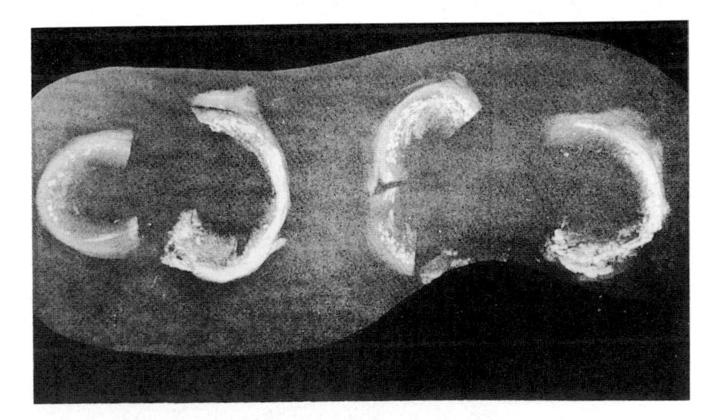

Fig. 10.30 Calcification of knee menisci. Radiograph of left and right menisci removed *post mortem* from a case of chondrocalcinosis. (× 0.5)

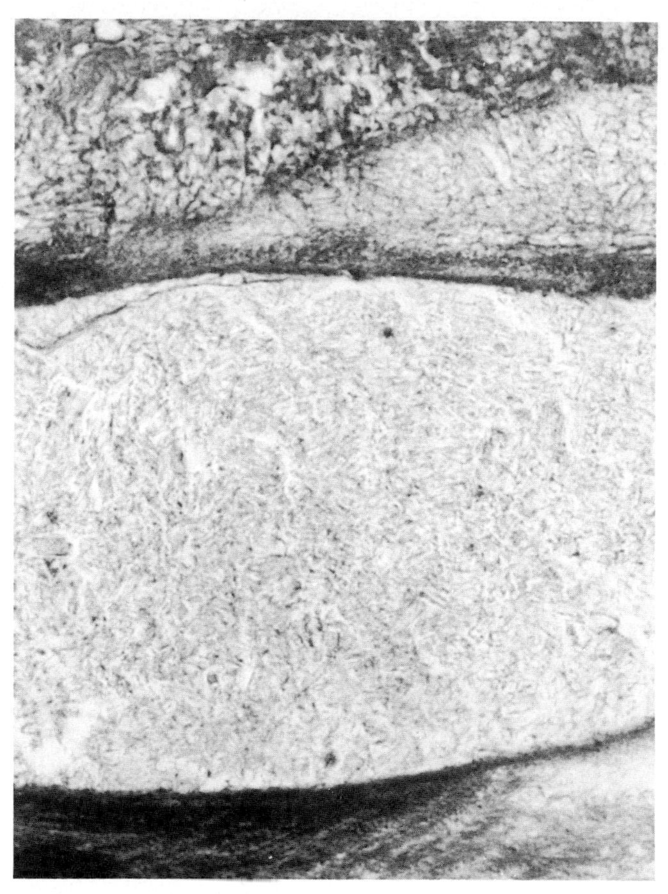

Fig. 10.31 Intervertebral disc chondrocalcinosis. There is an extensive but ill-defined aggregate of calcium pyrophosphate dihydrate crystals within the annulus fibrosus of the disc. Deposits of this kind may be chance findings when laminectomy is undertaken for the relief of intervertebral disc protrusion or prolapse. (HE × 500.)

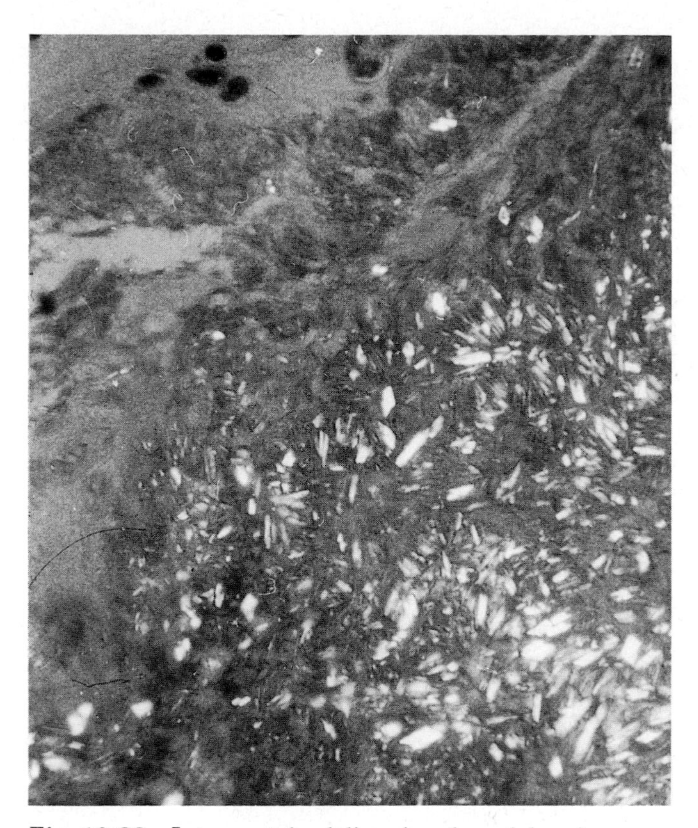

Fig. 10.32 Intervertebral disc chondrocalcinosis. The crystal aggregates within the disc are doubly refractile and of characteristic shape. (HE × 500.)

plasma cells may preponderate. Haemosiderin-containing cells may be seen. In a minority (20 per cent) of cases, virtually identical groups of crystals lie within dense tissue of low vascularity and there is little cellular reaction. This category of response is typical of joints which show the changes of osteoarthrosis and in which a secondary inflammatory reaction has led to the fibrosis recognized in chronic osteoarthrosis. Small clusters of crystals are intimately related to compact collagen and the orientation of the crystal clusters follows the collagen crimp. Scanning electron microscopy and electron probe X-ray microanalysis of the capsular and periarticular crystalline foci confirm both the nature and the location of the crystals.

Amyloid The coexistence of amyloid with calcium pyrophosphate deposition is sufficienly frequent to suggest the possibility that the phenomena may be related. Amyloid was identified in the hip joint capsular tissue of six cases of

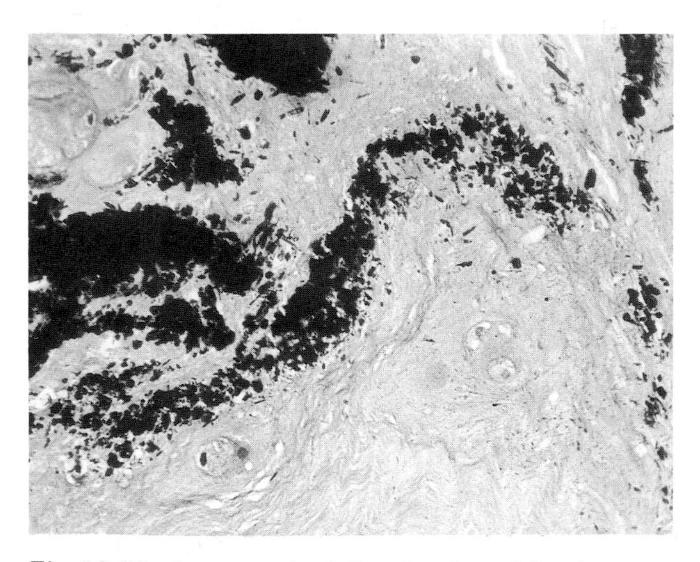

Fig. 10.33 Intervertebral disc chondrocalcinosis. The site of the accumulations of calcium pyrophosphate dihydrate crystals within the annulus of the disc is clearly shown by stains for calcium. (von Kossa × 238.)

57 subjected to hip arthroplasty: there were CPPD deposits in the immediate vicinity.[225] Similar observations, reported by these authors, have been made on eight knee joints; they also revealed amyloid coincident with CPPD.

Calcium pyrophosphate dihydrate deposition disease of individual joints

Wrist and carpometacarpal joints Here, CPPD deposition disease (chondrocalcinosis) is a frequent feature of pseudogout.[188] The cartilage is the most common site; the hyaline cartilage of the intercarpal and carpometacarpal joints is also affected. Subchondral sclerosis with discrete subchondral cysts can occur in the absence of chondrocalcinosis. The cysts may be secondary to the escape of synovial fluid into the subjacent bone or to subchondral bone necrosis; both processes may occur together[189] and pathological fracture has been described through such a large cyst.[229]

Hip joint In the hip joint, the femoral capital cartilagenous changes have been studied more closely than the acetabular. Synovitis and synovial chondrocalcinosis are recognized and cortical erosion is a feature of changes that, in other respects, replicate osteoarthrosis. It is possible that aseptic necrosis[120] contributes to the rapid destruction of the hip. How this is associated with CPPD deposition disease is not known. There are analogous changes in the knee.[118]

Temporomandibular joint arthropathy Temporomandibular joint arthropathy due to calcium pyrophosphate is extremely uncommon. In the case described by Pritzker *et al.*,[181] osteoclastic bone destruction with remodelling and new bone formation surrounded a mass of CPPD crystals but there was little cellular reaction. The absence of hyaline cartilage from the adult temporomandibular joint gave particular significance to the crystal deposits. The initial crystal deposition was likely to have been synovial or bony. In analogy to the CPPD deposition disease found by peripheral joint aspiration in the same patient, the temporomandibular arthritis was presumed also to have been due to CPPD deposition.[78]

Neuropathic joints Chondrocalcinosis in association with tabetic arthropathy is characterized by CPPD deposition[100,196] and a synergistic interaction between these disorders has been suggested but not proved. CPPD crystals also accumulate in non-tabetic, mechanically unstable joints,[219] confirming a relationship between physical microtrauma and crystal deposition.

Experimental studies

Crystalline deposits of CPPD can be produced experimentally in rabbits.[59a] CPPD deposition disease has also been identified in rhesus monkeys[194] including *Macaca mulatta*[187] and *Macaca sylvanus*.[105] In *M. sylvanus*, vertebral hyperostosis and intervertebral disc CPPD deposits were investigated *in vitro*. Amorphous and orthorhombic calcium pyrophosphate tetrahydrates are precursors in the formation of triclinic- and monoclinic CPPD.[150] The air pouch model of Selye[215] can be regarded as a facsimile synovium to study the tissue responses to CPPD. Mononuclear cells, and, to a lesser extent polymorphs, phagocytose the CPPD crystals which later become embedded in the mock synovial pouch lining, suggesting the sequence which may occur in the true synovium.[222]

Other calcium phosphate deposition diseases
(Table 10.8)

A distinct category of disease is characterized by the deposition of the basic calcium phosphates.[47,84,130] Complex factors regulate normal and abnormal calcification. They are outlined on p. 107 *et seq*. Why normal structures such as the fibrocartilagenous menisci of the knee joint are calcified in some species, such as the mouse[122,237] and not others, is not clear. Calcification, the deposition of insoluble calcium phosphate (Tables 10.8 and 10.9) (Fig. 10.34) in tissues, centres on matrix vesicles. There is often mitochondrial involvement. Calcification subsequently extends

Table 10.8

Some calcium-containing crystals deposited in human articular cartilage (modified from Dieppe *et al.*[48])

	Phosphorous/ calcium ratio
*Calcium hydroxyapatite $Ca_5OH(PO_4)_3.H_2O$	0.46
Calcium pyrophosphate dihydrate $Ca_2P_2O_7.2H_2O$	0.77
Calcium phosphate dihydrate (orthophosphate) (Brushite) $CaHPO_4.2H_2O$	0.77

* Strictly, hydroxyapatite is $Ca_{10}(PO_4)_6(OH)_2$. However, there is lack of precision in terminology due to incomplete understanding of bone structure. $Ca_5OH(PO_4)_3.H_2O$ is correctly termed calcium hydroxylapatite.

Table 10.9

Matrix vesicles in calcific disease (after Anderson[8]).

Class	Disease	Mitochondrial involvement	Extracellular membrane vesicle involvement
Cardiovascular			
	Atherosclerosis	–	+
	Medial sclerosis	+	+
	Valvular disease	–	+
Osteoarticular			
	Osteoarthrosis, 'apatitic'	–	+
	Calcifying tendinitis	–	+
	Myositis ossificans	–	+
	Osteosarcoma	–	+
	Chondrosarcoma	–	+
Renal			
	Metastatic calcification	+	+
Non-skeletal neoplasia		+	+
Miscellaneous			
	Tympanosclerosis	–	+
	Calcinosis cutis (calcergy)	–	+
	Failed blood pumps	–	+
	Failed intrauterine contraceptive devices	+	+
	Dental plaque and calculus	+	+

in relation to matrix components, particularly the type II collagen of cartilage.

Abnormal calcification

Abnormal calcification is[9] recognized in a wide variety of diseases, including many of those of connective tissue; to some, the term 'calcinosis' is applied. A combination of factors determines the deposition of calcium-containing crystals.[20a] They include a raised concentration of precipitating ions and the formation of specific nucleators. The inappropriate formation of insoluble mineral may occur when calcium or phosphate levels are raised: it is then said to be *metastatic*. Alternatively, calcification may occur in diseased tissues when the levels of extracellular calcium and phosphate are normal. This process is *dystrophic* calcification.

Abnormal calcification is essentially a cellular phenomenon.[7] Under abnormal circumstances there may be a disturbance not only of the energy-dependent mechanism that maintains a concentration of calcium within the cytoplasm much lower than that of the extracellular fluid, but also of the high normal concentration of calcium within mitochondria. If a mitochondrion is so loaded with calcium phosphate that its energy-producing processes are impaired, a vicous circle is established and the exclusion of calcium from the cell and thus, in turn, from the mitochondrion itself, is no longer practicable. The formation of electron-dense calcium phosphate opacities within heart muscle mitochondria, for example, is an early indication of irreversible ischaemic cell injury.

Metastatic calcification

Metastatic calcification is liable to occur when serum and extracellular fluid levels of calcium or of phosphate are abnormally high. The process is exemplified by the tissue changes in hyperparathyroidism and in hypervitaminosis D. Hypercalcaemia, however, is not, by itself, a sufficient cause for metastatic calcification: a high level of plasma calcium is recognized[165a] in disorders such as myeloma (lymphotoxin), metastatic cancer (parathormone- related protein and transforming growth factor-α) and sarcoidosis (vitamin D), but metastatic calcification is not constant or generalized. Although adenosine triphosphate and pyrophosphate prevent early crystallization, once calcium hydroxyapatite has formed, there is a tendency for crystallization to continue in the extracellular matrix.

Dystrophic calcification

Dystrophic calcification occurs, for example, at sites of tissue injury or necrosis. It is thought that matrix vesicles invariably act as niduses for calcium hydroxyapatite crystal growth. Dystrophic calcification may occur around joints (calcific periarthritis). The deposits may be symptomless or cause inflammation. Calcium also accumulates in altered soft tissues after local mechanical stress, as in 'tennis thumb', and in destructive forms of neurogenic arthropathy and infection. The tendency to calcium deposition may be familial, as in tumoral calcinosis (see p. 405) and in some examples of calcific periarthritis.

Calcification of a dystrophic character is a frequent feature of the generalized connective tissue diseases. It is of particular importance in progressive systemic sclerosis (see Chapter 17) where calcinosis cutis and circumscripta are recognized; in polymyositis/dermatomyositis (see Chapter 16); in systemic lupus erythematosus (see Chapter 14); and in relapsing polychondritis (see p. 711). Other sites where dystrophic calcification is common include: the ar-

Fig. 10.34 Crystals of calcium phosphate.
(SEM × 700.)

terial media in Mönckeburg's sclerosis, where calcification occurs at the periphery of elastic fibres;[111] the aortic valve in calcific aortic disease: here, calcification occurs, with advancing age, in matrix vesicles derived from mesenchymal cells; and in neoplasms such as breast carcinoma where concentric, lamellar deposits of hydroxyapatite form. Although intracellular structures may nucleate the formation of hydroxyapatite such as that of the psammoma bodies in ovarian neoplasms and the Michaelis-Gutmann body in malakoplakia,[140] intracellular mineralization also begins within the blind intracellular lumina that occur in adenocarcinomata. The high frequency of dystrophic calcification in malignant mesenchymal neoplasms is also attributable to their matrix vesicle content.

New techniques for the analysis of calcific deposits have begun to show that the material is not uniform. It had been accepted that in the subcutaneous calcific deposits of the CREST subcategory of progressive systemic sclerosis (see p. 695), the mineral was in the form of classical calcium hydroxyapatite. It now appears that the deposits are highly heterogeneous and in a nonstoichiometric, carbonated, calcium deficient, apatitic, solid phase. They coexist with dense globules that present a poorly organized, more or less amorphous phase, or as microcrystalline tricalcium phosphate, admixed with scattered calcium hydroxyapatite crystals.[44] Future work is likely to demonstrate comparable heterogeneity in other connective tissue mineral deposits.

Calcifying tenosynovitis

The deposition of calcium hydroxyapatite crystals in the rotator cuff of the shoulder joint, especially in the tendon of the supraspinatus muscle, 10–20 mm proximal to its insertion into the greater tuberosity, is a common cause of shoulder pain.[80,146,173] The disorder is relatively frequent in young adults. Identical symptoms can result from calcification of other tendons such as that of the pectoralis minor. Surgical exploration enables a picture to be formed of the sequence of microscopic events. In an unusual category (the 'Milwaukee' shoulder), McCarty *et al.*[135] identified a tear of the rotator cuff in seven of the eight shoulder joints in four elderly women. The synovial fluid contained radioopaque calcium hydroxyapatite crystals (Table 10.10), activated collagenase and neutral proteinase but it is not known whether these are constant findings (Fig. 10.35).

Table 10.10

Crystals identified in the Milwaukee shoulder syndrome[137]

Calcium hydroxyapatite
Octacalcium phosphate
Tricalcium phosphate (Whitlockite)

Calcification may begin, not in the tendon itself but in nearby, more vascular tissue.[80] These authors found that the hyperplastic synovium and peritendinous tissues contained many round psammoma-like foci of calcification. The pectoralis minor tendon revealed focal fibrillation and fragmentation. Circumscribed granular or globular proteinaceous zones were surrounded by epithelial cells, foreign-body giant-cells and lymphocytes accompanying a meshwork of young blood vessels. The 50–150 μm diameter globular structures were laminated, stained blue or brown with a trichrome stain but negatively for calcium. However other reports have identified calcium salts within the tendinous radioopacities. Larger lesions revealed central cavitation and a fibrous margin.

The evidence suggests that an ischaemic and traumatic lesion might predispose to secondary, peritendinous calcification. The role of the supraspinatus tendon insertion (see p. 101) is not known. Local hypoxia is presumably the

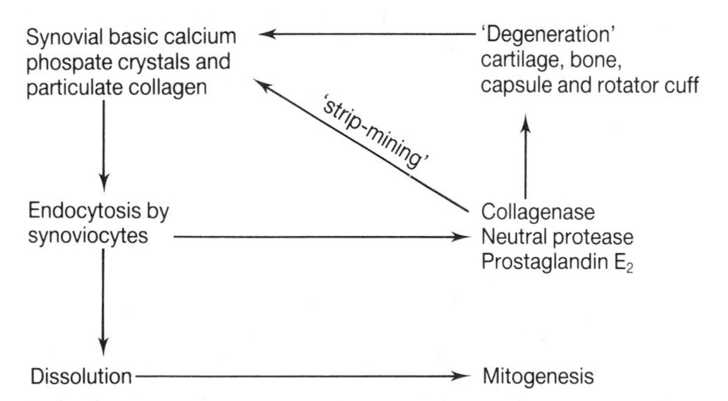

Fig. 10.35 Connective tissue injury in the Milwaukee syndrome.
Hypothetical scheme for the pathogenesis of the tissue lesions of this form of calcium hydroxyapatite deposition disease. Redrawn from McCarty and Halverson.[130]

result of an impaired arterial blood supply and is caused by repetitive mechanical trauma. Fibrocartilagenous metaplasia is invoked.[244,245] Transmission electron microscopic studies (Fig. 10.36) show calcium crystals in matrix vesicles among collagen fibres.[202] The calcium deposits are insoluble calcium hydroxyapatite. Whether reabsorption of the deposits can occur, with reconstitution of a collagenous zone, is unclear.[244,245] Synovitis and bursitis are sequelae but polymorph infiltration appears uncommon.[203]

Tumoral calcinosis (calcifying collagenolysis)

From time to time, young healthy persons, with no history of trauma, form increasingly large, radioopaque, multilobulated masses of calcium-containing material within the subcutaneous tissues overlying the shoulder, elbow or hip joints.[99,117,155,242] The masses are pockets of white, milk-like fluid; the loculi are arranged in clusters. Usually, there is no systemic disturbance of calcium metabolism and no visceral, metastatic deposition of calcium elsewhere. A familial predisposition is sometimes demonstrable. Negro races are susceptible. The masses increase to sizes in excess of 200 mm in diameter and weight 500 g but remain firm and are not tender. Smaller masses may form in the tissues near the joints of the vertebral column and pelvis, the ribs, wrists and feet. Carpal tunnel obstruction may result. Inadequate surgical excision can leave residual fistulae or excite infection but the possibility exists of effective medical treatment with agents such as aluminium hydroxide which evoke phosphorus deprivation.[45a]

The mineral deposits are of calcium hydroxyapatite.[21] Transmission electron microscopy shows needle-like crystals within mononuclear cells and extracellularly, not associated with collagen. Electron probe X-ray microanalysis[83] demonstrates only calcium and phosphate with calcium: phosphate ratios of 1.54 ± 0.03. The electron diffraction patterns resemble those of calcium hydroxyapatite, evidence confirmed by X-ray diffraction analysis. The total lipid is 1–10 per cent of the demineralized dry weight, the uronic acid 1–6 μg/mg dry weight. No abnormal collagen types are found and the collagen Hyp (see Chapter 4) content in demineralized material is 2–9 μg/mg.

The histological features of early cases show small, cystic fluid-containing spaces traversed by fine fibrous trabeculae.[242] Within the fluid are masses of 3 μm diameter, round, calcified bodies and larger laminated granules. The granules stain by the von Kossa method; the decalcified residue reacts with the periodic acid–Schiff technique but weakly with Alcian blue. Within the cysts, among the granules, are larger irregular masses. The cyst edges include many mononuclear phagocytes, occasional foreign-body giant-cells, some lymphocytes, plasma cells, eosinophil granulocytes and lipid-containing histiocytes. The

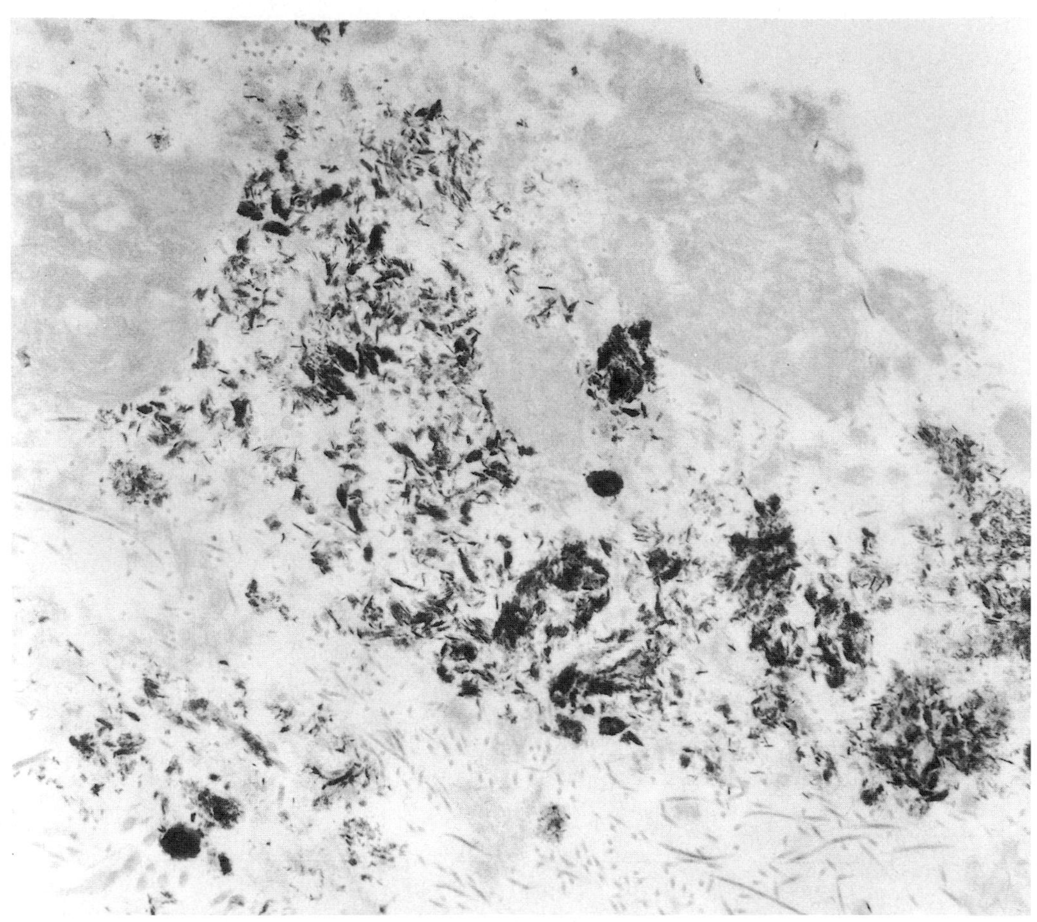

Fig. 10.36 Crystals of calcium hydroxyapatite in the Milwaukee syndrome. (TEM × 9200.) Courtesy of Professor Paul Dieppe.

giant cells and histiocytes give evidence of alkaline phosphatase activity. The most active lesions are highly vascular. Extension into nearby adipose tissue and skeletal muscle is recognized.

In older, less cellular lesions, the 'cysts' no longer contain fluid. The granules of calcified material become larger and amorphous, less fully mineralized, and collagen-containing. The irregular-shaped masses within the cysts are also larger and compacted, the surrounding fibrous tissue denser, less cellular and less vascular. A role for proteolysis has been suggested and the disorder has been called calcifying collagenolysis.[242] Whether the multinucleated giant cells of tumoral calcinosis are osteoclasts remains speculative.

Experimental soft-tissue calcification

The factors that inhibit the deposition of calcium phosphates from solution *in vivo* can be overcome experimentally.[138,184a] The procedures include the induction of calciphylaxis and calcergy.[217] These artefacts are forms of metastatic and dystrophic mineralization. They do not appear to have exact analogues in human disease.

Calciphylaxis

Calciphylaxis[218] is a condition in which the preliminary administration of an agent such as parathyroid hormone or a vitamin D 'sensitizes' an animal to the subsequent action of an agent, such as iron dextran, or to mechanical trauma, leading to local or systemic calcification. In *local caliphylaxis*, subcutaneous challenge of a sensitized rat with, say, 50 μg ferric chloride results within 2–3 days in the heavy local deposition of calcium hydroxyapatite at the site of the iron injection. There is a critical period of 24 h between sensitization and challenge before which calcification does not begin. The nature of the local response is similar whatever the identity of the sensitizing compound. In *systemic calciphylaxis*, an agent such as ferric chloride is injected into the jugular vein of an animal similarly pretreated. Calcium is deposited in the thyroid and parathyroid glands. Other systemically administered compounds cause calcification in different organs. Further variants of the syndrome can be produced by the additional, simultaneous action of locally or systemically administered mast cell-discharging agents such as dextran. Chronic overdosage of a rat with some vitamin D derivatives leads to changes that resemble premature ageing and that include widespread arterial and cartilagenous calcification. Curiously, this state, accompanied by loss of weight and connective tissue atrophy, can be prevented by challenging agents such as aluminium chloride.

Calcergy Calcergy results from the topical application of compounds such as lead acetate which will cause local calcification without any previous 'sensitization' by systemic agents that modulate mineralization. The effects may be *local* as when lead acetate is given by subcutaneous injection, or *systemic*, as when lead acetate is given intravenously. Mast cell-discharging compounds such as dextran modify these responses either when given locally or systemically.

Calcium hydroxyapatite and articular disease (Fig. 10.34)

Apatite crystals (Table 10.10) often accumulate in the articular cartilage, menisci and tendons of synovial joints and in the discs of fibrocartilaginous joints. Calcium hydroxyapatite crystals are exceedingly small (0.01–1.00 μm long) and generally beyond the resolving power of the light microscope. This is one reason why their significance in soft tissue-, as opposed to bone pathology, has been overlooked. However, when crystal aggregates (approximately 0.1–10.0 μm diameter) are found or individual crystals retrieved, scanning electron microscopy, transmission electron microscopy and electron probe X-ray microanalysis allow their identification. Larger, smooth-surfaced bodies such as those found in tumoral calcinosis (see p. 405) can also be examined and identified by these techniques.

The presence of calcium hydroxyapatite crystals in midzone (zone II–IV) hyaline articular cartilage, in relation to matrix vesicles, has prompted the suggestion that the crystals influence the onset of osteoarthrosis (see Chapter 22).[4,5] Calcium hydroxyapatite crystals may also accumulate in synovial tissues and appear capable of promoting soft-tissue inflammation. Experimentally, calcium hydroxyapatite crystals can be phagocytosed both by macrophages and by synovial cells in culture.[95] The crystals exert a mitogenic effect on synovial fibroblasts,[39] and can provoke a matrix in which normal and abnormal chondrocytes can express collagen and proteoglycan synthesis *in vitro*.[68,121a,123]

Hyperoxaluria

Secondary hyperoxaluria

Hyperoxaluria is a recognized feature of some cases of chronic renal failure. The crystals of calcium oxalate may accumulate in connective tissues, including synovial joints.

Costochondral and articular cartilages are also sites for this form of crystal deposition disease. For pathological diagnosis, percutaneous bone biopsy is recommended. The crystal aggregates are round and globular; they show a slight yellow colouration and are doubly refractile. Individual crystals are rhomboidal; they are arranged in rosette-like, radial arrays. The deposits stain positively with the von Kossa technique for calcium but only when other calcium salts have precipitated simultaneously. Polarized light microscopy and X-ray diffraction permit positive identification.

Hyperoxaluria has been recognized in patients with fat malabsorption, in some after jejunoilial bypass operations, and in others with chronic inflammatory bowel disease, disorders in which there is a recognized tendency to develop polyarthritis (see p. 786). The increased absorption of oxalate from the gut is a consequence of defective binding to calcium. Occasionally, gout-like attacks of acute arthritis have been described but arthropathy[88] usually results from periarticular tophi in soft tissues.[206]

Primary hyperoxaluria

In heritable, primary hyperoxaluria, there is continuous excess synthesis and excess excretion of oxalic acid. Renal calculi and nephrocalcinosis are early features; they culminate in progressive renal excretory failure with death by the age of 20 years.[256]

Xanthinuria

Very rarely, some individuals have a genetically determined deficiency of xanthine oxidase.[96] At least 45 examples of this deficiency have been described. Serum uric acid levels are low but arthropathy is an associated clinical sign in 7 per cent of cases. Since xanthine crystals injected experimentally can produce acute arthritis, it is possible that the presence of xanthine crystals in the synovia can cause inflammation, 'xanthine gout'. Xanthine and hypoxanthine crystals have been demonstrated in the uncommon, associated myopathy but, so far, these crystals have not been proved to be present in synovial fluid or synovial tissue.[96] There is an analogy with the muscle aggregates of xanthine in cases of gout treated with xanthine oxidase inhibitors.[250]

Cholesterol arthropathy

Cholesterol is an important, normal component of cell membranes; it exists in the plasma in the low density lipoprotein fraction. Nevertheless, there is very little cholesterol in normal synovial fluid. Occasionally the synovial fluid of patients with rheumatoid arthritis, ankylosing spondylitis or other inflammatory joint diseases comes to contain a great excess, sufficient to cause the synovial fluid to be turbid, viscous and yellow-white.[56]

Exceptionally, synovial fluid cholesterol may be crystalline. This phenomenon is described in patients with osteoarthrosis and persistent effusion.[60] The crystals are identified by polarized light microscopy: they are rhomboidal, notched and 10–50 μm in diameter or needle-shaped and only 1–20 μm in size. The characteristic morphology of synovial fluid cholesterol crystals can be established by scanning electron microscopy but X-ray diffraction is necessary for positive identification. Larger clusters (tophi) may form.[61]

The explanation for cholesterol accumulation in synovial joints is incomplete; the crystals either excite or are associated with the excitation of an inflammatory reaction. Cholesterol crystals injected into the knee joints of rabbits provoke inflammation:[59a,180] the crystals persist for long periods within the synovium and cholesterol granulomata, hyperplasia of synovial cells and fibrosis are consequences.

Disorders of bone related to disease of non-calcified connective tissue

Because of the overlap between the diseases of calcified and non-calcified connective tissues, brief accounts are given of some of the relevant disorders of bone that may affect joints. In each section, the reader is referred to reviews which provide more extensive discussion. The articular consequences of hyperparathyroidism are described on p. 416.

Osteomalacia

Bone is formed continuously throughout normal life, replacing the tissue lost as a result of resorption and catabolism. Replacement is a dual mechanism. Osteoid, bone matrix, synthesized by osteoblasts, is impregnated by insoluble calcium hydroxyapatite in a process catalysed by osteoblast alkaline phosphatase. When mineralization is incomplete, usually because of a deficiency of the active metabolites of vitamin D, the accumulation of excess, non-mineralized osteoid constitutes the disorder, osteomalacia.[36,151,191,247] A deficiency of phosphate may also cause osteomalacia, but chronic deficiency of calcium generally results in secondary hyperparathyroidism.

Osteomalacia is often present in elderly persons living alone for long periods on deficient diets; it is also a regular complication of intestinal malabsorption, chronic renal tubular disease and tropical malnutrition. In established, severe or advanced osteomalacia, ill-defined bone pain and tenderness accompany weakness and impair muscular movement. The diagnosis of osteomalacia can only be confirmed by bone biopsy; the sample is frequently obtained from the iliac crest (Fig. 10.37).[156,177] The extent of the calcification front can be examined after *in vitro* or *in vivo* tetracycline labelling.[139] The relative volume of tissue represented by non-mineralized matrix may be measured by histomorphometric techniques.[190] An increase in the volume of osteoid coupled with a decrease in the extent of the calcification front confirms a diagnosis of osteomalacia. The *in vivo* administration of two tetracycline labels, separated by a known time interval before bone biopsy, allows a mineralization rate to be calculated. Decrease in this rate is an early and sensitive indicator of impaired mineralization. Treatment, for example, of elderly persons suffering from defective vitamin D and calcium intake, can be rapidly curative.

Osteomalacia, like osteoporosis, can contribute to and exaggerate the severity of articular disease. The effects of osteomalacia on the non-mineralized connective tissues are limited. In the long bones, zones of fibroosseous disorganization and repair comprise Looser's zones. They may be of mechanical origin. There is little surrounding reaction. In vertebral bone, osteomalacia impairs strength and contributes to intraosseous protrusion of the nuclei pulposus. The relative contribution made by osteomalacia and by osteoporosis to vertebral bone disease is often difficult to determine. Osteoporosis is common among the elderly in whom osteomalacia is also often recognized; the influence of modest degrees of osteomalacia may be slight.

When vitamin D deficiency develops in early childhood, the defective mineralization of epiphyseal zones of provisional calcification disorganizes endochondral ossification and the more complex but analogous disease, rickets, is manifest. The immediate effects of rickets are recognized in childhood and adolescence; the late consequences impair the structure and function of the calcified connective tissues throughout life. The musculoskeletal symptoms and signs of rickets are fully recorded.[151]

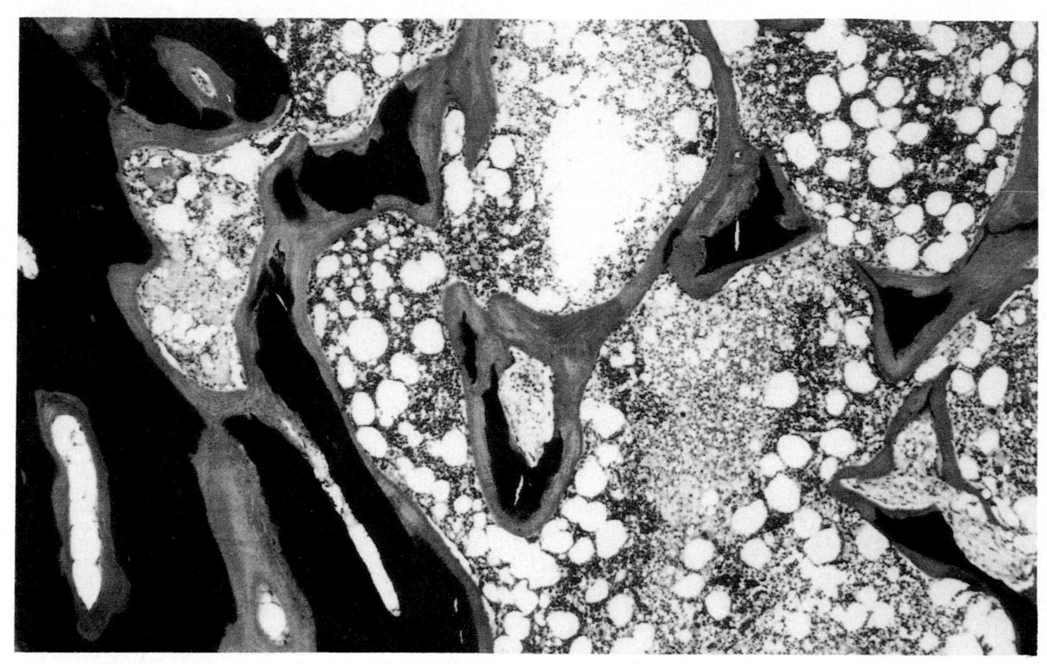

Fig. 10.37 Osteomalacia. Synovial and fibrous joints may be prejudiced by the bone lesions of this disorder. In this figure, bone of cortex is shown (at left), cancellous, medullary bone (at right). Mineralized matrix appears black, unmineralized osteoid grey. The appearances are those of severe osteomalacia. (von Kossa HE × 144.)

Hypophosphataemia

Persistently low serum phosphate levels result in a form of osteomalacia. Hypophosphataemia (with hyperphosphaturia) may be due to failure of renal tubular reabsorption of phosphate. This is accompanied by aminoaciduria in the Fanconi syndrome. Individuals with hypophosphataemic osteomalacia/rickets are usually replete with vitamin D and the disorder is, therefore, resistant to vitamin D treatment. Hypophosphataemia may, very rarely, arise *de novo* in association with certain neoplasms. This phenomenon results in 'oncogenic' osteomalacia. The hypophosphataemia may be very profound and the osteomalacia severe. The associated neoplasms are frequently vascular in type and display the features of haemangiopericytoma. It is possible that the neoplasm produces a chemical factor which selectively inhibits renal tubular reabsorption of phosphate but the precise pathogenetic mechanisms are mysterious.[142]

Hypophosphataemia also occurs in intestinal malabsorption and as the result of prolonged ingestion of antacids (phosphate-binding effect). The symptoms referable to the musculoskeletal system can simulate ankylosing spondylitis, calcifying tendinitis and myopathy.[164] They respond readily to treatment.

Osteoporosis

In osteoporosis, the amount of bone tissue per unit volume of anatomical bone is reduced.[191,260] There is a small net imbalance between the processes of bone resorption and formation so that, with time, there is an aggregate loss and the quantity of bone in the whole skeleton gradually diminishes.[247] The bone that is formed is, however, normally and fully mineralized unless a further disorder of bone, such as osteomalacia, is coincidental. In uncomplicated cases, the trabeculae of cancellous bone seen in two-dimensional representations are always abbreviated, often slender and may be diminished in number. Cortical bone is thinned. Osteoporosis is a common disease of the elderly[259a] and is rare in the young[94]; by definition, osteoporosis is therefore a frequent association of systemic connective tissue diseases such as rheumatoid arthritis in older women. In many cases, consequently, the pathological features of infective, immunological, metabolic and degenerative diseases of the synovial joints and systemic connective tissues, are complicated by osteoporosis particularly when there has been evidence of maternal osteoporosis.[214a]

The diagnosis of osteoporosis is suspected on the basis of clinical X-ray. This procedure is, however, of low sensitivity and some 25 per cent of the bone mass is lost before osteoporosis is recognizable. However, new techniques such as dual photon absorptiometry offer the possibility of recognizing changes in bone density of as little as ± 1 per cent. The established lesions, seen at necropsy, include the presence of an open meshwork of porous bone (Fig. 10.38); compression fractures of vertebrae, some of which assume a characteristic fish-tail deformity (Fig. 10.39); kyphosis; and increasing scoliosis. Rib-bone fractures readily and the strength and density of the limb bones is reduced. The

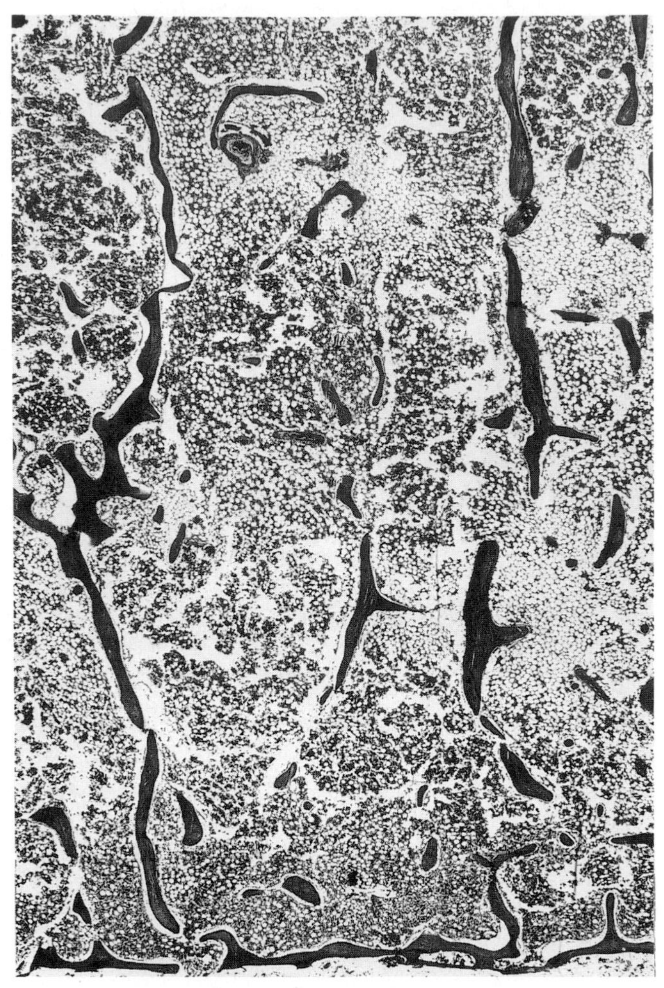

Fig. 10.38 Osteoporosis.
In theory, osteoporotic bone should be less elastic than normal bone and therefore less capable of absorbing the forces transmitted during articular load-bearing. However, osteoporotic bone appears often to sustain microfracture. Since there is a normal capacity for fracture healing, in the absence of other metabolic disease, these microfractures repair, restoring bone elasticity and predisposing to osteoarthrosis of the adjacent synovial joints. (HE × 10.)

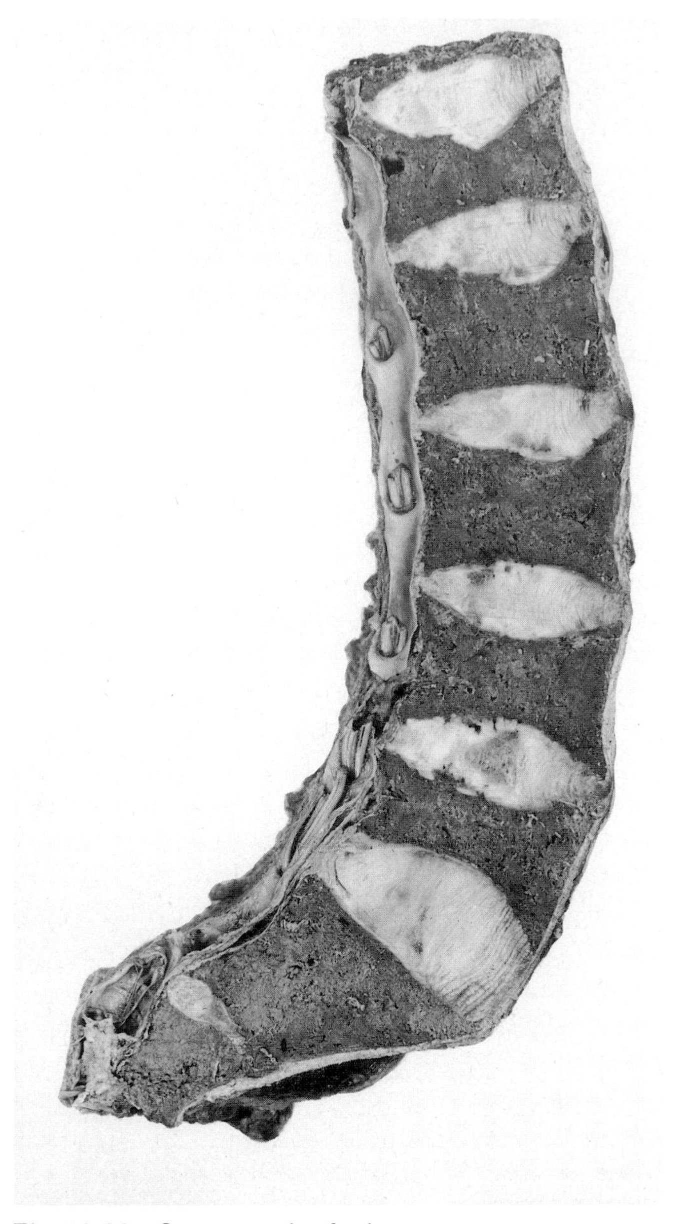

Fig. 10.39 Osteoporosis of spine.
The vertebral bodies have become biconcave. The intervertebral discs protrude into the porotic bone. The shape of the fourth and fifth lumbar vertebrae resembles a 'fish-tail'.

diminished density of the vertebral bones predisposes to intraosseous prolapse of residual nuclei pulposus so that Schmorl's nodes (see p. 936) often complicate osteoporosis.

Paget's disease of bone

Pain, deformity and fracture complicate established Paget's disease.[35] The articular (Chapter 22) and rheumatic manifestations have been surveyed.[67,6] In the large limb joints, epiphyseal or juxtaarticular bone may be directly involved. Alternatively, osteoarthrosis may coexist with Paget's disease and, under these circumstances, pagetoid change may affect articular bone. Gout is an occasional accompaniment. In the spine, pain may be attributed to lumbosacral Paget's disease alone but other causes of back pain such as ankylosing spondylitis and psoriatic arthropathy can coexist.[35] There may be pressure upon the foramina of the vertebrae so that neurological and vascular changes ensue.

Fluorosis

Whereas small amounts of fluoride in drinking water, of the order of one part per million, exert an important influence in stabilizing bone mineral crystals and in preventing dental caries, large amounts of fluoride, ingested over long periods, lead to a generalized bone disease.[54,247] The end results superficially resemble the late stages of disseminated ankylosing hyperostosis (see p. 947).

Geographical fluorosis

Excess fluoride is ingested in drinking water in geographical zones where water sources are drawn from rocks with a high fluoride content. The incidence and severity of geographical fluorosis have received particular attention in India[221,223,226] and Tanganyika. Grey, brown-grey or more darkly mottled zones appear in the adult teeth. Deciduous teeth are seldom affected, presumably because the duration of exposure required to induce clinical fluorosis is roughly proportional to the fluoride content of the drinking water and to the volume of water ingested. The condition is thus most likely to be seen in zones where high temperatures combine with manual labour to stimulate a large daily intake of water with a high fluoride content. The concentration of fluoride in such water sources may be up to 12 parts per million.

Occupational fluorosis

The problem of occupational fluorosis was first identified in Denmark.[162,163] Industrial fluorosis was investigated in this country by the Medical Research Council[1] with special reference to the influence of aluminium fluoride on human and animal populations living near smelting plants. The disease in poultry is comparable to that in mammals. Workers exposed chronically to the dust of cryolite, a sodium aluminium fluoride used in the manufacture of aluminium, were found to display 'a curious whiteness and density of (X-ray) shadows of the ribs, clavicles and cervical vertebrae, a loss of the normal sharp vertebral outlines and extensive calcification of the intervertebral ligaments'. The disease affects the vertebral column, causing pain, stiffness, rigidity and ultimately, ankylosis. Limb joints are also disordered; the distribution of lesions is apparently influenced by occupational stress to the skeleton (Figs 10.40 and 10.41). Osteosclerosis, with cortical and periosteal new bone formation, leads to a spinal rigidity with marginal osteophyte formation and calcification within the paravertebral ligaments resembling that seen in long-standing ankylosing spondylitis (Fig. 19.7). By contrast with the latter disease, in which ankylosis is accompanied by osteoporosis, the 'bamboo' spine of fluorosis is accompanied by an extreme, generalized form of osteosclerosis. There is commonly a form of associated osteoarthrosis with widespread osteophytosis and lipping. In the fingers, the resulting osteophytes do indeed resemble those of osteoarthrosis. The osteosclerotic process also affects skull sutures and the inner ear, leading to VIIIth nerve deafness.

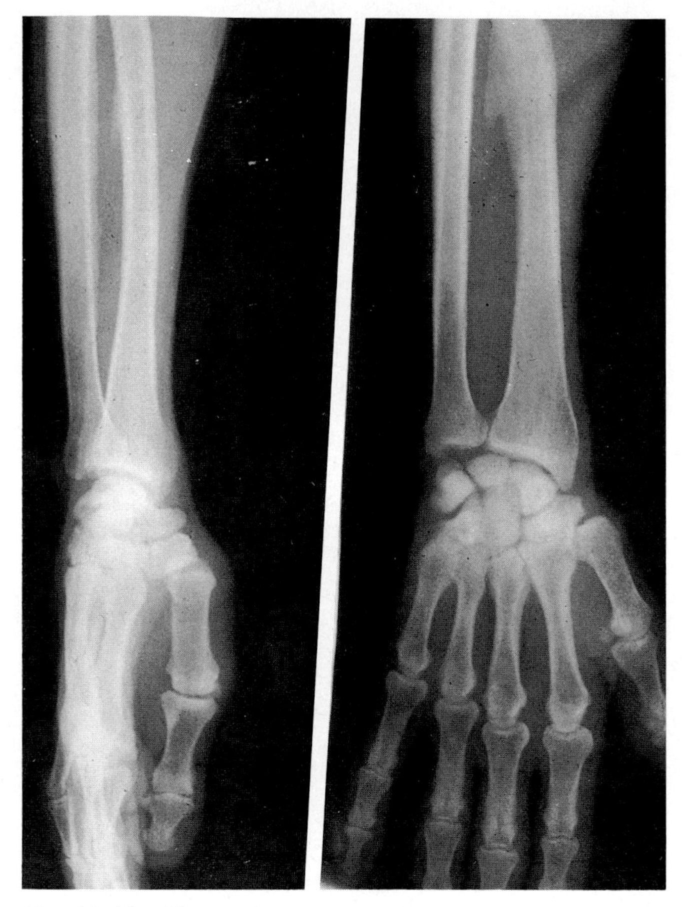

Fig. 10.40 Fluorosis.
The radiological changes, typical of endemic fluorosis, centre on periosteal new bone formation. The new bone is dense and thick so that the outlines of these forearm and hand bones appear unusually clear.

The biochemical disorder

Although blood fluoride levels are low, affected persons continue to excrete excess urinary fluoride for some years after leaving the zone of fluoride-laden water supplies. Raised serum alkaline phosphatase levels have been reported, presumably reflecting the new bone formation occurring in osteosclerotic tissues, while analysis of the bones themselves confirms the anticipated high fluoride

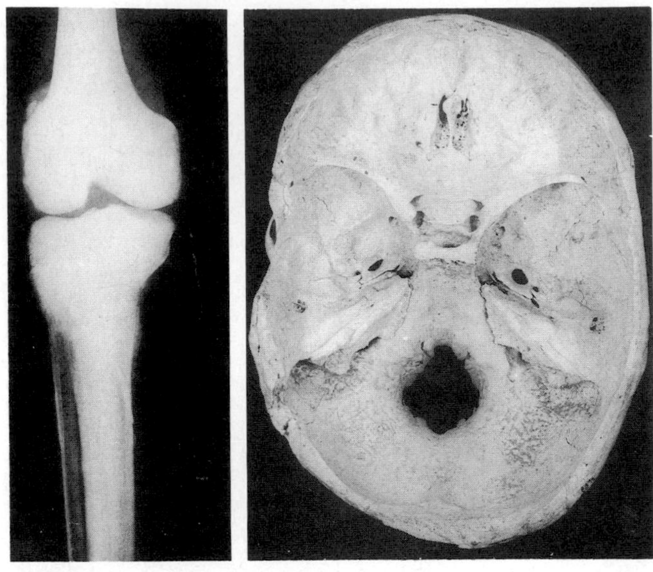

Fig. 10.41 Endemic fluorosis.
At left, radiograph of part of lower limb showing increased radioopacity of tibia. At right, thick, heavy bone of skull. No diploe remains and osteophytes protrude from the margins of the foramen magnum.

content. Sodium fluoride can increase the rate of proliferation of bone cells in culture.[63] Fluoride also increases the alkaline phosphatase content of bone cells and embryonic bone and enhances the growth and mineralization of the latter. In fluoride bone disease, a direct effect on osteoblasts therefore appears probable, complementing the cytotoxic effect of fluoride on bone resorbing cells.[184] There are other well-documented actions. After absorption, fluorine exchanges with hydroxide in the hydroxyapatite (bone mineral) lattice. Fluoride is therefore distributed in the skeleton in the same manner as calcium hydroxyapatite.[247] The incorporation of fluoride in the lattice of new hydroxyapatite crystals may protect against the acute toxic effects of very high concentrations of this element. Since the crystal produced by substituting fluoride for hydroxyl ions is larger than that of non-substituted hydroxyapatite, and therefore more stable, fluoride is thought to have a comparable stabilizing action on bone and tooth mineral.[247]

Pathological characteristics

The undesirable effects of fluoride on bone were recognized experimentally long before the human disease was identified.[23] Histologically, there are alternating zones of hypomineralization and of hypermineralization in dentine and enamel: of diminished mineralization and of increased calcification in cartilage; and of decreased mineralization within osteons, trabeculae and the margins of osteocytic lacunae, with an increased extent of heavily mineralized cement lines and osteon margins, in bone.[148] Although, therefore, hypermineralization is characteristic, zones of defective calcification may be sufficient to lead to pseudofractures resembling those of osteomalacia; and excess osteoid seams may be identified in iliac crest biopsies. In suspected cases, electron probe X-ray microanalysis allows the fluoride to be identified and measured.

The final state of the skeleton is bizarre. There is widespread, excess mineral deposition in the bones of the axial skeleton, the margins of the intervertebral discs, the tendons and ligaments. New bone formation is stimulated, osteophytes appear and the marginal involvement of diarthrodial joints leads to impaired movement. The changes in the intervertebral discs simulate syndesmophytes but the disc lesions of fluorosis are distinguishable from those of ankylosing spondylitis by the predominance of osteosclerosis as opposed to osteoporosis. Hyperostosis of the tibia, which may be confused with syphilis or Paget's disease, accompanies the vertebral disorder.

Other metabolic disease

Copper deficiency

Copper is an essential co-factor for enzymes including lysyl oxidase; 90 per cent of plasma copper is present in the α_2-globulin, caeruloplasmin; the element is stored in and excreted from the liver. Copper deficiency[45,85] exerts effects that are species-specific. Anaemia, neutropenia and osteoporosis are recognized in the human and injury to the brain and arteries is characteristic of Menkes' steely-hair syndrome (see p. 355). In sheep, by contrast, the effects are neurological. In those species in which copper deficiency culminates in arterial rupture, there is a close pathological similarity both to the Marfan syndrome (see p. 340) and to the effects of the propionitriles (lathyrism) (see p. 413). The vascular effects of copper deficiency in pigs have been closely investigated by Danks.[45] Dissecting aneurysm is one result. In dogs, osteoporosis results from diminished osteoblast activity.[183]

Manganese and Mseleni disease

Mseleni disease is an arthropathy indigenous to Northern Zululand.[51] The mode of inheritance has been sought but there is no evidence of a genetic origin. By contrast with conventional osteoarthrosis (see Chapter 22), the articular cartilage of the femoral head is covered by an array of degenerate, fibrillated cartilage and regenerated, new hyaline cartilage. The cause of Mseleni disease has been attributed to manganese deficiency.

Bismuth arthropathy

The excessive or prolonged administration of bismuth compounds may culminate in encephalopathy. The cerebral disorder, in turn, is likely to be associated with shoulder joint arthropathy.[26] There is osteolysis due to osteonecrosis of parts of the head of the humerus, sometimes with cartilage destruction. These changes are accompanied by remodelling of the humeral head. The pathogenesis of bismuth arthropathy is not clear but there is a resemblance to the joint changes which are rare accompaniments of other metal intoxications.

Lathyrism

The ingestion of large quantities of the seeds of certain varieties of sweet pea, particularly *Lathyrus sativus, L. cercera* and *L. clymenum,* produces a neurological syndrome in which leg cramp and stiffness may be followed quite suddenly by irreversible spastic paralysis progressing to complete paraplegia. Prolonged consumption of these vegetables is only necessary under famine conditions; the disease is therefore one of countries like India and North Africa where grain crops are often inadequate.[230]

Pathogenesis

Important responses are produced in the experimental animal and in cultured tissues by chemical extracts of these peas.[252] Numerous animal species, ranging from the toad to the monkey[216] are susceptible to these reactions but there is a wide range of sensitivity among different animals to the action of any single lathyrogen.

The reaction to Lathyrus seeds has been clarified by analysing the response of the rat. This animal is immune to the toxic effects of *L. sativus* and *L. cicera,* both toxic for man, but is highly susceptible to the actions of *L. odoratus,* the garden sweet pea. The toxic effects of *L. odoratus* in the rat were first described by Geiger, Steenbock and Parsons.[73] Other studies followed.[179,178] An analogy was drawn between the skeletal changes in rats fed *L. odoratus* peas and those of human adolescent kyphoscoliosis, intervertebral disc nucleus herniation and osteoarthrosis. The development of inguinal hernia and of dissecting aortic aneurysm suggested that the connective tissue reaction was highly diverse. Dissecting aneurysms, limb and skeletal deformities resembled the changes of Marfan's syndrome (see p. 340) in humans.

Biochemical changes

The toxic factor in Lathyrus seeds was water-and-alcohol-soluble but resistant to heat. McKay et al.[145] and Schilling and Strong[205] separated and purified a compound, β(N-β-L-glutamyl)-aminopropionitrile. A derivative, β-aminopropionitrile, was also an effective cause of bone lesions, aortic dissection and paraplegia in the rat.[12] Menzies and Mills[157] suggested that the basic change in the rat aorta and bones was the accumulation of excess intercellular matrix. The material was thought to be chondroitin sulphate proteoglycan. It was postulated that the toxin inhibited an enzyme concerned with normal connective tissue metabolism. A consequent defect was one of impaired cross-linking between collagen fibres.[125,127,154]

Pathological characteristics

The effects of a diet containing a high proportion of *L. odoratus* peas is precisely similar to the results of administering the toxic factor β(N-β-l-glutamyl)-aminopropionitrile or one of the simpler propionitrile derivatives.[179] There are longer-term degenerative changes in cartilage, laxity of the capsular and ligamentous insertions and dissecting aortic aneurysm. In observations made on auricular chondrocytes in culture, the changes in elastic cartilage were shown by inhibition of elastin cross-linkages. The formation of elastic fibres was diminished but only in the presence of relatively large amounts of β(N-β-l-glutamyl)-aminopropionitrile.[165]

There are four phases of microscopic disorder:[157] an increase in intercellular matrix; the disruption of connective tissue fibres; haemorrhage; and repair. The matrix comes to contain excess chondroitin sulphate proteoglycan. Reticular fibres are disorganized; they become irregular and disintegrated. Rupture of the aortic intima occurs with dissection. Where degeneration of the reticular aortic skeleton continues without rupture, an irregular network of fine elastic fibres appears in conjunction with collagen formation and scarring.

In bone, there is disturbance of the zones of active growth, particularly in the epiphyseal and periosteal regions.[179,157] The changes are more rapid in younger animals. After two weeks, there is an increase in the number of cells in the cartilage columns which become tall and irregular. Six weeks later, clumps of cartilage cells have replaced the irregular columns. As many as 2–10 cells may be found in each cartilagenous lacuna. New bone trabeculae are disorderly and among them are islands of chondroblasts. Later, with the accumulation of more proteoglycan, the orderly structure of the bone is wholly lost (Fig. 10.42). Similar changes occur within the fibrocartilage of the intervertebral discs.

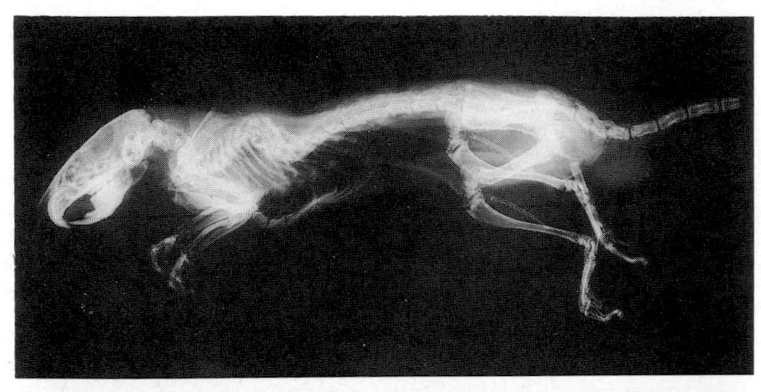

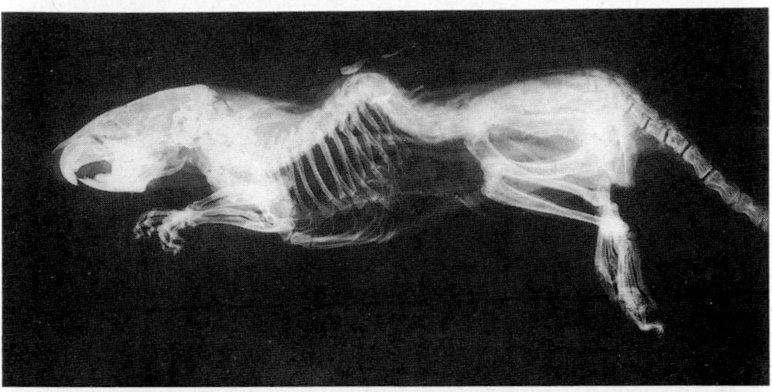

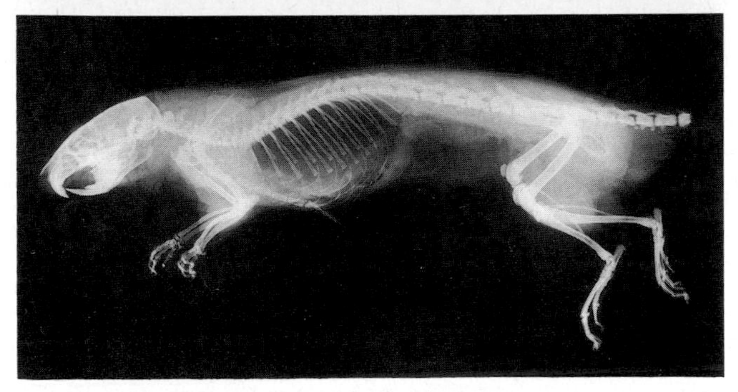

Fig. 10.42 Experimental lathyrism.
The effects on the skeleton of the young rat of β-amino propionitrile are shown. The compound may be fed or injected. At top, moderate skeletal deformity; at centre, severe deformity; and at bottom, normal, control rat skeleton.

Scurvy

A chronic deficiency in the diet of man of ascorbic acid, results in the state of scurvy. There is a fault in collagen stability and structure leading to a profound effect on wound healing. Bleeding is a feature.

Under physiological conditions, the enzyme gulonolactone oxidase is required to catalyse the conversion of L-gulonolactone to L-ascorbic acid. Man, monkeys and guinea-pigs lack this enzyme which is, however, present in other species for whom ascorbic acid is not therefore an essential dietary factor. In the absence of ascorbic acid, abnormal collagen synthesis is inevitable in man since this substance is necessary as a co-factor in the hydroxylation of collagenous Pro and Lys (see p. 174). There is a failure of hydroxylation of peptidylproline in the collagen of certain peptidyllysine residues. The collagen formed under these adverse circumstances is underhydroxylated and unstable; it is prematurely degraded, a state contributing *inter alia* to defective wound healing.[126]

The consequences of scurvy for the connective tissues are highly adverse.[92,246] The integrity of the extracellular connective tissue matrix in all parts of the body is impaired by defective collagen synthesis and its instability. There are severe defects in bone strength and growth[101] and a susceptibility to infection. Bone and muscle pain are consequences of local periosteal and perivascular haemorrhage. They impair movement. Intraarticular haemorrhage is encountered in the knee and ankle joints.

In modern times, scurvy has ceased to be an overt problem in Western countries. Little has been added to pathological knowledge since the extensive studies by Wolbach and Maddock[257] but the role of ascorbic acid in connective tissue metabolism continues to attract much interest.

Endocrine arthropathy

All connective tissues are sensitive to the direct and indirect effects of anterior pituitary, thyroid, adrenocortical, sex, pancreatic and other hormones. Consequently, a large variety of endocrine diseases, ranging from acromegaly to diabetes mellitus, is complicated by abnormal connective structure and function. For the same reasons, the connective tissue system is vulnerable in disorders of the midbrain and hypothalamus; and abnormalities of connective tissue cell function may be anticipated when there is defective expression of hormone receptors and of the intracellular pathways that they modulate.

Among the connective tissues, the synovial joints appear particularly prone to endocrine-mediated disease. Emphasis is therefore placed on the endocrine arthropathies.[104]

Joint disease in hyperpituitarism

Because of the responsiveness of cartilage to growth hormone, it is not surprising that in hyperpituitarism there is cartilage as well as skeletal overgrowth just as cartilage and bone atrophy or hypoplasia may be anticipated in hypopituitarism. The joint changes in acromegaly are characteristic.[103] They are also reversible.[116] The clinical features received attention in the classical paper of Marie[152] and were subject to detailed comment by Atkinson[11] who reviewed all the reported cases of the disease. Because the disease is very uncommon, and because of advances in clinical endocrinology, the opportunity to investigate acromegalic joints and connective tissue is now most exceptional. However, the pathological features of acromegalic joint disease were fully described in the older literature[55,101] and it has long been recognized that the synovial joints in acromegaly may be both large and painful.[103]

Kellgren, Ball and Tutton[109] described the articular and limb changes in 25 cases and gave an account of the pathological features in three necropsies; a more recent comment was that of Good.[77] The author has had the opportunity for a pathological study of one such case and a detailed report of another was given by Remagen.[186] Kellgren and his colleagues considered that there were two clinical forms of acromegalic arthropathy. The most common variety showed excessive joint mobility, a complaint of instability, a thickening of synovial and periarticular connective tissues and recurrent effusions. Less commonly, there occurred large osteophytic bony outgrowths and deformity of bone ends, leading to limitation of movement and affecting many joints, usually over a period of ten or more years.

The principal systemic disturbance in acromegaly is an overgrowth of connective tissue. The character of this hyperplasia of the connective tissue system varies in different anatomical locations. The change is quantitative rather than qualitative and the water content, the tropocollagen content and the structure of the connective tissue are normal. There are few modern analyses of the biochemistry of the connective tissues. Among the articular and periarticular adipose tissues there is a replacement of fat cells by fibrous tissue. Articular hyaline cartilage is increased in thickness and altered in character. A form of degeneration, distinct from that encountered in osteoarthrosis, is common. Superficial cartilage layers remain intact while more deeply situated layers lose their hyaline structure, becoming 'fibrillar'. Loss of cartilage may occur but there is an undermining of the margins of the ulcerated zone by contrast with the shallower excavations of osteoarthrosis. Comparable changes affecting the intra- and periarticular tissue of large joints, such as the knee, lead to an increase in joint space assessed radiologically. The synovial tissue of bursae is hypertrophic and fibrotic.

Bone structure is altered. In the large joints, the bones undergo gradual remodelling. The diameter of the shaft of long bones is increased. In the small bones of the hands and feet, a marginal formation of osteophytic bony outgrowths at the articular margins accompanies localized bone atrophy affecting the terminal parts of the phalanges (Fig. 10.43). An 'arrowhead' deformity ensues. The phalangeal shafts narrow and resemble pipestems. A proportion of patients develops osteoporosis accentuating the appearances described above. An incidental complicating osteoarthrosis alters and exaggerates the remodelling of the phalangeal margins.

The mechanisms by which excess growth hormone causes the variegated connective tissue changes of acromegaly is via the somatomedins,[10] insulin-like growth factors. These agents directly stimulate chondrocyte division, proteoglycan synthesis and collagen formation (p. 26). They also provoke appositional bone growth. In

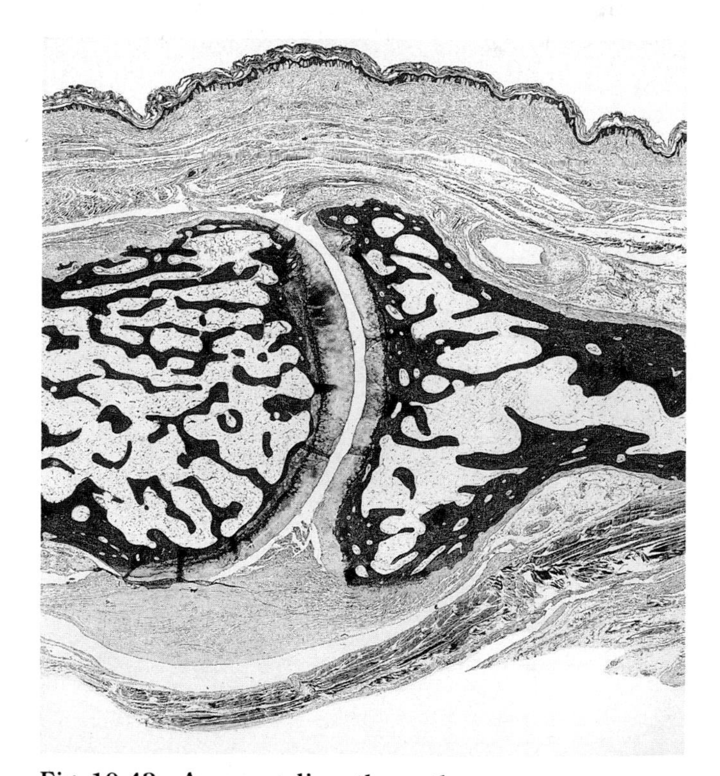

Fig. 10.43 Acromegalic arthropathy.
Distal interphalangeal joint from patient with pituitary adenoma. There is high bone density, some increase in articular cartilage thickness of the proximal end of the distal phalanx, and a marginal formation of new lamellar bone at the periphery of the proximal phalanx. The appearances are those of 'arrow-head' deformity. (HE × 8.)

acromegaly, the plasma concentrations of the somatomedins are raised, a change in contradistinction to the fall anticipated with advancing age. However, Johanson[104] has emphasized that the endocrine disorders in acromegalic arthropathy are complex; there is often hypothyroidism and this defect, together with altered levels of the sex hormones, complicates the cell and tissue responses to growth hormone excess.

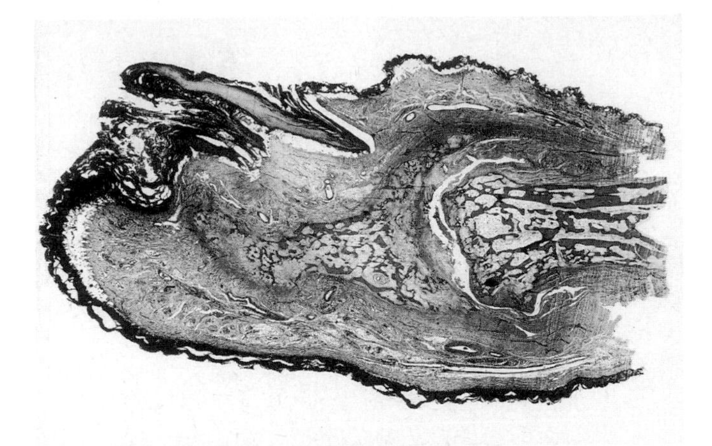

Fig. 10.44 Hyperparathyroidism.
The distal phalanx of this finger from a case of secondary hyperparathyroidism, is extensively remodelled. Compact cortical bone is almost entirely absent and the cancellous bone of the shaft is replaced by vascular fibrous tissue. There is new periosteal woven bone formation and extensive destruction of the articular cartilage. Chondrocalcinosis may coexist but the changes due to calcium pyrophosphate dihydrate deposition disease can be differentiated. (HE × 4.)

Joint disease in hypothyroidism

Articular and bone disease are commonplace in congenital and juvenile cretinism and may develop in acquired myxoedema.[16,20,104] When the quantity or mode of action of thyroid hormone is impaired, both chondrocyte 'degeneration' at epiphyseal growth zones and matrix mineralization are accelerated, whereas chondrocyte proliferation is slowed. The epiphyseal plate becomes thickened; growth is retarded. Analogous changes in articular chondrocyte function are likely. The process of vascularization of the calcified, zone V of articular cartilage may be activated and here, and at sites of tendinous and ligamentous insertion into bone, remodelling of articular structure takes place. There is a predisposition to osteoarthrosis.

Changes also occur in the non-cartilagenous connective tissues. An excess of hyaluronate accumulates, for example, in the subcutaneous tissues. The synovial villi enlarge and may be found to contain crystalline deposits of calcium pyrophosphate (p. 393). These changes, together with alterations in the amounts, organization and mechanical properties of nearby bone, determine that symptomatic articular disease in untreated hypothyroidism is frequent. The articular changes may be accompanied by joint swelling and muscle stiffness. There is often a non-inflammatory synovial effusion. The carpal tunnel syndrome is an easily recognizable complication of hypothyroid joint disease of the upper limb but the knees and small joints of the hands and feet account for much of the symptomatic disease.

Joint disease in hyperparathyroidism

Three categories of articular disease may complicate untreated primary hyperparathyroidism.[104] In a first type of arthropathy, joint disease is the result of subchondral bone disorganization.[30] Bone reabsorption prejudices the integrity and the mechanical characteristics of articular cartilage and, in established hyperparathyroidism can cause extensive destruction of the small joints of fingers (Figs 10.44, 10.45 and 10.46) and toes. The appearance of such lesions related to larger joints, particularly the knee, may lead to an erroneous diagnosis of a polyarthritis such as rheumatoid arthritis. The erosions of bone that are recognized in late, untreated hyperparathyroidism are foci of osteoclastic bone reabsorption combined with the synthesis of a highly vascular connective tissue matrix that has a rudimentary myxoid (p. 306) appearance. In the larger, weight-bearing- and in the intervertebral joints, mechanical loads can lead to crush fracture; as a result of such injuries, traumatic synovitis and secondary osteoarthrosis may develop.[30] Second, prolonged hypercalcaemia may lead to metastatic calcification of non-osseous connective tissues. The deposits of calcium hydroxyapatite affect, for example, articular cartilage and synovial tissue.

In a third sequence, an association between primary hyperparathyroidism and gout has been identified. The underlying cause is likely to be renal dysfunction attributable to prolonged hypercalcaemia. The articular changes that follow are first, subchondral tophus formation; second, urate-induced cartilage injury; and third, synovitis.

Hyperparathyroidism can complicate any of the systemic and local connective tissue diseases. Hyperuricaemia and gout[212] may occur in males in whom parathyroid adenoma is subsequently diagnosed. Patients with rheumatoid arthritis may develop hyperparathyroidism and a complex clinical syndrome can evolve. Thus, in one instance, an elderly woman had been found to have rheumatoid arthritis in middle age. Ten years before her death, a parathyroid adenoma was identified but surgical removal was not

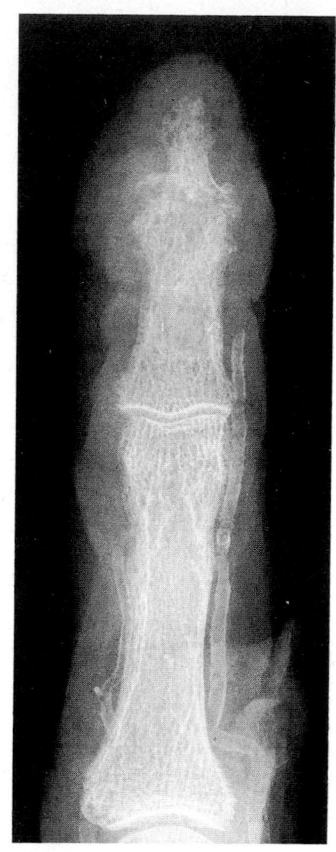

Fig. 10.45 Hyperparathyroidism.
Radiograph of finger from patient known to have had untreated secondary hyperparathyroidism. Note calcification of digital arteries. The distal interphalangeal joint is grossly disorganized. There is a dorsal fracture of the margin of the distal phalanx; reorganization and repair are accompanied by destruction of the articular surfaces.

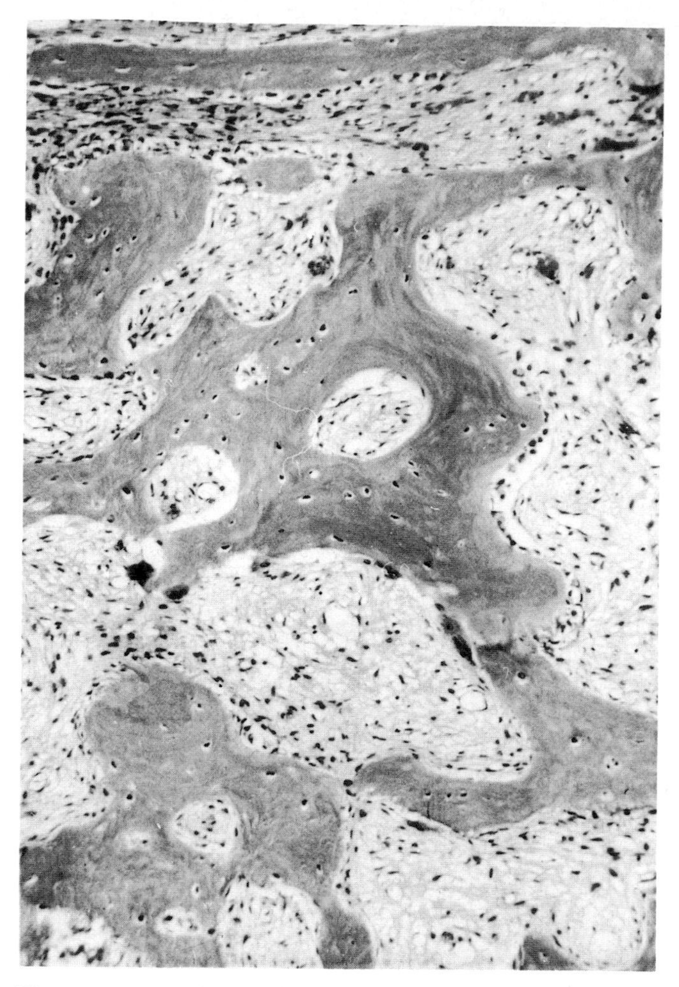

Fig. 10.46 Hyperparathyroidism.
In this specimen of bone from a site adjoining an articular surface, there is vigorous osteoclastic activity with lacunar bone reabsorption. Considerable osteoblastic activity accompanies the presence in the bone marrow, of an excess of loose, vascular connective tissue. (HE × 83.)

advised because of the presence of rheumatic aortic and mitral valve disease. At the time of her death, a 15 g parathyroid adenoma was found and the hands displayed rheumatoid arthritis both of the metacarpophalangeal joints and metastatic calcification of the digital arteries.

Joint disease in diabetes mellitus

Diabetes mellitus is very common and the coincidental occurrence of osteoarthrosis, rheumatoid arthritis and the spondyloarthropathies is well recognized. Insulin, the insulin-like growth factors (somatomedins)[10] and growth hormone closely influence chondrocyte metabolism so that particular forms of diabetic arthropathy might be anticipated. Diabetic limbs are susceptible to the results of ischaemia and bacterial infection (p. 733) and the articular tissues are often exposed to the effects of these complications. Diabetics are also prone to a form of neurogenic arthropathy (p. 906).

Three varieties of diabetic joint disease have attracted attention. The contribution of endocrine factors to the pathogenesis of osteoarthrosis is considered in Chapter 22. Ankylosing hyperostosis occurs with unexpectedly high frequency in diabetes mellitus; it is described on p. 947. Finally, an unexplained progressive articular contracture has been recognized in young, insulin-dependent diabetics;[40] there is finger weakness and restricted joint mobility.[197]

REFERENCES

1. Agate J N. Industrial fluorosis. MRC Memorandum No. 22. London: HM Stationery Office, 1949.

2. Agudelo C A, Weinberger A, Schumacher H R, Turner R, Molina J. Definitive diagnosis of gout by identification of urate crystals in asymptomatic metatarso-phalangeal joints. *Arthritis and Rheumatism* 1979; **22**: 559–60.

3. Agudelo C A, Schumacher H R. The synovitis of acute gouty arthritis. A light and electron microscopic study. *Human Pathology* 1973; **4**: 265–79.

3a. Alarcon G S, Reveille J D. Gouty arthritis of the axial skeleton including the sacroiliac joints. *Archives of Internal Medicine* 1987; **147**: 2018–9.

4. Ali S Y, Griffiths S. New types of calcium phosphate crystals in arthritic cartilage. *Seminars in Arthritis and Rheumatism* 1981; **11**: 124–6.

5. Ali S Y, Griffiths S. Formation of calcium phosphate crystals in normal and osteoarthritic cartilage. *Annals of the Rheumatic Diseases* 1983; **42**; Suppl 1: 45–8.

6. Altman R D, Collins B. Musculoskeletal manifestations of Paget's disease of bone. *Arthritis and Rheumatism* 1980; **23**: 1121–7.

7. Anderson H C. Calcific diseases—a concept. *Archives of Pathology and Laboratory Medicine* 1983; **107**: 341–8.

8. Anderson H C. Mineralization by matrix vesicles. *Scanning Electron Microscopy* 1984; Part 2: 953–64.

9. Anderson H C. Mechanisms of pathologic calcification. *Rheumatic Diseases Clinics of North America* 1988; **14**: 303–19.

10. Ashton I K, Francis M J. An assay for plasma somatomedin: [3H] thymidine incorporation by isolated rabbit chondrocytes. *Journal of Endocrinology* 1977; **74**: 205–12.

11. Atkinson F R B. *Acromegaly*. London: John Bale, Sons and Danielsson, 1932, p. 50.

12. Bachuber T E, Lalich J J, Angevine D M, Schilling E D, Strong F M. Lathyrus factor activity of beta-amino propionitrile and related compounds. *Proceedings of the Society for Experimental Biology and Medicine N.Y.* 1955; **89**: 294–7.

13. Ball J. New knowledge of intervertebral disc disease. *Journal of Clinical Pathology* 1978; **31**: Supplement (Royal College of Pathologists) **12**: 200–4.

14. Bardin T, Schumacher H R, Lansaman J, Rothfuss S, Dryll A. Transmission electron microscopic identification of silicon-containing particles in synovial fluid: potential confusion with calcium pyrophosphate dihydrate and apatite crystals. *Annals of the Rheumatic Diseases* 1984; **43**: 624–7.

15. Becker M A, Seegmiller J E. Recent advances in the identification of enzyme abnormalities underlying excessive purine synthesis in man. *Arthritis and Rheumatism* 1975; **18**: 687–94.

16. Berkheiser S W. Adult hypothyroidism: report of advanced case with autopsy study. *Journal of Clinical Endocrinology* 1955; **15**: 44–53.

17. Berkowitz D. Gout, hyperlipidemia and diabetes interrelationships. *Journal of the American Medical Association* 1966; **197**: 77–80.

18. Bjelle A O. The glycosaminoglycans of articular cartilage in calcium pyrophosphate dihydrate (CPPD) crystal deposition disease. *Calcified Tissue Research* 1973; **12**: 37–46.

19. Bjelle A O, Sundström B K G. An ultrastructural study of the articular cartilage in calcium pyrophosphate dihydrate (CPPD) crystal deposition disease (chondrocalcinosis articularis). *Calcified Tissue Research* 1975; **19**: 63–71.

20. Bland J H, Frymoyer J W. Rheumatic syndrome of myxedema. *New England Journal of Medicine* 1970; **282**: 1171–4.

20a. Boskey A L, Bullough P G, Vigorita V, Di Carlo E. Calcium–acidic phospholipid–phosphate complexes in human hydroxyapatite-containing pathologic deposits. *American Journal of Pathology* 1988; **133**: 22–9.

21. Boskey A L, Vigorita V J, Sencer O, Stuchin S A, Lane J M. Chemical, microscopic and ultrastructural characterization of the mineral deposits in tumoral calcinosis. *Clinical Orthopaedics and Related Research* 1983; **178**: 258–69.

22. Boss G R, Seegmiller J E. Hyperuricemia and gout: classification, complications and management. *New England Journal of Medicine* 1979; **300**: 1459–68.

23. Brandl J, Tapeiner H. Ueber die Ablagerung der Fluorverbindungen im Organismus, nach Futterung mit Fluornatrium. *Zeitschrift für Biologie*, München und Leipzig, 1891; **10**: 518–39.

24. Brown W E, Gregory T M. Calcium pyrophosphate crystal chemistry. *Arthritis and Rheumatism* 1976; **19**: 446–63.

25. Buchanan W W, Klinenberg J R, Seegmiller J E. The inflammatory response to injected microcrystalline monosodium urate in normal, hyperuricemic, gouty and uremic subjects. *Arthritis and Rheumatism* 1965; **8**: 361–7.

26. Buge A, Hubault A, Rancurel G. Les arthropathies de l'intoxication par le bismuth. *Revue de Rhumatisme et des Maladies Osteoarticulaire* 1975; **42**: 721–9.

27. Bunn C W. *Chemical Crystallography* (2nd edition). Oxford: Clarendon Press, 1961.

28. Burch T A, O'Brien W M, Need R, Kurland L T. Hyperuricaemia and gout in the Mariana Islands. *Annals of the Rheumatic Diseases* 1966; **25**: 114–16.

29. Bywaters E G L. Gout in the time and person of George IV: a case history. *Annals of the Rheumatic Diseases* 1962; **21**: 325–38.

30. Bywaters E G L, Dixon A St J, Scott J T. Joint lesions of hyperparathyroidism. *Annals of the Rheumatic Diseases* 1963; **22**: 171–87.

31. Cameron H U, Fornasier V L, Macnab I. Pyrophosphate arthropathy. *American Journal of Clinical Pathology* 1975; **63**: 192–8.

32. Cannon P J, Stason W B, Demartini F E, Sommers S C, Laragh J L. Hyperuricemia in primary and renal hypertension. *New England Journal of Medicine* 1966; **275**: 457–64.

33. Caswell A M, Guilland-Cumming D F, Hearn P R. Pathogenesis of chondrocalcinosis and pseudogout. Metabolism of inorganic pyrophosphate and production of calcium pyrophosphate dihydrate crystals. *Annals of the Rheumatic Diseases* 1983; **42**; Suppl 1: 27–37.

34. Caswell A M, McGuire M K B, Russell R G G. Studies of pyrophosphate metabolism in relation to chondrocalcinosis. *Annals of the Rheumatic Diseases* 1983; **42**; Suppl 1: 98–9.

35. Cawley M I D. Complications of Paget's disease of bone. *Gerontology* 1983; **29**: 276–87.

36. Chalmers J. Osteomalacia. A review of 93 cases. *Journal of the Royal College of Surgeons of Edinburgh* 1968; **13**: 255–75.

37. Chang Y H, Gralla E J. Suppression of urate crystal-induced canine joint inflammation by heterologous anti-polymorphonuclear leukocyte serum. *Arthritis and Rheumatism* 1968; **11**: 145–50.

38. Cherian P V, Schumacher H R. Immunochemical and ultrastructural characterization of serum proteins associated with monosodium urate crystals (MSU) in synovial fluid cells from patients with gout. *Ultrastructural Pathology* 1986; **10**: 209–19.

39. Cheung H S, Story M T, McCarty D J. Mitogenic effects of hydroxyapatite and calcium pyrophosphate dihydrate crystals on cultured mammalian cells. *Arthritis and Rheumatism* 1984; **27**: 668–74.

40. Choulot J J, Saint-Martin J. Contractures articulaires évolutives. Une complication méconnue du diabète insulino-dépendant. *Nouvelle Presse Médicale* 1980; **9**: 515–17.

41. Copeman W S C. *A Short History of the Gout and the Rheumatic Diseases.* Berkeley: University of California Press, 1964.

42. Cross A D, Jones R A. *An Introduction to Practical Infra Red Spectroscopy* (3rd edition). London: Butterworths, 1969.

43. Currey H L F. Significance of radiological calcification of joint cartilage. *Annals of the Rheumatic Diseases* 1966; **25**: 295–306.

44. Dalculsi G, Fauré G, Kerebel B. Electron microscopy and microanalysis of a subcutaneous heterotopic calcification. *Calcified Tissue International* 1983; **35**: 723–7.

45. Danks D M. Copper deficiency in humans. In: Ciba Foundation Symposium 79: *Biological Roles of Copper.* Amsterdam: Excerpta Medica, 1980: **209**.

45a. Davies M, Clements M R, Mawer E B, Freemont A J, Tumoral calcinosis: clinical and metabolic response to phosphorus deprivation. *Quarterly Journal of Medicine* 1987; **NS63**: 493–503.

46. Dieppe P A. New knowledge of chondrocalcinosis. *Journal of Clinical Pathology* 1978; **31**; Supplement (Royal College of Pathologists) **12**: 214–22.

47. Dieppe P A, Calvert P. *Crystals and Joint Disease.* London: Chapman and Hall, 1983.

48. Dieppe P A, Crocker P, Huskisson E C, Willoughby D A. Apatite deposition disease. A new arthropathy. *Lancet* 1976; **i**: 266–9.

49. Dieppe P, Hornby J, Swan A, Hutton C, Preece A. Laboratory handling of crystals. *Annals of the Rheumatic Diseases* 1983; **42**; Suppl 1: 60–3.

50. Dodds W J, Steinbach H L. Gout associated with calcification of cartilage. *New England Journal of Medicine* 1966; **275**: 745–9.

51. du Toit G T. Hip disease of Mseleni. *Clinical Orthopaedics and Related Research* 1979; **141**: 223–8.

52. Eade A W T, Swannell A J, Williamson N. Pyrophosphate arthropathy in hypophosphatasia. *Annals of the Rheumatic Diseases* 1981; **40**: 164–70.

53. Editorial. Crystals in joints. *Lancet* 1980; **i**: 1006–7.

54. Epker B N. A quantitative microscopic study of bone-remodelling and balance in a human with skeletal fluorosis. *Clinical Orthopaedics and Related Research* 1967; **55**: 87–93.

55. Erdheim J. *Die Lebensvorgänge im normalen Knorpel und seine Wucherung bei Akromegalie.* Berlin: Julius Springer, 1931.

55a. Eshel G, Gulik A, Halperin N, Avrahami E, Schumacher H R, McCarty D J, Caspi D. Hereditary chondrocalcinosis in an Ashkenazi Jewish family. *Annals of the Rheumatic Diseases* 1990; **49**: 528–30.

56. Ettlinger R E, Hunder G G. Synovial effusions containing cholesterol crystals. Report of 12 patients and review. *Mayo Clinic Proceedings* 1979; **54**: 366–74.

57. Faires J S, McCarty D J. Acute arthritis in man and dog after intrasynovial injection of sodium urate crystals. *Lancet* 1962a; **ii**: 682–4.

58. Faires J S, McCarty D J. Acute synovitis in normal joints of man and dog produced by injections of microcrystalline sodium urate, calcium oxalate and corticosteroid esters. *Arthritis and Rheumatism* 1962b; **5**: 295–6.

59. Fallet G H, Vischer T L, Micheli A. Chondrocalcinosis. *Schweizerische Medizinische Wochenschrift* 1982; **112**: 888–97.

59a. Fam A G, Schumacher H R. Crystal-induced joint inflammation. In: Greenwald R A, Diamond H S, eds. *CRC Handbook of Animal Models for the Rheumatic Diseases* (Volume II). Boca Raton, Florida: CRC Press Inc, 1988: 99–125.

60. Fam A G, Pritzker K P H, Cheng P-Y, Little A H. Cholesterol crystals in osteoarthritic joint effusions. *Journal of Rheumatology* 1981; **8**: 273–80.

61. Fam A G, Sugai M, Gertner E, Lewis A. Cholesterol tophus. *Arthritis and Rheumatism* 1983; **26**: 1525–8.

62. Fam A G, Topp J R, Stein H B, Little A H. Clinical and roentgenographic aspects of pseudogout: a study of 50 cases and a review. *Canadian Medical Association Journal* 1981; **124**: 545–51.

63. Farley J R, Wergedal J E, Baylink D J. Fluoride directly stimulates proliferation and alkaline phosphatase activity of bone forming cells. *Science* 1983; **222**: 330–2.

64. Fellström B, Lindsjö M, Danielson B G, Ljunghall S, Wikstrom B G. Binding of glycosaminoglycans to sodium urate and uric acid crystals. *Clinical Science* 1986; **71**: 61–4.

65. Ferrerroca O, Brancos M A, Franco M, Sabate C, Querol J R. Massive articular chondrocalcinosis—its occurrence with calcium pyrophosphate crystal deposits in nucleus pulposus. *Archives of Pathology and Laboratory Medicine* 1982; **106**: 352–4.

66. Fiechtner J J, Simkin P A. Urate spherulites in gouty synovia. *Journal of the American Medical Association* 1981; **245**: 1533–6.

67. Franck W A, Bress N M, Singer F R, Krane S M. Rheumatic manifestations of Paget's disease of bone. *American Journal of Medicine* 1974; **56**: 592–603.

68. Gardner D L, Lameletie J, Lawton D M, Wilson N H F. Response of cultured chondrocytes to porous ceramic: light- and low temperature scanning electron microscopic studies. *Journal of Pathology* 1987; **152**: 189A.

69. Garrod A B. On the blood and effused fluids of gout, rheumatism, and Bright's disease. *Transactions of the Medico-Chirurgical Society of Edinburgh* 1854; **37**: 49–60.

70. Garrod A B. A Treatise on Gout and Rheumatic Gout (rheumatoid arthritis) (3rd edition). London: Longmans Green, 1876.

71. Gaucher A, Fauré G, Netter P, Malaman B, Steinmetz J. Identification of microcrystals in synovial fluids by combined

scanning electron microscopy and X-ray diffraction: application to triclinic calcium pyrophosphate dihydrate. *Biomedicine* 1977; **27**: 242–4.

72. Gaucher A, Fauré G, Netter P, Pourel J, Duheille J. Identification des cristaux observés dans les arthropathies destructices de la chondrocalcinose. *Revue du Rhumatisme* 1977; **44**: 407–14.

73. Geiger B J, Steenbock H, Parsons H T. Lathyrism in the rat. *Journal of Nutrition* 1933; **6**: 427–42.

74. Gerster J C, Baud C A, Lagier R, Boussina I, Fallet G H. Tendon calcifications in chondrocalcinosis. A clinical radiologic, histologic and crystallographic study. *Arthritis and Rheumatism* 1977; **20**: 717–22.

75. Ginsberg M H, Kozin F, O'Malley M, McCarty D J. Release of platelet constituents by monosodium urate crystals. *Journal of Clinical Investigation* 1977; **60**: 999–1007.

76. Glatt M, Dieppe P, Willoughby D. Crystal-induced inflammation, enzyme release and the effects of drugs in the rat pleural space. *Journal of Rheumatology* 1979; **6**: 251–8.

77. Good A E. Acromegalic arthropathy. A case report. *Arthritis and Rheumatism* 1964; **7**: 65–74.

78. Good A E, Upton L G. Acute temporomandibular arthritis in a patient with bruxism and calcium pyrophosphate deposition disease. *Arthritis and Rheumatism* 1982; **25**: 353–5.

79. Gordon R B, Counsilman A C, Cross S M C, Emmerson B T. Purine synthesis de novo in lymphocytes from patients with gout. *Clinical Science* 1982; **63**: 429–35.

80. Gravanis M B, Gaffney E F. Idiopathic calcifying tenosynovitis: histopathologic features and possible pathogenesis. *American Journal of Surgical Pathology* 1983; **7**: 357–61.

81. Green D, Arsever L, Grumet K A, Ratnoff O D. Classic gout in Hageman factor (factor XII) deficiency. *Archives of Internal Medicine* 1982; **142**: 1556–7.

82. Hall A P. Barry P E, Dawber T R, McNamara P M. Epidemiology of gout and hyperuricemia: a long-term population study. *American Journal of Medicine* 1967; **42**: 27–37.

83. Hall T, Echlin P, Kaufman R. *Microprobe Analysis as Applied to Cells and Tissues*. London: Academic Press, 1974.

84. Halverson P B, McCarty D J. Basic calcium phosphate (apatite, octacalcium phosphate, tricalcium phosphate) crystal deposition diseases. In: D J McCarty, ed, *Arthritis and Allied Conditions: A Textbook of Rheumatology* (eleventh edition). Philadelphia: Lea & Febiger, 1989: 1737–55.

85. Hambridge K M. Zinc and chromium in human nutrition. *Journal of Human Nutrition* 1978; **32**: 99–110.

86. Hartung E F. Symposium on gout: historical considerations. *Metabolism* 1957; **6**: 196–208.

87. Hasselbacher P. Binding of IgG and complement protein by monosodium urate monohydrate and other crystals. *Journal of Laboratory and Clinical Medicine* 1979; **94**: 532–41.

88. Hasselbacher P. Stimulation of synovial fibroblasts by calcium oxalate and monosodium urate monohydrate. A mechanism of connective tissue degradation in oxalosis and gout. *Journal of Laboratory and Clinical Medicine* 1982; **100**: 977–85.

89. Hasselbacher P, Schumacher H R. Localisation of immunoglobulin in gouty tophi by immunohistology and on the surface of monosodium urate crystals (MSU) by immunoagglutination. *Arthritis and Rheumatism* 1976; **19**: 802.

90. Hauge M, Harvald B. Heredity in gout and hyperuricemia. *Acta Medica Scandinavica* 1955; **152**: 247–57.

91. Healey L A, Skeith M D, Decker J L. Hyperuricemia in Filipinos: interaction of heredity and environment. *American Journal of Human Genetics* 1967; **19**: 81–5.

92. Hess A F. *Scurvy: Past and Present*. Philadelphia: J B Lippincott & Co., 1920.

93. Hess R J, Martin J H. Pyarthrosis complicating gout. *Journal of the American Medical Association* 1971; **218**: 592–3.

94. Hills E, Dunstan C R, Wong S Y P, Evans R A. Bone histology in young adult osteoporosis. *Journal of Clinical Pathology* 1989; **42**: 391–7.

95. Hirsch R S, Smith K, Vernon-Roberts B. A morphological study of macrophage and synovial cell interactions with hydroxyapatite crystals. *Annals of the Rheumatic Diseases* 1985; **44**: 844–51.

96. Holmes E W, Wyngaarden J B. Hereditary xanthinuria. In: Schriver C R, Beaudet A L, Sly W S, Valle D, eds, *The Metabolic Basis of Inherited Disease* (6th edition). New York: McGraw-Hill Book Company, 1989: 1085–94.

97. Howell D S, Martell-Pelletier J, Pelletier J P, Morales S, Muniz O. NTP pyrophosphohydrolase in human chondrocalcinotic and osteoarthritic cartilage: II: Further studies on histologic and subcellular distribution. *Arthritis and Rheumatism* 1984; **27**: 193–9.

98. Inagaki K, Morioka S, Yokoyama H. Development of urate crystals in tophus. Report II. *Yokahama Medical Bulletin* 1969; **20**: 171–5.

99. Inclan A, Leon P, Camejo M G. Tumoral calcinosis. *Journal of the American Medical Association* 1943; **121**: 490–5.

100. Jacobelli S, McCarty D J, Silcox D C, Mall J C. Calcium pyrophosphate dihydrate crystal deposition in neuropathic joints. *Annals of Internal Medicine* 1973; **79**: 340–7.

101. Jaffe H L. *Metabolic, Degenerative and Inflammatory Diseases of Bones and Joints*. Philadelphia: Lea & Febiger, 1972.

102. Janssen T, Rayan G M. Gouty tenosynovitis and compression neuropathy of the median nerve. *Clinical Orthopaedics and Related Research* 1987; **216**: 203–6.

103. Johanson N A, Vigorita V J, Goldman A B, Salvati E A. Acromegalic arthropathy of the hip. *Clinical Orthopaedics and Related Research* 1983; **173**: 130–9.

104. Johanson N A. Endocrine arthropathies. *Clinics in Rheumatic Diseases* 1985; **11**: 297–323.

105. Kandel R A, Renlund R C, Cheng P T, Rapley W A, Mehren K G, Pritzker K P H. Calcium pyrophosphate dihydrate crystal deposition disease with concurrent vertebral hyperostosis in a Barbary ape. *Arthritis and Rheumatism* 1983; **26**: 682–7.

106. Katz W A. Deposition of urate crystals in gout: altered connective tissue metabolism. *Arthritis and Rheumatism* 1975; **18**: 751–6.

107. Kellermeyer R W, Breckenridge R T. The inflammatory process in acute gouty arthritis. I. Activation of Hageman factor by sodium urate crystals. *Journal of Laboratory and Clinical Medicine* 1965; **65**: 307–15.

108. Kelley W N, Fox I H, Palella T D. Gout and related disorders of purine metabolism. In: Kelley W N, Harris E D,

Ruddy S, Sledge C B, eds, *Textbook of Rheumatology* (3rd edition). Philadelphia: W B Saunders & Co., 1989: 1395–448.

109. Kellgren J H, Ball J, Tutton G K. The articular and other limb changes in acromegaly. *Quarterly Journal of Medicine* 1952; **21**: 405–24.

110. Kernodle G W, Allen N B. Acute gout presenting in the manubriosternal joint. *Arthritis and Rheumatism* 1986; **29**: 570–2.

111. Kim K M, Valigorsky J M, Mergner W F, Jones R T, Pendergrass R F, Trump B F. Aging changes in the human aortic valve in relation to dystrophic calcification. *Human Pathology* 1976; **7**: 47–60.

112. Kippen I, Klinenberg J R, Weinberger A, Wilcox W. Factors affecting urate solubility *in vitro*. *Annals of Rheumatic Diseases* 1974; **33**: 313–7.

113. Kjellstrand C M, Campbell D C, von Hartisch B, Buselmeier T. Hyperuricaemic acute renal failure. *Archives of Internal Medicine* 1974; **133**: 349–59.

114. Kohn N N, Hughes R E, McCarty D J, Faires J S. The significance of calcium pyrophosphate crystals in the synovial fluid of arthritic patients: the 'pseudogout syndrome'. II. Identification of crystals. *Annals of Internal Medicine* 1962; **56**: 738–45.

115. Kozin F, McCarty D J. Protein adsorption to monosodium urate, calcium pyrophosphate dihydrate and silica crystals. Relationship to the pathogenesis of crystal-induced inflammation. *Arthritis and Rheumatism* 1976; **19**: 433–8.

116. Lacks S, Jacobs R P. Acromegalic arthropathy—a reversible rheumatic disease. *Journal of Rheumatology* 1986; **13**: 634–6.

117. Lafferty F W, Reynolds E S, Pearson O H. Tumoral calcinosis: a metabolic disease of obscure etiology. *American Journal of Medicine* 1965; **38**: 105–18.

118. Lagier R. Femoral cortical erosions and osteoarthrosis of the knee with chondrocalcinosis. *Fortschrift auf dem Gebiete der Röntgenstrahlen und der Nuklearmedizin* 1974; **120**: 460–7.

119. Lagier R, Boivin G, Lacotte D, Gerster J C. Histological study of a case of recurrent olecranon bursitis with mixed calcium pyrophosphate dihydrate and apatite crystal deposits. *Virchows Archiv. A Pathological Anatomy and Histopathology* 1985; **405**: 453–61.

120. Lagier R, Martin E, Radi I. Étude anatomo-radiologique d'une 'coxarthrose destructirce rapide' avec omarthrose et chondrocalcinose. *Revue de Rheumatisme* 1971; **38**: 317–23.

121. Lagier R, Wildi E. Fréquence de la chondrocalcinose dans une série de 1000 disques intervertébraux excisés chirurgicalement. *Revue du Rhumatisme* 1979; **46**: 303–7.

121a. Lamelettie M D J. The growth of chondrocytes *in vitro* on porous hydroxyapatite ceramic. BSc Thesis, University of Manchester, 1987.

122. Lawton D M, Bartley C J, Gardner D L. X-ray microanalysis of mouse knee meniscus. *Institute of Physics Conference Series* 1988: **93**: 587–8.

123. Lawton D M, Lamalettie M D J, Gardner D L. Biocompatibility of hydroxyapatite ceramic: response of chondrocytes in a test system using low temperature scanning electron microscopy. *Journal of Dentistry* 1989; **17**: 21–7.

124. Leaney B J, Calvert J M. Tophaceous gout producing spinal cord compression. *Journal of Neurosurgery* 1983; **58**: 580–2.

125. Levene C I. Collagen as a tensile component in the developing chick aorta. *British Journal of Experimental Pathology* 1961; **42**: 89–94.

126. Levene C I. Diseases of the collagen molecule. *Journal of Clinical Pathology* 1978; **31**; Supplement (Royal College of Pathologists) **12**: 82–94.

127. Levene C I, Gross J. Alterations in state of molecular aggregation of collagen induced in chick embryos by β-aminopropionitrile (lathyrus factor). *Journal of Experimental Medicine* 1959; **110**: 771–90.

128. Levinson D J. Clinical gout and the pathogenesis of hyperuricaemia. In: McCarty D J, ed. *Arthritis and Allied Conditions: A Textbook of Rheumatology* (eleventh edition). Philadelphia: Lea & Febiger, 1989: 1645–76.

129. Lichtenstein L, Scott H W, Levin M H. Pathologic changes in gout: survey of eleven necropsied cases. *American Journal of Pathology* 1956; **32**: 871–95.

130. McCarty J, Halverson P. Basic calcium phosphate (apatite, octacalcium phosphate, tricalcium phosphate) crystal deposition diseases. In McCarty, D J. ed. *Arthritis and Allied Conditions: A Textbook of Rheumatology* (10th edition), Philadelphia: Lea & Febiger, 1985; 1547–64.

131. McCarty D J. Crystal-induced inflammation of the joints. *Annual Review of Medicine* 1970; **21**: 357–66.

132. McCarty D J. Calcium pyrophosphate dihydrate crystal deposition disease—1975. *Arthritis and Rheumatism* 1976; **19**: 275–85.

133. McCarty D J. Crystals, joints and consternation. *Annals of the Rheumatic Diseases* 1983; **42**: 243–53.

133a. McCarty D J. Pathogenesis and treatment of crystal-induced inflammation. In; McCarty D J, ed. *Arthritis and Allied Conditions: A Textbook of Rheumatology* (10th edition), Philadelphia: Lea & Febiger, 1494–514.

134. McCarty D J, Gatter R A. Pseudogout syndrome (articular chondrocalcinosis). *Bulletin on the Rheumatic Diseases* 1964; **14**: 331–4.

135. McCarty D J, Halverson P B, Carrera G F, Brewer B J, Kozin F. 'Milwaukee Shoulder'—association of microspheroids containing hydroxyapatite crystals, active collagenase, and neutral protease with rotator cuff defects. *Arthritis and Rheumatism* 1981; **24**: 464–73.

136. McCarty D J, Hollander J L. Identification of urate crystals in gouty synovial fluid. *Annals of Internal Medicine* 1961; **56**: 452–70.

137. McCarty D J, Lehr J R, Halverson P B. Crystal populations in human synovial fluid—identification of apatite, octacalcium phosphate, and tricalcium phosphate. *Arthritis and Rheumatism* 1983; **26**: 1220–4.

138. McClure J. Local calcergy. A histological, histochemical and electron-probe X-ray analytical study. *Experientia* 1980; **36**: 1102–3.

139. McClure J. Demonstration of calcification fronts by *in vivo* and *in vitro* tetracycline labelling. *Journal of Clinical Pathology* 1982; **3**: 1278–82.

140. McClure J. Malakoplakia. *Journal of Pathology* 1983; **140**: 275–330.

141. McClure J, Smith P S. Calcium pyrophosphate dihydrate deposition in the intervertebral discs in a case of Wilson's disease. *Journal of Clinical Pathology* 1983; **36**: 764–68.

142. McClure J, Smith P S. Oncogenic osteomalacia. *Journal of Clinical Pathology* 1987; **40**: 446–53.

143. McClure J, Smith P S, Gramp A A. Calcium pyrophosphate dihydrate (CPPD) deposition in ochronotic arthropathy. *Journal of Clinical Pathology* 1983; **36**: 894–902.

144. McGee-Russell S M. Histochemical methods for calcium. *Journal of Histochemistry and Cytochemistry* 1958; **6**: 22–42.

145. McKay G F, Lalich J J, Schilling E D, Strong F M. A crystalline 'Lathyrus factor' from *Lathyrus odoratus*. *Archives of Biochemistry* 1954; **52**: 313–22.

146. McKendry R J R, Uhthoff H K, Sarkar K, Hyslop P St G. Calcifying tendinitis of the shoulder: prognostic value of clinical, histologic and radiologic features in 57 surgically treated cases. *Journal of Rheumatology* 1982; **9**: 75–80.

147. Magid S K, Gray G E, Anand A. Spinal cord compression by tophi in a patient with chronic polyarthritis: case report and literature review. *Arthritis and Rheumatism* 1981; **24**: 1431–4.

148. Malcolm A S, Storey E. Osteofluorosis in the rabbit: microradiographic studies. *Pathology* 1971; **3**: 39–51.

149. Mandel N, Mandel G. Structures of crystals that provoke inflammation. *Advances in Inflammation Research* (volume 5). New York: Raven Press, 1982.

150. Mandel N S, Mandel G S, Carroll D J, Halverson P B. Calcium pyrophosphate crystal deposition—an *in vitro* study using a gelatin matrix model. *Arthritis and Rheumatism* 1984; **27**: 789–96.

151. Mankin H J. Rickets, osteomalacia and renal osteodystrophy. *Journal of Bone and Joint Surgery* 1974; **56**: 101–28.

152. Marie P, Sur deux cas d'acromégalie; hypertrophie singulière, non-congénitale, des extrémités supérieures, inférieure et céphalique. *Revue de Médicine* 1886; **6**: 297–333.

153. Markel S F, Hart W R. Arthropathy in calcium pyrophosphate dihydrate crystal deposition disease. *Archives of Pathology and Laboratory Medicine* 1982; **106**: 529–33.

154. Martin G R, Gross J, Piez K A, Lewis M S. On the intramolecular cross-linking of collagen in lathyritic rats. *Biochimica et Biophysica Acta* 1961; **53**: 599–601.

155. Medsger T A. Tumoral calcinosis. In: McCarty D J, ed, *Arthritis and Allied Conditions* (10th edition). Philadelphia: Lea & Febiger, 1985: 1025.

156. Melsen F, Melsen B, Mosekilde L. An evaluation of the quantitative parameters applied in bone histology. *Acta Pathologica Microbiologica Scandinavica* (Section A) 1978; **86**: 63–9.

157. Menzies D W, Mills K W. The aortic and skeletal lesions of lathyrism in rats on a diet of sweet pea. *Journal of Pathology and Bacteriology* 1957; **73**: 223–7.

158. Moens C, Moens D, Moens P. Prevalence of monosodium urate and calcium pyrophosphate dihydrate crystals in postmortem knee synovial fluid. *Arthritis and Rheumatism* 1985; **28**: 1319–20.

159. Mohr W. Gelenkkrankheiten. Diagnostik und Pathogenese makroskopischer und histologischer Strukturveränderungen. Stuttgart: Georg Thieme Verlag, 1984.

160. Mohr W, Hersener J, Wilke W, Weinland G, Beneke G. Pseudogicht (chondrokalzinose). *Zeitschrift für Rheumatologie* 1974; **33**: 107–29.

161. Mohr W, Oehler K, Hersener J, Wilke W. Chondrocalcinose der Zwischenwirbelscheiben. *Zeitschrift für Rheumatologie* 1979; **38**: 11–26.

162. Møller P F, Gudjonsson S V. Massive fluorosis of bones and ligaments. *Acta Radiologica* (Stockholm) 1932; **13**: 269–94.

163. Møller P F, Gudjonsson S V. The Classic: massive fluorosis of bones and ligaments. *Clinical Orthopaedics and Related Research* 1967; **55**: 5–15.

163a. Moscowitz R W. Diseases associated with the deposition of calcium pyrophosphate or hydroxyapatite. In: Kelley W N, Harris E D, Ruddy S, Sledge C B, eds. *Textbook of Rheumatology* (3rd edition). Philadelphia, London: W B Saunders Company, 1989: 1449–67.

164. Moser C R, Fessel W J. Rheumatic manifestations of hypophosphatemia. *Archives of Internal Medicine* 1974; **134**: 674–8.

165. Moskalewski S, Langeveld C H, Scherft J P. Influence of beta-aminopropionitrile (BAPN) on cell growth and elastic fiber formation in cultures of auricular chondrocytes. *Experientia* 1983; **39**: 1147–8.

165a. Mundy G R. Hypercalcaemia of malignancy revisited. *Journal of Clinical Investigation* 1988; **82**: 1–6.

166. Muniz O, Pelletier J-P, Martell-Pelletier J, Morales S, Howell D. NTP pyrophosphohydrolase in human chondrocalcinotic and osteoarthritic cartilage: I. Some biochemical characteristics. *Arthritis and Rheumatism* 1984; **27**: 186–92.

167. Nagahashi M. Scanning electron microscopic observations of calcium pyrophosphate crystals of joint tissues and synovial fluid. *Japanese Journal of Genetics* 1979; **53**: 793–805.

168. Nriagu J O. Saturnine gout among Roman aristocrats: did lead poisoning contribute to the Fall of the Empire? *New England Journal of Medicine* 1983; **308**: 660–3.

169. O'Brien W, Burch T A, Bunim J J. Genetics of hyperuricaemia in Blackfeet and Pima Indians. *Annals of Rheumatic Diseases* 1966; **25**: 117–19.

169a. Ortman B L, Pack L L. Aseptic loosening of a total hip prothesis secondary to tophaceous gout. *Journal of Bone and Joint Surgery* 1987; **69-A**: 1096–9.

170. Owen D S Jnr, Toone E, Irby R. Coexistent rheumatoid arthritis and chronic tophaceous gout. *Journal of the American Medical Association* 1966; **197**: 123–6.

171. Paik C H, Alavi I, Dunea G, Weiner L. Thalassemia and gouty arthritis. *Journal of the American Medical Association* 1970; **213**: 296.

172. Paul H, Reginato A, Schumacher H R. Morphological characteristics of monosodium urate: a transmission electron microscope study of intact natural and synthetic crystals. *Annals of the Rheumatic Diseases* 1983; **42**: 75–81.

173. Pedersen H E, Key J A. Pathology of calcareous tendinitis and subdeltoid bursitis. *Archives of Surgery* 1951; **62**: 50–63.

174. Perricone E, Brandt K D. Enhancement of urate solubility by connective tissue. I. Effect of proteoglycan aggregates and buffer cation. *Arthritis and Rheumatism* 1978; **21**: 453–60.

175. Perricone E, Brandt K D. Enhancement of urate solubility by connective tissue. II. Inhibition of sodium urate crystallisation by cation exchange. *Annals of the Rheumatic Diseases* 1979; **38**: 467–70.

176. Phelps P. Polymorphonuclear leukocyte motility *in vitro*: IV. Colchicine inhibition of chemotactic activity formation after phagocytosis of urate crystals. *Arthritis and Rheumatism* 1970; **13**: 1–10.

177. Phillips H B, Owen-Jones S, Chandler B. Quantitative histology of bone: a computerized method of measuring the total mineral content of bone. *Calcified Tissue Research* 1978; **26**: 85–9.

178. Ponseti I V, Baird W A. Scoliosis and dissecting aneurysm of aorta in rats fed with *Lathyrus odoratus* seeds. *American Journal of Pathology* 1952; **28**: 1059–77.

179. Ponseti I V, Shepherd R S. Lesions of the skeleton and of other mesodermal tissues in rats fed sweet-pea (*Lathyrus odoratus*) seeds. *Journal of Bone and Joint Surgery* 1954; **36**A: 1031–58.

180. Pritzker K P H, Fam A G, Omar S A, Gertzbein S D. Experimental cholesterol arthropathy. *Journal of Rheumatology* 1981; **8**: 281–90.

181. Pritzker K P H, Phillips H, Luk S C, Koven I, Kiss A, Houpt J B. Pseudotumor of temporomandibular joint: destructive calcium pyrophosphate dihydrate arthropathy. *Journal of Rheumatology* 1976; **3**: 70–80.

182. Pritzker K P H, Zahn C E, Nyburg S C, Luk S C, Houpt J B. The ultrastructure of urate crystals in gout. *Journal of Rheumatology* 1978; **5**: 7–18.

183. Read R A, Kent G N, Price R I. Inhibition of osteoblast function in the osteoporosis of copper deficiency in dogs. *Calcified Tissue International* 1988; **42**; suppl: A10.

184. Ream L J, Hull D L. Scott J N, Pendergrass P B, Fluoride ingestion during multiple pregnancies and lactations: microscopic observations on bone of the rat. *Virchows Archiv B Cell. Pathology* 1983; **44**: 35–44.

184a. Reginato A J, Schumacher H R. Spontaneous and experimental articular and periarticular calcification. In: Greenwald R A, Diamond H S, eds. *CRC Handbook of Animal Models for the Rheumatic Diseases* (Volume II). Boca Raton, Florida: CRC Press Inc., 1988: 127–48.

185. Reginato A J, Schumacher H R, Martinez V A. The articular cartilage in familial chondrocalcinosis. Light and electron microscopic study. *Arthritis and Rheumatism* 1974; **17**: 977–92.

186. Remagen W. New findings in acromegaly and subsequent arthrosis. *Virchows Archiv für pathologische Anatomie und Physiologie und für klinische Medizin* 1965; **340**: 8–24.

187. Renlund R C, Pritzker K P H, Cheng P T, Kessler M J. Rhesus monkeys (Macaca mulatta) as a model for calcium pyrophosphate dihydrate crystal deposition disease. *Journal of Medical Primatology* 1986; **15**: 11–16.

188. Resnick D, Niwayama G, Goergen T, Utsinger P D, Shapiro R. The incidence and specificity of pyrophosphate arthropathy. *XIVth International Congress of Rheumatology* 1977; Abstract 444.

189. Resnick D, Niwayama G, Goergen T G, Utsinger P D, Shapiro R F, Haselwood D H, Wiesner K B. Clinical radiographic and pathologic abnormalities in calcium pyrophosphate dihydrate deposition disease (CPPD): pseudogout. *Radiology* 1977; **122**: 1–15.

190. Revell P A. Histomorphometry of bone. *Journal of Clinical Pathology* 1983; **36**: 1323–31.

191. Revell P A. *Pathology of Bone*. New York: Springer-Verlag, 1986.

192. Reynolds P P, Knapp M J, Baraf H S B, Holmes E W. Moonshine and lead – relationship to the pathogenesis of hyperuricaemia in gout. *Arthritis and Rheumatism* 1983; **26**: 1057–64.

193. Rizzoli A J, Trujeque L, Bankhurst A D. The coexistence of gout and rheumatoid arthritis: case reports and a review of the literature. *Journal of Rheumatology* 1980; **7**: 316–24.

194. Roberts E D, Baskin G B, Watson E, Henk W G, Shelton T C. Animal models of human disease: calcium pyrophosphate deposition disease (CPPD) in non-human primates. *American Journal of Pathology* 1984; **116**: 359–61.

195. Rodnan G P. A gallery of gout—being a miscellany of prints and caricatures from the 16th century to the present day. *Arthritis and Rheumatism* 1961; **4**: 27–46.

196. Rondier J, Cayla J, Guiraudon C, Charpentier Y le. Arthropathie tabetique et chondrocalcinose articulaire. *Revue de Rhumatism* 1977; **44**: 671–4.

197. Rosenbloom A L, Silverstein A H, Lezotte D C, Riley W J, Maclaren N K. Limited joint mobility in diabetes mellitus of childhood: natural history and relationship to growth impairment. *Journal of Pediatrics* 1982; **101**: 874–8.

198. Rothschild B M, Sienknecht C W, Kaplan S B, Spindler J S. Sickle cell disease associated with uric acid deposition disease. *Annals of the Rheumatic Diseases* 1980; **39**: 392–5.

199. Rubinow A, Sonnenblick M. Amyloidosis secondary to polyarticular gout. *Arthritis and Rheumatism* 1981; **24**: 1425–7.

200. Russell I J, Papaioannou C, McDuffie F C. Complement activation in human serum by urate crystals: evidence for a role for IgG. *Clinical Research* 1979; **27**: 335A.

201. Ryan L M, McCarty D J. Calcium pyrophosphate crystal deposition disease; pseudogout; articular chondrocalcinosis. In: McCarty D J, ed, *Arthritis and Allied Conditions: A Textbook of Rheumatology* (eleventh edition). Philadelphia: Lea & Febiger, 1989: 1711–36.

201a. Ryan L M, McCarty D J. Calcium pyrophosphate crystal deposition disease; pseudogout; articular chondrocalcinosis. In: McCarty D J, ed. *Arthritis and Allied Conditions. A Textbook of Rheumatology* (10th edition). Philadelphia: Lea and Febiger, 1985: 1515–46.

202. Sarkar K, Uhthoff H K. Ultrastructural localisation of calcium in calcifying tendinitis. *Archives of Pathology and Laboratory Medicine* 1978; **102**: 266–9.

203. Sarkar K, Uhthoff H K. Ultrastructure of the subacromial bursa in painful shoulder syndromes. *Virchows Archiv A* (Pathological Anatomy and Histopathology) 1983; **400**: 107–17.

204. Scheele K W. Examen chemicum calculi urinarii. *Opuscula* 1776; **2**: 73.

205. Schilling E D, Strong F M. Isolation, structure and synthesis of a lathyrus factor from *L. odoratus*. *Journal of the American Chemical Society* 1954; **76**: 2848.

206. Schmidt K L, Leber H W, Schutterie G. Arthropathy in primary oxalosis—crystal synovitis or osteopathy? *Deutsche Medizinische Wochenschrift* 1981; **106**: 19–22.

207. Schumacher H R. The synovitis of pseudogout: electron microscopic observations. *Arthritis and Rheumatism* 1968; **11**: 426–35.

208. Schumacher H R. Pathology of the synovial membrane in

gout: light and electron microscopic studies. Interpretation of crystals in electron micrographs. *Arthritis and Rheumatism* 1975; **18**: 771–82.

209. Scott J T. Gout. In: Scott J T, ed, *Copeman's Textbook of the Rheumatic Diseases* (5th edition). Edinburgh: Churchill Livingstone, 1978a: 647–91.

210. Scott J T. New knowledge of the pathogenesis of gout. *Journal of Clinical Pathology* 1978b; **31**: Supplement (Royal College of Pathologists) **12**: 205–13.

211. Scott J T. Gout. In: Scott J T, ed, *Copeman's Textbook of the Rheumatic Diseases* (6th edition). Edinburgh: Churchill Livingstone, 1986: 883–937.

212. Scott J T, Dixon A St J, Bywaters E G L. Association of hyperuricaemia and gout with hyperparathyroidism. *British Medical Journal* 1964; **11**: 1070–3.

213. Seegmiller J E. Human aberrations of purine metabolism and their significance for rheumatology. *Annals of the Rheumatic Diseases* 1980; **39**: 103–17.

214. Seegmiller J E, Rosenbloom F M, Kelley W N. Enzyme defect associated with a sex-linked human neurological disorder and excessive purine synthesis. *Science* 1967; **155**: 1682–4.

214a. Seeman E, Hopper J L, Bach L A, Cooper M E, Parkinson E, McKay J, Jerums G. Reduced bone mass in daughters of women with osteoporosis. *New England Journal of Medicine* 1989; **320**: 554–8.

215. Selye H. Use of 'granuloma pouch' technic in the study of antiphlogistic corticoids. *Proceedings of the Society for Experimental Biology and Medicine N.Y.* 1953; **82**: 328–33.

216. Selye H. Lathyrism. *Revue Canadienne Biologique* 1957; **16**: 1–82.

217. Selye H, Berczi I. The present status of calciphylaxis and calcergy. *Clinical Orthopaedics and Related Research* 1970; **69**: 28–54.

218. Selye H, Gentile G, Veileux R. An experimental model of calcareous subdeltoid bursitis induced by calciphylaxis. *Arthritis and Rheumatism* 1962; **5**: 219–25.

219. Settas L, Doherty M, Dieppe P. Localised chondrocalcinosis in unstable joints. *British Medical Journal* 1982; **285**: 175–6.

220. Shirahama T, Cohen A S. Ultrastructural evidence for leakage of lysosomal contents after phagocytosis of monosodium urate crystals. A mechanism of gouty inflammation. *American Journal of Pathology* 1974; **76**: 501–12.

221. Siddiqui A H. Fluorosis in Nalgonda district, Hyderabad-Deccan. *British Medical Journal* 1955; **11**: 1408–13.

222. Sin Y M, Sedgwick A D, Moore A, Willoughby D A. Studies on the clearance of calcium pyrophosphate crystals from facsimile synovium. *Annals of the Rheumatic Diseases* 1984; **43**: 487–92.

223. Singh A, Dass R, Hayreh S S, Jolly S S. Skeletal changes in endemic fluorosis. *Journal of Bone and Joint Surgery* 1962; **44**-B: 806–15.

224. Sokoloff L. The pathology of gout. *Metabolism* 1957; **6**: 230–41.

224a. Sokoloff L, Varma A A. Chondrocalcinosis in surgically resected joints. *Arthritis and Rheumatism* 1988; **31**: 750–6.

225. Sorensen K H, Teglbjaerg P S, Ladefoged C, Christensen H E. Pyrophosphate arthritis with local amyloid deposition. *Acta Orthopaedica Scandinavica* 1981; **52**: 129–33.

226. Srikantia S G, Siddiqui A H. Metabolic studies in skeletal fluorosis. *Clinical Science* 1965; **28**: 477–85.

227. Stavric B, Johnson W J, Grice H C. Uric acid nephropathy: an experimental model. *Proceedings of the Society for Experimental Biology and Medicine N.Y.* 1969; **130**: 512–16.

228. Stecher R M, Hersh A H, Solomon W M. The heredity of gout and its relationship to familial hyperuricaemia. *Annals of Internal Medicine* 1949; **31**: 595–614.

229. Stern P, Weinberg S. Pathological fracture of the radius through a cyst caused by pyrophosphate arthropathy—report of a case. *Journal of Bone and Joint Surgery* 1981; **63**-A: 1487–8.

230. Stockman R. Lathyrism. *Edinburgh Medical Journal* 1917; **19**: 277–307.

231. Stockman A, Darlington L G, Scott J T. Frequency of chondrocalcinosis of the knees and avascular necrosis of the femoral head in gout: a controlled study. *Annals of the Rheumatic Diseases* 1980; **39**: 7–11.

232. Sydenham T. Differentiation of gout from rheumatism. Tractatus de podagra et hydropei. London: G. Kettilby, 1683.

233. Taillard W, Lagier R. Pseudo-spondylolisthesis et chondrocalcinose. *Revue de Chirurgie Orthopedique* 1977; **63**: 149–56.

234. Talbott J H, Altmann R D, Yu T F. Gouty arthritis masquerading as rheumatoid arthritis or vice versa. *Seminars in Arthritis and Rheumatism* 1978; **8**: 77–114.

235. Talbott J H, Terplan K L. The kidney in gout. *Medicine* 1960; **39**: 405–67.

236. Talbott J H, Yu T F. *Gout and Uric Acid Metabolism*. New York: Grune and Stratton, 1976.

237. Tanaka A, O'Sullivan F X, Koopman U J, Gay S. Etiopathogenesis of rheumatoid arthritis-like disease in MRL/l mice: II. ultrastructural basis of joint destruction. *Journal of Rheumatology* 1988; **15**: 10–16.

238. Terkeltaub R, Ginsberg M H. Molecular mechanisms of gouty arthritis: new insights into an old disease. *Survey and Synthesis of Pathology Research* 1984; **3**: 386–96.

239. Terkeltaub R A, Ginsberg M H, McCarty D J. Pathogenesis and treatment of crystal-induced inflammation. In: McCarty D J, ed, *Arthritis and Allied Conditions: A Textbook of Rheumatology* (11th edition). Philadelphia: Lea & Febiger, 1989: 1691–710.

240. Terkeltaub R, Smeltzer D, Curtiss L K, Ginsberg M H. Low density lipoprotein inhibits the physical interaction of phlogistic crystals and inflammatory cells. *Arthritis and Rheumatism* 1986; **29**: 363–70.

241. Thompson G R, Ming Ting Y, Riggs G A, Fenn H E, Denning R M. Calcific tendinitis and soft-tissue calcification resembling gout. *Journal of the American Medical Association* 1968; **203**: 464–72.

242. Thomson J G. Calcifying collagenolysis (tumoural calcinosis). *British Journal of Radiology* 1966; **39**: 526–32.

243. Tse R L, Phelps P. Polymorphonuclear leukocyte motility *in vitro*. V. Release of chemotactic activity following phagocytosis of calcium pyrophosphate crystals, diamond dust, and urate crystals. *Journal of Laboratory and Clinical Medicine* 1970; **76**: 403–15.

244. Uhthoff H K. Calcifying tendinitis, an active, cell mediated calcification. *Virchows Archiv. A Pathological Anatomy and Histo-*

pathology 1975; **366**: 51–8.

245. Uhthoff H K, Sarkar K, Maynard J A. Calcifying tendinitis: a new concept of its pathogenesis. *Clinical Orthopaedics and Related Research* 1976; **118**: 164–8.

246. van Wersch H J. *Scurvy as a Skeletal Disease.* Utrecht: Dekker and van de Vegt, NV, 1954.

247. Vaughan J. *The Physiology of Bone* (3rd edition). Oxford: Clarendon Press, 1981.

247a. Vernon-Roberts B. Synovial fluid and its examination. In: Scott J T, ed. *Copeman's Textbook of the Rheumatic Diseases* (6th edition). Edinburgh, London: Churchill-Livingstone, 1986: 251–77.

248. Wall B, Agudelo C A, Tesser J R P, Mountz J, Holt D, Turner R A. An autopsy study of the prevalence of monosodium urate and calcium pyrophosphate dihydrate crystal deposition in first metatarsophalangeal joints. *Arthritis and Rheumatism* 1983; **26**: 1522–4.

249. Wallace S L, Bernstein D. Relationship between gout and the kidney. *Metabolism* 1963; **12**: 440–6.

250. Watts R W E, Scott J T, Chalmers R A, Bitensky L, Chayen J. Microscopic studies on skeletal muscle in gout patients treated with allopurinol. *Quarterly Journal of Medicine* 1971; **40**: 1–14.

251. Watts R W E. Hyperuricaemia: some biochemical aspects. *Proceedings of the Royal Society of Medicine* 1969; **62**: 853–7.

252. Weaver A L, Spittell J A. Lathyrism. *Mayo Clinic Proceedings* 1964; **39**: 485–9.

253. Webb J, Champion D G, Frecker A S, Robinson R G. Calcium pyrophosphate crystal synovitis with articular chondrocalcinosis ('Pseudogout' syndrome). *Medical Journal of Australia* 1970; **1**: 461–5.

254. Weissman G. Gout—molecular basis for inflammation caused by interaction of monosodium urate with liposomes and lysozymes. *Nature* 1972; **240**: 1–6.

255. Whitehouse F W, Cleary W J. Diabetes mellitus in patients with gout. *Journal of the American Medical Association* 1966; **197**: 73–6.

256. Williams H E, Smith L H. Primary hyperoxaluria. In: Stanbury J B, Wyngaarden J B, Fredrickson D S, Goldstein J L, Brown M S, eds, *The Metabolic Basis of Inherited Disease* (5th edition). New York: McGraw-Hill, 1983: 204–28.

257. Wolbach S B, Maddock C L. Cortisone and matrix formation in experimental scorbutus and repair therefrom with contributions to pathology of experimental scorbutus. *Archives of Pathology* 1952; **53**: 54–69.

258. Wollaston W H. On gouty and urinary concretions. *Philosophical Transactions of the Royal Society of London* 1797; **87**: 386–400.

259. Wolman M. On the use of polarized light in pathology. *Pathology Annual* (volume 5). New York: Appleton-Century-Crofts, 1970: 381.

259a. Wong S Y P, Kariks J, Evans R A, Dunstan C R, Hills E. The effect of age on bone composition and viability in the femoral head. *Journal of Bone and Joint Surgery* 1985; **67**-A: 274–83.

259b. Woolf A D, Dieppe P A. Mediators of crystal-induced inflammation in the joint. *British Medical Bulletin* 1987; **43**: 429–44.

260. Woolf A D, Dixon A St J. *Osteoporosis: a Clinical Guide.* London: Martin Dunitz, 1988.

261. Wyngaarden J B, Kelley W N. Gout. In: Stanbury J B, Wyngaarden J B, Fredrickson D S, Goldstein J L, Brown M S, eds, *The Metabolic Basis of Inherited Disease* (5th edition). New York: McGraw-Hill Book Company, 1983: 1043–114.

262. Zaharopoulos P, Wong J Y. Identification of crystals in joint fluids. *Acta Cytologica* 1980; **24**: 197–202.

263. Žitňan D, Sităj S. Calcification multiples du cartilage articulaire. Ninth International Congress on Rheumatic Diseases, Toronto. 1957.

264. Žitňan D, Sităj S. Mnolsopocetna familiarta kalcifikaaz artikularynch chrupiek. *Bratiplavské lekárske listy* 1958; **38**: 217–28.

265. Žitňan D, Sităj S. Natural course of articular chondrocalcinosis. *Arthritis and Rheumatism* 1976; **19** (Suppl): 363–90.

Chapter 11

Amyloid and Amyloidosis

Amyloid is a generic name given to classes of chemically distinct but physically similar proteins deposited in slowly increasing amounts in the extracellular matrix. The result is a group of disorders given the collective name, amyloidosis.[17,45,46,50,60,91,107]

The most common class of amyloid related to connective tissue disease is the secondary amyloid encountered in rheumatoid arthritis[42] (see Chapter 13). Older patients with rheumatoid arthritis may also accumulate senile amyloid. Juvenile chronic polyarthritis and ankylosing spondylitis are occasionally complicated by amyloidosis but because of their relative infrequency, amyloidosis in these disorders is a rare pathological problem. Amyloidosis has also been described with dermatomyositis,[44] psoriatic arthropathy,[6,94] Reiter's disease[81] and, rarely, systemic lupus erythematosus.[58] In addition, amyloid is now identified with increasing frequency in the articular cartilages, discs, menisci and ligaments of otherwise normal, aged people. Amyloid accumulates in the hyaline cartilages in myelomatosis but renewed interest in connective tissue amyloidosis has come from its recognition in patients treated by haemodialysis for chronic renal failure.

The amyloids (Table 11.1)[22,91] are fibrillar proteins laid down between vascular endothelial cells, their supporting collagen and nearby basement membranes, in the intercellular substance between the medial muscle cells of larger arteries and in aggregates within synovial, cartilagenous, subcutaneous, endocrine, intermuscular and other tissues (Figs 11.1, 11.2, 11.3). There is usually a second, associated, non-fibrillar protein, the component-P (Table 11.2). In addition to their fibrillar structure, the amyloids have two further characteristics. First, it is clear that amyloid proteins assume a physical form described as a β-pleated sheet; the amino acid chains are arranged perpendicular to the long axis of the fibril. Second, amyloids are weakly birefringent but display stronger apple-green double refractility after staining with the cotton dye, Congo red. The physical organization of amyloid fibrils accounts for these various properties and is so uniform that a proposal was made that amyloidosis should be called β-fibrillosis.[45]

Historically, amyloid was recognized as 'lardaceous disease':[96] it had been seen by earlier writers but not distinguished from fatty change. Virchow[113] recommended the designation 'amyloid' because the deposits reacted like starch and were similar to cellulose; before Virchow's demonstration, 'amyloid' had been used to denote a substance identified by the treatment of cellulose with sulphuric acid.[92] The cellulose nature of amyloid was soon disproved and Friedreich and Kekule[39] finally established that the amyloids were albuminous, i.e. that they were proteins. In practice, it was customary to demonstrate macroscopic deposits of amyloid in tissues by the *post-mortem* application of iodine and sulphuric acid: the test yielded a characteristic mahogany-brown colour.

Amyloid fibrils: classification

Amyloid fibrils can be isolated in large quantities after the homogenization of tissue samples in fluids of low ionic strength: opalescent colloidal solutions are formed. X-ray diffraction can then be used to show the β-pleated sheet fibrillar protein structure.[8,32]

The formation of amyloids is believed to occur in stages.

Table 11.1

Amyloidosis and amyloid deposits with special reference to connective tissue locations (modified from Cohen and Connors[21] and Glenner and Page[46])

Amyloid fibril or clinical designation*		Classification	Sites of deposition	Biochemical composition when known
Systemic				
Acquired				
	'Primary'† Idiopathic or with immunocyte dyscrasias such as myelomatosis or monoclonal gammopathy (IDA)	AL	Cartilage	$A\varkappa_{III}$, $A\lambda_{VI}$ etc.
	'Secondary' Resulting from chronic connective tissue disease such as rheumatoid arthritis, chronic infection, Crohn's disease	AA	Kidney, intestine, artery, adrenal, pituitary, spleen	A-A prototype
	Haemodialysis-associated	AH	Carpal tunnel, bone	A-β_2-microglobulin-like
	Experimental	AA		
Heredofamilial				
	Experimental Senile amyloid in senescence-accelerated (SAM) mouse	AS_{sam}		Apo A-II like
	Neuropathic Portuguese	AF_p		A-prealbumin
	Japanese	AF_j		A-prealbumin
	Icelandic	AF_i		A-cystatin C (γ-protein)
	Non-neuropathic			
	Familial cardiomyopathy	AF_c	Heart	A-prealbumin
	Familial Mediterranean	AA		A-A
Non-systemic				
Organ-limited				
	Immunocyte-derived		Respiratory tract	
	Senile			
	Cardiac	AS_{c_1} AS_{c_2}	Heart, synovia, cartilage, ligaments, menisci	A-prealbumin A-protein name
	Cerebral (Alzheimer's disease)	AS_{b_1}	Brain	A-protein name (β-protein)
Localized				
	Cutaneous	AD	Skin: lichen or nodular	
	Endocrine organ-associated			
	Medullary carcinoma of thyroid	AE_t	Within neoplasm	A-calcitonin
	Insulinoma	AE_i	Within neoplasm	A-insulin
	Repeated localized insulin injection	AE_i	At sites of repeated injection	A-insulin
	Atrial natriuretic peptide	AE_a	Atrial wall	

* Amyloid serum proteins, the precursors of some amyloids, are designated with S before the class, e.g. SAA is the precursor of AA. Amyloid P component (AP) and its precursor, SAP, are classed separately.

† Glenner points out that in IDA, the systemic amyloidosis is 'secondary' to the underlying immunocyte-derived abnormality; he therefore discards the term 'secondary amyloidosis' and substitutes the expression 'reactive systemic amyloidosis'.

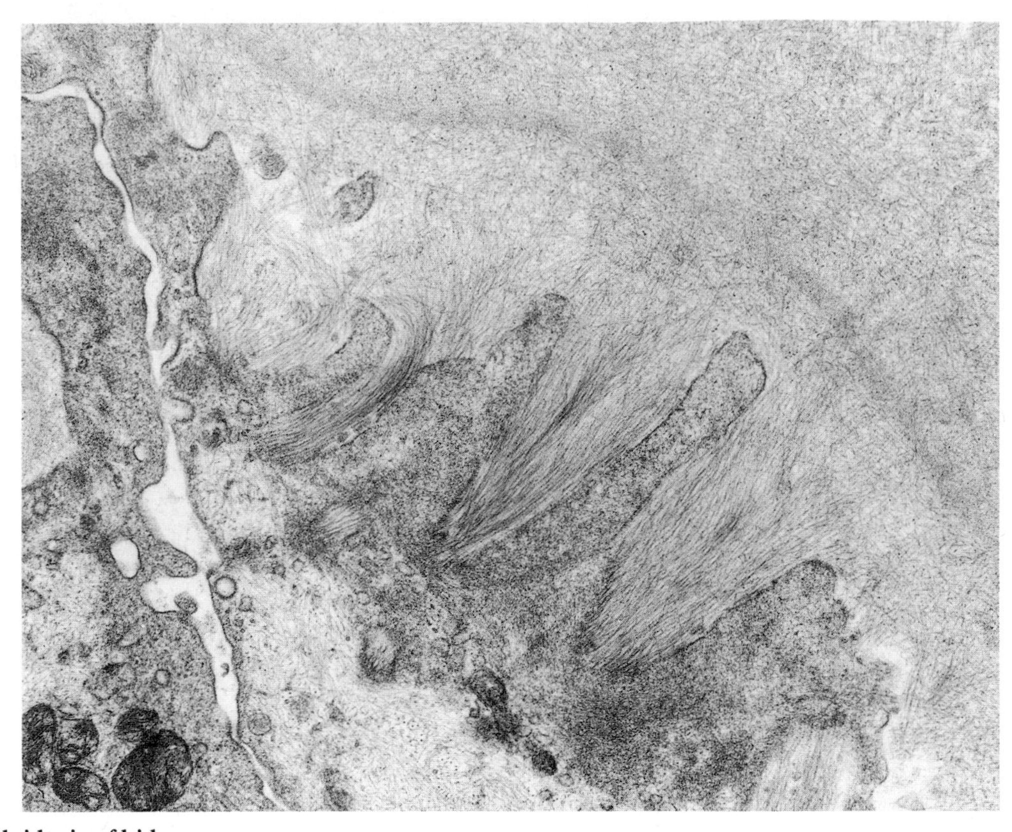

Fig. 11.1 Amyloidosis of kidney.
Numerous beaded fibrils of amyloid extend around and within the basement membrane and occupy much of the subepithelial, mesangial connective tissue matrix. (TEM × 29 500.)

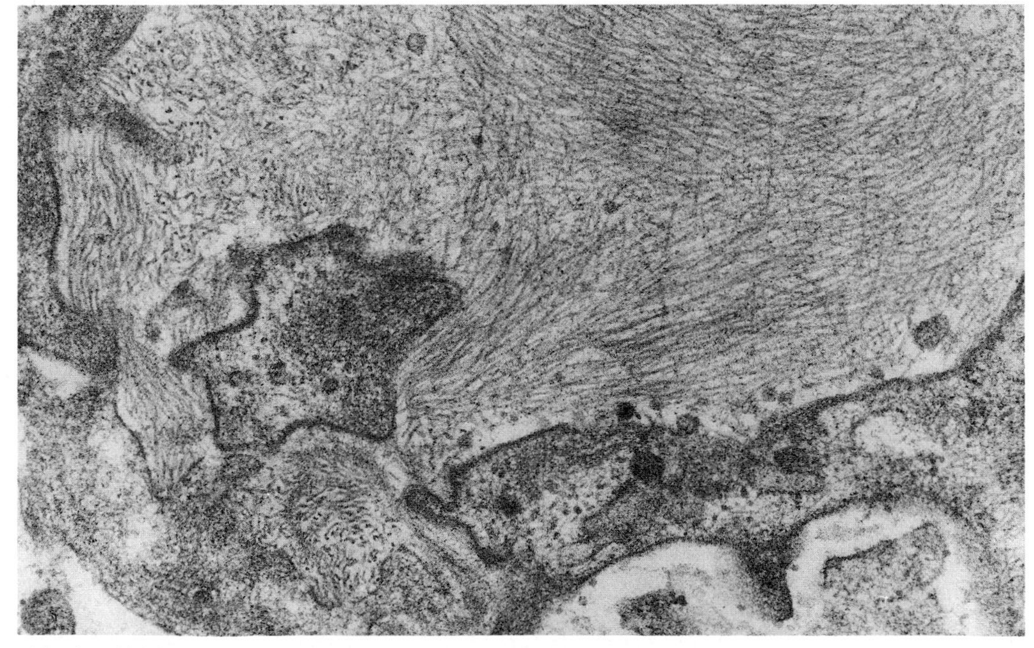

Fig. 11.2 Amyloidosis of kidney.
The beaded amyloid fibrils lie between the cytoplasmic processes of a renal glomerular endothelial cell. (TEM × 44 600.)

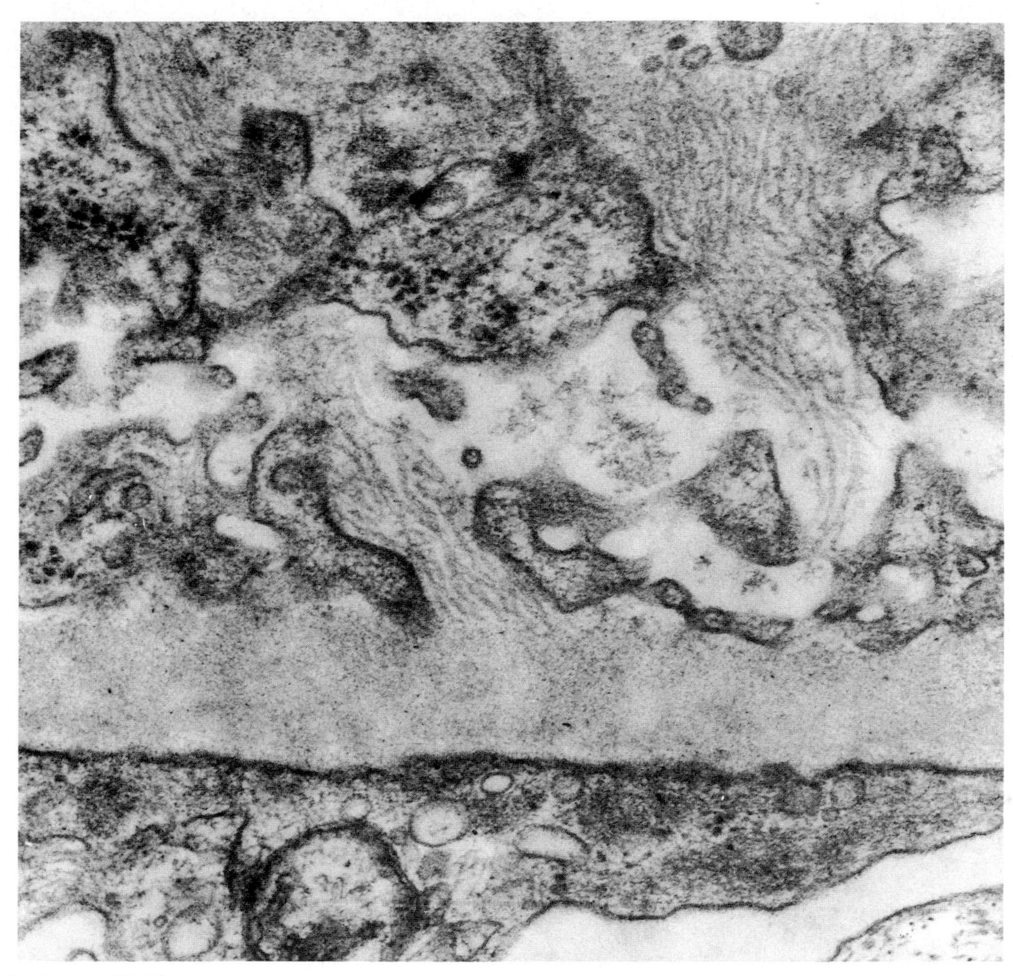

Fig. 11.3 Amyloidosis of kidney.
In this figure, amyloid fibrils lie between the processes of a glomerular epithelial cell; they display a beaded structure. There is a periodicity of 47 nm; the fibrils are 21 nm thick. A suggestion has been advanced that the fibrils are, in fact, tubular. (TEM × 52 000.)

Table 11.2
Properties of amyloid components[50]

	Fibrils	P-component
Proportion of amyloid	~ 95 per cent	~ 5 per cent
TEM structure	Fine branching fibrils of variable length, 10–15 nm diameter	Laterally stacked rods of variable length; ring of globular subunits in cross-section
Protein	Almost entirely	~ 90 per cent
Carbohydrate	Very little: glycosaminoglycan is associated	~ 10 per cent
Solubility	In water	In saline
Antigenicity	Weak	Strong
Physical properties	Polymerizes under physiological conditions	Calcium-dependent binding to polyanions

For each class, there are distinct precursor molecules: in some instances, the precursors are serum proteins which may be normal constituents such as β_2-microglobulin; other precursors are abnormal reactants such as serum amyloid A, an acute-phase protein formed quickly as inflammation begins.

The amyloids tend to give rise to or be associated with individual clinical syndromes. They can therefore be classified by their clinical behaviour. However, classification is also possible by their tendency to systemic or local deposition, by their acquired or heredofamilial origin, and, above all, by the identity of the fibrillar protein (Table 11.1). Seven amyloid classes can now be designated by the letter A (amyloid) followed by a second letter which may be L, A, H, S, F, D or E indicating the nature of the fibrillar protein. Subclasses are shown by a subscript letter and number. The chemical composition of several of the most important classes of amyloid is known in detail as are the physical and ultrastructural characteristics both of the fibrillar protein and of the constant P-component (see below). Much of an amyloid deposit *in vivo* is water. There is a small carbohydrate moiety not now thought to be glycosaminoglycan and a little immunoglobulin that is bound to the amyloid non-specifically.

In this account, the seven amyloid classes are outlined briefly; a description is then given of the accumulation of amyloids in the connective tissues.

Systemic amyloidosis

(Table 11.1)

Systemic amyloid may accumulate because of heredofamilial factors or be acquired, on the basis of environmental disorders. When there is a demonstrable cause for acquired amyloidosis, it is said to be 'secondary'; when no cause has been found, amyloidosis remains 'primary' (idiopathic).

Acquired amyloidosis

Primary acquired systemic amyloidosis: amyloid AL The proteins of amyloid AL are derived from homogeneous immunoglobulin light-chains or NH_2-terminal fragments of them.[57] Enzymatic cleavage seems to make the light-chains prone to fibril formation. There is no unique light-chain but the variable region of λ_{VI} is particularly fibrillogenic. λ chains assume a β-pleated structure more readily than $\varkappa$, although $\varkappa_{III}$ is also amyloidogenic. The accumulation of amyloid is often unexplained ('primary') but there are frequently large numbers of bone marrow plasma cells in affected individuals. In other patients, the synthesis of the light-chain amyloid precursor is associated with myelomatosis or with another immunocyte dyscrasia such as lymphoma, monoclonal gammopathy or macroglobulinaemia. Experimentally, it is interesting that proteolytic enzymes such as trypsin, pepsin and some lysosomal proteases are capable of changing selected immunoglobulin light-chains *in vitro* into microfibrils with the transmission electron microscopic appearance of amyloid microfibrils (see p. 433).

Secondary acquired systemic amyloidosis: amyloid AA (Figs 11.4 and 11.5) Amyloid may be secondary to immunological, infectious or reactive disorders of connective tissue such as rheumatoid arthritis

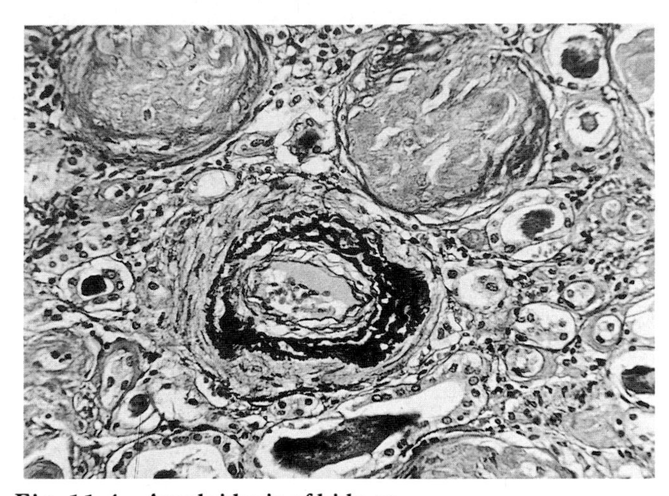

Fig. 11.4 Amyloidosis of kidney.
In this case of rheumatoid arthritis, the accumulation of amyloid A has advanced to a stage of glomerular obliteration. The deposits of amyloid are accompanied by the formation of collagen. Amyloid nephrosclerosis has resulted and the patient has developed end-stage renal failure with uraemia. (Elastic-van Gieson × 180.)

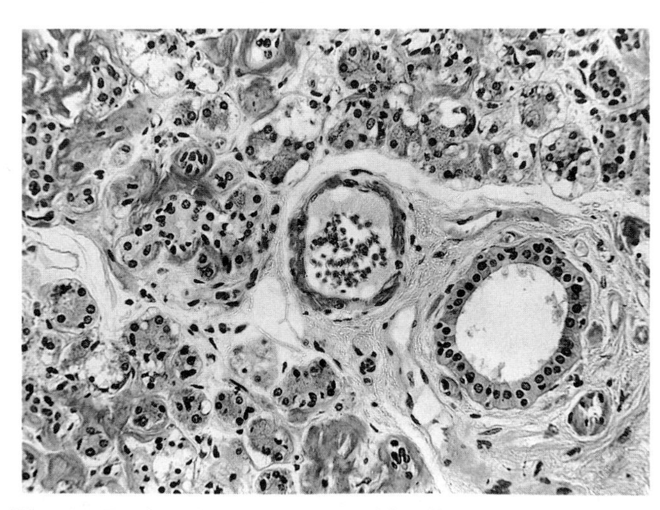

Fig. 11.5 Amyloidosis of parotid salivary gland.
Amyloid lies within the wall of an arteriole (centre) and between the acini and ducts of the gland parenchyma. (HE × 170.)

(see Chapter 13), tuberculosis (see p. 743) and ankylosing spondylitis (see p. 768). Amyloid A protein, the fibrillar component, is derived by enzymatic cleavage from a circulating precursor, serum amyloid A-related protein. Serum amyloid A, a soluble α-β globulin, mol. wt 16–18 kD, can be isolated with the high density lipoproteins and with albumin: it is a complex of which one part, SAAL (apo SAA), is a lower molecular weight (approximately 1.2 kD) polypeptide containing all the amyloid A protein-related determinants. Serum amyloid A has been shown to be an acute-phase reactant but elevation of the serum concentration of this protein in response to inflammation does not necessarily lead to amyloid deposition in the tissues. Amyloid A also forms the fibrils of the deposits in familial Mediterranean fever (see p. 368) and those of most forms of induced experimental amyloidosis (see p. 439).

Amyloid A protein is consequently the most common form of amyloid fibril. It comprises 76 amino acid residues that have been fully sequenced. The molecular weight is approximately 8.5 kD and the amino acid sequence is identical with that of part of the larger serum amyloid A molecule. Although there is no precise resemblance to any other naturally occurring protein, there is homology between the amyloid A proteins obtained from different species just as there is similarity but not identity between the AA amyloids recovered from subjects with different diseases. Such chemical variability as there is lies in the C-terminal and, to a lesser extent, in the N-terminal ends of the molecule.

Haemodialysis-related amyloidosis: amyloid AH[49]

The proteins of AH fibrils are derived from β_2-microglobulin, a small protein of molecular weight 118 kD. β_2-microglobulin is present in most body fluids and is associated on the plasma membrane with the histocompatibility antigens (see Chapter 5). The molecule of β_2-microglobulin exhibits homology with the constant-region domains of immunoglobulin heavy-chains. Normally, β_2-microglobulin is excreted by glomerular filtration and reabsorbed and catabolysed by the proximal renal tubule. Under circumstances of chronic renal failure, treated by haemodialysis, β_2-microglobulin accumulates: conventional dialysis membranes are impermeable to the molecule. As a result, β_2-microglobulin, which apparently has the molecular characteristics necessary for assembly in a β-pleated sheet conformation, accumulates in selected tissues.

Acquired experimental amyloidosis: amyloid AA (see p. 439)

Amyloid of class AA can be produced in many animal species. The mouse has been employed in most studies but rabbits, hamsters, guinea-pigs, monkeys, mink and chickens are among species capable of amyloid formation. Many avian species form amyloid in old age.[28,29]

Heredofamilial amyloidosis

Senescence-accelerated mouse amyloidosis: amyloid AS

This form of genetically determined systemic amyloidosis has attracted growing attention.[101,102]

Dominant heredofamilial amyloidosis: amyloid AF

There are Portuguese (AF_p) and Japanese (AF_j) subclasses of neuropathic heredofamilial amyloidosis. The fibril protein in each is a variant of prealbumin in which methionine is substituted for valine in position 30. The genetic mutation can be recognized in early life, before evidence of the disease, usually associated with polyneuropathy, becomes evident. There are other variants of the altered prealbumin. Amyloidosis of the heart (AF_c) may also be heritable.[37] In preliminary evidence, the fibril appeared to contain methionine substituted for leucine in position 110.[57] It is interesting that the amyloid fibril of senile cardiac amyloidosis, an organ-limited form of the disease, also shows partial sequence homology with normal prealbumin. It appears that this variety of prealbumin-derived amyloid has an affinity for the heart. In a third (Icelandic) AF_i (cerebral) form of the dominant category of heritable amyloidosis, the leptomeninges and meningeal blood vessels contain fibrils, the polypeptides of which have amino acid homology with fragments of β-trace protein (cystatin C). The protein, a proteinase inhibitor, exists in certain neuroendocrine cells. Individuals with this form of amyloidosis die prematurely from cerebral haemorrhage.

Recessive heredofamilial amyloidosis: amyloid AA

The most important complication of familial Mediterranean fever, better called recurrent polyserositis (see p. 368) is amyloidosis. The amyloid fibril is AA. Although arthritis is frequent in familial Mediterranean fever, it is not yet known whether amyloid is deposited in the synovia and cartilages as it is in skeletal muscle.

Non-systemic amyloidosis

Amyloid deposits in this category may be limited to single organs or to selected localized parenchymal sites.

Organ-limited amyloidosis

Immunocyte-derived, organ-limited amyloidosis

The chemical nature of the amyloid nodules that form, for example, in the larynx or trachea, is not known. The amyloid of widespread pulmonary amyloidosis is not well understood.

Senile, organ-limited amyloidosis:[27,59,79,80] **amyloid AS** In advanced old age, amyloid aggregates in the heart,[13] the central nervous system,[83,85,89] aorta[61] and cerebral blood vessels,[2,85] and in the hyaline cartilages and other connective tissues. In cardiac amyloid the fibrillar protein (AS_c) is similar to prealbumin.[26] Cardiac dysfunction, perhaps related to small coronary artery amyloidosis, may result (Fig. 11.6).[95,105] A new form of cardiac amyloid has been detected in the sewing ring of an autologous fascia lata graft and in porcine valve bioprostheses.[47] Amyloid has also been identified in the tissue of aortic and mitral valves excised surgically.[70] Amyloid may, moreover, be deposited in the aorta itself. In cerebral amyloid, AS_b, the protein accumulates in Alzheimer's disease in the neuritic plaques, in the neurofibrillary tangles and in small blood vessels. In each site, the amyloids react with an antiprealbumin antiserum[104] and are a β-trace protein. Amyloid P-component may not be present. However, the fibrillar structure is identical to that of the systemic amyloids. The limited studies of connective tissue cartilagenous, discal, meniscal and ligamentous amyloid that have been made show that the deposits do not have the histochemical and immunocytochemical properties of amyloid AL: whether they are AA or AS is not established.

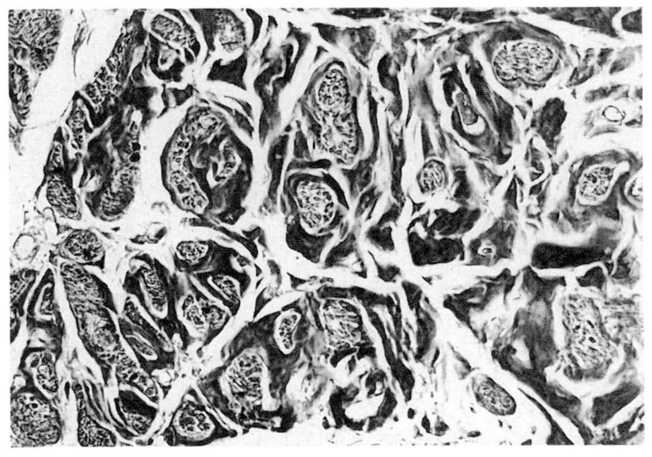

Fig. 11.6 Cardiac amyloid.
Substantial, structureless deposits of amyloid extend between and among the cardiac cell fasciculi. (HE × 377.)

Localized amyloidosis

Localized cutaneous amyloidosis: amyloid AD The skin may be a site for deposits of primary or myeloma-associated (AL) amyloid and the aggregates are common in clinically uninvolved skin;[98] a comparable situation is observed in secondary (AA) amyloidosis. However, distinct cutaneous amyloid aggregates are detectable in some cutaneous lesions: they do not react for AA, AL, prealbumin or fibronectin[10] and are designated amyloid AD. They may bind to elastic material[118] which has long been known to have an affinity for Congo red.

Localized endocrine-associated amyloidosis: amyloid AE Amyloid was found to be a characteristic, intrinsic component of medullary carcinoma of the thyroid.[51] Subsequently, insulinomata and other APUD* cell neoplasms have also been shown to be capable of amyloid deposition.[90] In the case of the amyloid identified in medullary (C-cell) carcinoma of the thyroid, the amyloid is calcitonin (see p. 107) or precalcitonin. Although amyloid has been found in localized pancreatic and pituitary deposits, its chemical nature in humans is not certain. However, the proteolytic digestion *in vitro* of insulin and glucagon gives rise to substances that resemble amyloid microfibrils. It is widely believed, therefore, that in humans neuroendocrine polypeptides other than calcitonin are likely to be sources for local amyloid deposits. Repeated local injections of insulin can indeed lead to amyloid fibril formation at the site of injection.[30] Atrial amyloid may comprise natriuretic polypeptide.

Amyloid component-P

Amyloid component-P, present in almost all amyloid classes, is a rod-shaped molecule formed of five doughnut-shaped subunits stacked one upon the other. Amyloid P-component is identical to a glycoprotein normally present in human serum, serum component-P. Serum component-P is a rapidly and continuously synthesized α-globulin. Amyloid component-P binds to fibrillar amyloid in a union dependent on calcium. Like C-reactive protein, an acute-phase protein, amyloid component-P is a pentraxin: both are similar ultrastructurally and in the sequence of the amino acids that constitute their primary molecular organization; they are nevertheless antigenically distinct. Anti-component-P antisera can be used in broad diagnostic tests for amyloid. Amyloid component-P is held to comprise 5–15 per cent of the mass of an amyloid.

Histochemistry and immunohistology of amyloid

In paraffin-embedded, haematoxylin and eosin-stained sections of fixed tissue, amyloid is an homogeneous,

* APUD: amine precursor uptake decarboxylase.

faintly eosinophilic material; no internal structure is resolved.[36,108] The depositis are only weakly birefringent. Small amounts can be overlooked, particularly when they coincide with the lesions of a connective tissue disease. In spite of claims to the contrary,[52] the spatial relationship to collagen and to reticular fibres is not constant; moreover, the tinctorial properties are different from those of collagen. It is essential therefore that, in diagnosis, special stains such as Congo red be employed;[119] many derive from those that were used in the textile industry to stain cellulose.

Metachromasia, a property of amyloid stained with methyl or crystal violet, may be evidence of the presence of associated glycosaminoglycan; it is abolished by methylation.[86] Metachromasia with toluidine blue is inconstant but is enhanced by previous partial digestion with pepsin. The violaceous, coloured material is partly removed by preliminary treatment with testicular hyaluronidase. There is a faintly positive response with the periodic acid-Schiff reagents, suggesting the presence of material containing diglycol links, perhaps glycoprotein. There is a correspondingly weak interaction of tissue amyloid with alcian blue and an increased uptake at pH 5.7 in magnesium chloride by comparison with a slight interaction at pH 2.6. The character of this result suggests the presence in amyloid of insoluble complexes between amyloid protein and polyanions that are probably carbohydrate, usually sulphated and perhaps heparan sulphate rather than sialic acid.

Congo red is best used in an alkaline technique.[93] Carnoy's fixative encourages intense staining but also the highest proportion of false-positive responses.[15] Differentiation in alcohol is important: it must be complete and the Congo red stain itself should be in saturated sodium chloride. The value of Congo red is increased by the use of polarized light, enhancing the natural birefringence of amyloid: the amyloid/dye complexes change colour from the salmon-pink characteristic of Congo red-stained tissue to a delicate apple-green.[31] This dichroic reaction allows extremely small amounts of amyloid to be identified and is the method preferred for the routine examination of biopsy specimens.[107] Sirius red stains amyloid a deep crimson but the dye/amyloid complex does not fluoresce in ultraviolet light[11] in spite of variable fluorescence in light of a longer, visible wavelength.

Fluorochromes have also been used with ultraviolet light to demonstrate amyloid. A pink fluorescence can indeed be shown after the application of Congo red. More vigorous fluorescence is obtained by the use of Phorwhite BBU,[115] an optical brightener for cellulose, and with thioflavine T,[111,112] a basic dye which has the disadvantage of staining many other negatively charged tissue components. When thioflavine T was compared with Congo red, Sirius red and crystal violet morphometrically, the thioflavine method was found to show the greatest variability, the least reproducibility.[64] All the stains differed in specificity and sensitivity and all were open to observer and other errors.

An important advance in understanding the staining properties of the amyloids came from Romhanyi's[97] analyses of the responses of these proteins to tryptic digestion. Amyloid sensitive to trypsin was found to be AA; amyloid resistant to trypsin was probably AL. Subsequently, Wright, Calkins and Humphrey[120] showed that tryptic digestion could be replaced by the simpler procedure of oxidizing the tissue with potassium permanganate. After brief potassium permanganate and sulphuric acid treatment, AA protein lost its affinity for Congo red and its birefringence; AL protein and AE amyloid retained their usual staining reactions. The potassium permanganate technique has found wide acceptance:[110] its specificity has been examined by comparing reactions with Congo red after potassium permanganate treatment, with those given by the same tissues tested by an unlabelled immunoperoxidase technique using rabbit antisera against human AA protein.[40]

The availability of antisera against the strongly immunogenic amyloid component-P has encouraged tests for amyloid by indirect immunofluorescence methods. Antiserum component-P reacts with all forms of amyloid but this technique has also revealed a more widespread localization of amyloid component-P than had been anticipated, e.g. in normal glomerular capillary basement membrane (see p. 50); in large vessels; and around elastic material. Antisera against amyloid proteins can be used to identify amyloid in tissue sections by light microscopy[62] and similar techniques can be employed to analyse amyloid by transmission electron microscopy.[74,75] In one study using antisera monospecific for immunoglobulin heavy-chains, κ- and λ-light-chains and AA protein, classification was possible in 44 of 50 cases: 20 were AL, 24 AA.[41] Protein-A gold preparations can be chosen for ultrastructural purposes: the gold-labelled deposits can be identified in thin sections and both the location and identity of the amyloid fibrils determined.[77]

Ultrastructure of amyloid

Amyloid fibrils

An identical fibrillar fine structure[23] is shared by the amyloids of men and of animals.[22,24] Recognition of the fibrils remains an important diagnostic criterion.[107] In thin sections of fixed tissue, the unit fibril is 7.5–10.0 nm in diameter. Although early work suggested the presence of a beaded structure and a double-stranded longitudinal arrangement, the present view is that there is a subunit filament of 2.5–3.5 nm in diameter extending within the 7.5–10.0 nm diameter fibril.[24] Two or more of the subunit filaments, occasionally crossing each other, form an amyloid fibril. The three-dimensional subunit amyloid fibril is consequently believed to be *either* two or more helical 2.5–3.5 nm filaments of long pitch; *or* three or several 2.5–3.5 nm filaments assembled in parallel to form a tube-like array. In amyloid fibril deposits, a ring-like structure having a diameter comparable to the amyloid fibril, may be the amyloid fibril in cross-section. There is indeed substantial evidence that tissue amyloid deposits are cylinders, not solid fibrils.[9]

Preparations of *isolated* amyloid macromolecules can be obtained from native material or from amyloid-like fibrils created *in vivo*. Native fibrils *in vitro* closely resemble those found in the tissue deposits, both in staining characteristics and in ultrastructure. High-resolution transmission electron microscopy after shadowing confirms the longitudinal subunit structure: the subunit, entitled a protofibril or filament, is 2.5–3.5 nm in diameter and two or more such filaments, occasionally twisted about each other, form the native amyloid fibril. Amyloid-like fibrils reconstituted from Bence–Jones proteins by peptic cleavage and precipitation of the variable segments, closely resemble the native amyloid fibril in ultrastructure. Protofibrils, thinner and filamentous, form first: they aggregate laterally as fibrils.

Amyloid component-P

Amyloid component-P was identified ultrastructurally as a rod-like structure, 10 nm in diameter, with a 4 nm periodic clear banding.[7] Each rod was shown to comprise an assembly of pentagonal discs with a central, empty core.[18] Amyloid component-P and amyloid fibrils often coexisted in the same preparation.[103] As rods, the side view of amyloid component-P indicated segments 2–3 nm wide and 1–1.5 nm apart. Face on, the units were doughnut-like structures with a 9–10 nm outer diameter, a 4 nm inner diameter. Within the outline of this doughnut, five globular subunits could often be detected. Amyloid component-P and C-reactive protein share more than half their amino acid structure, i.e. they display more than 50 per cent homology. Amyloid component-P is seen by transmission electron microscopy to exist as paired discs (dimers) but C-reactive protein exists as single discs (monomers).

Amyloidosis of articular tissues

Amyloid may accumulate in synovial tissue, in articular hyaline and fibrocartilages, in capsular and periarticular tissues and in ligaments.[109] There are four circumstances under which this may occur: 1. when amyloid AL forms in the 'primary' disease; 2. when amyloid AA is deposited in the course of immunological connective tissue disorders such as rheumatoid arthritis or in reactive diseases such as ankylosing spondylitis; 3. in old age, when asymptomatic AS or unidentified amyloid fibrils are widespread; and 4. when amyloid AH results from long-term haemodialysis.

Articular amyloid in myeloma and in primary amyloidosis

Amyloid arthropathy in myelomatosis[33] has attracted much attention since its early descriptions. A large mass may form in the soft tissues and within the cartilage whence it may extend into the joint space or juxtaarticular bone (Fig. 11.7). The identity of the AL deposits can be established by the use of immunoperoxidase and immunogold/silver techniques. The frequency of amyloid AL arthropathy has been recorded as approximately 5 per cent of myeloma cases.[53] In an unique account, intracytoplasmic synovial amyloid fibrils were described.[38] Macrophages (see p. 38) play a key part in amyloidogenesis and it is possible that synovial A cells are important in synthesizing articular amyloid.[53]

In an early report, two forms of primary articular amyloidosis were recognized.[116] In the first, more common variety, both synovia and periarticular tissues were infiltrated by amyloid that was so extensive that palpable subcutaneous nodules, joint swelling, limited movement and hypertrophied synovia were identified.[88] In the second, bone adjacent to a joint was replaced by amyloid[72] resulting in much swelling and in pathological fracture. Thirty-nine examples were reviewed by Wiernik:[116] their mean age was 53 years. The joint involvement was usually symmetrical, affecting large as well as small articulations. In 24 cases, the frequency of articular amyloidosis was, in order: shoulders, knees, wrists, metacarpophalangeal or interphalangeal, elbows and hips. The spine, ankle and acromioclavicular joints were less often affected. The carpal tunnel syndrome proved to be a frequent complication of wrist joint disease.

Synovia

Synovial amyloid causes a stiffening and pallor of the tissues. The deposits stain with Congo red, display birefring-

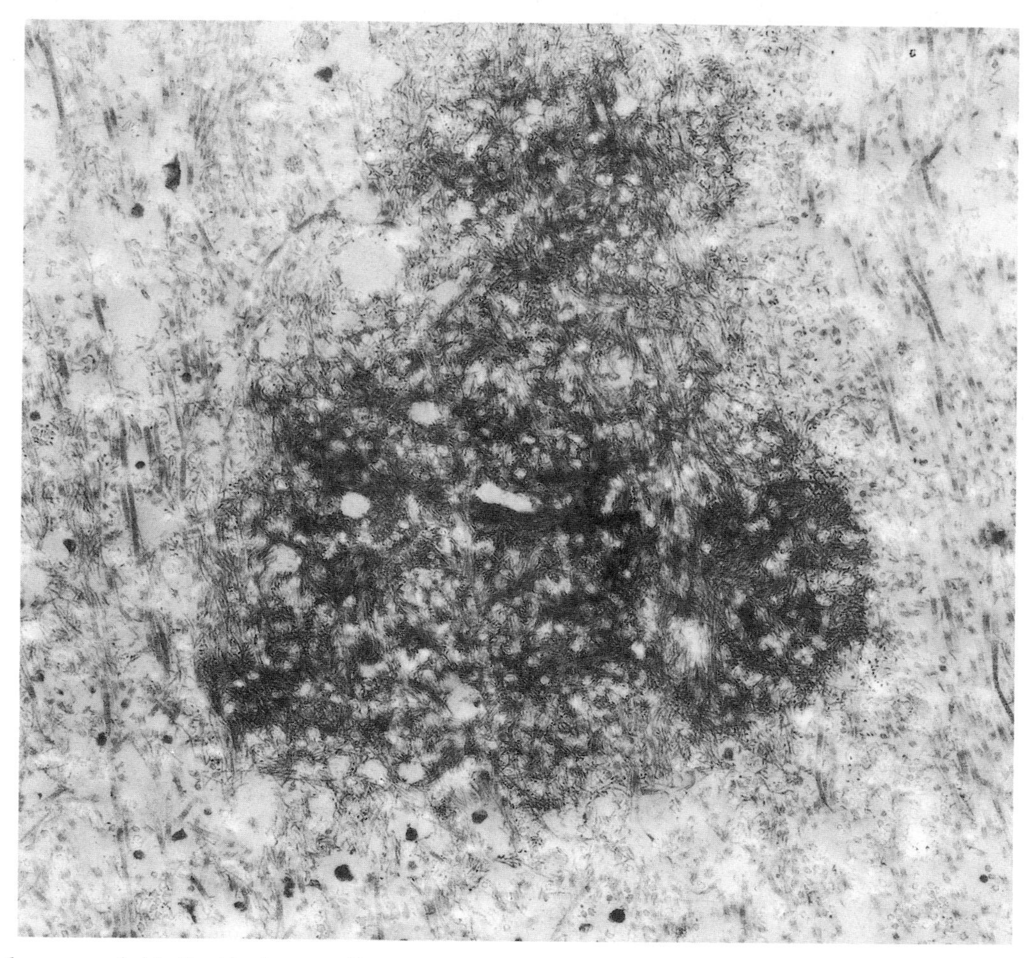

Fig. 11.7 Myeloma amyloid of articular cartilage.
In this case of myelomatosis, λ-chain myeloma protein accumulated in the extracellular matrix. The fibrils, aggregating as β-pleated sheets, reacted histochemically and immunologically as 'primary' amyloid. (TEM × 14 940.)

ence and dichroism and the characteristic microfibrillar structure by electron microscopy; they lie between and among synoviocytes. Uncertainty exists whether the amyloid is secreted by synoviocytes, deposited from the synovial fluid in which aggregated masses of amyloid can be found,[116] sometimes with synovial villous fragments, or passed to the extracellular tissue spaces from a capillary ultrafiltrate.[114] The synovial fluid can be employed in diagnosis.[73] In the case of myeloma described by French,[38] type A synoviocytes contained amyloid microfibrils. Whether this material had been phagocytosed from the large extracellular accumulations or represented synovial cell amyloid secretion remained unclear. The synovial amyloid deposits resisted potassium permanganate oxidation and were believed therefore to be of immunoglobulin origin. All forms of myeloma, including the IgD[88] and pure Bence–Jones varieties, have now been shown to be capable at times of evoking this form of synovial deposit.[53]

Cartilage

Cartilagenous amyloid deposits, linear or globular,[14] appear parallel to the surface in cartilage zones I and II but radially or obliquely, as 'strings of beads', in deeper zones. Around chondrocytes, are amyloid-free pericellular zones, an observation conflicting with that of Hickling *et al.*[53] The cells themselves contain no amyloid. Thin sections confirm the identity and disposition of these deposits; they are mainly confined to hyaline cartilage. Intervertebral disc fibrocartilage is involved with advancing age.[114]

Articular amyloid in systemic connective tissue disease

Although secondary, systemic amyloid A is frequent in chronic rheumatoid arthritis, synovial deposits in this disease (see Chapter 12), in ankylosing spondylitis (see Chapter 19) and in the other systemic connective tissue disease are exceptional.

Articular amyloid in ageing

Amyloidosis of synovial joints

With advancing age, small deposits of amyloid can often be recognized in the articular cartilages of normal elderly persons.[48] Comparable changes are found in senescent mice.[102] Ladefoged and Christensen[68] investigated 30 necropsies of mean age 72.7 years in a hospital-selected series: 93 per cent could be shown to have amyloid in the superficial zones of the hyaline articular cartilage (Fig. 11.8), while 43 per cent had amyloid in the articular capsule. In both sites, the quantity of amyloid increased with age; there was a significant correlation between the amount and the age. Material with the staining and ultrastructural characteristics of amyloid is indeed frequently present in normal human joints.[66]

These capsular and cartilage deposits are usually symptomless.[82] In 91 unselected individuals, of mean age 70 years, the overall frequency of articular amyloid in the sternoclavicular and hip joints, was 56 per cent.[48] Deposits in the former joint were much commoner than in the latter. The amyloid aggregates were microfibrillar. There was no association with previous connective tissue disease and only two patients had generalized amyloidosis. Strangely, the sternoclavicular deposits were largely confined to the fibrocartilagenous discs and the capsule was affected in only one instance. The amyloid deposits were resistant to potassium permanganate oxidation. In contrast to the frequency of involvement of the synovial joints, islands of amyloid were only found in the heart of 12 per cent of this group of aged patients, most often in the interatrial septum. Cartilagenous amyloidosis, it is suggested, should be classed as a form of localized disease.[48]

Relationship of amyloidosis to osteoarthrosis

Christensen and Sorensen[20] made the surprising observation that amyloid can be found in the articular capsular connective tissue in many patients subjected to arthro-

a

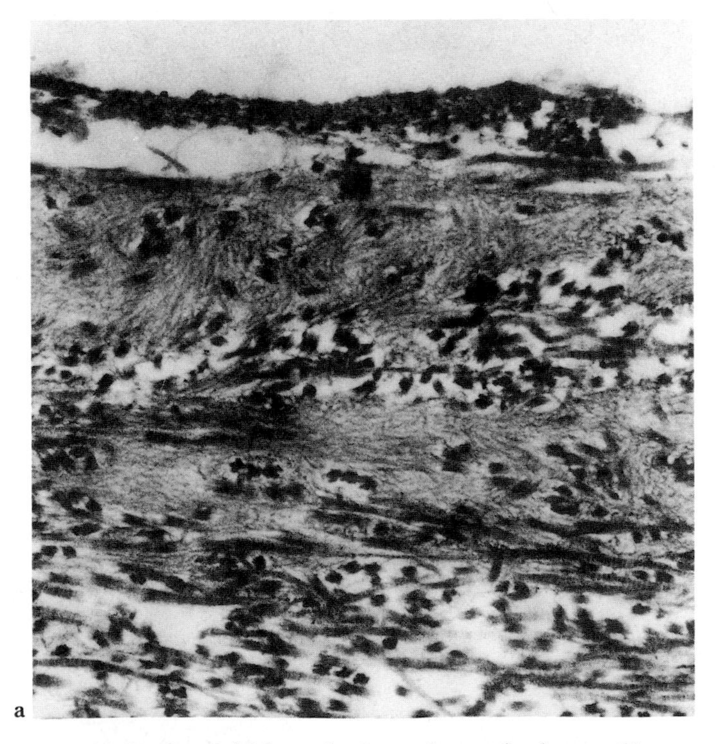

b

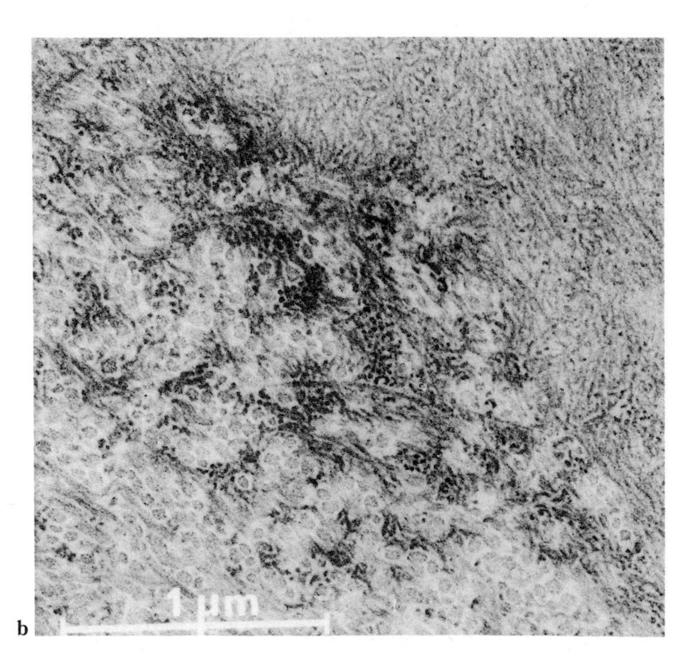

Fig. 11.8 Amyloid deposits in ageing articular cartilage.
Articular cartilage from aged synovial joint cartilage and fibrocartilage as well as tissue from periarticular ligaments, is often found to contain fibrillar material with the properties of amyloid. Tests for amyloid A tend to be negative but the identity of this form of amyloid is not yet certain; it is probably best described as 'senile'. The aggregates appear by light microscopy as linear streaks or pericellularly. **a.** Amyloid fibrils interwoven among collagen fibrils at articular surface. **b.** Fibrillar amyloid situated more deeply in zone I cartilage. (TEM **a.** × 26 000; **b.** × 35 000.)

plasty for osteoarthrosis. The deposits are 'congophilic' and the Congo red-stained material is birefringent and exhibits positive dichroism. Ladefoged[66] confirmed the presence of amyloid in 65 per cent of 116 osteoarthrosic femoral heads from patients of mean age 61 years who had not had primary amyloidosis, myeloma or tuberculosis. Only one patient suffered from rheumatoid arthritis. Seven cases demonstrated massive amyloid deposits, 28 moderate deposits. However, there was no correlation between femoral head deformity and the degree of amyloid deposition; no attempt was made to relate cartilage amyloidosis to early fibrillation. Articular capsular amyloid was found to be more common in males than females[71] but there was no difference between the amounts deposited in the right or the left joint capsules or between the amounts found in the load-bearing and non-load-bearing parts of the femoral head cartilage. With few exceptions, when amyloid was identified in the joint capsule, it was always recognizable in the cartilage. However, cartilage often contained demonstrable amyloid when the capsule did not. These observations were confirmed.[34] In osteoarthrosic joints, the amyloid deposits were sometimes near sites of calcium pyrophosphate deposition (p. 885).

Amyloidosis of fibrocartilagenous joints

Amyloid is deposited in the intervertebral discs of rapidly ageing, SAM mice (p. 439).[101] The aggregates are found throughout the spine. These observations draw attention to the few reports of human intervertebral disc amyloidosis. One example, with calcification of the disc, was described[3] and may have been an instance of coexistent amyloidosis and chondrocalcinosis (see p. 401). Systematic surveys, however, suggest that intervertebral disc amyloid in the aged is relatively common.[69,114] Takeda *et al.*[106] investigated 84 patients: 44 had demonstrable amyloid deposits. The youngest was aged 27 but the frequency of amyloidosis rose abruptly in the fourth decade. In my own experience, amyloidosis appears commonplace in small aggregates in sacroiliac joints where it can be detected in the cartilagenous and non-cartilagenous parts of the articulation as well as in adjacent ligaments.

Amyloidosis of menisci

Studies made in this laboratory on knee joint menisci from 17 patients subjected to arthroplasty for rheumatoid arthritis, have demonstrated superficial amyloid deposits in a large proportion of cases, sometimes associated with chondrocalcinosis. These observations entirely confirm those of Ladefoged.[67] The amyloid deposits were not related to the secondary AA aggregates of rheumatoid disease.

Hereditary amyloidosis and joint diseases

Articular deposits in hereditary amyloidosis are very unusual. In a rare example, Eyanson and Benson[35] reported the development of erosive, inflammatory arthropathy in nine patients from two kindreds with hereditary amyloidosis. Unlike cases of primary AL or myeloma arthropathy, there were no large deposits within synovial tissues but inflammatory cartilagenous erosion was evident.

Haemodialysis amyloid

In 1984, Charra *et al.*[19] gave preliminary reports of amyloid detected during the surgical decompression of the carpal tunnel in patients with renal failure maintained for long periods on haemodialysis (Figs. 11.9 and 11.10). Their accounts were substantiated by Munoz-Gomez.[87] In none of the affected individuals was there any factor predisposing to amyloidosis. The mean duration of haemodialysis was 8.6 years. The amyloid fibrils were 9–10 nm thick.[84] The amyloid protein was identified as β_2-microglobulin[12,43] and the presence of this molecule in the amyloid deposits within the carpal tunnel tenosynovium[5,16] demonstrated by the use of an anti-β_2-microglobulin antibody.[12,78] The provision of special hiflux dialysis membranes now makes it possible to minimize the frequency of amyloid AH in future populations of haemodialysis patients.

Following the recognition of β_2-microglobulin residues that showed congophilia and a microfibrillar structure, fibrils were successfully prepared artificially, the first demonstration of amyloidogenesis *in vitro* with the precursor molecules intact.[25,49] That β_2-microglobulin amyloid, haemodialysis amyloid, was first observed in the carpal tunnel synovium[5,16] an observation which is probably only a reflection of the vulnerability of this anatomical site to materials that occupy the tunnel space. For similar, mechanical reasons bone containing β_2-microglobulin or other amyloid deposits is prone to pathological fracture.

Articular amyloidosis and crystal deposition

Amyloid and calcium pyrophosphate dihydrate (see p. 401) may coincide.

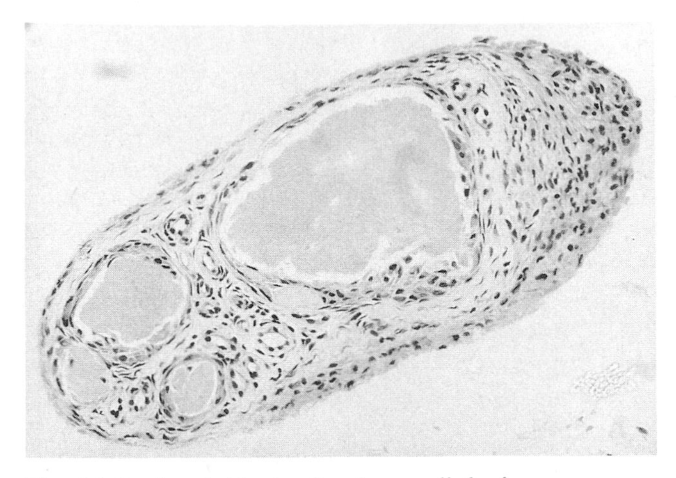

Fig. 11.9 Amyloidosis after haemodialysis.
Tenosynovium from carpal tunnel in patient who had been maintained on long-term haemodialysis for chronic renal failure. Amyloid is uniform, grey material. Note macrophage polykaryon at left margin of amyloid aggregate. (HE × 72.) Courtesy of Professor John McClure.

Pathogenesis
(Fig. 11.11)

Amyloidosis[21] has long been associated with procedures used to enhance antibody production and with disorders in which inflammation and infection are prolonged. Amyloid, it has been suspected, may be related to an alteration or abnormality in immunoglobulin production. Immunoglobulin and complement are components of some amyloid deposits, and this view accords with ideas on the nature and pathogenesis of both inflammatory connective tissue diseases such as rheumatoid arthritis, in which amyloidosis is common, and with the nature of myeloma and of the proteins secreted by myeloma cells. Why, in the latter circumstances, synovial amyloid is found more often than in the former, and why some individuals with chronic inflammatory disease form synovial amyloid while others do not, remains unexplained. The possibility that amyloidogenesis may occur in two phases has been widely discussed.

In a first phase, there are raised IgG levels. In parenthesis, it should be noted that amyloid in humans can occur in cases of agammaglobulinaemia. Altered B-cell function may be involved. Alternatively, a defect in T-suppressor

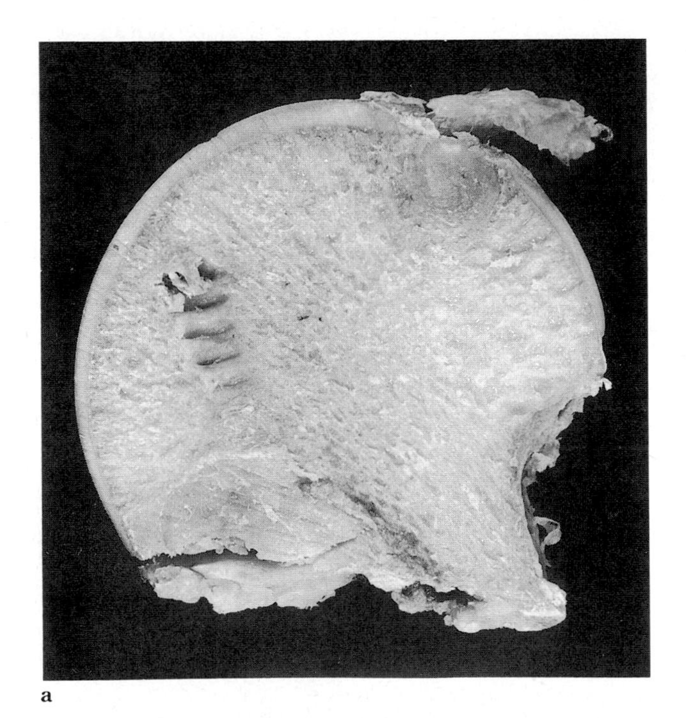

a

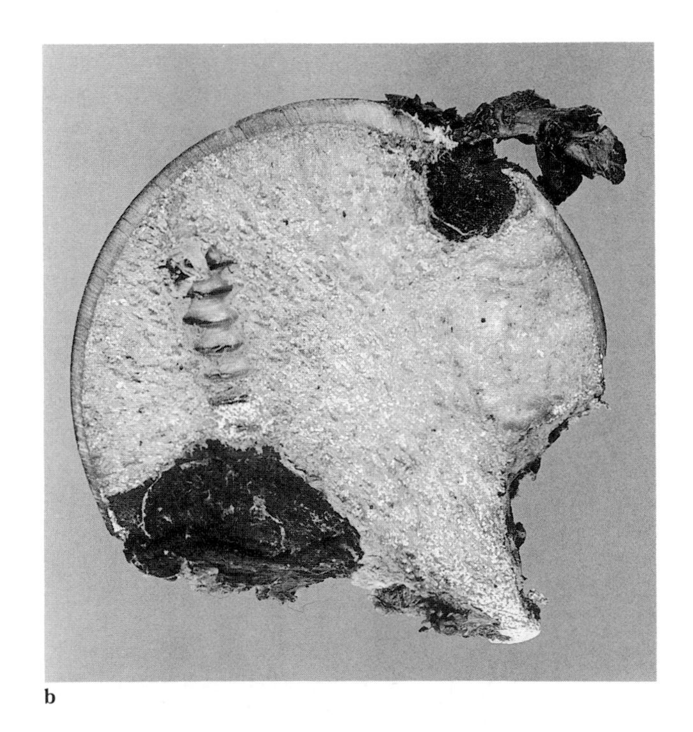

b

Fig. 11.10 Amyloid after haemodialysis.
Femoral head from male aged 58 years, treated for 11 years by haemodialysis for chronic renal failure. Evidence of aluminium bone disease was identified. Patient sustained fracture of femoral neck. **a.** Femoral head bone, seen in this figure, contains islands of non-osseous material (upper right, near ligamentum teres; and lower left). (× 1.2.) **b.** Iodine and sulphuric acid test for amyloid reveals dark, mahogany-brown coloration, indicative of amyloid. (× 1.2.)

cells may permit an escape of B-cell function from normal control. Thus, a particular set of lymphoid cells may be activated and there is evidence of both the depletion of T-cell populations during amyloid formation and of impaired cell-mediated immune responsiveness. The development of tolerance to a particular antigen during amyloid induction may be a part explanation for the altered protein synthesis.

In a second phase, the mononuclear macrophage is implicated. It acts as a common pathway, via which a large number of factors ranging from prolonged inflammation to excessive quantities of circulating endocrine polypeptides, may cause the synthesis of amyloid fibrils.

Amyloidosis of animals

The pathogenesis of amyloid can be investigated in animals which spontaneously develop the disorder. Alternatively, a wide variety of procedures can be used to cause amyloidosis in experimental animals and amyloidogenesis can also be investigated in tissue culture.[56]

At least 21 species in addition to humans are known to develop amyloidosis spontaneously.[1,54] Amyloidosis is, for example, an important disorder in birds.[28,29] Since the first account of mice with age-associated amyloidosis and cystic renal disease, many other reports of spontaneous murine amyloidosis have been published.[99] The amyloid of the senescence accelerated mouse (SAM) has attracted much interest; the fibrillar AS_{sam} protein has a molecular weight of 5.2 kD. Although much is known of the sites where AS_{sam} is laid down, there are few reports of its localization to articular cartilage, fibrocartilage, ligament or synovium.

Experimental production of amyloidosis

Many techniques have been developed that lead to the reproducible formation of amyloid in the laboratory animal.[1,65] The most frequently used animals have been the mouse,[4] hamster, rabbit and guinea-pig. Agents employed to provoke amyloidogenesis have included casein,[1] injected or fed, infection with *Leishmania* spp.,[100] adjuvants, bacterial antigens, endotoxins,[55,56] and parabiosis.[76,117]

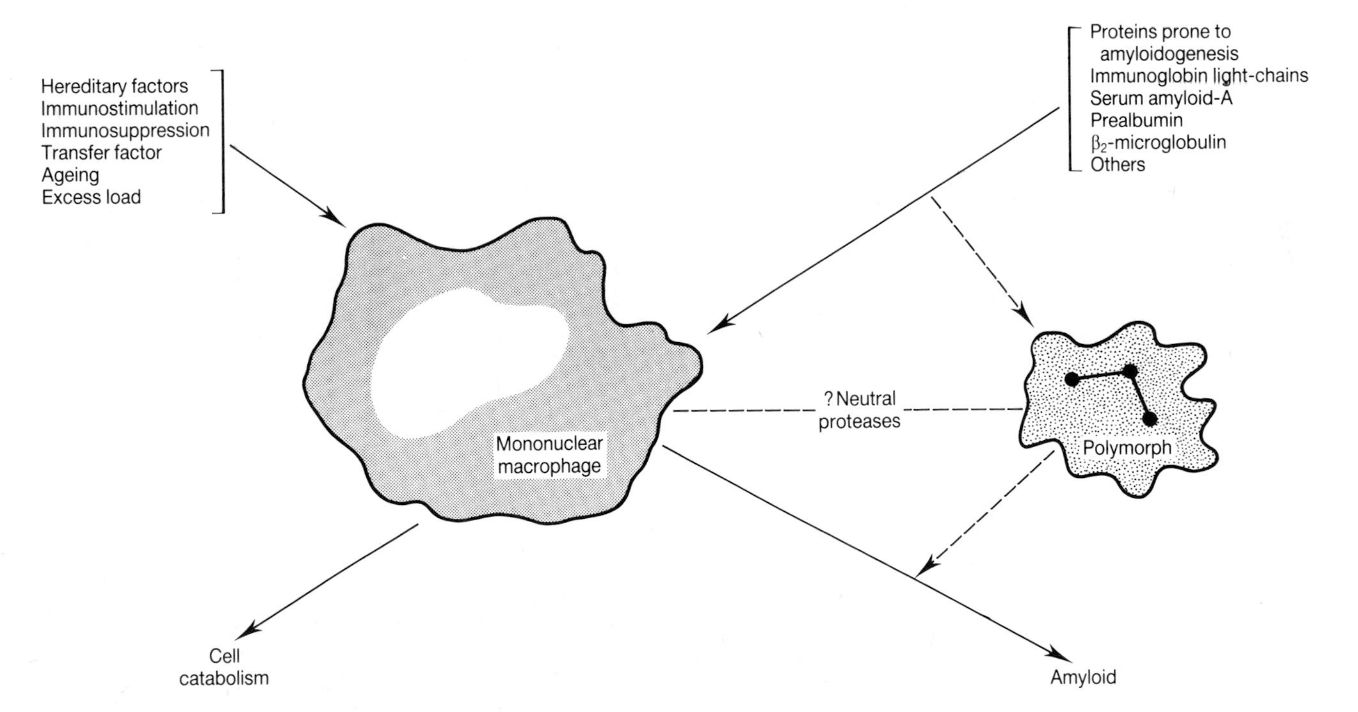

Fig. 11.11 Biphasic hypothesis for origin of amyloid.
In many instances, amyloid appears to be due to abnormal processing of normal proteins which are sometimes present in excess, e.g. when they are produced by functioning endocrine neoplasms or given parenterally (insulin). Defect may be inherited or acquired. Process centres on the macrophage. Factors provoking amyloidogenesis include sustained inflammatory responses, immunosuppression and ageing. Amyloid synthesis can be accelerated by low concentrations of serum protease inhibitors and by 'transfer factors' that may be cytokines. Modified from Cathcart and Ignaczak.[17]

Amyloid fibrillogenesis

In the case of amyloid L protein, proteolytic cleavage of the paraprotein L chains produces fragments that can polymerize as β-pleated sheets.[91] Whole L chains can occasionally behave similarly and the aggregation of these proteins as insoluble fibrillar masses in the connective tissues is readily understood. In the case of amyloid A protein, an origin can be traced to the precursor acute-phase protein, serum amyloid A. However, other factors are necessary for AA deposition; they include raised levels of serum amyloid A and the presence of an 'amyloid-enhancing factor'. Elevated amounts of serum amyloid A can be induced by administering to mice supernatants from macrophages cultured in the presence of lipopolysaccharide.[63] Amyloid-enhancing factor always appears experimentally before AA deposits can be seen by light microscopy; the factor may be a nucleoprotein. In a disorder such as rheumatoid arthritis (see Chapter 13), it is therefore reasonable to propose a sequence of three coincident processes: first, inflammation, with the synthesis of cytokines such as interleukin-1 which stimulate the hepatic production of apo-serum amyloid A, itself associated in the circulation with high density lipoprotein to form a serum amyloid A complex; second, amyloid-enhancing factor which appears when inflammation is chronic; and third, further factors which are not yet understood but which may include heparin and/or heparan sulphate. Glycosaminoglycan and AA deposition are synchronous and probably coincidental.

REFERENCES

1. Abruzzo J L, Gross A F, Christian C L. Studies on experimental amyloidosis. *British Journal of Experimental Pathology* 1966; **47**: 52–9.
2. Ackerman R H, Richardson E P, Heros R C. Amyloid arteriopathy of cerebral vessels. *New England Journal of Medicine* 1982; **307**: 1507–14.
3. Ballou S P, Khan M A, Kushner I. Diffuse intervertebral disc calcification in primary amyloidosis. *Annals of Internal Medicine* 1976; **85**: 616–17.
4. Barth W F, Willerson J T, Asofsky R, Sheagren J N, Wolff S M. Experimental murine amyloid. III. Amyloidosis induced with endotoxins. *Arthritis and Rheumatism* 1969; **12**: 615–26.
5. Bergada E, Montoliv J, Bonal J, Subias R, Lopez Redret J, Revert L. Carpal tunnel syndrome (CTS) with local and articular amyloid deposits in haemodialysis patients. *Kidney International* 1984; **26**: 579.
6. Berger P A. Amyloidosis—a complication of pustular psoriasis. *British Medical Journal* 1969; **2**: 351–3.
7. Bladen H A, Nylen M U, Glenner G G. The ultrastructure of human amyloid as revealed by the negative staining technique. *Journal of Ultrastructure Research* 1966; **14**: 449–59.
8. Bonar L, Cohen A S, Skinner M M. Characterization of the amyloid fibril as a cross-β protein. *Proceedings of the Society for Experimental Biology and Medicine N.Y.* 1969; **131**: 1373–5.
9. Bourgeois, N, Buyssens N, Goovaerts G. Ultrastructural appearance of amyloid. *Ultrastructural Pathology* 1987; **11**: 67–76.
10. Breathnach S M, Bhogal B, Debeer F C, Black M M, Pepys M B. Primary localized cutaneous amyloidosis—dermal amyloid deposits do not bind antibodies to amyloid-A protein, prealbumin or fibronectin. *British Journal of Dermatology* 1982; **107**: 453–60.
11. Brigger D, Muckle T J. Comparison of Sirius red and Congo red as stains for amyloid in animal tissues. *Journal of Histochemistry and Cytochemistry* 1975; **23**: 84–8.
12. Bruckner F E, Burke M, Pereira R S, Eisinger A J, Bending M, Kwan J, Osman A K, Watson B. Synovial amyloid in chronic haemodialysis contains beta-2-microglobulin. *Annals of the Rheumatic Diseases* 1987; **46**: 634–7.
13. Buerger L, Braunstein H. Senile cardiac amyloidosis. *American Journal of Medicine* 1960; **28**: 357–67.
14. Bywaters E G L, Dorling J. Amyloid deposits in articular cartilage. *Annals of the Rheumatic Diseases* 1970; **29**: 294–306.
15. Carson F L, Kingley W B. Nonamyloid green birefringence following Congo red staining. *Archives of Pathology and Laboratory Medicine* 1980; **104**: 333–5.
16. Cary N R, Sethi D, Brown E A, Erhardt C C, Woodrow D F, Gower P E. Dialysis arthropathy: amyloid or iron? *British Medical Journal* 1986; **293**: 1392–4.
17. Cathcart E S, Ignaczak T F. Amyloidosis. In: Kelley W N, Harris E D, Ruddy S, Sledge C B, eds, *Textbook of Rheumatology* (2nd edition). Philadelphia: W B Saunders & Co., 1985: 1469–87.
18. Cathcart E S, Wollheim F A, Cohen A S. Plasma protein constituents of amyloid fibrils. *Journal of Immunology* 1967; **99**: 376–85.
19. Charra B, Calemard E, Uzan M, Terrat J C, Vanel T, Laurent G. Carpal tunnel syndrome, shoulder pain and amyloid deposits in long-term haemodialysis patients. *Kidney International* 1984; **26**: 549.
20. Christensen H E, Sorensen K H. Local amyloid formation of capsula fibrosa in arthrosis coxae. *Acta Pathologica et Microbiologica Scandinavica*, Section A 1972; **80**: 128–31.
21. Cohen A S, Connors L H. The pathogenesis and biochemistry of amyloidosis. *Journal of Pathology* 1987; **151**: 1–10.
22. Cohen A S, Shirohama T, Sipe J D, Skinner M. Amyloid proteins, precursor, mediator and enhancer. *Laboratory Investigation* 1983; **48**: 1–4.
23. Cohen A S, Calkins E. Electron microscopic observations on a fibrous component in amyloid of diverse origins. *Nature* 1959; **183**: 1202–3.
24. Cohen A S, Shirahama T, Skinner M. Electron microscopy of amyloid. In: Harris J R, ed, *Electron Microscopy of Proteins.* London: Academic Press, 1982; vol. 3.
25. Connors L H, Shirahama T, Skinner M, Fenves A, Cohen

A S. *In vitro* formation of amyloid fibrils from intact beta-2 microglobulin. *Biochemical and Biophysical Research Communications* 1985; **131**: 1063–8.

26. Cornwell G G, Westermark P, Natvig J B, Murdoch W. Senile cardiac amyloid—evidence that fibrils contain a protein immunologically related to prealbumin. *Immunology* 1981; **44**: 447–52.

27. Cornwell G G, Westermark P. Senile amyloidosis: a protean manifestation of the aging process. *Journal of Clinical Pathology* 1980; **33**: 1146–52.

28. Cowan D F. Avian amyloidosis. I. General incidence in zoo birds. *Pathologia Veterinaria* 1968a; **5**: 51–8.

29. Cowan D F. Avian amyloidosis. II. Incidence and contributing factors in the family Anatidae. *Pathologia Veterinaria* 1968b; **5**: 59–66.

30. Dische F E, Wernstedt C, Westermark G T, Westermark P, Pepys M B, Rennie J A, Gilbey S G, Watkins P J. Insulin as an amyloid-fibril protein at sites of repeated insulin injections in a diabetic patient. *Diabetologia* 1988; **31**: 158–61.

31. Divry P, Florkin M. Sur les propriétés optiques de l'amyloide. *Comptes Rendus des Sciences et Mémoires de la Societé de Biologie* 1927; **97**: 1808–10.

32. Eanes E D, Glenner G G. X-ray diffraction studies on amyloid filaments. *Journal of Histochemistry and Cytochemistry* 1968; **16**: 673–7.

33. Editorial. Amyloid arthropathy. *Annals of the Rheumatic Diseases* 1985; **44**: 727–8.

34. Egan M S, Goldenberg D L, Cohen A S, Segal D. The association of amyloid deposits and osteoarthritis. *Arthritis and Rheumatism* 1982; **25**: 204–8.

35. Eyanson S, Benson M D. Erosive arthritis in hereditary amyloidosis. *Arthritis and Rheumatism* 1983; **26**: 1145–9.

36. Francis R J. Amyloid. In: Bancroft J D, Stevens A, eds, *Theory and Practice of Histological Techniques* (3rd edition). Edinburgh: Churchill Livingstone, 1990: 155–75.

37. Frederiksen T E, Gotzsche H, Harboe N, Kiaer W, Mellemgaard K. Familial primary amyloidosis with severe amyloid heart disease. *American Journal of Medicine* 1962; **33**: 328–48.

38. French B T. Amyloid arthropathy in myelomatosis—intracytoplasmic synovial deposition. *Histopathology* 1980; **4**: 21–8.

39. Friedreich N, Kekule A. Zür amyloidfrage. *Virchows Archiv für Pathologische Anatomie und Physiologie und für Klinische Medizin* 1859; **16**: 50–65.

40. Fujihara S. Differentiation of amyloid fibril proteins in tissue sections—two simple and reliable histological methods applied to fifty-one cases of systemic amyloidosis. *Acta Pathologica Japonica* 1982; **32**: 771–82.

41. Gallo G R, Feiner H D, Chuba J V, Beneck D, Marion P, Cohen D H. Characterization of tissue amyloid by immunofluorescence microscopy. *Clinical Immunology and Immunopathology* 1986; **39**: 479–90.

42. Gardner D L. *Pathology of the Connective Tissue Diseases*. London: Edward Arnold, 1965: 272–90.

43. Gejyo F, Odani S, Yamada T, Honma N, Saito H, Suzuki Y, Nakagawa Y, Kobayashi H, Maruyama Y, Hirasawa Y, Suzuki M, Arakawa M. Beta-2 microglobulin: a new form of amyloid protein associated with chronic haemodialysis. *Kidney* 1986; **30**: 385–90.

44. Gelderman A H, Levine R A, Arndt K A. Dermatomyositis complicated by generalized amyloidosis. *New England Journal of Medicine* 1962; **267**: 858–61.

45. Glenner G G. Amyloid deposits and amyloidosis. *New England Journal of Medicine* 1980; **302**: 1283–92; 1333–43.

46. Glenner G G, Page D L. Amyloid, amyloidosis and amyloidogenesis. In: Richter G W, Epstein M A, eds, *International Review of Experimental Pathology*. New York: Academic Press, 1976; **15**: 1–81.

47. Goffin Y A, Gruys E, Sorenson G D, Wellens F. Amyloid deposits in bioprosthetic cardiac valves after long-term implantation in man: a new localization of amyloidosis. *American Journal of Pathology* 1984; **114**: 431–2.

48. Goffin Y A, Thoua Y, Potvilege P R. Microdeposition of amyloid in the joints. *Annals of the Rheumatic Diseases* 1981; **40**: 27–33.

49. Gorevic P D, Casey T T, Stone W J, Diraimondo C R, Prelli F C, Frangione B. Beta-2 microglobulin is an amyloidogenic protein in man. *Journal of Clinical Investigation* 1985; **76**: 2425–9.

50. Gorevic P D, Franklin E C. Amyloidosis. *Annual Review of Medicine* 1981; **32**: 261–71.

51. Hazard J B. The C cells (parafollicular cells) of the thyroid gland and medullary thyroid carcinoma: a review. *American Journal of Pathology* 1977; **88**: 213–50.

52. Heller H, Missmahl H P, Sohar E, Gafni J. Amyloidosis: its differentiation into perireticulin and pericollagen types. *Journal of Pathology and Bacteriology* 1964; **88**: 15–34.

53. Hickling P, Wilkins M, Newman G R, Pritchard M H, Jessop J, Whittaker J, Nuki G. A study of amyloid arthropathy in multiple myeloma. *Quarterly Journal of Medicine* 1981; **50**: 417–34.

54. Higuchi K, Takeda T. Animal models. In: Marrink J, Van Rijswijk M H, eds, *Amyloidosis*. Dordrecht: Martinus Nijhoff Publishers, 1986: 283–91.

55. Hoffman J S, Benditt E P. Changes in high density lipoprotein content following endotoxin administration in the mouse. Formation of serum amyloid protein-rich subfractions. *Journal of Biological Chemistry* 1982a; **257**: 10510–17.

56. Hoffman J S, Benditt E P. Secretion of serum amyloid protein and assembly of serum amyloid protein-rich high density lipoprotein in primary mouse hepatocyte culture. *Journal of Biological Chemistry* 1982b; **257**: 10518–22.

57. Husby G, Sletten K. Chemical and clinical classification of amyloidosis 1985. *Scandinavian Journal of Immunology* 1986; **23**: 253–65.

58. Huston D P, McAdam K P, Balow J E, Bass R, DeLellis R A. Amyloidosis in systemic lupus erythematosus. *American Journal of Medicine* 1981; **70**: 320–3.

59. Ishii T, Hosoda Y, Ikegami N, Shimada H. Senile amyloid deposition. *Journal of Pathology* 1983; **139**: 1–22.

60. Isobe T, Araki S, Uchino F, Kito S, Tsubara E, eds. *Amyloid and Amyloidosis*. New York, London: Plenum Press, 1988.

61. Iwata T, Kamei T, Uchino F, Mimaya H, Yanagaki T, Etoh H. Pathological study on amyloidosis—relationship of amyloid deposits in the aorta to aging. *Acta Pathologica Japonica* 1978; **28**: 193–203.

62. Kaa C A van de, Hol P R, Huber J, Linke R P, Kooiker C J, Gruys E. Diagnosis of the type of amyloid in paraffin wax embedded tissue sections using antisera against human and animal amyloid proteins. *Virchows Archiv A: Pathological Anatomy and Histopathology* 1986; **408**: 649–64.

63. Kisilevsky R, Snow A D, Subrahmanyan L, Boudreau L, Tan R. What factors are necessary for the induction of AA amyloidosis? In: Marrink J, Van Rijswijk M H, ed, *Amyloidosis.* Dordrecht: Martinus Nijhoff Publishers, 1986: 301–10.

64. Kosma V M, Collan Y, Kulju T, Aalto M L, Jantunen E, Karhunen J, Selkainaho K. Reproducibility of morphometric measurements of amyloid after various staining methods. *Analytical and Quantitative Cytology and Histology* 1985; **7**: 267–70.

65. Kuczynski M H. Edwin Goldmanns Untersuchungen über cellulare vergänge in gefolge des verdaungsprozesses auf Grund nachgelässener Präparate dargestellt und durch neue versuche ergäntz. *Virchows Archiv für Pathologische Anatomie und Physiologie und für Klinische Medizin* 1922; **239**: 185–302.

66. Ladefoged C. Amyloid deposits in human hip joints. A macroscopic, light and polarization microscopic and electron microscopic study of congophilic substance with green dichroism in hip joints. *Acta Pathologica Microbiologica et Immunologica Scandinavica*, Section A 1982; **90**: 5–10.

67. Ladefoged C. Amyloid deposits in the knee joint at autopsy. *Annals of the Rheumatic Diseases* 1986; **45**: 668–72.

68. Ladefoged C, Christensen H E. Congophilic substance with green dichroism in hip joint autopsy material. *Acta Pathologica et Microbiologica Scandinavica* 1980; **88**: 55–8.

69. Ladefoged C, Fedders O, Petersen O F. Amyloid in intervertebral discs: a histopathological investigation of surgical material from 100 consecutive operations on herniated discs. *Annals of the Rheumatic Diseases* 1986; **45**: 239–43.

70. Ladefoged C, Rohr N. Amyloid deposits in aortic and mitral valves. A clinicopathological investigation of material from 100 consecutive heart valve operations. *Virchows Archivs A* (Pathological Anatomy and Histopathology) 1984; **404**: 301–12.

71. Ladefoged C H R, Christensen H E, Sorensen K H. Amyloid in osteoarthritic hip joints. *Acta Orthopaedica Scandinavica* 1982; **53**: 587–90.

72. Lai K N, Chan K W, Siu D L S, Wong C C, Yeung D. Pathologic hip fractures secondary to amyloidoma. Case report and review of the literature. *American Journal of Medicine* 1984; **77**: 937–43.

73. Lakhampal S, Li C Y, Gerty M A, Kyle R A, Hunder G G. Synovial fluid analysis for diagnosis of amyloid arthropathy. *Arthritis and Rheumatism* 1987; **30**: 419–23.

74. Levo Y, Livni N, Laufer A. Diagnosis and classification of amyloidosis by an immunohistological method. *Pathology, Research and Practice* 1982; **175**: 373–9.

75. Linke R P. Monoclonal antibodies against amyloid fibril protein AA. Production, specificity, and use for immunohistochemical localization and classification of AA-type amyloidosis. *Journal of Histochemistry and Cytochemistry* 1984; **32**: 322–32.

76. Linke R P, Kuni H. Ubertragungsversuche der Amyloidose durch parabiose homozgoter Mäuse. *Virchows Archiv für Pathologische Anatomie und Physiologie und für Klinische Medizin* 1969; **346**: 89–102.

77. Linke R P, Nathrath W B J, Wilson P D. Immuno-electron microscopic identification and classification of amyloid in tissue sections by the postembedding protein-A gold method. *Ultrastructural Pathology* 1983; **4**: 1–7.

78. McClure J, Bartley C J, Ackrill P. Carpal tunnel syndrome caused by amyloid containing beta-2-microglobulin: a new amyloid and a complication of long-term haemodialysis. *Annals of the Rheumatic Diseases* 1986; **45**: 1007–11.

79. McKeown F. Heart disease in old age. *Journal of Clinical Pathology* 1963; **16**: 532–7.

80. McKeown F. *Pathology of the Aged.* London: Butterworths, 1965; **56**: 180–2.

81. Miller L D, Brown E C, Arnett F C. Amyloidosis in Reiter's Syndrome. *Journal of Rheumatology* 1979; **6**: 225–31.

82. Mohr W. Zür klinischen Aussage histologischer untersuchungsverfahren bei rheumatischen krankheiten. *Therapiewoche* 1978; **28**: 5848–64.

83. Morel F, Wildi E. General and cellular pathochemistry of senile and presenile alterations of the brain. *Proceedings of the First International Congress of Neuropathology* 1952; **2**: 347.

84. Morita T, Suzuki M, Kamimura A, Hirasawa Y. Amyloidosis of a possible new type in patients receiving long-term haemodialysis. *Archives of Pathology and Laboratory Medicine* 1985; **109**: 1029–32.

85. Mountjoy C Q, Tomlinson B E, Gibson P H. Amyloid and senile plaques and cerebral blood vessels. A semi-quantitative investigation of a possible relationship. *Journal of the Neurological Sciences* 1982; **57**: 89–103.

86. Mowry R W, Scott J E. Observations on the basophilia of amyloids. *Histochemie* 1967; **10**: 8–32.

87. Munoz-Gomez J, Bergada-Barado F, Gomez-Perez R, Llopart-Buisan E, Subias-Sobrevia E, Rotes-Querol J, Sole-Arques M. Amyloid arthropathy in patients undergoing periodic haemodialysis for chronic renal failure: a new complication. *Annals of the Rheumatic Diseases* 1985; **44**: 729–33.

88. Nashel D J, Widerlite L W, Pekin T J. IgD myeloma with amyloid arthropathy. *American Journal of Medicine* 1973; **55**: 426–30.

89. Pantelakis S. Un type particulier d'angiopathie senile du système nerveux central: l'angiopathie congophile: topographie et frequence. *Monatsschrift für Psychiatrie und Neurologie* 1954; **128**: 219–56.

90. Pearse A G E. *Histochemistry: Theoretical and Applied* (volume 2, 4th edition). Edinburgh: Churchill Livingstone, 1985: 576–88.

91. Pepys M B, Baltz M L. Amyloidosis. In Scott J T, ed, *Copeman's Textbook of the Rheumatic Diseases* (6th edition). Edinburgh: Churchill Livingstone, 1986: 1024–53.

92. Puchtler H, Sweat F. A review of early concepts of amyloid in context with contemporary chemical literature from 1839 to 1859. *Journal of Histochemistry and Cytochemistry* 1966; **14**: 123–34.

93. Puchtler H, Sweat F, Levine M. On the binding of Congo red by amyloid. *Journal of Histochemistry and Cytochemistry* 1962; **10**: 355–64.

94. Qureshi M S, Sandle G I, Kelly J K, Fox H. Amyloidosis complicating psoriatic arthritis. *British Medical Journal* 1977; **2**: 302.

95. Roberts W C, Waller B F. Cardiac amyloidosis and throm-

boembolism causing dysfunction. Analysis of 54 necropsy patients. *American Journal of Cardiology* 1983; **52**: 137–46.

96. Rokitansky C. *Handbuch der Pathologischen Anatomie.* Vienna: Braumuller und Seidel, 1842: **3**, 311; 384; 424.

97. Romhanyi G. Differences in ultrastructural organisation of amyloid as revealed by sensitivity or resistance to induced proteolysis. *Virchows Archiv A* (Pathological Anatomy and Histopathology) 1972; **357**: 29–52.

98. Rubinow A, Cohen A S. Skin involvement in generalized amyloidosis. *Annals of Internal Medicine* 1978; **88**: 781–5.

99. Scheinberg M A, Cathcart E S, Eastcott J W, Skinner M, Benson M. Shirahama T, Bennett M. The SJL/J mouse: a new model for spontaneous age-associated amyloidosis. 1. Morphologic and immunochemical aspects. *Laboratory Investigation* 1976; **35**: 47–54.

100. Shibolet S, Merker H J, Sohar E, Gafni J, Heller H. Cellular proliferation during the development of amyloid: electron microscopic observations on the kidneys of Leishmania-infected hamsters. *British Journal of Experimental Pathology* 1967; **48**: 244–9.

101. Shimizu K, Ishii M, Yamamuro T, Takeshita S, Hosokawa M, Takeda T. Amyloid deposition in intervertebral discs of senescence-accelerated mouse. *Arthritis and Rheumatism* 1982; **25**: 710–12.

102. Shimizu K, Kasai R, Yamamuro T, Hosokawa M, Takeshita S, Takeda T. Amyloid deposition in the articular structures of AKR senescent mice. *Arthritis and Rheumatism* 1981; **24**: 1540–3.

103. Shirahama T, Cohen A S. High-resolution electron microscopic analysis of the amyloid fibril. *Journal of Cell Biology* 1967; **33**: 679–708.

104. Shirahama T, Skinner M, Westermark P, Rubinow A, Cohen A S, Brun A, Kemper T L. Senile cerebral amyloid: prealbumin as a common constituent in the neuritic plaque, in the neurofibrillary tangle, and in the microangiopathic lesion. *American Journal of Pathology* 1982; **107**: 41–50.

105. Smith T J, Kyle R A, Lie J T. Clinical significance of histopathologic patterns of cardiac amyloidosis. *Mayo Clinic Proceedings* 1984; **59**: 547–55.

106. Takeda T, Sanada H, Ischii M, Matsushita M, Yamamuro T, Shimitzu K, Hosokawa M. Age-associated amyloid deposition in surgically-removed herniated intervertebral discs. *Arthritis and Rheumatism* 1984; **27**: 1063–6.

107. Tribe C R, Mackenzie J C. Amyloidosis. In: Bacon P A, Hadler N M, eds, *The Kidney and Rheumatic Disease.* London: Butterworths, 1982: 297–322.

108. Tribe C R, Perry V A. Diagnosis of amyloidosis. Association of Clinical Pathologists (London). 1979; Broadsheet 92.

109. Uehlinger E. Die Amyloid-Arthropathie. *Wissenschaftliche Zeitschrift der Friedrich Schiller-Universität,* Jena, *Mathematische-naturwissenschaftliche Reihe* 1970; **20**: 336–42.

110. van Rijswijk M H, van Heusden C W G J. The potassium permanganate method: a reliable method for differentiating amyloid AA from other forms of amyloid in routine laboratory practice. *American Journal of Pathology* 1979; **97**: 43–58.

111. Vassar P S, Culling C F A. Fluorescent stains, with special reference to amyloid and connective tissues. *Archives of Pathology* 1959; **68**: 487–98.

112. Vassar P S, Culling C F A. Fluorescent amyloid staining of casts in myeloma nephrosis. *Archives of Pathology* 1962; **73**: 59–63.

113. Virchow R. Ueber einen Gehirn und Rückenmark des Menschen auf gefundene Substanz mit der chemischen Reaction der Cellulose. *Virchows Archiv für Pathologische Anatomie und Physiologie und für Klinische Medizin* 1854; **6**: 135–8.

114. Wagner T, Mohr W. Age distribution of amyloid in the intervertebral discs. *Annals of the Rheumatic Diseases* 1984; **43**: 663–4.

115. Waldrop F S, Puchtler H, Valentine L S. Fluorescence microscopy of amyloid. *Archives of Pathology* 1973; **95**: 37–41.

116. Wiernik P H. Amyloid joint disease. *Medicine* 1972; **51**: 465–79.

117. Williams G. Amyloidosis in parabiotic mice. *Journal of Pathology and Bacteriology* 1964; **88**: 35–41.

118. Winkelman R K, Peters M S, Venencie P Y. Amyloid elastosis: a new cutaneous and systemic pattern of amyloidosis. *Archives of Dermatology* 1985; **121**: 498–502.

119. Wolman M. Amyloid, its nature and molecular structure. Comparison of a new toluidine blue polarized light method with traditional procedures. *Laboratory Investigation* 1971; **25**: 104–10.

120. Wright J R, Calkins E, Humphrey R L. Potassium permanganate reaction in amyloidosis. A histologic method to assist in differentiating forms of this disease. *Laboratory Investigation* 1977; **36**: 274–81.

Chapter 12

Rheumatoid Arthritis: Cell and Tissue Pathology

Rheumatoid arthritis is a worldwide life-shortening systemic disease unique to man. It is dominated by symmetrical aseptic polyarthritis. There is a genetic predisposition (see p. 243) but the cause(s) of rheumatoid arthritis, more common in women than men, remains unknown. Occasionally, children are affected. The mechanisms of tissue injury are complex and not yet fully understood. The immune system is disturbed, with incompletely explained alterations in cell-mediated events, particularly in the synovial joints, and abnormalities of humoral immunity. Articular cartilage is degraded by enzymes released from activated mononuclear phagocytes responding to cytokines. Autoantibodies are formed against immunoglobulin, collagen and proteoglycan, and tissues in many parts of the body are disordered as a result of the deposition of immune complexes.

The pathological changes of rheumatoid arthritis centre upon the 264 synovial joints (see p. 128). The synoviochondral margins are the sites of acute inflammation. Visceral lesions may be encountered in any part of the body ('rheumatoid disease') but subcutaneous granulomata and disordered haematopoiesis are highly characteristic. The most typical pattern is of slowly advancing disease with progressive disability and insidious, systemic illness. A few cases, particularly those that are seropositive, are rapidly progressive or fulminating; deformity quickly ensues. In many other instances, there is a fluctuating course. Rarely, the disease spontaneously remits. There is no specific treatment but deformity can be minimized, suffering reduced and disability lessened by a wide range of drugs, and by an increasing number of prophylactic and remedial surgical procedures. Many of these regimens are associated with individual, iatrogenic complications.

The pathological changes of rheumatoid arthritis merge with, and may be indistinguishable from, those of the other visceral connective tissue diseases, particularly systemic lupus erythematosus, vasculitis, progressive systemic sclerosis and polymyositis/dermatomyositis. There is an analogy with mixed connective tissue disease. Cases of rheumatoid arthritis that display features of another systemic connective tissue disease are said to 'overlap'. There is a pathological basis for this phenomenon. Klemperer, Pollack and Baehr[226] ('diffuse collagen disease') and Pagel and Treip[318] ('viscerocutaneous collagenosis') recognized that the tissue changes of systemic lupus erythematosus, progressive systemic sclerosis, polymyositis/dermatomyositis and of vasculitis could, from time to time, be found in rheumatoid arthritis. There is no clear association with rheumatic fever although, in Third World countries, both diseases remain common and may coexist.

History

Rheumatoid arthritis was not known to antiquity and has not been demonstrated conclusively in human skeletons disinterred by archaeologists. Non-calcified tissues are destroyed by putrefaction and there is no substantial evidence of rheumatoid arthritis in mummified or tanned remains. Circumstantial evidence from portraits painted by Rubens has been used to argue that rheumatoid arthritis existed in the seventeenth century.[12] Nevertheless, there is no record of rheumatoid arthritis in the anatomical collection of John Hunter[20] and the first clear descriptions of the skeletal lesions were those of Adams.[1,2] Later, rheumatoid arthritis emerged as a growing problem in ageing Western populations to take the place of tuberculosis, poliomyelitis and other major epidemic and endemic infections. Extensive accounts of the pathological anatomy of rheumatoid arthritis were recorded.[71,121,128,142,227,228,411] The aetiology and pathogenesis of rheumatoid arthritis have been fully discussed[169a,459a] but hypotheses to explain its natural history continue to be propounded.

Clinical definition

There is no single diagnostic sign or symptom of rheumatoid arthritis: the disease is identified by the coincidence of a variety of characteristic changes[372,423] (Figs 12.1, 12.2, 12.3). To achieve uniformity in diagnostic practice, it became necessary to formulate criteria that could be used internationally. The American Rheumatism Association (ARA) (now the American College of Rheumatology)[355] made proposals for this purpose (Table 12.1). It was recommended that all studies of rheumatoid arthritis should define patients in terms of these characteristics so that valid comparisons between different populations would be possible. The 1958 Revised Criteria proved to have limitations; further revisions were therefore made[14] (Table 12.2). The diagnosis of rheumatoid arthritis can be made with 91 per cent sensitivity if a patient fulfils four of these seven new criteria. Alternatively, a diagnosis with 86 per cent sensitivity can be made by using only two of the seven criteria. These are: 1. recognition of 'arthritis', i.e. soft-tissue swelling in at least 3 of 14 candidate joint groups; and 2. *either* radiographic changes *or* the presence in the serum of a significantly raised anti-immunoglobulin (rheumatoid factor) titre.

Histological criteria are no longer regarded as essential for the clinical diagnosis of rheumatoid arthritis, a situation analogous with that in systemic lupus erythematosus (see Chapter 14) and osteoarthrosis (see p. 568). This change does not demonstrate a decline of interest in understanding the histogenesis of rheumatoid arthritis. Rather, the new criteria reflect the most sensitive and specific tests found to be of merit in clinical practice, not those of greatest value in unravelling the natural history of the disease. However, they create an impression that it may be less necessary to

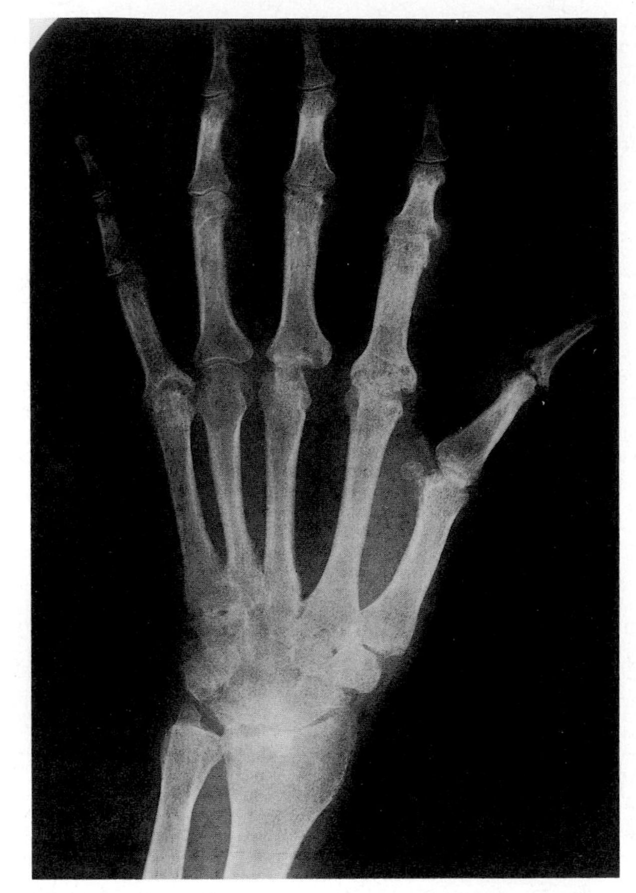

Fig. 12.1 Radiograph of hand in rheumatoid arthritis. Left hand radiograph showing severe rheumatoid arthritis with extensive erosive arthropathy involving the metacarpophalangeal and radioulnar joints. Courtesy of Department of Diagnostic Radiology, University of Manchester.

define the tissue changes in rheumatoid arthritis than the radiological (Fig. 12.4), molecular or immunological. This is misleading since only a knowledge of *all* aspects of rheumatoid arthritis can lead to its full understanding. A factor contributing to the decline of tissue studies is the cumbersome nature of classical diagnostic histopathological procedures (see p. 478). There is a possibility that new techniques such as those of fibreoptic arthroscopy, automated morphometry, cytofluorimetry and confocal microscopy may redress this situation.

Pathological changes

Any of the synovial joints may be the initial site for rheumatoid arthritis but, commonly, the disease first attacks the peripheral joints of the limbs[142,145,192,402] (Fig. 12.5). Syno-

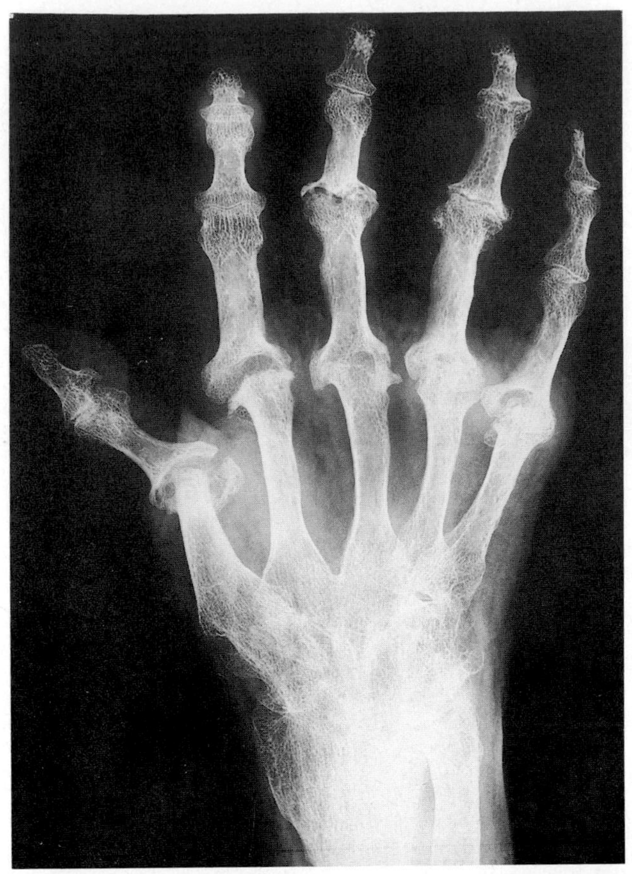

Fig. 12.2 Radiograph of hand in advanced rheumatoid arthritis.
Note the frequent subluxation of metacarpophalangeal joints. Bone adjoining first metacarpophalangeal joint is fractured. Joint space reduction is accompanied by remodelling of periarticular bone with osteophyte formation, part of secondary osteoarthrosis.

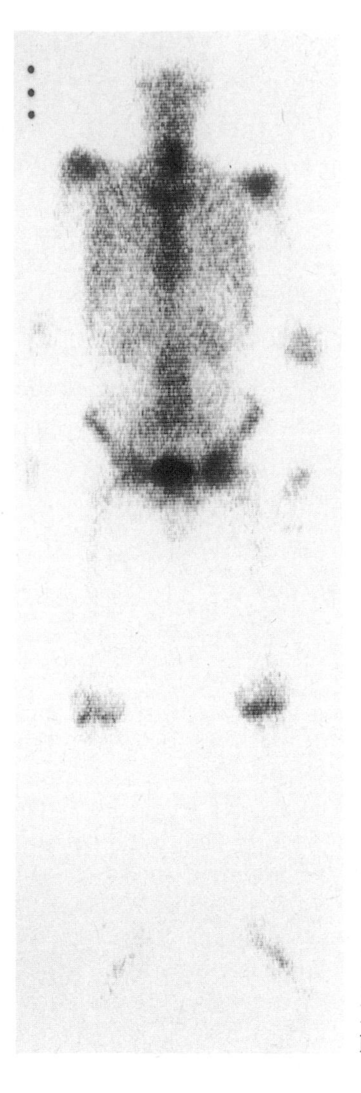

Fig. 12.3 Isotope scan in rheumatoid arthritis.
Note increased uptake in left hip and knee joints.

Table 12.1

1958 criteria for the diagnosis of rheumatoid arthritis[355]

1. Morning stiffness
2. Swelling of a joint
3. Swelling of another joint
4. Pain on movement or tenderness in a joint
5. Symmetrical swelling
6. Rheumatoid nodule
7. Rheumatoid factor
8. Radiographic changes
9. Mucin clot (in synovial fluid)
10. Synovial biopsy
11. Nodule biopsy

'Classical RA' = 7 of 11 criteria
'Definite RA' = 5 of 11 criteria
'Probable RA' = 3 of 11 criteria

Table 12.2

1987 criteria for the diagnosis of rheumatoid arthritis[14]

1. Morning stiffness in or around joints for at least one hour.
2. Soft-tissue joint swelling observed by the physician in at least 3 of 14 joint groups. The groups are: right or left MCP,* PIP, wrist, elbow, knee, ankle, MTP.
3. Soft-tissue swelling in a hand joint (MCP, PIP or wrist).
4. Symmetrical swelling of one joint area in (2) above.
5. Rheumatoid nodule.
6. Presence of a significant titre of serum rheumatoid factor by a method positive in $<$ 5 per cent of the normal population.
7. Radiographic changes of wrists and/or hands: erosions or juxtaarticular osteoporosis.

Rheumatoid arthritis is diagnosed when four of the seven criteria are met.

*MCP = metacarpophalangeal; PIP = proximal interphalangeal; MTP = metatarsophalangeal.

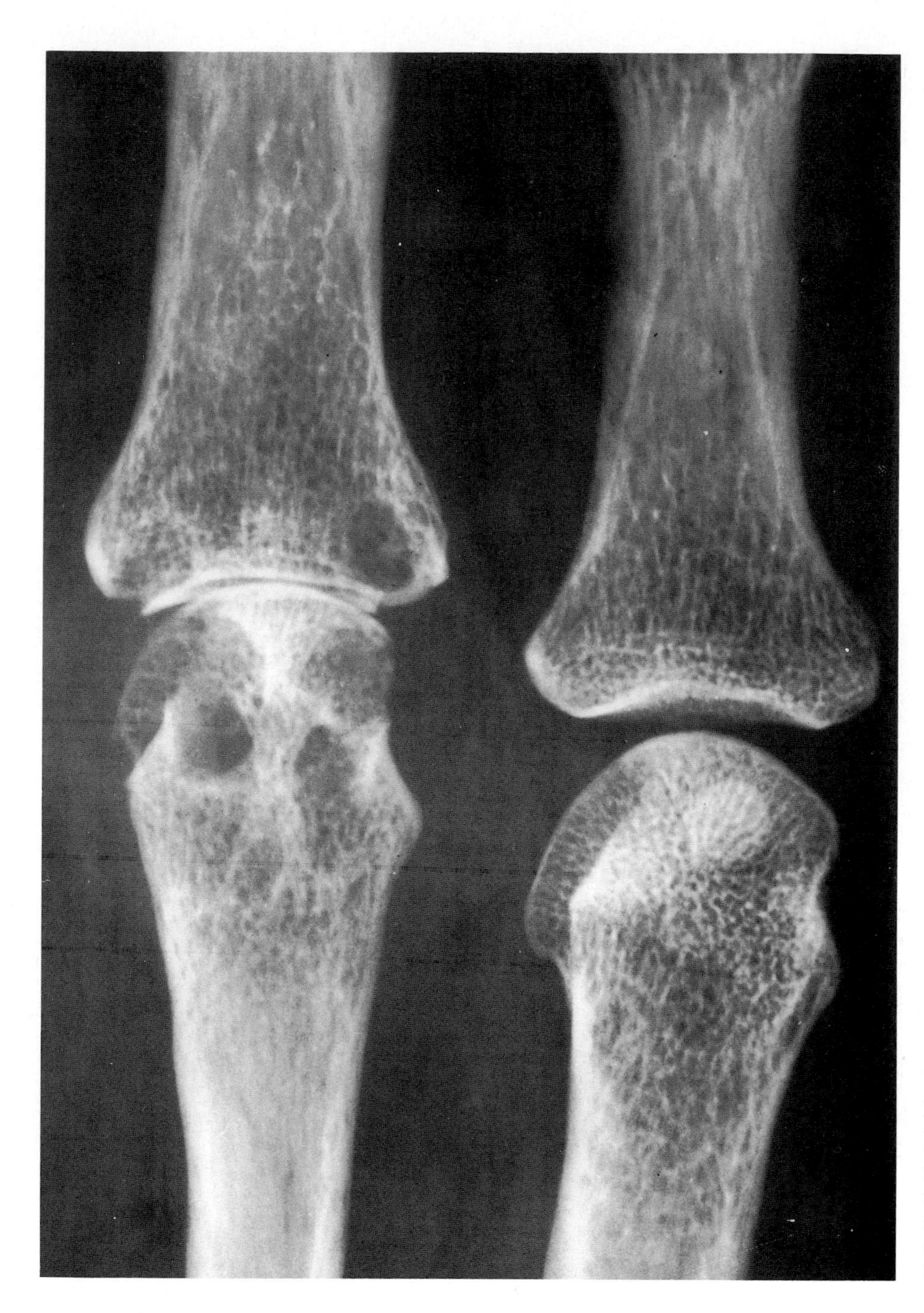

Fig. 12.4 Microfocal radiograph of metacarpophalangeal joints in rheumatoid arthritis. Rheumatoid joint at left, normal control at right. Note loss of joint space and localized osteoporosis. Several pseudocysts are present. Microfocal radiography offers up to 5 × enlargement of details of joint and bone structure with enhanced precision of diagnosis. Courtesy of Dr C Buckland-Wright.

vial tissue is the principal target. The symphyses and synarthroses are not sites of primary inflammatory disease. The metacarpophalangeal and metatarsophalangeal and, to a lesser degree, the proximal interphalangeal articulations are symmetrically affected. Hyaline cartilage, which is avascular, is progressively destroyed. Several or many joints are simultaneously inflamed, and in continuing, decreasing order of frequency, rheumatoid arthritis disorganizes the wrists, ankles, knees, shoulders, elbows and hips. Polyarthritis also involves the synovial joints of the spine and the cervical neurocentral joints. The temporomandibular, cricoarytenoid and cricothyroid joints and those of the middle ear are attacked regularly. The synovia of tendon sheaths, joint recesses and bursae are also inflamed. The articular capsules, tendons, ligaments, menisci, labra and discs are secondarily implicated. Nearby bone is directly reabsorbed and there is often osteoporosis, not limited to parts of the skeleton adjoining foci of arthritis.

a

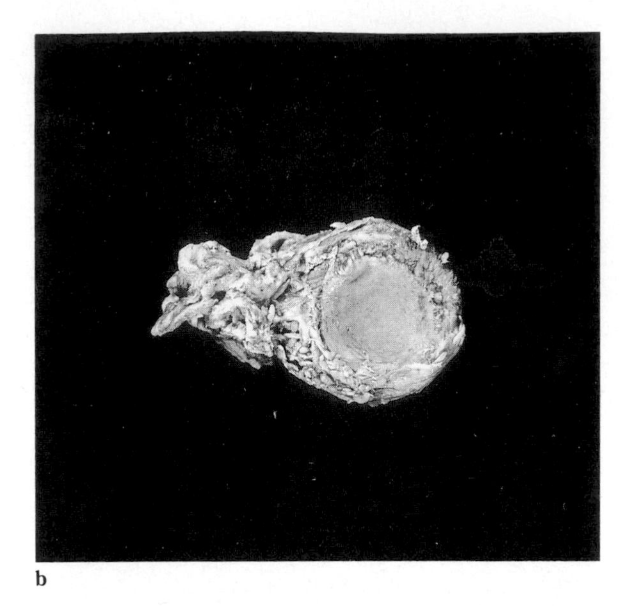

b

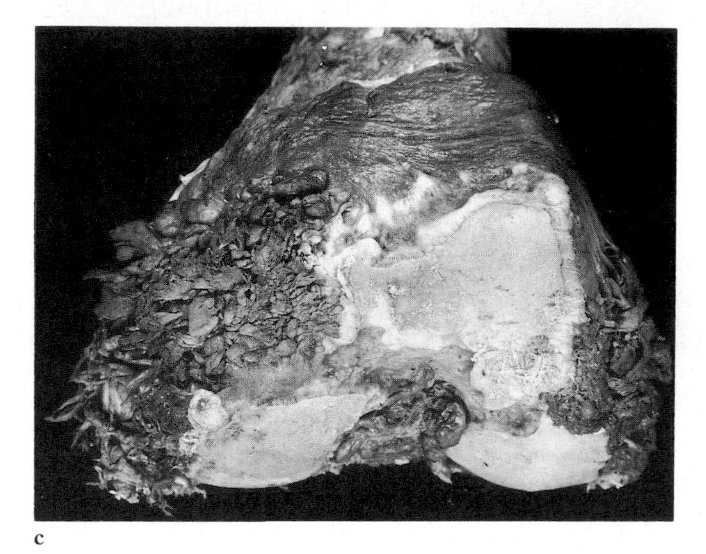

c

Fig. 12.5 Anatomical progression of rheumatoid arthritis in two cases.
a. Anterior surface of lower end of humerus of female aged 60 years with disease of two years' duration. Rough irregular pannus covers the margin of the capitulum and the entire articular surface of the trochlea; **b.** Head of radius of same case. Rim of marginal pannus has replaced half of the cartilage surface; **c.** Lower end of femur of male case. Note hypertrophic synovial villi, marginal erosion of cartilage of condyles and secondary osteophytosis at edge of denuded patellar surface of femur.

Synovial disease

Macroscopic changes

The pathological anatomy of rheumatoid arthritis of the hip, knee and finger has been fully described.[71,121,142,372,401] The extent of the disease can be measured accurately by techniques such as microfocal radiography[49] (Fig. 12.4) and magnetic resonance imaging (see p. 162). In early arthritis, the synovia are swollen and red. Synovial processes protrude from the opened joint. An excess of sterile, thin, opalescent or yellow-orange synovial fluid escapes. The synoviochondral junction is delineated by a congested, vascular tissue, the pannus (L. cloth), that extends onto and begins to replace the cartilage periphery (see Figs 12.5 and 6). Synovial villi adhere loosely to the cartilage surface; the adhesions become coextensive with the vascular tissue at the articular margins. However, the appearances depend upon the techniques of examination and the joint anatomy. Thus, ambulatory arthroscopy of mobile patients gives a distinctive perspective in joints with discs or menisci[69a] and cartilage destruction is not seen unless there is meniscal degeneration.

The opposed, roughened synovia and cartilage are loosely joined by strands of fibrin. Opaque islands of fibrin sit upon the synovial surfaces and 'rice' bodies (see p. 451) lie within the synovial fluid. Organization begins. Much of the joint space is eventually obliterated by fibrous ankylosis but the persistence of a large exudate is a factor contributing to the delayed formation of adhesions. There is concurrent inflammation of the synovia of the tendon sheaths and bursae (see p. 471) and changes in muscle strength and contractility.

With time, the rough, granular and dull-red surface of the synovial membrane becomes less swollen, and the villous processes less prominent. However, remission and exacerbation are usual so that the appearances in a single joint change. The affected articulations in an individual patient display a varied pattern of disease severity. A faint, orange-brown colouration of the synovia recalls the florid synovial processes of villonodular synovitis and the discoloured synovia of haemosiderosis (see Chapter 24). Synovial villous atrophy follows and the inflamed synovia give way to a pale, thinned and fibrous membrane. Unsuspected silent bacterial infection is not uncommon but may be revealed only when a joint is opened surgically or at autopsy.

Remission may occur spontaneously or be encouraged by treatment. When inflammation subsides, however, anatomical signs of indolent tissue destruction persist. In the knee joint, for example, the menisci, subject to degradative processes originating in their marginal synovia, are often found to have been extensively destroyed. In the same way, tendons may have ruptured, ligaments weakened, discs eroded. The chronic active inflammatory process also destroys bone such as that of the cervical vertebrae (see p. 475). There is little evidence that articular cartilage can repair but joints can be effectively reconstituted by surgical intervention (Fig. 12.6). Synovectomy has been used to alleviate disease by the excision of the affected tissue but a 'neosynovium' forms, recurrence is possible, and the value of classical synovectomy is now seriously questioned.[261]

Microscopic changes

(Fig. 12.7)

The disease is long-lasting and the early changes are not necessarily comparable with those recognized in the active, chronic disorder.[184,194] A characteristic is the conspicuous persistence of hyperplastic layers of synovial cells which may proliferate during episodes of exacerbation.[122,123] The influences of drug treatment must always be considered. The effects of rheumatoid arthritis on the synovium of a large joint such as the knee are not uniform and there are substantial differences between the synovial appearances in different parts of the same joint.[195] Whether there is corresponding variation in synovial reaction between the different joints of a single individual is less certain since multiple joints are seldom subject to simultaneous biopsy. However, the symmetrical pattern of the arthritis in a classical case suggests that the extent and severity of the histological changes in, say, the left and right second metacarpophalangeal joints are likely to be closely similar. The significance of individual microscopic changes for the pathological diagnosis of rheumatoid arthritis is considered on p. 478.

The early lesions of rheumatoid arthritis have been explored by needle-, arthroscopic- and open biopsy.[386] A sequence of histological disorders underlies the cardinal signs of inflammation. Redness, pain, heat, swelling and altered function are outward manifestations of capillary and venular dilatation, a slowing of blood flow through the delicate vascular loops of the synoviochondral margin and a relative reduction of axial blood flow. Oedema follows with surface and incorporated deposits of fibrin, foci of necrosis, the presence of increasing numbers of large, poorly differentiated mesenchymal cells and the disorganization of collagen fibre bundles. The margination of polymorphs is seen with the active migration of these cells and, subsequently, of macrophages, through intercellular endothelial junctions. Although the superficial synovium contains type IV collagen, there is no subsynoviocytic basement membrane and polymorphs pass quickly from the blood vessels of the areolar synovial connective tissue into the joint spaces and synovial fluid. They are followed by increasing numbers of lymphocytes and mononuclear macrophages. The identity of these cells is discussed on p. 455 *et seq.*

Thus the initial, acute inflammatory response is succeeded by a mononuclear cell infiltrate, largely of plasma cells and lymphocytes; within two to six months lymphocytic foci may become prominent.[236] Synovial cell hyperplasia is usual and multinucleated giant cells frequent.[348] Occasionally, early microscopic lesions are present which simulate those of rheumatoid subcutaneous granulomata (see p. 487). There is great variation in the severity of early microscopic change from case to case[118,236] and within joint compartments[85,218] but the extent of this variation may be much less when studies are confined to early cases not treated with disease-modifying drugs such as gold salts, penicillamine, antimalarials, corticosteroids or other immunosuppressive agents.[353]

Exudation

Fibrin (Fig. 12.8) The inflammatory cellular exudate is accompanied by an extravascular accumulation of plasma proteins. Fibrinogen escapes from the small blood vessels. Polymerized fibrin aggregates, on synovial surfaces, form eosinophilic laminae, strands and masses within the joint recesses, upon the surfaces of other structures such as ligaments and on the articular cartilage itself. The eosinophilic exudate is still sometimes termed 'fibrinoid' (see p. 294). Much fibrin is soon incorporated into the synovial villi and quickly covered by synovial cells. The quantity of fibrin declines linearly with time. The presence of fibrin in active, chronic synovitis is, of course, not pathognomonic of

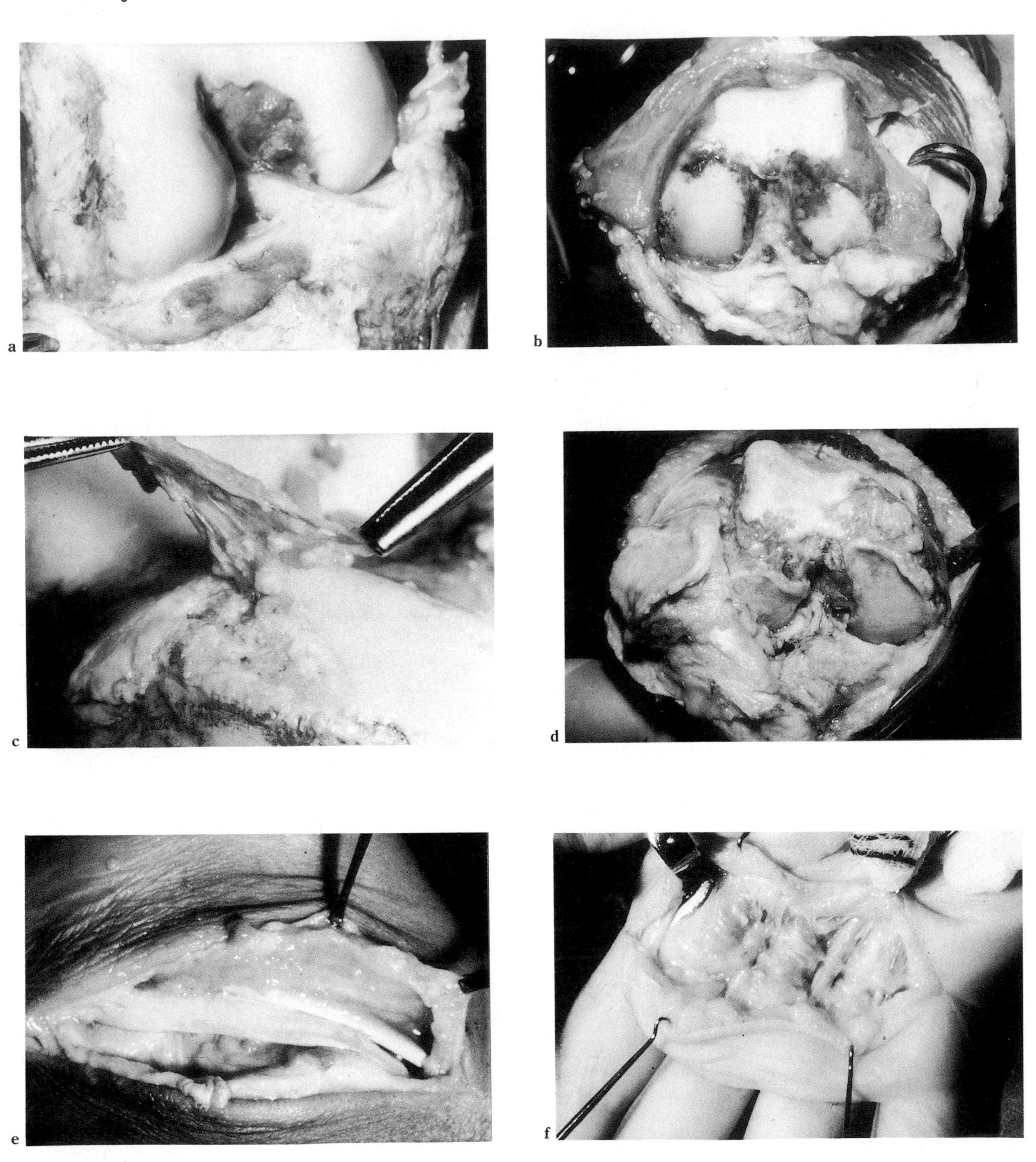

Fig. 12.6 Rheumatoid arthritis—sequence of pathological changes seen at surgery.
a. Hyperaemia of intercondylar synovium and of synovium adjoining tibial condyles; **b.** Extension of pannus around femoral condyles with progressive cartilage destruction; **c.** Exuberant synovial tissue adhering to femoral condylar surface. Note patchy cartilage destruction and marginal osteophytosis; **d.** More extensive loss of femoral condylar cartilage and erosion of patellar surface of femur; **e.** Rheumatoid synovitis of tendinous synovium; **f.** Rheumatoid synovitis of synovium of flexor tendons of palm. Courtesy of Mr W A Souter.

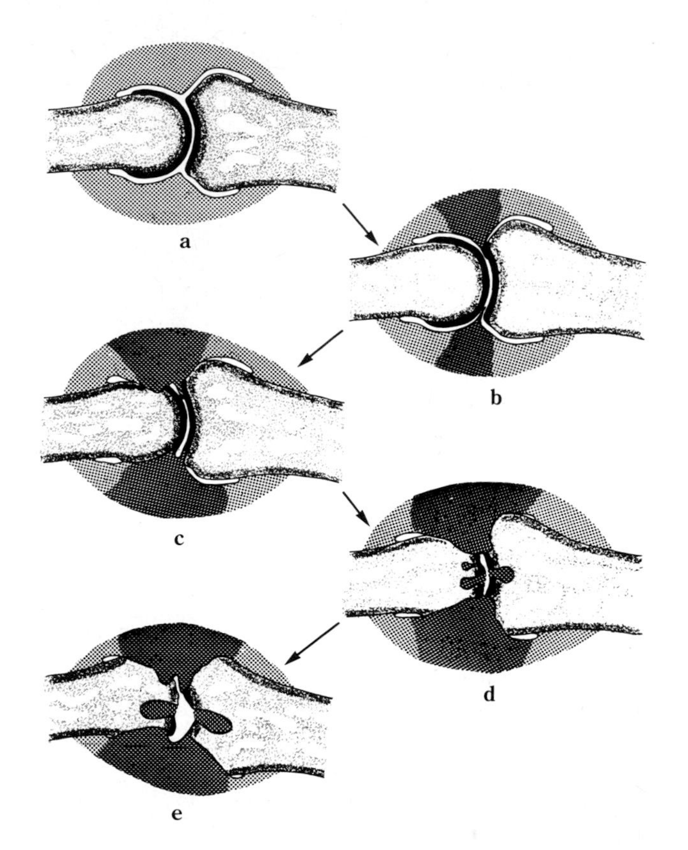

Fig. 12.7 Rheumatoid arthritis: evolution of microscopic changes.
a. Normal interphalangeal synovial joint. Articular cartilage is shown in black, compact bone in heavy stippling, cancellous bone in light stippling, and synovial and periarticular tissue in cross-hatching. The joint space between opposing cartilage surfaces is relatively greater than would be the case in life. **b.** Early rheumatoid synovitis. Inflammation, indicated by dense cross-hatching, has led to synovial swelling and hyperaemia. The synovial processes adhere to the cartilage surfaces at the joint periphery. **c.** The marginal destruction of cartilage, granulation tissue formation and the osteoclastic reabsorption of nearby bone, have caused the loss of tissue that can be detected radiologically as 'erosion'. **d.** Granulation tissue, formed in continuity with the inflamed synovia, has continued to replace cartilage surfaces and has extended through them to form radiolucent zones that are pseudocysts. The extent of the synovitis and of the fascitis, tendinitis and cellulitis is now much greater. Adhesions have formed between the joint surfaces. **e.** Laxity of ligaments and capsule, with muscle atrophy and tendinitis, have encouraged subluxation. Little or no normal articular cartilage remains. Secondary osteoarthrosis has developed and there is osteoporosis. Fibrous ankylosis is likely. Shortening of the finger or toe is observed.

rheumatoid arthritis: fibrinous exudates form in the seronegative arthritides, in systemic lupus erythematosus and in bacterial arthritis but the quantities differ. In chronic bacterial arthritis (see p. 743), as in rheumatoid arthritis, the amount of fibrin may be sufficient to cause fibrin aggregation and the formation of 'rice bodies' (see below).

Occasionally, rheumatoid patients develop an antibody against Factor VIIIc, leading to acquired haemophilia (see p. 967). The specificity resembles that of lupus anticoagulant (see p. 572).[337] There are indeed resemblances between the synovia in rheumatoid arthritis and this tissue in haemophilia but coagulation defects are rarely regarded as a cause of intra-articular bleeding in rheumatoid synovitis.

Rice bodies When rheumatoid joints are aspirated through wide-bore needles after saline lavage, more than two-thirds of the aspirated fluids contain ovoid, flat or thread-like rice bodies.[329] Many rice bodies may be present in a single joint; in chronic, active disease their formation is episodic. The rice bodies range from 2 to 7 mm in diameter. The young rice body has a rich content of polymerized fibrin but transmission electron microscopy shows that only part of the structure is this substance. There is much protein–polysaccharide and some lipid.[7] Rice bodies are common in synovial fluid aspirates from seropositive cases but rare in seronegative arthritis in which, like systemic lupus erythematosus, the quantity of fibrin is less. There is, however, no correlation with the titre of rheumatoid factors. In seropositive rheumatoid arthritis, the presence of rice bodies is not related to the severity of clinical or radiological disease but the aspiration and removal of the bodies effects clinical improvement. With time, rice bodies undergo organization. Fibronectin is present and its distribution is identical to that of the contained fibrin.

Infiltration

Polymorphs The first class of cell to accumulate in the inflamed synovium is the neutrophil polymorph. The implication of polymorphs in rheumatoid arthritis synovitis is established. The number of polymorphs in active rheumatoid synovial fluid is initially high (see p. 463) but quickly declines unless there is persistent, latent infection or the coexistence of crystals. Many of the polymorphs are 'ragocytes', containing immune complexes. Polymorph enzymes are activated and released into the synovial fluid. Whether polymorphs simply reflect the sustained inflammatory response in rheumatoid arthritis or whether they play a significant part in articular injury is disputed. When immune complexes lodge in vessel walls, activating complement, there is little doubt that the lesions of vasculitis are directly attributable to polymorphs. In the case of articular cartilage in rheumatoid arthritis it is reasonable to propose as Mohr and Wessinghage[285] have done, that the polymorphs at the cartilage margin contribute to matrix degradation, but this view is not held by Bromley *et al.*,[43] who prefer to invoke the actions of mast cells.

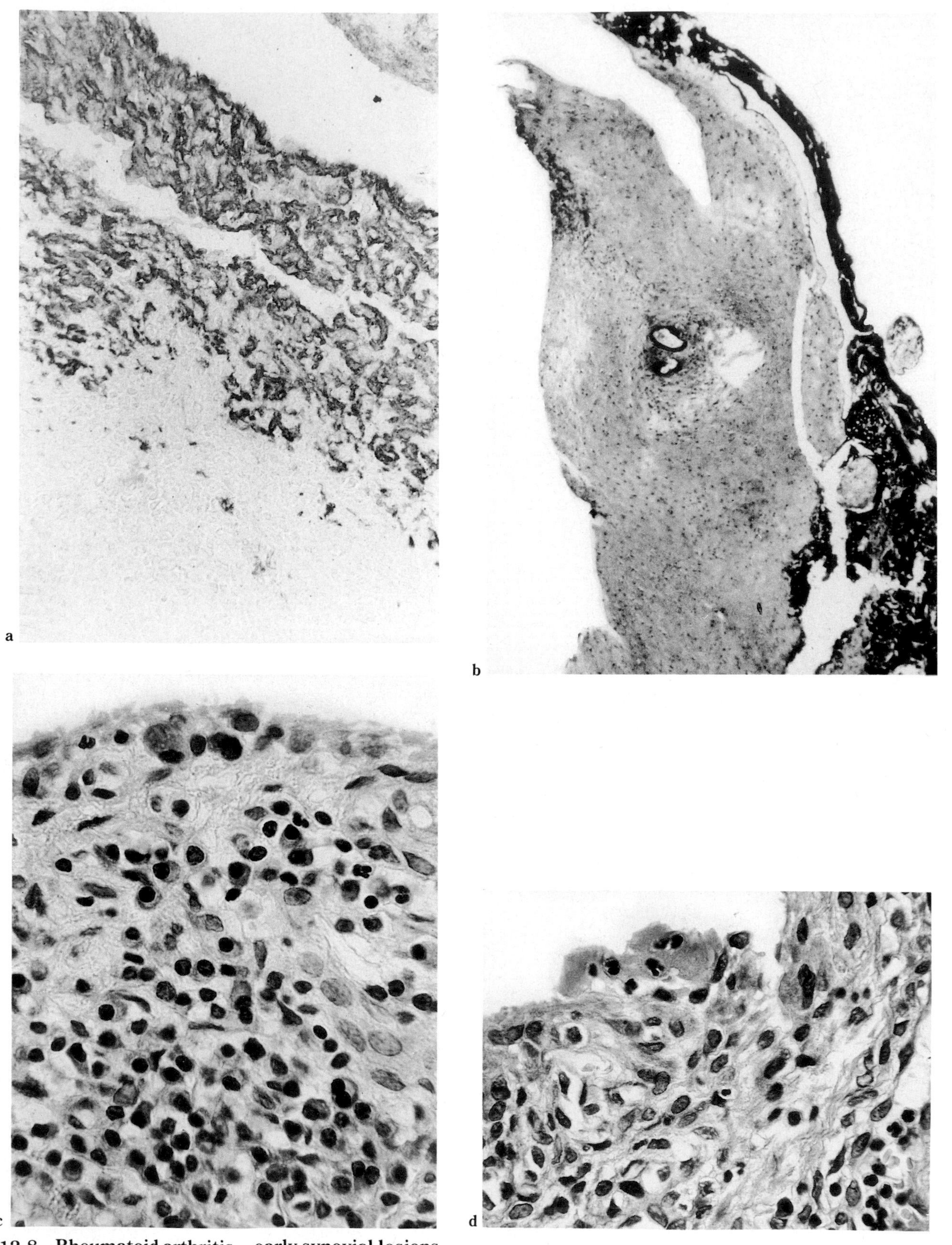

Fig. 12.8 Rheumatoid arthritis—early synovial lesions.
a. Fibrin at synovial surface. (PAP × 80.) **b.** Synovial necrotizing arteriolitis. This rare lesion is seen within a synovial villus that is almost wholly covered by fibrin. (HE × 60.) **c.** Extensive infiltrate of lymphocytes is accompanied by smaller numbers of plasma cells, mononuclear macrophages and occasional polymorphs. (HE × 480.) **d.** Focus of fibrin at synovial surface. Small numbers of polymorphs accompany early synovial cell hyperplasia and mononuclear macrophage infiltrate of subsynoviocytic tissue. (HE × 480.)

Mononuclear cells Characteristically, the subsynoviocytic connective tissues of the inflamed joints and of the tendon sheaths, bursal synovia and periarticular structures in the established disease, are extensively infiltrated by lymphocytes, plasma cells and macrophages (Fig. 12.9).

The high frequency in active, chronic rheumatoid synovitis of lymphocytes, plasma cells and mononuclear phagocytes and the occasional presence of lymphoid follicles, sometimes with germinal centres, suggested that the rheumatoid synovium could be regarded as an antibody-forming, ectopic lymphoid organ;[455] some of its functions appeared to be those of the reticuloendothelial system, others reflected the interplay of immunoregulatory cells with mononuclear macrophages. There are similarities between the lymphocyte populations of rheumatoid lymph nodes (p. 505) and the rheumatoid synovium.[456]

The numbers of individual cell phenotypes present varies with the duration of the disease and forms of therapy. Since the structure of the synovium of joints such as the knee differs markedly from one part to another, considerable caution is necessary in making generalizations regarding the cell populations. Little is known of the extent and degree to which these populations vary between smaller, peripheral joints such as the metacarpophalangeal and larger limb joints, such as the hip, within an individual patient. In addition, there is surprisingly little information regarding the cell populations of normal, human synovial joints. In many published studies, the control data is obtained from osteoarthrosic synovium in which rheumatoid-like changes are now well recognized (see Chapter 22).

The cellular characteristics of the rheumatoid synovium were initially studied in a series of nine early cases by Schumacher and Kitridou.[373] Even in early disease, the polymorph infiltrate (see p. 451) had already been succeeded by a sparse population of lymphocytes and occasional plasma cells. Vascular congestion, red cell extravasation and venulitis were indices of active inflammation but no lymphoid follicles had formed. The leucocyte aggregates were perivascular. In further, early work, increased numbers of lymphocytes were found to migrate from the synovial venules[230] which were of high endothelial type[458] and the synovium could be categorized into distinct zones which were plasma-cell rich, lymphocyte rich, or transitional.[200] On the basis of histochemical enzyme reactions, B-cells appeared to occupy the centre of the occasional lymphoid follicles but plasma cells were scattered more diffusely throughout the synovium and surrounded the follicles.[234] Subsequent studies extended these observations greatly and it came to be realized that the proportion of T lymphocytes was greater, of B lymphocytes less than had been suspected.[319]

Lymphocytes[294,460] The relative ease with which blood and synovial fluid samples can be collected, compared with synovial tissue, has determined that much of the data on the behaviour of rheumatoid lymphocytes has been derived from the former. However, any attempt to decipher the changes in this cell population in the synovial joints, must clearly take account of the cellular traffic that takes place continually between the thymus, blood, lymphoid tissues and synovium. The histopathological data are incomplete.

1. *Peripheral blood* The degree to which the mononuclear cell leucocyte population of the peripheral blood mirrors, is determined by, influences or regulates the corresponding cell population of the rheumatoid synovium, has been debated. Although there is some direct evidence that the immune function of the peripheral blood mononuclear cells reflects events in the synovium, assessed at biopsy,[270] it should not be assumed that there is either an immediate qualitative or quantitative relationship between the cellular activities of these two territories.

Normally, the proportion of T lymphocytes is approximately 80 per cent. The proportion of these cells that are activated in terms of the expression of major histocompatibility complex (MHC) class II antigens is less than 3 per cent. In rheumatoid arthritis, as in systemic lupus erythematosus, in some infections, and after active immunization, the percentage of MHC class II antigen-positive T-cells increases, and in rheumatoid arthritis may comprise as much as 15 per cent of the lymphocyte population. The absolute number of lymphocytes may remain remarkably constant. When MHC class II antigens are expressed, the molecules are actively synthesized by the blood lymphocyte; they are not simply adsorbed. The origin of the circulating activated lymphocytes is not certain but it is possible that they derive from the synovial tissues where approximately 40 per cent of the T-cells are MHC class II antigen-positive.

2. *Synovial fluid* The lymphocytes of the synovial fluid (see p. 463) fall into three main categories. Many are T-cells. The proportion of activated, E-rosette-forming, MHC class II-positive cells is increased above the levels found in the blood of matched, control material. While in some instances either mainly the T helper (T_H) or T suppressor (T_S) cells may be activated, in some series both populations are MHC class II positive. The proportion of B-cells is inexplicably lower in the synovial fluid than in the blood. This observation is confirmed in eluates prepared from enzymically digested synovial tissue samples from which B-cells and plasma cells may, however, be lost during elution. The low B-cell proportions may reflect either the rapid transformation of these cells to plasma cells or the decreased tendency of synovial B-cells to recirculate.

The proportion of synovial fluid T_S/Leu 2a-cells is in-

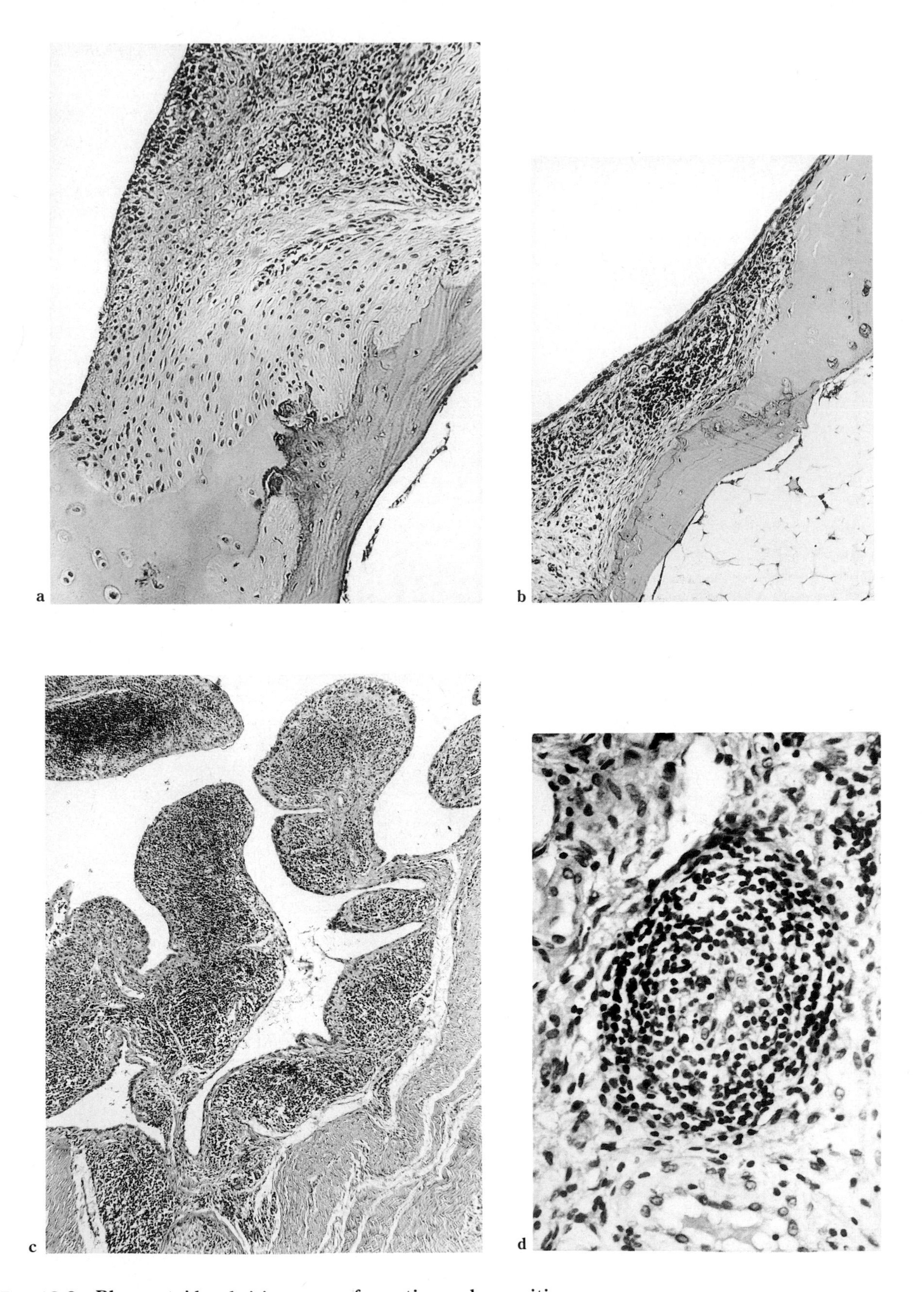

Fig. 12.9 Rheumatoid arthritis: pannus formation and synovitis.
a. Articular cartilage (bottom left) is incompletely replaced by cellular and vascular granulation tissue (HE × 120.) **b.** Granulation tissue (centre) extends across subchondral bony end plate (centre left) to replace articular cartilage (top right). (HE × 96.) **c.** Synovial villous hyperplasia. Observe extensive cellular infiltrate. (HE × 60.) **d.** Small lymphoid follicle with rudimentary germinal centre formation in synovium. (HE × 240.)

creased or comparable to that in the blood, the proportion of T_H/Leu 3a-cells normal or decreased. Thus the synovial fluid T_H:T_S-cell ratio is lower than that of the blood. Interpretation of these observations is not straightforward since the fluids from cases of ankylosing spondylitis, juvenile (non-rheumatoid) polyarthritis and psoriatic arthropathy show similar aberrations. There is reason to suspect that the altered distribution of the synovial fluid T-cell subsets may be influenced more by the duration of synovial disease than by its cause; in early non-infective arthritis, the T_H:T_S ratio may be higher (4.0) than in the blood (1.9) whereas in chronic rheumatoid and non-rheumatoid synovitis the ratio is lower than that of the blood.

There is a third synovial fluid lymphocyte population. These small cells are devoid of surface immunoglobulin and often have no Fc receptors. They are null cells. In rheumatoid synovial fluid, null cells may comprise up to 32 per cent of the lymphocyte population. The null cell population is thought to include natural killer cells and lymphocytes mediating antibody-dependent cytotoxicity.

3. *Synovium* Extreme caution is necessary in interpreting the cell populations of the rheumatoid synovium since features such as perivascular T-cell accumulation and the phenotypic heterogeneity of the extrinsic and intrinsic cells may be found in normal synovia.[252] The majority of the lymphocytes in the subsynoviocytic connective tissue, are T-cells (Fig. 12.10). Seventy to 95 per cent, eluted by DNase, are E-rosette-forming although the proportions vary in different parts of a single biopsy sample. Thus, the majority are activated. Tests made by indirect immunofluorescence and histochemical procedures confirm the T-cell lineage. Natural killer (NK) cells are sparse: they form part of the null-cell population. Nevertheless, natural killer cells may influence the immunological response of the rheumatoid synovium since they are thought to regulate immunoglobulin production and their infrequency in the subsynoviocytic rheumatoid tissue may explain, in part, the abundant production of immunoglobulins by this tissue. Although many plasma cells are present there is only a small proportion of B lymphocytes.

As Zvaifler and Silver[460] emphasize, lymphocyte phenotypic expression and cell function may not equate. However, tests by monoclonal antibodies and other procedures categorize as many as 40 per cent of the synovial tissue (subsynoviocytic) lymphocytes as activated in terms of the expression of MHC class II antigens and other markers. Whereas, in the normal peripheral blood, the T_H/Leu 3a:T_S/Leu 2a ratio is approximately 2:1, in the subsynoviocytic connective tissue in rheumatoid arthritis, this ratio has been shown to vary from 4:1 to 14:1.

The sites at which these varying proportions of T-cells are found in rheumatoid synovial lymphoid tissue are of importance.[238] When germinal centres are recognizable, their distribution recalls the localization of lymphocytes in the process of T lymphocyte-dependent, B-cell activation (Fig. 12.11). The sites at which they occur invites comparison with the distribution of the same cells in rheumatoid lymph nodes (p. 505). It is worth recalling, as Zvaifler and Silver[460] advise, that T_H/Leu 3a, helper/inducer cells are normally prolific in the blood and in tonsillar, paracortical and intestinal lamina propria tissues, as well as in the thymic medulla. They are not present in normal synovial tissues. T_S/Leu 2a, suppressor/cytotoxic cells predominate in the normal bone marrow and gut epithelium. In the lymph node paracortex, where antibody secreting B-cells are located, T_H/Leu 3a cells are juxtaposed to MHC class III HLA-DR-positive, ATPase-positive dendritic cells. In rheumatoid arthritis, in addition to the normal sites, T_H/Leu 3a cells are found concentrated in parts of the synovium with lymphoid follicles, particularly the rare follicles that have germinal centres. T_S/Leu 2a cells are scanty in the follicles of rheumatoid synovia and are located peripherally, a distribution different from their situation in rheumatoid lymph nodes (see p. 505) where they are interspersed among interdigitating and T_H/Leu 3a cells.

Mononuclear phagocytes These cells comprise the principal non-lymphocytic subsynoviocytic population. They play crucial parts in the immunopathological, metabolic and inflammatory processes in the rheumatoid synovium. The majority are MHC class II DR antigen-positive.[196,319] In histological sections and in eluates prepared from enzymically degraded synovial tissue, macrophages form approximately 5–15 per cent of the released cells; they are identifiable both by their content of enzymes such as lysozyme and naphthyl acetate esterase and by the expression of markers such as Mac387 (anti-calgranulin). Macrophages are distributed diffusely but are not identifiable in the germinal centres of the rare lymphoid follicles that have these structures. Macrophages often appear to be in contact with T lymphocytes, particularly with T-suppressor or T-cytotoxic cells.

Dendritic macrophages[213,333] Within the rheumatoid subsynoviocytic tissue, there is a further, minor population of so-called dendritic macrophages. Their surface antigenic structure has been carefully studied.[426] They are of stellate form. They are neither phagocytic nor of a lymphocytic lineage. They resemble the interdigitating cells of the rheumatoid lymph nodes (see p. 507) and their appearances and behaviour recall those of the blood dendritic cells. Dendritic macrophages, while relatively frequent in the rheumatoid synovium, are said not to be identifiable in normal synovial tissue or in osteoarthrosic synovium. The function of the dendritic macrophage, in analogy to the

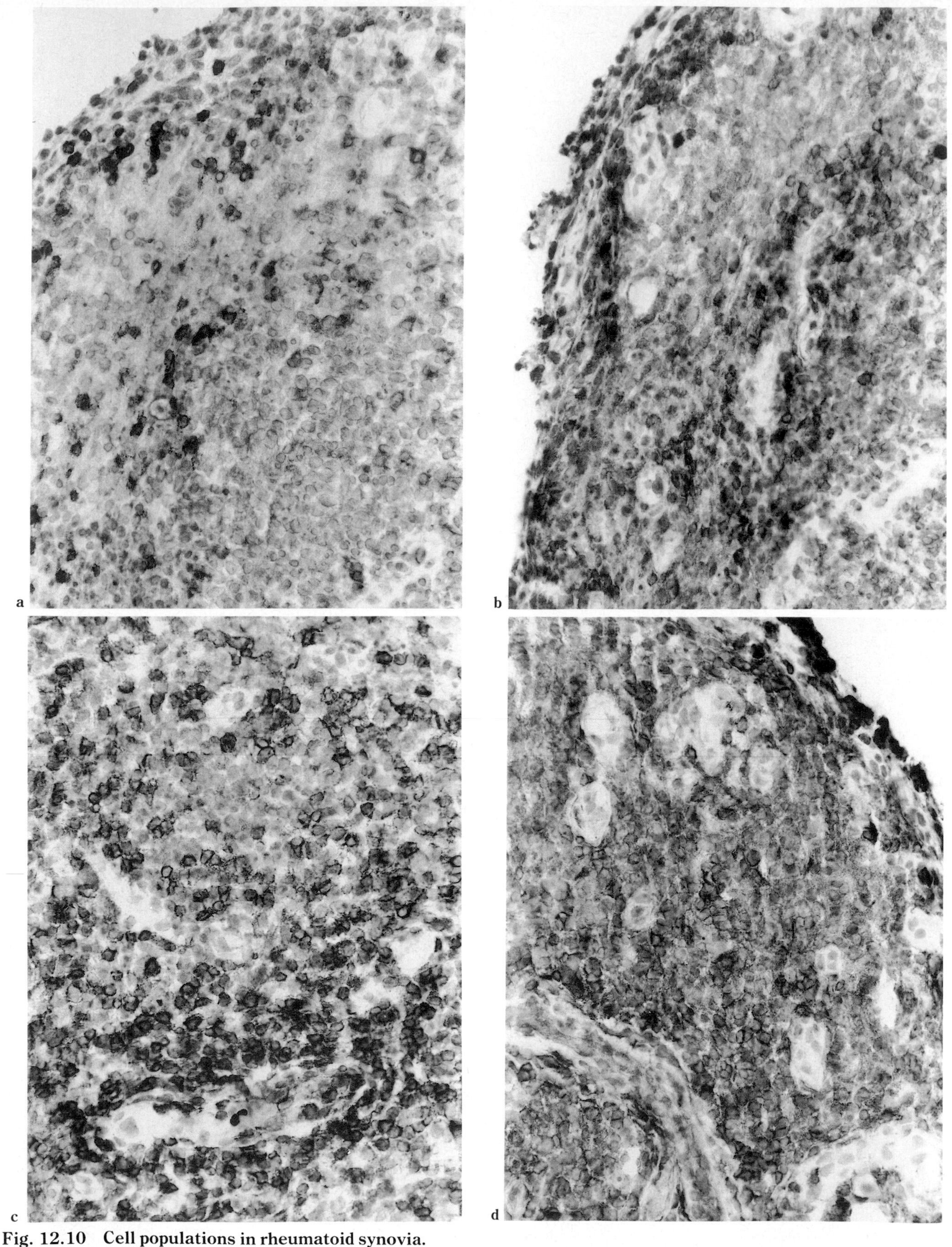

Fig. 12.10 Cell populations in rheumatoid synovia.
a. Small numbers of Leu2a-positive cytotoxic/suppressor T_S-lymphocytes beneath synoviocytic layer and at margin of lymphoid follicle. **b.** Scattered population of helper/inducer T_H-lymphocytes positive for Leu3a, often near small blood vessels. **c.** Monoclonal antibody defined mature T-cells identified by Leu 4. Again, a perivascular situation appears frequent. **d.** Presence of cells expressing HLD-DR antigens is indicated by dark-appearing reaction product. Reactivity is greatest in relation to synoviocytes (top right) but scattered positive mononuclear cells are discerned among the subsynoviocytic population. (PAP × 330.) Photographed from preparations lent by Dr Madeleine Rooney.

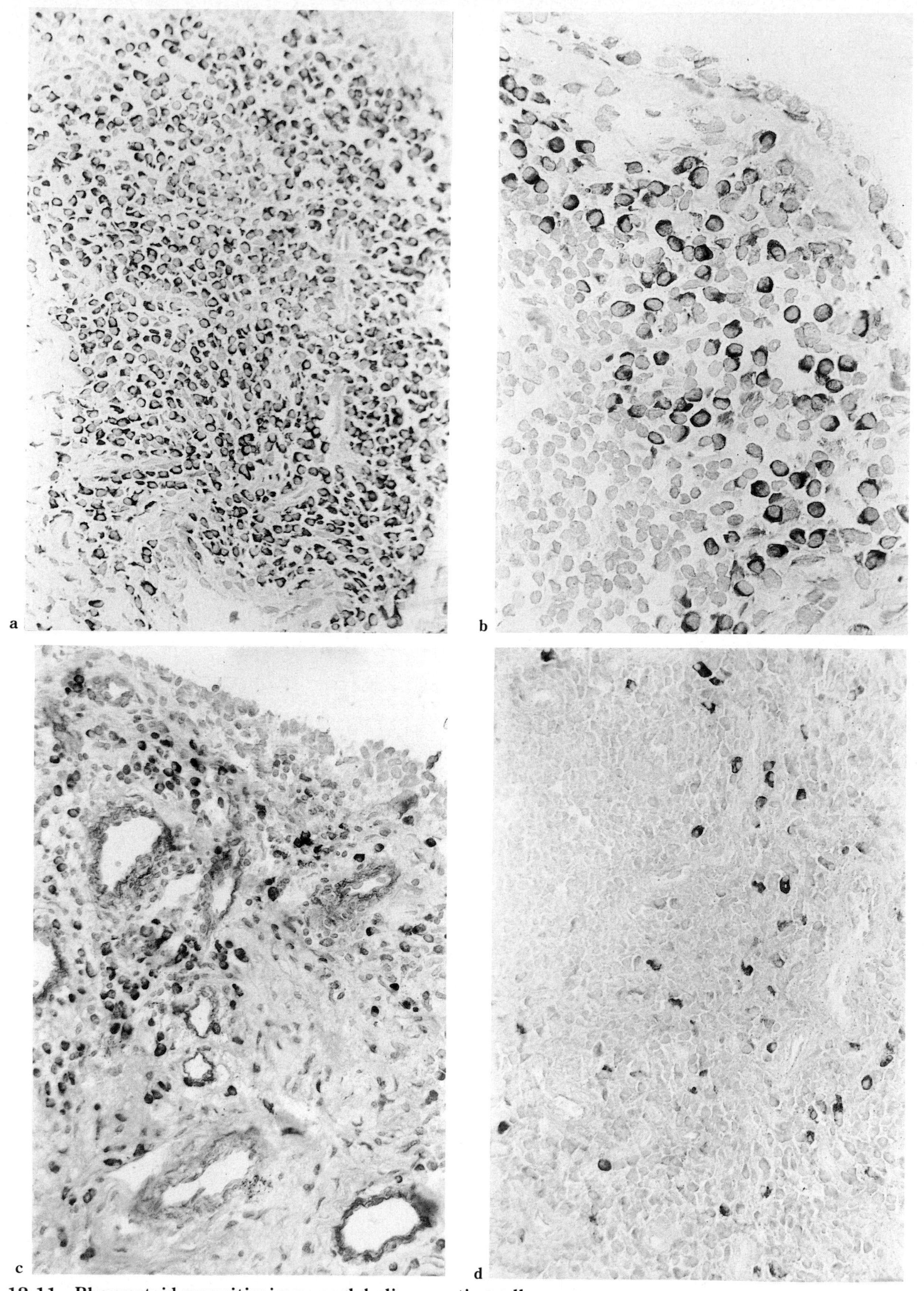

Fig. 12.11 Rheumatoid synovitis: immunoglobulin-secreting cells.
a. IgG is most frequent demonstrable intracellular immunoglobulin. Clusters of IgG-containing plasma cells are often observed. (PAP × 250.) **b.** Beneath synovial cell layer (at top) are large numbers of IgG-positive plasma cells, arranged around the periphery of a lymphoid follicle. (PAP × 480.) **c.** Cells reacting for IgM are less frequent. Here they lie around small subsynoviocytic blood vessels. (PAP × 250.) **d.** Small but significant numbers of IgA-reacting plasma cells are scattered throughout the synovium. (PAP × 250.)

lymph node interdigitating cells, is believed to be the presentation of antigen to T lymphocytes. Many are MHC class II (Ia)-positive, particularly in perivascular areas where they are often in contact with activated MHC class II-positive T-lymphocytes.

Plasma cells The many plasma cells of the rheumatoid synovium secrete immunoglobulin;[136] those of the transitional zone are particularly active and often adjoin macrophages.[199] Overall, plasma cells synthesizing IgG predominate. However, 90 per cent of the cells synthesizing IgM secrete rheumatoid factor compared to 50–60 per cent of those synthesizing IgG.[452] Only 10 per cent of IgA plasma cells are rheumatoid factor-positive. Rheumatoid plasma cells display higher rates of immunoglobulin synthesis than those in non-rheumatoid synovia.[138] When synovia from a variety of articular diseases were compared it was found that only rheumatoid tissue contained more than 10 per cent of plasma cells reacting for μ heavy-chains.[136] When the subclasses of immunoglobulin were considered, 81 per cent of synovial cell IgG plasma cells were IgG_1, 4 per cent IgG_2, 14 per cent IgG_3 and 0.9 per cent IgG_4.[167] In a minority of synovia, the population of IgG_3 cells is selectively increased. The reactivity of synovial rheumatoid factors is related to uncharacterized antigenic determinants in immunoglobulin Fc; the evidence suggests that the major determinant lies in the CH3 domain of the IgG_3 molecule.[350]

Functions of synovial mononuclear cells On the basis of the observed sites of macrophages and lymphocytes in the rheumatoid synovium, Janossy *et al.*[204] suggested that the dendritic macrophages of the synovial lymphoid follicular tissue were engaged in presenting processed autologous or extrinsic antigen to the T_H/Leu 3a cells, provoking T_H-cell proliferation and activation. Subsequently, these T_H-cells would induce B-cells to assume the morphology and functions of plasma cells, with increasingly vigorous immunoglobulin synthesis. The absence or exclusion of T_S/Leu 2a cells from the follicles could be held to imply a failure of a suppressor mechanism that would otherwise modulate or control the immunostimulatory process.[333a]

Other evidence also suggests that selected, impaired T-cell function can contribute to the sequence of inflammatory reactions in rheumatoid arthritis. Under artificial, *in vitro*, conditions, interleukin (IL)-2 production by rheumatoid synovial lymphocytes is decreased and the proliferation of lymphoblasts in response to IL-2 impaired.[74] Rheumatoid lymphocytes produce less interferon-γ after IL-2 treatment than normal cells although IL-2 strongly enhances NK-cell activity.

The observations on which the hypothesis of Janossy *et al.*[204] was constructed have not always been confirmed. Lindblad and Hedfors[251] found the presence of lymphoid foci with dendritic cells expressing class II HLA-DR antigens not to be specific for rheumatoid arthritis or any other disease but merely to reflect sites showing maximal inflammatory activity. Again, Young *et al.*[453] identified lymphoid follicles in only a minority of rheumatoid synovia, an observation with which the majority agree.[118,142,348] The inconstant presence of lymphoid follicles is not attributable to sampling. Within their synovial specimens, Young *et al.*[453] found that the T-cell lymphocytic infiltrate was usually diffuse, lacked germinal centres and included T_H- and T_S-cells in approximately equal numbers. Forre *et al.*[130,131] identified no increase in OKT4-cells, no reduction of OKT8-cells.

From these varied results, Zvaifler and Silver[460] deduce that because (as in Young's observations) the distinctive distribution patterns of immunocompetent cells in rheumatoid synovia are, as Young *et al.*[453] say, unique to individual patients, there may be distinct immunopathological mechanisms in different patients or subsets of patients. Alternatively, the discrepancies in the results observed by Janossy *et al.*[204] and Young *et al.*[453] may be attributable to differences in sampling, the selection of populations or the non-random effects of therapy.

Mast cells[81,83,161] Systematic counts reveal that the number of stainable mast cells is increased within the subsynoviocytic connective tissue in active, chronic rheumatoid arthritis[83] and in juvenile chronic polyarthritis.[17] The role of these cells in active synovitis is discussed on pp. 536 *et seq*. It is likely that they play a part in the destruction of marginal articular cartilage by rheumatoid synovial tissue.[43] Because mast cell numbers are also raised in the synovium in osteoarthrosis,[135] it appears unlikely that mastocytosis is a characteristic confined to rheumatoid arthritis.

Granulomata; iron; tissue debris; amyloid

Granulomata An inconstant feature of rheumatoid synovitis is the presence in the subsynoviocytic connective tissue, of sterile microscopic granulomata resembling lesions often found in the subcutaneous tissues and viscera (see p. 485). The marginal palisaded histiocytes are of synovial A cell origin. The cells of both the microscopic granuloma and the hemigranuloma display the same reactivity for MHC class II HLA-DR antigens, leucocyte common antigen and a macrophage marker, Mac387 (calgranulin), as do those of the subcutaneous granuloma (see p. 485).[16] The central, necrotic zone may form by the segregation, between hypertrophied, inflamed villi, of an inflammatory exudate. However, it is more probable that the lesions are a consequence of histotoxic effects caused by the local arrest of immune complexes.

Iron The synovium contains excess, insoluble iron present both as ferritin and as haemosiderin[36,300] (Fig. 12.12). Haemosiderin lies free within the extracellular matrix, accumulating after minor episodes of intra-articular or intrasynovial bleeding and within mononuclear phagocytes. Ferritin is present both in the synoviocyte cytoplasm and in the lysosomes of type A cells as well as in subsynoviocytic macrophages. Macrophages migrate towards iron deposits and may be surrounded by clusters of lymphocytes (see Fig. 12.12). The large content of synovial iron, easily recognized by special stains, has been measured. Levels as high as in haemochromatosis or in pigmented villonodular synovitis have been detected[379] but the quantity does not correlate with serum or synovial fluid iron levels. Iron is retained, paradoxically, in the face of a characteristic normocytic, hypochromic or normochromic anaemia (see p. 503). The extent of the iron deposits has been related to the degree of anaemia and to the duration and extent of the arthritis.

The presence of iron in synovial tissue is not confined to rheumatoid arthritis. Iron is abundant following trauma; in haemophilia, haemosiderosis and haemochromatosis; in pigmented villonodular synovitis; in psoriatic arthritis; in gout; and in some instances of ankylosing spondylitis. Iron dextran injected experimentally is also taken up by synovial cells when it is segregated in siderosomes.[155a]

Tissue debris and amyloid Rheumatoid synovial tissue occasionally includes fragments of cartilage derived from nearby zones of degradation.

Exceptionally, amyloid is identified in the superficial, subsynoviocytic zone. The accumulation of amyloid in synovial tissues is discussed in Chapter 11, p. 434.

Proliferation

Synoviocyte hypertrophy and hyperplasia The enlarged, congested and prominent synovial villi (Fig. 12.9) are delineated by a synoviocytic layer in which the cells manifest both hypertrophy and hyperplasia. Explanations for the evident increase in synoviocyte size rest on speculation that the cells display increased funtional activity. The mechanisms of synoviocyte hyperplasia are discussed on p. 541.

In morphological terms, synovial cell hyperplasia is a particular property of type A and 'intermediate' cells[118] (Fig. 12.12). The result of the proliferation is the formation of a three-, four- or many-layered synoviocytic 'intima'. However, there are wide differences in response in different parts of a single joint and between different joints. Few counts have been made of the synovial cell population either by morphometric techniques applied to tissue sections or by flow cytometric methods on disaggregated samples. The absence or rarity of mitotic figures has led to debate on the nature of the synovial cell response and the arguments are assembled on p. 541 *et seq.*

It is possible that increased cellularity is influenced by an enhanced capacity of rheumatoid synoviocytes to bind together. Adherent stellate synovial cells in culture form fibronectin and this glycoprotein may play a part in the pathogenesis of rheumatoid arthritis.[378] The stellate cells are probably fibroblastic. Complexes between the actin cytoskeleton of type B synoviocytes and the fibronectin of the extracellular matrix constitute foci which may be the adhesion sites for normal synovial cells.[392] The complexes probably contribute both to pannus formation, and to the synoviocyte responses. They are demonstrable by high resolution replication, immunofluorescence and immunoelectron microscopy.

On the basis of their structure, behaviour, histochemical reactions and surface antigen expression, the hyperplastic synoviocytes in rheumatoid arthritis were at first categorized into three classes.[50,332] The first class was said to be monocyte/phagocyte; they expressed Ia antigens, macrophage antigens and resembled macrophages. The second class also expressed much surface MHC class II antigen but was not phagocytic and had no IgFc receptors. It was devoid of antigens characterizing B-cells, T-cells, monocytes or fibroblasts. The third class was of non-phagocytic cells that resembled fibroblasts and expressed neither MHC class II antigenic determinants nor macrophage surface antigens.

However, these views have been modified. Thus, the majority of both the type A (macrophage-like) and type B (fibroblast-like) cells now appear to express both leucocyte- and macrophage markers[15,16] and MHC class II antigenic determinants[385] but these gene products are expressed differentially. Thus, a macrophage-like population manufactures DR- and DQ-gene products, whereas a fibroblast-like population expresses DR- but not DQ.[51] In cultures, the latter phenotype can be induced by interferon-γ.

The presence of fibronectin on cartilage surfaces in rheumatoid joints has been confirmed;[378,384] it may be formed by polymorphs.[278] Its presence may modulate the interaction between the molecules of the synovium and the articular cartilage[241] and it is therefore of particular interest that the concentration of fibronectin is diminished at the synoviochondral junction itself.[387]

Multinucleated giant cells (see also p. 541) (see Fig. 12.12) Within and beneath the hyperplastic synoviocytic layer, multinucleate, 40–50 μm diameter giant cells are recognized, sometimes in considerable numbers. The giant

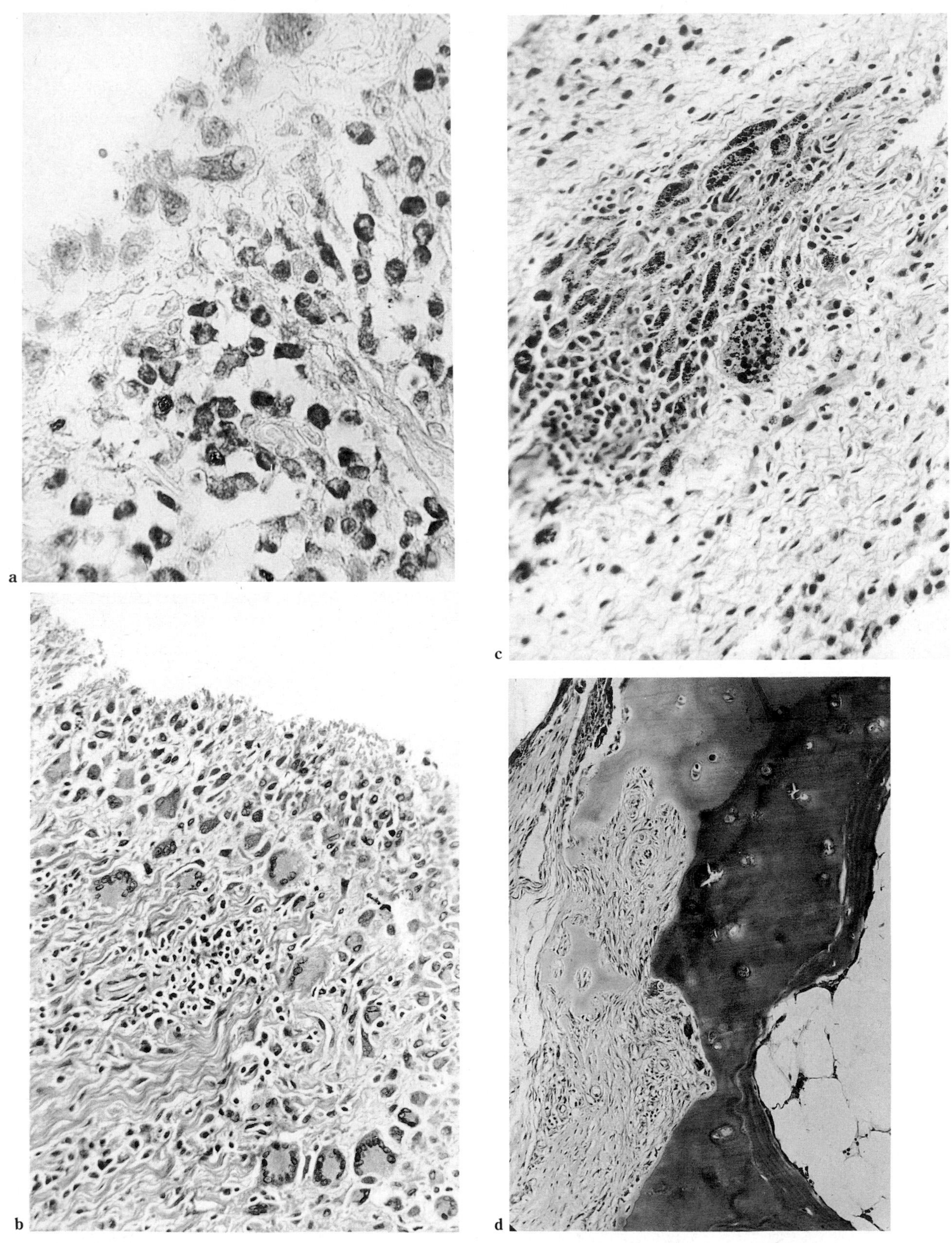

Fig. 12.12 Rheumatoid synovitis: synoviocyte hyperplasia; giant cells; haemosiderin; bone erosion.
a. Hyperplasia of synoviocytes (at top) with numerous plasma cells in subsynoviocytic connective tissue. (Methyl green pyronin × 600.) **b.** Synovial giant-cell formation. Individual stages in the evolution of these multinucleate cells can be recognized. Their presence is characteristic but not pathognomonic of rheumatoid synovitis. (HE × 140.) **c.** Synovial haemosiderin deposits. Mononuclear macrophages are packed with haemosiderin (black granules) and are incompletely surrounded by a cluster of lymphocytes. (HE × 240.) **d.** Marginal osteoclastic bone reabsorption. There is replacement of articular cartilage (at top) and of subchondral bony end-plate (centre): (HE × 96.)

cells are more frequent in rheumatoid arthritis with severe radiological changes[301] but their presence does not correlate with the titre of rheumatoid factor. However, none was present in nine seronegative cases.[166] The morphology of the giant cells is characteristic (see Fig. 12.12). There are 8–12 peripheral nuclei but the cell size and nuclear arrangement are distinct from those of gout and foreign body reactions, from Touton and mulberry cells and from the Langhans cell.

Vascular changes (see also p. 491) Small blood vessels play a crucial part in the initiation of rheumatoid synovitis (see Chapter 7) and in the regulation of cell migration to and from the synoviochondral margins. The basement membranes of postcapillary venules and capillaries of the transitional (see p. 69) subsynoviocytic tissue, and to a much smaller degree, of the lymphocyte-rich areas, are thickened.[274b] One explanation for this thickening, which is likely to influence the migration of leucocytes into the synovium, is the action of mononuclear macrophages and histiocytes in which the rheumatoid synovium abounds.[232]

The composition of the synovial tissue cellular infiltrate is also influenced by the height of the endothelium of the cells of the postcapillary venules.[458] In lymphocyte-rich areas, as in lymph nodes, the presence of tall endothelial cells accompanies lymphocyte migration. Simmling-Annefeld and Fassbender[390] proposed that the 'point of attack' in rheumatoid synovia was the capillary in which platelets accumulate. The capillary endothelium, they thought, underwent both 'progressive and retrogressive transformation', critical but non-specific parts of the exudative process. As in the case of mesenchymal cells, the responses of the endothelial cell and pericyte could be tabulated. There was, for example, extension and enlargement of the Golgi complex. It is difficult to relate these proposals to other studies of the rheumatoid synovium, particularly since it is widely believed that the postcapillary venule, not the capillary, holds the key to the processes of acute inflammation.

Repair and fibrosis Active inflammation commonly persists for a few or many years. Clinical remissions occur spontaneously. They are transient but ultimately the disease appears to subside. However, the histological evidence does not necessarily conform with the clinical. Evidence from necropsy studies of joints previously analysed by biopsy confirms that active disease may persist microscopically even when there is clinical remission. Synoviocyte hyperplasia, a subsynoviocytic lymphocytic infiltrate, increased numbers of mast cells, the accumulation of haemosiderin, the retention within the synovial tissues of fibrin slowly undergoing organization and fibrosis extending to the para-articular capsular and soft connective tissues, are the main findings. Vasculitis is exceptional. In contrast to the viscera, active synovial arteritis is seldom identified (see p. 491).

Cartilage destruction and the extension of pannus, accompany the formation of fibrous adhesions that bind together the opposing articular surfaces. As a result of ligamentous laxity and of the zonal destruction of tendons and the secondary osteoarthrosis and bone reabsorption and remodelling that accompany the sustained inflammation, the skeletal muscles that act across the joint cause first temporary and then permanent deformity. These established changes range from simple ulnar deviation at the carpometacarpal and wrist joints, to swan neck finger deformity (see p. 471), 'main-en-griffe', dislocation at interphalangeal joints, shortening of the fingers and, finally, with maturation of the new fibrous tissue, fibrous ankylosis. There are numerous variations upon this theme within individual joints (see p. 474). Particular pathological problems are posed by the cervical spine, temporomandibular joints, the elbows and ankles.

Ultrastructure

Scanning electron microscopy (Fig. 12.13) The hypertrophic synovial villi are readily demonstrable by scanning electron microscopy.[139,142,151] Their surfaces are, however, often obscured by a muddy deposit of fibrin. Nevertheless, leucocytes can be identified on the synovia. Rice bodies may be imaged. Synovial fluid sediments can be viewed by the same technique and fragments of collagen fibres and unidentified granular particles may be present.[175] The cartilage surfaces are partly detached.[198] At the partially disrupted surfaces of the peripheral cartilage, residual clusters of chondrocytes, with their nearby matrix, protrude into the joint space as the cartilage is degraded.[175] The appearances recall those caused experimentally by enzymic digestion.[311]

Transmission electron microscopy (Figs 12.14, 12.15)[23,34,35,189,308] Transmission electron microscopy has helped to elucidate both the synovial ultrastructure[92,156,157] and its functional disorders. The 20–25 nm periodicity of the fibrin strands resting on the synovia is confirmed, as are the increased height and number of synovial cells. Synovial A cells bear numerous filopodia and include coated vesicles, cell debris and lysosomes. The morphology of these lysosomes is very varied: some contain iron. The rough endoplasmic reticulum of the synovial cells is accompanied by dilatation of the cisternae. Golgi complexes are few and small, the large vacuoles scanty. In B cells, lipid droplets are frequent. Fine filaments may be abundant, replacing organelles and the accumulation of the beaded microfila-

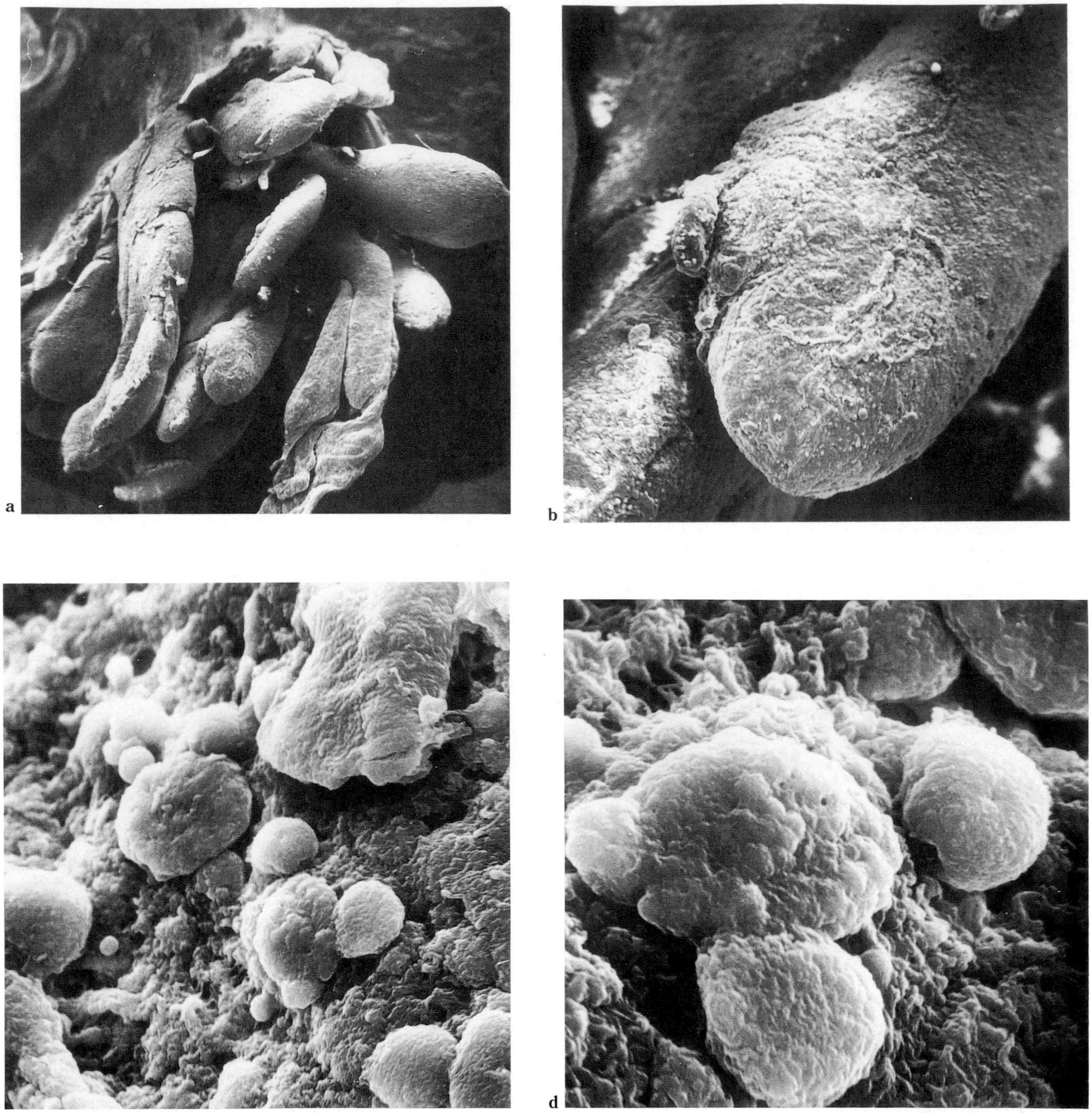

a

b

c

d

Fig. 12.13 Rheumatoid synovia: surface ultrastructure.
a. Hypertrophy of synovial villi with much increased surface area. **b.** Higher-power view of single villus. **c.** Polymorphs and lymphocytes lie upon synovial surface which is partially obscured by adherent fibrin (at right). **d.** Detail of three inflammatory cells resting upon synovial surface. (SEM. **a**: × 20; **b**: × 90; **c**: × 1 900; **d**: × 4425.)

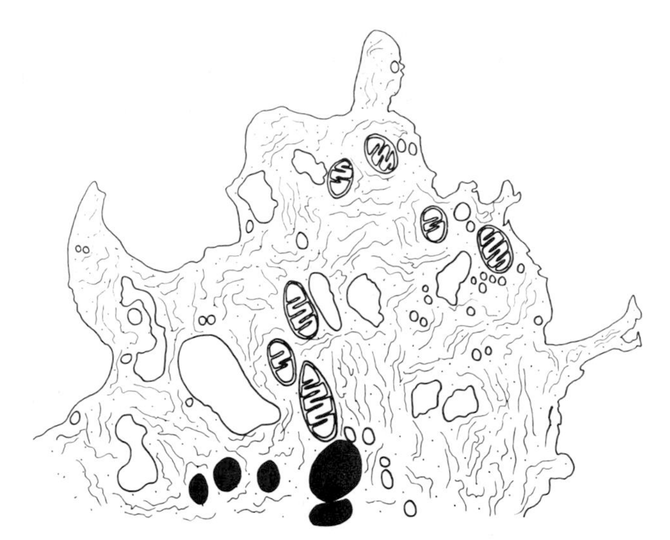

Fig. 12.14 Rheumatoid synovia: ultrastructure. Diagram of A(M) cell in rheumatoid synovium. Dense bodies are phagosomes. Cytoplasmic vacuoles are numerous and villous processes extend into joint cavity. Endoplasmic reticulum is sparse. From an electron micrograph of Dr A J Palfrey.

ments of amyloid has occasionally been described.

The ultrastructural changes of the rheumatoid synovium include an alteration in the ratio of A to B cells, with relatively more B cells having much rough endoplasmic reticulum, many pinocytotic vesicles and numerous intermediate filaments and microfilaments but fewer A cells of increased phagocytic activity. There is also an increase in the proportion of intermediate cells; heightened cellularity; and a change in the structure of intracellular organelles. The pronounced increase in intermediate filaments and microfilaments and the proliferation of pinocytotic vesicles and of rough endoplasmic reticulum is accompanied by the extracellular accumulation, near the synovial cells, of microfibrillar masses but not of intermediate filaments. Hollywell *et al.*[190] confirmed earlier evidence[292] that these phenomena were in no sense specific to or diagnostic of rheumatoid arthritis and the same may be true of aberrations in the fibronectin–actin adhesion sites.[392] The changes attributable to rheumatoid arthritis are shared with several other forms of arthritis including Reiter's syndrome (see Chapter 19), Crohn's disease (see p. 785) and Whipples' disease (see p. 788), each seronegative.

Synovial fluid

The synovial fluid in active chronic rheumatoid arthritis is increased in amount. It displays a series of alterations shared with other arthritides (see Table 12.3).[134,424] The

Table 12.3

Synovial fluid in rheumatoid arthritis[354]

	Normal	Definite rheumatoid arthritis*
Appearance	Clear	Clear-turbid
Clot	0	1.4–0.5
RBCs/l	0.160	2.6–5.9 × 10⁶
WBCs/l	0.063	13.8–16.8 × 10⁶
White cell types %		
Polymorphs	7	63–66
Lymphocytes	25	11–17
Monocytes	63	24–15
Relative viscosity at 38°C	235	11.9–14.5
Sugar mg/ml		
synovial fluid	Almost	78–58
serum	equal	95–91
Total protein, g/l	17.2	47.5–52.4
(exclusive of mucin)		
Albumin	10.2	31.1–26.7
Globulin	0.5	17.9–27.1
Mucin N_2	1.04	0.65–0.68
Mucin glucosamine	0.87	0.48–0.43
Type of mucin precipitate	G	P

G: tight, ropy clump in a clear solution.
P: small friable masses in a cloudy solution.
*In each instance, the first figure is for definite (American Rheumatism Association 1958 criteria) rheumatoid arthritis with a synovial fluid effusion for less than 6 weeks; the second for definite rheumatoid arthritis with a synovial fluid effusion for 52 weeks.

composition and amount of the synovial fluid reflect the activity of the systemic and local disease and the changes that occur in the blood. The analysis of the synovial fluid, by itself, does not permit the diagnosis of rheumatoid arthritis but is of value in monitoring the response to treatment and in recognizing other coincidental disorders such as the crystal arthropathies.[316] Microbiological tests are possible. They identify the bacteria and fungi (see Chapter 18) that complicate rheumatoid arthritis.

Cells

There is an early excess of polymorphs (Table 12.3), particularly in cases complicated by bacterial infection, and an increased number of mononuclear macrophages and of lymphocytes.

Polymorphs In unstained, wet preparations, more than 65 per cent of the polymorphs contain intracytoplasmic inclusions, believed to be immune complexes (see p. 238). These cells are 'ragocytes' (see p. 96). A high proportion

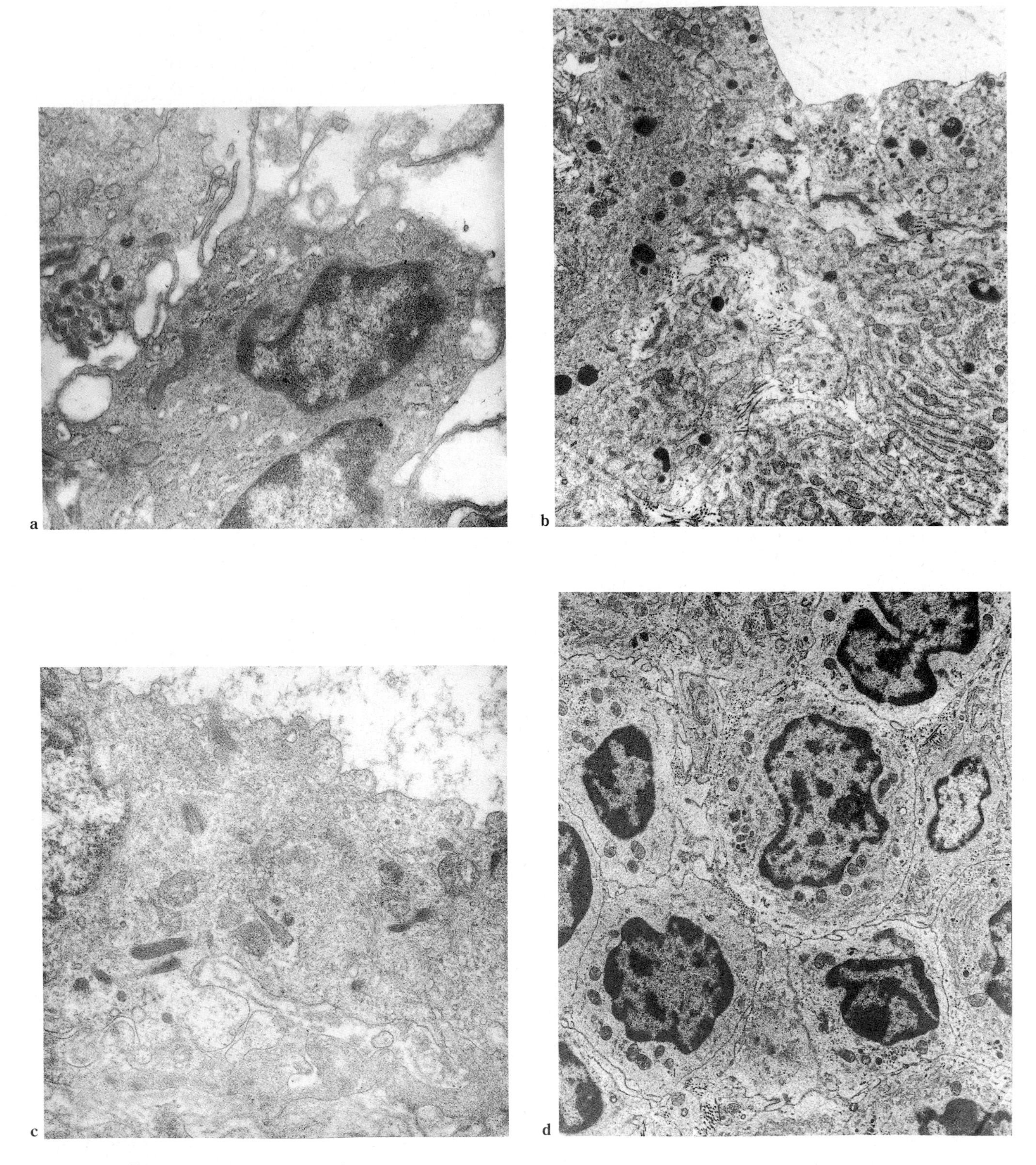

Fig. 12.15 Rheumatoid synovitis: ultrastructure.
a. Type A synoviocyte. Note villous, cytoplasmic processes, macrophage-like nucleus, and occasional, dense phagolysosomes. **b.** Part of adjoining synoviocyte. Cell at right has abundant endoplasmic reticulum characteristic of type B cell. **c.** Indeterminate synovial surface cells. Plasma membranes (at bottom) outline extracellular matrix structures within which fibrillar collagen and proteoglycan can be identified. **d.** Cluster of mononuclear macrophages and large lymphocytes in synovial lymphoid follicle. (TEM **a.** × 4 600; **b.** × 7 200; **c.** × 5 200; **d.** × 2 600.)

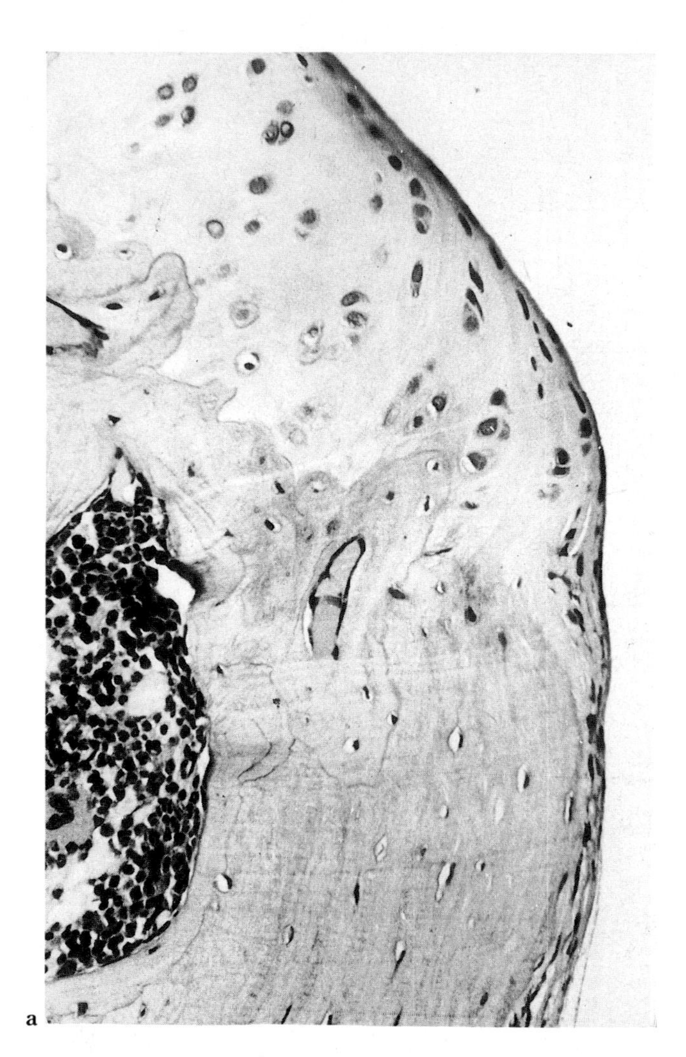

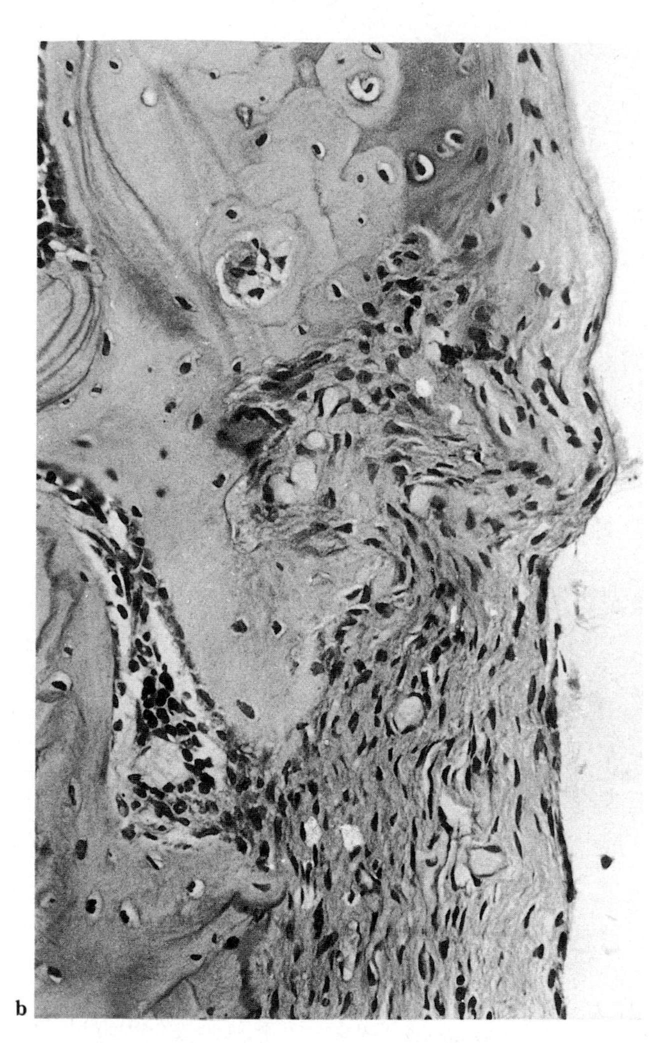

Fig. 12.16 Cartilage destruction in experimental joint disease.
Mechanism of cartilage loss and bone erosion in rheumatoid arthritis is analogous to results obtained from rats subjected to injections of the inflammatory agent turpentine. **a.** Normal control. Note hyaline articular cartilage (top), bone (bottom) and intervening calcified cartilage. **b.** Identical site 13 days after induction of experimental synovitis. Vascular granulation tissue merges with and replaces non-calcified and calcified articular cartilage. Pannus extends across articular cartilage surface (top). (HE × 300.)

of ragocytes, an index of the immune complex status and seropositivity, suggests a poor clinical outcome. IgG, IgM, complement and rheumatoid factors may be detected in these cells but the presence of IgG or complement is not confined to rheumatoid arthritis.[459] The polymorphs may also contain fragments of DNA derived from ingested leucocytes. The appearances may resemble those of the lupus erythematosus (LE) cell (see Fig. 14.2). Judged by the expression of complement C3 receptor and granulocyte function antigens 1 and 2, polymorph activation appears to occur within the joint.[116] Serial studies in a single joint suggest a similarity of cell type response over periods that include clinical relapses, whatever the treatment.[93]

Mononuclear macrophages and lymphocytes (see also p. 38) There are many mononuclear cells but they are less numerous than polymorphs. Chondrocytes are occasionally present but may not be easy to differentiate. Rheumatoid arthritis can be distinguished from non-inflammatory and degenerative joint diseases by the identification of subsets of synovial fluid mononuclear cells.[331] The majority of the many lymphocytes are T-cells. Many are activated. The remaining mononuclear cells include dendritic cells with the phenotype of antigen-presenting interdigitating cells. The proportion of both types of cell does not differ between seropositive and seronegative cases. The ratio of T:B cells, approximately

5:1, closely resembles that of the blood, but there are differences in the T_H:T_S ratios. They are approximately 3:1 for blood and approximately 1.5:1 for synovial fluid. The numbers of T_H-cells are decreased, those of the T_S-cells increased. The preponderance of T_S-cells does not parallel the higher proportions of T_H-cells in the synovial tissue (see p. 455) where T_H-cells may be 'trapped'.

The proportion of activated T_S-cells, indexed by the expression of MHC class II HLA-DR antigens, is approximately 40 per cent. The high numbers of these cells results in a diminished reaction to phytohaemagglutinin or concanavillin A and a low responsiveness of synovial fluid mononuclear cells in the allogeneic and autologous mixed lymphocyte reactions. Direct tests of synovial fluid T-cell function reveal little evidence of true helper or suppressor activity. The capacity of synovial fluid T-cells to induce or inhibit B-cell differentiation is negligible compared to those of the blood which are essentially normal. Moreover, there is low NK cell activity.[73]

Macromolecules

Fibrin There is much fibrin. Fibrinogen escapes from synovial venules during early, acute inflammation. There is also considerable Hagemann factor activity. The extrinsic coagulation cascade is activated. Little plasmin activity is recognized. The removal of fibrin coagula is deficient. In a few cases, acquired haemophilia is accompanied by anti-Factor VIIIC antibody.

Fibronectin (see also pp. 51 and 207) The level[207] of fibronectin is high compared to the plasma concentration.[63] Much fibronectin is found around synoviocytes probably because of the raised plasma levels.[68] The glycoprotein is detected between and around the synovial lining cells, not within their cytoplasm, and there may be a mechanism for fibronectin concentration. Fibronectin is incorporated into synovial fluid cryoproteins. The extent of this phenomenon correlates with the concentrations of complement component 3.[258]

Chondronectin Chondronectin (see also pp. 53 and 209) is present in concentrations correlating with albumin and fibrinogen. It is derived from the plasma[62] and its presence is an indicator of cartilage destruction.

Proteoglycan

Hyaluronate The viscosity of the synovial fluid is low as a result of a reduction in the amount and degree of polymerization of hyaluronate. There are simultaneous alterations in the quantities of lubricin (see p. 211) and other glycoproteins. Impaired mechanical function and diminished efficiency of lubrication are among the consequences. Low hyaluronate levels may impede cartilage repair (see p. 78).

Other proteoglycans The content of proteoglycan is inversely proportional to the degree of radiological evidence of joint disorganization.[367] In cases of corresponding severity some with, some without X-ray evidence of joint destruction, the proteoglycan concentrations are higher in the latter than in the former: proteoglycan levels, measured by ELISA, reflect degrees of cartilage degradation. They are, however, lower in rheumatoid arthritis than in juvenile rheumatoid arthritis.[366] Serial assays enable the progression of the joint disease to be measured.[368]

There are differences in proteoglycan concentrations in acute as opposed to chronic disease. Thus, the levels of keratan sulphate and total sulphated glycosaminoglycan are raised in patients with active, acute rheumatoid arthritis, just as they are in other forms of active arthritis.[342] Proteoglycan concentrations in patients with chronic disease, whether active or not, are not raised.

Immunoglobulins A variety of antibodies is found (see pp. 233 and 245).[136] The most important are listed on pp. 245 *et seq*; they include rheumatoid factors (see p. 233).[33] However, anti-collagens, anti-cytokeratins, anti-α_2-microglobulin and heterophil antibodies have also been detected. Antibodies to denatured collagen can be synthesized preferentially in rheumatoid synovial fluid.[358] The presence of antiviral antibodies draws attention to the possible aetiological role of virus (see p. 532).

Immune complexes of varying size are common. The frequency and quantity of immune complexes is greater than in the serum. IgG and IgM immune complexes predominate. Early immune complexes, unlike those that form later, are not autoreactive. They are not formed of rheumatoid factors and the contribution of rheumatoid factors to pathogenesis is still questioned. However, it is strongly suspected that immune complexes contribute to later tissue injury.

Complement

Serum complement levels are normal or raised in rheumatoid arthritis but synovial fluid levels are low by comparison with those in non-inflammatory arthropathy. These low concentrations reflect local complement consumption, one result of the presence in the synovial fluid of immune complexes. Synovial cell lysis and death may be mediated by complement. The low values of synovial fluid complement are associated with signs of complement activation by both complement pathways and complement degradation is increased. The presence of C3 and C4 breakdown products supports this view but interpretation has proved difficult since complement components can be formed by the synovial tissue locally. That complement activation and consumption reflect antibody-mediated synovial inflammation is shown by evidence that the level of synovial fluid complement degradation products correlates with the severity of synovitis.[99]

Cytokines

Interleukin (IL)-1 (p. 291) is present in synovial exudates; it may be induced by tumour necrosis factor (TNF) and this cytokine is detectable in symptomatic joints.[96a] The synovial fluid in rheumatoid arthritis and osteoarthrosis also displays bone-resorbing activity.[9] IL-1 activity is greater in rheumatoid arthritic than in osteoarthrosic synovial fluid but, in the latter, bone-resorbing activity is also relatively high. Whether this activity is of an osteoclast-stimulating cytokine is unclear; it is not IL-1 itself. IL-2 is present in rheumatoid synovial fluid, activating T-cells in a manner not seen in gout or traumatic arthritis. However, the quantity of IL-2 is deficient. There are low concentrations in synovial tissue. Stimulated rheumatoid synovial mononuclear cells secrete less IL-2 than normal and the responsiveness of rheumatoid lymphocytes to IL-2 is decreased. In response to IL-1, insufficient IL-2 may be formed to enable the promotion of lymphocytic immune reactivity.

Enzymes

Lysosomal hydrolases and proteases, oxidoreductases, phosphatase and collagenase are among the many enzymes present. The majority originate in the synovium (see p. 537). Superoxide dismutase and catalase, enzymes that scavenge oxygen anions and hydrogen peroxide (H_2O_2) (see p. 535), are absent from the cell-free synovial fluid supernate.

Inflammatory mediators

Inflammatory mediators (p. 284) including prostaglandins and leukotrienes, platelet activating factor and histamine are among those present (p. 533). Large amounts of prostaglandin E_2 and of collagenase are produced by adherent stellate rheumatoid synovial cells in culture. There is a pronounced increase in the quantities of PGE_2 and of the collagenase formed when these cells are exposed to a monocyte-derived factor.

Small molecules

Trace elements The concentrations of copper, iron, zinc and aluminium are raised.[306] The elevated synovial fluid copper is accompanied by a high synovial fluid caeruloplasmin level but non-caeruloplasmin-bound copper is not detectable.[447] The elevated zinc content correlates with total synovial fluid leucocyte numbers; the serum zinc level is normal.[41]

Other small molecules The technique of H^1 nuclear magnetic resonance (see p. 99) allows the simultaneous measurement of many other small molecular species. By this means, it has been found that triglyceride and creatinine levels can be indices of inflammatory activity.[444]

Cartilage destruction

Cartilage/synovium interface (Fig. 12.16)

There are three patterns of transition between the cellular pannus and articular cartilage.[231,381] In the first, a mononuclear cell infiltrate forms at the cartilage margin and endothelial cell buds and small vascular channels begin to penetrate the degraded cartilage matrix. In the second, mononuclear cells extend more deeply into the cartilage.[383] They are accompanied by increasing proportions of cells with a fibroblastic morphology. In the third, a vascular granulation tissue is seen to have formed, overlying and sometimes extending beneath the residual cartilage. Cartilage destruction (see p. 533) continues to be mediated by the mononuclear cell infiltrate. Repair begins but the fibrotic reaction is clearly a secondary response.

The mononuclear cells of the cartilage–synovium interface are mainly mononuclear macrophages and lymphocytes. Many transformed T-cells are present and transmission electron microscopy demonstrates the close association between these cells and the macrophages.[200] In cases of a non-erosive, quiescent character, mast cells (see p. 536) have been identified[43] but these cells are almost certainly present throughout the life-cycle of the disease. The proportions of cell types at the cartilage margin have been estimated as approximately: macrophages 40 per cent, fibroblasts 36 per cent, mast cells 12 per cent, and polymorphs 8 per cent.

The mechanisms of cartilage destruction are discussed on p. 533 *et seq.*

Light microscopy

The marginal degradation of cartilage is a much earlier phenomenon than is suspected clinically. Analogous changes occur in experimental disease (Figs 12.16 and 12.17). The first sign of hyaline cartilage injury is diminished metachromatic staining on exposure to polycationic dyes (see p. 544). The diminished stainability is first recognized at the joint periphery, in the matrix adjoining the

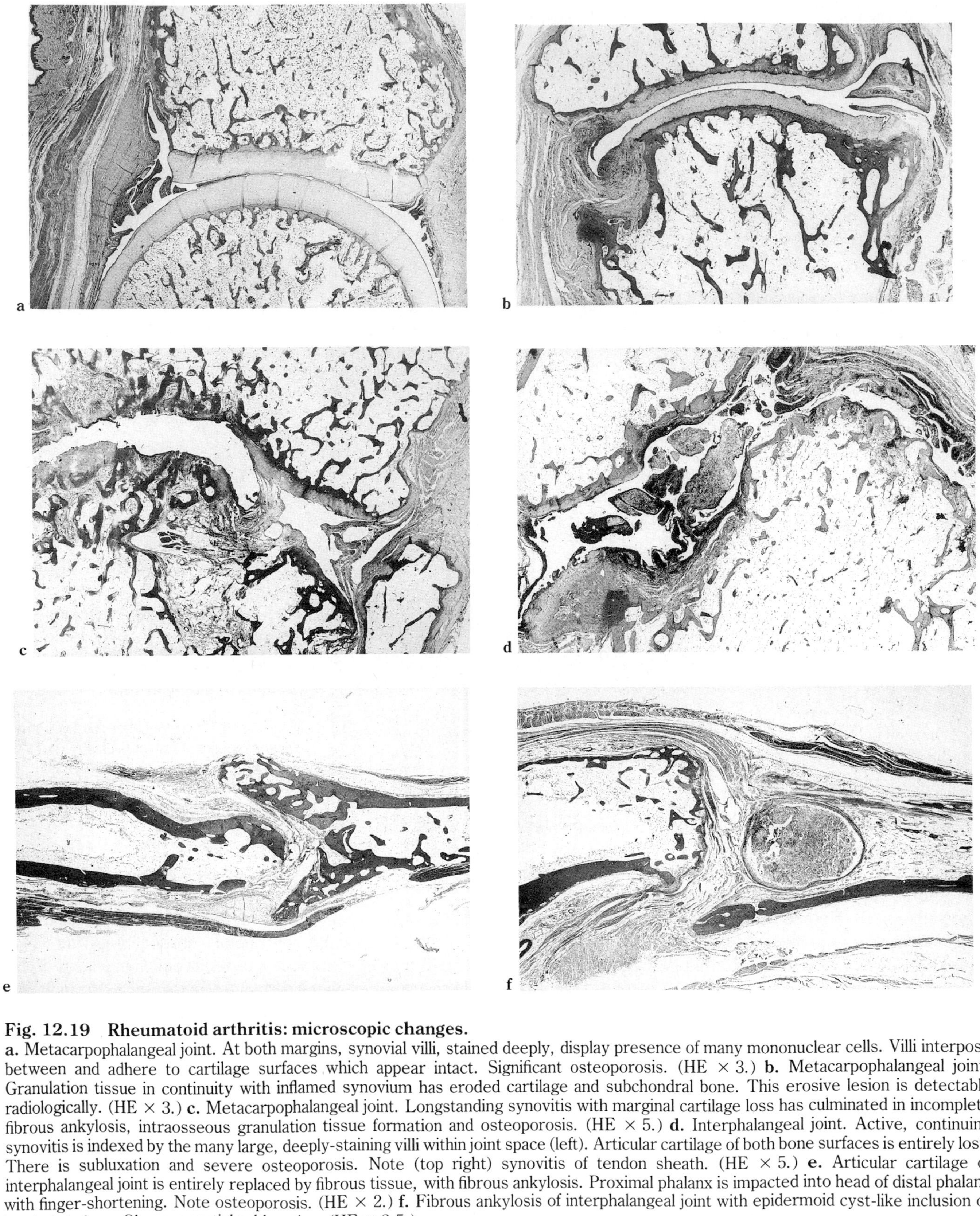

Fig. 12.19 Rheumatoid arthritis: microscopic changes.
a. Metacarpophalangeal joint. At both margins, synovial villi, stained deeply, display presence of many mononuclear cells. Villi interpose between and adhere to cartilage surfaces which appear intact. Significant osteoporosis. (HE × 3.) **b.** Metacarpophalangeal joint. Granulation tissue in continuity with inflamed synovium has eroded cartilage and subchondral bone. This erosive lesion is detectable radiologically. (HE × 3.) **c.** Metacarpophalangeal joint. Longstanding synovitis with marginal cartilage loss has culminated in incomplete fibrous ankylosis, intraosseous granulation tissue formation and osteoporosis. (HE × 5.) **d.** Interphalangeal joint. Active, continuing synovitis is indexed by the many large, deeply-staining villi within joint space (left). Articular cartilage of both bone surfaces is entirely lost. There is subluxation and severe osteoporosis. Note (top right) synovitis of tendon sheath. (HE × 5.) **e.** Articular cartilage of interphalangeal joint is entirely replaced by fibrous tissue, with fibrous ankylosis. Proximal phalanx is impacted into head of distal phalanx with finger-shortening. Note osteoporosis. (HE × 2.) **f.** Fibrous ankylosis of interphalangeal joint with epidermoid cyst-like inclusion of head of phalanx. Observe partial subluxation. (HE × 2.5.)

mal mechanical stresses that facilitate a vicious circle of destruction.

It is exceptional for the pathological examination of a whole rheumatoid joint to be possible before the disease has been present for many weeks or months. In one unusual instance, the precise duration and exact anatomical consequences of the disease could be recorded (see Fig. 12.5). Clinical diagnosis is, however, usually made after the inflammatory and destructive processes have long been established. When arthrotomy or autopsy allow the entire joint to be explored, the characteristic synovitis and pannus, with bone erosion, are usually seen to be complicated by a sequence of associated local tissue changes exaggerated or complemented by osteoporosis, osteoarthrosis, and fracture. They are exacerbated by progressive mechanical dysfunction.

Many of these anatomical sequelae of rheumatoid arthritis are the immediate results of joint disease. Others are determined by rheumatoid lesions of the tendons and ligaments and of dense connective tissues such as the fascia, aponeuroses, discs, menisci or labra.[142] The sequence of events is exemplified in the reports of E M Smith *et al.*[395] and R J Smith and Kaplan.[397] Nearly all aspects of finger joint deformity can be explained by the action of forces transmitted by the flexor tendons of the hand, actions amplified by ulnar dislocation of the extensor tendons. Fibrosis and contracture, once present, accentuate deformity. Soft-tissue changes such as periarticular fibrosis and intrinsic muscle contracture are secondary phenomena.[395] The head of a metacarpal, eroded by granulation tissue and softened by osteoporosis, fails to support the base of the terminal phalanx. Joint stability becomes increasingly dependent on the integrity of contiguous capsular, tendinous and ligamentous connective tissue.[397] However, these tissues are themselves prejudiced. The stretched collateral ligaments, impaired by granulation tissue, degenerate, a change often accompanied by tendon rupture.

Both the extent to which these anatomical changes develop and their significance vary greatly in proportion to the severity of the joint disease, the response to drug and symptomatic treatment, the influence of palliative or remedial surgery and the complexity of the structure of individual joints. Because of advances in surgical techniques, there is a particularly detailed understanding of the anatomical changes in the hip, knee, finger and elbow joints.[3]

Juxta-articular tissues

The passive accumulation of IgG in superficial hyaline cartilage[77a] is an association of rheumatoid synovitis but there is a much greater accumulation of immunoglobulin in related structures such as menisci, capsules and ligaments.

Tendons, ligaments and menisci

The tendon sheath synovia are common sites for active inflammation.[142,215] The frequency of tendon disease is approximately 40 per cent. The majority of tendon lesions are encountered in relation to the long flexor muscles of the fingers but the finger extensor tendons, those crossing the wrist joint, and the infrapatellar tendon[344] are among others in which the pathological lesions of rheumatoid arthritis have been noted. One consequence of rheumatoid tendinitis is 'spontaneous' rupture. However, rupture may also follow the local injection of corticosteroids.[26]

Tendon sheath lesions (Fig. 12.20) are irregularly distributed. The synovia are infiltrated by inflammatory cells resembling those in the synovia of the joint (see p. 451).[197] Occasionally, there is extensive replacement of the tenon and peritenon by granulation tissue. Fibrinoid is present. The appearances, which may include foci of necrosis, may resemble those of the subcutaneous granuloma (see p. 487). A sequence has been proposed in which oedema is followed by the reduction of the compact collagen-rich tendon to a swollen, homogeneous, granular mass; 'spontaneous' rupture;[344] the weakening of support to nearby joints; and an entrapment of tendons and of the median nerve causing a 'carpal tunnel syndrome'.[317] Rupture is due to attrition of the tendon over bony spurs; destruction of the tendon by granulation tissue; or tendon necrosis attributable to compression or endarterial intimal hyperplasia of the vincular blood vessels.[419]

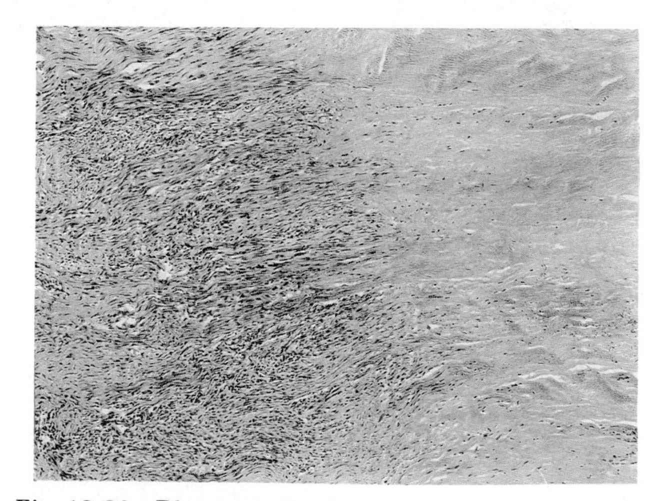

Fig. 12.20 Rheumatoid arthritis: tendon necrosis. Necrotic zone (top right) is separated from intact tendinous tissue by distinct sharp margin which may represent delineation of vascular territory. Note infiltrate of lymphocytes. There is a distant resemblance to rheumatoid granuloma. Tendon rupture is one possible consequence. (HE × 50.)

Menisci
(Fig. 12.21)

The avascular knee joint menisci are circumscribed by vascular synovial tissue which forms a recess at the superior and inferior margins of each meniscus. This tissue is the target for rheumatoid synovitis. As a result, the menisci are destroyed in a centripetal manner. Delicate endothelial cell processes, the harbingers of vascular syncytia, extend into the meniscal fibrocartilage. They are followed by a mononuclear cell infiltrate. Ultimately, the menisci are extensively destroyed and their excision becomes a necessary part of synovectomy with knee-joint arthroplasty.

Bursae and synovial cysts

Bursae The lining synovia are the sites of an inflammatory reaction that does not differ in essence from the synovitis of the joint itself (see p. 448). The bursal wall is formed of dense collagen. There is a chronically inflamed, synovium-like lining and a lymphocytic and plasmacytic infiltrate. The sterile contents include a viscous or thinner fluid and much cell debris. Lymphoid follicles are rarely present.[427] The cellular lining does not display the villous hypertrophy of true rheumatoid synovitis although variable quantities of haemosiderin within the wall indicate that the escape of blood into the affected space has occurred.

Cysts Communication between a bursa and the nearby joint may be reduced to a valve-like opening which may become occluded. A cyst forms and gradually increases in volume as fluid is retained. A very large size may be reached; cysts of the leg have been described with contents of 1–2 litres of fluid. Rheumatoid cysts are most commonly found in relation to the posterior (popliteal) aspect of the knee joint (Baker's cyst).[97,191,221,323,343] However, comparable structures develop at the shoulder,[203,380] elbow,[97,247] hip[65,249] and ankle[376] joints.

Rheumatoid cysts attract attention for cosmetic or mechanical reasons. Occasionally, they become infected. Pressure may be exerted upon nearby veins or nerves.[65] Rheumatoid cysts of the hip have been known to compress the urinary bladder.[80,430] A 'cyst' may be very long. High intracyst pressures are generated when the limb is fixed and rupture may suddenly occur. The signs of rupture of a popliteal (Baker's) cyst resemble those of synovial rupture,[97] and simulate thrombophlebitis.

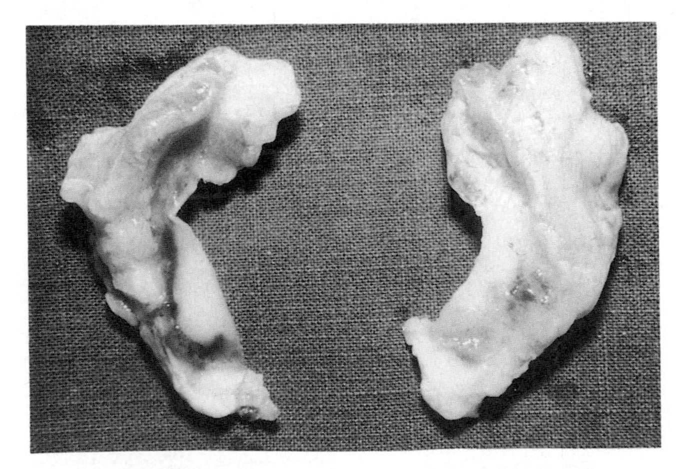

Fig. 12.21 Rheumatoid arthritis: knee joint menisci. Menisci are frequently damaged or destroyed by rheumatoid synovitis and are often removed at arthroplasty. In this Figure note extension of inflamed synovial tissue into and over the surface of the medial meniscus (right). Lateral meniscus (left) is less severely affected.

Bone disease

With the onset of arthritis, bone structure in rheumatoid arthritis quickly becomes abnormal.[361] The disorder is at first localized; vigorous remodelling ensues. Ultimately, however, bone disease becomes generalized. In juvenile rheumatoid arthritis, dwarfism may result. Although clinical and radiological signs of bone changes in rheumatoid arthritis have long been recorded, it has proved difficult to obtain finite evidence of their nature and extent. Contributory reasons for this difficulty have been the need to differentiate clearly rheumatoid arthritic bone anomalies from diseases such as the osteoporosis associated with ageing and the high frequency of iatrogenic medical and surgical changes in bone. Further difficulties have been insensitive techniques for the measurement of bone structure and function and a failure to standardize the site and form of samples taken for analysis. These problems have now been largely overcome.

Occasionally, attention is drawn to bone disorder in rheumatoid arthritis by single or multiple[279] pathological fractures which may be obscured clinically by the pain caused by the arthritis.[443] In other cases, bone rarefaction is observed locally during the radiography of inflamed joints and has been recorded in systematic surveys of the spine and limbs.[365] In these 164 patients, 'osteoporosis' was common in men and women whether treated with corticosteroids or not. Osteoporosis was particularly frequent in women aged over 50 years in whom, however, the 16 per cent frequency of spinal fracture was little greater than the frequency among non-rheumatoid arthritic North American women of comparable age. Conventional radiographic and densitometric techniques have proved insensitive guides to bone mass but photon absorptiometry offers the opportunity of reducing sources of diagnostic error.[299] In an early small series of 26 rheumatoid women matched with controls, there was no preferential loss of trabecular bone and no selective change at metaphyseal as opposed to diaphyseal sites. However, decreased bone rigidity was demonstrable and the decrease was exaggerated by corticosteroid treatment. More recent investigations, measuring total body calcium by neutron activation analysis and cortical bone areas, have demonstrated a reduction of bone mass in rheumatoid patients treated with non-steroidal anti-inflammatory drugs (NSAIDs), and a further reduction in those given low doses of corticosteroids.[345] There is also accelerated bone turnover suggesting an overall increase in bone metabolism as part of the rheumatoid process, not explicable either by immobilization or by associated metabolic bone disease.[338]

The pathological changes in rheumatoid bone are first, those demonstrable by conventional microscopic techniques, and second, those recognized only by labelling and morphometric methods. In general terms, these categories correspond to local, inflammatory or mechanical abnormalities, and to generalized 'metabolic' disease, respectively.

Local bone disease

Local disease may be erosive or porotic.

There are five varieties of erosive bone disorder.[286,287] They are: 1. marginal; 2. compressive erosions; 3. superficial surface erosions; 4. a subchondral, reparative, vascular response with chondroclastic activity and a lymphoplasmacytic infiltrate;[450] and 5. pseudocyst formation. Compressive (pressure) erosions appear to be biomechanical phenomena. Examples include the erosion of posterior rib bone, and *protrusio acetabulae*. Neither pressure erosions nor surface reabsorption are well understood histologically.

Other forms of local bone disorder encountered from time to time in rheumatoid arthritis are bone necrosis, attributable to steroid therapy, osteomyelitis, and abscess.

Marginal erosion

As rheumatoid synovitis advances, bone adjoining the synoviochondral junction is reabsorbed by osteoclasts.[45] Osteoclastic bone reabsorption is occasionally accompanied by chondroclastic reabsorption of zone V, calcified cartilage. These multinucleated cells confine their activities to margins of calcified tissue devoid of osteoid and of non-mineralized cartilage. There is an initial phase in which the non-mineralized lamina is degraded by osteoblastic or macrophage collagenase (see p. 32). Osteoclasts leave distinctive 'footprints', the Howship lacunae; they provide one measure of resorptive activity and can be recognized by scanning electron microscopy.[110] The process of marginal erosion (Fig. 12.19) is under parathormone and calcitonin control; it is susceptible to regulation by phosphonate. There is a variable degree of active inflammation, bone resorption and granulation tissue/repair in different parts of individual joints.[108]

Pseudocysts

Pseudocysts are radiolucent foci often found beneath or near the articular cartilage in joints affected by rheumatoid synovitis.[142,237] The alternative term 'geode' has been used.[207] However, 'geode' is a geological description meaning 'a large cavity in rock, lined with crystals that are free to grow inwards' and its use in the description of connective tissue disease is ambiguous.[144] Pseudocysts are found at any part of the joint surface.

The 'cystic' appearance is misleading since these structures have a solid, not a fluid content. Cruickshank, McLeod and Shearer[90] gave a detailed histological description. The pseudocyst is a zone of loose, vascular connective tissue extending in continuity with the granulation tissue (pannus) that covers or replaces the articular surface (see Fig. 12.9). Pseudocysts usually range in size from 1 mm to more than 5 mm in diameter but may be as large as 50 mm.[162,267] They are of irregular shape and may be multiple. Very rarely, pseudocyst formation without joint disease has been claimed.[291] The chondro-osseous junction and the subchondral bony end plates are insidiously destroyed. Bone trabeculae are reabsorbed by osteoclasts. The vascular connective tissue forms an intraosseous island in which lymphocytes and plasma cells accompany numerous fibroblasts and buds of capillary endothelium. Islands of residual cartilage and bone lie within the loose granulation tissue of the cyst. The quantity and nature of this matrix varies; it may be richly collagenous or formed of a loose, myxoid material replete with proteoglycan. However, new bone formation begins and, with time, the pseudocyst may be delineated by a radio-opaque border of circumferentially orientated, sclerotic trabecular bone.

Lesions similar to rheumatoid pseudocysts have been described in haemophiliac arthropathy (see p. 967), and Charcot's neurogenic arthropathy (see p. 906). Gouty tophi are distinguished by the presence of monosodium biurate (see p. 390) and the presence of many macrophages, some multinucleated. The cystic lesions of osteoarthrosis (see Chapter 22) arise during bone remodelling and may not be in continuity with the joint cavity; there is less active evidence of inflammation.

The pathogenesis of rheumatoid pseudocysts is controversial. Bone reabsorption, mediated both by osteoclasts and by the proteases of the rheumatoid granulation tissue, sets the scene. The extension of granulation tissue within the bone is enhanced by osteoporosis. The possibility that the pseudocyst is a form of rheumatoid granuloma[266,267] finds little support. However, pseudocyst formation can be precipitated by mechanical and hydrodynamic factors. Very high pressures can be generated in large inflamed joints by flexion.[97,205,206] The pressures are sufficiently high to impede the synovial circulation. Knee joint capsules, distended with fluid, rupture at pressures below those measured *in vivo* in rheumatoid arthritis. These high intra-articular pressures can propel synovial granulation tissue through the articular surface into porotic bone.

Localized osteoporosis This is a common association of active, chronic rheumatoid arthritis.[142] There is an increased turnover of juxta-articular bone.[383] Histomorphometry reveals an increased active osteoid surface but also increased osteoclastic bone reabsorption promoted by factors such as prostaglandin E and osteoclast activating factor (OAF)*.

Osteonecrosis Radiological evidence of bone necrosis is rare both in non-steroid treated rheumatoid arthritis and in cases given intravenous pulsed methylprednisolone.[442] There are few systematic pathological studies.

Generalized bone disease

Osteoporosis Osteoporosis is commonplace. The clinical evidence is variegated. Rarely, there is massive osteolysis.[276] Whether osteoporosis is related to rheumatoid arthritis directly is difficult to establish because of the influences of diminished exercise, bed rest, corticosteroid therapy and nutritional imbalance. Paradoxically, an increase in bone metabolism has been described as part of the rheumatoid process.[338] One marker of bone formation is γ-carboxyglutamic acid-containing protein (BGP), a plasma component. The levels of BGP in men and women with rheumatoid arthritis are lower than those of control patients. Whether rheumatoid arthritis alone, as opposed to corticosteroid treatment, is partly responsible for this change is suspected but not proven.[438] However, total body calcium is reduced, particularly in women. The reduction is exaggerated by corticosteroids.[345] Thus, the extent and cause(s) of generalized osteoporosis in rheumatoid arthritis remain controversial.

Early studies with tetracycline labelling and fluorescence microscopy[109] were equivocal. They demonstrated active demineralization. More recently, conventional histological and morphometric techniques have confirmed the presence of osteoporosis.[304] Measurements of trabecular bone mineral content by quantitative computed tomography indicated an increased prevalence of osteoporosis in 7 per cent of 88 patients with non-steroid-treated rheumatoid arthritis.[75] The bone mineral content, 135.8 ± 32.8 mg/ml potassium phosphate in the rheumatoid patients, was significantly lower than in 105 control patients (151 ± 32.1 mg/ml). Potassium phosphate in solution has attenuation properties similar to calcium hydroxyapatite. The difference from normal was most marked in younger patients. However, the frequency of osteoporosis appeared equally common among patients with osteoarthrosis.[304]

The cause(s) of the osteoporosis in rheumatoid arthritis have been clarified by morphometry. When steroid therapy was excluded, iliac crest biopsy revealed a low mean trabecular bone volume in rheumatoid female patients aged 34 to 50 years; a low mean trabecular bone volume was also suggested in males.[277] The mean trabecular plate thickness was low at all ages examined (34–80 years) but the mean trabecular plate density (an index of trabecular number) and the mean trabecular plate separation showed no age-related change. It was concluded that rheumatoid arthritis is associated with premature bone loss attributable to trabecular thinning. The predominant mechanism of bone loss appears to be reduced bone formation at the bone remodelling unit level.[76] One explanation may be the polypeptide inhibitor of osteoblast proliferation.[177] Another may be the presence in widened haversian canals of mononuclear inflammatory cells.[417]

Osteomalacia Osteomalacia has also been suspected. Signs of vitamin D deficiency were detected by iliac crest biopsy and morphometry in 20 of 29 Dutch rheumatoid patients, all receiving dietary vitamins and exposed to sunlight;[399] none had signs of malabsorption. Conflicting results were obtained in London where rheumatoid arthritis was not accompanied by osteomalacia.[304] In western England, again, as in Glasgow,[339] osteomalacia was commonplace. The differences between these series centre upon the criteria used for selecting patients. The significance of evidence that synovial fluid cells, in cases with high cell counts, can synthesize 1,25-dihydroxy-vitamin D3[176] remains unclear.

Regional joint disease

Any or all of the 264 synovial joints may be sequentially or simultaneously affected by rheumatoid arthritis. There are always associated changes in joint capsules, tendons, ligaments, soft connective tissue and blood vessels. Whereas the most common clinical problems encountered are attributable to disease of the limb joints, synovitis of the joints of the mandible, larynx and spine poses a threat to survival and demands special attention.

* The properties and amino acid composition of OAF are closely similar to those of other cytokines, particularly IL-1.

Arthritis of spinal joints

Although secondary disorders such as staphylococcal osteomyelitis may occur,[263] the high frequency and serious implications of primary rheumatoid arthritis of spinal joints has attracted particular interest.[22] There have been important surgical advances.[225] Garrod[150] was aware of cervical disease, the most common and most important form of rheumatoid spinal joint disease, but the possibility that the lumbar and other parts of the vertebral column could be affected by rheumatoid arthritis was recognized much more recently.[242,243]

Cervical spine[256,257]

The upper cervical spine is most vulnerable.[22] There are three main anatomical forms of rheumatoid cervical spine abnormality (Fig. 12.22): atlantoaxial subluxation, atlantoaxial impaction and subaxial subluxation.

Atlantoaxial subluxation (Fig. 12.22)

In atlantoaxial subluxation, erosive synovitis of the atlantoaxial, atlanto-odontoid and atlanto-occipital joints, together with chronic, destructive inflammation of the synovium-lined bursa between the odontoid process and the transverse ligament,[57] results in displacement of the atlas (C1) upon the axis (C2). Displacement is most commonly anteriorly (11–46 per cent of *post mortem* cases; 19–71 per cent of all rheumatoid cases).[256] It is a result of injury to or destruction of the transverse, alar and/or apical ligaments[119] (Fig. 12.24). Posterior atlantoaxial subluxation[364] is much less frequent; it constituted 6.7 per cent of cases of atlantoaxial subluxation in one series and may lead to spinal cord compression or vertebral artery injury. Lateral atlantoaxial subluxation may contribute to as many as 21 per cent of all cases. Superior subluxation is a term that has been applied to overriding of the atlas on the axis, an abnormality that has been encountered in cases with both posterior and lateral displacement of C1.[435]

Atlantoaxial impaction (Fig. 12.23)

In atlantoaxial impaction, a process of vertical subluxation, the skull settles onto the atlas, the atlas onto the axis. The extent of this process of pseudobasilar invagination is measured by plotting a line (McGregor's line) from the hard palate to the occiput in a lateral radiograph. Atlantoaxial impaction is diagnosed when the zenith of the odontoid process is 8 mm above the palato-occipital line in men and 9.7 mm in women. The principal tissue changes, in contrast to those of the common anterior atlantoaxial subluxation, are the erosion and loss of bone in relation to the atlanto-occipital and atlantoaxial joints.

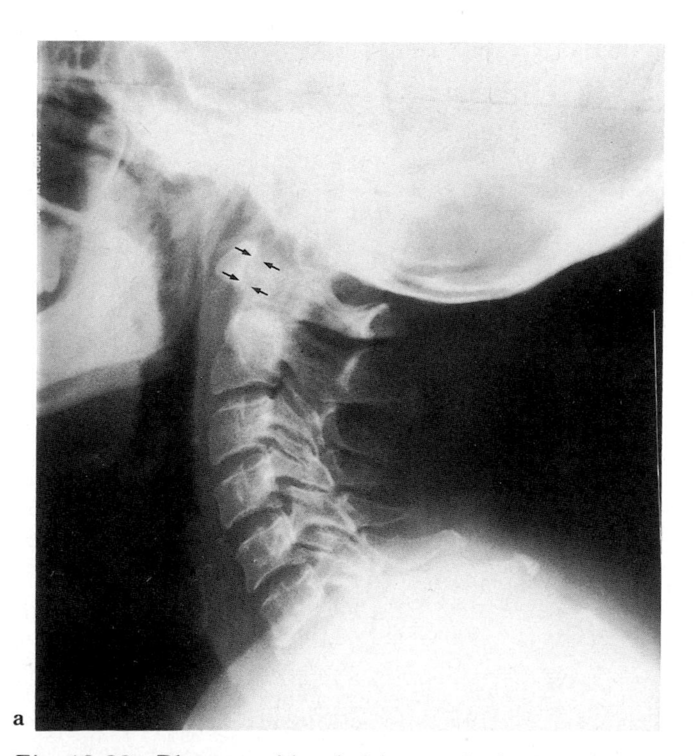

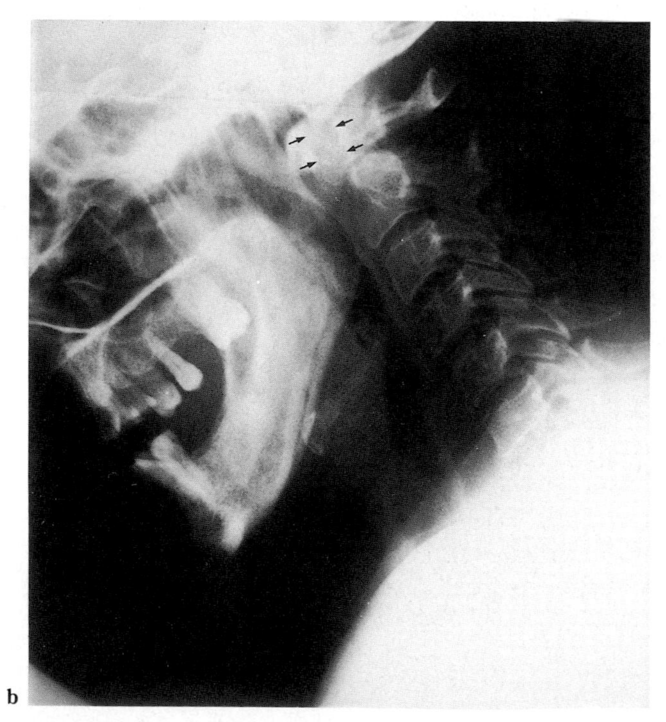

Fig. 12.22 Rheumatoid arthritis: cervical spine.
Radiographs of lateral cervical spine: **a.** in extension; **b.** in flexion. Note subluxation of the atlantoaxial joint (arrowed) in **b.** Courtesy of the Department of Diagnostic Radiology, University of Manchester.

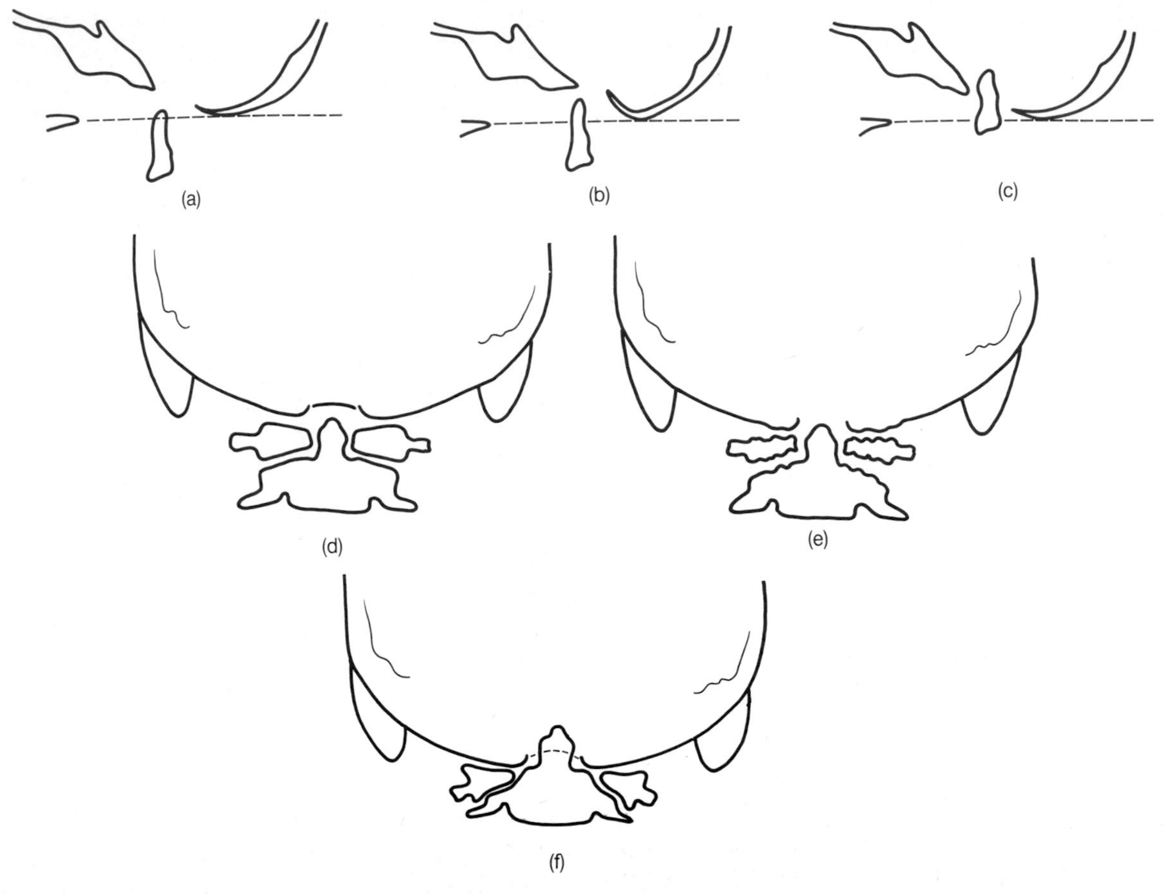

Fig. 12.23 Rheumatoid arthritis: cervical arthritis with basilar invagination and atlantoaxial impaction. **a.** Normal. Note position of odontoid process in relationship to the foramen magnum. McGregor's line is indicated. **b.** Basilar invagination. **c.** Atlantoaxial impaction. **d.** Mechanism of development of atlantoaxial impaction. Normal structure. **e.** Bone and connective tissue erosion. Severe bone loss mainly at C1/C2. **f.** Skull and C1 have settled onto C2. Odontoid protrudes into foramen magnum. Redrawn from Weissman.[437]

Subaxial subluxation In subaxial subluxation the displacements are found at one or more cervical vertebral levels. Subaxial subluxation frequently accompanies atlantoaxial disease and is recognizable in 10–20 per cent of patients with rheumatoid arthritis.[256] The disc spaces are narrowed and there is erosion of the end plates without osteophytosis.

Neurological complications Much attention has been given to the neurological complications of cervical arthritis. The changes are those of cervical myelopathy.[271] The likelihood of compression of the spinal cord, occasionally fatal,[441] is related to the degree of deformity (Fig. 12.24).[437]

Where there is odontoid herniation, ischaemic atrophy of the medullary pyramids, with loss of myelinated pyramidal tract fibres and astrocytosis, accompanies atrophy of paraspinal muscles. The walls of veins and some arteries of the subarachnoid space around the brain stem and cervical cord display a low-grade vasculitis.

Where there is atlantoaxial subluxation, a depression may be seen on the upper part of the cervical spinal cord, with necrosis of a variable extent of the cord segments. Among the histological results in those surviving long periods of incomplete subluxation are denervation of the diaphragm. There is intimal fibromuscular hyperplasia and stenosis of the anterior spinal arteries. Transverse depression of the dura mater and spinal cord; cervical nerve root atrophy with selective neuronal loss or astrocytosis; and necrosis of the anterior and/or posterior gray horns, are among the neuropathological complications.

Thoracic spine Rheumatoid synovitis of the thoracic spine is more common than had been believed.[186] Synovitis of the costovertebral joints extends to cause discitis and intervertebral disc destruction.[55,56] Bone erosion and disintegration of the disc margins is followed by fibrosis, narrowing of the interdisc spaces, posterolateral bone sclerosis and, occasionally, ankylosis. The adjacent apophyseal joints are, however, spared. The intervertebral disc lesions may accompany or be followed by degenerative intervertebral disc disease, calcium pyrophosphate deposition disease (see p. 393), osteoporosis and/or ankylosing hyperostosis.

Lumbar spine Radiological studies[242] drew attention to previously unsuspected changes in the lumbar spine. There were osteoporosis (see p. 409), subluxation, disc narrowing without vertebral osteophytosis, and apophyseal joint erosions. The lumbar changes, more common in males than females, were associated with seropositivity. Some of the lesions are closely similar to those in the thoracic spine.[186] Subluxation is attributable to apophyseal joint synovitis and erosion. However, there is a less likely suggestion that this change is the result of rheumatoid inflammation of the disc margins themselves. Disc space is lost and nearby bone becomes sclerotic. Discitis and bone sclerosis may be the direct result of rheumatoid joint disease but they can also be interpreted as reactions to subluxation, the consequence of instability of the posterior segments of the spine.[55]

Aural arthritis

Both the ball-and-socket incudostapedial joint and the saddle-shaped incudomalleal joint are susceptible to rheumatoid synovitis. Whether arthritis of these joints accounts for middle-ear deafness is still not certain. In two incudostapedial joints examined through the courtesy of Professor I Friedmann, synovial tissue was no longer present, but granulation tissue of low vascularity extended beneath the articular cartilage of the lenticular process of the incus, suggesting that rheumatoid synovitis had developed.

Temporomandibular arthritis[307]

Since the temporomandibular joints always function synchronously, disorganization of one inevitably disturbs all movements of the mandible. Temporomandibular rheumatoid synovitis is a severely disabling condition which may be unilateral or bilateral.[77,142] Franks[133] surveyed 100 cases of adult rheumatoid arthritis and found X-ray evidence of disease of the temporomandibular joint(s) in 86 per cent. Arthritis may develop in the course both of the adult and of the juvenile disease.[61] The progression of juvenile chronic polyarthritis can lead to delayed mandibular growth and in severe cases, facial deformity ('bird-face': mandibular micrognathia) is a possible result. In its advanced form, when nutrition is threatened, palliative surgery for temporomandibular arthritis may be necessary no matter how poor the general state of the patient.

Anatomical access to the temporomandibular joints is not easily negotiated. There are few pathological studies. The synovial tissue of the lower compartment is first affected by rheumatoid arthritis but the upper is soon involved. A pannus covers the disorganized articular surfaces and the intra-articular disc. Fibrous adhesions form followed by bone destruction; the condyle may be wholly destroyed and ankylosis may become complete.

Laryngeal arthritis

It has long been known that the larynx is susceptible to 'rheumatism'[264] but histological studies of the intrinsic joints of the larynx are more recent.[147,165,288,289] Ankylosis of the cricoarytenoid joint is one consequence. Rheumatoid pharyngitis and upper oesophageal ulceration[289] may develop (Fig. 12.25). The oesophageal lesion may be associated with amyloid vasculopathy. The histological changes in the small laryngeal joints closely resemble those encountered in the joints of the limbs (Fig. 12.26). Synovitis is followed by a marginal replacement of articular cartilage by granulation tissue infiltrated by plasma cells and lymphocytes. Periarthritis and myositis may be present but the suggestion that laryngeal dysfunction attributable to laryngeal neuropathy is due to rheumatoid vasculitis[449] has not been confirmed. Rarely, tracheal stenosis in rheumatoid arthritis may be due, not to rheumatoid disease, but to relapsing polychondritis.[352]

Manubriosternal arthritis

X-ray studies confirm the high frequency with which this joint is affected in rheumatoid arthritis. Its involvement is of particular interest because, like the symphysis pubis, which is not implicated in rheumatoid arthritis, the central part of the structure is fibrocartilagenous. This component is absorbed in approximately 30 per cent of normal persons; the structure comes to resemble a synovial joint. Presumably, only those whose manubriosternal joints have undergone this frequent but unexplained change are prone to manubriosternal rheumatoid synovitis. Pathologically, fibrous ankylosis, bone erosion and adjacent osteosclerosis are described.[220] Subluxation of the joint may occur, a phenomenon which may be associated with cervical and thoracic spinal disease. The forces generated by longstanding kyphosis are assumed to precipitate this displacement.[217]

Shoulder joint arthritis

Because of increasing incapacity among the elderly, surgical correction of shoulder joint disease is becoming more common. There are few systematic studies of the pathological material.

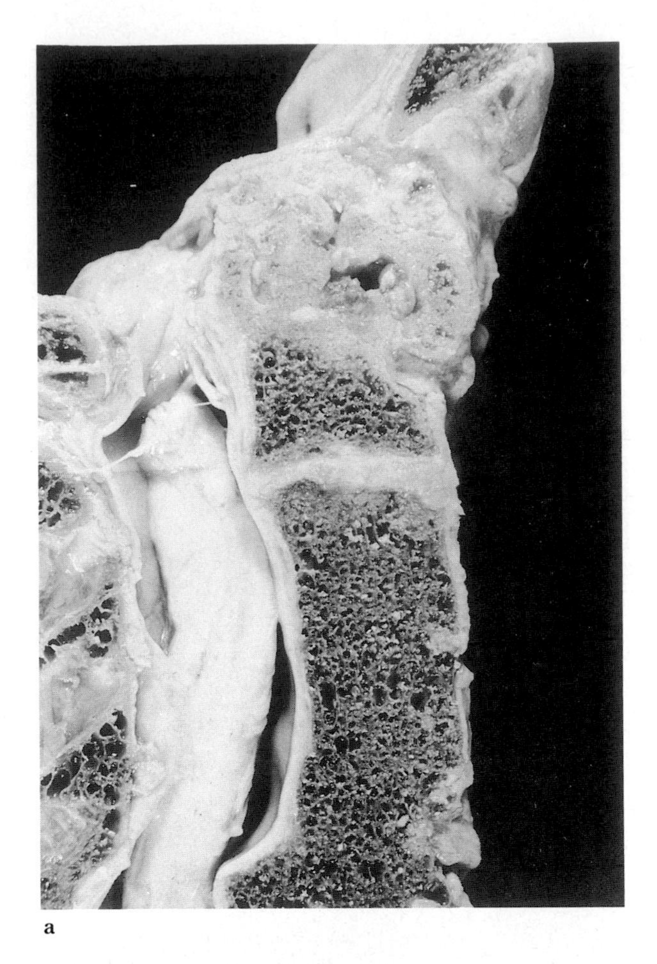

a

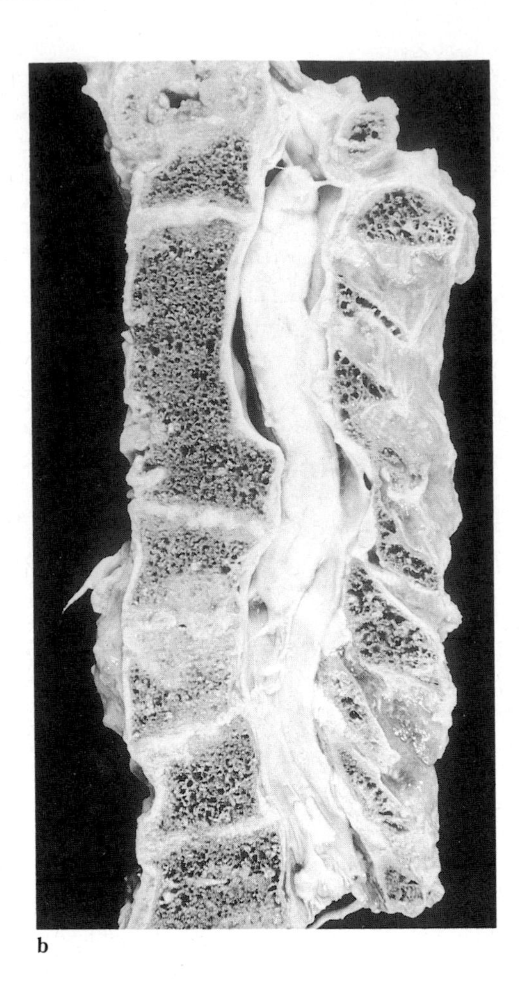

b

Fig. 12.24 Rheumatoid arthritis: spine.
a. Destruction of the median atlantoaxial joint and of the adjoining anterior arch of the atlas (C1) by rheumatoid granulation tissue. Part of the displaced, disintegrated odontoid process lies anterior to the medulla and immediately below the foramen magnum. **b.** Fusion of the bodies of C3, C4 and C5 has followed destruction of the corresponding main intervertebral joints. The C5/C6 and C7/C8 discs are relatively well preserved but the C6/C7 disc is the focus for rheumatoid spondylodiscitis originating in the inflamed synovial, neurocentral, joints.

Acromioclavicular joint

Arthritis of this joint is often associated with glenohumeral shoulder joint disease.[324] Eighty-five per cent of 49 rheumatoid patients with shoulder pain were found to have radiological evidence of acromioclavicular joint disorder. At arthrotomy, the joints were occupied by hypertrophic synovial tissue. Bone reabsorption led to the tapering of the clavicular ends of both aspects of the joint. The articular surfaces were usually wholly destroyed.

Elbow arthritis

Elbow joint arthroplasty is now practised with success and material from the initial operation and, very occasionally, from revision procedures, is becoming available.

Wrist arthritis

The wrist is commonly affected by rheumatoid synovitis.[67a] Instability may lead to dorsal dislocation or palmar displacement of the distal end of the ulna.[371] There are few systematic pathological studies but intraosseous pseudocyst formation may be conspicuous.[237]

Diagnostic biopsy

Few subjects in surgical pathology have proved so intractable as the problem of synovial biopsy in rheumatoid arthritis.[261] It is now generally accepted that, judged by the classical techniques of paraffin-section morphology, no single light-microscopic feature or combination of features is

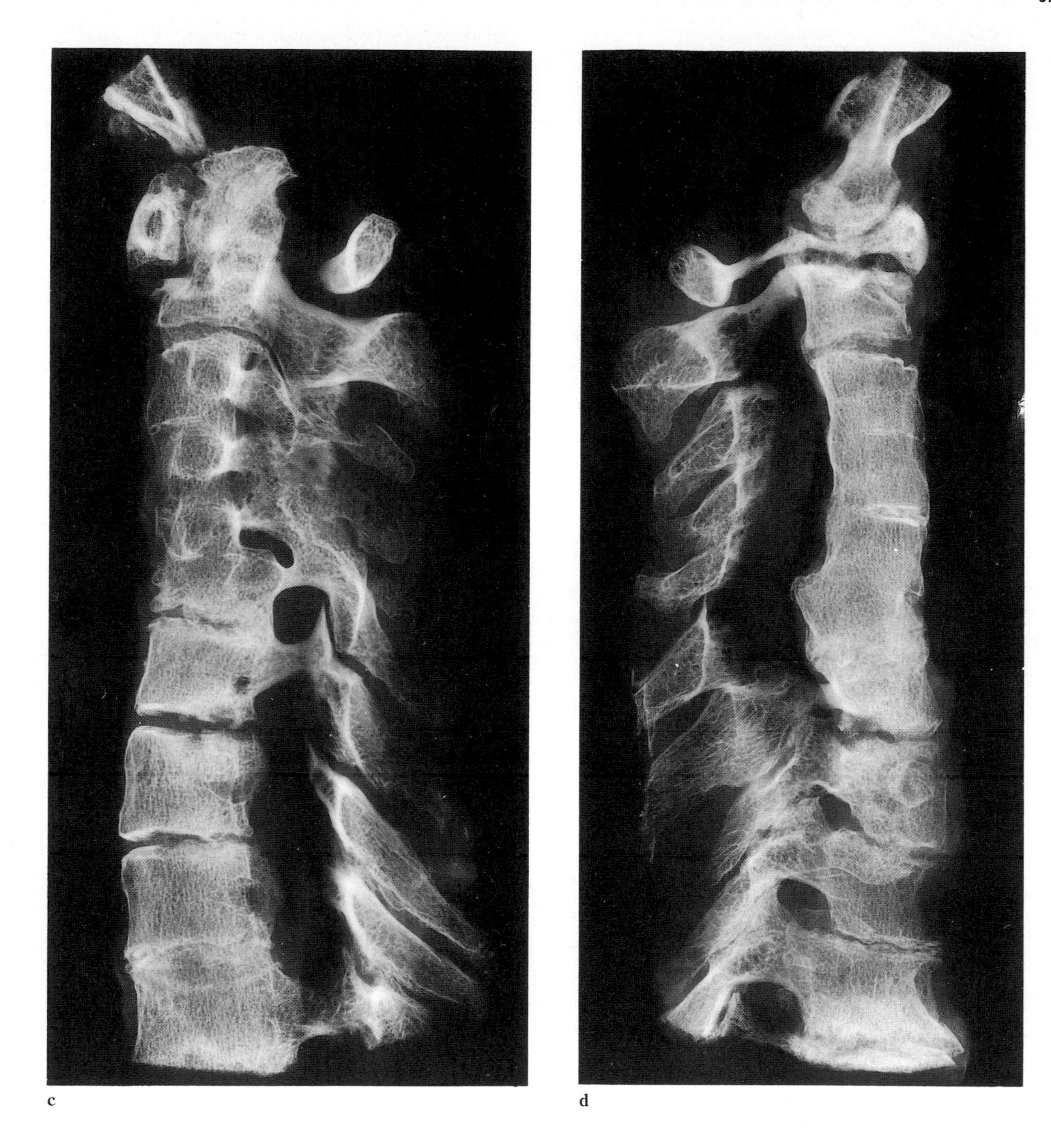

c d

Fig. 12.24—*cont.*
c. Radiograph of slab block from *post mortem* specimen divided sagittally. Note base of skull, margin of atlas (C1) and odontoid process of axis (C2). There is loss of clarity of images of the apophyseal joints together with intervertebral disc degeneration and apophyseal joint destruction of the lower cervical and upper thoracic intervertebral joints. **d.** Compare with Figure 12.24c. Occipital condyles rest on atlas (C1). There is fusion and loss of intervertebral disc tissue from the C3/C4, C4/C5, C5/C6 and C6/C7 articulations. Anterior osteophytes, associated with intervertebral disc degeneration give sharply angulated, 'cheese-cutter' anterior margins to main intervertebral joints between C8 and T1, T1 and T2 and T2 and T3. Courtesy of Professor René Lagier.

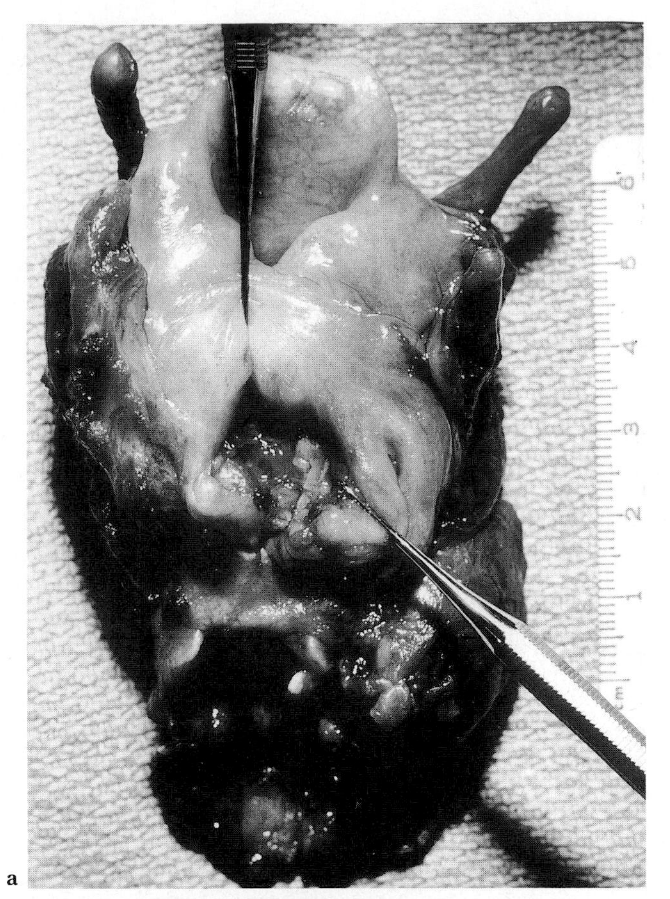

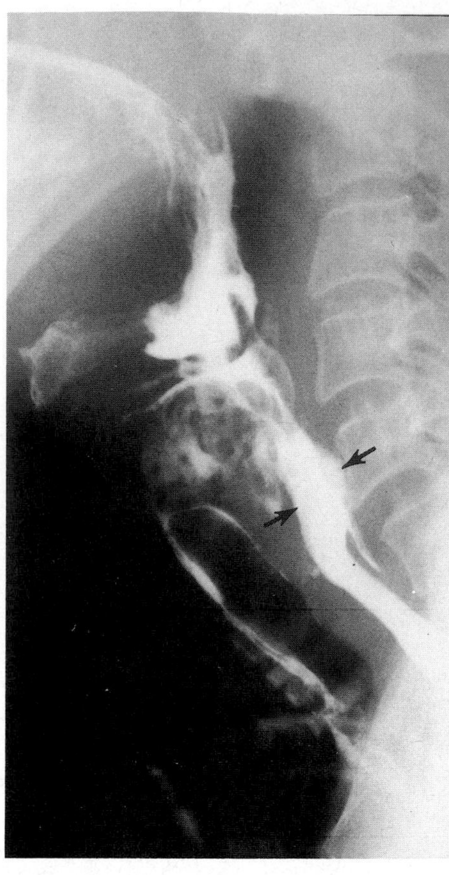

Fig. 12.25 Rheumatoid arthritis of larynx.
a. Inflammatory process has extended from intrinsic joints of larynx to erode laryngeal and oesophageal walls, creating laryngooesophageal fistula. **b.** Site of communication outlined by barium. Courtesy of Dr W W Montgomery.

pathognomonic[118,348] although the presence of an aseptic rheumatoid granuloma or hemigranuloma is unlikely to be attributable to any other disease. Few of the changes or combinations of changes seen in rheumatoid arthritis are confined to this disorder. Thus, a fibrinous exudate is common in many forms of arthritis and is, for example, characteristic of tuberculosis. A strong case has been argued for the role of polymorphs in causing cartilage breakdown in active rheumatoid arthritis[285] (see p. 535) but these cells accumulate in the majority of forms of acute synovitis and are in no way pathognomonic of rheumatoid arthritis. Synovial cell hyperplasia follows all surgical operations that open the joint cavity; and multinucleate giant cells are seen in a considerable variety of synovial diseases.[404a] Equally, sparse lymphocytic and plasmacytic infiltrates follow trauma and are a feature of viral infections such as mumps. Mast cells appear with increased frequency in all forms of synovitis with a vascular component and are recognized in osteoarthrosis.[135,143] Lymphoid follicles, with or without germinal centres, are unusual in synovial disease other than rheumatoid arthritis; but those with germinal centres are uncommon in rheumatoid arthritis itself. Considerable attention has been given to the proliferation of mesenchymal, subsynoviocytic fibroblast-like cells in established, chronic rheumatoid arthritis. Whatever their origin, few believe that they are pathognomonic of rheumatoid arthritis. Necrotizing vasculitis is very rare indeed and has seldom been reported in studies of polyarteritis nodosa or the other systemic vasculitides. Little help is offered in diagnosis by the finding of increased synovial vascularity and the thick-walled vessels so often seen in samples of the synovia from the lower limbs are normal features of the synovia from hip and knee.

Accumulations of haemosiderin may succeed any traumatic or surgical episode and many infective diseases. The particles of cartilage and bone that tend to occur in advanced disease reflect secondary disruption of the hard connective tissues and the progress of secondary osteoarthrosis, rather than rheumatoid arthritis itself. New collagen synthesis is an early association of fibroblast proliferation but this fibrosis is a consequence of virtually every form of severe or sustained joint injury, infection or irritation. Whether the

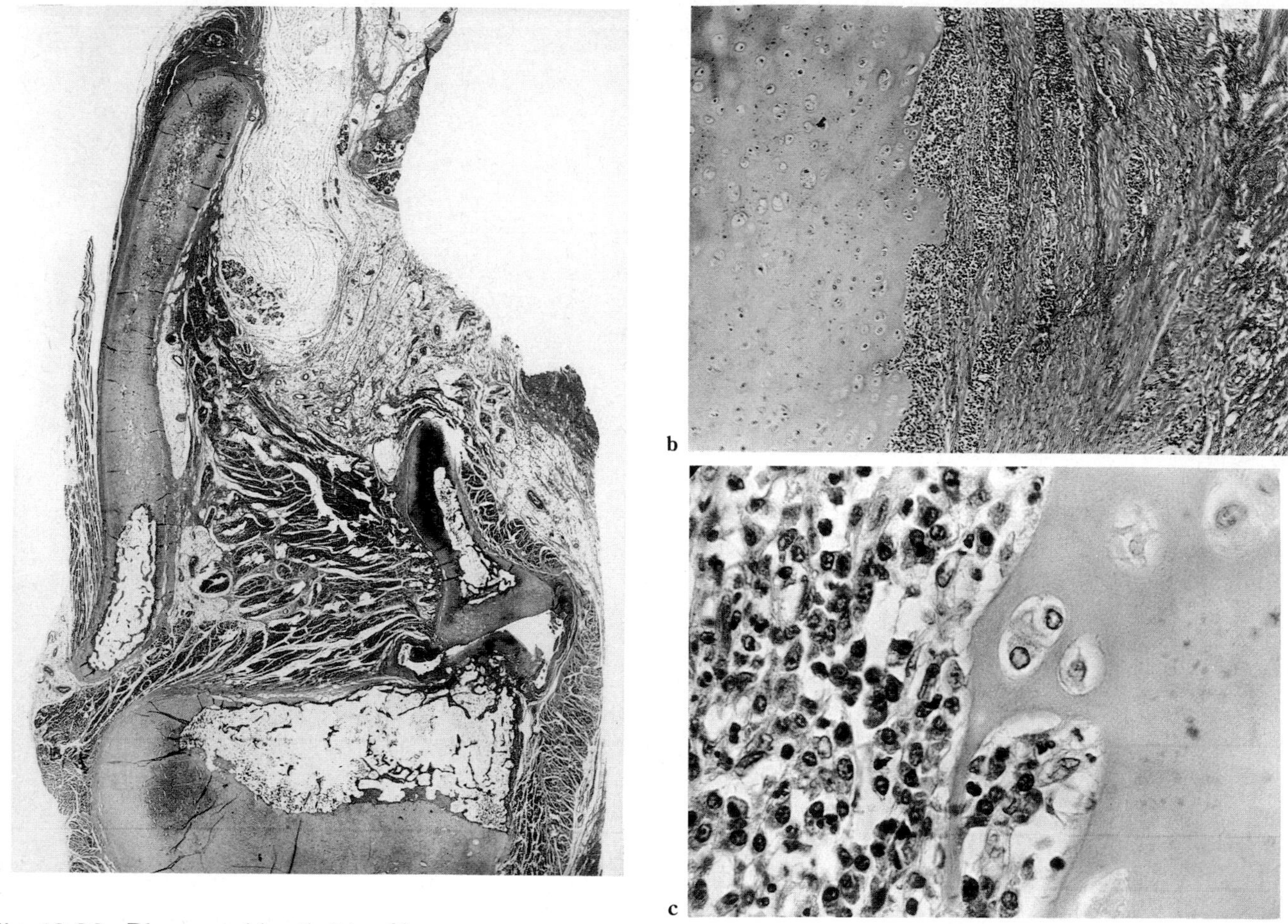

Fig. 12.26 Rheumatoid arthritis of larynx.
a. Whole laryngeal section in median plane. Section bisects thyroid, cricoid and arytenoid cartilages. Note margins of arytenoid cartilage (centre right) where inflammatory cell infiltration of synovia (dark grey-black) marks site of active, severe rheumatoid synovitis. Note also foci of ectopic formations of osteoporotic bone in each cartilage, a common age change. **b.** Margin of cricoid cartilage. Observe direct erosion of cartilage, without mediation of chondroclasts, by numerous lymphocytes. (HE × 60.) **c.** High-power view of **b.** At least one half of the cells present are plasma cells. Cartilage matrix degradation precedes chondrocyte death.

fibrosis occurs largely within the subsynoviocytic tissue or extends, in advanced cases, to implicate the capsule and periarticular connective tissues, is a matter of duration and severity, not of specificity.

Conventional microscopy

To overcome the problems of histological diagnosis posed by conventional subjective techniques, extensive retrospective[79,163,357,404,404a] and a few prospective[410] studies have been made of large numbers of synovial samples. Prospectively, the degree of inflammatory activity, judged by synovial fluid cytology, correlates with synovial tissue activity ('actual activity'); but a correlation with synovial cell proliferation ('basic activity') does not allow an assessment of clinical disease activity.[410] Retrospectively, Goldenberg and Cohen[163] found that, in rheumatoid synovia sampled by needle, only the presence of lymphoid follicles was specific although the degree and intensity of synovial cell hyperplasia and the extent of macrophage infiltration were characteristic. Soren[404,404a] could not determine either a definite or a probable diagnosis of rheumatoid arthritis on the basis of the presence of lymphoid follicles since these follicles were present nearly as often in other forms of synovitis as in rheumatoid arthritis. Nor could he confirm the specificity of the multinucleate giant cells described by Collins[71] and by Grimley and Sokoloff.[166] Soren stated that the diagnostic value of necrotizing or active vasculitis was questionable and agreed with Fassbender[121] and Lindner[253] that, although the specificity of synovial granulomata was undisputed, their contribution to diagnosis was small because of their rarity (approximately 2 per cent of biopsies). Hemigranulomata, with free surface fibrin or fibrinoid, were also of limited diagnostic value, partly because they were identified with equal frequency in osteoarthrosis as in rheumatoid arthritis and with *greater* frequency in ankylosing spondylitis. The presence of large fibroblasts, the mesenchymal cell of Fassbender,[121] seemed to be the most frequently observed and diagnostically useful synovial change. The only condition in which fibroblast proliferation appeared even more often was psoriatic arthropathy. Since rheumatoid synovial fibroblast proliferation is inhibited by corticosteroids, the higher frequency of this change in psoriatic arthropathy was, however, attributed by Soren to the much more frequent use of these drugs in rheumatoid arthritis.

Soren's[404a] (Table 12.4) work was extended in a detailed analysis of 393 joint operations on patients aged 2 to 76 years, seen at New York University Medical Center over a 15-year period.[79] Cases considered comprised 127 of definite or classical rheumatoid arthritis, 47 of possible or probable disease and 23 of juvenile rheumatoid arthritis. Each specimen was scored for 37 histological features and the number of cases displaying each feature calculated. The statistical significance of each feature was then tested as a means of differentiating between rheumatoid synovitis and seven other categories of disease. The results of these enquiries* are given in full by Cooper *et al.*[79] and by Rosenberger *et al.*[357] Their conclusions could be reached because each diagnosis had been established previously by clinical criteria. No account was taken of transmission electron microscopy or immunopathological features in the synovia. The results were often exotic: for example, the number of occasions when, say, a distinction of enterocolitic arthropathy from, say, osteoarthrosis rests upon synovial biopsy alone seems likely to be small. Nevertheless, the principle that detailed histological categorization can assist differential synovial diagnosis appeared to be established. Moreover,[79] the *degree* of agreement between the clinical and the discriminant function histological diagnosis could be derived mathematically; the weight and signi-

* Among the most valuable criteria influencing the diagnosis of definite or classical (1958) rheumatoid arthritis were: vs psoriatic arthropathy, oedema (+), fibrosis (−)*; vs Reiter's disease, fibrosis (−); vs ankylosing spondylitis, organizing fibrin (−), synovial cell hypertrophy (−), proliferating fibroblasts (−), diffuse lymphocytic infiltration (−), diffuse histiocytic infiltration (−), fresh haemorrhages (+), vascular thrombosis (+); vs enterocolitic arthropathy, fibrocytic hypercellularity (−), proliferating fibroblasts (−), focal lymphocytic aggregates (−), fibrosis (−); vs osteoarthrosis, 16 negative features, including synovial cell hyperplasia, diffuse lymphocytic infiltration and fibrosis; hyperaemia (+) and severe oedema (+); vs post-traumatic synovitis, 16 negative features, and four positive (capillary proliferation, hyperaemia, slight synovial cell proliferation and severe oedema).

(+) = significantly frequent association; (−) = significantly infrequent association.

Table 12.4

Histopathological features of synovitis[404a]

1. Hypertrophy of synoviocytes.
2. Hyperplasia of synoviocytes.
3. Giant cells in the synovial membrane.
4. Ulcer of the synovial membrane.
5. Fresh fibrin on the synovial membrane.
6. Organized fibrin on the synovial membrane.
7. Fibrin in the synovial membrane.
8. Hyperaemia of the synovial membrane.
9. Oedema of the synovial membrane.
10. Inflammatory cells in the walls of synovial blood vessels: (a) arterioles, (b) venules.
11. Inflammatory cells in perivascular infiltrates in the synovial membrane.
12. Inflammatory cells: (a) polymorphonuclear leucocytes, (b) lymphocytes, (c) plasmacytes, (d) histiocytes in focal accumulations, in the synovial membrane.
13. Inflammatory cells: (a) polymorphonuclear leucocytes, (b) lymphocytes, (c) plasmacytes, (d) histiocytes in diffuse infiltrates, in the synovial membrane.
14. Inflammatory cells: (a) lymphocytes, (b) plasmacytes in nodules (follicles), in the synovial membrane.
15. Haemorrhages in the synovial membrane.
16. Haemosiderin in the synovial membrane.
17. Fibroblasts in the synovial membrane.
18. Increased fibrocytes of the synovial membrane.
19. Proliferating capillaries of the synovial membrane.
20. Increased blood vessels of the synovial membrane.
21. Villi of the synovial membrane.
22. Thick-walled blood vessels in the synovial membrane.
23. Necrosis and/or inflammation of blood vessels of the synovial membrane.
24. Degeneration of collagen fibres of the synovial membrane.
25. Necrosis of the synovial membrane.
26. Hemigranulomata on the synovial membrane.
27. Granulomata in the synovial membrane.
28. Fragments of cartilage and/or bone on and in the synovial membrane: (a) with inflammatory reaction, (b) without inflammatory reaction.
29. Lipid deposits in the synovial membrane.
30. Fibrosis of the synovial membrane.

Phases of inflammation:

Acute	Hyperaemia Oedema Polymorphs	
Subacute	A transition from acute to chronic or healing	
Chronic	Lymphocytic Plasmacytic Histiocytic Fibrosis	Infiltration

Table 12.5

Gradations of inflammatory activity and severity used in assessing synovitis[404a]

Activity of inflammation (Graded 0–3)	
0	Essentially normal (or scar). No inflammatory changes except rare lymphocytes or plasmacytes.
1	Lymphocytes or plasmacytes in focal or diffuse infiltrates.
2	Hypertrophy of synoviocytes in addition to activity of grade 1.
3	Fibrin on synovial membrane and/or ulcer of membrane in addition to activity of grade 2.
Severity of inflammation	
1	Mild inflammatory reaction focally or throughout section.
2	Mild inflammatory reaction in some areas, moderate or marked inflammation in others.
3	Marked inflammatory reaction throughout section.

ficance of the discrimination could therefore be calculated. By applying the technique of linear discriminant function analysis it was consequently possible, with each disease, to construct a database. Prospectively, the likelihood of a new diagnosis could then be calculated from the frequency of the new histopathological findings. Any pathologist with an adequate database could, in theory, derive the same assistance from the microscopic features of a new case scored against an original analysis of his selected and correlated clinical diagnoses.[357]

In analogous studies, Geiler and Emmrich[153] compared the synovial abnormalities of the hand, finger and knee of 13 patients with rheumatoid arthritis of 6 months duration, with those of 16 established late cases conforming to the 1958 American Rheumatism Association criteria for definite disease (see Table 12.1). Thirty-two histological features were used to grade the early reaction. There were five early features: increased glycosaminoglycan constituting a lamina of mucoid (p. 306) change in the outer synovial tissue; vasculitis or perivasculitis; slight plasmacytosis and lymphocytosis; little fibrin and few polymorphs; and an *absence* of fibrosis and hyaline change. Synovial cell hyperplasia (see p. 541) was much less pronounced than in later cases and particular importance was attached to the common presence of immunocompetent cells (see p. 246). It is suspected that, in established rheumatoid arthritis, there are alterations in the redox balance and other indices of cellular metabolic activity.[54,183] However, there is little published evidence to show that recognition of changes in synovial cellular respiration and metabolism in the earliest stages of rheumatoid arthritis facilitates diagnosis.

Morphometry

Entirely different statistical studies have been made, based on the morphometry of individual features of the synovium in rheumatoid arthritis.[409] They have been considerably extended by Kennedy *et al.*[218,219] in enquiries made to discriminate between rheumatoid arthritis and osteoarthrosis. In 11 synovia from patients with longstanding classical or definite rheumatoid arthritis coming to surgery, Kennedy *et al.*[219] found villus formation, synoviocyte hyperplasia and a widespread lymphocyte, plasma cell and macrophage infiltrate. In the rheumatoid synovia, there were two main patterns of T-cells: diffuse, with a predominance of T_S (CD8)-cells; and aggregates with a predominance of T_H (CD4)-cells (see p. 247). Free synovia contained B-cells reacting for IgM, with pericellular IgG. There were numerous plasma cells between the lymphocyte aggregates and in discrete foci. Many cells in and between the aggregates were MHC class II HLA-DR-positive but tests for activated lymphocytes and macrophages were more often positive in cells in the interaggregate areas.

When the nuclear density of these preparations was assessed, a higher but varied density of cells was demonstrable in rheumatoid synovium in contrast to osteoarthrosis. In cell aggregates, and in the more superficial ('upper') and deeper ('lower') synovia, there were significantly more nucleated cells in rheumatoid arthritis than in osteoarthrosic synovia. The T_H-, macrophage/activated lymphocyte and plasma cell populations of rheumatoid synovia were entirely different in rheumatoid arthritis compared with osteoarthrosic biopsies. In the superficial layer, there were more T_S- than T_H-cells and in the lymphocytic aggregates there was a predominance of T_H (CD4)-cells. Macrophage/activated lymphocytes were numerous in the superficial layer. IgG plasma cells were present throughout the synovia; IgA plasma cells were more numerous in the superficial layer than IgM plasma cells.

Techniques of biopsy

In order to allow for the variability of cellular reaction in a single sample, it has been proposed that at least 2.5 mm^2 of synovium (approximately 12 microscopic fields at $\times$ 250) requires to be analysed.[218] This extent of tissue can be obtained from the three to five specimens taken at a single arthroscopic examination. Arthroscopy is less convenient than the relatively simple 'office' technique of needle biopsy. However, needle biopsy has serious limitations. The

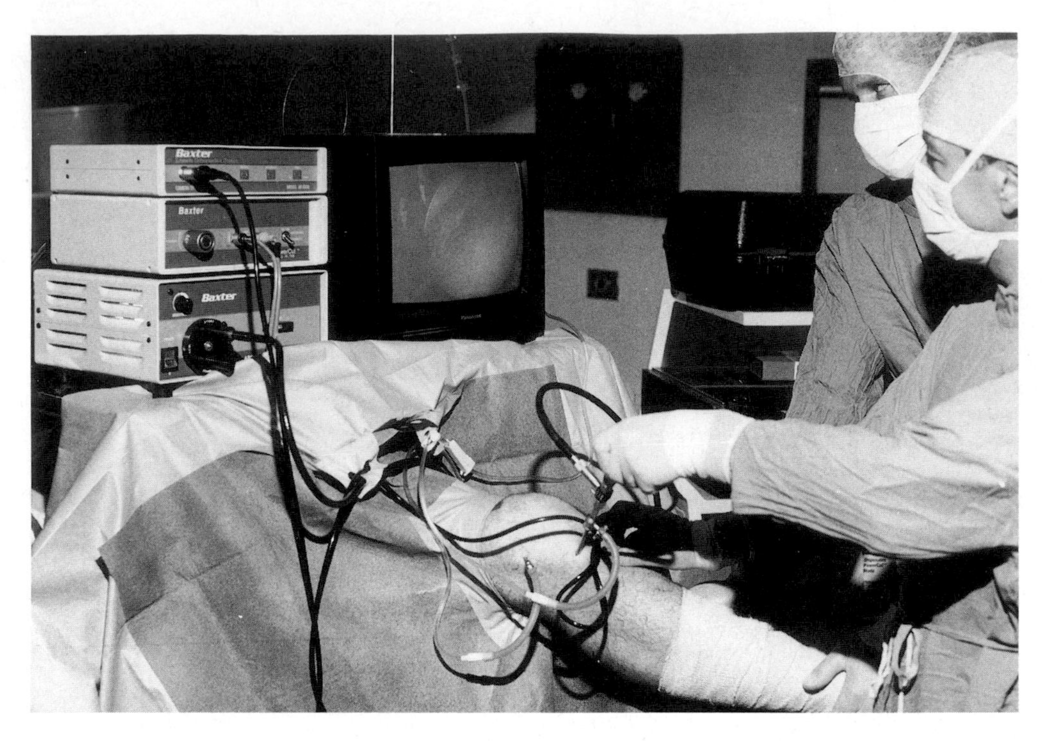

Fig. 12.27 Fibreoptic arthroscopy.
Fibreoptic arthroscopes allow exploration of the whole of the internal surfaces of large joints such as the knee. A high intensity xenon lamp provides intraarticular illumination. The operative field is displayed on a high-resolution colour monitor (at centre). Images can be recorded on video tape for future discussion. Separate small incisions provide access for instruments such as 5 mm diameter rotary cutter. Clarity of field of view is ensured by continual irrigation with isotonic fluid. Among procedures now made possible by this technique are biopsy of sites selected under direct vision, debridement of meniscal fibrocartilage, synovectomy, reconstructive surgery and grafting. Pathologists can view the video recording of the operation before surveying the microscopic preparations. Resected fibrocartilage or bone may be submitted in numerous small fragments that recall those obtained by transurethral prostatectomy. Specimens taken under these conditions remain suitable for immunocytochemistry and ultrastructural analysis. Courtesy of Mr R Strachan and Baxter Laboratories Inc.

biopsy is 'blind'; there can be no correlation between the gross appearances of the synovium and the microscopic; important parts of the joint may be inaccessible; and, because of these restrictions, histological interpretations are particularly prone to errors of omission. Moreover, as stated above, there is a majority view that rheumatoid arthritis affects different parts of a single large joint such as the knee to different degrees.[195] Open exploration at arthrotomy has continued to be advantageous to the surgeon and to the pathologist; because of the enhanced probability of accurate diagnosis, it has often appeared advantageous to the patient.

The situation has, however, changed again. A new generation of fibreoptic arthroscopes (Fig. 12.27), with continuous lavage, high intensity illumination, video-amplification and a range of sophisticated instruments, now offers a solution to some of these difficulties. Previously inaccessible parts of the joint can be readily seen and samples taken during arthroscopic synovectomy[69a] under good visual conditions. There is no reason why the pathologist should not survey video recordings of the operation on his own monitor while assessing the tissue sections so that a direct correlation of the *in vivo* appearances of the joint and the microscopic changes is practicable.

Extra-articular lesions

Patients with rheumatoid polyarthritis are suffering from a systemic 'rheumatoid disease'. Anaemia, lymphadenopathy, pulmonary, vascular and skin disorders are encountered regularly. Pathological studies show that extra-articular and visceral lesions are frequent.[38] Early investigators draw attention to abnormalities of the cardiovascular system but Christie[66] and Sinclair and

Cruickshank[391] described more widespread organ involvement. The presence of visceral lesions reflects a high titre of rheumatoid factor and heralds rapid progression and an adverse prognosis.

Coincidental disease

The characteristic age and sex distribution of rheumatoid arthritis in Western populations means that the pathological features of many other common disorders are likely to coincide with this systemic connective tissue disease. Prominent among these coincidental abnormalities are cancer, atherosclerosis and osteoporosis. Primary and metastatic soft- and hard (calcified) tissue neoplasms occur in rheumatoid arthritis with the same frequency as in the general population although therapeutic immunosuppression predisposes to lymphoma (see p. 555). There is particular interest in concurrent diseases that cause clinical symptoms and signs similar to those of rheumatoid arthritis. Chronic tophaceous gout, for example, occurs periodically in patients with rheumatoid arthritis;[316] by contrast, there is a negative correlation with calcium pyrophosphate deposition disease.[98]

Skin; granulomata

Skin

Immunoglobulin M and complement components are often demonstrable in and around cutaneous blood vessels in the uninvolved skin of patients with rheumatoid arthritis. The presence of these deposits correlates with an increased prevalence of vasculitis and of visceral disease. It has been suggested that immune complexes can cause injury to blood vessels without provoking clinical signs, a suggestion supported by the detection of perivascular immunoglobulin in seropositive patients with circulating immune complexes. Manifest vasculitis can then be provoked by the local release of histamine. Conversely, the disappearance of IgM deposits from the skin correlates with the regression of signs of systemic disease and improvement in the arthritis.[439]

Granulomata

With the exception of anaemia (see p. 503), the most frequent extra-articular rheumatoid lesion is a sterile, subcutaneous or visceral granuloma, the so-called rheumatoid nodule.[70,71,457] Although such granulomata usually form in the course of polyarticular disease, they may appear months or years before the onset of arthritis.[141] Subcutaneous granulomata are identified in approximately 20 per cent of cases of rheumatoid arthritis. The majority are seropositive. The granulomata are a frequent association of severe, progressive cases among whom other visceral lesions such as vasculitis, pulmonary fibrosis and serositis abound; they are rare in children with active, chronic polyarthritis in whom IgM rheumatoid factor is not usually formed. The structure of a granuloma (Fig. 12.28) is the only feature of the disease that can be said to be pathognomonic of rheumatoid arthritis. The nature, causes and life-cycle of the granuloma therefore appear to be of great importance in attempts to unravel the natural history of rheumatoid disease.

In children, rheumatoid-like nodules, distinguishable from those of granuloma annulare and rheumatic fever, have been recognized in patients in whom there was apparently no subsequent evidence of rheumatoid arthritis. However, this view is now questioned since cases are emerging where the interval between the identification of a childhood nodule and the onset of rheumatoid arthritis[315] has been as much as 50 years. The terms 'anarthritic rheumatoid arthritis' and 'anarthritic rheumatoid granuloma' are used to describe cases in which joint disease has not yet appeared.

There is a spectrum of abnormality. Benign variants of the rheumatoid granuloma have been identified. In 'rheumatoid nodulosis',[448] high rheumatoid factor titres and multiple granulomata are associated with episodic, low-grade synovitis and subchondral cystic lesions of the small bones of the hands and feet. The syndrome is most common in middle-aged men who are seropositive but have a low ESR. The prognosis is unexpectedly good.[448]

The common sites of rheumatoid granulomata are the subcutaneous tissues of skin exposed to frequent mechanical stresses, and of skin overlying bony prominences. Typical anatomical situations are the extensor aspect of the forearm, the occiput, the olecranon, the spine, the scapulae, the knee, the ischial regions, the lateral margins of the fingers of the dominant hand and tissue overlying the femoral tuberosities.

Macroscopic appearances Rheumatoid granulomata range from 1–2 mm to 20–30 mm in diameter; they are insidious in onset and increase slowly in size unless infection or ulceration supervene. There is a firm surrounding margin of fibrous tissue, binding the granuloma to nearby structures but the centre of the granuloma is soft, necrotic and often fluid. The margins, yellow-white due to a lipochrome pigment, become less vascular and increasingly collagenous with time.

Small, transient intracutaneous rheumatoid-like nodules are much less common than the indolent, larger subcutaneous lesions.[38] The intracutaneous nodules appear

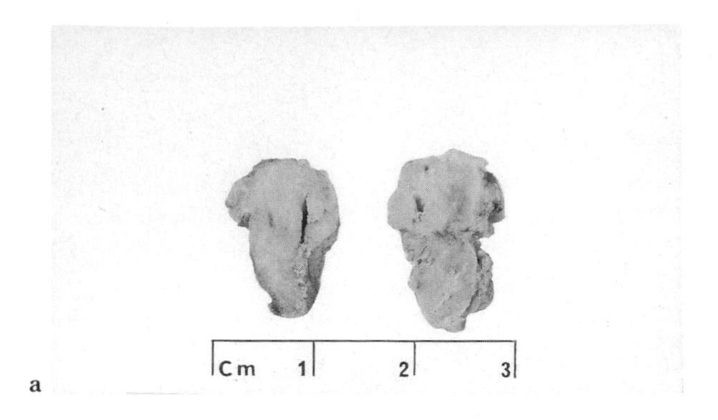

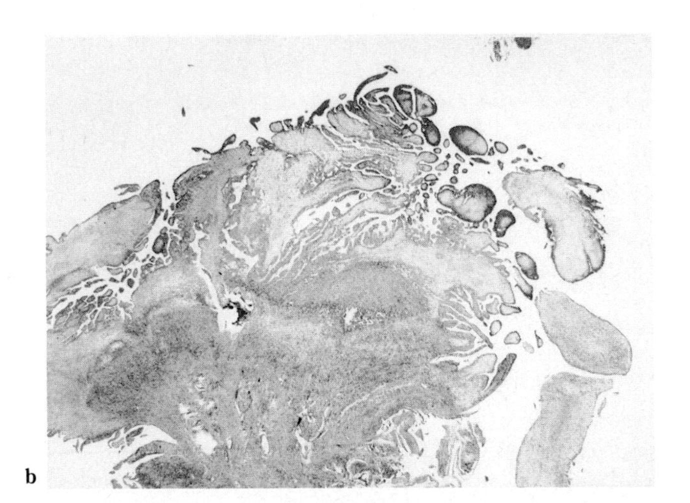

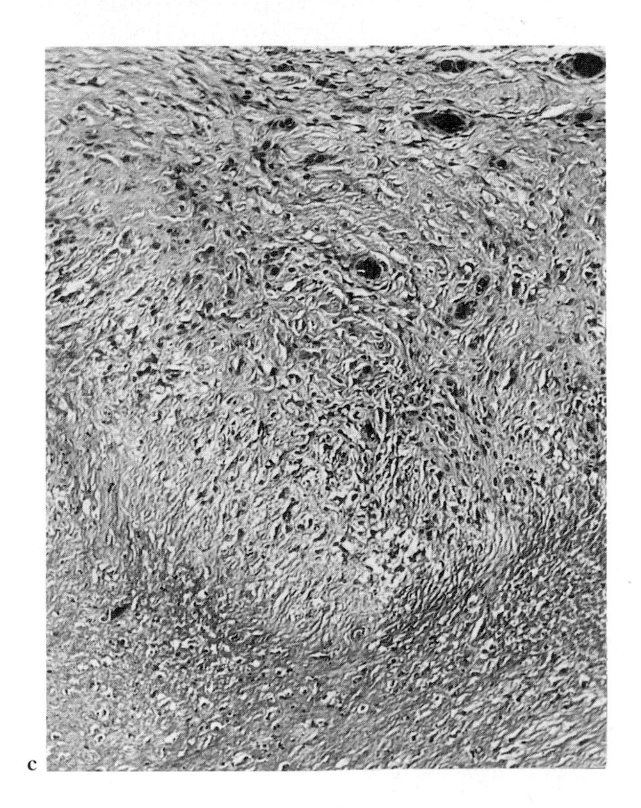

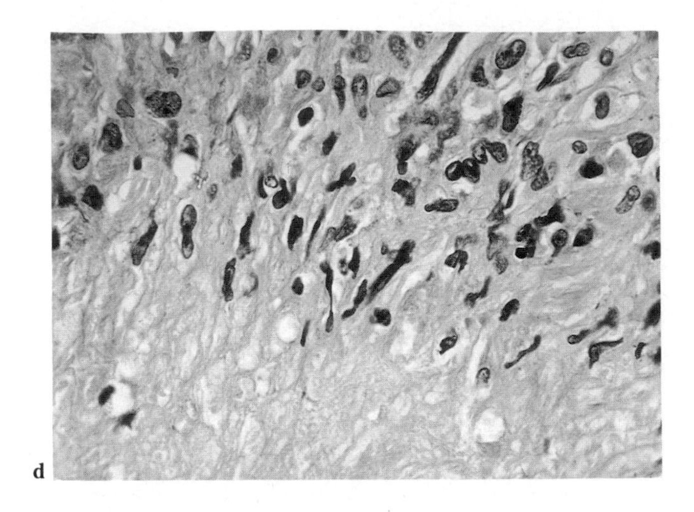

Fig. 12.28 Rheumatoid arthritis: granulomata.
a. Subcutaneous granuloma. Note cleft where fluid contents have escaped. **b.** Synovial granuloma. Extending almost horizontally across centre of field are gently curved outlines of elongated, flattened granuloma upper, central part of which is in continuity with synovial cavity, the phenomenon of hemigranuloma. Note grey-black outlines of villous processes attributable to presence of fibrin. **c.** Margin of subcutaneous granuloma. Note serpiginous edge separating necrotic centre (bottom) from external vascular fibrous tissue (top). **d.** High power view of field shown in **c.** **e.** Anarthritic rheumatoid nodule. The extensive fibrinous exudate creates a resemblance to Churg–Strauss granuloma. (HE **b:** × 11; **c:** × 100; **d:** × 600; **e:** × 140.)

particularly within the skin of manual workers and disappear with rest. They may also develop within the skin around an olecranon subcutaneous bursa. There is a relationship to vasculitis[38] and the evanescent nature of the intracutaneous granuloma recalls the behaviour of the analogous but little understood nodule of rheumatic fever (see p. 805).

Microscopic appearances The light-microscopic features have been described in detail.[139a, 142, 402] The initial response is the formation of a focus of proliferating capillaries in a matrix of undifferented mononuclear cells and fibroblasts. Necrosis begins at the edge of this granulation tissue. The granulomata become foci of circumscribed necrosis with serpiginous margins formed of elongated mononuclear cells often arranged in a palisade manner (see Fig. 12.28). These radially disposed cells are histiocytic in origin although they are said not to take up neutral red; they are potent sources of proteases that evoke continued tissue degradation and an extension of the necrotic zone. Intermediate zone cells contain ferritin.[393] Multinucleate Touton-type giant cells are occasionally seen at the granuloma margins. These cells may be a response to the lipid that is generated by cell necrosis and that can be shown by both light and transmission electron microscopy to be present within the pallisaded histiocytes. The giant cells are morphologically distinct from the giant cells of the rheumatoid synovium (see p. 459).

Transmission electron microscopic studies confirm that the central core of a granuloma comprises cell and nuclear debris, a meshwork of fibrin and the residue of partly degraded collagen fibrils, fat, other plasma proteins and proteoglycans.[69] There is little calcium and no evidence of the presence of bacteria, mycoplasmas, spirochaetes, chlamydia or viruses. Gieseking[158] and Gieseking, Baumer and Backmann[159] claim that mononuclear phagocytes are quite uncommon and that the cells which comprise the granuloma wall are proliferating, vascular parietal cells. There are (they say) two types. The first ('myoid lyocytes') contain clusters of myofilaments and are of smooth muscle origin, the second ('endothelial pinocytes') occur mainly in the outer part of the granuloma and are of endothelial origin.

Immunocytochemistry Three histological stages in the evolution of rheumatoid granulomata are recognized: acute, granulomatous and fibrotic.[139a] In the necrotic *centre* of the granulomata, strongly positive reactions with anti-IgG and anti-fibrinogen are given in stage 1, almost equally strong reactions in stage 2 but weak reactions in stage 3. The IgG located in the necrotic centre exists both by itself and in the form of IgG/IgM immune complexes. Aggregated IgG can be demonstrated as granules and as 'bumpy' deposits occupying distended spaces between the residual collagen fibrils. Denatured IgG can be shown to be capable of binding labelled IgM rheumatoid factor and complement. Plasma cells within the granulation tissue react strongly with anti-IgG and anti-IgM, very strongly with anti-aggregated IgG but weakly with anti-IgA and not at all for the β_{Ic} component of complement or for fibrinogen. The palisaded mononuclear cells do not react for IgG. However, responses similar to those of the granuloma itself are given by nearby tissues. Thus adjacent blood vessels contain IgG and fibrin. Fibrotic nodules give few immunocytochemical reactions.

The identity of the mononuclear palisaded cells at the *margin* of the rheumatoid granuloma and of the surrounding inflammatory cells has been clarified by immunocytochemical techniques.[16, 105] There are interesting similarities to the cytology of the synovium (see p. 459). The inner palisaded cells which resemble mononuclear macrophages are MHC class II HLA-DR antigen- and leucocyte common antigen-positive. Many are also strongly positive for α_1-antitrypsin, and for lysozyme[4] as well as for macrophage/granulocyte markers. A few of the palisaded cells, especially at the nodule margins, are plasma cells; IgG-secreting cells are more numerous than IgM or IgA.

The majority of the mononuclear cells *around* the nodule are MHC class II HLA-DR-positive macrophages and interdigitating reticulum cells. They are present close to nearby blood vessels. The proportion of cells reacting as macrophages is greatest at the pallisaded layer, least at the outer, perivascular edge. Like the radially arrayed cells of the granuloma margin, the perinodular cells are often positive for macrophage/granulocyte markers, for α_1-antitrypsin and for antilysozyme. A minority of the cells around the granuloma are B-cells, plasma cells and MHC class II HLA-DR-positive T-cells.

Synovial granulomata Rheumatoid granulomata of microscopic dimensions are occasionally identified in the subsynoviocytic connective tissues examined at biopsy. Their rarity in this location determines that their presence or absence contributes little in synovial diagnosis (see p. 478). Sometimes this lesion is near the free synovial surface and, when incomplete, forms an ulcer (see p. 449) termed a hemigranuloma.

Larger granulomata have been found, for example, within the olecranon bursa. Here, as in sites such as the sacral, ischial, gluteal and femoral trochanteric regions, rupture or ulceration may occur. Granulomata may also form within the richly collagenous but poorly vascularized tissue of tendons (see p. 101). Tendinous necrosis predisposes to rupture. Alternatively, the development of a granuloma within a flexor tendon, within or near the common flexor tendon sheath and beneath the fascia that constitute the carpal tunnel, can lead to entrapment.

Visceral granulomata Structurally similar granulomata are identified from time to time in a wide variety of sites. The lung and pleura attract attention because nodules, which may be confluent, can be mistaken for bronchogenic carcinoma. Rarely, laryngeal granuloma has been described.[137] In coal-workers with pneumoconiosis, the nodules[305] (see p. 485) may form before polyarthritis is manifest. At the base of heart valves, the presence of a rheumatoid granuloma can cause partial or complete heart block (see p. 491). Granulomata have also been identified in the myocardium and pericardium.[313] Central nervous system granulomata may form within the choroid plexus[224] and can compress the brain or spinal cord.[255] The sclera,[127] the bridge of the nose and antehelix of the ear[179] and the oral[393], laryngeal and retropharyngeal tissues are other uncommon sites. Ureteral stenosis is one consequence of a pelvic granuloma.[3] The non-rheumatoid granulomata recognized in transurethral prostatic resections are usually much smaller.[327] When rheumatoid granulomata occur in bone, the presence of primary or metastatic bone tumour may be suggested.[193,290] The bone lesion may be cystic and communicate with the synovial joint cavity by a narrow canal.

Diagnosis Many lesions require consideration in the differential diagnosis of rheumatoid subcutaneous granuloma,[214] fewer in the diagnosis of visceral granuloma. The subcutaneous lesions can be distinguished from those of infections such as staphylococcal abscess, from tuberculosis and leprosy, from complicated trauma where foreign bodies remain *in situ*, and from local, but deeper, hypersensitivity reactions such as erythema nodosum (p. 653). Granuloma annulare is a diagnostic problem in children. Gouty tophi display a predilection for sites such as the olecranon bursa and the presence of a sarcoid reaction to materials such as beryllium, or microorganisms such as brucella may sometimes be suspected. Rheumatoid granulomata can simulate benign and malignant neoplasms[179,193] and basal cell carcinoma[179] is among the numerous conditions which may resemble the rheumatoid lesion. The granulomata may also be confused with epithelial inclusion cysts.[446] Parasitic worm infestations such as the subcutaneous reaction to *Onchocerca* can mimic rheumatoid granulomata in West Africa. Rare but difficult problems arise when the rheumatoid granuloma is anarthritic (see p. 485): serological tests are negative in the analogous local condition of Churg–Strauss granuloma (see p. 639).

Histologically, therefore, rheumatoid granulomata have features in common with a wide variety of disorders (Table 12.6). Kaye, Kaye and Bobrove[214] give a simple classification of clinical entities associated with the presence of true rheumatoid-type nodules (Table 12.6).

Table 12.6

Clinical entities associated with rheumatoid granulomata[214]

Adult	
I	Associated with rheumatoid arthritis
a.	Rheumatoid factor present
b.	Rheumatoid factor absent
II	Associated with features of rheumatoid arthritis
a.	Nodulosis—with musculoskeletal symptoms
b.	Rheumatoid lung disease but no musculoskeletal symptoms
III	Not associated with rheumatoid symptoms
a.	Subcutaneous
b.	Pulmonary
c.	Cardiac and visceral
Child	
IV	Associated with juvenile chronic polyarthritis
a.	Rheumatoid factor present
b.	Rheumatoid factor absent
V	Not associated with rheumatic symptoms (benign rheumatoid nodules of childhood)

Experimental reproduction Many attempts have been made to reproduce rheumatoid granulomata experimentally. The regular occurrence of granulomata in the viscera weakens, but does not destroy, the argument that these lesions are largely mechanical in origin; it strengthens the role attributed to vasculitis, probably as a consequence of immune complex deposition. Particulate material such as bentonite and polymerized fibrin and liquids such as autologous blood and synovial fluid have been tested in normal and rheumatoid human and in animal tissues. None has elicited the formation of true rheumatoid granulomata. Analogues have been examined, among them the subcutaneous air sac (see p. 301) and the collagen sponge (see p. 301) but these 'test-beds' have not explained why islands of immune complexes should provoke regionally selective subcutaneous and visceral granulomata so commonly.

Foreign but not autologous fibrin implanted subcutaneously in rabbits provokes inflammation: the response creates a lesion resembling but not replicating the human rheumatoid nodule. There is a vasculitis that is of interest because of the meticulous demonstration by Sokoloff[400a] that, in man, the rheumatoid granuloma may be anatomically related to a focus of arteritis. Against this view is the evidence that in young rheumatoid nodules, of less than seven days duration, vasculitis of the smaller subcutaneous vessels, indexed by immune complex deposition, is not more frequent in cases with younger than with older nodules.[341]

A second implant of fibrin in a sensitized rabbit results in an accelerated reaction. In guinea-pigs, similarly, prior sensitization leads to the evolution of a persistent granuloma. Prolonged local responses of this kind are not evoked when soluble antigen is substituted for fibrin and the presence of a depot of antigen seems to be a prerequisite for the development of reactions morphologically comparable with the rheumatoid lesion. The reabsorption of a depot of insoluble antigen can also be delayed by local sensitization, the result of interference by antibody with the activity of macrophages during the attempted phagocytosis of insoluble antigen. Macrophage migration to the site of foreign antigen can be impaired by cytokines. This evidence suggests that excess local antigen is ineffectively phagocytosed both because of humoral and of cell-mediated mechanisms. Antibody-mediated lymphocytoxicity is impaired in rheumatoid synovitis and the mononuclear cells of rheumatoid granulomata may behave similarly.

Pathogenesis There are analogies with the nature and behaviour of the rheumatoid synovium (p. 449). On the basis of the morphological, immunological and cytochemical evidence, Ziff[457] supported the attractive, but not proven, hypothesis that nodule formation, in susceptible sites, originates with mechanically induced small blood vessel injury, catalysing the local accumulation in injured vessels of rheumatoid factor immune complexes. These complexes activate mononuclear macrophages. The number of macrophages is amplified by the liberation of chemotactic factors, including cytokines such as IL-1, released from the initially activated cells. A procoagulant, derived from the macrophages, catalyses the polymerization of fibrinogen and the resultant fibrin is seen as fibrinoid (p. 294). Tissue necrosis results from the mutual actions of neutral proteases, collagenase and other enzymes: they derive from the activated mononuclear cells. Chemotactic factors are also held responsible for the assembly of macrophages around the necrotic nidus, where oxygen partial pressures are presumably low. These chemotactic factors include transforming growth factor, tumour necrosis factor, granulocyte macrophage colony-stimulating factor and fibronectin. Macrophage receptors interact both with the polymerized fibrin and with the fibronectin that is deposited at the vascular margin of the nodule where oxygen partial pressures are presumably high.

Chronicity One explanation for the chronicity and indolence of rheumatoid granulomata may be the inability of the host to deal with large fibrin aggregates in which antigenic determinants have been so altered that anti-fibrin antibody has formed, resulting in the deposition locally of fibrin –antifibrin antibody immune complexes. The necrotic central material of the granuloma is not reabsorbed by the reparative response of vascular granulation tissue, as it would be in, say, the organization of an infarct. Vascular proliferation is induced by agents including tumour necrosis factor–alpha and transforming growth factor–beta. The local response to the zone of necrosis, perhaps the result of focal vasculitis, recalls the reaction to a focus of caseous necrosis in tuberculosis. It therefore appears that, whatever the initiating cause of rheumatoid granuloma, a mechanism of humoral and cell-mediated vasculitis overrides a reparative response and determines the persistence of the indolent zone of low-grade, active inflammation.

Involution Rheumatoid granulomata can involute spontaneously. The surrounding inflammatory cell exudate diminishes. An increased proportion of multinucleated giant cells forms in the palisaded cell layer where the residual cells lose their radial polarity and become progressively vacuolated. At this stage fibrosis is characteristic but the central zone retains a fluid content.[127] The establishment of a fistula is a serious event in the natural history of a subcutaneous granuloma. Bacteria may gain access. Treatment with corticosteroids accelerates involution. However, prolonged corticosteroid therapy can set the scene for the entry to an ulcerated granuloma or to the site of its excision, of antibiotic-resistant bacteria, particularly staphylococci. Bacterial and mycotic septicaemia and endocarditis may result.[148]

Cardiovascular system

The heart may be directly affected by rheumatoid arthritis. The arteries are frequently implicated. Venous disease is usually secondary.

Heart

The cardiac abnormalities encountered in rheumatoid arthritis may be primary or secondary.[201] There is the expected frequency of coincidental atherosclerosis, hypertension, calcific valvular disease and other abnormalities. All parts of the heart have been shown to be the occasional sites of rheumatoid lesions. The early accounts were summarized by Sokoloff[400] and by Gardner.[142] The prevalence of cardiac disease attributable to rheumatoid arthritis may be as high as 5–10 per cent of cases but the majority of patients who die with rheumatoid cardiac lesions do not come to necropsy. The pericardium/epicardium, myocardium, endocardium and valves, and the coronary blood vessels are all at risk. The heart may also be affected by disorders such as amyloidosis (p. 430) and bacterial or mycotic infection associated with rheumatoid arthritis or its treatment.

Pericardium Pericarditis is the most common abnormality[334] and in hospital necropsies, was recognized in 37 of 150 cases;[142] 15 of these displayed acute pericarditis, often fibrinous, while 22 examples were of chronic inflammation, often with fibrosis, fibrous adhesions or obliteration of the pericardiacal sac (Fig. 12.29). Either form may be however localized. The development of effusion is usually of little functional significance but cardiac tamponade has been described[420] and the development of constriction is an occasional reason for pericardiectomy.[25,219a,240] Microscopically, the lesions found in acute and in chronic rheumatoid pericarditis are not pathognomonic unless there are coincident rheumatoid granulomata. The structure of the pericardial granuloma closely resembles that of the common subcutaneous (see p. 485) and of the rare synovial (see p. 487) lesion. There are, however, few records of diagnosis by pericardial biopsy. The time of onset of pericarditis in rheumatoid arthritis is not clearly related to the duration of the disease although it is more probable during exacerbations.[201] Like the granuloma of rheumatoid arthritis, pericarditis may precede arthritis. However, patients with rheumatoid arthritis are prone to infection and it may be suspected that some instances of pericarditis (and of pleuritis) are manifestations of opportunistic, infectious, iatrogenic disease.

Myocardium A focal interstitial, non-specific myocarditis with lymphocytic and plasmacytic infiltration has been detected in 10–20 per cent of autopsy cases. Myocardial ischaemia, with necrosis and fibrosis, may also be the result of a coronary artery vasculitis.[414] Myocarditis may be extensive, resembling viral or Fiedler's myocarditis. A distinction from myocarditis precipitated by hypersensitivity responses to drugs used in therapy is not practicable although eosinophilia may suggest the latter diagnosis. When myocarditis coincides with granulomatous pericarditis in a seropositive case with systemic arteritis, there is a high probability that the myocardial lesion also is of rheumatoid origin. When myocardial granulomata are found they resemble those that form beneath the skin and in sites such as the pericardium and pleura.[374]

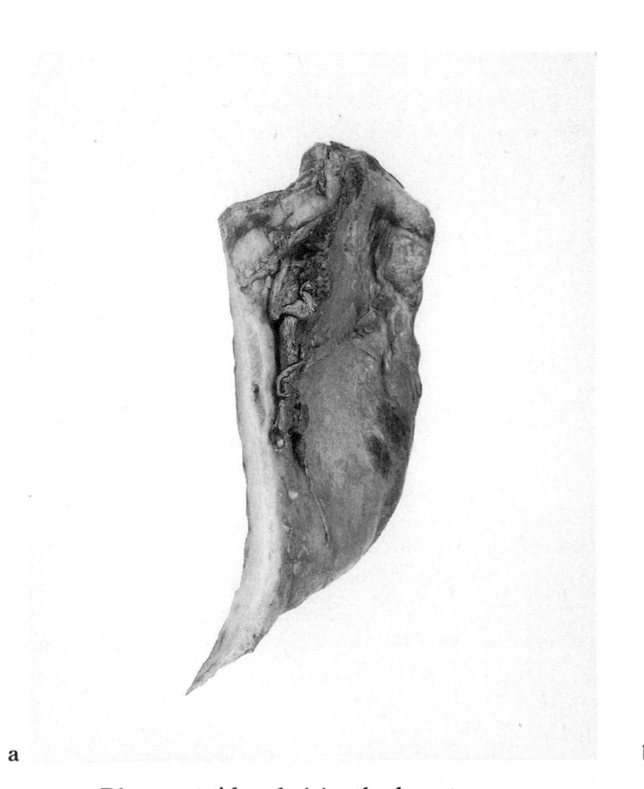

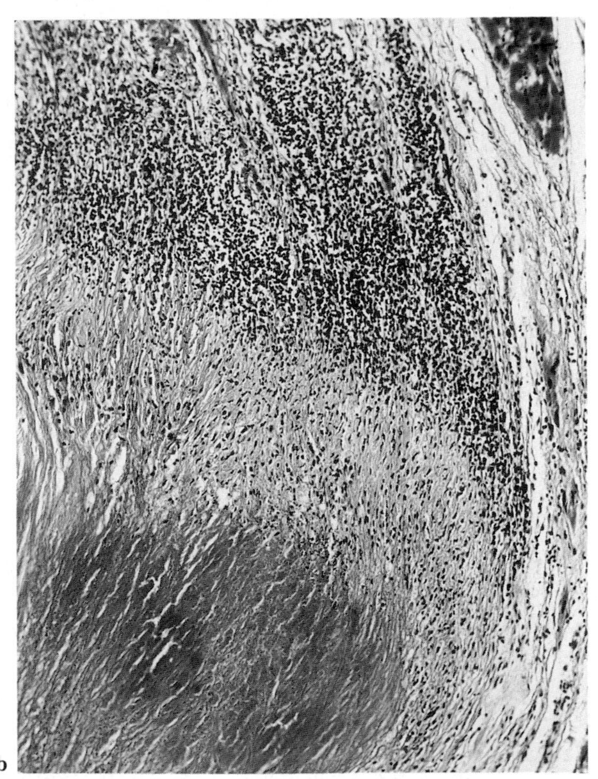

Fig. 12.29 Rheumatoid arthritis: the heart.
a. Constrictive pericarditis. Pericardial sac was excised surgically. Pale thickened parietal pericardium found to be infiltrated widely by lymphocytes and plasma cells. Much fibrous tissue present with many small blood vessels. **b.** Cardiac granuloma. Appearance of granuloma resembles that of subcutaneous, pulmonary and other similar nodules. Here the necrotic centre of the granuloma (bottom) lies at the base of the mitral valve. There is a margin of lymphocytes with a smaller number of mononuclear macrophages. (HE × 60.) Courtesy of Professor Bruce Cruickshank and the Editor, *Journal of Pathology and Bacteriology*.[88]

Endocardium Valvular lesions occur. They are usually of an indeterminate character. There is a lymphocytic, plasmacytic and histiocytic infiltrate. Vascularization takes place; red cells may escape and haemosiderin accumulates. Fibrosis develops: focal or diffuse thickening results with scarring and contracture, stenosis or incompetence. In approximately 3 per cent of cases, the valvular lesions are granulomata (see Fig. 12.29), most often affecting the aortic and mitral valves and readily identifiable as rheumatoid.[245] The granulomata are often small and do not cause anatomical deformity of a valve or valve ring; however, they are sometimes sufficiently large to replace much of the base of the valve and the connective tissue structure. All four valves are occasionally affected simultaneously.[351] In spite of the size of the granulomatous lesions, the endocardial surface is not usually disturbed and thrombi and verrucous vegetations tend not to form.[201]

Aortic valve Aortic valve disease is of particular interest because of the susceptibility of the aortic valve to incidental infection, to rheumatic endocarditis and to the lesions of ankylosing spondylitis (see Chapter 19) with which there may be an overlap.[347] Incompetence is the result of granulomata[351] or of non-granulomatous inflammation. The scarred, distorted valve can be replaced surgically[451] and valve prolapse may necessitate emergency surgery.[250] In rare instances, a congenital deformity, such as bicuspid aortic valve, underlies the inflammatory rheumatoid reaction[382] and may predispose to it.

Mitral valve Mitral valve lesions resemble, and are usually accompanied by, those of the aortic valve. Mitral disease is of lesser functional significance[201] and true granulomata have rarely been identified at surgery. The mitral lesion may lead to stenosis, incompetence or both, or the valve may retain a normal gross appearance. As with aortic valve disease, there is little relationship between the frequency of the clinical disease and the frequency of lesions found on microscopic study.

It is suggested[201] that there may be a negative correlation between rheumatoid valvular disease and bacterial endocarditis, perhaps because endocardial thrombosis is uncommon in the former.

Conduction defects occur periodically either because of granulomata affecting the conducting tissue of the interventricular septum or bundle of His, or because of a more diffuse inflammatory reaction.[314] Conduction defects may also appear either because there is (secondary) cardiac amyloidosis[421] or on the basis of coronary arteritis (see p. 492).[245] Complete heart block is rare. It is a complication of established, erosive rheumatoid arthritis with nodule formation.[5] In an analysis of 28 cases, the usual pathological finding was indeed a granuloma in or near the atrioventricular node or bundle of His. Complete heart block may coincide with aortic regurgitation[170] and culminate in infarction of viscera.[445]

Blood vessels

Vascular abnormalities play a large part in the pathogenesis of rheumatoid arthritis and the visceral and cutaneous complications of this disorder are frequently the result of vascular insufficiency.

Vasculitis is a particular association of extra-articular, visceral disease in erosive arthritis accompanied by high titres of IgM rheumatoid factor.[58] Blood vessels[375] of all sizes are susceptible but the most common forms of rheumatoid vascular disease affect the medium-size, muscular distributing arteries, the small arteries of the extremities and the venules. Many of the features of rheumatoid vasculitis are indistinguishable from and overlap with those of other forms of vasculopathy.[67] The reader is referred to Chapter 15. In a survey in which the clinical diagnostic criteria, response to therapy, prognosis, clinical laboratory and histopathological features were assessed over a five-year period, Scott *et al.*[375] showed that rectal biopsy, undertaken in 35 cases, was of particular value in diagnosis. Histological confirmation was also sought, and obtained, by skin (19 cases), kidney, ovary, appendix, lung or coronary artery (five cases) biopsy; and at necropsy (12 cases). Levels of IgG rheumatoid factor, anti-complement activity, and low levels of C4 were the best clinical markers for the presence of systemic vasculitis.

Arteries The mean age of patients with rheumatoid arthritis determines that these patients often have coincidental atherosclerosis and medial arterial calcification. An associated giant-cell arteritis is less likely. The possibility, suggested by the finding of low plasma α_2-globulin levels, that atheroma was less common in rheumatoid arthritis than normal, has not been substantiated.

Three principal forms of rheumatoid arterial disease are recognized: 1. intimal endarterial proliferation; 2. necrotizing arteritis; and 3. a low-grade, subacute arteritis. Granulomatous arteritis[347] is very uncommon. The diagnosis of arterial disease may only be made at autopsy or when surgery is undertaken for limb or visceral ischaemia. However, rectal biopsy (see above) is a safe, reliable procedure and enabled the recognition of necrotizing arteritis in eight cases among the 30 positive results that Scott *et al.*[375] obtained from 44 biopsies in 50 patients.

Intimal endarterial proliferation (Fig. 12.30) Peripheral vascular insufficiency in the limbs of those with rheumatoid arthritis is a well-recognized clinical problem.[58] Intimal, fibromuscular hyperplasia, with defective perfu-

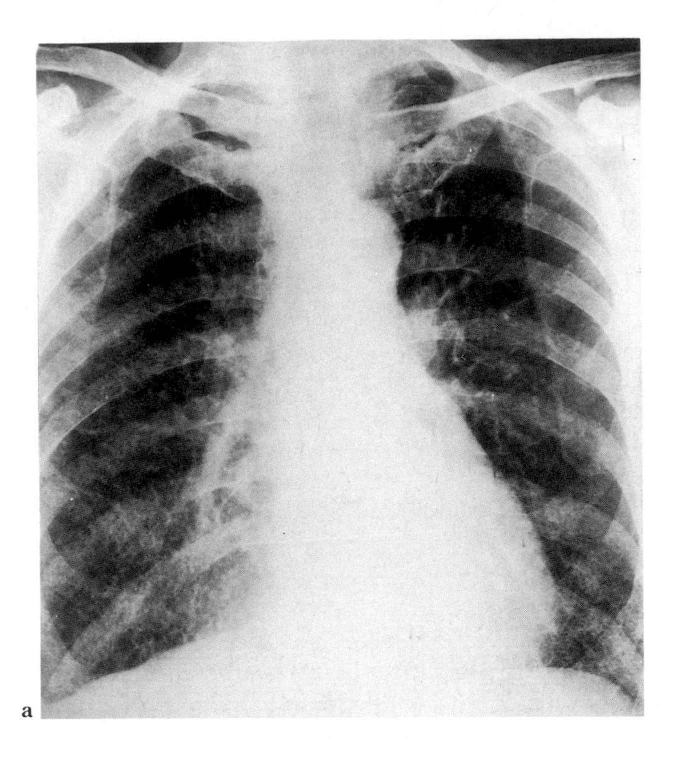

Fig. 12.33 Rheumatoid arthritis: the lung.
a. Radiograph of lung showing interstitial pneumonitis of right lower lobe with honeycomb appearance. **b.** Lung in long-standing rheumatoid arthritis. Lower part of upper lobe and entire lower lobe show widespread consolidation, fibrosis and cyst-like, honeycomb appearance.

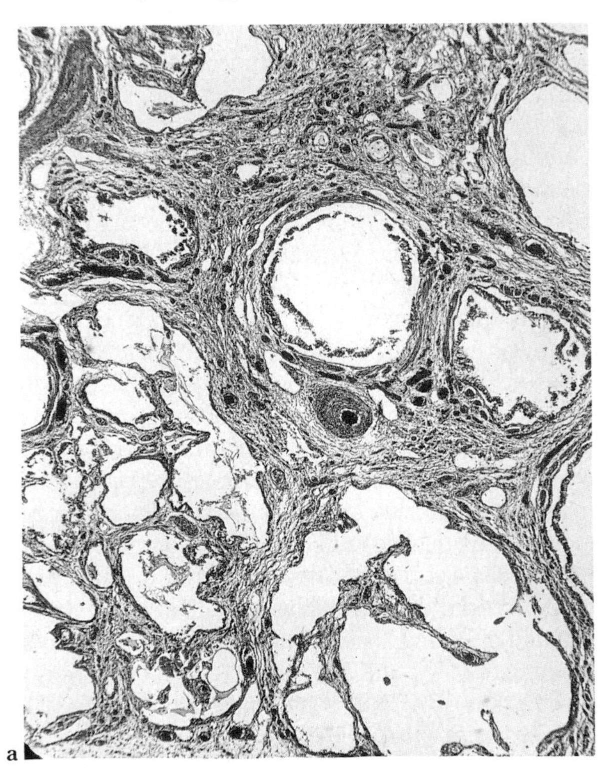

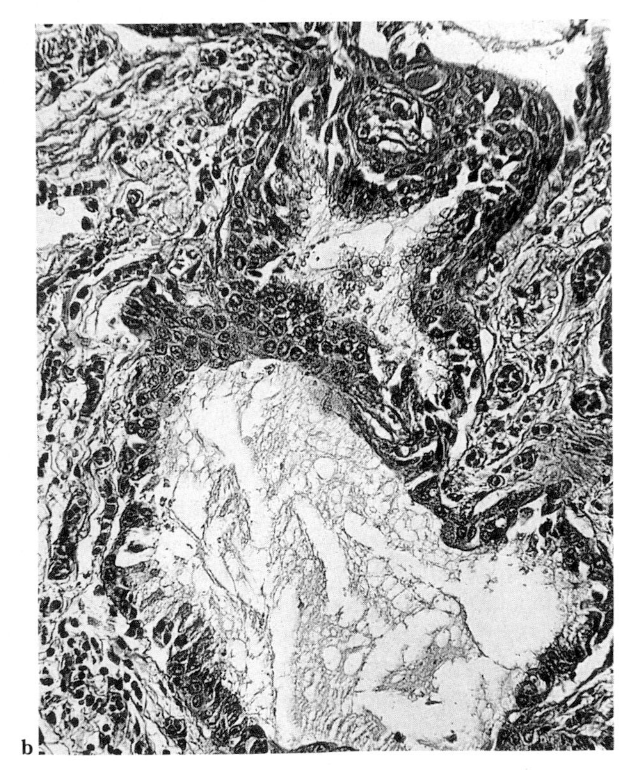

Fig. 12.34. Rheumatoid arthritis: interstitial pneumonitis with honeycomb lung appearance.
a. Enlarged air spaces, with destruction of interalveolar septa accompany dilated bronchioles lined by cuboidal epithelium. A pulmonary arterial branch (centre) is partly occluded by intimal fibromuscular hyperplasia. **b.** Residual air spaces are lined by hyperplastic cuboidal alveolar or low columnar bronchiolar epithelium. The appearances resemble those seen after lung injury and, occasionally, in systemic sclerosis. (HE. **a**: × 25; **b**: × 250.)

via polymorph collagenase. An enzyme of polymorph origin has been shown in bronchioalveolar lavage fluids.[434] The enzyme, in the small group of 10 patients tested by these authors, is apparently present entirely in active form.

An immune cell-mediated alveolitis is likely therefore to be the precursor of rheumatoid-associated pulmonary fibrosis.[356] The diffuse interstitial fibrosis that marks the end-point of this progressive disorder is not confined to rheumatoid arthritis but is identified periodically in patients with other systemic connective tissue diseases such as progressive systemic sclerosis and systemic lupus erythematosus. The interstitial fibrosis is also found in a variety of other states all of which have in common the likelihood of prolonged deposition, within the alveolar septa, of immune complexes that can activate complement and generate an insidious, widespread inflammatory reaction.

Pulmonary and pleural nodules Pulmonary granulomata[142,168] often develop close to a pleural surface but may be intrapulmonary (Fig. 12.35). Both the pleura-related lesions and those remote from a pleura are histologically closely similar to the subcutaneous granuloma (see p. 485). Needle biopsy may establish diagnosis effectively but this is an uncertain test. So important is it to establish the precise nature of a lung nodule or mass that thoracotomy is to be preferred. Occasionally, a pleura-related granuloma can be shown to be intimately associated with a focus of pulmonary vasculitis.[13] In rare instances, a rheumatoid nodule may form in the wall of a bronchus.[208] Carcinoma of the lung is suspected in differential diagnosis. Seropositivity is characteristic.

Pleural and pulmonary granulomata appear as discrete rounded, faintly radio-opaque, intrathoracic shadows. They may number as many as 20.[422] They vary in size from 10–20 to 70 mm in diameter and may be single or multiple. Their formation is unrelated to pneumoconiosis[359] and is not confined to men, in whom the granulomata are, however, twice as frequent as in women. The granulomata often coincide, in seropositive patients, with the formation of nodules subcutaneously and in the viscera.

Pneumoconiosis and Caplan's syndrome Caplan[59,60] recognized that seropositive coal workers with

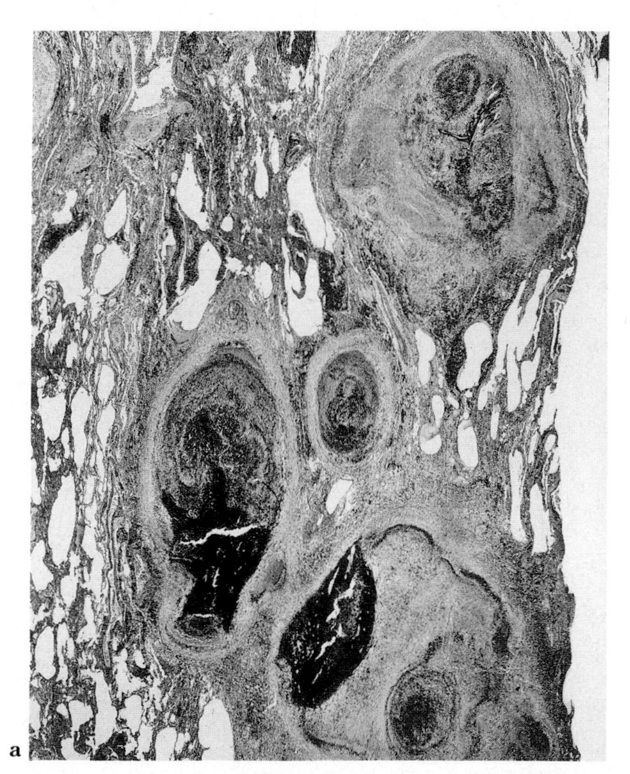

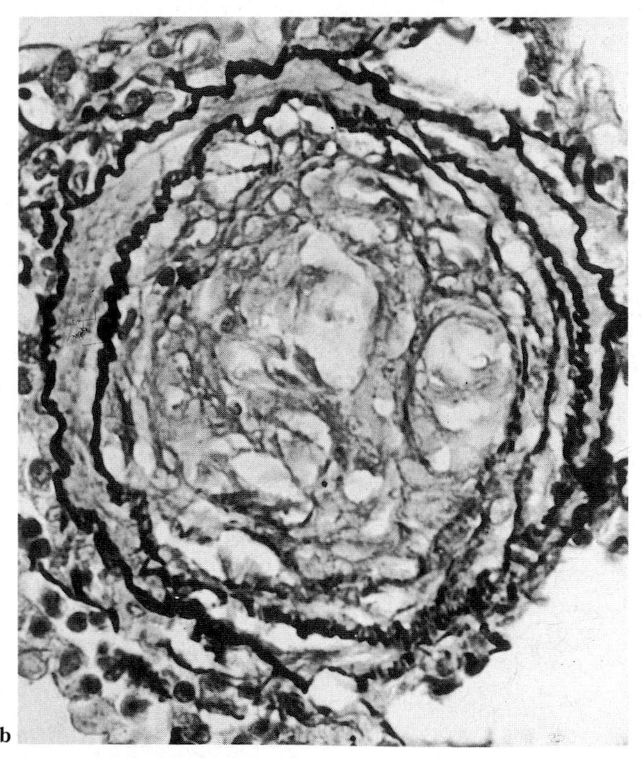

Fig. 12.35 Rheumatoid arthritis: pulmonary granulomata (Caplan's syndrome).
a. Contiguous rheumatoid granulomata, laden with carbon pigment, coalesce within this emphysematous and fibrotic lung from a coalminer with pneumoconiosis. Diagnosis of rheumatoid granuloma is confirmed by recognition of necrotic central zones within the nodules. (HE × 7.) **b.** Pulmonary arterial branch in lung of young woman with pulmonary hypertension and Raynaud's phenomenon. Thrombosis and recanalization accompany reduplication of the prominent elastic laminae. (Weigert's elastic—van Gieson × 375.)

rheumatoid arthritis tended to form discrete intrapulmonary rheumatoid granulomata (see Fig. 12.35), sometimes before the onset of the polyarthritis. Rheumatoid granulomata may also develop in workers exposed to gold, asbestos and dolomite.[11] The common precipitating factor is thought to be the silicate content of the ores[28] from which these substances are taken. By the same token, those seropositive individuals exposed to silicates in file-making and boiler scaling and in the manufacture of grinding wheels[330,407,422] were also prone to rheumatoid pulmonary nodule formation.

The histological features of Caplan's granulomata have been very fully described in texts on pulmonary pathology.[185]. But the mechanism of origin of rheumatoid lung nodules is no better understood than the aetiology of subcutaneous granulomata (see p. 485). However, it is highly likely that quartz particles, with protein adsorbed to their surfaces, can exert a toxic action on pulmonary macrophages, disorganizing the T-cell-mediated immune response to autologous or exogenous antigens. Localization of the granulomata, with the immobilization of silicate-containing macrophages in peribronchiolar foci, appears to have much in common with the genesis of silicotic nodules. Necrosis of the centres of pulmonary rheumatoid granulomata is likely to be due to the activation of macrophage proteases.[229]

Pulmonary vasculopathy Necrotizing arteritis and intimal arterial and arteriolar proliferation are the principal lesions. The latter is a frequent feature of diffuse interstitial pulmonary fibrosis with 'honeycomb-lung' change. Necrotizing vasculitis is usually multifocal but may be localized.[13] Arteriolar intimal fibromuscular proliferation may be a rare cause of idiopathic pulmonary hypertension (see Fig. 12.35).[146] Pulmonary vascular disease in rheumatoid arthritis can complicate treatment and has been identified in patients receiving renal dialysis for amyloid nephrosclerosis.

Infection[428]Patients with rheumatoid arthritis, whether subjected to treatment with corticosteroids or cytotoxic drugs or not, exhibit a lower resistance to bacterial infection than normal (see p. 531). It is not surprising, therefore, that they should be prone to chronic bronchopulmonary infection. Fistulae, arising at the sites of subpleural nodules, are effective portals of entry for organisms causing empyema, a common complication of longstanding rheumatoid arthritis.[209] Empyema has also been described as a complication of jugular venous catheterization.[10]

The bronchi may be the site of bronchiectasis although the evidence for this diagnosis is usually radiological rather than histological. In one series, 16 cases of 516 were affected. In 13, bronchiectasis preceded arthritis. Episodes of bronchopneumonia were common, and 22 per cent of the males and 4.5 per cent of the females in Walker's[428] study had chronic bronchitis, often commencing before the arthritis.

Less common pulmonary diseases

Bronchiolitis obliterans Obliterative bronchiolitis, a lesion most often resulting from the inhalation of toxic fumes, is one consequence of penicillamine therapy for rheumatoid arthritis.[322] Obliterative bronchiolitis of a constrictive type can also be found by meticulous injection studies and by serial section, without the mucosal gland hypertrophy or excess mucus secretion of chronic bronchitis, and without a history of exposure to chemical irritants or evidence of virus infection. The occluded bronchioles are thickened by granulation tissue. The inflammatory infiltrate, which is predominantly lymphocytic, may extend to the peribronchiolar tissues. Rheumatoid patients with obstructive airways disease are unusually non-smokers. Obstruction may be the result of peribronchiolar lymphoplasmacytic infiltration, destroying small airways in the same manner as the parotid salivary gland tissue is destroyed in Sjögren's syndrome.[27,129] Subsequently, the affected bronchioles may be wholly obliterated by collagenous connective tissue.[27] However, the cause of some cases remains uncertain.[152] The role of virus is suspected.

Eosinophilic pneumonitis Peripheral blood eosinophilia may complicate rheumatoid lung disease and be associated with pneumothorax.[82] Rarely, a more acute hypereosinophilic syndrome with vasculitis and thrombosis, may prove fatal.[188] A chronic eosinophilic pneumonia may coexist with bronchiolitis obliterans[78] but it is not known whether this is a coincidental association. The cavitation of rheumatoid pulmonary nodules may be a cause of pneumothorax and it is of interest that eosinophilia may coexist.[330]

Upper lobe cavitation syndrome Ankylosing spondylitis is not infrequently associated with an unusual form of upper lobe fibrotic and cavitating lung disease (see p. 779). Although a similar process may be suspected in rheumatoid arthritis, the cavitation of sterile granulomatous nodules is more frequent.[262] The histological appearances of cavitating rheumatoid pulmonary nodules is closely similar to those of the subcutaneous nodule. However, less is known of the phenotype of the cells comprising the lung lesion. The differential diagnosis of cavitating lung lesions has been discussed fully.[235]

Bronchocentric granulomatosis In the small number of cases of bronchocentric granulomatosis that have been reported in rheumatoid arthritis, the bronchi and bronchiolar lesions have generally been regarded as coincidental. Rarely, however a leucocytoclastic vasculitis (see p. 628) has been encountered in a seronegative rheumatoid patient with a bronchocentric, destructive process.[182]

Drug reactions The response to penicillamine is mentioned on p. 552. Pulmonary disease may complicate the administration of gold and methotrexate.

Urinary system

The age distribution of rheumatoid patients determines that many have coincidental urinary tract disease before the onset of arthritis while others develop hypertensive arteriolosclerosis, infection, neoplasia or calculus during the long course of the connective tissue disease (Fig. 12.36). The possibility that there might be a distinct form of rheumatoid nephropathy arose because of the widespread use of increasingly refined techniques for assessing biopsy material and also because of the need to distinguish spontaneous from iatrogenic lesions. Studies of renal function suggest that renal abnormalities are more frequent than in a control population[94] and measures of renal damage such as the excretion of N-acetyl-β-D-glucosaminidase correlate with rheumatoid disease activity.

Autopsy evidence

The published autopsy evidence is not easy to interpret. Many patients receive drugs with a potential for renal injury. There is also a need to define histological criteria precisely to allow comparisons between different series of cases. Disease patterns, and the definitions used to describe them, change continually. The early literature was reviewed.[142] Lawson and McLean,[244] for example, found renal lesions in 72 per cent of their cases. In compiling a table (Table 12.8) of the disorders itemized in three series, nephrosclerosis, chronic pyelonephritis, interstitial fibrosis and end-stage kidney are grouped together since it is unclear how they had been defined. They appear to comprise more than 90 per cent of the abnormalities identified at autopsy in the affected kidney. The frequency of renal vasculitis and of hydronephrosis and calculus varies in different populations and may have altered with time. The importance of renal amyloidosis (see Fig. 12.36) (see Chapter 11) remains high; there is no wholly effective treatment. Interest then centres on the remaining, much less common forms of renal disorder, particularly on glomerular abnormalities. Whether there is an entity which can be defined as 'rheumatoid renal disease' remains controversial.

Biopsy evidence

Mahallaway and Sabour[268] were among the first to postulate the existence of 'rheumatoid kidney'. Brun *et al.*[46]

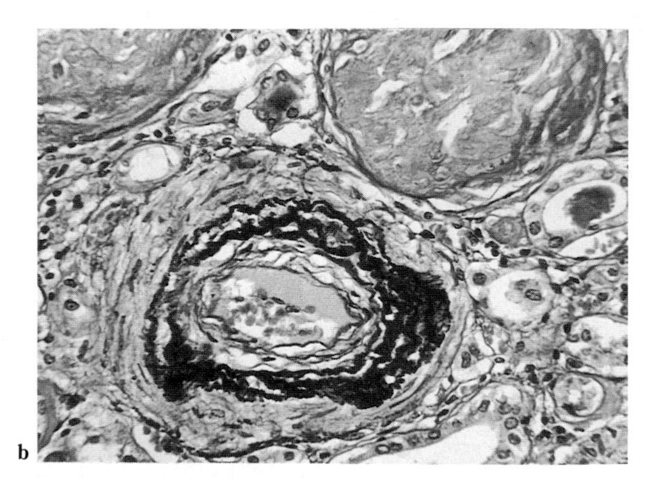

Fig. 12.36 Rheumatoid arthritis: renal disease.
a. A glomerulus is transected in the plane of the distal convoluted tubule and juxtaglomerular apparatus. There is fibrinoid change attributable to coincidental accelerated hypertension. Appearances simulate those of systemic sclerosis. **b.** Amyloid nephrosclerosis in kidney from a patient with chronic renal failure. Amyloid has provoked the formation of much new collagen within the obsolete glomeruli (at top) and the wall of an interlobular artery (centre) which displays excess elastic material and reduced lumen. (Weigert's–van Gieson × 240.)

Table 12.8

Frequency of renal disease in rheumatoid arthritis—assessed at autopsy

	Gardner[142] (148 cases) (%)	Ramirez[340] (76 cases) (%)	Boers *et al.*[40] (132 cases) (%)
Pyelonephritis	39 (26)	6 (7) acute	
Nephrosclerosis		41 (54)	118 (90)
End-stage kidney		35 (46)	11 (8)
Interstitial fibrosis		33 (43)	
Renal vasculitis	2 (1)	7 (8)	8 (6)
Hydronephrosis	4 (10)		4 (3)
Amyloidosis	17 (11)	5 (6)	14 (11)
Malformation	3 (2)		13 (10)
Infarct	2 (1)	10 (13)	5 (4)
Calculi	2 (1)	4 (5)	7 (5)
Papillary necrosis	11 (7)	3 (3)	2 (2)
Pyonephrosis	2 (1)		1 (1)
Perinephric abscess			
Glomerular disease	1 (1)		
Focal			5 (8)
Membranous			9 (7)
Mesangial		5 (6)	27 (38)
Proliferative		3 (4)	11* (14)
Hyalinization		34 (46)	
Tubular disease	2 (1)		
Tubulointerstitial nephritis			5 (3)
Necrosis			2 (2)
Tumours	0	0	9 (7)

*Six with vasculitis.
There is little doubt that the main differences shown between these series related to the increasing attention paid in recent years to microscopic disease.

could not confirm the identity of this disorder in their survey of 32 rheumatoid patients. The renal lesions found were highly heterogeneous. Pasternack *et al.*[321] took the opposite view and held that the local glomerulitis and interstitial tissue changes found in young patients with a recent onset of rheumatoid arthritis and no history of heavy or high analgesic abuse 'made the existence of rheumatoid kidney disease at least probable'. Subsequent investigations, with increasingly sophisticated histological methods, have tended to support this opinion. Akikusa *et al.*,[6] for example, identified five examples of glomerulonephritis unrelated to drug therapy, in the course of rheumatoid arthritis in female patients aged 36 to 61 years. Although angiitis was present in one case and a second displayed a systemic lupus erythematosus 'overlap' syndrome, three were accepted as rheumatoid glomerulonephritis. The pathogenesis was related to immune complex deposition; unlike systemic lupus erythematosus, rheumatoid factors are not 'protective'.[181] Necrotizing, crescentic, mesangioproliferative, mesangiocapillary and membranous glomerulonephritis were identified. In support of this evidence, and of their earlier work, Pasternack and his colleagues[180] demonstrated that a mild mesangial 'glomerulopathy' could be found in 12 of 39 rheumatoid patients, one of whom had amyloidosis, nine membranous glomerulonephritis, and two mesangial glomerulopathy together with extraglomerular amyloidosis. The mesangial lesions almost invariably contained immunoglobulin and complement. However, it was conceded that rheumatoid mesangial glomerulopathy 'may not represent a distinct clinicopathological entity'. In further studies Kuznetsky *et al.*[239] concluded that necrotizing glomerulonephritis is part of the spectrum of rheumatoid glomerular lesions; while Higuchi *et al.*[187] reported an example of membranous glomerulonephritis in a 16-year-old girl who had not been given gold or penicillamine. Although this syndrome is rare, they claim that rheumatoid arthritis is causally related to the renal lesion. It certainly appears that rheumatoid arthritis may present as a nephrotic syndrome without other cause for the renal disease.[388]

A large literature attests to the part anti-inflammatory and anti-rheumatic drugs play in causing renal disease in rheumatoid arthritis, both in adult and in juvenile arthritis.[8] Before its ban, phenacetin had long been associated with the development of non-obstructive pyelonephritis.[244] Although renal papillary necrosis was provoked by compounds containing this drug, corticosteroids could be protective. Renal papillary necrosis has also been attributed to non-steroidal anti-inflammatory drugs (NSAIDs) including phenylbutazone, indomethacin, ibuprofen and other analogous substances.[202] Salicylates may impair renal tubular function[52] but greater importance attaches to the development of membranous glomerulopathy in those given gold.[363] The development of gold nephropathy may be related to an absence of serum IgM rheumatoid factor.[394] There is a genetic predisposition to this response (see p. 531).

Alimentary system

Salivary glands

Enlargement of the salivary and lacrimal glands is an inconstant feature of Sjögren's syndrome (see p. 529). Alternatively, the glands may be fibrotic or atrophic. The acini may be displaced and occupied by a lymphocytic infiltrate with smaller proportions of plasma cells or they may be degenerate, containing inspissated eosinophilic material, or necrotic. Subsequent to epithelial cell injury, fatty and fibrous tissue forms. Analogous changes occur in the oesophageal glands.[100]

Principal salivary glands The mononuclear cells that occupy the salivary glands in Sjögren's syndrome are an index of an autoimmune reaction directed against a heterogeneous range of antigens.[398] Some of the humoral antibodies formed are against salivary gland and duct cell components. They are associated in the serum with anti-smooth muscle, anti-skeletal muscle, anti-liver, anti-gastric parietal cell and antiendocrine gland antibodies. The serum anti-salivary duct antibodies are of high specificity. Their presence tends to correlate inversely with the degree of focal lymphocytic infiltration. However, their formation may be an epiphenomenon.[440] The lymphocytic infiltrate includes many T_H-cells; the proportion of T_S-cells is decreased.[296] Uncontrolled clonal expansion may occur and there is a propensity for lymphoma to arise in the Sjögren syndrome. An intermediate phase, pseudolymphoma, with progressive lymphadenopathy involving the salivary glands, is recognized. The phenotype of the implicated lymphocytes is usually T-cell.

The interpretation of the structural changes in the salivary glands in rheumatoid arthritis takes account of the age-related alterations to which all organs are subject. Nowhere is this more important than in the alimentary tract where atrophy is a normal corollary of age. Drummond and Chisholm[104] found that in labial salivary glands, the acinar volume declined with age; simultaneously, the ductal and connective tissue volumes increased while the vascular volume remained unchanged. They measured these parameters in 36 rheumatoid patients, 18 of whom had Sjögren's syndrome. There was an overall reduction in salivary gland acinar volume but a significantly greater reduction in patients with Sjögren's syndrome. The volumes of connective tissue increased correspondingly but those of the ductal and vascular tissues did not differ from control values.

Labial salivary glands The proportions of T-cells and of B-cells in the labial salivary glands have been determined and compared with the proportions of mononuclear macrophages and mast cells.[415] The numbers of both T- and of B-cells were raised, those of the macrophages lowered. The number of mast cells was not altered and did not correlate with T-cell or B-cell counts. However, the proportions of mast cells were directly related to the degree of fibrosis and fatty tissue. This evidence indicated that in labial salivary glands fibrofatty change was an age-related, not a rheumatoid, phenomenon.

Mouth and oesophagus

There may be mucosal atrophy. Rheumatoid granulomata are rare (see p. 485). Although amyloid is not infrequently deposited in the tongue, mouth and lip, these sites are seldom used for diagnostic biopsy for which the rectum is preferred. One of the arguments in favour of this preference is the evidence that rectal biopsy permits vasculopathy (see p. 491) to be diagnosed as well as amyloid. However, it should be remembered that arteritis may affect oral and oesophageal vessels as well as intestinal. In both locations, vasculitis may coexist with amyloidosis.

Stomach and intestine

As many as 38 per cent of patients with rheumatoid arthritis have peptic ulcers but the number may be no greater than in osteoarthrosic controls.[269] An increased frequency of peptic ulcer in rheumatoid arthritic (as well as osteoarthrosic) patients has been attributed, without convincing histological evidence, to the administration of NSAIDs. One factor may be a heritable defect in mucus glycopeptide. Experimentally, a low proportion of high molecular weight mucus glycoprotein predisposes to gastric ulcer.[19] The high frequency of peptic ulcer was emphasized by Farah *et al.*;[120] 36 per cent of 185 patients with rheumatoid arthritis, but only 29 per cent of 45 patients with non-rheumatoid rheumatic disorders, had peptic ulcers, determined by endoscopy. The relative frequency of gastric ulcer was much higher in rheumatoid patients.

Chronic superficial gastritis, present in 30 per cent of 59 cases, and chronic atrophic gastritis, identified in 62.5 per cent, are also frequent.[272] However, as Doube and Collins[102] emphasize, they are common, age-related findings in the non-rheumatoid population. There are no histological features in these forms of gastritis nor in the chronic atrophic gastritis of Sjögren's syndrome that distinguish rheumatoid from non-rheumatoid patients. From time to time, rheumatoid patients with chronic atrophic gastritis develop megaloblastic anaemia.

Intestine

Although there is no direct evidence that the intestinal tract is intrinsically abnormal in rheumatoid arthritis,[102] the gut may play an important part in the immunopathogenesis of some cases which are identifiable by the demonstration of raised levels of IgA rheumatoid factor and wheat protein IgG. Gluten appears to be a common immunizing antigen.[312] Intestinal function may be impaired[325] and systematic studies of the duodenojejunal mucosa have shown villous atrophy of varying degrees more commonly in seronegative than in seropositive cases.[155] In the rare and exceptional cases with total villous atrophy, signs of arthritis disappear on a gluten-free diet, suggesting a relationship to latent forms of coeliac disease.

Arteritis, with ischaemic colitis or intestinal infarction, and amyloid deposition are well-recognized histological disorders in rheumatoid arthritis. Bowel infarction may also result from a mesenteric arterial proliferative endarteritis[260] similar to that often found in digital vessels (see p. 491). Non-specific inflammatory changes, with glandular atrophy and lymphocytic infiltration, are often detected in both small and large gut. Although a change in the small intestinal flora has been invoked in the pathogenesis of MHC class I HLA-B27-associated (reactive) arthropathies (see Chapter 19), there is little pathological data to show that a change in the structure or ultrastructure of the gut in rheumatoid arthritis contributes to this phenomenon. Analogous transmission light- and electron microscopic changes have been found in rectal mucosal cells.[412] The transmission electron microscopic changes include the accumulation of iron. However, there is no functional defect and the abnormalities may be iatrogenic.

Liver

Liver disease in rheumatoid arthritis is held to be uncommon but the liver is enlarged in approximately 10 per cent of patients[431]—an observation substantiated by scintiscan (see Chapter 3) results. There is frequent evidence of biochemical abnormality.[280] The extent to which liver disorder is attributable directly to the rheumatoid arthritis or to the numerous drugs given in treatment, is debated.[413] Few rheumatoid patients with pathological evidence of liver disease remain untreated with compounds that may be hepatoxic. Gold, salicylates, phenylbutazone, corticosteroids and azathioprine can cause hepatic abnormalities but penicillamine and the NSAIDs spare the liver.

Chemical disorders

Other tests of liver function are often altered because of the serum protein changes that are part of the rheumatoid disorder; markers of hepatocellular damage such as serum ornithine carbamyl transferase may be elevated although the possible influence on these tests of drugs such as gold and phenylbutazone must be considered. The serum alkaline phosphatase levels may also be raised: the elevation is of the liver isoenzyme. Since this rise may be accompanied by increased levels of other enzyme activities such as that of γ-glutamyltranspeptidase, it has been suggested that they correlate with rheumatoid activity[413] even though they do not reflect the severity of histological changes in the liver parenchyma.

Morphological changes

Histological abnormalities are regularly but not consistently observed.[280] Liver disease may contribute to death. One representative study of 31 patients with 'classical' or 'definite' rheumatoid arthritis[280] showed that 15 had hepatomegaly and/or splenomegaly. Of these, only three had normal liver function tests. Of the 31 livers, four were histologically normal, and 23 showed non-specific abnormalities such as fatty change and mononuclear cell infiltration. Only four revealed definite, structural disease: primary biliary cirrhosis, chronic active hepatitis, amyloidosis and alcoholic cirrhosis. In a further survey, of 26 patients, fatty liver (7), passive venous congestion (5), haemosiderosis (1) and cirrhosis (1) were recognized[95] but no correlation could be shown between the activity of the arthritis, the duration of the disease, and the extent of histological liver disorder. Amyloid was not identified. Primary biliary cirrhosis is indeed periodically encountered[114] and this disorder may coincide both with Sjögren's syndrome in which there is a raised frequency of liver disease and with autoimmune thyroiditis. Nodular, non-cirrhotic liver disease occurs not only in

Felty's syndrome but also may be recognized in the presence of portal hypertension with relatively slight splenomegaly and anaemia but not leucopenia.[171]

Haematopoietic system
(Fig. 12.37)

Erythrocytes[64,173,298]

The early evidence relating to the nature of the anaemia of rheumatoid arthritis is contained in a long series of reports by Duthie and his colleagues. Their papers are reviewed by Mowat[297] and by Gardner.[142] A moderate normocytic, hypochromic anaemia is one of the most characteristic features of rheumatoid arthritis. The cause of the anaemia is still not fully understood. Occasionally, there is true iron deficiency, judged by serum ferritin levels,[37] but this is not usual and may reflect coexistent disease, such as persistent bleeding from the gastric mucosa. Under these circumstances, there may be impaired mucosal iron absorption,[432] an observation at variance with the results of Benn *et al.*[29] When iron absorption is measured directly, uptake is normal in those with replete iron stores but paradoxically low in those with depleted stores.

Mucosal iron transfer is low in the former group. There is a low serum iron and low total iron-binding capacity. Iron is cleared quickly from the plasma although exogenous iron transferrin is incorporated into red blood cells only to a normal or reduced degree.

It appears therefore that iron, freely available in the diet, is absorbed normally but used inadequately for the synthesis of haemoglobin. The body contains ample reserves of iron in the bone marrow, spleen and liver. Indeed, the mean total splenic iron may be above normal. Iron is also sequestered within the hypertrophic synovial villi (see p. 459); the quantity retained may be a factor in determining anaemia.[300] Haemosiderin lies free within the marrow interstices and in mononuclear phagocytes. Ferritin is also present and may be found not only in macrophages but also in plasma cells. There is additional limited evidence of an increased rate of red blood cell destruction but the reduction in red cell survival is slight. Normal red cells transfused to rheumatoid patients survive normally but the autologous red cells of a patient with active disease have a diminished life span. One hypothesis advanced to explain the anaemia of rheumatoid arthritis therefore proposes that the red blood cell production rate fails to compensate for the slight but persistent increase in the rate of red cell destruction.

A defect in the release of iron from the reticuloendothelial system may be an important factor in the impaired synthesis of haemoglobin. When ineffective and effective haem synthesis were measured, it was found that the total erythroid haem turnover was decreased in rheumatoid arthritis but the haem turnover due to ineffective erythropoiesis was much increased.[362] The marrow iron supply was measured and distinguished from non-erythroid iron turnover. In rheumatoid arthritis, marrow iron turnover is normal. However, with the exception of iron-deficient individuals, ineffective iron turnover is less than normal.[64] In iron-deficient persons, there is a clear difference: ineffective iron turnover is much increased.

There may be other inherent abnormalities in the responsiveness of bone marrow erythropoietic precursor cells. Thus, there is an inverse relationship between immunoreactive erythropoietin and haemoglobin concentrations in anaemic rheumatoid patients. Erythropoietin synthesis may be defective but the explanation for this deficiency is not clear. The evidence for renal disease sufficient to account for this abnormality is debated (see p. 499). The erythropoietic response to the anaemia of rheumatoid arthritis is blunted.[18] Interleukin-1 is a humoral inhibitor of erythropoiesis; it may play a part in the development of the anaemia.[275] The overall proportions of red cell, while cell and platelet precursors is nevertheless normal. A reduced proportion of marrow erythroblasts has been described but this is not constant. It remains possible that erythroblast behaviour is selectively impaired. It should also be noted that bone marrow blood vessels are susceptible to arteritis and to amyloidosis[49a,416] and that these disorders may affect haematopoiesis.

A further contribution to the anaemia of rheumatoid arthritis may be the suppression, perhaps mediated by an immune mechanism, of the cell proliferative capacity of the bone marrow. Cell culture techniques have been used to assess the number of erythroid colonies formed by circulating progenitor cells.[173] The numbers of erythroid colonies correlates inversely with serum IgM and rheumatoid factor concentrations. A role for immune suppression is thus indirectly demonstrated. The nature of the factor(s) suppressing erythropoiesis is speculative. Adherent macrophages co-cultured with erythroid progenitor cells enhance the growth of bone marrow-forming colonies and this mechanism operates in anaemic rheumatoid patients as in normal subjects.[346] Mononuclear phagocytes themselves, therefore, do not exert a direct inhibitory effect on erythropoiesis. However, there may be an unexplained deficiency of IL-4.

Leucocytes

With the exception of the Felty syndrome (see p. 529) in which splenomegaly accompanies or causes hyperplenism, there is no constant white cell defect in rheumatoid arthritis. Leucopenia or agranulocytosis may, of course, result

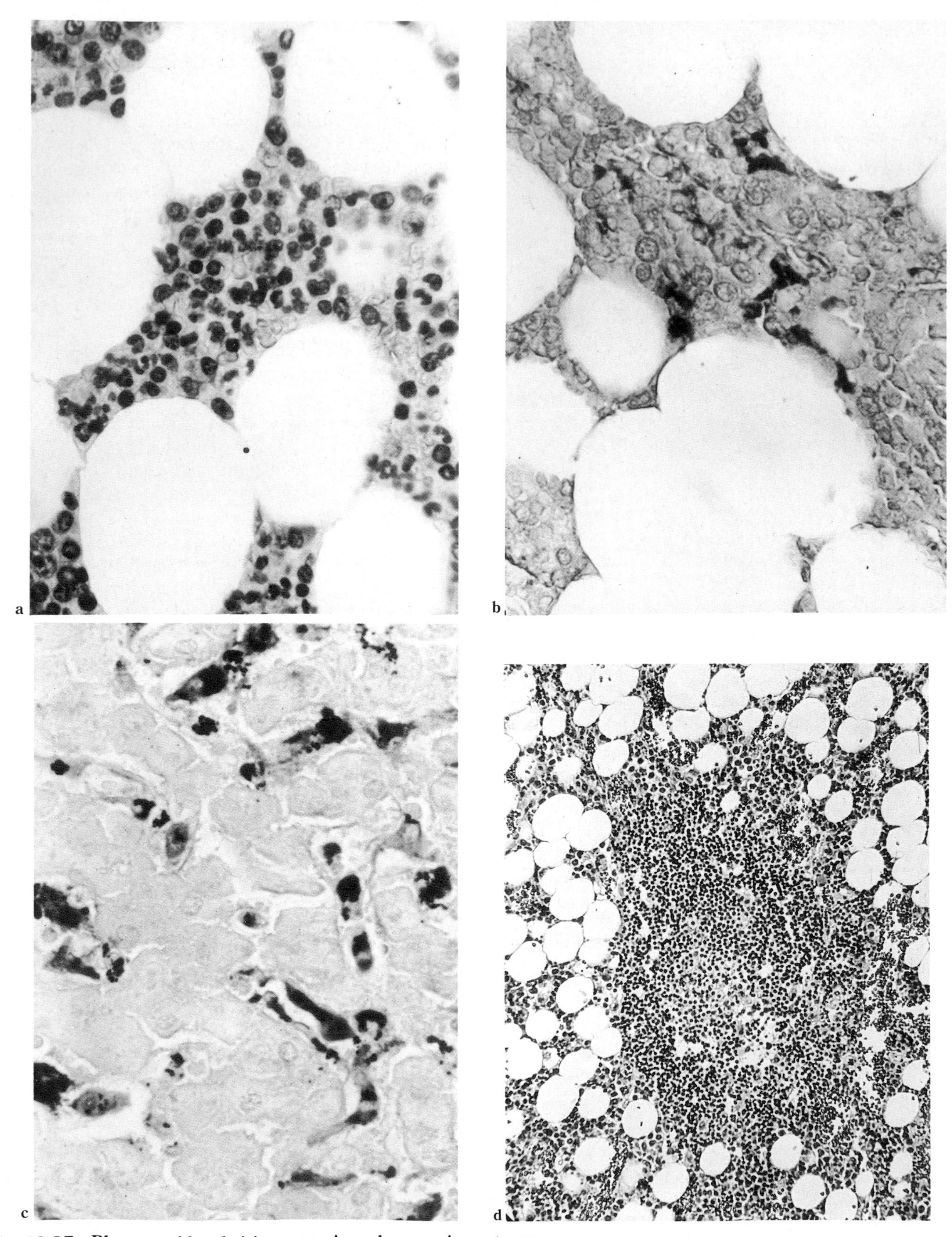

Fig. 12.37 Rheumatoid arthritis: anaemia and excess iron storage.
a. The hyperchromic or normochromic, normocytic anaemia of rheumatoid arthritis is accompanied by a bone marrow in which red cell, white cell and platelet precursors are present in normal proportions. **b.** Bone marrow contains at least normal quantities of stainable iron. **c.** Iron is stored in excess in the cells of the reticuloendothelial cells. In this liver, iron dextran given therapeutically is retained in an analogous distribution within Kupffer cells. **d.** Lymphoid follicles are recognized with increased frequency in the bone marrow. (**a**: HE × 380; **b**: Prussian blue reaction × 380; **c**: Prussian blue reaction × 380; **d**: HE × 140.)

from adverse drug reactions (see p. 552). Rheumatoid arthritis is one explanation for the presence in the bone marrow of lymphoid follicles, sometimes with germinal centres.[124] There is also usually an excess of marrow plasma cells the numbers of which are proportional to the levels of serum immunoglobulin.

The leucocytes in rheumatoid arthritis may display morphological differences from normal. Scanning electron microscopic studies suggest that peripheral blood polymorphs are often highly polarized. Transmission electron microscopy demonstrates the presence of many phagocytic vacuoles, a few electron-dense primary and secondary granules. Collectively, the electron microscopic changes resemble those recognized in normal polymorphs on exposure to C5a *in vitro*, during adherence to endothelial cells *in vivo* or during phagocytosis.[259]

Lymphoreticular system

A generalized disturbance of the immune system is characteristic of active rheumatoid arthritis and this is reflected in remarkable changes in the histology and cytology of the lymphoid tissues, particularly the lymph nodes, spleen and bone marrow. However, it should not at once be assumed that these aberrations are directly attributable to dysfunction of the cells mediating immune responsiveness. Subjects with rheumatoid arthritis often have other, coincidental diseases and they are unusually susceptible to infection (see p. 531). The cellular and tissue characteristics of the lymphoid organs in rheumatoid arthritis, like the disorders of the synovial tissues, mirror those of the blood and lymph. In any histological description, the observer is handicapped by the need to offer two-dimensional descriptions of continuously changing, three-dimensional processes. This limitation and those of sampling, contribute to differences in the interpretation of the lymphoid tissue in rheumatoid arthritis that are not yet resolved.

Lymph nodes

Generalized enlargement of the lymph nodes may be detected in three of every four rheumatoid patients[101, 172,295,310,418] and may be an early feature of the disease.[216] It is not confined to nodes draining inflamed joints and the distribution of the enlarged nodes is not a close reflection of the pattern of regional arthropathy.[295] The para-aortic, axillary, cervical and mediastinal nodes are among those commonly affected. The most conspicuous, and constant feature of rheumatoid lymphadenopathy is follicular hyperplasia (Fig. 12.38).[248] The size of the secondary follicles and their germinal centres is sometimes so great that the presence of follicular lymphoma may be suspected. Occasionally, patients with rheumatoid lymphadenopathy have been subjected to radiotherapy because of this interpretation. The histological features of the regional lymph nodes in juvenile rheumatoid arthritis and in Felty's syndrome are qualitatively similar to those in conventional adult rheumatoid arthritis.

Cellular changes Although the germinal centres are much enlarged, usually extending into the lymph node medulla, their mean area is no greater than that of cases with non-rheumatoid, non-neoplastic follicular hyperplasia.[233] The follicles are distinct, separate and round. It is of interest that the mean diameter of lymph node follicles in rheumatoid arthritis is also not greater than that of follicles in the lymph nodes of non-rheumatoid patients[233] although the area of lymph node cortices occupied by the secondary follicles is proportionately raised. There is little evidence within the germinal centres of heightened cell replication. Mitotic figures, tingible body macrophages and centroblasts are few although plasma cells and Russell bodies are recognized regularly.

The paracortices are usually extensive and well-defined. The sinuses are conspicuous. Within them, polymorphs are relatively frequent: they are more numerous in patients with evidence of active systemic disease.

The distribution of B-lymphoid cells in the germinal centres and of non-B lymphoid cells in the interfollicular areas conforms with the patterns seen in a reactive node.[216] Light- and heavy-chains of the major immunoglobulin classes show a polyclonal distribution, a valuable feature in distinguishing the appearances from those of follicular lymphoma. Although the pattern of reactivity with antibodies for immunoglobulin light- and heavy-chains is similar to that found in the nodes of patients with follicular hyperplasia but without rheumatoid arthritis, the strength of the binding, judged by an immunoperoxidase technique, is stronger in the former. Immunoglobulin is also bound in reticular patterns in the germinal centres and demonstrable within some paracortical histiocytes. The germinal centres contain IgG- and IgM- but few IgA-specific B cells, the lymphocytic coronas surrounding the germinal centres IgM- and IgD-cells. The reticular distribution of germinal centre immunoglobulin may reflect passive adsorption, a function of high extracellular immunoglobulin concentrations; alternatively, it may represent an active concentration of immunoglobulin by antigen-presenting cells. That the B cell immunoglobulin is locally produced is supported by evidence that a high proportion of coronal lymphocytes react for $\varkappa_{III}$ is a light chain epitope associated with antibody synthesis.[233]

As in the case of rheumatoid synovial tissue, very many of the paracortical and follicular cells in rheumatoid lymph

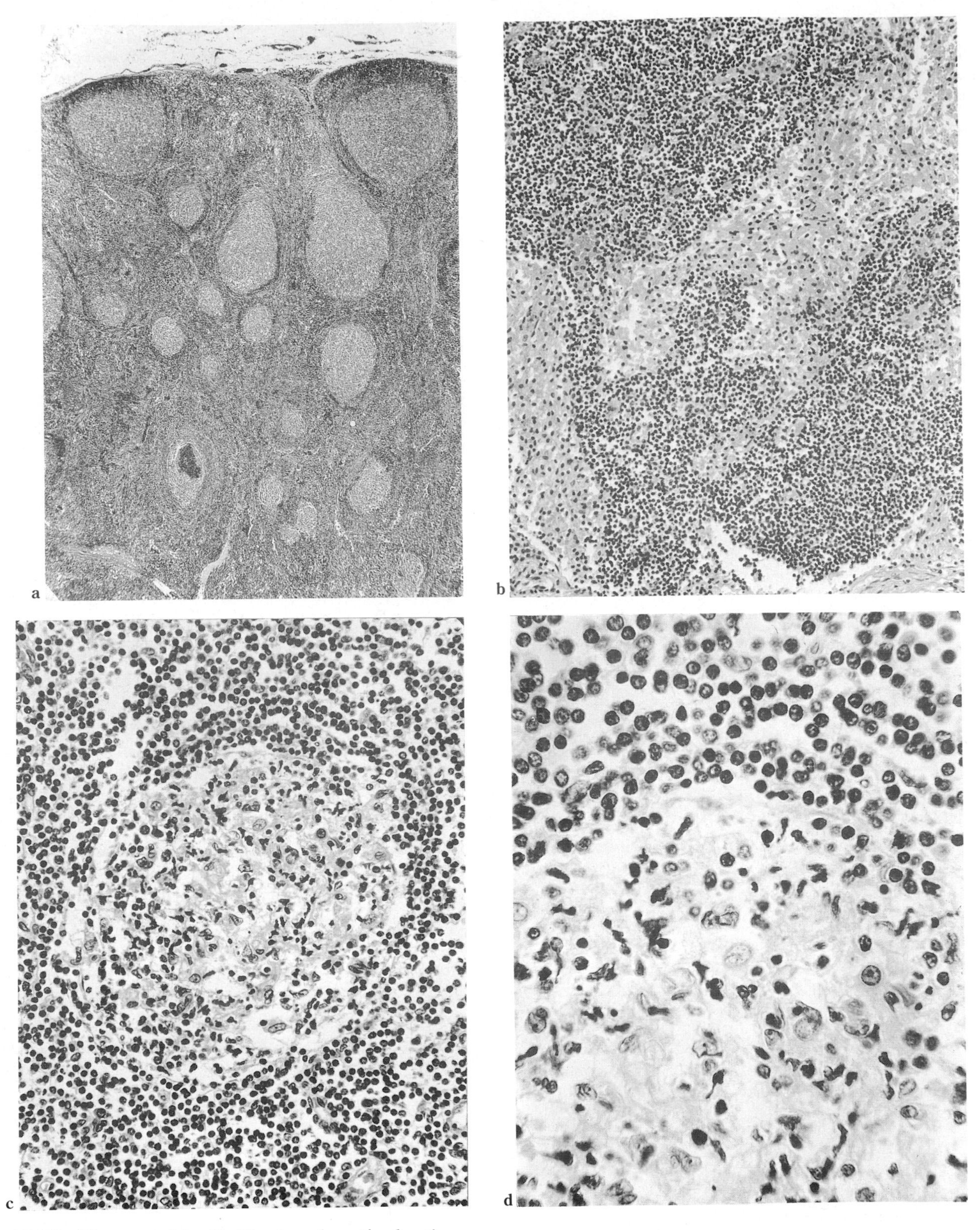

Fig. 12.38 Rheumatoid arthritis: lymphoreticular tissue.
a. In this enlarged axilary lymph node, many prominent germinal centres are present. Occasionally their size and distribution may suggest the diagnosis of follicular lymphoma. **b.** Sinus cell hyperplasia in a fatal case of rheumatoid arthritis. **c.** Lymphoid follicle with germinal centre from axillary lymph node. Among the lymphoblasts is much nuclear debris. **d.** High-power view of field shown in **c.** (HE **a**: × 25; **b**: × 140; **c**: × 250; **d**: × 500.)

nodes express MHC class II HLA-DR determinants. In the paracortex, another similarity is the high frequency of interdigitating (dendritic) cells, the number of which suggests highly active processes of antigen presentation. The role of T-cell populations in the subsequent immune responses is still being clarified. T_H-cells are commonplace in the germinal centres of lymph nodes from both rheumatoid and non-rheumatoid patients, although T_S-cells are relatively much more frequent in the lymphoid germinal centres in rheumatoid arthritics. It has been claimed that there is a diminution in the T_H:T_S-cell ratio (approximately 2:1)[233] in comparison with a ratio of approximately 3.8:1 in the synovia[218,219]. However, other observers find a smaller difference between the lymph node cortex and the synovial tissues. In spite of the relative scarcity of T_S-cells in rheumatoid lymph node germinal centres compared with the numbers of T_H-cells, there are many more T_S-cells in rheumatoid than in non-rheumatoid cases. One consequence may be the low germinal centre proliferative activity described above. Against this view is evidence that the proportions of T_S-cells are not related to measurements of germinal centre cell proliferation such as numbers of mitotic figures. It does appear likely, however, that cases of disease of high activity have lower T_S-cell numbers, higher T_H:T_S-cell ratios.

In the lymph node germinal centres, the proportion of cells expressing IL-2 receptors is increased in rheumatoid arthritis. The paracortex, however, contains reduced numbers of activated, IL-2-positive lymphocytes. These observations require to be considered in relation to the increased activation and IL-2 receptor expression by peripheral blood lymphocytes that, however, manufacture low levels of IL-2. The low numbers of activated paracortical lymphocytes may be an index of the increased migration of these cells towards synovial tissues, particularly in patients with active synovitis.

Lymphoma The relationship of rheumatoid arthritis to lymphoma is the subject of debate. Although the increased incidence of lymphoma in Sjögren's syndrome (see p. 529) is established, the true influence of rheumatoid arthritis on the occurrence of lymphoma remains contentious. Multiple myeloma, 'reticulum cell sarcoma', chronic lymphocytic leukaemia, Waldenström's macroglobulinaemia, and the hyperviscosity syndrome may coexist with rheumatoid arthritis and histiocytic medullary reticulosis, and myelofibrosis can give rise to a rheumatoid-like polyarthritis.[84] A number of reports suggest that lymphoma is more common than would be expected by chance. An eight-fold increase in the incidence of 'reticuloendothelial system' cancer was shown in one series.[335] There is indeed an increased frequency of lymphoproliferative (and of non-lymphoproliferative) cancer in patients treated for prolonged periods with immunosuppressive drugs such as cyclophosphamide[211] and azathioprine but the increase is slight, is not dose-related and is not site-specific.[389]

Immune deficiency States of immunological deficiency and autoallergic disease occur more often with rheumatoid arthritis than can be ascribed to chance alone, both in individuals and in families. Acquired and spontaneous immunological deficiency is believed to predispose to malignant lymphomata and, simultaneously, to rheumatoid arthritis. When there is disordered protein synthesis, paraproteinaemia may antedate a rheumatoid-like polyarthritis by several years, perhaps because the prolonged antigenic stimulus ultimately provokes a rheumatoid response.

Mechanisms of lymph node change The lymph nodes in rheumatoid arthritis display many of the features of acute, high-antigen-dose stimulation in rodents together with others common in chronic local antigen stimulation.[106] Only the sustained outer, cortical plasma cell differentiation is distinctive in rheumatoid arthritis; it persists in association with the prominent germinal centres. Thus, in the guinea-pig, mitogenic lymphokines injected into the peripheral lymphatics of the ear caused rapid paracortical cell division and germinal centre proliferation in regional lymph nodes. In the rabbit, regional lymph node antibody production was depressed by using, as antigen, protein to which iodine-125 was bound. The evidence from the experimental guinea-pig suggested that an autopharmacological response to a cytokine can cause a rheumatoid-like lymph node follicular hyperplasia leading in man, as in the antigen-primed animal, to accelerated B-cell maturation, presumably as a result of overresponsiveness of T-cells or their overstimulation. The evidence from the rabbit experiment indicated that interference with an antibody feedback mechanism during antigenic stimulation of a lymph node can provoke an abnormally prolonged T-cell reaction with unrestricted production of plasmacytes, as in rheumatoid arthritis. Thus, the regional lymph nodes in rheumatoid arthritis may 'lose sight of' the original antigenic stimulus with a resultant defective regulation of lymphocyte activation in the paracortex where there is a continual passage of recirculating lymphocytes across high vascular endothelium.

Spleen

Enlargement of the spleen is characteristic although the degree of enlargement, readily measured pathologically,[149] is often insufficient to be detectable clinically. The increased splenic mass results from histological changes that closely mirror those of the regional lymph nodes. The increased quantity of splenic tissue is accompanied by abnormal phagocytic and lymphoreticular activity. Evidence for enhanced phagocytosis is given by the increased rate at which particulate material, high molecular weight dyes and oil/glycerine emulsions are cleared from the circulating blood. It is likely that this mechanism contributes to the splenic retention of excess iron which is characteristic of rheumatoid arthritis; to the anaemia and leucopenia of Felty's syndrome; and perhaps to splenic amyloidosis (see p. 551).

It is also reasonable to propose that the spleen influences the rate of clearance of circulating immune complexes which are characteristic of rheumatoid arthritis (see p. 245). The results of measurements of the clearance from the blood of heat-damaged red blood cells were inclusive but tests with IgG-coated red cells demonstrated defective clearance. Nevertheless, the rates of red cell clearance did not correlate with levels of circulating immune complexes.[126] Thus, the contribution of splenic dysfunction to the lesions of rheumatoid arthritis provoked by immune complexes remains uncertain.

Nervous system

The brain, cord and peripheral nerves are the sites of lesions some of which are primary, some secondary and some the consequences of treatment. The neuropathological sequelae have been fully described.[223]

Encephalopathy

Encephalopathy[222] can result from intracranial vasculitis, from the presence of intracranial granulomata and so-called rheumatoid meningitis, from choroid plexus immune complex deposition, from secondary disorders such as amyloidosis and bacterial infection, or from the administration of drugs such as corticosteroids.

Large tracts of cerebral rheumatoid granulation tissue may form in individuals with quiescent joint disease who are virtually asymptomatic.[406] Active, sometimes necrotizing vasculitis, is detected histologically.[142] Aneurysms may form. Infarction of brain territories results. Meningeal disease may, however, occur in the absence of parenchymal disorder.[222] Rheumatoid cerebral granulomata are rare but have been identified within the dura mater, the falx cerebri, the cerebral parenchyma, the choroid plexus and the wall of the vertebral artery. The preferential localization of intracranial granulomata to sites such as the dura, rich in collagen, reflects the vascularity of this tissue and its susceptibility to immune complex deposition, rather than the chemical identity of the connective tissue. The relative frequency with which the choroid plexus is affected is an index of the susceptibility of this vascular structure to the deposition of soluble complexes of a particular size. Cerebrovascular amyloid may accumulate. The anterior pituitary gland is also a frequent site for amyloid deposition in long-standing rheumatoid arthritis.

The cerebral tissues are susceptible to opportunistic infections by bacteria, viruses, fungi and protozoa. Corticosteroids and other immunosuppressive drugs permit the onset of meningitis and abscess may result from local disease or as a consequence of pyaemia.

Myelopathy

Myelopathy can be a direct consequence of arthritis but much more often it is the secondary result of articular and bone disease of the vertebral column. Large masses of granulation tissue may form.[408] Rarely, spinal vasculitis has been detected and the vertebral artery walls have been a site of rheumatoid granuloma. Spinal dural granulomata are also recognized.

Rheumatoid arthritis of the cervical spine (see p. 475) threatens the integrity of the brain stem and spinal cord. Potentially fatal compression of the brain stem follows vertical subluxation of the odontoid process and the arterial supply to the brain stem is compromised.[396] When lower cervical segments are affected,[91] there is angulation and indentation of the anterior surface of the cord. Anterior horn necrosis is conspicuous: the destruction may extend to the posterior horns, the ventral part of the posterior columns and the medial part of the lateral columns. There is a pattern of compression distinct from that resulting from the cervical spondylosis of osteoarthrosis.[223] Where there is brain stem damage, vascular thrombosis of arteries such as the vertebral, right superior cerebellar and right anterior, inferior cerebellar may be implicated. Less is known of disease of the spinal nerve roots.

Peripheral neuropathy

Peripheral neuropathy[86,303] may be inflammatory, ischaemic or compressive. It may result from amyloidosis. Pathological signs of inflammation related to nerve trunks are more common than clinical. Vascular changes may be mild, with endarterial fibrosis, or may resemble those of polyarteritis nodosa.[326] Wallerian degeneration of myelinated and of 'non-myelinated' fibres has been recorded but segmental demyelination is also encountered. Reasonably extensive arteritis is necessary for the development of necrosis of the affected central fascicles of the mid-upper arm and mid-thigh.[223]

Flexor tendon synovitis is one cause of a median nerve entrapment neuropathy recognized in the carpal tunnel syndrome but other nerves such as the posterior tibial may be similarly affected. Overall, entrapment neuropathy may occur in as many as 45 per cent of cases. Large, myelinated fibres appear most vulnerable.

The autonomic nervous system may be implicated. The thermoregulatory and heating responses are diminished in areas related to cutaneous sensory impairment.[30] Lesions may be pre- or postganglionic and postural hypotension and gustatory sweating are recorded.[111]

Myopathy[223,349]

Muscle weakness in rheumatoid arthritis is often disproportionate to the severity of the disease and muscular dystrophy may be simulated.[403] Very occasionally, myasthenia gravis and rheumatoid arthritis occur within one family. Skeletal muscle granulomata are rare. When rheumatoid vasculitis affects muscle, the muscular changes are identical to those produced by polyarteritis nodosa.[274a] Only in recent years has it been possible to bring together techniques for the selective biopsy of affected muscle, identifying, by electromyography, sites to be sampled, and using transmission electron microscopy and histochemical methods for the microscopic assessment of the state of individual muscle cells. Early reports[86] drew attention to the frequency of nerve and skeletal muscle lesions in rheumatoid arthritis (Fig. 12.39); recent papers have systemically applied modern methods to the classification of the muscle disorder.

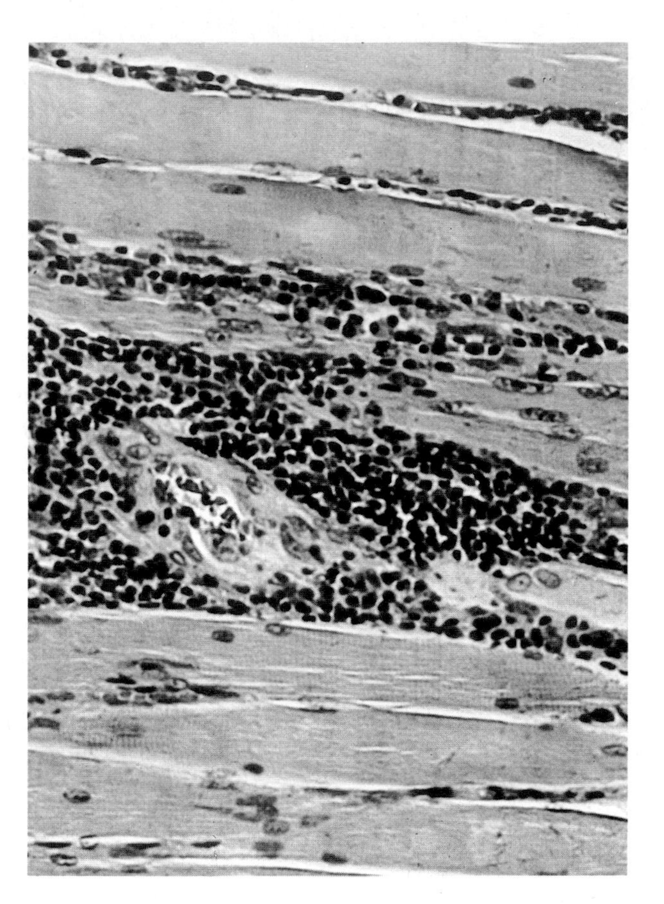

Fig. 12.39 Rheumatoid arthritis: skeletal muscle. A cluster of mature lymphocytes surrounds an intramuscular arteriole in the deltoid muscle. In this case, death was attributable to systemic candidiasis. (HE × 270.)

With the exception of an important class of individuals in whom corticosteroids have been used,[174] six groups of cases can be identified by microscopy and histochemistry[349] (Table 12.9). These authors express surprise that, in their series of 16 patients, there was no demonstrable relationship between the functional state and the activity of the arthritis assessed clinically, and the extent or type of muscle disorder assessed microscopically. Even when electromyography is used to identify an exact point for biopsy,[174] there remains the common problem in pathology that the extent of the tissue taken for microscopy is minute compared to the mass of affected muscle. The microscopic assessments made on 10 sections of quadriceps muscle, for example, make judgements on approximately 1×10^{-6} of the muscle mass. The chances against the random biopsy being representative of the state of a group of disorders that are themselves often focal and zonal, must necessarily be high. With this background, it is perhaps surprising that Reza and Verity[349] were able to delineate so clearly their categories of myopathy. However, their microscopic definitions form a basis for future studies of skeletal muscle change in rheumatoid arthritis. Thus, in *minimal lesion myopathy*, intact muscle mass but raised creatine phosphokinase levels were accompanied by very slight muscle cell

Table 12.9

Features of rheumatoid myopathy

Morphological abnormalities of skeletal muscle in rheumatoid arthritis[349]	Groups of cases of rheumatoid myopathy assessed histologically[174]
1. Minimal lesion myopathy	1. Connective tissue disease in muscle, e.g. myositis, polyarteritis nodosa
2. Inflammatory myopathy	2. Muscle cachexia
3. Type II muscle cell atrophy	3. Peripheral neuropathy
4. Neurogenic atrophy	4. Steroid myopathy
5. Overlap syndromes	
6. Vasculitis	

atrophy, perifibre lymphocytosis and rare, angular, atrophic cells the appearance of which suggested single cell denervation, perhaps inflammatory. In *inflammatory myopathy*, the lymphocytic aggregates described by Sokoloff *et al.*,[403] by Cruickshank[86] and by Haslock *et al.*[174] were present with little endomysial fibrosis and rare, reduced nicotine-adenine dinucleotide tetrazolium-negative atrophic, angular cells. In *type II (dark, fast)* fibre atrophy, the disease discriminates against this cell, with a loss of the normal type I/type II mosaicism so readily seen in myosin adenine triphosphatase preparations. In the *neurogenic syndromes*, there was fascicular disruption and increased endomysial fibroplasia with vascular prominence, perivascular inflammation and focal eccentric vasculitis. These are severe cases of seropostive rheumatoid arthritis with visceral disease and nodulosis, sometimes provoked to activity by a change in corticosteroid dosage. The *cross-over syndromes* displayed variation in muscle cell size, preferential type II cell atrophy and rare, angular atrophic cells. Myositis and synovitis coincided but the features of childhood polymyositis were not seen. Finally, the cases with predominant *vasculitis* might present with muscle weakness. There was a resemblance to polyarteritis nodosa and affected vessels could be identified. The response to corticosteroids, as in polymyalgia (p. 642), could be dramatic and beneficial.

Discriminating diagnosis on the basis of muscle microscopy now lies in the hands of those specializing in the pathology of muscle disease (Chapter 16). The advice of these laboratories should be taken before a case of rheumatoid arthritis with muscle weakness or atrophy is subjected to biopsy. Electron microscopy has contributed less to the understanding of rheumatoid myopathy than histochemistry, but in special circumstances, such as the definition of morphological change in muscle spindles,[265] the ultrastructural findings can prove to be of critical importance. In the spindles, the components most resistant to structural changes appear to be the nerve endings.

Ocular system

Ocular disease is a common and important feature of rheumatoid arthritis[178] and the drugs used in treatment may damage the eye. The relationship between eye disease and rheumatoid disease is, however, debatable so that Hazelman and Watson[178] state that 'the eyes are often the first tissues to be involved in rheumatoid arthritis' and 'the pathological changes run hand-in-hand with those in the joints' whereas Pisko *et al.*,[328] investigating 97 rheumatoid patients on NSAIDs, found a 31.9 per cent prevalence of ocular disease and a 24.5 per cent/year incidence in which none of the ocular lesions were definitely related either to the arthritis or to therapy.

Keratoconjunctivitis sicca may accompany severe corneal disease and forms part of Sjögren's syndrome (see p. 529). Biopsy of minor salivary glands reveals the characteristic changes (see p. 501). Practolol can provoke analogous disorder and the antinuclear antibodies formed in these patients may damage the conjunctiva leading to vascular and fibrotic changes resembling those that occur in practolol peritonitis (see p. 306). Corneal ulceration may develop alone or with scleritis, and perforation may occur. Immune complexes are capable of causing such damage as are proteolytic enzymes.[283] The retina may be injured by the administration of substantial amounts of chloroquine,[369] and vision can be seriously impaired. Retinal vasculitis is an occasional complication of rheumatoid arthritis[273] and chiasmal neuropathy may be a consequence of rheumatoid pachymeningitis.[436]

Endocrine system

Thyroid

A relationship has long been suspected between the autoimmune disease, Hashimoto's thyroiditis, and rheumatoid arthritis (Fig. 12.40a): rheumatoid arthritis displayed a higher prevalence of thyroid disease in 34 'unselected' cases of thyroiditis than in a control group. Moreover, thyroid autoantibodies were demonstrable more often in an unselected group of women with rheumatoid arthritis than in a control series.[48] Nevertheless, when the records of 521 US patients with rheumatoid arthritis were searched for evidence of thyroiditis and of diabetes mellitus, no difference was found between the real and expected numbers of persons with these diseases.[254] Anatomical studies have not revealed any particular thyroid lesion in rheumatoid arthritis although the thyroid arteries and arterioles are affected by the deposition of amyloid as frequently as are those of the adrenals (Fig. 12.40b), in which necrotizing arteritis is periodically discovered.

Pituitary

The anterior pituitary has been said to contain an excess of basophil cells with Schiff-positive granules at the poles of the nuclei, but this work has not been extended with modern cytochemical and immunological techniques. There is no longer any reason to suppose that failed adaptation to 'stress', mediated by the pituitary–adrenal axis, plays any

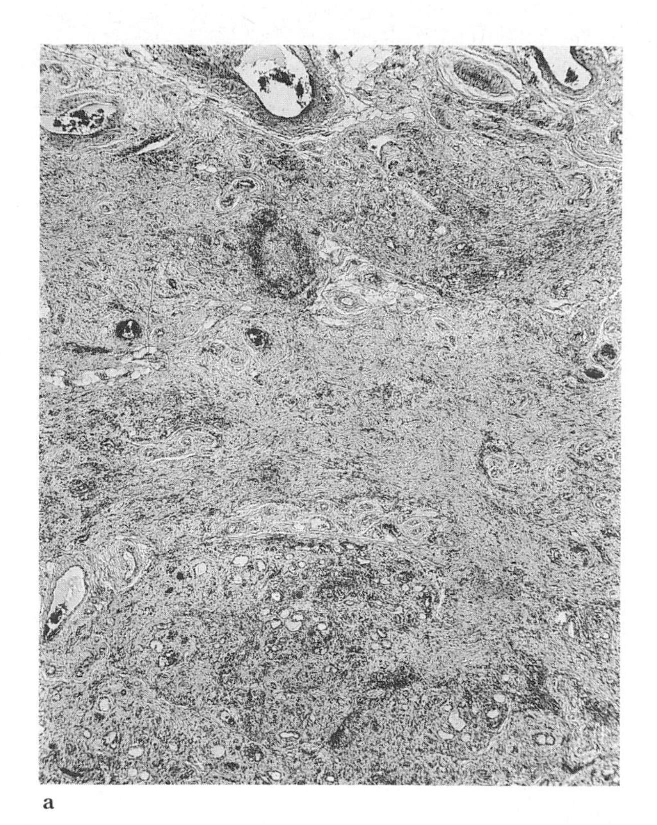
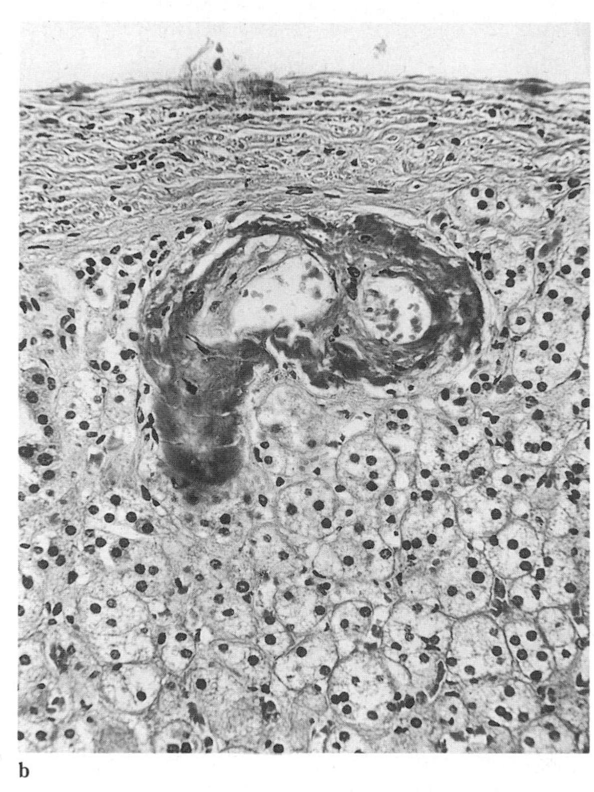

a b

Fig. 12.40 Rheumatoid arthritis: thyroid and adrenal glands.
a. Widespread fibrosis and lymphocytic infiltration of this thyroid gland accompany hypothyroidism. The appearances, of active chronic thyroiditis, may constitute part of an autoimmune syndrome linked with rheumatoid arthritis but analogous changes are frequent in the aged. **b.** Unless corticosteroid or ACTH treatment has been practiced, the adrenal glands remain of normal size. However, vasculitis, and in this case, amyloid, affecting a cortical arteriole, may be conspicuous. (**a**: HE × 25; **b**: × 250.)

primary part in the pathogenesis of rheumatoid arthritis. However, the adrenals and pituitary are affected by prolonged administration of corticosteroids so that adrenal cortical hypoplasia is frequently found at necropsy. The vessels of the anterior pituitary are also common sites for amyloidosis.

Gonads

Few systematic studies of the histological structure of the gonads in rheumatoid arthritis have been attempted although the testis is susceptible to the ischaemic changes caused by necrotizing arteritis.

Pancreas

In spite of claims by Bely (personal communication), there is no substantive evidence of a rheumatoid pancreatic disease. An unusual case of the insulin autoimmune syndrome in a patient with two small insulinomas and concurrent rheumatoid arthritis has been described.[140]

REFERENCES

1. Adams R. *Illustrations of the Effects of Rheumatic Gout or Chronic Rheumatic Arthritis on all the Articulations: with descriptive and explanatory statements*. London: John Churchill, 1857a.
2. Adams R. *A Treatise on Rheumatic Gout or Chronic Rheumatic Arthritis of all the Joints*. London: John Churchill, 1857b.
3. Adelson G L, Saypol D C, Walker A N. Ureteral stenosis secondary to retroperitoneal rheumatoid nodules. *Journal of Urology* 1982; **127**: 124–5.
4. Aherne M J, Bacon P A, Blake D R, Gallagher P J, Jones D B, Morris C J, Potter A R. Immunohistochemical findings in rheumatoid nodules. *Virchows Archiv* (Pathological Anatomy and Histopathology) 1985; **407**: 191–202.
5. Aherne M, Lever J V, Cosh J. Complete heart block in

rheumatoid arthritis. *Annals of the Rheumatic Diseases* 1983; **42**: 389–97.

6. Akikusa B, Irabu N, Kamei K, Tsuchida H, Kondo Y. Glomerulonephritis in patients with rheumatoid arthritis (RA). Report of five cases and review of the literature. *Acta Pathologica Scandinavica* 1986; **36**: 235–52.

7. Albrecht M, Marinette G V, Jacox R F, Vaughan J H. A biochemical and electron microscopy study of rice bodies from rheumatoid patients. *Arthritis and Rheumatism* 1965; **8**: 1053–63.

8. Allen R C, Petty R E, Lirenman D S, Malleson P N, Laxer R M. Renal papillary necrosis in children with chronic arthritis. *American Journal of Diseases of Children* 1986: **140**: 20–2.

9. Alwan W H, Dieppe P A, Elson C J, Bradfield J W B. Bone resorbing activity in synovial fluids in destructive osteoarthritis and rheumatoid arthritis. *Annals of the Rheumatic Diseases* 1988; **47**: 198–205.

10. Anthony V B, Sahn S A. Empyema after internal jugular catheterisation in rheumatoid arthritis. *Thorax* 1981; **36**: 958–9.

11. Antilla S, Sutinen S, Paakko P, Finell B. Rheumatoid pneumoconiosis in a dolomite worker: a light- and electron microscopic, and X-ray microanalytical study. *British Journal of Diseases of the Chest* 1984; **78**: 195–200.

12. Appelboom T, de Boelpaepe C, Ehrlich G E, Famaey J P. Rubens and the question of antiquity of rheumatoid arthritis. *Journal of the American Medical Association* 1981; **245**: 483–6.

13. Armstrong J G, Steele R H. Localised pulmonary arteritis in rheumatoid disease. *Thorax* 1982; **37**: 313–14.

14. Arnett F C, Edworthy S M, Block D A *et al.* The American Rheumatism Association 1987 criteria for the classification of rheumatoid arthritis. *Arthritis and Rheumatism* 1988; **31**: 315–23.

15. Athanasou N A, Quinn J, Heryet A, Puddle B, Woods C G, McGee J O'D. The immunohistology of synovial lining cells in normal and inflamed synovium. *Journal of Pathology* 1988; **155**: 133–42.

16. Athanasou N A, Quinn J, Woods C G, McGee J O'D. Immunohistology of rheumatoid nodules and rheumatoid synovium. *Annals of the Rheumatic Diseases* 1988; **47**: 398–403.

17. Athseya B M, Moser G, Schumacher H R, Hanson V, Dahms B, Thompson D M. Role of basophils and mast cells in juvenile rheumatoid arthritis. In Pepys J, Edwards A M, eds, *The Mast Cell: its Role in Health and Disease*. London: Pitman, 1979: 127–36.

18. Baer A M N, Dessypris E N, Goldwasser E, Krantz S B. Blunted erythropoietin response to anaemia in rheumatoid arthritis. *British Journal of Haematology* 1987; **66**: 559–64.

19. Bagshaw P F, Munster D J, Wilson J G. Molecular weight of gastric mucosa glycoprotein is a determinant of the degree of subsequent aspirin induced chronic gastric ulceration in the rat. *Gut* 1987; **28**: 287–93.

20. Baillie M. *The Morbid Anatomy of some of the Most Important Parts of the Human Body*. London: J Johnson, 1793.

21. Balbi B, Cosulich E, Risso A, Sacco O, Balzano E, Rossi G A. The interstitial lung disease associated with rheumatoid arthritis: evidence for imbalance of helper T-lymphocyte subpopulations at sites of disease activity. *Bulletin Europèene Physiolopathologie Réspiratoire* 1987; **23**: 241–7.

22. Ball J, Sharp J. Rheumatoid arthritis of the spine. In Hill A G S, ed, *Modern Trends in Rheumatology*. New York: Appleton-Century-Crofts, 1971: 117–38.

23. Barland P, Novokoff A B, Hamerman D. Fine structure and cytochemistry of the rheumatoid synovial membrane, with special reference to lysosomes. *American Journal of Pathology* 1964; **44**: 853–66.

24. Barrie H J. Histologic changes in rheumatoid disease of the metacarpal and metatarsal heads as seen in surgical material. *Journal of Rheumatology* 1981; **8**: 246–57.

25. Batley W J, Uddin J, Kelly H G. Rheumatoid arthritis complicated by constrictive pericarditis: report of a case treated successfully by pericardiectomy. *Canadian Medical Association Journal* 1969; **100**: 863–6.

26. Bedi S S, Ellis W. Spontaneous rupture of the calcaneal tendon in rheumatoid arthritis after local steroid injection. *Annals of the Rheumatic Diseases* 1970; **29**: 494–5.

27. Begin R, Masse S, Cantin A, Menard H-A, Bureau M-A. Airway disease in a subset of non-smoking rheumatoid patients. Characterization of the disease and evidence for an autoimmune pathogenesis. *American Journal of Medicine* 1982; **72**: 743–50.

28. Benedek T G. Rheumatoid pneumoconiosis. Documentation of onset and pathogenic considerations. *American Journal of Medicine* 1973; **55**: 515–24.

29. Benn H-P, Drews J, Randzio G, Jensen J M, Loffler H. Does active rheumatoid arthritis affect intestinal iron absorption? *Annals of the Rheumatic Diseases* 1988; **47**: 144–9.

30. Bennett P H, Scott J T. Autonomic neuropathy in rheumatoid arthritis. *Annals of the Rheumatic Diseases* 1965; **24**: 161–8.

31. Berg W B van den, Joosten L A B, Putte L B A van de, Zwarts W A. Electrical charge and joint inflammation. Suppression of cationic aBSA-induced arthritis with a competitive polycation. *American Journal of Pathology* 1987; **127**: 15–16.

32. Berg W B van den, Putte L B A van de, Zwarts W A, Joosten L A B. Electrical charge of the antigen determines intraarticular antigen handling and chronicity of arthritis in mice. *Journal of Clinical Investigation* 1984; **74**: 1850–9.

33. Bernstein R M. Humoral autoimmunity in systemic rheumatic disease—a review. *Journal of the Royal College of Physicians of London*, 1990; **24**: 18–25.

34. Bierther M, Wagner R. Elektonmikroskopische Untersuchungen des Pannus bei chronischer Polyarthritis. *Zeitschrift für Rheumatologie* 1974; **33**: 32–42.

35. Bierther M F W, Wegner K W. Elektronenmikroskopische Untersuchungen synovialer Gefassveranderungen bei chronische Polyarthritis. *Zeitschrift für Rheumaforschung* 1971; **30**: 214–22.

36. Blake D R, Hall N D, Bacon P A, Dieppe P A, Halliwell B, Gutteridge J N C. The importance of iron in rheumatoid disease. *Lancet* 1981; **ii**: 1142–4.

37. Blake D R, Scott D G I, Eastham E J. Rashid H. Assessment of iron deficiency in rheumatoid arthritis. *British Medical Journal* 1980; **280**: 527.

38. Bluestone R, Bacon P A, eds, Extraarticular manifesta-

tions of rheumatoid arthritis. *Clinics in the Rheumatic Diseases*. Philadelphia, London: W B Saunders Company, 1977: 3(3).

39. Boddington M M, Spriggs A I, Morton J A, Mowat A G. Cytodiagnosis of rheumatoid pleural effusions. *Journal of Clinical Pathology* 1971; **24**: 95–106.

40. Boers M, Croonen A M, Dijkmans B A C, Breedveld F C, Eulderink F, Cats A, Weening J J. Renal findings in rheumatoid arthritis: clinical aspects of 132 necropsies. *Annals of the Rheumatic Diseases* 1987; **46**: 658–63.

41. Bonebrake R A, McCall J T, Hunder G G, Polley H F. Zinc accumulation in synovial fluid. *Mayo Clinic Proceedings* 1972; **47**: 746–50.

42. Bromley M, Bertfield H, Evanson J M, Woolley D E. Bidirectional erosion of cartilage in the rheumatoid knee joint. *Annals of the Rheumatic Diseases* 1985; **44**: 676–81.

43. Bromley M, Fisher W D, Woolley D E. Mast cells at sites of cartilage erosion in the rheumatoid joint. *Annals of the Rheumatic Diseases* 1984; **43**: 76–9.

44. Bromley M, Woolley D E. Histopathology of the rheumatoid lesion. Identification of cell types at sites of cartilage erosion. *Arthritis and Rheumatism* 1984; **27**: 857–63.

45. Bromley M, Woolley D E. Chondroclasts and osteoclasts at subchondral sites of erosion in the rheumatoid joint. *Arthritis and Rheumatism* 1984; **27**: 968–75.

46. Brun C, Olsen T S, Raaschou F, Sorensen A W S. Renal biopsy in rheumatoid arthritis. *Nephron* 1965; **2**: 65–81.

47. Brunk J R, Drash E C, Swineford O. Rheumatoid pleuritis successfully treated with decortication. Report of a case and review of the literature. *American Journal of the Medical Sciences* 1966; **251**: 545–51.

48. Buchanan W W, Crooks J, Alexander W D, Koutras D A, Wayne E J, Gray K G. Association of Hashimoto's thyroiditis and rheumatoid arthritis. *Lancet* 1961; **i**: 245–8.

49. Buckland-Wright J C, Carmichael I, Walker S R. Quantitative microfocal radiography accurately detects joint changes in rheumatoid arthritis. *Annals of the Rheumatic Diseases* 1986; **45**: 379–83.

49a. Burkhardt R. Histomorphologische Untersuchungen über die Rolle des Knochenmarkes bei rheumatischen Krankheiten. *Zeitschrift für gesamte experimentelle Medizin* 1967; **143**: 1–66.

50. Burmester G R, Dimitriu-Bona A, Waters S J, Winchester R J. Identification of three major synovial lining cell populations by monoclonal antibodies directed to Ia antigens and antigens associated with monocytes-macrophages and fibroblasts. *Scandinavian Journal of Immunology* 1983; **17**: 69–82.

51. Burmester G R, Jahn B, Rohwer P, Zacher J, Winchester R J, Kalden J R. Differential expression of Ia antigens by rheumatoid synovial lining cells. *Journal of Clinical Investigation* 1987; **80**: 595–604.

52. Burry H C, Dieppe P A, Bresnihan F B, Brown C. Salicylates and renal function in rheumatoid arthritis. *British Medical Journal* 1976; **i**: 613–15.

53. Burt R W, Berenson M M, Samuelson C O, Cathey W J. Rheumatoid vasculitis of the colon presenting as pancolitis. *Digestive Diseases and Sciences* 1983; **28**: 183–8.

54. Butcher R G, Chayen J. Unbalanced production of reduced nicotinamide–adenine dinucleotide phosphate in rheumatoid synovial lining cells. *Biochemical Journal* 1971; **124**: 19P.

55. Bywaters E G L. The pathology of the spine. In Sokoloff L, ed, *The Joints and Synovial Fluid* (volume 2). New York: Academic Press, 1980: 428–547.

56. Bywaters E G L. Thoracic intervertebral discitis in rheumatoid arthritis due to costovertebral joint involvement. *Rheumatology International* 1981; **1**: 83–97.

57. Bywaters E G L. Rheumatoid and other diseases of the cervical interspinous bursae, and changes in the spinous processes. *Annals of the Rheumatic Diseases* 1982; **41**: 360–70.

58. Bywaters E G L, Scott J T. The natural history of vascular lesions in rheumatoid arthritis. *Journal of Chronic Diseases* 1963; **16**: 905–14.

59. Caplan A. Certain unusual radiological appearances in the chests of coal-miners suffering from rheumatoid arthritis. *Thorax* 1953; **8**: 29–37.

60. Caplan A, Payne R B, Withey J L. A broader concept of Caplan's syndrome related to rheumatoid factors. *Thorax* 1962; **17**: 205–12.

60a. Carlsson G E, Kopp S, Oberg T. Arthritis and allied diseases of the temporomandibular joint. In: Carlsson G E, Zarg G A, eds, *The Temporomandibular Joints – Function and Dysfunction*. Copenhagen: Munksgaard.

61. Carlsson G E, Zarg G A, eds, *The Temporomandibular Joints – Function and Dysfunction*. Copenhagen: Munksgaard, 1978.

62. Carsons S, Horn V J. Chondronectin in human synovial fluid. *Annals of the Rheumatic Diseases* 1988; **47**: 797–800.

63. Carsons S, Mosesson M W, Diamond H S. Detection and quantitation of fibronectin in synovial fluid from patients with rheumatic disease. *Arthritis and Rheumatism* 1981; **24**: 1261–7.

63a. Cathcart E S, Hayes K C, Gonnerman W A, Lazzari A A, Frantzblau C. Experimental arthritis in a non-human primate. I. Induction by bovine type II collagen. *Laboratory Investigation* 1986; **54**: 26–31.

64. Cavill I, Bentley D P. Erythropoiesis in the anaemia of rheumatoid arthritis. *British Journal of Haematology* 1982; **50**: 583–90.

65. Chilton C P, Darke S C. External iliac venous compression by a giant iliopsoas rheumatoid bursa. *British Journal of Surgery* 1980; **67**: 641.

66. Christie G S. The general changes in rheumatoid arthritis. In: King E S J, Lowe T E, Cox L B, eds, *Studies in Pathology*. Presented to Peter MacCullum. Victoria: Melbourne University Press, 1950: 133–44.

67. Chumbley L C, Harrison E G, Jr, De Remee R A. Allergic granulomatosis and angiitis (Churg–Strauss syndrome). Report and analysis of 30 cases. *Mayo Clinic Proceedings* 1977; **52**: 477.

67a. Clayton M L, Ferlic D C. The wrist in rheumatoid arthritis. *Clinical Orthopaedics and Related Research* 1975; **106**: 192–7.

68. Clemmensen I, Holund B, Andersen R B. Fibrin and fibronectin in rheumatoid synovial membrane and rheumatoid synovial fluid. *Arthritis and Rheumatism* 1983; **26**: 479–85.

69. Cochrane W, Davies D V, Dorling J, Bywaters E G L.

Ultramicroscopic structure of the rheumatoid nodule. *Annals of the Rheumatic Diseases* 1964; **23**: 345–63.

69a. Cohen S, Jones R. An evaluation of the efficacy of arthroscopic synovectomy of the knee in rheumatoid arthritis: 12–24-month results. *Journal of Rheumatology* 1987; **14**: 452–5.

70. Collins D H. The subcutaneous nodule of rheumatoid arthritis. *Journal of Pathology and Bacteriology* 1937; **45**: 97–115.

71. Collins D H. *The Pathology of Articular and Spinal Diseases*. London: Edward Arnold, 1949.

72. Colofiore J R, Schwartz E R. Monensin stimulation of arylsulfatase B activity in human chondrocytes. *Journal of Orthopaedic Research* 1986; **4**: 273–80.

73. Combe B, Pope R M, Darnell B, Kincaid W, Talal N. Regulation of natural killer cell activity by macrophages in the rheumatoid joint and peripheral blood. *Journal of Immunology* 1984; **133**: 709–13.

74. Combe B, Pope R M, Fischbach M, Darnell B, Baron S, Talal N. Interleukin-2 in rheumatoid arthritis: production of and response to interleukin-2 in rheumatoid synovial fluid, synovial tissue and peripheral blood. *Clinical and Experimental Immunology* 1985; **59**: 520–8.

75. Compston J E, Crawley E O, Evans C, O'Sullivan M M. Spinal trabecular bone mineral content in patients with non-steroid treated rheumatoid arthritis. *Annals of the Rheumatic Diseases* 1988; **47**: 660–4.

76. Compston J E, Vedi S, Mellish R W E, Croucher P, O'Sullivan M M. Reduced bone formation in non-steroid treated patients with rheumatoid arthritis. *Annals of the Rheumatic Diseases* 1989; **48**: 483–7.

76a. Constable T J, McConkey B, Paton A. The cause of death in rheumatoid arthritis. *Annals of the Rheumatic Diseases* 1978; **37**: 569.

77. Cook H P. Bilateral ankylosis of the temporomandibular joints following rheumatoid arthritis. *Proceedings of the Royal Society of Medicine* 1958; **51**: 694–6.

77a. Cooke T D, Hurd E R, Jasin H E, Bienenstock J, Ziff M. Identification of immunoglobulins and complement in rheumatoid articular collagenous tissues. *Arthritis and Rheumatism* 1975; **18**: 541–51.

78. Cooney T P. Interrelationship of chronic eosinophilic pneumonia, bronchiolitis obliterans, and rheumatoid disease: a hypothesis. *Journal of Clinical Pathology* 1981; **34**: 129–37.

79. Cooper N S, Soren A, McEwen C, Rosenberger J L. Diagnostic specificity of synovial lesions. *Human Pathology* 1981; **12**: 314–28.

80. Coventry M B, Polley H F, Weiner A D. Rheumatoid synovial cyst of the hip; report of three cases. *Journal of Bone and Joint Surgery* 1959; **41–A**: 721–30.

81. Crisp A J. Mast cells in rheumatoid arthritis. *Journal of the Royal Society of Medicine* 1984; **77**: 450–1.

82. Crisp A J, Armstrong R D, Grahame R, Dussek J E. Rheumatoid lung disease, pneumothorax, and eosinophilia. *Annals of the Rheumatic Diseases* 1982; **41**: 137–40.

83. Crisp A J, Chapman C M, Kirkham S A, Schiller A L, Krane S. Articular mastocytosis in rheumatoid arthritis. *Arthritis and Rheumatism* 1984; **27**: 845–51.

84. Crow J, Gumpel J M. Histiocytic medullary reticulosis presenting as rheumatoid arthritis. *Proceedings of the Royal Society of Medicine* 1977; **70**: 632–4.

85. Cruickshank B. Interpretation of multiple biopsies of synovial tissue in rheumatic diseases. *Annals of the Rheumatic Diseases* 1952; **11**: 137–45.

86. Cruickshank B. Focal lesions in skeletal muscles and peripheral nerves in rheumatoid arthritis and other conditions. *Journal of Pathology and Bacteriology* 1952; **64**: 21–32.

87. Cruickshank B. The arteritis of rheumatoid arthritis. *Annals of the Rheumatic Diseases* 1954; **13**: 136–46.

88. Cruickshank B. Heart lesions in rheumatoid disease. *Journal of Pathology and Bacteriology* 1958; **76**: 223–40.

89. Cruickshank B. Interstitial pneumonia and its consequences in rheumatoid disease. *British Journal of Diseases of the Chest* 1959: **53**: 226–40.

90. Cruickshank B, McLeod J G, Shearer W S. Subarticular pseudocysts in rheumatoid arthritis. *Journal of the Faculty of Radiologists* 1954; **5**: 218–26.

91. Davidson R C, Horn J R, Herndon J H, Grin O D. Brainstem compression in rheumatoid arthritis. *Journal of the American Medical Association* 1977; **238**: 2633–4.

92. Davies D V, Palfrey J. The fine structure of normal and rheumatoid synovial membrane. In Hill A G S, ed, *Modern Trends in Rheumatology* (2). London: Butterworths, 1971: 1–20.

93. Davis M J, Denton J, Freemont A J, Holt P J L. Comparison of serial synovial fluid cytology in rheumatoid arthritis: delineation of subgroups with prognostic implications. *Annals of the Rheumatic Diseases* 1988; **47**: 559–62.

93a. Denman A M. A viral aetiology for juvenile chronic arthritis? *British Journal of Rheumatology* 1988; **27**: 169–70.

94. Dieppe P A, Doyle D V, Burry H C, Tucker S M. Renal disease in rheumatoid arthritis. *British Medical Journal* 1976; **I**: 611–12.

95. Dietrichson O, From A, Christoffersen P, Juhl E. Morphological changes in liver biopsies from patients with rheumatoid arthritis. *Scandinavian Journal of Rheumatology* 1976; **5**: 65–9.

96. Di Giovine F S, Symons J A, Duff G W. Interleukin 1 and tumour necrosis factor in the pathogenesis of septic arthritis. In: Calabro J J, Dick W C, eds, *Infections and Arthritis*. Dordrecht: Kluwer Academic Publishers, 1989: 121–32.

96a. Di Giovine F S, Nuki G, Duff G W. Tumour necrosis factor in synovial exudates. *Annals of the Rheumatic Diseases* 1988; **47**: 768–72.

97. Dixon A St J, Grant C. Acute synovial rupture in rheumatoid arthritis: clinical and experimental observations, *Lancet* 1964; **i**: 742–5.

98. Doherty M, Dieppe P, Watt I. Low incidence of calcium pyrophosphate dihydrate crystal deposition in rheumatoid arthritis, with modification of radiographic features in coexistent disease. *Arthritis and Rheumatism* 1984; **27**: 1002–9.

99. Doherty M, Richards N, Hornby J, Powell R. Relation between synovial fluid C3 degradation products and local joint inflammation in rheumatoid arthritis, osteoarthritis, and crystal associated arthropathy. *Annals of the Rheumatic Diseases* 1988; **47**: 190–7.

100. Doig J A, Whaley K, Dick W C, Nuki G, Williamson J,

Buchanan W W. Otolaryngological aspects of Sjögren's syndrome. *British Medical Journal* 1971; **4**: 460–4.

101. Dorfman R F, Warnke R. Lymphadenopathy simulating the malignant lymphomas. *Human Pathology* 1974; **5**: 519–50.

102. Doube A, Collins A J. Is the gut intrinsically abnormal in rheumatoid arthritis? *Annals of the Rheumatic Diseases* 1988; **47**: 617–19.

103. Dreher R. Origin of synovial type A cells during inflammation. An experimental approach. *Immunobiology* 1982; **161**: 232.

104. Drummond J R, Chisholm D M. A quantitative histological study of human labial salivary glands in rheumatoid arthritis patients with and without Sjögren's syndrome. *IRCS Medical Science* 1986; **14**: 118–19.

105. Duke O L, Hobbs S, Panayi G S, Poulter L W, Rasker J J, Janossy G. A combined immunohistological and histochemical analysis of lymphocyte and macrophage subpopulations in the rheumatoid nodule. *Clinical and Experimental Immunology* 1984; **56**: 239–46.

106. Dumonde D C, Kelly R H, Morley J. Lymphoid and microvascular dysfunction in experimental models of rheumatoid inflammation. In Dumonde D C, ed, *Infection and Immunology in the Rheumatic Diseases*. Oxford: Blackwell, 1976: 375–403.

107. Dumonde D C, Steward M W, Brown K A. The role of microbial infection in rheumatic disease. In: Scott J T, ed, *Copeman's Textbook of the Rheumatic Diseases* (sixth edition). Edinburgh: Churchill Livingstone, 1986: 411–67.

108. Duncan H. Cellular mechanisms of bone damage and repair in the arthritic joint. *Journal of Rheumatology* 1983 (Supplement 11) **10**: 29–37.

109. Duncan H, Frost H M, Villanueva A R, Sigler J W. The osteoporosis of rheumatoid arthritis. *Arthritis and Rheumatism* 1965; **8**: 943–54.

110. Duncan H, Venkatasubramaniam K V, Riddle J M, Pitchford W C, Mathews C H E. Bone erosions in rheumatoid arthritis—a scanning, fluorescence and light microscope study. *Micron* 1981; **12**: 287–88.

111. Edmonds M E, Jones T C, Saunders W A, Sturrock R D. Autonomic neuropathy in rheumatoid arthritis. *British Medical Journal* 1979; **2**: 173–7.

112. Ellman P, Ball R E. Rheumatoid disease with joint and pulmonary manifestations. *British Medical Journal* 1948; **ii**: 816–20.

113. Ellman P, Parkes Weber F, Goodier T E W. A contribution to the pathology of Sjögren's disease. *Quarterly Journal of Medicine* 1951; **20**: 33–42.

114. Ellman M H, Weis M J, Spellberg M A. Liver disease in rheumatoid arthritis. *American Journal of Gastroenterology* 1974; **62**: 46–53.

115. Elson C J, Scott D G I, Blake D R, Bacon P A, Holt P D J. Complement-activating rheumatoid-factor-containing complexes in patients with rheumatoid vasculitis. *Annals of the Rheumatic Diseases* 1983; **42**: 147–50.

116. Emery P, Lopez A F, Burns G F, Vadas M A. Synovial fluid neutrophils of patients with rheumatoid arthritis have membrane antigen changes that reflect activation. *Annals of the Rheumatic Diseases* 1988; **47**: 34–9.

117. Engel U, Aru A, Francis D. Rheumatoid pleurisy. Specificity of cytological findings. *Acta Pathologica et Microbiologica Scandinavica* (Section A) 1986; **94**: 53–6.

118. Eulderink F. The synovial biopsy. In: Berry, C L, ed, *Bone and Joint Disease*. New York: Springer-Verlag, 1982: 26–72.

119. Eulderink F, Meijer K A E. Pathology of the cervical spine in rheumatoid arthritis. A controlled study of 44 spines. *Journal of Pathology* 1976; **120**: 91–108.

120. Farah D, Sturrock R D, Russell R I. Peptic ulcer in rheumatoid arthritis. *Annals of the Rheumatic Diseases* 1988; **47**: 478–80.

121. Fassbender H G. *The Pathology of Rheumatic Diseases* (translated by G Loewi). Berlin: Springer-Verlag, 1975.

122. Fassbender H G. Histomorphological basis of articular cartilage destruction in rheumatoid arthritis. *Collagen and Related Research* 1983; **3**: 141–55.

123. Fassbender H G, Simmling-Annefeld M. The potential aggressiveness of synovial tissue in rheumatoid arthritis. *Journal of Pathology* 1983; **139**: 399–406.

124. Faulkner-Jones B E, Howie A J, Boughton B J, Franklin I M. Lymphoid aggregates in bone marrow: study of eventual outcome. *Journal of Clinical Pathology* 1988; **41**: 768–75.

125. Faurschou P, Francis D, Faarup P. Thoracoscopic, histological and clinical findings in nine cases of rheumatoid pleural effusion. *Thorax* 1985; **40**: 371–5.

126. Fields T R, Gerardi E N, Ghebrehiwet B, Bennett R S, Lawley T J, Hall R P, Plotz P H, Karsh J R, Frank M M, Hamburger M I. Reticuloendothelial system Fc-receptor function in rheumatoid arthritis. *Journal of Rheumatology* 1983; **10**: 550–7.

127. Fienberg R, Colpoys F L. The involution of rheumatoid nodules treated with cortisone and of non-treated rheumatoid nodules. *American Journal of Pathology* 1951; **27**: 925–49.

127a. Firestein G S, Zvaifler N J. How important are T cells in chronic rheumatoid arthritis? *Arthritis and Rheumatism* 1990; **33**: 768–73.

128. Fischer A G T. *Chronic (Non-Tuberculous) Arthritis*. London: H K Lewis, 1929.

129. Forman M B, Zwi S, Gear A J, Kallenbach J, Wing J. Severe airway obstruction associated with rheumatoid arthritis and Sjögren's syndrome. *South African Medical Journal* 1982; **61**: 674–6.

130. Førre O, Thoen J, Dobloug J H *et al.* Detection of T lymphocyte subpopulations in the peripheral blood and the synovium of patients with rheumatoid arthritis and juvenile rheumatoid arthritis using monoclonal antibodies. *Scandinavian Journal of Immunology* 1982; **5**: 221–6.

131. Førre O, Thoen J, Lea T, Dobloug J H. Mellbye O J, Natvig J B, Pahle J, Solheim B G. *In situ* characterization of mononuclear cells in rheumatoid tissues, using monoclonal antibodies. *Scandinavian Journal of Immunology* 1982; **16**: 315–19.

132. Frank S T, Weg J G, Harklerod L E, Fitch R F. Pulmonary dysfunction in rheumatoid disease. *Chest* 1973; **63**: 27–34.

133. Franks A S T. Temporomandibular joint in adult rheumatoid arthritis. A comparative evaluation of 100 cases. *Annals of the Rheumatic Diseases* 1969; **28**: 139–45.

134. Freemont A J. Synovial fluid findings early in traumatic arthritis. *Journal of Rheumatology* 1988; **15**: 881–2.

135. Freemont A J, Denton J. Disease distribution of synovial fluid mast cells and cytophagocytic mononuclear cells in inflammatory arthritis. *Annals of the Rheumatic Diseases* 1985; **44**: 312–15.

136. Freemont A J, Rutley C. Distribution of immunoglobulin heavy chains in diseased synovia. *Journal of Clinical Pathology* 1986; **39**: 731–5.

137. Friedmann B A. Rheumatoid nodules of the larynx. *Archives of Otolaryngology* 1975; **101**: 361–3.

138. Fritz P, Hoenes J, Sall H J, Mischlinski M, Tuezek H V, Wegner G. Immunophotometric analysis of immunoglobulin synthesis and deposition in synovial membranes of patients with rheumatoid arthritis. *Acta Histochemica (Jena)* 1987; **34**: 109–14 (Suppl).

139. Fujita T, Inoue H, Kodama T. Scanning electron microscopy of the normal and rheumatoid synovial membranes. *Archivum Histologicum Japonicum* 1968; **29**: 511–22.

139a. Fukase M, Koizumi F, Wakaki K. Histopathological analysis of sixteen subcutaneous rheumatoid nodules. *Acta Pathologica Japonica* 1980; **30**: 871–82.

140. Fushimi H, Tsukuda S, Hanafusa T, Matsuyuki Y, Nishikawa M, Ishihara S, Kanao K. A case of insulin autoimmune syndrome associated with small insulinomas and rheumatoid arthritis. *Endocrinologica Japonica* 1980; **27**: 679–87.

141. Ganda O P. Caplan H I. Rheumatoid disease without joint involvement. *Journal of the American Medical Association* 1974; **228**: 338–9.

142. Gardner D L. Pathology of Rheumatoid Arthritis. London: Edward Arnold, 1972.

143. Gardner D L. The nature and causes of osteoarthrosis. *British Medical Journal* 1983; **286**: 418–24.

144. Gardner D L. Geode: crystal-containing cavity, lymph space, or pseudocyst? *Annals of the Rheumatic Diseases* 1985; **44**: 569–70.

145. Gardner D L. Pathology of rheumatoid arthritis. In: Scott J T, ed, *Copeman's Textbook of the Rheumatic Diseases* (6th edition). Edinburgh, London: Churchill Livingstone, 1986: 604–52.

146. Gardner D L, Duthie J J R, MacLeod J, Allan W S A. Pulmonary hypertension in rheumatoid arthritis: report of a case with intimal sclerosis of the pulmonary and digital arteries. *Scottish Medical Journal* 1957; **2**: 183–8.

147. Gardner D L, Holmes F. Anaesthetic and postoperative hazards in rheumatoid arthritis. *British Journal of Anaesthetics* 1961; **33**: 258–64.

148. Gardner D L. Krieg A F, Chapnick R. Fatal systemic fungus disease in rheumatoid arthritis with cardiac and pulmonary mycotic and rheumatoid granulomata. *InterAmerican Archives of Rheumatology* 1962; **5**: 561–86.

149. Gardner D L, Roy L M H. Tissue iron and the reticuloendothelial system in rheumatoid arthritis. *Annals of the Rheumatic Diseases* 1961; **20**: 258–64.

150. Garrod A E. *A Treatise on Rheumatism and Rheumatoid Arthritis*. London: Griffin, 1890.

151. Gaucher A, Faure G, Netter P, Pourel J, Duheille J. Etude en microscopie électronique à balayage de la membrane synoviale de la polyarthrite rhumatoïde. *Semaines des Hôpitaux de Paris* 1977; **53**: 2095–100.

152. Geddes D M, Corrin B, Brewerton D A, Davies R J, Turner-Warwick M. Progressive airway obliteration in adults and its association with rheumatoid disease. *Quarterly Journal of Medicine* 1977; **184**: 427–44.

153. Geiler G, Emmrich J. Veranderungen an der Synovialmembran bei Fruhfallen der rheumatoiden Arthritis. *Zeitschrift für Rheumatologie* 1980; **39**: 33–45.

154. Geisinger K M, Vance R P, Prater T, Semble E, Pisko E J. Rheumatoid pleural effusion, A transmission and scanning electron microscopic evaluation. *Acta Cytologica* 1985; **29**: 239–47.

155. Gendre J-P, Luboinski J, Prier A, Camus J-P, Quintrec Y le. Anomalies de la muqueuse jéjunale et polyarthrite rhumatoide: 30 cas. *Gastroénterologie Clinique et Biologique* 1982; **6**: 772–5.

155a. Ghadially F N. Siderosomes, haemosiderin and ferritin. In: *Ultrastructural Pathology of the Cell and Matrix* (third edition). London: Butterworths, 1988: 636–43.

156. Ghadially F N, Roy S. Ultrastructure of synovial membrane in rheumatoid arthritis. *Annals of the Rheumatic Diseases* 1967; **26**: 426–43.

157. Ghadially F N, Roy S. *Ultrastructure of Synovial Joints in Health and Disease*. London: Butterworths, 1969.

158. Gieseking R. Das feinmikroskopische Bild des Rheumatismus nodosus. *Beitrage für pathologische Anatomie* 1969; **138**: 292–320.

159. Gieseking R, Baumer A, Backmann L. Elektronoptische Untersuchungen an Granulomen des Rheumatismus nodosus. *Zeitschrift für Rheumaforschung* 1969; **28**: 165–75.

160. Glynn L E. The chronicity of inflammation and its significance in rheumatoid arthritis. *Annals of the Rheumatic Diseases* 1968; **27**: 105–21.

161. Godfrey H P, Ilardi C, Engber W, Graziano F M. Quantitation of human synovial mast cells in rheumatoid arthritis and other rheumatic diseases. *Arthritis and Rheumatism* 1984; **27**: 852–6.

162. Gohel V, Dalinka M K. Edeiken J. Giant rheumatoid pseudocyst. A case report. *Clinical Orthopaedics and Related Research* 1972; **88**: 151–3.

163. Goldenberg D L, Cohen A S. Synovial membrane histopathology in the differential diagnosis of rheumatoid arthritis, gout, pseudogout, systemic lupus erythematosus, infectious arthritis and degenerative joint disease. *Medicine* 1978; **57**: 239–52.

164. Golds E E, Cooke T D, Poole A R. Immune regulation of collagenase secretion in rheumatoid and osteoarthritic synovial cell cultures. *Collagen and Related Research* 1983; **3**: 125–40.

165. Gresham G A, Kellaway T D. Rheumatoid disease in the larynx and lung. *Annals of the Rheumatic Diseases* 1958; **17**: 286–92.

166. Grimley P M, Sokoloff L. Synovial giant cells in rheumatoid arthritis. *American Journal of Pathology* 1966; **49**: 931–54.

167. Haber P L, Kubakawa H, Koopman W J. Immunoglobulin subclass distribution of synovial plasma cells in rheumatoid

arthritis determined by use of monocloncal anti-subclass antibodies. *Clinical Immunology and Immunopathology* 1985; **35**: 346–51.

168. Hardt H von der. The special histopathological feature of 'Caplan's syndrome'. *Beitrage für Pathologie* 1970; **142**: 114–23.

169. Harper E. Collagenases. *Annual Review of Biochemistry* 1980; **49**: 1063–78.

169a. Harris E D. Pathogenesis of rheumatoid arthritis. In: Kelley W N, Harris E D, Ruddy S, Sledge C B, eds, *Textbook of Rheumatology* (second edition). W B Saunders & Co., 1985: 886–915.

170. Harris M. Rheumatoid heart disease with complete heart block. *Journal of Clinical Pathology* 1970; **23**: 623–6.

171. Harris M, Rash R M, Dymock I W. Nodular, non-cirrhotic liver associated with portal hypertension in a patient with rheumatoid arthritis. *Journal in Clinical Pathology* 1974; **27**: 963–6.

172. Hart F D. Extraarticular manifestations of rheumatoid arthritis. *British Medical Journal* 1969; **3**: 131–6.

173. Harvey A R, Clarke B J, Chui D H K, Kean W F, Buchanan W W. Anemia associated with rheumatoid disease. Inverse correlation between erythropoiesis and both IgM and rheumatoid factor levels. *Arthritis and Rheumatism* 1983; **26**: 28–34.

174. Haslock D I. Wright V, Harriman D G F. Neuromuscular disorders in rheumatoid arthritis. A motor-point muscle biopsy study. *Quarterly Journal of Medicine* 1970; **39**: 335–8.

175. Hayashi T. Stereoscopic ultrastructure of the rheumatoid synovial membranes. *Ryumachi* 1976; **16**: 35–56.

176. Hayes M E, Denton J, Freemont A J, Mawer E B. Synthesis of the active metabolite of vitamin D, 1,25$(OH)_2D_3$, by synovial fluid macrophages in arthritic diseases. *Annals of the Rheumatic Diseases* 1989; **48**: 723–9.

177. Hazelton R A, Vedam R, Masci P P, Whitaker A N. Partial purification and characterisation of a synovial fluid inhibitor of osteoblasts. *Annals of Rheumatic Diseases* 1990; **49**: 121–4.

178. Hazleman B L, Watson P G. Ocular complications of rheumatoid arthritis. In: Bluestone R, Bacon P A, eds, Extraarticular manifestations of rheumatoid arthritis. *Clinics in Rheumatic Diseases* 1977; **3.3**: 501–26.

179. Healey L A, Wilske K R, Sagebiel R W. Rheumatoid nodules simulating basal-cell carcinoma. *New England Journal of Medicine* 1967; **277**: 7–9.

180. Helin H, Korpela M, Mustonen J, Pasternack A. Mild mesangial glomerulopathy—a frequent finding in rheumatoid arthritis patients with hematuria or proteinuria. *Nephron* 1986a; **42**: 224–30.

181. Helin H, Korpela M, Mustonen J, Pasternack A. Rheumatoid factor in rheumatoid arthritis associated renal disease and in lupus nephritis. *Annals of the Rheumatic Diseases* 1986b; **45**: 508–11.

182. Hellems S O, Kanner R E, Renzetti A D. Bronchocentric granulomatosis associated with rheumatoid arthritis. *Chest* 1983; **83**: 831–2.

183. Henderson B. The biochemistry of the human synovial lining with special reference to alterations in metabolism in rheumatoid arthritis. *Pathology, Research and Practice* 1981; **172**: 1–24.

184. Henderson B, Edwards J C W. *The Synovial Lining in Health and Disease*. London: Chapman and Hall, 1987: 255–6; 316–25.

185. Heppleston A G. Environmental lung disease. In: Thurlbeck W M, ed. *Pathology of the Lung*. Stuttgart, New York: Georg Thieme, 1988: 591–685.

186. Heywood A W B, Meyers O L. Rheumatoid arthritis of the thoracic and lumbar spine. *Journal of Bone and Joint Surgery* 1986; **68**-B: 362–8.

187. Higuchi A, Suzuki Y, Okada T. Membranous glomerulonephritis in rheumatoid arthritis unassociated with gold or penicillamine treatment. *Annals of the Rheumatic Diseases* 1987; **46**: 488–90.

188. Hillerdal G, Marjanovic B, Aberg H. Rheumatoid arthritis, immune complex disease, and hypereosinophilic syndrome. *Acta Medica Scandinavica* 1979; **206**: 429–32.

189. Hirohata K, Kobayashi I. Fine structure of the synovial tissues in rheumatoid arthritis. *Kobe Journal of Medical Science* 1964; **10**: 195–225.

190. Hollywell C, Morris C J, Farr M, Walton K W. Ultrastructure of synovial changes in rheumatoid disease and in seronegative inflammatory arthropathies. *Virchows Archiv. A Pathological Anatomy and Histopathology* 1983; **400**: 345–55.

191. Hooper J C, Brookler M. Popliteal cysts and their rupture in the rheumatoid arthritic simulating thrombophlebitis. *Medical Journal of Australia* 1971; **1**: 1371–3.

192. Hough A J, Sokoloff L. Pathology: In: Utsinger P D, Zvaifler N J, Erhlich G E, eds, *Rheumatoid Arthritis: Etiology, Diagnosis, Management.* Philadelphia: J B Lippincott & Co., 1985: 49–69.

193. Hunder G G, Ward L E, Ivins J C. Rheumatoid granulomatous lesion simulating malignancy in the head and neck of the femur. *Mayo Clinic Proceedings* 1963; **40**: 766–70.

194. Huth F, Soren A, Klein W. Structure of synovial membrane in rheumatoid arthritis. *Current Topics in Pathology* 1972; **56**: 55–78.

195. Hutton C W, Hinton C, Dieppe P A. Intra-articular variation of synovial changes in knee arthritis: biopsy study comparing changes in patellofemoral synovium and the medial tibiofemoral synovium. *British Journal of Rheumatology* 1987; **26**: 5–8.

196. Igushi T, Kurosaka M, Ziff M. Electronmicroscopic study of HLA-DR and monocyte/macrophage staining cells in the rheumatoid synovial membrane. *Arthritis and Rheumatism* 1986; **29**: 600–13.

197. Inoue H, Julkunen H, Oka M, Vainio K. Scanning electron microscopic studies of extensor tendon degeneration in rheumatoid arthritis. *Acta Rheumatologica Scandinavica* 1970; **16**: 311–18.

198. Inoue H, Kodama T, Fujita T. Scanning electron microscopy of normal and rheumatoid articular cartilages. *Archivum Histologicum Japonicum* 1969; **30**: 425–35.

199. Ishikawa H, Hirohata K. An immunoelectron-microscopic study of the immunoglobulin-synthesizing cells of the rheumatoid synovial membrane. *Clinical Orthopaedics and Related Research* 1984; **185**: 290–9.

200. Ishikawa H, Ziff M. Electron microscopic observations of immunoreactive cells in the rheumatoid synovial membrane. *Arthritis and Rheumatism* 1976; **19**: 1–14.

269. Malone D E, McCormick P A, Daly L *et al.* Peptic ulcer in rheumatoid arthritis—intrinsic or related to drug therapy? *British Journal of Rheumatology* 1986; **25**: 342–4.

270. Malone D G, Wahl S M, Tsokos M, Cattell H, Decker J L, Wilder R L. Immune Function in severe, active rheumatoid arthritis. A relationship between peripheral blood mononuclear cell proliferation to soluble antigens and synovial tissue immunohistologic characteristics. *Journal of Clinical Investigation* 1984; **74**: 1173–85.

271. Manz H J, Luessenhop A J, Robertson D M. Cervical myelopathology due to atlantoaxial and subaxial subluxation in rheumatoid arthritis. *Archives of Pathology and Laboratory Medicine* 1983; **107**: 94–8.

272. Marcolongo R, Bagel P F, Montagnami M. Gastrointestinal involvement in rheumatoid arthritis: a biopsy study. *Journal of Rheumatology* 1979; **6**: 426–40.

273. Martin M F R, Scott D G I, Dieppe P A, Gilbert C, Easty D L. Retinal vasculitis in rheumatoid arthritis. *British Medical Journal* 1981; **282**: 1745–6.

274. Martinet Y, Haslam P L, Turner-Warwick M. Clinical significance of circulating immune complexes in 'lone' cryptogenic fibrosing alveolitis and those associated with connective tissue disorders. *Clinical Allergy* 1984; **14**: 491–7.

274a. Matsubara S, Mair W G P. Ultrastructural changes of skeletal muscles in polyarteritis nodosa and in arteritis associated with rheumatoid arthritis. *Acta Neuropathologica* 1980; **50**: 169–74.

274b. Matsubara T, Ziff M. Basement membrane thickening of postcapillary venules and capillaries in rheumatoid synovium. *Arthritis and Rheumatism* 1987; **30**: 18–30.

275. Maury C P J, Andersson L C, Teppo A-M, Partanen S, Juvonen E. Mechanism of anaemia in rheumatoid arthritis: demonstration of raised interleukin-1-beta concentrations in anaemic patients and of interleukin-1-mediated suppression of normal erythropoiesis and proliferation of human erythroleukaemia (HLE) cells *in vitro. Annals of the Rheumatic Diseases* 1988; **47**: 972–8.

276. Mbuyi-Muamba J M, Dequeker J, Burssens A. Massive osteolysis in a case of rheumatoid arthritis: clinical, histologic and biochemical findings. *Metabolic Bone Disease and Related Research* 1983; **5**: 101–5.

277. Mellish R W E, O'Sullivan M M, Garrahan N J, Compston J E. Iliac crest trabecular bone mass and structure in patients with non-steroid treated rheumatoid arthritis. *Annals of the Rheumatic Diseases* 1987; **46**: 830–6.

278. Menard C, Beaulieu A D, Audette M, Corbeil J, Latulippe L. Studies on fibronectin in inflammatory vs non-inflammatory polymorphonuclear leucocytes of patients with rheumatoid arthritis. II. Synthesis and release of fibronectin *in vitro. Clinical and Experimental Immunology* 1985; **60**: 347–54.

279. Miller B, Markheim H R, Towbin M N. Multiple stress fractures in rheumatoid arthritis. *Journal of Bone and Joint Surgery* 1967; **49**-A: 1408–14.

280. Mills P R, MacSween R N M, Dick W C, More I A, Watkinson G. Liver disease in rheumatoid arthritis. *Scottish Medical Journal* 1980; **25**: 18–22.

281. Mitchell N, Shepard N. The ultrastructure of articular cartilage in rheumatoid arthritis. *Journal of Bone and Joint Surgery* 1970; **52**-A: 1405–23.

282. Mitchell N S, Shepard N. Changes in proteoglycan and collagen in cartilage in rheumatoid arthritis. A study by light and electron microscopy. *Journal of Bone and Joint Surgery* 1978; **60**-A: 349–54.

283. Mohos S C, Wagner B M. Damage to collagen in corneal immune injury. Observation of connective tissue structure. *Archives of Pathology* 1969; **88**: 3–20.

284. Mohr W. Gelenkkrankheiten. Diagnostik und Pathogenese makroskopischer und histologischer Strukturveranderungen. Stuttgart: Georg Thieme Verlag, 1984.

285. Mohr W, Wessinghage D. The relationship between polymorphonuclear granulocytes and cartilage destruction in rheumatoid arthritis. *Zeitschrift für Rheumaforschung* 1978; **37**: 81–6.

286. Monsees B, Destouet J M, Murphy W A, Resnick D. Pressure erosions of bone in rheumatoid arthritis: a subject review. *Radiology* 1985; **155**: 53–9.

287. Monsees B, Murphy W A. Pressure erosions: a pattern of bone resorption in rheumatoid arthritis. *Arthritis and Rheumatism* 1985; **28**: 820–4.

288. Montgomery W W. Cricoarytenoid arthritis. *Laryngoscope* 1963; **73**: 801–36.

289. Montgomery W W, Goodman M L. Rheumatoid cricoarytenoid arthritis complicated by upper esophageal ulceration. *Annals of Otology* 1980; **89**: 6–8.

290. Morales-Piga A, Elena-Ibanez A, Zea-Mendoza A, Rocamora-Ripoll A, Belran-Gutierez J. Rheumatoid nodulosis: report of a case with evidence of intraosseous rheumatoid granuloma. *Arthritis and Rheumatism* 1986; **29**: 1278–83.

291. Morrey B F. Rheumatoid pseudocyst (geode) of the femoral neck without apparent joint involvement. *Mayo Clinic Proceedings* 1987; **62**: 407–11.

292. Morris C J, Farr M, Hollywell C A, Hawkins C F, Scott D L, Walton K W. Ultrastructure of the synovial membrane in seronegative inflammatory arthropathies. *Journal of the Royal Society of Medicine* 1983; **76**: 27–31.

293. Morris P B, Imber M J, Heinsimer J A, Hlatky M A, Reimer K A. Rheumatoid arthritis and coronary arteritis. *American Journal of Cardiology* 1986; **57**: 689–90.

294. Morrow J, Isenberg D. *Autoimmune Rheumatic Disease.* Oxford: Blackwell Scientific Publications, 1987.

295. Motulsky A G, Weinberg S, Saphir O, Rosenberg E. Lymph nodes in rheumatoid arthritis. *Archives of Internal Medicine* 1952; **90**: 660–76.

296. Moutsopoulos H M, Fauci A S. Immunoregulation in Sjögren's syndrome: influence of serum factors on T-cell subpopulations. *Journal of Clinical Investigation* 1980; **65**: 519–28.

297. Mowat A G. Hematologic abnormalities in rheumatoid arthritis. *Seminars in Arthritis and Rheumatism* 1971; **1**: 195–219.

298. Mowat A G. Connective tissue diseases. *Clinics in Haematology* 1972; **1**: 573–94.

299. Mueller M N, Jurist J M. Skeletal status in rheumatoid arthritis. A preliminary report. *Arthritis and Rheumatism* 1973; **16**: 66–70.

300. Muirden K D. The anaemia of rheumatoid arthritis: the

significance of iron deposits in the synovial membrane. *Australian Annals of Medicine* 1970a; **19**: 97–104.

301. Muirden K D. Giant cells, cartilage and bone fragments within rheumatoid synovial membrane: clinicopathological correlations. *Australian Annals of Medicine* 1970b; **19**: 105–10.

302. Natali P G, Ashton M. Soluble deoxyribonucleoprotein (sNP) complexes in the pleural effusion of rheumatoid arthritis. *Clinical Immunology and Immunopathology* 1978; **9**: 229–35.

303. Neumark T, Dombay M, Gaspardy G. Ultrastructural studies in rheumatoid polyneuropathy. *Acta Morphologica Academica Hungarica* 1979; **27**: 205–20.

304. Ng K C, Revell P A, Beer M, Boucher B J, Cohen R D, Currey H L F. Incidence of metabolic bone disease in rheumatoid arthritis and osteoarthritis. *Annals of the Rheumatic Diseases* 1984; **43**: 370–7.

305. Niedobitek F von. Zur Morphologie und Pathogenese des Caplan-Syndroms. *Zeitschrift für Rheumaforschung* 1969; **28**: 177–91.

306. Niedermeier W, Griggs J H. Trace metal composition of synovial fluid and blood serum of patients with rheumatoid arthritis. *Journal of Chronic Diseases* 1971; **23**: 527–36.

307. Norman J E de B, Bramley P. *A Textbook and Colour Atlas of the Temporomandibular Joint: Diseases, Disorders, Surgery*. London: Wolfe Medical Publications, 1990: 77–83.

308. Norton W L, Ziff M. Electron microscopic observations on the rheumatoid synovial membrane. *Arthritis and Rheumatism* 1966; **9**: 589–610.

309. Nosanchuk J S, Naylor B. A unique cytologic picture in pleural fluid from patients with rheumatoid arthritis. *American Journal of Clinical Pathology* 1968; **50**: 330–5.

310. Nosanchuk J S, Schnitzer B. Follicular hyperplasia in lymph nodes from patients with rheumatoid arthritis. A clinicopathologic study. *Cancer* 1969; **24**: 343–54.

311. O'Connor P, Brereton J, Gardner D L. Hyaline articular cartilage dissected by papain: light and scanning electron microscopy and micromechanical studies. *Annals of the Rheumatic Diseases* 1984; **43**: 320–6.

312. O'Farrelly C, Marten D, Melcher D, McDougall B, Price R, Goldstein A J, Sherwood R, Fernandes L. Association between villous atrophy in rheumatoid arthritis and a rheumatoid factor and gliadin-specific IgG. *Lancet* 1988; **ii**: 819–22.

312a. Ohno O, Cooke T D. Electron microscopic morphology of immunoglobulin aggregates and their interactions in rheumatoid articular collagenous tissues. Arthritis and Rheumatism 1978; **21**: 516–27.

313. Ojeda V J, Stuckey B G A, Owen E T, Walters M N-I. Cardiac rheumatoid nodules. *Medical Journal of Australia 1986;* **144**: 92–3.

314. Okada Y, Nakanishi I, Kajikawa K, Kawasaki S. An autopsy case of rheumatoid arthritis with an involvement of the cardiac conduction system. *Japanese Circulation Journal* 1983; **47**: 671–6.

315. Olive A, Maymo J, Lloreta J, Corominas J, Carbonell J. Evolution of benign rheumatoid nodules into rheumatoid arthritis after 50 years. *Annals of the Rheumatic Diseases* 1987; **46**: 624–5.

316. Owen D S, Toone E, Irby R. Coexistent rheumatoid arthritis and chronic tophaceous gout. *Journal of the American Medical Association* 1966; **197**: 953–6.

317. Page J W. Spontaneous tendon rupture and cervical vertebral subluxation in patients with rheumatoid arthritis. *Journal of the Michigan Medical Society* 1961; **60**: 888–92.

318. Pagel W, Treip C S. Viscero-cutaneous collagenosis. A study of the intermediate forms of dermatomysitis, scleroderma and disseminated lupus erythematosus. *Journal of Clinical Pathology* 1955; **8**: 1–18.

319. Palmer D G, Hogg N, Revell P A. Lymphocytes, polymorphonuclear leukocytes, macrophages and platelets in synovium involved by rheumatoid arthritis. A study with monoclonal antibodies. *Pathology* 1986; **18**: 431–7.

320. Panayi G S, Wooley P, Batchelor J R. Genetic basis of rheumatoid disease: HLA antigens, disease manifestations, and toxic reactions to drugs. *British Medical Journal* 1978; **ii**: 1326–8.

321. Pasternack A, Wegelius O, Makisara P. Renal biopsy in rheumatoid arthritis. *Acta Medica Scandinavica* 1967; **182**: 591–5.

322. Penny W J, Knight R K, Rees A M, Thomas A L, Smith A P. Obliterative bronchiolitis in rheumatoid arthritis. *Annals of the Rheumatic Diseases* 1982; **41**: 469–72.

323. Perri J A, Rodnan G P, Mankin H J. Giant synovial cysts of the calf in patients with rheumatoid arthritis. *Journal of Bone and Joint Surgery* 1968; **50**-A: 709–19.

324. Petersson C J. The acromioclavicular joint in rheumatoid arthritis. *Clinical Orthopaedics and Related Research* 1987; **223**: 86–93.

325. Pettersson T, Wegelius O, Skrifvars B. Gastro-intestinal disturbances in patients with severe rheumatoid arthritis. *Acta Medica Scandinavica* 1970; **188**: 139–44.

326. Peyonnard J-M, Charron L, Beaudet F, Couture F. Vasculitic neuropathy in rheumatoid disease and Sjögren syndrome. *Neurology* 1982; **32**: 839–45.

327. Pieterse A S, Aarons I, Jose J S. Focal prostatic granulomas. Rheumatoid like—probably iatrogenic in origin. *Pathology* 1984; **16**: 174–7.

328. Pisko E J, Turner R A, Yeatts R P, Burch P G, McKinley P H, Collins R L. Ocular pathology, tear production, and tear lysozyme in rheumatoid arthritis. *Journal of Rheumatology* 1982; **9**: 708–11.

329. Popert A J, Scott D L, Wainwright A C, Walton K W, Williamson N, Chapman J H. Frequency of occurrence, mode of development, and significance of rice bodies in rheumatoid joints. *Annals of the Rheumatic Diseases* 1982; **41**: 109–17.

330. Portner M M, Gracie W A. Rheumatoid lung disease with cavitary nodules, pneumothorax and eosinophilia. *New England Journal of Medicine* 1966; **275**: 697–700.

331. Poulter L W, Al-Shakarchi H A A, Campbell E D R, Goldstein A J, Richardson A T. Immunocytology of synovial fluid cells may be of diagnostic and prognostic value in arthritis. *Annals of the Rheumatic Diseases* 1986; **45**: 584–90.

332. Poulter L W, Duke O, Hobbs S, Janossy G, Panayi G. Histochemical discrimination of HLA-DR positive cell populations in normal and arthritic synovial lining. *Clinical and Experimental Immunology* 1982; **48**: 381–8.

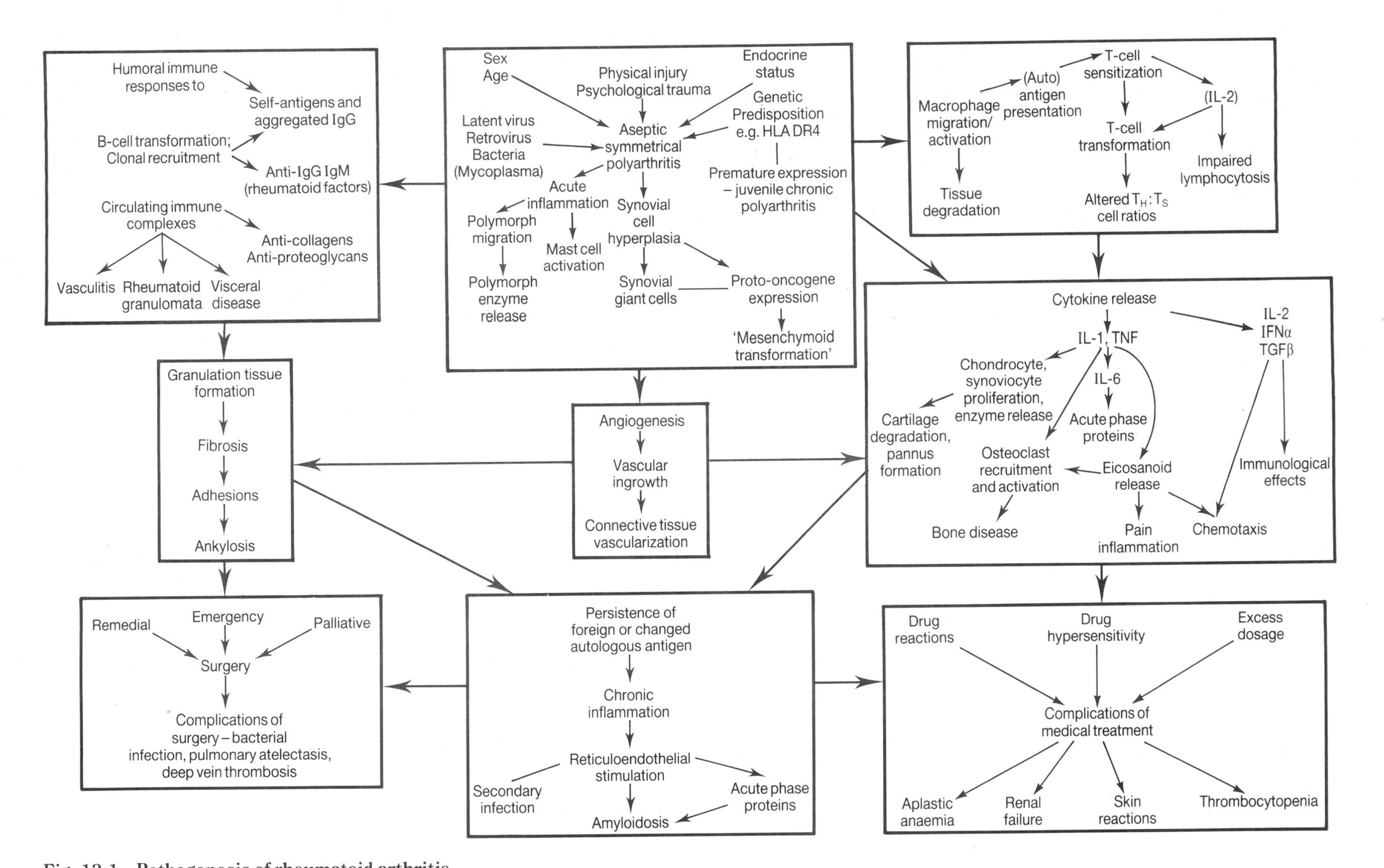

Fig. 13.1 Pathogenesis of rheumatoid arthritis.
This diagram indicates in simplified form factors known to be involved in the natural history (pathogenesis) of rheumatoid arthritis.

Role of polymorphs

Large numbers of polymorphs are present in the synovial fluid and synovial tissue in early rheumatoid arthritis (see p. 451). They are closely concerned with cartilage degradation.[231] That this view is disputed[42a] may reflect differences in the selection of cases for study and in their drug treatment. Polymorphs accumulate in and around the blood vessels in foci of rheumatoid vasculitis and necrosis and the capacity of these cells to injure tissues is not limited to the synovium. They are a common feature of rheumatoid lymphadenitis.

Polymorphs migrate quickly from the synovial blood vessels into the synovium and synovial fluid and towards the cartilage margins. Migration may be more rapid than normal.[304] Paradoxically, circulating polymorphs migrate slowly, a reflection of frequent interaction with immune complexes.[44] The movement of polymorphs in the synovium is influenced by factors in the synovial fluid other than immune complexes. One chemotactic factor is a C5a-like molecule of approximately 14 kD.[250] However, the rate of mobilization is much influenced by drug treatment;[304] it does not correlate with rheumatoid factor titres or total serum complement.

Subpopulations of blood polymorphs in rheumatoid arthritis vary in surface charge, electrophoretic mobility and adhesiveness to surfaces. Cells of low surface charge and high adhesiveness are frequent,[45] although both rheumatoid and normal polymorphs adhere equally to cultured porcine endothelium. In contrast to their increased adhesiveness to test surfaces, rheumatoid synovial fluid polymorphs have deficient phagocytic properties, one factor that may contribute to the high frequency of systemic and latent articular bacterial infection. The metabolic state of these cells is decreased but this may be a consequence of the diminished phagocytic activity, not a cause.[15] The mechanism leading to impaired phagocytosis is intrinsic. Also it does not correlate with the amounts of serum immune complexes or rheumatoid factors.[16]

Many other factors influence polymorph behaviour. Colony-stimulating factors and tumour necrosis factor enhance polymorph activity. There are consequential changes in polymorph surface antigens. Complement receptor 3(CR3) and the granulocyte function antigens (GFA) 1 and 2 are increasingly expressed, but the polymorph antigen GpIIb-IIIa, a protein with fibrinogen and fibronectin receptor activity expression, is apparently decreased.[105] The capacity of polymorphs to bind to cartilage, to degrade proteoglycan and to inhibit proteoglycan synthesis are potentiated by heat-aggregated IgG. In view of the frequency with which rheumatoid joints are silently infected with staphylococci and other commensals, it is therefore of interest that these properties of polymorphs are enhanced by culture-medium conditioned by mononuclear macrophages stimulated by coexistent *Staphylococcus aureus*.[22]

Influence of free radicals

When polymorphs are committed to the phagocytosis of immune complexes, a conformational change occurs in the plasma membrane and there is a characteristic respiratory burst. Free radicals derived from oxygen are generated. They include singlet oxygen (O), superoxide ($\dot{O}$) and hydroxyl (OH) radicals. They are accompanied by peroxide (H_2O_2).[32] One result is injury to cell membranes including those of the polymorphs themselves. Enzyme inhibitors can be prejudiced. Activated proteases are freed into the nearby tissues. Hydroxyl radicals are especially damaging to immunoglobulin, DNA, phospholipid and glycoprotein. Lipid peroxidation products form. In rheumatoid arthritis, when low levels of antioxidant such as ascorbate persist, tissues are particularly prone to injury.[282,283]

Tissue iron (see p. 458) plays an important part in activating molecular oxygen. At iron storage,[126] singlet oxygen radicals are generated, impairing immune complex clearance by macrophages and provoking tissue injury. The incorporation of ferric iron into ferritin, with the conversion of ferrous to ferric ions, involves the activation of molecular oxygen. The level of synovial fluid ferritin in rheumatoid arthritis is high and in proportion to indices of mononuclear and macrophage stimulation and of disease activity such as rheumatoid factor, complement and immune complex titres.[32,33] However, there is sufficient intra-articular ferritin-bound iron to provoke tissue destruction by free radicals.[31] It was therefore thought probable that oxidative radical reactions within the rheumatoid synovium would lead to the formation of hydroxyl radicals and to lipid peroxidation, with cell- and organelle membrane disruption and this has proved to be the case.

The potential for tissue injury displayed by the excess iron storage of rheumatoid synovia is demonstrated further by the tendency for synovial lymphocytes to move towards these stores. They, and macrophages, have receptors for iron-binding proteins. Such a mechanism may be one explanation for the clinical observation that iron-dextran given intravenously may cause the formation of low molecular weight iron-chelates, resulting in a synovial 'flare'.[361]

Polymorph enzymes

Polymorphs are thought to contribute to early cartilage injury.[231] They attack cartilage directly and indirectly. Moving towards the cartilage margins, they initiate a sequence of enzymic pathways that act to destroy matrix proteoglycan, proteoglycan–collagen links and, subsequently, collagen itself.[15,16] Macrophages participate in corresponding, but later, processes.

A first mechanism of polymorph matrix injury is by lysosomal enzyme activation; the enzymes are stored within the azurophilic granules. The lysosomal membranes of rheumatoid synoviocytes are abnormally labile, perhaps because of the presence in type A cells of excess iron. Iron-laden cells may discharge lysosomal enzymes more readily than normal. The phagocytosis of fragments of partially degraded cartilage is followed by intracellular connective tissue digestion. However, polymorph enzymes also diffuse into the nearby matrix. Activation takes place in the pericellular environment. High levels of activated acid hydrolases and proteinases have been shown in the synovial fluid.[155] The view that these enzymes contribute to tissue injury in rheumatoid arthritis is supported by the finding in synoviocytes of increased numbers of dense bodies, vacuoles and phagolysosomes.

A second mechanism invokes the polymorph specific granules. These specific granules are sources of an elastase, the most important polymorph enzyme capable of collagen digestion.[225,226,291] Immunoreactive enzyme has been shown within the polymorphs in rheumatoid synovium[226] and a similar enzyme is demonstrable in the articular cartilage of rabbits with antigen-induced arthritis (see p. 543).[291] Cathepsins B, D and G are also present and a neutral metalloproteinase has been identified.[287]

Role of mast cells

There is much speculation concerning the contribution made by mast cells (p. 36) to the pathogenesis of rheumatoid arthritis.[155,355] The evidence is circumstantial. It is probable that mast cell hyperplasia contributes to cartilage loss and perpetuation of inflammation in a secondary, coincidental, not in a primary manner. It remains to be determined whether drugs preventing mast cell degranulation can alleviate rheumatoid synovitis.

Mast cells contribute to three crucial processes: the release of inflammatory mediators; the promotion of new blood vessel formation; and generalized bone reabsorption.[79,80] The number of synovial fluid mast cells correlates strongly and positively with the synovial fluid histamine content.[218] Added to cultures of adherent synovial cells from human rheumatoid sources, dog mastocytoma mast cells provoke large increases in prostaglandin E synthesis and in collagenase production, both agents capable of degrading cartilage.[371] The synovial connective tissue mast cells respond to anti-IgE, compound 48/80 and calcium ionophore[144] by the release of histamine.[219] They can presumably undergo this response *in vivo*. Histamine itself can stimulate prostaglandin E production by rheumatoid synovial cells[335] and this is a further mechanism by which cartilage degradation could be caused.

In support of the significance of mast cell proliferation in rheumatoid inflammation, it has been found that mast cell counts are directly related to the intensity of clinical signs of arthritis; they are assumed to play active parts in mediating both inflammation and the cartilage degradation. Expressed as mast cells per synovial blood vessel, the proportions of mast cells is greater in patients with active than with inactive rheumatoid arthritis.[133a] The number of mast cells also correlates positively with an index of lymphocytic infiltration, with the proportion of T_H-cells and with the degree of plasma cell infiltration,[218,220] but negatively with the extent of incorporated fibrin. Many rheumatoid synovial fluids contain mast cells but their coincidence with cytophagocytic monocytes is not confined to rheumatoid arthritis and is also characteristic of the reactive arthritides (see Chapter 19).[120]

Other mechanisms of cartilage destruction

Synovitis precedes the centripetal degradation of the hyaline articular cartilage.[289a] Degradation takes place only in inflamed joints. A cycle of destruction is initiated beginning at the synoviochondral junctions, largely sparing cartilage not directly in contact with vascular tissue.[58]

Studies of the sequence of changes of rheumatoid arthritis in man, of experimental models and of cartilage in culture[319] suggest three ways in which cartilage is destroyed. Each centres on abnormal cell behaviour, and each invokes enzyme activation. The responsible enzymes may degrade collagen or proteoglycan or both categories of macromolecule together.[363] The enzymes may act extracellularly or within cells. Disorganization may be effected in a twin- or multiphase sequence, the action of enzyme(s) of broader specificity creating changes that allow the more intimate, selective action of a second class of catalyst.

It may be assumed that neither active enzymes nor enzyme-inhibitor complexes are carried to the synoviochondral junction in the blood stream. Local cell populations are the sources of the active enzyme(s). The incriminated cells could be: 1. the endothelial cells of the small vessels of the synoviochondral fringe (the process of neovascularization is discussed on pp. 295 and 540); 2. blood-derived polymorphs at the cartilage margin; 3. macrophages accumulating from the blood stream and histiocytes sited locally; 4. synoviocytes; 5. chondrocytes; and 6. bone cells, particularly osteoclasts. The contribution of polymorphs and of mast cells is considered on p. 535 *et seq.*

Role of macrophages

Macrophage-derived enzymes that can degrade cartilage collagen in rheumatoid arthritis include specific collagenases; acid cathepsins; neutral proteases; and others that may further degrade collagen that is already partially broken down (Figs 13.2 and 13.3).[120a,121] Enzymes that may degrade proteoglycan include those disorganizing multisubunit aggregates; those degrading proteoglycan core protein; and those degrading the glycosaminoglycan side-chains of the proteoglycan subunit. The influence of enzymes on collagen and proteoglycan degradation is considered further on pp. 184 and 202, respectively. There is evidence to show that enzyme activation is regulated by the removal of inhibitor(s) such as the human cartilage inhibitor of Lesjak and Ghosh.[209] The activities of such enzymes may, of course, also be modulated by agents such as cortisol.[320] Interleukin-1 can stimulate the release of latent metalloproteinase from human articular cartilage in culture[55] and it appears reasonable to argue a corresponding activation of this enzyme not only in ageing cartilage, as these authors suggest, but also in a contributory role in the cartilage loss of osteoarthrosis (see Chapter 22) and rheumatoid arthritis.[134a]

Role of synoviocytes

Synoviocytes themselves may synthesize and release latent neutral metalloproteases,[249] particularly collagenase.[108] A small basically charged calcium-dependent enzyme of molecular weight 33 kD has been identified in these circumstances. Immunoreactive enzyme has been shown to be present at the cartilage–pannus junction but activity is restricted at the time tissue is excised, becoming widespread, however, after short-term culture of the material.[369] This implies the release of collagenase inhibitor(s). The balance between collagenase and inhibitor production may hold one key to the extent and rapidity of cartilage destruction.[213]

Role of cytokines

The role of cytokines in rheumatoid synovitis is not yet fully understood.[8,11] That cytokines are released is beyond dispute. Interleukin-1, granulocyte macrophage colony-stimulating factor (GM-CSF), tumour necrosis factor-alpha (TNF-α)[86] and platelet-derived growth factor (PDGF) are all produced by rheumatoid synovia and the actions of IL-1 and TNF are affected locally by IL-6 which induces immunoglobulin production and the synthesis of acute phase proteins. Interleukin-1 and TNF stimulate the production of neutral proteases, including a collagenase, both by synovial fibroblasts (synovial B cells) and by chondrocytes adjoining the synoviochondral margin. Granulocyte macrophage colony-stimulating factor amplifies the cellular interactions by enhancing HLA-DR expression on macrophages and other antigen-presenting cells. This cytokine, and others, almost certainly exercise part of their influence by the additional release of inhibitory molecules so that a state of dynamic molecular interaction is achieved. One example of an inhibitor molecule is the 22 kD IL-1 inhibitor released from cultured human monocytes, a molecule with considerable amino acid sequence homology to IL-1β. There is less certainty about the mechanisms by which macrophages are

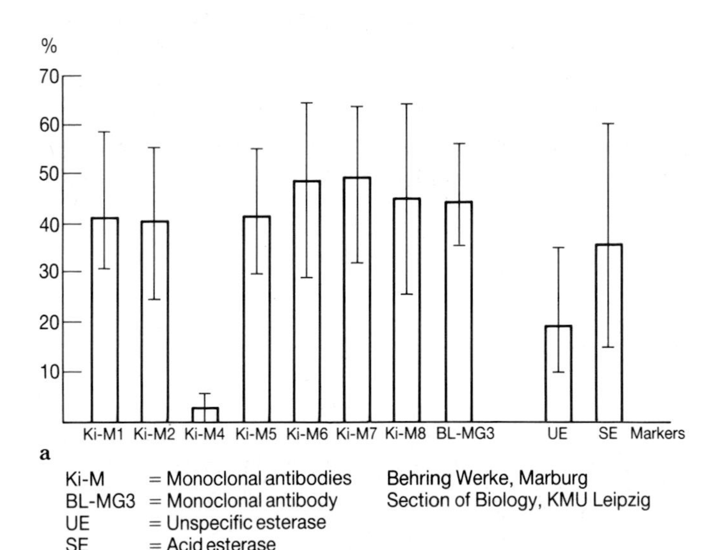

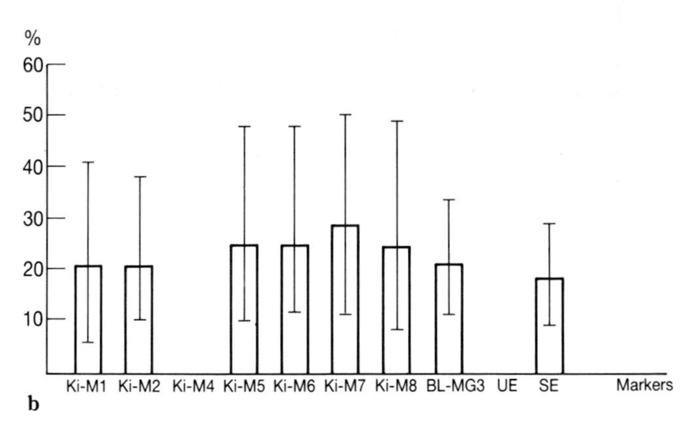

Fig. 13.2 Rheumatoid arthritis: distribution of mononuclear macrophages.
a. Mean percentage of synoviocytes from 21 cases of rheumatoid arthritis scored, as percentage, by monoclonal and enzymatic macrophage markers. **b.** Proportions of cells expressed as percentages, in diffuse lymphatic infiltrates of synovial tissue in 20 cases of rheumatoid arthritis assessed by monoclonal and enzymatic markers for mononuclear macrophages. Courtesy of Dr G Geiler.

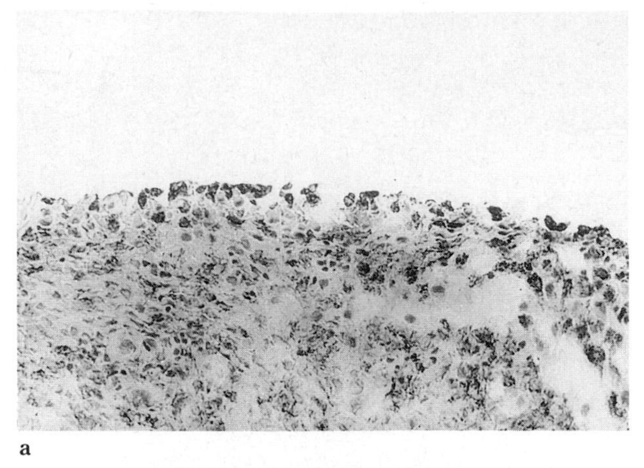
a

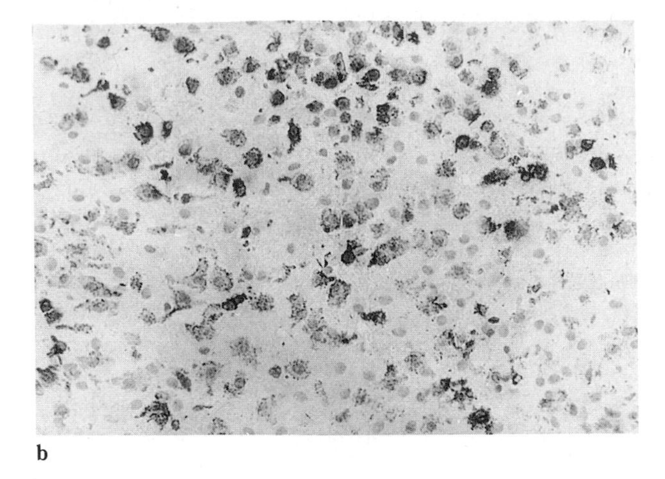
b

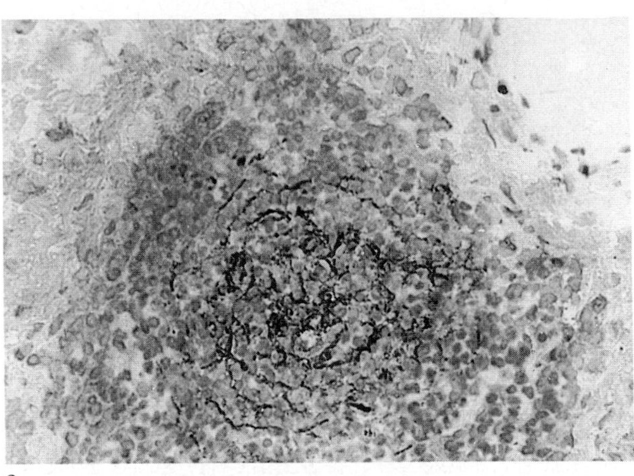
c

Fig. 13.3 Rheumatoid arthritis: distribution of cells identified as mononuclear macrophages. a. Synoviocytic layer (Ki-M8 × 140); **b.** Synoviocytic layer (Ki-M6 × 90); **c.** Synovial lymphoid follicle (Ki-M4 × 230). Courtesy of Dr G Geiler.

stimulated and it is unclear whether 'lymphokines' exist, or exist in excess, in rheumatoid synovia. The possibility remains that preactivated macrophages secrete cytokines without continuous induction or are susceptible to autocrine or paracrine stimulation.[11]

There is substantial experimental evidence that chondrocytes can degrade enzymatically their own pericellular and the adjacent intercellular matrices to an abnormal degree. The sequence may contribute to human disease. The reaction is an amplification of one of the cycles of enzyme-mediated responses that modulate normal cartilage homeostasis by a dynamic balance of catabolism and anabolism. Whether imbalance of such autodegradative pathways contributes to cartilage loss in rheumatoid arthritis has, so far, limited direct support but the abnormal sequence is believed to be involved in the pathogenesis of osteoarthrosis (see Chapter 22).

The evidence on which this concept is based stems from the experiments of Fell[111] (p. 88). Fell and Jubb[112] found that live synovial tissue in culture caused the breakdown of both the living and dead cartilage with which it was in contact. When similar experiments were made in which the synovium and the cartilage were not in contact, only the living cartilage was degraded. The data demonstrated unequivocally that the live chondrocytes were active in causing the degradation of their own pericellular matrix. It also showed that live synovial tissue exerts two actions on cartilage: the first, a direct influence by secreted proteinases; the second, an indirect action determined by live chondrocytes.[88,88a]

Interleukins The nature of the signals passed to live chondrocytes from the non-contiguous synovial cells excited intense interest. The messenger molecules proved to be a family of approximately 20 kD molecular weight proteins designated catabolins. Similar small proteins were shown to be secreted by activated macrophages,[175] by lymphocytes[289] and by synovial fibroblasts.[267] It became clear that 'catabolins' were cytokines[288] (see p. 291) (Fig. 13.4). They displayed the properties of interleukin (IL)-1.[288] Their role in provoking chondrocyte-mediated cartilage matrix degradation is summarized on p. 88 *et seq*.

With these provisos the *in vitro* and *in vivo* degradation of cartilage in rheumatoid arthritis by chondrocytes can now,

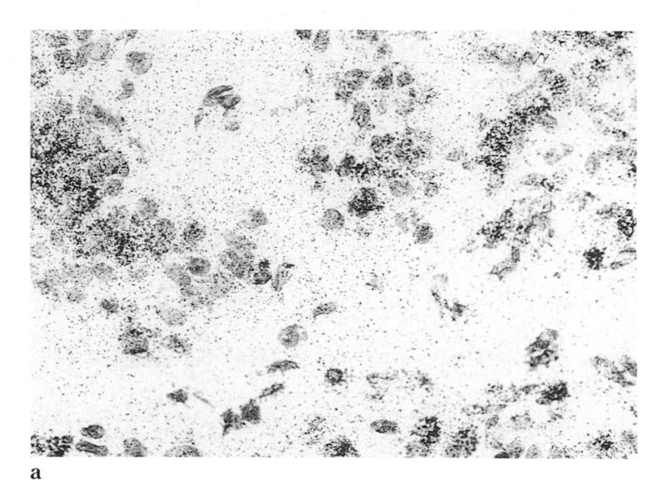

a

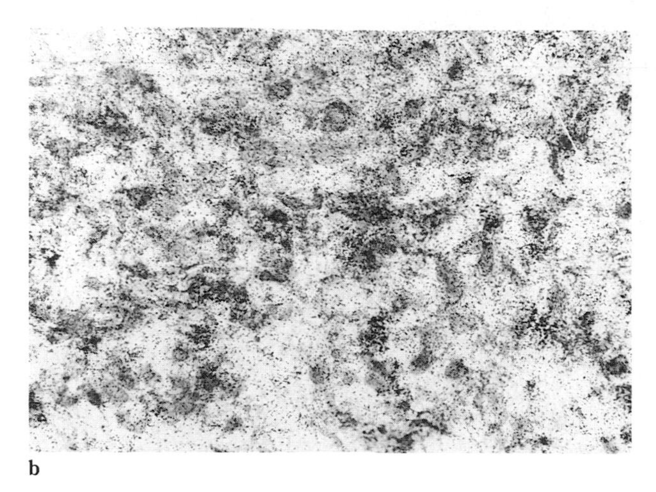

b

Fig. 13.4 Rheumatoid arthritis: cytokine synthesis.
a. Interleukin-1; **b.** Tumour necrosis factor. Alpha naphthyl esterase for mononuclear macrophages. *In situ* hybridization for cytokines. (× 400.) Courtesy of Professor G Duff.

therefore, be viewed as one of the many disorders of cytokines that have been identified in this disease.[63] The cytokines are inducible peptides with receptor-dependent biological actions that may be exerted on a wide range of target cells.[93] The mechanisms triggering IL-1 expression remain obscure. It appears likely that IL-1 can participate in inhibiting cartilage proteoglycan synthesis;[286] it may also retard chondrocyte mitosis *in vivo*.[63] It is clear that IL-1 regulates chondrocyte metabolism and cell-mediated matrix degradation[135,246] and it is established that it increases the rate of (pig) cartilage proteoglycan degradation *in vitro*,[258,273] a sequence enhanced by fibroblast growth factor. There are undoubtedly other influences on cartilage. For example, IL-1 activates chondrocytes and synoviocytes *in vitro* to produce prostaglandin E_2 and plasminogen activator[214] as well as collagen and proteoglycan-degrading enzymes[259] and perhaps other extracellular enzymes.[103]

Although the evidence directly relating IL-1 activity to cartilage injury in rheumatoid arthritis *in vivo* is indirect, there is now little doubt that synovial IL-1 contributes to the degradation.[43] But many of the other important, early tissue changes of rheumatoid arthritis are also mediated or caused by IL-1.[93,94] IL-1 plays a crucial role in immunoregulation (see p. 228) and the intra-articular injection of a partially purified synovial catabolin/IL-1 preparation causes acute synovitis in rabbits with a lymphocytic and plasmacytic infiltrate.[89] Recombinant IL-1 exerts a similar effect. IL-1 also leads to the early onset and increased incidence in mice of collagen arthritis (see p. 549).[186] Moreover, IL-1 gene expression is activated in rheumatoid synovial cells *in vivo*. IL-1 gene expression is not constitutive. Finally, high levels of IL-1 peptides are demonstrable in synovial exudates in rheumatoid arthritis and in plasma; the levels correlate with disease activity[98] and are inversely related to the synovial fluid concentration of proteoglycan.[294] Other tissue changes in rheumatoid arthritis are provoked by the cytokines TNF, IL-2 and IFN-γ. Some may be sequential.

Tumour necrosis factors Tumour necrosis factors (TNF) have been implicated in inflammatory joint disease for reasons analogous with those given for IL-1.[86] The evidence in relation to *septic* arthritis is given by Di Giovine, Symons and Duff.[86a] Tetta *et al.*[338] demonstrated high concentrations of TNF in the serum of patients with severe rheumatoid arthritis and TNF was often detected in the synovial fluid. Tumour necrosis factor stimulates the resorption and cellular interactions within rheumatoid synovial joints, although the cartilage-degrading activity is not always related to the degree of synovial cellularity. TNF has other effects. Both TNF-α and TNF-β (but not all-*trans*-retinoic acid) stimulate hyaluronate synthesis by synovial B cells;[51] hyaluronate formation is increased in rheumatoid arthritis. However, TNF activity in the rheumatoid synovial fluid may be low; it may be masked in cytolytic assays by an inhibitory macromolecule.[241]

Role of chondrocytes

Chondrocytes respond to changes in the composition of the pericellular matrix by adjusting the amount and quality of proteoglycan they secrete. It is therefore not wholly surprising to find that chondrocytes have a vigorous ability to degrade their pericellular matrix by the activation of a number of enzymes and this sequence has been discussed on p. 88.

Lysosomal enzymes Cartilage matrix degradation of the kind caused by the vitamin A alcohol retinol or by IL-1, depends upon the metabolic activity of the chondrocytes. There is a close association between cartilage matrix loss and lysosomal enzyme synthesis and secretion. An acid proteinase, identified as cathepsin D, was discovered but other lysosomal enzymes have also been demonstrated. Some such as aryl sulphatases A and B, display increased activities in osteoarthrosis (see Chapter 22)[68a] but have no demonstrable role in the cartilage changes of rheumatoid arthritis. Acid phosphatase and β-glucuronidase activity is particularly high in the superficial tangential cartilage zone in normal bovine material.[296]

The presence of cathepsin D, the most important lysosomal aspartate protease, was confirmed in rabbit ear and embryonic chick cartilage[362] and in human and monkey articular cartilages;[4,5] the enzyme is, however, active at pH 5.0 not at pH 7.2 and it appears likely that its actions are confined to the intracellular digestion of matrix components that have been released by other mechanisms before endocytosis. It has been confirmed that the inhibition of cathepsin D immunologically or chemically does not decrease the rate of proteoglycan degradation in a chick embryo limb culture system.[154] Indeed, the majority of lysosomal proteases have acid pH optima, incompatible with an extracellular location. Cathepsin B, for example, present in human articular cartilage,[23] displays maximum activity at pH 6.0. There is, therefore, still uncertainty about the identity and sequence of the enzymes activated *in vivo* and no proof that they contribute to cartilage loss in rheumatoid arthritis.

Neutral proteases For these reasons, interest was aroused by neutral proteases of the kind demonstrable in growth-plate epiphyseal cartilage[102] and in osteoarthrosic cartilage. Evidence derived from catabolin (IL-1)-induced pig cartilage degradation showed the presence of at least two types of protease, one probably cathepsin D, the other a neutral metalloenzyme.[345] Neutral proteases have long been known in isolated chondrocytes in culture[293] and a neutral metalloprotease was extracted from human articular cartilage.[292]

Role of lymphocytes

The synovial tissue in rheumatoid arthritis is often dominated by an infiltrate of T lymphocytes. These cells exert cytotoxic effects both directly (NK cells) and after sensitization (see p. 245). It is now apparent that, in addition, T lymphocytes secrete a proteoglycanase capable of breaking down rabbit articular cartilage *in vitro*.[178] Similar reactivity may influence human rheumatoid cartilage *in vivo*.

The role of IL-2 in promoting the multiplication and activities of T lymphocytes is mentioned on pp. 228 and 539. Activated T-cells produce both IL-2 and specific cell surface receptors (IL-2r) for this growth factor. IL-2 secretion is defective in rheumatoid arthritis and there is evidence that levels of soluble IL-2 receptor correlate with disease activity.[331] Serum IL-2 receptor levels may be a guide to the clinical state.[365]

Mechanisms of vascularization

Understanding of the relationship between blood flow and the mechanism of marginal cartilage loss in rheumatoid arthritis remains incomplete (Fig. 12.9). The integrity of the cartilage matrix, upon which depends its essential biomechanical properties, is a function of avascularity. Normal hyaline cartilage inhibits the ingrowth of small blood vessels. At sites of inflammation, where blood flow increases and the new formation of small vessels is promoted, cartilage matrix is degraded and lost. Much, if not all, of this loss is the result of the catabolic enzymes derived directly from polymorphs, macrophages or synovial cells, indirectly from chondrocytes. However, the inflamed synovium in some cases of rheumatoid arthritis produces an angiogenesis factor[47] which catalyses new vessel formation. The role of this factor in facilitating cartilage loss in rheumatoid arthritis is not clear[46] but there are obvious analogies with the angiogenic factors formed during the growth of experimental tumours.[316] Tumour angiogenesis factors may derive from mast cells and one, at least, can be blocked by protamine, an inhibitor of the heparin that is believed to be the responsible mast cell product.[202] Mast cells are numerous in experimental models of inflammatory arthritis[145] and in rheumatoid synovia (p. 458): they may exert a critical function in regulating synovial angiogenesis and hyaline cartilage breakdown in rheumatoid arthritis.

A further consequence of the increased regional blood flow at sites of active rheumatoid synovitis, is bone reabsorption. In older persons with active rheumatoid arthritis, the limb blood flow may be prejudiced by common, coincidental vascular diseases such as atherosclerosis. This does not prevent very active rheumatoid inflammation. Peripheral circulatory insufficiency in rheumatoid arthritis is also attributable to the endarterial, fibromuscular proliferation characteristic of this disease. However, the experimental interruption of blood flow to the limbs of animals with adjuvant arthritis delays, but does not materially alter, the natural history of the disease (Gardner, unpublished observations).

Mechanisms of cell proliferation

Synoviocyte hyperplasia

The cause of synoviocyte hyperplasia in rheumatoid arthritis, the frequency of overt cell division and the source of the additional synoviocytes are debated.[158] Mitotic figures are very rarely seen. This observation alone does not preclude cell division. However, no increased proportion of labelled cells is demonstrable by ^{3}H-thymidine autoradiography.[230] DNA determinations of Feulgen-stained preparations confirm this evidence[73] and further corroboration has been obtained by the application to the hyperplastic synovium of the monoclonal antibody, Ki67 which recognizes one phase of the cycle of cell division. The few labelled cells demonstrated in this way do not bear macrophage markers.[205,222]

It therefore appears that the cells of the hyperplastic rheumatoid synovium divide infrequently. One explanation is that during the acute phase of onset of rheumatoid arthritis, synoviocytes multiply rapidly but then settle down as a stable population in which further cell division is very infrequent. An alternative explanation, supported by Henderson, Revell and Edwards[158] is that cell recruitment continues, not by division but by the repopulation of the synovium with cells derived from the bone marrow. Two experiments support this view. The first demonstrated that many more guinea-pig synovial cells were labelled when animals with hypersensitivity arthritis were given ^{3}H-thymidine systemically than when this nucleotide was given intra-articularly.[92] The second study employed bone marrow cells with giant granules from beige mice. When these cells were given to irradiated histocompatible animals, the marker cells were subsequently identified as type A synoviocytes.[101] Finally, it remains possible that the recruitment of cells to the hyperplastic synovium is from precursors in the subsynoviocytic layer. Evidence to support this view, which does not exclude a bone marrow origin for the precursors, has come from ^{3}H-thymidine autoradiographic studies of the synovium in rabbits.[165,166]

Giant-cell formation

The ultrastructural features of the rheumatoid synovial giant cell (see p. 459) are those of the synovial A cell (macrophage).[141] Among these features are elaborately ruffled, intertwining, slender filopodia, granular endoplasmic reticulum, some smooth-surfaced vesicles and cytoplasmic filaments, and sparse microtubules. Single membrane-bounded cytoplasmic granules, when large and perinuclear, correspond to the brightly periodic acid-Schiff-staining forms seen by light microscopy. Smaller granules resemble the cytoplasmic dense bodies of normal synoviocytes and some activated macrophages.

The superficial giant-cells are distinctive[68,142] but they are not pathognomonic of rheumatoid arthritis[72] since they have been found in psoriatic arthropathy, ankylosing spondylitis, osteoarthrosis, and posttraumatic arthritis.[72,315] They are present in classical as well as in possible/probable (American Rheumatism Association, 1958) rheumatoid arthritis, definite/classical rheumatoid arthritis and in juvenile rheumatoid arthritis but not in Reiter's disease or enteropathic arthropathy. The deeper giant-cells are said to be found in rheumatoid arthritis alone.[142] The presence of rheumatoid giant-cells within the internal part of the synoviocytic layer conforms with the suggestion that they originate in type A synoviocytes. The demonstration that the giant-cell is phagocytic supports this view. There is lysosomal acid phosphatase activity and ingested erythrocytes may be seen within the giant-cell cytoplasm, rarely, however, in large numbers. It is highly probable that the synovial giant-cells are of the mononuclear phagocytic lineage. They react for the MHC class II antigen HLA-DR[14] and for leucocyte common antigen; many also react positively for a macrophage/granulocyte antigen, Mac387 (calgranulin). There is defective cell division. The persistence of an unidentified virus may be one explanation for giant-cell formation but macrophage fusion, as in tuberculosis, has also been considered.

Role of oncogenes

Cell proliferation in the subsynoviocytic connective tissue may be influenced by proto-oncogenes. The orderly expression of cellular proto-oncogenes is necessary for normal growth and differentiation. Many proteins encoded in cellular proto-oncogenes regulate enzyme activation, receptor activity and membrane function. Cellular proto-oncogenes are, however, relatively quiescent, expressing their regulatory and other functions only during defined phases of cell division and differentiation. When malignant changes are induced, point mutations, oncogene translocation, the loss of regulator sequences and amplification are among the mechanisms leading to transformation. Factors such as platelet-derived growth factor, epidermal growth factor and insulin-like growth factor play important parts in the regulation of the growth and development of normal mesenchymal cells such as fibroblasts (see p. 18). Platelet-derived growth factor B chain is encoded in the cellular oncogene *c-sis*, epidermal growth factor receptor in *c-erb-B*.

It therefore appears possible that disordered oncogene expression may play a part in the proliferative lesions of systemic connective tissue disease. However, it is a considerable misrepresentation of the observed changes to describe the rheumatoid arthritis synoviocytic or subsynoviocytic mesenchymal connective tissue as 'cancer-like'[306] and the concept of 'mesenchymoid transformation'[109] remains controversial. Nevertheless, cellular proto-oncogenes have been identified in these tissues and an abnormality of proto-oncogene expression is clearly one way in which disordered mesenchymal cell growth and differentiation in rheumatoid arthritis may be caused. The pattern of this expression has similarities to that of activated T-cells (see p. 228). *C-myc*, *c-fos* and *c-ras* are increasingly expressed whereas *c-myb* and *c-abl* are not.[192] Rheumatoid synovia contain a metalloproteinase with homology for a rat protein induced by a proto-oncogene.[122] This enzyme can activate latent collagenase; it can also degrade the non-collagenous matrix. It is possible that the human enzyme is induced in the same way.

Rheumatoid-like disease in animals

Models offer the possibility of analysing pathological mechanisms in isolation, under defined conditions, and of testing new therapeutic drugs and surgical procedures. Models of rheumatoid arthritis have therefore been much used in searching for the causes and cures for rheumatoid arthritis. There is no definitive evidence that rheumatoid arthritis occurs naturally in animals other than man but many spontaneous and induced animal diseases present features sufficiently similar to those of rheumatoid arthritis to allow them to be used with profit. The exhibition of pathological lesions in an animal disease similar to those of human rheumatoid arthritis does not establish that the animal has rheumatoid arthritis. But models are analogues, not replicas, and they can be employed effectively to dissect the pathogenesis of rheumatoid arthritis even though they are not 'identikits'. One limitation in the use of animal models remains the extreme difficulty of studying primates. An alternative to the study of animal models of rheumatoid arthritis is the choice of cell- and organ culture systems. They offer access to many aspects of disordered biology. They cannot, however, provide complete analogues of inflammatory processes since they have no vascular circulation.

Spontaneous rheumatoid-like disease

Many forms of infective and postinfective arthritis of animals display histological, immunological and chemical features like those of rheumatoid arthritis. Among those reviewed in other chapters are the arthritides attributed to erysipelothrix, the mycoplasmas and caprine arthritis/encephalitis virus (CAEV) (see Chapter 18).

The synovitis of the Pekinese *dog* is morphologically similar to that of human rheumatoid arthritis. In a further group of animals, selected by eliminating those with evidence of osteoarthrosis, infection or systemic lupus erythematosus, microscopic changes were identified resembling those of rheumatoid arthritis.[299] There was ultrastructural evidence of microvascular injury and tubuloreticular structures. Crystalline arrays of tubules resembling those found in neoplasia and in viral disease were present in synovial endothelial cells.

In the *cat*, a chronic progressive arthritis, not caused by identifiable bacteria or mycoplasmas, is aetiologically related to feline leukaemia virus and feline syncitia-forming virus infections.[263] One form of this cat disease, the more common, resembles Reiter's disease: there is osteopenia and periosteal new bone formation. The less common, deforming variety closely resembles rheumatoid arthritis. However, arthritis, confined to males, cannot be reproduced by the inoculation either of cell-free synovial tissue preparations from diseased animals or of tissue culture fluid containing leukaemia virus or syncitia-forming isolates.

In the *mouse*, the heritable disease of the MRL-*lpr/lpr* strain[148] has been advocated as a model of rheumatoid arthritis[252] (see p. 601).

In the captive gorilla, Browne described a rheumatoid arthritis-like state[48] but no rheumatoid-like disease has been identified in other *apes or monkeys*[52] nor can human rheumatoid arthritis be transmitted to them.[215]

Experimental induction of rheumatoid-like disease

Very many methods have been proposed for the experimental induction of rheumatoid arthritis. There is a voluminous literature. For this reason, the reader is referred to reviews, such as those of Sokoloff;[314] Goldings and Jasin,[134] Currey[82] and Greenwald and Diamond.[138a] The present account is confined to six groups of animal models of particular interest. As examples of contemporary work on experimental analogues of rheumatoid arthritis, adjuvant arthritis and anti-collagen arthritis are reviewed in some detail.

Immune complex and antigen arthritis

Immune complex arthritis Immune complex-mediated synovitis is one characteristic sign of serum sickness and a considerable literature records its investigation. It has been established that the formation of immune complexes in rheumatoid arthritis may precede cartilage degradation; the phenomena are spatially associated.[71] The complexes localize selectively in collagenous articular tissues. Confirmation of this close relationship between chondrocyte injury and the pinocytosis of immune complexes was obtained in rabbits in which an Arthus-type hypersensitivity arthritis had been produced[329] by the intraarticular challenge of animals sensitized to bovine serum albumin. The localization of immune complexes was identified in superficial cartilage by a peroxidase–antiperoxidase technique in rabbits sensitized and challenged with either bovine serum albumin or ferritin. Collagen/anticollagen complexes have also been employed directly, by intraarticular injection, to cause sustained arthritis.[312] These studies preceded a series of particular interest in which the intraarticular injection of human α_2-macroglobulin-collagenase complexes were found surprisingly to induce a more severe arthritis in rabbits than the enzyme alone.

Antigen arthritis Antigen arthritis is a term restricted in practice to a form of experimental hypersensitivity in which soluble antigen is injected locally into one or more joints of an animal previously sensitized to the antigen.[95] Provided sufficient time has elapsed after sensitization to allow adequate titres of circulating antibody to form, an Arthus-type response occurs within the synovia of the injected joint, rapidly leading to synovitis and secondary cartilage destruction.[3,95] Adjuvant is given with the antigen to boost the primary response to the sensitizing agent.

Tissue responses Bilateral challenging injections of antigen cause symmetrical chronic arthritis. The responses can be monitored by isotopic scans.[279] The histological changes have been described.[82,156] Synovial villous inflammatory congestion and hypertrophy with synoviocyte hyperplasia are followed by lymphocytic infiltration and lymphoid follicle formation. A pannus forms and cartilage is destroyed centripetally. Periarticular structures are implicated. There is great variability in the form of joint disease caused by antigen, depending on the challenging dose. With very large amounts of antigen, cartilage necrosis is evident; with lower doses, a milder, sustained disease results.[167] These differential effects relate to the differential tissue handling of antigen. Immunocytochemical methods help to clarify the nature of the articular response. Immunoglobulin and antigen are demonstrable as granules in the articular connective tissue and are thought to be components of immune complexes. Classical macrophages are present in the superficial synovium. Dendritic cells and T_H-cells occur mainly in clusters around small synovial blood vessels.[87] Both humoral and cell-mediated responses are implicated.

Hyperplasia of synoviocytes and of subsynoviocytic connective tissue cells, is detected by ^{3}HTdR autoradiography as early as three days after challenge.[156] The deeper cells may contribute to hyperplasia of the synovial intima. Accompanying this proliferative process, are synoviocyte metabolic changes that resemble those of rheumatoid arthritis. Increased activities of lysosomal hydrolases are among those recorded.

The products of antigen arthritis, like those caused by the injection of zymosan, damage bone and (in the case of antigen arthritis) bone marrow, as well as cartilage. Fibronectin co-distributes with the fibrinous exudate of antigen arthritis[301] and this distribution of fibronectin correlates with the fibrosis that characterizes the older lesions.

The distribution of antigen is not confined to the joint itself and the retention of antigen provides the key to the evolution of the prolonged disease (p. 468). The electrical charge of the antigen determines antigen distribution and persistence.[350] Arthritis caused by a positively charged antigen, amidated bovine serum albumin, can indeed be suppressed by the non-immunogenic polycationic protein, protamine chloride.[349]

Antigen arthritis can be modified by synovectomy, ameliorated by drug treatment and X-irradiation and exacerbated by intravenous antigen administration. The synovium that regenerates after synovectomy retains a diminished capacity to respond to intra-articular antigen. By contrast, flares of active inflammation can be provoked by the intravenous injection of small doses of antigen weeks after the onset of the arthritis.[348] The flare is not complement-dependent but is abolished by antilymphocyte serum. The degree of this articular responsiveness to intravenous injection of antigen is related not only to the capacity of small amounts of antigen to reach the synovia but to the hyperreactivity of the synovia of sensitized animals.[207] Flares caused by intravenous injection[271] are dependent upon the presence of MHC class II antigen, suggesting that an interaction between antigen-presenting cells and T lymphocytes is critical to the pathogenesis of the synovitis.

Antigen-arthritis is accompanied by a marked decrease in cartilage proteoglycan in the affected joints. The mechanism is one of inhibition of chondrocyte proteoglycan synthesis.[290] In addition, however, anti-proteoglycan antibodies are formed as a secondary consequence of cartilage injury. The larger proportion recognize a part of the proteoglycan molecule containing core protein and associated keratan sulphate.[372]

Cell wall arthritis

Bacterial products such as lipopolysaccharide effectively cause cartilage matrix degradation in organ culture, perhaps by the direct stimulation of chondrocytes. Bacterial components may therefore play a part in the genesis of septic arthritis and in joint disorders such as yersinial Reiter's disease that follow intestinal infection (p. 764). The responses are influenced by cyclic adenosine monophosphate.[26] Cell-free extracts of group A streptococci, injected intraarticularly, in rabbits, can promote a self-perpetuating, *local* arthritis. Moreover, cell wall polymers derived from group A streptococci, as well as *Salmonella typhimurium* lipopolysaccharide, can provoke local inflammation with smaller fragments of purified peptidoglycan–polysaccharide leading to oedema and transient arthritis. Larger fragments cause not oedema but an acute arthritis that becomes chronic.[107]

Of even greater interest is the observation that an acute, peripheral synovitis followed by chronic, remittent *polyarthritis* can be the result in rats of a single intraperitoneal or intravenous injection of a sterile aqueous suspension of group A streptococcal cell wall fragments.[81,360] The cyclical disease evolves over four to six months with synovial inflammation, cartilage destruction, cartilage and bone erosion, loss of function and ankylosis. The mechanisms underlying this bacterial cell wall-mediated arthritis are not yet wholly understood. The responsiveness of Sprague–Dawley rats to heat-killed streptococci and their sonicated fragments displays streptococcal group specificity. Whereas whole group A streptococci fail to induce arthritis, whole group B organisms regularly cause arthritis but only after a six- to eight-day latent period.[168,317a] By contrast, both sonicated group A and group B streptococcal cell wall products lead to arthritis after a latent period of only 24 h. Group D streptococcal products are inactive. Fragments of unrelated *E. coli* can reactivate arthritis caused by *Strep. pyogenes* cell walls, suggesting a mechanism for chronicity.[351]

Different rat strains exhibit differing degrees of susceptibility. The injected bacterial wall fragments are widely disseminated after injection and are demonstrable many weeks later within macrophages at sites of inflammation and within the spleen and liver. Resistance to degradation relates to arthritogenicity. The immune response to the bacterial products is multifactorial. There is a very short, variable, incubation period and the alternate complement pathway can be activated. All animals forming high levels of anti-peptidoglycan antibody develop severe, chronic arthritis but there is no correlation between anti-group A polysaccharide antibodies and joint disease, although the titre of these antibodies is 10 to 100 times greater than that of anti-peptidoglycan.[138] Arthritis can be induced in animals subjected to neonatal thymectomy, and delayed hypersensitivity is not demonstrable suggesting that cell-mediated immunity against bacterial cell wall antigens is not of pathogenetic significance.[168] It appears that the degree of macrophage cytotoxicity determines the severity and pattern of disease. The size of the injected peptidoglycan polysaccharide fragments is also crucial. Fragments with a molecular weight less than 5×10^6 kD are not arthritogenic but can promote increased endothelial permeability.[62] Large fragments (mol. wt approximately 500×10^6 kD) provoke little acute inflammation but lead to chronic joint disease. Small fragments exhibit the converse response.

Cell wall extracts, peptidoglycan or defined peptidoglycan subunits can cause acute joint inflammation in some strains of mouse after intravenous injection. DBA/IJ and $(\text{BA LB/C} \times \text{DBA/IJ})_{F_1}$ strains are particularly susceptible.[193] Anti-peptidoglycan antibodies are formed but do not seem responsible for the disease. Nevertheless complement components are implicated.

Adjuvant arthritis

Stoerk, Bielinski and Budzilovich[324] noted the anomolous development of polyarthritis in rats injected with splenic tissue homogenized in Freund's complete adjuvant (FCA)*. Independently, Pearson[261,261a] showed an identical reaction to skeletal muscle homogenized in FCA; surprisingly, the same polyarthritic response occurred when the aqueous phase, in which muscle tissue was incorporated, was omitted (Fig. 13.5).

Induction Adjuvant disease[358] is a systemic and articular reaction to FCA; the arthritis is a highly reproducible form of synovitis, of limited duration, widely used to test anti-inflammatory compounds. Adjuvant arthritis is believed to be unique to the rat although there is a hint that FCA can cause polyarthritis in the rhesus monkey. The inductive procedure is very precise. FCA is injected once into the dermis or directly into a regional lymph node.[243] Local inflammation occurs. After a latent period of 9–17 days, systemic illness and polyarthritis are manifest. The optimum quantity of injected mycobacteria averages 0.05 mg. The site of injection may vary but the route of injection must be intradermal and the injected material must have access to the regional lymph nodes.

The natural history of experimental FCA disease has been fully described. At the site of the intradermal injection, acute inflammation reaches a peak after two days. The histological changes centre on clusters of oil droplets: there is a vigorous, sterile polymorph response. A delayed, asymmetrical polyarthritis then begins in the limbs and tail (Fig. 13.5). The degree of inflammation is greatest in the

* Freund's complete adjuvant (FCA) is a suspension of finely ground, killed mycobacterial particles in a mineral oil. Nocardial or other acid/alcohol-fast microorganisms can be substituted for mycobacteria. Freund's incomplete adjuvant (ICFA) comprises oil only. Water-soluble compounds, such as antigens, can be homogenized with the oils to provide material for injection.

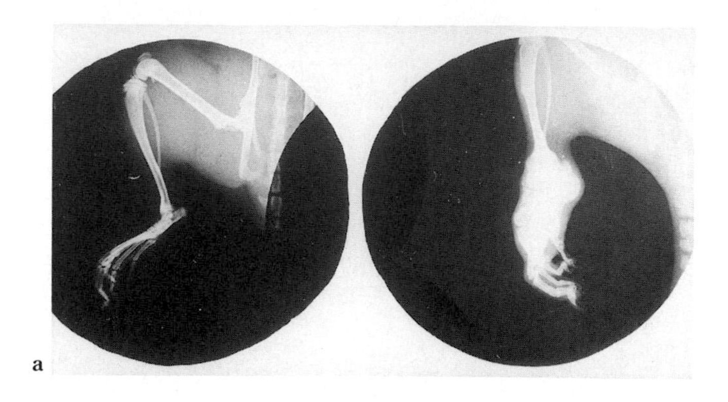

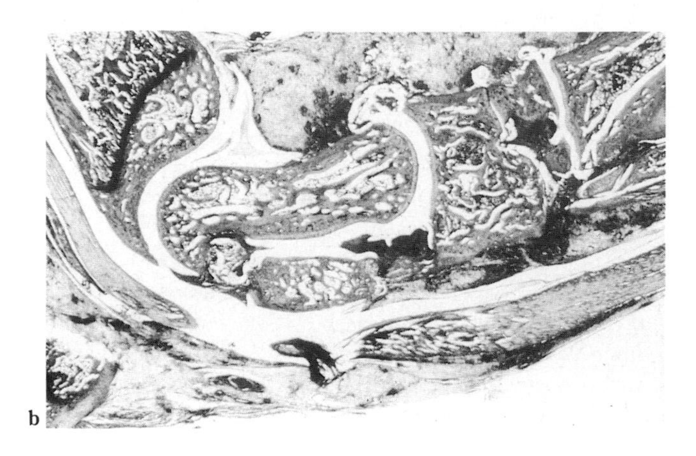

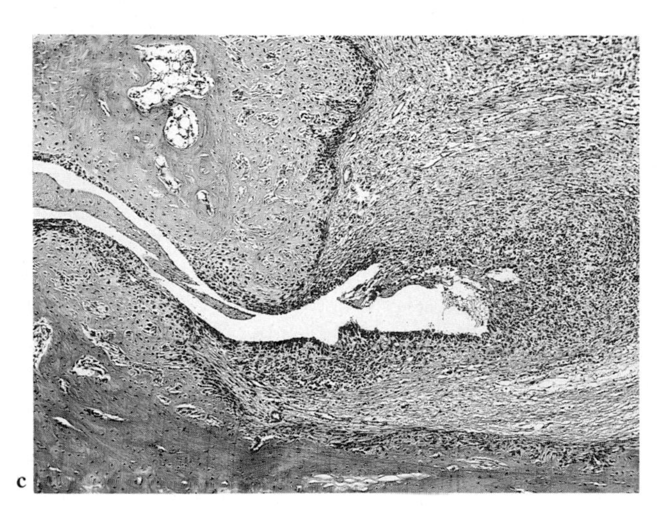

Fig. 13.5 Adjuvant arthritis.
a. Radiograph illustrating new bone formation four weeks after intradermal injection of adjuvant (right). Normal control is seen at left. **b.** Synovitis of the tarsal joints. The intense lymphocyte-rich synovial reactions are demonstrated as deep grey-black zones at tarsal joint margins. **c.** Active synovitis 25 days after intradermal injection of Freund's adjuvant. Already, the extensive mononuclear cell infiltrate of the synovium is accompanied by periosteal new woven bone formation which culminates in the appearances seen in Fig. 13.5a. (HE **b**: × 12; **c**: × 45.)

ankles, tarsal and metatarsophalangeal joints. The response is flitting in character and, in irregular sequence, the joints become red, tender and swollen. Microscopically, there is a delayed, secondary response with synovitis and periarticular inflammation in which polymorphs are numerous, mast cells relatively frequent. Mononuclear phagocytes and lymphocytes soon predominate. As the process extends, cartilage and bone are directly destroyed. Osteoclastic bone resorption is extremely active; it is accompanied by prolific, new, woven bone formation. After four weeks or so from the time of the initial injection, inflammation begins to subside. Joint destruction is now extensive and eventually bony ankylosis converts the entire foot into a radioopaque, osseous mass.

Animals with FCA arthritis display systemic illness, lose weight and become cachectic. However, there is no immediate increase in mortality. There are extraarticular lesions and the viscera are affected. Inflammatory nodules form on the ears and tail. Tendinitis, retinitis, conjunctivitis, iridocyclitis and diarrhoea are common. The extraarticular lesions resemble those seen in the articular tissues. Mononuclear cell infiltration, fibrin, and focal necrosis are recognized. The nodules of ear, tail, tendons and genitalia show a comparable perivascular cell aggregation. They are not, however, replicas of rheumatoid nodules. Fibroblast proliferation culminates in tendinous adhesions and fibrosis, causing contractures.

Cell wall reactivity of bacteria and other organisms Many microorganisms other than *Myc. tuberculosis* can be used to induce adjuvant arthritis. Of 16 dilipidated strains of mycobacteria, only *Myc. leprae* and *Myc. paratuberculosis/Johnei* were non-arthritogenic and non-pathogenic.[357] It has long been known that *Nocardia* and *Corynebacterium rubrum* can replace mycobacteria in FCA arthritis: the three species share a common cell wall structure. However, other microorganisms can also be arthritogenic and *C. diphtheriae*, *Streptomyces lavendulae* and *Streptomyces fradiae* as well as *Lactobacillus plantarum* have this property.

The choice of oils used in FCA influences the incidence of the arthritis following injection: squalene and methyl isostearate are particularly effective.

Susceptibility There are important differences in the susceptibility of different strains of rat to FCA. Lewis and Sprague-Dawley animals respond reproducibly: the manifestations of the disease are less variable in the former.[281]

The Holtzman strain reacts vigorously, the Buffalo strain weakly. Differences in susceptibility can be achieved by establishing inbred lines of a Wistar/Sprague-Dawley cross.[216] The capacity of germ-free animals constantly to develop severe joint disease shows that a bacterial flora is not necessary for the development of arthritis.

Various strains of conventional (non-germ-free) rats develop distinct clinical forms of adjuvant disease, emphasizing not only the importance of genetic differences but also the part played by environmental factors.[236] In the DA(RT1a) rat strain, the capacity to respond to FCA appears to be controlled by an autosomal, dominant gene locus, Ar, linked to the major histocompatibility complex genes with which it displays a high degree of recombination. Although the mechanism of action of the Ar gene locus is not known, it relates to immunological reactivity against self-antigen.

Measurement of response

Many techniques have been used to assay the severity of the adjuvant response. The most obvious depend on measuring foot-pad thickness, temperature and the accumulation of an isotopic label. The clustering of cells at a site of challenge by adjuvant or antigen can be measured directly by injecting 5-iodo-2′-deoxyuridine-iodine-125. Indirectly, the early soft-tissue lesions can be evaluated by magnetic resonance imaging (see p. 162)[337] and the extent of the joint changes by microfocal radiography.

Transfer

Adjuvant arthritis cannot be reproduced by the passive administration of pooled sera from sensitized rats nor can it be transferred to syngeneic animals by serum. Antibody does not play a significant part in the onset of adjuvant arthritis. However, the disease is transmissable by syngeneic thoracic duct lymphocytes.

Prevention

FCA arthritis can readily be prevented or its onset aborted by excision of the regional lymph nodes up to seven days after the initial injection; by thymectomy; or by whole-body irradiation. In each instance, the mode of prevention is by interfering with access of adjuvant to the immune system or by abrogating immune responsiveness. Adjuvant arthritis can also be inhibited by hypophysectomy or the administration of bromocriptine: manifestation of the disease requires prolactin and/or growth hormone. By contrast, arthritis is inhibited by adrenocorticotrophic hormone. The disease can be blocked by the promotion of mild, local peritoneal inflammation, e.g. by silica particles.[40] Concurrent parasitic infestations have the same effect. Peripheral nerve or spinal paralysis modify but do not prevent FCA arthritis.

Of great importance was the early observation that rats became resistant to adjuvant arthritis after pretreatment with lipopolysaccharides in ICFA but not in saline. Mycobacterial wax D in saline was protective whereas the same material in FCA caused arthritis. Protein antigens added to the wax D before inoculation were inhibitory.[262] When dead mycobacteria were injected intraperitoneally during the incubation period, the onset of arthritis was also inhibited; by contrast, PPD (purified protein derivative of *Myc. tuberculosis*) accentuated the lesions. The protective effect of mycobacteria was dose-related.[76] There was no limitation to the dissemination of mycobacterial antigen.

Treatment

Adjuvant arthritis can be modified and the severity of the inflammation lessened by a wide range of antiinflammatory drugs. One example is attributable to the effects of the prostaglandin synthetase inhibitor, Naproxen.[2] The reader is referred to the large pharmacological literature. Compounds affecting immune responsiveness suppress or prevent the arthritis. Early observations demonstrated that rubidomycin (daunorubicin) given within five days of the adjuvant, diminished or prevented the secondary polyarthritis. Anti-lymphocyte serum and anti-lymphocyte globulin have similar effects. In a recent study, the synthetic immunoregulatory thiabenzimidazole, tilomisole, has been found to have both antiinflammatory and immunomodulatory activities when tested in Lewis rats.[128]

The part played by bone resorption and new bone formation in determining final deformity is great. It is therefore noteworthy that a diphosphonate, 4-(chlorophenyl) thiomethylene diphosphonic acid, reduces the severity of the long-term adjuvant arthritic changes.[19] Because of the distant possibility that adjuvant arthritis could be the result of activation of latent virus, there is interest in the evidence that antiviral agents like Statolon, a broad-spectrum antiviral agent which induces interferon formation, can inhibit the development of adjuvant arthritis. Pyran copolymer, another interferon inducer, has a similar effect. However, these actions may be upon components of the protein synthetic mechanism other than virus.

Nature of adjuvant arthritis

Early evidence strongly suggested that FCA arthritis was a form of hypersensitivity response to an antigen associated with the injected mycobacteria. The view that FCA arthritis was caused by activation of a latent mycoplasmal or viral infection has now been largely abandoned, in spite of the effectiveness of some antiviral agents in therapy. The evidence favouring an immune response centred on the preventative role of immunosuppression, the need for lymph node and thymic integrity, the capacity to transmit the disease by lymphocyte suspensions and the blocking actions of Bacille Calmette et Guérin (BCG) or purified protein derivative (PPD). That FCA arthritis is a response to mycobacterial antigen seemed at first less likely when it became clear that not only mycobacterial lipoprotein but also cell wall peptidoglycan and finally natural or synthetic muramyl dipeptide (MDP) could substitute for the whole microorganism in the adjuvant, sometimes in aqueous emulsion. However, new evidence now strongly suggests that the disease mechanism can be attributed to cross-reactivity between host and microbial antigens.[298] There is structural homology between components of *Myc. tuberculosis* and cartilage macromolecules, particularly proteoglycans (p. 531).[353] Thus, T-cells that are arthritogenic in rats recognize an epitope formed by the amino acids at positions 180–188 in the sequence of a *Myc. bovis* BCG antigen and this antigen induces resistance to subsequent attempts to invoke adjuvant arthritis.[353] There is limited homology between the 180–188 sequence and rat proteoglycan link protein but not chicken or rat core protein. Adjuvant arthritis is triggered by activating T-cells with a specificity for self-antigen, a response regulated in part by MHC-linked genes. Cytokines are central to this activation.[129]

Muramyl dipeptide arthritis Simple, chemically defined molecules such as MDP were found to substitute for mycobacteria as adjuvants. The glycopeptide, N-acetyl muramyl-L-alanyl-D-isoglutamine, had been used to enhance immune responses and may have a role as an anti-tumour agent.[60] The presence of the D-isoglutamine residue is essential for the prolongation both of humoral- and of cell-mediated reactions.[200] It was then established that mycobacterial wax D, bacterial cell walls of other species, their peptidoglycans and fragments of the peptidoglycans were arthritogenic in the same manner as Freund's adjuvant. It became clear that a water-in-oil emulsion of MDP was itself arthritogenic[239] in rats. When immunodeficient rats were used, an aqueous form of lipid-conjugated MDP was effective.[196] Congenitally athymic nude (*rnu/rnu*) rats did not develop arthritis after the administration of MDP but *rnu/+* rats were highly susceptible.[195] A saline solution of MDP was capable of provoking a form of arthritis in inbred Lewis rats.[375] Analogues of MDP with adjuvant activity were arthritogenic after a single intravenous injection in BALB/c mice,[194] and the introduction of a stearoyl group into the C-9 hydroxyl group of MDP enhanced and prolonged the joint lesions. Adjuvant-active aqueous MDP analogues also caused polyarthritis in both euthymic and athymic ('nude') rats after single or multiple systemic injections.[197]

As in the case of classical Freund's adjuvant arthritis, it is still not certain whether MDP arthritis is attributable to an immune response to the glycopeptide; to cross-reactivity against a connective tissue component, possibly on the basis of molecular mimicry; or to an autoimmune response against a tissue component such as type II collagen or keratan sulphate proteoglycan which has been altered by the initial reaction to MDP. The availability of this simple, highly effective arthritogenic agent has however accelerated progress in understanding experimental adjuvant arthritis.

Collagen arthritis[328,356]

Humoral immune responses to collagen are commonplace in rheumatoid arthritis; they are however, also recognized in systemic lupus erythematosus, progressive systemic sclerosis, relapsing polychondritis and other diseases and may be one reaction to any injury or to surgical operations upon joints.[233–235] It is established that anticollagen antibodies develop in naturally occurring canine joint diseases and contribute to the formation of collagen–anticollagen immune complexes which are pathogenic.[20] Early studies, based on the detection of anti-type II collagen, revealed antibody in 3 to 71 per cent of cases of rheumatoid arthritis. With new techniques and in the light of information on the 4 or more collagen types present in cartilage (see Chapter 4) (II, VI, IX and XI), it can now be concluded that antibodies to a variety of native and denatured collagens occur in a proportion of patients with rheumatoid arthritis. Antibodies to denatured collagen are more widespread than antibodies to native collagens; antibodies to denatured type II collagen are the most common. Those patients with antibodies to native type II collagen appear to be a subset of rheumatoid patients:[191] an association with the inheritance of the MHC class II HLA genotypes DR2, DR3 and DR7 has been claimed, and a lack of association with DR4. By contrast, there is an association between the development of antibodies to native type II collagen and HLA-DR4. Antibodies formed against type XI collagen may in some instances be responding to the $\alpha1$(II) chain of type II collagen with which the $\alpha3$(XI) chain of type XI collagen has similarities.

Although biotinylated native type II collagen can bind to lymphocytes and plasma cells of the rheumatoid synovium, suggesting local anti-collagen antibody formation,[189] the significance of anti-collagen antibodies for the initiation and prolongation of rheumatoid arthritis remains uncertain. The reaction to collagen may be an immune response to a foreign antigen with which a native collagen shares epitopes or expresses homology. Alternatively the production of anti-collagen antibodies may be a sequel to arthritis provoked by another mechanism such as virus which may act by breaking tolerance. A further possibility is that anti-collagen antibodies are formed because of polyclonal B-cell activation or immune dysregulation, perhaps also caused by virus. Whether the anti-collagen responses initiate connective tissue lesions in rheumatoid arthritis is therefore unlikely but it seems probable that the pathological manifestations of the disease can be amplified, the course prolonged, if anti-collagen reactions are initiated.

It is therefore of much interest that native type II collagen, given by intradermal injection in FCA or ICFA, can induce sterile polyarthritis (collagen arthritis) in the rat.[77,342] A similar reaction has been caused in the mouse (see p. 549)[74] and non-human primate (see p. 550).[77] In the mouse, type IX and XI collagens are also arthritogenic.[36]

The conditions to be met for the successful production of collagen-arthritis are exacting; they help to explain its pathogenesis. The collagen must be native. Denatured type II collagen or collagen peptides are ineffective and native type I or type III collagens do not provoke a response. However, heterologous rather than homologous native type II collagen can be used. ICFA can be employed with the collagen so that the presence of mycobacteria or analogous components is not essential, emphasizing one of many features distinguishing collagen arthritis from adjuvant arthritis (see p. 544). However, the addition of synthetic MDP to the sensitizing injection greatly increases both the incidence of arthritis in the injected rats and the severity of the response.[57] The reactivity of individual rat strains to type II collagen is strongly influenced genetically. WF, LEW and DA strains are among the high-responders. The capacity to cause arthritis may be distinct from the severity of the immune response provoked by the injection of ICFA with type II collagen. This distinction is also shown in limited studies made with the reactions to 'minor' collagens. The collagen-arthritis response is under the control of multiple genes, RTI-linked and non-linked.[140] Resistance to collagen-arthritis often correlates with the expression of MHC class II (Ia) antigens.

Development of polyarthritis

Fourteen to 60 days after the injection of type II collagen, many sensitized rats suddenly develop an arthritis which increases in severity to reach a peak after four days and which persists for five to eight weeks. The peak onset is at 20 days. The disease may be bilateral or unilateral, but the hind limbs are mainly involved, the fore limbs occasionally. The ankle, tarsal and interphalangeal joints are most often affected. As the disease declines, permanent deformity culminates in bony ankylosis. The gross changes closely recall those of rat adjuvant arthritis. The progress of the inflammation can be monitored by local measurements of blood flow, swelling and temperature but serial radiography may prove valuable.[173] Unlike adjuvant arthritis, the spine is not affected and there are no visceral changes. However, a small proportion of animals develop focal inflammation of the ears (see below) displaying features in common with relapsing polychondritis (see p. 711).

Histological changes

Splenic and lymph node hypertrophy precede the arthritis. Synovial oedema, polymorph and extensive mononuclear cell infiltration are followed by synovial cell hyperplasia and synovial villous hypertrophy, cartilage destruction with pannus formation, fibrosis, periosteal new bone formation and bony ankylosis. Long after active inflammation has subsided, the synovial mononuclear infiltrate persists. A suggestion was advanced that there were two stages in the development of the synovial disorder, the first hyperplastic, with fibrin exudation, the second cellular with exudation. Many accept that there is a sufficient resemblance to rheumatoid arthritis to allow the model to be used in the experimental study of the latter disease. However, the extent of cartilage microscar formation[84] and the extremely active bone response are features that distinguish the rat and the human disorders. In collagen arthritis, as in human rheumatoid arthritis, anti-Ia (MHC class II)-reactive cells accumulate in the early pannus, persisting during the subsequent phases of cartilage erosion.[163] The many lymphocytes present in the early stages of active synovitis are T_H-cells, those seen in nodes later in the disease mainly T_S-cells. There are smaller numbers of B-cell and plasmacytes in this later phase.

Humoral immune responses

Rats with adjuvant and with collagen arthritis[341] develop evidence both of humoral and of cell-mediated immunity to type II collagen. Following sensitization, immune responses to collagen may be detectable in the absence of arthritis. There is no evidence that collagen arthritis represents the activation of latent infection. However, the provocative antigen, collagen type II, demonstrably persists for long periods. The suggestion that collagen may itself act as an adjuvant has received little support. It is possible to transfer collagen-arthritis to non-immunized recipient animals with serum from rats with collagen arthritis. The agent conferring this capacity is an IgG anti-collagen antibody that can be inhibited by the native type II collagen used for immunization. One of the arthritogenic determinants lies in the 3/4 region from the N-terminus of type II collagen.[161] The antibody, present in donor rats for at least three weeks, strongly cross-reacts with homologous type II collagen, not with denatured collagen; it localizes to the articular surfaces in recipient rats.[326]

Arthritic rats showed greater lymph node cell reactivity than non-arthritic rats.[247] Rabbit antibody against bovine native type II collagen did not lead to rat collagen arthritis in spite of the observation that this antibody was deposited on articular surfaces. Neither C4 nor C3 were detectable at this site and a failure to activate complement was suggested as the reason for the non-appearance of arthritis.[276] This evidence is incomplete.

The arthritis transferred by rat anticollagen antibody to *rnu/rnu* (athymic nude) rats is enhanced and prolonged by comparison with the effect of the same serum in *rnu*/+ rats,[332] evidence suggesting that anti-type II collagen antibody is not the only factor determining the onset of arthritis. The absence of functional T-cells alone does not inhibit the onset of the disease.[39] However, heterologous anti-idiotypic antisera directed towards antitype II collagen antibodies specifically suppress the development of collagen arthritis.[12]

Cell-mediated immune responses The lymph node cells from rats developing collagen-arthritis produce a leucocyte inhibitory factor and a lymphokine which is arthritogenic and is secreted in culture by the T-cells. Evidence has been adduced that a heterologous anti-T-cell antiserum can decrease the incidence of collagen arthritis.[39] Nevertheless, the prevailing view remains that anticollagen antibodies bear a closer relationship to the development of collagen arthritis than does the cell mediated immunity.[356]

Enhancement of collagen arthritis Analogues of retinoic acid can enhance the severity of collagen arthritis perhaps through an effect on collagenase production or PGE_2 secretion. Of immediate relevance to views on the pathogenesis of rheumatoid arthritis is evidence that rat cytomegalovirus infection increases the severity of collagen arthritis.

Suppression of collagen arthritis The immune status of recipient rats is crucial in determining their immune and arthritic responses to type II collagen. The concept that collagen arthritis is mediated by an immune response to type II collagen is supported by the finding that intravenous pretreatment injection with native soluble type II collagen reduced both the incidence of arthritis and IgM-, IgG- and cell-mediated immune responses.[78] Adjuvant arthritis is not suppressed in this way. The suppression of collagen-arthritis appears to be determined by a peptide near the COOH-terminus of type II collagen. Extending these observations, Phadke, Fouts and Parrish[266] showed that rats exposed to native type II collagen 7–10 days *after* immunization with native type II collagen displayed a lowered incidence of arthritis and of anti-collagen antibody levels but no diminution in the cell-mediated immune response. Rats with adjuvant arthritis also showed a reduced arthritic response.

Anti-collagen antibodies given before immunization with type II collagen prevent the development of collagen arthritis.[106] It is of interest that the intravenous injection of purified anti-collagen IgG itself causes a transient arthritis: a later, second injection is ineffective. A further indication that impaired B-cell responses can reduce the susceptibility to collagen arthritis is provided by preimmunization with anti-μ serum:[153] Splenic cells coupled to native type II collagen and given by intravenous injection up to 14 days before immunization with type II collagen are also suppressive. Sensitization to collagen, measured by IgG antibody titres, is decreased but haemagglutinating antibody and measurements of cell-mediated hypersensitivity are not, suggesting the importance of a particular class or subclass of autoantibodies in initiating collagen arthritis. Spleen and lymph node cells from rats immunized by type II collagen inhibit collagen arthritis but not the antibody response to collagen.[50] Arthritis can also be suppressed by antigen-specific T suppressor cells.[200a]

Other influences Anti-inflammatory drugs, such as indomethacin, inhibit the signs both of collagen and of adjuvant arthritis. D-penicillamine is without effect, although cyclophosphamide is suppressive as is polymorph depletion. Dimethyl sulfoxide is effective when given by intraperitoneal injection in toxic doses. Complement depletion delays arthritis without decreasing anticollagen antibody titres. Cyclosporin A given early suppresses arthritis but, given after the preclinical phase, enhances the subsequent disease.

Collagen arthritis in mice Mice of inbred strains aged more than seven weeks develop polyarthritis after immunization with heterologous or homologous native type II collagen.[74] Types IX and XI collagens are also arthritogenic but the incidence and severity of the polyarthritis is less than after type II collagen.[234] Chick sternal collagen can be used.[327] A booster intraperitonal injection of collagen is required approximately 21 days after the initial dose. The time course of the disease is different from the rat. Polyarthritis begins suddenly after 19–122 days (average 28–42 days); the maximum severity is soon reached. The incidence of the disease is high. The arthritogenic antigenic determinant rests in the region of the type II collagen molecule represented by CB peptide II.[336] There is genetic variation in responsiveness, influenced by the strain from which the collagen is derived.[367] $H\text{-}2^q(I)$ haplotype mice, for example, mount high anti-type II collagen antibody responses but do not always develop collagen arthritis. $H\text{-}2^r$ mice require higher doses of heterologous collagen than $H\text{-}2^q$.

A humoral immune response to type II collagen is essential for the development of mouse collagen arthritis. Immunization with homologous rather than heterologous type II collagen results in disease which may be clinically similar to rheumatoid arthritis;[36] however, the histological changes are slighter, and there is no bone erosion. Like the rat, mouse collagen arthritis can be transferred by the passive intravenous injection of high levels of anti-type II collagen antibody from animals with collagen arthritis. Previous administration of native type II collagen is protective.

Polyclonal anti-type II collagen antibody or small amounts of the corresponding monoclonal antibody behave similarly. The disease can, moreover, be suppressed by pretreatment with antisera against specific MHC class II (Ia) antigens.[368] Interestingly, a mild polyarthritis can be transferred by giving a human serum IgG fraction from a patient with seronegative rheumatoid arthritis.

A cellular immune response reaches a maximum two weeks after immunization. It is not specific to type II collagen: type I and denatured type II collagen evoke comparable proliferative responses when cultured with lymphocytes from animals immunized with native type II collagen. However, antigen-specific T_s-cells can ameliorate the severity of mouse collagen arthritis.[200a] These cells depress both *in vitro* cell-mediated immune responses to type II collagen and anti-type II collagen antibody levels.

Important autoimmune reactions in humans and mice are influenced, in a non-uniform manner, by sex. In NZB/NZW mouse lupus-like disease (see p. 600) for example, oestrogens potentiate but testosterone suppresses antibody production and the development of disease whereas, in the BXSB mouse, males are susceptible. In the case of collagen arthritis, male DBA/1 mice are more susceptible than female; oophorectomy renders female mice as prone as male.[162] Pregnancy in DBA/1 mice may lead to transient remission of arthritis but not long-term protection.[354]

Collagen arthritis in non-human primates

Squirrel monkeys (*Saimiri sciureus*, Letica) develop a mainly symmetrical polyarthritis after immunization with native bovine type II collagen; some die but others undergo remission.[59] The shoulders, hips and spine are spared. Anti-type II collagen antibody titres reach moderately high levels but there is no evidence of rheumatoid factor formation. There is periarticular, as well as articular, inflammation but fewer lymphocytes and plasma cells and more macrophages than in rheumatoid arthritis. Cebus monkeys (*Cebus albifrons*, Barranquilla) do not respond to native type II collagen and squirrel monkeys given FCA alone fail to develop arthritis.

Closely similar results were reported by Yoo *et al.*;[373] they injected the rhesus monkeys, *Macaca mulatta* and *Macaca fasicularia* with bovine type II collagen and, after several booster injections, all animals developed arthritis seven weeks following immunization. The mononuclear cell synovitis was characterized by giant cell formation. High levels of anti-type II collagen antibodies were recorded. These studies were extended by Rubin *et al.*[284] The use of booster injections of type II collagen encouraged more severe, persisting arthritis in all animals and evidence was gained of cell-mediated as well as of humoral immunity to bovine type II collagen. Seven animals given FCA alone did not respond.

There is therefore substantive evidence that monkeys can develop collagen arthritis in the same way as the rat and mouse. There is not, however, any proof that the experimental disease is a replica of rheumatoid arthritis in man.

Proteoglycan arthritis

The possibility that arthritis may be initiated or at least sustained by immune reactions against cartilage components other than collagen, has long excited interest. Rats and rabbits have been injected intraarticularly with suspensions of cartilage particles, causing inflammation, cartilage degradation and signs of delayed hypersensitivity. Anticartilage extract antisera have been given in a similar way, leading to local arthritis. However, there is greater contemporary interest in the antigenic properties of individual proteoglycans rather than in those of whole cartilage extracts or anticartilage antisera. The proposal that nasal and articular cartilage proteoglycans could cause a sterile, humoral antibody- and cell-mediated synovitis in rabbits and dogs is attributable to Glant and Olah[131,133] and Glant.[130] In the view of these authors, reimmunized animals develop a local immune-mediated inflammatory response against proteoglycan, releasing enzymes that degrade cartilage. Further antigenic determinants are unmasked on the proteoglycan molecules, leading to autoantibody formation and the onset of a self-perpetuating autoimmune reaction. In an extension of these studies Glant and his colleagues[132] used chondroitinase ABC-digested fetal human cartilage proteoglycan to cause polyarthritis and spinal disease ('ankylosing spondylitis') in female BALB/c mice. Animals given proteoglycan without FCA did not usually develop arthritis and expressed lower levels of immune response. The development of polyarthritis and spondylitis proved to be strain-specific but not haplotype specific. Cytotoxicity to mouse chondrocytes and sometimes rheumatoid factor formation; immune complex deposits in joint tissues and kidneys; and the formation of antimouse type II collagen antibodies, were associated phenomena.[227]

Porcine clostridial arthritis

Scandinavian studies have shown that aseptic polyarthritis in pigs can be provoked by a dietary change.[221] When fish meal was added to the feeds of eight-week old animals, clinical joint disease developed after approximately five days. Particular attention has been directed to the change in the intestinal flora which follows the dietary manipulation. The principal feature is an increase in the proportions of atypical type A *Clostridium perfringens*. Anti-*Clostridium perfringens* antibodies are formed and there are raised levels of serum immunoglobulins. One explanation for the delayed polyarthritis may therefore be the formation, and deposition in synovia, of circulating immune complexes. Alternatively, the syndrome may be analogous with Reiter's disease (see Chapter 19) in which either cross-reacting antibacterial antibodies or an immune response characteristic of particular MHC class I (HLA-B) determinants have been invoked.

The synovial lesions comprise an inflammatory cell exudate, lymphocytic infiltration, synovial cell hyperplasia, granulation tissue formation and synovial villous hypertrophy. Cartilage and bone erosion are frequent but not constant. There is evidence of systemic disease, with rheumatoid-like subcutaneous granulomata and, occasionally, a proliferative glomerulonephritis.

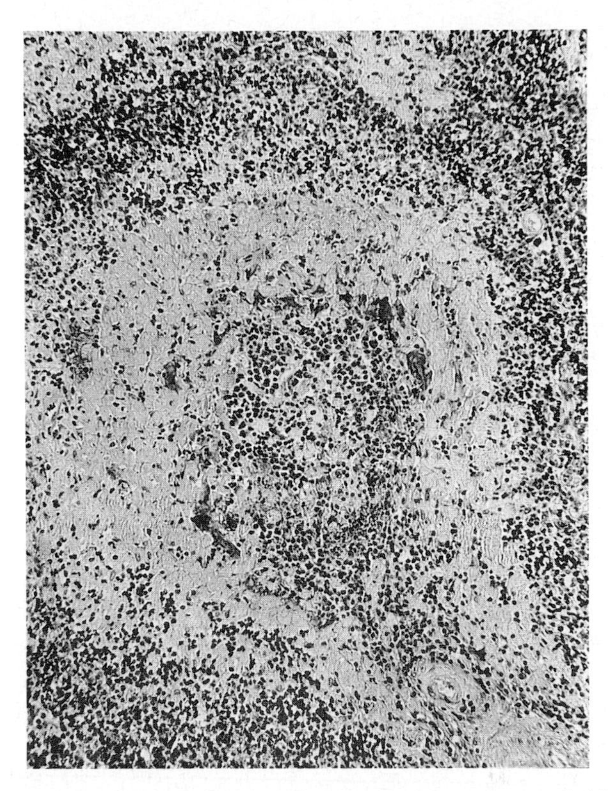

Fig. 13.6 Rheumatoid arthritis: lymph node amyloidosis.
Rheumatoid arthritis remains the commonest cause of 'secondary' amyloidosis in the Western world (see Chapter 11). (HE × 130.)

Complications and prognosis

In recent years, the variety of treatments for rheumatoid arthritis has increased greatly: there has been a corresponding, if not proportional increase in the diversity of complications, many inconvenient, some dangerous, a few fatal.

Amyloidosis

The deposition of amyloid AA (Chapter 11) of a 'secondary' type is a common finding in patients with chronic rheumatoid arthritis. There is reduced amyloid A-degrading activity.[356a] In biopsy series, amyloidosis is estimated to occur in 5–10 per cent of cases. However, *post mortem* studies revealed amyloid in 15.9 per cent of 385 patients and in 11.7 per cent of 145 patients.[123a] Lender and Wolf[208] identified amyloid in 15 per cent of 54 instances but Wegelius suggested that the frequency of the syndrome was 3–5 per cent.[356a] A cumulative incidence of 4 per cent was found in 1272 cases of juvenile chronic polyarthritis.[9] In ankylosing spondylitis there were three examples of amyloidosis among 48 subjects[176] and amyloid is periodically encountered in association with psoriatic arthropathy (see p. 781).

The pattern of organ involvement shown by secondary amyloidosis in rheumatoid arthritis is visceral.[123a,343] In the elderly, there is a high probability of concurrent senile amyloidosis.[123a] Amyloid AA is exceptional in rheumatoid synovia and cartilage but this amyloid is frequently deposited within small blood vessels in the kidney, spleen, adrenals (Fig. 12.40), lymph nodes (Fig. 13.6), gastrointestinal tract, oesophagus, liver and skin. Indeed, almost every viscus including the anterior pituitary, the salivary glands and the blood vessels of the bone marrow may be affected.[63a] Cardiac amyloid may be sufficient to precipitate heart failure. One of the most important sites for amyloid deposition in rheumatoid arthritis is in the renal glomeruli. There is a secondary synthesis of new collagen. Renal excretory failure is an inevitable result of this amyloid nephrosclerosis. The renal deposits may assume a characteristic ultrastructural form with radially arranged fibrils in groups extending from the epithelial aspect of the capillary wall.[85] These spicules are indices of rapid deposition and a fulminant course, a progression more probable in AL amyloidosis than in AA. There is no evidence that the treatment of rheumatoid arthritis with corticosteroids influences the prevalence of amyloidosis[123] and in an age-matched and sex-matched series of cases, Ozdemir *et al.*[255] also found no difference in the frequency of amyloidosis between treated and untreated cases. In spite of therapeutic tests of compounds such as dimethyl sulphoxide, renal amyloid is resistant to drug therapy; dialysis or transplantation offer the only prospects of survival.

Effects of treatment

Medical treatment

Gold[134a] It has long been known that gold, in the form of Myocrisin, accumulates in the tissues of patients with rheumatoid arthritis treated with the salt, sodium aurothiomaleate. Injected experimentally into the rabbit knee, myochrysine causes regressive and destructive changes in the synovial membrane.[127] Cells with electron dense gold bodies can be recognized in the synoviocytes and similar bodies are seen within the articular chondrocytes up to 14 months after injection.[127] In a human case, the deposition of gold after 12 years' therapy was associated with black pigmentation of the synovium,[270] a phenomenon thought to be due to the high concentration of carrier gold-197 in the preparation containing the unstable isotope gold-198. Blue-grey discolouration of the face, neck and hand skin has also been recognized in longstanding cases of rheumatoid arthritis,[24] a condition analogous with argyria and with plumbism. Electron microscopy confirms the presence of gold-coated granules in the cells of both skin and synovium.

An inherited predisposition to adverse reactions to gold is increasingly well understood (p. 531).[256a] The systemic effects of gold have long been documented.[30,134a]

More recently, the use of oral gold, for example, as Auranofin, has attracted attention. The systemic complications resemble those of the parenteral compound. When gold is combined with another slow-acting antirheumatic drug such as hydroxychloroquine,[302] the number of adverse reactions is significantly increased. Of 49 patients given intramuscular gold with a placebo, 17 withdrew from treatment, 10 because of adverse reactions including skin, oral, renal and marrow disorders. Among 52 given gold and hydroxychloroquinine, 18 withdrew for similar reasons.

Penicillamine Not only does a slow-acting compound such as penicillamine not stop the progression of joint disease, but the drug itself may provoke a number of characteristic pathological responses (p. 499).[30a]

Corticosteroids There is a very large literature.[27,123,364]

Non-steroidal anti-inflammatory drugs Aspects of tissue responses to non-steroidal anti-inflammatory drugs (NSAIDs) have been extensively reviewed.[219,364]

Incidental infection The possibility that rheumatoid arthritis is the direct or indirect result of viral or bacterial infection has been closely investigated for 50 years. These investigations have been complicated by evidence suggesting that untreated cases of rheumatoid arthritis are more susceptible to opportunistic and incidental infection than normal. This increased susceptibility is, of course, amplified by the prolonged use of agents that impair immune responsiveness, depress bone marrow or lymphoreticular function or injure tissue directly. The frequency of infection among hospital patients with rheumatoid arthritis is greater than among control patients in the same hospital.[257] Very often, the coincidental infections are themselves synovial.[100,181] Haemolytic *Staphylococcus aureus* is the organism most often recognized[277] but streptococci, pneumococci, pasteurella, haemophilus, proteus and coliforms are among aerobic organisms that have infected hip, wrist and knee joints.[179] *Serratia marcescens* is a rare cause.[91] There are now an increasing number of recorded instances of anaerobic infections: Ziment[375a] reported 47 examples, and *Bacteroides fragilis*[285] and *Bacteriodes melaninogenicus*[90] are among the responsible agents. The anaerobic infections are often silent and indolent and multiple joints may be implicated. Rising antibody titres may assist diagnosis. A number of instances of mycotic or fungal-like infections have supervened in rheumatoid arthritis, generally when a patient has been treated with corticosteroids. Thus, Gardner *et al.*[124] described (Fig. 13.7) articular nocardiosis (*Nocardia asteroides*) complicating rheumatoid arthritis (Fig. 13.7), and Graham and Frost[136] reported infection of a rheumatoid knee with *Candida guilliermondi*.

Surgical treatment

Surgery is[25,70] undertaken for the diagnosis and for the further treatment of rheumatoid arthritis.[56,75] Pathological problems relevant to the natural history of the disease may arise from both categories of procedure. However, the increasingly widespread adoption of arthroscopic synovectomy has led to a reduced morbidity and to diminished hospital stay and rehabilitation period while retaining the possibility of effective synovial ablation.[67]

Synovial tissue is excised early in rheumatoid arthritis to control the development of the disease or, later, for remedial and functional reasons.[96,317] In an analogous way, synovia can be ablated (synoviorthesis) by chemical agents,[38] by the local injection of radioactive isotopes,[146,206] or even by the external exposure of joints to ionizing radiation (see p. 553). Experimentally, synovectomy of the rabbit knee is followed by the regeneration of a complete new synovial lining layer.[185,280] It has been claimed that this neosynovium arises from mesenchymal cells of the subjacent bone but an origin from residual synoviocytes or a repopulation by bone marrow-derived cells appear possible. The cells of the capsule are inert. Regeneration is delayed experimentally when hydrocortisone is given simultaneously but this treatment pro-

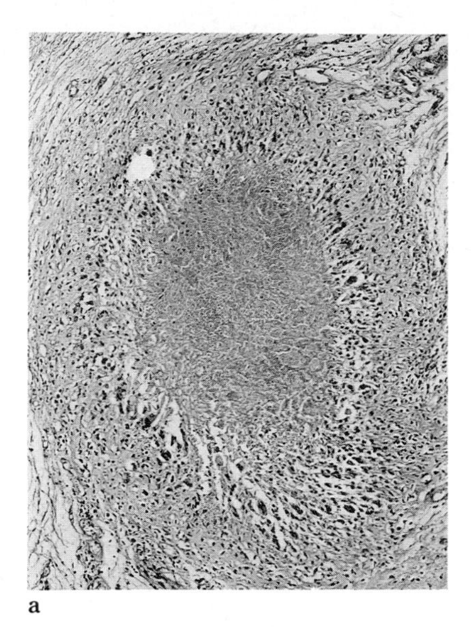
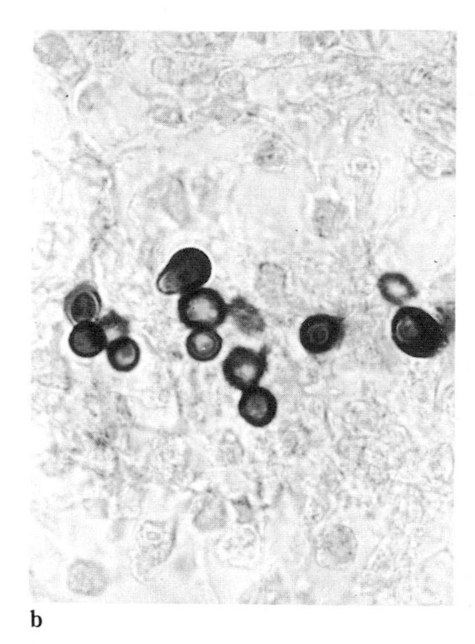
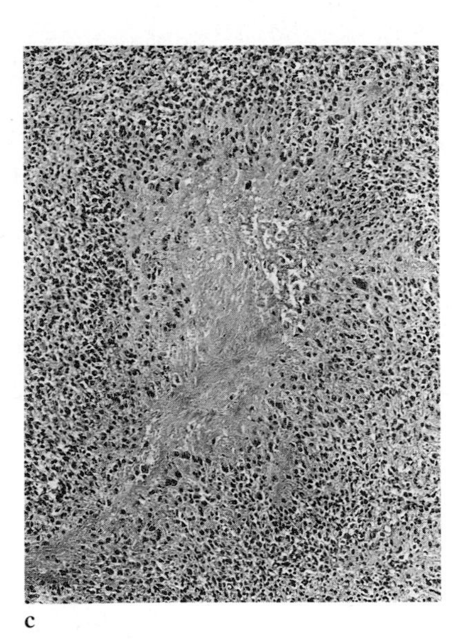

a b c

Fig. 13.7 Rheumatoid arthritis: systemic fungal infection.
In this patient, systemic candidiasis was a sequel to antibiotic-resistant staphylococcal infection. **a.** Subpericardial mycotic granuloma near tricuspid valve. **b.** *Candida albicans* demonstrated in granuloma. **c.** Subendocardial granuloma from the same patient. In this lesion, a rheumatic, not a mycotic, origin is suspected. (**a**, **c**: HE × 65; **b**: PAS × 875.)

tects against the cartilage fibrillation and chondrocyte death that can be a late sequel.[29]

After three years, the regenerated synovium is indistinguishable histologically from that of non-operated joints. Nevertheless, 5–18 months after surgery there is wide variation in the state of the tissue, some parts showing fibrosis, others active synovitis.[260] Later, an increase in perivascular plasma cells and lymphocytes accompanies synovial cell hypertrophy[238] and the synovial tissue manifest lysosomal hydrolase activity is slightly diminished. Although the regenerated tissue is less permeable to plasma proteins than the non-operated synovium, there are unchanged amounts of complement components C3 and C4. Early and late synovectomy have also been used effectively in the control of juvenile rheumatoid arthritis.[182] As in the adult, a wide range of joints, from the proximal interphalangeal and metacarpophalangeal to the knee and wrist,[65] has been approached.

Total joint replacement by arthroplasty is now widely practised in the late treatment of the secondary osteoarthrosis that complicates rheumatoid arthritis. Hips, knees and elbows[204,317] are replaced with increasing frequency and finger joints have been successfully replaced first by metal, more recently by polymeric materials including silicone and polypropylene. Ceramic (aluminium silicate) implants have also been introduced. The tissue reactions[114a] to joint prostheses and to the products of the wear and corrosion of prostheses have been discussed by Revell.[69,275] They are reviewed in Chapter 24 (see p. 959).

Among the complications of these operations that pose pathological problems are bacterial infection; loosening of the bone/acrylic interface without infection; and adverse tissue reactions to polymethyl methacrylate (acrylic bone cement), polyethylene and metals.[228] A reaction to silicone elastomer (silastic) is also recognized and silicone implants may not only elicit new connective tissue[201] but may yield fragments of silicone that are carried to regional lymph nodes. Here, a reaction simulating neoplasia can be evoked as long as eight years after arthroplasty.[143] Histologically, the silicone can be readily identified as irregular refractile particles within clusters of multinucleate giant cells.[188] The local response may be complicated histologically by the presence of particles of bone. Barium sulphate, employed to identify acrylic bone cement radiologically, is seen within the cement. Electron probe X-ray microanalysis of tissue from revised arthroplasties may therefore be required to distinguish iron, cobalt, barium, chromium and other elements from calcium and phosphate.

Irradiation

The reader is referred to Chapter 24. Local, X-irradiation was used to treat regional disease of joints such as the knee but was abandoned because of complications such as bone necrosis. Whole-body irradiation is seldom advocated but 'total lymphoid irradiation' in a manner adopted for the treatment of Hodgkin's disease, has been tested. There are frequent complications.[159] Among 11 patients in this small

Table 13.2

Mean age at death of 142 hospital cases* of rheumatoid arthritis examined at necropsy between 1929–66[142]

	Number	Mean age
Males	44	59.0
Mean age at death of whole male population		67.0†
Females	98	63.0
Mean age at death of whole female population		72.8†
Female/male sex ratio 2.2/1.0		
Of 106 cases examined *post mortem* 1950–66		
Males	31	58.7
Females	75	63.4

*Includes 4 cases of Still's disease, mean age at death 26 years.
†Registrar-General for Scotland, 1964.

series were examples of acute bacterial arthritis; death due to renal failure with amyloidosis; empyema; and Kaposi's sarcoma.

Causes of death

The causes of death[66] and age at death in rheumatoid arthritis have continued to attract interest (Tables 13.2 and 13.3, p. 444).[152,300] When rheumatoid arthritis is the immediate and proximate cause of death, cervical vertebral subluxation or dislocation, necrotizing renal papillitis, systemic vasculitis, cardiac granulomata, mesenteric arterial thrombosis and amyloidosis may be fatal. There is an increased susceptibility to infection[123a,257] especially by opportunistic organisms that emerge after immunosuppression or surgery. The complications of rheumatoid arthritis that follow prolonged immobilization include osteoporosis, urinary tract infection and deep venous thrombosis. Early clinical and epidemiological evidence suggested that there was a modest reduction in life expectancy in patients with rheumatoid arthritis.[69,97,123a,346] However, at a time when more people were living to a greater age, the number of deaths attributable to rheumatoid arthritis appeared to be diminishing,[366] and increased life span did not seem to be increasing the number of cases.

Among rheumatoid patients in Massachusetts deaths from all causes were increased by 86 per cent.[232] The peak mortality for UK females was in the 70–79 year decade, that for males approximately 5 years earlier.[366] The degree to which rheumatoid arthritis itself influences life expectancy has been increasingly questioned. Among the many difficulties in making this judgement is the bias introduced by selection of patients who have been hospitalized for established rheumatoid arthritis. Patients with rheumatoid arthritis display increased mortality from circulatory disease[232] and the detailed investigation of 500 males and 500 females revealed that deaths in a three-year period were approximately twice those in a control group. Cardiovascular and renal disease, and infection were the main immediate causes.[172] In a full study over an 11-year period, this raised mortality was again confirmed but it was demonstrated that the increased mortality came from the 75 per cent fraction of cases who had been in hospital during the 13 years before initial case selection.[6] Thus, there is a view that the apparently abbreviated life span in rheumatoid arthritis may be the result of the collection into the records of a subset of patients likely to develop severe disease and, likely, therefore, to be subjected to aggressive treatment. This view accords with evidence derived from a prospective study[272] of 100 cases of definite or classical (1958 American Rheumatism Association criteria) disease, fol-

Table 13.3

Records of four groups giving causes of death and age at death in rheumatoid arthritis[340a]

	Reah[274]	Boers *et al.*[35a]	Constable[169]	Rasker and Cosh[272]
A: Number of deaths	80	111	100	43
B: Mean age at death (years)	63	63	63	69
C: Deaths attributed to rheumatoid arthritis (%)	—	21	16	21
D: Deaths to which rheumatoid arthritis contributed (%)	—	13	28	16
C & D combined (%)	42.5	34	44	37
E: Deaths not related to rheumatoid arthritis (%)	57.5	66	56	63

The reader is referred to Scott.[374a]

lowed over 18 years. Forty-three patients died during that period and rheumatoid arthritis caused death in nine, mainly from vascular disease, amyloidosis (renal failure) and infection. Rheumatoid arthritis contributed to death in a further seven cases, and infection and drug treatment were important factors in these fatalities. The 16 patients who died from causes directly or indirectly attributable to rheumatoid arthritis developed the disease at an earlier age, and died at an earlier age, than the 27 cases of rheumatoid arthritis in whom death was not directly or indirectly attributable to the disease. The younger group with the poorer prognosis displayed a worse American Rheumatism Association grading after one year and a worse functional capacity than those who survived longer.

Many facts remained unclear and Abbruzzo,[1] drawing attention to the restriction of Rasker and Cosh's[272] analysis to those with definite/classic (1958) disease one year after onset, commented that: 'Rheumatoid arthritis, on balance, is an essentially benign, non-fatal disease'. However, Abbruzzo conceded that this observation obscured the existence of a 'high-risk subgroup' in which there is an increased mortality e.g. from cardiovascular and infectious disease.

There are inaccuracies in data recording and disease classification[170,171] but the main thrust of the debate is made clear from the evidence of Prior, Symmons *et al.*;[269] Mutru, Laakso, Isomaki and Koota;[237] and Symmons, Prior *et al.*[330] A reduced life expectancy, suggested by the autopsy data of Gardner,[123a] is confirmed. Mortality is often attributable to the rheumatoid disease itself.[303] In 489 consecutive patients with definite/classic (American Rheumatism Association 1958) disease followed over 11.2 years, cohort analysis indicated a three-fold increase in overall mortality in comparison with age- and sex-specific rates for the general England and Wales population. There was an excess of deaths from circulatory, respiratory and musculoskeletal disease in those first seen early in the disease, a small excess of cases of malignant disease and alimentary disorders, and an excess of deaths from infection, alimentary and genitourinary disease in those referred more than five years after the onset of rheumatoid arthritis.[268,269] The relative risk overall is 2.6 in UK men, 3.4 in UK women[330] although initially women may have a milder disease, the prognosis being affected by 'hormonal status'. These results accord with those of Mutru *et al.*[237] for a Finnish population and those of Mitchell, Spitz, Bloch, McShane and Fries[229] for a Canadian population. The evidence from 13 surveys was admirably summarized by Scott.[300] The data are however at variance with figures, derived from death certification, for Australian deaths recorded by the Australian Bureau of Statistics.[359]

There is particular interest in the deaths attributable to renal disease. Chronic nephritis, renal infection and amyloidosis are the most frequent causes of death in this category.[203] Amyloidosis was the cause of death in 5.8 per cent of 500 males and in 12.8 per cent of 500 females in this large series. Men with rheumatoid arthritis died from amyloidosis after a mean disease duration of 16.1 years, women after 21.6 years. The corresponding figures for renal deaths were 14.8 and 19.2 years. When amyloidosis was excluded, renal failure determined death in 9.1 per cent of the males and 15.5 per cent of the females in comparison with a frequency of 0.7 per cent among 500 male controls selected from the general population and of 0.0 per cent among 500 female controls.

Interest has also centred on the possibility that there is a raised frequency of malignant neoplastic disease in rheumatoid arthritis. There is strong evidence of an association between rheumatoid arthritis and lymphoproliferative cancer,[268] although this link does not appear to be attributable to drug treatment. However, the increased use of cyclophosphamide has brought to light a well-recognized complication of this substituted nitrogen mustard, the development of bladder cancer. When 119 cases of rheumatoid arthritis treated with cyclophosphamide were compared with 119 matched rheumatoid controls, 37 examples of malignant disease (six of the bladder, eight of the skin) were found in 29 patients in the former group, 16 examples in 16 patients in the latter. The difference was significant and has persisted over 13 years.[18]

REFERENCES

1. Abruzzo J L. Rheumatoid arthritis and mortality. *Arthritis and Rheumatism* 1982; **25**: 1020–3.

2. Ackerman N R, Rooks W H, Shott L, Genant H, Maloney P, West E. Effects of naproxen on connective tissue changes in the adjuvant arthritic rat. *Arthritis and Rheumatism* 1979; **22**: 1365–74.

3. Alexander I S, Gardner D L, Skelton-Stroud P N. Synovial response of *Papio cynocephalus* to exogenous antigen: histological and immunoperoxidase observations. *Annals of the Rheumatic Diseases* 1983; **42**: 448–51.

4. Ali S Y, Evans L. Enzymatic degradation of cartilage in osteoarthritis. *Federation Proceedings* 1973; **32**: 1494–8.

5. Ali S Y, Evans L. Studies on the cathepsins in elastic cartilage. *Biochemical Journal* 1969; **112**: 427–33.

6. Allebeck P, Ahlbom A, Allander E. Increased mortality among persons with rheumatoid arthritis, but where RA does not appear on death certificate. *Scandinavian Journal of Rheumatology* 1981; **10**: 301–6.

7. Alspaugh M A, Tan E M. Serum antibody in rheumatoid arthritis reactive with a cell-associated antigen. Demonstration by

precipitation and immunofluorescence. *Arthritis and Rheumatism* 1976; **19**: 711–19.

8. Alwan W H, Dieppe P A, Elson C J, Bradfield J W B. Bone resorbing activity in synovial fluids in destructive osteoarthritis and rheumatoid arthritis. *Annals of the Rheumatic Diseases* 1988; **47**: 198–205.

9. Ansell B M. *Rheumatic Disorders in Childhood.* London: Butterworth, 1980.

10. Ansell B M, Bywaters E G L. Juvenile chronic arthritis. In: Scott J T, ed, *Copeman's Textbook of the Rheumatic Diseases* (6th edition). Edinburgh: Churchill Livingstone, 1986: 1100–24.

11. Arend W P, Dayer J-M. Cytokines and cytokine inhibitors or antagonists in rheumatoid arthritis. *Arthritis and Rheumatism* 1990; **33**; 305–15.

12. Anita C, Kaibara N, Jingushi S, Takagishi K, Hote Kebuchi T, Arai K. Suppression of collagen arthritis in rats by heterologous anti-idiotype antisera against anticollagen antibodies. *Clinical Immunology and Immunopathology* 1987; **43**: 374–81.

13. Asherson R A, Muncey F, Pambakian H, Brostoff J, Hughes G R V. Sjögren's syndrome and fibrosing alveolitis complicated by pulmonary lymphoma. *Annals of the Rheumatic Diseases* 1987; **46**: 701–5.

14. Athanasou N A, Quinn J, Woods C G, McGee J O'D. Immunohistology of rheumatoid nodules and rheumatoid synovium. *Annals of the Rheumatic Diseases* 1988; **47**: 398–403.

15. Attia W M, Shams A H, Ali M K H, Clark H W, Brown T McP, Bellanti J A. Studies of phagocytic cell function in rheumatoid arthritis. 1. Phagocytic and metabolic activities of neutrophils. *Annals of Allergy* 1982; **48**: 279–82.

16. Attia W M, Shams A H, Ali M K H, Jang L W, Clark H W, Brown T McP, Bellanti J A. Studies of phagocytic functions in rheumatoid arthritis. II. Effects of serum factors on phagocytic and metabolic activities of neutrophils. *Annals of Allergy* 1982; **48**: 283–7.

17. Baker D G, Dayer J-M, Roelke M, Schumacher H R, Krane S M. Rheumatoid synovial cell morphologic changes induced by a mononuclear cell factor in culture. *Arthritis and Rheumatism* 1983; **26**: 8–14.

18. Baker G L, Kahl L E, Zee B C, Stolzer B L, Agarwal A K, Medgser T A. Malignancy following treatment of rheumatoid arthritis with cyclophosphamide. Long-term case-control follow-up study. *American Journal of Medicine* 1987; **83**: 1–9.

19. Barbier A, Breliere J C, Remandet B, Roncucci R. Studies on the chronic phase of adjuvant arthritis: effect of SR 41319, a new diphosphonate. *Annals of the Rheumatic Diseases* 1986; **45**: 67–74.

20. Bari A S M, Carter S D, Bell S C, Morgan K, Bennett D. Anti-type II collagen antibody in naturally occurring canine joint diseases. *British Journal of Rheumatology* 1989; **28**: 480–6.

21. Barnes C G, Turnbull A L, Vernon-Roberts B. Felty's syndrome. A clinical and pathological survey of 21 patients and their response to treatment. *Annals of the Rheumatic Diseases* 1971; **30**: 359–74.

22. Bates E J, Kowanko I C, Ferrante A. Conditioned medium from stimulated mononuclear leucocytes potentiates the ability of human neutrophils to damage human articular cartilage. *Annals of the Rheumatic Diseases* 1988; **47**: 1–9.

23. Bayliss M T, Ali S Y. Studies on cathepsin B in articular cartilage. *Biochemical Journal* 1978; **171**: 149–54.

24. Beckett V L, Doyle J A, Hadley G A, Spear K L. Chrysiasis resulting from gold therapy in rheumatoid arthritis: identification of gold by X-ray microanalysis. *Mayo Clinic Proceedings* 1982; **57**: 773–7.

25. Beddow F H, ed. *The Surgical Management of Rheumatoid Arthritis.* John Wright & Sons, Bristol, 1988.

26. Bednar M S, Hubard J R, Steinberg J J, Broner F A, Sledge C B. Cyclic AMP-regulating agents inhibit endotoxin-mediated cartilage degradation. *Biochemical Journal* 1987; **244**: 63–8.

27. Behrens T W, Goodwin J S. Glucocorticoids. In: McCarty D J, ed, *Arthritis and Allied Conditions. A Textbook of Rheumatology* (11th edition). Philadelphia, London: Lea and Febiger, 1989: 613–6.

28. Bennett J C. The etiology of rheumatoid arthritis. In: Kelley W N, Harris E D, Ruddy S, Sledge C B, eds, *Textbook of Rheumatology* (second edition). Philadelphia: W B Saunders & Company, 1985: 879–86.

29. Bentley G, Kreutner A, Ferguson A B. Synovial regeneration and articular cartilage changes after synovectomy in normal and steroid-treated rabbits. *Journal of Bone and Joint Surgery* 1975; **57**-B: 454–62.

29a. Berg W B van den, Joosten L A B, Putte L B A van de, Zwarts W A. Electrical charge and joint inflammation. Suppression of cationic aBDA-induced arthritis with a competitive polycation. *American Journal of Pathology* 1987; **127**: 15–16.

29b. Berg W B van den, Putte L B A van de, Zwarts W A, Joosten L A B. Electrical charge of the antigen determines intraarticular antigen handling and chronicity of arthritis in mice. *Journal of Clinical Investigation* 1984; **74**: 1850–9.

30. Bernstein R M. Humoral autoimmunity in systemic rheumatic disease—a review. *Journal of the Royal College of Physicians of London*, 1990; **24**: 18–25.

30a. Berry H. Penicillamine, gold and antimalarials. In: Scott J T, ed., *Copeman's Textbook of the Rheumatic Diseases* (6th edition). Edinburgh, London, Melbourne, New York: Churchill Livingstone, 1986: 52534.

31. Biemond P, Swaak A J G, Eijk H G van, Koster J F. Intra-articular ferritin bound iron in rheumatoid arthritis. A factor that increases oxygen free radical-induced tissue destruction. *Arthritis and Rheumatism* 1986; **29**: 1187–93.

32. Blake D R, Gallagher P J, Potter A R, Bell M J, Bacon P A. The effect of synovial iron on the progression of rheumatoid disease. A histologic assessment of patients with early rheumatoid synovitis. *Arthritis and Rheumatism* 1984; **27**: 495–501.

33. Blake D R, Hall N D, Bacon P A, Dieppe P A, Halliwell B, Gutteridge J N C. The importance of iron in rheumatoid disease. *Lancet* 1981; **ii**: 1142–4.

34. Blendis L M, Ansell I D, Lloyd-Jones K, Hamilton E, Williams R. Liver in Felty's syndrome. *British Medical Journal* 1970; **i**: 131–5.

35. Bloch K J, Buchanan W W, Wohl M J, Bunim J J. Sjögren's syndrome: a clinical, pathological and serological study of 62 cases. *Medicine* 1965; **44**: 187–231.

35a. Boers M, Croonen A M, Dijkmans B A C, Breeveld F C, Eulderink F, Cats A, Weening J J. Renal findings in rheumatoid arthritis: clinical aspects of 132 necropsies. *Annals of the Rheumatic Diseases* 1987; **46**: 658–63.

36. Boissier M C, Feng X Z, Carlioz A, Roudier R, Fournier C. Experimental autoimmune arthritis in mice. I. Homologous type II collagen is responsible for self-perpetuating chronic polyarthritis. *Annals of the Rheumatic Diseases* 1987; **46**; 691–700.

37. Boissier M C, Ghiocchia G, Ronziere M C, Herbage D, Fournier C. Arthritogenicity of minor cartilage collagens (types IX and XI) in mice. *Arthritis and Rheumatism* 1990; **33**: 1–8.

38. Boussina I, Lagier R, Ott H, Fallet G. Radiological opacities of the knee after intra-articular injections of osmic acid. *Radiologica Clinica* 1976; **45**: 417–24.

39. Brahn E, Trentham D E. Effect of antithymocyte serum on collagen arthritis in rats: evidence that T-cells are involved in its pathogenesis. *Cellular Immunology* 1984; **86**: 421–8.

40. Bramm E, Binderup I, Arrigoni-Martelli E. Inhibition of adjuvant arthritis by intraperitoneal administration of low doses of silica. *Agents and Actions* 1980; **60**: 435–8.

41. Brewer E J, Bass J, Baum J L, Cassidy J T, Fink C, Jacobs J, Hanson V, Levinson J E, Schaller J, Stillman J S. Current proposed revision of JRA criteria. *Arthritis and Rheumatism* 1977; **20**: 195–9.

42. Brewerton D A. Causes of arthritis. *Lancet* 1988; **ii**: 1063–6.

42a. Bromley M, Fisher W D, Woolley D E. Mast cells at sites of cartilage erosion in the rheumatoid joint. *Annals of the Rheumatic Diseases* 1984; **43**: 76–9.

43. Brown M F, Hazleman B L, Dingle J T, Dandy D J, Murley A H G. Production of cartilage degrading activity by human synovial tissues. *Annals of the Rheumatic Diseases* 1987; **46**: 319–23.

44. Brown K A, McCarthy D, Perry J D, Dumonde D C. Reduction of the surface charge of blood polymorphonuclear cells by rheumatoid sera and heat-induced aggregated human IgG(HAGG). *Annals of the Rheumatic Diseases* 1988; **47**: 359–63.

45. Brown K A, Perry J D, Black C, Dumonde D C. Identification by cell electrophoresis of a subpopulation of polymorphonuclear cells which is increased in patients with rheumatoid arthritis and certain other rheumatological disorders. *Annals of the Rheumatic Diseases* 1988; **47**: 353–8.

46. Brown R A, Tomlinson I W, Hill C R, Weiss J B, Phillips P, Kumar S. Relationship of angiogenesis factor in synovial fluid to various joint diseases. *Annals of the Rheumatic Diseases* 1983; **42**: 301–7.

47. Brown R A, Weiss J B, Tomlinson I W, Phillips P, Kumar S. Angiogenic factor from synovial fluid resembling that from tumours. *Lancet* 1980; **i**: 682–5.

48. Browne T McP, Clark H W, Bailey J A, Gray C W. A mechanistic approach to treatment of rheumatoid-type arthritis naturally occurring in a gorilla. *Transactions of the American Clinical and Climatological Association* 1970; **82**: 227–47.

49. Burkhardt R. Histomorphologische Untersuchungen uber die Rolle des Knochenmarkes bei rheumatischen Krankheiten. *Zeitschrift für gesamte experimentelle Medizin* 1967; **143**: 1–66.

50. Burrai I, Henderson B, Knight S C, Staines N A. Suppression of collagen type II-induced arthritis by transfer of lymphoid cells from rats immunized with collagen. *Clinical and Experimental Immunology* 1985; **61**: 368–72.

51. Butler D M, Vitti G F, Leizer T, Hamilton J A. Stimulation of the hyaluronic acid levels of human synovial fluid fibroblasts by recombinant human tumor necrosis factor alpha, tumor necrosis factor beta (lymphotoxin), interleukin-1 alpha, and interleukin-1 beta. *Arthritis and Rheumatism* 1988; **31**: 1281–9.

52. Bywaters E G L. Observations on chronic polyarthritis in monkeys. *Journal of the Royal Society of Medicine* 1981b; **74**: 794–9.

53. Bywaters E G L, Ansell B M. Monoarticular arthritis in children. *Annals of the Rheumatic Diseases* 1965; **24**; 116–22.

54. Bywaters E G L, Edmonds J. Biopsies and tissue diagnosis in the rheumatic diseases. *Clinics in Rheumatic Diseases* 1975; **2**: 179–210.

55. Campbell I K, Roughley P J, Mort J S. The action of human articular-cartilage metalloproteinase on proteoglycan and link protein. Similarities between products of degradation *in situ* and *in vitro*. *Biochemical Journal* 1986; **237**: 117–22.

56. Cappell H A, Kelly I G. Surgical management of rheumatic diseases. *Annals of the Rheumatic Diseases* 1990; **49** (Suppl 2): 823–82.

57. Carlson R P, Blazek E M, Dakto L J, Lewis A J. Humoral and cellular immunologic responses in collagen-induced arthritis in rats: their correlation with severity of arthritis. *Journal of Immunopharmacology* 1984; **6**: 379–88.

58. Carlsson G E, Kopp S, Oberg T. Arthritis and allied diseases of the temporomandibular joint. In: Carlsson G E, Zarg G A, eds, *The Temporomandibular Joints – Function and Dysfunction*. Copenhagen: Munksgaard, 1978.

59. Cathcart E S, Hayes K C, Gonnerman W A, Lazzari A A, Frantzblau C. Experimental arthritis in a non-human primate. I. Induction by bovine type II collagen. *Laboratory Investigation* 1986; **54**: 26–31.

60. Chedid L, Lederer E. Past, present and future of the synthetic immunoadjuvant MDP and its analogs. *Biochemical Pharmacology* 1978; **27**: 2183–6.

61. Cherian P V, Schumacher H R. Immunoelectronmicroscopic characterization of intracellular inclusions in synovial fluid cells of patients with rheumatoid arthritis. *Ultrastructural Pathology* 1983; **5**: 15–27.

62. Chetty C, Klapper D G, Schwab J H. Soluble peptidoglycan-polysaccharide fragments of the bacterial cell wall induce acute inflammation. *Infection and Immunity* 1982; **38**: 1010–19.

63. Chin J E, Lin Y. Effects of recombinant human interleukin-1 on rabbit articular chondrocytes. Stimulation of prostanoid release and inhibition of cell growth. *Arthritis and Rheumatism* 1988; **31**: 1290–6.

63a. Chumbley L C, Peacock O S. Amyloidosis of the conjunctiva – an unusual complication of trachoma: a case report. *South African Medical Journal* 1977; **52**: 897–8.

64. Clasener H A L, Biersteker P J. Significance of diphtheroids isolated from synovial membranes of patients with rheumatoid arthritis. *Lancet* 1969; *ii*: 1031–3.

65. Clayton M L, Ferlic D C. The wrist in rheumatoid arthritis. *Clinical Orthopaedics and Related Research* 1975; **106**: 192–7.

66. Cobb S, Anderson F, Bauer W. Length of life and cause of death in rheumatoid arthritis. *New England Journal of Medicine* 1953; **249**: 553–6.

67. Cohen S, Jones R. An evaluation of the efficacy of arthroscopic synovectomy of the knee in rheumatoid arthritis: 12–24-month results. *Journal of Rheumatology* 1987; **14**: 452–5.

68. Collins D H. *The Pathology of Articular and Spinal Diseases*. London: Edward Arnold, 1949.

68a. Colofiore J R, Schwartz E R. Monensin stimulation of arylsulfatase B activity in human chondrocytes. *Journal of Orthopaedic Research* 1986; **4**: 273–80.

69. Constable T J, McConkey B, Paton A. The cause of death in rheumatoid arthritis. *Annals of the Rheumatic Diseases* 1978; **37**: 569.

70. Cooke T D V, Chir B. A scientific basis for surgery in rheumatoid arthritis. *Clinical Orthopaedics and Related Research* 1986; **208**: 20–4.

71. Cooke T D, Hurd E R, Jasin H E, Bienenstock J, Ziff M. Identification of immunoglobulins and complement in rheumatoid articular collagenous tissues. *Arthritis and Rheumatism* 1975; **18**: 541–51.

72. Cooper N S, Soren A, McEwen C, Rosenberger J L. Diagnostic specificity of synovial lesions. *Human Pathology* 1981; **12**: 314–28.

73. Coulton L A, Henderson B, Chayen J. The assessment of DNA-synthetic activity. *Histochemistry* 1981; **72**: 91–9.

74. Courtenay J S, Dallman M J, Dayan A D, Martin A, Mosedale B. Immunisation against heterologous type II collagen induces arthritis in mice. *Nature* 1980; **283**: 666–8.

75. Coventry M B, Polley H F, Weiner A D. Rheumatoid synovial cyst of the hip; report of three cases. *Journal of Bone and Joint Surgery* 1959; **41A**: 721–30.

76. Cozine W S, Stanfield A B, Stephens C A L, Mazur M T. Adjuvant disease—the paradox of prevention and induction with complete Freund's adjuvant. *Proceedings of the Society for Experimental Biology and Medicine N.Y.* 1972; **141**: 911–14.

77. Cremer M. Type-II collagen-induced arthritis in rats. In: Greenwald R A, Diamond H S, *CRC Handbook of Animal Models for the Rheumatic Diseases* (volume I). Boca Raton, Florida: CRC Press Inc., 1988: 17–27.

78. Cremer M A, Hernandez A D, Townes A S, Stuart J M, Kang A H. Collagen-induced arthritis in rats: antigen-specific suppression of arthritis and immunity by intravenously injected native type II collagen. *Journal of Immunology* 1983; **131**: 2995–3000.

79. Crisp A J. Mast cells in rheumatoid arthritis. *Journal of the Royal Society of Medicine* 1984; **77**: 450–1.

80. Crisp A J, Chapman C M, Kirkham S A, Schiller A L, Krane S. Articular mastocytosis in rheumatoid arthritis. *Arthritis and Rheumatism* 1984; **27**: 845–51.

81. Cromartie W J, Craddock J H, Schwab H J, Anderle S K, Yang C. Arthritis in rats after systemic injection of streptococcal cells or cell walls. *Journal of Experimental Medicine* 1977; **146**: 1585–1602.

82. Currey H L F. Animal models of arthritis. In: Scott J T, ed, *Copeman's Textbook of the Rheumatic Diseases* (sixth edition). Edinburgh: Churchill-Livingstone, 1986: 468–85.

82a. Deighton C M, Walker D T. The familial nature of rheumatoid arthritis. *Annals of the Rheumatic Diseases* 1991; **50**: 62–5.

83. Denman A M. A viral aetiology for juvenile chronic arthritis? *British Journal of Rheumatology* 1988; **27**: 169–70.

84. DeSimone D P, Parsons D B, Johnson K E, Jacobs R P. Type II collagen-induced arthritis. A morphologic and biochemical study of articular cartilage. *Arthritis and Rheumatism* 1983; **26**: 1245–58.

85. Dickman S H, Churg J, Kahn T. Morphologic and clinical correlates in renal amyloidosis. *Human Pathology* 1981; **12**: 160–9.

86. Di Giovine F S, Nuki G, Duff G W. Tumour necrosis factor in synovial exudates. *Annals of the Rheumatic Diseases* 1988; **47**: 768–72.

86a. Di Giovine F S, Symons J A, Duff G W. Interleukin I and tumour necrosis factor in the pathogenesis of septic arthritis. In: Calabro J J, Dick W C, eds, *Infections and Arthritis*. Dordrecht: Kluwer Academic Publishers, 1989: 121–32.

87. Dijkstra C D, Dopp E A, Vogels I M C, Noorden C J F van. Macrophages and dendritic cells in antigen-induced arthritis. An immunohistochemical study using cryostat sections of the whole knee joint of rat. *Scandinavian Journal of Immunology* 1987; **26**: 513–23.

88. Dingle J T. Articular damage in arthritis and its control. *Annals of Internal Medicine* 1978; **88**: 821–6.

88a. Dingle J T. Catabolin—a cartilage catabolic factor from synovium. *Clinical Orthopaedics and Related Research* 1981; **156**: 219–31.

89. Dingle J T, Page Thomas D P, King B, Bard D R. In vivo studies of articular tissue damage mediated by catabolin/interleukin 1. *Annals of the Rheumatic Diseases* 1987; **46**: 527–33.

90. Dodd M J, Griffiths I D, Freeman R. Pyogenic arthritis due to Bacterioides complicating rheumatoid arthritis. *Annals of the Rheumatic Diseases* 1982; **41**: 248–9.

91. Dowart B B, Abrutyn E, Schumacher H R. Serratia arthritis. Medical eradication of infection in a patient with rheumatoid arthritis. *Journal of the American Medical Association* 1973; **225**: 1642–3.

92. Dreher R. Origin of synovial type A cells during inflammation. An experimental approach. *Immunobiology* 1982; **161**: 232–45.

93. Duff G W. Interleukins and clinical medicine. In Sheppard M C, ed, *Advanced Medicine* (volume 24). London: Baillière Tindall, 1988: 237–46.

94. Duff G W. Peptide regulating factors in non-malignant disease. *Lancet* 1989; **i**: 1432–5.

95. Dumonde D C, Glynn L E. The production of arthritis in rabbits by an immunological reaction to fibrin. *British Journal of Experimental Pathology* 1962; **43**: 373–83.

96. Duthie R B, Bentley G, eds. *Mercer's Orthopaedic Surgery* (8th edition). London: Edward Arnold, 1983.

97. Duthie J J R, Brown P E, Truelove L H, Baragar F D, Lawrie A J. Course and prognosis in rheumatoid arthritis: a further report. *Annals of the Rheumatic Diseases* 1964; **23**: 193–204.

98. Eastgate J A, Symons J A, Wood N C, Grinlinton F M, Di Giovine F S, Duff G W. Correlation of plasma interleukin-1 levels with disease activity in rheumatoid arthritis. *Lancet* 1988; **ii**: 706–9.

99. Eden W van, Holoshitz J, Cohen I. Antigenic mimicry between mycobacteria and cartilage proteoglycans: the model of adjuvant arthritis. *Concepts in Immunopathology* 1987; **4**: 144–70.

100. Editorial. Septic arthritis in rheumatoid disease. *British Medical Journal* 1976; **ii**: 1089–90.

101. Edwards J C W, Willoughby D A. Demonstration of bone marrow-derived cells in synovial lining by means of giant in-

tracellular granules as genetic markers. *Annals of the Rheumatic Diseases* 1982; **41**: 177–82.

102. Ehrlich M G, Armstrong A L, Neuman R G, Davis M W, Mankin H J. Patterns of proteoglycan degradation by a neutral protease from human growth-plate epiphyseal cartilage. *Journal of Bone and Joint Surgery* 1982; **64**-A: 1350–4.

103. Elford P R, Meats J E, Sherrard R M, Russell R G G. Partial purification of a factor from human synovium that possesses interleukin-1, chondrocyte stimulating and catabolin-like activities. *FEBS Letters* 1985; **179**: 247–51.

104. Ellman P, Cudkowicz L, Elwood J S. Therapy of 'Felty's syndrome'. *Annals of the Rheumatic Diseases* 1955; **14**: 84–9.

104a. Ellman P, Parkes Weber F, Goodier T E W. A contribution to the pathology of Sjögren's disease. *Quarterly Journal of Medicine* 1951; **20**: 33–42.

105. Emery P, Lopez A F, Burns G F, Vadas M A. Synovial fluid neutrophils of patients with rheumatoid arthritis have membrane antigen changes that reflect activation. *Annals of the Rheumatic Diseases* 1988; **47**: 34–9.

106. Englert M, McReynolds R A, Landes M J, Oronsky A L, Kerwar S S. Pretreatment of rats with anticollagen IgG renders them resistant to active type II collagen arthritis. *Cellular Immunology* 1985; **90**: 258–66.

107. Esser R E, Anderle S K, Chetty C, Stimpson S A, Cromartie W J, Schwab J H. Comparison of inflammatory reactions induced by intraarticular injection of bacterial cell wall polymers. *American Journal of Pathology* 1986; **122**: 323–34.

107a. Eulderink F. The synovial bipsy. In Berry C L, ed, *Bone and Joint Disease*. New York: Springer-Verlag, 1982: 26–72.

107b. Eulderink F, Meijer K A E. Pathology of the cervical spine in rheumatoid arthritis. A controlled study of 44 spines. *Journal of Pathology* 1976; **120**: 91–108.

108. Evanson J M, Jeffrey J J, Krane S M. Studies on collagenase from rheumatoid synovium in tissue culture. *Journal of Clinical Investigation* 1968; **47**: 2639–51.

109. Fassbender H G. *The Pathology of Rheumatic Diseases* (translated by G Loewi). Berlin: Springer-Verlag, 1975.

110. Fava-de-Moraes F, Friedman H, Egami M I, Bevilacqua E M A F, Cossermelli W. Histochemical study of labial salivary glands in Sjögren's syndrome. *Journal of Oral Pathology* 1978; **7**: 135–42.

111. Fell H B. The Strangeways Research Laboratory and cellular interactions. In: Dingle J T, Gordon J L, eds, *Cellular Interactions*. Amsterdam: Elsevier, North/Holland Biomedical Press, 1981: 1–14.

112. Fell H B, Jubb R W. The effect of synovial tissue on the breakdown of articular cartilage grown in organ culture. *Arthritis and Rheumatism* 1977; **20**: 1359–71.

113. Felty A R. Chronic arthritis in the adult, associated with splenomegaly and leuopenia. A report of 5 cases of an unusual clinical syndrome. *Bulletin of the Johns Hopkins Hospital* 1924; **35**: 16–20.

114. Ferrel P B, Aitcheson C T, Pearson G R, Tan E M. Seroepidemiological study of relationships between Epstein–Barr virus and rheumatoid arthritis. *Journal of Clinical Investigation* 1981; **67**: 681–7.

114a. Fitzgerald R H Jr, Kelly P J. Total joint arthroplasty. Biologic causes of failure. *Mayo Clinic Proceedings* 1979; **54**: 590–6.

115. Fong S, Carson D A, Vaughan J H. Rheumatoid factor. In: Gupta S, Talal N, eds, *Immunology of Rheumatic Diseases*. New York: Plenum Medical Book Company, 1985: 167–96.

116. Fox R, Chilton T, Rhodes G, Vaughan J H. Lack of reactivity of rheumatoid arthritis synovial membrane DNA with cloned Epstein–Barr virus DNA probes. *Journal of Immunology* 1986; **137**: 498–501.

117. Fox R, Sportsman R, Rhodes G, Luka J, Pearson G, Vaughan J. Rheumatoid arthritis synovial membrane contains a 62,000-molecular weight protein that shares an antigenic epitope with the Epstein–Barr virus-encoded associated nuclear antigen. *Journal of Clinical Investigation* 1986; **77**: 1539–47.

118. Fox R I, Robinson C A, Curd J G, Kozin F, Howell F V. Sjögren's syndrome. Proposed criteria for classification. *Arthritis and Rheumatism* 1986; **29**: 577–85.

119. Freemont A J. Molecules controlling lymphocyte endothelial interactions in lymph nodes are produced in vessels of inflamed synovium. *Annals of the Rheumatic Diseases* 1987; **46**: 924–8.

120. Freemont A J, Denton J. Disease distribution of synovial fluid mast cells and cytophagocytic mononuclear cells in inflammatory arthritis. *Annals of the Rheumatic Diseases* 1985; **44**: 312–15.

120a. Friedmann B A. Rheumatoid nodules of the larynx. *Archives of Otolaryngology* 1975; **101**: 361–3.

121. Fukase M, Koizumi F, Wakaki K. Histopathological analysis of sixteen subcutaneous rheumatoid nodules. *Acta Pathologica Japonica* 1980; **30**: 871–82.

122. Fulton R, Forrest D, McFarlane R, Onions D, Neil J C. Retroviral transduction of T-cell antigen receptor beta-chain and *myc* genes. *Nature* 1987; **326**: 190–4.

123. Gardner D L. Amyloidosis in rheumatoid arthritis treated with hormones. *Annals of the Rheumatic Diseases* 1962; **21**: 298–9.

123a. Gardner D L. *The Pathology of Rheumatoid Arthritis*. London: Edward Arnold, 1972.

123b. Gardner D L. Geode: crystal-containing cavity, lymph space, or pseudocyst? *Annals of the Rheumatic Diseases* 1985; **44**: 569–70.

124. Gardner D L. Krieg A F, Chapnick R. Fatal systemic fungus disease in rheumatoid arthritis with cardiac and pulmonary mycotic and rheumatoid granulomata. *InterAmerican Archives of Rheumatology* 1962; **5**: 561–86.

125. Gaston J S H, Rickinson A B, Yao Q Y, Epstein M A. The abnormal cytotoxic T-cell response to Epstein–Barr virus in rheumatoid arthritis is correlated with disease activity and occurs in other arthropathies. *Annals of the Rheumatic Diseases* 1986; **45**: 932–6.

126. Ghadially F N, ed. Siderosomes, haemosiderin and ferritin. *Ultrastructural Pathology of the Cell and Matrix* (third edition). London: Butterworths, 1988: 636–43.

127. Ghadially F N, Lalonde J-M A, Thomas I, Massey K I. Long-term effects of myochrysine on the synovial membrane and aurosomes. *Journal of Pathology* 1978; **125**: 219–24.

128. Gilman S C, Carlson R P, Daniels J F, Datko L, Berner P R, Chang J, Lewis A J. Immunological abnormalities in rats with

adjuvant-induced arthritis—II. Effect of antiarthritic therapy on immune function in relation to disease development. *International Journal of Immunopharmacology* 1987; **9**: 9–16.

129. Gilman S C, Daniels J F, Wilson R E, Carlson R P, Lewis A J. Lymphoid abnormalities in rats with adjuvant-induced arthritis. I. Mitogen responsiveness and lymphokine synthesis. *Annals of the Rheumatic Diseases* 1984; **43**: 847–55.

130. Glant T T. Induction of cartilage degradation in experimental arthritis produced by allogeneic and xenogeneic proteoglycan antigens. *Connective Tissue Research* 1982; **9**: 137–44.

131. Glant T T, Csongor J, Szucs T. Immunopathologic role of proteoglycan antigens in rheumatoid joint disease. *Scandinavian Journal of Immunology* 1980; **11**: 247–52.

132. Glant T T, Mikecz K, Arzoumanian A, Poole A R. Proteoglycan-induced arthritis in BALB/c mice. Clinical features and histopathology. *Arthritis and Rheumatism* 1987; **30**: 201–12.

133. Glant T T, Olah I. Experimental arthritis produced by proteoglycan antigens in rabbits. *Scandinavian Journal of Rheumatology* 1980; **9**: 271–9.

133a. Godfrey H P, Ilardi C, Engber W, Graziano F M. Quantitation of human synovial mast cells in rheumatoid arthritis and other rheumatic diseases. *Arthritis and Rheumatism* 1984; **27**: 852–6.

134. Goldings E A, Jasin H E. Arthritis and autoimmunity in animals. In: McCarty D J, ed, *Arthritis and Allied Conditions. A Textbook of Rheumatology* (11th edition). Philadelphia, London: Lea and Febiger, 1989: 465–81.

134a. Gordon D A. Gold compounds. In: Kelley D A, Harris E D, Ruddy S, Sledge C B eds., *Textbook of Rheumatology* (3rd edition). Philadelphia, London: WB Saunders Company, 1989: 804–23.

134b. Golds E E, Cooke T D, Poole A R. Immune regulation of collagenase secretion in rheumatoid and osteoarthritic synovial cell cultures. *Collagen and Related Research* 1983: **3**: 125–40.

135. Gowen M, Wood D D, Ihrie E J, Meats J E, Russell R G G. Stimulation by human interleukin-1 of cartilage breakdown and production of collagenase and proteoglycanase by human chondrocytes but not by human osteoblasts *in vitro*. *Biochimica et Biophysica Acta* 1984; **797**: 186–93.

136. Graham D R, Frost H M. *Candida guilliermondi* infection of the knee complicating rheumatoid arthritis: a case report. *Arthritis and Rheumatism* 1973; **16**: 272–7.

137. Grahame R, Armstrong R, Simmons N A, Mims C A, Wilton J M A, Laurent R. Isolation of rubella virus from synovial fluid in five cases of seronegative arthritis. *Lancet* 1981; **ii**: 649–51.

138. Greenblatt J J, Hunter N, Schwab J H. Antibody response to streptococcal cell wall antigens associated with experimental arthritis in rats. *Clinical and Experimental Immunology* 1980; **42**: 450–7.

138a. Greenwald R A, Diamond H S, eds. *CRC Handbook of Animal Models for the Rheumatic Diseases*. Boca Raton, Florida: CRC Press Inc., 1988.

139. Grennan D M, Dyer P A. Immunogenetics and rheumatoid arthritis. *Immunology Today* 1988; **9**: 33–5.

140. Griffiths M M, DeWitt C W. Genetic control of collagen-induced arthritis in rats: the immune response to type II collagen among susceptible and resistant strains and evidence for multiple gene control. *Journal of Immunology* 1984; **132**: 2830–6.

141. Grimley P M. Rheumatoid arthritis: ultrastructure of the synovium. *Annals of Internal Medicine* 1967; **66**: 623–4.

142. Grimley P M, Sokoloff L. Synovial giant cells in rheumatoid arthritis. *American Journal of Pathology* 1966; **49**: 931–54.

143. Groff G D, Schned A R, Taylor T H. Silicone-induced adenopathy eight years after metacarpophalangeal arthroplasty. *Arthritis and Rheumatism* 1981; **24**: 1578–81.

144. Gruber B, Poznansky M, Boss E, Partin J, Gorevic P, Kaplan A P. Characterization and functional studies of rheumatoid synovial mast cells. Activation by secretagogues, anti-IgE, and a histamine-releasing lymphokine. *Arthritis and Rheumatism* 1986; **29**: 944–55.

145. Gryfe A, Sanders P M, Gardner D L. The mast cell in early rat adjuvant arthritis. *Annals of the Rheumatic Diseases* 1971; **30**: 24–30.

146. Gumpel J M Radiosynovorthesis. *Clinics in the Rheumatic Diseases* 1978; **4**: 311–26.

147. Gupta S. Lymphocyte subpopulations. Phenotypic expression and functions in health and rheumatic diseases. In: Gupta S, Talal N, eds, *Immunology of Rheumatic Diseases*. New York: Plenum Medical Book Company, 1985: 21–83.

148. Hang L, Izui S, Theofilopoulos A N, Dixon F J. Suppression of transferred BXSB male SLE disease by female spleen cells. *Journal of Immunology* 1982; **128**: 1805–8.

149. Harley J B. Autoantibodies in Sjögren's syndrome. In: Talal N, ed., *Sjögren's Syndrome: a Model for Understanding Autoimmunity*. London, San Diego, New York: Academic Press, 1989: 75–86.

150. Harris E D. Pathogenesis of rheumatoid arthritis. In: Kelley W N, Harris E D, Ruddy S, Sledge C B, eds, *Textbook of Rheumatology* (third edition). W B Saunders & Co., 1989: 905–42.

151. Hart H, Marmion B P. Rubella virus and rheumatoid arthritis. *Annals of the Rheumatic Diseases* 1977; **36**: 3–12.

152. Hazes J M W, Silman A J. Review of UK data on the rheumatic diseases – 2. Rheumatoid arthritis. *British Journal of Rheumatology* 1990; **29**: 310–2.

153. Helfgott S M, Bazin H, Dessein A, Trentham D E. Suppressive effects of anti-μ serum on the development of collagen arthritis in rats. *Clinical Immunology and Immunopathology* 1984; **31**: 403–11.

154. Hembry R M, Knight C G, Dingle J T, Barrett A J. Evidence that extracellular cathepsin D is not responsible for the resorption of cartilage matrix in culture. *Biochimica et Biophysica Acta* 1982; **714**: 307–12.

155. Henderson B, Edwards J C W. *The Synovial Lining in Health and Disease*. London: Chapman and Hall, 1987: 255–6; 316–25.

156. Henderson B, Glynn L E, Chayen J. Cell division in the synovial lining in experimental allergic arthritis: proliferation of cells during the development of chronic arthritis. *Annals of the Rheumatic Diseases* 1982; **41**: 275–81.

157. Henderson B, Pettipher E R, Higgs G A. Mediators of rheumatoid arthritis. *British Medical Bulletin* 1987; **43**: 415–28.

158. Henderson B, Revell P A, Edwards J C W. Synovial lining cell hyperplasia in rheumatoid arthritis; dogma and fact. *Annals of the Rheumatic Diseases* 1988; **47**: 348–9.

159. Herbst M, Fritz H, Sauer R. Total lymphoid irradiation of intractable rheumatoid arthritis. *British Journal of Radiology* 1986; **59**: 1203–7.

160. Hinzpeter E N, Naumann G, Bartelheimer H K. Ocular histopathology in Still's disease. *Ophthalmology Research* 1971; **2**: 16–24.

161. Hirofuji T, Kakimoto K, Hori H, Nagai Y, Saisho K, Sumiyoshi A, Koga T. Characterization of monoclonal antibody specific for human type II collagen: possible implication in collagen-induced arthritis. *Clinical and Experimental Immunology* 1985; **62**: 159–66.

162. Holmdahl R, Klareskog L, Rubin K, Bjork J, Smedegard G, Jonsson R, Andersson M. Role of T lymphocytes in murine collagen induced arthritis. *Agents and Actions* 1986; **19**: 295–305.

163. Holmdahl R, Rubin K, Klareskog L, Dencker L, Gustafson G, Larsson E. Appearance of different lymphoid cells in synovial tissue and in peripheral blood during the course of collagen II-induced arthritis in rats. *Scandinavian Journal of Immunology* 1985; **21**: 197–204.

164. Holoshitz J, Klajman A, Drucker I, Lapidot Z, Yaretzky A, Frenkel A, van Eden W, Cohen I R. T lymphocytes of rheumatoid arthritis patients show augmented reactivity to a fraction of mycobacteria cross-reactive with cartilage. *Lancet* 1986; **ii**: 305–9.

165. Howat D W. Possible origin of synovial lining cell hyperplasia in rheumatoid arthritis. *Journal of the Royal Society of Medicine* 1987; **80**: 477–8.

166. Howat D W, Glynn L E, Bitensky L, Chayen J. The origin of the apparent synovial lining cell hyperplasia in rheumatoid arthritis: evidence for a deep stem cell. *British Journal of Experimental Pathology* 1987; **68**: 259–66.

167. Howson P, Shepard N, Mitchell N. The antigen induced arthritis model: the relevance of the method of induction to its use as a model of human disease. *Journal of Rheumatology* 1986; **13**: 379–90.

168. Hunter N, Anderle S K, Brown R R, Dalldorf F G, Clark R L, Cromartie W J, Schwab J H. Cell-mediated immune response during experimental arthritis induced in rats with streptococcal cell walls. *Clinical and Experimental Immunology* 1980; **42**: 441–9.

169. Hurst N P, Nuki G, Wallington T. Functional defects of monocyte C3b receptor-mediated phagocytosis in rheumatoid arthritis (RA): evidence for an association with the appearance of a circulating population of non-specific esterase-negative mononuclear phagocytes. *Annals of the Rheumatic Diseases* 1983; **42**: 487–93.

170. Ingemar B, Lindahl B. The causal sequence on death certificates: errors affecting the reliability of mortality statistics for rheumatoid arthritis. *Journal of Chronic Diseases* 1985; **38**: 47–57.

171. Ingemar B, Lindahl B, Allander E. Problems in the classification of cause of death diagnoses affecting the reliability of mortality statistics for rheumatoid arthritis. *Journal of Chronic Diseases* 1985; **28**: 409–18.

171a. Ilonen J, Reijonen H, Arvilommi H, Jokinen I, Mottenen T, Hannonen P. HLA-DR antigens and HLA-DQ beta chain polymorphism in susceptibility to rheumatoid arthritis. *Annals of the Rheumatic Diseases* 1990; **49**: 494–6.

172. Isomaki H A, Mutru O, Koota K. Death rate and causes of death in patients with rheumatoid arthritis. *Scandinavian Journal of Rheumatology* 1975; **4**: 205–8.

173. Jamieson T W, De Smet A A, Cremer M A, Kage K L, Lindsley H B. Collagen-induced arthritis in rats. Assessment by serial magnification radiography. *Investigative Radiology* 1985; **20**: 324–30.

174. Jasin H E. Autoantibody specificities of immune complexes sequestered in articular cartilage of patients with rheumatoid arthritis and osteoarthritis. *Arthritis and Rheumatism* 1985; **28**: 241–8.

175. Jasin H E, Dingle J T. Human mononuclear cell factors mediate cartilage matrix degradation through chondrocyte activation. *Journal of Clinical Investigation* 1981; **68**: 571–81.

176. Jayson M I V, Salmon P R, Harrison W. Amyloidosis in ankylosing spondylitis. *Rheumatology and Physical Medicine* 1971; **11**: 78–82.

177. Jones V E, Jacoby R K, Cowley P J, Warren C. Immune complexes in early arthritis. II. Immune complex constituents are synthesized in the synovium before rheumatoid factors. *Clinical and Experimental Immunology* 1982; **49**: 31–40.

178. Kammer G M, Sapolsky A I, Malemud C J. Secretion of an articular cartilage proteoglycan-degrading enzyme activity by murine T lymphocytes *in vitro*. *Journal of Clinical Investigation* 1985; **76**: 395–402.

179. Karten I. Septic arthritis complicating rheumatoid arthritis. *Annals of Internal Medicine* 1969; **70**: 1147–58.

180. Kataaha P K, Mortazavi-Milani S M, Russell G and Holborow E J. Anti-intermediate filament antibodies, antikeratin antibody, and antiperinuclear factor in rheumatoid arthritis and infectious mononucleosis. *Annals of the Rheumatic Diseases* 1985; **44**: 446–9.

181. Kellgren J H, Ball J, Fairbrother R W, Barnes K L. Suppurative arthritis complicating rheumatoid arthritis. *British Medical Journal* 1958; **i**: 1193–200.

182. Kampner S L, Ferguson A B Jr. Efficacy of synovectomy in juvenile rheumatoid arthritis. *Clinical Orthopaedics and Related Research* 1972; **88**: 94–109.

183. Kennedy W P U, Partridge R E H, Matthews M B. Rheumatoid pericarditis with cardiac failure treated by pericardiectomy. *British Heart Journal* 1966; **28**: 602–8.

184. Kessler H S. A laboratory model for Sjögren's syndrome. *American Journal of Pathology* 1968; **58**: 671–85.

185. Key J A. The reformation of synovial membrane in the knees of rabbits after synovectomy. *Journal of Bone and Joint Surgery* 1925; **7**: 793.

186. Killar L M, Dunn C J. Interleukin-1 potentiates the development of collagen-induced arthritis in mice. *Clinical Science* 1989; **76**: 535–8.

187. Kingsley G, Pitzalis C, Panayi G S. Immunogenetic and cellular immune mechanisms in rheumatoid arthritis: relevance to new therapeutic strategies. *British Journal of Rheumatology* 1990; **29**: 58–64.

188. Kircher T. Silicone lymphadenopathy: a complication of silicone elastomer finger joint prostheses. *Human Pathology* 1980; **11**: 240–4.

189. Klareskog L, Rubin K, Holmdahl R. Binding of collagen type II to rheumatoid synovial cells. *Scandinavian Journal of Immunology* 1986; **24**: 705–14.

190. Klein W, Rosenbauer K A, Rupprecht L, Krämer J, Huth F. Beitrag zur Kenntnis der juvenilen monartikulären rheumatoiden Arthritis. *Virchows Archiv Abteilung A: Pathologische Anatomie* 1972; **357**: 359–68.

191. Klimiuk P S, Clague R B, Grennan D M, Dyer P A, Smeaton I, Harris R. Autoimmunity to native type II collagen—a distinct genetic subset of rheumatoid arthritis. *Journal of Rheumatology* 1985; **12**: 865–70.

192. Klinman D M, Mushinski J F, Honda M, Ishigatsubo Y, Mountz J D, Raveche E S, Steinberg A D. Oncogene expression in autoimmune and normal peripheral blood mononuclear cells. *Journal of Experimental Medicine* 1986; **163**: 1292–1307.

193. Koga T, Kakimoto K, Hirofuji T, Kotani S, Ohkuni H, Watanabe K, Okada N, Okada H, Sumiyoshi A, Saisho K. Acute joint inflammation in mice after systemic injection of the cell wall, its peptidoglycan, and chemically defined peptidoglycan subunits from various bacteria. *Infection and Immunity* 1985; **50**: 27–34.

194. Koga T, Kakimoto K, Hirofuji T, Kotani S, Sumiyoshi T, Saisho K. Muramyl dipeptide induces acute joint inflammation in the mouse. *Microbiological Immunology* 1986; **30**: 717–23.

195. Kohashi O, Aihara K, Ozawa A, Kotani S, Azuma I. New model of a synthetic adjuvant, N-acetylmuramyl-L-alanyl-D-isoglutamine-induced arthritis. Clinical and histologic studies in euthymic nude and euthymic rats. *Laboratory Investigation* 1982; **47**: 27–36.

196. Kohashi O, Kohashi Y, Kotani S, Osawa A. A new model of experimental arthritis induced by an aqueous form of synthetic adjuvant in immunodeficient rats (SHR and nude rats). *The Ryumachi* 1981; **21** (Suppl.): 149–56.

197. Kohashi O, Kohashi Y, Shigematsu N, Ozawa A, Kotani S. Acute and chronic polyarthritis induced by an aqueous form of 6-O-acyl and N-acyl derivatives of N-acetylmuramyl-alanyl-D-isoglutamine in euthymic rats and athymic nude rats. *Laboratory Investigation* 1986; **55**: 337–46.

198. Konttinen Y T, Bergroth V, Kunnamo I, Haapasaari J. The value of biopsy in patients with monoarticular juvenile rheumatoid arthritis of recent onset. *Arthritis and Rheumatism*, 1986; **29**: 47–53.

199. Korn J H. Cellular and biochemical interactions in the rheumatoid joint. In: Utsinger P D, Zvaifler N J, Ehrlich G E, eds, *Rheumatoid Arthritis: Etiology, Diagnosis, Management*. Philadelphia: J B Lippincott & Co., 1985: 90–132.

200. Kotani S, Watanabe Y, Kinoshita F, Shimono T, Morisaki I, Shiba T, Kusomoto S, Tarumi Y, Ikenaka K. Immunoadjuvant activities of synthetic N-acetyl-muramyl-peptides or -amino acids. *Biken Journal* 1975; **18**: 105–11.

200a. Kresina T F. Suppression of collagen arthritis by $Ly1^-2^+$ antigen-specific T suppressor cells. *Annals of the New York Academy of Sciences* 1986; **475**: 350–2.

201. Ksander G A, Vistnes L M. Collagen and glycosaminoglycans in capsules around silicone implants. *Journal of Surgical Research* 1981; **31**: 433–9.

202. Kuettner K E, Pauli B U. Inhibition of neovascularization by a cartilage factor. In: Nugent J, O'Connor M, eds, *Development of the Vascular System*. London: Pitman, 1983: 163–73.

203. Laakso M, Mutru O, Isomaki H, Koota K. Mortality from amyloidosis and renal diseases in patients with rheumatoid arthritis. *Annals of the Rheumatic Diseases* 1986; **45**: 663–7.

204. Laine V, Vainio K. The elbow in rheumatoid arthritis. In: Hymans W, Paul W D, Herschel H, eds, *Early Synovectomy in Rheumatoid Arthritis*. Amsterdam: Excerpta Medica, 1969: 112.

205. Lalor P A, Mapp P I, Hall P A, Revell P A. Proliferative activity of cells in the synovium as demonstrated by a monoclonal antibody, Ki67. *Rheumatology International* 1987; **7**: 183–6.

206. Lee P. The efficacy and safety of radiosynovectomy. *Journal of Rheumatology* 1982; **9**: 165–8.

207. Lems J W, Berg W B van den, Putte L B A van de, Zwarts W A. Flare of antigen-induced arthritis in mice after intravenous challenge. Kinetics of antigen in the circulation and localization of antigen in the arthritic and non-inflamed joint. *Arthritis and Rheumatism* 1986; **29**: 665–74.

208. Lender M, Wolf E. Incidence of amyloidosis in rheumatoid arthritis. *Schriftenreihe für Wasser, Boden und Luftig* 1972; **1**: 109–12.

209. Lesjak M S, Ghosh P. Polypeptide proteinase inhibitor from human articular cartilage. *Biochimica et Biophysica Acta* 1984; **789**: 266–77.

210. Lowther D A, Gillard G C, Baxter E, Handley C J, Rich K A. Carageenin-induced arthritis. III. Proteolytic enzymes present in rabbit knee joints after a single intraarticular injection of carageenin. *Arthritis and Rheumatism* 1976; **19**: 1287–94.

211. Luder A S, Naphtali V, Porat E B, Lahat N. Still's disease associated with adenovirus infection and defect in adenovirus directed natural killing. *Annals of the Rheumatic Diseases* 1989; **48**: 781–6.

212. Lukes R J, Tindle B H. Immunoblastic lymphadenopathy. A hyperimmune entity resembling Hodgkin's disease. *New England Journal of Medicine* 1975; **292**: 1–8.

213. McGuire M K B, Meats J E, Ebsworth N M, Murphy G, Reynolds J J, Russell R G G. Messenger function of prostaglandins in cell to cell interactions and control of proteinase activity in the rheumatoid joint. *International Journal of Immunopharmacology* 1982; **4**: 91–102.

214. McGuire-Goldring M B, Meats J E, Wood D D, Ihrie E J, Ebsworth N M, Russell R G G. *In vitro* activation of human chondrocytes and synoviocytes by a human interleukin-1-like factor. *Arthritis and Rheumatism* 1984; **27**: 654–62.

215. McKay J M K, Sim A K, McCormick J N, Marmion B P, McCraw A P, Duthie J J R, Gardner D L. Aetiology of rheumatoid arthritis: an attempt to transmit an infective agent from patients with rheumatoid arthritis to baboons. *Annals of the Rheumatic Diseases* 1983; 443–7.

216. MacKenzie A R, Sibley P R, White B P. Resistance and susceptibility to the induction of rat adjuvant disease. Diverging susceptibility and severity achieved by selective breeding. *British Journal of Experimental Pathology* 1979; **60**: 507–12.

217. Male D, Young A, Pilkington C, Sutherland S, Roitt I M. Antibodies to EB virus- and cytomegalovirus-induced antigens in early rheumatoid disease. *Clinical and Experimental Immunology* 1982; **50**: 341–6.

218. Malone D G, Irani A-M, Schwartz L B, Barrett K M, Metcalfe D D. Mast cell numbers and histamine levels in synovial fluid from patients with diverse arthritides. *Arthritis and Rheumatism* 1986; **29**: 956–63.

219. Malone D E, McCormick P A, Daly L *et al.* Peptic ulcer in

rheumatoid arthritis—intrinsic or related to drug therapy? *British Journal of Rheumatology* 1986; **25**: 342–4.

220. Malone D G, Wilder R L, Saavedra-Delgado A M, Metcalfe D D. Mast cell numbers in rheumatoid synovial tissues. *Arthritis and Rheumatism* 1987; **30**: 130–7.

221. Mansson R, Norberg B, Olhagen B, Bjorklund N-E. Arthritis in pigs induced by dietary factors. Microbiological, clinical and histological studies. *Clinical and Experimental Immunology* 1971; **9**: 677–93.

222. Mapp P I, Revell P A. Ultrastructural localisation of muramidase in the human synovial membrane. *Annals of the Rheumatic Diseases* 1987; **46**: 30–7.

223. Marmion B P. Infection, autoimmunity and rheumatoid arthritis. *Clinics in Rheumatic Diseases* 1978; **4**: 565–86.

224. Meats J E, McGuire M B, Russell R G. Human synovium releases a factor which stimulates chondrocyte production of PGE and plasminogen activator. *Nature* 1980; **286**: 891–2.

225. Menninger H, Burkhardt H, Lambusch M, Rissotto R. Proteolytische Ablaufe im Knorpelstoffwechsel und ihre Beeinflussung durch Antirheumatika. *Arzneimittel Forschung* 1982; **32**: 1376–81.

226. Menninger H, Putzier R, Mohr W, Tillmann K. Granulocyte elastase at the site of cartilage erosion by rheumatoid tissue. *Zeitschrift für Rheumaforschung* 1980; **39**: 145–56.

227. Mikecz K, Glant T, Poole A R. Immunity to cartilage proteoglycans in BALB/c mice with progressive polyarthritis and ankylosing spondylitis induced by injection of human cartilage proteoglycan. *Arthritis and Rheumatism* 1987; **30**: 306–18.

228. Mirra J M, Marder R A, Amstutz H C. The pathology of failed total joint arthroplasty. *Clinical Orthopaedics and Related Research* 1982; **170**: 175–83.

229. Mitchell D M, Spitz P W, Young D Y, Bloch D A, McShane D J, Fries J F. Survival, prognosis and causes of death in rheumatoid arthritis. *Arthritis and Rheumatism* 1986; **29**: 706–14.

230. Mohr W, Beneke G, Mohing W. Proliferation of synovial lining cells and fibroblasts. *Annals of the Rheumatic Diseases* 1975; **34**: 219–24.

231. Mohr W, Wessinghage D. The relationship between polymorphonuclear granulocytes and cartilage destruction in rheumatoid arthritis. *Zeitschrift für Rheumaforschung* 1978; **37**: 81–6.

232. Monson R R, Hall A P. Mortality among arthritics. *Journal of Chronic Diseases* 1976; **29**: 459–67.

233. Morgan K. What do anti-collagen antibodies mean? *Annals of the Rheumatic Disease* 1990; **49**: 62–5.

234. Morgan K, Buckee C, Collins I, Ayad S, Clague R B, Holt P J L. Antibodies to type II and XI collagens: evidence for the formation of antigen specific as well as cross-reacting antibodies in patients with rheumatoid arthritis. *Annals of the Rheumatic Diseases* 1988; **47**: 1008–13.

235. Morgan K, Clague R B, Collins I, Ayad S, Phinn S D, Holt P J L. Incidence of antibodies to native and denatured cartilage collagens (types II, IX, and XI) and to type I collagen in rheumatoid arthritis. *Annals of the Rheumatic Diseases* 1987; **46**: 902–7.

235a. Morrow J, Isenberg D. *Autoimmune Rheumatic Disease*. Oxford: Blackwell Scientific Publications, 1987.

236. Muir V Y, Dumonde D C. Different strains of rats develop different clinical forms of adjuvant disease. *Annals of the Rheumatic Diseases* 1982; **41**: 538–43.

237. Mutru O, Laakso M, Isomaki H, Koota K. Ten year mortality and causes of death in patients with rheumatoid arthritis. *British Medical Journal* 1985; **290**: 1797–9.

238. Myllala T, Peltonen L, Puranen J, Korhonen L K. Consequences of synovectomy of the knee joint – clinical, histopathological and enzymatic changes and changes in 2 components of complement. *Annals of the Rheumatic Diseases* 1983; **42**: 28–35.

239. Nagao S, Tanaka A. Muramyl dipeptide-induced adjuvant arthritis. *Infection and Immunity* 1980; **28**: 624–6.

240. Nagase H, Cawston T E, Silva M de, Barrett A J. Identification of plasma kallikrein as an activator of latent collagenase in rheumatoid synovial fluid. *Biochimica et Biophysica Acta* 1982; **702**: 133–42.

241. Neale M L, Williams B D, Matthews N. Tumour necrosis factor activity in joint fluids from rheumatoid arthritis patients. *British Journal of Rheumatology* 1989; **28**: 104–8.

242. Nepom G T, Hansen J A, Nepom B S. The molecular basis for HLA Class II associations with rheumatoid arthritis. *Journal of Clinical Immunology* 1987; **7**: 1–7.

243. Newbold B B. Role of the lymph nodes in adjuvant-induced arthritis in rats. *Annals of the Rheumatic Diseases* 1964; **23**: 392–6.

244. Ng K C, Brown K A, Perry J D, Holborow E J. Anti-RANA antibody: a marker for seronegative and seropositive rheumatoid arthritis. *Lancet* 1980; **i**: 447–9.

245. Norval M, Marmion B P. Attempts to identify viruses in rheumatoid synovial cells. *Annals of the Rheumatic Diseases* 1976; **35**: 106–13.

246. O'Byrne E M, Schroder H C, Goldberg R L. Catabolin/interleukin-1 regulation of cartilage and chondrocyte metabolism. *Agents and Actions* 1987; **21**: 341–4.

247. Ofosu-Appiah W A, Morgan K, Holt P J L. Native type II collagen-induced arthritis in the rat. Studies of the humoral response to collagen at the cellular level. *Journal of Rheumatology* 1984; **11**: 432–7.

248. Ohno O, Cooke T D. Electron microscopic morphology of immunoglobulin aggregates and their interactions in rheumatoid articular collagenous tissues. *Arthritis and Rheumatism* 1978; **21**: 516–27.

249. Okada Y, Nagase H, Harris E D Jr. A multi-substrate metalloproteinase from rheumatoid synovial cells. *Transactions of the Association of American Physicians* 1986; **99**: 143–53.

250. Okamoto T, Ueda K, Kambara T, Kutsuna T. Chemotactic activity for polymorphonuclear and mononuclear leukocytes in rheumatoid synovial fluids. *Acta Pathologica Japonica* 1986; **36**: 1109–22.

251. Ollier W, Thomson W, Welch S, De Lange G G, Silman A. Chromosome 14 markers in rheumatoid arthritis. *Annals of the Rheumatic Diseases* 1988; **47**: 843–8.

252. O'Sullivan F X, Fassbender H G, Gay S, Koopman W J. Etiopathogenesis of the rheumatoid arthritis-like disease in MRL/l mice. I. The histomorphologic basis of joint destruction. *Arthritis and Rheumatism* 1985; **28**: 529–36.

253. Osung O A, Chandra M, Holborow E J. Antibody to intermediate filaments of the cytoskeleton in rheumatoid arthritis. *Annals of the Rheumatic Diseases* 1982; **41**: 69–73.

254. Ottenhoff T H M, Torres P, Aguas J T da las, Fernandez

able factors in cartilage which inhibit invasion by vascularized mesenchyme. *Laboratory Investigation* 1975; **32**: 217–22.

317. Souter W A. Surgical management of the rheumatoid elbow. In: Beddow F M, ed., *The Surgical Management of Rheumatoid Arthritis*. London, Boston: Wright, 1988: 69–82.

317a. Spitznagel J K, Goodrum K J, Warejcka D J. Rat arthritis due to whole group B streptococci. Clinical and histopathologic features, compared with groups A and D. *American Journal of Pathology* 1983; **112**; 37–47.

318. Stastny P. Association of the B-cell alloantigen DRw4 with rheumatoid arthritis. *New England Journal of Medicine* 1978; **298**: 869–71.

319. Steinberg J J, Hubbard J R, Sledge C B. *In vitro* models of cartilage degradation and repair. In: Otterness I *et al.* eds, *Advances in Inflammation Research* (volume 11). New York: Raven Press, 1986: 215–41.

320. Steinberg J J, Tsukamoto S, Sledge C B. Breakdown of cartilage proteoglycan in a tissue culture model of rheumatoid arthritis. *Biochimica et Biophysica Acta* 1983; **757**: 47–58.

320a. Steinberg J, Sledge C B, Noble J, Stirrat C R. A tissue-culture model of cartilage breakdown in rheumatoid arthritis. Quantitative aspects of proteoglycan release. *Biochemical Journal* 1978; **180**: 403–12.

321. Stewart S M, Alexander W R M, Duthie J J R. Isolation of diphtheroid bacilli from synovial membrane and fluid in rheumatoid arthritis. *Annals of the Rheumatic Diseases* 1969; **28**: 477–87.

322. Stewart S M, Duthie J J R, MacKay J M K, Marmion B P, Alexander W R M. Mycoplasmas and rheumatoid arthritis. *Annals of the Rheumatic Diseases* 1974; **33**: 346–52.

323. Stierle G, Brown K A, Rainsford S G, Smith C A, Hamerman D, Stierle H E, Dumonde D C. Parvovirus associated antigen in the synovial membrane of patients with rheumatoid arthritis. *Annals of the Rheumatic Diseases* 1987; **46**: 219–23.

324. Stoerk H, Bielinski T C, Budzilovich T. Chronic polyarthritis in rats injected with spleen in adjuvants. *American Journal of Pathology* 1954; **30**: 616.

325. Stovell P B, Ahuja S C, Inglis A E. Pseudarthrosis of the proximal femoral epiphysis in juvenile rheumatoid arthritis. *Journal of Bone and Joint Surgery* 1975; **57**-A: 860–1.

326. Stuart J M, Tomoda K, Yoo T J, Townes A S, Kang A H. Serum transfer of collagen-induced arthritis. II. Identification and localization of autoantibody to type II collagen in donor and recipient rats. *Arthritis and Rheumatism* 1983; **26**: 10: 1237–44.

327. Stuart J M, Townes A S, Kang A H. Nature and specificity of the immune response to collagen in type II collagen-induced arthritis in mice. *Journal of Clinical Investigation* 1982; **69**: 673–83.

328. Stuart J M, Townes A S, Kang A H. Collagen autoimmune arthritis. *Annual Review of Immunology* 1984; **2**: 199–218.

329. Sumi M, Maeda M, Cooke T D V. Deleterious interactions of immune complexes with tibial cartilage of antigen-induced arthritic rabbits. II. Chondrocyte degradation. *Clinical Orthopaedics and Related Research* 1986; **212**: 260–74.

330. Symmons D P M, Prior P, Scott D L, Brown R, Hawkins C F. Factors influencing mortality in rheumatoid arthritis. *Journal of Chronic Diseases* 1986; **39**: 137–45.

331. Symons J A, Wood N C, Di Giovine F S, Duff G W. Soluble IL-2 receptor in rheumatoid arthritis. Correlation with disease activity, IL-1 and IL-2 inhibition. *Journal of Immunology* 1988; **141**: 2612–18.

332. Takagishi K, Kaibara N, Hotokebuchi T, Arita C, Morinaga M, Arai K. Serum transfer of collagen arthritis in congenitally athymic nude rats. *Journal of Immunology* 1985; **134**: 3864–7.

333. Talal N (ed.) *Sjögren's Syndrome: A Model for Understanding Autoimmunity*. 2nd International Symposium, 1989. London: Academic Press.

334. Taylor D J, Woolley D E. Evidence for both histamine H_1 and H_2 receptors on human articular chondrocytes. *Annals of the Rheumatic Diseases* 1987; **46**: 431–5.

335. Taylor D J, Yoffe J R, Brown D M, Woolley D E. Histamine stimulates prostaglandin E production by rheumatoid synovial cells and human articular chondrocytes in culture. *Arthritis and Rheumatism* 1986; **29**: 160–5.

336. Terato K, Cremer M A, Hasty K A, Kang A H, Hasty D L, Townes A S. Physiocochemical and immunological studies of the renatured alpha 1(II) chains and isolated cyanogen bromide peptides of type II collagen. Collagen and Related Research 1985; **5**: 469–80.

337. Terrier F, Hricak H, Revel D, Alpers C E, Reinhold C E, Levine J, Genant H K. Magnetic resonance imaging and spectroscopy of the periarticular inflammatory soft-tissue changes in experimental arthritis of the rat. *Investigative Radiology* 1985; **20**: 813–23.

338. Tetta C, Camussi G, Modena V, Vittorio C D, Baglioni C. Tumour necrosis factor in serum and synovial fluid of patients with active and severe rheumatoid arthritis. *Annals of the Rheumatic Diseases* 1990; **49**: 665–7.

339. Theofilopoulos A N, Dixon F J. Murine models of systemic lupus erythematosus. *Advances in Immunology* 1985; **37**: 269–390.

340. Thorne C, Urowitz M B. Long-term outcome in Felty's syndrome. *Annals of the Rheumatic Diseases* 1982; **41**: 486–9.

341. Trentham D E, Townes A S, Kang A H, David J R. Humoral and cellular sensitivity to collagen in type II collagen-induced arthritis in rats. *Journal of Clinical Investigation* 1978; **61**: 89–96.

342. Trentham D E, Townes A S, Kang A H. Autoimmunity to type II collagen: an experimental model of arthritis. *Journal of Experimental Medicine* 1977; **146**: 857–68.

343. Tribe C R. Amyloidosis in rheumatoid arthritis. In Hill A G S, ed, *Modern Trends in Rheumatology* (volume 1). London: Butterworths, 1966: 121–38.

344. Tribe C R, Mackenzie J C. Amyloidosis. In Bacon P A, Hadler N M, eds, *The Kidney and Rheumatic Disease*. London: Butterworths, 1982: 297–322.

345. Tyler J A. Chondrocyte-mediated depletion of articular cartilage proteoglycans *in vitro*. *Biochemical Journal* 1985; **225**: 493–507.

346. Uddin J, Kraus A S, Kelly H G. Survivorship and death in rheumatoid arthritis. *Arthritis and Rheumatism* 1970; **13**: 125–30.

347. Utsinger P D, Zvaifler N J, Weiner S B. Etiology. In: Untsinger P D, Zvaifler N J, Ehrlich G E, eds, *Rheumatoid Arthritis: Etiology, Diagnosis, Management*. Philadelphia: J B Lippincott & Co., 1985: 21–48.

348. Van de Putte L B A, Lens J W, Van den Berg W B,

Kruijsen M W M. Exacerbation of antigen-induced arthritis after challenge with intravenous antigen. *Immunology* 1983; **49**: 161–7.

349. Van den Berg W B, Joosten L A B, Van de Putte L B A, Zwarts W A. Electrical charge and joint inflammation. Suppression of cationic aBSA-induced arthritis with a competitive polycation. *American Journal of Pathology* 1978; **127**: 15–26.

350. Van den Berg W B, Van de Putte L B A, Zwarts W A, Joosten L A B. Electrical charge of the antigen determines intraarticular antigen handling and chronicity of arthritis in mice. *Journal of Clinical Investigation* 1984; **74**: 1850–9.

351. Van den Broek M F, Van den Berg W B, Van de Putte L B A, Severijnen A J. Streptococcal cell wall-induced arthritis and flare-up reaction in mice induced by homologous or heterologous cell walls. *American Journal of Pathology* 1988; **133**: 139–49.

352. Van Eden W, Holoshitz J, Cohen I. Antigenic mimicry between mycobacteria and cartilage proteoglycans: the model of adjuvant arthritis. *Concepts in Immunopathology* 1987; **4**: 144–70.

353. Van Eden W, Thole J E R, Van der Zee R, Noordzij A, Van Embden J D A, Hensen E J, Cohen I R. Cloning of the mycobacterial epitope recognized by T lymphocytes in adjuvant arthritis. *Nature* 1988; **331**: 171–3.

354. Waites G T, White A. Effect of pregnancy on collagen-induced arthritis in mice. *Clinical and Experimental Immunology* 1987; **67**: 467–76.

355. Wasserman S I. The mast cell and synovial inflammation. Or, what's a nice cell like you doing in a joint like this? *Arthritis and Rheumatism* 1984; **27**: 841–4.

356. Watson W C, Townes A S, Kang A H. Immunopathogenic mechanisms in type II collagen autoimmune arthritis. *Pathological and Immunopathological Research* 1986; **5**: 297–304.

356a. Wegelius O, Teppo A-M, Maury C P J. Reduced amyloid A-degrading activity in serum in amyloidosis associated with rheumatoid arthritis. *British Medical Journal* 1982; **284**: 617–18.

357. Whitehouse M W. Rat polyarthritis: induction with adjuvants constituted with mycobacteria (and oils) from the environment. *Journal of Rheumatology* 1982; **9**: 494–501.

358. Whitehouse M W. Adjuvant-induced polyarthritis in rats. In: Greenwald R A, Diamond H S, eds, *CRC Handbook of Animal Models for the Rheumatic Diseases* (volume I). Boca Raton, Florida: CRC Press Inc., 1988: 3–16.

359. Wicks I P, Moore J, Fleming A. Australian mortality statistics for rheumatoid arthritis 1950–81: analysis of death certificate data. *Annals of the Rheumatic Diseases* 1988; **47**: 563–9.

360. Wilder R L. Streptococcal cell wall-induced polyarthritis in rats. In: Greenwald R A, Diamond H S, eds, *CRC Handbook of Animal Models for the Rheumatic Diseases* (volume I). Boca Raton, Florida: CRC Press Inc., 1988: 33–40.

361. Winyard P G, Blake D R, Chirico S, Gutteridge J M C, Lunec J. Mechanism of exacerbation of rheumatoid synovitis by total-dose iron-dextran infusion: *in vivo* demonstration of iron-promoted oxidant stress. *Lancet* 1987; **i**: 69–72.

362. Woessner J F. Cartilage cathepsin D and its action on matrix components. *Federation Proceedings* 1973; **32**: 1485–8.

363. Woessner J F, Howell D S. The enzymatic degradation of connective tissue matrices. In: Owen, R, Goodfellow J, Bullough P, eds, *Scientific Foundations of Orthopaedics and Traumatology*. Edinburgh: Churchill-Livingstone, 1980: 232–41.

364. Wolfe F. 50 years of antirheumatic therapy: the prognosis of rheumatoid arthritis. *Journal of Rheumatology* 1990; **22**: 24–32 (Suppl).

365. Wood N C, Symons J A, Duff G W. Serum interleukin-2-receptor in rheumatoid arthritis: a prognostic indicator of disease activity? *Journal of Autoimmunity* 1988; **1**: 353–61.

366. Wood P H N. Recent trends in sickness absence and mortality. Statistical appendix to Digest of Data on the Rheumatic Diseases. *Annals of the Rheumatic Diseases* 1970; **29**: 324–9.

367. Wooley P H, Luthra H S, Krco C J, Stuart J M, David C S. Type II collagen-induced arthritis in mice. II. Passive transfer and suppression by intravenous injection of anti-type II collagen antibody or free native type II collagen. *Arthritis and Rheumatism* 1984; **27**: 1010–17.

368. Wooley P H, Luthra H S, Lafuse W P, Huse A, Stuart J M, David C S. Type II collagen-induced arthritis in mice. III. Suppression of arthritis by using monoclonal and polyclonal anti-Ia antisera. *Journal of Immunology* 1985; **134**: 2366–74.

369. Woolley D E, Tetlow L C, Evanson J M. Collagenase immunolocalization studies of rheumatoid and malignant tissues. In: Woolley D E, Evanson J M, eds, *Collagenase in Normal and Pathological Tissues*. Chichester: John Wiley and Sons, 1980: 105–25.

370. Wynne-Roberts C R, Anderson C H, Turano A M, Baron M. Light- and electron-microscopic findings of juvenile rheumatoid arthritis synovium: comparison with normal juvenile synovium. *Seminars in Arthritis and Rheumatism* 1978; **7**: 287–302.

371. Yoffe J R, Taylor D J, Woolley D E. Mast cell products stimulate collagenase and prostaglandin E production by cultures of adherent rheumatoid synovial cells. *Biochemical and Biophysical Research Communications* 1984; **122**: 270–6.

372. Yoo J U, Kresina T F, Malemud C J, Goldberg V M. Epitopes of proteoglycans eliciting an antiproteoglycan response in chronic immune synovitis. *Proceedings of the National Academy of Sciences of the United States of America* 1987; **84**: 832–6.

373. Yoo T J, Stuart J M, Takeda T, Sudo N, Floyd R A, Ishibe T, Olson G, Orchik D, Shea J J, Kang A H. Induction of type II collagen autoimmune arthritis and ear disease in monkey. *Annals of the New York Academy of Sciences* 1986; **475**: 341–2.

374. Youinou P, Goff P le, Colaco C B, Thivolet J, Tater D, Viac J, Shipley M. Antikeratin antibodies in serum and synovial fluid show specificity for rheumatoid arthritis in a study of connective tissue diseases. *Annals of the Rheumatic Diseases* 1985; **44**: 450–4.

375. Zidek Z, Masek K, Jiricka Z. Arthritogenic activity of a synthetic immunoadjuvant, muramyl dipeptide. *Infection and Immunity* 1982; **35**: 674–9.

375a. Ziment I, Davis A, Finegold S M. Joint infection by anaerobic bacteria: a case report and review of the literature. *Arthritis and Rheumatism* 1969; **12**: 627–35.

376. Zvaifler N J. Etiology and pathogenesis of rheumatoid arthritis. In: McCarty D J, ed, Arthritis and allied conditions. *A Textbook of Rheumatology* (11th edition). Philadelphia: J B Lippincott & Co., 1989: 659–73.

Chapter 14
Systemic Lupus Erythematosus

Systemic lupus erythematosus (SLE)[213,297] is a not infrequent, progressive connective tissue disorder in which, characteristically, there is the anomalous production of antibodies to an amazingly broad spectrum of autoantigens. The disease, which displays significant differences in racial distribution, affects females many times more often than males. The majority of organs are involved but the synovial joints, skin (Fig. 14.1), kidneys, brain, blood vessels and heart are particularly vulnerable. With increased clinical awareness and the introduction of diagnostic laboratory tests of enhanced sensitivity, milder forms of the disorder have been recognized in recent years with much increased frequency. There has been a corresponding improvement in prognosis only partly attributable to better forms of drug therapy or to the use of plasmapheresis or thoracic duct drainage.[146] Whereas untreated SLE was formerly regarded as potentially fatal, the estimated five-year survival rate in some series has improved to approximately 98 per cent.[113] The cause of SLE remains unknown but there may be a hereditary predisposition to unidentified virus(es) or to the action of sensitizing drugs or chemicals.

Systemic lupus erythematosus often begins between the ages of 15 and 25 years but an onset in early childhood is occasional. However, approximately 15 per cent of patients develop the disease after the age of 60 years, usually with an insidious onset and a relatively high frequency of interstitial lung disease. In the UK, the prevalence of SLE is one in 2000 women (G V R Hughes, personal communication). The female:male ratio is 9:1. In the USA, the incidence for the whole population has been found to be higher: in one report it was 7.6 cases per 100 000, the prevalence 1 case in 1969. In some parts of the USA, when Blacks only are considered,[93] the prevalence may be as high as 1 individual per 245 Black females aged 15 to 64.

Definition and classification

Clinical classification is on the basis of four or more of the 11 most useful diagnostic criteria agreed by a Subcommittee of the American Rheumatism Association (American College of Rheumatology)[279] (Table 14.1). It is of interest that, in the deliberations leading to the formulation of these 11 criteria, biopsies of skin and kidney were eliminated because they were seldom performed. Yet renal biopsy offers very high diagnostic sensitivity and extreme specificity, and skin biopsy high sensitivity and very high specificity.

Overall,[279] diagnostic sensitivity and specificity are both 96 per cent. Equal accuracy was obtainable by using the frequently performed tests for serum antinuclear antibody together with a long list of 19 individual, variable criteria. However, this option was regarded as too cumbersome for practical application. The number of patients escaping correct classification when the 1982 criteria are used is small. Serum complement determinations, although helpful in measuring disease activity, do not aid diagnosis or classification.

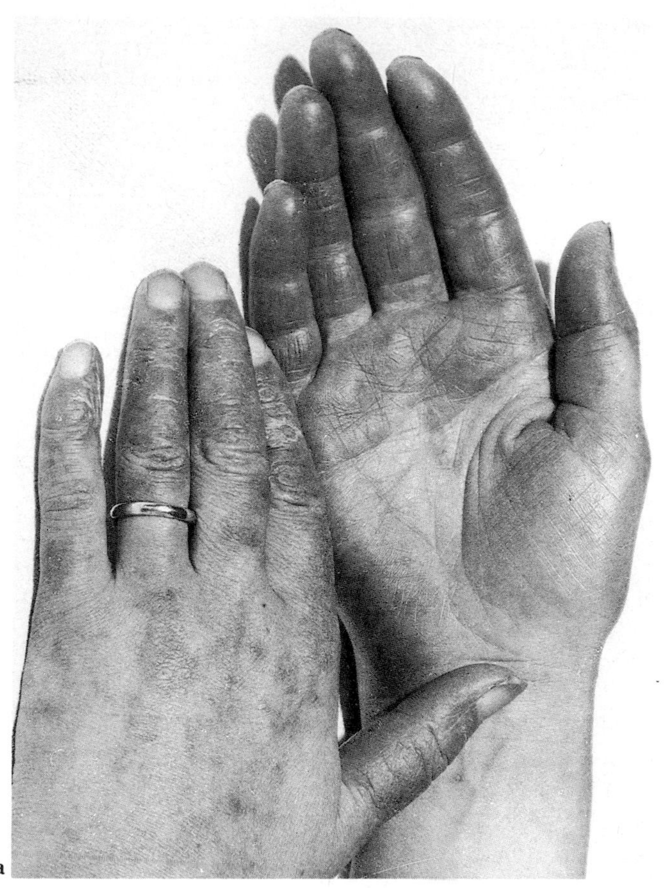

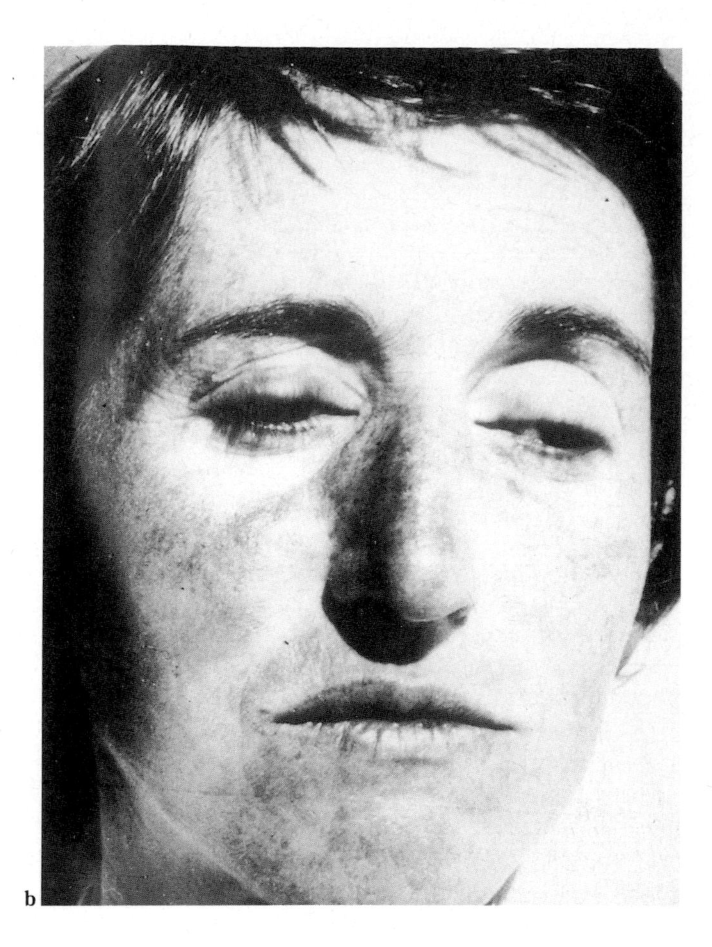

Fig. 14.1 Skin lesions in SLE.
a. In this young woman, there is a patchy erythematous eruption of the dorsum of the hand accompanied by peripheral congestion and a suggestion of finger clubbing. **b.** In this older patient, a scaling, erythematous rash extends across the bridge of the nose and involves both malar regions.

Table 14.1
Criteria for the classification (diagnosis) of SLE*

Criterion	Relative sensitivity	Relative specificity
Antinuclear antibody	1	10
Arthritis	2	11
Immunological disorder	3	7
Haematological disorder	4	8
Malar rash	5	= 3
Serositis	6	9
Renal disorder	7	6
Photosensitivity	8	= 3
Oral ulceration	9	= 3
Neurological disorder	10	2
Discoid rash	11	1

*1982 revised criteria[279] for the classification of SLE. Skin and renal biopsy are excluded since, in the USA, they are seldom performed. To identify a person with SLE, the individual is required to display four or more of the 11 criteria, serially or simultaneously.

History

Descriptions of lupus erythematosus (LE) exist in the writings of Hippocrates.[142] The modern name was adopted by Cazenave.[52] Lupus erythematosus was thought to be an unusual, localized, chronic skin disease, occasionally severe. It gradually emerged that LE could exist as a visceral disease (*systemic* lupus erythematosus—SLE), independent of skin changes. Kaposi[163] drew attention to an occasional widespread, and sometimes fatal, acute systemic disorder and described the pathological changes. For many years there was confusion between discoid LE and lupus vulgaris.

The visceral changes were discussed by Osler[233,234,235] who drew attention to arthritis and neurological disease, and by Jadassohn[153] who recognized the diffuse distribution of the disorder. Libman and Sacks,[193] describing a particular form of endocarditis in SLE, helped to establish its unique identity. Their observations were extended by Gross.[116] He identified haematoxylin bodies (see p. 572) and numerous minute myocardial infarcts. The widespread nature of the endothelial changes found in the blood vessels of many parts of the body impressed Baehr, Klemperer and Schifrin;[20] they suggested that hypersensitivity to an unknown antigen might be one explanation for SLE, a view supported by Gross.[116] However, Klemperer, Pollack and Baehr[175] revised their earlier views: they emphasized the diffuse occurrence of zones of collagen necrosis and the frequency of fibrinoid degeneration (see p. 294) and collagen sclerosis (see p. 301), and concluded that acute SLE was best viewed as a primary disorder of collagenous connective tissue, to be called 'diffuse collagen disease'. Later,[177] these authors advanced the new hypothesis that SLE and progressive systemic sclerosis (see Chapter 17) were both hypersensitivity disorders of the connective tissue system, i.e. of collagenous tissues. These views were discussed by Klemperer.[172] The thesis was extended by the discovery of the LE cell phenomenon;[127] by the recognition that patients could give false-positive serological tests for syphilis; by the demonstration of abnormal circulating antibodies; by the electron microscopical analysis of the renal changes;[90] and by evidence demonstrating that the abnormal antibodies were directed against components of the patient's own cells or tissues, i.e. that they were autoantibodies. The pathogenesis of SLE has been fully discussed by Zvaifler and Woods[309,312] and the literature analysed by Dubois[76,77] and Wallace and Dubois.[297] A unique, personal view of the disease was given by Ropes.[255]

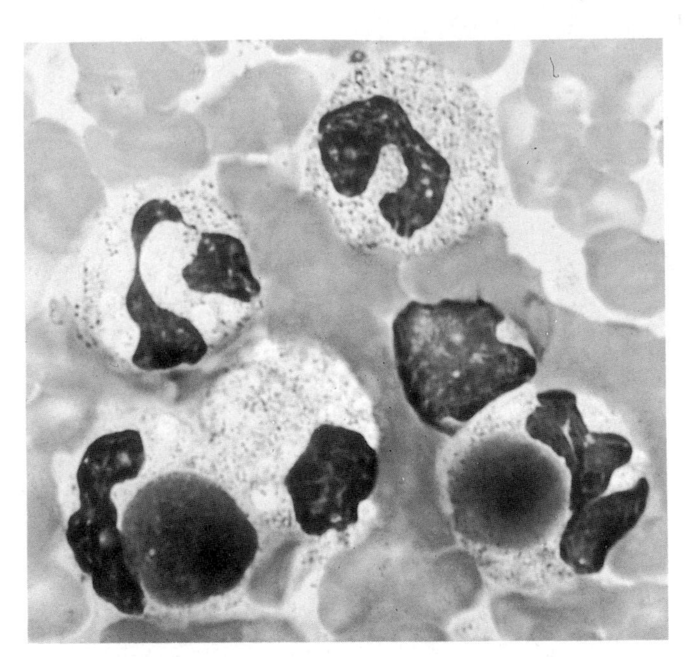

Fig. 14.2 Lupus erythematosus (LE) cell test—lupus cell phenomenon.
It was shown by Hargraves, Richmond and Morton[127] that, *in vitro*, live blood leucocytes vigorously phagocytosed the effete nuclei and nuclear material that is always present in the circulation. The response is attributable to the presence in the plasma, and in serum prepared from clotted blood, of antinuclear antibodies. In one sense, the 'LE cell phenomenon' as it was called, is artefactual; there is little evidence that it occurs *in vivo* in spite of the obvious morphological similarity between 'LE cells' and the basophilic bodies detected in histological preparations of, for example, the kidneys. The artefact was shown to be reproducible and became the basis for an early laboratory test for the diagnosis of SLE; the test still has a screening value although its sensitivity is low. The lupus erythematosus cell test can be undertaken with live leucocytes from the patient herself, with leucocytes from other individuals or with animal cells.

In this Figure, large, amorphous uniformly basophilic 'bodies' lie within the cytoplasm of polymorphs; their nuclei are displaced to one side. The cells form part of a cluster (Leishmann × 1200).

Clinical presentation

The variety of organ involvement is matched only by the diversity of the symptoms and signs with which the disease originates.[257,261,262,265a] There is an early arthritis or arthralgia. Evidence of systemic illness begins months or years after the onset of joint disease which, in young adults, is generally diagnosed as rheumatoid arthritis and, in children, as juvenile rheumatoid arthritis or rheumatic fever. An erythematous facial rash (discoid lupus) confined to the malar regions, explains the old confusion with lupus vulgaris. Severely affected patients, febrile and anaemic, may be found to have renal disease with albuminuria and haematuria. The signs of diffuse glomerulonephritis are confirmed. The spleen is slightly enlarged and pleuritis and pericarditis become evident.

Assessments have been made of the most useful laboratory tests in the diagnosis and management of SLE (Fig. 14.2). Patients with cerebral manifestations and thrombocytopenia are the most difficult to assess. No single test distinguishes reliably between cases grouped as severely active, moderately active and inactive, but inactive disease can be differentiated from active by assays of circulating immune complexes, platelet counts and the erythrocyte sedimentation rate.[218] Measurements of circulating immune complexes, double-stranded (ds-) DNA binding, lymphocyte numbers and CH50, the consumption of complement, allow severely active cases to be separated from less active. The further analysis of anti-dsDNA antibodies can increase the sensitivity of these tests. Anti-dsDNA antibodies are heterogeneous (Fig. 14.2). They differ in (sub)class, specificity, complement-fixing ability and avidity. High avidity anti-dsDNA antibodies mirror high disease activity. Low avidity anti-dsDNA antibodies are of secondary importance:[228] they are detectable in a category of older (>45 years) female (95 per cent) patients with a low incidence of renal disease.

In spite of differences in sensitivity, it has been suggested that a set of laboratory tests can be constructed that might provide diagnostic criteria as specific as those of the original (1971)[16] preliminary criteria for SLE.[219] The choice of tests for anti-DNA antibodies, for anti-Sm antibody (p. 599) and for immune deposits at the dermal–epidermal junction of uninvolved ('normal') skin, assists the recognition of cases of SLE. Eighty per cent have abnormal values for at least two of the three tests and anti-red blood cell antibodies are present in 76 per cent of the affected patients. Other, closely similar studies suggest that the disease state is best reflected by antinuclear antibody, anti-native DNA antibody, low complement component C3 and C4 levels, circulating immune complexes and cold lymphocytoxins.[91] In some patients there are periods, however, when no abnormal immune phenomena can be detected.

Pathological changes

Unit lesions

Just as a microscopic tubercle can be regarded as the structural unit of tuberculosis, so a series of 'unit lesions' (Table 14.2) is morphologically characteristic of SLE. In varying combinations and permutations, these microscopic disorders, very largely the result of immune complex deposition, determine the organ changes of the disease. It is not clear why their frequency varies in different cases or why some viscera, e.g. kidney, synovial joint, brain, appear so much more vulnerable than others. A simple explanation for the varied organ involvement attributes the frequency of microscopic disease to differential organ blood flow.

Table 14.2

The 'unit' light- and electron microscopic lesions of SLE

Light microscopic lesions	
Vasculopathy	Basophilic bodies
Fibrinoid	Necrosis
Thrombosis	Collagen sclerosis
Basement membrane change	Granulomata
Ultrastructural lesions	
Subendothelial, Subepithelial	Electron-dense deposits
Intracellular, Extracellular	Virus-like or organized deposits

Vasculopathy

Vasculopathy is very frequent. Larger vessels may develop arteritis (see p. 582), accounting for skin and mouth ulcers, nervous system disease[40] and, occasionally, accelerated hypertension. Intestinal infarction and gangrene are complications of mesenteric arteritis[36] but focal vasculitis may affect any viscus from the ovaries and testes to the synovia and aortic vasa vasorum: the coronary arteries are susceptible.[180] Smaller arteries and the penicilliary arteries of the spleen are affected by onion-skin collagen sclerosis (Fig. 14.3). Capillaries display endothelial cell hyperplasia

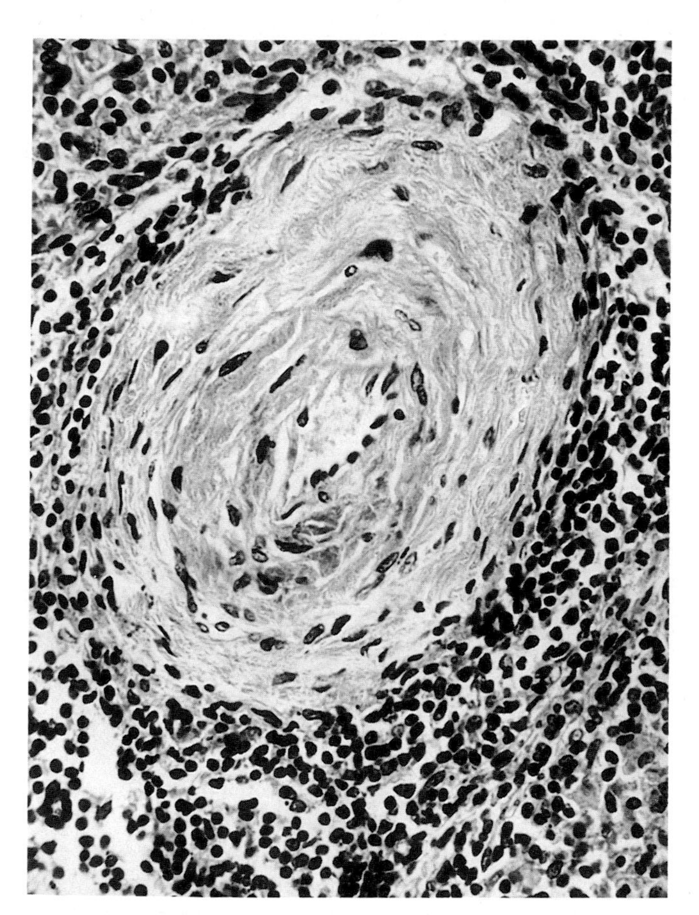

Fig. 14.3 Penicilliary artery of spleen in SLE. The artery is circumscribed by a series of concentric laminae of collagenous connective tissue within which the nuclei of several fibrocytes can be seen. The appearances, those of onion-skin fibrosis, are distinguished from the normal periarterial connective tissue quantitatively. For the diagnosis of 'onion-skin' change, there should be at least seven laminae of fibrous tissue. The cause of this change is not known but it can be described as a form of collagen sclerosis. (HE × 390.)

and it is significant that in addition to the presence of endothelial cell virus-like inclusions[226] (see p. 582), anti-endothelial antibody may be present in the serum which may contain immune complexes capable of binding to endothelial cells.[55] Vasculitis has contributed to protein-losing nephropathy and has caused massive colonic haemorrhage.[136] In the pulmonary circulation, fibromuscular endarterial thickening, sometimes with thrombosis, is occasional.[102,296]

Fibrinoid

In SLE, fibrinoid (p. 294) is a faintly basophilic material: it contains DNA[174] and protein although the presence not of DNA but of plasma protein was found by others.[220] Fibrinoid is suceptible to tryptic digestion.[295] In myocardial and vascular lesions, in the glomeruli and in the dermis, anti-fibrin antibody confirmed the presence of fibrin[105], an observation that was not, however, always substantiated.[292,293]

Thrombosis

Many forms of venous and arterial thrombosis are common in SLE. The thrombotic tendency may contribute to the onset of fibrinoid change.[162] Deep-vein thrombi form, a complication affecting renal veins and the inferior vena cava, whether or not haemodialysis has been undertaken. Of a series of factors associated with coagulation, lupus anticoagulant, a substance inhibiting components of the coagulation cascade, has the strongest association with thrombosis in SLE patients.[133] Glomerular capillary thrombosis is of particular importance (see p. 587) and thrombotic microangiopathy (TTM) (see p. 660) is occasionally recognized. The frequency of thrombosis may culminate in the excessive consumption of fibrinogen, a change that regularly accompanies haemorrhagic phenomena (see p. 590). However, these and other coagulation defects[43] also reflect the presence of the lupus anticoagulant.[195]

The lupus anticoagulant is an antibody. Its properties[43] may be associated in part with IgG or IgM, or both. A cofactor, similar to complement, is required. The lupus anticoagulant antibody is directed against acidic phospholipids involved in the formation of prothrombin activator. The result of the presence of lupus anticoagulant is to cause gross disorders in tests of coagulation. That these abnormal properties do not often promote haemorrhagic phenomena is attributable to the retention of the overriding actions of the platelets.

Lupus anticoagulant is closely similar to the anticardiolipin antibodies the presence of which is demonstrable by Venereal Disease Research Laboratory (VDRL) and Cardiolipin Wasserman Reaction (CWR) reactions. Anti-cardiolipin antibodies and lupus anticoagulant comprise distinct antibody subgroups.[203] Biological false-positive tests for syphilis (STS) are often given by sera from patients with SLE because of this similarity. However, the frequency of positive anti-cardiolipin tests made by sensitive techniques such as solid-phase radioimmunoassay and enzyme-linked immunosorbent assay (ELISA) is much greater than the frequency of positive VDRL tests.

There is also evidence for the release of platelet-activating factor by basophil polymorphs on exposure to DNA[48] and circulating platelet aggregating immune complexes exist.[164] There is increased Factor VIII complex activity[14] with a diminished release of plasminogen activator after venous occlusion, changes attributable to endothelial damage by immune complexes. The presence of raised levels of circulating immune complexes has been proposed as one explanation for thrombotic microangiopathy (see Chapter 15).[53]

Basement membrane changes

Basement membrane changes are characteristic (Fig. 14.4). Prominence of glomerular capillary basement membranes is frequent and contributes to the wire-loop lesion of membranous nephropathy (see p. 586). Occasionally, a similar change is seen in the renal tubular basement membranes. There is now no doubt that, although basement membrane permeability is often abnormal, the thickening detected by light microscopy[175] is attributable to the presence of immune complex deposits adjacent to the basement membrane, beneath the endothelial cell plasma membranes. Anti-basement membrane antibodies are demonstrable; when present, they are not nephrotoxic.

Basophilic (haematoxyphilic) bodies

Basophilic bodies (Fig. 14.5) are pathognomonic although their recognition has been claimed in other connective disorders, including polymyositis.[210] Gross[116] was the first to observe clumps and pockets of haematoxylin-stained bodies of spindle shape.[64] Haematoxylin-staining basophilic masses were noted in necrotic lymph nodes[106] and it became clear that these structures, although inconstant, were highly characteristic of SLE.[171] Their composition and origin has excited much interest: like the LE cell, their formation sheds light on the fundamental mechanisms of tissue injury in SLE. Basophilic bodies are most commonly found in the glomeruli and beneath the endocardium, but they have been identified in all tissues that are sites of SLE lesions. The bodies are approximately 5–15 μm in diameter. The small size demands close scrutiny by light- or electron microscopy since they must be distinguished from the nuclear debris formed in karyorrhexis, from late nor-

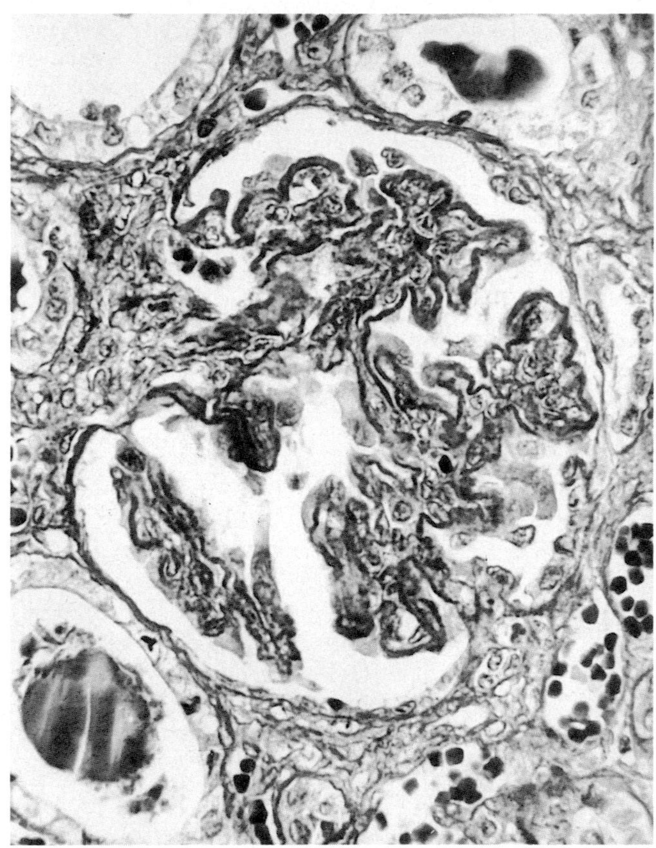

Fig. 14.4 Basement membrane change in SLE.
From a girl aged 14 years with rapidly advancing SLE and lupus nephropathy. There is conspicuous thickening of the renal glomerular basement membranes; their outlines stand out as 'wire loops'. The patient showed the clinical signs of the nephrotic syndrome and proteinaceous casts are seen in the renal tubules. (HE × 500.)

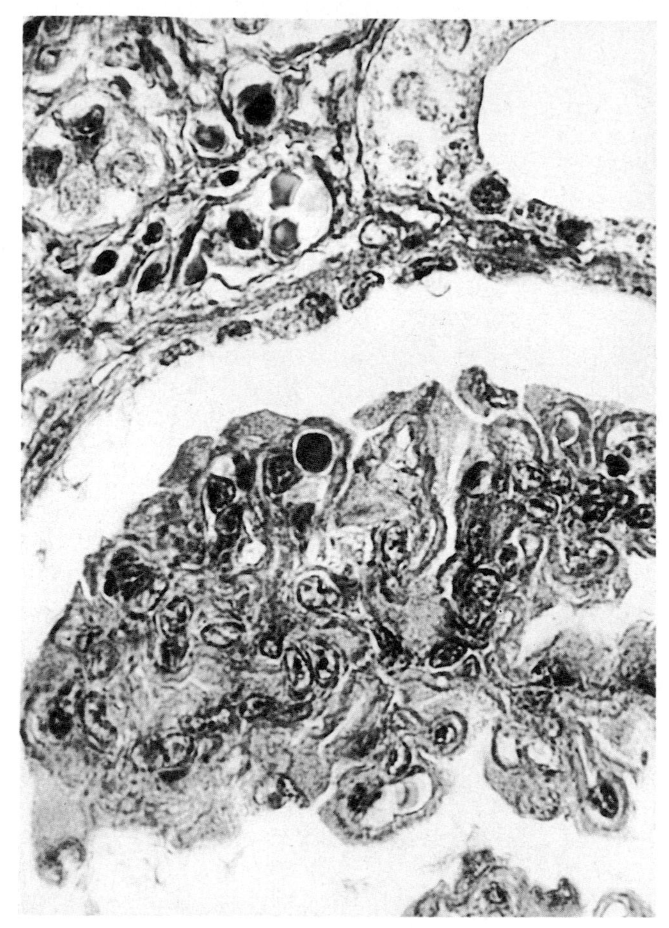

Fig. 14.5 Renal basophilic body formation in SLE.
In the centre of this field, near the margin of the glomerulus, is a black-appearing, rounded body; it is situated in a glomerulus which is affected by focal glomerulonephritis. The microscopic body is slightly larger than a red blood cell. In the stained preparation the body was deeply basophilic (haematoxyphilic). These uncommon structures are composed of degraded nuclear material. (HE × 825.)

moblasts, from fragments of macrophages and other cells, from preparative and stain artefacts and from saprophytic and parasitic protozoa and fungi.

Basophilic bodies are well-defined, round, ovoid or polygonal; purple-red or pale red-blue when stained with haematoxylin and eosin; devoid of organized structure and membrane material; free from cytoplasm; and less dense than normal, nearby nuclei.[310] The bodies may be single, as in a glomerulus, or multiple and clustered. Particular difficulty may be encountered in identifying basophilic bodies in zones of necrosis but tests showing the absence of calcium, Gram-negativity, the presence of DNA and the loss of organized structure, are of value in achieving recognition. The Feulgen stain is positive[174] and the depth of staining may be measured by integrating microdensitometry. RNA is not present. It is possible that the DNA is depolymerized. Nuclear histone is proportionately less than in normal nuclei but there is much extranuclear protein and some carbohydrate, not present in lupus erythematosus (LE) cells.[106] The additional, exogenous protein is likely to be immunoglobulin,[292] almost certainly antinuclear antibody; the basophilic bodies are believed to be nucleoprotein–antibody complexes. There is an analogy with the immunoglobulin found as a stippled band or as larger masses at the epidermodermal junction (see p. 580).[280]

The relative infrequency of basophilic bodies has meant that there have been few electron microscopic studies. Faith and Trump[89] reported the presence of both cytoplasmic and nuclear material. After staining, the basophilic body is highly electron dense[215] and contains cytoplasmic derivatives such as vesicles, vacuoles and granules. A transition from degenerate nuclei is not clear, although Grishman and Churg[115] believed that basophilic bodies were of nuclear origin. Ordonez and Gomez[230] demonstrated that the bodies included both dense polymorph chromatin and immune complex material. In the glomerulus, their location is usually in or near mesangial cells.[115,215]

Necrosis

Necrosis is a consequence of vascular disease and the extent of the necrotic lesion is in proportion to the size of the territory supplied by or derived from the affected vessel. Mesenteric arteritis, for example, may precipitate intestinal infarction. In smaller vessels, such as glomerular capillaries, it is not always clear whether segmental tuft necrosis is the result of an inflammatory response to immune complex deposition or to capillary thrombosis; the processes may be interrelated. Large territories of infarction in lymph nodes (Fig. 14.6) are often accompanied by the aggregation, in peripheral sinusoids, of basophilic DNA-containing masses. The cause(s) of the central necrosis of the rheumatoid arthritis-type granulomata that develop in a small proportion of cases of SLE remain unknown: immune complex deposition with vasculitis is suspected.

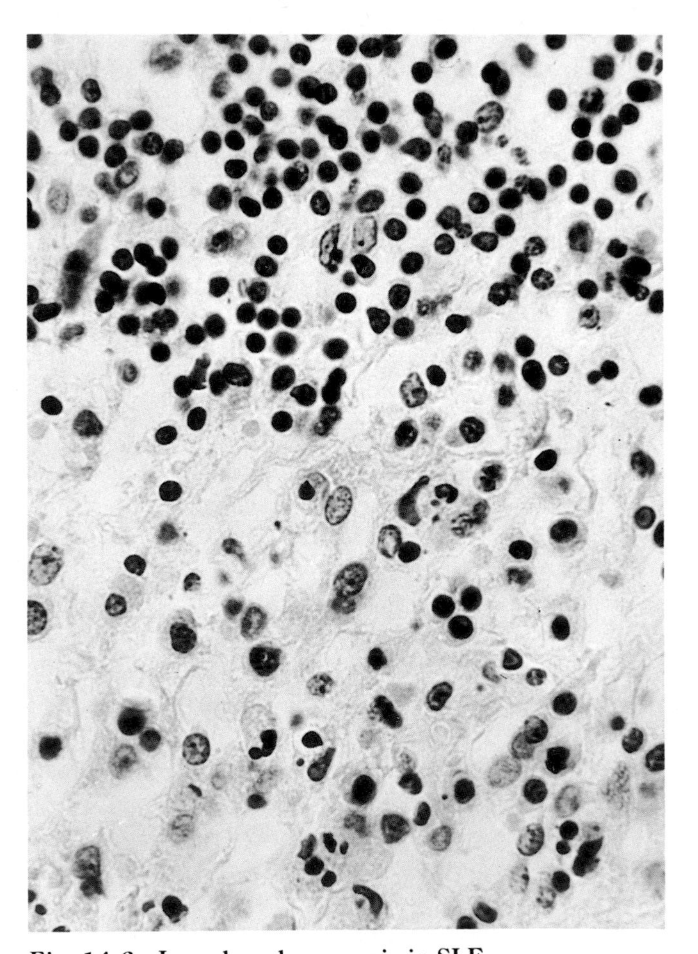

Fig. 14.6 . Lymph node necrosis in SLE.
The upper third of this field appears as normal lymphoid tissue; the lower two-thirds has undergone almost complete necrosis. The mechanism of this response is uncertain but may represent an interaction between lymphocytes and antilymphocytic antibodies. (HE × 560.)

Collagen sclerosis

Collagen sclerosis is a term given to excess local collagen formation in SLE: it has been applied to the multilayer, onion skin-like change that encapsulates the splenic penicilliary arteries. The process of fibrosis (see Chapter 8) is closely related: it may occur at sites of old infarction, in the lungs (see p. 583), around subcutaneous granulomata and, as fibromuscular hyperplasia, within small systemic and pulmonary arteries.

Collagen sclerosis of glomeruli is the end-result of the irreversible forms of lupus nephropathy (see p. 586).

Granulomata

Granulomata with a central zone of epithelioid cells surrounded by numerous histiocytes among which were foci of fibrinoid,[64] were a central part of a theory explaining SLE on the basis of hypersensitivity, advanced by Teilum.[283] Neither giant-cells nor eosinophils were present. Although Pollack[249] described granulomata in the splenic capsule, oesophagus and skeletal muscle, Cruickshank[65] recorded only one palmar tendon sheath granuloma and found granulomata around sclerosed vessels or trabeculae in only two of 18 spleens. Necrosis, fibrinoid and basophilic bodies were not seen. Subcutaneous nodules of the knee and elbow formed in a case of rheumatoid arthritis during the presentation of lupus erythematosus-like signs including glomerulonephritis and a positive antinuclear antibody test.[161] In six further cases conforming with the initial diagnostic criteria, single or multiple nodules developed near the elbow or finger joints[123]: three were microscopically of rheumatoid character. In nine further examples of classical SLE, all with polyarthritis, single or multiple rheumatoid-type granulomata formed; seven were at the elbow and in two of these there were Achilles tendon granulomata.

Although early accounts are difficult to interpret, it therefore appears that rheumatoid-like subcutaneous or visceral granulomata form in only 5–10 per cent of cases of definite SLE.[103] That such a high proportion of the described cases presented clinical features of the SLE/rheumatoid arthritis overlap syndrome, emphasizes the difficulty of establishing that granulomata are a true part of classical SLE.

Electron-dense deposits

These deposits, recognized in glomerular tufts and at epidermodermal junctions, are immune complexes or immunoglobulin: the presence of complement is a positive index of activity. Immunoelectron microscopy can be used to establish whether an electron-dense material is nucleic acid, nucleoprotein, antibody or antigen. In light-microscopic practice, immunofluorescence and now, more conveniently, immunoperoxidase and immunogold techniques, can achieve the same result.

Virus-like bodies

Three varieties of particulate material have attracted attention; their presence is not confined to SLE but they are more common in this than in other diseases. The recognition of these ultrastructural aggregates may therefore offer limited diagnostic help although they are not believed to be of aetiological significance.

Myxoviruses Interwoven, tubular,[121] myxovirus-like structures were recognized in glomerular endothelial cells in lupus erythematosus nephropathy.[77] The particles, measuring 23 × 100 nm, were of RNA. Similar inclusions have been found in skin, endothelial and lymphoid cells, and in polymorphs, histiocytes, fibrocytes, and even extracellularly. They have also been seen in lung, liver, muscle and all other tissues involved in this disease. Attempts to show the presence of myxovirus antigen have been unsuccessful: serological and cultural tests have proved negative. Therefore, in spite of their resemblance to the inclusions of subacute sclerosing panencephalitis, it has been concluded that the inclusions are non-specific products of cell injury: they offer diagnostic guidance but are not pathognomonic. Even if ultimately they prove to be of viral origin, and they are much larger than the known myxoviruses, they may be coincidental consequences of SLE rather than causal agents.

Tumour viruses Particles resembling C-type oncornaviruses have been found in human as they have in New Zealand (NZB) mouse tissue (see p. 599). Since normal human sera have low natural levels of anti-mammalian C-type and other retrovirus antibodies, the significance of these intracellular particles is uncertain. It is probable that they are opportunistic.

Reoviruses Rarely, organized extracellular deposits have been found in human SLE and in NZB mouse renal disease. The deposits measure approximately 1.0 × 10–15.0 nm. There is a resemblance to myxo- and reovirus structure but it is more likely that the material is a collection of phospholipid membranes. The fingerprint pattern of the deposits is identified between renal basement membranes and endothelial cells. These membrane assemblies are characteristic of lupus erythematosus; they are not found in other diseases.

Tissues and viscera

The unit lesions of SLE are expressed with varying frequency in the viscera and supporting tissues. The ubiquity of organ involvement is characteristic and recalls the widespread nature of tissue involvement in untreated syphilis.

Osteoarticular system

Joints Objective signs of arthritis and subjective complaints of arthralgia, together, are the most frequent early evidence of SLE:[198] they are present in 90–95 per cent of cases. Articular disease, often diagnosed as rheumatoid arthritis, may be recognized up to five or more years before the characteristic features of SLE appear; in other instances, arthritis develops during the course of SLE. The joint disorder is symmetrical, remitting and polyarticular, usually beginning in the proximal interphalangeal, metacarpophalangeal, knee and wrist articulations; less frequently in the ankles, elbows and shoulders. In longstanding cases, Jaccoud's non-erosive deforming arthropathy (see p. 805) may persist. The mechanism of elbow contracture is distinct from the inflammatory sequence of the more common joint disorders. Atlantoaxial subluxation may develop[19] but the synovial joints of the spine are generally spared. However, SLE and ankylosing spondylitis may coexist.[229] Temporomandibular arthritis is occasionally recognized but intrinsic joints of the larynx and those of the middle ear are not affected. The arthritis of SLE may be transient but can progress to fibrous ankylosis, subluxation and deformity. These changes, influenced by the common involvement of tendons, ligaments and soft, periarticular tissues are much less severe than in rheumatoid arthritis. Juxtaarticular bone erosion is unusual but avascular bone necrosis (see p. 577) is itself an important cause of chronic articular disease in SLE.

Synovial fluid The synovial fluid (p. 89) of affected joints is only slightly increased in volume, clear or slightly turbid, sterile and of moderate-to-high viscosity. A fibrin coagulum may form. There is considerable mucin clot formation. Leucocyte counts are less than 3×10^6/litre; mononuclear cells predominate but lymphocytes are much commoner than mononuclear phagocytes. Polymorphs may contain inclusions similar to those of the synovial fluid cells in rheumatoid arthritis (see Chapter 12). As in rheumatoid arthritis, but in contrast with all other connective tissue diseases, complement levels are low. Lupus erythematosus cells may be identified and antinuclear factor can sometimes be shown to be present in undiluted specimens.[185] However, this change has been detected in rheumatoid arthritis and in Sjögren's syndrome and is not specific for SLE.

Synovium Microscopic evidence of synovitis is usual but, as in rheumatoid arthritis, may be detected in patients in whom there is no clinical sign of arthritis or in whom the disease is in remission. In a classical account of the histological features of the metacarpophalangeal, proximal interphalangeal, knee and ankle joints from 10 patients examined *post mortem*, Cruickshank[63] identified a thick layer of fibrin-like material on or under the synovial surfaces: the material was Schiff-positive, bright red with the picro–Mallory method, deep violet with phosphotungstic acid–haematoxylin but not metachromatic. There was reduction in the number of synoviocytes which were sometimes absent so that the fibrin-like material rested upon subsynoviocytic connective tissue. Occasionally, there was synovial cell necrosis. Basophilic bodies (see p. 572) of small size were described in eleven of the fourteen joints studied: they were most numerous among synoviocytes but could also be found in deeper tissue and, rarely, were exclusively present at this site.

Inflammation was usually of slight degree; the most severe change was a sparse lympho- and plasmacytic infiltrate. Oedema was exceptional, polymorphs rare, and macrophages absent. Fibrosis of the deeper tissues was present in six cases. Vasculitis was seen in only one joint, from a patient who displayed the visceral lesions both of SLE and of polyarteritis nodosa. Intimal fibromuscular proliferation was recognized in the small arteries or veins of four other patients.

These results contrast with those obtained by needle biopsy.[185] Superficial deposits of fibrin-like material were seen on the synovial surfaces of four patients subjected to biopsy in an investigation of twenty-five cases. However, this material did not stain for fibrin and did not display the ultrastructural features of fibrin. Synovial cell proliferation was focal or diffuse and included A, B and intermediate-type cells (see p. 32). Inflammation of varied intensity, usually perivascular but sometimes diffuse, was common: the predominant cells were mononuclear but three specimens included polymorphs, and in one patient, inflammatory cells infiltrated vessel walls to constitute a vasculitis. Another specimen showed small-vessel fibrinoid necrosis. Necrotic cells, nuclear debris and amorphous haematoxyphilic material were scattered in the synoviocytic and subsynoviocytic layers but there were no objects larger than nuclei. Vascular laminae were occasionally obliterated by platelets and fibrin-like material. Vascular endothelial cell inclusions of several kinds were seen by electron microscopy. They comprised: aggregates of 24 nm diameter microtubules resembling paramyxovirus; extracellular, larger masses of similar material, near nuclear debris; inclusions with the electron density of nuclear chromatin within venular endothelial cells; and, (in one case) subendothelial, electron-dense material resembling that of lupus nephritis (see p. 588).

Some of the microscopic changes described by Cruickshank[63] were probably *post mortem* artefacts. There is also the difficulty that in his paper no account is given of the clinical criteria used to select cases; there is no record of the severity of the disease nor of the clinical laboratory features and no description of treatment. The results reported by Cruickshank[63] and by Labowitz and Schumacher[185] are divergent and there is therefore scope for more extensive, controlled studies of the synovium from treated and untreated cases, categorized according to the 1982 American Rheumatism Association criteria.[279] Morphometric techniques have not yet been applied to this problem.

Incidental bacterial infection of the synovial joints in SLE occurs periodically, just as it does in rheumatoid arthritis. Salmonella, gonococci and other septic microorganisms have been implicated. No report of dialysis-related synovial amyloidosis (see p. 437) in SLE has yet appeared.

Articular cartilage Few investigations have been made. Erosive, ankylosing, destructive arthritis is not characteristic and the opportunity for cartilage biopsy is exceptional. The joint space may appear narrowed radiologically[111] but account must be taken of the possible presence of the avascular bone necrosis (see p. 577), and secondary osteocartilagenous sequestration which may be a consequence of corticosteroid therapy or renal dialysis. Furthermore, the criteria now applied to the selection of cases and their definitive diagnosis are distinct from, and more precise than, those applicable before 1982, and earlier authors may have been inexact in their exclusion of putative cases of rheumatoid arthritis or of overlap syndromes. Cruickshank[63] described articular cartilage damage in five of six finger joints; in three there was marginal erosion with thick layers of granulation tissue on the cartilagenous surface or on the subchondral bone where this had been exposed by cartilage loss. The microscopic changes were similar to, but less severe than, those of rheumatoid arthritis. Foci of inflammation were seen in adjacent bone but no comment was made on the presence or absence of osteoclastic bone reabsorption or chondroclastic activity.

Tendon sheaths Tenosynovitis occurs in 8 per cent of patients. Rarely, the Achilles and patellar tendons rupture as they do in rheumatoid arthritis. The possibility that rupture is influenced by local or systemic corticosteroid treatment must be considered. In one of the few reports of the histological changes in the tendon sheaths, Cruickshank[63] found fibrin-like material on or beneath the synovial surface in six of seven cases. The number of synoviocytes was reduced but necrosis was not observed.

However, inflammation was more severe than in the articular synovia, and disruption of the collagen fibre bundles of the tendon was present in one instance where tenosynovitis was particularly acute. Lymphocytes and plasma cells were numerous, neutrophil and eosinophil polymorphs occasional. In four cases there was slight synovial fibrosis. Arteriolar intimal fibromuscular hyperplasia was present in two. One case displayed a 2 mm thick nodule of mesenchymal synovial cells which occupied the synovial space between the tendon and its synovium: the nodule (see p. 485) did not have a necrotic centre and was not bounded by fibrous tissue. Basophilic bodies containing Feulgen-positive DNA were present in the tissue from two patients.

Bone Bone disease in SLE usually takes the form of osteoporosis and avascular necrosis (Fig. 14.7). However, a high prevalence of skeletal cystic lesions had been recognized radiologically. There is a higher concentration of C-reactive protein in these patients than in those without multiple cyst formation.[184]

Osteoporosis, perhaps the result of cytokines formed by cells within the synovium,[281] is usually juxtaarticular. Nevertheless, the long-term administration of corticosteroids and the predominantly female population determine that systemic osteoporosis is also common. Chronic renal failure (see p. 588), with the need for dialysis, is responsible for other varieties of renal bone disease including osteomalacia, osteitis fibrosa and hyperostosis (Fig. 14.7b).

Avascular necrosis of bone[253] is commonplace in SLE and is a major cause of chronic joint disease.[179] Dubois and Cozen[78] reported the association. In 90 per cent of cases, the signs of avascular necrosis are polyarticular[68,96] and, in one unusual example, bone relating to 13 joints was affected. Evidence of bone disorder is usually symmetrical and is characteristic of early disease in young patients with evidence of systemic vascular abnormalities and subjected to large amounts of corticosteroid. One third of the patients required orthopaedic surgery. The histological changes of avascular bone necrosis in SLE[51] are not known to differ from those in other forms of idiopathic avascular bone necrosis.[253]

In Klippel *et al.'s*[179] series of 375 cases studied between 1962 and 1977, 31 (8 per cent) developed avascular necrosis of bone. During the same period, 0.4 per cent of 710

a

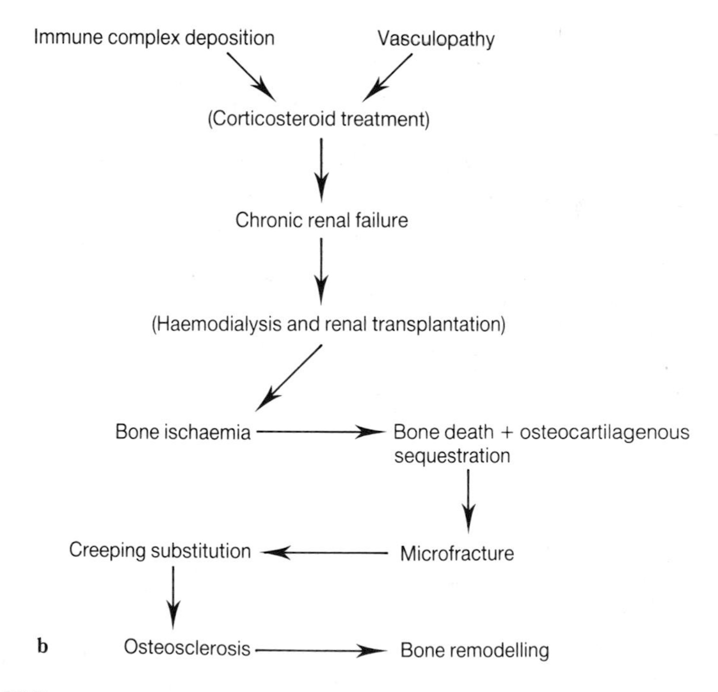

b

Fig. 14.7 Avascular necrosis of bone of femoral head in SLE.
a. A focus of underlying bone necrosis has led to cartilage disruption near the ligamentum teres. Although the administration of corticosteroids and the use of haemodialysis, for renal failure, are commonplace in the treatment of SLE, bone necrosis occurs not infrequently in the absence of treatment and may affect multiple bone sites. Synovial joints such as the hip may be implicated as a result of the intraarticular bone lesion. (Natural size.) **b.** Pathogenesis of change associated with bone necrosis in SLE.

cases of rheumatoid arthritis were found to have bone necrosis. Radiographic changes in SLE patients with symptoms referrable to joints and bone were identified in most instances (Table 14.3) but these signs were also found in 26 per cent of a control, asymptomatic group of SLE patients.

The likely causes of avascular bone necrosis in SLE include therapeutic steroids, fat embolism and ischaemia. In the two cases examined by Siemsen, Brook and Meister[271] and by Labowitz and Schumacher[185] respectively, vasculitis was not present in bone excised at biopsy but in both instances vasculitis was detected in the nearby synovium and muscle. The possibility that avascular bone necrosis in SLE is a direct consequence of vascular disease remains high but the role of corticosteroids is still suspect.

Table 14.3

Frequency of avascular bone necrosis in SLE[179]

Sites	Number of symptomatic patients identified by X-ray	X-ray evidence of avascular bone necrosis	
		Bilateral	Unilateral
Femoral head	30	17	7
Femoral condyle	30	12	7
Talus	24	7	2
Humeral head	24	5	
Carpus	24	5	1
Radial head	17	2	2
Metacarpal/metatarsal	24	1	1

Small numbers of asymptomatic patients were also found to have X-ray evidence of avascular bone necrosis.

Muscle Muscle pain is common in SLE; tenderness is less frequent and weakness unusual. Atrophy, flaccid paralysis, absence of reflexes and elevated serum creatinine phosphokinase levels have all been described.[255]

The results of autopsy studies were discussed by Cruickshank.[66] Perivascular lymphocytic infiltrates, and, less often, myolysis with a more extensive cellular infiltration, had been recognized by Klemperer, Pollack and Baehr.[175] Whether they were distinguishable from the changes in other connective tissue diseases by simple morphological criteria, as Steiner and Chason[276] believed, is unlikely: these authors described a polymorph and lymphocytic infiltrate with few plasma cells. Degenerative muscle cell change may be frequent.[196] Vasculitis was occasional but venulitis constant.

Early observers found muscle biopsy (see p. 679) of little value in diagnosis. However, hydropic change, sarcolemmal nuclear proliferation and perivenous mononuclear cell clusters enabled Pearson and Yamazaki[243] to distinguish SLE skeletal muscle biopsies from the change in non-SLE patients. Although vacuolar myopathy is recognized in untreated patients, this lesion, with swollen sarcolemmal nuclei and nuclei sited centrally within a vacuolar space, may also be provoked by corticosteroid or by chloroquine treatment.

In an early ultrastructural morphometric investigation of muscle biopsies, material from cases of SLE was compared with blocks from cases of progressive systemic sclerosis, rheumatoid arthritis, tuberculosis and from normal individuals.[227] In SLE there were fewer capillaries per unit area of muscle. However, there was a raised mean vascular diameter and an increase in the thickness of the basement membrane. The latter measurement was positively correlated with the use of corticosteroids in treatment. There were no basement membrane-related electron-dense deposits. These observations have been greatly extended.[95]

In a systematic biopsy study of 19 patients with SLE, not categorized according to the American Rheumatism Association (1982)[279] criteria for SLE but comprising 15 females and 4 males, 18 Caucasians and 1 Black, Oxenhandler *et al.*[237] made detailed histological, enzyme histochemical and direct immunofluorescence studies of deltoid, quadriceps femoris or other skeletal muscle blocks. Five clinically involved individuals were found to have an inflammatory myopathy, two with perivascular atrophy and one with vasculitis: one biopsy revealed perivascular atrophy without inflammation; two had neurogenic injury; one neurogenic mixed inflammatory myopathy; and four showed minimal histological changes. However, clinically uninvolved subjects displayed essentially similar abnormalities.

The results of enzyme tests based on the assessment of myofibrillar ATPase at pH 9.4 and on reactions for succinate and NADH dehydrogenases, showed that skeletal muscle in eight cases displayed type 1 fibre predominance; in six cases selective type 2 atrophy; in three cases fibre-blocking; and, in one case, type 1 fibre atrophy. The pathogenic significance of type 1 fibre predominance is unknown. Selective type 2 fibre atrophy is a non-specific phenomenon associated both with disease and corticosteroid administration. It is also encountered in other connective tissue disorders including polymyalgia rheumatica (see p. 642) and in myasthenia gravis, an observation that draws attention to the occasional coexistence of SLE and myasthenia. This circumstance is marked by the presence of two distinct populations of anti-DNA and anti-acetylcholine receptor antibodies.[27] Attention has therefore been directed to the immunopathological changes. IgG deposition around sarcolemmal basement membrane was found in 13 cases;

myofibrillar-related IgG in 5; vascular immunoglobulin or complement in 5; and IgG-containing globules in 10. Basement membrane-related immunoglobulin is analogous to the deposits found in the skin (see p. 580). Myofibrillar immunoglobulin deposition recalls the distribution of some viral antigens and may affect muscle membrane function. Immunoglobulin and complement have been recognized in apparently normal blood vessels in skin, kidney and skeletal muscle. The absence of microscopic injury at these foci of immune complex deposition is not yet explained.

The interpretation of these various muscle changes at present shows no correlation either with clinical evidence of muscle disease or with the diagnosis of SLE. Although the muscle abnormalities recognized in this analysis are thought to be restricted to SLE or to related connective tissue disease, there is no obvious explanation for their presence in terms of pathogenesis.

Skin

Some form of skin disease occurs in 70–85 per cent of cases of SLE.[103,257] Although a butterfly facial rash developed in 73 per cent of 142 cases,[255] with a less characteristic rash in a further 8 per cent, cutaneous lesions are often initially lacking.[190] In 20 per cent of patients, these changes never develop. In one in five patients with SLE, the cutaneous disorder is a well-defined discoid lesion which may be the first, local, evidence of a disease that is to become systemic.

The most characteristic skin lesion is an erythematous rash, with oedema, developing in a butterfly, facial distribution, affecting both malar regions and extending across the bridge of the nose. A maculopapular rash of other parts of the body is commonplace. The lesions of discoid lupus, referred to previously, may be present as long as 25 years before the systemic nature of lupus erythematosus is discovered. Vasculitic disorders, with splinter haemorrhages, hyperpigmentation, alopecia, purpura and mucosal ulceration of the palate or trachea are among the variegated changes that appear periodically. Purpura is often attributable to thrombocytopenia but is sometimes a response to vasculitis.

The cutaneous lesions of discoid lupus (Fig. 14.8) are fully described in textbooks on dermatopathology.[190] Characteristically, there is hyperkeratosis with keratotic plugging; atrophy of those layers of the epidermis with nucleated cells (the stratum malpighii); hydropic change of the basal cells; a focal and unevenly distributed lymphocytic infiltrate, usually around dermal appendages and blood vessels; oedema, dilatation of small blood vessels; and a limited extravascular extravasation of red blood cells.

The more acute skin changes of non-discoid, erythema-

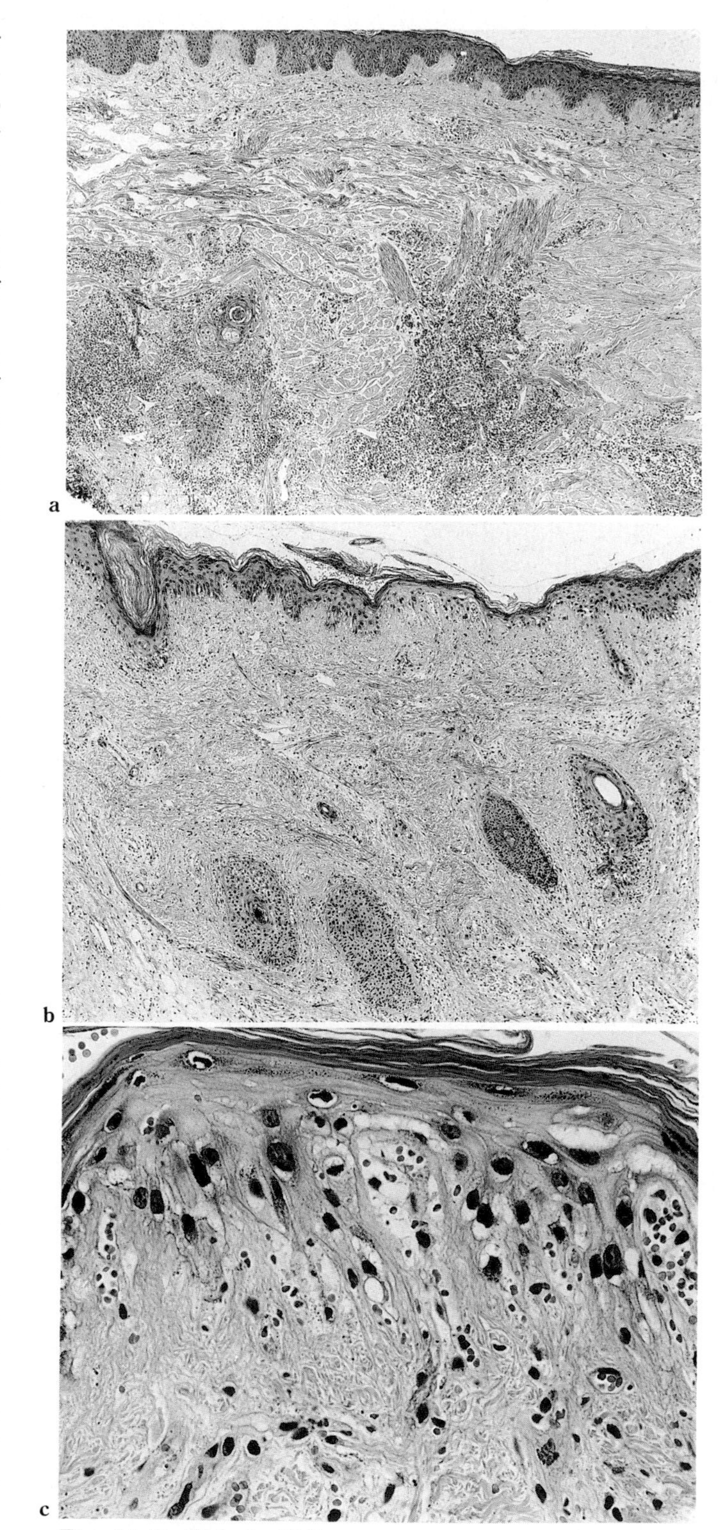

Fig. 14.8 Skin in SLE.
a. There is slight hyperkeratosis. A mononuclear cell infiltrate surrounds the blood vessels, nerve fibre bundles, glands and erector pili muscles. **b.** The epidermis at right is atrophic. At left, there is hyperkeratosis and follicular plugging. **c.** In this field, pronounced basal cell hydropic degeneration is seen. (Courtesy of Dr Philip McKee.) (**a:** HE × 40; **b:** HE × 40; **c:** HE × 220.)

tous type may be limited in extent: they are characteristic of SLE but not pathognomonic. They are distinguished from the changes of established discoid lupus quantitatively, not qualitatively, and are, in general, more acute, less destructive and reversible. There is fibrinoid change of the collagenous, dermal connective tissue. The foci of fibrinoid are periodic acid-Schiff-positive.

It is important to observe that the skin lesions may be only of a non-specific, acute or chronic inflammatory nature. In one case in four, no histological evidence of cutaneous disease is detectable by biopsy or at necropsy. Perivascular inflammatory cell infiltration is identified in 15 per cent of cases; subendothelial fibromuscular intimal hyperplasia of arterioles in 8 per cent. However, fibrinoid change of the walls of these small skin vessels is rare (4 per cent) and necrotizing vasculitis is exceptional.

The mechanism of skin injury in SLE has attracted much interest. Gamma-globulins were shown by direct immunofluorescence to be present in relation to the basement membrane that lies deep to the basal cell layer of the epidermis.[41] The immunoglobulins were present whether or not the basal cells were the site of the characteristic vacuolation: they were indicators of cell injury rather than immediate causes. The appearances were said to be those of a 'lupus band'. Subsequently, it was demonstrated[61] that the clinically normal skin of patients with SLE was affected in the same way. This phenomenon is of diagnostic significance[190,245] and has been used to assess disease activity.[259] The presence of skin immunoglobulin and complement deposits correlates well with serum C3 levels but poorly with most of the criteria for active nephritis.[37,50]

The direct immunofluorescence test for IgG, IgM and complement is performed on cryostat sections of skin taken at excisional biopsy. A lupus-band of positive immunofluorescence, thought to indicate the site of immune complex accumulation, is found in the skin lesions in 90–95 per cent of cases of both systemic and discoid lupus erythematosus (Fig. 14.9). There are very few false-positive reactions. The deposits in discoid lupus erythematosus are coarse and granular in comparison with those in SLE; they may not be present in some early lesions; in scarred tissue; or where there has been long-term, local treatment with corticosteroids. Sun-exposed skin gives a higher frequency of positive reactions than covered skin.

The cutaneous lesions and immune complex deposits have been investigated by transmission electron microscopy and the sub-basement membrane site of the immune complexes confirmed: the complexes lie within the intercellular matrix and upon collagen fibres. It is easy to understand that the basal cells appear to be the immediate target for chemical mediators released or activated by the local binding of complement to immune complexes. The tubular, viral-like structures (see p. 575) found within

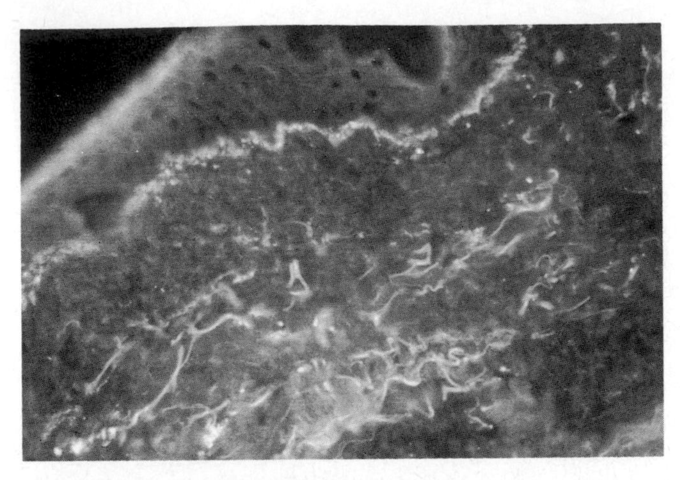

Fig. 14.9 Immunoglobulin in relation to epidermal basement membrane in SLE.
An antiimmunoglobulin antiserum has been used in an immunofluorescence test to show the presence of immunoglobulin at the site of the subepidermal basement membrane (top). Note natural fluorescence of dermal elastic material (lower part of field). Courtesy of Professor R McKie. (Immunofluorescence × 140.)

dermal vascular endothelial cells are probably not viral but appear to represent a cell reaction product.[190]

Subcutaneous or periarticular granulomatous nodules (see p. 577) are found in 5–7 per cent of patients. Similar nodules have been described in the larynx. The microscopic structure of the nodules is not recognizably distinct from that of rheumatoid granulomata (see pp. 485–489). The nodules may be transient, slowly regressing as the patient responds to treatment but increasing again in size with relapse. In their behaviour, therefore, the subcutaneous nodules of SLE appear less indolent than those of rheumatoid arthritis and less prone to persistence, ulceration and secondary infection.

Mucosal lesions develop; they are analogous to those of the skin. The ulcers of the trachea are attributable to the inflammatory response to immune complex deposition: there is a mononuclear cell infiltrate, many basophilic bodies and speckled nuclear staining by immunofluorescent tests for serum antinuclear antibodies.

Heart and blood vessels

The classical cardiovascular lesions of *untreated* SLE were described and analysed between 1923 and 1941.[117,175] Much has been learnt of their ultrastructure and of the immunological changes associated with them. Since the early descriptions of endocarditis[193] and vascular disease[20] it has been clear that cardiac and blood vessel disease (see p. 571) is central to the pathology of SLE.[112,132,135,152,255]

Heart[39,62] The pericardium, myocardium, endocardium and blood vessels are all affected.[64]

Pericardium Clinical signs of pericarditis are found in at least 25 per cent of patients but evidence of pericardial disease is detected *post mortem* in more than 75 per cent of cases with classical SLE. Constrictive pericarditis has also been recognized in instances of drug-induced SLE (see p. 594).[18,38] Particularly when treatment has controlled the progress of SLE, the pericardial sac may be found to have been obliterated by dense adhesions. Localized adhesions form in 16 per cent of cases and obliteration of the pericardial space occurs in 25 per cent. Cardiac tamponade is rare. Microscopically, active low-grade inflammation is characterized by a sparse inflammatory cell infiltrate with rare foci of fibrinoid. When inflammation is more severe, a widespread mononuclear and, occasionally, polymorph or histiocytic response has been described. Haemorrhage is unusual and a fibrinous exudate inconstant. The small blood vessels display the changes described below. Basophilic bodies can be detected very occasionally.

Myocardium Clinically, myocardial disease is less common than pericardial but anatomical changes are recognized much more frequently *post mortem*. Recent and old myocardial infarction is identified in a few cases and, from time-to-time, coronary artery vasculitis can be shown to be an associated if not causal disorder. Microscopically, small foci of perivascular myocarditis or more widespread, diffuse inflammation are seen.[255] There is local fibrinoid change both in the walls of small myocardial vessels, near these vessels (33 per cent) and in connective tissue septa. In Rope's[255] series, fibrinoid was found *post mortem* in nine of 58 cases, Aschoff bodies, presumably coincidental, in two. Although healing with fibrosis is commonplace, focal atrophy and necrosis of cardiac muscle cells is only occasional. Instances of staphylococcal abscess are attributable to the immunosuppressive effects of corticosteroids or to cytotoxic drug treatment.

Endocardium Prospective echocardiographic studies reveal that clinically important valvular lesions are frequent and may require surgical intervention.[99] Among 74 patients, there were seven with Libman–Sacks endocarditis, six with rigid thickened valves with stenosis, regurgitation or both; and five with other forms of valve disease without dysfunction, a total of 24 per cent. The frequency of valvular lesions found *post mortem* is, not unexpectedly, higher and has been estimated to be 35–42 per cent. Anatomical abnormalities are largely confined to the mitral valve. Of the remaining cases the tricuspid valve is next in order of frequency followed by the aortic (Fig. 14.10) and pulmonary.

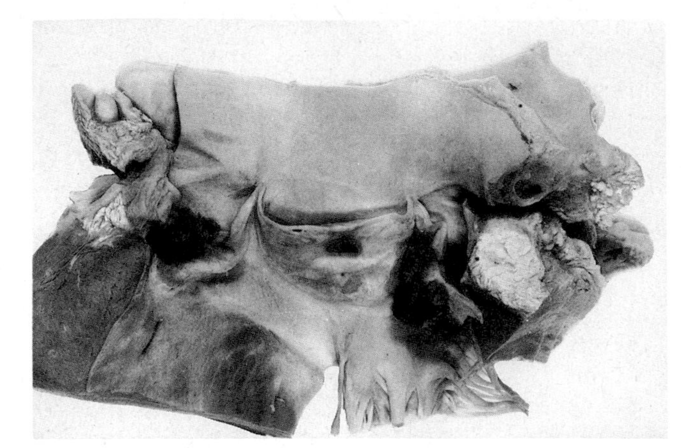

Fig. 14.10 Aortic valve in SLE carditis.
Verrucous vegetations protrude from the surfaces of the aortic valve cusps. In this case, of a 20-year-old female, death occurred from renal failure two years after the development of a facial rash. (× 0.70.)

Although verrucous vegetations of the form described by Libman and Sacks[193] are recognized on either aspect of the mitral valve,[64] they are predominantly on the ventricular aspect.[255] Pocket lesions[116] develop as the compact vegetations extend from the mitral valve on to the chordae tendineae; they come to lie between the mitral valve leaflets and the mural endocardium. Thickening of the valve or annulus or of the chordae tendineae is sometimes found without the presence of vegetations. Mitral stenosis is rare, aortic insufficiency occasional.[270] The chordae tendineae may themselves be a site for fibrinoid change and may rupture.[168] Mitral stenosis and aortic insufficiency, with thickened, rolled, calcified cusps, are very unusual and valvulitis sufficient to cause clinical disease very uncommon. Exceptionally, fulminant SLE has resulted in acute mitral and aortic valve incompetence.[31] Sinoatrial node dysfunction is implicated in many SLE patients who develop arrhythmias or conduction defects; however, the sinoatrial node is seldom abnormal structurally.[154] Those lesions that occur are focal and inflammatory and related to the nodal artery. Neonatal lupus has been found to coexist with congenital heart block.[133]

Microscopy reveals fibrinoid change of the subendocardial connective tissue, myocytic hyperplasia and, occasionally, basophilic bodies (Fig. 14.11). Subsequent fibrosis is less extensive than in rheumatic endocarditis but small numbers of capillaries often extend into the affected tissue.

Bacterial endocarditis is superimposed on established lupus endocarditis in 10 per cent of necropsy cases; mycotic and other infections are more probable when treatment with corticosteroid and cytotoxic drugs is maintained.

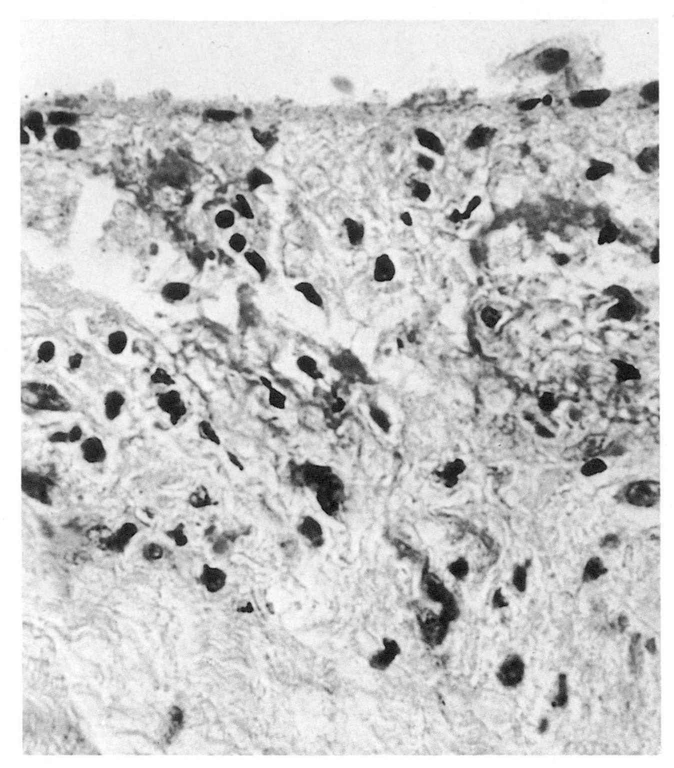

Fig. 14.11 Endocarditis in SLE.
Beneath the endocardial surface (at top) there are small numbers of mononuclear cells, principally macrophages, and very occasional polymorphs. Irregular-shaped basophilic aggregates ('bodies') accompany the presence of material staining as 'fibrinoid'. (HE × 560.)

Blood vessels

Arteries Atheroma is encountered with the low frequency anticipated in a young, female population. The arterial lesions in SLE are largely confined to small muscular arteries and arterioles.[7] Small foci of fibrinoid are identified as an early change; extensive, segmental arterial or arteriolar disease with necrosis and subendothelial fibromuscular hyperplasia recall the lesions of polyarteritis and of accelerated (malignant) hypertension, respectively. They tend to affect smaller arterial branches than those disordered in progressive systemic sclerosis. There is an associated lymphocytic and plasmacytic infiltrate and the inflammatory cells may become pyknotic. Basophilic bodies are seen. Thrombosis is unusual. The acute necrotizing vasculitis of SLE progresses to fibrosis and healing, although injury to the internal elastic lamina is characteristic of cases where microaneurysm results.

Any of the organs and tissues may be occasional sites for arterial disease. The changes are common in the kidney (see p. 584). A diffuse, florid vasculitis affecting the arteries and arterioles of all parts of the brain has been responsible for death in fulminating SLE provoked by photoexposure.[40] Jejunal necrosis and perforation have been described, and necrotizing arteritis, with aneurysm of the hepatic artery, has caused fatal haemorrhage into the liver.[241] It is relevant to the subsequent discussion that IgG, complement and fibrin were found in these vascular lesions.

Veins Venulitis as well as arteritis is a recognized but unusual feature of some SLE lesions.[196] Veins are affected less often. Deep-vein thrombosis is commonplace. In an unusual case, membranous obstruction of the inferior vena cava above the hepatic veins, relieved by a renal vein–right atrial interposition graft, was thought to have released gut-derived antigen into the systemic circulation, leading to SLE.[307] Alternatively, immune hyperactivity might have resulted from an abnormal abdominal venous circulation.

Pulmonary blood vessels Pulmonary hypertension, attributable to interstitial pulmonary fibrosis, primary pulmonary arteriolitis or arteriolar fibromuscular hyperplasia, has been encountered in exceptional cases.[16,44,102,135,255] Raynaud's phenomenon may coexist.[296]

Ultrastructure

Finely granular and 'organized' electron-dense deposits can be found in the walls of small blood vessels.[114] In one case, organized, crystal-like deposits, not fibrin, lay beneath glomerular basement membranes and subendothelially. Small deposits were present in the heart between collagen and elastic fibres and between collagen fibres themselves. Similar deposits were found in the subsynoviocytic tissue and at the epidermodermal junction. In a second case, comparable deposits were present in the mesangium and along the basement membranes, around the aorta and in the pericardium. The deposits were distinct from paramyxovirus tubules.[119,120] The partly crystallized material appeared to correspond to that shown by immunofluorescence and was of a wire-loop character.

Endothelial cell inclusions (see p. 575) were found in muscle, skin, glomerular and peritubular capillaries.[226] At first, it appeared possible that they were myxovirus or paramyxovirus tubules (see p. 575). Another explanation for the occurrence of vascular injury is the possibility that virus promotes the formation of cross-reacting antibodies. It is therefore of interest that IgG anti-endothelial cell antibody was found in SLE sera.[55] The antigen was shown, by *in vivo* culture tests, to be able to bind to endothelial cells.

Immunopathology Immunological mechanisms resulting in cardiac and vascular disease in SLE are most likely to originate in the lodgement of immune complexes in cardiac tissue or in vessel walls, with the initiation of complement-mediated inflammation. Thus, immune complex deposition may be responsible for the uncommon complication of coronary vasculitis. Immune reactants, including IgG, CIq and fibrin, have been identified in inflamed, and in a different distribution, in non-inflamed coronary arteries.[180] In an investigation of the immunological phenomena in 10 cases of severe SLE with cardiac disease, direct immunofluorescence was used to locate immunoglobulins G, M and E, complement components C1q, C3, C4, C5, fibrin, properdin and α_2-macroglobulin.[32] Deposits of these proteins were present predominantly in the walls of myocardial or pericardial blood vessels, but, in one case with Libman–Sacks endocarditis, immunoglobulin and complement were demonstrated in the valve stroma and vegetations. The deposits were more diffuse than the microscopic inflammatory foci with which they were associated. Nevertheless, the cardiac lesions of SLE are thought to be related to immune complex deposition, particularly in patients with persistently elevated anti-dsDNA titres and in those with clinical evidence of disease activity.

Respiratory system

Clinical signs of lung and pleural disease are very common and have been recognized since Osler's[235] description (Fig. 14.12). Radiological signs of pleural effusion, for example, were present in 55 per cent of 133 patients in one series[255] and pulmonary opacities in 56 per cent. There have been many reports[283] and reviews[134] but the central issue for pathologists is this: Which lesions of the lung and pleura in SLE are directly attributable to the disease and which are caused by secondary or opportunistic infection, by renal or cardiac failure, or by treatment?

The lesions attributed to SLE include acute alveolitis, alveolar haemorrhage, alveolar wall necrosis, oedema, hyaline membrane, interstitial pneumonitis, interstitial fibrosis, fibrosing alveolitis,[291] vasculitis, intimal vascular proliferation, pleuritis and pleural effusion. When alternative explanations for these changes were sought in an analysis of all patients listed in the files of the Johns Hopkins Hospital,[134] only the following disorders could be attributed directly to SLE with confidence: interstitial fibrosis (5); vasculitis (2); haematoxylin bodies (1); interstitial pneumonitis (11); pleuritis (22); alveolar wall necrosis (1); pleural effusion (3); alveolar haemorrhage (2); and oedema (3).

The remaining lung abnormalities such as hyaline membrane, encountered periodically in SLE, were believed to be due variously to intercurrent infection, congestive heart failure, increased intracranial pressure, renal failure, aspiration, or oxygen toxicity.

The explanation for lupus pneumonitis is unclear. Tubuloreticular inclusions (see p. 575) have been found in pulmonary capillary endothelial cells.[197] Granular immunoglobulins G and M, and complement component C1q

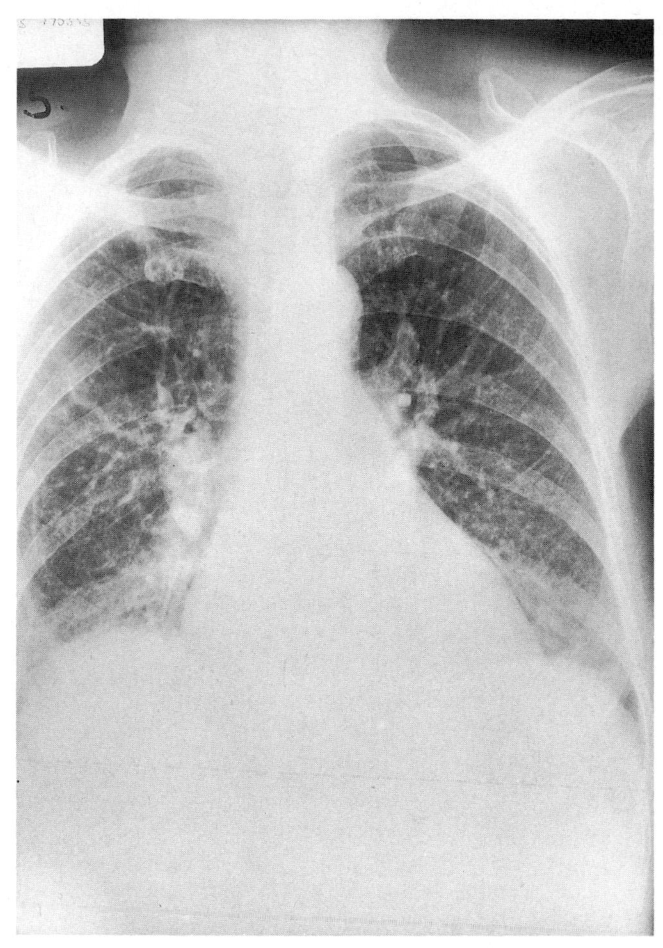

Fig. 14.12 Radiograph of lung in SLE. There is diffuse oedema with prominence of the pulmonary arteries. The interlobar septa can be seen and there is slight cardiac enlargement. The appearances are not specific to SLE.

deposits in interalveolar septa and circulating immune complexes, have led to a view that alveolitis and alveolar wall necrosis may be attributable to complement-mediated inflammation. Pleural fluid complement levels are low. However, the immunofluorescent findings do not correlate with light-microscopic changes[247] and they may result from lung injury secondary to infection, infarction or anaemia. As Haupt *et al.*[134] comment, loss of alveolar membrane integrity may have allowed a non-specific deposition of immunoglobulin and complement.

The conventional light-microscopy of the pleura and lungs in SLE was reviewed by Cruickshank.[65] The pneumonitis may be associated with a basophilic, mucinous oedema of the alveolar walls.[22] There is thickening of alveolar septa and a mononuclear cell infiltrate in some, with alveolar necrosis in others. A moderate plasma cell infiltrate, slight histiocytic desquamation, and acute ulcerative and focal necrosis of the alveoli and bronchioles are seen.[84] Alveolar haemorrhage is an uncommon[82,191] but important factor contributing to death. Churg *et al.*[54] believed that there is an association with immune complex deposition. The interstitial fibrosis of SLE may display a close similarity to the changes of progressive systemic sclerosis: oesophageal dysfunction and Raynaud's phenomenon may coexist. Obliterative bronchiolitis[169] is now recognized, as it is in rheumatoid arthritis. Obliterative pulmonary arterial disease[4] is another shared entity. Pulmonary vasculitis (see p. 656) is rare. The common terminal bronchopneumonia of SLE cannot be regarded as specific to this disorder and pulmonary amyloid[224] is a curiosity.

Urinary system

Kidney The problems of renal disease dominate the pathology of SLE. Although many cases of SLE are not subjected to biopsy, it is believed that the kidney is invariably the site of immunoglobulin, complement and/or immune complex deposition. The frequency with which signs of renal disease are found varies widely: for example, in Ropes'[255] series of 142 patients, 73 per cent had albuminuria in the absence of infection, 40 per cent had haematuria, 68 per cent casts and 47 per cent elevated blood urea levels at some time. The corresponding figures for another series[260] were (within six months of diagnosis): 54, 35, 30 and 67 per cent, respectively; and (at diagnosis or during the course of the disease) 78, 42, 30 and 59 per cent, respectively.

Pathological lesions Some cases have no clinical evidence of renal involvement but are found to have histological signs of glomerulonephritis;[205] the type of renal lesion cannot be deduced from clinical evidence alone. At the opposite end of the diagnostic spectrum are cases of putative idiopathic or post-streptococcal glomerulonephritis, subsequently shown to be examples of late-onset SLE.[2] Renal disease is therefore very common. It is an adverse feature[35] although the prognosis of even severe lupus nephritis has continued to improve[47,216] as drug regimens have been modified. Renal failure has become less common while opportunistic infection has emerged as the single largest cause of death (see p. 603). The renal vascular lesions have been described in increasing detail.[165]

Anatomically,[132,221] there are no diagnostic features. Depending on the stage of the disease, the kidneys may appear normal, congested, pale and large or finely scarred and atrophic. A microcystic appearance may follow prolonged dialysis. Nephrocalcinosis is sometimes detectable.

A so-called parenchymatous nephritis with albuminuria has been observed[79] and each of the four cases of endocarditis investigated by Libman and Sacks[193] (see p. 581) had albuminuria with casts. The renal microscopic changes were recorded by Klemperer *et al.*[173,175] Early accounts were reviewed by Cruickshank[64–66] but his analyses have been overtaken by the advances made during the past 15 years. Cruickshank[65] recorded basement membrane thickening; the 'wire loop' appearance of glomerular capillaries; focal glomerular necrosis; and proliferative glomerulonephritis. Early electron microscopic observations[58] suggested that basement membrane thickening was due to subendothelial fibrinoid. At sites of focal glomerular necrosis, basophilic bodies were often seen. Tubular and interstitial abnormalities were thought to occur only in advanced disease but it became clear that diffuse glomerulonephritis was much the most common renal lesion.[23,250]

Table 14.4

WHO classification of lupus nephritis[24,35]

Class I	*Normal kidney* No changes detectable by light microscopy, transmission electron microscopy or immunofluorescence
Class II	*Mesangial changes alone* a. Minimal mesangial alteration b. Mesangial glomerulitis
Class III	*Focal and segmental proliferative glomerulonephritis*
Class IV	*Diffuse, proliferative glomerulonephritis* (including membranoproliferative glomerulonephritis)
Class V	*Membranous glomerulonephritis*
Class VI	*Diffuse sclerosis*

It is not possible to predict the histological severity of lupus nephritis. Diagnostic biopsy is therefore often necessary. Repeat biopsy is frequently desirable, to explain significant clinical change or to search for reasons for a deficient response to treatment. Although needle biopsy is the most common procedure, open exploration with wedge biopsy is practiced in some centres and offers a more adequate sample for microscopy.

Classification of lupus nephritis The most common and most important renal abnormalities are: mesangial nephritis; focal, proliferative glomerulonephritis; diffuse proliferative glomerulonephritis; and membranous nephritis.[258] However, the kidney may show no light-microscopic changes or disease may have advanced to cause a diffuse sclerosis in which more than one pattern of abnormality is seen at biopsy. The varieties of nephritis were classified by the World Health Organization[24,35] (Table 14.4) and this classification, often modified, is widely used[47,151] (Fig. 14.13).

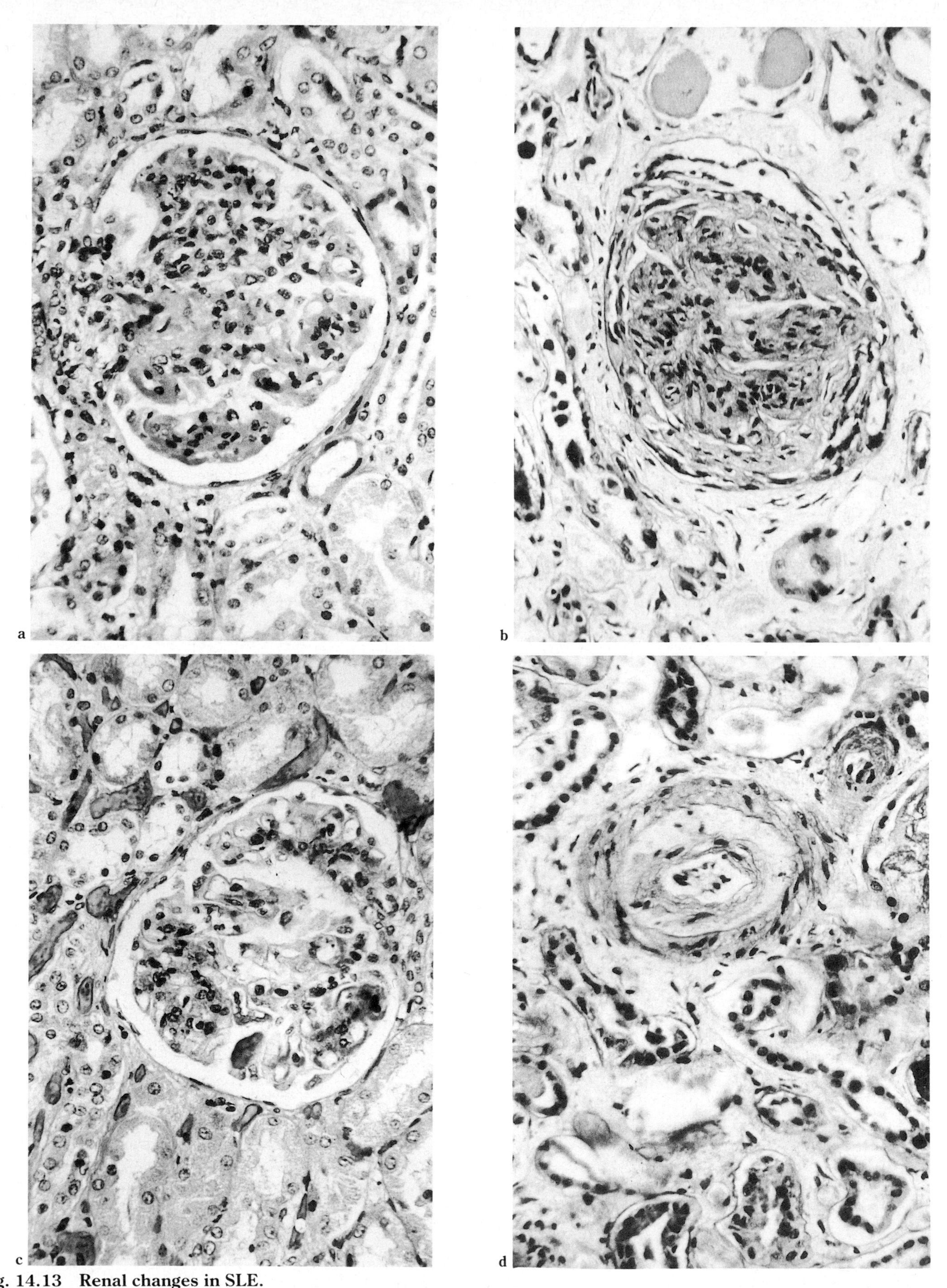

Fig. 14.13 Renal changes in SLE.
a. Acute proliferative glomerulonephritis; **b.** Progressive glomerulonephritis with capsular epithelial cell proliferation; **c.** Focal glomerulonephritis, with fibrinoid; **d.** Renal arteriolar subendothelial cell connective tissue and smooth muscle cell hyperplasia. (HE × 280.)

Class I—normal kidney: In some cases, light microscopy may show no abnormality. However, immunoglobulin, complement or immune complex deposition is almost invariably found by immunofluorescence or electron microscopy. The problem may be quantitative: in an example quoted by Dunnill[79] only 1 in 30 glomeruli showed light-microscopic changes whereas all displayed immunofluorescence and electron microscopical abnormalities. Light-microscopy of conventional 5 μm paraffin sections is a less sensitive test for lupus nephropathy than the examination of 1 μm plastic-embedded sections.

Class II (17 per cent)—mesangial changes alone: In class IIa, light-microscopy reveals no abnormality. However, deposits of immune reactants are found by immunofluorescence within the mesangium and dense deposits are detectable in this site after conventional staining for electron microscopy. In class IIb, mesangial glomerulitis is recognized when segmental or global, focal or diffuse hypercellularity is confined to the mesangium, or when there is an increased matrix with a widening of the mesangial stalk. No changes are seen in the peripheral glomerular loop. Immune reactants are found within the mesangium only by immunofluorescence and electron microscopy, and renal tubular, vascular and interstitial changes are minimal or absent. Immunoglobulin A may predominate.[151]

Class III (10 per cent)—focal and segmental, proliferative glomerulonephritis: This is characteristic. By definition, fewer than 50 per cent of glomeruli are affected. There are localized segments of glomerular, nuclear crowding due to intra- and extracapillary cell proliferation. Capillary lumina may be obliterated or occupied by hyaline thrombi and there are often islands of necrosis in which cell disintegration is accompanied by fibrinoid. Polymorphs infiltrate the affected sites. Mesangial changes similar to those of class II disease are present. Immunofluorescence confirms the presence of many coarse deposits of mesangial immunoglobulin and complement: granular deposits may lie on the basement membrane of peripheral capillaries. The presence of the deposits is confirmed by electron microscopy. Tubular abnormalities and interstitial inflammation are seen: they are focal and periglomerular. The structural changes of this form of lupus nephritis are similar to those of Henoch–Schönlein disease and those of a wide range of experimental focal glomerulopathies.[100] The recognition of the active segmental disease is an indication for vigorous treatment but progression to renal failure is rare.[79]

Class IV (approximately 50 per cent)—diffuse proliferative glomerulonephritis: By definition, not less than 50 per cent of glomeruli are affected and often the disease is wholly diffuse. The degree of glomerular involvement varies but the whole tuft may be affected. The generalized increase in cellularity, the obliteration of capillary lumina and foci of fibrinoid resemble the segmental change of class III lupus nephritis but the diffuse disorder involves a greater proportion of the glomerular surfaces. Within some tufts there is localized thickening of the basement membranes causing a 'wire-loop' appearance. Subendothelial deposits of IgG and C3 are shown by immunofluorescence: IgA and IgM may also be detectable. This class of lupus nephritis is taken to include membranoproliferative disease. Immunoglobulins are identified within the mesangium and there is prominent mesangial cell proliferation, circumferential extension of the mesangial matrix, basement membrane reduplication and lobulation of glomerular tufts. In this subcategory, glomerular necrosis is slight or absent. Previously, the prognosis of this class of lupus nephritis was very poor. However, continuing improvements in drug therapy now determine a five-year survival of approximately 30 per cent.

Class V (15 per cent)—membranous glomerulonephritis: In many cases of lupus nephritis, there is diffuse involvement of all glomeruli which are affected uniformly by a capillary thickening indistinguishable from that of idiopathic membranous glomerulonephritis. The thickened wall is eosinophilic and its hyaline appearance simulates the structure of amyloid and of some instances of diabetic glomerulosclerosis. Klemperer, Pollack and Baehr[176] and others believed that fibrinoid was present. There is no excess glomerular cellularity. In thin sections, stained with a silver impregnation technique, spikes of basement membrane-like material protrude between subepithelial granular deposits which are shown by immunofluorescence microscopy to be largely IgG[151] and C3, and which are believed to be small, soluble immune complexes. The discrete deposits are readily seen by electron microscopy: some are mesangial. Subendothelial deposits are few or absent and tend to be near the mesangial stalk. Tubular interstitial changes are less than in lupus renal disease of classes III and IV. The response to treatment is often positive but the membranous glomerulonephritic process remits in only one-third of cases; untreated, renal failure ensues.

Class VI—diffuse sclerosis: As the renal disease of SLE progresses, an increasing proportion of glomeruli become sclerotic and collagenous. A variegated histological picture emerges and more than one pattern (class) of change may be seen in a single biopsy. Glomerular sclerosis is an adverse sign of disease activity. Glomerular thrombosis is one factor leading to sclerosis and it is therefore of interest that capillary thrombi (see p. 572) are rare in mesangial nephropathy, uncommon in membranous disease but relatively frequent (approximately 50 per cent) in diffuse and focal glomerulonephritis. The development of thrombotic microvascular disease may be associated with the presence of lupus anticoagulant.[167]

In 15 series of cases reported over an 11-year period[47,151] the principal categories of histological change found in patients with renal disease varied substantially, partly due to the selection of cases and partly due to racial differences. The frequency of minimal, focal and mesangial disease ranged from 10 per cent to 70 per cent; of membranous nephropathy from 6 per cent to 28 per cent; and of diffuse proliferative disease from 20 per cent to 81 per cent. The respective mean figures were 30 per cent, 15 per cent and 55 per cent; the Japanese results were particularly varied.

Table 14.5

Renal pathology scoring system[17]

Activity index	Chronicity index
Glomerular abnormalities	
1. Cellular proliferation	1. Glomerular sclerosis
2. Fibrinoid necrosis; karyorrhexis	2. Fibrous crescents
3. Cellular crescents	
4. Hyaline thrombi, wire loops	
5. Leucocyte infiltration	
Tubulointerstitial abnormalities	
1. Mononuclear cell infiltration	1. Interstitial fibrosis
	2. Tubular atrophy

Because of the importance of contrasting the response of different cases to treatment, and of being able to judge the progress of individual cases, a scoring system (Table 14.5) for nephritic lesions has been proposed; it will be of interest to examine its contribution in comparative studies of new therapeutic regimens.

Biopsy and prognosis The value of renal biopsy in the diagnosis of obscure cases of renal disease is beyond question. The relative contribution towards the diagnosis of SLE made by light-microscopy, immunofluorescent microscopy and electron microscopy is less clear. In Western practice, no renal investigation of a case of suspected SLE can now be considered complete without the use of all three methods. Yet electron microscopy contributes to the precision of diagnosis in only 1 case in 10 of renal disease. Nevertheless, to omit ultrastructural studies in an undiagnosed case could be to invite legal difficulty.

There is, nevertheless, a wide difference of opinion regarding the indications for renal biopsy in SLE.[202] The material, examined by light-, immunofluorescence and electron microscopy should be classified (see p. 584) and 'activity' should be assessed (Table 14.5). A scoring system has been advocated and morphometric analyses are becoming practicable on a routine basis. Even when such careful studies are made, it appears that 10–12 per cent of biopsies display atypical features. The distinction between mild (focal) and severe (diffuse) proliferative glomerulonephritis is not well defined, and transitions from one form of disease to another occur sufficiently often to make it difficult to forecast the clinical course. Membranous glomerulonephritis is an exception to this generalization. Not surprisingly, there are limitations to the value of classifying renal disease, for prognostic purposes, in an individual patient. However, the recognition of histological signs of disease activity appear to offer some guidance to the rate of progression of the disease and its response to treatment.

The problem was addressed critically by Whiting-O'Keefe *et al.*[304] They demonstrated that, whereas a straightforward histological classification of renal disease did not add significantly to the value of a clinical assessment in predicting change in renal function over a 12-month period, both the measurement of the proportion of sclerotic glomeruli and the determination by transmission electron microscopy of the extent of subendothelial deposits added to the predictive power of the clinical assessment.

A further study[17] incorporated histological evidence of activity and chronicity into the examination of 102 patients entering a randomized therapeutic trial. While there was a modest increased risk for the development of end-stage renal disease in those with diffuse proliferative or membranoproliferative glomerulonephritis, semiquantitative scores for activity and chronicity indices (Table 14.5) enabled the identification of subgroups with relatively high rates of renal failure.

In the light of these reports, it is not surprising to learn that, in two investigations of SLE with onset at ages 5–65 years (71 patients) and of SLE with onset before 20 years (36 patients), there was no correlation between the initial histopathological classification and the prognosis.[47,216] Nor did conventional renal histopathology provide an accurate basis for assessing fetal survival in women with SLE.[71]

One, earlier view of the value of histological classification was given by Appel *et al.*[15] These authors denied that any single light, transmission electron or immunofluorescence microscopic feature offered greater accuracy than class in predicting prognosis. The outcome for class IIa and IIb (mesangial) lesions was favourable. Those with class III (focal, segmental) disease often progressed to class IV and displayed an equally poor prognosis. Cases of histological class V (membranous) disorder displayed a better prognosis than proliferative disease: as Donadio *et al.*[75] had demonstrated, they pursued a relatively benign course. In this latter series, 6 of 28 patients died of cardiovascular disease. Although proteinuria persisted, renal function was not prejudiced by steroid therapy.

Glomerular sclerosis remains an important cause of renal failure. One precipitating factor may be thrombosis; another is necrosis. When 105 renal biopsies were searched for fibrin thrombi, it was therefore of interest that 31 of 63 biopsies in diffuse and focal proliferative nephritis contained thrombi, whereas thrombi were present in only 3 of 15 biopsies in membranous nephropathy and in none of 21 with mainly mesangial changes.[160] A striking association has been observed between the presence of the circulating (lupus) anticoagulant (LAC) (see p. 572)[43] and the presence of glomerular capillary thrombi, assessed on renal biopsy. If biopsy is repeated, it is evident that late glomerulosclerosis is much more common if an initial biopsy shows evidence of thrombosis. The circulating anticoagulant and glomerular thrombi indicate severe disease activity.[160]

In end-stage renal failure resulting from proliferative glomerulonephritis or from widespread glomerular fibrosis, peritoneal or haemodialysis is the mainstay of treatment. The prognosis of cases of SLE nephropathy treated by haemodialysis is comparable to that of patients with end-stage renal disease due to disorders other than SLE.[60] During 12 years, six of the 28 patients treated by these authors died, mainly among those receiving high doses of prednisone. During dialysis, disease activity is minimal and, in 14 dialysed patients, treated for more than three months, no deaths were attributable directly to active SLE. Morbidity was due largely to infection and cardiovascular insufficiency.[156]

Renal transplantation has been used successfully in treating intractable lupus erythematosus nephropathy. In one case, transplantation to a standard site was successful in spite of the presence of inferior vena cava thrombosis.[299] Glomerulonephritis is not known to develop in the grafted organ although immune complexes remain in the circulation.[10]

Immunopathology Systematic analyses of renal biopsy material gave characteristic results.[141] However, there were geographical and racial differences. Thus, IgG was predominant in the capillaries of class V Japanese cases[151] but IgA predominated in the mesangium of class II cases. In 47 Malaysian cases, 43 displayed hepatitis B surface antigen (HbsAg) in the glomerular immune complex.[194] When the sites of complement deposition in Ishikura's cases were examined in detail, it was found that the complex of C5b, C6, C7, C8 and C9, the membrane attack complex (MAC), was localized to both glomerular and peritubular regions. The MAC and immune complexes displayed similar localities in glomeruli but immune complexes were rarely seen in a peritubular location.[33] By contrast with classical immune complex disease, the circulation in cases of immune complex lupus nephritis contained excess antibody. Extraglomerular immunoglobulin and complement deposits were frequent.[151]

The immunoglobulin deposits in the glomeruli and renal vasculature include antinuclear and anti-DNA antibodies. They may be derived from immune complexes formed in the circulation but it is also possible that anti-DNA antibody reacts *in situ* with renal cell nuclear DNA. The anti-DNA antibody and DNA antigen can be eluted from the kidney. High serum levels of free native anti-DNA antibody correlate with the onset of diffuse proliferative glomerulonephritis[289] and the antibody is of high avidity. There is a suggestion that both in humans and in the mouse, nephritis is associated with a conserved anti-DNA antibody idiotype.[302]

At sites of glomerular fibrinoid, not only IgG, C1q and C3 but also the membrane attack complex (MAC) are present: the MAC may directly mediate tissue injury.[33] Some of the immune complexes deposited in the kidney contain viral antigens such as those of the C-type RNA HEL-12 virus[239] and immunoglobulin eluted from the kidney may react with cells infected with this virus.

Ultrastructure Ultrastructural studies were integral to the demonstration of the cell and basement membrane changes in the kidney (Fig. 14.14). In addition, they allowed the recognition of the intra- and extracellular aggregates (see p. 575) that were thought at first to be virus. In one survey, Ishikura *et al.*[151] found intracellular microtubular structures in glomerular endothelial cells in 70 per cent of cases; both glomerular and interstitial endothelia were affected. Extracellular tubular structures were reported in 32 per cent of renal biopsies in SLE; they were much more

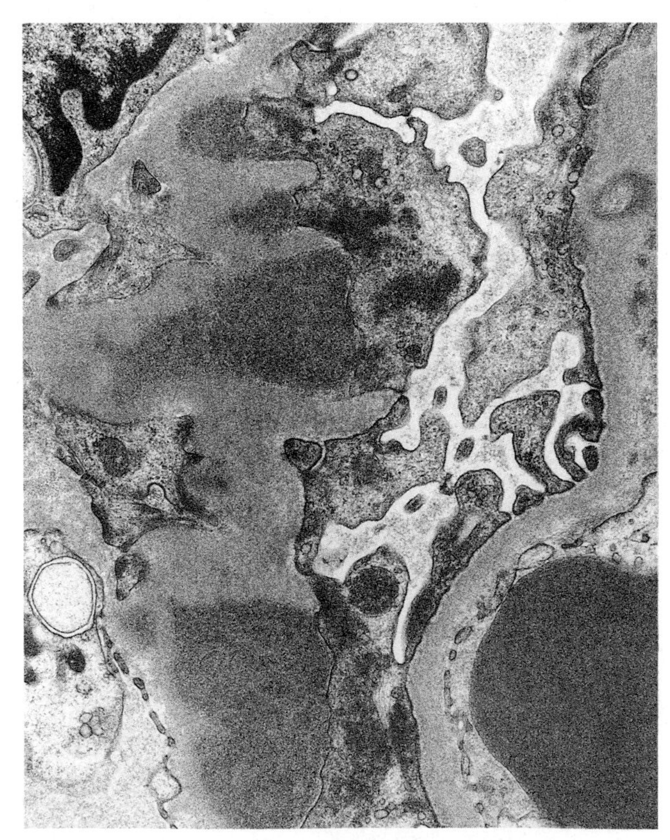

Fig. 14.14 Ultrastructural changes in SLE of kidney. Electron-opaque immune complexes are seen (at centre), mainly on the subepithelial aspect of a glomerular basement membrane. Note red blood cell (bottom right) within glomerular capillary and epithelial, urinary space (centre). (Courtesy of Dr Alan Currey.) (TEM × 21 500.)

frequent in other forms of glomerulopathy.[155] Spherical microparticles were seen in 46 per cent of the 80 cases of this series, most often in membranous nephropathy. Organized deposits with straight, parallel microfilaments were rare. However, a third category of extracellular electron microscopic dense deposit with a fingerprint pattern of phospholipid membranes (see p. 575) was not recognized.

The transmission electron microscopic study of the subepithelial deposits in renal SLE reveals two groups of cases:[266] the first, type I, are regular, homogeneous, and diffusely distributed in each glomerulus: they are seen in class V (membranous glomerulonephritis) cases; the second, type II, are irregular, larger, variably electron-dense and associated with class IV (proliferative glomerulonephritis) cases. The different deposits may correlate with prognosis and the response to therapy.

Scanning electron microscopic investigations have been few. The cells of the kidney can be released from thawed biopsies and the residual glomerular and tubular basement membranes surveyed.[301] Glomerular changes include: epimembranous, crater-like deformities, some with immune complex material; moth-eaten basement membranes; and secondary basement membrane formation within capillaries, abnormalities that are not recognized in tubular basement membranes.

Urinary bladder There may be a reduction in the capacity of the bladder and an increase in the thickness of the irregular walls[232] with hydronephrosis and ureteric dilatation. Bladder disease may be a primary manifestation of SLE but there are few histological studies.

Alimentary system

Gastrointestinal tract Involvement of the gastrointestinal tract in SLE was not mentioned by Klemperer *et al.*[175] However, gastrointestinal lesions are now well recognized[65,143,257] but are less constant than renal and neurological disease. Many symptoms and signs have been attributed to treatment with azathioprine, chloroquine or salicylates. Anorexia, nausea, vomiting or diarrhoea occur in 50 per cent of cases; abdominal pain and haemorrhage are less constant. Peptic ulceration was recognized in 7 of 35 patients in one series.[255] Oral ulceration and Sjögren's syndrome (see Chapter 13) are occasionally identified.

Two-thirds of SLE patients have evidence of serosal inflammation such as perisplenitis or perihepatitis; adhesions form. Ascites occurs: low levels of ascitic fluid complement have been noted and immune complex deposition observed. There is a characteristic but inconstant small-vessel arteritis;[143] larger arteries are sometimes affected in a manner analogous to polyarteritis nodosa, and venulitis is recorded.[136] Among the consequences of vasculitis have been gastritis, jejunal ulceration and perforation, colitis, and haemorrhage. However, understanding of the nature of these lesions is incomplete, and their distinction from common gut disorders such as ulcerative colitis not always clear-cut. Vasculitis causes local ischaemia. As in rheumatoid arthritis, intestinal vascular disease, mucosal ulceration, oedema, haemorrhage,[136] infarction and perforation may develop in sequence: the depth and extent of the lesions is related to the size of the affected arteries. There may be concomitant arteritis of other tissues and organs, particularly the skin.[311] Mesenteric arteritis, identified radiographically, has been confirmed at surgery[248] but arteriography does not always provide evidence of this disorder. Rothfield[257] reported three deaths from intestinal perforation among 365 patients; Zizic *et al.*[311] four deaths from colonic perforation among 15 deaths in 197 patients.

Dysphagia is occasional and, as in rheumatoid arthritis, arteritis may underlie oesophageal ulceration[132] and extend to the larynx.[255] Oesophageal motor dysfunction has suggested an association with systemic sclerosis or an overlap syndrome.

The coexistence of ulcerative colitis and SLE has been claimed, but in one large series the systemic disease was believed to be drug (sulphonamide)-linked[7] and the association has not been proven. Intestinal malabsorption is another rare feature of SLE: it has been attributed to mucosal injury, vasculitis, atherosclerosis and mucosal ischaemia but the pathological aspects of the syndrome are not well understood.[143] The coexistence of SLE and the Canada–Cronkhite syndrome (see p. 658) of gastrointestinal polyposis has been encountered.[182]

Liver Whether *clinically significant* liver disease occurs in SLE is still debated.[183,209] The liver is often enlarged.[255] Although jaundice is rare, objective tests of liver function frequently reveal abnormalities and high serum glutamate oxaloacetate transaminase levels are characteristic. When a set of five tests was used to assess 206 patients, 43 met strict diagnostic criteria for liver involvement. The criteria were: an abnormal liver biopsy and a two-fold or greater increase in at least four measurements of serum total bilirubin, serum glutamate oxaloacetate and pyruvate transaminases (SGOT and SGPT), lactate dehydrogenase and alkaline phosphatase. At least two different tests had to be abnormal for hepatic involvement to be confirmed.[264] The administration of salicylates has a recognizable, adverse effect on liver function in patients with active disease: 7 of 16 patients receiving 50 mg/kg day developed raised serum transaminase levels within two weeks.[268]

The earlier literature contains many references to liver lesions.[64] It has become clear that signs of liver disease sufficient to warrant biopsy are indeed very common.[209]

Klemperer, Pollack and Baehr[175] encountered periphepatitis and focal necrosis with portal vein thrombosis, and centrilobular necrosis, fatty change and hepatocyte atrophy were among the ill-defined changes recognized.

The papers of Joske and King[158] and of Mackay, Taft and Cowling[200] were important stimuli to unravelling the hepatic lesions: lupus erythematosus (LE) cells were found in preparations made from the serum of patients with chronic, active viral hepatitis. The term 'lupoid hepatitis' was introduced to describe the syndrome and caused considerable confusion. It was by no means clear whether the immunological changes were evidence that lupoid hepatitis was a separate entity from the hepatic lesions of SLE, as Mackay *et al.* believed, or whether so-called lupoid hepatitis represented an autoimmune process superimposed on the cell lesions caused in the liver by virus.[201]

A careful analysis has helped to clarify the situation.[118] Eighty-three patients with a miscellany of liver diseases were investigated prospectively, 12 patients with 'lupoid' hepatitis retrospectively. Three assays for dsDNA, the lupus band test (see p. 580) and circulating immune complexes were used to search for proof of SLE; enzyme and other techniques were chosen to analyse liver function. Patients with various forms of liver disease had anti-dsDNA, as examined by older methods and occasionally gave positive lupus band, antinuclear antibody and immune complex tests. However, evidence of an essential immunological distinction between patients with SLE and those with chronic, active 'autoimmune' hepatitis was demonstrated by the observation that the former had specific anti-dsDNA antibody, using the *Crithidia luciliae* assay, and gave strong lupus skin band tests, usually with the presence of two or more immunoglobulin classes.

In treated SLE, postnecrotic cirrhosis may occur and primary biliary cirrhosis has been described.[28] Nevertheless, liver disease is not a cause of death. Where hepatic abnormalities are encountered *post mortem*, cardiac failure, infection, transfusional haemosiderosis, arteritis and the presence of basophilic bodies contribute to the histological picture. Naturally, the more careful the study the more likely the finding of hepatic abnormalities. Even so, the claim that liver involvement in SLE is more common than previously recognized and that severe and even fatal liver disease can occur, could be regarded as exaggerated.[264] Three of the four cirrhotic patients examined by these authors revealed a peculiar form of canalicular cast with biliary cholestasis, findings that might be difficult to distinguish from the effects of drug treatment. In this series only 3 of 238 patients died from liver failure. The most common histological diagnoses in the most severe cases with disease proven by liver biopsy were: granulomatous hepatitis (7) (two with features of chronic active hepatitis); cirrhosis (4) (three with cholestasis); acute hepatitis (1); chronic active hepatitis (4); chronic persistent hepatitis (3); primary biliary cirrhosis (1); normal (1); and candidiasis (1). These studies were made on a population of cases of SLE selected by the preliminary (1974), not by the revised (1982) criteria for the disease.

Among other, less frequent hepatic disorders encountered in SLE are: hepatic venocclusive disease of uncertain origin,[240] nodular regenerative hyperplasia[170] without vasculitis, and the presence of hepatocyte cytoplasmic concentric membranous bodies, a possible ultrastructural index of increased protein synthesis.[269] The occlusion of small hepatic veins has been related to the titre of lupus anticoagulant[223] (see p. 572).

Pancreas Pancreatitis may be secondary to SLE and has been attributed to vasculitis.[76] In three examples, Pollack, Grove, Kark *et al.*[249a] described vasculitis, uraemia and platelet thrombi. With advances in therapy and the widespread use of cytotoxic agents and steroids, known to cause pancreatitis, the significance of pancreatic disease in SLE has become more difficult to evaluate.

Haematopoietic system

Anaemia is frequent: it is usually normocytic and hypochromic although iron deficiency is unusual. However, in 10–15 per cent of patients the anaemia is haemolytic. Anti-red blood cell antibodies are detected and Coombs tests are often positive. Leucopenia is also common and anti-white cell antibodies may be present in the plasma. Leucopenia is characteristic. The proportion of neutrophil granulocytes is normal and increases may coincide with episodes of infection—an observation that must be reconciled with the increased susceptibility of SLE patients to microbial disease whether or not steroid therapy has been used.[274] Infection remains a major cause of death (see p. 603). The erythrocyte sedimentation rate is elevated and may be very high. This elevation may reflect disease activity but can indicate concomitant, opportunistic infection. C-reactive protein levels are elevated only moderately in active cases but are substantially raised when there is intercurrent infection.[244] The haematological disorders of SLE are discussed in detail in texts on clinical rheumatology.[166,198,267]

Particular interest centres on disorders of coagulation. Haemorrhagic phenomena are characteristic. As part explanation, demonstrable thrombocytopenia has been implicated, perhaps attributable to antiplatelet antibodies. However, many more sera contain antiplatelet antibodies than the numbers of cases of thrombocytopenia and a state of compensated thrombocytolysis has been proposed. This is disputed.[43] There is likely also to be a hypoaggregatability of platelets.

Thrombotic episodes are common and widespread, both

in the venous[85] and arterial[131] circulations. Thrombosis in the deep or superficial leg veins, in the renal, portal and hepatic veins, and in the inferior vena cava is recognized.[129] In the arterial circulation, thrombi occur in the cerebral, retinal, coronary and limb vessels. A particular feature is placental vessel thrombosis leading to infarction and fetal death.[72]

A full explanation for the frequent occurrence of thrombosis is lacking. However, thrombosis, fetal death and thrombocytopenia are clearly associated with the presence in the circulation of antibodies that bind cardiolipin and other phospholipids (see p. 572). It is highly probable that these antibodies are similar to, if not identical with, lupus anticoagulant (LAC). Gamma-globulin from patients with LAC inhibits prostacyclin formation (see Chapter 7). Low prostacyclin levels can promote platelet aggregation and thus thrombosis, and LAC can also inhibit prekallikrein activity and fibrinolysis and may bind platelet membrane phospholipid. Both phenomena can cause thrombosis.

Lymphoreticular system and thymus

The spleen and lymph nodes are characteristically affected by SLE so that splenomegaly and lymphadenopathy are found in approximately 50 per cent of all untreated cases.

Spleen The early literature is summarized by Cruickshank.[65] Perisplenitis is frequent; splenic atrophy and hyposplenism are encountered.[73] There is concentric, onion-skin, perivascular collagen deposition (see p. 574) around the penicilliary arteries. Since laminae of collagen are always present in these sites in normal spleens, it is necessary to stipulate that onion-skin collagenization must comprise at least seven concentric layers of connective tissue to be considered admissable evidence of SLE. Like the lymph nodes, single or multiple foci of necrosis of varying sizes are encountered but the aggregation of basophilic bodies seems to be less conspicuous than in the lymph nodes, perhaps because of splenic macrophage degradative activity. The presence of an inflammatory reaction may be an index of vasculopathy rather than of primary lymphoid necrosis. When there is an haemolytic anaemia, splenic fibrosis is likely to accompany haemosiderosis and there may be a corresponding hyperplasia of splenic macrophages.

Lymph nodes Enlargement of the lymph nodes was found clinically in 78 per cent of the patients surveyed by Ropes[255] and in a corresponding proportion of the 58 patients in her series examined *post mortem*. Cervical, axillary, mesenteric, inguinal and tracheobronchial node involvement was commonplace. Opaque, yellow foci of necrosis (Fig. 14.15) were thought by early observers to be tuberculous and of course, treated or untreated, many patients with SLE die from bacterial infection. However, the necrotic foci in SLE evoke no inflammatory reaction. Nearby cortical sinuses are occasionally packed with the aggregates of basophilic, degraded nuclear material which is the summation of the nuclear changes caused by antinuclear antibody but no polymorph response is seen. Klemperer, Pollack and Baehr[175] confirmed the usual absence of acid alcohol-fast bacilli and analogous studies were reported by Fox and Rosahn[98] and by Harvey *et al.*[132] In 12 *post mortem* cases, Cruickshank[64] found follicular hyperplasia in only one, but sinus cell hyperplasia with cell injury and karyorrhexis was common as was plasmacytosis. Basophilic bodies were encountered in only one case, in spite of Cooper's[59] comment that these structures are more likely to be identified in the kidney and lymph node than elsewhere. Like the kidney, lung, synovium and skin, lymph node cells have been found to contain tubular inclusions which may be viral or may represent ultramicroscopic

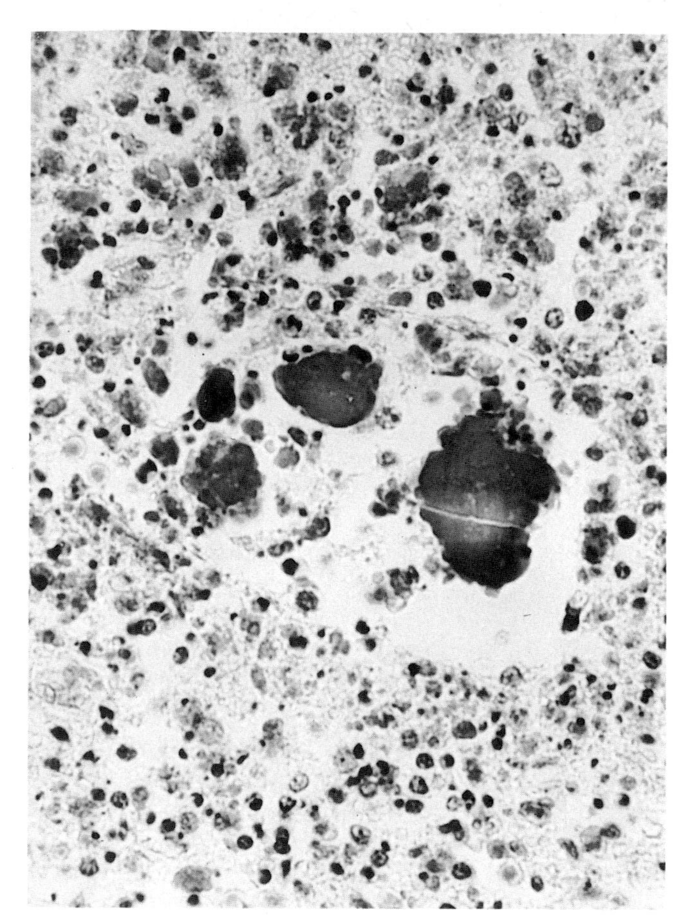

Fig. 14.15 Lymph node in SLE.
Aggregated basophilic bodies in lymph node sinus. The material is of nuclear origin. Compare with Fig. 14.6. The blue-staining nuclear debris is anatomically related to the surrounding lymph node necrosis. (HE × 595.)

evidence of cell injury. Lymphocytes and macrophages may be involved, locally or within the circulation.

It is inevitable that the role of treatment be considered when the lymph nodes are examined in SLE. After steroid treatment had become widespread, Moore, Weisberger and Bowerfind[211,212] reported that follicular hyperplasia was common. However, Cruickshank[66] commented that neither cortisone nor adrenocorticotrophic hormone influenced the microscopic lymph node lesions. Cytotoxic and immunosuppressive drugs have been widely and successfully employed in the preparation of patients for renal transplantation but there is limited evidence that these compounds exert beneficial long-term effects on the changes of the lymphoid tissues. An early account[132] confirmed that lymphadenopathy disappeared within one to three weeks, while Dubois and Arterberry[77] reported regression of lymphadenopathy in 409 cases within one week —some enlargement persisted despite improvement in the clinical condition of these patients. Non-Hodgkin's lymphoma occurs irrespective of immunosuppressive treatment[3] but a rare association of SLE-like disorders in older males is the development of immunoblastic lymphoma,[278] a possible consequence of immunosuppressive therapy or simply an unusual reflection of aberrant control of T-cell function.

Thymus The evidence relating disorders of the immune mechanism and of tolerance to the origin of SLE have naturally drawn attention to the role of the thymus. The thymus has been found to be small.[126] There is cortical atrophy, medullary spindle epithelial cell aggregates, increased cystic change and scanty epithelial Hassall's corpuscles. There are a few plasma cells but only occasional lymphoid follicular germinal centres.[110,204] Systemic lupus erythematosus may follow thymectomy for myasthenia gravis[45] and evidence of defective T_S-cell function has been obtained, partially reversible by thymic hormone. The disease can also be a rare association of thymoma[56] and in the presence of thymoma there may be evidence of functional thymic deficiency. Thymectomy does not lead to improvement in the activity of the SLE.

Nervous system

Neuropsychiatric disorders are common.[1,128] In addition to non-organic psychological disorders, organic brain syndromes and seizures are the most common abnormalities: together, these three groups of disease are recognized in 50 per cent of most clinical series. There is evidence that the persistent circulation of dsDNA occurs specifically in patients with vasculitis and central nervous system (CNS) involvement[277] although this finding is inconstant. Seizures, usually of grand mal character, were often terminal before the availability of corticosteroids. The neuropsychiatric changes of SLE are unpredictable; several levels of the nervous system may be affected simultaneously. Visual defects and cranial nerve lesions are encountered but peripheral neuropathies are much less frequent. Guillain-Barré syndrome, transverse myelopathy, and movement disorders are occasional but it is important to note that some of the neurological disorders of SLE are attributable to mycotic or bacterial opportunistic infection, to uraemia, to steroid therapy or to hypertension. Computed tomography (CT) and magnetic resonance imaging (MRI)[294] help to define the sites of the organic brain lesions. Basal ganglionic calcification, for example, can be located.[225] Because of the suspected part played by foci of defective brain metabolism in causing neurological disease, it is not surprising that oxygen-15 scans can assist in the delineation of minor abnormalities.

Much of the pathological evidence rests upon the *post mortem* studies of Johnson and Richardson,[157] Ropes[255] and Ellis and Verity:[87] 19 of the patients described by Ropes[255] were included in the more detailed series of 24 reported by Johnson and Richardson.[157] The pathological changes centre on ischaemic or anoxic cell loss, infarction, haemorrhage and vasculopathy. The arterioles and small arteries may display mural fibrinoid change, intimal cell proliferation, luminal fibrin thrombi or, rarely, a true vasculitis. Aneurysm is occasional, attributable to vasculitis, and may rupture causing subarachnoid haemorrhage and cerebral infarction.[97] The characteristic lesions of thrombotic microangiopathy (see p. 660) have been recognized.

The pathogenesis of the transient, functional disturbances of neurones in SLE, of the demyelinating lesions, and of the vasculopathies are likely to be different. When the contributions made by uraemia, hypertension, corticosteroid therapy and infection are excluded, attention is necessarily focused on the role of the small blood vessels. The thrombotic phenomena may be associated with the properties of lupus anticoagulant (LAC) (see p. 572). There is a suggested correlation between the presence of infarct-like lesions, shown by magnetic resonance imaging (see p. 162) and serum LAC but the anticoagulant does not appear to have a direct pathogenic role in diffuse SLE encephalopathy.[94]

It is far from certain whether any of the focal vascular lesions of the brain in SLE are attributable to immune complex deposition. Surprisingly, the thin-walled blood vessels of the choroid plexus do not show microscopic abnormalities resembling those of the kidney.[87] Nevertheless, the deposition of immunoglobulin and sometimes of complement has been demonstrated in the choroid plexuses of a few fatal cases of cerebral SLE. Immunoglobulin has, however, been identified in the choroid plexus of control patients and it is therefore possible that tissue injury, detectable at light-microscopic level, may depend upon changes in the selective permeability of choroidal vessels to complement components.

Not surprisingly, it has proved difficult to obtain precise correlation between the clinical signs of cerebral lupus and the sites of the randomly scattered foci of cell injury or loss. It is clear, however, that cases with generalized seizures and cranial nerve defects tend to display cortical and brain-stem infarcts. Vascular abnormalities can account for transient foci of ischaemia, for microinfarcts or for larger zones of necrosis. It is probable that vasoactive substances are also liberated locally. One mechanism is the abnormal binding of phospholipids to endothelial cell membranes with the decreased release of arachidonic acid (see Chapter 7), lowered prostacyclin levels and a tendency to platelet aggregation and thrombosis.

Antineuronal and antiglial cell antibodies are detectable in the serum and cerebrospinal fluid; the former may be responsible for focal demyelination directly. However, their specficity and avidity vary widely and glycolipids as well as phospholipids may be disorganized selectively. The latter may contain lymphocytotoxic antibody.[305] It is self-evident that extraneural antibody formed, for example, in response to viral antigen, may gain access to the brain if the blood–brain barrier is prejudiced. Alternatively, stimuli to antibody formation may be in response to primary neuronal injury with the processing of antigen by mononuclear macrophages external to the central nervous system. Some light may be shed on these questions by further studies of murine lupus (see p. 598) since immunoglobulin and complement have been found in the choroid plexuses of older, surviving NZB/NZW mice (see p. 600).[187]

Ocular system

Although conjunctivitis, episcleritis, subconjunctival lesions and eyelid lesions are among the clinical features recognized in large series of cases, there are few published reports of the pathology of the eye.[64] Just as the brain is impaired by the ischaemic consequences of vascular disease, so periodically the retina is affected by retinal arteritis, arteriolar occlusion and thrombosis. An association with Sjögren's syndrome (see Chapter 13) and keratoconjunctivitis sicca is recognized.

Endocrine system

With rare exceptions, endocrine dysfunction in SLE is the result of treatment. It is not necessary to reiterate the effects of corticosteroids on adrenal cortical and anterior pituitary function. Although pancreatic fibrosis, occasionally with lymphocytic infiltration, diffuse acinar ectasia, and vasculitis were seen histologically,[255] pancreatitis was not encountered in this series, in contrast to the cases quoted by Dubois and Arterberry[76] and by Wallace and Dubois.[297] Diabetes mellitus is not more common than in the general population. However, attention has been drawn to the instance of a child in whom the onset of diabetes mellitus was followed by the development of SLE two years later:[231] her cells expressed the MHC class I antigens HLA A2, Aw24, B8 and B27 which may have predisposed to a multiple endocrine adenopathy syndrome. The mechanisms by which autoimmune thyroiditis can occasionally be associated with both antinuclear factor formation and the onset of SLE is not understood[303] but the inheritance of particular HLA determinants may indicate one way in which an heritable predisposition can operate.

Associated syndromes

The Sjögren (sicca) syndrome (p. 529) is a thread linking clinical 'overlap' syndromes in which signs of more than one systemic connective tissue disease coexist. The relationship of some of these cases to so-called 'mixed connective tissue disease' (see p. 709) is not clear. It has long been known that the histological[238] features of more than one connective tissue disease may coexist but the evidence for overlap syndromes is mainly immunological and clinical rather than morphological.

Rheumatoid arthritis, systemic sclerosis, polymyositis and necrotizing vasculitis may each merge with SLE. SLE may coexist with Sjögren's syndrome,[6,11] sometimes in the presence of hepatic cirrhosis.[181] The incidence of keratoconjunctivitis sicca in SLE is high, particularly when there is arthritis. Biopsy of the minor salivary glands of 50 cases of SLE[6] showed that 47 were abnormal: there was inflammatory cell infiltration (43); glandular atrophy (28); and ductular change (38). In 31 cases, plasma cells predominated: in the remainder, lymphocytes outnumbered plasma cells but in only one case were no plasma cells present. Overall, there was no correlation between the duration of the disease or its activity or treatment, and the minor salivary gland histological abnormalities. The signs of Sjögren's syndrome are common in some cases in which the lesions of rheumatoid arthritis and SLE coexist. The proportion of cases with antisalivary duct cell antibody is approximately the same in rheumatoid arthritis and SLE. One reason for the diagnosis of an overlap syndrome is the finding of a positive lupus erythematosus cell test in a patient with rheumatoid arthritis.[29] Deforming, erosive arthritis is rarely associated with definite SLE and seropositivity: rheumatoid arthritis cases with a positive lupus erythematosus cell test have a guarded prognosis.

Drug-induced systemic lupus erythematosus[140]

In 1954 Perry and Schroeder,[246] and Dustan, Taylor, Corcoran and Page[81] described a lupus erythematosus-like syndrome that followed the treatment of severe cases of hypertension with hydralazine (l-hydrazinophthalazine). Facial skin rashes, malaise, fever, anaemia and splenomegaly were followed by evidence of renal disease and plasma protein changes. A small proportion of lupus erythematosus cell tests was positive. Attempts to reproduce the syndrome in dogs[57,100] were only partially successful, and the hypertensive rat[101] was insensitive to lupus-like change.

The onset of serositis draws attention to the development of the drug-induced disease. The principal clinical findings include polyarthralgia and arthritis, with pleuritis and pericarditis. Renal and cerebral signs are rare. Ninety-five per cent of patients display antinuclear antibody formation but anti-dsDNA antibody is usually absent. Although the clinical manifestations are reversible on withdrawal of the drug, the serum lupus erythematosus (antinuclear) factor persists for some months.

It is now believed that the adverse response to hydralazine[46] is a genetic characteristic associated with the inheritance of the MHC class II antigen HLA-DR4.[25] It is more probable in women than men. The syndrome is accompanied by immune complex deposition in tissues, a tendency increased by the inhibition of C4.[272] The role of complement system protein C4 is itself genetically determined.[273] Hydralazine, procainamide[108,186] and isoniazid, drugs with chemical and metabolic similarities to hydralazine and a propensity to provoke a SLE-like disease, are acetylated by the hepatic N-acetyltransferase system. Enzyme activity is genetically determined. Individuals with the Mendelian dominant characteristic acetylate quickly (fast acetylators); the minority, who are homozygous for the recessive allele, are slow acetylators and tend to develop a mild SLE-like disease if given one of these drugs over a prolonged period. In the case of hydralazine[207] and procainamide, slow acetylators are prone to develop antinuclear antibodies, and antinuclear factor formation can be provoked in mice by isoniazid or hydralazine.[49]

The drug-activated SLE syndrome is influenced by the amount and duration of the compound given: it is essentially a pharmacological response (Table 14.6). There is no evidence for an association between acetylator phenotype and the development of idiopathic SLE.[21] The nuclear histone epitopes reacting with sera from cases of drug-induced SLE are no less restricted than those reacting with sera from spontaneous cases.[107] Altered DNA conformations such as (left-handed) Z-DNA, are more immunogenic than (right-handed) B-DNA.[288] Procainamide and hydralazine increase the rate of transition of B- to Z-DNA. They promote the aggregation of calf-thymus DNA and it is therefore possible that these compounds may activate SLE in part by actions upon native DNA.

Table 14.6

Drug-induced SLE

Drugs in which an association with a lupus-like syndrome is proven:[140]	
Hydralazine	Methyldopa
Procainamide	Chlorpromazine
Isoniazide	Quinidine
Drugs probably associated with a lupus-like syndrome:	
Anticonvulsant drugs	Sulphasalazine
Antithyroid agents	Beta-blockers
Penicillamine	Lithium
Drugs possibly associated with a lupus-like syndrome	
Paraminosalicylic acid	Penicillin
Oestrogens	Griseofulvin
Gold salts	Reserpine
	Tetracycline
Drugs activating SLE by a dose-related effect in slow acetylators, with antinuclear antibody formation:[5]	
Hydralazine	
Procainamide	
Isoniazid	
Anticonvulsants e.g. diphenylhydantoin	
Chlorpromazine	
Some drugs activate SLE by an allergic, non-dose-related effect irrespective of acetylator phenotype, with little antinuclear antibody formation:	
Aminosalicylic acid	Penicillin
D-penicillamine	Phenylbutazone
Griseofulvin	Practolol
L-dopa	Propylthiouracil
Methyldopa	Quinidine
Methysergide	Tetracycline
Oral contraceptives	Folazimide

Aetiology and pathogenesis

Much has been learnt of systemic lupus erythematosus from studies of animal models (see pp. 250 and 597).

Systemic lupus erythematosus can be said to be a syndrome. It is a heterogeneous disorder with an inconstant phenotypic expression. It seems possible that a number of different causes may take effect against a variable genetic background, to result in a multifacetted disease state. Although it was thought that SLE was primarily a vascular disorder, detailed studies of the histological lesions led to the suggestion that the illness was a systemic, mesenchymal anomaly,[70] probably a diffuse disease of collagenous connective tissue.[177] As Woods and Zvaifler[309] point out, any putative cause must explain the persistence of SLE, the associated immunological abnormalities, and the mechanisms of tissue injury.

Virus

Since the first demonstration by transmission electron microscopy of virus-like inclusions in tissue cells in SLE,[120] it seemed increasingly probable that SLE would be shown to be of infective origin. Patients frequently have raised antibody titres against measles and rubella viruses and early electron micrographs displayed structures resembling paramyxovirus, the family of single-stranded (ss) RNA viruses to which the measles virus belongs. However, sera from SLE patients normally also contain antibodies against Epstein–Barr (EB) and other DNA viruses, as well as antibodies against a number of RNA viruses including mumps and poliovirus. These findings may represent a broad, underlying, altered immune reactivity to various exogenous agents and their significance would be naturally enhanced if the virus(es) could be isolated from SLE tissue. However, there has been little success in attempts to recover virus directly although a type C oncornavirus has been retrieved, an agent of the category implicated in dog SLE (see p. 598) as well as in many neoplasms. Viral antigens have been demonstrated in human SLE tissue even if they have not been recovered from them. Methods such as immunofluorescence have been employed in attempted viral identification. Thus, the leucocytes of some SLE patients contain a C-type RNA virus antigen and SLE kidney may have similar deposits although retroviruses of the human T-cell leukaemia family are not detectable.[34] A gp-70-like virus antigen has been described on peripheral SLE lymphocytes but the presence of this agent, which resembles baboon endogenous virus, has not been confirmed. There is no evidence that human T lymphocyte virus (HTLV) or human immunodeficiency virus (HIV-1) retroviruses participate in the pathogenesis of SLE but antibodies against the p17 and p24 core proteins of HIV-1 have been found in a significant proportion of patients many of whom may have circulating retroviral mRNA.

Heredity

Whereas SLE is found in about one person in 2×10^3 female members of the population, in relatives of those with the disease the prevalence is about 1 in 200. Abnormalities of the immune mechanism, such as the formation of antinuclear antibody, are also much more common in relatives of those with SLE than in the general population. The problem remains that, in human SLE as in animal disorders such as murine NZB/NZW disease (see p. 600), an increased familial incidence may simply reflect the vertical, e.g. transplacental, passage of an infectious agent such as a C-type oncornavirus.

Additional evidence for the role of heredity in SLE comes from twin studies: approximately 50 per cent of monozygotic twins are concordant. There is also a high frequency of SLE in certain races. On average, SLE is three times more common in Blacks than in Caucasians. The evidence associating the inheritance of major histocompatibility complex antigens with SLE is discussed by Walport, Black and Batchelor[298] (see Chapter 5). The haplotype A1, B8, DR3 is implicated as is DR2.

Some light has been shed on susceptibility to SLE by those forms of the disorder associated with a inherited deficiency of the 'early' components of complement such as C2 (see p. 244), and with the administration of drugs such as hydralazine and procainamide (see p. 594). Heritable C2 deficiency, a circumstance contributing to immune complex disease,[265] is associated with a tendency to develop both systemic and discoid lupus erythematosus. However, the observation that acquired, as opposed to inherited, C2 deficiency also predisposes to SLE, suggests that the common link is not the inheritance but any significant complement-mediated impairment of host response to pathogenic microorganisms. In the case of drug-related SLE, the common factor is the inheritance of a tendency for 'slow' activity by the enzyme N-acetyl transferase which is responsible for conjugating and detoxicating drugs such as hydralazine. This form of enzyme inheritance is a marker for disease predisposition. Additional genetic factors outside the MHC of somatic chromosome 6 may also be important in disease susceptibility. One example is the expression of the erythrocyte C3b receptor that may lead to impaired handling of circulating immune complexes.[306]

Sex

Systemic lupus erythematosus is a disease of mature females: it is much less common in the young and old. The role of oestrogens in permitting the disease, supported by evidence from murine NZB/NZW LE in which androgens have been shown to be protective (see p. 603) is further substantiated by the association of SLE with Klinefelter's syndrome. Zvaifler[312] suggests that oestrogens may increase anti-DNA antibody synthesis, cause a switch to the formation of nephrotoxic IgG rather than IgM and diminish immune complex clearance.

Environment

The relatively high frequency of SLE among Black populations cannot be explained by factors such as malnutrition, exposure to sunlight, crowding and socioeconomic status which are common to analogous Caucasian populations. However, the role of environmental agents of an infective nature is suggested by the demonstration that dogs in households with cases of SLE may develop comparable antibody changes (see p. 598). The transmission of a causal microbial agent could, of course, be from man to dog or vice versa. It has also been observed that staff handling SLE sera form more antinuclear, anti-dsDNA and lymphocytotoxic antibody than control persons.

Immunopathology

Dominating the disorders of SLE and intimately associated with its pathogenesis are abnormalities of the immune mechanism.[254] The immunopathology of SLE is further discussed in Chapter 5. Although patients with SLE respond normally to exogenous antigen, characteristically they form many autoantibodies and display defective T-cell functions. Explanations are therefore sought for anomalous B- and T-cell behaviour: the defects are closely related to genetic predisposition and may be provoked by virus.

Abnormal humoral immunity

In SLE, antibodies are acquired passively[42] or formed inappropriately against nuclear, cytoplasmic, nucleolar and other antigens. Antinuclear antibodies include anti-dsDNA,[122] antihistone and antistructural (non-histone) proteins (Table 14.7). There is enhanced binding of sera from patients with SLE to the left-handed Z-DNA form of polynucleotides.[287,288] Anticytoplasmic antibodies include antibodies to common antigens such as rRNP, ssRNA and transfer RNA (tRNA). Antibodies are also formed against tissue-specific antigens such as the microsomes of gastric parietal, thyroid epithelial and adrenal cortical cells, and lupus anticoagulant (LAC) (see p. 572) has attracted growing interest.

One possibility is that autoantibodies are formed when tolerance fails because clones of autoreactive cells arise on account of somatic mutation. A cause of somatic mutation could be viral infection. Such an infection cannot offer the whole explanation for this phenomenon which is clearly related to genetic predisposition to the disease; nor can it

Table 14.7

Non organ-specific antibodies in SLE[213]

Antibody	Association	Frequency (per cent)
Antinuclear antibodies		
Anti-dsDNA	Highly specific for SLE	90
Anti-ssDNA	Non-specific: found in other diseases	60
Anti-nRNP	Low titre in SLE. High titre suggests mixed connective tissue disease (p. 709)	40
Anti-Sm	Commoner in Black races: may reflect disease activity	25
Anti-La (SSB)	Sjögrens syndrome	15
Anti-SL	Fever and lymphadenopathy	8
Anti-PCNA	No specific associations	3
Anti-DNA-histone	In drug-induced SLE, 95 per cent	50
Anticytoplasmic antibodies		
Anti-Ro (SSA)	Lupus erythematosus negative for antinuclear antibody and Sjögren's syndrome	40

dsDNA = double stranded DNA; ssDNA = single stranded DNA; nRNP = nuclear ribonucleoprotein; Sm = an acidic glycoprotein; La(SSB) = a ribonucleoprotein; anti-SL = anti-systemic sclerosis/systemic lupus erythematosus antibody; PCNA = proliferating cell nuclear Ag; Ro(SSA) = a small cytoplasmic ribonuclear protein present in liver, spleen, kidney, lung and lymph node.

explain why antibodies are formed only to certain 'self' tissue antigens. Other explanations include the escape of a small proportion of the lymphocytes formed daily from the mechanisms that regulate diurnal immune reactivity, either because of a fault in T_S-cell function or because of a deficiency of the anti-idiotypic antibodies that are directed against T-cell receptor sites (see p. 250). A further possibility is the activation, later in life, of maternal lymphoid cells that have reached the fetus early in development but remained dormant. Reacting in adult life to agents such as virus, these cells could then provoke a reaction similar to a graft-versus-host response. An additional possibility is a change in macrophage cell membranes of the kind known to be caused by C-type retroviruses. Tissue antigens and macromolecules released continually during the processes of daily cell renewal are processed by the mononuclear phagocyte system. Normally, there is no association with the macrophage cell surface MHC class II molecules: such an association is necessary to allow T_H-cells to respond to antigen. Viruses or drugs reactive at lipophilic membranes could cause this association. The genetic basis of such a response is in part suggested by the observation that macrophage surface MHC class II molecules are coded by MHC genes.

Human leucocytes have a receptor for DNA. The function of this receptor is the 'internalization' and degradation of exogenous DNA. Evidence that the DNA-receptor is functionally defective in more than 90 per cent of a population of patients with SLE and mixed connective tissue disease supports the concept of a virus-induced change in membrane function.[30] Alternatively, membrane alterations could be brought about by anti-leucocyte antibody.

Abnormal cell-mediated immunity

Tests of skin reactivity and of leucocyte migration inhibition suggest that depressed cell-mediated immunity (see p. 250) may also contribute to the pathogenesis of SLE.[13,144,159] There is a diminished appearance of mononuclear macrophages in experimental inflammatory human skin tests.[242] In part, cell-mediated immune hyporeactivity may be attributed to lymphocytopenia. However, there are also important qualitative changes in T-cell reactivity: many defects are known. There are fewer T-cells than normal, defective delayed hypersensitivity reactions and abnormal responses to T-cell mitogens. Of particular interest is the evidence showing impaired T_H-cell function and poor primary antibody responses. Some of the lost T-cells appear to be those capable of becoming suppressor cells. This loss of suppressor cells is the most readily detectable T-cell dysfunction in patients with active SLE but without gross lymphopenia.[69]

The abnormalities of central immune regulation were thought to involve at least two distinct classes of gene.[208] More recently, severely impaired natural killer (NK) and killer (K) cell activity have also been found, changes that could contribute to the pathogenesis of SLE by disturbing mechanisms responsible for the elimination of virus-infected cells.[88] The lowered numbers of NK cells, analysed by means of a monoclonal antibody, HNK-1, suggested that many of the NK cells with the HNK-1 marker are abnormal and functionally immature.[83] There is certainly an intrinsic defect of T-cell function.[150] A cause of the altered T4+/T8+ T-cell ratios (see p. 250) in SLE could be the activity of anti-T-cell antibody.[214] Since interferon (IFN)-γ and interleukin (IL)-2 take part in the expression of the cytotoxic activity of lymphocytes, it is of interest that IL-2 but not interferon-γ can potentiate or fully restore the deficient cytotoxic effector function of peripheral mononuclear cells.[290]

In summary, the present evidence indicates that SLE is dominated by B-cell hyperactivity. The cause of this phenomenon remains obscure; there are genetic and environmental factors. Critical but unsolved questions are whether the autoantibodies in SLE arise from normally 'prohibited' B-cells by abnormal stimulation; because of an intrinsic defect in B stem-cells; or by aberrant immunoregulation. The concomitant T-cell defects appear to be secondary to anti-lymphocyte antibody production. Complement deficiencies are also important in the genesis of SLE though the mechanisms of susceptibility to the disorder have still to be elucidated. The syndrome of drug-related SLE (see p. 594) is an example of the pronounced influence of chemical and pharmacological agents on the induction of SLE-related immune disturbance in genetically predisposed individuals.[189]

Systemic lupus erythematosus-like disease in animals

Much has been learnt of the pathogenesis of human SLE from animal models. Immunological aspects of these models are discussed in Chapter 5. Diseases that resemble human SLE more or less closely arise spontaneously or can be induced in monkeys, dogs, rabbits and mice;[149,275] those in the New Zealand mouse are of particular importance. Attempts have been made to reproduce drug-induced SLE in dogs and rats (see p. 594) but the disorders caused in this way have attracted less interest than those arising spontaneously on a genetic basis (Fig. 14.16).

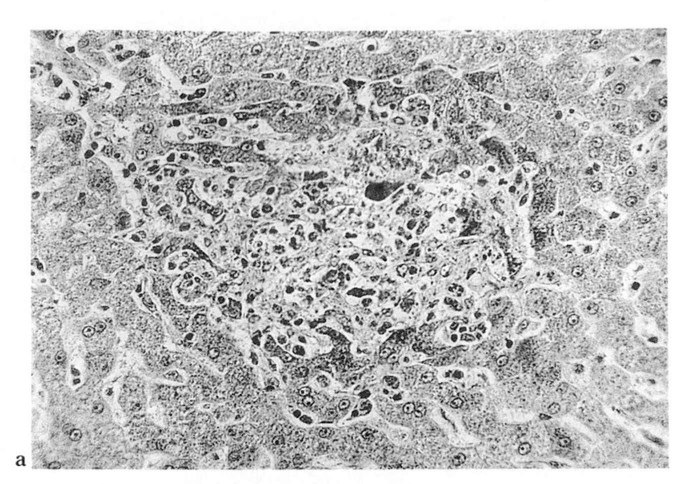
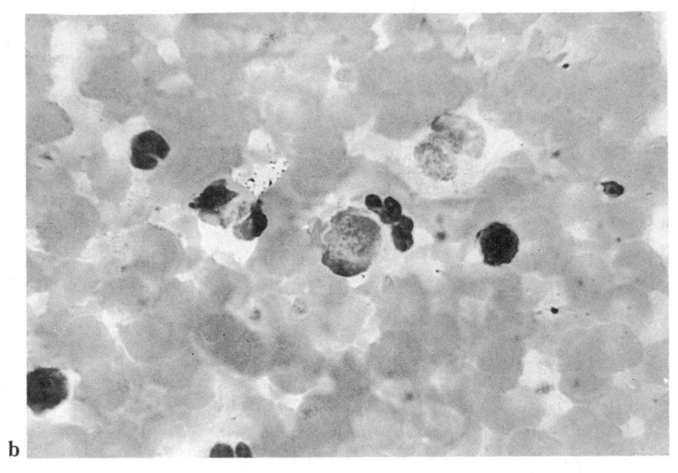

Fig. 14.16 Experimental hydralazine syndrome.
a. Focal hepatocyte necrosis following the administration of large doses of hydralazine to a normotensive dog for six months. (HE × 140.)
b. In a lupus erythematosus cell test (see Fig. 14.2), polymorphs have phagocytosed nuclear material which, however, retains identifiable microstructure and shows the changes of the artefactual so-called 'Tart' cell. (Giemsa × 580.)

Non-human primates

Few searches for naturally occurring primate systemic lupus erythematosus have been made. However, some female cynomologous macaques (*Macaca fascicularis*) from Indonesia, fed with a diet containing alfalfa sprouts, occasionally became anaemic, formed antinuclear antibody and/or anti-dsDNA antibody and displayed low serum complement levels.[206] Alfalfa seeds cause similar effects. A main component of alfalfa seeds and sprouts is L-canavanine sulphate. This agent may exert a direct toxic effect on monkey tissues. However, there appears to be considerable variability in response and an inherited predisposition to alfalfa toxicity has been proposed. In theory, L-canavanine could substitute for arginine in histone and impair genome function or be incorporated with protein in reactions catalysed by arginyl-transfer RNA synthetase. Either proposal could contribute to the formation of the autoantibodies identified in this strange syndrome. L-canaline, formed from L-canavanine by arginase, could act as an antimetabolite, inactivating enzymes that require vitamin B_6 as co-factor.

Simian virus

That virus may provoke human SLE has been suggested by the demonstration of viral antigen, related to baboon endogenous virus, on the circulating blood lymphocytes of normal individuals and of patients with SLE. However, this observation could not be validated by Hogg and Zvaifler (quoted by Zvaifler[312]) when they used serum, carefully adsorbed for non-viral tissue antigen, to examine a control clinical series.

Dog

A number of dog breeds develop autoimmune disorders with characteristics in common with SLE: poodles, cocker spaniels, a fox terrier, German Shepherd dogs,[145] a wire-haired terrier and a mongrel were found to be affected.[192] An association has been shown with the inheritance of the MHC class I antigen, DLA-A7.[282] The animals had autoimmune haemolytic anaemia, thrombocytopenic purpura and a membranous glomerulonephritis. Antinuclear antibodies were formed and lupus erythematosus cell tests were positive. When antinuclear antibody-positive and -negative dogs were housed together, the serological abnormalities were shown to be transmissible. Cell-free extracts of spleens from dogs with this lupus-like syndrome could convey serological abnormalities to other species.[192] The possibility that C-type oncornaviruses are implicated in the canine lupus syndrome, a suggestion coming from the recognition of virus-like structures in transmission electron microscopic sections, led to epidemiological enquiries. It seemed possible that dogs with SLE-like disease could be a threat to human households. However, an investigation of 83 members of 23 households exposed to 19 dogs with high titre antinuclear antibodies (ANA) revealed no differences in the levels of ANA, anti-DNA, anti-RNA, antilymphocyte antibody, rheumatoid factor or immunoglobulin compared with 50 members of 18 control households matched for dog age, sex and veterinary observer.[252]

Mouse

Considerable insight into the nature of human SLE has come from investigations of analogous disorders in mice.[12,26,109,217,285] The enquiries that began these experiments can be traced to the emergence of the New Zealand black (NZB) mouse, selected for breeding and found to be prone to an autoimmune haemolytic anaemia.[137,138,139] A considerable number of strains of mice susceptible to SLE-like disease is now known (Table 14.8). The disorders which they develop, and the analogous human abnormalities, are indicated in Table 14.9. It is clear that there is no single responsible autoimmune gene.

New Zealand black (NZB) mice

Male and female animals are likely to develop autoimmune haemolytic anaemia. Signs of anaemia, which is accompanied by reticulocytosis and hepatosplenomegaly, begin at approximately 9 months of age and by 12 months almost all animals have red blood cell-bound antibody, detectable in a direct Coombs test. Anti-ss- and -dsDNA antibodies are also recognized, together with antibodies for other nuclear epitopes and natural thymocytotoxic antibody (NTA). There are circulating immune complexes.

In addition to haemolytic anaemia, two further characteristics dominate the life-cycle of this disease-ridden strain of mouse. There is hyperplasia of the thymic medulla and of germinal centres in the spleen and lymph node cortices. Widespread plasmocytosis of the splenic medulla and lymph nodes is characteristic. Lymphoma (or pseudolymphoma) develops in 10–20 per cent of animals[217] but immunosuppression by compounds such as azathioprine increases the incidence of thymic lymphoma. Renal disease also commonly develops. A membranous glomerulonephritis leads to proteinuria. Immune complexes are located beneath glomerular capillary basement membranes. Proliferative features emerge; azotaemia develops and the animals die in renal failure.

The immunological abnormalities underlying NZB-disease centre on B-cell hyperactivity. There is an accelerated maturation of B-cell bone marrow precursors. Ly-1 positive B-cells are responsible for the polyclonal secretion of IgM autoantibodies. The role of thymic lymphocytes in causing or permitting B-cell hyperactivity is suspected. There are demonstrable abnormalities in a wide range of T-cell functions including: anomalous responses to interleukin-2; the atypical generation of T_S-cells; a susceptibility to the induction of tolerance; and abnormalities in reactions to mitogens and in the autologous mixed lymphocyte reaction. That there is an abnormality in the secretion of the

Table 14.8

Mouse models of systemic lupus erythematosus[109,217]

Strain of mouse	Haplotype	Sex with most severe disease	Mean time to death (months)	Pathological features	Autoantibodies	T_H-cell dependence	B-cell abnormality	Accelerating factor
NZB	$H\text{-}2^d$	M & F	15.5 (Male) 14 (Female)	Haemolytic anaemia Membranous glomerulonephritis Lymphoid hyperplasia Lymphoma	Anti-red blood cell Excess IgM production Natural thymocytotoxic	+	Yes	Environmental
(NZBx NZW)F_1	$H\text{-}2^{d/2}$	F	15 (Male) 9 (Female)	Membranoproliferative glomerulonephritis Sjögren's syndrome	Anti-native DNA Anti-gp 70	++	Yes	Oestrogen
MRL/MP-*lpr*/*lpr*	$H\text{-}2^k$	F	6 (Male) 5 (Female)	Membranoproliferative glomerulonephritis Lymphadenopathy Rheumatoid-like arthritis Vasculitis	Anti-native DNA Anti-Sm* Rheumatoid factor Anti-gp 70	++++	No	Autosomal recessive *lpr* gene
BXSB	$H\text{-}2^b$	M	5 (Male) 20 (Female)	Proliferative glomerulonephritis Degenerative coronary artery disease	Anti-native DNA Anti-red blood cell	++	Yes	Y chromosome (not hormonal)
Moth-eaten	$H\text{-}2^b$	M & F	1	Alopecia (hair loss) Glomerulonephritis Infection Immunosuppression	Anti-DNA Anti-red blood cells Natural thymocytotoxic			
Palmerston-North	$H\text{-}2^q$	F	1.1	Polyarteritis Immune complex nephritis	Anti-DNA			
Swan	$H\text{-}2^k$	M & F	18	Glomerulonephritis (mild)	Anti-DNA			

* An extractable nuclear antigen (ENA) test measures antibodies to various non-DNA and nucleic acids.[217] The ENA test is of value in the diagnosis of mixed connective tissue disease (see p. 709). The antigens are sensitive to RNase and trypsin: they are indeed ribonucleoprotein. Sm antigen (named after the initial patient, Smith), is closely related but is resistant to RNase and trypsin.

Table 14.9

Analogies between human and murine systemic lupus erythematosus[275]

Human disorder	Mouse strain	Mouse disorder
Familial incidence Concordance in identical twins	NZB	Inheritance of autoimmune disease
Discordance between non-identical twins		Influence of environmental factors
Female preponderance HLA-linkage	(NZB × NZW)F_1	Exacerbation by oestrogen H-2 linkage
Rare male forms of SLE	BXSB	Male sex-linked inheritance
B-cell hyperactivity	NZB BXSB MRL/l	B-cell hyperactivity
Relatives of patients with SLE have signs of autoimmunity without overt disease	Congenic + Recombinant inbred lines	Many autoimmune traits inherited independently
Dominant inheritance of antinuclear and anti-ssDNA antibodies	Backcross offspring	Involvement of dominant, recessive and modifying genes

autoanti-idiotypic antibodies regulating anti-red blood cell antibody production has been proposed.

New Zealand black/white mice

The F_1 progeny of NZB and NZW mice develop a more severe form of autoimmune disease. Females are much more susceptible than males. An immune complex membranoproliferative glomerulonephritis begins to cause clinical signs within 6 months and culminates in death in females by 8 to 10 months of age, and in males in the second year and often by 15 months.

Pathology The renal disorders of NZB/W mice have been said to provide a good model of human SLE glomerulonephritis; they dominate the pathology of this mouse disorder although it is of extreme interest that NZB/W mice also sustain lymphocytic infiltration of the lacrimal and salivary glands, providing an analogy with Sjögren's syndrome (see p. 529). The renal abnormalities accompany the increased formation of antinuclear antibody. DNA–antiDNA immune complexes and complement are deposited both in the mesangium and in relation to glomerular capillary basement membranes. The quantities of deposited complexes increase, collagenization of glomeruli begins and ultimately culminates in renal excretory failure with azotaemia.

Immune mechanisms The immunological abnormalities of NZB/W mice invoke humoral- and cell-mediated disorder.[286]

There is hypergammaglobulinaemia. There may be an underlying xenotropic (Table 14.10, p. 603) type C retrovirus infection. High titres of antibodies form against nuclear antigens, particularly ss- and dsDNA, RNA and synthetic polynucleotides. The great range of antibodies formed is explicable by the presence of cross-reacting idiotypes (see p. 235). It is of importance that antibodies to gp70 (the glycoprotein envelope of the endogenous type C retrovirus (see p. 595) are also formed. Initially, antibodies are IgM. With increasing age, IgG comes to be the principal anti-DNA isotype—a change accounting for the enhanced glomerular inflammatory response. Circulating immune complexes appear, and include gp70. Natural thymocytotoxic antibody is also demonstrable and contributes to the disordered humoral immunity. Although B-cell hyperactivity underlies these responses, these cells exert additional effects independent of T-cell regulation. The effects include high levels of spontaneous IgM production, even in neonatal NZB and NZB/W animals.

There are many abnormalities of T-cell function.[9] The importance of T-cell behaviour in the pathogenesis of NZB/W mouse disease has been emphasized by the demonstration that treatment with a monoclonal anti-L3T4 antibody, specific for T_H-cells, prevents the onset of autoimmunity.[308]

Early thymic atrophy contributes to a deficiency of both thymic hormones and cytokines which regulate the maturation and behaviour of T-cells. Thus, there may be diminished T_S-cell function. There is a reduced formation of interleukin-2 and a diminished lymphocyte responsiveness to this cytokine. Whether the impaired capacity of macrophages to clear circulating immune complexes is intrinsic and influenced by interleukin-1, for example, or whether it is attributable to 'reticuloendothelial blockade' is unclear. In NZB/W mice, natural killer (NK) cells also display impaired functions and are less able than usual to kill chosen targets.

Virus The recognition of virus-like structures in renal (see p. 588) and other tissues in human SLE, and the identification of type C RNA viruses in the tissues of NZB mice, drew attention to the possibility that virus could be the cause of NZB/W disease (Table 14.9). The xenotropic type C retrovirus is integrated into the mouse genome. Nevertheless, autoantibodies to these viruses are detected in autoimmune mice. The importance of these antibodies, in the context of NZB/W disease, lies in their role in the formation of immune complexes that may contribute to renal disease. However, retrovirus infection is not a whole explanation for the autoimmune state: the gene coding for high levels of virus expression is transmissible to SWR mice. The hybrid animal remains wholly free of evidence of autoimmune disease in spite of prolific virus production.

MRL mice

The MRL/MP-*lpr*/*lpr* (MRL/*lpr*) mouse[12] rapidly develops an autoimmune syndrome with immune complex glomerulonephritis, lymphoproliferation ('*lpr*') and, occasionally (10–20 per cent), a rheumatoid-arthritis-like articular disorder.[236] Necrotizing arteritis, alopecia and skin lesions associated with vasculitis,[8] are additional features which determine that this mouse strain provides a valuable model of at least two human connective tissue diseases. A further strain, the MRL/MP-$^{++}$, develops a less severe autoimmune syndrome without the lymphoproliferation of the MRL/*lpr* mouse.

Pathology The accelerated membranoproliferative glomerulonephritis of MRL/*lpr* mice is associated with anti-DNA antibody formation. The lymphoproliferation is indexed by great enlargement of the axillary and cervical lymph nodes and by splenomegaly. The arteries that are sites of necrotizing lesions, display fibrinoid change (see p. 572) and are infiltrated by polymorphs. Vessels in the kidneys, heart and gonads are susceptible. The articular disease of MRL/*lpr* mice is unique: although only a small proportion of animals show overt signs of arthritis, more than 75 per cent have ultrastructural evidence of rheumatoid-like joint lesions.[125] There is a significant correlation between the development of erosive arthritis and serum IgM rheumatoid factor titres.[125]

Immune mechanisms The immunological features of MRL/*lpr* disease centre on the rapid expression of autoantibody formation, hypergammaglobulinaemia, circulating immune complex formation, thymocytotoxic antibody formation and thymic involution. The autoantibodies include anti-ss and anti-dsDNA. However, anti-Sm* is also formed. The capacity of T-cell subsets to exercise immunoregulation is seriously impaired. The lymphocytes identified in the enlarged nodes are weakly staining Ly-1 cells. They are insensitive to suppressor signals from the Ly-123 cell that regulates feedback inhibition. There is defective interleukin-2 production and the mouse T-cells are unresponsive to endogenous interleukin-2.[9,217] Many macrophages are activated, as indicated by their expression of MHC class II HLA antigens. This characteristic, a feature of rheumatoid arthritis (see p. 227), indicates an abnormally high rate of antigen presentation.

Genetics The genetics of MRL/*lpr* mice are simpler than those of the NZB mouse. The haplocyte (H-2^k) determines that cell transfer can be undertaken: the *lpr* gene is a single locus autosomal recessive gene that acts only as an accelerating factor. When the *lpr* gene is transferred to mice with no potential for autoimmune disease, anti-DNA antibody and rheumatoid factor are formed, lymphoproliferation develops and the animals die prematurely. MRL disease is T-cell dependent. T-cell-derived differentiation factors (cytokines) are formed by MRL lymphocytes; these factors are believed to promote the activation and maturation of adjacent B-cells, catalysing IgG autoantibody secretion.

BXSB mice

This recombinant inbred strain was derived from crossing C57B1/6J females with SB/le males.[222] The male BXSB mice manifest signs of autoimmunity early and die within five to seven months. The females survive much longer with a low-grade autoimmune disorder.

Pathology The principal pathological features of BXSB disease are a Coombs-positive haemolytic anaemia; a rapidly progressive membranoproliferative glomerulonephritis

* An extractable nuclear antigen (ENA) test measures antibodies to various non-DNA proteins and nucleic acids.[217] The ENA test is of value in the diagnosis of mixed connective tissue disease (see p. 709). The antigens are sensitive to RNase and trypsin: they are indeed ribonucleoprotein. Sm antigen (named after the initial patient, Smith), is closely related but is resistant to RNase and trypsin.

with immune complex deposition; and a 'degenerative' arterial disease, particularly of the coronary arteries. The coronary artery disease in (NZW × BXSB)F_1 males is constant and accounts for death by 5 months. The vascular changes have been attributed to low circulating levels of immune complexes. They contrast with the necrotizing polyarteritis of MRL/*lpr* mice.

Immune mechanisms Underlying BXSB disease is a Y-chromosome-linked gene *not* influenced by male hormones. The as yet unidentified gene linked to the BXSB chromosome has been designated Yaa (Y chromosome-linked autoimmune acceleration).[152a] Among the antibodies formed are: rheumatoid factor, natural thymocytotoxic antibody, anti-DNA, and antired blood cells. There are circulating immune complexes, hypocomplementaemia but only low levels of anti-gp 70 (see p. 600). Thymic atrophy occurs and the abnormal B-cell function of these animals is exacerbated by thymectomy. T_H-cells exercise a role and anti-T-cell antibodies delay the onset of the SLE-like syndrome. The available evidence suggests that pathogenesis is, however, principally related to an inherited abnormality of B-cell function. There is defective tolerance induction at the B-cell level.[124]

Moth-eaten mice

These animals are the consequence of a mutation in the C57B/6J strain. They develop an autoimmune, SLE-like disorder. Anti-DNA, natural thymocytotrophic and anti-red blood cell antibodies are formed. Immune complexes appear in the circulation and glomerulonephritis develops. The life-span is very brief since the mice are immunodeficient and highly susceptible to infection, restricting their value in experimental studies. The regulation of B-cell function is defective.

Palmerston North (PN) mice

The original observation in this genetically distinct strain was of polyarteritis. Continued inbreeding has revealed an autoimmune, SLE-like condition. Female mice are more severely affected than males.

Pathology An immune complex glomerulonephritis develops. Glomerular abnormalities appear very early. Within one month of birth, thickened glomerular basement membranes and segmental glomerular hypercellularity are described. Fibrinoid change and crescent formation are seen. Renal deposits of IgG, IgM and complement component C3 are recognized. They become worse and renal failure occurs. Evidence of arteritis is detected in the kidneys but vascular disease is not confined to these organs and changes are found in vessels in the lymph nodes, thymus, ovaries and spleen. Thymic hyperplasia is succeeded by atrophy but lymph nodes are enlarged throughout life and malignant neoplasms, particularly lymphoma, are recognized in 14 per cent of animals.

Immune mechanisms The immunological abnormalities are expressed early. There is hypergammaglobulinaemia. Anti-DNA antibodies may be found at birth and, by two months, nearly two-thirds of female animals display DNA-binding antibody. Anti-red blood cell and anti-thymocyte antibodies are found. The enhanced B-cell activity is likely to be related to defective T-cell function. There are many splenic lymphocytes with reduced levels of Ly-1 antigen: they display a weak responsiveness to H-2 compatible cells *in vivo*.

Virus The PN strain does not express murine leukaemia virus-related gp-70, infectious ecotropic or xenotropic type C retroviruses.

Swan mice (Swiss antinuclear) mice

Antinuclear antibodies form and result in a mild glomerulonephritis manifest at approximately ten months of age. Anti-DNA antibodies appear and thymic atrophy occurs but the SLE-like disease that results is slow to develop and of mild character.

Pathogenesis of mouse systemic lupus erythematosus

Heredity

Intensive enquiry has failed to reveal a common genetic denominator for the three main forms of mouse disease. The *H2* genes are dissimilar. There is no single, shared genotype and cross-mating does not increase the disease frequency. It is of interest that a haplotype such as H2d may be shared by mice with no signs of autoimmunity. In every case, however, the inheritance of the immunodeficiency *xid* gene prevents or retards the SLE-like disease. One hypothesis therefore places the origin of the mouse syndromes in genetic aberrations of early B-cell function, manifest early in life by one of a number of possible predisposing genes. In the mouse as in man, the stem-cell defect, which may be in B-cells or in T-cell regulation of B-cell responses, or both,[275] sets the scene for abnormally frequent and excessive antibody reactions to a wide variety of endogenous or 'self' antigens to which, normally, there is immunological tolerance. Interestingly, the humoral responses of the immune system to exogenous antigen, such as those of bacteria and viruses, remain normal. The propensity for abnormal antibody to form against self-antigen is accelerated by oestrogens but slowed by androgens; it is also inhibited by diets restricted in phospholipids.[92]

Virus

Particular interest has concentrated on the finding that, throughout life, there are high concentrations of type C virus-like (ecotropic) particles in the tissues of affected mice together with demonstrable murine viral antigens (p. 600). Other, xenotropic viruses are also present. The glomerular lesions include antibody to virus-related antigen, especially to the glycoprotein envelope of gp-70 retrovirus (see p. 600) and there is a suspicion that this agent is of particular importance. Nevertheless, just as there is no constant, demonstrable genetic anomaly, so no single viral agent or combination of viruses can explain wholly the lupus-like tissue responses in mouse disease. Dixon's[74] evidence demonstrated disparate associations (Table 14.10) and drew attention to the observation that non-autoimmune murine strains such as NZW, AKR and BALB/C also carry ecotropic and xenotropic viral agents. There is no consistent relationship between the development of murine autoimmune lupus-like disease and the demonstrable presence of any single viral agent or combination of viruses.[312]

Table 14.10

Viruses in murine lupus erythematosus-like disease

Mouse disease	Viruses* present
NZB/NZW disease	Ecotropic; xenotropic
MRL *lpr/lpr* disease	Ecotropic; no xenotropic
BXSB disease	No ecotropic; occasional xenotropic

*Ecotropic viruses can only infect and replicate in the cells of the species in which they exist.

Xenotropic viruses cannot infect cells of their parent species but multiply in heterologous cells.

Complications and prognosis

The severity of SLE appears to have decreased spontaneously during the years since the introduction of cortisone. However, acute disease and sudden death still occur. An increased incidence of renal disease, arthropathy and neuropsychiatric disorder is evident but spectacular advances in the control of the renal abnormalities have changed the prognosis and the mean survival time has lengthened.[255] Nevertheless, treatment may, itself, cause iatrogenic disorders.[178] Occasionally, the hazards of treatment exceed those of the disease:[86] prolonged bed rest introduces the dangers of deep-vein thrombosis (see p. 582) and osteopenia. Improperly regulated blood transfusion may transmit hepatitis B or C or human immunodeficiency viruses. Infection remains a notorious hazard of SLE whether immunosuppression has been used or not. Bacterial,[274] mycotic[300] and viral disease are the largest categories. Pregnancy can exacerbate active SLE and oral contraceptives may have the same effect. Surgical operations should be avoided[255] although SLE can itself provoke lesions, such as peritonitis, which demand intervention. The adverse effects of antimalarial therapy, and the complications of treatment with aspirin, non-steroidal antiinflammatory drugs, corticosteroids and immunosuppressive agents are discussed by Lanham and Hughes[188] and by Hughes,[146] respectively.

In spite of a remarkable improvement in survival, SLE still has a high mortality rate.[256] Twenty per cent of cases die within 10 years of diagnosis and the mean age at death is 34 years. The most common causes of death are the systemic lesions of SLE itself, particularly of the kidney, and infection. In a 10-year period these categories accounted for 31 and 33 per cent of deaths, respectively; more than half the fatal systemic lesions were renal. When haemodialysis has been used in treating renal disease, deaths attributable to renal disorder become uncommon although the length of survival is not influenced.

Effective treatment with prednisone or with a prednisone/azathioprine combination, for example, preserves renal function, and in one series gave a five-year survival rate of 82 per cent. The second main cause of death, as in the West Indian study reported by Harris *et al.*[130] is infection. When cyclosphosphamide is used, systemic lupus erythematosus undergoes significant amelioration and the number of T_3(CD3)-, T_4(CD4)-, T_8(CD8)- and B-cells declines progressively. The B-cell changes appear transient.[199]

Less common causes of death include central nervous system disease, especially 'seizures'; pulmonary, cardiovascular, haematological and gastrointestinal lesions; incidental cerebrovascular accidents and pulmonary embolism; gastrointestinal bleeding; and myocardial infarction. Some of these disorders are iatrogenic; many are infective. Together, they contribute 17 per cent of fatalities in SLE. A further 9 per cent of individuals die from diseases unrelated to SLE and 13 per cent from unknown causes.[256] There is no evidence of a predisposition to neoplasia although instances of lymphoma are recorded (see p. 592). Among them is Burkitts' lymphoma.[251] Amyloidosis A is rare.[148,284] Active SLE may persist in spite of treatment or may reappear. Under these circumstances, disorders such as myocardial infarction, thought to be infrequent, emerge as important factors contributing to death,[263] and vascular disease, especially coronary artery atherosclerosis, assumes unexpected significance.

REFERENCES

1. Adelman D C, Saltiel E, Klinenberg J R. The neuropsychiatric manifestations of systemic lupus erythematosus: an overview. *Seminars in Arthritis and Rheumatism* 1986; **15**: 185–99.
2. Adu D, Williams D G, Taube D, Vilches A R, Turner D R, Cameron J S, Ogg C S. Late onset systemic lupus erythematosus and lupus-like disease in patients with apparent idiopathic glomerulonephritis. *Quarterly Journal of Medicine* 1983; **52**: 471–88.
3. Agudelo C A, Schumacher H R, Glick J H, Molina J. Non-Hodgkin's lymphoma in SLE: report of 4 cases with ultrastructural studies in 2. *Journal of Rheumatology* 1981; **8**: 69–78.
4. Aitchison J D, Wynn Williams A. Pulmonary changes in disseminated lupus erythematosus. *Annals of the Rheumatic Diseases* 1956; **15**: 26–32.
5. Alarçon-Segovia D. Drug-induced systemic lupus erythematosus and related syndromes. *Clinics in Rheumatic Diseases* 1975; **1**: 573–82.
6. Alarçon-Segovia D, Ibanez G, Velazquez-Forero F, Hernandez-Ortiz J, Gonzalez-Jimenez Y. Sjögren's syndrome in systemic lupus erythematosus. Clinical and subclinical manifestations. *Annals of Internal Medicine* 1974; **81**: 577–83.
7. Alarçon-Segovia D, Osmundson P J. Peripheral vascular syndromes associated with systemic lupus erythematosus. *Annals of Internal Medicine* 1965; **62**: 907–19.
8. Alexander E L, Moyer C, Travlos G S, R J B, Murphy E D. Two histopathologic types of inflammatory vascular disease in MRL/Mp autoimmune mice. Model for human vasculitis in connective tissue disease. *Arthritis and Rheumatism* 1985; **28**: 1146–55.
9. Altman A, Theofilopoulos A N, Weiner R, Katz D H, Dixon F J. Analysis of T-cell function in autoimmune murine strains. Defects in production and responsiveness to interleukin-2. *Journal of Experimental Medicine* 1981; **154**: 791–808.
10. Amend W J, Vincenti F, Feduska N J, Salvatierra D, Johnston W H, Jackson J, Tilney N, Garovoy M, Burwell E L. Recurrent systemic lupus erythematosus involving renal allografts. *Annals of Internal Medicine* 1981; **94**: 444–8.
11. Andonopoulos A P, Skopouli F N, Dimou G S, Drosos A A, Moutsopoulos H M. Sjögren syndrome in systemic lupus erythematosus. *Journal of Rheumatology* 1990; **17**: 201–4.
12. Andrews B A, Eisenberg R A, Theofilopoulos A N, Izui S, Wilson C B, McCohahey P J, Murphy E D, Roths J B, Dixon F J. Spontaneous murine lupus-like syndromes: clinical and immunopathological manifestations in several strains. *Journal of Experimental Medicine* 1978; **148**: 1198–215.
13. Andrianakos A A, Tsichlis P N, Merikas E G, Marketos S G, Sharp J T, Merikas G E. Cell-mediated immunity in systemic lupus erythematosus. *Clinical and Experimental Immunology* 1977; **30**: 89–96.
14. Angles-Cano E, Sultan Y, Clauvel J P. Predisposing factors to thrombosis in systemic lupus erythematosus: possible relation to endothelial cell damage. *Journal of Laboratory and Clinical Medicine* 1979; **94**: 312–23.
15. Appel G B, Silva F G, Pirani C L, Meltzer J I, Estes D. Renal involvement in systemic lupus erythematosus (SLE): a study of 56 patients emphasizing histologic classification. *Medicine* 1978; **57**: 371–410.
16. American Rheumatism Association 1971 criteria for SLE. See: Cohen A S, Reynolds W E, Franklin E C *et al.* Preliminary criteria for the classification of systemic lupus erythematosus. *Bulletin of the Rheumatic Diseases* 1971; **21**: 643–8.
17. Austin H A, Mueng L R, Joyce K M, Antonovych T A, Kullick M E, Klippel J H, Decker J L, Balow J E. Prognostic factors in lupus nephritis: Contribution of renal histologic data. *American Journal of Medicine* 1983; **75**: 382–91.
18. Aylward P E, Tonkin A M, Bune A. Cardiac tamponade in hydralazine-induced systemic lupus erythematosus. *Australian and New Zealand Journal of Medicine* 1982; **12**: 546.
19. Babini S M, Cocco J A M, Babini J C, Delasota M, Arturi A, Marcos J C. Atlantoaxial subluxation in systemic lupus erythematosus—further evidence of tendinous alterations. *Journal of Rheumatology* 1990; **17**: 173–7.
20. Baehr G, Klemperer P, Schifrin A. A diffuse disease of the peripheral circulation (usually associated with lupus erythematosus and endocarditis). *Transactions of the Association of American Physicians* 1935; **50**: 139–55.
21. Baer A N, Woosley R L, Pincus T. Further evidence for the lack of association between acetylator phenotype and systemic lupus erythematosus. *Arthritis and Rheumatism* 1986; **29**: 508–14.
22. Baggenstoss A H. Symposium on systemic lupus erythematosus: visceral lesions in disseminated lupus erythematosus. *Proceedings of the Mayo Clinic* 1952; **27**: 412–9.
23. Baldwin D S, Lowenstein J, Rothfield N F, Gallo G, McClusky R T. The clinical course of proliferative and membranous forms of lupus nephritis. *Annals of Internal Medicine* 1970; **73**: 929–42.
24. Barba L, Pawlowski I, Brentjens J R, Andres A. Diagnostic immunopathology of the kidney biopsy in rheumatic diseases. *Human Pathology* 1983; **14**: 290–304.
25. Batchelor J R, Welsh K I, Tinoco R M, Dollery C T, Hughes G R V, Bernstein R, Ryan P, Naish P F, Aber G M, Bing R F, Russell G I. Hydralazine-induced systemic lupus erythematosus: influence of HLA-DR and sex on susceptibility. *Lancet* 1980; **i**: 1107–9.
26. Bearer E, Gershwin M E, Castles J J. Lupus – insights from experimental models. *Clinical and Experimental Rheumatology* 1986; **4**: 161–8.

27. Benchetrit E, Pollack A, Flusser D, Rubinow A. Coexistence of systemic lupus erythematosus and myasthenia gravis —two distinct populations of anti-DNA and anti-acetylcholine receptor antibodies. *Clinical and Experimental Rheumatology* 1990; **8**: 71–4.

28. Benner E J, Gourley R T, Cooper R A, Benson J A Jr. Chronic active hepatitis with lupus nephritis. *Annals of Internal Medicine* 1968; **68**: 405–13.

29. Bennett R M. Mixed connective tissue disease and other overlap syndromes. In: Kelley W N, Harris E D, Ruddy S, Sledge C B, eds, *Textbook of Rheumatology* (3rd edition). Philadelphia, London: W B Saunders & Co., 1989: 1147–65.

30. Bennett R M, Peller J S, Merritt M M. Defective DNA receptor function in systemic lupus erythematosus and related diseases: evidence for an autoantibody influencing cell physiology. *Lancet* 1986; **i**: 186–8.

31. Benotti J R, Sataline L R, Sloss L J, Cohn L H. Aortic and mitral insufficiency complicating fulminant systemic lupus erythematosus. *Chest* 1984; **86**: 140–3.

32. Bidani A K, Roberts J L. Schwartz M M, Lewis E J. Immunopathology of cardiac lesions in fatal systemic lupus erythematosus. *American Journal of Medicine* 1980; **69**: 849–58.

33. Biesecker G, Katz S, Koffler D. Renal localization of the membrane attack complex in systemic lupus erythematosus nephritis. *Journal of Experimental Medicine* 1981; **154**: 1779–94.

34. Boumpas D T, Popovic M, Mann D L, Balow J E, Tsokos G C. Type-C retroviruses of the human T-cell leukaemia family are not evident in patients with systemic lupus erythematosus. *Arthritis and Rheumatism* 1986; **29**: 185–8.

35. Boyce N W, Holdsworth S R, Thomson N M, Atkins R C. Renal involvement in systemic lupus erythematosus. *Medical Journal of Australia* 1984; **140**: 775–9.

36. Brown C H, Scoulon P J, Haserick J R. Mesenteric arteritis with perforation of the jejunum in a patient with systemic lupus erythematosus. Report of a case. *Cleveland Clinic Quarterly* 1964; **31**: 169–78.

37. Brown M M, Yount W J. Skin immunopathology in systemic lupus erythematosus. *Journal of the American Medical Association* 1980; **243**: 38–42.

38. Browning C A, Bishop R L, Heilpern J, Singh J B, Spodick D H. Accelerated constrictive pericarditis in procainamide-induced systemic lupus erythematosus. *American Journal of Cardiology* 1984; **53**: 376–7.

39. Bulkley B H, Roberts W C. The heart in systemic lupus erythematosus and the changes induced in it by corticosteroid therapy. A study of 36 necropsy patients. *American Journal of Medicine* 1975; **58**: 243–64.

40. Bunning R D, Laureno R, Barth W F. Florid central nervous system vasculitis in a fatal case of SLE. *Journal of Rheumatology* 1982; **9**: 735–8.

41. Burnham T K, Neblett T R, Fine G. Application of fluorescent antibody technic to the investigation of lupus erythematosus and various dermatoses. *Journal of Investigative Dermatology* 1963; **41**: 451–6.

42. Buyon J P. Neonatal lupus and congenital complete heart block—manifestations of passively acquired autoimmunity. *Clinical and Experimental Rheumatology* 1989; **7**: S199–S203.

43. Byron M A. The clotting defect in SLE. *Clinics in Rheumatic Diseases* 1982; **8**: 137–51.

44. Bywaters E G L. Collagen disease. *Transactions of the Medical Society of London* 1959; **75**: 149–52.

45. Calabrese L H, Bach J F, Currie T, Bidt D, Clough J, Krakauer R S. Development of systemic lupus erythematosus after thymectomy for myasthenia gravis—studies of suppressor cell function. *Archives of Internal Medicine* 1981; **141**: 253–5.

46. Cameron H A, Ramsay L E. The lupus syndrome induced by hydralazine: a common complication with low-dose treatment. *British Medical Journal* 1984; **289**: 410–2.

47. Cameron J S, Turner D R, Ogg C S, Williams D G, Lessof M H, Chantler C, Leibowitz S. Systemic lupus with nephritis: a long-term study. *Quarterly Journal of Medicine* 1979; **189**: 1–24.

48. Camussi G, Tetta C, Coda R, Benveniste J. Release of platelet-activating factor in human pathology. I. Evidence for the occurrence of basophil degranulation and release of platelet-activating factor in systemic lupus erythematosus. *Laboratory Investigation* 1981; **44**: 241–51.

49. Cannat A, Seligmann M. Induction by isoniazid and hydralazine of antinuclear factors in mice. *Clinical and Experimental Immunology* 1968; **3**: 99–105.

50. Caperton E M, Bean S F, Dick F R. Immunofluorescent skin test in systemic lupus erythematosus. *Journal of the American Medical Association* 1972; **222**: 935–7.

51. Catto M. Ischaemia of bone. *Journal of Clinical Pathology* 1977; **30**, suppl. 11 (Royal College of Pathologists): 78–93.

52. Cazenave P L A. Des principales formes du lupus et de son traitement. *Gazette des Hôpitaux Civils et Militaires* 3me séries 1850; **2**: 383.

53. Cecere F A, Yoshinoya S, Pope R M. Fatal thrombotic thrombocytopenic purpura in a patient with systemic lupus erythematosus: relationship to circulating immune complexes. *Arthritis and Rheumatism* 1981; **24**: 550–3.

54. Churg A, Franklin W, Chan K L, Kopp E, Carrington C B. Pulmonary hemorrhage and immune-complex deposition in the lung—complications in a patient with systemic lupus erythematosus. *Archives of Pathology and Laboratory Medicine* 1980; **104**: 388–91.

55. Cines D B, Lyss A P, Reeber M, Bina M, Dehoratius R J. Presence of complement-fixing antiendothelial cell antibodies in systemic lupus erythematosus. *Journal of Clinical Investigation* 1984; **73**: 611–25.

56. Claudy A L, Touraine J L, Schmitt D, Viac J, Moreau X. Thymoma and lupus erythematosus. *Thymus* 1983; **5**: 209–22.

57. Comens P. Experimental hydralazine disease and its similarity to disseminated lupus erythematosus. *Journal of Laboratory and Clinical Medicine* 1956; **47**: 444–54.

58. Comerford F R, Cohen A S. The nephropathy of systemic lupus erythematosus. An assessment by clinical, light- and electron microscopic criteria. *Medicine* 1967; **46**: 425–73.

59. Cooper N S. The role of histopathology. In: Cohen A S, ed, *Laboratory Diagnostic Procedures in the Rheumatic Diseases*. Boston: Little, Brown & Co., 1967: 266–303.

60. Coplon N S, Diskin C J, Petersen J, Swenson R S. The long-term clinical course of systemic lupus erythematosus in

end-stage renal disorder. *New England Journal of Medicine* 1983; **308**: 186–90.

61. Cormane R H. 'Bound' globulin in the skin of patients with chronic discoid lupus erythematosus and systemic lupus erythematosus. *Lancet* 1964; **ii**: 534–5.

62. Crozier I G, Li E, Milne M J, Nicholls M G. Cardiac involvement in systemic lupus erythematosus detected by echocardiography. *American Journal of Cardiology* 1990; **65**: 1145–8.

63. Cruickshank B. Lesions of joints and tendon sheaths in systemic lupus erythematosus. *Annals of the Rheumatic Diseases* 1959; **18**: 111–9.

64. Cruickshank B. The basic pattern of tissue damage and pathology of systemic lupus erythematosus. In: Dubois E L, ed, *Lupus Erythematosus*. Los Angeles: University of Southern California Press, 1965.

65. Cruickshank B. The basic pattern of tissue damage and pathology of systemic lupus erythematosus. In: Dubois E L, ed, *Lupus Erythematosus* (2nd edition). Los Angeles: University of Southern California Press, 1976: 12–71.

66. Cruickshank B. The basic pattern of tissue damage and pathology of systemic lupus erythematosus. In: Dubois E L, ed, *Lupus Erythematosus* (3rd edition). Philadelphia: Lea and Febiger, 1987: 53–104.

67. Cruickshank B, Hill A G S. Histochemical identification of connective-tissue antigen in rat. *Journal of Pathology and Bacteriology* 1953; **66**: 283–9.

68. Darlington L G. Osteonecrosis at multiple sites in a patient with systemic lupus erythematosus. *Annals of the Rheumatic Diseases* 1985; **44**: 65–6.

69. Decker J L, Steinberg A D, Reinertsen J L, Plotz P H, Balow J E. Klippe J H. Systemic lupus erythematosus—evolving concepts. *Annals of Internal Medicine* 1979; **91**: 587–604.

70. Denzer B S, Blumenthal S. Acute lupus erythematosus disseminatus. *American Journal of Diseases of Children* 1937; **53**: 525–40.

71. Devoe L D, Loy G L, Spargo B H. Renal histology and pregnancy performance in systemic lupus erythematosus. *Clinical and Experimental Hypertension*—Part B: Hypertension in Pregnancy 1983; **2**: 325–40.

72. De Wolf F, Carreras L O, Moerman P. Decidual vasculopathy and extensive placental infarction in a patient with thromboembolic accidents, recurrent fetal loss and a lupus anticoagulant. *American Journal of Obstetrics and Gynecology* 1982; **142**: 829–34.

73. Dillon A M, Stein H B, English R A. Splenic atrophy in systemic lupus erythematosus. *Annals of Internal Medicine* 1982; **96**: 40–3.

74. Dixon F J. Rous-Whipple Lecture: The pathogenesis of murine systemic lupus erythematosus. *American Journal of Pathology* 1979; **97**: 9–16.

75. Donadio J V, Holley K E, Wagoner R D, Ferguson R H, McDuffie F C. Treatment of lupus nephritis with prednisone and combined prednisone and azathioprine. *Annals of Internal Medicine* 1972; **77**: 829–35.

76. Dubois E L. *Lupus Erythematosus*. Los Angeles: University of Southern California Press, 1965.

77. Dubois E L, Arterberry J D. Etiology of discoid and systemic lupus erythematosus. In: Dubois E L, ed, *Lupus Erythematosus* (2nd edition). Los Angeles: University of Southern California Press, 1976: 90–123.

78. Dubois E L, Cozen L. Avascular (aseptic) bone necrosis associated with systemic lupus erythematosus. *Journal of the American Medical Association* 1960; **174**: 966–71.

79. Dunnill M S. *Pathological Basis of Renal Disease* (2nd edition). London, Philadelphia: Bailliere, Tindall, 1984.

80. Dunnill M S. Pleural mesothelioma (editorial). *European Journal of Respiratory Diseases* 1984; **65**: 159–61.

81. Dustan H P. Taylor R D, Corcoran A C, Page I H. Rheumatic and febrile syndrome during prolonged hydralazine treatment. *Journal of the American Medical Association* 1954; **154**: 23–9.

82. Eagen J W, Memoli V A, Roberts J L, Matthew G R, Schwartz M M, Lewis E J. Pulmonary hemorrhage in systemic lupus erythematosus. *Medicine* 1978; **57**: 545–60.

83. Egan M L, Mendelsohn S L, Abo T, Balch C M. Natural killer cells in systemic lupus erythematosus—abnormal numbers and functional immaturity of HNK + cells. *Arthritis and Rheumatism* 1983; **26**: 623–9.

84. Eisenberg H, Dubois E L, Sherwin R P, Balchum O J. Diffuse interstitial lung disease in systemic lupus erythematosus. *Annals of Internal Medicine* 1973; **79**: 37–45.

85. Elias M, Eldor A. Thromboembolism in patients with the 'lupus'-type circulating anticoagulant. *Archives of Internal Medicine* 1984; **144**: 510–5.

86. Elliott R W, Essenhigh D M, Morley A R. Cyclophosphamide treatment of systemic lupus erythematosus: risk of bladder cancer exceeds benefit. *British Medical Journal* 1982; **284**: 1160–1.

87. Ellis S G, Verity M A. Central nervous system involvement in systemic lupus erythematosus: a review of neuropathologic findings in 57 cases, 1955–57. *Seminars in Arthritis and Rheumatism* 1979; **8**: 212–21.

88. Ewan P W, Barrett H M, Pusey C D. Defective natural killer (NK) and killer (K) cell function in systemic lupus erythematosus. *Journal of Clinical and Laboratory Immunology* 1983; **10**: 71–6.

89. Faith G C, Trump B F. The glomerular capillary wall in human kidney disease, acute glomerulonephritis, systemic lupus erythematosus and preeclampsia-eclampsia and a review. *Laboratory Investigation* 1966; **15**: 1682–1718.

90. Farquhar M G, Verrier R L, Good R A. An electron microscopic study of the glomerulus in nephrosis, glomerulonephritis and lupus erythematosus. *Journal of Experimental Medicine* 1957; **106**: 649–60.

91. Fellman B, Grob P J. Systemischer lupus erythematosus —Klinik und Laboratorium. *Schweizerische Medizinische Wochenschrift* 1979; **109**: 1237–50.

92. Fernandes G, Talal N. SLE: hormones and diet. *Clinical and Experimental Rheumatology* 1986; **4**: 183–5.

93. Fessel W J. Systemic lupus erythematosus in the community: incidence, prevalence, outcome and first symptoms: the high prevalence in Black women. *Archives of Internal Medicine* 1974; **134**: 1027–35.

94. Fields R A, Sibbitt W L, Toubbeh H, Bankhurst A D.

Neuropsychiatric lupus erythematosus, cerebral infarctions and anticardiolipin antibodies. *Annals of the Rheumatic Diseases* 1990; **49**: 114–7.

95. Finol H J, Montagnani S, Marquez A, Deoca I M, Muller B. Ultrastructural pathology of skeletal muscle in systemic lupus erythematosus. *Journal of Rheumatology* 1990; **17**: 210–9.

96. Fishel B, Caspi D, Eventov I, Avrahami E, Yaron M. Multiple osteonecrotic lesions in systemic lupus erythematosus. *Journal of Rheumatology* 1987; **14**: 601–4.

97. Fody E P, Netsky M G, Mrak R E. Subarachnoid spinal hemorrhage in a case of systemic lupus erythematosus. *Archives of Neurology* 1980; **37**: 173–4.

98. Fox R A, Rosahn P D. Lymph nodes in disseminated lupus erythematosus. *American Journal of Pathology* 1943; **19**: 73–99.

99. Galve E, Candell-Riera J, Pigrau C, Permanyer-Miralda G, Garcia-Del-Castillo H, Soler-Soler J. Prevalence, morphologic types, and evolution of cardiac valvular disease in systemic lupus erythematosus. *New England Journal of Medicine* 1988; **319**: 817–23.

100. Gardner D L. The response of the dog to oral L-hydrazinophthalazine (hydralazine). *British Journal of Experimental Pathology* 1957; **38**: 227–35.

101. Gardner D L. The effect of hydralazine (L-hydrazinophthalazine) on the kidneys of rats treated with cortexone. *British Journal of Experimental Pathology* 1958; **39**: 552–6.

102. Gardner D L, Duthie J J R, Macleod J, Allan W S A. Pulmonary hypertension in rheumatoid arthritis: report of a case with intimal sclerosis of the pulmonary and digital arteries. *Scottish Medical Journal* 1957; **2**: 183–8.

103. Gilliam J N, Sontheimer R D. Skin manifestations of SLE. *Clinics in Rheumatic Diseases* 1982; **8**: 207–18.

104. Ginzler A M, Fox T T. Disseminated lupus erythematosus: cutaneous manifestations of systemic disease (Libman-Sacks). Report of a case. *Archives of Internal Medicine* 1940; **65**: 26–50.

105. Gitlin D, Craig J M, Janeway J A. Studies on the nature of fibrinoid in the collagen diseases. *American Journal of Pathology* 1957; **33**: 55–77.

106. Godman G C, Deitch A D, Klemperer P. The composition of the LE and haematoxylin bodies of systemic lupus erythematosus. *American Journal of Pathology* 1958; **34**: 1–23.

107. Gohill J, Cary P D, Couppez M, Fritzler M K. Antibodies from patients with drug-induced idiopathic lupus erythematosus react with epitopes restricted to the amino and carboxyl termini of histone. *Journal of Immunology* 1985; **135**: 3116–21.

108. Goldberg M J, Husain M, Wajszczuk W, Rubenfire M. Procainamide-induced lupus erythematosus pericarditis encountered during coronary bypass surgery. *American Journal of Medicine* 1980; **69**: 159–62.

109. Goldings E A, Jasin H E. Arthritis and autoimmunity in animals. In: McCarty D J, ed, *Arthritis and Allied Conditions: A Textbook of Rheumatology* (11th edition). 1989. Philadelphia: Lea & Febiger, 1989: 465–81.

110. Goldstein G, Mackay I R. Contrasting abnormalities in the thymus in systemic lupus erythematosus and myasthenia gravis: a quantitative histological study. *Australian Journal of Experimental Biological and Medical Science* 1965; **43**: 381–90.

111. Gould D M, Daves M L. Roentgenologic findings in systemic lupus erythematosus. Analysis of 100 cases. *Journal of Chronic Diseases* 1955; **2**: 136–45.

112. Griffith G C, Vural I L. Acute and subacute disseminated lupus erythematosus: correlation of clinical and postmortem findings in eighteen cases. *Circulation* 1951; **3**: 492.

113. Grigor R, Edmonds J, Lewkonia R, Bresnihan B, Hughes G R V. Systemic lupus erythematosus. A prospective analysis. *Annals of the Rheumatic Diseases* 1978; **37**: 121–8.

114. Grishman E, Churg J. Connective tissue in systemic lupus erythematosus. *Archives of Pathology* 1971; **91**: 156–67.

115. Grishman E, Churg J. Ultrastructure of hematoxylin bodies in systemic lupus erythematosus. *Archives of Pathology and Laboratory Medicine* 1979; **103**: 573–6.

116. Gross L. The heart in atypical verucous endocarditis (Libman-Sacks). *Contributions to the Medical Sciences in Honor of Dr Emanuel Libman by his Pupils, Friends and Colleagues*. New York: The International Press, 1932; **2**: 527–50.

117. Gross L. Cardiac lesions in Libman-Sacks disease with consideration of its relationship to acute diffuse lupus erythematosus. *American Journal of Pathology* 1940; **16**: 375–407.

118. Gurian L E, Rogoff T M, Ware A J, Jordan R E, Combes B, Gilliam J N. The immunologic diagnosis of chronic active 'autoimmune' hepatitis: distinction from systemic lupus erythematosus. *Hepatology* 1985; **5**: 397–402.

119. Gyorkey F, Min K W, Gyorkey P. Myxovirus-like structures in human collagen diseases. *Arthritis and Rheumatism* 1969; **12**: 300.

120. Gyorkey F, Min K W, Sinkovics J G, Gyorkey P. Systemic lupus erythematosus and myxovirus. *New England Journal of Medicine* 1969; **280**: 333.

121. Haas J E, Yunis E J. Tubular inclusions of systemic lupus erythematosus. Ultrastructural observations regarding their possible viral nature. *Experimental and Molecular Pathology* 1970; **12**: 257–63.

122. Hahn B H. Antibodies to DNA. Epiphenomena or pathogens? In: Gupta S, Talal N, eds, *Immunology of Rheumatic Diseases*. New York: Plenum Medical Book Company, 1985: 221–35.

123. Hahn B H, Yardley J H, Stevens M B. 'Rheumatoid' nodules in systemic lupus erythematosus. *Annals of Internal Medicine* 1970; **72**: 49–58.

124. Hang L, Izui S, Theofilopoulos A N, Dixon F J. Suppression of transferred BXSB male SLE disease by female spleen cells. *Journal of Immunology* 1982; **128**: 1805–8.

125. Hang L, Theofilopoulos A N, Dixon F J. A spontaneous rheumatoid-arthritis like disease in MRL/l mice. *Journal of Experimental Medicine* 1982; **155**: 1690–701.

126. Hare W, MacKay I. Thymic size in systemic lupus erythematosus. *Archives of Internal Medicine* 1969; **124**: 60–3.

127. Hargraves M M, Richmond H, Morton R. Presentation of two bone marrow elements: the 'Tart' cell and the 'L.E.' cell. *Proceedings of the Mayo Clinic* 1948; **23**: 25–8.

128. Harris E N, Hughes G R V. Cerebral disease in systemic lupus erythematosus. *Seminars in Immunopathology* 1985; **8**: 251–66.

129. Harris E N, Hughes G R V, Gharavi A E. Anticardiolipin

antibodies and the lupus anticoagulant. *Clinical and Experimental Rheumatology* 1986; **4**: 187–90.

130. Harris E N, Williams E, Shah D J, De Ceulaer K. Mortality of Jamaican patients with systemic lupus erythematosus. *British Journal of Rheumatology* 1989; **28**: 113–7.

131. Hart R G, Miller V T, Coule B M, Bril V. Cerebral infarction associated with lupus anticoagulant—preliminary report. *Stroke* 1984; **15**: 114–8.

132. Harvey A McG, Shulman L E, Tumulty P H, Conley C L, Shoenrich E H. Systemic lupus erythematosus. *Medicine* 1954; **33**: 291–437.

133. Hasselaar P, Derksen R H W M, Blokzojl L, Hessing M, Nieuwenhuis H K, Bouma B N, Degroot P G. Risk factors for thrombosis in lupus patients. *Annals of the Rheumatic Diseases* 1989; **48**: 933–40.

134. Haupt H M, Moore G W, Hutchins G M. The lung in systemic lupus erythematosus: analysis of the pathologic changes in 120 patients. *American Journal of Medicine* 1981; **71**: 791–8.

135. Hejtmancik M R, Wright J C, Quint R, Jennings F L. The cardiovascular manifestations of systemic lupus erythematosus. *American Heart Journal* 1964; **68**: 119–30.

136. Helliwell T R, Flook D, Whitworth J, Day D W. Arteritis and venulitis in systemic lupus erythematosus resulting in massive lower intestinal haemorrhage. *Histopathology* 1985; **9**: 1103–13.

137. Helyer B J, Howie J B. Positive lupus erythematosus tests in a cross-bred strain of mice, NZB/BL-NZY/BL. *Proceedings of the University of Otago (New Zealand) Medical School* 1961; **39**: 3–4.

138. Helyer B J, Howie J B. Renal disease associated with positive lupus erythematosus tests in a cross-bred strain of mice. *Nature* 1963a; **197**: 197.

139. Helyer B J, Howie J B. Spontaneous autoimmune disease in NZB/BL mice. *British Journal of Haematology* 1963b; **9**: 119–31.

140. Hess E. Drug-related lupus. *New England Journal of Medicine* 1988; **318**: 1460–2.

141. Hill G S, Hinglais N, Tron F, Bach J F. Systemic lupus erythematosus. Morphologic correlations with immunologic and clinical data at the time of biopsy. *American Journal of Medicine* 1978; **64**: 61–79.

142. Hippocrates (460–375 BC). *Works,* with English translation, Jones W H S and Withington E T, eds. London: Heinemann, 1923–31.

143. Hoffman B I, Katz W A. The gastrointestinal manifestations of systemic lupus erythematosus: a review of the literature. *Seminars in Arthritis and Rheumatism* 1980; **9**: 237–47.

144. Horwitz D A. Cellular immunity and immunoregulation in systemic lupus erythematosus. In: McCarty D T, ed, *Arthritis and Allied Conditions: A Textbook of Rheumatology* (11th edition). Philadelphia: Lea & Febiger, 1989: 1055–67.

145. Hubert B, Teichner M, Fournel C, Monier J C. Spontaneous familial systemic lupus erythematosus in a canine breeding colony. *Journal of Comparative Pathology* 1988; **98**: 81–9.

146. Hughes G R V. The treatment of SLE: the case for conservative management. *Clinics in Rheumatic Diseases* 1982; **8**: 299–313.

147. Hughes G V R. *Connective Tissue Diseases* (3rd edition). Oxford: Blackwell Scientific Publications, 1987.

148. Huston D P, McAdam K P W J, Balow J E, Bass R, Delellis R A. Amyloidosis in systemic lupus erythematosus. *American Journal of Medicine* 1981; **70**: 320–4.

149. Huston D P, Steinberg A D. Animal models of human systemic lupus erythematosus. *Yale Journal of Biology and Medicine* 1979; **52**: 289–305.

150. Indiveri F, Scudeletti M, Pierri I, Traverso A, Cerri C, Ferrone S. PHA-T cells in systemic lupus erythematosus and in rheumatoid arthritis: abnormalities in HLA class II antigen induction and in autologous mixed lymphocyte reactions. *Cellular Immunology* 1986; **97**: 197–203.

151. Ishikura H, Yoshiki T, Yamaguchi J, Kondo N, Tateno M, Aizawa M, Itoh T. Lupus nephritis—clinicopathology and immunopathology of 80 biopsy cases. *Acta Pathologica Japonica* 1984; **34**: 1087–98.

152. Ito M, Kagiyama Y, Omura I, Hiramatsu Y, Kurata E, Kanaya S, Ito S, Fujino T, Kusaba T, Jimi S. Cardiovascular manifestations in systemic lupus erythematosus. *Japanese Circulation Journal* 1979; **43**: 985–94.

152a. Izui S. Autoimmune accelerating genes *lpr* and *Yaa,* in murine systemic lupus erythematosus. *Autoimmunity* 1990; **6**: 113–29.

153. Jadassohn J. *Handbuch der Haut und Geschlechtskrankheiten* (quoted by Harvey A M *et al.*[132]). Berlin: J Springer, 1927–37.

154. James T N, Rupe C E, Monto R W. Pathology of the cardiac conduction system in systemic lupus erythematosus. *Annals of Internal Medicine* 1965; **63**: 402–10.

155. Jao W, Lao I O, Sreekanth S, Abrahams C. Unusual intraglomerular structures in renal glomerular diseases. *Laboratory Investigation* 1980; **42**: 126.

156. Jarrett M P, Santhanam S, Greco F del. The clinical course of end-stage renal disease in systemic lupus erythematosus. *Archives of Internal Medicine* 1983; **143**: 1353–60.

157. Johnson R T, Richardson E P. The neurological manifestations of systemic lupus erythematosus: a clinicopathological study of 24 cases and a review of the literature. *Medicine* 1968; **47**: 337–69.

158. Joske R A, King W E. The 'L.E.-cell' phenomenon in active chronic viral hepatitis. *Lancet* 1955; **ii**: 477–80.

159. Kammer G M, Stein R L. T-Lymphocyte immune dysfunctions and systemic lupus erythematosus. *Journal of Laboratory and Clinical Medicine* 1990; **115**: 273–82.

160. Kant K S, Pollak V E, Weiss M A, Glueck H I, Miller M A, Hess E V. Glomerular thrombosis in systemic lupus erythematosus: prevalence and significance. *Medicine* 1981; **60**: 71–86.

161. Kantor G L, Bickel Y B, Barnett E V. Coexistence of systemic lupus erythematosus and rheumatoid arthritis. Report of a case and review of the literature with clinical, pathologic and serologic observations. *American Journal of Medicine* 1969; **47**: 433–44.

162. Kaplan A P. The intrinsic coagulation, fibrinolytic and kinin-forming pathways of man. In: Kelley W N, Harris E D, Ruddy S, Sledge C B, eds, *Textbook of Rheumatology* (2nd edition). Philadelphia: W B Saunders & Co., 1985: 95–114.

163. Kaposi M K. Neue Beiträge zur Kenntniss des Lupus erythematosus. *Archiv für Dermatologie und Syphilis* 1872; **4**: 36–78.

164. Kasai N, Parbtani A, Cameron J S, Yewdall V, Shepherd Z P, Verroust P. Platelet-aggregating immune complexes and intraplatelet serotonin in idiopathic glomerulonephritis and systemic lupus. *Clinical and Experimental Immunology* 1981; **43**: 64–72.

165. Katz S M, Korn S, Umlas S L, Dehoratius R J. Renal vascular lesions in systemic lupus erythematosus. *Annals of Clinical and Laboratory Science* 1990; **20**: 147–53.

166. Kelley W N, Harris E D, Ruddy S, Sledge C B, eds, *Textbook of Rheumatology* (3rd edition). Philadelphia, London: W B Saunders Company.

167. Kincaid-Smith P, Nicholls K. Renal thrombotic microvascular disease associated with lupus anticoagulant. *Nephron* 1990; **54**: 285–8.

168. Kinney E L, Wynn J, Ward S, Babb J D, Wineshaffer C, Zelis R. Ruptured chordae tendiniae: its association with systemic lupus erythematosus. *Archives of Pathology and Laboratory Medicine* 1980; **104**: 595–6.

169. Kinney W W, Angelillo V A. Bronchiolitis in systemic lupus erythematosus. *Chest* 1982; **82**: 646–8.

170. Klemp P, Timme A H, Sayers G M. Systemic lupus erythematosus and nodular regenerative hyperplasia of the liver. *Annals of the Rheumatic Diseases* 1986; **45**: 167–70.

171. Klemperer P. The concept of collagen diseases. *American Journal of Pathology* 1950; **26**: 505–19.

172. Klemperer P. Significance of intermediate substances of connective tissue in human disease. *Harvey Lectures* 1955; **49**: 100–23.

173. Klemperer P. Concept of connective-tissue disease. *Circulation* 1962; **25**: 869.

174. Klemperer P, Gueft B, Lee S L, Leuchtenberger C, Pollister A W. Cytochemical changes in acute lupus erythematosus. *Archives of Pathology* 1950; **49**: 503–16.

175. Klemperer P, Pollack A D, Baehr G. Pathology of disseminated lupus erythematosus. *Archives of Pathology* 1941; **32**: 569–631.

176. Klemperer P, Pollack A D, Baehr G. On the nature of acute lupus erythematosus. *New York State Journal of Medicine* 1942; **42**: 2225–6.

177. Klemperer P, Pollack A D, Baehr G. Diffuse collagen disease, acute disseminated lupus erythematosus and diffuse scleroderma. *Journal of the American Medical Association* 1942; **119**: 331–2.

178. Klippel J H. Systemic lupus erythematosus—treatment-related complications superimposed on chronic disease. *Journal of the American Medical Association* 1990; **263**: 1812–5.

179. Klippel J H, Gerber L H, Pollak L, Decker J L. Avascular necrosis in systemic lupus erythematosus—silent symmetric osteonecroses. *American Journal of Medicine* 1979; **67**: 83–7.

180. Korbet S M, Schwartz M M, Lewis E J. Immune complex deposition and coronary vasculitis in systemic lupus erythematosus. Report of two cases. *American Journal of Medicine* 1984; **77**: 141–6.

181. Krook H. Liver cirrhosis in patients with a lupus erythematosus-like syndrome. *Acta Medica Scandinavica* 1961; **169**: 713–26.

182. Kubo T, Hirose S, Aoki S, Kaji T, Kitagawa M. Canada-Cronkite syndrome associated with systemic lupus erythematosus. *Archives of Internal Medicine* 1986; **146**: 995–6.

183. Kushimoto K, Nagasawa K, Ueda A, Mayumi T, Ishii Y, Yamauchi Y, Tada Y, Tsukamoto H, Kusaba T, Niho Y. Liver abnormalities and liver membrane autoantibodies in systemic lupus erythematosus. *Annals of the Rheumatic Diseases* 1989; **48**: 946–52.

184. Laasonen L, Gripenberg M, Leskinen R, Skrifars B, Edgren J. A subset of systemic lupus erythematosus with progressive cystic bone lesions. *Annals of the Rheumatic Diseases* 1990; **49**: 118–20.

185. Labowitz R, Schumacher H R Jnr. Articular manifestations of systemic lupus erythematosus. *Annals of Internal Medicine* 1971; **74**: 911–21.

186. Ladd A T. Procainamide-induced lupus erythematosus. *New England Journal of Medicine* 1962; **267**: 1357–8.

187. Lampert P W, Oldstone M B A. Host immunoglobulin G and complement deposits in the choroid plexus during spontaneous immune complex disease. *Science* 1973; **180**: 408–10.

188. Lanham J G, Hughes G R V. Antimalarial therapy in SLE. *Clinics in Rheumatic Diseases* 1982; **8**: 279–98.

189. Lee S L, Chase P H. Drug-induced systemic lupus erythematosus. A critical review. *Seminars in Arthritis and Rheumatism* 1975; **5**: 83–103.

190. Lever W F, Schaumburg-Lever G. *Histopathology of the Skin* (6th edition). Philadelphia: J B Lippincott & Co., 1983.

191. Lewis E J, Schurr P H, Busch G J, Galvanek E, Merrill J P. Immunopathologic features of a patient with glomerulonephritis and pulmonary hemorrhage. *American Journal of Medicine* 1973; **54**: 507–13.

192. Lewis R M, Schwartz R, Henry W B Jnr. Canine systemic lupus erythematosus. *Blood* 1965; **25**: 143–60.

193. Libman E, Sacks B. A hitherto undescribed form of valvular and mural endocarditis. *Archives of Internal Medicine* 1924; **33**: 701–38.

194. Looi L M, Prathap K. Hepatitis B virus surface antigen in glomerular immune complex deposits of patients with systemic lupus erythematosus. *Histopathology* 1982; **6**: 141–8.

195. Love P E, Santoro S A. Antiphospholipid antibodies. Anticardiolipin and the lupus anticoagulant in systemic lupus erythematosus (SLE) and in non-SLE disorders. Prevalence and clinical significance. *Annals of Internal Medicine* 1990; **112**: 682–98.

196. Lowman E W, Slocumb C H. The peripheral vascular lesions of lupus erythematosus. *Annals of Internal Medicine* 1952; **36**: 1206–16.

197. Lyon M G, Bewtra C, Kenik J G, Hurley J A. Tubuloreticular inclusions in systemic lupus pneumonitis: report of a case and review of the literature. *Archives of Pathology and Laboratory Medicine* 1984; **108**: 599–600.

198. McCarty D J, ed. *Arthritis and Allied Conditions. A Textbook of Rheumatology* (11th edition). Philadelphia: Lea & Febiger, 1989.

199. McCune W J, Golbus J, Zeldes W, Bohlke P, Dunne R,

Fox D A. Clinical and immunologic effects of monthly administration of intravenous cyclophosphamide in severe systemic lupus erythematosus. *New England Journal of Medicine* 1988; **318**: 1423–31.

200. Mackay I R, Taft L I, Cowling D C. Lupoid hepatitis. *Lancet* 1956; **ii**: 1323–6.

201. Mackay I R, Wood I J. Lupoid hepatitis: a comparison of 22 cases with other types of chronic liver disease. *Quarterly Journal of Medicine NS* 1962; **31**: 485–507.

202. Mcluskey R T. The value of the renal biopsy in lupus nephritis. *Arthritis and Rheumatism* 1982; **25**: 867–75.

203. McNeil N P, Chesterman C N, Krilis S A. Anticardiolipin antibodies and lupus anticoagulants comprise separate antibody subgroups with different phospholipid binding characteristics. *British Journal of Haematology* 1989; **73**: 506–13.

204. MacSween R N, Anderson J R, Milne J A. Histological appearances of the thymus in systemic lupus erythematosus and rheumatoid arthritis. *Journal of Pathology and Bacteriology* 1967; **93**: 611–19.

205. Mahajan S K, Ordónez N G, Feitelson P J, Lim V S, Spargo B H, Katz A I. Lupus nephropathy without clinical renal involvement. *Medicine* 1977; **56**: 493–502.

206. Malinow M R, Bardana E J Jnr, Pirofsky B, Craig S, McLaughlin P. Systemic lupus erythematosus-like syndrome in monkeys fed alfalfa sprouts: role of a nonprotein amino acid. *Science* 1982; **216**: 415–6.

207. Mansilla-Tinoco R, Harland S J, Ryan P J, Bernstein R M, Dollery C T, Hughes G R V, Bulpitt C J, Morgan A, Jones J M. Hydralazine, antinuclear antibodies and the lupus syndrome. *British Medical Journal* 1982; **284**: 936–8.

208. Miller K B, Schwartz R S. Familial abnormalities of suppressor-cell function in systemic lupus erythematosus. *New England Journal of Medicine* 1979; **301**: 803–9.

209. Miller M H, Urowitz M B, Gladman D D, Blendis L M. The liver in systemic lupus erythematosus. *Quarterly Journal of Medicine* 1984; **53**: 401–9.

210. Mitsuhashi T, Hojo K, Kanayama Y. Atypical collagen disease. Polymyositis, prominent hematoxylin body formation and some manifestations of progressive systemic sclerosis. *Acta Pathologica Japonica* 1980; **30**: 1019–35.

211. Moore R D, Weisberger A S, Bowerfind E S. Histochemical studies of lymph nodes in disseminated lupus erythematosus. *Archives of Pathology* 1956; **62**: 472–8.

212. Moore R D, Weisberger A S, Bowerfind E S Jr. An evaluation of lymphadenopathy in systemic disease. *Archives of Internal Medicine* 1957; **99**: 751–9.

213. Morgan S H, Hughes G R V. Connective tissue disorders. *Medicine International* 1984; **2**: 397–408.

214. Morimoto C, Reinherz E L, Distaso J A, Steinberg A D, Schlossman S F. Relationship between systemic lupus erythematosus T cell subsets, anti-T cell antibodies and T-cell functions. *Journal of Clinical Investigation* 1984; **73**: 689–700.

215. Morita M, Sakaguchi H. Ultrastructure of renal glomerular hematoxylin bodies. *Ultrastructural Pathology* 1984; **7**: 13–19.

216. Morris M C, Cameron J S, Chantler C, Turner D R. Systemic lupus erythematosus with nephritis. *Archives of Disease in Childhood* 1981; **56**: 779–83.

217. Morrow J, Isenberg D. *Autoimmune Rheumatic Diseases*. Oxford: Blackwell Scientific Publications, 1987: 48–147.

218. Morrow W J W, Isenberg D A, Toddpokropek A, Parry H F, Snaith M L. Useful laboratory measurements in the management of systemic lupus erythematosus. *Quarterly Journal of Medicine* 1982; **51**: 125–38.

219. Moses S, Barland P. Laboratory criteria for a diagnosis of systemic lupus erythematosus. *Journal of the American Medical Association* 1979; **242**: 1039–43.

220. Movat H Z, More R H. The nature and origin of fibrinoid. *American Journal of Clinical Pathology* 1957; **28**: 331–53.

221. Muehrcke R C, Kark R M, Pirani C L, Pollack V E. *Lupus Nephritis*. Baltimore: Williams and Wilkins, 1957.

222. Murphy E D, Roths J B. A Y chromosome associated factor in strain BXSB producing accelerated autoimmunity and lymphoproliferation. *Arthritis and Rheumatism* 1979; **22**: 1188–94.

223. Nakamura H, Uehara H, Okada T, Kambe H, Kimura Y, Ito H, Hayashi E, Yamamoto H, Kishimoto S. Occlusion of small hepatic veins associated with systemic lupus erythematosus with the lupus anticoagulant and anti-cardiolipin antibody. *Hepato-Gastroenterology* 1989; **36**: 393–7.

224. Nomura S, Kumagai N, Kanoh T, Uchino H, Kurihara J. Pulmonary amyloidosis associated with systemic lupus erythematosus. *Arthritis and Rheumatism* 1986; **29**: 680–2.

225. Nordstrom D M, West S G, Andersen P A. Basal ganglia calcification in central nervous system lupus erythematosus. *Arthritis and Rheumatism* 1985; **28**: 1412–6.

226. Norton W L. Endothelial inclusions in active lesions of systemic lupus erythematosus. *Journal of Laboratory and Clinical Medicine* 1969; **74**: 369–79.

227. Norton W L, Hurd E R, Lewis D C, Ziff M. Evidence of microvascular injury in scleroderma and systemic lupus erythematosus: quantitative study of the microvascular bed. *Journal of Laboratory and Clinical Medicine* 1968; **71**: 919–33.

228. Nossent J C, Huysen V, Smeenk R J T, Swaak A J G. Low avidity antibodies to dsDNA as a diagnostic tool. *Annals of the Rheumatic Diseases* 1989; **48**: 748–52.

229. Olivieri I, Gemignani G, Balagi M, Pasquariello A, Gremignai G, Pasero G. Concomitant systemic lupus erythematosus and ankylosing spondylitis. *Annals of the Rheumatic Diseases* 1990; **49**: 323–4.

230. Ordonez N G, Gomez L G. The ultrastructure of glomerular haematoxylin bodies. *Journal of Pathology* 1981; **135**: 259–65.

231. O'Regan S W. HLA-B8, autoimmune polyendocrinopathy and systemic lupus erythematosus. *Canadian Medical Association Journal* 1979; **121**: 1168–9.

232. Orth R W, Weisman M H, Cohen A H, Talner L B, Nachtsheim D, Zvaifler N J. Lupus cystitis: primary bladder manifestations of systemic lupus erythematosus. *Annals of Internal Medicine* 1983; **98**: 323–6.

233. Osler W. On the visceral complications of erythema exudativum multiforme. *American Journal of Medical Science* 1895; **110**: 629–46.

234. Osler W. On the visceral manifestations of the erythema

group of skin diseases. *Transactions of the Association of American Physicians* 1903; **18**: 599–624.

235. Osler W. On the visceral manifestations of the erythema group of skin diseases. *American Journal of Medical Science* 1904; **127**: 1–23.

236. O'Sullivan F X, Fassbender H G, Gay S, Koopman W J. Etiopathogenesis of the rheumatoid arthritis-like disease in MRL/l mice. I. The histomorphologic basis of joint destruction. *Arthritis and Rheumatism* 1985; **28**: 529–36.

237. Oxenhandler R, Hart M N, Bickel J, Scearce D, Durham J, Irvin W. Pathologic features of muscle in systemic lupus erythematosus—a biopsy series with comparative clinical and immunopathologic observations. *Human Pathology* 1982; **13**: 745–57.

238. Pagel W, Treip C S. Viscerocutaneous collagenosis: study of intermediate forms of dermatomyositis, scleroderma and disseminated lupus erythematosus. *Journal of Clinical Pathology* 1955; **8**: 1–18.

239. Panem S, Ordonez N G, Katz A I, Spargo B J, Kirsten M D. Viral immune complexes in systemic lupus erythematosus. *Laboratory Investigation* 1978; **39**: 413–20.

240. Pappas S C, Malone D G, Rabin L, Hoofnagle J H, Jones E A. Hepatic veno-occlusive disease in a patient with systemic lupus erythematosus. *Arthritis and Rheumatism* 1984; **27**: 104–8.

241. Paronetto F, Deppisch L, Tuchman L R. Lupus erythematosus with fatal hemorrhage into the liver and lesions resembling those of periarteritis nodosa and malignant hypertension. *American Journal of Medicine* 1964; **36**: 948–55.

242. Passero F C, Myers P. Decreased numbers of monocytes in inflammatory exudates in SLE. *Journal of Rheumatology* 1981; **8**: 62–8.

243. Pearson C M, Yamazaki J N. Vacuolar myopathy in systemic lupus erythematosus. *American Journal of Clinical Pathology* 1958; **29**: 455–63.

244. Pepys M B, Lanham J G, De Beer F C. C-reactive protein in SLE. *Clinics in Rheumatic Diseases* 1982; **8**: 91–103.

245. Percy J S, Smyth C J. The immunofluorescent skin test in systemic lupus erythematosus. *Journal of the American Medical Association* 1969; **208**: 485–8.

246. Perry H M, Schroeder H A. Syndrome simulating collagen disease caused by hydralazine (apresoline). *Journal of the American Medical Association* 1954; **154**: 670–3.

247. Pertschuk L P, Moccia L F, Rosen Y, Lyons H, Marino C M, Rashford A A, Wollschlager C M. Acute pulmonary complications in systemic lupus erythematosus. Immunofluorescence and light microscopic study. *American Journal of Clinical Pathology* 1977; **68**: 553–7.

248. Phillips J C, Howland W J. Mesenteric arteritis in systemic lupus erythematosus. *Journal of the American Medical Association* 1968; **206**: 1569–70.

249. Pollack A D. Some observations on the pathology of systemic lupus erythematosus. *Journal of the Mount Sinai Hospital of New York* 1959; **26**: 224–40.

249a. Pollack V E, Grove W J, Kark R M, Muehrcke R C, Pirani C L, Steck I E. Systemic lupus erythematosus simulating acute surgical condition of the abdomen. *New England Journal of Medicine* 1958; **259**: 258–66.

250. Pollak V E, Pirani C L, Schwartz F D. The natural history of the renal manifestations of systemic lupus erythematosus. *Journal of Laboratory and Clinical Medicine* 1964; **63**: 537–50.

251. Posner M A, Gloster E S, Bonagura V R, Valacer D J, Ilowite N T. Burkitt's lymphoma in a patient with systemic lupus erythematosus. *Journal of Rheumatology* 1990; **17**: 380–2.

252. Reinertsen J L, Raslow R A, Klippel J H, Hurvitz A I, Lewis R M, Rothfield N F, Zvaifler N J, Steinberg A D, Decker J L. An epidemiologic study of households exposed to canine systemic lupus erythematosus. *Arthritis and Rheumatism* 1980; **23**: 546–68.

253. Revell P A. *Pathology of Bone*. Berlin: Springer Verlag, 1986: 223–8.

254. Roitt I M. *Essential Immunology* (6th edition). Oxford: Blackwell Scientific Publications, 1988: 238–53.

255. Ropes M W. *Systemic Lupus Erythematosus*. Cambridge, Massachusetts: Harvard University Press, 1976.

256. Rosner S, Ginzler E M, Diamond H S. A multicentre study of outcome in systemic lupus erythematosus: two causes of death. *Arthritis and Rheumatism* 1982; **256**: 612–7.

257. Rothfield N. Clinical features of systemic lupus erythematosus. In: Kelley W N, Harris E D, Ruddy S, Sledge C B, eds, *Textbook of Rheumatology*. Philadelphia, London: W B Saunders, 1981: 1106–32.

258. Rothfield N F. Clinical features of systemic lupus erythematosus. In: Kelley W N, Harris E D, Ruddy S, Sledge C B, eds, *Textbook of Rheumatology* (2nd edition). Philadelphia: W B Saunders & Co., 1985; 1070–97.

259. Rothfield N, Marino C. Studies of repeat skin biopsies of nonlesional skin in patients with systemic lupus erythematosus. *Arthritis and Rheumatism* 1982; **25**: 624–30.

260. Rothfield N F. Systemic lupus erythematosus. In: Katz W A, ed, *Rheumatic Diseases: Diagnosis and Management*. Philadelphia: J B Lippincott & Co., 1977; 765–76.

261. Rothfield N F. Systemic lupus erythematosus: clinical aspects and treatment. In: McCarty D J, ed, *Arthritis and Allied Conditions: A Textbook of Rheumatology* (10th edition). Philadelphia: Lea & Febiger, 1985; 927–8.

262. Rothfield N. Systemic lupus erythematosus: clinical aspects and treatment. In: McCarty D J, ed, *Arthritis and Allied Conditions: A Textbook of Rheumatology* (11th edition). Philadelphia: Lea & Febiger, 1989; 1022–48.

263. Rubin L A, Urowitz M B, Gladman D D. Mortality in systemic lupus erythematosus—the bimodal pattern revisited. *Quarterly Journal of Medicine* 1985; **55**: 87–98.

264. Runyon B A, Labrecque D R, Anuras S. The spectrum of liver disease in systemic lupus erythematosus—report of 33 histologically-proved cases and review of the literature. *American Journal of Medicine* 1980; **69**: 187–94.

265. Schifferli J A, Ng Y C, Peters D K. The role of complement and its receptor in the elimination of immune complexes. *New England Journal of Medicine* 1986; **315**: 488–95.

265a. Schur P H. Clinical features of SLE. In: Kelley W N, Harris E D, Ruddy S, Sledge C B, eds, *Textbook of Rheumatology*

(3rd edition). Philadelphia, London: W B Saunders Company, 1989: 1101–29.

266. Schwartz M M, Roberts J L, Lewis E J. Subepithelial electron-dense deposits in proliferative glomerulonephritis of systemic lupus erythematosus. *Ultrastructural Pathology* 1982; **3**: 105–18.

267. Scott J T, ed. *Copeman's Textbook of the Rheumatic Diseases* (6th edition). Edinburgh: Churchill-Livingstone, 1986.

268. Seaman W E, Plotz P H. Effect of aspirin on liver tests in patients with RA or SLE and in normal volunteers. *Arthritis and Rheumatism* 1976; **19**: 155–60.

269. Shapiro S H, Wessley Z, Lipper S. Concentric membranous bodies in hepatocytes from a patient with systemic lupus erythematosus. *Ultrastructural Pathology* 1985; **8**: 241–7.

270. Shulman H J, Christian C L. Aortic insufficiency in systemic lupus erythematosus. *Arthritis and Rheumatism* 1969; **12**: 138–45.

271. Siemsen J K, Brook J, Meister L. Lupus erythematosus and avascular bone necrosis: a clinical study of three cases and review of the literature. *Arthritis and Rheumatism* 1962; **5**: 492–501.

272. Sim E, Gill E W, Sim R B. Drugs that induce systemic lupus erythematosus inhibit complement component C4. *Lancet* 1984; **ii**: 422–4.

273. Speirs C, Fielder A H L, Chapel H, Davey N J, Batchelor J R. Complement system protein C4 and susceptibility to hydralazine-induced systemic lupus erythematosus. *Lancet* 1989; **i**: 922–4.

274. Staples P J, Gerding D N, Decker J L, Gordon R S. Incidence of infection in systemic lupus erythematosus. *Arthritis and Rheumatism* 1974; **17**: 1–10.

275. Steinberg A D, Raveche E S, Laskin C A, Smith H R, Santoro T, Miller M L, Plotz P H. Systemic lupus erythematosus: insights from animal models. *Annals of Internal Medicine* 1984; **100**: 714–27.

276. Steiner G, Chason J L. Differential diagnosis of rheumatoid arthritis by biopsy of muscle. *American Journal of Clinical Pathology* 1948; **18**: 931–9.

277. Steinman C R. Circulating DNA in systemic lupus erythematosus—association with central nervous system involvement and systemic vasculitis. *American Journal of Medicine* 1979; **67**: 429–35.

278. Stevanovic G, Cramer A D, Taylor C R, Lukes R J. Immunoblastic sarcoma in patients with systemic lupus erythematosus-like disorders. *Archives of Pathology and Laboratory Medicine* 1983; **107**: 589–92.

279. Tan E M, Cohen A S, Fries J F, Masi A T, McShane D J, Rothfield N F, Schaller J G, Talal N, Winchester R J. The 1982 revised criteria for the classification of systemic lupus erythematosus. *Arthritis and Rheumatism* 1982; **25**: 1271–7.

280. Tan E M, Kunkel H G. An immunofluorescent study of the skin lesions in systemic lupus erythematosus. *Arthritis and Rheumatism* 1966; **9**: 37–45.

281. Tanaka Y, Watanabe K, Suzuki H, Saito K, Oda S, Suzuki H, Eto S, Yamashita U. Spontaneous production of bone-resorbing lymphokines by B-cells in patients with systemic lupus erythematosus. *Journal of Clinical Immunology* 1989; **9**: 415–20.

282. Teichner M, Krumbacher K, Doxiadis I, Doxiadis G, Fournel C, Rigal D, Monier J C, Grossewilde H. Systemic lupus erythematosus in dogs. Association to the major histocompatibility complex class-I antigen DLA-A7. *Clinical Immunology and Immunopathology* 1990; **55**: 255–62.

283. Teilum G. Miliary epithelioid-cell granulomas in lupus erythematosus disseminatus. *Acta Pathologica et Microbiologica Scandinavica* 1945; **22**: 73–9.

284. Teilum G. Pathogenesis of amyloidosis in the light of recent cytochemical investigations. *Acta Rheumatologica Scandinavica* 1957; **3**: 164–8.

285. Theofilopoulos A N, Dixon F J. Murine models of systemic lupus erythematosus. *Advances in Immunology* 1985; **37**: 269–390.

286. Theofilopoulos A N, Singer P A, Kofler R, Kono D H, Duchosal M A, Balderas R S. B-cell and T-cell antigen receptor repertoires in lupus arthritis murine models. *Springer Seminars in Immunopathology* 1989; **11**: 335–68.

287. Thomas T J, Meryhew N L, Messner R P. Enhanced binding of lupus sera to the polyamine-induced left-handed Z-DNA form of polynucleotides. *Arthritis and Rheumatism* 1990; **33**: 356–65.

288. Thomas T J, Messner R P. A left-handed (z) conformation of poly(dA-dC)poly(dG-dT) induced by polyamines. *Nucleic Acids Research* 1986; **14**: 6721–33.

289. Tron F, Bach J F. Relationship between antibodies to native DNA and glomerulonephritis in systemic lupus erythematosus. *Clinical and Experimental Immunology* 1977; **28**: 426–32.

290. Tsokos G C, Smith P L, Christian C B, Lipnick R N, Balow J E, Djeu J Y. Interleukin-2 restores the depressed allogeneic cell-mediated lympholysis and natural killer cell activity in patients with systemic lupus erythematosus. *Clinical Immunology and Immunopathology* 1985; **34**: 379–86.

291. Turner-Warwick M. Connective tissue disorders and the lung. *Australian and New Zealand Journal of Medicine* 1986; **16**: 257–62.

292. Vasquez J J, Dixon F J. Immunohistochemical study of lesions in rheumatic fever, systemic lupus erythematosus and rheumatoid arthritis. *Laboratory Investigation* 1957; **6**: 205–17.

293. Vasquez J J, Dixon F J. Immunohistochemical analysis of lesions associated with fibrinoid change. *Archives of Pathology* 1958; **66**: 504–17.

294. Vermers M, Bernstein R M, Bydder G M, Steiner R E, Young I R, Hughes G V R. Nuclear magnetic resonance (NMR) imaging of the brain in systemic lupus erythematosus. *Journal of Computer Assisted Tomography* 1983; **7**: 461–7.

295. Wagner B M. Histochemical studies of fibrinoid substances and other abnormal tissue proteins. III: Proteolysis of fibrinoids. *Journal of the Mount Sinai Hospital of New York* 1957; **24**: 1323–30.

296. Wakaki K, Koizumi F, Fukase M. Vascular lesions in systemic lupus erythematosus (SLE) with pulmonary hypertension. *Acta Pathologica Japonica* 1984; **34**: 593–604.

297. Wallace D J, Dubois E L. *Dubois' Lupus Erythematosus* (3rd edition). Philadelphia: Lea & Febiger, 1987.

298. Walport M J, Black C M, Batchelor J R. The im-

munogenetics of SLE. *Clinics in Rheumatic Diseases* 1982; **8**: 3–21.

299. Waltzer W C, Zincke H, Sterioff S. Renal transplantation —its use in a patient with systemic lupus erythematosus and complete occlusion of inferior vena cava. *Archives of Surgery* 1980; **115**: 987–8.

300. Watson J I, Mandl M A J, Rose B. Disseminated histoplasmosis occurring in association with systemic lupus erythematosus. *Canadian Medical Association Journal* 1968; **99**: 958–62.

301. Weidner N, Lorentz W B. Scanning electron microscopy of the acellular glomerular and tubular basement membrane in lupus nephritis. *American Journal of Clinical Pathology* 1986; **85**: 135–45.

302. Weisbart R H, Noritake D T, Wong A L, Chan G, Kacena A, Colburn K K. A conserved anti-DNA idiotype associated with nephritis in murine and human systemic lupus erythematosus. *Journal of Immunology* 1990; **144**: 2653–8.

303. White R G, Bass B H, Williams E. Lymphadenoid goitre and the syndrome of systemic lupus erythematosus. *Lancet* 1961; **i**: 368–73.

304. Whiting-O'Keefe Q, Henke J E, Shearn M A, Hopper J, Biava C G, Epstein W V. The information content from renal biopsy in systemic lupus erythematosus—stepwise linear regression analysis. *Annals of Internal Medicine* 1982; **96**: 718–22.

305. Williams G W, Bluestein H G, Steinberg A D. Brain-reactive lymphocytotoxic antibody in the cerebrospinal fluid of patients with systemic lupus erythematosus: correlation with central nervous system involvement. *Clinical Immunology and Immunopathology* 1981; **18**: 126–32.

306. Wilson J G, Wong W W, Schur P H, Fearon D T. Mode of inheritance of decreased C3b receptors on erythrocytes of patients with systemic lupus erythmatosus. *New England Journal of Medicine* 1982; **307**: 981–86.

307. Wilson J H P, Haalebos M M P, Laméris J S, Urk H van, Esch B van. Membranous obstruction of the inferior vena cava and systemic lupus erythematosus. *Digestion* 1983; **27**: 111–15.

308. Wofsy D, Seaman W E. Successful treatment of autoimmunity in NZB/NZW F_1 mice with monoclonal antibody to L3T4. *Journal of Experimental Medicine* 1985; **161**: 378–91.

309. Woods V L, Zvaifler N J. Pathogenesis of systemic lupus erythematosus. In: Kelley W N, Harris E D, Ruddy S, Sledge C B, eds, *Textbook of Rheumatology* (3rd edition). Philadelphia, London: W B Saunders Company, 1989: 1077–100.

310. Worthington J W, Baggenstoss A H, Hargraves M M. Significance of haematoxylin bodies in the necropsy diagnosis of systemic lupus erythematosus. *American Journal of Pathology* 1959; **35**: 955–69.

311. Zizic T M, Shulman L E, Stevens M B. Colonic perforations in systemic lupus erythematosus. *Medicine* 1975; **54**: 411–26.

312. Zvaifler N J. Etiology and pathogenesis of systemic lupus erythematosus. In: Kelley W N, Harris E D, Ruddy S, Sledge C B, eds, *Textbook of Rheumatology* (1st edition). Philadelphia: W B Saunders & Co., 1981: 1079–105.

Chapter 15
Vasculitis and Vasculopathy

Introduction and classification

Many of the most dramatic forms of systemic connective tissue disease are characterized by vascular disorders (vasculopathy). Arterial lesions are common, for example, in rheumatoid arthritis, systemic lupus erythematosus, polymyositis/dermatomyositis and systemic sclerosis. However, vascular diseases also occur as syndromes sufficiently characteristic to be considered in their own right. The majority of these vascular syndromes are inflammatory. Polyarteritis nodosa (PAN) is one example. There is often clinical and pathological overlap between the idiopathic ('primary') forms of inflammatory vasculopathy and those associated with systemic connective tissue disease, leading to difficulties in nomenclature and diagnosis.

Vasculopathy is a general term for the vascular diseases: they may be inherited or acquired. Vasculitis[76] is inflammation of vessel walls, usually,[390] or often,[334] with evidence of vessel wall necrosis. There is no wholly satisfactory classification because several forms of vessel may be affected in one disease (Table 15.1), because of overlap between different diseases, because understanding of cause and pathogenesis is frequently limited, and because the diverse signs are infrequently characteristic, rarely pathognomonic.[188a]

Table 15.1

Examples of differential susceptibility to disease of parts of the vascular system

Aorta and large conducting arteries	Takayasu's syndrome Ankylosing spondylitis Syphilis Atheromatous inflammation
Distributing, musculoelastic arteries	Rheumatoid vasculitis Polyarteritis nodosa Bacterial infection Rheumatic fever
Smaller muscular arteries	Progressive systemic sclerosis Diabetes mellitus Serum hepatitis
Resistance arterioles	Progressive systemic sclerosis
Capillaries	Systemic lupus erythematosus
Venules	Henoch–Schönlein purpura Hypersensitivity vasculitis Haemorrhagic virus infections
Sinusoids	Thrombotic microangiopathy
Small veins	Schistosomiasis
Large veins	Bacterial infection Trauma Thrombophlebitis

In this chapter, the diseases confined to blood vessel walls and considered by McCluskey and Fienberg[235] to be 'primary' (idiopathic), are discussed together with those such as Wegener's granulomatosis and Kawasaki disease where the primary tissue changes extend beyond the blood vessels. The chapter also reviews other conditions such as Takayasu's vasculopathy where the disorder may be non-inflammatory. Heritable, degenerative, traumatic and neoplastic vasculopathies are largely excluded but some, such as pseudoxanthoma elasticum and homocystinuria, are considered in Chapter 9.

In an important, early approach, Zeek[410] classified vasculitis[66] as: 1. small-vessel, hypersensitivity vasculitis (angiitis); 2. visceral, muscular artery disease, periarteritis nodosa; 3. rheumatic arteritis; 4. allergic, asthma-associated granulomatous angiitis; and 5. temporal arteritis. Zeek's work was modified by Alarçon-Segovia and Brown[5] to include vasculitis secondary to the systemic connective tissue diseases[115] and a number of debatable, related syndromes. Fauci, Haynes and Katz[92] devised a widely-used, extended classification. Many authors[390] find it convenient to exclude disorders such as giant-cell (temporal) arteritis and microbial vasculitis but any comprehensive approach, for example that of McCluskey and Fienberg[235] (Table 15.2), must offer scope for the inclusion of all relevant abnormalities. The most recent and most extensive approach to classification is that of the American College of Rheumatology.[222a]

There are other pathological possibilities. The vasculopathies can be classified by: 1. the topography and morphology of the affected vessel(s); 2. the aetiological agent, if known; 3. the histological appearances of the lesion(s); and 4. the time-course of the disease (Table 15.3). A first step in classification, therefore, could be *anatomical*. In any particular example of vasculopathy, a clear definition of the site, size and nature of the affected vessel is required. There are wide regional differences in susceptibility (see p. 617). Large, distributing; medium-sized, musculoelastic; small or very small (resistance) arteries and arterioles are periodically affected. Capillaries or sinusoids may be the primary site of disease; or venules, or smaller or larger veins may be implicated. A second step is *aetiological*. An inherited defect may be known or suspected, as in the Marfan syndrome (see p. 340) or pseudoxanthoma elasticum (see p. 355). The environmental cause of the disease may, for example, be, bacterial, as in opportunistic infection caused by *Pseudomonas pyocyaneus* (see p. 734); injury due to heat or cold (see p. 954); or the lodgement in a vessel of immune complexes (see p. 238). A third step is *histological* analysis, employing light- and transmission electron microscopic techniques; immunofluorescence and/or perox-

Table 15.2

Classification of vasculitis/vasculopathy[235]

I	*Primary vasculitides of unknown cause* (in which lesions, excluding secondary, ischaemic effects, are restricted to blood vessels)
	1. Polyarteritis nodosa
	2. Hypersensitivity vasculitis
	i. Cutaneous, i.e. confined to skin
	ii. Cutaneous and visceral
	iii. Hypocomplementaemic
	3. Unclassified or (overlap) with features of polyarteritis nodosa and hypersensitivity vasculitis
	4. Henoch–Schönlein purpura
	5. Essential mixed cryoglobulinaemia
	6. Giant cell arteritis
	7. Localized poly(peri)arteritis nodosa
II	*Connective tissue diseases in which vasculitis/vasculopathy periodically occurs*
	1. Rheumatoid arthritis
	2. Systemic lupus erythematosus
	3. Polymyositis/dermatomyositis
	4. Progressive systemic sclerosis
	5. Rheumatic fever
	6. Sjögren's syndrome
III	*Diseases with known causative agents in which vasculitis sometimes occurs*
	1. Viral infections such as hepatitis B (various forms of lesion, e.g. polyarteritis nodosa)
	2. Rickettsial infection
	3. Bacterial infection (e.g. opportunistic Pseudomonas infection)
	4. Drug addiction (e.g. to amphetamine)
	5. Drug reactions: hypersensitivity vasculitis
IV	*Wegener's (pathergic) granulomatosis*
V	*Allergic granulomatosis of Churg and Strauss*
VI	*Other forms of vasculopathy*
	Traumatic
	Heat/cold injury
	Chemical injury
	Irradiation
	Hypertensive
VII	*Kawasaki disease*
VIII	*Behçet's disease*
IX	*Takayasu's disease*
X	*Thrombotic microangiopathy*

idase–antiperoxidase or immunogold methods to search for antigens or antibodies present in affected vessels; enzyme- and molecular histochemical analyses to define functional changes and, occasionally, radiographic or autoradiographic methods. A fourth step is to plot the *time course* of the disease, if possible studying the anatomical and histological features in sequence.

Polyarteritis nodosa

Polyarteritis nodosa (PAN) is an uncommon, progressive, non-suppurative inflammatory syndrome affecting visceral, muscular arteries and arterioles. Polyarteritis nodosa is not a single entity: identical forms of focal vasculitis occur in systemic lupus erythematosus, rheumatoid arthritis or, occasionally, in hepatitis B infection.[235] The necrosis of scattered, discrete vascular segments leads to thrombosis and tissue ischaemia: the resultant infarcts cause a clinical condition as varied as the distribution of the arterial lesions. Aneurysms form: they are usually small.

In the United States, the incidence of PAN was found to be $0.7/10^5$ persons, the prevalence 6.3×10^5 and the mortality rate approximately $0.2/10^5$.[62] PAN is twice as common in males as in females and often begins in early middle age, although individuals of any age may be affected and cases under the age of 1 year have been recorded[86] (see p. 651). Formerly, a fatal outcome was usual but the progress of the disease can now be greatly modified by treatment with agents such as corticosteroids or cytotoxic and immunosuppressive drugs (see p. 628).

Historical background

The first report was that of Matani;[248] the coronary arteries were affected. The case of Rokitansky[323] was re-examined[83] and was the first in which a microscopic study was made. Early interest was aroused by the finding of multiple small aneurysms. With the paper of Küssmaul and Maier,[201] it was realized that the widespread distribution of the lesions accounted for the varied clinical findings. These authors introduced the term 'periarteritis nodosa' but it was 12 years before further examples were described.[32] With time,[198] increased accuracy in diagnosis showed that the disease was less rare than had been supposed; only 40 examples were annotated by Lamb[206] but by 1938, Boyd had collected 395 reports[145] and Mowrey and Lundberg[266] described 230 cases from the Armed Forces Institute of Pathology files, all but three post-1939. The term poly-

Table 15.3

Classification of the vasculopathies

This is a proposed classification of the vasculopathies, based on topography, histology and chronology. The terms are, of course, not mutually exclusive so that, for example, acute arteritis of small muscular arteries caused by the deposition of HBs antigen-antibody immune complexes may be accompanied, in another part of the vascular system, by reparative fibrosis of focal lesions, caused by the same mechanisms, in larger, distributing arteries.

At any point in time, the vessels affected by a vasculopathy may show one or more of the histological lesions listed below, or a combination of them. The Table is intended also to emphasize how diseases of blood vessels vary simultaneously in distribution, severity, structure and sequence.

In individual patients, several or many vasculopathic processes may co-exist., In an extreme example, a patient with rheumatoid arthritis may have developed 1. skin lesions due to hypersensitivity to ampicillin, and may have 2. rheumatoid necrotizing vasculitis of the viscera, 3. arterial amyloidosis, 4. limb gangrene due to rheumatoid endarterial fibromuscular occlusion, 5. atherosclerosis and 6. opportunistic *Pseudomonas pyocyaneus* vasculitis attributable to immunosuppression by corticosteroids.

Codes may be assigned to particular cases by using the numbers beside each term in a manner similar to the SNOMED system. A particular advantage of this system is to allow systematic filing and accurate comparisons between the series of cases from different laboratories.

Topography	Aetiology	Chronology	Histology
1. Aorta; large elastic, conducting arteries	1. Heritable disease	1. Acute	1. Oedema
2. Musculoelastic, distributing arteries	2. Acquired disease	2. Subacute	2. Plasmatic vasculosis
3. Small, muscular arteries	i Physical injury	3. Chronic	3. Fibrinoid
4. Resistance arterioles	ii Trauma	(The suggested time scales are relative and different scales clearly apply in individual diseases)	4. Tissue necrosis
5. Capillaries	iii Chemical injury		5. Karyorhexis/karyolysis
6. Venules	iv Metabolic/endocrine		6. Polymorph (Infiltrates)
7. Sinusoids	v Inflammatory		a. Neutrophil (Infiltrates)
8. Small veins	vi Immunological:		b. Eosinophil (Infiltrates)
9. Large veins	a. Antibody-mediated		7. Macrophage (Infiltrates)
	b. Cell-mediated		8. Plasma cell (Infiltrates)
	vii Infective:		9. Lymphocyte (Infiltrates)
	a. Virus		10. Giant cell (Infiltrates)
	b. Rickettsia		11. Mast cell (Infiltrates)
	c. Bacteria		12. Elastolysis
	d. Protozoa		13. Endothelial proliferation
	e. Metazoa		14. Thrombosis
	f. Fungi		15. Fibrosis/scar

arteritis nodosa came to represent the nature of the disease more exactly. The suggestion that arteritis might be a complication of sulphonamide therapy[316] and the experimental demonstration that foreign protein given intravenously to rabbits could cause similar lesions,[315,317–319] greatly increased interest in the nature of the human disease.

Clinical features

Polyarteritis nodosa (PAN) is a disseminated, febrile disorder with anaemia and a raised erythrocyte sedimentation rate. The signs are those of multiple, disseminated visceral infarction or tissue ischaemia: they mimic those of other systemic diseases of connective tissues.[341] Overlap syndromes are frequent[334] and it may be this very difficulty of interpreting the clinical disorder in terms of the conventional body systems which first raises a suspicion of the true diagnosis.[94,170,320,333] The sensitivity and specificity of classification have been enhanced by the recommendations of the American College of Rheumatology,[224a] based on the 'growth' of a classification tree (Tables 15.4 and 15.5).[37b]

Abdominal complaints are common. Pain, signs of intestinal obstruction, bleeding, diarrhoea or vomiting may introduce the clinical course. In selected populations there may be underlying sickle-cell disease.[243] All parts of the gastrointestinal tract may be affected: each of the viscera can sustain infarcts. Hepatic necrosis and fibrosis, cholecystitis or appendicitis are encountered and an association with retroperitoneal fibrosis is described.[164] Cardiac failure may follow myocardial ischaemia caused by coronary arteritis

Table 15.4

1990 ACR criteria for the classification of polyarteritis nodosa (traditional format).*
Modified from Lightfoot *et al.*[224a]

Definition
1. Loss of 4 kg or more of body weight since illness began, not due to dieting or other factors
2. Mottled livedo reticular pattern over the skin of portions of the extremities or torso
3. Pain or tenderness of the testicles, not due to infection, trauma, or other causes
4. Diffuse myalgias (excluding shoulder and hip girdle) or weakness of muscles or tenderness of leg muscles
5. Development of mononeuropathy, multiple mononeuropathies, or polyneuropathy
6. Development of hypertension with the diastolic BP higher than 90 mm Hg
7. Elevation of BUN >40 mg/dl or creatinine >1.5 mg/dl, not due to dehydration or obstruction
8. Presence of hepatitis B surface antigen or antibody in serum
9. Arteriogram showing aneurysms or occlusions of the visceral arteries, not due to arteriosclerosis, fibromuscular dysplasia, or other noninflammatory causes
10. Histologic changes showing the presence of granulocytes or granulocytes and mononuclear leukocytes in the artery wall

*For classification purposes, a patient shall be said to have polyarteritis nodosa if at least three of these 10 criteria are present. The presence of any three or more criteria yields a sensitivity of 82.2 per cent and a specificity of 86.6 per cent.

Table 15.5

1990 ACR classification tree criteria for polyarteritis nodosa (PAN). Modified from Lightfoot *et al.*[224a]

PAN subsets	Number of patients PAN/ non-PAN	Percentage correctly classified	Percentage PAN patients in subset	Non-PAN subsets	Number of patients PAN/ non-PAN	Percentage correctly classified	Percentage non-PAN patients in subset
Neuropathy and weight loss >6.5 kg; negative arteriogram and artery biopsy	10/19	34	8	Negative arteriogram and no neuropathy	8/555	99	81
Positive artery biopsy and neuropathy; negative arteriogram	13/10	57	11	Neuropathy; negative arteriogram and artery biopsy; weight loss ≤6.5 kg	7/53	88	8
Positive arteriogram, male sex; normal AST and ALT	6/3	67	5	Positive arteriogram; normal AST and ALT; female sex	0/7	100	1
Positive arteriogram and abnormal AST or ALT	74/42	64	63				

The classification tree yields a sensitivity of 87.3 per cent and a specificity of 89.3 per cent.

and, in childhood, polyarteritis is one of the rare causes of myocardial infarction. In the much more frequent adult cases, the onset of angina pectoris is evidence of coronary arteritis.

Renal dysfunction may be an initial sign and renal failure remains an important cause of death (see p. 628). Albuminuria and haematuria are detected in patients with lumbar pain and fever. In other instances, raised blood pressure directs attention to the renal system. Renal biopsy can then provide the first definitive evidence of PAN. In such cases, progressive glomerular destruction with microscopic lesions[70] may induce a state resembling glomerulonephritis; alternatively, a rapid and progressive decline with azotaemia and uraemia accompanies repetitive renal infarction.

New diagnostic laboratory tests have been developed. For example, raised levels of Factor VIII-related antigen are characteristic of systemic necrotizing vasculitis of the form found in PAN, rheumatoid arthritis and Wegener's granulomatosis,[402] but not of small vessel vasculitis of the kind characteristic of cutaneous vasculitis, Henoch–Schönlein purpura and some cases of rheumatoid arthritis. The diagnosis and management of microscopic polyarteritis as well as Wegener's granuloma, can be aided by the measurement of auto-anti-polymorph cytoplasmic antibodies[228,330] and the capacity to monitor such indices may assist studies of pathogenesis.

Peripheral neuritis with predominantly sensory or motor involvement is another early sign of the disease. A disturbance of the central nervous system is indicated by single or recurrent episodes of cerebral infarction. The effects of such infarcts, which may occur in specific sites, such as the optic pathway or in the white and grey cortical masses, range from the abrupt onset of blindness or of hemiplegia to ill-defined evidence of personality change or memory defects. Without electromyography, polymyositis in PAN may be difficult to distinguish from peripheral neuritis. However, widespread, unexplained muscle disease may occasionally be caused by the direct involvement of small arteries supplying skeletal muscle.

The skin and cutaneous tissues often show evidence of vascular disease; a legion of superficially distinct cutaneous lesions has been described.[217]

Pulmonary disease in PAN causes fleeting lung shadows occurring at intervals, varying widely in size and distribution and accompanied by pleuritis and effusion. Evidence was advanced for the existence of a group of patients with severe and overriding involvement of the respiratory system.[325] In these cases, pulmonary disease, perhaps with asthma, may precede by many months the occurrence of other visceral lesions. Such cases comprise approximately 30 per cent of patients but there is a further rare category in which PAN presents with involvement of the upper respiratory tract. This upper respiratory PAN is distinguishable microscopically from Wegener's granulomatosis and from the Churg–Strauss allergic syndrome by the absence of extra- or perivascular granulomata, and by the anatomical distribution of the lesions.

Pathological changes

(Table 15.6)

The widespread nature of the vascular lesions determines the distribution of the visceral and tissue changes: they are random and unpredictable, an observation thought to be related to their pathogenesis (see p. 626). Zones of infarction are common in the kidney, heart, liver and gut (see below) but, when small arteries and arteries are mainly affected, more subtle perfusion defects become evident so that a cardiomyopathy,[29] renal excretory failure, or memory defects may reflect territorial impairment of the heart, kidney or brain, respectively. Pulmonary ischaemic lesions occur, but by contrast with Wegener's granulomatosis they represent relatively uncommon and minor parts of the syndrome. In all, few organs are exempt and lesions of the peripheral nervous system, the endocrine organs, skeletal muscle, gut and skin are among those often found pathologically.[293]

Table 15.6

Frequency of organ involvement in polyarteritis nodosa

Organ	Involvement (per cent)
Kidney	85
Heart	76
Liver	62
Gastrointestinal tract	51
Jejunum	37
Ilium	27
Mesentery	24
Colon	20
rectosigmoid	10
Duodenum	10
Gall bladder	10
Appendix	7
Muscle	39
Pancreas	35
Testis	33
Peripheral nerve	32
Central nervous system	27
Skin	20

Many cases of PAN were originally diagnosed on the basis of the macroscopic recognition of multiple, small arterial aneurysms, often on coronary, mesenteric or renal arteries. With recourse to microscopy as an adjunct to diagnosis it was realized that the majority of instances of PAN showed only microscopic changes. However, there is no evidence that cases of polyarteritis with purely microscopic lesions differ fundamentally from those which present with macroscopic changes. The microscopic changes in the different viscera in PAN are similar, and it is convenient to consider first the structure of typical arterial lesions, and second, how these lesions affect individual organs and systems.

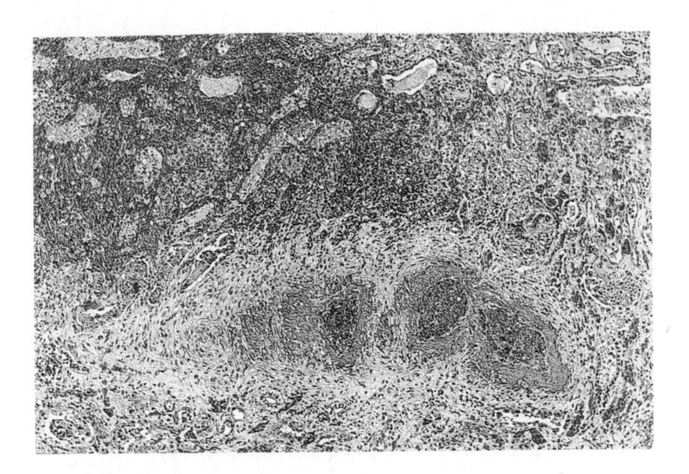

Fig. 15.1 Polyarteritis nodosa.
Right: a branch of a renal arcuate artery is cut longitudinally: it is the site of acute necrotizing inflammation. The arterial wall is infiltrated by polymorphs, macrophages and some lymphocytes, and the vascular lumen is occupied by recent thrombus. **Left**: a zone of recent infarction—the result of the arterial obstruction. (HE × 70.)

Arterial disease

There is random, focal, segmental necrosis of medium-sized, muscular, distributing visceral arteries (Figs 15.1 and 15.2). Muscular arteries are affected in many parts of

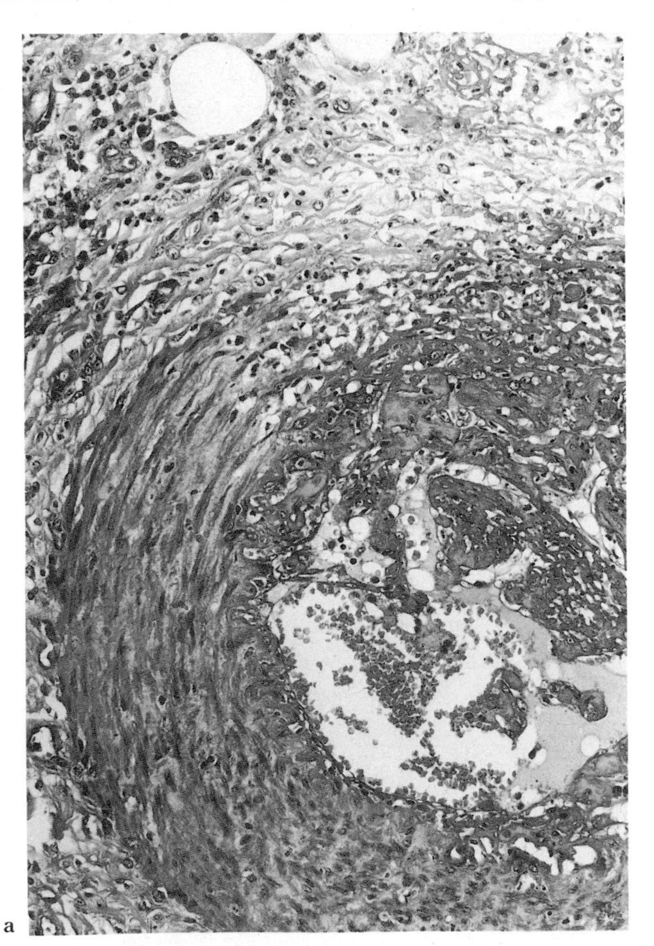

a

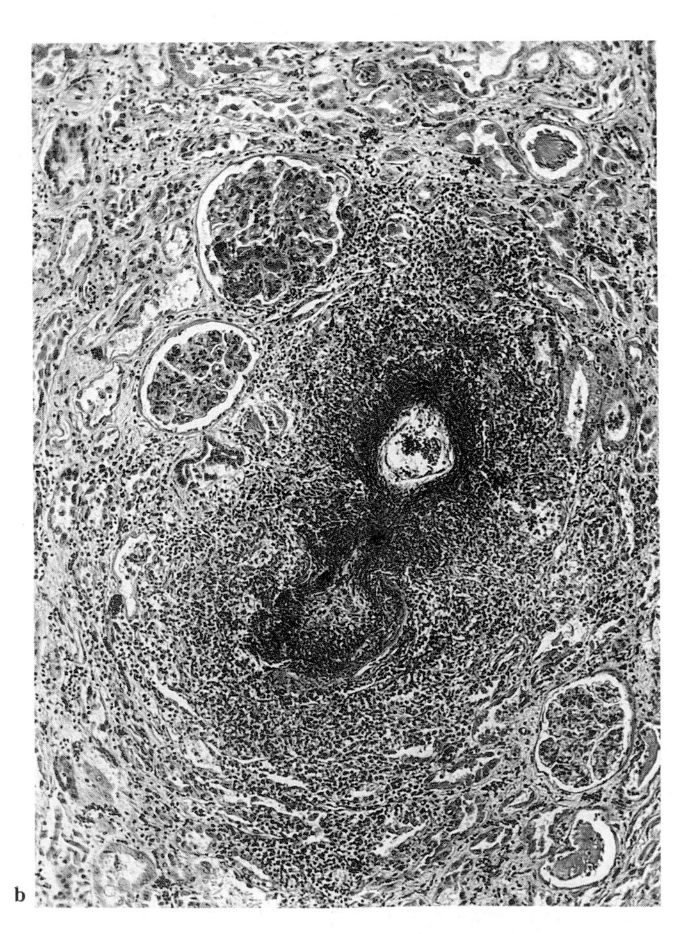

b

Fig. 15.2 Polyarteritis nodosa.
a. Segmental inflammation and cell necrosis affect the entire thickness of this medium-size, distributing artery. (HE × 285.) **b.** In the centre of the field, a renal interlobular artery is transected tangentially: the arterial wall is necrotic and shows the changes still widely described as fibrinoid. There is a vigorous surrounding inflammatory reaction and the lower part of the artery contains a recent thrombus. (HE × 60.)

the body. The lesions occur in clusters, develop in irregular sequence and, individually, evolve with time, passing, through stages of onset, progression and healing.

The earliest change is swelling and oedema of the muscular media. The appearances are well seen in rapidly frozen material, sectioned by cryostat. Artefacts may be avoided in tissue fixed by perfusion but this is seldom practicable. An eosinophilic, hyaline substance, derived from plasma proteins, accumulates in the subendothelial intima. The changes are those of plasmatic (fibrinous) vasculosis[212,213] or insudation.[326,405] Characteristically, swelling and oedema precede necrosis of the medial muscle of a chord of the vessel. The chord may include the whole of the circumference or only a part, but typically extends to include the whole thickness. Damage to smooth muscle and to the elastic laminae disrupts the vessel. The elastic laminae are destroyed when a cellular infiltrate permeates the artery, extending to the adventitia. The infiltrate includes neutrophil polymorphs and macrophages. With time, more mononuclear cells appear: they include lymphocytes and plasma cells. In a smaller proportion of cases, particularly in those with lung involvement, eosinophil polymorphs are relatively common. Cells may become very numerous and may mask the structure of the whole of one coat of the vessel. As the internal elastic laminae are disrupted, a focal loss of tensile strength in the vessel wall determines the points at which aneurysms form. Aneurysms appear simultaneously but inconstantly in a number of visceral sites or in crops, at intervals: their development heralds an advanced stage of the process. Nevertheless, severe disease is frequent in the absence of aneurysm. Disturbance of the fibrinolytic mechanism promotes fibrin and platelet deposition on the endothelial surface. Thrombosis occurs: its presence is common in the advancing disease.

With time, arterial healing begins, in different viscera at different times. The thrombus which occupies the arterial lumen is penetrated by vascular, mesenchymal tissue. Recanalization occurs. Injured zones in the media and elastic are replaced by collagen which is an inadequate substitute for the normal media (Fig. 15.3); it tends to give way in the face of the arterial pulse, with secondary aneurysm formation. In exceptional cases a perivascular granuloma develops: there is a poorly defined aggregate of macrophages and fibroblasts forming a mass which may be mistaken for an aneurysm. However, extra- and perivascular granuloma is a sign of allergic granulomatosis (see p. 639).

There are many cases of PAN with only microscopic lesions. Platt and Davson[308] made a study of the renal changes in 14; there were two categories according to whether or not severe and widespread glomerular lesions were present. In the former, the kidneys were large, pale but without architectural disturbance. There was patchy

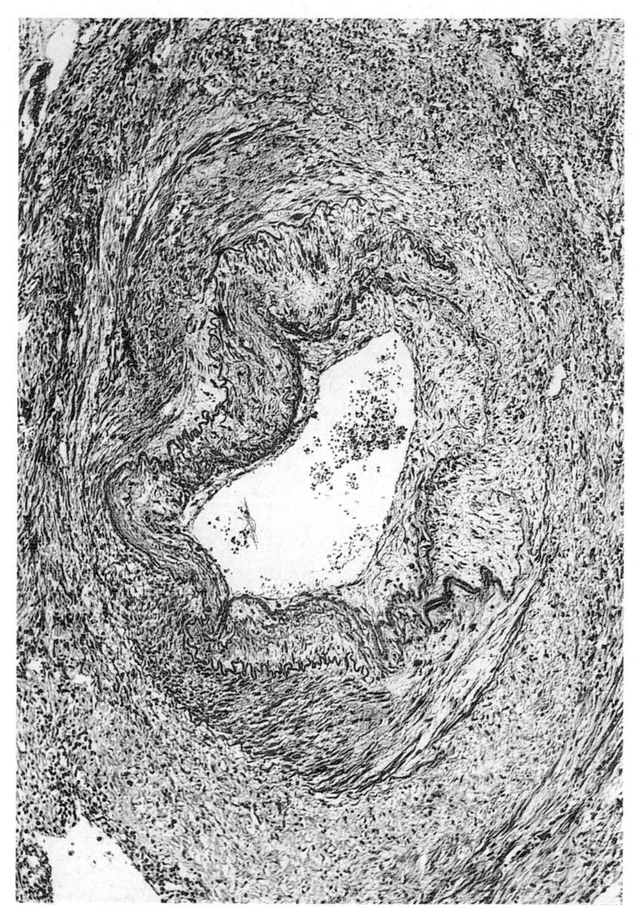

Fig. 15.3 Healed polyarteritis nodosa.
One segment of this small muscular artery has been severely damaged by earlier necrotizing vasculitis. The internal elastic lamina has been interrupted (at right) and here the vascular wall is largely replaced by organized fibrous tissue. It is at such sites that aneurysm formation is most probable. (Weigert–van Gieson × 80.)[147]

fibrinoid in glomerular tufts with occasional epithelial crescent formation and tuft fibrosis. Inflammatory cells surrounded the involved glomeruli and were present in the interstitial tissue. In a minority of the cases the renal arterial branches were affected. These cases of microscopic renal polyarteritis nodosa could be distinguished from focal embolic nephritis and from accelerated (malignant) hypertension.[70] Acute, proliferative and membranous glomerulonephritis were excluded on the basis of the histological characteristics.

Organ changes

Urinary system

Classical arteritis with aneurysm[201] is seen less often than microscopic segmental necrotizing glomerulitis[70] or focal, proliferative glomerulonephritis.[323] There are no clinical differences between those in whom the disease is confined to the glomerulus and those with histologically proven arteritis in other renal sites.[337] The most serious renal manifestation is rapidly progressive glomerulonephritis.[273]

The classical disease, of larger renal arteries and their branches, is often identified by the presence of single, or more usually by multiple, zones of infarction (Fig. 15.4); diagnostic help in identifying infarcts can be given by computed tomographic scans which can identify the wedge-shaped areas of decreased density.[309] When the disease has been present for some weeks, healed and recent infarcts may be identified side-by-side. Again, when the disease has been long-lasting and slowly progressive or has been controlled by the administration of corticosteroids or cytotoxic drugs, numerous small cortical scars are found and the variegated granular appearance of the kidney surface may resemble that of chronic glomerulonephritis, chronic pyelonephritis or diabetic or amyloid nephrosclerosis. An absolute distinction between these conditions is not possible simply on the basis of a gross examination. Occasionally, renal arterial thrombosis or aneurysm may identify PAN but aneurysms commonly become so large that they are recognizable by arteriography; they may develop quickly.[99]

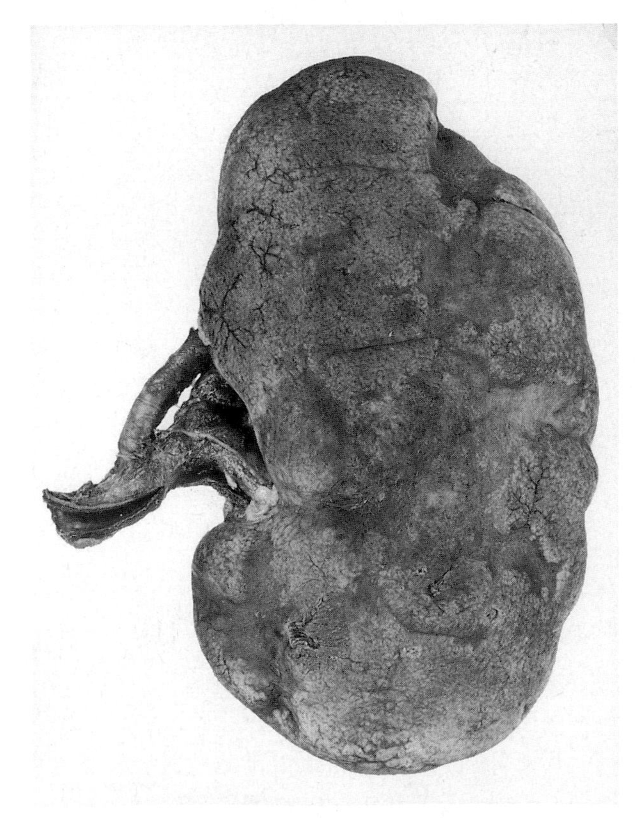

Fig. 15.4 Kidney: polyarteritis nodosa.
Impaired blood flow through branches of the renal artery has, over a period of months, led to a series of infarcts shown as zones of cortical atrophy. In the absence of recognizable aneurysm, the appearances are indistinguishable from those of other disorders such as multiple embolic infarction.

The microscopic disease[70] takes the form of glomerulonephritis (Fig. 15.5). Rarely, an association with Goodpasture's syndrome suggests the role of antibasement membrane antibody.[199] There is often crescent formation. Although necrosis and disintegration of the capsular epithelium may be seen, the glomerular tuft displays little cellular proliferation. A mononuclear but not polymorph cell infiltrate is occasional. Fibrin is demonstrable but immunofluorescence tests for immunoglobulin and immune complexes yield inconstant results and there are no dense electron microscopic deposits. Ultimately, collagenization with glomerular sclerosis, tubular atrophy and interstitial fibrosis come to be the signs of 'end-stage' renal failure which may, rarely, be brought about by extensive segmental infarction.

The microscopic renal changes in PAN may be simulated by those of accelerated hypertension, progressive systemic sclerosis (see Chapter 17), thrombotic thrombocytopenic microangiopathy (see p. 660) or the syndrome of haemolytic uraemia. Clinical evidence distinguishes the analogous lesions found in postpartum renal failure. At renal biopsy, arteritis is detected in only 34 per cent of cases.[337] The distinction of crescentic glomerular changes from other forms of glomerulonephritis is best made by assessing the extent of the glomerular lesion and by the results of immunofluorescence and transmission electron microscopic studies.[274]

Arteritis may occasionally affect the ureter,[252] bladder, testis and epididymis.

Cardiovascular system

Multiple aneurysms are encountered along the primary and smaller branches of the coronary arteries (Fig. 15.6). The presence of these aneurysms[304] may precipitate coronary artery thrombosis. Myocardial infarction is consequently one cause of death in PAN. Rarely, coronary artery aneurysm may rupture leading to cardiac tamponade. Schrader, Hochman and Bulkley[331] studied 36 necropsy cases with the clinical diagnosis of PAN. The majority had cardiac disease; 50 per cent displayed active or healed coronary arteritis. The most typical, severe lesions were of the small, immediately subepicardial vessels: there was occasional, incipient microaneurysm formation. Overall, the arteritis was less acute in the heart than in other viscera although interstitial myocarditis was occasional as were gross infarction[312] and fibrinous and fibrous pericarditis. The presence of widespread small artery disease may contribute to a non-hypertensive form of cardiomyopathy.[29]

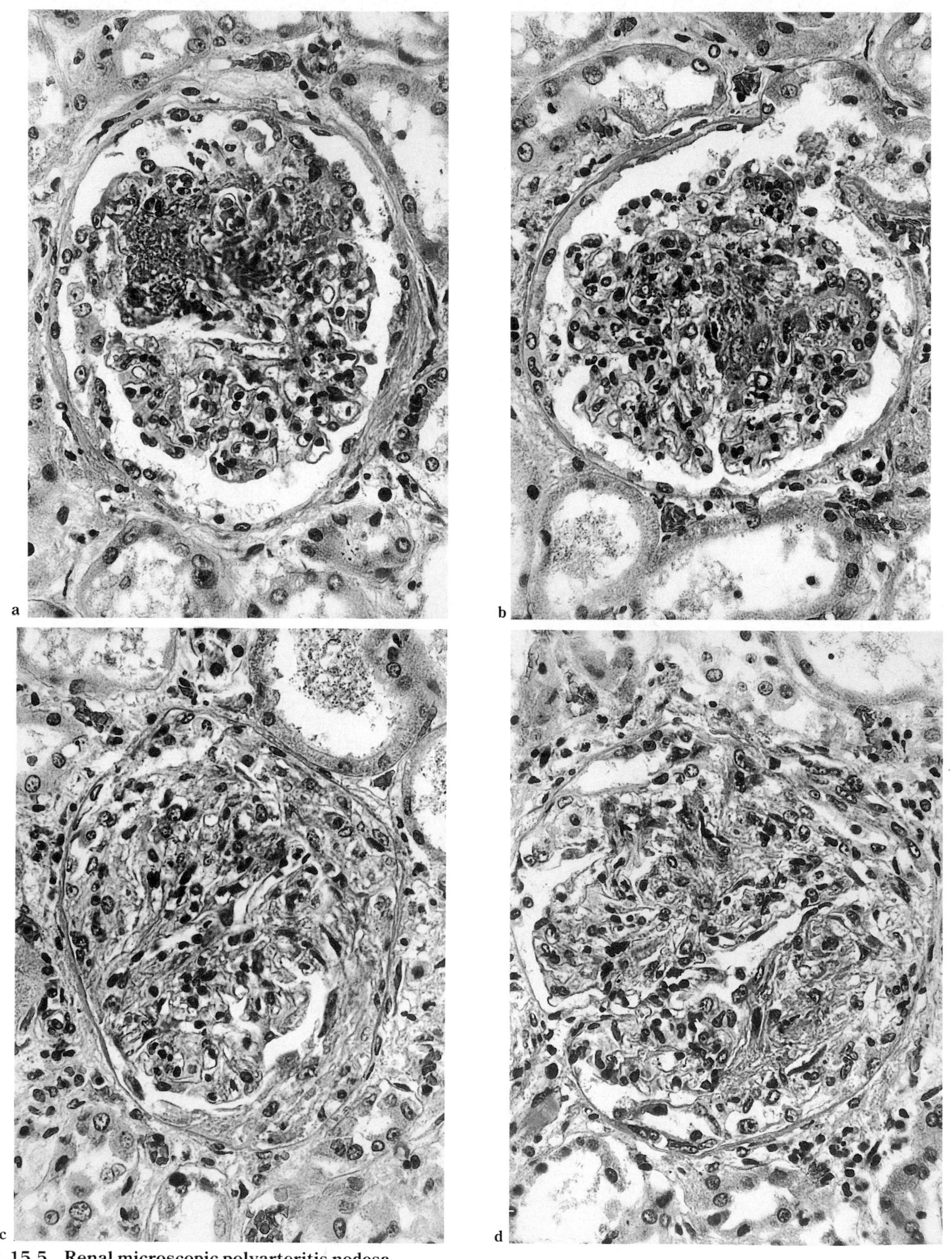

Fig. 15.5 Renal microscopic polyarteritis nodosa.
These fields illustrate sequential phases in the development of the glomerular lesions. **a.** One half of the glomerulus is the site of acute inflammation with fibrinoid change; **b.** similar quadrantic lesion in another glomerulus; **c.** healing of the lesion in the upper part of this glomerulus is accompanied by proliferation of the capsular epithelium; **d.** the formation of much collagen is evident so that a significant proportion of the glomerular tuft is functionless. (HE × 325.)

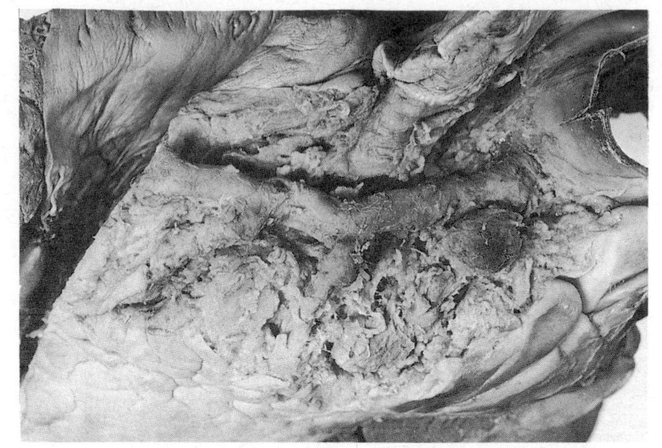

Fig. 15.6 Heart: coronary artery polyarteritis nodosa.
The circumflex branch of the left coronary artery runs horizontally across this field. Near the right margin, a 10 mm-diameter aneurysm has formed at the site of a focus of necrotizing arteritis. (Natural size.)

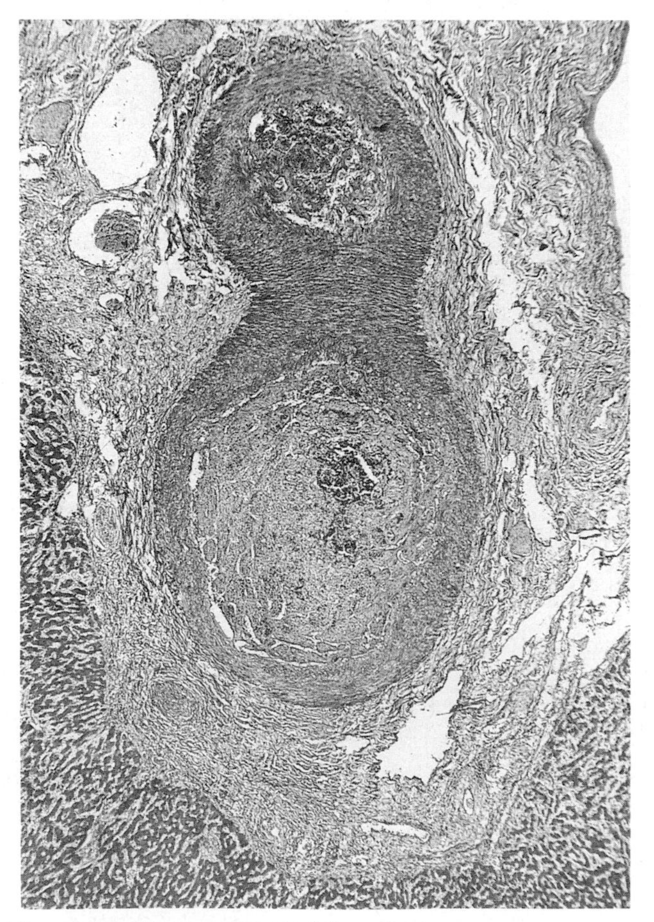

Fig. 15.7 Liver: polyarteritis nodosa.
A branch of the hepatic artery is divided tangentially. The inferior part of the vessel is thinned at a site of earlier vasculitis. Here, the arterial lumen is now occupied largely by organized thrombus. The new vascular lumen and that of the superior portion of the artery are both occluded by recent thrombus. (HE × 30.)

Alimentary system

Polyarteritis nodosa (Fig. 15.7) remains the single, most common known cause of hepatic arterial infarction.[37a,240] Infarcts are earlier signs of PAN than are aneurysms.[241] There is a rare association with nodular regenerative hyperplasia.[271] The gallbladder has long been known as a site at which PAN may be expressed.[139]

Gastrointestinal tract

In classical PAN, the characteristic finding is the presence of multiple small aneurysms, usually less than 10 mm in diameter, on the course of the mesenteric arteries and along the serosal border of the intestine. Occasionally these vessels rupture. The recognition of microscopic vasculitis may be difficult. At colonoscopy, non-specific inflammation or ulceration are suggestive changes as they are at rectal biopsy but overt arteritis is only identified in a minority of instances[42] and the diagnosis of gastrointestinal arteritis continues to depend heavily on clinical evidence of infarction obtained by arteriography and CT scanning (see p. 161).

Respiratory system

In the classical syndrome of PAN the lungs are not commonly affected. A survey[325] of 111 cases from nine medical schools suggested that, where the lungs were significantly involved, there were other distinctive features of the disease. In their 14 patients, in whom respiratory illness initiated and dominated the clinical picture, there were large numbers of eosinophil polymorphs in the blood and tissues, a feature not common in the usual visceral syndrome. There was a tendency for the arterial lesions to be accompanied by a granulomatous reparative reaction. In this reaction a segment of the wall of affected arteries is infiltrated and replaced by histiocytes and other mononuclear cells which are co-extensive with a periadventitial reaction. This reaction is of a similar low-grade, inflammatory nature. The work of Rose and Spencer has been overtaken by that of Liebow[224] (see p. 656). The group of cases in which lung lesions are frequent is distinguishable on both clinical and histological grounds from Wegener's granulomatosis in which sinus, nasal, pharyngeal and upper respiratory signs predominate.

Musculoskeletal system

The development of an aneurysm in the subcutaneous tissue and skeletal muscles occasionally produces a palpable swelling which can be explored at the time of biopsy. The skeletal muscle changes can be identified by electron microscopy.[249]

Bone

Subperiosteal and endosteal arteries may be the site of arteritis and subperiosteal new bone formation is one consequence:[251] 17 cases were reported by these authors. Gross discrepancies in leg thickness can result. Dermal, neural and muscular vasculitis often coexist but the limb changes may be dissociated from visceral disease.

Synovium

There are few reports in this country of synovial arteritis and it is noteworthy that the small synovial arteries are usually not affected by the vasculitis of rheumatoid arthritis, systemic lupus erythematosus and progressive systemic sclerosis. However, rheumatoid synovial vasculitis appears to be common in Eastern Europe and isolated cases of solitary synovial arteritis are now appearing in the UK.

Nervous system

The presence of cerebral aneurysms or of microscopic arteritic lesions in the central nervous system leads to zones of infarction which vary in extent with the size of the affected vessel and with the presence or absence of an adequate collateral circulation (Fig. 15.8). Multiple zones of infarction of different ages may be encountered at necropsy and similar lesions in the course of the spinal cord or peripheral nerves may cause a clinical neuropathy of a complex and variable kind. The vascular lesion of arteritic peripheral neuropathy are associated with a Wallerian-like degeneration[378] and transmission electron microscopic studies show unusually swollen axons and organelle accumulations. Non-myelinated fibres are also injured. Quantitative estimations fail to show selective vulnerability of large- or small-diameter myelinated fibre groups.

Skin

Subcutaneous necrotizing vasculitis in PAN is uncommon and cannot be differentiated histologically from the lesions of rheumatoid vasculitis (see p. 491). There is frequent clinical confusion with hypersensitivity vasculitis (see p. 628).[349]

Diagnosis

Angiography and computed tomographic scans provide useful, preliminary indices of vessel lesions.[309] Skin, skeletal muscle, kidney, testis, rectal mucosal and sural nerve biopsy can then give valuable diagnostic information.

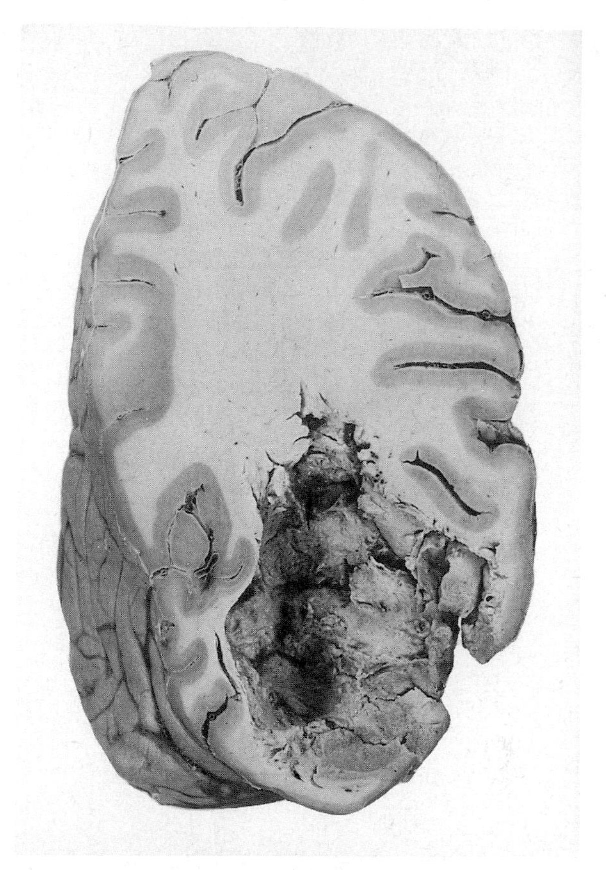

Fig. 15.8 Brain: polyarteritis nodosa.
The inferior part of this cerebral hemisphere is the site of a massive recent infarct, the consequence of acute necrotizing vasculitis of a major branch of the circle of Willis.

Usually, small arteries and veins are obtained. The site selected is chosen on the basis of clinical and imaging information. However, because of the high frequency of kidney involvement, renal biopsy is the single most useful test. In the absence of recognizable clinical lesions, blind muscle or skin biopsy is not helpful. The entire extent of a muscle block should be examined in semiserial section; even so, the samples are usually devoid of identifiable lesions. In one series, 15 of 24 skin biopsies, 12 of 22 skeletal muscle biopsies, 4 of 6 testicular biopsies and 5 of 10 renal biopsies showed evidence of vasculitis.[58] These results can be improved upon if more than one site or tissue is sampled, if multiple biopsies of different sites are performed and, particularly, if there is a demonstrable lesion such as a small aneurysm, that can be excised.

An absolute histological distinction between the vasculitis of PAN and of systemic connective diseases such as rheumatoid arthritis and systemic lupus erythematosus is seldom possible. Biopsy, alone, may not therefore be decisive. The lesions may be acute, necrotizing, visceral as well as superficial, and widespread. The clinical histories and serological tests, for example, for rheumatoid factor and antinuclear factor, may be helpful. Nevertheless, cases are regularly encountered in which long-standing seropositive rheumatoid patients die in renal failure with visceral cutaneous vasculitis, and vasculitis having a PAN-like character. Radnai[313a] believes that, although the differences between the arteritis of rheumatoid arthritis and of PAN may only be quantitative, a distinction *is* possible. In the acute form, there is more proliferation of connective tissue in the vessel wall in PAN than in rheumatoid arthritis; in the healing chronic phase, there is more fibrosis and a greater tendency to intimal proliferation in the former than in the latter.

Pathogenesis

Heredity

There is no known genetic predisposition to the syndrome of PAN although occasional cases have been recorded where a chromosomal translocation has been associated with macroglobulinaemia and arterial lesions said to be those of polyarteritis.[39] In the important instances of arteritis associated with serum hepatitis B antigenaemia (see p. 725), no association with any single HLA locus has been shown.

Infection

A considerable number of viruses, many bacteria, and some protozoa, worms and fungi can lodge in vessel walls and provoke arteritis. Small vascular disease is also commonplace in rickettsial disorders such as typhus.

Arteries are indeed particularly susceptible to viral infection; viraemia is characteristic of the incubation period of most viral diseases. Although it is likely that many viral lesions are promoted by immune complex formation rather than by direct infection (see pp. 238 and 722), there are established instances where true viral arteritis occurs. For example, virus is involved in the pathogenesis of at least three well-known animal forms of vasculitis: equine arteritis,[157] NZB/NZW mouse disease (see p. 600),[362] and Aleutian disease of blue mink.[2,310] The virus of equine viral arteritis (EAV), a small, enveloped, RNA virus and a member of the *Togaviridae*, is able to cause endothelial cell injury after localizing and proliferating in the arterial intima. Medial muscle cell necrosis ensues. Much is known of EAV:[372] it is the only member of the genus *arterivirus*.[373] Hepatitis A virus can cause glomerulonephritis and arteritis in the marmoset.[264] It appears likely that cytomegalovirus[255] and rubella virus injure arteries in the same, direct way. Necrotizing vasculitis, with disseminated intravascular coagulation, is one result of disseminated neonatal herpes simplex infection.[303]

A spontaneously occurring form of PAN was recognized in control outbred Palmerston North mice during a preliminary investigation of NZB/B1 mice (see p. 602); the natural history of this mouse PAN was described in detail.[394] Virus-like particles were identified in the glomerular epithelial cells. The strongest evidence of virus as a cause of human PAN is, however, in relation to hepatitis B virus infection.[335,336] Hepatitis B virus may, in theory, directly infect and damage vascular endothelial and intimal cells. Of patients with classical PAN, 30–50 per cent are believed to have been infected with hepatitis B virus[127,370] and approximately 1 per cent of all cases of serum hepatitis develop PAN. Patients undergoing long-term dialysis seem particularly susceptible.[74] During the early viraemic phases of serum hepatitis, immune complexes of hepatitis B (HB) antigen and anti-HBs antibody have been recognized in the circulation[176] and deposits of serum hepatitis B antigen, immunoglobulin and complement have been found in foci of arterial disease.[258] They are thought to provoke inflammation. The size and composition of the immune complexes is critical. Thus, the complexes demonstrable during the early serum-sickness-like phase of hepatitis B infection contain complement-fixing IgG1 and IgG3 subtypes; complexes in comparable patients without the syndrome do not. Serum hepatitis B antigen complexes disappear from the healed vascular lesions.

Although bacteraemia is not a characteristic of most bacterial infections, recurrent systemic bacterial vasculitis has been described after *Streptococcus pyogenes* infection.[231] *Salmonella typhi*, *S. typhimurium*, *Brucella sp.*, *Mycobacterium tuberculosis* and *Streptococcus viridans* are among the organisms that cause comparable lesions. The role of *Pseudomonas pyocyaneus* has attracted attention in immunocompromised subjects. Thrombosis, aneurysm and ischaemic disease may result. However, the lesions of classical PAN in man are sterile and direct bacterial infection cannot account for the widespread, focal disease.

The nematode *Strongyloides stercoralis*[382] and the fungus *Candida albicans*[116] and *Coccidioides*[397] are among the multicellular agents causing arteritis directly.

Other antigens and haptens

Drugs as antigens or haptens have also been implicated mainly because of experimental evidence.[274] Neoplastic antigens can provoke immune complex formation and may cause vasculitis: hairy-cell leukaemia is one example.[80] There is little support for the role of autoimmunity: anti-artery antibodies are not demonstrable. Nevertheless,

White and Grollman[400] were able to produce a necrotizing arteritis in the rat by the injection of suspensions of homologous arterial tissue in Freund's adjuvant. They demonstrated by the use of fluorescein conjugates that antiartery antibodies were located in the damaged vascular segments.

Role of immune complexes

The factors determining circulating immune complex formation, and the propensity for certain complexes to localize in vascular walls are discussed by Christian and Sergent[48] (see Chapter 5). The inflammatory phenomena attributable to immune complex deposition are described on pp. 238 and 533. Immune complexes are of varying size, and as they form, their size increases. Complexes formed in the region of great antigen excess are too small to be trapped within the arterial intima. The complexes that are likely to be most damaging are those formed in the region of slight antigen excess. The larger complexes formed in time, in the region of antibody excess, are phagocytosed and quickly removed by the reticuloendothelial system. Further factors determining the efficiency of complement fixation, and thus the likelihood of vasculitis, are the immunoglobulin subtype in the complex, the scale of host antibody production and local vascular factors, including the IgE-mediated platelet release of histamine (see p. 283).

It is now suspected that circulating immune complexes are responsible for the majority of cases of PAN but firm evidence to support this view is actually slight. The origin of this hypothesis is found in studies of serum sickness, a form of type III hypersensitivity (see p. 227) mediated by immune complexes. That the vasculitis of serum sickness closely resembled the lesions of PAN was remarked by Clark and Kaplan[54] and by Rich.[315] Rich described a group of cases in which PAN followed the prophylactic use of horse serum or of sulphonamides. Subsequently,[316] he obtained evidence that the repeated injection of horse serum or of horse serum with sulphonamides into rabbits produced a necrotizing vasculitis closely resembling PAN in man.[316] These results were confirmed using albumin as an antigen;[151,166] γ-globulin, fibrinogen and albumin were located in the vascular lesions.[125,203,254,298,299] Subsequently, the pathogenesis of experimental serum sickness was defined in detail.[56,57,72,120] After the intravenous injection into rabbits of large amounts of bovine serum albumin (BSA), circulating BSA/antiBSA immune complexes were found in the circulation.[126] Their presence coincided with the onset of arteritis and glomerulonephritis. At arterial bifurcations, subendothelial rabbit IgG, C3 and BSA were located near the internal elastic lamina. It appeared that the widespread, apparently hazhazard location of the lesions in PAN could be attributed to the capacity of immune complexes to localize at certain sites and to the response of the arterial wall to this localization. Once arteritis is active, the identification of a causal antigen is difficult because antigen, host immunoglobulin and complement can be degraded by the inflammatory response.[126] Experimentally, rabbit arteritis can be blocked by the deliberate removal of the animals' complement or polymorphs.

The demonstration of circulating immune complexes in some cases of PAN, and the microscopic evidence showing the association of immune complexes with sites of vascular injury and inflammation, does not exclude the possibility that some examples of the disease are mediated by delayed, not immediate, hypersensitivity. Reactions with vascular smooth muscle are among the early events.[267]

Cryoglobulins

The role of cryoglobulins has been considered. Among 53 patients,[58] a small number of whom displayed advancing immune complex disease with cryoglobulinaemia, HBs antigen or lowered complement levels, the spectrum of severity of the disease was considerable; 55 per cent survived five years. There was no association between cryoglobulin, HBs antigen and lowered complement levels, and either distinctive clinical features or prognosis. Intestinal and renal disease affected the prognosis adversely. Of 22 patients who died, 10 had active vasculitis that led to death or was present at death; in 4, the activity of the vasculitis could not be ascertained; and in 8, the vasculitis was inactive.

Hypertension

Among the other possible causes of PAN is raised blood pressure. Patients with PAN often have systemic hypertension. Arteritis is also a frequent association of experimental hypertension.[114] The necrotizing arteritis of acute rat hypertension, for example, is of abrupt onset[117] and develops when blood pressure has reached a high, sustained level.[345] The arteritic response is apparently independent of the precise cause of hypertension since virtually indistinguishable visceral arterial lesions with medial and elastic necrosis, acute inflammation and aneurysm formation are found in renal-, adrenal–regeneration-, salt- and deoxycortone hypertension, and in hypertension caused by perinephritis.[393] Similar arterial lesions occur spontaneously in ageing animals and may be induced by immunizing procedures irrespective of blood pressure changes.[393] Polyarteritis nodosa in the rat, therefore, is a single type of histological response attributable to several causes. Immunosuppression does not prevent experimental hypertensive vasculitis.[118] It appears likely that the development of systemic hypertension in patients with PAN is an incidental result of interference with blood flow to zones of renal cortex. Because of the experimental evidence, the possibility cannot be entirely excluded that in cases with severe hypertension injury to small vessels may lead to the liberation of antigens from damaged smooth muscle initiating an autoimmune response analogous with the reactions that occur in rheumatoid arthritis and systemic lupus erythematosus.

Prognosis

The response to treatment with corticosteroids, cytotoxic compounds or a combination of these drugs has greatly improved the outlook for patients with PAN.[58] The prognosis is most closely related to the size of the largest vessel involved and the degree of end-organ damage.[76] However, life-threatening disease can result from predominantly small artery or arteriolar injury: organ perfusion is threatened, a particular hazard in renal disease. The assertion that corticosteroids could exacerbate arteritis[16,192] was not confirmed. Among 130 patients followed between 1946 and 1962 the five-year survival was 13 per cent for untreated patients and 48 per cent for those treated intensively with corticosteroids or adrenocorticotrophic hormone.[104] In a survey of 53 patients followed for at least two years, the majority of those dying directly from the vascular disease were within the first year; after 12 months, myocardial infarction and cerebrovascular accidents accounted for the majority of deaths.[58] Fifty-five per cent survived five years. The prognosis is now better still, and among 34 patients with microscopic polyarteritis, all with focal necrotizing glomerulonephritis, 65 per cent survived five years, a figure reduced to 55 per cent when those with renal vasculitis alone were considered.[329] Renal involvement is therefore still an important determinant of mortality.[76] In the presence of renal disease, there is a one- to two-year survival of 30–50 per cent, a five-year survival of 30–40 per cent. This compares with five-year survivals of approximately 50 per cent of all cases of PAN when steroids are given, or approximately 80 per cent when immunosuppressive drugs are given as well.

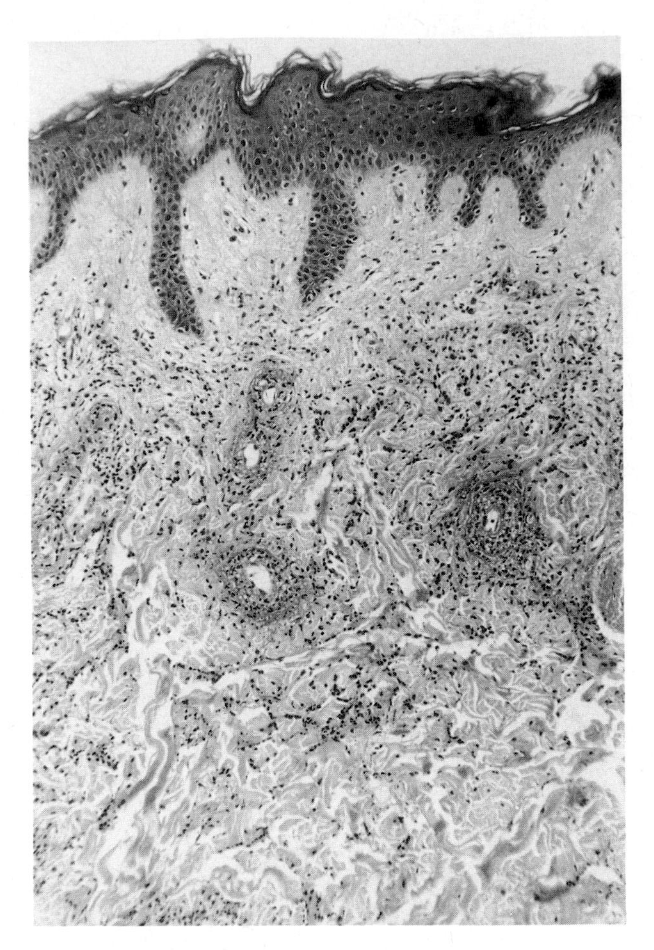

Fig. 15.9 Skin: hypersensitivity vasculitis.
Within the dermis, at the margin of the subcutaneous connective tissue, the walls of several venules appear thickened by the acute inflammatory reaction that is taking place within and around them. (HE × 33.)

Hypersensitivity vasculitis

In this poorly understood group of disorders, immune complexes are thought to be responsible for acute inflammation which begins in the walls of postcapillary venules. The size of the immune complexes and the reason they lodge at this site are not well understood.

Hypersensitivity vasculitis (Fig. 15.9)[15,94,333] was differentiated from other forms of vasculopathy by Zeek.[411] The most conspicuous part of the pathological process is mural infiltration of affected vessels by polymorphs and the subsequent disintegration of these cells, leucocytoclasis. The effects of hypersensitivity vasculitis are in proportion to the extent to which venous return from an affected tissue or organ is impeded. The most commonly recognized lesions that result are in the skin;[350] 'palpable purpura' is frequent. Consequently, the condition has attracted particular interest among dermatologists who have given much thought to its pathogenesis and classification.[66,399,401] Other organs are, however, affected.[62]

Particular forms of small vessel vasculitis include Henoch–Schönlein purpura, hypocomplementaemic vasculitis and the vasculitis of mixed cryoglobulinaemia; in each, there are histological similarities but other, distinguishing features.

Clinical presentation

The clinical classification of hypersensitivity vasculitis has been facilitated by the criteria agreed by the American College of Rheumatology[41a] (Tables 15.7 and 15.8).

A history of drug therapy or bacterial infection can sometimes be obtained. There is a sudden appearance of flat, erythematous skin macules which become papular.[62] The legs and lower back are susceptible but the hands and forearms, the face and upper trunk are sometimes affected. Evidence of systemic involvement includes signs referable to the lungs, kidney, heart and central nervous system. Eosinophilia, a raised erythrocyte sedimentation rate (ESR) and anaemia are frequent. Ill-defined pulmonary infiltrates, pleural effusion, pericarditis, peripheral neuropathy and encephalopathy are also recognized.[126] The rapid onset of renal failure or cardiac ischaemia may cause early death. When there is prompt diagnosis, withdrawal of the responsible antigen can allow recovery. The cutaneous necrotizing purpuric and ulcerating lesions are accompanied by malaise, fever, arthralgia and myalgia. A non-destructive arthritis is occasionally recognized.[346] Closely similar changes are found from time to time in the small vessels in disorders such as hepatitis B, systemic lupus erythematosus, rheumatoid arthritis, Sjögren's syndrome and Wegener's granulomatosis: the final common pathway by which the venules are injured is assumed to be the same.

Pathological changes

Much of the evidence relates to the skin[216,217] (Fig. 15.10). There is an infiltration of dermal venules by polymorphs which extend into the surrounding tissues. Both superficial papillary dermis and deeper vessels may be affected. There may be associated fibrinoid within the injured vessel wall but its recognition is not a prerequisite for diagnosis. The walls of affected venules become necrotic and local haemorrhage is the cause of clinical purpura and the explanation for the aggregates of haemosiderin residues at affected dermal sites. With time, the severity of the reaction subsides; polymorph nuclear debris remains but a macrophage response develops with polymorphs and mononuclear cells in varying proportions. In some instances the early cellular infiltrate is lymphocytic rather than granulocytic, and this may suggest the operation of a delayed rather than of an immediate, hypersensitivity response (see p. 227). The syndrome is as variegated as the causes.[234]

It had been thought that, particularly where there was renal involvement, small arteries and capillaries could be affected.[411] Now, however, the majority describe hypersensitivity vasculitis as a disorder solely or usually of postcapillary venules[62,66] (Fig. 15.11). Evidence to confirm this view comes from the study of 1 μm or thin, transmission electron microscopic sections.[235,351]

Visceral lesions occur but their extent and frequency is much less well understood than are those of the skin. The affected vessels are again the small veins or venules (Figs 15.10 and 15.11). Since glomerulonephritis or renal vasculitis occur rarely except in particular, distinct forms of the syndrome such as Henoch–Schönlein purpura or mixed cryoglobulinaemia (see below), it is not necessary to invoke renal capillary or arterial disease. McCluskey and Fienberg[235] suggest, rather, that the group of cases originally described by Zeek,[410,411] and which included examples with renal disease, probably corresponded to the

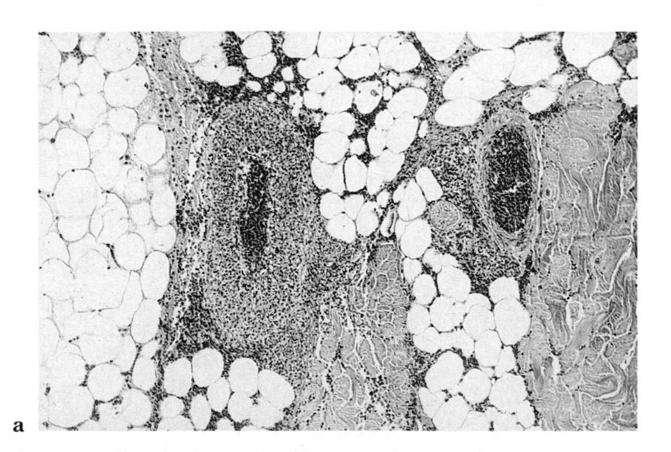

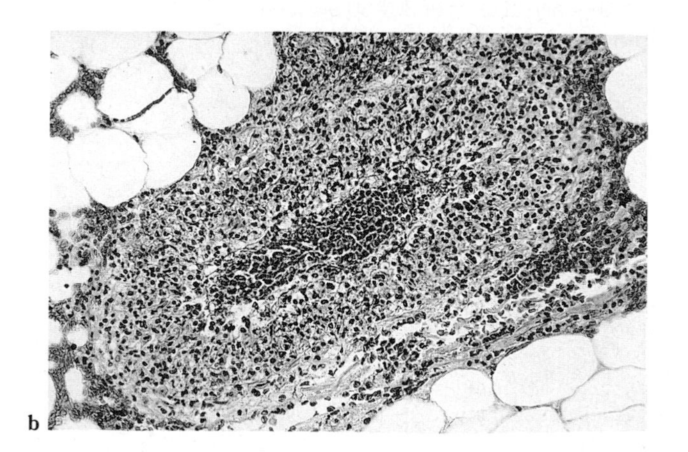

Fig. 15.10 Skin: hypersensitivity vasculitis.
a. The wall of the small subcutaneous vein (right) is incompletely surrounded by an acute inflammatory cellular exudate; a similar reaction entirely occupies the wall of the larger vessel seen in the left part of this field. (HE × 58.) **b.** The small vein seen in the left part of **a** is shown at higher magnification. (HE × 125.)

Table 15.7

1990 ACR criteria for the classification of hypersensitivity vasculitis (traditional format). Modified from Calabrese *et al.*[41a]

Definition of criteria
1. Development of symptoms after the age of 16 years
2. Medication was taken at the onset of symptoms that may have been a precipitating factor
3. Slightly elevated purpuric rash over one or more areas of the skin; does not blanch with pressure and is not related to thrombocytopenia
4. Flat and raised lesions of various sizes over one or more areas of the skin
5. Histologic changes showing granulocytes in a perivascular location

The presence of three or more criteria yields a sensitivity of 71.0 per cent and a specificity of 83.9 per cent.

Table 15.8

1990 classification tree criteria for hypersensitivity vasculitis (HSV).* Modified from Calabrese *et al.*[41a]

HSV subsets	Number of patients HSV/ non-HSV	Percentage correctly classified	Percentage HSV patients in subset	Non-HSV subsets	Number of patients HSV/ non-HSV	Percentage correctly classified	Percentage non-HSV patients in subset
Purpura; history of medication at disease onset	31/32	49	33	Palpable purpura without medication at disease onset, age ≤16 at disease onset	0/45	100	6
Purpura; no history of medication at disease onset; age >16 at disease onset	28/84	25	30	No purpura or maculopapular rash; no granulocytes in vessel wall	6/453	99	63
Maculopapular rash and biopsy demonstrating eosinophils in a venule or arteriole; no purpura	7/8	47	7.5	Maculopapular rash; no purpura; no eosinophils in venule or arteriole	14/64	82	9
Biopsy demonstrating granulocytes in the wall of a venule or arteriole; absence of purpura and maculopapular rash	7/28	20	7.5				

*The classification tree yields a sensitivity of 78.5 per cent and a specificity of 78.7 per cent.

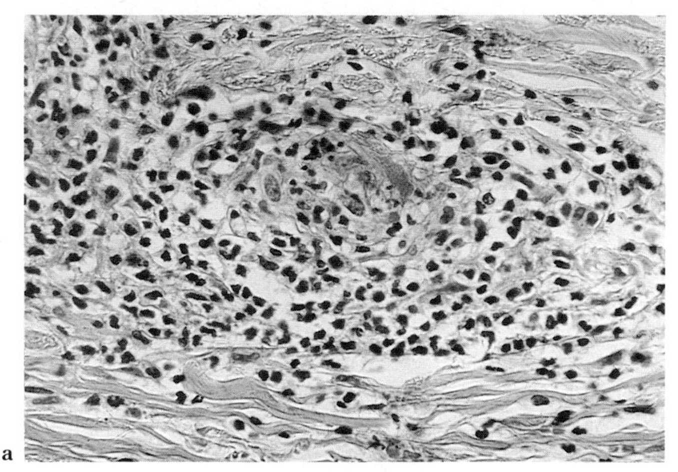

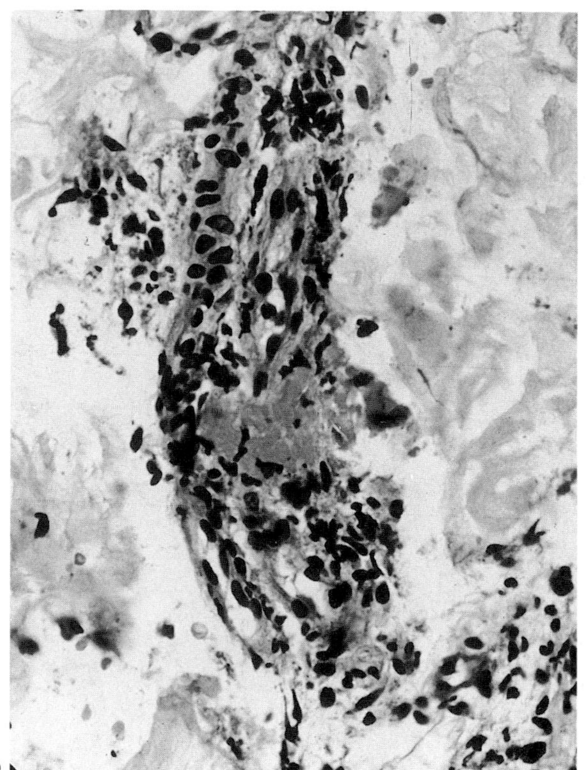

Fig. 15.11 Skin: hypersensitivity vasculitis.
a. Among the numerous cells comprising the inflammatory exudate which extends within and beyond the wall of this venule are the pyknotic nuclei of many polymorphs. They are accompanied by numerous mononuclear macrophages and smaller numbers of lymphocytes. (HE × 160.) **b.** The extent of the leucocytoclastic disruption of polymorphs in this dermal venule is greater than of the vessel shown in **a**. (HE × 400.)

microscopic form of PAN described by Davson, Ball and Platt[70].

The picture is not, however, clear-cut and in a systematic study of the microscopic changes in 30 subjects with a history of drug-related vasculitis, Mullick *et al.*[268] demonstrated that the vascular responses included arterioles, capillaries, venules and small veins. Occasionally small arteries were involved but larger, musculoelastic vessels were spared. Eosinophils were prominent and vessel wall infiltration the rule; a perivascular infiltrate alone was not regarded as a sufficient basis for diagnosis. Fibrinoid was never detected and tissue necrosis was not present nor was there thrombosis. Pre-exudative lesions were not present and sites of healing, with fibrosis and aneurysms, were not recognized.

Electron microscopy reveals interendothelial cell gaps and increased pinocytotic activity. There is recognizable basement membrane thickening and a coating of collagen fibres by fibrin. Immunofluorescence confirms the deposition of fibrin in venules; the demonstration of immunoglobulin (Ig) and C3 is occasionally perceived in nearby, intact skin. Although IgM is the most frequently detected Ig, IgG and IgA may also be present (Table 15.9). Immunofluorescence may also permit the local demonstration of bacterial antigens including those of streptococci, staphylococci and mycobacteria, united as immune complexes with immunoglobulin and complement.

Table 15.9

Direct immunofluorescence microscopy of skin lesions in hypersensitivity vasculitis[239]

Disease	Antisera				
	IgG	IgA	IgM	C3	Fibrin
Systemic	1/16	4/16	10/16	12/16	12/16
Drug-related	0/4	2/4	4/4	4/4	4/4
Idiopathic	2/17	4/17	12/17	14/17	13/17
Total	3/37	15/37	27/37	30/37	29/37

Aetiology and pathogenesis

A wide range of causes has been invoked (Table 15.10).

Infections are often provocative; the most important is with the β-*haemolytic streptococcus* although *Staphylococcus aureus* and *Mycobacterium leprae* have been implicated. In lepromatous leprosy, the cutaneous nodular lesions are initiated by a necrotizing angiitis provoked by chemotherapy. Drugs, including thiazide diuretics, phenytoin,[114,407] ephedrine,[403] antibiotics such as penicillin, and chemotherapeutic agents such as the sulphonamides, are also potent causes (Table 15.11). Tetracycline toxicity is occasionally provocative.[191]

The injection of foreign serum has been known to precipitate cutaneous vasculitis as have foods such as chocolate,

Table 15.10

Some causes of hypersensitivity (leucocytoclastic) vasculitis

1. Associated with chronic, coexistent immunologically related diseases:
    - Rheumatoid arthritis
    - Systemic lupus erythematosus
    - Sjögren's syndrome
    - Lymphoproliferative disorders including lymphoma
2. Associated with precipitating event:
    - Administration of drugs
    - Bacterial infection
3. Uncertain aetiology:
    - Associated with serological changes
        - Cryoglobulinaemia
        - Hypergammaglobulinaemia
        - C2 deficiency
    - Of characteristic clinical appearance
        - Nodular vasculitis/palpable purpura
        - Erythema elevatum diutinum
        - Henoch-Schönlein syndrome
        - Urticaria
4. Idiopathic

tomatoes, milk and fruit. In one unusual case, vasculitis was diagnosed in a 17-year-old boy treated for nine years with allopurinol for partial hypoxanthine-guanine phosphoribosyl transferase deficiency[391] (see p. 383).

Miscellaneous causes include hypergammaglobulinaemia in Sjögren's syndrome; systemic lupus erythematosus; lymphocytic lymphoma; hairy-cell leukaemia;[89] myeloproliferative disorders;[229] angioimmunoblastic lymphadenopathy;[10] immunoblastic lymphoma;[262] and C2 deficiency.

Table 15.11

Drugs associated with 30 cases of hypersensitivity vasculitis[268]

Allopurinol	Indocin
Ampicillin	Oxyphenbutazone
Bromide	Penicillin
Chlorothiazide	Potassium iodide
Chlorpropamide	Quinidine
Colchicine	Spironolactone
Dextran	Sulphonamide
Diphenylhydantoin	Tetracycline
Griseofulvin	Trimethadione

The proximate cause of the vasculitis is suspected to be immune complex deposition[305] although it may be difficult to obtain proof of this process by immunomicroscopic techniques. Where the lesions include greater proportions of lymphocytes, the role of delayed hypersensitivity has been proposed.[351] Low serum complement levels have been linked to polymorph infiltration, and normal complement levels to a lymphocytic response.[351] Polymorph disintegration releases lysosomal proteases and free radicals.[261]

Henoch–Schönlein purpura

Henoch–Schönlein purpura[8,112] can be regarded as a special form of hypersensitivity vasculitis.[62] The syndrome is most frequent in children. Sensitization to microbial, food or other antigens is suspected and the onset, often in the early months of the year, is marked by the development of purpura, oedema, synovitis and abdominal pain attributable to intussusception, intestinal haemorrhage (Fig. 15.12) or perforation. Renal disease is detected in approximately 50 per cent of cases: it is often slight but chronic renal disease may evolve.

The dermal lesions are those of leucocytoclastic venulitis; the renal lesions affect both venules and glomerular capillaries. Immunoglobulin and complement deposits are recognized. Similar venular disease is recognized in the small intestinal wall. Little is known of the synovitis. The glomeruli usually display only slight diffuse mesangial hypercellularity but occasionally a more severe response,

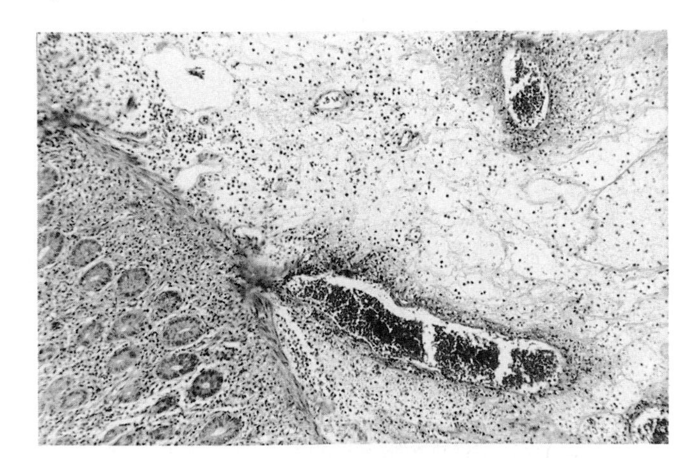

Fig. 15.12 Small intestine: Henoch–Schönlein vasculitis.
Two submucosal venules are the site of a vigorous inflammatory reaction: the circulation of the intestinal wall is prejudiced and the loose submucosal connective tissue is intensely oedematous. (HE × 22.)

with widespread capsular epithelial cell proliferation and crescent formation, is found on percutaneous biopsy.

Serum complement levels remain normal. Nevertheless, immune complexes are likely to be the cause of tissue injury. Immunoglobulin and complement deposits can be shown in affected vessels: the immunoglobulin is mainly IgA and it is suspected that complement is activated by the alternate pathway (see p. 238). The disease is self-limiting. However, impaired renal function may persist, particularly in older persons.

Hypocomplementaemic vasculitis

Hypocomplementaemic vasculitis was recognized when transient urticarial skin lesions in young adult women were found to be attributable to immune complex-mediated leucocytoclastic vasculitis.[237] As in Henoch–Schönlein purpura, arthralgia and abdominal pain occur but there is neither intussusception nor haemorrhage. The articular changes recall those of systemic lupus erythematosus. Renal dysfunction, with haematuria and proteinuria, is encountered and chronic obstructive airways disease is frequent.

Histologically, the skin changes, again, are those of leucocytoclastic vasculitis. Renal lesions are of slight severity: there is a mild membranoproliferative glomerulonephritis with electron-dense, lumpy-bumpy deposits on the epithelial aspect of the glomerular capillaries. Glomerular hypercellularity is not accompanied by crescent formation or necrosis. There are similarities to the lesions of systemic lupus erythematosus but neither serum antinuclear antibodies nor anti-DNA antibodies are formed. However, serum factors may be found which react both with C1q and monoclonal rheumatoid factor.

Mixed cryoglobulinaemia

Mixed cryoglobulinaemia[230,256] is dominated by the occurrence of cutaneous vasculitis. Visceral vascular disease also occurs. An illustrative case was recently described in detail:[321] a 33-year-old woman who had suffered from viral hepatitis while subject to drug-abuse had been treated both for gonorrhoea and genital herpes. She developed a purpuric, pretibial rash and arthralgia. A skin biopsy showed leucocytoclastic vasculitis; the serum contained IgM (κ) and IgG (κ and λ) cryoglobulins, indicating the most common form of (type II) cryoglobulinaemia (Table 15.12). Because of the association between hepatitis B and arteritis, it is important to note that a link between cryoglobulinaemia and hepatitis B infection is possible.[220]

Table 15.12

Mixed cryoglobulinaemia

Associated with	Cryoprecipitate
Type I	
Waldenström's macroglobulinaemia or with multiple myeloma	Homogeneous M component
Type II	
Vasculitis or other connective tissue disease	Monoclonal IgM, usually kappa, complexed with heterogeneous IgG
Type III	
Vasculitis or other connective tissue disease	IgG and IgM: no monoclonal component

Wegener's granulomatosis

In this very uncommon, strange disorder of young to middle-aged individuals, vasculitis and granulomata of the upper respiratory tract and lungs accompany ulceration of the nose and paranasal sinuses and pulmonary infiltrates.[62] Subsequently, focal glomerulonephritis develops. There may be other visceral lesions.[259] The aetiology is unknown although circulating immune complexes, the result of an humoral hypersensitivity reaction towards unidentified antigens, are thought to account for both the vasculitis and the renal disorder. Untreated, the disease is quickly fatal but there is a dramatically beneficial response to cytotoxic therapy.[93] Rarely, the condition begins in pregnancy.[363]

The first recognition of this form of vasculitis is attributable to Klinger[198] and to Rössle;[326] that the cases constituted a syndrome in their own right was proposed by Wegener[387,388,389] and accepted by Godman and Churg.[128] The classification and diagnosis of Wegener's granulomatosis may be improved by the adoption of the 1990 ACR criteria[209a] (Tables 15.13 and 15.14).

Clinical presentation

The onset is insidious. The disease affects previously healthy persons.[25] There is a suggestion that HLA-DR2 is inherited with increased frequency in Wegener's granulomatosis but not in PAN or the Churg–Strauss syndrome.[81] There is no familial association although the disease has

Table 15.13

1990 ACR criteria for the classification of Wegener's granulomatosis (traditional format).* Modified from Leavitt *et al.*[209a]

Definition of criteria
1. Development of painful or painless oral ulcers or purulent or bloody nasal discharge
2. Chest radiograph showing the presence of nodules, fixed infiltrates, or cavities
3. Microhaematuria (>5 red blood cells per high power field) or red cell casts in urine sediment
4. Histologic changes showing granulomatous inflammation within the wall of an artery or in the perivascular or extravascular area (artery or arteriole)

*For the purposes of classification, a patient shall be said to have Wegener's granulomatosis if at least two of these four criteria are present. The presence of any two or more criteria yields a sensitivity of 88.2 per cent and a specificity of 92.0 per cent.

Table 15.14

1990 classification tree criteria for Wegener's granulomatosis (WG).* Modified from Leavitt *et al.*[209a]

WG subsets	Number of patients WG/ non-WG	Percentage correctly classified	Percentage WG patients in subset	Non-WG subsets	Number of patients WG/ non-WG	Percentage correctly classified	Percentage non-WG patients in subset
Nasal or oral inflammation and active urinary sediment; absence of granulomatous inflammation and normal chest radiograph	8/23	23	9.5	Absence of granulomatous inflammation and absence of nasal or oral inflammation	8/598	99	83
Abnormal chest radiograph and nasal or oral inflammation; absence of granulomatous inflammation	8/8	50	9.5	Nasal or oral inflammation; negative findings for biopsy, chest radiograph, and urinary sediment	1/47	98	7
Granulomatous inflammation on biopsy and abnormal chest radiograph; absence of nasal or oral inflammation	13/8	62	15	Granulomatous inflammation on biopsy; absence of nasal or oral inflammation and normal chest radiograph	2/31	94	4
Granulomatous inflammation on biopsy and nasal or oral inflammation	45/7	87	53				

*The classification tree yields a sensitivity of 87.1 per cent and a specificity of 93.6 per cent.

been reported in two sisters.[270] In one large series, the mean age was 40.6 years, the mean time from onset to diagnosis 8.3 months.[93] The initial signs and symptoms are pulmonary, upper respiratory or articular. In the cases of Fauci *et al.*, the presenting evidence was of pulmonary infiltrates (71 per cent), sinusitis (67 per cent), arthritis or arthralgia (44 per cent), fever (34 per cent), otitis (25 per cent) and cough (34 per cent). The arthritis may be erosive.[178] Few patients present with evidence of renal disease but, in the absence of treatment, one in three patients with glomerulonephritis and uraemia progress to end-stage renal failure. The breast may be affected.[184] Otorrhoea, deafness and ulceration of the gums are other signs[209] but it is the concentration of changes in the upper respiratory tract which is distinctive. They are entirely different from those of midline neoplasia.[236,257,260] Auricular chondritis and saddle-nose deformity may simulate relapsing polychondritis[130] and isolated laryngeal disease has been recorded.[156]

Clinical laboratory tests provide few characteristic diagnostic findings. Leucopenia, a feature of pulmonary lymphomatoid granulomatosis (see p. 657), is not characteristic. A normocytic, normochromic anaemia is present. The erythrocyte sedimentation rate is very high. Positive tests for rheumatoid factor are often detected, usually of low titre. Serum IgA and IgM, but not IgE levels, may be elevated. Antinuclear factors are not recognized; however, circulating immune complexes can frequently be found and the recognition of antipolymorph cytoplasm antibodies[149] can provide diagnostic help (Fig. 15.13).

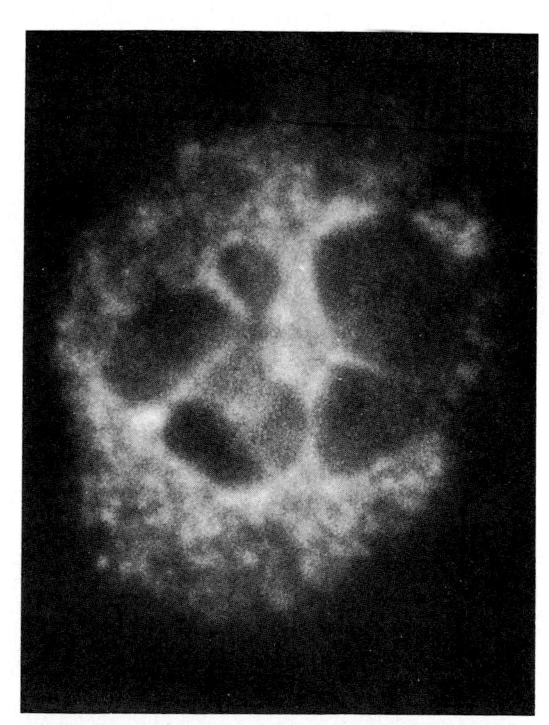

Fig. 15.13 Detection of antineutrophil cytoplasm antibodies in serum in Wegener's granulomatosis. Antineutrophil cytoplasm antibodies shown by indirect immunofluorescence testing on cytospin preparation of polymorphs. Rabbit antihuman IgG conjugated to fluorescein isothiocyanate. Reproduced courtesy of Mr R. Simpson.

Pathological changes

The triad described by Wegener[387] comprises: 1. necrotizing granulomata of the upper respiratory tract, particularly the air sinuses; 2. a necrotizing vasculitis affecting both pulmonary arteries and veins and selected extrapulmonary sites such as the kidney; and 3. a focal glomerulonephritis. A close relationship with the allergic granulomatosis and angiitis of Churg and Strauss[52] (see p. 639) is evident but there is neither eosinophilia nor an association with asthma.

Wegener's granulomatosis is not a form of primary vasculitis.[235] Much of the tissue destruction is attributable to extravascular granulomata, distinct from blood vessels. The presence of these granulomata is sufficient to allow a histological diagnosis; the certainty of diagnosis is strengthened if vasculitis is also found.

In Wegener's granulomatosis, pulmonary and systemic arteries, veins and capillaries are sites for an inflammatory reaction whereas, in polyarteritis nodosa (see p. 620), only the systemic arteries are affected.[332] Liebow[224] identified a 'limited' form of Wegener's granulomatosis with established lung lesions identical to those of the classical syndrome but neither upper respiratory tract involvement nor glomerulonephritis. Nevertheless, this limited syndrome may present both focal renal granulomata and other extrapulmonary lesions.

The gross pathological features (Table 15.15) of

Table 15.15

Frequency of organ involvement in Wegener's granulomatosis[93]

Organ or system	Number of cases	Per cent of cases
Lung	80	94
Paranasal sinus	77	91
Kidney	72	85
Joint	57	67
Nose/nasopharynx	54	64
Ear	52	61
Eye	49	58
Skin	38	45
Nervous system	19	22
Heart	10	12

Wegener's granulomatosis were summarized by Godman and Churg[128] in a paper describing seven new cases and reviewing 22 previously reported and by Walton[384] who added 10 cases and analysed 46 reported by others. Conn and Hunder[62] summarize contemporary views. The lungs, kidneys and spleen are most frequently involved but other viscera are affected from time to time. In the lungs, pale, grey-white zones of necrosis and consolidation contribute toward the shadows seen radiologically and are entirely distinct from the red-purple zones of infarction which result when blood vessels at the margins of granulomata become inflamed and subsequently thrombose.[210] In occasional cases, the pulmonary granulomata are smaller and give a miliary appearance which it is difficult to distinguish grossly from tuberculosis. There is macroscopic evidence of active inflammation with ulceration and infection of the respiratory mucosa of the paranasal sinuses which is thickened and hyperaemic. Bone destruction may be evident and the granulomatous reaction may extend from the sinuses into the skull. Ulcerating lesions are also encountered in the mouth, pharynx, palate, larynx,[367] trachea and skin. In the viscera, the lesions most commonly found are discrete granulomata.

The histological lesions are granulomata localized first to the paranasal sinuses and bronchi. The granulomatous zones mimic tuberculous infection of the paranasal sinuses, a resemblance exaggerated by the presence of occasional multinucleated, foreign body giant cells and by the coexistence of repair and coagulative, central necrosis. Giant cells are equally characteristic of the pulmonary granulomata and, occurring in relation to involved arteries, may resemble a giant-cell arteritis. In the mouth, analogous ulcerating lesions are present, part of a necrotizing stomatitis; occasionally they penetrate the entire thickness of the mucosa. The small arteries and arterioles in such zones are the sites for necrotizing inflammatory reactions.

The course of the disease is remorseless and progressive; it is punctuated by remittent fever with advancement of the ulcerating nasal lesions on the one hand, and zonal pulmonary consolidation, the sign of infarction, with thrombosis and the accumulation of fibrinoid in the necrotic zones, on the other.

Respiratory system

Upper respiratory tract

Vasculitis of venules or arterioles, with nearby granulomata, are initially localized to the paranasal sinuses, palate, nasopharynx, mouth or pharynx. Larger arteries and veins may be implicated; mural fibrinoid, with endovascular thrombosis, is frequent.

Biopsy of the nasal sinuses or nasopharynx often fails to show granulomatous vasculitis and the only recognizable change may be a non-specific, acute or chronic inflammatory reaction with necrosis and ulceration. In other parts of the upper respiratory tract there may be evidence of either vasculitis (Fig. 15.14) or granulomata but seldom of both. Occasionally, disease of the upper airways is sufficiently severe to cause erosion of the walls of a sinus; however, neither palatal perforation nor erosion, with fistula, of the skin of the nose or face is encountered—an important feature distinguishing Wegener's granulomatosis from local neoplastic disease.[93] However, collapse of the cartilage of the nose not infrequently leads to saddle-nose deformity, a feature of other systemic connective tissue disorders, particularly relapsing polychondritis (see p. 711).

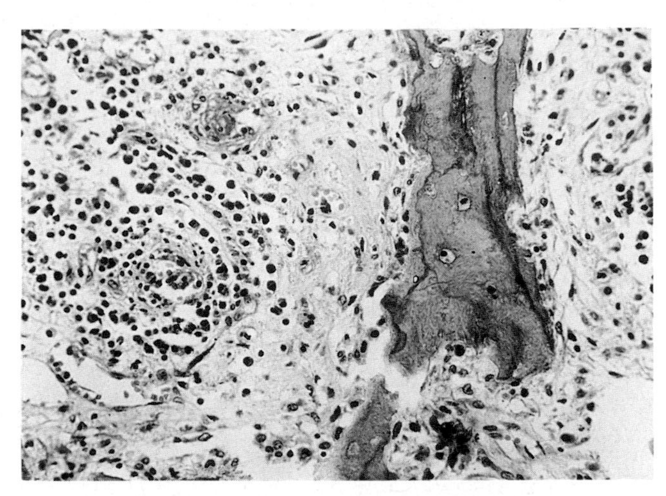

Fig. 15.14 Frontal air sinus: Wegener's granulomatosis.
In the upper left part of the field the wall of an arteriole is occupied by fibrinoid. At lower left centre, a second vessel is the focus for an aggregate of plasma cells, mononuclear macrophages and a small number of lymphocytes. The entire field is involved in a low-grade, granulomatous inflammatory reaction which is threatening the integrity of a bone trabecula (at right). (HE × 180.)

Lung

In one series, 94 per cent of patients had lung disease.[93] There are circumscribed, multiple, bilateral zones of inflammatory cell infiltration within which cavitation may occur. An overlying pleuritis, accompanied by a pleural exudate, reflects a response to the lung disorder. Pulmonary calcification and hilar lymphadenopathy are not present and the zones of inflammation, firm, solid and airless, may resemble or be accompanied by infarction. The underlying small-vessel vasculitis is not confined to pulmonary arterial branches: venules are also implicated.[313] Affected vessels are the site of a segmental infiltrate of lymphocytes and macrophages with small numbers of neutrophil and rare eosinophil polymorphs. One consequence is endovascular mural thrombosis, leading to segmental or lobular pulmonary infarction (Fig. 15.15). Another result is selective vessel-wall necrosis easily recognized by the loss of elastic material the destruction of which often evokes multinucleated giant-cell formation. In zones of pulmonary infarction, the appearances of coagulative necrosis simulate tuberculous caseation. Where the destructive granulomatous process has extended to involve bronchial walls, there is mucosal ulceration, a feature frequent in cases in which there is either incomplete response to treatment or relapse. Adjoining sites of lung infiltration are segments of atelectasis. Pneumothorax may be a result of a subpleural, cavitating mass.[179]

Biopsy is useful in confirming the diagnosis. Immunofluorescence microscopy has shown the presence of IgG and of complement as granular deposits in lung tissue[339] but direct evidence of immune complex deposition is inconstant. Thoracotomy, with open biopsy, establishes the presence of granulomatous vasculitis and its sequelae in two-thirds of cases. Sometimes, granulomata may be seen without vasculitis[93] or vasculitis without granulomata: in these cases, there are usually vasculitic lesions in other organs. In established examples, recognizable by other criteria, endobronchial disease may be present so that bronchial biopsy may be positive. Generally, however, this diagnostic procedure is negative in early, debatable cases.

Electron microscopic investigations of lung tissue obtained at biopsy have thrown light on the pathogenesis of the pulmonary lesions.[73] Early intravascular lysis of leucocytes in a case of 'limited' Wegener's granulomatosis was followed by platelet aggregation and fibrin deposition. Endothelial cell necrosis followed, with vascular occlusion, mucosal necrosis and death of pneumocytes. Macrophage immigration and fibroblast proliferation followed, a sequence which, in this case, provided a preliminary to successful cyclophosphamide therapy.

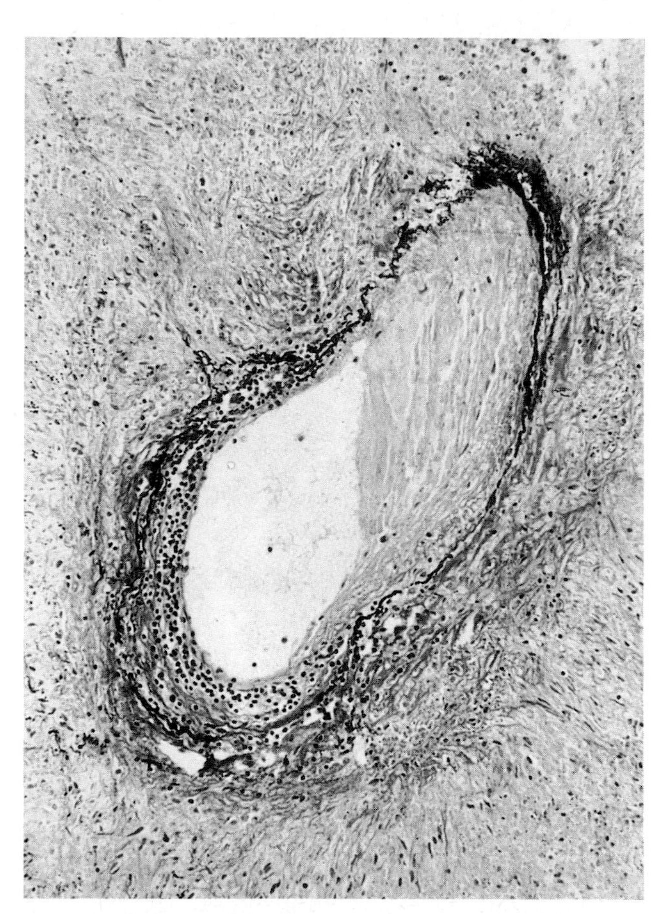

Fig. 15.15 Lung: Wegener's granulomatosis.
There is an acute necrotizing arteritis indicated by an inflammatory cell infiltrate of polymorphs and mononuclear macrophages but few eosinophil granulocytes. A small pulmonary artery has been cut longitudinally. The elastic lamina has been extensively destroyed. The lower part of the lumen is occupied by organizing thrombus and the effect on the pulmonary circulation is shown by the surrounding zone of infarction. (HE × 150.)

Urinary system

There is usually a focal necrotizing glomerulitis, thought by Godman and Churg[128] to resemble focal endocarditic glomerulonephritis. However, renal biopsy has shown that there is a wide range of abnormality extending from mild focal and segmental glomerulonephritis to fulminating diffuse necrotizing disease.[93] The progression of untreated renal disease may be very rapid. Early diagnosis and the institution of effective treatment are priorities.[306] Arteritis and renal granulomata appear to be rare but vascular lesions are identified more often at necropsy. Renal papillary necrosis has been described.[386] Transmission electron microscopy is likely to confirm the presence of subendothelial deposits of electron-dense material[276] in mild proliferative glomerulonephritis. In the latter case, which responded well to cytotoxic therapy with azathioprine, there were epithelial microvilli, slight foot process fusion and a marginal increase in mesangial matrix and collagen. Virus-like bodies were identified.

Osteoarticular system

Joints Two-thirds of patients in the series of Fauci *et al.*[93] had joint disorders; of these, 42 per cent had an arthritis which was polyarticular, never deforming, and affecting mainly the knees and ankles. The joint lesions may simulate rheumatoid arthritis and occasionally can present before the onset of respiratory tract disorder.[311]

Skeletal muscle

Very occasionally, non-specific inflammation, with a vasculitis of small arteries not attributable to drugs, accompanies non-specific inflammation and myopathy.

Cardiovascular system

Pericarditis and, less often, myocarditis, are occasional. Congestive cardiomyopathy was detected in rare instances,[93] but could not be attributed to cyclophosphamide with certainty. The atrioventricular node may be implicated directly, resulting in conduction defects.[7]

Skin

Clinically, papules, vesicles, purpura, ulceration or nodule formation are recognized.[93] The underlying lesions extend from a non-specific acute or chronic inflammation to a necrotizing granulomatous vasculitis.

Nervous system

Brain Both central and peripheral nervous system signs and symptoms are quite inconstant: mononeuritis multiplex, cranial nerve disease, syncope and diabetes insipidus are among the variegated lesions that have been reported.[93] Rarely, cerebral or hypothalamic signs predominate.[152]

Eyes Among the diverse lesions are conjunctivitis, episcleritis, uveitis, optic nerve vasculitis, retinal artery occlusion, nasolacrimal duct obstruction and proptosis.[93] In one case in three of this series, retro-orbital masses were responsible for the proptosis which resisted treatment. Where biopsy was made, acute and chronic inflammation were found to have occurred, with or without granulomatous vasculitis.

Other sites

Among other sites where the lesions of Wegener's granulomatosis have been described are: the parotid gland, the mastoid bone, the larynx, the thyroid gland, the breast, the cervical vertebra, the peritoneum, the liver, the spleen (Fig. 15.16), the vocal cords, and the tympanic membrane.[93] Inflammatory bowel disease may be simulated;[348] rectal and oral biopsy facilitate diagnosis.

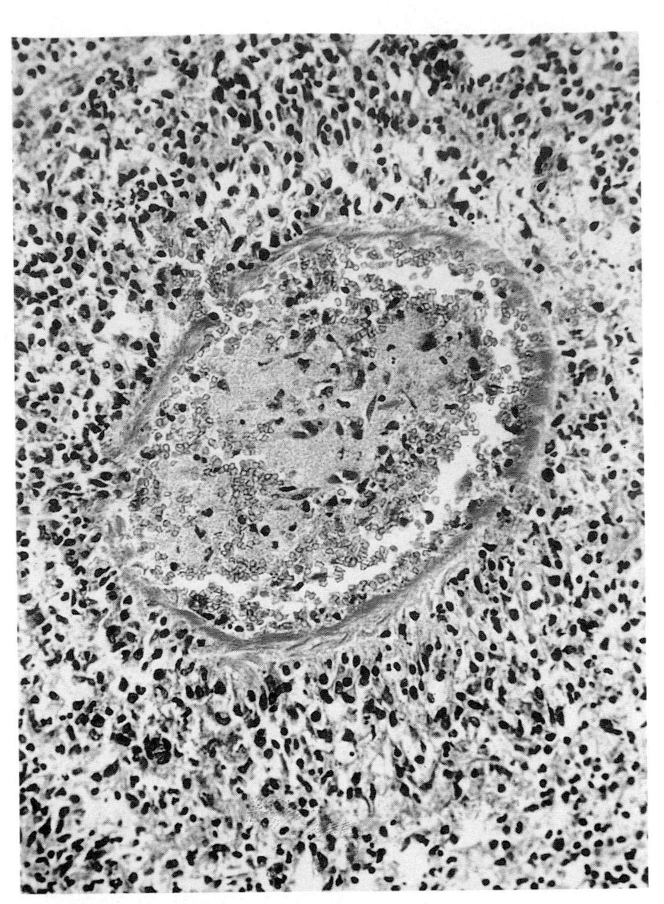

Fig. 15.16 Spleen: Wegener's granulomatosis. The media of this splenic artery is necrotic: the wall is replaced by an exudate which has the appearance of fibrinoid. The elastic lamina has been destroyed and the lumen of the artery is occluded by organizing thrombus. (HE × 250.)

Aetiology and pathogenesis

The Churg–Strauss syndrome and Wegener's granulomatosis may be members of a continuous spectrum of disorder,[128] a view, however, challenged by Liebow[224] (Table 15.16). The lung has a restricted repertoire of responses to injury and it should not be concluded that apparently similar conditions are necessarily variants of the same disorder. In Wegener's granulomatosis, the origin of disease in the upper respiratory tract or lungs, and evidence of the presence of increased titres of circulating immune complexes, together suggest that an unidentified antigen may gain access to a sensitized individual via the airways. The role of an Arthus phenomenon localized to the bronchi has been proposed[95] and the pulmonary deposition of IgG and complement has been reported.[339] Little is known, however, of the detailed mechanisms by which the granulomata of Wegener's disease begin. Since they occur

Table 15.16

Liebow's view of granulomatous lung disease with vasculitis[224]

Classical Wegener's
Limited angiitis and granulomatosis of Wegener type—no upper respiratory tract involvement and no renal disease: but may have focal renal granulomata and other, extrapulmonary lesions
Lymphomatoid granulomatosis
Necrotizing sarcoid angiitis and granulomatosis
Bronchocentric granulomatosis

in the absence of vasculitis, it seems possible that two distinct mechanisms play a part and that respiratory tract-borne antigens and blood-borne immune complexes create independent but related lesions. There is little to support a hereditary, racial or occupational predisposition.

Renal disease is delayed and secondary. There is substantial evidence to demonstrate glomerular and renal vascular deposits of antibody and complement: immune complex deposition is presumed, the consequence of circulating complexes.[324] The suggestion has been made that renal immune complexes, which are transient, are quickly cleared by phagocytosis. Alternatively, the renal and vascular lesions may both be consequences of viral infection or of the formation of antibasement membrane antibodies.[383]

Prognosis

Accounts of the symptomatology, pathology and management of two patients in whom the diagnosis of Wegener's granulomatosis was debated, help to explain problems of diagnosis and treatment.[124,218] In a typical case, a 44-year-old White business man complained of cough, haemoptysis, frontal headache and chest pain; he died within a few days. In another case a 59-year-old commercial fisherman who had had severe arthralgia for five years, quickly developed more severe joint pain with cough and haemoptysis. He had been treated for syphilis many years previously.

The mean survival time for untreated cases was five months: 82 per cent of patients died within one year, 90 per cent within two years. Renal failure with uraemia was a frequent cause of death; in other instances opportunistic or secondary infection developed. Those with inadequately treated, severe, systemic vasculitis and renal disease had a poor outcome; those with mainly respiratory disease had a better prognosis.[37,227]

The use of corticosteroids prolonged survival but it was the advent of nitrogen mustard[87] and other cytotoxic drugs that heralded a new era of treatment. Immunosuppressive agents including azathioprine,[6] cyclophosphamide,[73] antilymphoblast serum[197] and chlorambucil,[238] with a variety of corticosteroids—have been successful in suppressing very active disease and in greatly prolonging survival.[9] Cyclophosphamide is the drug of choice, and in 79 of 85 cases (93 per cent) complete remission was induced.[93] In only six of these 85 cases was death attributable to active granulomatosis, and there is evidence that granulomatous lesions, such as those of the brain, may be not only suppressed but 'eradicated'.[152]

One benefit of cytotoxic therapy may be to permit renal transplantation in cases where renal failure has occurred.[197] Transplantation is not necessarily accompanied by the risk of disease developing in the transplanted kidney.[374]

Allergic granulomatosis (Churg–Strauss phenomenon)

Churg and Strauss[52] described a clinical syndrome of asthma, fever and eosinophilia, together with symptoms of vascular embarrassment in various organ systems. The histological features in 13 cases were those of widespread vascular lesions 'of the type seen in periarteritis nodosa', with alterations in the vessel wall and in extravascular connective tissue. The changes included a necrotic eosinophilic exudate, 'fibrinoid', and an extravascular granulomatous proliferation of epithelioid and giant cells, an 'allergic granuloma'. These changes, Churg and Strauss believed, constituted an entity apart from classical polyarteritis nodosa, 15 cases of which, without asthma, did not display extravascular granulomata or granulomatous vascular changes. There was no reference to the work of Wegener[387,388] (see p. 633) but later Godman and Churg[128] and Churg[51] stated or implied that allergic granulomatosis, with tissue eosinophilia, closely resembled Wegener's granulomatosis which, it appeared, might also be attributed to hypersensitivity.

Few additional series of cases of the Churg–Strauss syndrome have been analysed. Those of Chumbley, Harrison and DeRemee[49] and of Finan and Winkelmann[97] are noteworthy. However, only six of the 30 cases reported by Chumbley and his colleagues were investigated *post mortem* and the studies of Winkelmann and Finan, who included 12 previously reported cases among their 27 patients, were centred largely on findings obtained by cutaneous biopsy. Twenty further cases provided the basis for Masi's analysis.[244a]

Clinical features

The clinical presentation of the syndrome was explicitly defined by Churg and Strauss[52] and has been described in detail by Lanham *et al.*[208] and in outline by Conn and Hunder.[62] It seems best to give a short summary of the original account.[52]

The patients were female rather than male, and aged 9 to 63 years. All had asthma beginning early or late in life and, on average three years before the terminal illness. Blood eosinophilia followed a period of increasingly severe asthma with fever. Untreated, the duration of the disease until death ranged from three months to five years. Recurrent episodes of pneumonia were sometimes indistinguishable from the findings of Loeffler's eosinophilic pneumonitis. Hypertension was frequent, cardiac failure characteristic. The cutaneous changes were those of non-thrombocytopenic purpura; erythematous, maculopapular or pustular rashes were seen but deep cutaneous or subcutaneous nodules[400] attracted special attention because, in five of seven cases, they provided the material for biopsy diagnosis. Lymphadenopathy, peripheral neuropathy, central nervous system symptoms and arthropathy were encountered. In one patient, articular signs led to the diagnosis of rheumatoid arthritis, a figure to be compared with the 26 cases of cutaneous extravascular necrotizing granuloma of Finan and Winkelmann[97] among whom four had rheumatoid arthritis, and two systemic lupus erythematosus.

Delineation of allergic granulomatosis may be aided by the use of the 1990 ACR criteria[244a] (Tables 15.17 and 15.18). There are now at least three assessments of the clinical syndrome. Some insist that diagnosis be restricted to the original criteria, and that necrotizing vasculitis, eosinophilic tissue infiltration and extravascular granulomata be demonstrated histologically before the condition is accepted. Others[208] believe that this syndrome, with asthma and eosinophilia (see Fig. 15.17), is readily recognized clinically and that rigid insistence on these histological characteristics has created the false impression of rarity. A further view questions whether the Churg–Strauss phenomenon is really distinct from Wegener's (localized) pathergic granulomatosis[390] and whether, indeed, it is a specific entity. Uncertainty regarding the nature of the Churg–Strauss syndrome has been made greater by the original authors. They[53] appear to have extended their diagnostic spectrum to include Loeffler's eosinophilic pneumonia and bronchocentric granulomatosis.[224]

Pathological changes

Many of the macroscopic, visceral changes are indistinguishable from those of the syndrome of polyarteritis nodosa (see p. 619). Infarcts, scars, haemorrhages and aneurysms are commonplace. Vascular disease may affect the lungs:

Table 15.17

1990 ACR criteria for the classification of Churg–Strauss syndrome (traditional format). Modified from Masi *et al.**[244a]

Criteria	Number of CSS† patients (n = 20)	Sensitivity (per cent)	Number of control patients (n = 787)	Specificity (per cent)
1. Asthma	19	100	782	96.3
2. Eosinophilia >10 per cent	20	95	708	96.6
3. Neuropathy, mono or poly	20	75	781	79.8
4. Pulmonary infiltrates, non-fixed	20	40	736	92.4
5. Paranasal sinus abnormality	14	85.7	366	79.3
6. Extravascular eosinophils	16	81.3	385	84.4

*For classification purposes, a patient shall be said to have Churg–Strauss syndrome if at least four of these six criteria are positive. The presence of four or more of these six criteria yields a sensitivity of 85 per cent and a specificity of 99.7 per cent.

†CSS = Churg–Strauss syndrome.

Table 15.18

1990 ACR classification tree criteria for Churg–Strauss syndrome. Modified from Masi *et al.*[244a]

CSS* subsets	Number of patients CSS/ non-CSS	Percentage correctly classified within subset	Percentage of total CSS patients in subset	Non-CSS subsets	Number of patients CSS/ non-CSS	Percentage correctly classified within subset	Percentage of total non-CSS patients in subset
Asthma and eosinophilia >10 per cent	18/2	90	90	No asthma or eosinophilia >10%	0/735	100	93.4
No asthma, eosinophilia >10 per cent, history of allergy	1/4	20	5	No asthma, but eosinophilia >10 per cent, no history of allergy	0/19	100	2.4
				Asthma without eosinophilia >10%	1/27	96.4	3.4

The sensitivity of this format of classification was 95 per cent and the specificity 99.2 per cent.
*CSS = Churg–Strauss syndrome.

there is pulmonary artery wall thickening and, sometimes, thrombosis. Lower lobe pulmonary consolidation accompanies the bronchial changes of asthma. The heart is severely affected with patchy myocardial scarring, ventricular hypertrophy, mural thrombi and pericarditis.

Microscopically, the arterial lesions are indeed those of polyarteritis nodosa. However, these focal lesions, with all stages of healing and fibrosis, are accompanied by a granulomatous response within the vessels. Inflammatory foci also occur in veins in the walls of which epithelioid cell granulomata are seen.

Extravascular connective tissue granulomata are characteristic: they are single, multiple or widespread (Figs 15.17 and 15.18). They comprise numerous eosinophil polymorphs with smaller numbers of macrophages and giant cells of either foreign body or Langhans type; multinucleate giant cells are more prominent as the disease advances. Plasma cells, lymphocytes and neutrophil polymorphs are also present. Although the heart, particularly the pericardium, is particularly severely affected, similar inflammatory foci are found in lung, bile duct, spleen, kidney, prostate,[408] ureter, perirenal fat, lymph node and skeletal muscle. The thymus may be affected severely and directly[181] but there is little to suggest that cell-mediated immunity is disordered. The syndrome has been identified in a woman with primary biliary cirrhosis and polychondritis:[61] in this case, the temporal arteries were affected by an arteritis in which the presence of fibrinoid distinguished the disorder from giant-cell arteritis (see p. 642).

The granulomatous nodules develop near small veins. The nodules measure 50 μm to 1 mm. Surrounding a core of necrotic cells and acidophilic collagen, are radially arranged macrophages and giant cells. The cells involved in the necrotic process are mainly eosinophil polymorphs with some macrophages and local tissue cells. Foci of eosinophils, at first intact, undergo necrosis. There is much eosinophil cationic protein and eosinophilic protein-x in these extravascular granulomata.[359] There is sometimes a resemblance to an epithelioid tubercle. The altered collagen, no longer doubly refractile, becomes increasingly acidophilic before losing its stainability. These fibre changes, which constitute what Churg and Strauss termed 'fibrinoid swelling', are most conspicuous where the cellular exudate is least, especially in chronic cases.

Immunopathology

Many but not all the reported cases were atopic; some had allergic rhinitis. Asthma was characteristic and many first-degree relatives were asthmatic. Hypocomplementaemia and circulating immune complexes were exceptional. Occasionally, there was a positive C1q-binding test, and IgM bound to complement and IgM were shown in several renal biopsies.[208] The latex test for rheumatoid factor was negative.

Serum IgE levels are raised but may return to normal as vasculitis remits.[208] Although this phenomenon has also been described in Wegener's granulomatosis and in polyarteritis nodosa, it remains possible that IgE-containing immune complexes play a part in the pathogenesis of the Churg–Strauss lesions.[200] Direct immunofluorescence tests on cutaneous granulomata revealed C3 or fibrin in six

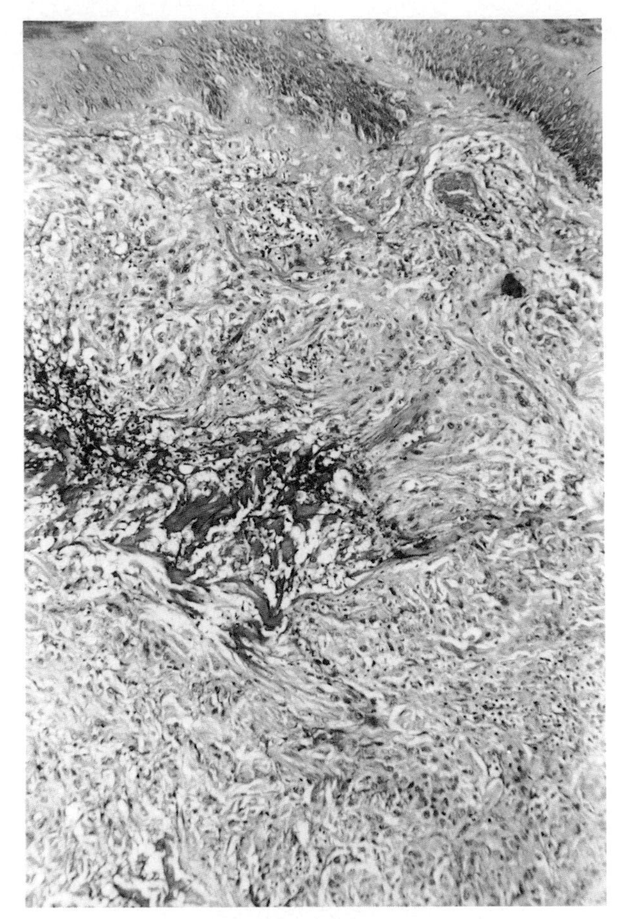

Fig. 15.17 Skin: allergic granulomatosis.[52]
Within this dermal tissue there is an ill-defined, stellate granuloma the centre of which is occupied by deeply basophilic (black-appearing) tissue debris. There is a surrounding cellular infiltrate among which are many polymorphs and eosinophil granulocytes. (HE × 33.)

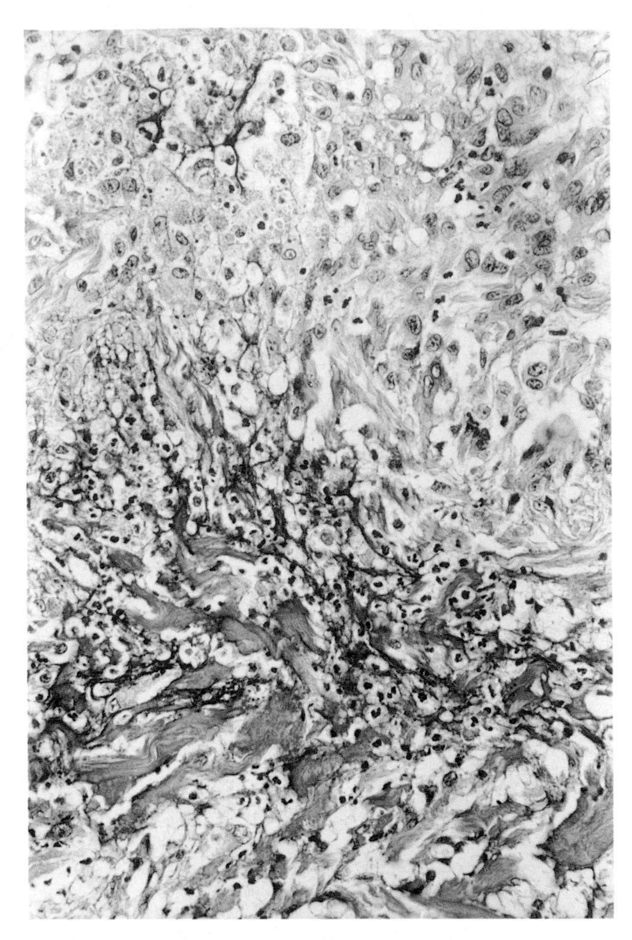

Fig. 15.18 Skin: allergic granulomatosis.
Detail from dermal lesion seen in Fig. 15.16. **Top**: the cellular reaction is of mononuclear macrophages; some display rudimentary giant-cell formation. **Bottom**: there are broad bundles of degenerate, hyaline-appearing collagen. They mingle with strands of basophilic exudate interspersed with polymorphs. (HE × 83.)

or eight cases, respectively. IgM or IgG were only occasionally identified.[97] Fibrosis and scar formation rather than resolution followed. Scars were prominent around blood vessels and throughout the heart.

Pathogenesis

The association with allergy is inescapable. The nature of the allergen remains uncertain and little is known of any genetic predisposition. The disease may be mediated by immune complexes that lodge in pulmonary and systemic vascular sites. IgE has been implicated, and immune complexes containing this reaginic antibody may be the proximate cause of the lesions. Phenytoin may provoke a systemic granulomatous vasculitis[111] and experimentally, an analogous granulomatous vasculitis has been caused in rats by glucan.[182]

Causes of death and prognosis

The response to corticosteroids is dramatic but occasional cases react best to azathioprine or to cyclophosphamide. Previously, survival was rare. Now the five-year survival may be as much as 62 per cent[49] and the mean survival time nine years. In another series of 16 cases, studied over six years, there was one death.[208] The causes of death in 50 cases are given in Table 15.19.

Giant-cell arteritis

Giant-cell arteritis is a frequent, but poorly understood, focal vasculitis encountered in elderly females and associ-

Table 15.19

Causes of death in 50 cases of allergic granulomatosis[208]

	Number	Per cent
Congestive cardiac failure/ myocardial infarction	24	48
Cerebral haemorrhage	8	16
Renal failure	9	18
Gastrointestinal tract perforation or haemorrhage	4	8
Status asthmaticus	4	8
Respiratory failure	1	2

ated with the clinical syndrome of 'polymyalgia rheumatica'.[135] Certain branches of the internal and external carotid arteries are particularly vulnerable but any visceral or peripheral artery may be affected. Inflammation of the superficial temporal artery, with giant cells, causes the clinical syndrome of 'temporal arteritis'.[85,132,153,172]

Historical aspects

Attention was drawn to the syndrome of temporal arteritis by Hutchinson[175] who wrote of a tightly fitting hat as a cause of headache. There is an early Arabian reference.[142] Horton and his colleagues rediscovered temporal arteritis and described the histological changes.[167,168,169]

Polymyalgia rheumatica (senile rheumatic gout[38]) was renamed by Barber.[17] It is now clear that temporal arteritis and polymyalgia rheumatica are local and systemic manifestations, respectively, of a disorder to the pathological features of which Gilmour[123] had given the descriptive term giant-cell arteritis.

Heredity

There is familial aggregation[221] and many reports of first-degree relatives with polymyalgia rheumatica or giant-cell arteritis.[136] An association with the inheritance of HLA-DR4 has been suggested.[26]

Frequency

Accurate information on the prevalence and incidence of the clinical syndromes of polymyalgia rheumatica and temporal arteritis is more easily obtained than is surgical or necropsy data on the frequency of giant-cell arteritis. Of 889 cases examined *post mortem* in which sections of the superficial temporal artery and the aorta were examined microscopically, 14 (1.6 per cent) had evidence of giant-cell arteritis.[291] Clinical signs of giant-cell arteritis were reported in 4.2 to 17.4 per 10^5 individuals aged over 50 years.[172] Increased awareness has led to more frequent diagnosis. In our ageing populations, further increases are anticipated. In one series[150] there was a female/male preponderance of 4:1, in another,[91] of 2:1. The mean age of the patients in the two series was 80 and 67 years, respectively.

Clinical features

Giant-cell arteritis[171a] causes three clinical syndromes[171,172,173] which reflect the vascular territories that are affected (Fig. 15.19, Tables 15.20 and 15.21).

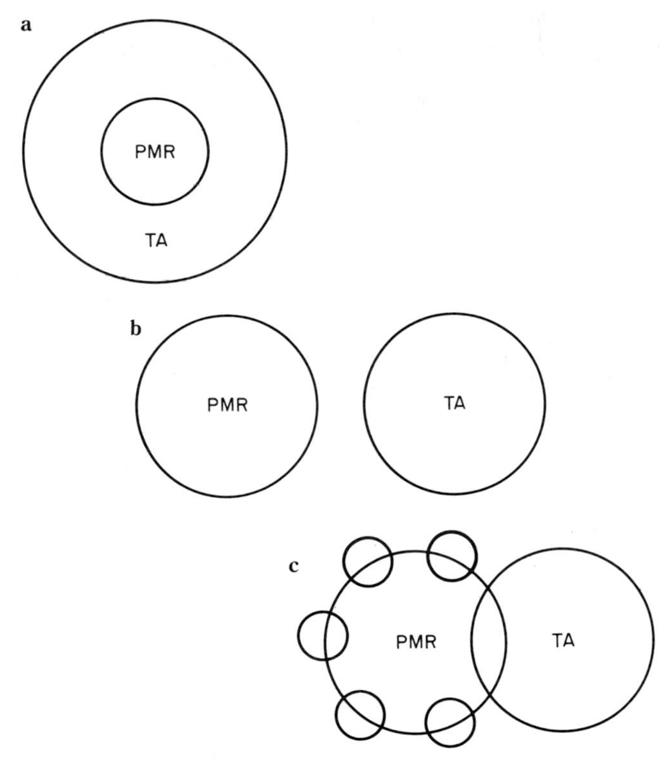

Fig. 15.19 Possible relationships of polymyalgia rheumatica (PMR) and temporal arteritis (TA).
a. The representation is incomplete; polymyalgia displays several features not found in temporal arteritis; **b.** the diagram does not account for similarities between the syndromes; **c.** overlap of temporal arteritis and polymyalgia demonstrating that several disorders may create the polymyalgia symptom complex. Those in the central overlap area display a high risk of developing the vascular sequelae of temporal arteritis. Redrawn from Goodman.[132]

Table 15.20

1990 ACR criteria for the classification of giant cell (temporal) arteritis (traditional format). Modified from Hunder *et al.*[171a]

Definition of criteria
1. Development of symptoms or findings beginning at age 50 or older
2. New onset of or new type of localized pain in the head
3. Temporal artery tenderness to palpation or decreased pulsation, unrelated to arteriosclerosis of cervical arteries
4. Erythrocyte sedimentation rate ≥50 mm/hour by the Westergren method
5. Biopsy specimen with artery showing vasculitis characterized by a predominance of mononuclear cell infiltration or granulomatous inflammation, usually with multinucleated giant cells

The presence of any three or more criteria yields a sensitivity of 93.5 per cent and a specificity of 91.2 per cent.

Table 15.21

1990 classification tree for giant cell (temporal) arteritis (GCA).* Modified from Hunder *et al.*[171a]

GCA subsets	Number of patients GCA/non-GCA	Percentage correctly classified	Percentage GCA patients in subset	Non-GCA subsets	Number of patients GCA/non-GCA	Percentage correctly classified	Percentage non-GCA patients in subset
Age ≥50 at disease onset and tenderness or decreased temporal artery pulsation	122/11	92	57	Vasculitis, with normal temporal arteries, no claudication, and negative biopsy	6/445	99	75
Age ≥50 at disease onset and artery biopsy showing mononuclear cell or granulomatous inflammation, with clinically normal temporal arteries	77/42	65	36	Vasculitis, age <50 at disease onset, claudication, normal temporal arteries, and biopsy without specified findings	0/3	100	<1
Age ≥50 at disease onset, claudication, normal temporal arteries and negative biopsy	5/2	71	2	Vasculitis, age <50 at disease onset, clinically normal temporal arteries, and biopsy with specified findings	2/83	98	14
				Vasculitis, with abnormal temporal arteries, age <50 at disease onset	2/7	78	1

*The classification tree yields a sensitivity of 95.3 per cent and a specificity of 90.7 per cent.

Polymyalgia rheumatica

An elderly woman, previously well, develops low-grade fever with weight loss. Arthralgia and myalgia are abrupt or insidious. Ill-defined pain and stiffness affect the neck, shoulder girdle and proximal parts of the arms. Malaise, fatigue and anxiety may persist for long periods before the diagnosis is established. Morning stiffness can simulate rheumatoid arthritis. Movement accentuates the pain which is common and severe at night. Pain inhibits movement but muscle atrophy and joint contractures contribute to inactivity. Signs of active synovitis are confirmed and the knee and sternoclavicular joints are often inflamed; synovitis of the shoulders and hips is occasional. There is a mild synovial fluid polymorph leucocytosis. A slight normocytic, hypochromic anaemia is usual. The erythrocyte sedimentation rate is very high but there is no leucocytosis. A diminished serum albumin and increases in α_2-globulin and fibrinogen are recorded. However, tests for antinuclear antibody and for rheumatoid factor are negative.

Temporal arteritis

The disease begins in old age with headache in the temporal region or with the rapid advance of impaired vision. The temporal arteries are often tender and nodular. The patient may be febrile: a raised erythrocyte sedimentation rate is evidence of the systemic nature of the disease. Examination reveals arterial abnormalities in other parts of the body; the blood pressure may be elevated but there is no consistent evidence of renal ischaemia. There is a beneficial response to small doses of corticosteroids but large doses may be needed quickly to prevent blindness.

Giant-cell arteritis syndrome

Although the symptoms and signs characteristic of temporal arteritis may be the first clinical evidence of this systemic disorder, arteritis is not confined to the territories of the external carotid arteries and widespread disorder is evident: jaw claudication and lingual Raynaud's phenomenon are indices of involvement of other cranial arteries.[68,301] In addition to complaints which simulate temporal arteritis and polymyalgia rheumatica, fever, malaise, claudication of the jaw, tender scalp nodules, and the aortic arch syndrome are identified. In one series of 25 patients followed over five years, six had concurrent autoimmune thyroid disease. Giant-cell arteritis may appear in an occult form[154] so that fever, anaemia, weight loss, claudication or neuropathy may be presenting signs of the disorder: it has been described in the very young when the role of virus is suspected.

Pathological changes

Unit lesions

The 'unit' lesions of giant-cell arteritis[147,148] centre on inflammation of all coats of the artery (Fig. 15.20). Histiocytes, lymphocytes, plasma cells and occasional neutrophil polymorphs are present in the adventitia. Eosinophil polymorphs are exceptional. The cell infiltrate is focal or of uneven distribution, reflecting a non-uniform inflammation of the media. Around these foci, fibronectin accumulates[46] and α_2-macroglobulin, lysozyme and Factor VIII are demonstrable. There is patchy loss of muscle cells by necrosis and a mononuclear phagocytic infiltrate. Giant, multinucleate cells are seen, particularly where there is demonstrable injury to, and loss of parts of the internal elastic

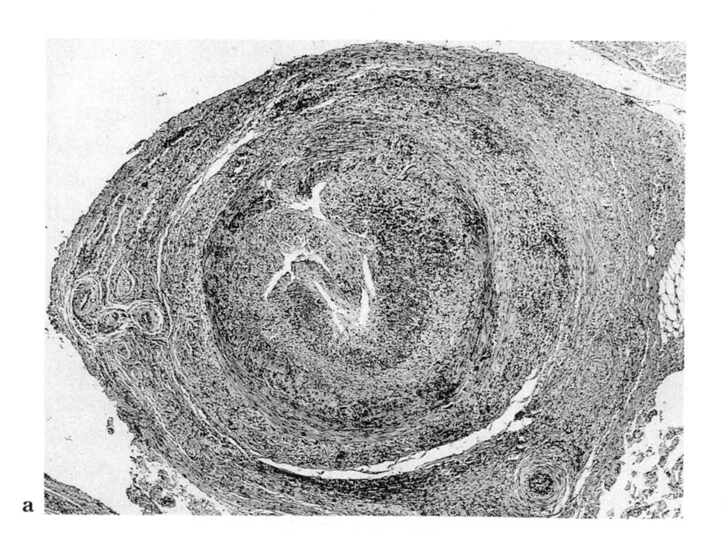

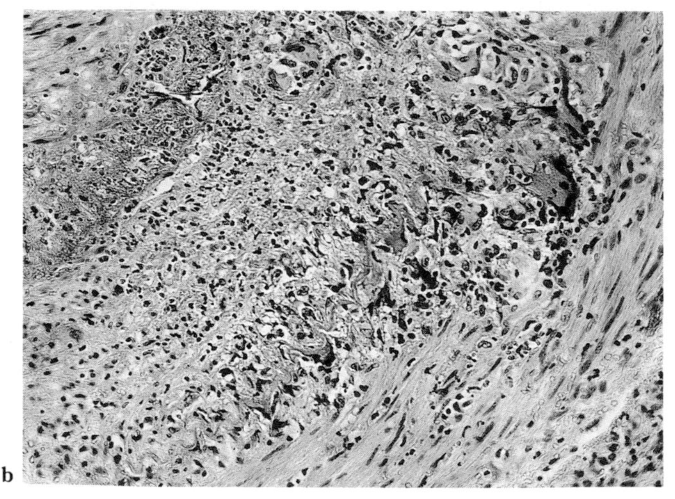

Fig. 15.20 Temporal artery: giant-cell arteritis.
a. There is extensive inflammatory cell infiltration of the vessel wall with involvement of the surrounding perivascular loose connective tissue. The arterial lumen is occupied by organizing thrombus. (HE $\times$ 45.) **b.** Higher-power view of same field. The internal elastic lamina runs diagonally across the field from lower left to upper right. Multinucleated giant-cells border the disintegrating internal elastic lamina. The arterial lumen (top left) is filled with thrombus. (HE $\times$ 150.)

Clinical presentation

The major clinical criteria[23,282,283] are: 1. recurrent aphthous mouth ulceration (99 per cent); 2. skin lesions including erythema nodosum-like eruptions, subcutaneous thrombophlebitis and hyperirritability of the skin (66 per cent); 3. eye lesions, including recurrent hypopyoniritis or iridocyclitis, and chorioretinitis (66 per cent); and 4. genital ulceration (80 per cent).

The minor criteria include arthritis or arthralgia (55 per cent); gastrointestinal pain and melaena; epididymitis; vascular lesions including thrombosis and aneurysm; central nervous system disorder, especially brain stem syndromes; meningoencephalomyelitis (22 per cent) and confusion. It is convenient to speak of 'complete' or 'incomplete' Behçet's syndrome depending on how many of the major criteria exist and on the presence or absence of ocular disease.[340] The vascular and neurological features are delayed; the latter, in particular, are ominous.

Pathological changes

As many as 60 necropsy reports are registered each year in Japan. There are numerous individual cases in the literature.[55,195,355,356] The retrospective analysis of 170 Japanese autopsies conducted between 1961 and 1976 offers an invaluable source of data.[205] These cases comprised 122 men and 48 women. The ages at autopsy ranged from less than 19 years to more than 70 years: the majority were in the fourth and fifth decades.

Skin and mucosae

Mucosal lesions were recorded in 88 of 170 necropsies.[205] The commonest lesions were genital ulcers (48), stomatitis and aphthous ulcers (24) and glossitis and lingual ulcers (11). There were cutaneous changes in 15 cases: 13 were 'dermatitis'.

The early oral and genital lesions include characteristic aphthous ulcers of the labial, buccal, gingival and lingual mucosae, sharply defined and with surrounding erythema. Pyoderma with larger or smaller pustular lesions may be spontaneous but similar lesions can be produced by puncturing 'sterile' skin with a needle; they tend to occur at sites of operation or incision. Erythema nodosum-like changes may coexist; in these, transmission electron microscopic evidence suggests that fat cells become detached from the basal lamina succeeded first by lymphocytic and then by macrophage infiltration. Fat cell lysis follows, promoting inflammation.[165]

Nervous system

Neurological changes were present in 58 of 170 necropsy cases.[205] The most frequent disorders were cerebrospinal demyelinization (23), multiple encephalomalacia (11) and perivascular cell infiltration and cerebral atrophy (5 each). Ophthalmic abnormalities (34) comprised: uveitis (23), optic nerve atrophy (4), and phthisis bulbi (3). There was one example of optic neuritis. Vasculitis or angiopathy underlie changes which can include meningitis, myelitis, brain stem syndromes and organic confusional states. In two cases dying after a neurological disorder, necropsy revealed diffuse meningoencephalitis with multiple foci of softening, demyelination and giant-cell accumulation in the pons, midbrain, cerebral peduncle and internal capsule.[355] Hypopyoniritis is the classical ocular disease.[100] However, episcleritis, conjunctivitis, keratitis, iridocyclitis, retinal thrombophlebitis and optic atrophy are common.[18] In Japan, Behçet's syndrome is a frequent cause of blindness. Benign intracranial hypertension with papilloedema has been described.[186]

Osteoarticular system

The synovial joints were not analysed in the report of Lakhanpal *et al.*[205] although erythema nodosum (2), polymyositis (1) and systemic lupus erythematosus (1) were mentioned briefly. There is increased interest in this aspect of the systemic disease; it is appreciated that a sterile arthritis can occur.[19,20,102,245,353,404] The synovial fluid may contain so many leucocytes that the appearances may mimic sepsis:[101] a response to colchicine can be shown, presumably because of the effect of this compound on polymorph chemotaxis. It is clearly essential to avoid confusion with gonococcal arthritis (see p. 736). The ultrastructural appearances of the synovium in Behçet's disease have been illustrated clearly.[131]

Arthropathy may be monoarticular or oligoarticular. Arthritis usually develops: the knees, ankles, wrists and elbows are most often affected. The small joints are rarely involved but the manubriosternal and sacroiliac articulations may be implicated. The synovitis is of limited extent, with involvement of only the superficial zone.[375] Of eight specimens from six patients, three confirmed as sterile, all save one were replaced by dense, inflamed, vascular granulation tissue composed of lymphocytes and a variable population of macrophages, neutrophil polymorphs and fibroblasts. A plasma cell infiltrate with lymphoid follicular formation was identified in only one. The three specimens that included part of the articular surface revealed pannus formation with marginal cartilage erosion. In spite of these changes the joints do not progress to fibrous ankylosis and deformity.

Articular morphology and function are preserved although subcutaneous nodules may form.[409]

Cardiovascular system

Cardiac abnormalities were present in 28 of 170 cases.[205] Cardiomegaly (12), endocarditis (4) and pericardial effusion were recorded as were numerous other disorders. Cardiac disease, with primary valvulitis, may result in aortic regurgitation.[60] Alternatively, congestive cardiac failure may develop without evident cause and Behçet's disease is therefore thought to be responsible for a form of congestive cardiomyopathy.[162] Arterial disease was very common: the most frequent abnormalities were abdominal aortic aneurysm (10), pulmonary artery thrombosis (5), and aortic atherosclerosis (4). In one 15-year-old boy, pulmonary arteritis accompanied right ventricular thrombosis.[13] However, many other associated diseases were noted including three cases of Takayasu's arteritis. Venous disorder was much less common than arterial and inferior vena-caval obstruction (4), oesophageal varices (3) and superior vena-caval thrombosis (2) were the most significant. Deep vein thrombosis of the legs and pelvis was surprisingly rare.

Multiple aneurysms of the aortic arch and the sinus of Valsalva have been found:[163] rupture is frequent. Abdominal aneurysm is characteristic.[226] Alternatively, there may be fusiform dilatation of the aorta[163] with aneurysms of the superior mesenteric artery and other vessels.

The fundamental lesions of Behçet's syndrome are therefore vascular.[18] The changes, particularly those of the arteries, recall those of other systemic connective tissue diseases and, indirectly, suggest explanations for pathogenesis. The frequency of vascular disease is high: 81 cases were found to be affected in a series of 800.[340] The category of Behçet's syndrome in which vascular disease predominates has been termed 'vasculoBehçet's disease'.[392] An appreciable proportion of cases with neurological disease (neuroBehçet's disease) are found to have lesions of small blood vessels as well; the latter may contribute to the former.

The histology of the arterial and venous lesions of Behçet's syndrome have been tabulated and described in detail.[109] The aorta and its main branches bear the brunt of the disease, with intimal fibromuscular thickening, loss of medial elastic tissue, adventitial fibrosis and lymphocytic infiltration of the vasa vasorum. Where small arteries are affected, 'punched-out' ulcers appear in the terminal ileum and ascending colon. Perforation occurs. The arterial disorder also includes intimal fibromuscular thickening with resulting zones of ischaemia and infarction. Small veins also are thick-walled and may thrombose. A thrombophlebitis characterizes the skin lesions, but, in large vessels, including the inferior vena cava, thrombosis is bland.

Respiratory system

Pulmonary disease is very frequent:[205,343] in one series 127 of 170 necropsy cases showed lung abnormalities. Pneumonia (66), pulmonary oedema (29) and pleuritis and pleural effusion (9) were the most frequent of many changes found. The so-called 'pneumonia' often presents with haemoptysis. The underlying disease is a granulomatous or necrotizing vasculitis (see p. 615).[77] Aneurysm may develop here, as it does in Wegener's syndrome.[30,45] Thrombosis, with pulmonary infarction, is one result. Cases with active lung disease can have circulating immune complexes. Arteritis is rarely seen in simple pulmonary thromboembolism and fibrinoid is unusual.

Alimentary system

Gastrointestinal tract

Of 170 cases examined at necropsy, 90 showed evidence of gastrointestinal disease[205] (Fig. 15.22). There were 29 examples of ileal ulcer of which

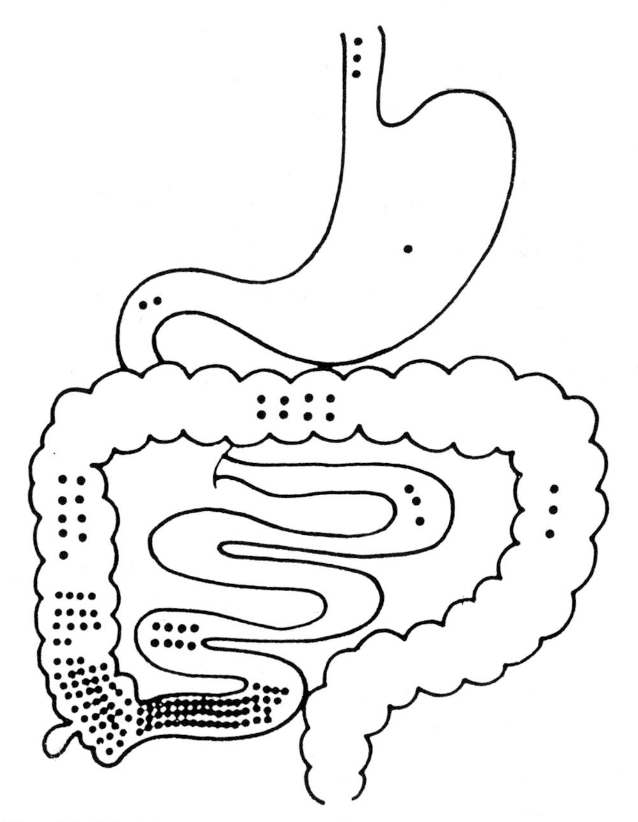

Fig. 15.22 Behçet's disease.
Location of gastrointestinal ulcers. Reproduced from Kasahara *et al.*[187]

17 had perforated; 15 of peritonitis; and 15 gastric ulcers. Many other disorders were noted but there was only one example each of necrotizing colitis, gastrointestinal haemorrhage and ulcerative colitis. Detailed histological studies of two cases of haemorrhagic diarrhoea showed that the basis of the inflammatory reaction that affected the whole thickness of the gut wall was a small-vessel vasculitis in which fibrinoid and leucocytoclasis were evident.[322] There is a good response to prednisone.[88] The use of labial biopsy has shown that latent Sjögren's syndrome is common in patients' with Behçet's disease[143] supporting the concept that autoimmunity plays a part in the pathogenesis of both disorders.

Intestinal disease is therefore frequent. Reviewing 136 cases coming to laparotomy, Kasahara *et al.*[187] found intestinal ulceration most often in the fourth and fifth decades: ulcers were especially common in the terminal ileum and caecum. Perforation was likely. Ulceration was localized or diffuse. Skip lesions were frequent and for this reason and because the stomach, duodenum and other parts of the alimentary tract were often affected, the preliminary diagnosis of Crohn's disease was sometimes made.

Liver, spleen and pancreas

Of 170 patients examined *post mortem*,[205] 58 had evidence of hepatobiliary disease, 37 of splenic disease and 6 of pancreatic disease. There were 21 instances of fatty liver, 18 of hepatic congestion and 6 of cirrhosis. There were no examples of hepatic infarction. However, hepatic vein thrombosis has been found as a cause of the Budd–Chiari syndrome.[395] Nineteen cases displayed 'splenitis', nine had splenic congestion but, curiously, only six were recorded as having splenomegaly.[195] There were only five instances of pancreatitis, in spite of the recognition of this sign as a frequent part of the disease spectrum.

Genitourinary system

Of 170 cases studied at necropsy, 60 revealed signs of renal disease but only six of genital disorders distinct from the mucosal changes listed (above). Pyelonephritis (23), cystitis (10) and nephrosclerosis (8) were the most common urinary abnormalities.[205] Biopsy studies do not accord wholly with these observations and in a prospective investigation of 10 cases with no symptoms of renal disease, morphological and immunofluorescent evidence of immune complex nephropathy was recognized in six of seven subjects.[159] Focal proliferative glomerulonephritis may lead to renal failure.[395]

Lymphoreticular system

There is very little evidence of lymphadenopathy and no consistent sign that Behçet's vasculitis is associated with a structural disorder of the immune system (see below).[71]

Other changes

Among other diseases observed *post mortem* in Lakhanpal's[205] series, were 13 miscellaneous neoplasms, some examples of bacterial, viral and fungal infection but no recorded cases of amyloidosis. However, in Middle Eastern cases, amyloid has been described as a complication of Behçet's syndrome.

Aetiology and pathogenesis

A genetic predisposition to Behçet's disease is thought to have been disseminated along the Silk Road followed by nomadic tribes between Europe and the Orient.[286] HLA-Bw51 may be implicated. There is also evidence of the inheritance of HLA-B5 and B27 with increased frequency and HLA-B5 appears to be associated with those cases with uveitis.

There are few clues to the nature of the immediate cause[1,75,211] although the racial clustering of cases in Turkey and Japan suggests the possible role of an infective agent, as in Kawasaki disease, or the operation of dietary factors or pollutants: guinea-pigs develop a Behçet-like state after prolonged treatment with organochemical compounds, pesticides or copper. Inclusion bodies have been described in the cells of hypopyon and oral exudates but no virus has yet been isolated. Nevertheless inoculations of human material can be used to cause mouse encephalitis.

Immunological abnormalities may influence the response to foreign antigen in food hypersensitivity; and T-cell reactivity may be impaired. However, the demonstration of changed humoral immunity shown by high levels of IgA, IgG, IgM, and C9, and the appearance of serum cryoprecipitates, especially when vasculitis is activated, are not necessarily specific and can be observed in uncomplicated recurrent aphthous stomatitis.[282,283] Immune complexes have been shown to be present in the sera: their level is related to disease activity and the titre is higher in neuro-ocular and arthritic forms of Behçet's disease than in the mucocutaneous forms.[219] The secretory component of IgA has been measured in the saliva and both the free and bound forms found to be deficient: the total protein concentration is normal.[1] Serum IgA levels are also normal but there is deficient IgA in the jejunal fluids. IgA-carrying B-cells are present in normal numbers and the evidence suggests an abnormality of the host defence mechanism at the mucosal surfaces.

The number of T-cells in the blood also appears normal[1] although this has been denied.[225] There may be a functional T-cell abnormality.[71] The mean proportion of T_H-cells is decreased, that of T_S-cells increased.[225] Cultures of T-cells from patients with Behçet's disease demonstrate the production of interferon-γ; the T-cells do not require the help of macrophages in secreting interferon-γ, leading to the view that an unknown causative agent such as a virus may stimulate the lymphocyte population *in vivo*.[106] An analogy with experimental adjuvant arthritis is suggested (see p. 544).

Prognosis and pathological response to treatment

The extent and nature of neurological and ocular disease determine prognosis. Their onset is ominous. The syndrome, in other respects, is benign.[79] There is no uniformly effective drug therapy but surgical intervention may be necessary, for example, in the treatment of aneurysm.

Kawasaki disease (mucocutaneous lymph node syndrome (MLNS or MCLS))

Kawasaki[189] and Kawasaki *et al.*[190] described an increasingly common, acute, febrile illness of Japanese children. The condition has continued to attract widespread interest, first, because it is now known to be ubiquitous, and second, because there is a mortality rate of 1.5–2.0 per cent, attributable to coronary artery vasculitis with aneurysm and to myocardial infarction. There is a close resemblance to the poorly understood condition of infantile polyarteritis nodosa and some believe that the two disorders are identical.[144]

Clinical presentation

Kawasaki disease presents with an erythematous rash, oral and pharyngeal lesions, a skin disorder, conjunctivitis and enlarged cervical lymph nodes. There is a resemblance to varieties of erythema multiforme such as the Stevens–Johnson syndrome. However, the development of myocarditis, pericarditis, aseptic meningitis, arthralgia or arthritis, enteritis and hepatitis are distinctive. Occasionally, pyoarthrosis develops.[35] Anaemia, polymorph leucocytosis and raised plasma IgE levels are among the laboratory findings. Initially, Kawasaki designated the disorder a 'mucocutaneous lymph node syndrome' (MCLS) but the failure of this term to recognize the importance of vasculitis and the uncertain significance of the lymphadenopathy cast doubt on the value of this designation.

A Japanese MCLS Committee[233] and, subsequently, the US Center for Disease Control[43,96] laid down criteria for the diagnosis of Kawasaki disease (Table 15.22) and these are widely accepted.

Kawasaki disease has continued to spread: many of the epidemiological features suggest that it is of an infectious nature. A Japanese survey revealed 80 cases in 1967 and 940 in 1973. By 1979, more than 24 000 Japanese cases had been reported.[253] Perhaps more significantly, there have now been three major Japanese epidemics in 1979, 1982 and 1985, with at least 15 000 cases in the 1982 outbreak. In 1985, the number of recorded cases was 4.5 times greater than in the interepidemic year of 1984.[406] There have now been North American epidemics and Kawasaki disease spares no ethnic group.

Initially, the acute disease was thought to be self-limiting. It was then realized that approximately 2 per cent of patients died during a delayed recovery phase. More recently, it has become clear that the late consequences of coronary arteritis may lead to death some years after the initial illness. Male mortality is 1.5 times higher than female.

Table 15.22

Clinical criteria for the diagnosis of Kawasaki disease

This table is based on the recommendations of the MCLS Research Committee of Japan[233] modified by the US Center for Disease Control.[43] The presence of five of the six criteria is required for diagnosis. Any finding listed under 3 or 4 is sufficient to establish that criterion.

1. Fever lasting five days or more
2. Bilateral conjunctivitis
3. Dry red lips with fissuring; strawberry tongue; oropharyngeal erythema
4. Palmar and plantar erythema; oedema with induration of hands and feet; membranous finger tip desquamation
5. Erythematous rash of trunk
6. Acute non-purulent cervical lymph node swelling, to diameters of 150 mm or more

Pathological changes

The lesions of Kawasaki disease usually resolve and even in those cases where coronary arteritis or aneurysm develop, there is often evidence of healing. Recently, there have been indications that Kawasaki disease may be less severe than was the case formerly but this may reflect increased perception and enhanced diagnostic sensitivity.[292]

Cardiovascular system

The pathology[365] of classical cases of Kawasaki disease is dominated by the severity, extent and progression of an arteritis which, although systemic, has a predilection for the coronary circulation. As in infantile polyarteritis nodosa, aortitis has been described and the carotid, subclavian and iliac vessels have been implicated. There is medial necrosis, periarterial inflammatory cell infiltration and thrombosis. Aneurysm frequently results. Although healing with fibrosis occurs over periods ranging from 4–6 months to 1–4 years, there is a significant risk that the aneurysm(s) will rupture, leading to sudden death. Acute myocardial infarction, which may be fatal, is a relatively common consequence of coronary artery thrombosis.

In an early study,[108] it was demonstrated that approximately 70 per cent of deaths occurred between 11 and 50 days after the onset of the febrile illness. The coronary artery lesions manifest specific times of onset, a selective localization and characteristic morphological features. Coronary arteritis was recognizable at 10 days; dilatation of the vessel was often found within the subsequent two days and the proximal parts of the main vessels were characteristically affected.[246]

The myocardial infarcts that complicate Kawasaki coronary arteritis usually occur within one year of the onset of the disease, but as many as 25 per cent take place later. About one quarter die during the initial attack and 16 per cent of the survivors have a second infarct.[188] In this series of 195 children, of whom 141 were male, 43 per cent continued to thrive. It is now possible, in selected cases, to undertake remedial surgery which may take the form of aortocoronary bypass, aneurysmectomy or combined procedures.[357] The disease may be restricted to one vessel and restoration of normal activity is practicable.

In cases where late death occurs suddenly from cardiac disease, there has been an opportunity to study the sequence of healing and organization of the cardiac and coronary artery lesions[364] and the early phases of healing were investigated in some detail by Fujiwara *et al.*[107] These authors analysed 69 autopsy cases, 38 acute, and defined the progression from panvasculitis with thrombosis to stenosis, recanalization and aneurysm. Small vessel angiitis accompanied the earlier lesions but was transient; 80 per cent of those cases examined in the phase of healing showed evidence of myocardial infarction.

Considerable light has been thrown on the changes in other systemic, non-coronary vessels by an investigation of 227 intercostal arteries from 17 fatal cases.[247] The proximal parts of the vessels were mainly affected. Dilatation or aneurysm formation was noted immediately beyond the origins of the vessels and smooth muscle necrosis accompanied an inflammatory cellular infiltration. Less common cardiac lesions, such as mitral insufficiency, occur.[196] The pathology of the cranial and peripheral nerves has attracted attention and muscle biopsy may show myonecrosis associated with circulating cryoglobulins and immune complexes.[161]

Lymphoreticular system

Less is known of the lymph nodes. Contrary to the statement of Melish,[253] diagnostic lymph node biopsy can be of value: the technique has been underused. Characteristic changes may be present in acute Kawasaki disease.[121] The changes include multiple foci of necrosis and microthrombi in the small blood vessels. In contrast to the widespread arteritis which affects, for example, coronary vessels, vasculitis is not a feature of lymph node vessels. Clearly, a wide spectrum of differential diagnosis must be considered, and among the disorders that require distinction from Kawasaki disease of lymph nodes are leptospirosis, thrombotic microangiopathy, systemic lupus erythematosus, tuberculosis and cat-scratch disease. Rickettsial infections may cause lymph node vasculopathy: in Rocky Mountain spotted fever, small arteries, veins and capillaries may be affected. Finally, the close similarity between Kawasaki disease and infantile polyarteritis nodosa suggests that the identity of the lymph node changes in the two disorders should be considered.

Immunopathological features

There is thymic involution and profound abnormalities of immunoregulation. The proportion of T_S-cells is reduced, that of T_H/MHC class II-positive cells increased. Enhanced synthesis of IgG and IgM is recognized.[215]

Aetiology and pathogenesis

Early pointers to possible allergic causes for Kawasaki disease included a tendency for children to have had eczema or nasopharyngitis; for BCG, smallpox, poliomyelitis or triple vaccine to have been administered; or for drugs such as those used in treating allergy, vomiting or the common cold to have been given to the pregnant mother. More recently, the immunological features of the disease and the probable role of infection have attracted interest. An abnormality of gastrointestinal immune regulation has been suggested by the finding of high levels of complement-fixing, circulating IgA-containing immune complexes.[287] There is thymic involution. Some preliminary experimental studies have failed to clarify the pathogenesis further. Epidemiological evidence strongly favours the hypothesis that a microbial agent is involved. However, studies of the role of rickettsiae and dust mites have been negative and recent evidence supporting the possibility that a retrovirus may be causally related to Kawasaki disease[342] has not been confirmed.[314] Nevertheless, filtrates from cultures of *Propionibacterium acnes* contain a cytopathogenic protein, to which antibodies increase in titre from age four years onwards in Kurume, Japan[369]—supporting the suggestion that this organism may have a role in promoting the characteristic three-yearly outbreaks of Kawasaki disease. Moreover, immunologically deficient mice and guinea-pigs injected with *Pseudomonas aeruginosa* develop a severe vasculitis with little evidence of inflammation but a tendency to aneurysm formation.[194] The response simulates the reaction of immunosuppressed patients to *Pseudomonas* spp.

Attempts to demonstrate a relationship to streptococcal infection are equally difficult to interpret. In another enquiry, nude BALB/C (*nu/nu*) mice and the heterozygous (*nu*/+) groups were maintained under specific pathogen-free and conventional conditions respectively. A β-haemolytic streptococcal vaccine was injected subcutaneously on five to ten occasions at weekly intervals. Between 15 and 60 days after the last injection, mitral valvulitis, together with myocarditis and endocarditis was present in both SPF nude and heterozygous mice. IgM and IgG were identified in the inflamed myocardium and in the arterial and periarterial zones. The relationship of this experimental syndrome to the pharyngitis-related MCLS remains uncertain.

Erythema nodosum

Erythema nodosum[31,84] is a form of subcutaneous vasculitis causing ill-defined, tender, red nodules of the skin of the anterior and lateral aspects of the lower legs. It is an increasingly uncommon disorder in Western countries, perhaps because the bacterial infections to which it is a hypersensitivity response, are themselves less frequent. Erythema nodosum was recognized by Willan;[396] an early account was that of Bury.[41]

Clinical presentation[19,263]

Groups of 10–50 mm diameter tender, dull-red, slightly raised nodules appear suddenly and without warning in the subcutaneous tissues; as the transient, ovoid nodules, which have firm centres but less sharply defined margins, fade and disappear they leave bruise-like, discoloured islands. Others take their place and a sequence of formation and regression extends over a six to eight week period. In one-third of cases, recurrence occurs. Systemic signs of illness are apparent. Fever and arthralgia or a symmetrical polyarthritis often develop.[18] The erythrocyte sedimentation rate and the white cell count are elevated, and occasionally conjunctivitis, myocarditis and erythema multiforme coexist. Comparable lesions, caused by enzymatic digestion of subcutaneous fat, occur rarely with obstructive and haemorrhagic pancreatitis.[263,269]

Pathological changes

It is generally agreed that erythema nodosum is primarily a vasculitis although the nature of the disease, which centres on and around small arteries and veins, has been debated.[98] The deeper dermis and subcutis display oedema, a polymorph infiltrate, and some congestion with the escape of erythrocytes into the loose extravascular tissue and local atrophy of fat cells (Figs 15.23 and 15.24). Although venular thrombosis is exceptional, the venous walls are the site of endothelial hyperplasia and a mild inflammatory cell infiltrate. There is a gradation in severity from acute phlebitis to a much milder response, and a lymphocytic infiltrate succeeds the acute reaction. In the later phase of the disease, mononuclear macrophages and multinucleate giant-cells are encountered. Occasionally, biopsy reveals the microscopic changes of giant-cell arteritis.[129] Plasma cells and eosinophils are exceptional but mast cells may be conspicuous. Arteriolitis is inconstant. Sarcoid-like, organized epithelioid cell granulomata are not seen but histiocytes can cluster around a characteristic central cleft, thought to be artefactual.

In the differential histological diagnosis, tuberculosis, sarcoidosis, brucellosis and metal or elemental granulomata are considered, while the cutaneous lesions of erythema multiforme, cutaneous necrotizing vasculitis and nodular vasculitis require exclusion. There are similarities with the

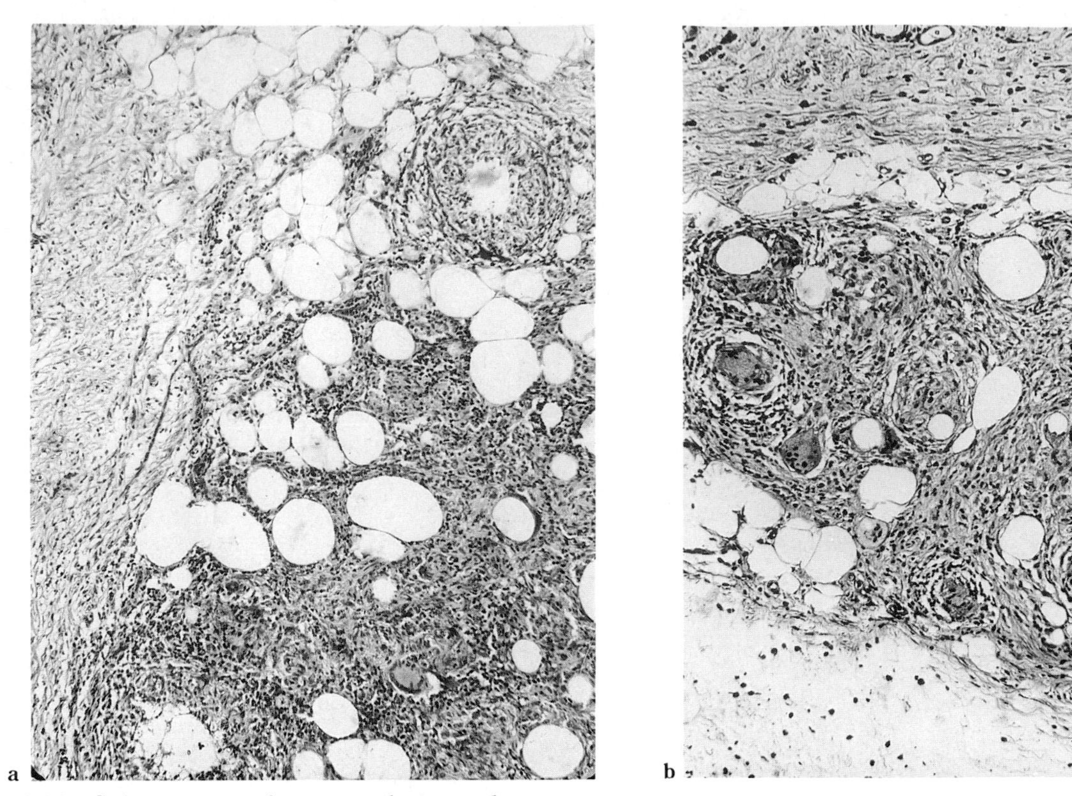

Fig. 15.23 Subcutaneous tissue: erythema nodosum.
a. The subcutaneous fat is infiltrated by mononuclear macrophages, polymorphs and lymphocytes. Occasional multinucleated giant cells are seen. The lesion is related to a low-grade vasculitis of small vessels such as the artery seen (top right). (HE × 90.) **b.** Compare with Fig. 15.23a. Numerous multinucleated giant cells are seen in relation to a zone of fat necrosis which has resulted from the vasculitis. (HE × 110.)

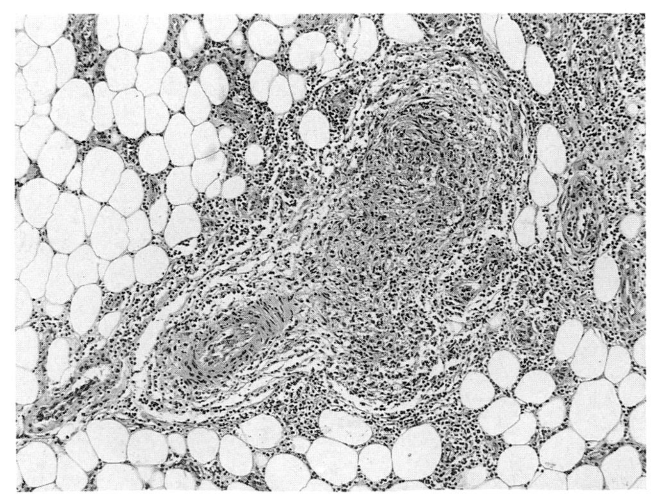

Fig. 15.24 Skin: nodular vasculitis.
Small, branching subcutaneous artery displaying inflammatory changes that are indistinguishable from those shown in Fig. 15.23. (HE × 70.)

cutaneous lesions of lepromatous leprosy[234] (see p. 746): transmission electron microscopic studies have been made of the subcutaneous erythema nodosum-like lesions[288] and *M. leprae* lie within deeper dermal and subcutaneous macrophages. Although angiitis was present in 56 per cent of 27 early lesions,[234] mastocytosis, glycosaminoglycan and reticular fibre increase and elastic fibre fragmentation suggest that this is not an adequate model on which to base an explanation of the pathogenesis of the idiopathic disease.[84]

Aetiology

Erythema nodosum is a common reaction to a miscellaneous group of agents, among which bacterial infection and drugs are important. Early in this century, tuberculosis was the main disease association; more recently, sarcoidosis in Scandinavia and streptococcal infection in the UK have become common antecedents. In west and south-west USA where coccidioidomycosis is endemic, this and other less common fungal infections are the most frequent causes. *Yersinia enterocolitica* and *Y. pseudotuberculosis* have been identified in Finnish series while sulphathiazole-treated lymphogranuloma venereum, hepatitis and amoebiasis are among the many other precedents of uncertain significance. Erythema nodosum leprosum is a characteristic lesion of lepromatous leprosy. Although the syndrome is not attributable to sulphone treatment, it appears when the mycobacteria are no longer viable. Ulcerative colitis and lymphoma are recognized associations of erythema nodosum but many other cases remain unexplained. Among the drugs thought to cause erythema nodosum are penicillin and the salicylates. Oral contraceptives have been incriminated.

Immunological aspects

Erythema nodosum is regarded as a hypersensitivity response. Acute-phase proteins including α_2-globulins are raised as are serum IgA and IgM; the former revert to normal as the rash fades. Elevated levels of complement and of cryofibrinogen have been described but no correlation has been found between erythema nodosum and specific humoral antibodies. That erythema nodosum may begin when delayed hypersensitivity reactions become positive, suggests an association with cell-mediated hypersensitivity. The onset of lymphocyte transformation, caused by exposure to the suspected antigen, and the presence of macrophage aggregating factor in the serum, support this view. Cultured lymphocytes from cases of erythema nodosum spontaneously release macrophage-activating factor during the active phases of the skin disease, not during recovery. Those cases in which immunofluorescence microscopy has shown complement and antigen in skin lesions have generally been found to display a nodular vasculitis. Conversely, the vessels in cases of nodular vasculitis have been shown to contain both immunoglobulin and bacterial antigens.[297]

Pathogenesis

The reaction of circulating antibody with local antigen to form immune complexes, as in the Arthus reaction, may account for the features of erythema nodosum leprosum but does not adequately explain the pathogenesis of the majority of cases. In contrast to nodular vasculitis, complement and immunoglobulin are not found locally. Cell-mediated reactions do not account for the clinical course, and cytokines, although sometimes demonstrable, are not provocative. As an alternative to an immune mechanism, disturbances non-specifically activating proteases and/or the Hageman pathway have been invoked.[84]

Erythema multiforme

In this disorder, a combination of target (iris), macular, papular, vesicular and bullous eruptions develops in acute episodes, sometimes with recurrence.[155,290] In rare instances the lesions become haemorrhagic. Occasionally there is evidence of an unusually severe systemic disturbance with an abrupt onset, fever, prostration and a widespread eruption becoming bullous and involving the mucous membranes. In these cases the term Stevens–Johnson syndrome is used. The benign form of erythema multiforme may be accompanied by arthralgia but, unlike classical erythema multiforme, polyarthritis is rare.

Pathological changes

In milder cases associated with macular and papular eruptions, epidermal spongiosus and intercellular oedema are accompanied by the presence in the dermis of a perivascular, inflammatory lymphocytic infiltrate.[217,289,351a] Occasional polymorphs and eosinophil granulocytes are seen. With more severe responses, the appearances resemble those of leucocytoclastic vasculitis (see p. 628). Endothelial cell injury and swelling, in small blood vessels, accompany the escape of red cells into the nearby connective tissue; a polymorph and eosinophil perivascular inflammatory cell infiltrate and the presence of nuclear debris are recognized. There may be bulla formation within the overlying epidermis the detached portion of which includes necrotic cells. The basal lamina remains on the floor of the bulla.

Aetiology and pathogenesis

The majority of cases appear to be hypersensitivity responses to foreign antigens.[351a] In approximately 15 per cent of patients the antigen is microbial: cases have accompanied Coxsackie virus, adenovirus and *Histoplasma* infection. Herpes simplex virus has been identified in some cases with recurrent skin disorders, and lesions resembling erythema multiforme have been produced by injecting herpes simplex antigen into the dermis of such patients. Low serum complement levels have suggested immune complex formation but no antiepidermal or antibasement membrane antibodies have been found locally by immunofluorescence. In many other instances (approximately 60 per cent) a chemical agent has been demonstrated; barbiturates, sulphonamides and penicillin have been implicated and a similar form of eruption has been recognized after therapeutic ionizing radiation. In a severe case observed personally, progressing to the Stevens–Johnson syndrome, phenylbutazone was also the cause of transient bone marrow suppression. Following withdrawal of the drug, the patient made a spontaneous recovery.

Pulmonary vasculitis

In pulmonary vasculitis[44,110] (Fig. 15.25), the vascular disease may be confined to the lung, as in limited Wegener's granulomatosis, or be part of a disseminated disorder. Classification offers problems but Chandler and Fulmer[44] describe three groups: those cases in which the lung is the principal organ involved; those in which the lung *may* be involved; and those in which the lung abnormality is only part of a systemic disorder. Several entities in which the lung may be implicated, such as Wegener's granulomatosis (see p. 633), the Churg–Strauss syndrome (see p. 639) and leucocytoclastic vasculitis (see p. 628), have already been described. Pulmonary involvement in the immunological disorders of the connective tissue system is discussed in appropriate sections of this volume, e.g. the lung disease of rheumatoid arthritis in Chapters 12 and 13; of systemic lupus erythematosus in Chapter 14. The present section is confined to the more important abnormalities not covered by these descriptions.

Classification

It is convenient to centre this account on the syndromes defined in Liebow's[224] review. The wider implications of pulmonary hypertension and pulmonary vascular disease are described in specialized works such as those of Harris and Heath.[146] Liebow[224] drew attention to a number of forms of pulmonary vasculitis with granulomatosis, a condition he defined as 'tissue necrosis with a peripheral, chronic cellular reaction not ascribable to occlusive lesions of the blood vessels'. There were five groups of cases (Table 15.23): Wegener's granulomatosis is described on p. 633.

Limited Wegener's granulomatosis

Eighty-five cases have been reviewed.[224] Wegener's granulomatosis can be 'limited' to the lung, Liebow believed,

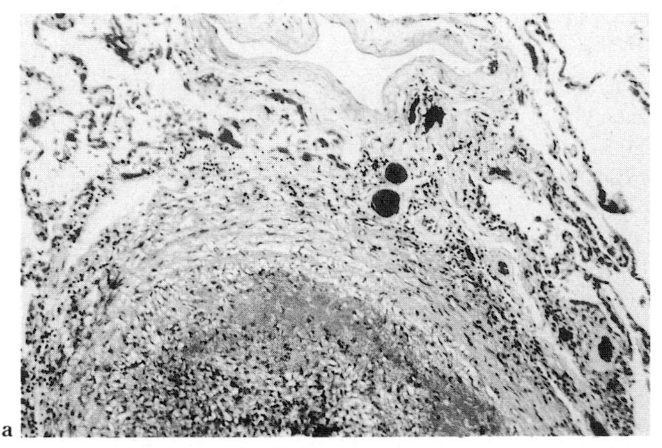

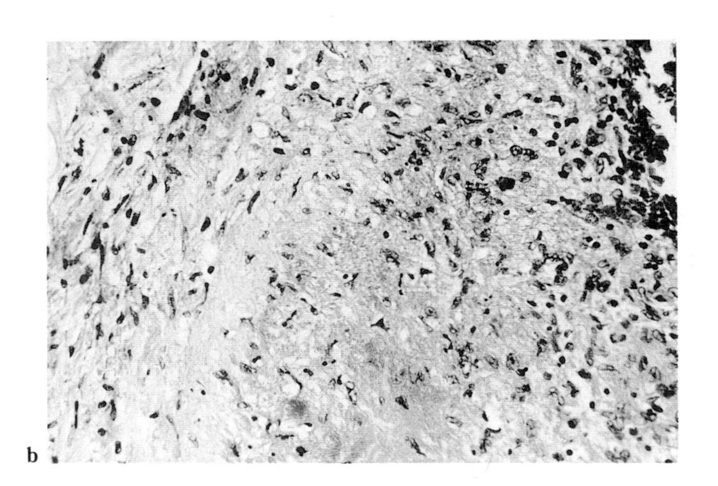

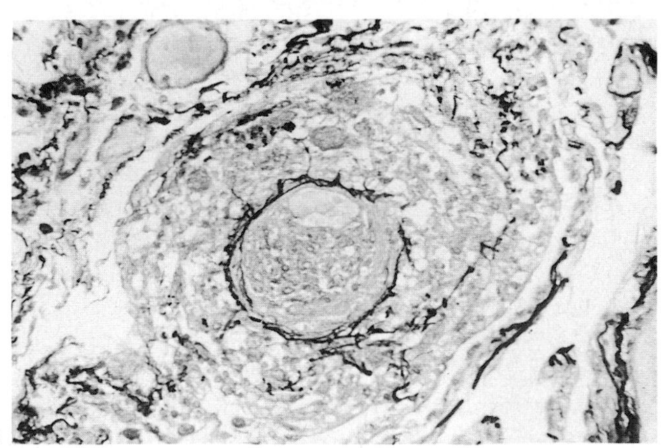

Fig. 15.25 Lung: necrotizing vasculitis.
a. A focal, segmental cellular infiltrate at the periphery of this small pulmonary artery is associated with a loss of integrity of the media and the presence, in the lumen, of recent thrombus. (HE × 113.) **b.** Compare with Fig. 15.25a. Portion of similar pulmonary artery, at higher magnification. In the lower, centre part of the field the disorganized vascular tissue has assumed the appearances of fibrinoid. (HE × 284.) **c.** The loss and disorganization of arterial elastic tissue is demonstrated. (Weigert–van Gieson × 284.)

Table 15.23

Pulmonary angiitis and granulomatosis[224]

Classical Wegener's granulomatosis
Limited angiitis and granulomatosis of the Wegener type
Lymphomatoid granulomatosis
Necrotizing sarcoid angiitis and granulomatosis
Bronchocentric granulomatosis

influencing prognosis beneficially. The 'limited' disorder, slightly more common in women than in men, is relatively more frequent than the complete triad. However, the distinction from the classical syndrome is not absolute and some cases of the 'limited' condition had lesions of other tissues such as the skin.

Lymphatoid granulomatosis

This is a lymphoproliferative disorder centred about blood vessels which are destroyed by the infiltration of lymphocytes, plasma cells and large mononuclear cells. Seventy-four cases have been reviewed.[224] Fibrinoid is not recognized; 10–15 per cent of cases progress to malignant lymphoma. Foci of necrosis within the infiltrate and pulmonary artery occlusion by the cell infiltrate or by thrombus, are characteristic. Adjacent bronchial lymph nodes are not involved. Nevertheless, there may be sufficient damage to the walls of bronchioles, with obstruction by granulation tissue, to lead to a misdiagnosis of bronchiolitis obliterans. Although survival is long and the effects of steroid therapy beneficial, the ultimate mortality, compounded by the occurrence of skin and central nervous system lesions, is very high.

Necrotizing 'sarcoid' vasculitis and granulomatosis

Necrotizing 'sarcoid' vasculitis with non-caseating granuloma and vessel necrosis and destruction is very uncommon. Eleven cases have been reviewed.[224] It is now considered to be part of the spectrum of pulmonary sarcoidosis.[50] The granulomata are sometimes single but often confluent; they are sterile. Foci of necrosis are seen. The vascular lesions are varied. Granulomata intrude on both arteries and veins: they extend into and through the walls interrupting the media, protruding beneath the endothelium and causing partial or total occlusion. Alternatively, there is a giant-cell arteritis. A third variety of vascular lesion comprises a limited lympho- and plasmacytic vasculitis with which necrosis is exceptional. Single or confluent granulomata occur beneath the bronchial epithelium and bronchioles can be obstructed by a polypoid-like mass formed by these lesions. However, conventional bronchiolitis obliterans may also be noted. The prognosis is good.

Bronchocentric granulomatosis

Bronchocentric granulomatosis differs from the other varieties of pulmonary granuloma described by Liebow:[224] the blood vessels are involved *incidentally*. Liebow[224] reviewed nine cases. The principal lesion is a plasma- and lymphocytic bronchial destruction, often with ulceration and necrosis leading to narrowing. Bronchiolar injury appears uniform with collapse and obliteration by a cellular infiltrate that includes giant cells.

Aetiology and pathogenesis

The proximate cause of vascular injury in most of the syndromes of acute pulmonary vasculitis is thought to be immune complex deposition.[182] Why the immune complexes aggregate selectively in different sites and why individual histological patterns result, is not certain. Nor is it clear why acute or prolonged inflammatory reactions develop (see Chapters 7 and 8). The hypersensitivity responses that result in immune complex formation are thought to derive from the secretion of antibodies against drugs, chemicals, immunoglobulins, bacteria, viruses and neoplastic cell proteins;[44] however, only some bacterial antigens and HBs antigen have been implicated with certainty.

The capacity to produce pulmonary vasculitis experimentally may throw further light on the disease mechanisms. In one study, in the rabbit, the intravenous injection of alveolar macrophages, obtained by lavage, led to pulmonary arteritis.[160] In another, glucan, a yeast cell-wall polysaccharide, was injected, causing a brisk, angiocentric granulomatous vascular reaction.[183]

Foreign body pulmonary vasculitis

Pulmonary vasculitis can result from the intravenous injection of drugs of addiction suspended in solution with insoluble particles.[11] Particles may also reach the lungs during therapeutic injection or transfusion.[119] Most episodes occur in adults but foreign-body pulmonary vasculitis can develop in children given contaminated intravenous fluids. Fatal pulmonary hypertension may result. In one infantile example, the particles were shown by electron probe X-ray microanalysis to contain either silicon and titanium, or talc;[34] the former provoked an histiocytic reaction, the latter granulomata. Analogous pulmonary arterial responses were observed during the study of pulmonary embolism and right ventricular infarction by the intravenous injection of glass beads in cats[40] or lycopodium spores in rabbits.[158] In the rat, a polyarteritis restricted to selected regions of the liver is one additional consequence.

Vasculitis in necrotizing enterocolitis

Immune complex vasculitis has now been identified as a feature of neonatal necrotizing enterocolitis.[137] Antigens such as milk pass the neonatal intestinal mucosa more easily in infants than in adults; and massive antigen diffusion may occur when the mucosal barrier is damaged by vascular insufficiency or ileus. Hypoxia, hyperviscosity plasticizers from polyvinyl chloride catheters, and stress are all factors invoked in the sequence,[137] and it is important that these views should be recalled when studying cases of unexplained adult enteritis in which evidence of vasculitis has been found.

Vasculitis and the Cronkhite–Canada syndrome

Rare instances have been described in which a generalized vasculitis of medium-sized and small arteries, without aneurysm, was accompanied by cutaneous anergy in individuals with the Cronkhite–Canada syndrome of gastrointestinal polyposis.[64,300]

Takayasu's disease

The Japanese ophthalmologist Takayasu[361] described visual impairment and absent carotid pulses in a young woman.[185] The disorder,[232,341,398] of uncertain cause, involved the origins of the branches of the aortic arch in an inflammatory reaction to be distinguished clinically from the effects of trauma, infection and hypersensitivity, and from the vasculitis of the systemic immunological diseases of the connective tissues.

Clinical presentation

Conn and Hunder[62] give a full account of the signs and symptoms. The sensitivity and specificity of classification may be enhanced by the use of the 1990 ACR criteria[9a] (Tables 15.24 and 15.25). Systemic, febrile illness may precede evidence of vascular obstruction ('pulseless disease') and arthralgia and low-grade arthritis are frequent. There follows decreased blood flow to the upper and lower parts of the body. Ischaemic limb ulcers may be accompanied by evidence of a new, collateral circulation. The face and brain may be affected so that facial skin and dental atrophy accompany signs of impaired central nervous system and retinal function.

Takayasu's arteritis is most frequent in the Orient and in Mexico. The symmetrical, bilateral obstruction to large aortic branches is often simulated by thromboembolism

Table 15.24

1990 ACR criteria for the classification of Takayasu arteritis (traditional format). Modified from Arend *et al.*[9a]

Definition of criteria
1. Development of symptoms or findings related to Takayasu arteritis at age ≤40 years
2. Development and worsening of fatigue and discomfort in muscles of one or more extremity while in use, especially the upper extremities
3. Decreased pulsation of one or both brachial arteries
4. Difference of >10 mm Hg in systolic blood pressure between arms
5. Bruit audible on auscultation over one or both subclavian arteries or abdominal aorta
6. Arteriographic narrowing or occlusion of the entire aorta, its primary branches, or large arteries in the proximal upper or lower extremities, not due to arteriosclerosis, fibromuscular dysplasia, or similar causes; changes usually focal or segmental

The presence of three or more of these criteria yields a sensitivity of 90.5 per cent and a specificity of 97.8 per cent.

Table 15.25

1990 ACR classification tree for Takayasu arteritis (TA). Modified from Arend *et al.*[9a]

TA subsets	Number of patients TA/ non-TA	Percentage correctly classified	Percentage of TA patients in subset	Non-TA subsets	Number of patients TA/ non-TA	Percentage correctly classified	Percentage non-TA patients in subset
Subclavian or aortic bruit and age ≤40 years at disease onset	50/7	88	79	Absence of subclavian or aortic bruit and normal brachial artery pulse	4/690	99	93
Decreased brachial artery pulse and BP difference >10 mm Hg (between arms)	6/1	86	10	Absence of subclavian or aortic bruit and BP difference <10 mm Hg (between arms)	1/11	92	1
Subclavian or aortic bruit and decreased brachial artery pulse	2/14	13	3	Subclavian or aortic bruit, age >40 years at disease onset, and normal brachial artery pulses	0/21	100	3

The classification tree yields a sensitivity of 92.1 per cent and a specificity of 97.0 per cent.

resulting from atherosclerosis, giant-cell arteritis or syphilis. Although 10–30-year-old females are said usually to be affected, symmetrical occlusion of the arteries to the upper extremities in middle-aged persons, and in particular males, is now not a rare presentation.[78,207]

Originally described in Japan, the distribution of this uncommon disorder is believed to be worldwide. Takayasu's syndrome may coexist with inflammatory bowel disorders such as Crohn's disease[103] and has been known for some years to be an occasional accompaniment to tuberculosis.[180]

Pathological changes

The findings in eight Thai cases were recorded by Vinijchaikul[377] and in 76 Japanese necropsies by Nasu.[272] Classically, arteritis is confined to the proximal parts of the branches of the aortic arch. The left carotid and subclavian arteries are affected more severely than the right. However, the ascending and descending thoracic and the abdominal aorta have been shown to be affected occasionally, as have the renal, femoral, iliac, pulmonary, coronary and mesenteric arteries.[328]

Samantray[328] described Takayasu's disease as a panarteritis, probably affecting first the adventitia and vasa vasorum, leading to secondary changes in the media in which loss of elastic tissue is characteristic. Proliferation of the intima and vascular stenosis occurs with occlusion or even aneurysm formation. Samantray[328] tabulated the microscopic changes in eight cases in whom biopsy was undertaken at surgery, and in two necropsy cases. In his series, plasma cells were few and lymphocytes of only modest frequency. Giant cells may be present. Daggett and Fallon[67] point out that in their 55-year-old man with claudication of the arms, the only important distinction is with giant-cell arteritis (see p. 642). In Takayasu's arteritis, intimal thickening and inflammation without marked medial involvement is characteristic. This contrasts with giant-cell arteritis. It is also notable that giant-cell arteritis involves the small muscular arteries of the head and neck, some affected in temporal arteritis, and that vascular occlusion of the large arteries in giant-cell arteritis is typical only in patients with widespread vascular disease.

Takayasu's disease is not confined to the arteries: the heart and kidneys may be implicated. One-third of cases have thickening and distortion of the aortic valve, and aortic insufficiency may develop. Occasionally, proliferative glomerulonephritis may be detected.[360] The recognition of circulating immune complexes has been described.[140,277] Antiaortic antibodies have been identified in the sera by some[371] but not all[12] investigators; whether these antibodies are those that lodge as immune complexes in the renal glomeruli is not clear. The role of cytotoxic T lymphocytes has been considered.[333]

Aetiology

The cause of Takayasu's syndrome is uncertain. There is a racial predisposition in Japanese, Thais and Koreans. An autoimmune disorder is proposed. Possible bases for this response are: 1. the association with the inheritance of HLA Bw52, present in 44 per cent of affected Japanese; 2. a putative link with DHO, a B-cell antigen; and 3. a recently described association with HLA DR4 and MB3.[379] Anti-artery antibodies have occasionally been identified[371] but whether they are cause or effect is not known. It is of interest that only 9 per cent of the group of North Americans analysed in this report, were Bw52-positive. Two per cent were of Korean origin. An association with tuberculosis has rarely[294] been shown but patients in India with Takayasu's arteritis are 46 times as likely to have had tuberculosis as the general population. In Mexico, the relationship may be even more common. An association with syphilis, previously reported, is now discounted and a similarity with the panaortitis of rheumatic fever is misleading.

Prognosis

Takayasu's arteritis responds poorly to corticosteroids. The prognosis is guarded: the arteritis tends to recur. Although surgical resection of the sites of occlusion or bypass grafting may relieve the signs of vascular obstruction,[14] the inflammation resumes. Amyloidosis may develop.[133] There is progressive involvement of the arterial system and eventually death.[67,177]

Thrombotic microangiopathy

Thrombotic, thrombocytopenic microangiopathy (TTM)[265] is a clinical syndrome (Moschowitz's syndrome) with characteristic histological features which may underlie or be associated with a number of different primary disorders. In infancy and early childhood, TTM may be a feature of the haemolytic–uraemic syndrome,[36] emphasizing the haematological changes implied in the title 'thrombotic microangiopathic haemolytic anaemia' introduced by Symmers.[358] In young adults, TTM is encountered periodically with *post partum* renal failure; in older persons with malignant hypertension, progressive systemic sclerosis, systemic lupus erythematosus[327] and during transplant rejection. TTM is characterized by a widespread disturbance of the vascular system with subendothelial fibrinous deposits, platelet thrombi and haemolysis. The disorder affects the systemic, precapillary arterioles and the signs are therefore widespread.

Clinical presentation

An acute onset[358] may lead to the diagnosis of neurological disease; there is pyrexia, musculoskeletal pain and meningism. The signs suggest meningitis or subarachnoid haemorrhage. Purpuric spots appear on the trunk and limbs and haemorrhages may occur from the viscera. Progressive haemolytic anaemia is found.[250] There is thrombocytopenia but no fault in the number, development or maturation of bone marrow megakaryocytes. Raised plasma globulin levels are detected with occasional false-positive serological tests for syphilis, positive lupus erythematosus cell tests and antiphospholipid antibodies.[27] Nevertheless, there is little evidence that the haemolytic anaemia is of immune origin. There are no detectable antiplatelet antibodies. Reticulocytosis and a highly cellular bone marrow accompany a low platelet count, a prolonged bleeding time and decreased prothrombin consumption.

The genetic predisposition to TTM is not well understood. However, the syndrome may be familial and husband and wife may be affected concomitantly. The disorder has been reported in young adult sisters,[105,275] in association with pregnancy. It is possible that these cases were instances of systemic lupus erythematosus with lupus anticoagulant formation (see p. 572).

Pathological changes

Visceral disease results from disseminated focal lesions of the precapillary segments of the systemic arterioles.[115,368] The brain, liver, spleen, lymph nodes, bone marrow and other viscera are involved. The pulmonary vessels are usually spared, the converse of the malignant carcinoid syndrome in which the smaller pulmonary arteries and arterioles show changes similar to TTM. Pale-staining, homogeneous, eosinophilic masses are found in the intima; they form small, crescentic occlusive lesions, distorting the arteriolar lumina (Fig. 15.26). Raised zones indent the vascular channels. The few cells within these zones are not

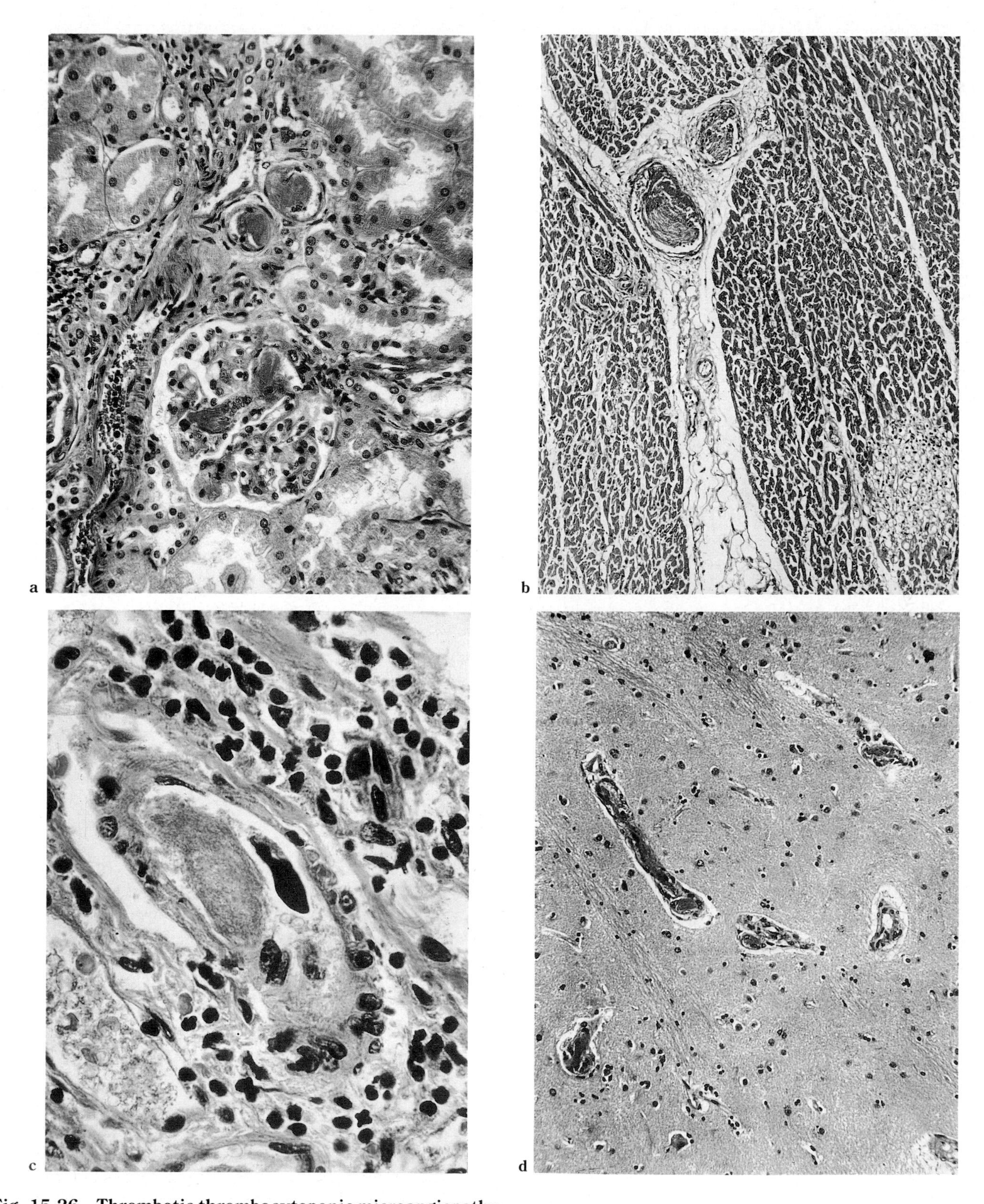

Fig. 15.26 Thrombotic thrombocytopenic microangiopathy.
a. Kidney: The small blood vessels (top centre) may be venules; they are occluded by well-formed platelet and fibrin thrombi. Note that similar material occupies part of the glomerular capillary tuft but that a long segment of arteriole (at left) is apparently unaffected. **b. Heart**: The small blood vessels that extend within the septa of intramyocardial connective tissue (at centre) appear also to be venules. They are obstructed by platelet and fibrin thrombi in the same manner as the vessels shown in Fig. 15.26a. **c. Liver**: A hepatic arteriole is blocked by platelet thrombus beside which lies a megakaryocyte. **d. Brain**: Venules within the cerebrum contain similar microthrombi to those shown in Figs 15.26a–c. (HE **a.** × 115; **b.** × 60; **c.** × 360; **d.** × 110.)

easily identifiable. The mural lesions are often the only signs of disease. However, thrombus may be superadded and the internal elastic lamina disrupted. When this occurs, microaneurysms form. Unlike the loss of elastic material in polyarteritis nodosa (see p. 620), there is no inflammatory cell response; elastic material loss is difficult to explain. The microvascular obstructions contribute to tissue ischaemia and promote infarction in those territories without a collateral circulation.[366]

The material found in the intima of affected vessels resembles platelet thrombus (Fig. 15.26). Mural thrombi quickly become covered by endothelial cells so that it is unlikely that the intimal deposit originates outside the vessel. It is probable that widespread platelet thrombi form on the endothelium of many visceral arterioles coinciding with and contributing to thrombocytopenia. The thrombi may then become incorporated in the intima by a covering of endothelium. There are few transmission electron microscopic studies. Conventional stains suggest that the deposit is a 'fibrinoid' (see p. 294)[354] composed of fibrin.[63]

Thrombotic microangiopathy is a common complication of antibasement membrane glomerulonephritis,[352] and may be found in approximately 75 per cent of patients with this disorder. The affected capillaries are often blocked by intravascular thrombi; red cells are fragmented (schistocytes) and there is a predominance of platelets or fibrin in the microthrombi. Subsequently, glomerular sclerosis may accompany arteriolosclerosis.

The diagnosis of TTM by biopsy is straightforward and safe; in view of the hazards of delayed treatment, biopsy is usually urgent. Bone marrow aspiration is preferred;[28] it is simpler than lymph node biopsy although both yield a high proportion of positive results. Alternatively, muscle, skin or synovia (Fig. 15.27) may be sampled but are less likely to reveal the diagnostic lesions.

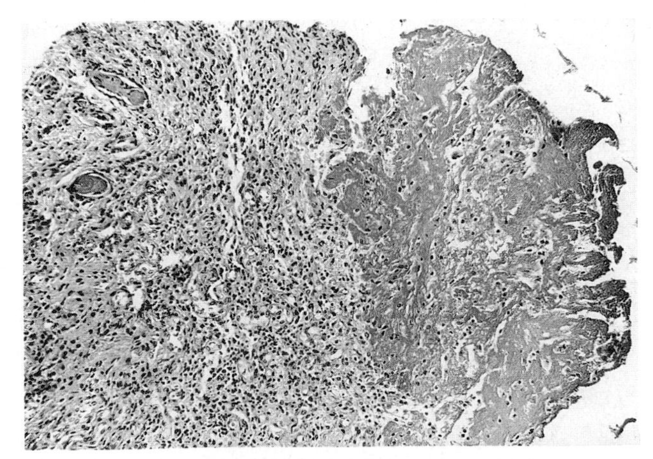

Fig. 15.27 Synovium: thrombotic microangiopathy. In this case of rheumatoid arthritis, microangiopathy was one consequence of treatment with phenylbutazone. **Left**: small synovial blood vessels, obstructed by platelet and fibrin thrombi, are seen. **Right**: fibrin has polymerized on the synovial surface in the course of the rheumatoid inflammatory response. (HE × 70.)

Aetiology and pathogenesis

No satisfactory explanation has been given for the aetiology of TTM. There is no evidence of a genetic predisposition. It is included in the connective tissue disorders (see p. 13) because of the occasional coexistence of systemic lupus erythematosus, polyarteritis nodosa, or glomerulonephritis; because of the (inconstantly) raised plasma globulin levels, sometimes with abnormal serological tests. There is a distant similarity to the vascular lesions in primary amyloidosis. TTM may indeed be an incidental end-result of a disturbance of the immune mechanism, the haemolytic anaemia one part of this underlying disease. Similarly, thrombocytopenia may be a result of a thrombotic syndrome similar to that accompanying the formation of lupus anticoagulant (see p. 572).

The difficulty in assessing the relative importance of possible aetiological mechanisms is illustrated by a case in which TTM was recognized in an elderly woman seven days after the commencement of treatment with phenylbutazone for rheumatoid arthritis. Hashimoto's thyroiditis was present. Vascular thrombi were demonstrable within the synovia of joints affected by the synovitis (Fig. 15.27) and in the abnormal thyroid, suggesting that TTM can coincide with autoimmune connective tissue disease and thyroid disease. The impaction within small blood vessels of platelets formed by an active bone marrow appeared to account for the thrombocytopenia and the clinical signs of the disease; the altered reaction to phenylbutazone may have been a manifestation of an abnormal immune response linked with rheumatoid arthritis or 'overlap' systemic lupus erythematosus.

Actinic vasculopathy

It was widely recognized that habitual exposure to the infrared radiation of domestic fires was a common cause of erythema *ab igne* but less attention had been given to the effects on the skin and the dermal blood vessels of chronic ultraviolet irradiation. The long-term effects on visceral arteries of X- and γ-radiation are, of course, fully documented.

One effect of ultraviolet radiation is the production of actinic granulomata, a reparative process that follows connective tissue radiation injury. There may be a relationship to granuloma annulare (see p. 488). The ultraviolet radiation injures elastic tissue and the dermal arteries undergo a form of intimal proliferation and elastosis that appears to be more severe on the superficial aspects of the affected vessels.[278,279,280] A relationship to temporal arteritis (see p. 642) has been proposed.

REFERENCES

1. Abdou N I, Schumacher H R, Colman R W, Sagawa A, Hebert J, Pascual E, Carroll E T, Miller M, South M A, Abdou N L. Behçet's disease—possible role of secretory component deficiency, synovial inclusions, and fibrinolytic abnormality in the various manifestations of the disease. *Journal of Laboratory and Clinical Medicine* 1978; **91**: 409–22.

2. Accinni L, Dixon F J. Degenerative vascular disease and myocardial infarction in mice with lupus-like syndrome. *American Journal of Pathology* 1979; **96**: 477–92.

3. Adamantiades B. Sur un cas d'iritis á hypopyon recidivante. *Annales d'Oculistique* 1931; **168**: 271.

4. Ainsworth R W, Gresham G A, Balmforth G F. Pathological changes in temporal arteries removed from unselected cadavers. *Journal of Clinical Pathology* 1961; **14**: 115–19.

5. Alarçon-Segovia D, Brown A L. Classification and etiologic aspects of necrotizing angiitides: an analytic approach to a confused subject with a critical review of the evidence for hypersensitivity in polyarteritis nodosa. *Proceedings of the Mayo Clinic* 1964; **39**: 205–22.

6. Aldo M A, Benson M D, Gomerford F R, Cohen A S. Treatment of Wegener's granulomatosis with immunosuppressive agents: description of renal ultrastructure. *Archives of Internal Medicine* 1970; **126**: 298–305.

7. Allen D C, Doherty C C, O'Reilly D P J. Pathology of the heart and cardiac conduction system in Wegener's granulomatosis. *British Heart Journal* 1984; **52**: 674–8.

8. Allen D M, Diamond L K, Howell D A. Anaphylactoid purpura in children (Schönlein–Henoch syndrome): Review with a follow-up of the renal complications. *AMA Journal of Diseases of Children* 1960; **99**: 833–54.

9. Appel G B, Gee B, Kashgarian M, Hayslett J P. Wegener's granulomatosis – clinical-pathologic correlations and long-term course. *American Journal of Kidney Diseases* 1981; **1**: 27–37.

9a. Arend W P, Michel B A, Bloch D A, Hunder G G, Calabrese L H, Edworthy S M, Fauci A S, Leavitt R Y, Lie J T, Lightfoot R W Jr, Masi A T, McShane D J, Mills J A, Stevens M B, Wallace S L, Zvaifler N J. The American College of Rheumatology 1990 criteria for the classification of Takayasu arteritis. *Arthritis and Rheumatism* 1990; **33**: 1129–34.

10. Arlet P, Laroche M, Delsol G, Seigneuric G, Duffant M, Le Tallec Y. Angioimmunoblastic lymphadenopathy with leucocytoclastic vasculitis of the skin. Two cases. *La Nouvelle Presse Médicale* 1982; **11**: 3713–16.

11. Arnett E N, Battle W E, Russo J V, Roberts W C. Intravenous injection of talc-containing drugs intended for oral use. A cause of pulmonary granulomatosis and pulmonary hypertension. *American Journal of Medicine* 1976; **60**: 711–18.

12. Asherson R A, Asherson G L, Schrire V. Immunological studies in arteritis of the aorta and great vessels. *British Medical Journal* 1968; **3**: 589–90.

13. Augarten A, Apter S, Theodor R. Right ventricular thrombus and pulmonary arteritis in Behcet's disease. *Israel Journal of Medical Sciences* 1987; **23**: 900–1.

14. Austen W G, Shaw R S. Surgical treatment of pulseless (Takayasu's) disease. *New England Journal of Medicine* 1963; **270**: 1228–31.

15. Bacon P A, Evolving concepts in vasculitis. *Quarterly Journal of Medicine* 1985; **57**: 609–10.

16. Baggenstoss A H, Schick R M, Polly H F. The effect of cortisone on the lesions of periarteritis nodosa. *American Journal of Pathology* 1951; **27**: 537–59.

17. Barber H S. Myalgic syndrome with constitutional effects: polymyalgia rheumatica. *Annals of the Rheumatic Diseases* 1957; **16**: 230–7.

18. Barnes C G. Miscellaneous and uncommon rheumatic conditions. In: Scott J T, ed, *Copeman's Textbook of the Rheumatic Diseases* (5th edition). Edinburgh, London, New York: Churchill-Livingstone 1978: 845–80.

19. Barnes C G. Other forms of arthritis: miscellaneous. In: Scott J T, ed, *Copeman's Textbook of the Rheumatic Diseases* (6th edition). Edinburgh: Churchill Livingstone, 1986a: 1229–30.

20. Barnes C G. Behçet's syndrome. In: Scott J T, ed, *Copeman's Textbook of the Rheumatic Diseases* (6th edition). Edinburgh: Churchill Livingstone, 1986b: 1231–4.

21. Behçet H. Über rezidivierende, aphthöse, durch ein Virus verursacht Geschwüre am Mund, am Auge und an den Genitalien. *Dermatologica Wochenschrift* 1937; **105**: 1152–7.

22. Behçet H. Some observations on the clinical picture of the so-called triple symptom complex. *Dermatologica* (Basel) 1940; **81**: 73–83.

23. Behçet's Disease Research Committee of Japan. Behçet's disease. *Japanese Journal of Ophthalmology* 1974; **18**: 291–4.

24. Bell D A, Mondschein M, Scully R E. Giant cell arteritis of the female genital tract: a report of three cases. *American Journal of Surgical Pathology* 1986; **10**: 696–701.

25. Bernhard J D, Mark E J. An 18-year-old man with cutaneous ulcers and bilateral pulmonary infiltrates. *New England Journal of Medicine* 1986; **314**: 1170–84.

26. Bignon J D, Barrier J, Soulillou J P, Martin P, Grolleau J Y. HLA DR4 and giant cell arteritis. *Tissue Antigens* 1984; **24**: 60–2.

27. Bird A G, Lendrum R, Asherson R A, Hughes G R V. Disseminated intravascular coagulation, antiphospholipid antibodies, and ischaemic necrosis of extremities. *Annals of the Rheumatic Diseases* 1987; **46**: 251–5.

28. Blecher T E, Raper A B. Early diagnosis of thrombotic microangiopathy by paraffin sections of aspirated bone-marrow. *Archives of Diseases of Childhood* 1967; **42**: 158–62.

29. Blétry O, Godeau P, Charpentier G, Guillevin L, Herreman G. Manifestations cardiaques de la périartérite noueuse. Frequence de la cardiomyopathie non-hypertensive. *Archives des*

Maladies du Coeur et des Vaisseaux, Paris 1980; **73**: 1027–36.

30. Blétry O, Grenier P, Jeannin L, Godeau P, Gourgon R. Behçet's disease with pulmonary artery aneurysms. *La Nouvelle Presse Médicale* 1981; **10**: 2813–16.

31. Blomgren S E. Erythema nodosum. *Seminars in Arthritis and Rheumatism* 1974; **4**: 1–24.

32. Bloomfield A L. Periarteritis nodosa. In: *A Bibliography of Internal Medicine: Selected Diseases*. Chicago: University of Chicago Press, 1960; 54–60.

33. Blumberg S, Giansiracusa D F, Docken W P, Kantrowitz F G. Recurrence of temporal arteritis. Clinical recurrence nine years after initial illness. *Journal of the American Medical Association* 1980; **244**: 1713.

34. Bowen J H, Woodard B H, Barton T K, Ingram P, Shelburne J D. Infantile pulmonary hypertension associated with foreign body vasculitis. *American Journal of Clinical Pathology* 1981; **75**: 609–14.

35. Bowler J, Dvaric D M, Roberts J M, Burke S W. Kawasaki syndrome presenting as pyarthrosis of the hip. A case report. *Journal of Bone and Joint Surgery* 1986; **68-A**: 467–8.

36. Brain M C. Microangiopathic hemolytic anemia. *Annual Review of Medicine* 1970; **21**: 133–44.

37. Brandwein S, Esdaile J, Danoff D, Tannenbaum H. Wegener's granulomatosis. *Archives of Internal Medicine* 1983; **143**: 476–9.

37a. Bras G, Brandt K H. Vascular disorders. In: MacSween R N M, Anthony P P, Scheuer P J, eds, *Pathology of the Liver* (2nd edition). Edinburgh, London: Churchill-Livingstone, 1987: 478–502.

37b. Breiman, L, Friedman J, Olshen R A, Stone C J. *Classification and Regression Trees*. Belmont CA: Wadsworth, 1984.

38. Bruce W. Senile rheumatic gout. *British Medical Journal* 1888; **ii**: 811–13.

39. Buchanan J G, Scott P J, McLachlan E M, Smith F, Richmond D E, North J D K. A chromosome translocation in association with periarteritis nodosa and macroglobulinemia. *American Journal of Medicine* 1967; **42**: 1003–10.

40. Büchner F. *Allgemeine Pathologie. Pathologie als Biologie und als Beiträg zur Lehre vom Menschens*. München-Berlin: Urban und Schwartzenberg, 1950.

41. Bury J S. A case of erythema with remarkable nodular thickenings and induration of the skin associated with intermittent albuminuria. *Illustrated Medical News* 1889; **3**: 145–8.

41a. Calabrese L H, Michel B A, Bloch D A, Arend W P, Edworthy S M, Fauci A S, Fries J F, Hunder G G, Leavitt R Y, Lie J T, Lightfoot R W Jr, Masi A T, McShane D J, Mills J A, Stevens M B, Wallace S L, Zflaifler N J. The American College of Rheumatology 1990 criteria for the classification of hypersensitivity vasculitis. *Arthritis and Rheumatism* 1990; **33**: 1108–13.

42. Camilleri M, Pusey C D, Chadwick V S, Rees A J. Gastrointestinal manifestations of systemic vasculitis. *Quarterly Journal of Medicine* 1983; **52**: 141–9.

43. Center for Disease Control. Multiple outbreaks of Kawasaki syndrome – United States. *MMWR* 1985; **34**: 33–5.

44. Chandler D B, Fulmer J D. Pulmonary vasculitis. *Lung* 1985; **163**: 257–73.

45. Charlier P, Blétry O, Grenier P, Jeannin L, Godeau P, Gourgon R. Maladie de Behçet avec aneurysmes arteriels pulmonaires. 2 observations. *Nouvelle Presse Médicale* 1981; **10**: 2813–16.

46. Chemnitz J, Christensen B C, Christoffersen P, Garbarsch C, Hansen T M, Lorenzen I. Giant cell arteritis. Histological, immunohistochemical and electronmicroscopic studies. *Acta Pathologica et Microbiologica Scandinavica* 1987; **95**: 251–62.

47. Chess J, Albert D M, Bhan A K, Paluck E I, Robinson N, Collins B, Kaynor B. Serologic and immunopathologic findings in temporal arteritis. *American Journal of Ophthalmology* 1983; **96**: 283–9.

48. Christian C L, Sergent J S. Vasculitis syndromes: clinical and experimental models. *American Journal of Medicine* 1976; **61**: 385–92.

49. Chumbley L C, Harrison E G, De Remee R A. Allergic granulomatosis and angiitis (Churg–Strauss syndrome). Report and analysis of 30 cases. *Mayo Clinic Proceedings* 1977; **52**: 477–84.

50. Churg A, Carrington C B, Gupta R. Necrotizing sarcoid granulomatosis. *Chest* 1979; **76**: 406–13.

51. Churg J. Allergic granulomatosis and granulomatous–vascular syndromes. *Annals of Allergy* 1963; **21**: 619–28.

52. Churg J, Strauss L. Allergic granulomatosis, allergic angiitis, and periarteritis nodosa. *American Journal of Pathology* 1951; **27**: 227–301.

53. Churg J, Strauss L. Interstitial eosinophilic pneumonitis, pleuritis, and angiitis. *New England Journal of Medicine* 1981; **304**: 611.

54. Clark E, Kaplan B I. Endocardial, arterial and other mesenchymal alterations associated with serum disease in man. *Archives of Pathology* 1937; **24**: 458–75.

55. Clinicopathological Conference, Postgraduate Medical School of London. Orogenital ulceration with phlebothrombosis (?Behçet's syndrome) complicated by osteomyelitis of lumbar spine and ruptured aorta. *British Medical Journal* 1965; **i**: 357–61.

56. Cochrane C G. Mechanisms involved in the deposition of immune complexes in tissues. *Journal of Experimental Medicine* 1971; **134**: Supplement 75–89.

57. Cochrane C G, Koffler D. Immune complex disease in experimental animals and man. *Advances in Immunology* 1973; **16**: 185–264.

58. Cohen R D, Conn D L, Ilstrup D M. Clinical features, prognosis and response to treatment in polyarteritis. *Mayo Clinic Proceedings* 1980; **55**: 146–55.

59. Cohle S D, Titus J L, Espinola A, Jachimczyk J A. Sudden unexpected death due to coronary giant cell arteritis. *Archives of Pathology and Laboratory Medicine* 1982; **106**: 171–2.

60. Comess K A, Zibelli L R, Gordon D, Fredrickson S R. Acute, severe aortic regurgitation in Behçet's syndrome. *Annals of Internal Medicine* 1983; **99**: 639–40.

61. Conn D L, Dickson E R, Carpenter H A. The association of Churg–Strauss vasculitis with temporal artery involvement, primary biliary cirrhosis and polychondritis in a single patient. *Journal of Rheumatology* 1982; **9**: 744–8.

62. Conn D L, Hunder G G. Vasculitis and related disorders. In: Kelley W N, Harris E D, Ruddy S, Sledge C B, eds, *Textbook of Rheumatology* (3rd edition). Philadelphia: W B Saunders & Co., 1989: 1167–99.

63. Craig J M, Gitlin D. The nature of the hyaline thrombi in thrombotic thrombocytopenic purpura. *American Journal of Pathology* 1957; **33**: 251–65.

64. Cronkhite L W, Canada W J. Generalized gastro-intestinal polyposis. An unusual syndrome of polyposis, pigmentation, alopecia and onychotrophia. *New England Journal of Medicine* 1955; **252**: 1011–15.

65. Cupps T R, Fauci A S. *The Vasculitides*. Philadelphia: W B Saunders & Co., 1981: 1–211.

66. Cupps T R, Springer R M, Fauci A S. Chronic, recurrent small-vessel cutaneous vasculitis. *Journal of the American Medical Association* 1982; **247**: 1994–8.

67. Daggett W M, Fallon J T. A 55-year-old man with intermittent claudication of the arms (Takayasu's arteritis). *New England Journal of Medicine* 1981; **305**: 1519–24.

68. Dare B, Byrne E. Giant cell arteritis: a five year review of biopsy-proven cases in a teaching hospital. *Medical Journal of Australia* 1980; **1**: 372–3.

69. Dasgupta B, Duke O, Kyle V, MacFarlane D G, Hazleman B L, Panayi G S. Antibodies to intermediate filaments in polymyalgia rheumatica and giant cell arteritis: a sequential study. *Annals of the Rheumatic Diseases* 1987; **46**: 746–9.

70. Davson J, Ball J, Platt R. The kidney in periarteritis nodosa. *Quarterly Journal of Medicine* 1948; **17**: 175–202.

71. de Vere-Tyndall A, Knight S, Burman S, Denman A M, Ansell B M. Lymphocyte responses in juvenile chronic arthritis and Behçet's disease—cell number requirements and effects of glucocorticosteroid therapy. *Clinical and Experimental Immunology* 1982; **50**: 549–54.

72. Dixon F J, Vasquez J J, Weigle W O, Cochrane C G. Pathogenesis of serum sickness. *Archives of Pathology* 1958; **65**: 18–28.

73. Donald D J, Edwards R L, McEvoy J D S. An ultrastructural study of the pathogenesis of tissue injury in limited Wegener's granulomatosis. *Pathology* 1976; **8**: 161–9.

74. Drueke T, Barbanel C, Jungers P, Digeon M, Poisson M, Brivet F, Trecan G, Feldmann G, Crosnier J, Bach J F. Hepatitis-B antigen-associated periarteritis nodosa in patients undergoing long-term hemodialysis. *American Journal of Medicine* 1980; **68**: 86–90.

75. Editorial. Behçet's disease. *British Medical Journal* 1978; **ii**: 234.

76. Editorial. Systemic vasculitis. *Lancet* 1985; **i**: 1252–4.

77. Efthimiou J, Johnston C, Spiro S G, Turner-Warwick M. Pulmonary disease in Behçet's syndrome. *Quarterly Journal of Medicine* 1986; **58**: 259–80.

78. Ehrlich G. Takayasu's arteritis and Still's disease. *Arthritis and Rheumatism* 1979; **12**: 1442.

79. Ehrlich G E. Intermittent and periodic arthritic syndromes. In: McCarty J D, ed, *Arthritis and Allied Conditions* (11th edition). Philadelphia: Lea & Febiger, 1989: 991–1009.

80. Elkon K B, Hughes G R V, Catovsky D, Claven J P, Dumont J, Seligmann M, Tannenbaum H, Esdaile J. Hairy-cell leukaemia with polyarteritis nodosa. *Lancet* 1979; **ii**: 280–2.

81. Elkon K B, Sutherland D C, Rees A J, Hughes G R V, Batchelor J R. HLA antigen frequencies in systemic vasculitis—increase in HLA-DR2 in Wegener's granulomatosis. *Arthritis and Rheumatism* 1983; **26**: 102–5.

82. Elling H, Skinhoj P, Elling P. Hepatitis B virus and polymyalgia rheumatica: a search for HBsAg, HBsAb, HBcAb, HBeAg and HBeAb. *Annals of the Rheumatic Diseases* 1980; **39**: 511–13.

83. Eppinger H. Pathogenesis (histogenesis) der aneurysmen einschliesslich des Aneurysma equi verminosum. *Archiv für Klinische Chirurgie* 1887; **35**: Suppl. 1: 1–563.

84. Epstein W L. Erythema nodosum. In: Samter M, ed, *Immunological Diseases* (4th edition). Boston: Little, Brown & Co., 1988: 1247–56.

85. Ettlinger R E, Hunder G G, Ward L E. Polymyalgia rheumatica and giant cell arteritis. *Annual Review of Medicine* 1978; **29**: 15–22.

86. Ettlinger R E, Nelson A M, Burke E C, Lie J T. Polyarteritis nodosa in childhood. A clinical pathologic study. *Arthritis and Rheumatism* 1979; **22**: 820–5.

87. Fahey E, Leonard E, Churg J, Godman G C. Wegener's granulomatosis. *American Journal of Medicine* 1954; **17**: 168–79.

88. Fallingborg J, Laustsen J. Colitis of Behçet's syndrome. *Acta Medica Scandinavica* 1984; **215**: 397–9.

89. Farcet J P, Wechsler J, Wirquin V, Divine M, Reyes F. Vaculitis in hairy-cell leukemia. *Archives of Internal Medicine* 1987; **147**: 660–4.

90. Fassbender H G, Annefeld M. Ultrastrukturelle veranderungen in der Skelettmuskulatur bei polymyalgia rheumatica. *Deutsche Medizinische Wochenschrift* 1986; **111**: 1799–1804.

91. Fauchald P, Rygvold O, Oystese B. Temporal arteritis and polymyalgia rheumatica. *Annals of Internal Medicine* 1972; **77**: 845–52.

92. Fauci A S, Haynes B F, Katz P. The spectrum of vasculitis: clinical pathologic, immunologic, and therapeutic considerations. *Annals of Internal Medicine* 1978; **89**: 660–76.

93. Fauci A S, Haynes B F, Katz P, Wolff S M. Wegener's granulomatosis: prospective clinical and therapeutic experiences with 85 patients for 21 years. *Annals of Internal Medicine* 1983; **98**: 76–85.

94. Fauci A S, Leavitt R Y. Vasculitis. In: McCarty D J, ed, *Arthritis and Allied Conditions: a Textbook of Rheumatology* (11th edition). Philadelphia: Lea & Febiger, 1989: 1166–88.

95. Fienberg R. Pathergic granulomatosis. *American Journal of Medicine* 1955; **19**: 829–31.

96. Fienberg R. Case records of the Massachusetts General Hospital: Case 43–1986. *New England Journal of Medicine* 1986; **315**: 1143–54.

97. Finan M C, Winkelmann R K. The cutaneous extravascular necrotizing granuloma (Churg–Strauss granuloma) and systemic disease: a review of 27 cases. *Medicine* 1983; **62**: 142–58.

98. Fine R M, Meltzer H D. Erythema nodosum: a form of allergic cutaneous vasculitis. *Southern Medical Journal* 1968; **61**: 680–86.

99. Fisher R G. Renal artery aneurysms in polyarteritis nodosa—a multiepisodic phenomenon. *American Journal of Roentgenology* 1981; **136**: 983–5.

100. France R, Buchanan R N, Wilson M W, Sheldon M B. Relapsing iritis with recurrent ulcers of the mouth and genitalia (Behçet's syndrome). Review: with report of additional case. *Medicine* 1951; **30**: 335–55.

101. Frayha R A. Arthropathy of Behçet's disease with marked synovial pleocytosis responsive to colchicine. *Arthritis*

and Rheumatism 1982; **25**: 235–6.

102. Frayha R A, Nasr F W. Erythema nodosum—arthropathy complex as an initial presentation of Behçet's disease. Report of five cases. *Journal of Rheumatology* 1978; **5**: 224–8.

103. Friedman C J, Tegtmeyer C J. Crohn's disease associated with Takayasu's arteritis. *Digestive Diseases and Sciences* 1979; **24**: 954–8.

104. Frohnert P P, Sheps S G. Long-term follow-up study of periarteritis nodosa. *American Journal of Medicine* 1967; **43**: 8–14.

105. Fuchs W E, George J N, Dotin L N, Sears D A. Thrombotic thrombocytopenic purpura. Occurrence two years apart during late pregnancy in two sisters. *Journal of the American Medical Association* 1976; **235**: 2126–7.

106. Fujii N, Minagawa T, Nakane A, Kato F, Ohno S. Spontaneous production of γ-interferon in cultures of T-lymphocytes obtained from patients with Behçet's disease. *Journal of Immunology* 1983; **130**: 1683–6.

107. Fujiwara H, Fujiwara T, Kao T-C, Ohshio G, Hamashima Y. Pathology of Kawasaki disease in the healed stage—relationships between typical and atypical cases of Kawasaki disease. *Acta Pathologica Japonica* 1986; **36**: 857–67.

108. Fujiwara H, Hamashima Y. Pathology of the heart in Kawasaki disease. *Pediatrics* 1978; **61**: 100–7.

109. Fukuda Y, Sakuma Y, Sumita M. Pathological studies of vascular changes in Behçet's disease. In: Shiokawa Y, ed, *Vascular Lesions of Collagen Diseases*. Baltimore: University Park Press, 1977: 212–25.

110. Fulmer J D, Kaltreider H B. The pulmonary vasculitides. *Chest* 1982; **82**: 615–24.

111. Gaffey C M, Chun B, Harvey J C, Manz H J. Phenytoin-induced systemic granulomatous vasculitis. *Archives of Pathology and Laboratory Medicine* 1986; **110**: 131–5.

112. Gairdner D. The Schönlein–Henoch syndrome (anaphylactoid purpura). *Quarterly Journal of Medicine NS* 1948; **17**: 95–122.

113. Gallagher P, Jones K. Immunohistochemical findings in cranial arteritis. *Arthritis and Rheumatism* 1982; **25**: 75–9.

114. Gardner D L. The relationship between intermittent hypotension and the prevention by hydralazine of acute vascular disease in rats with steroid hypertension. *British Journal of Experimental Pathology* 1960; **41**: 60–71.

115. Gardner D L. *Pathology of the Connective Tissue Diseases*. London: Edward Arnold, 1965: 191–202.

116. Gardner D L, Krieg A F, Chapnick R. Fatal systemic fungus disease in rheumatoid arthritis with cardiac and pulmonary mycotic and rheumatoid granulomata. *Archives of InterAmerican Rheumatology* 1962; **5**: 561–86.

117. Gardner D L, Laing C P. Measurement of enzyme activity of isolated small arteries in early rat hypertension. *Journal of Pathology and Bacteriology* 1965; **90**: 399–406.

118. Gardner D L, Quagliata F, Drossman M, Kalish M, Schimmer B. Attempted prevention of arteriolar lesions in accelerated rat hypertension by immunosuppression. *British Journal of Experimental Pathology* 1970; **51**: 242–52.

119. Garvan J M, Gunner B W. Particulate contamination of intravenous fluids. *British Journal of Clinical Practice* 1971; **25**: 119–21.

120. Germuth F G, Flanagan C, Montenegro M R. The relationships between the chemical nature of the antigen, antigen dosage, rate of antibody synthesis and the occurrence of arteritis and glomerulonephritis in experimental hypersensitivity. *Bulletin of the Johns Hopkins Hospital* 1957; **101**: 149–69.

121. Giesker D W, Krause P J, Pastuszak W T, Hine P, Forouhar F A. Lymph node biopsy for early diagnosis in Kawasaki disease. *American Journal of Surgical Pathology* 1982; **6**: 493–501.

122. Gillanders L A, Strachan R W, Blair D W. Temporal arteriography: a new technique for the investigation of giant cell arteritis and polymyalgia rheumatica. *Annals of the Rheumatic Diseases* 1969; **28**: 267–9.

123. Gilmour J R. Giant-cell chronic arteritis. *Journal of Pathology and Bacteriology* 1941; **53**: 263–77.

124. Ginns L C, Mark E J. Pulmonary densities, arthralgia, and renal disorder. *New England Journal of Medicine* 1981; **304**: 958–66.

125. Gitlin D, Craig J M, Janeway C A. Studies on the nature of fibrinoid in the collagen diseases. *American Journal of Pathology* 1957; **33**: 55–77.

126. Gocke D J, Healey L A. Polyarteritis and other primary vasculitides. In: Samter M, ed, *Immunological Diseases* (3rd edition). Boston: Little, Brown & Co., 1978: 1077–90.

127. Gocke D J, Hsu K, Morgan G, Bombardieri S, Lockshin M, Christian C L. Association between polyarteritis and Australia antigen. *Lancet* 1970; **ii**: 1149–53.

128. Godman G C, Churg J. Wegener's granulomatosis: pathology and review of the literature. *Archives of Pathology* 1954; **58**: 533–53.

129. Goldberg J W, Lee M L, Sajjad S M. Giant cell arteritis of the skin simulating erythema nodosum. *Annals of the Rheumatic Diseases* 1987; **46**: 706–8.

130. Goldenberg D L, Goodman M L. Case records of the Massachusetts General Hospital. A 43-year-old woman with a progressive saddle-nose deformity. *New England Journal of Medicine* 1985; **312**: 1695–703.

131. Gonzalez T, Ravina M, Martin-Herrera A, Gantes M A, Y Diaz-Flores L. Synovial ultrastructure in Behçet's syndrome. *Morfología Normal et Patológica*. Sec. B. 1982; **6**: 35–48.

132. Goodman B W. Temporal arteritis. *American Journal of Medicine* 1979; **67**: 839–52.

133. Graham A N, Delahunt B, Renouf J J, Austad W I. Takayasu's disease associated with generalized amyloidosis. *Australian and New Zealand Journal of Medicine* 1985; **15**: 343–5.

134. Graham E, Holland A, Avery A, Russell R W R. Prognosis in giant cell arteritis. *British Medical Journal* 1981; **282**: 269–71.

135. Graham J R, Fienberg R. Case records of the Massachusetts General Hospital. A 70 year-old woman with fever of unknown origin. *New England Journal of Medicine* 1986; **315**: 631–9.

136. Granato J E, Abben R P, May W S. Familial association of giant cell arteritis—a case report and brief review. *Archives of Internal Medicine* 1981; **141**: 115–17.

137. Gray E S, Lloyd D J, Miller S S, Davidson A I, Balch N J, Horne C H W. Evidence for an immune complex vasculitis in

neonatal necrotising enterocolitis. *Journal of Clinical Pathology* 1981; **34**: 759–63.

138. Greene G M, Lain D, Sherwin R M, Wilson J E, McManus B M. Giant cell arteritis of the legs. Clinical isolation of severe disease with gangrene and amputations. *American Journal of Medicine* 1986; **81**: 727–33.

139. Gruber G B. Zür Frage der Periarteritis nodosa mit besonderer Berücksichtigung der Gallenblasen—und Nieren—Beteiligung. *Virchows Archiv für Pathologische Anatomie und Physiologie und für Klinische Medizin* 1925; **258**: 441–501.

140. Gyotoku Y, Kakiuchi T, Nomaka Y, Saito Y, Ito I, Murao S. Immune complexes in Takayasu's arteritis. *Clinical and Experimental Immunology* 1981; **45**: 246–52.

141. Hall S, Persellin S, Lie J T, O'Brien P C, Kurland L T, Hunder G G. The therapeutic impact of temporal artery biopsy. *Lancet* 1983; **ii**: 1217–20.

142. Hamilton C R Jr, Shelley W M, Tumulty P A. Giant cell arteritis: including temporal arteritis and polymyalgia rheumatica. *Medicine* 1971; **50**: 1–28.

143. Hamza M, Cammoun M, Ayed H B. Lip biopsy in Behçet's disease. *Nouvelle Presse Médicale* 1981; **10**: 2748.

144. Hanson V. Systemic lupus erythematosus, dermatomyositis, scleroderma and vasculitides in children. In: Kelley W N, Harris E D, Ruddy S, Sledge C B, eds, *Textbook of Rheumatology* (2nd edition). Philadelphia: W B Saunders & Co., 1985: 1293–1313.

145. Harkavy J. *Vascular Allergy and its Systemic Manifestations*. London: Butterworths, 1963: 170–86.

146. Harris P, Heath D. *The Human Pulmonary Circulation: Its Form and Function in Health and Disease* (3rd edition). Edinburgh: Churchill Livingstone, 1986.

147. Harrison C V. Giant-cell or temporal arteritis: a review. *Journal of Clinical Pathology* 1947–8; **1**: 197–211.

148. Harrison C V. Giant cell or temporal arteritis. In: Hadfield G, ed, *Recent Advances in Pathology* (6th edition). London: Churchill, 1953: 187–8.

149. Harrison D J, Simpson R, Neary C, Wathen C G. Renal biopsy and antineutrophil antibodies in the diagnosis and assessment of Wegener's granuloma. *British Journal of Diseases of the Chest* 1988; **82**: 398–404.

150. Hauser W A, Ferguson R H, Holley K E, Kurland L T. Temporal arteritis in Rochester, Minnesota, 1951–1967. *Mayo Clinic Proceedings* 1971; **46**: 597–602.

151. Hawn C V, Janeway C A. Histological and serological consequences in experimental hypersensitivity. *Journal of Experimental Medicine* 1947; **85**: 571–90.

152. Haynes B F, Fauci A S. Diabetes insipidus associated with Wegener's granulomatosis successfully treated with cyclophosphamide. *New England Journal of Medicine* 1978; **299**: 764.

153. Hazleman B. Polymyalgia rheumatica and giant cell arteritis. In: Scott J T, ed, *Copeman's Textbook of Rheumatic Diseases* (6th edition). Edinburgh: Churchill Livingstone, 1986: 1278–91.

154. Healey L A, Wilske K R. Presentation of occult giant cell arteritis. *Arthritis and Rheumatism* 1980; **23**: 641–3.

155. Hebra F. Das umschriebene Eczem. Eczema marginatum. In: Virchow R, ed, *Handbuch der Speziellen Pathologie und Therapie*. Erlangen: Enke, 1860; **3**: 361–3.

156. Hellman D, Laing T, Petri M, Jacobs D, Crumley R, Stulbarg M. Wegener's granulomatosis: isolated involvement of the trachea and larynx. *Annals of the Rheumatic Diseases* 1987; **46**: 628–31.

157. Henson J B, Crawford T B. The pathogenesis of virus-induced arterial disease—Aleutian disease and equine viral arteritis. *Advances in Cardiology* 1974; **13**: 183–91.

158. Herbertson B M. Patchy necrosis of the myocardium of rabbits after anaphylactic shock and after experimental pulmonary embolism. *Journal of Pathology and Bacteriology* 1953; **66**: 211–22.

159. Herreman G, Beaufils H, Godeau P, Cassou B, Wechsler B, Boujean J, Chomette G. Behçet's syndrome and renal involvement—a histological and immunofluorescent study of 11 renal biopsies. *American Journal of Medical Sciences* 1982; **284**: 10–17.

160. Hicken P, Hogg J C, Pare P D. Experimental pulmonary arteritis. *Investigative Radiology* 1980; **15**: 299–307.

161. Hicks J T, Korenyi-Both A, Utsinger P D, Baran E M, McLaughlin G E. Neuromuscular and immunochemical abnormalities in an adult man with Kawasaki disease. *Annals of Internal Medicine* 1982; **96**: 607–9.

162. Higashihara M, Mori M, Takeuchi A, Ogita T, Miyamoto T, Okimoto T. Myocarditis in Behçet's disease—a case report and review of the literature. *Journal of Rheumatology* 1982; **9**: 630–3.

163. Hills E A. Behçet's syndrome with aortic aneurysms. *British Medical Journal* 1967; **4**: 152–4.

164. Hollingworth P, Denman A M. Retroperitoneal fibrosis and polyarteritis nodosa. *Journal of the Royal Society of Medicine* 1980; **73**: 61–4.

165. Honma T, Bang D, Saito T, Nakagawa S, Ueki H, Lee S. Ultrastructure of lymphocyte-mediated fat-cell lysis in erythema nodosum-like lesions of Behçet's syndrome. *Archives of Dermatology* 1987; **123**: 1650–4.

166. Hopps H C, Wissler R W. Experimental production of generalised arteritis and periarteritis (periarteritis nodosa). *Journal of Laboratory and Clinical Medicine* 1946; **31**: 939–7.

167. Horton B T, Magath B T. Arteritis of the temporal vessels: report of seven cases. *Proceedings of the Staff Meetings of the Mayo Clinic* 1937; **12**: 548–53.

168. Horton B T, Magath B T, Brown G E. An undescribed form of arteritis of the temporal vessels. *Proceedings of the Staff Meetings of the Mayo Clinic* 1932; **7**: 700–1.

169. Horton B T, Magath B T, Brown G E. Arteritis of the temporal vessels. *Archives of Internal Medicine* 1934; **53**: 400–9.

170. Hughes G R V. *Connective Tissue Diseases* (3rd edition). Oxford: Blackwell Scientific Publications, 1987: 200–18.

171. Hunder G G, Allen G L. The relationship between polymyalgia rheumatica and temporal arteritis. *Geriatrics* 1973; **28**: 134–42.

171a. Hunder G G, Bloch D A, Michel B A, Stevens M B, Arend W P, Calabrese L H, Edworthy S M, Fauci A S, Leavitt R Y, Lie J T, Lightfoot R W Jr, Masi A T, McShane D J, Mills J A, Wallace S L, Zvaifler N J. The American College of Rheumatology 1990 criteria for the classification of giant cell arteritis. *Arthritis and Rheumatism* 1990; **33**: 1122–8.

172. Hunder G G, Hazleman B L. Giant cell arteritis and polymyalgia rheumatica. In: Kelley W N, Harris E D, Ruddy S,

Sledge C B, eds, *Textbook of Rheumatology* (3rd edition). Philadelphia London Toronto: W B Saunders & Co., 1989: 1200–8.

173. Huston K A, Hunder G G. Giant cell (cranial) arteritis—a clinical review. *American Heart Journal* 1980; **100**: 99–107.

174. Huston K A, Hunder G G, Lie J T, Kennedy R H, Elveback L R. Temporal arteritis. A 25-year epidemiologic, clinical and pathologic study. *Annals of Internal Medicine* 1978; **88**: 162–7.

175. Hutchinson J. Diseases of the arteries. *Archives of Surgery* London 1890; **1**: 323–44.

176. Inman R D, McDougal J S, Redecha P B, Lochshin M D, Stevens C E, Christian C L. Isolation and characterisation of circulating immune complexes in patients with hepatitis B systemic vasculitis. *Clinical Immunology and Immunopathology* 1981; **21**: 364–74.

177. Ishikawa K. Natural history and classification of occlusive thrombo-aortopathy (Takayasu's disease). *Circulation* 1978; **57**: 27–35.

178. Jacobs R P, Moore M, Brower A. Wegener's granulomatosis presenting with erosive arthritis. *Arthritis and Rheumatism* 1987; **30**; 943–6.

179. Jaspan T, Davison A M, Walker W C. Spontaneous pneumothorax in Wegener's granulomatosis. *Thorax* 1982; **37**: 774–5.

180. Jellinek H, Littman I, Sülé E, Földi M, Máthé Z. Takayasu's disease and tuberculosis. *Acta Medica Academiae Scientiarum Hungaricae* 1960; **16**: 3–17.

181. Jessurun J, Azevedo M, Saldana M. Allergic angiitis and granulomatosis (Churg–Strauss syndrome): report of case with massive thymic involvement in a nonasthmatic patient. *Human Pathology* 1986; **17**: 637–9.

182. Johnson K J, Glovsky M, Schrier D. Pulmonary granulomatous vasculitis. Pulmonary granulomatous vasculitis induced in rats by treatment with glucan. *American Journal of Pathology* 1984; **114**: 515–6.

183. Johnson K J, Wilson B S, Till G O, Ward P A. Acute lung injury in rats caused by immunoglobulin A immune complexes. *Journal of Clinical Investigation* 1984; **74**: 358–69.

184. Jordan J M, Rowe W T, Allen N B. Wegener's granulomatosis involving the breast. Report of three cases and review of the literature. *American Journal of Medicine* 1987; **83**: 159–64.

185. Judge R, Currier R, Gracie W, Figley M. Takayasu's arteritis and the aortic arch syndrome. *American Journal of Medicine* 1962; **32**: 379–92.

186. Kalbian V V, Challis M T. Behçet's disease. Report of twelve cases with three manifesting as papilledema. *American Journal of Medicine* 1970; **49**: 823–30.

187. Kasahara Y, Tanaka S, Nishino M, Umemura H, Shiraha S, Kuyama T. Intestinal involvement in Behçet's disease: review of 136 surgical cases in the Japanese literature. *Diseases of the Colon and Rectum* 1981; **24**: 103–6.

188. Kato H, Ichinose E, Kawasaki T. Myocardial infarction in Kawasaki disease: clinical analyses in 195 cases. *Journal of Pediatrics* 1986; **108**: 923–7.

188a. Katz P, Fauci A S. Systemic vasculitis. In: Samter M, ed, *Immunological Diseases* (4th edition). Boston, Toronto: Little Brown & Co., 1988: 1417–35.

189. Kawasaki T. Acute febrile mucocutaneous syndrome with lymphoid involvement with specific desquamation of the fingers and toes in children. *Japanese Journal of Allergy* 1967; **16**: 178–222.

190. Kawasaki T, Kosaki F, Okawa S. A new infantile acute febrile mucocutaneous lymph node syndrome (MLNS) prevailing in Japan. *Pediatrics* 1974; **54**: 271–6.

191. Kelly J R, Andolsek K. Tetracycline toxicity presenting as generalized vasculitis. *Southern Medical Journal* 1978; **71**: 961–3.

192. Kemper J W, Baggenstoss A H, Slocumb C H. The relationship of therapy with cortisone to the incidence of vascular lesions in rheumatoid arthritis. *Annals of Internal Medicine* 1957; **46**: 831–51.

193. Kennedy L J Jr, Mitchinson M J. Giant cell arteritis with myositis and myocarditis. *California Medicine* 1971; **115**: 84–7.

194. Keren G, Wolman M. Can pseudomonas infection in experimental animals mimic Kawasaki's disease? *Journal of Infection* 1984; **9**: 22–9.

195. Kiernan T J, Gillan J, Murray J P, McCarthy C F D. Behçet's disease and splenomegaly. *British Medical Journal* 1978; **ii**: 1340–1.

196. Kitamura S, Kawashima Y, Kawachi K, Harima R, Ihara K, Nakano S, Shimazaki Y, Mori T. Severe mitral regurgitation due to coronary arteritis of mucocutaneous lymph node syndrome—a new surgical entity. *Journal of Thoracic and Cardiovascular Surgery* 1980; **80**: 629–36.

197. Kjellstrand C M, Simmons R L, Uranga V M, Buselmeier T J, Najarian J S. Acute fulminant Wegener's granulomatosis. Therapy with immunosuppression, hemodialysis and renal transplantation. *Archives of Internal Medicine* 1974; **134**: 40–3.

198. Klinger H. Grenzformen der Periarteritis nodosa. *Frankfurter Zeitschrift für Pathologie* 1931; **42**: 455–80.

199. Kondo N, Tateno M, Yamaguchi J, Yoshiki T, Itoh T, Kawashima N, Kataoka K. Immunopathological studies of an autopsy case with Goodpasture's syndrome and systemic necrotizing angiitis. *Acta Pathologica Japonica* 1986; **36**: 595–604.

200. Krapf F E, Manger B K, Kalden J R, Immunkomplexe bei Allergien: Nachweis von IgE-haltigen Immunkomplexen bei der Vaskulitis Churg–Strauss. *Klinische Wochenschrift* 1986; **64**: 563–9.

201. Küssmaul A, Maier R. Ueber eine bisher noch nicht beschrieben eigenthümliche Arterienerkrankung (Periarteritis nodosa), die mit Morbus Brightii und rapid fortschreitender allgemeiner Muskellähmung einhergeht. *Deutsch Archiv für Klinische Medizin* 1866; **1**: 484–518.

202. Kyle V, Dutoit S H, Elias-Jones J, Hazleman B. Giant cell arteritis of myometrial and axillary arteries and polymyalgia rheumatica. *Annals of the Rheumatic Diseases* 1987; **46**: 256–8.

203. Lachmann P J, Muller-Eberhard H J, Kunkel H G, Paronetto F. The localization of *in vivo* bound complement in tissue sections. *Journal of Experimental Medicine* 1962; **115**: 63–82.

204. Ladanyi M, Fraser R S. Pulmonary involvement in giant cell arteritis. *Archives of Pathology and Laboratory Medicine* 1987; **111**: 1178–80.

205. Lakhanpal S, Tani K, Lie J T, Katoh K, Ishigatsubo Y, Ohokubo T. Pathologic features of Behçet's syndrome: a review of Japanese autopsy registry data. *Human Pathology* 1985; **16**: 790–5.

206. Lamb A R. Periarteritis nodosa—a clinical and pathological review of the disease. *Archives of Internal Medicine* 1914; **14**: 481–516.

207. Lande A, Rossi P. The value of total aortography in the diagnosis of Takayasu's arteritis. *Radiology* 1975; **114**: 287–97.

208. Lanham J G, Elkon K B, Pusey C D, Hughes G R V. Systemic vasculitis with asthma and eosinophilia. A clinical approach to the Churg–Strauss syndrome. *Medicine* 1984; **63**: 65–81.

209. Largiadèr A. Wegenersche Granulomatose. *Schweizerische Medizinische Wochenschrift* 1964: **94**: 1000–3.

209a. Leavitt R Y, Fauci A S, Bloch D A, Michel B A, Hunder G G, Arend W P, Calabrese L H, Fries J F, Lie J T, Lightfoot R W, Masi A T, McShane D J, Mills J A, Stevens M B, Wallace S L, Zvaifler N J. The American College of Rheumatology 1990 criteria for the classification of Wegener's granulomatosis. *Arthritis and Rheumatism* 1990; **33**: 1101–7.

210. Leggatt P O, Walton E W. Wegener's granulomatosis. *Thorax* 1956; **11**: 94–100.

211. Lehner T. Behçet's syndrome and autoimmunity. *British Medical Journal* 1967; **i**: 465–7.

212. Lendrum A C, Fraser D S, Slidders W. Further observations on the age changes in extravascular fibrin. *Nederlands Tijdschrift voor Geneeskunde* 1964; **108**: 2373.

213. Lendrum A C, Fraser D D, Slidders W, Henderson R. Studies on the character and staining of fibrin. *Journal of Clinical Pathology* 1962; **15**: 401–13.

214. Leong A-S-Y, Alp M H. Hepatocellular disease in the giant cell arteritis/polymyalgia rheumatica syndrome. *Annals of the Rheumatic Diseases* 1981; **40**: 92–5.

215. Leung D Y M, Chu E T, Wood N, Grady S, Meade R, Geha R S. Immunoregulatory T-cell abnormalities in mucocutaneous lymph node syndrome. *Journal of Immunology* 1983; **130**: 2002–4.

216. Lever W F, Schaumberg-Lever G. *Histopathology of the Skin* (5th edition). Philadelphia, Toronto: J B Lippincott & Co., 1975: 160–3.

217. Lever W F, Schaumberg-Lever G. *Histopathology of the Skin* (6th edition). Philadelphia: J B Lippincott & Co., 1983: 167–70.

218. Levine B W, Mark E J. Reticulonodular pulmonary infiltrate during treatment for Wegener's granulomatosis. *New England Journal of Medicine* 1978; **298**: 729–36.

219. Levinsky R J, Lehner T. Circulating soluble immune complexes in recurrent oral ulceration and Behçet's syndrome. *Clinical and Experimental Immunology* 1978; **32**: 193–8.

220. Levo Y, Gorevic P D, Kassab H, Zucker-Franklin D, Franklin E C. Association between hepatitis B virus and essential mixed cryoglobulinemia. *New England Journal of Medicine* 1977; **296**: 1501–4.

221. Liang G C, Simkin P A, Hunder G G, Wilske K R, Healey L A. Familial aggregation of polymyalgia rheumatica and giant cell arteritis. *Arthritis and Rheumatism* 1975; **17**: 19–24.

222. Liang G C, Simkin P A, Mannik M. Immunoglobulins in temporal arteries: an immunofluorescent study. *Annals of Internal Medicine* 1974; **81**: 19–24.

222a. Lie J T and members and consultants of the American College of Rheumatology subcommittee on classification of vasculitis. Illustrated histopathologic classification criteria for selected vasculitis syndromes. *Arthritis and Rheumatism* 1990; **33**: 1074–87.

223. Lie J T, Failoni D D, Davis D C. Temporal arteritis with giant cell aortitis, coronary arteritis and myocardial infarction. *Archives of Pathology and Laboratory Medicine* 1986; **110**: 857–60.

224. Liebow A A. The J Burns Amberson Lecture—pulmonary angiitis and granulomatosis. *American Review of Respiratory Disease* 1973; **108**: 1–18.

224a. Lightfoot R W Jr, Michel B A, Bloch D A, Hunder G G, Zvaifler N J, McShane D J, Arend W P, Calabrese L H, Leavitt R Y, Lie J T, Masi A T, Mills J A, Stevens M B, Wallace S L. The American College of Rheumatology 1990 criteria for the classification of polyarteritis nodosa. *Arthritis and Rheumatism* 1990; **33**: 1088–93.

225. Lim S D, Haw C R, Kim N I, Fusaro R M. Abnormalities of T-cell subsets in Behçet's syndrome. *Archives of Dermatology* 1983; **119**: 307–10.

226. Little A G, Zarins C K. Abdominal aortic aneurysm and Behçet's disease. *Surgery* 1982; **91**: 359–62.

227. Littlejohn G O, Ryan P J, Holdsworth S R. Wegener's granulomatosis: clinical features and outcome in seventeen patients. *Australian and New Zealand Journal of Medicine* 1985; **15**: 241–5.

228. Lockwood C M, Bakes D, Jones S, Whitaker K B, Moss D W, Savage C O S. Association of alkaline phosphatase with an autoantigen recognised by circulating anti-neutrophil antibodies in systemic vasculitis. *Lancet* 1987; **ii**: 716–21.

229. Longley S, Caldwell J R, Panush R S. Paraneoplastic vasculitis. Unique syndrome of cutaneous angiitis and arthritis associated with myeloproliferative disorders. *American Journal of Medicine* 1986; **80**: 1027–30.

230. Lo Spalluto J, Dorward B, Miller W J Jr, Ziff M. Cryoglobulinemia based on interaction between a gamma macroglobulin and 7S gamma globulin. *American Journal of Medicine* 1962; **32**: 142–7.

231. Lucas S B, Moxham J. Recurrent vasculitis associated with beta-haemolytic streptococcal infections. *British Medical Journal* 1978; **1**: 1323.

232. Lupi-Herrera E, Sanchez-Torres G, Marcushamer J, Mispireta J, Horwitz S, Espino Vela J. Takayasu's arteritis: clinical study of 107 cases. *American Heart Journal* 1977; **93**: 94–103.

233. MCLS Research Committee of Japan. *Diagnostic Guideline of Infantile Acute Febrile Muco-cutaneous Lymphnode Syndrome (MCLS)* (1st edition 1970, revised 1974). Tokyo: Institute of Public Health.

234. Mabalay M C, Helwig E B, Tolentine J G, Binford C H. The histopathology and histochemistry of erythema nodosum leprosum. *International Journal of Leprosy* 1965; **33**: 28–49.

235. McCluskey R T, Fienberg R. Vasculitis in primary vasculitides, granulomatoses and connective tissue diseases. *Human Pathology* 1983; **14**: 305–15.

236. McDonald T J, DeRemee R A. Wegener's granulomatosis. *Laryngoscope* 1983; **93**: 220–36.

237. McDuffie F C, Sams W M Jr, MacDonald J E, Andreini P H, Conn D L, Samayoa E A. Hypocomplementaemia with

308. Platt R, Davson J. A clinical and pathological study of renal disease. Part II. Diseases other than nephritis. *Quarterly Journal of Medicine NS* 1950; **19**: 33–56.

309. Pope T L, Buschi A J, Moore T S, Williamson B R J, Brenbridge A N A G. CT features of renal polyarteritis nodosa. *American Journal of Roentgenology* 1981; **136**: 986–7.

310. Porter D D, Larsen A E, Porter H G. The pathogenesis of Aleutian disease of mink: III. Immune complex arteritis. *American Journal of Pathology* 1973; **71**: 331–4.

311. Pritchard M H, Gow P J. Wegener's granulomatosis presenting as rheumatoid arthritis (two cases). *Proceedings of the Royal Society of Medicine* 1976; **69**: 501–4.

312. Przybojewski J Z. Polyarteritis nodosa in the adult: report of a case with repeated myocardial infarction and a review of cardiac involvement. *South African Medical Journal* 1981; **60**: 512–18.

313. Przybojewski J Z, Maritz F. Pulmonary arteriovenous fistulas: a case presentation and review of the literature. *South African Medical Journal* 1980; **57**: 366–73.

313a. Radnai B. Comparative morphology of small vessel lesions in rheumatoid arthritis and periarteritis nodosa. *Acta Morphologica Academie Scientiarum Hungaricae* 1969; **17**: 69–79.

314. Rauch A M, Fultz P N, Kalyanaeaman V S. Retrovirus serology and Kawasaki syndrome. *Lancet* 1987; **i**: 1431.

315. Rich A R. The role of hypersensitivity in periarteritis nodosa; as indicated by seven cases developing during serum sickness and sulfonamide therapy. *Bulletin of the Johns Hopkins Hospital* 1942; **71**: 123–35.

316. Rich A R. Hypersensitivity in disease with especial reference to periarteritis nodosa, rheumatic fever, disseminated lupus erythematosus and rheumatoid arthritis. *Harvey Lectures, Series XLII: 1947*. Lancaster Pennsylvania: Science Press, 1948: 106–47.

317. Rich A R, Gregory J E. The experimental demonstration that periarteritis nodosa is a manifestation of hypersensitivity. *Bulletin of the Johns Hopkins Hospital* 1943a; **72**: 65–82.

318. Rich A R, Gregory J E, Experimental evidence that lesions with the basic characteristics of rheumatoid carditis can result from anaphylactic hypersensitivity. *Bulletin of the Johns Hopkins Hospital* 1943b; **73**: 239–64.

319. Rich A R, Gregory J E. Further experimental cardiac lesions of the rheumatic type produced by anaphylactic hypersensitivity. *Bulletin of the Johns Hopkins Hospital* 1944; **75**: 115–34.

320. Richardson J. *Connective Tissue Disorders*. Oxford: Blackwell, 1963: 104–24.

321. Robinson D R, Kirkham S E. A 33-year old woman with cutaneous vasculitis, arthralgia, and intermittent bloody diarrhea. *New England Journal of Medicine* 1984; **311**: 904–11.

322. Roge J, Fabre M, Durand B, Durand J, Benichou J, Paillas J, Roge F. Colitis associated with Behçet's disease—report of 2 cases with intestinal vasculitis. *Gastroentérologie Clinique et Biologique* 1982; **6**: 872–8.

323. Ronco P, Mougenot B, Kanfer A, Verroust P, Mignon F. Atteintes rénales des angéites nécrosantes. *Revue de Medecine Interne (Paris)* 1989; **10**: 227–34.

324. Ronco P, Verroust P, Mignon F, Kourilsky O, Vanhille P, Meyrier A, Mery J P, Morel-Maroger L. Immunopathological studies of polyarteritis nodosa and Wegener's granulomatosis—a report of 43 patients with 51 renal biopsies. *Quarterly Journal of Medicine* 1983; **NS52**: 212–23.

325. Rose G A, Spencer H. Polyarteritis nodosa. *Quarterly Journal of Medicine* 1957; **26**: 43–81.

326. Rössle R. Zum Formenkreis der rheumatischen Gewebsveranderungen, mit besondere Berücksichtigung der rheumatischen Gefässentzündungen. *Virchows Archiv für Pathologische Anatomie und Physiologie und für Klinische Medizin* 1933; **288**: 780–832.

327. Rothfield N F. Thrombotic thrombocytopenic purpura. In: McCarty D J, ed, *Arthritis and Allied Conditions: A Textbook of Rheumatology* (10th edition). Philadelphia: Lea & Febiger, 1985: 919.

328. Samantray S K. Takayasu's arteritis—a study of 45 cases. *Australian and New Zealand Journal of Medicine* 1978; **8**: 68–74.

329. Savage C O S, Winearls C G, Evans D J, Rees A J, Lockwood C M. Microscopic polyarteritis: presentation, pathology and prognosis. *Quarterly Journal of Medicine* 1985; **NS56**: 467–83.

330. Savage C O S, Winearls C G, Jones S, Marshall P D, Lockwood C M. Prospective study of radioimmunoassay for antibodies against neutrophil cytoplasm in diagnosis of systemic vasculitis. *Lancet* 1987; **i**: 1389–93.

331. Schrader M L, Hochman J S, Bulkley B H. The heart in polyarteritis nodosa: a clinicopathologic study. *American Heart Journal* 1985; **109**: 1353–9.

332. Schuh D, Herrman W R. Zür Morphologie der Vaskulitis bei der Wegenerschen Granulomatose. *Angiologica* 1969; **6**: 339–53.

333. Scott D G I. Vasculitis. In: *Copeman's Textbook of the Rheumatic Diseases* (6th edition). Edinburgh: Churchill Livingstone, 1986: 1292–324.

334. Scott D G I, Bacon P A, Elliott P J, Tribe C R, Wallington T B. Systemic vasculitis in a district general hospital 1972–1980; clinical and laboratory features, classification and prognosis of 80 cases. *Quarterly Journal of Medicine* 1982; **51**: 292–311.

335. Sergent J S, Lockshin M D, Christian C L, Gocke D J. Vasculitis with hepatitis B antigenemia: long-term observations in nine patients. *Medicine* (Baltimore) 1976; **55**: 1–18.

336. Sergent J S, Lockshin M D, Gocke D J, Christian C L. Polyarteritis nodosa with and without persistent hepatitis-associated antigen: long-term course in 21 patients. *Arthritis and Rheumatism* 1973; **16**: 568–9.

337. Serra A, Cameron J S, Turner D R, Hartley B, Ogg C S, Neild G H, Williams D G, Taube D, Brown C B, Hicks J A. Vasculitis affecting the kidney—presentation, histopathology and long-term outcome. *Quarterly Journal of Medicine* 1984; **53**: 181–207.

338. Sewell J R, Allison D J, Tarin D, Hughes G R V. Combined temporal arteriography and selective biopsy in suspected giant cell arteritis. *Annals of the Rheumatic Diseases* 1980; **39**: 124–8.

339. Shasby D M, Schwartz M I, Forstot J Z, Theofilopoulos A N, Kassan S S. Pulmonary immune complex deposition in Wegener's granulomatosis. *Chest* 1982; **81**: 338–40.

340. Shimizu T. Behçet's disease: a systemic inflammatory disease. In: Shiokawa Y, ed, *Vascular Lesions of Collagen Diseases*

and Related Conditions. Baltimore: University Park Press, 1977: 201–11.

341. Shiokawa Y, ed, *Vascular Lesions of Collagen Diseases and Related Conditions*. Baltimore: University Park Press, 1977.

342. Shulman S T, Rowley A. Does Kawasaki disease have a retroviral aetiology? *Lancet* 1986; **ii**: 545–6.

343. Slavin R E, De Groot W J. Pathology of the lung in Behçet's disease—case report and review of the literature. *American Journal of Surgical Pathology* 1981; **5**: 779–88.

344. Smith A J, Kyle V, Cawston T E, Hazleman B L. Isolation and analysis of immune complexes from sera of patients with polymyalgia rheumatica and giant cell arteritis. *Annals of the Rheumatic Diseases* 1987; **46**: 468–74.

345. Smith C C, Zeek P M, McGuire J. Periarteritis nodosa in experimental hypertensive rats and dogs. *American Journal of Pathology* 1944; **20**: 721–32.

346. Smukler N M, Schumacher H R. Chronic nondestructive arthritis associated with cutaneous polyarteritis. *Arthritis and Rheumatism* 1977; **20**: 1114–20.

347. Sofferman R A, Burlington M D. Lingual infarction in cranial arteritis. *Journal of the American Medical Association* 1980; **243**: 2422–3.

348. Sokol R J, Farrell M K, McAdams A J. An unusual presentation of Wegener's granulomatosis mimicking inflammatory bowel disease. *Gastroenterology* 1984; **87**: 426–32.

349. Soter N A. Clinical presentations and mechanisms of necrotizing angiitis of the skin. *Journal of Investigative Dermatology* 1976; **67**: 354–9.

350. Soter N A, Austen K F. Cutaneous necrotizing angiitis. In: Samter M, ed, *Immunological Diseases* (4th edition). Boston: Little Brown & Co., 1988: 1267–80.

351. Soter N A, Mihm M C, Gigli I, Dvorak H F, Austen K F. Two distinct cellular patterns in cutaneous necrotizing angiitis. *Journal of Investigative Dermatology* 1976; **66**: 344–50.

351a. Soter N A, Wuepper K D. Erythema multiforme and Stevens–Johnson syndrome. In: Samter M, ed, *Immunological Diseases* (4th edition). Boston: Little, Brown and Co., 1988: 1257–66.

352. Stave G M, Croker B P. Thrombotic microangiopathy in anti-glomerular basement membrane glomerulonephritis. *Archives of Pathology and Laboratory Medicine* 1984; **108**: 747–51.

353. Strachan R W, Wigzell F W. Polyarthritis in Behçet's multiple symptom complex. *Annals of Rheumatic Diseases* 1963; **22**: 26–35.

354. Stuart A E, MacGregor-Robertson G. Thrombotic thrombocytopenic purpura. A hyperergic micro-angiopathy. *Lancet* 1956; **i**: 475–9.

355. Sugihara H, Muto Y, Tsuchiyama H. Neuro-Behçet's syndrome: report of two autopsy cases. *Acta Pathologica Japonica* 1969; **19**: 95–101.

356. Sulheim O, Dalgaard J B, Andersen S Ry. Behçet's syndrome: report of case with complete autopsy performed. *Acta Pathologica et Microbiologica Scandinavica* 1959; **45**: 145–58.

357. Suma K, Takeuchi Y, Shiroma K, Tsuji T, Inoue K, Yoshikawa T, Koyama Y, Narumi J, Asai T, Kusakawa S. Early and late postoperative studies in coronary arterial lesions resulting from Kawasaki's disease in children. *Journal of Thoracic and Cardiovascular Surgery* 1982; **84**: 224–9.

358. Symmers W St C. Thrombotic microangiopathic haemolytic anaemia (thrombotic microangiopathy). *British Medical Journal* 1952; **ii**: 897–903.

359. Tai P-C, Holt M E, Denny P, Gibbs A R, Williams B D, Spry C J F. Deposition of eosinophil cationic protein in granulomas in allergic granulomatosis and vasculitis: the Churg–Strauss syndrome. *British Medical Journal* 1984; **289**: 400–2.

360. Takagi M, Ikeda T, Kimura K, Saito Y, Ishii M, Takeda T, Murao S. Renal histological studies in patients with Takayasu's arteritis—report of 3 cases. *Nephron* 1984; **36**: 68–73.

361. Takayasu M. Case with unusual changes of the central vessels of the retina. *Acta Societatis Ophthalmologicae Japonicae* 1908; **12**: 554.

362. Talal N, Steinberg A D. The pathogenesis of autoimmunity in New Zealand black mice. *Current Topics in Microbiology and Immunology* 1974; **64**: 79–103.

363. Talbot S F, Main D M, Levinson A I. Wegener's granulomatosis: first report of a case with onset during pregnancy. *Arthritis and Rheumatism* 1984; **27**: 109–12.

364. Tanaka N, Naoè S, Masuda H, Ueno T. Pathological study of sequelae of Kawasaki disease (MCLS)—with special reference to the heart and coronary arterial lesions. *Acta Pathologica Japonica* 1986; **36**: 1513–27.

365. Tanaka N, Sekimoto K, Fukushima T, Tokita H, Veno T, Naoe S. Pathological study of fatal mucocutaneous lymph node syndrome cases of Kawasaki disease: relationship with infantile polyarteritis nodosa. In: Shiokawa Y, ed, *Vascular Lesions of Collagen Diseases and Related Conditions*. Baltimore: University Park Press, 1977: 296–307.

366. Tapp E, Geary C G, Dawson D W. Thrombotic microangiopathy with macroscopic infarction. *Journal of Pathology* 1969; **97**: 711–17.

367. Thompson E A, Burman H J, Dolgopol V B, Bruyning E H. Wegener's granulomatosis with unusual pathologic findings. *New York State Journal of Medicine* 1963; **63**: 3565–9.

368. Thomson D, Gardner D L. Thrombotic microangiopathy in rheumatoid arthritis. *Scottish Medical Journal* 1969; **14**: 190–3.

369. Tomita S, Kato H, Fujimoto T, Inoue O, Koga Y, Kuriya N. Cytopathogenic protein in filtrates from cultures of *Propionibacterium acnes* isolated from patients with Kawasaki disease. *British Medical Journal* 1987; **295**: 1229–32.

370. Trepo C, Thirolet J. Hepatitis associated antigen and periarteritis nodosa (PAN). *Vox Sanguinis* 1972; **19**: 410–11.

371. Ueda H, Saito Y, Ito L, Yamaguchi H, Takeda T, Morooka S. Further immunological studies of aortitis syndrome. *Japanese Heart Journal* 1971; **12**: 1–21.

372. Van Berlo M F, Horzinet M C, Van der Zeijst B A M. Equine arteritis virus-infected cells contain six polyadenylated virus-specific RNAs. *Virology* 1982; **118**: 345–52.

373. Van Berlo M F, Rottier P J M, Spaan W J M, Horzinek M C. Equine arteritis virus-induced polypeptide synthesis. *Journal of General Virology* 1986; **67**: 1543–9.

374. Van Ypersele de Strihou C, Pirson Y, Vandenbroucke J-M, Alexandre G P J. Haemodialysis and transplantation in Wegener's granulomatosis. *British Medical Journal* 1979; **ii**: 93–4.

375. Vernon-Roberts B, Barnes C G, Revell P A. Synovial pathology in Behçet's syndrome. *Annals of the Rheumatic Diseases* 1978; **37**: 139–45.

376. Vilaseca J, Gonzalez A, Cid M C, Lopez-Vivancos J, Ortega A. Clinical usefulness of temporal artery biopsy. *Annals of the Rheumatic Diseases* 1987; **46**: 282–5.

377. Vinijchaikul K. Primary arteritis of the aorta and its main branches (Takayasu's arteriopathy). *American Journal of Medicine* 1967; **43**: 15–27.

378. Vital A, Vital C. Polyarteritis nodosa and peripheral neuropathy. Ultrastructural study of 13 cases. *Acta Neuropathologica* 1985; **67**: 136–41.

379. Volkman D J, Mana D L, Fauci A S. Association between Takayasu's arteritis and B-cell alloantigen in North Americans. *New Zealand Journal of Medicine* 1982; **306**: 464–5.

380. von Knorring J, Wasastjerna C. Liver involvement in polymyalgia rheumatica. *Scandinavian Journal of Rheumatology* 1976; **5**: 197–204.

381. von Rokitansky K. Ueber einige der wichtigsten Erkrankungen der Arterien. *Denkschrift der Käiserliche Akademie der Wissenschaft der Wien* 1852; **4**: 49–121.

382. Wachter R M, Burke A M, MacGregor R R. *Strongyloides stercoralis* hyperinfection masquerading as cerebral vasculitis. *Archives of Neurology* 1984; **41**: 1213–16.

383. Wahls T L, Bonsib S M, Schuster V L. Coexistent Wegener's granulomatosis and anti-glomerular basement membrane disease. *Human Pathology* 1987; **18**: 202–5.

384. Walton E W. Giant-cell granuloma of the respiratory tract (Wegener's granulomatosis). *British Medical Journal* 1958; **ii**: 265–70.

385. Walton J N. Some diseases of muscle. *Lancet* 1964; **i**: 447–52.

386. Watanabe T, Nagafuchi Y, Yoshikawa Y, Toyoshima H. Renal papillary necrosis associated with Wegener's granulomatosis. *Human Pathology* 1983; **14**: 551–7.

387. Wegener F. Über generalisierte aseptische Gefässerkränkungen. *Verhandlungen der Deutschen Gesellschaft für Pathologie* 1936; **29**: 202–10.

388. Wegener F. Über eine eigenartige rhinogene Granulomatose mit besonder Beteiligung des Arteriensystems und der Nieren. *Beiträge zur Pathologischen Anatomie und Physiologie* 1939; **102**: 36–68.

389. Wegener F. About the so-called Wegener's granulomatosis with special reference to the generalised vascular lesions. *Morgagni* 1968; **1**: 5–21.

390. Weinblatt M E, Fienberg R. A 31-year old asthmatic woman with rapidly progressive multisystem disease. *New England Journal of Medicine* 1987; **316**: 1139–47.

391. Weiss E B, Forman P, Rosenthal I M. Allopurinol-induced arteritis in partial HGPRTase deficiency—atypical seizure manifestations. *Archives of Internal Medicine* 1978; **138**: 1743–4.

392. Werner B. Das klinische Bild des Morbus Behçet aus Japan. *Medizinische Monatschrift* 1967; **21**: 78–82.

393. White F N, Grollman A. Experimental periarteritis nodosa in the rat. *Archives of Pathology* 1964; **78**: 31–6.

394. Wigley R D, Couchman K G, Maule R. Polyarteritis nodosa: the natural history of a spontaneously occurring model in outbred mice. *Australian Annals of Medicine* 1970; **19**: 319–27.

395. Wilkey D, Yocum D E, Oberley T D, Sundstrom W R, Karl L. Budd–Chiari syndrome and renal failure in Behçet disease. Report of a case and review of the literature. *American Journal of Medicine* 1983; **75**: 541–50.

396. Willan R. *On Cutaneous Diseases*. London: 1808; Vol. 1: 483.

397. Winer L H. Histopathology of the nodose lesion of acute coccidioidomycosis. *Archives of Dermatology and Syphilology* 1950; **61**: 1010–24.

398. Winkelmann R K. Granulomatous Vasculitis. In: Wolff K, Winkelmann R K, eds, *Vasculitis*. London: Lloyd-Luke (Medical Books) Ltd, 1980: 228–41.

399. Winkelmann R K, Ditto W B. Cutaneous and visceral syndromes of necrotizing or 'allergic' angiitis: study of 38 cases. *Medicine* 1964; **43**: 59–89.

400. Winkelmann R K, Buechner S A, Powell F C, Banks P M. The T-lymphocyte and cutaneous Churg–Strauss granuloma. *Acta Dermato-Venereologica (Stockholm)* 1983; **63**: 199–204.

401. Wolff K, Winkelmann R K, eds, *Vasculitis*. London: Lloyd-Luke, 1980.

402. Woolf A D, Wakerley G, Wallington T B, Scott D G I, Dieppe P A. Factor VIII related antigen in the assessment of vasculitis. *Annals of the Rheumatic Diseases* 1987; **46**: 441–7.

403. Wooten M R, Khangure M S, Murphy M J. Intracerebral hemorrhage and vasculitis related to ephedrine abuse. *Annals of Neurology* 1983; **13**: 337–40.

404. Wright V. The arthritis of ulcerative colitis, Reiter's disease and Behçet's syndrome. In: Hill A G S, ed, *Modern Trends in Rheumatology*. London: Butterworths, 1966: 337–46.

405. Yamaguchi H, Nakajima S, Torikata C, Takeuchi H, Kageyama K. Studies on the morphological changes of artery caused by exudation—initial changes of arteritis. *Acta Pathologica Japonica* 1972; **22**: 441–55.

406. Yanagawa H, Nakamura Y, Kawasaki T, Shigematsu I. Nationwide epidemic of Kawasaki disease in Japan during winter of 1985–1986. *Lancet* 1986; **ii**: 1138–9.

407. Yermakov V M, Hitti I F, Sutton A L, Khein S M, Flint A, Moss J A. Necrotizing vasculitis associated with diphenylhydantoin. *Human Pathology* 1983; **14**: 182–3.

408. Yonker R A, Katz P. Necrotizing granulomatous vasculitis with eosinophilic infiltrates limited to the prostate. Case report and review of the literature. *American Journal of Medicine* 1984; **77**: 362–4.

409. Yurdakul S, Yaxici H, Tuzuner N, Aytac S, Muftuoglu A U. Case Report. Olecranon nodules in a case of Behçet's disease. *Annals of the Rheumatic Diseases* 1981; **40**: 182–4.

410. Zeek P M. Periarteritis nodosa; a critical review. *American Journal of Clinical Pathology* 1952; **22**: 777–90.

411. Zeek P M. Periarteritis nodosa and other forms of necrotizing angiitis. *New England Journal of Medicine* 1953; **248**: 764–72.

Chapter 16
Polymyositis and Dermatomyositis

The clinical complex of polymyositis–dermatomyositis (PM–DM) is a heterogeneous group of acquired muscle diseases (myopathies) in which muscle cell (myofibre) injury and aseptic inflammation lead to weakness.[22,25,67] A wide range of heritable and acquired, non-inflammatory muscle disease is often considered in differential clinical diagnosis. The present account is centred on the pathology of adult polymyositis.

Myositis, in its broadest sense, implies muscle inflammation which may be focal or widespread, acute or prolonged, primary or secondary.[19] In theory, all forms of striated and non-striated muscle may be subject to myositis. In practice, the term is usually restricted to non-cardiac, striated muscle. Inflammation of the myocardium is designated 'myocarditis' and inflammation of involuntary non-striated muscles, such as those of the uterus and gastrointestinal tract, is seldom differentiated from the inflammatory processes affecting the viscus as a whole.

The PM–DM complex may present as a disease entity or it may appear with systemic connective tissue diseases, such as systemic lupus erythematosus or progressive systemic sclerosis. Dermatomyositis, in particular, may mark the advance of a cancer. Alternatively, the PM–DM complex may be a manifestation of treatment with drugs such as D-penicillamine given for rheumatoid arthritis or it may appear in the course of the hydralazine syndrome (see p. 594).

Historical background

Problems of definition (see below) delayed recognition of the PM–DM complex. An example of PM–DM, acceptable by modern criteria, was reported by Wagner.[65] Unverricht[64] described the details of a case which he had observed seven years previously and, in 1891, introduced the term 'dermatomyositis' to replace the older title 'polymyositis acuta progressiva'. The early reports were collated by Steiner[59] and many modern surveys have been made.[12,13,15,21,21a,22]

Definition

Polymyositis–dermatomyositis is a series of clinical syndromes which have nosological, serological and biochemical features in common. Definitions of the PM–DM complex have aimed primarily at satisfying the need for diagnosis. The purpose of these definitions is to enable the selective choice of treatment and to provide a sound basis for controlled therapeutic trials. Five main categories of disorder were identified within the complex[11,12] (Table 16.1). An alternative, broader classification has additional, clinical value[22] (Table 16.2). Within each of the five categories, histological, ultrastructural, histochemical and immunopathological phenomena offer help with diagnosis. However, some of these changes are shared between two or

Table 16.1

Classification of polymyositis/dermatomyositis[11,12]

Group	Diagnostic/prognostic entity	Histological changes	Histochemical changes	Immunopathological features
I	Primary idiopathic adult polymyositis (33%)	Loss of myofibrils Phagocytosis of necrotic muscle cells (> 75%) Ultrastructural evidence of cell necrosis (see text) Inflammation with perivascular macrophage and lymphocyte infiltration Muscle cell atrophy Fibrosis	Loss of myofibrillar ATPase activity Increased perimysial alkaline phosphatase, related to fibrosis Loss of oxidative enzyme activities Regenerating cells show raised alkaline phosphatase and oxidative enzyme activities	Immunoglobulin and complement accumulation in injured muscle cells T-cell anti-muscle response Raised T_H-cell proportions
II	Primary idiopathic adult dermatomyositis (33%)	Muscle changes similar to polymyositis Perifasicular muscle atrophy Necrosis and phagocytosis of muscle cells May be little perifasicular or perivascular inflammation	As in group I	Diffuse T-cell endomysial infiltration Activated macrophages Many B-cells High T4/T8 lymphocyte ratios T4 cells located near B-cells
III	Polymyositis/dermatomyositis with cancer (5–8%)	As in class I	As in group I	
IV	Childhood dermatomyositis or, rarely, polymyositis (8–22%)	Perifasicular muscle cell atrophy In 30%, muscle necrosis Occasionally, lymphocyte and macrophage infiltration Vasculopathy	Not well established	Increased T-cell proportions Mononuclear macrophages, HLA-DR+ve
V	Polymyositis associated with other systemic connective tissue disorders (15–30%)	Similar to those of adult polymyositis/ dermatomyositis	As in group I	See Chapters 12, 14 *et seq.*

more of the clinical groups and the disorders that underlie the abnormalities are not yet well understood. Consequently, a true pathological classification of the PM–DM complex, based on molecular and cellular disorders, is not practicable. This difficulty is compounded by the infrequency of cases and the heterogeneity of the complex. Finally, the use of necropsy studies, so valuable in establishing baselines for other systemic connective tissue diseases, is limited because death during the active phases of PM–DM is uncommon and few patients die in hospital.

Clinical features

The entire complex of non-localized, non-suppurative muscle inflammatory disorders is the PM–DM complex.[37] A distinction from muscular dystrophy is important because of the beneficial response of PM–DM to steroid treatment. Myasthenia gravis, granulomatous myositis and inclusion-body myositis are related disorders not included in the PM–DM complex and not discussed here.

Polymyositis[15] is an acquired primary, non-suppurative, inflammatory disorder in which degenerative changes of the lower and upper proximal limb muscles coincide with disorders of the muscles regulating neck and respiratory movements. In one-third of cases, only skeletal muscle is affected; in the remaining cases, other viscera such as the heart are involved.[36] Of 16 patients in this cardiac series, seven had congestive cardiac failure, four with microscopic myocarditis, and five presented evidence of coronary vasculitis, intimal thickening or medial calcific disease.

The associations of polymyositis include systemic connective tissue diseases such as rheumatoid arthritis,

Table 16.2

Additional classes of polymyositis/dermatomyositis included in revised (clinical) classification[22]

Group	Diagnostic entity	Histological changes
VI	Polymyositis with monoclonal gammopathy	As in adult polymyositis; may show inclusion bodies
VII	Polymyositis with AIDS	As in adult polymyositis. Viral antigen in T_H cells surrounding muscle fibres
VIII	Eosinophilic polymyositis; polymyositis in eosinophilic fasciitis	Fasciitis; eosinophilic granulocytes present in ~ 50%. Occasionally muscle necrosis with eosinophilia (see p. 708)
IX	Putative polymyositis/dermatomyositis	A clinical syndrome. Myopathy without microscopic evidence of myositis
X	Polymyositis with other systemic disease	Changes are those of: Crohn's disease; vasculitis; sarcoidosis; gluten enteropathy; chronic graft-versus-host disease; jejunoileal bypass surgery; Behçet's disease (see p. 647); myasthenia gravis; acne fulminans; lipoatrophy; Hashimoto's disease; Waldenström's macroglobulinaemia; Kawasaki disease (see p. 651)
XI	Polymyositis/dermatomyositis induced by drugs	Syndrome has been described after treatment with D-penicillamine, cimetidine, penicillin, ipecac, sulphonamides, procainamide, hydralazine
XII	Fascioscapulohumeral dystrophy with inflammation	May show inflammation resembling polymyositis
XIII	Inclusion body myositis	Frozen section necessary to show eosinophilic muscle cell inclusions and vacuoles outlined by basophilic material
XIV	Benign acute myositis of childhood	Muscle cell necrosis with leucocytic infiltration
XV	Postviral acute myositis in adults	Focal myositis with muscle cell phagocytosis
XVI	Postviral fatigue syndrome	Because serum enzymes remain normal, histological changes probably confined to atrophy
XVII	Polymyositis with agammaglobulinaemia	Interstitial inflammation with severe fasciitis
XVIII	Fungal and bacterial myositis	Changes are those of staphylococcal, clostridial, mycobacterial and other infections
XIX	Parasitic myositis	Changes are those of toxoplasmal, trypanosomal, sarcosporidial infestation or those due to cestodes or nematodes

systemic lupus erythematosus, progressive systemic sclerosis and Sjögren's syndrome;[55] neoplasia;[44] virus infection; and, particularly in the rare childhood disorder, vasculitis. A large variety of immunological abnormalities has been identified.[8] When polymyositis is accompanied by a skin rash, as it is in one-third of cases, the term 'dermatomyositis' is preferred.

Group I: Polymyositis

The most frequent presenting symptom is weakness of the proximal limb muscles. The legs and pelvic girdle are affected initially in 30–50 per cent of cases; the arms and shoulder girdle in 25–35 per cent.[21,21a] The degree of loss of muscle strength is proportionally less than the degree of late muscle atrophy. The patient is ill, swallows with difficulty and has lost weight. The disease develops insidiously over a period of months or even years but an acute onset is occasional. Muscle tenderness is frequent but there is neither skin rash nor calcinosis and arthropathy is rare. Contractures develop in the advanced disease. The most valuable objective diagnostic tests for polymyositis are electromyography, the laboratory measurement of creatine phosphokinase activity, and muscle biopsy.[15,21,51,58]

The erythrocyte sedimentation rate is raised. Hypergammaglobulinaemia is inconstant. Creatine is excreted in excess in the urine. In rare, acute instances myoglobinuria is present. Creatinine phosphokinase levels are raised on average 10 times above normal.[15] The lupus erythematosus cell test is sometimes positive and antinuclear antibody titres can be high. Anti-DNA antibodies are seldom found but other antibodies are of diagnostic importance (Table 16.3). The presence of antimuscle antibody is described.

Table 16.3

Diagnostic serological changes in autoantibodies in adult polymyositis/dermatomyositis[37]

	Group (number)				
Antigen	1 (31)	2 (20)	3 (5)	5 (18)	Total (74)
ANA† (per cent)	5 (16)	5 (24)	1 (17)	13 (72)	24 (32)
nRNP†	1	0	0	6 (33)	7 (9)
Sm‡	0	0	0	4 (22)	5 (5)
Jo-1‡	8/17(47)	0/11	0/3	2/13(15)	10/44(23)
Other antibodies of diagnostic value[26]					
PM-1					
Mi-1					
Ku*					

*High specificity and predictive value
†Low specificity and predictive value
‡High predictive value for interstitial lung disease in PM (Jo-1) or SLE (Sm).

Group II: Dermatomyositis

In most cases which develop dermatomyositis, a skin rash precedes muscle weakness. Photosensitization may be a precipitating factor. The symptoms and signs of muscle disease are indistinguishable from those of polymyositis but the extent of skin and muscle involvement is highly variable. Flat, bright-red or violaceous plaques form on convex surfaces, especially those of the nose, cheeks and chest; pain and swelling precede desquammation and even ulceration. The changes may merge with those of progressive systemic sclerosis (see Chapter 17) and calcinosis[59a] is then possible. The clinical laboratory tests yield results similar to those of polymyositis; the findings at muscle biopsy, in particular, are identical. Skin biopsy (see p. 682) is of value in diagnosis.

Group III: Polymyositis and/or dermatomyositis with neoplasia

In as many as 15 per cent of cases of PM–DM, the syndrome coincides with the development of neoplastic disease.[4,7,12] The significance of this observation and its possible relationship to understanding of the pathogenesis of PM–DM, are discussed on p. 689. The clinical and pathological features of Group III cases and the results of other laboratory investigations differ from other cases of PM–DM only with respect to the phenomena caused by the growth of the cancer.

Group IV: Childhood dermatomyositis
(Fig. 16.1)

Although rare, dermatomyositis is relatively twice as frequent as polymyositis in children.[20] Many of the clinical and laboratory findings resemble those of the adult disease but the presence of vasculitis clearly distinguishes the pathological features from all other forms of idiopathic polymyositis and dermatomyositis. Intestinal perforation may occur just as it does in rheumatoid arthritis and polyarteritis nodosa. In spite of the finding that this change and gastrointestinal haemorrhage cause death in 25 per cent of cases, the prognosis is much better than in the adult disease. Muscle changes in childhood dermatomyositis include extensive endothelial injury to cells lining arterioles, capillaries and veins. Platelet thrombi form.[6,20] Perhaps because of an impaired vascular supply to the subcutaneous tissues, calcinosis is more common than in adult dermatomyositis.

Group V: Dermatomyositis/polymyositis overlap group

By definition, cases in this category display the clinical features both of polymyositis or dermatomyositis *and* those of a systemic immunological disorder of connective tissue such as progressive systemic sclerosis,[10] rheumatoid arthritis or systemic lupus erythematosus; an association with the former is most common. It is often observed that the disability from the connective tissue disease overshadows the muscle disease due to PM–DM; the converse is exceptional.[15] The evidence of myositis may be obscured by signs closely resembling those of classical PM–DM. However, atrophy of type II muscle cells is more frequent in the overlap cases, and vasculitis and motor neurone denervation are common.

Pathological changes

Unless otherwise stated, the descriptions that follow are restricted to classical PM–DM.

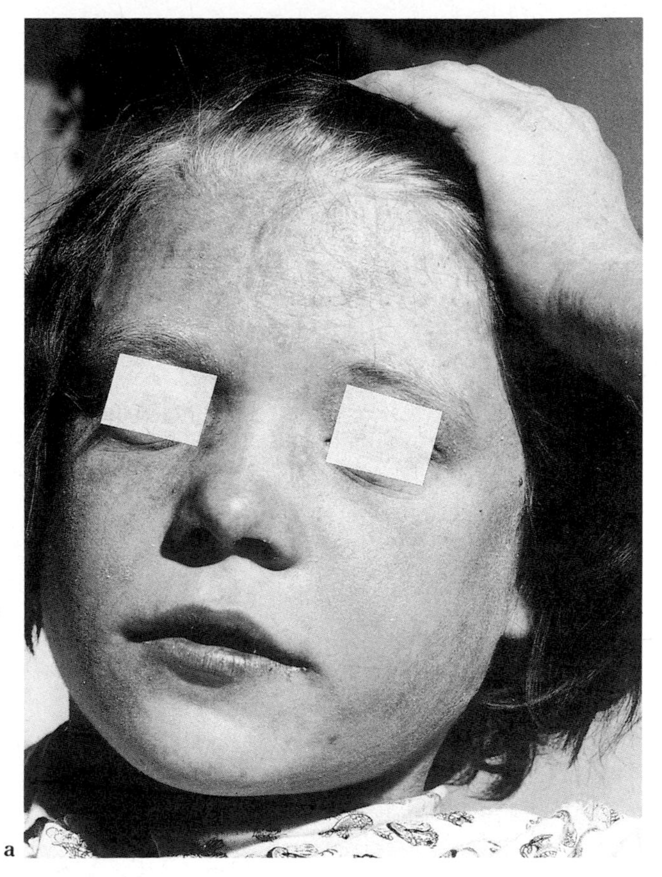

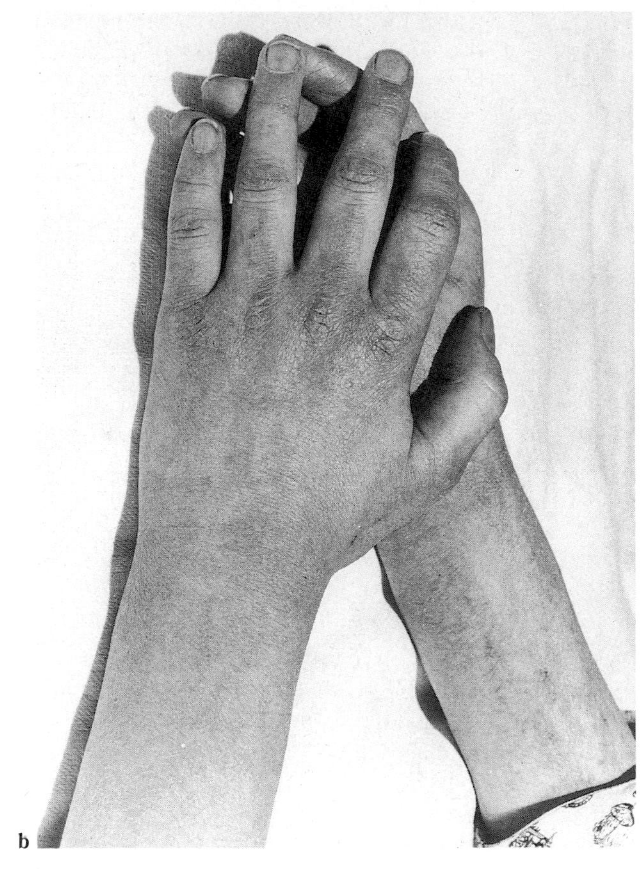

Fig. 16.1 Dermatomyositis in childhood.
a. The characteristic facies of dermatomyositis in a young child. There is a violaceous skin hue and periorbital oedema. **b.** The hands show linear erythema best seen on the extensor surfaces of the fingers.

Biopsy

Biopsy is an important contribution to diagnosis.

Histological assessments in cases of myositis are invariably made with a full knowledge of clinical, electromyographic, serological and clinical chemical investigations. The histopathological criteria are carefully specified by Anderson.[2] The disorder does not affect all parts of a muscle uniformly or to the same degree. A frequent difficulty with muscle biopsy, therefore, is the identification of a site likely to be affected histologically. A large limb muscle is preferred. Electromyography may aid the selection of an appropriate site: muscle that is partly but not wholly weakened or atrophic is recommended, and the biopsy may need to be repeated. However, care must be taken to avoid sites where electromyographic needles have been inserted or where muscle cell injury may have been caused by injections or by previous biopsy. Needle biopsy is rapid but the specimens are small and not always easily orientated; they may not be representative. Open biopsy avoids these difficulties but may create a scar.

Anderson[2] advocates removing three small cylinders of skeletal muscle. One block provides a specimen for transmission electron microscopy; a second is transfixed by pins and frozen rapidly; the third is stored for biochemical analysis. The material is divided into parts that are prepared for paraffin section (polarized light microscopy and staining), for cryostat section (immunofluorescence and histochemistry), for transmission electron microscopy, and for chemical analysis. Because the muscle disorders are focal and of random distribution, almost the whole of each block should be examined in semiserial section. Blocks for paraffin section are fixed in cold buffered formaldehyde after the trimmed block has been secured to a cardboard or wooden surface. The tissue is stained by haematoxylin and eosin, Masson's trichrome, phosphotungstic acid haematoxylin, periodic acid–Schiff and silver impregnation techniques before examination by conventional and polarized light microscopy. Small blocks adhered to metal stubs are quenched in pentane, butane or isohexane chilled with liquid nitrogen or nitrogen slush, before the preparation of thin (~2–4 μm), cryostat sections for the identification of fluorescent labels or for enzyme histochemistry. The blocks for conventional transmission electron microscopy are fixed in cold, buffered glutaraldehyde.

Histopathology

Because PM–DM is a focal disorder and because the severity of the tissue reactions varies in different sites, a single biopsy specimen may show no characteristic change. In classical, adult PM–DM a combination of lesions rather than a single abnormality is usual. Attempts have been made to associate individual lesions with the disease,[31] or with particular clinical symptoms and signs, but this has not proved easy.

Skeletal muscle

The affected muscles are pale; they may be soft and flaccid or, later, firm and fibrotic. In either instance, a loss of mass is evident. In longstanding cases, zones of calcification may be detected within affected muscles. The muscular changes are not confined wholly to the proximal parts of the limbs, in which, however, the effects are particularly severe. Lesions may also be encountered in the pharyngeal and lingual musculature. The striated muscle of the heart is much less severely involved, although, in longstanding cases, foci of fibrosis appear to replace cardiac muscle cells by a process analogous to that described in progressive systemic sclerosis (see Chapter 17).

The most common diagnostic changes in adult PM–DM are: a non-specific muscle cell degeneration; necrosis with phagocytosis; inflammation; atrophy; regeneration; and fibrosis. The occurrence of vasculitis, with resulting ischaemia, is common only in childhood myositis;[25] arteritis in the adult PM–DM syndrome is indicative of the coincidence of polyarteritis nodosa, rheumatoid arthritis or other systemic connective tissue disease.

The initial, degenerative changes of skeletal muscle cells and their subsequent, selective necrosis, may be limited to very small numbers of cells, the components of a single fasciculus, or they may encompass much larger structural units. The muscle cells of a motor unit, supplied by a single motor neurone, share common histochemical reactions, but the proportions of cell types varies greatly between different muscles. The differentiation between these cell types begins to be lost in PM–DM as cell degeneration advances. Initially, the degenerative changes may amount to only a loss of striations, hyalinization or a change in uniformity of staining. Subsequently, muscle cells become waxy, hyalinized and swollen. There is vacuolation due to fat accumulation. Disruption of cells leads to the occasional finding of empty sarcolemmal sheaths the nuclei of which may be considerably increased in number.

Biopsy often reveals evidence of an inflammatory cell response accompanying muscle cell necrosis (Figs 16.2, 16.3 and 16.4). There is local oedema but no fibrinoid change. Macrophages accumulate around the injured or effete cells; they also accumulate near venules from which, it can be assumed, they have migrated. Evidence of active inflammation is, however, often restricted to a lymphocytosis. Plasma cells are infrequent. Dawkins[25] reminds us that the presence of polymorphs or of eosinophil granulocytes or of epithelioid cell granulomata, should suggest other forms of bacterial, parasitic or traumatic myopathy. It is natural that the inflammatory cell response should be most evident at the margins of affected fascicles nearest to small blood vessels.

Selective injury to individual muscle cells or to those of a

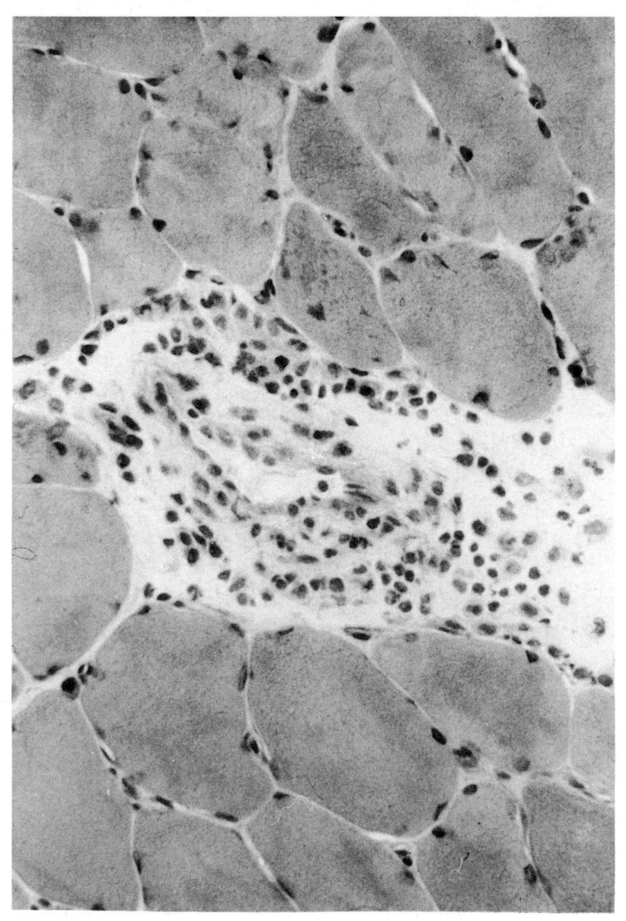

Fig. 16.2 Skeletal muscle: focal myositis.
Surrounding a venule sited among perifascicular connective tissue, there are small numbers of lymphocytes, a few macrophages and occasional polymorphs. The nearby muscle cell ('fibres') display some variability in cross-sectional area—an index of early atrophy in response to impaired blood flow. (HE × 330.)

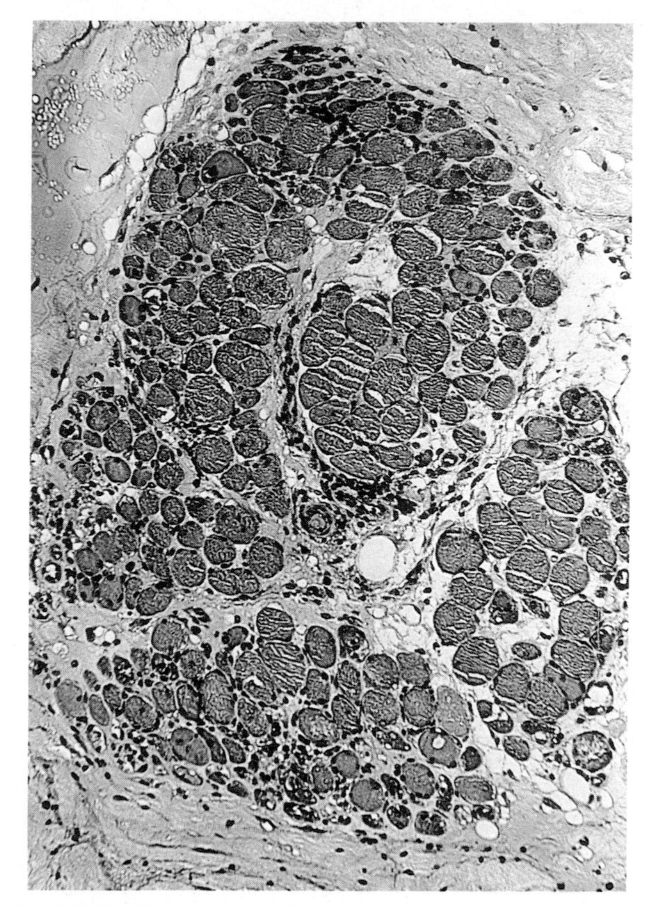

Fig. 16.3 Skeletal muscle: myositis.
Biopsy of rectus femoris muscle. Muscle cells ('fibres') are cut transversely. Within the perimysial connective tissue are numerous lymphocytes and macrophages. There is muscle cell atrophy. A reduction in cell numbers shows that some have been lost. By themselves, these changes are insufficient to permit a diagnosis of polymyositis but they exclude most forms of non-inflammatory myopathy. (HE × 110.)

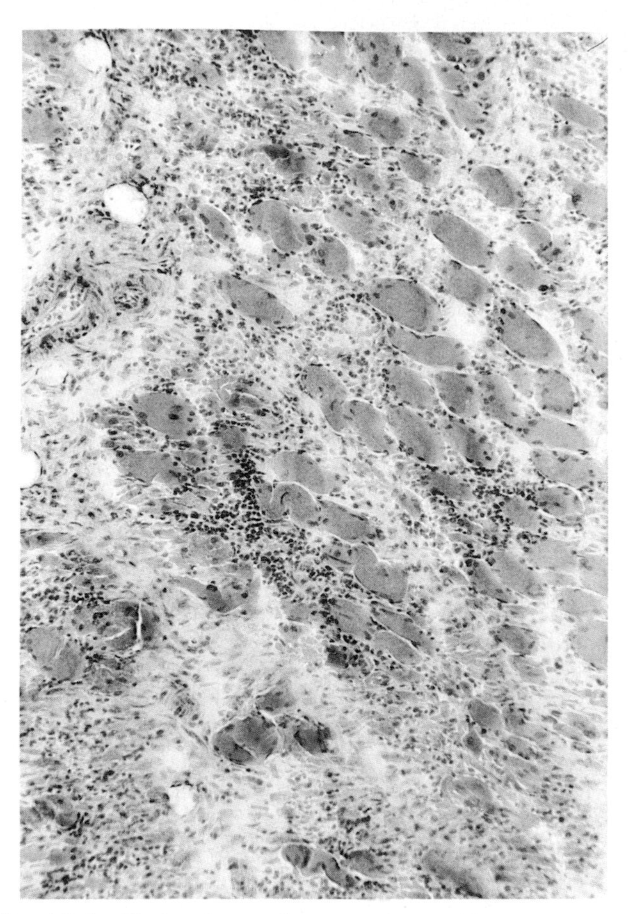

Fig. 16.4 Skeletal muscle: severe myositis.
Female aged 68 years with hyperthyroidism. Myositis of modest degree is not rare in hyperthyroidism but, in the present case, was unusually severe. There is widespread loss of muscle fibres, foci of lymphocytes and macrophages and an insidious, early replacement fibrosis. (HE × 280.)

single fasciculus, culminates in fibre atrophy (Figs 16.5, 16.6 and 16.7). The fibres become smaller and of abnormally varied size when viewed in transverse section. Occasional muscle cell nuclei are pyknotic. Particularly in childhood dermatomyositis, atrophy may be most pronounced at the periphery of individual fascicles, perhaps because of a differential reduction in blood supply. The small fibres may be contrasted with those undergoing atrophy because of denervation. Denervated muscle cells become abnormally angulated in cross-section but retain high oxidative enzyme activity. With time, the atrophic muscle cells in PM–DM disappear. Their loss is attributable to cell lysis. Their place is taken by adipose tissue.[1,21a,50,68] Considerable muscle cell regeneration may occur. Focal proliferation of sarcolemmal nuclei is frequent although mitotic figures are rarely recognized. RNA synthesis, an index of protein formation in regenerating muscle cells, is shown in groups of these cells by the increased sarcoplasmic basophilia after staining with haematoxylin and eosin. The inflammatory reaction lessens as time advances and corticosteroid therapy takes effect. A progressive, replacement fibrosis begins, advancing to the contractures and deformities which are late features of the disease.

In classical PM–DM, the contribution of vascular disease to pathogenesis is usually thought to be slight. Within affected muscles, the small arteries and arterioles, particularly those of the immediate precapillary regions, may, however, display a concentric, smudgy, eosinophilic intimal swelling which recalls the changes seen in the arteriolar intimal disorder of accelerated hypertension or the appearances in thrombotic microangiopathy (see p. 660). Indeed, microthrombi may be superadded, leading to vascular occlusion. Muscle infarction is very unusual.

The most severe histological changes are encountered in group III and group IV myositis: those of group II and V

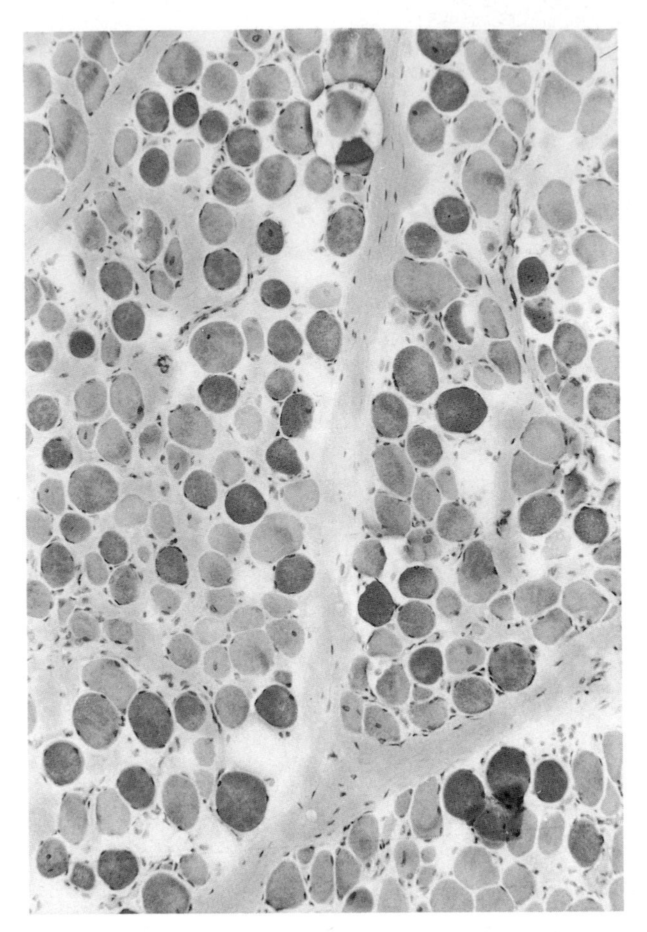

Fig. 16.5 Skeletal muscle: Duchenne muscular dystrophy.
In this, the commonest form of progressive, inherited wasting disease of skeletal muscle, there is muscle fibre necrosis, hyalinization and regeneration. There is no specific treatment for this sex-linked recessive disorder of males but the recent discovery of a characteristic, inherited protein, dystrophin, raises the possibility of new therapeutic approaches. (HE × 33.)

tend to be banal.[51] The histopathological examination of the skin in group II (dermatomyositis) is however less rewarding than the study of skeletal muscle.

In differential diagnosis, the non-inflammatory myopathies including denervation disorder, neuromuscular junction disease, muscular dystrophy, dystrophia myotonica, myasthenia and glycogen storage disease can often be distinguished clinically. Differential histopathological diagnosis is with the other forms of inflammatory muscle disease: they include infection, secondary polymyositis, giant-cell arteritis/myositis, and trauma.

Skin

When the skin is affected, there is epidermal atrophy (Fig. 16.8). The epidermal rete pegs are lost or flattened[48] (Fig. 16.8). By contrast, the collagenous connective tissue of the dermis is thickened and oedematous while individual collagen fibres merge together, losing their identity (Fig. 16.9). The dermis is infiltrated by varying numbers of mononuclear cells, often in aggregates; among them are large and small lymphocytes, histiocytes, plasma cells and occasional eosinophil granulocytes. The monocytic cells are often perivascular in location. The tinctorial characteristics of collagen change; the fibres become deeply eosinophilic and abnormally straight. These changes, which are usually progressive, may remit but tend to culminate in a rigid, inelastic structure covered with thin, atrophic epidermis resembling the skin of progressive systemic sclerosis (see Chapter 17). Telangiectatic blood vessels are seen and the basal epidermal cell layer becomes excessively pigmented. Deposition of calcium may occur. Erythematous oedema, sometimes facial, recalls a resemblance to systemic lupus erythematosus, while the subsequent fibrosis may also result in changes resembling those of progressive systemic sclerosis. In occasional cases distinction between these conditions on a microscopic basis is impossible.

Synovial joints

The joints become inflamed in one case in three. The arthritis may progress to erosion. The joint disease may coincide with or precede symptomatic muscle disease, and Schumacher *et al.*[57] recorded the surprising observation that all patients in his group with articular disease also had pulmonary disorders.[16] The hands, wrists and knees are most conspicuously involved. The synovial fluid is only slightly increased in volume. The fluid is of almost normal viscosity, is devoid of crystals and contains a modest number of cells of which the great majority are mononuclear.[57] Synovial biopsy reveals a small-to-moderate amount of fibrin upon the synovial surface with occasional focal loss of the cells of the synoviocytic layer. The remaining cells display minimal hyperplasia by contrast with rheumatoid arthritis. Scattered perivascular lymphocytes are seen with occasional plasma cells, perivascular fibrosis and hyalinization of small blood vessels. Electron microscopic studies have been made of very few synovial biopsies: they have revealed tubuloreticular structures in synovial vascular endothelium indistinguishable from those characteristic of systemic lupus erythematosus (see p. 578) and seen occasionally in other connective tissue diseases.

Bone

Osteoporosis is described but is, in any event, common in older women whether or not PM–DM is present. The effect of steroids on bone structure in PM–DM is not well documented.

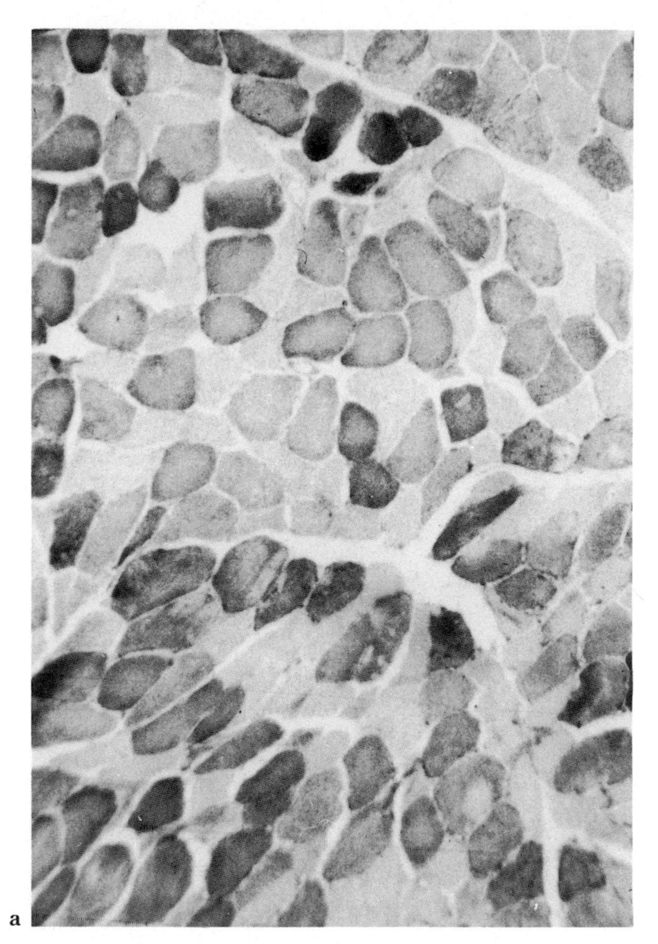

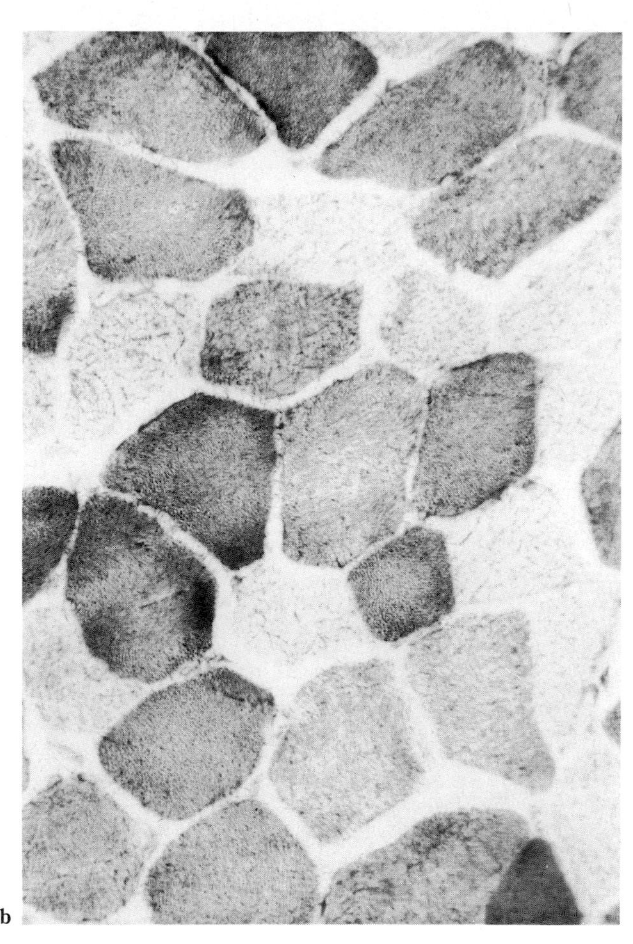

Fig. 16.6 Skeletal muscle: polymyositis and breast carcinoma.
a. The transected muscle fibres of this case are of variable size and there is an abnormal, patchy demonstration of oxidative enzyme activity. The normal, clearcut delineation of type I from type II muscle cells ('fibres') is lost. There is fibre atrophy but, in this section, no evidence of active inflammation. (NAD diaphorase × 113.) **b.** Higher power view of part of same specimen. The difference in degree of enzyme activity between the type I and type II fibres is blurred. (NAD diaphorase × 284.)

Nervous system

Neurological disturbances punctuate the history of some of the reported cases of this disease but it is not always certain whether the evidence of cerebral dysfunction is attributable to polymyositis or to associated treatment with drugs like adrenocorticotrophic hormone.[27]

Cardiovascular system

In addition to the cardiac disorders mentioned on p. 676 *et seq*, perivascular lesions in the alimentary tract have been described; they are much less frequent than in progressive systemic sclerosis. The effect of the disease on arteries and arterioles, with intimal fibrosis and superadded thrombosis, is held to account for occasional instances of gastrointestinal ulceration and haemorrhage; here again, the possible role of therapy must be considered. Dysphagia may be the result of oesophageal fibrosis.

Respiratory system

The changes merge with those encountered in progressive systemic sclerosis.[29] Pulmonary hypertension with cor pulmonale has been described[16] and in such cases the hypertension may be due to diffuse fibrosis culminating in episodes of secondary infection. Rapidly progressive fibrosing alveolitis is a rare complication.[30] The lung changes are likely to be exacerbated because of the defective operation of the respiratory muscles. Indeed, it remains possible that many of the pulmonary lesions are secondary to mechanically impaired ventilation. The laryngeal and pharyngeal muscles may be affected with

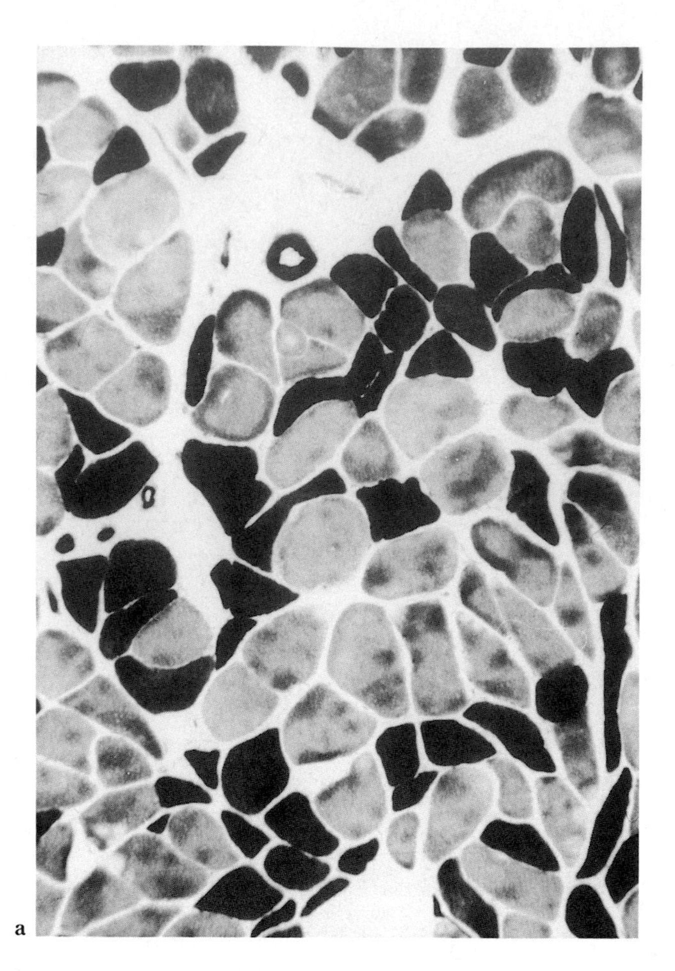

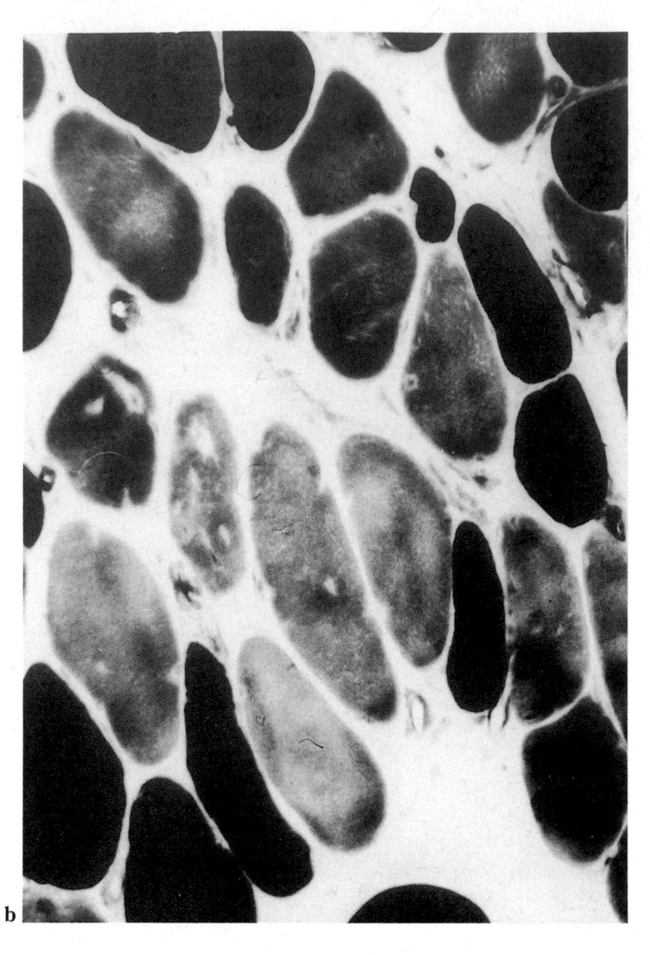

Fig. 16.7 Skeletal muscle: polymyositis and breast carcinoma.
a. Same case as Fig. 16.6. Myosin ATPase activity is high in type II fibres, low in type I. However, the distinction is less clear than in normal muscle. Moreover, within individual transected muscle fibres the enzyme reaction is uneven. There is partial atrophy of individual muscle cell ('fibres') but, in this field, no evidence of inflammation. (Myosin ATPase × 113.) **b.** Higher power view of part of same section. Vacuolation of some fibres; variable size, indicative of atrophy; and patchy and inconstant ATPase activity, are among the features observed microscopically. (Myosin ATPase × 284.)

weakness, defective coordination, hoarseness and the accidental inhalation of irritants.

Lymphoreticular system

Enlargement of the organs of this system may occur, comparable with that of rheumatoid arthritis (see p. 505).

Urinary system

Involvement of the renal system is also recognized and renal failure may play an important part in the fatal termination of the disease. The histological changes resemble those of progressive systemic sclerosis.

Childhood disease

There is evidence that in children, polymyositis displays a number of characteristic clinical and pathological features[5] and that it can be regarded as an entity distinct from adult PM–DM. The principal changes are in relation to blood vessels in the skin, gastrointestinal tract, muscle, fat and small nerves. Perivascular inflammatory cell infiltrates, accompanied by arteritis and phlebitis, are followed by intimal fibromuscular hyperplasia of the arteries and veins, vascular thrombotic occlusion, and infarction of related tissues. Vascular disease with ischaemic infarcts appears especially prominent in the gut, muscles and small nerves. The calcific material that is so characteristic of the muscle lesions in children is found by electron microscopy to be bone mineral, calcium hydroxyapatite.

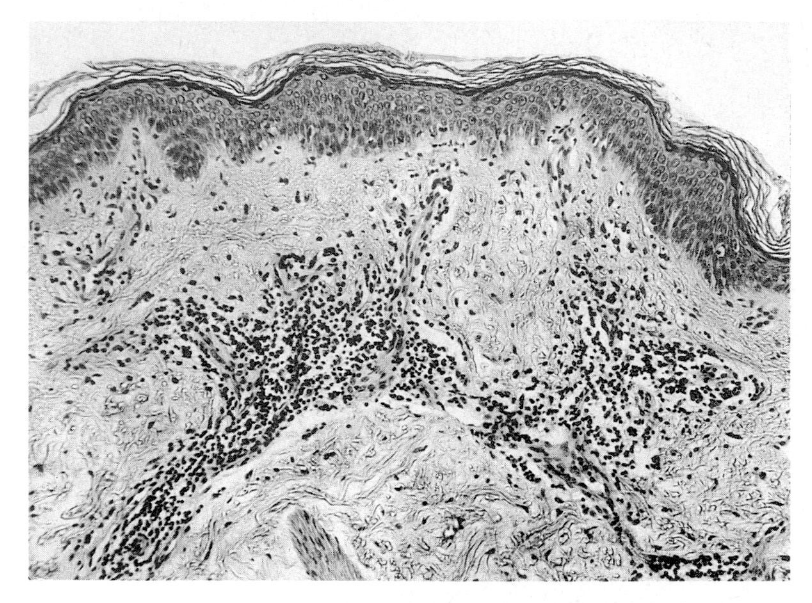

Fig. 16.8 Skin: dermatomyositis.
Epidermal atrophy is accompanied by liquefaction degeneration of scattered basal cells. There are many lymphocytic foci, particularly around dermal appendages. The histological distinction from the changes of early systemic sclerosis is not absolute. (HE × 120.)

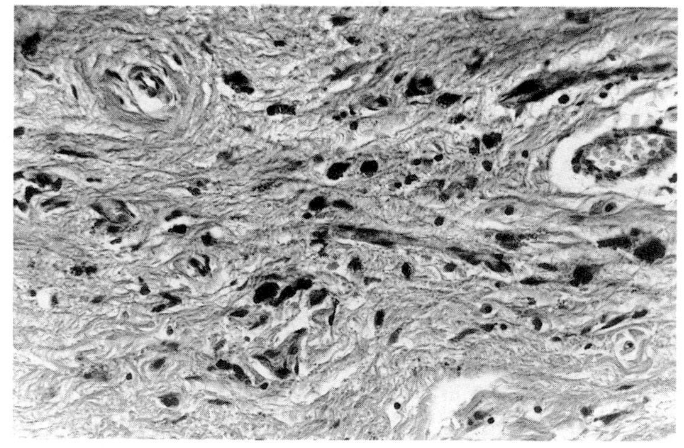

Fig. 16.9 Subcutaneous tissue: dermatomyositis.
Within the subcutaneous tissue there is a slight excess of mononuclear macrophages, related to small blood vessels. There is less new, collagen-rich connective tissue than in progressive systemic sclerosis. Small numbers of mast cells can be identified. (HE × 120.)

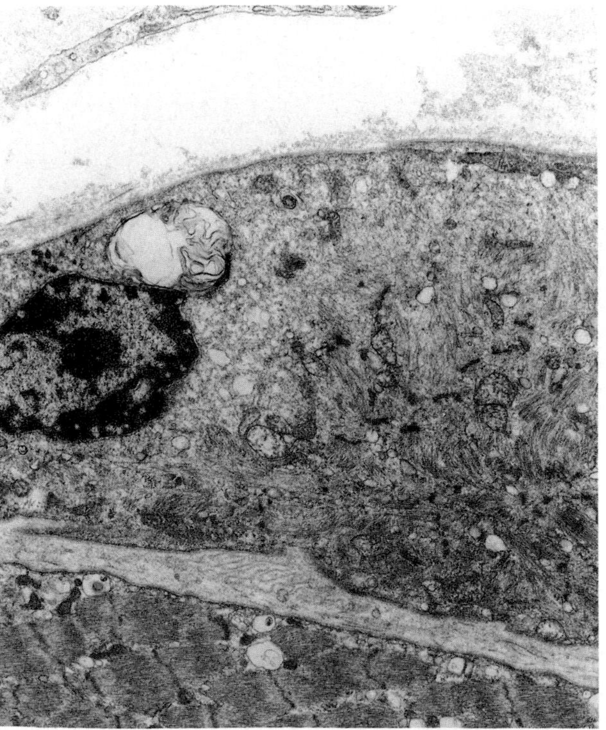

Fig. 16.10 Skeletal muscle: polymyositis.
An atrophic cell ('fibre') with disarray of the myofibrils. The small membranous whorl is of lysosomal origin, derived from the sarcoplasmic reticulum. (TEM × 7920.) Courtesy of Dr Janice R Anderson.

Ultrastructure

Polymyositis

Examination of skeletal muscle by electron microscopy[51] confirms the nature of the changes seen by light microscopy. The sequence of muscle cell injury and necrosis, inflammation, cell infiltration, muscle cell regeneration and fibrosis, is mirrored in a series of well-recognized fine structural changes which are illustrated by Carpenter and Karpati,[18] Anderson[2] and Carpenter[17] (Figs 16.10 and 16.11).

In adult polymyositis, the necrotic muscle cells are widely scattered. Some cells are regenerating. Plasma cells, mononuclear macrophages and $T_{C/S}$ (CD8) lymphocytes surround necrotic cells and indent the plasma membrane of others that are viable. Mononuclear macrophages are seen. Muscle cells may contain electron-dense material within vacuoles which have a double membrane; some of this material is derived from phospholipid membranes and may come to form spheromembranous bodies. There is a variable degree of muscle cell disorganization.

The small, punctate bodies, similar to those of apoptosis but less refractile than lipid, are aggregates of double-membrane bounded, dense structures. Some of this material is glycogen or of mitochondrial origin but much is structureless and unidentified. Comparable material is seen between the sarcolemma and basement membrane or within the basement membrane, remote from the muscle cells, a location which suggests a removal from the muscle cell by exocytosis.[17] Basal laminae are often reduplicated but basal laminar regeneration is often incomplete as evinced by the presence of woven bundles of T-tubular material. Muscle cell nuclei are occasionally pyknotic.

All stages of muscle cell regeneration may be recognized. The presence of numerous polyribosomes and of sparse, new myofibrils are indices of this process. There is a tendency for clumping of the sarcolemma and an overcontraction of myofibrils. Slighter degrees of muscle damage are indexed by Z-disc streaming and lysis, changes accom-

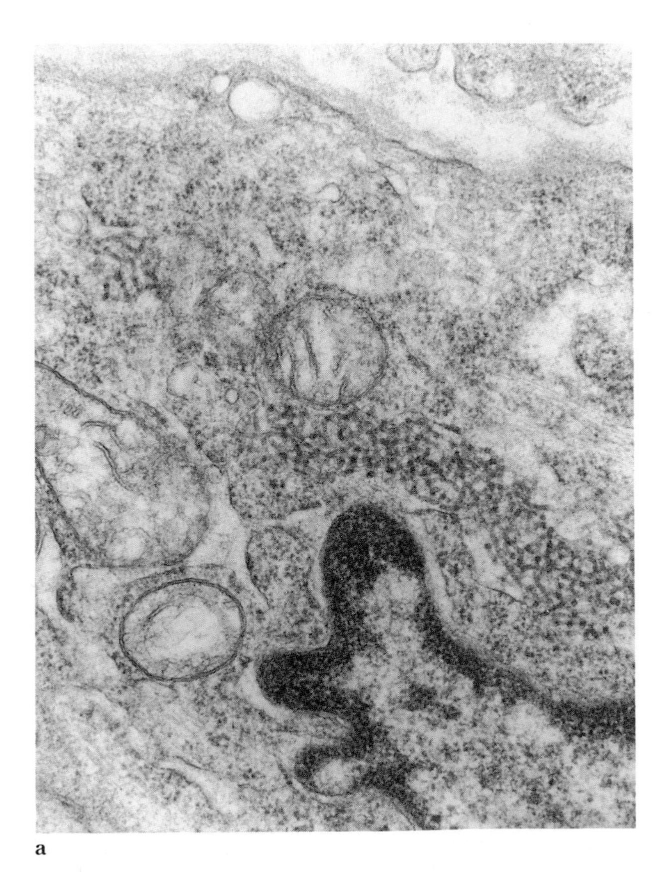

a

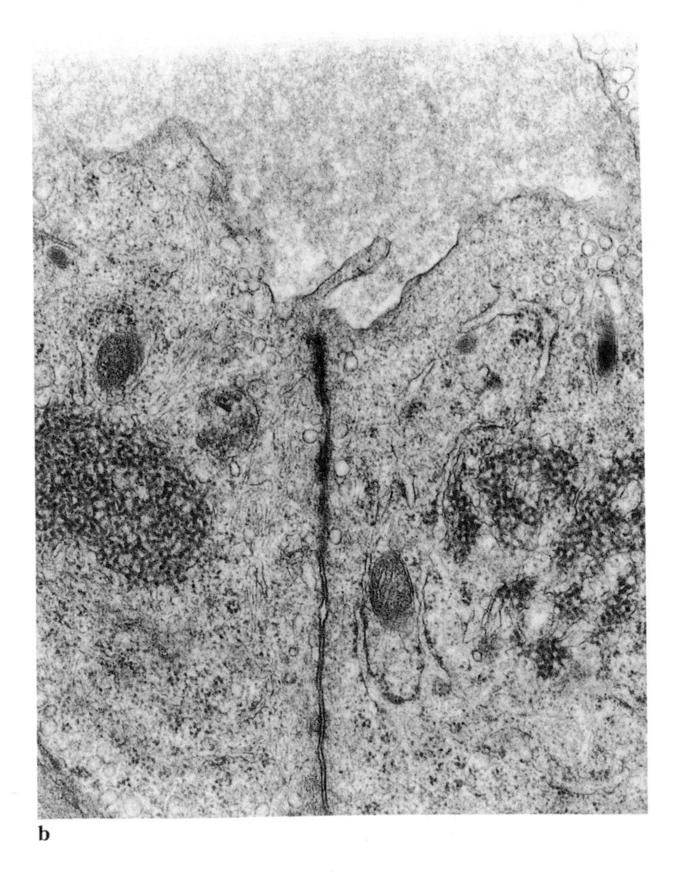

b

Fig. 16.11 Skeletal muscle: polymyositis.
a. Tubular arrays in endothelial cells of capillary. These structures are believed to be products of cell metabolism. They are closely similar to arrays found in cells in SLE and are almost certainly not viral components. **b.** Adult with necrotizing myopathy and skin rash. Note tubular arrays in adjacent endothelial cells. (TEM: **a.** × 37 800; **b.** × 30 300.) Courtesy of Dr Janice R. Anderson.

panied by the presence of dense sarcoplasmic bodies and meshworks of microfibrillar or tubular material that may be of T-tubular origin.

Dermatomyositis

The presence of perifascicular, muscle cell atrophy is probably determined by the pattern of blood flow. Arterial disease is focal. There is evidence that the afferent arteriolar circulation is first to the centre of muscle fasicles, with the perifasicular territory remote from this supply. There is a diminished capillary network, with a reduced proportion of capillaries per muscle cell. Endothelial cell hyperplasia accompanies capillary necrosis. The muscle cell changes in cases in which dermatomyositis accompanies the growth of a malignant neoplasm do not differ from those in primary, idiopathic adult PM–DM.

Histochemistry
(Table 16.4)

In PM–DM the histochemical changes vary with the stage and state of activity of the disease. Some enzyme activities are influenced by steroid therapy, an effect indexed by the accumulation of lipid droplets in muscle cells (Fig. 16.12). Necrotic cells clearly lose demonstrable oxidative- and, subsequently, hydrolytic enzyme activity. Atrophy may be selective so that changes in either type I or type II cells are demonstrable but the degree of type II change is affected by steroid treatment. Anderson[2] advocates the comparison of histochemical changes in muscle biopsies taken before and after treatment is begun.

Table 16.4

Enzyme activities of normal skeletal muscle

	Type I cells (so-called fibres)	Type II cells (so-called fibres)
ATPase, pH 9	Light (low activity)	Dark (high activity)
Oxidative enzymes	Dark	Light
Phosphorylase	Light	Dark

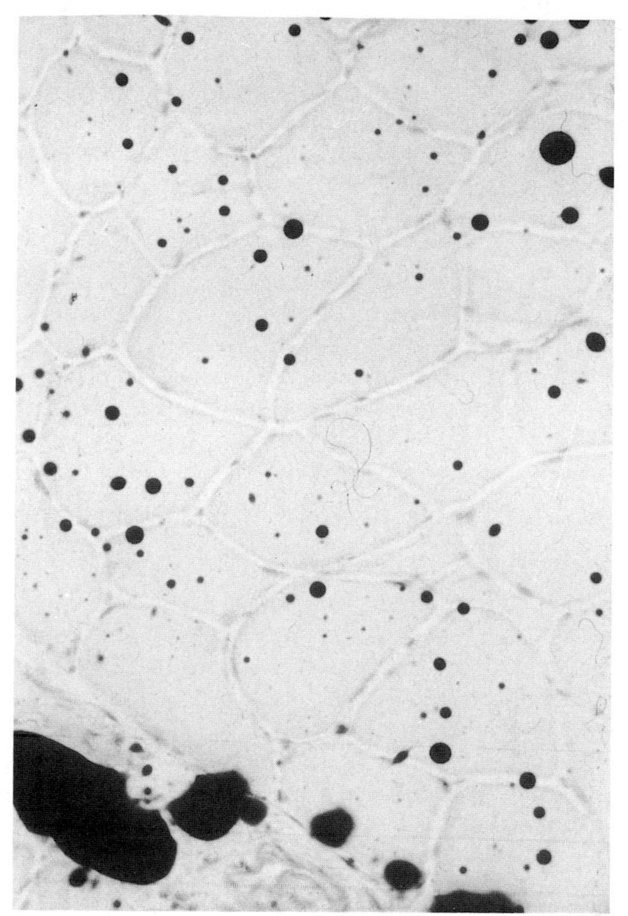

Fig. 16.12 Polymyositis: lipid droplet accumulation. The presence of lipid droplets, demonstrated by an oil red O reaction on a cryostat section, is a feature of metabolic, endocrine and alcoholic myopathies. In biopsies from patients with polymyositis, a positive reaction may be an indication of the effects of corticosteroid therapy. (Oil red O × 280.)

Immunopathology

The reader is referred to Chapter 5.

The PM–DM complex is a miscellany of disorders. No single group of immunopathological changes can be held characteristic of all the disorders within the complex. This heterogeneity, and the probable involvement of a series of aetiological agents, determine that it is not yet possible to differentiate members of the clinical PM–DM complex solely on the basis of immunohistological criteria. It is highly likely that the ability to discriminate between the main classes of disorder in the PM–DM complex (Tables 16.1 and 16.2) will, however, increase in the near future as antigens are refined and an increasing range of monoclonal antibodies becomes available.

Muscle

Humoral mechanisms (Table 16.3) Although the findings are not wholly consistent, four categories of change can be recognized.[26] Direct immunofluorescence shows immunoglobulin, complement and fibrin within necrotic muscle cells. However, intact cells from occasional patients with high serum anti-ribonucleoprotein antibody titres may also contain immunoglobulin, suggesting that the binding of immunoglobulin is an initiating event in cell injury. Synchronous identification of immunoglobulin, complement and fibrin is widely held to indicate that a local antibody–antigen reaction is responsible for the inflammation found at that site. Alternatively, the simultaneous demonstration of immunoglobulin and complement may simply mark a locus at which immune complexes accumulate.

The recognition of serum antinuclear antibodies is of much less significance in PM–DM than in systemic lupus erythematosus or mixed connective tissue disease. Nevertheless, antinuclear antibodies can impair cell staining by anti-immunoglobulin conjugates, drawing attention to the possibility that the localization of these conjugates may exert a pathogenetic role. Complement component C9 has been localized to both dead and viable muscle cells. The part that this component plays in muscle cell injury remains uncertain. The localization of immunoglobulin at the sarcolemma is not a phenomenon specific to PM–DM.

The presence of immunoglobulin in small blood vessels is frequent in systemic connective tissue diseases other than PM–DM. When immunoglobulin and complement coincide with fibrin in a focus of vasculitis of larger vessels, as may occasionally be found in PM–DM, it is reasonable to associate them with the cause of vascular injury. However, there is little to indicate that immune complex-mediated vasculitis contributes to any form of PM–DM other than childhood dermatomyositis.

Late in the disease, fibrosis becomes evident. Muscle biopsy material from 22 cases of polymyositis showed increased staining with antibodies against collagen types I, III, IV and V[52]. It is possible that an autoimmune mechanism underlies this collagen accumulation. Alternatively, an increased synthesis of several collagen types may be stimulated by immune complex deposition. Serum levels of the amino–terminal propeptide of type III collagen are elevated.[47]

Cell-mediated mechanisms The foci of inflammation that characterize the muscle lesions of PM–DM include varying numbers of lymphocytes. In areas adjacent to infiltrating leucocytes or where muscle cell damage is evident, MHC class II HLA antigens are expressed, apparently by the release of interferons-α and β.[38] Many of the cells are $T_{C/S}$ (CD8), cytotoxic cells although increased proportions of T_H (CD4)-cells have also been described. However, the suggestion that an antimuscle cell, T-cell-mediated hypersensitivity response typifies the pathogenesis of PM–DM is not established. In dermatomyositis, high proportions of T_H (CD4) cells have been found near B-cells.

Skin

In contrast to systemic lupus erythematosus (see p. 580), non-involved skin contains no demonstrable immunoglobulin deposits. Direct immunofluorescence is generally negative.[25] Affected skin in dermatomyositis contains both immunoglobulin and complement deposits. The demonstration of these junctional deposits is inconstant. The presence of antinuclear antibody is sometimes suggested by the immunofluorescent demonstration in the epidermis of speckled nuclear staining.[26]

Aetiology and pathogenesis

Inheritance

(see p. 244)

Although the PM–DM complex is not manifest simply on the basis of an inherited trait, there is evidence that some forms of PM–DM are associated with the inheritance of particular HLA antigens. This evidence relates to defects in immune regulation and a role for autoimmune disorder is suspected, mediated by both humoral and cellular mechanisms.[8] HLA-B8 and HLA-DR3 are associated with some subtypes of polymyositis and dermatomyositis.[26] These antigens may reflect an association with the A1, Cw7, B8, C4AQ0, C4B1, BfS, DR3 supratype,[25] an inheritance that marks genes modulating the immune response to a variety of antigens and that is associated with a number of other autoimmune diseases.

Immune disorders

By contrast with other systemic connective tissue diseases, the serum immunoglobulin levels are often low. However, the PM–DM complex is heterogeneous and there are both associations between juvenile dermatomyositis and hypogammaglobulinaemia and between the vasculopathy of juvenile dermatomyositis, and hypergammaglobulinaemia and immune complex formation. The serum complement levels in adult PM–DM are normal or raised but, occasionally, they are diminished. Like systemic lupus erythematosus (see p. 570), dermatomyositis may be accompanied by complement component C2 deficiency (see p. 244).

A variety of autoantibodies of diagnostic value has been

identified (Table 16.3). Whether these antibodies play a part in the pathogenesis of the skeletal muscle and other lesions is less clear. The antibodies are directed against components of skeletal muscle, cell nuclei and other epitopes. The significance of the anti-skeletal muscle antibodies has been questioned since similar antibodies are often present in the serum of patients with other wasting diseases of skeletal muscle.

Circulating mononuclear cells in PM–DM may display anti-muscle cytotoxic activity. Lymphocytes are transformed on exposure to skeletal muscle antigen. Circulating lymphocytes are toxic to skeletal muscle in culture[39] and tests for macrophage migration inhibition may be positive. A role for cell-mediated hypersensitivity in the pathogenesis of PM–DM is therefore strongly suspected but the mechanisms by which hypersensitivity injures muscle are not entirely clear. The autologous mixed lymphocyte reaction is impaired[54] although concanavalin A-induced suppression is normal. The part that $T_{C/S}$ (CD8)-cells play in causing the lesions of PM–DM is under investigation. One explanation for the foci of lymphocytes found in active polymyositis is the apparent migration of $T_{C/S}$ (CD8)-cells towards sites of muscle injury where they are found to accumulate.

Role of neoplasia

One of the most interesting associations, mentioned on p. 678, is between dermatomyositis and cancer.[7] Artefacts such as pyomyositis can be misleading[63] but there are many instances where dermatomyositis, in particular, has coincided with the diagnosis of neoplastic disease. Carcinomata of the bronchus, breast, colon or prostate are the most frequent accompaniments, but thymoma,[41] 'lymphosarcoma',[3] Hodgkin's disease[28] and malignant melanoma[61] are among the many other associations. Neoplastic disease may precede and appear to precipitate PM–DM whereas effective treatment of a neoplasm may alleviate the syndrome.[56]

The evidence associating PM–DM with cancer has been questioned.[12] Valuable new data which help to balance the argument, derive from a controlled retrospective study of 71 patients with PM–DM matched with 71 patients with non-PM–DM rheumatic disease and 71 patients with no inflammatory disorders.[44] These authors also studied the incidence of new cancers developing in a group of patients with PM–DM followed prospectively.

The results demonstrated that 15 of 71 PM–DM cases had antecedent or concurrent cancer compared with 4 of 71 control cases with other 'rheumatic' diseases and 1 of 71 control cases with non-inflammatory muscle disease. This survey confirmed a strong association between PM–DM and malignant neoplasia but suggested that the association was similar for polymyositis and dermatomyositis and for male and female patients. There was both an increased frequency of cancer diagnosed before the recognition of PM/DM, and evidence that most of the neoplasms presented at or about the same time of entry into the investigation. There was, however, no unexpected increase in the number of malignant neoplasms developing in patients followed prospectively.

On the basis of this evidence, the suggestion has emerged that neoplastic antigens may be initiating factors for PM–DM, perhaps due to the sharing of an antigenic constituent between muscle and neoplastic cells. A reaction with both antigens is postulated when the appropriate antibodies are liberated. The importance of an immune mechanism in these cases of dermatomyositis is supported by the demonstration of a specific local immediate hypersensitivity reaction when neoplastic extracts obtained from a patient with this disease are injected intradermally into her own skin.

Virus infections

The role of viruses in the causation of PM–DM is not established (see p. 690). Studies of viruses in pathogenesis are complicated by the frequent finding of virus-like particles in tissue examined at biopsy. However, virus has seldom been isolated from such cases. In an exceptional instance, Tang *et al.*[62] identified Coxsackie virus A9 in the skeletal muscle of a child with long-standing muscle weakness. Inclusion body myositis is also likely to be of viral origin.

It remains possible that viral infection in PM–DM may be a direct cause of muscle cell injury.[14] In a recent case where polymyositis developed in a young male seropositive for both human T-cell lymphotrophic virus (HTLV)-1 and human immunodeficiency virus (HIV), direct infection of injured muscle by HTLV-1 (but not by HIV) was demonstrated by *in situ* hybridization and immunocytochemistry.[69] Polymyositis is also an occasional result of influenza infection. While some enteroviruses are neurotropic, others can cause human myocarditis, epidemic pleurodynia (Bornholm disease) or myopathy in newborn mice. Earlier reports of the detection of intranuclear and intracytoplasmic filamentous nuclear aggregates, resembling picorna- or myxoviruses, are of uncertain significance since virus-like particles are present in muscle in other disorders and even in normal muscle. Coxsackie virus A2 and an echovirus have been isolated from the stools of chronic and acute cases of dermatomyositis and polymyositis, respectively, but no

Coxsackie virus has been isolated from the skeletal muscle itself in classical cases of polymyositis.

An alternative explanation for muscle cell injury is virus-induced autoimmunity. Among the antibodies detected in the five groups of PM–DM, were antinuclear antibody and anti-ribonucleoprotein[37] (Table 16.3). Anti-ss DNA titres were normal but the majority of cases had antibody to calf thymic nuclear antigens, particularly Jo-1, Mi, PM/Scl and Ku. The possibility has arisen that anti-Jo-1 may be the result of a virus-induced antigen change. Jo-1 antigen itself is histidyl-tRNA synthetase.[46] One way in which virus can provoke an autoimmune response is by molecular mimicry. Histidyl-tRNA, alanyl-tRNA and threonyl-tRNA are common autoantigens in PM–DM and anti-Jo-1 antibody can bind to small fragments of histidyl-tRNA. A search of the National Biomedical Research Foundation database revealed six matches to histidyl-tRNA synthetase with homology greater than or equal to eight of nine amino acids but 30 matches with this degree of homology to alanyl-tRNA. More than 50 per cent of these matches were with viral proteins including influenza virus, E3 virus and adenovirus.[66] Since Jo-1-autoantigen displays considerable homology with tropomyosin and keratin, it appeared likely that cross-reactions would occur if the alanyl-tRNA amino acid sequences formed the epitope.

Protozoal and metazoal infestation

For many years, attention has been drawn to the possibility that polymyositis and/or dermatomyositis may be linked to infestation with *Toxoplasma gondii*. Magid and Kagen[43] found no increased frequency of anti-*T. gondii* IgM antibody in immunofluorescence tests but Quilis and Damjanov[53] interpreted their light- and transmission electron microscopy and immunofluorescent studies as evidence of immune complex tissue injury related to *T. gondii* infection. The possible association with *T. gondii* was reviewed by Behan *et al.*;[9] they suggested that reactivation of a latent infection in an immunocompromised host was an acceptable explanation for these conflicting views. However, antiprotozoal chemotherapy does not cure PM–DM.

Animal models

A number of forms of PM–DM occur naturally in animals. Others have been reproduced experimentally.

The study of two groups of animal models of PM–DM has thrown light on aspects of the syndrome, particularly upon non-immune, multifocal muscle cell necrosis and upon auto-immune inflammatory muscle cell injury.[40] Among the first group are nutritional myopathies such as the lesions caused by vitamin E deficiency, Syrian hamster (genetic) myopathy, and the infectious disorders attributable to toxoplasmosis or to virus infection. In the second group are a series of spontaneous canine diseases, including 'pure' polymyositis, polymyositis with cancer, and eosinophilic myositis. Allergic myonecrosis, graft-versus-host disease, and allergic myositis are other analogous experimental disorders resulting from hypersensitivity, sometimes of allergic character. In the present account, attention will be confined to virus infection, Syrian hamster myopathy, the genetic dermatomyositis of collie dogs and Shetland sheep dogs, and allergic myositis.

Virus infection[69a]

One model of virus myositis is attributable to the simian AIDS virus.[40] Fifty per cent of recipient animals develop focal muscle cell necrosis, selective type II cell atrophy and inflammation. The possibility exists that this form of myositis is due to the inappropriate cytopathic action of lymphocytes themselves or to opportunistic infection. However, no microorganism other than the simian AIDS virus has been recovered from the muscle cells.

Syrian hamster myopathy

This is an inherited myopathy, transmitted as an autosomal recessive characteristic. The lesions are focal and inflammatory but affect both cardiac and skeletal muscle. The onset is approximately 10 weeks after birth. A perinuclear 'halo' around muscle cell nuclei displaces the contractile proteins. A sustained plasma- and lymphocytic reaction develops. The reason for the activation of an anti-muscle cell cytotoxic response in these animals is unknown.

Dog models of PM–DM

Polymyositis has long been known to occur in German shepherd dogs and in boxers, and a polymyositis-like disease was associated with systemic lupus erythematosus (see p. 598) in a standard poodle.[42] Histological changes resembling PM–DM may also accompany canine leukaemia and nasopharyngeal carcinoma. Both acute and, less often, chronic eosinophilic myositis have been identified in dogs. Particular interest has centred on a form of dermatomyositis found in collie dogs and Shetland sheep dogs.[32,33,33a,35] Many of the earlier lesions are of the skin: they include vesicle and pustule formation with ulceration of face, lips and ears. There is a dermal infiltrate of polymorphs, lymphocytes, mast cells and macrophages. The development of bilaterally symmetrical generalized muscle atrophy is recognized between 13 and 19 weeks of age and high levels of serum IgG accompany circulating immune complexes. There is evidence that a predisposition to the disorder is inherited as an autosomal dominant characteristic. *Post mortem* studies of 20 young animals and 10 neonates suggested that the most severe muscle lesions were in the temporalis, masseter, flexor digitorum, superficialis and gastrocnemius muscles.[34] Both type I and type II muscle cells were affected. Cutaneous and vascular lesions were also most severe in the head and extremities. No viruses were isolated from muscle but the endothelial cells of muscle capillaries contained crystalline arrays of virus-like structure.

Experimental immune myositis

Muscle cell injury has been studied after allogeneic transplantation. A graft-versus-host reaction results in cell-mediated muscle destruction. Early attempts were made to produce an experimental myositis by an anti-muscle immune response. Homogenized muscle preparations were injected parenterally with Freund's complete adjuvant (p. 544). One consequence was the coincidental discovery of adjuvant arthritis[49,60] (see p. 544), Subsequently, methods for the reproduction of allergic myositis in rats and guinea-pigs were perfected.[23,40] Heterologous muscle preparations with adjuvant were injected repeatedly into the foot-pad dermis. An acute, focal myositis resulted, closely resembling the histological lesions of PM–DM. Like Freund's adjuvant arthritis of rats (see p. 544). experimental allergic myositis can be transferred passively by preparations of cytotoxic lymphocytes but not by serum. Transmission electron microscopic changes include cell injury with necrosis, loss of Z bands and filaments and organelle and microsomal disintegration. It has been claimed that antigens from myofibrils provoke allergic myositis more potently than other microsomal preparations.[45]

REFERENCES

1. Adams R D, Denny-Brown D, Pearson C M. *Diseases of Muscle: A Study in Pathology* (2nd edition). New York: Hoeben, 1962.

2. Anderson J R. *Atlas of Skeletal Muscle Pathology*. Lancaster: MTP Press Ltd, 1985.

3. Arapakis G, Jordanoglou J. A case of lymphosarcoma associated with myopathy. *British Medical Journal* 1964; **2**: 32–3.

4. Arundell F D, Wilkinson R D, Haserick J R. Dermatomyositis and malignant neoplasm in adults. *Archives of Dermatology* 1960; **82**: 772–5.

5. Banker B Q, Victor M. Dermatomyositis (systemic angiopathy) of childhood. *Medicine* 1966; **45**: 261–89.

6. Banker B Q. Dermatomyositis of childhood: ultrastructural alterations of muscle and intramuscular blood vessels. *Journal of Neuropathology and Experimental Neurology* 1975; **34**: 46–75.

7. Barnes B E. Dermatomyositis and malignancy. A review of the literature. *Annals of Internal Medicine* 1976; **84**: 68–76.

8. Behan W M H, Behan P O. Immunological features of polymyositis/dermatomyositis. *Springer's Seminars in Immunopathology* 1985; **8**: 267–93.

9. Behan W M H, Behan P O, Draper I T, Williams H. Does toxoplasma cause polymyositis? Report of a case of polymyositis associated with toxoplasmosis and a critical review of the literature. *Acta Neuropathologica* 1983; **61**: 246–52.

10. Bjelle A, Henrikson K G, Hofer P A. Polymyositis in eosinophilic fasciitis. *European Neurology* 1980; **19**: 128–37.

11. Bohan A. Clinical presentation and diagnosis of polymyositis and dermatomyositis. In: Dalakas M C, ed, *Polymyositis and Dermatomyositis*. London: Butterworths, 1988: 19–36.

12. Bohan A, Peter J B. Polymyositis and dermatomyositis. *New England Journal of Medicine* 1975; **292**: 344–7; 403–7.

13. Bohan A, Peter J B, Bowman R L, Pearson C M. A computer-assisted analysis of 153 patients with polymyositis and dermatomyositis. *Medicine* 1977; **56**: 255–86.

14. Bowles N E, Richardson P J, Olsen E G J, Archard L C. Detection of coxsackie-B-virus specific RNA sequences in myocardial biopsy samples from patients with myocarditis and dilated cardiomyopathy. *Lancet* 1986; **i**: 1120–2.

15. Bradley W G. Inflammatory diseases of muscle. In: Kelley W N, Harris E D, Ruddy S, Sledge C B, eds, *Textbook of Rheumatology* (2nd edition). Philadelphia: W B Saunders & Co., 1985: 1225–45.

16. Caldwell I W, Aitchison J D. Pulmonary hypertension in dermatomyositis. *British Heart Journal* 1956; **18**: 273–6.

17. Carpenter S. Resin histology and electron microscopy in inflammatory myopathies. In Dalakas M C, ed, *Polymyositis and Dermatomyositis*. London: Butterworths, 1988: 195–215.

18. Carpenter S, Karpati G. The major inflammatory myopathies of unknown cause. In: Sommers S C, Rosen P R, eds, *Pathology Annual* (volume 16). New York: Appleton-Century-Crofts, 1981: 205–37.

19. Carpenter S, Karpati C. *Pathology of Skeletal Muscle*. Edinburgh: Churchill Livingstone, 1984: 416–740.

20. Carpenter S, Karpati G, Rothman S, Watters G. The childhood type of dermatomyositis. *Neurology* 1976; **26**: 952–62.

21. Currie S. Polymyositis and related disorders. In: Walton J, ed, *Disorders of Voluntary Muscles* (4th edition). New York: Churchill Livingstone, 1981: 545–7.

21a. Currie S. the inflammatory myopathies. In: Walton J, ed,

Diseases of Voluntary Muscle (5th edition). Edinburgh, London: Churchill Livingstone, 1988: 588–610.

22. Dalakas M C, ed, *Polymyositis and Dermatomyositis*. London: Butterworths, 1988.

23. Dawkins R L. Experimental myositis associated with hypersensitivity to muscle. *Journal of Pathology and Bacteriology* 1965; **90**: 619–25.

24. Dawkins R L. Experimental autoallergic myositis, polymyositis and myasthenia gravis. Autoimmune muscle disease associated with immunodeficiency and neoplasia. *Clinical and Experimental Immunology* 1975; **21**: 185–201.

25. Dawkins R L. Muscle disease. In: Holborow E J, Reeves W G, eds, *Immunology in Medicine: a Comprehensive Guide to Clinical Immunology* (2nd edition). New York: Academic Press, 1983: 840–51.

26. Dawkins, R L, Garlepp M J. Immunopathology of polymyositis and dermatomyositis. In: Dalakas M C, ed, *Polymyositis and Dermatomyositis*. London: Butterworths, 1988: 85–96.

27. Domzalski C A, Morgan V C. Dermatomyositis: diagnostic features and therapeutic pitfalls. *American Journal of Medicine* 1955; **19**: 370–82.

28. Dowsett R J, Wong R L, Robert N J, Abeles M. Dermatomyositis and Hodgkin's disease. Case report and review of the literature. *American Journal of Medicine* 1986; **80**: 719–23.

29. Ellman P. Pulmonary manifestations in the systemic collagen diseases. *Postgraduate Medical Journal* 1956; **32**: 370–87.

30. Fergusson R J, Davidson N McD, Nuki G, Crompton G K. Dermatomyositis and rapidly progressive fibrosing alveolitis. *Thorax* 1983; **38**: 71–2.

31. Greenfield J G, Shy G M, Alvord E C, Berg L. *An Atlas of Muscle Pathology in Neuromuscular Diseases*. Edinburgh: Livingstone, 1957: 32–7.

32. Hargis A M, Haupt K H, Hegreberg G A, Prieur D J, Moore M P. Familial canine dermatomyositis: Initial characterization of the cutaneous and muscular lesions. *American Journal of Pathology* 1984; **116**: 234–44.

33. Hargis A M, Haupt K H, Prieur D J, Moore M P. Animal models of human disease: dermatomyositis: familial canine dermatomyositis. *American Journal of Pathology* 1985; **120**: 323–5.

33a. Hargis A M, Prieur D J. Familial canine dermatomyositis. In: Greenwald R A, Diamond H S, eds *CRC Handbook of Animal Models for the Rheumatic Diseases*, volume I, 1988. Boca Raton, Florida: CRC Press Inc: 157–67.

34. Hargis A M, Prieur D J, Haupt K H, Collier L L, Evermann J F, Ladiges W C. Postmortem findings in four litters of dogs with familial canine dermatomyositis. *American Journal of Pathology* 1986; **123**: 480–96.

35. Haupt K H, Hargis A M, Prieur D J, Hegreberg G A, Moore M P. Familial canine dermatomyositis. Preliminary studies on a canine model of human dermatomyositis. *Federation Proceedings* 1984; **43**: 708.

36. Haupt H M, Hutchins G M. The heart and cardiac conduction system in polymyositis–dermatomyositis—a clinicopathologic study of 16 autopsied patients. *American Journal of Cardiology* 1982; **50**: 998–1006.

37. Hochberg M C, Feldman D, Stevens M B. Adult onset polymyositis/dermatomyositis: an analysis of clinical and laboratory features and survival in 76 patients with a review of the literature. *Seminars in Arthritis and Rheumatism* 1986; **15**: 168–78.

38. Isenberg D A, Rowe D, Shearer M, Novick D, Beverley P C L. Localization of interferons and interleukin-2 in polymyositis and muscular dystrophy. *Clinical and Experimental Immunology* 1986; **63**: 450–8.

39. Kakulas B A. Destruction of differentiated muscle cultures by sensitized lymphoid cells. *Journal of Pathology and Bacteriology* 1966; **91**: 495–503.

40. Kakulas B A. Animal models of polymyositis and dermatomyositis. In: Dalakas M C, ed, *Polymyositis and Dermatomyositis*. London: Butterworths, 1988: 133–54.

41. Klein J J, Gottlieb A J, Mones R J, Appel S H, Osserman K E. Thymoma and polymyositis. Onset of myasthenia gravis after thymectomy: report of two cases. *Archives of Internal Medicine* 1964; **113**: 142–52.

42. Krum S H, Cordiner G H, Anderson B C, Holliday T A. Polymyositis and polyarthritis associated with systemic lupus erythematosus in a dog. *Journal of the American Veterinary Association* 1977; **170**: 61–4.

43. Magid S K, Kagen L J. Serologic evidence for acute toxoplasmosis in polymyositis-dermatomyositis. Increased frequency of specific anti-toxoplasma IgM antibodies. *American Journal of Medicine* 1983; **75**: 313–20.

44. Manchul L A, Jin A, Pritchard K I, Tenenbaum J, Boyd N F, Lee P, Germanson T, Gordon D A. The frequency of malignant neoplasms in patients with polymyositis–dermatomyositis. *Archives of Internal Medicine* 1985; **145**: 1835–9.

45. Manghani D, Partridge T, Sloper J C, Smith P. Role of myofibrillar antigens in the pathogenesis of experimental myositis, with particular reference to lymphocyte sensitization, the transfer of the disease by lymphocytes from animals with experimental myositis to cultured muscle cells. In: Bradley W G, Gardner-Medwin D, Walton J N, eds, *Recent Advances in Myology*. Amsterdam: Excerpta Medica, 1975: 387–406.

46. Matthews M B, Bernstein R M. Myositis autoantibody inhibits histidyl-tRNA synthetase: a model for autoimmunity. *Nature* 1983; **304**: 177–9.

47. Myllyla R, Myllyla V, Tolonen U, Kivirikko K I. Changes in collagen metabolism in diseased muscle. I. Biochemical studies. *Archives of Neurology* 1982; **39**: 752–5.

48. Pagel W, Woolf A L, Asher R. Histological observations on dermatomyositis. *Journal of Pathology and Bacteriology* 1949; **61**: 403–11.

49. Pearson C M. Development of arthritis, periarthritis and periostitis in rats given adjuvants. *Proceedings of the Society for Experimental Biology and Medicine of New York* 1956; **91**: 95–101.

50. Pearson C M. Polymyositis and related disorders. In: Walton J N, ed, *Disorders of Voluntary Muscle*. Boston: Little, Brown & Company, 1964.

51. Pearson C M. Polymyositis and dermatomyositis. In: Samter M, ed, *Immunological Diseases* (3rd edition). Boston: Little, Brown & Co. 1978: 1091–108.

52. Peltonen L, Myllala R, Tolonen U, Myllala V V. Changes in collagen metabolism in diseased muscle. 2: Immunohistochemical studies. *Archives of Neurology* 1982; **39**: 756–9.

53. Quilis M R, Damjanov I. Dermatomyositis as an immunolo-

gic complication of toxoplasmosis. *Acta Neuropathologica* 1982; **58**: 183–6.

54. Ransohoff R M, Dustoon M M. Impaired autologous mixed lymphocyte reaction with normal concanavalin A—induced suppression in adult polymyositis/dermatomyositis. *Clinical and Experimental Immunology* 1983; **53**: 67–75.

55. Rosenberg N L, Carry M R, Ringel S P. Association of inflammatory myopathies with other connective tissue disorders and malignancies. In: Dalakas M C, ed, *Polymyositis and Dermatomyositis*. London: Butterworths, 1988: 37–69.

56. Scherbel A L, McCormack L J, Mackenzie A H, Atdjian M. Association of certain connective tissue syndromes and malignant disease. *Postgraduate Medicine* 1964; **35**: 619–28.

57. Schumacher H R, Schimmer B, Gordon G V, Bookspan M A, Brogadir S, Dorwart B B. Articular manifestations of polymyositis and dermatomyositis. *American Journal of Medicine* 1979; **67**: 287–92.

58. Steigerwald J C. Polymyositis – dermatomyositis: the inflammatory myopathies. In: Samter M, ed, *Immunological Diseases* (4th edition). Boston, Toronto: Little, Brown & Co., 1988: 1437–57.

59. Steiner W R. Dermatomyositis, with report of a case which presented a rare muscle anomaly but once described in man. *Journal of Experimental Medicine* 1903; **6**: 407–42.

59a. Storen G. Dermatomyositis with calcinosis universalis. *Nordisk Medicin* 1956; **55**: 472–3.

60. Stoerk H C, Bielinski T C, Budzilovich T. Chronic polyarthritis in rats injected with spleen in adjuvants. *American Journal of Pathology* 1954; **30**: 616.

61. Sunnenberg T D, Kitchens C S. Dermatomyositis associated with malignant melanoma: parallel occurrence, remission, and relapse of the two processes in a patient. *Cancer* 1983; **51**: 2157–8.

62. Tang T T, Sedmak G V, Siegesmund K A, McCreadie S R. Chronic myopathy associated with Coxsackie virus type A9: a combined electron microscopical and viral isolation study. *New England Journal of Medicine* 1975; **292**: 608–11.

63. Tucker R E, Winter W G, del Valle C, Uematsu A, Libke R. Polymyositis mimicking malignant tumour. *Journal of Bone and Joint Surgery* 1978; **60-A**: 701–3.

64. Unverricht H. Polymyositis acuta progressiva. *Zeitschrift für Klinische Medizin* 1887; **12**: 533–49.

65. Wagner E L. Fall einer seltenen Muskel—krankheit. *Archiv der Heilkunde* 1863; **4**: 282–3.

66. Walker E J, Jeffrey P D. Polymyositis and molecular mimicry, a mechanism of autoimmunity. *Lancet* 1986; **ii**: 605–7.

67. Walton J. *Disorders of Voluntary Muscle*. Edinburgh: Livingstone, 1981.

68. Walton J N, Adams R D. *Polymyositis*. Edinburgh: Livingstone, 1958.

69. Wiley C A, Nerenberg M, Cros D, Soto-Aguilar M C. HTLV-I polymyositis in a patient also infected with the human immunodeficiency virus. *New England Journal of Medicine* 1989; **320**: 992–5.

69a. Ytterberg S R, Schnitzer T J. Coxsackievirus B1-induced murine polymyositis. In: Greenwald R A, Diamond H S, eds, *CRC Handbook of Animal Models for the Rheumatic Diseases* (volume 1). Boca Raton, Florida: CRC Press Inc., 1988: 147–56.

Chapter 17

Systemic Sclerosis and Allied Disorders

Systemic sclerosis is a generalized connective tissue disease of unknown cause, affecting dark- as well as light-skinned races, three times as common in women as in men, and characterized by the excess formation, locally or systemically, of collagen-rich fibrous tissue. The incidence is approximately 6 cases per million population per year.[45]

Definition[13,85,86,101,119]

The most constant feature of systemic sclerosis is scleroderma, a broad term describing hardening of the skin.[192] There may or may not be associated systemic disease. Connective tissue-rich structures, such as the synovial joints and skeletal muscle, are often affected. There are frequent visceral lesions, particularly of the alimentary tract, lungs, heart and kidney. Occasionally, these visceral disorders occur in the absence of scleroderma. In its most characteristic, insidiously advancing form, the syndrome is termed 'progressive systemic sclerosis' (PSS). There are, however, localized varieties of scleroderma. One example is 'morphoea', a scleroderma-like lesion affecting single, or occasionally multiple, skin areas. There are other variants, such as the CREST syndrome (p. 695), dominated by vascular, oesophageal and calcific changes, and related disorders such as acroosteolysis and eosinophilic fasciitis. Clinical and laboratory features may be shared with systemic lupus erythematosus, rheumatoid arthritis and polymyositis–dermatomyositis, creating overlap syndromes. Moreover, features of systemic sclerosis may be encountered in mixed connective tissue disease (see p. 709). In all, therefore, there is a spectrum of abnormalities ranging from systemic to limited local disorder.[156] A classification is given in Table 17.1.

History

Historical aspects of systemic sclerosis were reviewed by Rodnan and Benedek.[153]

Lewin and Heller[104] attribute the first description of the generalized skin disease to Zacutus Lusitanus[108] and the

Table 17.1

Classification of systemic sclerosis[119]

Systemic sclerosis
With diffuse scleroderma
Diffuse, symmetrical, skin involvement
Rapid progression
Early visceral involvement
With CREST syndrome (see below)
Restricted skin involvement
Delayed visceral disease
Calcinosis
Telangiectasia
With 'overlap'
Associated features of systemic lupus erythematosus, rheumatoid arthritis or polymyositis/dermatomyositis
Localized cutaneous disease
Morphoea
Single or multiple, discrete skin plaques
Linear morphoea
With or without melorheostosis
Eosinophilic fasciitis (p. 708)

introduction of the name scleroderma to Gintrac.[60] An early account was that of Curzio.[38] A classical description of the various forms of the skin lesion was provided by Hutchinson.[79] Since the work of Steven[174] it has been widely accepted that the disease may affect many viscera in addition to the skin; his report included a clear morbid anatomical description of the visceral lesions in a single case. The name 'progressive systemic sclerosis' (PSS) was proposed by Goetz.[61] Enormous compilations of similar cases have been made[5,99,104] but the pathological literature[41,93,123] is relatively scant. The paper of Klemperer, Pollack and Baehr[95] was a landmark in the interpretation of the nature of progressive systemic sclerosis. In recent years there has been great interest in the visceral lesions.[100] Progressive systemic sclerosis merges clinically and morphologically with the other generalized connective tissue diseases, particularly myositis, arteritis, systemic lupus erythematosus and rheumatoid arthritis. The natural history of progressive systemic sclerosis is likely to be shaped if not determined by humoral and cell-mediated hypersensitivity (see p. 251).[119]

Clinical features

Systemic sclerosis is an uncommon and unusual disorder, of unknown aetiology, more frequent with advancing age, and characterized by a combination of indolent skin, vascular and visceral lesions.[166a] Tightening of the involved skin, arthritis, dysphagia, dyspnoea, renal dysfunction and peripheral vascular disease are signs which indicate the sites of the most frequent lesions. In localized zones of skin disease a central pale area corresponding to dermal thickening is surrounded by a margin of vascular stasis which gives the skin a violaceous colour. Scleroderma (morphoea) remains localized in this way. Cutaneous atrophy succeeds the dermal thickening and is often accompanied by excessive, patchy, localized pigmentation, particularly in the skin of the limbs. Telangiectatic spots accompany these changes and there may be a loss of hair in the affected zones.

There is much evidence that abnormal microvascular function and impaired peripheral blood flow contribute to, or initiate, the skin lesions and limb ischaemia of progressive systemic sclerosis.[134] Blood flow can be measured by the clearance of isotopes such as xenon–133.[102,130] Small vessels can be visualized by capillaroscopy or by video microscopy after the injection of fluorescent dyes[16] and temperature changes recorded.

With the passage of time, the cutaneous lesions become more dramatic. Loss of facial expression, with a thin, pinched nose and tightly drawn mouth (Fig. 17.1), are accompanied by progressive sensitivity of the limbs to cold. Raynaud's phenomenon appears. The tissues of fingers and toes begin to display evidence of ischaemia and small, superficial ulcers develop in zones of digital pallor (Fig. 17.2). The tissues of the affected limbs waste and calcium is frequently deposited within them. There is, however, no evidence of a generalized disorder of calcium metabolism. The skin of the fingers becomes tightened and with the digital ulceration there occurs an insidious atrophy of the distal phalangeal bones. The ultimate result may be cold, claw-like hands, superficially ulcerated and with digital amputation, susceptible to infection.

A particular, unusual, limited form of systemic sclerosis has been defined, the CREST syndrome,[193] in which *calcinosis* accompanies *Raynaud*'s phenomenon, *(o)esophageal* dysfunction, *sclerodactyly* and *telangiectasia*. It appears probable that progressive arterial insufficiency accounts both for the Raynaud's phenomenon and, in turn, for the atrophy and calcification.[84]

Involvement of the gastrointestinal tract is frequent. Oesophageal muscle contraction is disorganized (Fig. 17.3); emptying is delayed and incomplete and there is difficulty in swallowing. Consequently, hiatus hernia is common and ulceration may occur. Irregular, uncoordin-

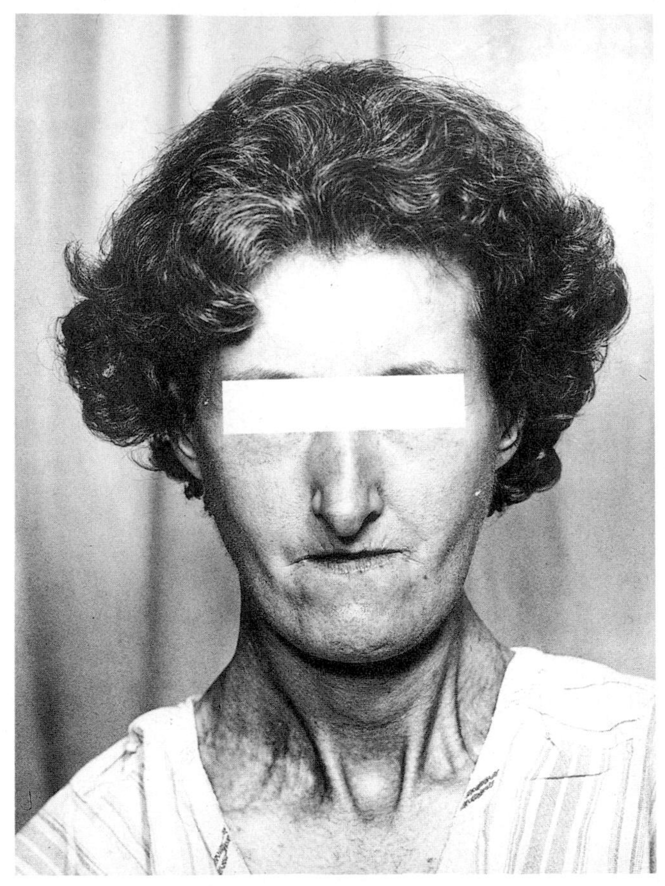

Fig. 17.1 Systemic sclerosis.
Note the indrawn margins of the mouth. The skin is atrophic, taut and constricted.

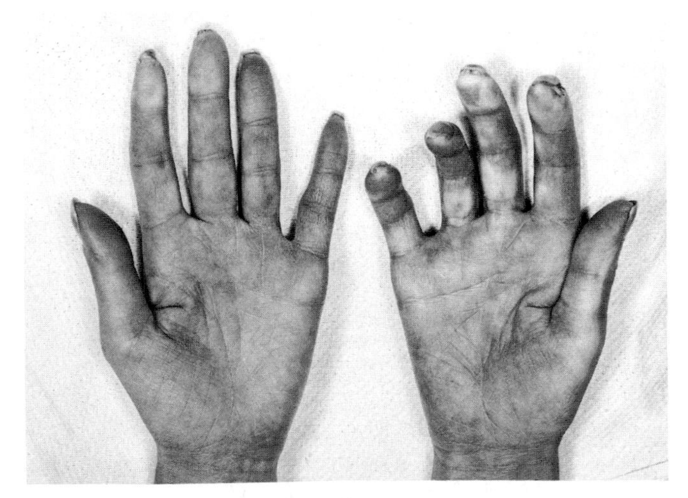

Fig. 17.2 Systemic sclerosis.
Peripheral vascular disease is manifested as Raynaud's phenomenon. There is patchy discoloration of the digits and incipient gangrene of the terminal parts of the fingers.

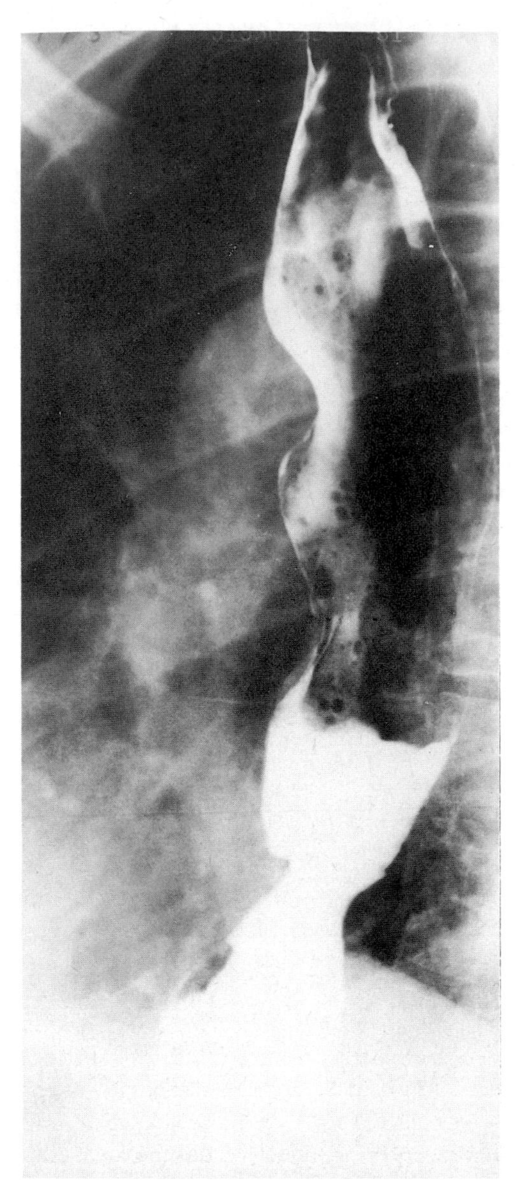

Fig. 17.3 Oesophagus in systemic sclerosis.
A barium swallow emphasizes the irregular and uncoordinated smooth muscle contractions that are the result of fibrosis and interference with the innervation of the oesophageal wall.

ated contractions of the oesophagus are recognized on screening. Patients may also present with malabsorption, atony, obstruction or ulceration of the small gut. Disturbances of liver function tests are described; jaundice is rare, but there are often changes in the plasma proteins. A raised γ-globulin level and increases in plasma fibrinogen are characteristic and as a result there is an increase in the erythrocyte sedimentation rate.

Dyspnoea is a late sign and is more likely to be evidence of lung than of heart disease. A progressively diminishing ventilatory capacity may be due both to constriction of the

Table 17.2

Criteria for systemic sclerosis[118]

Number of cases	264
Major criterion	
Proximal scleroderma	239
Minor criteria	
Sclerodactyly	19
Digital pitting scars	15
Bibasilar pulmonary fibrosis	8

Proximal scleroderma is the single major criterion

In the absence of proximal scleroderma Sclerodactyly Digital pitting scars of finger tips or loss of substance of the distal finger pad Bibasilar pulmonary fibrosis	Contribute as minor criteria

The major and two or more minor criteria were found in 97 per cent of cases of systemic sclerosis by comparison with 2 per cent in cases of systemic lupus erythematosus, polymyositis/dermatomyositis or Raynaud's syndrome.

skin of the upper trunk and to damage to lung tissue succeeded by repetitive infection. Myocardial changes are shown by the signs of ischaemic heart disease. Involvement of the kidney is frequent but signs of renal disease are late. Occasionally, patients with systemic sclerosis present with the nephrotic syndrome but, in the majority, renal involvement is recognized clinically only when terminal signs of renal failure appear. A curious association is the periodic but uncommon development of carcinoma of the lung, occasionally before the connective tissue disease is manifest. Forty-four cases were analysed in one review.[175]

There is no specific diagnostic clinical test except skin and renal biopsy.[192] The sensitized sheep cell agglutination (Rose–Waaler) test (see p. 233) is often positive and antinuclear factors may be detected in the serum. The establishment of agreed clinical criteria for systemic sclerosis is still a matter for debate.[119] In one approach, data from retrospective studies led to a scoring system of value. In another, the American Rheumatism Association* Scleroderma Criteria Cooperative Study evolved a simple preliminary classification prospectively (Table 17.2) in which scleroderma of the skin proximal to the metacarpophalangeal or metatarsophalangeal joints but including other parts of the limbs, face, neck or trunk, is the sole major criterion for clinical diagnosis.[118]

* Now the American College of Rheumatology.

Pathological changes

It is convenient to consider the pathological changes[93,119,150] of systemic sclerosis as unit lesions, affecting the organs and viscera to varying degrees.

Unit lesions

The unit lesions are vascular disease; epithelial atrophy; and fibrosis. Individually, these changes are not pathognomonic, but their organ distribution and pattern in systemic sclerosis are characteristic (Table 17.1) and they culminate in disease of the skin, gastrointestinal tract, lungs, kidneys, skeletal muscle and pericardium (Table 17.3). Although myocardial and pleural disease is also commonplace, a controlled retrospective study[41] of 58 patients with systemic sclerosis at necropsy in Baltimore during the years 1948 to 1966 confirmed that these disorders were often present in matched controls.

Table 17.3

Organs principally involved in systemic sclerosis[41]

Site	Prevalence in systemic sclerosis (per cent)	Excess prevalence over controls (per cent)
Skin	98	98
Oesophagus	74	74
Lung	81	59
Kidney	58	49
Small gut	48	46
Pericardium	53	41
Skeletal muscle	41	41
Large gut	39	39
Pleura	81	29
Myocardium	81	26

Skin

Skin changes predominate and are usually the presenting lesions.[103] The disease may indeed be confined to the skin (morphoea) (Fig. 17.4). In the first instance, the signs are those of a low-grade inflammatory response in which a dermal, perivascular lymphocytic infiltrate accompanies oedema; there is swelling of the collagen fibre bundles

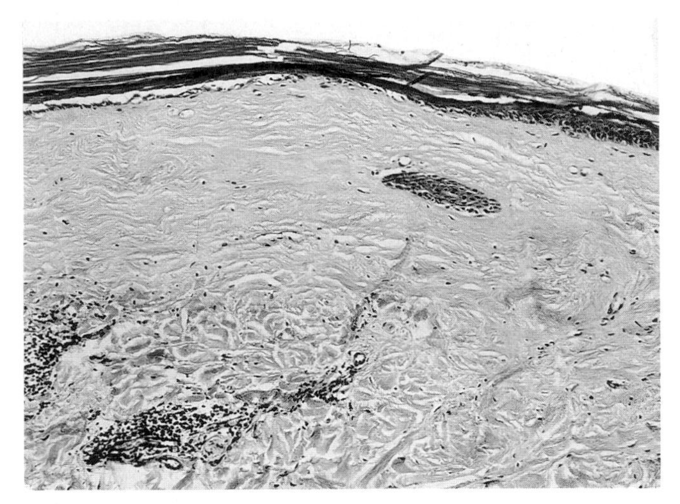

Fig. 17.4 Skin in scleroderma.
Atrophy of the epidermal rete pegs is accompanied by slight hyperkeratosis. Dermal appendages are not yet wholly lost and there are scattered foci of lymphocytes (bottom left). (HE × 70.) walled/ (HE × 20.)

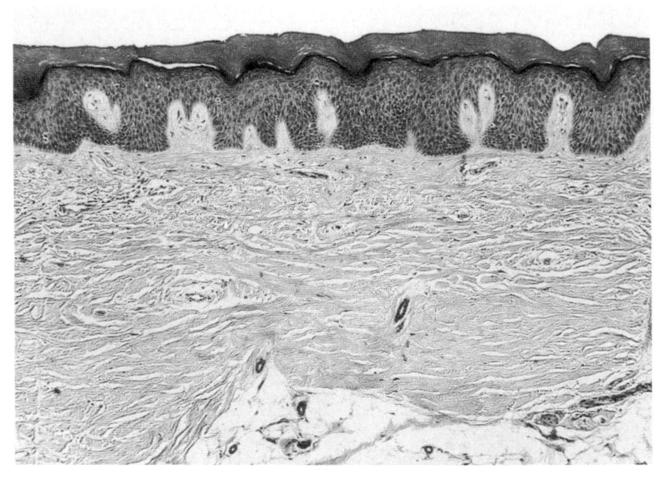

Fig. 17.6 Localized scleroderma (morphoea).
Keratin formation continues. There is limited epidermal atrophy but an almost total loss of dermal appendages and a great increase in the quantity of dermal collagenous connective tissue. (HE × 60.)

which lose their normal staining characteristics to become homogeneous and increasingly eosinophilic. Subcutaneous fat cells atrophy and inflammatory cells extend among them. There is a corresponding reduction in elastic tissue and a progressive increase in dermal collagen, fibrosis, which results in the thickening and immobility seen clinically (Fig. 17.5). The mass of collagen formed is disproportionate to the number of recognizable fibroblasts which gradually diminishes. An early increase in mast cells, particularly those of the reticular dermis, suggests that they may promote the fibrosis[68] (see p. 304). With the evolution of

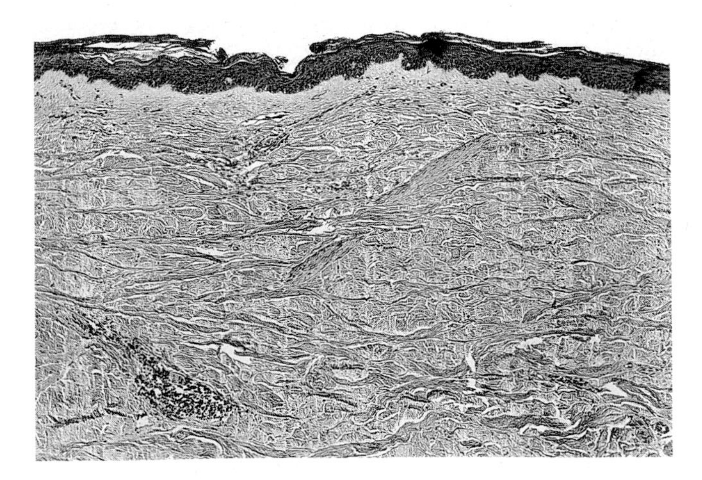

Fig. 17.5 Skin in scleroderma.
Widening of the epidermal rete processes and continued keratinization are features of this specimen in which overall thinning of the skin is associated with a relative increase in the proportions of dermal collagen. (HE × 40.)

the systemic disease, zones of fibrinoid (see p. 294) are sometimes seen within this connective tissue. There is a progressive loss of dermal appendages and of dermal and subcutaneous fat (Fig. 17.6). Rarely, subcutaneous nodules resembling those of rheumatoid arthritis are recognized.[19]

The small dermal blood vessels show an increase in intimal cellularity; the intima comes to contain much additional matrix material ('mucoid change', p. 306) which stains blue with haematoxylin and eosin. The appearances resemble those of the primitive mesenchyme, of the haematoxyphil substance of the Aschoff nodule (see p. 798), and of the intercellular matrix in myxomata and in localized myxoedema. The overlying epidermis is thin and atrophic. There is a variable increase in the amount of melanin in the basal layer of the epidermis; telangiectatic blood vessels appear as small clusters of dilated, superficial venules and capillaries. Particularly in the distal phalangeal soft tissues (Fig. 17.2) but sometimes in other places such as the periarticular soft tissues of the knee joint, dystrophic deposits of calcium occur. The deposits may ulcerate through the skin and calcific debris can escape onto the skin surface.

Viscera

The visceral lesions of systemic sclerosis are distributed in the territories of the purinergic 'third' nervous system,[23] in analogy with the distribution of the principal visceral lesions in Chagas' disease. In this South American protozoal disorder, the heart, oesophagus and colon suffer most.

Alimentary system

The organs of the alimentary system are important targets.[62] The oesophageal changes were described by Rake;[147] there is a focal loss of smooth muscle[41] and a replacement by fibrous tissue (Fig. 17.7). These changes, and the associated muscular incoordination, may be attributed to lesions of nerves, of small blood vessels or of smooth muscle (Fig. 17.7). Reflux oesophagitis with ulceration, erosion and inflammation, is common. However, the transmission electron microscopic appearances are distinct from those of hiatus hernia with reflux oesophagitis and comprise a characteristic pattern of disorder.[164] In seven of these author's cases there was fibrosis; this was severe in five. There were 'adequate' numbers of normal nerve fibres but pronounced changes in smooth muscle cells. Within them were thick, dense bodies (5/7); thick dense plaques (7/7); and an increase in cytoplasmic organelles (5/7). The most specific changes were in the capillary endothelial cells and basement membranes. The endothelial cells were often swollen (5/7); basement membrane lamination was present in six of seven cases and platelet thrombi were common (4/7). Disruption of capillary wall integrity offered an explanation for vascular insufficiency, muscle atrophy and fibrosis.

Rarely, the lower part of the oesophagus is lined by metaplastic columnar epithelium, the formation of which may be a result of systemic sclerosis.[24] High oesophageal strictures and ulceration are also sequelae. Gastric carcinoma and the gastric lesions of systemic sclerosis may very exceptionally be found to coexist.[160]

In the duodenal loop, atrophy, dilatation and fibrosis have been observed. In the small gut[120] a focal replacement of smooth muscle by collagen is recognized. Ultramicroscopic changes in the mucosa accompanied by alterations in absorption have been studied by electron microscopy: peroral jejunal biopsy reveals mucosal villous atrophy in a minority of cases. When passive intestinal permeability is altered, it is, therefore, attributable to other factors, such as bacterial overgrowth.[32] In the large gut, muscle atrophy may result in herniation of the mucosa, diverticulum formation and disturbances of the normal pattern of folds and segments. Dilatation and fibrosis develop. In rare cases, progressive myofibroblastic proliferation of the intima of mesenteric arterial branches may culminate in gangrene of the colon.

No specific liver disease is recognized; nodular regenerative hyperplasia is a very unusual association.[165] There are microvascular changes which may be responsible for hepatocyte disorder, and pancreatic necrosis has been associated with vascular insufficiency in the same way.[2] The gall bladder may be fibrotic.[35]

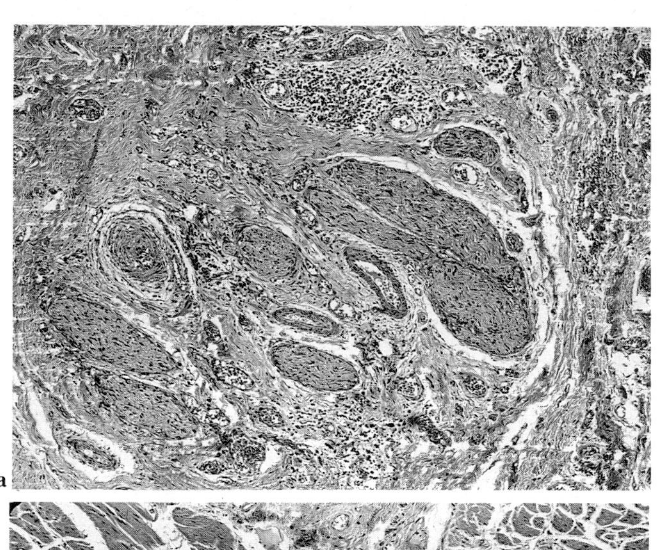

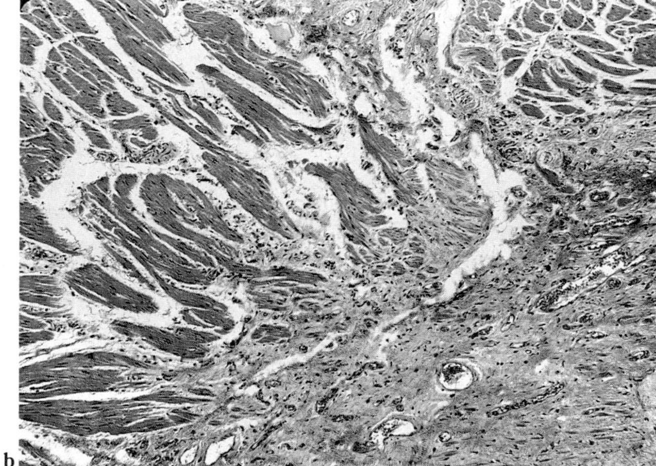

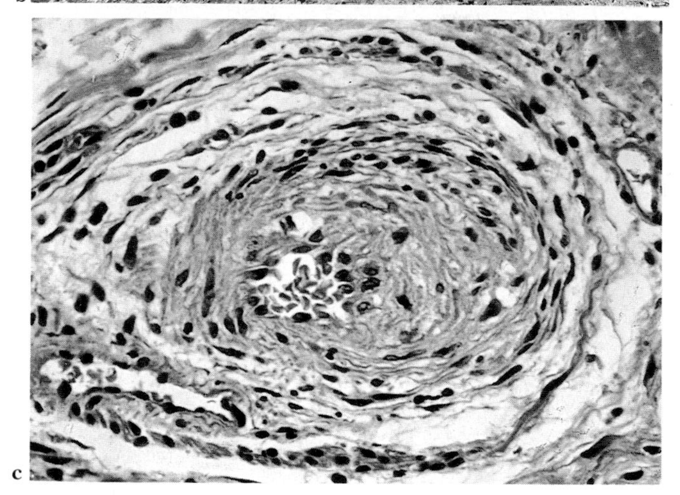

Fig. 17.7 Oesophagus in systemic sclerosis.
a. Neuromuscular complex. The walls of the small arteries are thickened. There is an excess of surrounding fibrous tissue which encases both the blood vessels and the adjacent nerve fibres. (HE × 60.) **b.** The muscular wall of the oesophagus is shown; much of the normal structure has been replaced by fibrous tissue (bottom right). (HE × 60.) **c.** Small artery in oesophageal wall. There is intimal fibromuscular hyperplasia and reduplication of the internal elastic lamina. The lumen appears diminished. (HE × 310.)

Respiratory system

The pulmonary lesions are well defined[26,34,46,186] Basement membrane thickening in pulmonary capillaries has been demonstrated by transmission electron microscopy[191] and contributes to defective gas interchange; it is accompanied by occlusive connective tissue hyperplasia of the pulmonary arteries and arterioles. There is a progressive and inexorable interstitial fibrosis. By contrast with idiopathic pulmonary fibrosis, the relative lung content of types I and III collagen is unaltered.[167] The consequence is a form of cystic change difficult to distinguish histologically from bronchiectasis but more widespread in distribution (Figs 17.8 and 17.9).[49,70] Associated infection is common and inflammation may play a part in the origin of the lung lesions.[136] The atrophic and fibrotic lung tissue heals incompletely. Calcification within fibrous foci is occasional. Residual clefts in the newly formed fibrous tissue come to be lined by regenerating alveolar cells which resemble the cuboidal cells of lung tissue proliferating after injury, or the cells of fetal lung. The hyperplastic process may be florid. The development of all forms of lung carcinoma has been described.[175] However, instances of alveolar cell carcinoma are unexpectedly frequent.[26,148]

The possibility of an occupational predisposition to systemic sclerosis arose when Erasmus[47] described 17 cases among underground gold miners in the Witwatersrand; a unique earlier account by Branwell[20] had drawn attention to systemis sclerosis in Scottish stone-masons. Rodnan and

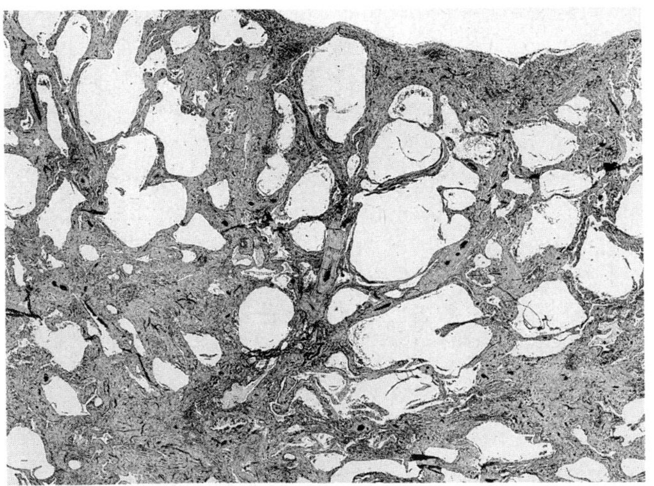

Fig. 17.8 Interstitial lung fibrosis in systemic sclerosis.
Throughout the lung, bordered by the pleural surface (top right), are numerous cystic spaces usually lined by low columnar or cuboidal epithelium. There is widespread loss of respiratory bronchioles, alveolar ducts and alveoli, and an excess of fibrous tissue. The changes comprise a variety of interstitial pulmonary fibrosis and have assumed the characteristics of 'honeycomb' lung. (HE × 4.2.)

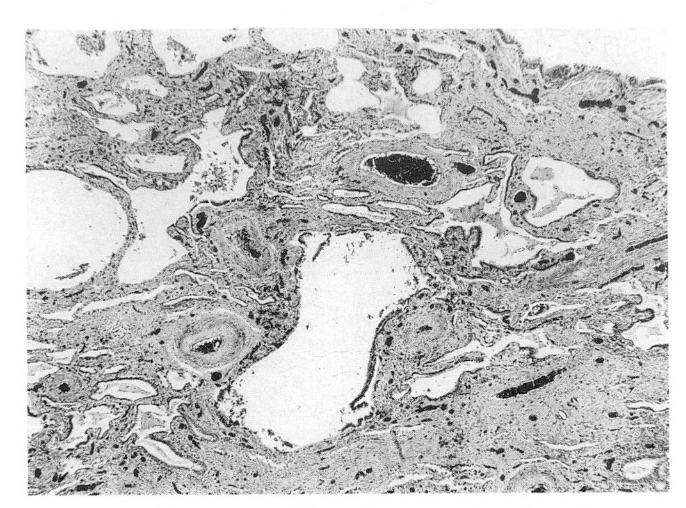

Fig. 17.9 'Honeycomb' lung in systemic sclerosis.
Among the greatly excessive fibrous tissue are cystic spaces often lined by low columnar epithelium. The small arteries are thick-walled. (HE × 20.)

his colleagues[154] subsequently reviewed their experience with over 150 cases of systemic sclerosis in Pittsburgh and discovered that 26 of 60 men (43 per cent) with this disease had either worked in coal mines or in analogous occupations with heavy exposure to siliceous dust. These observations, confirmed by others,[66] raised the possibility that, as in the case of the Caplan syndrome in rheumatoid arthritis (see p. 497), there exists a subpopulation in whom genetic predisposition permits fibrogenic dust to evoke a selectively severe pulmonary reaction. It may transpire that pulmonary systemic sclerosis can develop in advance of the skin disease, in analogy with Caplan's syndrome, but this is not yet known.

Urinary system

The renal lesions centre upon the blood vessels. The smaller muscular arteries are the site for disorders of the endothelial cells, basement membrane and intima. Renal disease has long been seen as an integral part of the syndrome.[126] Although haemodialysis clearly influences the outcome, renal failure is the cause of death in a substantial proportion of fatal cases.[101]

Acute renal disease in systemic sclerosis is characterized by arterial and arteriolar changes that resemble closely those of accelerated (malignant) hypertension[85] (Figs 17.10 and 17.11). The initial arterial lesions may include foci of fibrinoid. Fibrinoid may also be encountered in the thickened walls of glomerular capillary loops while glomerular lesions resembling those of the microscopic form of polyarteritis nodosa (p. 621) have been described (Fig. 15.5). Frequently, there is intimal fibromuscular hyperplasia with

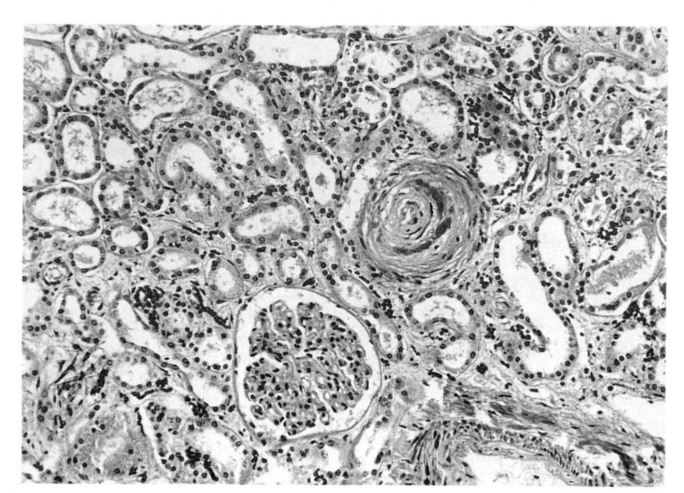

Fig. 17.10 Kidney in systemic sclerosis.
Close to a glomerulus, an interlobular artery is almost occluded by intense intimal fibromuscular hyperplasia; the dark material is fibrinoid. (HE × 75.)

an accumulation of basophilic extracellular matrix, the so-called 'mucoid' change (p. 306). Two features help in distinguishing these changes in systemic sclerosis from those of systemic hypertension. First, the changes of systemic sclerosis affect interlobular arteries and larger divisions of the arcuate arteries to a greater extent than in accelerated hypertension in which the acute lesions are largely confined to the afferent glomerular arterioles and those parts of the interlobular arteries immediately proximal to them; second, there may be a low-grade inflammatory and fibrous reaction around the arteries in systemic sclerosis, perhaps indicative of an underlying immune complex-

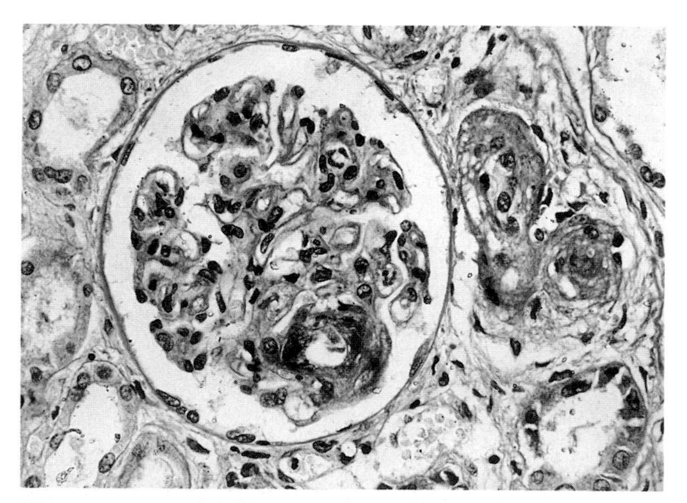

Fig. 17.11 Kidney in systemic sclerosis.
The afferent glomerular arteriole (centre right) and one loop of the glomerular capillary bed (lower centre) are distorted by the presence of fibrinoid. The appearances resemble those encountered in accelerated ('malignant') hypertension. (HE × 210.)

mediated cause (see p. 627). The renal vascular lesions of systemic sclerosis can themselves *lead to* severe hypertension[180] and arteriolar fibrinoid change may be one result.

In differentiating these features from those of accelerated hypertension with which systemic sclerosis may coexist,[180] it should be noted that similar changes can also be encountered in microangiopathic haemolytic anaemia,[42] in the haemolytic uraemic syndrome of children (in whom systemic sclerosis is very rare), and in subacute allograft rejection. Renal transplantation may be attempted in the treatment of systemic sclerosis and the histological question may arise: Has disease recurred in the transplanted kidney? As in the malignant phase of systemic hypertension, glomerular fibrinoid is identified.

Accompanying the slowly advancing forms of systemic sclerosis, with progressive skin disease, there are analogous changes in renal connective tissue. Basement membranes thicken and excess fibrous tissue forms. Both collagen type I and type III accumulate but there is a greater formation of collagen type III, indicated by an enhanced degree of immunofluorescent staining for this protein.[12] The renal vascular lesions of systemic sclerosis, which include thrombosis of narrowed vessels, simultaneously cause cortical zones of microscopic ischaemia and infarction. The result of focal cortical infarction, the scars which follow and the obstructive arteriolar lesions, is a reduction in the mass of the convoluted tubules supplied by these vessels. Fibrosis leads to a fine granular renal surface. The combination of focal renal cortical scarring and of the intervening atrophy of nephrons is to produce a small, granular kidney. The external configuration of the kidney may be indistinguishable from the appearances in chronic pyelonephritis, polyarteritis nodosa, amyloid nephrosclerosis, and in cases of systemic lupus erythematosus with end-stage renal disease.

The evidence that many patients with systemic sclerosis have circulating autoantibodies, including antinuclear factor and rheumatoid factor[7,51,52,114,149] has led to the suggestion that the renal lesions are attributable to immunological phenomena (see p. 251). Immunofluorescence microscopy of *post mortem* material revealed focal aggregates of immunoglobulin M and complement in the walls of renal arterioles and small arteries and in the region of glomerular basement membranes.[115] However, it was uncertain whether such aggregates were trapped fortuitously or selectively or whether their presence was the result of specific local antigen–antibody reactions. It was also unclear whether the immune complexes exerted local tissue-damage, i.e. were pathogenic, or whether their presence was coincidental. The investigations by McCoy *et al.*[111] of renal tissue from 11 specimens from seven patients with systemic sclerosis at biopsy, necropsy or nephrectomy (Figs 17.12–17.15), and by Lapenas, Rodnan and Cavallo[97]

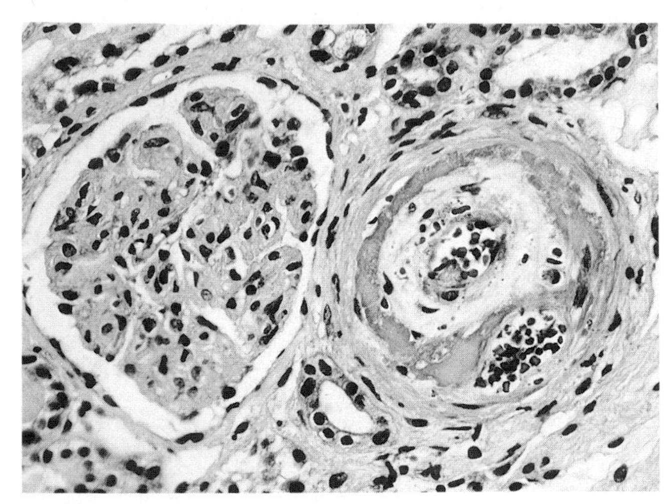

Fig. 17.12 Kidney in systemic sclerosis.
A 24-year-old female. Biopsy of kidney 10 weeks before transplantation. Compare with Fig. 17.13. (HE × 200.) Courtesy of Dr R C McCoy and the Editor *Laboratory Investigation.*[111]

on a further 16 patients, helped to clarify these questions. Diffuse deposits of immunoglobulin M and complement component Clq were bound to the intima of interlobular and arcuate arteries which showed fibromucoid changes on light microscopy and much intimal fibrillar and matrix material but no immune complex deposits by electron microscopy. It appeared that the immunoglobulin M and complement represented the interaction of complement-fixing antibody and antigen: fibrinogen was almost always present, and the antibody was thought to be antiglobulin i.e. of rheumatoid factor type. Characteristic vascular lesions recurred in an allograft after transplantation and antinuclear factor and anti-immunoglobulins could be eluted from the excised graft.

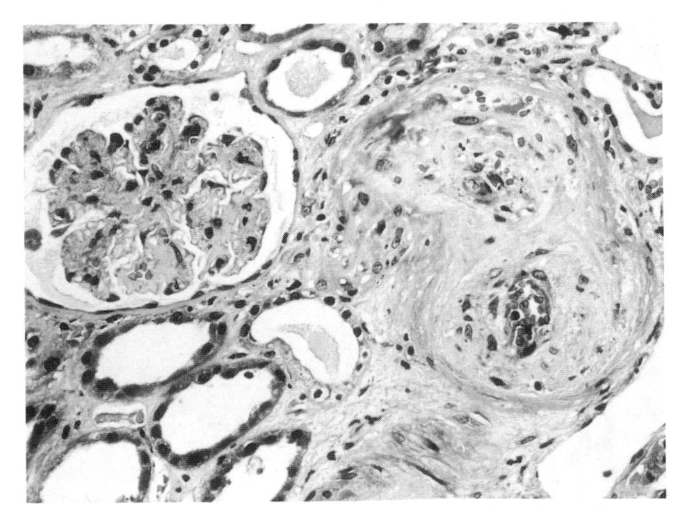

Fig. 17.13 Kidney transplant in systemic sclerosis.
Compare with Fig. 17.12. Biopsy of allograft from same case. The fibromuscular intimal proliferation, fibrinoid change and extravasation of red blood cells are closely similar to those seen in the recipient before transplantation. (HE × 160.) Courtesy of Dr R C McCoy and Editor *Laboratory Investigation.*[111]

Cardiovascular system

Blood vessels Peripheral vascular lesions in systemic sclerosis are very common. It is likely that they mediate most of the characteristics of the disease.[84] The most important are those which lead to digital and to renal vascular insufficiency. The nature of the early lesions of the digital, limb and larger arteries in systemic sclerosis is not certain but there is evidence that one mechanism may be the formation of antibodies binding to smooth muscle and to elastic material, accompanied by complement activation.[48] Nailfold biopsy offers the possibility of correlated vascular and connective tissue changes[178] while dermal and muscle changes can be readily compared.[133] The cuticles contain globular, eosinophilic, periodic-acid Schiff-positive deposits of serum proteins associated with parakeratosis and raised epithelial cell mitotic activity.[178] There is capillary ectasia and endothelial cell swelling and proliferation with diminished numbers of dermal capillaries and nerve bundles. The capillary density in the CREST syndrome (see p. 695) tends to be higher than in those patients with rapidly advancing systemic disease. Digital arteries of larger size are, however, also affected and their disordered function is closely related to the occurrence of Raynaud's phenomenon. The abnormal digital artery blood flow (Fig. 17.16) revealed by arteriography (see p. 159) is extensively documented.[39] The principal change seen *post mortem* is intimal fibromuscular hyperplasia.[159] Larger, or proximal vessels may be implicated. Severe narrowing is likely to be accompanied by thrombosis, organization and recanalization, lesions which produce appearances similar to those encountered in occasional cases of rheumatoid arthritis (Fig. 12.30). They are distinguished from thromboangiitis obliterans by the absence of venous and neural involvement. In less common instances, the digital arteries and their branches may be the site for an arteritis which varies in intensity from subacute to acute and which may also recall the arteritis of rheumatoid arthritis (see p. 491). In other cases, the arterial changes simulate the lesions of polyarteritis nodosa[179] (see p. 620). Rarely, there is an association with giant-cell arteritis.[142]

Heart

Although the clinical signs and symptoms of cardiac disease in systemic sclerosis are much less common than histological studies suggest, evidence of myocardial disorder is readily obtained when sensitive tests of blood flow are made. In one series, 77 per cent of cases displayed myocar-

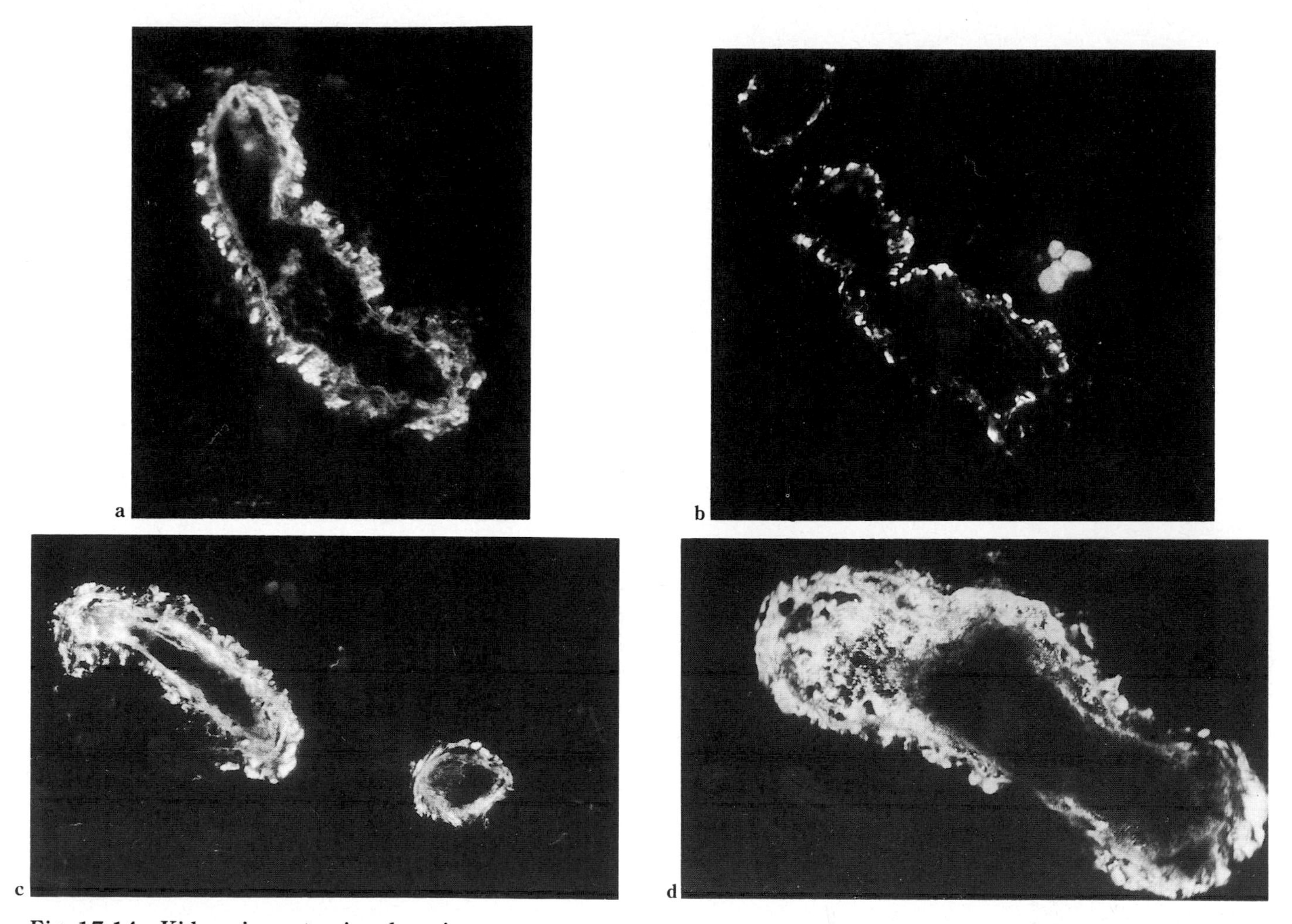

Fig. 17.14 Kidney in systemic sclerosis.
a. Immunoglobulin M (IgM) localized to the media of small interlobular artery. (Immunofluorescence × 250.) **b.** Localization of complement component C3 in pattern similar to that shown in Fig. 17.14a. (Immunofluorescence × 250.) **c.** IgM in interlobular artery of renal biopsy from patient before transplantation. Compare with Figs 17.12 and 17.14d. (Immunofluorescence × 250.) **d.** Compare with Fig. 17.14c. Artery from allograft examined by biopsy after transplantation. IgM is located in vessel wall in same pattern as in host kidney. (Immunofluorescence × 250.) Courtesy of Dr R C McCoy and the Editor *Laboratory Investigation.*[111]

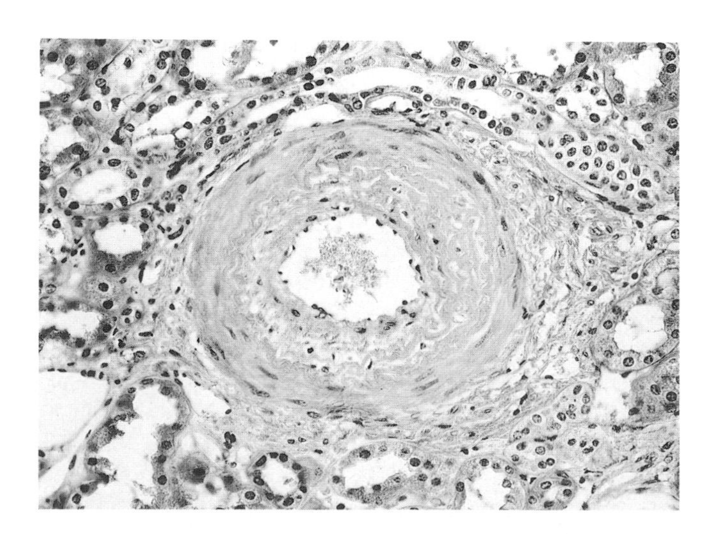

Fig. 17.15 Arterial disease in systemic sclerosis.
Large interlobular artery showing myofibroblastic intimal proliferation and reduplication of the internal elastic lamina. (HE × 246.) Courtesy of Dr R C McCoy and the Editor *Laboratory Investigation.*[111]

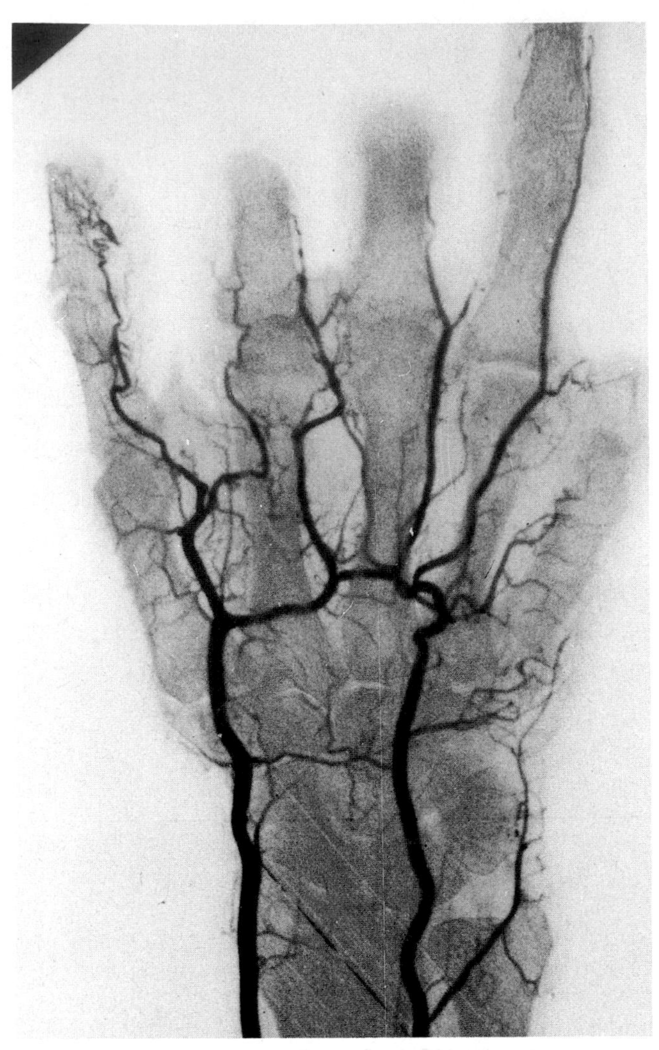

Fig. 17.16 Peripheral arteries in systemic sclerosis. A digital angiogram shows the occlusion of digital vessels associated with ischaemic lesions of the form seen in Fig. 17.2. Courtesy of Professor M I V Jayson.

dial vascular dysfunction.[58] Attention was first drawn to the cardiac lesion of systemic sclerosis by Weiss *et al.*[187]

Myocardium In the myocardium a main characteristic is the premature formation in 12 to 81 per cent of cases[17] of excess myocardial fibrous tissue,[41] which is often focal[163] (Fig. 17.17). In some instances, there is a more subtle pericellular fibrosis.[166] One explanation for the cardiac fibrosis suggests that it is a primary disorder, part of the widespread visceral collagenization of this disease. It is now widely accepted that the myocardial changes are, however, secondary to the same form of vascular and microvascular lesions that affect all the target organs.[100] Endothelial cell injury of coronary arteries and arterioles is characteristic: in cases of sudden death, the arteries display narrowing,

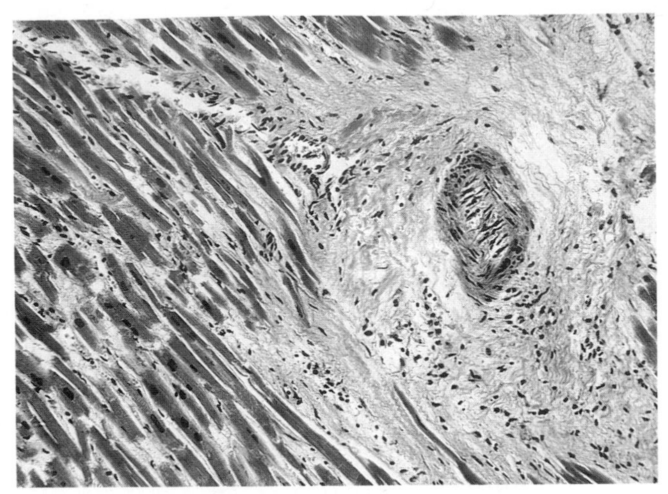

Fig. 17.17 Myocardium in systemic sclerosis. Excess collagenous connective tissue is distributed around a small coronary arterial branch. There is a focal loss of cardiac muscle cells and a sparse lymphocytic infiltrate. (HE × 105.)

fibrosis and fibrinoid change.[83] Concentric intimal fibromuscular hyperplasia has been seen in 17 per cent of cases.[41]

Endocardium In the endocardium, valvular fibrosis, with changes resembling those of healed rheumatic endocarditis and a verrucous endocarditis simulating systemic lupus erythematosus, have been recorded. Aortic valvulitis may accompany panaortitis.[162] The valve may perforate.

Pericardium There is often low-grade pericarditis with fibrosis;[17] these changes have been reported in 33–72 per cent of cases and may be identified in the absence of myocardial disease. Constriction is rare.

Osteoarticular system

Synovial joints

The joints are the site of inflammatory disease in approximately 60 per cent of cases.[6] The clinical signs, particularly limitation of movement, may be confused with those caused by the dermal fibrosis.[106] There are few histological reports. A study of 29 cases demonstrated a diffuse or focal lymphocytic and plasma cell synovial infiltrate.[151] Fibrin may be present. Unlike rheumatoid arthritis, there is little tendency to pannus formation, to bone reabsorption or to fibrous ankylosis, although periarticular demineralization is recognized by X-ray. Later, intense synovial fibrosis is accompanied by synovial cell atrophy and vascular obliteration. These changes closely resemble those of the definitive skin lesions, and contribute, as they do in tendon sheaths, to the peculiar stiffness, limited movement, and leathery crepitus of this form of 'stiff man' syndrome. An analagous form of joint disease is one result of *in utero* hypoxia.[125a] Bony ankylosis, particularly of the small peripheral joints, has been described but larger joints such as the hip may be similarly affected.[80] Other forms of joint disease have been recorded and a mutilating neurogenic arthropathy with subluxation and bone destruction has been encountered in a case of the CREST syndrome.[90]

Skeletal muscle

Involvement of the skeletal muscle in systemic sclerosis may lead to a mistaken diagnosis of polymyositis–dermatomyositis[185] (see Chapter 16), a disease with which many pathological characteristics are shared. In addition to muscle atrophy, which may affect the tongue, and the limitation of movement attributable to the constricting skin lesions, focal lymphorrhages occur in the perivascular and interstitial connective tissue planes. Occasionally there is evidence of focal myositis, of muscle cell atrophy and of replacement fibrosis. Ultrastructural studies have shown the narrowing of capillary lumina with endothelial cell proliferation.[120]

Central nervous system

Whether the brain is affected primarily by systemic sclerosis is conjectural: there is no pathological support for this suggestion.[41] Behavioural disorders are described,[194] attributed to cerebritis, but the effects of therapy must be remembered. As in systemic lupus erythematosus, high doses of corticosteroids may themselves cause the psychosis of hyperadrenocorticalism. Cranial nerve lesions and in particular, trigeminal sensory neuropathy have been encountered[176] but the pathological mechanisms are speculative. The VIIth, VIIIth and IXth cranial nerves may also be affected. The site of involvement is often peripheral but a reduced number of trigeminal sensory roots has been identified surgically, embedded in a thickened arachnoid.[75] Where retinopathy is recognized in systemic sclerosis, there is evidence that the light- and transmission electron microscopic appearances in the retina are indistinguishable from those of accelerated hypertension.[3]

Endocrine system

Disorders such as hypothyroidism have drawn attention to the relative frequency of signs of autoimmunity. Thyroid fibrosis was recognized in 14 per cent of 56 cases of systemic sclerosis by comparison with 2 per cent of controls.[63]

Aetiology

The cause(s) of systemic sclerosis and of localized scleroderma and morphoea remain unknown. However, there are a number of syndromes, caused by physical or chemical agents, which have sufficient similarity to systemic sclerosis to throw some light on the mechanisms by which the vascular connective tissue may respond in this disorder.

Infection

There is no firm support for the role of an infective agent.

Chemical agents

Particular interest has been directed towards the role of chemical agents; there is nothing to support any feature in common with retroperitoneal fibrosis (see p. 304). However, there is an association of a triad of symptoms and signs (Raynaud's phenomenon, scleroderma-like skin lesions and acroosteolysis) with chronic, occupational exposure to vinyl chloride. Vinyl chloride is the basis of the polyvinyl chloride polymer industry. Vinyl chloride is carcinogenic, promoting angiosarcoma of the liver; it has been shown to be capable of causing scleroderma-like changes in dermal blood vessels.[116] Another agent, known particularly for its ability to cause fibrosis, is bleomycin; it is employed in the chemotherapy of lung cancer and may provoke reversible skin lesions resembling those of systemic sclerosis.[53] There is also interest in the dermal fibrosis and fibrous myopathy produced by the analgesic pentazocine[158] and in the scleroderma-like changes encountered in children with phenylketonuria.[81]

Adjuvants

A reversible adjuvant disease has been described in which systemic disorders resembling those of systemic sclerosis, rheumatoid arthritis or systemic lupus erythematosus have been induced in women by the introduction of substances like paraffin and silicone in augmentation, cosmetic procedures.[96,125]

The suspected association between occupational exposure to silicates and the pulmonary fibrosis of systemic sclerosis is discussed above (see p. 700).

Pathogenesis

Against a background of a heritable predisposition which contributes to a pattern of abnormal humoral and cell-mediated immune responsiveness, there develop peripheral and visceral microcirculatory disorders accompanied by, or causing, excess connective tissue synthesis.

Heredity

The familial clustering of rare cases has drawn attention to the role of inheritance. Early observations demonstrated an association between HLA-B8 and rapidly progressive, widespread disease.[77] More recently, it has been established that there is a raised incidence of the haplotype A1-CW7-B8-Bfs-C4AQ0-B1-DR3. However, there are geographical/racial differences.[14] DRw3 may be more frequent in patients who develop the CREST syndrome. Overall, it can be concluded that there is evidence of a weak genetic influence on the origin of systemic sclerosis: one consequence could be the emergence of clones of T-cells reactive to autologous endothelium, collagen and RNA.

Immunity

(See Chapter 5)

On the basis of the concept summarized on p. 251, it appears reasonable to propose that clones of T-cells with immunity for previously unrecognized tissue-associated antigens, such as collagen, cutaneous connective tissue or RNA, become established. The defect in T_S-cell or enhanced T_H-cell function that renders this possible may be inherited, as suggested; however, the disorder may be acquired. In the same way, clones of B-cells can emerge with specificity for nuclear and/or nucleolar antigens. Under any of these circumstances, activated T-cells can react with tissue-associated antigens, releasing lymphokines. Either these cytokines or cytotoxic T-cells can then induce microvascular changes such as endothelial injury, increased vascular permeability or intimal myofibroblast proliferation. Cytokines chemotactic for fibroblasts attract neighbouring cells which accumulate at local T-cell reaction sites; simultaneously, fibroblast growth is initiated by the fibroblast growth factors (p. 302). Other cytokines provoke collagen synthesis by fibroblasts (see p. 18). Mononuclear macrophages, stimulated by cytokines, release interleukin-1, causing fibroblast mitosis and the increased synthesis of proteoglycan, collagenase and fibronectin. There is enhanced spontaneous interleukin-1 production.[187a] The entire sequence enhances matrix formation, remodelling and fibrosis (see Chapter 8). Such interactions could be perpetuated if, as has been suggested, the putative antigen(s) were fibroblast-associated antigens such as collagen.

Vascular disease

Evidence of circulatory impairment in systemic sclerosis is so frequent that it is natural to ask whether this is fundamentally not a vascular disorder.[84] The obstruction to digital arterial blood flow correlates poorly with the severity of scleroderma but there is mounting evidence that a microcirculatory abnormality underlies not only the dermal but also the visceral lesions. It is not immune complex-mediated.

Microcirculatory injury centres on endothelial cell proliferation and basement membrane thickening. Endothelial cells may be injured by serum factors. Raised levels of Factor VIII/von Willebrand factor antigen, of circulating platelet aggregates and β-thromboglobulin are indirect indices of endothelial cell disorder. Fibrin deposits may be cleared inefficiently. Blood viscosity is increased and serotonin metabolism impaired. Together, these disorders offer acceptable explanations for the circulatory abnormalities of systemic sclerosis.

Fibrosis

Following a series of early changes, some inflammatory, systemic sclerosis is dominated by fibrosis (p. 301) and excess collagen synthesis.[157] Calcinosis (p. 402) may follow. Isolated dermal fibroblasts in culture export types I, III and V collagen, dermatan sulphate proteoglycan and fibronectin at an accelerated rate[145] although cell growth and morphological characteristics are normal. Their behaviour, influenced genetically, may be a response to the impaired microcirculation: it is not thought to be a primary manifestation of the disease. Serum factors and mast cells promote fibrosis: they may be activated by ischaemia.

There is intense interest in the possibility that dermal fibroblasts (see p. 16) may synthesize excess and/or abnormal collagen and proteoglycan, partly for genetic reasons, but partly in response to local abnormalities of the circulation.[22] The collagen accumulation may also be enhanced immunologically (see p. 251)[27] and in response to mast cell activity (see p. 304). Transmission electron microscopic studies show that there is an abnormally large number of thin 20–40 nm diameter collagen fibrils with an incomplete cross-banding pattern and embryonic 'beaded' filaments.[69] Early analyses suggested that the concentration of collagen measured chemically[57,129] and the concentration of 'soluble dermal protein' were normal. Nevertheless, the concentration of collagen-bound hexosamines, but not hexoses, was increased, an observation suggesting abnormal assembly, cross-linking, or conformation of collagen fibrils, rather than altered collagen synthesis. In the initial stages of the disease there is excess deposition of type III collagen,[55] whereas in the later, fibrotic stage type I only or types I and III are revealed. The type III probably represents reticular fibrils.[56] The ratio of type I to type III procollagen, may, however, be normal.[182]

Investigations of the skin lesions led to extensive studies of dermal fibroblasts in culture. Some observers, surprisingly, failed to demonstrate increased collagen synthesis.[143] However, other, detailed observations showed that cells from systemic sclerosis skin accumulated more collagenase-sensitive protein and labelled Hyp than normal when monolayers from the corresponding portion of normal and abnormal biopsies were compared.[22] An increased response to ascorbic acid was demonstrable. The fibroblasts also secreted excess glycosaminoglycan.[21] Further insight into the disorder of dermal collagen synthesis has been gained by studying the effects of therapeutic compounds on fibroblast cell lines. An oestrogen, cyclofenil, had no influence on glycosaminoglycan or collagen synthesis.[94] By contrast, increased collagen synthesis could be inhibited by recombinant interferon-γ in a dose-dependent response.[161] The inhibitory effect was exerted both upon type I and upon type III procollagen mRNA.[88] Moreover, the increased urinary excretion of high molecular weight peptides derived from young collagen and of glycosaminoglycan was lessened by treatment with compounds such as penicillamine[15] which increases the proportion of soluble dermal collagen.[144]

A distinct approach raises the possibility that, in the systemic sclerosis dermis, populations of fibroblasts are selected that are induced by the influence of circulating growth factors to synthesize abnormally large amounts of collagen.[18] Inhibition of DNA synthesis and thus of collagen mRNA is impaired.[98] The net effect is the defective regulation of connective tissue synthesis.[187b]

Experimental models of systemic sclerosis

Although there is no exact replica of the human disease, a number of animal models have thrown light on aspects of pathogenesis.

The tight-skin (TSK) mouse

The TSK mouse strain,[87a] isolated in 1967 and described in 1976,[65] forms excessive quantities of dermal and subcutaneous collagen. The disorder is transmitted as an autosomal dominant characteristic. There is an increase in the proportion of soluble collagen. The dermis is of increased thickness and tensile strength and adheres unusually firmly to the subcutis.[135] By comparison with normal mice, the skin collagen contains fibres that are extremely wide and have very irregular cross-sectional profiles.[89] The average skin collagen content is approximately 2.5 times that of normal animals on a dry weight basis; the DNA content is not abnormal. This unusual genetic model illuminates mechanisms of skin collagen synthesis but there is no underlying immunological anomaly, no microvascular disorder and no inflammatory response so that it is an incomplete replica of systemic sclerosis.

Avian scleroderma

Bernier (quoted by van de Water and Gershwin[183]) described an inherited disorder of young white leghorn chickens in which necrosis of the comb was followed by polyarthritis, skin swelling and induration. Although the oesophagus, gut, lung, kidney, heart and testis were often abnormal, the lesions were distinct from those of human systemic sclerosis and neither in the viscera nor in the skin was there evidence of vascular disease. However, the recognition of serum extractable nuclear antigen, anti-DNA, anticytoplasmic antibodies and speckled antinuclear antibodies, suggested some features in common with systemic connective tissue disease in man, and the avian disorder continues to attract interest.[183,183a]

Prognosis

There is no satisfactory, specific treatment. Gradations of systemic sclerosis occur, and a simple classification, based on the extent of the skin changes, and on the presence of visceral lesions, can be helpful in prognosis.[5] Survival rates are best for younger, female patients without visceral involvement, poorest for older, male patients with acute, diffuse visceral disorder. The mean age at death for systemic sclerosis in the USA in 1977 was: for White males, 57.7 years, for White females, 58.2 years; for non-White males, 50.0 years, for non-White females, 49.3 years.[73]

The POEMS syndrome

The POEMS syndrome[4,171] is a rare variety of plasma cell dyscrasia with polyneuropathy, organomegaly, endocrinopathy, M protein and skin lesions: it is encountered principally in the Japanese. There is anasarca. Fam, Rubenstein and Cowan[50] emphasized the similarity to systemic sclerosis and other multisystem diseases in a case which presented with a solitary immunoglobulin A-secreting osteosclerotic myeloma of the sacrum.

The pathogenesis of the POEMS syndrome is obscure: it is possible that, as in amyloid neuropathy (see p. 431), the λ-M-proteins may be responsible for neural changes and organomegaly. There are features that bring to mind the endocrine properties of osteoid osteoma and of neoplastic osteomalacia[110] and are reminiscent of generalized (polyostotic) fibrous dysplasia. Histological studies have failed to provide evidence of any consistent underlying microangiopathy.

Eosinophilic fasciitis

Clinical features

In 1974 Shulman[172] drew attention to a 'diffuse fasciitis with eosinophilia' (eosinophilic fasciitis) in which painful swelling and induration of the skin[170] and soft tissues of the upper and/or of the lower extremities, and perhaps of the trunk, were followed quickly by the development of joint contractures due to involvement of the nearby tissues.[25,59,127] Although eosinophilic fasciitis has been said to be a variant of systemic sclerosis,[76] with which it may coexist,[36] Raynaud's phenomenon and visceral disease are absent.

Pathological changes

Histological studies of biopsy material reveal great local thickening of fascia due to collagen accumulation.[1] The fascia contains lymphocytes and plasma cells. In 19 of 53 cases the dermis as well as the fascia was involved[127] and in 33 of 53 cases, the skeletal muscle was affected. Indeed, the inflammatory lesions may be more severe in skeletal muscle than in fascia.[92] The epidermis is spared. In those cases with dermal involvement, there is a sparse lymphocytic and plasma cell infiltrate, with occasional eosinophils. Although the dermal collagen and appendices are usually normal, there is thickening and collagenization of the subdermal (subcutaneous) fascia. Perivascular localization of mononuclear cells is not conspicuous. However, vascular endothelial cell proliferation accompanies an infiltration of the vessel wall by inflammatory cells, with occasional obliteration of the lumen; lymphocytes and plasma cells are also encountered in moderate excess in the underlying muscle and muscular fascia. The description of instances of granulomatous vasculitis (see p. 633) raises the possibility that eosinophilic fasciitis should be classified as a vasculopathy.[105]

Immunofluorescent investigations of 14 biopsies have been reported[127] (Table 17.4). There is not yet a consensus of opinion on the significance of these findings. Transmission electron microscopic investigations confirm the increased number of small, irregularly periodic connective tissue fibres distinct from nearby, normal collagen. An inflammatory cell infiltrate of lymphocytes, histiocytes and occasional plasma cells and thickening of the vascular basement membrane are demonstrable.[37]

The cause of eosinophilic fasciitis remains unknown.

Table 17.4

Immunocytochemical studies of immunoglobulin deposits in eosinophilic fasciitis[127]

Location	Deposits
Deep fascia and septa of underlying muscle[152,155]	IgG, C3
Dermal epidermal junction	IgM
Collagen bundles[9]	IgG
Around blood vessels in deep dermis and fascia[59]	IgM C3 (trace)
Basement membrane[1,107,127,170]	IgG IgM IgG, C3, C4 No immunofluorescence staining

However, the illness is often preceded by an episode of unusually severe physical exertion. Hypergammaglobulinaemia (IgG) and the detection of immunoglobulin and C3 in the inflamed deep fascia, all suggest that a humoral immune mechanism is involved.[152,155] Concurrence with sarcoidosis is coincidental.[33] The disorder appears to be self-limiting with a tendency to remission. Symptomatic relief is given by corticosteroids.

Toxic oil and eosinophilia–myalgia syndromes

Further light has been shed on the nature of eosinophilic fasciitis by epidemics of two new diseases with similar clinical features and pathological findings. Both conditions have been associated with the accidental or deliberate ingestion of dietary contaminants or additives.

Toxic oil syndrome[92a] The use in cooking of industrial grade denatured rapeseed oil in place of olive oil led, in 1981, to the sudden outbreak in Spain of an illness in which the early signs of toxicity were followed by the later development of a scleroderma-like or progressive neuromuscular disease. By June 1982 there had been 19 828 cases with 315 deaths. The early occurrence of fever, headache, dyspnoea, cough, skin rashes, muscle pain and eosinophilia was followed after one month by oedema of the extremities, after a further one to two months by an inexorable neuropathy with signs resembling those of systemic sclerosis, including joint contractures. When lung disease was recognized, immunoglobulin E levels tended to be elevated. The serum creatine kinase levels were normal but the aldolase levels were elevated. Many of the recorded histopathological characteristics resembled those of eosinophilic fasciitis and the eosinophilia–myalgia syndrome (see below).

Eosinophilia–myalgia syndrome[163a] More recently, L-tryptophane has been linked to the onset of a further eosinophilic fasciitis-like disorder.[172a] L-Tryptophane has been used safely for many years as a mild sedative, anxiolytic or antidepressive drug. Suddenly, in late 1989, cases began to appear of patients with severe, diffuse but mainly proximal muscle myalgia, fever, cough, dyspnoea, skin rash and oedema. An analogy was drawn with the species-specific pulmonary disease caused by α, L-tryptophane in cattle.[26a] Anaemia, the presence of anomalous antibodies, normal creatine kinase but raised aldolase levels and pronounced eosinophilia, were features recalling those of the toxic oil syndrome. By February 1990, 1305 cases had been recorded; there were 15 deaths.

Among the causes of death were a Guillain-Barré-like syndrome of ascending polyneuropathy. The most characteristic histopathological signs were a 'non-specific', subacute inflammatory cell infiltrate of the transfascial epimysial and perimysial connective tissues.[22a] Plasma cells, lymphocytes and histiocytes were seen in equal numbers.

Particular significance attaches to the location of these changes; to new blood vessel formation in the epimysial tissues; to proteoglycan neosynthesis; to collagen deposition in the perimysial connective tissue; to the dermatomyositis-like changes in the peripheral zones without myofibre degeneration; and to the perimysial venulitis.

Mixed connective tissue (anti-ENA) disease

Mixed connective tissue disease is an immunological entity. Sharp *et al.*[168] described a group of patients with signs and symptoms like those of rheumatoid arthritis, systemic lupus erythematosus, systemic sclerosis or polymyositis who, they thought, were suffering from a 'distinct rheumatic disease'. All the patients possessed antibodies to a specific nuclear antigen that was susceptible to digestion by RNAase, i.e. that was an extractable nuclear antigen (ENA). The patients with anti-ENA, particularly those with only high titre anti-ENA, were clinically different from those with systemic lupus erythematosus, i.e. those with anti-DNA. Twenty-two of the original patients were reexamined in 1980.[131]

The concept of mixed connective tissue disease is controversial. It is particularly difficult for tissue pathologists for three reasons: first, histopathologists are wholly familiar with the clinical 'overlap syndromes' in which the signs and symptoms of a patient with systemic lupus erythematosus include some that are indistinguishable from, say, rheumatoid arthritis or systemic sclerosis; they are also aware that, as Pagel and Treip[138] showed, a patient dying with rheumatoid arthritis may display microscopic evidence of vasculitis just as a patient dying with systemic lupus erythematosus may show signs of synovitis. Second, they are aware that the formation of a particular class of antibody, e.g. IgG rheumatoid factor is, by itself, insufficient grounds for postulating the existence of an 'antirheumatoid factor disease'. Third, they are faced with the semantic difficulty that mixed connective tissue disease may be interpreted to mean 'disease of a mixed connective tissue' such as synovium, which often contains loose and dense connective tissue and adipose tissue and always has a synoviocytic layer and blood vessels. It is as though a class of patients was postulated, all synthesizing anticardiolipin antibody and therefore all suffering from 'anticardiolipin antibody disease'. But few would accept that the standard serological tests for syphilis are positive only in *Treponema pallidum* infection.

Clinical features

There is no single, unique diagnostic criterion. There is a 8:1, female:male sex ratio but no racial predominance in Blacks or Caucasians. The majority of cases are identified in the age group 10–30 years. No characteristic genotype is known. The clinical signs often resemble systemic lupus erythematosus, systemic sclerosis or polymyositis. There is fever, fatigue, weight loss and, almost always, skin disorders, particularly Raynaud's phenomenon, with swollen hands. The sicca syndrome is recognized. There is arthralgia and often arthritis: the changes may resemble those of psoriatic arthritis mutilans or rheumatoid arthritis. Rheumatoid arthritis-like nodules have been seen. However, the clinical signs of mixed connective tissue disease are sometimes sufficiently distinctive to allow the recognition of cases in advance of antiribonucleoprotein antibody formation.

Immunological changes
(Table 17.5)

Many cases of systemic lupus erythematosus and other systemic connective tissue diseases have antibodies reacting with ENA but the RNAase sensitivity of the ENA of mixed connective tissue disease is, so far, unique. Nuclear chromatin has multiple components including DNA, RNA histones and acidic proteins. Antinuclear antibodies are correspondingly diverse. One nuclear antigen is a glycoprotein and the anti-m antibody of systemic lupus erythematosus reacts with this antigen. Other antibodies react with a second distinctive nuclear antigen, a ribonucleoprotein designated Mo. This antigen is now known to be identical with the ENA of Sharp *et al.*[168]

Clinically, free serum ribonucleoprotein has been found in mixed connective tissue disease and circulating immune complexes have been identified. Anti-ssDNA antibodies are frequent. There may be hyperglobulinaemia, hypocomplementaemia, positive tests for rheumatoid factor and lymphocytotoxic antibody. The standard serological tests for syphilis may be falsely positive. In skin biopsies, epidermal nuclei may respond positively to immunofluorescent tests for antiribonucleoprotein antibody.

Table 17.5

Immunopathological features of mixed connective tissue disease[8]

Hypergammaglobulinaemia
Very high titres of anti-RNP antibody, susceptible to RNAase
Immune complex deposition in renal tissue
Circulating immune complexes
Hypocomplementaemia
Lymphocytotoxic antibody
Lymphoplasmacytic infiltration of synovium, intestine, heart, liver, skeletal muscle, salivary gland, lung

Pathological changes

There are few pathological reports of cases in which all the organs have been fully examined microscopically. Of 15 children (mean age 10.7 years, range 4 to 6 years) four died.[173] The histopathological changes recalled those of systemic sclerosis but there was less fibrosis. A tendency to intimal thickening of the coronary, pulmonary, and renal arteries and the aorta was observed with hyaline replacement of muscle in the gastrointestinal tract. In mixed connective tissue disease, renal disorders are less severe than in systemic lupus erythematosus although as many as 25 per cent of cases of mixed connective tissue disease have renal disease. The findings are usually those of membranous glomerulonephritis.[11,139] Immune complex clearance may be ineffective or the immune complexes themselves may differ in reactivity with Fc receptors or display an increased rate of Fc receptor generation. Immunofluorescence microscopy demonstrates IgG, C2 and C4 in granular glomerular deposits; electron microscopy confirms that the deposits are subepithelial. Occasionally, endarterial intimal fibromuscular hyperplasia of low cellularity is seen and severe hypertension can result. It is likely that the renal disorders are influenced by the number and activity of Kupffer and splenic sinusoidal cells. Little is known of the morphological state of these cells or of the histopathology of the lymph nodes.

Respiratory system

These include interstitial lung disease;[42a,189] pleurisy, sometimes with effusion; small zones of lower segmental consolidation; interstitial fibrosis; and pulmonary hypertension with arteriolar intimal hyperplasia or plexogenic angiopathy. Opportunistic infections may complicate treatment and can be fatal.

Cardiovascular system

Cardiac disease includes pericarditis, and myocarditis,[188] perhaps with heart block. There is an association between congenital complete heart block and maternal connective tissue disease: antibodies to the ribonucleoprotein antigen Ro(SS-A) are present in the majority of these mothers. Aortic insufficiency is a rare finding although verrucous endocarditis has not yet been recognized. Microscopically, the changes include perivascular lymphocytic and mononuclear macrophage infiltration; and the replacement of cardiac muscle cells by fibrous tissue.

Osteoarticular system

Synovitis is common[10] and the anatomical changes may simulate rheumatoid arthritis or psoriatic arthropathy. Tenosynovitis has been described and rheumatoid-like nodules are occasional. The development of marginal erosions, like those of rheumatoid arthritis, distinguishes the arthritis of mixed connective tissue disease from that of systemic lupus erythematosus (see Chapter 14). In both disorders aseptic necrosis of bone is encountered.

Skeletal muscle

Muscle cell degeneration with a mononuclear cell infiltrate accompanies IgG and IgM sites in intermysial blood vessels and around the plasmalemma.[137] There is no vasculitis. The appearances must be differentiated from those produced by steroid treatment. Occasionally the changes are indistinguishable from polymyositis.

Alimentary system

In the alimentary tract, patterns of altered oesophageal motility are found resembling those of systemic sclerosis;[132] little is known of the histological appearances. Intestinal vasculitis and malabsorption are described and chronic active hepatitis may be an associated change.

Central nervous system

In contrast to systemic lupus erythematosus, cerebral disease is not a serious prognostic feature; psychosis and convulsions are very uncommon. Cranial nerve (V nerve) neuropathy,[184] vascular disease, aseptic meningitis and transverse myelitis are among the disorders reported.

Pathogenesis

Prognosis

The cause of mixed connective tissue disease remains unknown: it may not be a single disease but, like rheumatoid arthritis or systemic lupus erythematosus, may represent a single form of immunological response to a variety of unidentified agents. There is no known genetic predisposition. The immunopathological features (Table 17.5) are strong indicators that mixed connective tissue disease, like systemic lupus erythematosus, is of autoimmune origin.

Mixed connective tissue disease responds well to steroids. However, the prognosis now appears less favourable than had at first been anticipated. The course may, nevertheless, be benign—the signs and symptoms resembling those of rheumatoid arthritis or systemic sclerosis rather than systemic lupus erythematosus, but renal, pulmonary and cardiac disease may be severe or fatal. Three relatively frequent causes of death (Table 17.6) distinguish mixed connective tissue disease from systemic lupus erythematosus: rapidly progressive pulmonary hypertension, myocarditis, and renovascular hypertension. Severe renal disease is infrequent but opportunistic infection, in part the consequence of steroid treatment, is common.

Table 17.6

Causes of death in mixed connective tissue disease[8]

Suicide	Pulmonary hypertension
Opportunistic infection	Renovascular hypertension
Myocarditis	Mesenteric vasculitis
Myocardial infarction	Pulmonary embolism
Gastrointestinal haemorrhage	Ruptured abdominal aortic aneurysm
Cerebral haemorrhage	'Shock' lung
	Unexplained ventricular fibrillation

Relapsing polychondritis

Definition
(Table 17.7)

Polychondritis[71] (polychondropathy,[82] relapsing polychondritis[109, 140]) is a rare, systemic disorder of the connective tissues of the ears, synovial and cartilagenous joints,

Table 17.7

Diagnostic clinical criteria for polychondritis[71]

An earlier definition (A) has been superseded (B)

A Recurrent inflammation involving two or more cartilagenous sites, at least one being of an organ of special sense, together with compatible diagnostic findings from biopsy of an affected cartilage.[43]

B 1. Bilateral auricular chondritis
2. Non-erosive, seronegative inflammatory polyarthritis
3. Nasal chondritis
4. Ocular inflammation
5. Respiratory tract chondritis
6. Cochlear and/or vestibular dysfunction
and
Confirmatory cartilage biopsy[109]

The definitive diagnosis is established when three or more of these clinical features are present; biopsy may then be desirable but is not essential.

nose, eye, upper respiratory tract and cardiovascular system. The cause of polychondritis remains uncertain but both humoral and cell-mediated immunological mechanisms play a part in its natural history. The effector pathway is the degradation of individual connective tissue components by enzymatic means.

Clinical features

The clinical presentation (Tables 17.7 and 17.8) is often dramatic, and the disease disabling, disfiguring and possibly fatal. Two classical examples have been analysed recently.[28] In one, there was severe, sterile inflammation of the cartilage-supported part of the left ear with pyrexia, leucocytosis and an elevated erythrocyte sedimentation rate. These changes were followed by respiratory stridor, diplopia and eye pain, tender nasal cartilage and compromise of the airway by glottic swelling and subglottic narrowing. In spite of eye and lung disease, there was a good response to substantial doses of corticosteroids. In other instances, arthropathy is the presenting sign, and aortic insufficiency, aneurysm, floppy mitral valve and pericarditis are among the cardiovascular disorders that appear in one case in four.

Systemic connective tissue diseases such as vasculitis, rheumatoid arthritis or systemic lupus erythematosus coexist with polychondritis, and it may not be easy to determine whether they are predisposing agencies or coincidental. When the features of Behçet's disease are present, the name MAGIC (mouth and genital ulcers with inflamed cartilage) (p. 715) syndrome has been proposed.[54] Relapsing polychondritis may also be a result of hydralazine (see p. 594).[40] A simultaneous destructive arthropathy may be present.[64] The differential clinical diagnosis of polychondritis is very wide and ranges from local trauma or infection as causes of the external ear disease; the Marfan syndrome, medial arterial mucoid change and syphilis as causes of the arterial changes; and vasculitis, Wegener's granuloma[113] and midline granuloma as agents of tracheal and bronchial cartilage collapse and of upper respiratory disorder. There can be multifocal neurological disorder.[190]

At least 300 cases have been reported. Males and females are affected equally; the majority occur between 20 and 60 years of age. Early clinical signs can be used to predict mortality: the 5-year and 10-year probabilities of survival after diagnosis in this series were 74 and 55 per cent, respectively.[122]

Table 17.8

Features of polychondritis presenting during the course of disease[71]

Disorder	Per cent
Inflammation of external ear	85
Arthropathy	76
synovial	68
costochondral	31
Inflammation of nasal cartilage	66
Inflammation of eye	59
Laryngotracheobronchial disease	51
Inner ear involvement	42
Cardiovascular disease	24
Skin disease	16

Pathological changes

Nine autopsies were listed by Dolan, Lemmon and Teitelbaum;[43] many others have been recorded[72,78,117,141] but in recent years the treatment of polychondritis with corticosteroids has often been effective and the material available for histological study has been limited to biopsy of the external ear. The morbid anatomical findings are often very diverse.

The ear, nose, tracheobronchial and laryngeal cartilages are soft, distorted, inflamed and collapsed. The ear is of misshapen, cauliflower form, the nose saddle-shaped and disfigured. The anatomical changes in the synovial joints are rarely mentioned. There may be pulmonary consolidation, collapse or bronchopneumonia. Aneurysms of the aorta and of its main branches are recognized in the ascending thoracic and abdominal parts[78] and the subclavian artery has been affected.[31] Aortic insufficiency, with dilatation of the valve ring, mitral and tricuspid insufficiency, floppy mitral valve, endocarditis and infarction are among other changes that have been described.[71] Vasculitis of the larger arteries and giant-cell arteritis, polyarteritis, Wegener's granuloma and Takayasu's disease are recorded but it is often difficult to determine whether these are integral parts of the polychondritis syndrome.

Ear

Acute inflammation of the external ear is characterized by a vigorous perichondritis. The small blood vessels are congested. The perichondrium is infiltrated by large mononuclear cells and by varying numbers of plasma cells and lymphocytes. There is marginal destruction of the substance of the auricular cartilage, cartilage necrosis, the formation of young, vascular reparative tissue and a progressive but remitting tissue disorganization (Fig. 17.18). The cartilage becomes fragmented into residual islands of elastic cartilage separated by bands of granulation tissue. The junction between basophilic, metachromatic cartilage matrix and the perichondrium is ill-defined.

Transmission electron microscopy[67,124,169] reveals chondrocyte hypertrophy and necrosis.[29] The cells contain many lysosomes and lipid droplets; they retain glycogen. Where necrosis has occurred, matrix vesicles, residual granules and cell debris can be seen and electron-dense, amorphous material may lie at the cartilage margin. Elastic fibres are fragmented[44] and stain poorly. Earlier reports emphasized defective staining of the amorphous central component and retention of a distinct peripheral margin of microfibrils.[67,169] However, Dryll *et al.*[44] reported various

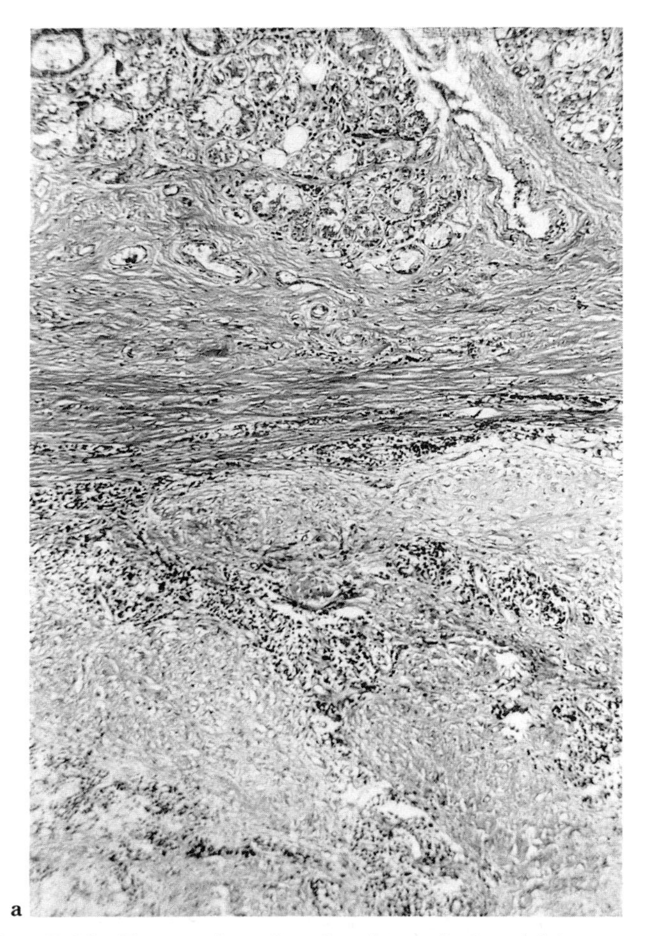

Fig. 17.18 External ear in relapsing polychondritis.
a. Margin of ear. The subcutaneous sweat glands and small blood vessels (at bottom) are sparsely infiltrated with lymphocytes. The perichondrial collagenous connective tissue (at centre) borders a zone of cartilagenous destruction (at top). The residual cartilage matrix is split into septa within which are numerous lymphocytes and plasma cells together with smaller numbers of polymorphs. (HE × 110.) **b.** Perichondrial inflammation is shown (at centre). The edge of the aural elastic cartilage is being directly destroyed by a cellular exudate in which polymorphs are prominent. (HE × 110.)

degrees of fragmentation ranging from the peripheral release of both fibrillar and amorphouse components to virtually complete fragmentation of the amorphous elastin and phagocytosis by chondrocytes of disintegrated fibrillar material.

Collagen fibrils appear less severely affected than elastin,[44] loose aggregates of banded material with a 90 nm periodicity corresponding to the sites of resorption of collagen associated with elastin fibres. These observations, in which advantage was taken technically of the enhanced contrast provided by tannic acid, appear to be of importance because of the suggestion (see below), derived from experimental studies, that an immunological reaction against collagen may be a primary feature of the human disease. The transmission electron microscopic data[44] does not support this concept.

Histochemistry

The published evidence is limited to the external ear. The loss of metachromatic matrix material is a conspicuous feature of early cartilage disease.[43,91] The loss is of glycosaminoglycan.[29,87] The Alcian blue-critical electrolyte concentration technique (see p. 54) confirms loss of all classes of glycosaminoglycan, particularly the most highly sulphated compounds.[78] There is a comparable depletion of glycosaminoglycan in parts of the aortic wall near the aneurysms, without regeneration. The nature of the change has been confirmed chemically.[78] There is little mention of polarized light microscopy or of elastic-van Gieson stains. There is no description of microassays of affected ear tissue for the amino acids characteristic of collagen or elastin nor, in the case of the matrix proteoglycan, descriptions of biochemical analyses.

Biopsy

The ear is the preferred site for biopsy. The availability of an adequate, representative sample of auricular cartilage can establish a definitive diagnosis.[181] Since the disease process may respond to corticosteroids and since cartilage destruction may remit, it is essential to obtain a sufficiently extensive tissue block without disfigurement. The ear lobule is not inflamed and should be avoided. Because of the possible need for special studies of the auricular connective tissue by immunofluorescence microscopy, by histochemistry and by transmission electron microscopy, parts of the biopsy should be quickly frozen; small portions should be fixed in buffered glutaraldehyde. Avidin–biotin complex demonstrations of anticartilage, anticollagen and antiproteoglycan antibodies may then be attempted on formalin-fixed, paraffin-embedded sections.

Aetiology and pathogenesis

There is no known genetic predisposition.

The precipitating agent in polychondritis may be chemical (e.g. alcoholism), immunological (e.g. autoimmune connective tissue disease), bacterial (e.g. diphtheria), viral (e.g. rubella), thermal, traumatic or radiation.

The presenting features of sterile, recurrent inflammation of cartilage, joints and vascular tissues are unique. Although tests for rheumatoid factor (17 per cent), antinuclear antibody (18 per cent), lupus erythematosus cells (8 per cent) and serological tests for syphilis (6 per cent) are sometimes positive, they are often explicable either by the coexistence of disorders such as rheumatoid arthritis and systemic lupus erythematosus, or are 'false-positives'. The demonstration that the urinary loss of glycosaminoglycans parallels disease activity[43,91] is no more than an index of tissue proteoglycan degradation—just as anaemia, leucocytosis and the raised ESR are markers of widespread, prolonged inflammation.

Of much greater interest is the demonstration that serum IgG, labelled with a fluorochrome, binds to the periphery of cartilage lacunae[43] and that anticartilage antibody is present in the patient's serum.[78] In some reports[72] anticartilage antibodies were not demonstrated but these authors found that human cartilage antigens were mitogenic when tested with peripheral blood lymphocytes. The serum may contain antibodies to type II collagen[121] but not to other collagen types or to proteoglycan; immune complexes may be present.[74] The presence in the serum of antinative collagen type II antibody lent weight to the view that specific injury to collagen, not a broader pattern of injury to other connective tissue constituents, might be the mechanism by which cartilage is injured. This concept has received support from experiments in which Sprague-Dawley rats were sensitized with native type II chick collagen, dissolved in acetic acid and emulsified in incomplete Freund's adjuvant.[112] Following sensitization, rats often (69 per cent) developed synovitis; small numbers of animals also displayed unilateral or bilateral foci of external ear inflammation accompanied by vasculitis. The lesions were 'multifocal, nodular chondritis', with cartilage necrosis and granulation tissue in which mononuclear cells (histiocytes) predominated. Whether or not ear lesions developed, circulating antinative type II collagen antibodies were present. When the ear lesions of two rats that had developed unilateral chondritis were tested by direct immunofluorescence, fibrinogen and C3 were found in the inflammatory exudate. Two of three rats with type II collagen-induced chondritis also developed cell-mediated hypersensitivity assessed by demonstrating leukocyte migration inhibition factor *post mortem*.

Although the experimental evidence is incomplete, by

analogy there is sufficient reason to speculate that autoimmunity to type II collagen could be the most important mechanism by which injury to hyaline, elastic and nonarticular cartilage is caused in human polychondritis. However, the view that anticollagen antibodies directly injury cartilage[112] is not easily reconciled with transmission electron microscopic studies of the human disease and does not explain its aetiology. Ultrastructural evidence suggests that, in the injured cartilages, proteoglycan degradation precedes collagen injury. This is the usual sequence in inflammatory articular disease. In experimental rabbit caragheenin arthritis, for example, loss of the metachromatic matrix is an early sign of advancing articular injury; and the labile, replaceable populations of proteoglycan are susceptible to early change in experimental rat, guinea-pig and dog joint disease. The mechanism of proteoglycan degradation and of any subsequent collagen injury is almost certainly enzymatic. In human disorders such as rheumatoid arthritis, the role of neutral collagenase has attracted interest (see p. 537), and the possibility of analogous actions in polychondritis, perhaps by the loss or inactivation of enzyme inhibitors, invites consideration. An early role for interleukin-1 in polychondritis awaits exploration. The contribution of T-cell dysfunction has been suspected and it is significant that decreased T_H-cell and increased T_S-cell proportions have been recorded.[30]

An interesting analogy is with the reversible, floppy ear syndrome produced in rabbits by the intravenous injection of active or inactive crude papain proteinase.[177] Within a few hours, the ear cartilage of injected rabbits loses its metachromasia: glycosaminoglycan is degraded, and proteoglycan altered. Water is lost and the mechanical properties of the ear cartilage change. The loss of glycosaminoglycan can be replicated by the effects of large doses of vitamin A. Both papain and vitamin A can degrade cartilage matrix by altering lysosomal membrane permeability, releasing activated hydrolytic enzymes which then act upon the adjacent cartilage matrix (see p. 536). Alternatively, of course, the mode of action of papain upon rabbit ear cartilage may be direct, very small quantities of active enzyme gaining access to the tissues. Since inactivated enzyme can cause rabbit ears to droop in the same way as active enzyme, this explanation seems unlikely. The analogy with the animal disorder therefore suggests that polychondritis in man may be a result of the systemic liberation of agents that attack lysosomal membranes.

In a fawn-hooded strain of rat, bilateral auricular chondritis appeared to develop spontaneously.[146] The disorder, which was familial, was not evidently related to the platelet storage pool deficiency from which these rats suffered.

Prognosis

Involvement of the laryngotracheobronchial tract is ominous and many cases die from respiratory disease; loss of the supporting structure of the air passages causes defective ventilation, sometimes with secondary bacterial infection. There is a need for early diagnosis because tracheal collapse can occur within two months.[128]

The MAGIC syndrome: concurrent mouth and genital ulcers with inflamed cartilage disorders

Firestein *et al.*[54] noted the frequency with which the features of relapsing polychondritis (the inflammatory destruction of cartilage) and Behçet's disease (genital and oral ulceration) coincided; they described five patients with these coexistent features. All had arthritis and perichondritis; three had vasculitis. The limited evidence from these cases led to a search for other signs of overlap. It was concluded that the association between polychondritis and Behçet's disease was particularly close and that a shared immunological abnormality was likely, possibly centred on a reaction to an antigen of elastic material or elastin.

REFERENCES

1. Abeles M, Belin D C, Zurier R B. Eosinophilic fasciitis. A clinicopathologic study. *Archives of Internal Medicine* 1979; **139**: 586–9.
2. Abraham A A, Joos A. Pancreatic necrosis in progressive systemic sclerosis. *Annals of the Rheumatic Diseases* 1980; **39**: 396–8.
3. Ashton N, Coomes E N, Garner A, Oliver D O. Retinopathy due to progressive systemic sclerosis. *Journal of Pathology and Bacteriology* 1968; **96**: 259–68.
4. Bardwick P A, Zvaifler N J, Gill G N, Newman D, Greenway G D, Resnick D L. Plasma cell dyscrasia with polyneuropathy, organomegaly, endocrinopathy, M protein, and skin changes: the POEMS syndrome: report on two cases and a review of the literature. *Medicine* 1980; **59**: 311–22.
5. Barnett A J. Scleroderma (progressive systemic sclerosis): progress and course based on a personal series of 118 cases. *Medical Journal of Australia* 1978; **2**: 129–34.
6. Baron M, Lee P, Keystone E C. The articular manifesta-

tions of progressive systemic sclerosis (scleroderma). *Annals of the Rheumatic Diseases* 1982; **41**: 147–52.

7. Beck J, Anderson J R, Gray K G, Rowell N R. Antinuclear and precipitation autoantibodies in progressive systemic sclerosis. *Lancet* 1963: **ii**: 1188–90.

8. Bennett R M. Mixed connective disease and other overlap syndromes. In: Kelley W N, Harris E D, Ruddy S, Sledge C B, eds, *Textbook of Rheumatology* (3rd edition). Philadelphia: W B Saunders & Co., 1989: 1147–65.

9. Bennett R M, Herron A, Keogh L. Eosinophilic fasciitis. Case report and review of the literature. *Annals of the Rheumatic Diseases* 1977; **36**: 354–9.

10. Bennett R M, O'Connell D J. The arthritis of mixed connective tissue disease. *Annals of the Rheumatic Diseases* 1978; **37**: 397–403.

11. Bennett R M, Spargo B H. Immune complex nephropathy in mixed connective tissue disease. *American Journal of Medicine* 1977; **63**: 534–41.

12. Black C M, Duance V C, Sims T J, Light N D. An investigation of the biochemical and histological changes in the collagen of the kidney and skeletal muscle in systemic sclerosis. *Collagen and Related Research* 1983; **3**: 231–44.

13. Black C M, Myers A R, eds, *Systemic Sclerosis*. New York and London: Gower Medical Publishing, 1985.

14. Black C M, Welsh K I, Maddison P J, Jayson M I V, Bernstein R M, Pereira R S, Batchelor R. HLA antigens in scleroderma. In: Black C M, Myers A R, eds, *Systemic Sclerosis*. New York and London: Gower Medical Publishing, 1985: 84–8.

15. Blumenkrantz N, Asboe-Hansen G. Variation of urinary acid glycosamino-glycans and collagen metabolite excretion with disease activity in generalized scleroderma. *Acta Dermato Venereologica* 1980; **60**: 39–43.

16. Bollinger A, Jager K, Siegenthaler W. Microangiopathy of progressive systemic sclerosis. Evaluation by dynamic fluorescence videomicroscopy. *Archives of Internal Medicine* 1986; **146**: 1541–5.

17. Botstein G R, Le Roy E C. Primary heart disease in systemic sclerosis (scleroderma)—advances in clinical and pathologic features, pathogenesis, and new therapeutic approaches. *American Heart Journal* 1981; **102**: 913–9.

18. Botstein G R, Sherer G K, Leroy E C. Fibroblast selection in scleroderma—an alternative model of fibrosis. *Arthritis and Rheumatism* 1982; **25**: 189–95.

19. Bourgeois P, Cywiner-Golenzer Ch, Lessana-Leibowitch M, Kahn M F, de Seze S. Subcutaneous and tendinous nodules in scleroderma. *Revue du Rheumatisme* 1976; **43**: 85–91.

20. Branwell B. Diffuse sclerodermia: its frequency; its occurrence in stonemasons; its treatment by fibrinolysin, elevations of temperature due to fibrinolysin injections. *Edinburgh Medical Journal* 1914; **12**: 387–401.

21. Buckingham R B, Prince R K, Rodnan G P. Progressive systemic sclerosis (PSS, scleroderma) dermal fibroblasts synthesize increased amounts of glycosaminoglycan. *Journal of Laboratory and Clinical Medicine* 1983; **101**: 659–69.

22. Buckingham R B, Prince R K, Rodnan G P, Taylor F. Increased collagen accumulation in dermal fibroblast cultures from patients with progressive systemic sclerosis (scleroderma). *Journal of Laboratory and Clinical Medicine* 1978; **92**: 5–21.

22a. Bulpitt K J, Verity M A, Clements P J, Paulus H E. Association of L-tryptophane and an illness resembling eosinophilic fasciitis. Clinical and histopathologic findings in four patients with eosinophilia-myalgia syndrome. *Arthritis and Rheumatism* 1990; **33**: 918–29.

23. Burnstock G. Purinergic nerves. *Pharmacological Reviews* 1972; **24**: 509–81.

24. Cameron A J, Payne W S. Barrett's esophagus occurring as a complication of scleroderma. *Mayo Clinic Proceedings* 1978; **53**: 612–5.

25. Caperton E M, Hathaway D E. Scleroderma with eosinophilia and hypergammaglobulinaemia. The Shulman syndrome. *Arthritis and Rheumatism* 1975; **18**: 391.

26. Caplan H. Honeycomb lungs and malignant pulmonary adenomatosis in scleroderma. *Thorax* 1959; **14**: 89–96.

26a. Carlson J R, Dyer I A, Johnson R J. Tryptophane-induced interstitial pulmonary emphysema in cattle. *American Journal of Veterinary Research* 1968; **29**: 1983–9.

27. Cathcart M K, Krakauer R S. Immunologic enhancement of collagen accumulation in progressive systemic sclerosis (PSS). *Clinical Immunology and Immunopathology* 1981; **21**: 128–33.

28. Center D, Goodman M L. Case records of the Massachusetts General Hospital; a 41 year old man with diffuse tracheal narrowing. *New England Journal of Medicine* 1984; **313**: 1530–7.

29. Charpentier Y Le, Chomette G, Baudion D, Lecler J P, Blétry O, Godeau P, Auriol M. Relapsing polychondritis. Histochemical, histoenzymological, immunofluorescent and ultrastructural studies of ear cartilage in three cases. *Pathologie Biologie* 1980; **28**: 509–15.

30. Check I J, Ellington E P, Moreland A. McKay M. T-helper suppressor cell imbalance in pyoderma gangrenosum, with relapsing polychondritis and corneal keratolysis. *American Journal of Clinical Pathology* 1983; **80**: 396–9.

31. Cipriano P R, Alonso D R, Baltaxe H A, Gay W A Jr, Smith J P. Multiple aortic aneurysms in relapsing polychondritis. *American Journal of Cardiology* 1976; **37**: 1097–1102.

32. Cobden I, Rothwell J, Axon A T R, Dixon M F, Lintott D J, Rowell N R. Small intestinal structure and passive permeability in systemic sclerosis. *Gut* 1980; **21**: 293–8.

33. Cohen M D, Allen G L, Ginsburg W W. Eosinophilic fasciitis and sarcoidosis: a case report. *Journal of Rheumatology* 1983; **10**: 347–9.

34. Collins D H, Darke C S, Dodge O G. Scleroderma with honeycomb lungs and bronchiolar carcinoma. *Journal of Pathology and Bacteriology* 1958; **76**: 531–40.

35. Copeman P W M, Medd W E. Diffuse systemic sclerosis with abnormal liver and gallbladder. *British Medical Journal* 1967; **3**: 353–4.

36. Coyle H E, Chapman R S. Eosinophilic fasciitis (Shulman syndrome) in association with morphoea and systemic sclerosis. *Acta Dermato Venereologica* 1980; **60**: 181–2.

37. Cramer S F, Kent L, Abramowsky C, Moskowitz R W. Eosinophilic fasciitis—immunopathology, ultrastructure, literature review, and consideration of its pathogenesis and relation to scleroderma. *Archives of Pathology and Laboratory Medicine* 1982; **106**: 85–91.

38. Curzio (also under Crusio) C. An account of an extraordin-

ary disease of the skin and its cure. Extracted from the Italian of Carlo Crusio, with a letter of the Abbe Nollet to Mr William Watson, by Robert Watson. *Philosophical Transactions of the Royal Society of London* 1754; **48**: 579–87.

39. Dabich L, Bookstein J J, Zweifler A, Zarafonetis C J D, Arbor A. Digital arteries in patients with scleroderma. *Archives of Internal Medicine* 1972; **130**: 708–14.

40. Dahlqvist A, Lundberg E, Ostberg Y. Hydralazine-induced relapsing polychondritis-like syndrome. *Acta Oto-laryngologica* 1983; **96**: 355–9.

41. D'Angelo W A, Fries J F, Masi A T, Shulman L E. Pathologic observations in systemic sclerosis (scleroderma). A study of fifty-eight autopsy cases and fifty-eight matched controls. *American Journal of Medicine* 1969; **46**: 428–40.

42. Dayer J M, Favre H, Rais M, Chatelenat F. Scleroderma, renal failure, malignant hypertension and microangiopathic hemolytic anemia. *Schweizerische Medizinische Wochenschrift* 1974; **104**: 864–7.

42a. Derderian S S, Tellis C J, Abbrecht P H, Welton R C, Rajagopal K R. Pulmonary involvement in mixed connective tissue disease. *Chest* 1985; **88**: 45–8.

43. Dolan D L, Lemmon G B Jr, Teitelbaum S L. Relapsing polychondritis. *American Journal of Medicine* 1966; **41**: 285–99.

44. Dryll A, Lansaman J, Meyer O, Bardin T, Ryckewaert A. Relapsing polychondritis—an ultrastructural study of elastic and collagen fibre degradation revealed by tannic acid. *Virchows Archives A Pathological Anatomy and Histopathology* 1981; **390**: 109–20.

45. Eason R J, Tan P L, Gow P J. Progressive systemic sclerosis in Auckland—a 10 year review with emphasis on prognostic features. *Australian and New Zealand Journal of Medicine* 1981; **11**: 657–62.

46. Ellman P. Pulmonary manifestations in the systemic collagen diseases. *Postgraduate Medical Journal* 1956; **32**: 370–87.

47. Erasmus L D. Scleroderma in gold miners on the Witwatersrand, with particular reference to pulmonary manifestations. *South African Journal of Laboratory and Clinical Medicine* 1957; **3**: 209–31.

48. Evans D J, Cashman S J, Walport M. Progressive systemic sclerosis: autoimmune arteriopathy. *Lancet* 1987; **i**: 480–2.

49. Evans M, Parker R A. Honeycomb lung and mitral stenosis in scleroderma. *Thorax* 1954; **9**: 154–8.

50. Fam A G, Rubenstein J D, Cowan D H. POEMS syndrome. Study of a patient with proteinuria, microangiopathic glomerulopathy, and renal enlargement. *Arthritis and Rheumatism* 1986; **29**: 233–41.

51. Fennell R H Jr, Maclachlan M J, Rodnan G P. The occurrence of antinuclear factors in the sera of relatives of patients with systemic rheumatic disease. *Arthritis and Rheumatism* 1962; **5**: 296.

52. Fennell R H, Rodnan G P, Vasquez J J. Variability of tissue-localizing properties of serum from patients with different disease states. *Laboratory Investigation* 1962; **2**: 24–31.

53. Finch W R, Buckingham R B, Rodnan G P, Prince R K, Winkelstein A. Scleroderma induced by bleomycin. In: Black C M, Myers A R, eds, *Systemic Sclerosis*. New York and London: Gower Medical Publishing, 1985: 114–21.

54. Firestein G S, Gruber H E, Weisman M H, Zvaifler N J, Barber J, O'Duffy J D. Mouth and genital ulcers with inflamed cartilage: MAGIC syndrome. Five patients with features of relapsing polychondritis and Behçet's disease. *American Journal of Medicine* 1985; **79**: 65–72.

55. Fleischmajer R, Gay S, Perlish J S, Cesarini J P. Immunoelectron microscopy of type III collagen in normal and scleroderma skin. *Journal of Investigative Dermatology* 1980; **75**: 189–91.

56. Fleischmajer R, Gay S, Meigel W N, Perlish J S. Collagen in the cellular and fibrotic stages of scleroderma. *Arthritis and Rheumatism* 1978; **21**: 418–28.

57. Fleischmajer R, Krol S. Chemical analysis of the dermis in scleroderma. *Proceedings of the Society for Experimental Biology and Medicine* 1967; **125**: 252–6.

58. Follansbee W P, Curtis E I, Medsger T A. Physiologic abnormalities of cardiac function in progressive systemic sclerosis with diffuse scleroderma. *New England Journal of Medicine* 1984; **310**: 142–8.

59. Fu T S, Soltani K, Sorensen L B. Eosinophilic fasciitis. *Journal of the American Medical Association* 1978; **240**: 451–3.

60. Gintrac E. Note sur la sclerodermie. *Revue médico-chirurgicale de Paris* 1847; **ii**: 263–7.

61. Goetz R H, Pathology of progressive systemic sclerosis (generalized scleroderma) with special reference to changes in viscera. *Clinical Proceedings* 1945; **4**: 337–92.

62. Goldgraber M B, Kirsner J B. Scleroderma of the gastrointestinal tract—a review. *Archives of Pathology* 1957; **64**: 255–65.

63. Gordon M B, Klein I, Dekker A, Rodnan G P, Medsger T A. Thyroid disease in progressive systemic sclerosis—increased frequency of glandular fibrosis and hypothyroidism. *Annals of Internal Medicine* 1981; **95**: 431–5.

64. Gouet D, Marechaud R, Nean J-Ph, Bontoux D, Sudre Y. Chronic atrophic polychondritis with destructive arthropathy. *La Presse Médicale* 1983; **12**: 1172–3.

65. Green M C, Sweet H O, Bunter L E. Tight-skin, a new mutation of the mouse causing excessive growth of connective tissue and skeleton. *American Journal of Pathology* 1976; **82**: 493–507.

66. Gunther G, Schuchardt E. Silikose und progressive sklerodermie. *Deutsche Medizinische Wochenschrift* 1970; **95**: 467–8.

67. Hashimoto K, Arkin C R, Kang A H. Relapsing polychondritis. An ultrastructural study. *Arthritis and Rheumatism* 1977; **20**: 91–9.

68. Hawkins R A, Claman H N, Clark R A F, Steigerwald J C. Increased dermal mast cell populations in progressive systemic sclerosis: a link in chronic fibrosis? *Annals of Internal Medicine* 1985; **102**: 182–6.

69. Hayes R L, Rodnan G P. The ultrastructure of skin in progressive systemic sclerosis (scleroderma): I. Dermal collagen fibers. *American Journal of Pathology* 1971; **63**: 433–40.

70. Heppleston A G. The pathology of honeycomb lung. *Thorax* 1956; **11**: 77–93.

71. Herman J H. Polychondritis. In: Kelley W N, Harris E N, Ruddy S, Sledge C B, eds, *Textbook of Rheumatology*. Philadelphia: W B Saunders & Co., 1989; 1513–22.

72. Herman J H, Dennis M V. Immunopathologic studies in relapsing polychondritis. *Journal of Clinical Investigation* 1973; **52**: 549–58.

73. Hochberg M C, Lopez-Acuna D, Gittelsohn A M. Mortal-

143. Perlish J S, Bashey R I, Stephens R E, Fleischmajer R. Connective tissue synthesis by cultured scleroderma fibroblasts 1. in vitro collagen synthesis by normal and scleroderma dermal fibroblasts. *Arthritis and Rheumatism* 1976; **19**: 891–901.

144. Pieraggi M T, Bonafe J L. Apprèciation de l'éffet de la D-penicillamine sur la synthèse et le maintien de la structure collagène, grace à la coloration histologique par le Rouge Sirius examinée en lumière polarisée. *Annales de Pathologie* 1982; **2**: 155–7.

145. Postlethwaite A E, Kang A H. Pathogenesis of progressive systemic sclerosis. *Journal of Laboratory and Clinical Medicine* 1984; **103**: 506–10.

146. Prieur D J, Young D M, Counts D F. Auricular chondritis in fawn-hooded rats: a spontaneous disorder resembling that induced by immunization with Type II collagen. *American Journal of Pathology* 1984; **116**: 69–76.

147. Rake G. On the pathology and pathogenesis of scleroderma. *Bulletin of the Johns Hopkins Hospital* 1931; **48**: 212–27.

148. Richards R L, Milne J A. Cancer of the lung in progressive systemic sclerosis. *Thorax* 1958; **13**: 238–45.

149. Ritchie R F. The clinical significance of titered autinuclear antibodies. *Arthritis and Rheumatism* 1967; **10**: 544–52.

150. Rivelis A L. Esclerosis sistemica progresiva—analisis de 66 cases. *Archives of InterAmerican Rheumatology* 1963; **6**: 496–524.

151. Rodnan G P. The nature of joint involvement in progressive systemic sclerosis (diffuse scleroderma). Clinical study and pathologic examination of synovium in twenty-nine patients. *Annals of Internal Medicine* 1962; **56**: 422–39.

152. Rodnan G P. Eosinophilic fasciitis. In: Samter M, ed, *Immunological Diseases* (3rd edition). Boston: Little, Brown and Company, 1978: 1127–41.

153. Rodnan G, Benedek T. An historical account of the study of progressive systemic sclerosis. *Annals of Internal Medicine* 1962; **57**: 305–19.

154. Rodnan G P, Benedek T G, Medsger T A, Cammarata R J. The association of progressive systemic sclerosis (scleroderma) with the Coal Miners' pneumoconiosis and other forms of silicosis. *Annals of Internal Medicine* 1967; **66**: 323–4.

155. Rodnan G P, Di-Bartolomeo A G, Medsger T A. Eosinophilic fasciitis. Report of seven cases of a newly recognised scleroderma-like syndrome. *Arthritis and Rheumatism* 1975; **18**: 422–3.

156. Rodnan G P, Jablonska S. Classification of systemic and localized scleroderma. In: Black C M, Myers A R, eds, *Systemic Sclerosis*. New York and London: Gower Medical Publishing, 1985: 3–6.

157. Rodnan G P, Lipinski E, Luksick J. Skin thickness and collagen content in progressive systemic sclerosis and localized scleroderma. *Arthritis and Rheumatism* 1979; **22**: 130–40.

158. Rodnan G P, Medsger T A. Pentazocine-induced fibrosis. In: Black C M, Myers A R, eds, *Systemic Sclerosis*. New York and London: Gower Medical Publishing, 1985: 122–4.

159. Rodnan G P, Myerowitz R L, Justh G O. Morphologic changes in the digital arteries of patients with progressive systemic sclerosis (scleroderma) and Raynaud's phenomenon. *Medicine* 1980; **59**: 393–408.

160. Roge J, Delavierre Ph., Durand H, Besançon-Lajeunesse L. Scleroderma et cancer d'estomac. *Semaine des Hôpitaux de Paris* 1971; **47**: 1211–13.

161. Rosenbloom J, Feldman G, Freundlich B, Jimenez S A. Inhibition of excessive scleroderma fibroblast collagen production by recombinant γ-interferon. Association with a coordinate decrease in types I and III procollagen messenger RNA levels. *Arthritis and Rheumatism* 1986; **29**: 851–6.

162. Roth L M, Kissane J M. Panaortitis and aortic valvulitis in progressive systemic sclerosis (scleroderma): report of case with perforation of an aortic cusp. *American Journal of Chemical Pathology* 1964; **41**: 287–96.

163. Rottenberg E N, Slocumb C H, Edwards J E. Cardiac and renal manifestations in progressive systemic scleroderma. *Proceedings of the Staff Meetings of the Mayo Clinic* 1959; **34**: 77–82a.

163a. Roubenhoff R, Coté T, Watson R, Levin M L, Hochberg M C. Eosinophilia-myalgia syndrome due to L-tryptophane ingestion. Report of four cases and review of the Maryland experience. *Arthritis and Rheumatism* 1990; **33**: 930–8.

164. Russell M L, Friesen D, Henderson R D, Hanna W M. Ultrastructure of the esophagus in scleroderma. *Arthritis and Rheumatism* 1982; **25**: 1117–23.

165. Russell M L, Kahn H J. Nodular regenerative hyperplasia of the liver associated with progressive systemic sclerosis: a case report with ultrastructural observation. *Journal of Rheumatology* 1983; **10**: 748–52.

166. Sackner M A, Akgun N, Kimbel P, Lewis D H. The pathophysiology of scleroderma involving the heart and respiratory system. *Annals of Internal Medicine* 1964; **60**: 611–30.

166a. Seibold J R. Scleroderma. In: Kelley W N, Harris E D, Ruddy, Sledge C B, eds, *Textbook of Rheumatology* (3rd edition). Philadelphia, London: W B Saunders Company, 1989: 1215–44.

167. Seyer J M, Kang A H, Rodnan G. Investigation of type I and type III collagens of the lung in progressive systemic sclerosis. *Arthritis and Rheumatism* 1981; **24**: 625–31.

168. Sharp G C, Irvin W S, Tan E M, Gould R G, Holman H R. Mixed connective tissue disease, an apparently distinct rheumatic disease syndrome associated with a specific antibody to an extractable nuclear antigen (ENA). *American Journal of Medicine* 1972; **52**: 148–59.

169. Shaul S R, Schumacher H R. Relapsing polychondritis. Electron microscopic study of ear cartilage. *Arthritis and Rheumatism* 1975; **18**: 617–25.

170. Shewmake S W, Lopez D A, McGlamory J C. The Shulman syndrome. *Archives of Dermatology* 1978; **114**: 556–9.

171. Shimpo S. Solitary myeloma causing polyneuritis and endocrine disturbances. *Japanese Journal of Clinical Medicine* 1968; **26**: 2444–56.

172. Shulman L E. Diffuse fasciitis with hypergammaglobulinaemia and eosinophilia: A new syndrome. *Journal of Rheumatology* 1974; **1** (Suppl.): 46.

172a. Shulman L E. The eosinophilia–myalgia syndrome associated with ingestion of L-tryptophane. *Arthritis and Rheumatism* 1990; **33**: 913–7.

173. Singsen B H, Swanson V L, Bernstein B H, Heuser E T, Hanson V, Landing B H. A histologic evaluation of mixed connective tissue disease in childhood. *American Journal of Medicine* 1980; **68**: 710–17.

174. Steven J L. A case of scleroderma leading to pronounced

hemiatrophy of the face, body and extremities with deformity and fibrous ankylosis of the joints after a lengthened period of superficial ulceration. *International Clinics* 1898; **7**: 195–202.

175. Talbott J H, Barrocas M. Carcinoma of the lung in progressive systemic sclerosis: a tabular review of the literature and a detailed report of the roentgenographic changes in two cases. *Seminars in Arthritis and Rheumatism* 1980; **9**: 191–217.

176. Teasdall R D, Frayha R A, Shulman L E. Cranial nerve involvement in systemic sclerosis (scleroderma): a report of 10 cases. *Medicine* 1980; **59**: 149–59.

177. Thomas L. Reversible collapse of rabbit ears after intravenous papain, and prevention of recovery by cortisone. *Journal of Experimental Medicine* 1956; **104**: 245–52.

178. Thompson R P, Harper F E, Maize J C, Ainsworth S K, Le Roy E C, Maricq H R. Nailfold biopsy in scleroderma and related disorders. *Arthritis and Rheumatism* 1984; **27**: 97–103.

179. Toth A, Alpert L I. Progressive systemic sclerosis terminating as periarteritis nodosa. *Archives of Pathology* 1971; **92**: 31–6.

180. Traub Y M, Shepiro A P, Rodnan G P, Medsger T A, McDonald R H, Steen V D, Psial T A, Tolchin S F. Hypertension and renal failure (scleroderma renal crisis) in progressive systemic sclerosis. Review of a 25-year experience with 68 cases. *Medicine* 1983; **62**: 335–52.

181. Trentham D E, Goodman M L. Relapsing polychondritis. *New England Journal of Medicine* 1982; **307**: 1631–9.

182. Uitto J, Bauer E A, Eisen A Z. Scleroderma. Increased biosynthesis of triple-helical type I and type III procollagens associated with unaltered expression of collagenase by skin fibroblasts in culture. *Journal of Clinical Investigation* 1979; **64**: 921–30.

183. Van der Water J, Gershwin M E. Avian scleroderma: an inherited fibrotic disease of white leghorn chickens resembling progressive systemic sclerosis. *American Journal of Pathology* 1985; **120**: 478–82.

183a. Van de Water J, Gershwin M E. Avian scleroderma. In: Greenwald R A, Diamond H S, eds, *CRC Handbook of Animal Models for the Rheumatic Diseases* (volume I). Boca Raton, Florida, CRC Press Inc., 1988: 195–204.

184. Vincent F M, van Houzen R N. Trigeminal sensory neuropathy and bilateral carpal tunnel syndrome—the initial manifestation of mixed connective tissue disease. *Journal of Neurology, Neurosurgery and Psychiatry* 1980; **43**: 458–60.

185. Walton J N, Gardner-Medwin D. The muscular dystrophies. In: Walton J N, ed, *Disorders of Voluntary Muscle* (5th edition). Edinburgh: Churchill Livingstone, 1988: 519–68.

186. Weaver A L, Divertie M B, Titus J L. The lung in scleroderma. *Mayo Clinic Proceedings* 1967; **42**: 754–66.

187. Weiss S, Stead E A Jr, Warren J V, Bailey B T. Scleroderma heart disease with a consideration of certain other visceral manifestations of scleroderma. *Archives of Internal Medicine* 1943; **71**: 749–76.

187a. Westacott C I, Whicker J T, Hutton C W, Dieppe P A. Increased spontaneous production of interleukin-l together with inhibitory activity in systemic sclerosis. *Clinical Science* 1988; **75**: 561–7.

187b. Whiteside T L, Ferrarini M, Hebda P, Buckingham R B. Heterogeneous synthetic phenotype of cloned scleroderma fibroblasts may be due to aberrant regulation in the synthesis of connective tissues. *Arthritis and Rheumatism* 1988; **31**: 1221–9.

188. Whitlow P L, Gilliam J N, Chubick A, Ziff M. Myocarditis in mixed connective tissue disease, association of myocarditis with antibody to nuclear ribonucleoprotein. *Arthritis and Rheumatism* 1980; **23**: 808–15.

189. Wiener-Kronish J P, Solinger A M, Warnock M L, Churg A, Ordonez N, Golden J A. Severe pulmonary involvement in mixed connective tissue disease. *American Review of Respiratory Diseases* 1981; **124**: 499–503.

190. Willis J, Atack E A, Kraag G. Relapsing polychondritis with multifocal neurological abnormalities. *Canadian Journal of Neurological Sciences* 1984; **11**: 402–4.

191. Wilson R J, Rodnan G P, Robin E D. An early pulmonary physiologic abnormality in progressive systemic sclerosis (diffuse scleroderma). *American Journal of Medicine* 1964; **36**: 361–9.

192. Winkelmann R K. Staging of scleroderma. In: Black C M, Myers A R, eds, *Systemic Sclerosis*. New York and London: Gower Medical Publishing, 1985: 24–8.

193. Winterbaur R H. Multiple telangiectasia, Raynaud's phenomenon and subcutaneous calcinosis: a syndrome mimicking hereditary hemorrhagic telangiectasia. *Bulletin of Johns Hopkins Hospital* 1964; **114**: 361–83.

194. Wise T N, Ginzler E M. Scleroderma cerebritis, an unusual manifestation of progressive systemic sclerosis. *Diseases of the Nervous System* 1975; **36**: 60–2.

Gall[262] investigated two cases and discovered synovial cell hyperplasia, congestion and a lymphocytic infiltrate. They detected HB_s antigen by direct immunofluorescence; viral particles were found by electron microscopy in synoviocytes and in the subsynovial vascular and connective tissues.

Parvovirus

Human parvovirus arthritis

The possibility has arisen that a parvovirus (HPV) infection may cause human diseases such as rheumatoid arthritis. A strain of parvovirus, RA-1, was isolated from human rheumatoid synovial tissue:[241,277,322] the virus was pathogenic for newborn mice. Subsequently, it was found that isolates from homogenates of human rheumatoid synovium often reacted with an anti-RA-1 virus antiserum.[53,291]

A distinct, B19 parvovirus arthritis has also been recognized.[265] In a series of six patients, two with early rheumatoid arthritis and four with inflammatory arthritis, sensitive assays for IgM, IgG and human parvovirus (HPV) antigen were used to search for this virus together with 'dot blot' hybridization for HPV DNA: the results confirmed the existence of recent HPV infection. Infection was not persistent but past HPV infection was more common in rheumatoid arthritis patients than in controls.[53] Little is known of the tissue reactions to HPV infection. The small joints of hands and feet are affected together with the elbows, wrists, knees and ankles, and nerve entrapment is among the early signs.

Aleutian disease of mink

Mink (*Mustela vison*), particularly those homozygous for the recessive gene that determines a blue–grey coat, are susceptible to a parvovirus which quickly provokes a grossly excessive secretion of antiviral immunoglobulin and often causes death with immune complex arteritis and glomerulonephritis.[234] The blue–grey coat is similar to that of the Aleutian blue fox, so that the affected mink are said to have Aleutian disease. The infective agent is the Aleutian disease virus (ADV). A similar agent infects ferrets but is less virulent. In mink of the Aleutian genotype, there is a 50 per cent mortality within 12 weeks. Non-Aleutian mink display progressive infection, persistent but non-progressive infection, or non-persistent and non-progressive infection. In the first category, there are high anti-ADV antibody titres, hypergammaglobulinaemia, immune complex lesions and a 50 per cent mortality in 30 weeks. In the second, there are low antibody titres and no immune complex lesions, while in the third, the virus infection ultimately resolves.

The tissue lesions of Aleutian disease centre on the plasma and lymphocytic response to ADV. There is widespread infiltration of many organs by these cells; the lymph nodes and spleen are much enlarged.[235] Animals with progressive infection display little clinical evidence of disease until a few days before their death when they exhibit the renal failure and haemorrhage or infarction which result from immune complex glomerulonephritis or arteritis, respectively. IgG and complement component C3 are deposited in a granular pattern on glomerular capillary walls. Antiviral antibody can be eluted from the affected kidneys. In low-grade disease, renal inflammation subsides and glomerular sclerosis evolves. The severity of arteritis ranges from acute necrotizing vasculitis with polymorph infiltration and fibrinoid change to a subacute arteritis with fibrosis marking the sites of elastic laminar and smooth muscle injury.

The pathogenesis of Aleutian disease of mink is not wholly understood. The lesions are not caused directly by the virus nor is virus inactivated by the binding of antibody to virions. The huge excess of antiviral antibody, reaching levels of approximately 50 g/l IgG, results in immune complex formation. One result is the onset of systemic renal and arterial disease. Another, perhaps offering an explanation for the persistence of virus, is the interaction of circulating immune complexes with T lymphocytes, impairing cell-mediated immunity. It is also possible that ADV replicates in T-cell subsets, diminishing T-cell responsiveness.

Enterovirus

Coxsackie virus arthritis

Polyarthritis is a rare manifestation of coxsackie virus B_2 and B_4 infections.[153] Occasionally, the duration of the arthritis in a young person may suggest the development of juvenile rheumatoid arthritis.

Echovirus arthritis

An instance of polyarthritis caused by Echovirus type 6, has been reported, resolving after nine days. In a further case, Echovirus type 9 was implicated.[32]

Alphavirus

Alphaviruses (arboviruses–arthropod-borne viruses) are intracellular parasites of a wide range of natural vertebrate hosts in Eastern Europe, North and South America, Africa, India, Asia and Australia.[194,280] They are transmissible by culicine and anopheline mosquitoes, by sandflies, ticks, midges, gnats and other insects, and, rarely, via milk. Febrile illness,[207] with or without a haemorrhagic state, may precede encephalitis; and in some of this large group of disorders arthritis is commonplace. There may be a genetic predisposition to infection.[105] In forms of epidemic polyarthritis, such as Australian Ross River fever, there is a characteristic synovitis.[137] Arthritis is an important feature of 'chikungunya' (Africa, Southeast Asia) and has also been reported as a result of infection by agents of the bunyavirus group.

Epidemic polyarthritis

Epidemic outbreaks of polyarthritis[10,75,76,210] caused by a group A alphavirus (arbovirus) transmitted by several species of mosquito[298] were described in New South Wales[221] in the Northern Territories of Australia[131] and in Fiji.[30] The sudden or gradual onset[103] of sore throat, headache, malaise, lethargy and other constitutional symptoms heralds arthritis of two or more joints that may be symmetrical or asymmetrical. Arthritis persists for 2–4 weeks and may be accompanied by inflammation of other periarticular joint tissues including the tendinous synovium, capsule and ligaments. Recovery is complete and there is no residual joint disease. The alphavirus has only once been recovered from the blood of a patient, never from the synovia. However, virus is recoverable from *Aedes*, *Culex* and *Mansonia* mosquitoes the bite of which precedes arthritis by 6–15 days. Fever is usually not accompanied by leucocytosis or by an elevated erythrocyte sedimentation rate (ESR). By contrast with rubella and hepatitis B virus arthritis, antiviral antibody is present during the stage of overt arthritis and there is neither depression of complement levels nor evidence of circulating immune complexes which are not therefore apparently implicated.

Chikungunya

Chikungunya ('He who walks bent up')[172,281a,284] was identified as an epidemic dengue-like disease in 1952;[192,246,250] it is transmitted by mosquitoes of the *Aedes* species from primary hosts including the baboon. Outbreaks occurred in Tanzania, other parts of Africa, southern India and southeast Asia. High fever and a maculopapular rash persist for only 5–7 days but the characteristically severe joint pain may continue for several months. A rheumatoid arthritis-like syndrome has been described[101] with the presence of rheumatoid factor, and four of five patients, with this syndrome were HLA-B27-positive. The arthralgia, unaccompanied by swelling, is usually polyarticular but occasionally only the knees are implicated. Regional lymphadenopathy is recognized. Virus is recoverable from the blood up to three days after the onset of symptoms, but the synovial fluid, which is bacteriologically sterile and contains mononuclear cells, and the synovial tissue, have not been examined for virus or surveyed microscopically.

O'nyong nyong

O'nyong nyong ('Joint crusher') is an endemic and sometimes epidemic virus infection transmitted by anopheline mosquitoes and described in Uganda.[130,271,326,327] O'nyong nyong closely resembles chikungunya. The onset of fever is sudden. Arthralgia is severe, lymphadenitis common and characteristic. Joint pain[284] is symmetrical and generalized and affects knees, elbows, wrists, fingers and ankles; heat, redness and swelling are not encountered. The attack lasts 4–7 days and leaves no residual joint disease. Virus has been isolated from the blood but there are no reports of synovial fluid analysis or of synovial biopsy.

Arthritis is encountered in *Mayaro* (Trinidad and Brazil) and *Sindbis* (Queensland, southern Africa and Asia) but descriptions of their pathological features are lacking.

Bunyamera

Arthralgia is frequent in these African mosquito-borne arbovirus diseases. It is not clear whether the response is infective or postinfective.

Rubivirus

Rubella arthritis

Arthritis and arthralgia are common features of rubella; rubella arthritis was recognized long before the causal agent had been identified.[6,44,84,114,164,229] A symmetrical, early morning inflammation of wrist, metacarpophalangeal, proximal interphalangeal and distal interphalangeal joints appears some days after the onset of the rash.[336] Indeed, the rash may fade before the arthralgia begins. Synovitis can present in the absence of a rash in a young adult living in a household where rubella has been diagnosed.

Pain and swelling predominate; redness is slight; and a similar involvement of the knees, wrists and other synovial joints is frequent. Arthritis and arthralgia may characterize individual epidemics of rubella suggesting either a predilection of individual viral types for joint tissue, or the operation of as yet unidentified environmental, predisposing factors. Females are affected with much greater relative frequency than men: indeed, 15–20 per cent of women with rubella

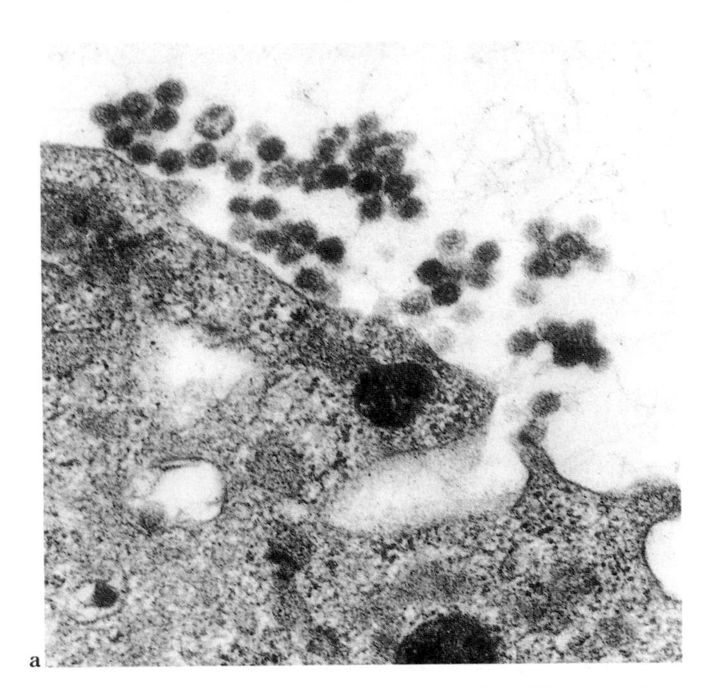

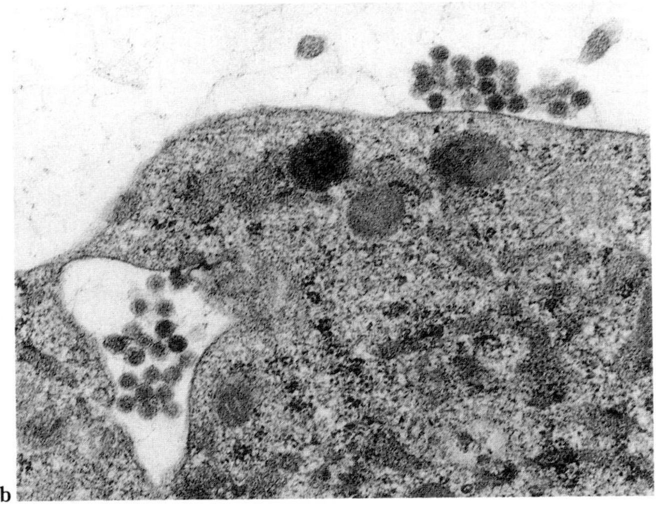

Fig. 18.5 Retrovirus at cell margin.
C-type retrovirus particles are seen budding from the plasma membrane of McCoy cells. (TEM: **a.** × 39 400; **b.** × 30 200.) Courtesy of Dr Alan Currey.

recovered after 12 months from synovial fluid cells, after 18 months from cell-free synovial fluid. There is continuous, restricted virus replication. Plasmacytosis is accompanied by the synthesis of much polyclonal IgG1 and IgM:[162,163] the concentration of the former, but not of the latter, is higher in the synovium than in the serum. In spite of the formation of agglutinating antibodies, neutralizing antibody is not secreted unless an adjuvant such as heat-killed *Mycobacterium tuberculosis* is injected with the CAEV virus.[219] CAEV disease is apparently the consequence of a defective immune response to the retrovirus: T-cell reactivity is promoted[72] but the behaviour of the increased proportions of B-cells is variable and unpredictable. There are clear differences from human rheumatoid arthritis in which the secretion of IgM anti-IgG is so common and in which the inheritance of HLA-D related antigens may set the scene for altered proportions of synovial T_H:T_S-cells.

Miscellaneous

Lymphogranuloma inguinale

Polyarthritis is a recognized complication. Wright and Logan[333] described involvement of the large joints of the lower limbs and of the wrists. A resemblance to rheumatoid arthritis has been suggested; the course, however, is long but benign with little microscopic evidence of joint damage. The synovial fluid remains sterile. The number of systematic pathological studies of the joints in this disease is small.

Erythema infectiosum

Although adults frequently develop arthralgia or arthritis of the wrists or knees, no causal agent has yet been recovered from the blood and other sites, and the synovial fluid and synovial tissue changes are not described or understood.

Endemic arthritis, possible viral

Self-limiting outbreaks of arthritis in which a viral origin is suspected have been described among the Navajo Indians,[211] in New Guinea[160] and in Nigeria.[127] In each instance, synovial biopsies were reported to have shown 'non-specific' changes.

Arthritis in the acquired immunodeficiency syndrome (AIDS)

The high frequency of opportunistic infections in AIDS involves the parasites *Pneumocystis carinii*, *Cryptosporidium* spp. and *Toxoplasma gondii*.[58] There is also an association with *Myc. avian intracellulare*, *Myc. tuberculosis*, campylobacter, shigella and salmonella; with candidiasis; with *Entamoeba histolytica* and *Giardia lamblia*; and with strongyloides.[339] Since these microorganisms, together with viruses such as herpes simplex and herpes zoster to which patients with AIDS are prone, include many known to parasitize synovial joints, it is to be anticipated that these agents will be found, with increasing frequency, to be responsible for opportunistic arthritis.

Rickettsial, coxsiellal and chlamydial arthritis

Rickettsiae,[48] coxsiellae[252] and chlamydiae,[255] like viruses, are obligate intracellular parasites. By contrast with the viruses, coxsiellae and chlamydiae have cell walls, divide by binary fission, contain RNA and DNA in bacteria-like ratios, and otherwise differ from bacteria only by their lack of certain enzymes necessary for extracellular life and multiplication. Rickettsiae and coxsiellae are natural inhabitants of arthropods: their parasitism of human cells is coincidental and exceptional. Chlamydiae, by contrast, are transmitted directly from man-to-man or from intermediary avian hosts: they have been established as a cause of polyarthritis in lambs[254] and are implicated in human disorders such as Reiter's disease (see p. 764).

In rickettsial and coxsiellal infection, arthritis or other connective tissue disease is extremely uncommon; but the chlamydiae have been associated with polyarthritis in two contexts.

First, in lymphogranuloma venereum, in which the infective agent is thought to be a member of the chlamydiae, the disease may be complicated by an active polyarthritis that affects the knees, ankles and wrists simultaneously, but which may adopt a migratory pattern reminiscent of rheumatic fever.[188] Lymphogranuloma arthritis, however, commonly assumes a chronic course, with persistent or recurrent effusions accompanied by little structural disorganization of the articular tissues. The causative organisms are not recoverable from the synovial fluid and it is possible that the arthritis is therefore of a hypersensitivity origin, due to immune complex deposition.

Second, chlamydiae have been invoked as causes of Reiter's disease[254] (see p. 764) and, less convincingly, of rheumatoid arthritis (see p. 532). The evidence is incomplete. In Reiter's disease the pathogenesis is more likely to be linked with infection indirectly. *Chlamydia trachomatis* was isolated from 5 of 16 men of a group of 531 who developed arthritis with 'non-specific' urethritis but from an identical proportion of those without arthritis; the difference was that 20 per cent of the series who were HLA-B27-positive developed arthritis compared to 2 per cent without this antigen.[169] There is therefore a genetic basis for this form of articular infection.

Mycoplasmal arthritis

In the 27 years since Sharp[267] reviewed the role of mycoplasmas in joint infection, important advances have been made in understanding mycoplasmal arthritis (Table 18.3). The mycoplasmas[136] are the smallest prokaryotes able to live and multiply in cell-free media. They have no cell wall and display exacting growth requirements. The colonies are minute and cultivation difficult. 'Classical' mycoplasmas are distinguished from a T ('tiny')-form that splits urea and is now called *Ureaplasma urealyticum*. Widely distributed in nature, mycoplasmas cause natural diseases of cattle, swine, sheep, goats, cats, rats, mice, and birds.[297] The diseases range from respiratory, urogenital and mammary infection to arthritis. It is frequently found that more than one mycoplasma can cause arthritis in a single species.

There are differences in species' responsiveness to experimental mycoplasma infection,[64] influenced by the route of infection. Joints may be infected directly or by blood-borne organisms carried, for example, from the lungs. Acute arthritis is characterized by the cardinal signs

Table 18.3

Mycoplasmas that cause arthritis[54]

Mycopl. arthritidis			
Rat	N	A	Conjunctivitis, urethritis, abscesses
Mouse	E	C	
Rabbit	E	C	
Mycopl. pulmonis			
Mouse	N	C	Pneumonia, reproductive tract disease
Rat	E	A	Pneumonia, reproductive tract disease
Rabbit	E	C	?Respiratory disease
Mycopl. hyorhinis			
Swine	N	C	Polyserositis
Mycopl. hyosynoviae			
Swine	N	C	
Mycopl. synoviae			
Fowl	N	C	Air sacculitis, hepatitis, endocarditis, anaemia
Mycopl. gallisepticum			
Fowl	N	C	Air sacculitis, arteritis of brain and heart
Mycopl. mycoides			
Cattle	N	C	Pleuropneumonia
Mycopl. bovis			
Cattle	N	C	Mastitis, reproductive tract disease
Mycopl. agalactiae			
Sheep	N	C	Mastitis and
Goats	N	C	conjunctivitis

N = natural disease; E = experimental disease; A = acute; C = chronic.

into rabbits.[118] The synovitis, of febrile polyarthritic (70 per cent) or monoarthritic (30 per cent) character, is marked by synovial fluid polymorph leucocytosis, frequently positive synovial fluid culture and occasional positive blood culture. The synovitis of gonococcal arthritis displays the features usual in bacterial joint infection.[47,302] Synovial cells are destroyed. There is synovial hyperaemia and a polymorph, and later, lymphocytic, macrophage and plasma cell infiltrate. Tenosynovitis may be a dominant characteristic.[302] This brisk, acute inflammatory reaction occasionally progresses to purulent arthritis with necrosis and dissolution of synovial, cartilagenous and capsular tissue. A complication shared by all forms of destructive bacterial arthritis is the subsequent occurrence of, first, fibrous and, later, bony ankylosis. This may cause diagnostic difficulty when gonococcal arthritis is confined to the sacroiliac joint.[243]

Staphylococcal arthritis

Staphylococci, in particular, *Staphylococcus aureus*, remain the most frequent cause of non-gonococcal suppurative arthritis.[258] In spite of the widespread use of antibiotics these organisms account for approximately 65 per cent of cases. Of 42 cases of septic arthritis, 25 were staphylococcal.[15] Trauma,[56] wound sepsis, endocarditis[187] and infection during immunosuppression or drug addiction are recognized predisposing states, as is neurogenic arthropathy.[201,251]

The knee, elbow, hip and shoulder are most often infected[15] but the sternoclavicular, metatarsophalangeal, wrist and ankle joints can be attacked. The infecting organism commonly comes from the patient's own skin or nose, or, in hospital infection, from skin or nasal carriers who are members of the hospital staff. Direct implantation may account for wound sepsis but arthritis is more likely to result from blood-borne organisms lodging in the synovia or metaphyseal bone. In the latter case, untreated osteomyelitis in children is the underlying disorder. Staphylococcal arthritis can be studied experimentally in rabbits following intraarticular injection.[161]

The pattern of histological response is described above. The phagocytosis and lysosomal digestion of staphylococci by synovial lining cells is less significant for the outcome of the infection than is the attack on the intraarticular microorganisms by polymorphs that enter the joint space early in inflammation. An index of polymorph behaviour is the rapid rise in the level of latent and active lysosomal proteases within the synovial fluid. These are enzymes that display a pronounced capacity for degrading articular cartilage and causing the permanent cartilage injury that is a feature of all the purulent and granulomatous synovial joint infections. It is axiomatic that progression to collagen loss is a preliminary to permanent cartilage disorganization, whereas transient controlled infection with a limited breakdown of proteoglycan is susceptible to repair.

The outcome of staphylococcal arthritis is determined by the pathogenicity and virulence of the organism, the effectiveness of selected early antibiotic treatment and by the use of appropriate surgical intervention. However, an important factor in the outcome of the infection is the level of natural and acquired immunity of the host (see Chapter 6). Humoral immunity, with the prompt secretion of IgM and IgG antibodies, results in the formation of immune complexes with one or other of the numerous staphylococcal antigens. Complexes lodge not only in the synovial fluid of the infected joint but also in the other joints and in relation to renal basal laminae. Complement levels fall and the systemic infection may be complicated by renal failure with immune complex glomerulonephritis. Immunity may be poor because of defective, non-specific host defence mechanisms. The capacity of polymorphs to destroy staphylococci intracellularly may be weakened by the action of drugs or ionizing radiation. The administration, locally or systemically, of immunosuppressive drugs lowers resistance, and in common articular diseases such as rheumatoid arthritis, large joints are sites for silent, unsuspected staphylococcal infection. Among the numerous states that may influence phagocytic function adversely and which may coexist with systemic connective tissue diseases, are: infancy, severe infection, rheumatoid arthritis, diabetes mellitus, the Chediak–Higashi syndrome, the 'lazy leucocyte syndrome', and some cases of agammaglobulinaemia. Chronic granulomatous disease may be contributory, while ethanol, salicylates and prednisolone may inhibit the adhesion of leucocytes to venular endothelium that is a preliminary to diapedesis, emigration and phagocytosis. Defective opsonization and complement component C3 and C3-inhibitor deficiencies also predispose to recurrent infection.[258]

Streptococcal arthritis

Streptococcal joint infection[121,151] usually originates by blood-borne dissemination of organisms from an infection of the skin or upper respiratory tract. However, direct trauma is another cause of streptococcal arthritis: very small puncture wounds are sufficient to initiate a dangerous infection with group A, β-haemolytic *Streptococcus pyogenes*. Arthritis due to group B streptococci is uncommon: at risk are pregnant women, diabetics, patients with carcinomatosis or cirrhosis and those on immunosuppressive regimens. The relevant literature is reviewed by Gaunt and Seal.[111] In one report, three of seven infected patients had undergone hip arthroplasty six months to six years previously and the operated hip was the site of infection. A further case had undergone splenectomy and a fifth case had neurogenic arthropathy with diabetes mellitus.[278] In some instances, the group B streptococcal infection may be aggressive, destroying several joints and causing permanent disability.[232] Group G β-haemolytic streptococcal infections are also very uncommon: septic arthritis results.[216] The disease tends to be polyarticular.[197] Streptococcal arthritis may be seen as an early sign of endocarditis and cases due to group B β-haemolytic and group D β-haemolytic organisms have been reported.[121] Analogous infections in the pig are attributable to *Streptococcus equisimilis*; the resemblance to *Erysipelothrix rhusiopathiae* infections (see p. 740) is apparent.

Pneumococcal arthritis

Arthritis has never been a frequent result of infection by *Streptococcus pneumoniae*. The reported cases, from 1888 to 1970, were reviewed.[306] Pneumococci were seen microscopically in the first patient described. The infection is often monoarticular. Destruction of articular structures may be followed, in the untreated patient, by injury to ligaments and extension of infection into periarticular muscles. The histological features are those of suppurative arthritis.

Brucella arthritis

Brucellosis is characteristically insidious in onset and behaviour; as a cause of unexplained pyrexia it is readily confused with enteric fever, tuberculosis and virus infection, including infectious mononucleosis.[258] Arthritis is an occasional complication. *Brucella melitensis* is the most frequent agent of human disease although *B. abortus* is responsible for the greatest proportion of cases in the USA. Accompanying the undulant fever there are often complaints of joint pain, sometimes simulating those of rheumatic fever, at other times localized to a single joint. Disseminated brucellosis may present as sternoclavicular[184] or sacroiliac[233] arthropathy. Spondylitis is a rare result of brucella infection[334] but true arthritis is uncommon.[133] A non-purulent infection has been recognized in the stifle joints of cattle.[36]

The organisms are carried from the site of infection in lymphatics to regional lymph nodes. From there, the small Gram-negative coccobacilli are conveyed by the bloodstream within infected mononuclear cells and are taken up by organs in which phagocytic reticuloendothelial cells are active. While the major organs of the reticuloendothelial system are infected in this way, almost any tissue may at some time be found to be the site for bacterial localization. Among the commonly infected sites are the bones and synovial tissue.

Infection may reach the synovia directly via the bloodstream but a preliminary osteomyelitis usually affects the bones of the spine, sacroiliac region or limbs. In spinal infection, the vertebral bone and the discs are destroyed. The characteristic local response to the intracellular multiplication of brucellae is the formation, within synovial tissues, of epithelioid cell follicles the appearances of which simulate the lesions of sarcoidosis (see p. 989). There is a close microscopic resemblance also to the granulomata of berylliosis and to the more extensively necrotic focal lesions of tuberculosis and of tularaemia. The opportunity to study these granulomata histologically is rare: first because of the low overall mortality of brucellosis and second, because of the need to avoid biopsy in the presence of suspected infection.

Haemophilus influenzae arthritis

Haemophilus influenzae arthritis has been regarded as a childhood disease second only to *Staphylococcus aureus* as a cause of suppurative arthritis.[34] However, increasing numbers of cases are now encountered in adults.[223] Hoaglund and Lord[141] added two cases to the six recorded in the literature. By 1986, 29 adult cases had been described:[34] 15 were monoarticular, 6 of the knee, and 22 of the 29 had predisposing causes such as alcoholic cirrhosis or a urinary tract infection (Table 18.4).

Meningococcal arthritis

Outbreaks of meningococcal meningitis are encountered among students, refugees and military personnel occupying crowded quarters, and among displaced and poverty-stricken populations. During epidemics of meningococcal meningitis, as many as 80 per cent of healthy troops may be carriers of *Neisseria meningitidis*. In interepidemic periods, among soldiers quartered in barracks or students in hostels, the carrier rate may be no more than 2 per cent. In any epidemic, instances of early mono- or polyarticular arthritis are encountered periodically. They arise during the bacteraemic phases of the disorder and can be shown to be due to meningococcal infection of the synovium. Immunization programmes may curtail manifest disease but do not necessarily prevent epidemics or reduce the frequency of joint infection which may be as high as 4 per cent.

The histological features of acute, early meningococcal arthritis cannot be readily distinguished from those of other forms of septic arthritis, such as that of gonococcal infection. The infection may be silent: purulent arthritis has been identified in synovial joints not recognized to be abnormal clinically. Bursitis and tenosynovitis are also encountered. Gram-negative diplococci can be identified within the numerous synovial fluid polymorphs: the organisms can be cultured, tested chemically and serotyped. Meningococcal septicaemia may culminate in adrenal haemorrhage and acute adrenal cortical insufficiency: the lesions are sterile and are thought to be analogues of the intravascular fibrin thrombi of the Shwartzman phenomenon. There is reason to suspect similar vascular changes as provocative factors in the synovitis of meningococcal infection, and a complication of the early disease is haemorrhage into the joints, a process analogous with the skin petechiae and ecchymoses.

In other patients, a delayed sterile arthritis develops 3–10 days after the onset of the illness. This response, which may be accompanied by cutaneous vasculitis, episcleritis or pericarditis, is characterized by low serum immune complement levels. It is thought that local, synovial immune-complex formation occurs, provoking a sterile postinfective arthritis.[128] An analogous synovitis can be caused in rabbits by the injection of purified meningococcal polysaccharide antigen into the knees of animals previously sensitized by the intravenous injection of heat-killed meningococci.

Erysipelothrix arthritis

Erysipelothrix rhusiopathiae, a small Gram-positive bacillus that occurs in chains, is a natural pathogen of domestic and wild animals and birds, and may infect man. The organism, a member of the family *Corynebacteriaceae*, may cause a chronic arthritis in pigs: the histological features resemble those of human rheumatoid arthritis.[274]

The acute infection, acute swine erysipelas, is transmissible by intravenous injection.[274] The acute lesions include skin haemorrhages and urticaria, pericarditis, lymph node eosinophilia and adrenal haemorrhages. Acute synovitis is characterized by vascular engorgement, lymphocytic infiltration and a serosanguineous effusion. Synovial villous hypertrophy develops: the villi are extensively infiltrated by lymphocytes and plasma cells; suppuration is exceptional. The changes are not pathognomonic.[304] The formation of a pannus originating at the synoviochondral junction accompanies cartilage erosion but cartilage destruction and absorption are limited. Fibrosis and granulation tissue formation occur and adhesions bind the opposing articular surfaces. Ankylosis of the tarsal and carpal joints is evident. The main intervertebral joints are affected and both osteophytosis and intervertebral ankylosis are recognized.

E. rhusiopathiae cannot be isolated from pigs with chronic arthritis. Like experimental rabbit mycoplasmal arthritis and a number of other animal and human arthritides, prolongation of the disease rests upon the persistence of non-infective antigen (p. 543) or upon a secondary, immunological disorder. The severity of the arthritis is related to levels of anti-*Erysipelothrix* antibody.[106] Transmission electron microscopic studies of the synovia of gnotobiotic or specific pathogen-free pigs, rats and mice[261] revealed coagulopathy, vascular necrosis and a fibrinous exudate preceding impaired connective tissue matrix synthesis and cartilage degradation. Immunofluorescence antifibrin tests demonstrated fibrin, IgG and C3 at sites of capillary thrombosis. With time, pannus formation and fibrosis ensued, becoming most severe 5–8 months after experimental infection.[260] Plasmacytosis with IgG formation was shown. The transmission electron microscopic changes included extensive subsynoviocytic cell proliferation, capillary endothelial cell thickening, subendothelial basement membrane broadening and splitting. Some of the appearances recall those described in human rheumatoid arthritis as mesenchymoid transformation.[92]

Anaerobic infections

There is increased interest in the role of anaerobic organisms as causes of joint infection (Table 18.7).[98,215] The

Table 18.7

Some anaerobic bacterial causes of arthritis

Bacteroides fragilis
Bacteroides sp.
Fusobacterium necropolorum
Fusobacterium sp.
Unidentified Gram-negative bacilli
Clostridium welchii
Clostridium sp.
Propionibacterium acnes
Anaerobic diphtheroids or corynebacteria
Peptococcus magnus
Peptococcus sp.
Peptostreptococcus sp. and anaerobic streptococci
Actinomyces sp.

classes of anaerobic bacteria responsible for septic arthritis were tabulated by Fitzgerald et al.[98] according to whether the underlying cause was postoperative, traumatic, or debilitating disease. Anaerobic infection may be caused directly by *Bacteroides fragilis* introduced into a joint in the course of steroid injections,[335] as a result of trauma[8] or occurring in the course of treatment for rheumatoid arthritis.[134]

Clostridia sp. have also been implicated. The most frequent explanation for *C. welchii* infection is trauma. For example, acute septic arthritis may result from lacerating or penetrating injury of the knee in children.[305] However, clostridial infection can follow intraarticular injections[266] and may even be an indirect consequence of adenocarcinoma of the intestinal tract in the absence of injury or treatment to the affected joint.[195]

Salmonella and Shigella infections

Salmonellae, known for their intestinal and visceral lesions, can directly infect joints: they also have a predilection for bone to which they are conveyed by the blood. Vasculitis is a feature of, for example, *S. typhimurium* infection. From bone, synovial joints are occasionally infected directly. The arthritis tends to occur in children and is monoarticular. *Yersinia enterocolitica* may be a cause of pyogenic arthritis.[338] Subacute polyarthritis is a complication of shigella dysentery; it follows 2–4 weeks after the active infection. An immune response to bacterial toxins develops by a mechanism similar to that following salmonella and yersinia infection.

Salmonellae, yersinia and shigellae are well-recognized causes of reactive arthritis (see p. 767). They have attracted growing interest because individuals inheriting HLA-B27 and other transplantation antigens are many times more likely to develop sterile polyarthritis after the infection than those who do not inherit this MHC class I antigen.

Spirochaetal infections

Congenital syphilis and tabes dorsalis are now rare in Western countries and leptospirosis and yaws constitute problems mainly in the Far East. The greatest interest is concentrated on the spirochaetal arthritis of Lyme disease.

Lyme arthritis

Lyme disease[285, 196a] is an endemic infectious multisystem disorder caused by *Borrelia burgdorferi* (Fig. 18.9), a spirochaete transmitted by the bite of a small ixodid tick.[40] The first American case is believed to have occurred in

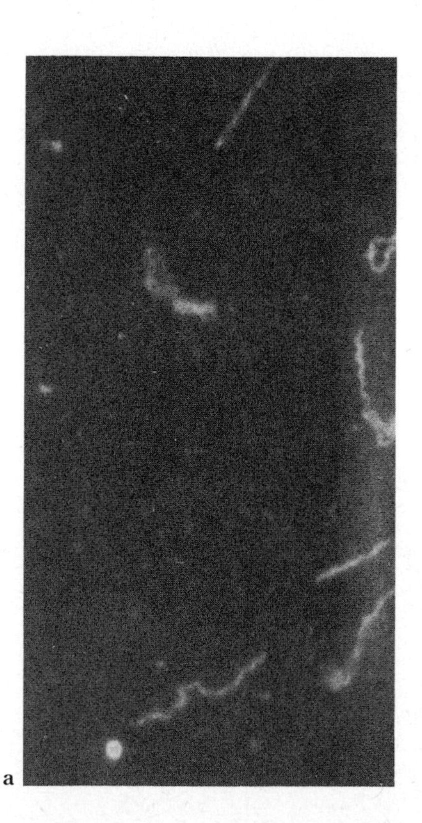

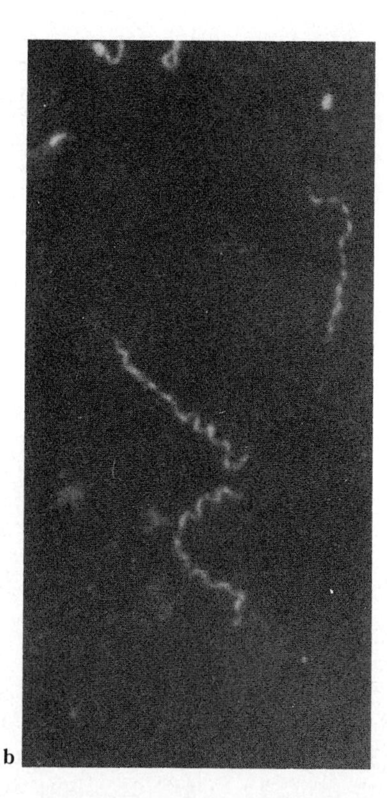

Fig. 18.9 Borrelia.
The spirochaetal organisms that cause Lyme arthritis are members of the family Borrelia. Members of this class of bacteria are shown in this figure. (× 1800.)

1965 but the disease had been known in Europe since 1913.[87,189] Fourteen European cases were described by Huaux *et al.*[152] Attention was drawn to the disorder in 1975 when an unusual geographical cluster of children with what was thought to be juvenile polyarthritis was identified in the Old Lyme area of Connecticut, USA.[286] Subsequently, new cases have been recognized in Europe and Australia.

The systemic disease, common in young people, usually begins during the season when *Ixodes dammini* is 'questing'. A macule or papule forms at the site of the arthropod bite and within 3–30 days a pathognomonic skin lesion develops: erythema chronicum migrans.[284] There is malaise, fever, headache and lymphadenopathy as the primary erythematous and secondary annular skin lesions appear. After a few weeks signs and symptoms of meningeal irritation are followed by a fluctuating meningoencephalitis with facial nerve paralysis and radiculoneuritis. Transient signs of heart block suggest cardiac disease in approximately 8 per cent of patients and hepato- and splenomegaly are among the other clinical signs.

Arthritis develops in 60 per cent of infected individuals; the onset may be as long as two years after the arthropod bite. Intermittent inflammation of one or more large limb joints, particularly the knee, is characteristic. Like serum sickness and, to an extent rheumatic fever, heat and swelling of the affected joint is disproportionate to pain and redness. In 10 per cent of patients, chronic inflammation may progress to joint destruction and secondary bone change. Although aspects of the disease, in particular the histological changes, resemble those of rheumatoid arthritis, there are important, clinical and immunogenetic differences. Lyme arthritis resembles the polyarthritis of Reiter's syndrome; complement levels are less diminished than in rheumatoid arthritis; tests for rheumatoid factor are seldom positive; and the HLA type is usually DRw2, as in systemic lupus erythematosus and multiple sclerosis, rather than Dw4, as in classical rheumatoid arthritis, or B27, as in Reiter's syndrome.

In active erythema chronicum migrans there is oedema of the dermal papillae and a heavy perivascular mononuclear infiltrate extending around dermal appendages and throughout all dermal layers.[284] There is epidermal hyperkeratization, intra- and extracellular oedema and haemosiderin deposition.

The synovium has been examined after synovectomy.[283] The hypertrophic synovium, of high vascularity, is infiltrated with lymphocytes and plasma cells. Two-thirds of the synovial lymphocytes are T-cells, one-fifth B-cells.[282] Synovial cell hyperplasia accompanies a polymorph infiltrate and surface deposits of fibrin. Lymphoid follicular aggregates form. An obliterative vasculopathy is frequent and, occasionally, *B. burgdorferi* can be demonstrated in or near the walls of these onion-skin like arterioles.[165] Pannus formation leads to marginal cartilage destruction. Culture of the affected synovial cells shows that they form large amounts of collagenase and of prostaglandin E_2 but small amounts of acid proteinases.

Lyme disease can be transmitted by intraperitoneal injection of *B. burgdorferi* into hamsters.[81] Spirochaetaemia develops and organisms can be recovered from many tissues ranging from the spleen to the eye. However, articular disease does not develop within nine months of inoculation and synovitis is not seen. The preferred host for immature *I. dammini* is the mouse *Peromycus leucopus*: it is not known whether these and the other vectors, such as deer, develop synovitis.

Syphilitic arthritis

There may be joint infection both in congenital and in acquired syphilis. The pathological changes can be classified as non-gummatous and gummatous. Very full descriptions of the classical disease are available.[158]

In congenital syphilis, osteochondritis and periostitis are accompanied by a low-grade inflammation of joint tissues with hyperaemia, swelling of synovia and a lymphocytic and plasma cell infiltrate. Occasional zones of necrosis may mimic the changes in the gummatous arthritis that is now rarely encountered in adults with tertiary syphilis. When joint swelling is seen in older children with congenital syphilis, there is little other evidence of inflammation and pain is conspicuously slight. The knees are affected, singly or together; to this now rare condition the term Clutton's joint is applied.[51] Histologically, the synovial villi in a typical case were found to be pink, thickened and gelatinous.[14] No spirochaetes were detected. The synovial tissues were oedematous, the synovial cells hyperplastic, and the synovial vessels thick-walled and hyalinized. Around them were clusters of plasma cells; moderate numbers of lymphocytes were scattered among the synovial villi. Corresponding to the synovial reaction there was thickening, oedema, round cell infiltration and increased vascularity of the periosteum. New bone formation was evident with increased density of pre-existing cortical bone.

In acquired syphilis, the disseminated infection which characterizes the secondary stages of the disease may be accompanied by a low-grade arthritis. Again, the knee joints are the most common site. The microscopic response resembles that described in congenital syphilis; it subsides spontaneously leaving no residual evidence of infection.

The features of Charcot's neurogenic arthropathy are described on p. 906 (Fig. 22.39).

Yaws arthritis

In children and adolescents, yaws caused by *Treponema pertenue* infects joint structures and the epiphyseal growth zones, producing changes similar to those encountered in congenital syphilis; 'sabre' deformity of the tibiae is one sequel.[138] Later in life, gummatous nodules are encountered in the periosteum near joints, particularly the elbows and knees. Microscopically, the periarticular nodules comprise a central zone of necrosis surrounded by dense connective tissue.

Leptospiral arthritis

Joint and muscle pains are common symptoms in leptospirosis icterrohaemorrhagica, caused by *Leptospira canicola*, and in rat bite fever caused by *Spirillum minus*. Arthritis and myocarditis have resulted from *L. pomona* infection. Joint involvement is also encountered in the infection caused by *Streptobacillus moniliformis* and in this context is interesting because of the use to which this organism has been put in studying experimental arthritis.

Very little is known of the histological changes in the joints in these diseases although the skeletal muscular lesions in leptospirosis have been described in detail both at necropsy and at biopsy. Focal acute myositis with muscle cell death is followed by the proliferation of sarcolemmal nuclei leading to healing and limited muscle cell regeneration.

Mycobacterial infections
(Table 18.8)

Actinomycotic arthritis is described on p. 749.

Tuberculous arthritis

Arthritis caused by *Mycobacterium tuberculosis*[12,158,171,183,205,220] is now very uncommon in European pathological practice; it is, however, not extinct and is encountered in three groups of patients: the aged, immigrants and the immunosuppressed. In two recent cases, tuberculous vertebral osteitis was mistakenly diagnosed as metastatic carcinoma in an 81-year-old woman with low back pain; and tuberculous synovitis of the wrist was erroneously diagnosed as traumatic tenosynovitis in a 25-year-old Malaysian student.

Extensive classical descriptions of tuberculous arthritis are contained in the *Handbuch*[183] and in modern works on orthopaedics.[82] The degree and frequency of the residual osteoarticular deformities are well represented in the classical pathological museums. The antiquity of the disease has been confirmed by palaeopathological studies. Impressions of the disorder in contemporary societies are best obtained

Table 18.8

Mycobacteria and spirochaetes known to cause arthritis

Acid-alcohol fast bacteria
Mycobacterium tuberculosis hominis
Mycobacterium tuberculosis bovis
Mycobacterium avium
Mycobacterium scrofulaceum
Mycobacterium kansasii
Mycobacterium fortuitum
Mycobacterium marinum
Mycobacterium szulgai
Mycobacterium triviale
Mycobacterium intracellulare
Mycobacterium leprae
Spirochaetes
Treponema pallidum
Borrelia burgdorferi
Streptobacillus moniliformis
Spirillum minis
Leptospira pomona

by studies in northeast Brazil, parts of southeast Asia and West Africa where endemic tuberculosis still abounds.

Tuberculous arthritis is a blood-borne infection disseminated from the lungs or intestine to the small blood vessels of the metaphyseal regions of growing long bones. Tuberculous osteitis (osteomyelitis) results. The joint is disorganized by the destruction of ephiphyseal cartilage or by the spread of infection through the articular soft tissues[183] (Figs 18.10 and 18.11). Occasionally, infection begins in the synovia. The thickened, hyperaemic synovial tissue comes to be covered by a grey-yellow exudate (Fig. 18.10) among which pale, white-yellow tubercles can be detected. The articular cartilage loses its normal translucency; portions of cartilage may be detached from nearby bone. Tuberculous granulation tissue may cover both articular cartilage and the synovial lining of the joint.

Histologically, connective tissue and bone are destroyed (Figs 18.12–18.15). The natural history of the disease is dominated by the onset of tuberculous arteritis with vascular thrombosis and ischaemia. Caseous necrosis is evident. Multinucleated giant-cells are present (Figs 18.12 and 18.13). Similar necrotic material, amorphous or structureless in appearance, lines the affected joint surfaces; within the hyperaemic synovial tissue are caseous foci bounded by epithelioid cells, lymphocytes and smaller numbers of giant cells. Where macrophages predominate, the granulomatous lesions of sarcoidosis or brucellosis may be simulated. Acid- and alcohol-fast bacilli can, however, be demonstrated by the Ziehl-Neelsen or auramine staining methods

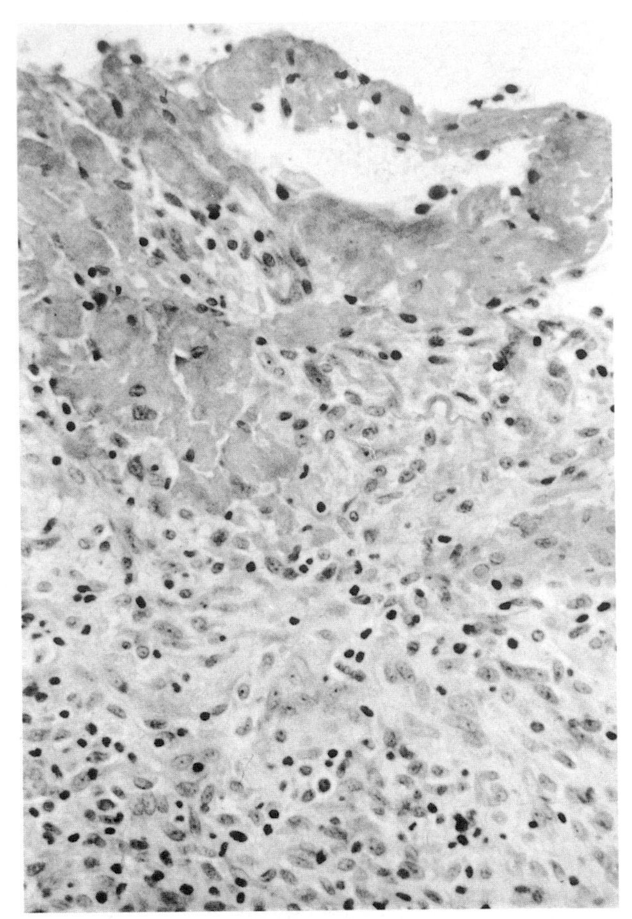

Fig. 18.14 Tuberculous synovitis.
The extensive fibrinous exudate (at top) and the presence of considerable numbers of lymphocytes and occasional polymorphs closely resembles appearances commonly seen in rheumatoid arthritis. (HE × 150.)

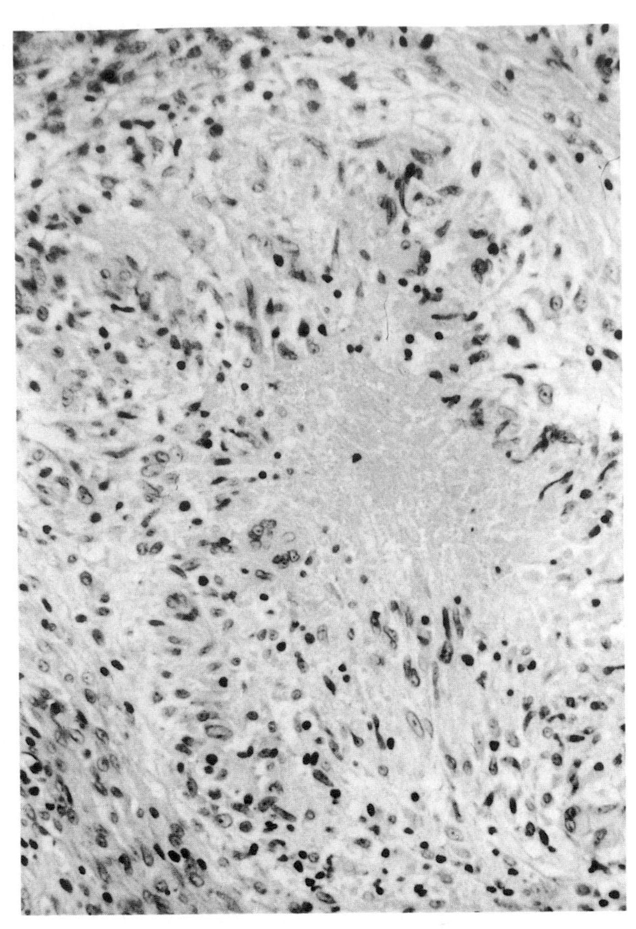

Fig. 18.15 Tuberculous synovitis.
In the centre of the field, a focus of caseous necrosis is outlined by arrays of elongated mononuclear macrophages together with small numbers of lymphocytes. The appearances recall those of rheumatoid synovial granuloma. (HE × 150.)

rheumatoid arthritis cross-reactivity to proteoglycan. Evidence to support this view was obtained by Ottenhoff *et al.*[230] who found high responsiveness to antigens specifically associated with *Myc. tuberculosis* but not to other mycobacterial antigens.

Atypical mycobacterial infections

It is well recognized that arthritis may be caused by 'saprophytic', 'anonymous' or unclassified bacteria related to *Mycobacterium tuberculosis* but differing in cultural, metabolic or colonial features.[181] Some gain access to the tissues as a result of minor trauma or are conveyed with foreign bodies: in others the mode of access remains uncertain but may be via pulmonary, gastrointestinal, skin or haematogenous routes. Infection may be opportunistic. Hoffman *et al.*[143] described an example of synovitis of the knee caused by *Myc. avium* in a man with systemic lupus erythematosus treated with prednisone and draw attention to 46 cases attributable to *Myc. scrofulaceum*, *Myc. kansasii*,[186] *Myc. fortuitum*, *Myc. marinum*, *Myc. szulgai*, *Myc. triviale*, *Myc. intracellulare* or unidentified mycobacterium sp. *Myc. marinum*, as its name suggests is likely to originate in the marine environment. Very little is known of the histological features in these 'new' infections (Figs 18.18 and 18.19) but the observed changes closely resemble those of the BCG reaction (Fig. 18.20).

Leprous arthritis

Infection with the obligate, very slowly multiplying intracellular parasite *Mycobacterium leprae*[27,70,157,168,244] results in a reactive syndrome the effects of which are determined mainly by the responses of the host. Where there is little cell-mediated immunity, lepromatous lesions of the skin, nasal mucosa, eye and lymph nodes develop;

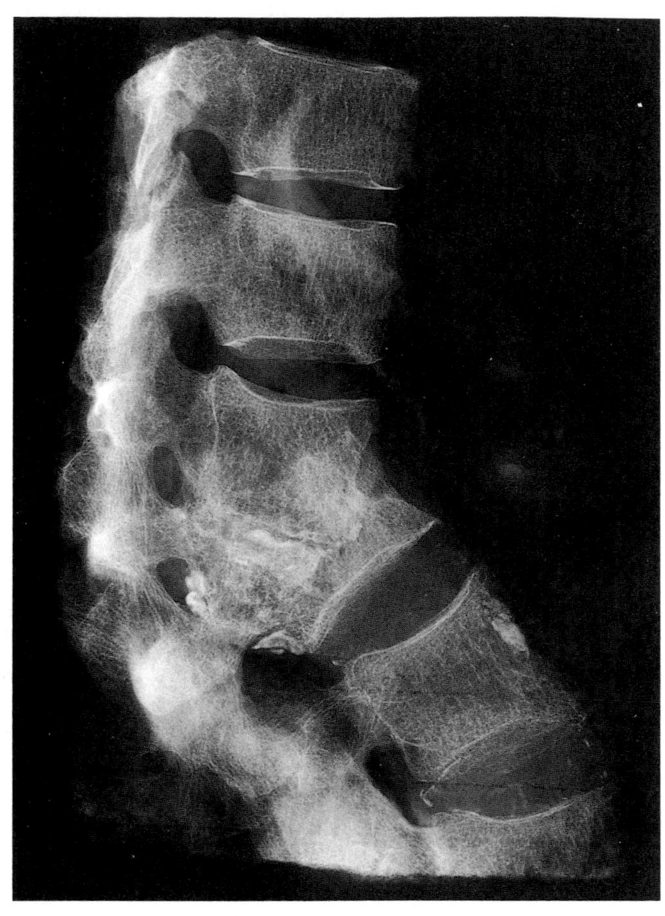

Fig. 18.16 Tuberculous vertebral osteitis.
Radiograph of specimen from case of healed tuberculous osteitis (Pott's disease). The cauda equina was spared. There is destruction of parts of two vertebral bodies and the intervening disc. New bone formation has led to the presence of a single bone mass replacing the injured tissues. (× ⅔.)

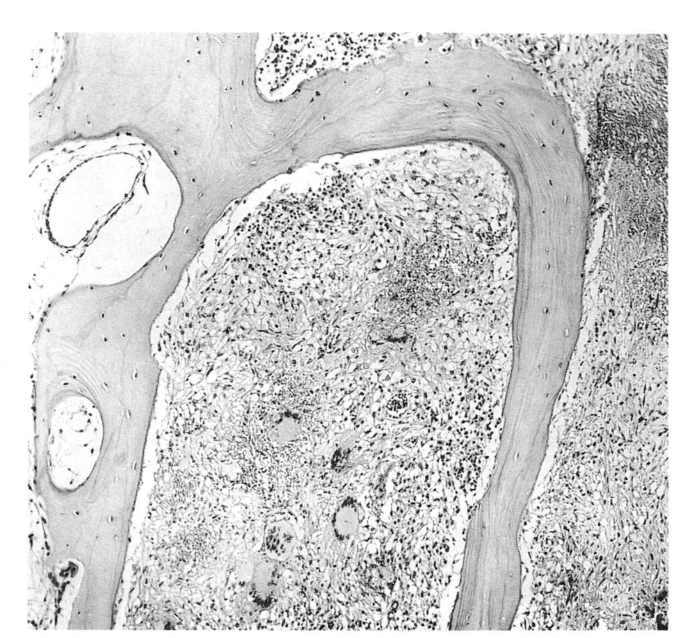

Fig. 18.17 Tuberculous vertebral osteitis.
Vertebral bone from male aged 80 years dying from miliary tuberculosis. Granulation tissue containing many lymphocytes and several multinucleated giant cells adjoins an island of caseous necrosis. (HE × 75.)

there may be circulating antibody and treatment can promote immune complex formation, vasculitis, erythema nodosum leprosum and synovitis. Late in the disease, amyloid may accumulate. Tuberculosis may coexist. Where a vigorous cell-mediated immune response occurs, the tuberculoid lesions are dominated by activated macrophages which, together with lymphocytes, constitute the tuberculoid granuloma. Nerve lesions result in sensory loss, so that muscle wasting, contractures, hand and foot ulcers and bone resorption culminate in a variety of secondary articular disorders.

Non-specific, 'rheumatic' clinical manifestations in leprosy may mimic those of the systemic connective tissue diseases, delaying diagnosis. However, there are at least three forms of synovitis directly attributable to leprosy: 1. *Myc. leprae* may directly infect the synovia;[146] 2. an immune complex synovitis syndrome may be part of the lepra reaction (reactional state) that is associated with the lepromatous disease; and 3. a secondary form of neuropathic arthropathy may complicate the neurological lesions of tuberculoid leprosy.

At the risk of some confusion with the nomenclature of hypersensitivity, leprologists now distinguish a type I lepra reaction encountered in borderline leprosy and involving a

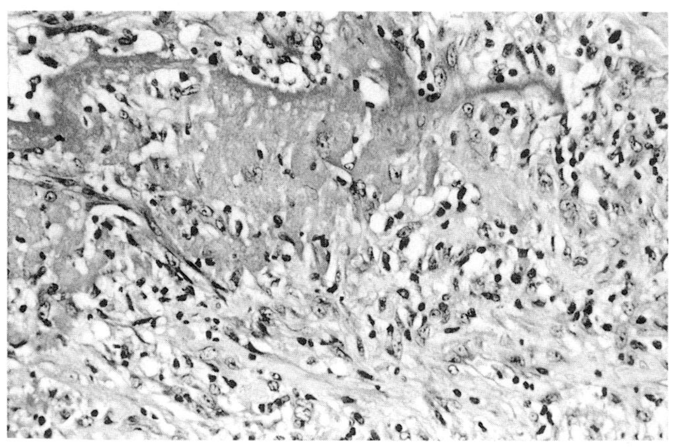

Fig. 18.18 Synovitis of wrist due to atypical mycobacterial infection.
Male aged 49 years with synovitis due *Mycobacterium kansasii*. Fibrin (dark grey) lies within granulation tissue in which there is a sparse lymphocytic infiltrate. (HE × 110.) Courtesy of Dr Peter Revell.

periostitis: a case report with histology of synovium. *Arthritis and Rheumatism* 1963; **6**: 341–8.

15. Argen R J, Wilson C H, Wood P. Suppurative arthritis: clinical features of 42 cases. *Archives of Internal Medicine* 1966; **117**: 661–6.

16. Armstrong R, Simmons N, Wilton J M A, Dyson M, Laurent R, Millis R, Mims C A. Chronic arthritis associated with the presence of intrasynovial rubella virus. *Annals of the Rheumatic Diseases* 1983; **42**: 2–13.

17. Atkin S L, Kamel M, El-Hady A M, El-Badawy S A, El-Ghobary A, Dick W C. Schistosomiasis and inflammatory polyarthritis: a clinical, radiological and laboratory study of 96 patients infected by *S. mansoni* with particular reference to the diarthrodial joints. *Quarterly Journal of Medicine* 1986; **NS59**: 479–87.

18. Backman A, Wallgren E I. On tuberculosis of bones and joints of BCG vaccinated children. *Acta Paediatrica* 1954; **43**: 252–8.

19. Banks K L, Jacobs C A, Michaels F H, Cheevers W P. Lentivirus infection augments concurrent antigen-induced arthritis. *Arthritis and Rheumatism* 1987; **30**: 1046–53.

20. Barden J A, Decker J L. Mycoplasma hyorhinis swine arthritis. I. Clinical and microbiologic features. *Arthritis and Rheumatism* 1971; **14**: 193–201.

21. Bayer A S. Arthritis associated with hepatitis: clinical and pathogenetic considerations. *Postgraduate Medicine* 1980; **67**: 175–8.

22. Bayer A S, Choi C, Tillman D B, Guze L B. Fungal arthritis V. cryptococcal and histoplasmal arthritis. *Seminars in Arthritis and Rheumatism* 1980; **9**: 218–27.

23. Bayer A S, Guze L B. Fungal arthritis. I. Candida arthritis: diagnostic and prognostic implications and therapeutic considerations. *Seminars in Arthritis and Rheumatism* 1978; **8**: 142–50.

24. Bayer A S, Guze L B. Fungal arthritis. II. Coccidioidal synovitis: clinical diagnostic, therapeutic and prognostic considerations. *Seminars in Arthritis and Rheumatism* 1979; **8**: 200–11.

25. Bayer A S, Scott V J, Guze L B. Fungal arthritis. IV. Blastomycotic arthritis. *Seminars in Arthritis and Rheumatism* 1979a; **9**: 145–51.

26. Bayer A S, Scott V J, Guze L B. Fungal arthritis. III: Sporotrichal arthritis. *Seminars in Arthritis and Rheumatism* 1979b; **9**: 66–74.

27. Beitzke H. Leprous disease of joints. In: Lubarsch O, Henke F, eds, *Handbuch der speziellen pathologischen Anatomie und Histologie*. Berlin: Springer Verlag, 1934; Volume 9, part II: 607–11.

28. Beitzke H. Seltene Mykosen der Knochen und Gelenke. In: Lubarsch O, Henke F, eds, *Handbuch der speziellen pathologischen Anatomie und Histologie*. Berlin: Springer Verlag, 1934: 612–34.

29. Bennett D, Taylor D J. Bacterial infective arthritis in the dog. *Journal of Small Animal Practice* 1988; **29**: 207–30.

30. Bennett N M, Cunningham A L, Fraser J R E, Speed B R. Epidemic polyarthritis acquired in Fiji. *Medical Journal of Australia* 1980; **1**: 316–7.

31. Bisbe J, Vilardell J, Valls M, Moreno M, Brancos M, Andrew J. Transient fungemia and candida arthritis due to Candida zeylanoides. *European Journal of Clinical Microbiology* 1987; **6**: 668–9.

32. Blotzer J W, Myers A R. Echovirus-associated polyarthritis. *Arthritis and Rheumatism* 1978; **21**: 978–81.

33. Blumberg S, Bienfang D, Kantrowitz F G. A possible association between influenza vaccination and small-vessel vasculitis. *Archives of Internal Medicine* 1980; **140**: 847–8.

34. Borenstein D G, Simon G L. *Hemophilus influenzae* septic arthritis in adults. A report of four cases and a review of the literature. *Medicine* 1986; **65**: 191–201.

35. Boudoulas O, Camisa C. *Nocardia asteroides* infection with dissemination to skin and joints. *Archives of Dermatology* 1985; **121**: 898–900.

36. Bracewell C D, Corbel M J. An association between arthritis and persistent serological reactions to Brucella abortus in cattle from apparently brucellosis-free herds. *Veterinary Record* 1980; **106**: 99–101.

37. Brassfield A L, Adams D S, Crawford T B, McGuire T C. Ultrastructure of arthritis induced by caprine retrovirus. *Arthritis and Rheumatism* 1982; **25**: 930–6.

38. Breckenridge R L, Buck L, Tooley E, Douglas G W. Listeria monocytogenes septic arthritis. *American Journal of Clinical Pathology* 1980; **73**: 140–1.

39. Bunning V K, Raybourne R B, Archer D L. Food-borne enterobacterial pathogens and rheumatoid disease. *Society for Applied Bacteriology Symposium Series* 1988; **17**: 87S–107S.

40. Burgdorfer W, Barbour A G, Hayes S F, Benach J L, Grunwaldt E, Davis J P. Lyme disease—a tick-borne spirochetosis? *Science* 1982; **216**: 1317–9.

41. Caranasos G J, Felker J R. Mumps arthritis. *Archives of Internal Medicine* 1967; **119**: 394–8.

42. Chambers R J, Bywaters E G L. Rubella synovitis. *Annals of the Rheumatic Diseases* 1963; **22**: 263–8.

43. Chandler F W, Kaplan W, Ajello L. *A Colour Atlas and Textbook of the Histopathology of Mycotic Diseases*. London: Wolfe Medical Publications, 1980.

44. Chantler J K, Ford D K, Tingle A J. Persistent rubella infection and rubella-associated arthritis. *Lancet* 1982; **i**: 1323–5.

45. Chantler J K, Tingle A J, Petty R E. Persistent rubella virus infection associated with chronic arthritis in children. *New England Journal of Medicine* 1985; **313**: 1177–23.

46. Cheevers W P, Roberson S, Klevjer-Anderson P, Crawford T B. Characterization of caprine arthritis-encephalitis virus: a retrovirus of goats. *Archives of Virology* 1981; **67**: 111–17.

47. Chiari H. Die eitrige Gelenkentzündungen. In: Lubarsch O, Henke F. eds, *Handbuch der speziellen pathologischen Anatomie und Histologie*. Berlin: Springer Verlag, 1934: 12–74.

48. Christie A B. *Infectious Diseases: Epidemiology and Clinical Practice* (3rd edition). Edinburgh: Churchill Livingstone, 1980: 775–99.

49. Clague H W, Harth M, Hellyer D, Morgan W K C. Septic arthritis due to *Nocardia asteroides* in association with pulmonary alveolar proteinosis. *Journal of Rheumatology* 1982; **9**: 469–72.

50. Clements J E, Narayan O, Cork L C. Biochemical characterization of the virus causing leukoencephalitis and arthritis in goats. *Journal of General Virology* 1980; **50**: 423–8.

51. Clutton H H. Symmetrical synovitis of the knee in hereditary syphilis. *Lancet* 1886; **i**: 391–3.

52. Cockshott P, MacGregor M. Osteomyelitis variolosa. *Quarterly Journal of Medicine* 1958; **27**: 369–87.

53. Cohen B J, Buckley M M, Clewley J P, Jones V E, Puttick A H, Jacoby R K. Human parvovirus infection in early rheumatoid and inflammatory arthritis. *Annals of the Rheumatic Diseases* 1986; **45**: 832–8.

54. Cole, B C, Cassell G H. Mycoplasma infections as models of chronic joint inflammation. *Arthritis and Rheumatism* 1979; **22**: 1375–81.

55. Collier W A. Infectious polyarthritis of rats. *Journal of Pathology and Bacteriology* 1939; **48**: 579–89.

56. Collins D H. Bacterial arthritis—specific infections of the joints. In: *The Pathology of Articular and Spinal Diseases*. London: Arnold, 1949: 130–49.

57. Cons F, Trevino A, Lavalle C. Septic arthritis due to *Nocardia brasiliensis*. *Journal of Rheumatology* 1985; **12**: 1019–21.

58. Cook G C. Opportunistic parasitic infections associated with the acquired immune deficiency syndrome (AIDS): parasitology, clinical presentation, diagnosis and management. *Quarterly Journal of Medicine* 1987; **65**: 967–83.

59. Cooper C, Cawley M I D. Bacterial arthritis in an English health district: a ten year review. *Annals of the Rheumatic Diseases* 1986; **45**: 458–63.

60. Cooper L Z, Ziring P R, Weiss H J, Matters B A, Krugman S. Transient arthritis after rubella vaccination. *American Journal of Diseases of Children* 1969; **118**: 218–25.

61. Cork L C, Hadlow W J, Crawford T B, Gorham J R, Piper R C. Infectious leukoencephalomyelitis of young goats. *Journal of Infectious Diseases* 1974; **129**: 134–41.

62. Crawford T B, Adams D S, Cheevers W P, Cork L C. Chronic arthritis in goats caused by a retrovirus. *Science* 1980; **207**: 997–9.

63. Crawford T B, Adams D S, Sande R D, Gorham J R, Henson J B. The connective tissue component of the caprine arthritis–encephalitis syndrome. *American Journal of Pathology* 1980; **100**: 443–54.

64. Currey H L F. Animal models of arthritis. In: Scott J T, ed, *Copeman's Textbook of the Rheumatic Diseases* (6th edition). Edinburgh, London: Churchill Livingstone, 1986: 468–85.

65. Curtis G D W, Newman R J, Slack M P E. Synovial fluid lactate and the diagnosis of septic arthritis. *Journal of Infection* 1983; **6**: 239–46.

66. Curtiss P H. Cartilage damage in septic arthritis. *Clinical Orthopaedics and Related Research* 1969; **64**: 87–90.

67. Curtiss P H. The pathophysiology of joint infections. *Clinical Orthopaedics and Related Research* 1973; **96**: 129–35.

68. Curtiss P H, Klein L. Destruction of articular cartilage in septic arthritis. II: *In vivo* studies. *Journal of Bone and Joint Surgery* 1965; **47**-A: 1595–604.

69. Dan M. Neonatal septic arthritis. *Israel Journal of Medical Sciences* 1983; **19**: 967–71.

70. David-Chausse J, Texier L, Dehais J, Bullier R, Louis-Joseph L. Manifestations articulaires au cour de deux cas de lèpre. *Bordeaux Mèdicale* 1978; **14**: 1183–90.

71. Davis W J, Larson H E, Simsarian J P, Parkmen P D, Meyer H M. Rubella immunity and resistance to infection. *Journal of the American Medical Association* 1971; **215**: 600–9.

72. DeMartini J C, Banks K L, Greenlee A, Adams D S, McGuire T C. Augmented T lymphocyte numbers in goats chronically infected with the retrovirus causing caprine arthritis-encephalitis. *American Journal of Veterinary Research* 1983; **44**: 2064–9.

73. Denman M. Viral aetiology of arthritis. *Bulletin on Rheumatic Diseases* 1987; series **2**: 1–4.

74. Deresinski S C, Stevens D A. Bone and joint coccidioidomycosis treated with miconazole. *American Review of Respiratory Disease* 1979; **120**: 1101–8.

75. Doherty R L, Gorman B M, Whitehead R H, Carley J G. Studies of epidemic polyarthritis: the significance of three Group A arboviruses isolated from mosquitoes in Queensland. *Australasian Annals of Medicine* 1964; **13**: 322–7.

76. Dowling P G. Epidemic polyarthritis. *Medical Journal of Australia* 1946; **1**: 245–6.

77. Drouhet E, Guilmet O, Kouvalchouk J-F, Chapman A, Ziza J-M, Laudet J, Brodaty D. First case of *Drechslera longirostrata* fungal infection in man. Spondylodiscitis as complication of endocarditis from valve prosthesis infection. Treatment with combined ketoconazole and amphotericin B. *La Nouvelle Presse Mèdicale* 1982; **11**: 3631–5.

77a. Dumonde D C, Steward M W. The role of infection in rheumatic disease. In: Scott J T, ed, *Copeman's Textbook of the Rheumatic Diseases* (5th edition). Edinburgh, London: Churchill Livingstone, 1978: 222–58.

78. Duncan J R, Ross R F. Fine structure of the synovial membrane in *Mycoplasma hyorhinis* arthritis of swine. *American Journal of Pathology* 1969; **57**: 171–86.

79. Dupont B, Drouhet E. Cutaneous, ocular, and osteoarticular candidiasis in heroin addicts: new clinical and therapeutic aspects in 38 patients. *Journal of Infectious Diseases* 1985; **152**: 577–91.

80. Duran H, Ferrandez L, Gomez-Castresana F, Lopez-Duran L, Mata P, Brandau D, Sanchez-Barbo A. Osseous hydatidosis. *Journal of Bone and Joint Surgery* 1978; **60**A: 685–90.

81. Duray P H, Johnson R C. The histopathology of experimentally infected hamsters with the Lyme disease spirochete, *Borrelia burgdorferi*. *Proceedings of the Society for Experimental Biology and Medicine* 1986; **181**: 263–9.

82. Duthie R B, Bentley G. Infections of the musculoskeletal system. In: *Mercer's Orthopaedic Surgery* (8th edition). London: Arnold, 1983: 466–552.

83. Duthie R B, Ferguson A B. *Mercer's Orthopaedic Surgery* (7th edition). London: Arnold, 1973: 487–530.

84. Editorial. Rubella synovitis. *British Medical Journal* 1963; **II**: 1547.

85. Editorial. Rheumatic manifestations in leprosy. *Lancet* 1981; **i**: 648–9.

86. Editorial. Rheumatoid arthritis and tuberculosis. *Lancet* 1986; **ii**: 321–2.

87. Editorial. Lyme disease in Europe. *Lancet* 1987; **ii**: 264–5.

88. Esposito A L, Gleckman R A. Acute polymicrobic septic arthritis in the adult: case report and literature review. *American Journal of the Medical Sciences* 1974; **267**: 251–4.

89. Esterley J R, Oppenheimer E H. Pathological lesions due to congenital rubella. *Archives of Pathology* 1969; **87**: 380–8.

90. Fainstein V, Gilmore C, Hopfer R L, Maksymiuk A, Bodey

studied by latex tests. *New England Journal of Medicine* 1958; **258**: 743–5.

165. Johnston Y E, Duray P H, Steere A C, Kashgarian M, Buza J, Malawista S E, Askenase P W. Lyme arthritis: spirochetes found in synovial microangiopathic lesions. *American Journal of Pathology* 1985; **118**: 26–34.

166. Jones R C, Goodwin R A. Histoplasmosis of bone. *American Journal of Medicine* 1981; **70**: 464–6.

167. Jones P G, Rolston K, Hopfer R L. Septic arthritis due to *Histoplasma capsulatum* in a leukaemic patient. *Annals of the Rheumatic Diseases* 1985; **44**: 128–9.

168. Karat A B A, Karat S, Job C K, Furness M A. Acute exudative arthritis in leprosy: a rheumatoid arthritis-like syndrome in association with erythema nodosum leprosum. *British Medical Journal* 1967; **II**: 770–2.

169. Keat A C, Maini R N, Nkwazi G C, Pegrum G D, Ridgway G L, Scott J T. Role of chlamydia trachomatis and HLA-B27 in sexually acquired reactive arthritis. *British Medical Journal* 1978; **1**: 605–7.

170. Keiser H, Ruben F L, Wolinsky E, Kushner I. Clinical forms of gonococcal arthritis. *New England Journal of Medicine* 1968; **279**: 234–40.

171. Kelly P J, Karlson A G. Granulomatous bacterial arthritis. *Clinical Orthopaedics and Related Research* 1973; **96**: 165–7.

172. Kennedy A C, Fleming J, Solomon L. Chikungunya viral arthropathy: a clinical description. *Journal of Rheumatology* 1980; **7**: 231–6.

173. Kennedy-Stoskopf S, Narayan O, Strandberg J D. The mammary gland as a target organ for infection with caprine arthritis encephalitis virus. *Journal of Comparative Pathology* 1985; **95**: 609–17.

174. Key J A, Large A M. Histoplasmosis of the knee. *Journal of Bone and Joint Surgery* 1942; **24**: 281–90.

175. Keystone E C, Cunningham A J, Metcalf A, Kennedy M, Quinn P A. Role of antibody in the protection of mice from arthritis induced by *Mycoplasma pulmonis*. *Clinical and Experimental Immunology* 1982; **47**: 253–9.

176. Keystone E C, Taylor-Robinson D, Osborn M F, Ling L, Pope C, Fornasier V. Effect of T-cell deficiency on the chronicity of arthritis induced in mice by *Mycoplasma pulmonis*. *Infection and Immunity* 1980; **27**: 192–6.

177. Kilroy A W, Schaffner W, Fleet W F, Lefkowitz L V, Karzon D T, Fenichel G N. Two syndromes following rubella immunization. *Journal of the American Medical Association* 1970; **214**: 2287–92.

178. Kirchhoff H, Heitmann J, Mielke H, Dubenkropp H, Schmidt R. Studies of polyarthritis caused by *Mycoplasma arthritidis* in rats. II. Serological investigation of rats experimentally infected with *Mycoplasma arthritidis* ISR 1. *Zentralblatt für Bakteriologie, Mikrobiologie, und Hygiene* 1983; **254**: 275–80.

179. Klevjer-Anderson P, Adams D S, Anderson L W, Banks K L, McGuire T C. A sequential study of virus expression in retrovirus-induced arthritis of goats. *Journal of General Virology* 1984; **65**: 1519–25.

180. Klevjer-Anderson P, Cheevers W P. Characterization of the infection of caprine synovial membrane cells by the retrovirus caprine arthritis-encephalitis virus. *Virology* 1981; **110**: 113–9.

181. Klinenberg J R, Grimley P M, Seegmiller J E. Destructive polyarthritis due to a photochromogenic mycobacterium. *New England Journal of Medicine* 1965; **272**: 190–3.

182. Kohn D F, Magill L S, Chinookoswong N. Localization of *Mycoplasma pulmonis* in cartilage. *Infection and Immunity* 1982; **35**: 730–3.

183. Konschegg T. Die Tuberkulose der Gelenke. In: Lubarsch O, Henke F, eds, *Handbuch der speziellen pathologischen Anatomie und Histologie*. Berlin: Springer Verlag, 1934: 438–68.

184. Lam K, Silverstein L M, Carlisle R J, Bayer A S. Disseminated brucellosis initially seen as sternoclavicular arthropathy. *Archives of Internal Medicine* 1982; **142**: 1193–6.

185. Langer H E, Bialek R, Mielke H, Klose J. Human dirofilariasis with reactive arthritis—case report and review of the literature. *Klinische Wochenschrift* 1987; **65**: 746–51.

186. Leader M, Revell P, Clarke G. Synovial infection with *Mycobacterium kansasii*. *Annals of the Rheumatic Diseases* 1984; **43**: 80–2.

187. LeFrock J, Mader J, Smith B, Carr B. Bone and joint infections caused by Gram-positive bacteria: treatment with cefotaxime. *Infection* 1985; **13**; Supplement 1: S50–S55.

188. Levy B. Arthritis in venereal disease, with particular reference to aetiology. *Medical Journal of Malaya* 1959; **5**: 42–58.

189. Lipschutz B. Ueber ein seltene Erythemform (Erythema chronicum migrans). *Archiv für Dermatologie und Syphilis* 1913; **118**: 349–56.

190. Lopez-Longo F J, Monteagudo I, Vaquero F J, Martinez Moreno J L M, Carreno L. Primary septic arthritis of the manubriosternal joint in a heroin user. *Clinical Orthopaedics and Related Research* 1986; **202**: 230–1.

191. Loveridge P. Legionnaires' disease and arthritis. *Canadian Medical Association Journal* 1981; **124**: 366–7.

192. Lumsden W H R. Epidemic of virus disease in southern province Tanganyika Territory in 1952–53: general description and epidemiology. *Transactions of the Royal Society of Tropical Medicine and Hygiene* 1955; **49**: 33–57.

193. McIntosh K. Recent advances in viral diagnosis. *Archives of Pathology and Laboratory Medicine* 1980; **104**: 30–4.

194. Mackenzie J S. *Viral Diseases in South-East Asia and the Western Pacific*. Sydney: Academic Press, 1982.

195. Macy N J, Lieber L, Habermann E T. Arthritis caused by *Clostridium septicum*—a case report and review of the literature. *Journal of Bone and Joint Surgery* 1986; **68**-A: 465–6.

196. Maisondieu P. Etude des manifèstations articulaires des oreillons. A propos de quelques observations. MD Thesis (No. 519) Paris. *Imprimèrie de la Faculté de Médecine* 1924; 23–4, 40–9, 72–3.

196a. Malawista S E. Lyme disease. In: McCarty D J, *Arthritis and Allied Conditions. A Textbook of Rheumatology* (11th edition). Philadelphia, London: Lea and Febiger, 1989: 1955–65.

197. March L M, Needs C J, Webb J. Streptococcus group G septic polyarthritis. *Australian and New Zealand Journal of Medicine* 1985; **15**: 647–9.

198. Marmion B P. A microbiologist's view of investigative rheumatology. In: Dumonde D C, ed, *Infection and Immunology in the Rheumatic Diseases*. Oxford: Blackwell Scientific Publications, 1976: 245–58.

199. Marmion B P, McKay J M. Rheumatoid arthritis and the virus hypothesis. In: Glynn L E, Schumberger H D, eds, *Experimental Models of Chronic Inflammatory Disease.* Bayer Symposium VI. New York: Springer Verlag, 1978; 188–211.

200. Marmor L, Peter J B. Candida arthritis of the knee joint. *Clinical Orthopaedics and Related Research* 1976; **118**: 133–5.

201. Martin J R, Root H S, Kim S O, Johnson L G. Staphylococcus suppurative arthritis occurring in neuropathic knee joints: a report of four cases with a discussion of the mechanisms involved. *Arthritis and Rheumatism* 1965; **8**: 389–402.

202. Matthews M, Shen F H, Lindner A, Sherrard D J. Septic arthritis in hemodialyzed patients. *Nephron* 1980; **25**: 87–91.

203. Mayer M E, Geiseler J, Harris B. Rapid diagnosis of septic arthritis by coagglutination. *Journal of Clinical Microbiology* 1983; **18**: 1424–6.

204. Melnick J L, ed, *Progress in Medical Virology.* Basel: S Karger AG, 1982: Volume 28.

205. Messner R P. Arthritis due to mycobacteria and fungi. In: McCarty D J, ed, *Arthritis and Allied Conditions* (10th edition). Philadelphia: Lea & Febiger, 1985: 1687–96.

206. Messner R P. Arthritis due to mycobacteria, fungi, and parasites. In: McCarthy D J, ed, *Arthritis and Allied Conditions. A Textbook of Rheumatology* (11th edition). Philadelphia, London: Lea and Febiger, 1989: 1925–37.

207. Miles J A R. Arthropod-borne virus diseases of man. *Australian Annals of Medicine* 1961; **10**: 317–26.

208. Mitchell H, Travers R, Barraclough D. Septic arthritis caused by *Pasteurella multocida. Medical Journal of Australia* 1982; **1**: 137.

209. Monto A S, Cavallaro J J, Whale E H. Frequency of arthralgia in women receiving one of three rubella vaccines. *Archives of Internal Medicine* 1970; **126**: 635–9.

210. Mudge P R, Aaskov J G. Epidemic polyarthritis in Australia 1980–1981. *Medical Journal of Australia* 1983; **70**: 269–73.

211. Muggia A L, Bennahum D A, Williams R C. Navajo arthritis—an unusual, acute, self-limited disease. *Arthritis and Rheumatism* 1971; **14**: 348–55.

212. Mulhern L M, Friday G A, Perri J A. Arthritis complicating varicella infection. *Pediatrics* 1971; **48**: 827–9.

213. Murray H W, Fialk M A, Roberts R B. Candida arthritis. A manifestation of disseminated candidiasis. *American Journal of Medicine* 1976; **60**: 587–94.

214. Myers A R. Septic arthritis caused by bacteria. In: Kelley W N, Harris E D, Ruddy S, Sledge CB, eds, *Textbook of Rheumatology* (2nd edition). Philadelphia: W B Saunders, 1985: 1507–27.

215. Nakata M M, Lewis R P. Anaerobic bacteria in bone and joint infections. *Reviews of Infectious Disease* 1984; **6** (Supplement 1): S165–70.

216. Nakata M M, Silvers J H, George W L. Group G streptococcal arthritis. *Archives of Internal Medicine* 1983; **143**: 1328–30.

217. Narayan O, Kennedy-Stoskopf S, Sheffer D, Griffin D E, Clements J E. Activation of caprine arthritis-encephalitis virus expression during maturation of monocytes to macrophages. *Infection and Immunity* 1983; **41**: 67–73.

218. Narayan O, Kennedy-Stoskopf S, Zink M C. Lentivirus–host interactions: lesions from visna and caprine arthritis–encephalitis viruses. *Annals of Neurology* 1988; **23** (supplement): S95–S100.

219. Narayan O, Sheffer D, Griffin D E, Clements J, Hess J. Lack of neutralizing antibodies to caprine arthritis–encephalitis lentivirus in persistently infected goats can be overcome by immunization with inactivated mycobacterium tuberculosis. *Journal of Virology* 1984; **49**: 349–55.

220. Nieberle K. Die Entstehung und Entwicklung der Tuberkulose der Haustiere. *Zeitschrift für Infektionskrankheiten, parasitären Krankeheiten und Hygiene der Haustiere* 1930; **38**: 219–32.

221. Nimmo J R. An unusual epidemic. *Medical Journal of Australia* 1928; **1**: 549–50.

222. Nitsche J F, Vaughan J H, Williams G, Curd J G. Septic sternoclavicular arthritis with *Pasteurella multocida* and *Streptococcus sanguis. Arthritis and Rheumatism* 1982; **25**: 467–9.

223. Norden C W, Sellers T F. *Hemophilus influenza* pyarthyrosis in an adult. *Journal of the American Medical Association* 1964; **189**: 694–5.

224. Norenberg D D, Bigley V, Virata R L, Liang G C. *Corynebacterium pyogenes* septic arthritis and plasma cell synovial infiltrate and monoclonal gammopathy. *Archives of Internal Medicine* 1978; **138**: 810–11.

225. Noyes F R, McCabe J D, Fekety R. Acute candida arthritis: report of a case and use of amphotericin B. *Journal of Bone and Joint Surgery* 1973; **55**A: 169–76.

226. Ogden J A. Pediatric osteomyelitis and septic arthritis: the pathology of neonatal disease. *Yale Journal of Biology and Medicine* 1979; **52**: 423–48.

227. Ogra P L, Herd K. Arthritis associated with induced rubella infection. *Journal of Immunology* 1971; **107**: 810–31.

228. Onion D K, Crumpacker C S, Gilliland B C. Arthritis of hepatitis associated with Australia antigen. *Annals of Internal Medicine* 1971; **75**: 29–33.

229. Osler W. *The Principles and Practice of Medicine* (6th edition). New York: Appleton & Co., 1906: 146.

230. Ottenhoff T H M, Torres P, de las Aguas J T, Fernandez R, Eden W van, de Vries R R P, Stanford J L. Evidence for an HLA-DR4-associated immune-response gene for *Mycobacterium tuberculosis. Lancet* 1986; **ii**: 310–3.

231. Petty B G, Sowa T, Charache P. Polymicrobial polyarticular septic arthritis. *Journal of the American Medical Association* 1983; **249**: 2069–72.

232. Pischel K D, Weisman M H, Cone R O. Unique features of group B streptococcal arthritis in adults. *Archives of Internal Medicine* 1985; **145**: 97–102.

233. Porat S, Shapiro M. Brucella arthritis of the sacroiliac joint. *Infection* 1984; **12**: 55/205–57/207.

234. Porter D D. Aleutian disease: a persistent parvovirus infection of mink with a maximal but ineffective host humoral immune response. *Progress in Medical Virology* 1986; **33**: 42–60.

235. Porter D D, Larsen A E, Porter H G. Aleutian disease of mink. In: Kunkel H G, Dixon F, eds, *Advances in Immunology* 1980; **29**: 261–86.

236. Pott P. Remarks on that kind of palsy of the lower limbs, which is frequently found to accompany a curvature of the spine. London: J Johnson, 1779.

237. Powell J M, Bass J W. Septic arthritis caused by *Kingella*

kingae. American Journal of Diseases of Children 1983; **137**: 974–6.

238. Quismorio F P, Jakes J T, Zarnow A J, Barber D, Kitridou R C. Septic arthritis due to *Arizona hinshawii. Journal of Rheumatology* 1983; **10**: 147–50.

239. Ray C G, Gall E P, Minnich L L, Roediger J, Benedetti C de, Corrigan J J. Acute polyarthritis associated with active Epstein–Barr virus infection. *Journal of the American Medical Association* 1982; **248**: 2990–3.

240. Raymond J, Bergeret M, Bargy F, Missenard G. Isolation of two strains of *Kingella kingae* associated with septic arthritis. *Journal of Clinical Microbiology* 1986; **24**: 1100–1.

241. Reid D M, Reid T M S, Brown T, Rennie J A N, Eastmond C J. Human parvovirus-associated arthritis: a clinical and laboratory description. *Lancet* 1985; **i**: 422–5.

242. Remafedi G, Muldoon R L. Acute monarticular arthritis caused by herpes simplex virus Type-1. *Pediatrics* 1983; **72**: 882–3.

243. Renney K M, Unusual case of gonococcal arthritis. *British Journal of Venereal Diseases* 1980; **56**: 35–6.

244. Riordan D C. The hand in leprosy. *Journal of Bone and Joint Surgery* 1960; **42**-A: 683–90.

245. Riou B, Bentata-Pessayre M, Krivitzky A, Delzant G. Arthrite sterno-costale á Candida chez un toxicomane. *Presse Médical* 1983; **12**: 364.

246. Robinson M C. Epidemic of virus disease in southern province Tanganyika Territory in 1952–53; clinical features. *Transactions of the Royal Society of Tropical Medicine and Hygiene* 1955; **49**: 28–32.

247. Robinson S C. Bacillus cereus septic arthritis following arthrography. *Clinical Orthopaedics and Related Research* 1979; **145**: 237–8.

248. Rosenbaum J, Lieberman D H, Katz W A. Moraxella infectious arthritis: first report in an adult. *Annals of the Rheumatic Diseases* 1980; **39**: 184–5.

249. Rosenthal J, Brandt K D, Wheat L J, Slama T G. Rheumatologic manifestations of histoplasmosis in the recent Indianapolis epidemic. *Arthritis and Rheumatism* 1983; **26**: 1065–70.

250. Ross R W. Newala epidemic; virus isolation, pathogenic properties and relationship to epidemic. *Journal of Hygiene* 1956; **54**: 177–91.

251. Rubinow A, Spark E C, Canoso J J. Septic arthritis in a Charcot joint. *Clinical Orthopaedics and Related Research* 1980; **147**: 203–6.

252. Saah A J, Hornick R B. Coxiella burnetii (Q fever). In: Mandell G L, Gordon Douglas R G Jr and Bennett J E, eds, *Principles and Practice of Infectious Diseases* (2nd edition). Chichester: John Wiley & Sons, 1985: 1088–90.

253. Salfield S. Filarial arthritis in the Sepik district of Papua New Guinea. *Medical Journal of Australia* 1975; **1**: 264–7.

254. Schachter J. Can chlamydial infections cause rheumatic disease? In: Dumonde D C, ed, *Infection and Immunology in the Rheumatic Diseases*. Oxford: Blackwell Scientific Publications, 1976: 151–7.

255. Schachter J. Chlamydial infections. *New England Journal of Medicine* 1978; **298**: 428–35, 490–5, 540–9.

256. Schaller J G. Arthritis and infections of bones and joints in children. *Pediatric Clinics of North America* 1977; **24**: 755–90.

257. Schmid F R. Bacterial arthritis. In: McCarty D J, ed, *Arthritis and Allied Conditions* (10th edition). Philadelphia: Lea & Febiger, 1985: 1662–85.

258. Schmid F R. Principles of diagnosis and treatment of bone and joint infections. In: McCarty D J, ed, *Arthritis and Allied Conditions. A Textbook of Rheumatology* (11th edition). Philadelphia, London: Lea and Febiger, 1989: 1863–91.

259. Schnitzer T J. Viral arthritis. In: Kelley W N, Harris E D, Ruddy S, Sledge C B, eds, *Textbook of Rheumatology* (3rd edition). Philadelphia: W B Saunders & Co., 1989: 1611–28.

260. Schultz L C, Drommer W, Seidler D, Ehard H, Leimbeck R, Weiss R. Experimental erysipelas in different species as a model for systemic connective tissue disease. II. The chronic phase with special reference to polyarthritis. *Beiträge zur Pathologie* 1975; **154**: 27–51.

261. Schultz L C, Drommer W, Seidler D, Ehard H, Mickcoitz G, Hertrampf B, Bohm K H. Experimental erysipelas in different species as a model for systemic connective tissue disease I. Systemic vascular processes during organ manifestation. *Beiträge zur Pathologie* 1975; **154**: 1–26.

262. Schumacher H R, Gall E P. Arthritis in acute hepatitis, and chronic active hepatitis. Pathology of the synovial membrane with evidence for the presence of Australia antigen in synovial membrane. *American Journal of Medicine* 1974; **57**: 655–64.

263. Schwartz J. What's new in mycotic bone and joint disease. *Pathology, Research and Practice* 1984; **178**: 617–34.

264. Seifert M H, Mathews J A, Phillips I, Gargan R A. Gas-liquid chromatography in diagnosis of pyogenic arthritis. *British Medical Journal* 1978; **II**: 1402–4.

265. Semble E L, Agudelo C A, Pegram P S. Human parvovirus B19 arthropathy in two adults after contact with childhood erythema infectiosum. *American Journal of Medicine* 1987; **83**: 560–2.

266. Seradge H, Anderson M G. Clostridial myonecrosis following intraarticular steroid injection. *Clinical Orthopaedics and Related Research* 1980; **147**: 207–9.

267. Sharp J T. The mycoplasmataceae (PPLO) as causes of joint infections. *Arthritis and Rheumatism* 1964; **7**: 437–42.

268. Sharp J T, Lisaky M D, Duffy J, Duncan M W. Infectious arthritis. *Archives of Internal Medicine* 1979; **139**: 1125–30.

269. Shelley W B. Herpetic arthritis associated with disseminated herpes simplex in a wrestler. *British Journal of Dermatology* 1980; **103**: 209–12.

270. Sherman L, Gazit A, Yaniv A, Dahlberg J E, Tronick S R. Nucleotide sequence analysis of the long terminal repeat of integrated caprine arthritis encephalitis virus. *Virus Research* 1986; **5**: 145–55.

271. Shore H. O'Nyong-nyong fever: an epidemic virus disease in East Africa. III. Some clinical and epidemiological observations in the northern province of Uganda. *Transactions of the Royal Society of Tropical Medicine and Hygiene* 1961; **55**: 361–73.

272. Shuper A, Mimouni M, Mukamel M, Varsano I. Varicella arthritis in a child. *Archives of Disease in Childhood* 1980; **55**: 568–9.

273. Sigal L H, Steere A C, Niederman J C. Symmetric polyarthritis associated with heterophile-negative infectious mononucleosis. *Arthritis and Rheumatism* 1983; **26**: 553–6.

274. Sikes D, Neher G M, Doyle L P. Studies on arthritis in

swine. I. Experimental erysipelas and chronic arthritis in swine. *American Journal of Veterinary Research* 1955; **16**: 349–66.

275. Silby H M, Farber R, O'Connell C J, Ascher J, Marine E J. Acute monoarticular arthritis after vaccination. Report of a case with isolation of vaccinia virus from synovial fluid. *Annals of Internal Medicine* 1965; **62**: 347–50.

276. Simpson M B, Merz G W, Kurlinski J P, Solomon M H. Opportunistic mycotic osteomyelitis: bone infections due to aspergillus and candida species. *Medicine* 1977; **56**: 475–82.

277. Simpson R W, McGinty L, Simon L, Smith C A, Godzeski C W, Boyd R J. Association of parvoviruses with rheumatoid arthritis of humans. *Science* 1984; **233**: 1425–8.

278. Small C B, Slater L N, Lowy F D, Small R D, Salvati E A, Casey J I. Group B streptococcal arthritis in adults. *American Journal of Medicine* 1984; **76**: 367–75.

279. Smith L R, Heaton C L. Actinomycosis presenting as Wegener's granulomatosis. *Journal of the American Medical Association* 1978; **240**: 247–8.

280. Spence L P, Thomas L. Application of haemagglutination and complement-fixation techniques to the identification and serological classification of arthropod-borne viruses; studies on Chikungunya and Mokande viruses. *Transactions of the Royal Society of Tropical Medicine and Hygiene* 1959; **53**: 248–55.

281. Spruance S L, Metcalf R, Smith C B, Griffiths M M, Ward J R. Chronic arthropathy associated with rubella vaccination. *Arthritis and Rheumatism* 1977; **20**: 741–7.

281a. Steere A C. Viral arthritis. In: McCarty D J (ed.) *Arthritis and Allied Conditions. A Textbook of Rheumatology* (11th edition). Philadelphia, London: Lea and Febiger, 1989: 1938–54.

282. Steere A C, Brinckerhoff C E, Miller D J, Drinker J, Harris E D, Malawista S E. Elevated levels of collagenase and prostaglandin E_2 from synovium associated with erosion of cartilage and bone in a patient with chronic Lyme arthritis. *Arthritis and Rheumatism* 1980; **23**: 591–9.

283. Steere A C, Gibofsky A, Patarroyo M E, Winchester R J, Hardin J A, Malawista S E. Chronic Lyme arthritis. Clinical and immunogenetic differentiation from rheumatoid arthritis. *Annals of Internal Medicine* 1979; **90**: 896–901.

284. Steere A C, Malawista S E, Viral arthritis. In: McCarty D J, ed, *Arthritis and Allied Conditions* (11th edition). Philadelphia: Lea & Febiger, 1985: 1697–712.

285. Steere A C, Malawista S E. Lyme Disease. In: McCarty D J, ed, *Arthritis and Allied Conditions* (10th edition). Philadelphia: Lea & Febiger, 1985: 1713–22.

286. Steere A C, Malawista S E, Hardin J A, Ruddy S, Askenase P W, Andiman W A. Erythema chronicum migrans and Lyme arthritis. *Annals of Internal Medicine* 1977; **86**: 685–98.

287. Steere A C, Malawista S E, Snydman D R, Shope R E, Andeman W A, Ross M R, Steele F M. Lyme arthritis. An epidemic of oligoarticular arthritis in children and adults in three Connecticut communities. *Arthritis and Rheumatism* 1977; **20**: 7–17.

288. Steigman A J. Rashes and arthropathy in viral hepatitis. *Mount Sinai Journal of Medicine* 1973; **40**: 752–7.

289. Steinitz M, Tamir S. Monoclonal rheumatoid factor produced in vitro by Epstein–Barr virus (EBV) cell line—a reagent to detect specific antibodies and weak cellular antigens. In: Baum S J, Ledney G D, Thierfelder S, eds, *Experimental Hematology Today*. Basel: S Karger AG, 1982: 211–17.

290. Steward M W, Katz F E, West N J. The role of low affinity antibody in immune complex disease. The quantity of anti-DNA antibodies in NZB/W F1 hybrid mice. *Clinical and Experimental Immunology* 1975; **21**: 121–30.

291. Stierle G, Brown K A, Rainsford S G, Smith C A, Hamerman D, Stierle H E, Dumonde D C. Parvovirus associated antigen in the synovial membrane of patients with rheumatoid arthritis. *Annals of the Rheumatic Diseases* 1987; **46**: 219–23.

292. Streifler J, Pitlik S, Garty M, Grosskopf I, Rosenfeld J B. Sternoclavicular arthritis and osteomyelitis due to *Pseudomonas aeruginosa*, not related to drug abuse. *Israel Journal of Medical Sciences* 1985; **21**: 458–9.

293. Stuckey M, Quinn P A, Gelfand E W. Identification of *Ureaplasma urealyticum* (T-strain *Mycoplasma*) in patient with polyarthritis. *Lancet* 1978; **ii**: 917–20.

294. Swaramappa M, Reddy C R R, Devi C S, Reddy A C, Reddy P K, Murthy D P. Acute Guinea-worm synovitis of the knee joint. *Journal of Bone and Joint Surgery* 1969; **51**A: 1324–30.

295. Tanaka H. *Mycoplasma pulmonis* arthritis in congenitally athymic (nude) mice. Clinical and biological features. *Microbiology and Immunology* 1979; **23**: 1055–65.

296. Taylor P W, Trueblood M C. Septic arthritis due to *Aerococcus viridans*. *Journal of Rheumatology* 1985; **12**: 1004–5.

297. Taylor-Robinson D, Taylor G. Do mycoplasmas cause rheumatic disease? In: Dumonde D C, ed, *Infection and Immunity in the Rheumatic Diseases*. Oxford, London, Edinburgh: Blackwell Scientific Publications, 1976.

298. Tesh R B. Arthritides caused by mosquito-borne viruses. *Annual Revue of Medicine* 1982; **33**: 31–40.

299. Tezner O. Ergebnisse der inneren Medizin und Kinderheilkunde. *Varicellen*. Berlin: Springer-Verlag, 1931: 476.

300. Thompson G R, Ferreyra A, Brackett R G. Acute arthritis complicating rubella vaccination. *Arthritis and Rheumatism* 1971; **14**: 19–26.

301. Thompson G R, Manshady B M, Weiss J J. Septic bursitis. *Journal of the American Medical Association* 1978; **240**: 2280–1.

302. Thompson S E, Jacobs N F, Zacarias F, Rein M F, Shulman J A. Gonococcal tenosynovitis-dermatitis and septic arthritis. *Journal of the American Medical Association* 1980; **244**: 1101–2.

303. Thornberry D K, Wheat L J, Brandt K D, Rosenthal J. Histoplasmosis presenting with joint pain and hilar adenopathy-pseudosarcoidosis. *Arthritis and Rheumatism* 1982; **25**: 1396–1402.

304. Tittiger F, Alexander D C. Studies on the bacterial flora of condemned portions from arthritic hogs. *Canadian Journal of Comparative Medicine* 1971; **35**: 244–8.

305. Torg J S, Lammot T R. Septic arthritis of the knee due to *Clostridium welchii*. *Journal of Bone and Joint Surgery* 1968; **50**A: 1233–6.

306. Torres J, Rathbun H K, Greenough W B. Pneumococcal arthritis: report of a case and review of the literature. *Johns Hopkins Medical Journal* 1973; **132**: 234–41.

307. Umber J, Chapman M W, Drutz D J. Candida pyoarthrosis. Report of a case and results of treatment with 5-

fluorocytosine. *Journal of Bone and Joint Surgery* 1974; **56**A: 1520–4.

308. van Pelt R W, Langham R F. Synovial fluid changes produced by infectious arthritis in cattle. *American Journal of Veterinary Research* 1968; **29**: 507–16.

309. Vergani D, Morgan-Capner P, Davies E T, Anderson A W, Tee D E H, Pattison J R. Joint symptoms, immune complexes and rubella. *Lancet* 1980; **ii**: 321–2.

310. Verinder D G. Septic arthritis due to *Mycoplasma hominis*. A case report and review of the literature. *Journal of Bone and Joint Surgery* 1978; **60-A**: 224.

311. Vincenti F, Amend W J, Feduska N J, Salvatierra O. Septic arthritis following renal transplantation. *Nephron* 1982; **30**: 253–6.

312. Waldvogel F A, Vasey H. Osteomyelitis: the past decade. *New England Journal of Medicine* 1980; **303**: 360–70.

313. Wall B A, Weinblatt M E, Darnall J J, Muss H. *Candida tropcalis* arthritis and bursitis. *Journal of the American Medical Association* 1982; **248**: 1098–9.

314. Walton K, Hilton R C, Sen R A. Pseudomonas arthritis treated with parenteral and intraarticular ceftazidime. *Annals of the Rheumatic Diseases* 1985; **44**: 499–500.

315. Waltzing P, Bloch-Michel H. The rosette phenomenon in filariasis and its use in the diagnosis of filarial arthritis. *La Nouvelle Presse Mèdicale* 1971; **79**: 2061–3.

316. Ward J R, Bishop B. Varicella arthritis. *Journal of the American Medical Association* 1970; **212**: 1954–6.

317. Washburn L R, Cole B C, Gelman M I, Ward J R. Chronic arthritis of rabbits induced by mycoplasmas. I. Clinical, microbiologic, and histologic features. *Arthritis and Rheumatism* 1980; **23**: 825–36.

317a. Washburn L R, Cole B C, Ward J R. Chronic arthritis of rabbits induced by mycoplasmas. II. Antibody response and the deposition of immune complexes. *Arthritis and Rheumatism* 1980; **23**: 837–45.

318. Washburn L R, Cole B C, Ward J R. Chronic arthritis of rabbits induced by mycoplasmas. III. Induction with nonviable *Mycoplasma arthritidis* antigen. *Arthritis and Rheumatism* 1982; **25**: 937–46.

319. Watanakunakorn C. *Serratia marcescens* osteomyelitis of the clavicle and sternoclavicular arthritis complicating infected indwelling subclavian vein catheter. *American Journal of Medicine* 1986; **80**: 753–4.

320. Wenzel R P, McCormick H J, Busch H J, Beam W E. Arthritis and viral hepatitis. *Archives of Internal Medicine* 1972; **130**: 770–7.

321. Wheeler J, Heffron W, Williams R. Migratory arthralgias and cutaneous lesions as confusing initial manifestations of gonorrhea. *American Journal of the Medical Sciences* 1970; **260**: 150–9.

322. White D G, Woolf A D, Mortimer P P, Cohen B J, Blake D R, Bacon P A. Human parvovirus arthropathy. *Lancet* 1985; **i**: 419–21.

323. Wilkerson R D, Taylor D C, Opal S M. *Nocardia asteroides* sepsis of the knee. *Clinical Orthopaedics and Related Research* 1985; **197**: 206–8.

324. Wilkins J, Leedom J M, Salvatore M A, Portnoy B. Clinical rubella with arthritis resulting from reinfection. *Annals of Internal Medicine* 1972; **77**: 930–2.

325. Wilkinson M C. A note on Poncet's tuberculous rheumatism. *Tubercle* 1967; **48**: 297–306.

326. Williams M C, Woodall J P, Gillett J D. O'Nyong-nyong fever: an epidemic virus disease in East Africa. VII. Virus isolations from man and serological studies up to July 1961. *Transactions of the Royal Society of Tropical Medicine and Hygiene* 1965; **59**: 186–97.

327. Williams M C, Woodall J P. Portersfield J S. O'Nyong-nyong fever: an epidemic virus disease in East Africa. V. Human antibody studies by plaque inhibition and other serological tests. *Transactions of the Royal Society of Tropical Medicine and Hygiene* 1962; **56**: 166–72.

328. Winn R E, Chase W F, Lauderdale P W, McCleskey F K. Septic arthritis involving *Capnocytophaga ochracea. Journal of Clinical Microbiology* 1984; **19**: 538–40.

329. Winter W G, Larson R K, Honnegar M M, Jacobsen D T, Pappagianis D, Huntington R W. Coccidioidal arthritis and its treatment. *Journal of Bone and Joint Surgery* 1975; **57**A: 1152–7.

330. Winter W G Jr, Larson R K, Zettas J P, Libke R. Coccidioidal spondylitis. *Journal of Bone and Joint Surgery* 1978; **60-A**: 240–4.

331. Woglom W H, Warren J. Pyogenic filterable agent in albino rat. *Journal of Experimental Medicine* 1938; **68**: 513–28.

332. Woo P, Panayi G S. Reactive arthritis due to infestation with *Giardia lamblia. Journal of Rheumatology* 1984; **11**: 719.

332a. Woods V L, Zvaifler N J. Pathogenesis of systemic lupus erythematosus. In: Kelley W N, Harris E D, Ruddy S, Sledge C B, eds, *Textbook of Rheumatology* (3rd edition). Philadelphia, London: W B Saunders Company, 1989: 1077–100.

333. Wright L T, Logan M. Osseous changes associated with lymphogranuloma venereum. *Archives of Surgery* 1939; **39**: 108–21.

334. Wynne H A, Lancaster R. Brucellosis as a rare cause of spondylitis. *Journal of the Royal Society of Medicine* 1985; **78**: 161–2.

335. Ximent I, Davis A, Finegold S M. Joint infection by anaerobic bacteria: a case report and review of the literature. *Arthritis and Rheumatism* 1969; **12**: 627–35.

336. Yanez J E, Thompson G R, Mikkelsen W M, Bartholomew L E. Rubella arthritis. *Annals of Internal Medicine* 1966; **64**: 772–7.

337. Yocum R C, McArthur J, Petty B G, Diehl A M, Moench T R. Septic arthritis caused by *Propionibacterium acnes. Journal of the American Medical Association* 1982; **248**: 1740–1.

338. Ziegler G, Euller I, Dellamonica P, Estesse H, Thyss J. Septic arthritis due to *Yersinia enterocolitica.* Two cases. *La Nouvelle Presse Mèdicale* 1980; **9**: 3097.

339. Zuckerman A J, Banatvala J E, Pattison J R. *Principles and Practice of Clinical Virology.* Chichester: John Wiley & Sons, 1987: 507–43.

340. Zvaifler N J, Woods V L. Etiology and pathogenesis of systemic lupus erythematosus. In: Kelley W N, Harris E D, Ruddy S, Sledge C B, eds, *Textbook of Rheumatology* (2nd edition). Philadelphia: W B Saunders & Co., 1985: 1042–69.

Chapter 19

Reactive and Postinfective Arthritis and Seronegative Spondyloarthropathy

The concept of '*reactive*' arthritis is used to describe sterile inflammatory joint disease associated with a recognizable infection but in which neither viable microorganisms nor their products can be identified in the joints or joint tissues.[3,67,200] Alternatively, arthritis can be said to be 'postinfective' if components of the microorganisms but not viable infective agents are recognized in affected joints.[67] In recent years, the distinction between 'reactive' and 'postinfective' arthritis has become increasingly difficult.[103] Meticulous searches, by immunofluorescence,[103] electron microscopic,[186] and other techniques, of the sterile synovial fluids and synovial tissues from cases of reactive arthritis have yielded evidence of *Yersinia enterocolitica* antigens (Figs 19.1 and 19.2),[103] *Chlamydia trachomatis*-like ultrastructural particles[186] and *Salmonella* lipopolysaccharide.[102] This evidence determines that cases of 'reactive' arthritis associated with these microorganisms should now be termed 'postinfective'. Moreover, the duration of the 'postinfective' stage appears to be much more protracted than had been suspected and bacterial antigens are identifiable within articular tissues for long periods after the initial infection. In the present account, the term 'reactive arthritis' is used subject to the implicit qualification that a number of syndromes within this category of disease are likely to be 'postinfective'.

'Reactive' arthritis[200] therefore remains an asymmetrical sterile inflammatory joint disease beginning 7–30 days after an extraarticular infection. The infection is attributable to chlamydia, mycoplasmas, bacteria or protozoa.[82] Susceptible individuals are usually male; they are clustered in families and often inherit the antigen HLA-B27 (see p. 243).[30,31,217] Following the initial infection, there is a notable failure to synthesize rheumatoid factors (see p. 233) so that patients with reactive arthritis are said to belong to a category of 'seronegative arthropathies'.[148,224] The peripheral articular disease attacks smaller numbers of joints than the symmetrical polyarthritis of rheumatoid arthritis (see Chapter 12); by contrast with the latter disorder, the axial skeleton is frequently affected.

The most common clinical form of reactive arthritis is Reiter's syndrome.[82a,124,189] However, three further forms of seronegative arthritis show conspicuous similarities to and overlap with Reiter's syndrome: ankylosing spondylitis, enteropathic arthropathy and psoriatic arthropathy.[147,149] Behçet's disease (which is discussed on p. 647) and juvenile rheumatoid arthritis (outlined in Chapter 12) also share features with the reactive arthritides and an analogous disorder has been described in cat-scratch disease which is now attributable to *Rothia dentocariosa.*[97]

The extreme frequency, in reactive arthritis and associated disorders, of spinal lesions, led to the inclusive description 'spondyloarthropathy' (spondylarthritis).[149] The

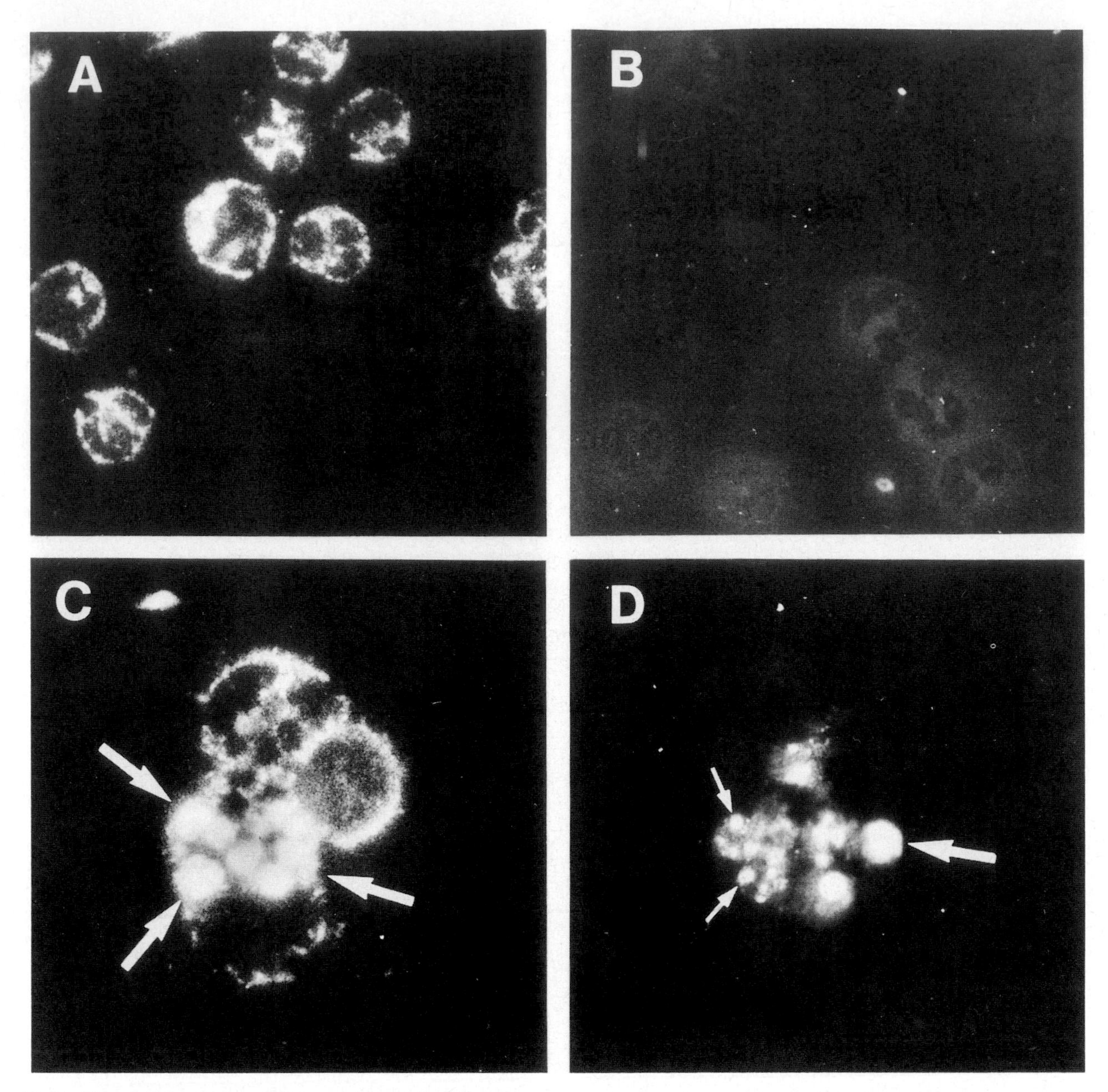

Fig. 19.1 *Yersinia enterocolitica* reactive arthritis.
Positive immunofluorescence staining with rabbit anti-*Y. enterocolitica* antiserum of synovial fluid cells from patient with arthritis triggered by *Y. enterocolitica* infection (A, C, D). Fig. 19.1B shows negative staining with yersinia-absorbed antiserum. Figs 19.1C and D demonstrate typical vacuoles (large arrows) with rod-shaped particles (small arrows). Culture of the synovium for *Y. enterocolitica* was negative. (A, B × 510; C, D × 750.) Reproduced from Granfors *et al.*[103] by courtesy of Dr K Granfors and the Editor, *New England Journal of Medicine.*

nature of these lesions almost certainly contains a clue to the unique characteristics of reactive and seronegative arthritis. It therefore appears logical to discuss the pathology and pathogenesis of these various disorders in a single chapter, in spite of the absence of absolute evidence that they are all reactive in nature.

Reiter's syndrome

Reiter's syndrome[44, 124, 171, 223] is a triad of non-gonococcal urethritis, arthritis and conjunctivitis.[18] There are early historical accounts[35] but acceptable definitions are recent. One agreed description is of 'an episode of peripheral

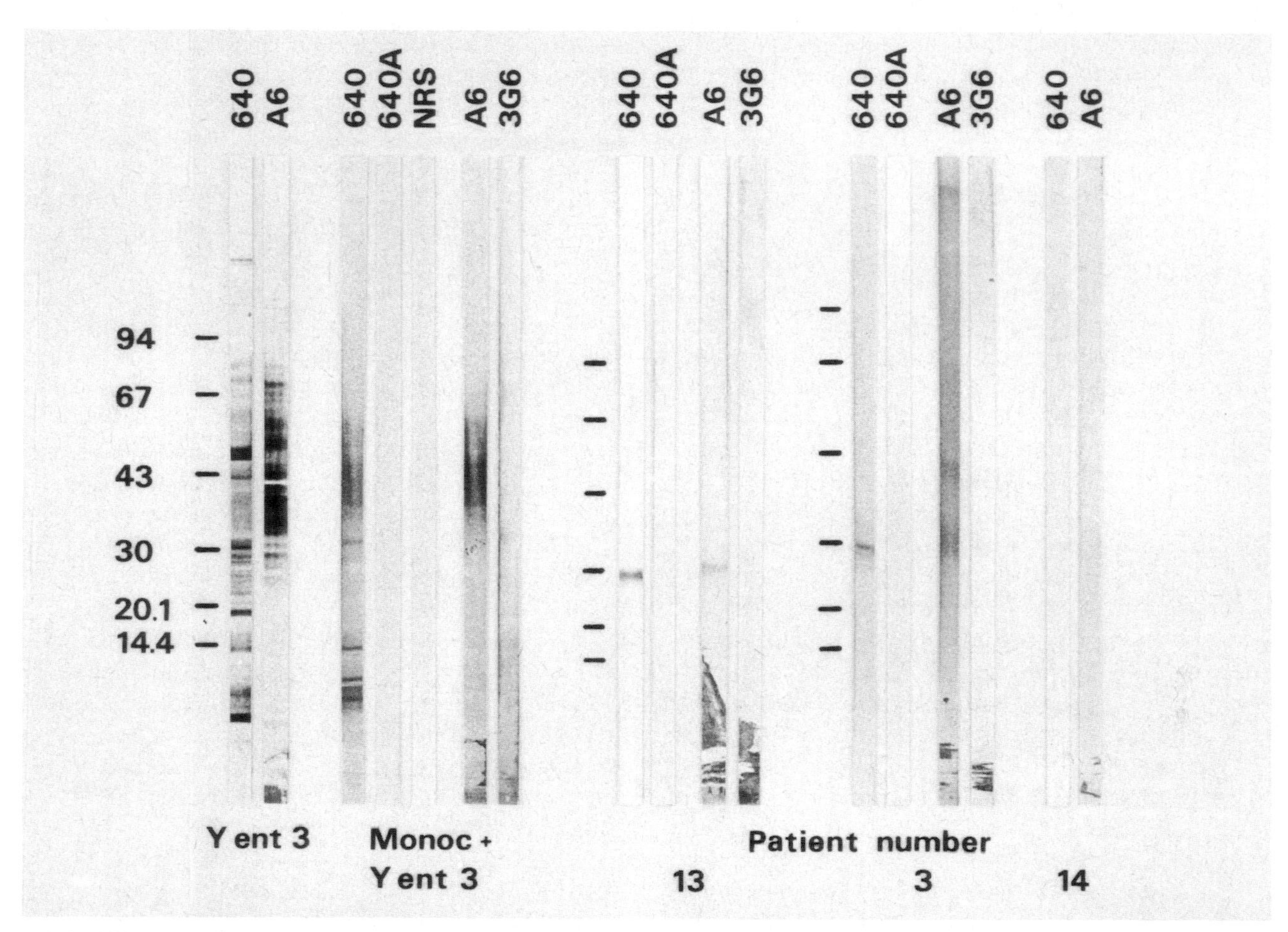

Fig. 19.2 Western blot of synovial fluid cells from cases of post-yersinia arthritis.
Positive (two patients) and negative (one patient) reactivity of solubilized synovial cells with rabbit anti-*Yersinia enterocolitica* and control antibodies. Y ent 3 = native yersinia control; Monoc+ Y ent 3 = yersinia control processed with mononuclear cells; 640 = rabbit antiserum to *Y. enterocolitica* 0:3;640A = absorbed rabbit antiserum to *Y. enterocolitica* 0:3; NRS = normal rabbit serum; A6 = monoclonal yersinia antibody; 3G6 = control monoclonal antibody. Reproduced from Granfors *et al.*[103] by courtesy of Dr K Granfors and the Editor, *New England Journal of Medicine.*

arthritis of more than one month's duration occurring in association with urethritis and/or cervicitis'.[216] European investigators were, however, convinced that an identical condition could develop after dysentery[163] so that the present concept is of a sterile asymmetrical arthritis following some days or weeks after an extraarticular infection by urethral *or* gut-associated microorganisms.

Historical aspects

Historical accounts[124] of the syndrome invoke conjunctivitis succeeding venereal infection[35] or dysentery[197] but the syndrome, unjustifiably, derives its eponymous designation from the descriptions of Reiter.[171] Interest in Reiter's syndrome grew following a huge Finnish epidemic of *Shigella flexneri* infection: there were 334 postdysenteric cases. In 1973, Brewerton *et al.*[33] and Schlosstein *et al.*[185] reported the association with HLA-B27.

Clinical features

Reiter's syndrome is relatively frequent: it is the most common form of inflammatory joint disease in young males. The syndrome occurs in females but the relative difficulty of confirming the genitourinary infection tends to conceal the true frequency of the disorder. Approximately 3 per cent of those with non-specific urethritis develop Reiter's disease; but 20 per cent of infected individuals who have inherited HLA-B27 manifest the arthropathy.[44] It is of great interest that this figure is similar to the proportion of HLA-positive individuals who develop ankylosing spondylitis (see p. 768). By contrast, only 1.5 per cent of those with shigella infections respond in this way.

The early clinical changes[44,82a] are polyarthritis; urethritis or cervicitis; back pain; eye disease; balanitis; heel pain; and tendinitis. Stomatitis, keratoderma blennorrhagica, nail lesions and aortitis are also sometimes present. The polyarthritis is asymmetrical and affects knees and ankles. However, the disorder may be monoarticular: it is often prolonged and recurrent. Among the tendinous insertions (entheses) that become inflamed are those of the heel and intercostal muscles. The ocular disorders include a sterile conjunctivitis, uveitis or iritis. Keratoderma characteristically affects the soles of the feet, the glans penis and the toes, but the skin response is occasionally widespread.

The late clinical features are spondyloarthropathy, sacroiliitis and a disorder indistinguishable from ankylosing spondylitis; severe ocular lesions such as optic neuritis; pericarditis and aortic incompetence; neuropathy; pleuritis; purpura; thrombophlebitis; and amyloidosis. There is no specific treatment (see below) and a study of 16 cases[212] showed that the prognosis was not good: 80 per cent of patients were likely to have arthritis five years after the onset of the disease.

The laboratory criteria of diagnostic value are: a raised erythrocyte sedimentation rate; anaemia; negative tests for rheumatoid factor; and synovial fluid changes. However, there is no single diagnostic test and HLA typing is of longer-term epidemiological rather than of short-term investigative value.

Pathological changes

Joints

Febrile, asymmetrical, mild or severe arthritis affects a variable number of synovial joints, particularly those of the lower limbs and, often, the toes and fingers. The early polyarthritis is variable rather than flitting so that new joints are attacked every few days. The congested, hyperaemic synovium resembles that in pyogenic infection. There is a brisk polymorph infiltrate of subsynovial connective tissues. The synovial fluid is turbid; it contains $2–5 \times 10^9$ polymorphs/litre in the early disease, greater proportions of lymphocytes later. The synovial fluid glucose is normal or only slightly reduced by contrast with the low glucose in pyogenic infection.

Early electron microscopic studies in five cases[158] confirmed the presence of vascular plugging with platelets and polymorphs; there were unidentified inclusions within mononuclear cells.[152,165] In nine cases with severe arthritis progressing to irreversible change described by Soren,[196] knee joint arthrotomy was performed one to nine years after joint disease began. All synovia were the sites of focal lymphocytic aggregates, perhaps an index of the duration of the disease. Eight cases showed hyperaemia, foci of histiocytes and fibroblast proliferation; seven had evidence of synovial cell hypertrophy and hyperplasia, five had progressed to villous synovial hypertrophy, and four displayed the presence of inflammatory cells in blood vessel walls and focal aggregates of plasma cells. Soren interpreted the relatively early plasmacytosis as suggestive of 'an immunologic response to the aetiologic agent in Reiter's syndrome'. In 12 patients in another study, 11 had synovial deposits of immunoglobulin or complement: in eight there was perivascular, in four interstitial IgM. C3 was identified in all cases around blood vessels, and in four interstitial staining for C3 was also confirmed.[11] The results supported the concept that immune complexes[154] play a part in the synovitis of Reiter's syndrome.

With the passage of time, the oedema and hyperaemia diminish, plasma cells and lymphocytes become fewer, and fibroblasts multiply. Granulation tissue forms and a pannus replaces marginal cartilage. The new connective tissue, increasing in density, forms adhesions; periostitis with periosteal new bone accompanies a marginal erosion of subchondral bone that can be identified radiologically. The ultimate result is chronic, erosive arthritis. As the activity of the synovial disease diminishes, the synovial changes come more closely to resemble those of rheumatoid arthritis but the electron microscopic appearances remain characteristic.[152]

Spondylitis frequently ensues[188]: 8 of 26 patients showed this change[98] which closely resembles the spinal disorder of ankylosing spondylitis (see p. 768). Sacroiliac disease is recognized in 20–30 per cent of cases. The histological abnormality which underlies the intervertebral and apophyseal joint disease and the insertional tendinitis of peripheral sites, such as the Achilles tendon, is an enthesopathy (see p. 101). Syndesmophytes (see p. 771) form and apophyseal ankylosis with paravertebral calcification and bridging are recognized. All patients affected in this way have bilateral radiological changes in the sacroiliac joints.

Extraarticular disease

Keratosis blenorrhagica affects the soles of the feet, the palms and the circumcised glans penis. There is hyperkeratosis, parakeratosis and acanthosis. The lesions closely resemble those of pustular psoriasis: they progress from erythematous macules less than 10 mm in diameter to papules with keratotic crusting. Keratosis is greater in the regions of greatest normal keratinization but the distribution of the lesions is distinct from that of psoriasis. The changes in the mucous membranes, including the urethra, uncircumcised glans penis and conjunctiva are similar to those of the cutaneous situations but lack evidence of keratosis.[108] Herpetiform lesions characterize the balanitis circinata; similar erosions derive from vesicles, papules or plaques in the buccal mucosa, urethra or bladder.

The ocular lesions tend to progress. Uveitis becomes severe. Iridocyclitis advances with intraocular haemorrhage, and optic neuritis may culminate in blindness. Aortic valve disease with incompetence has been described; it is indistinguishable from the aortic lesion of ankylosing spondylitis (see p. 779). In the early, active disease, pericarditis may occur. Necropsy records are few.[151] However, cardiac lesions have been recorded.[65] Aortic insufficiency has been found to be a feature, particularly in those individuals with spondylitis and the whole Reiter's triad[175,226] but many with aortic disease do not display the entire syndrome.

Aetiology and pathogenesis

The search for the cause of Reiter's syndrome has concentrated on three fields of investigation: the genetic predisposition; the immediate, effective agent; and experimental reproduction.

Heredity

The demonstration of a strong association with the inheritance of HLA-B27[33] began a new era of enquiry. The significance of the HLA-antigen, in immunity is discussed in Chapter 5 (see p. 241). Brewerton *et al.*[32] had shown that more than 95 per cent of patients with ankylosing spondylitis (see p. 768) possessed HLA-B27 and it was a short step to establish that there was a striking relationship between the inheritance of this histocompatability antigen and the occurrence of Reiter's syndrome and other seronegative spondyloarthropathies.

Infection

The role for an infective agent is strongly suspected but it has been widely believed that any organism causing the arthritis of Reiter's syndrome does not do so by direct invasion of articular tissues. Until recently, no synovial cell inclusions had been seen. The synovia are the target for a sterile inflammatory reaction that is assumed to be mediated by an immunological mechanism. One reason for this supposition is that the synovitis follows some days or weeks after the urethritis, conjunctivitis or enteritis.

Sexually acquired reactive arthritis In postvenereal Reiter's syndrome (sexually acquired reactive arthritis—SARA), two classes of microorganism have been implicated.

Non-specific urethritis is commonly attributable to *Chlamydia trachomatis* and evidence of this genitourinary infection can be obtained, at the onset of arthritis, in approximately 50 per cent of cases.[126,182] However, chlamydia can rarely be grown from the synovial tissues or synovial fluid so that the evidence for direct synovial involvement is limited although elevated titres of antichlamydial antibodies are often demonstrable. Indeed, patients with postvenereal Reiter's syndrome have higher titres of antichlamydial antibodies than those with uncomplicated urethritis, and the occurrence of arthritis is strongly associated with the development of a humoral immune response to chlamydial antigens. Using a monoclonal antibody to *C. trachomatis*, Keat *et al.*[125] have now demonstrated typical chlamydial elementary bodies in the synovial tissue or synovial fluid cells in five of eight patients with SARA and comparable observations have been made by Schumacher *et al.*[186] who employed both electron microscopic and immunoperoxidase localization techniques. They believe that the synovitis of Reiter's syndrome results directly from the presence of these chlamydial elementary bodies, an observation that has implications for treatment as well as pathogenesis. All the cases with synovial elementary bodies had high antichlamydial antibody titres. Moreover, mouse pneumonitis agent (MoPn), a *Chlamydia trachomatis* biovar, can be used to cause an inflammatory synovitis in mice.[117] The development of this arthritis depends strongly on previous immunization with the inactive microbial agent.

Ureaplasma urealyticum, known to be a cause of infective arthritis (see p. 731), has also been isolated from the genital tracts of a small number of patients with postvenereal Reiter's syndrome. The organism, like chlamydia, can cause urethritis but the mechanism by which it leads to reactive arthritis is not yet well understood. In experiments in which patients' synovial (but not blood) lymphocytes were tested against microbial antigens, lymphocytes from postvenereal cases of Reiter's syndrome were stimulated by ureaplasmal or by chlamydial antigen, or by both demonstrating that synovial lymphocytes register the cause of reactive arthritis and that they, and humoral antibodies, may be implicated in its pathogenesis.[83]

Postgastrointestinal infection When Reiter's syndrome follows a gastrointestinal infection, the responsible organism has usually been *Yersinia enterocolitica, Shigella dysenteriae, Shigella flexneri,* or *Salmonella typhimurium.* However, *Salmonella muenchen*[167] and organisms causing antibiotic-associated colitis have been implicated[168] and *Campylobacter jejuni*[71], *Clostridium difficile*[76,136] and *Giardia lamblia*[219] may also be responsible. The mechanism by which microbial infection provokes a delayed polyarthritis is not fully understood. The response is associated with the inheritance of HLA-B27 (see below). The possibility that, as in rheumatic fever (see p. 808), particular strains of arthritogenic microorganisms are involved, cannot be excluded.

The available evidence is conveniently summarized in relation to *Y. enterocolitica*. It is rare for the bacterium to be recovered directly from the synovial fluid[227] and even the late isolation of the organism from the stools may be difficult. Until recently evidence to prove the yersinial nature of a bowel infection has been largely serological therefore. Now, however, *Yersinia enterocolitica* antigens have been identified in the synovial fluid cells from infected joints.[103] The formation of IgA, IgG and IgM antibodies takes place; the persistence of IgA and IgG correlates with the occurrence of arthritis.[105,199] There is a relatively strong IgA antibody response, apparently due to chronic stimulation of the intestinal lymphoid tissue.[104] One way in which the antiyersinial antibodies could cause synovitis would be by the formation of immune complexes which would then lodge in synovial tissue. Immune complexes are indeed formed in recent yersiniosis and IgM complexes persist in higher concentration in those with arthritis than in those without. Moreover, the complexes can sometimes be found in the synovial fluid.[130] However, synovial fluid complement levels are high and the similarity of the complement/protein ratios to those of the synovial fluid in osteoarthritis or in traumatic synovitis argues against a mechanism such as that postulated in rheumatoid anthritis in which immune complexes bind and activate complement. An alternative mechanism could be the development of cross-reactivity, one explanation for which would be homology between parts of the component molecules of the yersinia and sequences in the polypeptide chains of synovial tissue components. There is little to support this view, although a closely similar concept has been widely canvassed with regard to the pathogenesis of ankylosing spondylitis (see p. 780).

Another possible pathway of synovial injury is cell-mediated hypersensitivity. In the synovial fluid of four cases of genitourinary Reiter's syndrome and one case of the postyersinial dysenteric disorder, the $T11^+$ T lymphocyte was the predominant synovial fluid mononuclear cell; in contrast to rheumatoid arthritis, $T4^+$ helper/inducer cells were more numerous than $T8^+$ cytotoxic/suppressor cells.[157] It is therefore of great interest that synovial (but not peripheral blood) lymphocytes from eight cases of enteric reactive arthritis were stimulated by salmonella, shigella and campylobacter antigens during tests in which rheumatoid synovial lymphocytes were minimally responsive.[83]

Salmonellae, including *S. typhi* and *S. paratyphi*, regularly cause septic arthritis[150] but *S. typhimurium* is peculiarly liable to provoke a reactive arthritis[132] and is regarded as individually arthritogenic. In analogy with the significance of *Yersinia enterocolitica* antigens in synovial fluid cells (see p. 94), salmonella lipopolysaccharide has now been recognized in similar cells.[102] Comparable responses occur after *Campylobacter jejuni*-induced infection.[71] In one case of Reiter's disease, the precipitating agent was thought be *Clostridium difficile* and analogous changes have been encountered after infection with *Leptospira* sp.[218] and *Giardia lamblia*.[219] There is experimental evidence to support the concept that individual mouse strain antibody responses to *S. typhimurium* are influenced by cross-reactivity to cell surface antigens.[220]

Evidence that throws light on the pathogenesis of gut-related Reiter's syndrome has also come from other animal studies. One of five strains of rat developed aseptic arthritis after intravenous challenge with *Yersinia enterocolitica* strain WA;[115] the persistence of infection supported the view that in the human disease the development of arthritis might be as much a reflection of prolonged antigenic stimulus as of any idiosyncrasy to the antigen itself. Many of the features of Reiter's disease are encountered in the rat after sensitization with Freund's adjuvant. In this model (see p. 544), acute polyarthritis of the hind limb joints, especially those of the feet, develops 9–17 days after an intradermal injection of Freund's complete adjuvent—i.e. of finely ground, heat-killed mycobacteria suspended uniformly in a light mineral oil. The response can also be elicited by nocardia preparations and by lipoprotein extracts of these organisms. It is known that the arthritogenic moiety is the cell wall component, muramyl dipeptide (see p. 547).

Ankylosing spondylitis

Ankylosing spondylitis[24,43,147] is a chronic, seronegative, inflammatory, systemic disease affecting at least 0.4 per cent of males in the UK but only 0.05 per cent of females, beginning insidiously or, occasionally, acutely, with pain in the sacroiliac, hip and sciatic regions. There is a very strong association with the inheritance of HLA-B27. The disease, much more common than previously suspected but often not recognized, involves the fibrocartilagenous joints of the axial skeleton and progresses with fibrosis and bony union to ankylosis. The sacroiliac joints which are both cartilagenous and synovial, are particularly vulnerable. The peripheral, diarthrodial joints are also affected and display features resembling those of rheumatoid arthritis. The osteoarticular changes of ankylosing spondylitis are unexpectedly common in persons with regional ileitis, ulcerative colitis and juvenile rheumatoid arthritis. In all forms of the disease, extraarticular and systemic lesions are likely. With time, involvement of the diarthrodial and of the fibrous joints of the vertebral column and ribs impairs thoracic mobility and predisposes to pulmonary infection or to traumatic paraplegia. Survival, however, is usually lengthy and incapacity seldom complete; the course of the disease is milder in women than in men.

Historical aspects

The history of ankylosing spondylitis has been wittily sketched by Bywaters[41] and cleverly outlined by Eudora Weltz (quoted by Smith[195]). There is convincing evidence of the existence of ankylosing spondylitis in Egyptian mummies as early as 2900 BC and other records in European neolithic communities. The first description of the pathology of the disease is attributable to Connor (1695) (Fig. 3, p. 4).[25] A further case was reported by Fagge.[75] Important contributions were made by Strumpell,[198] Bechterew[19,19a] and Marie[142] in relation to the clinical features of the disease but a comparatively small number of systematic pathological studies has been undertaken. In New York, 311 patients were said to have been seen between 1928 and 1954[26] but no records are available from the 10 000 necropsies performed during the same period.

Heredity

There is an increased familial prevalence and it had long been suspected that there might be a hereditary predisposition. However, the independent demonstration by Brewerton *et al.*[32] and by Schlosstein *et al.*[185] that the histocompatibility antigen HLA-B27 (see p. 243) was present in 72 of 75 and in 35 of 40 of their two groups of Caucasian patients, respectively, compared with an incidence of this antigen of 7 per cent in the general population, was unexpected. Moreover, 32 of 60 first-degree relatives were HLA-B27-positive.[30,31,32]

Whether the inheritance of HLA-B27 is causally related to ankylosing spondylitis or is coincidental is contraversial. One view, discussed further on p. 781, is that the sterochemical configuration of an antigenic determinant on the HLA-B27 complex is associated with the disease, perhaps because of homology with a cross-reacting component of arthritogenic bacteria such as *Klebsiella pneumoniae*.* That 5 per cent of Caucasians with ankylosing spondylitis are HLA-negative is not yet explained. There are also conspicuous racial differences in the frequency of ankylosing spondylitis which is uncommon among Negroes and Japanese, very common among some North American Indians. The development in HLA-positive male children of a seronegative arthropathy with enthesopathy has drawn attention to the occasional occurrence of premature ankylosing spondylitis.[178,183] Ankylosing spondylitis, the clinical and pathological features of which may overlap with other seronegative arthropathies, may coincide with diseases such as systemic lupus erythematosus in which the inheritance is not of HLA-B27 but, for example, of HLA-A1; B8; DR2 and DR3.[155]

* In UK terminology, *Klebsiella pneumoniae* is distinguished from *Kl. aerogenes*. This distinction is not made in the USA. However, *Kl. pneumoniae* is likely to be the approved name.

Clinical features

Ankylosing spondylitis is a disease of young adult males.[112] The sex ratio is variously recorded as between 10:1 (male to female) and 4:1. The mean age of onset is approximately 27 years and the peak incidence is in the third decade. A predictable pattern of clinical behaviour emerges during the first ten years[49] (Table 19.1). The disease commonly begins with pain in the lower back of sacroiliac distribution.

Table 19.1

Modified New York criteria for ankylosing spondylitis[209]

Diagnostic criteria
1. *Clinical*
   a. Low back pain and stiffness for more than three months which improves with exercise but is not relieved by rest.
   b. Limitation of motion of the lumbar spine in both the sagittal and frontal planes.
   c. Limitation of chest expansion relative to normal values, corrected for age and sex.
2. *Radiological*
   Sacroiliitis grade 2 bilaterally or sacroiliitis grade 3–4 unilaterally.

Grading criteria
1. Definite ankylosing spondylitis if the radiological criterion is associated with at least one clinical criterion.
2. Probable ankylosing spondylitis if
   a. Three clinical criteria present
   b. The radiological criterion is present without any signs or symptoms satisfying the clinical criteria. (Other causes of sacroiliitis to be considered.)

Pain is most severe during the early course of the disease and diminishes with time as fibrosis and ankylosis advance and inflammation regresses. Ankylosing spondylitis is a common cause for unexplained low back pain in young men but the prevalence of the disease in non-hospital populations is difficult to estimate. The frequency of ankylosing spondylitis in general population groups in temperate climates is likely to be greater than has been suspected.

Less common than early pain in the sacroiliac region is pain in other parts of the pelvis such as the hip. In a minority of cases the disease begins as a peripheral polyarthritis closely resembling rheumatoid arthritis. Ankylosing spondylitis is distinguished by the absence of systemic disease and anaemia, a relatively low erythrocyte sedimentation rate, the characteristic age and sex pattern and negative tests for rheumatoid factor. There is also pain and tenderness of the sternomanubrial and sternoclavicular joints, tender ischial tuberosities and tender heels: the hips, knees, ankles and feet are particularly affected and temporomandibular arthritis has been recorded as an early symptom. Subcutaneous nodules are rare. Transient but later persistent pain in the gluteal regions may be sufficiently sudden and severe to suggest crush fracture of the spine, disc prolapse, trauma or infection. However, a gradual onset with deep pain extending to the iliac crest, down to the knee, the trochanter or the femoral triangle, is more usual. Low lumbar backache, early morning stiffness and restricted intercostal movement require treatment. Hart[112] emphasized the significance of initial thoracic and cervical spinal symptoms (14/184). Cervical stiffness may be early or late.

The insidious progression of ankylosis means that trauma is hazardous and the stiff, rigid spine and ankylosed synovial joints predispose to fracture, fracture dislocation and tetraplegia. There are characteristic radiological changes.[164] Local trauma invokes local exacerbations of pain and it is accepted that injury and physical insult account for a proportion of the lesions subsequently seen histologically. Ankylosing spondylitis is a systemic disease and early non-granulomatous anterior uveitis (iridocyclitis), aortic incompetence with reflux, upper lobe pulmonary fibrosis and a raised erythrocyte sedimentation rate draw attention to the extent of the disorder just as amyloidosis, usually with renal deposition, is a regular, if uncommon, cause of death.

Pathological changes

Ankylosing spondylitis is a systemic, non-suppurative inflammatory disease.[14,15,39] The widespread lesions[145] affect the entheses; the synarthroses; the synovial joints; skeletal muscle; subcutaneous tissue; and the eye, heart, aorta, kidney and lung. Understanding of the pathology of the early lesions has been handicapped by the rarity of any justification for biopsy and by the infrequency with which cases come to *post mortem* examination in hospitals with osteoarticular laboratories.

There are two primary phenomena: enthesopathy and synovitis; and two associated disorders: enchondral ossification and trauma. There are also consequential changes, particularly amyloidosis, and secondary iatrogenic diseases, such as myelocytic leukaemia.

Ligamentous (enthesopathy) and articular-capsular lesions

The 'unit' lesion, i.e. the fundamental pathological change most characteristic of ankylosing spondylitis, is inflammation centred on the entheses.[156] These structural and functional components are the sites at which ligaments such as the anterior longitudinal ligament of the spine, are inserted into or take origin from bone. They have large vascular beds, unlike nearby ligamentous or fibrocartilagenous tissues, and are sites of relatively high connective tissue metabolic activity, rendering them vulnerable to the effects of inflammation.

The histological features of enthesitis were demonstrated by Ball.[13–15] At the enthesis, there is vascular congestion and a lymphocytic, plasmacytic though less often polymorph cellular infiltrate. Although the adjacent bone is the site of an osteitis, there is never evidence of bone necrosis. Whether osteoporosis is present or not does not influence the severity of the enthesopathy. Granulation tissue formation heralds repair and the insidious replacement of chondrified and calcified parts of the ligamentous insertions. Fibrosis follows, leading to functional loss and immobility. Reactive new bone formation begins. There are analogous changes in the joint capsules. The articular and bone defects are replaced firstly by fibrous tissue, then by bony tissue. The site of connective tissue injury becomes attached to the eroded ligamentous end, forming a new enthesis above the original bony surface. In this way, a small spur of reactive, cancellous bone is produced, subsequently to be remodelled and substituted by mature lamellar bone. Repair can occur in very small erosions, and the early inflammatory disease is transient. Why it should initiate bony ankylosis is not understood. There are paradoxes: the enthesopathic erosion of bone at tendinous insertions into the patellar and tibial tuberosities is rare compared with the high frequency of involvement of the spine and calcaneum.[4]

Cartilagenous and fibrous joint disease

The cartilagenous and fibrous joints are affected directly and by the occurrence of nearby enthesitis.[7,15] The articulations affected most severely are the main intervertebral joints, the cartilagenous component of the sacroiliac joints, and the manubriosternal joints.

Intervertebral joints Active inflammation begins in the attachments of the fibres of the annulus fibrosus to the vertebral body, the enthesis in the vascular peripheral tissue of the intervertebral disc itself and within adjacent bone (Fig. 19.3). The inflammatory changes closely resemble those described on p. 770. The common intervertebral ligament is unaffected but inflammatory cells may extend in the loose connective tissue between the outer part of the annulus and this ligament. Active inflammatory 'erosion' is often confined to a small part of the margin of the vertebra. Erosion may be uni- or bilateral. The reparative process generates new bone; one result is a remodelling or 'squaring' of the vertebral outline, the main cause of which is the underlying osteitis.[7,8] Widely regarded as late manifestations of the disease, osteitis with erosion may in fact be an early sign recognizable by biopsy.[144]

Bone formation is therefore seen to occur within the anatomical confines of the intervertebral disc margins (Figs 19.4 and 19.5). Fusion of the vertebrae occurs. The bony processes which cause this fusion are syndesmophytes. Extending throughout the spine, their presence results in the anatomical change of 'bamboo spine'. The radiological appearances (Figs 19.6, 19.7 and 19.8) are pathognomonic; they and the underlying bamboo spine are both readily distinguishable from ankylosing hyperostosis (see p. 947), fluorosis (see p. 411), osteophytosis and the effects of infection and injury.

The process of cartilagenous calcification and the evolution of syndesmophytosis can be traced by microradiography and by tetracycline labelling.[84] Although the bony bridges may, in the first place, be only 2–5 mm wide, ankylotic bone forms in spinal segments that are in other respects normal. Non-marginal syndesmophytes may also be found. Inflammation of nearby structures such as the cartilage endplate is not seen. The calcification and ossification of cartilage, the vascular ingrowth and consequential ossification of the disc[40,42,73] constitute a spondylodiscitis

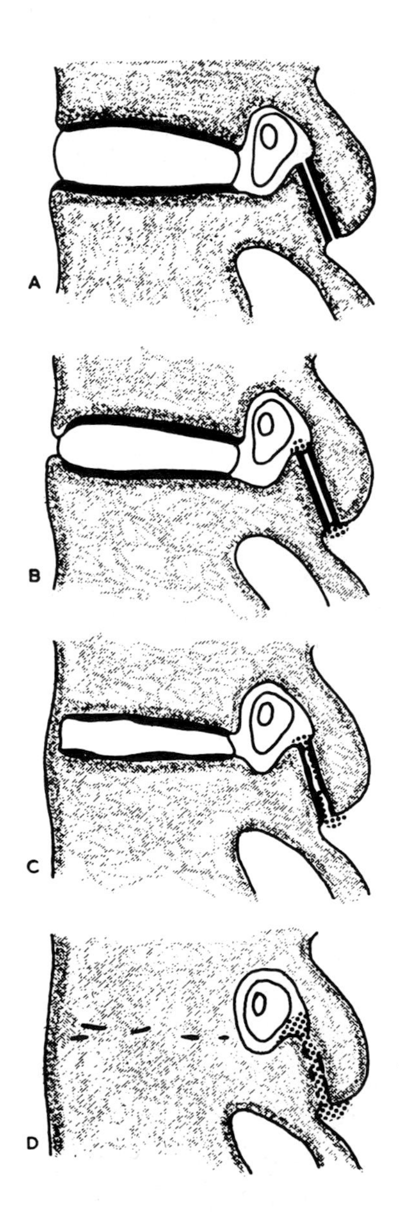

Fig. 19.3 Ankylosing spondylitis: evolution of intervertebral and apophyseal joint disease.
A. Normal vertebrae with main intervertebral and one apophyseal (posterolateral) joint. The former is fibrocartilagenous (a syndesmosis), the latter synovial (a diarthrosis). The intervertebral canal transmits nerve roots and blood vessels in close relation to the two joints. **B.** Early changes in ankylosing spondylitis take the form of enthesitis, an ill-defined osteitis at the anterolateral margins of the intervertebral discs, and a low-grade synovitis of the apophyseal and of the other vertebral, synovial joints (see Fig. 19.14). **C.** As the disease slowly advances, there is a marginal loss of the hyaline cartilage that links the intervertebral disc with the underlying bone; and new fibrous tissue and bone formation at the anterior and lateral margins of the discs, within the anatomical configuration of the ligaments that bind the discs and bones together. Synovitis develops in the (synovial) apophyseal and costoclavicular, joints. As the synovitis progresses, changes occur that closely resemble those of rheumatoid arthritis. **D.** In the late stages of the disease, there is complete destruction of the intervertebral joints. New bone replaces both the hyaline- and the fibrocartilages. Since the changes are within the anatomical contours mentioned in **B.** (above), the vertebral bodies retain their overall shape by contrast with, for example, the alterations to vertebral body contours that occur in osteophytosis and in ankylosing hyperostosis. Within the apophyseal and the other synovial joints, there is progressive hyaline cartilage destruction and fibrous ankylosis.

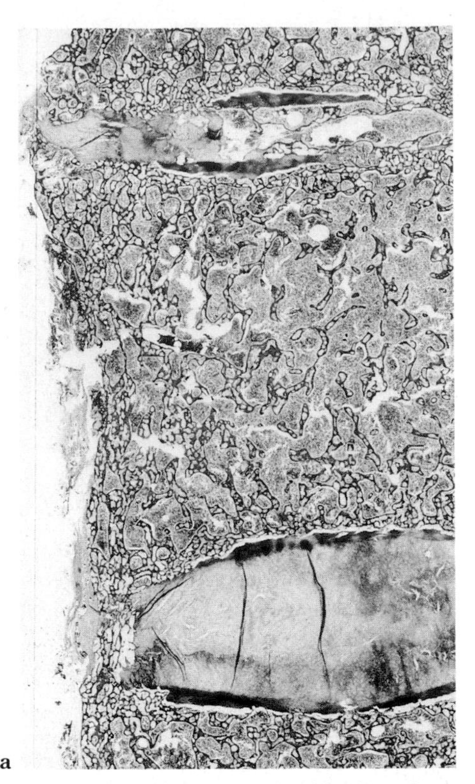

a

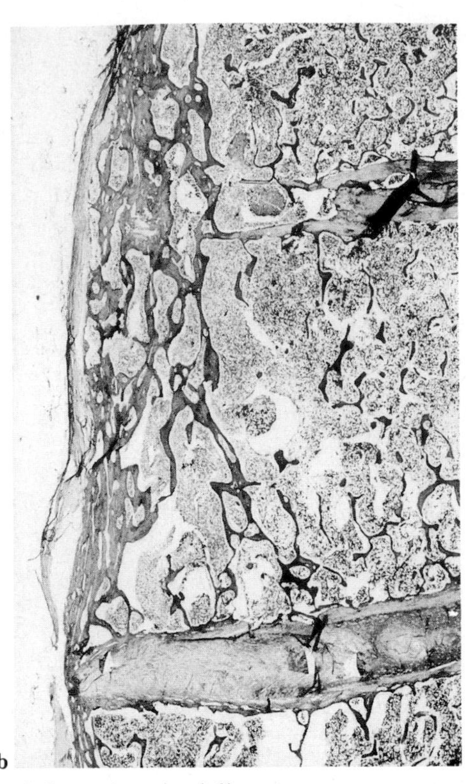

b

Fig. 19.4 Ankylosing spondylitis: vertebral lesions in established thoracic spinal disease.
a. Little remains of the osseocartilagenous endplates of the upper disc which is undergoing early ossification. At anterior margin (left), new bone encroaches on disc periphery. Lower disc is less severely affected but anterior margin is occupied by syndesmophyte. Note moderately severe osteoporosis. **b.** Almost continuous lamina of new, syndesmophyic bone unites adjoining vertebral bodies. Density of new bone contrasts with nearby, osteoporotic vertebral bone. (HE × 2.)

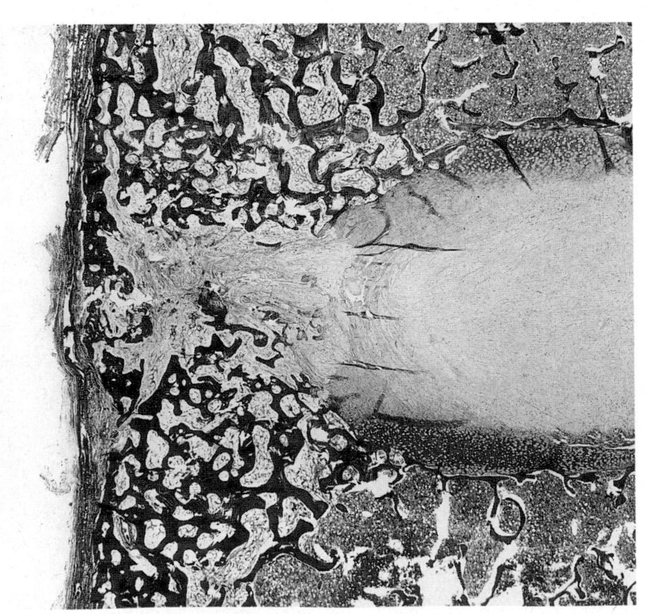

Fig. 19.5 Ankylosing spondylitis: intervertebral joint.
In this sagittal section, the anterior parts of two adjacent vertebral bodies are seen to be united by a bridge of fibrous tissue and new bone, which has formed within the anatomical outlines of the disc. The structure is a syndesmophyte. (HE × 4.)

the most severe forms of which occur in patients with advanced ankylosis. These lesions are believed to be traumatic. A destructive process affecting the entire disc–bone interface can be a result of fracture of the neural arch or undisplaced fracture of an entire spinal segment.[225] The spondylodiscitis is not infective. Segmental fracture is a complication of spinal osteoporosis and of lack of disc resilience and segmental variations in the extent and density of ankylosis at different spinal levels. The fractures are horizontal through the ankylosed disc and posterior arch: their late recognition may account for their identification as spondylodiscitis although very severe destruction in non-ankylosed segments can occur without preliminary fracture.

Cawley *et al.*[51] described the histological changes in spondylodiscitis. There are *localized* anterior rim lesions in the kyphotic thoracic segments, with replacement of the rims and of the disc by granulation tissue. The focal microscopic lesions of lumbar rims without disc disease are not wholly understood but may be consequences of simple trauma in the osteoporotic bone of a spondylitic spine. There are also severe lesions of the *whole* disc–bone interface: they appear similar whether or not there is neural arch fracture. Eroded disc tissue is replaced by granulation

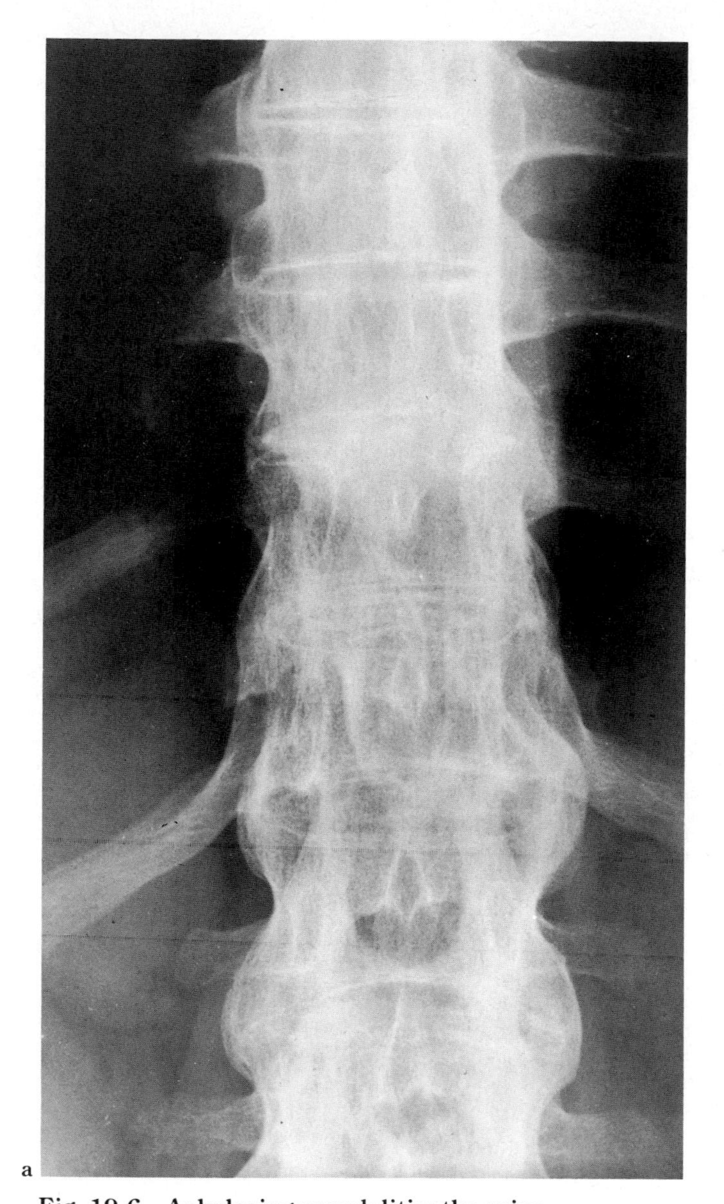
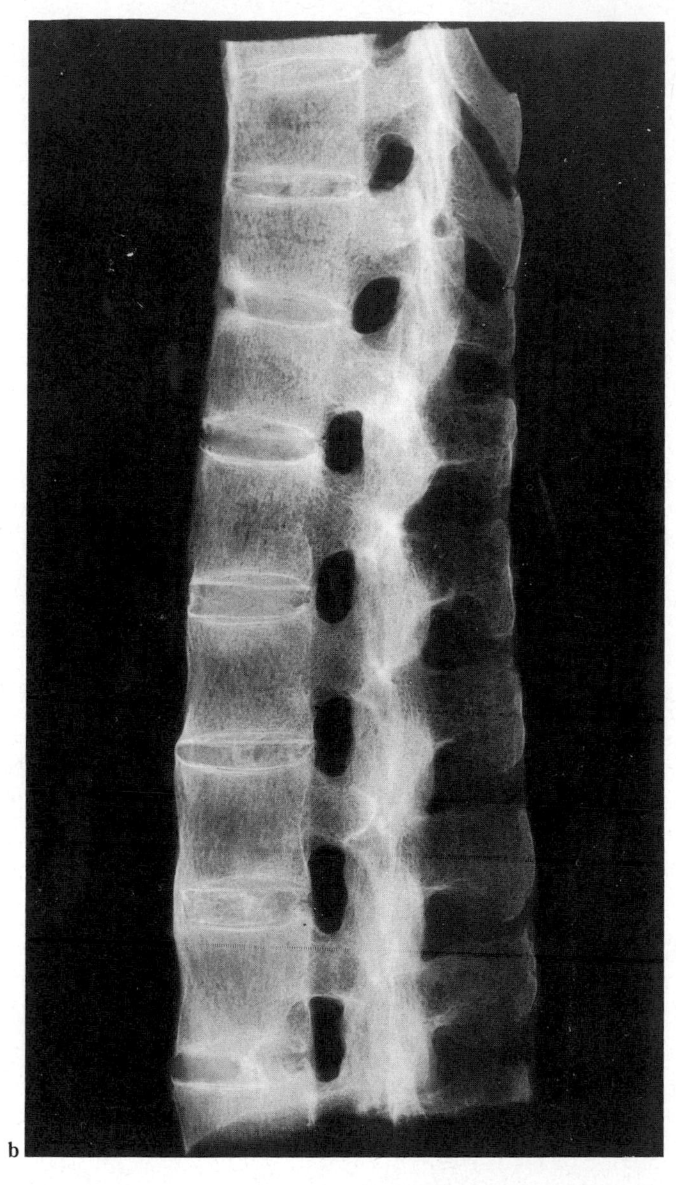

Fig. 19.6 Ankylosing spondylitis: the spine.
a. Posteroanterior view, clinical radiograph. The margins of the main intervertebral joints are united by new bone formed within the disc margins, the process of syndesmophytosis. **b.** Lateral view of anatomical specimen. The outlines of the intervertebral discs are clearly seen. The anterior margins of the lower thoracic and upper lumbar discs are replaced by bone. The posterior disc margins are obscured. In relation to the apophyseal joints, hyperostosis has succeeded synovitis and the joint spaces are lost. Fig. 19.6b by courtesy of Professor Bruce Cruickshank and the Editor, *Journal of Pathology*.

tissue among which are bone fragments and islands of fibrocartilage. Although there is active bone remodelling, with appositional bone sclerosis in adjoining vertebral bone, there is little reactive bone in the eroded disc margin.

The distribution of the intervertebral joint disease in ankylosing spondylitis is widespread and multifocal. It is said that the synovial atlantoaxial joints are spared. Consequently, no neurological complications ensue unless trauma provokes atlantoaxial dislocation and fracture.[15] It is interesting to contrast this observation with the comments of Fam and Cruickshank[77] in their case of psoriatic cervical arthritis (see p. 781).

Manubriosternal joint This articulation fuses in normal adult life in 10 per cent of individuals: investigations are difficult. Twenty-three to forty-seven per cent of manubriosternal joints are ankylosed in ankylosing spondylitis (Fig. 19.9); a peripheral bar of bone bridges the joint space[13] and complete synostosis is frequent. Inflammatory erosion, the result of enthesitis, is detected in one quarter

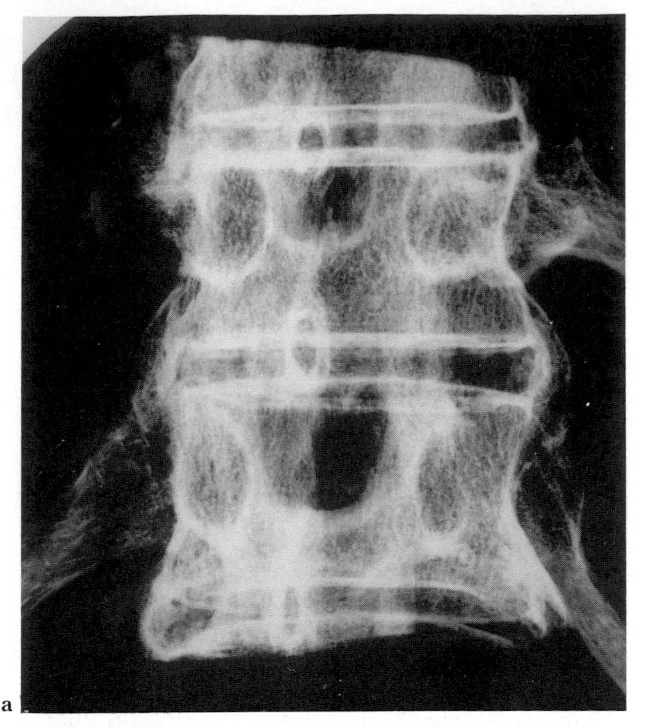
a

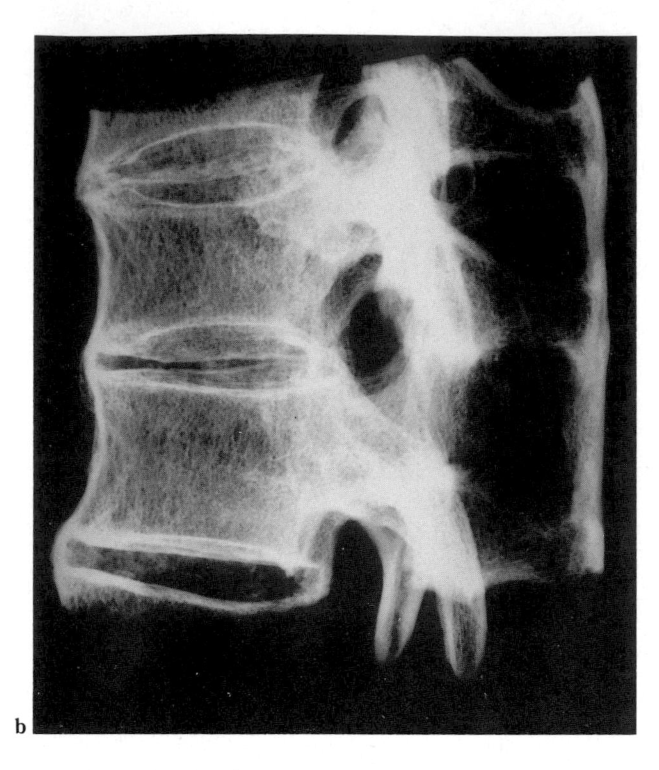
b

Fig. 19.7 Ankylosing spondylitis: the spine.
a. Radiograph of anatomical specimen for comparison with Fig. 19.6a. Detail of marginal, intervertebral syndesmophytes. Note outlines of costovertebral joints and osteoporosis of vertebral bodies. **b.** Further anatomical specimen. Lateral radiograph. There is new bone formation within the margins of the intervertebral discs and ankylosis of the apophyseal joints. Osteoporosis is associated with partial collapse of the upper vertebra.

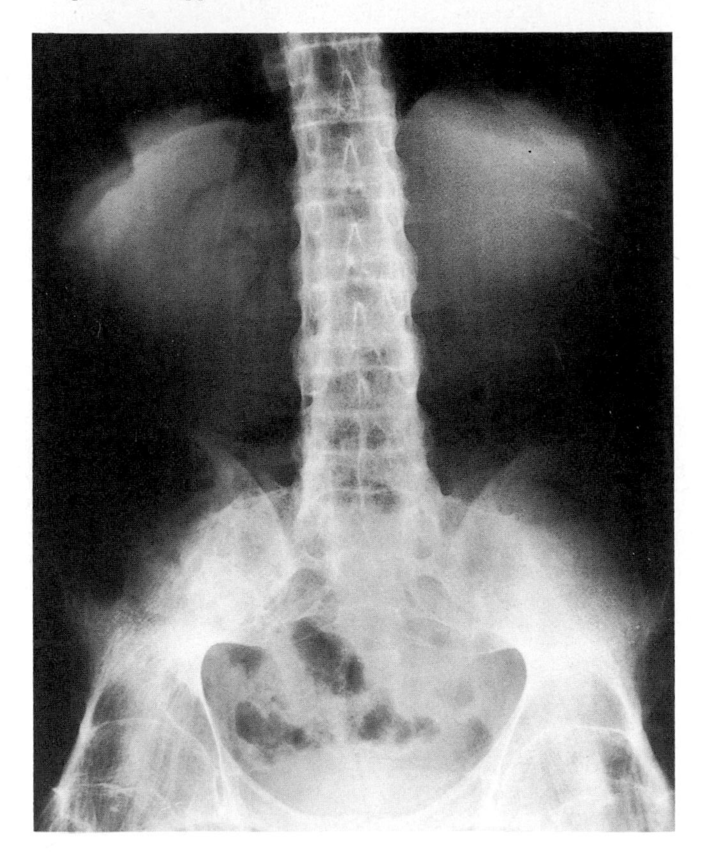

Fig. 19.8 Ankylosing spondylitis: the spine.
Late-stage ankylosing spondylitis of lumbosacral spine and pelvis showing bony ankylosis of the hip and sacroiliac joints and of the spine. Syndesmophytes unite the contiguous vertebral bodies at all levels, causing a 'bamboo-spine' appearance. Courtesy of the Department of Diagnostic Radiology, University of Manchester.

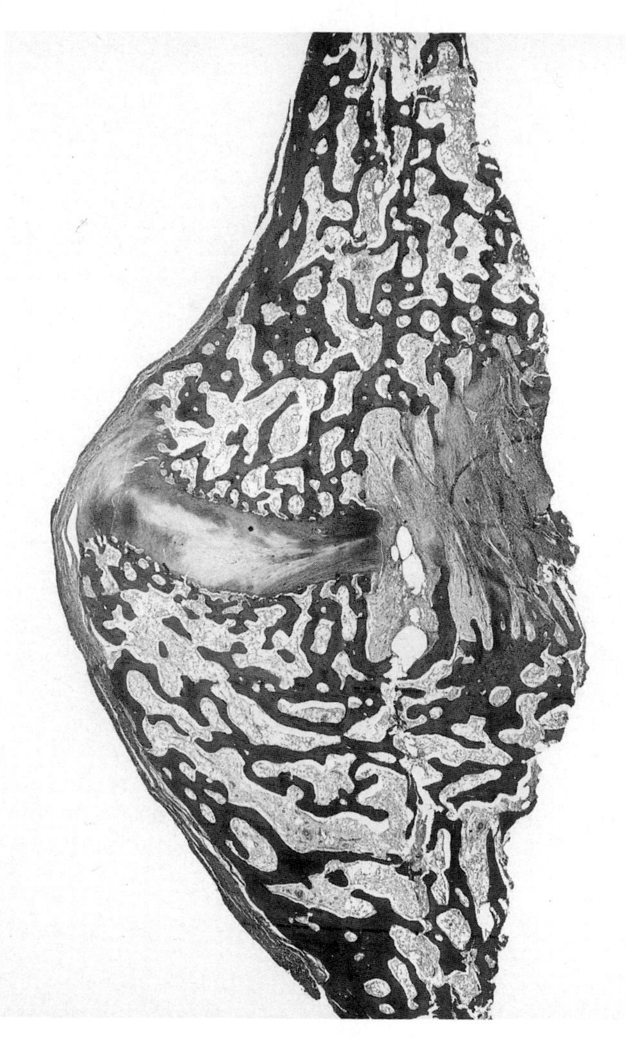

Fig. 19.9 Ankylosing spondylitis: manubriosternal joint.
Sagittal section of manubriosternal joint with anterior margin to left. Much of the fibrocartilage and adjacent bone is replaced by granulation tissue. There is slight, nearby hyperostosis. (HE × 5.)

of those subjects imaged radiologically and this enthesopathy may explain the osteitis identified by Cruickshank.[62] However, synovial tissue sometimes adjoins the manubriosternal joint and synovitis could explain erosion. Moreover, stress fracture can occur and complicate the histological findings.

Symphysis pubis Erosion of this joint in pregnancy and with pelvic or urinary infection[110] and comparable changes provoked by athletic training[111] can confuse the interpretation of the severe changes which may occur in the uncommon female ankylosing spondylitic patients.[172] Focal osteitis with erosion and joint narrowing and bone sclerosis have been recognized.[62]

Synovial disease

Synovitis both of the peripheral and of the axial synovial joints is almost invariable.

Peripheral synovial joints Larger joints are particularly at risk but smaller articulations, including the cricoarytenoid[134] and ossicular,[141] can be implicated.

Early accounts[56,86,106] of the synovial changes were quoted by Cruickshank.[61] He described the sequence of histological phenomena in the limb joints from 12 patients at necropsy or biopsy (Figs. 19.10 and 19.11). The microscopic changes closely resembled those of active rheumatoid arthritis but inflammatory cell infiltration was less[121] and fewer cells were present. There were rare zones of necrosis. Villous hypertrophy and synovial cell hyperplasia with congestion and oedema accompanied lymphocytic, plasmacytic and histiocytic infiltration. The cells were both diffusely and focally arranged, as in rheumatoid arthritis, and were often perivascular in distribution. These observations, which are comparable to those of Soren,[196] were confirmed and extended by Revell and Mayston.[173] Among 14 specimens aged 8 to 42 years, duration 6 months to 25 years, the main histological changes are shown in Table 19.2. In 10 cases the proportions of synovial cells staining for IgG, IgA and IgM were 78.6–88.2, 0.3–3.0

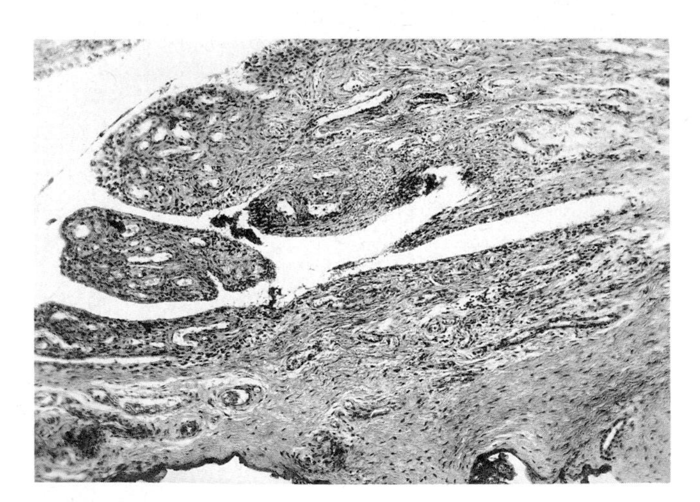

Fig. 19.10 Ankylosing spondylitis: hip joint synovitis.
Within the hypertrophic synovial villi there are small numbers of lymphocytes and mononuclear macrophages, constituting a low-grade inflammatory reaction. There is a slight excess of collagenous fibrous tissue. The appearances are indistinguishable from those of rheumatoid synovitis. (HE × 37.)

Fig. 19.11 Ankylosing spondylitis: chronic synovitis. Beneath an attenuated synovial cell lining (at top) is dense, collagenous connective tissue interspersed more deeply with scattered lymphocytes. (HE × 85.)

and 7.4–16.8, respectively; corresponding figures for 12 cases of rheumatoid arthritis were 49.2–92.2, 1.4–14.6 and 5.4–37.6, respectively. The predominance of IgG-positive cells in both diseases was evident.

The synovial fluid (see Chapter 2) contains fewer polymorphs, and more lymphocytes than the synovial fluid in classical rheumatoid arthritis.[121,127] The large synovial mononuclear cell is a monocyte-derived macrophage.[202] Some of the synovial fluid polymorphs, contain cytoplasmic vacuoles in which are Ig and C3; however, synovial fluid complement is not lowered.

As in rheumatoid arthritis, the role of iron in perpetuating if not in initiating synovitis has been considered. An exacerbation of ankylosing spondylitic synovitis provoked by intravenous iron dextran was reported by Canton, Downs and Abruzzo.[48] Feltesius, Lindh *et al.*[80] demonstrated that the peripheral blood polymorphs, and platelets in ankylosing spondylitis contained increased amounts of iron measured by nuclear microprobe; the polymorph stores were raised to a much greater level than those of the platelets, suggesting that synovial collagenase and prostaglandin synthesis and release could be enhanced and toxic free radicals liberated.

Table 19.2

Histological changes in synovium in ankylosing spondylitis[173]

Histological changes	Moderate	Severe
Synovial lining cell hyperplasia	7	1
Surface fibrin	5	1
Incorporated fibrin	3	0
Iron deposition	3	0
Diffuse lymphoid cells	5	2
Lymphoid cell aggregates	4	2
Fibrosis	3	0
Perivascular fibrosis	4	0
Increased vascularity	9	0
Perivascular lymphocytes	4	2

With the passage of time, small synovial blood vessels come to show endarterial fibrosis but this may be a result of the now obsolete practice of radiotherapy: nearby focal haemorrhage, with haemosiderin deposition, is common. Inflammation, which is less erosive than rheumatoid arthritis, is long-lasting, with round cell infiltration. Granulation tissue formation, with fibrosis, becomes prominent as time passes (Fig. 19.12) and this tissue, in continuity with the

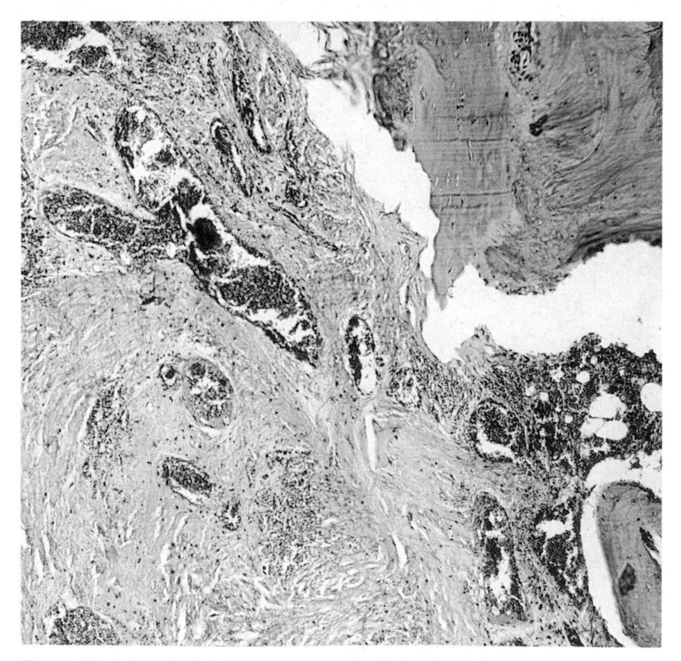

Fig. 19.12 Ankylosing spondylitis: hip joint. Extensive fibrosis has occurred within the periarticular synovial tissue; it coincides with the degradation and loss of articular cartilage. The vascular granulation tissue adjoins underlying bone. The residual synovial tissue contains many lymphocytes and mononuclear macrophages. (HE × 40.)

synovia, replaces and spreads over marginal articular cartilage zones. Fibrous tissue substitutes for cartilage (Fig. 19.13) and is also found in nearby bone which is progressively destroyed. Osteoporosis complicates local bone destruction. The final stages of the disease include extensive fibrous ankylosis but, unlike rheumatoid arthritis, they extend to bony ankylosis with diminishing evidence of inflammation. When the larger diarthrodial joints are affected, secondary osteoarthrosis is recognized. The fibrotic changes culminate in ankylosis.

Axial synovial joints Disease of the axial articulations is characteristic and, together with the fibrocartilagenous joint lesions, accounts for much of the disability of ankylosing spondylitis. The apophyseal, costovertebral and sacroiliac joints are most frequently affected, and ankylosing spondylitic patients are peculiarly susceptible to sacroiliac joint disease.[86,93,106]

In the case of the *apophyseal joints*, synovitis is accompanied by capsular inflammation, enthesitis. Continued inflammation culminates in fibrous ankylosis and synchondrosis. The joint space is obliterated and enchondral ossification follows, resulting in bony ankylosis. Ossification also occurs within the joint capsule. Ball[13] believes that the cartilage change represents a reaction of the cartilage–bone junction to the relief of mechanical stress. Synostosis

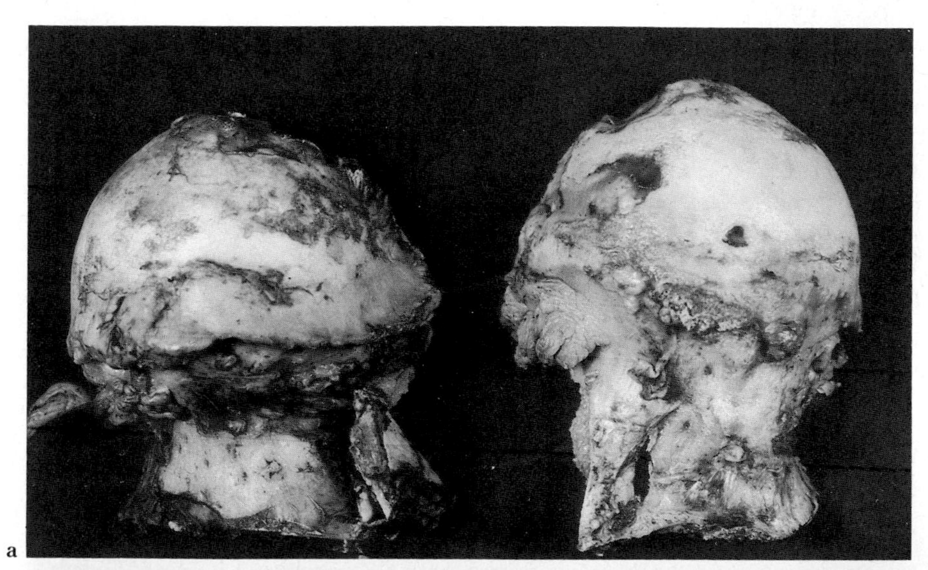
a

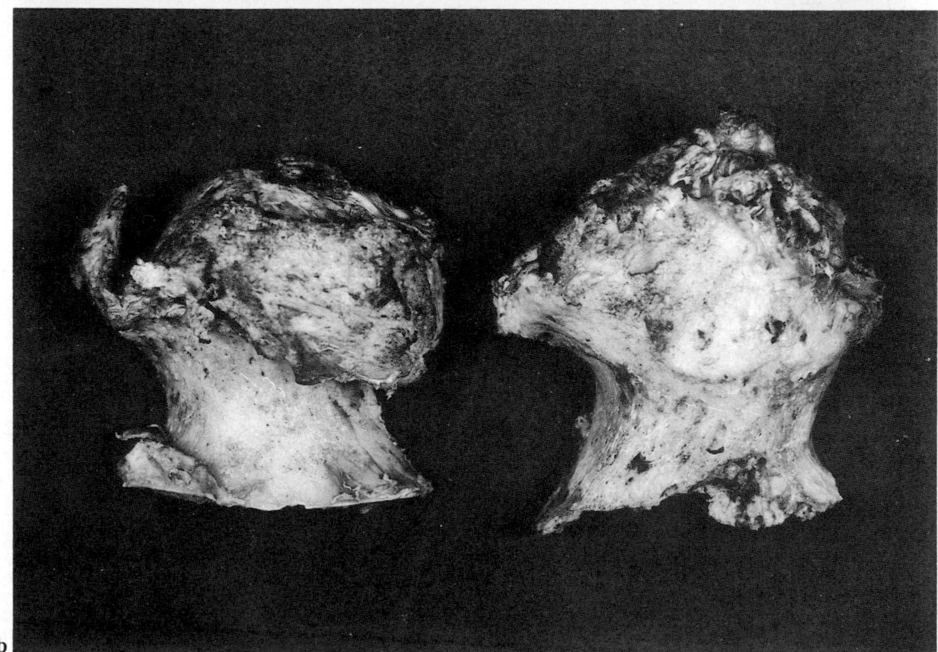
b

Fig. 19.13 Ankylosing spondylitis: hip joint.
a. Articular cartilage surfaces have been partly denuded by the extension of pannus both from the synovial margins and through defects in the principal regions of load-bearing cartilage. **b.** More advanced changes: compare with Fig. 19.13a. Secondary osteoarthrosis has developed and is accompanied by osteophytosis and bone remodelling so that the femoral head has assumed a flattened, mushroom-like shape (HE × 0.8.)

occurs but the anatomical contours of the joints are preserved. Apophyseal bony ankylosis is almost always accompanied by intervertebral disc ankylosis. It is interesting that costovertebral ankylosis is also accompanied by ankylosis of other components of the same spinal segment.

The fibrous, and later bony, ankylosis of the *costovertebral* (Fig. 19.14) *costoclavicular* and *costotransverse joints* severely limit thoracic movement; respiration becomes largely diaphragmatic. A similar process in the apophyseal joints, almost invariably associated with intervertebral, fibrocartilagenous spinal joint disease, results in dorsal kyphosis and immobility, occasionally so extreme that the patient comes to face the ground while standing upright. Modern surgical techniques have, however, greatly advanced and the correction of many of the classical deformities of ankylosing spondylitis is now possible.

Relatively few histological as opposed to radiological, studies have been made of the *sacroiliac joints* (Fig. 19.15). Synovitis is described in two papers.[66,93] Cruickshank[62] refers to sacroiliac osteitis and Schilling[184a] to fibroosteitis; but subchondral osteitis is a disputed change. The anatomical contours of the affected joints are preserved; an early change is circumferential ankylosis, the result of capsular and of enchondral ossification. The iliac progression is more severe than the sacral, accounting for much erosion.

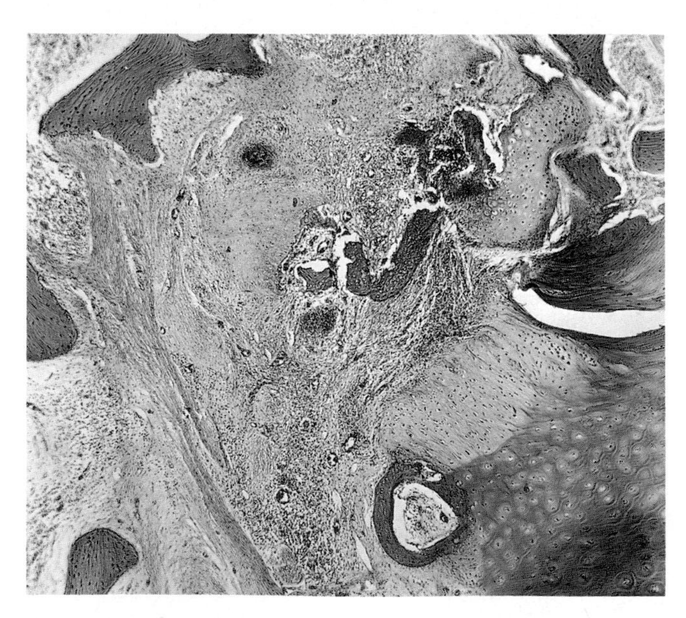

Fig. 19.14 Ankylosing spondylitis: costovertebral joint.
The costal cartilage (lower right) is united to the sternum by granulation tissue within which there are many lymphocytes arranged focally and diffusely. At centre, metaplastic cartilage and bone formation is taking place within the fibrovascular tissue. (HE × 22.) Courtesy of Professor Bruce Cruickshank and the Editor, *Annals of the Rheumatic Diseases*.

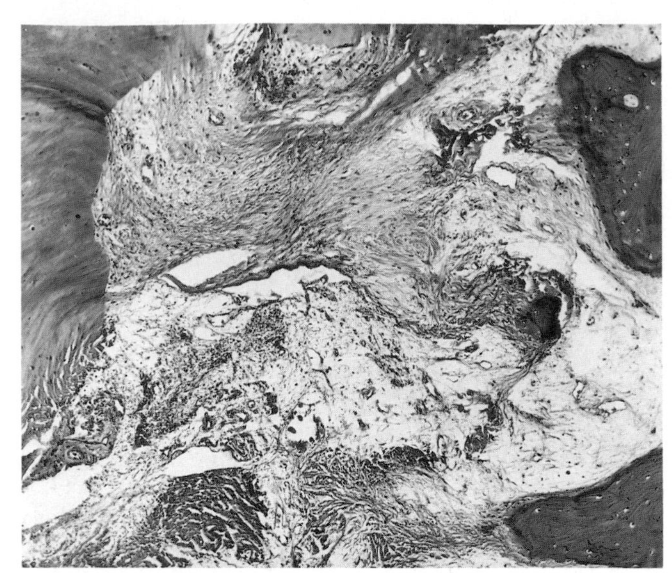

Fig. 19.15 Ankylosing spondylitis: sacroiliac joint.
The iliac cartilage (at left) and the adjacent bone have been extensively replaced by vascular granulation tissue the formation of which has followed an earlier inflammatory response. The sacroiliac joint is both synovial and cartilagenous and this reaction has occurred at the point where the two forms of joint structure merge. (HE × 54.)

Skeletal muscle

Severe muscle changes are recognized. It has been suggested that the abnormalities are neuropathic.[22] Among the histological and histochemical disorders found by these authors in three patients, one of whom was female, were 'moth-eaten' and target or targetoid type 1 fibres, and a loss of both type 2 fibres and of type 2 motor neurones.

Bursae

A synovitis has been described. Histologically, the appearances are identical to those of the synovial joints.[129]

Subcutaneous tissue

Subcutaneous nodules may occasionally form. Histologically, they are indistinguishable from those of rheumatoid arthritis.[179] In the young men described by these authors, rheumatoid factor was present in the serum and the possibility that rheumatoid arthritis and ankylosing spondylitis coexisted must be considered as in a case described by Eulderink (personal communication).

Extraskeletal disease

Ophthalmic system Unilateral acute anterior nongranulomatous uveitis may be a presenting sign of ankylosing spondylitis; it is associated not only with the classical disease but also with 'complicated' spondylitis. Uveitis may recur but tends to be self-limited. The histological changes have not been described and the reason for their occurrence is not understood.

Cardiovascular system Bernstein and Brock[23] and Graham and Smyth[101] identified cardiac disease in a small proportion of long-standing cases of ankylosing spondylitis: the lesions they recognized were almost always isolated aortic incompetence. Two such cases were found among 222 spondylitic patients examined by Ansell, Bywaters and Doniach[6] but the frequency of aortic valve disease is thought to be nearer 10 per cent[15,128].

In their early investigations Clark and Bauer[52] described a peculiar form of aortitis in patients with so-called rheumatoid arthritis of the spine and peripheral joints; Bauer, Clark and Kulka[17] and Clark *et al.*[53] confirmed these observations. In 519 spondylitic patients, Graham and Smythe[101] found 24 with aortic valvular insufficiency. Three of these cases were examined *post mortem*; the findings in the remaining cases and in two others, not included in the original series, revealed: 1. marked dilatation of the aortic valve ring, causing the aortic incompetence; 2. aneurysmal ballooning and dilatation of the sinus of Valsalva; 3. thickening, shortening and calcification of the aortic valve leaflets which sagged into the cavity of the left ventricle; 4. thickening of the rolled free margins of the aortic valve cusps; 5. adhesions and partial fusion of the commissures in some cases, but free, non-adherent commissures in others; 6. thickening of the intima of the sinus of Valsalva and of the regions of the aortic valve ring; and 7. chronic fibrous obliteration of the pericardial cavity.

Microscopically, there were scattered foci of collagen and of elastic tissue destruction in the vicinity of the valve ring with vascularization, round cell infiltration and the presence of occasional polymorphs. The overlying intima revealed fibrosis and a sparse chronic inflammatory cell infiltrate. The aortic valve cusps displayed fibrous thickening and zones of calcification. There was slight fibromuscular thickening of the intramural arterial myocardial branches of the coronary arteries. These lesions were considered to be the late stages of the more acute inflammatory changes of the aortic valve and proximal aorta observed by Clark, Kulka and Bauer,[53] by Schilder *et al.*[184] and by Valaitis *et al.*[207] In each of the five cases of Graham and Smythe[101], the mitral and other cardiac valves were normal. Apart from atherosclerosis, no lesion extended into the aorta beyond the valve ring. In particular, there were no Aschoff bodies.

A further analysis[128] discovered cardiovascular lesions in 14 per cent of 97 patients. In four necropsies, there was heart block (see below), extensive myocardial fibrosis but no vascular lesions and no inflammatory cell infiltrates (1); isolated aortic incompetence, with active aortitis or with calcification of the valve cusps (2 and 3); and aortic insufficiency, active aortitis *and* heart block (4). In both patients with active aortitis, an inflammatory reaction was localized to the root and proximal part of the ascending aorta. Microscopically, there were endarterial fibrosis and perivascular lymphocytic and plasma cell infiltration.

Heart block is therefore a recognized feature of cardiac disease in ankylosing spondylitis just as it is in rheumatoid arthritis. Julkunen[120] and Julkunen and Luomanmaki[122] reported that 12 of 250 patients with ankylosing spondylitis had transient or permanent first degree atrioventricular block; one case had right bundle branch block and four had third degree heart block. Weed *et al.*[211] concluded that 8 per cent of ankylosing spondylitic patients without aortic insufficiency had conduction defects compared to a figure of 0.5 per cent for a normal population. *Post mortem* study of the first of Weed's two cases showed a dilated aortic valve, slight diffuse thickening of the valve, but no commissural fusion. Microscopically, there was continuity of a fibrotic and endarteritic process from the base of the aortic valve to the apex of the muscular septum: the penetrating part of the atrioventricular bundle was almost completely replaced by fibrous tissue in continuity with that present in the aortic root. By contrast, Brewerton *et al.*[34] found the standard electrocardiograms to be normal in 73 of 74 patients. Nevertheless, early diastolic left ventricular abnormalities were identified by echocardiography in 16 of 30 men. In a parallel study, searching for the cause of these disorders, these authors reviewed the histology of 28 hearts from ankylosing spondylitic patients without hypertension, severe coronary artery disease or valvular anomaly. Only a mild, diffuse increase in interstitial connective tissue was seen: there was neither inflammatory change nor amyloidosis.

Arteries Vasculitis in ankylosing spondylitis is rare. G. Ball and Hathaway[12] described a controversial case in which the extent and activity of an arteritis was debatable and in which the joints were not investigated microscopically. Paloheimo *et al.*[161] confirmed that ankylosing spondylitis could coexist with Takayasu's arteritis; it is notable that three of their four reported cases were female.

Respiratory system It is now recognized that pulmonary function in cases of ankylosing spondylitis without symptoms of lung disease, may be defective[79] and that in longstanding ankylosing spondylitis the lungs may be sites for an unusual, non-tuberculous form of fibrosis.[47,119,208]

Although pulmonary tuberculosis (10 per cent), bronchiectasis (5 per cent) and chronic bronchitis and/or emphysema (17 per cent), are no more common *post mortem* in ankylosing spondylitis than in an unselected UK population,[64] an association with tuberculosis has been widely canvassed.[26,36] Cruickshank's earlier figures had revealed tuberculosis (10 per cent), bronchiectasis (1 per cent), and chronic bronchitis (8 per cent), and historically, such cases were often referred for antituberculous treatment. There is no association with earlier radiotherapy.

Commonly, the upper lobe disease is recognized many years after the onset of ankylosing spondylitis. There is a fine, unilateral soft atypical fibrous infiltrate that is progressive and destructive. Bilateral disease may ensue. Occasionally, a secondary fungal infection with organisms such as *Aspergillus* sp. develops.[143] Bronchocentric granulomatosis can coexist[176] and bronchiectasis may be superadded. Lobectomy may be complicated by bronchopleural fistula.

The pulmonary lesions are believed to be similar to those of the aorta. In four cases, needle biopsy revealed interalveolar fibrosis with degeneration of collagen.[119] In two cases (in this series of seven) in which right upper lobectomy was performed, fibrosis, hyalinized connective tissue and the presence of cavities lined with 'non-specific inflammatory tissue' were found in one, interalveolar fibrosis with 'degeneration' and elastic fragmentation in the other. The disorder is regarded as a non-specific pneumonitis in which lymphocytic and plasma cell infiltration progress to fibrosis, pleural thickening, bronchiectasis and cyst formation.[14]

Urinary system Although amyloidosis (see Chapter 11) is a relatively frequent cause of death in ankylosing spondylitis,[20,39] other renal disorders are uncommon. The majority of changes are vascular and coincidental.[64,135] Examples of IgA nephropathy are, however, now regularly reported together with instances of mesangioproliferative glomerulonephritis and membranous nephropathy.[192] These authors identified five cases of IgA nephropathy among 116 patients surveyed clinically, raising questions relating the immune responses to gut-associated microorganisms (see p. 786).

Aetiology and pathogenesis

The cause of ankylosing spondylitis remains uncertain. Although ankylosing spondylitis-like disorders occur in the dog[193] and non-human primate[2] and in the gorilla in which they are accompanied by the inheritance of HLA-B27, animal models of ankylosing spondylitis have not yet clarified its pathogenesis. In man, ankylosing spondylitis is associated with a defect in the mononuclear cell response to Epstein–Barr virus similar to that encountered in rheumatoid arthritis patients. It is not known whether this is a secondary phenomenon.[174]

However, the most important clues come from the similarity between the spondyloarthropathies that characterize both ankylosing spondylitis and reactive arthritis, and the overwhelming evidence of a genetic predisposition, particularly in males. Outbreaks of salmonella, shigella and yersinia intestinal infection were followed by an unexpectedly large number of cases of reactive arthritis[163] among patients with enteritis who were HLA-B27-positive. It was natural to search for evidence of a link between a delayed, immune response to bacterial antigen and the connective tissue disease.

One form of response could be the consequence of antibacterial antibodies, formed in genetically susceptible persons during subclinical intestinal infection. A disease-specific antigen might exist in circulating, pathogenic immune complexes.[37] Immunoglobulin A has been implicated and cutaneous deposits of this class of antibody are one index of its possible role in pathogenesis.[55] Raised titres of antiklebsiella IgA antibodies are demonstrable in active ankylosing spondylitis[58] but the degree of disease activity and the IgA titres do not correlate so that a non-specific release of IgA into the circulation is proposed, without fundamental significance for the origin of the spondyloarthropathy. Antibody titres to the D-Ala-D-Ala moiety of peptidoglycan, an important cell wall component of most bacteria and the arthritogenic element in Freund's adjuvant (see Chapter 13), are elevated in ankylosing spondylitis as they are in Reiter's syndrome but not in rheumatoid arthritis or osteoarthrosis,[162] supporting the possibility that similar antibodies are causally related to the reactive spondyloarthropathies.

Two further concepts have emerged. The first implicates the immune response to enteropathic bacteria in direct cross-reactions with HLA-B27. *Klebsiella pneumoniae* has attracted particular interest.[60] This organism was isolated more often from the faeces of patients with active disease than from those whose disease was inactive.[69] The same faecal organism was associated with acute anterior uveitis in ankylosing spondylitic subjects.[70] It seemed possible that the cross-reactivity between articular connective tissue and bacterial molecular components might be on the basis of molecular mimicry, a mechanism implicated in other connective tissue disorders such as polymyositis–dermatomyositis (see Chapter 16). Cross-reactivity between anti-HLA-B27$^+$ lymphocyte antisera and *Klebsiella pneumoniae* was shown[68,213] and anti-*Kl. pneumoniae* antisera displayed lymphocytotoxic activity against HLA-B27+ lymphocytes from ankylosing spondylitis-positive patients. Human monospecific antisera for HLA-27 also bound to *Kl. pneumoniae* extracts.[10] There therefore seemed some reason to believe that cross-reactivity between antibodies or cells directed against *Kl. pneumoniae* might directly react with HLA-B27 antigens,[204] provoking or modifying the initial inflammatory responses in ankylosing spondylitis. The evidence remains controversial and a variety of views have been expressed. Cameron *et al.*,[45] for example, failed to find specific cross-reactivity between *Kl. pneumoniae* antibody and peripheral mononuclear cells from HLA-B27-positive ankylosing spondylitic patients.[45]

Alternative views have been advanced. One proposal envisaged the possibility that antigens in the outer membrane of *Kl. pneumoniae* could *modify* HLA-B27.[90] These authors considered that a modified HLA-B27 complex could induce effector cells that would injure targets such as synovium and cartilage, leading to ankylosing spondylitis. The mechanism might invoke Ir-type genes (see p. 243); an antigen-binding receptor might have a 2-domain structure, the constant portion being the HLA-B27 molecule, the variable region an idiotypic structure specific for some *Kl. pneumoniae* idiotopes (see p. 243). It is certainly clear that cross-reactive antisera, prepared against certain bacteria, can discriminate between ankylosing spondylitis-positive and -negative patients.[91] The reactivity could be attributed to a *modifying factor* that has been found in, and purified from, bacterial cell walls.[91] This well-characterized factor cross-reacts with the HLA-associated cell surface structure on ankylosing spondylitic lymphocytes and other cells. Alternatively, this 'factor' may simply include the consecutive amino acid residues 188–193 in *Kl. pneumoniae* and those of HLA-B27.1 (72–77) with which they are known to be homologous.[187] However, this cannot offer a complete explanation for HLA-B27/*Kl. pneumoniae* cross-reactivity since only 29 per cent of ankylosing spondylitic sera bind to the synthesized peptide sequence representing residues 69–84 of HLA-B27.1. Schwimmbek *et al.*[187] therefore suggest that an autoimmune response against HLA-B27.1 may be the pathogenetic mechanism in *subsets* of ankylosing spondylitis and Reiter's syndrome patients. The reaction may indeed be against *Kl. pneumoniae* component(s) which clearly share limited sequence homology with HLA-B27.

It is also possible that plasmids, harboured by *Klebsiella* sp. may play a part in modifying cells bearing HLA-B27, setting the scene for the development of ankylosing spondylitis.[46] The cross-reacting molecular determinant shared by organisms such as *Klebsiella* sp, *Salmonella* sp., *Escherichia coli* and *Campylobacter* sp. may be related to early events in pathogenesis.[166] Such an antigen has been found to be shared by three strains of arthritogenic bacteria.[159] However, there are conflicting views and Georgopoulos *et al.*[94] found no reproducible cross-reactions between *Klebsiella* sp. antigens and cells from HLA-B27 donors with ankylosing spondylitis. This evidence contrasts with that of Geczy, *et al.*[92] who demonstrated that cytotoxic T-lymphocytes can recognize certain bacterial antigens associated with HLA-B27, setting the scene for the initial stages of ankylosing spondylitis.

Prognosis

Curiously, life expectation is not shortened although the quality of life is impaired.[26] In a large, early series[63,64] 3 per cent of patients died from leukaemia and 2 per cent from aplastic anaemia, the consequences of X-ray therapy.[146] Malignant skeletal neoplasms have followed radiotherapy[215] and skin cancers have been described.[169] Treatment with phenylbutazone was reported to cause agranulocytosis, thrombocytopenia or aplastic anaemia and the use of other analgesic and antiinflammatory drugs has invoked dyspepsia, melaena and haematemesis from gastric erosion or ulceration. However, the most important common hazard of treatment appears to be the ankylosis encouraged by excessively prolonged bed rest.

Psoriatic arthropathy

Psoriatic arthropathy (psoriatic arthritis) was an 'arthritis confined to the distal joints, associated with psoriasis';[16] it is now viewed as 'an inflammatory polyarthritis associated with psoriasis, usually having a negative sensitized sheep cell agglutination test (SCAT)'.[222] There is a wide spectrum of disorder[144a] and Wright and Moll[224] distinguished five broad clinical groups:

1. Distal interphalangeal joints predominantly affected
2. A severely deforming variety with widespread ankylosis, arthritis mutilans (5 per cent)
3. Symmetrical polyarthritis indistinguishable from seronegative rheumatoid arthritis
4. Asymmetrical, mono- or oligoarthritis
5. Ankylosing spondylitis (5 per cent).

Clinical features

Psoriasis and rheumatoid arthritis are both common disorders: the former associated with the inheritance of HLA-B13, -B3 and other antigens, the latter with HLA-D-related antigens. However, there is no longer any doubt that the arthritis of psoriasis and rheumatoid arthritis are distinct entities.[123,221] Psoriatic joint disease shares many clinical features with other seronegative disorders particularly ankylosing spondylitis, Reiter's disease and enteropathic arthropathy. Because of the frequency with which the spinal joints are affected, one category of HLA-B27-positive psoriatic arthropathy, with ankylosing spondylitis and Reiter's disease, may be designated as spondyloarthropathy. Psoriasis occurs two or three times more often in patients with arthritis than in the general population; but the estimates of arthritis in psoriatic patients vary widely. Approximately 0.1 per cent of the whole population has psoriatic arthritis; a male predominance is suggested. The peak age of onset of arthritis is between 36 and 45 years.

Pathological changes

Articular disease

There are few anatomical observations since there are rare opportunities to examine the tissues from cases of psoriatic arthropathy *postmortem*. The microscopic changes closely resemble those of rheumatoid arthritis.[16,21,190,224] The non-specificity of the histological appearances has often been emphasized[59] and more recent arthroscopic studies have also failed to differentiate between the characteristics of the two diseases.[118] However, systematic, semiquantitative observations, even of small numbers of cases, have begun to suggest measurable differences by which psoriatic arthritis might be identifiable[196] and it is hoped that similar methods, combined with immunolocalization techniques (see below) may introduce an element of useful diagnostic and prognostic discrimination.

In early psoriatic synovitis, there is hyperaemia, a limited fibrinous exudate and a marginal, erosive loss of cortical bone near articular margins. Cartilage loss is followed by replacement fibrosis but bony ankylosis is exceptional. In the small proportion of cases advancing to a severe, destructive arthropathy, the 'mutilans' group, the extent of the joint lesions is distinctive (Fig 19.16).

Soren[196] reviewed 13 accessions (Table 19.3). Where the duration of disease was less than five years, the number of specimens was so small that there was difficulty in identifying histological features that could be said to be characteristic (Figs 19.17 and 19.18). There was more success with cases of longer duration and synoviocyte 'histiocytic' infiltrates, and fibrosis[78,190] were among criteria of particular significance. A 'degenerative alteration of capillaries and smaller blood vessels, sometimes with focal or annular necrosis of the subendothelial layer and media', could not be related to age. Soren remarked on the smaller number of proliferating fibroblasts than in rheumatoid arthritis; he also confirmed the absence of granulomata or hemigranulomata (see Chapter 12). With time, in cases where gross destruction of the joint develops, bone resorption and deformity of the middle phalanges of both fingers and toes are found to be associated with less pronounced resorption of the proximal end of these bones; cup-like deformity results.

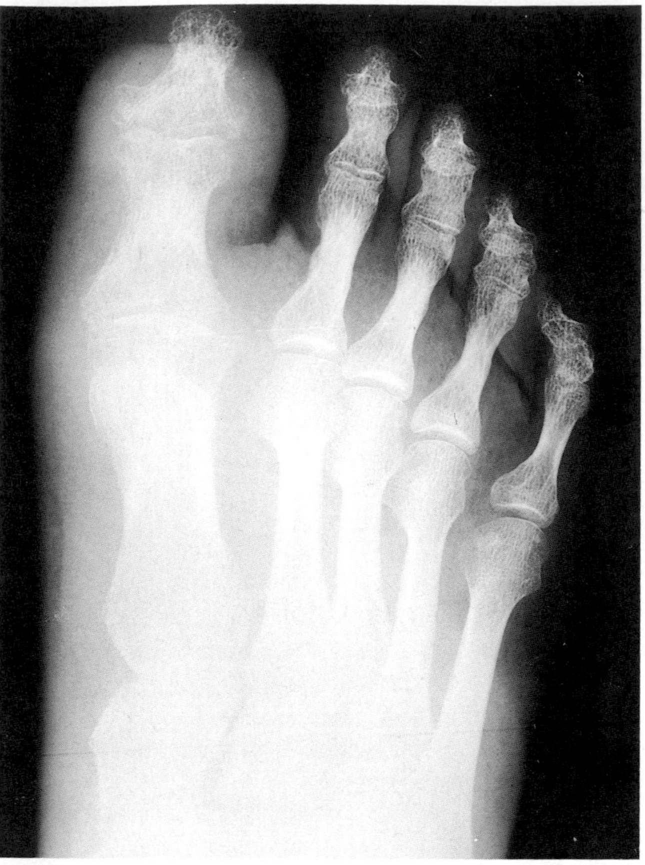

Fig. 19.16 Psoriatic arthropathy: arthritis mutilans of the foot.
A particularly severe and deforming arthritis has caused advanced radiological change in the distal metatarsophalangeal joints. There is irregular loss of joint space and nearby osteoporosis.

Occasional transmission electron microscopic studies of the synovia have been made;[74] they reveal conspicuous vascular endothelial cell swelling, inflammatory cell infiltration, thickening of the vascular walls and subsynoviocytic fibrosis. Synovial villi are not prominent; the A and B cells of the hyperplastic synovial cell layer are not abnormal and no electron-dense deposits are identified. The absence of any particular synovial lining cell reaction confirms the observations of Fassbender,[78] of Soren[196] and of Cooper *et al.*[57]

The synovial vasculature has attracted interest. In the larger, lower limb joints, small synovial arteries and the arterioles are normally thick-walled. In patients with psoriatic arthritis, dermal capillaries were 'meandering' and had tight terminal convolutions,[170] phenomena not observed in rheumatoid arthritis or systemic lupus erythematosus. There is, moreover, a brisker response to reactive hyperaemia in the small arteries of the phalanges in psoriasis and there have been attempts to relate these functional anomalies to the ultrastructural abnormalities observed by Espinoza *et al.*[74]

Table 19.3

Microscopic change in the synovium in 11 cases of psoriatic arthropathy of more than five years' duration[196]

Microscopic changes*	Case incidence (per cent)
Synovial villous hypertrophy	77
Synoviocyte hypertrophy	85
Synoviocyte hyperplasia	77
Organized surface fibrin	54
Hyperaemia	77
Oedema	69
Perivascular inflammatory cells	62
Foci of	
lymphocytes	77
plasma cells	62
histiocytes	77
Diffuse collections of	
lymphocytes	54
plasma cells	54
histiocytes	54
Synovial fibroblasts	85
Increased fibrocytes	92
Fibrosis	62

*Only features identified in more than 50 per cent of specimens are shown.

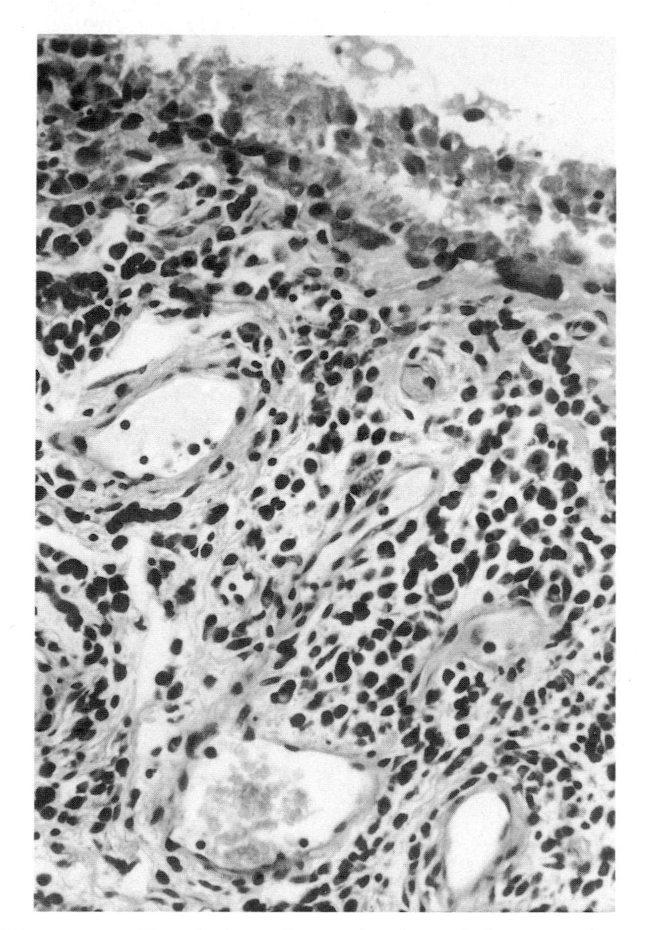

Fig. 19.17 Psoriatic arthropathy: knee joint synovium. In this biopsy, the synoviocytes (at top) are seen to have undergone moderate hyperplasia while the subsynoviocytic connective tissue, in which there are several dilatated venules but no free fibrin or haemosiderin is infiltrated widely by numerous plasma cells and smaller numbers of mononuclear macrophages. (HE × 385.)

Small numbers of studies have been made of the immunopathological changes. Although circulating immune complexes are identifiable in approximately 40 per cent of patients,[131] their concentration is not in proportion to the presence or absence of synovitis. Immunoglobulins and complement have been localized to the synovium[203] and IgG–anti-IgG immune complexes have been eluted. Nevertheless, synovial fluid complement levels are not lowered.[201] The synovial lymphocytes are mainly T-cells.[29] Some react for immunoglobulin but rheumatoid factor-secreting cells are not recognizable. Unlike rheumatoid arthritis, there appear to be no anomalies of peripheral blood lymphocytes and anticollagen antibodies have not been found in the serum.

There is very little published evidence of the state of the spinal and other axial joints in those cases of psoriatic arthritis with spondylitis. Marginal bone overgrowth occurs at the entheses. In a case of unusual interest, Fam and Cruickshank[77] described gross destruction of the C3/C4 and C5/C6 intervertebral discs in a 68-year-old man who had had psoriasis for 12 years. The skin disease began 10 years after the onset of a deforming arthritis. Death followed the development of tetraplegia: the C5 vertebra had shifted 5 mm anteriorly in relation to C6. Histological study showed erosive disease, with replacement of the discs by granulation tissue containing foci of plasma cells and lymphocytes. Apophyseal joints were similarly affected and there was much new bone at the anterior margins of the lower cervical vertebrae.

Psoriatic arthritis may develop in childhood.[191,194] Small hand and foot joints and the sacroiliac joints are particularly vulnerable; there is often tendon sheath involvement, occasionally iridocyclitis. Very little is known of the pathological features of the juvenile disease.

Systemic disease

Upper lobe pulmonary disease in psoriatic arthritis has been described much less often than in ankylosing spondylitis. The characteristically asymmetrical sacroiliac joint disease

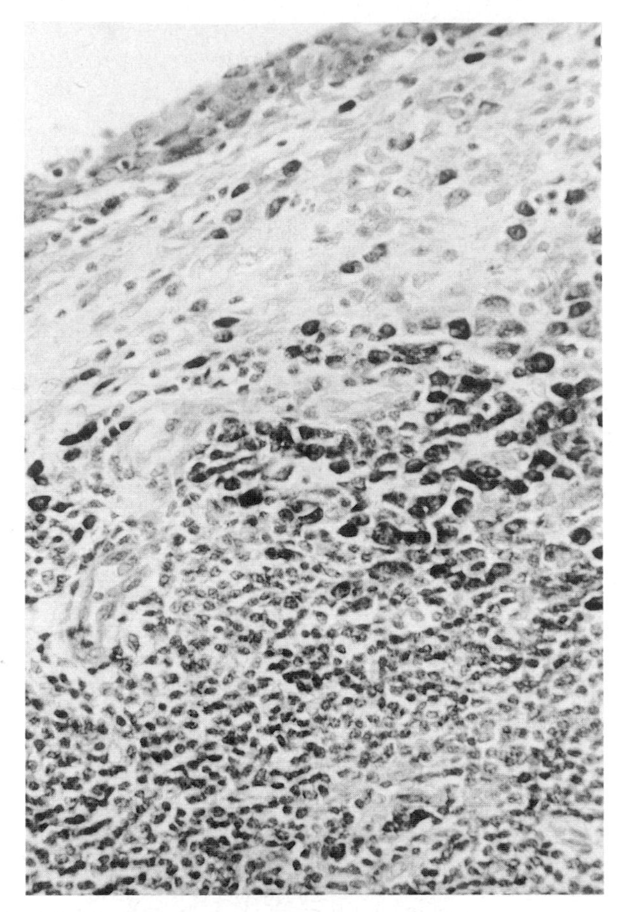

Fig. 19.18 Psoriatic arthropathy: knee joint synovium. Compare with Fig. 19.17. In the lower half of the field, a large lymphoid follicle is bounded by a more superficial marginal infiltrate of differentiated plasma cells. There is slight synoviocytic hyperplasia. (HE × 385.)

and the non-uniform, osteophyte-like syndesmophytes of psoriatic spinal arthritis attracted attention in the case of a man who developed non-granulomatous upper lobe pulmonary fibrosis 30 years after the onset of psoriasis:[107] *Aspergillus* sp. infection was superimposed on *Streptococcus pneumoniae* pneumonia.

Aortic regurgitation may develop in psoriatic arthritis as in ankylosing spondylitis; the thickened aortic cusps and rolled-up appearances with fibrosis are indistinguishable from those of the carditis of ankylosing spondylitis.[153] Amyloidosis (Fig 19.19) may complicate psoriatic arthritis just as it does ankylosing spondylitis and in rare instances, amyloid has been detected in the psoriatic synovium.[87]

Aetiology and pathogenesis

Genetic and environmental factors are operative. In patients with uncomplicated psoriasis, HLA-B13, -B37, Cw6 and other antigens are often recognized; in those who develop spondylitis, there is an association with HLA-B27.[29a] In childhood, psoriasis can be provoked by bacterial infection, e.g. by streptococci. Antistreptococcal toxins and enzymes have been found to be more frequent in patients with psoriatic arthritis than in control patients, and it is logical to speculate that the synovial response may be reactive, as in Reiter's disease, or due to cross-reactivity or molecular mimicry, as suggested for ankylosing spondylitis and rheumatic fever.

In young adults, the skin lesions may be triggered by trauma in the same way that joint destruction in other forms of seronegative spondyloarthropathy can be promoted.[160] The skin lesions of psoriasis can be evoked by mechanical, physical or chemical injury, the Koebner phenomenon. It has been suggested that a similar pattern of events may initiate synovitis or enthesopathy but this has not been established. There seems little histological similarity between the parakeratotic lesions of the psoriatic plaque and the inflammatory changes in the synovial joints.

Prognosis

Wright and Moll[224] and Wright[222] believed that the prognosis for psoriatic arthropathy was better than that for rheumatoid arthritis. However, Gladman *et al.*[96] distinguished three categories of patient rather different from those of Wright and Moll and suggested a greater degree of long-term disability. The categories were those with:

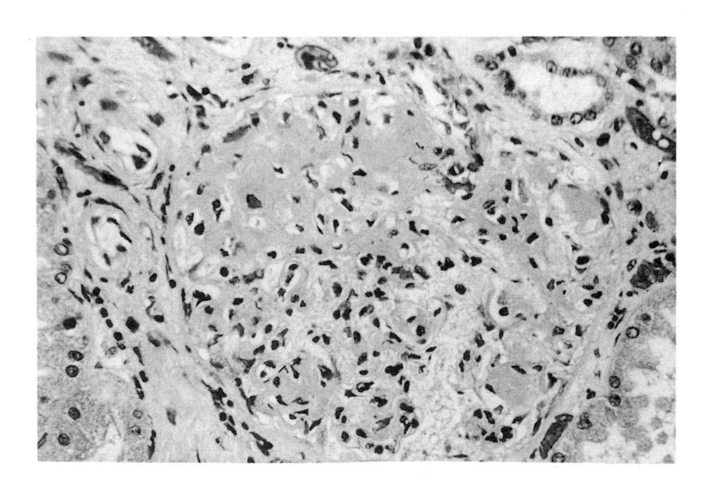

Fig. 19.19 Psoriatic arthropathy: renal amyloidosis. Much of this glomerulus is occupied by accumulations of congophilic amyloid. The chemical identity of the amyloid in psoriatic arthropathy, an uncommon complication of this chronic disease, is not yet established. (HE × 284.)

1. A deforming, erosive arthropathy (40 per cent of 220 cases)
2. Five or more deformed joints (17 per cent)
3. Substantial disability (11 per cent).

There is radiological progression of synovitis, and the segmental formation of syndesmophytes[109] but limited evidence of clinical progression.[72] For the pathologist, this radiological data is of interest but there is little information that sheds light on either the microscopic progression of the disease or the relationship, if any, between clinical progression, clinical prognosis and the sequence of the tissue changes.

Enteropathic arthropathy

Inflammatory disease of the gastrointestinal system is associated with sterile seronegative arthritis more often than can be ascribed to chance.[4a,81,100,113,218a] Ulcerative colitis, for example, was found as long ago as 1929 to be complicated with unexpected frequency by arthritis. Interest in the nature and cause of the joint diseases that parallel gastrointestinal inflammation has, however, accelerated because of studies of the pathogenesis of reactive arthritis (see pp. 763 and 767) and of the role of the HLA antigens in some of these disorders. Antibiotic treatment[180] and intestinal surgery[210] may also lead to arthritis.

Ulcerative colitis

Ten to 15 per cent of patients with ulcerative colitis develop joint disease, a proportion much greater than that of rheumatoid arthritis in a general population.[1,100,116] The distinction from rheumatoid arthritis is emphasized by the frequency with which arthritis coincides with exacerbations of the intestinal disorder; the patients are seronegative and subcutaneous granulomata very rarely form. Although peripheral joint disease is common and begins approximately six years after the onset of colitis, the anatomical sites affected extend to those of the spine. Spondylitis develops and displays features similar to those of ankylosing spondylitis (see p. 768) and to the spinal changes that succeed Crohn's disease (see below).[139] There is a wide range of ages but in the series of McEwen *et al.*[139] the age at onset of the intestinal disorder in 38 cases was 27.0 years and of the development of spondylitis 27.5 years. The male : female ratio was 15:4. In order of frequency, the affected joints in these cases were the hip (32 per cent) and shoulder (29 per cent). The spine and sacroiliac joint were involved in 39 per cent compared with 34 per cent in ankylosing spondylitis, 10 per cent in psoriasis and 9 per cent in Reiter's syndrome. The distal interphalangeal joints were entirely spared. Excluding the joints peripheral to the hips and shoulders, of which on average less than one was affected per patient, permanent joint disease could be confirmed in only two joints, averaging 0.7/patient. These figures were in striking contrast with those for 39 patients with spondylitis and Reiter's disease of whom, on average 5.2 and 2.9 joints/patient were affected respectively, 4.1 and 2.1 permanently.

The onset of the arthritis is sudden. The disease is migratory and often limited to one or to a few joints, usually of the lower limb. Rheumatoid factor titres are not raised and antinuclear factors are not found. Arthritis is more likely to complicate chronic than acute fulminating colitis and to be asssociated in women with extensive rather than limited bowel disease. Patients with erythema nodosum, oral ulceration and uveitis are more likely to develop arthritis than those without these signs. Unlike the ankylosing spondylitis that is unexpectedly often an association of ulcerative colitis, HLA-B27 is inherited with normal frequency.

There is a synovial fluid polymorph leucocytosis: complement levels are normal. The histological features[138,140] of occasional cases have been recorded.[196] In four biopsies there was synovial cell and villous hypertrophy, synovial hyperaemia, and a diffuse histiocytic infiltrate; in three there was synovial oedema, the presence of inflammatory cells in the walls of blood vessels, and a diffuse lymphocytic and plasma cell infiltrate; and in two there was synovial cell hyperplasia, perivascular inflammatory cells, focal accumulations of histiocytes, synovial membrane 'haemorrhages' and an increased number of fibrocytes. Cartilage 'erosion' is slight, the outcome good with little residual limitation of function due to fibrous adhesion and contracture. The synovial response to colectomy is occasionally beneficial.[137]

Crohn's disease

Now known to affect any part of the gastrointestinal tract, Crohn's disease[9,85] is no longer so clearly differentiated from ulcerative colitis as had been the case (Fig. 19.20). It is therefore of great interest that the clinical signs of the peripheral arthritis that may complicate Crohn's disease are indistinguishable from those associated with ulcerative colitis; the frequency of the peripheral arthritis in the former (about 20 per cent) is, however, somewhat greater than in the latter (about 9–11.5 per cent).

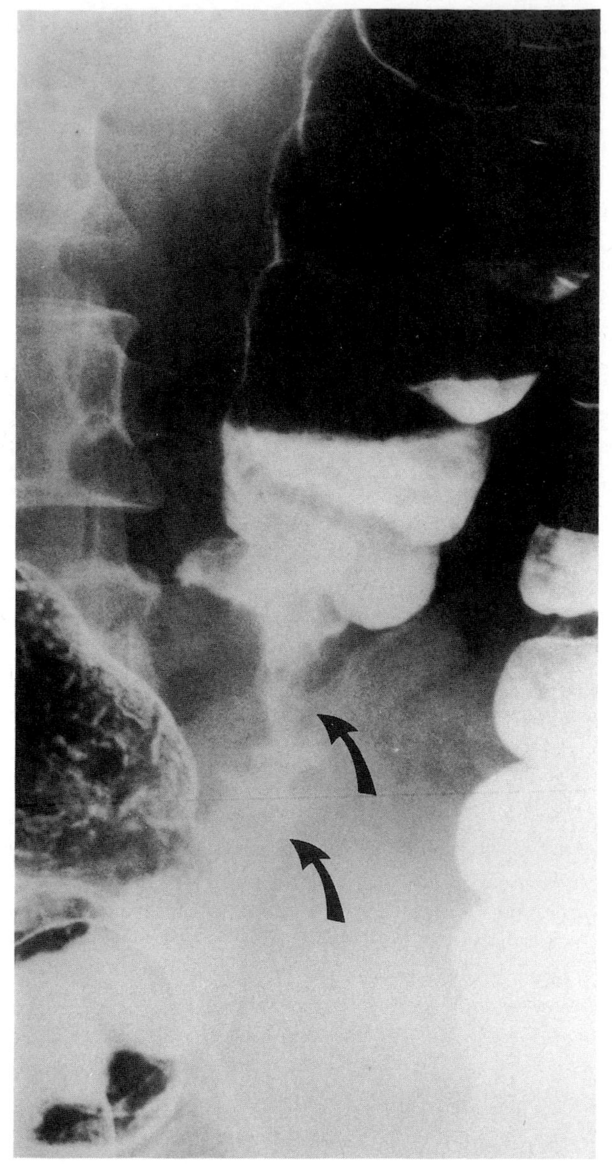

Fig. 19.20 Crohn's disease.
Double-contrast barium enema in a patient who had previously undergone a right hemicolectomy and small bowel resection for Crohn's disease, subsequently developing a recurrence at the anastomotic site (arrowed). Courtesy of the Department of Diagnostic Radiology, University of Manchester.

Soren[196] describes the synovial changes in two cases. Many of the features closely resemble those of the arthritis of ulcerative colitis, with moderate synovial cell hypertrophy and slight hyperplasia, oedema, hyperaemia and a mononuclear cell infiltrate. However, the synovial response may be granulomatous, resembling the intestinal lesions of Crohn's disease.[5] Soren[196] comments that whereas in ulcerative colitis there is a much greater proportion of lymphocytes, in Crohn's disease plasma cells predominate throughout the subsynoviocytic tissue. Because of the small number of cases described, the significance of these differences for the pathogenesis of these forms of enteropathic arthropathy is uncertain.

It is possible that new light may be shed on the synovial disorder of ulcerative colitis and of Crohn's disease when immunofluorescence, immonunoperoxidase, transmission electron microscopic and other analytical techniques have been fully applied. Meanwhile, it is a reasonable assumption that the synovitis is caused in the same manner as that following intestinal bypass (see below) in which a change in the flora of all or parts of the gut results in access of bacterial antigens to the vascular system, antibody formation and the circulation of immune complexes that lodge in synovial vessels, activating complement and causing inflammation.

The granulomata encountered in the synovium, muscle and bone in exceptional cases of Crohn's disease resemble those of the gastrointestinal tract.

Intestinal disease and ankylosing spondylitis

Ankylosing spondylitis affects approximately 50 per 1×10^5 individuals in the general population; 95 per cent of cases are male. In ulcerative colitis and in Crohn's disease this frequency is greatly increased so that either ulcerative colitis or Crohn's disease coexist with ankylosing spondylitis in approximately 5.0–6.9 per cent of instances. When the importance of HLA-B27 in ankylosing spondylitis was discovered attention turned naturally to intestinal disease. It was discovered that approximately 4.5 per cent of patients with inflammatory gastrointestinal disease had HLA-B27-positive ankylosing spondylitis, and approximately 1.5 per cent had HLA-B27-negative ankylosing spondylitis. Thus, both HLA-B27-positive and HLA-B27-negative ankylosing spondylitis are 100–300 times more prevalent in patients with inflammatory intestinal disease than expected by chance.

The clinical features of ankylosing spondylitis in intestinal disease are not distinguishable from the random disorder although the hips, shoulders and knees are affected slightly more often, and the male predominance is less conspicuous. The pathological characteristics are held to be identical.

Arthritis and intestinal bypass operations

The surgical procedures of jejunocolic and jejunoileal bypass achieved popularity for the correction of extreme obesity during the years 1965 to 1975 and it is estimated that more than 1×10^5 operations have been performed in the USA.[54] Many complications resulted from the operation; they included metabolic bone disease and infection but an arthritis–dermatitis syndrome attracted particular interest. Arthritis developed in approximately 30 per cent of patients subjected to jejunocolostomy and approximately 10 per cent of those treated by jejunoileostomy.[210] The joint disease was often polyarticular and symmetrical, beginning as soon as two or as late as 33 months after surgery. In order of frequency, the knee, metacarpophalangeal, shoulder, wrist, interphalangeal, elbow, ankle and spine were affected but many other joints were at risk.[54] Tests for rheumatoid factor were negative. Among the associated disorders were tenosynovitis, pleuritis and pericarditis, Raynaud's phenomenon and myalgia.[95] Occasionally, the arthritis–dermatitis syndrome coincided with rheumatoid arthritis.

The histological appearances shown by biopsy reveal a non-specific, inflammatory response (Figs 19.21, 19.22 and 19.23). Crystals are not present. IgA and complement are deposited in the synovium. There is a similarity to the synovitis of ulcerative colitis and Crohn's disease and the presence of a mainly lymphocytic synovial infiltrate suggests the influence of a cell-mediated immune response as part of reaction to the profound change in intestinal flora.

The natural history of bypass arthropathy was clarified by the recognition in the serum of cryprotein complexes that included IgG, IgM, IgA, C3, C4, C5 and C3-activator fragment of the properdin complex, and IgG antibodies against both *Escherichia coli* and *Bacteroides fragilis*. It is suggested that an abnormal systemic absorption of bacterial antigen, provoked by a change in bacterial flora and pH, can lead to synovial inflammation; and that antibodies, against these antigens form immune complexes, lodge in synovial vessels and activate C3, 4, 5 by the classical and alternate pathways. The synovial fluid contains 1.2–5.8 white blood cells $\times 10^9$/litre of which between 25 and 75 per cent are polymorphs.[206]

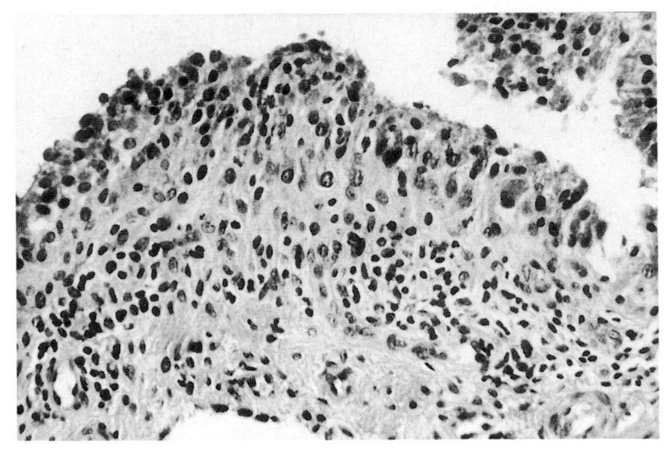

Fig. 19.22 Enteropathic arthropathy: knee joint synovium.
Compare with Fig. 19.16. The hyperplastic synovial cell layer (at top) is infiltrated by polymorphs. Many lymphocytes and fewer plasma cells and mononuclear macrophages lie more deeply. At bottom, there are a dilated venule and (right) an intact arteriole. (HE × 200.)

Postjejunoileostomy patients occasionally developed mixed cryoglobulinaemia and polyarthritis with dermatitis and renal insufficiency: IgG and *E. coli* antigens were found to be deposited along glomerular capillary walls.[205] In view of the debatable but stimulating evidence of cross-reactivity between *Klebsiella pneumoniae* and lymphocytes of HLA-B27-positive individuals with ankylosing spondylitis, it is intriguing to learn that anti-IgG antibodies and anti-*Klebsiella* sp. antibodies were identifiable in the serum

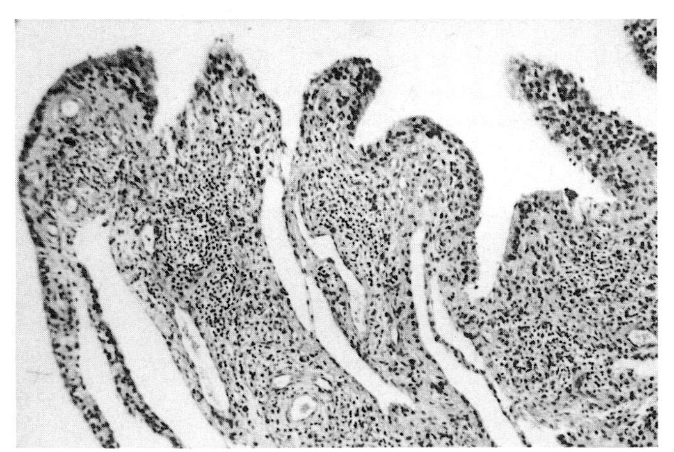

Fig. 19.21 Enteropathic arthropathy: knee joint synovium.
The large but attenuated synovial villi contain an extensive infiltrate of mononuclear cells. There is slight synovial cell hyperplasia. (HE × 80.)

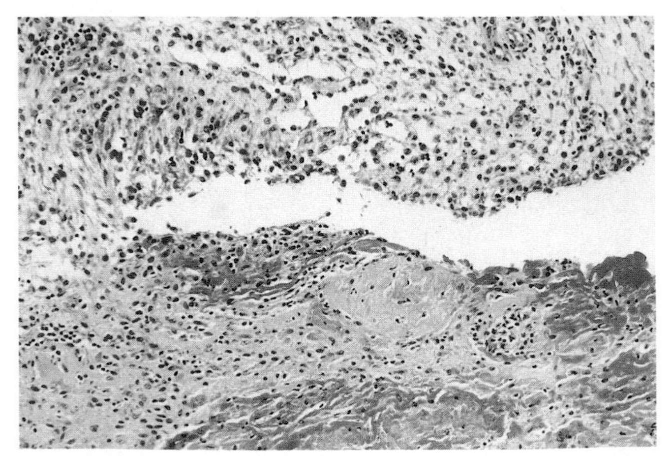

Fig. 19.23 Enteropathic arthropathy: knee joint synovium.
In this part of the same biopsy used to provide Figs 19.21 and 22, abundant polymerized fibrin lies upon and is being incorporated into the synovium. (HE × 90.)

cryoglobulin and in a renal biopsy eluate of a patient who developed postjejunoileal bypass arthritis undertaken for the relief of obesity.[88]

Arthritis in Whipple's disease

Whipple[214] described a syndrome in which weight loss, steatorrhoea and abdominal signs accompany polyarthritis.[50,133] The migratory polyarthritis, usually bilateral, affects ankles, knees, shoulders, elbows and fingers.[100] Arthritis does not herald intestinal symptoms and may vanish when the latter develop. Ten males are affected for every female. The syndrome[99] is a very uncommon disorder of middle-aged men; there is familial clustering. Fever, skin pigmentation and inflammation of pleura, lung, pericardium and thyroid accompany lymphadenopathy; central nervous system disease may develop with little evidence of intestinal abnormality or it may occur during a relapse after antibiotic treatment. No associated HLA antigen inheritance is known.

Sixty-five to 90 per cent of patients with Whipple's disease develop arthritis or arthralgia; many present with an arthropathy before intestinal disease is recognized and Whipple's disease should be suspected in otherwise undiagnosed articular disorders. Although inflammation may persist, articular cartilagenous erosion is slight and the arthritis responds to antibiotic treatment, as do the other manifestations of the disease, without fibrosis or deformity.

The disorder is associated with, and assumed to be caused by, the accumulation of periodic acid-Schiff-positive bacterial residues inside granulomatous aggregates of foamy macrophages in the small intestinal mucosa. These cells can be found by peroral biopsy of the jejunum but are also identified in a wide variety of other sites including the synovium, lymph nodes, heart, liver, spleen and brain, and in peritoneal aspirates and in renal granulomata. Subcutaneous granulomata are uncommon in Whipple's disease but may show the same cell aggregates.[99] The bacteria can be identified by electron microscopy even when the results of periodic acid-Schiff staining are in dispute (Fig. 19.24).[114]

Synovial biopsy may determine early diagnosis[181] although the eventual explanation for intestinal malabsorp-

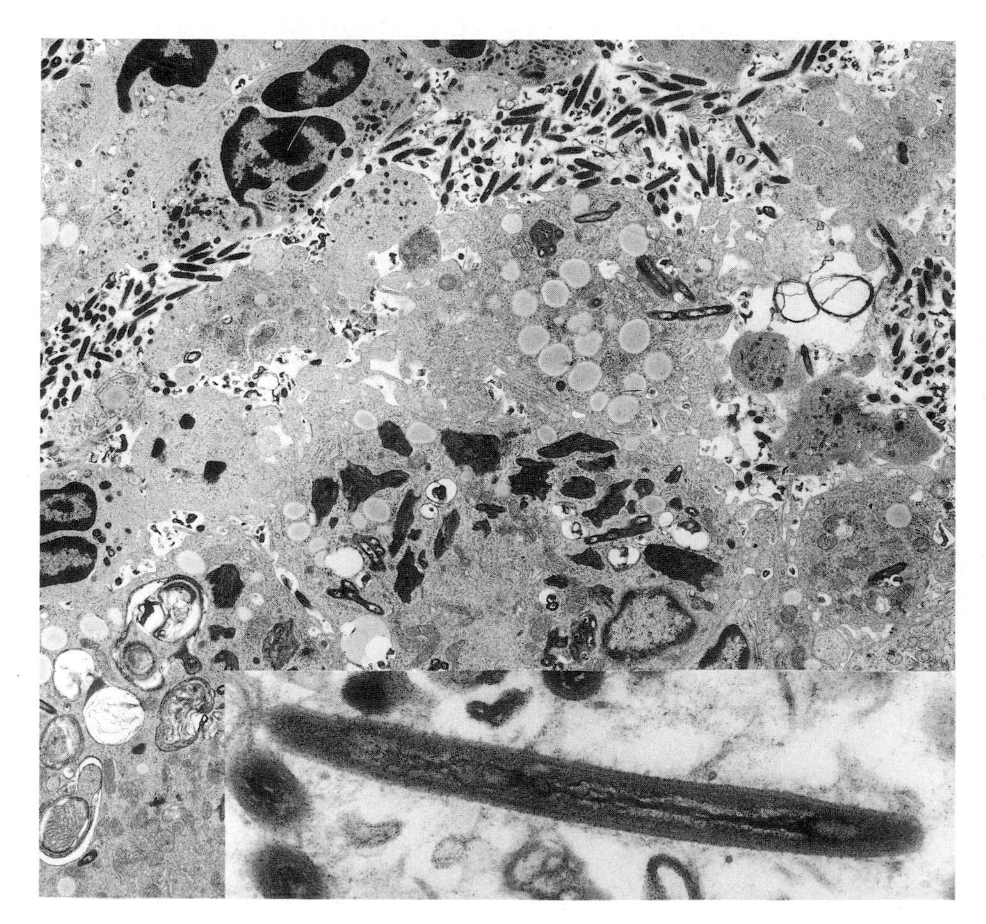

Fig. 19.24 Whipple's disease: intestinal mucosa. **a.** Within some of the mononuclear macrophages that lie in the superficial mucosa, are a small number of bacteria the presence of which is thought to account for the changes of Whipple's disease. Similar bacteria have been identified in the synovia. (TEM × 3950.) **b.** *Inset* Detail of bacterium shown in Fig. 19.24a. (TEM × 53 000.)

tion is likely to come from jejunal biopsy. The synovial macrophages are periodic acid-Schiff-positive; transmission electron microscopy reveals the presence in these cells of vacuoles containing partly degraded cell debris and of bacteria[114] that are poorly preserved and impossible to categorize.

There is a synovial fluid polymorph leucocytosis; the fluid is sterile. An association exists between Whipple's disease and sacroiliitis and spondylitis. However, authentic examples of Whipple's disease with ankylosing spondylitis are rare.

Arthritis with antibiotic-induced pseudomembranous enterocolitis

Antibiotics such as lincomycin and clindamycin cause dramatic changes in gut flora that are expressed as fulminating and sometimes fatal enterocolitis, a condition first recognized during treatment with 'broad-spectrum' tetracycline derivatives.[89] Analogous disorders may succeed the use of some penicillins.

Arthritis was described as a complication of clindamycin-induced arthritis by Rollins and Moeller[177] and a similar condition following the administration of ampicillin was recorded by Rothschild, Masi and June.[180] The arthritis was monoarticular and gonococcal arthritis was considered in a differential diagnosis. The excess synovial fluid contained many polymorphs and much protein. Synovial biopsy is inadvisable in these circumstances so that little is known of the synovial histology. The cause of antibiotic-induced enterocolitis is now widely believed to be *Clostridium difficile* which proliferates as the normal intestinal flora is destroyed.[27,76] Whether the gut disorder is attributable to this anaerobe or to its exotoxin is not known; nor is it certain whether the exotoxin can exert any influence on the synovia. It appears more probable that the synovitis is of a reactive character. The onset is usually 2–3 weeks after antibiotic treatment has begun. However, neither blood nor synovial fluid humoral- or cell-mediated immune mechanisms have been very fully studied.

Arthritis in coeliac disease

Arthritis may be a prominent diagnostic sign in gluten-sensitive enteropathy.[28] Little is known of the synovial histopathology. The most commonly affected joints are the lumbar spine, hip, knee and shoulder. Patients may be HLA-B27-positive but other antigens, including HLA A1, B8 and DR3 may also be inherited.

REFERENCES

1. Acheson E D. An association between ulcerative colitis, regional enteritis and ankylosing spondylitis. *Quarterly Journal of Medicine* 1960; **29**: 489–9.

2. Adams R F, Flinn G S, Douglas M. Ankylosing spondylitis in a nonhuman primate: a monkey tale. *Arthritis and Rheumatism* 1987; **30**: 956–7.

3. Ahvonen P, Sievers K, Aho K. Arthritis associated with Yersinia enterocolitica infection. *Acta Rheumatologica Scandinavica* 1969; **15**: 232–53.

4. Albert J, Lagier R. Enthesopathic erosive lesions of patella and tibial tuberosity in juvenile ankylosing spondylitis. Anatomico-radiological study of a case with tibial bursitis. ROFO: *Fortschritte auf dem Gebiete der Röntgenstrahlen und der Nuklearmedizin* 1983; **139**: 544–8.

4a. Aldo-Benson M A. Enteropathic arthritis. In: McCarty D J, ed, *Arthritis and Allied Conditions. A Textbook of Rheumatology* (11th edition). Philadelphia, London: Lea and Febiger, 1989: 972–9.

5. Al-Hadidi S, Khatib G, Chhatwal P, Khatib R. Granulomatous arthritis in Crohn's disease. *Arthritis and Rheumatism* 1984; **27**: 1061–2.

6. Ansell B M, Bywaters E G L, Doniach I. The aortic lesions of ankylosing spondylitis. *British Heart Journal* 1958; **20**: 507–15.

7. Aufdermaur M. Spondylitis ankylosans. In: Doerr W, Seifert G, eds, *Spezielle Pathologische Anatomie* (volume 18). Berlin: Springer Verlag, 1984.

8. Aufdermaur M. Pathogenese der Gelenkverknocherung bei Spondylitis ankylosans. *Therapiewoche* 1985; **35**: 2614–8.

9. Austad W R, Thompson G R, Joseph R R. Regional enteritis presenting as acute arthritis. *Michigan Medicine* 1968; **67**: 324–9.

10. Avakian H, Welsh J, Ebringer A, Entwistle C C. Ankylosing spondylitis, HLA-B27 and klebsiella. II. Crossreactivity studies with human tissue typing sera. *British Journal of Experimental Pathology* 1980; **61**: 92–6.

11. Baldassare A R, Weiss T D, Tsai C C, Arthur R E, Moore T L, Zuckner J. Immunoprotein deposition in synovial tissue in Reiter's syndrome. *Annals of the Rheumatic Diseases* 1981; **40**: 281–5.

11a. Ball G V. Ankylosing spondylitis. In: McCarty D J, ed, *Arthritis and Allied Conditions. A Textbook of Rheumatology* (11th edition). Philadelphia, London: Lea and Febiger, 1989: 934–43.

12. Ball G V, Hathaway B. Ankylosing spondylitis with widespread arteritis. *Arthritis and Rheumatism* 1966; **9**: 737–45.

13. Ball J. Enthesopathy of rheumatoid and ankylosing spondylitis. *Annals of the Rheumatic Diseases* 1971; **30**: 213–23.

14. Ball J. Articular pathology of ankylosing spondylitis. *Clinic-*

al Orthopaedics and Related Research 1979; **143**: 30–7.

15. Ball J. Pathology and pathogenesis. In: Moll JHM, ed, *Ankylosing Spondylitis*. Edinburgh: Churchill Livingstone, 1980: 96–112.

16. Bauer W, Bennett G A, Zeller J W. Pathology of joint lesions in patients with psoriasis and arthritis. *Transactions of the Association of American Physicians* 1941; **56**: 349–52.

17. Bauer W, Clark W S, Kulka J P. Aortitis and aortic endocarditis, an unrecognized manifestation of rheumatoid arthritis. *Annals of the Rheumatic Diseases* 1951; **10**: 470–1.

18. Bauer W, Engleman E P. A syndrome of unknown etiology characterized by urethritis, conjunctivitis, and arthritis (so-called Reiter's disease). *Transactions of the Association of American Physicians* 1942; **57**: 307–13.

19. Bechterew W von. Oderevenielost pozvonochika s iskrevlenigem yevo, kak osobaya forma zabolievanya. *Vrach* 1892; **13**: 899–903.

19a. Bechterew W von. Steifigkeit der Wirbelsäule und ihre Verkrümmung als besondere Erkrankungsform. *Neurologisches Centralblatt* 1893; **12**: 426–34.

20. Benedek T G, Zawadzki A. Ankylosing spondylitis with ulcerative colitis and amyloidosis. Report of a case and review of the literature. *American Journal of Medicine* 1966; **40**: 431–439.

21. Bennett R M. Psoriatic arthritis. In: McCarty D H, ed, *Arthritis and Allied Conditions* (11th edition). Philadelphia: Lea and Febiger, 1989: 954–71.

22. Berman L, Isaacs H, Pickering A. Structural abnormalities of muscle tissue in ankylosing spondylitis. *South African Medical Journal* 1976; **50**: 1238–40.

23. Bernstein L, Brock O J. Cardiac complications in spondylarthritis ankylopoietica. *Acta Medica Scandinavica* 1949; **135**: 185–94.

24. Bluestone R. Ankylosing spondylitis. In: McCarty D J, ed, *Arthritis and Allied Conditions*. (10th edition). Philadelphia: Lea & Febiger, 1985: 819–40.

25. Blumberg B S, Blumberg J L. Bernard Connor (1666–1698) and his contribution to the pathology of ankylosing spondylitis. *Journal of Historical Medicine* 1958; **13**: 349–66.

26. Blumberg B S, Ragan C. The natural history of rheumatoid spondylitis. *Medicine* 1956; **35**: 1–31.

27. Bolton R P, Wood G M, Losowsky M S. Acute arthritis associated with *Clostridium difficile* colitis. *British Medical Journal* 1981; **283**: 1023–4.

28. Bourne J T, Kumar P, Huskisson E C, Mageed R, Unsworth D J, Wojtulewski J A. Arthritis and coeliac disease. *Annals of the Rheumatic Diseases* 1985; **44**: 592–8.

29. Braathen L R, Fyrand O, Mellbye O J. Predominance of cells with T-markers in the lymphocytic infiltrates of synovial tissue in psoriatic arthritis. *Scandinavian Journal of Rheumatology* 1979; **8**: 75–80.

29a. Brewerton D A. HL-A 27 in Reiter's disease and psoriatic arthropathy. *International Journal of Dermatology* 1975; **14**: 39–40.

30. Brewerton D A. HLA-B27 and the inheritance of susceptibility to rheumatic disease. *Arthritis and Rheumatism* 1976; **19**: 656–68.

31. Brewerton D A. Inherited susceptibility to rheumatic disease. *Journal of the Royal Society of Medicine* 1978; **71**: 331–8.

32. Brewerton D A, Caffrey M, Hart F D, James D C O, Nicholls A, Sturrock R D. Ankylosing spondylitis and HL-A 27. *Lancet* 1973b; **ii**: 904–7.

33. Brewerton D A, Caffrey M, Nicholls A, Walters D, James D C O. Reiter's disease and HLA 27. *Lancet* 1973a; **ii**: 996–8.

34. Brewerton D A, Gibson D G, Goddard D H, Jones T J, Moore R B, Pease C T, Revell P A, Shapiro L M, Swettenham K V. The myocardium in ankylosing spondylitis. A clinical, echocardiographic, and histopathological study. *Lancet* 1987; **ii**: 995–8.

35. Brodie B C. *Pathological and Surgical Observations on Diseases of the Joints*. London: Longman, Hurst, Rees, Orme and Brown, 1818: 55.

36. Brown W M, Doll R. Mortality from cancer and other causes after radiotherapy for ankylosing spondylitis. *British Medical Journal* 1965; **I**: 1327–32.

37. Bruneau C, Bonin H. Evidence for a disease specific antigen in circulating immune complexes in ankylosing spondylitis. *Clinical and Experimental Immunology* 1983; **53**: 529–35.

38. Burbacher C R, Weiland A H. Gonorrheal arthritis. *Journal of the Florida Medical Association* 1938; **24**: 433–5.

39. Bywaters E G L. A case of early ankylosing spondylitis with fatal secondary amyloidosis. *British Medical Journal* 1968; **i**: 412–16.

40. Bywaters E G L. The early lesions of ankylosing spondylitis (a demonstration to the Heberden Society). *Annals of the Rheumatic Diseases* 1969; **28**: 330.

41. Bywaters E G L. Historical aspects of ankylosing spondylitis. *Rheumatology and Rehabilitation* 1979; **18**: 197–203.

42. Bywaters E G L. The pathology of the spine. In: Sokoloff L, ed, *The Joints and Synovial Fluid*. New York: Academic Press, 1980; **2**: 427–47.

43. Calin A. Ankylosing spondylitis. In: Kelley W N, Harris E D, Ruddy S, Sledge C B. *Textbook of Rheumatology* (3rd edition). Philadelphia: W B Saunders & Co., 1989: 1021–37.

44. Calin A. Reiter's syndrome. In: Kelly W N, Harris E D, Ruddy S, Sledge C B, eds, *Textbook of Rheumatology* (3rd edition). Philadelphia: W B Saunders & Co., 1989: 1038–52.

45. Cameron F H, Russell P J, Easter J F, Wakefield D, March L. Failure of *Klebsiella pneumoniae* antibodies to cross-react with peripheral blood mononuclear cells from patients with ankylosing spondylitis. *Arthritis and Rheumatism* 1987; **30**: 300–5.

46. Cameron F H, Russell P J, Sullivan J, Geczy A F. Is a klebsiella plasmid involved in the aetiology of ankylosing spondylitis in HLA-B27-positive individuals? *Molecular Immunology* 1983; **20**: 563–6.

47. Campbell A H, MacDonald C B. Upper lobe fibrosis associated with ankylosing spondylitis. *British Journal of Diseases of the Chest* 1965; **59**: 90–101.

48. Cantor R I, Downs G E, Abruzzo J L. Acute exacerbation of ankylosing spondylitis after an iron dextran infusion. *Annals of Internal Medicine* 1972; **77**: 933–4.

49. Carette S, Graham D, Little H, Rubenstein J, Rosen P. The natural disease course of ankylosing spondylitis. *Arthritis and Rheumatism* 1983; **26**: 186–90.

50. Caughey D E, Bywaters E G L. The arthritis of Whipple's disease. *Annals of the Rheumatic Diseases* 1963; **22**: 327–35.

51. Cawley M I D, Chalmers T M, Kellgren J H, Ball J.

Destructive lesions of vertebral bodies in ankylosing spondylitis. *Annals of the Rheumatic Diseases* 1972; **31**: 345–58.

52. Clark W S, Bauer W. Cardiac changes in rheumatoid arthritis. *Annals of the Rheumatic Diseases* 1948; **7**: 39–40.

53. Clark W S, Kulka P, Bauer W. Rheumatoid aortitis with aortic regurgitation. An unusual manifestation of rheumatoid arthritis (including spondylitis). *American Journal of Medicine* 1957; **22**: 580–92.

54. Clarke J, Weiner S R, Bassett L W, Utsinger P D. Bypass disease. *Clinical and Experimental Rheumatology* 1987; **5**: 275–87.

55. Collado A, Sanmarti R, Bielsa I, Castel T, Kanterewicz E, Canete J D, Brancos M A, Rotes-Querol J. Immunoglobulin A in the skin of patients with ankylosing spondylitis. *Annals of the Rheumatic Diseases* 1988; **47**: 1004–7.

56. Collins D H. *The Pathology of Articular and Spinal Diseases.* London: Arnold, 1949: 313–27.

57. Cooper N S, Soren A, McEwen C, Rosenberger J L. Diagnostic specificity of synovial lesions. *Human Pathology* 1981; **12**: 314–28.

58. Cooper R, Fraser S M, Sturrock R D, Gemmell C G. Raised titres of anti-Klebsiella IgA in ankylosing spondylitis, rheumatoid arthritis, and inflammatory bowel disease. *British Medical Journal* 1988; **296**: 1432–4.

59. Coste F, Solnica J. La polyarthrite psoriasique. *Revue Française d'Études Cliniques et Biologiques* 1966; **11**: 578–99.

60. Cowling P, Ebringer R, Cawdell D, Ishii M, Ebringer A. C-reactive protein, ESR, and klebsiella in ankylosing spondylitis. *Annals of the Rheumatic Diseases* 1980; **39**: 45–9.

61. Cruickshank B. Histopathology of diarthrodial joints in ankylosing spondylitis. *Annals of the Rheumatic Diseases* 1951; **10**: 393–404.

62. Cruickshank B. Lesions of cartilaginous joints in ankylosing spondylitis. *Journal of Pathology and Bacteriology* 1956; **71**: 73–84.

63. Cruickshank B. Pathology of ankylosing spondylitis. *Bulletin on the Rheumatic Diseases* 1960; **10**: 211–14.

64. Cruickshank B. Pathology of ankylosing spondylitis. *Clinical Orthopaedics and Related Research* 1971; **74**: 43–58.

65. Csonka G W, Litchfield J W, Oates J K, Willcox R R. Cardiac lesions in Reiter's disease. *British Medical Journal* 1961; **i**: 243–7.

66. Dihlmann W, Lindenfelser R, Selberg W. Sakroiliakale histomorphologie der ankylosierenden spondylitis als Beiträg zur Therapie. *Deutsche Medizinische Wochenschrift* 1977; **102**: 129–32.

67. Dumonde D C, ed, *Infection and Immunology in the Rheumatic Diseases.* Oxford: Blackwell Scientific Publications, 1976: 95–6.

68. Ebringer A, Cowling P, Ngwa Suh N, James D C O, Ebringer R W. Crossreactivity between *Klebsiella aerogenes* species and B27 lymphocyte antigens as an aetiological factor in ankylosing spondylitis. In: Dausset J, Svejgaard A, eds, *HLA and Disease.* Paris: Inserm, 1976; **56**: 27.

69. Ebringer R W, Cawdell D R, Cowling P, Ebringer A. Sequential studies in ankylosing spondylitis: association of *Klebsiella pneumoniae* with active disease. *Annals of the Rheumatic Diseases* 1978; **37**: 484.

70. Ebringer R W, Cawdell D, Ebringer A. Klebsiella pneumoniae and acute anterior uveitis in ankylosing spondylitis. *British Medical Journal* 1979; **i**: 383.

71. Ebright J R, Ryan L M. Acute erosive reactive arthritis associated with *Campylobacter jejuni*-induced colitis. *American Journal of Medicine* 1984; **76**: 321–3.

72. Editorial. Prognosis of psoriatic arthritis. *Lancet* 1988; **ii**: 375–6.

73. Engfeldt B, Romanus R, Ydén S. Histological studies of pelvo-spondylitis ossificans (ankylosing spondylitis) correlated with clinical and radiological findings. *Annals of the Rheumatic Diseases* 1954; **13**: 219–28.

74. Espinoza L R, Vasey F B, Espinoza CG, Bocanegra T S, Germain B F. Vascular changes in psoriatic synovium. A light and electron microscopic study. *Arthritis and Rheumatism* 1982; **25**: 677–84.

75. Fagge C H. A case of simple synostosis of the ribs to the vertebrae, and of the arches and articular processes of the vertebrae themselves, and also of one hip joint. *Transactions of the Pathological Society of London* 1877; **28**: 201–8.

76. Fairweather S D, Youngs D, George R H, Burdon D W, Keighley M R B. Arthritis in pseudomembranous colitis associated with an antibody to *Clostridium difficile* toxin. *Journal of the Royal Society of Medicine* 1980; **73**: 524–5.

77. Fam A G, Cruickshank B. Subaxial cervical subluxation and cord compression in psoriatic spondylitis. *Arthritis and Rheumatism* 1982: **25**: 101–6.

78. Fassbender H G. *Pathology of Rheumatic Diseases.* Berlin: Springer-Verlag, 1975: 245–58.

79. Feltelius N, Hedenström H, Hillerdal G, Hällgren R. Pulmonary involvement in ankylosing spondylitis. *Annals of the Rheumatic Diseases* 1986; 45: 736–40.

80. Feltelius N, Lindh U, Venge P, Hallgren R. Ankylosing spondylitis: a chronic inflammatory disease with iron overload in granulocytes and platelets. *Annals of the Rheumatic Diseases* 1986; **45**: 827–31.

81. Ferguson R H. Enteropathic arthritis. In: McCarty D J, ed, *Arthritis and Allied Conditions* (10th edition). Philadelphia: Lea & Febiger, 1985: 867–73.

82. Firestein G S, Zvaifler N J. Reactive arthritis. *Annual Review of Medicine* 1987; **38**: 351–60.

82a. Ford D K. Reiter's syndrome: reactive arthritis. In: McCarty D J, ed, *Arthritis and Allied Conditions: A Textbook of Rheumatology* (11th edition). Philadelphia, London: Lea and Febiger, 1989: 944–53.

83. Ford D K, da Roza D M, Schultzer M. The specificity of synovial mononuclear cell responses to microbiological antigens in Reiter's syndrome. *Journal of Rheumatology* 1982; **9**: 561–7.

84. Francois R J. Some pathological features of ankylosing spondylitis as revealed by microradiography and tetracycline labelling. *Clinical Rheumatology* 1981; **1**: 23–9.

85. Frayha R, Stevens M B, Bayless T M. Destructive monarthritis and granulomatous synovitis as the presenting manifestations of Crohn's disease. *Johns Hopkins Medical Journal* 1975; **137**: 151–5.

86. Freund E. A contribution to the pathogenesis of spondylitis ankylopoietica. *Edinburgh Medical Journal* 1942; **49**: 91–109.

87. Friedman R, Agus B, Ames E. Amyloid arthropathy in a

patient with psoriasis and amyloidosis. *Arthritis and Rheumatism* 1981; **24**: 1320–3.

88. Gamble C N, Kimchi A, Depner T A, Christensen D. Immune complex glomerulonephritis and dermal vasculitis following intestinal bypass for morbid obesity. *American Journal of Clinical Pathology* 1982; **77**: 347–52.

89. Gardner D L. Aureomycin-resistant staphylococcal enterocolitis. Report of two fatal cases. *Lancet* 1953; **ii**: 1236–8.

90. Geczy A F, Alexander K, Bashir H V, Edmonds J P, Upfold L, Sullivan J. HLA-B27, klebsiella and ankylosing spondylitis: biological and chemical studies. *Immunological Reviews* 1983; **70**: 23–50.

91. Geczy A F, van Leeuwen A, van Rood J J, Ivanyi P, Breur B S, Cats A. Blind confirmation in Leiden of Geczy factor on the cells of Dutch patients with ankylosing spondylitis. *Human Immunology* 1986; **17**: 239–45.

92. Geczy A F, McGuigan L E, Sullivan J S, Edmonds J P. Cytotoxic T lymphocytes against disease-associated determinant(s) in ankylosing spondylitis. *Journal of Experimental Medicine* 1986; **164**: 932–7.

93. Geiler G. Die spondylarthritis ankylopoietica aus pathologisch—anatomischer Sicht. *Deutsche Medizinische Wochenschrift* 1969; **94**: 185–8.

94. Georgopoulos K, Dick W C, Goodacre A, Pain R H. A reinvestigation of the cross-reactivity between klebsiella and HLA-B27 in the aetiology of ankylosing spondylitis. *Clinical and Experimental Immunology* 1985; **62**: 662–71.

95. Ginsberg J, Quismorio F P, De Wind L T, Mongan E S. Musculoskeletal symptoms after jejuno-ileal shunt surgery for intractable obesity. Clinical and immunologic studies. *American Journal of Medicine* 1979; **67**: 443–8.

96. Gladmann D D, Shuckett R, Russell M L, Thorne J C, Schachter R K. Psoriatic arthritis (PSA)—an analysis of 220 patients. *Quarterly Journal of Medicine* 1987; NS**62**: 127–41.

97. Goddard N J, Golding D N. Cat-scratch disease presenting with arthropathy of the ankles. *Journal of the Royal Society of Medicine* 1989; **82**: 499–500.

98. Good A E. Involvement of the back in Reiter's syndrome. *Annals of Internal Medicine* 1962; **57**: 44–59.

99. Good A E, Beals T F, Simmons J L, Ibrahim M A. A subcutaneous nodule with Whipple's disease. Key to early diagnosis? *Arthritis and Rheumatism* 1980; **23**: 856–9.

100. Good A E, Utsinger P D. Enteropathic arthritis. In: Kelley W N, Harris E D, Ruddy S, Sledge C B, eds, *Textbook of Rheumatology* (2nd edition). Philadelphia: W B Saunders & Co., 1985: 1031–41.

101. Graham D C, Smythe H A. The carditis and aortitis of ankylosing spondylitis. *Bulletin of Rheumatic Diseases* 1958; **9**: 171–4.

102. Granfors K, Jalkanen S, Lindberg A A, Maki-Ikola O, von Essen R, Lahesmaa-Rantala R, Isomaki H, Saario R, Arnold W J, Toivanen A. Salmonella lipopolysaccharide in synovial cells from patients with reactive arthritis. *Lancet* 1990; **335**: 685–8.

103. Granfors K, Jalkanen S, von Essen R, Lahesmaa-Rantala R, Isomaki O, Pekkola-Heino K, Merilahti-Palo R, Saario R, Isomaki H, Toivanen A. Yersinia antigens in synovial-fluid cells from patients with reactive arthritis. *New England Journal of Medicine* 1989; **320**: 216–21.

104. Granfors K, Toivanen A. IgA-anti-yersinia antibodies in yersinia triggered reactive arthritis. *Annals of the Rheumatic Diseases* 1986; **45**: 561–5.

105. Granfors K, Viljanen M, Tiilikainen A, Toivanen A. Persistence of IgM, IgG, and IgA antibodies to yersinia in yersinia arthritis. *Journal of Infectious Diseases* 1980; **141**: 424–9.

106. Güntz E. Beiträg zur pathologischen Anatomie der Spondylarthritis ankylopoetica. *Fortschrift für Rontgenstrahlung* 1933; **47**: 683–93.

107. Guzman L R, Gall E P, Pitt M, Lull G. Psoriatic spondylitis: association with advanced nongranulomatous upper lobe pulmonary fibrosis. *Journal of the American Medical Association* 1978; **239**: 1416–7.

108. Hancock J A H. Surface manifestations of Reiter's disease in the male. *British Journal of Venereal Diseases* 1960; **36**: 36–9.

109. Hanly J G, Russell M L, Gladman D D. Psoriatic spondyloarthropathy: a long-term prospective study. *Annals of the Rheumatic Diseases* 1988; **47**: 386–93.

110. Harris N H. Lesions of the symphysis pubis in women. *British Medical Journal* 1974; **4**: 209–11.

111. Harris N H, Murray R O. Lesions of the symphysis pubis in athletes. *British Medical Journal* 1974; **4**: 211–4.

112. Hart F D. Clinical features and complications. In: Moll J M H, ed, *Ankylosing Spondylitis.* Edinburgh: Churchill Livingstone, 1980: 53–68.

113. Haslock I. Enteropathic arthritis. In: Scott J T, ed, *Copeman's Textbook of the Rheumatic Diseases* (6th edition) Edinburgh: Churchill Livingstone, 1986: 806–18.

114. Hawkins C F, Farr M, Morris C J, Hoare A M, Williamson N. Detection by electron microscope of rod-shaped organisms in synovial membrane from a patient with the arthritis of Whipple's disease. *Annals of the Rheumatic Diseases* 1976; **35**: 502–9.

115. Hill J L, Yong Z, Laheji K, Kono D H, Yu D T Y. Experimental animal models of yersinia infection and yersinia-induced arthritis. *Contributions to Microbiology and Immunology* 1987; **9**: 228–32.

116. Hochberg M C, Feinstein R S, Moser R L, Ryan M J. Colitic arthritis. *Johns Hopkins Medical Journal* 1982; **151**: 173–80.

117. Hough A J, Rank R G. Induction of arthritis in C57B1/6 mice by chlamydial antigen. Effect of prior immunization or infection. *American Journal of Pathology* 1988; **130**: 163–72.

118. Jayson M I V, Henderson D R F. Arthroscopy in the diagnosis of inflammatory joint disease. *Rheumatology and Rehabilitation* 1973; **12**: 195–7.

119. Jessamine A G. Upper lung lobe fibrosis in ankylosing spondylitis. *Canadian Medical Association Journal* 1968; **98**: 25–9.

120. Julkunen H. Rheumatoid spondylitis—clinical and laboratory study of 149 cases compared with 182 cases of rheumatoid arthritis. *Acta Rheumatologica Scandinavica* 1962; suppl. **4**: 1–110.

121. Julkunen H. Synovial inflammatory cell reaction in chronic arthritis. *Acta Rheumatologica Scandinavica* 1966; **12**: 188–96.

122. Julkunen H, Luomanmaki K. Complete heart block in rheumatoid (ankylosing) spondylitis. *Acta Medica Scandinavica* 1964; **176**: 401–5.

123. Kammer G M, Soter N A, Gibson D J, Schur P H.

Psoriatic arthritis: a clinical immunologic and HLA study of 100 patients. *Seminars in Arthritis and Rheumatism* 1979; **9**: 75–97.

124. Keat A. Reiter's syndrome and reactive arthritis in perspective. *New England Journal of Medicine* 1983; **309**: 1606–15.

125. Keat A, Thomas B, Dixey J, Osborn M, Sonnex C, Taylor-Robinson D. *Chlamydia trachomatis* and reactive arthritis: the missing link. *Lancet* 1987; **i**: 72–4.

126. Keat A C, Thomas B J, Taylor-Robinson D, Pegrum G D, Maini R N, Scott J T. Evidence of *Chlamydia trachomatis* infection in sexually acquired reactive arthritis. *Annals of the Rheumatic Diseases* 1980; **39**: 431–7.

127. Kendall M J, Farr M, Meynell M J, Hawkins C F. Synovial fluid in ankylosing spondylitis. *Annals of the Rheumatic Diseases* 1973; **32**: 487–92.

128. Kinsella T D, Johnson L G, Sutherland R I. Cardiovascular manifestations of ankylosing spondylitis. *Canadian Medical Association Journal* 1974; **111**: 1309–11.

129. Lagier R, Albert J. Bilateral deep infrapatellar bursitis associated with tibial tuberosity enthesopathy in a case of juvenile ankylosing spondylitis. *Rheumatology International* 1985; **5**: 187–190.

130. Lahesmaa-Rantala R, Granfors K, Isomaki H, Toivanen A. Yersinia specific immune complexes in the synovial fluid of patients with yersinia triggered reactive arthritis. *Annals of the Rheumatic Diseases* 1987; **46**: 510–4.

131. Laurent M R, Panayi G S, Shepperd P. Circulating immune complexes, serum immunoglobulins, and acute phase proteins in psoriasis and psoriatic arthritis. *Annals of the Rheumatic Diseases* 1981; **40**: 66–9.

132. Lemaire V, Ryckewaert A. Rhumatisme post-salmonellien. *Presse Médicale* 1978; **7**: 2239–40.

133. Le Vine M E, Dobbins W O, III. Joint changes in Whipple's disease. *Seminars in Arthritis and Rheumatism* 1973; **3**: 79–93.

134. Libby D M, Schley W S, Smith J P. Cricoarytenoid arthritis in ankylosing spondylitis. A cause of acute respiratory failure and cor pulmonale. *Chest* 1981; **80**: 641–3.

135. Linder E, Pasternack A. Immunofluorescence studies on kidney biopsies in ankylosing spondylitis. *Acta Pathologica et Microbiologica Scandinavica* B, 1970; **78**: 517–25.

136. McCluskey J, Riley T V, Owen E T, Langlands D R. Reactive arthritis associated with *Clostridium difficile*. *Australian and New Zealand Journal of Medicine* 1982; **12**: 535–7.

137. McCulloch D K, Fraser D M, Turner A L. Arthritis preceding fulminant ulcerative colitis and responding to colectomy. *British Medical Journal* 1980; **281**: 839.

138. McEwen C. Arthritis accompanying ulcerative colitis. *Clinical Orthopaedics and Related Research* 1968; **57**: 9–17.

139. McEwen C, Ditata D, Lingg C, Porini A, Good A, Rankin T. Ankylosing spondylitis and spondylitis accompanying ulcerative colitis, regional enteritis, psoriasis and Reiter's disease. *Arthritis and Rheumatism* 1971; **14**: 291–318.

140. McEwen C, Lingg C, Kirsner J B, Spencer J A. Arthritis accompanying ulcerative colitis. *American Journal of Medicine* 1962; **33**: 923–41.

141. Magaro M, Ceresia G, Frustaci A. Arthritis of the middle ear in ankylosing spondylitis. *Annals of the Rheumatic Diseases* 1984; **43**: 658–9.

142. Marie P. Sur la Spondylose Rhizomelique. *Revue de Mèdicine* Paris 1898; **18**: 285–315.

143. Martelli N A, Rosenberg M, Olmedo G. Espondilitis anquilopoyetica, retraccion de ambos lobulos superiores y micetoma pulmonar. *Medicina* 1983; **43**: 425–32.

144. Menkes C J, Kahan A, Feldmann J-L, Vinh T. Spondylite cervicale destructice. Manifestation initiale exceptionelle de la spondylarthrite ankylosante. *La Presse Mèdicale* 1983; **12**: 227–9.

144a. Michet C J, Conn D L. Psoriatic arthritis. In: Kelley W N, Harris E D, Ruddy S, Sledge C B, eds, *Textbook of Rheumatology* (3rd edition). Philadelphia: W B Saunders, 1989: 1053–63.

145. Mohr W. Gelenkkrankheiten. *Diagnostik und Pathogenese makroskopischer und histologischer Strukturveranderungen*. Stuttgart: Georg Thieme Verlag, 1984.

146. Mole R H, Major I R. Myeloid leukaemia frequency after protracted exposure to ionizing radiation: experimental confirmation of the flat dose-response found in ankylosing spondylitis after a single treatment course with X-rays. *Leukaemia Research* 1983; **7**: 295–300.

147. Moll J M H. *Ankylosing Spondylitis*. Edinburgh: Churchill Livingstone, 1980.

148. Moll J M H. Seronegative arthropathies. *Journal of the Royal Society of Medicine* 1983; **76**: 445–7.

149. Moll J M H, Haslock I, Macrae I F, Wright V. Associations between ankylosing spondylitis, psoriatic arthritis, Reiter's disease, the intestinal arthropathies, and Behçet's syndrome. *Medicine* 1974; **53**: 343–64.

150. Molyneux E, French G. Salmonella joint infection in Malawian children. *Journal of Infection* 1982; **4**: 131–38.

151. Mori K, Zak F G. Reiter's syndrome (with complete autopsy). *Acta Dermatologica-Venerealogica* (Stockholm) 1960; **40**: 362–7.

152. Morris C J, Farr M, Hollywell C A, Hawkins C F, Scott D L, Walton K W. Ultrastructure of the synovial membrane in seronegative inflammatory arthropathies. *Journal of the Royal Society of Medicine* 1983; **76**: 27–31.

153. Muna W F, Roller D H, Craft J, Shaw R K, Ross A M. Psoriatic arthritis and aortic regurgitation. *Journal of the American Medical Association* 1980; **244**: 363–4.

154. Munthe E. Relationship between IgG complexes and anti-IgG antibodies in rheumatoid arthritis. *Acta Rheumatologica Scandinavica* 1970; **16**: 240–56.

155. Nashel D J, Leonard A, Mann D L, Guccion J G, Katz A L, Sliwinski A J. Ankylosing spondylitis and systemic lupus erythematosus. *Archives of Internal Medicine* 1982; **142**: 1227–8.

156. Niepel G A, Kostka D, Kopecky S, Manca S. Enthesopathy. *Acta Rheumatologica et Balneologica Pistiniana* 1966; **1**: 28–64.

157. Nordstrom D, Konttinen Y T, Bergroth V, Leirisalo-Repo M. Synovial fluid cells in Reiter's syndrome. *Annals of the Rheumatic Diseases* 1985; **44**: 852–6.

158. Norton W L, Lewis D, Ziff M. Light and electron microscopic observations on the synovitis of Reiter's disease. *Arthritis and Rheumatism* 1966; **9**: 747–57.

159. Ogasawara M, Kobayashi S, Hill J L, Kono D H, Yu D T Y. Rabbit antisera against three different bacteria which can induce

reactive arthritis: analysis by ELISA, immunoprecipitation and Western blot. *Immunology* 1985; **54**: 665–76.

160. Olivieri I, Gherardi S, Bini C, Trippi D, Ciompi M L, Pasero G. Trauma and seronegative spondyloarthropathy: rapid joint destruction in peripheral arthritis triggered by physical injury. *Annals of the Rheumatic Diseases* 1988; **47**: 73–6.

161. Paloheimo J A, Julkunen H, Siltanen P, Kajander A. Takayasu's arteritis and ankylosing spondylitis: report of four cases. *Acta Medica Scandinavica* 1966; **179**: 77–85.

162. Park H, Schumacher H R, Zeiger A R, Rosenbaum J T. Antibodies to peptidoglycan in patients with spondylarthritis: a clue to disease aetiology? *Annals of the Rheumatic Diseases* 1984; **43**: 725–8.

163. Paronen I. Reiter's disease—a study of 344 cases observed in Finland. *Acta Medica Scandinavica* 1948; 131: suppl. **212**: 1–112.

164. Patton J T. Differential diagnosis of inflammatory spondylitis. *Skeletal Radiology* 1976; **1**: 77–85.

165. Pekin T J, Malinin T I, Zvaifler N J. Unusual synovial fluid findings in Reiter's syndrome. *Annals of Internal Medicine* 1967; **66**: 677–84.

166. Prendergast J K, Sullivan J S, Geczy A, Upfold L I, Edmonds J P, Bashir H V, Reiss-Levy E. Possible role of enteric organisms in the pathogenesis of ankylosing spondylitis and other seronegative arthropathies. *Infection and Immunity* 1983; **41**: 935–41.

167. Puddey I B. Reiter's syndrome associated with *Salmonella muenchen* infection. *Australian and New Zealand Journal of Medicine* 1982a; **12**: 290–1.

168. Puddey I B. Reiter's syndrome following antibiotic-associated colitis. *Australian and New Zealand Journal of Medicine* 1982b; **12**: 292–3.

169. Rampling R P, Lambert H E. Multiple basal cell carcinomas in ankylosing spondylitis treated with X-ray therapy. *British Journal of Radiology* 1985; **58**: 178–81.

170. Redisch W. Capillaroscopic observations in rheumatic diseases. *Annals of the Rheumatic Diseases* 1970; **29**: 244–53.

171. Reiter H. Ueber eine bisher unerkannte spirochäteninfektion (Spirochaetosis arthritica). *Deutsche Medizinische Wochenschrift* 1916; **42**: 1535–6.

172. Resnick D, Dwosh I L, Goergen T G, Shapiro R P, Utsinger P D, Wiesner K B, Bryan R L. Clinical and radiographic abnormalities in ankylosing spondylitis; a comparison of men and women. *Radiology* 1976; **119**: 293–7.

173. Revell P A, Mayston V. Histopathology of the synovial membrane of peripheral joints in ankylosing spondylitis. *Annals of the Rheumatic Diseases* 1982; **41**: 579–86.

174. Robinson S, Panayi G S. Deficient control of in vitro Epstein–Barr virus infection in patients with ankylosing spondylitis. *Annals of the Rheumatic Diseases* 1986; **45**: 974–7.

175. Rodnan G P, Benedek T G, Shaver J A, Fennell R H. Reiter's syndrome and aortic insufficiency. *Journal of the American Medical Association* 1964; **189**: 889–94.

176. Rohatgi P K, Turrisi B C. Bronchocentric granulomatosis and ankylosing spondylitis. *Thorax* 1984; **39**: 317–8.

177. Rollins D, Moeller D. Acute migratory polyarthritis associated with antibiotic-induced pseudomembranous colitis. *American Journal of Gastroenterology* 1976; **65**: 353–6.

178. Rosenberg A M, Petty R E. A syndrome of seronegative enthesopathy and arthropathy in children. *Arthritis and Rheumatism* 1982; **25**: 1041–7.

179. Rosenthal S H, Lidsky M D, Sharp J T. Arthritis with nodules following ankylosing spondylitis. *Journal of the American Medical Association* 1968; **206**: 2893–4.

180. Rothschild B M, Masi M T, June P L. Arthritis associated with ampicillin colitis. *Archives of Internal Medicine* 1977; **137**: 1605–6.

181. Rubinow A, Canoso J J, Goldenberg D L, Cohen A S. Synovial fluid and synovial membrane pathology in Whipple's disease. *Arthritis and Rheumatism* 1976; **19**: 820.

182. Schachter J. Chlamydial infections. *New England Journal of Medicine* 1978; **298**: 428–35; 490–5; 540–9.

183. Schaller J, Bitnum S, Wedgwood R J. Ankylosing spondylitis with childhood onset. *Journal of Pediatrics* 1969; **74**: 505–16.

184. Schilder D P, Harvey W P, Hufnagel C. Rheumatoid spondylitis and aortic insufficiency. *New England Journal of Medicine* 1956; **255**: 11–17.

184a. Schilling F von. Röntgenmorphologische Befund bei der spondylitis ankylopoietica. *Zeitschrift für Rheumaforschung* 1969; **1**: 33.

185. Schlosstein L, Terasaki P I, Bluestone R, Pearson C M. High association of an HL-A antigen, W27, with ankylosing spondylitis. *New England Journal of Medicine* 1973; **288**: 704–6.

186. Schumacher H R, Magge S, Cherian P V, Sleckman J, Rothfuss S, Clayburn G, Sieck M. Light and electron microscopic studies on the synovial membrane in Reiter's syndrome. Immunocytochemical identification of chlamydial antigen in patients with early disease. *Arthritis and Rheumatism* 1988; **31**: 937–46.

187. Schwimmbeck P L, Yu D T Y, Oldstone M B A. Autoantibodies to HLA B27 in the sera of HLA B27 patients with ankylosing spondylitis and Reiter's syndrome. Molecular mimicry with *Klebsiella pneumoniae* as potential mechanism of autoimmune disease. *Journal of Experimental Medicine* 1987; **166**: 173–81.

188. Sharp J. Differential diagnosis of ankylosing spondylitis. *British Medical Journal* 1957; **i**: 975–8.

189. Sharp J T. Reiter's syndrome (reactive arthritis). In McCarty D J, ed, *Arthritis and Allied Conditions.* (10th edition) Philadelphia: Lea & Febiger 1985: 841–9.

190. Sherman M S. Psoriatic arthritis—observations on the clinical, roentgenographic and pathological changes. *Journal of Bone and Joint Surgery* 1952; **34A**: 831–52.

191. Shore A, Ansell B M. Juvenile psoriatic arthritis—an analysis of 60 cases. *Journal of Pediatrics* 1982; **100**: 529–35.

192. Shu K-H, Lian J-D, Yang Y-F, Lu Y-S, Wang J-Y, Lan J-L, Chou G. Glomerulonephritis in ankylosing spondylitis. *Clinical Nephrology* 1986; **25**: 169–74.

193. Sikes D, Hayes F A, Prestwood A K, Smith J F. Ankylosing spondylitis and polyarthritis of the dog: physiopathologic changes of tissues. *American Journal of Veterinary Research* 1970; **31**: 703–12.

194. Sills E M. Psoriatic arthritis in childhood. *Johns Hopkins Medical Journal* 1980; **146**: 49–53.

195. Smith R D. The petrified man. *Journal of Rheumatology* 1983; **10**: 106.

196. Soren A. *Histodiagnosis and Clinical Correlation of*

Rheumatoid and other Synovitis. Philadelphia: J B Lippincott & Co., 1978.

197. Stoll M. Quoted by Huette in 'De l'arthrite dysenterique.' *Archives générales de Médicine* 1869; **14**: 129–31.

198. Strumpell E A G G. Bemerkung uber chronische ankylosierende Entzundung der Wirbelsaule und der Huftgelenke. *Deutsche Zeitschrift für Nervenheilkungen* 1897; **11**: 338–42.

199. Toivanen A, Lahesmaa-Rantala R, Vuento R, Granfors K. Association of persisting IgA response with yersinia triggered reactive arthritis: a study on 104 patients. *Annals of the Rheumatic Diseases* 1987; **46**: 898–901.

200. Toivanen A, Toivanen P, eds, *Reactive Arthritis*. Boca Raton, Florida: CRC Press, 1988.

201. Townes A S, Sowa J M. Complement in synovial fluid. *Johns Hopkins Medical Journal* 1970; **127**: 23–37.

202. Traycoff R B, Pascual E, Schumacher H R. Mononuclear cells in human synovial fluid. Identification of lymphoblasts in rheumatoid arthritis. *Arthritis and Rheumatism* 1976; **19**: 743–8.

203. Ullman S. Deposits of complement and immunoglobulins in dermal and synovial vessels in psoriasis. *Acta Dermatologica et Venerealogica* 1978; **58**: 272–3.

204. Upfold L I, Sullivan J S, Geczy A F. Biochemical studies on a factor isolated from klebsiella K43-BTS1 that cross reacts with cells from HLA-B27 positive patients with ankylosing spondylitis. *Human Immunology* 1986; **17**: 224–38.

205. Utsinger P D. Systemic immune complex disease following intestinal bypass surgery. *Journal of the American Academy of Dermatology* 1980; **2**: 488–95.

206. Utsinger P D, Farber N, Shapiro R F, Ely P H, McLaughlin G E, Wiesner K B. Clinical and immunologic study of the post-intestinal bypass arthritis dermatitis syndrome. *Arthritis and Rheumatism* 1978; **21**: 599.

207. Valaitis J, Pilz C G, Montgomery M M. Aortitis with aortic valve insufficiency in rheumatoid arthritis. *Archives of Pathology* 1957; **63**: 207–12.

208. Vale J A, Pickering J G, Scott G W. Ankylosing spondylitis and upper lobe fibrosis and cavitation. *Guy's Hospital Reports* 1974; **123**: 97–119.

209. Van der Linden S, Valkenburg H A, Cats A. Evaluation of diagnostic criteria for ankylosing spondylitis. A proposal for modification of the New York Criteria. *Arthritis and Rheumatism* 1984; **27**: 361–8.

210. Wands J R, La Mont J T, Mann E, Isselbacher K J. Arthritis associated with intestinal bypass procedure for morbid obesity. Complement activation and characterization of circulating cryoproteins. *New England Journal of Medicine* 1976; **294**: 121–4.

211. Weed C L, Kulander B G, Mazzarella J A, Decker J L. Heart block in ankylosing spondylitis. *Archives of Internal Medicine* 1966; **117**: 800–6.

212. Weinberger H W, Ropes M W, Kulka J P, Bauer W. Reiter's syndrome, clinical and pathological observations—a long-term study of 16 cases. *Medicine* 1962; **41**: 35–91.

213. Welsh J, Avakian H, Cowling P, Ebringer A, Wooley P, Panayi G, Ebringer R W. Ankylosing spondylitis, HLA-B27 and klebsiella. I: Crossreactivity studies with rabbit antisera. *British Journal of Experimental Pathology* 1980; **61**: 85–91.

214. Whipple G H. A hitherto undescribed disease characterized anatomically by deposits of fat and fatty acids in the intestinal and mesenteric lymphatic tissues. *Johns Hopkins Hospital Bulletin* 1907; **18**: 382–91.

215. Wick R R, Gössner W. Follow-up study of late effects in 224-Ra treated ankylosing spondylitis patients. *Health Physics* 1983; **44**: 187–95.

216. Willkens R F, Arnett F C, Bitter T, Calin A, Fisher L, Ford D K, Good A E, Masi A T. Reiter's syndrome: evaluation of preliminary criteria for definite disease. *Arthritis and Rheumatism* 1981; **24**: 844–9.

217. Winchester R J. The major histocompatibility complex. In: Kelley W N, Harris E D, Ruddy S, Sledge C B, eds, *Textbook of Rheumatology* (3rd edition). Philadelphia, London: W B Saunders Company, 1989: 101–37.

218. Winter R J D, Richardson A, Lehner M J, Hoffbrand B I. Lung abscess and reactive arthritis: rare complications of leptospirosis. *British Medical Journal* 1984; **288**: 448–9.

218a. Wollheim F A. Enteropathic arthritis. In: Kelley W N, Harris E D, Ruddy S, Sledge C B, eds, *Textbook of Rheumatology* (3rd edition). Philadelphia, London: W B Saunders, 1989: 1064–75.

219. Woo P, Panayi G. Reactive arthritis due to infestation with *Giardia lamblia*. *Journal of Rheumatology* 1984; **11**: 719.

220. Wooley P H, Ebringer A. Crossreactivity as a factor in the immune response to *Salmonella typhimurium* in CBA and BALB/C mice. *Journal of Medical Microbiology* 1980; **13**: 11–17.

221. Wright V. Psoriatic arthritis. In: Kelley W N, Harris E D, Ruddy S, Sledge C B, eds, *Textbook of Rheumatology* (2nd edition). Philadelphia: W B Saunders & Co, 1985.

222. Wright V. Psoriatic arthritis. In: Scott J T, ed, *Copeman's Textbook of the Rheumatic Diseases* (6th edition). Edinburgh: Churchill Livingstone, 1986; 775–86.

223. Wright V. Reiter's disease. In: Scott J T, ed, *Copeman's Textbook of the Rheumatic Diseases* (6th edition). Edinburgh: Churchill Livingstone, 1986; 787–805.

224. Wright V, Moll J M H. *Seronegative Polyarthritis*. Amsterdam: North Holland, 1976.

225. Yan A C M C, Chan R N W. Stress fractures of the fused dorso-lumbar spine in ankylosing spondylitis. *Journal of Bone and Joint Surgery* 1974; **56–B**: 681–7.

226. Yates D B, Scott J T. Cardiac valvular disease in chronic inflammatory disorders of connective tissue. *Annals of the Rheumatic Diseases* 1975; **34**: 321–5.

227. Ziegler G, Euller L, Dellamonica P, Etesse H, Thyss J. Arthrites septiques à *Yersinia enterocolitica*. Deux observations. *La Nouvelle Presse Mèdicale* 1980; **9**: 3097.

Chapter 20

Rheumatic Fever

Definition

Rheumatic fever is a systemic febrile illness that is an occasional, late reaction to pharyngitis caused by group A, β-haemolytic steptococci.[14,15,24,61] Any streptococcal M serotype may be implicated but a few are most frequent.

The prevalence of rheumatic fever varies greatly in different continents. Rare in most Western countries, the disease is still commonplace in the Third World. The incidence of rheumatic fever remains unrelated to the effective use of antibiotics in therapy and prophylaxis. The prevalence of rheumatic fever began to decrease early in the present century, many years before chemotherapeutic and antibiotic treatment was introduced. The lowest incidence rates in the United States were reached between 1971 and 1981,[122] but during the 1980s increased incidence rates of up to eight-fold were recorded. They remain unexplained.

Rheumatic fever is characterized by pancarditis; acute, non-suppurative, transient polyarthritis; chorea; erythema marginatum; and subcutaneous nodule formation. Rheumatic fever is a consequence only of nasopharanygeal infection; it does not complicate wound sepsis, cellulitis or other streptococcal disease.[155] There are differences in the humoral responses to streptococci causing skin infection and those leading to rheumatic fever.[17,74] Underlying these differences is a hereditary predisposition to rheumatic fever which may also influence the capacity of T-lymphocytes to mediate tissue injury. The pathogenesis of the disease, however, is not yet fully understood. Nevertheless, rheumatic fever can be prevented by antibiotics and treated effectively with salicylates.

History

There have been many reviews.[19,24,142] A classical, clinical description was that of Sydenham.[145] The valvular, cardiac lesions were recognized by Lancici[87] and related to the clinical signs of rheumatic fever by Pulteney.[125] Bouillaud[20,21] gave a full account of endocarditis and Baillie[7] mentioned the changes observed by Pitcairn in the heart of a patient dying from valvular disease.[108] Although focal cardiac lesions had been seen by Romberg,[131] their specificity for rheumatic fever was first demonstrated by Aschoff.[5] The cyclical development of the Aschoff body was elucidated by Gross and Ehrlich;[63,64] Klinge[83] agreed that connective tissue injury preceded cell proliferation.

The significance of nasopharyngitis was appreciated by Fowler.[47] Poynton and Paine[123] believed that the disease was the result of infection with a mysterious *Streptococcus rheumaticus* but the nature of this organism was never proved. A constant association between rheumatic fever and conventional streptococcal infection was suspected and

Glover[58] showed that waves of nasopharyngitis did indeed invariably precede outbreaks of rheumatic fever by intervals that were always 2–3 weeks. Glover's evidence was substantiated by the identification of the responsible streptococci and the demonstration, in affected persons, of a rise in antistreptolysin O (ASO) titres.[29,32]

Clinical presentation

The clinical face of acute rheumatic fever is dominated by synovitis, the pathological by pancarditis.[8,24,147] Rheumatic fever was said to 'lick the joints but bite the heart'.[21,96] However, the severity of the disease has waned remarkably during the past 50 years. The many instances of rheumatic heart disease found in the absence of a clinical history strongly suggest that 'silent' cases occur and that cardiac lesions can develop without polyarthritis.[43]

Rheumatic fever begins abruptly in a child or young adult recently recovered from group A, β-haemolytic streptococcal pharyngitis. Two to three weeks (mean 18.7 days) after the onset of the infection, and irrespective of whether lymphadenitis or scarlatina has complicated the bacterial disease, high fever, headache, joint pain and swelling develop. There is an elevated erythrocyte sedimentation rate, raised plasma viscosity, a mild hypochromic anaemia not responsive to iron therapy and a polymorph leucocytosis. Plasma acute-phase reactants such as C-reactive protein increase sharply.

The articular disease is characteristic although similar forms of arthritis may occur after yersinia (see p. 768) or neisseria (see p. 736) infections. Diagnosis may be assisted by radionuclide imaging (see p. 159).[162] The knees, ankles, wrists and elbows are affected in succession: exquisitely painful inflammation begins as quickly in one as it subsides in another, the faint red colour of the tender, swollen joints differing from the pale but warm and swollen articulations of acute rheumatoid arthritis (see p. 445). Occasionally, the distinction between rheumatic fever and rheumatoid arthritis is less clear and prolonged pain and disability occur with non-progressive, correctable ulnar deviation.[24]

In the young child with rheumatic fever, polyarthritis is relatively less frequent than in the adult patient in whom, by contrast, carditis occurs less often than in the child. Carditis is recognized in 90–92 per cent of those less than 3 years of age, in 50 per cent aged 3–6 years, in 32 per cent of those aged 14–17 years and in 15 per cent of adult cases. The diagnostic criteria of Duckett Jones[14,147] are widely used (Table 20.1).[33]

Table 20.1

Jones criteria (revised) for guidance in the diagnosis of rheumatic fever[33]

Major manifestations
- Carditis
- Polyarthritis
- Chorea
- Erythema marginatum
- Subcutaneous nodules

Minor manifestations

Clinical:
- Arthralgia
- Fever
- Previous rheumatic fever or rheumatic heart disease

Laboratory:
- Acute-phase reactions
  - Erythrocyte sedimentation rate
  - C-reactive protein
  - Leucocytosis
- Electrocardiogram
  - Prolonged P-R interval

PLUS

Supporting evidence of preceding streptococcal infection such as increased ASO titre or other evidence of streptococcal antibodies; positive throat culture for group A streptococci; recent scarlet fever.

The presence of two major, or of one major and two minor criteria, indicates a high probability of the presence of rheumatic fever if supported by evidence of a preceding streptococcal infection. The absence of the latter should make the diagnosis doubtful, except in situations in which rheumatic fever is first discovered after a long latent period from the antecedent infection, as in Syndenham's chorea or low-grade carditis.

Non-rheumatic causes of chorea must be excluded. Note: raised ASO titres can be caused by non-group A streptococcal infections and by streptococcal skin infections.

Pathological changes

The lesions of rheumatic fever are widely disseminated in the cardiovascular system, the synovial joints, the subcutaneous tissues, the brain and the serous membranes. It is convenient to consider first, the characteristic 'unit' lesions of rheumatic fever, second the tissue and organ changes that evolve from these lesions.

The 'unit' lesions

Interstitial haematoxyphil substance

Interstitial haematoxyphil substance (IHS) (chromotropic substance), the 'mucoid oedema' of the older literature, can be identified in the endocardium, subendocardium and myocardium of the atrial appendages even when Aschoff nodules (see below) are not recognized. IHS is thought to be glycosaminoglycan: it is metachromatic with toluidine blue, thionin and polychrome methylene blue, stains with Hale's dialysed iron and with alcian blue but is only faintly periodic acid-Schiff-positive.[82] Similar material can be found in the valves of hearts in the absence of rheumatic carditis but not in the atria and ventricles. The significance of IHS in relation to rheumatic carditis is obscure. There are few studies with modern immunocytochemical techniques.

Fibrinoid

Fibrinoid (see p. 294) is a characteristic microscopic feature of rheumatic fever:[60] uncertainty concerning its nature has meant that some writers[89] avoid the term 'fibrinoid' and substitute 'altered collagen'. Lannigan showed, indeed, that the material stains for collagen but not for fibrin. Others cling to the term 'fibrinoid' and emphasize that the early structural changes within the connective tissue include matrix oedema, collagen fragmentation and a lymphocytic, histiocytic and neutrophil and eosinophil polymorph infiltration.

The tissue appearance called fibrinoid (see p. 294) is identifiable in the majority of the acute rheumatic fever lesions: it is present in the margins of inflamed synovial tissue, in the pericardial and pleural exudates, in the intramyocardial Aschoff nodule, in affected blood vessels and in the transient subcutaneous nodule. Zones of swollen, fused, eosinophilic 'altered collagen' occur, surrounded by IHS in which there are lymphocytes or macrophages.[89] The foci of fibrinoid appear as small islands or in long attenuated accumulations. Where the eosinophilic collagen retains its outline, the term 'reticular Aschoff body' has been used. Gamma-globulin is present.[94,95]

Cell infiltration

Cell infiltration is characteristic. The cells that are attracted to foci of inflammation in the heart and synovium early in the disease include neutrophils and some eosinophil polymorphs. However, lymphocytes with a few histiocytes and some plasma cells soon predominate. With time, larger cells become more frequent; some are multinucleate. Their arrangement and shape are similar in cardiac and in synovial tissues but, in the subcutaneous nodule, histiocytes are common, multinucleate giant-cell formation infrequent. With time, the nuclear chromatin of the larger cells seen in fixed, paraffin-embedded sections becomes condensed. Mitotic activity is infrequent.

Granulomata

In cardiac tissue, Aschoff[5] described and illustrated (Figs 20.1, 20.2) minute (submiliary) aggregates of cells in the intramuscular connective tissue. He concluded that these lesions were pathognomonic of rheumatic fever. *'Aschoff nodules'* ('bodies') are found immediately beneath the endocardium, within loose connective tissue, in the intermuscular septa, particularly of the left side of the heart, and within the normally avascular tissue of the cardiac valves. The nodules evolve through phases of formation, growth and regression, a so-called life-cycle.[63,64] Much additional evidence was gained by the study of left atrial biopsies at mitral valvulotomy.[89,104,121] In the first place there was a minute zone of IHS and altered collagen within the oedematous connective tissue near small blood vessels (Fig. 20.2). Fibrinoid appeared and a mesenchymal cellular infiltrate was recognized. Fibrin was demonstrable.[91,92]

The cellular population of the Aschoff body has excited

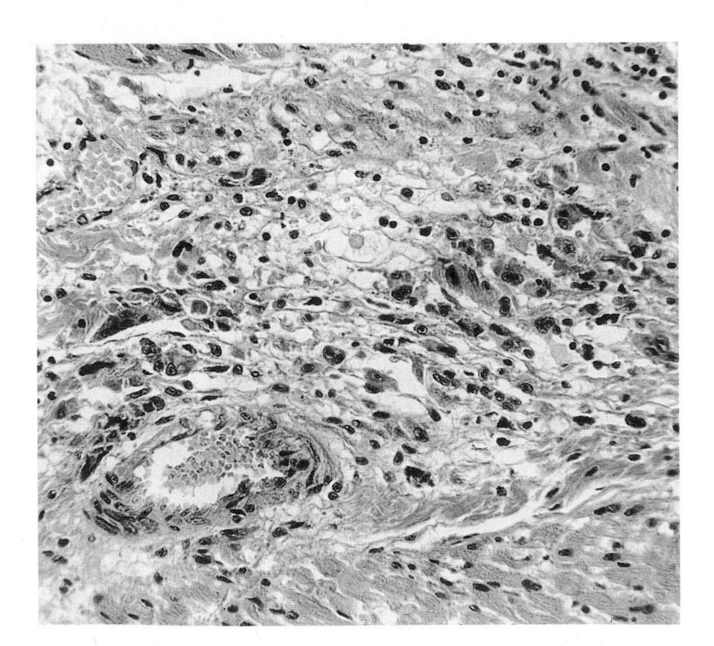

Fig. 20.1 Aschoff body in rheumatic fever.
Interstitial myocardial connective tissue is seen between cardiac muscle bundles (top left and lower right). Within the connective tissue are many mononuclear macrophages. The larger cells may be termed Anitschkoff myocytes: some are multinucleated. The cellular focus lies near a coronary arteriole. Lymphocytes are also present; polymorphs, are characteristically rare. The blood vessel is intact. Compare with Figs 20.4 and 20.5. (HE × 195.)

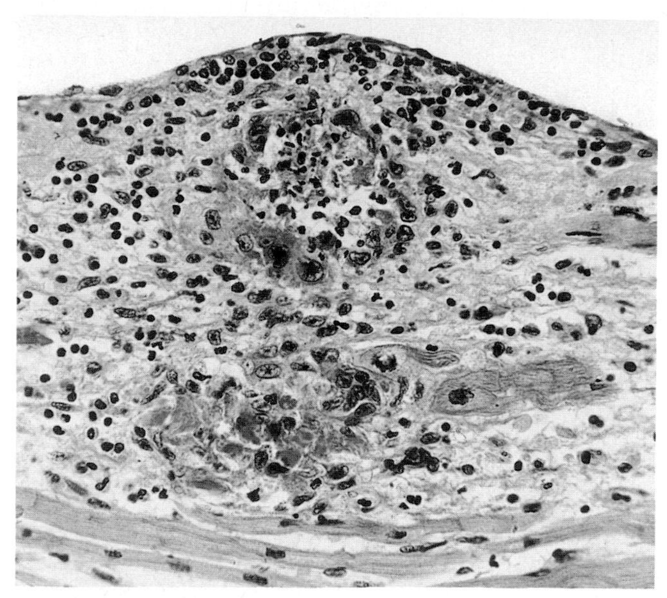

Fig. 20.2 Rheumatic vasculitis.
Mononuclear cells, mainly lymphocytes, lie immediately beneath the endocardium (at top) and cluster around and within a disrupted arteriole (above) and a disorganized group of cardiac muscle cells (below). (HE × 240.)

controversy. Lymphocytes are seen and there are small numbers of plasma cells with infrequent neutrophil and even rarer eosinophil polymorphs. The most characteristic cell is large and mononuclear and has a basophilic cytoplasm. It is distinguished with difficulty from a histiocyte. However, the nucleus in cross-section is elongated and spindle-shaped and resembles an owl's eye. In longitudinal section, the owl-eye nuclei are serpiginous and have fine processes projecting to the nuclear membrane: the term 'caterpillar' nucleus is therefore used.

For historical reasons, based on observations of the responses of myocardial tissue to foreign bodies, the large caterpillar nuclei cells of the Aschoff body were called myocytes, Anitschkow myocytes[3] or Aschoff cells. There is evidence that the Anitschkow cell is of mesenchymal origin;[154] it may become fibroblastic, secrete collagen and remain as a fibrocyte. It has always appeared likely that Anitschkow cells are histiocytes or mononuclear macrophages. They certainly display acid phosphatase activity[38] and have ultrastructural features that recall those of macrophages. However, an alternative view, vigorously argued by Murphy and Becker,[111] has been that the Anitschkow cell is of muscle origin.[110] The cells contain contractile protein[9] and may attain a very large size and become multinucleated. There are still other views concerning the nature of these cells. Thus, Pienaar and Price[120] believed, from experimental electron microscopic evidence, that Anitschkow cells were derived from pericytes or endothelial cells. Others have suggested an origin from nerves[71] or lymphatics.

Early electron microscopic investigations of the cardiac nodule[90,91,93,94,95] confirmed that the multinucleated cells had an electron-dense cytoplasm with abundant endoplasmic reticulum and occasional small vacuoles; there were long, thin cytoplasmic processes sometimes connected with adjacent cells. Inside these processes, clumps of granular material were enclosed. The fine structural identity of mast cells, plasma cells and monocytes could be recognized but the presence of peripheral collagen fibrils with a 64 nm periodicity was only occasional.

Early histochemical studies of cardiac nodules[90,91,93,94,95] suggested that the fibrinoid of the Aschoff nodule was derived from degenerate collagen. No fibrin was found. Since that time, immunofluorescence and electron microscopic techniques for detecting fibrin have greatly increased in specificity, sensitivity and ease; and there is wide support for the view that small quantities of fibrin and other proteins derived from plasma are present within the nodule. With time, the quantity of fibrin diminishes; fibroblasts replace less mature mononuclear, mesenchymal cells and there is a corresponding but limited formation of new collagen. A loss of metachromasia of the intercellular matrix, demonstrable with cationic dyes, accompanies ageing of the nodule. Giant cells, myocytes and histocytes become fewer. There is no permanent change in the structure of arterioles or venules at the site of the lesion. The evolution of the lesion is complete, it is believed, when it is finally converted to a minute scar of low cellularity, a view that is difficult to prove because of the problem of identifying such small foci.

The Aschoff body (see p. 798) is specific to rheumatic fever. However, its appearance is delayed until 3–4 weeks after the onset of the disease. Moreover, rheumatic fever carditis can occur in the absence of Aschoff nodules and rheumatic fever inflammation is not confined to the heart any more than Aschoff bodies are to the myocardium. Aschoff bodies have long been known to be present in biopsy samples of atrial appendages examined at cardiotomy in patients in whom there was no evidence whatsoever of active disease. Consequently, those who adopt the Aschoff body as the only valid evidence for the experimental reproduction of rheumatic fever are probably applying criteria that are too narrow.

Fibrosis

Fibrosis (see p. 301) with scar formation is the common end-result of pericardial and of endocardial and valvular inflammation and accounts for the ultimate valvular deformity. Paradoxically, the synovial, pulmonary and subcutaneous lesions do not behave in this way and there is no evidence that the inflammatory foci of the central nervous system culminate in perivascular or extracerebral fibrosis or in cerebral gliosis. The synovial changes are particularly puzzling. There is intensely active inflammation with a cellular exudate similar to that encountered in the cardiac lesions; yet synovial replacement fibrosis and fibrous ankylosis do not occur. Instead, the resolution of the inflammatory reaction is so rapid and so complete that proteolytic enzymic activation must be assumed and lysosomal hydrolases invoked to explain the degradation and removal of the products of inflammation.

Cardiovascular system

Rheumatic fever affects all parts of the heart (pancarditis) and may injure the aorta, the small arteries and the arterioles.[156a] Venules are implicated as components of the inflammatory reactions; veins are affected incidentally when cardiac failure is accompanied by venous thrombosis.

Heart

Rheumatic fever provokes a pancarditis in which the pericardium, myocardium and endocardium are affected. The advent of valvulotomy and of atrial biopsy led to rapid advances in understanding of the endomyocardial lesions of rheumatic fever. Rheumatic carditis has also been recognized by endomyocardial biopsy performed during cardiac catheterization.[151]

The significance of rheumatic fever as a cause of disability and death is determined by the extent to which the valves and myocardium are injured. It is convenient to describe primary changes in the heart during acute rheumatic fever; and late secondary effects upon cardiac structure and function.

Primary acute rheumatic carditis

Acute rheumatic fever causes pancarditis.[35] The most characteristic lesions are those of the myocardium and of the valvular endocardium. Occasionally, cardiac failure may develop within 2–6 weeks of the onset of the polyarthritis. At necropsy, the heart is enlarged because of ventricular dilatation: the left ventricle is particularly vulnerable. The ventricular muscle is pale and soft but there is little hypertrophy.

Pericarditis Acute fibrinous pericarditis masks the external cardiac surface (Fig. 20.3). Within the pericardial sac is an excess of slightly turbid fluid containing flakes of fibrin but no blood. Cardiac tamponade is not caused. When the volume of the inflammatory exudate is small, the pericardial

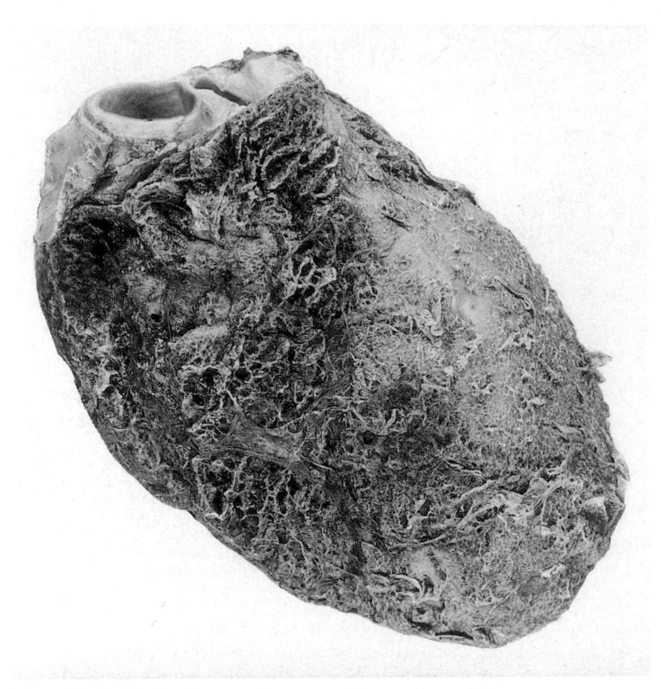

Fig. 20.3 Fibrinous rheumatic pericarditis.
The pericardial surface is covered by aggregates of polymerized fibrin, an index of the increased permeability of the pericardial venules in acute rheumatic fever. (× 0.5.)

and epicardial surfaces are loosely bound together by fibrin so that, when they are separated artefactually, a characteristic 'bread-and-butter' surface structure remains (Fig. 20.3). Microscopically, the appearances are those of acute inflammation. Subserosal blood vessels are congested. Around them, and beneath the surface, a few polymorphs and some lymphocytes, plasma cells and histocytes, lie among the oedematous loose connective tissue. Acute vasculitis may affect the small arteries and arterioles.

Rheumatic pericarditis is a much brisker response than the subacute reaction encountered in rheumatoid arthritis but there may be little to distinguish the changes in rheumatic fever from the pericarditis of systemic lupus erythematosus. The changes of uraemic pericarditis in rheumatoid arthritis with amyloid nephropathy may be simulated but immunopathological methods allow a distinction to be made.[118] Rheumatic pericarditis apparently resolves more slowly than rheumatic synovitis; however, residual fibrosis may bind together the opposing epicardial surfaces. Nevertheless constrictive pericardial fibrosis does not ensue.

Endocarditis and myocarditis Endocarditis and myocarditis (Figs 20.4, 20.5) entirely overshadow rheumatic pericarditis in importance. Common to both sites of injury are the chromotropic change, fibrinoid and cellular infiltrate described above (see p. 798). The most characteristic lesion is the Aschoff body (see p. 799). Myocarditis

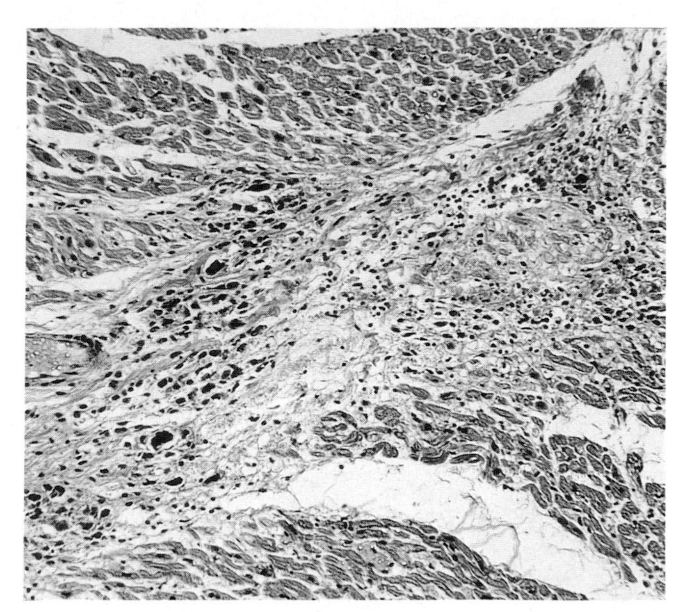

Fig. 20.4 Intramyocardial Aschoff body.
The presence of these focal cell clusters within the cardiac connective tissue demonstrates that the heart has, at some time, been subject to the injurious processes of rheumatic fever. The Aschoff nodule is no longer accepted as a sign of active clinical disease. (HE × 120.)

is frequent in acute and in recurrent, subclinical rheumatic carditis; and late myocardial failure is a characteristic sequel to rheumatic valvular disease. Aschoff bodies, both of classical and of a reticular variety, are encountered within the intermuscular septa of the atrial myocardium, often near small blood vessels; they are less frequent within the ventricular myocardium. Becker and Murphy[9] believed that cardiac valvular connective tissue contained scattered smooth muscle cells and those of the subepicardial tissue show changes resembling those of the intramyocardial Aschoff body.[9,109] Atrial endocardial Aschoff bodies contained smooth muscle type actomyosin: cells and fragments of atrial myocardial Aschoff bodies contained striated muscle-type actomyosin. It has been observed that interstitial haematoxyphil substance, condensed around Aschoff nodules, may extend into the surrounding tissues[89], and sometimes around the whole circumference of the atrial appendage, within the myocardial septa and between cardiac muscle cells. Immunofluorescence studies suggest that antibodies against streptococcal cell wall antigens, and in particular against M-associated protein, is bound to the sarcolemma of heart muscle cells. In spite of the role that this antibody may play in impairing cardiac muscle cell function, there is no corresponding myocyte injury detectable by light microscopy.

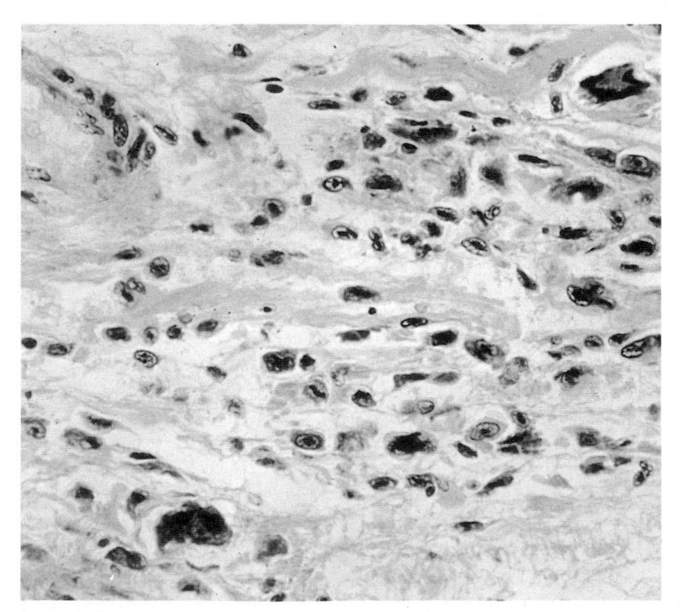

Fig. 20.5 Cellular responses in intramyocardial Aschoff body.
Two multinucleated cells and others with a characteristic caterpillar arrangement of nuclear chromatin, are seen. (HE × 350.)

Secondary chronic rheumatic heart disease

Rheumatic carditis may represent a single, severe episode of cell and connective tissue injury; or episodes of focal inflammation may recur and remit over prolonged periods of time, particularly when, in the absence of prophylatic antibiotic cover, pharyngeal streptococcal infection recurs. Aschoff bodies can be identified on the exposed surfaces of atrial appendices.[28] The appearances suggest that the cells are mesenchymal. With each episode, clusters of Aschoff bodies form, evolve, and undergo fibrosis and scarring. Since Aschoff bodies are very small, scattered, intermuscular lesions, mainly of the left atria and ventricle, it is not realistic to attribute chronic cardiac failure to their presence alone. There is no persistent, chronic injury to cardiac muscle cells themselves: these cells may, however, be directly and severally injured in recurrent episodes of rheumatic fever. The principal cause of longstanding cardiac muscle cell disorder is the functional disturbance of scarred and contracted cardiac valves, leading ultimately to heart failure.[35]

Chronic endocardial disease

The microscopic inflammatory foci which so commonly appear in the cardiac valves are distinguished from the response characteristic of the Aschoff nodule itself by the fact that, normally, the valves have no intrinsic blood supply. The oedematous valvular connective tissue comes to contain interstitial haematoxyphil substance and fibrinoid can be demonstrated. There is an infiltrate of histiocytes and lymphocytes but no smooth or cardiac muscle cells are present. In general configuration, the exudative and proliferative valvular lesion mimics the Aschoff nodule. A disturbance of the overlying endothelial surface is produced,[135] presumably activating a tissue thromboplastin and interrupting normal fibrinolysis; the loss of endothelium and the exposure of basement membrane collagen, activate platelets. Platelets and fibrin are laid down as minute, sterile, verrucous vegetations, largely confined to the lines of closure of the cardiac valves apparently because endothelial loss is precipitated mechanically over zones of abnormal valve tissue. The morphology of the aortic and mitral valvular lesions has been studied by scanning electron microscopy;[135] acute lesions with fibrin and platelet deposits, have been identified.

With time, the inflamed valve and the superimposed vegetations undergo a series of regressive changes leading to granulation tissue formation, fibrosis, scarring and contracture. The normally avascular valve is permeated by capillaries growing in from the deeper valve margins. With the capillary buds come proliferating fibroblasts. The granulation tissue extends to the overlying vegetations, and valve and vegetations are slowly but remorselessly converted to fibrous scar tissue. In time, increasing quantities of collagen are formed; vascularity diminishes but small blood vessels do not disappear. The collagen shrinks with age; valve deformity results (Fig. 20.6).

Endocarditis of the valves is the key to the subsequent, late development of cardiac failure following upon valvular stenosis and incompetence. The focal inflammatory reaction affects the valves of the left side of the heart particularly. The mitral valve is most commonly attacked; the aortic valve suffers next in frequency, while combined mitral and

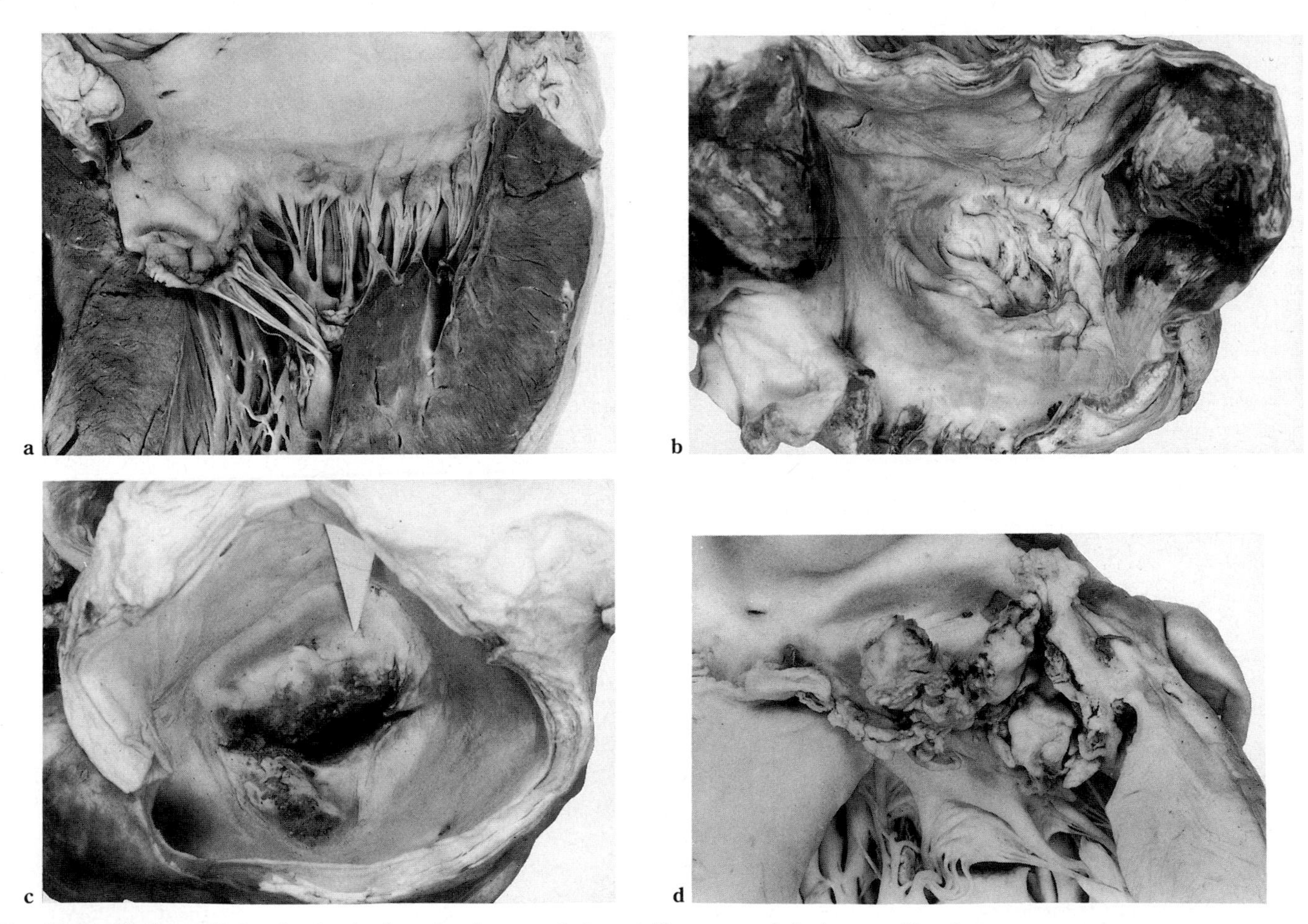

Fig. 20.6 Endocardial valvular lesions in rheumatic heart disease and their complications.
a. Acute rheumatic endocarditis. Along the line of closure of the mitral valve is a thin, discrete row of pale, pink-grey vegetations many of which are only 1–2 mm in diameter. **b.** Established mitral stenosis. Upper aspect of stenotic mitral valve. Stenosis is the culmination of the acute inflammatory process depicted in **a.** A large, dark grey-appearing mass of thrombus lies in the atrial appendage, one result of atrial fibrillation. **c.** Surgical relief of mitral stenosis. In early practice, the stenosed valve was split by valvulotomy, shown here. The insertion of an artificial valve is now preferred. **d.** Infective endocarditis. The infection of valves that have been scarred and vascularized by the processes of rheumatic endocarditis is still common even in countries where acute rheumatic fever has become a rarity. *Streptococcus viridans* is implicated as are other bacteria; inspite of antibiotic treatment, mortality remains as high as 30 per cent.

aortic valve lesions come third in order of prevalence. In approximately 5 per cent of cases, there is evidence of tricuspid endocarditis; the pulmonary valve is very rarely disturbed.

The effects of these microscopic changes on the structure and function of the cardiac valves are eventually disastrous (Fig. 20.6). Fibrosis, with subsequent contracture, affects the mitral, aortic and tricuspid valve; the chordae tendineae; and the posterior wall of the left atrium. The severity of the resulting symptoms and signs depends as much on the capacity of the myocardium to adapt by hypertrophy to work against increasing resistance as on the precise degree of valvular stenosis.

Mitral valve Adhesions bind the opposing leaflets of the mitral valve together, particularly at critical sites such as the insertions of the thickest and strongest chordae tendineae. The valve edges also become joined *between* these insertions and the commissures; the fibrous adhesions are usually weaker here than at the sites of insertion. The valve orifice is gradually converted to a small, stenotic opening with rigid margins. The whole valve structure may assume an inflexible, inelastic, immobile 'button-hole' configuration. Shortening, by contracture, of the principal chordae, pulls the margins of the scarred valve downwards, creating a 'funnel' deformity. Rigidity and irregularity of the mitral valve and of the nearby endocardial and perivalvular tissues are now exaggerated by the dystrophic deposition of calcium salts (see p. 403) which may be confined to zones deep to the valve ring but which may project at the edges of the residual valve orifice or upon the surfaces of the deformed valve leaflets.

Aortic valve The aortic valve alone may be a site of rheumatic valvulitis but more often it is affected synchronously with involvement of the mitral valve. Fibrosis results from intrinsic changes in the valve structure, the result of focal inflammation, and from the evolution of the small verrucous vegetations that form when thrombosis occurs at the zones of contact, at valve margins and commissures. A vicious circle is established in which scarring causes diminished mobility of the normally delicate aortic cusps; diminished mobility and the presence of organizing thrombi lead to an extension of scar tissue; and the entire process is complicated and exaggerated by dystrophic calcification. The final state is that of stenosis and functional insufficiency.

Chronic myocardial disease Myocardial disease commonly occurs as a result of the valvular defects. There is dilatation, hypertrophy and atrophy, in varying proportions.[35] Failure of the myocardium, with congestion, is influenced by an inconstant and variable quantity of scar tissue. Nevertheless, scarring[50] as a sequel to Aschoff body formation can, by itself, have little mechanical influence on cardiac function: the Aschoff nodes are too few and too small. However, additional insidious changes in myocardial cell function occur and provoke atrial fibrillation, atrial flutter and partial heart block. Their onset is due to ectopic foci of excitation at foci of injury that have been demonstrated by recent immunological studies. The use of serial cardiac sections to correlate the electrocardiographic and the structural alterations in the conducting mechanism, has been advocated.[49]

The right atrial muscle may be affected more severely than expected. In chronic rheumatic heart disease, enzyme histochemical studies for succinate dehydrogenase and adenosine triphosphatase failed to show muscle cell degeneration in atrial biopsies although there was ultrastructural evidence of muscle disease.[153] Transmission electron microscopic studies of Indian patients, among whom tricuspid valve disease is frequent, often reveal cardiac myofibre disorders.[153]

Because of the decline in incidence of rheumatic fever in Western countries and increased interest in disorders such as 'floppy mitral valve' that may have an heritable origin (see p. 347), there is a need to distinguish between these disorders quickly and reliably. There is good reason to suggest that macroscopic examination is sufficient.[10] The postrheumatic mitral valve is fibrotic and firm with thickened, fused leaflets and commissures. Funnel-shaped deformity, interchordal fusion and rigidity are characteristic and calcification may occur at any site. The floppy valve, by contrast, shows laxity of leaflets with dome-like deformities extending above the annulus. There is thinning and attenuation of leaflets which may rupture. There is no commissural fusion and the valve remains soft and flexible.

Blood vessels

Small vessels mediate the inflammatory responses of the synovia, heart, subcutaneous tissue and nervous system; however, the aorta and some arteries may be affected primarily.

The arterial lesions of acute rheumatic fever are widespread;[57,137] their systemic distribution is similar to that of the arteritis of polyarteritis nodosa (see p. 620) and changes are consequently found in the cutaneous, muscular, coronary, pulmonary, pancreatic, gonadal and cerebral vessels. The lesions take the form of a zonal alteration in connective tissue structure. Altered vascular permeability is suggested by the finding of swollen, basophilic endothelial cells beneath which eosinophilic proteinaceous material may collect. The media appears to be involved secondarily. There is an accumulation within the media, first, of interstitial haematoxyphil substance and subsequently, of fibrinoid.

With the appearance of fibrinoid, the connective tissue skeleton of the artery is progressively disorganized. Muscle cell necrosis may be detected while, accompanying this response, mononuclear cells accumulate. The entire thickness of the vessel wall is implicated as in polyarteritis nodosa but the associated inflammatory reaction is usually of a milder character and is confined to a sparse histiocytic, lymphocytic and plasma cell infiltrate (Fig. 20.2). Occasionally a disseminated necrotizing vasculitis is encountered in a fatal case of rheumatic fever[5] and in such cases arterial thrombosis with subsequent recanalization may be a sequel.

A tendency to incomplete, localized, mural thrombosis leads to the deposition, at sites of active vasculitis, of small marginal thrombi. The thrombi, formed of platelets and fibrin in the first instance, are quickly covered with endothelial cells and remain as verrucous projections within the vessel lumen, at the site of the inflammatory response. Here, their appearance excites comparison with the analogous but non-inflammatory changes encountered in smaller vessels in thrombotic microangiopathy (see p. 660). They are similar to the valvular endocardial vegetations described above (see p. 801). Arteritis was envoked as one explanation for the rheumatoid granuloma; I know of no evidence to suggest that the analogous, acute lesion in rheumatic fever is the result of rheumatic vasculitis.

Particular attention has been paid to *cardiac* vascular lesions in Japanese work.[50] Active rheumatic fever angiitis may include muscle destruction and intimal hyperplasia as well as thrombosis; panangiitis is characteristic. In their series of 41 cases examined *post mortem*, Fukuda and Okada[49] found seven with angiitis, most frequently in patients with mitral regurgitation.

In the aorta, raised, dull red-brown ridges may be encountered upon the endothelium above the aortic valve.[116] They are composed of hyperplastic intimal connective tissue and, probably, smooth muscle cells; new capillaries can be recognized. The intimal changes, more usually microscopic, overlie medial zones where fibrosis has succeeded ischaemic necrosis and where the fibrous foci are surrounded by histiocytes, lymphocytes and regenerating myocytes. Degeneration of medial muscle with loss of elastic is apparently the consequence of a rheumatic periaortitis in which the simultaneous presence of a vasculitis of the vasa vasorum with fibrinoid, oedema and cellular infiltration combine to produce zones of connective tissue change resembling those encountered in the cardiac Aschoff body. The effects of these limited medial aortic changes are slight and insufficient to lead to aneurysm formation.

Osteoarticular system

The affected diarthrodial joints are the sites for a severe acute, transient, sterile synovitis.[24,31,45,52,83] Jaffe,[72] Sissons,[136] and Catto[25] devote only a few lines to the description of the synovial joints in rheumatic fever, the first author solely in the context of the differential diagnosis of secondary syphilis, the latter two without a text reference. No evidence of overt infection is recognizable and no bacteria have been identified within or isolated from the synovial tissues. Attempts to implicate virus have been made without success. A more chronic, low-grade synovitis with swelling of the metacarpophalangeal and proximal-interphalangeal joints does not respond to salicylates and may mimic rheumatoid arthritis or juvenile rheumatoid arthritis.[24]

The periarticular tissues are red and swollen. The joint capsule is distended by a modest quantity of turbid fluid that contains many neutrophil polymorphs and smaller numbers of mononuclear cells. As many as 50×10^6 cells/litre may be present. The synovial fluid is viscous and contains 30–50 g protein/litre. The acutely inflammed synovial villi are oedematous and hyperaemic (Fig. 20.7). Fibrin is exuded and adheres to and partially masks the synovial surfaces: it can

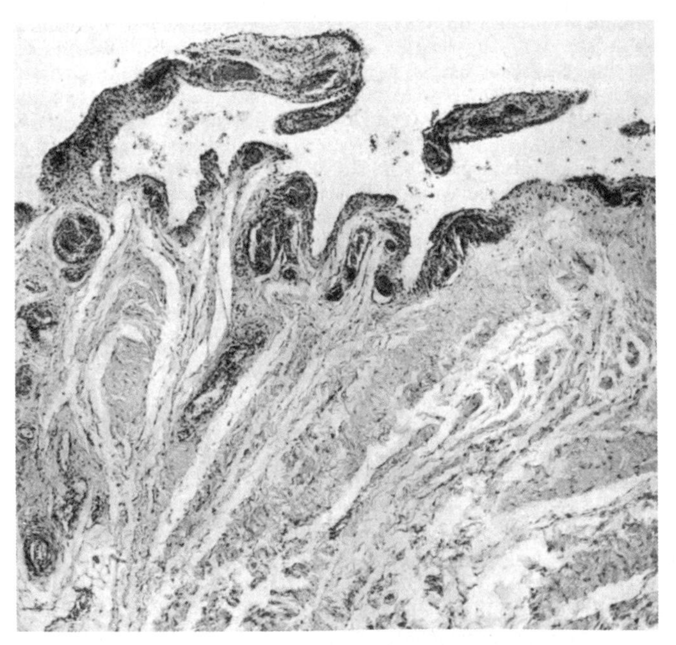

Fig. 20.7 Synovitis in rheumatic fever.
Synovial tissue obtained from the shoulder joint four days before death in the third week of a recrudescence of rheumatic fever. There is vascular congestion and oedema of the synovial connective tissue. Many polymorphs are recognizable in the subsynoviocytic layer. (HE × 40.) From Collins D H, *Pathology of Articular and Spinal Disease*[31] reproduced by permission of the publishers.

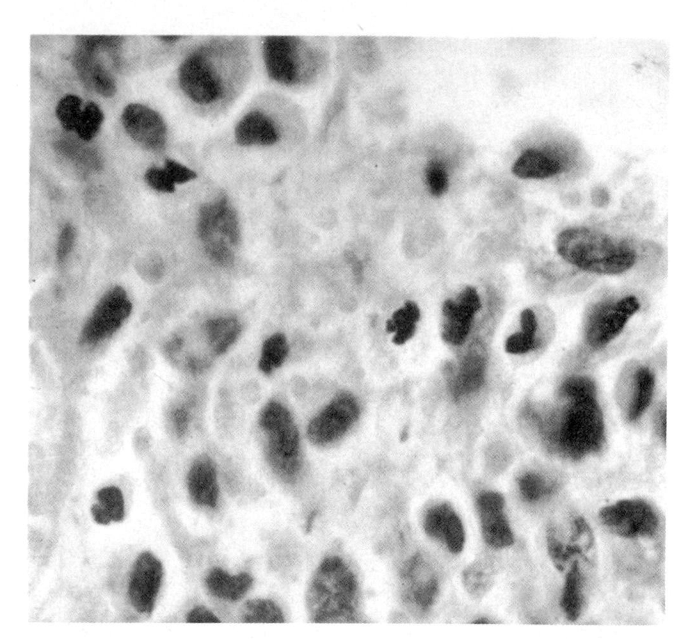

Fig. 20.8 Acute rheumatic synovitis.
Higher-power view of field from tissue illustrated in Fig. 20.7. Numerous polymorphs infiltrate between the type A and B cells. (HE × 1000.) From Collins D H, *Pathology of Articular and Spinal Diseases.*[31] Reproduced by permission of the publishers.

be seen by scanning electron microscopy and identified by transmission electron microscopy or by selective staining. The circulus vasculosus of the synoviochondral junction is prominent but, by contrast with the early pannus of rheumatoid arthritis, there is no marginal degradation of cartilage.

The older literature describes the finding of identical synovial changes in the finger, toe, ankle, wrist, knee, elbow, shoulder and hip joints.[83] Synoviocytes may be lost beneath fibrin deposits. It is not certain whether fibrin is incorporated into the synovial tissue, as is commonly the case in rheumatoid arthritis. Although a very early polymorph infiltrate is recognized, the synovia are soon infiltrated with mononuclear phagocytes. Later, lymphocytes and plasma cells become more frequent. However these cells are never as common as in overt rheumatoid arthritis, while polymorphs remain numerous (Fig. 20.8). Synovial cell hyperplasia is provoked and the thickness of the lining layer increases. Generally, signs of inflammation subside very quickly and completely, although persistent inflammation may be a characteristic of cases in those countries where rheumatic fever and rheumatoid arthritis are both common and the diseases coincide. Nevertheless, small vessels may remain congested and the synovial tissue mildly oedematous. Neither necrosis nor nodule (see p. 798) formation occur. Perivascular aggregates of inflammatory cells are detected; vasculitis, vascular thrombosis and intimal fibromuscular hyperplasia are, however, very unusual. There is no tendency for synovial villi to become attached to the surfaces of articular cartilage and none for pannus formation to evolve. There is no evidence to show that the arthritis of rheumatic fever ever progresses to that of rheumatoid arthritis.

If immediate resolution does not take place, macrophages engulf the residual fibrin producing a local microscopic lesion that resembles the Aschoff body. However, as Fassbender[45] points out, whereas the compact cardiac musculature contributes to the circumscribed character of the cardiac nodule, the looser character of much synovial tissue ensures a more open structure, less sharply defined. Multinucleate giant-cell forms are thought to be of macrophage origin. A sequence of regressive changes follows: salicylate or corticosteroid treatment has a dramatic effect on the cardinal signs of inflammation but the other microscopic features resolve less readily. Macrophages, lymphocytes and plasma cells decrease in number but some lymphocytes may persist for long periods. There is a mild stimulus to new collagen synthesis but fibroblast activity is slight and residual fibrosis minimal. Bone erosions do not develop but there is evidence that the inflammatory reaction may extend to involve the periosteum, and here fibrinoid has been described.[83] Foci of fibrinoid, in these circumstances, are most frequent over bony surfaces such as the olecranon, patella and calcaneum. Granulomata, distantly resembling those of the rheumatoid nodule, may form within the periarticular connective tissues and within tendon sheaths and fascia, independent of the transient arthritic response.

Although the migratory polyarthritis of rheumatic fever almost always subsides within a few days of onset, and although the disappearance of the histological signs of inflammation is remarkably complete, a more sustained reaction is occasionally seen. In this there is persistence of the signs of inflammation but a limited infiltration of the connective tissues of the joint with mononuclear cells including plasma cells, lymphocytes and macrophages.

Skin

Subcutaneous rheumatic fever nodules (granulomata)[13,24] (Fig. 20.9) are found in the more severe cases, occasionally in the absence of a history of rheumatic fever. They are small oval, spherical or irregular-shaped structures within the tissues overlying bony prominences such as the elbow and occiput. The nodules, ranging in size from 4 to 25 mm in diameter, are painless and not tender; they may be readily moved in the planes of the connective tissues but are quite commonly attached to the capsule and pericapsular structures of a nearby joint. In contrast to the persistent, indolent nodules of rheumatoid arthritis,[13] the nodules of rheumatic fever have a transient existence, the duration of their evolution apparently corresponding approximately to that of the acute synovial polyarthritis. Thus, rheumatic nodules emerge and resolve within a space of days; there is no tendency to ulceration of the overlying skin or to sinus formation. Scar tissue does not result.

The histological appearance of the rheumatic fever nodule is characteristic but not pathognomonic[31,39,100] (Fig. 20.9). The structure is not encapsulated and the margins merge with those of the surrounding loose connective tissue. There is an irregular central zone of disorganization: material with the staining characteristics of fibrinoid is detectable and there is matrix proteoglycan degradation but no collagen loss. Bordering this central zone is a pallisade-like layer of mononuclear cells, principally histiocytes, with their long axes orientated in a radial manner recalling the structure of the rheumatoid nodule and suggesting the influence of positive or negative chemotaxis. Multinucleate cells are uncommon but moderate numbers of plasma cells, neutrophil polymorphs and lymphocytes are seen at the periphery; eosinophil polymorphs are rare. Few immunocytochemical studies have been done.

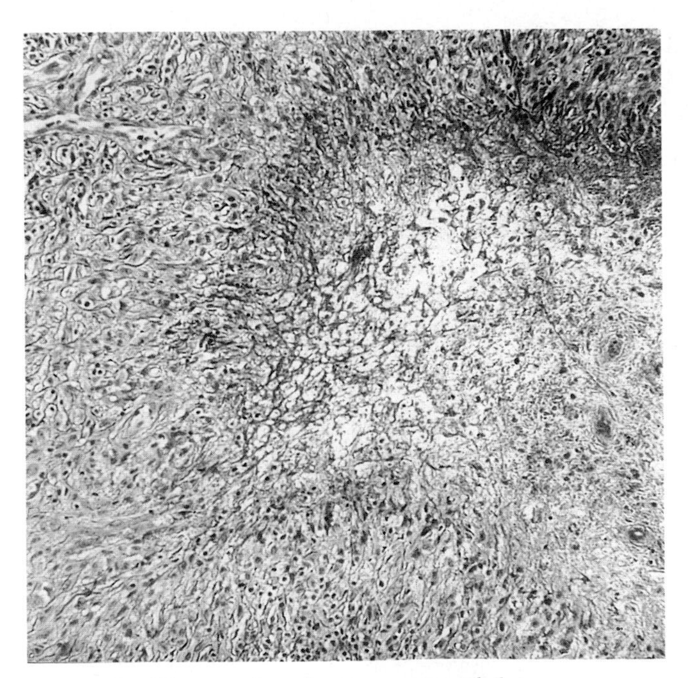

Fig. 20.9 Rheumatic subcutaneous nodule.
A central zone of ill-defined, eosinophilic tissue debris is bounded by strands of fibrin which extend into the nearby vascular connective tissue. The shape and organization of the evanescent nodule resembles that of focal infections such as invasive candidiasis (see p. 752) and there are morphological similarities to the rheumatoid nodule (see p. 485) which is distinguished by its indolence and persistence. The nodule of rheumatic fever is followed neither by superinfection nor scarring. (HE × 110.)

The general appearance of the rheumatic subcutaneous nodule has a distant resemblance to the Aschoff body and it is certainly reasonable to draw an analogy with that structure. In children, the microscopic appearance and location of the nodules resemble those of granuloma annulare (see p. 488). On account of the brief life-cycle of the rheumatic nodule and its rarity, and a disinclination to undertake a needless biopsy, the European pathologist now seldom has an opportunity to examine a proven example of rheumatic fever granuloma. However, the existence of the nodule is frequently recalled when the differential diagnosis of unexplained dermal and subcutaneous granulomata are discussed.

Respiratory system

There is a widely held view that specific abnormalities develop in pulmonary structure and function both in acute rheumatic fever and in the recurrent disease.[140] The cause of *rheumatic pneumonitis* is unknown but there is speculation that the lesions result from a form of hypersensitivity. Accounts of the lung disorder in rheumatic fever are traced historically to Cheadle,[27] but have become fewer as the disease has diminished in frequency.[57,65,66,129]

The inflammatory pulmonary lesions in acute rheumatic fever are a combination of an inconstant vasculitis, focal alveolar injury, necrosis and haemorrhage; and the presence of much intraalveolar oedema fluid that may coalesce on alveolar surfaces. Moderate numbers of mononuclear macrophages come to lie within the alveoli the walls of which are lined by a compact, eosinophilic lamina formed of exuded fibrin[102] and probably, of other proteins of plasma origin. The appearances of the alveoli resemble those encountered in hyaline membrane disease of the newborn,[140] in the lungs of patients with severe hypertension treated with anti-hypertensive agents, and in uraemic oedema.[41] The consequences of this widespread microscopical change, the exact mechanism of which is not clear, is pulmonary consolidation. There is a proliferation of type II pneumocytes.

The heavy, congested, dark, red-purple lungs collapse incompletely after removal, retaining their shape and, resisting distortion. They have a rubber-like consistency. There is pleuritis with a fibrinous exidate on the visceral pleural surfaces. The lower lobes are more severely affected that the upper: there is oedema and segmental zones of haemorrhage, but infarction cannot be identified unless, in longer standing cases, thromboembolism has complicated cardiac failure.

Rheumatic pneumonitis is most common in the lower lobes of the lungs: they become heavy, firm, of elastic texture and, on section, deep purple-red. The shape is retained in a manner suggestive of pneumonitic consolidation. Much inflammatory fluid is present and can be demonstrated by pressure upon the lungs. Scattered upon the congested cut surfaces are zones of more intense haemorrhage. The conspicuous red-brown confluent colouration of viral pneumonitis is not seen but distinction from viral pneumonia, on anatomical grounds, may be difficult.

Resolution of the pneumonitis is usually complete: however, where the connective tissue microskeleton of the lung is destroyed in zones of necrosis, repair is accompanied by scar formation. Occasionally, organization with fibrosis follow exudation; and loose connective tissue plugs come to partly fill alveolar ducts, a phenomenon not specific to rheumatic pneumonitis. The pulmonary changes of rheumatic fever are not easily distinguished from prolonged pulmonary oedema in left ventricular cardiac failure, and in uraemia. The presence of fibrinoid within the walls of pulmonary arterial branches and of endarterial fibrosis may aid diagnosis.

The pulmonary vascular changes of rheumatic fever should be distinguished from those which accompany the pulmonary hypertension of mitral valve contracture.[98] Where pulmonary hypertension is established, the smooth muscle cells of the pulmonary arteries and arterioles are hypertrophic. When pressures in the pulmonary artery circulation are known to have been particularly high or to have risen very quickly, a necrotizing arteriolitis may be recognized. The vasculitis may be part of a hypersensitivity response attributable either to antibody cross-reacting with haemolytic streptococci, or to immune complex deposition. In occasional cases of long-standing mitral stenosis, circumscribed zones of intraalveolar bone formation are encountered (Fig. 20.10).[44] These zones are found in the lower parts of the lungs within alveoli and alveolar ducts. There is no inflammatory or fibrous tissue reaction but the bony nodules are generally regarded as a late consequence of a focal rheumatic pneumonitis. The shape of the nodules conforms with that of the fibrinous casts in uraemic lungs.[98]

Central nervous system[161]

Chorea is now a rare manifestation of rheumatic fever, and its frequency continues to decline. The onset of chorea is late; although it may appear with carditis, it may develop in isolation. Little is known of the pathological changes that underlie the strange, purposeless movements, the muscular weakness and the emotional disturbance; and there is no explanation of why chorea is almost exclusively confined to females when the onset is postpubertal and why, in women, it is so rare in adults.

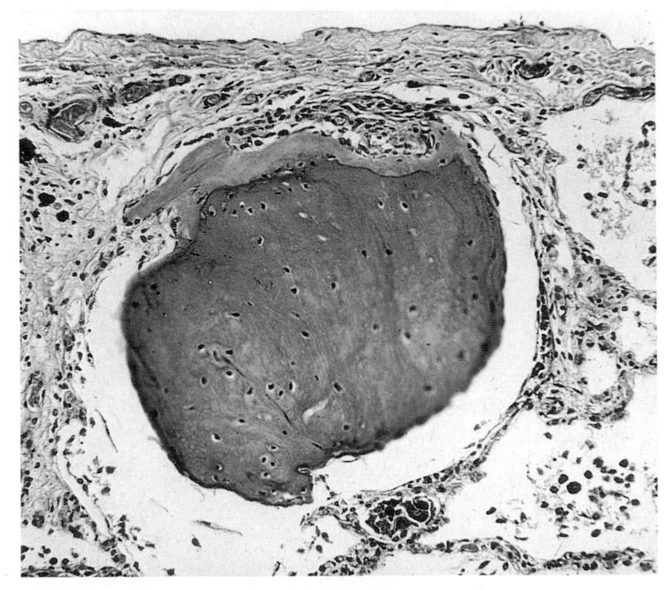

Fig. 20.10 The lung in rheumatic mitral stenosis. Sustained pulmonary hypertension may culminate in focal, intrapulmonary bone formation. (HE × 120.)

Stollerman[142] reminds us of the reasons for the uncertainty surrounding the pathological features of chorea. Chorea is rarely fatal. Some who die of carditis may have involvement of the central nervous system without chorea. No single central nervous system site is consistently affected; Aschoff bodies are not found within the brain;[34,113] and it has not proved possible to correlate the clinical findings with the pathological changes. The neuropathological lesions are not conspicuous histologically: they include arteritis, cellular degeneration, perivascular round-cell infiltration and occasional petechial haemorrhages distributed widely in the cortex, cerebellum and basal ganglia.[23,82] However, there is evidence of a direct interaction between antigenic sites in cerebral neurones and antistreptococcal antibodies since the sera of approximately 50 per cent of children who develop Sydenham's chorea give immunofluorescent staining that can be adsorbed with streptococcal membranes.[130]

Urinary system

An obliterating, rheumatic endarteritis of medium-size and small renal cortical arteries may lead to multiple zones of infarction and, subsequently, to a rheumatic nephrosclerosis.[22] However, the relationship between acute post-streptococcal glomerulonephritis and rheumatic fever remains controversial and the distinction between the respective renal lesions uncertain. The conditions may coincide.[11] To clarify the position, 22 renal biopsies on patients with acute rheumatic fever were surveyed.[62] In all, half of the patients had glomerular lesions, usually slight and often detectable only by transmission electron microscopy. The lesions included basement membrane thickening, electron-dense deposits and increased mesangial cells. They displayed considerable variability.[55]

Aetiology and pathogenesis
(Fig. 20.11)

Microorganisms

The association of rheumatic fever with preceding nasopharyngeal infection with a group A, β-haemolytic streptococcus is established beyond reasonable doubt.[1,69,132,143] The disease does not follow infection by the same group of organisms at other sites. The bacteria often persist in the nasopharynx after the onset of rheumatic fever which is characteristically 2–3 weeks (*Pope*: 3–4 weeks[122]) after the onset of either first or recurrent infections. The remaining organs of the body, including the heart and synovial fluid, are always sterile. Although overt nasopharyngeal infection is characteristic, varying proportions of infections that precede rheumatic fever are clinically silent (subclinical) or are not recognized. Thus, in recent US outbreaks, 45–75 per cent of cases gave no history of nasopharyngitis and only 15 per cent had been treated actively for this condition.[122] Nevertheless, it is clear that that streptococcoal components initiate tissue injury during the preliminary 2–4 week postinfective period although streptococcal antigens have never been recognized in affected tissues. Because of recent experience with the identification of yersinia, chlamydia and salmonella antigens in 'reactive' arthritis (see p. 764), caution should be exercised in stating that this observation is definitive.

The first proof of a relationship between rheumatic fever and streptococcal infection was epidemiological (see p. 811). Streptococcal virulence is strongly associated with the component M protein (Table 20.2)—a surface molecule which affords protection against phagocytosis.[81a] Serotyping by means of anti-M protein antisera is a potent epidemiological tool. With this technique, it has been demonstrated that there are at least 75 serotypes of group A streptococcus. Types 5, 6, 18 and 24 are most frequently associated with rheumatic fever and in recent North American outbreaks, types 5, 14, 18 and 24 were implicated in one, type 18 in another.[122] When an outbreak occurs in an institution such as a school or in a military barracks, a single serotype is usually responsible but in sporadic cases no particular serotype can be implicated.

Because of the involvement of small numbers of serotypes of group A streptococci in the initiation of rheumatic fever, it seems likely that there are 'rheumatogenic' strains of these organisms. They are distinct from the bacteria that provoke acute glomerulonephritis and which are said to be 'nephritogenic'.[18,142] Rheumatogenic strains have large quantities of stable M protein, substantial hyaluronate capsules and properties characteristically induced by teichoic acid.[134] These properties are likely to contribute to a

Table 20.2

Cell components of haemolytic streptococci

Component	Antigenicity	Biological properties	Pathological effects
Hyaluronate	–	Virulence factor, impedes phagocytosis	None known in men; measured in complicated streptococcal infection
M protein	+	Type specific antigen; Virulence factor, impedes phagocytosis	Type-specific hypersensitivity reactions in man
M-associated protein	+	Non-type specific antigen, cross-reactive with cardiac tissue	Non-type-specific immunotoxic reactions in man
C carbohydrate	+	Group-specific antigen, possibly cross-reactive with heart valves	None (in isolated form)
Peptidoglycan	+	Common antigen, cross-reactive with peptidoglycan of other bacteria	Endotoxin-like activity; inflammation
Cell membrane	+	Group-specific antigens, cross-reactive with cardiac and renal tissues	Transient nephritis in experimental animals

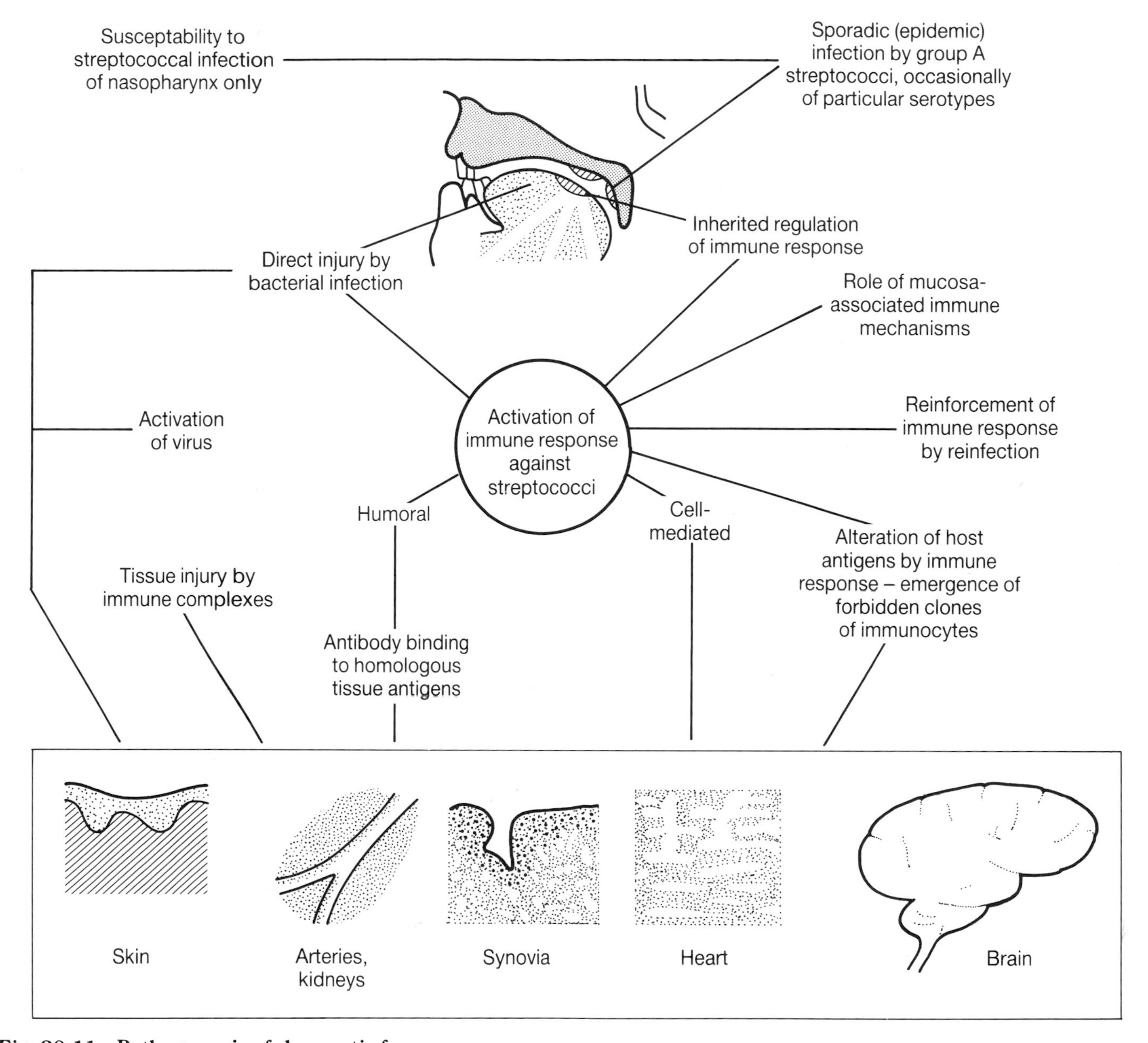

Fig. 20.11 Pathogenesis of rheumatic fever.
Suggested mechanisms thought to operate in the development of the disease.

tendency for strains causing rheumatic fever to adhere tenaciously to pharyngeal cells for long periods.[133] The remote possibility that an initial infection by streptococci activates a latent virus infection which then, in turn, attacks the synovia, heart and other tissues has received little support despite the finding of virus particles in rheumatic lesions.[132] The well-known capacity for coxsackie and other viruses to cause myocarditis does not appear relevant although, clearly, such an infection could account for the formation of anticardiac antibodies, the presence of which is characteristic of rheumatic fever.

The chemical and biological characteristics of group A streptococci have been analysed in great detail.[69] Their cellular and extracellular products are increasingly well understood (Tables 20.2 and 20.3). A peptide extracted from type M protein contains the myosin cross-reactive epitope and anti-M5 antibodies cross-react with epitopes on human heart muscle cell sarcolemma and myosin. The biologically active amino terminal moieties of proteins M5, 6 and 24 display mutual homology; they are not, however, homologous with equivalent components of the nephritogenic organisms, serotypes M1 and M49.

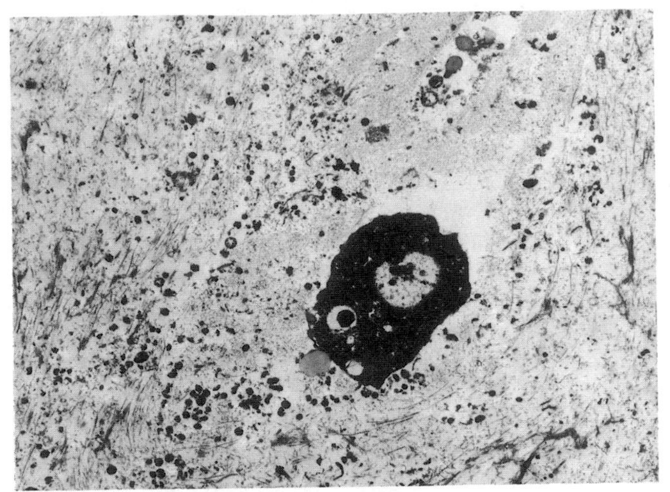

Fig. 21.2 Ageing articular cartilage.
Residual structure of effete chondrocyte from zone II hyaline cartilage of 64-year-old female. (TEM × 3962.)

and enzymic protein anabolism, and a shift to the altered use of energy sources. The maintenance of energy production by glycolysis, at the expense of nucleic acid synthesis, may be a price paid for energy conservation.

Mitochondria

Mitochondria become fewer with advancing age, an index of a diminishing capacity for oxidative phosphorylation.

Microtubules and cytoplasmic filaments

Adult ageing chondrocytes contain pressure-sensitive, approximately 25 nm diameter microtubules composed in part of tubulin. Microtubules subserve important functions in relation to the mitotic spindle and the chondrocyte cilium. They have a special significance in relation to cell shape and intracellular organization but are concerned also with the movement of materials and with collagen secretion. Closely related to lipid droplets (see p. 22), chondrocytes often display hierarchies of these cytoplasmic filaments which confer structural rigidity and are prominent in elastic cartilage chondrocytes. Normally, cytoplasmic filaments, which include a population of 5–8 mm diameter actin filaments and of non-actin, non-tubulin 10 mm diameter filaments, form a minor constituent of the cell. With advancing age, large masses or whorls of beaded vimentin microfilaments accumulate[37] perhaps because of the decline or obstruction of normal metabolic pathways.

Lipid

The content of *intracellular* lipid decreases after middle age as cartilage metabolism diminishes. The size of the lipid droplets does not alter, either in human hyaline or elastic cartilage. However, there is a slight, age-related increase in total and *extracellular* lipid throughout mature, adult life. The lipid may represent the residue of dead or apoptotic chondrocytes but it may also comprise matrix vesicles (see p. 108) or be derived from the synovial fluid. The possible contribution of matrix vesicles to the early phases of osteoarthrosis is considered in Chapter 22. Superficial zone I cartilage contains much more extracellular lipid than the deeper zones. Although the quantity rises with age, there is no obvious association with the accumulation of calcium salts.

Ageing of the extracellular matrix

The extent of the delicate pericellular matrix increases with age. There are collagen fibres of increasing thickness. It is now thought likely that the quantity of the remaining extracellular matrix also increases, contributing to the decline in cell density (see p. 826). Significant changes in the quality of the extracellular matrix are also commonplace with advancing age.[106] Some are attributable to changes in the content of small molecules, free ions and/or water; many are the result of alterations in the amounts, size and structure of proteoglycan, or of changes in the cross-linking, organization and disposition of fibrous type II collagen. In the final analysis, all these changes are the direct or indirect consequence of age-related changes in the metabolic processes of articular chondrocytes.

Proteoglycans

The physical attributes of hyaline articular cartilage, and, presumably, of the many other connective tissues rich in proteoglycan, are determined critically by their water content (p. 827). The elasticity of ageing articular cartilage, for example, is impaired. This change strongly suggests that, with age, the collagen meshwork that normally regulates the extent to which the proteoglycan gel can expand, becomes less effective (see p. 824). In addition, many age-related changes in the proteoglycans themselves have now been shown (Fig 21.3). The total quantity of proteoglycan in ageing human cartilage changes very little.[12] However, there are important alterations in synthesis, largely affecting glycosaminoglycans and confined to the periods of growth and maturation. The chondroitin sulphates diminish in relative amount and in the deep zones (III, IVA, IVb) are replaced to an increasing degree by keratan sulphate. The size of core proteins diminishes. Fragments accumulate that are isolated hyaluronate-binding regions. There is ultimately a depletion of the large glycosaminoglycan-rich

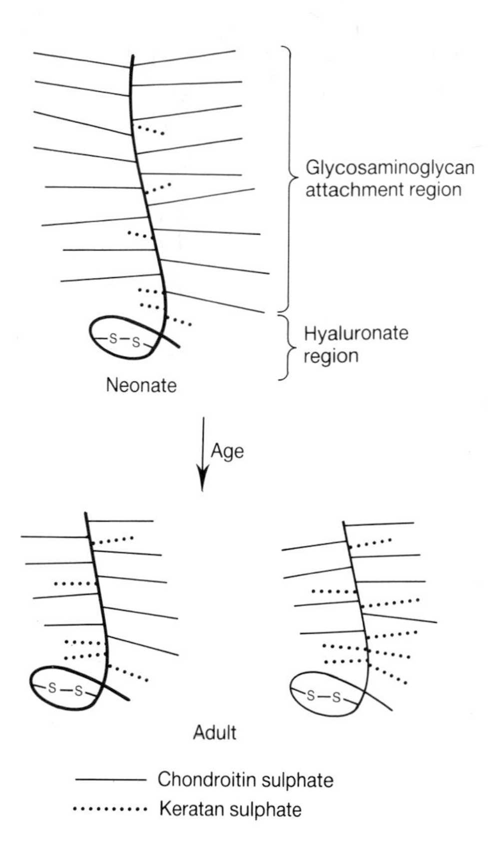

Fig. 21.3 Age changes in cartilage proteoglycan subunits.
Central protein core bears chondroitin sulphate and keratan sulphate chains. The following changes occur with age: size of subunits diminishes; number and size of chondroitin sulphate chains decrease; number and size of keratan sulphate chains increase; size of core protein decreases; and varieties of core protein evolve with globular S–S bridged hyaluronate-binding regions. Redrawn from Roughley and Mort courtesy of the authors and the Editor, *Clinical Science*.[93]

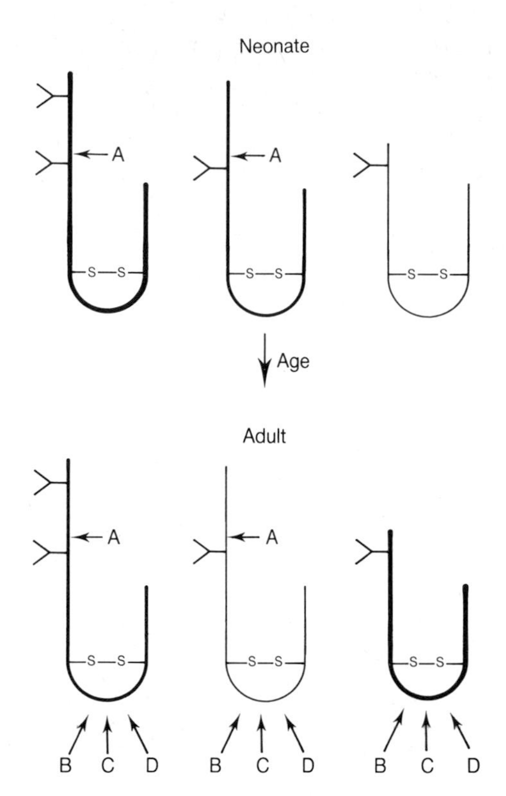

Fig. 21.4 Age changes in cartilage link proteins.
Degrees of glycosylation near N-terminus of protein backbones vary. There is extensive intramolecular S-S bridging. Three link proteins are present both in the adult and neonate. Larger link proteins differ in degree and/or type of glycosylation. Smallest are derived from larger by proteolytic cleaving near N-terminus (A). Relative abundance of various components changes as age advances: changes are indicated by varied thickness of lines. Redrawn from Roughley and Mort courtesy of the authors and the Editor, *Clinical Science*.[93]

proteoglycans, essential for normal cartilage behaviour, a progressive sequence that is likely to contribute to the destructive changes of osteoarthrosis (see Chapter 22).

The structure of proteoglycan subunits alters[93] (Fig. 21.3). There is a decrease in the number and size of the chondroitin sulphate chains and an increase in 6-relative to 4-sulphation. An increase in the number and size of keratan sulphate chains is recognized together with a decrease in the number of 0-linked oligosaccharide chains per core protein, the amino acid composition of which alters. The hyaluronate-binding region is unaltered. The net result is the predominance of smaller, less glycosylated proteoglycan subunits. Roughley and Mort[93] claim three possible explanations for these changes: 1. variation in core protein gene expression; 2. varied activity of post-translational enzymes; or 3. modification of proteoglycan by proteolysis after its secretion. It is likely that all three mechanisms are active.

Studies of *sheep* nasal material, suggest that the increase in keratan sulphate proteoglycan is due, not to the number of keratan sulphate chains but to an increase in the length of individual chains.[102] It has long been known that there are alterations in the length of *human* glycosaminoglycan side-chains and their composition.[27] Simultaneously, there is a decline in chondroitin sulphate, with a decrease in chain number and length, and an increase in keratan sulphate, with an increase in chain number and length. The result is that proteoglycan subunits become smaller with age although they retain their aggregatability.

In very early *bovine* life, there is a prompt increase in cartilage proteoglycan content relative to collagen.[49] By two years of age, there are increasing proportions of a distinct, small keratan sulphate proteoglycan, lesser proportions of a large chondroitin sulphate proteoglycan. In subsequent ageing, the proportion of subpopulations of proteoglycans and all molecular parameters, remain constant. However, the articular cartilage proteoglycans from large *human* joints differ in composition. Among those of low buoyant density, small proteoglycans, not derived from large monomers, are the major component in young (4, 11 years) cartilage. In old (70, 73 years) cartilage, fragments of large monomers containing keratan sulphate and hyaluronate-binding region are the main component.[97]

Proteoglycan aggregates The structure of the aggregating proteoglycans of *human* articular cartilage changes throughout fetal, postnatal and adolescent life[92] (Fig. 21.5). Synthetic and degradative processes both alter. The aggregating proteoglycan subunits are most abundant at birth. Their quantity then decreases, reaching a plateau at the onset of skeletal maturation. In postnatal human articular cartilage, there are at least two populations of aggregating proteoglycans. A small keratan sulphate-rich, chondroitin sulphate-deficient population appears from age four years and the large, single population of polydisperse chondroitin sulphate and keratan sulphate proteoglycans decreases from age one year.[112] When young *calves* and older *steers* are compared, the aggregated and non-aggregated monomers from calf cartilage are found to be longer and less variable in length than those from the older animal.[15] Steer proteoglycan aggregates are shorter and have fewer monomers. Studies of the age-related decrease in *human* proteoglycan aggregate size demonstrate that an increase in cartilage hyaluronate content is accompanied by a decrease in the size of the hyaluronate chains.[45] Since there is no age related change in the size of newly synthesized hyaluronate, modification of the hyaluronate chain may take place in the extracellular matrix.

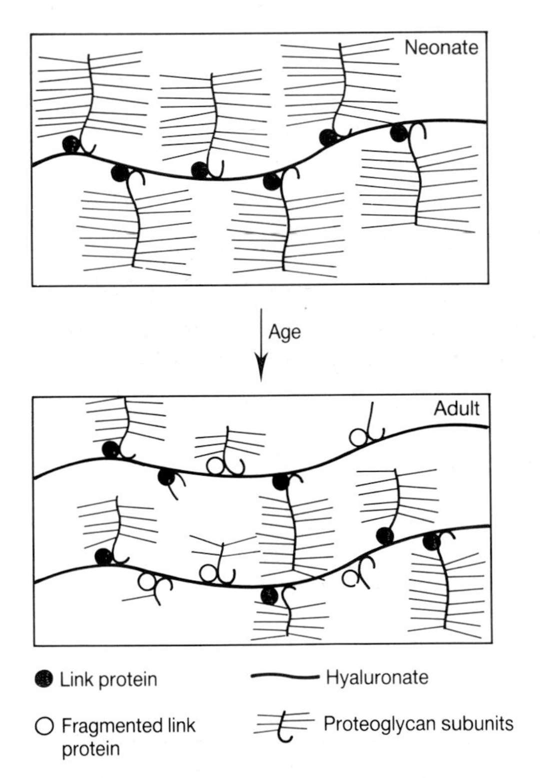

Fig. 21.5 Age changes in cartilage extracellular matrix organization.
Proteoglycan aggregates are shown. They consist of central hyaluronate filaments to which are attached proteoglycan subunits associated with link protein. Note large proteoglycan subunits and intact link proteins in neonate, smaller subunits and fragmented link proteins in adult. Proteolytic cleavage in glycosaminoglycan attachment regions, attributable to proteolytic cleavage of proteoglycan, creates molecules that range in size from intact subunits to hyaluronate-binding regions. Higher concentration of hyaluronate in adult is reflected in the increased closeness of the aggregates. Redrawn from Roughley and Mort courtesy of the authors and the Editor, *Clinical Science*.[93]

Link proteins (Fig. 21.4) Alterations in link proteins with age have also been detected.[82] The smaller of the three main link proteins identified in *human* articular cartilage may be derived by proteolysis. During ageing, further proteolysis is apparent.[81] With advancing age, multiple components are found, additional to the 48, 44, and 41 kD molecular weight proteins present at all ages. The age changes are attributable both to altered synthesis and increased degradation. In the articular cartilage of *rabbits* there is also a decrease with age in the capacity of chondrocytes to synthesize link proteins. This progressive change accompanies a decline in the capacity of chondrocytes to proliferate in culture.[89]

Proteoglycan degradation and altered antigenicity Changes in proteoglycan degradation (see p. 202) occur throughout life; they are attributable to altered chondrocyte behaviour and advance with ageing. The principal alterations are due to proteinase activity (see below). Another cause of the age-related changes in *human* cartilage proteoglycans can be damage mediated by hydrogen peroxide.[91] One effect of these changes may be an alteration of antigenic structure. In *rabbits*, adult proteoglycans express all fetal antigenic determinants and have others not identified on fetal proteoglycans.[21] These age-related changes in antigenic structure may underlie the emergence of an antiproteoglycan autoimmune state.

Collagens

As age advances, there are characteristic changes in the size and organization of articular cartilage collagen fibres.[107] There is an overall increase in fibre diameter[67] attributable to the aggregation together of individual fibres. One reason for this sequence may be a rise in the numbers of insoluble cross-links; another, a reduction in proteoglycan monomer size. These age-associated changes culminate in the appearance of large amianthoid (asbestos-like) fibres (Fig.

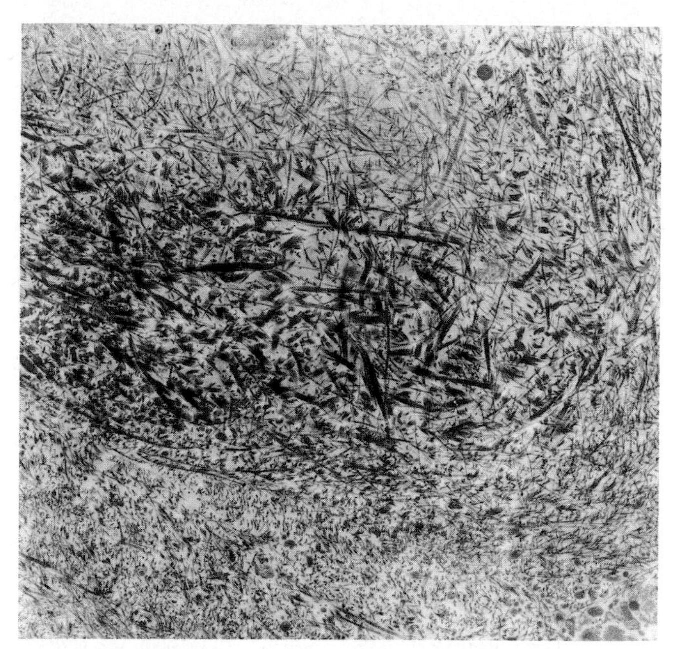

Fig. 21.6 Ageing articular cartilage.
Condensation of abnormally large ('amianthoid') collagen fibres in cartilage matrix. Focus constitutes a microscar. Reasons for microscar formation are not clear. (TEM × 6600.)

21.6) in which individual collagen fibres are arranged in parallel. The formation of these fibres with advancing age is a result both of collagen fibre remodelling and of alterations in proteoglycan size and composition.

Collagen types A slight change in the proportions of collagen types in senescence has been suggested. Type II collagen increases from 82 to 97 per cent with advancing age[83] but type 11 (1α, 2α, 3α) collagen decreases. High molecular weight collagen disappears with ageing, although it reappears in repair fibrocartilage. In osteoarthrosis, the 'minor' collagens increase in degenerate tissue, in new osteophyte-surface fibrocartilage (type X) and in reparative cartilage.

Collagen cross-links[60] The mature hydroxypyridinium cross-linking amino acids of collagen, 5–10 times greater in cartilage than in bone, reach a maximum by 10–15 years of age, staying in essentially the same range throughout life.[29] The ratio of the two variants of the mature cross-link, hydroxylysylpyridinoline and lysylpyridinoline, is always greater than 10:1 in cartilage compared to bone (3.5:1) and does not vary with age in adult life. The borohydride-reducible ketoamine cross-links change with age in a reciprocal manner so that the ratio of hydroxypyridinium, to borohydride reducible cross-links is age-related. The borohydride-reducible cross-links disappear from cartilage by the age of 10–15 years.

Protease activity

Explanations for the changes in proteoglycan content, monomer size and aggregatability that have been identified in advancing age in human cartilage, include the existence of elevated protease activity and the presence of diminished levels of protease inhibitors. Thus, a comparison of the non-weight-bearing with the weight-bearing parts of tibial condylar cartilages reveals a rise in neutral metalloenzyme activity levels in older (> 50 year) cartilage, an elevation that increases as surface roughening advances.[68] Interestingly, neutral metalloenzyme proteoglycan-degrading activity is higher in non-weight-bearing than in weight-bearing areas but the overall proteoglycan-degrading enzyme activity is higher in the superficial than in the deep zones of fibrillated tissue. When this change is overt, the large molecular weight proteinase inhibitor, α_1-trypsin inhibitor, is present in increased amounts, the low molecular weight collagenase inhibitor (Sephadex G75 included fractions) in reduced amounts.[38] However, the picture is not entirely clear and paradoxically the low molecular weight collagenase inhibitor fractions appear slightly elevated in intact but aged cartilage.

Metabolism

Evidence has slowly accumulated that adult cartilage uses less oxygen and has a higher glycolytic activity than immature cartilage. Whether such changes continue throughout life, representing an aspect of ageing, is less certain. Care is necessary in taking into account the contribution of diminishing cellularity to alterations in signs of integrated cartilage metabolism. Although there is a differential decline with ageing in the activity of different groups of respiratory enzymes, a reduction in oxygen uptake and an increase in glycolysis, there is no substantial evidence of an age-related decrease in sulphate incorporation or protein synthesis. The alterations in proteoglycan size and composition may indeed indicate the selective heightening of some aspects of the metabolism of individual chondrocytes. An increase in alkaline phosphatase activity accompanies these alterations in the cartilage of *rabbits*.[84]

Calcification

With advancing age, dystrophic calcification occurs at random in human non-articular cartilages such as those of the tracheal, laryngeal, bronchial and costal cartilages. The deposits are predominantly calcium hydroxyapatite but calcium pyrophosphosphate dihydrate and basic calcium phosphates (see Chapter 10) are frequently found in aged fibrocartilagenous tissue and in intervertebral discs. Occasionally, the mineralization is metastatic. Calcification is commonly succeeded by ossification in the non-articular cartilages of the larynx, trachea and ribs. Some light has been shed on these changes by experimental studies with small mammals.

The concentration of elements such as sodium, phosphorus, sulphur, chlorine and calcium in the nuclei and cytoplasm of young and old *rat* tibial articular chondrocytes differs little when these elements are assessed semiquantitatively by electron-probe X-ray microanalysis.[78] However, the concentrations in an intracellular location and in the extracellular matrix differ significantly and change with age. Thus, the extracellular matrix of older animals contains more calcium and sulphur but less potassium and chlorine than that of younger animals. There is also a measurable difference in sodium, potassium and chlorine between the extracellular matrix of tibial articular as opposed to xiphisternal cartilage.

In ICR mice, 'spontaneous' osteoarthrosis is common in old age. Matrix vesicles adjoining chondrocytes in the mandibular condylar cartilage of these animals react positively for acid phosphatase but not for alkaline phosphatase and are devoid of calcium complexes.[62] It remains uncertain therefore whether the vesicles are involved in age-related as opposed to osteoarthrosis-provoking mineralization. The lack of calcium is confirmed by electron energy loss spectroscopy, an observation contrasting with the presence of calcium in matrix vesicles in mineralizing zones at all ages. In accordance with the observations of Ali[2–5] the matrix vesicles may characterize a degenerative process which is, however, not necessarily associated with calcification.[63]

Amyloids

At least four forms of amyloid (see Chapter 11) may be identified in ageing human cartilage. Much the most common is an amyloid which displays the characteristic green dichroism and birefringence of all amyloids but which is not amyloid A: it is not oxidized by potassium permanganate and does not interact with antiamyloid A antisera. The frequency of this age-related amyloid, which may be a form of amyloid S, is high,[56] and it can be identified in capsular and ligamentous tissues as well as in cartilage. In the sacroiliac joints, it can be detected in almost all aged joints.[34] Very much less often, the ageing articular cartilages come to include amyloid A, amyloid L or amyloid H. These glycoproteins are recognized in very occasional cases of rheumatoid arthritis, in multiple myeloma and in chronic haemodialysis patients respectively (p. 437).

Changes with age in the organization of connective tissues

Cellularity

The reader is referred to Chapters 1 and 2.

For the reasons given above, discussion centres on cartilagenous tissues. The cellularity of cartilage, high in fetal and in neonatal tissue, declines with time.[100] However, the changes differ according to whether the tissue is articular or non-articular, whether it is hyaline, fibrous or elastic. Moreover, altered cellularity is not constant within one organ and differs in different parts of a single cartilage. There is a reduction in the cell density of non-articular, hyaline, elastic and fibrocartilages, and in human costal cartilage the diminution in cellularity after age 30 may be as much as 25 per cent.[99] The evidence relating to hyaline cartilage is incomplete and sometimes conflicting.

Cellularity of hyaline articular cartilage

Particular care is necessary in interpreting and comparing results obtained by different authors because of the increasingly wide range of techniques used in establishing cell numbers which have sometimes been expressed per unit area, sometimes per unit volume and sometimes as cell numbers per unit area of overlying cartilage surface. Much of the older, classical work was conducted on dehydrated tissue, prepared as stained, paraffin-embedded sections.[99] The established principles of morphometry were not always observed and sampling errors were encountered. In other instances, the problems of shrinkage and dehydration were reduced by the use of rapidly frozen human cartilage blocks, prepared as cryostat sections.[89a, 89b] The artefacts introduced by failure to protect matrix macromolecules are now well-known (p. 23) and can be taken into account. Nevertheless, the additional possibility of making cell counts on fully hydrated, fixed or unfixed human cartilage slices, using the techniques of scanning optical microscopy, is a reminder that even the results of counts made after low temperature preparative procedures, may have to be re-examined.

When older results are considered, it appears that there is no overall decline in the cell density of humeral head and femoral condylar cartilage between the ages of 25 and 85 years if the entire thickness of the tissue is assessed. When a single zone is considered, however, there may be significant changes so that in the superficial zones (zone I and II) of the femoral condyles, there is a reduction in cell density which accompanies an increase in cell density of the deeper zones. This pattern is selective anatomically: it is not recognized in the cartilage of the femoral head where a decrease in cell density in the superficial zones is sufficient to account for an age-related fall in the cellularity of the whole tissue.[108] Femoral head cartilage cellularity decreases by up to 35 per cent between the fourth and ninth decades of life.[107a,108] Stockwell[100] believes that these various results can be reconciled by taking into account the demonstration that the cellularity of zone I femoral head and femoral condylar cartilages (but not apparently the humeral head) declines differentially. The decrease in zone I femoral head cartilage is enough to account for the diminished cellularity of the whole cartilage whereas, in the femoral condyle, there is a slight increase in cellularity of the deeper zones, balancing the reduction observed in zone I.

In the case of the human femoral condyle, investigations of *post mortem* specimens collected from 77 subjects aged 72 ± 15 years, confirmed a decreasing cell density, expressed as cells/mm^2, as a function of age.[89a,89b] The counts were made on 10 μm cryostat sections. Diminishing densities, approximating overall to half the total initial figures, were recognized in each of the four main cartilage zones. Since the initial cell densities were greatest in zone I cartilage, the greatest absolute reduction in cell numbers occurred in this zone. With the methods used by these authors, techniques open to criticism because of the established likelihood of ice-crystal formation and subsequent cell/matrix disruption on thawing, the observation that there were many empty 'lacunae' (see p. 23) in the preparations is not wholly surprising. The mean diameter of these 'lacunae' did not change but Quintero and his colleagues showed[89a] that the number of empty lacunae increased with age. The validity of this result was supported by the finding that the ageing cells displayed decreased alcianophilia and a reduced uptake of $^{35}SO_4$.

The cellularity of the apical and posterior (non-weight bearing) regions of the femoral condyles has been compared and related to age.[78a,79,89a,89b] Fibrillated cartilage has a lower cell density than non-fibrillated. However, normal-appearing cartilage from joints with osteoarthrosis displays a lower cell density than cartilage from intact joints. With increasing age, there is a raised number of cell clusters (clones) (clones/mm^2) but a decreased number of cells/clone. Fissured apical femoral condylar load-bearing cartilage has a slightly lower density of cell clusters, with a lower density of clonal cells and of the number of chondrocytes/cluster, than fissured posterior condylar cartilage.

It is not yet known whether the cellularity of the much thinner cartilage of the large, congruent joints, such as the ankle, or of the small condylar joints, such as those of the fingers and toes, declines with age. There is a reciprocal relationship between cell density and cartilage thickness which is evident when analogous joints of large and small mammals are compared.[96,96a] This is a scale effect and it is therefore likely that substantial changes in cell density will be recognized when the hyaline cartilages of human distal interphalangeal and metacarpophalangeal joints, for example, are measured and compared with the hip and knee. These cell changes are reflected in altered integrated metabolic activity.[86]

The proposal[100] that overt fibrillation can entirely explain the reduced cell density of ageing human zone I articular cartilage is difficult to accept since age-related (as opposed to preosteoarthrosic) fibrillation is of a highly selective distribution (see Chapter 22). Greater attention to changes in the differential amounts of the cellular- and extracellular matrix components is required. Investigations in this direction may reconcile the growing morphometric evidence of age-related increases in cartilage thickness and the accumulating biochemical evidence of age-related alteration in proteoglycan size (see p. 822) and collagen-cross-linking (see p. 824). Chondrocyte loss by conventional mechanisms of necrosis, with pyknosis and karyorrhexis, can be recognized occasionally by light- and transmission electron microscopy. However, cell death by apoptosis may complicate the interpretation of the decline in cell numbers although few studies have been made to assess its importance.

There is an analogous reduction in the numbers of synovial intimal cells.[50] This change includes a diminished proportion of type B cells, a relative increase in type A cells. Atrophic cells become apparent, with few organelles. There is reduced synovial vascularity and an increased fibrous tissue component.

Water content

With advancing age, the water content (p. 75) of mature connective tissue is thought to change. In whole thickness femoral head cartilage, for example, the water content has been found to decline from approximately 76 per cent at age 15 years to approximately 69 per cent at age 70 years[106] with a greater decrease in the deeper zones than in the superficial. An increase of fixed charge density is observed in parallel with the reduction in water content. However, it is not established that these results can be extrapolated to other joints, and the identification of entirely normal material in the aged is fraught with practical difficulties. Moreover, the drying of stored material is slowed but not abolished at −20°C, the temperature often used to store cartilage before analysis.

Thickness

Hyaline cartilage

The older proposal that cartilage becomes thinner with age[10] is likely to have stemmed from studies of fibrillated, osteoarthrosic specimens in which bone exposure was the result of a breaking away of cartilage. More recent evidence is against the view that ageing alone leads to thinning. The mean thickness of ageing human humeral cartilage[70] and of patellar cartilage[96], does not decrease with time. Nevertheless, there have been few systematic measurements of the thickness of hyaline cartilage in even the most important synovial joints in man; and fewer still that have taken proper account of the asymmetry of the major articulating surfaces. When the human hip joint was investigated, measurements of magnified lateromedial radiographs revealed that the thickest part of the femoral head cartilage lay 20° anterior to the zenith; the entire cartilage increased in thickness with age between the ages of 20 and 50 years, but the greatest increase was of the thickest part.[8]

Calcified cartilage

The proportion of viable cells declines with time. There are corresponding alterations in the role of matrix vesicles.[63] The thickness of the calcified zone (zone V) of articular cartilage at first remains constant although there are wide local variations about a mean of 134 μm.[40] However, as senility advances there is a decrease in the thickness of this zone.[59] The change in thickness is influenced by endochondral ossification which leads to new subchondral bone formation and a thinning of the zone V cartilage. There is a simultaneous incorporation of the non-calcified articular cartilage into the calcified zone by advancement of the calcification front. The tidemark is an index of this front; it becomes duplicated. Theoretically, this age-related change in the tidemark should preface an increase in the thickness of the zone V calcified cartilage. That it does not, and that thinning occurs, indicates that increased endochondral ossification at the deeper aspect of the zone of calcification is relatively greater than advancing calcification at the superficial margin. It is likely that this active process may influence the altered geometry of the articular surfaces in advancing age.[16,30] The mechanical properties of synovial joint cartilage are much influenced by the junctions between the deepest, calcified zone V and the bone end-plate, and between the calcified and non-calcified cartilage; these interfaces are sites for the propagation of microfractures in osteoarthrosis (p. 859).

Surface structure

The appearances of mammalian cartilage surfaces[31] have been investigated by very many gross and microscopic techniques (Figs 21.7–21.10). They are summarized in Chapter 3. Anatomical observations[35] serve to show that, in general, the surfaces of mature, adult human hip and knee joints deteriorate with time. The relationship of these changes to osteoarthrosis is discussed in Chapter 22. Other joints have been studied in detail, particularly in small mammals.[95]

En face microscopic enquiries confirm that the third order features seen on all exposed, moist, non-loaded cartilage bearing surfaces from disarticulated human and other mammalian joints are the superficial representations of underlying, zone I chondrocytes (see Chapter 3). The number of these features declines linearly with age but their diameter and height/depth increase.[64] These trends parallel changes in the molecular composition of the extracellular matrix and precede a series of alterations in colour, roughness and matrix organization that is readily demonstrable by techniques such as scanning electron microscopy.[36] As they progress, a pattern of enhanced,

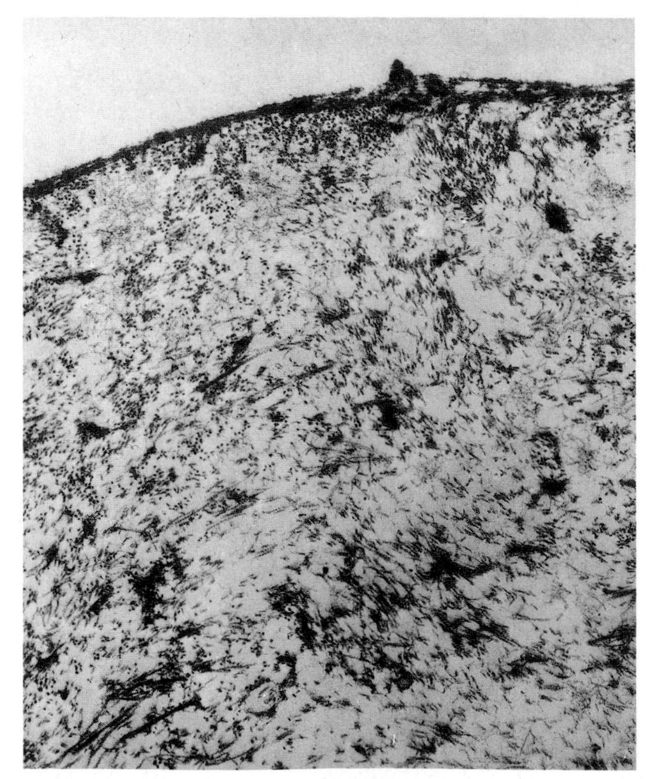

Fig. 21.7 Ageing and disorganization of articular cartilage.
Bearing surface and zone I matrix of hyaline articular cartilage from 64-year-old female. Collagen fibres are disordered and widely separated. (TEM × 11 000.)

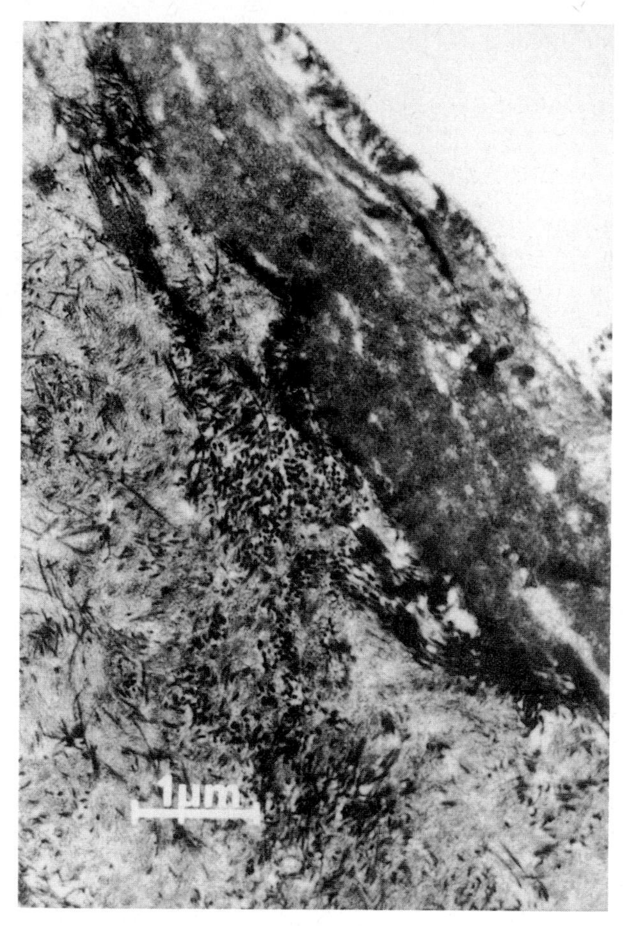

Fig. 21.8 Ageing and disorganization of cartilage surface.
Electron-dense material forms incomplete lamina. There is close resemblance to deposits found characteristically on bearing surfaces in osteoarthrosis (see Chapter 22). Nature of deposit, which should be distinguished from normal ultrastructural lamina (see Chapter 2), remains uncertain: it is neither fibrin nor cell debris nor amyloid. The lamina may, however, be lipoprotein or glycoprotein. (TEM × 11 000.)

selective roughening appears. The appearances are termed 'fibrillation'. The histological characteristics of fibrillation are described on p. 858. When this structural change is recognized only microscopically, it may be said to be 'minimal' (covert). When it can be seen naked eye, it is 'overt'. The early naked eye recognition of these areas is helped by first gently painting the fresh surface with a suspension in saline of Indian ink.[31,71] These authors carefully define descriptive terms of use in analysing cartilage surface age changes[51] (Table 21.1) and applicable in the investigation of osteoarthrosis (see Chapter 22).

The changes of cartilage surfaces with age can usefully be defined at four levels of resolution: anatomical, light microscopic, ultrastructural and molecular.

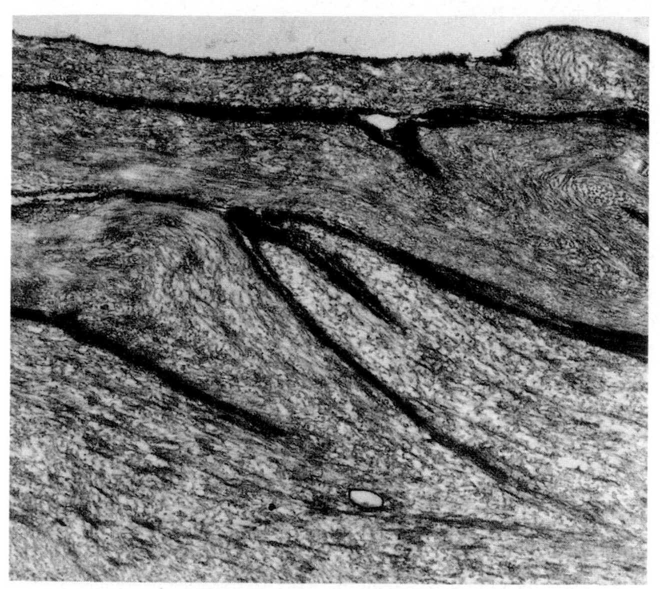

Fig. 21.9 Ageing and ultrastructural disorganization of articular cartilage.
Femoral condyle from 77-year-old female. Fine splits have formed at the bearing surface: they are delineated by electron-dense margins and are mainly tangential to the surface. The formation of these microscopic clefts constitutes a threat to the integrity of the whole cartilage mass since, from them, cracks are likely to be propagated through the entire cartilage thickness. (TEM × 6700.)

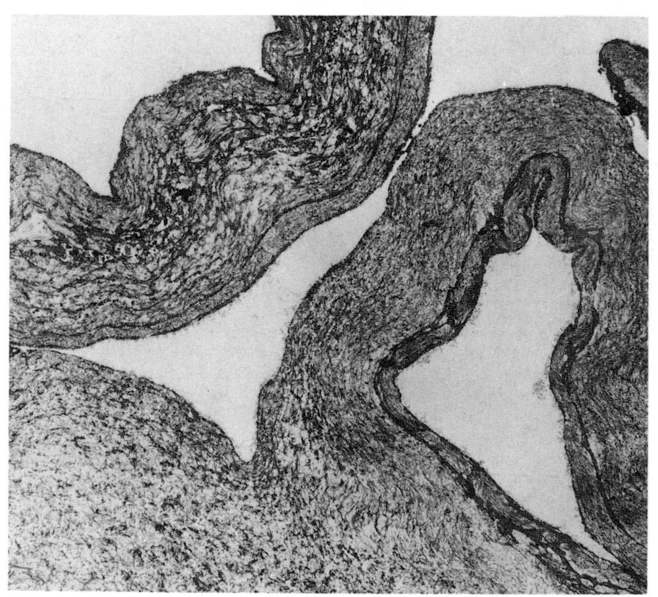

Fig. 21.10 Ageing and ultrastructural disorganization of articular cartilage.
Femoral condyle of 64-year-old male. As age-related disruption of cartilage surfaces continues, splits and fissures lead to detachment of collagen fibre bundles, the ends of which become free. Collagen fibre bundles undergo crimping. At a higher level of resolution, transmission electron microscopy demonstrates presence of lipid lamellae which assume a whorled convoluted structure similar to that seen in some forms of experimental osteoarthrosis. (TEM × 1048.)

Table 21.1

Types of articular cartilage lesions seen in joints of a Liverpool population[31]

Non-destructive or potentially destructive	Cartilage not thinned Regressive changes Minimal fibrillation (various *en face* patterns) 'Ravines' Overt fibrillation Tiny splits at uncalcified/calcified interface
Destructive	Cartilage thinned *Overt fibrillation with deep splitting ('splitting') *Horizontal splitting at uncalcified/calcified interface ('gouging') *Smoother-surfaced destructive thinning ('grinding') Localized rounded or elongated defect
	Full-thickness loss of uncalcified cartilage *Area of exposed calcified cartilage and/or bone
	Loss of bony height beneath exposed bone *Abrasive wear of exposed bone with necrosis of superficial osteocytes *Small segment of non-septic bone necrosis ?Bony collapse from microfractures in viable bone deep to surface *Subarticular bone 'cysts' large enough to suggest possibility of 'collapse'
Reparative	Re-covering of an exposed bone surface by new non-osseous layer (+ destructive changes in new non-osseous surface tissue)
Peripheral remodelling	Covering of cartilage surface by new non-osseous tissue ('pannus') Replacement of old cartilage by new non-osseous tissue Intracartilagenous ossification (can lead to 'expansile remodelling') Peripheral bony outgrowths ('osteophytic lipping')
Remodelling of cartilage base	Active ossification at chondro-osseous junction Duplication of tidemark (tide line) ?Thickening of calcified zone
Other lesions	Modulation of chondrocytes to 'fibroblasts', for example

*Indicates where progressive osteoarthrosic changes occur subsequently.

Anatomy The colour of cartilage alters with time. Young cartilage retains a delicate, blue-white translucency; old cartilage comes to be opaque and yellow-brown due to the accumulation of a 10 kD molecular weight acid glycoprotein.[105] There is no evidence that this pigment, unlike that of ochronosis (see p. 351), affects the mechanical qualities of the ageing tissue. Older cartilage appears less smooth than young. The extent of the surface roughening with age was examined in detail by Freeman and Meachim.[31] Using a dissecting microscope and Polaroid camera, Meachim recorded the surface topography and morphology of the patellofemoral joint cartilage,[28,80] the shoulder and hip joint cartilages[77] and of the lateral tibial plateau cartilage[71,78a] (Fig. 21.11). Meachim also investigated the ankle joint.[73] The extent of intact patellar cartilage and of areas of minimal fibrillation (Table 21.1), of overt fibrillation, of bone exposure and of osteophytosis were mapped and measured. Other features such as 'parallel line' patterns of minimal fibrillation, 'ravines', smooth-surfaced destructive thinning, peripheral fibrous covering and localized incomplete defects were variously recorded for individual joints (Table 21.1).

Analogous investigations were made by Byers and his colleagues and described in a series of unique but complex papers.[18,19,19a,19b,20] Byers[17] set out to test the hypothesis that *limited* degenerative changes of the hip joint are related to ageing and that, independently, there are *progressive* (degenerative) changes that can be regarded as a disease process (osteoarthrosis).

Byers[18] surveyed macroscopically 375 right femoral heads, 363 right acetabula and 154 left femoral heads obtained randomly at *post mortem* on London hospital patients. The results centre on a series of integrated schematic diagrams less sophisticated (but also less comprehensible) than the mercatorial projections used to good effect by Meachim[77] (Figs 21.12 and 21.13). They provide evidence that there are indeed two statistically independent classes of degenerative cartilage change on the femoral head. Analogous changes were present on the acetabulum where, however, they could not be shown to be independent variables.

Limited progression In the case of the alterations 'of limited progression', the argument[18] rested on the recognition of two groups of surface change that, at whatever age, had seldom advanced anatomically to full thickness cartilage loss and bone exposure. In the first group, the alterations affected an area lying below the fovea; a crescentic area starting above the fovea and extending posteriorly; a roughly triangular area anterior to the fovea; and a narrow zone, adjacent to the femoral head margin, extending over the superior capital surface from approximately the mid-

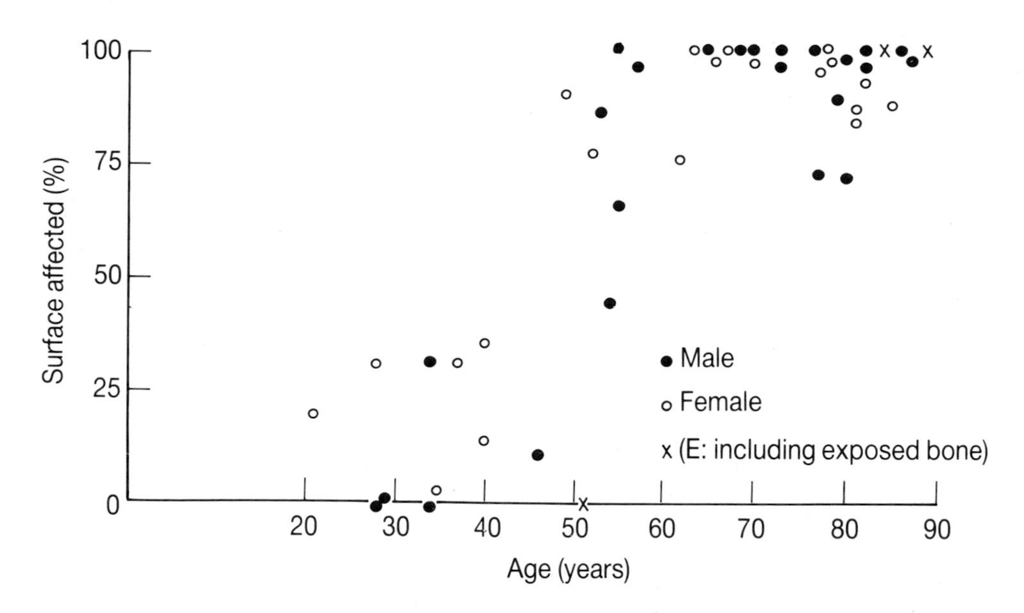

Fig. 21.11 Overt fibrillation of tibial plateau.
Proportion of bare, non-meniscus-covered area of left lateral human tibial plateau affected by overt fibrillation of hyaline articular cartilage. Redrawn from Meachim by permission of Dr George Meachim and the Editor, *Journal of Anatomy*.[74]

point posteriorly to the mid-point anteriorly, divided into two equal regions (Fig. 21.14).

The five *macroscopic* lesions identified in the areas comprising the first group were (a) a fine granularity; (b) a superficial fraying, fissuring or flaking; (c) cartilage loss; (d) ossification at the site of cartilage loss; and (e) bone exposure at this site. The lesions of the second group were parafoveal, that is, contiguous with but usually either in front of or behind, or above or below the fovea (Fig. 21.14). They comprised clearly demarcated, depressed areas of cartilage with a smooth, soft base beneath which ossification was often present.

According to their individual extent, the lesions constituting the two groups were each placed in one of five 'stages of development', i. e. of extent.

Progression The term progression was chosen by Byers[18] to delineate an uncommon class of femoral head disorganization that was distinguishable anatomically from the changes of 'limited progression' and that was of an advanced character. With one exception, 'progressive' changes were antero-superior and outside, i. e. beyond, the sites affected by the so-called lesions 'of limited progression'. In advanced cases, there were additional, inferior changes. The 'earliest' (i.e. least) disorder was the presence of a finely granular surface in a clearly demarcated circular area. The disorder appeared superficial but persisted (it was proposed) until cartilage loss occurred. Coarse fibrillation was not seen. 'Later', i. e. in other cases, the classical lesions of established osteoarthrosis (Chapter 22) were identified. It is important, in considering the significance of Byers'[20, 115] views, to note that no cases of an intermediate character were recognized.

These controversial observations were extended histologically.[19] The gross lesions of 'limited progression' were shown microscopically to be: fraying; fibrillation, splitting; the formation of a fibrous surface layer; cellular resorption; and ossification. By contrast, 'progressive' lesions were identified solely as fibrillation (p. 858). Subsequently, the significance of the interrelationship between the different alterations in the hip joint was examined mathematically.[19a] The independent character (in arithmetic terms) of 'limited' and 'progressive' alterations was confirmed. A strong statistical association between the occurrence of 'progressive' cartilage change and osteophytosis was shown and between (pseudo)cyst formation and osteophytosis.

Finally, the validity of the concept that two categories of hyaline articular lesion, 'progressive' and 'of limited progression' exist in Western populations, was tested by means of an investigation of the joints of the feet. Nine articular surfaces were examined in each of 54 amputated feet in a group of patients aged 10–79 years.[115] Approximately 60 per cent of joint surfaces had acquired lesions by the third decade; the lesions were usually peripheral. Advanced lesions were very uncommon. Bone exposure was identified on only three surfaces, each of the first metatarsophalangeal joints; these three instances were in

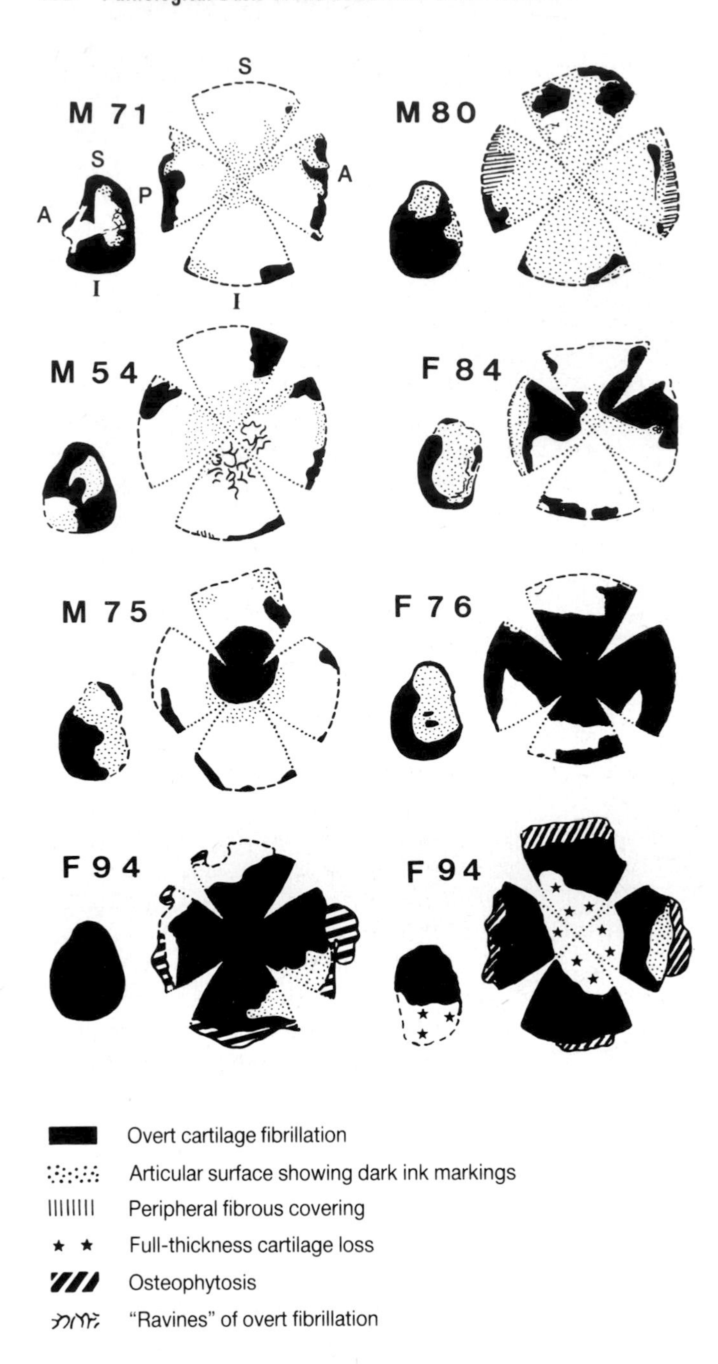

Fig. 21.12 Anatomical changes on the cartilage surfaces of the ageing human shoulder joint.
a. Representative selection of mercatorial projections of the shoulder joint from eight patients aged 71 to 94 years. In each instance, the glenoid is shown to the left of the (projected) humeral head. Anterior (A), posterior (P), superior (S) and inferior (I) aspects of both joint surfaces are indicated. The key shows the symbols used in depicting the extent and nature of the cartilage changes recorded in the opened shoulder joint. Reproduced by courtesy of Dr G Meachim and the Editor, *Journal of Anatomy*.[77]

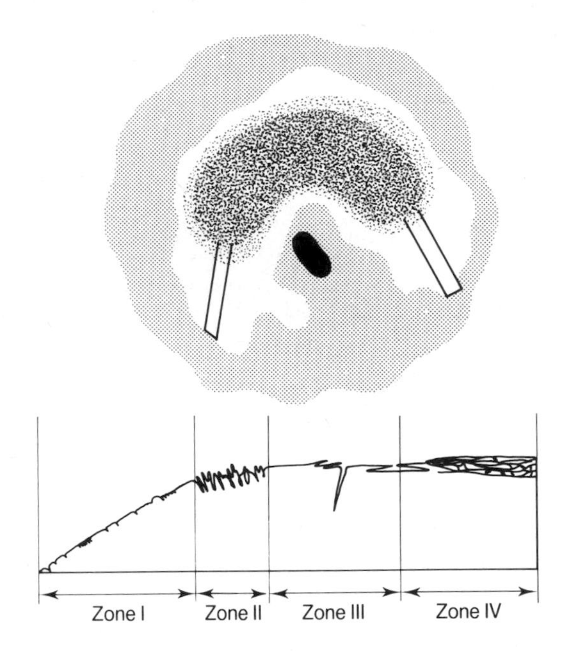

Fig. 21.13 Distribution of femoral head lesions. Rectangular areas demonstrate sites from which blocks were taken. Below, extent of lesions of cartilage zones indicated as height and shape of graph lines. Modified from Byers courtesy of Dr Paul Byers and the Editor, *Seminars in Arthritis and Rheumatism*.[20]

patients aged 50–59 years. It has been established that patients with femoral neck fracture and Asian individuals, two populations believed to be relatively immune to osteoarthrosis (p. 881), display hip joint cartilage changes of limited progression but very rarely progressive lesions. These observations, together with the results of the foot study, were held once more to confirm the independent nature and significance of the two classes of disorder. In spite of this reaffirmation, Byers'[18,19,19b] views have not escaped criticism.

Results at variance with those obtained by Byers for the hip joint,[20] have been reported for the patellar and trochlear (patellar surface of femur) aspects of the knee.[79] These authors examined the knee joints of 57 females and 63 males over the age of 50 years (Table 21.4) *post mortem*. The frequency and severity of the lesions increased with age. The femoral lesions were more common than the patellar but the latter were more common and more severe in women than men.

The weakness of Byers' case lies, of course, in the extrapolation from a (large) series of single selected anatomical end-points (the anatomical features recorded in each individual case), to a theory of 'progression'. There is, in fact, no firm scientific evidence in Byers' papers that any of the changes of 'limited progression' were not capable of

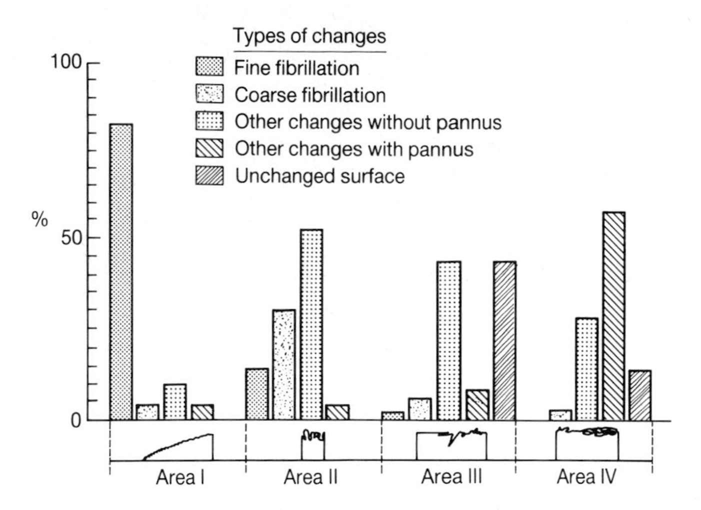

Fig. 21.14 Distribution of forms of femoral head lesions.
Differential distribution of cartilage changes. Note, for example, that area I changes are mostly 'fine fibrillation', area II changes mostly 'other changes without pannus', area III features (where there is much unchanged surface) often 'unchanged surface' without pannus, area IV changes very frequently 'other changes with pannus'. Modified from Byers *et al.* courtesy of Dr Paul Byers and the Editor, *Seminars in Arthritis and Rheumatism.*[20]

'progression' or that any of the 'progressive' lesions were not limited.

The strength of Byers'[18,19,19b] case is the great attention to detailed anatomical change; the demonstration of a relationship between the frequency of structural hip joint cartilage change and age; and the further support given to the widely held view that age-related hyaline articular cartilage disorganization is distinct from, but may coexist with, osteoarthrosis.

It is of interest to compare these observations with those of Meachim.[70,72–75,77]

Some of Meachim's results are summarized in Tables 21.2 and 21.3. Areas of fibrillation present in the second decade developed particularly at the joint periphery and on certain other characteristic sites such as the mid-zone of the inferomedial segment of the femoral head and the mid-zone of the acetabular surface. With advancing age, but varying individually, fibrillation spread tangentially and involved an increasing proportion of the total area of each surface. Bone exposure, by the vertical progression of these age-related changes, was uncommon in the left shoulder (Fig. 21.12) or hip, even in the very old, a finding that contrasts with the earlier recognition of this lesion at patellofemoral and radiohumeral articulations.

Age-related fibrillation of the patellofemoral joint surfaces was neither uniform topographically nor in the time sequence of development so that overt fibrillation was found to affect different sites in a characteristic order (Tables 21.2 and 21.3). Overt fibrillation, for example, was first seen to develop in the medial strip of the medial patellar facet and at the articular periphery. Overt fibrillation of the ankle was common and often recognizable in young adults. It tended to become more extensive with advancing age, initially and especially affecting the cartilage periphery and the boundaries between the central and malleolar territories. The centre was spared. By contrast, minimal ankle joint cartilage fibrillation occurred at any site, advancing in the centre with age. Full-thickness cartilage loss was rare. Superficial fraying and splitting of the cartilage of the left lateral tibial plateau were normal findings but, particularly in younger adults, were often accompanied by large areas of intact cartilage. The bare area and the meniscus-covered posterior segment were more susceptible to overt fibrilla-

Table 21.2

Typical order in which forms of fibrillation affect patellar sites according to degree of overt fibrillation on patellar surface as a whole[77]

Order affected			
1. 'Area a' medial facet	2. Patellar periphery	3. Region of transverse ridge	4. Lateral facet and 'area b' medial facet
Overt	Overt	Parallel linear ± intact	Parallel linear ± intact
Overt	Overt	Ravines and/or parallel linear	Parallel linear
Overt	Overt	Overt	Parallel linear ± overt
Overt	Overt	Overt	Overt
Overt	Overt ± bone exposure	Bone exposure	Bone exposure

Table 21.3

Typical order in which forms of fibrillation affect femoral sites according to degree of overt fibrillation on opposing femoral surface as whole[77]

Order affected				
1. 'Area c' medial condyle; curved ridge on lateral condyle; femoral groove periphery (excluding lateral margin)	2. Medial facet of groove	3. Mid-line region of groove	4. Medial part of lateral facet of groove	5. Lateral part of lateral facet of groove (including lateral margin)
Overt	Parallel linear or intact	Parallel linear or intact	Parallel linear or intact	Intact
Overt	Parallel linear ± overt	Overt	Parallel linear ± overt	Intact and/or parallel linear and/or irregular minimal
Overt	Overt	Overt	Overt	Intact and/or parallel linear and/or irregular minimal
Overt	Overt	Overt	Overt	Overt
Overt	Overt	Overt ± bone exposure	Overt ± bone exposure	Overt ± bone exposure

tion than the meniscus-covered lateral and anterior segments. There was a relationship between the proportion of the bare area affected by overt fibrillation and age.

Table 21.4

Lesions identified in 120 patients aged 50 years or more on the patellar and trochlear cartilages[79]

Stage I:	Non-extended fissures		
II:	Fissures extending over > 25 per cent of the articular surface		
III:	Fissures associated with small but deep cartilage ulcerations		
IV:	Deep, extended cartilage ulceration exposing subchondral bone.		
Patellar lesions: Mostly bilateral and symmetrical (93.3 per cent)	Stage I		26.2 per cent
	II		22.5 per cent
	III		27.1 per cent
	IV		17.5 per cent
Deep extended ulcerations: (stage IV)	Age	< 60 years	1.9 per cent
		60–70	8.0 per cent
		70–80	13.6 per cent
		> 80	38.9 per cent

Light microscopy[57,58] (Table 21.1) The light microscopic changes of fibrillation are described on p. 85. They do not differ qualitatively from the fibrillation that constitutes the earliest microscopic lesion of osteoarthrosis (see Chapter 22). It is convenient to adopt Freeman and Meachim's[31] classification of the microscopic lesions although preferable to regard them not as different 'types' of lesion[32] but as different qualitative and quantitative local expressions of age change. It is of interest to compare this classification with the categories used (see p. 858) to categorize the microscopic changes of osteoarthrosis.

Normal By incident light microscopy, a surface may appear normal after painting with Indian ink. Nevertheless, other analytical techniques reveal quantitative histological changes related to age.

Minimal fibrillation Indian ink may be retained superficially in reticular or dentate patterns so that, on perpendicular section, there is very superficial fraying or splitting of the surface zones. Tangential splits develop and microscopic flaps of detached cartilage are seen. Some may be visible to the naked eye. The patterns of splitting may be aligned parallel to directional surface movements or to the direction of the main superficial collagen fibre bundles.

Ravines Deeper splits that may be parallel with, or at right angles to, the superficial splits of minimal fibrillation.

Thinning Thinning is a change that may be detected by the naked eye and that is measurable.

Overt fibrillation When the retention of Indian ink occurs in confluent surfaces areas, there is a more pronounced disintegration of the articular surface, many flaps of loosened tissue and a random or orderly arrangement of this change. Overt fibrillation must be distinguished from the marginal encroachment of fibrous tissue onto or replacing articular hyaline cartilage.

Horizontal splitting[76] Splits often run tangentially within superficial cartilage layers. In addition, Meachim and Bentley[76] emphasized the frequency of the horizontal splits that result, at all ages, between the uncalcified and the calcified cartilage layers of, for example, the patella. It is suggested that they are the result of shearing damage sustained during life. These intracartilagenous splits contrast with the osteoarthrosic splitting away of the whole cartilage from the bone that follow osteonecrosis and bone infarction (see Chapter 22).

Ultrastructure The ultrastructural appearances of fibrillation[87] are described on p. 850 and exemplified by Figs 21.7 to 21.10.

Molecular changes The molecular changes in cartilage structure with age are described on p. 822. The reader is also referred to Chapter 4.

Vascularity

The larger limb arteries supplying the joints are prone to atherosclerosis and plaques are particularly likely to form at foci of repetitive mechanical stress and turbulence. Much less is known of changes in the small arteries and arterioles of the connective tissues. During early life, the marrow contacts made by subchondral vessels become fewer, a change which may impair chondrocyte metabolism. In advanced old age, it is proposed that thickening of the capillary basement membrane is attributable to an endothelial 'reaction response' consequent on modifications in blood flow through arterioles and venules. These changes may be caused by 'senile involution of the perivascular connective tissues',[24] particularly of the perimicrovascular proteoglycans.

Elastic cartilage

Much is known of the changes in the organization of elastic tissues with age[90] but understanding of the senescence of elastic cartilage is limited. Elastogenesis and the calcification of elastic tissue have been important topics for study.[55] The excess formation of elastic material in breast carcinoma and alterations with age in the elastic material of the dermis[104] have also been objects of intensive enquiry. In breast carcinoma[46] the tumour cells secrete a cytokine which promotes elastin synthesis by fibroblasts and by smooth muscle cells; it also stimulates cell proliferation in fibroblast culture. In the ageing skin, the altered mechanical properties are attributable to a disordered supermolecular organization of the elastic fibre network with a reduction in rubber-like reversible deformability. It is reasonable to postulate similar phenomena in elastic cartilage.

Bone

The total quantity of compact bone declines with age at the rate of approximately 3 per cent/decade from age 40, in both sexes.[69] After the menopause the rate of decrease rises to approximately 9 per cent/decade. Trabecular bone mass also diminishes. However, the scatter of results is great. An average loss of 6–8 per cent/decade beginning in the third and fourth decades is suggested, in both sexes. The implications for the age distribution of pathological fracture are clear. The growing frequency, among ageing Western populations, of fracture of the femoral neck is particularly serious and these fractures now account for a significant proportion of all hip joint arthroplasties. The tissues of a few of the elderly women treated in this way are found to display coexistent osteomalacia; some also have bone necrosis, commonly in associaton with osteoarthrosis. The relationship between osteoporosis and osteoarthrosis is further considered in Chapter 22.

Changes with age in the mechanical properties of cartilage

The mechanical properties of cartilage (Chapter 6) change with age.[52,111]

Compression

Mechanical tests of ageing cartilage have been made in compression and in tension. Tests in compression reveal that, under peak loads, deformation of the femoral head cartilage in whole loaded hip joints is not uniform; compliance increases directly with age in the period 28–85 years.[6] Creep compliance does not increase correspondingly. By contrast, tests of cartilage from the superior surface of the femoral head made with an indentor show no alteration of Young's modulus with age.[7] Testing of the whole joint enables the detection of fluid expressed within the first 30 s of loading, a change not detectable by indentation procedures. The demonstration that older cartilage is generally softer and more readily deformed than younger accords with measurements obtained by a confined compression creep test.[9] In these experiments, made with human patellar cartilage, the intrinsic equilibrium modulus and the permeability to fluid flow were highly correlated with the water content. As the water content rose, the matrix became softer and more permeable. In parenthesis, it should be noted that these authors found only a marginal decrease in the equilibrium modulus with increasing age and surface 'degeneration' and permeability was not significantly correlated with age or 'degeneration'. It remains difficult to reconcile these observations with the accumulating evidence of age-related alterations in proteoglycan monomer size and proteoglycan aggregatability (see p. 822).

Tension

In parallel with altered compliance the tensile stiffness, fracture strength and fatigue resistance all deteriorate.[52,53] The changes are differential so that the tensile strength of zone I cartilage declines with age after the third decade whereas the deterioration in tensile strength of deep zone (zones III and IV) cartilage is continuous, throughout life.[53] There are analogous changes in the tensile strength of tendon.[42] These characteristics are linearly related to collagen content and it appears possible that the diminished tensile strength of ageing cartilage may be a consequence of a change in collagen alignment; cross-linking, theoretically, should increase. There is a change in the water content of mid-zone cartilage and increased collagen extensibility is a prerequisite for the implied increase in proteoglycan domains.

On the basis of these various results it can be postulated that an increase in hyaline articular cartilage total water content, resulting from collagen change and involving proteoglycan expansion, leads to a vicious circle of increased compliance and diminished fatigue strength. At first sight, they are at variance with data suggesting that cartilage water content declines with age.[106] Ageing cartilage responds to the normal diurnal peak loads of, for example,

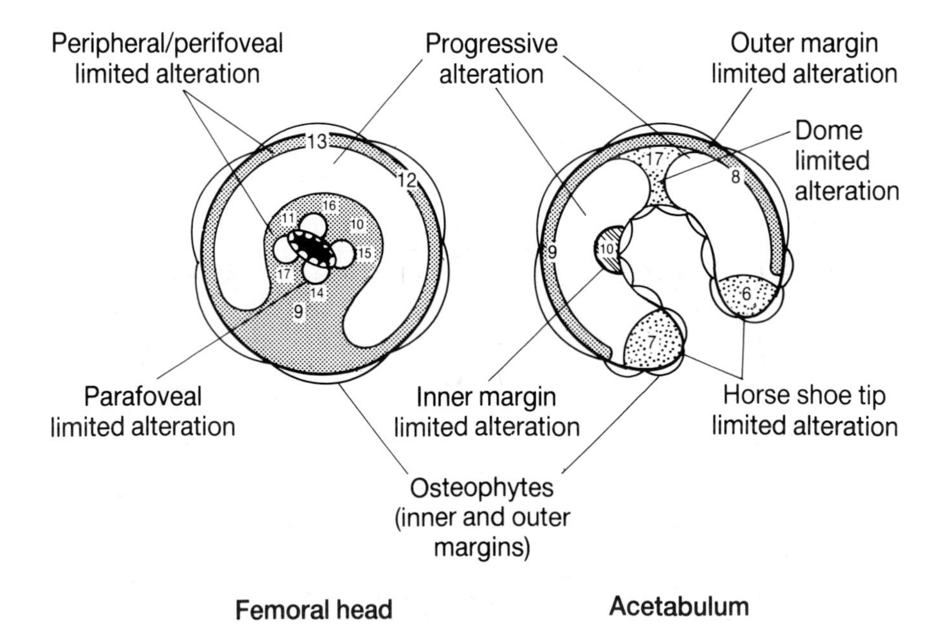

Fig. 21.15 Age-related and non-age-related lesions of hip joint.
On femoral head, note, for example, sites of 'limited alteration peripheral and parafoveal' lesions, age-related but not anatomically progressive, by comparison with semilunar distribution of sites of 'progressive alteration', not age-related but anatomically unrestricted and advancing to cartilage loss. Analogous changes are recognisable on the acetabular surfaces. Modified from Byers *et al.* courtesy of Dr Paul Byers and the Editor, *Seminars in Arthritis and Rheumatism.*[20]

walking, by increased deformation. Eventually, these normal forces could be expected to bring about mechanical failure and collagen fracture.[32] Accelerated breakdown, minimal and, later, overt fibrillation ensue. In this way, there may develop the age-related changes so common from the onset of adult life[18,70] (Figs 21.13–21.16).

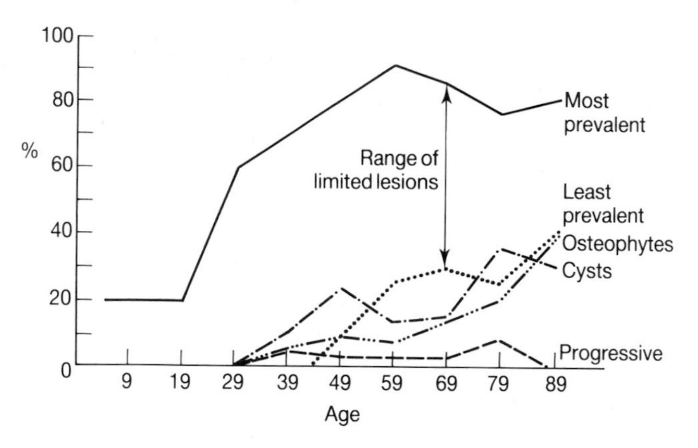

Fig. 21.16 Age prevalence of cartilage lesions of different extent.
Lesions of limited anatomical extent are very frequent, often from an early age. Those that are anatomically 'progressive', are relatively uncommon. Modified from Byers courtesy of Dr Paul Byers and the Editor, *Seminars in Arthritis and Rheumatism*.[20]

Changes with age in the structure, function and clinical behaviour of joints

Much less is known of the structure and function of whole, ageing joints than of their components. The available evidence suggests that the efficiency of movement and of load-bearing decline with advancing age.

Age and joint structure

The customary, normal age-changes in the manubriosternal and sacroiliac[34,98] joints have been summarized in Chapter 3. Other joints undergo comparable alterations. Thus, the articular cartilage of the acromial end of the clavicle is covered by hyaline cartilage from birth until approximately 16 years of age.[103] After age 17 years, the surface becomes fibrocartilagenous. A comparable change occurs in the structure of the cartilage of the acromion process of the scapula but it is effected approximately seven years later.

Age and joint mobility

By the age of 79 years, many elderly persons have restricted movement of knee (20 per cent), hip (66 per cent) and other joints.[101] Those who had been sedentary are more disabled than those whose occupation had been physically strenuous. To examine the influence of age on the joints, active function has been assessed by measuring shoulder joint flexibility. One thousand elderly persons were studied; all lived in their own homes. Overall, there was approximately 30° less mobility than in younger individuals.[11] Women retained less flexibility than men, with a diminution of abduction of approximately 10°/decade (Table 21.5). Multiple regression analysis revealed that the effects of age on shoulder abduction were attributable to declining health, strength and customary use. The contribution made by customary use was most marked in those with a recognized disability. By contrast with these measurements of active abduction, age and knee position did not appear to affect the passive stiffness of the ankle joints in normal women.[22] Analogous studies have been made in rhesus monkeys (*Macca mulatta*).[25] The results demonstrated an age-dependent loss of passive mobility.

Table 21.5

Reduction with age in range of abduction at shoulder joint[11]

Age	Male	Female
65–74 years	129 ± 14°	124 ± 19°
> 74 years	121 ± 19°	114 ± 22°

Normal threshold for adequate function is approximately 120°.

Age and arthritis[23,44]

Although common disorders such as rheumatoid arthritis increase in frequency with age, there is little evidence that age, *per se*, predisposes to systemic connective tissue disease. The controversial question of the relationship between age-related articular cartilage fibrillation and osteoarthrosis is discussed on pp. 830 and 886.

REFERENCES

1. Adolphe M, Ronot X, Jaffray P, Hecquet C, Fontagne J, Lechat P. Effects of donor's age on growth kinetics of rabbit articular chondrocytes in culture. *Mechanisms of Ageing and Development* 1983; **23**: 191–8.

2. Ali S Y. Calcification of cartilage. In: Hall B K, ed, *Cartilage Structure and Function.* New York: Academic Press, 1983: 343 –78.

3. Ali S Y. Apatite-type crystal deposition in arthritic cartilage. *Scanning Electron Microscopy* 1985; **4**: 1555–56.

4. Ali S Y, Griffiths S. New types of calcium phosphate crystals in arthritic cartilage. *Seminars in Arthritis and Rheumatism* 1981; **11**: 124–6.

5. Ali S Y, Griffiths S. Formation of calcium phosphate crystals in normal and osteoarthritic cartilage. *Annals of the Rheumatic Diseases* 1983; **42** Suppl 1: 45–8.

6. Armstrong C G, Bahrani A S, Gardner D L. *In vitro* measurement of articular cartilage deformations in the intact human hip joint under load. *Journal of Bone and Joint Surgery* 1979; **61-A**: 744–55.

7. Armstrong C G, Bahrani A S, Gardner D L. Changes in the deformational behaviour of human hip cartilage with age. *Journal of Biomechanical Engineering* 1980; **102**: 214–220.

8. Armstrong C G, Gardner D L. Thickness and distribution of human femoral head articular cartilage. *Annals of the Rheumatic Diseases* 1977; **36**: 407–412.

9. Armstrong C G, Mow V C. Variations in the intrinsic mechanical properties of human articular cartilage with age, degeneration and water content. *Journal of Bone and Joint Surgery* 1982; **64-A**: 88–94.

10. Barnett C H, Cochrane W, Palfrey A J. Age changes in articular cartilage of rabbits. *Annals of the Rheumatic Diseases* 1963; **22**: 389–400.

11. Bassey E J, Morgan K, Dallosso H M, Ebrahim S B. Flexibility of the shoulder joint measured as range of abduction in a large representative sample of men and women over 65 years of age. *European Journal of Applied Physiology and Occupational Physiology* 1989; **58**: 353–60.

12. Bayliss M T, Ali S Y. Age-related changes in the composition and structure of human articular cartilage proteoglycans. *Biochemical Journal* 1978; **176**: 683–693.

13. Bittles A H, Collins K J. *The Biology of Human Ageing.* Society for the Study of Human Biology Symposium Series. Cambridge: Cambridge University Press, 1986.

14. Brocklehurst J C. *Textbook of Geriatric Medicine and Gerontology* (4th edition) Edinburgh: Churchill Livingstone, 1991.

15. Buckwalter J A, Kuettner K E, Thonar E J. Age-related changes in articular cartilage proteoglycans: electron microscopic studies. *Journal of Orthopaedic Research* 1985; **3**: 251–7.

16. Bullough P, Goodfellow J, O'Connor J. The relationship between degenerative changes and load-bearing in the human hip. *Journal of Bone and Joint Surgery* 1973; **55-B**: 746–58.

17. Byers P D. Theoretical medicine: the pathogenesis of osteoarthritis. *Annals of the Rheumatic Diseases* 1988; **47**: 258–9.

18. Byers P D, Contepomi C A, Farkas T A. A *post-mortem* study of the hip joint including the prevalence of the features of the right side. *Annals of the Rheumatic Diseases* 1970; **29**: 15–31.

19. Byers P D, Contepomi C A, Farkas T A. *Postmortem* study of the hip joint. II. Histological basis for limited and progressive cartilage alterations. *Annals of the Rheumatic Diseases* 1976; **35**: 114–21.

19a. Byers P D, Contepomi C A, Farkas T A. *Post-mortem* study of the hip joint. III. Correlations between observations. *Annals of the Rheumatic Diseases* 1976; **35**: 122–6.

19b. Byers P D, Hoaglund F T, Purewal G S, Yau A C M C. Articular cartilage changes in Caucasian and Asian hip joints. *Annals of the Rheumatic Diseases* 1974; **33**: 157–61.

20. Byers P D, Pringle J, Oztop F, Fernley H N, Brown M A, Davison W. Observations on osteoarthrosis of the hip. *Seminars in Arthritis and Rheumatism* 1977; **6**: 277–303.

21. Champion B R, Reiner A, Roughley P J, Poole A R. Age-related changes in the antigenicity of human articular cartilage proteoglycans. *Collagen and Related Research* 1982; **2**: 45–50.

22. Chesworth B M, Vandervoort A A. Age and passive ankle stiffness in healthy women. *Physical Therapy* 1989; **69**: 217–24.

23. Christian C L. Arthritis in the elderly. *Medical Clinics of North America* 1982; **66**: 1047–52.

24. Curri S. Histochemical aspects of the ageing of connective tissue. *Phlebologie* 1986; **39**: 791–4.

25. De Rouseau C J. Ageing in the musculoskeletal system of rhesus monkeys: II. Degenerative joint disease. *American Journal of Physical Anthropology* 1985; **67**: 177–84.

26. Dominice J, Levasseur C, Larno S, Ronot X, Adolphe M. Age-related changes in rabbit articular chondrocytes. *Mechanisms of Ageing and Development* 1986–87; **37**: 231–40.

27. Elliott R J, Gardner D L. Changes with age in the glycosaminoglycans of human articular cartilage. *Annals of the Rheumatic Diseases* 1979; **38**: 371–7.

28. Emery I H, Meachim G. Surface morphology and topography of patellofemoral cartilage fibrillation in Liverpool necropsies. *Journal of Anatomy* 1973; **116**: 103–120.

29. Eyre D R, Dickson I R, Van Ness K. Collagen cross-linking in human bone and articular cartilage. Age-related changes in the content of mature hydroxypyridinium residues. *Biochemical Journal* 1988; **252**: 495–500.

30. Freeman M A R. The pathogenesis of osteoarthrosis: an hypothesis. In: Apley A G, ed, *Modern Trend in Orthopaedics* (6th edition). London: Butterworths, 1972: 40–94.

31. Freeman M A R, Meachim G. Ageing and degeneration. In: Freeman M A R, ed, *Adult Articular Cartilage* (2nd edition). Tunbridge Wells: Pitman Medical, 1979: 487–543.

32. Freeman M A R, Todd R C, Pirie C J. The role of fatigue in the pathogenesis of senile femoral neck fractures. *Journal of Bone and Joint Surgery* 1975; **56-B**: 698–702.

33. Gardner D L, Ageing of articular cartilage. In: Brocklehurst J G, ed, *Textbook of Geriatic Medicine and Gerontology* (4th edition). Edinburgh, London: Churchill-Livingstone, 1991: 792–812.

34. Gardner D L, Brereton J, Hollinshead M. Changes with age

in the human sacroiliac joint cartilages and chondrocytes. *European Journal of Experimental Musculoskeletal Research* 1991 (submitted).

35. Gardner D L, Mazuryk R, O'Connor P, Orford C R. Anatomical changes and pathogenesis of osteoarthrosis in man, with particular reference to the hip and knee joints. In: Lott D J, Jasani M K, Birdwood G F B, eds, *Studies in Osteoarthrosis: Pathogenesis, Intervention and Assessment.* Chichester: John Wiley & Sons, 1987: 21–48.

36. Gattone V H-2nd, Saul F P, O'Connor B L, McNamara M C. Scanning electron microscopic study of age-related surface changes in rat femoral head articular cartilage. *Journal of Submicroscopic Cytology* 1982; **14**: 99–106.

37. Ghadially F N. *Ultrastructural pathology of the cell and matrix. A Text and Atlas of Physiological and Pathological Alterations in the Fine Structures of Cellular and Extracellular Components* (3rd edition). London: Butterworths, 1988: 890–4.

38. Ghosh P, Andrews J L, Osborne R A, Lesjak M S. Variation with ageing and degeneration of the serine and cysteine proteinase inhibitors of human articular cartilage. *Agents and Actions* (Supplement) 1986; **18**: 69–81.

39. Goldman R, Rockstein M. *The Physiology and Pathology of Human Aging.* New York: Academic Press, 1979.

40. Green W T, Martin G N, Eanes E D, Sokoloff L. Microradiographic study of the calcified layer of articular cartilage. *Archives of Pathology* 1970; **90**: 151–8.

41. Hall D A. *The Ageing of Connective Tissue.* New York: Academic Press, 1976.

42. Haut R C. Age-dependent influence of strain rate on the tensile failure of rat-tail tendon. *Journal of Biomechanical Engineering* 1983; **105**: 296–9.

43. Hayflick L. Recent advances in the cell biology of ageing. *Mechanisms of Ageing and Development* 1980; **14**: 59–79.

44. Holden G. Age and arthritis. *Journal of the Royal Society of Medicine* 1982; **75**: 389–94.

45. Holmes M W A, Bayliss M T, Muir H. Hyaluronic acid in human articular cartilage. Age-related changes in content and size. *Biochemical Journal* 1988; **250**: 435–41.

46. Hornebeck W, Derouette J C, Brechemier D, Adnet J J, Robert L. Elastogenesis and elastinolytic activity in human breast cancer. *Biomedicine* 1977; **26**: 48–52.

47. Hough A J, Webber R J. Ageing phenomena and osteoarthritis: cause or coincidence? Claude P Brown Memorial Lecture. *Annals of Clinical and Laboratory Science* 1986; **16**: 502–10.

48. Huber-Bruning O, Wilbrink B, Vernooij J E, Bijlsma J W, den Otter W, Huber J. Contrasting reactivity of young and old human cartilage to mononuclear cell factors. *Journal of Rheumatology* 1986; **13**: 1191–2.

49. Inerot S, Heinegård D. Bovine tracheal cartilage proteoglycans. Variations in structure and composition with age. *Collagen and Related Research* 1983; **3**: 245–62.

50. Jilani M, Ghadially F N. An ultrastructural study of age-associated changes in the rabbit synovial membrane. *Journal of Anatomy* 1986; **146**: 201–15.

51. Kahn A R, Kahane J C. India ink pinprick assessment of age-related changes in the cricoarytenoid joint (CAJ) articular surfaces. *Journal of Speech and Hearing Research* 1986; **29**: 536–43.

52. Kempson G E. Mechanical properties of articular cartilage. In: Freeman M A R, ed, *Adult Articular Cartilage* (2nd edition). Tunbridge Wells: Pitman Medical 1979: 333–414.

53. Kempson G E. Relationship between the tensile properties of articular cartilage from the human knee and age. *Annals of the Rheumatic Diseases* 1982; **41**: 508–11.

54. Kirkwood T B L, Holliday R. Ageing as a consequence of natural selection. In: Bittles A H, Collins K J, ed, *The Biology of Human Ageing.* Society for the Study of Human Biology, Symposium Series (volume 25). Cambridge: Cambridge University Press 1986: 1–16.

55. Labat-Robert J, Robert L. Ageing of the extracellular matrix and its pathology. *Experimental Gerontology* 1988; **23**: 5–18.

56. Ladefoged C. Amyloid deposits in human hip joints. A macroscopic, light and polarization microscopic and electron microscopic study of congophilic substance with green dichroism in hip joints. *Acta Pathologica, Microbiologica et Immunologica Scandinavica* 1982; **90**: 5–10.

57. Lagier R. Anatomo-pathologie du vieillissment osteo-articulaire. *Lyon Mèdicale* 1971; **226**: 737–54.

58. Lagier R. An anatomicopathological approach to the study of articular ageing. *Fortschrift für Röntgenstrählung* 1976; **124**: 564–70.

59. Lane L B, Bullough P G. Age-related changes in the thickness of the calcified zone and the number of tidemarks in adult articular cartilage. *Journal of Bone and Joint Surgery* 1980; **62–B**: 372–75.

60. Light N D and Bailey A J. Changes in crosslinking during ageing in bovine tendon collagen. *FEBS Letters* 1979; **97**: 183–8.

61. Lindner J, Grasedyck K. Age-dependent changes in connective tissue. *Arzneimittelforschung* 1982; **32**: 1384–96.

62. Livne E, von der Mark K, Silbermann M. Morphologic and cytochemical changes in maturing and osteoarthritic articular cartilage in the temporomandibular joint of mice. *Arthritis and Rheumatism* 1985; **28**: 1027–38.

63. Livne E, Oliver C, Leapman R D, Rosenberg L C, Poole A R, Silbermann M. Age-related changes in the role of matrix vesicles in the mandibular condylar cartilage. *Journal of Anatomy* 1987; **150**: 61–74.

64. Longmore R B, Gardner D L. The surface structure of ageing human articular cartilage: a study by reflected light interference microscopy (RLIM). *Journal of Anatomy* 1978; **126**: 353–65.

65. McKeown E F. *Pathology of the Aged.* London: Butterworths, 1965: 282–307.

66. McKeown E F. De Senectute. The F E Williams Lecture. *Journal of the Royal College of Physicians* 1975; **10**: 79–99.

67. Mallinger R, Stockinger L. Amianthoid (asbestoid) transformation: electron microscopical studies on aging human costal cartilage. *American Journal of Anatomy* 1988; **181**: 23–32.

67a. Maroudas A, Kuettner K, eds. *Methods in Cartilage Research.* London, San Diego, New York: Academic Press, 1990.

68. Martel-Pelletier J, Pelletier J-P. Neutral metalloproteinases and age-related changes in human articular cartilage. *Annals of the Rheumatic Diseases* 1987; **46**: 363–9.

69. Mazess R B. On ageing bone loss. *Clinical Orthopaedics and Related Research* 1982; **165**: 239–52.

70. Meachim G. Effect of age on the thickness of adult articular cartilage at the shoulder joint. *Annals of the Rheumatic Diseases* 1971; **30**: 43–6.

71. Meachim G. Light microscopy of Indian ink preparations of fibrillated cartilage. *Annals of the Rheumatic Diseases* 1972; **31**: 457–64.

72. Meachim G. Articular cartilage lesions in osteoarthritis of the femoral head. *Journal of Pathology* 1972; **107**: 199–210.

73. Meachim G. Cartilage fibrillation at the ankle joint in Liverpool necropsies. *Journal of Anatomy* 1975; **119**: 601–10.

74. Meachim G. Cartilage fibrillation on the lateral tibial plateau in Liverpool necropsies. *Journal of Anatomy* 1976; **121**: 97–106.

75. Meachim G. Age-related degeneration of patellar articular cartilage. *Journal of Anatomy* 1982; **134**: 365–72.

76. Meachim G, Bentley G. Horizontal splitting in patellar articular cartilage. *Arthritis and Rheumatism* 1978; **21**: 669–74.

77. Meachim G, Emery I H. Cartilage fibrillation in shoulder and hip joints in Liverpool necropsies. *Journal of Anatomy* 1973; **116**: 161–79.

78. Middleton J F, Hunt S, Oates K. Electron probe X-ray microanalysis of the composition of hyaline articular and non-articular cartilage in young and aged rats. *Cell and Tissue Research* 1988; **253**: 469–75.

78a. Mitrovic D, Borda-Iriarte O, Naveau B, Stankovic A, Uyzan M, Quintero M, Ryckewaert A. Résultats de l'éxamen autopsique des cartilages des genoux chez 120 sujets décédés en milieu hospitalier. II. Articulation fémoro-tibiale. *Revue du Rhumatisme et des Maladies Osteo-articulaires* 1989; **56**: 505–10.

79. Mitrovic D, Quintero, M. Stankovic A, Ryckewaert A. Cell density of adult human femoral condylar articular cartilage. *Laboratory Investigation* 1983; **49**: 309–16.

80. Mitrovic D, Stankovic A, Borda-Iriarte O, Quintero M. Ryckewaert A. Résultats de l'éxamen autopsique des cartilages des genoux chez 120 sujets décédés en milieu hospitalier. I. Articulation fémoro-patellaire. *Revue du Rhumatisme et des Maladies Osteo-articulaires* 1987; **54**: 15–21.

81. Mort J S, Caterson B, Poole A R, Roughley P J. The origin of human cartilage proteoglycan link-protein heterogeneity and fragmentation during ageing. *Biochemical Journal* 1985; **232**: 805–12.

82. Mort J S, Poole A R, Roughley P J. Age-related changes in the structure of proteoglycan link proteins present in normal human articular cartilage. *Biochemical Journal* 1983; **214**: 269–72.

83. Nemeth-Csoka M N, Meszaros T. Minor collagens in arthrotic human cartilage. Changes in content of 1α, 2α, 3α and M-collagen with age and in osteoarthrosis. *Acta Orthopaedica Scandinavica* 1983; **54**: 613–9.

84. Ohta N, Kawai N, Kawaji W, Hirano H. Changes in alkaline phosphatase activity in rabbit articular cartilage associated with ageing and joint contracture. *Histochemistry* 1983; **77**: 417–422.

85. Oryschak A F, Ghadially F N, Bhatnagar R. Nuclear fibrous lamina in the chondrocytes of articular cartilage. *Journal of Anatomy* 1974; **118**: 511–5.

86. Pataki A, Ruttner J R, Abt K. Age-related histochemical and histological changes in the knee-joint cartilage of C57B1 mice and their significance for the pathogenesis of osteoarthrosis. I. Oxidative enzymes. *Experimental Cell Biology* 1980; **48**: 329–48.

87. Pidd J G, Gardner D L, Adams M E. Ultrastructual changes in the femoral condylar cartilage of mature American foxhounds following transection of the anterior cruciate ligament. *Journal of Rheumatology* 1988; **15**: 663–9.

88. Pieraggi M T, Julian M, Bouissou H. Fibroblast changes in human ageing. *Virchows Archiv A Pathological Anatomy and Histopathology* 1984; **402**: 275–87.

89. Plaas A H, Sandy J D. Age-related decrease in the link-stability of proteoglycan aggregates formed by articular chondrocytes. *Biochemical Journal* 1984; **220**: 337–40.

89a. Quintero M, Mitrovic D R, Stankovic A, de Sege S, Miravet, L, Ryckewaert A. Aspects cellulaires du vieillissement du cartilage articulaire. I. Cartilage condylieu à surface normale, prélevé dans les genoux normaux. *Revue du Rhumatisme* 1984; **51**: 375–9.

89b. Quintero M, Mitrovic D R, Stankovic A, de Ség̀e S, Miravet L, Ryckewaert A. Aspects celentaires du vieillissement du cartilage articulaire. II. Cartilage condylieu à surface fisurée prélevé dans les genoux 'hormaux' et 'arthrosiques'. *Revue du Rhumatisme* 1984; **51**: 445–9.

90. Robert L, Jacob M P, Frances C, Godeau G, Hornebeck W. Interaction between elastin and elastases and its role in the ageing of the arterial wall, skin and other connective tissues. A review. *Mechanisms of Ageing and Development* 1984; **28**: 155–66.

91. Roberts C R, Mort J S, Roughley P J. Treatment of cartilage proteoglycan aggregate with hydrogen peroxide. Relationship between observed degradation products and those that occur naturally during ageing. *Biochemical Journal* 1987; **247**: 349–57.

92. Roughley P J. Structural changes in the proteoglycans of human articular cartilage during ageing. *Journal of Rheumatology* 1987; **14**: (special number) 14–5.

93. Roughley P J, Mort J S. Ageing and the aggregating proteoglycans of human articular cartilage. *Clinical Science* 1986; **71**: 337–44.

94. Schofield J D, Weightman B. New knowledge of connective tissue ageing. *Journal of Clinical Pathology* 1978; **31**: supplement (Royal College of Pathologists) 12: 174–180.

95. Silbermann M, Levine E. Age-related degenerative changes in the mouse mandibular joint. *Journal of Anatomy* 1979; **129**: 507–20.

96. Simon W H. Scale effects in animal joints. I. Articular cartilage thickness and compressive stress. *Arthritis and Rheumatism* 1970; **13**: 244–56.

96a. Simon W H. Scale effects in animal joints. II. Thickness and elasticity in the deformability of articular cartilage. *Arthritis and Rheumatism* 1971; **14**: 493–502.

97. Stanescu V, Chaminade F, Muriel M P. Age-related changes in small proteoglycans of low buoyant density of human articular cartilage. *Connective Tissue Research* 1988; **17**: 239–52.

98. Stewart T D. Pathological changes in ageing sacroiliac joints. A study of dissecting-room skeletons. *Clinical Orthopaedics and Related Research* 1984; **183**: 188–96.

99. Stockwell R A. The cell density of human articular and costal cartilage. *Journal of Anatomy* 1967; **101**: 753–63.

100. Stockwell R A. *Biology of Cartilage Cells.* Cambridge: Cambridge University Press, 1979: 245–7.

101. Svanborg A. Practical and functional consequences of ageing. *Gerontology* 1988; **34**: suppl. i: 11–5.

102. Theocharis D A, Kalpaxis D L, Tsiganos C P. Cartilage keratan sulphate; changes in chain length with ageing. *Biochimica et Biophysica Acta* 1985; **841**: 131–4.

103. Tiurina T V. Age-related characteristics of the human acromioclavicular joint. *Arkhiv Anatomii, Gistologii i Embryologii (Leningrad)* 1985; **89**: 75–81.

104. Uitto J. Connective tissue biochemistry of the ageing dermis. Age-related alterations in collagen and elastin. *Dermatological Clinics* 1986; **4**: 433–46.

105. Van der Korst J K, Willekens F L H, Lansink A G W, Henrichs A M A. Age-associated glycopeptide pigment in human costal cartilage. *American Journal of Pathology* 1977; **89**: 605–620.

106. Venn M F. Variation of chemical composition with age in human femoral head cartilage. *Annals of the Rheumatic Diseases* 1978; **37**: 168–174.

107. Verzar F. Aging of the collagen fiber. In: Hall D A, Jackson D S, eds, *International Review of Connective Tissue Research* (volume 2). New York: Academic Press, 1964: 244–300.

107a. Vignon E, Arlot M, Vignon G. Le vieillissement du cartilage de la tete femorale humaine. Etude macroscopique de 42 pieces. *Lyon Médicale* 1973; **229**: 661–9.

108. Vignon E, Arlot M, Vignon G. A study of the cell density of the femoral head cartilage in relation to age. *Revue du Rhumatisme* 1976; **43**: 403–405.

109. Viidik A. Age-related changes in connective tissues. In: Viidik A, ed, *Lectures on Gerontology 1: On the Biology of Ageing* Part A. London: Academic Press 1982: 173–212.

110. Vogel H. *Connective Tissue and Ageing.* Amsterdam: Excerpta Medica, 1973.

111. Vogel H G. Influence of maturation and ageing on mechanical and biochemical properties of connective tissue in rats. *Mechanisms of Ageing and Development* 1980; **14**: 283–92.

112. Webber C, Glant T T, Roughley P J, Poole A R. The identification and characterization of two populations of aggregating proteoglycans of high buoyant density isolated from postnatal human articular cartilages of different ages. *Biochemical Journal* 1987; **248**: 735–40.

113. Weiss A, Livne E, Bernheim J, Silbermann M. Structural and metabolic changes characterizing the ageing of various cartilages in mice. *Mechanisms of Ageing and Development* 1986; **35**: 145–60.

114. Young J Z. Ageing and senescence. In: *An Introduction to the Study of Man.* Oxford: Oxford University Press, 1974: 287–302.

115. Youngman H R, Cooper F, Byers P. Prevalence of cartilage lesions in foot joints: a test of the concept of limited and progressive lesions. *Annals of the Rheumatic Diseases* 1987; **46**: 515–9.

Chapter 22
Osteoarthrosis and Allied Diseases

Introduction and definition

Osteoarthrosis[317] is an ubiquitous class of non-inflammatory synovial joint abnormality in which progressive cartilage disorder and bone change culminate in joint malfunction, mechanical failure and symptomatic disease.

The designation 'osteoarthritis' is preferred in North America.[183,191] European investigators believe that the suffix *-itis* implies, incorrectly, that a primary event in 'osteoarthritis' may be inflammatory. They therefore use 'osteoarthrosis' to mean 'degenerative disease of synovial joints'. Neither term indicates any full understanding of the nature of the disorder. Indeed, it is a measure of our ignorance that it is still necessary to reiterate that osteoarthrosis, like cardiomyopathy or encephalopathy, is a syndrome or complex, not a disease entity. To establish uniformity of usage, rationalize diagnosis and facilitate controlled therapeutic trials, significant efforts have been directed towards establishing an internationally acceptable working definition of the clinical syndrome of osteoarthrosis. Thus, osteoarthrosis (osteoarthritis) is said to be:

> . . . a heterogeneous group of conditions that lead to joint symptoms and signs which are associated with defective integrity of articular cartilage, in addition to related changes in the underlying bone and at joint margins. Although articular cartilage is poorly innervated and defects in cartilage are not, in themselves, symptomatic, a clinical syndrome, which often includes pain, may evolve from such defects.[14]

There is no agreed pathological definition. Meachim and Brooke[293] state that: 'osteoarthrosis is a diagnostic label for degenerative disease of synovial joints'. Hough and Sokoloff[183] define osteoarthritis as 'an inherently non-inflammatory disorder of movable joints characterized by deterioration and abrasion of articular cartilage as well as by formation of new bone at the joint surfaces'.

In the present volume, the connective tissues are viewed as a single system, synovial joints as bone-cartilage-synovial fluid-cartilage-bone continua. On this basis osteoarthrosis is 'the aggregate of the degenerative disorders of the cells and matrix of the connective tissue of synovial joints that develop with time, and for whatever reason, in excess of those attributable solely to senescence'.

The variety of these definitions clearly indicates the difficulties faced in offering an account of the pathology of 'osteoarthrosis'. Although there are sound clinical reasons for considering that osteoarthrosis is a category of disease affecting all freely movable (diarthrodial) joints,[183] biological criteria suggest the desirability of placing degenerative diseases of cartilagenous and fibrocartilagenous joints, the synchondroses and syndesmoses, in a separate group. They are therefore considered in Chapter 23. Osteoarthrosis of spinal synovial joints and degenerative disease of the main, fibrocartilagenous intervertebral joints commonly coexist but these structures are structurally and functionally different. The pathological changes present distinctive features. Degenerative disease of the lateral and median atlantoaxial, atlantooccipital, neurocentral, apophyseal, costovertebral and sacroiliac joints is *osteoarthrosis*; degenerative disease of the main intervertebral joints is *spondylosis*.

History

There are many reviews.[75,99,120,181,331,366,438,484] Classical accounts date from the time of Hippocrates[180] (BC 460–375) and Aretaeus (AD 81–138).[19] Osteoarthrosis is known to have existed, therefore, from the earliest days of civilized man and its imprints have indeed been detected in Egyptian mummy prehistory.[399] However, pathological understanding of the nature of osteoarthrosis is modern. The gross anatomy of the syndrome was described by John Hunter[199] but Virchow[477] and Charcot[69] caused confusion because they viewed osteoarthrosis and rheumatoid arthritis as a single entity which they called 'arthritis deformans'. Adams[5] recognized the pathological features of rheumatoid arthritis but it was not until the early years of this century that osteoarthrosis and rheumatoid arthritis were clearly differentiated.[30,181,331]

Clinical presentation

Criteria and classification

In many instances, the cause(s) of a particular example of osteoarthrosis is known and the disorder is 'secondary'. In others, no cause or causes are discernible and the syndrome remains 'idiopathic' (Table 22.1). Part of the confusion that still surrounds the investigation of osteoarthrosis can be attributed to a lack of precision in defining the form of disease under study. Osteoarthrosis following trauma is 'chronic post-traumatic osteoarthropathy'; the osteoarthrosis that develops late in rheumatoid arthritis is 'chronic rheumatoid osteoarthropathy'; and the articular syndrome that succeeds experimental procedures, such as the aseptic surgical division of a canine cruciate ligament, is 'postcruciate ligament section arthropathy'. In the study of cardiac disorders, many important truths have come from the investigation of 'cardiac failure' but a full understanding of the mechanisms of a particular example of heart disease usually rests on the exact definition of the cause of the syndrome, whether it be coronary artery insufficiency, systemic hypertension, hyperthyroidism, and so on. A comparable view may be taken of osteoarthrosis. In broad terms, 'osteoarthrosis' is 'synovial joint failure'. To comprehend the nature and pathogenesis of individual examples of end-stage synovial joint failure (Fig. 22.1), it is essential to analyse the precise mechanisms by which it originates. Although the terminal, anatomical results may be indistinguishable, the origins and pathogenesis of the osteoarthrosis which is 'postdysplastic hip joint arthropathy', for example, are by no means necessarily the same as those which lead to the occupational, infective or endocrine arthropathies of the same joint.

In cellular terms, any form of osteoarthrosis detectable clinically is necessarily of an advanced or terminal nature. The classification of osteoarthrosis and the formulation of diagnostic criteria[14,16] rest upon late manifestations of cartilage and bone disorder. The large proportion of asymptomatic patients, the non-specific nature of much early symptomatology and the lack of diagnostic laboratory tests compound this problem. Open biopsy is rarely practised for the diagnosis of osteoarthrosis but diagnostic fibreoptic arthroscopy is now widely undertaken. When the results of this procedure and of others[15] such as magnetic resonance imaging and microfocal radiography are available, considerably more than 40 per cent of patients found to have the structural and functional changes of osteoarthrosis have no subjective complaints. At the conclusion of a detailed study[16] involving 12 North American centres, it was decided that 'no single set of classification criteria could satisfy

Table 22.1

Classification for subsets of osteoarthrosis[14]

I. IDIOPATHIC
  A. *Localized*
    1. *Hands:* e.g. Heberden's and Bouchard's nodes (nodal), erosive interphalangeal arthritis (non-nodal)
    2. *Feet:* e.g. hallux valgus, hallux rigidus
    3. *Knee:*
      a) Medial compartment
      b) Lateral compartment
      c) Patellofemoral compartment (e.g. chondromalacia)
    4. *Hip:*
      a) Eccentric (superior)
      b) Concentric (axial, medial)
      c) Diffuse (coxae senilis)
    5. *Spine* (particularly cervical and lumbar)
      a) Apophyseal
      b) Intervertebral (disc)*
      c) Spondylosis (osteophytes)*
      d) Ligamentous (hyperostosis) (Forestier's disease: DISH†)*
    6. *Other single sites:* e.g. shoulder, sacroiliac, temporomandibular, acromioclavicular, wrist, ankle
  B. *Generalized:* includes 3 or more areas listed above
    1. Small (peripheral) and spine*
    2. Large (central) and spine*
    3. Mixed (peripheral and central) and spine*

II. SECONDARY
  A. *Post-traumatic*
  B. *Congenital or developmental disease*
    1. *Localized*
      a) Hip disease e.g. Legg—Calvé—Perthes, congenital hip dislocation
      b) Mechanical and local factors e.g. unequal leg length, obesity, hypermobility syndromes
    2. *Generalized*
      a) Bone dysplasias e.g. epiphyseal dysplasia
      b) Metabolic disease e.g. ochronosis, haemachromatosis, Gaucher's disease, Ehler's—Danlos syndrome
  C. *Calcium deposition disease*
    1. Calcium pyrophosphate deposition disease
    2. Apatite arthropathy
    3. Destructive arthropathy (shoulder, knee)
  D. *Other bone and joint disorders*
    e.g. rheumatoid arthritis, urate arthropathy (gout), avascular bone necrosis, septic arthritis, Paget's disease, osteochondritis
  E. *Other diseases*
    1. Endocrine: e.g. diabetes mellitus, acromegaly, hypothyroidism, hyperparathyroidism
    2. Neuropathic
    3. Miscellaneous: e.g. Kashin–Beck disease, caisson disease, frostbite

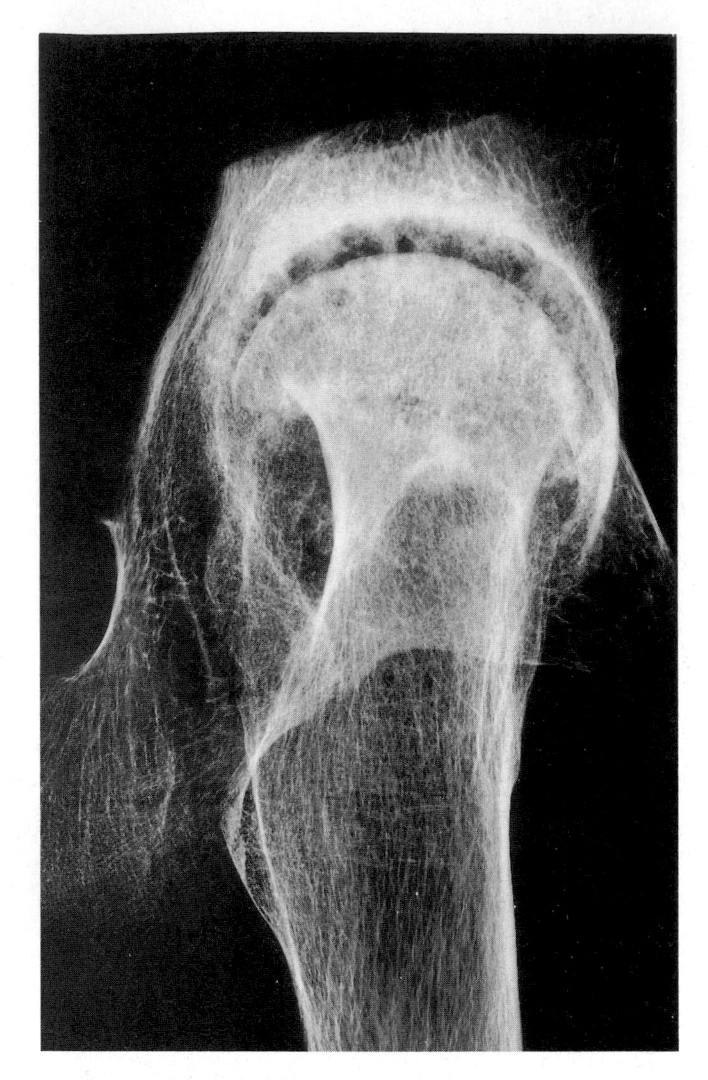

Fig. 22.1 Osteoarthrosis of the hip joint.
A single anteroposterior radiograph, assessed for joint space narrowing and cyst formation, yields very great sensitivity in assessing osteoarthrosis clinically.[15] In this conventional radiograph, three principal features are shown: osteosclerosis of bone adjoining the articular surfaces of the acetabulum and femoral head, multiple pseudocyst formation, and generalized osteoporosis (osteopenia). The two-dimensional film gives little information about the intercartilagenous space, the true 'joint space', although the appearances suggest an increase in the interosseous space.

*Since the principal articulations of the spine, the main intervertebral joints, are fibrous and not synovial, degenerative disease of these joints is not classified in the present text as osteoarthrosis and the reader is referred to Chapter 23.

†DISH = diffuse idiopathic skeletal hyperostosis.

all circumstances to which the criteria for osteoarthrosis of the knee [for example] would be applied'. Separate sets of classification criteria were therefore designed to be used under different circumstances (Table 22.2). They were developed by recursive partitioning after initial data had been evaluated by a 'Delphi' technique of opinion sampling.[13d,14] An example of a clinical classification tree for knee osteoarthrosis (Fig. 22.2) devised by these authors is shown in Fig. 22.3.

Table 22.2

Relative ranking, in order of importance, of clinical, laboratory and radiographic criteria for the identification of osteoarthrosis[16]

History
1. Pain
2. Age > 50 years
3. Decreased function
4. Swelling
5. Stiffness for < 30 min

Physical findings
1. Crepitus
2. Bony enlargement
3. Limitation of motion
4. Instability
5. Tenderness

Laboratory findings
1. Normal erythrocyte sedimentation rate
2. Non-inflammatory synovial fluid findings
3. Negative or very low rheumatoid factor titres

Radiographic parameters
1. Osteophytes
2. Narrowing (of interosseous space)*
3. Cysts
4. Varus
5. Valgus
6. Chondrocalcinosis

*The term 'joint space' is open to misinterpretation and it is essential to distinguish the intercartilagenous (identified by contrast arthrography) from the interosseous space, recognizable by simple X-radiography.

The most common forms of secondary osteoarthrosis are attributable to major or repetitive, minor injury; rheumatoid arthritis; infection; nutritional, endocrine and metabolic osteoarticular disease; and malformation (Table 22.3). The pathology of secondary osteoarthrosis is best approached from knowledge of the individual, underlying abnormalities. Rheumatoid arthritis (see Chapter 12) is one example. In idiopathic osteoarthrosis the cause of the disorder is not demonstrable. Much of the large, present effort devoted to understanding osteoarthrosis is directed towards this second class of disease (see p. 881 and 896).

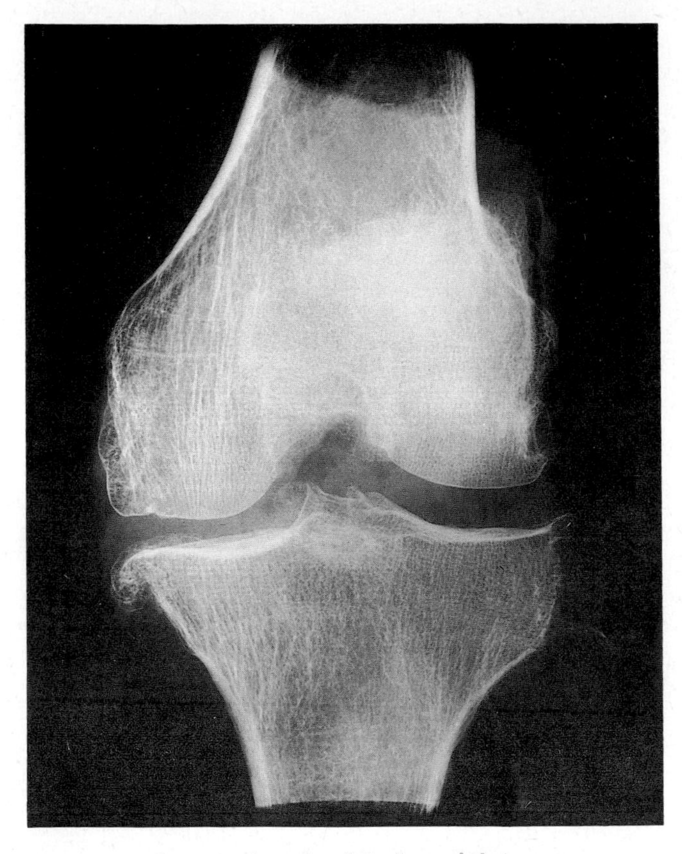

Fig. 22.2 Osteoarthrosis of the knee joint.
An anteroposterior radiograph of the knee has been a classical procedure for assessing radiographic progression of osteoarthrosis clinically.[15] In this *post mortem* specimen, the wide separation of the opposed femoral and tibial condyles suggests relative integrity of the articular cartilages of both bones. However, there is osteophytic lipping of all surfaces and substantial osteophytes are seen, especially at the edge of the medial condyles. In contrast with the appearances shown in Fig. 22.1, no pseudocysts are present.

Prevalence

Osteoarthrosis is one of the most common and most disabling disorders of Western society. Because of its high frequency, osteoarthrosis contributes greatly to the large drug bills paid by ageing populations; for the same reason, osteoarthrosis imposes heavy loads on surgical resources. In Third World countries, where populations do not so often

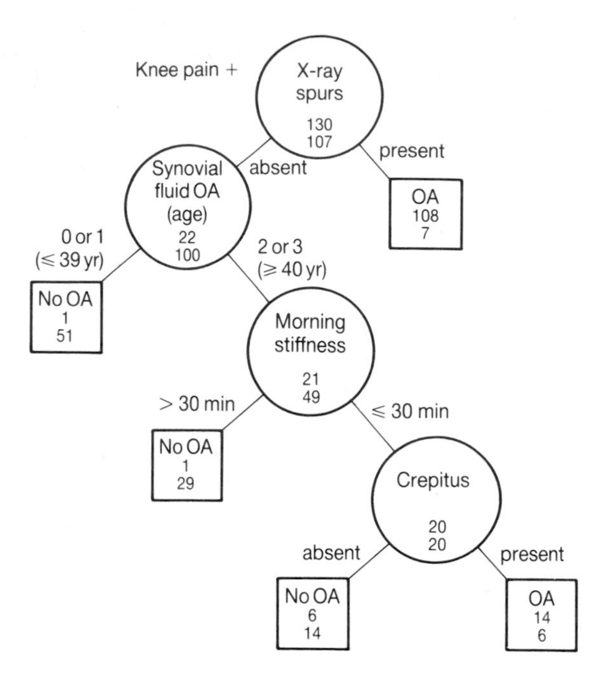

Fig. 22.3 Osteoarthrosis (osteoarthritis) of the knee: classification tree for clinical, radiological and laboratory (non-histopathological) criteria.
Studies of the development of classification criteria for osteoarthritis (osteoarthrosis) have revealed serious deficiencies in some previously used methods. Classification trees were therefore 'grown'. The methodology is summarized by Altman *et al.*;[14] the details are too extensive to be included in this text. In the example shown here,[14] 237 patients with knee joint pain are seen to have radiological evidence of spurs (osteophytes). When the nine features of synovial fluid held to be characteristic of osteoarthrosis were taken into account, 22 of those with knee pain but without radiological evidence of osteophytosis were found to have these synovial fluid changes while 100 did not. Again, when the 122 patients with knee pain but without osteophytes were analysed in terms of the presence of none or one versus two or three of the diagnostic synovial fluid criteria, 51 of those below the age of 39 years were regarded as not having osteoarthrosis; one patient had the disease. In turn, in those above the age of 40 years, the presence of early morning stiffness was used to extend the tree further, and so on. The overall sensitivity of this elegant but complex approach to the diagnostic classification of osteoarthrosis is 94 per cent, the overall selectivity 88 per cent. It is hoped that the use of fibreoptic arthroscopy will allow for the useful introduction of histological criteria that remain, meanwhile, variable and of uncertain significance (p. 856). Redrawn from Altman *et al.*[14]

survive to ages at which osteoarthrosis becomes symptomatic, the position is quite different, and osteoarthrosis, like many cancers, is a much less important cause of disability.

The prevalence of established, pathologically advanced idiopathic osteoarthrosis can be assessed indirectly by the use of medical histories, physical examination and radiography[390], and X-ray and other imaging techniques are of great value in differentiating individual forms of

Table 22.3

Disorders and circumstances predisposing to osteoarthrosis in man. Occasionally, the disorders fall into more than one category[29]

Malformation:	Chondrodystrophy*
	Osteogenesis imperfecta*
	Epiphyseal hip dysplasia
	Acetabular protrusion*
	Congenital hip dysplasia*
	Slipped femoral head epiphysis*
Endocrine:	Acromegaly
	Diabetes mellitus*
	Hypothyroidism
Metabolic:	Chondrocalcinosis
	Kashin–Beck disease
	Renal bone disease
	Gaucher's disease
	Ochronosis*
	Mucopolysaccharidoses*
Occupational:	Coal mining
	Ballet dancing
Sport:	Football
	Water skiing
	Jogging
	Parachuting
Injury:	Fracture at joint surfaces
Bone necrosis:	Drug induced
	Post-dialytic
	Caisson disease
	Thiemann's disease
Surgical procedures:	Meniscectomy
	Malunited fracture
Blood disease:	Haemoglobinopathies*
Coagulopathies:	Haemophilia*
Inflammatory:	Rheumatoid arthritis*
Drug use or abuse:	Antiinflammatory and antirheumatic, e.g. indomethacin

*Disorders that are inherited or have a major genetic predisposition.

osteoarthrosis.[165,464] Indeed, many of the early, pioneering investigations of the epidemiology of osteoarthrosis centred on the application of X-radiography to selected populations.[247] Thus, in an urban-rural population of the North of England, Lawrence, Bremner and Bier[248] found the prevalence of osteoarthrosis, judged by clinical and radiographic tests, to be 52 per cent in males, 51 per cent in females. Excluding minimal clinical/radiographic disease, the figures were 19 and 22 per cent, respectively. Five or more joint groups were affected in 0.5 per cent of males, 1.8 per cent of females. Osteoarthrosis was shown to be the most common cause of lost working time in an urban female population,[216] and the third most common cause in an urban male population. The expectation of life has risen during the past 40 years and there is little reason to suspect a diminution of these high figures in contemporary societies. Thus, in a population of 1405 subjects aged 63–94 years, radiographic evidence of osteoarthrosis increased from 27 per cent in those aged less than 70 years, to 44 per cent in those aged more than 80 years.[118]

The prevalence of presymptomatic osteoarthrosis, judged objectively by radiography,[118] in large joints such as the knee, increases with age to a greater degree in women than in men. There is also a less well understood category of 'generalized' osteoarthrosis, predominantly of older women. This uncommon form of osteoarthrosis, little studied pathologically, displays an inherited predisposition and is accompanied by local and systemic signs of inflammation.[217] The distal interphalangeal joints are particularly susceptible and here osteophytes form. They are still called Heberden's nodes.

The prevalence of the molecular and microscopic changes which precede asymptomatic osteoarthrosis cannot yet be assessed. There are no specific laboratory diagnostic tests for the disorder although criteria such as the measurement of proteoglycan degradation products[460] (see p. 202) may index the progress of cartilage injury. The identification in the serum or synovial fluid of unique proteins, enzymes[231] or polypeptides characteristic of the connective tissue responses in osteoarthrosis is likely to assist future diagnosis and facilitate the conduct of therapeutic trials. Vigorous searches for such 'markers' are in progress.

Osteoarthrosis is an ubiquitous disorder of mammals and has been identified in many other extant and prehistoric vertebrates[438] (Fig. 22.4). The disorder is commonplace in

Fig. 22.4 Osteoarthrosis in extinct vertebrates. *Triceratops*, a rhinoceros-like ornithischian dinosaur of the Upper Cretacean period. The presence of marginal osteophytes at the main bearing surfaces of the limb joints, together with eburnation, strongly suggests that this large (7 m long) creature, like other dinosaurs, was susceptible to osteoarthrosis. Courtesy of the Natural History Museum.

domestic animals, such as dogs, pigs and horses, and investigations of osteoarthrosis in these species have shed important light on the nature of the syndrome (see p. 896). In man, racial differences have been detected. In Hong Kong Chinese, for example, hip joint osteoarthrosis is less common than among Caucasian populations. Hereditary factors (see p. 883) now attract much attention.

Pathological changes

Osteoarthrosis is a conglomerate group of abnormalities of connective tissue cell function.[168] Cartilage and bone cell behaviour is aberrant. Influenced to varying degrees by genetic information, the principal consequences are the synthesis, export and assembly of hierarchies of matrix macromolecules which fail prematurely.

One result of the defective cell function which underlies osteoarthrosis is the release into the blood and/or synovial fluid of products of macromolecular degradation such as keratan sulphate[497] or specific antigens that can be identified by monoclonal[460] or polyclonal antibodies.[173] Unique non-proteoglycan matrix glycoproteins, including a collagen,[264] exemplify the 'markers' that are sought. Other indices include the intermediate filaments tubulin, vimentin and desmin synthesized by mesenchymal cells; S-100, a calcium-binding protein; and glycoproteins such as fibronectin and chondronectin that have surface-interactive characteristics. There is evidence that each of these antigens may be expressed abnormally in osteoarthrosis leaving little doubt that chondrocyte function is perverted. Much less is known of osteoblast, osteocyte and osteoclast behaviour. As a corollary, it is clear that the degradation of articular cartilage in osteoarthrosis is indirectly attributable to cytokines such as interleukin-1[369] (see p. 88) and that the repair and reconstitution of mesenchymal tissue in osteoarthrosic cartilage is modulated by agents such as transforming growth factor β and the fibroblast growth factors.[169]

The collection of standardized samples of cartilage with early matrix disorganization can be facilitated by the preliminary painting of a bearing surface with a dilute, isotonic suspension of Indian ink*.[198,290] However, it is self-evident that the earliest pathological phenomena of osteoarthrosis are not accessible to the techniques of morphological analysis: only the methods of biophysics, biochemistry and molecular biology allow them to be examined directly. To enable such approaches, the biologist requires material

* A colloidal solution of carbon particles. Commercial preparations may contain phenol as a preservative.

from an individual with established disease which has not yet advanced to a degree of disintegration recognizable by simple inspection.

There is, therefore a Catch-22 situation: the study of the earliest changes of osteoarthrosis by the methods of cellular and molecular pathology is impracticable until the disease is established; the disease can only be said to be established long after the earliest alterations in cell behaviour have been translated into organizational changes in connective tissue that are reflected in identifiable, micro- and macroscopic abnormalities.

For these several reasons, little is known of the earliest, molecular disorders in human osteoarthrosis and the corpus of knowledge assembled about the initial abnormalities has come largely from studies of spontaneous or induced disease in animals (see p. 896). One approach to the resolution of this problem is to design tests of preosteoarthrosic cell behaviour that can be applied to fresh living human tissue predicted to be in a preosteoarthrosic state on the basis of genetic markers or secreted macromolecules. Another approach, less sensitive but of possible value to those seeking to study the biological, cellular and ultrastructural disorders of early osteoarthrosis, is to investigate fresh, random samples of human connective tissue by biophysical procedures such as the measurement of tissue permeability or micromechanical behaviour, allowing tissue identified as preosteoarthrosic then to be assessed for cell abnormality and molecular disorder.

One way in which to identify hyaline cartilage surfaces that have undergone the macromolecular changes of osteoarthrosis but in which disintegration has not advanced to a point at which microscopic abnormalities can be seen, is to use a simple physicochemical test that is analagous with the imbibition of contrast medium during radiography.[445a] Gardner *et al.*[135] have demonstrated that the organic dye light green GS (molecular weight 790.34) diffuses through excised human tibial condylar cartilage at a rate conveniently measurable by scanning optical microscopic morphometry. The rate of diffusion of the dye increases geometrically with the degree of cartilage surface disintegration (fibrillation—see p. 858). Three contiguous, parallel blocks of 'fresh' cartilage are prepared by cryostat section. One sample is examined by cryostat section to assess the degree of microscopic fibrillation; the second is tested simultaneously with light green GS. When the results of this measurement of dye diffusion rate are known, the third block can be used for ultrastructural, chemical or immunological analysis or for micromechanical assessment with the reasonable certainty that the cartilage is in a preosteoarthrosic state.

There are numerous full accounts of the macroscopic and microscopic phenomena in osteoarthrosis, both in the older[438] and more recent literature.[29,136,183,293,310] Many authorities find it convenient to divide descriptions of the pathology of the syndrome into 'early', 'intermediate' and 'late'. However, these arbitrary time scales have little foundation in systematic observation. It is more logical to divide descriptions of osteoarthrosis into 'molecular', 'ultrastructural', 'microscopic' and 'anatomical' according to an ascending scale of morphological defect. There is usually no way in which the chronology of the lesions can be confirmed directly.

Molecular changes[44,168]

Detailed studies have been made of the chemical composition of osteoarthrosic cartilage and of its metabolism;[182,268,273,320,323] the reader is referred to these sources for full accounts of the human and animal disorder. There are very few analyses of the chemistry and metabolism of osteoarthrosic bone. In the case of the human syndrome, the selection of osteoarthrosic cartilage for chemical analysis has inevitably rested on late clinical criteria or upon insensitive techniques such as the Indian ink painting of cartilage surfaces (see p. 834). The study of naturally occurring animal disease is constrained in the same way. In experimental animal studies, chemical investigations are, of course, possible from the moment the syndrome is initiated (Table 22.4).

Water

The water content of osteoarthrosic femoral head cartilage is increased[275,277] and an increased avidity for water has been demonstrated.[278] When human femoral condylar cartilage was analysed, the degree of the increased water content depended upon or reflected the anatomical extent of the cartilage disruption[44] (fibrillation, see p. 858). When fibrillation was confined to superficial cartilage, the water content ranged from 82 per cent in the 350 μm slice nearest the surface to 70 per cent in the 350 μm slice nearest the bone. In normal specimens, the corresponding figures were 78 and 68.5 per cent, while in osteoarthrosic cartilage with a smooth surface the values were 77 and 67.5 per cent, respectively. When fibrillation extended deeply, the water content was 83 per cent in superficial cartilage, 75 per cent in deep cartilage. Deeply extending fibrillation was accompanied by a decreased rate of glycosaminoglycan synthesis and a decreased glycosaminoglycan content; when fibrillation remained superficial, neither aspect of glycosaminoglycan metabolism was measurably altered. In the canine model of osteoarthrosis which follows surgical division of a cruciate ligament, an early increase in the hydration both of the tibial condylar hyaline cartilage, and of the correspond-

Table 22.4

Biochemical changes in articular cartilages in dog arthropathy caused by surgical section of cranial (anterior) cruciate ligament ('ACL disease')[3,62,63,262,263,323,407]

Water content:	Early increase of both hyaline and fibrocartilages before histological changes
Dry weight:	Increased
Mass:	Increased
Collagens:	Synthesis of type II collagen increased tenfold within 2 weeks of ACL section. New collagen less glycosylated than normal. Collagen/non-collagenous protein ratio elevated. No formation of type I collagen. Changes in 'minor' collagens, particularly type IX, suspected but not proved
Proteoglycans:	No consistent change in proteoglycan content Increased proportion of proteoglycans characteristic of immature cartilage[3] Linear increase with time in rate of proteoglycan synthesis, *in vivo* and *in vitro* Increased chondroitin sulphate content relative to keratan sulphate, with longer proteoglycan chondroitin sulphate side-chains Increased turnover of proteoglycan; rapid breakdown, with active release of $\sim 1 \times 10^6$ kD molecules binding to hyaluronate less well than normal[350] No change in aggregatability Molecules of normal structure retained but degraded proteoglycans quickly diffuse out. Active breakdown of core protein
Unidentified proteins:	Synthesis of new proteins recognized
Glycoproteins:	Loss of staining. 550 kD glycoprotein present[119]
Enzyme activities:	Activation/secretion of neutral proteases leading to proteoglycan loss[360,361] Increased cartilage and synovial fluid collagenase activity, correlating with degree of synovitis[361] and reaching a peak 4 weeks after aseptic surgery Increased cartilage and synovial fluid hydrolase activity[413]

ing meniscal fibrocartilage, precedes microscopic changes in cartilage structure.[323]

The altered water content of osteoarthrosic cartilage is attributed to a progressive fault in the capacity of the cartilage collagen microskeleton to retain proteoglycan in normal compression. At first, the increase in cartilage water was difficult to reconcile with evidence that, in the established human disorder, proteoglycans were depleted. However, experimental data in early experimental dog disease revealed no overall early change in total proteoglycan content. Later, when collagen degradation affects the restraining collagen fibrillar meshwork, the loss of abnormal proteoglycan advances quickly so that the water content declines.

DNA

It is clear that DNA increases focally in osteoarthrosic cartilage when chondrocyte clusters form (see p. 856); confirmation of this observation may be made by microdensitometry. Later, when fibrillation (see p. 858) is established, the number of cells in the superficial zones of hyman hyaline articular cartilage becomes difficult to evaluate; the measurement of DNA unreliable. In the anterior cruciate ligament model of osteoarthrosis, a significant increase in cell density in the tibial condyle is detectable within 2–4 weeks of surgery and there is a corresponding increase in cartilage DNA. Late in the human and animal disorder, as the matrix is degraded during fibrillation and fragmentation, the content of DNA declines in proportion to cartilage loss.

Proteoglycans

The changes in the human disorder are still not fully understood; those of animal models are better validated. Many studies have confirmed a progressive loss of the matrix glycosaminoglycans that are demonstrable by cationic dyes such as safranin O and alcian blue.[273] There is a corresponding, measurable decrease in fixed charge density,[469] an alteration which provides an index of a large series of progressive alterations in proteoglycan composition. An increase in proteoglycan chondroitin-4-sulphate and a decrease in keratan sulphate[495] appear to represent the reversion of chondrocyte synthesis and secretion to a less mature form, changes mirrored by the altered cell morphology (Fig. 22.7, p. 852). In the dog cruciate ligament model, this change in proteoglycan content[3] is first identified in the central, load-bearing tibial cartilage but it is anatomically progressive. A sequence of alterations in proteoglycan side-chains has been confirmed in naturally occurring canine osteoarthrosis.

As one result of altered chondrocyte behaviour, osteoarthrosic proteoglycans initially become larger. Their secretion is delayed. Early evidence suggested that these macromolecules were more extractable from osteoarthrosic than from normal cartilage and that a higher than normal proportion existed as non-aggregated subunits. However, in canine cartilage, in animals subjected to cruciate ligament section, the interaction with hyaluronate remains essentially unaffected.[323] *In vitro*, the synthesis of link protein in response to RNA from osteoarthrosic chondrocytes appears normal[461] and the physiological function of this glycoprotein *in vivo* is probable.

Proteoglycan biosynthesis changes. In both human osteoarthrosic cartilage and in cartilage from the cruciate ligament dog model, $^{35}SO_4$-sulphate incorporation is increased in excess of the rise in cell numbers.[402] *In vitro*, this heightened activity persists in the absence of endocrine or humoral agents. Initially, proteoglycans are small, cannot react with hyaluronate and are not increasingly catabolysed. Later, however, there is exaggerated proteoglycan loss. Degradative activity is mediated by chondrocytes and the role of a variety of enzymes catalysing matrix breakdown has been shown (p. 892).[363] The enzymes include lysosomal acid proteases, cathepsins D and B, a neutral proteoglycanase and a collagenase (see Chapter 4). The induction of abnormal enzymes has also been suggested. The mechanism by which these enzymes are induced and activated invokes the release of cytokines, particularly of interleukin-1. The nature of this action is considered in Chapter 2 and again in Chapter 12.

Collagens

Although a defect in the conversion of procollagen to collagen has been identified in dogs with spontaneous osteoarthrosis (p. 897),[300] there is no early, consistent change in collagen content either per unit weight of cartilage or per unit DNA.[272] Late in the disease, as cartilage is progressively lost, the total quantity of collagen in all affected cartilages declines. No constant change of collagen phenotype from type II to type I has been shown.[156,195] Whether significant alterations occur in the proportions of type VI, IX and XI collagens is suggested[329] but not proven; there is particular interest in the quantities and organization of type IX collagen. Early in the experimental canine model of osteoarthrosis that follows cruciate ligament section, the incorporation of ^{3}H-proline into type II collagen is greatly increased. The rate of collagen synthesis is also considerably increased in proportion to a concomitant increase in non-collagenous protein.[67,323] The increased rate of collagen synthesis is also evident in human osteoarthrosic cartilage.[252] The increment may be as much as five times normal. It is of interest, in view of observations on the repair of cartilage in osteoarthrosis, that the new collagen formed retains the type II phenotype.[122]

Fibronectin

Fibronectin is one of a number of glycoproteins found in increased amount in dissociative extracts of osteoarthrosic articular cartilage.[50,301] The largest concentrations are in zone I (surface zone) cartilage. The glycoprotein is released as part of the amorphous material which accumulates around osteoarthrosic chondrocytes and cells secreting fibronectin are common in human osteoarthrosis.[385] Localization of fibronectin within the surface of residual femoral head cartilage demonstrates that the amounts remaining vary considerably.[210] Analogous observations in the naturally occurring osteoarthrosis of Labrador retriever dogs shows the amount of fibronectin to be as much as 20 times normal.[50,52,53,499]

Ultrastructural changes

Transmission electron microscopy

The selection of material, and the small size of the samples, mean that considerable care is necessary in interpreting disorders identified by electron microscopy.[136,308,346,429,486–488,502] The coexistence of age changes and the effects of drug treatment have always to be considered.

In human material, the changes found in ageing and in osteoarthrosic cartilage were said at first to be indistinguishable.[400] However, claims of this kind were often based on the study of very limited amounts of tissue.

It is now generally agreed that an electron-dense material, with no identifiable ultrastructure, forms on the cartilage surfaces, merging with and replacing the superficial, fine surface lamina (see p. 70) (Fig. 22.5). The electron-dense material is thought to be derived from effete chondrocytes and synovial fluid macromolecules but it is likely to include both the fibronectin which is identifiable at these surfaces and materials such as fibrin which escape from synovial blood vessels when an associated secondary inflammation (see p. 876) occurs. The random arrangement of zone I and zone II collagen is lost and the quantity of interfibrillar electron lucent matrix increased (Fig. 22.6). Collagen fibres come to have a much wider and more variable range of diameter than normal. In addition to this wide diversity of size, which may be from 30 to 450 nm,[486] there are occasional amianthoid fibres, up to 700 nm in thickness.[142] Similar large fibres are however found in normal ageing cartilage:[182,429] they display only the major 67 nm periodicity. Transmission electron microscopy also

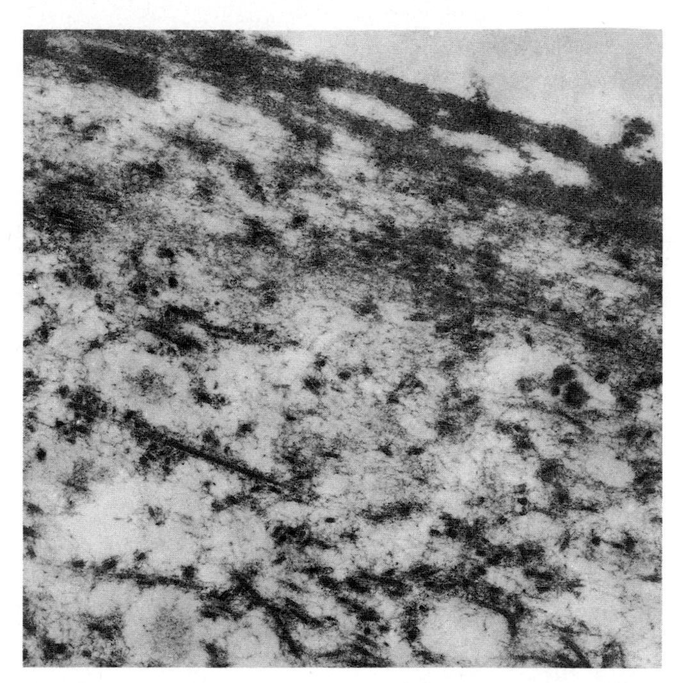

Fig. 22.5 Experimental canine osteoarthrosis following section of an anterior cruciate ligament (ACL arthropathy).
Twenty-four weeks after the operation, electron-dense material is incorporated in the ultrastructural surface lamina which forms a continuum between the synovial fluid and the uppermost lamina of the fibre-rich part of this zone I (superficial zone) cartilage. The type II collagen fibre-bundles are widely separated allowing the expansion and increased hydration of the nearby proteoglycans. (TEM × 24 225.)

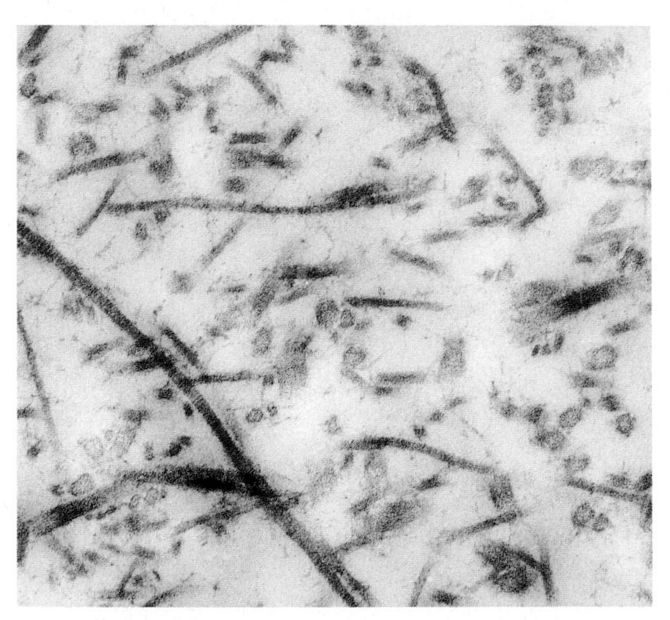

Fig. 22.6 Experimental canine anterior cruciate ligament arthropathy.
At the same time interval as in Fig. 22.4, the collagen fibre bundles of the zone II matrix are even more widely separated. A disorganization of the collagen–proteoglycan interrelationship seems probable. (TEM × 51 700.)

reveals the presence of occasional 'microscars' where the disappearance of chondrocytes by lysis has led to a focus of collagen fibre condensation. Microscars, like amianthoid fibres,[182,429] are features of advancing age: their presence may be associated with the increased tendency in ageing for collagen to develop more, and stronger, cross-links, leading to an increasingly crystalline structure.

The alterations of proteoglycan structure, amount and arrangement that are thought, on chemical evidence, to be crucial to the cartilage matrix disorganization of osteoarthrosis, are much more difficult to demonstrate by transmission electron microscopy than alterations in the robust, insoluble collagen.[429] Nevertheless, labels such as ruthenium red allow proteoglycan/collagen associations to be defined, as do electron-dense polycationic dyes such as cupromeronic blue that can be used in critical electrolyte concentration techniques.[415,346] Early evidence with these labelling and staining methods confirms the disruption of proteoglycan/collagen association sites (see p. 55). A reduction in proteoglycan stainability is characteristic of the early disease.

Many chondrocytes in all zones have been found to have pericellular spaces bounded by an amorphous peripheral rim.[308] The technical difficulties in interpreting these appearances are discussed in Chapter 1 (p. 23). The pericellular matrices contain increased quantities of proteoglycan. On this basis, morphologically distinct clones of chondrocytes have been recognized.[486] One class has an intensely staining pericellular halo, the other has not. Around cells of the first variety, the pericellular matrix comprises mature circumferential, closely packed collagen fibres together with many small beaded fibrils. There is an increase in matrix electron density, many matrix vesicles and lipid accumulations. The chondrocyte nuclei are ovoid and compact with numerous perinuclear, intracellular intermediate filaments. However, rough endoplasmic reticulum, Golgi components and mitochondria are identified. Around cells of the second variety, the matrix is restricted and compact. These chondrocytes contain fewer cytoplasmic filaments and lysosomes but lipid droplets and glycogen granules are recognizable. Centrioles may be seen but whether their presence is an indirect index of mitotic activity remains uncertain: dividing cells are not detected.

With the morphological progression of the disease, evidence of chondrocyte degradation becomes more frequent. Myelin figures and chromatin condensates appear. With time, cell remnants provide the only evidence of irreversible chondrocyte injury. Chondrocyte loss coincides with the progression of the matrix degradation that appears by light microscopy as fibrillation (see p. 858). The rate of cartilage matrix destruction exceeds that of repair which is focal and incomplete.

A proportion of superficial chondrocytes sustain irreversible injury early in osteoarthrosis; some die or undergo apoptosis. Those chondrocytes that are not lost early from zones I and II are large, more ovoid than normal, and retain a fibroblast-like fine structure. The continued presence of much rough endoplasmic reticulum, a large Golgi apparatus, many ribosomes and cytoplasmic filaments, conforming with isotopic labelling and other metabolic analyses, supports the view that, in early osteoarthrosis, chondrocyte metabolism and activity are increased above normal (Figs 22.7 and 22.8). Provided the nuclear membrane remains intact, and the intracellular filaments occupy less than half of the cytoplasmic volume, there is no reduction of $^{35}SO_4$-sulphate, ^{3}H-proline or ^{14}C-glucosamine incorporation, assessed by autoradiography; once the nucleolemma is disrupted, synthetic activities decrease markedly.

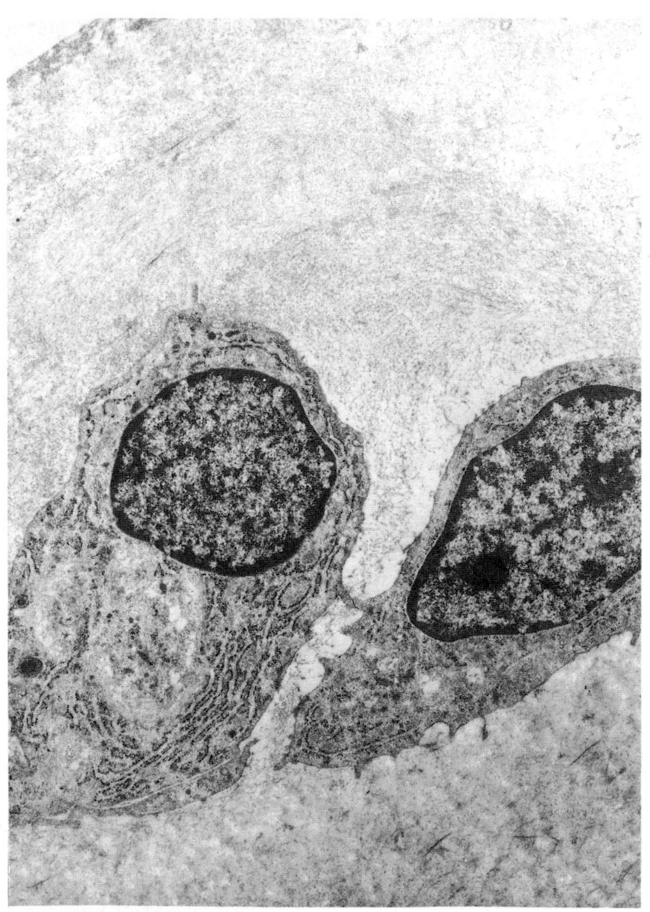

Fig. 22.7 Experimental canine anterior cruciate ligament arthropathy.
In this model of osteoarthrosis, there is an early and progressive change in the numbers and morphology of articular chondrocytes. The abnormalities, particularly severe in the cells of zone I, are illustrated in this preparation made 24 weeks after operation. Close to the articular surface (top left) is a pair of ovoid chondrocytes. In this situation, chondrocytes are normally single and ellipsoidal. (TEM × 8880.)

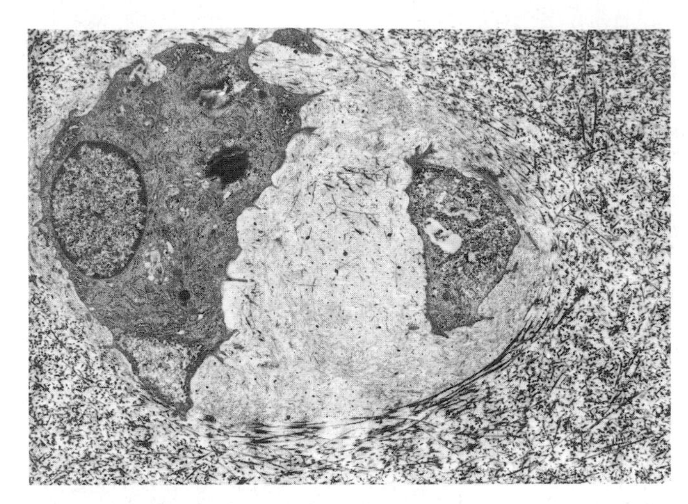

Fig. 22.8 Experimental canine anterior cruciate ligament arthropathy.
As cartilage disorganization becomes established, an excess of pale-appearing pericellular matrix forms between the apposed chondrocytes of zones II and III. The electron-lucent matrix contains finely fibrillar material but it is not known whether this is type II collagen, another type of fibrillar collagen or another material. (TEM × 3000.)

As osteoarthrosis advances, the articular cartilage comes to contain numerous matrix vesicles.[9,12] These vesicles (see p. 108) are associated with the presence of ultramicroscopic clusters of calcium hydroxyapatite crystals (see p. 402). It has been proposed that the crystals form and grow because of changes in the enzymatic regulation of local calcium and phosphate fluxes. In turn, the crystal niduses become focal cartilage dyshomogeneities; they are points at which mechanical stresses are concentrated, possible causes of microfracture and progressive cartilage disruption.

The ultrastructural appearances of cartilage in dogs with spontaneous osteoarthrosis have been described;[160] those of experimental canine post-cruciate ligament section arthropathy are outlined on pp. 899 *et seq.*

Scanning electron microscopy

Scanning electron microscopic techniques confirmed the presence and severity of the ultrastructural surface changes of early osteoarthrosis.[205,382,383,502] Preliminary studies were handicapped technically by practices such as the air-drying, acetone drying and critical-point drying of cartilage specimens but 'optimum' techniques[58] have become more general and artefacts less common. Redler[382] and Minns and Steven[305] summarize the observed osteoarthrosic changes. There is surface disruption, with tortuous, nodular collagen fibres and fibre bundles, fissures, craters and a peeling away of laminae of superficial, zone I fibres[205] (Fig. 22.9). The subsurface fibrillar meshwork of relatively small collagen fibrils coalesces into an ill-defined material; ultimately, cartilage loss culminates in exposure of calcified cartilage and, later, bone.

Low-temperature scanning electron microscopy

The early ultrastructural surface changes of human and of experimental dog and mouse osteoarthrosis were demon-

Fig. 22.9 Osteoarthrosis of human tibial condyle; conventional scanning electron microscopy.
a. This low-power micrograph depicts approximately 1 mm^2 of the articular surface of a microscopically fibrillated tibial condylar cartilage surface. **b.** Closer examination reveals that zone I collagen fibre bundles are disrupted and often fractured. Synovial fluid cells present at the surface include a small number of erythrocytes released during dissection. **c.** In other parts of the same joint surface, the collagen fibre bundles retain their alignment; separation of the bundles is beginning and crimping can be seen (top right). **d.** In this higher-power view, the periodicity of the partially crimped collagen fibre bundles can be seen. It is possible to suspect the beginning of a separation of the individual fibres that comprise the bundles. (SEM **a**: × 100; **b**: × 500; **c**: × 500; **d**: × 2000.)

strated by low-temperature scanning electron microscopy.[130,137,338] In humans, the gradual loss of the pericellular matrix leaves conspicuous, radiating perichondrocytic fibril meshworks[338] (Figs 22.10 and 22.11). The appearances resemble those produced when controlled enzymatic degradation is deliberately employed to allow systematic examination of the residual cells and deeper matrix.[337] The technique reveals early pericellular proteoglycan and water loss which evidently precede the surface collagen disorganization of advancing fibrillation.

Conventional and low-temperature scanning electron microscopic methods offer limited resolution. Neither this technique nor scanning transmission electron microscopy are ideal approaches to the surfaces of bulk material. Consequently, the possibility of preparing platinum/carbon replicas of rapidly frozen osteoarthrosic cartilage blocks and of the analysis of these replicas by transmission electron microscopy has attracted interest. Replicas of the fully hydrated, proteoglycan-containing surfaces can be studied at high resolution and it is therefore possible, with this method, to search for intimate details of macromolecular disorganization in very early experimental osteoarthrosis.[137,137a]

Confocal scanning optical microscopy

A renaissance in light-microscopy has centred on scanning optical instruments.[133,421] Images of morphologically intact osteoarthrosic cartilage have been obtained by optical sectioning and three-dimensional analyses of chondrocyte changes in osteoarthrosis have begun (Fig. 22.12). Tandem scanning microscopy offers somewhat similar opportunities.

Other methods

Advances in technique have allowed other electron microscopic investigations of osteoarthrosic cartilage to be made. Freeze-etching was used by Spycher, Moore and Ruttner[446] to prepare replicas of the affected surfaces of human femoral heads. Chondrocytes were investigated. A pericellular corona contained membrane-bounded vesicles. These structures were thought to be derived from the plasma membrane of the chondrocytes but the relationship of the vesicles to the onset of osteoarthrosis remained obscure.

One of the immediate effects of any significant mechanic-

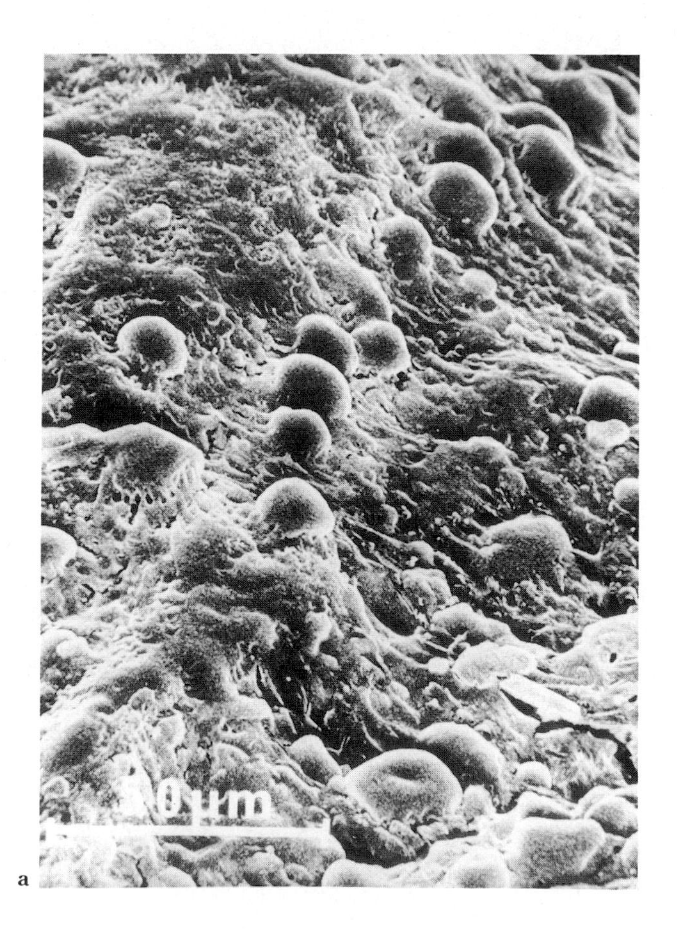

a

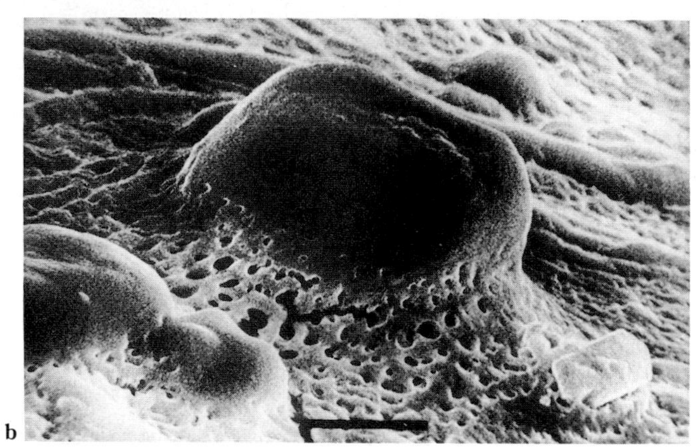

b

Fig. 22.10 Osteoarthrosis of human tibial condyle; low temperature scanning electron microscopy.
Fresh cartilage blocks were rapidly frozen in nitrogen slush at −210 °C, transferred to the microscope in the cold state without exposure to the ambient atmosphere and examined at temperatures significantly better than −140 °C. They were therefore unfixed, virtually fully hydrated and in an *in vivo*-like state. **a.** At low magnification, an area of microscopic fibrillation displays single chondrocytes and cell clusters protruding from the bearing surface following the pericellular degradation and loss of cartilage matrix. Residual collagen fibre bundles can be detected. **b.** At higher magnification, it is clear that around the chondrocyte or chondrocyte cluster, a loss of matrix proteoglycan leaves a residual laciform meshwork of collagen fibre bundles that continue tenuously to prevent the cell(s) from release. (**a**: × 760; **b**: × 2000.)

Fig. 22.11 Experimental canine anterior cruciate ligament arthropathy; low temperature scanning electron microscopy.
Eight weeks after cruciate ligament section, fissures and troughs were detectable in the surface of the unfixed, fully hydrated medial femoral condyle. One explanation for the development of these features is a form of plastic deformation; another is crack propagation. (× 1980.)

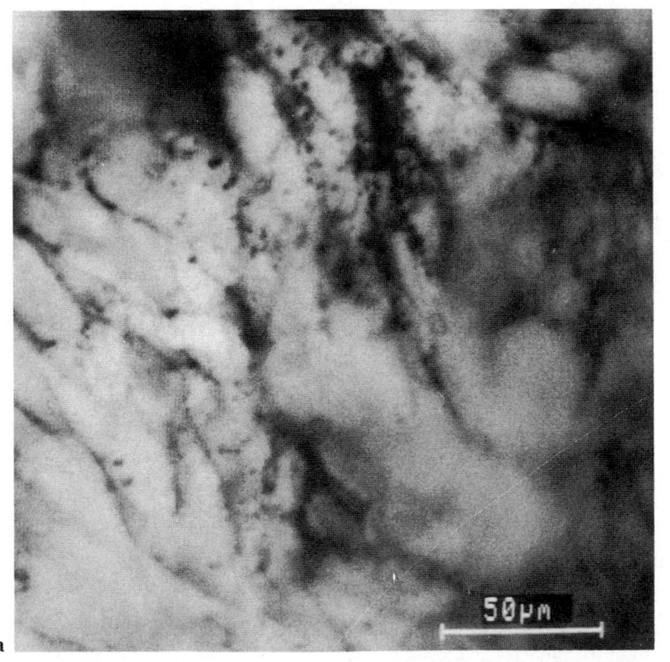

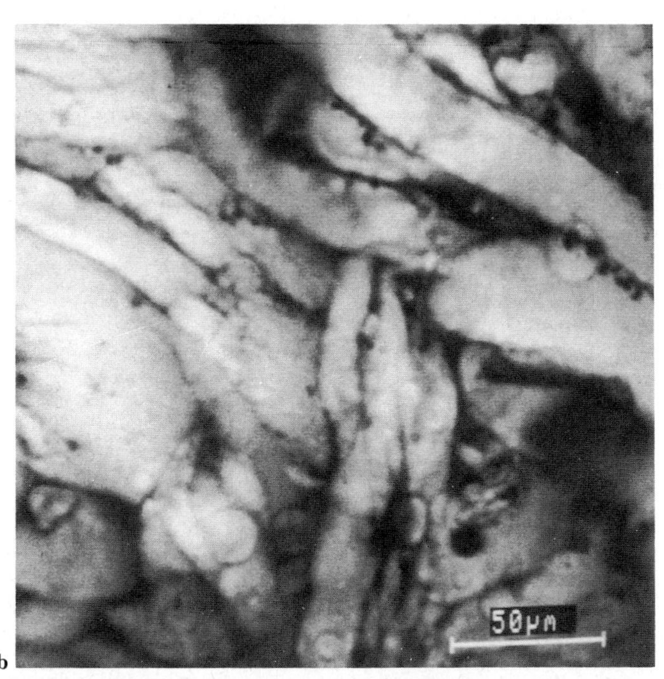

Fig. 22.12 Fibrillation of bearing surface of medial tibial condyle; confocal scanning optical microscopy.
Serial optical sections tangential to the fibrillated surface. In spite of their resemblance to the electron optical images derived from the amplification of secondary electrons during scanning electron microscopy, these figures are, in fact, representations of photons reflected from narrow planes of focus. (Scanning laser microscopy × 400.)

al agent on articular cartilage surfaces must be upon the ultramicroscopic lamina (see pp. 142–143) which lies between and is coextensive with the synovial fluid and the most superficial part of the zone I cartilage. The surface lamina is negatively charged. The density of these charges declines with age[246] and this decline may exacerbate the macromolecular surface disorganization that culminates in the microscopic fibrillation of osteoarthrosis.

Light microscopic changes

Whichever the joint or anatomical site, and irrespective of whether the joint is frequently exposed to gravitational as well as to intrinsic loads ('weight-bearing') or largely to non-gravitational forces ('non-weight bearing'), a series of characteristic, interrelated, microscopically recognizable changes follows the macromolecular disorders described on pp. 848 and 849. Articular cartilage, bone, periarticular structures and the synovia are all affected.

(a) Cartilage

The volume of cartilage matrix[99] greatly exceeds that of the chondrocytes; the extracellular matrix is relatively abundant and it is not surprising that it has attracted much interest. Although matrix disorder has been studied more fully than chondrocyte dysfunction, osteoarthrosis is a cellular disease and it is logical to comment first on the chondrocyte population.

Chondrocyte proliferation

The early structural changes of osteoarthrosis are characterized by a conspicuous increase in chondrocyte numbers, a change which is reproducible experimentally. Mitotic figures are not seen in random human specimens but ^{3}H-thymidine given to rabbits, the patellar groove cartilage of which has been excised, demonstrates enhanced incorporation of the label not only in the residual traumatized cartilage but also in the cartilages of the sham-operated or non-operated joints.[174] Analogous changes have been identified in the cartilages of dogs subject to surgical section of an anterior cruciate ligament (see p. 898).

In human osteoarthrosic cartilage, a remarkable process of proliferation of the cells remaining between the fibrillary clefts begins and clones of chondrocytes appear as microscopic cell islands (Fig. 22.13). Eight, sixteen or more cells constitute each island (as many as 200 have been counted) and they closely adjoin cleft (see p. 858) margins. Mitotic figures are absent and tests made with monoclonal antibodies such as Ki67 which recognize cells in the cycle of division, have proved inconclusive. However, the demonstration that the affected tissue contains many more cells per unit volume than normal and that there is an increased tissue DNA content, clearly implies either cell division or cell recruitment from another, extraarticular site. There is no observational or experimental evidence to support the concept of cell recruitment, neither is there evidence that the increased cell density is attributable to matrix loss. Indeed, in early experimental dog postcruciate ligament section arthropathy, the mean cartilage thickness increases. Amitotic chondrocyte division is thought improbable. In spite of the poor capacity of hyaline articular cartilage for repair by regeneration, it must be assumed therefore that there is, after all, enhanced cell division, a view supported by experimental evidence that shows an early increase in DNA content in postcruciate ligament section arthropathy (see p. 849). In this form of canine osteoarthrosis the increase in chondrocyte numbers is very early. Although hypercellularity takes place without apparent mitotic activity, it is accompanied by conspicuous changes in cell morphology and organization (see p. 857), suggesting a reversion to an inappropriately undifferentiated structure. These alterations parallel the increased synthesis of macromolecules characteristic of the early disease. It appears probable that comparable changes occur in early human osteoarthrosis. Any subsequent, selective death of chondrocytes is likely to be ischaemic; it cannot account for the initial biochemical and biomechanical phenomena.

The stimulus to, and mechanisms of, chondrocyte proliferation are poorly understood.[103] A mechanism similar to that by which keratinocytes form clusters without cell division has been considered.[266] A role for growth factors (see p. 28) is assumed. That chondrocyte clusters arise simply because the cells come closer to the synovial fluid, the major source of the oxygen and metabolites required for chondrocyte survival,[75] seems unlikely. Normal hyaline cartilage not in continuity with the vascular synovium or bone marrow has a poor capacity for repair.

Chondrocyte degradation and loss The few detailed analyses of chondrocyte change in established, overt osteoarthrosis stress the occurrence of cell death in superficial zone I cartilage.[293] Disorganization and loss of zone I and zone II cartilage exposes chondrocytes to abnormal environmental stresses; surface cells shrink and die, those of the deeper zones III and IVa becoming at first swollen and clumped. Much of this work has been upon the numerous examples of advanced, stage 2, 3 or 4 (see p. 869) femoral head disease made available by the increasingly common operation of hip joint arthroplasty. In such surgical excision specimens, Meachim and Brooke[293] readily admit that the 'rows of empty osteocyte lacunae' that they detect

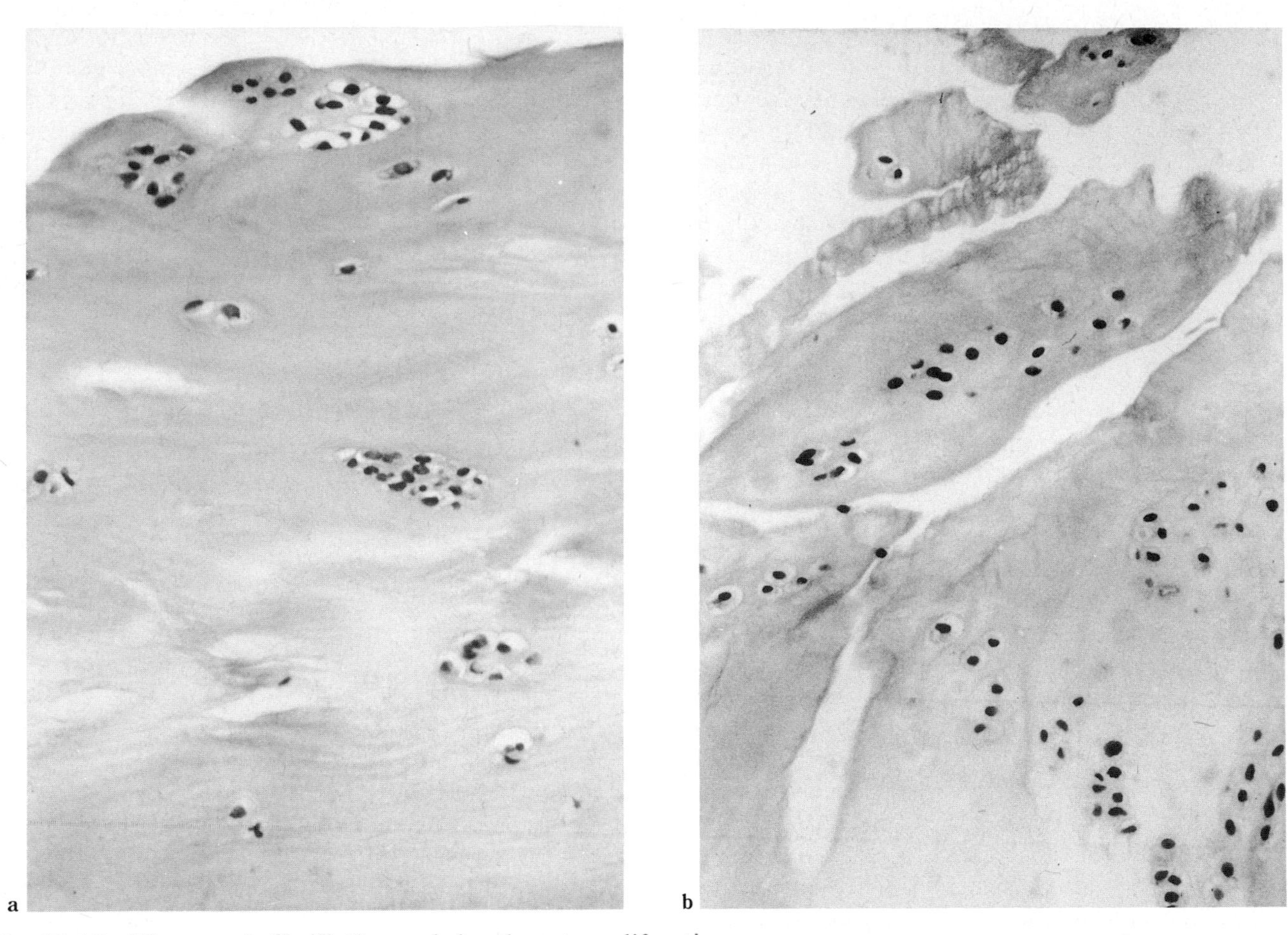

Fig. 22.13 Microscopic fibrillation and chondrocyte proliferation.
a. Tangential clefts, microfractures, are appearing within the cartilage. In advance of this process, zone I (superficial zone) chondrocytes have increased in numbers and appear in clusters that are often termed clones on the assumption that they have originated from the multiplication of a single cell. There is, however, no evidence of mitotic activity. In spite of a suggestion that ^{3}H-thymidine is incorporated at an increased rate, it remains possible that the new cells have been recruited from outside the cartilage. (HE × 280.) **b.** In this field, tangential and vertical clefts have developed. Where collagen fibre bundles have become partially detached (top left), they appear crimped. Several chondrocyte clusters have formed in the matrix islands remaining between the clefts; when cartilage amyloid is present, this glycoprotein can often be detected at the margins of these fissures. (HE × 280.)

may be associated with cytotoxic, avascular or mechanically induced tissue necrosis. When, however, microscopic study is confined to joints less prone to bone necrosis, there is much less evidence of chondrocyte injury or death.[435] The significance of empty chondrocyte chondrons (see p. 23), the spaces defined by the pericellular collagenous capsule, is, indeed, open to question. When formalin-fixed paraffin-embedded blocks of decalcified, dehydrated articular material are prepared as haematoxylin and eosin-stained sections, conventional light microscopy reveals varying numbers of empty chondrons in the majority of specimens from aged subjects. When formalin-fixed but fully hydrated 2–300 μm-thick slices of the same cartilage are examined by confocal light microscopic cell counts, the cell density is higher, the number of empty spaces less.

Cartilage swelling It was often claimed that the most sensitive index of articular hyaline cartilage abnormality recognizable by light-microscopy in osteoarthrosis was a focal, basophilic swelling of the extracellular matrix.[36] Hough and Sokoloff[183] describe this change as 'chondromucoid softening'. Small 'blisters' are said to appear deep to the tangential zone I lamina of collagen fibre bundles.[75] These 'blisters' burst, causing shallow defects in the cartilage. The defects have frayed edges. However, 'chondromucoid softening' and 'surface blistering' are subjective criteria. Objectively, the most delicate light-microscopic changes resemble those identified early in experimental osteoarthrosis; they include increased hydration, collagen fibre disorganization and chondrocyte atypia (see p. 898).

Collagen disorganization

Meachim and Brooke[293] propose an early separation of collagen fibres, accompanying and perhaps initiating the matrix swelling that is permitted when proteoglycans, no longer constrained by collagen, are enabled to expand. These early collagen abnormalities are central to theories which attribute osteoarthrosic fibrillation (see below) to a reduction in the tensile strength of the cartilage collagen fibre bundles,[126,293,320] and thus to a fault in the regulation of proteoglycan compression. One explanation for a progressive failure of collagen–proteoglycan links is a postulated defect in the quality or organization of type IX collagen.

Loss of metachromasia

The systematic examination of ageing human tibial condyles reveals a highly variable staining intensity with metachromatic polycationic stains.[136] It is difficult to be certain when a low concentration of glycosaminoglycans, indexed, for example, by weak toluidine blue stainability, can be taken to represent early, prefibrillary osteoarthrosis. As the lesions of osteoarthrosis progress morphologically, with chondrocyte proliferation and fibrillation, reduced matrix stainability becomes more conspicuous. There is a significant inverse correlation between the severity of the histological process and polysaccharide concentration.[273] The loss of metachromasia and a decrease in safranin O and in alcian blue reactivity, are accepted as indices of the progressive leaching out of newly synthesized, but abnormal, proteoglycans from their sites within the prejudiced collagen meshwork. All serious attempts to grade the microscopic changes of osteoarthrosis are obliged to take account of these changes.[273]

Fibrillation (Fig. 22.14)

The accentuated microscopic roughening of the cartilage surface, and the formation of both small tangential and oblique and vertical clefts into and through the cartilage matrix is a secondary phenomenon termed fibrillation. There is splitting of the matrix parallel to the surface, in the axis of the collagen fibre bundles; horizontal flakes form and begin to be detached. The separation and splitting of cartilage components extends more deeply but begins to assume a tangential and then a vertical orientation. Small blisters are said to appear deep to the tangential surface fibre bundles; the blisters burst, producing shallow defects in the cartilage with frayed edges. The splitting process takes place focally, in preferred cartilage zones and islands where loss or change in matrix proteoglycan or in their component glycosaminoglycans is demonstrable as a reduction of metachromatic staining.

Fibrillation (Fig. 22.14) is readily identified by conventional microscopy ('microscopic fibrillation'[125]) and is regarded as a hallmark of established osteoarthrosis. There are two drawbacks to this viewpoint. First, it is clear that by the time fibrillation is recognizable, the initial cytogenetic and molecular changes of osteoarthrosis are long established; and second, fibrillation is not an unique structural change since an indistinguishable microscopic phenomenon is commonplace in cartilage senescence (see Chapter 21). Fibrillation is a form of crack propagation; it is determined by well-established mechanical principles (see Chapter 6). The superficial, tangential microscopic cracks are usually in alignment with zone I collagen fibre bundles. As the cracks extend, the free ends of disrupted zone I collagen bundles lie at the cartilage surfaces. Devoid of restraint but subject to intrinsic microscopic and molecular forces, the surface collagen bundles assume a 'crimped' pattern (Fig. 22.9); they come to resemble small, coiled ropes.[330]

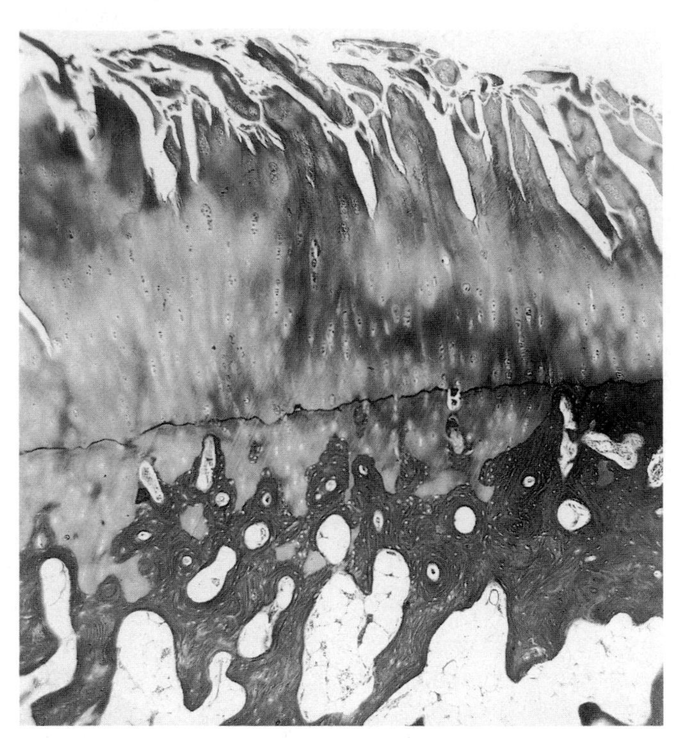

Fig. 22.14 Microscopic fibrillation and early osteosclerosis.
Tangential and vertical clefts have been propagated downwards from the articular surface; the longest crack extends to within a short distance of the tide-line.
Whether the microfractures that constitute this process of fibrillation extend at random or in an axis determined either by differential matrix composition or by the location of the chondrons, is not yet certain. (HE × 37.)

Fibrillation advances focally; many cartilage areas are spared but, in affected parts, splitting gradually becomes deeper. The diurnal imposition of shearing forces determines that separation of the cartilage matrix components extends to follow the orientation of the zone II and zone III collagen fibre bundles.[296] The continued softening of additional islands of cartilage leads to a redistribution and local concentration of stresses; lateral support for residual, intact cartilage is reduced. The lessened local restraint results in disproportionately large deformations under load, enlarging the defects. The clefts ('splits') become less horizontal and more vertical. Finally, they impinge on the calcified cartilage and on subchondral bone. Horizontal splits develop at the tide-line. Flakes of cartilage are then detached from the calcified cartilage.

The progressive nature of osteoarthrosic fibrillation (Fig. 22.15) is widely held to distinguish the lesion from the non-progressive lesions of ageing (see Chapter 21). It is important to record that the presence of fibrillation is often complicated by other concurrent disease. Amyloid is frequently detected, particularly in the superficial (zone I) cartilage of elderly subjects (see p. 436). In one series of 116 hip joint capsules, 28 per cent contained amyloid.[240] Amyloid deposits, identified by both Congo red dichroism and an affinity for anti-amyloid P component antibody, tend to be within the cartilage matrix, at the margins of surface cracks. Whether amyloidosis contributes to or even causes crack propagation is uncertain. There can be little doubt that its presence prejudices the mechanical integrity of the cartilage matrix.

Advancing fibrillation and the cumulative loss of cartilage fragments results in an overall thinning of this tissue. The anatomical distribution of preferred areas of osteoarthrosic fibrillation is discussed below and in Chapter 21. The freed particles of cartilage enter the synovial fluid and are effectively phagocytosed by the synovia (see p. 36) although larger fragments may persist in the joint space. Fragments are recognizable by fibreoptic arthroscopy and in synovial aspirates. Ferrography allows the magnetic separation of the particles and is a sensitive if little-used technique for detecting early cartilage breakdown.[111]

An understanding of the shape and direction assumed by the splits of fibrillation has been helped by studies in which a sharp, round pin has been used to prick the articular surfaces.[198] When the normal articular surface is punctured in this way, the elongated split that extends beneath the round puncture hole is in a direction parallel with the main collagen fibre bundles near the joint surface so that a 'split pattern' that is normal for each surface is obtained (Fig. 6.4).[294,335,336] The naturally occurring, early fibrillation of senescence is at first in the form of horizontal cracks and then vertical clefts, parallel to the lines of the split pattern.[296] Although deeper extensions of the vertical splits are influenced by the collagen fibre pattern, it also seems

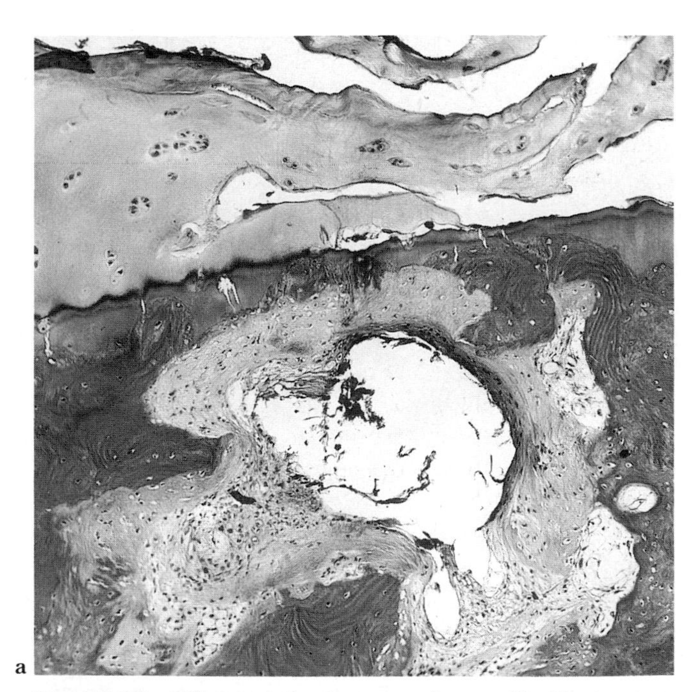

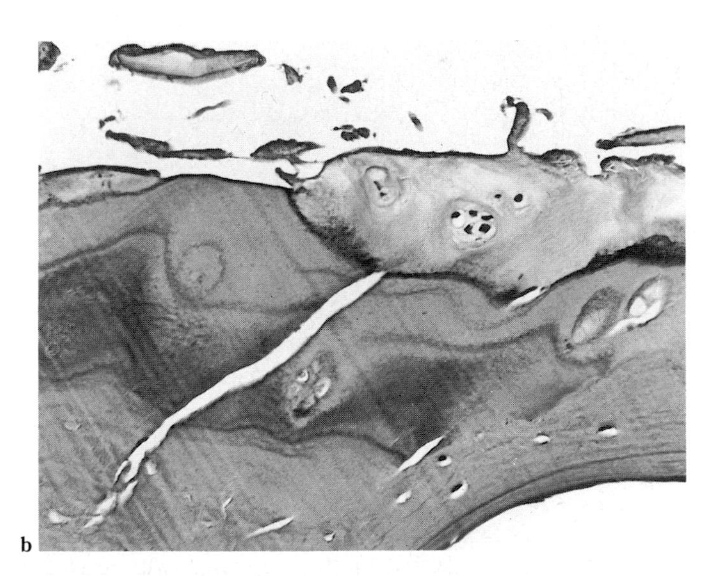

Fig. 22.15 Microscopic changes of overt fibrillation.
a. Less than half of the cartilage-bearing surface shown in this micrograph is covered by articular cartilage. The clefts that constitute the hallmark of microscopic fibrillation have extended to the bone in the right side of this field and fragments of cartilage have broken away, exposing the subchondral bony endplate. New bone has formed but within the marrow spaces and a change in the nature of the mesenchymal tissue has led to the development of a small radiotranslucent pseudocyst. **b.** Detail from same case. Cracks through the articular cartilage have been propagated as far as the tide-line which can be seen (upper top left) to form a small part of the free bearing surface. A small chondrocyte cluster lies within the residual, zone IV cartilage, and an oblique microfracture has extended downwards from the tide-line, through the calcified cartilage, into the hyperostotic bony endplate. The explanation for the origin of this oblique fracture must lie in shearing forces exerted at the tide-line. (HE **a.** × 60; **b.** × 170.)

possible that the direction and extent of the splits may be determined partly by the situation of chondrons (see p. 23). These units may act to concentrate shear stresses. Horizontal splits occur at the tidemark, where the interface of two different structures, the uncalcified and calcified cartilages, gives material discontinuity. The distinct elastic moduli determine shearing at the interface.

There are obvious changes in the overall mechanical properties of the affected cartilage. Osteoarthrosic cartilage is less able than normal to resist the loss of water under load. This phenomenon is attributable not to changes in proteoglycans that result in altered osmotic pressure, but to decreased fixed change density.[279] Direct mechanical testing demonstrates that osteoarthrosic femoral head cartilage is weaker in tension and softer in compression.[393] The thickness of the cartilage at the zenith is decreased and the bone density increased. Loss of proteoglycan (measured by fixed charge density) does not appear to be the initial event preceding osteoarthrosis and compressive stiffness does not depend on proteoglycan. Explanations were sought for the primary structural events in femoral head osteoarthrosis. Tests were made with a microcompression technique, employing artefactual notches to examine the propagation of cartilage tearing/splitting. They showed that two structural characteristics crucially compromise cartilage integrity. They were diminution of the strength of first, interfibril crosslinks and second, individual fibrils.[46]

Crystal deposition Crystal aggregates are not uncommon in the superficial, zone I cartilage of elderly persons sustaining subcapital femoral neck fracture.[449] The relationship of these crystals to osteoarthrosis is debated.

Calcium pyrophosphate Osteoarthrosic cartilage contains raised quantities of calcium, phosphorus and magnesium,[368] not usually in crystalline forms. Substantial amounts of pyrophosphate in particular may be produced by the abnormal cartilage.[190] The pyrophosphate is derived from frequent cell division and matrix synthesis or from remodelling calcified sites, accounting for much of the pyrophosphate found in osteoarthrosic synovial fluids.

Deposits of crystalline calcium pyrophosphate dihydrate are frequently identified in and among shreds of fibrillated osteoarthrosic cartilage.[39] Both chondrocalcinosis (see p. 393) and osteoarthrosis are increasingly common with advancing age and their coincidence is not unexpected. Nevertheless, the possibility remains that the pyrophosphate crystals may exacerbate fibrillation both mechanically and by promoting the activation of chondrocyte proteases through the expression of synoviocyte interleukin-1 (see p. 88).

Calcium hydroxyapatite Calcium hydroxyapatite crystals have been identified in osteoarthrosic synovial fluid and similar crystals have been found repeatedly in the hyaline cartilages of synovial joints in osteoarthrosis.[91,92] Complementing these observations is the demonstration that both the chondrocytes and matrix vesicles in regions close to the tidemark display alkaline phosphatase activity.[384] Many matrix vesicles are present in the extracellular matrix of osteoarthrosic cartilage and their presence is specifically related to phosphatase activity and the initiation of mineralization in the region of the tide-mark.

Electron dense, lipid-containing 'cellular debris' was found by early workers in the perichondrocytic matrix of rabbit cartilage[8a] and identical structures were recognized in osteoarthrosic human cartilage.[289] This extracellular lipid was shown to be identical with the matrix vesicles of epiphyseal cartilage[11] and in zone IV/V human articular cartilage such vesicles are associated with the initiation of calcium hydroxyapatite crystal formation in and near the tide-mark. The crystals may be of least three different forms: fine (deep zone), dense ('cuboid') (superficial zone) and needle-shaped (surface).

Crystals of calcium hydroxyapatite are detected easily within the cartilages of the femoral head[341] and tibial condyles. The extent of the hydroxyapatite deposits has been measured.[449] The femoral head deposits are more frequent in osteoarthrosic (41 per cent of 54 subjects aged 65.2 ± 8.6 years) than in control tissue (6.0 per cent).[341] The presence of hydroxyapatite crystals within the cartilage may be an index of their causal role in the pathogenesis of osteoarthrosis (see p. 893). Alternatively, the crystals may be no more than a reflection of the advancing abrasive wear of exposed bone. It seems reasonable to suppose as Gordon, Villanueva, Schumacher and Gohel[158] have done, that calcium hydroxyapatite and calcium pyrophosphate crystals may contribute, together with bone fragments, to both cartilage degradation in osteoarthrosis and to the onset and prolongation of synovitis.

The significance of ultrastructural aggregates of these crystals and their relationship to matrix vesicles is discussed further on pp. 107 and 893. Ohira and Ishikawa[341] rarely found matrix vesicles in foci of calcification.

Lipid deposition The small quantities of lipid present in superficial (zone I) cartilage include surfactants (see p. 22) and arachidonic acid. Their relationship to osteoarthrosis is not certain. Chondrocytes in osteoarthrosic cartilage contain raised amounts of triglyceride and complex lipids but this material may be present before microscopic fibrillation is detected. It should be recalled that increased intracellular lipid is recognized in senescence (see p. 822). The excess is likely to be an index of impaired chondrocyte metabolism and may be attributable to the coincidental ischaemia which, in sites such as the femoral head, may lead to superficial chondrocyte and deeper bone necrosis.

Recognition of osteoarthrosic changes in small cartilage samples The increasing sophistication of diagnostic techniques such as the resection of osteophytes by fibreoptic arthroscopy, means that very small but numerous fragments of cartilage now often reach the histopathologist. Comparable small samples may be available for biochemical analysis. Under these circumstances, it is not possible to grade the lesions of osteoarthrosis by classical criteria such as those of Mankin *et al.*[273] which have proved so useful in the study of whole joints in experimental animals, for example. The cytological analysis of individual chondrocytes[307] or of discrete, separated chondrocyte clusters offers one solution to this dilemma. These microscopic samples can be tested for the presence of proteins such as S100 the expression of which is increased in fibrillated cartilage chondrocytes and for fibronectin (see p. 850).

The problems of sampling have also been addressed by Sachs, Goldberg *et al.*[404] Three forms of cartilage change were recognizable. In the first, metachromasia was absent from perichondronal sites, forming pale halos among a normal-staining interterritorial matrix; the chondrocytes were spheroidal, evenly distributed, and single or paired. In the second, perichondronal metachromatic staining was intense but reduced interterritorially; the chondrocytes were spheroidal, arranged in clusters or columns commonly with clones of 3–12 cells. In the third, metachromatic staining was inconstant and irregular, the chondrocytes small and stellate, or larger and evenly dispersed. None of the small tissue samples prepared for organ culture from the hip and knee cartilages of normal joints displayed the first, second or third change whereas cartilage samples from patients with disabling osteoarthrosis often revealed more than one or even all three of the categories of abnormality. This analysis also emphasizes the impossibility of distinguishing the extent of tissue structure disorganization from surface appearances alone and the errors that can arise when composite samples are chosen for chemical or micromechanical testing.

(b) Bone

The relationships between the structure and function of osteoarthrosic cartilage and osteoarthrosic bone are intimate: neither tissue can usefully be analysed without the other. Three interrelated patterns of bone behaviour require consideration: first, the contribution of bone to the genesis of cartilage disintegration and joint failure; second, the influence on osteoarthrosis of primary but coincidental bone disease; and third, the changes in bone caused by osteoarthrosis.

Role of bone in the natural history of osteoarthrosis The reader is referred to pp. 890 *et seq.* where the contribution of bone to the pathogenesis of osteoarthrosis is more fully discussed.

Cortical and cancellous bone The earliest observations of human and prehominid skeletons immediately demonstrated the prevalence of bone disease as an integral part of articular degeneration and early studies of the pathological anatomy of osteoarthrosis (see p. 869) confirmed the impression that cartilage and bone lesions are inextricably associated. Mechanical factors transmitted by or derived from bone have therefore always seemed likely to be of importance in the initiation if not in the perpetuation of osteoarthrosis. The large limb bones, of very great mass by comparison with the articular cartilages, respond elastically to diurnal impact and static loading, absorbing a large proportion of the applied forces. Articular cartilage, although of higher elasticity, absorbs a relatively small proportion of these normal forces.[371,372,373,375]

Tide-line, calcified cartilage and chondrosseous junction It appears likely that the zone of calcified cartilage, zone V, acts to attenuate the shearing forces which tend to disrupt the interface between cartilage and bone. There are changes with age in the tide-line, a mineralization front; in the thickness of the calcified cartilage; and in the bone-forming front at the bone-calcified cartilage junction (see Chapter 21). Analogous disorders occur with and predispose to osteoarthrosis; they are not fully understood. Reduplication of the calcification front that constitutes the tide-line is a feature of senescence; it is particularly obvious at sites of advancing fibrillation and is characterized by an extensive deposition of calcium hydroxyapatite crystals in the territorial matrix of nearby deep zone (zone V) chondrocytes. The interface between calcified and non-calcified cartilage is a plane of material non-homogeneity and structural weakness. Whether physical changes in the reduplicated tide-line contribute to an alteration in the physical properties of this interface is a matter of speculation—few direct measurements have been made—but the occurrence of small, horizontally propagating microfractures is a common microscopic finding.

Osteoarthrosis and coincidental bone disease Osteoarthrosis is increasingly frequent as age advances and many of the bone diseases common in old age are coincidental findings. Metastatic and primary malignant bone neoplasms may extend to osteoarthrosic bone and cartilage and the changes of Paget's disease not infrequently merge with those of osteoarthrosic bone sclerosis, particularly in the vicinity of the hip joint.

Bone necrosis Bone necrosis is encountered in approximately 6 per cent of femoral heads excised during the surgical treatment of osteoarthrosis.[203,498] The appearances are those of infarction. Infarction occurs in sites of osteosclerosis (eburnation) so that it is secondary to the changes of osteoarthrosis, not a cause of them. Since eburnation is a consequence of active, highly vascular new bone formation, this is not surprising.

Osteoporosis It has been widely claimed that osteoarthrosis, for example, of the hip joint, is significantly less common in subjects with osteoporosis than in those without osteopenia. Osteoarthrosis and postmenopausal osteoporosis in women appear anthropometrically different.[90] The infrequency of osteoarthrosis in osteoporotic individuals has been used to support the argument that the mechanical attributes of normal bone contribute to bone sclerosis which has been held always to precede cartilage disorder (see below). That osteoarthrosis and osteoporosis seldom occur together is, however, no longer a tenable view. Indeed, there is evidence that osteoarthrosis develops as often in osteoporotic patients as in the general population.[175] Osteoporosis does not protect against the development of osteoarthrosis,[175] although it is still held that systemic osteoporosis is rare in those with idiopathic osteoarthrosis of the hip whereas local (disuse) osteoporosis may accompany the joint disease. It is possible that the new techniques which have been introduced for the study of bone in osteoporosis[138] may shed additional light on this controversy but these methods have not been adequately tested in osteoarthrosis.

Osteoarthrosis and other metabolic bone diseases Osteomalacia remains a common problem in the aged and is often identified in those who sustain fractures such as those of the femoral neck or shaft. There is evidence (see p. 474) that osteomalacia is unexpectedly frequent in persons with rheumatoid arthritis, a common cause of secondary osteoarthrosis, but the relationship of osteomalacia to idiopathic osteoarthrosis is not well understood.

Pagetic arthropathy (see p. 410) (Fig. 22.16) Paget's disease of bone, common in the elderly, becomes increasingly frequent as age advances. The vertebrae, pelvis and long limb bones of the leg are particularly prone to this unexplained disorder of bone formation and reabsorption.[208] Nearby joints, particularly the hip, knee and apophyseal, are susceptible to a form of disorganization resembling osteoarthrosis (Fig. 22.16). Although the joint disease has been termed Pagetic 'arthritis'[166] it seems reasonable to attribute the cartilage changes, with replacement of the subchondral bone by soft vascular tissue, a loss of support for articular cartilage and a collapse of foci of cartilage into bone defects, to the physical deficiencies in bone structure. True osteoarthrosic defects, such as progressive fibrillation and pseudocyst formation, may accompany Pagetic arthropathy but it is often difficult to be certain whether the osteoarthrosic disorder is secondary to Pagetic bone change or coincidental with it. In terms of Radin's hypotheses (see p. 891), it can be argued that the increased density and elasticity of Pagetic bone, with a diminished capacity to absorb forces imposed upon the limb joints during walking, predispose to cartilage disruption just as they do to fracture. Whether the control of Paget's disease by compounds such as calcitonin or diphosphonates ameliorates the associated arthropathy or not remains uncertain.

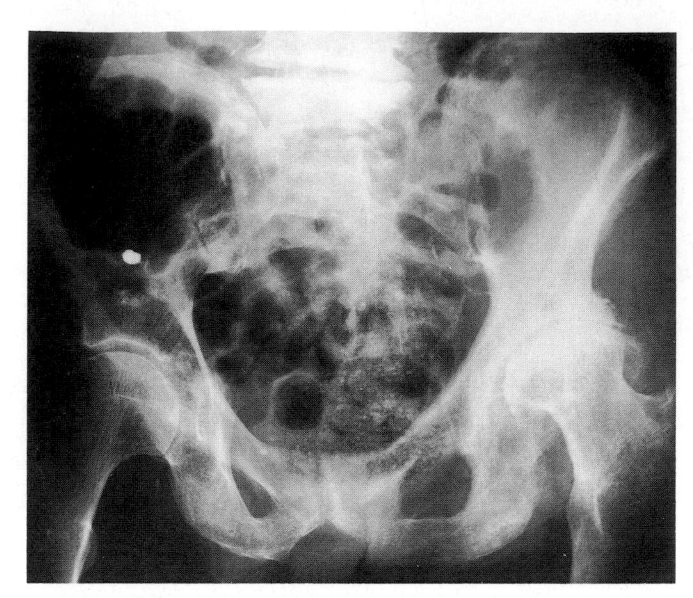

Fig. 22.16 Pagetic arthropathy.
In this elderly patient, the bone of the left acetabula and femoral head is patchily radioopaque; the interosseous space of the hip joint greatly reduced. The right hip joint serves as a radiographic control. The bone changes are those of active, severe Paget's disease; the articular disorder is a form of secondary osteoarthrosis and an osteosarcoma is present.

Bone changes resulting from osteoarthrosis (Figs 22.17–22.20) As fibrillation advances and cartilage splits (microfractures) extend more deeply, an active osteoblastic response begins within the subchondral bone. New appositional bone formation takes place both at the margins of the bone end-plate and on adjacent trabeculae. The result is a variety of localized hyperostosis which progresses to form a zone of compact bone in which the haversian systems are sustained by a high blood supply. The mechanism which initiates new bone formation in

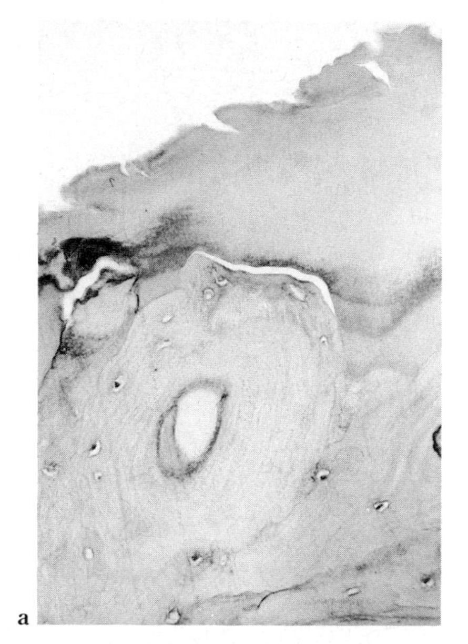
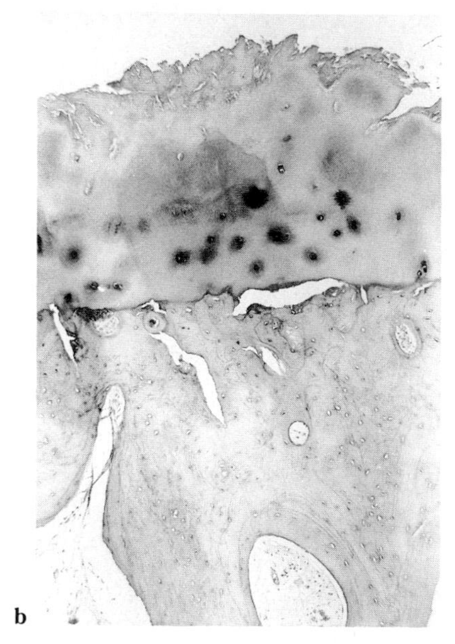
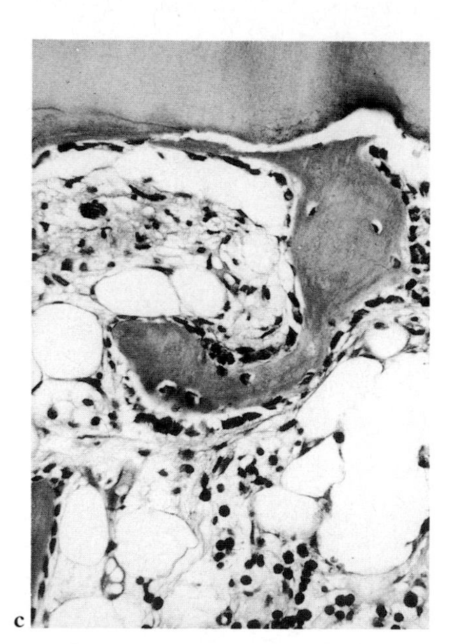

Fig. 22.17 Microfracture at the junction between the calcified and non-calcified cartilage.
a. Near an area where loss of cartilage has resulted in bone exposure, a microscopic crack has extended along the interface that is formed by the tide-line between the non-calcified hyaline cartilage (above) and the very narrow lamina of calcified cartilage (below). Viable osteocytes lie within the lacunae of the underlying hyperostotic bone. **b.** In the centre of this bearing surface, both horizontal and vertical microfractures can be seen. New, appositional bone formation occurs in advance of cartilage loss and it is interesting to compare the relatively mild fibrillation of the cartilage seen in this section with the hyperostosis of the underlying bone. **c.** Detail from bone adjoining part shown in previous Figures. Active new bone formation is taking place beneath the articular cartilage. At first the apparently woven character of the bone trabecula that is surrounded by numerous osteoblasts suggests that it is a new island of young bone. However, the horizontal microfracture that separates the island from the tide-line demonstrates that the trabecula has broken away from the deep margin of the bony endplate. The zone of calcified cartilage is so thin as to be unrecognizable. (HE **a:** × 65; **b:** × 16; **c:** × 160.)

advance of complete, whole-thickness cartilage loss is not fully understood. Mechanical stimuli, translated as chemical messages, may be conveyed from the increasingly denuded cartilage to subchondral osteoblasts by cytokines.

When the viability of osteocytes was tested by determining the activity of lactate dehydrogenase, 16 of 25 patients with idiopathic osteoarthrosis displayed non-viable cells in the centres of many femoral head bone trabeculae.[498] Bone death[500] is therefore a common feature of idiopathic osteoarthrosis but whether it is a cause of or a result of the disorder is uncertain (see p. 862). The interpretation of data on osteocyte death requires care: analytical methods may provoke osteocyte loss and cell death may be induced by cortiscosteroid therapy.

In the absence of osteomalacia, the hyperostotic bone of osteoarthrosis retains a normal mineral content. Fazzalari, Darracott and Vernon-Roberts,[117] for example, were able to confirm that there was no difference between the mineral content of male and female osteoarthrosic femoral neck bone as opposed to bone from patients with non-pathological fracture (Fig. 22.21). However, subjects with osteoarthrosis form less bone surface than patients with fractured femoral neck for the same mineralized bone volume (Fig. 22.22). It is of interest that trabecular thickness and spacing do not correlate with age.[116]

The increased density of the bone beneath osteoarthrosic cartilage may be a local response to the repair of microfractures or it may be secondary to the cartilage disorder. In osteoarthrosis-prone STR/ORT mice (see p. 903), as in man, the amount of bone in the medial tibial epiphysis is greater than in the lateral. The quantity of bone in young STR/ORT mice becomes greater than in control CBA animals but the increase does not explain the differential occurrence of osteoarthrosis: there is a wide time-span in the onset of cartilage degeneration and some mice develop a high bone density without osteoarthrosis. It has been suggested that the bone sclerosis of murine osteoarthrosis is attributable to diminished osteoclasis.

Eburnated bone is composed of an exceptionally dense, lamellar matrix; the cell density is low. Although measurements have not been made of the subjacent trabeculae of the cancellous bone, there is an impression that the cancellous structure is not correspondingly altered. The new, subchondral bone, in part woven, in part lamellar, is less strong than nearby, normal bone. For this reason, vertical or tangential microfractures are seen extending through

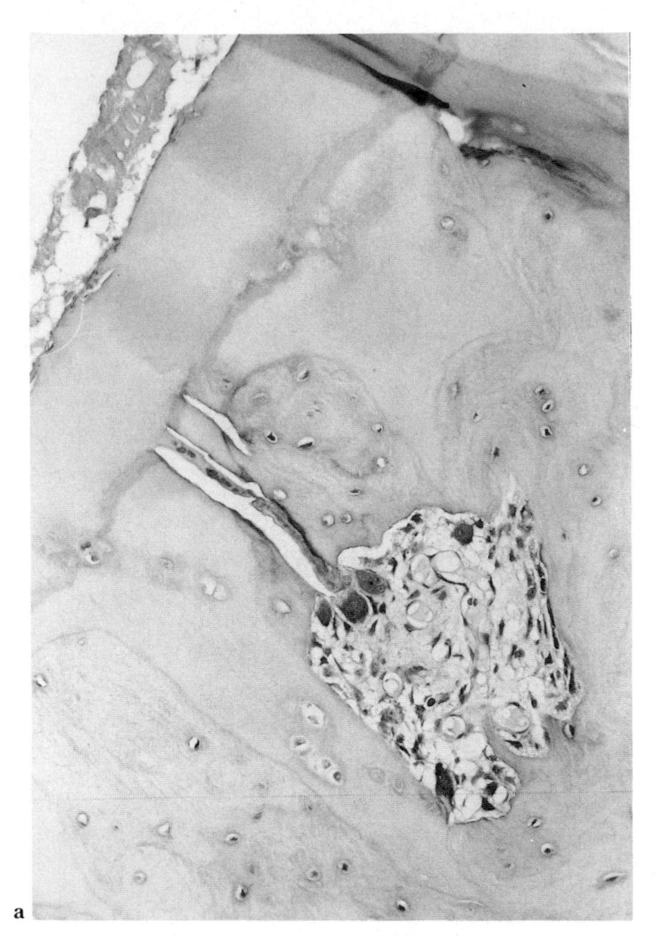

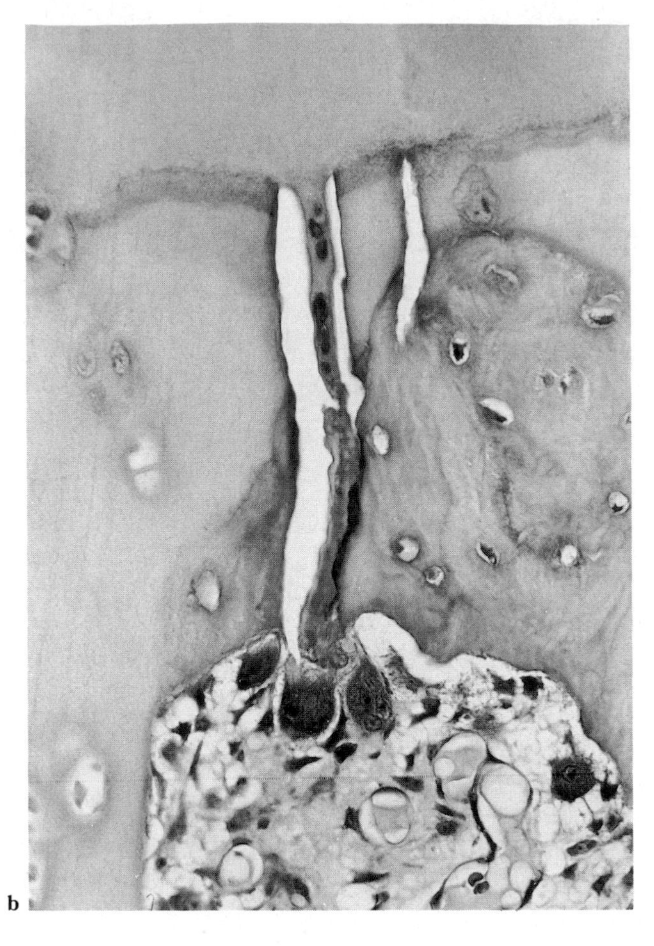

Fig. 22.18 Osteoclast response to microfracture.
a. When microfractures that have originated at the tide-line penetrate to the deep bone margin exposing an osteoid seam to the collagenolytic activity of osteoblasts, osteoclasts are likely to be recruited. Their function is to initiate remodelling at the fracture edge. In this field, the relationship of a microfracture to the surrounding osteosclerotic bone is clearly shown. **b.** Detail of same field. A secondary consequence of osteoclast activation in foci of the kind shown here, is the initiation of a change in marrow structure which leads to pseudocyst formation. Note in this micrograph how a finger-like process of hyaline cartilage forms the left-hand margin of the microfracture, while sclerotic bone forms the right-hand margin. Shear stresses at this interface may have contributed to fracture. (HE **a**: × 120; **b**: × 260.)

part of or, much less often, the whole of the thickness of the remodelled bone end-plate. It is not unusual to see focal osteoblastic[405] and osteoclastic, cellular reactions at the deep margin of microfractures: the occurrence of a fracture is a clear local stimulus to focal reparative bone remodelling.

Measurements of $^{99}Tc^{m}$-polyphosphate and of $^{99}Tc^{m}$-diphosphonate uptake demonstrate that these compounds accumulate, in osteoarthrosic femoral heads, not only in denuded weight-bearing areas, but in pseudocyst walls and in osteophytes, at the osteochondral junctions.[73] However, there is no generalized increase in bone synthesis. Patients with osteoarthrosis do not display an abnormally high 'mineral' density, assessed by gamma-ray attenuation in forearm bones or by the radiographic thickness of the second metacarpal bone.[8] Whether this observation will be confirmed by the more sensitive techniques of computed tomography (CT) scanning and photon absorptiometry is not yet known. New bone formation beneath the bearing surfaces is accompanied by bone remodelling here and at the articular periphery. Osteoclastic bone reabsorption and new lamellar bone formation take place at the deep margins of the bone end-plate and at the margins of nearby bone trabeculae. The bone formation culminates in local bone sclerosis, detectable radiologically. It is a paradox that this common response can occur in the osteoporotic bone of aged persons. Some have suggested that the stimulus to new bone formation may be the extension of fibrillation.[75] However, bone sclerosis is regularly seen deep to cartilage devoid of penetrating fissures. Extensive cartilage loss may overlie intact bone. The rate of growth of bone at reminer-

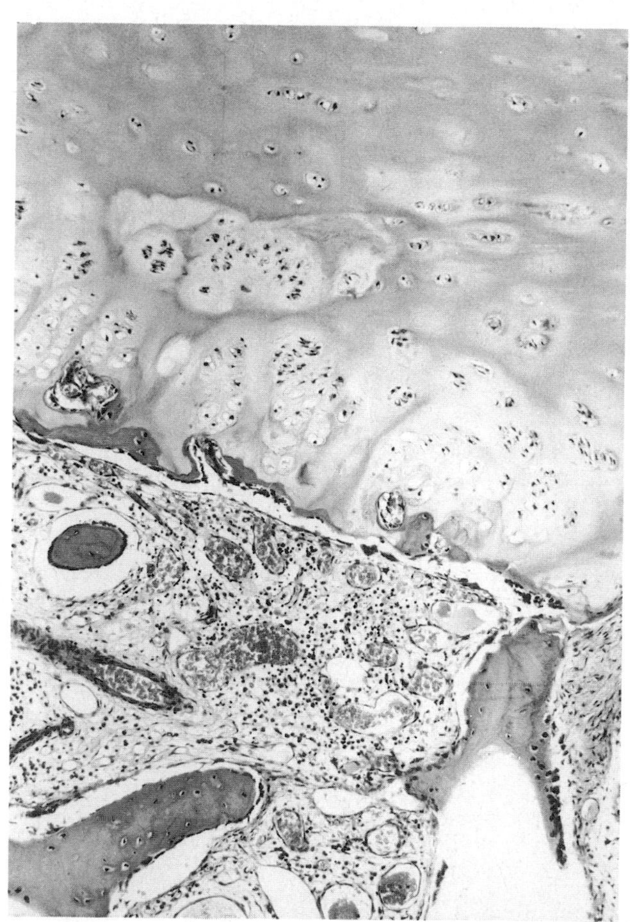

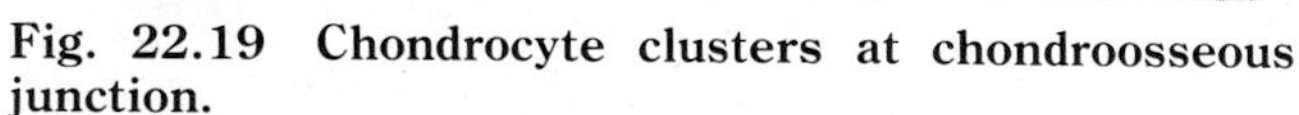

Fig. 22.19 Chondrocyte clusters at chondroosseous junction.

Fig. 22.13 demonstrates that focal increases in chondrocyte numbers, for whatever reason, commonly take place within zone I, before or concurrent with early fibrillation. Exceptionally, a similar process of chondrocyte recruitment occurs at the cartilage-bone junction. In this section, the bony endplate is very thin, the zone V calcified cartilage absent. The presence of small sinusoidal blood vessels extending into the articular cartilage from its deeper aspect confirms that enchondral ossification is occurring; whether this process is related to the formation of chondrocyte clusters is uncertain. (HE × 120.)

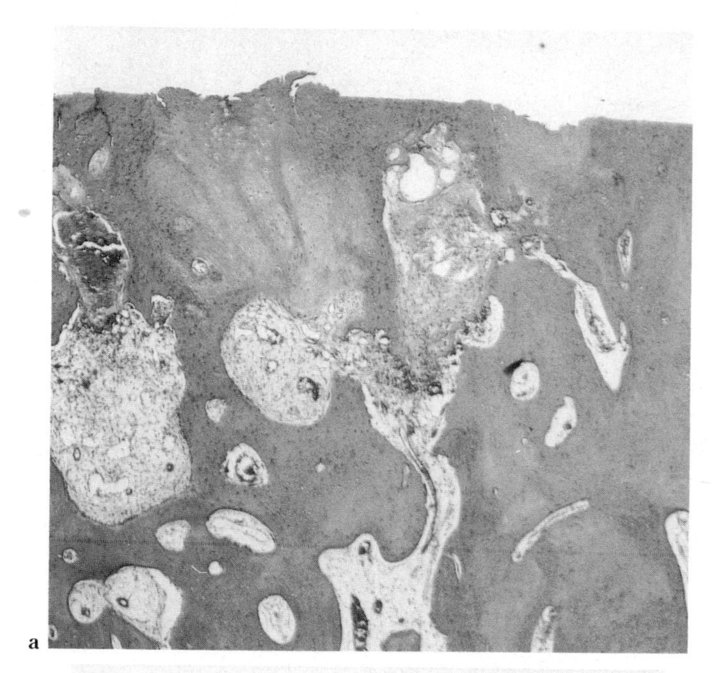

a

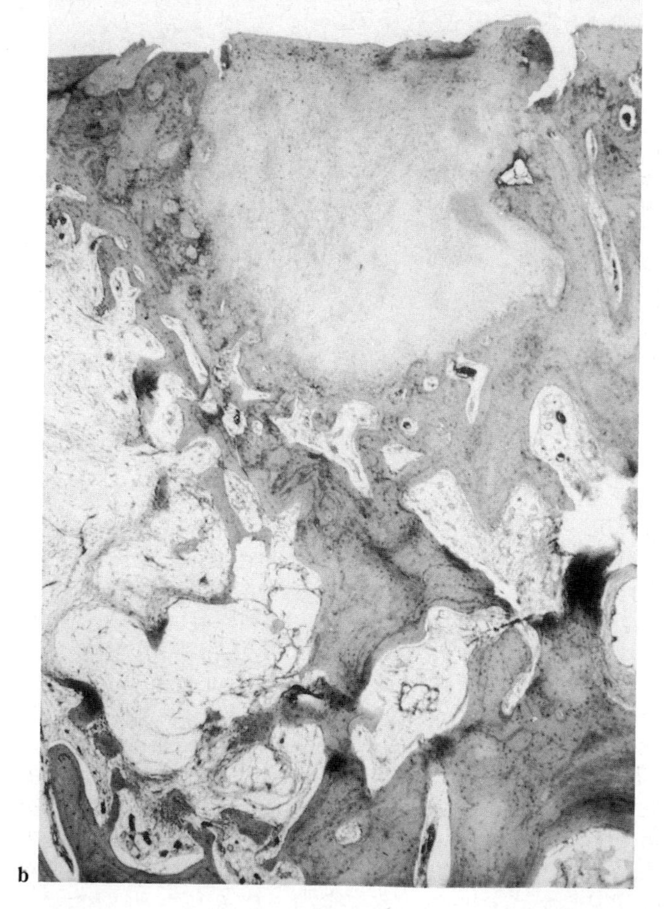

b

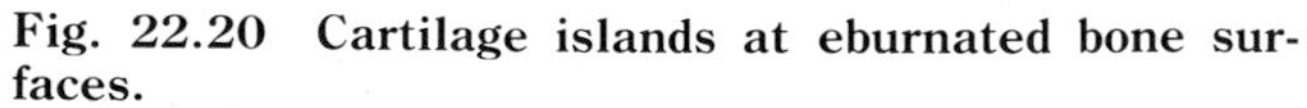

Fig. 22.20 Cartilage islands at eburnated bone surfaces.

a. The left central part of this osteosclerotic bone, from which all hyaline cartilage has been lost, is interrupted by the presence of an island of cartilage in continuity with a zone of fibrovascular marrow tissue. Although such islands are often described as 'plugs', suggesting that they have been inserted physically into the bone, the islands may represent residual hyaline cartilage or metaplastic tissue. **b.** Detail of island of hyaline cartilage within osteosclerotic bone in case of anatomical grade IV osteoarthrosis. (HE **a:** × 20; **b:** × 110.)

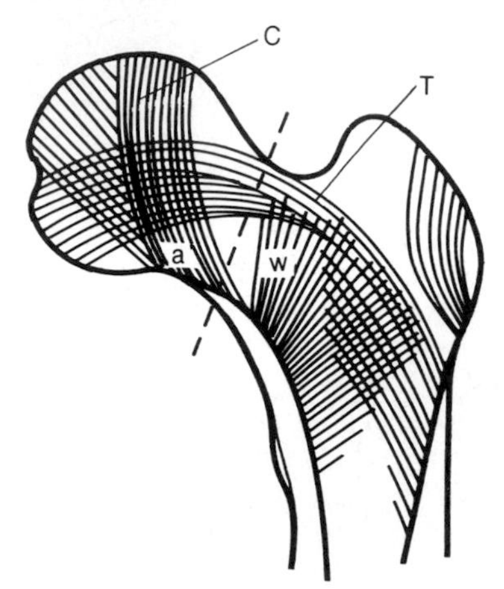

Fig. 22.21 Site of bone sampling in comparative studies of osteoarthrosis.[116]
Head and neck of femur demonstrating principal regions of compressive (C) and tensile (T) stress. a. Site chosen by Fazzalari *et al.* for histoquantitation. (W = Ward's triangle.) Redrawn from Fazzalari, Darracott and Vernon-Roberts.[116]

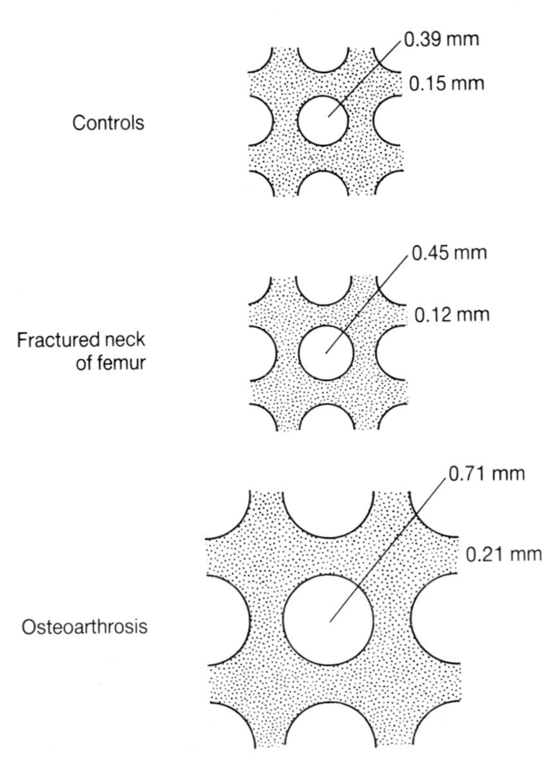

Fig. 22.22 Changes in trabecular bone structure in a selected stress region of the femoral neck.[116]
No significant difference was observed in the mineral content of bone from the femoral necks of 75 cases of osteoarthrosis treated by arthroplasty, and 18 cases of transcervical fracture of the femoral neck. There were 66 control cases. However, in terms of the site selected for measurement, osteoarthrosic cases generated less trabecular surface than bone from cases of fracture for the same amount of mineralized bone volume. The diagram displays the relationships between trabecular bone from the three groups of cases. Modified and redrawn from Fazzalari, Darracott and Vernon-Roberts.[116]

alization fronts is approximately 1 μm/day and a disparity between the extent of cartilage loss and bone formation is to be expected.

Zone V calcified cartilage diminishes as reabsorption occurs; osteoclastic activity is seen. The vascularity of the subchondral bone is high. New vessels form, an index of the low grade inflammatory reaction that can develop at this site (see p. 892). There is a sparse plasma cell, lymphocytic and mononuclear macrophage infiltrate. Fibroblasts multiply. Young granulation tissue evolves, possibly a reaction to the horizontal microfractures (cracks) at the tide-line. Granulation tissue also forms in the wake of osteonecrosis, between the calcified cartilage and the bone end plate.

The abnormally dense but physically weak hyperostotic bone, exposed by cartilage loss, is susceptible to abrasive wear. The hyperostotic bone is haphazardly interrupted by the presence of islands of cartilage but whether this tissue is residual or metaplastic is not always clear. Fibrous tissue may also be present. Excoriated by weight-bearing and movement, fragments of bone break off; limited parts of the bone marrow are directly exposed to the joint cavity. Bone is a poor substitute for cartilage as a bearing material; there are many blood vessels mediating the release of inflammatory mediators. Bone also contains nerve endings. Inflammatory changes and pain are the expected results.

Remodelling; osteophytosis (Fig. 22.23) Osteophyte formation is a common but poorly understood accompaniment of osteoarthrosis; it is characteristic of the syndrome but not pathognomonic. Osteophytosis may occur in the absence of overt osteoarthrosis and vice versa. It is believed that progressive cartilage loss leads to a redistribution of a local concentration of compressive and shear stresses. Lateral support for residual intact cartilage is reduced and the lessened local restraint results in disproportionately large deformities under load, extending the defect. A modelling response in marginal bone is promoted. The response culminates in new cartilage and bone formation and the growth of osteophytes that increase articular contact areas and enlarge the stress-bearing surfaces.[127]

The origins of the islands of bone that form excrescences at so many arthrosic articular margins lie in the capacity of periosteal osteoprogenitor cells to react to mechanical, physical and chemical stress. In the dog cruciate ligament model of osteoarthrosis (see p. 898), minute islands of dividing chondrocytes form microscopic chondrophytes within 3–4 weeks of surgery. Few comparable studies have been made of young human joints but it can be surmised that chondrophytosis precedes marginal bone formation. It appears, however, that bone growth is usually initiated directly, at the margin that lies deep to the vascular synoviochondral junction. Periosteal bone growth at the surface is accompanied by the remodelling of cortical and trabecular

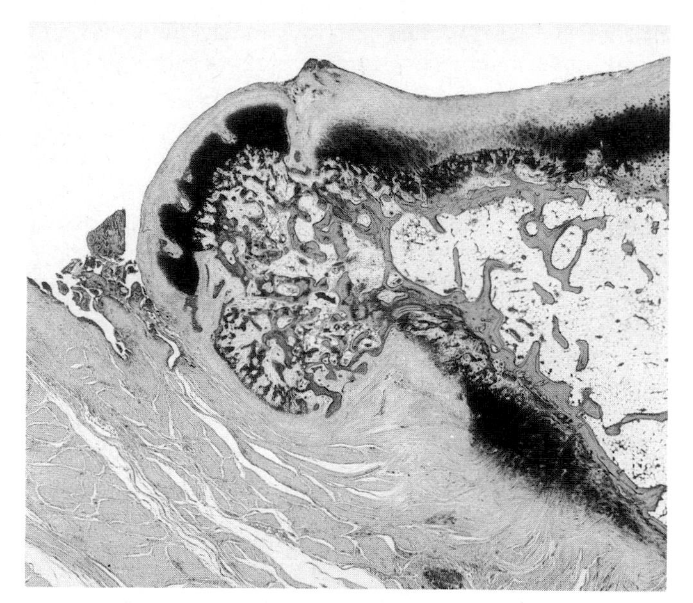

Fig. 22.23 Osteophyte formation.
An island of new bone has formed at the synoviochondral margin; it is separated from the nearby ligamentous and subsynovial connective tissues by a very thin cortical lamina. The focus of new bone is capped by a well-defined layer of collagen-rich cartilage ('fibrocartilage') which has, however, a high content of stainable proteoglycan. (Toluidine blue × 8.)

bone at the deeper margins. There is no tide-line. With time, the islands of bone thus formed appear as rows of smoothly contoured excrescences at the articular edges, growing into synovial cavities and recesses; others are more orderly uniform processes that lead to 'lipping'. Unlike the enthesetic reaction of ankylosing spondylitis, osteophytosis does not necessarily begin within ligamentous or tendinous insertions although it may do so; in this way the reactions are distinct from the exophytic bone formation of vertebral hyperostosis (see p. 947) and of those exostoses that arise, from trauma, for example, on the surface of long bones.

With time, marginal osteophytes come to be covered by a new formation of cartilage. This cartilage has a high content of type II collagen, arranged in substantial fibre bundles orientated radially. The structure is that of a hyaline cartilage unusually rich in collagen fibre bundles; in spite of the dearth of type I collagen it could be said to be fibrocartilage. There is much less metachromatic matrix than in the residual hyaline cartilage. The osteophytic processes, covered by fibrous or fibrocartilagenous tissue, extend into the synovial cavity in a direction that reflects the normal contour of the surfaces from which they develop. Collins[75] described shelf-like extensions at the edges of flat surfaces, curved lips round the concave surfaces, and osteophytes that turned backwards towards the bone shaft, at the edges of convex surfaces. The cumulative effect is to extend and exaggerate the contours of the articulating planes, a response to abnormal forces and to the local pressures of the joint capsule upon the zones of new bone growth. Pressure exerted by contact with opposing articular surfaces, by the intraarticular fibrocartilagenous menisci and by capsular ligaments, determines the alterations in cell growth and secretory patterns that lead ultimately to osteophyte formation. There is experimental evidence to show how mesenchymal cells differentiate in this way in a form of metaplasia unrelated to the proximity of the epiphysis.

A close relationship exists between the chondroosseous hyperplasia of osteophytosis and the concurrent but much more expansive process of remodelling of the inferomedial part of the femoral head, for example, that can culminate in grotesque, mushroom-shape deformity (Fig. 22.28) or in the elbow in enormous new bone formations. However, the processes are distinct and may develop largely independently. Often, there are mirror image changes at opposing joint surfaces and the two or more articulating structures come to interlock in a bizarre way, regulated by the new islands and rings of bone. New bone formation at the margins of subchondral bone is frequently accompanied by vigorous osteoclastic bone reabsorption. When the deep zone (zone V) of calcified cartilage is thin or absent, as it is in the face of excessive cartilage loading (see p. 888), there is a similar reabsorption of the exposed, deep hyaline cartilage surfaces. The responsible cells are designated chondroclasts (see p. 32). The processes of microscopic remodelling continue insidiously over long periods of time, contributing to the changed contours of affected subchondral bone but to a lesser degree than the analogous processes occurring synchronously at the synoviochondral, articular margins.

Pseudocysts Irregularly distributed within subarticular bone, often in zones of osteosclerosis, are varying numbers of round or ovoid islands of loose connective tissue which replace bone trabeculae and which have a 'myxoid' appearance, with stellate mesenchymal cells situated among a loose metachromatic connective tissue matrix (Fig. 22.24). Fluid is seldom present and the designation 'cyst' is derived from the property of radiolucency. Histologically, these structures are 'pseudocysts' in analogy with those recognized in rheumatoid arthritis (see p. 473). The margins of a 'pseudocyst' may be formed by inert, flattened connective tissue cells but osteoclastic bone reabsorption and osteoblastic new bone formation are both often seen.[293] Pseudocysts range in size from 5–8 mm to 10–20 mm; they are particularly common beneath the bearing surfaces of osteoarthrosic femoral heads. They are encountered less often in abnormal knee, shoulder and apophyseal joints. As in the case of the osteosclerosis with which pseudocysts

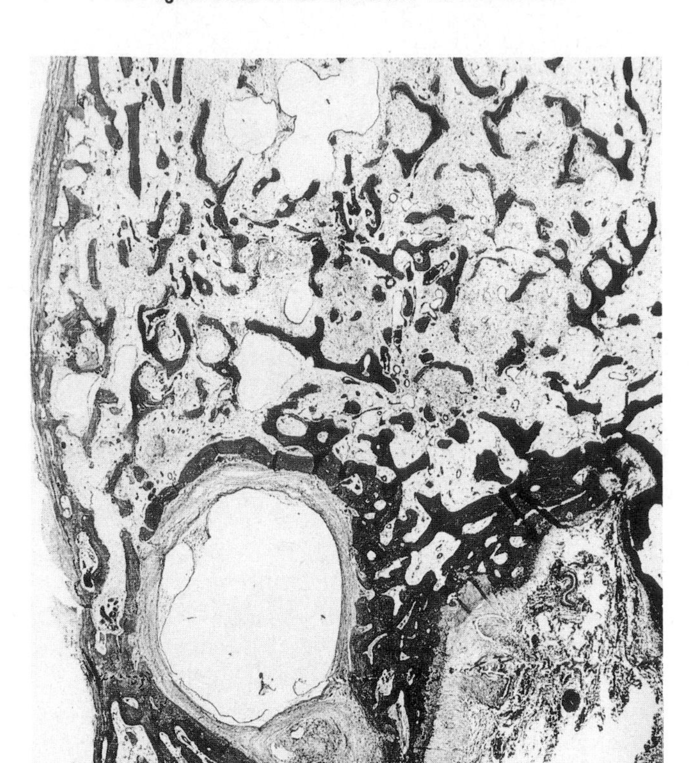

Fig. 22.24 Radius: pseudocyst formation.
Within the styloid process of the radius, and surrounded by a zone of sclerotic bone, lies a thin-walled pseudocyst the softer mesenchymal connective tissue contents of which have been lost. At the inferior, medial margin of the pseudocyst, a broad band of fibrous material represents a tangential section through a connective tissue core which joins the pseudocyst wall with the fibrillated hyaline cartilage covering the radial surface. (HE × 9.)

tend to be associated, pseudocysts are particularly frequent at sites where gravitational loads are greatest and may form before cartilage loss has led to a reduction of the interosseous space.

The pseudocysts seen, for example, within femoral heads affected by grade IV osteoarthrosis, are bounded by a thin lamina of compact bone resting upon orderly, normal circumferential trabeculae. Although osteoarthrosis may be associated with untreated gout and, very frequently in older persons, with chondrocalcinosis, the osteoarthrosic pseudocyst is devoid of crystals. As in rheumatoid arthritis, it seems likely that mechanical factors explain the development of these structures: porotic bone yields before abnormal forces concentrated at foci of material weakness in sites determined by the location of a cartilage defect and by changed bone geometry.

Other tissues

The ravages of osteoarthrosis are 'panarticular'. It is unrealistic to expect cartilage to be lost, bone eburnated and excoriated, and mechanical stability to be impaired without changes in articular function and, consequently, the insidious onset of microscopic lesions of menisci, discs, ligaments and tendons.

Fibrocartilaginous menisci, such as those of the knee, undergo age-related disorder (Fig. 2.18) in the same way as does hyaline cartilage. Osteoarthrosis, however, does not influence meniscal structure until a late stage of the disease; advanced cartilagenous disorganization, sometimes with secondary synovitis, may develop in the presence of intact, aged menisci. Grade IV osteoarthrosis, however, is often accompanied by meniscal degeneration (Fig. 2.18) or loss. In these cases, severe isolated monarticular osteoarthrosis, confined to a single part of opposing surfaces in the multifaceted joint, may reflect early local injury to a meniscus, initiating secondary osteoarthrosis. There is an analogy with the effects of meniscectomy after which cartilage degradation is accelerated (p. 84).

As osteoarthrosis advances, the fibrocartilagenous menisci as well as the hyaline cartilages become fibrillated and degenerate to a degree in excess of 'normal' age change (see p. 82). The free edges of the knee menisci break up, liberating fragments of fibrocartilage into the joint spaces above and below the menisci. There are frequent, comparable and extensive changes in the discs of the temporomandibular joints and it can be assumed that similar alterations are possible in the discs of other joints such as the inferior radioulnar. Tendons and ligaments also imperceptibly alter and may atrophy but little is known of these histological phenomena.

The synovial abnormalities of osteoarthrosis are discussed on pp. 874 *et seq.*

Reparative changes

The capacity of cartilage for repair is discussed in Chapter 2. Repair can be facilitated by ensuring that cartilage defects are in continuity with bone faults that expose marrow blood vessels or with the vascular synovium. The capacity of bone for repair is common knowledge.

The reparative changes of human osteoarthrosis were reviewed by Meachim and Osbourne[297] and by Meachim and Roberts.[298] Femoral heads excised surgically for osteoarthrosis displayed: exposed bone; loose, fibrous connective tissue; tissue intermediate in type between fibrous tissue and fibrocartilage; chondroid tissue; reparative fibrocartilage; unclassified tissue; and parts of the original hyaline cartilage. Some surface discontinuities are ruptured pseudocysts; other gaps are plugged by fibrous or fibrocartilagenous tissue that may extend 2–5 mm into the subjacent bone. Where islands of cartilage are found within subchondral bone, they may represent the extension of proliferating chondrocytes through discontinuities in the defective calcified cartilage and bone end-plate. Alternatively, they can be identified as a form of metaplasia, analogues of the marginal chondrophytes that are occasional precursors of osteophytosis. The reparative, surface fibrocartilagenous tissue may itself undergo fibrillation during the progression of osteoarthrosic destruction.

New fibrous-, cartilagenous- and bone tissue frequently forms during the series of processes described above. However, these new tissues are seldom in continuity with the main sites of cartilage loss and contribute little to its arrest or mitigation. The extent and frequency of reparative processes in osteoarthrosis excite very great interest because of the possibility that these phenomena can be induced or encouraged therapeutically (see p. 893). Established procedures such as osteotomy can give symptomatic relief in longstanding osteoarthrosis but the proposal that systemic or local drug therapy can stimulate the new formation of cartilage remains contentious.

The process of repair in osteoarthrosic knees can be assisted by implants of materials such as carbon fibre.[304] Meachim and Brooke[293] have considered the analagous situation posed by the response of the osteocartilagenous tissues to osteotomy. They conclude that the key to the new formation of cartilage lies in the persistence of islands of chondroid or 'myxoid' tissue within discontinuities in the exposed, hyperostotic bone. Cartilage cells grow outwards from these foci, secreting a new extracellular matrix. In untreated individuals, they propose that this process is inhibited by continued abrasive wear. After osteotomy, altered alignments and changed local mechanical conditions create a more favourable environment for the tissue.

These views remain controversial. They contrast with the observation that the spontaneous repair of excised cartilage defects is improbable unless the deficiency extends deeply as far as the bone marrow (see Chapter 2). The key to both new cartilage formation and bone remodelling after osteotomy may lie in molecules such as angiogenesis and growth factors secreted as a result of a complex series of local cytokine-mediated phenomena promoted by the tissue injuries of the surgical procedure (see p. 78).

Anatomical changes
(Figs 22.25 and 22.26)

The consequence of the molecular, cellular and microscopic abnormalities described above is the development of a series of progressive, overt anatomical lesions which are usually confined to a single joint but which may be polyarticular or even, occasionally, generalized (see p. 847). On the basis of Byers' analysis[55] and of Meachim's accounts[288,289,293] of the development of fibrillation with age, the overt lesions of osteoarthrosis can be said to be the sum of 1. the areas of fibrillation in situations where age alone would not invoke cartilage surface degradation; and 2. the lesions exceeding in severity those that occur with age alone, whatever their location.

Extensive accounts have been given of the anatomical osteoarthrosic changes of the major synovial joints.[29,75,120,136,207,243,366,422,438] The anatomical disorders of the lower limb joints have received much more attention

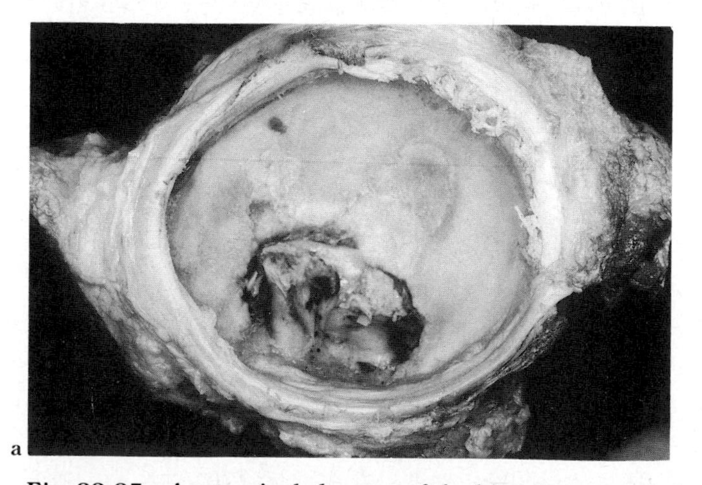
a

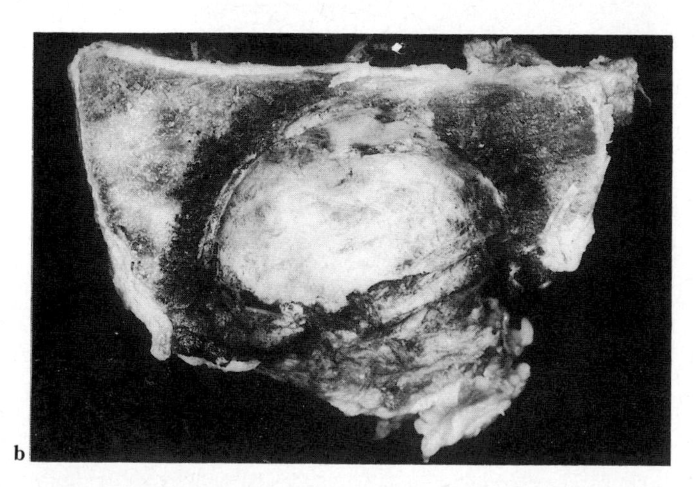
b

Fig. 22.25 Anatomical changes of the hip joint: acetabulum.
a. Note the circumferential acetabular labrum with the central zone at which the ligamentum teres originates. An ill-defined area of overt fibrillation extends in a biradiate fashion from the perifoveolar location. **b.** Acetabulum bisected and painted with Indian ink. There is overt, patchy, ill-defined fibrillation which is most conspicuous at the acetabular margin.

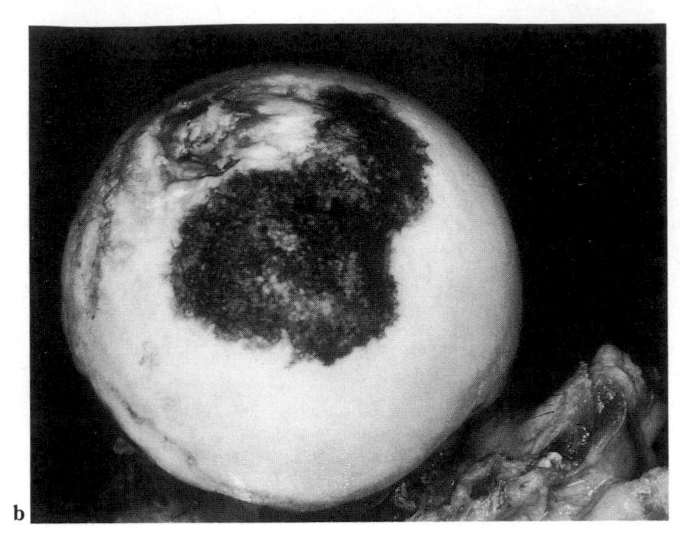

Fig. 22.26 Anatomical changes of the hip joint: femoral head.
a. Femoral head from same patient as Fig. 22.25a. An ill-defined area of overt fibrillation adjoins the fovea, occupying much of the superior, load-bearing surface. **b.** Same specimen as in **a** after painting with Indian ink. The extent of the area of overt fibrillation is now readily seen; it can be distinguished from a perifoveolar area of age-related fibrillation.

than the upper, the large joints more than the small. Some articulations such as the atlantooccipital, vertebral apophyseal, sacroiliac, carpal and tarsal have been neglected. Biopsy is infrequent and they are often inaccessible at necropsy. Material for *post mortem* study is difficult to obtain.

The importance of continuing the task of defining the gross structural features of this common human disorder has been very clearly put by Bullough.[48] The synovial joints are complex and highly heterogeneous structures. Their susceptibility to osteoarthrosis is not uniform. The known difference in the extent and prevalence of osteoarthrosis of the hip and shoulder joints could reasonably be predicted. Less clear are answers to questions such as: Is hip joint disease more severe in the left or right hip joints? Is there any relationship, as might be expected mechanically, between osteoarthrosis of the right hip, knee and ankle, or between, say, the right acetabulum and the right lateral and medial femoral condyles? Answers to these enquiries are not easily found and there is little evidence that they are being sought. Many years ago Heine[176] (Table 21.5) noted that osteoarthrosis began early in the knee and great toe but, by the age of 60 years, was unexpectedly common in the acromioclavicular joint. The significance of this observation has never been explained.

The gross pathological diagnosis of osteoarthrosis is best determined by assessing the site and degree of disorder in a particular synovial joint against a baseline of structure derived from analyses of the normal, ageing population. Such surveys have been made systematically for the hip joint by Byers *et al.*[55] and Meachim and Emery;[295] for the knee by Meachim;[292] for the ankle by Meachim;[291] and for the shoulder by Meachim and Emery.[295] Collins[75] (Table 22.6) modified the anatomical classification of osteoarthrosis adopted by Heine[176] and by Fischer[120] and recommended the use of four grades of severity. A simple classification of this kind remains of value.

Grade I. The certain distinction of grade I osteoarthrosis from the limited fibrillation and changes of age, demands meticulous examination of the whole surface of a joint, and a careful comparison of the pattern of change with that of the normal distribution of age change (see Chapter 21) established in a control population.[126] The sensitivity by which grade I change is recognized is enhanced by preliminary painting of the fresh (unfixed) cartilage surface with Indian ink. The femoral head may be considered as an example. Grade I osteoarthrosis is present when significant naked-eye roughening (fibrillation) is present in areas additional to those such as the peripheral or perifoveolar regions (see p. 830) subject to age-related fibrillation alone.[55] Among the susceptible areas are the anterosuperior and inferomedial parts of the femoral head. Shallow pits and grooves are also seen with small blisters beneath the cartilage surface. The blisters 'burst', causing shallow defects in the cartilage with frayed edges. There is no area of whole thickness cartilage loss, no recognizable marginal osteophytosis and no detectable synovial inflammation. Contrary to Collins'[75] view, grade I osteoarthritis may also be diagnosed when progressive loss in excess of age-related change affects areas such as the circumference of the femoral head or

Table 22.5[176]

Frequency of macroscopic evidence of osteoarthrosis identified *post mortem*

Age groups (years)	Knee		Shoulder		Hip		Elbow		Great toe		Acromio-clavicular		Sternoclavicular	
	A (per cent)	B (per cent)	A (per cent)	B (per cent)	A (per cent)	B (per cent)	A (per cent)	B (per cent)	A (per cent)	B (per cent)	A (per cent)	B (per cent)	A (per cent)	B (per cent)
15–19	3.1	0.0	0.0	0.0	0.0	0.0	0.0	0.0	0.0	0.0	0.0	0.0	0.0	0.0
20–29	9.2	0.0	0.8	0.0	0.8	0.0	1.7	0.0	9.6	0.0	4.4	0.0	0.0	0.0
30–39	48.1	1.0	2.0	0.0	7.8	0.0	18.0	0.0	27.0	0.0	20.0	0.0	2.0	0.0
40–49	74.0	0.8	4.7	0.0	16.7	0.8	26.9	0.0	35.7	0.0	35.9	0.8	6.2	0.0
50–59	87.1	2.6	14.3	0.7	44.8	0.7	61.7	1.3	60.5	7.9	73.0	1.3	16.1	0.0
60–69	92.6	12.0	22.1	2.6	60.0	2.7	73.7	5.2	72.6	13.7	91.0	9.6	36.3	1.1
70–79	97.5	33.3	44.1	9.7	76.1	12.2	87.4	10.5	78.1	18.2	100.0	12.3	40.0	1.5
80–95	100.0	39.4	60.0	15.7	89.4	16.7	95.8	15.5	84.6	24.6	95.7	11.4	62.0	7.0

A = macroscopic evidence of osteoarthrosis of all grades; B = macroscopic evidence of osteoarthrosis of grades III and IV.

Table 22.6

Osteoarthrosis in man: pathological grades[75]

I	Patches of fibrillation or softening in central areas of articular cartilage—i.e. those not displaying simple age changes
II	Fibrillation more pronounced; early marginal chondro-osteophytosis with related synovial cell hyperplasia
III	Changes more severe; commencing exposure of subarticular bone and more generalized synovial disease. Osteophytosis
IV	Extensive cartilage loss and bone exposure; eburnation and bone grooving. Destruction of intraarticular ligaments. Fibrosis or atrophy of synovial fringes. Limb shortening, subluxation

the perifoveolar region where age change (non-progressive change) normally prevails.

Grade II: With progression of the cartilage disorganization, there are larger areas of fibrillation affecting parts of the articular surface in excess of those where age-related fibrillation is customary (see p. 830). On the lower end of the femur, for example, the intercondylar, patellar surface is often roughened. With time the main, anterior and inferior parts of the medial condyle and, subsequently, the lateral condyle show increasingly large areas of fibrillation. The lateral condyle is affected later, the posterior surfaces of the condyles very infrequently. Fibrillation in the affected sites is deeper and cartilage loss more common than in grade I. However, bone is not exposed. At the chondrosynovial junctions new cartilage and bone formation is evident but the osteochondrophytes are small and inconspicuous. Evidence of juxtaarticular synovitis is recognized. Bone remodelling is not yet detected.

Grade III: Grade III osteoarthrosis is present when at least one area of whole thickness cartilage loss is recognized, with bone exposure (Figs 22.27–22.31). By the time the bone surface becomes weight-bearing, through cartilage loss, there has been subchondral new bone formation. The thickened, new bone is whiter than normal cartilage; this change to an ivory-like structure is described as eburnation. Around the most severely affected areas, where cartilage loss and bone exposure appear, the less advanced changes of fibrillation may be seen.

Osteophytes now form at the synovial margins: bone remodelling, brought about by lacunar osteoclastic reabsorption and, possibly by osteocytic osteolysis, begins to alter the anatomical contours of the joints. Bone sclerosis is detectable radiologically. Where bone exposure is long-standing, abrasive wear of one bone surface upon another leads to the formation of a smooth, white surface. However, linear scarification disrupts this surface structure so that grooves and ridges form in the axes of preferred movements. The abrasive process transects the haversian systems. Linear and focal crushing, crumbling and loss of abraded bone, with trabecular collapse

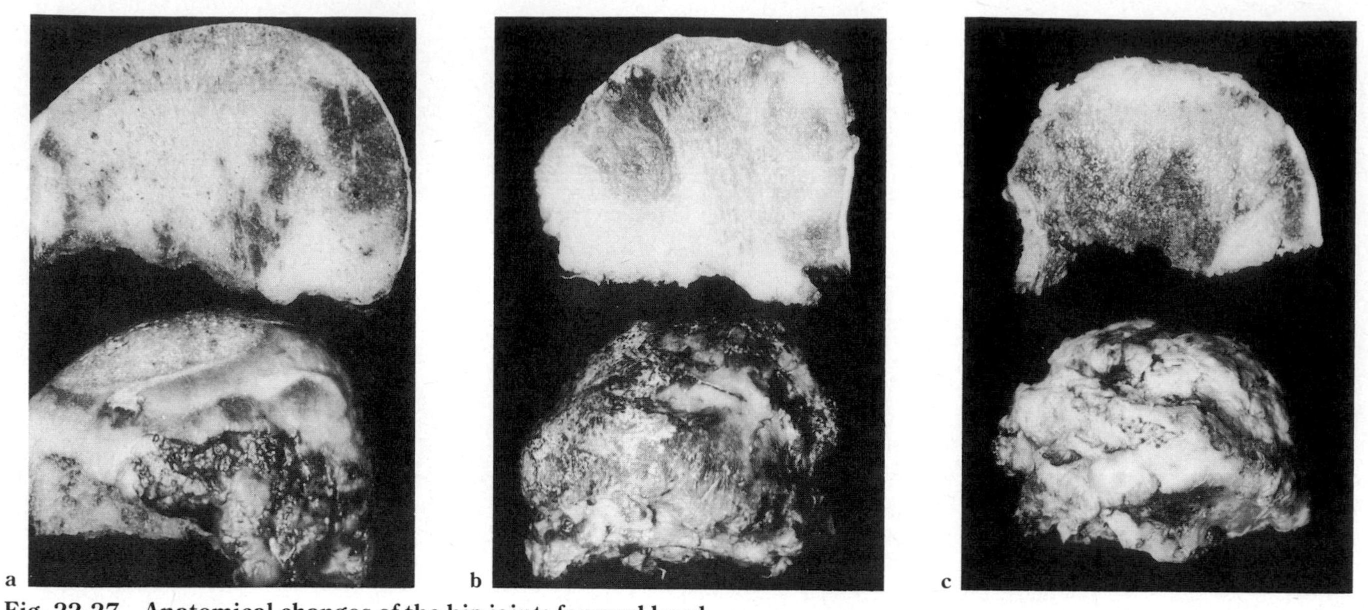

Fig. 22.27 Anatomical changes of the hip joint: femoral head.
a. Articular cartilage has been lost from approximately half of the bearing surface of the femoral head. At top, the extent of the loss of cartilage can be seen in section; at bottom, the many osteophytic processes that adjoin the abnormal surface can be seen *en face*. **b.** In this specimen, section of the femoral head (at top) reveals two large pseudocysts surrounded by eburnated, dense bone. At bottom, the degree of remodelling of the affected bearing surface is demonstrated. **c.** Cartilage and bone have been lost from the upper left part of this femoral head. There is marginal lipping and the remodelling, together with osteophytosis, contributes to the mushroom-like shape of the bone.

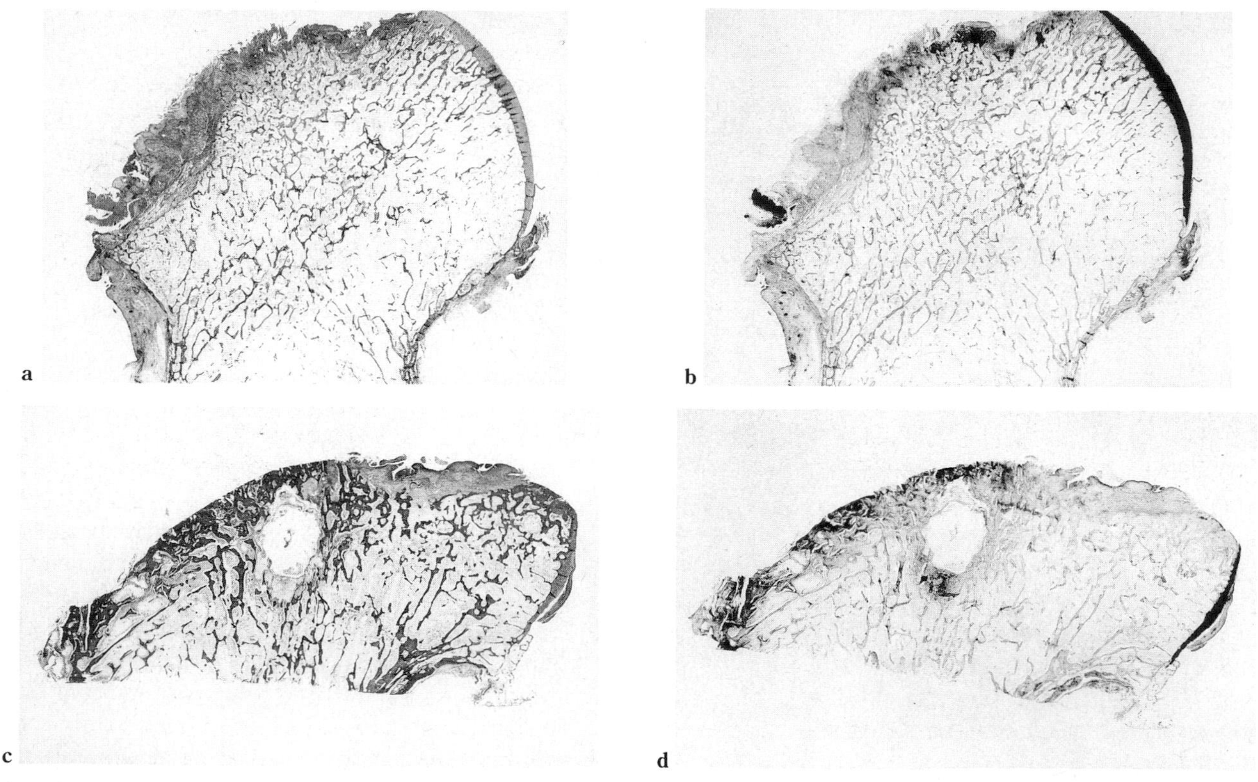

Fig. 22.28 Anatomical changes of the hip joint: femoral head destruction and remodelling; cartilage matrix loss.
a. In this specimen, cut in an oblique, coronal plane, the severe destruction of approximately one-third of the head of the femur suggests the operation of factors such as bone necrosis or infarction of which there was no residual evidence. There is osteopenia. **b.** In this section from the same block, stainable matrix proteoglycan is now restricted to residual cartilage in the upper right part of the field. The integrity of this cartilage is in conspicuous contrast with the changes seen *top left*. **c.** In this instance, remodelling is accompanied by osteosclerosis, pseudocyst formation and a mushroom-like deformity of the contours of the femoral head. **d.** In this section from the same block as in **c**, stainable matrix proteoglycan is limited to a small marginal zone of cartilage seen at right. (**a:** HE × 1.5; **b:** toluidine blue × 1.5; **c:** HE × 1.5; **d:** toluidine blue × 1.5.)

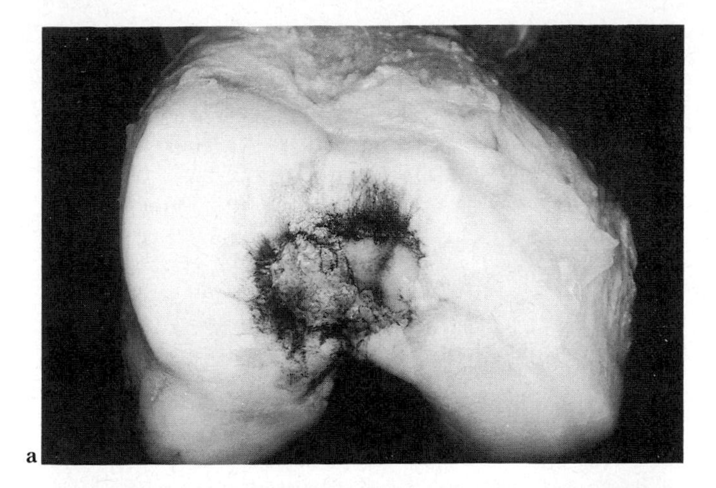

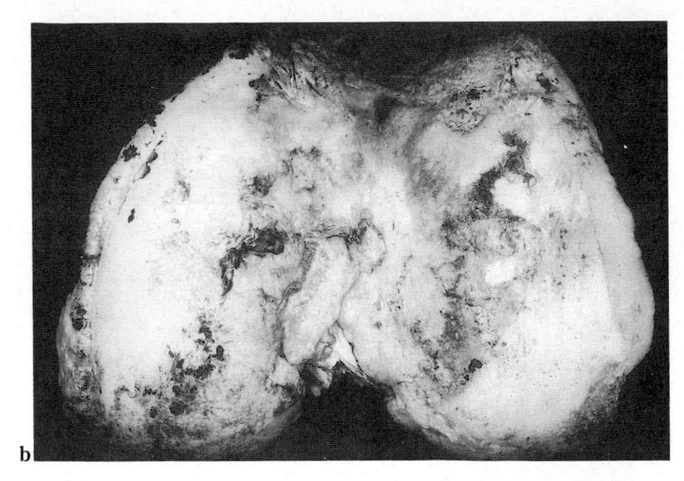

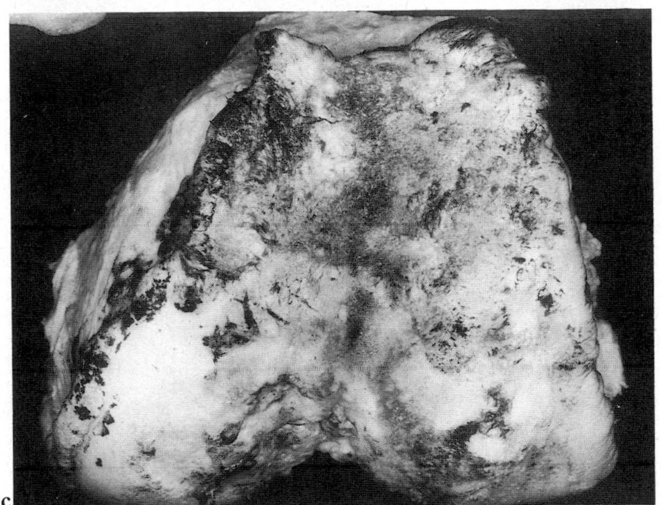

Fig. 22.29 Anatomical changes of the knee joint: femoral condyles.
a. Lower end of femur from a female aged 68 years. Overt fibrillation of part of the patellar surface has advanced to incomplete cartilage loss. Main condylar-bearing surfaces are spared. **b.** Lower end of femur from a female aged 87 years. Indian ink painting reveals that the principal load-bearing surfaces of the femoral condyles are relatively intact whereas the articular margins show both overt fibrillation and lipping (osteophytosis). **c.** From same patient as in **b.** Specimen has been rotated anteriorly to demonstrate the more widespread fibrillation, with cartilage loss and osteophytosis which particularly affect the patellar groove.

and a loss of bone height, culminate in the formation of irregularly shaped islands of pale-red, vascular bone marrow that appear prominently on the bearing surfaces. A shallow band of necrotic bone is often recognized at the exposed osseous surface but, periodically, secondary infarction of a larger zone occurs. The shape of this zone of necrotic bone is determined by the local vascular territory. Low-grade inflammatory reactions in the synovia advance, and synovial fibrosis and thickening, but no fibrous ankylosis, are present. Subchondral pseudocysts become visible beneath the articular surface contours.

Grade IV: There is now whole thickness loss of cartilage from large areas of the bearing surfaces. Eburnation, bone scarification, marrow exposure and a gross disorganization of the greater part of the articular surfaces are usual. Remodelling is active: there are changes in the anatomical contours so that opposing, articular profiles of, for example, the femorotibial joint, are adapted and moulded to the opposite contour. At the hip joint, the femoral head undergoes a process of expansive remodelling. Loss of bone from the central parts of the bearing surface is accompanied by new bone and fibrocartilage formation at the margins. The laterally expanded but flattened surface assumes a mushroom-like shape. A 'tilt deformity' may be simulated but the shift of the centre of the femoral head is apparent, not real. Among the explanations advanced to account for this remodelling process are a prime role for osteophytes; the stimulation of osteochondrogenesis by wear fragments in the synovial fluid; or a compensation for the mechanical disorders provoked by wear and osteophytosis, leading to a form of artificial osteotomy.

At the synoviochondral junction, osteophytes grow larger (Fig. 22.29) and may form grotesque, disabling configurations: one affected bearing surface is actively shaped to the shape of the opposing surface. Capsular fibrosis, low-grade synovial cell hyperplasia, the separation of fragments of cartilage (and bone) and their uptake by synovial phagocytes are among the variegated changes that characterize this, the most severe anatomical form of osteoarthrosis. The interosseous ('joint') space is reduced. The vascularity of the affected joints is much increased and is a reason for the increased isotope uptake in the subchondral regions that is demonstrable clinically.

Synovial changes

The histopathological changes in the synovia of idiopathic osteoarthrosis have excited controversy. Debate centres on three questions: Are the synovia abnormal? What are the nature and the extent of the abnormalities? and Are the abnormalities unique to or characteristic of the syndrome of idiopathic osteoarthrosis?

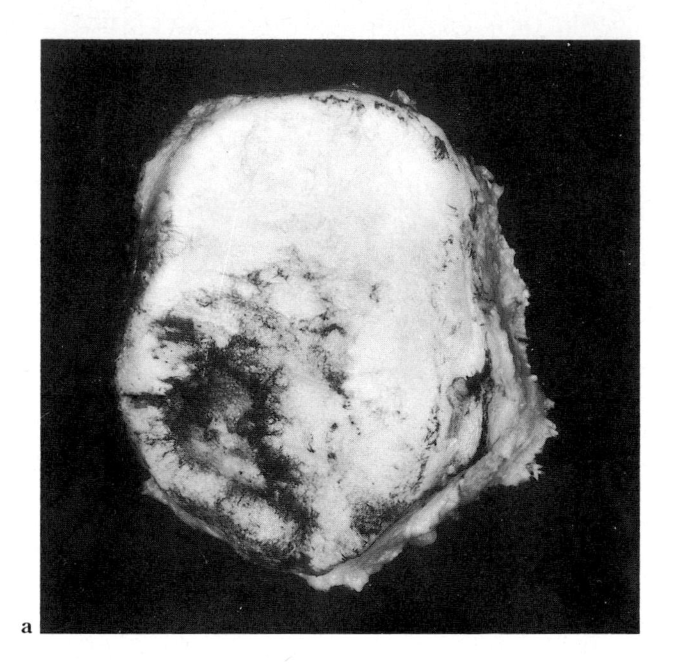

a

Fig. 22.30 Anatomical changes of the knee joint: patellae.
a. Fibrillation and cartilage loss are much more severe on the medial than on the lateral aspect of the patellar bearing surfaces. **b.** Complete cartilage loss from part of the patellar surface. Marginal osteophytosis and remodelling characterize this severe disorder. **c.** Same specimen as in **b** after painting with Indian ink.

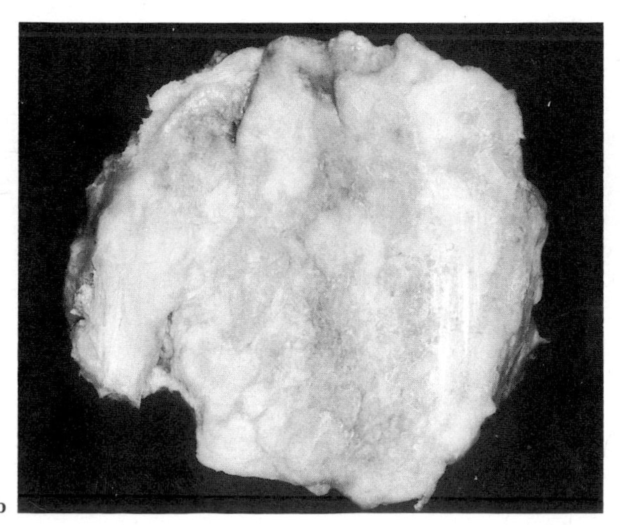

b

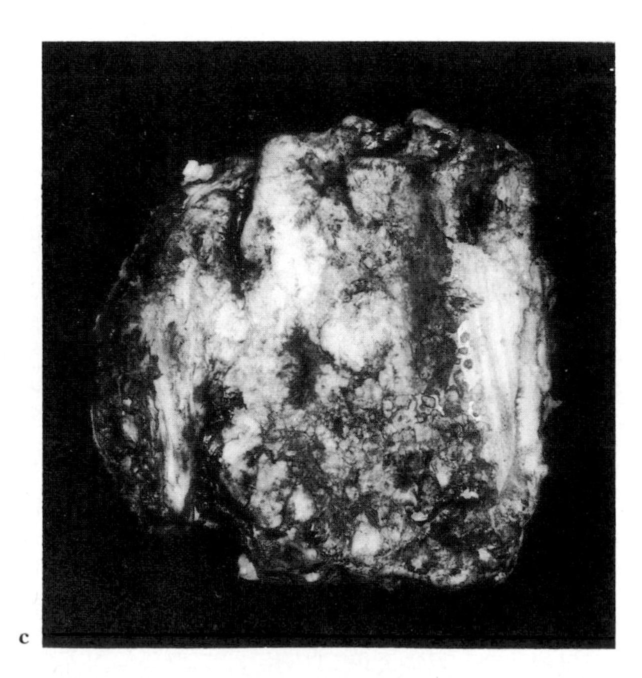

c

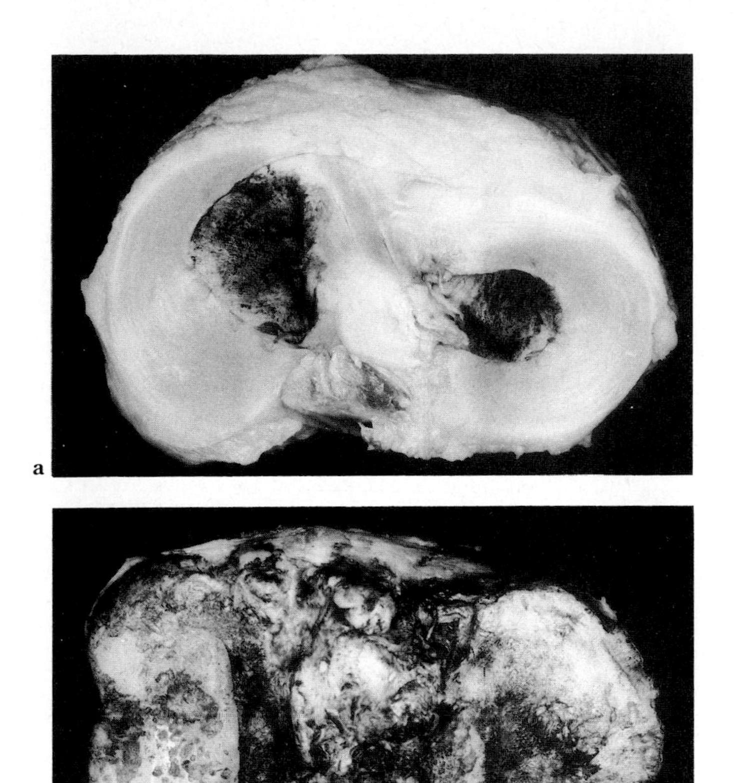

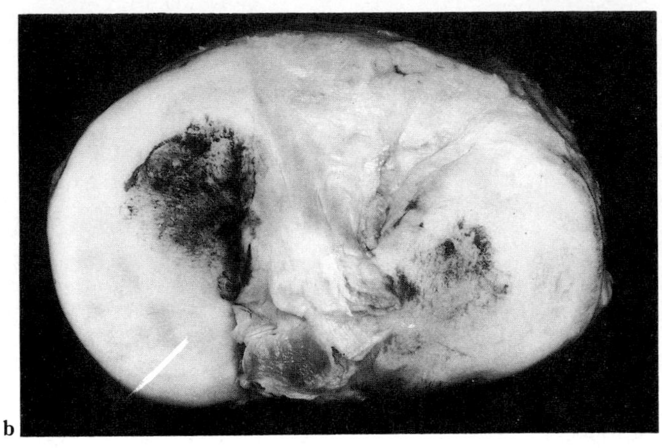

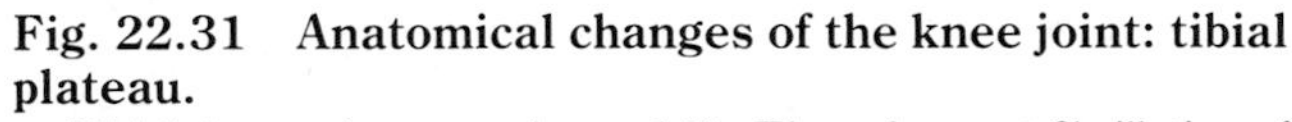

Fig. 22.31 Anatomical changes of the knee joint: tibial plateau.
a. Tibial plateau from a male aged 39. There is overt fibrillation of the central parts of both the medial and lateral condyles; the distribution of the roughened areas is enhanced by Indian ink painting. **b.** Same specimen. The menisci have been removed, confirming the integrity of the submeniscal cartilage. **c.** Tibial plateau from a female aged 85. Although there is extreme disorganization of the bearing surfaces of both condyles, with widespread cartilage loss and osteosclerosis, the severity of the changes is clearly greater in the central parts of the condyles than in the lateral.

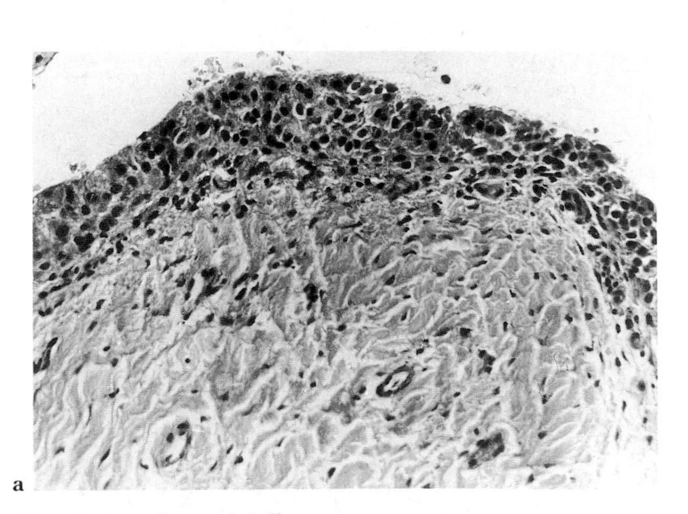

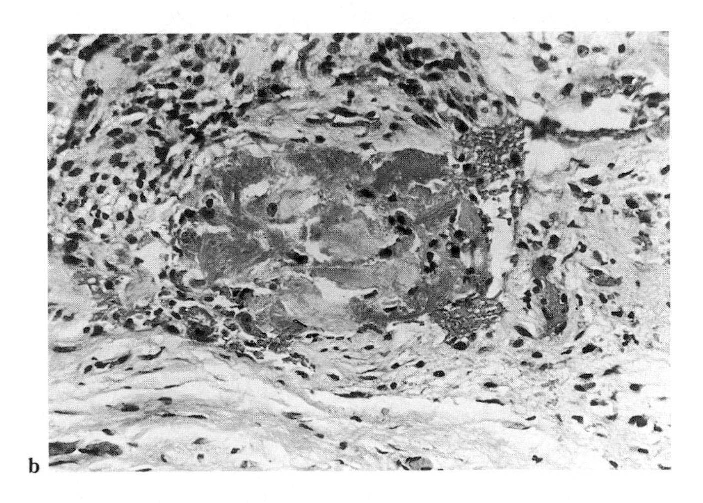

Fig. 22.32 Synovial disease.
a. Synoviocyte hyperplasia is a common secondary feature of osteoarthrosic synovial inflammation. Mitotic figures are infrequent and multinucleate giant cells recognized much less often than in rheumatoid synovitis. **b.** As osteoarthrosis progresses, cartilage disintegration and bone excoriation result in the liberation of particles of both materials into the synovial space. Here, a cluster of cartilage fragments lies grouped within a thin lamina of fibrous tissue; evidently, a chronic granulomatous inflammatory reaction has been elicited. (**a:** HE × 180; **b:** × 180.)

Inflammatory phenomena

Early accounts described a histological structure which varied little from normal.[75,398] Widely disparate accounts suggested abnormalities which differed by anatomical site so that synoviocyte atrophy and synovial fibrosis were features of hip joint disease, synoviocyte hyperplasia and hypertrophy those of knee joint osteoarthrosis.[200] Inflammatory changes of the kind seen in rheumatoid arthritis were stated not to be present. As investigations became more searching, it was realized that one or more cardinal signs of active inflammation were not rare clinically in cases with 'inflammatory osteoarthritis'. Some, a severely involved subset, demonstrated marginal joint erosions.[463]

Retrospective microscopic enquiries, often on patients who had been subject to arthroplasty for advanced disease, showed that 'moderate inflammatory changes'[151] were more frequent than had been anticipated (Figs 22.32 and 22.33). Indeed, it proved possible to demonstrate active synovitis in relatively large proportions of biopsy specimens collected over a 15-year period[81] (Table 22.7) although polymorphs and eosinophil granulocytes were rare. These observations were used, just as they were for rheumatoid arthritis, post-traumatic synovitis and juvenile chronic polyarthritis, to examine the statistical relationships between pairs of diseases. Nineteen histological features were identified which were less (usually) or more (seldom) often significantly frequent in osteoarthrosis than in definite classic rheumatoid arthritis defined by the 1958 American Rheumatism Association criteria (see p. 445). When linear discriminant functions were derived for the five most populous groups, there was agreed differentiation in 77.6 per cent of instances between the histopathological diagnosis of rheumatoid arthritis and 'osteoarthritis'.[395] In general, the

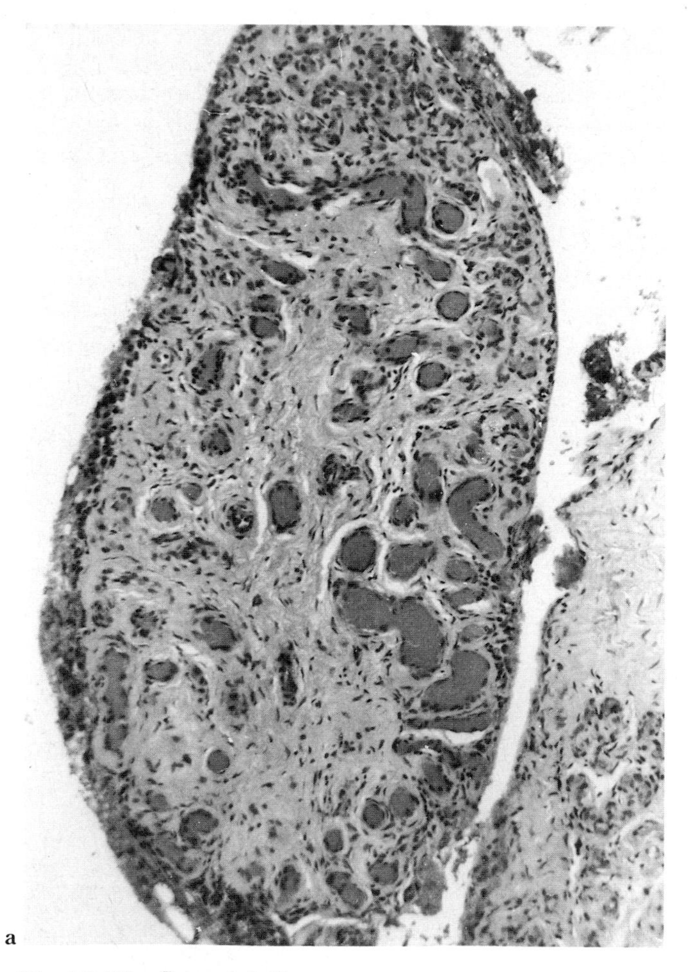
a

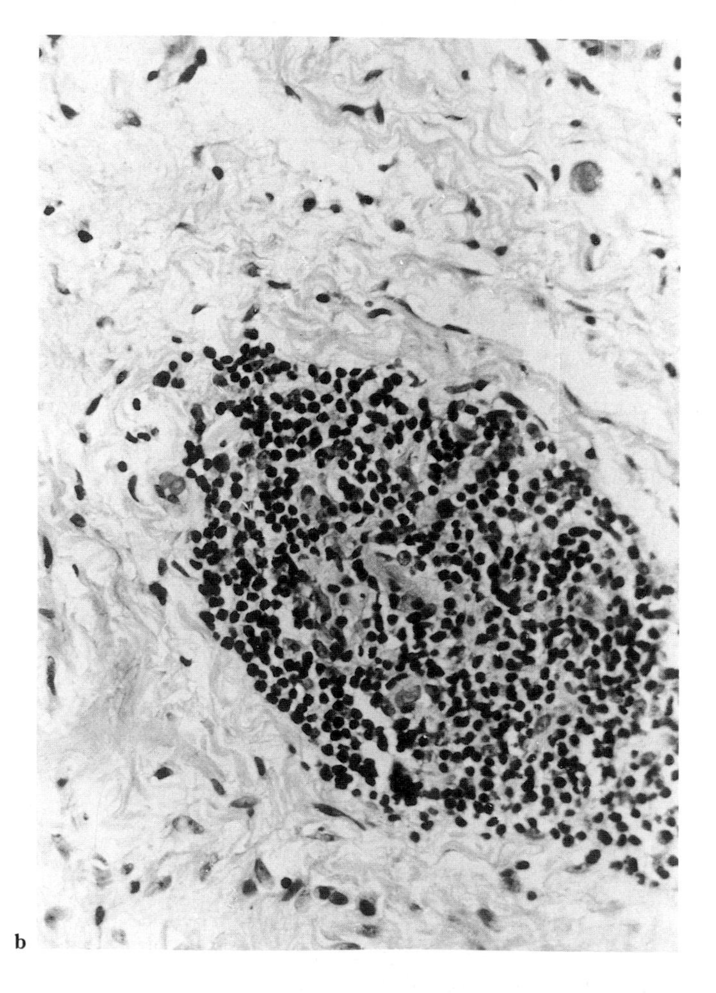
b

Fig. 22.33 Synovial disease.
a. At the margin of this femoral head, the synovium is hypervascular. There is an increased number of arterioles with many venules and a corresponding proliferation of capillaries. The increased vascularity of the osteoarthrosic synovium contributes to the high concentration of isotopes given during scanning procedures; new bone formation, with high vascularity, is a further reason for the increased uptake. **b.** One of the most characteristic features of the synovitis which is found adjoining cartilage margins at selected parts of joints such as the knee, is a focal, perivascular lymphocytic aggregation. (**a:** × 120; **b:** × 180.)

Table 22.7

Most frequent histological changes indicative of active chronic synovitis in 74 cases of joint disease examined over a 15-year period in New York[81]

Histological change	Frequency (per cent)
Hyperaemia	86.5
Foci of lymphocytes	81.1
Foci of histiocytes	77.0
Synoviocyte hyperplasia	63.5
Synoviocyte hypertrophy	68.9
Foci of plasmacytes	50.0
Perivascular inflammatory cells	50.0

histological changes conformed with those that had been described by Soren.[444,445]

Prospective studies were at first confined to cases treated by arthroplasty for severe deformity or disability. Arnoldi, Reiman and Bretlau,[25] on the basis of 24 cases of osteoarthrosis subject to hip joint surgery, concluded that histological signs of synovitis in osteoarthrosis are indeed common. They claimed that 'synovitis was a very early feature of osteoarthrosis: a preliminary proliferative phase with oedema and vascular congestion was followed by later fibrosis.'[387] The cartilagenous changes in the five 'adequate' specimens examined postoperatively by Goldenberg, Egan and Cohen[152] were 'mild'. Nevertheless, three of these five cases displayed 'synovial membrane inflammation' in terms of synovial cell hyperplasia, vascular change or leucocytic infiltration, and it was concluded that synovial inflammation was often present.

It is therefore widely accepted that there is 'an element of active synovitis' in a proportion of patients with idiopathic osteoarthritis,[293] a conclusion in accord with my own study of 80 synovial biopsies and with the work of Revell, Mayston, Lalor and Mapp.[391] In Meachim and Brooke's[293] series of 52 cases, lympho-plasmacytic infiltration, a broad index of sustained inflammation, was minimal or absent in 18 cases, slight in 24, and of moderate intensity in 10 (19 per cent). These results agree with general experience. The problem remains: What causes these changes and why are they inconstant?

At first it seemed sufficient to postulate that synovitis in osteoarthrosis was attributable to the fragments of cartilage and bone released into the joint cavity in advanced disease. The debris includes calcium hydroxyapatite crystals which are phlogistic.[96] However, the possibility that crystals such as those of calcium pyrophosphate (see p. 393), or calcium hydroxyapatite (see p. 402) and/or detritic fragments of bone or cartilage evoke synovitis in osteoarthrosis, for example by the release of prostaglandins, neutral proteases or collagenase, has been contradicted by evidence that there is no specific relation between the presence of inflammatory synovial infiltration and that of detritus or crystals.[391]

Immunopathology

It then emerged that whereas the articular cartilage in 16 per cent of cases of osteoarthrosis secondary to mechanical causes could be shown to include immunoglobulins and complement,[78–80] more than 50 per cent of specimens of cartilage from cases of idiopathic osteoarthrosis also reacted for these proteins. When these studies were extended to the synovia, immunofluorescence reactions for IgA, IgG and IgM and complement component 3 were positive in substantial proportions of patients, mainly women, with (largely) polyarticular idiopathic osteoarthrosis (Table 22.8). The high frequency with which immunoglobulin A was demonstrable is of particular interest. It appeared unlikely that the humoral immunological phenomena shown in osteoarthrosis were primary, pathogenic changes.[78] There was no correlation between the extent of the deposits of immunoglobulins and complement in cartilage and the severity of the synovial changes. The cartilage deposits appeared indeed to be coincidental epiphenomena. The patchy, localized distribution of deposits indicated their early occurrence and their degradation with the passage of time.

Cell-mediated responses to proteoglycan and to collagen are less frequent in osteoarthrosis than in rheumatoid arthritis. However, it remains possible that these reac-

Table 22.8

Relationship between grade of severity of synovial microscopic change, joint site and immunofluorescent identification of immunoglobulins and complement[78]

Joint	Number	Histological grade*	IgA	IgG	IgM	C
Knee	29	5.5	19	16	9	15
Hip	24	6.0	20	18	13	20
Foot	9	4.0	8	8	4	5
Hand	3	7.0	3	2	1	3

*Maximum score 15, reflecting four grades of synovial cell hyperplasia, villous hypertrophy, and lympho/plasmacytic and macrophage infiltration.

of particular interest that even before cartilage loss is complete in a vertical plane, the underlying bone has thickened by lamellar appositional new bone formation to such an extent that, on surface inspection, it may resemble ivory on account of its pale colour and opacity (see p. 871). Bone, eburnated in this way, is an extremely inefficient substitute for hyaline cartilage as a load-bearing material: it is vascular and contains nerve endings. With movement, excoriation occurs and the coarsely scarred, white bone surface breaks away, releasing fragments into the synovium and exposing the vascular channels of the subjacent marrow. Cartilage loss in grade IV hip joint disease is usually extensive upon the femoral head, less complete upon the acetabulum.

Hip joint osteoarthrosis is accelerated by the presence of concurrent infection, by continued occupational trauma, by the presence of crystals such as those of calcium pyrophosphate, by coexistent causes of inflammation such as rheumatoid arthritis, and by metabolic or endocrine dysfunction. It is of interest that antiinflammatory agents, such as indomethacin, have been shown to exacerbate hip osteoarthrosis, perhaps by allowing tissue destruction to advance silently, leading to the need for early arthroplasty.[379] Hip joint osteoarthrosis is very common in Paget's disease; it is possible that the deformities of the femoral neck and shaft predispose to osteoarthrosis.

Knee

This complex articulation is composed of two groups of bearing surfaces: the femorotibial, itself in lateral and medial parts, and the patellofemoral. Valgus and varus deformities predispose to osteoarthrosis. A pattern of normal, age-related fibrillation, common but not anatomically progressive, has been suggested,[36,292] but more severe, progressive lesions are also found.

Whereas the loads borne by the femoral head are distributed over a significant part of the main bearing surface, those of the convex femoral condyles are carried by areas that, in the normal, erect adult, may be no more than 200 mm^2. The areas affected by progressive osteoarthrosis are correspondingly confined to particular parts of the joint. The patella is first prejudiced; the patellar groove of the femur, the medial parts of the tibial condyles, i.e. those not covered by the menisci, the anterior part of the medial femoral condyle and the corresponding part of the lateral condyle, are affected in this order of frequency. The posterior parts of the femoral condyles, adjoining the posterior recess, are affected slightly: here, little weight is borne by the fully flexed knee, and, indeed, in full flexion, the forces exerted upon the joint tend to separate the articular surfaces.

Finger

The distal interphalangeal joints of older persons are insidiously affected but other finger joints, including the proximal interphalangeal and the metacarpophalangeal, can also be the site of osteoarthrosis and of osteophytosis. Marginal osteophytes form at the proximal end of the terminal phalanx; the osteophytes are most prominent dorsally. The deformity constitutes one variety of Heberden's node. In older men, Heberden's nodes are thought to be of mechanical origin; they progress little and cause relatively slight disability. Nodes are exceptional in paralysed limbs.[124] In women after the menopause, and particularly over the age of 60 years, the nodes increase progressively in size and may come to affect several fingers. In these women, the development of the osteoarthrosic deformity of Heberden's nodes is influenced by heredity; affected individuals simultaneously develop osteoarthrosis of other joints such as the apophyseal, knee and first tarsometatarsal. The condition is then termed generalized osteoarthrosis (GOA) (see p. 847); occasionally there is evidence of systemic illness with an elevated erythrocyte sedimentation rate.

Microscopically, the site of origin of Heberden's nodes is closely related to the insertions of tendons, within which they may grow, and to the attachments of the joint capsule, tendinous expansions and vinculae. The initial step in the growth of a node is the local, paraarticular formation, from the periosteium, of an island of compact, 'myxoid' mesenchymal tissue. With time, chondroid transformation occurs and the prominence comes to be, first, a chondrophyte, and then, by enchondral change, an osteophyte. The circumferential bony prominences, cancellous centrally but bounded by a lamina of compact bone, are capped by fibrocartilage; they are exactly analogous to the larger osteophytes of joints, such as the knee and elbow, and to the osteophytes of the apophyseal joints of the spine. However, the bony outgrowth may have no cartilagenous cap and the remaining articular structures of the joint may retain an entirely normal appearance.

Toe

In addition to its susceptibility to gout, the first metatarsophalangeal joint is prone to osteoarthrosis; in one classical series only the knee was affected more often.[176] One form of osteoarthrosis of the metatarsophalangeal joint comprises hallux rigidus: central, articular cartilage is progressively destroyed by fibrillation and subchondral, hypertrophic new bone formation contributes to disability and deformity. Another common and comparable deformity of the first metatarsophalangeal joint is hallux valgus. Compression and distortion by ill-fitting shoes create an abduction deformity that become permanent. There is secondary pain and disability, and bone hypertrophy accompanies remodelling and, sometimes, the formation of new articular facets.

Spine

By definition, osteoarthrosis only affects the freely moving, synovial joints. The pathological features of apophyseal osteoarthrosis are described on p. 931. In a proportion of individuals, further occasional neurocentral (Luschka) joints, of synovial character, develop at the lateral margins of the cervical fibrocartilagenous intervertebral joints. Here, the late onset of osteoarthrosis can provoke nerve root or nerve compression, leading to osteophyte formation and nerve entrapment.[66]

Temporomandibular
(Fig. 2.27b)

Osteoarthrosis increases in frequency with age in this, as in many other synovial joints but is very unusual before the age of 40 years.[60] There are analogies with the sternoclavicular joint in which the fibrocartilagenous disc is disrupted.[225,422] Whether there is a genetic predisposition to temporomandibular osteoarthrosis is not clear, nor is it known whether age-related changes in the cellular and molecular units of the joint lead to cartilage disorganization and loss. Mechanical stress, particularly upon the lateral part of the joint during mastication, exerts the greatest deformation. Loss of molar support is associated with temporomandibular osteoarthrosis. Consequently, it is deduced that this loss of support, with increased functional load, is one of the factors responsible for the development of osteoarthrosis of this joint. Muscular hyperactivity (bruxism), unilateral chewing, deformity after trauma and congenital defects are additional or alternative causative factors.

The pattern of cartilage injury and loss in osteoarthrosis of the temporomandibular joint corresponds closely to that recognized in other affected joints. At the joint margins, increased bone vascularity evokes new bone formation and osteophytosis: simultaneously, there may be reorganization and cyst formation. Remodelling of the lateroposterior part of the temporal eminence and of the lateroanterior part of the condyle, follow. The intraarticular disc has little or no capacity for remodelling and repair. The increased compressive forces that predispose to fibrillation lead on to thinning of the fibroelastic disc, to cell necrosis, to degradation of the intercellular matrix and ultimately, to perforation.[60] There is increased vascularity of the affected disc, suggesting repair; and disorganization of one disc can place abnormal loads on the other, promoting bilateral osteoarthrosis.

Temporomandibular joint osteoarthrosis may be early and superficial or established and deep; the changes are focal, affecting less than the whole surface. The lateral parts and the temporal component are especially vulnerable, the medial parts and the supporting condyle relatively spared. Remodelling of bone is possible without subsequent osteoarthrosis[60] and secondary inflammation may occur. Concomitant calcium pyrophosphate dihydrate crystal deposition has been suggested radiographically.[61]

Aetiology and pathogenesis

The importance of discovering the precise nature of the biological disorders that culminate in clinical osteoarthrosis is self-evident. It is possible that chemical agents will be discovered that will allow chondrocytes in early osteoarthrosis to repopulate the depleted cartilage matrix with new macromolecules. Early diagnosis will be necessary to make this suggestion realistic. Whole cartilage grafts[101] or chondrocyte cultures cannot take the place of lost tissue and the great advances in arthroplasty for the late, established disease have concealed the need for effective medical treatment. Genetic, social and epidemiological surveys may identify population groups prone to primary osteoarthrosis. If so, new life-styles, changed diets, exercise programmes and drug supplements could encourage whole populations to embark on regimens to minimize the risk or delay the onset of osteoarthrosis. It may become necessary to think not only of individual sufferers but of susceptible societies. Lengthy and large preventative programmes may be considered but so much suffering results from osteoarthrosis that new approaches to prophylaxis will be considered carefully.

Secondary osteoarthrosis is one result of a wide variety of hereditary (see Chapter 9) and environmental agents.[37,42,125,129,132,168,184,186,187,192,312,438] Table 22.3 indicates their diversity. They include mechanical and traumatic (see Chapter 24), infective (see Chapter 18), immunological (see Chapters 12–17), inflammatory, endocrine and metabolic (see Chapter 10) causes. The pathological end-stages of the syndrome may give little indication of the identity of the initiating factors.

The cause(s) of 'idiopathic' osteoarthrosis are necess-

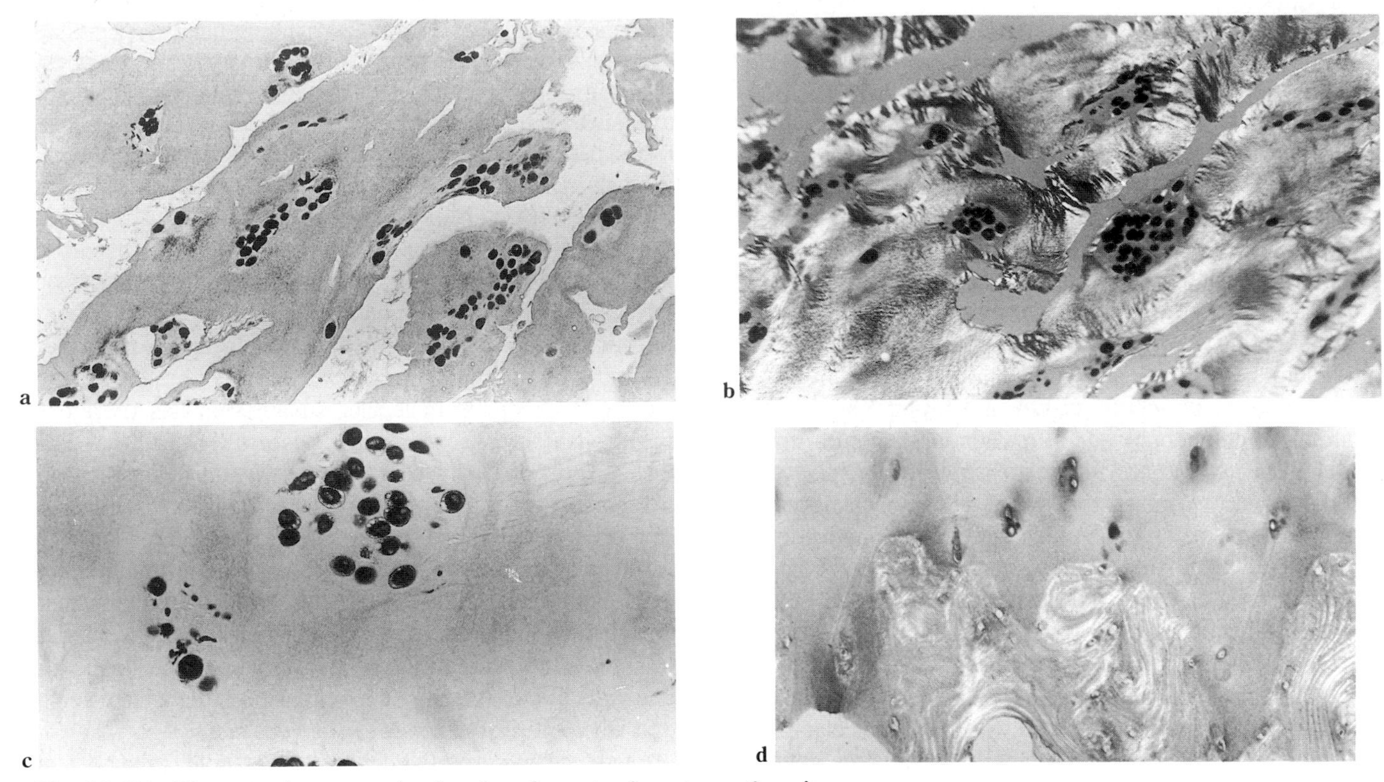

Fig. 22.35 Phenotypic expression by chondrocytes in osteoarthrosis.
a. Identification of S100 in chondrocyte cell clusters in superficial fibrillated cartilage. (ABC × 150.) **b.** Expression of vimentin demonstrated by application to fibrillated cartilage of monoclonal antibody against vimentin. (Polarized light, ABC × 150.) **c.** Monoclonal antibody confirms expression of tubulin by cells of chondrocyte clusters. (ABC × 300.) **d.** Osteoarthrosic cartilage contains excess fibronectin. Here, antifibronectin antibody demonstrates presence of this matrix glycoprotein in pericellular matrix of deep (zone 4b) zone cartilage. (ABC × 50.)

tomedin synthesis, release or action. The complex role of growth hormone includes the regulation of cartilage acid hydrolases[194] and neutral proteases but, again, there is little indication that the endocrine control of these systems is down-regulated in osteoarthrosis.

If acromegalic arthopathy (see p. 415), and the joint disorders that accompany gigantism, diabetes mellitus*, myxoedema and hyperthyroidism are accepted as examples of osteoarthrosis, then there is *a priori* evidence for suspecting that disordered chondrocyte growth and division may underlie a proportion of cases of 'idiopathic' osteoarthrosis. Much of the direct evidence has come from animal studies. In pituitary dwarf (dw/dw) mice, for example, cartilage disorder and secondary osteoarthrosis are characteristic.[423] The cartilage changes can in part be reversed by growth hormone.

Much less is known of the behaviour of bone cells. Osteoblast responsiveness appears to influence the surgical response to arthroplasty[405] but the part played in this sequence by growth factors and cytokines, and the relationship of this reactivity to that of bone change in the underlying articular disease, remain unclear.

* Diabetics do not have a noticeably higher frequency of osteoarthrosis. However, young diabetics display a stiff finger syndrome.

Molecular disorders

The biochemical literature has been extensively reviewed (p. 848).[320,321,322,323]

Evidence for the development of molecular or metabolic changes in cartilage that may predispose to osteoarthrosis, perhaps by permitting the normal forces exerted during daily movements to provoke excess deformation, has come from histochemical and biochemical studies. Histochemical reactions show the loss, from superficial cartilage zones, of metachromatic staining intensity,[75,273,274] an index of proteoglycan integrity.[74] The reduction is proportional to the severity of the disorder. Experimental procedures used to simulate natural joint disease such as division of the anterior cruciate ligament in the dog also cause a very early and extensive loss of matrix staining. It has been believed therefore that, early in osteoarthrosis, there is degradation or loss of proteoglycans or their components, perhaps with an abrupt alteration in the rate and extent of proteoglycan synthesis and turnover.

A conspicuous change in osteoarthrosic cartilage is in water content. Normally, the water content of hyaline cartilage is greatest in the surface zones, least in the deep zones.[469] In human osteoarthrosic cartilage there is an increase in water[275] (see p. 848), a change also found very soon after the experimental induction of joint disease.[262]

At first it might seem difficult to reconcile the demonstrable loss of cartilage proteoglycans in osteoarthrosis with the increase in water: as components of proteoglycans, glycosaminoglycans are the single main determinant of cartilage water, holding much water within proteoglycan domains; the loss of proteoglycan, it might be anticipated, would result in a decreased water content. One explanation lies in the fact that the tendency for cartilage to swell as proteoglycans retain ever greater quantities of water (the pressure gradient) is normally balanced, at each level, by tension in the network of cartilage collagen.[276] Any disruption of the collagen will encourage swelling in proportion to the content of proteoglycan within the disrupted network. Cartilage surface zones normally have the lowest concentration of proteoglycan and lose proteoglycan first as osteoarthrosis develops: those zones which have highly organized dense collagen laminae, swell least. The middle zones, with a high proteoglycan content, imbibe water quickly as the meshwork of collagen loses its restraining force. Swelling is considerable. The deepest zones contain much proteoglycan, loss of which is a later phenomenon; deep-zone collagen is also preserved intact until a late period in the evolution of osteoarthrosis and swelling is normally restrained. Whatever the degree of swelling, it is essential to remember that the crucial cartilage qualities of ionic equilibria, fluid flow and swelling pressure, with consequent elasticity of the tissue and its responses to compressive loads, are all determined primarily by the proteoglycan content, a disturbance of which remains the single most important abnormality in osteoarthrosis.[32,436]

It is natural to search for evidence of altered collagen quality, content or arrangement in osteoarthrosis. Although the scanning electron microscope is a crude tool with which to seek for changes in collagen fibrils or fibre bundles, there is some evidence that the arrangement in human osteoarthrosis and in experimental dog joint disease is altered.[205] The rate of collagen synthesis is increased in osteoarthrosic human femoral heads in comparison with normal.[274] The increased rate is in proportion to the severity of the disease. There is no change in the collagen phenotype which remains type II. Nevertheless, an increased accumulation of type VI collagen is suspected.[263] In experimental dog osteoarthrosis, associated with hip dysplasia, procollagen may accumulate suggesting that there is a partial defect in the conversion of procollagen to collagen.[300] The accumulation of this procollagen might interfere with normal collagen fibril structure and function, weakening the cartilage and permitting water retention, abnormal swelling and material weakness.

Abnormal materials

The possible contribution made by aggregates of crystals is considered on pp. 852 and 893. Deposits within cartilage of pigments, metabolites or other materials such as amyloid seem likely to alter mechanical properties and thus to predispose to osteoarthrosis. As cartilage ages, a yellow-brown pigment is deposited (see pp. 78 and 830). However, the degree of pigmentation measured in costal cartilage does not correlate with knee osteoarthrosis. The pigment is an acid glycoprotein (see p. 830). The inherited disorder of ochronosis (see p. 351) also precipitates premature cartilage breakdown, an effect that results from impregnation of cartilage by an oxidized, polymeric, melanin-like pigment derived from accumulated homogentisic acid. The deep brown discolouration of the synovia in haemosiderosis, haemophilia (p. 967) and haemochromatosis (see p. 369) is not mirrored by a corresponding pigmentation of the cartilage matrix. In haemosiderosis, the cartilage remains intact. Iron is taken up by chondrocytes in haemophilia but the cells seem able to retain metabolic activity in the face of this iron load. The mechanism of cartilage breakdown in haemophilia is likely to be related to synovial inflammation and to be mediated by agents that activate proteases.

Amyloid deposits in the articular cartilage of older people have been found to be more common than suspected.[150] These aggregates must alter the mechanical properties of cartilage. It is therefore not surprising that amyloid is seen in a substantial proportion of knee and hip cartilages[239,241] examined after arthroplasty for osteoarthrosis.[104] The deposits in knee joint cartilage are often near the cartilage surfaces, and calcium pyrophosphate crystals may lie among amyloid microfibrils. Amyloid complicates a rare neuropathic joint disease associated with macroglobulinaemia.[416]

Abnormal geometry

The geometric structure of a synovial joint[22] may be disturbed either because of defective growth, dysplasia, or because of injury or destruction. Alterations in joint geometry[223] are likely to promote cartilage breakdown and reactive bone change. The necrosis of chondrocytes can leave residual lacunae that can act to concentrate stress and to initiate cracks if the defects are longer than the critical crack length and of appropriate orientation. Geometric differences may explain the two forms of degenerative lesion recognized by Byers *et al.*[55,56] on femoral heads and on acetabula: both may have an identical cause. Whereas, in some anatomical parts of the joint, loading aggregates stress concentration and leads to progressive cartilage injury (osteoarthrosis), others are free from loading stress and deteriorate slowly.

Mechanical change

Mechanical disorders that range from fracture to congenital dysplasia, lead to secondary osteoarthrosis. Selected populations, such as ballet dancers, are prone to osteoarthrosis of vulnerable joints such as the metatarsophalangeal.[18] A factor contributing to osteoarthrosis was thought to be obesity. However, obesity by itself does not appear to induce or provoke osteoarthrosis although this is debated.[153,409,442,465]

Mechanical failure of the composite structure of cartilage must occur either in the fibre network, in the hydrated proteoglycan matrix or in, or between, both.[219,220] The mechanical disorder may reflect the intrinsic properties of the whole material[339] or of its constituents;[345] the geometry of the joint and the local characteristics of resultant stresses; and the molecular, biochemical and mechanical history of the tissue.

A difficulty in testing osteoarthrosic cartilage mechanically is the obvious need to disorganize the cadaver joint to prepare samples.[353] Within these limitations, some results have been obtained and Kempson *et al.*[221] demonstrated a decrease in cartilage stiffness that correlated with osteoarthrosic grading. Adjacent, normal appearing cartilage was abnormally soft on indentation and displayed reduced tensile stiffness and strength. When whole normal hip joints were tested[20,21] there was a greatly increased compressive compliance with age. The changes in compressive properties were explicable by the reported increase in water[276] (see p. 885).

The mechanical breakdown of cartilage in osteoarthrosis may result from: 1. the abnormal loading of abnormal cartilage (as in congenital dislocation); 2. the normal loading of abnormal cartilage, for example, following multiple microfracture; or, 3. the abnormal loading of normal cartilage. A fourth mechanism is the failure of joint lubrication. However, there is now little support for the view that a primary defect in joint lubrication can cause osteoarthrosis (Table 22.9). Whether a failure of the secretion or synthesis of the recently demonstrated glycoprotein lubricin (see p. 211)[452] can explain a predisposition to osteoarthrosis is not known but does not find favour.

The evidence demonstrating that compressive compliance increases with age[21,23] suggests that critical fracture strain is more likely to be exceeded in older cartilage, particularly if there is an accumulation of residual strain with repeated cycling; an accumulation of permanent strains rather than microfractures may be the cause of fatigue failure. This permanent strain may be due to changes in the proteoglycan or collagen or in the binding between them.

A widely promoted theory accounting for mechanical breakdown of articular cartilage in osteoarthrosis is tensile failure of the collagen fibres, either because of a diminished tensile strength and stiffness with age[221] and/or because of an accumulation of fatigue damage with repeated cycling, with a reduction in the fatigue resistance.[485] This hypothesis is based on the evidence that: 1. both the increase in the mechanical tensile properties and fibrillation are age-associated; 2. fibrillation must involve failure of fibres, with the formation of clefts and splits; 3. cartilage adjacent to fibrillation has reduced tensile properties; 4. the increase in thickness and water content of mildly fibrillated cartilage suggests abnormally distensible fibres; and 5. fibrillation may be preceded by abnormal separation of collagen fibres, with fragmentation and disorientation.[347]

The roughened, fibrillated cartilage in osteoarthrosis is subject to increased wear: there is abrasion and loss of tissue. Cartilage in which there is overt fibrillation and softening displays increased deformability. The contact areas between opposed surfaces are consequently increased under small loads. When cartilage thickness is reduced to less than 25 per cent normal function is not possible.[485]

Analogous problems are presented by associated fibrocartilage. The integrity of the knee joint menisci is a corollary of normal joint function. Removal of a meniscus, a surgical procedure formerly recommended for medial compartmental tears, can lead to irreversible cartilage damage.[206,236] Equally, excision of part or the whole of a meniscus is the basis for a variety of experimental models of osteoarthrosis[303] (see p. 902). Considerable protection is afforded to the human knee if a rim of meniscal tissue is allowed to remain at meniscectomy[49] and meniscal repair rather than removal is increasingly practised.

Diminished movement

Immobilization[178] Very long periods of immobility drastically and permanently change synovial joint structure.[6] The causes, prevention and treatment of contracture and ankylosis are recorded in a large orthopaedic literature. Experimental investigations have demonstrated very early alterations in hyaline cartilage after enforced immobility. There are analogous changes in the fibrous connective tissues.[6] The cartilage changes are similar to those identified in some forms of experimental osteoarthrosis to the pathogenesis of which they are clearly relevant.

Humans There is less opportunity to make controlled studies of human than of animal joints and it is essential to emphasize that the effects of immobility differ at different ages and between different joints. The small number of clinical investigations has centred on the large lower limb joints, usually the knee, of patients in whom compound fracture, incomplete fracture fixation or infection has required long periods in plaster. The joints of diseased limbs have been investigated after amputation. For obvious reasons, there are few records of immobilized normal joints. The opportunity to make such studies occasionally arises when amputation (Fig. 9.33, p. 366) is undertaken for congenital deformity or paralysis, or at *post mortem* after stroke, paraplegia or quadriplegia.

In joints from patients immobilized in plaster, ankylosis is accompanied by the replacement of articular cartilage by enchondral ossification. Bony ankylosis begins but may not be conspicuous radiographically.[110] The sequence of microscopic changes varies according to whether the articular surfaces are held in direct opposition or not.

In knees in which the surfaces are immobilized in a non-opposed position, fibrofatty, areolar synovial tissue extends into the joint space, merges with tissue extending onto the opposed articular surfaces, and obliterates the joint cavity, establishing fibrous ankylosis. Articular cartilage is reabsorbed and, in places, the new vascular fibrous tissue becomes coextensive with the bone marrow.

When the articular surfaces are directly opposed, fibrillation (see p. 858) of superficial (zone I) cartilage begins, with cystic defects in this and in deep zone cartilage. The 'cysts' contain a reparative mesenchymal tissue. The opposed surfaces become rigidly fused by progressively mature fibrofatty tissue. The deep zone cartilage and subchondral bone are traversed by increasingly dense fibrous tissue, a process culminating in intraarticular fibrosis and in fibrous and, ultimately, bony ankylosis.

Experimental animals[177] In principle, the effects of deliberately immobilizing an animal joint for very long periods closely resemble those recognized in humans. It is also clear that periodic or continuous short-term immobilization (more than 30 days) induces cartilage degeneration.[470]

The changes resulting from prolonged immobilization by plaster casts, compression devices, internal splints, nerve section and combinations of these techniques have been recorded in rats,[456] rabbits,[147] and monkeys.[406] The resulting changes depended on whether the bearing surfaces were opposed and on whether the limbs were immobilized in flexion[443] or extension.[245] The responses of immobilization vary according to the maturity of the animal. Early studies in mature rabbits, typified by those of Sood,[443] demonstrated thinning and increased cellularity of the non-mineralized hyaline cartilage of the patellar groove of the femur with increased cellularity and reduced matrix stainability. Prolonged remobilization reversed these changes. The macroscopic, radiographic, isotopic, histological,[245] scanning electron[59] and transmission electron microscopic changes resembled those of naturally occurring osteoarthrosis sufficiently strongly to suggest that immobilization provides an acceptable model of this disease.

Macroscopically, the cartilage both of immobilized opposed and of mobile, non-opposed (non-loaded) cartilage becomes thinner. With time, the cartilage assumes a yellow colour and appears matt. There are concomitant changes in bone. Technetium-99m-MDP bone scintigraphy offers a non-invasive means of recognizing early joint disease.[464] With time, the increasingly vascular bone becomes thicker. Using a semiquantitative method based on stereomicroscopy of scanning electron microscopic specimens, it is apparent that all rabbit knee articular surfaces deteriorate during six weeks of immobilization, whether or not the surfaces are continuously compressed.[177] Surface smoothness and evenness remain constant but splitting is apparent. It is confined to the medial tibial condyle.

Microscopically, the changes are more severe in rabbit joints immobilized in extension than in flexion. Matrix stainability declines within two days.[462] Surface disorganization leading to fibrillation, is evident within seven days.[177,211] Chondrocyte numbers increase in the superficial zone (zone I). More deeply, the columnar arrangement of the lacunae is lost. Foci of chondrocyte hyperplasia are seen and chondroosteophytes form.

When three-dimensional parameters are measured after eight weeks' immobilization in extension, an increase in cell density, but decrease in cell size, is found in the superficial zone cartilage of lateral tibial condylar cartilage. However, there is an increase in cell size in deep zone cartilage.[357] The contralateral, non-immobilized knee cartilage is also abnormal.

The microscopic and ultrastructural changes are accompanied by a decrease in matrix chondroitin sulphate, a feature common to all species tested. Diminished proteoglycan content in immobilized *sheep* or *dog* joints reflects a lowered rate of synthesis, indexed by diminished incorporation of ^{35}S-sulphate whereas immobilized adult *rabbit* joints display increased ^{35}S-sulphate incorporation.[471–473] In the dog aspirin (p. 894) aggravates the deleterious effects of immobilization.[352] There is increased proteoglycan extractability, perhaps due to impaired aggregation. The changed proteoglycan content is accompanied by a rise in cartilage hydration. These phenomena,[318] like the increased proteoglycan extractability, are analogous to the properties of immature cartilage and of cartilage from dogs with experimental osteoarthrosis[263,264] or from older humans.[204] One explanation for the species differences in proteoglycan invokes the effects of repetitive minor injury.[228] In support of this concept is the evidence, in the immobilized adult rabbit joint, of synoviocyte hyperplasia and of mononuclear cell infiltration, changes analogous to those recognized in the canine stifle joint up to three months after aseptic division of an anterior cruciate ligament has caused osteoarthrosis-like disease.[131] The presence of mononuclear synovial cell infiltration recalls the effects brought about by catabolin/interleukin-1 *in vitro* and in arthritides such as rheumatoid arthritis. There is, therefore, a possibility that the very rapid early loss of proteoglycan from adult rabbit immobilized cartilage can be a consequence of matrix degradation attributable to chondrocyte proteases activated by synovial-derived cytokines.

In contrast to the rapid changes in the quality and quantity of proteoglycans in immobilized joint cartilage, there is little alteration in collagen content. However, there is an enhanced collagen turnover; raised hydroxyproline labelling in immobilized rabbit joints[471] is paralleled by heightened prolyl hydroxylase activity in dog joints.[453]

The mechanical consequences of prolonged joint immobilization are softening and an increased vulnerability to injury of the chondrocytes[349] which are incompletely protected against adverse stresses. Collagen loss is a very late change. When it occurs, unlike proteoglycan, it is not reversible by simple procedures such as remobilization.

Remobilization Provided that conditions allow moderate exercise after a period of immobilization and do not impose severe mechanical stresses, the early thinning of articular cartilage and loss of proteoglycan are entirely reversible in rabbits[443] and in rats.[112] The response of dogs is less clear-cut. Palmoski *et al.*[355] found restoration of the proteoglycan population. Kiviranta *et al.*[229] made similar observations but noted selective, continued cartilage thinning affecting limited regions of the knee joint. The reformation of proteoglycans was particularly evident in the patellofemoral and tibial condylar regions, but not in the femoral condyle. In the dog, again, there was perpetuation of a thinning of the calcified (zone V) cartilage. It appears highly likely that these differences are attributable to differences in species, age, and loading conditions. It may be hazardous to extrapolate from the rat, rabbit and dog directly to humans.

Increased, excessive or abnormal loading

It is necessary to make a distinction between continuous and intermittent loads, between sustained loads and sudden impacts, and between small and large loads. The relationship between major and minor trauma and the origins of osteoarthrosis are discussed on pp. 889 and 890. Joint hypermobility may be significant.[159]

The effects of sustained alterations in load-bearing have been studied experimentally. Stance or gait have been altered by operation, amputation or casting.[354,455] The consequential drawbacks of a change of gait have been avoided in a study of the effects of increased gravitational load brought about by centrifugation. In general, there is an increase in cartilage thickness associated with chondrocyte hypertrophy. Superficial cartilage lesions, confirmed in rabbits, may be limited to certain species. The cartilage proteoglycan content may be slightly raised (dogs) or lowered (rabbits). In animals with a single immobilized, limb, this increase is detected in the contralateral joint; the increase may be as much as 25–35 per cent in less than three months. The cartilage content of chondroitin sulphate may be lowered (dogs, rabbit) or raised (sheep); in the former the keratan sulphate:chondroitin sulphate ratios are elevated. Divergent evidence regarding the rates of proteoglycan synthesis suggest that these may or may not[437] be raised in the more heavily loaded cartilages. There is no significant evidence of an elevated collagen content but the demonstration of decreased glycosaminoglycan extractability has suggested that proteoglycan/collagen interactions may be strengthened in response to increased load bearing.[154,155]

Kiviranta[228] suggests that the mechanisms by which cartilage responds to increased sustained loads is mediated by cAMP and cGMP, known to influence the part played by insulin-like growth factors (somatomedins) in increasing glycosaminoglycan synthesis. There are concomitant changes in calcium and in the prostaglandins.

Occupation Occupational predisposition to osteoarthrosis has long been recognized. The occupations in which osteoarthrosis is especially common are, however, unusual and certain sportsmen are vulnerable.[2] The elbows of some groups of coal miners, the toes of ballet dancers[45] and the knees of footballers are thought to be particularly susceptible. The evidence, however, is conflicting. Whereas the footballers of Leeds are believed not to be prone to osteoarthrosis of the knee, those of Vejle[232] are vulnerable to hip joint disease. The susceptibility of different lower limb joints differs significantly. Thus, Adams[2] found osteoarthrosis of the knee to be uncommon. His radiographic study of the ankle joint supported the same conclusion. When comparable investigations were extended to the hip joint, however, osteoarthrosis was found to be significantly more frequent than in control, non-football-playing patients.[232]

Parachutists[325] sustain many minor and some major orthopaedic injuries but do not develop accelerated osteoarthrosis. Under certain circumstances, running (see below) can be beneficial[349] but excessive exercise on hard surfaces may be harmful. In the case of ballet dancers, osteoarthrosis was found to affect the metatarsophalangeal joints of 54 per cent of 44 retired dancers but degenerative changes were not confined to these joints and six individuals had hip joint disease, six knee osteophytosis, four tibiofemoral osteoarthrosis and four patellofemoral osteoarthrosis.[18]

Osteoarthrosis and sports injuries: effects of running

The frequency and severity of osteoarthrosis in those who have habitually played games or taken part in sports that expose synovial joints to repetitive excess loads and to traumatic injury, has excited controversy.[2] Rugby football players and those practising judo or water-skiing, are widely suspected to be particularly at risk. However, firm evidence is relatively slight.

In the case of habitual, long-distance running, a controlled study of 41 individuals aged 50 to 72 years showed that there was a substantially increased bone mineral content, estimated by computed tomographic scanning, but no raised frequency of clinical or radiological signs of osteoarthrosis.[244] Essentially the same conclusions were reached in a further study of lower limb disorder.[356a]

Experimental studies of running

Early observations suggested that sustained periods of running by experimental animals led to increased cartilage thickness, increased cell numbers and protection against degradative lesions.[230] By contrast, very long, excessively strenuous exercise caused decreased cartilage stainability.

When these changes were investigated in more detail, it appeared that moderate, walking exercise by dogs over a six-month period was insufficient to alter the activity of chondrocyte oxidoreductases.[226] However, prolonged treadmill exercise by rabbits led to a raised cartilage proteoglycan content,[454,455] altered collagen metabolism[453] and to enlarged zone I nuclei.[358] The effects of obligatory treadmill running on the femoral condylar cartilage of dog legs that had been immobilized for six weeks, with consequential increase in water and decreased thickness and proteoglycan, was highly adverse. Voluntary, unrestrained walking on the other hand, reversed the disorders caused by immobilization.[349,470,493] As Kiviranta[228] points out, it is the severity of loading which determines the cartilage responses, particularly in the sedentary animal.

As in spontaneous[204] and experimental[316] degenerative joint disease, the mean proteoglycan size diminished[468] and proteoglycan aggregatability lessened.[351] The effects of running on collagen are evidently less. The collagen content of rabbit[455] and dog[403] cartilage is not increased and moderate exercise does not change the rate of collagen synthesis.[454] Moderate running exercise after 10 weeks of training, for one hour daily at 4 km/hour, on a treadmill inclined upwards at 15°, for 15 weeks, led to increased non-calcified cartilage thickness but to no change in the thickness of the calcified cartilage.[230] The thickened cartilage displayed a raised glycosaminoglycan content, to a greater extent on the medial than on the lateral femoral condyles. By contrast, much more strenuous running (daily on an inclined treadmill at 15° for 15 weeks, followed by 20 km/day on the same plane for 15 weeks) caused a reduction in thickness of the medial femoral condyle—a change accompanied by diminished glycosaminoglycan content, especially in zone I cartilage.[229a]

Major trauma: post-traumatic osteoarthrosis

Some aspects of traumatic arthropathy are outlined in Chapter 24 (see p. 955) where the immediate effects of direct traumatic injury to a joint are summarized. Post-traumatic osteoarthrosis is the combination of the long-term effects of one or more episodes of sudden, external violence upon the organism.[242] The lesions may originate primarily in bone or in cartilage. They culminate in osteoarthrosic remodelling of which the salient characteristics are: 1. more or less complete deterioration in cartilage structure and function; 2. intrinsic remodelling of subjacent bone, with osteosclerosis and sometimes cyst formation; and 3. extrinsic architectural change with marginal osteophytes and low-grade synovitis. There are corresponding radiological changes.

The sequence of osteocartilagenous injury culminates in local mechanical dysfunction; cartilage degradation; and secondary osteoarthrosis. The accompanying low-grade synovitis may be provoked by the cumulative effects of cartilage and/or bone fragments lodged within the joint, and cartilage degeneration is promoted if foreign bodies or septic arthritis have complicated injury.

The likelihood of osteoarthrosis developing after fracture is influenced by whether the fracture passes through the joint; the degree of angular deformity; the extent of soft tissue damage; the ensuing laxity of the joint; the coincidence with generalized osteoarthrosis; the involvement of nearby articular blood vessels; and, perhaps, the degree and duration of immobilization. Lower limb joints are more susceptible to post-traumatic osteoarthrosis than upper ones.[498a]

The frequency of secondary osteoarthrosis after intra-articular fracture of the leg is difficult to quantify but is estimated to be approximately 10 per cent of both proximal and distal components if a lower limb joint is implicated.

Chondromalacia patellae may be a special form of post-traumatic, lower limb osteoarthrosis (see p. 879). An alternative cause of postfracture osteoarthrosic change is mechanical. Comparable results to those of post-traumatic osteoarthrosis are encountered after osteonecrosis. Judged by the effects of injury on earning capacity and social history, approximately one-third of cases with condylar fractures of the knee and about one-quarter of those with fracture–dislocation of the hip, including central subluxation and acetabular injury, are awarded compensation for the original injury in Switzerland.[68] Not surprisingly, the severity of post-traumatic osteoarthrosis is related to both the severity of injury and to the type of fracture. Thus 52.4 per cent (11 cases) of comminuted fracture of the femoral head were compensated, compared with 43.3 per cent (13 cases) of compression fracture of the tibial plateau.

When trauma to the hip is sufficiently severe to cause dislocation or acetabular fracture, the consequent arthropathy may be compounded by avascular necrosis of the femoral head, sciatic nerve injury, disruption of the acetabulum or severe musculoskeletal injury.[172] The ankle and calcaneonavicular joints are also vulnerable,[359] as are the patella, medial malleolus, talus, calcaneum and the metatarsal heads.[115] Fissuring of the articular cartilages, fibrillation, and the propagation of microfractures of the chondroosseous junction, are some of the characteristic microscopic changes. The precise configuration of the primary, anatomical joint surfaces influences the frequency of post-traumatic osteoarthrosis. Osteoarthrosis is for example, more frequent in patients with a concave talar profile, a characteristic of youth, than in those with a less concave talar surface, a feature of older persons.[392] Fracture dislocation therefore exerts particularly adverse effects on the ankle joints of younger persons, in terms of post-traumatic osteoarthrosis.

Minor trauma[271]

The possibility that a sustained disturbance of a delicately adjusted mechanism regulating repair after trivial, day-to-day mechanical injury can account for the aggregate injury of osteoarthrosis must be considered. This concept receives support from the demonstration that only hyaluronate, from among a number of macromolecular substances extracted from or analogous to those in cartilage, reduces chondrocyte metabolism of ^{35}S-sulphate, probably by interacting with proteoglycan at the cell surface.[490,491] Since hyaluronate is a critical component of normal cartilage matrix, there is evidently an intrinsic feed-back mechanism for regulating matrix synthesis. Perhaps a disorder of chondrocytes may explain an insidious breakdown of this system. Many attempts have been made to study the effects of deliberate experimental injury.[271]

Superficial lacerations or incisions (see p. 78) are poor models of normal mechanical insults and single or repetitive blunt injuries are closer to the style of trauma encountered in daily life. Nevertheless, investigations of superficial laceration and incision[288,458] have shown how chondrocytes quickly react by a burst of mitotic activity, enhanced matrix synthesis, and increased catabolic enzyme activity. The changes persist for a few days only. There is little or no further progress towards healing of the injury by the replacement of the lost cartilage and there is no proof that the unhealed injuries progress over long periods to osteoarthrosis-like change.

The response to deeper cartilage injury (see p. 79) extending to bone, and creating haematoma formation and a granulation tissue reaction, is similar to that in other vascular tissues. There is little resemblance to natural osteoarthrosis, and marginally, no cartilage resurfacing. The resulting fibrous tissue is gradually changed to a cartilage-like tissue, becoming fibrocartilaginous. This new tissue persists indefinitely. The slightly roughened surface is depressed below the level of the surrounding intact cartilage which remains intact and free from osteoarthrosis. The extent to which repair occurs is influenced by the size of the defect and by the extent to which the healing limb is mobilized. Continuous passive movement over four weeks encourages rapid healing in deeply injured rabbit cartilage with the production of a tissue that resembles hyaline more closely than fibrocartilage, a response that may be further encouraged by an induced electrical field.

Continuous compression

Frequent (habitual) loading of cartilage from sites such as the femoral condyles maintains proteoglycan concentrations at higher levels than in cartilage.[437] The effects of continuous compression are largely artefactual. However, they show that mild, sustained experimental compression causes the appearance of chondrocyte hyperplasia followed by nuclear degradation, diminished matrix staining, the ingrowth of blood vessels into calcified and marginal zones, indentation of the opposed cartilage surfaces and deformity.[147] The severity of cartilage deterioration appears to be in proportion to the duration rather than to the magnitude of the cartilage compression.[161]

Influence of bone

The intimate functional relationship between the cartilaginous and bone components of synovial joints and their anatomical continuity suggests that the origins of osteoarthrosis should be sought in both tissues. Whereas the cartilage of large limb joints is very susceptible to the damaging effects of loading (see pp. 267 and 888), it is resistant to shear stress and abrasion. By contrast, bone, of lower elasticity, is present in such large masses that it absorbs much of the forces sustained during impulsive loading, sparing cartilage. Thus, the main force of impact sustained on walking downstairs, running or jumping is transmitted to adjacent bone which is strong and elastic yet deformable. Cartilage is spared the adverse effects of repeated impact. It has therefore been argued that an increase in the stiffness of bone, with a decrease in its capacity to attenuate peak dynamic forces in the form of impulsive loads, could initiate cartilage damage.[374] This hypothesis has attracted widespread interest and is supported by an impressive array of experimental studies, outlined below. Against the hypothesis is the failure to find any consistent bone change in osteoarthrosis that would be necessary to result in increased stiffness, and evidence from direct studies of circumferentially unrestrained articular cartilage, that this tissue can survive impact loads of up to 25 kN/m^2. Less than 13 kN is sufficient to fracture a femur.[389] Expressed in another way, if the contact area in impact loading is greater than 500 mm^2, then a compressive load sufficient to fracture a femur should not damage articular cartilage.

Contribution of bone to the genesis of osteoarthrosis[374]

Joints (see Chapter 3) are bearings. They are subject to two principal forms of stress, shear (from articulation) and longitudinal (from weight-bearing). The greatest proportion of the longitudinal loads borne by the limbs is impulsive. The attenuation of these loads had been attributed to the elastic and viscoelastic properties of cartilage and synovial fluid. However, it began to be suspected that limb bones and soft-tissues were of greater significance.[227] Although not as elastic as cartilage and synovial fluid, bone and the nearby non-mineralized connective tissues exist in relatively very large amounts. They are consequently regarded as highly effective shock absorbers.[373]

In a series of stimulating and provocative studies, Radin and his colleagues observed a consistent relationship, in human *post mortem* material, between early articular cartilage degeneration and mechanical stiffening of the subchondral bone. One cause of such stiffening in this material is, of course, the increased osteogenesis which acompanies the hyperostosis and osteophytosis of advancing osteoarthrosis.[71]

Since articular cartilage is highly resistant to abrasive wear caused by the friction of the opposed surfaces, and since it will not wear out even if the lubricating properties of the synovial fluid are extinguished,[373] it appeared logical to consider the role of bone and the adjacent tissues in the genesis of cartilage degeneration. Articular cartilage, Radin suggested, is highly susceptible to the effects of the impulsive loads which accompany most human diurnal movements.

To examine this hypothesis, experiments were made in guinea-pigs, rabbits, sheep and cattle. In guinea-pigs, in which skeletal immaturity judged by the non-closure of epiphyses, persists throughout life, impulsive loads equivalent to the body weight were applied for brief daily periods. Degenerative changes in knee joint articular cartilage were caused; these changes, including the early loss of stainable glycosaminoglycan, were invariably accompanied by mechanical stiffening of the subchondral bone. There was no immediate explanation for this change but bone stiffening appeared to precede cartilage degradation. The tests were extended to skeletally mature rabbits.[372] Daily, one-hour periods of repetitive 60 cpm body weight loading were employed. Again, measured bone stiffening preceded microscopic cartilage injury. The reason for the bone stiffening appeared to be the occurrence and healing of microfractures of the subchondral, cancellous bone trabeculae. The pathogenesis of these changes was examined in detail.

The repetitive application of loads of $1.5 \times$ body weight in 40 cpm cycles for 40 min daily, a procedure judged to be equivalent to hopping, was sufficient to cause an increase of 20 per cent in subchondral bone stiffness.[371] Within seven days, there was a reduction in superficial zone (zone I) cartilage proteoglycan, confirmed by a 17 per cent fall in hexosamine, an increase in ^{3}H-thymidine incorporation, increases in proteoglycan and protein synthesis but no demonstrable change in cartilage collagen content. An essentially similar but progressive sequence followed 20 days of testing. When fewer (15 days) impulsive loads were employed for even shorter (20 min) periods, there was still a demonstrable 20 per cent increase in bone stiffness but only a slight increase in cartilage cell numbers. These features accompanied a significant rise in protein and proteoglycan *formation* and in ^{3}H-thymidine incorporation but no change in proteoglycan (hexosamine) or collagen (hydroxyproline) *content*. The changes were those of very early metabolic adjustment. When attenuated studies were adopted, using intermittent four- and five-day loading cycles each of 20 min/day, each separated by nine-day rest periods, bone stiffness still increased but there was no decline in cartilage hexosamine content. Cell division rose only slightly but proteoglycan synthesis was much enhanced. When a similar experiment was followed by a 28-day rest period, bone stiffness was shown to return to normal although a slight increase in protein and proteoglycan content persisted, as did a mild rise in sulphate incorporation. The hexosamine (proteoglycan) content fell by 21 per cent.

These results clearly demonstrated the complex sequence of events which follows significantly increased, repetitive impulsive loading of a major articulation. The experiments were then extended to the phenomena of wear, by studies of bovine metatarsophalangeal joints.[375] Stiffening by 27 per cent, of excised subchondral bone, by the introduction of methyl methacrylate, significantly increased the degree of wear *in vitro*; by contrast, preliminary stiffening of cartilage by exposure to glutaraldehyde decreased wear. It was deduced that articular cartilage wear depends on both the properties of the cartilage and on the properties of bone, and that molecular change in osteoarthrosic cartilage, such as the increased water content (see p. 885), are indices of macromolecular composition, not prime causes of susceptibility to wear.

The view that osteoporosis and osteoarthrosis were mutually exclusive appeared to hold difficulties for Radin's concepts. Radin then postulated an increased frequency of microfracture in the trabeculae of porotic bone. However, the role of microfracture has been systematically assessed and evidence obtained that these fractures are much less common in osteoarthrosic human limb bones than would be required to sustain Radin's theories.[117] When the femoral heads from 67 patients subjected to arthroplasty because of severe osteoarthrosis were compared with those of 66 controls, there was in fact a smaller number of microfractures in the former than in the latter. The structure of the macerated bone allowed the recognition of two groups, sclerotic and porotic, and the former was divisible into two subgroups. In one of these subgroups, microfractures became less frequent as the percentage of mineralized bone increased.

There is experimental evidence to support this conclusion. Although the subchondral bone of osteoarthrosis-prone STR/ORT mice is thicker than in non-osteoarthrosic control CBA/ORT mice, bone thickening is not the cause of the osteoarthrosic changes.[483]

Blood flow and angiogenesis

It has long been suspected that altered arterial and venous blood flow may play a part in the origin of osteoarthrosis.[28,75,299,344] Although isotopic scanning techniques often reveal increased blood flow at sites of osteoarthrosis, it now seems most probable that this is an index of the increased metabolic activity of cartilage and bone cells in active disease and of the associated secondary synovitis, rather than a primary disorder. However, the precise explanation for the raised vascularity of the articular tissues continues to demand attention. One explanation is the synthesis and release of heparin-binding protein growth factors[238] resembling tumour growth factor.[123] Another is the action of a low molecular weight non-protein endothelial cell-stimulating angiogenic factor (ESAF).[47] The action of this small molecule may be by the activation of procollagenase and progelatinase, allowing new vascular buds to extend into a partially degraded matrix. Inhibitors of collagenase and of gelatinase may be able to inhibit ESAF. Platelet factor IV has the ability to inhibit collagenase and angiogenesis. The possibility remains that ESAF may also promote cartilage calcification and ossification by mechanisms similar to those that operate during endochondral ossification. Antagonists of ESAF may therefore not only diminish cartilage vascularization and degradation but also may delay new bone growth and osteophytosis.

The possibility that delayed emptying and congestion of the intraosseous veins, attributable to increased peripheral resistance, contributes to hip joint osteoarthrosis, still merits consideration. The evidence has been reviewed by Kiaer.[224] Alterations of bone morphology follow the experimental ligation of peripheral veins but the lesions differ from those of osteoarthrosis. Osteotomy gives symptomatic relief in osteoarthrosis and leads to morphological changes but these observations are only admissible evidence that venous obstruction contributes to the onset of osteoarthrosis if they are supported by experimental data. This support is given by evidence that the new bone formation of osteophytes is an early manifestation of experimental osteoarthrosis.[280,281] There is good reason to suspect a substantially raised articular blood flow as one explanation for the osteogenesis.[146]

Endogenous degradation

A wide variety of external agents ranging from trauma and immobilization to inflammatory mechanisms associated with crystal deposition, are capable of activating chondrocytic enzymes, leading to matrix degradation.[93] The possibility that these enzymes are responsible for initiating matrix breakdown in idiopathic osteoarthrosis has attracted growing attention.

Early studies of the actions of crude papain strongly suggested that the depletion of aural and costal cartilage matrix proteoglycan could result from intravenous enzymes.[265,457] Direct measurements of cathepsin D activity in human osteoarthrosic cartilage indicated that this index was significantly raised[13,17,408] and a concept emerged that enzymatic degradation could be a final common pathway in osteoarthrosis, whatever the prime cause.[40,184a,188] Human osteoarthrosic cartilage expressed high thiol proteinase cathepsin B activity,[31] particularly in the superficial zone (zone I). But the first evidence implicating an enzyme in the extralysosomal degradation of collagen, as opposed to proteoglycan, was the demonstration of a collagenase which was present in cartilage bound to a trypsin-sensitive inhibitor.[107] The correlation between phosphatase levels in subchondral bone and the severity of osteoarthrosis indicated parallel heightened changes in bone metabolism.[388]

Glynn[149] proposed that the enzymes degrading cartilage might originate in synovial A cells; they could, he surmised, be neutralized by inhibitors originating in B cells, a view in accord with the evidence that synovial inflammation in osteoarthrosis is more frequent than had been expected (see p. 876). Although synovial fluid from osteoarthrosic knees has a diminished capacity to inhibit proteolysis[327], no significant relationship between the degree of synovial inflammation and the activity of the lysosomal enzymes cathepsin D and acid phosphatase was demonstrable.[140] Nevertheless, in inflamed osteoarthrosic synovia, the total and active neutral metallocollagenolytic enzyme activity and the total and active neutral metalloproteoglycan degrading enzyme activities were all significantly elevated,[282] changes accompanied by the raised activities of serine collagenolytic enzyme activity and of serine proteoglycan-degrading activity. The main agents responsible for the degradation of osteoarthrosic cartilage were clearly collagenase(s) and proteoglycanase(s). The presence of these enzymes was confirmed in osteoarthrosic human cartilage.[105] Ethylene diamine tetraacetic acid (EDTA), a chelator of metal cations of the kind acting as enzyme inhibitors, was effective in the treatment of a rabbit model of osteoarthrosis[105], and the levels of human osteoarthrosic cartilage collagenolytic activity were shown to correlate with the severity of tibial condylar disease.[362] In the case of the proteoglycan-degrading, neutral metalloproteinases, enzyme activity appeared to alter proteoglycan structure in at least two ways: first, by a reduction in the ability of proteoglycan to form aggregates with hyaluronate, an action attributable to cleavage of core protein near the hyaluronate-binding globular domain, and second, limited proteolytic cleavage of the remainder of the core protein.[285]

Inflammation and osteoarthrosis

Osteoarthrosic joints often become inflamed. In primary generalized osteoarthrosis, there is systemic illness with a raised erythrocyte sedimentation rate. When the synovia are examined at the time of hip arthroplasty for idiopathic osteoarthrosis, they show low-grade inflammation with a slight mononuclear cell infiltrate, a little fibrin and areas of synovial villous hypertrophy.[25,151,444] The excess mast cells is a secondary phenomenon. These findings have been confirmed by electron microscopy. Since fragments of cartilage and bone are often found within the inflamed synovia, it seemed reasonable to attribute the inflammation to the release of the particles into the joint cavity and to their uptake by the synovia.

That there may be an alternative explanation has been suggested by the finding of bone mineral crystals, calcium hydroxyapatite, within the synovia. Apatite crystals, needle-shaped and 100 nm long, lie at the limit of resolution of the polarizing light microscope. The crystals are often in clusters. Apatite crystals, as opposed to fragments of bone, have only attracted attention since scanning electron microscopy has been used to survey synovial fluid and transmission electron microscopy to analyse synovial sections. Although the apatite crystals themselves can obviously be derived from the breakdown of osteoarthrosic cartilage, where they have been identified by some[8a,9,10] but not all investigators in relation to matrix vesicles, the possibility that their presence can initiate endogenous cartilage degradation in osteoarthrosis commands respect.[97]

Protection and treatment

Diminished movement

There is a low incidence of osteoarthrosis in hip and knee joints controlled by muscles weakened by anterior poliomyelitis.[148] This evidence appears to support the view that rest or immobility protects against osteoarthrosis by removing mechanical stimuli that lead to cartilage degradation.

However, the factors that determine whether diminished or increased mechanical stimuli exaggerate or protect against cartilage degradation and bone formation are numerous and complex. They are reviewed on pp. 886–889.

Drug treatment

A substantial proportion of the surgical and medical resources of the Western world is now directed to the palliative treatment of established osteoarthrosis. Each of these classes of therapy is plagued by pathological side-effects and complications,[48a] the details of which are beyond the scope of this text. The reader is referred to Ghosh.[144] Some of the tissue reactions to surgical procedures such as arthroplasty[167,172a] are outlined in Chapter 24. The pathological complications of drug treatment are recorded in a voluminous literature.

'To develop drugs specifically for osteoarthrosis . . . the thinking that has gone into the development of antirheumatic drugs should be discarded.'[320] The available evidence emphasizes the early, non-inflammatory changes of osteoarthrosic cartilage matrix. Nevertheless, pain and secondary inflammation are late results of these earlier cartilage changes and controlled trials to determine the best regimen of palliative analgesic and antiinflammatory treatment continue to be reported.[286] Whether there can be true restorative treatment in the sense of cartilage reorganization remains debatable.[89] It seemed possible that aspirin could protect against the development of the cartilage defects that are likely to follow surgery of disorders such as recurrent patellar dislocation.[72] More surprising was evidence of the suppression by sodium salicylate of the augmented glycosaminoglycan synthesis that follows surgical division of the anterior cruciate ligament in the dog,[354] observations confirmed by the demonstration that indomethacin also decreased net glycosaminoglycan synthesis in cartilage after surgical section of this cruciate ligament.[353a] This treatment does not appear to reverse the defective aggregation of the proteoglycan or the defect in proteoglycan-collagen interaction. Nevertheless, the efficiency of treatment of one induced abnormality has encouraged the further search for agents that will restore chondrocyte function to normal at an early stage of the disease. It may also prove possible to increase the resistance of cartilage surfaces to enzymatic degradation.[446a]

Here, selective aspects of the non-surgical treatment of osteoarthrosis are considered. For the pathologist, there is particular importance in learning whether an individual form of treatment has not only alleviated subjective symptoms but has initiated or catalysed reparative processes in cartilage and bone. In this respect, the available objective information on human disease is very limited although the results of many studies on experimental animal models of osteoarthrosis have been published.[475,476a]

Glycosaminoglycan polysulphate

This substance blocks catabolic processes in cartilage. The actions are upon neutral proteases and upon lysosomal hydrolases. Glycosaminoglycan polysulphate (GAGPS) has been tested extensively in animal models of osteoarthrosis as well as in the human disease.[1]

Using the form of rabbit joint disorder that follows partial medial meniscectomy, Howell, Manicourt, Muniz *et al.*[189] demonstrated that GAGPS greatly reduced both the histological indices of secondary osteoarthrosis and the extent of the lesions. Subsequently, Howell, Carreno, Pelletier and Muniz[188] confirmed that the activities of neutral metalloproteases were higher in the cartilage of animals not so treated. Indices of proteoglycan content were restored to normal when GAGPS was used therapeutically or prophylactically. Analogous results were obtained when GAGPS was tested by intramuscular injection into rabbits with the joint disease caused by immobilization (see p. 887).[154,155] In further studies Hannan, Ghosh *et al.*[170] found that systemic (subcutaneous) administration of GAGPS gave partial protection against the proteoglycan changes induced in dogs by bilateral medial meniscectomy.

The results[155] were in general analogous to those obtained when pentosan polysulphate (SP54) was given intramuscularly into rabbits with the joint lesions provoked by immobilization[154] and with those derived from studies of C57-Black mice with heritable osteoarthrosis (p. 903) to which GAGPS was given for long periods by subcutaneous injection.[108] GAGPS is also effective in alleviating the physical and chemical disorders of radial carpal articular cartilage caused in horses by partial or full thickness incisions, or by the intraarticular injection of iodoacetate, respectively. The equine lesions display less fibrillation and erosion after GAGPS than in untreated animals and fewer chondrocytes die.[501]

Non-steroidal antiinflammatory drugs

Few would seriously question the evidence that salicylates and numerous other non-steroidal antiinflammatory compounds (NSAIDs)[476a] benefit the symptoms and signs of osteoarthrosis, although the comparative efficacy of different drugs can only be determined by properly controlled clinical trials.[286] However, evidence that any NSAID exerts a fundamental effect on the cellular processes of osteoarthrosis, on the subsequent macromolecular disorganization or upon the limited reparative processes that accompany the syndrome is a subject for vigorous debate. Much of the evidence has, of necessity, derived from the study of experimental animal models of the disease.

When NSAIDs were tested in inflamed air pouches,[418] implanted with minced fragments of autologous cartilage, the inflammatory cell response was found to be reduced but the loss of proteoglycan from the (disaggregated) cartilage particles was not prevented.[417] The compound D-penicillamine, an agent which can enhance inflammation but inhibit cartilage degradation, had the converse effect. In whole animals with surgically induced arthropathy, different results may be anticipated and the abnormality may be made worse by acetylsalicyclic acid. When the available data were reviewed,[43] it appeared that salicylates and some other NSAIDs did, in fact, suppress proteoglycan synthesis both in normal and in degenerating articular cartilage.[353a] The possibility of induced degradation has also been raised.[283]

Clinical suspicions that some at least, of the NSAIDs may have damaging effects on cartilage have centred on indomethacin, a potent inhibitor of prostaglandin synthesis, and a syndrome of accelerated osteoarthrosis, attributable to indomethacin, has been postulated (Fig. 22.36). Although there is little direct microscopic evidence to show that indomethacin, unlike salicylates, directly damages cartilage or bone, it has been established that patients treated with this NSAID come to hip joint arthroplasty sooner than a control group given the weak inhibitor of prostaglandin synthesis, azapropazone.[379] The synovia in the indomethacin-treated patients contained less prostaglandin E_2, thromboxane B_2 and 5-hydroxyeicosatetraenoic acid (see Chapter 7) than those given azapropazone. It could therefore be argued that the lower proteoglycan content of the articular cartilage in those given indomethacin may be a direct rather than an indirect effect of this drug.

Corticosteroids

Although cartilage is susceptible to the actions of corticosteroids[428] so that the intraarticular injection of these agents can provoke cartilage degeneration[36a] and even a steroid arthropathy,[302] corticosteroids can be chondroprotective.[494] Models of synovial joint inflammation such as the granuloma pouch[418] have been used to test the effects of glucocorticoids. Much appears to depend on the dose and route of administration. In the model employed by Williams and Brandt[494] intraarticular iodoacetate (see p. 904) acts directly to cause chondrocyte injury or death, loss of safranin-O stainable matrix, fibrillation and subsequent osteophytosis. The dose-dependent action of the steroid triamcinolone that they showed to protect cartilage from the action of iodoacetate, is likely to be exerted on chondrocyte metabolism.

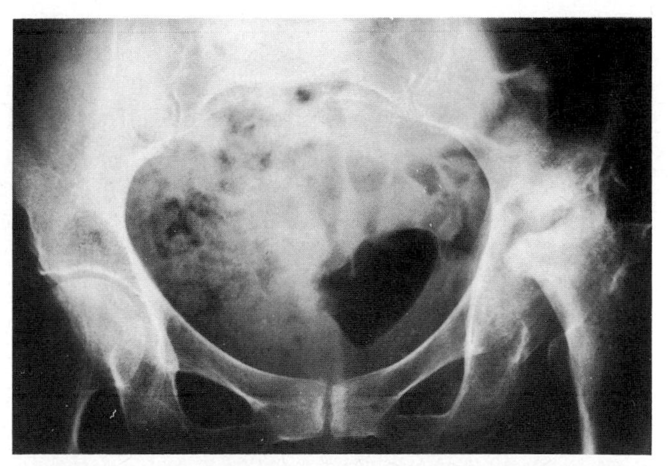

Fig. 22.36 The influence of drug treatment on the progression of the disease.
It is accepted that local corticosteroid injections may exacerbate joint destruction. Whether systemic corticosteroids and nonsteroidal antiinflammatory drugs have a comparable effect is debated. Indomethacin accelerates hip joint osteoarthrosis in the sense that prolonged administration of the drug results in a requirement for palliative surgery sooner than is the case in patients not treated in this way.[379] This radiograph is one of a series showing that a patient treated with indomethacin exhibited rapidly progressive and severe joint destruction. The right hip joint provides an intact radiographic control.

Antioestrogens

Extensive enquiries have demonstrated the role of sex hormones in the pathogenesis of osteoarthrosis in mice[424,425] (p. 903). It has seemed logical to test the ameliorative effects of antioestrogens. When tamoxifen, for example, was given by intramuscular injection to rabbits previously subjected to partial medial meniscectomy, the severity of cartilage injury but not of osteophyte formation was reduced.[396]

Hyaluronate

The role of synthetic synovial fluid substitutes is mentioned on p. 99. Hyaluronate is not now believed to be the main component responsible for the unique lubricating qualities of synovial fluid. However, the macromolecule has other biological properties that suggest it could benefit joints that are the site of osteoarthrosis.[27a,451a] In particular, there is evidence that hyaluronate may enhance the limited process of cartilage repair which accompanies the advancing lesions of osteoarthrosis. Sodium hyaluronate may alleviate the symptoms of temporomandibular arthritis, for example, to much the same degree as a corticosteroid.[234] The mechanisms of action are not yet fully understood: isotonic sterile solutions of high viscosity may influence reparative processes more significantly than those (approximately 0.1 per cent weight/volume) with a viscosity and molecular weight equal to or less than that of normal synovial fluid. When tests were made to examine the effects of intraarticular sodium hyaluronate on inflammatory mediators in osteoarthrosis and rheumatoid arthritis, cAMP was found to increase, and prostaglandin E_2 to decrease as the synovial fluid volume declined in response to treatment.[370]

Rumalon

No substance has aroused greater interest in the treatment of osteoarthrosis than the extract of bone marrow and cartilage termed Rumalon. Many claims have been made that repeated intramuscular injection of this biological preparation can alleviate the symptoms of osteoarthrosis[478] but neither the mode of action of the complex nor its efficacy is yet agreed. Silberberg and Hasler[431] showed that a cartilage-bone marrow extract stimulated chondrocyte hypertrophy and organelle development in C57B mice but the molecules to which these cells respond are not yet characterized. The effect of Rumalon on the pathogenesis of osteoarthrosis is obscure.

Oncotherapeutic agents

Studies of *post mortem* knee joint cartilage from 18 patients with a wide variety of malignant neoplasms, treated by alkylating agents, antimetabolites, plant alkaloids, antitumour antibiotics, nitrosoureas, random synthetic agents and/or corticosteroids, strongly suggested that the cartilage surface, tidemark and calcified cartilage remained intact.[283] However, the surprising observation was made that in comparison with a series of control *post mortem* cases, there was increased cartilage cellularity, decreased safranin-O stainability, depleted pericellular staining and disorganization of the superficial zone collagen meshwork. These changes were among a range of disorders that, it was claimed, resembled those of early osteoarthrosis. Correspondingly, the proteoglycan content was diminished, the DNA content increased. There were raised neutral collagenolytic and metalloproteoglycan-degrading enzyme activities and the active form of a neutral collagenolytic enzyme was also elevated.

The results of this unusual study are unexpected; their significance is not yet clear. The tissue lesions, with increased chondrocyte numbers, are the converse of those expected and have a bearing on the unresolved problem of why cell numbers in osteoarthrosic cartilage are raised (see p. 856). However, uncontrolled factors such as the duration and extent of immobilization may have influenced cartilage behaviour independently of chemotherapy and further enquiries are necessary.

Osteoarthrosis in non-humans

In theory, the best approach to understanding of the cell and tissue changes in human osteoarthrosis would be the analysis of material taken at the onset of the disease. However, early osteoarthrosis is symptomless. At later stages, residual, unaffected cartilage and bone are often found in parts of joints subject to arthrotomy but there are sound ethical and biological reasons for not touching this cartilage. At arthroplasty, the remains of the articular surfaces is frequently so severely disorganized that little can be learnt from it.

Alternative approaches are the investigation of naturally occurring osteoarthrosis in other mammals or the production of artefactual disease by deliberate insults to the internal or external environment of joints. Many techniques have been explored (Table 22.10). They have ranged from freezing and burning, to dropping on hard floors; the injection locally of acids, alkalis or fixative; interruption of the blood supply; and the application of enzymes. The early literature was reviewed by Gardner.[128]

Recently, there has been some standardization of approach. The greatest effort has centred on dogs and rabbits because enough cartilage is available for extensive biochemical analysis. Surgical procedures[313] are preferred: they have been used to cause elective, controlled impairment of joint function in mixed and closed colony groups.

Since the first days of anatomical observation it has been apparent that all adult mammals are susceptible to osteoarthrosis. Less is known of the other chordates.[467] In addition to naturally occurring disease, many species have been used to induce models of osteoarthrosis in the reasonable expectation that the pathogenesis of the model would help understanding of spontaneous osteoarthrosis. Horses, pigs, hens, dogs, sheep, rabbits, rats, guinea-pigs, mice, hamsters and monkeys are among the classes used for this purpose. Here, it is convenient to outline aspects of naturally occurring and experimental animal disease which have proved particularly valuable during the past decade. There is a very large literature beginning with the work of Redfern[381] and the reader is referred to reviews by Adams and Billingham,[4] Schwartz,[413] Van Sickle[466] and Moskowitz *et al.*[313]

Primates

There is limited evidence. Naturally occurring osteoarthrosis varies in frequency according to the environmental conditions. In 169 rhesus monkeys maintained in a zoo, only one example of spontaneous osteoarthrosis was identified.[57] Comparable results were obtained by Kessler, Turnquist, Pritzker and London.[223] They found radiographic evidence of osteoarthrosis in two of 13 (15 per cent) rhesus monkeys (*Macaca mulatta*) raised and maintained in cages. By contrast, there was a much higher frequency (12 of 15 animals—80 per cent) of such changes in formerly free-ranging rhesus macaques. Osteoarthrosis has been sought but rarely found in baboons (*Papio cynocephalus*) under investigation for other experimental reasons.[7,267] Direct attempts to induce osteoarthrosis in non-human primates have been even fewer but Lufti[255] was able to cause medial degenerative disease by resecting the meniscus from the right knees of a group of 12 young grivet monkeys (*Circopithecus aethiops aethiops*). The descriptions by Stecher[447] of a group of gorillas held in captivity are among the rare reports of osteoarthrosis in apes.

Other mammals

Horses

Lameness due to osteoarthrosis is a serious problem in young racehorses and a common abnormality in older workhorses.[377] Relatively little is known of the systematic pathology of the equine disease although some synovial joints, like the sacroiliac, have been studied in great detail. Many attempts have been made to validate the effects of intraarticular agents such as hyaluronate, on equine joint function and synovial fluid characteristics but there are few controlled experiments in which treated and untreated cartilage has been assessed, chemically, histologically and mechanically.[27a,179]

Pigs

Young Landrace pigs frequently develop a locomotor disorder, 'leg weakness', associated with a local, primarily non-inflammatory disturbance of epiphyseal plate endochondral ossification ('osteochondrosis'). A proportion of the affected animals progress to proliferative, non-inflammatory lesions of the articular cartilage surfaces or of the deeper zone cartilage ('arthrosis').[162,163] The affected joints are mainly the elbow and stifle (31.3 per cent), lumbar intervertebral (28.1 per cent) and hip (15.6 per cent). By the age of 18 months, the lesions of osteochondrosis tend to regress, to repair or to progress to arthrosis.

Yorkshire pigs, which have shorter bodies, narrower hindquarters, longer femurs and differently shaped stifle joints, less often develop leg weakness, osteochondrosis and arthrosis when bred under the same conditions of feeding and housing. They grow more slowly. It therefore appeared that genetic differences culminating in different skeletal proportions rather than the mechanical effects of differential growth rates and body weights, were principally responsible for the leg disease. Nevertheless, experimental overloading led to reversible osteochondrosis. Dietary changes in calcium, phosphate and protein intake altered the angle between the femoral condyles and the femoral head but not the incidence of bone and joint lesions.

The mechanisms of porcine osteochondrosis and arthrosis are likely to be closely similar to those identified in the degenerate cartilage of immobilized rabbit limbs (see p. 887) or in the legs of dogs following cruciate ligament section (see p. 898). There is a decreased cartilage proteoglycan content, with an increased proportion of small monomers, a loss of or damage to core protein, and a diminished quantity of hyaluronate.[328]

Dogs

The dog has proved to be a particularly valuable species in which to study the pathogenesis of osteoarthrosis.

Canine hip dysplasia

A number of strains of large dog are prone to heritable osteoarthrosis. Hip[256,259] and shoulder[466] joints are particularly affected. Lines of the Labrador Retriever have been bred that display a high frequency of hip dysplasia and, in some colonies, approximately 90 per cent of animals come to have osteoarthrosis.[259] German Shepherd dogs are also prone to the disease.

Canine hip dysplasia is initially a radiographic diagnosis; there is subluxation, remodelling of the femoral head and neck, a shallow acetabulum and changes in the acetabular rim. In one study, as large a proportion as 40 per cent of young and older dogs had multiple joint involvement.[343] Pathological changes are often found in hip joints that are radiographically intact. Focal cartilage lesions accompany fraying of the round ligament and enlargement of the joint capsule.[257]

The early hip joint disease includes a low-grade synovitis and increased volumes of synovial fluid.[259] The soft cartilage displays loss of superficial zone proteoglycan, scaling and fibrillation. There is an increase in chondrocyte numbers, especially in zone I, and in advanced disease, chondrocyte clusters are common.

The biochemical changes in canine hip dysplasia comprise reduced proteoglycan, increased water, but no significant change in DNA. The turnover of proteoglycan is raised, that of DNA diminished.[258] The transmission electron microscopic evidence of a loosening of superficial collagen fibres accompanies the presence of an electron-dense, superficial amorphous material.[496] There is an early loss of the ultrastructural surface lamina (see p. 70) and a decrease in the proportion of type A synoviocytes. Although the total collagen content of the degenerate cartilage is not thought to be changed,[258] measurements of hydroxy ^{14}C-proline accumulation in immature dogs demonstrate decreased collagen deposition *in vivo* and *in vitro*.[51] The diminished collagen deposition is accompanied by increased collagenolytic activity *in vitro*.[51] In cartilage from older animals, there is an increased amount of a high molecular weight collagen species not detected in normal animals. This collagenous protein is a procollagen and its presence suggests a defect, genetically determined, in the conversion of procollagen to collagen (p. 174).

Anterior cruciate ligament disease A factor precipitating osteoarthrosis in the dog is 'spontaneous' rupture of an anterior cruciate ligament.[24,41] One consequence of the rupture and of the resultant joint disease is the formation of antibodies to the type I collagen of the cruciate ligaments and to the type II collagen of the disordered articular cartilage.[332] Large animals are especially susceptible to anterior cruciate ligament rupture. Prompt surgical repair of the ligament is effective in the prevention of osteoarthrosis and early postoperative mobilization can facilitate repair.[340,365] For these reasons, and because the major joints of large animals provide adequate material for chemical analysis, microscopic study and for physical testing, the experimental aseptic division of a canine anterior cruciate ligament has become a standard procedure in the investigation of degenerative joint disease. Strictly, anterior cruciate ligament section results in a disease that is a model of that which that follows spontaneous anterior cruciate ligament rupture. But, as Toynbee emphasizes in his *History of Civilisations*, a model need only be an analogue, not a replica. It is sufficient if the model throws light on the original phenomenon and experimental post-anterior cruciate ligament section canine stifle joint disease (ACL disease) has proved to be of great value in stimulating ideas on the nature of human idiopathic osteoarthrosis.[209,323]

To divide a dog anterior cruciate ligament, a technique of open operation was devised by Magnuson[270] and exploited by Marshall,[280] Marshall and Olsson,[281] Schwartz,[412] Schwartz, Oh and Leveille,[414] Vignon *et al*,[476] Orford *et al.*,[348] and Pidd, Gardner and Adams.[364] An alternative procedure, in which the anterior cruciate ligament is divided by blind, closed operation,[367] was used by McDevitt and Muir,[262] McDevitt, Gilbertson and Muir,[260] Eyre *et al.*,[113] Stockwell, Billingham and Muir,[450,451] Gardner *et al.*,[131,134] Sandy *et al.*,[407] Lipowitz *et al.*[251] and Pelletier *et al.*[360,361] The open operation closely resembles the surgical approach selected when a spontaneously ruptured cruciate ligament is repaired.[41,114] The precise form of surgery adopted to divide the anterior cruciate ligament is important since it influences the extent of articular soft-tissue injury. The anterior cruciate ligament is covered by a delicate synovial membrane, section of which inevitably causes some intraarticular bleeding.

Histological changes (Figs 22.37 and 22.38) The histological changes following anterior cruciate ligament section therefore begin in the synovium. There is low-grade, acute sterile inflammation which may persist for two to 12 weeks postoperatively.[131,251,360,361] Lymphocytic foci remain within the subsynoviocytic connective tissue: they are often near aggregates of haemosiderin. Increased numbers of mast cells are seen.[134] The rise in the number of mast cells becomes significant at four weeks and remains elevated for a further 12 weeks. The superficial zone cartilage stains less deeply than normal[229a] for glycosaminoglycan. Zone I chondrocytes lose their ellipsoidal shape and become ovoid. Zone II cells are round and often paired. There is thought to be an increase in cell density[450] but this is disputed and Vignon *et al.*,[474,476] who used nitric acid to decalcify their blocks, reported a decreased number of cells and the presence of empty lacunae. Cartilage DNA has been found to be normal but DNA synthesis, estimated by autoradiography after ^{3}H-thymidine incorporation, is elevated,[412] an observation not confirmed by Pelletier *et al.*[360] Chondrocyte clusters form in response to early cartilage surface splitting which may be vertical in tibial cartilage, tangential in femoral.[476] After the first few weeks, the changes in cell numbers and morphology are overtaken by new matrix synthesis. The cartilage increases in amount and thickness while the ratio of cell numbers to matrix volume returns to normal. The structural changes settle at a new plateau of cell activity.[136]

Anatomical consequences The anatomical consequences begin with postoperative joint swelling: there is an increased volume of orange, opaque synovial fluid. Indian ink painting reveals cartilage surface fibrillation earlier on the medial tibial than on the lateral tibial or femoral condylar surfaces. Periarticular fibrous thickening develops and, within eight weeks, islands of chondrocytes, chondrophytes, form at the subperiosteal juxtaarticular margins at sites where vascular osteophytes develop subsequently.[146,280] There is a demonstrable increase in vascularity at the articular margins.[344] An increase in subchondral bone density is recognized. Cartilage loss advances and bone becomes eburnated. Meniscal fibrillation is seen. The severed cruciate ligament is calcified or resorbed and the synovia remain vascular, thickened, granular or villous.

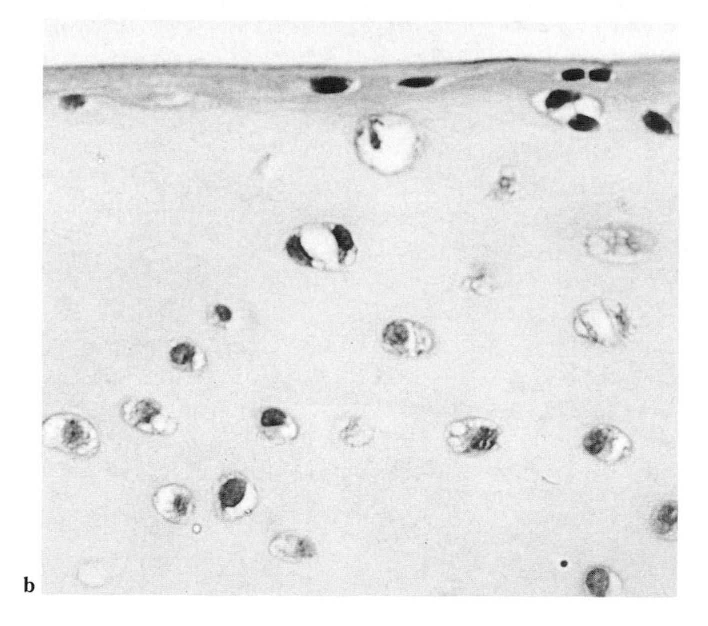

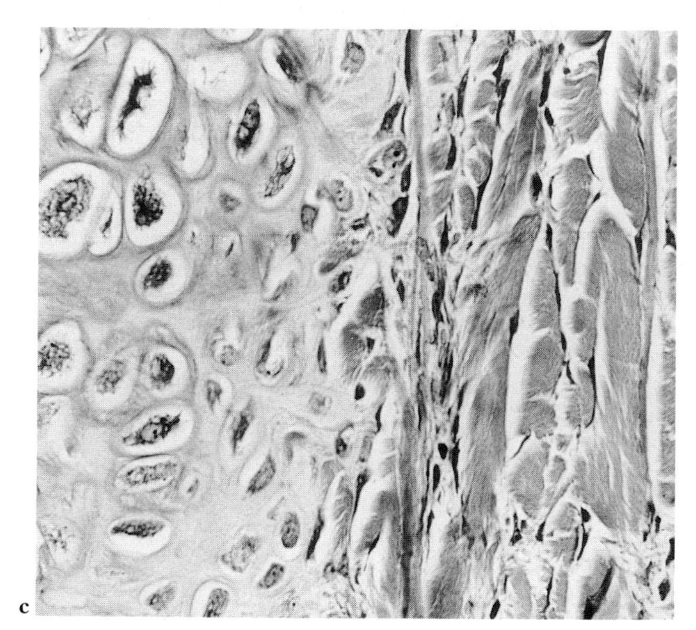

Fig. 22.37 Experimental joint disease.
A reproducible form of arthropathy can be caused by the aseptic surgical division of a canine anterior cruciate ligament. **a.** There is a conspicuous and relatively early increase in the cell density of the femoral and medial tibial condylar cartilages. **b.** Immediately beneath the zone I superficial lamina of compact collagen fibre bundles, a series of changes in chondrocyte morphology reflect their altered metabolism and secretory characteristics. The cells, normally ovoid, become more rounded; they lie in pairs in a site where cells are normally single; and between the nuclei of two such cells, in a single chondron, a pale-staining zone represents the accumulation of the finely fibrillar material seen in Fig. 22.8. **c.** At the synoviochondral margin, a multiplication of chondrocytes within the periosteal tissue leads to the development of cartilage islands (chondrophytes), the precursors of osteophytes. In this situation, therefore, osteophyte formation is by enchondral ossification. (HE **a:** × 125; **b:** × 500; **c:** × 320.)

Mechanics The operated joint becomes unstable.[24] There is a reduction in tensile stiffness of the superficial zone tibial condylar cartilage demonstrable three weeks after anterior cruciate ligament section; it is progressive.[326] The change in tensile stiffness was recognizable in deep zone cartilage by 23 weeks[326] and was indexed by a reduction in the tensile modulus. The cartilage becomes softer within two weeks of surgery[476] and microtests made in compression reveal increased, selective deformability of the midzone cartilage.[339]

Ultrastructure Ultrastructural studies[347,361,451] demonstrate matrix disorganization. There is increased separation of superficial collagen fibre bundles which are less well orientated than normal. Chondrocytes display much rough endoplasmic reticulum. Zone I cells become ovoid, and in

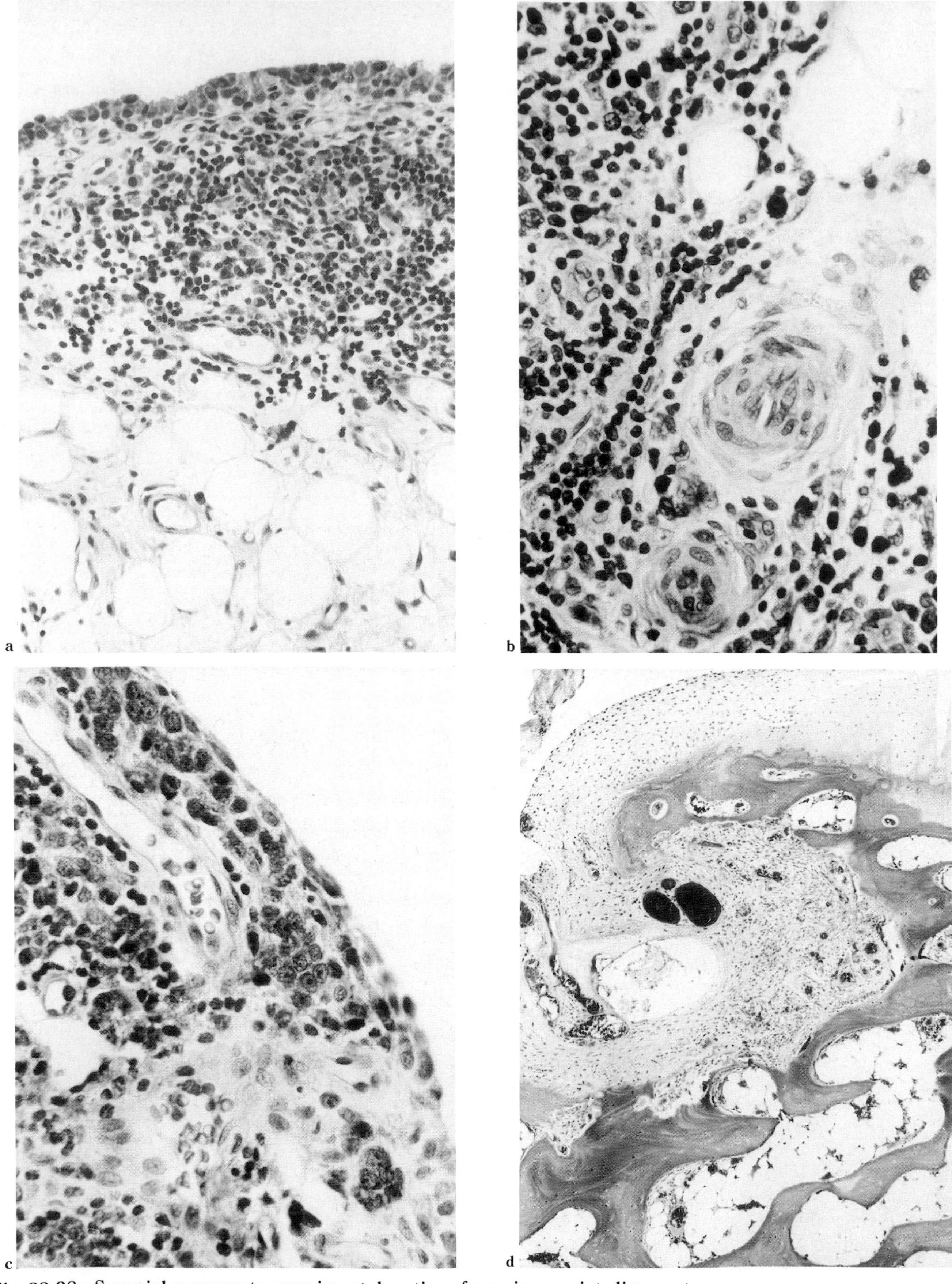

Fig. 22.38 Synovial response to experimental section of a canine cruciate ligament.
a. Three months after the surgical procedure, evidence of a significant diffuse lymphocytic infiltrate persists. There is little residual synovial cell hyperplasia. **b.** Lymphocyte aggregates are often in relation to subsynoviocytic blood vessels. **c.** There is a persistent increase in synovial vascularity. Some of the new vessels lie within 30 μm of the synovial surface. **d.** At the synoviochondral junction, osteoclast-mediated lacunar bone reabsorption and osteoblastic new bone formation are indices of remodelling at the site of osteophyte formation. (**a:** × 120; **b:** × 260; **c:** × 260; **d:** × 30.)

both zones I and II paired cells are often arranged in enlarged, electron-lucent lacunae. The surface is fissured by 2–12 weeks, depending on the cartilage studied, and a deposit of unidentified electron-dense material appears on or in the cartilage surface, replacing the normal, ultra-structural surface lamina.[347]

Replicas of free parts of normal dog femoral condyles made at −196°C (77 K) show that, whereas much of the surface is covered by an amorphous lamina, some parts are not obscured and have a collagen-rich structure; between the fibres, the gentle convexities of expanded, hydrated proteoglycan domains can be visualized.[137a] Early in the course of the post-anterior cruciate ligament section disease, this structure changes and around surface chondrocytes appear halos of matrix disorganization resembling those seen by conventional scanning electron microscopy at the free surfaces of normal cartilage that have been partially digested by papain.[337]

When normal dog articular cartilage surfaces are viewed in the unfixed, hydrated state by low-temperature scanning electron microscopy, a hierarchy of orderly surface features is seen.[137b] After anterior cruciate ligament section, cracks and fissures appear.[338] The process of crack propagation is a form of plastic deformation. Ultimately, by 12–16 weeks, patterns of crazing form. The evidence, obtained at −180°C (93 K) helps to confirm that the disorganization seen by conventional transmission- and scanning electron microscopy is not artefactual.

Biochemical changes These are summarized in Table 22.4.

Neurogenic arthropathy

Dogs[333,334] have also been subjected to ganglionectomy or nerve section[103] to provoke neurogenic arthropathy (see p. 906). In some experiments, cartilage degradation quickly followed surgery; in others[333] ganglionectomy alone was without affect but after subsequent transection of the anterior cruciate ligament severe disease developed rapidly.

Scarification of the supracondylar cartilage[288,458]

Scarification was used to demonstrate that there is cartilage repair even when the subchondral bone is not breached. The effects of laceration after division of an anterior cruciate ligament have been considered.[351] Further aspects of cartilage repair are considered in Chapter 2.

Partial and total meniscectomy

Partial meniscectomy leads to osteoarthrosis in the dog as it does in the rabbit.[84] The degree of cartilage change is related to the amount of meniscus removed.

Bilateral medial meniscectomy

In greyhounds, this procedure has been used to study the biochemical effects exerted by impaired joint stability on articular cartilage composition.[145] There are similarities to the results obtained in the beagle dog and Canadian foxhound following anterior cruciate ligament section.[364]

Other forms of experimental dog disease

Osteoarthrosis has been caused in dogs as in rabbits by the intraarticular injection of papain.[411] Osteophytosis results from the intraarticular injection of homogenates of autogenous costal cartilage.[71]

Cats

Osteoarthrosis is a much less common and severe clinical problem in cats than in dogs. Consequently, the natural history of osteoarthrosis in the cat has received much less attention than has the disease in dogs. Maturation changes have been identified on the surfaces of the articular cartilage[143] but the relationship of these changes to osteoarthrosis is uncertain.

Rabbits

An enormous variety of procedures has been used during the past 100 years to promote cartilage degeneration in rabbits. Although it has been convenient to term the resultant injuries 'osteoarthrosis',[435] the experimental lesions can often be designated more exactly 'traumatic arthropathy'.

Rabbit joints have been heated or frozen.[434] Foreign materials such as polypropylene[309] or talc[141] have been implanted in the knee joint. Cell-free extracts of homologous cartilage have been given intravenously[65] and cartilage degradation has been provoked by the intraarticular knee joint injection of papain[82,83,324] and by the injection of hydrocortisone.[70]

The early changes have been imaged by scintigraphy.[464]

Mechanical procedures have been tested with equal enthusiasm. Rabbit joints have been immobilized (see p. 887), compressed in flexion or extension,[85] compressed and immobilized[459] or simply struck once[139] or repeatedly.[373] However, as the sophistication of views on the nature of osteoarthrosis advanced, it began to appear that, in the rabbit as in the dog, surgical manipulations capable of causing limited, measurable joint instability might prove of the greatest interest. For this reason, the medial collateral ligament was divided,[420] the medial meniscus excised wholly[38,235,419] or in part.[314] The articular cartilages of the patellar and femoral patellar groove have been scarified.[288]

In analogy with the dog (see p. 898), the posterior cruciate ligament has been sectioned[86] or the patella excised.[137c] Complex, combined procedures have been advocated and that of Hulth[196,197] used by Bohr[38] comprises the removal of the medial meniscus and the section of both cruciate ligaments. The procedure causes extreme joint instability.

The models of joint disease which occur after total or partial meniscectomy, with or without simultaneous division of the collateral or cruciate ligaments have attracted particular recent attention. They are reviewed by Adams and Billingham.[4] Morphologically, the surface changes that develop in models in which collateral or cruciate ligaments are divided are more severe than those in which partial meniscectomy alone is undertaken or in which a meniscal tear is simulated. Histologically, there is chondrocyte degeneration and death, fissuring, fibrillation, flaking, whole thickness cartilage loss and, first, tibial, then later, femoral osteophytosis. Safranin-O proteoglycan-staining diminishes and cell cloning appears. Biochemically, fewer studies have been made than in the dog anterior cruciate ligament model.[106]

Effects of partial meniscectomy

In addition to the histological changes detailed above, osteochondrophyte formation is prominent.[315] The osteochondrophytes increase in size with time. There is a brisk marginal synovial response with cell proliferation and heightened vascularity[315] but the vascular changes do not precede osteophytosis. The cause of osteophyte formation appears to be mechanical force exerted abnormally at the site of non-calcified soft-tissue attachments. There is IgG and C3 deposition at the cartilage surfaces.[319] Chemical analyses demonstrate increased cartilage protein and glycosaminoglycan synthesis in parallel with the degenerative lesions.[287] The early increase in chondrocyte numbers is accompanied by raised protein synthesis but glycosaminoglycan synthesis remains constant throughout a 12-week period of observation.[316] As in the case of the rabbit model used by Butler *et al.*,[54] Steinetz[448] and Columbo *et al.*,[76,77] agents such as glycosaminoglycan polysulphate (see p. 894), active against proteinases, are effective in lowering cartilage netural metalloprotease activity.[188] There is a corresponding improvement in the scale of histological cartilage degradation.

Effects of anterior lateral meniscectomy with section of the fibular collateral and sesamoid ligaments

This complex procedure[77] causes a predictable and reproducible form of degenerative joint disease with fibrillation, cartilage loss, chondrocyte cluster formation, osteophytosis, chondrocyte loss, and proteoglycan depletion. The model has proved of value in assessing the protective effects of antirheumatic drugs,[76] some of which have been given intraarticularly.[54] The cartilage from animals submitted to this combined procedure, contains an increased amount of fibronectin[50] (see pp. 51 and 207), a phenomenon shared with the cartilage of Labrador Retriever dogs with hip dysplasia (see p. 897) and believed to be a general property of osteoarthrosis.[52]

Effects of resection of the medial meniscus with division of the medial collateral and both cruciate ligaments[197]

This procedure causes severe joint instability but has value as a model in the study of osteoarthrosis. Raised levels of the lysosomal enzymes arylsulfatase and β − glucuronidase have been identified. The elevated enzyme activities appear to be earlier and brisker in synovial than in cartilagenous tissues, indirect evidence of the substantial postoperative inflammatory reaction to which the joints of these animals are subject.

Guinea-pigs

To induce reproducible models of osteoarthrosis that could be the basis of pathogenetic and therapeutic studies, guinea-pigs have been subjected to surgical procedures analogous to those used in the dog, rabbit and rat. Guinea-pigs have also been maintained on dietary regimens that modify osteoarthrosis, and they have been injected locally with agents like iodoacetate that damage cartilage directly.[493]

Surgical procedures have been of two categories, intraarticular ('invasive') and extraarticular. Those of the former have included division of the anterior cruciate and the major part of the medial collateral ligament, with or without partial meniscectomy[414] or meniscectomy alone.[34,35]

The latter category has excited especial interest. Arsever and Bole[26] found progressive changes, identical with those of osteoarthrosis, 10–24 weeks after unilateral resection of a segment of the gluteal muscles at their sacral origin, after intrapatellar ligament tendotomy or after a combined operation. The mild gait abnormality caused by these procedures is not accompanied by trauma to the hind-limb joints. There is no synovitis. However, the contralateral, non-operated knee develops degenerative changes but at a later time. The results offer an interesting contrast to those obtained by surgical manipulations within the joint when the rapidity of onset of the cartilage lesions, like those of the cruciate ligament dog model (see p. 898), reflect the inflammation that complicates surgical trauma.

Rats

There is no suitable strain of rat with heritable osteoarthrosis and the small size of the limb joints determines that surgery is difficult. The collection of samples of cartilage sufficiently large for chemical and histological analysis *and* for physical testing is demanding. Nevertheless, studies have been made of the effects on articular cartilage of dividing an anterior cruciate ligament. Other rat studies have centred on nutritional or chemical techniques for causing articular disorders. Propylthiouracil leads to disturbed endochondral ossification,[87] and articular cartilage is affected secondarily. Fasting has comparable and very rapid effects.[88] Insulin-like growth factor (somatomedin) activity declines during the fasting.[311] In an entirely different category of work, rats have been injected intraarticularly with enzymes such as crude papain.[270] Disorganization of the joint is accompanied by conspicuous pseudocyst formation.

Hamsters

The Chinese hamster develops juvenile diabetes, a disorder associated with disordered cartilage metabolism and studies have been made of the joints in this animal.[430] In diabetic animals, femoral head chondrocytes display atrophy of the endoplasmic reticulum, abnormal mitochondria and large lipid inclusions. The relationship of these changes to those of the cartilage in human diabetes mellitus is uncertain.

Mice

Because of their ease of handling and short life-span, many studies have been made of genetic and environmental contributions to osteoarthrosis in mice.[432] Some examples of these investigations are considered. Different strains of mice develop osteoarthrosis with widely differing frequency. STR/IN mice are susceptible,[426] as are C57/B1 animals.[380] Degenerative changes commonly appear in the STR/ORT substrain as age advances, infrequently in CBA/ORT animals. Together the STR/IN strain and the STR/ORT substrain have been used to investigate the important question of the relationship between obesity and osteoarthrosis. There is no correlation between the body weight of STR/IN mice and the incidence of osteoarthrosis,[441] even when a relative ('ponderal') index is adopted.[442] Walton[482] made a more direct approach to the quantitation of the adipose tissue of STR/ORT mice. After plotting body weight against age, over a period of 18 months, the subcutaneous and intraabdominal adipose tissue was weighed. Again, there proved to be no correlation between the frequency of osteoarthrosis and adiposity.

STR/IN mice This strain of mice displays a high susceptibility to osteoarthrosis. Underlying this is a form of dyschondrogenesis[426] in which focal cartilage necrosis and a disorder of the epiphyseal growth zone accompany a focal failure of cartilage reabsorption, neovascularization and chondrophytosis. In the STR/ORT substrain, tibial cartilage fissures. Fibrillation advances with the formation of deep concavities overlying eburnated bone but femoral condylar changes are less advanced.[479–481]

Cartilage injury in the male STR/ORT mouse has been attributed to impaired chondrocyte respiration, judged by the proportion of cells manifesting lactate dehydrogenase activity.[13a–c] There is a parallel time course between the recognition of inactive cells and evidence of cartilage damage, although the former precedes the latter by 4–5 weeks. In young STR/ORT mice the glucose-6-phosphate dehydrogenase (G-6-PDH) activity of the lateral tibial cartilage was greater but more variable than that in control mice of the CBA/HT6 strain.[98] Relative to the lateral cartilage, the activity of G-6-PDH decreased with age. The decline was detectable long before the development of the histological changes characteristic of this form of murine osteoarthrosis, suggesting an early alteration in chondrocyte metabolism at least in part determined genetically.

Osteosclerosis develops in ageing STR/ORT mice just as it does in human osteoarthrosis. Development of osteosclerosis itself does not cause articular degeneration,[483] although bone sclerosis and cartilage degradation are closely related chronologically. There is no increase in osteoblastic activity so that, in this model, osteosclerosis may be due to decreased osteoclastic bony reabsorption.

C57/B1 mice This strain of mice develops osteoarthrosis spontaneously and has been employed to examine the relationship of cartilage degeneration to age (see Chapter 21). With advancing age, there is a focal loss of chondrocyte respiratory enzyme activity and an increase in regressive changes in chondrocyte nuclei which become decreasingly basophilic.[380] The life-cycle of chondrocytes of this strain and of Jax 6, DBA2, C3H and 129 were analysed by transmission electron microscopy.[433] Cell death was followed by ossification or replacement by fibrillar scars. Of particular interest was the observation that, with advancing age, superficial zone chondrocytes came to resemble those of younger mid-zones, an observation recalling the cytological phenomena that quickly follow anterior cruciate ligament division in the dog (see p. 898). C57/B mice have been of particular value in assessing the effects of anti-inflammatory drugs on cartilagenous lesions.[492]

C57B1 Jax 6 mice This strain has also found a place in determining the influence of endocrine changes on cartilage degenerative disease. Somatotrophin promotes chondrocyte breakdown and microscar formation in very young animals,[427] although the hormone is able to correct the defects of chondrocytes found in hypopituitary dwarf (dw/dw) mice.[423] Oestrogen given early inhibits the development of osteoarthrosis in C57BL/Jax 6 mice; later in life, it modifies the articular lesions.[424] By contrast, testosterone increases the incidence of osteoarthrosis in female mice.[425]

When consideration is given to the pathogenesis of osteoarthrosis of specialized synovial joints such as the temporomandibular or sacroiliac, it is logical to turn to analogous joints in animals. Thus, the temporomandibular joint of ICR mice has provided a valuable model.[253]

Birds and other animals

Hens

There are strong reasons why the experimental induction of osteoarthrosis should not be attempted in avian species. The bone and joint structure is radically different from that of the mammals; the quantity of material available for chemical analysis and physical testing is very small; and the analogy between bipedal birds and bipedal primates is false.

Nevertheless, results that are both valuable and unexpected have emerged from a sustained study in which progressive knee disorganization, with cartilage degradation, regularly develops 2–3 months after the intraarticular injection of sodium iodoacetate.[214] The reproducibility of this model, which is strictly a direct, chemically induced cartilage injury, not an osteoarthrosis, has determined its particular suitability for testing antirheumatic and chondroprotective drugs.[213] The glycosaminoglycan-peptide complex 'Rumalon' is among those shown to be beneficial. However, in the course of this study, it became clear that a number of therapeutic corticosteroids and other nonsteroidal, antiinflammatory drugs (NSAIDs) exerted inhibitory effects on connective tissue anabolism,[212] an observation in accord with the now substantial evidence that indomethacin potentiates osteoarthrosic cartilage breakdown.[379]

Other animals

Severe osteoarthrosic changes have been found in the Nile hippopotamus (*Hippopotamus amphibius*)[205a] and there appears to be no reason why osteoarthrosis should not be caused experimentally in any form of vertebrate under appropriate circumstances.

Kashin–Beck disease

Kashin–Beck disease is a form of generalized osteoarthrosis limited to selected geographical areas of Asia; the alternative name, Urov disease, indicates that the condition was originally recognized among settlers living by the Urov river.[440] The disease is not congenital but may be present in childhood. Kashin–Beck disease derives its eponymous designation from the original accounts of Kashin[215,215a] and of Beck.[33,329a] Until recently the disease attracted little interest in Western countries.

The disease affects young children of 6–14 years of age. The earlier the onset, the more severe the clinical signs. The onset is insidious. Vague joint and muscular pains exaggerated by effort are accompanied by fatigue and symptoms of mild but generalized illness. Joint effusions are not encountered but the bone and cartilagenous changes which underlie the eventual deformity simulate joint swelling. Radiological appearances include thinning of cortical bone with a destruction and reabsorption of joint surfaces and a premature closing of epiphyses. The nearby muscles display slight but progressive wasting. The erythrocyte sedimentation rate is not raised, and there is neither anaemia nor leucocytosis. With time, there is progressive disability, impaired limb movement, acrocyanosis and paraesthesiae. At this stage, the evidence of limitation in epiphyseal growth becomes clear and the limb bones, as in achondroplasia, become relatively short for the age of the patient. When the disease has been present for some years, the deformity is likely to be greater and spondyloarthrosis is added to the syndrome. The stature is small but there is no evidence of a disturbance of calcium or of phosphate metabolism and none of a deficiency of calcium absorption or utilization.

Pathological characteristics

In young persons there is osteochondrosis: the zones of epiphyseal necrosis impair growth and osteochondromatous bodies are liberated into the joint, simulating the effects of osteochondritis dissecans (see p. 958). Changes are limited to the synovial joints and periarticular structures and to hyaline cartilage. The disturbance is confined to zones in which epiphyseal bone growth is continuing. The zone of provisional calcification is abbreviated and rarified; later, it contains excess fibrous tissue, becomes dense and wider and forms an irregular line reminiscent of that encountered in rickets. At the same time, the osseocartilagenous tissue between the epiphysis and the diaphysis becomes widened. In older, adult individuals, fibrillation and loss of articular cartilage lead to the exposure of eburnated bone and marginal osteophytosis. The changes are therefore those of secondary osteoarthrosis. Ankylosis seldom develops. Any disorder which occurs in the synovial tissues appears to be secondary to the bone deformities. Flattening of the affected vertebrae may result in disc degeneration; in younger persons, together with the limb deformity, this may produce a moderate adolescent kyphoscoliosis.

Pathogenesis

The cause remains uncertain.[440] There is a selective geographical distribution pointing to racial or hereditary predisposition. However, infective, chemical, toxic, nutritional and endocrine factors have also been implicated.

Excess dietary iron

In 1939, Hiyeda[180a,b] suggested that chronic excessive dietary intake of iron was responsible for Kashin–Beck disease in a manner analogous to haemosiderosis of the Bantu. Although this suggestion is supported by the improvement which follows the removal of populations to non-endemic areas and by the fact that older persons entering an area for the first time are not affected, cirrhosis of the liver, an almost invariable concomitant of Bantu haemosiderosis, is not encountered. Rabbits fed a high-iron diet sustain ulcerating cartilagenous lesions and fibrosis of synovial joints when, concurrent with the feeding, the joints are repeatedly traumatized. Since repeated trauma plays no part in Kashin–Beck disease, this experimental work is of uncertain significance. Moreover, only indirect evidence of synovial iron deposition is available in the human disorder and chondrocyte iron uptake does not appear likely to cause the cartilage and bone destruction characteristic of Kashin–Beck disease.

Selenium deficiency

Selenium is an essential trace element for humans.[489] The selenoenzyme glutathione peroxidase prevents peroxidative cell damage; it destroys peroxide generated during dismutation. Chronic selenium deficiency causes decreased tissue glutathione peroxidase activity. About 70 μg selenium/day is needed to maintain normal body levels. Sodium selenite has a therapeutic effect in patients with Kashin–Beck disease and dietary supplements are prophylactic.

Mycotoxicosis

Kashin–Beck disease has also been attributed to the prolonged ingestion of wheat contaminated by the fungus *Fusaria sporotrischiella*. It was thought that the fungus acted indirectly by liberating toxic amines from wheat protein. Such an action was demonstrated in albino rats and puppies. The liberated amines caused vasoconstriction of small intraosseous arteries. There is evidence that the use of grain from uncontaminated crops can eradicate Kashin–Beck disease.

Mseleni disease

The related topic of Mseleni disease[102] is discussed in Chapter 10.

Chondromalacia patellae

In young persons, particularly in males who have sustained an injury to the knee, forcing the patella against the medial condyle of the femur, a premature softening of the patellar articular cartilage may develop. The disorder is termed chondromalacia patellae.[439] There is disability, pain, effusion, impaired movement and crepitus.

The macroscopic and microscopic appearances of the patellar cartilage are indistinguishable from those of osteoarthrosis.[439] Selective fibrillation of the patellar cartilage is, of course, a normal, age-related change.[109] In a preliminary ultrastructural study of chondromalacic cartilage, there was dedifferentiation of many zone I and II chondrocytes to fibroblast-like cells, while the structure of deeper zone III and IV cells suggested increased protein synthesis and secretion.[503] The territorial matrix appeared increasingly fibrous. Systematic studies of material from 12 young patients were in accord with the concept that chondromalacic changes are precipitated by mechanical overload.[342] Swelling of the zone I, superficial matrix accompanied disorganization of the collagen fibre network. These abnormalities together with the presence at the surface of extending fissures and of an electron-dense material, closely recalled those of idiopathic osteoarthrosis. Chondrocyte clusters were conspicuous; individual chondrocytes contained raised number of organelles. The migration of fibroblast-like cells across the cartilage surface may have represented a form of reparative reaction.

The causes of chondromalacia patellae have been much debated. A similar disorder is encountered among long-distance runners; and hereditary and systemic factors are implicated. Chondromalacia patellae has been regarded as a separate entity from idiopathic osteoarthrosis.[1a] Nevertheless, there is the same increase in bone stiffness that is held to underlie the progressive cartilage lesions of this disease (see p. 890).

The pathological changes are reminiscent of the knee joint osteoarthrosis of STR/ORT mice[480] (see p. 903), a heritable disorder of ageing male animals in which patellar dislocation is characteristic. Mechanical factors are certainly operative.[157] The role of cathepsins (see p. 892) has been proposed as an alternative agency but the reasons for the anatomically selective destruction of cartilage by enzymes that might be expected to degrade many other, additional parts of the knee joint, cannot be explained on this basis.

Neurogenic arthropathy

Whereas motor paralysis protects synovial joints against osteoarthrosis, the afferent inervation is essential to prevent cumulative injury from daily mechanical insults. Any disease which impairs or destroys nerves or fibre tracts mediating the sensations of position, pain and movement is likely to lead quickly to cartilage degradation whether or not the joint is affected by the disease directly. The condition is one of neurogenic arthropathy.[193,306,401] Diabetes mellitus,[376] peripheral nerve injury, macroglobulinaemia with peripheral amyloid neuropathy[416] amyloidosis, spina bifida with peripheral nerve involvement, congenital insensitivity to pain, hereditary sensory neuropathy, leprosy and cord lesions such as those of tabes dorsalis and syringomyelia are among the many known causes. There is an analogy with analgesic ('indomethacin hip') arthropathy, the nature of which continues to excite controversy (see p. 895).[94,379]

Tabetic arthropathy

Neurogenic joint disease is seen in its classical form in tabes dorsalis when the osteoarthrosis is termed Charcot's arthropathy.[69] Tabes is now rare in Western countries; tabetic arthropathy is stated to occur in about 5 per cent of instances of tabes dorsalis. Generally, only one joint is affected but as many as 6 or 7 have been implicated.

Syringomyelia

The arthropathy of syringomyelia is a relatively frequent feature of a disease which, however, is even rarer than tabes dorsalis. Syringomyelia results in porosities of the cervical segments of the spinal cord and the arthropathy, which accompanies loss of pain and temperature sensation, is consequently of upper limb distribution.

Pathological changes

The joint changes in neurogenic arthropathy (Figs 22.39 and 22.40) are distinguished from those of other forms of osteoarthrosis by their rapid progression and intensity. In the early stages, injury may initiate the process. Thereafter, the course of this quickly progressive degenerative disease is punctuated by episodes of trauma with effusion, sometimes fracture, haemorrhage and ultimately repair. The end-result is a deformed and distorted articulation in which osteophytic and post-traumatic remodelling combine to caricature normal structure.

Tabetic joint disease is largely confined to the lower limbs where the knee joint is usually involved, with the ankle, tarsus and hip following in this order of frequency. The pathological changes in the fully established case are extreme.[207] They are dominated by silent, pathological fractures. X-ray reveals bone destruction and fragmentation. There is subchondral bone reabsorption, periarticular new bone formation and the presence of intraarticular chondroosseous particles. Chondrocalcinosis is an occasional, associated disorder.[394] Intercurrent suppurative infection may complicate this disorganization. The joints are much enlarged due to the presence of many bone fragments and to healing at articular ends in malaligned positions. There are marginal exostoses and periosteal osteophytes. Fracture fragments are incorporated in the joint capsule and this capsule and the periosteal articular tissues are thickened or ossified. Fracture fragments may ultimately separate from the bone ends. The fragments tend to be absorbed while the bone ends are eroded and destroyed.

The articular capsule is fibrotic, the synovium thickened and intraarticular fibrin and fibrous adhesions are present. Many fragments of bone and cartilage are present in the subsynovial tissues: some may be viable and grow as chondroosteophytic islands. Metaplastic ossification may occur at tendinous insertions. There is complete or extensive loss of hyaline articular cartilage; the bone ends, where they persist, are covered by fibrous tissue or fibrocartilage. Alternatively, white, polished or excoriated, eburnated bone may constitute the articular surface.

Pathogenesis

The normal joint is protected by reflexes that guard against the adverse affects of overextension, excess load, impact and inappropriate movement. The joint in neurogenic arthropathy has diminished protection. Impairment of the afferent innervation is generally held to predispose to the quickly cumulative effects of repeated, diurnal micro-

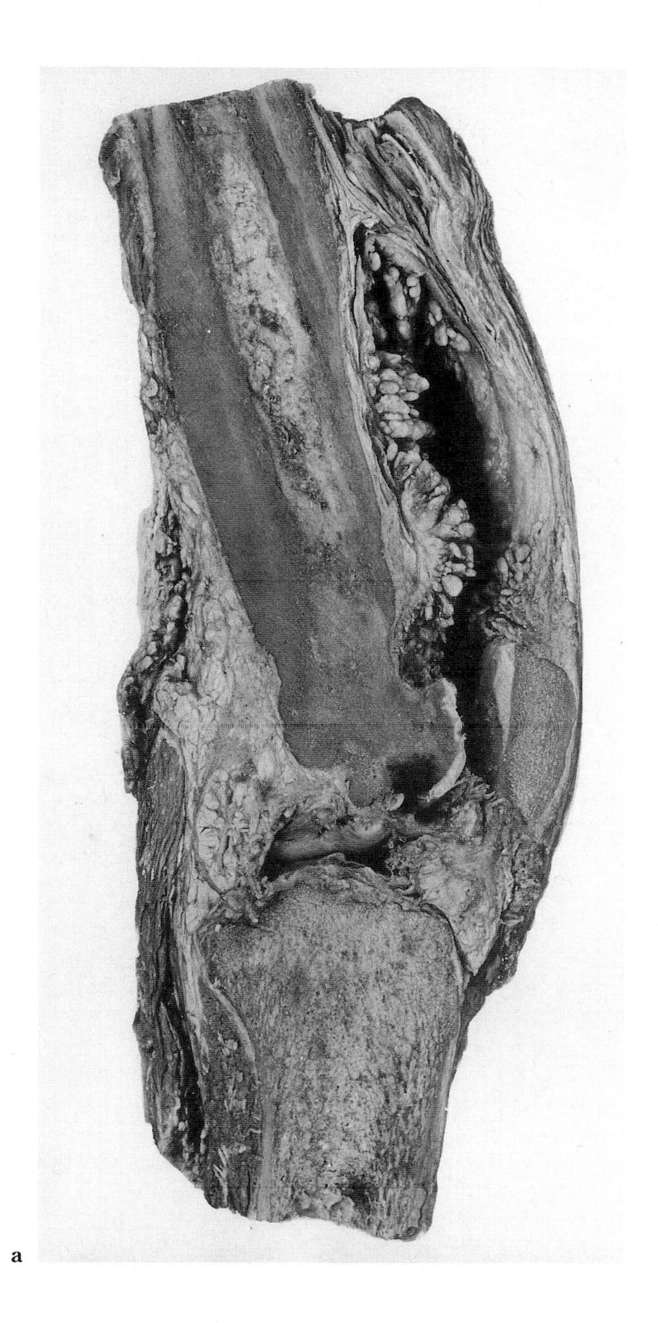

a

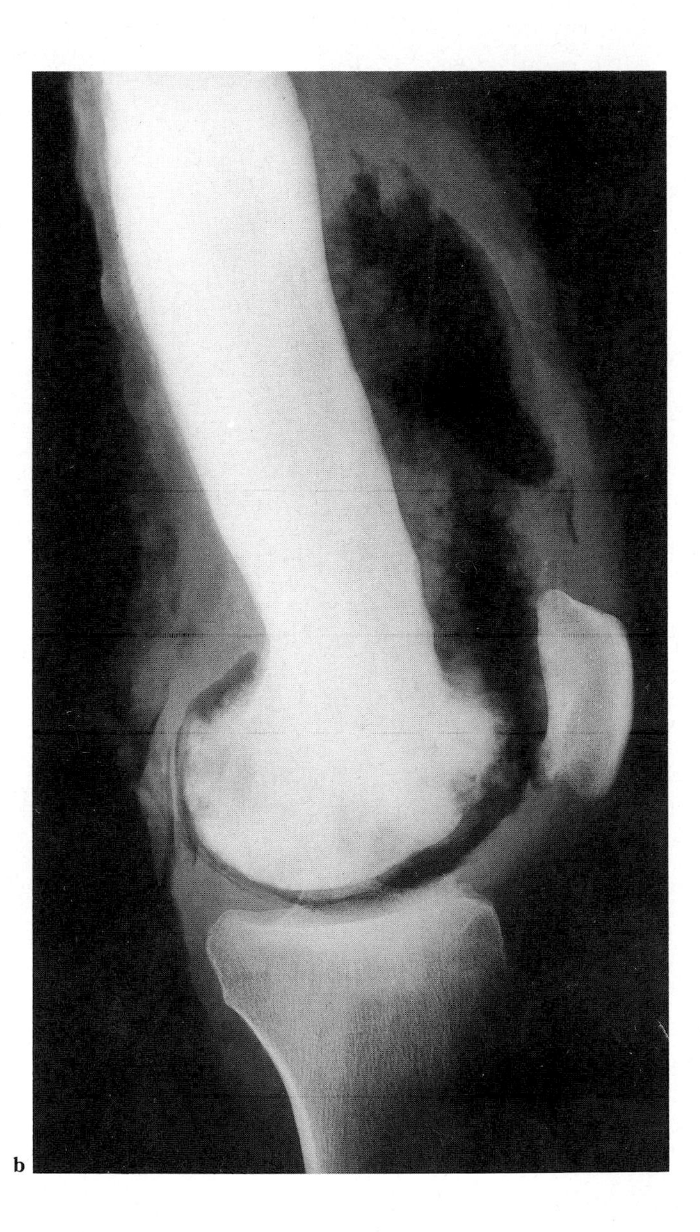

b

Fig. 22.39 Charcot's neurogenic arthropathy.
a. In this patient, a great increase in bone density is associated with the presence of an unusually large quadriceps bursa and severe secondary osteoarthrosis. The rapid progression of the joint disease is one feature of tabes dorsalis. **b.** A radiograph of the specimen shown in Fig. 22.38a illustrates the bone changes.

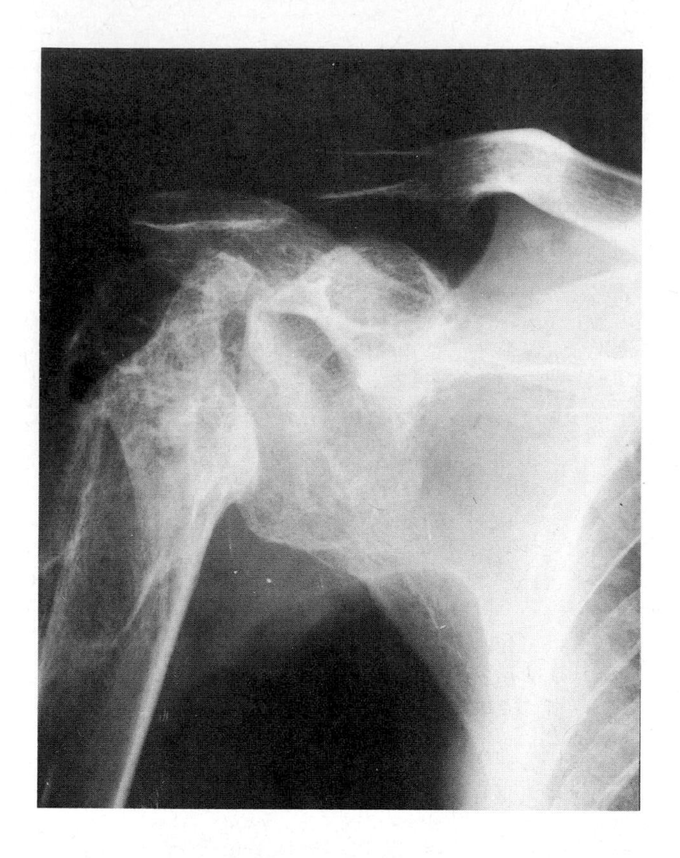

trauma. A similar mechanism is believed to obtain in the experimental animal. Dividing the sensory component of the sciatic nerve in the dog may rapidly result in hind-limb osteoarthrosis (see p. 901). In cats, there is an associated osteosclerosis and this draws attention to the contribution to neurogenic bone and joint disease of changes in the vascular supply to the limb. Neurally initiated vascular reflexes increase local blood flow in the osteoarthropathy of diabetes mellitus[100] and similar mechanisms may operate in other forms of neurogenic disease.

Fig. 22.40 Neurogenic arthropathy in syringomyelia. The shoulder is grossly disorganized. There is a loss of articular cartilage. Bone remodelling is accompanied by osteoporosis.

REFERENCES

1a. Abernethy P J, Townsend P R, Rose R M, Radin E L. Is chondromalacia patellae a separate clinical entity? *Journal of Bone and Joint Surgery* 1978; **60-B**: 205–10.

1. Adam M, Krabcova M, Musilova J, Pesakova V, Brettshneider I, Deyl Z. Contribution to the mode of action of glycosaminoglycan-polysulphate (GAGPS) upon human osteoarthrotic cartilage. *Arzneimittel Forschung* 1980; **30**: 1730–2.

2. Adams I D. Osteoarthrosis and sport. In: Wright ed, *Clinics in Rheumatic Diseases*. Philadelphia: W B Saunders & Co., 1976; **2**: 523–41.

3. Adams M E, Grant M D, Ho A. Cartilage proteoglycan changes in experimental canine osteoarthritis. *Journal of Rheumatology* 1987; **14**: 107–9 (suppl 14).

4. Adams M E, Billingham M E J. Animal models of degenerative joint disease. In: Berry C L, ed, *Current Topics in Pathology Volume 71: Bone and Joint Disease*. Berlin: Springer-Verlag 1982: 265–297.

5. Adams R. A Treatise on Rheumatic Gout, or Chronic Rheumatic Arthritis, of all the Joints. London: John Churchill, 1857.

6. Akeson W H, Amiel D, Able M F, Garfin S R, Woo SL-Y. Effects of immobilisation on joints. *Clinical Orthopaedics and Related Research* 1987; **219**: 28–37.

7. Alexander I S, Gardner D L, Skelton-Stroud P N. Synovial response of *Papio cynocephalus* to exogenous antigen: histological and immunoperoxidase observations. *Annals of the Rheumatic Diseases* 1983; **42**: 448–51.

8. Alhava E M, Kettunen K, Karjalainen P. Bone mineral in patients with osteoarthrosis of the hip. *Acta Orthopaedica Scandinavica* 1975; **46**: 709–15.

8a. Ali S Y. Matrix vesicles and apatite nodules in arthritic cartilage. In: Willoughby D A *et al.*, eds, *Perspectives in Inflammation*. Lancaster: MTP Press, 1977: 211–23.

9. Ali S Y. Mineral-containing matrix vesicles in human osteoarthrotic cartilage. In: Nuki G. *The Aetiopathogenesis of Osteoarthrosis* London: Pitman Medical, 1980: 105–16.

10. Ali S Y, Griffiths S. New types of calcium phosphate crystals in arthritic cartilage. *Seminars in Arthritis and Rheumatism* 1981; **11**: Suppl. 1: 124–6.

11. Ali S Y. Apatite-type crystal deposition in arthritic cartilage. *Scanning Electron Microscopy* 1985; **4**: 1555–66.

12. Ali S Y, Bayliss M T. Enzymic changes in human osteoarthritic cartilage. In: Ali S Y, Elves M W, Leaback D H eds; *Normal and Osteoarthritic Articular Cartilage* 1974. London: Institute of Orthopaedics, 1974: 189–201.

13. Ali S Y, Evans L. Enzymic degradation of cartilage in osteoarthritis. *Federation Proceedings* 1973; **32**: 1494–8.

13a. Altman F P. Metabolic changes associated with murine osteoarthritis. In: Pattison J R, Bitensky L, Chayen J, eds, *Quantitative Cytochemistry and its Applications*. London: Academic Press, 1979: 99–104.

13b. Altman F P. A metabolic dysfunction in early murine osteoarthritis. *Annals of the Rheumatic Diseases* 1981; **40**: 303–6.

13c. Altman F P. Enzyme histochemical changes in murine osteoarthritis. *Histochemistry* 1982; **74**: 43–8.

13d. Altman R, Alarcón G, Applecrouth D *et al.* The American College of Rheumatology criteria for the classification and reporting of osteoarthritis of the hip. *Arthritis and Rheumatism* 1991; **34**: 505–14

14. Altman R, Asch E, Bloch D, Bole G, Borenstein D, Brandt K, Christy W, Cooke T D, Greenward R, Hochbert M, Howell D, Kaplan D, Koopman W, Longley S, Mankin H, McShane D J, Medsger T, Meenan R, Mikkelsen W, Moskowitz R, Murphy W, Rothschild B, Segal M, Sokoloff L, Wolfe F. Development of criteria for the classification and reporting of osteoarthritis—classification of osteoarthritis of the knee. *Arthritis and Rheumatism* 1986; **29**: 1039–49.

15. Altman R D, Fries J F, Bloch D A, Carstens J, Cooke T D, Genant H, Gofton P, Groth H, McShane D J, Murphy W A, Sharp J T, Spitz P, Williams C A, Wolfe F. Radiographic assessment of progression in osteoarthritis. *Arthritis and Rheumatism* 1987; **30**: 1214–25.

16. Altman R D, Meenan R F, Hochberg M C, Bole G C, Brandt K, Cooke T D V, Greenwald R A, Howell D S, Kaplan D, Koopman W J, Mankin H, Mikkelsen W M, Moskowitz R, Sokoloff L. An approach to developing criteria for the clinical diagnosis and classification of osteoarthritis—a status report of the American Rheumatism Association Diagnostic Subcommittee on Osteoarthritis. *Journal of Rheumatology* 1983; **10**: 180–3.

17. Altman R D, Pita J L, Howell D S. Degradation of proteoglycans in human osteoarthritic cartilage. *Arthritis and Rheumatism* 1973; **16**: 179–85.

18. Andersson S, Nilsson B, Hessel T, Saraste M, Noren A, Stevens-Andersson A, Rydholm D. Degenerative joint disease in ballet dancers. *Clinical Orthopaedics and Related Research* 1989; **238**: 233–6.

19. Aretaeus the Cappadocian [ca. 70AD]. On arthritis. Translated by Adams F, ed, *The Extant Works of Aretaeus*. London: New Sydenham Society, 1856.

20. Armstrong C G, Bahrani A S, Gardner D L. *In vitro* measurement of articular cartilage deformations in the intact human hip joint under load. *Journal of Bone and Joint Surgery* 1979; **61-A**: 744–55.

21. Armstrong C G, Bahrani A S, Gardner D L. Changes in the deformational behaviour of human hip cartilage with age. *Journal of Biomechanical Engineering* 1980; **102**: 214–20.

22. Armstrong C G, Gardner D L. Thickness and distribution of femoral head articular cartilage. Changes with age. *Annals of the Rheumatic Diseases* 1977; **36**: 407–12.

23. Armstrong C G, Mow V C. Biomechanics of normal and osteoarthrotic articular cartilage. In: Staub L R, Wilson P D, *Clinical Trends in Orthopedics*. New York: Thieme-Stratton, 1982: 189–97.

24. Arnoczky S P, Marshall J L. The cruciate ligaments of the canine stifle: an anatomical and functional analysis. *American Journal of Veterinary Research* 1977; **38**: 1807–14.

25. Arnoldi C C, Reiman I, Bretlau P. The synovial membrane in human coxarthrosis: light- and electron microscopic studies. *Clinical Orthopaedics and Related Research* 1980; **148**: 213–20.

26. Arsever C L, Bole G G. Experimental osteoarthritis induced by selective myectomy and tendotomy. *Arthritis and Rheumatism* 1986; **29**: 251–61.

27. Ash P, Francis M J O. Response of isolated rabbit articular and epiphyseal chondrocytes to rat liver somatomedin. *Journal of Endocrinology* 1975; **66**: 71–8.

27a. Auer J A, Fackelman G E, Gingerich D A, Felter A W. Effect of hyaluronic acid in naturally occurring and experimentally induced osteoarthritis. *American Journal of Veterinary Research* 1980; **41**: 568–74.

28. Axhausen G. Über die Entstehung der Randwulste bei der Arthritis deformans. *Virchows Archiv für pathologische Anatomie und Physiologie* 1925; **255**: 144–71.

29. Ball J. In: Scott J T, ed. *Copeman's Textbook of the Rheumatic Diseases* (6th edition). Edinburgh: Churchill Livingstone, 1986: 821–45.

30. Bannatyne G A. *Rheumatoid Arthritis: its Pathology, Morbid Anatomy and Treatment* (4th edition). Bristol: John Wright, 1906.

31. Bayliss M T, Ali S Y. Studies on cathepsin B in human articular cartilage. *Biochemical Journal* 1978; **171**: 149–54.

32. Bayliss M T, Venn M, Maroudas A, Ali S Y. Structure of proteoglycans from different layers of human articular cartilage. *Biochemical Journal* 1983; **209**: 387–400.

33. Beck E B. To the problem of disforming endemic osteoarthritis in the Baikal district. *Russian Physician* 1906: **3**: 74–5.

34. Bendele A M. Progressive chronic osteoarthritis in femorotibial joints of partial medial meniscectomized guinea-pigs. *Veterinary Pathology* 1987; **24**: 444–8.

35. Bendele A M, White S L. Early histopathologic and ultrastructural alterations in femorotibial joints of partial medial meniscectomized guinea pigs. *Veterinary Pathology* 1987; **24**: 436–43.

36. Bennett G A, Waine H, Bauer W. *Changes in the Knee Joint at Various Ages*. Boston: Commonwealth Fund, 1942.

36a. Bentley G. Articular cartilage studies and osteoarthrosis. *Annals of the Royal College of Surgeons of England* 1975; **57**: 86–100.

37. Bland J H. The reversibility of osteoarthritis: a review. *American Journal of Medicine* 1983; **74**: 16–26.

38. Bohr H. Experimental osteoarthritis in the rabbit knee joint. *Acta Orthopaedica Scandinavica* 1976; **47**: 558–65.

39. Boivin G, Lagier R. An ultrastructural study of articular chondrocalcinosis in cases of knee osteoarthritis. *Virchow's Archives B* (Pathological anatomy) 1983; **400**: 13–29.

40. Bollet A J, Nance J L. Biochemical findings in normal and osteoarthritic articular cartilage. II. Chondroitin sulfate concentration and chainlength, water, and ash content. *Journal of Clinical Investigation* 1966; **45**: 1170–7.

41. Bradney I W. Treatment of osteoarthritis of the femorotibial joint of the dog by synovectomy and debridement, and repair of the ruptured anterior cruciate ligament. *Journal of Small Animal Practice* 1979; **20**: 197–207.

42. Brandt K D. Pathogenesis of Osteoarthritis. In: Kelley W N, Harris E D, Ruddy S, Sledge C B, eds, *Textbook of Rheumatology*. Philadelphia: W B Saunders & Co., 1981: 1457–70.

43. Brandt K D, Palmoski M J. Effects of salicylates and other non-steroidal anti-inflammatory drugs on articular cartilage. *American Journal of Medicine* 1984; **77(1A)**: 65–9.

44. Brocklehurst R, Bayliss M T, Maroudas A, Coysh H L, Freeman M A R, Revell P A, Ali S Y. The composition of normal

and osteoarthritic articular cartilage from human knee joints. *Journal of Bone and Joint Surgery* 1984; **66-A**: 95–106.

45. Brodelius, A. Osteoarthrosis of the talar joints in footballers and ballet dancers. *Acta Orthopaedica Scandinavica* 1961; **30**: 309–314.

46. Broom N D. The altered biomechanical state of human femoral head osteoarthritic articular cartilage. *Arthritis and Rheumatism* 1984; **27**: 1028–39.

47. Brown R A, Weiss J B. Neovascularisation and its role in the osteoarthritic process. *Annals of the Rheumatic Diseases* 1988; **47**: 881–5.

48. Bullough P C. Understanding osteoarthritis. The value of anatomical studies. *Journal of Rheumatology* 1987; **14**: 189–90.

48a. Bullough P G, DiCarlo E F, Hansraj K K, Neves M C. Pathologic studies of total joint replacement. *Orthopaedic Clinics of North America* 1988; **19**: 611–25.

49. Burr D B, Radin E L. Meniscal function and the importance of meniscal regeneration in preventing late medial compartment osteoarthrosis. *Clinical Orthopaedics and Related Research* 1982; **171**: 121–6.

50. Burton Wurster N, Butler, M, Harter S, Colombo C, Quintavalla J, Swartzendruber D, Arsenis C, Lust G. Presence of fibronectin in articular cartilage in two animal models of osteoarthritis. *Journal of Rheumatology* 1986; **13**: 175–82.

51. Burton Wurster N, Hui-Chou C, Greisen H A, Lust G. Reduced deposition of collagen in the degenerated articular cartilage of dogs with degenerative joint disease. *Biochimica et Biophysica Acta* 1982; **718**: 74–84.

52. Burton Wurster N, Lust G. Synthesis of fibronectin in normal and osteoarthritic articular cartilage. *Biochimica et Biophysica Acta* 1984; **800**: 52–8.

53. Burton Wurster, Lust G. Deposition of fibronectin in articular cartilage of canine osteoarthritic joints. *American Journal of Veterinary Research* 1985; **46**: 2542–5.

54. Butler M, Colombo C, Hickman L, O'Byrne E, Steele R, Steinetz B, Quintavalla J, Yokoyama N. A new model of osteoarthritis in rabbits. III. Evaluation of anti-osteoarthritic effects of selected drugs administered intraarticularly. *Arthritis and Rheumatism* 1983; **26**: 1380–6.

55. Byers P D, Contepomi C A, Farkas T A. A post mortem study of the hip joint including the prevalence of the features of the right side. *Annals of the Rheumatic Diseases* 1970; **29**: 15–31.

56. Byers P D, Pringle J, Oztop F, Fernley H N, Brown M A, Davison W. Observations on osteoarthrosis of the hip. *Seminars in Arthritis and Rheumatism* 1977; **6**: 277–303.

57. Bywaters E G L. Observations on chronic polyarthritis in monkeys. *Journal of the Royal Society of Medicine* 1981; **74**: 794–799.

58. Cameron C H S, Gardner D L, Longmore R B. The preparation of human articular cartilage for scanning electron microscopy. *Journal of Microscopy* 1976; **108**: 1–12.

59. Candolin T, Videman T. Surface changes in the articular cartilage of rabbit knees during immobilisation. A scanning electron microscopic study of experimental osteoarthritis. *Acta Pathologica et Microbiologica Scandinavica* 1980; **88**: 291–7.

60. Carlsson G E. Mandicular dysfunction and temporomandibular joint pathosis. *Journal of Prosthetic Dentistry* 1980; **43**: 658–6.

61. Carlsson G E, Hassler O, Oberg T. Microradiographic study of human temporomandibular discs obtained at autopsy. *Journal of Oral Pathology* 1973; **2**: 265–71.

62. Carney S L, Bayliss M T, Norman J M, Muir H. Electrophoresis of ^{35}S-labelled proteoglycans on polyacrylamide—agarose composite gels and their visualization by fluorography. *Analytical Biochemistry* 1986; **156**: 38–44.

63. Carney S L, Billingham M E J, Muir H, Sandy J D. Structure of newly synthesised (^{35}S)-proteoglycans and (^{35}S)-proteoglycan turnover products of cartilage explant cultures from dogs with experimental osteoarthritis. *Journal of Orthopaedic Research* 1985; **3**: 140–147.

64. Carroll G J. Spectrophotometric measurement of proteoglycans in osteoarthritic synovial fluid. *Annals of the Rheumatic Diseases* 1987; **46**: 375–9.

65. Caruso I, Mantellini P. L'arthrosi sperimentale (rapporti con le artropatie degenerative dell'uomo. *Reumatismo* 1967; **19**: 1–16.

66. Cave A J E, Griffiths J D, Whitely M M. Osteoarthritis deformans of the Luschka joints. *Lancet* 1955; **i**: 176–9.

67. Chaminade F, Stanescu V, Stanescu R, Maroteaux P, Peyron J G. Noncollagenous proteins in cartilage of normal subjects and patients with degenerative joint disease. *Arthritis and Rheumatism* 1982; **25**: 1078–83.

68. Chapchal G. Posttraumatic osteoarthritis after injury of the knee and hip joint. In: Chapchal G, ed., *Reconstructive Surgery and Traumatology* Basel: Karger, 1978; **16**: 87–94.

69. Charcot J M. *Leçones sur les Maladies des Vieillards et les Maladies Chroniques*. Paris: A. Delahaye, 1867.

70. Cherney D D, Baxter W D, DiDio L J A. Synovial membrane of the rabbit knee during an induced degenerative arthropathy. *Acta Anatomica* 1970; **75**: 225–47.

71. Chrisman O D, Fessel J M, Southwick W O. Experimental production of synovitis and marginal articular exostoses in the knee joints of dogs. *Yale Journal of Biology and Medicine* 1965; **37**: 409–12.

72. Chrisman O D, Snook G A. Studies on the protective effect of aspirin against degeneration of human articular cartilage. *Clinical Orthopaedics and Related Research* 1968; **56**: 77–8.

73. Christensen S B, Arnoldi C C. Distribution of ^{99m}Tc-phosphate compounds in osteoarthritic femoral heads. *Journal of Bone and Joint Surgery* 1980; **62-A**: 90–6.

74. Christensen S B, Reimann I. Differential histochemical staining of glycosaminoglycans in the matrix of osteoarthritic cartilage. *Acta Pathologica et Microbiologica Scandinavica* 1980A; **88**: 61–8.

75. Collins D H. *Pathology of Articular and Spinal Diseases.* London: Edward Arnold, 1949: 74–115.

76. Colombo C, Butler M, Hickman L, Selwyn M, Chart J, Steinetz B. A new model of osteoarthritis in rabbits. II. Evaluation of anti-osteoarthritic effects of selected antirheumatic drugs administered systemically. *Arthritis and Rheumatism* 1983; **26**: 1132–9.

77. Colombo C, Butler M, O'Byrne E, Hickman L, Swartzendruber D, Selwyn M, Steinetz B. A new model of osteoarthritis in rabbits. I. Development of knee joint pathology following lateral meniscectomy and section of the fibular collateral and sesamoid ligaments. *Arthritis and Rheumatism* 1983; **26**: 875–86.

78. Cooke T D V. Immune pathology in polyarticular osteoarthritis. *Clinical Orthopaedics and Related Research* 1986; **213**: 41–9.

79. Cooke T D V, Bennett E L, Ohno O. The deposition of immunoglobulins and complement components in osteoarthritic cartilage. *International Orthopaedics* 1980; **4**: 211–7.

80. Cooke T D V, Bennett E L, Wright L, Wyllie J. Relationships of immune deposits in osteoarthritic cartilage to disease site, pattern and synovial reaction. In: Peyron J G, ed., *Epidemiology of Osteoarthritis* Paris: Geigy, 1980: 113.

81. Cooper N S, Soren A, McEwen C, Rosenberger J L. Diagnostic specificity of synovial lesions. *Human Pathology* 1981; **12**: 314–28.

82. Coulais Y, Marcelon G, Cros J, Guiraud R. Etude d'un modele experimental d'arthrose. I. Induction et etude ultrastructurale. *Pathologie et Biologie* 1983; **31**: 577–82.

83. Coulais Y, Marcelon G, Cros J, Guiraud R. Etude d'un modele experimental d'arthrose. II. Etude biochimique du collagene et des proteoglycanes. *Pathologie et Biologie* 1984; **32**: 23–8.

84. Cox J S, Nye C E, Schaefer W W, Woodstein I J. The degenerative effects of partial and total resection of the medial meniscus in dogs' knees. *Clinical Orthopaedics and Related Research* 1975; **109**: 178–83.

85. Crelin E S, Southwick W O. Changes induced by sustained pressure in the knee joint articular cartilage of adult rabbits. *Anatomical Record* 1964; **149**: 113–33.

86. Davis W, Moskowitz R W. Degenerative joint changes following posterior cruciate ligament section in the rabbit. *Clinical Orthopaedics and Related Research* 1973; **93**: 307–12.

87. Dearden L C, Mosier H D. Growth retardation and subsequent recovery of the rat tibia, a histochemical, light and electron microscopic study. I. After propylthiouracil treatment. *Growth* 1974a; **38**: 253–75.

88. Dearden L C, Mosier H D. Growth retardation and subsequent recovery of the rat tibia, a histochemical, light and electron microscopic study. II. After fasting. *Growth* 1974b; **38**: 277–94.

89. Denko C W. Treatment of osteoarthritis with Rumalon R. *Arthritis and Rheumatism* 1978; **21**: 495–6.

90. Dequeker J, Goris P, Uytterhoeven R. Osteoporosis and osteoarthritis (osteoarthrosis). Anthropometric distinctions. *Journal of the American Medical Association* 1983; **249**: 1448–51.

91. Dieppe P A, Calvert P. *Crystals and Joint Disease*. London: Chapman and Hall, 1983: 207–10.

92. Dieppe P A, Doyle D V, Huskisson E C, Willoughby D A, Crocker P R. Mixed crystal deposition disease and osteoarthritis. *British Medical Journal* 1978; **I**: 150.

93. Dodge G R, Poole A R. Immunohistochemical detection and immunochemical analysis of type II collagen degradation in human normal, rheumatoid, and osteoarthritic articular cartilage and in explants of bovine articular cartilage cultured with interleukin 1. *Journal of Clinical Investigation* 1989; **83**: 647–61.

94. Doherty M, Holt M, MacMillan P, Watt I, Dieppe P. A reappraisal of 'analgesic hip'. *Annals of the Rheumatic Diseases* 1986; **45**: 272–6.

95. Doherty M, Watt I, Dieppe P. Influence of primary generalised osteoarthritis on development of secondary osteoarthritis. *Lancet* 1983; **ii**: 8–11.

96. Doyle D V. Tissue calcification and inflammation in osteoarthritis. *Journal of Pathology* 1982; **136**: 199–216.

97. Doyle D V, Dieppe P A, Crocker P R, Willoughby D A. Mixed crystal deposition in an osteoarthritic joint. *Journal of Pathology* 1977; **123**: 1–4.

98. Dunham J, Chambers M G, Jasani M K, Bitensky L, Chayen J. Changes in the orientation of proteoglycans during the early development of natural murine osteoarthritis. *Journal of Orthopaedic Research* 1988; **8**: 101–4.

99. Ecker A. Über Abnutzung und Zerstorung der Gelenkknorpel. *Archiv für Physiologie und Heilkunde* 1843; **2**: 325–48.

100. Edelman S V, Kosofsky E M, Paul R A, Kozak G P. Neuro-osteoarthropathy (Charcot's joint) in diabetes following revascularization surgery. *Archives of Internal Medicine* 1987; **147**: 1504–8.

101. Editorial. Limb salvage surgery. *Lancet* 1988; **ii**: 662–3.

102. Editorial. Mseleni disease. Lancet 1985; **ii**: 483–484.

103. Egar M, Wallace H, Singer M. Partial denervation effects on limb cartilage regeneration. *Anatomy and Embryology* 1982; **164**: 221–8.

104. Egen M S, Goldenberg D L, Cohen A S, Segal D. The association of amyloid deposits and osteoarthritis. *Arthritis and Rheumatism* 1982; **25**: 204–8.

105. Ehrlich M G. Degradative enzyme systems in osteoarthritic cartilage. *Journal of Orthopaedic Research* 1985; **3**: 170–84.

106. Ehrlich M G, Mankin H J, Jones H, Grossman A, Crispen C, Ancona D. Biochemical confirmation of an experimental model for osteoarthritis. *Journal of Bone and Joint Surgery* 1975: **57-A**: 392–6.

107. Ehrlich M G, Mankin H J, Jones H, Wright R, Crispen C, Vigliani G. Collagenase and collagenase inhibitors in osteoarthritic and normal human cartilage. *Journal of Clinical Investigation* 1977; **59**: 226–33.

108. Elling H. Histopathologische untersuchungen zür Beeinflussung der genetisch fixierten Arthrose der 'C57-black-Maus' durch Glycosaminoglykanpolysufat. *Arzneimittel Forschung* 1987; **37**: 940–3.

109. Emery I H and Meachim G. Surface morphology and topography of patello-femoral cartilage fibrillation in Liverpool necropsies. *Journal of Anatomy* 1973; **116**: 103–20.

110. Enneking W F, Horowitz M. The intra-articular effects of immobilization on the human knee. *Journal of Bone and Joint Surgery* 1972; **54-A**: 973–85.

111. Evans C H, Mears D C, Stanitski C L. Ferrographic analysis of wear in human joints—evaluation by comparison with arthroscopic examination of symptomatic knees. *Journal of Bone and Joint Surgery* 1982; **64-B**: 572–8.

112. Evans E B, Eggers G W N, Butler J K, Blumel J. Experimental immobilization and remobilization of rat knee joints. *Journal of Bone and Joint Surgery* 1960; **42-A**: 737–58.

113. Eyre D R, McDevitt C R, Billingham M E J, Muir H. Biosynthesis of collagen and other matrix proteins of articular cartilage in experimental osteoarthrosis. *Biochemical Journal* 1980; **188**: 823–37.

114. Fackelman G E. Tendon surgery. *Veterinary Clinics of North America* 1983; **5**: 381–90.

115. Farkas T A, Reffy A, Frenyo S. Microlesions of articular cartilage as a possible cause for post-traumatic osteoarthritis.

Archiv für Orthopadische und Unfall-Chirurgie 1975; **81**: 279–84.

116. Fazzalari N L, Darracott J, Vernon-Roberts B. Histomorphometric changes in the trabecular structure of a selected stress region in the femur in patients with osteoarthritis and fracture of the femoral neck. *Bone* 1985; **6**: 125–33.

117. Fazzarlari N L, Vernon-Roberts B, Darracott J. Osteoarthritis of the hip. Possible protective and causative roles of trabecular microfractures in the head of the femur. *Clinical Orthopaedics and Related Research* 1987; **216**: 224–33.

118. Felson D T, Naimark A, Anderson J, Kazis L, Castelli W, Meenan R F. The prevalence of knee osteoarthritis in the elderly. The Framingham osteoarthritis study. *Arthritis and Rheumatism* 1987; **30**: 914–8.

119. Fife R S. Alterations in a cartilage matrix glycoprotein in canine osteoarthritis. *Arthritis and Rheumatism* 1986; **29**: 1493–1500.

120. Fischer A G T. *Chronic (non-tuberculous) Arthritis.* London: H K Lewis, 1929.

121. Fitton-Jackson S. Environmental control of macromolecular synthesis in cartilage and bone: morphogenetic response to hyaluronidase. *Proceedings of the Royal Society of London* (Biology) 1970; **175**: 405–53.

122. Floman Y, Eyre D R, Glimcher M J. Induction of osteoarthrosis in the rabbit knee joint: biochemical studies on the articular cartilage. *Clinical Orthopaedics and Related Research* 1980; **147**: 278–88.

123. Folkman J, Merler E, Abernathy C, Williams G. Isolation of a tumor factor responsible for angiogenesis. *Journal of Experimental Medicine* 1971; **133**: 275–88.

124. Francon F, Diaz R, Combey P, Lathoud J. Heberden's nodes in organic hemiplegia. *Progress in Medicine* 1987; **98**: 5–10.

125. Freeman M A R. The pathogenesis of primary osteoarthrosis: an hypothesis. *Modern Trends in Orthopaedics* 1972; **6**: 4–94.

126. Freeman M A R, Meachim G. Ageing and degeneration. In: Freeman M A R, ed., *Adult Articular Cartilage* (2nd edition). Tunbridge Wells: Pitman Medical, 1979: 487–543.

127. Fukubayashi T and Kurosawa H. The contact area and pressure distribution pattern of the knee. *Acta Orthopaedica Scandanavica* 1980; **51**: 871.

128. Gardner D L. The experimental production of arthritis: a review. *Annals of the Rheumatic Diseases* 1960; **19**: 297–317.

129. Gardner D L. The nature and causes of osteoarthrosis. *British Medical Journal* 1983; **286**: 418–24.

130. Gardner D L. Methods for the selection and collection of hyaline articular cartilage samples. In: Maroudas A, Kuettner K, eds, *Methods Used in Research On Cartilaginous Tissues* New York: Academic Press, 1990.

131. Gardner D L, Bradley W A, O'Connor P, Orford C R, Brereton J D. Synovitis after surgical division of the anterior cruciate ligament of the dog. *Clinical and Experimental Rheumatology* 1984; **2**: 11–15.

132. Gardner D L, Elliot R J, Armstrong C G, Longmore R B. The relationship between age, thickness, surface structure, compliance and composition of human femoral head articular cartilage. In: Nuki G, ed, *The Aetiopathogenesis of Osteoarthritis*. Tunbridge Wells: Pitman Medical, 1980: 65–83.

133. Gardner D L, Elliot D, Simpson R. Rapid bone morphometry with blocks, not sections. Application of confocal scanning microscopy to the diagnosis of metabolic bone disease. *Journal of Pathology* 1990; **160**: 160A.

134. Gardner D L, Farragar B, Butterworth K. Synovial mastocytosis following aspectic division of an anterior cruciate ligament in the dog. In preparation.

135. Gardner D L, McArthur S D, Cunningham D S. A new quantitative optical method for the identification of covert osteoarthrosis. *Journal of Pathology* 1991; **163**: 171A.

136. Gardner D L, Mazuryk R, O'Connor P. Orford C R. Anatomical changes and pathogenesis of OA in man, with particular reference to the hip and knee joints. In: Lott D J, Jasani M K, Birdwood G F B, eds, *Studies in Osteoarthrosis: Pathogenesis, Intervention and Assessment.* Chichester: John Wiley & Sons, 1987: 21–48.

137. Gardner D L, Oates K, Lawton D M, Pidd J G, Middleton J F S. Methods for the study of cartilage by low temperature scanning electron microscopy and related techniques. In: Maroudas A, Kuettner K, eds, *Methods Used in Research on Cartilaginous Tissues.* New York: Academic Press, 1990: 63–7.

137a. Gardner D L, O'Connor P, Middleton J F S, Oates K, Orford C R. An investigation by transmission electron microscopy of freeze replicas of dog articular cartilage surfaces: the fibre-rich surface structure. *Journal of Anatomy* 1983; **137**: 573–82.

137b. Gardner D L, O'Connor P, Oates K. Low-temperature scanning electron microscopy of dog and guinea-pig hyaline articular cartilage. *Journal of Anatomy* 1981; **132**: 267–82.

137c. Garr E L, Moskowtiz R W, Davis W. Degenerative changes following experimental patellectomy in the rabbit. *Clinical Orthopaedics and Related Research* 1973; **92**: 296–304.

138. Garrahan N J, Mellish R W E, Compston J E. A new method for the two-dimensional analysis of bone structure in human iliac crest biopsies. *Journal of Microscopy* 1986; **142**: 341–9.

139 Gédéon P, Mazières B, Ficat P. Un nouveau modele d'arthrose experimentale: la contusion du cartilage. Ètude experimentale et clinique. *Revue du Rhumatisme* 1978; **45**: 401–8.

140. Gedikoglu O, Bayliss M T, Ali S Y, Tuncer I. Biochemical and histological changes in osteoarthritic synovial membrane. *Annals of the Rheumatic Diseases* 1986; **45**: 289–92.

141. Gershuni D H, Amiel D, Gonsalves M and Akeson W H. The biochemical response of rabbit articular cartilage matrix to an induced talcum synovitis. *Acta Orthopaedia Scandinavica* 1981; **52**: 599–603.

142. Ghadially F N, Lalonde J-M A, Yong N K. Ultrastructure of amianthoid fibers in osteoarthrotic cartilage. *Virchow's Archiv B Cell Pathology* 1979; **31**: 81–6.

143. Ghadially F N, Moshurchak E M, Ghadially J A. A maturation change in the surface of cat articular cartilage detected by the scanning electron microscope. *Journal of Anatomy* 1978; **125**: 349–60.

144. Ghosh P. Anti-rheumatic drugs and cartilage. *Bailliere's Clinical Rheumatology* 1988; **2**: 309–38.

145. Ghosh P, Sutherland J M, Taylor T K F, Bellenger C R, Pettit G D. The effect of bilateral medial meniscectomy on

articular cartilage of the hip joint. *Journal of Rheumatology* 1984; **11**: 197–201.

146. Gilbertson E M M. The development of periarticular osteophytes in experimentally induced osteoarthritis in the dog. *Annals of the Rheumatic Diseases* 1975; **34**: 12–25.

147. Ginsberg J M, Eyring E J, Curtiss P H. Continuous compression of rabbit articular cartilage producing loss of hydroxyproline before loss of hexosamine. *Journal of Bone and Joint Surgery* 1969; **51-A**: 467–74.

148. Glynn J H, Sutherland I D, Walker G F, Young A C. Low incidence of osteoarthrosis in hip and knee after anterior poliomyelitis: a late review. *British Medical Journal* 1966; **ii**: 739–42.

149. Glynn L E. Primary lesion in osteoarthrosis. *Lancet* 1977; **i**: 574–75.

150. Goffin Y A, Thona Y, Potvliege P R. Microdeposition of amyloid in the joints. *Annals of the Rheumatic Diseases* 1981; **40**: 27–33.

151. Goldenberg D L, Cohen A S. Synovial membrane histopathology in the differential diagnosis of rheumatoid arthritis, gout, pseudogout, systemic lupus erythematosus, infectious arthritis and degenerative joint disease. *Medicine* 1978; **57**: 239–52.

152. Goldenberg D L, Egan M S, Cohen A S. Inflammatory synovitis in degenerative joint disease. *Journal of Rheumatology* 1982; **9**: 204–9.

153. Goldin R H, McAdam L, Louie J S. Clinical and radiological survey of the incidence of osteoarthrosis among obese patients. *Annals of the Rheumatic Diseases* 1976; **35**: 349–53.

154. Golding J C, Ghosh P. Drugs for osteoarthrosis. I. The effects of pentosan polysulphate (SP54) on the degradation and loss of proteoglycans from articular cartilage in a model of osteoarthrosis induced in the rabbit knee joint by immobilization. *Current Therapeutic Research* 1983a; **33**: 173–84.

155. Golding J C, Ghosh P. Drugs for osteoarthrosis II. The effects of a glycosaminoglycan polysulphate ester (Arteparon) on proteoglycan aggregation and loss from articular cartilage of immobilized rabbit knee joints. *Current Therapeutic Research* 1983b; **34**: 67–80.

156. Goldwasser M, Astley T, Rest M van der, Glorieux F H. Analysis of the type of collagen present in osteoarthritic human cartilage. *Clinical Orthopaedics and Related Research* 1982; **187**: 298–302.

157. Goodfellow J W, Hungerford D S, Woods C. Patellofemoral mechanics and pathology. II. Chondromalacia patellae. *Journal of Bone and Joint Surgery* 1976; **58-B**: 291–9.

158. Gordon G V, Villanueva T, Schumacher H R, Gohel V. Autopsy study correlating degree of osteoarthritis, synovitis and evidence of articular calcification. *Journal of Rheumatology*, 1984; **11**: 681–6.

159. Grahame R. How often, when and how does joint hypermobility lead to osteoarthritis? *British Journal of Rheumatology* 1989; **28**: 320.

160. Greisen H A, Summers B A, Lust G. Ultrastructure of the articular cartilage and synovium in the early stages of degenerative joint disease in canine hip joints. *American Journal of Veterinary Research*, 1982; **43**: 1963–71.

161. Gritzka T I, Fry L R, Cheesman R L, Lavigne A. Deterioration of articular cartilage caused by continuous compression in a moving rabbit joint. *Journal of Bone and Joint Surgery* 1973; **55-A**: 1698–720.

162. Grondalen T. Osteochondrosis and arthrosis in pigs. *Acta Veterinaria Scandinavica* 1974a; **15**: 1–25.

163. Grondalen T. Osteochondrosis, arthrosis and leg weakness in pigs. *Nord Vet* 1974b; **26**: 534–37.

164. Gysen P, Franchimont P. Radioimmunoassay of proteoglycans. *Journal of Radioimmunoassay* 1984; **5**: 221–43.

165. Hackenbroch M H Jr, Bruns H, Widenmayer W. Beitrag zur ätiologie der Coxarthrosen. Katamnestische Beurteilung von 976 Coxarthrosen nach radiologischen und klinischen Gesichtpunkten. *Archives of Orthopaedic and Traumatic Surgery* (München) 1979; **95**: 275–83.

166. Hadjipavlou A, Lander P, Srolovitz H. Pagetic arthritis. Pathophysiology and management. *Clinical Orthopaedics and Related Research* 1986; **208**: 15–9.

167. Haines J F, Noble J. Revision arthroplasty of the knee: two problem knees. *Journal of the Royal College of Surgeons of Edinburgh* 1986; **31**: 255–7.

168. Hamerman D. The biology of osteoarthritis. *New England Journal of Medicine* 1989; **32**: 1322–30.

169. Hamerman D, Klagsbrun M. Osteoarthritis. Emerging evidence for cell interactions in the breakdown and remodelling of cartilage. *American Journal of Medicine* 1985; **78**: 495–9.

170. Hannan N, Ghosh P, Bellenger C, Taylor T. Systemic administration of glycosaminoglycan polysulphate (Arteparon) provides partial protection from damage produced by meniscectomy in the canine. *Journal of Orthopaedic Research* 1987; **5**: 47–59.

171. Harris F. Hormonal control of growth. Owen R, Goodfellow J, Bullough P: *Scientific Foundations of Orthopaedics and Traumatology* Edinburgh: Churchill Livingstone, 1980: 193–5.

172. Harris W H. Traumatic arthritis of the hip after dislocation and acetabular fractures: treatment by mold arthroplasty. An end-result study using a new method of result evaluation. *Journal of Bone and Joint Surgery* 1969; **51A**: 737–55.

172a. Harris W H, Sledge C B. Total hip and knee replacement. *New England Journal of Medicine* 1990; **323**: 725–31, 801–7.

173. Hascall V C, Glant T T. Proteoglycan epitopes as potential markers of normal and pathologic cartilage metabolism. *Arthritis and Rheumatism* 1987; **30**: 586–8.

174. Havdrup T, Telhag H. Mitosis of chondrocytes in normal adult joint cartilage. *Clinical Orthopaedics and Related Research* 1980; **153**: 248–52.

175. Healey J H, Vigorita V J, Lane J M. The coexistence and characteristics of osteoarthritis and osteoporosis. *Journal of Bone and Joint Surgery* 1985; **67-A**: 586–92.

176. Heine J. Über die Arthritis deformans. *Virchows Archiv für pathologische Anatomie und Physiologie und für klinische Medizin*, 1926; **260**: 521–663.

177. Helminen H J, Jurvelin J, Kuusela T, Heikkilä R, Kiviranta I, Tammi M. Effect of immobilization for six weeks on rabbit knee articular surfaces as assessed by the semiquantitative stereomicroscopic method. *Acta Anatomica* 1983; **115**: 327–35.

178. Helminen H J, Kiviranta I, Säämänen A-M, Tammi M, Paukkonen K, Jurvelin J, eds. *Joint loading. Biology and Health of Articular Structures*. Bristol: John Wright, 1987.

179. Hilbert B J, Rowley G, Antonas K N, McGill C A, Reynoldson J A, Hawkins C D. Changes in the synovium after the intra-articular injection of sodium hyaluronate into normal horse joints and after arthrotomy and experimental cartilage damage. *Australian Veterinary Journal* 1985; **62**: 182–4.

180. Hippocrates, 460–360 BC. Oeuvres complete d'Hippocrate. Traduction nouvelle avec le texte grec en regard . . . Par E Littré. 10 volumes. Paris: J B Bailliere, 1839–61.

180a. Hiyeda K. On cause of endemic diseases prevailing in Manchoukuo (Kaschin–Beck's disease, so-called Kokusan disease, and endemic goitre). *Transactiones Societatis pathologicae Japonicae* 1939; **29**: 325–32.

180b. Hiyeda K. Cause of Kaschin–Beck's disease. *Japanese Journal of Medical Sciences, Part V (Pathology)* 1939; **4**: 91–106.

181. Hoffa A, Wollenberg G A. *Arthritis deformans und sogennante Gelenkrheumatismus. Eine röntgenologische und anatomische Studie.* Stuttgart: F. Enke., 1908.

182. Hough A J, Mottram F C, Sokoloff L. The collagenous nature of amianthoid degeneration of human costal cartilage. *American Journal of Pathology* 1973; **73**: 201–10.

183. Hough A J, Sokoloff L. Pathology of osteoarthritis. In: McCarty D J, ed., *Arthritis and Allied Conditions: A Textbook of Rheumatology*, (11th edition). Philadelphia: Lea & Febiger, 1989: 1571–91.

184. Hough A J, Webber R J. Ageing phenomena and osteoarthritis: cause or coincidence? Claude P Brown Memorial Lecture. *Annals of Clinical and Laboratory Science* 1986; **16**: 502–10.

184a. Howell D S. Degradative enzymes in osteoarthritic human articular cartilage. *Arthritis and Rheumatism* 1975; **18**: 167–77.

185. Howell D. S. Etiopathogenesis of osteoarthritis. In: Moskowitz R W, Howell D S, Goldberg V M, Mankin H J, eds, *Osteoarthritis: Diagnosis and Management.* Philadelphia: W B Saunders & Co., 1984: 129–46.

186. Howell D S. Pathogenesis of osteoarthritis. *American Journal of Medicine* 1986; **80** (suppl. 4B): 24–8.

187. Howell D S. Etiopathogenesis of osteoarthritis. In: McCarty D J, *Arthritis and Allied Conditions* (11th edition). Philadelphia: Lea & Febiger, 1989: 1595–1604.

188. Howell D S, Carreno M R, Pelletier J-P, Muniz O E. Articular cartilage breakdown in a lapine model of osteoarthritis: action of glycosaminoglycan polysulfate ester (GAGPS) on proteoglycan degrading enzyme activity, hexuronate, and cell counts. *Clinical Orthopaedics and Related Research* 1986; **213**: 69–76.

189. Howell D S, Manicourt D H, Muniz O E, Tornero G, Carreno M R. Action of Arteparon, a neutral protease inhibitor on erosions in a rabbit model of osteoarthrosis. *Arthritis and Rheumatism* 1984; **27**: S41.

190. Howell D S, Muniz O, Pita J C, Enis J E. Pyrophosphate release of osteoarthritis cartilage incubates. *Arthritis and Rheumatism* 1976; **19**: 488–94.

191. Howell D S, Talbott J H, eds, Osteoarthritis Symposium. *Seminars in Arthritis and Rheumatism* 1981; **11**: 1–149.

192. Howell D S, Woessner J F, Jimenez S, Seda H, Schumacher H R. A view on the pathogenesis of osteoarthritis. *Bulletin on the Rheumatic Diseases* 1978–9; **29**: 996–1001.

193. Hubault A. Arthropathies nerveuses. *Revue du Neurologie* 1982; **138**: 1009–17.

194. Hubbard J R, Liberti J P. Regulation of cartilage acid hydrolases by growth hormone. *Endocrinology* 1982; **110**: 1483–8.

195. Hui-Chou C S, Lust G. The type of collagen made by the articular cartilage in joints of dogs with degenerative joint disease. *Collagen and Related Research* 1982; **2**: 245–56.

196. Hulth A. Experimental osteoarthritis. *Acta Orthopaedica Scandinavica* 1982; **53**: 1–6.

197. Hulth A, Lindberg L, Telhag H. Experimental osteoarthritis in rabbits. Preliminary report. *Acta Orthopaedica Scandinavica* 1970; **41**: 522–30.

198. Hultkranz J W. Über die Spaltrichtungen der Gelenkknorpel. *Verhandlungen der anatomische Gesellschaft, Jena* 1898; **12**: 248–56.

199. Hunter J. Manuscript 54, 1759. (In the possession of the Royal College of Surgeons of England.)

200. Huth F, Soren A, Rosenbauer K A, Klein W. Fine structural changes of the synovial membrane in arthrosis deformans. *Archives of Pathology* 1973; **359**: 201–11.

201. Hutton C W. Generalised osteoarthritis: an evolutionary problem? *Lancet* 1987; **ii**: 1463–5.

202. Hutton C W. Osteoarthritis—the cause not result of joint failure. *Annals of the Rheumatic Diseases* 1989; **48**: 958–61.

203. Ilardi C F, Sokoloff L. Secondary osteonecrosis in osteoarthritis of the femoral head. *Human Pathology* 1984; **15**: 79–83.

204. Inerot S, Heinegard D, Andell L, Olsson S-E. Articular cartilage proteoglycans in aging and osteoarthritis. *Biochemical Journal* 1978; **169**: 143–56.

205. Inoue H. Alterations in the collagen framework of osteoarthritic cartilage and subchondral bone. *International Orthopaedics* 1981; **5**: 47–52.

205a. Jack S W, Thacker H L. Degenerative joint disease in a Nile hippopotamus. *Journal of the American Veterinary Association* 1985; **187**: 1235.

206. Jackson J P. Degenerative changes in the knee after meniscectomy. *British Medical Journal* 1968; **ii**: 525–7.

207. Jaffe H L. Degenerative joint disease. *Metabolic, Degenerative, and Inflammatory Diseases of Bones and Joints.* Philadelphia: Lea & Febiger, 1972: 735–78.

208. Jaffe H L. Paget's disease. In: Jaffe H L, *Metabolic Degenerative, and Inflammatory Diseases of Bones and Joints.* Philadelphia: Lea & Febiger, 1972: 240–271.

209. Johnson R G. Transection of the canine anterior cruciate ligament: a concise review of experience with this model of degenerative joint disease. *Experimental Pathology* 1986; **30**: 209–213.

210. Jones K L, Brown M, Ali S Y, Brown R A. An immunohistochemical study of fibronectin in human osteoarthritic and disease-free articular cartilage. *Annals of the Rheumatic Diseases* 1987; **46**: 809–15.

211. Jurvelin J, Helminen H J, Lauritsalo S, Kiviranta I, Säämänen A-M, Paukkonen K, Tammi M. Influences of joint immobilisation and running exercise on articular cartilage surfaces of young rabbits. *Acta Anatomica* 1985; **122**: 62–8.

212. Kalbhen D A. The inhibitory effects of steroidal and non-steroidal antirheumatic drugs on articular cartilage in osteoar-

throsis and its counteraction by a biological GAG–peptide complex (Rumalon). *Zeitschrift für Rheumatologie* 1982; **41**: 202–11.

213. Kalbhen D A. New aspects on pharmacotherapy of osteoarthrosis. In: Simon L, Loyau G, eds, *L'Arthrose: Perspectives et Realites*. Paris: Masson S A, 1986: 255–62.

214. Kalbhen D A. Chemical model of osteoarthritis—a pharmacological evaluation. *Journal of Rheumatology* 1987; **14** (Suppl 14): 130–1.

215. Kashin N I. The information of the spreading of goitre and cretinism within the limits of the Russian Empire. *Moscow Medical Newspaper* 1861; **5–7**: 39–51.

215a. Kashin N I. The goitre and cretinism in the limits of Russia and outside. *The Records of the Scientific Medical Society of the Eastern Siberia, Irkusk,* 1868–9: 1–206.

216. Kellgren J H, Lawrence J S, Aitken-Swann J. Rheumatic complaints in an urban population. *Annals of the Rheumatic Diseases* 1953; **12**: 5–15.

217. Kellgren J H, Lawrence J S, Bier F. Genetic factors in generalized osteoarthrosis. *Annals of the Rheumatic Diseases* 1963; **22**: 237–55.

218. Kemp S F, Hintz R L. The action of somatomedin on glycosaminoglycan synthesis in cultured chick chondrocytes. *Endocrinology* 1980; **106**: 744–9.

219. Kempson G E. Mechanical properties of articular cartilage. In: Freeman M A R, ed, *Adult Articular Cartilage* (2nd edition). Tunbridge Wells: Pitman Medical, 1979a: 333–414.

220. Kempson G E. Mechanical properties of articular cartilage and their relationship to matrix degradation. *Biochemical Journal* 1979b; **183**: 111–13.

221. Kempson G E. Relationship between the tensile properties of articular cartilage from the human knee and age. *Annals of the Rheumatic Diseases* 1982; **41**: 508–11.

222. Kennedy T D, Plater-Zyberk C, Partridge T A, Woodrow D F, Maini R N. Morphometric comparison of synovium from patients with osteoarthritis and rheumatoid arthritis. *Journal of Clinical Pathology* 1988; **41**: 847–52.

223. Kessler M J, Turnquist J E, Pritzker K P H, London W T. Reduction of passive extension and radiographic evidence of degenerative knee joint diseases in cage-raised and free-ranging aged rhesus monkeys (Macaca mulatta). *Journal of Medical Primatology* 1986; **15**: 1–9.

224. Kiaer T. The intraosseous circulation and pathogenesis of osteoarthritis. *Medical Science Research* 1987; **15**: 759–63.

225. Kier R, Wain S L, Apple J, Martinez S. Osteoarthritis of the sternoclavicular joint. Radiographic features and pathologic correlation. *Investigative Radiology* 1986; **21**: 227–33.

226. Kincaid S A, Van Sickle D C. Effects of exercise on the histochemical changes of articular chondrocytes in adult dogs. *American Journal of Veterinary Research* 1982; **43**: 1218–26.

227. Kincaid S A, Van Sickle D C. Bone morphology and postnatal osteogenesis. Potential for disease. *Veterinary Clinics of North America (Small Animal Practice)* 1983; **13**: 3–17.

228. Kiviranta I. Joint loading influences on the articular cartilage of young dogs. *Quantitative Histochemical Studies on Matrix Carbohydrates*. Kuopio: University of Kuopio Publications, 1987.

229. Kiviranta I, Jurvelin J, Tammi M, Säämänen A M, Helminen H J. Weight bearing controls glycosaminoglycan concentration and articular cartilage thickness in the knee joints of young beagle dogs. *Arthritis and Rheumatism* 1987; **30**: 801–9.

229a. Kiviranta I, Tammi M, Jurvelin J, Helminen H J. Topographical variation of glycosaminoglycan content and cartilage thickness in canine knee (stifle) joint cartilage. Application of the microspectrophotometric method. *Journal of Anatomy* 1987; **150**: 265–76.

230. Kiviranta I, Tammi M, Jurvelin J, Säämänen A-M, Helminen H J. Moderate running exercise augments glycosaminoglycans and thickness of articular cartilage in the knee joint of young beagle dogs. *Journal of Orthopaedic Research* 1988; **6**: 188–95.

231. Kleesick K, Reinards R, Okusi J, Wolf B, Greiling H. UDP-D-xylose: proteoglycan core protein beta-D-xylosyl-transferase: a new marker of cartilage destruction in chronic joint diseases. *Journal of Clinical Chemistry and Clinical Biochemistry* 1987; **25**: 473–81.

232. Klunder K B, Rud B, Hansen J. Osteoarthritis of the hip and knee joint in retired football players. *Acta Orthopaedica Scandinavica* 1980; **51**: 925–8.

233. Knowlton R G, Katzenstein P L, Moskowitz R W, Weaver E J, Malemud C J, Pathria M N, Jimenez S A, Prockop D J. Genetic linkage of a polymorphism in the type II procollagen gene (COL2A1) to primary osteoarthritis associated with mild chondrodysplasia. *New England Journal of Medicine* 1990; **322**: 526–30.

234. Kopp S, Carlsson G E, Haraldson T, Wenneberg B. Long-term effect of intra-articular injections of sodium hyaluronate and corticosteroid on temporomandibular joint arthritis. *Journal of Oral and Maxillofacial Surgery* 1987; **45**: 929–35.

235. Korkala O, Karaharju E, Gronblad M, Aalto K. Articular cartilage after meniscectomy. Rabbit knees studied with the scanning electron microscope. *Acta Orthopaedica Scandinavica* 1984; **55**: 2737.

236. Krause W, Pope M, Johnson R, Wilder D. Mechanical changes in the knee after meniscectomy. *Journal of Bone and Joint Surgery* 1976; **58-A**: 599–604.

237. Kresina T F, Malemud C J, Moskowitz R W. Analysis of osteoarthritic cartilage using monoclonal antibodies reactive with rabbit proteoglycan. *Arthritis and Rheumatism* 1986; **29**: 863–87.

238. Kuettner K E, Pauli B U. Inhibition of neovascularization by a cartilage factor. *Ciba Foundation Symposium* 1983; **100**: 163–73.

239. Ladefoged C. Amyloid in osteoarthritic hip joints. A pathoanatomical and histological investigation of femoral head cartilage. *Acta Orthopaedica Scandinavica* 1982; **53**: 581–6.

240. Ladefoged C. Amyloid in osteoarthritic hip joints: deposits in relation to chondromatosis, pyrophosphate, and inflammatory cell infiltrate in the synovial membrane and fibrous capsule. *Annals of the Rheumatic Diseases* 1983; **42**: 659–64.

241. Ladefoged C, Christensen H E, Sorensen K H. Amyloid in osteoarthritic hip joints. Depositions in cartilage and capsule. Semiquantitative aspects. *Acta Orthopaedica Scandinavica* 1982; **53**: 587–90.

242. Lagier R. Anatomo-pathology of posttraumatic arthroses of the lower extremities. *Zeitschrift für Unfallmedizin und Berufskrankheiten* 1975; **68**: 3–21.

243. Lagier R. Anatomo-pathogie de la coxarthrose a l'usage du medicin practicien. *Journal Médicale de Strasbourg (Europa Medica)* 1978; **9**: 115–9.

244. Lane N E, Bloch D A, Jones H H, Marshall W H, Wood P D, Fries J F. Long-distance running, bone density, and osteoarthritis. *Journal of the American Medical Association* 1986; **255**: 1147–1151.

245. Langenskiöld A, Michelsson J-E, Videman T. Osteoarthritis of the knee in the rabbit produced by immobilization. *Acta Orthopaedica Scandinavica* 1979; **51**: 1–14.

246. Laver-Rudich Z, Silbermann M. Cartilage surface charge. A possible determinant in ageing and osteoarthritic processes. *Arthritis and Rheumatism* 1985; **28**: 660–70.

247. Lawrence J S. The epidemiology of degenerative joint disease: occupational and ergonomic aspects. In: Helminen H J, Kiviranta I, Säämänen A-M, Tammi M, Paukkonen K, Jurvelin J, eds, *Joint Loading: Biology and Health of Articular Structures.* Bristol: John Wright, 1987: 316–51.

248. Lawrence J S, Bremner J M, Bier F. Osteoarthrosis. Prevalence in the population and relationship between symptoms and X-ray changes. *Annals of the Rheumatic Diseases* 1966; **25**: 1–24.

249. Lawrence J S, de Graeff R, Laine V A I. *The Epidemiology of Chronic Rheumatism* Oxford: Blackwell, 1962.

250. Lindblad S, Hedfors E. Arthroscopic and immunohistologic characterization of knee joint synovitis in osteoarthritis. *Arthritis and Rheumatism* 1987; **30**: 1081–8.

251. Lipowitz A J, Wong P L, Stevens J B. Synovial membrane changes after experimental transection of the cranial cruciate ligament in dogs. *American Journal of Veterinary Research* 1985; **46**: 1166–70.

252. Lippiello L, Hall D, Mankin H J. Collagen synthesis in normal and osteoarthritic human cartilage. *Journal of Clinical Investigation* 1977; **59**: 593–600.

253. Livne E, Von der Mark K, Silbermann M. Morphologic and cytochemical changes in maturing and osteoarthritic cartilage in the temporomandibular joint of mice. *Arthritis and Rheumatism* 1985; **28**: 1027–38.

254. Lourie J A. Is there an association between ABO blood groups and primary osteoarthrosis of the hip? *Annals of Human Biology* 1983; **10**: 381–4.

255. Lufti A M. Morphological changes in the articular cartilage after meniscectomy. An experimental study in the monkey. *Journal of Bone and Joint Surgery* 1975; **57-B**: 525–8.

256. Lust G, Beilman W T, Dueland D, Farrell P W. Intra-articular volume and hip joint instability in dogs with hip dysplasia. *Journal of Bone and Joint Surgery* 1980; **62-A**: 576–82.

257. Lust G, Geary J C, Sheffy B E. Development of hip dysplasia in dogs. *American Journal of Veterinary Research* 1973; **34**: 87–91.

258. Lust G, Miller D R. Biochemical changes in canine osteoarthrosis. In: Nuki G, ed, *The Aetiopathogenesis of Osteoarthrosis*. Tunbridge Wells: Pitman Medical, 1980: 47–51.

259. Lust G, Summers B A. Early asymptomatic stage of degenerative joint disease in canine hip joints. *American Journal of Veterinary Research* 1981; **42**: 1849–55.

260. McDevitt C A, Gilbertson E, Muir H. An experimental model of osteoarthritis: early morphological and biochemical changes. *Journal of Bone and Joint Surgery* 1977; **59-B**: 24–35.

261. McDevitt C A, Miller R R. Biochemistry, cell biology and immunology of osteoarthritis. *Current Opinions in Rheumatology* 1989; **3**: 303–14.

262. McDevitt C A, Muir H. Biochemical changes in the cartilage of the knee in experimental and natural osteoarthritis in the dog. *Journal of Bone and Joint Surgery* 1976; **58-B**: 94–101.

263. McDevitt C A, Pahl J A, Ayad S, Miller R R, Uratsoji M, Andrish J T. Experimental osteoarthritic articular cartilage is enriched in guanidine soluble type VI collagen. *Biochemical and Biophysical Research Communications* 1988; **157**: 250–5.

264. McDevitt C A, Sidorsky K D, Arnoczky S P, Mitchell M, Amadio P C, Cox M J, Warren R F. A protein enriched in extracts of experimental canine and human osteoarthritic cartilage. In: Sen A, Thornhill T, ed, *Development and Diseases of Cartilage and Bone Matrix*. New York: Alan R Liss, 1987: 275–80.

265. McElligott T F, Potter J L. Increased fixation of sulfur[35] by cartilage *in vitro* following depletion of the matrix by intravenous papain. *Journal of Experimental Medicine* 1960; **112**: 743–50.

266. McGuire J, Fedarko N. Keratinocytes form multi-cell clusters in the absence of cell division or collagen synthesis; tunicamycin interferes with cell cluster formation. In: Seiji M, Bernstein I A, eds, *Current Problems in Dermatology*, volume 11: *Normal and Abnormal Epidermal Differentiation.* Basel: Karger 1983: 83–96.

267. McKay J M K, Sim A M, McCormick J N, Marmion B P, McCraw A P, Duthie J J R, Gardner D L. Aetiology of rheumatoid arthritis: an attempt to transmit an infective agent from patients with rheumatoid arthritis to baboons. *Annals of the Rheumatic Diseases* 1983; **42**: 443–7.

268. McKibbin B, Maroudas A. Nutrition and metabolism. In: Freeman M A R, ed, *Adult Articular Cartilage* (2nd edition). Tunbridge Wells: Pitman Medical, 1979: 461–486.

269. Magnuson P B. Joint debridement: surgical treatment of degenerative arthritis. *Surgery, Gynaecology and Obstetrics* 1941; **73**: 1–9.

270. Magyar E, Talerman A, Wouters H W. Bone cysts in experimentally induced arthritis and osteoarthritis in rats. *Archivum Chirurgicum Nederlandicum* 1974; **26**: 233–42.

271. Mankin H J. The response of articular cartilage to mechanical injury. *Journal of Bone and Joint Surgery* 1982; **64-A**: 460–71.

272. Mankin H J, Brandt K D. Biochemistry and metabolism of cartilage in osteoarthritis. In: Moskowitz R W, Howell D S, Goldberg V M, Mankin H J, eds, *Osteoarthritis: Diagnosis and Management.* Philadelphia: W B Saunders & Co., 1984: 43–79.

273. Mankin H J, Dorfman H, Lippiello L, Zarins A. Biochemical and metabolic abnormalities in articular cartilage from osteoarthritic human hips. II. Correlation of morphology with biochemical and metabolic data. *Journal of Bone and Joint Surgery* 1971; **53-A**: 523–37.

274. Mankin H J, Johnson M E, Lippiello L. Biochemical and metabolic abnormalities in articular cartilage from osteoarthritic human hips. III. Distribution and metabolism of amino sugar-containing macromolecules. *Journal of Bone and Joint Surgery* 1981; **63-A**: 131–9.

275. Mankin H J, Thrasher A Z. Water content and binding in normal and osteoarthritic human cartilage. *Journal of Bone and Joint Surgery* 1975; **57-A**: 76–80.

276. Maroudas A. Physicochemical properties of articular cartilage. In: Freeman M A R, ed, *Adult Articular Cartilage* (2nd edition). Tunbridge Wells: Pitman Medical, 1979: 215–90.

277. Maroudas A, Schneiderman R. 'Free' and 'exchangeable' or 'trapped' and 'non-exchangeable' water in cartilage. *Journal of Orthopaedic Research* 1987; **5**: 133–8.

278. Maroudas A, Venn M F. Chemical composition and swelling of normal and osteoarthrotic femoral head cartilage. II. Swelling. *Annals of the Rheumatic Diseases* 1977; **36**: 399–406.

279. Maroudas A, Ziv I, Weisman N, Venn M. Studies of hydration and swelling pressure in normal and osteoarthritic cartilage. *Biorheology* 1985; **22**: 159–69.

280. Marshall J L. Periarticular osteophytes: initiation and formation in the knee of the dog. *Clinical Orthopaedics and Related Research* 1969; **62**: 37–47.

281. Marshall J L, Olsson S-E. Instability of the knee. A long-term experimental study in dogs. *Journal of Bone and Joint Surgery* 1971; **53-A**: 1561–70.

282. Martel-Pelletier J, Cloutier J M, Pelletier J P. Neutral proteases in human osteoarthritic synovium. *Arthritis and Rheumatism* 1986; **29**: 1112–21.

283. Martel-Pelletier J, Pelletier J-P. Degradative changes in human articular cartilage induced by chemotherapeutic agents. *Journal of Rheumatology* 1986; **13**: 164–74.

284. Martel-Pelletier J, Pelletier J-P, Cloutier J-M, Malemud C J. Newly synthesized and endogenous proteoglycans in human osteoarthritic knee cartilage. *Journal of Rheumatology* 1987; **14**: 321–8.

285. Martel-Pelletier J, Pelletier J-P, Malemud C J. Activation of neutral metalloprotease in human osteoarthritic knee cartilage: evidence for degradation in the core protein of sulphated proteoglycan. *Annals of the Rheumatic Diseases* 1988; **47**: 801–8.

286. Martinez-Lavin M, Holman K I, Smyth C J, Vaughan J H. A comparison of naproxen, indomethacin and aspirin in osteoarthritis. *Journal of Rheumatology* 1980; **7**: 711–16.

286a. Matucci-Cerinic M, Lombardi A, Lotti T, Pignone A, Senesi C, Buzzi R, Aglietti P, Partsch G, Cagnoni M. Fibrinolytic activity in the synovial membrane of osteoarthritis. *British Journal of Rheumatology* 1990; **29**: 249–53.

287. Mayor M B, Moskowitz R W. Metabolic studies in experimentally-induced degenerative joint disease in the rabbit. *Journal of Rheumatology* 1974; **1**: 17–23.

288. Meachim G. The effect of scarification on articular cartilage in the rabbit. *Journal of Bone and Joint Surgery* 1963; **45-B**: 150–61.

289. Meachim G. Age changes in articular cartilage. *Clinical Orthopaedics and Related Research* 1969; **64**: 33–44.

290. Meachim G. Light microscopy of Indian ink preparations of fibrillated cartilage. *Annals of the Rheumatic Diseases* 1972; **31**: 457–64.

291. Meachim G. Cartilage fibrillation at the ankle joint in Liverpool necropsies. *Journal of Anatomy* 1975; **119**: 601–10.

292. Meachim, G. Cartilage fibrillation on the lateral tibial plateau in Liverpool necropsies. *Journal of Anatomy* 1976; **121**: 97–106.

293. Meachim G, Brooke G. The pathology of osteoarthritis. In: Moskowitz R W, Howell D S, Goldberg V M, Mankin H J. *Osteoarthritis: Diagnosis and Management*. Philadelphia: W B Saunders & Co., 1984: 29–42.

294. Meachim G, Denham D, Emery I H, Wilkinson P H. Collagen alignments and artificial splits at the surface of human articular cartilage. *Journal of Anatomy* 1974; **118**: 101–118.

295. Meachim G, Emery I H. Cartilage fibrillation in shoulder and hip joints in Liverpool necropsies. *Journal of Anatomy* 1973; **116**: 161–79.

296. Meachim G, Fergie I A. Morphological patterns of articular cartilage fibrillation. *Journal of Pathology* 1975; **115**: 231–40.

297. Meachim G, Osborne G V. Repair at the femoral articular surface in osteo-arthritis of the hip. *Journal of Pathology* 1970; **102**: 1–8.

298. Meachim G, Roberts C. Repair of the joint surface from subarticular tissue in the rabbit knee. *Journal of Anatomy* 1971; **109**: 317–27.

299. Meachim G, Stockwell R A. The Matrix. In: Freeman M A R, ed, *Adult Articular Cartilage* (2nd edition). Tunbridge Wells: Pitman Medical 1979: 1–67.

300. Miller D, Lust G. Accumulation of procollagen in the degenerative articular cartilage of dogs with osteoarthritis. *Biochimica et Biophysica Acta* 1979; **583**: 218–31.

301. Miller D R, Mankin H J, Shoji H, D'Ambrosia R D. Identification of fibronectin in preparations of osteoarthritic human cartilage. *Connective Tissue Research* 1984; **12**: 267–75.

302. Miller W T, Restifo R A. Steroid arthropathy. *Radiology* 1966; **86**: 652–7.

303. Minns R J, Muckle D S. The role of the meniscus in an instability model for osteoarthritis in the rabbit knee. *British Journal of Experimental Pathology* 1982; **63**: 18–24.

304. Minns R J, Muckle D S, Donkin J E. The repair of osteochondral defects in osteoarthritic rabbit knees by the use of carbon fibre. *Biomaterials* 1982; **3**: 81–86.

305. Minns R J, Steven F S. The collagen fibril organization in human articular cartilage. *Journal of Anatomy* 1977; **123**: 437–57.

306. Mitchell J K. On a new practice in acute and chronic rheumatism. *American Journal of Medical Science*, 1831; **8**: 55–64.

307. Mitchell M L, Sokoloff L. A method for cytologic examination of cartilaginous lesions. *Archives of Pathology and Laboratory Medicine* 1987; **111**: 342–5.

308. Mitchell N, Shepard N. Pericellular proteoglycan concentrations in early degenerative arthritis. *Arthritis and Rheumatism* 1981; **24**: 958–64.

309. Mitrovic D, Garcia F, Guillermet V, Garcia G, Darmon N, Ryckewaert A. La réaction des tissus articulaires à la présence d'un corps étranger implanté dans le genom du lapin. *Comptes Rendus de l'Academie Scientifique de Paris*, 1982; **295**: 321–4.

310. Mohr W. *Gelenkkrankheiten. Diagnostik und Pathogenese makrokopischer und histologische Strukturveranderungen*. New York: George Thieme Verlag, 1984: 173–202.

311. Mosier H D, Dearden L C, Jansons R A, Hill R R. Cartilage sulfation during catch-up growth after fasting in rats. *Endocrinology* 1978; **102**: 386–92.

312. Moskowtiz R W. Cartilage and osteoarthritis: current concepts. *Journal of Rheumatology*, 1977; **4**: 329–31.

313. Moskowitz R W. Experimental models of osteoarthritis. In: Moskowitz R W, Howell D S, Goldberg V M, Mankin H J, eds,

Osteoarthrosis: Diagnosis and Management. Philadelphia: W B Saunders & Co., 1984: 109–28.

314. Moskowitz R W, Davis W, Sammaco J, Martens M, Baker J, Mayor M, Burstein A H, Frankel V H. Experimentally induced degenerative joint lesions following partial meniscectomy in the rabbit. *Arthritis and Rheumatism* 1973; **16**: 397–405.

315. Moskowitz R W, Goldberg V M. Studies of osteophyte pathogenesis in experimentally induced osteoarthritis. *Journal of Rheumatology* 1987; **14**: 311–20.

316. Moskowitz R W, Goldberg V M, Malemud C J. Metabolic responses of cartilage in experimentally induced osteoarthritis. *Annals of the Rheumatic Diseases* 1981; **40**: 584–92.

317. Moskowitz R W, Howell D S, Goldberg V M, Mankin H J, eds, *Osteoarthritis: Diagnosis and Management.* Philadelphia: W B Saunders & Co., 1984.

318. Moskowitz R W, Howell D S, Goldberg V M, Muniz D, Pita J C. Cartilage proteoglycan alterations in an experimentally induced model of rabbit osteoarthritis. *Arthritis and Rheumatism* 1979; **22**: 155–63.

319. Moskowitz R W, Kresina R F. Immunofluorescent analysis of experimental osteoarthritic cartilage and synovium—evidence for selective deposition of immunoglobulin and complement in cartilaginous tissues. *Journal of Rheumatology* 1986; **13**: 391–6.

320. Muir H. Molecular approach to the understanding of osteoarthrosis. *Annals of the Rheumatic Diseases* 1977; **36**: 199–208.

321. Muir H M. Biochemistry. In: Freeman M A R, ed, *Adult Articular Cartilage* (2nd edition). Tunbridge Wells: Pitman Medical, 1979: 145–214.

322. Muir H M. The chemistry of the ground substance of joint cartilage. In: Sokoloff L, ed, *The Joints and Synovial Fluid.* New York: Academic Press, 1980: 27–94.

323. Muir H M, Carney S L. Pathological and biochemical changes in cartilage and other tissues of the canine knee resulting from induced joint instability. In: Helminen H J, Kiviranta I, Säämänen A-M, Tammi M, Paukkonen K, Jurvelin J, eds, *Joint Loading. Biology and Health of Articular Structures.* Bristol: John Wright, 1987: 47–63.

324. Murray D G. Experimentally induced arthritis using intraarticular papain. *Arthritis and Rheumatism* 1964; **7**: 211–19.

325. Murray-Leslie C F, Lintoll D J, Wright V. The knee and ankles in sport and veteran military parachutists. *Annals of the Rheumatic Diseases* 1977; **36**: 327–31.

326. Myers E, Hardingham T E, Billingham M E J, Muir H. Changes in the tensile and compressive properties of articular cartilage in a canine model of osteoarthritis. *Transactions of the 32nd meeting of the Orthopaedic Research Society* 1986; **11**: 231.

327. Nahir A M, Glynn L E, Bitensky L, Chayen J. Decreased inhibition of proteolysis by osteoarthritic synovial fluids. *British Journal of Experimental Pathology* 1986; **67**: 453–9.

328. Nakano T, Thompson J R, Aherne F X. Cartilage proteoglycans from normal and osteochondrotic porcine joints. *Canadian Journal of Comparative Medicine* 1985; **49**: 219–26.

329. Nemeth-Csoka M, Meszaros T. Minor collagens in arthrotic human cartilage. Changes in content of 1-α, 2-α, 3-α and M-collagen with age and in osteoarthrosis. *Acta Orthopaedica Scandinavica* 1983; **54**: 613–9.

329a. Nesterov A I. The clinical course of Kashin–Beck disease. *Arthritis and Rheumatism* 1964; **7**: 29–40.

330. Nicholls S P, Gathercole L J, Keller A, Shah J S. Crimping in rat tail tendon collagen: morphology and transverse mechanical anisotropy. *International Journal of Biological Macromolecules* 1983; **5**: 283–8.

331. Nichols E H, Richardson F L. Arthritis deformans. *Journal of Medical Research* 1909; **21**: 149–222.

332. Niebauer G W, Wolf B, Bashey R I, Newton C D. Antibodies to canine collagen types I and II in dogs with spontaneous cruciate ligament rupture and osteoarthritis. *Arthritis and Rheumatism* 1987; **30**: 319–27.

333. O'Connor B L, Palmoski M J, Brandt K D. Neurogenic acceleration of degenerative joint lesions. *Journal of Bone and Joint Surgery* 1985; **67-A**: 562–72.

334. O'Connor B L, Woodbury P. The primary articular nerves to the dog knee. *Journal of Anatomy* 1982; **134**: 563–72.

335. O'Connor P, Bland C, Bjelle A, Gardner D L. Production of split patterns on the articular cartilage surfaces of rats. *Journal of Pathology* 1980; **130**: 15–21.

336. O'Connor P, Bland C, Gardner D L. Fine structure of artificial splits in femoral condylar cartilage of the rat: a scanning electron microscopic study. *Journal of Pathology* 1980; **132**: 169–79.

337. O'Connor P, Brereton J, Gardner D L. Hyaline articular cartilage dissected by papain: light- and scanning electron microscopy and micromechanical studies. *Annals of the Rheumatic Diseases* 1984; **43**: 320–6.

338. O'Connor P, Oates K, Gardner D L, Middleton J F S, Orford C R, Brereton J D. Low temperature and conventional scanning electron microscopic observations of dog femoral condylar cartilage surface after anterior cruciate ligament division. *Annals of the Rheumatic Diseases* 1985; **44**: 321–7.

339. O'Connor P, Orford C R, Gardner D L. Differential response to compressive loads of zones of canine hyaline articular cartilage: micromechanical, light- and electron microscopic studies. *Annals of the Rheumatic Diseases* 1988; **47**: 414–20.

340. Ogata K, Whiteside L A, Andersen D A. The intra-articular effect of various postoperative managements following knee ligament repair. An experimental study in dogs. *Clinical Orthopaedics and Related Research* 1980; **150**: 271–6.

341. Ohira T, Ishikawa K. Hydroxyapatite deposition in osteoarthritic articular cartilage of the proximal femoral head. *Arthritis and Rheumatism* 1987; **30**: 651–60.

342. Ohno O, Naito J, Iguchi T, Ishikawa H, Hirohata K, Cooke T D V. An electron microscopic study of early pathology in chondromalacia of the patella. *Journal of Bone and Joint Surgery* 1988; **70-A**: 883–99.

343. Olsewski J M, Lust G, Rendano V T, Summers B A. Degenerative joint disease: multiple joint involvement in young and mature dogs. *American Journal of Veterinary Research* 1983; **44**: 130–8.

344. Ono K. The microvasculature of dog knee joints in normal and experimental arthritis. *Journal of the Japanese Orthopaedic Association* 1978; **52**: 401–11.

345. Orford C R, Gardner D L. Proteoglycan association with collagen d band in hyaline articular cartilage. *Connective Tissue Research* 1984; **12**: 345–48.

346. Orford C R, Gardner D L. Ultrastructural histochemistry of the surface lamina of normal articular cartilage. *Histochemical Journal* 1985; **17**: 222–33.

347. Orford C R, Gardner D L, O'Connor P. Ultrastructural changes in dog femoral condylar cartilage following anterior cruciate ligament section. *Journal of Anatomy* 1983; **137**: 653–63.

348. Orford C R, Gardner D L, O'Connor P, Bates G, Swallow J J, Brito-Babapulle L A P. Ultrastructural alterations in glycosaminoglycans of dog femoral condylar cartilage after surgical division of an anterior cruciate ligament; a study with cupromeronic blue in a critical electrolyte concentration technique. *Journal of Anatomy* 1986; **148**: 233–44.

349. Palmoski M J, Brandt K D. Running inhibits the reversal of atrophic changes in canine knee cartilage after removal of a leg cast. *Arthritis and Rheumatism* 1981; **24**: 1329–37.

350. Palmoski M J, Brandt K D. Hyaluronate binding by proteoglycans: comparison of mildly and severely osteoarthritic regions of human femoral head cartilage. *Clinica Chimica Acta* 1981; **70**: 87–95.

351. Palmoski M J, Brandt K D. Proteoglycan aggregation in injured articular cartilage. A comparison of healing lacerated cartilage with osteoarthritic cartilage. *Journal of Rheumatology* 1982; **9**: 189–97.

352. Palmoski M J, Brandt K D. Aspirin aggravates the degeneration of canine joint cartilage caused by immobilization. *Arthritis and Rheumatism* 1982; **25**: 1333–42.

353. Palmoski M J, Brandt K D. Effects of static and cyclic compressive loading on articular cartilage plugs *in vitro*. *Arthritis and Rheumatism* 1984; **27**: 675–81.

353a. Palmoski M J, Brandt K D. Proteoglycan depletion, rather than fibrillation, determines the effects of salicylate and indomethacin on osteoarthritic cartilage. *Arthritis and Rheumatism* 1985; **28**: 548–53.

354. Palmoski M J, Colyer R A, Brandt K D. Marked suppression by salicylate of the augmented proteoglycan synthesis in osteoarthritic cartilage. *Arthritis and Rheumatism* 1980; **23**: 83–91.

355. Palmoski M J, Perricone E, Brandt K D. Development and reversal of a proteoglycan defect in normal canine knee cartilage after immobilisation. *Arthritis and Rheumatism* 1979; **22**: 508–17.

356. Palotie A, Vaisanen P, Ott J, Ryhanen L, Elima K, Vikkula M, Cheah K, Vuorio E, Peltonen L. Predisposition to familial osteoarthrosis linked to type II collagen gene. *Lancet* 1989; **i**: 924–7.

356a. Panush R S, Schmidt C, Caldwell J R, Edwards N L, Longley S, Yonker R, Webster E, Nauman J, Stork J, Pettersson H. Is running associated with degenerative joint disease? *Journal of the American Medical Association* 1986; **255**: 1152–4.

357. Paukkonen K, Jurvelin J, Helminen H J. Effects of immobilization on the articular cartilage in young rabbits. A quantitative light microscopic stereological study. *Clinical Orthopaedics and Related Research* 1986; **206**: 270–80.

358. Paukkonen K, Selkäinaho K, Jurvelin J, Kiviranta I, Helminen H J. Cells and nuclei of articular cartilage chondrocytes in young rabbits enlarged after non-strenuous physical exercise. *Journal of Anatomy* 1985; **142**: 13–20.

359. Pelet D, Reichen A. Post traumatic arthrosis of the ankle joint following calcaneal fractures. *Zeitschrift für Unfallmedizin und Berufskrankheiten* 1975; **68**: 44–46.

360. Pelletier J P, Martel-Pelletier J. Cartilage degradation by neutral proteoglycanases in experimental osteoarthritis—suppression by steroids. *Arthritis and Rheumatism* 1985; **28**: 1393–1401.

361. Pelletier J P, Martel-Pelletier J, Altman R D, Ghandhur-Mnaymneh L, Howell D S, Woessner J F Jr. Collagenolytic activity and collagen matrix breakdown of the articular cartilage in the Pond–Nuki dog model of osteoarthritis. *Arthritis and Rheumatism* 1983; **26**: 866–74.

362. Pelletier J P, Martel-Pelletier J, Howell D S, Ghandhur-Mnaymneh L, Enis J E, Woessner J F. Collagenase and collagenolytic activity in human osteoarthritic cartilage. *Arthritis and Rheumatism* 1983; **26**: 63–8.

363. Pelletier J-P, Martel-Pelletier J, Malemud C J. Proteoglycans from experimental osteoarthritic cartilage: degradation by neutral metalloproteases. *Journal of Rheumatology* 1987; **14** (Suppl. 14): 113–5.

364. Pidd J, Gardner D L, Adams M E. Ultrastructural changes in the femoral condylar cartilage of mature American foxhounds following transsection of the anterior cruciate ligament. *Journal of Rheumatology* 1988; **15**: 662–9.

365. Piper T L, Whiteside L A. Early mobilization after knee ligament repair in dogs. An experimental study. *Clinical Orthopaedics and Related Research* 1980; **150**: 277–82.

366. Pommer G. *Mikroskopische Untersuchungen über Gelenkgicht*. Jena: Fischer, 1929.

367. Pond M J, Nuki G. Experimentally induced osteoarthritis in the dog. *Annals of the Rheumatic Diseases* 1973; **32**: 387–8.

368. Pritzker K P H, Châtauvert J M D, Grynpas M D. Osteoarthritic cartilage contains increased calcium, magnesium and phosphorus. *Journal of Rheumatology* 1987; **14**: 806–10.

369. Pujol J-P, Loyau G. Interleukin-1 and osteoarthritis. *Life Sciences* 1987; **41**: 1187–98.

370. Punzi L, Schiavon F, Cavasin F, Ramonda R, Gambari P F, Todesco S. The influence of intra-articular hyaluronic acid on PGE_2 and cAMP of synovial fluid. *Clinical and Experimental Rheumatology* 1989; **7**: 247–50.

371. Radin E L, Ehrlich M G, Chernack R, Abernethy P, Paul I L, Rose R M. Effect of repetitive impulse loading on the knee joints of rabbits. *Clinical Orthopaedics and Related Research* 1978; **131**: 288–93.

372. Radin E L, Parker H G, Pugh J W, Steinberg R S, Paul I L, Rose R M. Response of joints to impact loading—III. Relationship between trabecular microfractures and cartilage degeneration. *Journal of Biomechanics* 1973; **6**: 51–7.

373. Radin E L, Paul I L. Response of joints to impact loading. I. *In vitro* wear. *Arthritis and Rheumatism* 1971; **14**: 356–62.

374. Radin E L, Rose R M. Role of subchondral bone in the initiation and progression of cartilage damage. *Clinical Orthopaedics and Related Research* 1986; **213**: 34–40.

375. Radin E L, Swann D A, Paul I L, McGrath P J. Factors influencing articular cartilage wear *in vitro*. *Arthritis and Rheumatism* 1982; **25**: 974–80.

376. Raju U B, Fine G, Partamian J O. Neuropathic neuroarthropathy (Charcot's joint). *Archives of Pathology and Laboratory Medicine* 1982; **106**: 349–51.

377. Raker C W, Baker R H, Wheaf J D. Pathophysiology of equine degenerative joint disease and lameness. *Proceedings of the American Association of Equine Practitioners* 1966; **12**: 229–52.

378. Rapin C H, Lagier R. Raised serum albumin in hip osteoarthrosis: a comparative study in women of some blood chemical parameters in ageing and in cases of femoral neck fractures, osteoporotic vertebral crush fractures, and hip osteoarthrosis. *Annals of the Rheumatic Diseases* 1988; **47**: 576–81.

379. Rashad S, Revell P, Hemingway A, Low F, Rainsford K, Walker F. Effect of non-steroidal anti-inflammatory drugs on the course of osteoarthritis. *Lancet* 1989; **ii**: 519–22.

380. Rataki A, Ruttner J R, Abt K. Age-related histochemical and histological changes in the knee-joint cartilage of C57B1 mice and their significance for the pathogenesis of osteoarthritis. *Experimental Cell Biology* 1980; **48**: 329–48.

381. Redfern P. *Anormal Nutrition in the Human Articular Cartilages, with Experimental Researches on the Lower Animals.* Edinburgh: Sutherland and Knox, 1850.

382. Redler I. A scanning electron microscopic study of human normal and osteoarthritic articular cartilage. *Clinical Orthopaedics and Related Research* 1974; **103**: 262–8.

383. Redler I, Zimny M L. Scanning electron microscopy of normal and abnormal articular cartilage and synovium. *Journal of Bone and Joint Surgery* 1970; **52-A**: 1395–1404.

384. Rees J A, Ali S Y. Ultrastructural localisation of alkaline phosphatase activity in osteoarthritic human articular cartilage. *Annals of the Rheumatic Diseases* 1988; **47**: 747–53.

385. Rees J A, Ali S Y, Brown R A. Ultrastructural localization of fibronectin in human osteoarthritic articular cartilage. *Annals of the Rheumatic Diseases* 1987; **46**: 816–22.

386. Reimann I. Pathological human synovial fluids, viscosity and boundary lubricating properties. *Clinical Orthopaedics and Related Research* 1976; **119**: 237–41.

387. Reimann I, Arnoldi C C, Nielsen O S. Permeability of synovial membrane to plasma proteins in human coxarthrosis. Relation to molecular size and histologic changes. *Clinical Orthopaedics and Related Research* 1980; **147**: 296–300.

388. Reimann I, Christensen S B. A histochemical study of alkaline and acid phosphatase activity in subchondral bone from osteoarthrotic human hips. *Clinical Orthopaedics and Related Research* 1979; **140**: 85–91.

389. Repo R U, Finlay J B. Survival of articular cartilage after controlled impact. *Journal of Bone and Joint Surgery* 1977; **59-A**: 1068–76.

390. Resnick D, Niwayama G. *Diagnosis of Bone and Joint Disorders* (2nd edition). Philadelphia, London: W B Saunders Company, 1988: 1365–479.

391. Revell P A, Mayston V, Lalor P, Mapp P. The synovial membrane in osteoarthritis: a histological study including the characterization of the cellular infiltrate present in inflammatory osteoarthritis using monoclonal antibodies. *Annals of the Rheumatic Diseases* 1988; **47**: 300–07.

392. Riede U N, Heitz Ph, Ruedi Th. Gelenkmechanische der Talusform auf die Biomechanik des oberen Sprunggelenkes. *Langenbecks Archiv für Chirurgie* 1971; **330**: 174–84.

393. Roberts S, Weightman B, Urban J, Chappell D. Mechanical and biochemical properties of human articular cartilage in osteoarthritic femoral heads and in autopsy specimens. *Journal of Bone and Joint Surgery* 1986; **68-B**: 278–88.

394. Rondier J, Cayla J, Guiraudon C, Charpentier Y Le. Arthropathie tabétique et chondrocalcinose articulaire. *Revue du Rhumatisme* 1977; **44**: 671–4.

395. Rosenberger J L, Cooper N S, Soren A, McEwen C. A statistical approach to the histopathologic diagnosis of synovitis. *Human Pathology* 1981; **12**: 329–37.

396. Rosner I A, Boja B A, Goldberg V M, Moskowitz R W. Tamoxifen therapy in experimental osteoarthritis. *Current Therapeutic Research* 1983; **34**: 409–14.

397. Roughley P J, White R J, Poole A R. Identification of a hyaluronic acid-binding protein that interferes with the preparation of high-buoyant density proteoglycan aggregates from adult human articular cartilage. *Biochemical Journal* 1985; **231**: 129–38.

398. Roy S. Ultrastructure of synovial membrane in osteoarthritis. *Annals of the Rheumatic Diseases* 1967; **26**: 517–27.

399. Ruffer M A. *Studies in the Palaeopathology of Egypt.* Chicago, Illinois: University of Chicago Press, 1921.

400. Rüttner J R, Spycher M A. Electron microscopic investigations on ageing and osteoarthrotic human cartilage. *Pathologia et Microbiologia* 1968; **31**: 14–24.

401. Ryckewaert A, Naveau B. Les affections ostéo-articulaire de J M Charcot à nos jours. *Revues du Neurologie* 1982; **138**: 997–1008.

402. Ryu J, Treadwell B V, Mankin H J. Biochemical and metabolic abnormalities in normal and osteoarthritic human articular cartilage. *Arthritis and Rheumatism* 1984; **27**: 49–57.

403. Säämänen A-M, Tammi M, Kiviranta I, Jurvelin J, Helminen H J. Maturation of proteoglycan matrix in articular cartilage under increased and decreased joint loading. A study in young rabbits. *Connective Tissue Research* 1987; **16**: 163–75.

404. Sachs B L, Goldberg V M, Getzy L L, Moskowitz R W, Malemud C J. A histopathologic differentiation of tissue types in human osteoarthrosis cartilage. *Journal of Rheumatology* 1982; **9**: 210–16.

405. Saito M, Saito S, Ohzono K, Ono K. The osteoblastic response to osteoarthritis of the hip. Its influence on the long-term results of arthroplasty. *Journal of Bone and Joint Surgery* 1987; **69-B**: 746–51.

405a. Sakkas L I, MacFarlane D G, Bird H, Welsh K I, Panayi G S. Association of osteoarthritis with homozygosity for a 5.8 Kb Taq I fragment of the alpha 1-antichymotrypsin gene. *British Journal of Rheumatology* 1990; **29**: 245–8.

406. Salter R B, Field P. The effects of continuous compression on living articular cartilage. An experimental investigation. *Journal of Bone and Joint Surgery* 1960; **42-A**: 31–49.

407. Sandy J D, Adams M E, Billingham M E J, Plaas A, Muir H. *In vivo* and *in vitro* stimulation of chondrocyte biosynthetic activity in early experimental osteoarthritis. *Arthritis and Rheumatism* 1984; **27**: 388–97.

408. Sapolsky A I, Altman R D, Howell D S. Cathepsin D activity in normal and osteoarthritic human cartilage. *Federation Proceedings* 1973; **32**: 1489–93.

409. Saville P D, Dixon J. Age and weight in osteoarthritis of the hip. *Arthritis and Rheumatism* 1968; **11**: 635–44.

410. Saxne T, Wollheim F A, Heinegård D, Pettersson H.

Difference in cartilage proteoglycan level in synovial fluid in early rheumatoid arthritis and reactive arthritis. *Lancet* 1985; **ii**: 127–8.

411. Scheck M, Sakovich L. Degenerative joint disease of the canine hip. Experimental production by multiple papain and prednisone injections. *Clinical Orthopaedics and Related Research* 1972; **86**: 115–20.

412. Schwartz E R. Metabolic response during early stages of surgically induced osteoarthritis in mature beagles. *Journal of Rheumatology* 1980; **7**: 788–800.

413. Schwartz E R, Grunwald R A. Experimental models of osteoarthritis. *Bulletin on the Rheumatic Diseases* 1979; **30**: 1030–3.

414. Schwartz E R, Oh W H, Leveille C R. Experimentally induced osteoarthritis in guinea-pigs. Metabolic responses in articular cartilage to developing pathology. *Arthritis and Rheumatism* 1981; **24**: 1345–55.

415. Scott J E, Orford C R. Dermatan sulphate-rich proteoglycan associates with rat tail tendon collagen at the **d** band in the gap region. *Biochemical Journal* 1981; **197**: 213–16.

416. Scott R B, Elmore SMcD, Brackett N C, Harris W O, Still W J S. Neuropathic joint disease (Charcot joints) in Waldenström's macroglobulinemia with amyloidosis. *American Journal of Medicine* 1973; **54**: 535–8.

417. Sedgwick A D, Moore A R, Sin Y M, Al-Duaij, Landon B, Willoughby D A. The effect of therapeutic agents on cartilage degradation *in vivo*. *Journal of Pharmaceutical Pharmacology* 1984; **36**: 709–10.

418. Selye H. Use of 'granuloma pouch' technique in the study of antiphlogistic corticoids. *Proceedings of the Society for Experimental Biology and Medicine* 1953; **82**: 328–33.

419. Shapiro F, Glimcher M J. Induction of osteoarthrosis in the rabbit knee joint. *Clinical Orthopaedics and Related Research* 1980; **147**: 287–95.

420. Shoji H, D'Ambrosia R D, Dabezies E J, Taddonio R F, Pendergrass J, Gristina A G. Articular cartilage and subchondral bone changes in an experimental osteoarthritis model. *Surgical Forum* 1978; **29**: 554–6.

421. Shotton D M. Confocal scanning optical microscopy and its applications for biological specimens. *Journal of Cell Science* 1989; **94**: 175–206.

422. Silberberg M, Frank E L, Jarrett S R, Silberberg R. Aging and osteoarthritis of the human sternoclavicular joint. *American Journal of Pathology* 1959; **35**: 851–65.

423. Silberberg M, Hasler M, Silberberg R. Articular cartilage of dwarf mice: submicroscopic effects of somatotrophin. *Pathologia et Microbiologia* 1966; **29**: 137–55.

424. Silberberg M, Silberberg R. Modifying action of estrogen on the evolution of osteoarthrosis in mice of different ages. *Endocrinology* 1963; **72**: 449–51.

425. Silberberg M, Silberberg R. Role of sex hormone in the pathogenesis of osteoarthrosis of mice. *Laboratory Investigation* 1963; **12**: 285–9.

426. Silberberg M, Silberberg R. Dyschondrogenesis and osteoarthrosis in mice. *Archives of Pathology* 1964; **77**: 519–24.

427. Silberberg M, Silberberg R, Hasler M. Ultrastructure of articular cartilage of mice treated with somatotrophin. *Journal of Bone and Joint Surgery* 1964; **46-A**: 766–80.

428. Silberberg M, Silberberg R, Hasler M. Fine structure of articular cartilage in mice receiving cortisone acetate. *Archives of Pathology* 1966; **82**: 569–82.

429. Silberberg R. Ultrastructure of articular cartilage in health and disease. *Clinical Orthopaedics and Related Research* 1968; **57**: 233–57.

430. Silberberg R, Gerritsen G, Hasler M. Articular cartilage of diabetic Chinese hamsters. *Archives of Pathology and Laboratory Medicine* 1976; **100**: 50–4.

431. Silberberg R, Hasler M. Response of articular cartilage of adult mice to administration of a cartilage-bone marrow extract. *Experimental Medicine and Surgery* 1968; **26**: 235–48.

432. Silberberg R, Silberberg M. Pathogenesis of osteoarthrosis. *Pathologica et Microbiologica* 1964; **27**: 447–57.

433. Silberberg R, Silberberg M, Feir D. Life cycle of articular cartilage cells: an electron microscope study of the hip joint of the mouse. *American Journal of Anatomy* 1964; **114**: 17–47.

434. Simon W H, Lane J M, Beller P. Pathogenesis of degenerative joint disease produced by *in vivo* freezing of rabbit articular cartilage. *Clinical Orthopaedics and Related Research* 1981; **155**: 259–68.

435. Simon W H, Richardson S, Herman W. Long-term effects of chondrocyte death on rabbit articular cartilage *in vivo*. *Journal of Bone and Joint Surgery* 1976; **58-A**: 517–26.

436. Simon W H, Wohl D L. Water content of equine articular cartilage: effects of enzymatic degradation and artificial fibrillation. *Connective Tissue Research* 1982; **9**: 227–32.

437. Slowman S D, Brandt K D. Composition and glycosaminoglycan metabolism of articular cartilage from habitually loaded and habitually unloaded sites. *Arthritis and Rheumatism* 1986; **29**: 88–94.

438. Sokoloff L. *The Biology of Degenerative Joint Disease*. Chicago: University of Chicago Press, 1969: 1–4.

439. Sokoloff L. Pathology and pathogenesis of osteoarthritis. In: McCarty D J, ed, *Arthritis and Allied Conditions* (9th edition). Philadelphia: Lea and Febiger, 1979: 1135.

440. Sokoloff L. Endemic forms of osteoarthritis. *Clinics in the Rheumatic Diseases* 1985; **11**: 187–202.

441. Sokoloff L, Crittenden L B, Yamamoto R S, Jay G E. The genetics of degenerative joint disease in mice. *Arthritis and Rheumatism* 1962; **5**: 531–45.

442. Sokoloff L, Micklesen O, Silverstein E, Jay G E Jr. Experimental obesity and osteoarthritis. *American Journal of Physiology* 1960; **198**: 765–70.

443. Sood S C. A study of the effects of experimental immobilisation on rabbit articular cartilage. *Journal of Anatomy* 1971; **108**: 497–507.

444. Soren A. *Histodiagnosis and clinical correlation of rheumatoid and other synovitis*. Philadelphia: J B Lippincott & Co., 1978: 116–22.

445. Soren A, Klein W, Hulth F. Microscopic comparison of the synovial changes in post-traumatic synovitis and osteoarthritis. *Clinical Orthopaedics and Related Research* 1976; **121**: 191–5.

445a. Spring M W, Buckland-Wright J C. Contrast medium imbibition in osteoarthritic cartilage. *British Journal of Radiology* 1990; **63**: 823–5.

446. Spycher M A, Moor H, Rüttner J R. Electron microscopic

investigations on ageing and osteoarthrotic human articular cartilage II. The fine structure of freeze-etched ageing hip joint cartilage. *Zeitschrift für Zellforschung und mikroskpische Anatomie* 1969; **98**: 512–24.

446a. Stanescu R, Stanescu V. *In vitro* protection of the articular surface by cross-linking agents. *Journal of Rheumatology* 1988; **15**: 1677–82.

447. Stecher R M. Osteoarthritis in the Gorilla. Description of a skeleton with involvement of the knee and the spine. *Laboratory Investigation* 1958; **7**: 445–57.

448. Steinetz B G, Colombo C, Butler M C, O'Byrne E, Steele R E. Animal models of osteoarthritis: possible applications in a drug development program. *Current Therapeutic Research* 1981; **20**: suppl. S60–S75.

449. Stockwell R A. Distribution of crystals in the superficial zone of elderly human articular cartilage of the femoral head in subcapital fracture. *Annals of the Rheumatic Diseases* 1990; **49**: 231–5.

450. Stockwell R A, Billingham M E. Early response of cartilage to abnormal factors as seen in the meniscus of the dog knee after cruciate ligament section. *Acta Biologica Hungarica* 1984; **35**: 281–91.

451. Stockwell R A, Billingham M E J, Muir H. Ultrastructural changes in articular cartilage after experimental section of the anterior cruciate ligament of the dog knee. *Journal of Anatomy* 1983; **136**: 425–39.

451a. Strachan R K, Smith P, Gardner D L. Hyaluronate in rheumatology and orthopaedics: Is there a role? *Annals of the Rheumatic Diseases* 1990; **49**: 949–52.

452. Swann D A. Structure and function of lubricin, the glycoprotein responsible for the boundary lubrication of articular cartilage. In: Franchimont P, ed, *Articular Synovium.* Basel: Karger, 1982: 45–58.

453. Tammi M, Kiviranta I, Peltonen L, Jurvelin J, Helminen H J. Effects of joint loading on articular cartilage collagen metabolism; assay of procollagen prolyl 4-hydroxylase and galactosylhydroxylysyl glucosyltransferase. *Connective Tissue Research* 1988; **17**: 199–206.

454. Tammi M, Paukkonen K, Kiviranta I, Jurvelin J, Säämänen A-M, Helminen H J. Joint loading induced alterations in articular cartilage. In: Helminen H J, Kiviranta I, Säämänen A-M, Tammi M, Paukkonen K, Jurvelin J, eds, *Joint Loading. Biology and Health of Articular Structures.* Bristol: John Wright, 1987: 64–88.

455. Tammi M, Säämänen A-M, Jauhininen A, Malminen O, Kiviranta I, Helminen H. Proteoglycan alterations in rabbit knee articular cartilage following physical excercise and immobilization. *Connective Tissue Research* 1983; **11**: 45–55.

456. Thaxter T H, Mann R A, Anderson C E. Degeneration of immobilized knee joints in rats. *Journal of Bone and Joint Surgery* 1965; **47-A**: 567–85.

457. Thomas L. Reversible collapse of rabbit ears after intravenous papain, and prevention of recovery by cortisone. *Journal of Experimental Medicine* 1956; **104**: 245–52.

458. Thompson R C. An experimental study of surface injury to articular cartilage and enzyme responses within the joint. *Clinical Orthopaedics and Related Research* 1975; **107**: 239–48.

459. Thompson R C, Bassett C A L. Histological observations on experimentally induced degeneration of articular cartilage. *Journal of Bone and Joint Surgery* 1970; **52-A**: 435–43.

460. Thonar E J, Lenz M E, Klintworth G K, Caterson B, Pachman L M, Glickman P, Katz R, Huff J, Kuettner K. Quantification of keratan sulphate in blood as a marker of cartilage catabolism. *Arthritis and Rheumatism* 1985; **28**: 1367–76.

461. Treadwell B V, Mankin D P, Ho P K, Mankin H J. Cell-free synthesis of cartilage proteins: partial identification of proteoglycan core and link proteins. *Biochemistry* 1980; **19**: 2269–75.

462. Troyer H. The effect of short-term immobilization on the rabbit knee joint cartilage. A histochemical study. *Clinical Orthopaedics and Related Research* 1975; **107**: 249–57.

463. Utsinger P D, Resnick D, Shapiro R F, Wiesner K B. Roentgenologic, immunologic and therapeutic study of erosive (inflammatory) osteoarthritis. *Archives of Internal Medicine* 1978; **138**: 693–7.

464. Vanharanta H, Kuusela T, Kiuru A. Early detection of developing osteoarthritis by scintigraphy: an experimental study on rabbits. *European Journal of Nuclear Medicine* 1984; **9**: 426–8.

465. Van Saase J C L M, Vandenbroucke J P, van Rommunde L K J, Valkenburg H A. Osteoarthritis and obesity in the general population. A relationship calling for an explanation. *Journal of Rheumatology* 1988; **15**: 1152.

466. Van Sickle D C. Experimental models of osteoarthritis. In: Scarpelli D G, Migaki G, eds, *Comparative Pathobiology of Major Age-Related Diseases: Current Status and Research Frontiers.* New York: Alan Liss, 1984: 175–88.

467. Van Sickle D C, Kincaid S A. Comparative arthrology. In: Sokoloff L, ed, *The Joints and Synovial Fluid* Volume 1. New York: Academic Press, 1978: 1–47.

468. Vasan N. Proteoglycans in normal and severely osteoarthritic human cartilage. *Biochemical Journal* 1980; **187**: 781–7.

469. Venn M, Maroudas A. Chemical composition and swelling of normal and osteoarthrotic femoral head cartilage. I. Chemical composition. *Annals of the Rheumatic Diseases* 1977; **36**: 121–9.

470. Videman T. Experimental osteoarthritis in the rabbit. Composition of different periods of repeated immobilization. *Acta Orthopaedica Scandinavica* 1982; **53**: 339–47.

471. Videman T, Eronen I, Candolin T. ^{3}H-Proline incorporation and hydroxyproline concentration in articular cartilage during the development of osteoarthritis caused by immobilisation. A study *in vivo* with rabbits. *Biochemical Journal,* 1981; **200**: 435–40.

472. Videman T, Eronen I, Friman C. Glycosaminoglycan metabolism in experimental osteoarthritis caused by immobilisation. *Acta Orthopaedica Scandinavica* 1981; **52**: 11–21.

473. Videman T, Eronen I, Friman C, Langenskiöld A. Glycosaminoglycan metabolism of the medial meniscus, the medial collateral ligament and the hip joint capsule in experimental osteoarthritis caused by immobilisation of the rabbit knee. *Acta Orthopaedica Scandinavica* 1979; **50**: 465–470.

474. Vignon E, Arlot M, Hartman D, Moyen B, Ville G. Hypertrophic repair of articular cartilage in experimental osteoarthrosis. *Annals of the Rheumatic Diseases* 1983; **42**: 82–8.

475. Vignon E, Bejui J, Hartman D J, Ville G, Vial B, Mattieu P. Études quantitative des lesions arthrosiques experimentelles chez le lapin. Interet pour l'etude des medications anti-

arthrosiques. *Revue du Rhumatisme et des Maladies Osteoarticulaire (Paris)* 1986; **53**: 649–52.

476. Vignon E, Hartman J D, Vignon G, Moyen B, Arlot M, Ville G. Cartilage destruction in experimentally induced osteoarthritis. *Journal of Rheumatology* 1984; **11**: 202–7.

476a. Vignon E, Vignon G. Méthodes d'evaluation des anti-inflammatoires non-steroidiens dans la polyarthrite rhumatoide. *Semaine des Hopitaux de Paris* 1983; **59**: 3180–2.

477. Virchow R. Zur Geschichte der Arthritis deformans. *Virchows Archiv für pathologisch Anatomie und Physiologie, und für klinische Medizin.* 1869; **47**: 298–303.

478. Wagenhauser F J, Amira A, Borrachero J, Brummer L, Clausen C, Winer J. Die Behandlung der Arthrosen mit Knorpel-Knochenmark-Extrakt. Ergenbnisse eines Multi-Centre Trials. *Schweizerische Medizinische Wochenschrift* 1986; **98**: 904–7.

479. Walton M. Degenerative joint disease in the mouse knee; radiological and morphological observations. *Journal of Pathology* 1977a; **123**: 97–107.

480. Walton M. Degenerative joint disease in the mouse knee; histological observations. *Journal of Pathology* 1977b; **123**: 109–22.

481. Walton M. Studies of degenerative joint disease in the mouse knee; scanning electron microscopy. *Journal of Pathology* 1977c; **123**: 211–17.

482. Walton M. Patella displacement and osteoarthrosis of the knee joint in mice. *Journal of Pathology* 1979; **127**: 165–72.

483. Walton M, Elves M W. Bone thickening in osteoarthrosis. Observations of an osteoarthrosis-prone strain of mouse. *Acta Orthopaedica Scandinavica* 1979; **50**: 501–6.

484. Weichselbaum A. Die senilen Veranderungen der Gelenke und deren Zusammenhang mit der Arthritis deformans. *Sitzungs Bericht der Akademie Wissenschaftliche Wien (mathematischenaturwissenschaftliche Klasse)* 1887; **75**: 193.

485. Weightman B, Kempson G E. Load carriage. In: Freeman, M A R, ed, *Adult Articular Cartilage* (2nd edition). Tunbridge Wells: Pitman Medical, 1979: 291–331.

486. Weiss C. Light and electron microscopic studies of osteoarthritic cartilage. In: Simon W H, ed, *The Human Joint in Health and Disease.* Philadelphia: University of Pennsylvania Press, 1978: 112–21.

487. Weiss C. Normal and osteoarthritic articular cartilage. *Orthopaedic Clinics of North America* 1979; **10**: 175–89.

488. Weiss C, Mirow S. An ultrastructural study of osteoarthritic changes in the articular cartilage of human knees. *Journal of Bone and Joint Surgery*, 1972; **54-A**: 954–72.

488a. Westacott C I, Whicher J T, Barnes I C, Thompson D, Swan A J, Dieppe P A. Synovial fluid concentrations of five different cytokines in rheumatic diseases. *Annals of the Rheumatic Diseases* 1990; **49**: 676–81.

489. Westermark T W. Consequences of low selenium intake for man. In: Bratter P, Schramel P, eds, *Trace Elements—Analytical Chemistry in Medicine and Biology* Volume 3. New York: Walter de Gruyter, 1984: 49–70.

490. Wiebkin O W, Muir H. The inhibition of sulphate incorporation in isolated adult chondrocytes by hyaluronic acid. *FEBS Letters* 1973; **37**: 42–6.

491. Wiebkin O W, Muir H. Influence of the cells on the pericellular environment. *Philosophical Transactions of the Royal Society of London B* 1975; **271**: 283–91.

492. Wilhelmi G, Faust R. Suitability of the C57 black mouse as an experimental animal for the study of skeletal changes due to ageing, with special reference to osteo-arthrosis and its response to tribenoside. *Pharmacology* 1976; **14**: 289–96.

493. Williams J M, Brandt K D. Temporary immobilization facilitates repair of chemically induced articular cartilage injury. *Journal of Anatomy* 1984; **138**: 435–46.

494. Williams J M, Brandt K D. Triamcinolone hexacetonide protects against fibrillation and osteophyte formation following chemically induced articular cartilage damage. *Arthritis and Rheumatism* 1985; **28**: 1267–74.

495. Williams J M, Katz R J, Childs D, Lenz M E, Thonar E J-M A. Keratan sulfate content in the superficial and deep layers of osteophytic and nonfibrillated human articular cartilage in osteoarthritis. *Calcified Tissue International* 1988; **42**: 162–6.

496. Wiltberger H, Lust G. Ultrastructure of canine articular cartilage: comparison of normal and degenerative (osteoarthritic) hip joints. *American Journal of Veterinary Research* 1975; **36**: 727–40.

497. Witter J, Roughley P J, Webber C, Roberts N, Keystone E, Poole A R. The immunologic detection and characterization of cartilage proteoglycan degradation products in synovial fluids of patients with arthritis. *Arthritis and Rheumatism* 1987; **30**: 519–29.

498. Wong SYP, Evans R A, Needs C, Dunstan C R, Hills E, Garvan J. The pathogenesis of osteoarthritis of the hip. Evidence for primary osteocyte death. *Clinical Orthopaedics and Related Research* 1987; **214**: 305–12.

498a. Wright V. Posttraumatic osteoarthritis – a medico-legal minefield. *British Journal of Rheumatology* 1990; **29**: 474–8.

499. Wurster N B, Lust G. Fibronectin in osteoarthritic canine articular cartilage. *Biochemical and Biophysical Research Communications* 1982; **109**: 1094–1101.

500. Yamazaki J. Experimental study on the development of aseptic necrosis of the femoral head – with a comparison of osteoarthritis of the hip in terms of collagen metabolism. *Hokkaido Igako Zasshi* 1985; **60**: 544–54.

501. Yovich J V, Trotter G W, McIlwraith C W, Norrdin R W. Effects of polysulphated glycosaminoglycan on chemical and physical defects in equine articular cartilage. *American Journal of Veterinary Research* 1987; **48**: 1407–14.

502. Zelander T. Ultrastructure of articular cartilage. *Zeitschrift für Zellforschung* 1959; **49**: 720.

503. Zimny M L and Redler J. An ultrastructural study of patellar chondromalacia in humans. *Journal of Bone and Joint Surgery* 1969; **51-A**: 1179–1190.

Chapter 23

Intervertebral Joint Disease

Introduction

For developmental, structural and functional reasons, the vertebral column occupies an unique place in pathology.[9a, 16a, 37a] On the one hand, the calcified and non-calcified connective tissues of the spine are susceptible to almost all those diseases which attack the hard- and soft-tissues of other parts.[6,50] On the other hand, the spine is uniquely affected by the inflammatory spondyloarthropathies (see Chapter 19) and by mechanical disorders such as intervertebral disc protrusion and prolapse. Because of their inseparably close relationships, the vertebral bones and non-calcified connective tissues are very often disorganized simultaneously by metastatic tumour deposits,[42] by bacterial infection, by trauma[79] and by acquired or heritable malformation. In the present Chapter, in which neoplastic disease[48] is not reviewed, the osseous components of the spine[37] are considered in so far as is necessary for a proper discussion of the non-mineralized tissues. Partly because of incomplete knowledge, and partly for convenience, much more attention is directed to the fibrocartilagenous intervertebral discs than to the spinal ligaments and tendons and their origins and insertions.

Historical aspects

Understanding of the pathology of the intervertebral joints is modern. Mixter and Barr[65] recognized the important contribution made by the intervertebral discs to the origin of 'lumbago' (intractable lower back pain) and 'sciatica' (intractable pain in the distribution of the sciatic nerve). It was realized that intervertebral discs could, wholly or in part, be displaced without major trauma. Displacement could precipitate nerve or nerve root pain, with or without motor disturbances or paraplegia, depending on the spinal segments(s) affected. Early reports came independently from Middleton and Teacher[64] in Glasgow, and from Goldthwait[36] in Boston. Virchow[92] had already described the anatomical finding of disc fibrocartilage herniation and Luschka[62] had recognized the posterior protrusion of disc tissue. Interest in the pathology of the intervertebral discs was stimulated by the monumental work of Schmorl and Junghans.[75,76] A corresponding growth in interest in clinical aspects of intervertebral disc lesions followed.[22] The majority of the disorders attributable to intervertebral disc abnormality were of the lower limbs. Disturbances of the lumbar discs received the greatest attention. More recently, it began to be appreciated that cervical intervertebral joints were also susceptible to inflammatory, mechanical and degenerative disease.[11,12]

The non-calcified connective tissues of the spine are much less accessible to pathological study than those of the limbs, and they have attracted correspondingly fewer investigations. Nevertheless, there is a considerable, classical literature to which detailed reference is made by Collins,[18] Gardner,[34] Jaffe,[49] and Bywaters.[14] There are many accounts of the pathology of spinal disease in older encyclopedias such as those of Henke and Lubarsch[42] and numerous examples of advanced forms of disorders of the axial skeleton are to be seen in the osteoarticular museums of New York, Zurich, Geneva, and the Royal Colleges of Surgeons of England and Edinburgh.

The joints of the spine are susceptible to different classes of disease the prevalence of which is closely related to the structure and function of the two categories of spinal articulation. However, each spinal segment functions as an integrated unit so that, for example, osteoarthrosis of the posterolateral (apophyseal-'facet') synovial joints is a factor in disorganization of the main fibrocartilagenous joints just as degenerative disease of the latter, with anatomical malposition and impaired movement, exacerbates secondary osteoarthrosis of the former.

Synovial joints of the spine

Each of the superior articular facets of the atlas articulates with a condyle of the occipital bone of the skull at a synovial joint. In the same way, the vertebral arches are apposed at a series of posterolateral ('facet') apophyseal joints which adjoin the intervertebral foramina and which, like the costovertebral joints, have a freely moving 'diarthrodial' function and spaces lined by a synovial membrane. The sacroliliac joints (see p. 133) are both synovial and cartilagenous.

Cartilagenous joints of the spine

The seven cervical, twelve thoracic, and five lumbar vertebrae articulate with each other and with the sacrum at a series of 23 fibrocartilagenous joints ('symphyses', see p. 126). By definition these articulations have no synovia; the fibrocartilages have no direct blood supply.

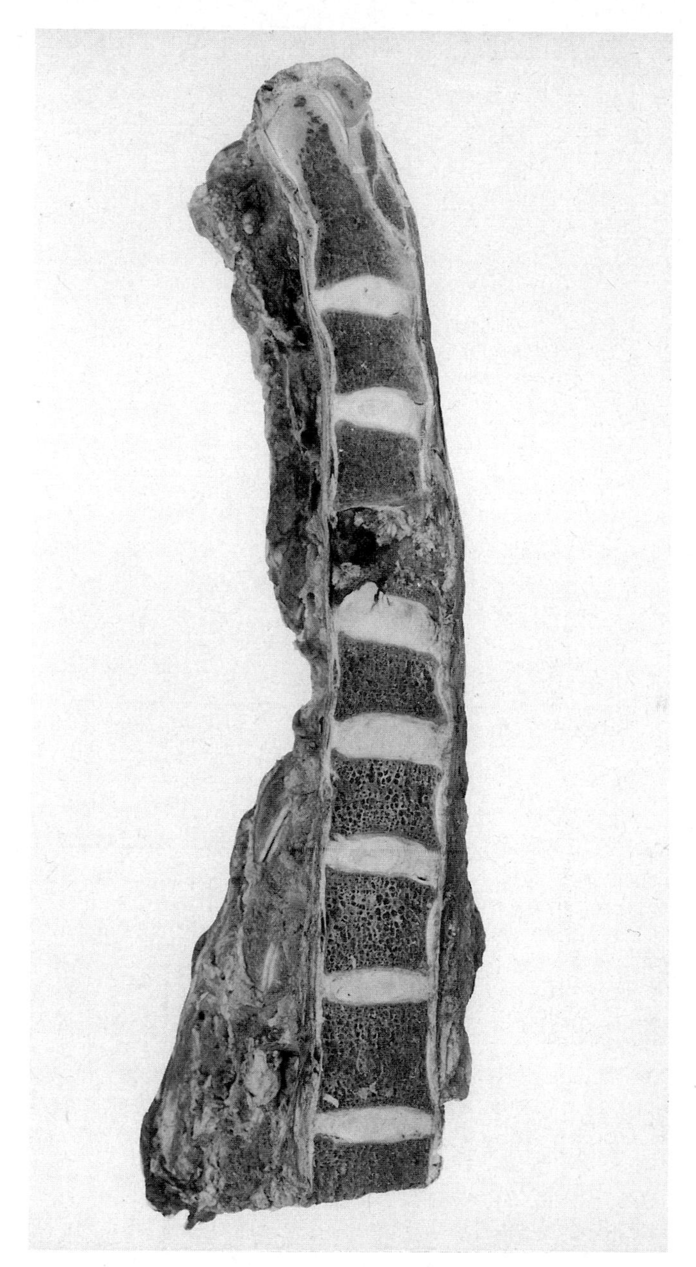

Fig. 23.1 Traumatic disintegration of intervertebral joint.
Falls in which the patient lands on the feet are often complicated by vertebral fracture. In this case, disruption of the body of C5 is accompanied by disintegration of the C4/C5 intervertebral disc. (× 0.5.)

Categories of spinal joint disease

For these reasons, the disorders of the non-calcified connective tissues of the spine can usefully be considered in

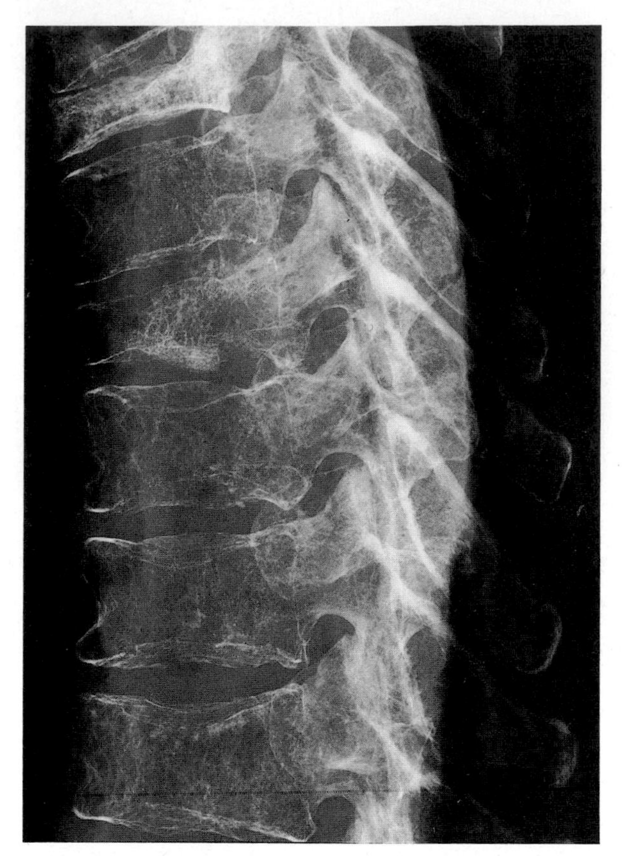

Fig. 23.2 Osteomalacia, vertebral collapse, osteophytosis and intervertebral joint disorganization.
An extremely severe, prolonged defect in bone mineralization complicated by collapse of vertebral bodies has indirectly contributed to the abnormal radiographic structure of a number of the principal, fibrocartilagenous, intervertebral joints.

Fig. 23.3 Selective erosion of vertebral bone.
Classically, selective erosion of the bone of the anterior surfaces of the thoracic vertebrae was attributable to aortic aneurysm.[50a] In the present instance, the cause was gastric ulcer of a thoracic stomach. As the radiograph shows, the non-mineralized intervertebral discs had undergone earlier age-related degeneration and disc collapse is associated with anterior angling of the disc space and osteophytic lipping of the vertebral margins.

two categories. The spinal synovial joints are the target for any or all of the hereditary and acquired, physical, chemical, immunological, inflammatory, endocrine, metabolic and neoplastic disorders to which the synovial joints of the appendicular skeleton are heir.[19] The pathological changes affecting the main fibrocartilagenous articulations are distinctive (Figs 23.1–23.3). The disease processes to which they are prone have much in common with prevalent disorders of the discs of the temporomandibular, sternoclavicular, inferior radioulnar and knee joints. However, the intervertebral fibrocartilagenous discs are complex structures; the pathological processes affecting them are more varied than those causing disorganization of the other, smaller discs.

There are three qualifications to these important generalizations. First, the close anatomical and functional relationships between the main fibrocartilagenous intervertebral- and the synovial apophyseal joints of, for example, the cervical spine, determine that the disorders of one *secondarily* influence the behaviour and structure of the other (Fig. 23.4).[27,28] Thus, cervical spondylosis is a compound series of abnormalities of both classes of articulation. Second, a particular set of adventitious neurocentral (Luschka) synovial joints may arise *within* the margins of the cervical intervertebral fibrocartilages.[40] Under these circumstances, the normally avascular tissues of the cervical intervertebral discs may be destroyed directly by inflammatory diseases such as rheumatoid arthritis, disorders which, by definition, are confined to vascular articular tissues. Third, there exists a class of disease, described in Chapter 19, of which ankylosing spondylitis is an example, in which the disease process affects *primarily* both synovial and fibrocartilagenous articulations.

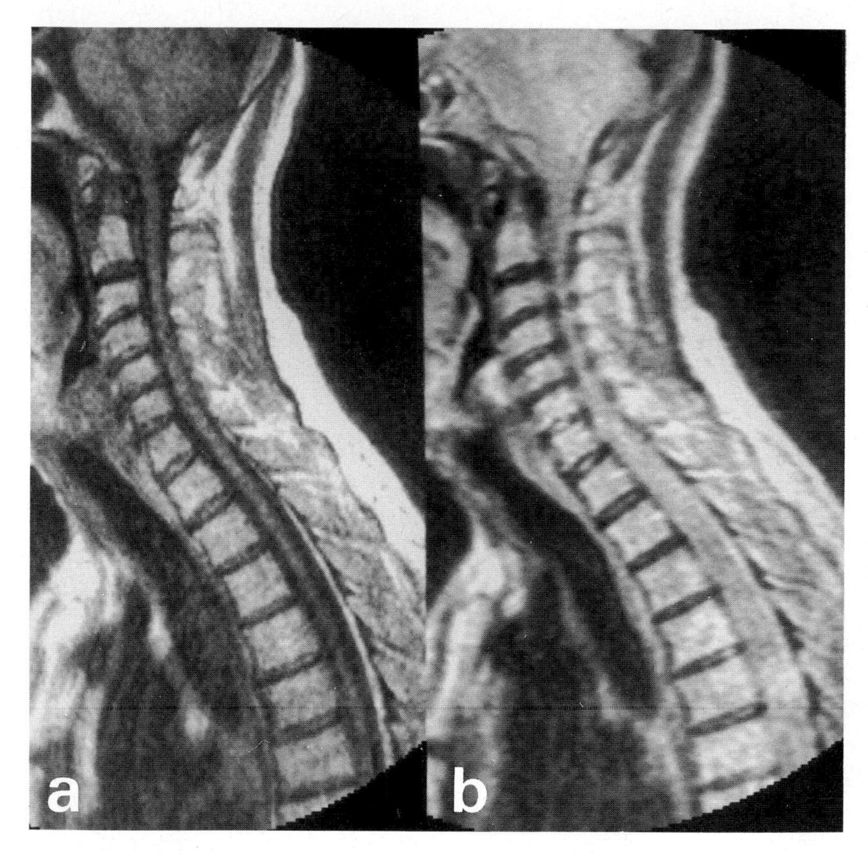

Fig. 23.4 Magnetic resonance imaging and spinal disease.
Magnetic resonance imaging (MRI) has revolutionized the recognition and interpretation of diseases of the intervertebral and apophyseal joints. In this instance, cervical spondylosis is demonstrated on a sagittal MRI scan in **a.** T_1-weighted and in **b.** T_2-weighted sequences. In **a**, the spinal cord, brain stem and cerebellum are well demonstrated by the low signal from the surrounding cerebrospinal fluid. In **b**, there is a high signal from the cerebrospinal fluid, demonstrating severe narrowing of the cervical part of the spinal canal. Reproduced by courtesy of the Department of Diagnostic Radiology, University of Manchester.

Heritable disease

Mucopolysaccharidoses

The most severe examples of mucopolysaccharidosis (p. 359) lead to disease of the axial skeleton. In the course of the Hürler syndrome (Table 4.5, p. 204) glycosaminoglycan accumulates in the connective tissues of the vertebral column. Spinal deformity ensues. Analogous abnormalities can be caused experimentally. Thus, malformation can be induced by treating mice with fluorodeoxycytidine.[54] Vertebral malformations include fusion (50 per cent), dysplasia, cleft, aplasia and hypoplasia. The form and degree of malformation correlate with observed changes in the amounts and nature of the vertebral column glycosaminoglycans.

I-cell disease

I-cell disease (see Chapter 4) is attributed to the diminished activity of UDP-N-acetylglucosamine: glycoprotein-N-acetylglucosaminylphosphotransferase. There is ovoid deformity of the vertebral bodies and anteroinferior beaking of the lower thoracic and lumbar vertebrae. The intervertebral discs may undergo degeneration and dystrophic calcification.[66]

Ochronosis

The heritable disorder and biochemical abnormalities encountered in alkaptonuria are described on pp. 351 *et seq.* As one result of a defect in the degradation of tyrosine, attributable to an inherited fault in the mechanism of action of homogenistic acid oxidase, homogenistic acid accumulates in many parts of the body, bound as an insoluble oxidation product to fibrous, fibrocartilagenous, hyaline cartilagenous and other connective tissues.

The intervertebral discs are particularly vulnerable. The accumulation of homogenistic acid is symptomless until adult life when its presence catalyses the degeneration and splitting of the intervertebral disc fibrocartilages. It is believed[14] that the accumulation of calcium hydroxyapatite follows these changes. Discal calcification is widespread and often extreme: parts of the degenerate, calcified discs may be protruded peripherally. The apophyseal joints are affected less severely,[59] partly, perhaps, because pigment accumulation is deep within the specialized hyaline articular cartilage (see p. 68). Secondary effects on spinal function are also exerted by the deposition of pigment within tendons and by the accumulation of connective tissue debris upon the surface of the principal spinal ligaments. Here, a low-grade, aseptic inflammatory reaction may be provoked, affecting both the ligaments and nearby joints.

Epiphyseal dysplasia

The spine is susceptible to a variety of malformations.[83a]

Spondyloepiphyseal dysplasia is one example of a series of heritable disorders (p. 332) the consequences of which create severe disturbances of spinal structure and function. Morquio's disease is another. The majority of forms of spondyloepiphyseal dysplasia find clinical expression in childhood; a number are recognized in adult life. Jaffe[49] discusses in detail the skeletal disease of (multiple) epiphysial dysplasia. Many parts of the body are affected but the vertebrae undergo a reduction in height. There are consequential effects on the growth and contours of the spine.[14]

Chondrodystrophia fetalis (achondroplasia)

The pathological features of the spine are outlined on p. 365 *et seq.*

Chemical, metabolic and nutritional disease

Crystal deposition

Dystrophic calcification of the intervertebral discs is a frequent and easily identified phenomenon.[29a] It may reach extreme forms in disorders such as alkaptonuria (see p. 351) but it is commonplace in degenerate discs in the absence of generalized metabolic disorder. The deposits are of dicalcium phosphate dihydrate.[17] However, it is now recognized that crystalline calcium pyrophosphate also frequently accumulates in intervertebral disc tissue (Fig. 23.2).[63] Similar deposits may complicate haemochromatosis.[16] Severe disc degeneration may follow.[80] The demonstration of calcium pyrophosphate crystals in biopsy material from the operation of discectomy may offer the first diagnostic indication of calcium pyrophosphate deposition disease[24] (see p. 393), of which primary hyperparathyroidism is one treatable cause.[7] Calcium pyrophosphate crystals may also aggregate at sites where repetitive intervertebral disc surgery has been performed.[24]

Amyloidosis

Amyloid has been recognized in articular hyaline cartilage with increasing frequency (see Chapter 11). In recent years the susceptibility of the fibrocartilagenous joints to amyloidosis has become evident. In an investigation of surgical biopsy material from 100 consecutive operations, Ladefoged, Fedders and Petersen[57] identified amyloid in 41. Tested with potassium permanganate, the amyloid retained its stainability with Congo red and its green dichroism; it was not amyloid A. It was also noted that crystalline calcium phosphate was present in the disc tissues in 26 instances. There was no topographical relationship to the amyloid deposits. By contrast, in six of eight cases where calcium pyrophosphate dihydrate crystals were found, the deposits were close to amyloid aggregates.

The cause of fibrocartilagenous amyloid accumulations is not known. It may be related to age. The suggestion has emerged that the amyloid is of a senile (A_S) variety (see p. 426). Takeda, Sanada *et al.*[83] investigated surgically removed intervertebral disc specimens from 84 patients. Fourty-four had potassium permanganate-resistant amyloid within the fibrocartilage. Statistical analysis suggested strongly that the presence of amyloid was an age-related phenomenon; it did not, of course, show that the fibrocartilagenous amyloidosis was due to senescence.

Lathyrism

Experimental lathyrism, caused by the administration of β-aminopropionitrile (see p. 413), is complicated by the onset of skeletal as well as cardiovascular abnormalities. Severe spinal defects ensue. They are described on p. 414.

Acromegaly

Whereas in pituitary gigantism the vertebral bodies and intervertebral discs undergo an appropriate and proportionate (but excessive) growth, in acromegaly there is a continuous growth of the cartilagenous endplates. A disproportionate increase in intervertebral disc size occurs and zones of ectopic calcification appear at the chondroosseous junctions adjoining sites of attachment of the anterior intervertebral ligament. There is a superficial resemblance to ankylosing hyperostosis. The vertebral bodies are excessively long and kyphosis and/or lordosis may ensue.

Degenerative disease

Intervertebral disc degeneration

With advancing age, the fibrocartilages of the human intervertebral discs undergo molecular changes that are the programmed consequences of cell senescence (Figs 23.5 and 23.6).[21,72] Analogous changes occur in other species.[2] There are alterations in the identity, distribution and organization of the collagens and in the content and composition of the proteoglycans (see p. 193).[77] Type III collagen may coexist with type I and II (see p. 126).[1,96] The water content falls and the physical characteristics of the disc tissue deteriorate.[46] A brown pigment accumulates. As a result of the interactions between these age-related phenomena and the diurnal loads to which each disc is subjected,[45a] the collagen fibre bundles begin to separate and/or to fracture. Clefts or cracks appear in three principal situations. In the lumbar vertebrae they are commonplace from early middle-age.

A first category of cleft is recognized, in sagittal sections of the spine, midway between the hyaline cartilagenous vertebral endplates and the centre of the disc.[89] The clefts extend. The upper and lower components tend to fuse, isolating the central part of the disc. Posterior and posterolateral extensions of the clefts are more common than anterior. The splits may reach as far as the posterior edge of the annulus fibrosus where neovascularization may be provoked. In Vernon-Roberts' view, the generation of new blood vessels at this margin supports the concept that extension is by physical tearing rather than by endogenous, proteolytic degradation.[88] However, increased understanding of angiogenic factors (see p. 295) suggests that they may play a part in this process.

A second category of cleft is circumferential (annular). These fissures, also, are more frequent posterolaterally than anteriorly. As the collagen fibre bundles separate, the planes of separation assume a radial direction. Like the central, planar clefts that lie parallel to the vertebral bone endplates, the circumferential clefts create a microenviron-

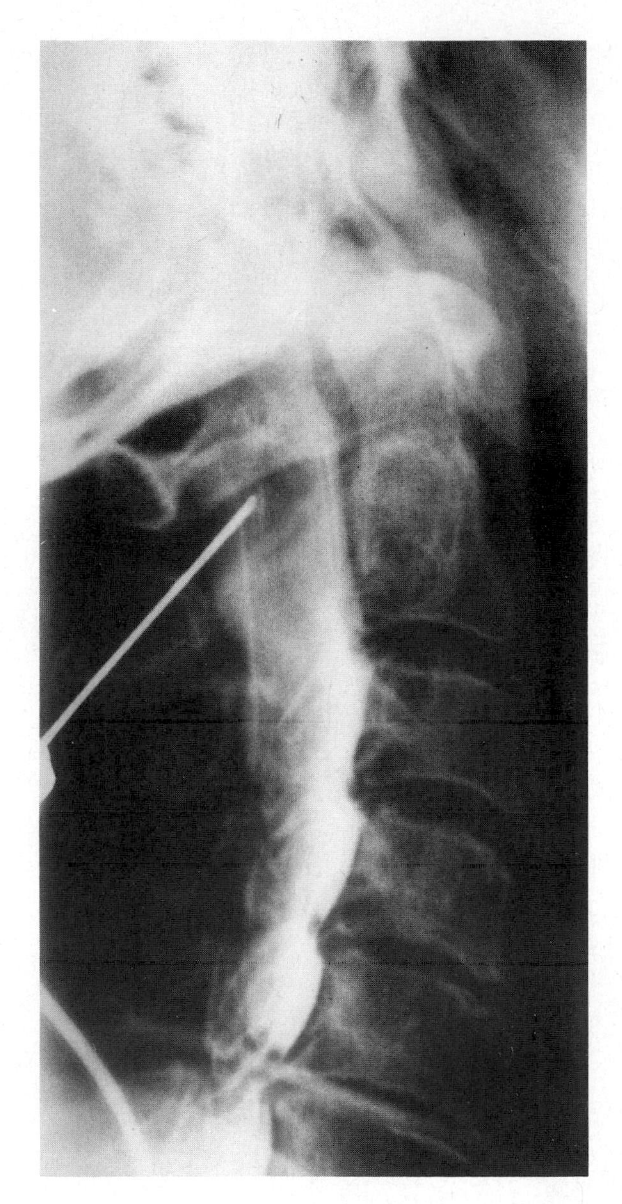

Fig. 23.5 Cervical and intervertebral disc disease: cervical myelogram.
Fifty-seven-year-old Caucasian female with a history of pain in the right shoulder and elbow of one year's duration. Cervical myelogram using water-soluble contrast medium. Lateral view, with patient in prone position. The spinal needle lies between neural arch of C1–2. The spinal cord is outlined by contrast medium in subarachnoid space. Note anterior indentation of column of contrast medium at C2–3, C3–4, C4–5 and, most severely, at C5–6. The reduced depth of the disc spaces at C4–5 and C5–6 is associated with marginal osteophyte formation and irregular disc margins. The use of myelography is associated with recognized hazards and, in Western countries, its place is increasingly taken by methods such as magnetic resonance imaging. Nevertheless, the images obtained by myelography facilitate understanding of the pathological anatomy of intervertebral disc disease.

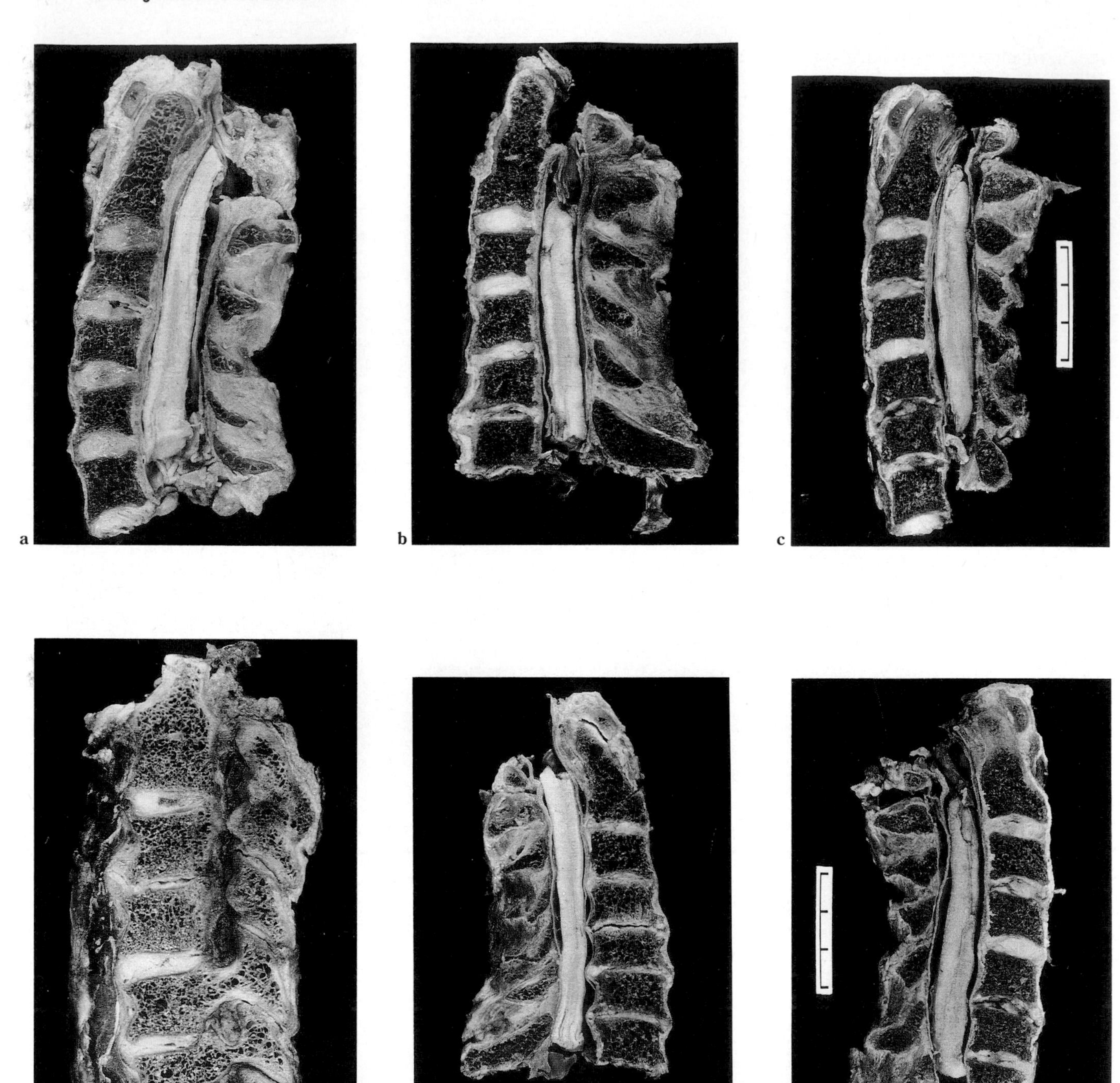

Fig. 23.6 Intervertebral disc degeneration.
Sagittal sections of cervical spines examined *post mortem*, indicating different sites of intervertebral disc disease and varied frequency of involvement of cervical discs. **a.** In this patient, the C3–4 disc has disintegrated. **b.** Here, the C5–6 intervertebral disc is degenerate; the change is associated with anterior osteophyte formation. **c.** In this example, the C3–4, C5–6 and C6–7 discs display various degrees of degeneration. **d.** Each of the cervical intervertebral discs seen in this block is abnormal; the greatest loss of intervertebral space is at the site of the C3–4 disc. **e.** There is severe disruption of the C4–5 intervertebral disc and the remaining cervical discs show only slightly less disorganization. **f.** Here, the principal abnormalities are seen at C3–4, C5–6 and C6–7.

ment in which compressive and shear stresses promote the protrusion (herniation) or prolapse of disc tissue posterolaterally or posteriorly (see p. 938).[45b]

In a systematic study of 117 specimens, Hilton, Ball and Benn[45] analysed the distribution and severity of anterior and posterior annular tears of the intervertebral discs of the lower thoracic and lumbar spine. The number of tears became greater with age. There was no significant difference between the frequency of tears in males and females. The presence of anterior and posterior disc tears at L4 and L5 did not indicate that there were tears at other spinal levels. In individuals aged 50 years or more, the number of anterior tears was greatest in the L2 and L1 discs, the number of posterior tears greatest in the L4 and L5 discs.

A third category of (microscopic) disc cleft develops close to the parts at which the annulus fibrosus is attached to the body of the vertebral bone.[43,76,91] The direction of these small clefts ('rim lesions') is at right angles to the annular collagen fibre bundles. The formation of new blood vessels may be stimulated here as at the posterior margins of the disc. Rim lesions[43] are apparently more common at the superior rather than at the inferior margins of the lumbar discs. There is an adjoining, concomitant osteosclerosis and an osteophytosis (see p. 866) that tends to be unilateral. In subjects aged more than 50 years, at least one vertebral rim is invariably prejudiced.[43]

The degradation and disorganization of intervertebral disc tissue are more frequent and more severe when the discs are also the site of vertical (Schmorl's node—see p. 936) or posterolateral prolapse.[89] This is to be expected since disc degeneration is the most important known factor predisposing to protrusion or prolapse. However, Vernon-Roberts also draws attention to the anomalous observation that single or multiple Schmorl's nodes can be recognized in a proportion of adolescent or young adult spines in which neither degeneration of disc tissue nor narrowing of intervertebral disc spaces is detected. In this category of individual, genetic predisposition, perhaps expressed as a fault in collagen synthesis or assembly;[5,26] defective bone structure or exposure to excessive mechanical stress (see Chapter 6) are among the explanations for disc prolapse. Where a discontinuity of the hyaline cartilage/vertebral bone endplate has been created, the opportunity is present for blood vessels to extend from the bone and bone marrow into the fibrocartilage of the disc. Vascularization creates the opportunity for calcification or even ossification. Foci of ossification may indeed come to exist within the disc tissue.

A secondary consequence of age-related dehydration of intervertebral disc collagen fibre bundles is a shrinkage of this tissue. The lateral dimensions cannot alter: fibrocartilage and bone are closely secured, so that a reduction takes place in the vertical height of the discs. As the processes of cleft formation, protrusion and prolapse (see p. 935) advance, the diurnal compressive loads sustained by the vertebral column culminate in an additional, lateral displacement ('bulging') of the disc, a sequence which helps to explain the paradox that dehydration and centrifugal protrusion can coincide.

The postural consequences for the spine, of serial intervertebral disc vertical shrinkage, are considered on p. 943. The opposing surfaces of the posterolateral facet joints are synchronously displaced so that the superior and inferior margins of the anterior facet override those of the posterior facet. One result is a reduction in the circumference of the closely adjacent intervertebral foramina with constriction of the nerve root canal. Another consequence may be enhanced osteosclerosis of the vertebral bone endplate.

Apophyseal joint osteoarthrosis (spondyloarthrosis)

The distinction of secondary from idiopathic osteoarthrosis is explained in Chapter 22. In the case of the apophyseal joints of the spine, there is good reason to suppose that both categories of disease are commonplace, although the histological features of the two classes of disorder merge microscopically. A resemblance to chondromalacia is not unexpected.[23]

Secondary osteoarthrosis of one or (more usually) several or many apophyseal joints (Figs 23.7–23.10) may result from congenital deformity; from acquired physical or mechanical disorders, including kyphosis and scoliosis; from major or repetitive minor injury; from infection; and from aseptic inflammatory disease, particularly rheumatoid arthritis. However, the greatest interest lies in the contribution to apophyseal joints osteoarthrosis made by intervertebral disc degeneration and protrusion or prolaspe.

The main fibrocartilagenous intervertebral joints and the posterolateral (apophyseal) synovial joints are so closely placed anatomically and so intimately related functionally, that significant disorganization of the former inevitably results in abnormality of the latter. Altered movement, increased or decreased load-bearing, immobilization, occupational or sports injury are among the factors known to cause osteoarthrosis of the appendicular synovial joints. They are likely to provoke apophyseal joint disease. But the same environmental factors exert considerable influences on intervertebral fibrocartilagenous joint function and the interrelationships of the disease processes affecting the two classes of articulation are evident.

Vernon-Roberts and Pirie[91] and Vernon-Roberts[89] believe that apophyseal joint osteoarthrosis exhibits macroscopic and microscopic differences from the classical features of osteoarthrosis recognized in large weight-bearing

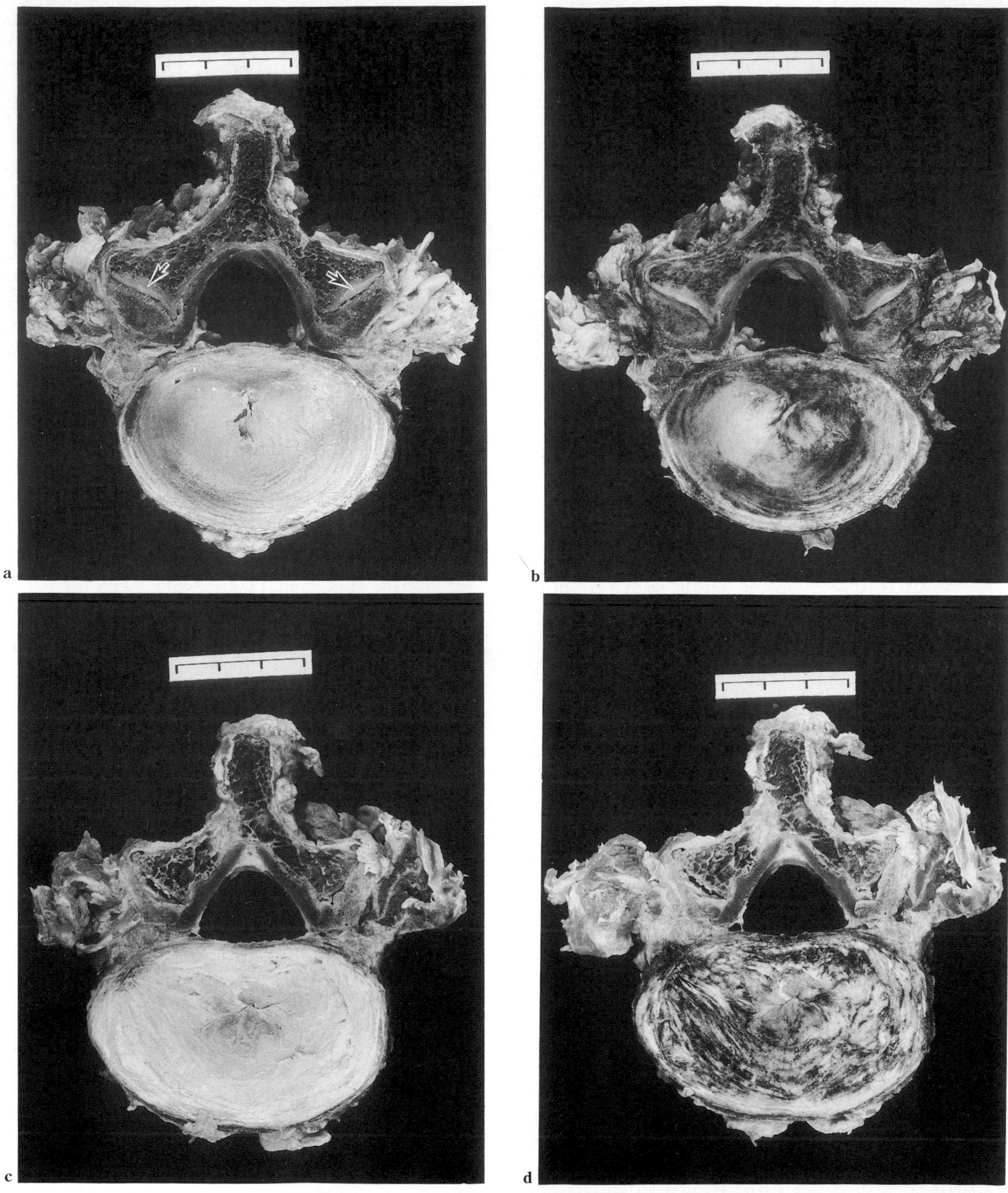

Fig. 23.7 Lumbar intervertebral disc degeneration and apophyseal joint osteoarthrosis.
Horizonal preparations. **a.** The site of the nucleus pulposus is identified near the centre of the disc. The remainder of the disc appears intact. Observe the thick hyaline cartilage of the posterior parts of the apophyseal joints. **b.** The same preparation after painting with dilute Indian ink. **c.** In this disc, posterolateral fissures extend from the disc centre towards the periphery. There is severe osteoarthrosis of the apophyseal joints. **d.** The same preparation after painting with dilute Indian ink. (× 0.75.)

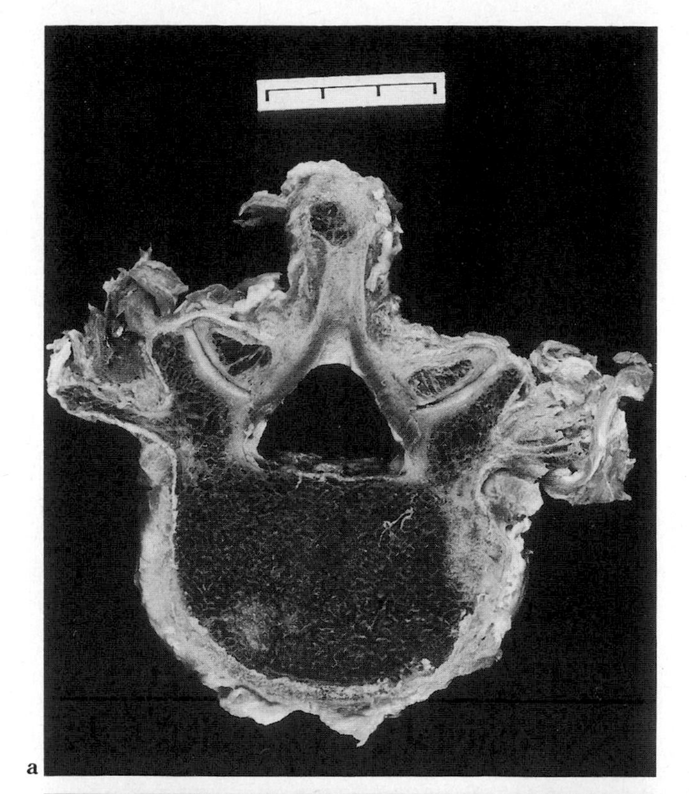

a

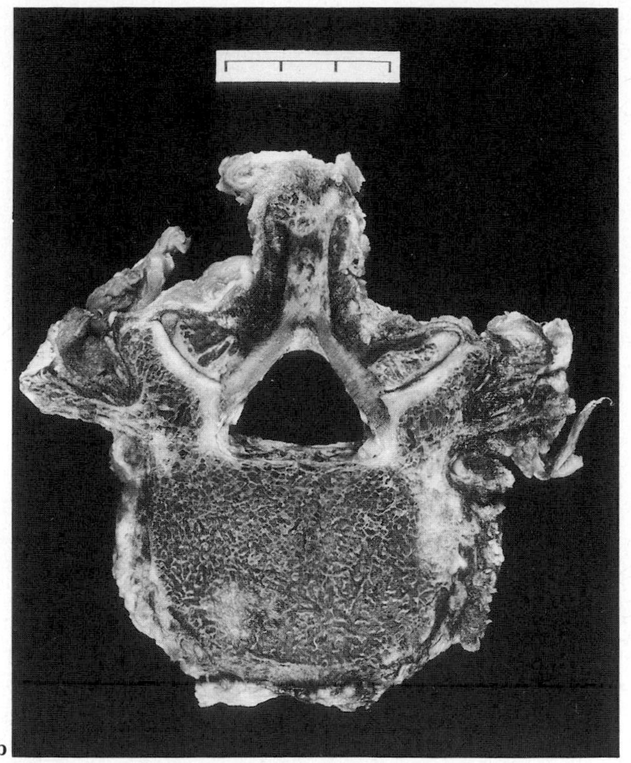

b

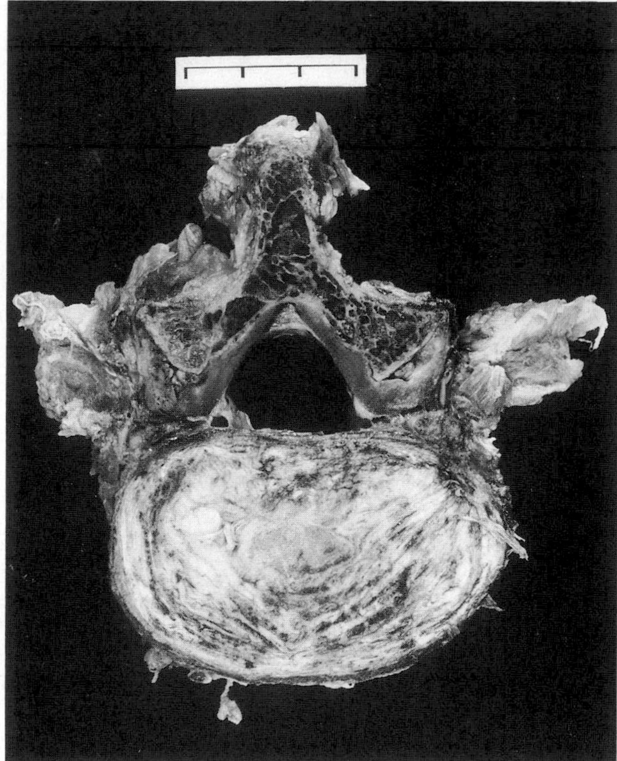

c

Fig. 23.8 Lumbar apophyseal joint osteoarthrosis. Horizontal preparations. Compare with Fig. 23.7. **a.** This lumbar vertebra has been divided at the level of the pedicles. The facet (apophyseal) joints are displayed. Note the loss of hyaline cartilage and dense subchondral bone of the anterior bearing surface. **b.** Similar preparation, after painting with dilute Indian ink. The facet joints are highlighted. **c.** Comparable sections made at the level of the intervertebral discs. There is some disc degeneration but the principal abnormality is the osteoarthrosic disintegration of the facet joints.

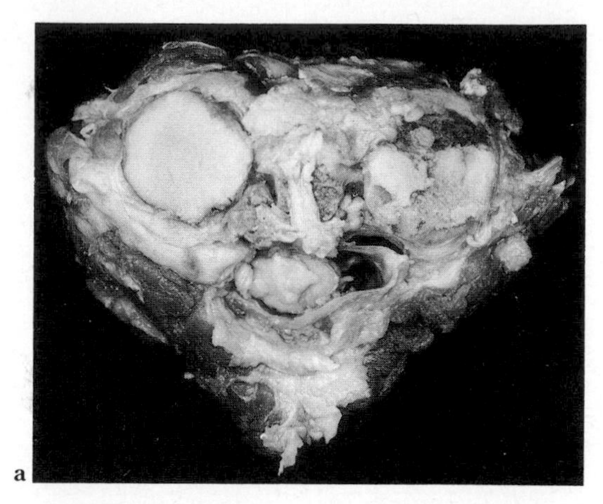

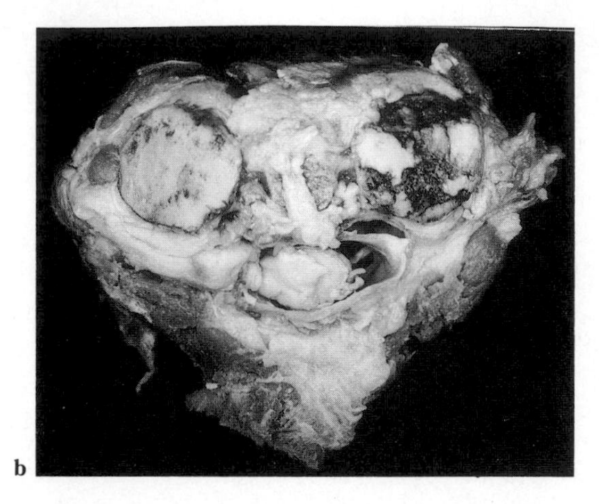

Fig. 23.9 Atlanto-occipital joint osteoarthrosis.
Changes analogous to those commonly seen in the apophyseal (facet) joints are often recognized in the synovial atlanto-occipital articulations. **a.** Hyaline cartilage of left bearing surface shows only circumferential fibrillation. The right surface is severely disorganized and has lost at least three-quarters of its cartilage. Bone is exposed. **b.** The same preparation after Indian ink painting. Enhanced contrast enables location of abnormal areas to be easily identified. (× 0.70.)

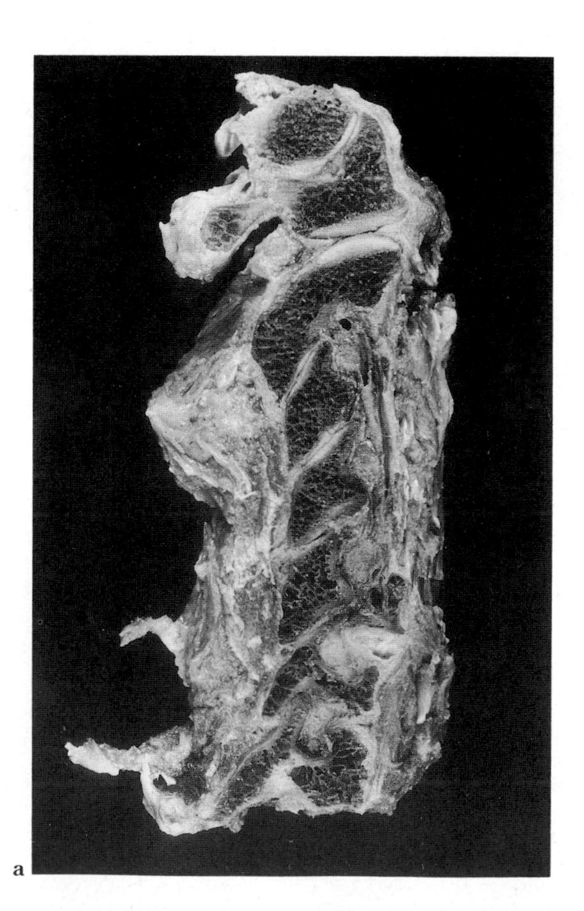

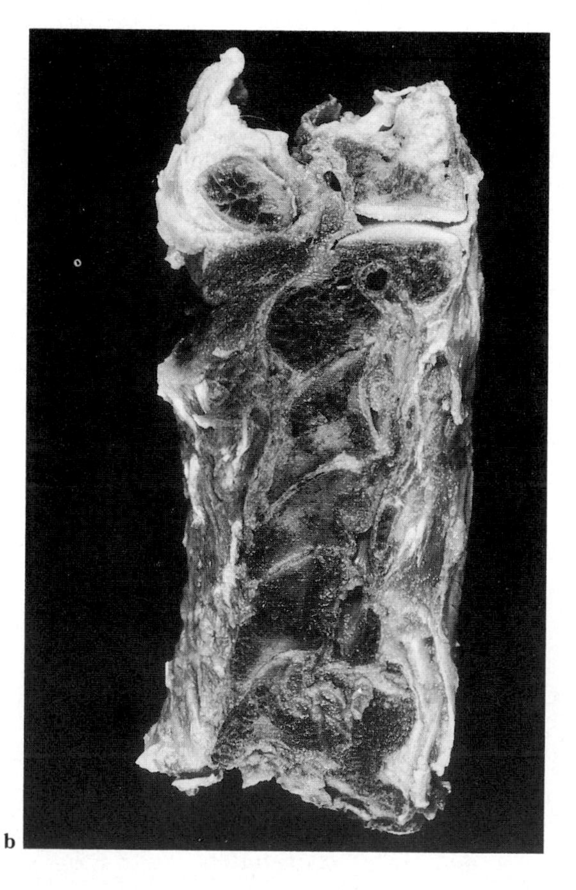

Fig. 23.10 Cervical spine apophyseal (facet) joint osteoarthrosis.
a. Paramedian, sagittal slab block. Different joints show varying degrees of cartilage loss and interosseous space narrowing. **b.** Compare with Fig. 23.10a. The extent of cartilage loss is, on average, greater.

joints such as the hip and knee. The hyaline cartilage surfaces of the vertebral facets may be retained largely intact even when osteophytes (see p. 943) have formed and subchondral osteosclerosis has developed. They concede, however, that the characteristic macroscopic and microscopic features of the classical disorder may also be present.

There are at least two possible explanations for these phenomena. First, it is important to note the specialized structure of the vertebral facets. The hyaline cartilage which covers their bearing surfaces includes a radially orientated array of collagen fibre bundles quite different from the collagenous microskeleton of, say, the femoral condyle. The differences observed by Vernon-Roberts and Pirie may be no more than a reflection of this specialized structure. Second, the apophyseal joints of many older persons may be the object simultaneously of idiopathic osteoarthrosis and of changes secondary to intervertebral disc degeneration. Under these circumstances, the resultant sequence of anatomical and histological change is likely to be distinct from the pattern attributable to, say, idiopathic osteoarthrosis alone. Apophyseal joint structures, like those of the intervertebral fibrocartilagenous joints, are of course, subject to age-related disorder (see p. 826).

Kashin–Beck disease

The pathological features of Kashin–Beck disease (epidemic endochondral dysostosis) are summarized on p. 904 *et seq.* Spinal abnormalities develop.[82]

Mechanical disorder

Direct trauma

Traumatic injuries of the fibrocartilagenous joints of the spine may be severe or trivial. Inevitably, major injury, discussed in texts on orthopaedic surgery,[47] disrupts the synovial apophyseal joints as well as the large articulations. The effects of trauma on synovial joints are considered further in Chapters 22 and 24. Minor mechanical disruption of the fibrocartilagenous joints is mentioned on p. 84.

The anatomical organization of the principal intervertebral joints, with fibrocartilagenous discs bound to nearby vertebral bone by extremely strong collagenous links, constitutes an articulation that is disrupted only in most unusual circumstances. Generally, when spinal fracture occurs in, for example, the victims of earthquakes, the vertebral bones themselves are broken rather than the fibrocartilaginous unions. Likewise, in compression fractures sustained by falls from great heights (Fig. 3.5), bone rather than fibrocartilage is impacted.

Indirect and passive injury: intervertebral disc protrusion and prolapse
(Figs 3.53 and 3.58)

In spite of their strength and the resistance of the concentric, circumferential and oblique collagen fibre bundles to both compressive, tensile and shear stresses, the nuclei pulposus of the intervertebral discs, particularly those of the large lumbar articulations, are commonly displaced upon or into adjacent tissues (Fig. 23.11).[3] Extraforaminal disc herniation may simulate retroperitoneal neoplasia.[22a]

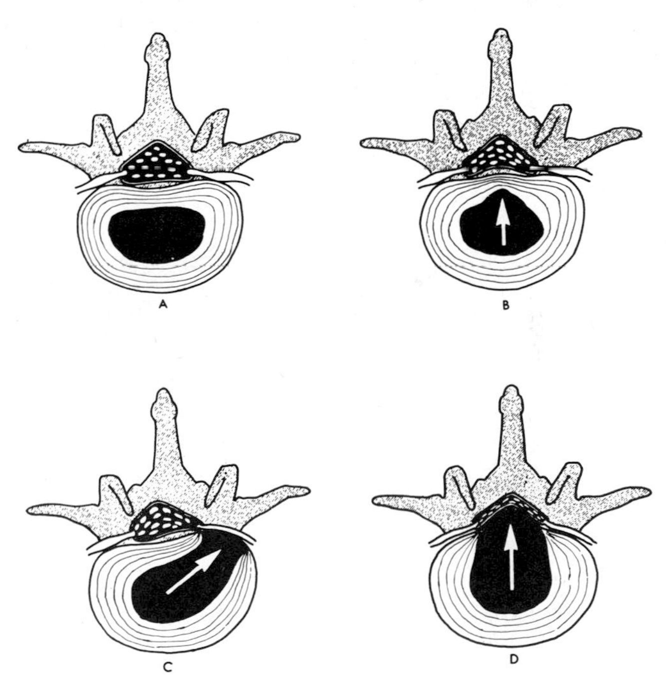

Fig. 23.11 Diagram of posterolateral and posterior protrusion and prolapse of lumbar intervertebral disc nuclei pulposus.
a. Normal relationships between annulus fibrosus, nucleus pulposus and spinal canal containing cauda equina. **b.** Posterior protrusion is slight; it is resisted by strong intervertebral ligaments. The effects are proportional to the degree of compression of nerve roots in the cauda equina. **c.** Posterolateral prolapse has taken place at focus of weakness of annulus. The nucleus is seen to compress a nerve root within the intervertebral foramen. The clinical signs are those of compression of the affected segment. **d.** In exceptional instances, the prolapse is central and the effects severe.

There are a number of reasons for displacement. First, loads imposed on otherwise normal intervertebral joints may be very high. Typically, loads sufficient to displace disc tissue are sustained during the sudden lifting of heavy weights in the flexed posture.[68] Second, for developmental and structural reasons, not all parts of each disc are equally strong. The posterior part of the annulus, and particularly the posterolateral chord, is attached to adjacent bone less firmly than the remainder; it is thinner than the rest of the disc and weaker. Displacement may occur in childhood[51,52,93] and is often recurrent.[25] Third, the integrity of the fibrocartilagenous disc and of the plate of hyaline cartilage which unites the disc with the nearby vertebral bone, depends upon the density, trabecular volume and adequacy of mineralization of this bone. Any sequence, such as osteoporosis, which prejudices bone strength inevitably increases the probability of intraosseous intervertebral disc displacement, even in response to normal, diurnal loads. Fourth, active support to the intervertebral joints is given by skeletal muscular activity, passive support by the strong spinal ligaments. Disuse and disease often progress to muscle atrophy and ligamentous laxity, diminishing the adequacy of the active, muscular protection afforded to the intervertebral joints during sudden, reflex movements.

Anterior displacement

The anterior and anterolateral parts of the lumbar intervertebral discs often protrude but seldom prolapse or rupture. The normal lumbar lordosis results in forces which are more severe on posterior parts of the lumbar discs; but postural change culminates in kyphosis, exaggerating compressive stresses on the anterior parts. The projection of these forces is facilitated by tilting of the vertebrae at the apophyseal joints and by rotation about the axis of the strong anterior longitudinal spinal ligament.[18,34] The nuclei pulposus are developmentally, chemically, histologically and mechanically distinct from the tissues of the annuli fibrosus. There is a myxoid structure, a higher water content and greater elasticity. Consequently, compressive and shear forces exerted upon the intervertebral discs are likely to cause displacement of the less deformable nuclear material into and through the more deformable and frequently open-textured annular tissue.

Experimental models of intervertebral disc degeneration have provided material for chemical and physical analysis. In the rabbit, anterior protrusion of the nucleus pulposus can be caused by a surgical incision into the disc.[61] An unexplained consequence of prolapse of material from the nuclei pulposus is embolism of the small blood vessels of the spinal cord, causing ischaemic myelopathy. There is evidence, from studies of mink, that such emboli may enter the pulmonary circulation.[39]

Vertical displacement (herniation; protrusion; prolapse

Displacement (herniation) of the nuclei of the intervertebral discs is commonly vertical (Fig. 23.12), much less often posterolateral, and rarely, posterior. In a random series of mature, adult cases studied *post mortem*, as many as 50 per cent of the vertebrae contained displaced nuclear tissue. Evidence of posterolateral displacement could be recognized in one instance in 10.

The occasional observations of earlier observers were extended and confirmed by Schmorl and Junghans.[75,76] Examining a series of 1000 spines *post mortem*, Schmorl recognized that portions of the intervertebral discs, often corresponding to the central or paracentral location of the nuclei, protruded more or less completely into the adjacent vertebral bone in 38 per cent of the specimens examined (Schmorl's nodes). Protrusion was more frequent in males (39.9 per cent) than females (34.3 per cent).[89] However, these figures are now believed to be underestimates. In anatomical studies of the thoracolumbar spine, Hilton, Ball and Benn[45] identified signs of nuclear protrusion in 76 per cent of cases, with a frequency that was approximately the same before and after the age of 50 years. It appeared that nuclear protrusion stemmed from a period of life when the myxoid tissue of the nucleus still retained a high water content and had not yet become collagenized. However, the high prevalence of nuclear protrusion in younger age groups also raised the question of heritable predisposition, perhaps in the form of genetically determined differences in discal collagens and/or proteoglycans. X-ray imaging is a less sensitive method for the detection of Schmorl's nuclear protrusions than dissection. The advent of magnetic resonance imaging has greatly increased the sensitivity with which protruded or prolapsed nuclei can be identified (see Chapter 3) and the frequency of clinically asymptomatic nuclear protrusion may be even higher than the figures obtained by Ball and his colleagues.

The development of Schmorl's nodes is not a new phenomenon. Comparing a thirteenth to sixteenth century Aberdeen population with a London population of the eighteenth to nineteenth centuries, Saluja *et al.*[74a] found 35.3 per cent of the former (35.9 per cent of males, 33.6 per cent of females) to show residual evidence of Schmorl's node formation compared with 11.4 per cent of the latter (18.4 per cent male, 1.8 per cent female). The reason for the unexpectedly low proportion of affected eighteenth to nineteenth century London individuals could not be explained by age.

Initially, nuclear discal tissue that is protruded vertically is recognized as a small island or 'node' of grey-white tissue, in continuity with the residual (lumbar) intervertebral disc but extending into the adjacent bone of the vertebral body.

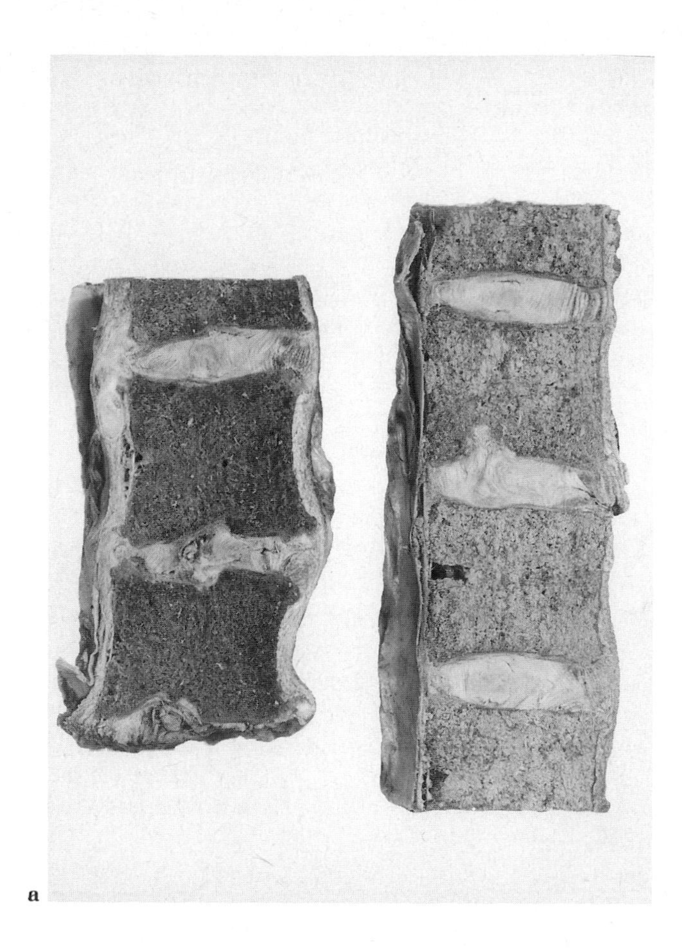

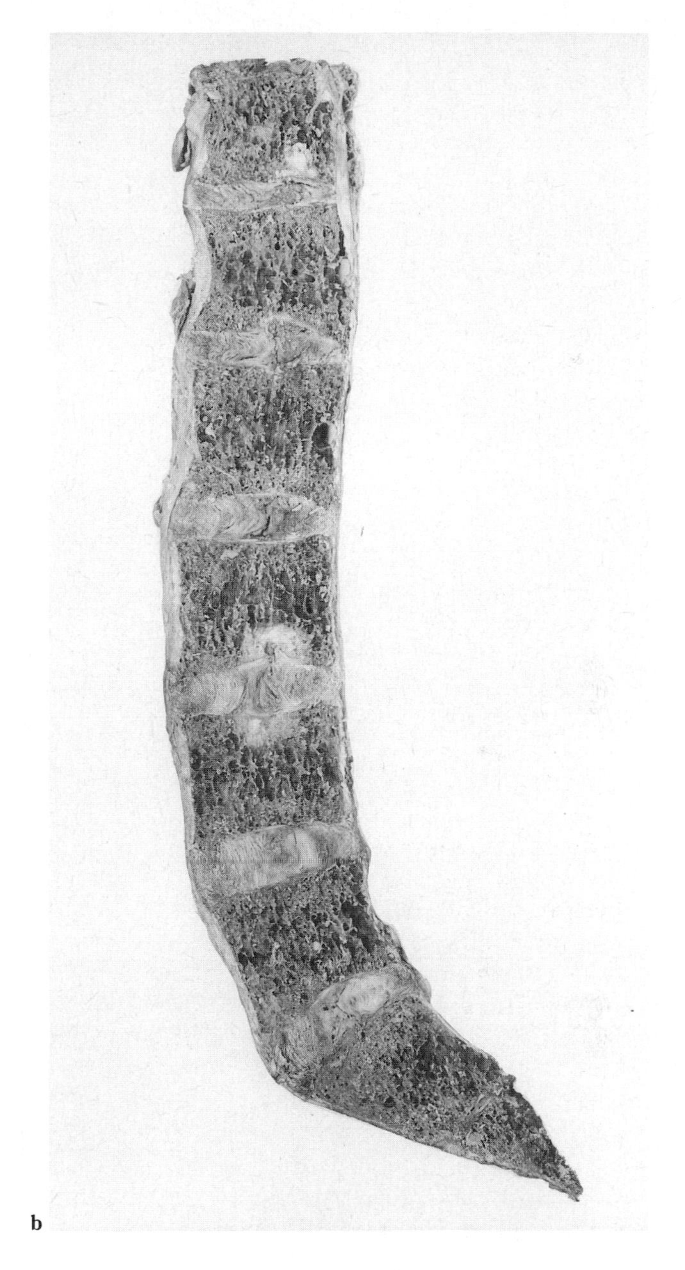

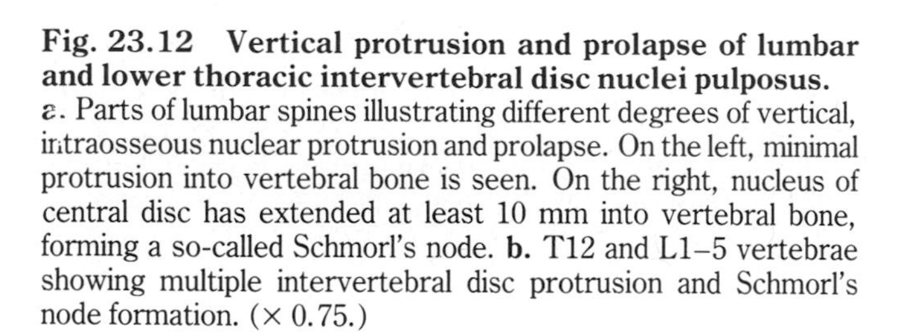

Fig. 23.12 Vertical protrusion and prolapse of lumbar and lower thoracic intervertebral disc nuclei pulposus. **a.** Parts of lumbar spines illustrating different degrees of vertical, intraosseous nuclear protrusion and prolapse. On the left, minimal protrusion into vertebral bone is seen. On the right, nucleus of central disc has extended at least 10 mm into vertebral bone, forming a so-called Schmorl's node. **b.** T12 and L1–5 vertebrae showing multiple intervertebral disc protrusion and Schmorl's node formation. (× 0.75.)

The protrusion may be upwards or downwards and herniated nuclear material is often seen to have extended simultaneously in both directions. A single, small vertical protrusion often displays a flattened, mushroom-like shape but bifid herniations are dumb-bell-like. Although the nuclei pulposus are often described as central, their location is usually off-centre and nuclear protrusion is frequently found to be closer to the posterior or posterolateral vertical axis of the disc than to the anterior. However, as Vernon-Roberts points out, small protrusions may be found in other parts of the disc.

The bone of the vertebrae into which vertical protrusions occur may be normal but, in the elderely female, is often porotic. Expelled into the trabecular bone, through the endplate, the nuclear tissue creates irregularly dispersed microfractures and comes to occupy bone marrow spaces the cells of which may undergo necrosis. A low-grade inflammatory reaction occurs (but see p. 939). As the process of herniation continues, more bone is displaced or resorbed until the forces promoting protrusion are balanced by those ensuring bone integrity. The formation of limited amounts of metaplastic fibrocartilage around the protruded 'node' is often followed by the limited synthesis of new bone. A cartilagenous and ultimately bony 'shell' encapsulates the protruded nucleus. New trabecular bone is formed by appositional growth but the continued operation of

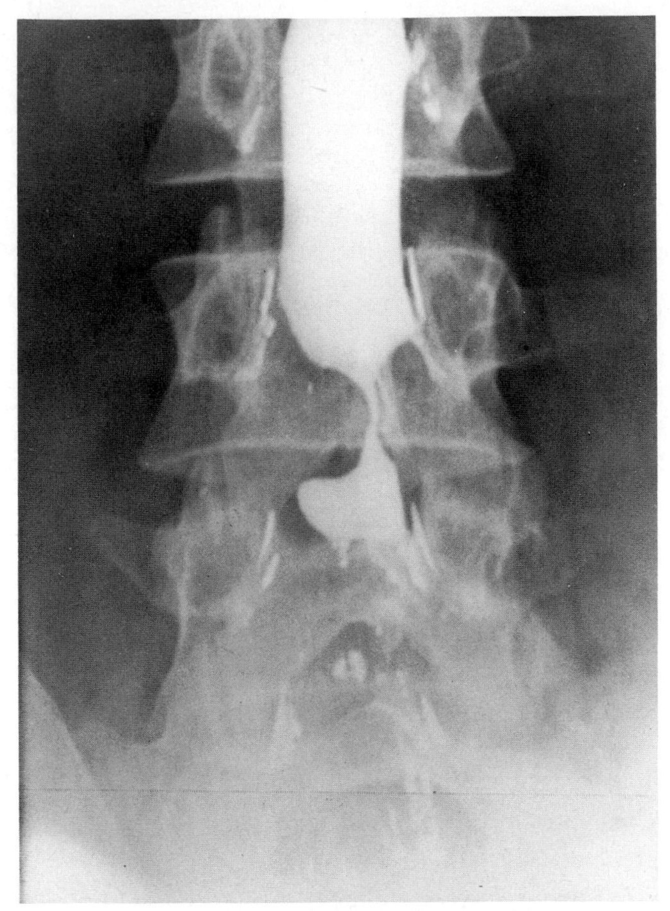

Fig. 23.13 Myelogram of intervertebral disc prolapse. Radio-opaque contrast medium has been introduced into the epidural space, in this now less favoured procedure. The space has failed to fill uniformly because of the presence of a massive prolapsed nucleus pulposus.

mechanical stress may be one reason why microfractures, often healed or healing, are so commonly seen.[90] The process of extrusion of nuclear tissue from the avascular intervertebral disc into the vascular bone promotes vascularization. New blood vessels grow centripetally, from the bone into the degenerate, residual disc tissue. Granulation tissue formation, collagen synthesis and fibrosis, calcification and ossification are the structural consequences.

Posterolateral displacement (Figs 3.53, 3.58 and 23.13)

The relative infrequency with which nuclear protrusion occurs in the median, sagittal plane is a consequence of the central position, between, and at, the posterior surfaces of the vertebrae, of the strong posterior longitudinal ligament. The lateral expansions of this ligament restrict protrusion to horizontal levels above and below the central plane. Although the posterior and posterolateral parts of the annulus fibrosus of the intervertebral disc are weaker than the lateral and anterior components, they are much less vulnerable to nuclear protrusion and prolapse than the chondroosseous endplates. The endplates are mechanically weak. However, it is likely that the greater vulnerability of these structures is also a result of the imposition of much larger and more frequent vertical compressive stresses than the forces imposed on the posterior parts of the disc.

The composition of the tissue protruded posterolaterally is controversial. The material has been said to comprise components of the nucleus in addition to portions of the annulus and adjacent hyaline cartilage endplate. This view,[76] based largely on *post mortem* enquiries, is in accord with the results of biopsy studies.[89] Surgical specimens, which often contain fragments of granulation tissue, cartilage or bone, may be largely of annular fibrocartilage, of nuclear myxoid connective tissue containing stellate notochord-like cells, or both. By contrast, it has been proposed that since the posterolateral collagen fibre bundles of the annulus can be displaced by the forcible extension of nuclear tissue, the protrusion is principally of the latter, without elements derived from a disintegrated annulus.[18] The answer may lie in the chronology of different protrusions. The more recent, acute herniation is usually of nuclear myxoid material which has extended quickly through and between the fibre bundles of the annulus. A more insidious, gradual extension of discal tissue may include granulation tissue, fibrous tissue and calcified or ossified material reflecting the process of repair by fibrosis.

Rather different views of the histological development of intervertebral disc herniation have been expressed by Yasuma *et al.*[97] and Lipson.[60] Yasuma described a sequence of change in which 'myxomatous' degeneration of the annulus fibrosus advanced with age. The collagenous fibre bundles 'reversed their usual direction'. Excised disc material examined at biopsy demonstrated the presence of annular tissue with myxomatous degeneration in the majority, while nuclear tissue was rare.

Clinical consequences of disc protrusion

The clinical consequence of protrusion of a lumbar nucleus is intractable pain. The spinal cord terminates at lumbar segments L1-2. Consequently, protrusion of intervertebral disc nuclei below this point exerts pressure upon the cauda equina and nerve roots, not upon the spinal cord. The pain is in a femoral nerve distribution when protrusion of nuclei from the higher lumbar discs compresses the L3 and L4 nerve roots. When pressure is exerted by nuclei protruding from the distal segments, the L4, L5 and S1 nerve roots are correspondingly irritated or compressed. The severity and duration of pain are much influenced by the extent of each protrusion and the degree to which its anatomical normality is restored by conservative treatment. Studies made *post mortem* with spines from children and young adults clearly demonstrate the rubber-like, reversible deformability of the normal nucleus pulposus (Fig. 3.8). Restoration to a normal site of the protruded older adult nucleus is much less likely. The advance of fibrosis, with new cartilage and bone formation at the sites of prolapse, eventually ensures that the symptoms of nerve root irritation decline although there may be persistent or permanent objective evidence of impaired sensory and motor neurone function.

Structural consequences of disc protrusion

The eventual, histological results of posterolateral intervertebral nuclear protrusion include neovascularization and fibrosis.[53] Scar tissue formation and collagenization of the protruded nuclear segment lead to a reduction in its size. The fibrotic portion becomes calcified or ossified. However, further episodes of nerve root irritation or compression are likely as adjacent malleable parts of the same nucleus are protruded in turn.

The anatomical consequences of protrusion and/or prolapse of the nuclei of the intervertebral discs include a reduction in the vertical distance between the vertebral bodies; an anterior inclination of the vertebral bodies, particularly in regions such as the lumbar where the normal anatomical alignment is lordotic not kyphotic; decreased spinal mobility with impaired rotational and flexural movements; in the case of posterolateral displacements, compression of nerve roots or nerves; the promotion of anterior and lateral new bone formation; and the imposition of abnormal mechanical shear stresses on the apophyseal joints, leading to premature secondary osteoarthrosis (see Chapter 22).[56] The formation of bone at the margins of posterolateral nuclear protrusions is a form of osteophytosis (see p. 943). The osteophytic processes are not large but their situation determines that they reduce the lumina of the intervertebral foramina, reproducing some of the effects caused by nuclear protrusion. The osteophytes are not sufficiently large to extend medially to constrict the spinal canal.

Spondylolisthesis

Spondylolisthesis describes the anterior displacement of one vertebra upon the vertebra below. It may follow severe injury but more often is recognized as a congenital defect resulting from excessive mobility of a spinal segment.[94] The underlying abnormality centres on either the intervertebral disc, or, more often, on the instability of the posterolateral, apophyseal joints and of their ligaments. Segmental pain may develop. It is a result of a stretching of the intervertebral ligaments.

Immunological disease

Humoral responses

It is established that, in inflammatory connective tissue disorders such as rheumatoid arthritis, anticollagen and antiproteoglycan antibodies are commonly identifiable. Whether they are pathogenic or merely epiphenomena is not yet certain. In cases where rheumatoid inflammation, for example, of the cervical neurocentral (Luschka) joints, contributes directly to cervical intervertebral disc degradation, the synthesis of antibodies against the intervertebral disc collagens and proteoglycans may also be anticipated. Since there is no certain way in which to discriminate between antiintervertebral disc collagen- and proteoglycan antibodies, and antibodies formed against analogous macromolecules undergoing degradation in the hyaline cartilages of the synovial joints, it has not yet been possible to differentiate between these processes. To achieve this understanding would seem particularly important in the seronegative spondyloarthropathies (see Chapter 19).

The presence of excess immunoglobulins M and G in the serum of patients recovering from clinical episodes of protrusion or prolapse of intervertebral disc(s) tissue has been demonstrated by Naylor *et al.*[69,70] The possibilities have therefore been raised that first, hypersensitivity mechanisms may contribute to the pathogenesis of disc protrusion or, second, that the raised antibody titres may simply reflect the intervertebral disc connective tissue injury sustained in the course of disc herniation or its treatment.[84]

Cell-mediated responses

Analogous studies have been made of cell-mediated responses. In patients with intervertebral disc protrusion, coming to surgery, evidence of delayed hypersensitivity to cellular and matrix (proteoglycan) antigens may accompany a raised IgM antibody titre. The role of delayed hypersensitivity as a cause or index of intervertebral disc protrusion has therefore been considered. The evidence for an active cell-mediated mechanism has been tested without success by Urovitz and Fornasier.[86] Necropsy and biopsy material from 108 Schmorl's nodes (see p. 936) and 110 intervertebral discs was searched for histological evidence of cellular infiltration. No infiltrates were found but granulation tissue and vascular invasion were seen in almost all specimens. It is relevant to observe that the autoimmune response by rabbits to nuclear pulposus antigen can evolve without clinical signs of disc disease.[10]

Transplantation of intervertebral discs

When the intervertebral joints of young, immature Sprague–Dawley rats are transplanted into non-functional sites, ankylosis tends to occur.[29] Ankylosis in the joints of the youngest animals is almost invariable in the proximal part of the tail, rare in the distal part unless this is curved. Ankylosis occurs through chondroid metaplasia of the intervertebral connective tissue. Cartilage is ultimately replaced by bone. Diminished metabolism and an absence of functional activity are regarded as the contributory factors. There have been few attempts to transplant human intervertebral discs and the main lessons from these animal studies may be in relation to the processes of fibrocartilagenous ankylosis and ossification in ankylosis spondylitis (see p. 768).

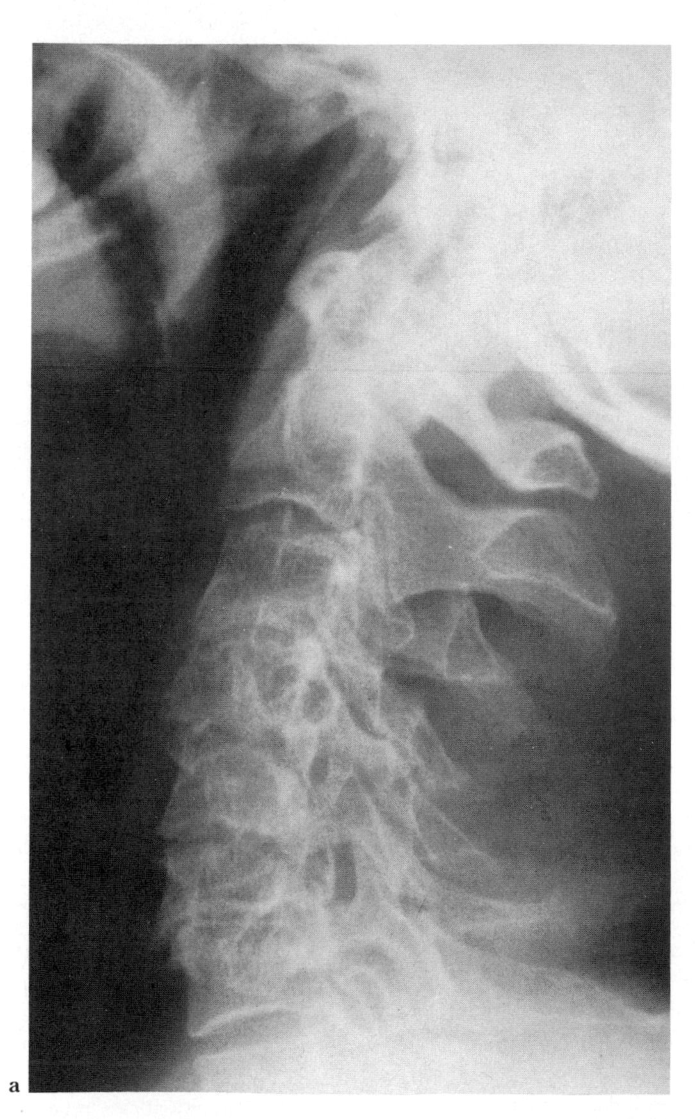

a

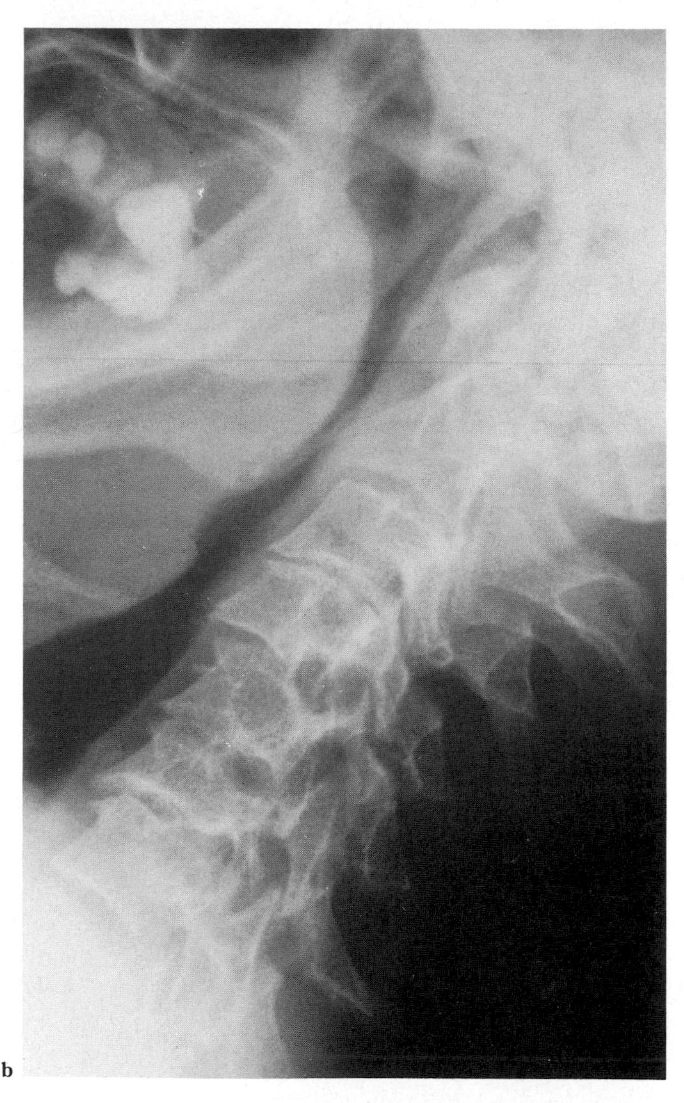

b

Fig. 23.14 Cervical spinal spondylosis with rheumatoid arthritis of the apophyseal (facet) joints and of the vertebral margins.
Fifty-nine-year-old Caucasian male. The spine is viewed **a.** in extension and **b.** in flexion. Degenerative intervertebral disc disease is shown by the reduction in disc spaces between all vertebrae from C3 to C7, osteophytic lipping of the vertebral bodies, and sclerosis of the vertebral bone margins. Rheumatoid changes comprise erosions of the margins of the apophyseal (facet) joints, most conspicuous at C4–5, together with erosion of the margins of the vertebral bodies at the same level. There is C4–5 vertebral subluxation.

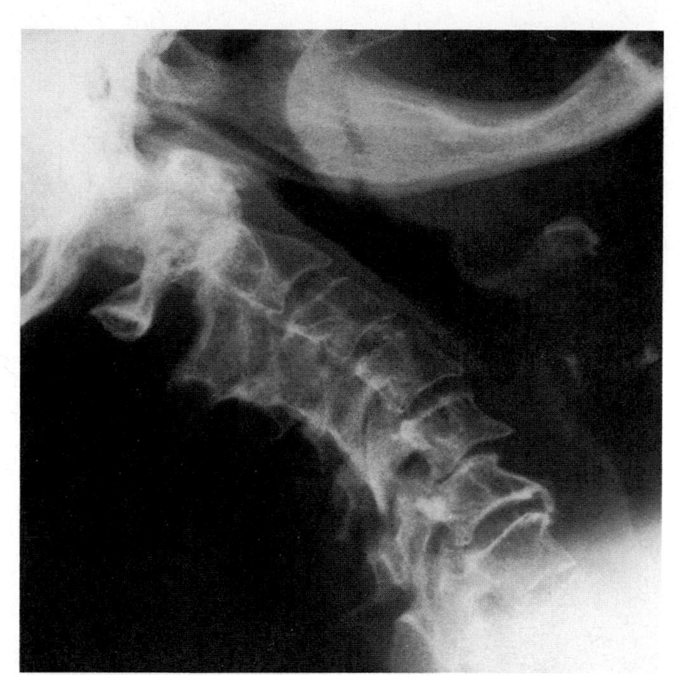

Fig. 23.15 Rheumatoid arthritis of cervical spine. Caucasian female aged 69 years with seronegative disease. A lateral view of the cervical spine shows profound demineralization, and anterior displacement of C1 on C2, and of C5 on C6. There is erosive destruction of all the apophyseal (facet) joints between C2 and C6.

Rheumatoid arthritis

Important aspects of rheumatoid arthritis of the spine[8] (Figs 23.14 and 23.15) are described on p. 475.

Ankylosing spondylitis and the seronegative spondyloarthropathies
(Table 23.1, Figs 3.41 and 19.3)

The principal pathological changes encountered in ankylosing spondylitis (Fig. 23.16) and the other seronegative spondyloarthropathies are described in Chapter 19.

Table 23.1

Pathological features of ankylosing hyperostosis and ankylosing spondylitis*

	Ankylosing hyperostosis	Ankylosing spondylitis
Association with HLA-B27	–	+++
Hereditary influence	±	+++
Sex ratio (F:M)	1:2	1:8
Age at onset (in years)	90% > 50	90% < 35
Inflammation (histologically and ESR)	Absent	Present at start and in active phases
Vertebral lesions	Bone apposition over full length	Bone erosion at vertebral rim leading to 'squaring' on X-ray followed by adjacent bone sclerosis
Osteophytes	Broad base, large, usually starting below vertebral rim	Slim syndesmophytes replace outer annulus fibrosus, starting at rim
Ankylosis	Arched bridges, mainly thoracic R > L	Flat bridges, ascending from lumbar R = L
Disc diameter	Increased by chondroid metaplasia of adjacent connective tissue	Reduced by destruction of outer annulus fibrosus
Intervertebral, costovertebral and sacroiliac joints	Intact	Inflammation soon followed by marginal ankylosis
Kyphosis	None (or slight)	Pronounced in 60%

* Modified from data of Professor Frits Eulderink.

Infection

Tuberculosis

In populations where social and preventive medical measures have controlled the transmission of *Mycobacterium tuberculosis*, immunoincompetent adolescents and adults are still susceptible to bone and joint tuberculosis. In Western societies, tuberculosis of the spine is seldom encountered except among first- and second-generation immigrant populations. However, from time-to-time tuberculous osteitis of vertebral bone is recognized in an elderly person in whom the presence of a solitary metastatic spinal tumour had been suspected. The principal intervertebral joint(s) contiguous with the infected bone are destroyed secondarily. Untreated, the results may include kyphoscoliosis, tuberculous meningitis, paraplegia and iliopsoas abscess. The pathological anatomy and histology of tuberculo-

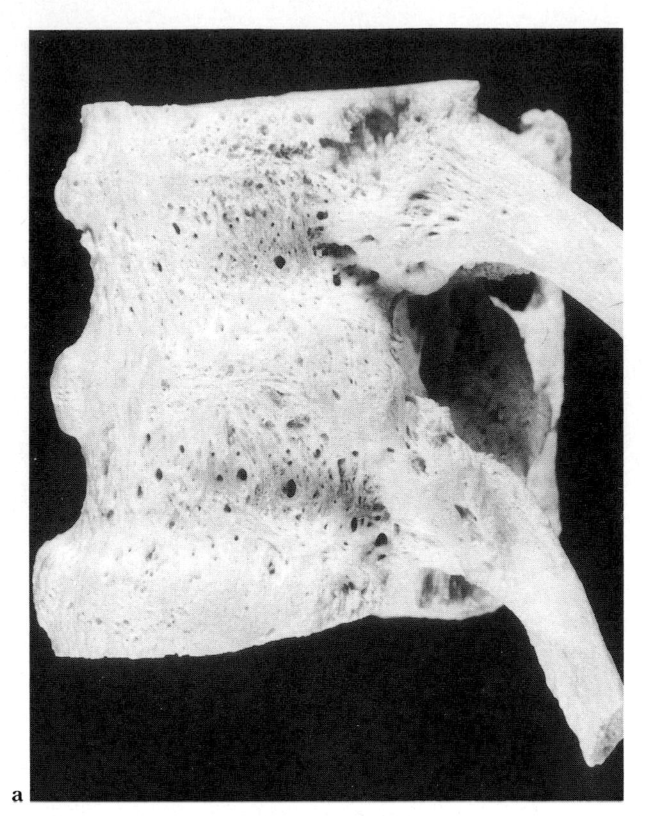

a

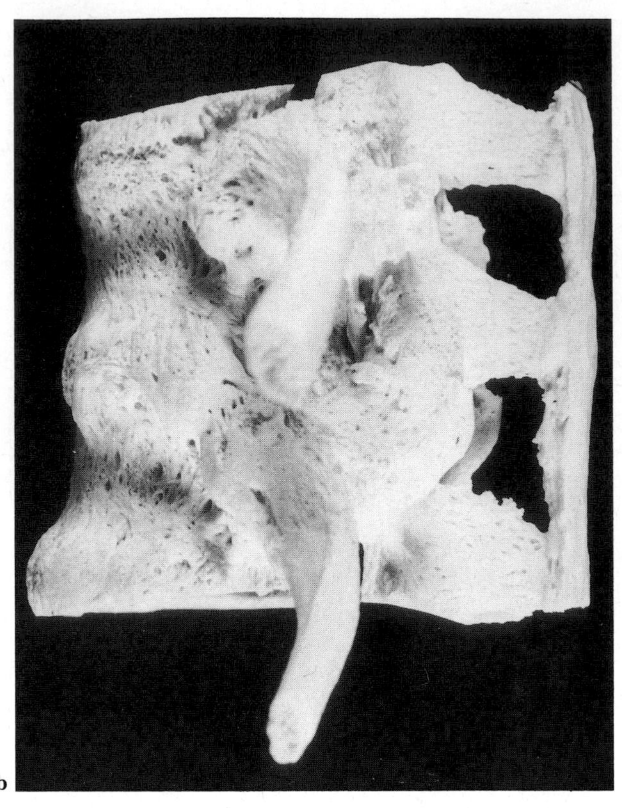

b

Fig. 23.16 Ankylosing spondylitis of thoracic spine.
a. Lateral aspect of the spine in the final phase of ankylosis. The entire osteoarticular structure of the vertebral column, including the components of both the synovial and the cartilagenous joints, has been converted into a rigid, calcified mass. **b.** Oblique view of same specimen. Reproduced by courtesy of Professor René Lagier.

sis of the intervertebral fibrocartilagenous joints are fully described in the texts listed on p. 743.[55] Following infection of the anterior metaphyseal regions of two or more contiguous vertebrae, a tuberculous osteomyelitis with giant cell systems, zones of caseous necrosis and granulation tissue formation, spreads directly through the vertebral bone cortex destroying disc tissue and promoting prolapse of the affected discs. Healing is associated with bony ankylosis.

Brucellosis

Bone and articular synovial tissue are destroyed insidiously by brucella (p. 739) and the vertebrae may be implicated. The cartilagenous intervertebral joints are affected less often than in tuberculosis. Microscopically, the non-necrotizing epithelioid cell granulomata of brucellosis closely resemble those of sarcoidosis. They are distinct from the caseating foci encountered in tuberculosis and tularaemia.

Other bacteria

Infection of the avascular intervertebral discs is a secondary result of focal, contiguous vertebral osteomyelitis. Intervertebral discitis may occur without demonstrable cause or may complicate direct, traumatic injury. Occasionally, infection follows surgical operations on the vertebrae or discs themselves. The use of acupuncture to treat the symptoms of prolapsed intervertebral disc (see p. 935) has been complicated by staphylococcal infection of an intervertebral disc space[38] and staphylococi are the most frequent causes of bacterial discitis. Periodically, however, other aerobic organisms including coliforms, and anaerobes such as *Bacteriodes fragilis* have been implicated. In childhood, vertebral osteomyelitis is uncommon, accounting for only 2 per cent of cases of osteomyelitis. The most common, infecting organism is, again, *Staphylococcus aureus* but a variety of other bacteria has been identified. There are, for example, three recorded cases of childhood intervertebral discitis attributed to *Kingella kingae* (see p. 733).[95]

Fungi

Infection of vertebral bone by blood-borne organisms such as *Candida albicans* is recognized in immunosuppressed or immunodeficient individuals. Periodically, the disease may extend to destroy intervertebral disc tissue and in one instance this sequence was followed by superinfection with *Staphylococcus aureus.*[71]

Response to therapeutic and other agents

The protrusion or prolapse of an intervertebral disc frequently causes neurological changes sufficiently severe to demand immediate or delayed, elective surgery. The diagnostic and orthopaedic aspects of these conditions are described fully in texts on imaging techniques[73] and on orthopaedic practice.[47] Magnetic resonance imaging plays an increasingly important part in these procedures (Fig. 23.4).

In addition to active surgical intervention in cases of intervertebral disc protrusion, active searches have continued for alternative, non-operative or medical forms of therapy. Two examples are briefly considered.

Chymopapain injection

As an alternative to surgical excision of intervertebral disc tissue for the relief of the symptoms of protrusion, the lysis of the nucleus pulposus by the direct injection of the plant proteinase, chymopapain, was proposed.[81] *In vitro*, there is a linear relationship between the amount of enzyme used and the quantity of glycosaminoglycan released. Tests made with nuclei pulposus *post mortem* are invalidated because intervertebral disc tissue quickly swells in isotonic solution leading to a leaching out of proteoglycan (see p. 126).[85] Consequently, the testing of enzyme injection techniques has been based largely on alternative materials such as bovine nasal cartilage.[13] There are few reports of the histological changes caused in human intervertebral discs by this form of treatment.

Effects of systemic glycosaminoglycan polysulphate

Glycosaminoglycan polysulphate (GAGPS) is one of an increasing number of compounds tested in the treatment of synovial joint osteoarthrosis (see p. 894). Glycosaminoglycan may be administered locally, into a synovial joint, or systemically. After experimental intramuscular injection, GAGPS is taken up by collagen-rich tissues such as those of the knee joint menisci.[4] The concentration achieved within intervertebral disc tissue is equivalent to that attained in the hyaline cartilage of the knee. Penetration is via the outer part of the annulus fibrosus. Fibronectin is among the components which may bind GAGPS, and the compound stabilizes the complexes that fibronectin may form with collagen. Little is known of the influence of systemic GAGPS on the histological structure of the human intervertebral disc.

Vertebral body osteophytosis (Table 23.2)

The presence of osteophytes (see p. 866) at the anterolateral margins of the vertebral bodies is an extremely common finding in older persons (Figs 23.17–23.22). A relationship to the presence of intervertebral disc degeneration is clear: disc degeneration and osteophytosis commonly occur together. However, the mechanisms of

Table 23.2

Pathological features of ankylosing hyperostosis and of spinal osteophytosis*

	Ankylosing hyperostosis	Spinal osteophytosis
Location	Principally thoracic	Principally lumbar
Contiguous levels affected	Invariable	Much less often
Intervertebral disc	Intact	Degenerate; frequent irregular endplates; nearby osteosclerosis
Posterior intervertebral joints	Intact	Osteoarthrosic
Osteophyte	Broad base; below vertebral rim; bridging	Less broad base; origin at vertebral rim
Paraspinal and extraspinal ossification	Frequent	Absent
Posterior longitudinal ligament	Often intact	Often affected

* Modified from data of Professor Frits Eulderink.

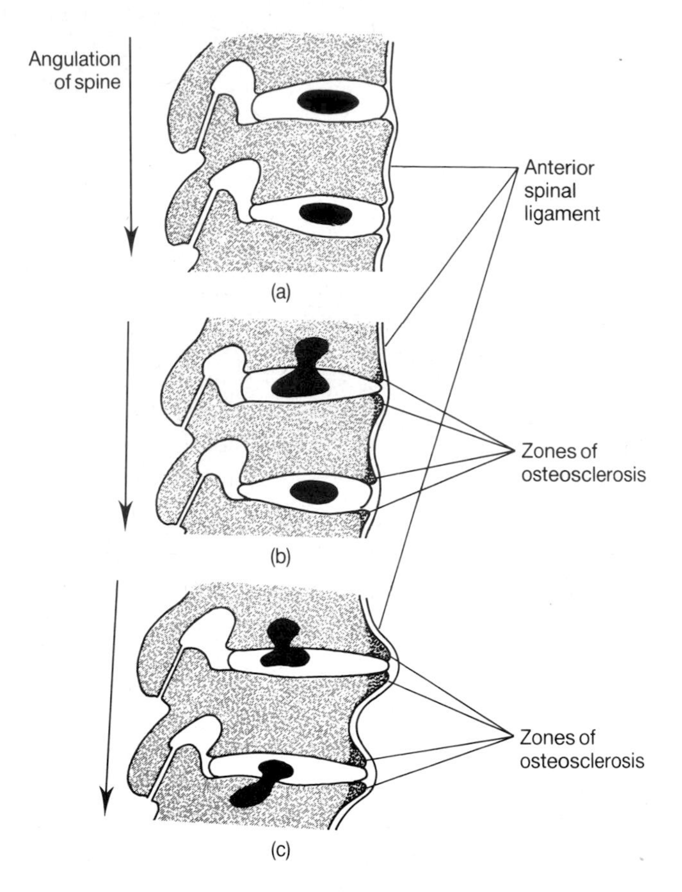

Fig. 23.17 Vertebral osteophytosis – diagram of sequence of disc changes.
a. Normal lumbar part of the vertebral column. The nuclei pulposus are shown in black. The anterior margin of the spine is bounded by the anterior common ligament. **b.** Vertical, upward prolapse of the nucleus pulposus of the upper intervertebral disc has taken place producing the anatomical appearance still known as a Schmorl's node. The disc margin protrudes slightly and new bone formation has been excited at the vertebral rim. **c.** More advanced changes proceeding from those shown in **b.** Schmorl's nodes have formed in adjacent vertebral bodies. The normal kyphotic shape of the spine is exaggerated, a change occurring occasionally in adolescence but commonly in old age. The intervertebral discs are narrowed; their margins protrude anteriorly. New bone formation at the vertebral rims culminates in osteophytosis.

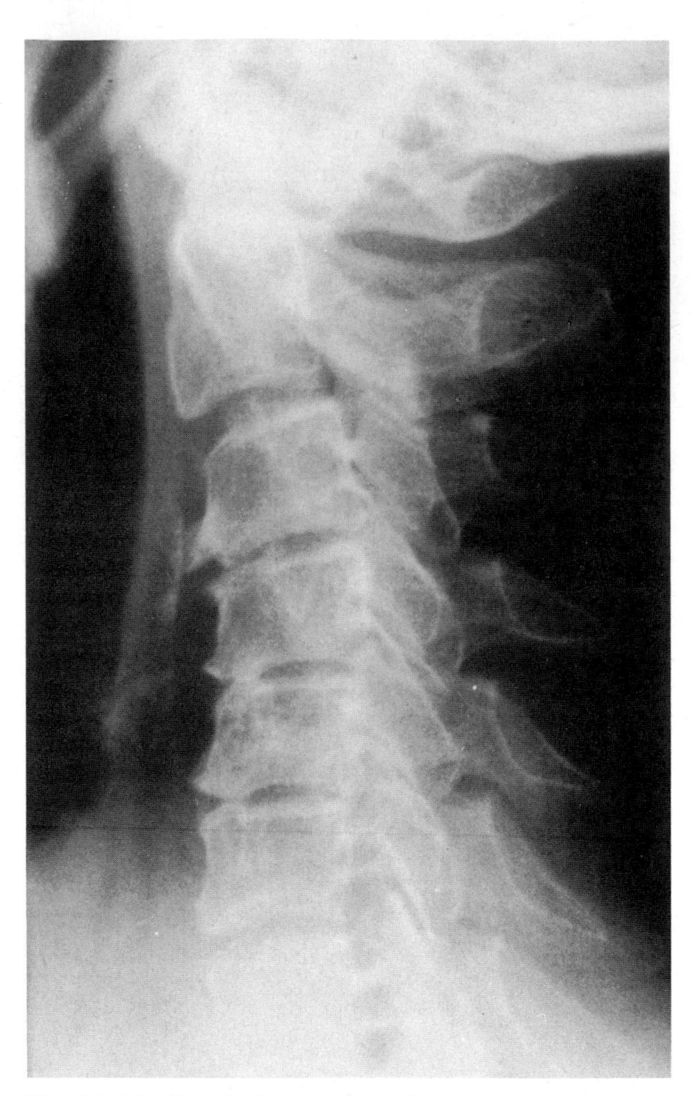

Fig. 23.18 Cervical osteophytosis.
Lateral cervical radiograph showing cervical osteophytes affecting the C3, 4, 5 and 6 vertebrae.

vertebral body osteophytosis are still not wholly understood.

Vertebral body osteophytes are often symmetrical. This feature and the absence of bony fusion between contiguous osteophytes distinguishes osteophytosis from ankylosing hyperostosis (see p. 947). The lumbar vertebrae often display larger and more extensive osteophytes than the cervical. However, all parts of the spine may be implicated. Nevertheless, generalized vertebral osteophytosis is not a component of systemic disease: it is an index of processes such as intervertebral disc degeneration, not of generalized osteoarthrosis (see p. 847). Much of the classical epidemiology of osteoarthrosis has, necessarily, been based on the X-ray assessment of bone structure. However, osteophytosis of the main intervertebral joints is not an index of spinal osteoarthrosis. The presence and severity of spinal osteoarthrosis can be judged by the presence of osteophytosis of the apophyseal joints, seen best in posterolateral radiographs.

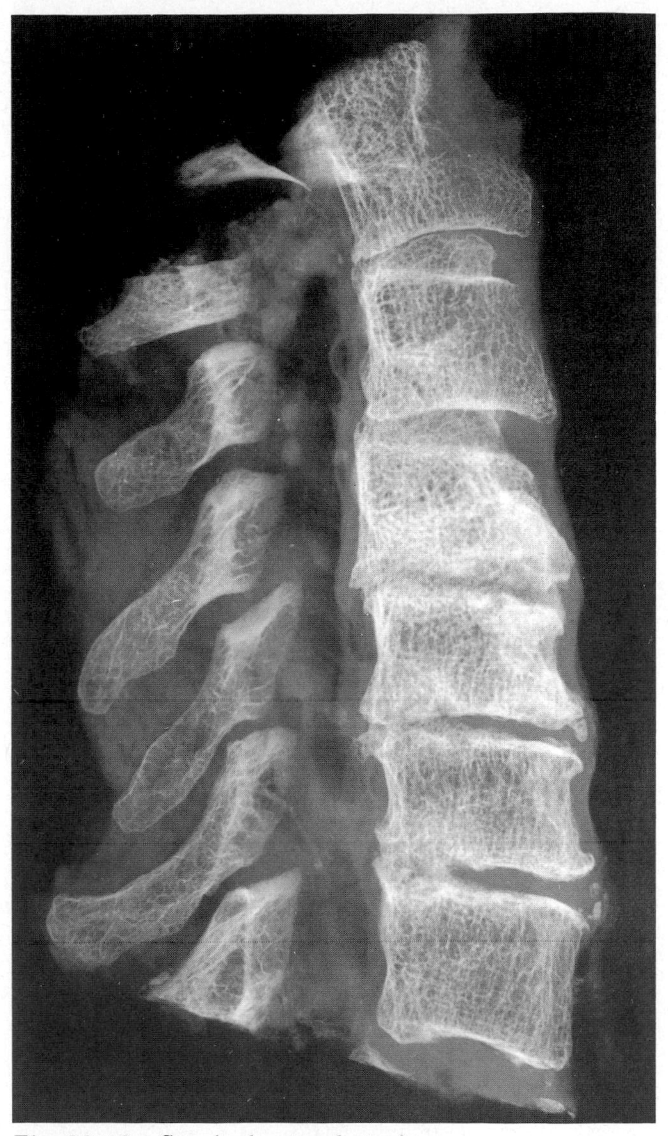

Fig. 23.19 Cervical osteophytosis.
Radiograph of a slab specimen. Disc degeneration and osteophytosis are particularly severe at the C4–5 level.

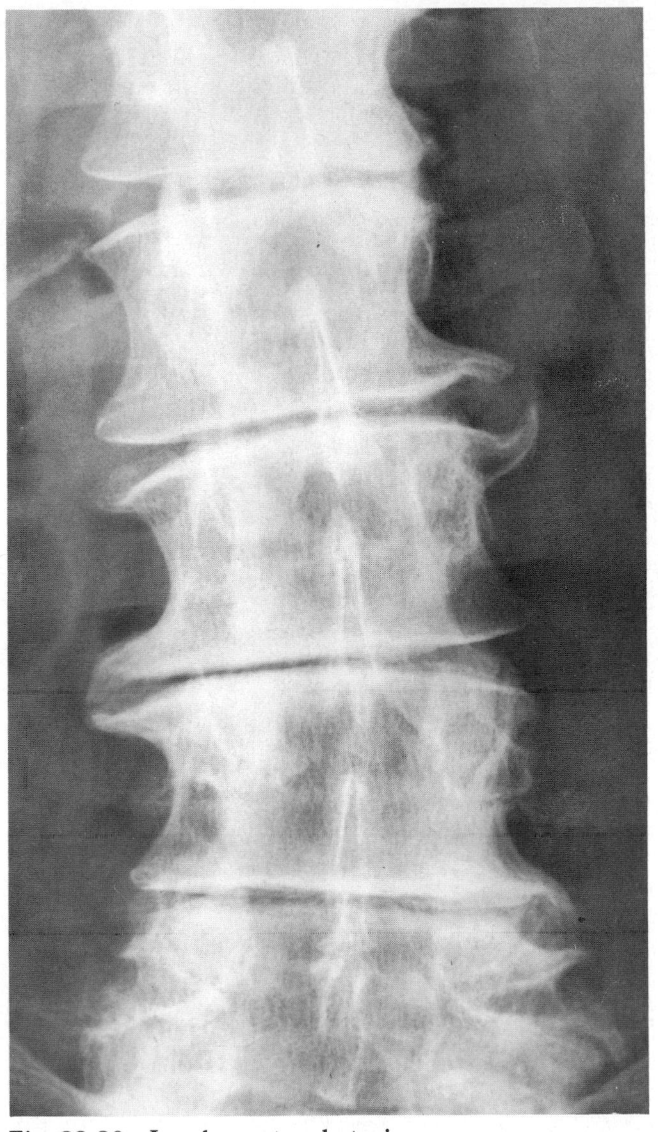

Fig. 23.20 Lumbar osteophytosis.
Compare with Fig. 23.18. Adjacent osteophytes lie close together but do not fuse.

It is widely accepted that the degree of vertebral body osteophytosis is proportional to the degree of reduction in the height of the vertebral bodies. This reduction (see p. 929) is an indirect measure of the extent of intervertebral disc degeneration. A process is therefore proposed in which disc degeneration leads to a progressive anterior tilting of the spine (Fig. 23.17).[18,89] The degenerate disc tissue, displaced anteriorly, raises and irritates the marginal vertebral body periosteum. New bone formation begins, confined largely to zones lateral to the anterior longitudinal ligament. Anterolateral osteophytes form.

An alternative hypothesis rests on the demonstration by Schmorl[75,76] of small tears in the periphery (rim) of the annulus fibrosus (see p. 126). This process culminates in the protrusion, anterolaterally, of discal tissue. In turn, periosteal osteoblasts are provoked to new bone synthesis and islands of bone are formed adjacent to the anterior longitudinal ligament. François[33] and others have demonstrated that cartilage neosynthesis or chondroid metaplasia is a preliminary to the formation of the osteophytes which are associated with rim lesions. This sequence is an attractive way of accounting for osteophytosis. The evidence brings together the processes observed in experimental osteophytosis of synovial joints such as the knee, in which

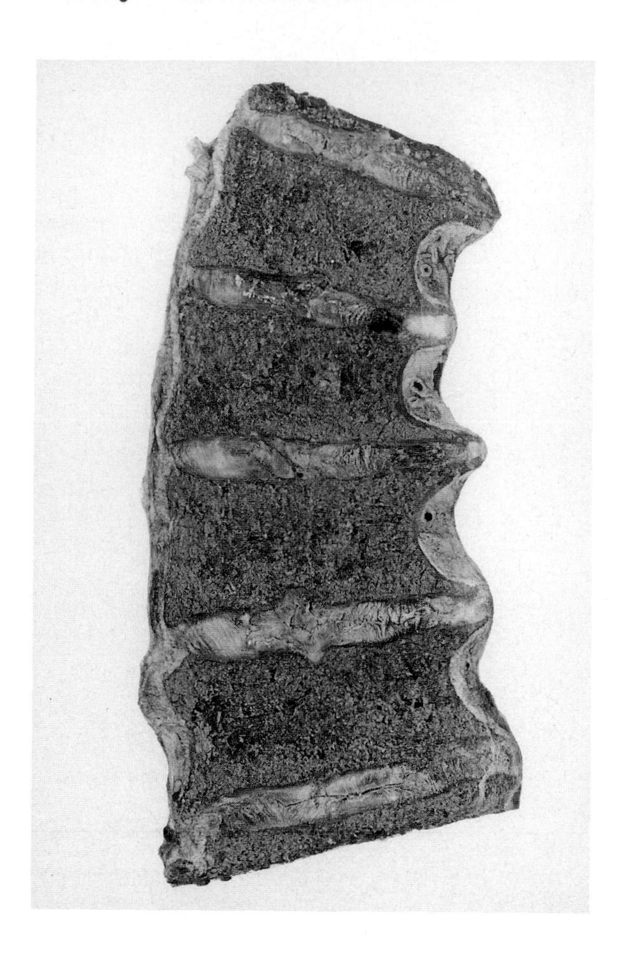

Fig. 23.21 Lumbar intervertebral disc degeneration and osteophytosis.
Slab block illustrating anatomical appearances corresponding to the radiograph in Fig. 23.19. There is early Schmorl's node formation.

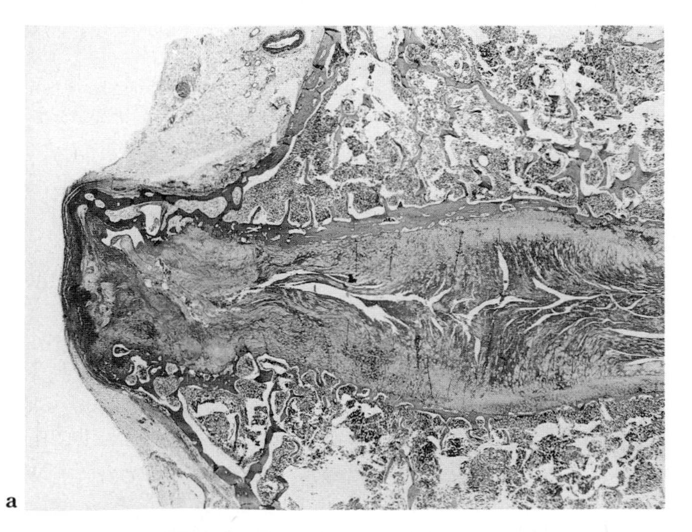
a

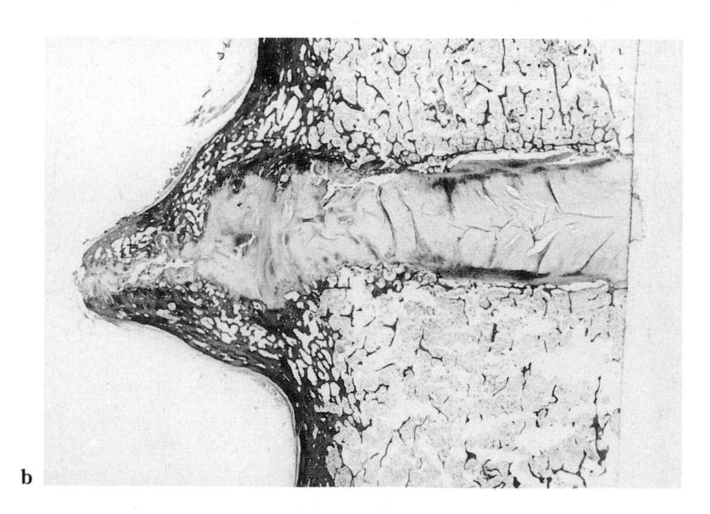
b

Fig. 23.22 Histological changes in osteophytosis.
a. New bone formation extends outwards from the vertebral rim. There is fissuring of the intervertebral disc and the substance of the disc protrudes laterally. The osteophytes are not in contact nor are they coextensive. (× 4.) **b.** Here, the osteophytes are contiguous but not in continuity. The appearances closely resemble those recognized in ankylosing hyperostosis. (× 2.5.)

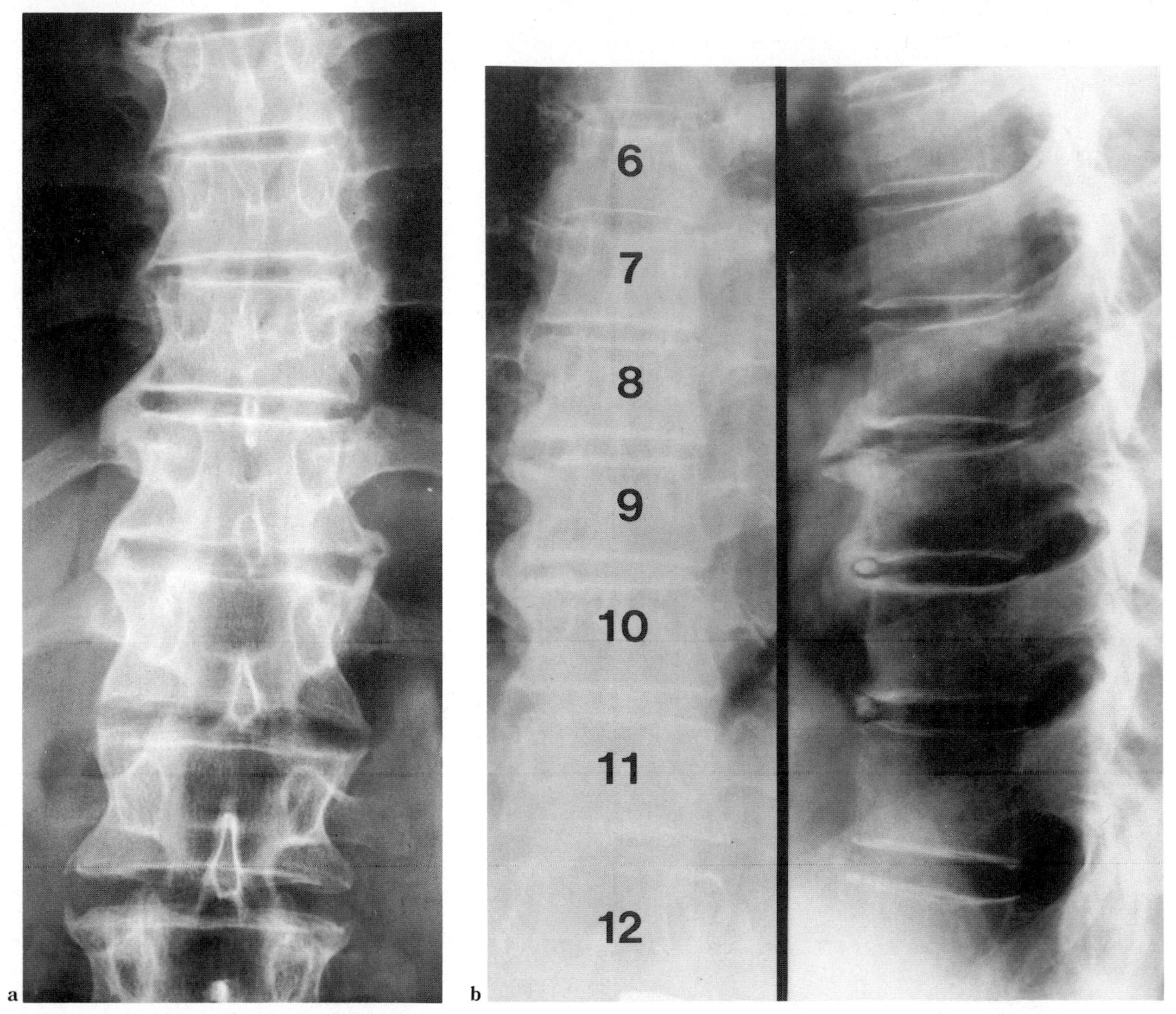

Fig. 23.23 Diffuse intervertebral spinal hyperostosis (DISH).
a. Ankylosing hyperostosis affecting the lower thoracic and upper lumbar vertebrae. **b.** Lateral view of osteophytic lipping of the anterior vertebral margins of T8, 9, 10 and 11. The changes are accompanied by fusion of the osteophytes. Reproduced by courtesy of Dr J T Patton.

the growth of chondrophytes is found to precede new bone formation (see p. 898), with those demonstrable in the spine.

Ankylosing hyperostosis

Early anatomical enquiries demonstrated that long, irregular bone outgrowths that immobilized two or more adjacent vertebrae were occasional findings in the spines of elderly subjects.[30–32,58] The condition[18] was clearly differentiated from ankylosis spondylitis (see Chapter 19) (Tables 23.1 and 23.2); it was unrelated either to spondyloarthrosis or osteoarthrosis.

Clinically, the disorders of ankylosing hyperostosis (Figs 23.23 and 23.24), most frequent in the elderly, are much commoner in males than females. The right side of the spine is affected more often than the left. There is an association with diabetes mellitus and, occasionally, with hypocalcaemia.[30] The condition has been recognized in animals including the horse, in extant skeletons as well as in those of extinct species; it can be reproduced in cats by the

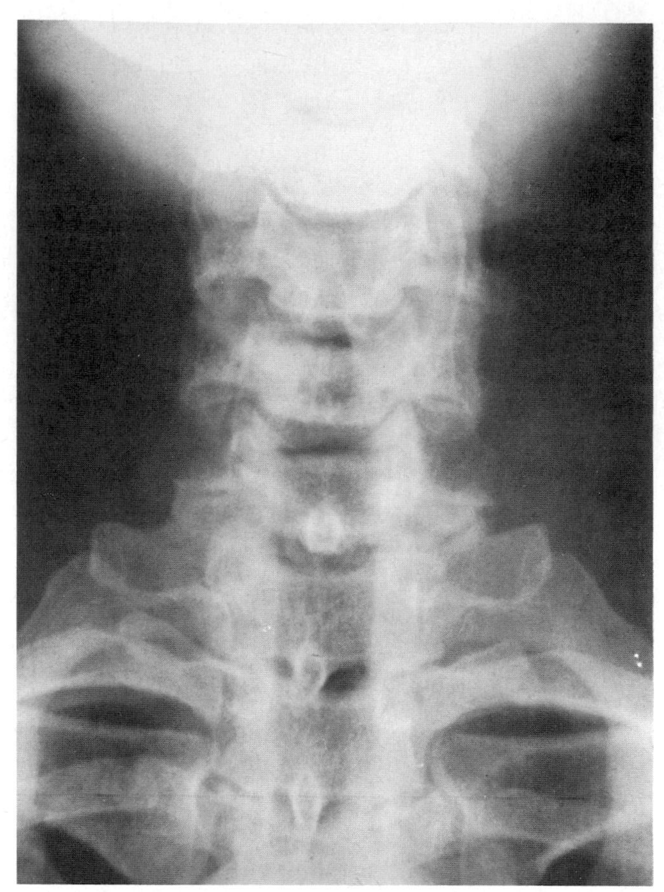

Fig. 23.24 Cervical hyperostosis.
There is widespread new, vertebral bone formation. The appearances merge with those of ankylosing hyperostosis (Forestier's disease) in which, however, the asymmetric deposition of new bone is often on the right and is principally thoracic.

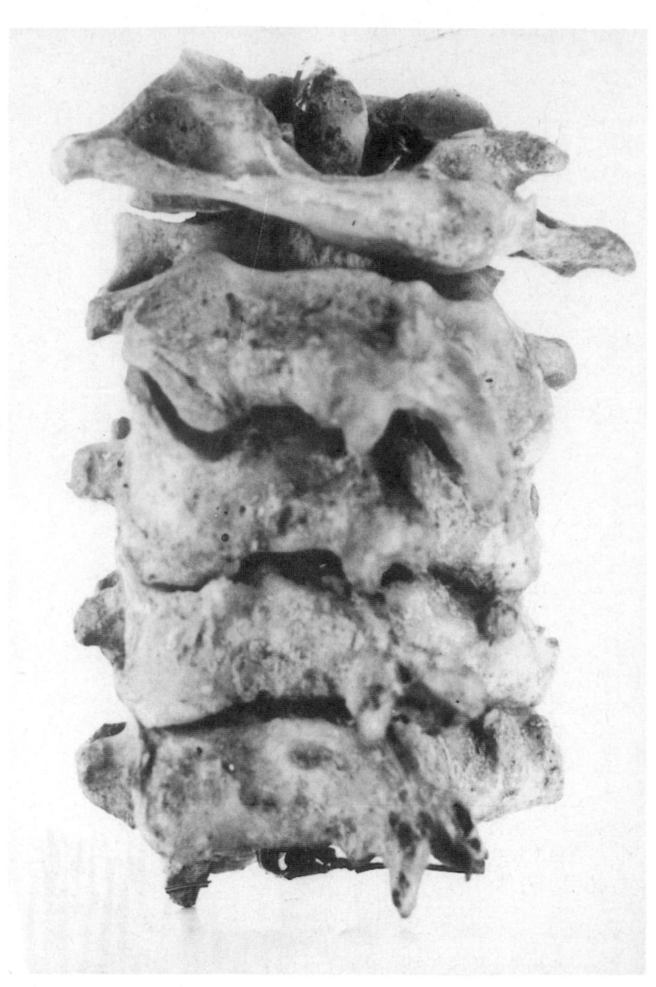

Fig. 23.25 Cervical vertebral hyperostosis.
Compare with Fig. 23.23. New bone 'drips' from the vertebral margins. The changes are asymmetric. Involvement of the posterior surfaces of the vertebral bodies distinguishes the appearances from those of ankylosing hyperostosis.

administration of vitamin A.[78] An analogy has been drawn with the bone changes of fluorosis (see p. 411) in which, however, the hyperostosis is generalized.

Radiologically, cervical and thoracic hyperostosis evolve through three stages. A slight, laminar thickening of the anterior aspects of the vertebrae with masking of the anterior surface of the discs, is followed by accentuated thickening and the formation of a bony spur[37] of candle-flame-like shape. Finally, fusion of opposed upper and lower spurs occurs. The syndesmophytic process formed in this way is much thicker than the intradiscal bone plates of ankylosing spondylitis (see p. 771).

Anatomically, discontinuous spurs extending asymmetrically outwards from the vertebral bodies are pathognomonic (Figs 23.25–23.27).[67] They frequently fuse and, in advanced cases, spread across the entire extent of the surface of the spine.[74] Microscopically, ossification takes the form of new, exophytic processes, not intraligamentous bone formation.[87] There are no comparable processes on the posterior walls of the vertebral bodies.

Osteochondrosis

The term osteochondrosis has been applied in the context both of animal synovial (see p. 897) and of human fibrocartilagenous joint disease. In humans, the process is one of erosion of the vertebral bone which constitutes the plate or plateau adjoining the intervertebral disc.[58] When a Schmorl's node (see p. 936) is identified, the progressive

Fig. 23.26 Ankylosing hyperostosis.
Note the relatively intact intervertebral discs within this slab block, and the anterior fusion of the osteophytic bony processes.

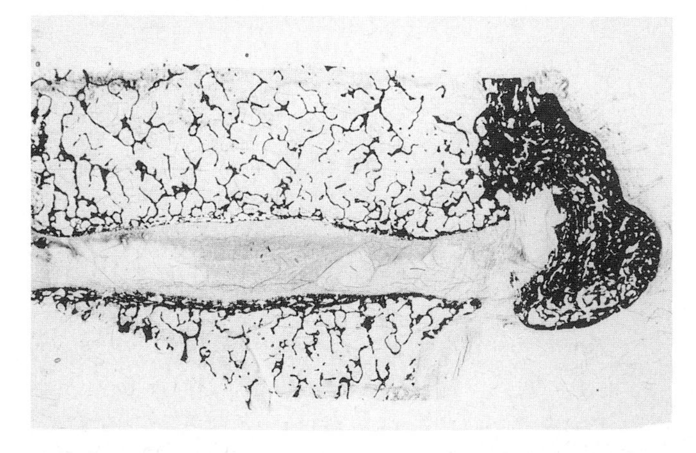

Fig. 23.27 Ankylosing hyperostosis.
Section through osteophytic processes that have extended beyond the margins of two adjacent thoracic vertebrae, fusing together and creating a densely ossified, radio-opaque mass.

degradation of endplate bone structure may be attributed to the associated intervertebral disc degeneration. The radiological changes bear similarities to those of infective discitis (see p. 941).

Bursitis

Cavities containing synovial fluid may form between the spinous processes of the lumbar vertebrae. Bursae are found when the interspinous distance is small compared to the height of the spine.[15] Bursitis may occur (Baastrup's syndrome). The pain which results can be relieved by bending forwards.

Synovial cysts

Encapsulated synovial cysts may appear as extradural masses.[41] They may be bilateral.[20] Lying beside a lumbar facet joint, for example, the cyst may cause extradural compression or spinal stenosis. Spinal synovial cysts are thought to result from the herniation of synovium through the facet joint capsule.[9]

Sacroiliac joint disease

The paired sacroiliac articulations (see p. 133) act as important bearings at the lower end of the vertebral column. The thick sacral cartilage, with its open meshwork of cancellous bone, and the thin iliac cartilage bound to a thicker, denser bone end plate, constitute a shock-absorbing system that acts to protect the central nervous system. The sacroiliac joints are susceptible to all the known disorders both of synovial and of fibrocartilagenous joints: rheumatoid synovitis, infection and trauma frequently disorganize the anterior sacral parts of the joints; osteoarthrosis, amyloidosis and degenerative disease are likely to often affect the posterior components (Fig. 23.28). Age-related and endrocrine change in the sacroiliac joints determine that the younger female joints remain mobile, particularly in pregnancy, until later middle-age. The male joints, by contrast, undergo early fibrous fusion so that by early middle-age no movement can be elicited by manipulation of a *post mortem* specimen nor can the opposed surfaces be distracted without gross force and/or incision.

Although the sacroiliac joints are components of the vertebropelvic axis, diseases of these structures are often

a

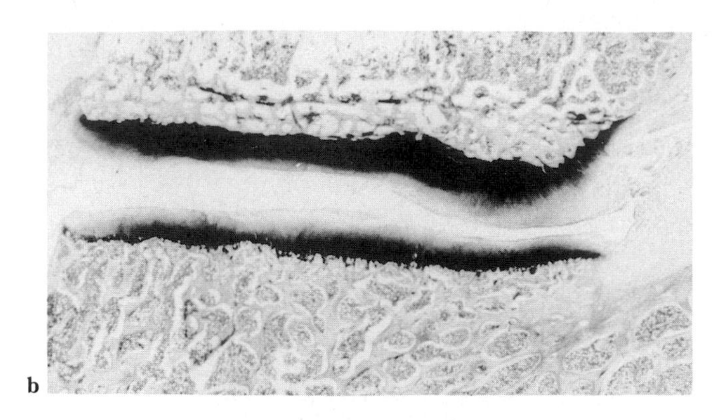
b

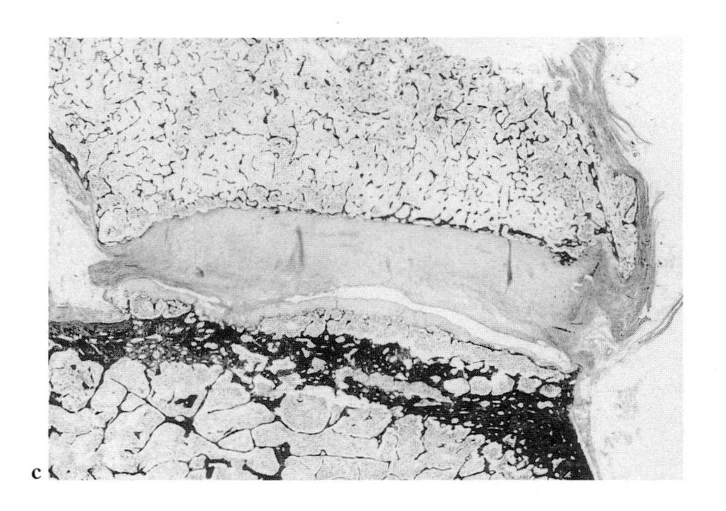
c

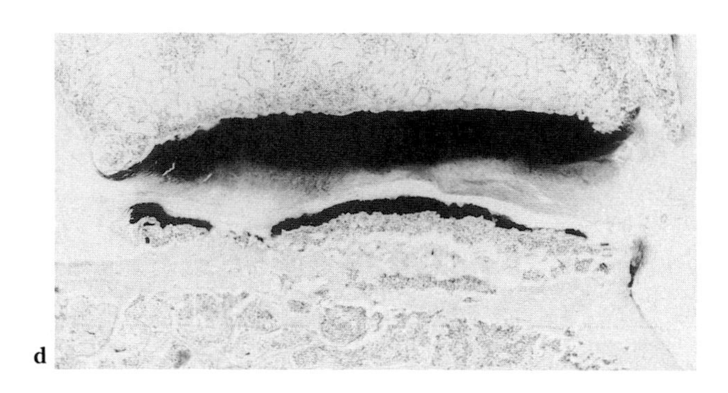
d

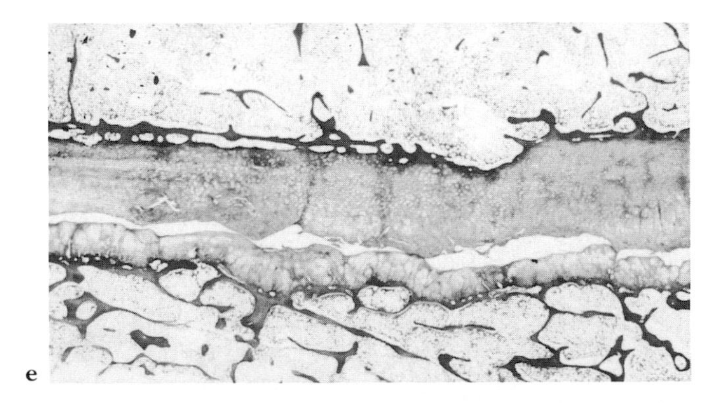
e

Fig. 23.28 Sacroiliac joint: variations in structure with age.
a. Note thick sacral cartilage (top) and much thinner iliac cartilage (bottom). Synovial lining of joint lies anteriorly (at right). By contrast with cartilage structure, iliac bone is dense; sacral bone has a more open, cancellous texture. In the old, amyloid is almost always found in the dense, ligamentous tissue (far right) and free, polymerized fibrin commonly lies in the joint space. At left, the joint is seen to become first cartilagenous, then fibrous. (HE × 2.) **b.** In this section, synovial tissue lies at right. Both cartilages are rich in metachromatic proteoglycan but, as is usual in osteoarthrosis, the superficial zones of the cartilages are depleted of this matrix material. (Toluidine blue × 2.5.) **c.** An oblique lamina of dense bone lies beneath the remaining tissue and the attenuated iliac cartilage. There is superficial fibrillation of the sacral cartilage adjoining a synovial 'bridge'. (HE × 1.5.) **d.** Focal absence of metachromatic matrix from the iliac cartilage marks the site where vascular, synovial pedicles traverse the sacroiliac joint space. (Toluidine blue × 1.5.) **e.** In this elderly subject, the osteoporotic bone structure highlights the hyaline sacral, the much more fibrous iliac cartilages. (HE × 4.2.)

only considered, by convention, in terms of the seronegative spondyloarthropathies.[14] One reason for apparent neglect of the pathology of the sacroiliac joints must be their anatomical inaccessibility.[35a] It is relatively difficult, time-consuming and inconvenient to examine them *post mortem* and biopsy studies, for obvious reasons, are extremely uncommon.

When systematic dissections were made of 30 paired sacroiliac joints, selecting material from male and female patients dying in hospital from diseases not directly related to the osteoarticular system, the wide spread of incidental sacroiliac bone disease was found to include metastatic carcinoma, leukaemia and chondrosarcoma. The synovia were occasional sites of rheumatoid synovitis while the articular cartilages were prone to degenerative change, osteoarthrosis and amyloidosis.[35] For reasons not related to previous traumatic injury, fibrin, identified by histochemical stains, transmission electron microscopy and by the application of an antifibrin monoclonal antibody, was found to be present in the synovial joint space in 15 of 17 instances.[35]

REFERENCES

1. Adam M, Deyl Z. Degenerated annulus fibrosus of the intervertebral disc contains collagen type III. *Annals of the Rheumatic Diseases* 1984; **43**: 258–63.

2. Adler J H, Schoenbaum M, Silberberg R. Early onset of disk degeneration and spondylosis in sand rats (*Psammomys obesus*). *Veterinary Pathology* 1983; **20**: 13–22.

3. Andrae R. Uber Knorpelnötchen am hinteren Ende der Wirbelbandscheiben im Bereiche des Spinalkanals. *Beitrage zür pathologische Anatomie und zür allgemeinen Pathologie (Jena)* 1929; **82**: 464–74.

4. Andrews J L, Sutherland J, Ghosh P. Distribution and binding of glycosaminoglycan polysulfate to intervertebral disc, knee joint articular cartilage and meniscus. *Arzneimittel Forschung* 1985; **35**: 144–8.

5. Ayad S, Weiss J B. Biochemistry of the intervertebral disc. In: Jayson M I V, ed, *The Lumbar Spine and Back Pain* (3rd edition). Edinburgh: Churchill Livingstone, 1987: 100–37.

6. Bailey R W, Sherk H H, Dunn E J, Fielding J W, Long D M, Ono K, Penning L, Stauffer E S (Editorial Committee of the Cervical Spine Research Society): *The Cervical Spine*. Philadelphia: J B Lippincott & Co., 1983.

7. Ball J. New knowledge of intervertebral disc disease. *Journal of Clinical Pathology* 1978; **31** suppl. (Royal College of Pathologists) **12**: 200–4.

8. Ball J, Sharp J. Rheumatoid arthritis of the cervical spine. In: Hill A G S, ed, *Modern Trends in Rheumatology*. London: Butterworth, 1971.

9. Bhushan C, Hodges F J III, Wityk J J. Synovial cysts (ganglia) of the lumbar spine simulating extradural mass. *Neuroradiology* 1979; **18**: 263–8.

9a. Bland J. *Disorders of the Cervical Spine*. Philadelphia, London: W B Saunders Company, 1987.

10. Bobechko W P, Hirsch C. Autoimmune response to nucleus pulposus in the rabbit. *Journal of Bone and Joint Surgery* 1965; **47**: 574–80.

11. Brain W R. Spondylosis: the known and the unknown. *Lancet* 1954; **i**: 687–93.

12. Brain W R, Northfield D, Wilkinson M. The neurological manifestations of cervical spondylosis. *Brain* 1952; **75**: 187–225.

13. Buttle D J, Tudor J, Barrett A J. Effect of X-ray contrast media on the action of chymopapain on the intervertebral disc: an *in vitro* study of cartilage degradation. *British Journal of Radiology* 1984; **57**: 475–7.

14. Bywaters E G L. The pathology of the spine. In: Sokoloff L, ed, *The Joints and Synovial Fluid* (volume II). London: Academic Press, 1980: 427–547.

15. Bywaters E G L, Evans S. The lumbar interspinous bursae and Baastrup's syndrome. *Rheumatology International* 1982; **2**: 87–96.

16. Bywaters E G L, Hamilton E B D, Williams R. The spine in idiopathic haemochromatosis. *Annals of the Rheumatic Diseases* 1971; **30**: 453–65.

16a. Camins M B, O'Leary P F. *The Lumbar Spine*. New York: Raven Press, 1987.

17. Chou C-W. Pathological studies on calcification of the intervertebral discs. *Journal of the Japanese Orthopaedic Association* 1982; **56**: 331–45.

18. Collins D H. *The Pathology of Articular and Spinal Diseases*. London: Edward Arnold, 1949.

19. Collins D H. Degenerative diseases. In: Nassim R, Burrows H J, eds, *Modern Trends in Diseases of the Vertebral Column*. London: Butterworth, 1959.

20. Conrad M R, Pitkethly D T. Bilateral synovial cysts creating spinal stenosis: CT diagnosis. *Journal of Computer Assisted Tomography* 1987; **11**: 196–7.

21. Crock H V. A reappraisal of intervertebral disc lesions. *Medical Journal of Australia* 1970; **1**: 983–9.

22. Dandy W E. Serious complications of ruptured intervertebral discs. *Journal of the American Medical Association* 1942; **119**: 474–7.

22a. Eckardt J J, Kaplan D D, Batzdorf U, Dawson E G. Extraforaminal disc herniation simulating a retroperitoneal neoplasm. Case report. *Journal of Bone and Joint Surgery* 1985; **67-A**; 1275–7.

23. Eisenstein S M, Parry C R. The lumbar facet arthrosis syndrome. Clinical presentation and articular surface changes. *Journal of Bone and Joint Surgery* 1987; **69-B**: 3–7.

24. Ellman M H, Vazques L T, Brown N L, Mandel N. Calcium pyrophosphate dihydrate deposition in lumbar disc fibrocartilage. *Journal of Rheumatology* 1981; **8**: 955–8.

25. Epstein J A, Lavine L S, Epstein B S. Recurrent herniation of the lumbar intervertebral disk. *Clinical Orthopaedics and Related Research* 1967; **52**: 169–78.

26. Eyring E J. The biochemistry and physiology of the intervertebral disc. *Clinical Orthopaedics and Related Research* 1969; **67**: 16–28.

27. Farfan H F, Cossette J W, Robertson G H, Wells R V, Kraus H. The effects of torsion on the lumbar intervertebral joints: the role of torsion in the production of disc degeneration. *Journal of Bone and Joint Surgery* 1970; **52-A**: 468–97.

28. Farfan H F, Sullivan J B. The relation of facet orientation to intervertebral disc failure. *Canadian Journal of Surgery* 1967; **10**: 179–85.

29. Feik S A, Storey E. Joint changes in transplanted caudal vertebrae. *Pathology* 1982; **14**: 139–47.

29a. Feinberg J, Boachiedjei O, Bullough P G, Boskey A L. The distribution of calcific deposits in intervertebral discs of the lumbosacral spine. *Clinical Orthopaedics and Related Research* 1990; **254**: 303–10.

30. Forestier J, Lagier R. Ankylosing hyperostosis of the spine. *Clinical Orthopaedics and Related Research* 1971; **74**: 65–83.

31. Forestier J, Lagier R, Certonciny A. Le concept d'hyperostose vertebrale ankylosante. Approche anatomo-radiologique. *Revue du Rhumatisme* 1969; **36**: 655–61.

32. Forestier J, Rotes-Querol J. Senile ankylosing hyperostosis of the spine. *Annals of the Rheumatic Diseases* 1950; **9**: 321–30.

33. François R J. *Le Rachis dans le Spondyloarthrite Ankylosante.* Brussels: Editions Arscia, 1975.

34. Gardner D L. *Pathology of the Connective Tissue Diseases* London: Edward Arnold, 1965.

35. Gardner D L, Hollinshead M, Brereton J. Comparative thickness, cell density and histochemistry of sacral and iliac cartilages in man. *British Journal of Rheumatology*, 1985; **24**: 90.

35a. Gardner D L, Vernon-Roberts B, Brereton J D, Hollinshead M S. Scanning electron microscopic observations and image analysis of human sacroiliac joints: age changes and their significance. *Journal of Pathology* 1984; **142**: A15.

36. Goldthwait J E. The lumbo-sacral articulation. *Boston Medical and Surgical Journal* 1911; **164**: 365–72.

37. Goobar J E, Clark G M. Sclerosis of the spinous processes and low back pain ('cock spur' disease). *Archives of InterAmerican Rheumatology* 1962; **5**: 587–96.

37a. Grahame R, ed. *Clinics in Rheumatic Disease* (volume 6), 1980.

38. Hadden W A, Swanson A J G. Spinal infection caused by acupuncture mimicking a prolapsed intervertebral disc. *Journal of Bone and Joint Surgery* 1982; **64-A**: 624–6.

39. Hadlow W J. Pulmonary emboli of nucleus pulposus accompanying degeneration of intervertebral disks in ranch mink. *Veterinary Pathology* 1982; **19**: 444–7.

40. Hayashi K, Yabuki T. Origin of the uncus and of Luschka's joint in the cervical spine. *Journal of Bone and Joint Surgery* 1985; **67-A**: 785–91.

41. Hemminghytt S, Daniels D L, Williams A L, Haughton V M. Intraspinal synovial cysts: natural history and diagnosis by CT. *Radiology* 1982; **145**: 375–6.

42. Henke F, Lubarsch O. Handbuch der speziellen pathologischen Anatomie und Histologie, 27 volumes. Berlin: J. Springer, 1924–39.

43. Hilton R C, Ball J. Vertebral rim lesions in the dorsolumbar spine. *Annals of the Rheumatic Diseases* 1984; **43**: 302–7.

44. Hilton R C, Ball J. Letter: Vertebral rim lesions in dorsolumbar spine. *Annals of the Rheumatic Diseases* 1984; **43**: 856–7.

45. Hilton R C, Ball J, Benn R T. Vertebral end-plate lesions (Schmorl's nodes) in the dorsolumbar spine. *Annals of the Rheumatic Diseases* 1976; **35**: 127–32.

45a. Hilton R C, Ball J, Benn R T. *In-vitro* mobility of the lumbar spine. *Annals of the Rheumatic Diseases* 1979; **38**: 378–83.

45b. Hilton R C, Ball J, Benn R T. Annular tears in the dorsolumbar spine. *Annals of the Rheumatic Diseases* 1980; **39**: 553–8.

46. Holm S H, Urban J P G. The intervertebral disc: factors contributing to its nutrition and matrix turnover. In: Helminen H J, Kiviranta I, Tammi M, Säämänen A-M, Paukkonen K, Jurvelin J, eds, *Joint Loading.* Bristol: John Wright, 1987: 187–224.

47. Hughes S. *Textbook of Orthopaedic Surgery.* Edinburgh: Churchill-Livingstone, 1987.

48. Jaffe H L. *Tumours and Tumorous Conditions of the Bones and Joints.* London: Henry Kimpton, 1958.

49. Jaffe H L. *Metabolic, Degenerative, and Inflammatory Diseases of Bone and Joints.* Philadelphia: Lea & Febiger, 1972.

50. Jayson M I V ed, *The Lumbar Spine and Back Pain* (3rd edition). Edinburgh: Churchill Livingstone, 1987.

50a. Jorgensen L, Lorentzen J E. Saccular aneurysm of the abdominal aorta with erosion of the lumbar spine. *Ugeskrift for Laeger (Copenhagen)* 1985; **147**: 1777–8.

51. Key I. Intervertebral disc lesions in children and adolescents. *Journal of Bone and Joint Surgery* 1950; **32-A**: 97–102.

52. King A B. Surgical removal of a ruptured intervertebral disc in early childhood. *Journal of Pediatrics* 1959; **55**: 57–62.

53. Kirkaldy-Willis W H, Wedge J T, Yong-Hing K, Reilly J. Pathology and pathogenesis of lumbar spondylosis and stenosis. *Spine* 1978; **3**: 319–28.

54. Kleinebrecht J, Svejcar J. Correlation of vertebral malformations with the synthesis and content of mucopolysaccharides during chondrogenesis. *Virchows Archives A Pathological Anatomy and Histology* 1979; **382**: 271–81.

55. Konschegg T. Die Tuberkulose der Gelenke. In: Lubarsch O, Henke F eds, *Handbuch der speziellen pathologischen Anatomie und Histologie* Berlin: Springer, 1934: 438–68.

56. Krayenbühl H, Zander E. Rupture of lumbar and cervical intervertebral disks. *Documenta Rheumatologica; English edition* (Basel) 1956; No. 1: 1–80.

57. Ladefoged C, Fedders O, Petersen O F. Amyloid in intervertebral discs: a histopathological investigation of surgical material from 100 consecutive operations on herniated discs. *Annals of the Rheumatic Diseases* 1986; **45**: 239–43.

58. Lagier R. L'hyperostose vertebrale en pathologie comparée. *Revue du Rhumatisme* 1979; **46**: 467–73.

59. Lagier R, Sit'aj S. Vertebral changes in ochronosis. *Annals of the Rheumatic Diseases* 1974; **33**: 86–92.

60. Lipson S J. Metaplastic proliferative fibrocartilage as one alternative concept to herniated intervertebral disc. *Spine* 1988; **13**: 1055–60.

61. Lipson S J, Muir H. Experimental intervertebral disc degeneration. Morphologic and proteoglycan changes over time. *Arthritis and Rheumatism* 1981; **24**: 12–21.

62. Luschka H von. Die Halbgelenke des Menschlichen Körpers. Eine Monographie. Berlin: Reimer, 1858.

63. McCarty D J, Fr, Gatter R A. Identification of calcium hydrogen phosphate dihydrate crystals in human fibrocartilage. *Nature* 1964; **201**: 391–2.

64. Middleton C S Teacher J H. Injury of the spinal cord due to rupture of an intervertebral disc during muscular effort. *Glasgow Medical Journal* 1911; **76**: 1–6.

65. Mixter W J, Barr J S. Rupture of intervertebral disc with involvement of the spinal canal. *New England Journal of Medicine* 1934; **211**: 210–5.

66. Mogle P, Amitai Y, Rotenberg M, Yatziv S. Calcification of intervertebral disks in I-cell disease. *European Journal of Pediatrics* 1986; **145**: 226–7.

67. Mohr W. Morphologie und Pathogenese der Spondylosis hyperostotica. In: Ott V R, ed, *Spondylosis Hyperostotica.* Stuttgart: Ferdinand Enke Verlag, 1982: 17–34.

68. Nachemson A. The load on lumbar discs in different positions of the body. *Clinical Orthopaedics and Related Research* 1966; **45**: 107–22.

69. Naylor A. The biophysical and biomechanical aspects of intervertebral disc herniation and degeneration. *Annals of the Royal College of Surgeons of England* 1962; **31**: 91–114.

70. Naylor A, Happey F, Turner R L, Shentall R D, West D C, Richardson C. Enzymic and immunological activity in the intervertebral disc. *Orthopaedic Clinics* 1975; **6**: 51–8.

71. Pohjola-Sintonen S, Ruutu P, Tallroth K. Hematogenous Candida spondylitis. A case report. *Acta Medica Scandinavica* 1984; **215**: 85–7.

72. Puschel J. Der Wassengehalf normaler und degenerierter Zurachenwirbelscheiben. *Beitrage für pathologische Anatomie* 1930; **84**: 123–30.

73. Resnick D, Niwayama G. *Diagnosis of Bone and Joint Disorders* (2nd edition). Philadelphia: W B Saunders & Co., 1988: 1480–615.

74. Revell P A, Pirie C J. The histopathology of ankylosing hyperostosis of the spine. *Rhumatologie* 1981; **33**: 99–104.

74a. Saluja G, Fitzpatrick K, Bruce M, Cross J. Schmorl's nodes (intravertebral herniations of intervertebral disc tissue) in two historic British populations. *Journal of Anatomy* 1986; **145**: 87–96.

75. Schmorl G, Junghans H. Die gesunde und kranke Wirbelsaule im Rontgenbild. *Fortschrift für Röntgenstrahlung* 1932; **43**: supplement.

76. Schmorl G, Junghans H. In: Besemann E F, transl, *The Human Spine in Health and Disease.* New York: Grune and Stratton, 1971.

77. Scott J E, Haigh M. Proteoglycan-collagen interactions in intervertebral disc. A chondroitin sulphate proteoglycan associates with collagen fibrils in rabbit annulus fibrosus at the **d-e** bands. *Bioscience Reports* 1986; **6**: 879–88.

78. Seawright A A, English P B, Gartner R J. Hypervitaminosis A and hyperostosis of the cat. *Nature* 1965; **206**: 1171–2.

79. Sims-Williams H, Jayson MIV, Baddeley H. Small spinal fractures in back pain patients. *Annals of the Rheumatic Diseases* 1978; **37**: 262–5.

80. Sit'aj S, Zitnan D. Spinal changes in chondrocalcinosis. *Proceedings of the Sixth European Congress of Rheumatology,* Prague, 1967.

81. Smith L, Brown J E. Treatment of lumbar intervertebral disc lesions by direct injection of chymopapain. *Journal of Bone and Joint Surgery* 1967; **49-B**: 502–19.

82. Takamori. *Kascin-Beck Disease* (volumes 1 and 2). Tokyo: Gifu University School of Medicine, 1968.

83. Takeda T, Sanada H, Ishii M, Matshushita M, Yamamuro T, Shimizu K, Hosokawa M. Age-associated amyloid deposition in surgically-removed herniated intervertebral discs. *Arthritis and Rheumatism* 1984; **27**: 1063–5.

83a. Theiler K. Vertebral malformation. *Advances in Anatomy, Embryology and Cell Biology* 1988; **112**: 1–99.

84. Thomas A M C, Afshar F. The microsurgical treatment of lumbar disc protrusion. Follow-up of 60 cases. *Journal of Bone and Joint Surgery* 1987; **69-B**: 696–8.

85. Urban J P, Maroudas A. Swelling of the intervertebral disc *in vitro. Connective Tissue Research* 1981; **9**: 1–10.

86. Urovitz E P M, Fornasier V L. Autoimmunity in degenerative disk disease. A histopathologic study. *Clinical Orthopaedics and Related Research* 1979; **142**: 215–8.

87. Van Lindhoudt D, François R J. Hyperostose vertebrale ankylosante apport de la microradiographie et de la microscopie de fluorescence à l'étude du rachis dorsal. *Rhumatologie* 1981; **33**: 89–98.

88. Venner R M, Crock H V. Clinical studies of isolated disc resorption in the lumbar spine. *Journal of Bone and Joint Surgery* 1981; **63-B**: 491–4.

89. Vernon-Roberts B. Pathology of intervertebral discs and apophyseal joints. In: Jayson M I V ed, *The Lumbar Spine and Back Pain* (3rd edition). Edinburgh: Churchill Livingstone, 1987: 35–55.

90. Vernon-Roberts B, Pirie C J. Healing trabecular microfractures in the bodies of lumbar vertebrae. *Annals of the Rheumatic Diseases* 1973; **32**: 406–12.

91. Vernon-Roberts B, Pirie C J. Degenerative changes in the intervertebral discs of the lumbar spine and their sequelae. *Rheumatology and Rehabilitation* 1977; **16**: 13–21.

92. Virchow R. *Untersuchung über die Entwicklung des Schadelgrundes im gesunden und krankhaften Zustand, und über den Einfluss desselben auf Schadelform, Gesichtsbildung und Gehirnbau.* Berlin: Georg Reimer, 1857.

93. Wahren H. Hernie des Nucleus pulposus bei einem 12 jahrigen Kind. *Acta Orthpaedica Scandinavica* 1939; **10**: 286–88.

94. Wiltse L L, Newman P H, Macnab I. Classification of spondylosis and spondylolisthesis. *Clinical Orthopaedics and Related Research* 1976; **117**: 23–9.

95. Woolfrey B F, Lally R T, Faville R J. Intervertebral diskitis caused by *Kingella kingae. American Journal of Clinical Pathology* 1986; **85**: 745–9.

96. Yasui N, Ono K, Yamaura I, Konomi H, Nagai Y. Immunohistochemical localization of types I, II and III collagens in the ossified posterior longitudinal ligament of the human cervical spine. *Calcified Tissue International* 1983; **35**: 159–63.

97. Yasuma T, Makino E, Saito S, Inui M. Histological development of intervertebral disc herniation. *Journal of Bone and Joint Surgery* 1986; **68-A**: 1066–72.

Chapter 24

Responses of Connective Tissues to Physical and Chemical Agents, Neoplasia, and to other Stimuli

In any systematic account of disease, there comes a point when classification falters either through lack of knowledge of the disorder or insufficient understanding of its aetiology and pathogenesis.[88] The connective tissue diseases are no exception. There are miscellaneous conditions attributable to physical and chemical agents. There are others in which the connective tissue system is responding to the growth of benign tumours, or to the spread and metastasis of malignant neoplasms. There remains a small subset of human diseases in which little of significance is known of their nature or cause(s) and in which their existence as diagnostic entities is questionable.

Physical and chemical agents

Every form of physical and chemical agency is capable of disorganizing the connective tissues. Among the physical agents that can effect such change are: heat or cold; ionizing radiation; violent acceleration or deceleration; tensile, compressive or sheer stress; missile injuries; and explosions in air, underground or underwater. The effects of many of these agents include fractures, strains, dislocations, evulsions, protrusions or prolapse and they find their place in textbooks of orthopaedics, physical medicine and sports medicine.

The non-mineralized connective tissues are susceptible to autolysis the phenomena of which are easily and quickly identified in vascular structures such as the synovia but which are much less conspicuous and of insidious, slower onset in avascular or hypovascular tissues such as articular cartilage. Cell death, with nuclear pyknosis and karyorrhexis or karyolysis, is recognizable in ischaemia of the vascular connective tissues. However, apoptosis may be responsible for cell loss or disorganization in the connective tissues of low vascularity.

Heat and cold

Burns and scalds

Burns of sufficient severity to destroy the epidermis, injure dermal connective tissue. In increasingly severe injury, tendons, ligaments, joint capsules and synovia may be directly affected by thermal injury. In one example, extensive burns, followed by keloid formation, were complicated by the late development of temporomandibular ankylosis with osteosclerosis of the condyloid process.[155]

Frostibite

Extreme environmental conditions encountered occupationally or during exploration, Arctic or Antarctic travel or at high altitude, are capable of damage to both articular and non-articular connective tissue and bone. Half of those affected by frostbite develop subsequent arthropathy.[171] The clinical and radiological disorders resemble secondary osteoarthrosis.[101] In children, epiphyseal fusion or defective epiphyseal growth may result in clinodactyly. Little is known of the histological changes. Experiments have been made to test the effects of cold on connective tissue. There are three phases of injury:[152] prenecrotic, necrotic and regenerative. They are associated with a complex sequence of changes in collagen, proteoglycan, elastic material and lipid, catabolism of which can be monitored by the urinary excretion of products such as Hypro and hexosamine. Intracellular ice crystal formation leads to shifts of water and electrolytes, and to cell injury. However, chondrocytes appear relatively resistant to cooling and the long-lasting effects of frostbite may be attributed to vascular endothelial change, platelet aggregation, thrombosis, and vascular insufficiency. Bone remodelling follows, possibly superimposed on bone necrosis.

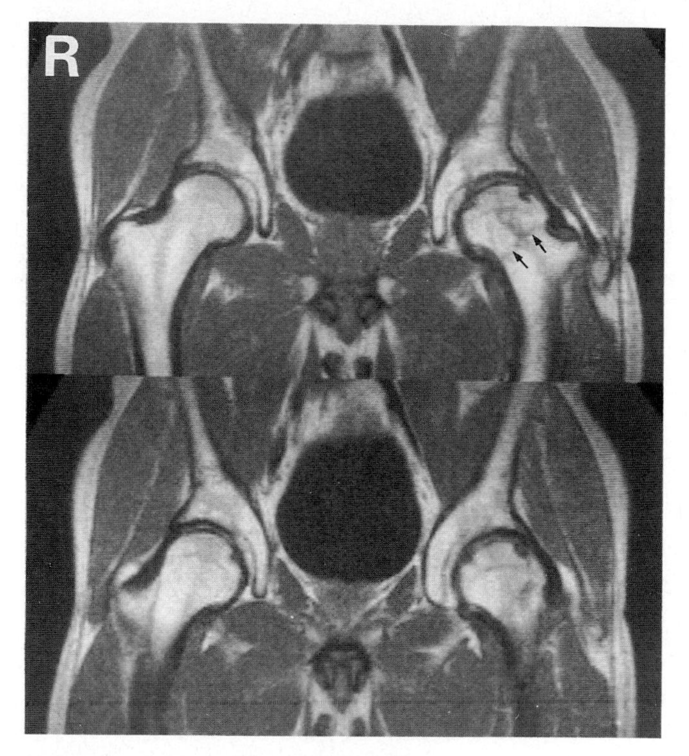

Fig. 24.1 Traumatic arthropathy.
Bone ischaemia of left femoral head secondary to fracture of the femoral neck on two contiguous T_1-weighted images. Note the irregular contour and reduced signals within the left femoral head and the site of the previous fracture (arrowed). The articular cartilage is well shown. Normal right hip for comparison. Courtesy of the Department of Diagnostic Radiology, University of Manchester.

Mechanical injury

The influence of excess or diminished mechanical loads on articular cartilage is considered in Chapter 22 (see p. 886) where the relationship between injury and osteoarthrosis is discussed (Fig. 24.1)[105] and the influence of spontaneous and induced instability[125] reviewed. The effects of scarification of cartilage[109] and the capacity to incorporate $^{35}SO_4$ after injury[110] have been examined experimentally.

Traumatic arthropathy

Blunt trauma, insufficient to disrupt an articular cartilage surface, is capable of causing chondrocyte disorganization,[151] a loss of ruthenium-red staining matrix[116,120] and a change in collagen fibre diameter.[29] There is an early increase in cartilage water. When injury is more severe, as in acute traumatic arthritis, it is followed by painful swelling of the joint accompanied by inflammatory effusion. The injury may be accompanied by intraarticular fracture, the healing of which then complicates and delays the resolution of the articular lesions.

Direct injury Within the joint itself there may be direct synovial injury with an acute synovitis in which the villi are swollen and hyperaemic. Liquid fat may be present in the synovial fluid.[137] Much triglyceride in the fluid suggests bone marrow leakage with bone injury but a low phospholipid content is an even more valuable indicator of trauma whether there is fracture or not. The histological response[162] may be disproportionate to the severity of the injury (Fig. 24.2). Fat necrosis occurs and may contribute to the chylous effusion.[177] The fluid secreted by the disorganized villi constitutes an effusion in which there are red blood cells, polymorphs and an excess of protein. Direct injury to articular cartilage or intraarticular fibrocartilages, such as those of the knee joint, is rare; indirect injury to these structures, particularly to the fibrocartilages, is common. In the injured hip, the acetabular fibrocartilage displays little response. The associated inflammatory reaction occurs at the margins of the joint surfaces and within synovial reflections at sites where fibrocartilage is attached. Cartilagenous debris within the knee can be identified by arthroscopy.

Healing and resolution of the inflammatory process are slow, perhaps on account of the low cellularity and low metabolic rate of some of the injured structures. Torn fibrocartilages heal poorly (see Chapter 2). Synovia become fibrotic. The metabolism of the articular cartilage itself is impaired. The injury of nearby tendinous and capsular tissue is frequent. The removal of necrotic material and the

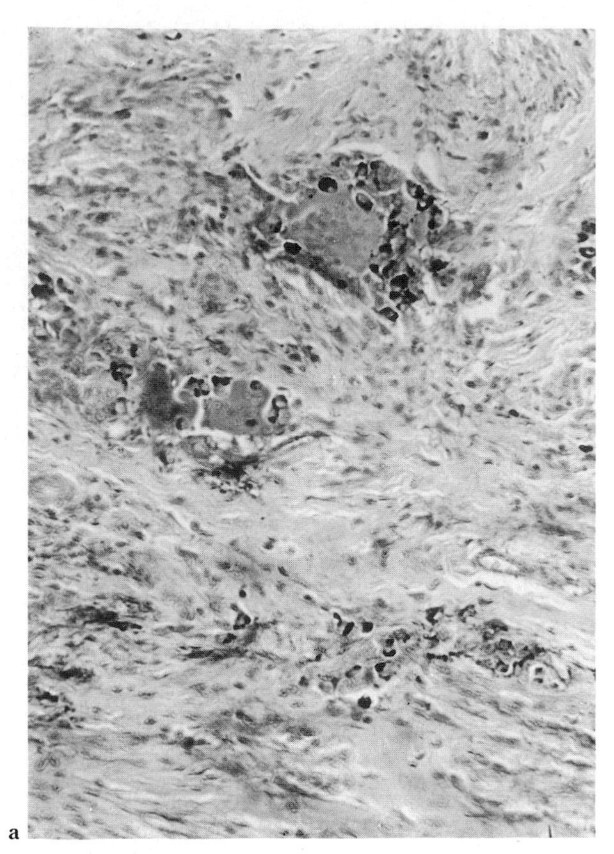

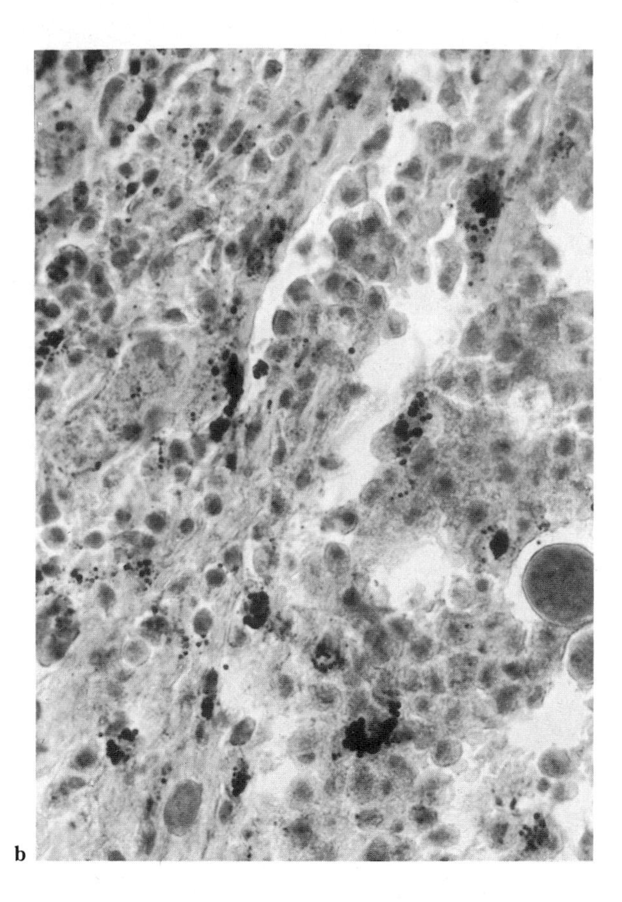

Fig. 24.2 Traumatic arthropathy.
Direct traumatic injury to knee joint synovial tissue in young male adult. **a.** Fat necrosis has elicited inflammatory reaction with multinucleated giant-cells in zones of reactive early fibrosis. Response is disproportionately severe in view of trivial clinical injury. Giant-cells contain stainable fat. **b.** Sudanophilic (black) fat droplets within cells at site of intense synovial cell proliferation. Note presence of mononuclear macrophages and plasma cells. (**a**: HE × 190; **b**: Sudan III × 500.)

replacement of displaced tissue are essential preliminaries to healing. Repeated episodes of trauma, infection, the presence of foreign or necrotic material and pathological changes, such as metastatic tumour, lead to a perpetuation of the arthritic changes and to delay in healing. The most frequent complication of repeated trauma is secondary osteoarthrosis.[114]

Vibration Although persistent exposure to low frequency vibration, as in the use of pneumatic drills, is widely recognized to cause vasospasm, the osteocartilagenous lesions that result are selective. For example, radiotranslucent bone 'cysts' are identified with increased frequency in the semilunate but not in other arm or wrist bones.[104] There is little doubt that bone change precipitates cartilagenous degradation.[138]

Battered child syndrome Fracture, subdural haematoma, bruising and soft-tissue swellings are characteristic; but failure to thrive may be a first sign of ill-treatment.[3] Since metaphyseal fractures are frequent, it is not surprising that irregular deformity of the distal humerus, for example, and parametaphyseal opacities should simulate articular disease.

Sports, road and occupational injury The synovial lesions encountered at arthrotomy in 35 cases with earlier sports, road or occupational injury were considered by Soren *et al.*[162] Light microscopy revealed slight synovial cell hyperplasia but rarely hypertrophy with a few multinucleate giant-cells. Perivascular concentric collagen proliferation was seen and haemosiderin was occasionally present. Twenty per cent of specimens included loose lymphocytic and plasma cell infiltrates, but polymorphs were rare. Scanning electron microscopy indicated that a fine, microvillous pattern still characterized the surface of residual synoviocytes and electron probe X-ray microanalysis confirmed that crystalline material contained iron. Transmission electron microscopy detected A, B and intermediate cells in early cases but after six months, siderosomes were no longer seen, although collagen was dense and abundant, and the synovial cell layer atrophic. Bony fusion

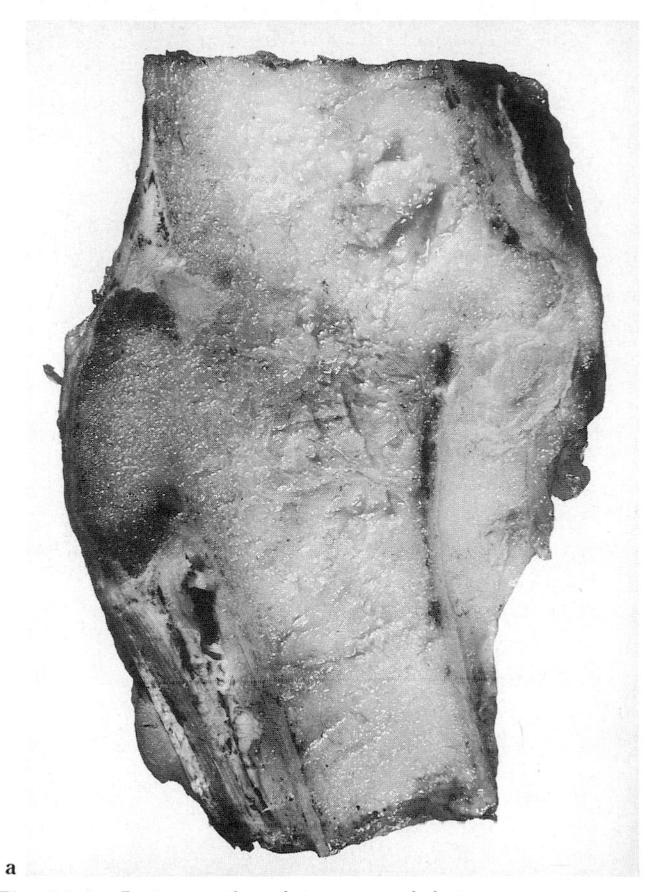

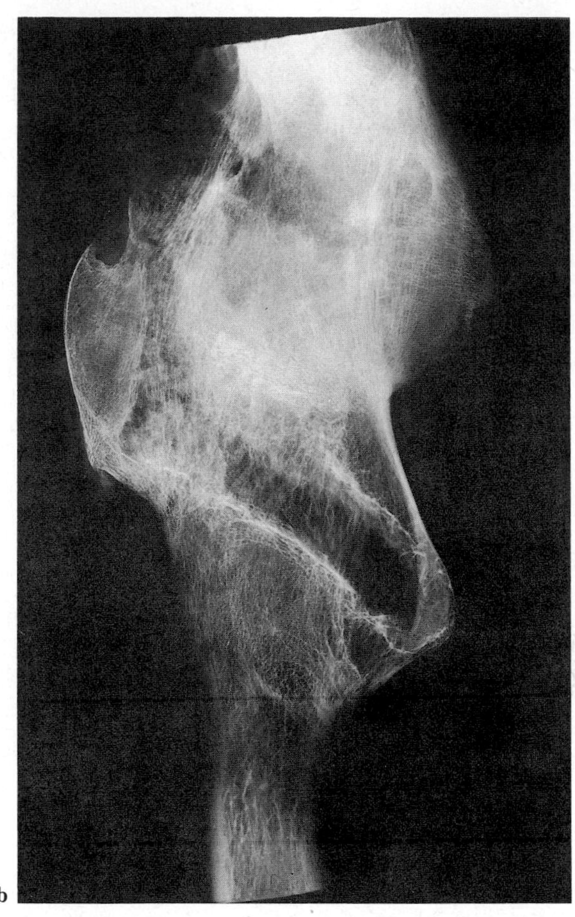

Fig. 24.3 Late results of trauma to joint.
a. Bony ankylosis after severe injury with fusion of patella, femur and tibia into single mass. **b.** X-ray of sagittal section of specimen to show coalescence of three bones seen in Fig. 24.3a.

of the knee (Fig. 24.3) and ankle, now rare in adults, are late consequences of the limb deformity that can result from road accidents in childhood. Contributory factors include extensive intraarticular fracture; prolonged immobilization; and chronic infection.

Sudeck's atrophy

Immobilization exerts a detrimental effect on articular cartilage and it has long been appreciated that the bone changes characteristic of Sudeck's post-traumatic atrophy[84,86] may be accompanied by cartilagenous and synovial disease. Pannus formation, erosive disease, fibrous and, occasionally, bony ankylosis, are identifiable: they simulate the disorders of primary generalized connective tissue disease.

Response to foreign materials

The parts played by monosodium biurate (see p. 383) calcium pyrophosphate dihydrate (see p. 393), calcium dihydrogen phosphate (see p. 402), calcium hydrooxyapatite (see p. 407), fluoride (see p. 411), iron (see p. 967) and bismuth (p. 413) are considered elsewhere. In addition osmium, thorium, gold[127,173,174] and yttrium may accumulate in synovia after therapeutic administration (see p. 552); and lead (plumbism), copper (Wilson's disease), silver (argyria), arsenic, beryllium, antimony, cobalt, nickel, cadmium and chromium can be shown to remain within the synovia after prolonged occupational ingestion, therapeutic administration or in the course of an inborn error of metabolism. Bismuth may be associated with an unusual shoulder joint disease. The accumulation of aluminium in cartilage, synovial fluid and synovial tissue has been described in patients on haemodialysis for chronic renal failure and subjected to hip joint arthroplasty following osteonecrosis of the femoral head. Aluminium phosphate is identifiable within synovial cell lysosomes. The tissue responses to the metals and polymers used in prosthetic surgery are described on p. 960 *et seq*.

Osteochondrosis

This term is applied to chondrocyte death, with proteoglycan loss, in the articular–epiphyseal cartilage complex. It constitutes a significant clinical disease in pigs[18] but may contribute to the onset of osteochondritis dissecans in man. The relationship of osteochondrosis to osteoarthrosis in swine is considered in Chapter 22 (see p. 897). One explanation for osteochondrosis is defective vascular perfusion of the affected territory.[180] There is an analogy with Legg–Calvé–Perthes disease.

Osteochondritis dissecans

Osteochondritis dissecans[8,73] is an infrequent disorder of the articular extremity of a limb, usually affecting the femoral condyle. It is a disease of young males aged 15–25 years. Familial cases have been described. The abnormality may be bilateral and, in these instances, other joints such as the elbow may be implicated. The radiological and magnetic resonance imaging changes are pathognomonic.

Characteristically, a fragment of necrotic bone together with its covering of articular cartilage, is first delimited and then extruded into the joint space. If the extruded particle is composed solely of cartilage, the term 'chondritis dissecans' has been suggested but both this term and the designation osteochondritis dissecans are imprecise since hyaline cartilage, normally avascular, cannot undergo primary inflammation.

Pathological features Macroscopically, it is difficult to define the lesion if the cartilage surface is uninterrupted. Generally, however, the osteochondritic fragment is recognized by an encircling surface grove; later, the affected cartilage is grossly irregular, soft and discoloured. When detachment is partial, the intraarticular body hangs from the joint margin. When detachment is complete, the elliptical structure lies within the joint constituting an approximately 25 × 5 × 15 mm diameter mass with a small but variable amount of subchondral bone. The free body may increase in size; it may also become encased in a thick layer of calcified cartilage and fibrocartilage to form a so-called 'joint mouse'.

Microscopically, two groups of change are recognized. First, there is complete or subtotal bone necrosis involving the marrow. The subchondral bone trabeculae are thickened; appositional new bone formation may precede the necrosis. The deeper surface of the body is covered by fibrous or fibrocartilagenous connective tissue or even by metaplastic bone. The defect in the condyle is also occupied by fibrous connective tissue and the hyaline cartilage surrounding the defect is fibrillated and undermined. Second, the synovia may contain microscopic particles of cartilage or of necrotic bone. There are signs of low-grade synovitis.

Pathogenesis[7] Loose bodies in the osteochondritic knee can be the result of severe or of minor trauma. However, an alternative view, which reconciles the poorly understood changes of osteochondrosis (cartilage necrosis) with the subsequent phenomena of osteochondritis dissecans, emphasizes the likely role of defective blood supply. However, ischaemic necrosis cannot always be confirmed when a loose body is examined microscopically. A further explanation is the mechanical contribution of an exceptionally long tibial spine or an aberrant location of the posterior cruciate ligament leading to abnormal loads upon parts of the medial femoral condyle. Fracture is also a suggested cause. An entirely different viewpoint[8] stresses the unevenness of ossification in childhood. The irregularities are generally self-correcting. However, if a more severe, focal, lasting defect results, subchondral bone sclerosis is seen to be accompanied by thickening of the overlying cartilage. Shear stresses (p. 262) at the margins of the focal defect result in separation of the cartilage from the deeper part. A tide-mark is often absent. Barrie[7,8] notes that the development of focal ossification within the separated (avascular) cartilage implies a tissue bridge to the (vascular) epiphysis. Fracture of this bridge provokes callus formation.

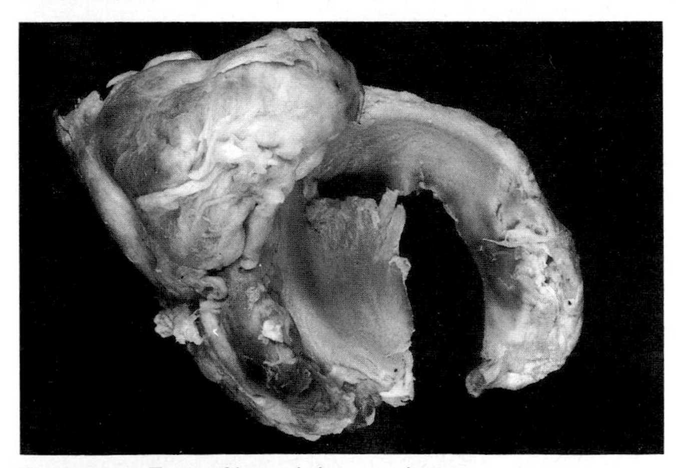

Fig. 24.4 Tear of knee joint meniscus.
Disruption by shear stress has torn posterolateral part of meniscus. Although anatomical shape is not easily defined, disc is probably right medial.

Injuries to discs and menisci

Injuries to discs

Mechanisms and consequences of physical disruption of discs are exemplified by the disorganization of intervertebral discs considered in Chapter 23.

Injuries to menisci
(Fig. 24.4)

The causes and results of meniscal injury are reviewed in terms of the tears of knee joint menisci that are a very common cause of internal derangement of this complex articulation. Located between the opposed femoral and tibial condyles, and attached to and co-extensive with the collateral ligaments, the menisci of the knee are susceptible to excessive movements in abduction, under load. Indirect mechanical injuries affect the medial meniscus very much more often than the lateral, an interesting correlate of the tendency for osteoarthrosis to prejudice the non-meniscal-covered hyaline cartilage of the medial condyle to a greater extent than the lateral. One reason why lateral menisci are less often torn than medial may be the less firm union of the former with the collateral ligament. Very occasionally, direct, impact injury to a medial meniscus may occur when the flexed knee is directly struck by an external object.

The close structural relationship between the medial meniscus and the medial collateral ligament determines that tears of the latter accompany injuries to the former. Bleeding occurs from the blood vessels supplying the ligament. A haematoma forms and becomes organized. Fibrous tissue is generated and the adjacent articular capsular connective tissue is thickened. The nearby synovium is disrupted and synovitis is accompanied by exudation.

The meniscus itself is incompletely or completely torn; there may be circumferential splitting but tears are usually oblique or transverse. The anterior attachment site, which is close to the insertion/origin of the anterior cruciate ligament, may be distracted. Like the response to ligament rupture,[123] it is thought likely that anticollagen antibodies form.

Menisci heal poorly. The process is considered further in Chapter 2 (p. 84) and is by substitutive fibrous, not by new fibrocartilage formation; the healing sequence is closely dependent on the contiguity of the synovial membrane. Large tears generally persist without healing but repair is catalysed by operations in which synovial tissue is approximated to the site of the tear. As Collins[23] points out, the original size or semilunar shape of the disc is seldom perfectly restored. Although a wedge of fibrous tissue, triangular in cross-section, remains to separate the femoral and tibial condyles, the separation is less complete than normal.

Surgery

A wide variety of materials has been used in prosthetic and reconstructive surgery.[60a] The majority are capable of initiating connective tissue inflammatory reactions (Table 24.1).[30a]

Table 24.1

Some materials used to study tissue ingrowth into implants

Materials
Aluminium oxide ceramic
Calcium hydroxyapatite ceramic
Wire mesh cylinders
Cigarette-paper cylinders
Silicate powders
Synthetic sponges
Caragheenan
Gauze
Cotton wool
Viscous cellulose sponges
Agar agar

Arthroplasty[37,93]

It is estimated that 20–30 000 operations for the replacement of joints by protheses are undertaken each year in the UK. The majority are successful but many revision operations are necessary. In the USA the combined total is many times greater. Among the early causes of failure are dislocation and infection. These complications[31a] may also occur later when wear, loosening of the components and implant dysfunction are additional difficulties. The commoner complications of arthroplasty are listed in Table 24.2. The pathological changes in the periprosthetic tissues are considered by Mirra, Marder and Amstutz.[119]

Table 24.2

Complications of arthroplasty

	Mechanical	Biological
Early:		
	Malposition of components	Skin necrosis
	Early dislocation	Infection (primary)
	Bone fracture	Haematoma
	Injury to nerves and blood bessels	Thromboembolism
		Cardiac arrest due to bone cement (polymethyl methacrylate)
Late:		
	Traumatic fracture around prosthesis	Non-union
	Fatigue fracture of component	Ectopic ossification
	Loosening	Metastatic infection
	Late dislocation	Mycotic aneurysm
		Reactivation of tuberculosis

The materials used for arthroplasty include metals, polymers (plastics and acrylics), silicones and ceramics. Ideally, these materials should be mechanically acceptable, susceptible to sterilization, and chemically inert. They should not excite inflammatory or hypersensitivity reactions and should not be oncogenic. However, no material yet discovered conforms absolutely with all these requirements and local reactions to their presence are an inevitable consequence of arthroplasty. The nature of these responses and methods that have proved of value in examining the tissue reactions pathologically were reviewed by Revell.[141,142]

The tissues available for pathological study are first, those at the site of operation and implant; and second, those recognized locally when failed implants are revised.[15a]

In the first category, the tissue reactions are largely with bone and bone marrow. Many are caused by the cement, polymethylmethacrylate, a self-curing polymer that retains the implant within the bone. Others are attributable to the implanted material itself. Since there is an increasing trend towards the insertion of prostheses held *in situ* without polymethylmethacrylate, there is growing interest in the bone reactions to cementless fixation.

In the second category, are the reactions of the non-osseous connective tissues examined at revision surgery.[30a] The present account is confined to this group; those that occur in relation to bone, at the site of the inserted prosthesis, are outlined by Revell.[142]

Tissue reactions to metals The classical pathological literature describes tissue and cell responses to a wide variety of metals that gain access to the connective tissues by chance, at surgery or during trauma. In orthopaedic practice the greatest interest is in cobalt–chrome steel alloy (Fig. 24.5) and in stainless steel. However, porous vitallium (a chrome-cobalt-molybdenum alloy),[17] nickel,[79] and titanium alloys have been studied closely. In military surgery, the interaction between lead and traumatized tissue has long been important (Fig. 24.7).

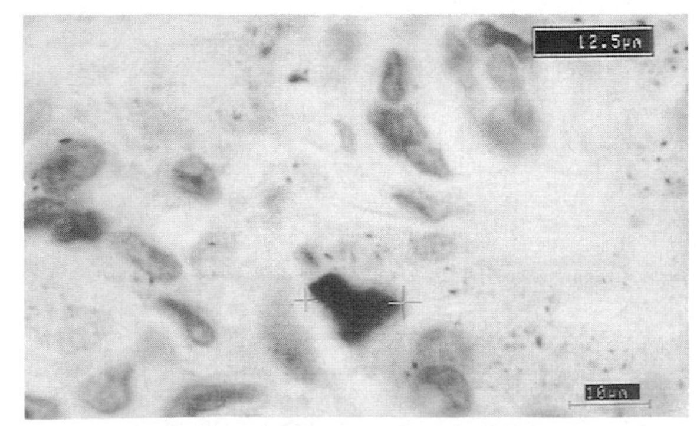

Fig. 24.5 Tissue reaction to cobalt–chrome steel. Histological preparation viewed by confocal scanning optical microscopy.

Cobalt–chrome alloy; stainless steel The synovia and adjoining soft connective tissue are discoloured black or grey-black. The coincidence of haemosiderin resulting from surgical intervention may add an orange-brown colouration. Not infrequently, the tissue adjoining a prosthesis may assume a soft, caseous appearance so that aerobic or anaerobic infection is suggested. Microscopically, the metal prosthesis is circumscribed by a loose aggregate of fibrin and granulation tissue external to which increasing quantities of fibrous tissue form a false capsule of varying extent. Mononuclear phagocytes containing fine, particulate metallic debris lie within the granulation tissue, often in groups or islands. Electron probe X-ray analysis is used to confirm the identity of the deposits.[85] Cells containing the alloy have a characteristic foamy or finely granular appearance. The metal is recognized as exceedingly small black particles of irregular shape. Many are so small that they lie at the limits of resolution of the light-microscope with which they are best detected by oil-immersion microscopy and polarized light. The intracellular particles display form birefringence. They can also be demonstrated by reflected light confocal microscopy and, after appropriate preparation, by scanning electron microscopy. Larger, metallic fragments are also often present. A low-grade vasculopathy, with lymphocytic vascular permeation, is seen but occasionally the presence of an hypersensitivity response is suggested by the recognition of a necrotizing vasculitis with fibrinoid change.

Revell[142] points out that the quantities of metals present at sites of prostheses vary: the largest amounts are found at sites where metal surfaces articulate; lesser amounts where metal adjoins bone; and the least where metal articulates with high density polyethylene.[175] The quantities of individual metals in the soft, periprosthetic tissues often do not correspond to those present in the original alloy and it is suspected that the differential solubility of, for example, cobalt and chromium, contributes to this discrepancy. Extensive studies have been made *in vitro* of the response of cells in culture to metallic particles.[172] Particulate cobalt–nickel and cobalt–chromium alloys damage murine macrophages *in vitro*, releasing lactate dehydrogenase and diminishing glucose-6-phosphate dehydrogenase activity.[139] By contrast, titanium and molybdenum appear better tolerated. Analogous observations have been made on the tissues of experimental animals.[68a]

Titanium Titanium as a titanium–aluminium–vanadium alloy has been used to manufacture static orthopaedic devices such as plates, pins and rods, and to form hip, and, less often, knee prostheses.[15,95] Fragments of titanium alloy can be liberated from these structures by abrasive wear, or by wear at the site of prosthetic fracture (fretting). Corrosion is not seen but the freeing of the alloy in solution is possible.

The consequence of titanium–aluminium–vanadium alloy liberation is the accumulation of the steel as grey-black particles within groups of connective tissue macrophages (Fig. 24.6).[2] Some particles lie free in the tissue. Like other small metallic tissue fragments released under similar circumstances, the particles of titanium show form birefringence. Their presence can be demonstrated by electron probe X-ray microanalysis (Fig. 24.6c) and the quantities measured by neutron activation analysis.[117] The question of whether the low-grade inflammatory and sustained fibrotic reactions seen at sites of titanium–aluminium–vanadium alloy accumulation are clinically significant is not resolved. The inflammatory response may be a result of T-cell-mediated delayed hypersensitivity. The blackened synovia attract immediate attention when revision arthroplasty or the removal of pins or plates are attempted. However,

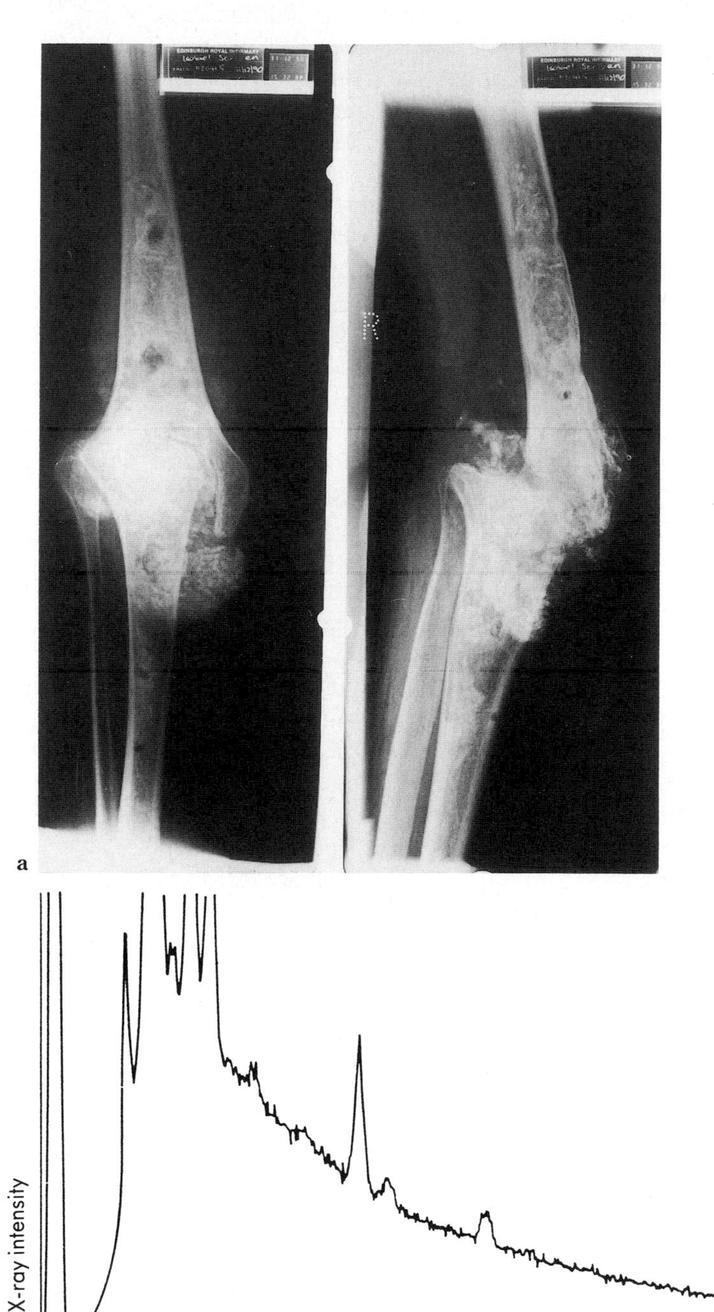

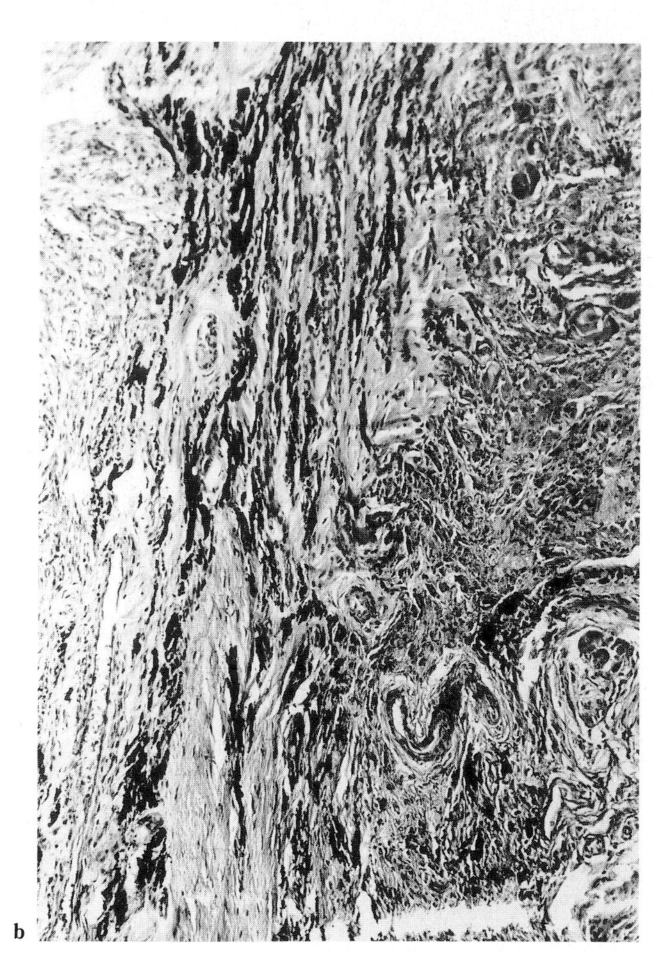

Fig. 24.6 Tissue reaction to titanium alloy.
a. Titanium–aluminium–vanadium alloy accumulations in periarticular tissues of the knee in a 76-year-old female from whom a rejected titanium alloy prosthesis had been removed. **b.** Titanium alloy aggregates in periarticular macrophages. There is often a lymphocytic response attributable to delayed hypersensitivity. (HE × 80.) **c.** Electron probe X-ray microanalysis of periarticular tissues at the site of a titanium alloy prosthesis. Note peaks for titanium K_{α} (4.5 keV) and K_{β} (4.93 keV). Dry tissue was aluminium coated. Reproduced by courtesy of Professor S Hughes and Dr K Oates.

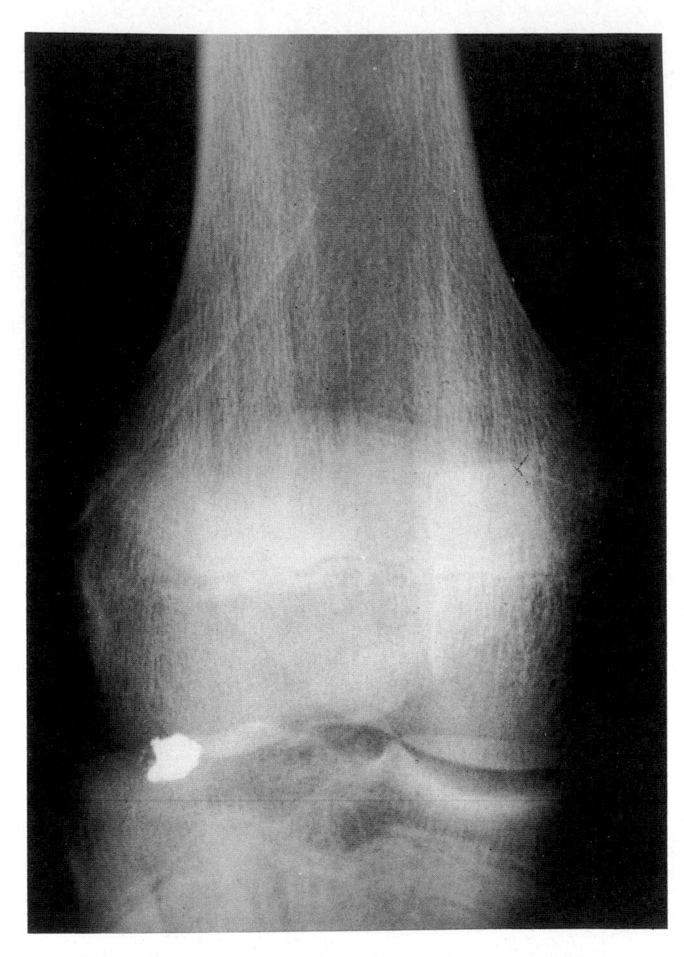

Fig. 24.7 Tissue reaction to lead.
Knee-capping injury caused by firing of revolver bullet at close range. Persistent presence of lead-containing missile in synovial tissue may provoke acute or chronic lead poisoning.

there is no direct correlation between the amount of titanium within the collagen-rich tissue and the measured amount of titanium, or between the amount of black-appearing deposit and the quantity of fibrous tissue.[117]

Lead Particles of lead and iron are seen at sites of bullet wounds and whole or fragmented bullets not infrequently lodge in joints (Fig. 24.7).[136] The missile remnants are commonly accompanied by pieces of bone, broken at impact. Lead is dissolved by synovial fluid[92] and acute or chronic lead toxicity may follow injury, particularly when the bullet is in or in direct contact with a space such as a synovial joint or bursa. Failure to remove particulate debris may cause lead synovitis[156] catalysing infection and predisposing to secondary osteoarthrosis.

Tissue reactions to polymers In many successful prostheses, a metal component articulates with a surface formed from an ultrahigh molecular weight polymer such as polyethylene.[68b,143] Alternative materials such as polyesters and polyformaldehyde have been tested and polyoxymethylene polyacetal is favoured in a Scandinavian hip prosthesis.[108] The amount of polyethylene debris is greater in convex- than concave-bearing surfaces. However, there are considerable individual differences in response between patients with identical prostheses.

Abrasive wear liberates fine and large particles of polyethylene into the connective tissues around a prosthesis.[68b] The material retains no stain: it is colourless and transparent but strongly bifringent. The particles, of all sizes, can be seen readily therefore by light-microscopy and often measure from 0.5 to 50 μm (Fig. 24.8 a, b). Occasionally very large shavings of polyethylene debris are encountered. Small particles are taken up by mononuclear phagocytes; larger particles elicit multinucleated giant-cell formation. In general, the connective tissues within which the polyethylene lies are collagenous[179] but necrosis may sometimes be evident. Fibrous polymers elicit comparable reactions (Fig. 24.8 c, d).

Tissue reactions to polymethyl methacrylate The self-curing cement polymethylmethacrylate has long been used to bond prostheses to bone. Within bone, mechanical trauma, blood vessel injury, the heat generated by the polymerizing material and its toxicity, lead to necrosis of nearby bone and bone marrow. Repair follows the demarcation of the necrotic focus by a zone of reactive hyperaemia; the reparative process extends over a period of months. A thin lamina of connective tissue of low cellularity defines the eventual limit of this repair tissue but remodelling of adjacent bone, with osteoclastic reabsorption, and the formation of new bone parallel to the cement surface, continue over a period of years. Remodelling is in accordance with Wolff's law so that the anatomical organization of the new bone, its amount and density conform with the tensile and compressive stresses (see Chapter 6) to which the bone is subjected.

It has long been known that particles of polymerized polymethylmethacrylate often lodge in soft connective tissue adjoining a prosthesis. The formation of a neosynovium may be initiated.[53] The significance of this synovia-like tissue is that it may be susceptible to inflammation, the mediators of which can initiate bone lysis and prosthetic loosening. Since polymethylmethacrylate is soluble in xylene, it is recognized in conventional paraffin sections only by the barium sulphate or other radioopaque material added to it before insertion, to allow radiographic control of the operative procedures. The barium salt comprises a finely granular, dispersed, weakly birefringent, grey material seen within the approximately 50–100 μm

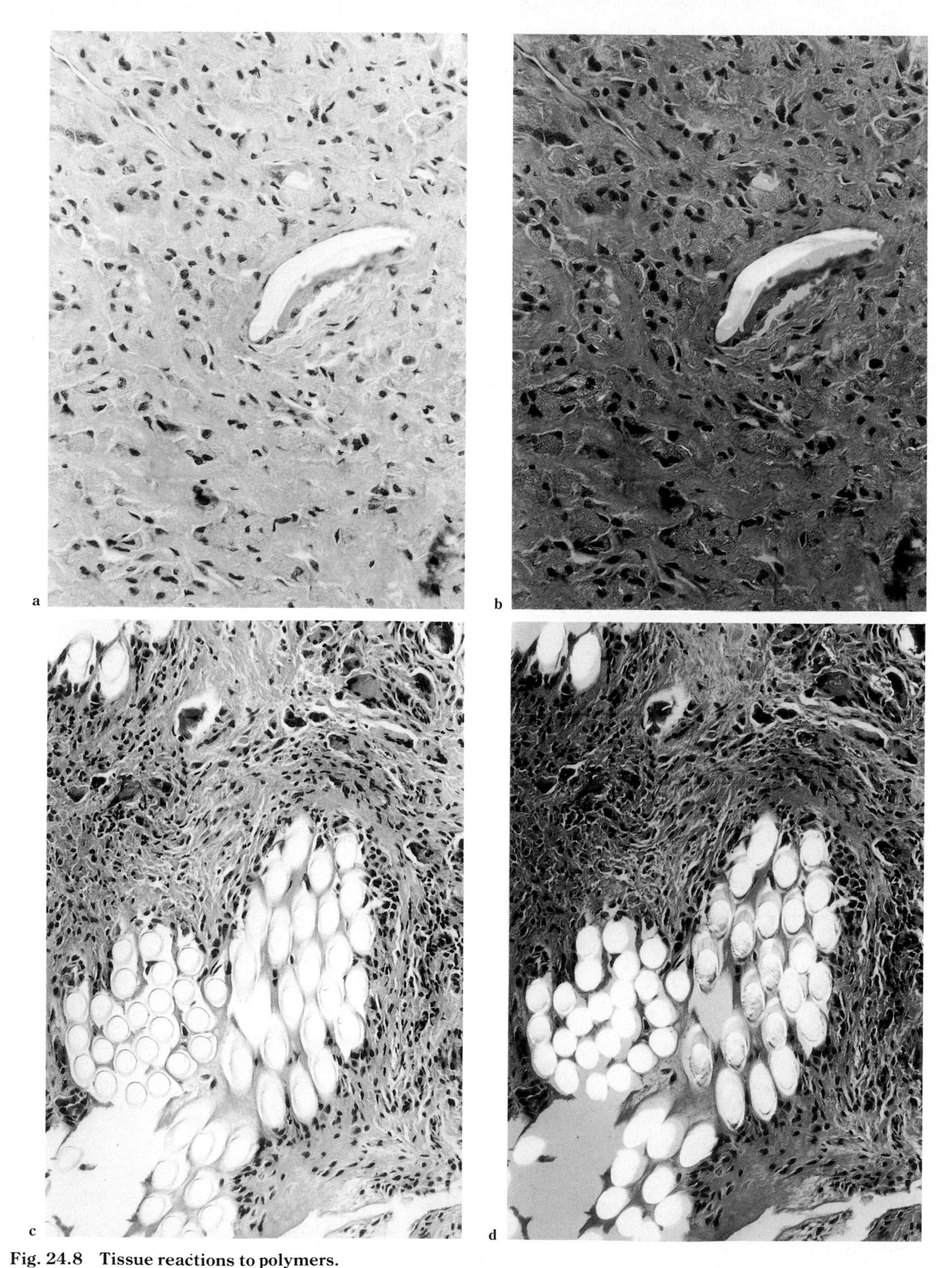

Fig. 24.8 Tissue reactions to polymers.
a. Debris composed of high-density polyethylene near site of hip arthroplasty. Fragments are circumscribed by macrophages fusing to form multinucleated giant cells. **b.** Same field, viewed in plane-polarized light. **c.** Repair of cruciate ligament injury by insertion of artificial fibrous polymer. Note persistence of polymer in knee joint and nearby collagen-rich fibrous tissue formed in response to the implant. **d.** Same field, viewed in plane-polarized light. (**a,b:** × 160; **c,d:** × 320.)

diameter empty tissue spaces (acrylic pearls)[175] from which polymethylmethacrylate has been dissolved. The granules are 1–2 μm in diameter. The direct identification of polymethylmethacrylate in tissue sections is possible in frozen sections stained by Sudan III or Oil Red O.[96] Polymethylmethacrylate may persist in mononuclear phagocytes to which the cement gives a frothy cytoplasmic appearance. Large pieces of polymethylmethacrylate provoke a multinucleated giant-cell reaction.

Pedley, Meachim and Gray[133] showed that polymethylmethacrylate particles can be positively identified by transmission electron microscopy; a biphasic, heterogeneous appearance results from an interaction between the acrylic and the epoxy resin used in embedding. These authors[111,113] were also able to distinguish the long-term responses to polymethylmethacrylate in rat and guinea-pig tissues.

Tissue reactions to carbon fibre Carbon, in various structural forms, has valuable properties as an implanted material.[1,61] Biocompatability is of a high order. Unfortunately, glassy carbon has low values of strain-to-fracture and strain energy: it is very brittle. Fibrous carbon also breaks easily.[130] The particles may be carried to lymph nodes. After surgical procedures near or in joints, carbon particles may lodge within the synovia or cause abrasive wear of articular surfaces. Synovitis may ensue. The histological changes that follow the successful implantation of carbon fibre are remarkably consistent.[118] A core of carbon fibre is concentrically enveloped by coherent laminae of fibroblasts and collagen. When the carbon fibres fragment, an easily recognizable pattern of black, thin rectangular particles is identified within mononuclear phagocytes.

Tissue reactions to bioglass and silica glass Bioglass[63] has direct bone- and tooth-bonding properties. It produces a calcium oxide-phosphate film at the interface between its silicon oxide-rich layer and the nearby tissue. Little inflammation is provoked by comparison with implants of silica glass;[71] there is slight fibrosis and no distinct boundary between bioglass and bone. When fragments of conventional glass or filaments of fibreglass gain access to connective tissues or joints, as they may do in accidents with windows or tableware, or occupationally in building or fibreglass workers or surfboard boat builders,[22] a vigorous local inflammatory reaction results. In one recent case of unexplained synovitis, the clue to diagnosis was provided by the patient's work: he was a glazier. Prolonged search revealed a single very small, weakly-doubly refractile glass fragment in the subsynoviocytic connective tissues.

Synovial fluid may come to contain particles rich in silicon. In glass-workers this is an occupational hazard. Glass fragments may, however, be introduced artefactually into synovial fluid samples during collection and processing. The fragments can be distinguished from crystals of natural and synthetic triclinic and monoclinic calcium pyrophosphate dihydrate, monosodium biurate and apatite conglomerates by means of scanning electron microscopy and electron probe X-ray microanalysis.[35] The artefactual particles have characteristic angled and rod shapes and display high silicon ($K\alpha$) but low calcium ($K\alpha_1$) peaks.

Tissue reactions to silicone Silicone elastomers (silicone 'rubbers') have found wide acceptance as substitutes for tissues and organs. They attract interest pathologically first, because *systemic* connective tissue disease occasionally develops as a late complication of silicone or paraffin injected into the tissues for cosmetic reasons,[16,82] and second, because silicone fragments cause synovitis and low-grade cellulitis.[13a,15b,146] The systemic responses may simulate progressive systemic sclerosis which is three times commoner than expected in women after cosmetic surgery; they may also mimic rheumatoid arthritis, systemic lupus erythematosis and polymyositis/dermatomyositis or they may take the form of an ill-defined adjuvant disease (see p. 544).

Silicone elastomers have important uses in finger and toe arthroplasty and in tendon repair. The adverse *local* connective tissue responses include infection, dislocation, fracture, synovitis and lymphadenopathy. The connective tissue and synovial reactions to silicones include low-grade, chronic inflammatory responses with multifocal, fibroblastic proliferation, and the presence of amorphous, pale, refractile but not birefringent material. Small pieces of silicone elastomer lie within mononuclear macrophages; larger pieces excite a new multinucleate, foreign-body giant-cell response or are surrounded by varying numbers of individual macrophages (Fig. 24.9).[182] The identity of the silicone deposits can be confirmed by electron probe X-ray microanalysis. Generally, an intact, smooth-surfaced silicone implant, for example in tendon transplantation in which silicone rods are used as stents, elicits little tissue reaction. A smooth, fibrous capsule forms[32] and the chemical composition of this tissue resembles that encountered at other sites of wound healing or hypertrophic scar (see p. 311).[81] More serious problems arise when silicone particles are released at sites of abrasive wear or prosthetic fracture. Some of these particles are carried to regional lymph nodes[78] and in rheumatoid patients, granulomata may be induced by the silicone.[57]

Tissue reactions to ceramics Porous ceramics such as aluminium oxide have been used to study connective tissue regeneration[178] and to investigate bone–ceramic

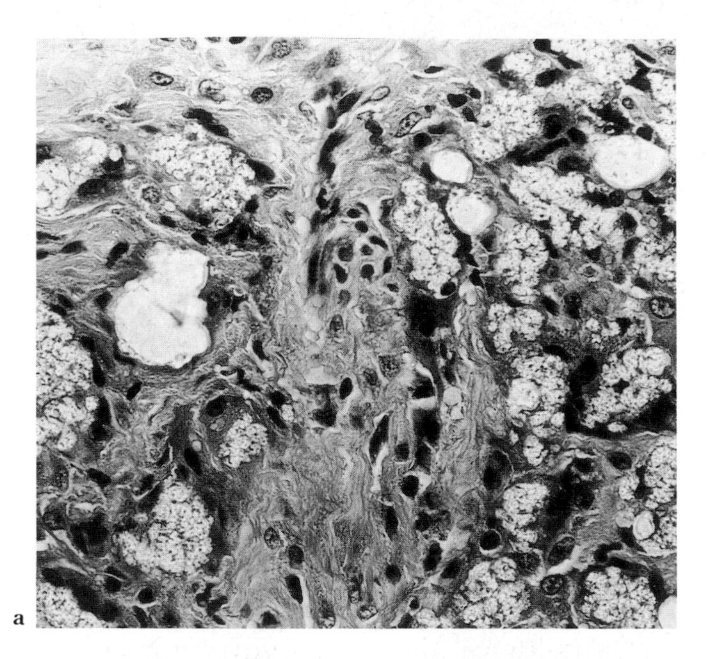
a

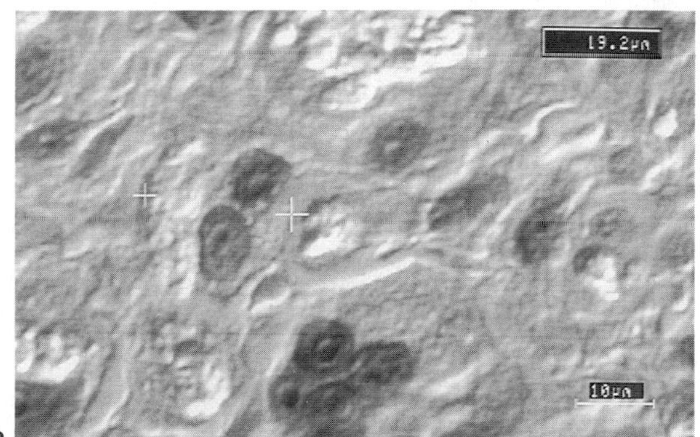
b

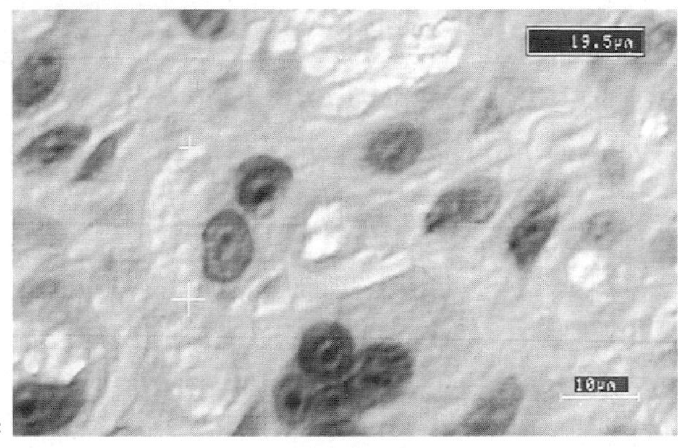
c

Fig. 24.9 Tissue reactions to silicone elastomer.
Silicone has been used to form part of prosthesis. With abrasive wear, fragments of silicone have broken away and can be seen within mononuclear macrophages. Note pale particles within phagocytes. (**a.** Transmitted white light, conventional mode × 320. **b.** 488 nm laser light, differential interference contrast mode × 900. **c.** 488 nm laser light, transmission mode × 900.)

interfaces.[44] The inert material causes fibrous tissue formation and there is a correlation between mechanical strength and the extent of collagen synthesis, the peak of which is only reached many days after experimental implantation into rats. Analogous reactions have been investigated after the implantation of a porous hydroxyapatite ceramic[19,43,89,90] and this material has proved particularly valuable for the *in vitro* three-dimensional study of chondrocytes.

Tissue reactions to other surgical materials Alternative, biocompatible materials are being continually sought and the product (adduct) created by reacting elastin with a fibrinogen derivative[4] is one example of the many being tested. A further composite, a collagen-like prosthetic material applicable to the repair of tendons and ligaments, is a polyester fibre created around an elastomeric core.[55]

Arthrodesis

With advances in techniques for arthroplasty and osteotomy, the use of arthrodesis remains the procedure of choice for a diminishing number of intractable orthopaedic problems. Among these are chronic pyoarthrosis with or without previous prosthetic arthroplasty, defects created by the resection of neoplasms, severe instability or chronic dislocation. The pathological features are initially those of the primary disease. Later, there is a local histological response to surgery, with new bone formation and remodelling.

Osteotomy

The reader is referred to texts on orthopaedic surgery. The correction of deformity by osteotomy relieves pain. The pathological complications include fracture and the osteoporosis and cartilage degradation that are responses to immobilization (see p. 886).

Transplantation

There have been important recent advances in the control of immunological rejection reactions and the treatment of connective tissue, articular and bone disease by allografts has entered a new phase.[30]

Joint transplantation

The principles were established early in this century.[94] The problems were summarized by Porter and Lance.[135] The transplants may be of whole limbs or of whole joints. In addition, parts of joints such as articular surfaces may be transplanted or bone grafts used during arthroplasty, e.g. to accommodate the problem of *protrusio acetabuli* (see p. 879).[66] 'Shells' of bone[15] can be used to support transplanted articular cartilage.

The immunology of joint and cartilage transplantation is reviewed by Elves,[31] that of bone grafting in an Editorial.[30] Some aspects of cartilage transplantation are mentioned in Chapter 2. Animal models have been developed for the study of osteoarticular transplants.[64]

Synovectomy

Ablation of the synovium by surgical, chemical or physical procedures has long seemed a worthwhile approach to the relief of pain and disability in disorders such as rheumatoid arthritis (see p. 552).[99] Fibreoptic endoscopes have made it practicable for rheumatologists to practice this procedure.[6] Although synovectomy is now less favoured,[100] the structural and functional consequences of synovectomy still require attention. There are special problems in haemophiliac arthropathy.[5]

Surgical synovectomy

Until recently, adequate access to the most important joints such as the knee was only provided by open operation. Now, however, fibreoptic arthroscopes allow access to all parts of many joints[72] during operations that are devoid of the morbidity that is associated with open synovectomy.[21] Among the joints subject to synovectomy in rheumatoid arthritis are the knee, elbow, shoulder and wrist. In theory, any accessible joint may be approached. Following synovectomy, a neosynovium comes to line the joint cavity. It is susceptible to the same disease process that characterizes the original disorder. Within 12 months of operation, rheumatoid synovitis, for example, may once again be evident arthroscopically[132] and the synovial fluid displays the elevated levels of lysosomal enzyme activity, and the complement component 3 and 4 levels[122] seen in the untreated disorder. Synovial A and B cells can be distinguished within five weeks after operation and regeneration of the excised synovium in rabbits subjected to synovectomy, capsulectomy and patellectomy is almost complete after 20 weeks (see Chapter 2).[147] Although synovectomy may harm articular cartilage in healthy animal joints, systemic corticosteroids diminish this hazard.[11] The presence of the synovium influences fibrocartilage regeneration after experimental meniscectomy.[77]

Chemical synovectomy

Chemicals such as osmium tetroxide which 'fix' (denature) proteins and antimitotic agents such as phosphoramides have been administered intraarticularly either to destroy the synovial cell populations in rheumatoid arthritis or to reduce the population of mononuclear cells that mediate the synovitis. The late results, with fibrosis and synoviocyte regeneration, are not dissimilar from those caused by surgical synovectomy. One consequence is the development of zones of radioopacity where osmium tetroxide has bound to articular adipose tissue,[14] promoting calcification. The margins of the synovial adipose tissue nearest to the joint cavity are blackened.

Radiation synovectomy

The attempt to control inflammatory joint disease by X-irradiation often proved disastrous. Bone necrosis was one complication. The early tests of colloidal thorium dioxide (thorotrast) were equally unfavourable and carried the late risk of malignant disease of the reticuloendothelial system. By contrast, radiosynoviothesis by the intraarticular injection of radiocolloids with short half-lives has proved to be relatively safe with overall complication rates of less than 4 per cent.[91] Among the isotopes tested have been gold (^{198}Au), yttrium (^{90}Y)[83] and rhenium (^{186}Re). For smaller joints erbium (^{169}Er) has been preferred.[58] One of the factors inhibiting the clinical use of intraarticular radiocolloids is the possibility of extraarticular dissemination of the isotope, with chromosomal damage to circulating lymphocytes and the uptake and concentration of the isotope in regional lymph nodes. However, dysprosium (^{165}Dy), in the form of dysprosium-165-ferric hydroxide macroaggregates, causes inguinal lymph node radiation levels significantly lower than those resulting from the other radiocolloids.[183]

Haemorrhage; haematopoietic disease

Disorganization of loose or compact non-mineralized connective tissue is often a result of haemorrhage within these tissues. There are many possible causes. In some instances, the connective tissues are prejudiced directly or indirectly by abnormalities of the haematopoietic system.

Haemorrhagic arthropathy

Intraarticular haemorrhage is usually traumatic but may come from synovial haemangioma or haemangiopericytoma. Haemorrhage into articular tissues may complicate heritable disorders of the coagulation mechanism and can result from the use of anticoagulants. Analogous states have been investigated after blood has been injected experimentally. Intraarticular bleeding is a feature of chronic inflammatory diseases such as rheumatoid arthritis, in which the quantity of iron retained by the synovium may contribute to the refractory anaemia of the disease; and more acute bleeding may occur repetitively in progressive renal failure with uraemia. In rheumatoid arthritis, and in uraemic arthropathy, the conglomerate papillary synovial processes simulate the appearances of pigmented villonodular synovitis (see p. 981).

Intraarticular bleeding Prolonged, repeated bleeding into a synovial joint prejudices cartilage integrity.

Biochemical and mechanical changes The stifle joint articular cartilage of mature mongrel dogs, for example, showed a loss of glycosaminoglycan after only four weeks of repeated intraarticular injections of blood.[24] Total collagen was not, however, significantly reduced until after 12 or more weeks of testing. Cartilage thickness change was inconstant. Compressive indentation tests revealed a correlation between vertical deformability and total hexosamine content, and between vertical deformability and total collagen content; but there was no correlation between shear resistance and these biochemical values.

Microscopic changes Sterile intraarticular bleeding or the repeated injection of blood[24] lead to synovial hypertrophy and synovial cell hyperplasia and to an infiltration of plasma cells and lymphocytes. The relationship between these effects and matrix degradation is not clear. Siderosomes appear in zone I, II and III chondrocytes and degenerative cell changes develop.[148,149] Siderosomes are also identified in nearby synoviocytes and are more numerous in type B than in type A synovial macrophages.[126] The numbers of siderosomes declines with time. In culture, the synovia liberate interleukins (see pp. 88 and 536) causing chondrocytes to degrade their immediate, pericellular matrix by proteolysis. A comparable indirect mechanism by which inflamed synovia catalyse cartilage breakdown in chronic haemarthrosis may be considered a possible alternative to direct injury caused by proteinases derived from synoviocytes or the inflammatory cells that infiltrate them.

Haemophiliac arthropathy[28,165] (Fig. 24.10) Haemophilia A and von Willebrand's disease result from heritable deficiences of clotting Factor VIII, the former of Factor VIII quality, the latter of Factor VIII quantity. In

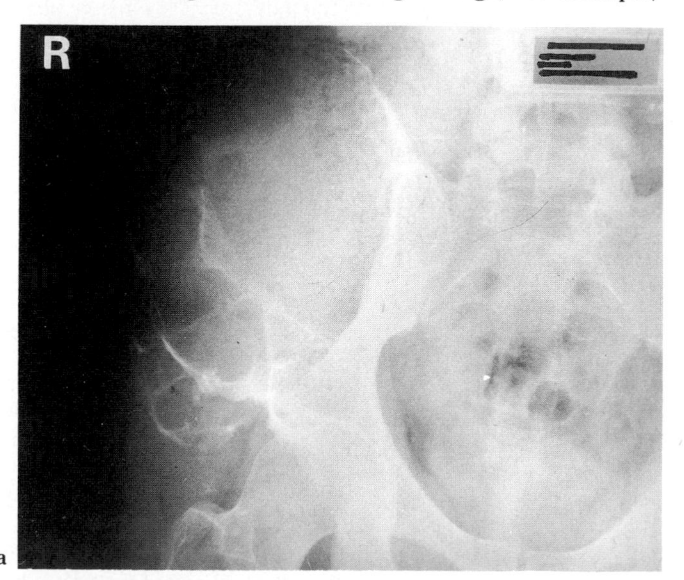

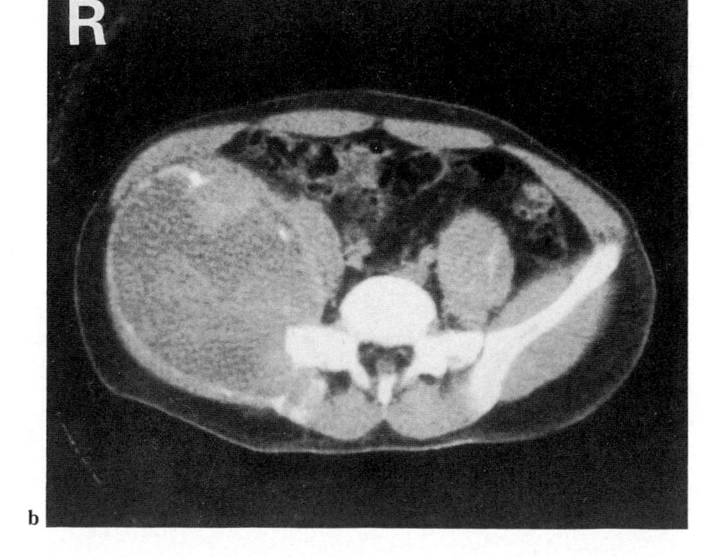

Fig. 24.10 Haemophiliac disease.
Haemophiliac pseudotumour of the right innominate bone. A large destructive lesion produces a soap-bubble pattern, an appearance that can mimic a malignant neoplasm. (**a**: plain radiograph; **b**: CT scan.) Courtesy of the Department of Diagnostic Radiology, University of Manchester.

haemophilia B, there is a deficiency of Factor IX. Haemophilias A and B are transmitted as sex-linked, recessive characteristics; von Willebrand's disease as an autosomal dominant trait. The deficiency in haemophilia A is the synthesis of an abnormal Factor VIII in which biological clotting activity is partly or wholly lost although the immunological identity of the material is retained. In von Willebrand's disease there is decreased Factor VIII synthesis.

Before the advent of iatrogenic HIV infection and the acquired immunodeficiency syndrome, chronic, deforming arthropathy was the most serious, non-fatal complication of more than 80 per cent of cases of haemophilia. It may occasionally be complicated by septic polyarthritis.[54,80,102] It is a rare consequence of von Willebrand's disease. The prophylactic use of natural or recombinant Factor VIII can greatly reduce the frequency of the arthropathy of

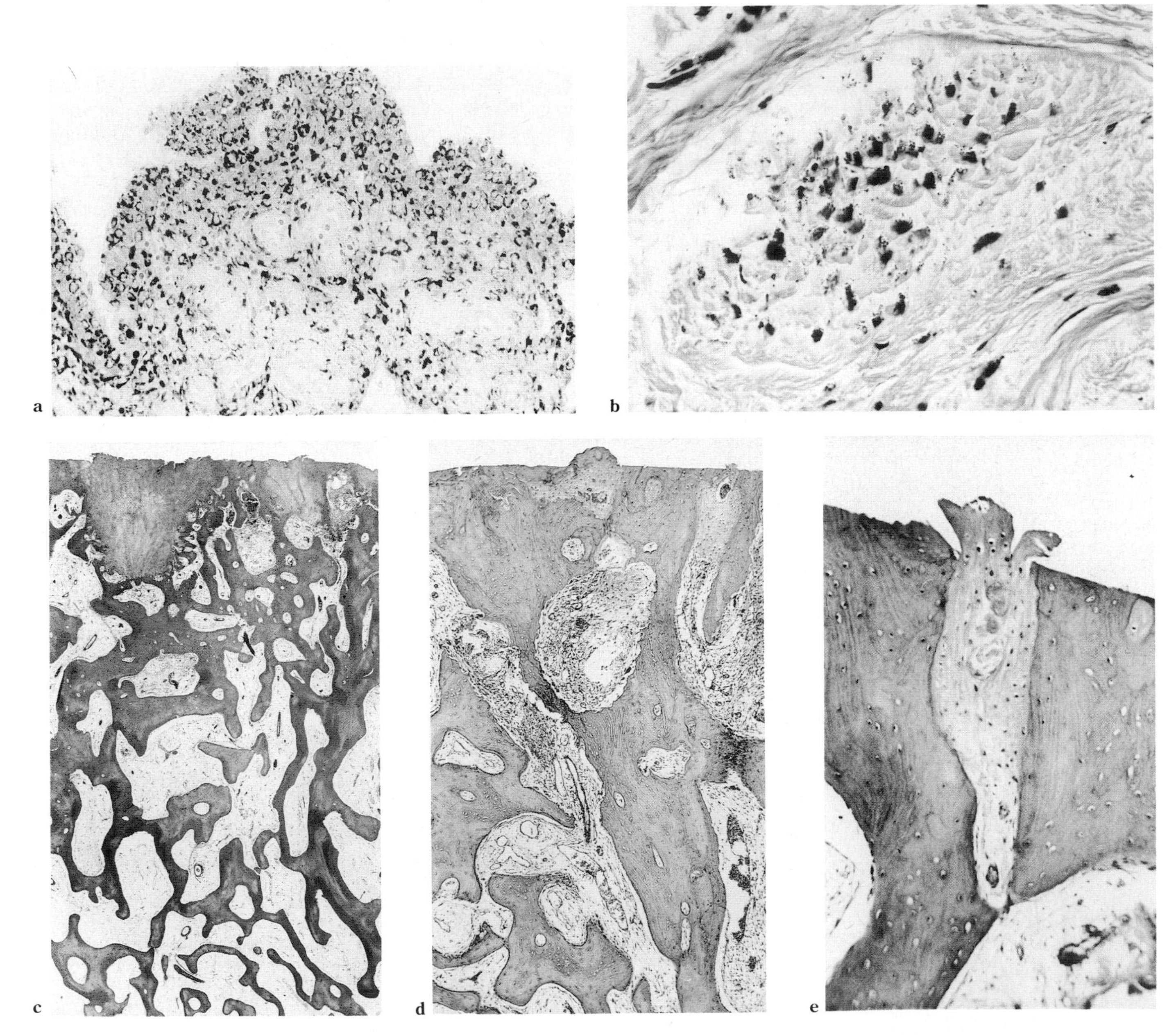

Fig. 24.11 Haemophiliac arthropathy.
a. Synovial tissue laden with haemosiderin. **b.** Higher-power view of part of same field. **c.** Severe, secondary osteoarthrosis. Note osteosclerosis and residual 'plugs' of cartilage at bearing surface. **d.** Further part of same tissue. Osteoclastic bone reabsorption is an index of the remodelling process. **e.** Detail of field adjoining **d.** Vertical microfracture is occupied by repair fibrocartilage of low vascularity. (**a**: Prussian blue reaction × 83; **b**: × 113; **c**: HE × 8; **d**: × 28; **e**: × 113.)

haemophilia A. The severity of the arthropathy of haemophilia is proportional to the degree of Factor VIII or Factor IX deficiency. Thus, 'spontaneous' intra- and periarticular haemorrhages are most likely in grade 4 haemophilia in which the activities of Factor VIII or Factor IX are less than 1 per cent of normal. It is notable that Factor XIII deficiency, a rare inborn error of 'fibrin stabilizing factor', may also be complicated by recurrent haemarthrosis and by arthropathy.[170]

Pathological changes (Figs 24.11–24.13) The arthropathy of haemophilia has been carefully studied anatomically[73,144] and experimentally.[67,149,160] It is of interest to compare the joint changes with those of various forms of secondary osteoarthrosis.[112,114,115] The primary disturbance is haemorrhage from synovial vessels into the joint cavity and periarticular soft tissues. Intraarticular haemorrhages may begin in the first two years of life and are recurrent. The knee, elbow and ankle joints are affected most frequently, the hip less often. Red blood cells are taken up by synovial A cells and by histiocytes. An inflammatory reaction within these vascular tissues follows; the response, sustained by repetitive haemorrhage, is more severe than the synovitis which follows intraarticular bleeding, for example, after surgery in non-haemophiliacs. In a single episode, venous or capillary bleeding from articular vessels is likely to be prolonged. The synovial spaces fill with blood. The joint becomes swollen, tender and immobile. Ultimately, the pressure within the joint comes to approach the pressure within the injured vessels

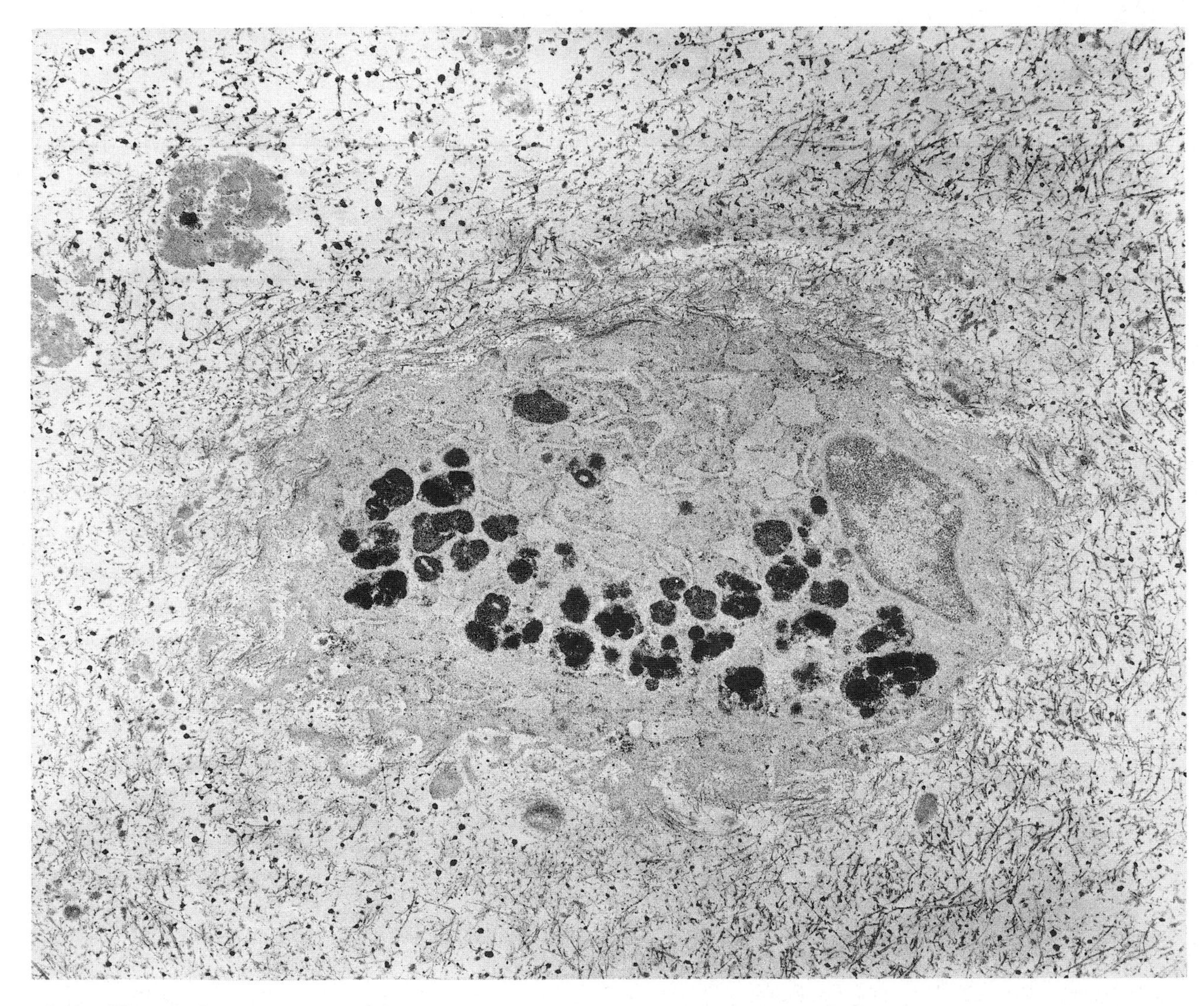

Fig. 24.12 Haemophiliac arthropathy.
Siderotic chondrocyte seen in cartilage of haemophilic joint. (Uranyl acetate—phosphotungstic acid × 16 060.) Courtesy of Professor A J Hough and the Editor, *Archives of Pathology*.[68]

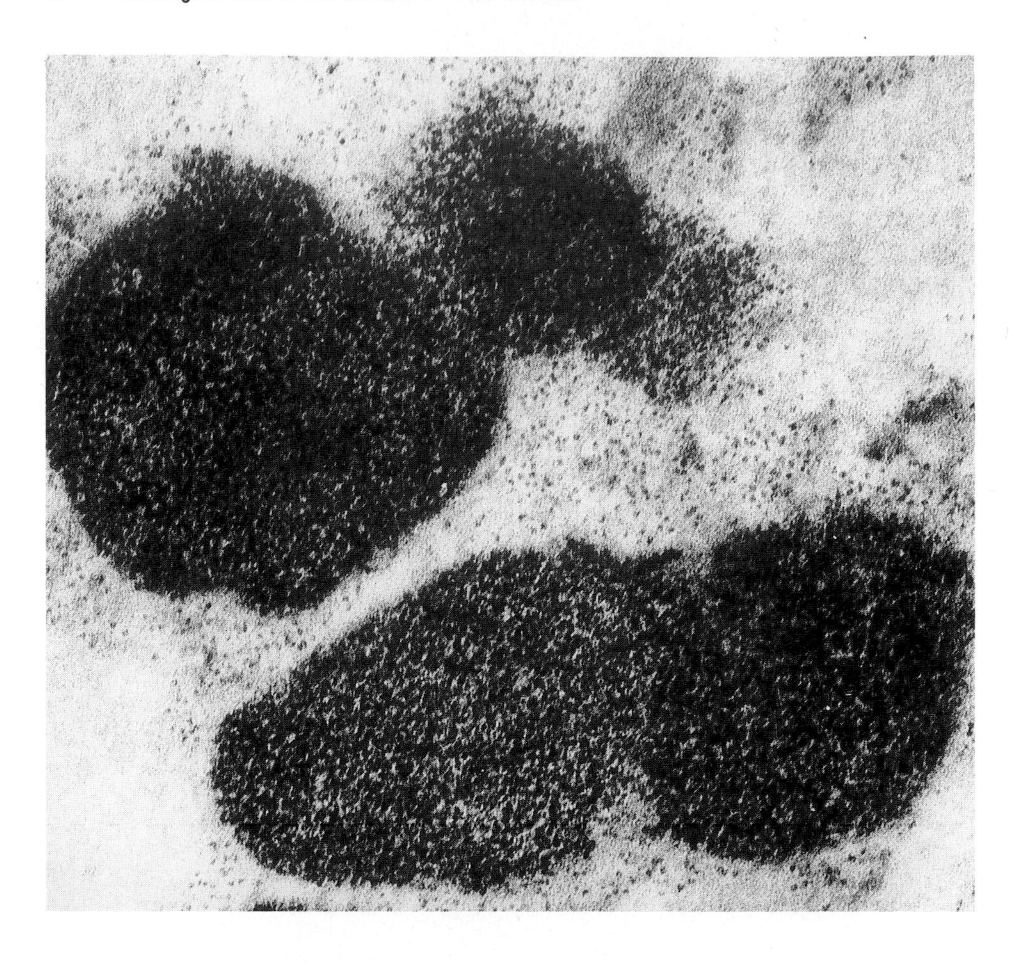

Fig. 24.13 Haemophiliac arthropathy.
Detail from Fig. 23.15. Siderosmomes in articular chondrocyte. (Uranyl acetate—lead citrate × 133 250.) Courtesy of Professor A J Hough and the Editor, *Archives of Pathology*.[68]

and bleeding ceases. Slow reabsorption of blood takes place. The synovial tissues are laden with haemosiderin and undergo hypertrophy; resolution is seldom complete.

Repeated episodes of bleeding lead to orange-brown discoloration of the hyperplastic synovia, evident from an early stage of the arthropathy. There is also an abundance of iron and iron-containing pigment in the capsular and periarticular tissues where repetitive haemorrhage is also likely. As time passes, the long-term effects of iron aggregation, recurrent low-grade synovitis and periarticular disease are seen to culminate in a progressive accumulation of collagen-rich fibrous tissue. Within the synovial joints, fibrous ankylosis and contracture are probable results. There are similarities to the anatomical changes which may result from inadequately treated bacterial arthritis (see Chapter 18). The morphological changes are exacerbated by each episode of intraarticular haemorrhage. Repetitive intraarticular bleeding from an early age encourages premature cartilage degradation and fibrillation. A form of secondary osteoarthrosis emerges. As fibrillation advances, the whole thickness loss of articular cartilage progresses to bone exposure and eburnation. Haemorrhages may extend beneath the cartilage and into the subchondral cancellous bone. The replacement of this blood and the disrupted bone and marrow by a loose, vascular mesenchymal tissue results in the formation of pseudocysts recognizable radiologically.

The advanced, destructive arthropathy characteristic of the untreated haemophilias is recorded in earlier reports, largely of necropsy material.[73,144] Among other pathological complications, are femoral neuropathy which may complicate bleeding into the iliopsoas muscle; Volkman's ischaemic contracture, which may succeed haemorrhage into the flexor muscles of the forearm; and pseudotumour of the femur or ilium which may simulate malignant bone neoplasia (Fig. 24.9).

Pathogenesis (Fig. 24.14)[101a,102] In spite of much research, the mechanisms of cartilage injury and of the subsequent fibrosis are not wholly clear.[101a] There are low levels of synovial tissue thromboplastin but this is a characteristic of the normal synovium.

Cartilage injury in haemophilia may be direct or indirect. Indirect mechanisms appear to be more important although haemorrhage into subchondral bone, with the formation of an organizing haematoma, can cause cartilage disorganization directly. Pressure changes that result from intraarticu-

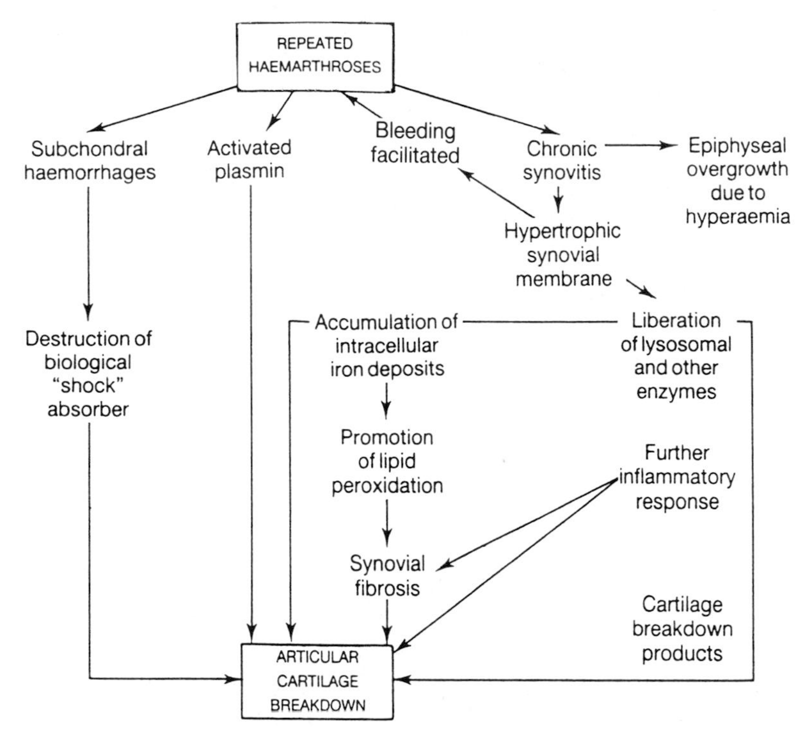

Fig. 24.14 Diagrammatic representation of pathogenesis of haemophiliac arthropathy.

lar effusion or haemorrhage can also impair cartilage metabolism.

Iron aggregates in haemophilia can provoke synovial cell lysosomal enzyme activation, catalysing cartilage degradation.[49,127] The iron deposits are highly iron-saturated ferritin and discrete, ovate particles of almost pure iron. Both may perpetuate inflammation by promoting lipid peroxidation.[121] Intraarticular haemorrhage also activates cartilage plasmin but whether this enzyme can degrade cartilage directly remains debatable.

To determine whether iron initiates cartilage injury directly by the binding of small amounts to matrix proteoglycan by chelation or indirectly by poisoning chondrocytes by iron salt formation, a study was made in five human and two canine cases.[68] In parenthesis, it had been suggested that the reason cartilage is not overtly pigmented in haemophilia and other chronic haemarthroses, is that much of the cartilage is lost by the time the joint is examined.

Two pigments were in fact present in residual cartilage. The first, haemosiderin, was detected by light-microscopy in the cytoplasm of a minority of chondrocytes: there was no stainable iron in the cartilage matrix. The second pigment was an extracellular, bilirubin-like material at the cartilage surface. Transmission electron miscroscopy (Figs 24.12 and 24.13) confirmed degenerative changes of chondrocytes and of their perichondral matrices. Membrane bounded siderosomes were present in the cells: the siderosomes were composed of ferritin or ferritin-like particles; and fine ferritin particles were dispersed in the cytoplasm but not in mitochondria. However, no iron was found in the cells from three early cases of the disease and none was present in the cartilage matrix of any of the seven subjects. The evidence argues strongly against a direct damaging effect of iron on the cartilage matrix and in favour of impaired or abnormal chondrocyte function.

Monolayer cultures of haemophiliac synovial cells grew fibroblast-like cells containing much pigment together with smaller, deeply-pigmented round cells.[102] Explants of synovium and of adherent cells secreted latent collagenase and neutral proteinase; lysozyme was also secreted by the explants but was not detected in the culture medium. The potential degradative activity of the synovium in haemophilia appeared to be confirmed. Recurrent haemorrhage without manifest inflammation seemed to be established as a sufficient explanation for cartilage breakdown. However, the participation of synovial interleukin-1 is likely and could play an important part in the activation of chondrocyte proteinases.[102]

Haemosiderosis (Fig. 24.15)

The synovial tissues have many of the properties of the reticuloendothelial system (see p. 36).[46] It is therefore not surprising that synovial cells phagocytose blood cells introduced into joints. They also accumulate particulate and insoluble

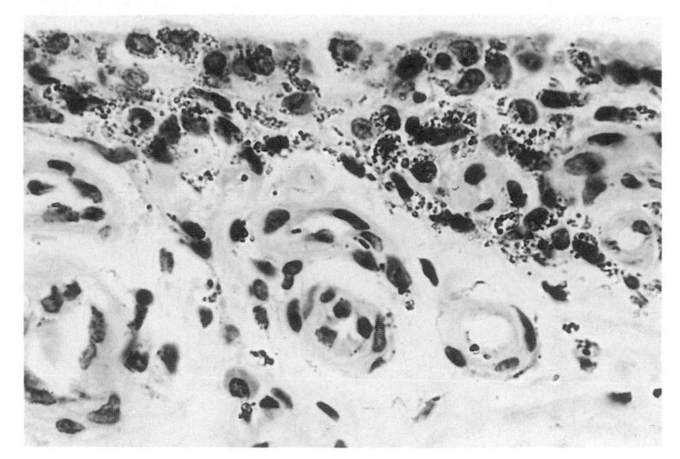

Fig. 24.15 Synovial haemosiderosis.
Numerous refractile deposits of haemosiderin are scattered throughout the synovial tissues of this boy aged 12 to whom 94 litres of blood had been transfused for the treatment of aplastic anaemia. (HE × 500.)

materials from the circulation. Red blood cells given by transfusion break down: the haemoglobin is degraded and haemosiderin forms. When a great deal of blood is transfused as, for example, in the treatment of intractable aplastic anaemia, much iron pigment accumulates in the form of haemosiderin, not only in the spleen, liver and bone marrow, but also in the type A synoviocytes and, progressively, in the subsynovial macrophages. Unlike the free iron of haemophiliac arthropathy, the accumulated haemosiderin is not inflammatory. There is no subsequent fibrotic response as in haemophilia, and no tendency for secondary calcification or chondrocalcinosis as in haemochromatosis (see Chapter 9). The microscopic synovial response in haemosiderosis is therefore slight.

Haemochromatosis Haemochromatosis is discussed in Chapter 9 (see p. 369).

Connective tissues and haematopoietic disease

Disorders of haematopoiesis, of haemoglobin structure and of the coagulation mechanism may cause articular and extraarticular connective tissue abnormalities. Articular disorders often result from intraarticular haemorrhage.

Leukaemia[97] The extensive involvement of haematopoietic and lymphoreticular tissues in chronic lymphocytic and in myelogenous leukaemia contributes to a high frequency of symptoms such as migratory polyarthritis, arthralgia, bone pain, and bone tenderness. In chronic lymphocytic leukaemia, tissue infiltration by small, relatively mature lymphocytes, is invariable; articular and periarticular tissues may be affected. T-cell leukaemia also can simulate seronegative arthritis.[159] Joint symptoms may, however, be the result of indirect change including the compression of nerve roots and haemorrhage into articular tissues. In other instances, rheumatoid factor may be detected and subcutaneous nodules form that must be distinguished from those of rheumatoid arthritis and rheumatic fever.

Myelofibrosis The synovial tissues are a rare site for extramedullary haematopoiesis. Megakaryocytes, erythroid and leucocytic precursors can be found within the synovial tissue (Fig. 24.16).[62] The clinical signs of a non-specific arthritis may draw attention to the disorder. Myelofibrosis and leukaemia periodically provoke secondary gout (see Chapter 10).

Lymphoma Bone pain, and the effects of nerve compression or infiltration, occasionally attract attention. The cells of Hodgkin's and non-Hodgkin's lymphomata[107] may infiltrate articular tissues directly but articular changes are more often attributable to disease of nearby bone. As in leukaemia, mono- and polyarticular symptoms and signs can simply reflect incidental infection or may result from haemorrhage, immunoglobulin deposition or therapy.

Haemoglobinopathy[154] Clinical signs of musculoskeletal disease are encountered periodically in the hereditary anomalies of haemoglobin structure.[33] Secondary gout may develop. Visceral and articular lesions are frequent in those with sickle-cell (haemoglobin AS) trait. Other inherited disorders of haemoglobin, some associated with sickling such as haemoglobin C disease and thalassaemia,[56] may be accompanied by articular and connective tissue disorders. In thalassaemia, many patients develop a specific osteoarthropathy as they approach the second and third decades of life. Although infection is a common precipitating factor and osteomyelitis a frequent complication, septic arthritis is rare.

In sickle-cell disease, the pathological lesions usually comprise bone involvement near joints. There is much additional bone marrow. Bone expansion accompanies cortical thinning and the presence of infrequent, coarse bone trabeculae. Vertebrae are indented centrally. Compression or compression fractures lead to kyphosis and lordosis. Osteophytosis and vertebral fusion are described.[154] Joint effusions are common. Some of the pathological changes are attributable to aseptic bone necrosis; multiple small bone marrow infarcts contribute to altered synovial haemodynamics. Secondary osteoarthrosis is a complication.

Schumacher[154] describes the microscopic findings at needle biopsy of the synovium. There is a focal proliferation of synovial cells and the presence of a small number of chronic inflammatory cells. Small vessel congestion is seen

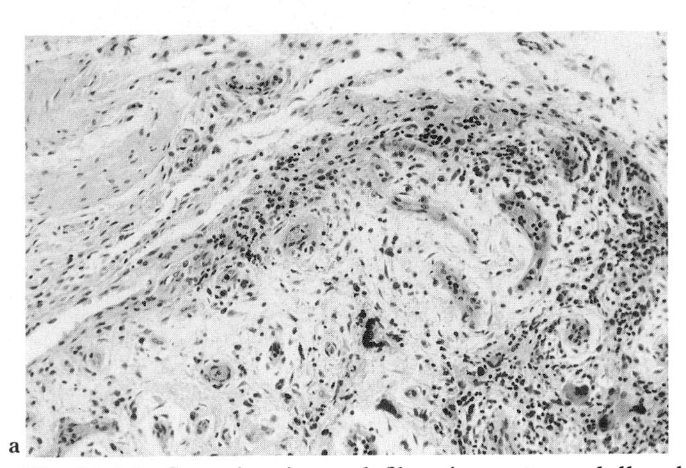
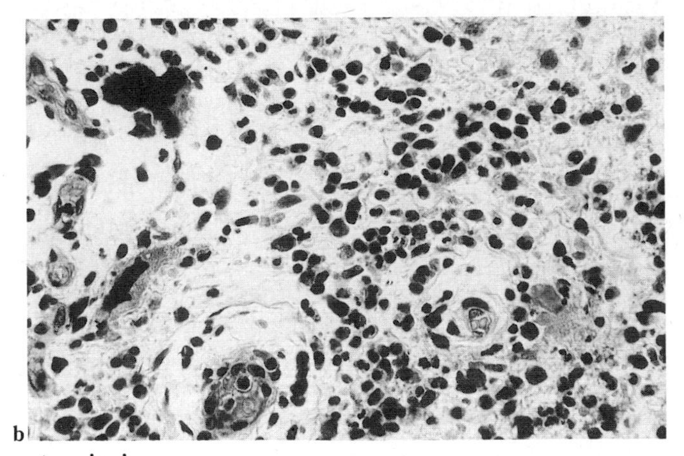

Fig. 24.16 Synovium in myelofibrosis—extramedullary haematopoiesis.
a. Islands of blood cell formation can be identified by the presence in lower, right part of field, of large megakaryoblasts. **b.** At higher magnification, identity of megakaryoblasts and red- and white cell precursors is confirmed. (**a**: HE × 110; **b**: × 284.)

and microvascular thrombosis is characteristic. Transmission electron microscopic evidence confirms the presence in synovial vessels of sickled red blood cells containing the tactoids of crystallized haemoglobin. Exceptional instances of chronic synovitis with secondary cartilage degradation have been encountered.[154a]

Polycythaemia vera In an analysis of 101 unselected patients, 14 were found to have secondary gout.[27] By contrast, the frequency of osteoarthrosis and of inflammatory polyarthritis did not differ from that of the general population. Fifty-eight per cent of the males and 42 per cent of the females had hyperuricaemia compared with expected normal frequencies of 4.6 per cent and 1.26 per cent, respectively.

Angioimmunoblastic lymphadenopathy

This is an immunoproliferative disease of unknown cause.[26] There are features in common with systemic lupus erythematosus (see Chapter 14). Most patients are elderly. They develop a systemic disease with a raised erythrocyte sedimentation rate, lymphadenopathy, cutaneous rash, hepatosplenomegaly and hypergammaglobulinaemia. From time-to-time, a seronegative arthralgia or arthritis symmetrically affect the elbows, wrists, knees or ankles. Occasionally, there is coexistent rheumatoid arthritis. Arthritis may precede the systemic illness or coincide with it. The lymph node changes are characteristic. There is an extensive proliferation of small blood vessels and an infiltrate of immunoblasts, plasmacytes, eosinophil granulocytes and some polymorphs. Skin lesions are also recognized but less is known of the synovial changes. It is assumed that they resemble those of the lymph nodes and skin. The synovial fluid is of increased volume, sterile but containing numerous leucocytes, sometimes with an excess of polymorphs.

Reactions to lipids

Intraarticular injection of lipid

Autologous, omental fat injected into rabbit knees is avidly taken up by synoviocytes. Lipid-laden cells pass to the subsynoviocytic tissue and, after seven days, no lipid remains in the joint space and little in the deeper tissue.[47] The chondrocytes also accumulate lipid although evidently there is a preliminary processing of the fat since it is the fatty acid rather than the glycerol moiety that is incorporated. These results are of interest because of the view that lipid within the most superficial cartilage zones may contribute to joint lubrication (see Chapter 6).

Fat can be injected into synovial joints in the form of liposomes, small phospholipid-bound sacs, and these sacs can be used to carry drugs selectively to the abnormal synovial membrane in rheumatoid arthritis.[158] Liposomes demonstrate a multilamellar structure in which concentric phospholipid bilayers are separated by water; they form spontaneously when a dried lipid film is hydrated. Liposomes can be employed as biodegradable vehicles for the administration of drugs such as corticosteroids, directly into synovial joints.[158] The therapeutic agent is quickly and specifically brought to the site at which its action is most effective since the liposomes, like red cell ghosts and many other small particles, are rapidly phagocytosed by the type A synovial cells.

Lipid in articular disease

From time-to-time, synovial fluid is found to contain fat either in the form of lipid droplets that can be identified microscopically or as increased fractions of free fatty acids, cholesterol or triglycerides. Fat may be liberated into the synovial fluid by direct trauma to the adipose tissue of the synovium.[12] Occasionally the fluid may be chylous. This unusual result of lymphatic obstruction can be a complication of systemic filariasis (see Chapter 18).[25]

Multicentric reticulohistiocytosis

Multicentric reticulohistiocytosis is a rare systemic disorder in which a destructive polyarthropathy, resembling rheumatoid arthritis, accompanies nodular skin and mucosal lesions and systemic disorder. In 25 per cent of cases, there is an associated malignant neoplasm.

The cutaneous lesions occupy the dermis and superficial subcutis. There is an infiltrate of mononuclear macrophages and very occasional lymphocytes which may extend to and merge with the epidermal basement membrane. The predominant cell population, however, is of large, sometimes multinucleate (giant) cells (megalocytes), 40–100-μm in diameter, with a conspicuously eosinophilic, finely granular cytoplasm and occasional pale-staining cytoplasmic vacuoles. The megalocytes contain diastase-resistant, periodic acid Schiff-positive material and lipid which adsorbs Sudan II and Oil Red O. The cells react positively for muramidase and α_1-antitrypsin[41] and are demonstrably histiocytic. The synovia contain islands of pale, foamy macrophages and, separately, small groups or cluster of megalocytes identical with those seen in the dermis (Fig. 24.17). They may display the same histochemical properties as the cutaneous giant-cells and are avidly phagocytic. Transmission electron microscopy enables the synovial cells to be classed as 'dark' (electron-dense) and 'light' (electron-lucent) (Fig. 24.18). The latter closely resemble

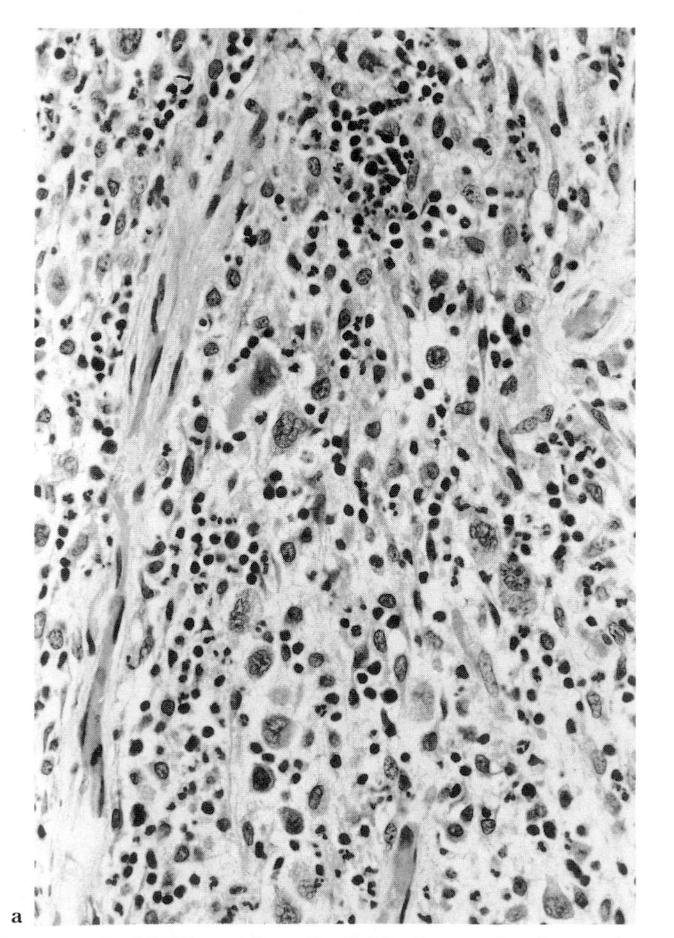

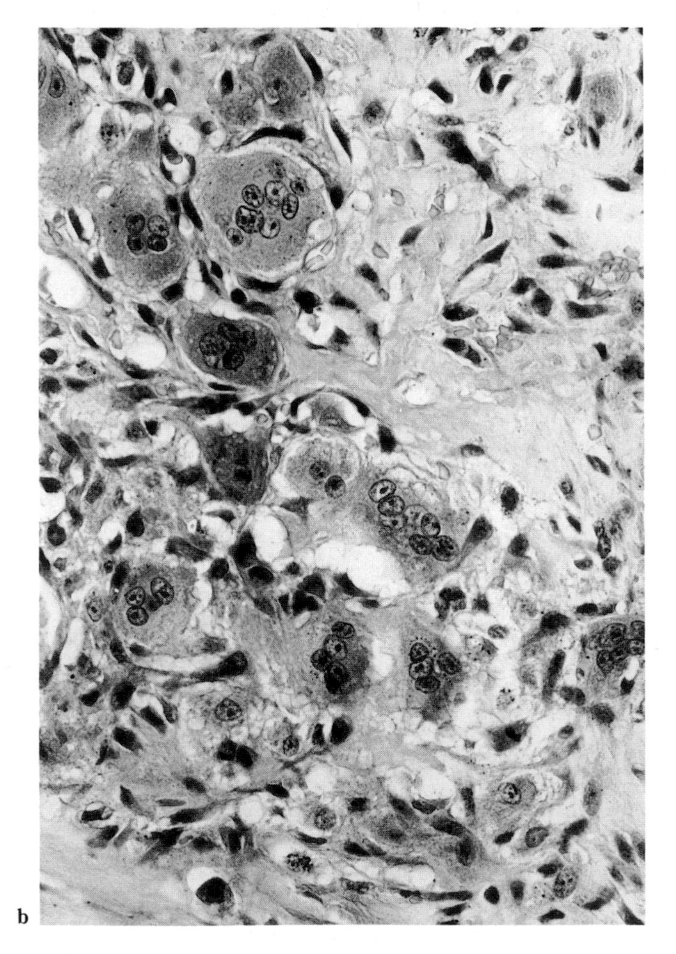

Fig. 24.17 Multicentric reticulohistiocytosis of synovium.
a. Active synovitis is indexed by brisk polymorph infiltrate. Note characteristic giant, multinucleated cells lying amongst fibrous stroma. (HE × 80.) **b.** Array of multinucleated giant cells indistinguishable from those found in the skin. (HE × 320.)

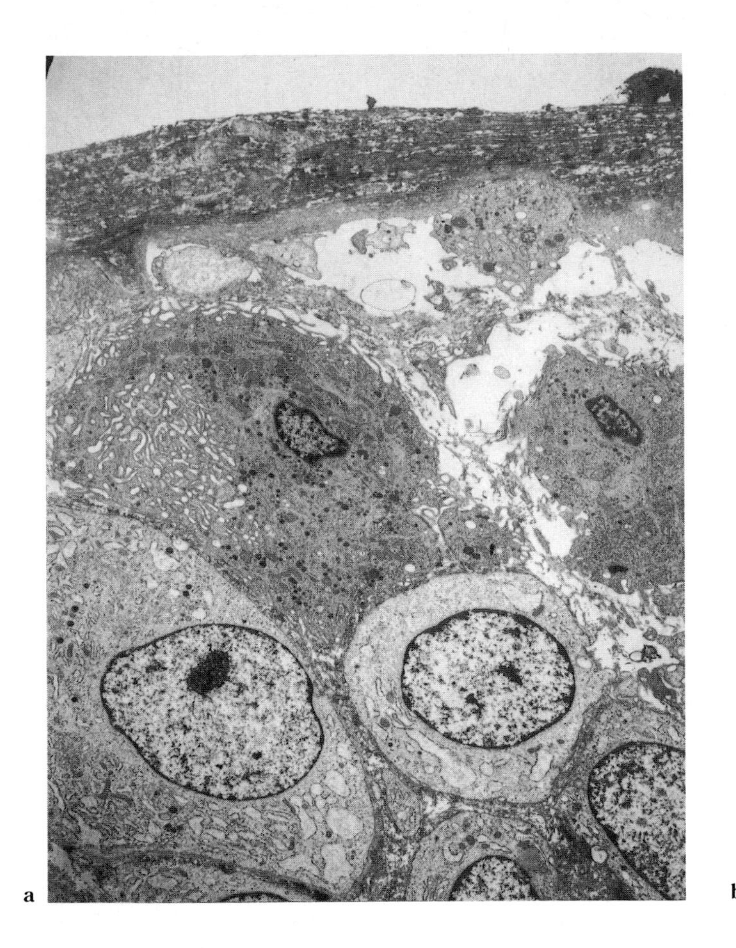

a

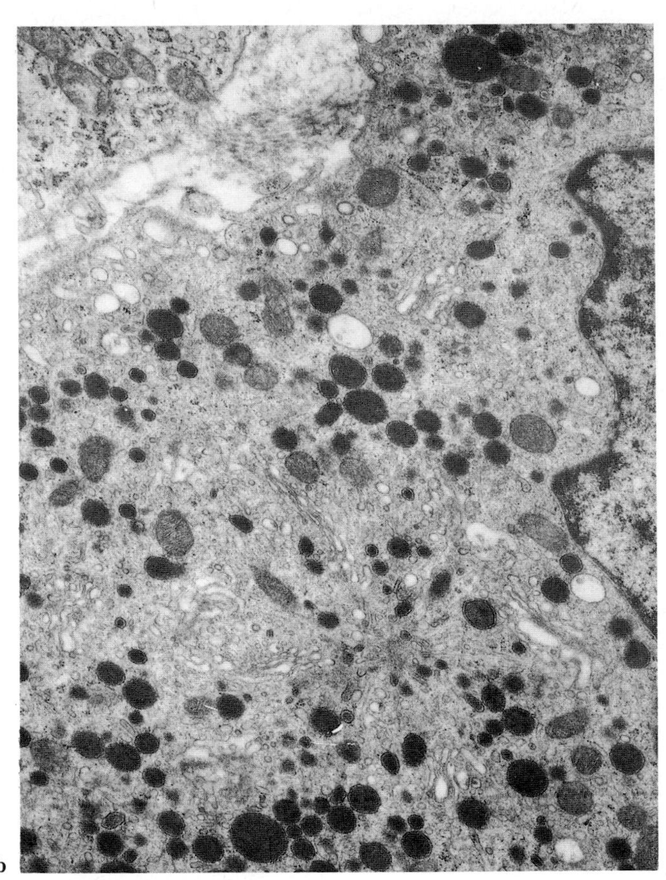

b

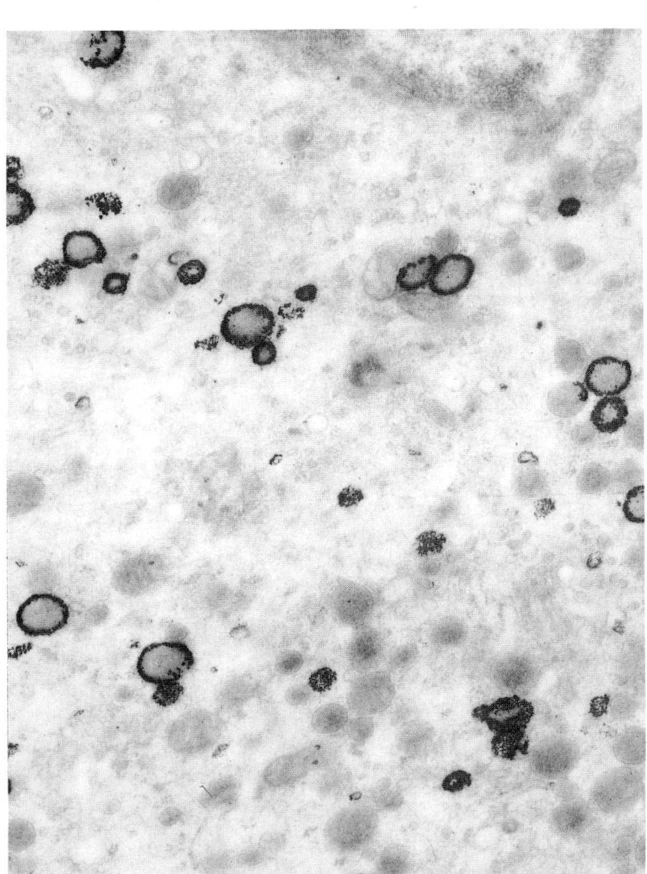

c

Fig. 24.18 Multicentric reticulohistiocytosis.
a. Surface of synovium with fibrinous exudate, an index of the associated destructive arthropathy. **b.** Cytoplasm of megalocyte with osmiophilic granules and large, stellate Golgi apparatus. **c.** Rims of osmiophilic granules are aryl sulphatase-positive. (TEM **a**: × 1310; **b**: × 14 040; **c**: × 20 075.) Courtesy of Dr A J Freemont and the Editor, *Annals of the Rheumatic Diseases.*[41]

type A synoviocytes. Their osmiophilic granules react for acid phosphatase and aryl sulphatase, the former in the centre of the granule, the latter in the periphery, a zone devoid of acid phosphatase activity.[41]

Accompanying the characteristic focal cellular synovial infiltrate is an inflammatory arthropathy in which a prolific fibrinous exudate is one hallmark of bone destruction and a secondary synovitis. Fragments of bone lie within the eosinophilic giant cells. Synoviocyte hyperplasia is evoked. The synovial fluid displays a low cell count; it contains a few bizarre, large macrophages, and megalocytes. These appearances, with poor mucin clot formation, are thought to be diagnostic of the disease.[41]

The nature and cause of multicentric reticulohistiocytosis remain unknown. It is no longer regarded as a lipidosis but is evidently an abnormality of tissue macrophage behaviour. The megalocytes appear to form by the fusion of the smaller histiocytes. Many patients, who are more often female than male and usually aged 50–70 years, display skin sensitivity to tuberculoprotein but mycobacterial residues are not demonstrable in the skin or articular lesions. Little is known of the genetics of multicentric reticulohistiocytosis but the possibility remains that there is both a racial and heritable predisposition.

Starch synovitis

The adverse, granulomatous tissue reactions to the magnesium silicate, talc are well recognized. Starch powder was chosen to replace talc, for example, as a lubricant for surgical gloves but starch, too, can cause adverse connective tissue responses when it gains access to body cavities. One reaction of this kind is synovitis. Starch may enter the joint or tendinous synovium accidentally, from surgeon's gloves, or deliberately during the attempted autoinjection of a narcotic.[42] The histological consequences include an inflammatory exudate comprising polymorphs, lymphocytes and macrophages; subsynoviocytic granulation tissue; and non-caseating granulomata including multinucleate giant-cells. These cells contain granules that stain positively with Lugol's iodine and display a Maltese cross structure when examined in polarized light. The granules react positively with the periodic acid-Schiff technique, before and after diastase digestion.

Bursae, ganglia, cysts
(Figs 24.19–24.22)

Bursae

Connective tissue spaces lined by synovial tissue are present near or in continuity with many normal synovial joints. With repeated episodes of minor injury or with the wear and tear which is occasionally determined by selected occupations, the outer connective tissue wall of the bursa becomes thickened and there is an increased secretion of synovial fluid. The knee is particularly susceptible and the prepatellar bursa is prone to inflammation. Occasionally, such bursae display the histological signs of specific diseases such as rheumatoid arthritis. When a bursa has sustained long-lasting and severe trauma the contents come to include increasing amounts of fibrin of inflammatory origin which may become organized into strands and networks of bizarre shape (Fig. 24.19b).

The connective tissue related to synovial joints other than the knee may, through chronic injury, undergo a

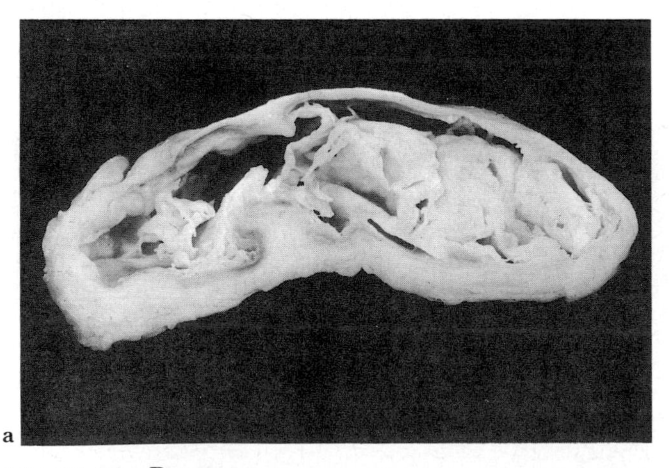
a

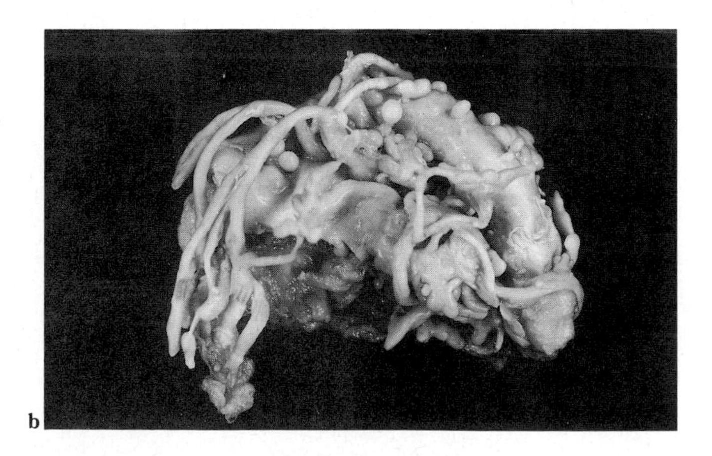
b

Fig. 24.19 Bursae.
a. Contents of this large bursa are soft, grey-white and homogeneous. They resemble the caseous material of a tuberculous abscess. The walls are compact, dense and fibrous. **b.** Bursa excised from knee joint. There are many ramifying, fibrous processes. (**a**: × 2; **b**: × 0.8.)

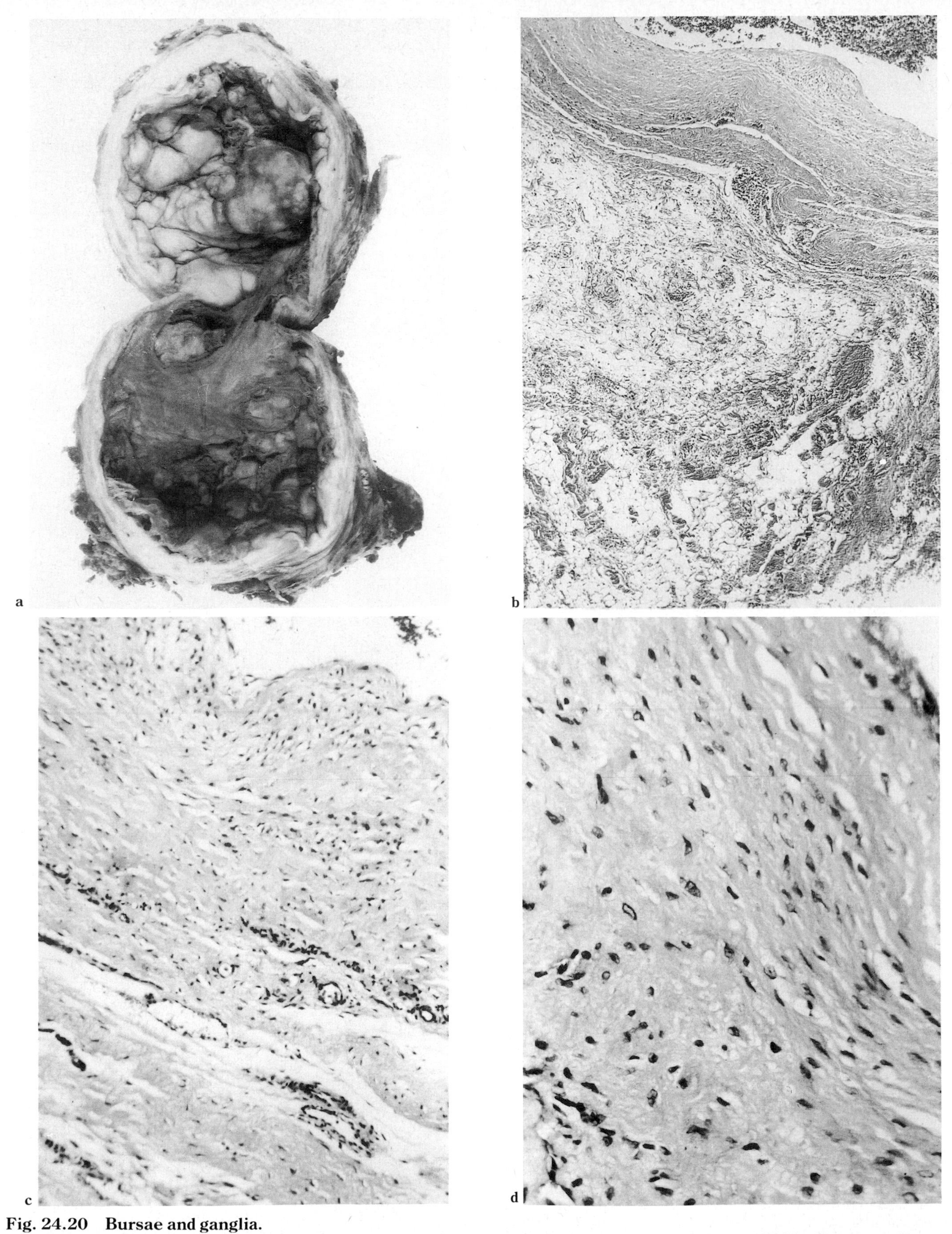

Fig. 24.20 Bursae and ganglia.
a. Prepatellar bursa: indolent swelling, 70 mm in diameter, present in prepatellar region of woman for 1 year. Wall is thick, dense fibrous tissue; internal lining surface has sessile, lobulated structure. **b.** Ganglion: wall of this circumscribed lesion into which bleeding has taken place, is composed of collagen-rich connective tissue. The mucoid material usually present has escaped. **c.** Higher power view of part of material shown in **b. d.** Further field from **b.** (**a**: × ½; **b**: HE × 35; **c**: × 113; **d**: × 284.) Fig. 24.20a is reproduced by permission of the President and Fellows of the Royal College of Surgeons of Edinburgh.

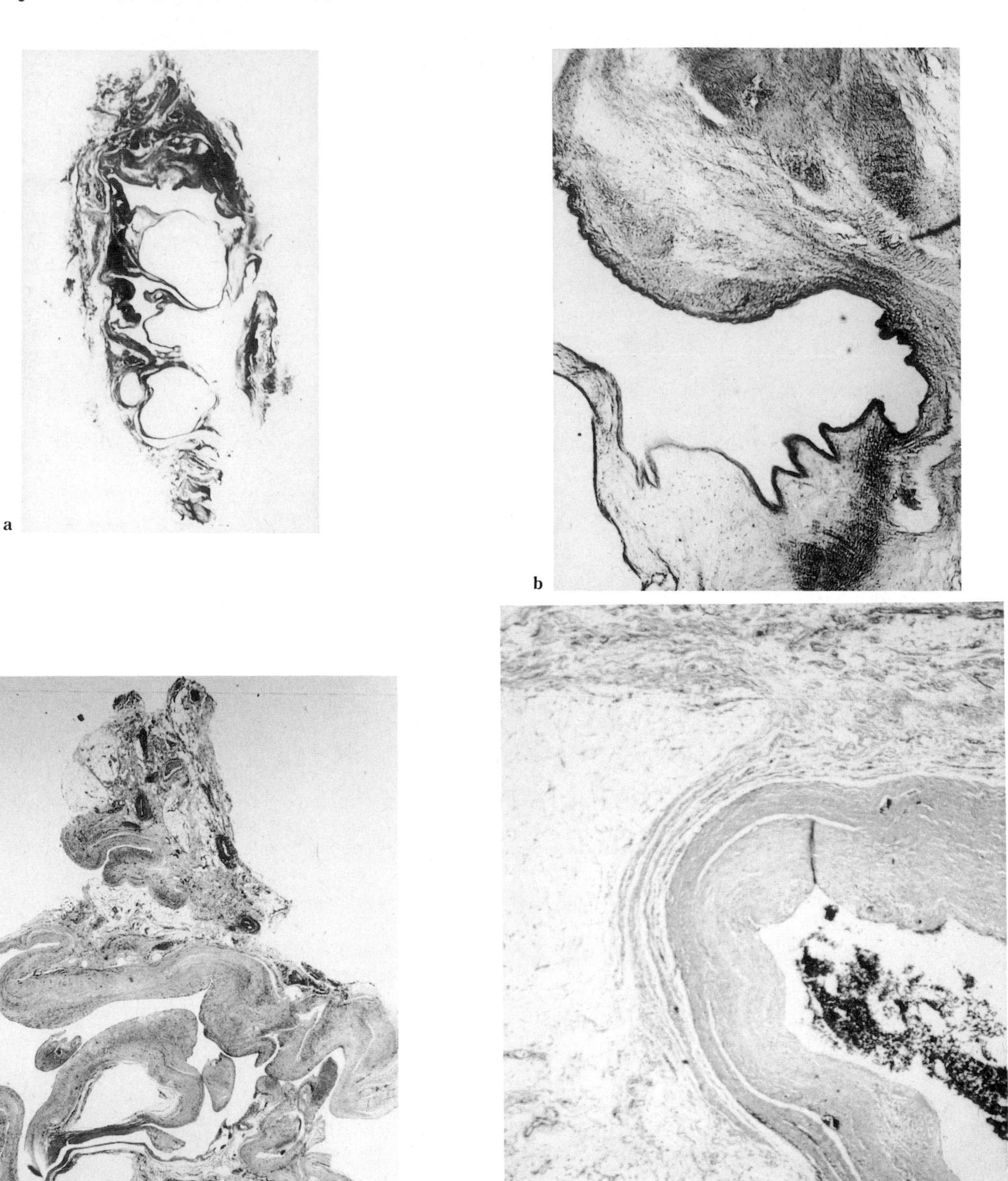

Fig. 24.21 Meniscal cyst; ganglion.
a. Increasing pain at medial side of knee followed a trivial injury in this male aged 39 years. Several meniscal cysts are present. They contained mucoid fluid. **b.** Cysts can be seen to be lined by flat (squamous) cells of synovial origin. **c.** Ganglion: cosmetic reasons and local discomfort drew attention to this fluctuant 24 mm diameter mass at the wrist joint of a 62-year-old male. The myxoid contents have escaped, to be replaced by a little blood. **d.** The wall is formed of fibrous tissue. (**a:** HE × 12; **b:** × 40; **c:** × 18; **d:** × 55.) Figs 24.21a and b are reproduced by permission of the President and Fellows of the Royal College of Surgeons of Edinburgh.

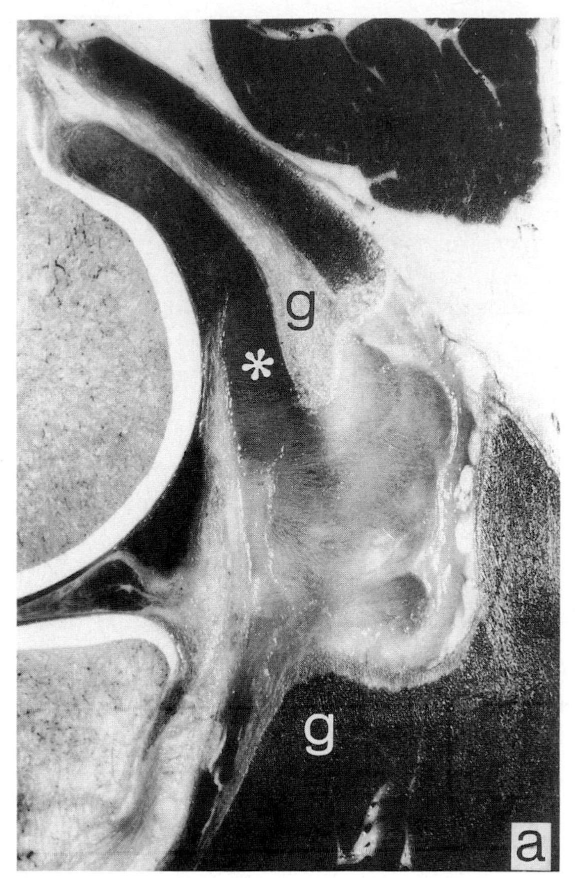

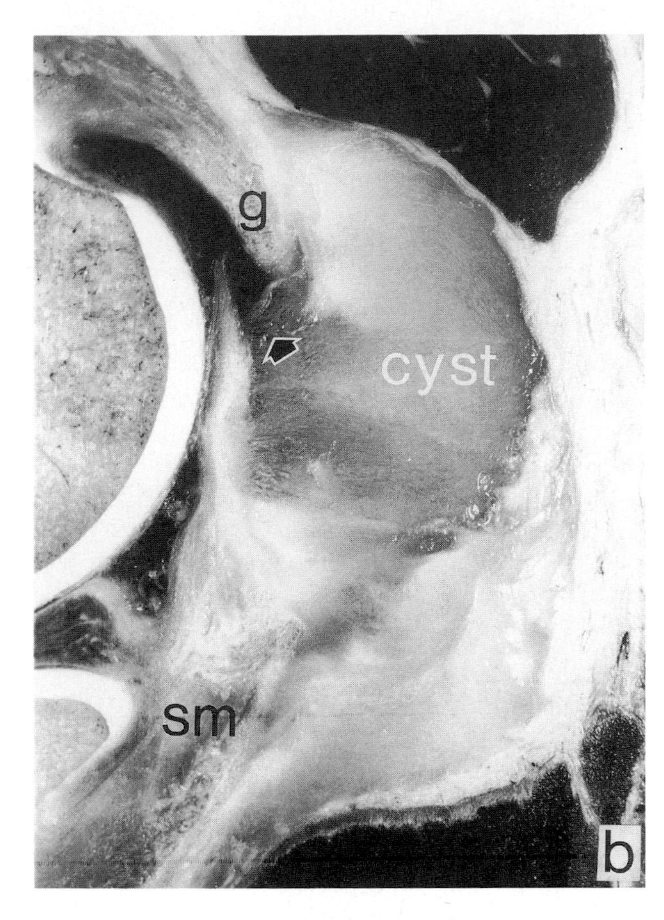

Fig. 24.22 Bursa of the knee.
Knee joint cavity communicates with gastrocnemiosemimembranosus bursa by transverse slit which opens on joint flexion. In this preparation, made by heavy duty cryomicrotome (LKB 2250 PMW, Bromma, Sweden), sagittal section shows **a.** lateral part of capsular opening and **b.** medial edge of capsular opening with large cyst displacing tendons. By courtesy of Dr W. Rauschning and the Editor, *Annals of the Rheumatic Diseases.*[140a]

sequence of events leading to new, adventitious bursa formation.[23] The local collagen fibres and ground substance become components of a focus of fibrinoid: the compact connective tissue material dissolves and a cystic structure appears containing a mixture of cell debris, extracellular fluid, altered ground substance and an inflammatory exudate. The fibrinogen coagulates and a curious laminated, sponge-like object is formed. A synovial lining may form around this nucleus and outside there occurs progressive thickening of the nearby connective tissue. Calcification of the adventitial bursal contents is rare but the bursa may become infected. Ultimately, fibrosis with obliteration of the sac may occur if mechanical irritation ceases.

Ganglia
(Figs 24.20b–d and 24.21c, d)

A ganglion (knot) is a common, slowly enlarging, 10 to 35 mm diameter, fluctuant uni- or multiloculated swelling found in adults on the extensor or flexor aspects of the wrist, near the ankle or in relation to the knee or other joints. Ganglia usually lie close to a joint capsule or tendon sheath; they are unsightly but not painful. Removal may also be considered to allow histological exclusion of neoplasia. The gross, microscopic and ultrastructural appearances of ganglia have been described in detail[46a] but their cause remains uncertain and their natural history is incompletely understood.

Dissection shows that a ganglion contains thick, viscous, mucinous fluid, often released during resection. There is seldom any evidence of continuity with a joint or tendon sheath synovial cavity.

Microscopically, the walls of the loculi of a ganglion are formed of compact fasicles of fibrillar collagen between which are elongated cells identified as myofibroblasts.[46a] Similar cells, bearing no apparent resemblance to synovial cells, form an incomplete lining for the cavities. Between individual cavities, a looser connective tissue contains small blood vessels. Neither haemosiderin nor lipid are present and multinucleated giant cells are not seen. Nearby bone is preserved intact and there is no tendency for a ganglion to compress or invade surrounding structures.

Transmission electron microscopy reveals the presence of myofibroblasts[46a] with a limited amount of extracellular matrix in which proteoglycan and fibrous collagen can be seen. Occasionally, mast cells and mononuclear macrophages are present. Understanding of the cause of ganglia rests heavily on knowledge of the source and nature of the contained fluid. The fluid is not synovial. Macromolecules within the nearby, connective tissue matrix may be secreted by the lining cells within which proteoglycan-like, electron-dense particles can be imaged. An alternative explanation proposes that the fluid contents of ganglia come from a process of 'degeneration' or depolymerization of the extracellular matrix but there have been few biochemical studies comparing the composition of ganglionic fluid with that of the normal extracellular matrix. A persistent, ganglion-like swelling can be caused by the periarticular injection of high molecular weight, high viscosity hyaluronate into rabbits and a local excess of hyaluronate secretion by cells provoked by trauma seems the most probable explanation for the formation of ganglia in humans.

Cysts[70]

(Fig. 24.21a, b)

Meniscal cysts From time-to-time, the extracellular matrix of parts of the lateral intra-articular knee joint meniscus accumulates a hyaluronate-rich, collagen-poor material. The defect is thought to be provoked by injury. Water is retained. The process of matrix change accompanies the circumferential formation of a condensate of collagenous fibrous tissue. A cystic cavity is formed. Synovial-like cells, of squamous character, come to line the cyst. The histological appearances of the contents of the cavity are myxoid (mucinoid); the cavities are frequently multiple and may be multiloculated. The accumulation of water results in an increase in size so that surgical excision of the cyst(s) becomes necessary.

Response of connective tissue to neoplasia

Desmoplastic response

Many of the most common carcinomata, particularly of bronchus, breast and gastrointestinal tract, form or cause to be formed, substantial quantities of collagenous connective tissue. The reader is referred to texts on the pathology of cancer.

Articular involvement by nearby tumours

The onset and extension of malignant neoplasms may be indicated by monarticular or polyarticular arthropathy, remote from the tumour site.[176] Periodically, there is direct joint involvement. The benign neoplasm, osteoid osteoma, is often associated with a highly vascular, hyperostotic reaction in which the adjacent bone is responsible for a zone of radioopacity surrounding the radiolucent nidus that constitutes the centre of the neoplasm. When the lesion is located near an articular surface, the mechanical and pharmacological properties of osteoid osteoma are responsible for cartilage degradation, and for secondary hypertrophic synovial changes that can be mistaken for tuberculosis or rheumatoid arthritis.

Primary malignant bone tumours also extend periodically directly into synovial joints.[134] Giant-cell tumour of bone may destroy articular cartilage directly. Figure 24.23

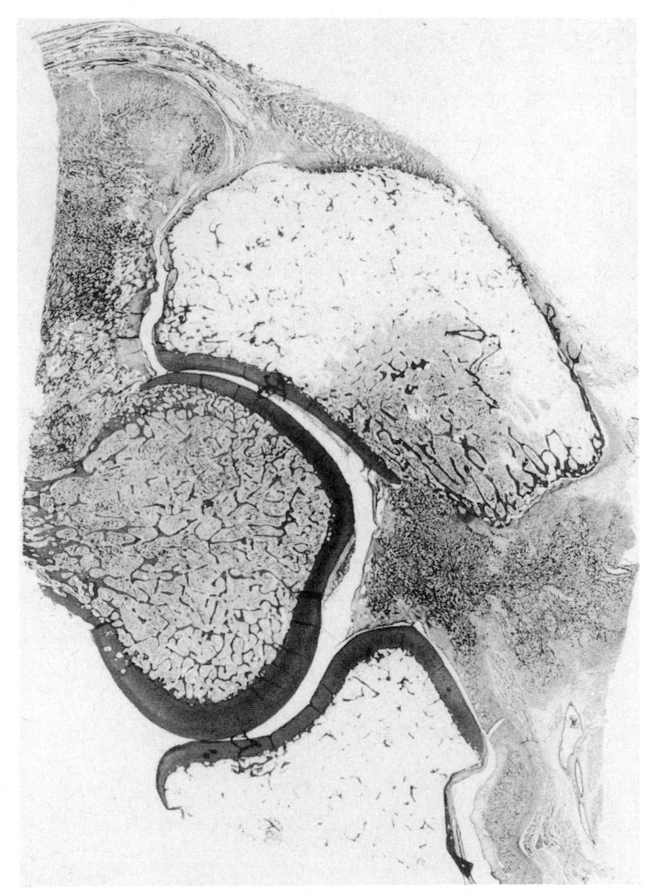

Fig. 24.23 Articular invasion by recurrent osteosarcoma.
Following chemotherapy, this osteosarcoma in a man aged 19 years recurred and extended down the humerus to involve the elbow joint. Note the dark-appearing tumour tissue within the synovia and the adjacent bones. (HE × 1.8.)

illustrates an osteosarcoma of the humerus infiltrating the elbow joint synovia. In a rare case, chondrosarcoma (Fig. 24.24a) was found to occupy the olecranon fossa following arthroplasty for chronic traumatic arthropathy associated with synovial chondromatosis (see p. 985).

Extraosseous connective tissue metastasis

The diversity of sites to which carcinomata such as those of the bronchus may metastasize is limitless. I have recently examined material from secondary deposits in a subungual finger tip and at the root of a molar tooth. However, the synovia are apparently spared and the reasons for this apparent immunity to metastasis have excited comment.

Apparent immunity of synovia to metastasis

Systematic search of the literature reveals only 30 reports of cases in which metastases from primary carcinomata, such as those of the bronchus and breast, have lodged in the synovia (Fig. 24.24b). Metastatic cancer cells, borne in the blood, may, of course, not reach the synovia. Alternatively, they may not be arrested within the joint. When every case is excluded in which the possibility of a intraosseous origin for the synovial neoplasm is suspected, no more than eight valid reports remain. To confirm or refute the significance of this observation, random, multiple synovial blocks have been examined from the knee joints of 29 patients dying with metastatic carcinomatosis. No microscopic secondary deposits have been found. To examine the possibility that the synoviocytes form substances that inhibit the growth of metastatic carcinoma cells lodging in the synovia, the supernates from synovial cell cultures were added to cultures of selected tumour cell lines.[166] No inhibition of tumour cell growth was measurable. The possibility remains that there is a mechanism of inhibition of tumour cell growth effective only *in vivo*.

Pigmented villonodular synovitis

Pigmented villonodular synovitis[74] is a localized, proliferative process found in synovial joints and tendon sheaths (Fig. 24.25). The knee and the tendon sheaths of the hand are most commonly affected[140] but the hip, foot, wrist and shoulder joints are also susceptible. In the tendon sheaths, the term 'giant-cell tumour of tendon sheath' is often used and the wide range of older names used to describe pigmented villonodular synovitis indicates the uncertainty regarding its nature and cause.[39]

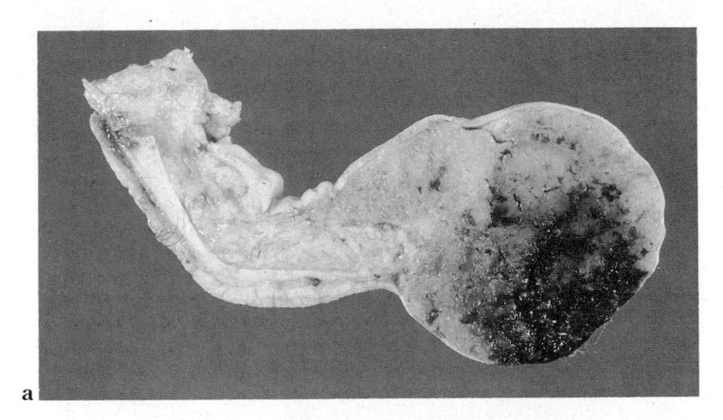

a

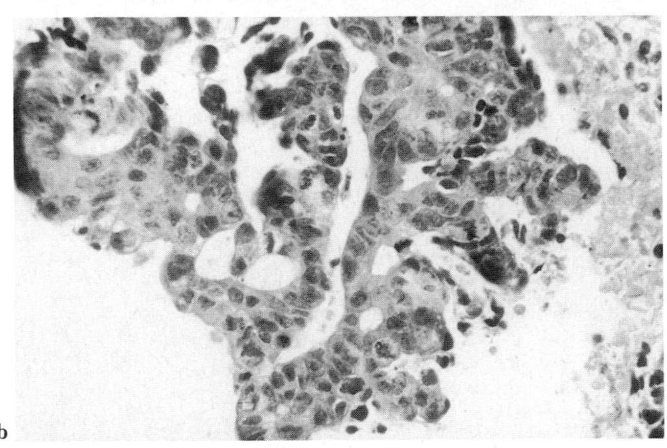

b

Fig. 24.24 Neoplastic destruction of synovial joint and synovial metastasis.
a. Primary chondrosarcoma of terminal phalanx of this 60-year-old man has destroyed distal interphalangeal joint. **b.** Carcinoma of colon metastatic to knee joint synovium in elderly female. Fig. 24.24a is reproduced by courtesy of Dr S W B Ewen; Fig. 24.24b by courtesy of Professor O. Myre-Jensen.

Clinical presentation

Pigmented villonodular synovitis is a monarticular arthropathy in which pain and swelling affect a knee or other joint in males or females in the age range 20–40 years. Rarely, more than one joint is involved. Exceptionally the disease is bilateral. Investigation of the synovial fluid reveals a brown-red, blood-stained liquid with slight polymorph leucocytosis. Radiologically, soft-tissue synovial nodularity may extend to include erosive, sclerotic or cystic bone lesions. Arteriograms confirm high vascularity.

Pathological changes

It is convenient to discriminate between local and diffuse forms of the disorder. In the former, a single localized nodule or mass which may be sessile or pedunculated, displays a variegated white-orange, or yellow-brown colour when the circumscribed lesion is bisected. The diffuse disorder simulates the more prolific forms of synovial

a

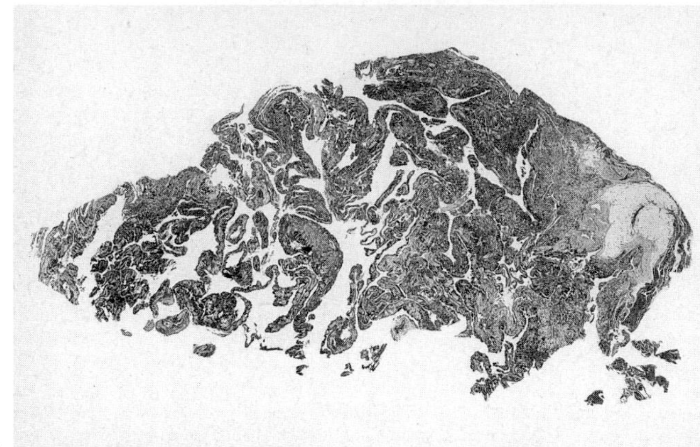

b

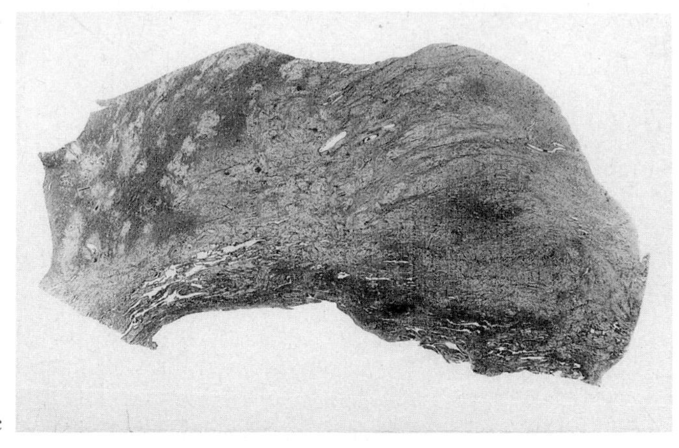

c

Fig. 24.25 Pigmented villonodular synovitis.
a. Opened knee joint is bounded by hypertrophic synovial villous processes and by many small, pedunculated nodules closely similar to those recognized in giant cell tumour of tendon sheath (xanthofibroma—benign fibrous histiocytoma). Tissue is orange-brown. **b.** Part of material excised from knee of 30-year-old male with diffuse pigmented villonodular synovitis. Loose meshwork of subsynoviocytic connective tissue is identical to that present in finer processes seen in **a. c.** In other more compact parts of specimen demonstrated in Fig. 24.25b, dense fibrous tissue is interspersed with clusters of lipid-containing cells. Much haemosiderin is present and there are many multinucleated giant cells. Fig. 24.25a is reproduced by permission of the President and Fellows of the Royal College of Surgeons of Edinburgh.

villous hypertrophy seen in florid cases of active, chronic rheumatoid arthritis in which intraarticular haemorrhage often occurs. The numerous, prominent, smooth-surfaced brown synovial processes of varied size and shape, protrude from the opened joint and affect all parts of the compartment. Many of the synovial processes are polypoidal and average 5–10 mm in diameter.

Microscopically (Fig. 24.26), the cells of the synoviocyte layer are found to be hyperplastic, the synovial lining layer three to seven or more cells in depth. Within these cells, type B synoviocytes have been said to be common, and type A cells difficult to identify.[48] Many of the entire cell population in this study contained lysosome-derived residual bodies, resulting from the phagocytosis of fibrin. Sideromes were numerous. Multinucleate synovial giant-cell formation is frequent. In the subsynoviocytic connective tissue, the presence of many macrophages, often containing iron and ingested red blood cells or their derivatives, is common. Other mononuclear cells are rich in the lipid which, with ingested iron pigments, gives the nodules of pigmented villonodular synovitis their characteristic colour. Mitotic figures are commonplace both among these mononuclear cells and in the synoviocyte population. Clefts lined by synovial-like cells form divisions between individual villi; they may also be present within the more solid parts of the nodules. There is a widespread hyaluronate- and collagen-rich stroma, and, particularly in the more cellular, less collagenous zones, many small blood vessels.

Bone adjoining the extending articular lesions of pigmented villonodular synovitis may be affected occasionally. It has been proposed that bone erosion, the extension of the cellular proliferative tissue and the formation of intraosseous, cyst-like lesions, is a pressure phenomenon. The relatively high frequency of bone lesions in the hip rather than the knee has been attributed to the few articular recesses in the former. The cells constituting the lesions of pigmented villonodular synovitis extend into bone through vascular foramina.

Pathogenesis

There are two main views on pathogenesis: first, that pigmented villonodular synovitis is a form of benign, recurrent neoplasm; second, that the disorder is inflammatory. Other concepts include the proposals that pigmented villonodular synovitis is a localized disease of lipid metabolism or that it results from trauma or repetitive haemarthrosis. There is no evidence of a genetic predisposition. Nodular forms of pigmented villonodular synovitis are prone to recur

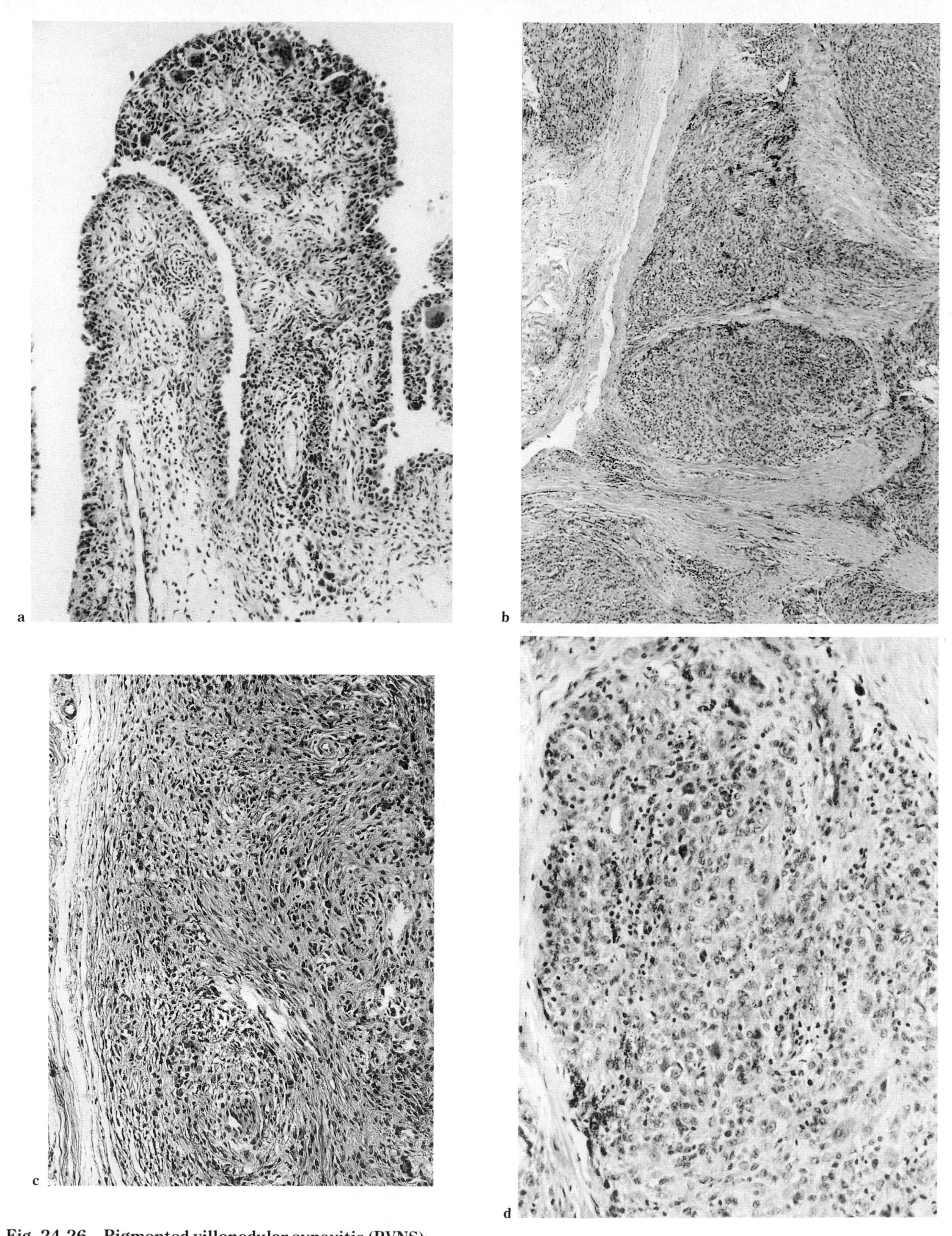

Fig. 24.26 Pigmented villonodular synovitis (PVNS).
a. Hypertrophic synovial villus. Note cellularity and high vascularity of villus with multinucleated giant cells at upper margins. **b.** Islands of mononuclear, macrophage-like cells within compact part of diffuse lesion in PVNS. Cellular foci which have neoplastic-like quality reminiscent of fibromatoses (see Chapter 8), are circumscribed by laminae of dense fibrous tissue. Much haemosiderin may be present. **c.** Whorls of mononuclear cells, many containing lidid, adjoin margin of compact lesion. **d.** Among the mononuclear cells, mitotic activity is frequent. An inflammatory component is indicated by the presence of scattered polymorphs. (HE **a**: × 125; **b**: × 50; **c**: × 120; **d**: × 160.)

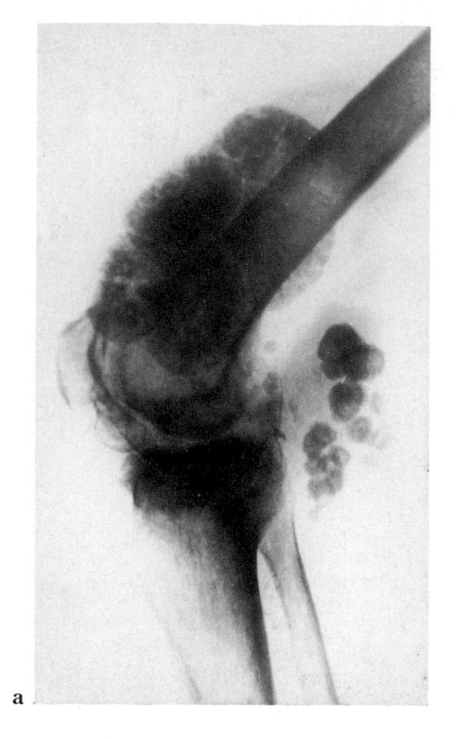

a

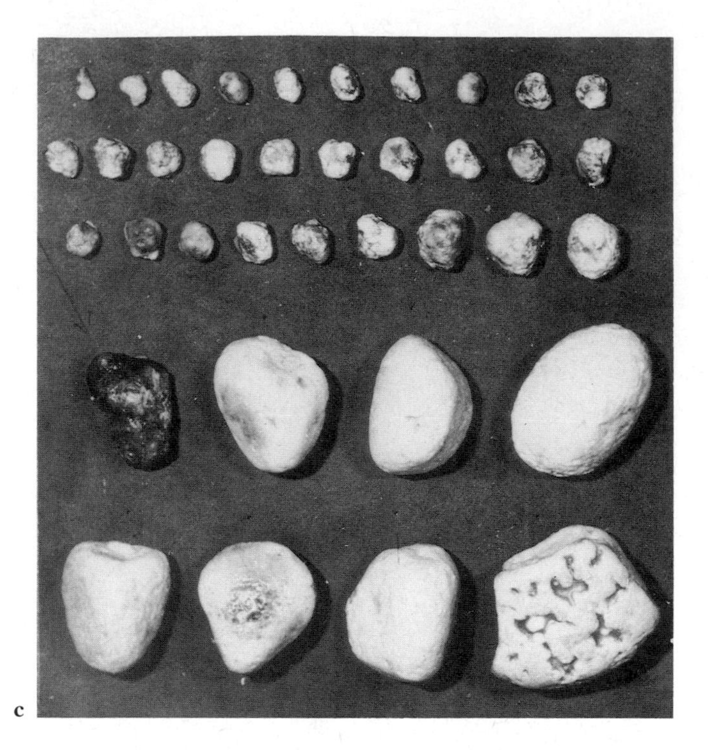

c

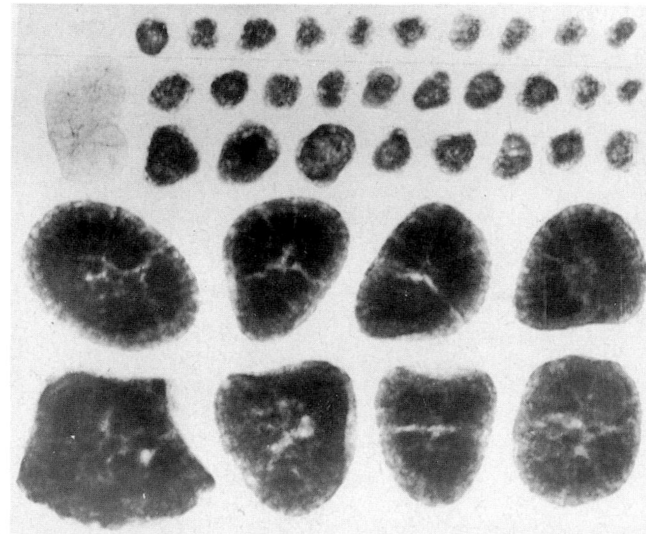

b

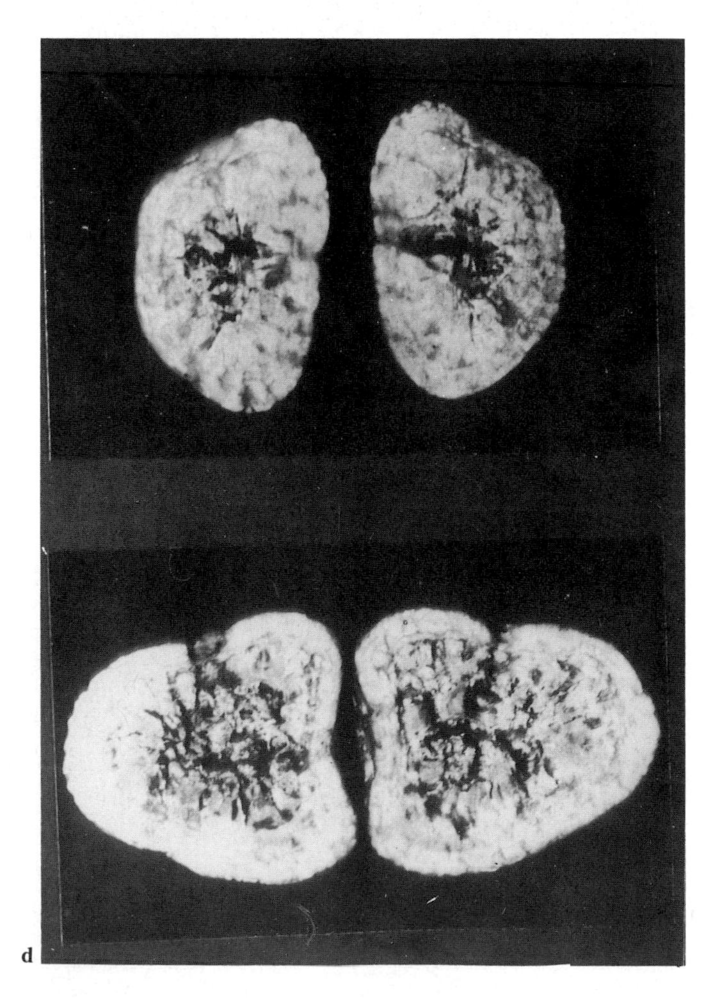

d

Fig. 24.27 Synovial osteochondromatosis.
a. Radioopaque islands of bony and cartilagenous tissue within knee joint. **b.** X-ray of synovial osteochondromatous foci after removal. **c.** External, smooth surfaces of islands of osseochondral tissue excised from knee joint. **d.** Internal structure of osteochondral nodules from knee joint synovia, demonstrating foci of bone within cartilage matrix. (**a**: × ¼; **b,c**: × ½; **d**: Natural size.) Reproduced by courtesy of the President and Fellows of the Royal College of Surgeons of Edinburgh.

and an analogy with the fibromatoses (see p. 313) may be considered.

The evidence in favour of an inflammatory disease includes the presence of connective tissue cell phagocytosis and of mononuclear cell proliferation, with haemosiderin aggregates and collagen and hyaluronate synthesis.[39] The alternative, older view, that pigmented villonodular synovitis is a benign neoplastic process, still finds support.[140] These authors emphasize the extension of the proliferating subsynovial mononuclear cells into nearby compact connective tissue and the enhanced mitotic activity of the recurrent lesions.

Synovial osteochondromatosis

In this uncommon condition, cartilage and/or bone islands develop within the subsynovial connective tissues (Figs 24.27 and 24.28). They form a series of pedunculated, intraarticular nodules. As with osteochondritis dissecans, young adult males between the ages of 17 and 40 are affected most commonly. The synovial villi in which cartilagenous or bony transformation occurs, often escape into the joint cavity where they form multiple loose bodies. The synovial tissue of bursae and of tendon sheaths may also be affected. Microscopically (Fig. 24.28), zones of hyaline cartilage, of cancellous bone or of both, are detected within a covering of compact fibrous or looser synovial tissue. Osteoarthrosis is likely to result if the loose bodies are not removed surgically; there is direct mechanical injury to articular cartilage.

Pathogenesis

It is widely believed that the foci of chondromatosis arise by a process of metaplasia but it is at least possible that chondrocytes can be lodged in the synovia after their

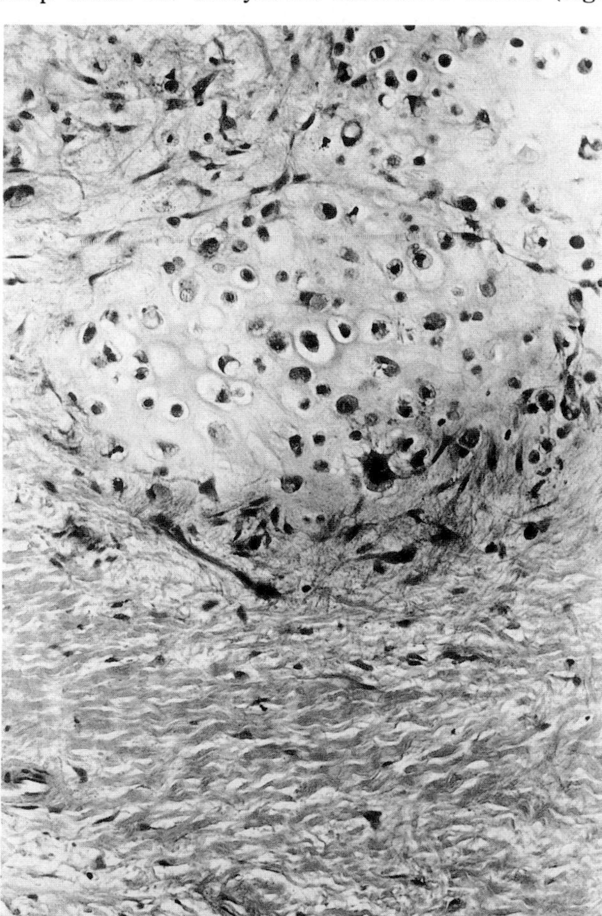

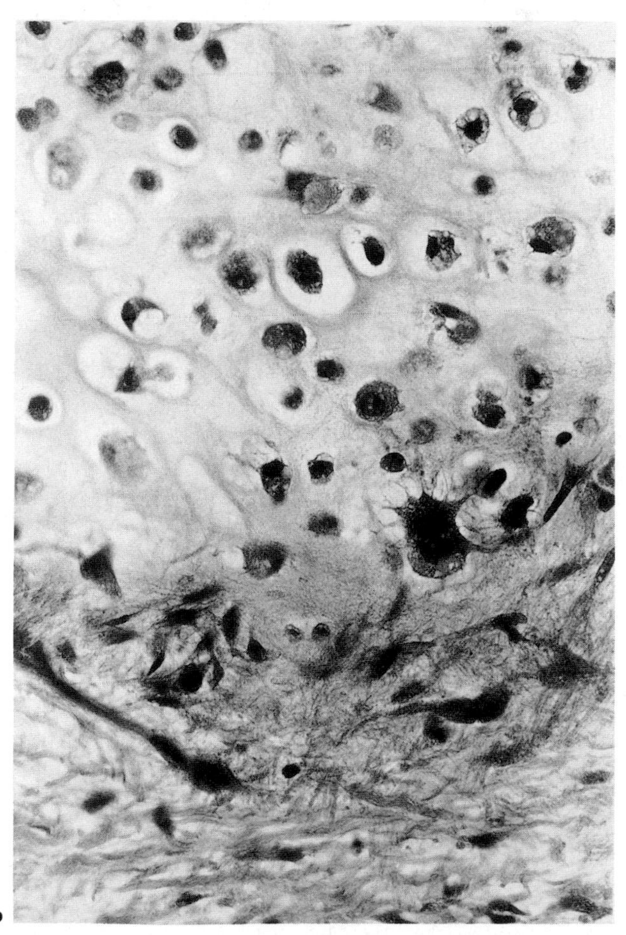

Fig. 24.28 Synovial chondromatosis.
a. Synovium from the elbow joint of a young female who sustained repeated injury during childhood and adolescence. The tissue displayed in this figure was examined at biopsy at age 15 years; 8 years later, elbow arthroplasty was performed. (HE × 160.) **b.** Higher power view of cells from the same specimen. In rare instances, cell density and disparity in cell and nuclear size are among the microscopic features that raise the possibility that synovial chondromatosis may precede chondrosarcoma. (HE × 320.)

release from cartilage surfaces. Chondrocytes can readily be grown in two- and three-dimensional culture (see p. 24) and, during the resection of chondrosarcoma, are easily introduced into the tissue along tracks of surgical exploration or excision. It is therefore not wholly surprising that islands of chondrocytes should be found occasionally multiplying within the synovia of patients in whom trauma or osteoarthrosis has led to osseochondral fragmentation. Under these circumstances, the term chondromatosis is applicable. Clearly, an identical response may be seen if the subsynoviocytic connective tissue undergoes cartilagenous (chondroid) metaplasia. Whichever origin is thought most likely, chondromatosis constitutes a benign, slowly progressive disorder. The islands of cartilage, sometimes partially calcified and periodically incorporating foci of bone, form osseochondral (osteocartilagenous) nodules, which may become detached from the synovia and lie free in a joint.

Complications

Synovial chondromatosis is generally monoarticular. The knee, elbow, hip and occasionally the ankle[9] are affected most often. There is a tendency for local recurrence after excision and the slowly growing islands of cartilage may advance into nearby bone. In diagnosis, the proliferation of chondrocytes and the varying degrees of their differentiation are occasionally causes for thought since, very rarely, chondrosarcoma has been found to evolve from chondromatosis.[60] The extraosseous origin of the lesion may point to the non-neoplastic nature of a progressive, chondroid articular mass. However, in a recent case, analogous with that of Hamilton *et al.*,[60] synovial chondromatosis in a girl aged 15 years advanced to invasive chondrosarcoma eight years later. The entire bone of the olecranon was occupied in a 'wall-to-wall' pattern by a grade 2 tumour.

Responses of connective tissue to other stimuli

Granuloma annulare

In this chronic skin disorder prevalent in children and young adults, groups of very small nodules are arranged in a circular or ovoid disposition on the extensor surfaces of the extremities (Fig. 24.29). The hands, fingers, feet, ankles and elbows are affected most commonly. The nodules are dermal; they elicit neither pain nor discomfort and there is little accompanying inflammatory reaction. Unlike the nodules of rheumatic fever, there is no association with a previous streptococcal, nasopharyngeal infection and no formation of antiimmunoglobulin autoantibodies. Arthritis does not precede or accompany nodule formation.

Microscopically, the dermal nodules have a general structure recalling that of the subcutaneous rheumatic (see Chapter 20) and rheumatoid (see Chapter 12) nodules. Granuloma annulare is distinguished from the former by its dermal location and compact structure and from the latter by its small size, isolated nature and superficial location. Mononuclear macrophages or histiocytes, lymphocytes and fibroblasts surround central foci of fibrinoid and the irregular margin recalls the serpiginous structure seen in rheuma-

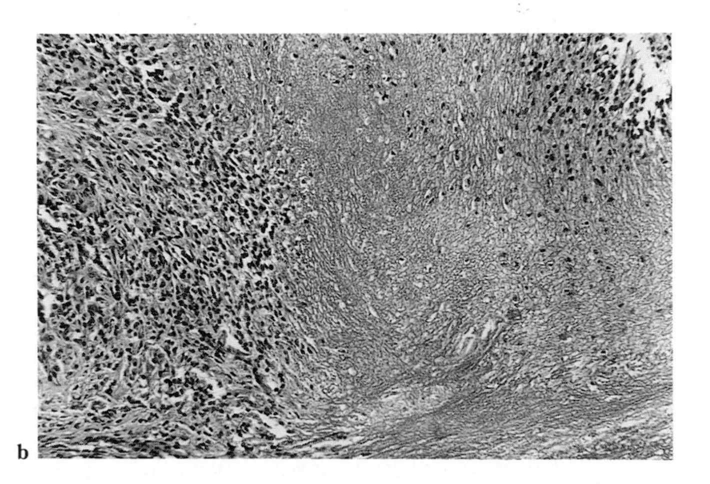

Fig. 24.29 Granuloma annulare.
a. Dermal nodule. Border of necrotic focus at left is formed by serpiginous zone of vascular connective tissue. There is a structural similarity to a small, isolated rheumatoid nodule but marginal pallisading of mononuclear cells is not recognized. **b.** Higher-power view of field seen in **a**. Necrotic focus (at right) is bounded by loose, vascular connective tissue within which is an infiltrate of mononuclear macrophages and lymphocytes. (HE **a**: × 60; **b**: × 105.)

toid granulomata. Fasciculi of intact but disordered collagen fibres, with some proteoglycan, lie among the fibrinoid.

The cause of granuloma annulare remains uncertain. An association with immediate and delayed hypersensitivity responses has been claimed and, from time-to-time, the disorder has been attributed to nearby, focal necrotizing immune complex arteritis or arteriolitis. Dermal reactions of a broadly similar nature are encountered in leprosy, necrobiosis lipoidica diabeticorum and xanthomata but these lesions are readily distinguished microscopically.

Hypertrophic osteoarthropathy

In exceptional cases of thoracic disease, and occasionally in extrathoracic disease, and in metastatic neoplasia,[181] clubbing of the fingers, periostitis and arthritis develop, constituting the rare condition of hypertrophic osteoarthropathy.[51,106,157] Usually, symptoms of hypertrophic osteoarthropathy precede signs of lung disease; the symptoms regress very quickly when the lung lesion is eradicated.

Clubbing of the fingers and toes is itself not confined to hypertrophic osteoarthropathy and is encountered in a wide variety of disorders ranging from cyanotic congenital heart disease and bacterial endocarditis to liver disease,[76] Crohn's disease and ulcerative colitis.[38] The morphological appearances of clubbed fingers were described by Bigler.[13] The concept of hypertrophic osteoarthropathy was advanced independently. The full syndrome of hypertrophic osteoarthropathy was identified much more recently and is now recognized in conditions ranging widely from cholestatic hepatic cirrhosis[69] to late-onset agammaglobulinaemia.[10]

Clinical presentation[157]

Bilateral, symmetrical painful swelling of the knees, ankles and metacarpophalangeal joints may simulate rheumatoid arthritis. The elbows, wrists and shoulders are less often affected. More males than females are implicated in an age range from 45–70 years. Swelling and tenderness may antedate clinical recognition of the underlying disease by as much as two or three years. In a typical case, the cause is a bronchial carcinoma but other neoplasms such as nasopharyngeal lymphoepithelioma in man,[59] and fibrosarcoma in the dog,[164] may result in identical changes. The swollen, distal joints display increased blood flow and isotopic scans reveal a localized uptake of nuclide in those parts of the skeleton where bone and articular changes are manifest. Radiologically, there is periosteal new bone formation with a radiolucent layer deep to the periosteal reactive bone. Tests reveal the presence of serum acute-phase reactants; occasionally there is a positive reaction for antinuclear factor but rarely for rheumatoid factor. Synovial fluid, present in excess, contains only a small increase of cells which may be mononuclear phagocytes with few polymorphs and lymphocytes. The synovial fluid is sterile and contains less than 5 per cent protein.

Pathological characteristics

Macroscopic changes

The affected terminal phalanges lose their normal contour. The angle between the nail plate and the dorsum of the finger is diminished and may ultimately become greater than 180°, with a dorsal convexity; this is the earliest evidence of finger clubbing. Bigler[13] described five grades of severity. In the mildest, there is little other than a change in the nail angle. In the most severe the finger becomes claw-shaped and the circumference much increased in the terminal part of the phalanx, giving a spoon-like or ball-shaped appearance. The overlying skin is atrophic, the colour cyanosed.

Bone

Increased blood flow and periosteal oedema accompany increased osteoblastic activity. New, appositional bone formation begins on the endosteal surface but the principal response is the formation of a layer of woven bone at the periosteal margin. Radiographically, new bone formation is recognized at the distal ends of the bones. If the underlying lesion is corrected, a reduction in peripheral blood flow may be detected within a few hours of operation and the more substantial changes in structure of soft connective tissue and of bone disappear within a few weeks.[40] The underlying microscopic disturbance is recognized most clearly in the nail bed; it is thickened to a varying degree and contains excess loose connective tissue and mesenchymal cells among which is a network of glomus-like arteriovenous channels. Groups of lymphocytes and eosinophil granulocytes may be present. Corresponding to these changes is periosteal oedema and thickening of the palmar surface. When a severe nail bed reaction is present, foci of primitive vascular mesenchymal tissue are also found in the palmar periosteum.

Synovia

A mild, non-specific synovitis develops with congestion, a perivascular mononuclear cell infiltrate and villous hypertrophy. Rarely, the affected synovia are infiltrated simultaneously by the carcinoma that constituted the underlying lung lesion.[34] Synovial changes analogous to those seen in the soft tissues of the clubbed digits and periosteum were recognized in cases of hypertrophic osteoarthropathy associated with bronchogenic carcinoma. In the presence of congenital cardiac disease and of cirrhosis, synovia displayed low-grade inflammation and, occasionally, calcification.[76] Although very slight synovial lymphocytic, plasmacytic and histiocytic infiltration was reported by Ropes and Bauer,[145] there was little evidence of inflammation among the synovial tissue of the eight patients with bronchial carcinoma reported by Schumacher.[153]

Ultrastructure Unexpectedly, electron-dense deposits were found in the walls of venules of five cases of hypertrophic pulmonary osteoarthropathy. Fibrin and necrotic cell debris was also present.[153] The electron-dense material had a resemblance to the deposits identified in the glomerular capillary walls in systemic lupus erythematosus but also in other disorders including carcinoma of the colon. The composition of the deposits in hypertrophic osteoarthropathy was not established; but comparable deposits are infrequent in inflammatory disease such as experimental antigen-induced arthritis, transient undiagnosed arthritis, rheumatoid arthritis and gouty arthritis.[153] As Schumacher points out, electron-dense synovial vascular deposits have not been seen in cases with non-inflammatory effusions.

Experimental study

Bone changes closely resembling those that occur naturally in human hypertrophic osteoarthropathy, can be produced in dogs by directing the inferior vena caval blood flow to the left atrium.[45] There is a marked fall in pO_2 and an increase in peripheral blood flow. The metacarpal bones develop periosteal vascularization and hypertrophy with the formation of woven bone containing foci of cartilage. Canine hypertrophic pulmonary osteoarthropathy can also be reproduced by provoking the development of squamous carcinoma chemically.[87] Subperiosteal new bone formation of metatarsal and tarsal bones, and of the tibia and fibula are described.

Aetiology and pathogenesis

It is generally accepted that the proliferation of bone and connective tissue which occurs on the periphery of the limbs is a result of increased local blood flow.[50] The increased flow is brought about by the opening up of arteriovenous communications similar to those of the glomus bodies. These vascular channels can be identified microscopically; the associated overgrowth of connective tissue may therefore be analogous to the unilateral limb hypertrophy which accompanies certain cases of congenital arteriovenous anomalies and hamartomata.

Two principal theories have been advanced to account for the dilatation of the vascular channels. The first takes note of the favourable response to vagotomy and the reduction of blood flow and locally elevated pulse pressure that follow excision of the underlying lung neoplasm. Vagal afferent impulses, transmitted centrally from the affected lung segment, are ended. Adrenergic block by drugs such as propranolol and atropine are beneficial and tend to support this theory; but the benefits of atropine have not been not confirmed. It is relevant to note the rare association of hypertrophic pulmonary osteoarthropathy with autonomic neuropathy.[150]

A second theory postulates the production, by tissue of mainly reticuloendothelial origin, of osteogenic substances able to provoke periosteal new bone growth. It is argued that these substances are normally inactivated by the pulmonary circulation. However, in the presence of a pulmonary arteriovenous shunt, osteogenic substances can escape into the systemic circulation. The theory has not been substantiated. Raised circulating levels of growth hormone have also been suggested but, like the increased oestrogen excretion suggested by Ginsburg and Brown,[52] have not been confirmed.

Granulomatous synovitis

Infection

The synovium, paraarticular tissues and bone may be at risk in tuberculosis; they are also attacked by leprosy, yaws and brucellosis. Pathological aspects of these disorders are reviewed in Chapter 18. Although tuberculosis of bones and joints is now uncommon in developed countries there is still a high incidence in first-generation immigrants and a raised frequency in their progeny. Like leprosy and, much less often, tertiary syphilis, tuberculous arthritis is, however, frequently encountered in China, India, Brazil, and Vietnam where pulmonary tuberculosis is still rife.

Synovial granulomata are periodically the result of local infection, after minor trauma, by atypical mycobacteria such as *Myc. marinum* and *Myc. kansasii* and several species have been implicated.[169] Occasionally, infection is opportunistic because of the prior use of systemic steriods or because of the development of the acquired immunodeficiency syndrome. Brucellosis also provokes local sarcoid-like epithelioid cell granulomata and sarcoidosis may affect the diarthrodial joints (see p. 989).

Plant thorn synovitis

Plant thorn synovitis[20,167] is an unusual local disorder that has attracted notoriety. The thorns enter the knees of children, characteristically in warm climates where plants such as the date and sentinel palm abound.[168] In the UK, injury sustained during the pruning of the blackthorn is an explanation for upper extremity arthritis. The heat, redness and tenderness of inflammation characterize the swollen joint. The synovia are oedematous and hypertrophic. No foreign body is seen until microscopy reveals thorns that can often be identified individually: they may be doubly refractile and periodic acid-Schiff-positive. The disorder of plant thorn synovitis enters the differential diagnosis of acute, monoarticular seronegative arthritis. It is salutary to recall that the presence of a single thorn may be a sufficient explanation for fatal tetanus.

Other causes of granulomatous synovitis[70]

When infection and the more obvious forms of granulomatous synovitis have been excluded, the possibility should be entertained that foreign bodies such as talc may reach the synovia as a result of intravenous injection in narcotic addicts (see p. 657).[129] The full explanation of the synovial and connective tissue response in such a patient may be complex because of susceptibility to endocardial thrombi, vasculitis, hepatitis B, AIDS, and trauma. Giant synovial cysts, such as those which occur in rheumatoid arthritis, may herniate into the calf and thigh and be accompanied by a sterile granulomatous synovitis.[70]

Granulomatous synovitis may develop in Crohn's disease.[65] In addition to the infective causes of granulomatous arthritis and connective tissue disease, numerous foreign bodies, chemical agents, adjuvants and extraneous materials may cause lesions of a granulomatous nature. The agents, of which many are described in earlier pages, range from beryllium and Freund's adjuvant, crystals, the particulate debris of accidents, the metallic and polymeric particles derived from arthropathy, to vegetable material accidentally implanted in a joint and suspensions used to convey local therapeutic agents.

Sarcoidosis

Clinical and immunological features

Sarcoidosis[75] (Figs 24.30 and 24.31) is an indolent, granulomatous disorder characterized by the formation in many

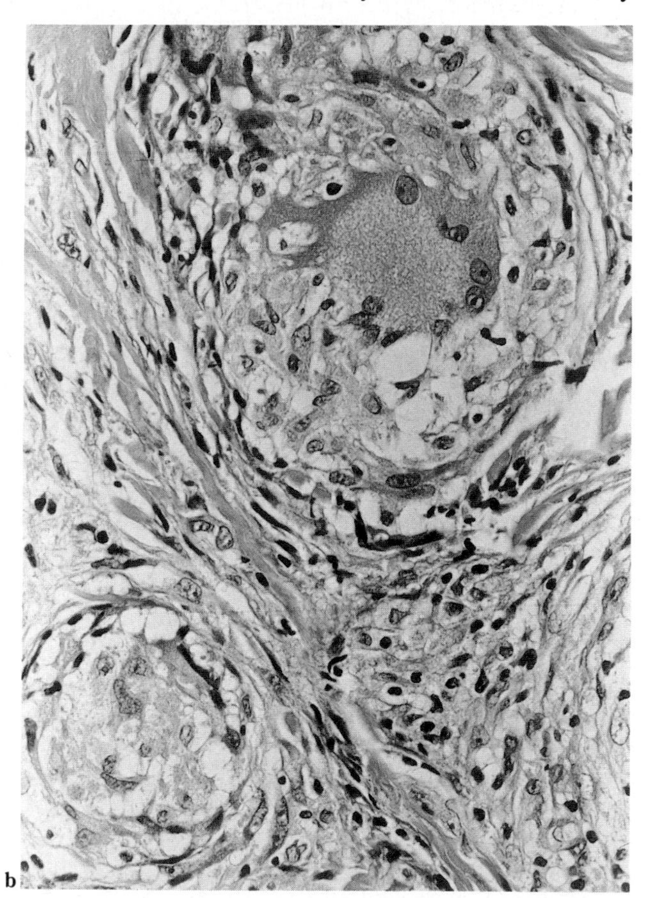

Fig. 24.30 Synovial sarcoidosis.
a. Epithelioid cell granuloma, devoid of caseation, and dominated by the presence of multinucleated giant cells with foamy cytoplasma and attenuated cytoplasmic processes. (HE × 320.) **b.** Similar field. Note foamy cytoplasm of giant cell (top centre). Compare with Fig. 24.31. (HE × 320.)

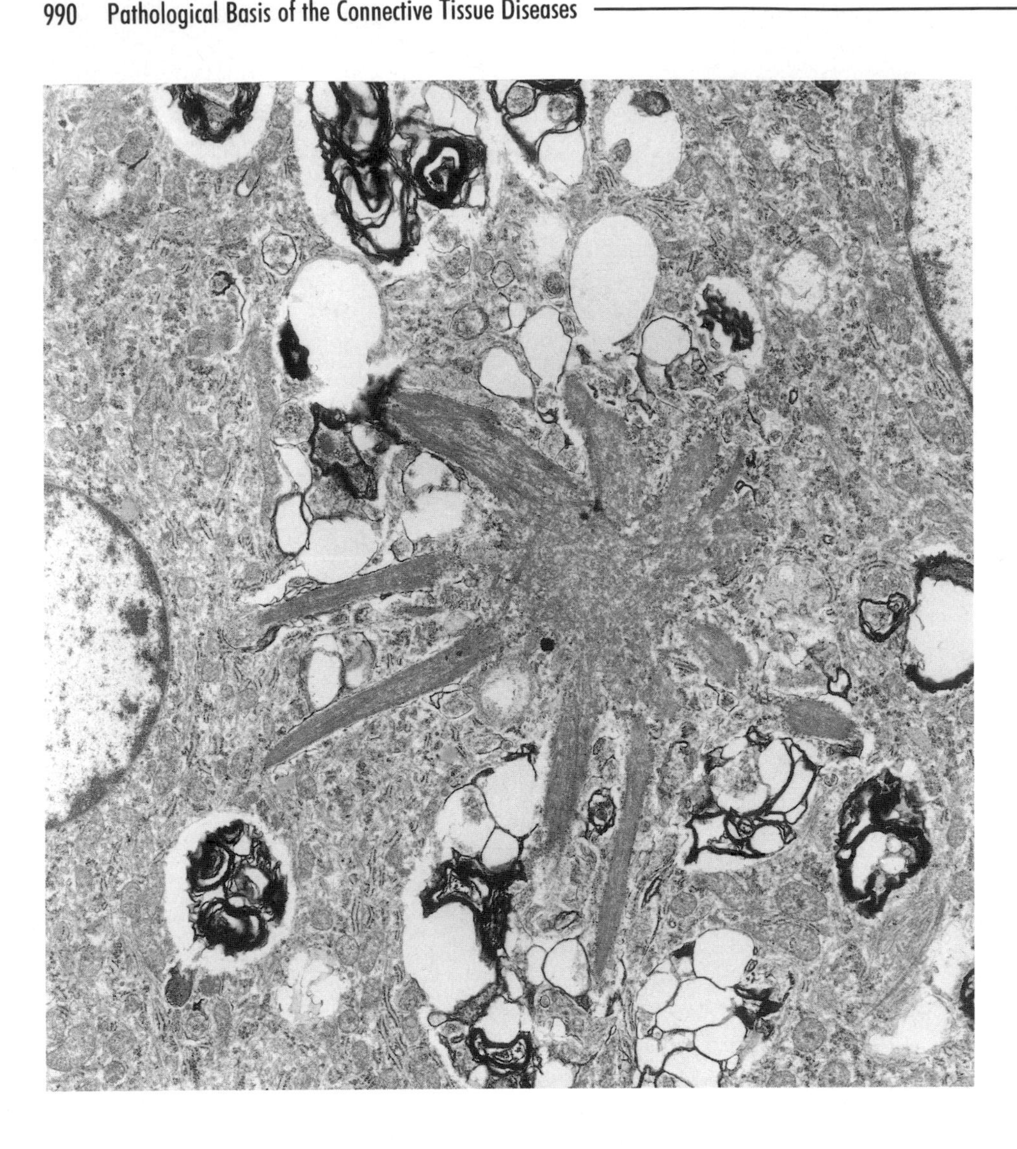

Fig. 24.31 Synovial sarcoidosis.
Asteroid body in multinucleate cell within which are many myelin bodies. (TEM × 8888.) Courtesy of Professor C J Kirkpatrick and Dr A S Curry and the Editor, *Ultrastructural Pathology.*[78a]

parts of the body but especially in the skin, lymph nodes, eyes and lung, of sterile, epithelioid cell, non-caseating cellular foci. Males and females are affected in equal proportion but there are relatively many more cases in those aged under 40 years than in older age groups. The HLA-B8 genotype is inherited with unexpectedly high frequency. The synovial joints and bone are periodically affected. Polyarthralgia,[6] bone pseudocysts and dactylitis are the most frequent results. Sarcoidosis mimics other disorders, including the chronic non-infective polyarthritides and tuberculosis. Erythema nodosum is an indication of hypersensitivity. The Kveim test is usually positive and the tuberculin test negative. Hypercalcaemia may be brief and limited or progressive, with nephrocalcinosis, renal calculus formation and uraemia as consequences. Hepatic acid phosphatase levels are raised and there may be elevated serum concentrations of IgG, IgA or IgM. A considerable proportion of cases is seronegative. Tests of monocyte phagocytic activity and of membrane receptor activity indicate increased cellular immune responsiveness. Immune complexes may be detected in the circulation. Sarcoidosis responds to corticosteroids given locally or systemically but histological confirmation of the diagnosis is mandatory.

Articular disease

The articular lesions have been studied in detail.[6,161,163] Arthritis is relatively more frequent in female patients than male. The majority of cases occur in adults but occasionally, sarcoidosis is responsible for polyarthritis in children.[124] Two categories of case are recognized: acute and chronic.

Little is known of the synovial lesions in acute sarcoid synovitis. The tendon sheaths display non-caseating epithelioid cell granulomata accompanied by multinucleate giant cell formation.[163] Where the synovial fluid has been examined, signs of inflammation are confirmed. Although the presence of sarcoid granulomata allows the diagnosis to be substantiated with some confidence, it is clearly essential to exclude the possibility that the granulomata are due to foreign bodies, mycobacteria or to brucella infection, or to chemical agents such as beryllium and zirconium.

Where chronic joint involvement is present, single or multiple joints may be affected. The examination of the synovial fluid reveals evidence of sterile inflammation. Sarcoid granulomata are often present in the synovia (Fig. 24.30). However, chronic sarcoid arthropathy may persist in the absence of granulomata.[128] In patients with predominantly extraarticular lesions such as erythema nodosum, thought to be of immune complex origin, the synovial response may be confined to a sparse lymphocytic infiltrate or may take the form of a periarticular reaction only.

Pathogenesis

Whether sarcoidosis is an immunological response to exogenous antigen is still unclear. In a recent case seen personally, sarcoidosis of bone and synovium developed in a middle-aged woman with myasthenia gravis. Whether both diseases are explicable on an autoimmune basis, remains uncertain. It is not known that sarcoidosis can provoke myasthenia. There remains a possibility that treatment given for myasthenia could have precipitated a sarcoid-like, epithelioid cell granulomatous reaction. The immunological changes in sarcoidosis, including elevated immunoglobulin levels, augmented antiviral antibody levels, diminished lymphocyte counts and lowered cell-mediated immune responsiveness, correlate with the extent, activity and severity of the disease. It is of interest that the proportion of T-cells is diminished. However, there is no single immunological abnormality that can explain the widespread tissue changes and the immunological abnormalities may therefore be secondary. Clearly, the susceptibility of the synovial tissues is much less than in rheumatoid polyarthritis.

Endometriosis

Painful joint swelling is, very rarely, attributable to extra-abdominal endometriosis.[131] Arborizing glandular elements are associated with limited nearby haemosiderin aggregates within macrophages. The glandular epithelium is tall columnar; the cells contain glycogen near a well-defined basement membrane. In spite of the infrequency of synovial endometriosis, it remains of importance in the differential diagnosis of biphasic synovial sarcoma.

Malacoplakia

In classical malacoplakia[98] (Gk: soft place), the urinary system is the site of a granulomatous process in which accumulations of macrophages have been attributed to defective lysis of microorganisms such as *E. coli* or *Staphylococcus aureus*. Malacoplakia may occur in patients with defective immunity and has been associated with lymphoma and carcinoma. In exceptional instances, malacoplakia develops in cutaneous or musculoskeletal sites. In one such case, periarticular malacoplakia evolved as a granulomatous mass that extended beneath the patellar tendon in a patient who had undergone renal allografting.[36] *Escherischia coli* septicaemia had been followed by the formation of a painful swelling of the lower leg succeeded by lymphadenopathy and the local culture of *E. coli*. The granulation tissue contained many large histiocytes within which were Michaelis–Gutmann bodies.

REFERENCES

1. Adams D, Williams D F. The response of bone to carbon–carbon composites. *Biomaterials* 1984; **5**: 59–64.
2. Agins H J, Alcock N W, Bansal M, Salvati E A, Wilson P D, Pellici P M, Bullough P G. Metallic wear in failed titanium-alloy hip replacements. *Journal of Bone and Joint Surgery* 1988; **70-A**: 347–56.
3. Akbarnia B, Torg J S, Kirkpatrick J, Sussman S. Manifestations of the battered-child syndrome. *Journal of Bone and Joint Surgery* 1974; **56-A**: 1159–66.
4. Aprahamian M, Lambert A, Balboni G, Lefebvre F, Schmitthaeusler R, Damge C, Rabaud M. A new reconstituted connective tissue matrix: preparation, biochemical, structural and mechanical studies. *Journal of Biomedical Materials Research* 1987; **21**: 965–77.
5. Arnold W D. Synovectomy. *Annals of the New York Academy of Sciences* 1975; **240**: 338–9.
6. Arnold W J, Kalunian K. Arthroscopic synovectomy by rheumatologists: time for a new look. *Arthritis and Rheumatism* 1989; **32**: 108–11.
7. Barrie H J. Hypothesis—a diagram of the form and origin of loose bodies in osteochondritis dissecans. *Journal of Rheumatology* 1984; **11**: 512–3.
8. Barrie H J. Osteochondritis dissecans 1887–1987. A centennial look at König's memorable phrase. *Journal of Bone and Joint Surgery* 1987; **69-B**: 693–5.
9. Bauer M, Jonsson K. Synovial chondromatosis of the ankle. *Fortschrift für Röntgenstrahlung* 1987; **146**: 548–50.
10. Beluffi G, Marseglia G L, Monafo V, Martini A, Re R,

Ugazio A G. Pulmonary hypertrophic osteoarthropathy in a child with late-onset agammaglobulinaemia. *European Journal of Pediatrics* 1982; **139**: 199–201.

11. Bentley G, Kreutner A, Ferguson A B. Synovial regeneration and articular cartilage changes after synovectomy in normal and steroid-treated rabbits. *Journal of Bone and Joint Surgery* 1975; **57-B**: 454–62.

12. Berk R N. Liquid fat in the-knee joint after trauma. *New England Journal of Medicine* 1967; **277**: 1411–2.

13. Bigler F C. The morphology of clubbing. *American Journal of Pathology* 1958; **34**: 237–61.

13a. Bogoch E R. Silicone synovitis. *Journal of Rheumatology* 1987; **14**: 1086–8.

14. Boussina I, Lagier R, Ott H, Fallet G H. Radiological opacities of the knee after intra-articular injections of osmic acid. *Radiologica Clinica* 1976; **45**: 417–24.

15. Branemark P-I, Hansson B-O, Adell R, Breine U, Lindstom J, Halle O, Ohman A. Osseointegrated implants in the treatment of the edentulous jaw. Experience from a 10-year period. *Scandinavian Journal of Plastic and Reconstructive Surgery* 1977; **11**: Supplement 16: 1–132.

15a. Bullough P G, DiCarlo E F, Hansraj K K, Neves M C. Pathologic studies of total joint replacement. *Orthopaedic Clinics of North America* 1988; **19**: 611–25.

15b. Byrd W E, Nunley J A. Bilateral silicone induced synovitis of the wrist. *Journal of Rheumatology* 1987; **14**: 1202–5.

16. Byron M A, Venning V A, Mowat A G. Post-mammoplasty human adjuvant disease. *British Journal of Rheumatology* 1984; **23**: 227–9.

17. Cameron H U, Pilliar R M, MacNab I. Porous vitallium in implant surgery. *Journal of Biomedical Materials Research* 1974; **8**: 283–9.

18. Carlson C S, Hilley H D, Henrikson C K, Meuten D J. The ultrastructure of osteochondrosis of the articular-epiphyseal cartilage complex in growing swine. *Calcified Tissue International* 1986; **38**: 44–51.

19. Cheung H S. *In vitro* cartilage formation on porous hydroxyapatite ceramic granules. *Cellular and Developmental Biology* 1985; **21**: 353–7.

20. Chow D, Cooke E D V, Feltis T. Thorn-induced synovitis. *Canadian Medical Association Journal* 1987; **136**: 1057–8.

21. Cleland L G, Treganza R, Dobson P. Arthroscopic synovectomy: a prospective study. *Journal of Rheumatology* 1987; **13**: 907–10.

22. Cleland L G, Vernon-Roberts B, Smith K. Fibre-glass induced synovitis. *Annals of the Rheumatic Diseases* 1984; **43**: 530–4.

23. Collins D H. *The Pathology of Articular and Spinal Diseases.* London: Edward Arnold, 1949.

24. Convery F R, Woo S L-Y, Akeson W H, Amiel D, Malcolm L L. Experimental hemarthrosis in the knee of the mature canine. *Arthritis and Rheumatism* 1976; **19**: 59–67.

25. Das G C, Sen S B. Chylous arthritis. *British Medical Journal* 1968; **2**: 27–9.

26. Davies P G, Fordham J N. Arthritis and angioimmunoblastic lymphadenopathy. *Annals of the Rheumatic Diseases* 1983; **42**: 516–8.

27. Denman A M, Szur L, Ansell B M. Joint complaints in polycythaemia vera. *Annals of the Rheumatic Diseases* 1964; **23**: 139–44.

28. DePalma A F. Haemophilic arthropathy. *Clinical Orthopaedics and Related Research* 1967; **52**: 145–65.

29. Donohue, J M, Buss D, Oegema T R, Thompson R C. The effects of indirect blunt trauma on adult canine articular cartilage. *Journal of Bone and Joint Surgery* 1983; **65-A**: 948–957.

30. Editorial. Limb salvage surgery. *Lancet* 1988; **ii**: 662–3.

30a. Editorial. Granulomatous reaction in total hip arthroplasty. *Lancet* 1990; **335**: 203–4.

31. Elves M W. The immunobiology of joints. In: Sokoloff L, ed, *The Joints and Synovial Fluid* (Volume I). New York: Academic Press, 1978: 331–406.

31a. Epps C H, ed. *Complications in Orthopaedic Surgery.* Philadelphia: J B Lippincott, 1978.

32. Eskeland G, Eskeland T, Hovig T, Teigland J. The ultrastructure of normal digital flexor tendon sheath and of the tissue formed around silicone and polyethylene implants in man. *Journal of Bone and Joint Surgery* 1977; **59-B**: 206–12.

33. Espinoza L R, Spilberg I, Osterland C K. Joint manifestations of sickle cell disease. *Medicine* 1974; **53**: 295–305.

34. Fam A G, Cross E G. Hypertrophic osteoarthropathy, phalangeal and synovial metastases associated with bronchogenic carcinoma. *Journal of Rheumatology* 1979; **6**: 680–6.

35. Faure G, Netter P, Bene M-C. Silicone-containing particles in synovial fluid: scanning electron microscopy coupled with analytical techniques allows an easy identification and differentiation from pathologically relevant crystals. *Annals of the Rheumatic Diseases* 1985; **44**: 148–50.

36. Fehring T K, Limbird T J, Brooks A L. Periarticular malakoplakia. *Clinical Orthopaedics and Related Research* 1984; **184**: 236–40.

37. Fifield R. New lives for painful hips. *New Scientist* 1987; **116**: 35–8.

38. Fischer D S, Singer D H, Feldman S M. Clubbing, a review, with emphasis on hereditary acropathy. *Medicine* 1964; **43**: 459–79.

39. Flandry F, Hughston J C. Current concepts review. Pigmented villonodular synovitis. *Journal of Bone and Joint Surgery* 1987; **69-A**: 942–9.

40. Frand M, Koren G, Rubinstein Z. Reversible hypertrophic osteoarthropathy associated with cyanotic congenital heart disease. *American Journal of Diseases of Childhood* 1982; **136**: 687–9.

41. Freemont A J, Jones C J P, Denton J. The synovium and synovial fluid in multicentric reticulohistiocytosis—a light-microscopic, electron microscopic and cytochemical analysis of one case. *Journal of Clinical Pathology* 1983; **36**: 860–6.

42. Freemont, A J, Porter M L, Tomlinson I, Clague R B, Jayson M I V. Starch synovitis. *Journal of Clinical Pathology* 1984; **37**: 990–2.

43. Gardner D L, Lamaletie J, Lawton D M, Wilson N H F. Response of cultured chondrocytes to porous ceramic: light—and low temperature scanning electron microscopic studies. *Journal of Pathology* 1987; **152**: 189A.

44. Gatti A M, Zaffe D, Poli G P, Galetti R. The evaluation of the interface between bone and a bioceramic dental material.

Journal of Biomedical Materials Research 1987; **21**: 1005–11.

45. Gerbode F, Birnstingl M, Braimbridge M. Experimental hypertrophic osteoarthropathy. *Surgery* 1966; **60**: 1030–5.

46. Ghadially F N. Overview article: the articular territory of the reticuloendothelial system. *Ultrastructural Pathology* 1980; **1**: 249–64.

46a. Ghadially F N. *Fine Structure of Synovial Joints: a Text and Atlas of the Ultrastructure of Normal and Pathological Tissues.* London, Boston: Butterworths, 1983: 307–25.

47. Ghadially F N, Janzen H K, Mehta P N. Synovial membrane in experimental lipoarthrosis. *Archives of Pathology* 1970; **89**: 291–301.

48. Ghadially F N, Lalonde J-M A, Dick C E. Ultrastructure of pigmented villonodular synovitis. *Journal of Pathology* 1979; **127**: 19–26.

49. Ghadially F N, Oryschak A F, Ailsby R L, Mehta P N. Electron probe X-ray analysis of siderosomes in haemarthrotic articular cartilage. *Virchows Archives B: Cellular Pathology* 1974; **16**: 43–9.

50. Ginsburg J. Observations on the peripheral circulation in hypertrophic pulmonary osteoarthropathy. *Quarterly Journal of Medicine NS* 1958; **27**: 335–52.

51. Ginsburg J. Hypertrophic pulmonary osteoarthropathy. *Postgraduate Medical Journal* 1963; **39**: 639–45.

52. Ginsburg J, Brown J B. Increased oestrogen excretion in hypertrophic pulmonary osteoarthropathy. *Lancet* 1961; **ii**: 1274–6.

53. Goldring S R, Jasty M, Roelke M S, Rourke C M, Bringhurst F R, Harris W H. Formation of a synovial-like membrane at the bone–cement interface. *Arthritis and Rheumatism* 1986; **29**: 836–42.

54. Goldsmith J C, Silberstein P T, Fromm R E, Walker D Y. Hemophilic arthropathy complicated by polyarticular septic arthritis. *Acta Haematologica* 1984; **71**: 121–3.

55. Goodship A E, Wilcock S A, Shah J S. The development of tissue around various prosthetic implants used as replacements for ligaments and tendons. *Clinical Orthopaedics and Related Research* 1985; **196**: 61–8.

56. Gratwick G M, Bullough P G, Bohne W H O, Markenson A L, Peterson C M. Thalassemic osteoarthropathy. *Annals of Internal Medicine* 1978; **88**: 494–501.

57. Groff F D, Schned A R, Taylor T H. Silicone-induced adenopathy eight years after metacarpophalangeal arthroplasty. *Arthritis and Rheumatism* 1981; **24**: 1578–81.

58. Gumpel J M. Radiosynoviorthesis. *Clinics in Rheumatic Diseases* 1978; **4**: 311–26.

59. Guthrie T Y, Hatfield H A. Hypertrophic pulmonary osteoarthropathy heralding relapse of lymphoepithelioma. *Archives of Otolaryngology* 1984; **110**: 552–3.

60. Hamilton A, Davis R I, Hayes D, Mollan R A B. Chondrosarcoma developing in synovial chondromatosis. *Journal of Bone and Joint Surgery* 1987; **69-B**: 137–40.

60a. Harris W B, Sledge C B. Total hip and total knee replacement. *New England Journal of Medicine* 1990; **323**: 725–31; 801–7.

61. Haubold A D, Shim H S, Bokros J C. Carbon in medical devices. In: Williams D F, ed, *Biocompatibility of Clinical Implant Materials.* Boca Raton: CRC Press, 1982: 3–42.

62. Heinicke M H, Zarrabi M H, Gorevic P D. Arthritis due to synovial involvement by extramedullary haematopoiesis in myelofibrosis with myeloid metaplasia. *Annals of the Rheumatic Diseases* 1983; **42**: 196–200.

63. Hench L L, Splinter R J, Allen W C, Greenlee T K. Bonding mechanisms at the interface of ceramic prosthetic materials. *Journal of Biomedical Materials Research* 1971; **2**: 117–41.

64. Henry, W B, Schachar N S, Wadsworth P L, Castronovo F P, Mankin H J. Feline model for the study of frozen hemijoint transplantation: qualitative and quantitative assessment of bone healing. *American Journal of Veterinary Research* 1985; **46**: 1714–20.

65. Hermans P J, Fievely M L, Descamps C L, Aupaix M A. Granulomatous synovitis and Crohn's disease. *Journal of Rheumatology* 1984; **11**: 710–2.

66. Hirst P, Esser M, Murphy J C M, Hardinge K. Bone grafting for protrusio acetabuli during total hip replacement. A review of the Wrightington method in 61 hips. *Journal of Bone and Joint Surgery* 1987; **69-B**: 229–233.

67. Hoaglund F T. Experimental hemarthrosis. The response of canine knees to injections of autologous blood. *Journal of Bone and Joint Surgery* 1967; **49-A**: 285–98.

68. Hough A J, Banfield W G, Sokoloff L. Cartilage in hemophilic arthropathy. *Archives of Pathology and Laboratory Medicine* 1976; **100**: 91–6.

68a. Howie D W, Vernon-Roberts B. The synovial response to intraarticular cobalt–chrome wear particles. *Clinical Orthopaedics and Related Research* 1988; **232**: 244–54.

68b. Howie D W, Vernon-Roberts B, Oakeshott R, Manthey B. A rat model of resorption of bone at the cement–bone interface in the presence of polyethylene wear particles. *Journal of Bone and Joint Surgery* 1988; **70**: 257–63.

69. Huaux J P, Geubel A, Maldague B, Michielsen P, Hemptinne B de, Otte J B, Nagant de Deuxchaisnes C. Hypertrophic osteoarthropathy related to end-stage cholestatic cirrhosis: reversal after live transplantation. *Annals of the Rheumatic Diseases* 1987; **46**: 342–5.

70. Iacono V, Gauvin G, Zimbler S. Giant synovial cyst of the calf and thigh in a patient with granulomatous synovitis. *Clinical Orthopaedics and Related Research* 1976; **115**: 220–4.

71. Ito G, Matsuda T, Inoue N, Kamegai T. A histological comparison of the tissue interface of bioglass and silica glass. *Journal of Biomedical Materials Research* 1987; **21**: 485–97.

72. Jackson R W. Current concepts review: arthroscopic surgery. *Journal of Bone and Joint Surgery* 1983; **65-A**: 416–20.

73. Jaffe H L. *Metabolic, Degenerative, and Inflammatory Disease of Bones and Joints.* Philadelphia: Lea & Febiger, 1972: 584–98; 721–34.

74. Jaffe H L, Lichtenstein L, Sutro C J. Pigmented villondular synovitis, bursitis, and tenosynovitis. A discussion of the synovial and bursal equivalents of the tenosynovial lesion commonly denoted as xanthoma, xanthogranuloma, giant cell tumour or the myeloplexoma of the tendon sheath, with consideration of this tendon sheath lesion itself. *Archives of Pathoology* 1941; **31**: 731–65.

75. James D G. The many faces of sarcoidosis. *Journal of the Irish Medical Association* 1974; **67**: 329–36.

76. Kieff E D, McCarty D R Jr. Hypertrophic osteoarthropathy

with arthritis and synovial calcification in a patient with alcoholic cirrhosis. *Arthritis and Rheumatism* 1969; **12**: 261–8.

77. Kim J-M, Moon M-s. Effect of synovectomy upon regeneration of meniscus in rabbits. *Clinical Orthopaedics and Related Research* 1979; **141**: 287–84.

78. Kircher T. Silicone lymphadenopathy. A complication of silicone elastomer finger joint prostheses. *Human Pathology* 1980; **11**: 240–4.

78a. Kirkpatrick C J, Curry A, Bisset D L. Light- and electron-microscopic studies on multinucleated giant cells in sarcoid granuloma; new aspects of asteroid and Schaumann bodies. *Ultrastructural Pathology* 1988; **12**: 581–97.

79. Kirkpatrick C J, Mohr W, Haferkamp O. The effects of nickel ions on articular chondrocyte growth in monolayer culture. *Research in Experimental Medicine* (Berlin) 1982; **181**: 259–64.

80. König F. Die Gelenkerkrankungen bei Blutern, mit besonderer Berucksichtigung der Diagnose. *Klinische Vortrage für Chirurgie* 1892; **11**: 233–42.

81. Ksander G A, Vistnes L M. Collagen and glycosaminoglycans in capsules around silicone implants. *Journal of Surgical Research* 1981; **31**: 433–9.

82. Kumagai Y, Shiokawa Y, Medsger T A, Rodnan G P. Clinical spectrum of connective tissue disease after cosmetic surgery. Observations on eighteen patients and a review of the Japanese literature. *Arthritis and Rheumatism* 1984; **27**: 1–12.

83. Kyle V, Hazleman B L, Wraight E P. Yttrium-90 therapy and ^{99m}Tc pertechnetate knee uptake measurements in the management of rheumatoid arthritis. *Annals of the Rheumatic Diseases* 1983; **42**: 132–7.

84. Lagier R. Post-traumatic Sudeck's dystrophy localized in the metatarso-phalangeal region. *Fortschrift für Rontgenstrahlung* 1983; **138**: 496–9.

85. Lagier R, Bertrand J. Phagocytosis of chromium during patellar osteoarthritic remodelling associated with a knee prosthesis. *Virchows Archiv A: Pathological Anatomy and Histology* 1979; **382**: 119–26.

86. Lagier R, Van Lindhoudt D. Articular changes due to disuse in Sudeck's atrophy. *International Orthopaedics* 1979; **3**: 1.

87. Lavi Y, Paladugu R R, Benfield J R. Hypertrophic pulmonary osteoarthropathy in experimental canine lung cancer. *Journal of Thoracic and Cardiovascular Surgery* 1982; **84**: 373–6.

88. Lawrence J S, de Graeff R, Laine V A I. *The Epidemiology of Chronic Rheumatism.* Oxford: Blackwell, 1962.

89. Lawton D M, Lamaletie M D J, Gardner D L. Biocompatibility of hydroxyapatite ceramic: response of chondrocytes in a test system using low temperature scanning electron microscopy. *Journal of Dentistry* 1989; **17**: 21–7.

90. Lawton D M, Lamaletie M D J, Gardner D L. Low temperature scanning electron microscopy of the superficial envelope of canine chondrocytes in culture. *Cell Biology: International Reports*, 1991; **15**: 47–54

91. Lee P. The efficacy and safety of radiosynovectomy. *Journal of Rheumatology* 1982; **9**: 165–8.

92. Leonard M H. The solution of lead by synovial fluid. *Clinical Orthopaedics and Related Research* 1969; **64**: 255–61.

93. Levy R N, Volz R G, Kaufer H, Mathews L S, Capozzi, Sturm P, Sherry H. Progress in arthritis surgery with special reference to the current status of total joint arthroplasty. *Clinical Orthopaedics and Related Research* 1985; **200**: 299–321.

94. Lexer E. Ueber Gelenktransplantation. *Medizinische Klinik* 1908; **4**: 817.

95. Linder L, Albrektsson T, Branemark P-I, Hansson H-A, Ivarsson B, Jonsson U, Lundstrom I. Electron microscopic analysis of the bone-titanium interface. *Acta Orthopaedica Scandinavica* 1983; **54**: 45–52.

96. Linder L, Carlsson A S. The bone-cement interface in hip arthroplasty. *Acta Orthopaedica Scandinavica* 1986; **57**: 495–500.

97. Luzar M J, Sharma H M. Leukaemia and arthritis: including reports on light, immunofluorescent, and electron microscopy of the synovium. *Journal of Rheumatology* 1983; **10**: 132–5.

98. McClure J. Malakoplakia. *Journal of Pathology* 1983; **140**: 275–330.

99. McEwen C. Early synovectomy in the treatment of rheumatoid arthritis. *New England Journal of Medicine* 1988; **279**: 420–1.

100. McEwen C. Multicentre evaluation of synovectomy in the treatment of rheumatoid arthritis. Report of results at the end of five years. *Journal of Rheumatology* 1988; **15**: 765–9.

101. McKendry R J R. Frostbite arthritis. *Canadian Medical Association Journal* 1981; **125**: 1128–30.

101a. Madhok R, Bennett D, Sturrock R D, Forbes C D. Mechanisms of joint damage in an experimental model of hemophilic arthritis. *Arthritis and Rheumatism* 1988; **31**: 1148–55.

102. Mainardi C L. Biochemical mechanisms of articular destruction. *Rheumatic Disease Clinics of North America* 1987; **13**: 215–33.

103. Mainardi C L, Levine P H, Werb Z, Harris E D. Proliferative synovitis in hemophilia. Biochemical and morphological observations. *Arthritis and Rheumatism* 1978; **21**: 137–44.

104. Malchaire J, Maldague B, Huberland J M, Croquet F. Bone and joint changes in the wrists and elbows and their association with hand and arm vibration exposure. *Annals of Occupational Hygiene* 1986; **30**: 461–8.

105. Mankin H J. The reaction of articular cartilage to injury and osteoarthritis. *New England Journal of Medicine* 1974; **291**: 1285–92; 1335–40.

106. Marie P. De l'osteo-arthropathie hypertrophiante pneumique. *Revues de Mèdicine* (Paris) 1890; **10**: 1–36.

107. Mariette X, Deroquancourt A, Dagay M F, Gisselbrecht C, Clauvel J P, Oksenhandler E. Monarthritis revealing non-Hodgkins T-cell lymphoma of the synovium. *Arthritis and Rheumatism* 1988; **31**: 571–2.

108. Mathieson E B, Lindgren J U, Reinholt F P, Sudmann E. Tissue reactions to wear products from polyacetal (Delrin) and UHMW polyethylene in total hip replacement. *Journal of Biomedical Materials Research* 1987; **21**: 459–66.

109. Meachim G. The effect of scarification on articular cartilage in the rabbit. *Journal of Bone and Joint Surgery* 1963; **45-B**: 50–61.

110. Meachim G. Sulphate metabolism of articular cartilage after surgical interference with the joint. *Annals of the Rheumatic Diseases* 1964; **23**: 372–80.

111. Meachim, G, Brooke G. The synovial response to intra-articular acrylic cement particles in guinea-pigs. *Biomaterials* 1984; **5**: 69–74.

112. Meachim G, Brooke G. The pathology of osteoarthritis. In: Moskowitz R, Howell D S, Goldberg V M, Mankin H J, eds, *Osteoarthritis: Diagnosis and Management*. Philadelphia, London: W B Saunders, 1984b: 29–42.

113. Meachim G, Brooke G, Pedley R B. The tissue response to acrylic particles implanted in animal muscle. *Biomaterials* 1982; **3**: 213–9.

114. Meachim G, Fergie I A. Morphological patterns of articular cartilage fibrillation. *Journal of Pathology* 1975; **115**: 231–40.

115. Meachim G, Ghadially F N, Collins D H. Regressive changes in the superficial layer of human articular cartilage. *Annals of the Rheumatic Diseases* 1965; **24**: 23.

116. Meachim G, Stockwell R A. The matrix. In: Freeman M A R, *Adult Articular Cartilage* (2nd edition). Tunbridge Wells: Pitman Medical, 1979: 1–67.

117. Meachim G, Williams D F. Changes in non-osseous tissue adjacent to titanium implants. *Journal of Biomedical Materials Research* 1973; **7**: 555–72.

118. Mendes D G, Angel D, Grishkan A, Boss J. Histological response to carbon fibre. *Journal of Bone and Joint Surgery* 1985; **67-B**: 645–9.

119. Mirra J M, Marder R A, Amstutz H C. The pathology of failed total joint arthroplasty. *Clinical Orthopaedics and Related Research* 1982; **170**: 175–83.

120. Mitchell N, Shepard N. Pericellular proteoglycan concentrations in early degenerative arthritis. *Arthritis and Rheumatism* 1981; **24**: 958–64.

121. Morris C J, Wainwright A C, Steven M M, Blake D R. The nature of iron deposits in haemophilic synovitis. *Virchows Archiv A (Pathological Anatomy and Histopathology)* 1984; **404**: 75–85.

122. Myllala T, Peltonen L, Puranen J, Korhonen L K. Consequences of synovectomy of the knee joint: clinical, histopathological, and enzymatic changes and changes in two components of complement. *Annals of the Rheumatic Diseases* 1983; **42**: 28–35.

123. Niebauer G W, Wolf B, Bashey R I, Newton C D. Antibodies to canine collagen types I and II in dogs with spontaneous cruciate ligament rupture and osteoarthritis. *Arthritis and Rheumatism* 1987; **30**: 319–27.

124. North A F, Fink C W, Gibson W M, Levison J E, Schucter S L, Howard W K, Johnson N H, Harris C. Sarcoid arthritis in children. *American Journal of Medicine* 1970; **48**: 449–55.

125. O'Connor B L, Visco D M, Heck D A, Myers S L, Brandt K D. Gait alterations in dogs after transection of the anterior cruciate ligament. *Arthritis and Rheumatism* 1989; **32**: 1142–7.

126. Okada Y, Nakanishi I, Munehiro C, Umeda S, Ichizen H, Masuda S. The presence of siderosomes in synovioblasts (B cells) of chronic spontaneous hemarthrosis. *Archives of Pathology and Laboratory Medicine* 1984; **108**: 968–72.

127. Oryschak A F, Ghadially F N. Aurosomes in rabbit articular cartilage. *Virchows Archiv: Abteilung B: Zellpathologie* 1974; **17**: 159–68.

128. Palmer D G, Schumacher H R. Synovitis with non-specific histological changes in synovium in chronic sarcoidosis. *Annals of the Rheumatic Diseases* 1984; **43**: 778–82.

129. Pare J A P, Fraser R G, Hogg J C, Howlett J G, Murphy S B. Pulmonary mainline granulomatosis: talcosis of intravenous methadone abuse. *Medicine* 1979; **58**: 229–39.

130. Parsons J R, Bhayani S, Alexander H, Weiss A B. Carbon fibre debris within the synovial joint. A time-dependent mechanical and histologic study. *Clinical Orthopaedics and Related Research* 1985: **196**: 69–76.

131. Patel V C, Samuels H, Abeles E, Hirjibehedin P F. Endometriosis at the knee. A case report. *Clinical Orthopaedics and Related Research* 1982; **171**: 140–44.

132. Patzakis M J, Mills D M, Bartholomew B A, Clayton M L, Smyth C J. A visual, histological, and enzymatic study of regenerating rheumatoid synovium in the synovectomized knee. *Journal of Bone and Joint Surgery* 1973; **55-A**: 287–300.

133. Pedley R B, Meachim G, Gray T. Identification of acrylic cement particles in tissues. *Annals of Biomedical Engineering* 1979; **7**: 319–28.

134. Pinstein M L, Sebes J I, Scott R L. Transarticular extension of chondrosarcoma. *American Journal of Roentgenology* 1984; **142**: 779–80.

135. Porter B B, Lance E M. Limb and joint transplantation. A review of research and clinical experience. *Clinical Orthopaedics and Related Research* 1974; **104**: 249–74.

136. Primm D D. Lead arthropathy—progressive destruction of a joint by a retained bullet. *Journal of Bone and Joint Surgery* 1984; **66-A**: 292–4.

137. Rabinowtiz J L, Gregg J R, Nixon J E. Lipid composition of the tissue of human knee joints. II. Synovial fluid in trauma. *Clinical Orthopaedics and Related Research* 1984; **190**: 292–8.

138. Radin E L, Rose R M. Role of subchondral bone in the initiation and progression of cartilage damage. *Clinical Orthopaedics and Related Research* 1986; **213**: 34–40.

139. Rae T. A study on the effects of particulate metals of orthopaedic interest on murine macrophages *in vitro*. *Journal of Bone and Joint Surgery* 1975; **57-B**: 444–50.

140. Rao A S, Vigorita V J. Pigmented villonodular synovitis (giant cell tumor of the tendon sheath and synovial membrane). A review of eighty-one cases. *Journal of Bone and Joint Surgery* 1984; **66-A**: 76–94.

140a. Rauschning W. Anatomy and function of the communication between knee joint and popliteal bursae. *Annals of the Rheumatic Diseases* 1980; **39**: 354–8.

141. Revell P A. Tissue reactions to joint prostheses and the products of wear and corosion. In Berry C L, ed, *Current Topics in Pathology*. Berlin: Springer-Verlag, 1982; **71**: 73–101.

142. Revell P A. Reactions of bone adjacent to joint prostheses and other implants. In: Revell P A, ed, *Pathology of Bone*. Berlin, Heidelberg, New York, Tokyo: Springer-Verlag, 1986: 217–23.

143. Revell P A, Weightman B, Freeman M A R, Vernon-Roberts B. The production and biology of polyethylene wear debris. *Archives of Traumatic and Orthopaedic Surgery* 1978; **91**: 167–81.

144. Rodnan G P, Brower T D, Hellstrom H R, Didisheim P, Lewis J H. Postmortem examination of an elderly severe hemophiliac, with observations on the pathologic findings in hemophilic joint disease. *Arthritis and Rheumatism* 1959; **2**: 152–61.

145. Ropes M W. Bauer W. *Synovial Fluid Changes in Joint Disease*. Cambridge, Massachusetts: Harvard University Press, 1963: 88.

146. Rosenthal D I, Rosenberg A E, Schiller A L, Smith R J.

Destructive arthritis due to silicone: a foreign-body reaction. *Radiology* 1983; **149**: 69–72.

147. Rosenthal R E, Oda J E, Lesker P A. Experimental synovectomy in the rabbit knee. Structural and functional considerations. *Clinical Orthopaedics and Related Research* 1972; **88**: 242–6.

148. Roy S. Ultrastructure of articular cartilage in experimental hemarthrosis. *Archives of Pathology* 1968; **86**: 69–76.

149. Roy S, Ghadially F N. Synovial membrane in experimentally produced chronic haemarthrosis. *Annals of the Rheumatic Diseases* 1969; **28**: 402–14.

150. Rudd A G, Nicholas D, Hodkinson H M. Autonomic neuropathy and hypertrophic osteoarthropathy in association with malignancy. *British Journal for Disease of the Chest* 1985; **79**: 396–9.

151. Sachs B L, Goldberg V M, Getzy L L, Moskowitz R W, Malemud C J. A histopathologic differentiation of tissue types in human osteoarthrosis cartilage. *Journal of Rheumatology* 1982; **9**: 210–16.

152. Sanyal S, Sengupta K P, Biswas S K, Pal N C. Connective tissue necrosis after cold injury. *Indian Journal of Medical Research* 1984; **79**: 554–65.

153. Schumacher H R. Articular manifestations of hypertrophic pulmonary osteoarthropathy in bronchogenic carcinoma. A clinical and pathologic study. *Arthritis and Rheumatism* 1976; **19**: 629–36.

154. Schumacher H R, Andrews R, McLoughlin G. Arthropathy in sickle cell disease. *Annals of Internal Medicine* 1973; **78**: 203–11.

154a. Schumacher H R, Dorwart B B, Bond J, Alavi A, Miller W. Chronic synovitis with early cartilage destruction in sickle cell disease. *Annals of the Rheumatic Diseases* 1977; **36**: 413–9.

155. Schwartz E E, Weiss W, Plotkin R. Ankylosis of the temporomandibular joint following burn. *Journal of the American Medical Association* 1976; **235**: 1477–8.

156. Sclafani S J A, Vuletin J C, Twersky J. Lead arthropathy: arthritis caused by retained intra-articular bullets. *Radiology* 1985; **156**: 299–302.

157. Segal A M, Mackenzie A H. Hypertrophic osteoarthropathy: a 10-year retrospective analysis. *Seminars in Arthritis and Rheumatism* 1982; **12**: 220–32.

158. Shaw I H, Knight C G, Page Thomas D P, Phillips N C, Dingle J T. Liposome-incorporated corticosteroids: I. The interaction of liposomal cortisol palmitate with inflammatory synovial membrane. *British Journal of Experimental Pathology* 1979; **60**: 142–50.

159. Soesbergen R M van, Feltkamp-Vroom T M, Feltkamp C A, Somers R, Beek W P van. T cell leukaemia presenting as chronic polyarthritis. *Arthritis and Rheumatism* 1982; **25**: 87–91.

160. Sokoloff L. Biochemical and physiological aspects of degenerative joint diseases with special reference to hemophilic arthropathy. *Annals of the New York Academy of Sciences* 1975: 285–90.

161. Sokoloff L, Bunim J J. Clinical and pathological studies of joint involvement in sarcoidosis. *New England Journal of Medicine* 1959; **260**: 842–7.

162. Soren A, Rosenbauer K A, Klein W, Huth F. Morphological examinations of so-called posttraumatic synovitis. *Beitrage für Pathologie* 1973; **150**: 11–30.

163. Spilberg I, Siltzbach L E, McEwen C. The arthritis of sarcoidosis. *Arthritis and Rheumatism* 1969; **12**: 126–37.

164. Stephens L C, Gleiser C A, Jardine J H. Primary pulmonary fibrosarcoma associated with *Spirocerca lupi* infection in a dog with hypertrophic pulmonary osteoarthropathy. *Journal of the American Veterinary Medical Association* 1983; **182**: 496–8.

165. Stevens M M, Yogarajah S, Madhok R, Forbes C D, Sturrock R D. Haemophilic arthritis. *Quarterly Journal of Medicine* 1986; **NS 58**: 181–97.

166. Strachan M W J, Gardner D L. Apparent immunity of synovial joints to metastasis: response of tumour cells to supernates from synovial cell cultures. *Journal of Pathology* 1989; **158**: 355A.

167. Stromqvist B, Edlund E, Lidgren L. A case of blackthorn synovitis. *Acta Orthopaedica Scandinavica* 1985; **56**: 342–3.

168. Sugarman M, Stobie D G, Quismorio F P, Terry R, Hanson V. Plant thorn synovitis. *Arthritis and Rheumatism* 1977; **20**: 1125–8.

169. Sutker W L, Lankford L L, Tompsett R. Granulomatous synovitis: the role of atypical mycobacteria. *Reviews of Infectious Diseases* 1979; **1**: 729–35.

170. Thakker S, McGehee W, Quismorio F P Jr. Arthropathy associated with Factor XIII deficiency. *Arthritis and Rheumatism* 1986; **29**: 808–11.

171. Tishler J M. The soft-tissue and bone changes in frostbite injuries. *Radiology* 1972; **102**: 511–3.

172. Vernon-Roberts B. Prosthetic implant reactions. *Australian and New Zealand Journal of Medicine* 1978; **8**: Supplement 1: 159–62.

173. Vernon-Roberts B. Action of gold salts on the inflammatory response and inflammatory cell function. *Journal of Rheumatology Supplement* 1979; **6**: 120–9.

174. Vernon-Roberts B, Dore J L, Jessop J D, Henderson W J. Selective concentration and localization of gold in macrophages of synovial and other tissues during and after chrysotherapy in rheumatoid patients. *Annals of the Rheumatic Diseases* 1976; **35**: 477–86.

175. Vernon-Roberts B, Freeman M A R. The tissue response to total joint replacement prostheses. In: Swanson S A V, Freeman M A R, eds, *Scientific Basis of Joint Replacement* Tunbridge Wells: Pitman Medical, 1977: 86–129.

176. Weinblatt M E, Karp G I. Monoarticular arthritis: early manifestation of a rhabdomyosarcoma. *Journal of Rheumatology* 1981; **8**: 685–8.

177. White R E, Wise C M, Agudelo C A. Post-traumatic chylous joint effusion. *Arthritis and Rheumatism* 1985; **11**: 1303–6.

178. Wie H, Beck E I. Mechanical properties and hydroxyproline content of connective tissue in porous ceramic implants. *Acta Pathologica et Microbiologica Scandinavica* 1979; **87**: 193–200.

179. Willert H G, Semlitsch M. Tissue reactions to plastic and metallic wear products of joint endoprostheses. In: Gschwend N, Dbrunner H U, eds, *Total Hip Prosthesis*. Bern: Huber, 1976: 205–39.

180. Woodard J C, Becker H N, Poulos P W Jr. Articular

cartilage blood vessels in swine osteochondrosis. *Veterinary Pathology* 1987; **24**: 118–23.

181. Yacoub M H, Simon G, Ohnsorge J. Hypertrophic pulmonary osteoarthropathy in association with pulmonary metastases from extrathoracic tumours. *Thorax* 1967; **22**: 226–31.

182. Yamashina M, Moatamed F. Peri-articular reactions to microscopic erosion of silicone-polymer implants. *American Journal of Surgical Pathology* 1985; **9**: 215–9.

183. Zalutsky M R, Venkatesan P P, English R J, Shortkroff S, Sledge C B, Adelstein S J. Radiation synovectomy with ^{165}Dy-FHMA: lymph node uptake and radiation dosimetry calculations. *International Journal of Nuclear Medicine and Biology* 1986; **12**: 457–65.

Subject Index